Physical Constants

Avogadro's number	$N_A = 6.022 \times 10^{23}\ \text{mol}^{-1}$
Boltzmann's constant	$k = 1.381 \times 10^{-23}$ J/K
Electron charge magnitude	$e = 1.602 \times 10^{-19}$ C
Gas constant	$R = 8.315$ J/mol·K
Gravitational constant	$G = 6.673 \times 10^{-11}\ \text{N}\cdot\text{m}^2/\text{kg}^2$
Permeability of free space	$\mu_0 = 4\pi \times 10^{-7}$ T·m/A
Permittivity of free space	$\epsilon_0 = 8.854 \times 10^{-12}\ \text{C}^2/\text{N}\cdot\text{m}^2$
Planck's constant	$h = 6.626 \times 10^{-34}$ J·s
Mass of electron	$m_e = 9.109 \times 10^{-31}$ kg
Mass of neutron	$m_n = 1.675 \times 10^{-27}$ kg
Mass of proton	$m_p = 1.673 \times 10^{-27}$ kg
Speed of light in vacuum	$c = 2.998 \times 10^8$ m/s

Useful Data

Air

Density (20°C, 1 atm)	1.20 kg/m^3
Molecular mass	29.0 g/mol
Speed of sound (20°C)	343 m/s

Water

Density (4°C)	1000 kg/m^3
Latent heat of fusion	3.35×10^5 J/kg
Latent heat of vaporization	2.26×10^6 J/kg
Specific heat	4190 J/kg·C°
Speed of sound (20°C)	1480 m/s

Earth

Mass	5.98×10^{24} kg
Radius (mean)	6.37×10^6 m
Distance from sun (mean)	1.50×10^{11} m

Moon

Mass	7.35×10^{22} kg
Radius (mean)	1.74×10^6 m
Distance from earth (mean)	3.84×10^8 m

Sun

Mass	1.99×10^{30} kg
Radius (mean)	6.96×10^8 m

Color Code

Vectors

Abstract →
Displacement →
Velocity →
Acceleration →
Force →
Electric Field →
Magnetic Field →

Electric current →
Electric field line →
Magnetic field line →
Light ray →
Real object or image ↑
Virtual object or image ↑

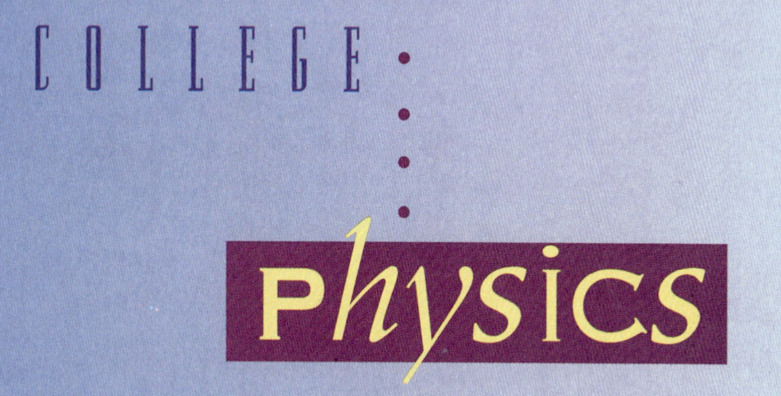
COLLEGE
PHYSICS

COLLEGE: PHYSICS

VINCENT P. COLETTA

Professor
Department of Physics
Loyola Marymount University
Los Angeles, California

illustrated

St. Louis Baltimore Berlin Boston Carlsbad Chicago London Madrid
Naples New York Philadelphia Sydney Tokyo Toronto

Publisher: James M. Smith
Executive Editor: Lloyd W. Black
Managing Editor: Judy H. Hauck
Project Manager: Mark Spann
Production Editor: Elizabeth Fathman
Manuscript Editor: Carl Masthay
Designer: David Zielinski
Manufacturing Managers: Kathy Grone, Theresa Fuchs
Art Developer: Cynthia Maciel
Illustrations: Precision Graphics, Network Graphics, Kurt Griffin, Inc.
Photo Researcher: Donata Dettbarn
Photographers: David Coletta, Patrick Watson

Printed in the United States of America
Composition by Graphic World
Color Separation by Color Dot Graphics, Inc.
Printing/binding by Von Hoffman Press, Inc.

Mosby-Year Book, Inc.
11830 Westline Industrial Drive
St. Louis, Missouri 63146

Library of Congress Cataloging-in-Publication Data

Coletta, Vincent.
College physics / Vincent Coletta.
p. cm.
Includes index.
ISBN 0-8016-7722-X
1. Physics. I. Title.
QC21.2.C625 1994
530—dc20
94-30904
CIP

95 96 97 98 99 / 9 8 7 6 5 4 3 2

Preface

Why Study Physics?

Imagine you have lived your entire life without being able to see the world in color. Imagine you see only black and white. Then imagine that someone gives you a way to see color for the first time. Before you experience color vision, however, you could not know what you were missing. You would have only the word of others to imagine how wonderful it could be. Seeing the world without any understanding of physics is like seeing the world without color. It is being blind to much of the beauty, richness, and depth of the physical universe. A good course in physics provides the means to open this universe to you, to see the world with the insight that physics provides. More specifically, knowledge of physics allows you:

To begin to appreciate the diverse phenomena of the world in a new, more unified way, to see a world governed by physical principles, and to understand how these principles serve as a foundation for understanding other sciences such as biology and chemistry.

To be able to apply the principles of physics to the solution of problems and to understand how physical principles have been used to solve enormous technical problems, opening up new vistas of experience of the physical universe undreamed of 50 or 100 years ago—space exploration, lasers, electron microscopes, computer memory chips, magnetic resonance imaging, and so on.

To wonder about the mysteries of the physical universe that remain and the vistas that will unfold within your lifetime.

You are very likely taking physics because it is required for your major, and you may rightly regard it as a challenge, a means for realizing your professional goals. But if you can be open to some of the broader educational objectives, you may find your physics course to be a lasting, enriching experience, and you may even find that such an attitude enhances your chance for success.

Goals of the Text

Designed for a 1-year course in college physics, using algebra and trigonometry, this text is the product of more than 20 years of experience teaching college physics. For more than 10 years I have worked to develop the best possible college physics textbook. The following are some general goals that have guided my work:

1. Use the most direct, concise language possible to convey ideas.
2. Use illustrations and photographs as effectively as possible to aid understanding.
3. Improve the explanation of difficult concepts.
4. Introduce abstract physical concepts wherever possible by appealing to common experiences that illustrate the concepts.
5. Present applications of physical principles—to biology, modern technology, sports and other everyday activities—in a way that clearly distinguishes applications from physical principles.
6. Present derivations in a way most likely to help, not hinder, understanding. Adapt both the style of derivations and placement of derivations to individual topics in such a way that the relationship of physical concepts is demonstrated in the clearest possible way.

7 Formulate new and interesting examples and problems, many drawn from real-life experiences, to make the college physics course more interesting for both students and professors.

Features

Organization

The overall organization of this text is very traditional. The following departures from tradition are intended to improve unity and coherence: a separate chapter on universal gravitation placed just after chapters 4 and 5 on Newton's laws and applications; a single chapter treating all mechanical waves, including sound, in an integrated way; an introduction to wave optics that begins with a qualitative overview of the wave properties of light, showing how diffraction and interference are related before each is discussed separately and quantitatively.

To allow the reader to quickly locate specific topics, I have made frequent use of subsection headings within sections of a chapter.

Explanations

The text offers particularly good explanations of many difficult concepts, including instantaneous speed, relative motion, universal gravitation, energy, Archimedes' principle, surface tension and capillarity, entropy, the eye and visual acuity, measurement of time in relativity, and wave-particle duality. A completely original feature is an elementary, quantitative discussion of optical coherence. The treatment of electricity and magnetism is unusually thorough and effective.

Applications

Numerous applications of physics to biology, technology, sports, and everyday life help motivate student interest. Every effort is made to distinguish applications from fundamental physical principles. For example, many applications are presented in examples and problems.

Each half of the book contains one extended biophysical application: in Chapter 11 the physics of fluids is applied to the human circulatory system, and in Chapter 25 optics is applied to the human eye. These unusually detailed biophysical applications are chosen for the richness of the results that follow from simple physical principles.

Illustrations and Photographs

Great care and meticulous attention to detail has been given to the development of full-color art that would realize the enormous potential of pictures to teach physics. The text contains approximately 1100 drawings and over 400 photographs, nearly all of which were planned as the manuscript was written, so that words and pictures work together to convey ideas.

Illustrations accompanying end-of-chapter problems are particularly plentiful in mechanics, where they serve to ease the student toward increasingly abstract thinking. Three-dimensional perspective drawings are used extensively, especially in the chapter on magnetism. In the chapters on optics, unusually careful ray diagrams are provided, for example, to show chromatic aberration and image formation by a microscope.

Illustrations often accompany examples. These illustrations are placed within the body of an example for easy reference. Often an example contains two illustrations, one relating to the formulation of the question, and a second relating to the solution (a free body diagram, for example).

Examples

Over 300 worked examples guide the student first to the solution of elementary problems and then to the solution of conceptually and/or mathematically more complex problems. A general problem-solving strategy is outlined in a special section preceding the problem set in Chapter 1. This strategy is then reinforced in the solution of examples throughout the book.

Particular attention is given to solution of "word problems" in kinematics. In Chapter 2 I introduce the technique of translating questions formulated in words to questions expressed in symbols. For example, the question "If a speed of 70 m/s is needed for a plane to leave the ground, how long a runway is required?" becomes "Find x when $v = 70$ m/s." This kind of translation from words to symbols is surprisingly effective in helping students overcome their difficulty with word problems.

Questions and Problems

The focus of student effort in a physics course is on problem solving. Therefore I have tried to make the over 2000 end-of-chapter questions and problems a strong feature of this text. They serve to build understanding of physical principles and to stimulate student interest in applications of physics to a wide variety of subjects, including biology and sports. Many reviewers have praised the originality and effectiveness of the problems.

The questions encourage students to build their qualitative, conceptual understanding of physics. Answers to odd-numbered questions are at the end of the questions in each chapter.

Problems are rated in difficulty by the number of stars appearing next to them: those with no stars are easiest, one-star problems are more difficult, and two-star problems are most difficult. Answers to odd-numbered problems are provided at the end of the book.

Historical Insights

Historical background is provided in introducing certain topics, for example, universal gravitation, electromagnetic waves, and atomic theory. These are areas in which history provides insight into the meaning of physical concepts by showing how these concepts evolved. A secondary goal is to provide insight into the process of discovery in physics, revealing it as a human activity. *In Perspective* essays give additional historical material.

Essays

Two kinds of essays are provided: 9 *In Perspective* historical essays and 12 *Closer Look* essays that involve physical concepts or applications. Both kinds of essays are intended to stimulate students to think about ideas beyond what is required for the course.

The *In Perspective* essays are mainly short biographies of physicists who have made some of the most important discoveries in physics: Galileo, Newton, Faraday, Einstein, Feynman, Hawking, and Curie. These are more than just a few paragraphs; they offer enough depth to humanize their subjects and sometimes to help understand what motivated their discoveries.

The *Closer Look* essays are discussions of physical principles and applications that encourage the student to think about subjects likely to arouse interest. For example, "Magic in the Sky" describes rare atmospheric optical effects such as the glory and Fata morgana. The "Energy to Run" explains, in terms of energy principles, why it is so much easier to ride a bicycle than it is to run at the same speed. "Electrical Effects in the Human Body" provides the biophysical basis for understanding why an electric shock that produces only a small electric current inside the body can nevertheless be lethal. "Biomagnetism" describes how magnetotectic bacteria have evolved in such a way as to take advantage of the earth's magnetic field. "Structure of the Retina and Color Sensitivity" describes the biophysics of the human eye. "General Relativity" shows how simple questions about relative motion led to a profound theory with amazing astronomical implications.

Ancillaries

The following ancillaries are available with this text:

Instructor's Manual with Complete Solutions

This manual presents answers and worked-out solutions to all problems in the text in a form that correlates closely with the text's approach to problem solving. Also included are sample course outlines.

Instructor's Manual on Disk

The Instructor's Manual is also available on IBM and Macintosh floppy disk.

Study Guide with Selected Solutions

Written by Martha Weller, the Study Guide contains worked-out solutions to 25% of the problems from the text, problem-solving tips and suggestions, and a chapter outline and summary of each chapter.

Testbank

Written by Paul Feldker, Ralph Barnett, and Larry Russell of St. Louis Community College, the printed testbank contains approximately 1100 quantitative problems and 400 concept-based questions. These problems and questions have been carefully prepared to complement the problem-solving style presented in the text.

ESATEST III Computerized Testbank

Developed by Engineering Software Associates, *ESATEST III* for Windows and Macintosh is the leading computerized testing system. Instructors can select, edit, add, or delete questions and construct and print tests and answer keys. The program has graphics capabilites, and users have the option of preparing algorithmically based questions. Ease of use and extensive printer support contribute to the software's popularity among educators.

Transparency Acetates

A total of 306 full-color transparency acetates contain nearly 700 of the high-quality renderings found in the text. The figures have been selected specifically for their instructional value. Images and labels are clear and large so that the acetates may be appreciated by students in the rear of large lecture halls.

ViewStudy™ CD-ROM for Physics

Another entry in the innovative Mosby ViewStudy™ series, this low-cost (free to adopters) CD-ROM contains a database of artwork and photographs and their accompanying captions taken from the text. Available for Windows or Macintosh, the disc permits users to access the images and their respective captions by topic or concept and by figure number indexed to the book. A built-in tool allows instructors to arrange and show images. Students can trace concepts visually and then print an image and its description in a note card format for easy review. Images can be exported for use in other applications, such as word processors and computerized testing systems.

Physics Quick Reference

This unique, pocket-sized supplement presents a section-by-section review of each chapter and includes key concepts and equations. It is shrink-wrapped free with each text.

Interactive Physics II™ Problems Set Disk

This problems set disk contains selected examples and problems from the text set up in the *Interactive Physics* environment. The disk is to be used with Knowledge Revolution's *Interactive Physics II,* a powerful, dynamic simulator of the physical behavior of objects. IBM and Macintosh versions are available.

f(g) Scholar™ Mosby Custom Version

Available for Windows and Macintosh, this unique software tool by Future Graph, Inc., allows science students to solve and analyze numerical problems of all types. Featuring an algebraic calculator with more than 300 built-in functions, a spreadsheet optimized for science, a graphing/charting module, and a full drawing package, *f(g) Scholar* also excels in preparing lab reports or homework assignments. The Mosby Custom Version offers the complete *f(g) Scholar* package with full documentation and a special Problems Disk that presents 50 problems from the text set up and worked out using the software. The Mosby Custom Version is also specially priced to be affordable to all students.

f(g) Scholar™ Evaluation Version

A full working version of this dynamic software tool is available free to instructors in either Windows or Macintosh format for evaluation.

f(g) Scholar™ Network Version and Mosby Custom Version Problems Disk

f(g) Scholar is available in network versions. Instructors please see your Mosby sales representative for more details. For those implementing a networked version of the software, the Problems Disk is available as a low-cost stand-alone that can be bundled with the text.

Physics Multimedia CD-ROM

Mosby offers the ultimate electronic learning tool for college physics. Included on the CD-ROM are the following features: the still-image database from the Viewstudy™ CD-ROM, a series of motion video demonstrations and animations that explore physics phenomena, the full contents of the *Interactive Physics II* Problems Set Disk, the complete *Study Guide with Selected Solu-*

tions, and the complete *Physics Quick Reference*. All features are accessible individually or through a concept map of the text. Available for both Windows and Macintosh.

Acknowledgments

The publication of an introductory college science textbook is a major undertaking, involving the efforts of many people, all of whom deserve thanks for their work.

First, I wish to thank the thousands of students who over the years have used this text in manuscript form and helped it take its final form. Their patience with a manuscript having cosmetically rough illustrations and few photographs will benefit others who will learn physics from the finished book.

Next I wish to express my gratitude for the efforts of the many reviewers who have taken the time to look at this book as it was being developed and who strongly influenced every feature of the book. Thanks to:

Stanley Bashkin, University of Arizona

Jay S. Bolemon, University of Central Florida

Louis H. Cadwell, Providence College

George Caviris, S.U.N.Y., Farmingdale

Robert W. Coakley, University of Southern Maine

Lawrence B. Coleman, University of California, Davis

Lattie F. Collins, East Tennessee State University

John Cooper, Auburn University

Donald A. Daavettila, Michigan Technological University

Miles J. Dresser, Washington State University, Pullman

Henry Fenichel, University of Cincinnati

Donald R. Franceschetti, Memphis State University

Philip W. Gash, California State University, Chico

Bernard S. Gerstman, Florida International University

Barry Gilbert, Rhode Island College

Joe S. Ham, Texas A & M University

Paul Happem, Philadelphia College of Pharmacy & Science

Hugh Hudson, University of Houston

Richard Imlay, Louisiana State University

Lawrence A. Kappers, University of Connecticut, Storrs

Paul L. Lee, California State University, Nothridge

Donald H. Lyons, University of Massachusetts, Boston

Rizwan Mahmoo, Slippery Rock University

Robert H. March, University of Wisconsin, Madison

James J. Merkel, University of Wisconsin, Eau Claire

Roger L. Morehouse, California State Polytechnic University, Pomona

J. Ronald Mowery, Harrisburg Area Community College

Darden Powers, Baylor University

Wayne F. Reed, Tulane University

Donald E. Rehfuss, San Diego State University

Joseph A. Schaefer, Loras College

Cindy Schwarz, Vassar College

Joseph Shinar, Iowa State University

Donald L. Sprague, Western Washington University

Fred J. Thomas, Sinclair Community College

Martha R. Weller, Middle Tennessee State University

John G. Wills, Indiana University

Richard L. Wolfson, Middlebury College

Lonnie L. VanZandt, Purdue University

George O. Zimmerman, Boston University

I have been most fortunate to have my book published by Mosby, a company of highly competent professionals with an uncompromising commitment to excellence in educational publishing. I wish to thank my publisher, Jim Smith, who shared my vision for this book and who has been very encouraging throughout its development and production. Jim is an ideal editor: knowledgable about the discipline (a surprisingly rare editorial quality), vitally interested in developing good books, and willing to make all the right decisions to achieve that goal. One of those decisions was to hire Lloyd Black as the new physics editor. Like Jim, Lloyd has a technical background in physics; together they form the nucleus of a powerful physics publishing team.

It is a pleasure to thank my managing editor, Judy Hauck, for her support and encouragement, her concern for pedagogical goals, and her creativity in searching for new ways to meet those goals. With the eye of an artist and the mind of a scientist, Judy contributed to the book in countless ways.

Thanks also to manuscript editor Dr. Carl Masthay for his meticulous attention to detail, for adding some insights stemming from his knowledge of astronomy, and for contributing an original photograph of the recent eclipse.

Thanks to project manager Mark Spann for overseeing the production process, with concern for both timeliness and quality, and with complete candor about what could or could not be done.

The organizational skill and unflappable good humor of production editor Liz Fathman were greatly appreciated in bringing the project to completion. Liz always got things done quickly and efficiently without ever letting those around her panic.

Less than 8 months before the final publication deadline, the photo program in this text was little more than an ambitious plan. Two people were mainly responsible for realization of that plan: photo researcher Donata Dettbarn and photographer David Coletta. Donata literally searched the world for the best possible photographs. No obstacle was too great for her, as she conducted the photo search with dedication and taste. I have thoroughly enjoyed working with her and sharing her appreciation for each new beautiful picture as it came in.

Over 100 of the photographs were shot by my son David. I thank him for his technical expertise, his artistic talent, and for his persistence in working on a project that was at times frustrating for him, as I kept insisting on getting just the right shot. Another who contributed both by researching photos and in providing his own photographs was Pat Watson, who I wish to thank for his efforts.

Thanks to art editor Cynthia Maciel for her creativity and organizational skill, for working with me to convey concepts through art as clearly as possible, and for suffering with me for over a year through the tedious job of checking and correcting illustrations. Thanks also to Cub Griffin for his quick, excellent work in making final additions to the art.

Thanks to Dave Zielinski for designing a beautiful layout for the book and for designing a cover that is both beautiful and meaningful.

The production of this book actually began years ago when I began using parts of the manuscript in my classes. Over the years, I was assisted in preparing the manuscript by dozens of student workers. I wish to thank all of them, especially Lorena Flores and Susanne Thomasson, who, after using the manuscript as students, each assisted me for 3 years—typing, drafting art, and laying out pages. The cheerful work and encouragement of all those students meant a lot to me at a time when the book was far from completion.

The Mosby marketing managers and sales representatives are a team of dedicated professionals. I am most grateful for their knowledge and enthusiasm as they begin presenting this book to the physics community. In particular, I wish to thank Anne McKeough (director of marketing), Cathy Bailey (marketing manager), Michael Weitz (product manager), and Janet Blanner (advertising manager).

Professor Martha Weller from Middle Tennessee State University deserves special thanks for her work on the solutions manual. Faced with the enormously tedious job of not only solving every problem in the book, but also communicating that solution in a form that would be helpful to students, she surpassed all expectations and produced solutions that are carefully detailed and insightful. Her solutions teach principles of physics and will be a strong aid to all who use them.

Thanks to my mentor, Professor Gerald Jones of the University of Notre Dame, who years ago convinced me of the need for a better college physics text and who was a model for me of what a physics professor should be.

I wish to thank my colleagues in physics and other disciplines who have been supportive, in particular Professor Emeritus Hanford Weckbach, S.J., who contributed so graciously of his time and knowledge of physics demonstrations and equipment, Professor John Bulman, who read several of the essays and made helpful suggestions, and Professor Madhu Amar, who helped with some of the photographs and even provided one mole of gold for a picture. Thanks also to my friend and former colleague, Dr. William Kaune, President of EMF Factors, who provided helpful technical information.

Thanks to Erianne Aichner for the many ways she has helped me over the years. She has been much more than a secretary. She has been a true friend, gladly going beyond the call of duty to help, and most importantly offering her continued encouragement. Thanks also to Janice Meichtry for her help with the steady stream of Federal Express packages that came through her office during production.

Thanks to my friends who have been so supportive, especially Sandy Patterson, Sarah de Heras, and Richard Harris. Thanks to my family, Mary, Paul, John, David, Annemarie, Catherine, Michelle, Tiffany, Melanie, Taylor, and Ryan, who have been both supportive and understanding of the demands of the book.

Finally, thanks to my parents, who instilled in me the love of knowledge and logical thinking that made this book possible, and to whom I dedicate this book.

Brief Contents

Introduction 1

Measurement and Units 6

In Perspective Essay: N-rays, Polywater, and Cold Fusion 14

1 Description of Motion 16

Problem solving strategy 36

2 Motion in a Straight Line 41

Closer Look Essay: Free Fall in Air 53

In Perspective Essay: Galileo Galilei 56

3 Motion in a Plane 64

4 Newton's Laws of Motion 87

5 Friction and Other Applications of Newton's Laws 118

Closer Look Essay: Microscopic Description of Friction 124

6 Gravitation 139

In Perspective Essay: Origins of the Theory of Universal Gravitation 152

In Perspective Essay: Isaac Newton 154

7 Energy 162

Closer Look Essay: The Energy to Run 184

8 Momentum 199

9 Rotation 214

10 Static Equilibrium 237

11 Fluids 257

12 Temperature and Kinetic Theory 295

13 Heat 320

14 Thermodynamics 338

15 Harmonic Motion 366

16 Mechanical Waves; Sound 386

Closer Look Essay: The Ear 418

17 The Electric Field 427

18 **Electric Potential 457**

19 **Electric Current 493**

Closer Look Essay: Superconductivity 501

20 **Direct Current Circuits 518**

Closer Look Essay: Electrical Effects in the Human Body 532

21 **Magnetism 550**

Closer Look Essay: Biomagnetism 574

22 **Electromagnetic Induction and AC Circuits 588**

In Perspective Essay: Michael Faraday 616

23 **Light 628**

24 **Geometrical Optics 662**

25 **The Eye and Optical Instruments 698**

Closer Look Essay: Structure of the Retina and Color Sensitivity 719

26 **Wave Optics 731**

Closer Look Essay: Magic in the Sky 754

27 **Relativity 756**

Closer Look Essay: General Relativity 784

In Perspective Essay: Albert Einstein 788

28 **Quantum Concepts 798**

In Perspective Essay: Richard Feynman 810

In Perspective Essay: Stephen Hawking 813

29 **The Atom 818**

Closer Look Essay: Lasers 841

Closer Look Essay: Semiconductors 845

30 **Nuclear Physics and Elementary Particles 854**

In Perspective Essay: Marie Curie 886

Appendix

A **Review of Mathematics A-1**

B **Gauss' Law B-0**

C **Models of Electrical Conduction in Metals C-1**

D **Selected Isotopes D-0**

E **Answers to Odd-Numbered Problems E-0**

Index

Contents

Introduction 1

Measurement and Units 6

In Perspective Essay: N-Rays, Polywater, and Cold Fusion 14

1 Description of Motion 16

1-1 Trajectory of a Particle 16
1-2 Speed 18
1-3 Displacement 22
1-4 Vector Algebra 23
1-5 Components of Vectors 27
1-6 Velocity 31
Problem Solving Strategy 36

2 Motion in a Straight Line 41

2-1 Acceleration in One Dimension 41
2-2 Linear Motion at Constant Acceleration 45
2-3 Free Fall 49
Closer Look Essay: Free Fall in Air 53
* 2-4 Graphical Analysis of Linear Motion 54
In Perspective Essay: Galileo Galilei 56

3 Motion in a Plane 64

3-1 Acceleration on a Curved Path 65
3-2 Projectile Motion 67
3-3 Circular Motion 72
3-4 Reference Frames and Relative Motion 76

4 Newton's Laws of Motion 87

4-1 Classical Mechanics 88
4-2 Force 88
4-3 Newton's First Law 90
4-4 Mass 91
4-5 Newton's Second and Third Laws 92
4-6 Force Laws 96
4-7 The Concept of Force 101
4-8 Applications of Newton's Laws of Motion 102

5 Friction and Other Applications of Newton's Laws 118

5-1 Friction 118
Closer Look Essay: Microscopic Description of Friction 124
5-2 Centripetal Force 125
5-3 Center of Mass 127

6 Gravitation 139

6-1 Universal Gravitation 139
6-2 Gravitational Attraction of the Earth 144
* 6-3 Noninertial Reference Frames 147
In Perspective Essay: Origins of the Theory of Universal Gravitation 152
In Perspective Essay: Isaac Newton 154

7 Energy 162

7-1 Work and Kinetic Energy 163
7-2 Gravitational Potential Energy; Constant Gravitational Force 169
7-3 Gravitational Potential Energy; Variable Gravitational Force 173
7-4 Spring Potential Energy; Conservation of Energy 177
7-5 Conservative and Non-Conservative Forces 180
7-6 Power 182
Closer Look Essay: The Energy to Run 184
* 7-7 Energy of a System of Particles 187

8 Momentum 199

8-1 Impulse and Linear Momentum 199
8-2 Momentum of a System of Particles; Conservation of Linear Momentum 202
8-3 Collisions and Kinetic Energy 204

9 Rotation 214

9-1 Description of Rotational Motion 214
9-2 Torque 219
9-3 Dynamics of Rotation About a Fixed Axis 221
9-4 Rotational Kinetic Energy 224
9-5 Angular Momentum 226
* 9-6 Energy Analysis of Running 228

10 Static Equilibrium 237

10-1 Conditions for Static Equilibrium 237
10-2 Center of Gravity 242
* 10-3 Stress and Strain 244

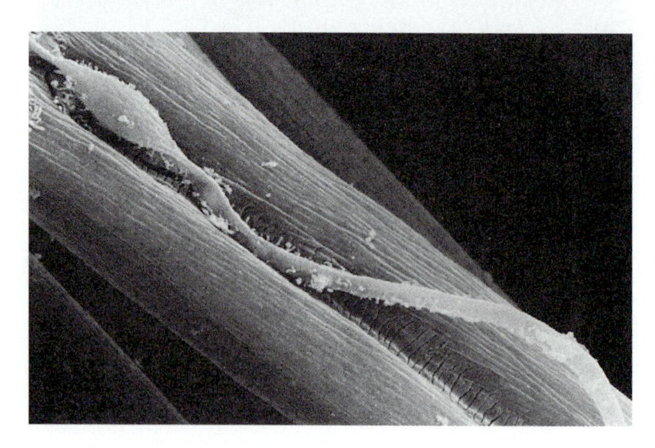

11 Fluids 257
11-1 Properties of Fluids 257
11-2 Pressure in a Fluid at Rest 259
11-3 Archimedes' Principle 265
* 11-4 Surface Tension and Capillarity 270
11-5 Fluid Dynamics; Bernoulli's Equation 273
* 11-6 Viscosity 279
* 11-7 Poiseuille's Law 282

12 Temperature and Kinetic Theory 295
12-1 Temperature Measurement 295
12-2 Ideal Gas Law 298
12-3 Kinetic Theory; Model of an Ideal Gas 303
* 12-4 Derivation of the Ideal Gas Law 306
* 12-5 Vapor Pressure and Humidity 309
12-6 Thermal Expansion 311

13 Heat 320
13-1 Definition of Heat 320
13-2 Calorimetry 321
13-3 Radiation 326
13-4 Convection 329
13-5 Conduction 330

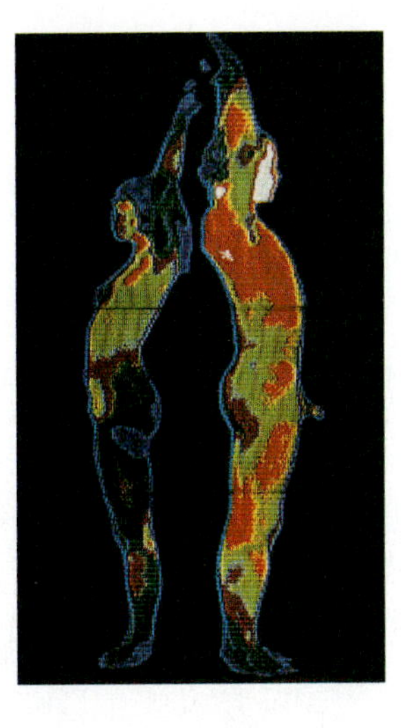

14 Thermodynamics 338
14-1 Thermodynamic Systems 339
14-2 First Law of Thermodynamics 341
14-3 Heat Engines and Refrigerators 346
14-4 Second Law of Thermodynamics 349
* 14-5 Human Metabolism 355

15 Harmonic Motion 366
15-1 Simple Harmonic Motion 366
15-2 Relationship Between SHM and Circular Motion 369
15-3 Mass and Spring 371
15-4 The Pendulum 374
15-5 Damped and Forced Oscillations 378

16 Mechanical Waves; Sound 386
16-1 Description of Waves 387
16-2 Wave Speed 392
16-3 Moving Sources and Observers: The Doppler Effect 395
16-4 Power and Intensity; The Decibel Scale 400
16-5 Time Dependence of the Displacement of a Particle of the Medium 404
16-6 Superposition of Waves; Beats; Standing Waves 406
Closer Look Essay: The Ear 418

17 The Electric Field 427

17-1 Electric Charge 427
17-2 Coulomb's Law 431
17-3 The Electric Field 434
17-4 Fields Produced by Continuous Distributions of Charge 440
17-5 Field Lines 444

18 Electric Potential 457

18-1 Electric Potential Energy and Electric Potential 458
18-2 Capacitance 468
18-3 Dielectrics 476
* 18-4 The Oscilloscope 478

19 Electric Current 493

19-1 Electric Current 493
19-2 Ohm's Law 495
19-3 Electric Power; Batteries and AC Sources 500
Closer Look Essay: Superconductivity 501
* 19-4 Electric Current and Ohm's Law on the Microscopic Level 509

20 Direct Current Circuits 518

20-1 Description of Circuits 519
20-2 Kirchhoff's Rules 520
20-3 Equivalent Resistance 523
* 20-4 Multiloop Circuits 526
20-5 Measurement of Current, Potential Difference, and Resistance 528
20-6 RC Circuits 529
Closer Look Essay: Electrical Effects in the Human Body 532
20-7 Electric Shock and Household Electricity 534

21 Magnetism 550

21-1 The Magnetic Field 550
21-2 Magnetic Forces on Current-Carrying Conductors 554
21-3 Motion of a Point Charge in a Magnetic Field 560
21-4 Magnetic Fields Produced by Electric Currents 563
21-5 Magnetic Fields Produced by Permanent Magnets 569
* 21-6 Magnetic Materials 572
Closer Look Essay: Biomagnetism 574

22 Electromagnetic Induction and AC Circuits 588
22-1 Faraday's Law 589
22-2 Inductance 597
* 22-3 Alternating Current Circuits 603
In Perspective Essay: Michael Faraday 616

23 Light 628
23-1 Electromagnetic Waves 629
23-2 The Nature of Light 632
23-3 The Propagation of Light 638
23-4 Reflection and Refraction 644

24 Geometrical Optics 662
24-1 Plane Mirrors 663
24-2 Spherical Mirrors 665
24-3 Lenses 676

25 The Eye and Optical Instruments 698
25-1 The Human Eye 699
25-2 The Magnifier 710
25-3 The Microscope 713
25-4 The Telescope 716
Closer Look Essay: Structure of the Retina and Color Sensitivity 719
25-5 Factors Limiting Visual Acuity 722

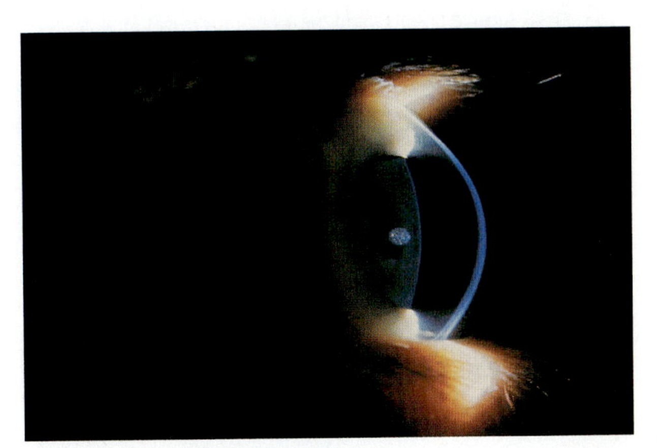

26 Wave Optics 731
26-1 Wave Properties of Light 732
26-2 Interference 736
26-3 Diffraction 741
26-4 Polarization 748
Closer Look Essay: Magic in the Sky 754

27 Relativity 765
27-1 Measurement of Time; Einstein's Postulates 766
27-2 Time Dilation 770
27-3 Length Contraction 776
27-4 Relative Velocity 779
27-5 Relativistic Mass and Energy 781
Closer Look Essay: General Relativity 784
In Perspective Essay: Albert Einstein 788

28 Quantum Concepts 798

28-1 Photons 799
28-2 Wave-Particle Duality 804
28-3 The Uncertainty Principle 807
In Perspective Essay: Richard Feynman 810
In Perspective Essay: Stephen Hawking 813

29 The Atom 818

29-1 Atomic Spectra and the Bohr Model of the Atom 819
29-2 Wave Properties of Electrons; Quantum Mechanics 830
29-3 Quantum Theory of Atomic Structure and Spectra; X-Rays 835
Closer Look Essay: Lasers 841
Closer Look Essay: Semiconductors 845

30 Nuclear Physics and Elementary Particles 854

30-1 Nuclear Structure 854
30-2 Radioactive Decay 861
30-3 Nuclear Reactions; Fission and Fusion 871
30-4 Biological Effects of Radiation 877
30-5 Elementary Particles 880
In Perspective Essay: Marie Curie 886

Appendix

A Review of Mathematics A-1

B Gauss' Law B-0

C Models of Electrical Conduction in Metals C-1

D Selected Isotopes D-0

E Answers to Odd-Numbered Problems E-0

Index I-1

Photo credits P-1

Introduction

Stonehenge

Why can some insects walk on water? What causes the beautiful colors in a soap bubble or an oil film? Is time travel in a time machine a scientific possibility? If you have ever wondered about such questions, you possess the most basic motivation for the study of physics: curiosity about physical phenomena. The origins of physics began with human curiosity in a prescientific age long ago. Two fundamental problems stimulated that curiosity: the **nature of motion** and the **composition of matter.**

Motion

Observing the stars was important in many early societies, for both practical and mystical reasons. In Egypt astronomers were able to predict the annual flooding of the Nile, which coincided with the first appearance each spring of the star Sirius. Astrology, a system of beliefs begun by the Babylonians and later developed by the Greeks, held that our personalities are affected by the position of the planets relative to the stars at the instant we are born. Ancient astronomers observed and charted the heavens. In England an early civilization produced Stonehenge, an arrangement of huge stones, which may have been used as a primitive observatory 1500 years before the birth of Christ.

Observation ultimately led to the attempt to organize astronomical data. In the second century, the Greek philosopher Ptolemy developed a system for describing the motion of the sun, moon, and planets about a stationary earth. Although earlier Greek astronomers had suggested that the earth moves about the sun, this idea was lost in antiquity, and Ptolemy's model was generally accepted until the middle of the sixteenth century. Then, in 1543, the Polish astronomer and mathematician Nicolaus Copernicus published *De Revolutionibus,* in which he challenged the Ptolemaic description of the universe and proclaimed that all the planets, including the earth, revolve about the sun. In the early seventeenth century, Galileo Galilei built one of the first telescopes and used it to observe the heavens. He is probably best known for his defense of the Copernican system and his ensuing controversy with the Catholic Church.

Of even greater significance for the development of physics, however, was Galileo's study of the nature of motion. He defined mathematically precise concepts with which to describe motion and performed simple experiments that led him to formulate the law of inertia.

Galileo was aware of the potential of the powerful new ideas he introduced. In *Two New Sciences* he wrote:

> The theorems set forth in this brief discussion, if they come into the hands of other investigators, will continually lead to wonderful new knowledge. It is conceivable that in such a manner a worthy treatment may be gradually extended to all the realms of nature.

Experimental and theoretical studies by many scientists led to a fundamental understanding of motion by 1687, when the book *Philosophiae Naturalis Principia Mathematica* was published in England. This monumental work by Isaac Newton expounded his system for explaining both celestial and terrestrial motion. Today this branch of physics, covered in the first 10 chapters of this book, is called "classical mechanics."

In his own time, the scientific effect* of Newton's ideas was to provide a complete solution to the problem of understanding planetary motion. In historical perspective, however, Newton's work assumes even more importance, in that it is the basis for later physical theories. For example, investigations into the nature of electric and magnetic phenomena are built upon classical mechanics.

Matter

The other ancient problem of physics was the relationship between bulk matter and the microscopic particles of which it is composed. About 400 B.C. the Greek philosopher Democritus speculated that all matter consists of indivisible particles which he called "atoms." What we call atoms today are *not* indivisible. But the possibility of finding the subatomic particles that truly are the basic building blocks of matter remains an intriguing question that has been only partially answered. Physicists continue to search for the fundamental particles. They also seek a unified understanding of the multiplicity of particles that have emerged from that search.

The general trend in our study of physics will be toward an increasingly microscopic view, since many of the questions that arise in a study of macroscopic phenomena have their answer on the atomic or molecular level. For example, we shall study classical mechanics—a macroscopic view—in Chapters 1 to 10 and then apply these concepts on the microscopic level to explain temperature and heat. Our study of electricity and magnetism (Chapters 17 through 22) will involve a description of the electron and proton, the basic charged particles within the atom. Chapters 29 and 30 will complete our gradual transition from a macroscopic description of matter to a particle description with a study of atoms, nuclei, and elementary particles.

---

*Newton's work had great influence on other areas of thought. Newtonian mechanics revealed a "clockwork universe," a world of intricate but orderly motion. This picture of the universe as an ordered whole influenced religious, philosophical, and political ideas.

Applications

Although the questions we ask in physics are typically motivated by curiosity rather than by practical considerations, discoveries in physics have often had a dramatic impact on technology. This in turn has had very obvious and pervasive effects on modern society. For example, our entire electronic technology began with fundamental questions about the nature of electricity and magnetism. Early experiments utilized simple, large-scale equipment. Today we have sophisticated electromagnetic communications systems and small, powerful computers utilizing tiny integrated circuits. These technologies are changing the way we live. The discovery of high-temperature superconductors in 1987 may lead to further change—practical electric cars, 300-mph magnetically levitated trains, smaller and more powerful computers using superconducting components, and controlled fusion as an abundant new source of energy. An educated person in this high-tech age should have some knowledge of the principles of physics underlying the new technologies, if only to make the technologies less mysterious.

A knowledge of physics is important for professional competence in many fields, since physics serves as a foundation for other sciences. For example, principles of quantum mechanics and atomic physics are used in chemistry to understand chemical bonding, and principles of electricity are used in biology to understand many biological processes, including nerve conduction. Thus a knowledge of physics is essential for chemists and biologists and also for physicians, dentists, engineers, geologists, architects, and physical therapists.

Aside from the very practical reasons for knowing some physics, there can be an esthetic value. Our appreciation of the world around us is enhanced when we view phenomena with the eyes of a physicist. Admittedly this is an acquired taste, and not everyone who takes a physics course will agree. But you may find pleasure in seeing physical principles applied to your environment—in understanding, for example, why you are pushed outward in a car as you turn a corner, or why a spinning baseball curves, or why the sky is blue.

Many of the examples and problems in this text describe applications of physics to everyday life. There are also more lengthy descriptions of applications, many involving biological systems. Newton's laws of motion, applied to the flow of fluids,

Eighteenth-century equipment used to study electricity.

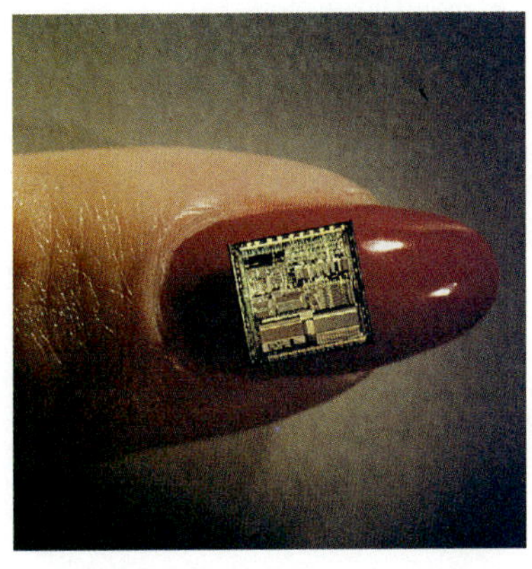

A modem integrated circuit containing hundreds of thousands of electrical components.

are used to understand the human circulatory system. Principles of electricity are used to explain household electricity and electric shock. Optical principles are applied both to the human eye and to the microscope. Quantum physics is used to understand lasers. Nuclear physics is applied to carbon dating of ancient archeological specimens.

Essays on various applications of physics appear throughout the text under the heading "A Closer Look." The topics of these essays are not commonly part of an introductory course and are included only to stimulate your interest. The text also contains "In Perspective" essays that give brief biographical sketches of physicists who have made some of the most important discoveries in physics.

Mathematics

Mathematics has an important role in the study of physics because the laws of physics are written in the language of mathematics; that is, they are expressions of precise relationships, set in a conceptual framework. As Galileo said nearly 400 years ago,

> Philosophy is written in this grand book, the universe, which stands continually open to our gaze. But the book cannot be understood unless one first learns to comprehend the language and reads the letters in which it is composed. It is written in the language of mathematics, and its characters are triangles, circles, and other geometric figures. . . . Without these one wanders about in a dark labyrinth.

Thus there are certain mathematical prerequisites for a study of physics, mainly high-school algebra and trigonometry. This text does not require a knowledge of calculus. Appendix A gives a brief review of the math you will need in your study of physics.

Derivations

Some physical laws are more fundamental than others. Derivations in the text show how the more fundamental laws lead to secondary ones, which are then applicable to particular kinds of problems. Although students often find derivations unappealing, they are an important part of physics. A physics book is not an encyclopedia. It certainly does not deal with all the detailed manifestations of physical law that occur in nature. Instead it provides a coherent description of fundamental physical laws and demonstrates the unity of physics—the relationship of physical laws—through derivations.

There is another, very practical reason for supplying derivations in a text, rather than simply describing results. If you thoroughly understand a derivation, you will also understand when the derived equation may be used. Obviously the intelligent use of equations requires that you know where they are applicable. Thus you should at least be aware of the assumptions made in the text when equations are derived.

Study Objectives

You should have two objectives in studying physics: (1) to obtain a **unified understanding of the fundamental principles of physics,** their areas of applicability, and their interrelationships, and (2) to develop the ability to **solve practical problems** using these principles. The two goals are complementary, and it is impossible to achieve one without the other.

When the answer to a problem is provided, it is sometimes possible to "solve" the problem by a trial-and-error process of formula juggling devoid of any real understanding. If you have no idea of why the principle used should apply, obtaining a correct numerical answer will not be of much value. Therefore the time you spend working on specific problems should be preceded by a careful reading of the text and lecture notes, so that you thoroughly understand the concepts.

A first course in physics is usually a great challenge and is often undertaken with considerable fear and anxiety. You may find it difficult to understand some of the concepts and impossible to solve certain problems. This is to be expected. It is definitely possible to learn from your mistakes in physics, and so it is important not to become discouraged. Like learning a new athletic skill or learning to play a musical instrument, learning physics requires practice. If you continue to work hard, your understanding will grow. You will find that you are able to understand many phenomena with a relatively small number of concepts. It can be very satisfying to see that the workings of both nature and technology can be explained in terms of just a few basic laws. It can also be satisfying to use physical laws to predict the outcome of a laboratory experiment and then verify your prediction by making measurements.

In this book I attempt to explain in the clearest possible way some of the mysteries of the physical universe. It is my hope that you will learn from it, and at the end of your physics course you will be able to look back on it as a positive experience.

Math phobic's nightmare

Measurement and Units

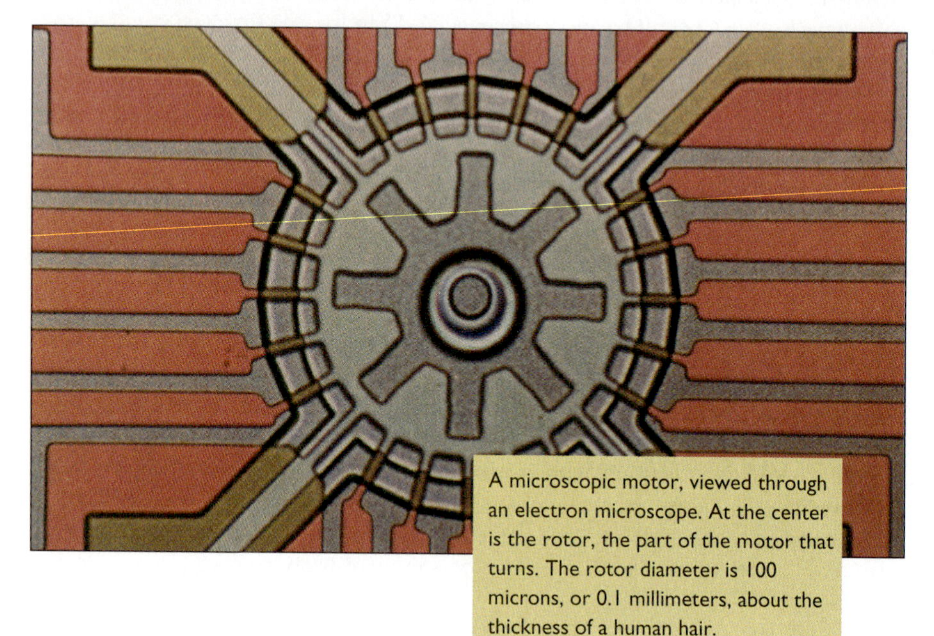

A microscopic motor, viewed through an electron microscope. At the center is the rotor, the part of the motor that turns. The rotor diameter is 100 microns, or 0.1 millimeters, about the thickness of a human hair.

Use of Units

Suppose you are driving with a friend and your gas tank is nearly empty. When you ask your friend how far it is to the nearest gas station, she responds, "About 3," with no indication of whether she means 3 blocks or 3 miles or perhaps some other measure of distance. Such a response is not helpful. Your friend gives you no information at all unless she gives the numerical value of the distance *and* the unit of length in terms of which that distance is measured—blocks, miles, whatever. As this example illustrates, it is important to express the numerical value of a quantity in terms of some unit. Distance or length may be expressed in units such as meters, feet, miles, or kilometers. Time may be expressed in units of seconds, hours, days, or years. Speed may be expressed in units of miles per hour, kilometers per hour, meters per second, and so on. Always remember to **include the unit in expressing the numerical value of any physical quantity.** Without the unit the number is meaningless.

Fundamental Quantities

All physical quantities can be defined in terms of a very small number of **fundamental physical quantities.** All the quantities studied in the first three chapters of this book can be defined in terms of just two fundamental quantities: **length** and **time.** For example, the speed of an object is defined as the distance traveled by the object divided by the elapsed time. In Chapter 4 we shall need to introduce a third fundamental quantity, **mass,** which is measured in units such as kilograms or grams.

We define the fundamental quantities by defining how we measure them. For instance, the length of an object is *defined* by comparing the object with multiples of some standard length, say, a meter. When we say that a basketball player is 2 meters tall, we mean that 2 vertical meter sticks, one on top of the other, will just reach from the floor to the top of the player's head. The time of any event is *defined* by measurement of the event's time on a clock, using standard units of time—hours, minutes, and seconds. We shall return to certain subtle questions concerning measurement of time in Chapter 27 when we study Einstein's theory of relativity.

Base and Derived Units

The units used to express fundamental quantities are called **base units,** and the units used to express all other quantities are called **derived units.** Thus meters, feet, seconds, and hours are all base units, since they are used to measure the two fundamental quantities length and time. The units miles per hour and meters per second are examples of derived units.

Base units are further characterized as being either primary or secondary. For each fundamental quantity, one base unit is designated the primary unit and all other units for that quantity are secondary. For measuring time, the second is the primary base unit, and minutes, hours, days, and so on are all secondary base units.

SI System of Units

For consistency and reproducibility of experimental results, it is important that all scientists use a standard system of units. The **Système International (SI)** is the system now used in most scientific work throughout the world. This system uses primary base units of meters, kilograms, and seconds for measurements of length, mass, and time respectively. In this book we shall use SI units primarily. Occasionally, however, in the early chapters, we shall also use the British system, in which length is measured in feet and force in pounds, since some of these units will be more familiar to you than the corresponding SI units.

Definition of the Second

The primary SI unit of time is the **second** (abbreviated s). Before 1960 the second was defined as a certain fraction of a day; that is, the second was defined as a fraction of the time required for one rotation of the earth. According to this definition, there are 86,400 seconds in a day.* A difficulty with this standard is that the earth's rate of rotation is not constant, and so a day is not a constant, reproducible standard of time. The earth's rotational rate experiences small random fluctuations from day to day. In addition there are seasonal variations and a gradual slowing down over the years. To take these facts into account, the second was defined, before 1960, as 1/86,400 of an average day during the year 1900.

---

*NOTE: 1 day = 24 hours, 1 hour = 60 minutes, and 1 minute = 60 s. So 1 day = (24) (60) (60) s = 86,400 s.

Like the other devices shown here, an atomic clock (bottom) is used to measure time.

We now use a more precise and reproducible standard to define the second—a standard consistent with the earlier definition. The second is now defined in terms of the radiation emitted by cesium atoms. Radiation is a periodic wave phenomenon. The time per cycle, called the period, is characteristic of the radiation's source. **The second is defined to be 9,192,631,770 periods of radiation emitted by cesium atoms under certain conditions.** The device used to measure time with cesium radiation is a large and elaborate device called an atomic clock. Atomic clocks are extremely accurate. Two of them will agree with one another to within 1 part in 10^{13}. An atomic clock is maintained by the National Institute of Standards and Technology.

Secondary units of time, such as minutes (min) and hours (h), are defined in terms of the second (1 min = 60 s, 1 h = 60 min = 3600 s).

Definition of the Meter

The primary SI unit of length is the **meter** (abbreviated m). Originally the meter was defined as one ten-millionth (10^{-7}) of the distance from the earth's equator to the North Pole. Later the meter was redefined to be the distance between two lines engraved on a certain bar made of a platinum-iridium alloy and carefully preserved in a French laboratory. The distance between the engraved lines was consistent with the older, less precise definition. Copies of the standard meter bar were distributed throughout the world.

In 1960 the meter was redefined as a certain multiple of the wavelength of the orange light emitted by krypton atoms under certain conditions. This newer atomic definition was again made consistent with the older definitions.

The most recent definition of the meter was made in 1983. By that time measurements of the speed of light had become so precise that their accuracy was limited by the precision of the krypton standard meter. The speed of light in a vacuum is a fundamental constant of nature. And since time could be measured on an atomic clock with much greater precision than distance could be measured, it made sense to turn the definition of the unit of length around, defining it in terms of speed and time. As of 1983 **the meter is defined to be the distance traveled by light in a vacuum during a time interval of 1/299,792,458 second.** So now, by definition, the speed of light is 299,792,458 m/s.

Names of Units

Some derived units in the SI system are given no special name. An example is the unit of speed, m/s. Other derived units are given special names. An example is the SI unit of force, the newton (abbreviated N), defined as 1 kg-m/s^2. The names of some SI derived units and their definitions are given on the inside front cover of this book.

Powers of Ten

Units that are powers-of-ten multiples of other units are often convenient to use, and so we use certain prefixes to denote those multiples. For example, *centi-* means a factor of 10^{-2}, *milli-* means a factor of 10^{-3}, and *kilo-* means a factor of 10^{+3}. Thus 1 centimeter (cm) = 10^{-2} m, 1 millimeter (mm) = 10^{-3} m, and 1 kilometer (km) = 10^{+3} m. The most commonly used powers-of-ten prefixes are listed on the inside front cover of this book.

Conversion of Units

It is often necessary to convert units from one system to another. For example, you may need to convert a distance given in miles to units of meters. To do this, you can use the conversion factor 1 mile = 1609 meters. A table of useful conversion factors is given on the inside front cover of this book.

EXAMPLE 1 Astronomical Distance

The star Sirius is about 8 light-years from earth. A light-year (abbreviated LY) is a unit of distance—the distance light travels in 1 year. One LY equals approximately 10^{16} m. Express the distance to Sirius in meters.

SOLUTION Since the ratio $\dfrac{10^{16}\text{ m}}{1\text{ LY}}$ equals 1, we can multiply the equation expressing distance in light-years by this factor without changing the equation. We then cancel units of light-years and find the distance in meters.

$$\text{distance} = 8\,\cancel{\text{LY}}\left(\frac{10^{16}\text{ m}}{1\,\cancel{\text{LY}}}\right) = 8 \times 10^{16}\text{ m}$$

EXAMPLE 2 Speed in SI Units

Express a speed of 60 miles per hour (mi/h) in meters per second.

SOLUTION We use the conversion factors 1 mi = 1609 m and 1 h = 3600 s, expressed as ratios, in such a way that when we multiply by these ratios we can cancel out miles and hours, leaving units of meters per second.

$$\text{speed} = \left(60\ \frac{\cancel{\text{mi}}}{\cancel{\text{h}}}\right)\left(\frac{1609\ \text{m}}{1\ \cancel{\text{mi}}}\right)\left(\frac{1\ \cancel{\text{h}}}{3600\ \text{s}}\right)$$
$$= 27\ \frac{\text{m}}{\text{s}}$$

Consistency of Units

In solving problems, we shall often use equations expressing relationships between various physical quantities. Algebraic symbols such as x and t are used to represent the physical quantities. Whenever we solve for the value of one quantity by substituting numerical values for other quantities in an equation, it is important to include the units along with the numerical values. The units are carried along in the calculation and treated as algebraic quantities. We then obtain from the calculation both the numerical answer and the correct units.

Using units in this way will alert you when you make certain common errors. For example, consider the following equation from Chapter 2:

$$x = \tfrac{1}{2}at^2$$

where x represents distance, a represents acceleration, and t represents time. Suppose we wish to calculate x at time $t = 5$ s, given an acceleration $a = 4$ m/s^2. We substitute these values into the equation and find:

$$x = \frac{1}{2}\left(4\ \frac{\text{m}}{\text{s}^2}\right)(5\ \text{s})^2$$
$$= \frac{1}{2}(4)(5)^2\left(\frac{\text{m}}{\text{s}^2}\right)(\text{s}^2)$$
$$= 50\ \text{m}$$

We obtain our answer in meters, a correct unit for distance.

Suppose we had mistakenly written the equation as $x = \tfrac{1}{2}at$, forgetting the exponent 2 on the t. When we substitute values into this incorrect equation, we obtain units for x of $\left(\frac{\text{m}}{\text{s}^2}\right)(\text{s}) = \frac{\text{m}}{\text{s}}$, which are clearly incorrect units for a distance. These incorrect units reveal that we have made some kind of mistake—either we have used an incorrect equation or we have substituted a quantity with incorrect units into a correct equation.

Of course, getting the units to come out correctly is no guarantee that you have not made some other kind of error. But at least it allows you to eliminate some kinds of errors.

Significant Figures

When you measure any physical quantity, there is always some uncertainty in the measured value. For example, if you measure the dimensions of a desk with a meter stick marked with smallest divisions of millimeters, your measurements may be accurate to the nearest millimeter but not to the nearest tenth of a millimeter. When you state the dimensions, you could explicitly indicate the uncertainty in your measurements. For example, you might measure the length of a desk to the nearest millimeter (or tenth of a centimeter) and express the desk's length as 98.6 ± 0.1 cm. This means that you believe the length to be between 98.5 cm and 98.7 cm.

In this text, we shall not explicitly indicate the uncertainty in a measured value. We shall, however, imply this uncertainty by the way we express a value. Saying that the length of the desk is 98.6 cm means that we have some confidence in the three measured digits, in other words, confidence that the true length differs from this number by no more than 0.1 cm. We say that there are three **significant figures** in this measurement. If you say that the length is 98.60 cm, giving four significant figures, you are implying that your measurement is accurate to four significant figures, in other words, that the uncertainty is no more than 0.01 cm. Since your measurement does not have this degree of precision, it would be misleading to state the result in this way. The measured length is 98.6 cm, *not* 98.60 cm.

Sometimes the number of significant figures is unclear. For example, if we say a certain distance is 400 m, are the two zeroes significant figures or are they included just to indicate the location of the decimal point? Do we mean that the uncertainty in distance is 1 meter? Using powers-of-ten notation avoids such ambiguity. For example, if we say the distance is 4.00×10^2 m, we give three significant figures, meaning that the uncertainty in distance is 0.01×10^2 m, or 1 m. But if we say the distance is 4.0×10^2 m, we give two significant figures, meaning that the uncertainty is 0.1×10^2 m, or 10 m.

Often it is tempting to state a result you have calculated with too many significant figures, simply because this is the way the numerical value appears on your calculator. For example, suppose you wish to calculate the area of a desk top. You measure a length of 98.6 cm and a width of 55.2 cm. You compute the rectangular area by multiplying these two numbers. Your calculator reads 5442.72. But each of your measurements is accurate to only three significant figures, and so the product is also accurate to only three significant figures. Thus you should round off the calculator reading and state the area as 5440 cm^2 or, better yet, 5.44×10^3 cm^2.

Now suppose you had measured the width of the desk with less precision than the length—measuring only to the nearest centimeter, so that the result is 55 cm. The length is known to three significant figures, but the width is known to only two significant figures. The less precise measurement is the limiting factor in the precision with which you can calculate the area. **When two or more numbers are multiplied or divided, the final answer should be given to a number of significant figures equal to the smallest number of significant figures in any of the numbers used in the calculation.** So, when you multiply a length of 98.6 cm times a width of 55 cm, do *not* state the area as 5423 cm^2, as indicated on your calculator. Rather, round off to two significant figures and state the area as 5400 cm^2 or as 5.4×10^3 cm^2.

When numbers are added or subtracted, the uncertainty in the calculated value is limited by the number having the greatest uncertainty. Thus, **when you add or subtract, the number of decimal places retained in the answer should equal the smallest number of decimal places in any of the quantities you add or subtract.** For example, the sum 12.25 m + 0.6 m + 44 m should *not* be written as 56.85 m but rather should be rounded off to 57 m.

Often numbers that appear in equations do not represent measured values and so are not subject to the rules for significant figures. For example, the numerical factor $\frac{1}{2}$ appears in the equation $x = \frac{1}{2}at^2$. This number is exact. There is no uncertainty in its value, and it places no limitation on the number of significant figures to which x can be calculated. If, for example, values for a and t are known to 3 significant figures, x may be calculated to 3 significant figures.

Order-of-Magnitude Estimates

It is often useful to estimate a number to the nearest power of ten. Such an estimate is called an **order-of-magnitude estimate.** Estimating is appropriate either when the available data do not permit any greater accuracy or when you don't need to know the number with any greater accuracy. Estimates can also be useful in checking the results of a more careful calculation, simply because it is so easy to calculate when you are working only with powers of ten.

For example, suppose we want to estimate the number of high schools in the United States. We could begin by estimating the country's high school age population. Since high school takes 4 years and an average lifetime is roughly 80 years, we might estimate that one person in 20 of the 200 million or so people in the country is of high school age. This gives a high-school-aged population of about 10 million, or 10^7. Of course this estimate is not very accurate, since not all age groups are equally represented in the population, and certainly not everyone of high school age is in high school. But 10^7 high school students is a reasonable order-of-magnitude estimate. Next we estimate that the average high school has an enrollment on the order of 10^3. This means that the number of high schools in the United States is on the order of $10^7/10^3 = 10^4$.

Problems

Standards of Length and Time

1 Two atomic clocks keep almost exactly the same time, but one runs faster than the other by 1 part in 10^{13}. How long would you have to wait before the clocks' readings differed by 1 s?

2 Before 1799 the legal standard of length in France was the foot of King Louis XIV. Since the king could not personally measure the length of everything with his foot, what was needed to make this standard unit of measure at all useful?

3 In 1983 the meter was redefined as the distance traveled by light in a vacuum during a time interval of 1/299,792,458 s, so that now the speed of light is exactly 299,792,458 m/s. Why wasn't the number rounded off so that the speed could be exactly 300,000,000 m/s?

Unit Conversion

4 How many picoseconds are in 1 h?

5 How many volts are in 30 megavolts?

6 How many milliamps are in 0.2 amp?

7 A picture has dimensions of 20 cm by 30 cm. Find the area in m^2.

8 A rectangular metal plate has dimensions of 8 cm by 5 cm by 3 mm. Find the plate's volume in m^3.

9 A football field is 100 yards long. Express this length in meters.

10 A room has dimensions of 5 m by 4 m. How many square yards of carpet are required to carpet the room?

11 Express the speed of light in units of mi/s.

12 You are driving on the Autobahn in Germany at a speed of 180 km/h. Express your speed in mi/h.

Consistency of Units

13 In the following equations, t is time in s, v is velocity in m/s, and a is acceleration in m/s^2:

$$v = at \qquad v = \frac{a^2}{t} \qquad v = at^2$$

Which of these equations is consistent with the units?

14 In the following equations, t is time in s, x is distance in m, v is velocity in m/s, and a is acceleration in m/s^2:

$$v^2 = 2ax \qquad v = \frac{a^2t^2}{x} \qquad v = \sqrt{xat}$$

Which of the equations is consistent with the units?

15 If you calculate v^2/a, where v is in m/s and a is in m/s^2, what units will your answer have?

Significant Figures

16 How many significant figures are in each of the following numbers: (a) 25.673; (b) 2200; (c) 2.200×10^3; (d) 3005; (e) 0.0043; (f) 4.30×10^{-3}?

17 How many significant figures are in each of the following numbers: (a) 165; (b) 500; (c) 5.00×10^2; (d) 40,001; (e) 0.0070; (f) 7.000×10^{-3}?

18 Round off each number in Problem 16 to two significant figures.

19 Round off each of the following quantities to three significant figures: (a) 5782 m; (b) 2.4751×10^5 s; (c) 3.822×10^{-3} kg; (d) 0.06231 m.

20 A nickel has a radius of 1.05 cm and a thickness of 1.5 mm. Find its volume in m^3.

21 A rectangular plot of land has dimensions of 865 m by 2234 m. How many acres is this? (1 acre = 43,560 square feet.)

22 Find the sum of the following distances: 4.65 m, 31.5 cm, 52.7 m.

23 Find the sum of the following masses: 21.6 kg, 230 kg, 55 g.

Order-of-magnitude Estimates

24 Estimate the total volume of water on earth.

25 Estimate the total volume of the earth's atmosphere.

26 Estimate the number of heartbeats in an average lifetime.

27 Estimate the total time you will spend during your lifetime waiting for traffic lights to change from red to green.

28 Estimate the total time you will spend during your lifetime waiting in line at the grocery store.

29 Estimate the total number of pediatricians in the United States.

30 Estimate the total number of teachers of college English composition courses in the United States.

31 Estimate the surface area of a water reservoir that is 10 m deep and big enough to supply the water needs of the Los Angeles area for 1 year.

32 (a) Estimate the maximum traffic capacity in one direction on an interstate highway in units of cars per minute. (b) Estimate how long it would take to evacuate a city of 1 million on one interstate highway.

33 Estimate the number of M&M's needed to fill a 1-liter (10^3 cm^3) bottle.

34 Suppose you are a visitor on another planet and observe the setting sun. You notice that your little finger, which is 1 cm wide, just covers the sun when you extend your arm out and hold your finger 1 m away from your eyes. The bottom edge of the sun begins to dip below the horizon, and 5 minutes later the sun completely disappears. Estimate the length of a day on the planet.

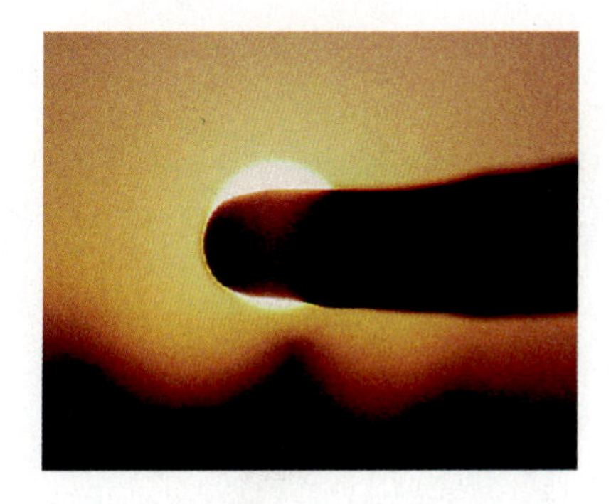

N-Rays, Polywater, and Cold Fusion

Most nonscientists believe that science advances inexorably—facts leading to theory, leading to further facts that allow refinement of the theory, and so on. It may appear that simple adherence to a cookbook like "scientific method" is a surefire route to success—good scientific law, like a good cake, coming from careful measurement and mixing of the right ingredients under just the right conditions. The history of science, however, teaches otherwise. Sometimes a scientist's insight leads to a brilliant and unexpected discovery, for example, Einstein's theory of relativity. And sometimes what appears at first to be a revolutionary discovery turns out to be a disappointing mistake.

Such mistakes are not always easy to recognize. An erroneous claim of an important discovery can arise from a desire to interpret facts in a way that confirms a scientist's own theories. It can be difficult for one to refute such a claim by attempting to reproduce an experiment because no two experiments are ever performed under exactly the same conditions. The scientist who claims to have observed a new effect can always say to another scientist who has repeated the experiment and not seen the effect: "But you have not carried out my experiment in exactly the way I did. You did not use reagents from the same source; you did not observe long enough or carefully enough; your instruments were faulty. . . ."

Our first example occurred long ago, in 1903—before, you might think, we knew as much as we know today. In that year, the French scientist René Blondlot, who had been working with the newly discovered X-rays, claimed to have found another new type of radiation. Blondlot named his rays "N-rays" after Nancy, the city in which he was working. He turned out paper after paper on the subject and became quite famous, drawing many well-known scientists into the ranks of his supporters and giving birth to an entire school of N-ray research.

Reports of failures to confirm Blondlot's findings began to appear, but defenders of Blondlot offered critiques of the challengers' experimental methods and powers of observation. Finally, in late 1904, Blondlot was visited by the American Robert Wood, professor of physics at Johns Hopkins University. Through some clever sleight of hand, Wood fooled Blondlot into claiming to observe N-rays under conditions in which, given Blondlot's own theories, they could not possibly have appeared. When Wood described his trick in a letter to *Nature*, a prestigious scientific journal, it was the beginning of the end for N-rays.

Our second example dates back only to the late 1960s. In those years, a Soviet scientist studying the properties of water discovered what he claimed to be a polymerized form of water. This so-called polywater, which had been produced by long and repeated heating of the water in an elaborate glassware apparatus, was observed to be a clear, plastic-like material. The discovery sent shock waves through the scientific community and was given apocalyptic coverage in the press, which claimed with some scientific support that the clumping of the water molecules into this undrinkable form could, if not contained, eventually spread to all the water in the world. The supposed polywater, after careful measurement and analysis, was finally unmasked as ordinary water containing certain impurities. This simple conclusion took nearly 7 years to reach! That much time was required before all the researchers who had failed to reproduce the experiment were finally believed.

Our third example is of still more recent vintage. In April 1989, chemists Stanley Pons and Martin Fleischmann called a press conference to announce the discovery of an amazing new process called "cold fusion." Fusion is the combining of light atomic nuclei to form heavier nuclei and is the process by which the sun generates energy. The process was believed to require sunlike conditions of extremely high temperature in order to overcome the electrical repulsion between the positively charged nuclei and get them close enough together so that they could fuse and release energy. Huge research efforts around the world have been devoted to controlling fusion in enormously large and expensive machines, producing high temperatures (Fig. A). Although steady progress has been made, these machines are not yet energy efficient; that is, they do not generate as much energy as is needed to operate them.

Pons and Fleischmann claimed to have bypassed the usual way of creating fusion at high temperatures by producing cold fusion in a small beaker containing heavy (deuterium-rich) water and a palladium coil carrying electricity (Fig. B). This process, if it had actually worked, would have provided virtually limitless low-cost energy and would have solved the world's energy problems.

Not surprisingly, government and industry were anxious to invest in this work, and other scientists around the

Fig. A Inside the Princeton fusion reactor.

Fig. B Cold fusion apparatus.

world were eager to reproduce the cold fusion effect. However, very few of the many labs that repeated the "cold fusion" experiment found any evidence that fusion was taking place. Pons and Fleischmann attacked their critics and maintained their claim. Lack of positive experimental results by others eventually took its toll. The general consensus today is that, whatever Pons and Fleischmann may have observed, it was not fusion. Like the proponents of N-rays and polywater before them, Pons and Fleischmann had been too caught up in visions of a grand discovery, and had abandoned the openness required to successfully probe the secrets of nature.

Fig. C Enormous energy is displayed in a solar flare, seen during a solar eclipse.

CHAPTER 1 Description of Motion

Motion is a basic phenomenon of nature. From early infancy we sense moving objects in our surroundings. The sensation of our own motion can provide powerful experiences—the "weightless" thrill of a rollercoaster ride or the backward shove we feel in a jet taking off or in an accelerating sports car. The study of motion in physics is motivated by a desire to transcend simple sensory awareness of motion—to quantify, to measure, to understand motion. In this chapter we begin such a study.

1-1 Trajectory of a Particle

The motion of a body can be recorded in various ways. A multiflash photograph made with a strobe light records a golf swing in Fig. 1-1. The duration of each flash is short enough that the golf club is effectively frozen at the instant of the flash. In this way we can see in a still photograph the motion that has occurred.

Fig. 1-1 Multiflash photograph of a golf swing.

If we wish to quantify motion, we must have some means of measuring the positions of the moving body at different times. In 1884 the artist Eadweard Muybridge studied human motion by taking a sequence of photographs of subjects moving near a background grid (Fig. 1-2). These photographic sequences provided a quantitative record of the path followed by each part of the moving body.

A smaller grid is sometimes used in an elementary physics laboratory to record the much simpler motion of a "projectile" (an object that has been thrown, or "projected"), as shown in Fig. 1-3. This multiflash photograph not only indicates the trajectory or path of the projectile, but also provides a record of its location at successive instants.

Fig. 1-2 Study of motion by the nineteenth-century photographer Muybridge.

In our description of motion, we shall frequently indicate the motion of a particle* by a sequence of dots corresponding to positions at equal time intervals.

Fig. 1-3 Projectile motion.

Cartesian Coordinates

The position of a particle is described mathematically by **Cartesian coordinates** x, y, and z. Consider the position of a particle at a particular time t. We may specify this position by giving the coordinates of the particle at this time: $x(t)$, $y(t)$, and $z(t)$, as indicated in Fig. 1-4. The coordinates are measured from the origin O and may be either positive or negative. For example, if the particle is to the left of the origin, $x(t)$ is negative, and if the particle is to the right of the origin, $x(t)$ is positive. Keep in mind that the coordinates depend on time, as our notation $x(t)$, $y(t)$, $z(t)$ implies. As a particle moves, its coordinates change.

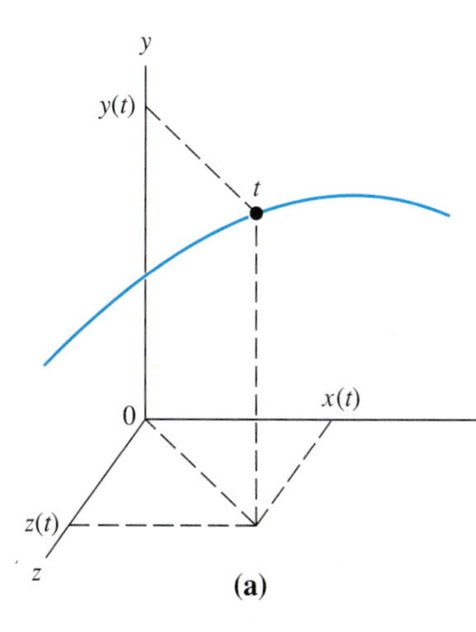

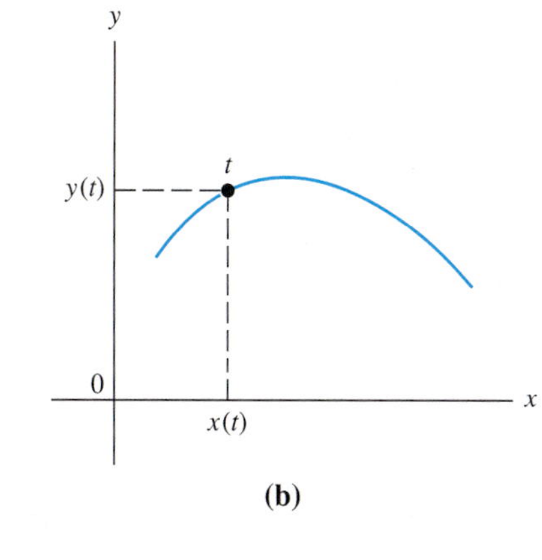

Fig. 1-4 (a) Cartesian coordinates $x(t)$, $y(t)$, $z(t)$ specify the position at time t of a particle moving along a three-dimensional path. **(b)** Cartesian coordinates $x(t)$, $y(t)$ specify the position at time t of a particle moving in a plane.

Path Length

Path length $s(t)$ is defined as the distance traveled along a trajectory in time t, measured from the starting point of the motion at $t = 0$ (Fig. 1-5). Unlike the Cartesian coordinates, s is always a positive quantity. Path length is the total distance covered, and so it can only increase with time as a particle moves.

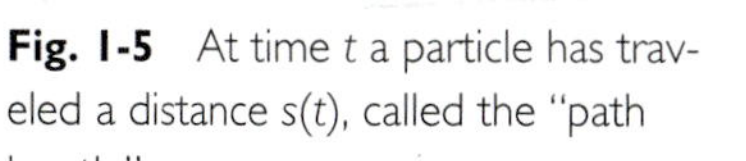

Fig. 1-5 At time t a particle has traveled a distance $s(t)$, called the "path length."

---

*By "particle" we mean an object whose size is small compared with the relevant dimensions of a problem. Whether a particular object can be treated as a particle, rather than as a collection of particles, depends on the problem under consideration as well as on the object. For example, the earth can be treated as a particle when we are studying its motion about the sun but not when we are considering the relative motion of any of its parts.

1-2 Speed

Uniform Motion

Fig. 1-6 A particle in uniform motion moves equal distances in equal time intervals.

Suppose a particle moves equal distances during equal time intervals, so that, as in Fig. 1-6, the dots representing the motion are equally spaced. Such motion is called "uniform motion." The particle's **speed,** denoted by v, is defined as the change in path length Δs divided by the corresponding change in time Δt:

$$v = \frac{\Delta s}{\Delta t} \qquad \text{(uniform motion)} \qquad (1\text{-}1)$$

If v is known, it may be used to find the distance traveled during any given time interval, since it follows from Eq. 1-1 that

$$\Delta s = v\,\Delta t \qquad \text{(uniform motion)} \qquad (1\text{-}2)$$

EXAMPLE 1 Traveling at Constant Speed

Let the dot sequence in Fig. 1-6 represent the motion of a car moving along a curved path. Let the spacing between the dots be 12.5 m and the corresponding time interval 0.500 s. (a) Compute the speed of the car. (b) If the car continues at this speed for 4.00 h, how far does it travel?

SOLUTION **(a)** Eq. 1-1 tells us that we can determine the car's speed if we know how far the car moves (Δs) in a given time interval (Δt). The problem statement tells us that the spacing between any two dots is 12.5 m and that the car takes 0.500 s to cover this distance. Thus we know both Δs and Δt and can use Eq. 1-1:

$$v = \frac{\Delta s}{\Delta t} = \frac{12.5 \text{ m}}{0.500 \text{ s}} = 25.0 \text{ m/s}$$

(b) Here we can use Eq. 1-2 because we know v and Δt and want to find Δs:

$$\Delta s = v\,\Delta t = (25.0 \text{ m/s})\,(4.00 \text{ h})$$

Notice, however, that we have two different units for time: seconds and hours. When working physics problems, you should always use consistent units within a problem; times should be all in seconds, say, or all in hours; distances should be all in meters or all in kilometers; and so forth.

To get consistent units here, we multiply by the conversion factor $\frac{3600 \text{ s}}{1 \text{ h}}$, since this factor is equal to 1 and allows us to cancel the unit h.

$$\Delta s = (25.0 \text{ m/s})\,(4.00 \text{ h})\left(\frac{3600 \text{ s}}{1 \text{ h}}\right) = 3.60 \times 10^5 \text{ m}$$

or, since 10^3 m = 1 km,

$$\Delta s = 3.60 \times 10^2 \text{ km}$$

Nonuniform Motion

Typically, motion is not uniform. You can see this in multi-image photographs of a person walking (Fig. 1-7). Dots of light indicate the instantaneous position of the hip, knee, and ankle at equal time intervals. Notice that the dots along the knee's path are not equally spaced and those along the ankle's path are even less uniformly spaced. Such uneven spacing tells us that the knee and ankle move at speeds that are not constant. In other words, their motion is nonuniform. Where the dots are very close together, the knee (or ankle) was moving slowly and so did not move far during the time interval. Where the dots are farther apart, the knee (or ankle) was moving quickly and so covered a greater distance.

Suppose we use $v = \Delta s/\Delta t$ to define speed for nonuniform as well as uniform motion. Unlike the case of uniform motion, the value of the speed for nonuniform motion depends on the particular time interval Δt for which the speed is computed. Furthermore, since the motion is not uniform during an arbitrary time interval, the ratio $\Delta s/\Delta t$ is an average property of the motion over this time interval. We refer to this ratio as **average speed,** denoted by $\bar{v}$:

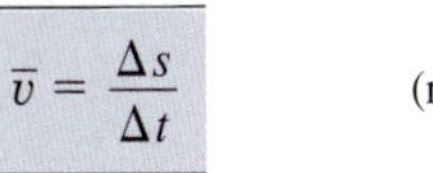

$$\bar{v} = \frac{\Delta s}{\Delta t} \qquad \text{(nonuniform motion)} \qquad (1\text{-}3)$$

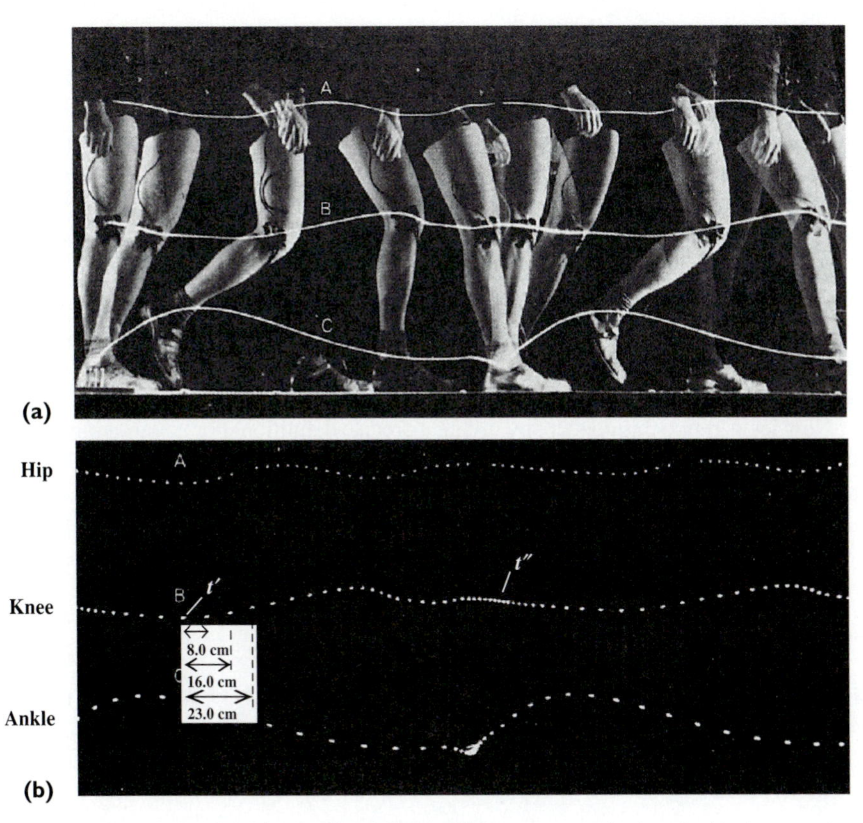

Fig. 1-7 In a study of walking by Gjon Mili, small lights were attached to the hip, knee, and ankle. The lower photograph was made by placing a rotating plate with a slit in front of the camera, so that the light reaching the camera was chopped and positions at equal time intervals were recorded. (After Napier J: *Sci Am* 216(4):56, 1967.) Mili's goal was to develop better artificial legs.

EXAMPLE 2 Average Speed in Traffic

In city traffic a car travels a distance of 15.0 km in 30.0 min. (a) Compute the car's average speed in kilometers per hour during this interval. (b) The driver estimates the time necessary to travel an additional 40.0 km, on the assumption that the average speed will remain the same. How long is this?

SOLUTION **(a)** Applying Eq. 1-3, we compute the average speed:

$$\bar{v} = \frac{\Delta s}{\Delta t} = \left(\frac{15.0 \text{ km}}{30.0 \text{ min}}\right)\left(\frac{60.0 \text{ min}}{1.00 \text{ h}}\right) = 30.0 \text{ km/h}$$

(b) We now wish to find the time interval Δt required to travel a distance $\Delta s = 40.0$ km at an average speed $\bar{v} = 30.0$ km/h. We apply Eq. 1-3 and solve for the unknown Δt in terms of the known quantities Δs and $\bar{v}$. Multiplying Eq. 1-3 by Δt and dividing by $\bar{v}$, we obtain

$$\Delta t = \frac{\Delta s}{\bar{v}} = \frac{40.0 \text{ km}}{30.0 \text{ km/h}} = 1.33 \text{ h}$$

Instantaneous Speed

What does it mean to say that an object is moving at a certain speed at some instant? For example, what does it mean to say that a car is traveling at a speed of 60 mi/h at the instant it passes a radar checkpoint? It does not necessarily mean that the car is traveling at a uniform rate. The car may be speeding up or, more likely, slowing down as it passes the radar unit. Of course, we can say that 60 mi/h is simply the reading of the car's speedometer or of the radar unit at some instant. But this does not explain the meaning of the number that is measured. Roughly speaking, we mean by the speed of the car at any instant the car's average speed over a very short time interval. Since 60 mi/h = 27 m/s = 2.7 m/0.1 s, a car moving at 60 mi/h will travel 2.7 m in the next tenth of a second. If this speed is maintained for 1.0 s, the car will travel 27 m, or if the same speed is maintained for 1.0 h, the car will travel 60 mi.

Now let us refine this concept by giving a precise definition. For any moving body, the ratio $\Delta s/\Delta t$, computed over shorter and shorter time intervals, eventually begins to approach a fixed numerical value, which we call the "limit" of the ratio. We define the instantaneous speed v to be this limiting value.

$$v = \lim_{\Delta t \to 0} \frac{\Delta s}{\Delta t} \qquad (1\text{-}4)$$

Read this equation: "v equals the limit as Δt approaches zero of $\Delta s/\Delta t$." The physical meaning of this definition is illustrated in the following example.

EXAMPLE 3 A Knee's Instantaneous Speed

(a) Estimate the maximum value of the instantaneous speed of the knee in Fig. 1-7. The time interval between adjacent images is 0.025 s. (b) Find the point at which the knee's speed is a minimum.

SOLUTION **(a)** The relatively wide spacing of the dots at the point marked t' in Fig. 1-7 tells us that the maximum speed occurs at approximately this point. The distances to the three adjacent points to the right of t' are given in the figure. We shall compute the ratio $\Delta s/\Delta t$ for each of the three intervals, beginning with the largest. The distance $\Delta s_1 = 23 \text{ cm} = 0.23 \text{ m}$ corresponds to a time interval of 3 times the flash interval of 0.025 s: $\Delta t_1 = 3(0.025 \text{ s}) = 0.075 \text{ s}$. Thus

$$\frac{\Delta s_1}{\Delta t_1} = \frac{0.23 \text{ m}}{0.075 \text{ s}} = 3.1 \text{ m/s}$$

The next smallest distance $\Delta s_2 = 16 \text{ cm} = 0.16 \text{ m}$ corresponds to twice the flash interval: $\Delta t_2 = 2(0.025 \text{ s}) = 0.050 \text{ s}$. We find

$$\frac{\Delta s_2}{\Delta t_2} = \frac{0.16 \text{ m}}{0.050 \text{ s}} = 3.2 \text{ m/s}$$

For the smallest interval, we find

$$\frac{\Delta s_3}{\Delta t_3} = \frac{0.080 \text{ m}}{0.025 \text{ s}} = 3.2 \text{ m/s}$$

To two significant figures, the last two values of $\Delta s/\Delta t$ are equal. If we were able to calculate the ratio $\Delta s/\Delta t$ for any value of Δt less than 0.025 s, we would get the same result. For example, if we could somehow accurately determine the distance traveled by the knee during the interval from t' to $t' + 0.010$ s, the ratio $\Delta s/\Delta t$ for this shorter time interval might be 3.19 m/s. But rounding off, we find this is again 3.2 m/s. No matter how small we make the time interval, the ratio will have this same value. Thus this is the **limiting value** of $\Delta s/\Delta t$, which is by definition the value of the instantaneous speed at t':

$$v(t') = 3.2 \text{ m/s}$$

If the motion at t' continued for 1.0 s, the knee would move a distance of 3.2 m; or if the motion continued at this same rate for 10 s, the knee would move 32 m. But of course the motion doesn't continue at a uniform rate. The fact that the dots get closer and closer together after time t' tells us that the knee is slowing down.

(b) The closeness of the dots at t'' tells us that the minimum speed occurs approximately at this point.

EXAMPLE 4 Estimating a Car's Instantaneous Speed

The wheels on a certain car are 70.0 cm in diameter. During a 0.150 s interval, they complete one revolution. (a) What is the car's average speed during this time interval in km/h and mi/h? (b) How is this value likely to compare with the instantaneous speed of the car during this time interval?

SOLUTION **(a)** As the wheel turns through one revolution, the car moves through a distance equal to the circumference of the wheel. The circumference of a circle of diameter d equals πd. Thus

$$\Delta s = \pi d = \pi(70.0 \text{ cm}) = 2.20 \times 10^2 \text{ cm} = 2.20 \text{ m}$$

Since $\Delta t = 0.150$ s, we have

$$\bar{v} = \frac{\Delta s}{\Delta t} = \left(\frac{2.20 \text{ m}}{0.150 \text{ s}}\right)\left(\frac{1 \text{ km}}{10^3 \text{ m}}\right)\left(\frac{3600 \text{ s}}{1 \text{ h}}\right) = 52.8 \text{ km/h}$$

or

$$\bar{v} = 52.8 \frac{\text{km}}{\text{h}}\left(\frac{1 \text{ mi}}{1.61 \text{ km}}\right) = 32.8 \text{ mi/h}$$

(b) This value represents the average of the instantaneous speed throughout the 0.150 s time interval. The question here is how much the instantaneous speed of the car is likely to vary over this time interval. Typically, cars never increase or decrease in speed by more than about 5 or 10 km/h in any 1 s interval (except during a collision). So during one revolution of the wheels (0.150 s), we can expect that the instantaneous speed will probably remain within about 1 km/h of the average speed. Thus the value we have computed is at least reasonably close to the limiting value of $\Delta s/\Delta t$, which is the instantaneous speed at the beginning of the time interval.

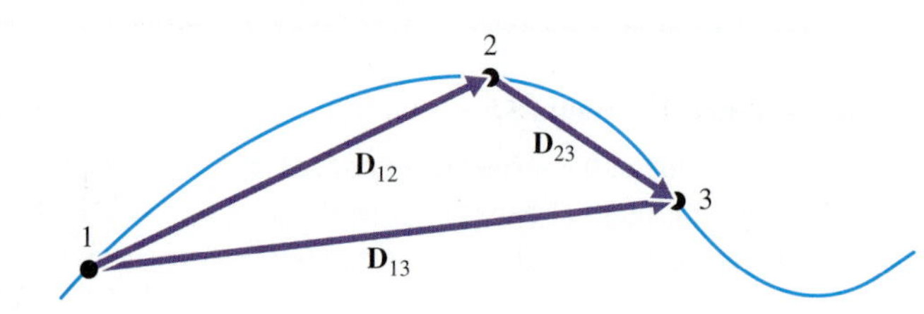

Fig. 1-8 Displacement $\mathbf{D}_{12}$ followed by displacement $\mathbf{D}_{23}$ results in a total displacement $\mathbf{D}_{13}$.

1-3 Displacement

The curved line in Fig. 1-8 represents the trajectory of a particle, with the particle's locations at three instants indicated by points 1, 2, and 3. The change in position of the particle from one point to another is called the **displacement** of the particle and is represented in the figure by an arrow connecting the two points. Fig. 1-8 shows three displacements: $\mathbf{D}_{12}$, $\mathbf{D}_{23}$, and $\mathbf{D}_{13}$. Notice that a displacement does not show the path taken by the particle; it gives only the change in position.

Addition of Two Displacements

Fig. 1-8 indicates a triangular relationship between $\mathbf{D}_{12}$, $\mathbf{D}_{23}$, and $\mathbf{D}_{13}$. This geometric relationship corresponds to the fact that displacement $\mathbf{D}_{12}$ followed by displacement $\mathbf{D}_{23}$ is equivalent to (in terms of initial and final positions) displacement $\mathbf{D}_{13}$. We shall represent this relationship by the equation

$$\mathbf{D}_{13} = \mathbf{D}_{12} + \mathbf{D}_{23} \tag{1-5}$$

We must emphasize, however, that **the + sign in this equation does not represent ordinary addition.** It represents the geometric relationship of the three displacements. Eq. 1-5 is just a shorthand for the triangular relationship shown in Fig. 1-8. Thus **the *length* of the displacement $\mathbf{D}_{13}$ is not equal to the sum of the lengths of $\mathbf{D}_{12}$ and $\mathbf{D}_{23}$** (except in the special case where $\mathbf{D}_{12}$ and $\mathbf{D}_{23}$ are parallel to each other and point in the same direction).

Addition of Three Displacements

This idea of displacement addition can be extended to a sequence of three or more displacements, as illustrated in Fig. 1-9.

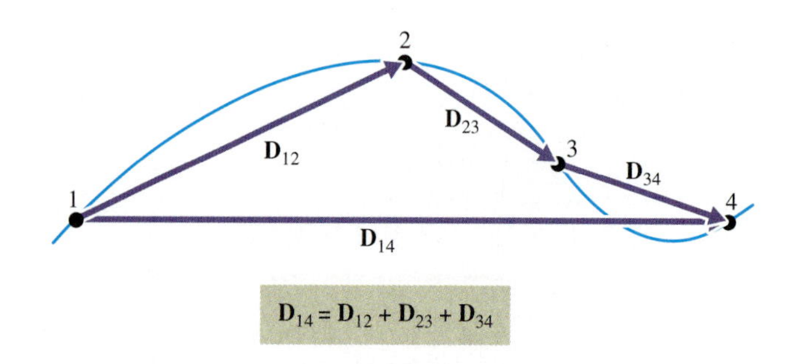

Fig. 1-9 Displacement $\mathbf{D}_{12}$ followed by displacements $\mathbf{D}_{23}$ and $\mathbf{D}_{34}$ results in a total displacement $\mathbf{D}_{14}$.

EXAMPLE 5 Displacement of a Golf Ball

A golfer takes three strokes to get the ball in the hole. The displacements, one for each stroke, are

$$\mathbf{D}_{12} = 200 \text{ m, north}$$

$$\mathbf{D}_{23} = 150 \text{ m, } 20° \text{ east of north}$$

$$\mathbf{D}_{34} = 20 \text{ m, east}$$

Find the single displacement that would have resulted in a hole-in-one.

SOLUTION Using a convenient scale, we draw the three displacements end to end, measuring the direction of $\mathbf{D}_{23}$ with a protractor (Fig. 1-10). We then find the single equivalent displacement $\mathbf{D}_{14}$ by drawing a straight line from the tail of $\mathbf{D}_{12}$ to the tip of $\mathbf{D}_{34}$ and measuring its direction and length (to scale).

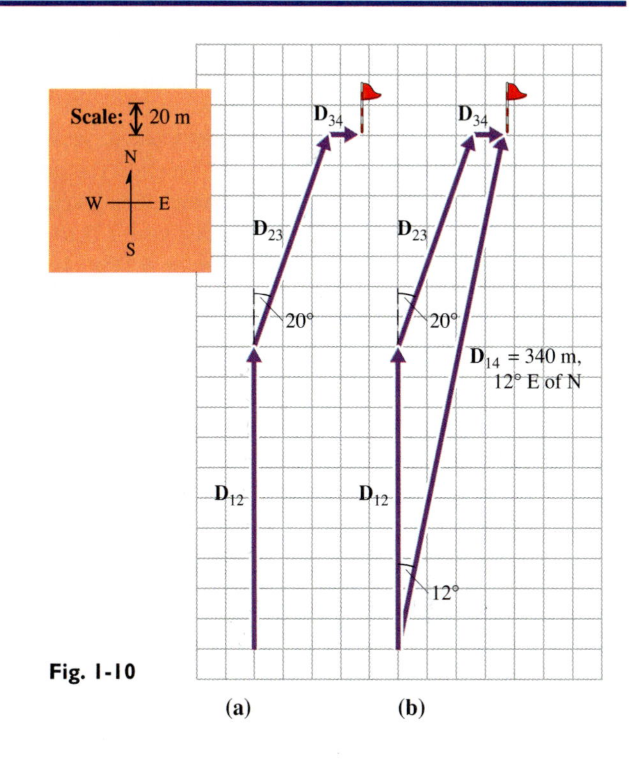

Fig. 1-10

1-4 Vector Algebra

Vectors and Scalars

Displacement is an example of a **vector, a physical quantity that has both magnitude and direction.** Examples of other vectors of importance in physics are velocity, acceleration, force, and electromagnetic fields. All these quantities have a direction as well as a magnitude. For example, the weight of an object is a force vector that is directed vertically downward. Quantities that have only magnitude are called **scalars.** Path length, speed, mass, and temperature are all examples of scalars. We shall denote vectors by boldface letters **A**, **B**, **C**, and so forth, and the magnitude of a vector **A** by $|\mathbf{A}|$ or simply by A. This section describes the algebra of vectors in a general way, in a discussion that is just an extension of what we did with displacements in the previous section.

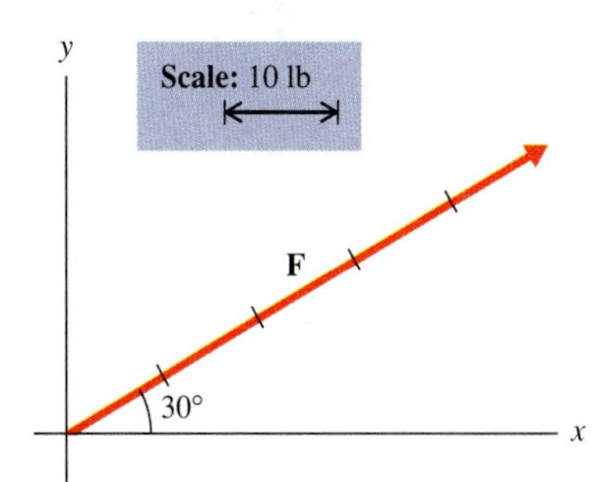

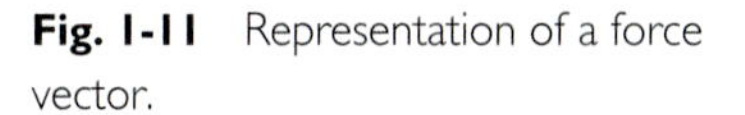

Fig. 1-11 Representation of a force vector.

Graphical Representation

We can represent any vector by drawing an arrow showing the direction of the vector. The arrow's length, drawn to some appropriate scale, indicates the magnitude of the vector. For example, suppose we wish to represent a 50 lb force* vector **F** that acts in a direction 30° above the x-axis. We can do this by first adopting a scale factor, say, 1.0 cm for every 10 lb, and then drawing a 5.0 cm arrow in a direction 30° above the x-axis (Fig. 1-11).

*Force is a concept that will be carefully defined in a later chapter. Here it is enough to think of a force as either a push or a pull in a specific direction. The magnitude of the push or pull may be measured in pounds.

Vector Addition

Vector addition is the mathematical operation of combining two vectors **A** and **B** to form a third "resultant" vector **R**, the "vector sum" of **A** and **B**. The vector sum represents the same kind of triangular relationship as in the addition of displacements, previously discussed.

To understand the significance of vector addition, we shall use the example of a particle being acted on by two forces (Fig. 1-12a). In adding the two vectors, we first draw vector **A** to scale and then draw vector **B**, also to scale, starting from the tip of vector **A** (Fig. 1-12b). The resultant **R** extends from the tail of **A** to the tip of **B**, as seen in the figure. The resultant force is important because the effect of the two forces **A** and **B** applied to the particle is found to be the same as that of the resultant force **R** (Fig. 1-12c).

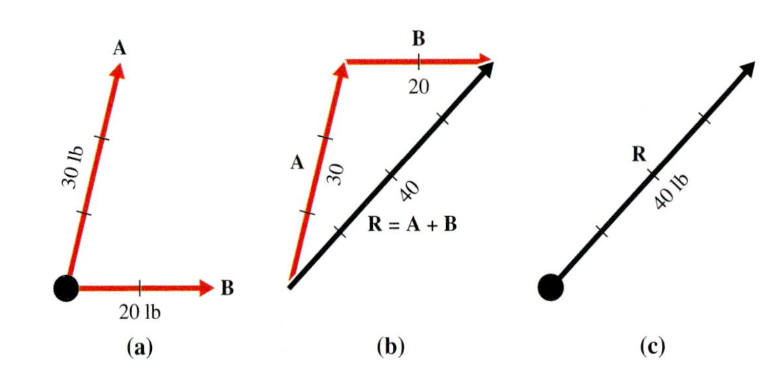

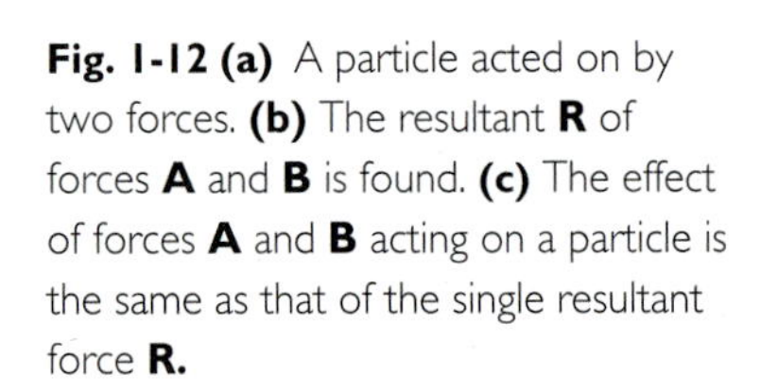
Fig. 1-12 (a) A particle acted on by two forces. **(b)** The resultant **R** of forces **A** and **B** is found. **(c)** The effect of forces **A** and **B** acting on a particle is the same as that of the single resultant force **R.**

The order of vector addition is unimportant, as shown in Fig. 1-13, where we first draw **B** and then draw **A**, starting from the tip of **B**. Thus

$$\mathbf{B} + \mathbf{A} = \mathbf{A} + \mathbf{B} \qquad (1\text{-}6)$$

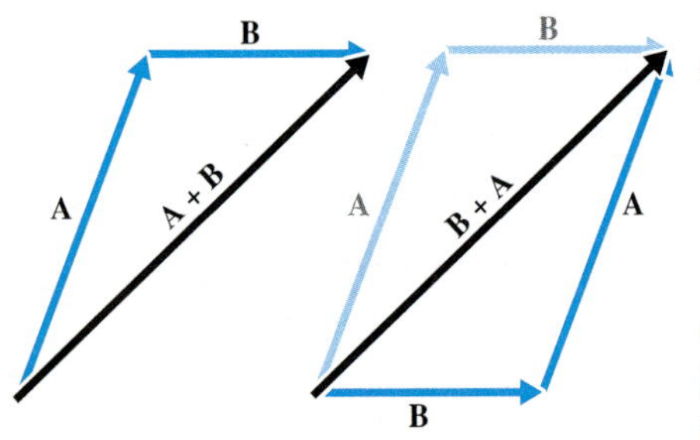

Fig. 1-13 The sum of two vectors **A** and **B** is the same, whether we draw **A** first and connect the tail of **B** to the tip of **A** (**A + B**) or draw **B** first and connect the tail of **A** to the tip of **B** (**B + A**).

The triangle formed by vector addition of two vectors implies that **the magnitude of the sum of two vectors is ordinarily less than the sum of the two magnitudes.** In fact, only if **A** and **B** point in the same direction will the magnitude of the resultant equal the sum of the two magnitudes (Fig. 1-14a). If **A** and **B** are oppositely directed, $|\mathbf{A} + \mathbf{B}|$ equals the difference in the lengths of the two vectors, as shown in Fig. 1-14b.

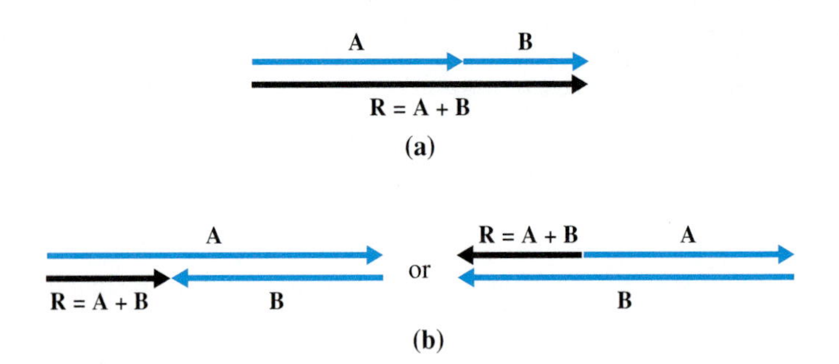

Fig. 1-14 (a) If vectors **A** and **B** are in the same direction, their vector sum has magnitude equal to the sum of the magnitudes of **A** and **B**. **(b)** If vectors **A** and **B** point in opposite directions, their vector sum has magnitude equal to the difference in the magnitudes of **A** and **B**.

EXAMPLE 6 Adding Two Vectors

A is a vector of magnitude 4 units pointing to the right. **B** is a vector of magnitude 3 units, whose direction we allow to vary. Find the magnitude of **A** + **B** if the angle between **A** and **B** is (a) 0°; (b) 90.0°; (c) 180°.

SOLUTION (a) An angle of 0° tells us that the two vectors point in exactly the same direction. Thus the magnitude of the resultant is merely the sum of the magnitudes of the two original vectors (Fig. 1-15a).

(b) Here we must use the Pythagorean theorem to determine the magnitude of **A** + **B** because this resultant vector is the hypotenuse of the right triangle formed by the two original vectors (Fig. 1-15b).

(c) The 180° angle tells us that the two vectors point in opposite directions, and so the magnitude of the resultant is the difference in length between them (Fig. 1-15c).

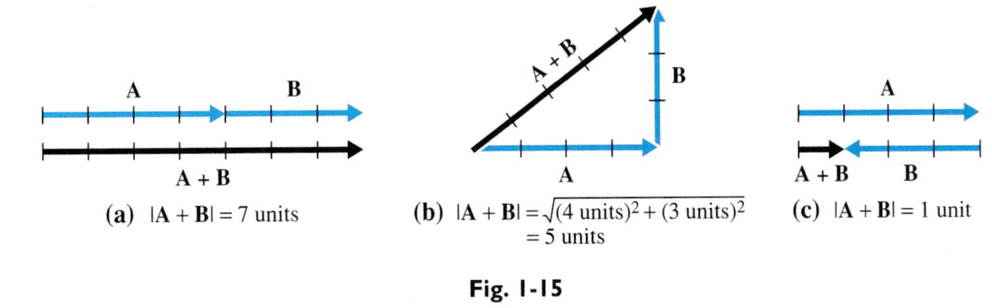

Fig. 1-15

Polygon Rule

To add three or more vectors, simply follow the rule for addition of two vectors and add the vectors to each other one by one. Fig. 1-16a shows this procedure, called the "polygon method": Place the vectors to be added tail to tip, preserving their length and direction. The sum, or resultant, is found by connecting the tail of the first arrow with the tip of the last arrow. Addition of vectors **A**, **B**, and **C** is represented by the vector equation

$$\mathbf{R} = \mathbf{A} + \mathbf{B} + \mathbf{C}$$

Fig. 1-16b uses three vectors to illustrate that the order of addition is unimportant, so that we may also write

$$\mathbf{R} = \mathbf{C} + \mathbf{A} + \mathbf{B}$$

or any other order of addition.

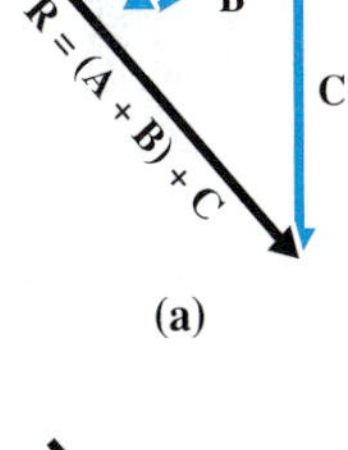

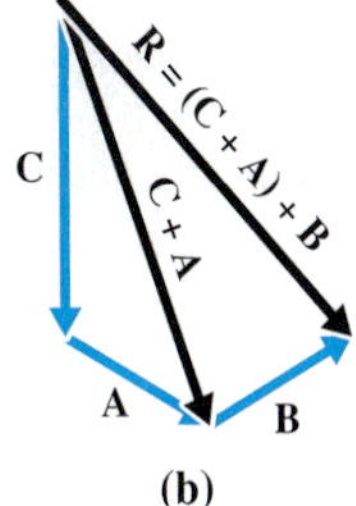

Fig. 1-16 The same resultant **R** is obtained by performing vector sums **(a) A + B + C** and **(b) C + A + B**.

Vector Subtraction

Fig. 1-17 Vector $-\mathbf{A}$ is directed opposite $\mathbf{A}$ but has the same magnitude as $\mathbf{A}$.

The negative of a vector A, is defined to be a vector of the same length as A, pointing in the opposite direction (Fig. 1-17). With this definition, we define **vector subtraction as addition of a negative vector:**

$$\mathbf{A} - \mathbf{B} = \mathbf{A} + (-\mathbf{B}) \tag{1-7}$$

Vector subtraction is illustrated in Fig. 1-18a. It follows from the definition of the negative vector that $\mathbf{A} - \mathbf{A} = \mathbf{0}$, as shown in Fig. 1-18b.

The manipulation of vector addition and subtraction is similar to ordinary algebra. Suppose that $\mathbf{R} = \mathbf{A} + \mathbf{B}$, where $\mathbf{R}$ and $\mathbf{A}$ are two vectors that are known, and you wish to find the vector $\mathbf{B}$. You can use vector subtraction to isolate the unknown vector $\mathbf{B}$:

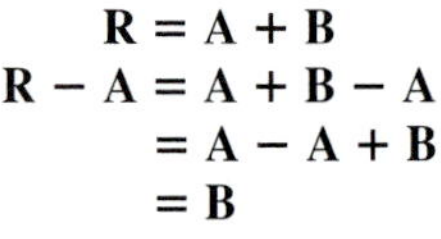

$$\begin{aligned} \mathbf{R} &= \mathbf{A} + \mathbf{B} \\ \mathbf{R} - \mathbf{A} &= \mathbf{A} + \mathbf{B} - \mathbf{A} \\ &= \mathbf{A} - \mathbf{A} + \mathbf{B} \\ &= \mathbf{B} \end{aligned}$$

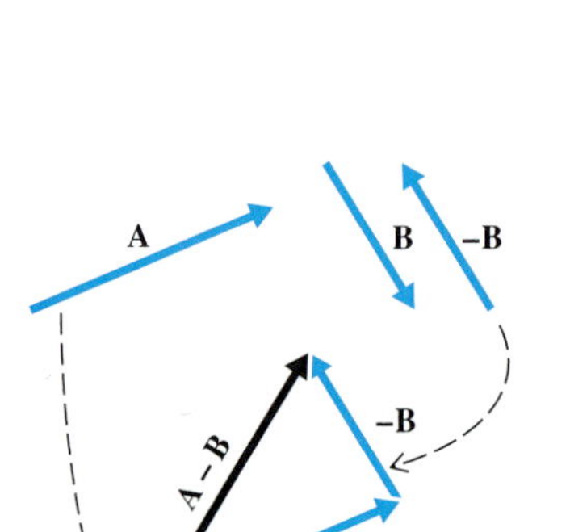

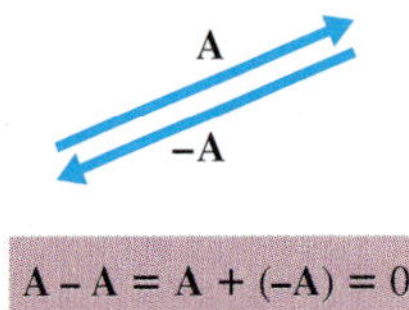

Fig. 1-18 Vector subtraction. **(a) A − B**; **(b) A − A = 0**.

EXAMPLE 7 Final Leg of a Trip

A person wishes to end a trip 50 km directly north of the starting point. The first leg of the trip results in a displacement of 20 km, 30° west of north. What final displacement is necessary?

SOLUTION We shall label the first displacement **A**, the second **B**, and the resultant **R**. Both **A** and **R** are given. The problem is to find **B**.

$$\mathbf{R} = 50\text{ km, north}$$

$$\mathbf{A} = 20\text{ km, }30^\circ\text{ west of north}$$

$$\mathbf{B} = ?$$

$$\mathbf{R} = \mathbf{A} + \mathbf{B}$$

Solving for **B**, we obtain

$$\mathbf{B} = \mathbf{R} - \mathbf{A}$$

The vector subtraction is carried out in Fig. 1-19a, where we find that **B** is directed 17° east of north and has a magnitude of 34 km.

Another way to solve this problem is to draw **A** and **R** from a common origin. Then **B** is constructed as the vector that must be added to **A** to give the resultant **R** (Fig. 1-19b).

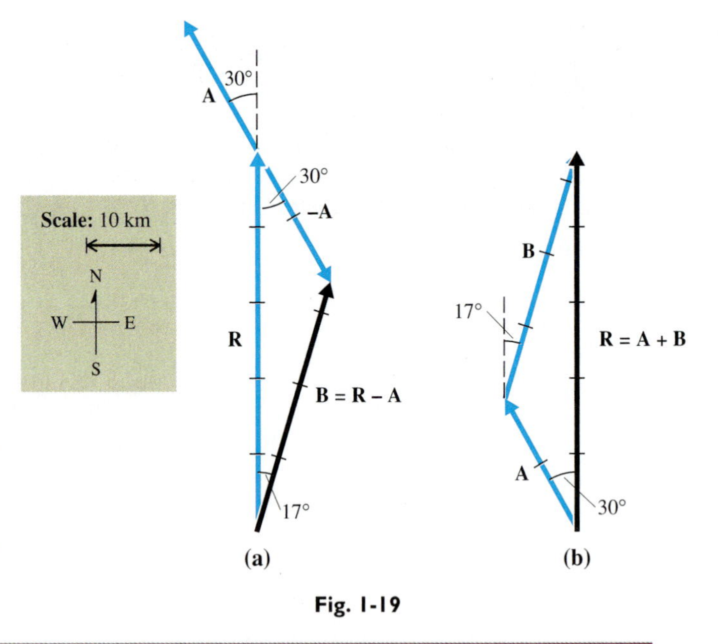

Fig. 1-19

Multiplication of a Vector by a Scalar

Multiplication of a vector by a positive scalar is defined to be a stretching (or shrinking) of the vector by the scalar factor. For example, the vector 2.5**A** is a vector parallel to **A** and 2.5 times as long (Fig. 1-20a). The vector 0.3**A** is a vector parallel to **A** and 0.3 times as long (Fig. 1-20b).

Multiplication of a vector by a negative scalar is defined as reversal of the vector's direction, as well as stretching of its length by the magnitude of the scalar (Fig. 1-20c).

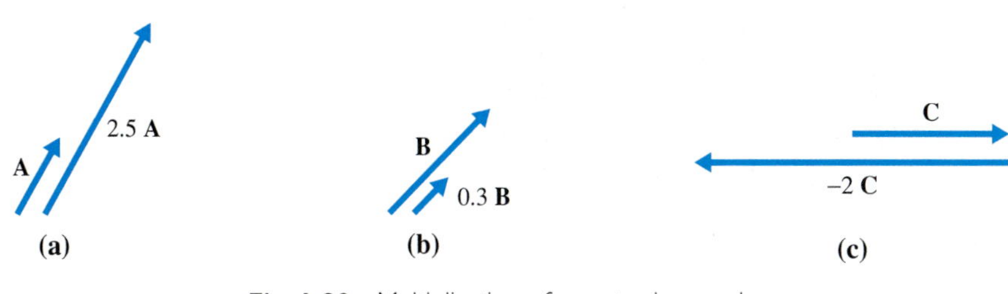

Fig. 1-20 Multiplication of a vector by a scalar.

1-5 Components of Vectors

If you draw a vector from the origin of a coordinate system, both its length and its direction are determined by specifying the coordinates of the point at the tip of the vector; these coordinates are called the **components** of the vector. The components A_x and A_y of a vector **A** are shown in Fig. 1-21. The x component of a vector is positive if the vector points to the right of the origin and negative if it points to the left of the origin. The y component of a vector is positive if the vector points above the origin and negative if it points below.

A vector may be specified either by its components or by its length and direction (as measured by an angle relative to a reference direction). Trigonometric functions are used to relate the two descriptions. See Appendix A for a review of trigonometry.

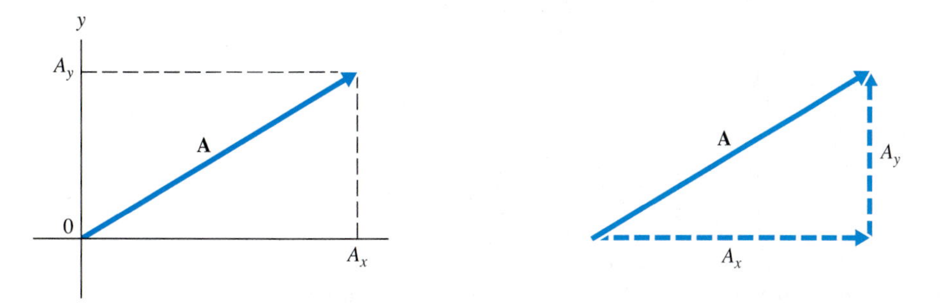

Fig. 1-21 Components of a vector.

EXAMPLE 8 Finding Components of Vectors

Find the x and y components of the vectors shown in Fig. 1-22.

SOLUTION The vector **A** is directed upward and to the right and so has positive x and y components. We can find the components by considering the right triangle formed by the vector and its components (Fig. 1-23a). Using the definitions of the cosine ($\cos\theta$ = adjacent side/hypotenuse) and the sine ($\sin\theta$ = opposite side/hypotenuse), we obtain

$$A_x = +A\cos 40.0° = +5.00(\cos 40.0°) = 3.83$$

$$A_y = +A\sin 40.0° = +5.00(\sin 40.0°) = 3.21$$

Components of **B** and **C** are found in a similar fashion. The direction of **B** indicates that it has a negative x component and a positive y component (Fig 1-23b), whereas **C** has a positive x component and a negative y component (Fig. 1-23c):

$$B_x = -B\sin 30.0° = -7.00(\sin 30.0°) = -3.50$$

$$B_y = +B\cos 30.0° = 7.00(\cos 30.0°) = 6.06$$

$$C_x = +C\sin 60.0° = 10.0(\sin 60.0°) = 8.66$$

$$C_y = -C\cos 60.0° = -10.0(\cos 60.0°) = -5.00$$

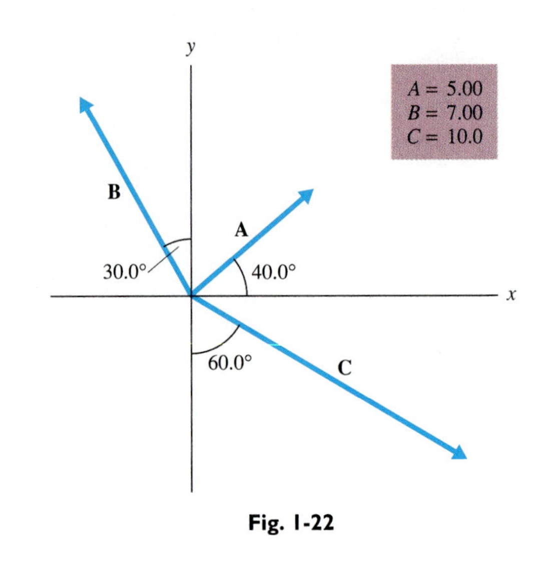

Fig. 1-22

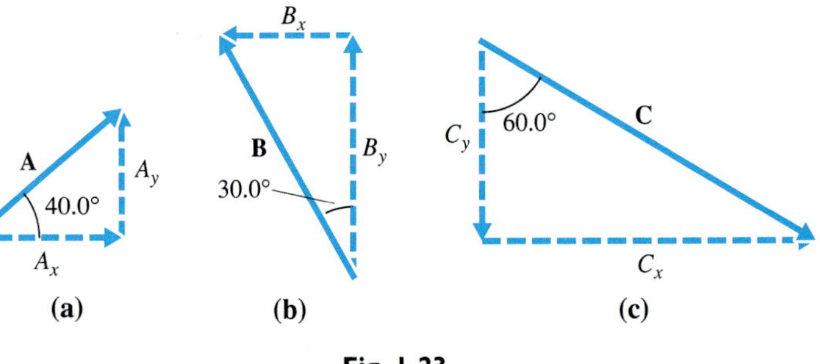

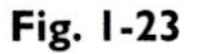
Fig. 1-23

EXAMPLE 9 Finding a Vector's Magnitude and Direction

A certain vector **A** has components $A_x = 5.00$, $A_y = -3.00$. Find the magnitude and direction of **A**.

SOLUTION The vector **A** is shown in Fig. 1-24. Its magnitude is found by applying the Pythagorean theorem:

$$A = \sqrt{A_x^2 + A_y^2} = \sqrt{5.00^2 + 3.00^2} = 5.83$$

From the definition of the tangent, we find that

$$\tan\theta = \frac{3.00}{5.00}$$

or

$$\theta = \arctan\frac{3.00}{5.00} = 31.0°$$

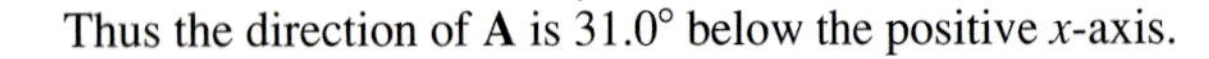
Thus the direction of **A** is 31.0° below the positive x-axis.

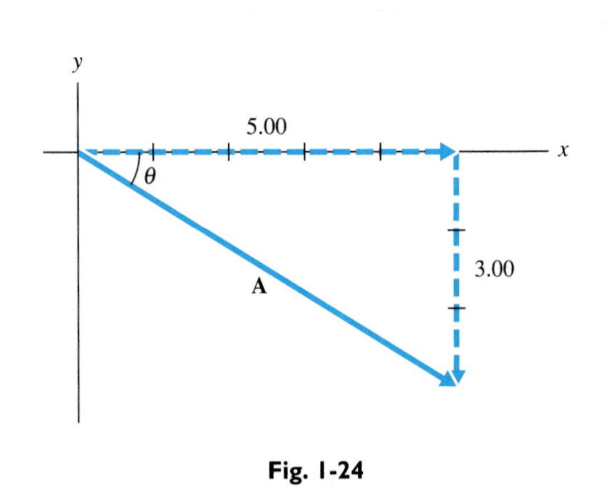

Fig. 1-24

Addition of Components

It is possible to add vectors nongraphically by *adding their components*. Consider two vectors **A** and **B** and their resultant **R** (Fig. 1-25). From the figure we see that

$$R_x = A_x + B_x$$
$$R_y = A_y + B_y \tag{1-8}$$

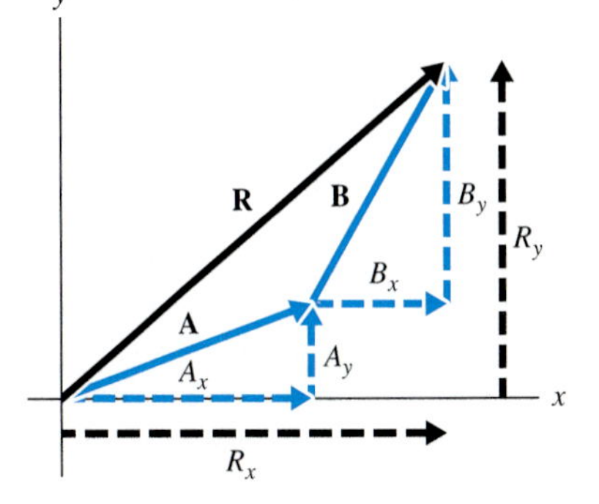

Fig. 1-25 The x component of a resultant vector **R** = **A** + **B** equals the sum of the x components of **A** and **B**, and the y component of **R** equals the sum of the y components of **A** and **B**.

Once you know the components of **R**, you can find its magnitude and direction.

For simplicity in Fig. 1-25, we have used two vectors having positive x and y components. The method is valid in general, however, even when the vectors have negative components, as in the following example.

EXAMPLE 10 Adding Two Vectors Using Components

Find the vector sum (resultant) of the two vectors shown in Fig. 1-26a.

SOLUTION From Eq. 1-8,

$$R_x = A_x + B_x = 3.00 + 2.00 = 5.00$$
$$R_y = A_y + B_y = 4.00 - 6.00 = -2.00$$

Since we need to determine the magnitude and direction of **R**, it is a good idea to first sketch the vector (Fig. 1-26b). **R** has magnitude

$$\sqrt{5.00^2 + 2.00^2} = 5.39$$

and $$\theta = \arctan \frac{2.00}{5.00} = 21.8°$$

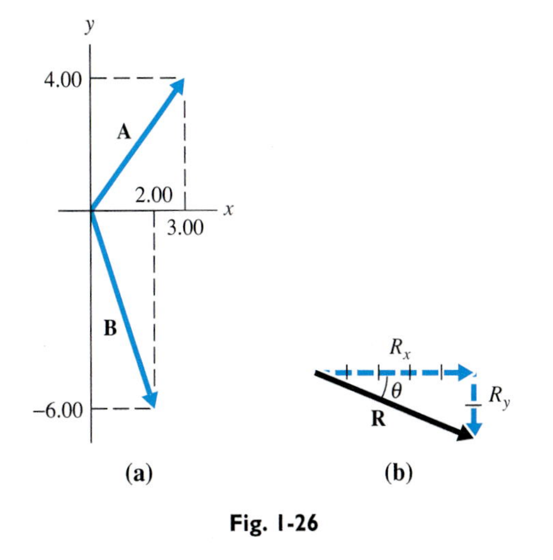

Fig. 1-26

This method is easily extended to addition of three or more vectors.

If $$\mathbf{R} = \mathbf{A} + \mathbf{B} + \mathbf{C} + \cdots$$

then

$$R_x = A_x + B_x + C_x + \cdots$$
$$R_y = A_y + B_y + C_y + \cdots \tag{1-9}$$

EXAMPLE 11 Adding Four Vectors

The force vectors in Fig. 1-27a have magnitudes $F_1 = 15.0$ lb, $F_2 = 20.0$ lb, $F_3 = 10.0$ lb, and $F_4 = 25.0$ lb. Find the resultant force.

SOLUTION The components of a vector depend on both the vector and the coordinate system chosen. We are free to choose any coordinate system. Here, to make our task as simple as possible, we choose a coordinate system with axes oriented so that two of the vectors ($\mathbf{F}_3$ and $\mathbf{F}_4$) lie along them (Fig. 1-27b). Then, from Eq. 1-9,

$$\begin{aligned} R_x &= F_{1x} + F_{2x} + F_{3x} + F_{4x} \\ &= (15.0 \text{ lb})(\cos 25.0°) + (20.0 \text{ lb})(\cos 45.0°) + 0 - 25.0 \text{ lb} \\ &= 2.74 \text{ lb} \end{aligned}$$

$$\begin{aligned} R_y &= F_{1y} + F_{2y} + F_{3y} + F_{4y} \\ &= (15.0 \text{ lb})(\sin 25.0°) + (20.0 \text{ lb})(\sin 45.0°) + 10.0 \text{ lb} + 0 \\ &= 30.5 \text{ lb} \end{aligned}$$

The resultant force **R** has magnitude

$$\begin{aligned} R &= \sqrt{R_x^2 + R_y^2} = \sqrt{(2.74 \text{ lb})^2 + (30.5 \text{ lb})^2} \\ &= 30.6 \text{ lb} \end{aligned}$$

The resultant makes an angle θ with the x-axis (Fig. 1-27c):

$$\theta = \arctan\left(\frac{30.5}{2.74}\right) = 84.9°$$

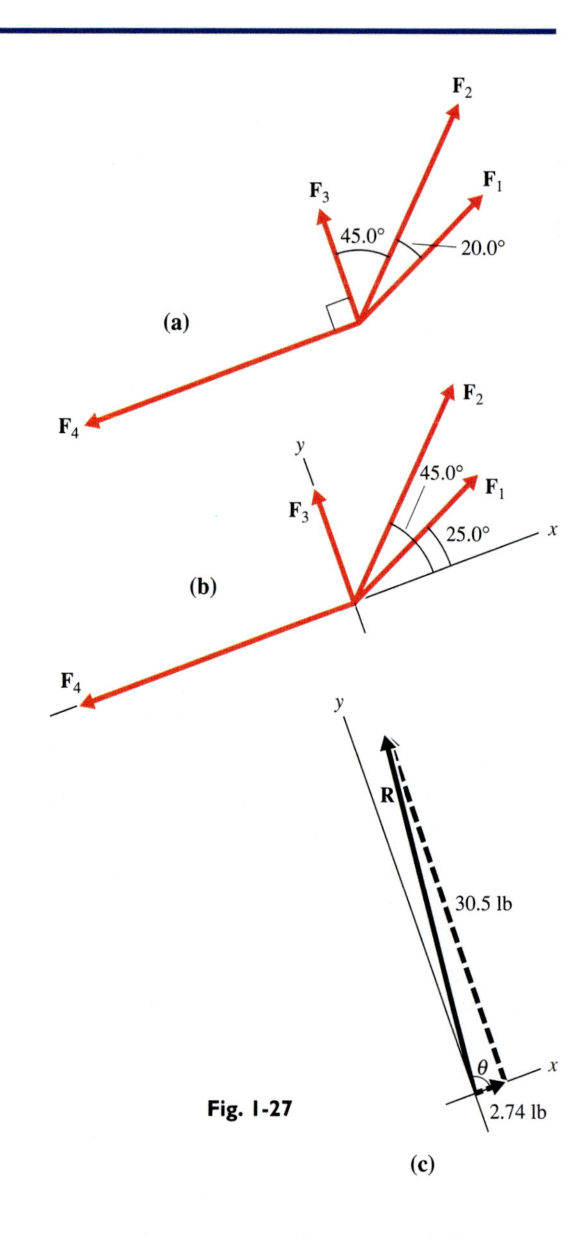

Fig. 1-27

The component method may also be applied to vector subtraction Let

$$\mathbf{R} = \mathbf{A} - \mathbf{B} = \mathbf{A} + (-\mathbf{B})$$

The x component of $-\mathbf{B}$ is $-B_x$, and the y component of $-\mathbf{B}$ is $-B_y$. Thus

$$\begin{aligned} R_x &= A_x + (-B_x) \\ &= A_x - B_x \end{aligned}$$

and

$$R_y = A_y - B_y$$

1-6 Velocity

Though we use the words "speed" and "velocity" interchangeably in everyday language, their precise meanings in physics are significantly different. **Speed is a scalar quantity measuring the rate of motion, but velocity is defined to be a vector quantity that gives the direction of motion as well as its rate.** For example, suppose a car is moving east at 50 km/h. The car's speed is 50 km/h, a scalar quantity. The car's velocity, on the other hand, is a vector, directed east and having a magnitude of 50 km/h.

Average Velocity

A body's average velocity $\overline{\mathbf{v}}$ over a time interval Δt is defined as its displacement **D** divided by Δt.

$$\overline{\mathbf{v}} = \frac{\mathbf{D}}{\Delta t} \qquad (1\text{-}10)$$

This definition means that average velocity is a vector of magnitude $\frac{|\mathbf{D}|}{\Delta t}$, pointing in the direction of the displacement **D**.

Fig. 1-28 shows the average velocity of a body over a certain interval Δt. Notice that the velocity vector $\overline{\mathbf{v}}$ is parallel to the displacement **D**. Fig. 1-28 also shows the components of both vectors. If we take the triangle formed by the displacement vector and its components (Δx and Δy) and divide each side of the triangle by the time interval Δt, we obtain the triangle formed by the velocity vector $\overline{\mathbf{v}}$ and its components ($\overline{v}_x$ and $\overline{v}_y$). Each velocity component ($\overline{v}_x$ or $\overline{v}_y$) equals the change in the corresponding coordinate (Δx or Δy), divided by Δt

$$\overline{v}_x = \frac{\Delta x}{\Delta t} \qquad (1\text{-}11)$$

$$\overline{v}_y = \frac{\Delta y}{\Delta t} \qquad (1\text{-}12)$$

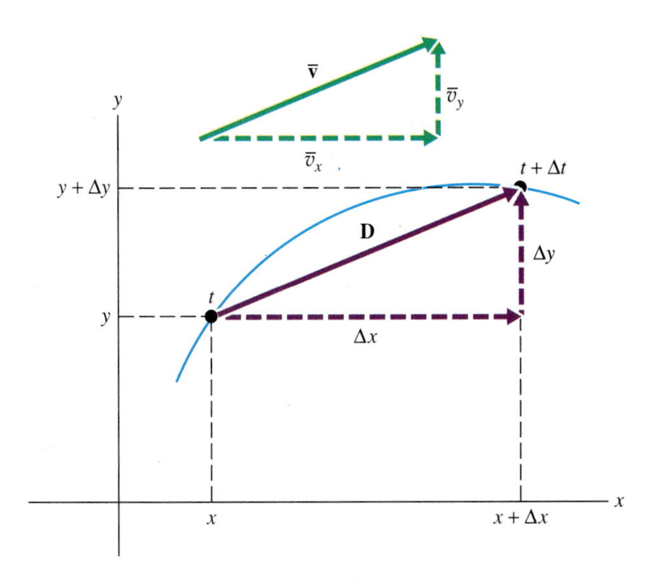

Fig. 1-28 The displacement and average velocity of a particle over a time interval Δt.

EXAMPLE 12 A Basketball's Average Velocity

A basketball rises 1.0 m and moves 4.0 m horizontally in its 0.80 s flight from a player's hands into a basket. Find the ball's average velocity.

SOLUTION Fig 1-29a shows the ball's trajectory. We find the ball's average velocity vector $\overline{\mathbf{v}}$ by first computing this vector's components $\overline{v}_x$ and $\overline{v}_y$, using Eqs. 1-11 and 1-12.

$$\overline{v}_x = \frac{\Delta x}{\Delta t} = \frac{4.0 \text{ m}}{0.80 \text{ s}} = 5.0 \text{ m/s}$$

$$\overline{v}_y = \frac{\Delta y}{\Delta t} = \frac{1.0 \text{ m}}{0.80 \text{ s}} = 1.3 \text{ m/s}$$

These components are shown in Fig. 1-29b, along with the vector $\overline{\mathbf{v}}$, whose magnitude $|\overline{\mathbf{v}}|$ and direction θ are found in the same way as for any other vector.

$$\begin{aligned} |\overline{\mathbf{v}}| &= \sqrt{\overline{v}_x^{\,2} + \overline{v}_y^{\,2}} \\ &= \sqrt{(5.0 \text{ m/s})^2 + (1.3 \text{ m/s})^2} \\ &= 5.2 \text{ m/s} \end{aligned}$$

$$\theta = \arctan\left(\frac{1.3 \text{ m/s}}{5.0 \text{ m/s}}\right) = 15^\circ$$

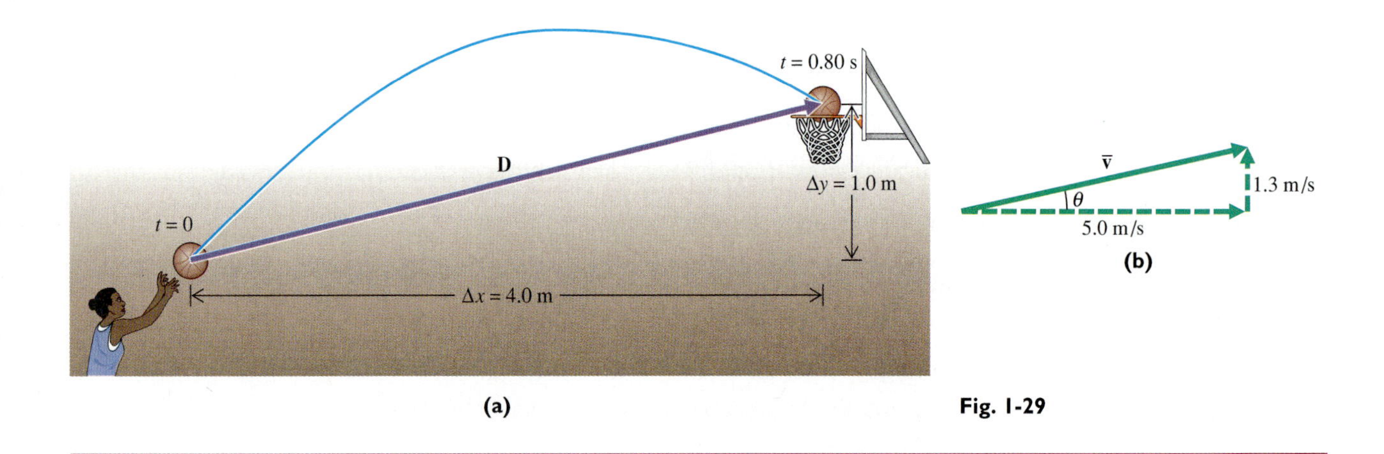

Fig. 1-29

Instantaneous Velocity

Instantaneous velocity is the name we give to **the vector that the average velocity approaches as** $\Delta t \to 0$. We shall see that the instantaneous velocity, denoted by **v**, is a vector **tangent to the trajectory** and having **magnitude equal to the instantaneous speed** v.* Instantaneous velocity gives both the direction of motion and the speed at an instant. Fig. 1-30 shows a car rounding a curve. The car's instantaneous velocity **v** continually changes, both in magnitude and direction.

*We anticipated this result when we used the symbol v for the instantaneous speed in section 1-2.

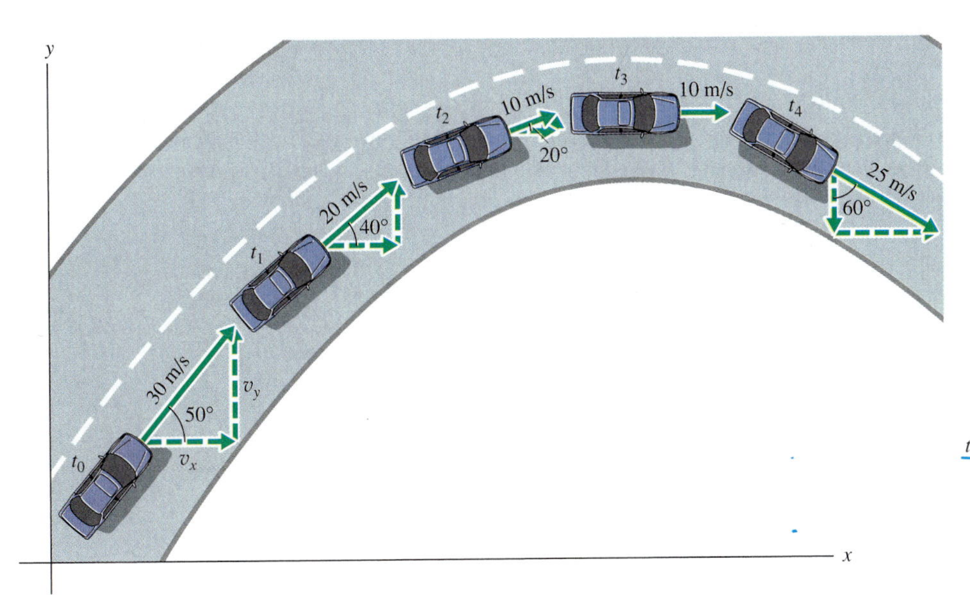

Fig. 1-30 A car rounds a curve. The car's instantaneous velocity is shown at different times.

When we calculate a body's average velocity over a very short time interval, we find that the velocity's magnitude is (very nearly) the body's instantaneous speed. Fig. 1-31 shows why this is so. In this figure average velocity is computed over diminishing time intervals Δt, starting with a fixed time t. We see that as Δt approaches 0, the displacement magnitude $|\mathbf{D}|$ approaches the path length Δs.

$$|\mathbf{D}| \to \Delta s \quad \text{as} \quad \Delta t \to 0$$

Dividing by Δt, we find that the average velocity magnitude $|\mathbf{D}|/\Delta t$ approaches the average speed.

$$\frac{|\mathbf{D}|}{\Delta t} \to \frac{\Delta s}{\Delta t} \quad \text{as} \quad \Delta t \to 0$$

But as $\Delta t \to 0$, $\dfrac{\Delta s}{\Delta t}$ becomes the instantaneous speed, v. So the expression above says that the average velocity vector's magnitude approaches the instantaneous speed as $\Delta t \to 0$. Fig. 1-31 also shows how, as $\Delta t \to 0$, the direction of the average velocity vector approaches the direction of the tangent to the trajectory.

Components of the instantaneous velocity vector, v_x and v_y, are shown in Fig. 1-30. These components are the values approached by the average velocity components as Δt approaches 0.

$$\bar{v}_x = \frac{\Delta x}{\Delta t} \to v_x \quad \text{as} \quad \Delta t \to 0 \tag{1-13}$$

$$\bar{v}_y = \frac{\Delta y}{\Delta t} \to v_y \quad \text{as} \quad \Delta t \to 0 \tag{1-14}$$

The component v_x measures motion along the x-axis, the rate of change of the x coordinate, whereas the component v_y measures motion along the y-axis, the rate of change of the y coordinate. Since the instantaneous speed v is the magnitude of the velocity vector, v is related to the vector's components by the Pythagorean theorem:

$$v = \sqrt{v_x^2 + v_y^2} \tag{1-15}$$

(a)

(b)

(c)

Tangent to trajectory

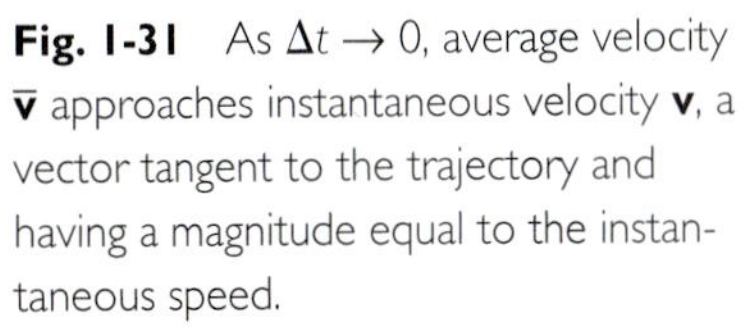

Fig. 1-31 As $\Delta t \to 0$, average velocity $\bar{\mathbf{v}}$ approaches instantaneous velocity $\mathbf{v}$, a vector tangent to the trajectory and having a magnitude equal to the instantaneous speed.

EXAMPLE 13 Estimating Where a Car Will Move

(a) In Fig. 1-30 find the car's velocity components v_x and v_y at time t_0. (b) About how far does the car travel in the x direction during the 0.10 s time interval after time t_0?

SOLUTION (a) From Fig. 1-30, we see that at time t_0, the instantaneous velocity vector has magnitude $v = 30$ m/s and makes an angle of 50° with the x-axis. Thus

$$v_x = (30 \text{ m/s})(\cos 50°) = 19 \text{ m/s}$$

$$v_y = (30 \text{ m/s})(\sin 50°) = 23 \text{ m/s}$$

(b) Since 0.10 s is a short time interval, it is reasonable to use v_x to estimate the distance Δx traveled over this time interval.

$$v_x \approx \frac{\Delta x}{\Delta t}$$

$$\Delta x = v_x \, \Delta t = (19 \text{ m/s})\,(0.10 \text{ s}) = 1.9 \text{ m}$$

CHAPTER 1 SUMMARY

A particle's position at any time t is specified by its Cartesian coordinates x, y, and z at that instant. If a particle is confined to a single plane, coordinates x and y suffice to specify position.

A particle that moves a distance Δs in a time interval Δt has an average speed $\bar{v}$ defined as the ratio $\Delta s/\Delta t$.

$$\bar{v} = \frac{\Delta s}{\Delta t}$$

A particle's instantaneous speed v at time t is the limiting value approached by the particle's average speed, computed over smaller and smaller time intervals, with the initial time in each computation held fixed at t.

$$v = \lim_{\Delta t \to 0} \frac{\Delta s}{\Delta t}$$

In the special case of uniform motion, a particle moves equal distances in equal time intervals, and so the ratio $\Delta s/\Delta t$ has the same value for any time interval. In this case, average speed is the same as instantaneous speed.

$$\bar{v} = v = \frac{\Delta s}{\Delta t} \qquad \text{(uniform motion)}$$

Given two vectors **A** and **B**, the vector sum **A** + **B** is found as shown in Fig. 1-12. Several vectors may be added in any order without changing the result. The vector difference **A** − **B** is found when one adds the vector −**B** to **A**, as shown in Fig. 1-18a. The magnitude of the sum of two vectors is not ordinarily equal to the sum of the magnitudes of the two vectors:

$$|\mathbf{A} + \mathbf{B}| \neq A + B \qquad \text{(unless } \mathbf{A} \text{ and } \mathbf{B} \text{ are in the same direction)}$$

A vector **A** may be expressed in terms of its components. For a vector in the xy plane, the vector's magnitude A is related to the components A_x and A_y by the equation

$$A = \sqrt{A_x^2 + A_y^2}$$

Addition and subtraction of vectors may be done by addition and subtraction of the respective components:

$$\mathbf{R} = \mathbf{A} + \mathbf{B} - \mathbf{C}$$

$$R_x = A_x + B_x - C_x$$

$$R_y = A_y + B_y - C_y$$

A particle's displacement **D** is a vector indicating the particle's change in position over some time interval Δt. The vector is an arrow drawn from the particle's initial position to its final position.

Average velocity $\bar{\mathbf{v}}$ is defined as displacement **D** divided by the time interval Δt over which the displacement occurred.

$$\bar{\mathbf{v}} = \frac{\mathbf{D}}{\Delta t}$$

and has components

$$\bar{v}_x = \frac{\Delta x}{\Delta t} \quad \text{and} \quad \bar{v}_y = \frac{\Delta y}{\Delta t}$$

Instantaneous velocity is the vector the average velocity approaches as the time interval approaches zero. The magnitude of the instantaneous velocity vector is instantaneous speed:

$$|\mathbf{v}| = v = \lim_{\Delta t \to 0} \frac{\Delta s}{\Delta t}$$

Questions

1 In Fig. 1-7 what is the approximate orientation of the lower leg when the ankle moves at minimum speed?

2 You make a trip from your home to the grocery store and then return home. What is your resultant displacement?

3 What is the direction of the displacement of the tip of an hour hand on a wall clock, as the time changes from 12:00 to 6:00?

4 Show how to add three vectors of equal magnitude and get a resultant of zero.

5 Which vector sum is different from the other three: (a) **A** − **B** + **C**; (b) −**B** + **C** + **A**; (c) **A** + **B** + **C**; (d) **C** + **A** − **B**.

6 Is it possible to add two vectors of unequal magnitude and get a resultant of zero?

7 If a particle has constant velocity, does it necessarily have constant speed?

8 If a particle has constant speed, does it necessarily have constant velocity?

9 If motion of a particle is along the x-axis, which statement is correct: (a) v_x is constant; (b) $v_y = v_z = 0$; (c) $v_x = 0$.

10 Is the magnitude of the average velocity vector always equal to the average speed?

11 For which of the times in Fig. 1-30 is the velocity component v_y greatest?

12 In Fig. 1-7 find the approximate point where the ankle's upward velocity component is greatest.

13 For which of the times in Fig. 1-30 is the velocity component v_y negative?

Answers to odd-numbered questions

1 vertical; **3** straight down; **5** (c); **7** yes; **9** (b); **11** t_0; **13** t_4

Problem-Solving Strategy

There is no single, simple approach to all physics problems, but there are a few useful general guidelines. First, try to **formulate the question** clearly. Often it helps to draw a sketch, so that you have a clear idea of what is happening. It is also useful to list all the relevant information given in the problem. Second, **determine the principles that apply** to the physical situation and will be useful in solving the problem. At this stage, you may find that you need to break the problem into smaller parts, which you can solve separately. As a third and final step, apply the principles and **work through the mathematics** to a quantitative solution. Very often the first two steps, rather than the third, cause the greatest difficulty. Solving physics problems is very much like solving "word problems" in elementary mathematics. Sometimes it is difficult to know where to start, but an important bonus derived from learning how to solve these problems is that you will develop a more analytical way of thinking, a skill that can be useful in areas outside of physics.

Students often remark that they can follow the instructor's solution perfectly but are totally lost when they try to solve a problem on their own. It is like being led through a maze. There are many wrong turns and dead ends, which lead nowhere. If you have a guide through the maze, it is easy to follow and you avoid all the wrong turns, but without that guide, your progress may be random and undirected. Learning physics is like learning to get through the maze without assistance. Once you have an understanding of physics, you will be like someone who can stand above the maze and see which path will lead to your goal. You will begin to recognize the dead-end paths without going down every one of them. Remember that this course is a learning experience. Don't be afraid to make a wrong turn. Just be sure to question it, so that you understand why it was wrong.

Problems (listed by sections)

1-2 Speed

Uniform motion

1 A cyclist moves at a constant speed of 15 mi/h. Express the speed in units of ft/s.

2 How long is required to travel 300 km along a freeway at a constant speed of (a) 50.0 km/h; (b) 60.0 km/h?

3 Light travels at a constant speed of 3.00×10^8 m/s, whereas sound travels through the air at a constant speed of 340 m/s.
(a) How long does it take for light to travel from a lightning stroke to an observer 1.00 km away?
(b) How long after the flash is seen is the thunder produced by the lightning heard?

4 The driver of a car is initially moving at a constant speed of 72.0 km/h when a traffic light turns red. If 0.500 s reaction time is required before the brakes can be applied, what is the distance in meters traveled by the car before it begins to slow down?

5 A person on earth communicating with an astronaut on the moon asks a question. How long must the person on earth wait for a response if the astronaut answers 1.00 s after the message is received? The moon is 3.84×10^5 km from the earth, and the speed of radio waves is 3.00×10^8 m/s.

★6 External stimuli are communicated to the brain by means of electrical signals propagating along nerve cells at a speed of approximately 30 m/s. Similarly, electrical messages are sent at the same speed from the brain along nerve cells to the muscles. Reflex actions are controlled by a relatively simple nerve circuit from a muscle to the spine and back to the muscle. Estimate the reflex time for a stimulus at the knee.

7 The world record in the 100 yard dash is 9.1 s. Estimate the world record in the 100 m dash.

Average speed

8 Compute the average speed in m/s of a runner who completes a mile in 4.00 min.

9 Blood circulating from the heart to the hands and back to the heart travels a distance of about 2.0 m in 40 s. Find the average speed of the blood.

10 You drive from Los Angeles to Portland, a distance of 1.6×10^3 km, in 30 h, including stops. What steady speed would have allowed you to arrive at the same time if you had driven at this speed nonstop?

11 During 5 successive 1.00 min intervals, a runner moves at the following constant speeds: 0.400 km/min, 0.240 km/min, 0.160 km/min, 0.160 km/min, and 0.320 km/min. Compute the total distance traveled and the average speed.

★12 A tortoise and a hare race over a course 1.00 km long. The tortoise moves at a constant speed of 2.00 m/s. The hare moves at a speed of 10.0 m/s for 60.0 s, rests for 10.0 min, and continues at 10.0 m/s for 40.0 s.
(a) Sketch s versus t for both the tortoise and the hare on the same graph.
(b) Who wins the race?
(c) What is the hare's average speed?

Instantaneous speed

13 A plane accelerates down a runway. Markers beside the runway are 10.0 m apart. The plane moves between two adjacent markers in 0.200 s. Estimate the instantaneous speed of the plane as it passes the first marker.

14 By analyzing a multiflash photograph of a golfer hitting a golf ball, one finds that the head of the club travels 50.0 cm in 0.0100 s, just before it strikes the ball. Find the approximate instantaneous speed of the club head at the instant of contact.

★15 Let $s = ct^3$, where $c = 1.0$ m/s^3. Compute the average speed over the time intervals $\Delta t_1 = 1.0 \times 10^{-1}$ s, $\Delta t_2 = 1.0 \times 10^{-2}$ s, and $\Delta t_3 = 1.0 \times 10^{-3}$ s, with all time intervals starting at the time $t = 1.0000$ s. What is the instantaneous speed at $t = 1.0000$ s?

★★16 Paradoxes relating to the nature of motion were formulated by the ancient Greeks. The paradoxes were not resolved because the Greeks had no clear understanding of concepts like instantaneous speed. One of Zeno's paradoxes can be stated as follows: a runner wishing to run 100 meters must first cover half that distance. But before he can cover the 50 meters, he must first travel 25 meters, and so on. It is clear that before the runner can travel 100 meters, or any finite distance, he must first move through an infinite number of shorter distances, and this he can never do in a finite time. Resolve this paradox.

1-3 Displacement

17 A traveler flies from Spokane to Atlanta by way of Los Angeles. Los Angeles is approximately 1600 km due south of Spokane, and Atlanta is approximately 3200 km due east of Los Angeles (Fig. 1-32). Find the traveler's total displacement.

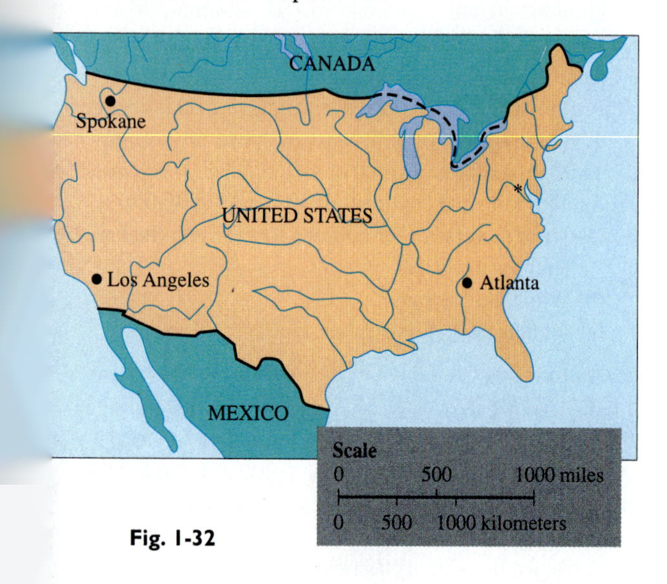

Fig. 1-32

18 A traveler first drives 20.0 km east, then 30.0 km southeast, and finally 10.0 km south. Find the traveler's total displacement.

1-4 Vector Algebra

Graphical vector addition

19 Four forces each have a magnitude of 50 lb. Their respective directions are north, south, east, and west. Find the resultant force.

20 A 40 lb force acts to the right, while a 100 lb force acts at an angle of 45° above the first. Construct the vector sum of the two forces graphically, using a convenient scale.

21 Electric fields are vector quantities whose magnitudes are measured in units of volts/meter (V/m). Find the resultant electric field when there are two fields, $\mathbf{E}_1$ and $\mathbf{E}_2$, where $\mathbf{E}_1$ is directed vertically upward and has magnitude 100 V/m and $\mathbf{E}_2$ is directed 45° to the left of $\mathbf{E}_1$ and has magnitude 150 V/m.

22 Find the resultant of the vectors shown in Fig. 1-33.

23 Find the resultant of the vectors shown in Fig. 1-34.

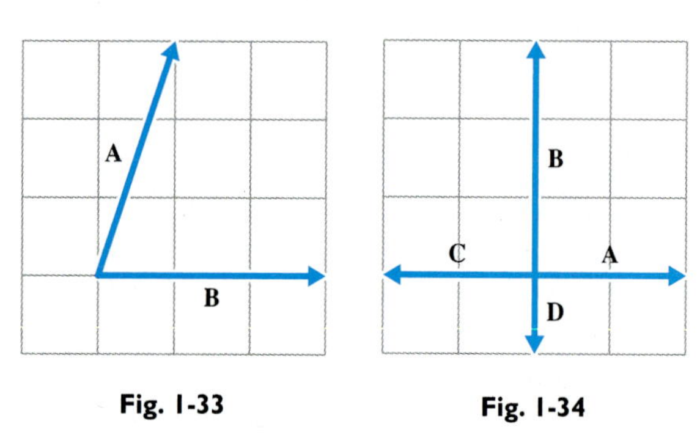

Fig. 1-33 **Fig. 1-34**

Graphical vector subtraction

24 A force of magnitude 20.0 lb directed toward the right is exerted on an object. What other force must be applied to the object so that the resultant force is zero?

25 The forces shown in Fig. 1-35 are applied to an object. Find the third force that will produce a zero resultant force.

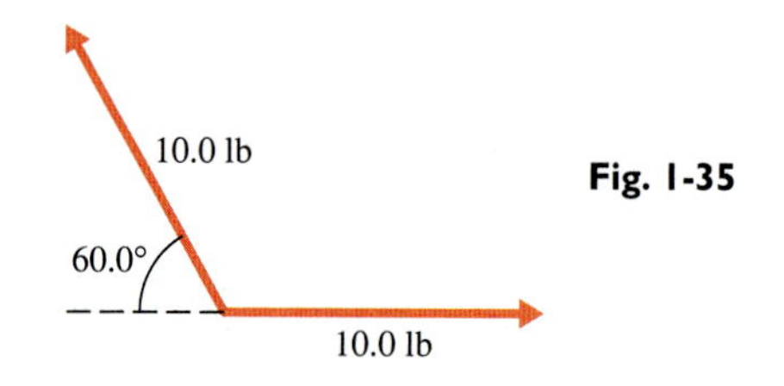

Fig. 1-35

26 A force $\mathbf{F}_1$ of magnitude 100 lb is directed toward the right. Find a second force $\mathbf{F}_2$ that will produce a resultant force of 200 lb to the left.

27 A force $\mathbf{F}_1$ of magnitude 100 lb is directed vertically upward. Find a second force $\mathbf{F}_2$ that will produce a resultant force of 141 lb, 45.0° to the right of $\mathbf{F}_1$.

1-5 Components of Vectors

Finding components

28 A certain vector $\mathbf{A}$, of magnitude 5.00 units, points in a direction 30.0° to the left of the negative y-axis. Find A_x and A_y.

29 Find the x and y components of the force vector shown in Fig. 1-36.

30 Find the x and y components of the force shown in Fig. 1-37.

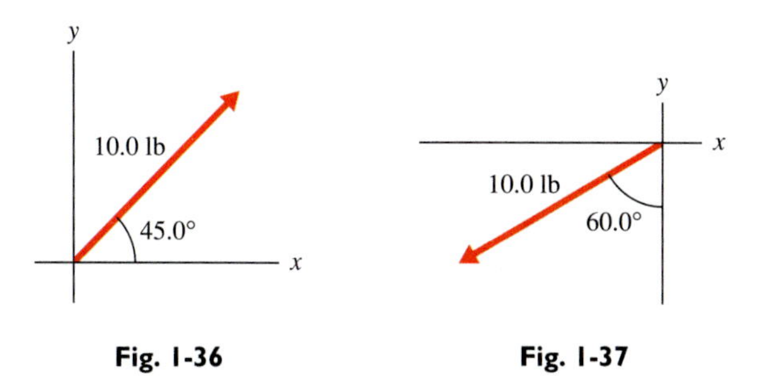

Fig. 1-36

Fig. 1-37

31 A force directed 50.0° below the positive x-axis has an x component of 5.00 lb. Find its y component.

Addition by components

32 **A** makes an angle of 30.0° above the positive x-axis, and **B** makes an angle of 45.0° below the negative x-axis. $A = 3.00$ units, and $B = 2.00$ units. Find **A** + **B** and **A** − **B**.

33 Three vectors **A**, **B**, and **C** have the following components: $A_x = 5.00$, $A_y = 7.00$; $B_x = 4.00$, $B_y = -3.00$; $C_x = -2.00$, $C_y = 5.00$.
(a) Find the x and y components of **A** + **B** + **C**.
(b) Find the magnitude and direction of **A** + **B** + **C**.

34 Find the resultant of the vectors shown in Fig. 1-38.

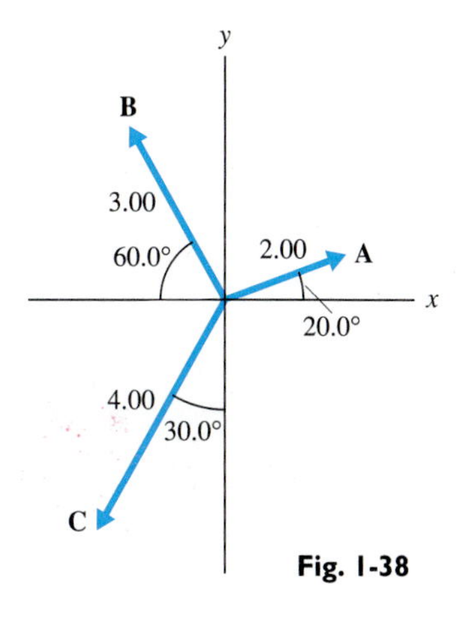

Fig. 1-38

35 A car traveling uphill experiences the forces indicated in Fig. 1-39. Find the resultant force on the car.

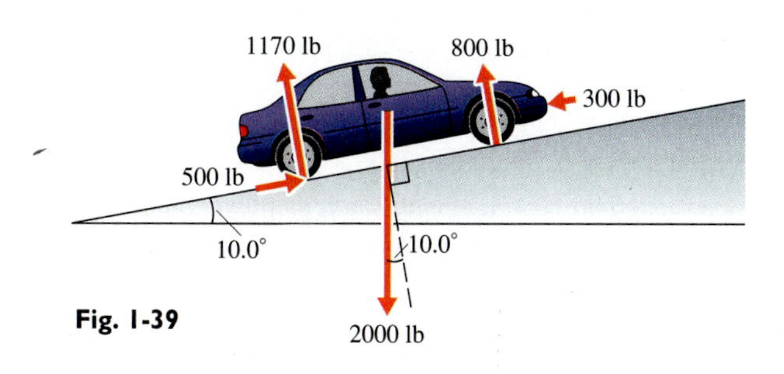

Fig. 1-39

36 Use the component method of vector addition to solve the problem in Ex. 7.

37 Find the resultant of the vectors shown in Fig. 1-40.

38 Find the resultant of the vectors shown in Fig. 1-41.

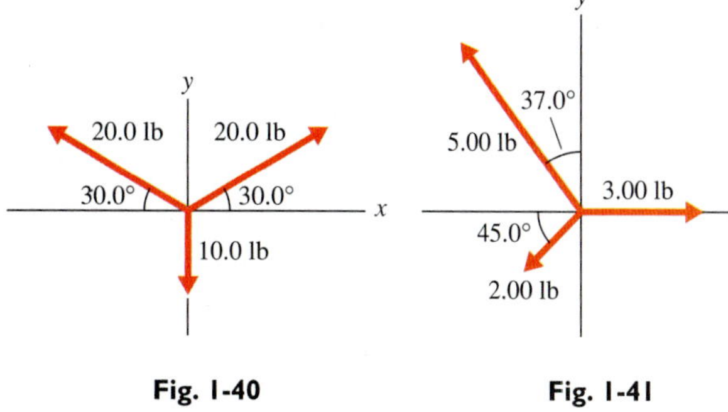

Fig. 1-40

Fig. 1-41

39 Find the additional force that must act on the dog in Fig. 1-42 in order to make the resultant force zero.

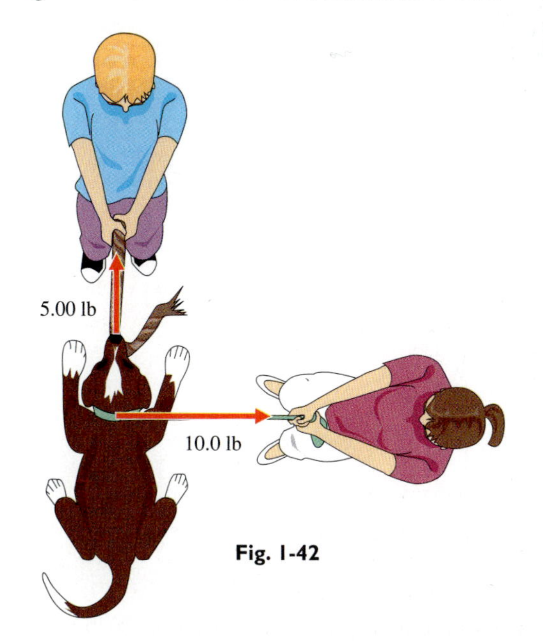

Fig. 1-42

★ **40** Find the smallest additional force that must be applied to the block in Fig. 1-43 in order to have the resultant force directed along the positive y-axis.

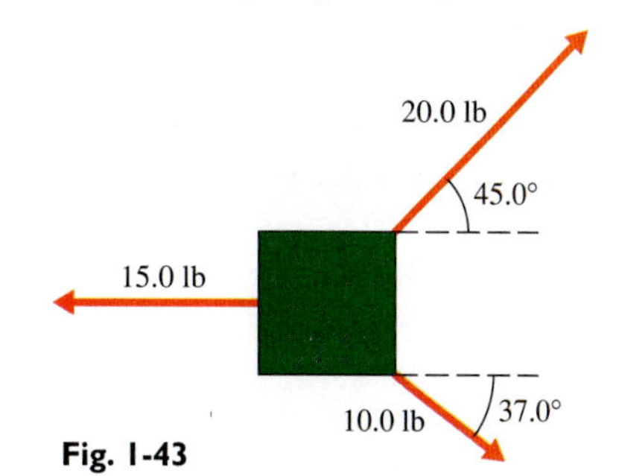

Fig. 1-43

41 Find the additional force that must be applied to the dog in Fig. 1-42 in order to make the resultant force 15.0 lb, directed toward the boy.

1-6 Velocity

Average velocity

42 At $t = 0$ the coordinates of a particle are $x = 0$, $y = 2.00$ m. At a time 1.00 s later, its coordinates are $x = 3.00$ m, $y = 4.00$ m. Find the components of the particle's average velocity during this time interval.

43 During a 10.0 s interval, the position of a particle changes from $x = 2.00$ m, $y = 4.00$ m to $x = 6.00$ m, $y = 4.00$ m. Find the particle's average velocity.

44 Find the x and y components of the average velocity of the particle whose position changes from $x = 3.00$ m, $y = 0$ to $x = 0$, $y = 4.00$ m over a 1.00 s interval.

45 In 1827 Robert Brown discovered that when a suspension of pollen in water is viewed under a microscope, the pollen particles are observed to move very erratically. In 1905 Einstein showed how this "Brownian motion" was a consequence of the random thermal motion of the water molecules that collide with the pollen. Fig. 1-44 shows a typical path of such a particle during a time interval of 0.0100 s. Compute the average velocity during this time interval. How does the magnitude of this vector compare with the average speed?

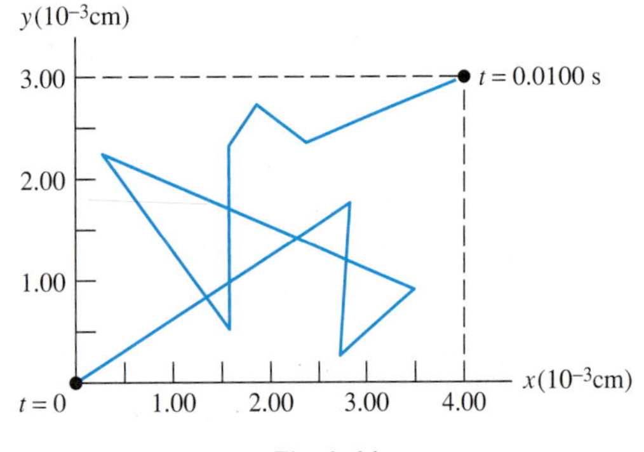

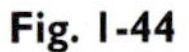
Fig. 1-44

Instantaneous velocity

46 A car travels around a curve at a constant speed of 80.0 km/h. What is the car's instantaneous velocity at an instant when it is headed west?

47 During a 5.00 s interval, a particle's coordinates change from $x = 10.0$ m, $y = 5.00$ m to $x = 30.0$ m, $y = -5.00$ m. Assuming the particle's velocity is constant, what will its coordinates be at the end of the next 5.00 s interval?

48 (a) In Fig. 1-30 find the car's velocity components at time t_1. (b) Estimate how far the car moves in the y direction from time t_1 to time $t_1 + 0.10$ s.

49 In Fig. 1-30 about how long does it take for the car's x coordinate to increase by 1.0 m from its value at time t_2?

Additional problems

★ **50** (a) A car travels at an average speed of 80.0 km/h for 1.00 h and then at an average speed of 40.0 km/h during a second hour. What is the car's average speed during the 2.00 h interval? (b) A car travels 1.00 km at an average speed of 80.0 km/h and then 1.00 km at an average speed of 40.0 km/h. What is the car's average speed during the 2.00 km interval?

51 In a tug-of-war, two teams pull on a rope in opposite directions. (a) If both teams exert a force of 500 lb on the rope, what is the resultant force on the rope? (b) If the team on the right exerts a force of 600 lb and the team on the left exerts 500 lb, what is the resultant force on the rope?

★ **52** A runner with a good awareness of her pace runs along a path of unknown length at a speed of 0.20 mi/min and then walks back to her starting point at a speed of 0.05 mi/min. She neglects to note her time for each part of her path but does measure the total round-trip time to be 50.0 min. How far has she run?

★ **53** A sleepy man drives a car along a straight section of highway at a constant speed of 88.0 km/h. His eyes begin to close, and his car moves at an angle of 5.00° relative to the road. For how long can his attention lapse before the car begins to move across the lane divider, originally 1.00 m from the side of the car?

★ **54** A man 2.00 m tall, walking near a street light at night, casts a 4.00 m long shadow directly in front of him. The lightpost is 6.00 m behind the man. (a) How tall is the lightpost? (b) If the man is moving away from the lightpost at the speed of 1.00 m/s, how fast does the tip of his shadow move?

55 A hockey player faces the goal, which is 10.0 m in front of him (due north). Instead of trying to make a direct shot, the player passes the puck to a teammate who is 6.00 m away in a direction 30.0° to the right of the direction to the goal. What final displacement is needed to score a goal?

★ **56** Two sprinters run a 100 m dash, with the winner finishing in 10.0 s. If the other runner was 1.00 m behind the winner at the finish, how long did it take her to run the 100 m?

★ **57** The cheetah is the fastest land animal, able to reach quickly a maximum speed of 30 m/s. But it is able to maintain this speed for only about 10 s. Suppose that the cheetah stalks a prey that has a maximum speed of 25 m/s but is able to maintain this speed for much longer than 10 s. If it is to catch its prey, what is the maximum distance the cheetah can be from the prey when the chase begins?

CHAPTER 2 Motion in a Straight Line

A fantasy spaceship on its way to the moon—based on the movie *A Mouse on the Moon.*

In the 1950s Peter Sellers' comedy *A Mouse on the Moon,* a scientist from a small imaginary country invents an unconventional spaceship and then challenges the United States and the Soviet Union in a race to the moon. Unlike a normal spaceship, which achieves a large velocity quickly and then coasts, this fictional spaceship leaves the earth very slowly but accelerates continuously, gradually increasing its speed throughout much of its flight to the moon. The effect of accelerating over such a long time is that the spaceship reaches an enormous final velocity and arrives first on the moon (see Problem 15). The motion of this spaceship is particularly simple because the ship moves in a straight line. In this chapter you will solve many real-life problems involving accelerated linear motion. For example, you may find the length of the shortest runway from which a jet can take off (Ex. 3), compute the height of a certain basketball player's vertical jump (Ex. 7), determine the thickness required of automobile air bags to prevent injury in a collision (Problem 14), or measure your reaction time (Problem 55). You will actually be able to predict the future in certain cases; that is, you can predict the future motion of bodies, based on a knowledge of their initial conditions.

2-1 Acceleration in One Dimension

For linear motion, only a single Cartesian coordinate is needed to specify the position of a particle. We shall use x for horizontal motion and y for vertical motion. The velocity vector of any particle moving in either the x direction or the y direction has only a single nonzero component, v_x or v_y. For motion along the x-axis, the average velocity is expressed in component form by Eq. 1-11:

$$\bar{v}_x = \frac{\Delta x}{\Delta t} \tag{2-1}$$

and the instantaneous velocity is expressed in component form by Eq. 1-15:

$$v_x = \underset{\Delta t \to 0}{\text{limit}} \frac{\Delta x}{\Delta t} \tag{2-2}$$

Acceleration

When the velocity of a body changes, we say the body "accelerates." For example, you can accelerate a car by applying pressure to the accelerator pedal (Fig. 2-1a). The car's acceleration is a measure of how rapidly its velocity changes. Applying the brakes is another way of changing a car's velocity, and so this braking action also accelerates the car (Fig. 2-1b). Since this kind of acceleration involves a decrease in velocity, it is often referred to as "deceleration."

Still another way of accelerating a body is to change its direction of motion rather than its speed (Fig. 2-1c). A change in the direction of motion means a change in the velocity *vector* and therefore an acceleration. A change in direction, however, means that the motion is not linear, and so we shall save the discussion of such acceleration for Chapter 3.

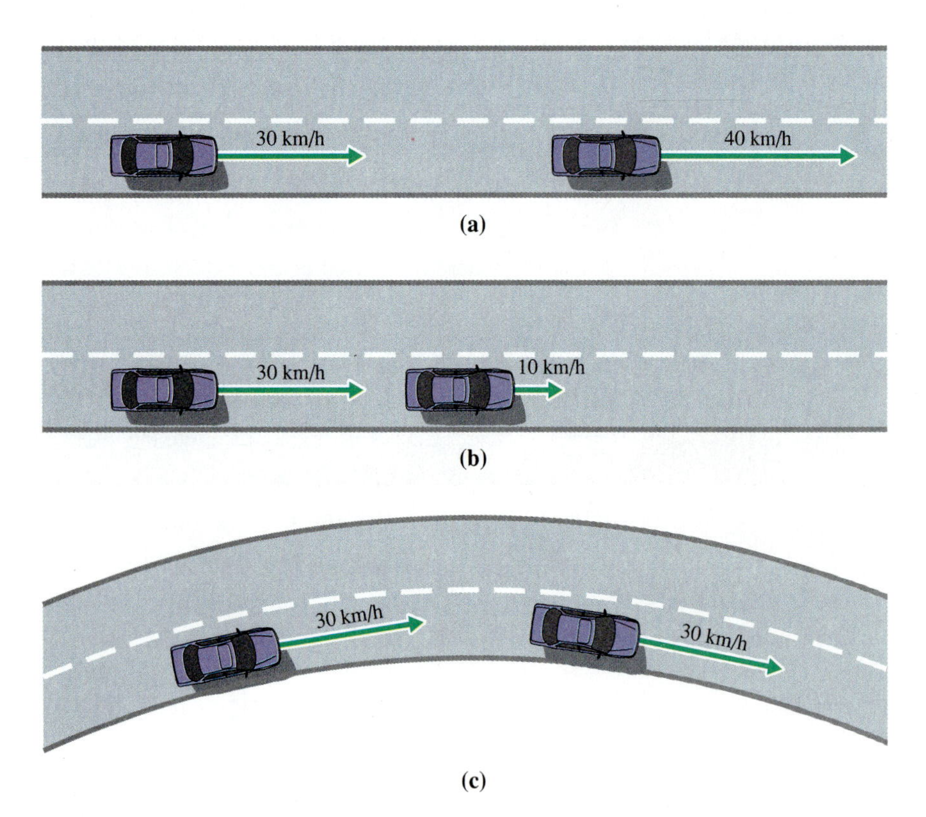

Fig. 2-1 Examples of a car experiencing acceleration, that is, a change in velocity.

The definition of acceleration parallels the definition of velocity: **acceleration is defined as the rate of change of velocity,** just as velocity is defined as the rate of change of position. We define the **average acceleration** $\bar{\mathbf{a}}$ during a time interval Δt as the velocity change $\Delta\mathbf{v}$ divided by Δt:

$$\bar{\mathbf{a}} = \frac{\Delta\mathbf{v}}{\Delta t} \tag{2-3}$$

Since acceleration is proportional to the change in velocity vector, acceleration is itself a vector. Fig. 2-2 shows how the acceleration vector associated with either an "accelerating" or "decelerating" car points in the direction of the change in velocity $\Delta\mathbf{v}$.

For motion along the x-axis, average acceleration is expressed in component form as

$$\bar{a}_x = \frac{\Delta v_x}{\Delta t} \tag{2-4}$$

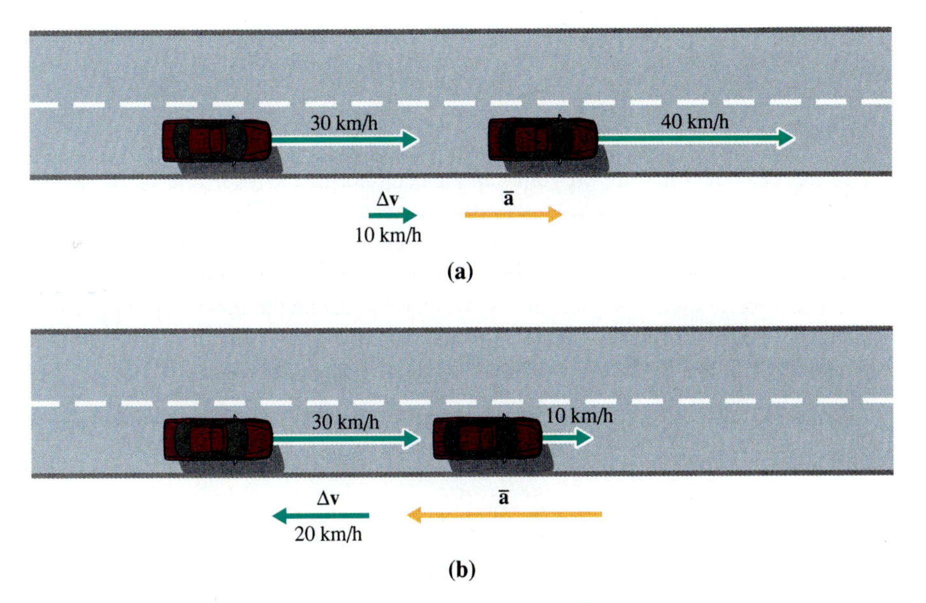

Fig. 2-2 Acceleration of a car that **(a)** accelerates or **(b)** decelerates.

EXAMPLE 1 Accelerating From Zero to Sixty

A car starts from rest and accelerates from 0 to 97.0 km/h (60.3 mi/h) in 9.00 s. Compute the car's average acceleration.

SOLUTION We apply Eq. 2-4, taking the x-axis to be the direction of motion.

$$\bar{a}_x = \frac{\Delta v_x}{\Delta t} = \frac{97.0 \text{ km/h} - 0}{9.00 \text{ s}}$$
$$= 10.8 \text{ km/h-s}$$

We read this answer as "10.8 kilometers per hour per second," which means that, on the average, the car's velocity increased by 10.8 km/h each second. A system of units more often used for acceleration is "meters per second per second" or "meters per second squared." Converting to these units, we find

$$\bar{a}_x = \left(10.8 \frac{\text{km}}{\text{h-s}}\right)\left(\frac{1000 \text{ m}}{1 \text{ km}}\right)\left(\frac{1 \text{ h}}{3600 \text{ s}}\right) = 3.00 \text{ m/s}^2$$

The meaning of this numerical value is that, on the average, the car's velocity increased 3.00 m/s each second, as illustrated in Fig. 2-3. (Although the figure is based on the assumption of a constant acceleration of 3.00 m/s², the car could have had acceleration that varied and simply averaged 3.00 m/s².)

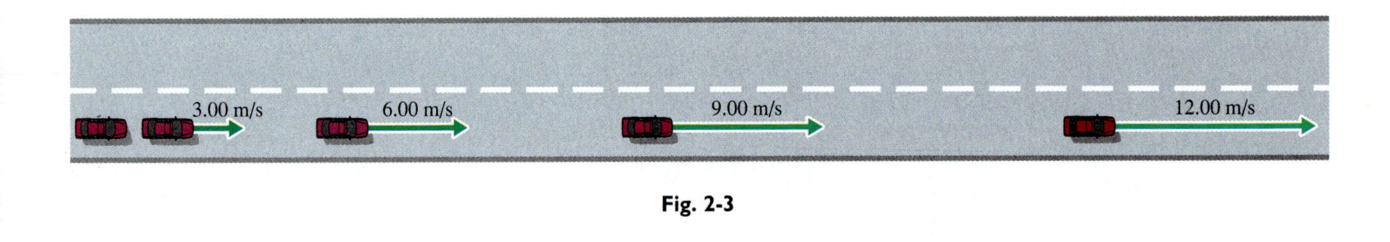

Fig. 2-3

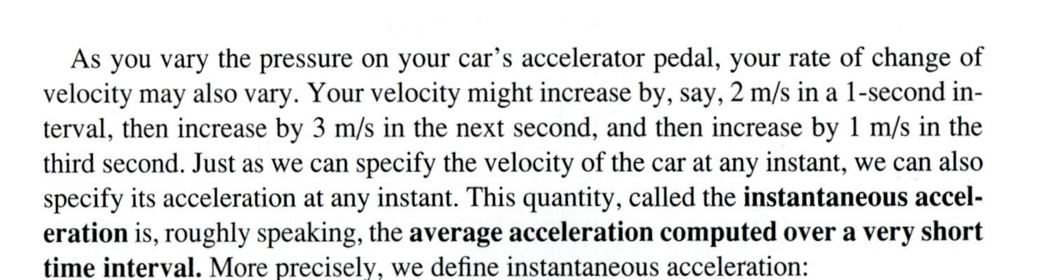

As you vary the pressure on your car's accelerator pedal, your rate of change of velocity may also vary. Your velocity might increase by, say, 2 m/s in a 1-second interval, then increase by 3 m/s in the next second, and then increase by 1 m/s in the third second. Just as we can specify the velocity of the car at any instant, we can also specify its acceleration at any instant. This quantity, called the **instantaneous acceleration** is, roughly speaking, the **average acceleration computed over a very short time interval.** More precisely, we define instantaneous acceleration:

$$\mathbf{a} = \lim_{\Delta t \to 0} \frac{\Delta \mathbf{v}}{\Delta t} \tag{2-5}$$

This expression is read "**a** equals the limit as Δt approaches zero of $\Delta\mathbf{v}/\Delta t$." It means that if we compute a body's average acceleration $\Delta\mathbf{v}/\Delta t$ over shorter and shorter time intervals, the result approaches a definite limit vector **a**.

For motion along the x-axis, we can express the instantaneous acceleration in component form:

$$a_x = \lim_{\Delta t \to 0} \frac{\Delta v_x}{\Delta t} \tag{2-6}$$

EXAMPLE 2 Braking

The driver of a car applies the brakes, causing the car to slow down. The instantaneous velocity at a certain instant is 10.0 m/s. The velocity 0.010 s later is 9.90 m/s. Estimate the car's instantaneous acceleration at the initial instant.

SOLUTION Since the time interval is so small, we can reasonably assume that the acceleration is approximately constant. Thus we can easily compute a_x, using Eq. 2-4:

$$a_x \approx \overline{a}_x = \frac{\Delta v_x}{\Delta t} = \frac{9.90\ \text{m/s} - 10.0\ \text{m/s}}{0.010\ \text{s}}$$
$$= -10\ \text{m/s}^2$$

The symbol $\approx$ means 'is approximately equal to'. We use it here instead of an equals sign because we are making an approximation when we say that the instantaneous acceleration is the same as the average acceleration.

The minus sign in the answer tells us that the acceleration is in the negative x direction, meaning that v_x is decreasing; that is, the car is slowing down.

2-2 Linear Motion at Constant Acceleration

When we know a body's acceleration and its initial velocity, it is possible to predict where that body will be at some future time and how fast it will be moving. In the special case of linear motion at constant acceleration, it is relatively easy to obtain algebraic expressions relating the position x, velocity v_x, acceleration a_x, and time t, as we shall show in this section. Using these equations we will be able to predict the position and velocity of an accelerated body at a given time. There are many practical situations in which a body moves in a straight line with constant acceleration, and so the equations we derive here will be used frequently in the solution of problems.

In order to obtain an expression for velocity v_x at time t, we apply the definition of acceleration along the x-axis (Eq. 2-5), using the symbol v_{x0} to express the initial value of velocity at time $t = 0$.

$$a_x = \frac{\Delta v_x}{\Delta t} = \frac{v_x - v_{x0}}{t - 0}$$

Solving for v_x, we obtain

$$v_x = v_{x0} + a_x t \tag{2-7}$$

From high-school algebra, you should recognize this equation as having the form $y = mx + b$, which is the equation for a straight line of slope m and y intercept b. Here the variables are t instead of x and v_x instead of y. The slope is a_x, and the v_x intercept is v_{x0}.

Notice that when $t = 0$, Eq. 2-7 reduces to $v_x = v_{x0}$, the value of the velocity at $t = 0$. Both v_{x0} and a_x are constants. If both are known, Eq. 2-7 can be used to find the velocity v_x at any time t. For example, if $v_{x0} = 0$ and $a_x = 2\ \text{m/s}^2$, Eq. 2-7 becomes

$$v_x = (2\ \text{m/s}^2)t$$

At time $t = 1$ s,

$$v_x = (2\ \text{m/s}^2)(1\ \text{s}) = 2\ \text{m/s}$$

and at $t = 10$ s,

$$v_x = (2\ \text{m/s}^2)(10\ \text{s}) = 20\ \text{m/s}$$

Next we shall find an expression for x as a function of t. We apply Eq. 2-1 for the average velocity v_x for the time interval from 0 to an arbitrary final time t. We let x_0 denote the initial position and x denote the position at time t:

$$\overline{v}_x = \frac{\Delta x}{\Delta t} = \frac{x - x_0}{t - 0}$$

Solving for x, we obtain

$$x = x_0 + \overline{v}_x t \tag{2-8}$$

When acceleration is constant, the average velocity can be computed as the average of the particle's initial velocity v_{x0} and its final velocity v_x:

$$\overline{v}_x = \tfrac{1}{2}(v_{x0} + v_x) \tag{2-9}$$

It is important to understand that this expression for average velocity is not valid in general. The definition of average velocity is $\Delta x/\Delta t$, and only when acceleration is constant is this "average velocity" equal to the numerical average of the initial and final values of velocity.

Substituting the expression for $\overline{v}_x$, given by Eq. 2-9, into Eq. 2-8, we obtain

$$x = x_0 + \tfrac{1}{2}(v_{x0} + v_x)\, t \tag{2-10}$$

Finally we substitute in this equation the expression for v_x given in Eq. 2-7 and obtain

$$x = x_0 + \tfrac{1}{2}(v_{x0} + v_{x0} + a_x t)t$$

or

$$x = x_0 + v_{x0}t + \tfrac{1}{2}a_x t^2 \tag{2-11}$$

Notice that when $t = 0$, this equation reduces to $x = x_0$. The quantities x_0, v_{x0}, and a_x are all constants. If all are known, Eq. 2-11 can be used to find the position x at any time t.

Eqs. 2-7 and 2-11 alone are sufficient to solve any problems in linear motion at constant acceleration. However, it is sometimes more convenient to use Eq. 2-10. If you know the initial and final velocities but not the acceleration, you can use Eq. 2-10 to find the final position x without bothering to calculate a_x.

We obtain another useful equation by eliminating t between Eqs. 2-7 and 2-10, so that we have a direct relationship between position and velocity. Solving Eq. 2-7 for t, we obtain

$$t = (v_x - v_{x0})/a_x$$

Then substituting this expression for t in Eq. 2-10, we get

$$\begin{aligned} x &= x_0 + \tfrac{1}{2}(v_{x0} + v_x)(v_x - v_{x0})/a_x \\ &= x_0 + \tfrac{1}{2}(v_x^2 - v_{x0}^2)/a_x \end{aligned}$$

Solving for v_x^2, we find

$$v_x^2 = v_{x0}^2 + 2a_x(x - x_0) \tag{2-12}$$

Eqs. 2-7, 2-10, 2-11, and 2-12 are called **equations of motion,** or **kinematic equations.** You will use them frequently to solve problems. A key step in these solutions is to formulate the question in terms of the symbols contained in the equations of motion. After this is done, deciding which equation or equations to use is usually straightforward. You should remember, however, that these particular equations of motion are valid only in the special case of **linear motion at constant acceleration.**

EXAMPLE 3 Minimum Length for a Runway

A jet plane beginning its takeoff moves down the runway at a constant acceleration of 4.00 m/s^2.

(a) Find the position and velocity of the plane 5.00 s after it begins to move.

(b) If a speed of 70.0 m/s is required for the plane to leave the ground, how long a runway is required?

SOLUTION Because the acceleration is constant, we can apply the equations of motion derived above.

(a) We take the origin of the x-axis to be the initial position of the plane, so that $x_0 = 0$ (Fig. 2-4).

It is useful to begin by listing all the data given in the problem:

$$a_x = 4.00 \text{ m/s}^2$$

$$v_{x0} = 0$$

$$x_0 = 0$$

The problem may be stated in terms of the symbols as follows:

Find x and v_x at $t = 5.00$ s.

Eqs. 2-7 and 2-11 are the obvious ones to use. When x_0 and v_{x0} are zero, these two equations reduce to

$$v_x = a_x t$$

and

$$x = \tfrac{1}{2} a_x t^2$$

At $t = 5.00$ s,

$$v_x = (4.00 \text{ m/s}^2)(5.00 \text{ s}) = 20.0 \text{ m/s}$$

$$x = \tfrac{1}{2}(4.00 \text{ m/s}^2)(5.00 \text{ s})^2 = 50.0 \text{ m}$$

(b) The problem here may be stated:

Find x when $v_x = 70.0$ m/s.

We shall use Eq. 2-12 because it contains the single unknown x, as well as a_x and v_x, which are known. With $v_{x0} = 0$ and $x_0 = 0$, Eq. 2-12 reduces to

$$v_x^2 = 2a_x x$$

Solving for x, we obtain

$$x = \frac{v_x^2}{2a_x} = \frac{(70.0 \text{ m/s})^2}{2(4.00 \text{ m/s}^2)} = 613 \text{ m}$$

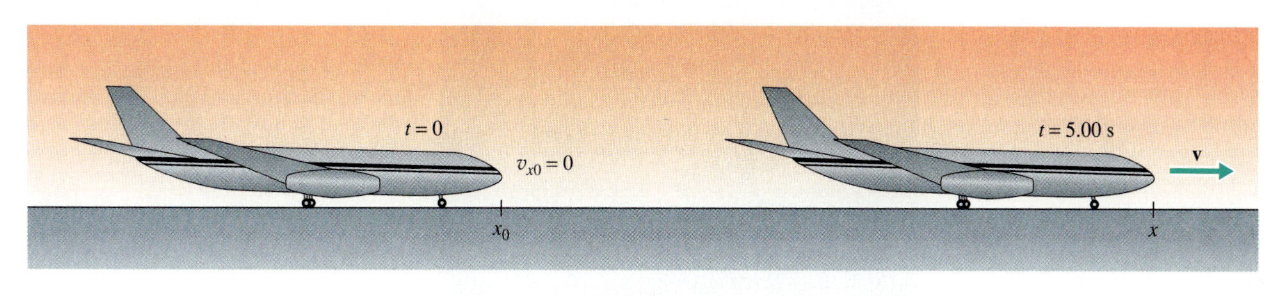

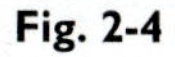

Fig. 2-4

EXAMPLE 4 Computing the Length of a Flight

Between takeoff and landing, an airplane in level flight **(1)** accelerates at a constant rate from 250 km/h to 950 km/h over a 15 min interval, **(2)** travels at a constant speed of 950 km/h for 25 min, **(3)** decelerates at a constant rate from 950 km/h to 250 km/h over a 15 min interval. Find the total distance traveled by the plane.

SOLUTION The motion consists of three parts, each having a different acceleration. In other words, a_x is not constant over the entire motion. However, a_x is constant *within each part,* and so we can apply our constant-acceleration equations if we solve the problem in three parts. Let x_1, x_2, and x_3 denote the distances traveled during the three parts of the motion. The sum $x_1 + x_2 + x_3$ is then the total distance. In each part, we shall let $x_0 = 0$.

(Part **1**) $v_{x0} = 250$ km/h

$v_x = 950$ km/h at $t = 15$ min

Find x at $t = 15$ min.

Eq. 2-10 gives

$$x_1 = \tfrac{1}{2}(v_{x0} + v_x)t$$

$$= \tfrac{1}{2}(250 \text{ km/h} + 950 \text{ km/h})(15 \text{ min})\left(\frac{1 \text{ h}}{60 \text{ min}}\right)$$

$$= 150 \text{ km}$$

(Part **2**) Velocity is constant in this interval.

$v_x = 950$ km/h

$a_x = 0$

Find x at $t = 25$ min.

Since v_x is constant, we can rearrange the defining equation for speed, $v_x = x/t$, to get

$$x_2 = v_x t$$

(Notice that both Eqs. 2-10 and 2-11 reduce to this simple form when $a_x = 0$, or, to put it another way, when $v_{x0} = v_x$.) Evaluating x_2, we obtain

$$x_2 = (950 \text{ km/h})\,(25 \text{ min})\left(\frac{1 \text{ h}}{60 \text{ min}}\right) = 400 \text{ km}$$

(Part **3**) $v_{x0} = 950$ km/h

$v_x = 250$ km/h at $t = 15$ min

Find x at $t = 15$ min.

Eq. 2-10 gives

$$x_3 = \tfrac{1}{2}(v_{x0} + v_x)t$$

$$= \tfrac{1}{2}(950 \text{ km/h} + 250 \text{ km/h})(0.250 \text{ h}) = 150 \text{ km}$$

Finally, we add the values obtained in parts 1, 2, and 3:

$$x_1 + x_2 + x_3 = 150 \text{ km} + 400 \text{ km} + 150 \text{ km} = 700 \text{ km}$$

Fig. 2-5 A feather and an apple falling freely in a vacuum accelerate at the same rate.

2-3 Free Fall

We are all familiar with the phenomenon we call "gravity," the earth's pull on bodies close to it. Release an object you are holding and it falls downward, pulled toward the earth by gravity. Galileo Galilei, whose life is discussed at the end of this chapter, was the first scientist to make a controlled study of the earth's gravity. Galileo discovered that gravity not only pulls all objects toward the earth, but also accelerates those objects while they are falling. In other words, the speed of any body falling freely continuously increases. Galileo determined that **when air resistance is negligible all bodies near the earth's surface fall with the same constant acceleration, denoted by g** (Fig. 2-5). The magnitude of this gravitational acceleration varies somewhat over the earth's surface, from 9.78 m/s^2 to 9.83 m/s^2, as indicated in Table 2-1. In problems and examples, we shall assume the value 9.80 m/s^2.

In this section we consider only freely falling bodies for which air resistance is negligible. The essay at the end of this section describes the effect of nonnegligible air resistance on falling bodies.

In describing the motion of a falling object, we shall use the y coordinate, choosing the upward direction as positive. Because the acceleration is constant, the equations derived in the previous section are again applicable once we make the following changes:

$$x \rightarrow y$$

$$v_x \rightarrow v_y$$

$$a_x \rightarrow a_y = -g$$

The equation $a_y = -g$ implies that the acceleration of gravity is in the negative y direction, in other words, downward. Eqs. 2-7, 2-10, 2-11, and 2-12 may now be written in terms of the y variables:

$$v_y = v_{y0} - gt \quad (2\text{-}13)$$

$$y = y_0 + \tfrac{1}{2}(v_{y0} + v_y)t \quad (2\text{-}14)$$

$$y = y_0 + v_{y0}t - \tfrac{1}{2}gt^2 \quad (2\text{-}15)$$

$$v_y^2 = v_{y0}^2 - 2g(y - y_0) \quad (2\text{-}16)$$

Table 2-1 Gravitational acceleration at various points on earth

Location	Latitude (north)	Elevation (m)	g (m/s^2)
Panama	9°	6	9.782
Honolulu	21°	6	9.789
New Orleans	30°	2	9.793
Atlanta	34°	324	9.795
San Francisco	38°	114	9.800
Denver	40°	1638	9.796
New York	41°	38	9.803
Rome	41°	49	9.803
Chicago	42°	182	9.803
Seattle	48°	58	9.807
London	51°	48	9.811
Copenhagen	56°	14	9.816

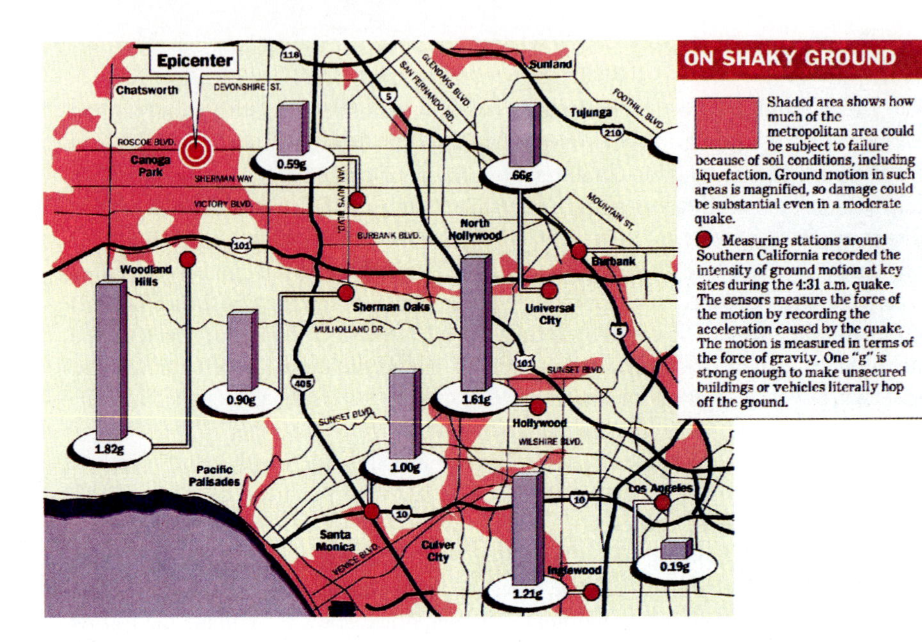

Acceleration is often measured in "g's," that is, as a multiple of gravitational acceleration g. This map shows local values of ground acceleration in g's in the Los Angeles area during the earthquake there on January 17, 1994.

EXAMPLE 5 Measuring a Cliff by Dropping a Rock

A rock is dropped from the top of a cliff and strikes the ground 4.00 s later. Find the position and velocity of the rock at 1.00 s intervals and determine the height of the cliff.

SOLUTION In symbols, our task is to find y and v_y for various values of t. Therefore we use Eqs. 2-15 and 2-13. Choosing the origin of the y-axis to be the top of the cliff, we set $y_0 = 0$. Also $v_{y0} = 0$, since the rock is dropped and not thrown. Our equations then reduce to

$$y = -\tfrac{1}{2}gt^2$$

$$v_y = -gt$$

$$t = 1.00\text{ s:}\quad y = -\tfrac{1}{2}(9.80\text{ m/s}^2)(1.00\text{ s})^2 = -4.90\text{ m}$$

$$v_y = -(9.80\text{ m/s}^2)(1.00\text{ s}) = -9.80\text{ m/s}$$

$$t = 2.00\text{ s:}\quad y = -\tfrac{1}{2}(9.80\text{ m/s}^2)(2.00\text{ s})^2 = -19.6\text{ m}$$

$$v_y = -(9.80\text{ m/s}^2)(2.00\text{ s}) = -19.6\text{ m/s}$$

$$t = 3.00\text{ s:}\quad y = -\tfrac{1}{2}(9.80\text{ m/s}^2)(3.00\text{ s})^2 = -44.1\text{ m}$$

$$v_y = -(9.80\text{ m/s}^2)(3.00\text{ s}) = -29.4\text{ m/s}$$

$$t = 4.00\text{ s:}\quad y = -\tfrac{1}{2}(9.80\text{ m/s}^2)(4.00\text{ s})^2 = -78.4\text{ m}$$

$$v_y = -(9.80\text{ m/s}^2)(4.00\text{ s}) = -39.2\text{ m/s}$$

The rock drops 78.4 m during its 4.00 s of free fall, and so this must be the height of the cliff. Fig. 2-6 shows the position and velocity of the rock at the times for which we computed them. Notice that the change in velocity during each 1.00 s interval is always −9.80 m/s.

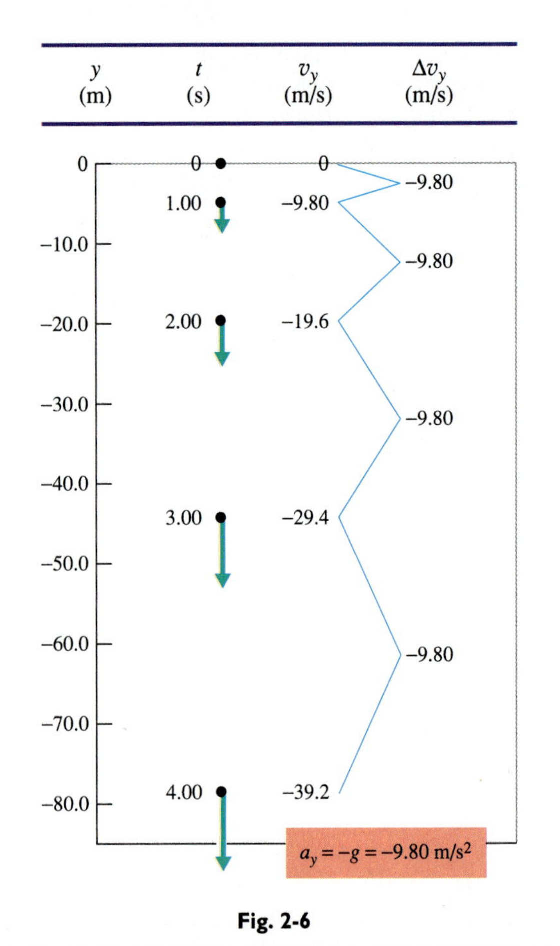

Fig. 2-6

The equations for free fall apply when an object is moving up as well as down. An object with an upward initial velocity will still experience a downward acceleration. This means that the velocity v_y will first be reduced to zero and then will become negative. Throughout the motion, the same equations apply since acceleration is the same throughout.

EXAMPLE 6 A Vertical Throw

A pitcher throws a baseball vertically upward with an initial speed of 15.0 m/s, releasing the ball when it is 2.00 m above the ground and then catching it at the same point.

(a) What is the maximum height reached by the ball?

(b) How long is the ball in the air?

SOLUTION **(a)** Fig. 2-7 shows a sketch of the ball's motion. The height of the ball is maximum at the instant the ball stops moving upward, that is, when $v_y = 0$. We state the problem in terms of our symbols as follows: Find y when $v_y = 0$. Eq. 2-16 provides the most direct method for finding y. Letting $y_0 = 0$, we have

$$v_y^2 = v_{y0}^2 - 2gy$$

$$0 = v_{y0}^2 - 2gy$$

$$y = \frac{v_{y0}^2}{2g} = \frac{(15.0\text{ m/s})^2}{2(9.80\text{ m/s}^2)} = 11.5\text{ m}$$

Since the origin was chosen 2.00 m above the ground, the maximum height is 13.5 m.

This result can also be obtained less directly if we first solve Eq. 2-13 for the time to reach the maximum height and then substitute this value of t into Eq. 2-15.

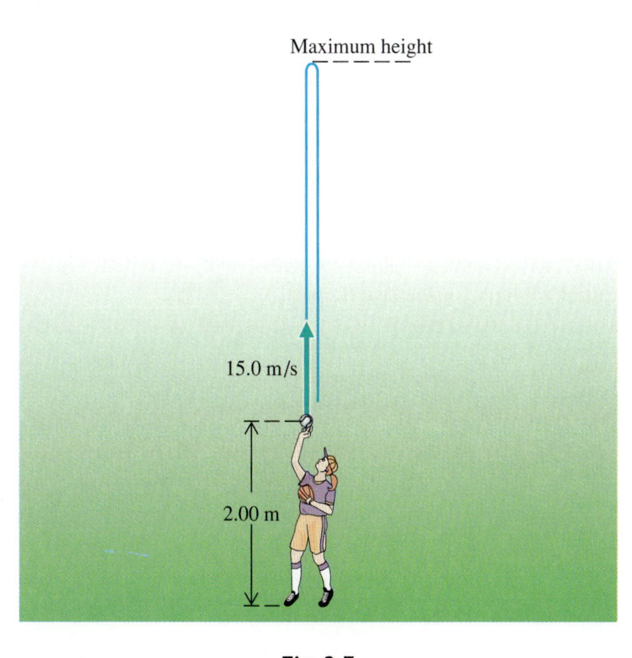

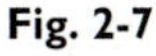

Fig. 2-7

(b) The time the ball is in the air is the time it takes to return to its starting point, $y = 0$, where it is caught. This problem may be stated: Find t when $y = 0$. We use Eq. 2-15:

$$y = y_0 + v_{y0}t - \tfrac{1}{2}gt^2$$

$$0 = 0 + v_{y0}t - \tfrac{1}{2}gt^2$$

One solution of this equation is $t = 0$, the initial time at which the ball is released. Here, though, we are interested in the nonzero value of t, the time when the ball returns to the origin. Solving for this value of t, we find

$$t = \frac{2v_{y0}}{g} = \frac{2(15.0\text{ m/s})}{9.80\text{ m/s}^2} = 3.06\text{ s}$$

Fig. 2-8 shows the baseball's position and velocity vector at 0.250 s intervals.

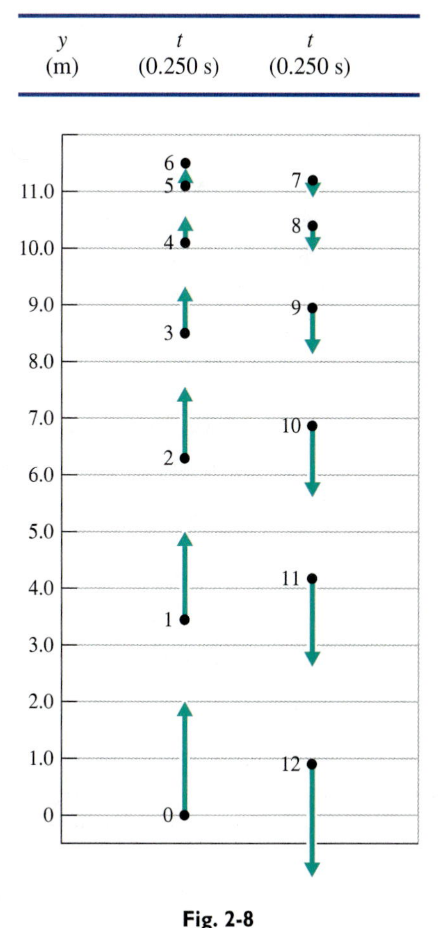

Fig. 2-8

EXAMPLE 7 Computing the Height of a Vertical Jump

Fig. 2-9 shows a basketball player jumping up to "slam dunk" a ball. Analyzing the motion, we see that the player first crouches and then springs upward, leaving the ground with some initial velocity and moving a short distance off the ground before falling back. Assume that his body remains straight from the instant it leaves the ground, so that each part of it undergoes the same motion. Then it is valid to treat the body as a particle with an upward initial velocity. That initial velocity is produced by the legs pushing the body upward.

Suppose that while the feet are in contact with the floor the upper body experiences a constant upward acceleration of magnitude 20.0 m/s² and is accelerated upward through a distance of 40.0 cm. Find the maximum height reached.

SOLUTION There are two parts to the motion, as illustrated in Fig. 2-10: **(1)** The upper body experiences a constant upward acceleration from the beginning of the crouch until the feet leave the floor; **(2)** the entire body experiences a constant downward gravitational acceleration from the instant the feet leave the floor until the maximum height is reached. (Of course, the body experiences the same gravitational acceleration as the feet return to the floor, but we are not concerned with that part of the motion here.)

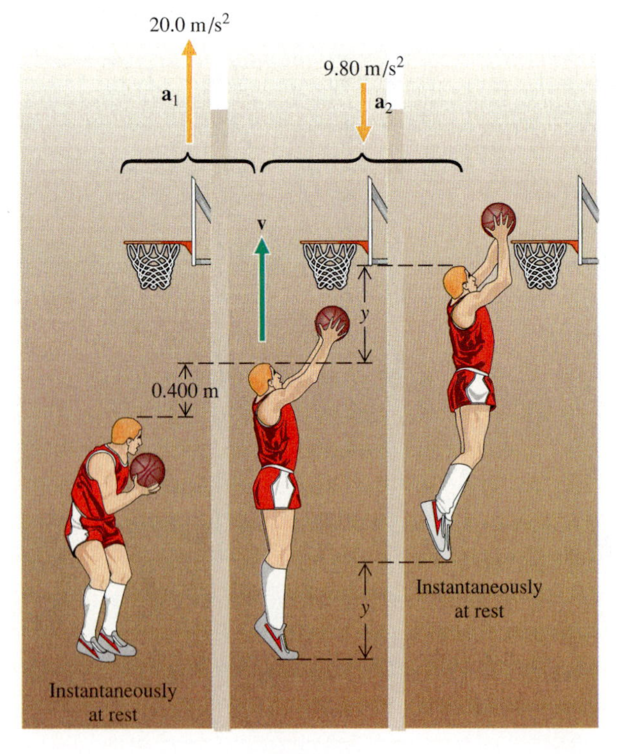

Fig. 2-10

Fig. 2-9

(Part **1**) First we must find the velocity of the body as the feet leave the floor; this value is the final velocity for the first part of the motion.

$y_0 = 0$

$v_{y0} = 0$

$a_y = +20.0 \text{ m/s}^2$, taking the y-axis as positive upward.

Find v_y when $y = 0.400$ m.

We can use a modified version of Eq. 2-16 if we set a_y equal to $+20.0 \text{ m/s}^2$ rather than $-g$:

$$v_y^2 = v_{y0}^2 + 2a_y(y - y_0)$$
$$= 0 + 2(20.0 \text{ m/s}^2)(0.400 \text{ m} - 0) = 16.0 \text{ m}^2/\text{s}^2$$

or

$$v_y = 4.00 \text{ m/s}$$

This final velocity for the first part of the motion is also the initial velocity for the second part.

(Part **2**) $v_{y0} = 4.00$ m/s

Now we wish to find y when $v_y = 0$. We apply Eq. 2-16:

$$v_y^2 = v_{y0}^2 - 2g(y - y_0)$$
$$0 = v_{y0}^2 - 2gy$$

Solving for y, we obtain

$$y = \frac{v_{y0}^2}{2g} = \frac{(4.00 \text{ m/s})^2}{2(9.80 \text{ m/s}^2)} = 0.816 \text{ m}$$

Free Fall In Air

When a falling body experiences significant air resistance, its motion cannot be described by the equations for free fall at constant gravitational acceleration. For some bodies the effect of air resistance is apparent. For example, a leaf falls to the ground much more slowly than predicted by the free-fall equations. The leaf does not accelerate downward at 9.8 m/s^2.

Parachutes are designed to make use of air resistance. A parachutist hits the ground with a safe impact velocity only because of the large air resistance provided by her parachute (Fig. 2-A).

Parachutes and leaves experience more air resistance than most other falling bodies because they have, for their weight, a relatively large surface area pushing against the air as they fall. Air resistance on a falling body depends on the ratio of the body's weight w to its cross-sectional area A. The smaller the ratio w/A, the greater the effect air resistance will have.

Fig. 2-A A parachute has much air resistance and therefore allows a safe jump.

The effect of air resistance on a falling body varies with the body's speed. A body experiences no air resistance as it just begins to fall. It is only after a body begins to move downward that air resistance can begin to have an effect. Air resistance increases as the falling body's speed increases. Any body, if it falls for a long enough time through the air, will eventually experience large air resistance. For example, suppose you drop a rock from the edge of a very high cliff. Initially, the rock will accelerate downward at 9.8 m/s^2. But as the rock's speed increases, air resistance will start to become an important factor. The rock will continue to accelerate downward, but its acceleration will gradually decrease because of air resistance. After 2 or 3 seconds, the effect of air resistance will be appreciable and downward acceleration will be somewhat less than 9.8 m/s^2. After 5 or 6 seconds, downward acceleration will be reduced to a value much less than 9.8 m/s^2. Eventually, the rock will have nearly zero acceleration; that is, it will continue to fall with nearly constant velocity. Any body falling through air for a long enough time will approach a limiting velocity, called **terminal velocity** (Fig. 2-B). The value of the terminal velocity, however, depends very much on the body. Using analysis of forces acting on a body falling through the air, it is possible to show that the body's terminal velocity v_T is given by

$$v_T = \sqrt{\frac{w}{CA}}^{\,*}$$

*A proof of this result is outlined in Problem 54 of Chapter 4.

where C is a number that depends on the body's shape.† Notice that $v_T \propto \sqrt{w/A}$. Thus a body with a small ratio of weight to cross-sectional area will have a small terminal velocity, which it will quickly approach. A leaf, for example, comes close to its terminal velocity after falling only a few centimeters.

A human falling through air typically has a terminal velocity of about 60 m/s. This is the approximate speed achieved by skydivers who leap from airplanes and descend for a time with closed parachutes. Of course, skydivers must open their parachutes well before reaching the

†A very streamlined body will have a relatively small value of C and will experience less air resistance than an unstreamlined body with the same ratio of weight to area.

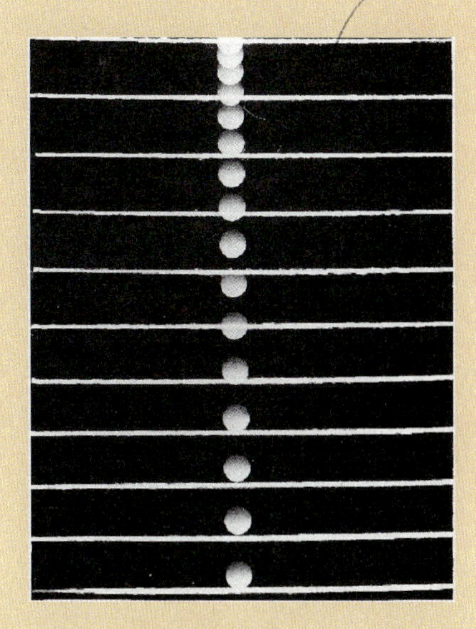

Fig. 2-B A multiflash photo of a Styrofoam ball falling through the air shows it approaching its terminal velocity, as indicated by the nearly equal spacing of the bottom images.

Continued.

A Closer Look

ground in order to reduce their terminal velocity to a lower, safer value.

Insects and other small creatures are not injured when they fall from great heights because their terminal velocities are quite low. Smaller bodies have a lower ratio of weight to area than larger bodies of the same shape. Hence small bodies tend to have small terminal velocities. To understand this more quantitatively, we can do a simple calculation relating terminal velocity for a human to the terminal velocity of an imaginary creature of human shape, but with all linear dimensions reduced to 1% of human dimensions. Weight is proportional to volume, and volume is proportional to the product of three linear dimensions, each of which is reduced by a factor of 10^{-2}. Thus both the weight and the volume of our imaginary creature are reduced from human values by a factor of $(10^{-2})^3 = 10^{-6}$. Area is the product of two linear dimensions, and so the area is reduced by a factor of $(10^{-2})^2 = 10^{-4}$. Since $v_T \propto \sqrt{\frac{w}{A}}$, v_T is reduced by a factor of $\sqrt{10^{-6}/10^{-4}} = 10^{-1}$.

Thus such a small creature would have a terminal velocity only 10% of a human's 60 m/s terminal velocity, or 6 m/s—small enough that our imaginary creature could jump off a tall building and probably avoid serious injury, since it would hit the ground at the same speed as a human jumping from a height of only 1.8 m.

*2-4 Graphical Analysis of Linear Motion

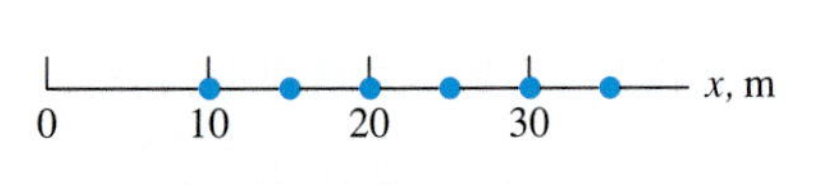

Fig. 2-11 The position of a particle is shown at 1-second intervals.

Suppose a particle moves in a straight line at a constant speed, as illustrated in Fig. 2-11 for a particle moving in the positive x direction at 5 m/s. For this kind of motion, the graph of x versus t is a straight line. As shown in Fig. 2-12a, the slope of this line is the particle's velocity v_x:

$$v_x = \frac{\Delta x}{\Delta t} = \text{slope of } x \text{ versus } t \text{ line} \qquad \text{(constant velocity)}$$

Since v_x is constant, the graph of v_x versus t is a horizontal line (Fig. 2-12b). Since there is no change in v_x, the acceleration $a_x = \frac{\Delta v_x}{\Delta t}$ equals zero at all times, as indicated in the graph of a_x versus t (Fig. 2-12c).

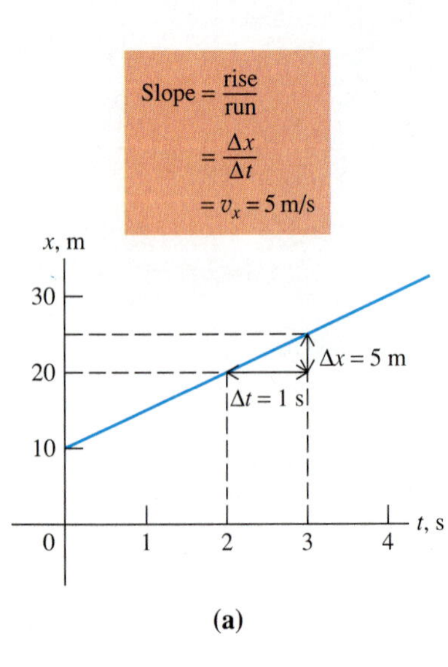

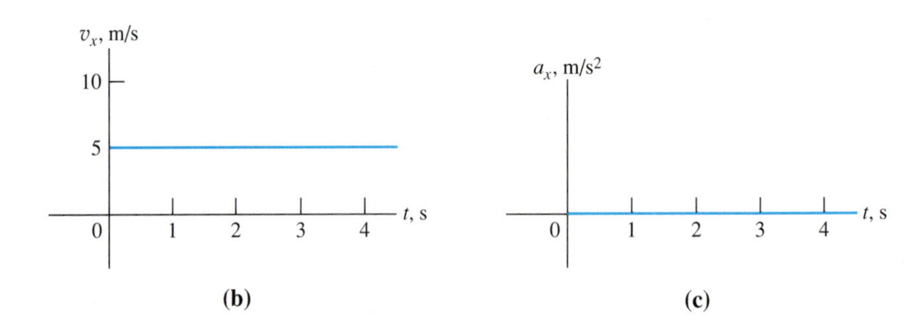

Fig. 2-12 Graphs of x, v_x, and a_x versus t for linear motion at constant velocity.

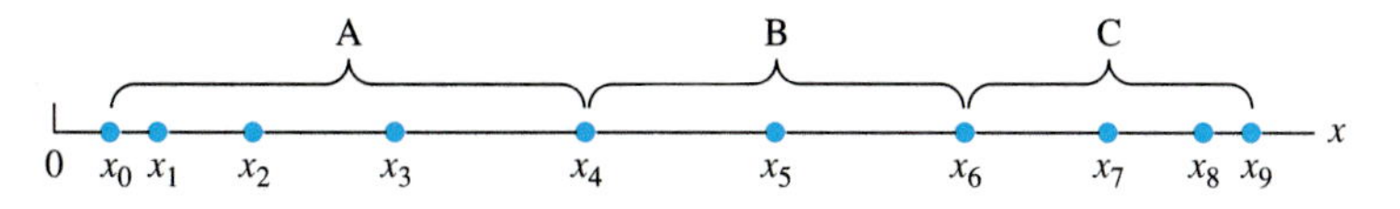

Fig. 2-13 Positions of a particle that (A) accelerates, (B) moves at constant velocity, and (C) decelerates.

Two points on the curve, t and $t + \Delta t$, are connected by a straight line of slope $\Delta x/\Delta t$. The line is nearly parallel to the tangent line, and so the slopes of the two lines are approximately equal. The approximation is improved by making Δt even smaller; as Δt approaches zero, $\Delta x/\Delta t$ approaches the slope of the tangent as its limiting value.

Next we consider the graphical interpretation of linear motion when velocity changes. To be specific, we can think of a car moving along a straight section of road. The sequence of dots in Fig. 2-13 indicates the position of the car at equal time intervals. Let $x_0, x_1, x_2, \ldots$ denote the values of the coordinate $x(t)$ at the corresponding times 0 s, 1 s, 2 s, ... The motion illustrated in Fig. 2-13 describes a car that, during interval A, starts from rest and moves to the right at increasing speed, then, during interval B, moves to the right at constant speed, and then slows down and comes to a stop during interval C.

This motion can also be indicated on a graph of x versus t, as shown in Fig. 2-14a, where the points have been connected by a smooth curve. The $x(t)$ curve slopes upward to the right, meaning that the x coordinate is increasing with time; that is, the car is moving in the positive x direction. A small part of this graph is shown as an inset to the figure on a greatly enlarged scale; from the inset we see that **the slope of a line tangent to the $x(t)$ curve at any point equals the value of the instantaneous velocity v_x at that point.**

For part A of the motion, the slope of the tangent to the $x(t)$ graph increases with t, which means that the instantaneous speed v_x is increasing, as indicated in Fig. 2-14b, where v_x is plotted versus t. For part B, Fig. 2-14a shows x versus t as a straight line, and Fig. 2-14b shows the corresponding constant velocity v_x—the slope of the straight line. For part C of the motion, the slope of the $x(t)$ curve in Fig. 2-14a gradually decreases to zero, corresponding to part C of Fig. 2-14b.

The acceleration a_x bears the same relationship to v_x as v_x bears to x; in other words, a_x is the instantaneous rate of change of v_x, just as v_x is the instantaneous rate of change of x. Thus it follows that the slope of the tangent to the v_x versus t curve at any point is the value of acceleration a_x at that point. For part A, the fact that v_x increases linearly with t means that it has a constant positive slope, and so Fig. 2-14c shows a positive constant acceleration for part A of the motion. In part B, where v_x is constant, the acceleration is zero—the slope of the horizontal line in Fig. 2-14b—and this zero acceleration is shown for part B of Fig. 2-14c. Finally, for part C, v_x decreases linearly with t and so has a constant negative slope, and this is shown in part C of Fig. 2-14c.

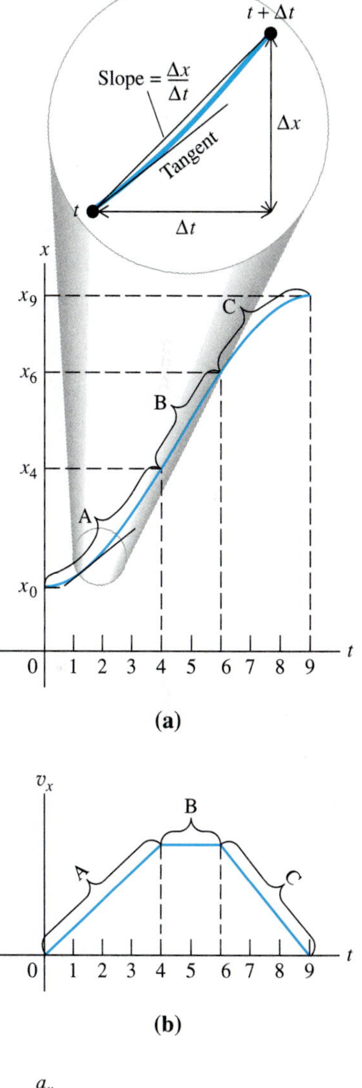

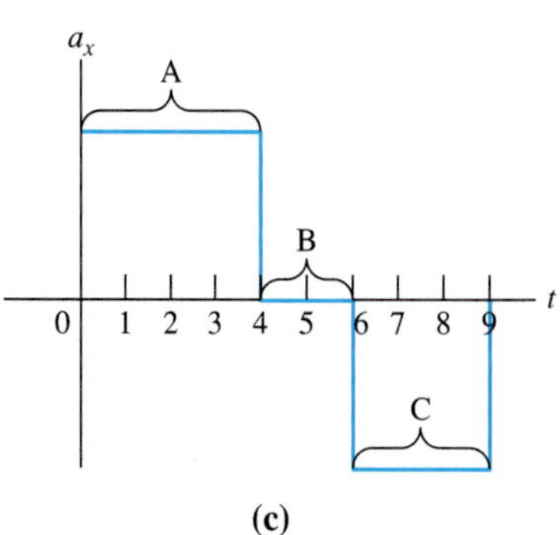

Fig. 2-14 Graphs of x, v_x, and a_x versus t for the motion shown in Fig. 2-13.

Galileo Galilei (1564-1642)

Galileo Galilei was a multifaceted Renaissance man, physically and intellectually strong, with a deep interest in mathematics and astronomy, a thorough knowledge of contemporary technology (clock-making and shipbuilding, for example), and an eloquent literary talent.

At the age of 17, Galileo entered the University of Pisa, intending to study medicine. Despite his obvious ability, his readings in mathematics and mechanics diverted him from his medical studies; he lost his scholarship and had to leave the university without a degree. He continued to work at home, studying pendulum motion and the pressure exerted on bodies submerged in water. The importance of his early work was eventually recognized, and he was appointed professor of mathematics at the University of Pisa at the age of 25. For the next 18 years Galileo continued his research and teaching, and his fame grew.

In Galileo's time, philosophy, science, and religion were all intertwined. Through the work of theologian Thomas Aquinas (1225–1274) the ancient philosophy of Aristotle (384–322 B.C.) had been reconciled with Christian theology and had come to have a central role in it. Followers of Aquinas, the Scholastics, were the dominant intellectual force of Galileo's day, and the philosophy of Aristotle was part of their accepted doctrine.

Part of Aristotle's philosophy was his physics, which in those days was called "philosophy of nature." This was very different from modern physics, in that it consisted mainly in describing the place of things in the ultimate plan of a purposeful universe. Thus a rock fell to earth because its nature was to be on the earth. This was natural motion, as opposed to the violent or forced motion of an object that was thrown or pushed. Another natural motion was that of the celestial bodies.

We would hardly call this body of beliefs a science of motion today, and indeed it was more of a system for classifying motion—part of Aristotle's larger philosophical system. Although his system was not quantitative for the most part, Aristotle did make a few quantitative statements, based apparently on casual observation and simple intuition. Even though they were wrong, these statements were useful in stimulating thought because of the importance attached to all of Aristotle's work.

Aristotle believed that heavier bodies fall to the earth faster than lighter bodies, an object 10 times as heavy as another dropping through a given distance in $\frac{1}{10}$ the time. This belief was surely based on the observation that lightweight objects, such as leaves and feathers, drop through the air more slowly than denser bodies do. But Aristotle had been uncharacteristically precise in formulating his simple general law. And in so doing, he was subject to being proved wrong.

As early as the fifth century John Philoponus had disproved Aristotle's statement by performing an experiment with two dense bodies, one 10 times as heavy as the other, and observing that they both struck the ground at virtually the same instant.* This experiment was again performed, and Aristotle's error rediscovered, by Simon Stevin in the sixteenth century. From time to time, other thinkers had challenged Aristotelian doctrine on the nature of motion. Despite the evidence, however, Aristotle's views were accepted by the Scholastics as part of the philosophical system that was so intertwined with Christian theology in the sixteenth and early seventeenth centuries.

Throughout his scientific career, Galileo fought a constant battle for the acceptance of the validity of scientific knowledge gained through experiment. It is hard for us to understand today the low status accorded to experimental observation in those days. There was no experimental science. On the one hand were the artisans and builders, who had practical goals, and on the other were the intellectuals, who dealt with the world of ideas—philosophy and mathematics. Galileo combined the talents of a practical man, an inventor, with the ability to formulate abstract concepts, characteristic of a philosopher. In applying these concepts to describe concrete physical

*It is a popular myth that Galileo first did this experiment, dropping two objects from the Leaning Tower of Pisa. Whether or not he did the experiment, he was certainly not the first to do it.

phenomena, he used the precise language of mathematics.

One of Galileo's most important contributions to science was his analysis of freely falling bodies. Great ingenuity was required in designing experiments, for there was no sufficiently accurate means of directly measuring free fall, as we can do quite easily today with a strobe light and a camera. The best time-keeping device Galileo could improvise was a water clock, here described in his famous *Dialogues:*

> For the measurement of time, we employed a large vessel of water placed in an elevated position; to the bottom of this vessel was soldered a pipe of small diameter giving a thin jet of water, which we collected in a small glass during the time of each descent . . . ; the water thus collected was weighed, after each descent, on a very accurate balance; the differences and ratios of these weights gave us the differences and ratios of times. . . .*

Instead of trying to study free fall directly, Galileo measured the time of descent for a ball rolling down an inclined plane. The incline had the effect of reducing gravitational acceleration, so that he could make accurate measurements with his water clock. He showed that, at various angles of inclination, the distance traveled was proportional to the square of the time, proving that this was motion at constant acceleration (see Eq. 2-11 with x_0 and v_{x0} set equal to 0). Since this result was valid for all the angles of inclination he was able to test, Galileo concluded that it should be true for a vertical incline as well, in other words for free fall. Furthermore, Galileo believed that in the absence of air—in a vacuum—*all* bodies, both heavy and light, would fall at the same rate. He was unable to prove this result, but he did make it plausible by discussing motion of various bodies in media of different densities—mercury, water, and air:

> Have you not observed that two bodies which fall in water, one with speed a hundred times as great as that of the other, will fall in air with speeds so nearly equal that one will not surpass the other by as much as one hundredth part? Thus, for example, an egg made of marble will descend in water one hundred times more rapidly than a hen's egg, while in air falling from a height of twenty cubits (30 ft) the one will fall short of the other by less than four finger-breadths.

After numerous other comparisons, Galileo concludes ". . . *then we are justified in believing it highly probable that in a vacuum all bodies would fall with the same speed.*" This sort of conceptualization was remarkable in that the vacuum pump had not yet been discovered and most thinkers believed that a vacuum was impossible. Galileo thought of a vacuum as a sort of limiting case, which might or might not be approachable in practice. If a perfect vacuum were attainable, however, free fall in it would be much simpler than free fall in air or any other media.

Perhaps even more important than the results Galileo obtained were the scientific methods he introduced:

1. He designed experiments allowing him to study the phenomenon of falling motion in the simplest way consistent with his technology.
2. His description was mathematically precise.
3. His conceptual model was idealized, in that it dealt with motion in the absence of friction and air resistance.
4. He was satisfied with a description of motion, leaving it to others to speculate over the causes of motion:

> The present does not seem to be the proper time to investigate the cause of the acceleration of natural motion concerning which various opinions have been expressed by various philosophers. . . . Now, all these fantasies, and others too, ought to be examined; but it is not really worth while. At present it is . . . [our purpose] merely to investigate and to demonstrate some of the properties of accelerated motion. . . .

Galileo gained fame for his wide-ranging scientific studies. To name only a few of his many accomplishments, he explained why large ships could not be built to the same scale as small boats (and the related problem of why large animals were necessarily shaped differently from smaller animals), he suggested that light might travel at a finite speed and attempted (unsuccessfully) to measure this speed, and after hearing of the invention of the telescope he made his own and used it to make fundamental discoveries in astronomy.

After his astronomical discoveries, Galileo devoted most of the last half of his life to promoting the Copernican theory of astronomy. This theory states that the earth spins on its axis and orbits the sun, in direct conflict with literal Biblical statements and with Aristotle, who had claimed that the sun and all other heav-

Continued.

*Except where otherwise noted, all quotations are from the Crew and de Salvio translation of *Dialogues Concerning Two New Sciences.*

enly bodies orbit the earth. Galileo was stubbornly committed to the Copernican theory. Although he was a religious man, he believed that the words of scripture should not be applied literally to scientific matters. He wrote of "academic philosophers . . . more in love with their doctrines than with Truth . . . publishing numerous writings full of vain arguments . . . and committing the grave error of adorning their own words with extracts from the Bible which were never meant to be interpreted literally.*"

Galileo fully realized that the work of earlier thinkers had been swallowed up by the dominant Scholastic force, and so he took his case outside the university and wrote in Italian rather than in the traditional academic Latin, apparently feeling that if his work was to have lasting impact, it must reach beyond a few biased scholars. He wrote in the appealing style of the Platonic dialogues, rejecting the more analytical style of Aristotle. His work makes good reading even today, and it was evidently successful, although at great personal cost.

Finally Galileo was brought before the Inquisition, forced to renounce his belief in the Copernican theory, and kept under house arrest for the last few years of his life. As an old man, on the verge of blindness, he returned to his early studies on forces and motion. He had a manuscript smuggled out of Italy, and the work *Dialogues Concerning Two New Sciences* was published in Holland shortly before his death. This book, written in the Platonic style, indicates that Galileo's spirit was unbroken by the ordeal of the Inquisition. He remained a free and independent thinker, one who has profoundly affected the world in which we live. His work was only the beginning of a powerful new method for studying nature, as he himself recognized: ". . . we may say the door is now opened, for the first time, to a new method fraught with numerous and wonderful results which in future years will command the attention of other minds."

*After Charon, *Cosmology*, p. 108.

CHAPTER 2 SUMMARY

For linear motion along the x-axis, average velocity $\bar{\mathbf{v}}$ and instantaneous velocity $\mathbf{v}$ are specified by the components $\bar{v}_x$ and v_x, where

$$\bar{v}_x = \frac{\Delta x}{\Delta t}$$

and
$$v_x = \lim_{\Delta t \to 0} \frac{\Delta x}{\Delta t}$$

Average acceleration $\bar{\mathbf{a}}$ and instantaneous acceleration $\mathbf{a}$ are respectively defined as

$$\bar{\mathbf{a}} = \frac{\Delta \mathbf{v}}{\Delta t}$$

and
$$\mathbf{a} = \lim_{\Delta t \to 0} \frac{\Delta \mathbf{v}}{\Delta t}$$

where the x components are expressed

$$\bar{a}_x = \frac{\Delta v_x}{\Delta t}$$

and
$$a_x = \lim_{\Delta t \to 0} \frac{\Delta v_x}{\Delta t}$$

If the x coordinate of a particle is plotted versus time, the particle's velocity v_x is the slope of the line tangent to the curve at any instant. If v_x is plotted versus time, the particle's acceleration a_x is the slope of the line tangent to the curve at any instant.

For linear motion at constant acceleration,

$$v_x = v_{x0} + a_x t$$

$$x = x_0 + \tfrac{1}{2}(v_{x0} + v_x)t$$

$$x = x_0 + v_{x0}t + \tfrac{1}{2}a_x t^2$$

$$v_x^2 = v_{x0}^2 + 2a_x(x - x_0)$$

For free fall in the absence of air resistance, taking the y-axis as positive upwards:

$$a_y = -g$$

$$v_y = v_{y0} - gt$$

$$y = y_0 + \tfrac{1}{2}(v_{y0} + v_y)t$$

$$y = y_0 + v_{y0}t - \tfrac{1}{2}gt^2$$

$$v_y^2 = v_{y0}^2 - 2g(y - y_0)$$

Questions

1 Which of the following cars has the greatest acceleration magnitude: *A*, whose speed increases from 0 to 2 m/s in 1 second; *B*, whose speed increases from 20 m/s to 23 m/s in 1 second; or *C*, whose speed decreases from 10 m/s to 5 m/s in 1 second? All cars move in a straight line.

2 A car is initially moving in reverse. The driver applies the brakes, slowing the car. What is the direction of the car's acceleration, relative to the car?

3 For the situation described in the last question, let the positive x direction be the forward direction. What are the signs of the x components of (a) velocity; (b) acceleration?

4 If v_x is negative and a_x is positive, is the speed increasing or decreasing?

5 Is it possible for an object with zero instantaneous velocity to have a nonzero instantaneous acceleration?

6 A person jumps upward, leaving the ground with the same initial velocity as an insect that is able to jump a vertical distance of 5 cm. Will the person jump a vertical distance less than, equal to, or greater than 5 cm?

7 For a body undergoing linear motion at constant acceleration, which sets of quantities would not be sufficient to determine the change in position $x - x_0$: (a) a_x, v_{x0}, t; (b) a_x, v_{x0}, v_x; (c) a_x, v_x, t; (d) $v_{x0}, v_x, \bar{v}_x$?

8 Suppose that three balls start from the same elevation with different initial velocities. Ball A starts from rest, ball B is thrown vertically upward at speed v, and ball C is thrown vertically downward at speed v. Which ball has the least speed as it hits the ground?

9 Consider an object that is initially at rest and then begins falling through the air. During what part of the object's motion will its acceleration be greatest, if it experiences significant air resistance?

Answers to odd-numbered questions

1 C; **3** (a) −, (b) +; **5** yes; **7** d; **9** the first part

Problems (listed by sections)

2-1 Acceleration in One Dimension

1 A sprinter starts from rest and accelerates to 10.0 m/s in a time interval of 1.00 s. Find the magnitude and direction of the runner's average acceleration $\bar{\mathbf{a}}$ during this interval.

2 A girl swims the length of a pool and back in 30.0 s. The pool is 20.0 m long, and the girl's speed is constant.
(a) Compute her speed v and her average velocity $\bar{\mathbf{v}}$.
(b) When is her acceleration not equal to zero?

3 A world champion drag racer starts from rest and accelerates over a quarter-mile drag strip in 5.80 s.
(a) Find the racer's average velocity $\bar{v}_x$ in m/s.
(b) The car's speed at the end of the strip is 402 km/h (250 mi/h). Find the average acceleration $\bar{a}_x$ in m/s^2.

4 A football player carrying the ball runs straight ahead at the line of scrimmage and directly into a wall of defensive linemen. The ball carrier has an initial speed of 8.00 m/s and is stopped in a time interval of 0.200 s. Find the magnitude and direction of his average acceleration.

5 A baseball player hits a line drive. Just before the ball is struck, it is moving east at a speed of 40.0 m/s (90 mi/h). Just after contact with the bat, 1.00×10^{-3} s later, the ball is moving west at a speed of 50.0 m/s (112 mi/h). Find the ball's average acceleration.

6 A car moves along a straight section of highway. Its speedometer readings at 1.0 s intervals are 50 km/h at $t = 0$, 54 km/h at $t = 1.0$ s, 56 km/h at $t = 2.0$ s, 70 km/h at $t = 3.0$ s.
(a) Find the magnitude of the average acceleration $\bar{a}_x$, in m/s^2, over the 3.0 s interval.
(b) Estimate the instantaneous acceleration, in m/s^2, at $t = 0$.

7 At 0.100 s intervals, the x coordinates of a particle are 0, 1.01 cm, 2.05 cm, and 3.12 cm. Estimate the instantaneous velocity and acceleration at $t = 0$.

8 A Ferrari F40, the world's fastest production car, can start from rest and achieve an average acceleration of 6.70 m/s^2 over a 4-second interval. How fast is the car moving after 4.00 s?

9 Find the constant acceleration of a particle whose velocity changes by 3.00 m/s every 0.100 s.

2-2 Linear Motion at Constant Acceleration

10 Find the position and velocity of a particle at $t =$ 2.00 s if the particle is initially moving east at a speed of 20.0 m/s and experiences an acceleration of magnitude 4.00 m/s^2, directed west.

11 An elevator is initially moving upward at a speed of 12.00 m/s. The elevator experiences a constant downward acceleration of magnitude 4.00 m/s^2 for 3.00 s.
(a) Find the magnitude and direction of the elevator's final velocity.
(b) How far did it move during the 3.00 s interval?

12 A paratrooper is initially falling downward at a speed of 30.0 m/s before her parachute opens. When it opens, she experiences an upward instantaneous acceleration of 75.0 m/s^2.
(a) If this acceleration remained constant, how much time would be required to reduce the paratrooper's speed to a safe 5.00 m/s? (Actually the acceleration is not constant in this case, but the equations of constant acceleration provide an easy estimate.)
(b) How far does the paratrooper fall during this time interval?

13 The brakes are applied to a car traveling on a dry, level highway. A typical value for the magnitude of the car's acceleration is 5.00 m/s^2. If the car's initial speed is 30.0 m/s, how long does it take to stop and how far does it travel, starting from the moment the brakes are applied?

14 A human being cannot tolerate acceleration of much more than a few hundred m/s^2 without sustaining serious permanent injury. If the acceleration is sustained for more than a few seconds, the maximum tolerable acceleration is much less. Suppose a car is initially moving at 30.0 m/s and then is quickly stopped in a collision.
(a) What is the minimum time over which the driver must be decelerated in order that he not experience an acceleration of magnitude greater than 1.00×10^2 m/s^2? Assume constant acceleration.
(b) What is the minimum distance the driver must be allowed to move forward during the deceleration? (Air bags are an improvement over seat belts because the distance the driver moves forward is greater with a compressible air bag than with a seat belt. This does not mean, however, that you would want to move forward in the car without air bags. Why not?)

15 Suppose that the fictional spaceship described at the beginning of this chapter accelerates at a constant rate of 0.100 m/s^2.
(a) How long would it take to reach the moon, 3.84×10^8 m away?
(b) What would be its final velocity?

16 A ball is thrown upward. It leaves the hand with a velocity of 15.0 m/s, having been accelerated through a distance of 0.500 m. Compute the ball's upward acceleration, assuming it to be constant.

17 A fullback preparing to carry the football starts from rest and accelerates straight ahead. He is handed the ball just before he reaches the line of scrimmage. Assume that the fullback accelerates uniformly (even during the handoff), reaching the line with a velocity of 8.00 m/s. If he takes 1.00 s to reach the line, how far behind it did he start?

★ **18** A block slides down an inclined plane, starting from rest and being pushed with a constant acceleration of 5.00 m/s^2 over a distance of 20.0 cm. It then decelerates at a constant rate of 1.00 m/s^2 (because of friction), until it again comes to rest. Find the total time the block is in motion.

★ **19** A cyclist starts at the top of a straight slope with an initial velocity of 3.00 m/s. Five seconds later, she is at the bottom of the incline, having traveled 50.0 m. Find her velocity at the bottom of the hill and her acceleration, assuming it to be constant.

20 A sprinter starts from rest and accelerates to her maximum speed of 10.0 m/s in a distance of 10.0 m.
(a) What was her acceleration, if you assume it to be constant?
(b) If this maximum speed is maintained for another 90.0 m, how long does it take her to run 100 m?

2-3 Free Fall

21 You throw a ball straight down from an apartment balcony to the ground below. The ball has an initial velocity of 5.00 m/s, directed downward, and it hits the ground 2.00 s after it is released. Find the height of the balcony.

22 You throw a ball vertically from an apartment balcony to the ground 15.0 m below. Find the ball's initial velocity if it hits the ground 1.00 s after you release it.

23 A boy tosses a football upward. If the football rises a vertical distance of 5.00 m and the boy catches it at the same point he released it, what is the velocity of the ball just before he catches it?

24 A girl drops a rock from the edge of a cliff and observes that it strikes the bottom 2.00 s later. How high is the cliff?

25 A rock is thrown vertically upward from the edge of a cliff. The rock reaches a maximum height of 20.0 m above the top of the cliff before falling to the base of the cliff, landing 6.00 s after it was thrown. How high is the cliff?

26 A ball thrown vertically upward has an upward velocity of 6.00 m/s at a point 12.0 m above where it was thrown. How long does the ball take to reach that point?

27 A flower pot on a balcony railing falls to the ground 10.0 m below. A cat directly beneath the pot sees it begin to fall. How long does the cat have to move?

28 A woman looks out her apartment window and sees a ball moving downward at 20.0 m/s. What were the ball's position and velocity 5.00 s earlier, if you assume it was in free fall for that long?

★ **29** Suppose that on earth you can throw a ball vertically upward a distance of 15.0 m. Given that the acceleration of gravity on the moon is 1.67 m/s^2, how high could you throw a ball on the moon?

★ **30** An arrow is shot vertically upward and then 2.00 s later passes the top of a tree 40.0 m high. How much longer will the arrow travel upward, and how high will it go?

★ **31** A rock is dropped from a treetop 19.6 m high, and then, 1.00 s later, a second rock is thrown down. With what initial velocity must the second rock be thrown if it is to reach the ground at the same time as the first?

★ **32** Show that an object thrown upward takes as long going up as coming down. HINT: Show that the time spent going up is v_{y0}/g and that the total time is $2v_{y0}/g$.

★ **33** A rocket is fired vertically upward from the surface of the earth at a constant acceleration of 20.0 m/s^2. After 30.0 s the fuel in the first stage of the rocket is expended, and the empty container is released and falls back to earth. Find the maximum height reached by the container.

*2-4 Graphical Analysis of Linear Motion

34 Interpreting Fig. 2-15 as a graph of v_x versus time, for which of the times t_1, t_2, t_3, t_4, t_5 is (a) v_x greatest; (b) v_x least; (c) a_x greatest; (d) a_x least?

35 Interpreting Fig. 2-15 as a graph of x versus time, for which of the times t_1, t_2, t_3, t_4, t_5, is (a) x greatest; (b) x least; (c) v_x greatest; (d) v_x least; (e) v greatest; (f) v least; (g) a_x greatest; (h) a_x least?

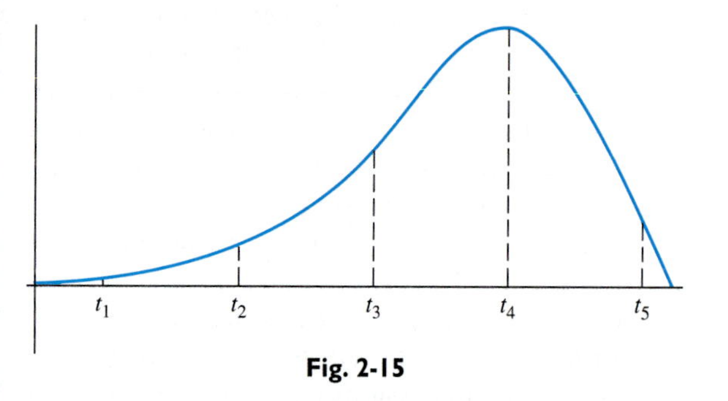

Fig. 2-15

36 Describe in words the motion represented in Fig. 2-16, and sketch the corresponding graph of a_x versus t.

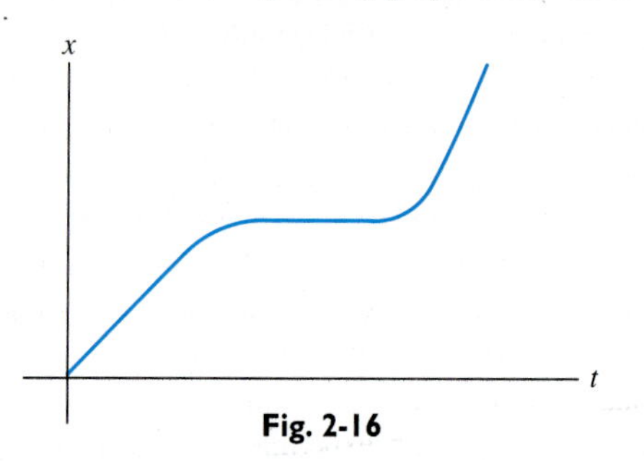

Fig. 2-16

37 Describe in words the motion represented in Fig. 2-17, and sketch the corresponding graph of a_x versus t.

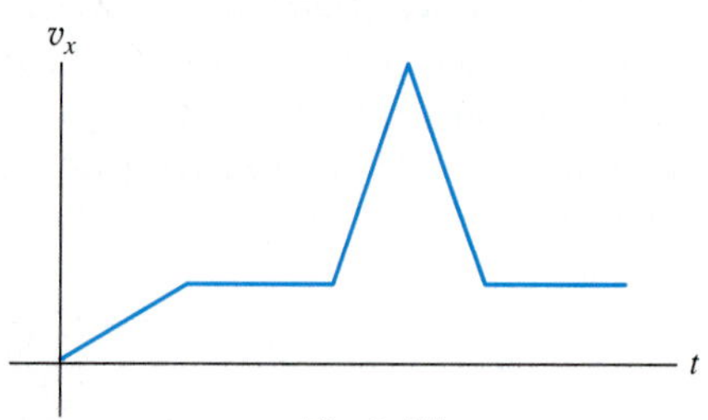

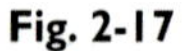

Fig. 2-17

Additional Problems

★ **38** Two cars leave a stoplight, one 0.500 s before the other. The second car accelerates with twice the acceleration of the first.

(a) How long after the first car begins to move does the second car pass it?

(b) What is the ratio of their speeds at that instant?

★ **39** A certain traffic light is yellow for 3.00 s before it turns red. As you approach the intersection, you must decide whether you can safely go through the yellow light before it turns red or whether you have time to stop before reaching the intersection. Suppose your reaction time is 0.500 s, you are initially moving at 20.0 m/s, and you can decelerate at a maximum rate of 5.00 m/s² when you step on the brakes and accelerate at a maximum rate of 3.00 m/s² when you step on the gas. (a) To come to a complete stop before reaching the intersection, what is the minimum distance you can be from the light when it turns yellow? (b) To pass the light before it turns red, what is the maximum distance you can be from the light when it turns yellow? (c) Repeat the calculations for parts (a) and (b) for an initial speed of 35.0 m/s, and find the range of distances for which you can neither stop nor get through the light in time.

40 After receiving a kickoff, football player "Rocket" Ishmael is running straight toward an opponent, who tries to tackle him. Each player moves at a speed of 9.00 m/s, and they are initially headed straight toward each other. To avoid being tackled, Ishmael accelerates laterally at 4.00 m/s², 0.500 s before they would otherwise make contact. If you assume his opponent has no time to change course, by how much do they miss?

★ **41** A child stands on a balcony directly above a friend 5.00 m below. Each throws a ball toward the other with the same initial speed of 10.0 m/s. If each catches a ball at the same instant, which child threw a ball first? How much later was the second ball thrown?

★ **42** A soccer player is about to attempt to score a goal by a penalty kick. The kicker is positioned a distance d directly in front of the goal keeper, who has a reaction time of 0.300 s and who can accelerate his arms and upper body to either side at 10.0 m/s². The kick will be aimed to pass 1.00 m to the side of the goal keeper's initial position. Find the maximum value of d, such that the goal keeper will be unable to intercept the ball, unless before the kick he correctly anticipates which way it is going. Assume that the soccer ball has a constant velocity of 30.0 m/s and that the goal keeper knows where the ball is going at the instant the ball is kicked.

★★ **43** A basketball player 2.00 m tall who is jumping vertically can accelerate upward at the rate of $2g$ until his feet leave the floor. Show that if the player begins his upward leap in a crouched position, so that the top of his head is a distance $2\text{ m} - d$ above the floor, his head will reach a maximum height of $2\text{ m} + 2d$.

★ **44** A mountain climber looks up a slope and sees an avalanche of boulders accelerating toward her. The boulders are initially 60.0 m above, moving at an initial speed of 2.00 m/s and accelerating at 4.00 m/s². The climber reacts in 0.500 s and begins to run down the slope. At what speed must she run in order to reach the bottom of the slope, 30.0 m below, before the boulders?

★★ **45** A cable car starts from rest and accelerates up a hill in San Francisco at a constant 2.00 m/s². A boy initially 15.0 m behind the car runs toward it at a constant speed of 8.00 m/s.

(a) How long does he take to reach the car?

(b) If the boy kept running past the car, when would the end of the car pass him again?

(c) What is the maximum acceleration the car should have in this problem if the boy is able to catch it?

★ **46** If a juggler requires 0.500 s to handle each object that is juggled, how high must the objects be thrown if there are (a) two objects? (b) three objects? (c) four objects?

47 A car traveling at 25.0 m/s crashes into a brick wall. The front 1.00 meter of the car crumples, and the driver comes to rest. Find the acceleration of the driver, assuming he is held in his seat by a seat belt and assuming that the acceleration is constant.

★★ **48** A boy exploring a cave drops a rock into a dark crevasse. He hears the rock strike the bottom 10.0 s after dropping it. The speed of sound is 330 m/s. Find the depth of the crevasse. HINT: Show that the time t_1 during which the rock drops is the solution to the quadratic equation

$$t_1^2 + 67.3t_1 - 673 = 0$$

★ **49** A runner travels 100 m in 10.0 s. Assume that it takes him 1.50 s to reach his top speed, which is then maintained for the rest of the race. Find his (a) average speed; (b) acceleration during the first 1.50 s; (c) maximum speed.

★ **50** The reaction time of the average automobile driver is 0.750 s. Assuming a maximum braking rate of $a_x = -6.00$ m/s², compute the minimum stopping distance for a car moving at an initial speed of 15.0 m/s (34 mi/h), 30.0 m/s (67 mi/h), and 45.0 m/s (100 mi/h).

★ **51** A baseball player preparing to steal second base takes a lead off first. His lead is limited by the pitcher, who can throw to first base before the player touches it if the lead is too large. What is the maximum lead the runner at first can take, if you assume the following: (1) as the pitcher begins to throw to first, the runner can begin to move toward the base 0.500 s before the ball is released; (2) the ball travels the 66.0 ft between the pitcher's mound and first base at a constant speed of 90.0 mi/h and the runner starts from rest and accelerates at 10.0 ft/s².

★ **52** A baseball player at first base is attempting to steal second. To be successful, he ordinarily must take a lead toward second base. How far must he stand off first base, given the following data: (1) the distance from first to second is 90.0 ft, the distance from the pitcher to the catcher is 60.0 ft, and the distance from the catcher to second base is 120 ft; (2) the baseball is thrown at a speed of 90.0 mi/h, and the runner runs at an average speed of 15.0 mi/h; (3) the runner begins to run 0.500 s before the ball is released by the pitcher, 1.50 s is required by the catcher to catch the ball and release his throw to second, and 0.500 s is required for the second baseman to make the tag.

53 In reacting to catch a glass that is about to fall off a table, you should be able to move your hand toward the glass with an acceleration considerably greater than 9.80 m/s². To test this assertion, try the following experiment. Move your hand back and forth over an interval of 20.0 cm as many times as you can in 10 s. Let N be the number of times your hand moves one way through this distance. Assume that your hand accelerates at a constant rate from rest to its maximum speed at the midpoint of the 20.0 cm interval. Show that your hand's acceleration is given by $a = (8.00 \times 10^{-3}\ \text{m/s}^2)N^2$. Compute a, assuming $N = 100$.

54 Suppose you are on a strange planet. When you flip a coin upward, it reaches a height 10.0 times as great as it would reach on earth. What is the value of gravitational acceleration on this planet?

55 Fig. 2-18 shows an experiment you can perform to measure your reaction time. When someone drops a ruler between your thumb and forefinger, you attempt to catch it. The ruler is released without warning and you react when you see the motion. By measuring how far the ruler falls before you catch it, you can determine how long it takes to react—your reaction time. Suppose the ruler drops 24.0 cm. What is your reaction time?

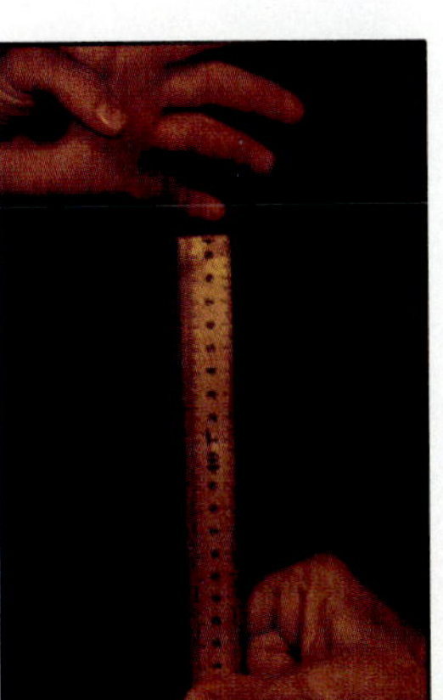

Fig. 2-18 Measuring reaction time.

CHAPTER 3 Motion in a Plane

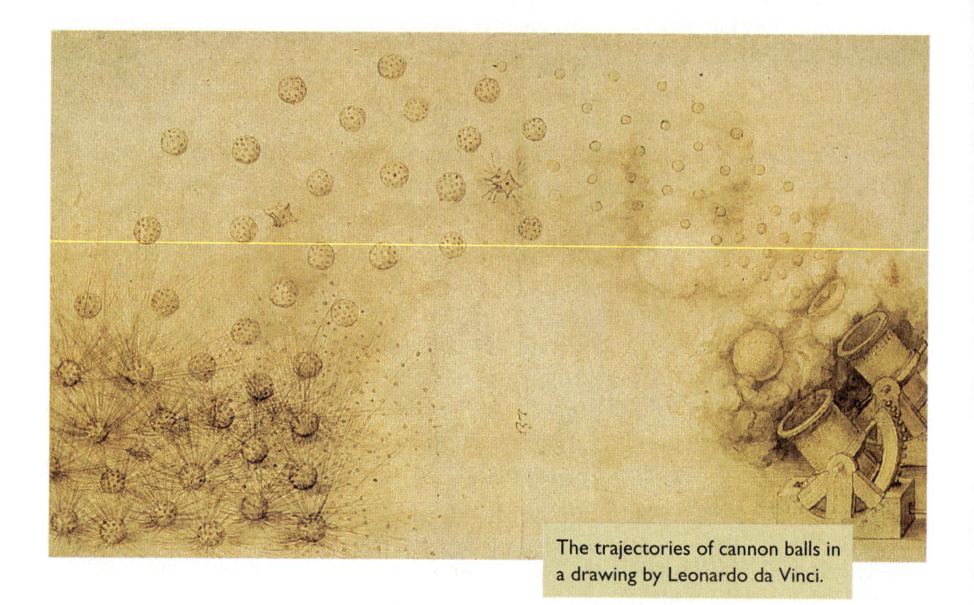
The trajectories of cannon balls in a drawing by Leonardo da Vinci.

In this chapter we shall generalize our discussion of accelerated motion to include nonlinear motion. For simplicity we shall limit our study to motion in a single plane. There are many examples of such motion: the trajectory of a baseball, football, or any other projectile is in a vertical plane, the trajectory of a car rounding a curve is in a horizontal plane, and the trajectory of an earth satellite is in a plane passing through the center of the earth.

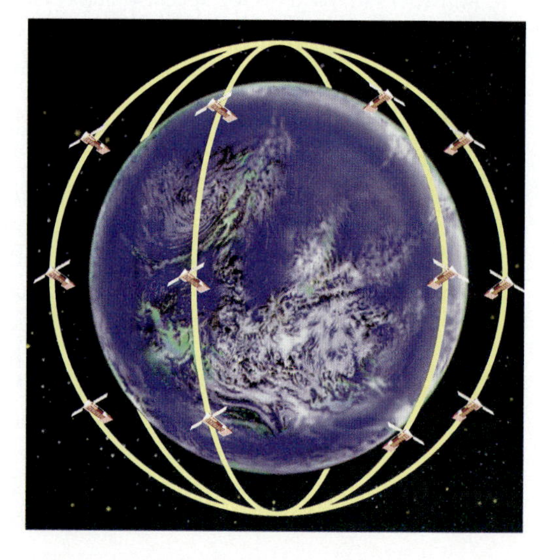
Each of the satellites in this proposed new satellite network moves in a plane. The satellites would provide a worldwide cellular telephone system, enabling subscribers to be reached at the same telephone number, no matter where they travel throughout the world.

3-1 Acceleration on a Curved Path

In Chapter 2 acceleration was defined as the rate of change of velocity. For linear motion, which is restricted to one unchanging direction, this definition implies that there is nonzero acceleration only when there is a change of speed. For nonlinear motion, however, the definition implies that there is acceleration even when the speed is constant, for the following reason: as a particle moves along a curved path, its velocity vector constantly changes direction. Since there is a change in the direction of the velocity vector, the particle is accelerated whether its speed changes or not.

In studying linear motion in Chapter 2, we were able to represent any motion as being either along the x-axis or along the y-axis. With motion in two dimensions, we need to be concerned with movement in both the x and y directions—in other words, with movement in the xy plane. We shall obtain an expression for average acceleration in the xy plane by applying the definition of average acceleration (Eq. 2-3):

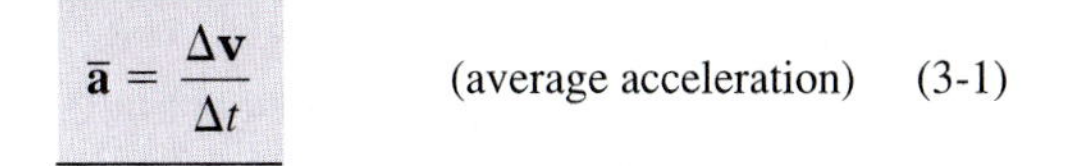

$$\bar{\mathbf{a}} = \frac{\Delta \mathbf{v}}{\Delta t} \qquad \text{(average acceleration)} \qquad (3\text{-}1)$$

In two dimensions, the change in velocity vector $\Delta\mathbf{v}$ in general has components Δv_x and Δv_y. The x and y components of the average acceleration vector are therefore the rates of change of the x and y components of velocity:

$$\bar{a}_x = \frac{\Delta v_x}{\Delta t} \qquad (3\text{-}2)$$

$$\bar{a}_y = \frac{\Delta v_y}{\Delta t} \qquad (3\text{-}3)$$

As we learned in Chapter 2, the instantaneous acceleration $\mathbf{a}$ is the limiting value of the average acceleration for a time interval approaching zero (Eq. 2-4):

$$\mathbf{a} = \underset{\Delta t \to 0}{\text{limit}} \frac{\Delta \mathbf{v}}{\Delta t} \qquad \text{(instantaneous acceleration)} \qquad (3\text{-}4)$$

In two dimensions, the instantaneous acceleration has components a_x and a_y, which are the respective limits of $\bar{a}_x$ and $\bar{a}_y$ for Δt approaching zero:

$$a_x = \underset{\Delta t \to 0}{\text{limit}} \frac{\Delta v_x}{\Delta t} \qquad (3\text{-}5)$$

$$a_y = \underset{\Delta t \to 0}{\text{limit}} \frac{\Delta v_y}{\Delta t} \qquad (3\text{-}6)$$

EXAMPLE 1 Accelerating on a Curve

A car initially is traveling east at a speed of 20.0 m/s. Then, 3.00 s later, having rounded a curve, the car is traveling 30.0° north of east at a speed of 25.0 m/s. Find the car's average acceleration during the 3.00 s interval.

SOLUTION In Fig. 3-1a we have sketched the motion of the car, with initial velocity **v** and final velocity **v′**. The corresponding change in velocity $\Delta\mathbf{v} = \mathbf{v}' - \mathbf{v}$ is indicated in Fig. 3-1b. Our knowns are: $\mathbf{v}$ = 20.0 m/s directed east, $\mathbf{v}'$ = 25.0 m/s directed 30.0° north of east, and Δt = 3.00 s. Our task is to find $\bar{\mathbf{a}}$ in terms of **v** and **v′**. We shall first find the x and y components of the car's average acceleration. Applying Eq. 3-2, we express the x component of acceleration in terms of the change in the x component of velocity:

$$\bar{a}_x = \frac{\Delta v_x}{\Delta t} = \frac{v_x' - v_x}{\Delta t}$$

From Fig. 3-1b we find

$$\bar{a}_x = \frac{v' \cos 30.0° - v}{\Delta t}$$

$$= \frac{(25.0 \text{ m/s})(\cos 30.0°) - 20.0 \text{ m/s}}{3.00\text{s}}$$

$$= 0.550 \text{ m/s}^2$$

Next we find the y component of acceleration, applying Eq. 3-3 and again using Fig. 3-1b:

$$\bar{a}_y = \frac{\Delta v_y}{\Delta t} = \frac{v_y' - v_y}{\Delta t}$$

$$= \frac{(25.0 \text{ m/s})(\sin 30.0°) - 0}{3.00 \text{ s}}$$

$$= 4.17 \text{ m/s}^2$$

The magnitude and direction of $\bar{\mathbf{a}}$ are found from its components in the usual way. First the magnitude:

$$|\bar{\mathbf{a}}| = \sqrt{\bar{a}_x^2 + \bar{a}_y^2} = \sqrt{(0.550 \text{ m/s}^2)^2 + (4.17 \text{ m/s}^2)^2}$$

$$= 4.21 \text{ m/s}^2$$

As indicated in Fig. 3-1c, the vector $\bar{\mathbf{a}}$ makes an angle θ with the x-axis, where

$$\theta = \arctan\left(\frac{\bar{a}_y}{\bar{a}_x}\right) = \arctan\left(\frac{4.17}{0.550}\right)$$

$$\approx 82°$$

The car experiences an average acceleration of 4.21 m/s² in a direction 82° north of east. This considerable acceleration is caused mainly by the change in direction (Δv_y) rather than by the change in speed. The same rate of change of speed on a straight road would produce an acceleration of only (25.0 m/s − 20.0 m/s)/3.00 s = 1.67 m/s².

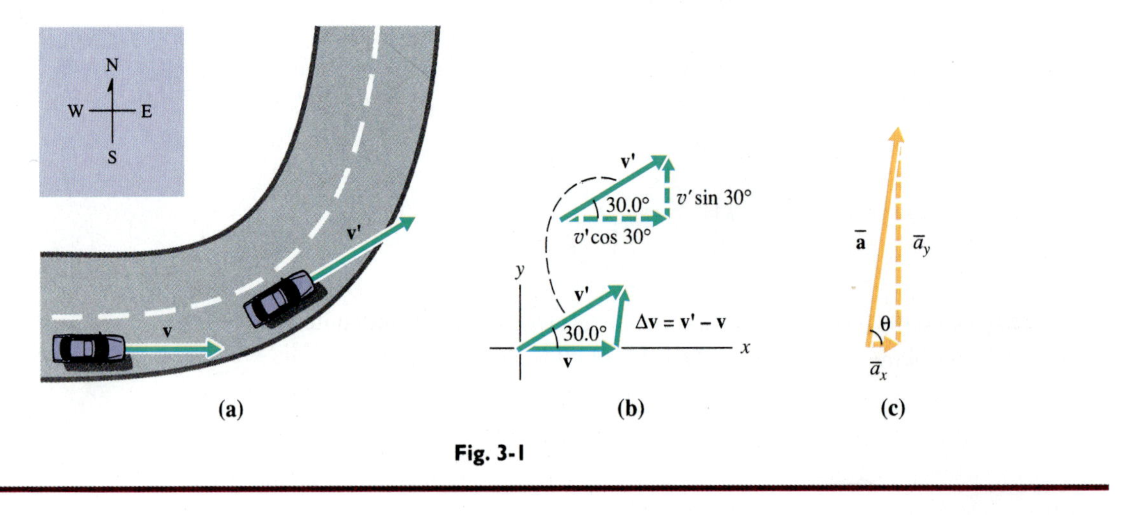

Fig. 3-1

If the x and y components of a particle's acceleration are known functions of time, we can find expressions for both the particle's velocity (v_x and v_y) and its position (x and y) as functions of time. In other words, we can predict the future motion of the particle. We shall see an example of this in the following section, for the particularly simple case of projectile motion.

3-2 Projectile Motion

In Section 2-3 we learned that, in the absence of air resistance, a freely falling object near the surface of the earth has constant acceleration. Whether an object falls from rest or is given some vertical initial velocity, the same free-fall equations apply.

Now suppose we attempt to describe the motion of an object that is thrown into the air with an initial velocity that is not vertically directed. The object is thrown, or *projected,* with an initial velocity vector $\mathbf{v}_0$ that makes some angle θ_0 with the horizontal, as in Fig. 3-2.

Fig. 3-2 Projectile motion.

An experimental fact that was first recognized by Galileo is that any projectile experiences exactly the same acceleration as a freely falling body does: its acceleration vector, denoted by $\mathbf{g}$, is a constant vector of magnitude 9.8 m/s^2 pointing in the downward direction.* Choosing our coordinate system as in Fig. 3-3, we have $a_x = 0$, $a_y = -g$, and $x_0 = y_0 = 0$, where x_0 and y_0 are the coordinates of the original position of the projectile. Both components of acceleration are constant. In Sections 2-2 and 2-3 we derived equations of motion for constant acceleration in either the x or the y direction. So we may apply the two sets of equations: Eqs. 2-7 and 2-11 for the x direction (with a_x and x_0 set equal to zero) and Eqs. 2-13 through 2-16 for the y direction (with y_0 set equal to zero):

$$\begin{aligned} &\text{(a) } a_x = 0 \\ &\text{(b) } v_x = v_{x0} \\ &\text{(c) } x = v_{x0}t \end{aligned} \tag{3-7}$$

$$\begin{aligned} &\text{(a) } a_y = -g \\ &\text{(b) } v_y = v_{y0} - gt \\ &\text{(c) } y = \tfrac{1}{2}(v_{y0} + v_y)t \\ &\text{(d) } y = v_{y0}t - \tfrac{1}{2}gt^2 \\ &\text{(e) } v_y^2 = v_{y0}^2 - 2gy \end{aligned} \tag{3-8}$$

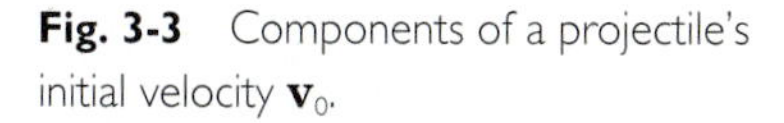

Fig. 3-3 Components of a projectile's initial velocity $\mathbf{v}_0$.

The components of the initial velocity vector, v_{x0} and v_{y0}, determine the entire motion of the projectile. In many sports, the skill of an athlete rests on his or her ability to impart the correct initial velocity to a ball, thereby determining where it will go. In basketball, for example, once the ball leaves a player's hand, its value of $\mathbf{v}_0$ is fixed and its motion is thereafter governed by Eqs. 3-7 and 3-8 (as long as its path is unobstructed). Whether the player makes the basket depends on the value of $\mathbf{v}_0$. In football, the crucial problem for the quarterback trying to complete a pass is to release the ball at the right time with the right initial velocity, so that it arrives downfield in the hands of the intended receiver.

*We assume here that air resistance is negligible and that the trajectory of the particle is very small compared with the radius of the earth.

EXAMPLE 2 Projecting a Marble Horizontally

A marble rolls along a table at a constant speed of 1.00 m/s and then falls off the edge of the table to the floor 1.00 m below. (a) How long does the marble take to reach the floor? (b) At what horizontal distance from the edge of the table does the marble land? (c) What is its velocity as it strikes the floor? (d) Indicate in a diagram the marble's position and velocity at 0.100 s intervals.

SOLUTION Projectile motion of the marble begins as it leaves the table (Fig. 3-4). Since the marble is initially moving horizontally, $v_{y0} = 0$ and $v_{x0} = 1.00$ m/s. In order to use Eqs. 3-7 and 3-8, we must take the origin to be at the edge of the table, so that $x_0 = y_0 = 0$. **(a)** This problem can be stated: Find t when $y = -1.00$ m. Because $v_{y0} = 0$, Eq. 3-8d reduces to

$$y = -\tfrac{1}{2}gt^2$$

Solving for t, we find

$$t = \sqrt{-2y/g} = \sqrt{(-2)(-1.00\text{ m})/(9.80\text{ m/s}^2)} = 0.452\text{ s}$$

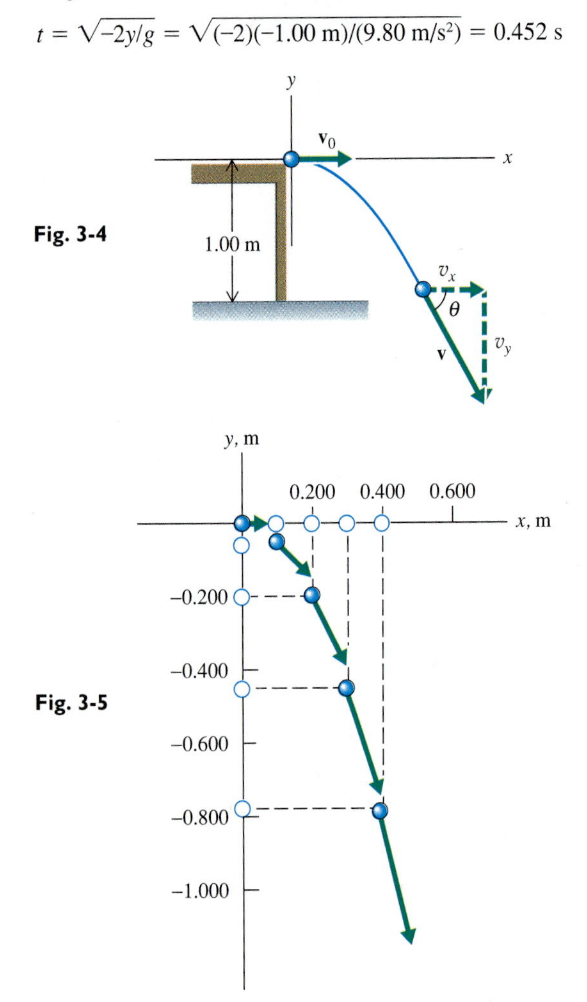

Fig. 3-4

Fig. 3-5

(b) Here we want the marble's x coordinate at the instant it strikes the floor; that is, we wish to find x when $t = 0.452$ s. Eq. 3-7 gives

$$x = v_{x0}t = (1.00\text{ m/s})(0.452\text{ s}) = 0.452\text{ m}$$

(c) Here we must find $\mathbf{v}$ at $t = 0.452$ s. The x component of velocity is constant throughout the motion (Eq. 3-7b):

$$v_x = v_{x0} = 1.00\text{ m/s}$$

We find the y component by using Eq. 3-8b:

$$v_y = v_{y0} - gt = 0 - (9.80\text{ m/s}^2)(0.452\text{ s}) = -4.43\text{ m/s}$$

Thus the velocity vector has magnitude

$$v = \sqrt{v_x^2 + v_y^2} = \sqrt{(1.00\text{ m/s})^2 + (-4.43\text{ m/s})^2}$$
$$= 4.54\text{ m/s}$$

and is directed at an angle θ below the horizontal, where

$$\theta = \arctan\left(\frac{|v_y|}{v_x}\right) = \arctan\left(\frac{4.43}{1.00}\right) = 77.3°$$

As the marble hits the floor, its velocity is 4.54 m/s directed 77.3° below the horizontal.

(d) The marble's coordinates and velocity components at times $t = 0$, 0.100 s, 0.200 s, 0.300 s, and 0.400 s are obtained by direct substitution into Eqs. 3-7b and c, and Eqs. 3-8b and d. The results were used to obtain Fig. 3-5. As illustrated in the figure, the motion in the x direction is that of a body moving at constant velocity, while the motion in the y direction is that of a freely falling body. What we see is the vector sum of these two effects. Fig. 3-6 shows the motion of a horizontally projected object.

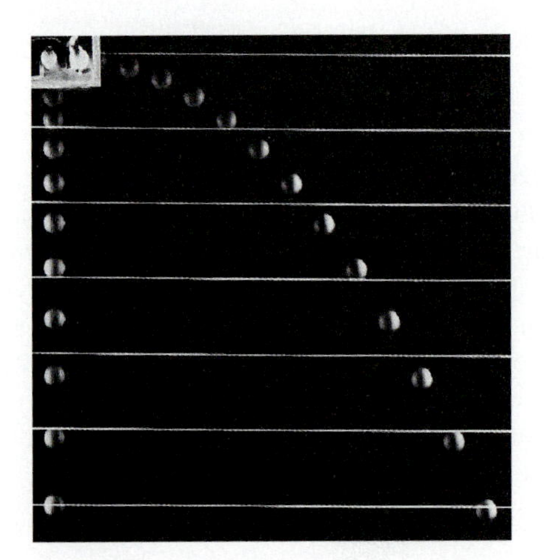

Fig. 3-6 One ball is released from rest, and at the same instant the other is given a horizontal initial velocity. Both balls are at the same elevation at any instant.

EXAMPLE 3 Throwing Too High

A quarterback, standing on his opponents' 35-yard line, throws a football directly downfield, releasing the ball at a height of 2.00 m above the ground with an initial velocity of 20.0 m/s, directed 30.0° above the horizontal.
(a) How long does it take for the ball to cross the goal line, 32.0 m (35 yards) from the point of release? (b) The ball is thrown too hard and so passes over the head of the intended receiver at the goal line. What is the ball's height above the ground as it crosses the goal line?

SOLUTION To better visualize the situation described here, we first sketch the trajectory (Fig. 3-7):
(a) The problem here is to find t when $x = 32.0$ m. We can use Eq. 3-7c ($x = v_{x0}t$), if we first find v_{x0}. From Fig. 3-7 we see that

$$v_{x0} = v_0 \cos \theta_0 = (20.0 \text{ m/s})(\cos 30.0°)$$
$$= 17.3 \text{ m/s}$$

Now we apply Eq. 3-7c and solve for t.

$$x = v_{x0}t$$

$$t = \frac{x}{v_{x0}} = \frac{32.0 \text{ m}}{17.3 \text{ m/s}} = 1.85 \text{ s}$$

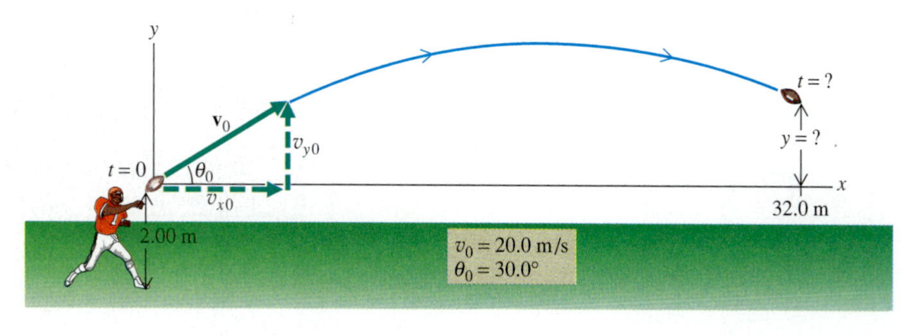

Fig. 3-7

(b) We want to find y when $x = 32.0$ m, or, since we have already found the time in part (a), we can state this: find y when $t = 1.85$ s. We apply Eq. 3-7d:

$$y = v_{y0}t - \tfrac{1}{2}gt^2,$$

where

$$v_{y0} = v_0 \sin \theta_0 = (20.0 \text{ m/s})(\sin 30.0°)$$
$$= 10.0 \text{ m/s}$$

Thus

$$y = (10.0 \text{ m/s})(1.85 \text{ s}) - \tfrac{1}{2}(9.80 \text{ m/s}^2)(1.85 \text{ s})^2$$
$$= 1.73 \text{ m}$$

Since $y = 0$ is 2.00 m above the ground, this means the ball is 3.73 m above the ground as it crosses the goal line—much too high to be caught at that point.

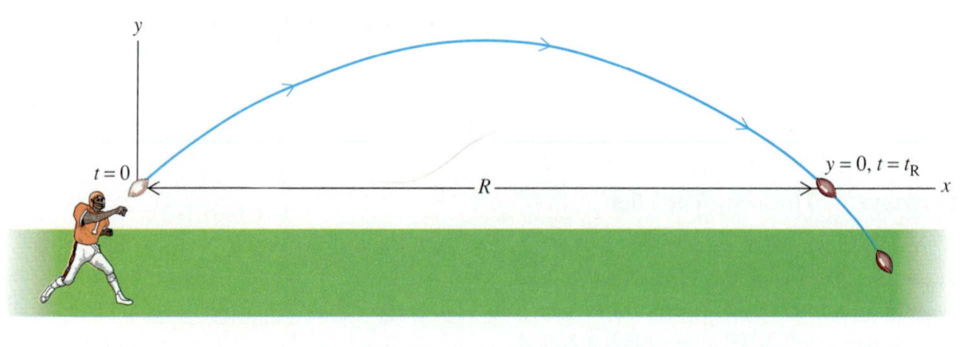

Fig. 3-8 The range of a projectile.

Although Eqs. 3-7 and 3-8 are sufficient to solve any problem in projectile motion, it is sometimes convenient to have other formulas indicating various aspects of the projectile's path. We shall find an expression for a projectile's horizontal range R, the horizontal distance traveled by the projectile before returning to its initial elevation ($y = 0$), illustrated in Fig. 3-8. And we shall find an expression for the time t_R the projectile takes to travel the distance R. We can find the time t_R at which the projectile returns to its initial elevation using Eq. 3-8d, setting $y = 0$.

$$y = 0 = v_{y0}t - \tfrac{1}{2}gt^2$$

The nonzero solution to this equation is the time t_R:

$$t_R = \frac{2v_{y0}}{g} \tag{3-9}$$

We can obtain an expression for the horizontal range by applying Eq. 3-7c.

$$x = v_{x0}t$$

Setting $x = R$ when $t = t_R$ and using Eq. 3-9 for t_R, we find

$$R = v_{x0}\left(\frac{2v_{y0}}{g}\right)$$

Next we substitute

$$v_{x0} = v_0 \cos\theta_0 \qquad \text{and} \qquad v_{y0} = v_0 \sin\theta_0$$

to obtain

$$R = \frac{2v_0^2 \sin\theta_0 \cos\theta_0}{g}$$

We can use the trigonometric identity $\sin 2\theta_0 = 2\sin\theta_0 \cos\theta_0$ to express this result more concisely:

$$R = \frac{v_0^2 \sin 2\theta_0}{g} \tag{3-10}$$

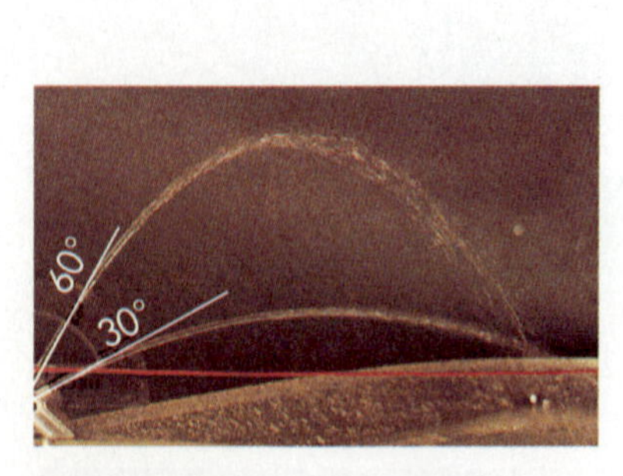

Fig. 3-9 Water is projected from two tubes at the same speed—from one at an angle of 30° and from the other at 60°. Why are the ranges equal?

The dependence of R on v_0^2 shows that a doubling of v_0 quadruples the range. For example, if a ball is thrown a distance of 20 m, doubling the initial speed will increase the range to 80 m. For a fixed value of v_0, R will be maximum at an angle $\theta_0 = 45°$, for which $\sin 2\theta_0 = \sin 90° = 1$.

EXAMPLE 4 Snowball Strategy

A clever strategy in a snowball fight is to throw two snowballs at your opponent in quick succession, the first one with a high trajectory and the second one with a lower trajectory and shorter time of flight, so that they both reach the target at the same instant. Suppose your opponent is 20.0 m away. You throw both snowballs with the same initial speed v_0, but θ_0 is 60.0° for the first snowball and 30.0° for the second. If they are both to reach their target at the same instant, how much time must elapse between the release of the two snowballs?

SOLUTION We need to find the time of flight for each snowball. The time t_R is determined by v_{y0}, the vertical component of initial velocity, according to Eq. 3-9:

$$t_R = \frac{2v_{y0}}{g} = \frac{2v_0 \sin \theta_0}{g}$$

To find t_R, we need to know, in addition to the initial angle θ_0 (a given), the initial speed v_0, which is not given. We can find v_0 by applying the range equation (Eq. 3-10):

$$R = \frac{v_0^2 \sin 2\theta_0}{g}$$

Solving for v_0, we obtain

$$v_0 = \sqrt{\frac{Rg}{\sin 2\theta_0}}$$

We obtain the same value for v_0 whether we use $\theta_0 = 30.0°$ or $\theta_0 = 60.0°$, since $\sin 2(30.0°) = \sin 2(60.0°)$:

$$v_0 = \sqrt{\frac{(20.0 \text{ m})(9.80 \text{ m/s}^2)}{\sin 60.0°}} = 15.0 \text{ m/s}$$

Now we can apply Eq. 3-9 and find t_R for each snowball.

$$t_R = \frac{2v_{y0}}{g} = \frac{2v_0 \sin \theta_0}{g}$$

For the first snowball,

$$t_R = \frac{2(15.0 \text{ m/s})(\sin 60.0°)}{9.80 \text{ m/s}^2} = 2.65 \text{ s}$$

For the second snowball,

$$t_R' = \frac{2(15.0 \text{ m/s})(\sin 30.0°)}{9.80 \text{ m/s}^2} = 1.53 \text{ s}$$

Thus you should wait a time Δt before making your second throw, where Δt is the difference in the times of flight:

$$\Delta t = t_R - t_R' = 2.65 \text{ s} - 1.53 \text{ s} = 1.12 \text{ s}$$

The trajectory equation describes the path of a projectile by relating its x and y coordinates. We derive this equation by solving Eq. 3-7c for t to obtain $t = x/v_{x0}$ and then substituting this expression for t into Eq. 3-8d:

$$y = v_{y0}t - \tfrac{1}{2}gt^2$$

$$= v_{y0}\left(\frac{x}{v_{x0}}\right) - \tfrac{1}{2}g\left(\frac{x}{v_{x0}}\right)^2$$

From Fig. 3-3 we see that $v_{y0}/v_{x0} = \tan \theta_0$. Substituting into the equation above, we obtain

$$y = (\tan \theta_0)x - \left(\frac{g}{2v_{x0}^2}\right)x^2 \qquad (3\text{-}11)$$

This equation has the form $y = ax + bx^2$, the general equation for a parabola. Thus a projectile has a parabolic trajectory (Fig. 3-10).

Fig. 3-10 Parabolic paths of projectiles.

EXAMPLE 5 Shooting With the Right Velocity

A basketball player shoots the ball at a hoop 3.00 m above the floor from a horizontal distance of 6.00 m from the center of the hoop. The ball leaves the player's hand 2.00 m above the floor at an angle of 45.0° with the horizontal. With what initial speed must the ball be shot in order to hit the center of the basket, without hitting the rim or backboard?

SOLUTION First we sketch the motion (Fig. 3-11). Once again we must take the origin to be at the initial position of the ball because the equations describing projectile motion were derived on the assumption $x_0 = y_0 = 0$. Our problem is to find v_0 such that $y = 1.00$ m when $x = 6.00$ m. We use the trajectory equation (Eq. 3-11) because it relates v_0 to quantities that are given:

$$y = (\tan \theta_0)x - \frac{gx^2}{2v_{x0}^2}$$

Substituting $v_{x0} = v_0 \cos \theta_0$, we have

$$1.00 \text{ m} = (\tan 45.0°)(6.00 \text{ m}) - \frac{(9.80 \text{ m/s}^2)(6.00 \text{ m})^2}{2v_0^2 \cos^2 45.0°}$$

Solving for v_0, we find

$$v_0 = 8.40 \text{ m/s}$$

Through experience a good shooter knows how to give the ball just this initial speed.

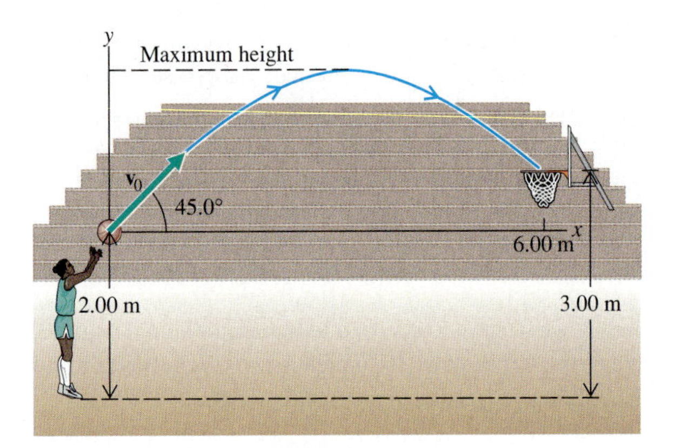

Fig. 3-11

3-3 Circular Motion

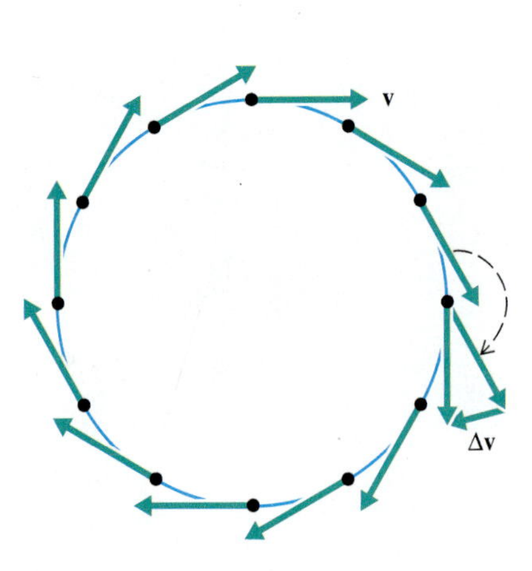

Fig. 3-12 A particle moving along a circular path at constant speed has a velocity **v** that changes direction. The velocity change $\Delta\mathbf{v}$ from one position to the next is constructed in the figure.

A particle moving along a circular path at constant speed is said to undergo uniform circular motion. Fig. 3-12 shows the position and the velocity vector at selected points for a particle moving in this way. The dots are equally spaced and the velocity vectors are of uniform length, indicating that the speed is constant. However, the fact that **v** changes direction as the particle moves along the circle implies that there is nonzero acceleration. The change in **v** from one point to the next is a vector, $\Delta\mathbf{v}$, pointing toward the inside of the curve, as indicated in Fig. 3-12. The particle's path is bending inward, and therefore its acceleration is inward.

Next we shall determine the direction and magnitude of the instantaneous acceleration of a particle in uniform circular motion, using Fig. 3-13. Fig. 3-13a shows the particle's velocity vector drawn tangent to the trajectory at two points P and P′. The particle moves from P to P′ along an arc of length Δs over a time interval Δt, and the velocity vector changes from **v** to $\mathbf{v}' = \mathbf{v} + \Delta\mathbf{v}$ during this time. The triangle formed by vectors **v**, **v**′, and $\Delta\mathbf{v}$ is constructed in the figure. We see that vector $\Delta\mathbf{v}$ is in the same direction as a line from the center of the arc Δs to point O at the center of the circle. The particle's average acceleration $\bar{\mathbf{a}} = \Delta\mathbf{v}/\Delta t$ is a vector proportional to $\Delta\mathbf{v}$, and so it points in the same direction as $\Delta\mathbf{v}$.

In Fig. 3-13b $\Delta\mathbf{v}$ is constructed for a shorter time interval Δt. Again vectors $\Delta\mathbf{v}$ and $\bar{\mathbf{a}}$ are directed toward the center of the circle. Because Δt is shorter, this second figure shows a smaller change in velocity, $\Delta\mathbf{v}$. However, the vector $\bar{\mathbf{a}} = \Delta\mathbf{v}/\Delta t$ has slightly greater magnitude than in the first figure. As the time interval Δt approaches zero, $\Delta\mathbf{v}$ also approaches zero, but the ratio $\Delta\mathbf{v}/\Delta t$ approaches as a limit the instantaneous acceleration **a**, shown in Fig. 3-13c. Notice that **a is directed toward the center of the circle,** perpendicular to **v**.

Next we wish to find an expression for the magnitude of **a**. In Fig. 3-13a the angle θ between the equal length vectors **v** and **v′** is the same as the angle θ between the two equal length lines OP and OP′, since **v** is perpendicular to OP and **v′** is perpendicular to OP′. (Remember that the velocity vector is always tangent to the trajectory and therefore is always perpendicular to the radius of the circle.) This means that the triangle formed by the velocity vectors is geometrically similar to the triangle OPP′. Therefore ratios of corresponding sides of these two triangles are equal. For very short time intervals, the length of line PP′ is approximately equal to the arc length Δs. Thus

$$\frac{|\Delta\mathbf{v}|}{v} \approx \frac{\Delta s}{r}$$

Multiplying both sides of this equation by $v/\Delta t$, we obtain

$$\frac{|\Delta\mathbf{v}|}{\Delta t} \approx \frac{v}{r}\frac{\Delta s}{\Delta t}$$

As Δt approaches 0, the equation becomes exact and $\Delta s/\Delta t$ approaches v, the instantaneous speed. Thus the limiting value of $\frac{|\Delta\mathbf{v}|}{\Delta t}$, the magnitude of the instantaneous acceleration **a**, is given by

$$a = \frac{v^2}{r} \quad \text{(uniform circular motion)} \quad (3\text{-}12)$$

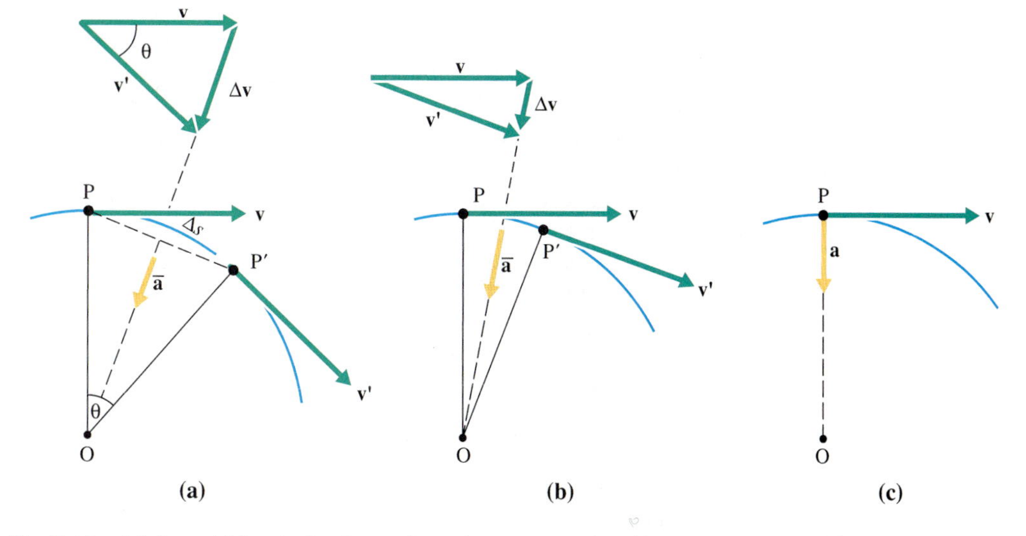

Fig. 3-13 **(a)** A particle's velocity change $\Delta\mathbf{v}$ and average acceleration $\bar{\mathbf{a}}$ are constructed for some time interval Δt. **(b)** $\Delta\mathbf{v}$ and $\bar{\mathbf{a}}$ are constructed for a shorter time interval. **(c)** Instantaneous acceleration, **a**.

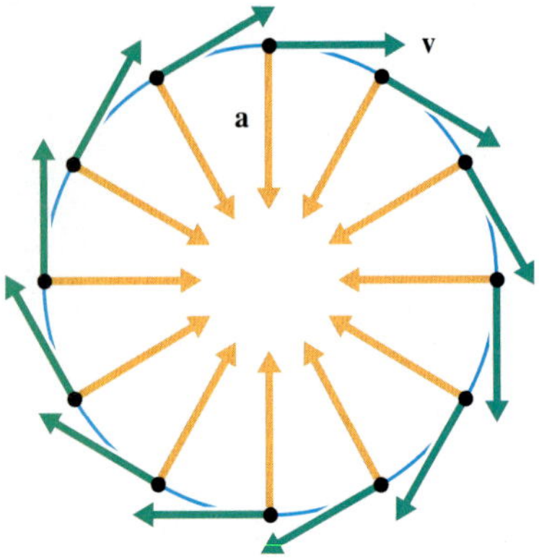

Fig. 3-14 Centripetal acceleration **a** is at any instant directed toward the center of the circular path, perpendicular to velocity **v**.

In Fig. 3-14 the illustration of Fig. 3-12 is repeated, this time with the acceleration vectors drawn in. The acceleration vector always points from the instantaneous position of the particle to the center of the circle. The term **centripetal acceleration** is used to describe **a**. "Centripetal" means 'directed toward the center'.

Note that although the magnitude of **a** is constant for uniform circular motion, the acceleration vector is not constant; it continuously changes direction—that direction always being toward the center of the circle.

As a physical example of uniform circular motion, suppose a rock is attached to a string and swung horizontally overhead along a circular path at a constant rate. The rock experiences centripetal acceleration; its acceleration vector points inward along the string (Fig. 3-15). We shall see in Chapter 4 that a force is required to produce an acceleration. The string maintains the circular path and its associated inward acceleration by exerting an inward pull (a force) on the rock. The greater the speed of the rock, the greater the acceleration and the greater the force provided by the string. A person swinging a rock in this way must pull harder on the string as the speed is increased. If at some instant the string is released, the rock will fly off tangent to the circle, in the direction of its instantaneous velocity.

Fig. 3-15 A rock swung overhead from a string moves along a circular path.

The effect of acceleration is felt by passengers in a car making a sharp turn. Suppose that a car moves at a speed of 10 m/s around a curve of radius 20 m. The car's centripetal acceleration $v^2/r = (10 \text{ m/s})^2/20 \text{ m} = 5 \text{ m/s}^2$, or approximately 0.5 g. Such a turn would cause a passenger to move relative to the car until something (perhaps the door) can provide a strong enough force to compel the passenger to follow the circular path. This force is directed inward, toward the center of the circle, that is, in the direction of the centripetal acceleration.

It is possible for a projectile to have a circular trajectory. If the initial velocity of a projectile is very large, so that its trajectory is not very small relative to the size of the earth, the description of projectile motion in Section 3.2 is no longer valid. If the initial velocity is great enough (on the order of thousands of meters per second), the projectile may become a satellite orbiting the earth (Fig. 3-16). For the special case of a circular orbit (achieved by giving the satellite just the right initial velocity), the satellite's speed is constant and its motion is uniform and circular. In this case the centripetal acceleration is equal to gravitational acceleration. This means that if a satellite could orbit the earth at an elevation of only a few kilometers above the surface, its centripetal acceleration would be 9.8 m/s², directed toward the center of the earth. Air resistance precludes such a low orbit, however. At a more realistic elevation of 200 km, a satellite would experience a centripetal acceleration of 9.2 m/s², the same acceleration that would be experienced by a body released from rest at this altitude (see Problem 26).

Fig. 3-16 In this drawing from the *Principia,* Newton showed how a particle projected horizontally from a high elevation might be given an initial velocity great enough to orbit the earth. For smaller values of v_0 the paths (D and E) are approximately parabolic, but if v_0 is sufficiently great, the particle *falls around the earth,* rather than into it. Thus Newton anticipated artificial satellites over 300 years ago.

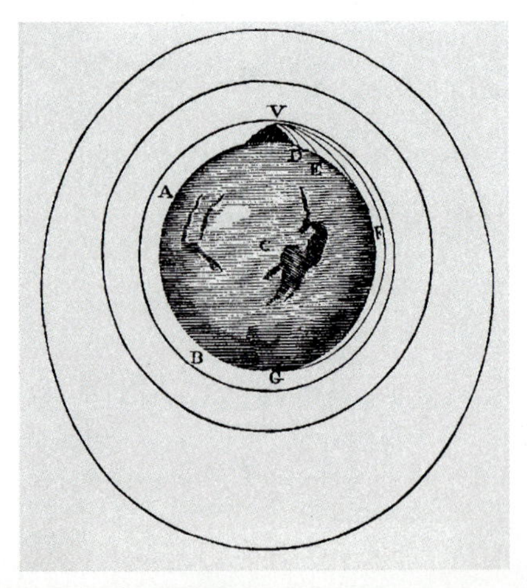

EXAMPLE 6 Swinging an Arm

Extend your arm straight up overhead and then swing it in a vertical plane, so that your hand follows a circular path. If you rotate your arm as fast as possible, you should be able to achieve a rate of 2.0 revolutions per second. Compute the acceleration of the blood in your fingertips at this rate, assuming your arm is 65 cm long.

SOLUTION As indicated in Fig. 3-17, the acceleration vector is directed toward the center of the circle, which is at your shoulder joint. The magnitude of the acceleration is given by Eq. 3-12.

$$a = \frac{v^2}{r}$$

To use this equation, we must know the speed v. We compute v by dividing the path length for one revolution ($2\pi r$, the circumference of the circle) by the time for one revolution, called the **period** and denoted by T:

$$v = \frac{2\pi r}{T}$$

Since the rate of rotation is 2.0 rev/s, the period $T = 0.50$ s. Combining the two preceding expressions, we obtain

$$a = \frac{(2\pi r/T)^2}{r} = \frac{4\pi^2 r}{T^2}$$

$$= \frac{4\pi^2(0.65 \text{ m})}{(0.50 \text{ s})^2} = 100 \text{ m/s}^2$$

This is approximately 10 g. Such an acceleration is quite painful.

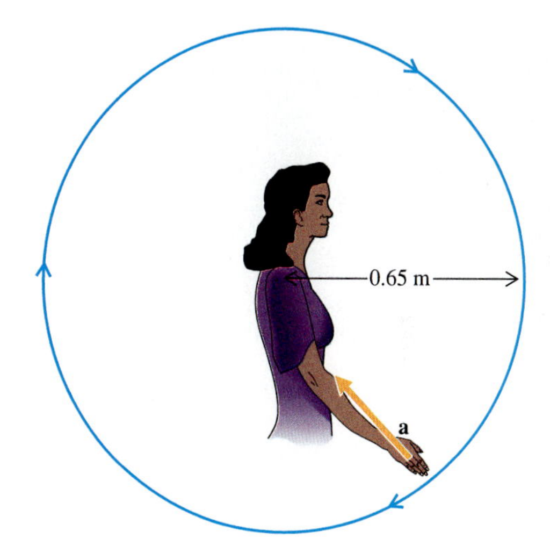

Fig. 3-17

Fig. 3-18 A car moves relative to the earth, and the earth moves relative to the car.

3-4 Reference Frames and Relative Motion

We shall sometimes refer to a coordinate system used in the description of motion as a "reference frame." It is not hard to imagine yourself standing at the origin of a reference frame and viewing a moving body.

Since we are earthbound, it is always tempting to think of the earth as stationary and to describe any motion from the earth's reference frame, that is, from a set of coordinate axes fixed with respect to the surface of the earth.* However, there is no fundamental reason to give this reference frame a uniquely privileged status. The earth moves relative to other bodies. Suppose, for example, you are in a moving car or train. As you look out the window, you see passing trees and buildings. Relative to you, these objects are moving. Indeed, the entire earth is moving relative to you (Fig. 3-18).

We shall often find it convenient to describe the motion of a body from a reference frame that is moving relative to the earth. For example, a moving plane or train serves as a natural reference frame for describing the motion of passengers inside.

It is sometimes useful to be able to go from one reference frame to another, that is, to have transformation equations that allow us to relate descriptions of motion in two different reference frames. Such relationships are of practical importance to an airplane pilot, for example. The pilot is obviously interested in the plane's motion relative to the earth. But the plane is moving relative to the *air,* and so, when determining the plane's position, the pilot must know how to compensate for the velocity of the air relative to the earth. We shall now derive the equations relating the relative motion of three bodies (earth, air, and plane in our example). Each of the three arbitrary moving bodies, labeled A, B, and C, are treated on equal footing. Each body moves relative to the other two.

Relative displacement is a displacement vector directed from one body to another. Let $\mathbf{D}_{AB}$ denote the displacement of A relative to B, that is, the displacement vector required to go from the position of B to the position of A (Fig. 3-19a). Using this notation, $\mathbf{D}_{AC}$ denotes the displacement of A relative to C, and $\mathbf{D}_{CB}$ denotes the displacement of C relative to B. These relative displacement vectors are related to each other. From Fig. 3-19b we see that vectors $\mathbf{D}_{AB}$, $\mathbf{D}_{AC}$, and $\mathbf{D}_{CB}$ form a vector addition triangle, with $\mathbf{D}_{AB}$ equal to the vector sum of the other two vectors:

$$\mathbf{D}_{AB} = \mathbf{D}_{AC} + \mathbf{D}_{CB} \tag{3-13}$$

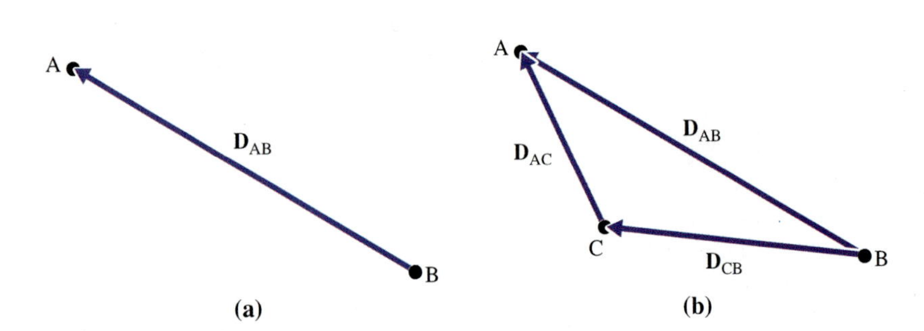

Fig. 3-19 **(a)** The displacement $\mathbf{D}_{AB}$ of A relative to B is a vector directed from B to A. **(b)** The relative displacements of any three bodies are related by a vector addition triangle.

*Historically, this led to the view that the earth was the center of the universe—the geocentric theory advocated by Aristotle.

As the bodies move, these relative displacement vectors change. **The change in relative displacement per unit time is called relative velocity.** The changes in relative displacement vectors are related in the same way as the relative displacements at any instant:

$$\Delta\mathbf{D}_{AB} = \Delta\mathbf{D}_{AC} + \Delta\mathbf{D}_{CB}$$

Dividing by the time interval Δt during which the displacement occurs, we obtain

$$\frac{\Delta\mathbf{D}_{AB}}{\Delta t} = \frac{\Delta\mathbf{D}_{AC}}{\Delta t} + \frac{\Delta\mathbf{D}_{CB}}{\Delta t}$$

Taking the limit of this expression as Δt approaches zero, we obtain a relationship between the relative velocity vectors $\mathbf{v}_{AB}$, $\mathbf{v}_{AC}$, $\mathbf{v}_{CB}$, each of which describes the velocity of one body relative to another:

$$\mathbf{v}_{AB} = \mathbf{v}_{AC} + \mathbf{v}_{CB} \tag{3-14}$$

This expression tells us that the velocity of A relative to B is the vector sum of the velocity of A relative to C and the velocity of C relative to B. The significant point to note about this equation is the relationship between the subscripts. (We may want to use subscripts other than A, B, and C to denote the bodies in relative motion, and we can, just as long as we always maintain the same relationship between subscripts.) The first subscript on the left side of the equation (A) is the same as the first subscript on the right side. The last subscript on the left side (B) is the same as the last subscript on the right side. The second and third subscripts on the right side are the same (C). Following this rule, we could write, for example, $\mathbf{v}_{XZ} = \mathbf{v}_{XY} + \mathbf{v}_{YZ}$.

Sometimes it is useful to relate $\mathbf{v}_{AB}$, the velocity of A relative to B, to $\mathbf{v}_{BA}$, the velocity of B relative to A. From Fig. 3-20 we see that

$$\mathbf{D}_{BA} = -\mathbf{D}_{AB}$$

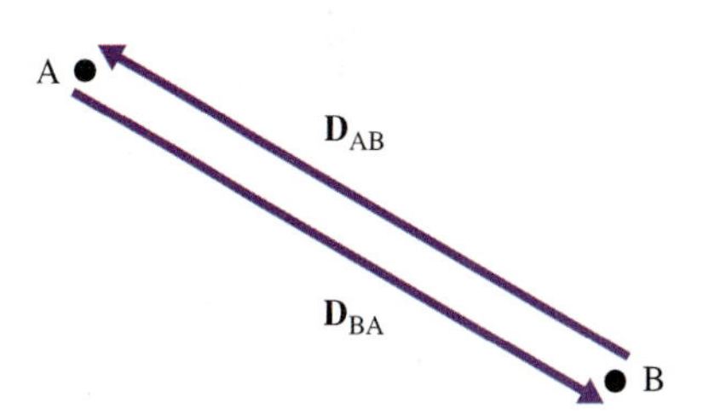

Fig. 3-20 The displacement of B relative to A ($\mathbf{D}_{BA}$) is a vector equal in magnitude but opposite in direction to the displacement of A relative to B ($\mathbf{D}_{AB}$).

and the change in relative displacement is given by

$$\Delta\mathbf{D}_{BA} = -\Delta\mathbf{D}_{AB}$$

Dividing by Δt, we have

$$\frac{\Delta\mathbf{D}_{BA}}{\Delta t} = \frac{-\Delta\mathbf{D}_{AB}}{\Delta t}$$

Taking the limit as Δt approaches zero, we obtain a second useful relationship between relative velocities:

$$\mathbf{v}_{BA} = -\mathbf{v}_{AB} \tag{3-15}$$

Thus each vector is the negative of the other; the relative velocity vectors have equal magnitudes but are oppositely directed. For example, if A is moving east at 20 m/s relative to B, then B is moving west at 20 m/s relative to A.

Application of the equations of relative motion requires only a careful labeling of the subscripts corresponding to the bodies and simple vector addition.

EXAMPLE 7 Navigating a Plane in a Crosswind

A pilot flying with an airspeed of 325 km/h wishes to fly due north in a 70.0 km/h wind blowing from east to west. In what direction should she head, and what is her speed relative to the earth?

SOLUTION First we assign a letter to each body: E, earth; P, plane; and A, air. Next we express the given information in terms of the relative velocity vectors:

$$\mathbf{v}_{AE} = 70.0 \text{ km/h, west}$$

$$|\mathbf{v}_{PA}| = 325 \text{ km/h}$$

$$\mathbf{v}_{PE} \text{ is directed north}$$

In this problem, we have been given one vector completely, only the magnitude of the second vector, and only the direction of the third vector. Our task is to find the direction of the second vector and the magnitude of the third. Using the relationship described in Eq. 3-14, we have

$$\mathbf{v}_{PE} = \mathbf{v}_{PA} + \mathbf{v}_{AE}$$

We can now use the vector triangle corresponding to this equation (Fig. 3-21) to obtain the desired information:

$$\theta = \arcsin\left(\frac{v_{AE}}{v_{PA}}\right) = \arcsin\left(\frac{70.0}{325}\right) = 12.4°$$

$$v_{PE} = \sqrt{v_{PA}{}^2 - v_{AE}{}^2} = \sqrt{(325 \text{ km/h})^2 - (70.0 \text{ km/h})^2} = 317 \text{ km/h}$$

To have her velocity relative to the earth directed north, the pilot must point the plane 12.4° east of north. The resulting speed relative to the ground has been reduced to 317 km/h. Fig. 3-22 indicates the position of the plane as seen from the earth at three successive instants.

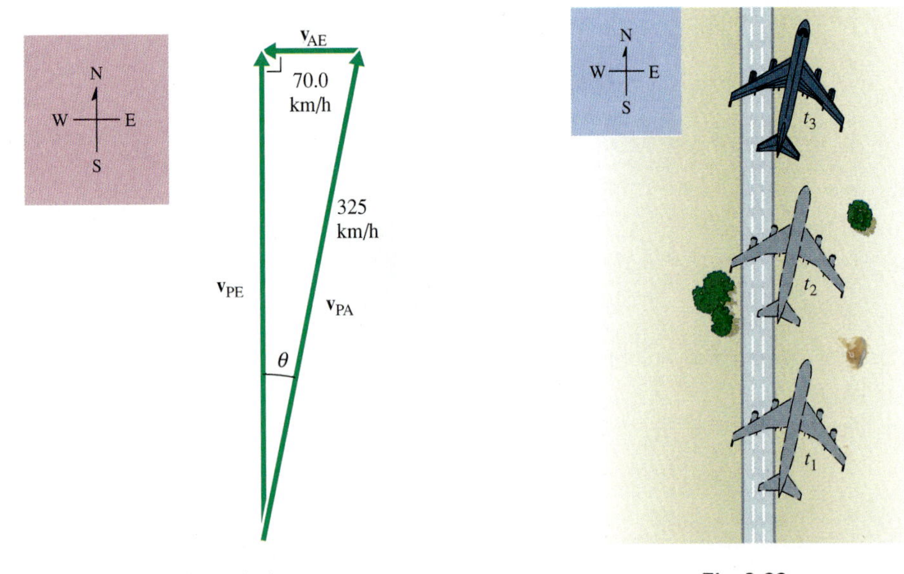

Fig. 3-21 Velocity $\mathbf{v}_{PE}$ is the vector sum of vectors $\mathbf{v}_{PA}$ and $\mathbf{v}_{AE}$.

Fig. 3-22

EXAMPLE 8 Walking on a Moving Sidewalk

An airline passenger late for a flight walks on an airport "moving sidewalk" at a speed of 5.00 km/h relative to the sidewalk, in the direction of its motion. The sidewalk is moving at 3.00 km/h relative to the ground and has a total length of 135 m. (a) What is the passenger's speed relative to the ground? (b) How long does it take him to reach the end of the sidewalk? (c) How much of the sidewalk has he covered by the time he reaches the end?

SOLUTION The situation is sketched in Fig. 3-23a. We assign a letter to each body in relative motion: P, passenger; S, sidewalk; G, ground. The relative velocities $\mathbf{v}_{PS}$ and $\mathbf{v}_{SG}$ are given:

$$\mathbf{v}_{PS} = 5.00 \text{ km/h, to the right}$$

$$\mathbf{v}_{SG} = 3.00 \text{ km/h, to the right}$$

(a) Here we must find the magnitude of the vector $\mathbf{v}_{PG}$, given the magnitude and direction of two other vectors. We find the velocity $\mathbf{v}_{PG}$ by using Eq. 3-14:

$$\mathbf{v}_{PG} = \mathbf{v}_{PS} + \mathbf{v}_{SG}$$

Here the vectors are parallel, and so the vector addition is quite simple (Fig. 3-23b). We add vectors by adding magnitudes:

$$\begin{aligned} v_{PG} &= v_{PS} + v_{SG} \\ &= 5.00 \text{ km/h} + 3.00 \text{ km/h} \\ &= 8.00 \text{ km/h} \end{aligned}$$

(b) The length of the sidewalk is 135 m, and so this is the distance Δx_G the passenger travels relative to the ground. So our problem is to find Δt when $\Delta x_G = 135$ m. The rate at which this distance along the ground is covered by the passenger is v_{PG}, where

$$v_{PG} = \frac{\Delta x_G}{\Delta t}$$

Therefore

$$\begin{aligned} \Delta t = \frac{\Delta x_G}{v_{PG}} &= \frac{135 \text{ m}}{8.00 \text{ km/h}\left(\dfrac{1.00 \text{ m/s}}{3.60 \text{ km/h}}\right)} \\ &= 60.8 \text{ s} \end{aligned}$$

(c) The problem here is to determine how much of the sidewalk's surface the passenger moves over. If he were standing still and not walking along the surface, he would cover none of it. Because he is moving relative to the surface at velocity $\mathbf{v}_{PS}$, he does move some distance Δx_S relative to the surface. The problem is to find Δx_S when $\Delta t = 60.8$ s, since we found in part (b) that this is the time interval during which he is on the moving sidewalk. His velocity relative to the sidewalk is $v_{PS} = \Delta x_S/\Delta t$, and so

$$\begin{aligned} \Delta x_S &= v_{PS}\,\Delta t \\ &= (5.00 \text{ km/h})\left(\frac{1.00 \text{ m/s}}{3.60 \text{ km/h}}\right)(60.8 \text{ s}) \\ &= 84.4 \text{ m} \end{aligned}$$

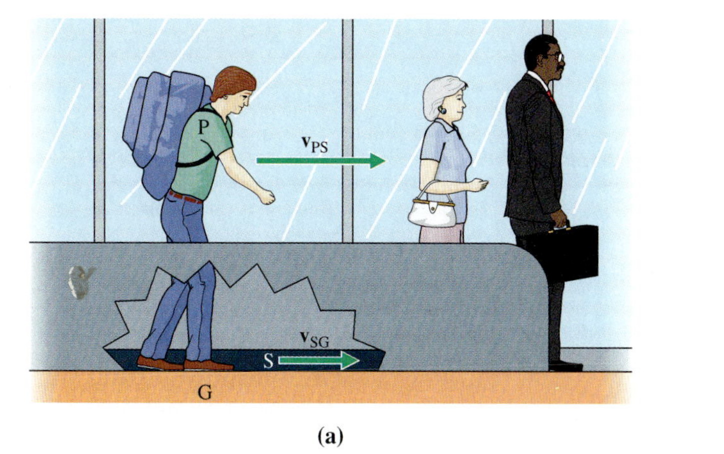

(a)

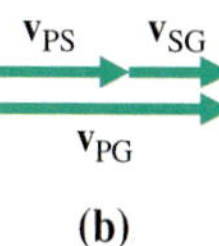

(b)

Fig. 3-23

EXAMPLE 9 Running in the Rain

Rain is falling vertically at a speed of 20.0 m/s. A woman runs through the rain at a speed of 5.00 m/s. (a) What is the velocity of the rain relative to the woman? (b) How far in front of her would an umbrella have to extend to keep the rain off if she holds the umbrella 1.50 m above her feet?

SOLUTION (a) We assign the following letters: W, woman; R, rain; and E, earth. We are given the relative velocities $\mathbf{v}_{RE}$ and $\mathbf{v}_{WE}$, which are drawn in Fig. 3-24a. We want to find the relative velocity $\mathbf{v}_{RW}$. Following the usual rule, we obtain an expression for $\mathbf{v}_{RW}$:

$$\mathbf{v}_{RW} = \mathbf{v}_{RE} + \mathbf{v}_{EW}$$

To perform the vector addition, we must first find the velocity $\mathbf{v}_{EW}$, which is not given. But $\mathbf{v}_{EW}$ is the negative of $\mathbf{v}_{WE}$, which is given. Performing the vector addition, we obtain the diagram shown in Fig. 3-24b. From the figure we find

$$v_{RW} = \sqrt{v_{RE}^2 + v_{EW}^2} = \sqrt{(20.0 \text{ m/s})^2 + (5.00 \text{ m/s})^2}$$
$$= 20.6 \text{ m/s}$$

and

$$\theta = \arctan\left(\frac{v_{EW}}{v_{RE}}\right) = \arctan\left(\frac{5.00 \text{ m/s}}{20.0 \text{ m/s}}\right)$$
$$= 14.0°$$

(b) The rain as seen by the running woman is shown in Fig. 3-24c. From the figure we see that the distance d the umbrella must extend is

$$d = (1.50 \text{ m})(\tan 14.0°)$$
$$= 0.375 \text{ m} = 37.5 \text{ cm}$$

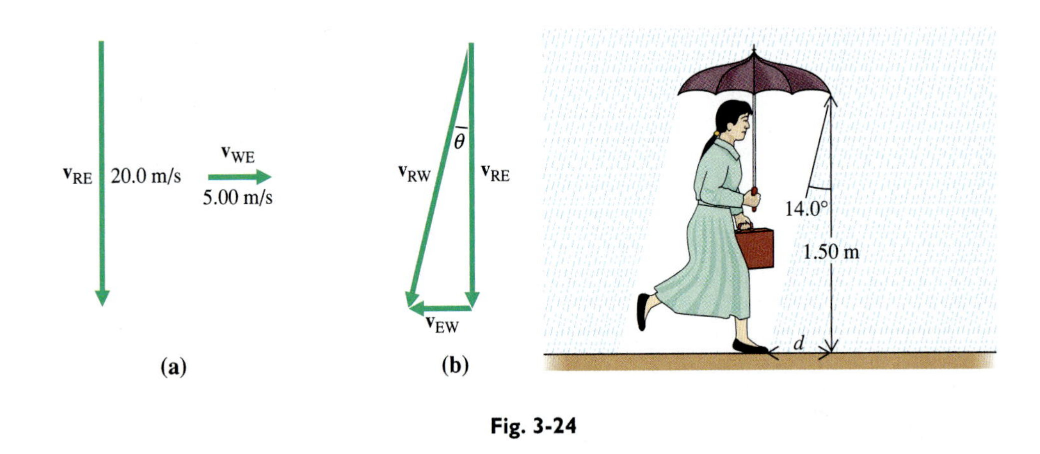

Fig. 3-24

CHAPTER 3 SUMMARY

For motion in a plane, average acceleration $\bar{\mathbf{a}}$, defined as $\Delta\mathbf{v}/\Delta t$, has Cartesian components

$$\bar{a}_x = \frac{\Delta v_x}{\Delta t} \quad \text{and} \quad \bar{a}_y = \frac{\Delta v_y}{\Delta t}$$

whereas instantaneous acceleration **a**, defined as $\lim_{\Delta t \to 0} \frac{\Delta \mathbf{v}}{\Delta t}$, has components

$$a_x = \lim_{\Delta t \to 0} \frac{\Delta v_x}{\Delta t} \quad \text{and} \quad a_y = \lim_{\Delta t \to 0} \frac{\Delta v_y}{\Delta t}$$

In the absence of air resistance, the motion of a projectile over a limited region near the earth's surface is described by the following equations for the projectile's x and y coordinates, and its velocity components v_x and v_y:

$$a_x = 0 \qquad a_y = -g$$

$$v_x = v_{x0} \qquad v_y = v_{y0} - gt$$

$$x = v_{x0}t \qquad y = \tfrac{1}{2}(v_{y0} + v_y)t$$

$$y = v_{y0}\, t - \tfrac{1}{2}gt^2$$

$$v_y^2 = v_{y0}^2 - 2gy$$

The x and y coordinates of a projectile are related at any instant by the equation

$$y = (\tan\theta_0)x - \left(\frac{g}{2v_{x0}^2}\right)x^2$$

The range R of a projectile is the horizontal distance it travels before falling to its original elevation and is given by

$$R = \frac{v_0^2 \sin 2\theta_0}{g}$$

A projectile travels the distance R in time

$$t_R = \frac{2v_{y0}}{g}$$

When a particle moves along a circular path of radius r at constant speed v, its acceleration, called centripetal acceleration, is directed toward the center of the circle and has magnitude

$$a = \frac{v^2}{r}$$

Relative velocity vectors are related by the equations

$$\mathbf{v}_{AB} = \mathbf{v}_{AC} + \mathbf{v}_{CB}$$

$$\mathbf{v}_{BA} = -\mathbf{v}_{AB}$$

Questions

1 Is an object moving along a curved path necessarily accelerated?

2 If an object moves along a linear path, is its acceleration necessarily zero?

3 If your speedometer reading is constant, does this necessarily mean your car is not accelerating?

4 The initial speed with which a ball is thrown is doubled, with the angle of projection fixed. Is the maximum height to which the ball rises doubled?

5 A person running away from you at a speed of 3 m/s throws a ball vertically upward (as seen by him), the ball leaving his hand with an initial velocity of 4 m/s (relative to him). What is the speed of the ball relative to you?

6 When riding a bicycle, is the force of the air resistance greater when riding with the wind or against the wind? Explain.

7 Would you expect a typical trajectory of a Ping-Pong ball to be a parabola? Explain.

8 While driving a car around a curve, an observer sees a road sign. Is the road sign moving with respect to the car? Is it accelerated with respect to the car?

Answers to Odd-Numbered Questions

1 yes; **3** no; **5** 5 m/s; **7** no

Problems (listed by sections)

3-1 Acceleration on a Curved Path

Average acceleration

1 A cyclist is initially traveling north, then turns 90.0° and moves west at the same speed. Find the direction of the cyclist's average acceleration.

2 A halfback is initially running south at a speed of 10.0 m/s, quickly cuts to his right, and 0.500 s later is running 60.0° west of south at 10.0 m/s. Find his average acceleration.

3 A pilot claims to have seen a UFO moving initially at a speed of about 440 m/s in an easterly direction and then, in a time interval of only 1.0 s, turning 45° and moving southeast at 440 m/s. Compute the UFO's average acceleration during the turn.

4 A particle is initially moving along the positive x-axis at a speed of 10.0 m/s. After 2.00 s, the particle is moving along the negative y-axis at a speed of 5.00 m/s. Find the x and y components of the particle's acceleration.

5 The initial and final velocities of a particle are shown in Fig. 3-25. Find the particle's average acceleration if the change in velocity takes place in a 10.0 s interval.

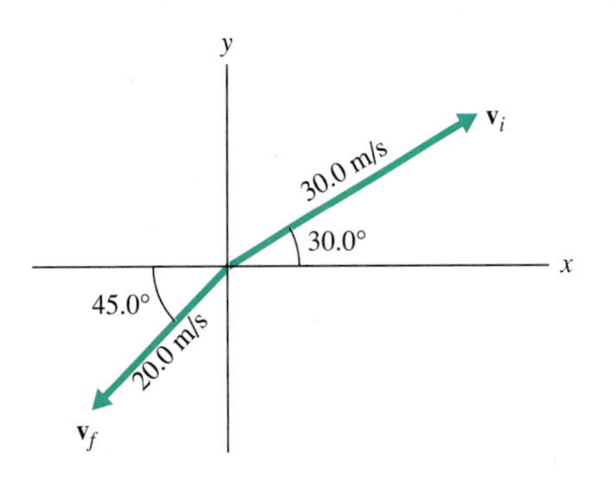

Fig. 3-25

Instantaneous acceleration

6 An aircraft initially flying north at 225 m/s turns toward the east and 0.100 s later is flying 0.0400° east of north at the same speed. Estimate the aircraft's instantaneous acceleration.

7 A baseball has a velocity of 44.0 m/s (98.4 mi/h), directed horizontally, as it is released by a pitcher. The ball's velocity 0.0100 s before it is released is 42.0 m/s, directed 3.00° above the horizontal. Estimate the ball's instantaneous acceleration just before it is released.

8 A child on a Ferris wheel is moving vertically upward at 5.00 m/s at one instant, and 0.100 s later is moving at 5.00 m/s at an angle of 86.0° above the horizontal. Estimate the child's instantaneous acceleration.

3-2 Projectile Motion

9 A water pistol aimed horizontally projects a stream of water with an initial speed of 5.00 m/s.

(a) How far does the water drop in moving 1.00 m horizontally?

(b) How far does it travel before dropping a vertical distance of 1.00 cm?

10 A beam of electrons in a television tube moves horizontally with a velocity of 1.00×10^7 m/s. How far will the electrons drop as they travel a horizontal distance of 20.0 cm?

11 Standing on a balcony, you throw your keys to a friend standing on the ground below. One second after you release the keys, they have an instantaneous velocity of 13.9 m/s, directed 45° below the horizontal. What initial velocity did you give them?

12 A baseball pitcher throws a pitch with an initial velocity of 44.0 m/s, directed horizontally. How far does the ball drop vertically by the time it crosses the plate 18.0 m away?

13 An archer wishes to shoot an arrow at a target at eye level a distance of 50.0 m away. If the initial speed imparted to the arrow is 70.0 m/s, what angle should the arrow make with the horizontal as it is being shot?

14 A fox fleeing from a hunter encounters a 0.800 m tall fence and attempts to jump it. The fox jumps with an initial velocity of 7.00 m/s at an angle of 45.0°, beginning the jump 2.00 m from the fence. By how much does the fox clear the fence? Treat the fox as a particle.

★**15** The object of the long jump is to launch oneself as a projectile and attain the maximum horizontal range (Fig. 3-26). Here we shall treat the long jumper as a particle even though the human body is fairly large compared to the size of the trajectory. Actually there is one point within the athlete's body, called the "center of mass" (to be studied in Chapter 5), that behaves as a projected particle. Our analysis of projectile motion implies that the long jumper should try to maximize v_0 and take off at an angle as close to 45.0° as possible. However, it is easier to get a large value of v_{x0} (by a running start) than it is to get a large value of v_{y0}; consequently θ_0 is usually much less than 45.0°. Suppose the jumper takes off with $v_{x0} = 9.00$ m/s and jumps with a value of v_{y0} sufficient to reach a vertical height of 1.00 m. Find v_0, θ_0, and the horizontal range. The world record, as of 1994, is 8.95 m.

Fig. 3-26

16 Suppose that a world-class long jumper jumped on the moon with the same initial velocity as that which produced a world record of 8.95 m on earth. What would be the lunar record for the long jump? Gravitational acceleration on the moon is 1.67 m/s².

17 In an article on the use of the sling as a weapon (Korfmann M: *Sci Am* 229:34, Oct. 1973), the author states that a skilled slinger can sling a rock a distance of about 400 m. What is the minimum speed the rock must have, when it leaves the sling, to travel exactly 400 m?

★**18** Prove the following:

(a) The maximum height of a projectile equals $v_{y0}^2/2g$.

(b) The time it takes a projectile to reach its maximum height equals v_{y0}/g.

(c) The time it takes a projectile to descend from its maximum height to its original elevation is the same as the time to ascend, v_{y0}/g.

(d) The y component of velocity is reversed when a projectile descends to its original elevation: $v_y = -v_{y0}$.

19 A football is kicked 60.0 meters. If the ball is in the air 5.00 s, with what initial velocity was it kicked?

20 A baseball player hits a home run over the left-field fence, which is 104 m from home plate. The ball is hit at a point 1.00 m directly above home plate, with an initial velocity directed 30.0° above the horizontal. By what distance does the baseball clear the 3.00 m high fence, if it passes over it 3.00 s after being hit?

3-3 Circular Motion

21 A runner moving at a constant speed of 10.0 m/s rounds a curve of radius 5.00 m. Compute the acceleration of the runner. Are these numbers realistic?

22 A large merry-go-round completes one revolution every 10.0 s. Compute the acceleration of a child seated on it, a distance of 6.00 m from its center.

23 In Problem 17, the initial velocity of the rock is produced by rotating the sling in a circle. What rate of rotation, in rev/s, is necessary to give the rock the required speed? Take the radius of the circle to be 1.50 m.

24 A certain centrifuge produces a centripetal acceleration of magnitude exactly $1000g$ at a point 10.0 cm from the axis of rotation. Find the number of revolutions per second.

25 Find the speed of a lunar orbiter in a circular orbit that is just above the surface of the moon, given that the orbiter's acceleration is equal to the moon's gravitational acceleration of 1.67 m/s². The radius of the moon is 1.74×10^6 m.

26 An artificial earth satellite has a circular orbit of radius 6.50×10^6 m (which means it is orbiting approximately 130 km above the surface of the earth) in an equatorial plane. The period T (the time required for one complete orbit) is 5.22×10^3 s (about 1.5 h).

(a) Compute the (constant) speed of the satellite.

(b) If the satellite is directly above the equator and traveling east at time t, find the average acceleration during the time interval from t to $t + T/40.0$.

(c) Find the satellite's instantaneous acceleration at time t.

3-4 Reference Frames and Relative Motion

27 Two cars, A and B, travel in the same direction on a straight section of highway. A has a speed of 70.0 km/h, and B a speed of 90.0 km/h (both relative to the earth).

(a) What is the speed of B relative to A?

(b) If A is initially 400 m in front of B, how long will it take for B to reach A?

28 A plane is headed due west with an air speed of 225 m/s. The wind blows south at 20.0 m/s. Find the velocity of the plane relative to the earth.

29 A plane is headed east with an air speed of 250.0 m/s. The wind blows southeast at 40.0 m/s. Find the velocity of the plane relative to the earth.

30 A man observes snow falling vertically when he is at rest, but when he runs through the falling snow at a speed of 6.00 m/s, it appears to be falling at an angle of 30.0° relative to the vertical. Find the speed of the snow relative to the earth.

⋆ **31** You are driving your car with a velocity of 20.0 m/s, north, approaching an intersection. Another car approaches the intersection with a velocity of 25.0 m/s, west (Fig. 3-27).

(a) Find the velocity of the other car relative to you.

(b) The cars are initially each 100.0 m from the intersection. Sketch the path of the other car as you see it.

32 A man can row a boat at a speed of 6.00 km/h in still water. If he is crossing a river where the current is 3.00 km/h, in what direction should his boat be headed if he wants to reach a point directly opposite his starting point?

⋆ **33** A river 100.0 m wide flows toward the south at 33.3 m/min. A girl on the west bank wishes to reach the east bank in the least possible time. She can swim 100.0 m in still water in 1.00 min.

(a) How long does it take her to cross the river?

(b) How far downstream does she travel?

(c) What is her velocity relative to land?

(d) What is the total distance she travels?

(e) In what direction must she swim if she wishes to travel straight across the river?

Additional problems

⋆ **34** In the game of darts, the player stands with feet behind a line 2.36 m from a dartboard, with the bull's-eye at eye level. Suppose you lean across the line, release a dart at eye level 1.80 m from the board, and hit the bull's-eye (Fig. 3-28). Find the initial velocity of the dart, if the maximum height of its trajectory is 1.00 cm above eye level.

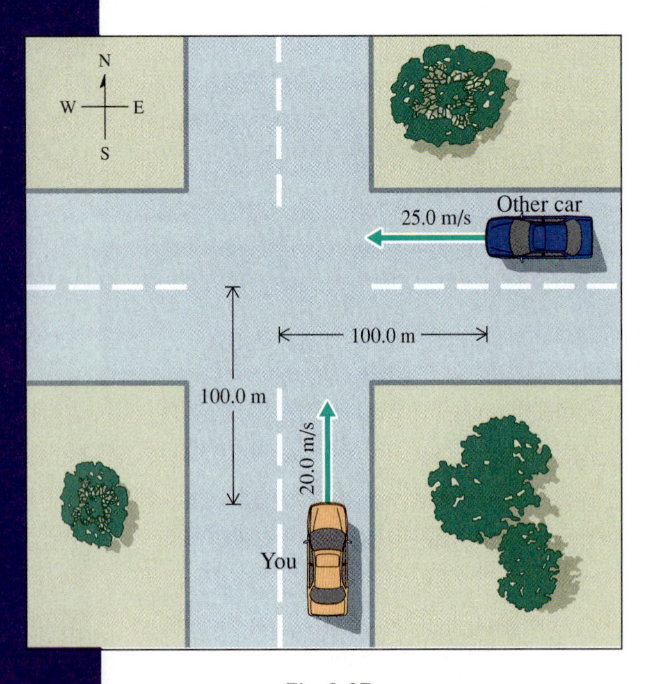

Fig. 3-27

Fig. 3-28

★★ **35** A football is to be thrown by a quarterback to a receiver who is running at a constant velocity of 10.0 m/s directly away from the quarterback, who intends for the ball to be caught a distance of 40.0 m away. At what distance should the receiver be from the quarterback when the ball is released? Assume the football is thrown at an initial angle of 45.0° and that it is caught at the same height at which it is released.

36 A particle requires 4.00 s to complete a circular path of radius 20.0 m. At $t = 0$, the particle is moving east and has instantaneous acceleration directed south. Find the particle's average acceleration from $t = 0$ to (a) $t = 4.00$ s; (b) $t = 2.00$ s; (c) $t = 1.00$ s. (d) Find the magnitude of its instantaneous acceleration at $t = 0$.

★ **37** In the shot put, a heavy lead weight—the "shot"—is given an initial velocity, starting from an initial elevation approximately equal to the shot putter's height, say, 1.90 m. If $v_0 = 8.00$ m/s, find the horizontal distance traveled by the shot for (a) $\theta_0 = 0°$; (b) $\theta_0 = 40.0°$; (c) $\theta_0 = 45.0°$.

★ **38** A tennis ball is struck at the base line of the court, 12.0 m from the net. The ball is given an initial velocity with a horizontal component equal to 24.0 m/s at an initial elevation of 1.00 m.

(a) What vertical component of initial velocity must be given to the ball, such that it barely clears the 1.00 m high net?

(b) How far beyond the net will the ball hit the ground?

39 The sun travels about the center of our galaxy in a nearly circular orbit of radius 2.5×10^{17} km in a period of about 2.0×10^{8} years. Compute the magnitudes of the velocity and acceleration of the sun relative to the center of the galaxy.

★ **40** The moon travels about the earth at uniform speed in a nearly circular orbit of radius 3.8×10^{5} km in a period of about 27 days. The earth travels about the sun at uniform speed in a circular orbit of radius 1.5×10^{8} km in a period of about 365 days. Compute the magnitudes of the velocity and acceleration of (a) the moon relative to the earth and (b) the earth relative to the sun. (c) What is the maximum acceleration of the moon relative to the sun, and during what phase of the moon does this occur? The orbits of the moon and earth are very nearly coplanar.

★ **41** The earth is approximately a sphere of radius 6.37×10^{3} km. A particle P at rest on the surface of the earth at 40.0° north latitude moves in a circular path as the earth rotates on its axis (Fig. 3-29). Compute the magnitudes of the particle's velocity and acceleration relative to the center of the earth.

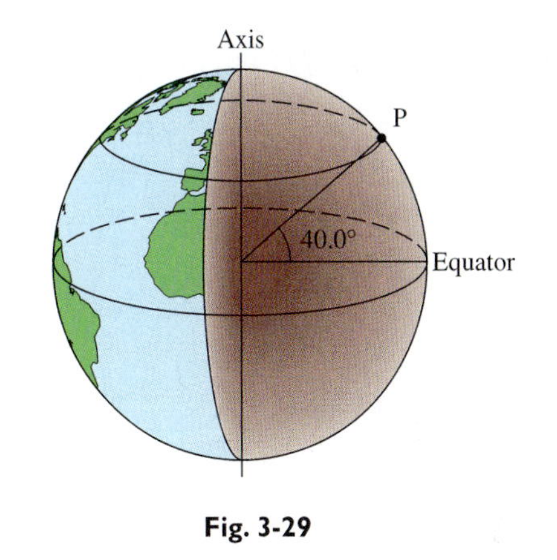

Fig. 3-29

★ **42** A rocket is fired at 40.0° north latitude with an initial velocity in an eastward direction. Its velocity relative to the center of the earth is of magnitude 11.0 km/s, just enough to escape the earth. Find the velocity of the rocket relative to the ground. What would be the advantage in firing the rocket from a site closer to the equator?

★ **43** A golfer must hit an approach shot to the green over a tree.

(a) What initial velocity must be imparted to the ball so that it will follow the trajectory indicated in Fig. 3-30?

(b) Find the horizontal distance d that the ball travels after it clears the tree before hitting the ground.

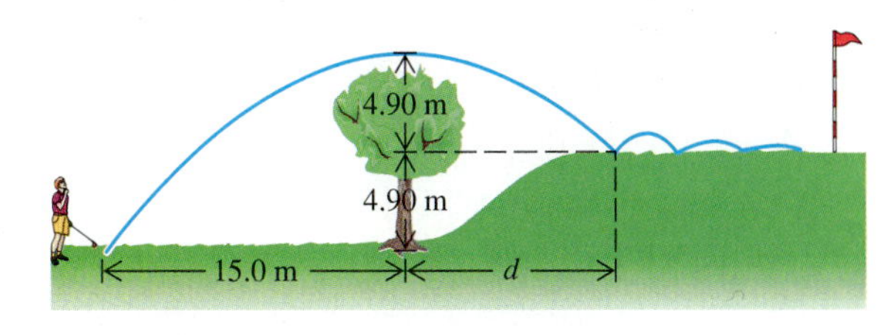

Fig. 3-30

★ **44** A point that is instantaneously on the top edge of an automobile tire moves in the forward direction at a speed of 40.0 m/s, as the car moves at a constant speed of 20.0 m/s.

(a) Find the velocity of the point relative to a passenger in the car.

(b) Find the centripetal acceleration of the point relative to the earth. The tire's radius is 35.0 cm.

★ **45** When you walk, your upper leg rotates about the hip joint, the knee describing an approximately circular arc relative to the hip. During its forward motion, the maximum speed of your knee relative to the ground is approximately twice the speed of your hip relative to the ground. What is the maximum centripetal acceleration of your knee when you are walking at a speed of 2 m/s, if the length of your upper leg is 0.5 m?

★ **46** A tennis ball is served at a height of 3.00 m with an initial horizontal component of velocity equal to 25.0 m/s.

(a) What should the vertical component of initial velocity be if the ball clears the 1.00 m high net, 12.0 m away?

(b) At what angle is the initial velocity vector below the horizontal?

(c) At what angle below the horizontal would the initial velocity vector be directed if there were no gravitational acceleration?

★★ **47** Airplane A is flying at a constant velocity of 1.20×10^2 m/s north. Airplane B is flying at a constant velocity of 1.60×10^2 m/s west.

(a) What is the velocity of B relative to A?

(b) If, at a certain instant, the pilot of A observes the pilot of B 75.0 m directly north, what will be the smallest distance between the pilots as they pass? (HINT: Consider the path of B as seen by A.)

★★ **48** Water from a garden hose has a maximum horizontal range of 10.0 m. The hose is used to put out a fire on a rooftop. Find the initial angle θ_0 that will allow the water to reach the greatest possible horizontal distance d from the edge of the roof, as shown in Fig. 3-31. (HINT: Consider first the angle θ_1.)

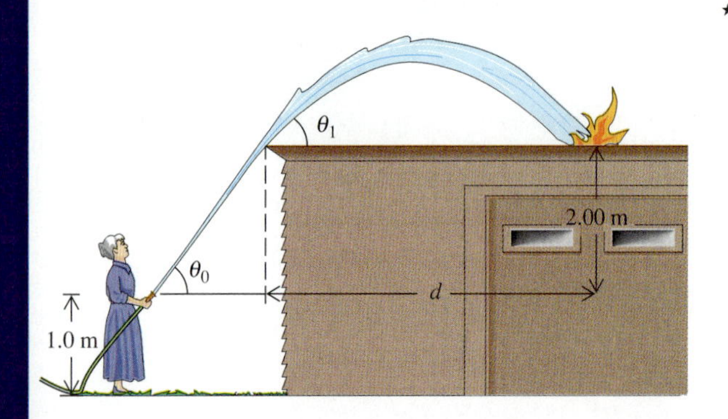

Fig. 3-31

★★ **49** Suppose you are standing on a 30.0° slope and kick a ball on the ground up the slope, giving the ball an initial velocity of 10.0 m/s, directed at an angle of 45.0° above the horizontal.

(a) At what distance from your feet will the ball strike the ground?

(b) Repeat your calculation with the initial velocity directed 50.0° above the horizontal.

★★ **50** A rock is thrown from the edge of a cliff to the ground 20.0 m below. The rock has an initial velocity of 15.0 m/s, directed 30.0° above the horizontal.

(a) How long does it take the rock to reach the ground?

(b) How far from the base of the cliff does the rock strike the ground?

(c) Find the velocity of the rock just before it strikes the ground.

★ **51** A car is found in swampy ground 100.0 m from the base of a cliff 40.0 m high. The car is headed directly away from the cliff. Find the car's initial velocity, assuming that it left the edge of the cliff with a horizontal initial velocity and that it did not roll after hitting the ground.

★ **52** A passenger in a car moving at 50.0 km/h looks out her side window and observes rain falling vertically. A man standing outside in the rain observes the rain falling at an angle of 30.0° with the vertical. Find the speed of the rain relative to the earth.

53 Suppose you want to leap from the top of a building to the top of an adjacent building of the same height, across a gap of 3.00 m. With what minimum initial velocity would you have to jump?

★ **54** A dart player stands 2.40 m from a dartboard and throws a dart, releasing it at eye level, which is also the level of the bull's-eye. The dart strikes the board 0.200 s after it is released, at a point 2.00 cm directly below the center of the bull's-eye.

(a) Find the x and y components of the dart's initial velocity.

(b) Find the angle that the dart's velocity vector makes with the horizontal as it strikes the board.

★ **55** A car traveling along a straight section of road at a speed of 100.0 km/h approaches a curve of radius 20.0 m. The driver must apply the brakes to round the curve. The car decelerates at the rate of 3.00 m/s^2 while the brakes are applied.

(a) If the centripetal acceleration of the car is not to exceed 4.00 m/s^2 as it rounds the curve, at what distance from the beginning of the curve must the brakes be applied?

(b) If the driver continues to apply the brakes as she begins to round the curve, what is the instantaneous acceleration of the car?

CHAPTER 4 Newton's Laws of Motion

The great meteor crater in Arizona

Psychics and fortune-tellers try to predict the future. Such predictions are rarely confirmed, however. There are simply too many unforeseeable circumstances to allow anyone to predict human affairs reliably. Yet it is sometimes possible to predict the future for mechanical systems. For example, we can predict the future course of a newly observed comet, using Newton's laws of motion. The eighteenth century French scientist Laplace believed that this predictive capacity of Newtonian mechanics could, in principle, be applied even to human events. He wrote:

> If an intellect were to know, for a given instant, all the forces that animate nature and the condition of all the objects that compose her, and were also capable of subjecting these data to analysis, then this intellect would encompass in a single formula the motions of the largest bodies in the universe as well as those of the smallest atom; nothing would be uncertain for this intellect, and the future as well as the past would be present before its eyes.

Although Laplace's belief turned out to be wrong, Newtonian mechanics does have a remarkable predictive capacity, as we shall see in this chapter.

In the three preceding chapters we described motion, using the concepts of velocity and acceleration. However, we have not yet discussed how the motion of a body results from forces acting on the body. In this chapter we shall begin our study of **dynamics,** that part of mechanics that relates the motion of a body to forces exerted on the body by its surroundings. We shall use Newton's laws of motion, together with several force laws, to describe and explain the connection between forces and motion.

4-1 Classical Mechanics

In 1687 Isaac Newton, whose life is described at the end of Chapter 6, published his great work *Philosophiae Naturalis Principia Mathematica*. In the *Principia* Newton established a complete conceptual and mathematical system for understanding motion. He formulated three general laws of motion and used them, along with his law of universal gravitational force, to solve the ancient problem of understanding the solar system. Starting from these laws, Newton was able to calculate planetary orbits precisely. He was also able to explain the behavior of comets and of ocean tides. Today, more than 300 years later, Newton's system of mechanics, called "classical mechanics," is still used to describe that part of nature most accessible to human observation.* Newton's laws are applied to an enormous variety of physical systems. For example, they are used to determine internal forces and stresses in the design of rigid structures; they are used to study the forces acting on and within the human body under various conditions; and they are used to calculate the engine thrust necessary to send a spacecraft to a given destination.

There are two general kinds of problems encountered in classical mechanics:

1 Given the acceleration of a particle, find the forces exerted on the particle by its physical environment. For example, determine the force of air resistance on a parachutist accelerating toward the earth at a given rate.

2 Given a particle's initial position and initial velocity and the forces exerted on it by its physical environment, determine the particle's subsequent motion. For example, given the location and velocity of a comet relative to the sun, determine the comet's position and velocity at any time in the future.

4-2 Force

As a first step in developing the concept, think of force as either a push or a pull exerted by one body on another. Historically the force concept developed from human pushes and pulls and the accompanying feeling of muscular exertion.

Anytime one body exerts a force on a second body, the body exerting the force also experiences a force, called a "reaction force" (Fig. 4-1).

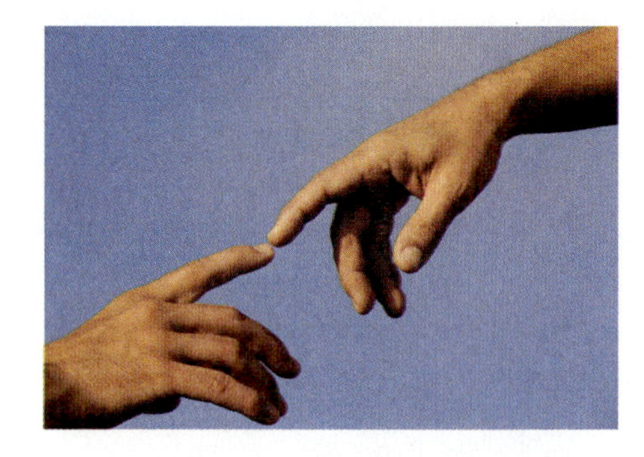

Fig. 4-1 Touching and being touched —action and reaction.

*Only in the twentieth century have Newton's laws failed in their ability to describe physical systems and then only in situations remote from everyday experience, as when an object is moving at nearly the speed of light or when the system is of atomic or subatomic dimensions. We shall study these domains of "modern physics" in Chapters 27 to 30. There we shall find that the laws of classical physics are superseded by the more general laws of relativity and quantum physics. However, we do not need to introduce the more difficult methods of modern physics into the solution of problems that can be successfully solved using classical physics. Furthermore, a thorough grounding in classical mechanics is essential to an understanding of modern physics.

The mutual **interaction** between two bodies is illustrated in Fig. 4-2 for several systems. Certain forces act only when the two bodies are touching, as in examples a, b, c, and f in Fig. 4-2. These are called **contact forces.** There are other forces, however, that act even when the interacting bodies are not touching. This **action at a distance** is easy to observe in the case of two permanent magnets (example d). The gravitational force is another example of a noncontact force. Near the surface of the earth, a force acts on any body, pulling it toward the center of the earth (example e). The body's weight is a measure of this attractive force.

Although contact is not necessary for there to be forces acting between two bodies, the strength of the interaction generally depends on how close to each other the two bodies are. Thus magnets must be fairly close to each other if they are to experience an observable mutual force, and a body must be somewhere in the vicinity of the earth to experience fully the earth's gravitational pull.

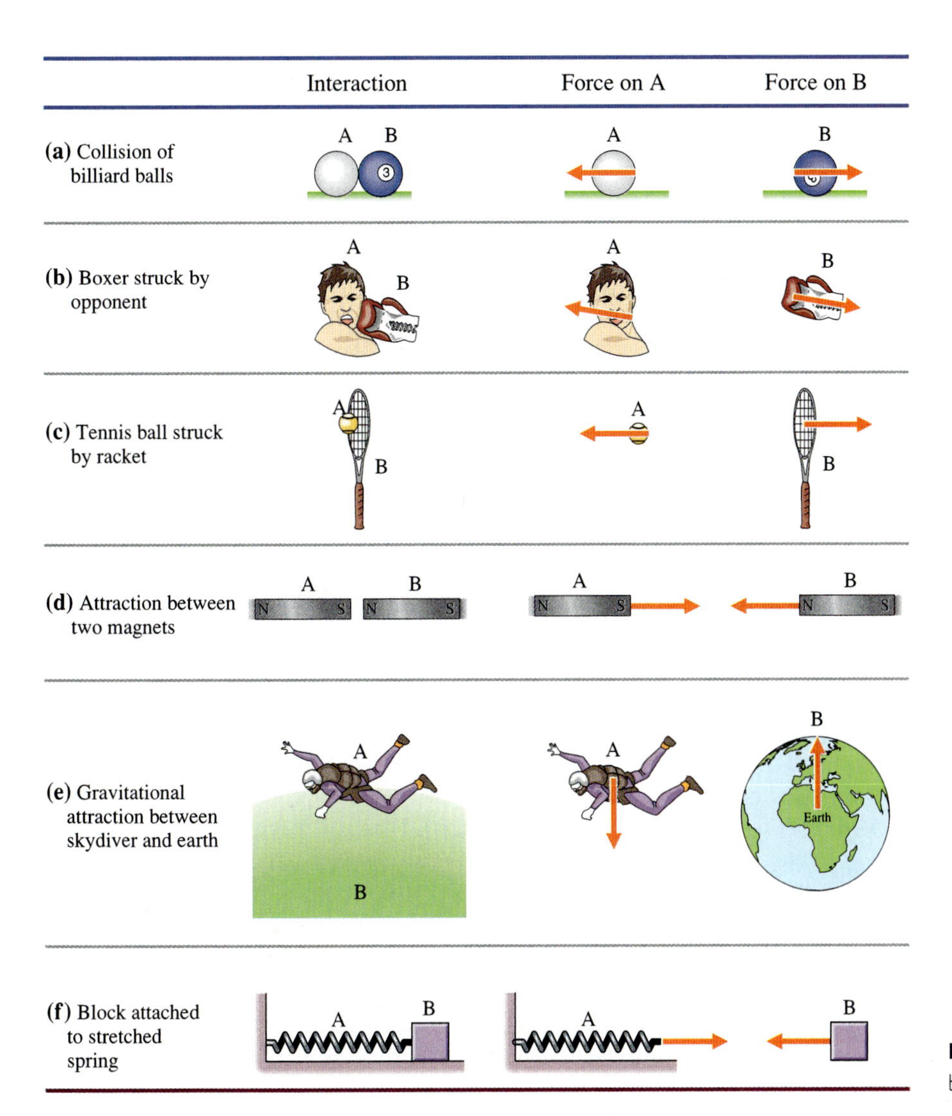

Fig. 4-2 Forces between interacting bodies.

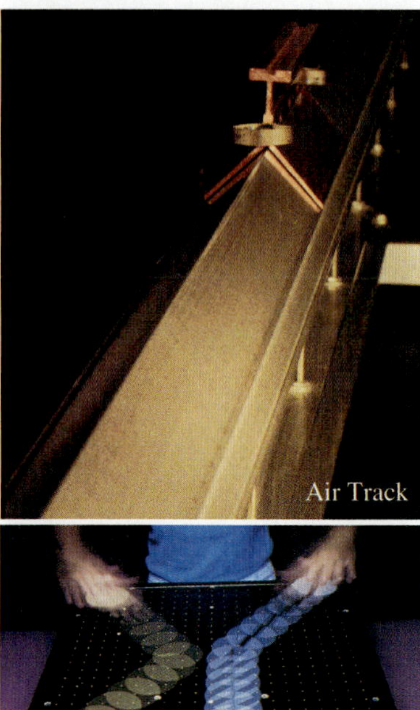

Fig. 4-3 An air track and an air table have surfaces with hundreds of tiny holes bored in them. Air is blown out through the holes, thereby allowing the "cars" to ride on a nearly frictionless cushion of air. Games such as air hockey utilize the same principle.

4-3 Newton's First Law

Newton's Statement

Newton's first law of motion states that *"Every body continues in its state of rest, or of uniform motion in a straight line, except when it is compelled to change that state by forces impressed upon it."* The tendency of a body to maintain its state of rest or of uniform motion in a straight line is called **inertia,** and the first law is sometimes called the **law of inertia.**

If a body either remains at rest or moves uniformly in a straight line, the body's velocity is constant and its acceleration is therefore zero. Thus another way of stating the first law is that: **a body will have zero acceleration if no forces act upon it.**

The first law implies that the effect of a force is to accelerate a body—to change its state of motion. This implication makes more precise our original notion of force as a push or a pull.

Galileo and Aristotle

The first law was partially formulated by Galileo when he was studying objects given an initial velocity on a smooth horizontal plane. Galileo observed that the smoother the surface, the farther an object travels before coming to rest. He concluded that, in the absence of friction, an object would travel forever, no force being necessary to maintain its motion.* Galileo's ideas were in sharp contrast to those of Aristotle, who believed that motion could not exist without the application of force. Aristotle's belief was doubtless derived from common experience, where friction is a factor and where an applied force is necessary to maintain motion by balancing the frictional force. For example, if you want to slide a book along the surface of your desk, you must continuously apply a force to the book in order to cancel the force of friction. Otherwise the book quickly comes to rest.

The air track and air table are devices for producing sliding motion with very little friction (Fig. 4-3), and so they approximate the ideal conditions envisioned by Galileo. So little friction is present on their surfaces that, once an object is given an initial velocity, it continues to move for a considerable time.

Inertial Reference Frames

Is Newton's first law valid for an observer in *any* reference frame? To answer this question, suppose that you are in outer space and observe an isolated body at rest. Another observer, who is accelerating with respect to you, views the same body and observes it to be accelerated. Since the body is isolated, there is nothing around to produce a force on it. Newton's first law is obviously satisfied for you, since both the force on the body and its acceleration equal zero. But for the other observer, Newton's first law is violated because the body appears to be accelerated without any force acting on it.

Whether Newton's first law is satisfied for any given observer depends on the reference frame of the observer. **A reference frame in which Newton's first law (the law of inertia) is satisfied is called an "inertial reference frame."**

Given one inertial reference frame, any other reference frame moving at constant velocity with respect to it is also inertial. In our example, if still another observer comes along, one who is moving at constant velocity relative to you rather than accelerating, she observes the isolated body moving at constant velocity. Newton's first law is satisfied in her reference frame as well as yours.

---

*Galileo believed that this ability to travel forever would be true for a perfectly smooth *circular* path around a perfectly spherical earth, rather than for a *straight-line* path. Descartes, a contemporary of Newton, was responsible for recognizing that this principle applies only to linear motion.

Can we name at least one physical reference frame that is inertial? Since the principle of inertia was formulated on earth, it is reasonable to assume that the earth itself is such an inertial frame. This turns out to be a good approximation in many cases but not exactly correct. The first law works in any reference frame with respect to which distant stars are either at rest or moving at constant velocity. It is a remarkable fact that one of the simplest laws of physics, discovered by observation and experiment on earth, is connected to the most distant matter in the universe. Because of the earth's daily rotation, points on the earth experience acceleration with respect to the stars, and so the earth's surface is not a truly inertial reference frame. However, the magnitude of this rotational acceleration is small—only about 0.03 m/s^2, as shown in Problem 41 of Chapter 3. Therefore, for most practical purposes we can ignore this small acceleration and **take the surface of the earth to be an inertial reference frame.**

It is not only Newton's first law that is valid in any inertial reference frame. It turns out that **all the laws of physics are valid in any inertial reference frame.**

4-4 Mass

Mass is a measure of the inertia of a body; that is, the mass of a body is a measure of the body's resistance to acceleration. Some bodies are harder to accelerate than others. Consider, for example, a bowling ball and a billiard ball, both initially at rest on a billiard table. If you strike the billiard ball with a cue stick, you can easily apply enough force to the ball to give it a significant velocity. The billiard ball is relatively easy to accelerate. Strike the bowling ball with the cue in the same way, however, and it will hardly move. To give the bowling ball the same acceleration you gave the billiard ball would require a much larger force. A bowling ball resists acceleration more than a billiard ball. A bowling ball has more mass than a billiard ball.

How do we quantify the concept of mass? Mass is a fundamental property of matter, just as length is a fundamental property of space (or of matter in space) and time is a fundamental property of existence. We define all these fundamental quantities by defining how we measure them. In the case of length and time, this quantification is familiar and accepted. Length is quantitatively defined when we establish a process for measuring the length of any body. Measurement of a body's length is a comparison between that length and multiples of some standard length, say, the meter. Time is quantified when we establish a process for measuring any time interval with respect to a standard unit of time. Measurement of a time interval is accomplished when we note the readings of a clock at the beginning and end of that interval.

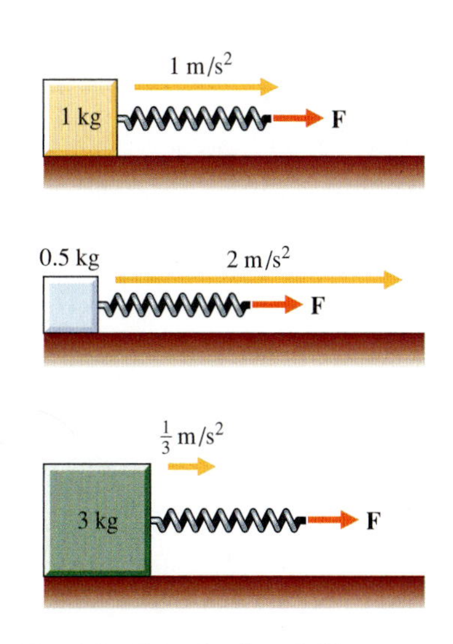

Fig. 4-4 Three bodies of different mass are accelerated by the same force.

Likewise the concept of mass can be made quantitative by reference to a standard mass. The scientific standard of mass, the standard kilogram, is a cylinder made of a very durable platinum-iridium alloy and kept in a sealed vault in Paris. Copies of this standard are in laboratories all over the world.

The mass of any object can be defined by the following experiment. Place a copy of the standard kilogram (abbreviated kg) on a frictionless surface and apply a force sufficient to give the kilogram an acceleration of 1 m/s^2 (Fig. 4-4). Next, apply this same force to any other body whose mass you want to determine.* The mass of the body is defined to be the inverse of the acceleration the body experiences under the action of this force. For example, if a body experiences an acceleration of 2 m/s^2, it has a mass of 0.5 kg by definition. If another body is accelerated at a rate of $\frac{1}{3}$ m/s^2 by the same force, it has a mass of 3 kg.

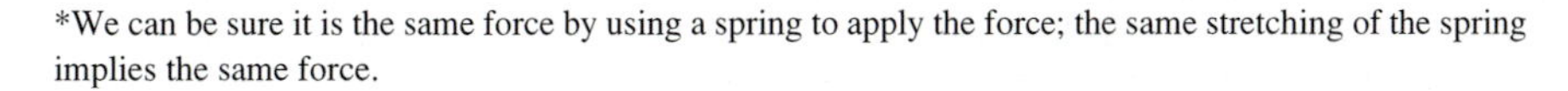

*We can be sure it is the same force by using a spring to apply the force; the same stretching of the spring implies the same force.

Mass is an additive property of matter. If a body of mass m_1 is attached to a body of mass m_2, the mass of the combination is $m_1 + m_2$. For example, if we place a 2 kg mass and a 3 kg mass together on an air track and apply the same force as before, we will observe an acceleration of $\frac{1}{5}$ m/s^2. This means that when we combine the 2 kg and 3 kg masses, we have a total of 5 kg.

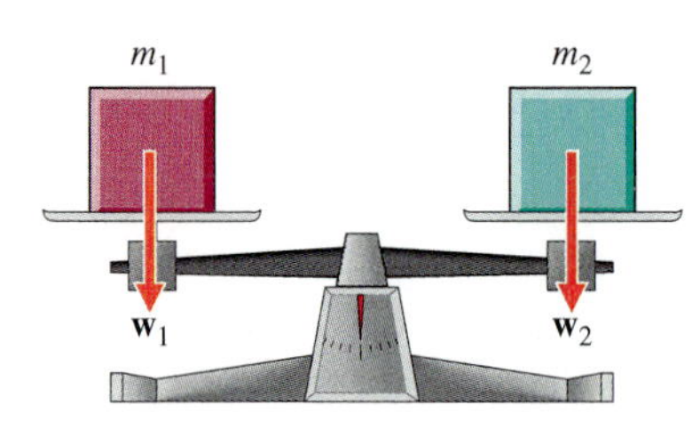

Fig. 4-5 A balance is used to measure mass.

We shall show in the next section that the weight of a body is proportional to its mass. This proportionality allows for a much easier method of measuring mass than the method used to define it. As a practical procedure, we can use an equal-arm balance to measure mass (Fig. 4-5). An unknown mass is balanced with multiples or submultiples of the standard mass. Balance is achieved when the forces acting on the two arms of the balance are equal. These forces are equal to the weights of the two masses. Equality of the weights implies equality of the masses.

Because weight and mass are proportional to each other, the two are often confused. It is important to distinguish clearly between them. Mass, a scalar quantity, is a measure of a body's inertia; weight, a vector quantity, is a measure of the earth's gravitational pull on the body.

4-5 Newton's Second and Third Laws

Second Law

The acceleration of a particle is determined by the resultant force acting on the particle. According to **Newton's second law** of motion, the acceleration is in the direction of the resultant force $\Sigma\, \mathbf{F}$ (Fig. 4-6), and the magnitude of the resultant force equals the product of mass times acceleration. Using vector notation, the second law is expressed

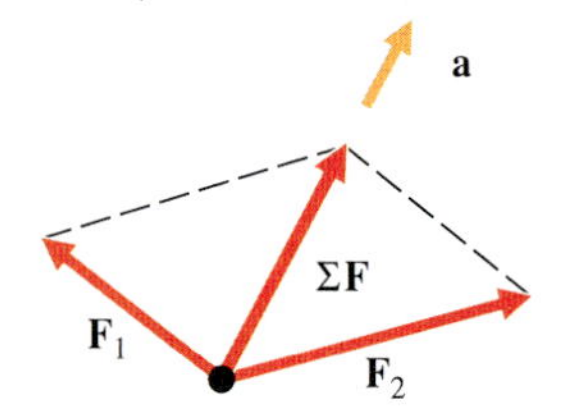

Fig. 4-6 A particle subject to two forces is accelerated in the direction of the resultant force $\Sigma\, \mathbf{F}$.

$$\Sigma\, \mathbf{F} = m\mathbf{a} \tag{4-1}$$

This vector equation implies that each component of the resultant force equals the mass times the corresponding component of acceleration. For forces in the xy plane, the second law in component form is written

$$\Sigma\, F_x = ma_x \qquad \Sigma\, F_y = ma_y$$

If we know the acceleration of a particle, we can use the second law to find the resultant force acting on the particle. On the other hand, if we know the forces acting on the particle, we can use the second law to find the particle's acceleration. We can then use the acceleration to predict the future motion of the particle.* When acceleration is the unknown, we may express the second law in the form

$$\mathbf{a} = \frac{\Sigma\, \mathbf{F}}{m} \tag{4-2}$$

or

$$a_x = \frac{\Sigma\, F_x}{m} \qquad a_y = \frac{\Sigma\, F_y}{m}$$

*The equations for the position and velocity of a particle undergoing linear motion at constant acceleration are an example of this (Section 2-2).

Units

The unit of force is obviously related to units of mass and acceleration by the second law ($\Sigma \mathbf{F} = m\mathbf{a}$). We define the newton, abbreviated N, to be the force that produces an acceleration of 1 m/s² when acting on a 1 kg mass. Thus

$$1 \text{ N} = 1 \text{ kg-m/s}^2 \tag{4-3}$$

The dyne is the force necessary to accelerate a 1 gram mass at the rate of 1 cm/s²:

$$\begin{aligned} 1 \text{ dyne} &= 1 \text{ g-cm/s}^2 = (10^{-3} \text{ kg})(10^{-2} \text{ m})/\text{s}^2 \\ &= 10^{-5} \text{ kg-m/s}^2 \\ &= 10^{-5} \text{ N} \end{aligned}$$

The pound is the unit of force in the British system. Although the pound may be defined independently, it is perhaps simplest to relate it to the newton:

$$1 \text{ lb} = 4.45 \text{ N} \tag{4-4}$$

The unit of mass in the British system is the slug. Since the British unit of acceleration is ft/s², we may write

$$1 \text{ slug} = (1 \text{ lb})/(1 \text{ ft/s}^2)$$

or

$$1 \text{ slug} = (4.45 \text{ N})/(0.305 \text{ m/s}^2) = 14.7 \text{ kg}$$

Table 4-1

System of units	Mass	Acceleration	Force
SI	kilogram	m/s²	N = kg-m/s²
cgs	gram	cm/s²	dyne = g-cm/s² (1 dyne = 10^{-5} N)
British	slug	ft/s²	lb = slug-ft/s² (1 lb = 4.45 N)

EXAMPLE 1 Computing the Force to Accelerate a Body

Find the force that must be exerted on a 0.500 kg air-track car to give it an acceleration of 3.00 m/s².

SOLUTION According to Newton's second law, the resultant force equals the product of the car's mass and its acceleration:

$$\Sigma \mathbf{F} = m\mathbf{a}$$

If we choose the x-axis along the track, we have only an x component of acceleration. Denoting the single horizontal force by **F**, we find its x component:

$$F_x = ma_x = (0.500 \text{ kg})(3.00 \text{ m/s}^2) = 1.50 \text{ N}$$

Third Law

Newton's third law of motion states that forces result from the mutual interaction of bodies and therefore always occur in pairs, as in Fig. 4-2. The third law states further that **these forces are always equal to each other in magnitude and opposite in direction.** Notice that this last statement does not mean that the forces cancel, since they do not act on the same body. The two forces involved in the third law *always* act on two different bodies. Failure to recognize this point is a common source of error in problem solving.

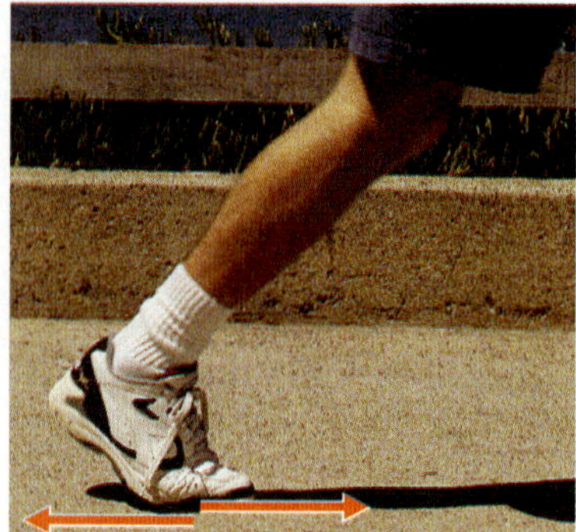

(a)

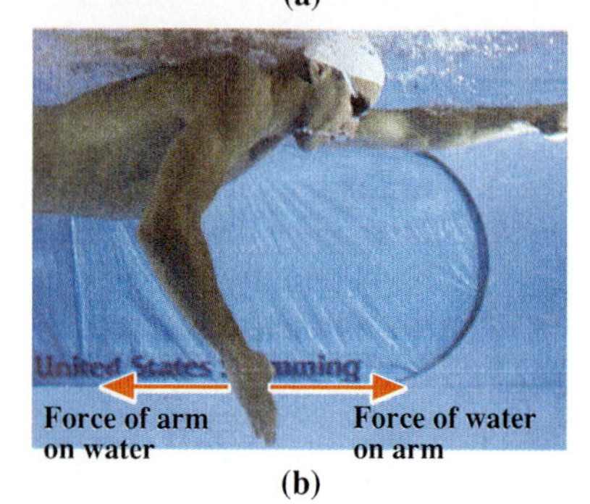

(b)

Fig. 4-7 Using Newton's third law to produce human motion.

The forces occurring in any interaction are often referred to as action and reaction forces. This terminology should not be misinterpreted. Neither force occurs before the other. Either force may be called the action force; the other is then called the reaction force. Action-reaction forces are shown in Fig. 4-2 for several systems.

The third law does not imply that the effect of the two forces will be the same. For example, when a rifle fires a bullet, the forces on the bullet and the rifle have equal magnitude, but the bullet, because of its much smaller mass, experiences a much greater acceleration than the rifle. Or when one boxer punches another in the face, the forces on the face and the fist are equal in magnitude, but the effects of the two forces are quite different.

Newton's third law is utilized in locomotion. For example, in walking you move forward by pushing one foot backward against the floor. The reaction force of the floor on your foot produces the forward acceleration of your body (Fig. 4-7a). In swimming, forward motion is provided primarily by your arms, which push the water backwards, thereby producing a reaction force of the water on your arms in the forward direction (Fig. 4-7b). The flight of birds is also based on this principle.*

*In analyzing the flight of birds, Leonardo da Vinci (1452–1519) recognized that when a bird's wings thrust against the air, the air pushes back on the wings and thereby supports the bird. But Leonardo's anticipation of Newton's third law as well as his other scientific discoveries had no influence on the development of science because they were unknown until hundreds of years later. Leonardo was so concerned about keeping his discoveries secret that he wrote in a mirror-image code, so that his words could be read only when seen in a mirror.

EXAMPLE 2 Pushing on a Wall

A standing person pushes against a wall with a horizontal force (Fig. 4-8). (a) Why doesn't the section of wall in contact with his hand move? (b) According to the third law, the person's horizontal push on the wall is accompanied by a reaction force on the person. Why doesn't he move away from the wall as a result of this reaction force?

Fig. 4-8

SOLUTION (a) The section of wall in contact with the hand does not move in response to the applied force because other forces are exerted on it by the other parts of the wall in contact with that section.* Since the wall doesn't move, the resultant of all forces must be zero, according to Newton's second law ($\Sigma \mathbf{F} = m\mathbf{a} = \mathbf{0}$).

*Actually there is a very slight movement of the surface when it is first pushed. As soon as the surface is slightly deformed, the surrounding parts of the wall begin to create a force opposing that exerted by the hand. If the wall's surface is soft (for example, cork) the deformation is readily observable.

Continued.

EXAMPLE 2—cont'd

(b) The wall certainly exerts an outward force on the hand (Fig. 4-9). If this force were unbalanced, the person would move away from the wall. Since the person is standing at rest, Newton's second law implies that the sum of the forces acting on the person must be zero. So there must be another force acting on the person, one that cancels the outward force of the wall. This other force on the person cannot be the reaction force to the wall's outward push. Remember action-reaction forces always act on different bodies.

The other force acting on the person is provided by the interaction between the feet and the floor. The feet must push out against the floor so that the floor will push back against the feet. As illustrated in Fig. 4-9, this pushing in opposite directions by the wall and floor produces a resultant force of zero. (If the person were on roller skates, $\mathbf{F}_2$ would be smaller than $\mathbf{F}_1$ and the person would move to the right.)

Fig. 4-9

EXAMPLE 3 Finding the Acceleration of a Body

Three astronauts, each of mass 70.0 kg, "float" in an orbiting space station and simultaneously exert forces on a block having a mass of 20.0 kg, as indicated in Fig. 4-10a. (a) Find the x and y components of the block's acceleration. (b) Find the instantaneous acceleration of the astronaut exerting the force $\mathbf{F}_1$.

SOLUTION **(a)** We apply the component form of Newton's second law to the block, in order to find a_x and a_y:

$$a_x = \frac{\Sigma F_x}{m}$$

From the figure, we find the x component of each force and then substitute into our acceleration equation:

$$a_x = \frac{F_{1x} + F_{2x} + F_{3x}}{m}$$

$$= \frac{(90.0 \text{ N})(\cos 30.0°) + 0 + (175 \text{ N})(\cos 45.0°)}{20.0 \text{ kg}}$$

$$= 10.1 \text{ m/s}^2$$

We obtain a_y in the same manner:

$$a_y = \frac{\Sigma F_y}{m} = \frac{F_{1y} + F_{2y} + F_{3y}}{m}$$

$$= \frac{(90.0 \text{ N})(\sin 30.0°) + 125 \text{ N} - (175 \text{ N})(\sin 45.0°)}{20.0 \text{ kg}}$$

$$= 2.31 \text{ m/s}^2$$

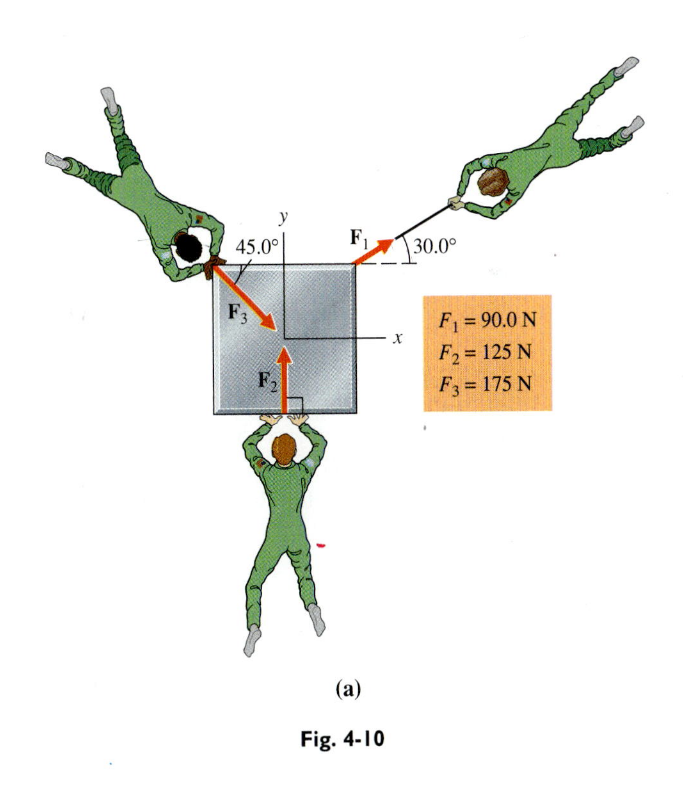

(a)

Fig. 4-10

Continued.

EXAMPLE 3—cont'd

(b) If we know all the forces acting on the astronaut exerting force $\mathbf{F}_1$, we can apply Newton's second law to find her acceleration. The only force acting on her is the reaction force to the force $\mathbf{F}_1$ she exerts on the block. According to Newton's third law, this reaction force $\mathbf{F}_1'$ is the negative of the force $\mathbf{F}_1$:

$$\mathbf{F}_1' = -\mathbf{F}_1$$

The force $\mathbf{F}_1'$ has the same magnitude as $\mathbf{F}_1$ and is directed 30.0° below the negative x-axis, as shown in Fig. 4-10b. To find the astronaut's acceleration $\mathbf{a}_1'$, we apply Newton's second law:

$$\mathbf{a}_1' = \frac{\Sigma\,\mathbf{F}}{m} = \frac{\mathbf{F}_1'}{m}$$

This vector equation implies that the astronaut's acceleration is in the same direction as the force $\mathbf{F}_1'$ and has magnitude equal to the magnitude of that force divided by the mass:

$$a_1' = \frac{F_1'}{m} = \frac{90.0\text{ N}}{70.0\text{ kg}}$$
$$= 1.29\text{ m/s}^2$$

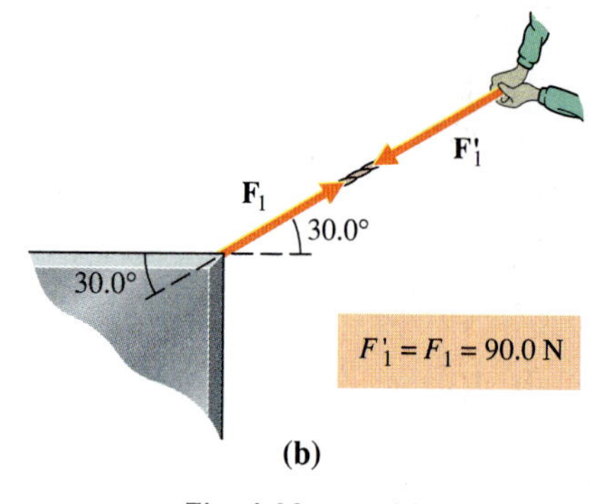

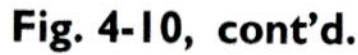

(b)

Fig. 4-10, cont'd.

4-6 Force Laws

A force law relates the force on a body to the body's surroundings. In this section we shall discuss several important force laws that will be useful in applying Newton's laws.

Weight on Earth

Perhaps the simplest of all force laws is the gravitational force law for a body of mass m near the surface of the earth. We can find an expression for this force by considering a body of mass m that is falling freely and experiencing negligible air resistance (Fig. 4-11). According to Newton's second law, the resultant force acting on any body equals the product of its mass and acceleration:

$$\Sigma\,\mathbf{F} = m\mathbf{a}$$

The falling body is subject only to the earth's gravitational force, which we refer to as the body's weight (denoted by $\mathbf{w}$); thus the resultant force equals the weight ($\Sigma\,\mathbf{F} = \mathbf{w}$). We know from experiment that, in the absence of air resistance, all freely falling bodies near the earth's surface experience the same acceleration $\mathbf{a} = \mathbf{g}$, as discussed in Section 2-3. Substituting the resultant force and acceleration into Newton's second law, we obtain an expression for the weight of a body of mass m on earth:

$$\mathbf{w} = m\mathbf{g} \tag{4-5}$$

Fig. 4-11 A freely falling body of mass m experiences acceleration $\mathbf{g}$.

Although we have derived this equation for a falling body, we may apply it quite generally to any body on or near the earth's surface. The gravitational force arises from the mutual interaction of the earth and the body. Whenever a body is close to the earth's surface, the body experiences a downward force **w**, equal to the product of its mass m and gravitational acceleration **g**. The same force acts *irrespective of the body's motion or of the presence of other forces*. This is our first example of a force law. It allows us to compute the gravitational force on a body (in other words, the body's weight), given its physical environment (on or near the surface of the earth). We shall use this force law frequently in solving problems.

EXAMPLE 4 Forces on a Man

Find the forces acting on a standing man whose mass is 90.0 kg.

SOLUTION According to Eq. 4-5, the man experiences a force **w** in the downward direction (the direction of **g**), and the magnitude of this force is

$$w = mg = (90.0 \text{ kg})(9.80 \text{ m/s}^2) = 882 \text{ N}$$

or

$$w = 882 \text{ N}\left(\frac{1.00 \text{ lb}}{4.45 \text{ N}}\right) = 198 \text{ lb}$$

Since the man is standing at rest, his acceleration is zero and so the second law implies that there must be another force to cancel the weight and produce a resultant force equal to zero, as shown in Fig. 4-12. This other force is produced by the contact between his feet and the surface on which he is standing. We denote this surface force by **S** and use the second law to solve for it:

$$\Sigma \mathbf{F} = m\mathbf{a} = \mathbf{0}$$

$$\mathbf{S} + \mathbf{w} = \mathbf{0}$$

Thus

$$\mathbf{S} = -\mathbf{w}$$

This equation says that the forces are oppositely directed and have equal magnitudes:

$$S = w = 882 \text{ N}$$

We could have just as easily solved this problem using Newton's second law in component form. Taking the positive y-axis in the upward direction, we have

$$\Sigma F_y = ma_y = 0$$

or

$$S - w = 0$$

Therefore

$$S = w = 882 \text{ N}$$

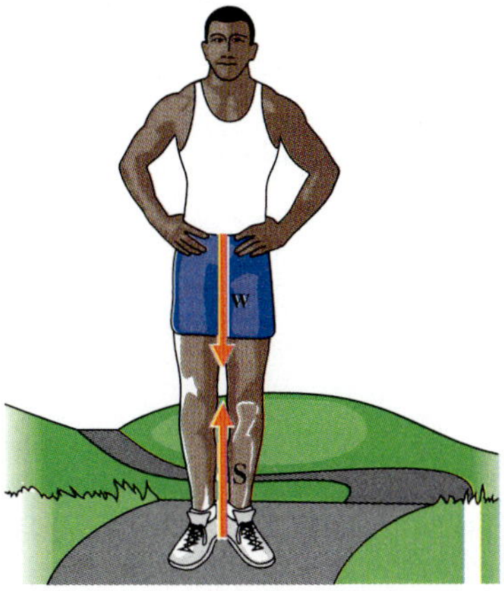

Fig. 4-12

The forces **S** and **w** are equal here because the man is stationary. It is possible for him to increase the force **S** by pushing down on the ground with a force greater than his weight. By Newton's third law, the upward force on his feet will then be greater. There would then be a resultant upward force of magnitude $S - w$, and the man would accelerate upward. In other words, by pushing on the ground with a force greater than his weight, the man can jump.

Variation of Weight on Earth

The value of g varies slightly from point to point on earth. (The variation arises from several factors, to be discussed in Chapter 6.) In particular, g is a function of latitude. For example, at the equator $g = 9.78\ \mathrm{m/s^2}$, at 40° north latitude $g = 9.80\ \mathrm{m/s^2}$, and at the North Pole $g = 9.83\ \mathrm{m/s^2}$. It follows from Eq. 4-5 ($\mathbf{w} = m\mathbf{g}$) that the weight of any object also varies slightly over the surface of the earth. The value of g is less at the equator than at the North Pole by $0.05\ \mathrm{m/s^2}$, which is about 0.5%. Thus the weight of a body is also 0.5% less at the equator than at the North Pole. If you weigh 1000 N (about 225 lb) at the North Pole, you can "lose" about 5 N, or 1 lb, by moving to the equator! You won't be any slimmer, though, because your mass is unchanged.

Fig. 4-13 *(Physics Today 41:9, Sept 1988.)*

Fundamental Forces

In light of the apparent diversity of forces one observes in nature, it is a wonderful fact that there are only four fundamental kinds of force:

Gravitational Force The force of gravity on earth is a special case of the gravitational interaction—that occurring between the earth and a body on or near its surface. We shall see in Chapter 6 that gravitation is a universal phenomenon; an attractive gravitational force acts between *any* two bodies anywhere in the universe.

Electromagnetic Force Magnetic forces and forces of static electricity are examples of the electromagnetic interaction, which acts between particles having electric charge. Electromagnetic forces are discussed in Chapters 17 to 22.

Nuclear Forces There are two fundamental nuclear forces: the **strong interaction** and the **weak interaction.** The strong interaction is responsible for the stability of the atomic nucleus, whereas the weak interaction is responsible for the type of radioactivity known as "beta decay." These forces are discussed in Chapter 30.

By the 1970s physicists had discovered that the electromagnetic and weak forces can be regarded as different manifestations of a single force, called the "electroweak force." Some physicists continue to work toward a further unification, developing "grand unified theories," which, if successful, will unify the strong force and the electroweak force. An even more ambitious goal is the unification of all the fundamental forces, including gravity.

The struggle to find unity in the forces of nature is an ongoing one. At one time electricity and magnetism were believed to be unrelated phenomena. As a result of discoveries in the nineteenth century, however, we now know that the force between electric charges and the force between magnets are special examples of a more general electromagnetic interaction. Viewed at the most fundamental level, maybe there really is only one force.

Derived Forces

All forces in nature can in principle be derived from one of the fundamental forces. In particular Eq. 4-5 ($\mathbf{w} = m\mathbf{g}$)can be derived from the general gravitational force law, as we shall show in Chapter 6.

All contact forces arise from electromagnetic interactions between the charged particles in the bodies making contact. For example, the collision of billiard balls, a boxer's punch, the pressure on a body submerged in water, and the frictional force on a car's tires all arise from electromagnetic forces acting between the interacting bodies. Even the forces holding matter together—atom to atom—are electromagnetic in origin.

Tension

Typically the forces acting between the parts of a solid body are complex, and so it would be very difficult to find a general force law for computing them. The special case of a flexible body, such as a rope or a string, is somewhat simpler. Again it would be difficult to find an expression for the magnitude of the forces acting within the string, since such an expression depends on particular qualities of the string. However, we can say something about the direction of these forces.

The fact that a string is flexible means that it bends when you push on it. In other words, a string cannot transmit a push. It can of course transmit a pull. The shape of the string adjusts itself so that this force acts along the string. Any section of a flexible rope or string exerts a force on any adjacent section. This force, called **"tension," is a pull tangent to the string** (Fig. 4-14).

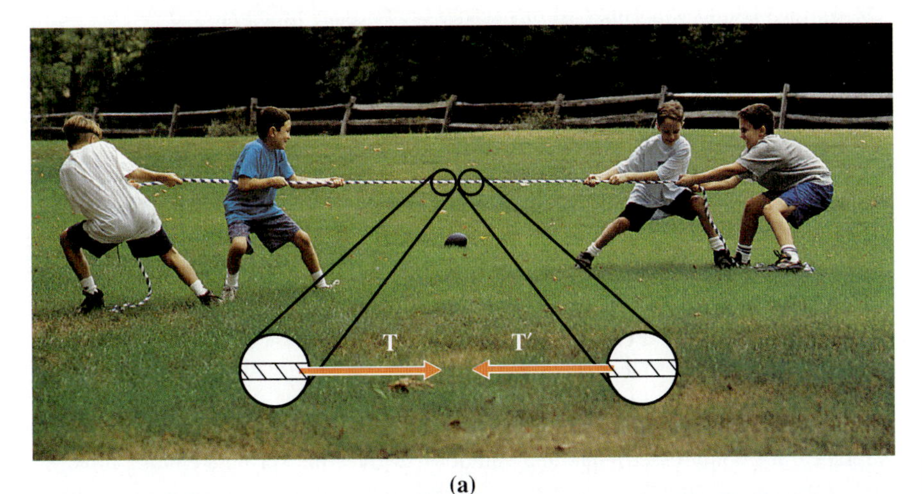

(a)

(b)

Fig. 4-14 **(a)** In a tug-of-war, the rope is under great tension, meaning that there is a large tension force exerted by any section of the rope on an adjacent section. **(b)** A much greater tension is present in the cables supporting the Golden Gate Bridge. At point P, the section to the right of P exerts a force **T** on the section to the left of P. (There is also a reaction force, not shown in the figure.)

The cable consists of 27,572 strands of flexible wire.

Spring Force

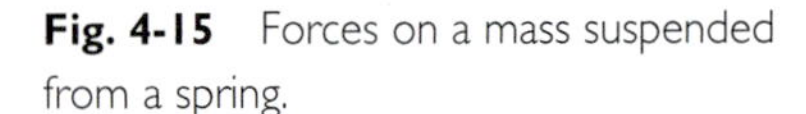

Fig. 4-15 Forces on a mass suspended from a spring.

The force exerted by a stretched spring is a particularly simple example of a contact force. When a spring is stretched, some of the adjacent molecules within the spring are pulled slightly farther apart from each other, and an attractive electromagnetic force attempts to pull them back to their original positions. Compression of a spring also produces a force in the spring. In this case, adjacent molecules are pushed together, and it is a repulsive electromagnetic force that is at work, attempting to push the molecules back to their original positions.

When an object hangs vertically at rest from a spring, Newton's second law predicts that the spring exerts a force **F** sufficient to cancel the object's weight (Fig. 4-15). Thus the force exerted by the spring is equal in magnitude to the weight supported. We can experimentally determine a force law for a stretched spring by hanging weights from the spring and measuring the corresponding stretch. When we do so, we find that most springs stretch or compress in direct proportion to the force applied to them, so long as the amount of stretching or compression is not too large. Put another way, the magnitude of the force **F** exerted by the spring is directly proportional to the spring's change in length $\Delta\ell$. This may be expressed

$$F = k\,\Delta\ell$$

where k is called the force constant of the spring. The force constant indicates the stiffness of the spring. The larger the value of k, the stiffer the spring, that is, the larger the force that must be applied to produce a given change in length $\Delta\ell$.

We can express the spring force law in a useful alternative form, a form that indicates the direction as well as the magnitude of the spring force. Consider the force **F** exerted on a block by a horizontal spring, as shown in Fig. 4-16. The origin of the x-axis is chosen at the position of the block for which the spring is relaxed (neither stretched nor compressed). The magnitude of x gives the spring's change in length ($\Delta\ell$) as it is either stretched or compressed. When x is positive, the spring is stretched and exerts a pull to the left, so that F_x is negative. When x is negative, the spring is compressed and exerts a push to the right so that F_x is positive. In either case, the sign of F_x is opposite the sign of x. Both the magnitude and the direction of the spring force are indicated by writing the force law in the form

$$F_x = -kx \tag{4-6}$$

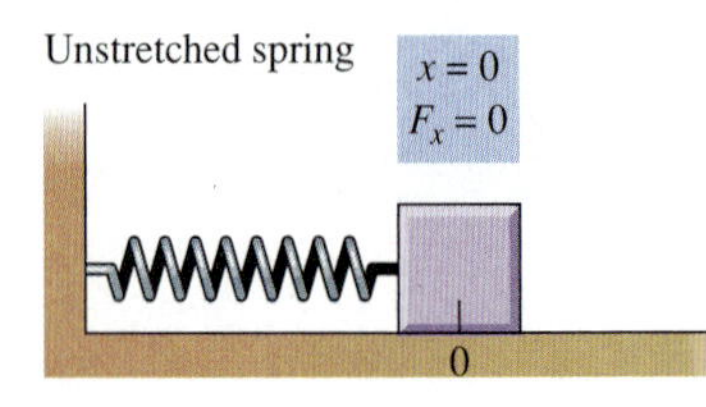

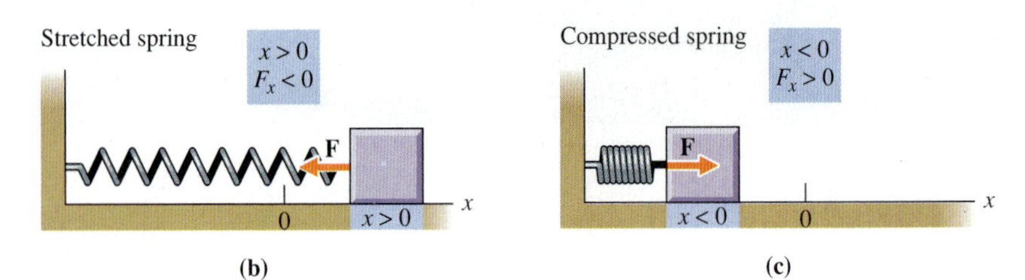

Fig. 4-16 The force **F** exerted by a spring on a block attached to the spring varies in magnitude and direction, depending on the compression or stretching of the spring.

EXAMPLE 5 Mass on a Spring

Find the instantaneous acceleration of a 1.00 kg mass suspended from a spring of force constant 5.00 N/cm, when the spring is stretched 10.0 cm. The mass is initially at rest.

SOLUTION The forces **F** and **w** acting on the mass are shown in Fig. 4-17. We take the y-axis to be positive in the upward direction. Using Eq. 4-6 with y substituted for x, we obtain the y component of the force exerted on the mass by the spring:

$$F_y = -ky = -(5.00\text{ N/cm})(-10.0\text{ cm}) = +50.0\text{ N}$$

The only other force acting on the mass is its weight **w**, which acts along the negative y-axis and has magnitude given by Eq. 4-5:

$$w = mg = (1.00\text{ kg})(9.80\text{ m/s}^2) = 9.80\text{ N}$$

The spring force exceeds the weight. Therefore the second law predicts an upward acceleration:

$$a_y = \frac{\Sigma F_y}{m} = \frac{50.0\text{ N} - 9.80\text{ N}}{1.00\text{ kg}}$$
$$= 40.2\text{ m/s}^2$$

The acceleration we have calculated is instantaneous, corresponding to a particular value of y. The acceleration is not a constant. As the mass moves, the length of the spring changes, and therefore the force the spring exerts also changes. The changing force produces changing acceleration.

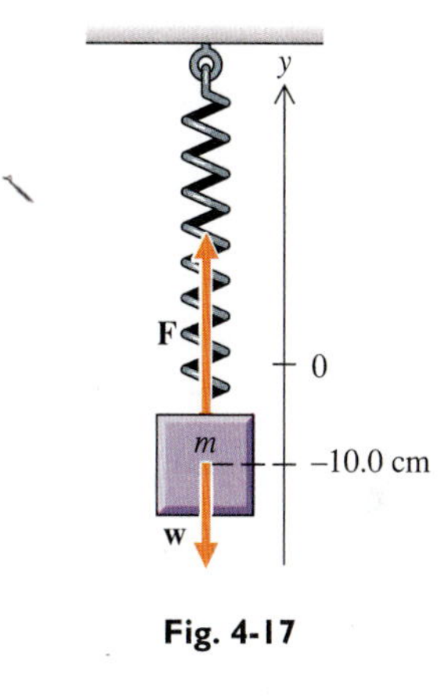

Fig. 4-17

4-7 The Concept of Force

Force is a subtle physical concept, one that developed over hundreds of years. It is therefore not surprising that understanding the precise nature of this concept requires some careful thought. We began our discussion of force in this chapter with the simple qualitative concept of a push or a pull. With our discussion of Newton's first and second laws of motion, we arrived at a refinement of the force concept as that which tends to produce acceleration. It is sometimes stated that force is "defined" by Newton's second law to be mass times acceleration. The difficulty with this kind of statement is that it leaves the impression that the second law is merely a definition* and therefore that it says nothing substantive about nature. But implicit in the second law is the idea that there are force laws—equations for computing the force on a body from knowledge of its physical environment.

The essential physical fact is that the interaction between an object and its physical environment can produce an acceleration of the object. If there were only one kind of force in nature, we could express the relationship between the acceleration and the environment directly and eliminate the concept of force. As it is, there are different kinds of interactions in nature and many different force laws.

Thus force is a useful intermediate concept—a unifying element in the logical structure of physics. The force concept is a way of relating the motion of a body to the body's surroundings. We think of a body experiencing "forces" produced by other bodies in the surroundings. Each force is a vector whose magnitude and direction can sometimes be computed from one of a number of force laws. When all the forces acting on the body are known, the second law can be used to find its acceleration.

*However, the second law is used to define the *unit* of force ($1\text{ N} = 1\text{ kg-m/s}^2$).

4-8 Applications of Newton's Laws of Motion

As stated at the beginning of this chapter, there are two kinds of problems in classical mechanics: (a) to find unknown forces acting on a body, given the body's acceleration, and (b) to predict the future motion of a body, given the body's initial position and velocity and the forces acting on it. For either kind of problem, we use Newton's second law ($\Sigma \mathbf{F} = m\mathbf{a}$). The following general strategy is useful for solving such problems:

1 **Choose a body** to which you will apply Newton's second law and isolate that body by drawing a diagram of it free of its physical surroundings. The body chosen is called a **free body,** and the diagram, which will include forces as described in step 2, is called a **free-body diagram.**

The free body may be a whole body, part of a body, or a collection of bodies. We may apply the second law to any system, as long as the acceleration **a** is the same for all parts of that system. In this case, the system behaves as a particle and Newton's second law is valid (see Problem 52). This condition on **a** is satisfied for any body at rest, such as in Ex. 4, where $\mathbf{a} = \mathbf{0}$ for all parts of the human body, or for any rigid body moving without rotating,* as for the mass on the spring in Ex. 5.

2 **Identify all the forces** exerted *on* the free body by objects in the surroundings, and draw these forces in the free-body diagram. Any object that is in contact with the free body will exert a force on it. In addition, there may be various noncontact forces: gravitational, electric, or magnetic. In this chapter the only noncontact force we shall need to consider is the weight of the object.

Do not include in the free-body diagram the forces exerted *by* the free body on the surroundings. Include only forces acting *on* the body.

Nor should you include forces acting *between* parts of the free body. Thus in Ex. 4 we considered the standing person as a particle, ignoring the human body's internal structure. Of course the body has various parts, each of which exerts forces on other parts. These are internal forces, however, and **only external forces should be included in the free-body diagram because only these forces determine the free body's acceleration.**

At times we may be interested in computing forces that are normally regarded as internal forces. We can compute such forces if we make an appropriate choice of the body to which we apply Newton's second law, so that these forces are external to the body. For example, we may find the tension in a rope by choosing a section of the rope as the free body, so that the tension is an external force.

3 Choose an inertial reference frame with convenient coordinate axes, **apply force laws,** and **apply Newton's second law** in component form. This step may require resolving force vectors into their components along the coordinate axes.

If the choice of the free body was a good one, there will be enough information to solve for the unknowns in the problem. Some problems may require the analysis of two or more related free-body diagrams.

---

*The particles of a rotating body do not all experience the same acceleration. However, in the next chapter we shall find that Newton's second law may still be used to describe the motion of a certain point—the center of mass of the body.

EXAMPLE 6 Finding the Tension in a Cable

A block of marble whose weight is 2.00×10^4 N is suspended from a cable supported by a crane (Fig. 4-18a). The cable's weight is 4.00×10^2 N. (a) Find the tension in the top and bottom of the cable when the block and cable are both at rest. (b) Find the tension in the top and bottom of the cable when the block is accelerating downward at the rate of 2.50 m/s^2.

SOLUTION (a) To find the tension in the top of the cable, choose as a free body the block and cable. Such a choice makes the tension $\mathbf{T}_1$ an external force. This tension force is the only contact force acting on the system. The only other external forces are the weight of the block $\mathbf{w}_b$ and the weight of the cable $\mathbf{w}_c$. The three external forces are shown in the free-body diagram of Fig. 4-18b.

Since the system is unaccelerated, we know from Newton's second law that the vector sum of the external forces equals zero. We choose our coordinate axes as indicated in Fig. 4-18b so that all forces lie along the y-axis and we need only apply the equation

$$\Sigma F_y = 0$$

From our free-body diagram, we see that $\mathbf{T}_1$ acts along the positive y-axis and $\mathbf{w}_c$ and $\mathbf{w}_b$ act along the negative y-axis. Thus

$$T_1 - w_c - w_b = 0$$

Solving for T_1, we obtain

$$T_1 = w_c + w_b = 4.00 \times 10^2 \text{ N} + 2.00 \times 10^4 \text{ N}$$
$$= 2.04 \times 10^4 \text{ N}$$

Next we find the tension in the bottom of the cable by choosing the block and a small section of cable at the bottom as the free body (Fig. 4-18c). The tension $\mathbf{T}_2$ and the weight $\mathbf{w}_b$ are the only external forces. Again we apply the second law:

$$\Sigma F_y = 0$$
$$T_2 - w_b = 0$$
$$T_2 = w_b = 2.00 \times 10^4 \text{ N}$$

The values for T_1 and T_2 are not surprising. Tension T_1 in the top of the cable balances the combined weight of the cable and block, whereas tension T_2 in the bottom of the cable balances the weight of the block alone.

(b) Here the block and cable have an acceleration $a_y = -2.50$ m/s^2. We shall apply Newton's second law and shall therefore need to find the masses of the block and the cable, m_b and m_c, using the force law $w = mg$:

$$m_b = \frac{w_b}{g} = \frac{2.00 \times 10^4 \text{ N}}{9.80 \text{ m/s}^2}$$
$$= 2040 \text{ kg}$$

$$m_c = \frac{w_c}{g} = \frac{4.00 \times 10^2 \text{ N}}{9.80 \text{ m/s}^2}$$
$$= 40.8 \text{ kg}$$

We again use the free body shown in Fig. 4-18b to solve for T_1, applying Newton's second law:

$$\Sigma F_y = ma_y$$
$$T_1 - w_c - w_b = (m_c + m_b)a_y$$
$$T_1 = w_c + w_b + (m_c + m_b)a_y$$
$$= 4.00 \times 10^2 \text{ N} + 2.00 \times 10^4 \text{ N} + (2040 \text{ kg} + 40.8 \text{ kg})(-2.50 \text{ m/s}^2)$$
$$= 1.52 \times 10^4 \text{ N}$$

And using Fig. 4-18c, we find the tension T_2:

$$\Sigma F_y = ma_y$$
$$T_2 - w_b = m_b a_y$$
$$T_2 = w_b + m_b a_y$$
$$= 2.00 \times 10^4 \text{ N} + (2040 \text{ kg})(-2.50 \text{ m/s}^2)$$
$$= 1.49 \times 10^4 \text{ N}$$

Notice that the tensions T_1 and T_2 are now less than the weights supported. The reason is that the weights are accelerating downward. If the acceleration were equal to $\mathbf{g}$, the tension forces would be zero.

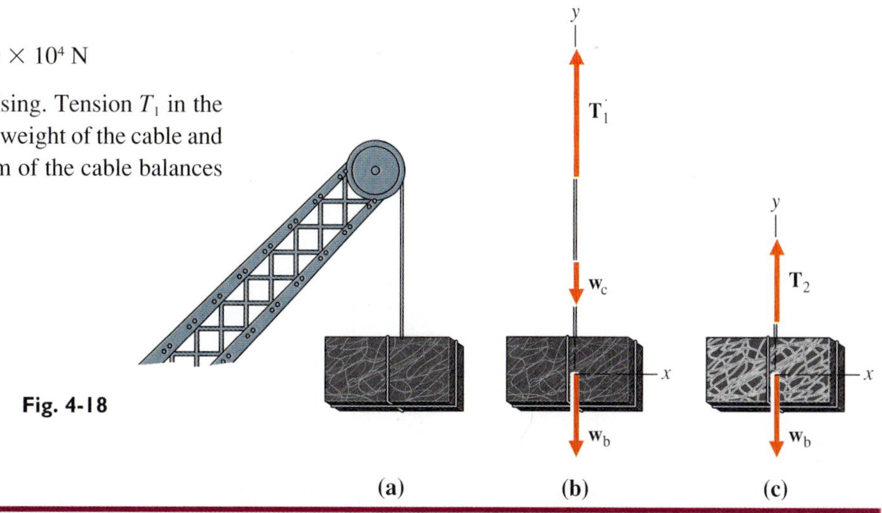

Fig. 4-18

It is a good approximation to ignore the mass of a cable, rope, or string whenever this mass is much less than other masses in a problem. The tension then is transmitted undiminished throughout, a fact that can be seen in the preceding example when we set m_c and w_c equal to zero. Then $T_1 = T_2$ in both parts a and b.

EXAMPLE 7 Forces on a Foot

Find the forces on each foot of a woman standing at rest if her weight of 575 N (129 lb) is evenly distributed between her two feet. Neglect the weight of the foot.

SOLUTION To solve for the forces exerted on the feet by the supporting surface, we choose the woman as the free body. The weight **w** and two equal contact forces **S** are the only external forces acting on the body (Fig. 4-19a). Applying Newton's second law, we solve for the magnitude of **S**:

$$\Sigma F_y = ma_y = 0$$

$$S + S - w = 0$$

$$S = \frac{w}{2} = \frac{575 \text{ N}}{2} = 288 \text{ N} \quad \text{(about 65 lb)}$$

We have found the force exerted on either foot by the supporting surface, but this is not the only force acting on the foot. In addition, the leg and upper body exert a downward force on the foot. To solve for this unknown force, we choose the foot alone as the free body and draw our free-body diagram (Fig. 4-19b), with two external forces: **S**, produced by the contact with the supporting surface, and **F**, produced by the contact with the leg and upper body. We neglect the weight of the foot, which is small compared with these other forces. Newton's second law implies that the two forces cancel:

$$\mathbf{F} = -\mathbf{S}$$

$$F = S = 288 \text{ N}$$

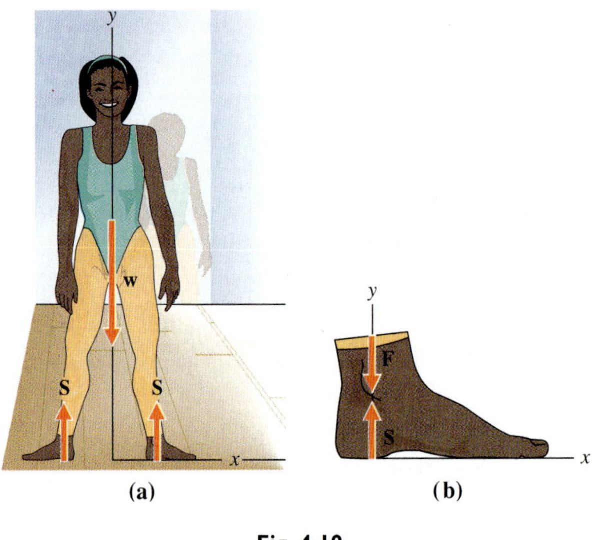

Fig. 4-19

The two forces **F** and **S** are both equal in magnitude to half the weight of the body. In other words, half the weight of the body pushes down on each foot and is supported by the surface. The two opposing forces **F** and **S** produce no acceleration of the foot, but they do cause some compression.

EXAMPLE 8 Forces on Accelerating Blocks

Two blocks are pushed along a frictionless horizontal surface by a constant 6.00 N force (Fig. 4-20). Find the acceleration of each block and the forces on it, given that $m_1 = 1.00$ kg and $m_2 = 2.00$ kg.

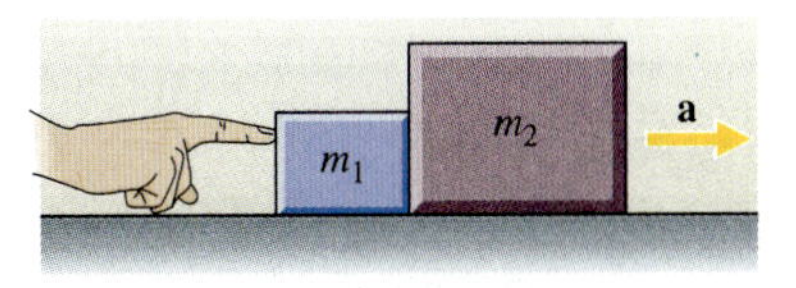

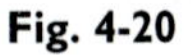

Fig. 4-20

SOLUTION We may choose m_1, m_2, or the combination of m_1 and m_2 as a free body, since m_1 and m_2 have a common acceleration. The three free-body diagrams are shown in Fig. 4-21. Notice that there is a contact force $\mathbf{F}_2$ exerted on m_2 by m_1 and a reaction force $\mathbf{F}_2'$ exerted on m_1 by m_2. This contact force must be smaller than the applied force $\mathbf{F}_1$; otherwise there would be no net force to provide for the acceleration of m_1.

In solving this problem, we begin with the free-body diagram of the two-block combination because in the other diagrams there are too many unknowns. There is no motion in the vertical direction, and thus $a_y = 0$ and the surface forces and weights cancel. We apply Newton's second law to the motion along the x-axis and solve for a_x:

$$\Sigma F_x = ma_x$$

$$F_1 = (m_1 + m_2)a_x$$

$$a_x = \frac{F_1}{m_1 + m_2} = \frac{6.00 \text{ N}}{1.00 \text{ kg} + 2.00 \text{ kg}}$$

$$= 2.00 \text{ m/s}^2$$

Now that we have found a_x, we may apply Newton's second law to m_2 and solve for the unknown $\mathbf{F}_2$:

$$\Sigma F_x = ma_x$$

$$F_2 = m_2 a_x = (2.00 \text{ kg})(2.00 \text{ m/s}^2)$$

$$= 4.00 \text{ N}$$

We can check this result by applying Newton's second law to m_1 and solving for a_x:

$$a_x = \frac{\Sigma F_x}{m}$$

$$= \frac{F_1 - F_2}{m} = \frac{6.00 \text{ N} - 4.00 \text{ N}}{1.00 \text{ kg}}$$

$$= 2.00 \text{ m/s}^2$$

This, of course, agrees with our previously computed value of a_x.

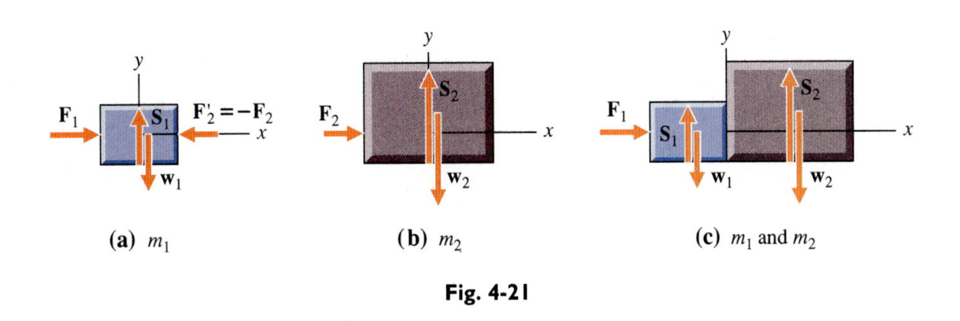

Fig. 4-21

EXAMPLE 9 Tension in Strings Supporting a Weight

A 10.0 N weight is supported at rest by string of negligible mass, as shown in Fig. 4-22a. Find the tension in each string.

SOLUTION The tension throughout the vertical string is obviously just equal to the weight supported—10.0 N. (You can prove this result by choosing the weight and any section of the vertical string as a free body and applying Newton's second law.)

The tension in the other two strings is not so obvious. We must choose a free body for which these forces are external forces and for which there is sufficient information to solve for the forces. In problems such as this, the right choice for the free body may not be apparent. There are many bodies one might choose but from which no information is gained—the ceiling, for example, or a section of one string. The useful free body here is either the knot where the three strings meet or the knot and some section of each string. The three tension forces are all external to the knot; they are shown resolved into vector components in the free-body diagram (Fig. 4-22b).

We already know that the tension T_3 is 10.0 N. We apply Newton's second law in component form to the knot:

$$\Sigma F_x = ma_x = 0$$

$$T_2 \cos 30.0° - T_1 \cos 45.0° = 0$$

This gives us a relationship between the two unknowns T_1 and T_2. A second equation relating T_1 and T_2 is obtained when we equate the sum of the y components of the forces to zero:

$$\Sigma F_y = ma_y = 0$$

$$T_1 \sin 45.0° + T_2 \sin 30.0° - T_3 = 0$$

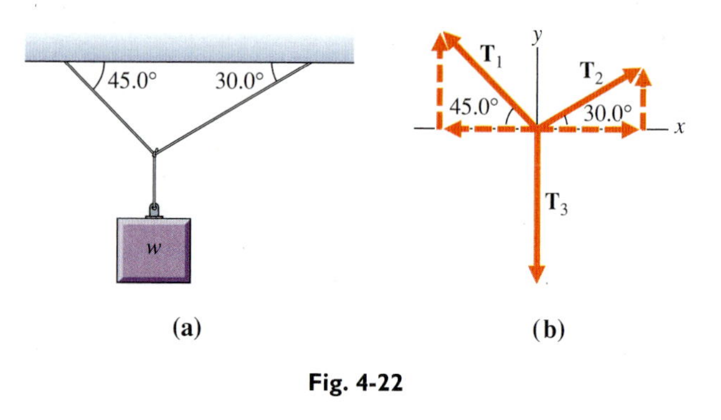

Fig. 4-22

Solving the two linear equations for the two unknowns in terms of T_3 and the angles, we obtain

$$T_1 = \frac{T_3}{\sin 45.0° + \cos 45.0° \tan 30.0°}$$

$$= \frac{10.0 \text{ N}}{\sin 45.0° + \cos 45.0° \tan 30.0°}$$

$$= 8.97 \text{ N}$$

$$T_2 = \frac{T_1 \cos 45.0°}{\cos 30.0°}$$

$$= \frac{(8.97 \text{ N})(\cos 45.0°)}{\cos 30.0°}$$

$$= 7.32 \text{ N}$$

When a flexible rope or string passes over a frictionless pulley of negligible mass, the tension is the same on both sides of the pulley. A frictionless and massless pulley changes the direction of the tension force but leaves the magnitude unchanged (see Problem 45 of Chapter 9).

EXAMPLE 10 Atwood's Machine

Two unequal masses, m_1 and $m_2 > m_1$, are suspended from opposite ends of a rope of negligible mass that passes over and is supported by a frictionless, stationary pulley of negligible mass (Fig. 4-23a). The greater mass m_2 will accelerate downward and the smaller mass m_1 will experience an acceleration of equal magnitude in the upward direction. By adjusting the values of m_1 and m_2, we can make the acceleration as small as we want. (This simple device is called "Atwood's machine.") Find expressions for the magnitude of the acceleration and the tension in the rope as functions of m_1 and m_2.

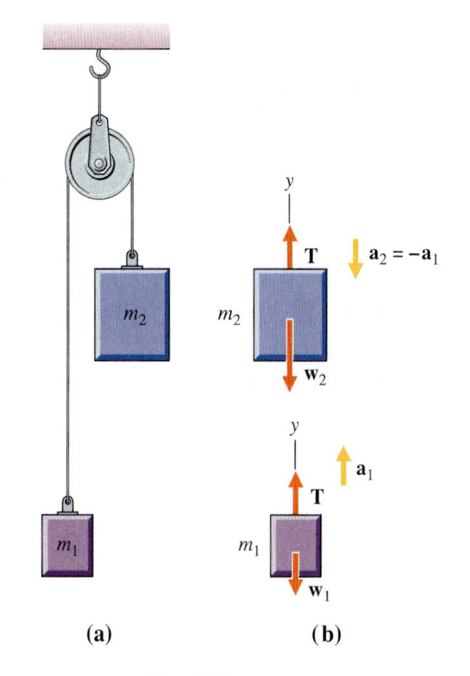

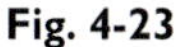
Fig. 4-23

SOLUTION First we choose as free bodies the two masses (Fig. 4-23b). The tension force on each mass is the same, and the two accelerations a_1 and a_2 are equal in magnitude and opposite in direction. This follows from the fact that when one mass moves a certain distance upward, the other moves the same distance downward in the same time interval.

We apply Newton's second law to each body:

$$\Sigma F_y = ma_y$$

With our choice of coordinate axes, $a_y = +a$ for m_1 and $a_y = -a$ for m_2. Thus we have

$$T - w_1 = m_1 a$$
$$T - w_2 = m_2(-a)$$

We have two linear equations with two unknowns, T and a. Solving for the unknowns in terms of the masses and weights, we obtain

$$a = \frac{w_2 - w_1}{m_2 + m_1} \qquad T = w_1 + \frac{m_1(w_2 - w_1)}{m_2 + m_1}$$

or, using $w = mg$,

$$a = \frac{m_2 - m_1}{m_2 + m_1} g \qquad T = \frac{2m_1 m_2}{m_2 + m_1} g$$

EXAMPLE 11 Instantaneous Force On a Runner's Foot

Fig. 4-24 shows a simplified model of a force platform used in biomechanical research to study the force exerted on the ground by the foot of a running person. Suppose that the platform has a mass of 5.0 kg and each of the four springs has a force constant of 1.0×10^6 N/m.

At some instant, the vertical springs are compressed 0.50 mm. At the same time, each horizontal spring differs from its relaxed length by 0.10 mm, the left spring compressed and the right spring stretched. The platform has a vertical component of acceleration of 5.0 m/s² in the upward direction and a horizontal component of 2.0 m/s² in the backward direction. Find the horizontal and vertical components of force on the platform.

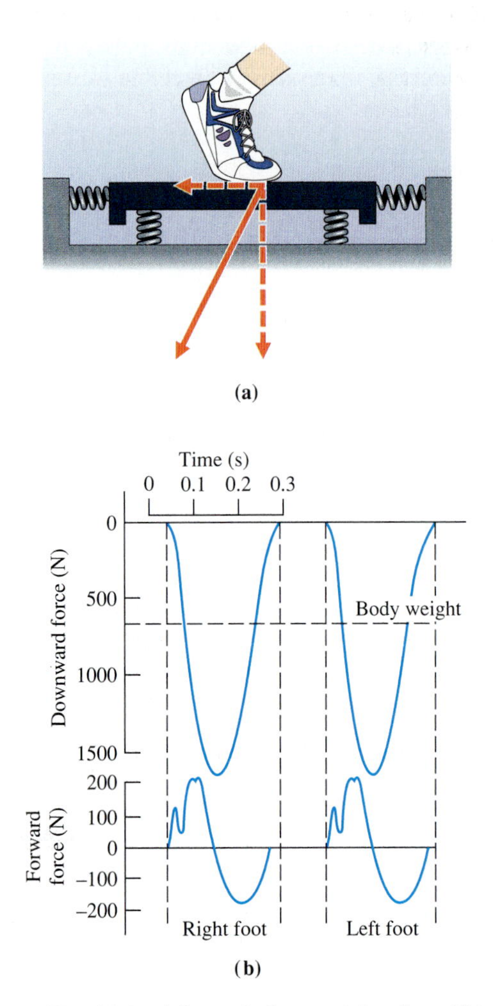

Fig. 4-24 A force platform and data for a 68 kg man running at 3.5 m/s. (From Alexander R McN: *Biomechanics*, New York, 1975, Halstead Press.)

SOLUTION We choose the platform as the free body and show in the free-body diagram the four forces exerted by the springs—$\mathbf{F}_1$, $\mathbf{F}_2$, $\mathbf{F}_3$, $\mathbf{F}_4$—the weight of the platform $\mathbf{w}$, and the force exerted by the foot, $\mathbf{F}_5$ (Fig. 4-25). We apply Newton's second law to the motion along the x-axis and solve for the horizontal component of $\mathbf{F}_5$:

$$\Sigma F_x = ma_x$$

$$F_3 + F_4 + F_{5x} = ma_x$$

$$F_{5x} = ma_x - F_3 - F_4$$

The two horizontal spring forces, $\mathbf{F}_3$ and $\mathbf{F}_4$, are of equal magnitude $k\,\Delta\ell$. Substituting into the last equation, we obtain

$$\begin{aligned} F_{5x} &= ma_x - 2k\,\Delta\ell \\ &= (5.0\text{ kg})(-2.0\text{ m/s}^2) - 2(1.0 \times 10^6\text{ N/m})(1.0 \times 10^{-4}\text{ m}) \\ &= -210\text{ N} \end{aligned}$$

Next we apply the second law to the motion along the y-axis:

$$\Sigma F_y = ma_y$$

$$F_1 + F_2 + F_{5y} - w = ma_y$$

$$F_{5y} = ma_y + w - F_1 - F_2$$

or, using $F_1 = F_2 = k\,\Delta\ell$ and $w = mg$,

$$\begin{aligned} F_{5y} &= ma_y + mg - 2k\,\Delta\ell \\ &= (5.0\text{ kg})(+5.0\text{ m/s}^2) + (5.0\text{ kg})(9.80\text{ m/s}^2) - 2(1.0 \times 10^6\text{ N/m})(5.0 \times 10^{-4}\text{ m}) \\ &= -930\text{ N} \end{aligned}$$

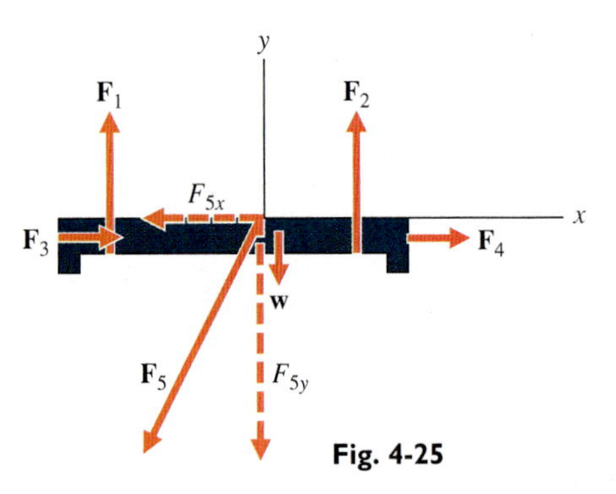

Fig. 4-25

At the instant considered, the foot exerts on the ground a backward force of 210 N (47 lb) and a downward force of 930 N (210 lb). The ground therefore exerts a reaction force of 210 N in the forward direction and 930 N in the upward direction. The graph in Fig. 4-24 indicates that the vertical force on the foot of a 68 kg (150 lb) runner reaches a maximum value of about 1700 N (380 lb), about 2.5 times the weight of the runner. This suggests why running on a hard surface without proper shoes can so easily lead to injuries. A well-cushioned heel on a running shoe reduces the maximum force on the foot as it hits the surface by lengthening the time of contact and thereby reducing the maximum instantaneous acceleration.

CHAPTER 4 SUMMARY

Mass is a measure of a body's inertia, that is, its tendency to resist acceleration. **Force** is the result of a mutual interaction between two bodies. Forces can sometimes be calculated from force laws. Each force on a body tends to accelerate the body in the direction of the force. This tendency may be opposed by the presence of other forces.

Newton's Laws of Motion

First law If no forces act on a body, the body continues in its state of rest or of uniform motion in a straight line.

Second law The resultant force on a body equals the product of the body's mass and acceleration.

$$\Sigma \mathbf{F} = m\mathbf{a}$$

or

$$\Sigma F_x = ma_x \qquad \Sigma F_y = ma_y$$

Third law Forces occur in action-reaction pairs—two forces equal in magnitude and opposite in direction, acting on two different bodies.

Force Laws

Weight The gravitational force of the earth on a body of mass m near the surface of the earth is its weight $\mathbf{w}$, where

$$\mathbf{w} = m\mathbf{g}$$

Tension The tension in a flexible body, such as a rope or string, is an attractive force between adjacent sections of the rope or string, and tangent to it.

Spring force The magnitude of the force exerted by a spring on an object is related to the change in length $\Delta\ell$ of the spring by the equation

$$F = k\,\Delta\ell$$

where k is the force constant of the spring, a measure of its stiffness. The x component of this force may be expressed as

$$F_x = -kx$$

In applying Newton's second law, any body or combination of bodies may be used as the free body as long as all parts of the free body have the same acceleration **a.** Then one may draw a free-body diagram, which shows all the external forces acting on the body. These forces determine the acceleration of the body, through the second law. In your analysis you must use an **inertial reference frame,** any reference frame in which Newton's first law is satisfied, that is, in which an isolated body is not accelerated. The laws of physics are valid only in inertial reference frames.

Questions

1 Aristotle had to invent an elaborate process in order to describe the motion of projectiles as forced motion. He argued that an arrow moves through the air by pushing aside the air, which then rushes around to the tail of the arrow and propels it forward. According to Newton, what force is needed to produce the horizontal component of the arrow's velocity?

2 You apply the brakes on your car, stopping suddenly, and are thrown forward. What force is responsible for your forward motion?

3 As viewed from the earth, a body is at rest. The same body is viewed by an observer on an escalator moving at a constant speed of 3 m/s. Are Newton's laws satisfied for this observer?

4 Suppose you are in a completely enclosed compartment in an airplane flying to an unknown destination. The walls are shielded so that you can't detect the earth's magnetic field, and you are not able to observe anything else outside the compartment. You place a ball on the floor, and it remains at rest.

(a) What can you conclude about the velocity of the plane?

(b) Is there any other experiment you could perform to determine the plane's speed or direction of motion? The compartment is a well-equipped physics laboratory.

5 Consider a planet the same size as earth, but one on which a day is much shorter than an earth day. Compared to the surface of the earth, would the surface of the planet be better or worse, as an approximation to an inertial reference frame?

6 Is it possible for an object to move along a curved path without any force acting on it?

7 You are pulling in a fish, using a fishing line that is very close to its breaking point. Should you (a) pull the fish in as quickly as possible or (b) pull the fish in slowly and without jerking the line?

8 If a horse tries to pull a cart, exerting a force on the cart in the forward direction, the cart will exert a backward force on the horse.

(a) Since these two forces are equal in magnitude and opposite in direction, is it not then impossible for the horse and cart to move?

(b) If the horse and cart together are considered as the free body, what other body exerts the force necessary to accelerate the free body forward?

9 In a tug of war, the winning team pulls on a rope of negligible mass and drags the losing team across a line. Does the winning team (a) pull harder on the rope than the losing team or (b) push harder on the ground than the losing team?

10 Blocks A and B collide on a frictionless horizontal surface. Block A, of mass 5 kg, experiences an instantaneous acceleration of 10 m/s^2 to the right, while block B experiences an instantaneous acceleration of 2 m/s^2 to the left. What is the mass of B?

11 You are an overweight space-age commuter, traveling from planet to planet and so experiencing varying values of g. Which way can you be sure the diet you are following is effective—by measuring (a) your weight on a spring scale, or (b) your mass on a balance?

12 If the earth's pull on a 40 N brick is 10 times as great as its pull on a 4 N book, when both are in free fall, why do they have the same acceleration?

13 An astronaut of mass m is in a spaceship accelerating vertically upward from the earth's surface with acceleration of magnitude a. The contact force exerted on the astronaut has a magnitude given by (a) mg; (b) ma; (c) $m(g + a)$; (d) $m(g - a)$; (e) $m(a - g)$.

14 A skydiver is observed to have a terminal speed of 55 m/s in a prone position and 80 m/s in a vertical position. Which of the following can be concluded from this observation?

(a) The force of gravity on the skydiver is less in the prone position than in the vertical position.

(b) The force of air resistance on the skydiver is proportional to the speed of the body.

(c) The force of air resistance is greater at 55 m/s in the prone position than at 80 m/s in the vertical position.

(d) The force of air resistance at 55 m/s in the prone position is the same as at 80 m/s in the vertical position.

(e) None of the above.

15 A heavy blanket hangs from a clothesline. Will the tension in the clothesline be greater if it sags a little or if it sags a lot?

16 In analyzing the forces on a halfback running with a football, a free-body diagram is drawn. If the player's entire body is the chosen free body, which of the following forces should not be drawn in the diagram: (a) his weight; (b) the force exerted on him by the ground; (c) the force he exerts on the ground; (d) the tension in the calf muscles?

Answers to Odd-Numbered Questions

1 None, since there is no horizontal acceleration; **3** yes; **5** worse; **7** b; **9** (a) no; (b) yes; **11** b; **13** c; **15** If it sags a little.

Problems (listed by sections)

4-4 Mass

1 The same force that gives the standard 1 kg mass an acceleration of 1.00 m/s² acts on a body, producing a horizontal acceleration of 1.00×10^{-2} m/s². No other horizontal force acts on the body. Find its mass in kg.

2 The same force that gives the standard 1 kg mass an acceleration of 1.00 m/s² acts first on body A, producing an acceleration of 0.500 m/s², and then on body B, producing an acceleration of 0.333 m/s². Find the acceleration produced when A and B are attached and the same force is applied.

4-5 Newton's Second and Third Laws

(Unless otherwise stated, all systems are assumed to be viewed from an inertial reference frame.)

3 Is the particle shown in Fig. 4-26 accelerated?

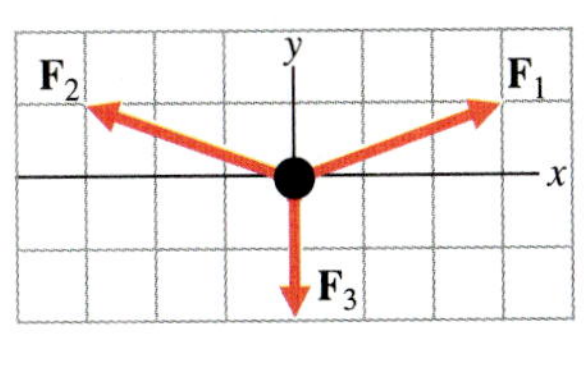

Fig. 4-26

4 Is the particle shown in Fig. 4-27 accelerated?

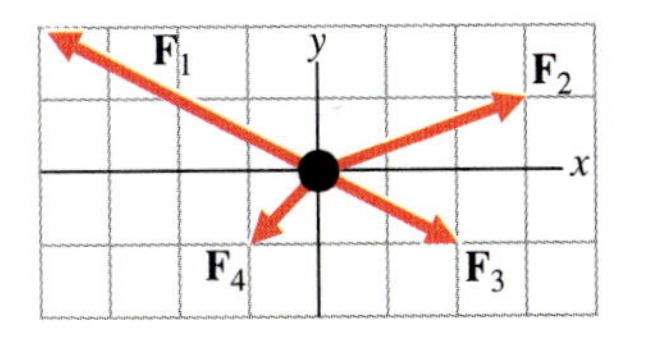

Fig. 4-27

5 The particle shown in Fig. 4-28 is at rest. Find the magnitudes of $\mathbf{F}_1$ and $\mathbf{F}_2$.

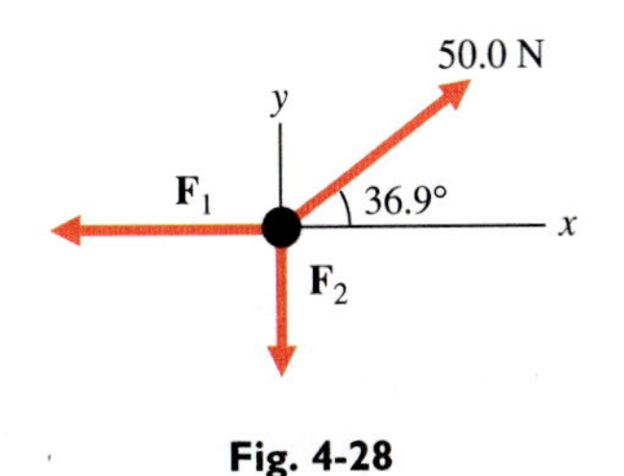

Fig. 4-28

6 The particle shown in Fig. 4-29 is at rest. Find the magnitude and direction of **F**.

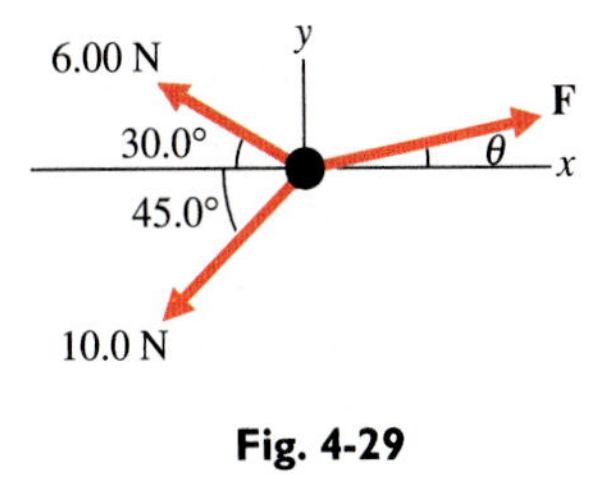

Fig. 4-29

7 A boat is pulled at constant velocity by the two forces shown in Fig. 4-30. Find the horizontal force exerted on the boat by the water.

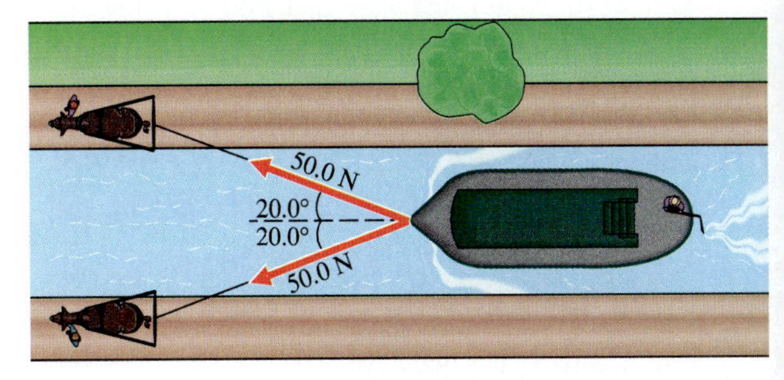

Fig. 4-30

8 A log is dragged along the ground at a constant speed by a force of 425 N at an angle of 45.0° above the horizontal. Find the horizontal component of force exerted by the ground on the log.

9 The canvas tarpaulin shown in Fig. 4-31 is stretched by horizontal forces applied by means of ropes. Find the x and y components of **F**.

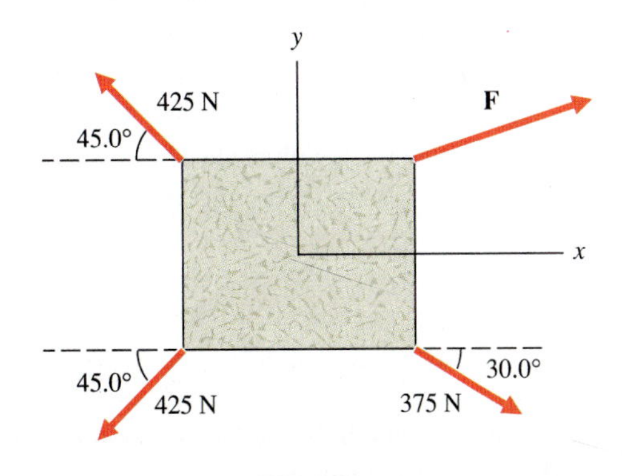

Fig. 4-31

10 A ball is released from rest in an elevator and falls 1.00 m to the floor in 0.400 s. Is the elevator an inertial reference frame?

11 Three children fight over a small stuffed animal of mass 0.200 kg, pulling with the forces indicated in Fig. 4-32. Find the instantaneous acceleration of the toy.

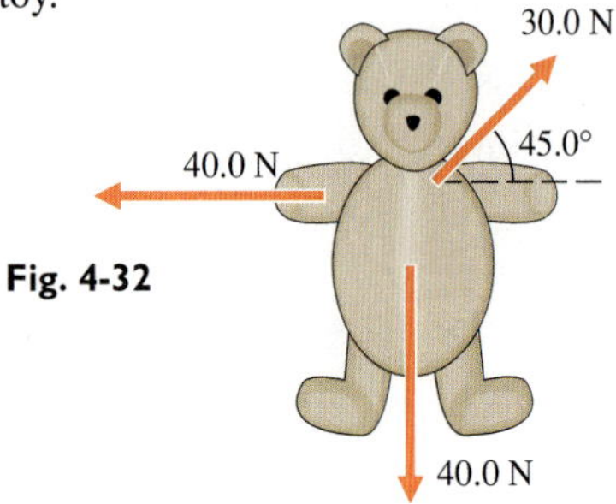

Fig. 4-32

12 Two hockey players strike a puck of mass 0.300 kg with their sticks simultaneously, exerting forces of 1.20×10^3 N, directed west, and 1.00×10^3 N, directed 30.0° east of north. Find the instantaneous acceleration of the puck.

13 A girl scout paddling a canoe pushes the water back with her paddle, exerting a backward force of 155 N on the water. Find the acceleration of the girl and the canoe if their combined mass is 90.0 kg.

14 A golf ball of mass 4.50×10^{-2} kg is struck by a club. Contact lasts 2.00×10^{-4} s, and the ball leaves the tee with a horizontal velocity of 50.0 m/s. Compute the average force the club exerts on the ball by finding its average acceleration.

15 A 3.00 kg mass is acted upon by four forces in the horizontal (xy) plane, as shown in Fig. 4-33. Find the acceleration of the mass.

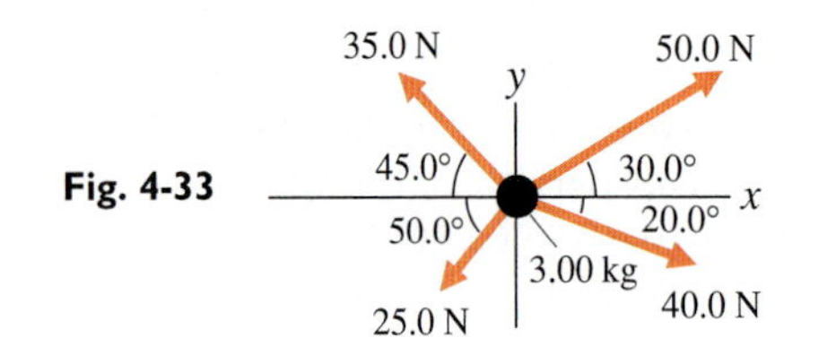

Fig. 4-33

★**16** A boxer stops a punch with his head. To approximate the force of the blow, treat the opponent's glove, hand, and forearm as a particle of mass 1.50 kg moving with an initial velocity of 20.0 m/s. Estimate the force exerted on the head if (a) the hand moves forward 10.0 cm while delivering the blow and then coming to rest; (b) the head is deliberately moved back during the punch so that the hand moves forward 20.0 cm while decelerating.

17 A boat and its passengers have a combined mass of 5.10×10^2 kg. The boat is coasting into a pier at a speed of 1.00 m/s. How great a force is required to bring the boat to rest in 1.00×10^{-2} s?

★**18** A 110 kg fullback runs at the line of scrimmage.
(a) Find the constant force that must be exerted on him to bring him to rest in a distance of 1.0 m in a time interval of 0.25 s.
(b) How fast was he running initially?

★**19** A car traveling initially at 50.0 km/h crashes into a brick wall. The front end of the car collapses, and the 70.0 kg driver, held in his seat by a shoulder harness, continues to move forward 1.00 m after the initial contact, decelerating at a constant rate. Find the horizontal force exerted on him by the seat harness.

4-6 Force Laws

Weight

20 (a) Compute your weight in N.
(b) Compute your mass in kg and in slugs.
(c) How much weight would you lose in going from the North Pole, where $g = 9.83$ m/s², to the equator, where $g = 9.78$ m/s², assuming no loss in mass?

21 (a) A 1.00 kg book is held stationary in the hand. Find the forces acting on the book and the reaction forces to each of these.
(b) The hand now exerts an upward force of 15.0 N on the book. Find the book's acceleration.
(c) As the book moves upward, the hand is quickly removed from the book. Find the forces on the book and its acceleration.

22 Find the vertical force exerted by the air on an airplane of mass 5.00×10^4 kg in level flight at constant velocity.

23 Just after opening a parachute of negligible mass, a parachutist of mass 90.0 kg experiences an instantaneous upward acceleration of 1.00 m/s². Find the force of the air on the parachute.

Tension

★**24** A small weight hangs from a string attached to the rearview mirror of a car accelerating at the rate of 1.00 m/s². What angle does the string make with the vertical?

25 In a tug of war, two teams pull on opposite ends of a rope, attempting to pull the other team across a dividing line. Team A accelerates toward team B at the rate of 0.100 m/s². Find all the horizontal forces acting on each team if the weight of each team is 1.00×10^4 N and the tension in the rope is 5.00×10^3 N.

Spring Force

26 The block in Fig. 4-34 rests on a frictionless surface. Find its instantaneous acceleration when the spring on the left is compressed 5.00 cm while the spring on the right is stretched 10.0 cm. Each spring has a force constant of 1.00×10^3 N/m.

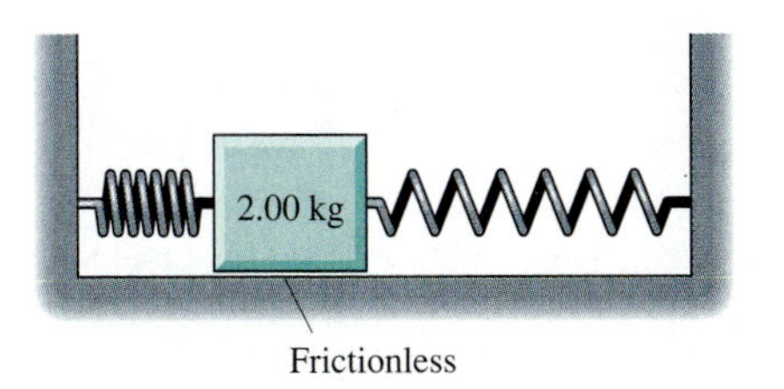

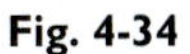
Fig. 4-34

★**27** When a 0.100 kg mass is suspended at rest from a certain spring, the spring stretches 4.00 cm. Find the instantaneous acceleration of the mass when it is raised 6.00 cm, compressing the spring 2.00 cm.

4-8 Applications of Newton's Laws of Motion

28 Find the tension in the ropes shown in Fig. 4-35 at points A, B, C, D, and E. The pulleys have negligible mass.

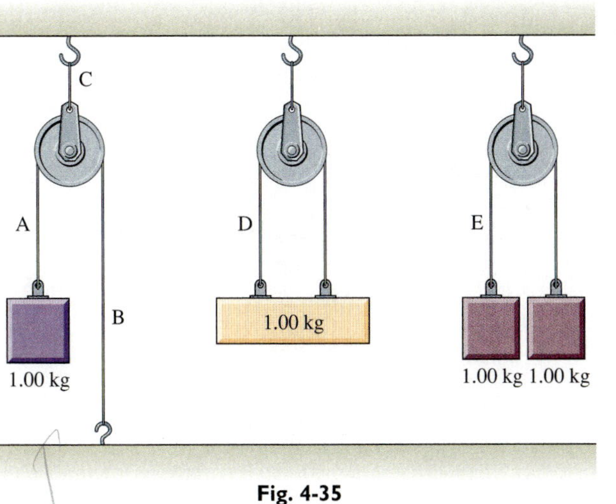

Fig. 4-35

29 Find the tension in each string in Fig. 4-36.

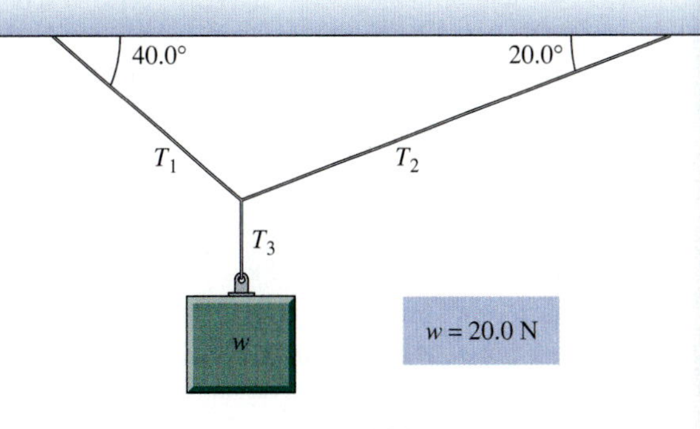

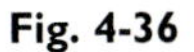
Fig. 4-36

★**30** Find the tension in each string in Fig. 4-37.

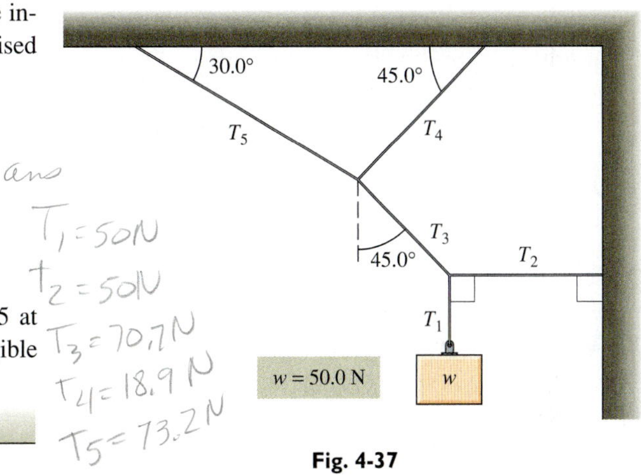

Fig. 4-37

31 A crate weighing 5.00×10^2 N is lifted at a slow, constant speed by ropes attached to the crate at A and B (Fig. 4-38). These two ropes are joined together at point C, and a single vertical rope supports the system.
(a) Find the tension T_1 in the vertical rope.
(b) Find the tensions T_2 and T_3 in the other ropes.

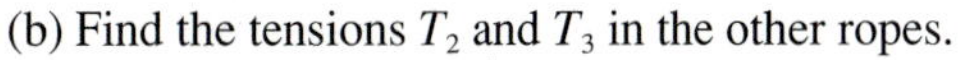

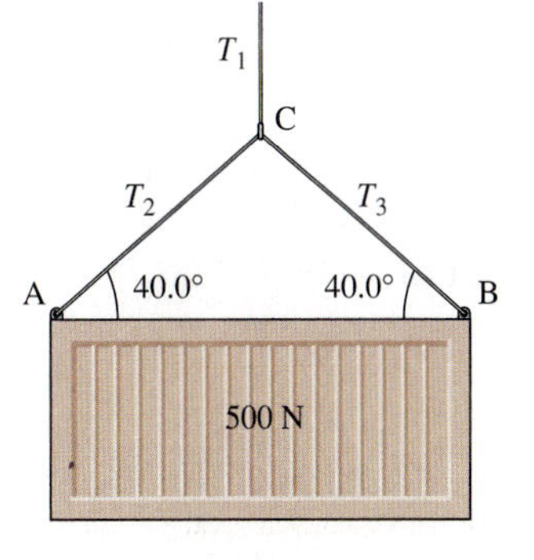

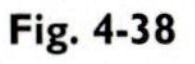
Fig. 4-38

32 A picture of width 40.0 cm, weighing 40.0 N, hangs from a nail by means of flexible wire attached to the sides of the picture frame. The midpoint of the wire passes over the nail, which is 3.00 cm higher than the points where the wire is attached to the frame. Find the tension in the wire.

33 Three blocks are suspended at rest by the system of strings and frictionless pulleys shown in Fig. 4-39. What are the weights w_1 and w_2?

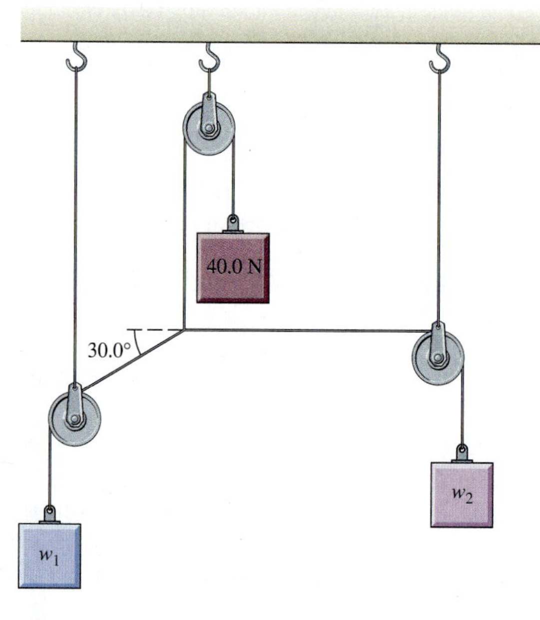

Fig. 4-39

★ **34** Find θ and w in Fig. 4-40, assuming that the arrangement is at rest.

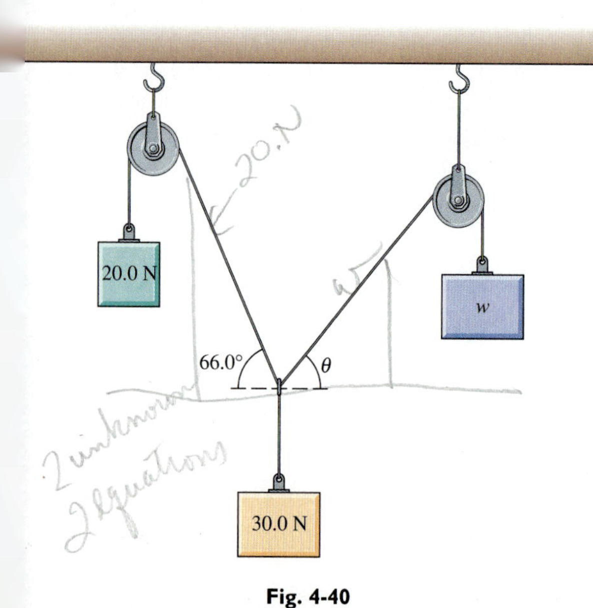

Fig. 4-40

35 A person weighing 710 N lies in a hammock supported on either end by ropes that are at angles of 45° and 30° with the horizontal (Fig. 4-41). Find the tension in the ropes.

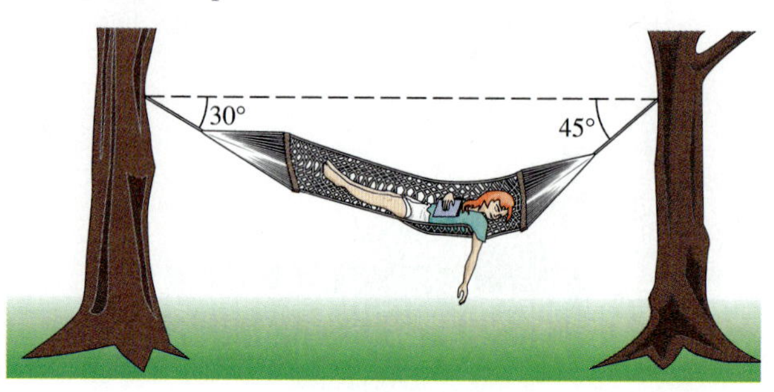

Fig. 4-41

36 Compute the acceleration of each mass in Fig. 4-23a and the tension in the rope. Let m_1 = 1.00 kg and m_2 = 2.00 kg.

37 Two blocks are connected by a string and are pulled vertically upward by a force of 165 N applied to the upper block, as shown in Fig. 4-42.

(a) Find the tension T in the string connecting the blocks.

(b) If the blocks start from rest, what is their velocity after having moved a distance of 10.0 cm?

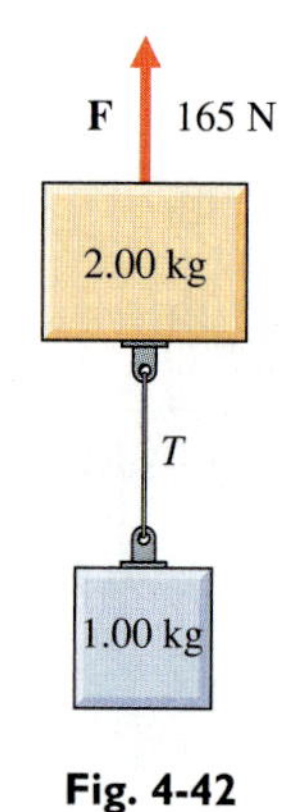

Fig. 4-42

★ **38** Two blocks are initially at rest on frictionless surfaces and are connected by a string that passes over a frictionless pulley (Fig. 4-43). Find the tension in the string.

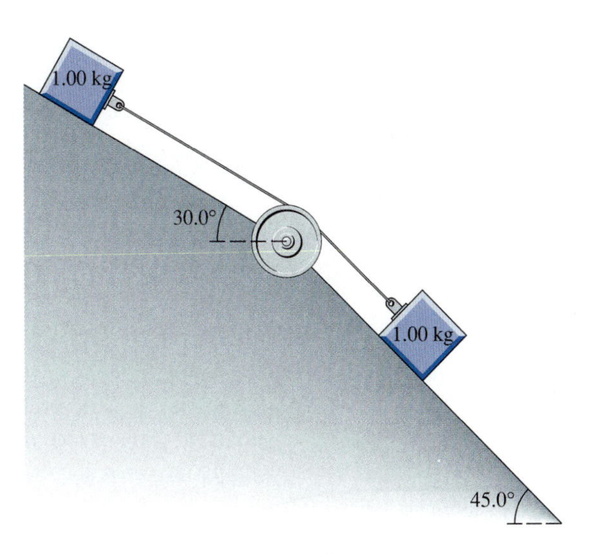

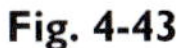
Fig. 4-43

★ **39** Two blocks connected by a string are on a horizontal frictionless surface. The blocks are connected to a hanging weight by means of a string that passes over a pulley (Fig. 4-44).

(a) Find the tension T in the string connecting the two blocks on the horizontal surface.

(b) How much time is required for the hanging weight to fall 10.0 cm if it starts from rest?

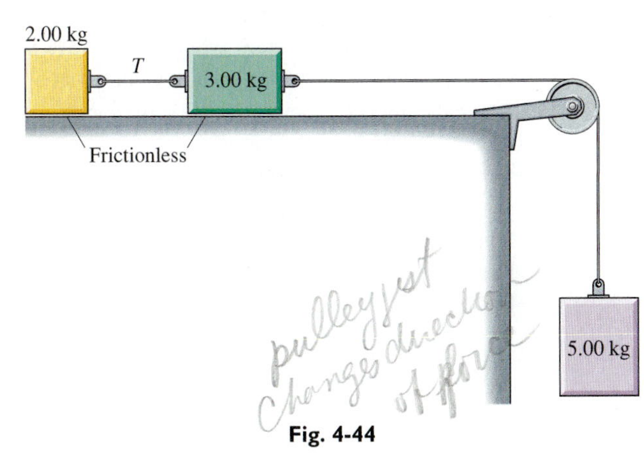

Fig. 4-44

★ **40** Find the acceleration of the 1.00 kg block in Fig. 4-45.

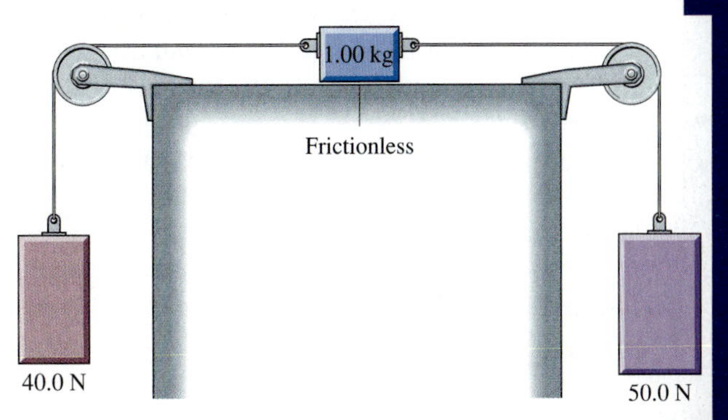

Fig. 4-45

★ **41** Two children of equal weight are suspended on opposite ends of a rope hanging over a pulley. Child A begins to slide down the rope, accelerating downward at a rate of 2.00 m/s^2. Find the direction and magnitude of child B's acceleration, assuming B doesn't slide.

42 A jet airplane has an instantaneous acceleration of 2.00 m/s^2 at an angle of 20.0° above the horizontal. Compute the horizontal and vertical components of force exerted on a 50.0 kg passenger by the airplane seat.

★★ **43** A boy weighing 4.00×10^2 N jumps from a height of 2.00 m to the ground below. Assume that the force of the ground on his feet is constant.

(a) Compute the force of the ground on his feet if he jumps stiff-legged, the ground compresses 2.00 cm, and the compression of tissue and bones is negligible.

(b) Compute the force his legs exert on his upper body (trunk, arms, and head), which weighs 2.50×10^2 N, under the conditions assumed above.

(c) Now suppose that his knees bend on impact, so that his trunk moves downward 40.0 cm during deceleration. Compute the force his legs exert on his upper body.

★ **44** A car is stuck in a mudhole. In order to move the car, the driver attaches one end of a rope to the car and the other end to a tree 10.0 m away, stretching the rope as much as possible (Fig. 4-46). The driver then applies a horizontal force of 4.00×10^2 N perpendicular to the rope at its midpoint. The rope stretches, with its center point moving 50.0 cm to the side as a result of the applied force. The car begins to move slowly. What is the tension in the rope?

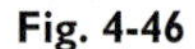

Fig. 4-46

Additional problems

★ **45** A painter on a platform raises herself by pulling on a rope connected to a system of pulleys (Fig. 4-47). If the painter and the platform combined weigh 1050 N, what force must she exert on the rope in order to raise herself slowly?

Fig. 4-47

★ **46** Three blocks, each having a mass of 1.00 kg, are connected by rigid rods of negligible mass and are supported by a frictionless surface. Forces $\mathbf{F}_1$ and $\mathbf{F}_2$, of magnitude 5.00 N and 10.0 N respectively, are applied to the ends of the blocks (Fig. 4-48). Find the forces acting on block B.

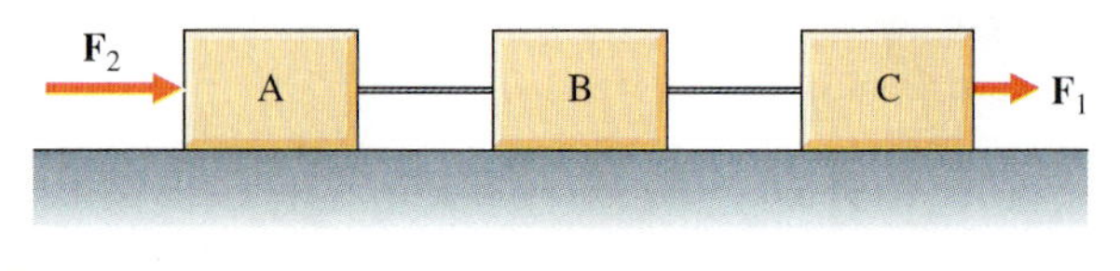

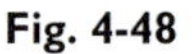

Fig. 4-48

★ **47** A football of mass 0.420 kg is thrown 60.0 m by a quarterback who imparts to it an initial velocity at an angle of 45.0° above the horizontal. If the quarterback moves his hand along an approximately linear path of length 40.0 cm while accelerating the football, what force does his hand exert on the ball, assuming the force to be constant?

48 A wet shirt weighing 5.00 N hangs from the center of a 10.0 m long clothesline, causing it to sag 5.00 cm below the horizontal. Find the tension in the line.

★ **49** A Ping-Pong ball is given an upward initial velocity. The force of air resistance causes the times of ascent and descent to be unequal. Which time is greater?

50 A basketball player stands in front of a basket and, without bending his knees, jumps straight up. The player weighs 1.00×10^3 N. His feet push downward on the floor with a constant force of 2.00×10^3 N for a time interval of 0.100 s, after which they leave the floor. Find (a) his acceleration while his feet are in contact with the floor; (b) his body's velocity as his feet leave the floor; (c) the maximum height he moves upward during the jump.

★★ **51** Each of the two identical springs in Fig. 4-49 has force constant $k = 1.00 \times 10^3$ N/m.

(a) Find the unstretched length of each spring.

(b) Find the instantaneous acceleration of the weight if it is pulled 10.0 cm lower and released.

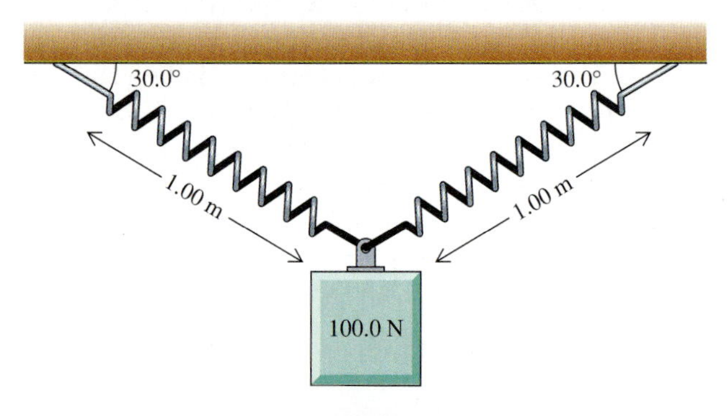

Fig. 4-49

★★ **52** Consider two particles, of mass m and m', having a common acceleration **a**, as shown in Fig. 4-50. These particles are subject to equal magnitude internal forces $\mathbf{F}_i$ and $\mathbf{F}'_i$, and to external forces $\mathbf{F}_e$ and $\mathbf{F}'_e$. Show that it follows from Newton's second law applied to each particle separately that we may apply the second law to the system of two particles if we use the net external force, the combined mass $m + m'$, and the common acceleration **a**.

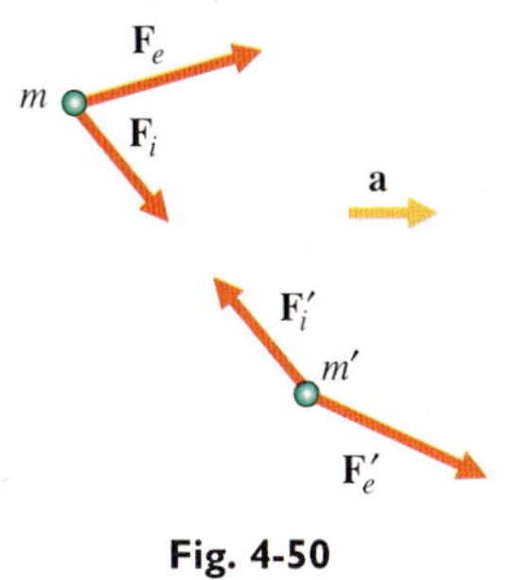

Fig. 4-50

★★ **53** Find the angles θ_1 and θ_2 in Fig. 4-51 if all the weights are at rest.

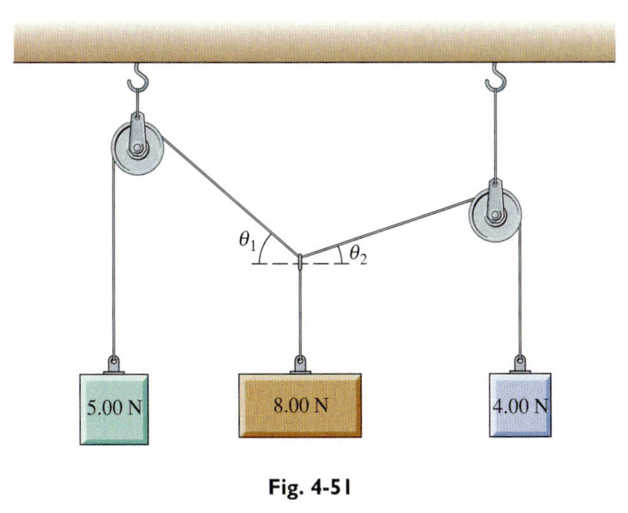

Fig. 4-51

★ **54** The force of air resistance **R** on a freely falling body is in a direction opposing the velocity and has magnitude approximately given by $R = CAv^2$, where A is the cross-sectional area of the body in the plane perpendicular to the motion, v is the speed, and C is a constant depending on body shape and air density. Show that if the y-axis is taken to be positive in the downward direction, a falling body experiences an acceleration $a_y = g(1 - CAv_y^2/w)$. Show that as a_y approaches zero, v_y approaches terminal velocity $v_T = \sqrt{w/CA}$.

★★ **55** A runner moving through the air experiences a force **R** because of the air (Fig. 4-52). This force, which is a function of the runner's velocity relative to the air, is approximately proportional to the square of the relative speed v_r. The magnitude of force **R** may be expressed as $R = CAv_r^2$, where A is the cross-sectional area of the body in the plane perpendicular to the motion. Suppose a runner first moves a distance D along the ground at constant velocity in the direction of a steady wind and then moves the same distance in the opposite direction at the same speed with respect to the ground. For both parts of the motion, express R in terms of the runner's speed v (relative to the ground) and the speed of the wind v_w. By how much does the average of these two values exceed the average magnitude of **R** in the absence of wind? This problem illustrates how wind generally produces a higher average value of air resistance on a runner, even though the runner runs with the wind the same distance he or she runs against the wind.

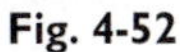

Fig. 4-52

★ **56** In Fig. 4-53 the mass of block A is 10.0 kg and that of block B is 15.0 kg. The pulley is massless and frictionless.

(a) What is the largest vertical force **F** that can be applied to the axle of the pulley if B is to remain on the floor?

(b) What will be the acceleration of A when this maximum force is applied?

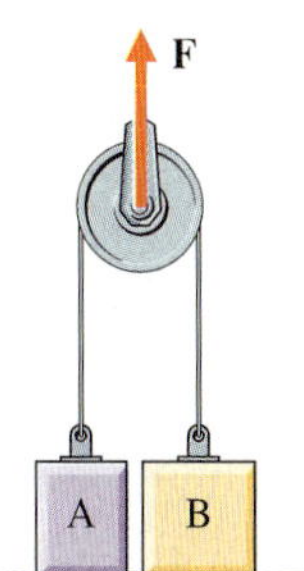

Fig. 4-53

CHAPTER 5

Friction and Other Applications of Newton's Laws

Imagine a skier descending a slope covered with fresh powder, gliding back and forth, leaving gracefully curving tracks in the snow. The skier experiences motion and forces that we will describe in this chapter. The skier's "center of mass" follows a curving path determined by the forces of gravity and friction. In this chapter we shall study friction, motion along circular arcs, and motion of a body's center of mass as applications of Newton's laws of motion.

5-1 Friction

Friction is a force acting between two surfaces, tending to prevent the surfaces from sliding over each other. In some cases friction is desirable. For example, most forms of transportation depend on friction. If a road is covered with ice, it is difficult to walk or drive a car on it because the frictional force between the ice and either your feet or the tires of your car may not be great enough to prevent slipping.

Often friction is not desirable. For example, it can cause machine parts to wear. For this reason, oil and other lubricants are used to reduce friction. Friction in human joints is very low because our bodies contain a natural lubricating system. Consequently, though our bones rub against each other at the joints as we move, bones do not normally wear out, even after many years of use.

We can begin to understand friction by using Newton's laws to analyze some simple experiments. As our first case, consider a book lying on a table. Since the book is at rest, the resultant force on it must be zero. So we know that the book's weight must be canceled by an opposing force **S** supplied by the table's surface* (Fig. 5-1a).

*This surface force arises from an elastic deformation of the surface. For a rigid surface, this deformation is so slight as to be undetectable by the casual observer. In the case of a heavy object on a nonrigid surface, however—a dumbbell on a rubber mat at the gym, for example—the effect is easy to see.

Now suppose you push very lightly on the book, applying a horizontal force **P** directed to the right (Fig. 5-1b). If this force is small enough, the book will remain at rest. To keep the book from moving, the surface must now exert a force that is directed to the left. In other words, there will be a horizontal component of **S** to cancel **P** and maintain a resultant force of zero. The vector component of **S** along the surface is called **friction** and is denoted by **f**. The vector component of **S** normal (perpendicular) to the surface is called the **normal force** and is denoted by **N**. In force diagrams, we usually draw **f** and **N**, the vector components of **S**, rather than **S** itself.

If you now slowly increase the horizontal force **P**, at first the book does not move. But when you make **P** large enough, the book begins to move. Once it is moving, the force needed to maintain the motion is smaller than the force needed to begin movement. The forces acting on the book for various values of **P** are shown in Fig. 5-2. In Fig. 5-2a to Fig. 5-2c, as **P** increases, **f** does also, and so equilibrium is maintained. In Fig. 5-2c, the force of friction has reached its maximum $\mathbf{f}_{max}$. When **P** exceeds $\mathbf{f}_{max}$, the book accelerates. Once the book is moving, the frictional force usually decreases to a value below $\mathbf{f}_{max}$; thus a force **P** only slightly greater than $\mathbf{f}_{max}$ can produce a significant acceleration (Fig. 5-2d). In Fig. 5-2e, the magnitude of **P** has been reduced so that it just balances the magnitude of the reduced frictional force; therefore the net force is again zero. In accordance with Newton's second law, the book now continues to move at constant velocity.

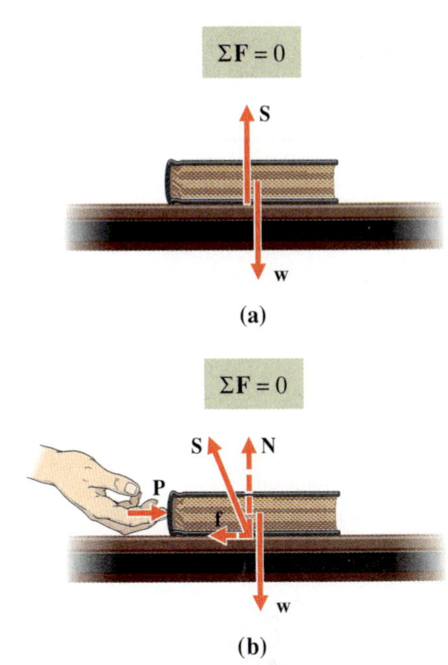

Fig. 5-1 **(a)** A surface supports a book, exerting a force **S** equal in magnitude to the book's weight **w.** **(b)** When you exert a force **P** on the book, the surface pushes back with a frictional force, **f**, that is tangent to the surface.

Friction between surfaces that slide over each other is called **kinetic friction,** whereas friction between surfaces at rest with respect to each other is called **static friction.** The kinetic frictional force between two surfaces is often less than the maximum static frictional force, as we have assumed in our description of the friction between book and table.

The maximum frictional force that the table can exert on the book depends on the two surfaces in contact: the kind of material each is made of, their smoothness, whether they are dry or lubricated, and so on. The force $\mathbf{f}_{max}$ also depends on the magnitude of the forces pressing the surfaces together. You can demonstrate this by repeating the experiment described above while someone presses down on the book. To move the book now, you must exert a greater horizontal force. The normal force also increases, as it must to balance the additional downward force as well as the weight of the book.

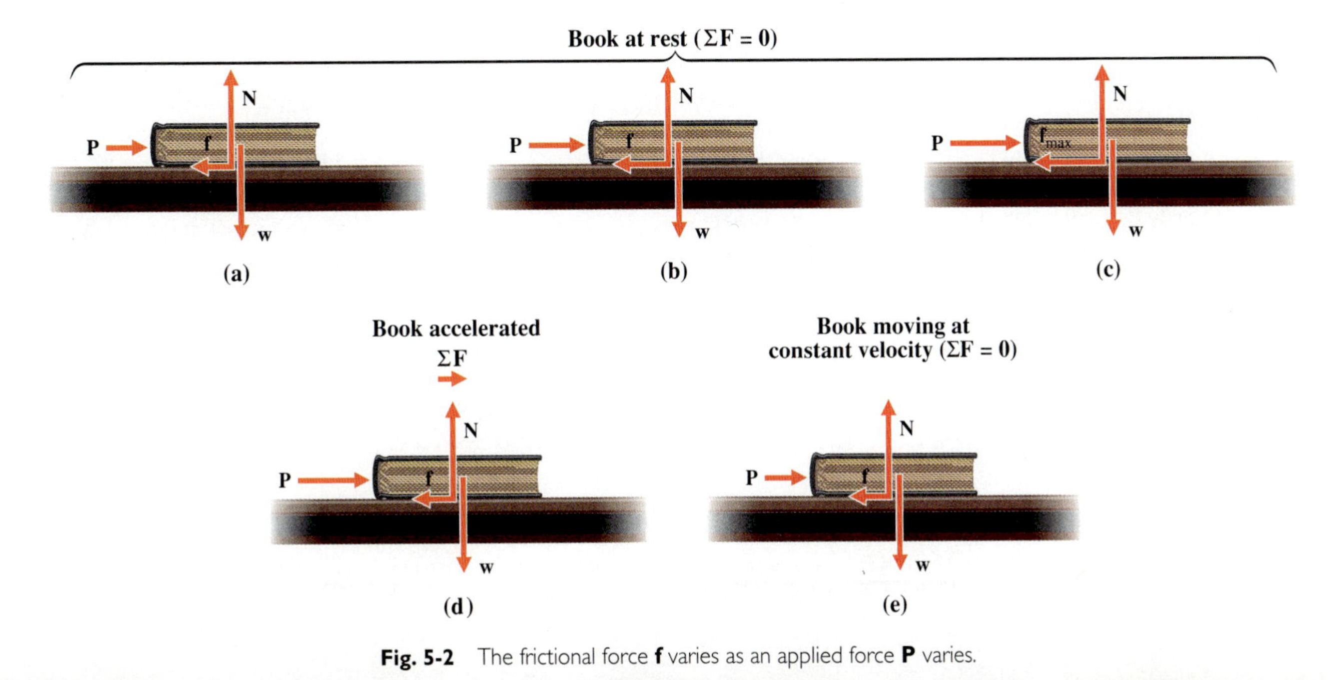

Fig. 5-2 The frictional force **f** varies as an applied force **P** varies.

Table 5-1 Coefficients of Static Friction

Surfaces	μ_s
Bone joints in mammals	0.002–0.01
Ski wax on snow, 0° C, dry	0.04
Ski wax on snow, 0° C, wet	0.1
Ski wax on snow, −20° C	0.2
Teflon on steel	0.04
Graphite on steel	0.1
Steel on steel, lubricated with motor oil	0.2
Steel on steel, clean	0.7
Brake material on cast iron	0.4
Wood on wood, wet	0.2
Wood on wood, dry	0.25–0.5
Wood on leather	0.3–0.4
Wood on brick	0.6
Glass on glass, clean	0.9
Rubber on solids	1–4

We find that **the maximum frictional force increases in direct proportion to the magnitude of the normal force.** The constant of proportionality is called the **coefficient of static friction** and denoted by μ_s, (μ is the Greek letter mu, pronounced /myou/), and so we can write

$$f_{max} = \mu_s N \qquad \text{(static friction)} \qquad (5\text{-}1)$$

The coefficient μ_s depends on the surfaces in contact and their condition. Some representative values of μ_s are given in Table 5-1.

The magnitude of the static frictional force **f** depends on the situation. Provided that its magnitude is less than f_{max}, **f** will always be such that the body remains at rest with a resultant force of zero. That is, if the external horizontal force trying to move the body has a magnitude less than f_{max}, then friction will be able to balance the external force, and the object will remain at rest.

The force of kinetic friction is also proportional to N, with a coefficient μ_k that is usually less than or equal to μ_s and depends on the sliding velocity:

$$f = \mu_k N \qquad \text{(sliding friction)} \qquad (5\text{-}2)$$

Eqs. 5-1 and 5-2 are the force laws that allow us to compute the force of friction whenever we know the normal force. They are not fundamental laws and they are not exact, but they are useful approximate expressions that work very well in most applications.

EXAMPLE 1 Forces on a Sliding Book

The book in Fig. 5-1 has a weight of 8.00 N. The coefficient of static friction between the book's surface and the table's surface is 0.400, and the coefficient of kinetic friction between the two surfaces is 0.300. (a) Find the magnitude of the minimum horizontal force **P** required to move the book, which is initially at rest. (b) Find the magnitude of the force **P** required to move the book at constant velocity.

SOLUTION (a) Fig. 5-2c shows the appropriate free-body diagram. Motion begins when the force **P** barely exceeds the maximum force of static friction, $\mathbf{f}_{max}$. So the minimum force required equals $\mathbf{f}_{max}$, the magnitude of which is found using Eq. 5-1 ($f_{max} = \mu_s N$).

$$P_{min} = f_{max} = \mu_s N$$

Inspecting our free-body diagram (Fig 5-2c), we see that the only vertical forces are **N** and **w**. Applying Newton's second law, we find

$$\Sigma F_y = ma_y = 0$$
$$N - w = 0$$
$$N = w = 8.00 \text{ N}$$

Inserting this result into the equation for P_{min}, we find

$$P_{min} = \mu_s N = (0.400)(8.00 \text{ N}) = 3.20 \text{ N}$$

(b) Here we use Fig. 5-2e as our free-body diagram. Since the book's velocity is constant, $\mathbf{a} = \mathbf{0}$. Thus

$$\Sigma F_x = ma_x = 0$$
$$P - f = 0$$
$$P = f$$

Since this is kinetic or sliding friction, we apply Eq. 5-2:

$$P = f = \mu_k N$$

As in part (a), $\Sigma F_y = 0$ and so forces **N** and **w** balance; that is, $N - w = 0$, or $N = w$. Inserting this into the preceding equation for P, we find

$$P = \mu_k w = (0.300)(8.00 \text{ N}) = 2.40 \text{ N}$$

EXAMPLE 2 Walking on Ice

A person tries to walk up a 15° incline on a street covered with ice (Fig. 5-3a). Is this possible if the coefficient of static friction between shoes and ice is 0.2?

SOLUTION The forces acting on the person are shown in Fig. 5-3b. We choose the y-axis perpendicular to the incline because with this choice of axes, the motion will be along the x-axis and there will be no component of acceleration along the y-axis. We need to investigate the x components of force in order to determine whether there can be a net force and therefore an acceleration along the positive x-axis. As each foot pushes backwards in the attempt to move forward, the ice surface exerts a forward frictional force on the foot. This force is opposed by the x component of the person's weight, $-w \sin \theta$. Thus

$$\Sigma F_x = f - w \sin \theta$$

This relation tells us that, in order for the person to move forward, f must be greater than $w \sin \theta$. This may or may not be possible, depending on the value of μ_s, since

$$f_{\max} = \mu_s N$$

The normal force is related to w and θ through the condition $\Sigma F_y = 0$ (since $a_y = 0$). The normal force just cancels the component of the weight perpendicular to the surface:

$$N = w \cos \theta$$

(It is a common error to assume that N always equals w. Notice that this is not true here.)

Combining the three preceding equations, we obtain an expression for the maximum value of the resultant force (ΣF_x):

$$(\Sigma F_x)_{\max} = \mu_s(w \cos \theta) - w \sin \theta$$
$$= (w \cos \theta)(\mu_s - \tan \theta)$$

Only if $\mu_s - \tan \theta$ is positive can $(\Sigma F_x)_{\max}$ be positive. Otherwise it will not be possible to walk up the incline. For this example, $\theta = 15°$. Thus

$$\mu_s - \tan \theta = 0.2 - \tan 15° = -0.07$$

So $(\Sigma F_x)_{\max}$ is negative, and it is impossible to go up the incline. In fact, it would be impossible even to stand on the icy incline.

(a)

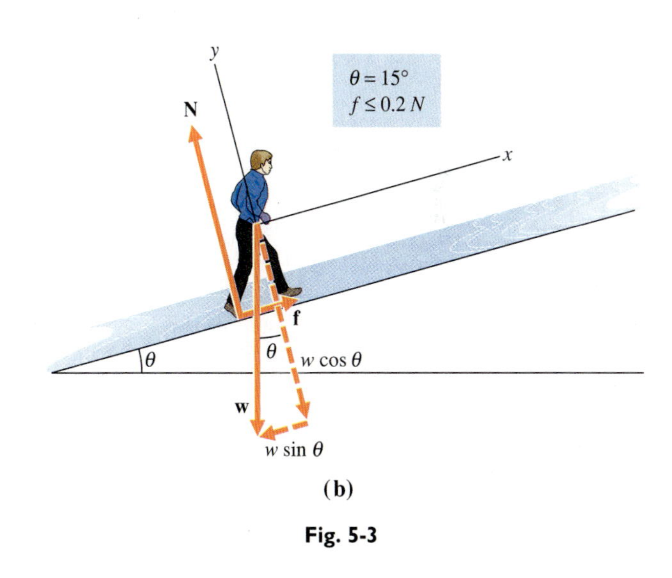

(b)

Fig. 5-3

EXAMPLE 3 Time to Complete a Ski Run

A skier starts with an initial velocity of 4.0 m/s, directed down a 22° slope. The slope is 130 m long, and μ_k is approximately independent of velocity and equal to 0.10. Find the skier's final velocity and the time she takes to reach the bottom. Neglect air resistance.

SOLUTION The first step is to find the skier's acceleration by considering the forces in the free-body diagram (Fig. 5-4).

$$a_x = \frac{\Sigma F_x}{m} = \frac{w \sin\theta - f}{m}$$

From Eq. 5-2, we know that

$$f = \mu_k N$$

Since $a_y = 0$, we have

$$\Sigma F_y = 0$$

$$N - w\cos\theta = 0$$

$$N = w\cos\theta$$

Combining these expressions for a_x, f, and N, we obtain

$$a_x = \frac{w\sin\theta - \mu_k(w\cos\theta)}{m}$$

or, using the relation $w = mg$,

$$a_x = g[\sin\theta - \mu_k(\cos\theta)]$$
$$= (9.8\ \text{m/s}^2)(\sin 22° - 0.10\cos 22°) = 2.8\ \text{m/s}^2$$

Now we may use the equations for linear motion at constant acceleration from Chapter 2. First we find the velocity v_x at the bottom of the slope:

$$v_x^2 = v_{x0}^2 + 2a_x x$$
$$= (4.0\ \text{m/s})^2 + 2(2.8\ \text{m/s}^2)(130\ \text{m}) = 740\ \text{m}^2/\text{s}^2$$
$$v_x = 27\ \text{m/s}$$

Next we relate the time t to the distance x, using Eq. 2-10:

$$x = \tfrac{1}{2}(v_{x0} + v_x)t$$

Solving for t, we find

$$t = \frac{2x}{v_{x0} + v_x} = \frac{2(130\ \text{m})}{4.0\ \text{m/s} + 27\ \text{m/s}}$$
$$= 8.4\ \text{s}$$

(Actually, air resistance would be a significant factor in this situation, and so the time required would be somewhat longer.)

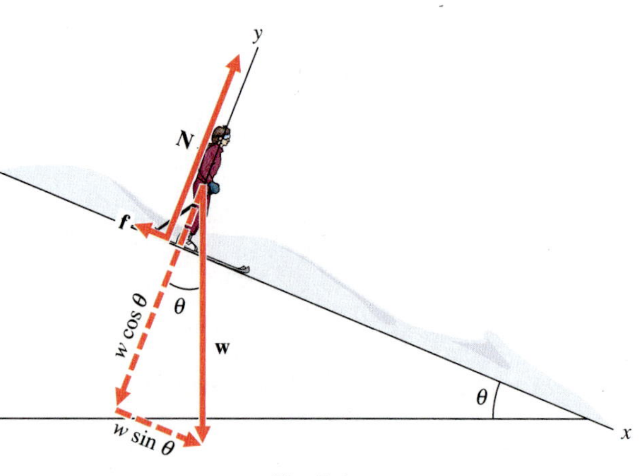

Fig. 5-4

EXAMPLE 4 Connected Blocks Accelerating Together

Determine how long the hanging block in Fig. 5-5a takes to reach the floor. The coefficient of kinetic friction between m_1 and the table is 0.40; the pulley is frictionless and has negligible mass; $m_1 = 1.0$ kg and $m_2 = 0.50$ kg.

SOLUTION First we must find the acceleration of m_2. We draw a separate free-body diagram for each block (Figs. 5-5b and c) and apply Newton's second law to each. The two blocks are coupled together by the string; therefore both have the same magnitudes of acceleration, a, and tension, T.

For m_1, both the acceleration $\mathbf{a}_1$ and the tension $\mathbf{T}_1$ are along the positive x-axis. Applying Newton's second law, we obtain

$$\Sigma F_x = m_1 a_{1x}$$

$$T - f = m_1 a \qquad (1)$$

But $f = \mu_k N = 0.40\,N$, and $N = w = m_1 g$ (since $\Sigma F_y = 0$). Thus substituting $f = 0.40 m_1 g$ into Eq. 1, we find

$$T - 0.40 m_1 g = m_1 a \qquad (2)$$

In applying Newton's second law to m_2, we choose the positive x-axis in the downward direction, so that $a_{2x} = +a$:

$$\Sigma F_x = m_2 a_{2x}$$

$$m_2 g - T = m_2 a \qquad (3)$$

Adding Eqs. 2 and 3, we find

$$m_2 g - 0.40 m_1 g = m_1 a + m_2 a$$

$$a = \frac{m_2 - 0.40 m_1}{m_1 + m_2}\, g$$

$$= \frac{0.50\text{ kg} - 0.40\text{ kg}}{1.0\text{ kg} + 0.50\text{ kg}}(9.8\text{ m/s}^2)$$

$$= 0.65\text{ m/s}^2$$

Finally, we use Eq. 2-11, which relates distance and time for linear motion at constant acceleration, and solve for t:

$$x = \tfrac{1}{2} a_x t^2$$

$$t = \sqrt{2x/a_x}$$

$$= \sqrt{2(1.0\text{ m})/(0.65\text{ m/s}^2)} = 1.8\text{ s}$$

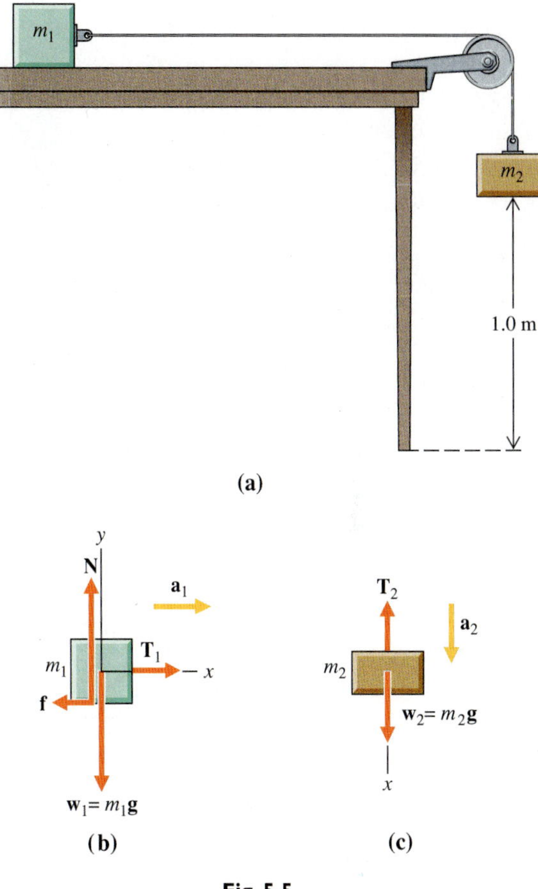

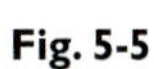

Fig. 5-5

Microscopic Description of Friction

Frictional forces arise from electromagnetic interaction between the molecules of the two surfaces in contact. We shall now describe a simple model that accounts for the main features of macroscopic friction. Fig. 5-A is a sketch of a highly magnified region of interaction between two surfaces for a block resting on a horizontal surface when no frictional force acts. As the blowup shows, the surfaces are in contact over only a small fraction of their total area. The total normal force on the upper surface does not act at a single point; instead, it is the sum of a large number of very small upward forces, acting at the various points of contact.

If we exert a horizontal force **P** on the block, attempting to slide it to the right, a frictional force **f** arises, which is directed to the left, opposing **P**. Like the normal force, this frictional force is also the resultant of many small forces, occurring only where the surfaces are in true microscopic contact. These microscopic frictional forces can be either attractive or repulsive, as indicated in Fig. 5-B.

The normal force changes when the microscopic area of contact between the two surfaces changes. If someone pushes down on the block, there will be contact between the two surfaces at more places, and the total microscopic area of contact will increase. This increased area of contact allows more molecules of the lower surface to exert upward forces, thereby contributing to a larger total normal force, which is necessary to balance the increased downward force. The increased area of contact also allows for an increased frictional force. Again, the reason is simply that there are more molecules in contact. The magnitudes of both **N** and $\mathbf{f}_{max}$ are approximately proportional to the number of molecules in close contact, or to the microscopic area of contact. Therefore f_{max} is approximately proportional to N.

Although the magnitude of the frictional force depends on the microscopic area of contact, it does not depend on the apparent, or macroscopic, area of contact. Thus the same force would be required to move any of the three identical blocks shown in Fig. 5-C, even though they have different orientations and therefore different apparent areas of contact with the table. The area of microscopic contact is the same for all three, since they have the same weight and must therefore have the same number of molecules collectively exerting the normal force that balances the weight.

$\mathbf{N} = \mathbf{N}_1 + \mathbf{N}_2 + \mathbf{N}_3 + \ldots$

Fig. 5-A The normal force is the resultant of a large number of forces arising from molecular interaction at points of microscopic contact between the two surfaces.

Fig. 5-B At A, where the surfaces are being pulled apart, the lower surface resists motion of the upper surface to the right by exerting an attractive molecular force $\mathbf{f}_A$, directed to the left. At R, where the surfaces are being pushed into each other, the lower surface resists motion of the upper surface to the right by exerting a repulsive force $\mathbf{f}_R$, also directed to the left. Both $\mathbf{f}_A$ and $\mathbf{f}_R$ contribute to the total frictional force.

Fig. 5-C The same force is required to move any of the three identical blocks, irrespective of orientation. In other words the force of friction is independent of the apparent area of contact.

A Closer Look

Polishing a rough surface will at first tend to reduce the frictional force because of a decrease in the number of places like R in Fig. 5-B. When two identical blocks have highly polished surfaces and are free of adsorbed gases, oils, and other impurities, however, the frictional force can be many times larger than the normal force. In this case, the surface irregularities are so reduced that a much larger number of the molecules on each surface are within range of the electromagnetic forces exerted by molecules on the other surface. Then the same forces that hold the parts of each block together internally begin to hold the two blocks to each other. The same effect is present in a stack of microscope slides (Fig. 5-D).

Lubricants greatly reduce friction between solids. The lubricant separates the surfaces of the solids beyond the range of their molecular forces, leaving only the weaker frictional force between lubricant and solid. The lubrication of human knee joints (Fig. 5-E) varies with need. When you are standing still, much of the lubricating synovial fluid is absorbed into the cartilage, so that friction between your bones is great enough to prevent slipping. When you are moving, the increased pressure on the cartilage squeezes out more synovial fluid, thereby reducing friction and allowing the bones to slide.

Fig. 5-D Slipping of these glass microscope slides is prevented by the strong frictional force acting between their very clean surfaces.

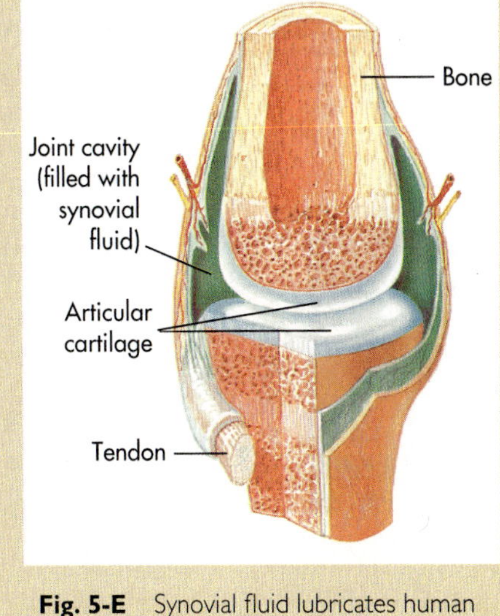

Fig. 5-E Synovial fluid lubricates human joints.

5-2 Centripetal Force

We saw in Chapter 3 that a body undergoing circular motion is continuously subject to an acceleration, called centripetal acceleration, directed toward the center of the circle. Because there is an acceleration, Newton's second law implies that there must be acting on the body a resultant force that is also directed toward the center of the circle. Using Eq. 3-12 ($a = v^2/r$), we obtain an expression for the magnitude of this net force acting on a body in circular motion:

$$|\Sigma \mathbf{F}| = ma$$

$$|\Sigma \mathbf{F}| = \frac{mv^2}{r} \tag{5-3}$$

Any net force that produces this centripetal acceleration is termed **centripetal force.**

Eq. 5-3 should not be confused with a force law. Force laws always express a particular kind of force (weight, spring force, friction) in terms of some physical object in the environment (the earth, a spring, a surface in contact). But centripetal force can be a force of any kind. Or it can result from a combination of different kinds of forces. In the examples and problems that follow, we shall see several kinds of centripetal forces: friction, normal force, tension, and weight.

EXAMPLE 5 Rounding a Curve at Maximum Speed

Find an expression for the maximum speed with which a car can travel without slipping around a level, circular curve of radius r.

SOLUTION The maximum speed of the car is determined by the maximum centripetal force that can be exerted on it. As Fig. 5-6 shows, the forces acting on the car are its weight and the surface force exerted by the road on the tires. The only force that acts in the horizontal plane is friction. Therefore the frictional force must be the centripetal force:

$$f = \frac{mv^2}{r}$$

Since there is to be no motion along the radius of the curve, this is a static frictional force, whose magnitude will vary as required to maintain the motion. The maximum frictional force determines the maximum velocity:

$$f_{max} = \frac{mv_{max}^2}{r}$$

But $f_{max} = \mu_s N$, and it follows from cancellation of the vertical components of force that $N = mg$, and therefore $f_{max} = \mu_s mg$. Thus

$$\mu_s mg = \frac{mv_{max}^2}{r}$$

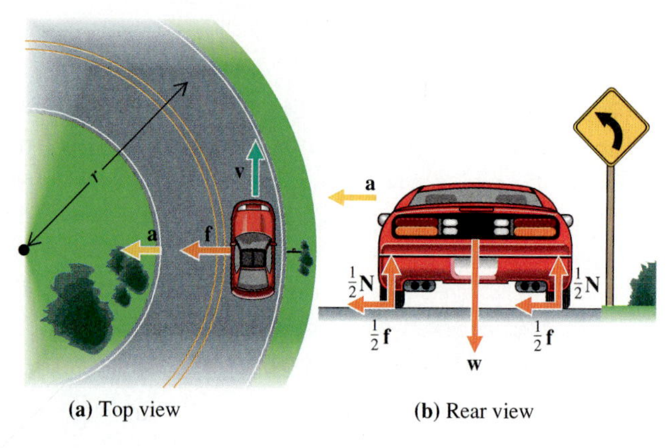

(a) Top view (b) Rear view

Fig. 5-6

Solving for v_{max}, we obtain

$$v_{max} = \sqrt{\mu_s gr}$$

If $\mu_s = 1.0$, a typical value for tires on a dry road, $v_{max} = \sqrt{gr}$. A curve of radius 10 m could be safely driven at a speed no greater than 9.9 m/s (22 mi/h), but a curve of radius 100 m could be traveled at speeds up to 31 m/s (70 mi/h) without any skidding.

EXAMPLE 6 Car Swinging on a Cable

One of the rides at an amusement park consists of small cars that are supported by steel cables and rotated in a circle at gradually increasing speed (Fig. 5-7a). As the speed increases, the cars rise. Find the speed of a car when its center is 5.00 m from the cable's point of support and the cable makes an angle of 35.0° with the vertical.

SOLUTION We choose the car as the free body. The only forces acting on the car are its weight **w** and the tension **T** in the cable (Fig. 5-7b). The component of **T** in the horizontal direction is the centripetal force:

$$T \sin \theta = \frac{mv^2}{r}$$

Because there is no more vertical motion once the angle θ is reached, the vertical forces now cancel:

$$T \cos \theta = mg$$

Dividing this equation into the preceding one, we obtain

$$\tan \theta = \frac{v^2}{gr}$$

But it is evident from Fig. 5-7a that r is itself a function of θ:

$$r = \ell \sin \theta$$

where ℓ is the length of the cable. Substituting this value for r into the equation above, we find

$$\tan \theta = \frac{v^2}{g\ell \sin \theta}$$

or

$$v = \sqrt{g\ell \sin\theta \tan\theta} = \sqrt{(9.80\,\text{m/s}^2)(5.00\,\text{m})(\sin 35.0°)(\tan 35.0°)} = 4.44\text{ m/s}$$

Fig. 5-7

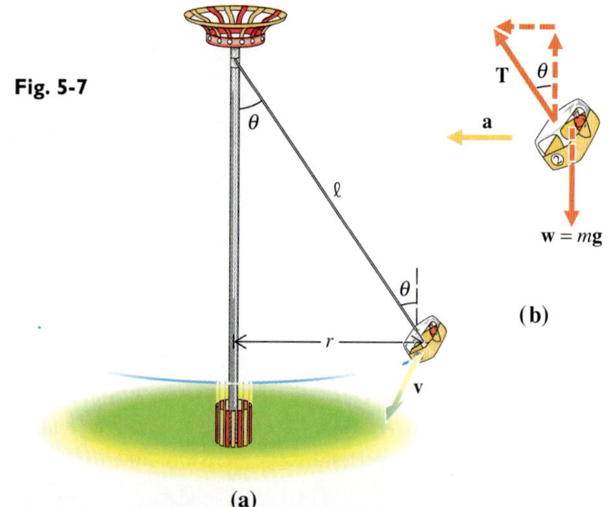

5-3 Center of Mass

A System Moving as a Single Particle

Newton's second law ($\Sigma\,\mathbf{F} = m\mathbf{a}$) applies to single particles and also to systems of particles in which the particles all have the same acceleration **a**. For example, when a car moves in a straight line, all parts of the car's body have the same velocity and acceleration. Another example is a ski jumper who, after leaving the ramp, maintains body and skis in a fixed, rigid position (Fig. 5-8). In both these cases, we can apply Newton's second law to the whole system because each system acts as a single particle.

Fig. 5-8 Each part of the system of skier and skis has the same velocity and acceleration, during the interval shown.

Motion of the Center of Mass

Even when the parts of a system have different velocities and accelerations, however, there is still one point in the system whose acceleration we may find by applying Newton's second law. This one point is called the **center of mass** of the system, and its acceleration is denoted by $\mathbf{a}_{cm}$. Thus

$$\Sigma\,\mathbf{F} = M\mathbf{a}_{cm} \tag{5-4}$$

where $\Sigma\,\mathbf{F}$ is the vector sum of the external forces acting on the system and M is the mass of the whole system. This equation implies that **the motion of the center of mass is the same as that of a particle of mass M subject to the same external forces as those acting on the system.** For example, suppose a wheel rolls along a horizontal frictionless surface with no resultant force on the wheel. According to Eq. 5-4, the wheel's center of mass (its geometric center) will experience no acceleration. Although all other parts of the wheel are accelerated because of rotation, the center of mass moves at constant velocity—just like a particle on which no force acts. Other examples of center-of-mass motion are shown in Fig. 5-9.

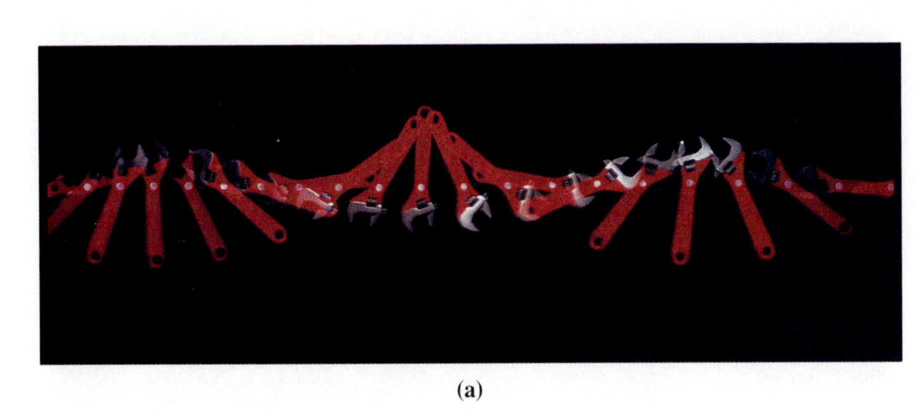

(a)

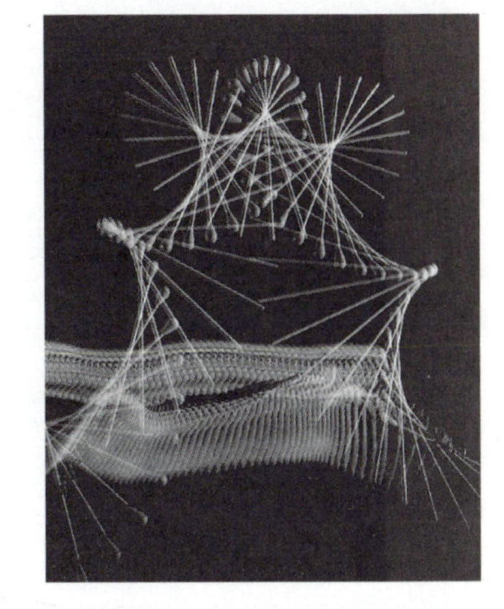

(b)

Fig. 5-9 **(a)** A rotating wrench slides along a frictionless horizontal surface. There is no resultant force on the wrench, and therefore its center of mass follows a linear path at constant speed. **(b)** A whirling baton tossed into the air experiences only the external force of gravity—in other words, its weight $\mathbf{W} = M\mathbf{g}$. Its center-of-mass acceleration $\mathbf{a}_{cm} = \mathbf{W}/M = M\mathbf{g}/M = \mathbf{g}$, and the center of mass follows the characteristic parabolic path of a simple projectile.

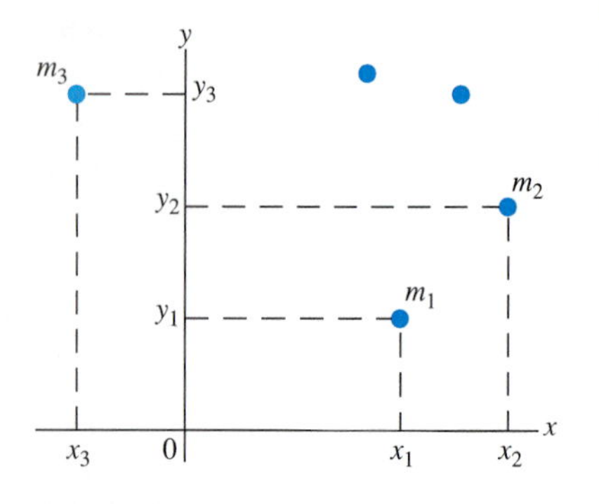

Fig. 5-10 Finding the center of mass of particles.

Definition of the Center of Mass

Next we shall define the center of mass and then, after computing the center of mass for some simple systems, we shall apply Eq. 5-4 (derived at the end of this section). For a system of n particles (Fig. 5-10), the coordinates of the center of mass relative to an arbitrary origin O is defined to be the following weighted average of the particles' coordinates:

$$x_{cm} = \frac{\Sigma\, mx}{\Sigma\, m} = \frac{m_1x_1 + m_2x_2 + \cdots + m_nx_n}{m_1 + m_2 + \cdots + m_n} \tag{5-5}$$

and

$$y_{cm} = \frac{\Sigma\, my}{\Sigma\, m} = \frac{m_1y_1 + m_2y_2 + \cdots + m_ny_n}{m_1 + m_2 + \cdots + m_n} \tag{5-6}$$

EXAMPLE 7 Center of Mass of Two Particles

Find the center of mass of a system made up of two particles A and B, 1.0 m apart, if each particle has a mass of 25 g.

SOLUTION If we choose as origin the position of particle A (Fig. 5-11a), we have

$$x_{cm} = \frac{\Sigma\, mx}{\Sigma\, m} = \frac{m_Ax_A + m_Bx_B}{m_A + m_B}$$
$$= \frac{(25\text{ g})(0) + (25\text{ g})(1.0\text{ m})}{25\text{ g} + 25\text{ g}} = 0.50\text{ m}$$

that is, the point midway between the two particles is the center of mass.

To see that the choice of origin is completely arbitrary, consider what happens when we choose the midpoint as our origin (Fig. 5-11b). In this case, we would compute x_{cm} as follows:

$$x_{cm} = \frac{m_Ax_A + m_Bx_B}{m_A + m_B} = \frac{(25\text{ g})(-0.50\text{ m}) + (25\text{ g})(+0.50\text{ m})}{25\text{ g} + 25\text{ g}}$$
$$= 0$$

Thus we see that the center of mass is the same physical point in the system, independent of the choice of origin. It is possible to prove this result in general for any system.

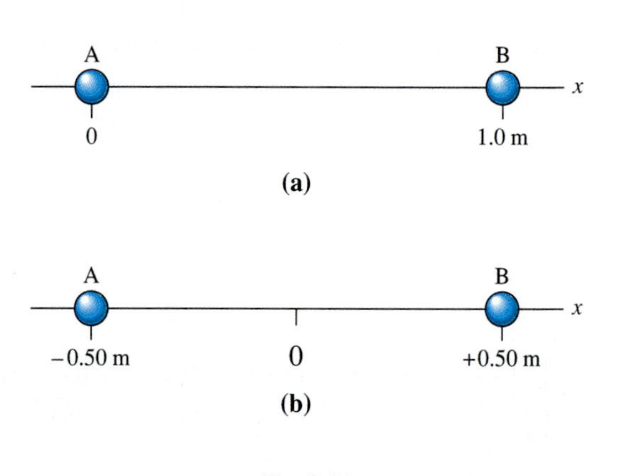

Fig. 5-11

Finding the center of mass of a homogeneous sphere or other symmetrical body is particularly simple. Fig. 5-12 shows three symmetrical bodies in which the center of symmetry is chosen as the origin. For each particle of mass m and coordinate $x = +d$, there is a second particle of mass m and coordinate $x = -d$. Thus the contributions of these two particles to $\Sigma\, mx$ cancel. Since the entire body can be divided into such pairs, $x_{cm} = 0$. We can apply the same argument to the y coordinates to show that $y_{cm} = 0$. Thus the center of mass of each body is at the origin—the center of symmetry.

Suppose you wish to find the center of mass of a system consisting of two or more extended bodies. If you know the center of mass of each body, you can find the center of mass of the system simply by treating each body of mass M as a particle of mass M located at the body's center of mass. For example, you can find the center of mass of the system of the earth and the moon by treating the earth as a particle located at the center of the earth and the moon as a particle located at the center of the moon. Problem 51 outlines a proof of this result.

In a rigid body, the center of mass is fixed with respect to the body. If a system of particles is not rigidly bound together, however, its center of mass will not necessarily be fixed with respect to any point in the system. For example, a human body is clearly not rigid. If a person stands upright, the center of mass is located within the midsection of the body. But if a person bends sufficiently at the waist, the center of mass can be outside the body, as we shall see in the following example.

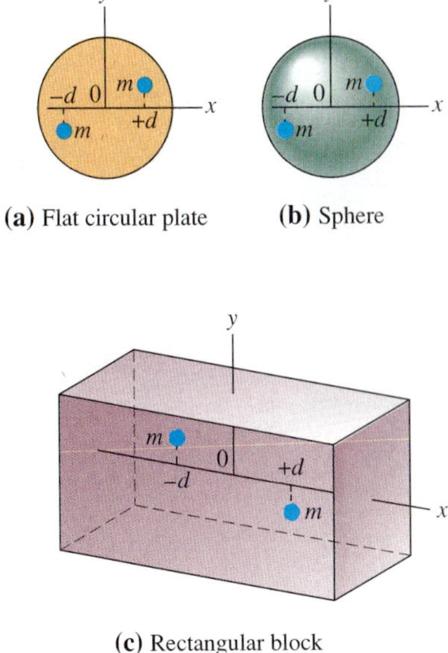

Fig. 5-12 Finding the center of mass of a symmetric body.

EXAMPLE 8 Center of Mass of a Sitting Person

The mass distribution of a sitting person can be roughly approximated by the rectangles shown in Fig. 5-13. Find the person's center of mass.

SOLUTION We find the center of mass of the entire body by treating each rectangle as though its mass were concentrated at its geometric center. The center of mass of the 65 kg rectangle is at point P, with coordinates $x = 10$ cm, $y = 45$ cm, and the center of mass of the 25 kg rectangle is at point P′ with coordinates $x = 60$ cm, $y = 6.0$ cm. Applying Eqs. 5-7 and 5-8 we find

$$x_{cm} = \frac{(65\text{ kg})(10\text{ cm}) + (25\text{ kg})(60\text{ cm})}{65\text{ kg} + 25\text{ kg}}$$

$$= 24\text{ cm}$$

$$y_{cm} = \frac{(65\text{ kg})(45\text{ cm}) + (25\text{ kg})(6.0\text{ cm})}{65\text{ kg} + 25\text{ kg}}$$

$$= 34\text{ cm}$$

These coordinates indicate that the center of mass, shown by the red dot in the figure, is outside the body.

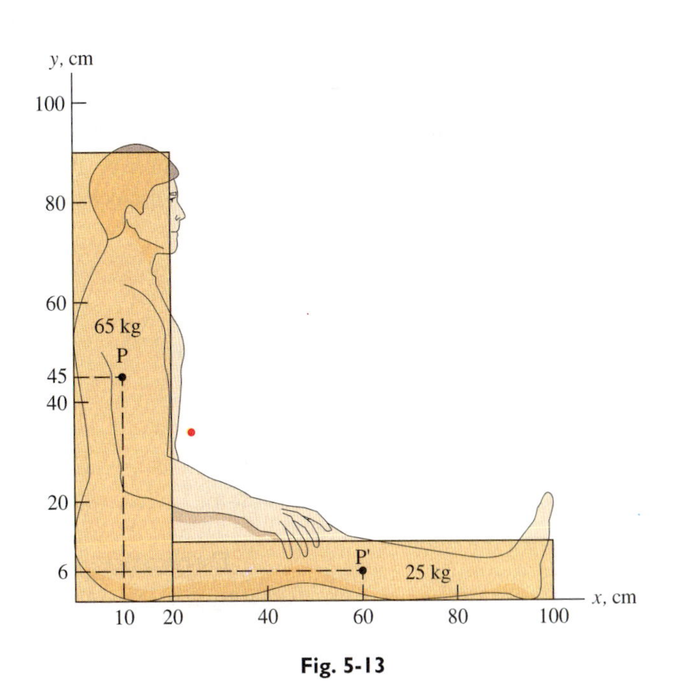

Fig. 5-13

Fig. 5-14 The high jumper's initial velocity determines the maximum height reached by the body's center of mass. To clear the greatest possible height for a given center-of-mass height, the jumper rolls over the bar, keeping as low as possible the parts of his body that are not directly over the bar at any instant.

The center of mass of a projectile follows a parabolic trajectory and is described by the equations developed in Chapter 3 for a single particle (if air resistance is negligible). A fireworks display is a beautiful demonstration of this effect. Immediately after exploding, the fireworks form a symmetric pattern about the center of mass. Since the forces on the fragments during the explosion are all internal forces, they cancel each other and have no effect on the motion of the center of mass, which will continue to follow an approximately parabolic trajectory, just as though there had been no explosion. The center of mass of a long jumper or a high jumper also follows a parabolic trajectory, though the shape of the body may change considerably during the jump, and consequently the position of the center of mass relative to parts of the body may also change (Fig. 5-14).

EXAMPLE 9 Two Skaters Meet

Two ice skaters, John (mass = 75 kg) and Michelle (mass = 55 kg), initially standing motionless on frictionless ice, 8.0 m apart, begin to pull on opposite ends of a rope (Fig. 5-15). How far does John move before he meets Michelle?

Fig. 5-15

SOLUTION Since no net external force acts on the system of John and Michelle, their center of mass, which is initially at rest, must remain at rest. When they meet, it must be at this point. We can use Eq. 5-7 to compute the initial distance of the center of mass from John:

$$x_{cm} = \frac{\Sigma\, mx}{\Sigma\, m}$$

$$= \frac{(75\text{ kg})(0) + (55\text{ kg})(8.0\text{ m})}{75\text{ kg} + 55\text{ kg}}$$

$$= 3.4\text{ m}$$

This is the distance John moves to meet Michelle.

Derivation of the Dynamics of the Center of Mass

Now we shall derive Eq. 5-4 ($\Sigma\, \mathbf{F} = M\mathbf{a}_{cm}$). We begin with the equation defining the x coordinate of the center of mass for an n particle system (Eq. 5-5):

$$x_{cm} = \frac{m_1x_1 + m_2x_2 + \cdots + m_nx_n}{m_1 + m_2 + \cdots + m_n} \tag{5-5}$$

Denoting the sum of the particle masses in the denominator by M, and multiplying this equation by M, we obtain

$$Mx_{cm} = m_1x_1 + m_2x_2 + \cdots + m_nx_n \tag{5-7}$$

When the particles move, their center of mass may move. The change in the center-of-mass x coordinate is related to the particles' changes in coordinates, Δx_1, $\Delta x_2, \ldots$, by an equation of the same form as the equation above.

$$M\,\Delta x_{cm} = m_1\,\Delta x_1 + m_2\,\Delta x_2 + \cdots + m_n\,\Delta x_n$$

Dividing this expression by Δt gives

$$M\frac{\Delta x_{cm}}{\Delta t} = m_1\frac{\Delta x_1}{\Delta t} + m_2\frac{\Delta x_2}{\Delta t} + \cdots + m_n\frac{\Delta x_n}{\Delta t}$$

In the limit as Δt approaches zero, this equation becomes a relationship between instantaneous velocities:

$$Mv_{cm,x} = m_1 v_{1x} + m_2 v_{2x} + \cdots + m_n v_{nx} \tag{5-8}$$

When the particles are accelerated, their center of mass may be accelerated. We can obtain an expression for the x component of center-of-mass acceleration $a_{cm,x}$ in terms of the particles' accelerations, $a_{1x}, a_{2x}, \ldots$, by repeating the operations that led from Eq. 5-7 to Eq. 5-8:

$$M\,\Delta v_{cm,x} = m_1\,\Delta v_{1x} + m_2\,\Delta v_{2x} + \cdots + m_n\,\Delta v_{nx}$$

$$M\frac{\Delta v_{cm,x}}{\Delta t} = m_1\frac{\Delta v_{1x}}{\Delta t} + m_2\frac{\Delta v_{2x}}{\Delta t} + \cdots + m_n\frac{\Delta v_{nx}}{\Delta t}$$

$$Ma_{cm,x} = m_1 a_{1x} + m_2 a_{2x} + \cdots + m_n a_{nx} \tag{5-9}$$

We can go through the same series of steps that led from Eq. 5-5 to Eq. 5-9, starting with the expression for the y coordinate of the center of mass (Eq. 5-6), leading to the following expression relating the y component of center-of-mass acceleration to the y components of the particles' accelerations:

$$Ma_{cm,y} = m_1 a_{1y} + m_2 a_{2y} + \cdots + m_n a_{ny} \tag{5-10}$$

This equation has the same form as Eq. 5-9. Eqs. 5-9 and 5-10 give the x and y components of the center-of-mass acceleration vector $\mathbf{a}_{cm}$. Together these equations are equivalent to the vector equation

$$M\mathbf{a}_{cm} = m_1\mathbf{a}_1 + m_2\mathbf{a}_2 + \cdots + m_n\mathbf{a}_n$$

We can apply Newton's second law to each particle. Thus in the preceding equation we may insert $m_1\mathbf{a}_1 = \Sigma\mathbf{F}_1$, $m_2\mathbf{a}_2 = \Sigma\mathbf{F}_2, \ldots, m_n\mathbf{a}_n = \Sigma\mathbf{F}_n$, where $\Sigma\mathbf{F}_1$, $\Sigma\mathbf{F}_2, \ldots, \Sigma\mathbf{F}_n$ are the resultant forces on the particles:

$$M\mathbf{a}_{cm} = \Sigma\mathbf{F}_1 + \Sigma\mathbf{F}_2 + \cdots + \Sigma\mathbf{F}_n \tag{5-11}$$

The right-hand side of this equation is the sum of all the forces acting on all of the particles in the system. It includes forces acting between the particles (internal forces) as well as forces exerted on the system from outside (external forces). From Newton's third law, we know that forces always occur in action-reaction pairs. Therefore, when we sum over all forces, the internal forces all cancel, leaving only the sum of the external forces, which we denote simply by $\Sigma\mathbf{F}$. And so Eq. 5-11 is equivalent to Eq. 5-4 ($\Sigma\mathbf{F} = M\mathbf{a}_{cm}$).

In the special case where all parts of the system have the same acceleration $\mathbf{a}$, this acceleration will also be the acceleration of the center of mass, and the equation $\Sigma\mathbf{F} = M\mathbf{a}_{cm}$ reduces to the statement of the second law for a single particle: $\Sigma\mathbf{F} = M\mathbf{a}$.

CHAPTER 5 SUMMARY

The magnitude of the force of static friction varies from zero to a maximum value proportional to the normal force pushing the two surfaces together:

$$f_{max} = \mu_s N$$

where μ_s is the coefficient of static friction. The force of kinetic friction is given by

$$f = \mu_k N$$

where μ_k is the coefficient of sliding friction.

Centripetal force is the force or combination of forces that produces the centripetal acceleration associated with circular motion. This resultant force is directed toward the center of the circle and has magnitude

$$|\Sigma \mathbf{F}| = \frac{mv^2}{r}$$

The location of the center of mass is defined by the equations

$$x_{cm} = \frac{\Sigma\, mx}{\Sigma\, m} \qquad y_{cm} = \frac{\Sigma\, my}{\Sigma\, m}$$

The motion of the center of mass of a system is determined by the equation

$$\Sigma\, \mathbf{F} = M\mathbf{a}_{cm}$$

Questions

1 (a) What is the minimum possible value of μ_s?
(b) Is there a finite maximum possible value of μ_s?

2 (a) What is the magnitude of the normal force acting on each of the glass slides shown in Fig. 5-D?
(b) What is the value of μ_s for these surfaces?

3 Which would require a greater force in order to be moved: a 10 N brick on a horizontal wooden surface or a 100 N steel block on a horizontal Teflon surface?

4 A book lies on a table. One end of the table is gradually raised, so that the surface is inclined. When the incline is great enough, the book begins to slide. Will sliding begin sooner if the book lies flat or stands on end, or does it happen at the same point in either case? Assume that the book does not tip over when it is standing on end.

5 You accelerate your car forward. What is the direction of the frictional force on a package resting on the floor of the car?

6 A sheet of paper is initially at rest beneath a book on a table. You jerk the paper quickly to the right, exerting a large force on it. Explain what happens to the book and why.

7 Your rear-wheel-drive car is stuck in snow. Would it be better to (a) remove a heavy object from the trunk to reduce the weight on the rear wheels or (b) place a heavy object in the trunk to increase the weight on the rear wheels?

8 Suppose you were on a perfectly frictionless ice pond. Would it be possible for you to walk or crawl very carefully to shore? Explain your answer.

9 You drive your car, first along level pavement and then up a hill.
(a) Will the maximum possible force of static friction increase, decrease, or remain the same as you leave the level part and start up the hill? Assume that the surface of the road is the same.
(b) Is it easier to spin your wheels on the hill or on the level part of the road?

10 What is the ratio of the maximum deceleration when braking to the maximum acceleration of a car with two-wheel-drive on level ground? Assume that the car's weight is equally distributed over the wheels and that when the car has sufficient power and good enough brakes, the only limitation on the rates of acceleration or deceleration is the friction that can be provided by the road.

11 As a car rounds the top of a hill on a straight section of highway, it moves along an approximately circular path in a vertical plane. What force or combination of forces provides the centripetal force necessary to keep the car on this path: (a) friction; (b) normal force; (c) weight; (d) air resistance; (e) weight minus normal force; (f) friction and weight?

12 Suppose you are riding a Ferris wheel at the amusement park. If the wheel rotates at a constant rate, is the force exerted on you by your seat greatest at the top, the bottom, or some point in between?

13 A tall old redwood tree, almost devoid of branches, tapers gradually as it rises from the ground. Is the center of mass of the tree closer to the top or the base?

14 A 3 m long javelin is thrown vertically upward, and the tip reaches a maximum height of 10 m. What is the maximum height of the javelin's center of mass, assuming the mass is uniformly distributed?

15 Where is the center of mass of a doughnut?

16 (a) In Example 9, will the place where John and Michelle meet be affected by how they pull on the rope, that is, by who pulls first and by how hard or how long each person pulls?
(b) Will the time it takes them to meet be affected by how they pull on the rope?

Answers to Odd-Numbered Questions

1 (a) 0; (b) no; **3** the 10 N brick; **5** forward; **7** b; **9** (a) decrease; (b) easier on the hill; **11** e; **13** base; **15** at the center of the hole

Problems (listed by sections)

5-1 Friction

1 Suppose the book in Fig. 5-2 weighs 9.0 N. To move the book from rest requires a minimum horizontal force **P** of magnitude 3.0 N. The book will keep moving at constant velocity if the magnitude of the force **P** is reduced to 2.0 N. Find the coefficients of static and kinetic friction between the book and the table.

2 Suppose that while moving into an apartment you move a refrigerator into place by sliding it across the floor. The refrigerator weighs 1500 N and the coefficient of static friction between the floor and the refrigerator is 0.45. What is the least force you could exert on the refrigerator to move it?

3 You are holding a bulletin board weighing 6.0 N in place against a wall while your friend secures it to the wall. To keep it from slipping, you apply a force perpendicular to the bulletin board, pressing it directly into the wall. How large must this force be if the coefficient of static friction with the wall is 0.40?

4 To push a certain box across a level floor at constant velocity requires that a horizontal force be applied to the box. The magnitude of this force is half as great as the vertical force required to lift the box. Find the coefficient of kinetic friction between box and floor.

5 A friend on skis stands still on level ground, covered with dry snow at 0° C. How hard would you have to push your friend in the forward, horizontal direction to move him if your friend weighs 800 N?

6 A wooden block weighing 5 N is at rest on top of a wooden desk. The coefficient of static friction between the block and the desk is 0.5. Find the minimum horizontal force required to move the block if (a) the block is being pressed into the surface with a 3 N force; (b) the block is partially supported by a vertical string under a tension of 3 N.

7 Suppose you are standing on skis, headed straight down a slight incline covered with dry snow at 0° C. What is the minimum angle of the incline that will cause you to slide?

8 When you try to walk too fast on a slippery floor, you slip because walking fast requires you to exert a greater frictional force against the floor than the surface allows. What is the maximum frictional force you can exert against the floor with one foot that bears your full weight of 560 N, if $\mu_s = 0.20$?

9 What force must be applied to push a carton weighing 250 N up a 15° incline, if the coefficient of kinetic friction is 0.40? Assume the force is applied parallel to the incline and the velocity is constant.

10 Two blocks, each of mass 1.0 kg, are pushed along the horizontal surface of a table by a horizontal force **P** of magnitude 9.8 N, directed to the right, as shown in Fig. 5-16. The blocks move together to the right at constant velocity. Find (a) the frictional force exerted on the lower block by the table; (b) the coefficient of kinetic friction between the surface of the block and the table; (c) the frictional force acting on the upper block.

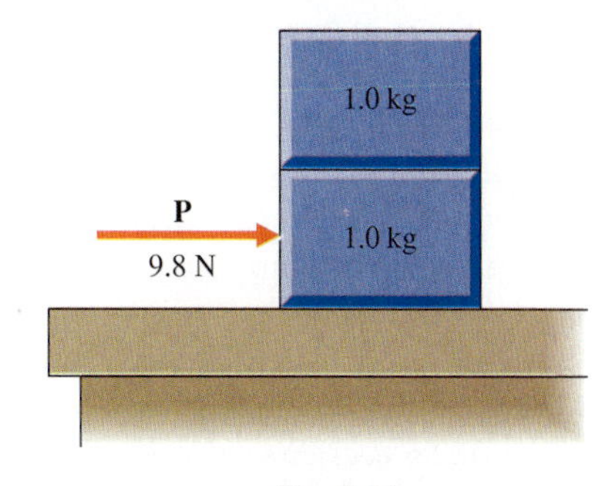

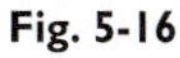
Fig. 5-16

11 In the previous problem, suppose that the magnitude of **P** is increased to 11.8 N, resulting in an acceleration of 1.0 m/s^2 of the two blocks to the right. Find (a) the frictional force exerted on the upper block by the lower block; (b) the frictional force exerted on the lower block by the upper block; (c) the frictional force exerted on the lower block by the table.

12 A skier of mass 64 kg skis straight down a 12° slope at constant velocity. Draw a free-body diagram of the skier with the various external forces acting on her. Include the force of air resistance, which is directed opposite the velocity.
(a) Find the value of the normal force.
(b) The force of air resistance has a magnitude of 75 N. Find the frictional force on the skis.
(c) What is the coefficient of kinetic friction?

13 A block weighing 10.0 N rests on a 30° inclined plane. Find (a) the normal force exerted by the plane on the block; (b) the frictional force exerted by the plane on the block; (c) the magnitude of the total force exerted by the plane on the block; (d) the normal force and the frictional force exerted by the block on the plane.

14 The coefficient of static friction can be measured experimentally in the following way. Place an object on a surface that is initially horizontal. Then incline the surface at an angle θ, which is gradually increased. When θ reaches some maximum value θ_{max}, the object will begin to slide. This angle is easily measured. The coefficient of static friction is then found from the equation $\mu_s = \tan \theta_{max}$.
(a) Prove this result by analyzing the forces acting on the block.
(b) What is θ_{max} for a piece of dry, clean glass resting on a glass surface?

15 A hockey puck is given an initial velocity of 40.0 m/s along the ice. Find the speed of the puck 1.00 s later if the coefficient of kinetic friction between puck and ice is 0.600. (HINT: The result is independent of the mass of the puck.)

16 A car of mass 975 kg coasts on a level road, starting with an initial speed of 15.0 m/s and coming to rest 15.0 s later. Find the frictional force acting on the car.

17 (a) Find the maximum rate of deceleration of a car on a dry, level road, assuming the coefficient of static friction between tires and pavement equals 1.0.
(b) Find the minimum stopping distance if the car has an initial speed of 25 m/s.
(c) Repeat parts a and b assuming a wet, oily surface for which $\mu_s = 0.10$.

18 Find the maximum acceleration of a car with two-wheel drive on a dry, level road with $\mu_s = 1.0$, assuming that the car's weight is equally distributed between the front and rear wheels.

19 A cardboard box slides along a wood floor with an initial velocity of 5.00 m/s. If the box comes to rest after traveling 3.00 m along the floor, what is the coefficient of kinetic friction between floor and box?

20 A small package rests on the horizontal dashboard of a car. If $\mu_s = 0.30$, what is the minimum acceleration of the car that will cause the package to slip off, assuming that the car is on a level road?

5-2 Centripetal Force

21 A 0.100 kg rock is attached to a 2.00 m long string and swung in a horizontal circle at a speed of 30.0 m/s. Find the tension in the string. Neglect the effect of gravity.

22 A 1.00 kg mass is attached to a spring of force constant 10.0 N/cm and placed on a frictionless surface. By how much will the spring stretch if the mass moves along a circular path of radius 0.500 m at a rate of 2.00 revolutions per second?

23 A passenger of mass 50.0 kg is in a car rounding a level curve of radius 100.0 m at a speed of 20.0 m/s.
(a) Assuming that friction is the only horizontal force acting on the passenger, find the frictional force.
(b) What would happen if $\mu_s = 0.3$?

★ 24 Find the angle at which a curve of radius r should be banked in order that a car moving at a speed v will not need any frictional force to round it (Fig. 5-17).

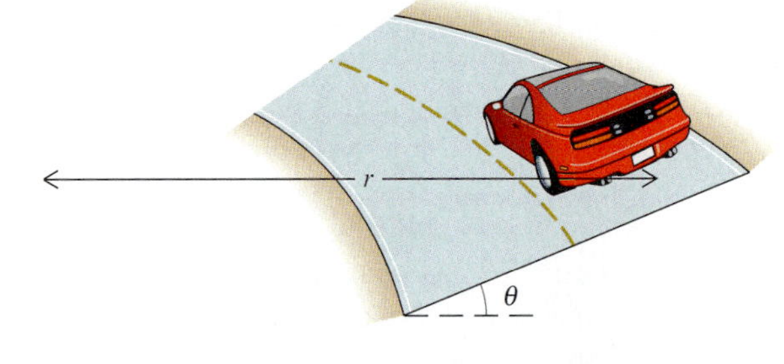

Fig. 5-17

25 A sprinter of mass 70.0 kg runs at a constant speed of 10.0 m/s in a straight line. Find the extra force the sprinter must exert on the ground to round a curve of radius 10.0 m.

26 A sling carrying a rock of mass 0.200 kg is swung in a vertical circle of radius 1.50 m at a speed of 62.6 m/s (sufficient to give a horizontal range of 400 m, as indicated in Problem 17, Chapter 3). Find the force the hand must exert on the sling when the rock is (a) at the top if its circular arc and (b) at the bottom of its circular arc.

27 A girl weighing 285 N is on a swing, supported by two chains of length 2.00 m. Find the tension in each chain when the chains are vertical and her speed is 3.00 m/s.

★ **28** As a car rounds the top of a hill at a speed of 20.0 m/s, it very briefly loses contact with the pavement. This section of the road has an approximately circular shape (Fig. 5-18). Find the radius r.

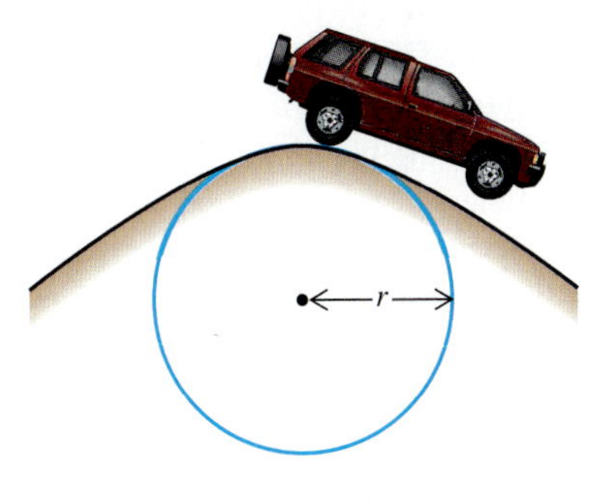

Fig. 5-18

5-3 Center of Mass

29 The mass of the earth is 5.98×10^{24} kg, the mass of the moon is 7.36×10^{22} kg, and the distance between the centers of the earth and the moon is 3.82×10^8 m. How far is the center of mass of the earth-moon system from the center of the earth?

30 Find the center of mass of the particles in Fig. 5-19, which are positioned at the corners of a square of edge length 30.0 cm.

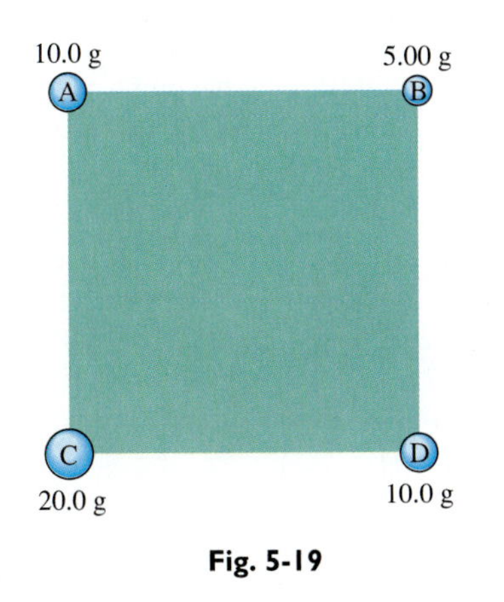

Fig. 5-19

31 Find the center of mass of the bricks shown in Fig. 5-20.

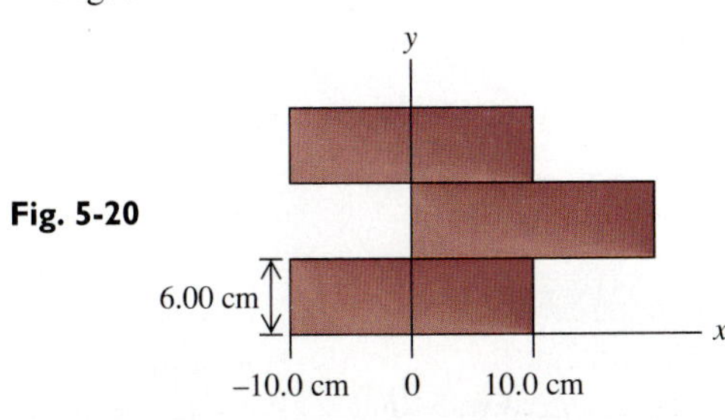

Fig. 5-20

32 A simple model of the mass distribution of a pole vaulter is given in Fig. 5-21. The pole vaulter is shown at the high point in the jump, at which point the midsection of his body is just over the bar (at the origin).
(a) Find the center-of-mass coordinates of the pole vaulter's body. Treat the rectangles as uniform distributions of mass. The center of mass of each rectangle is indicated by a dot, next to which is the mass of the rectangle.
(b) If the center of mass is at an elevation of 5.10 m, what is the height of the bar?

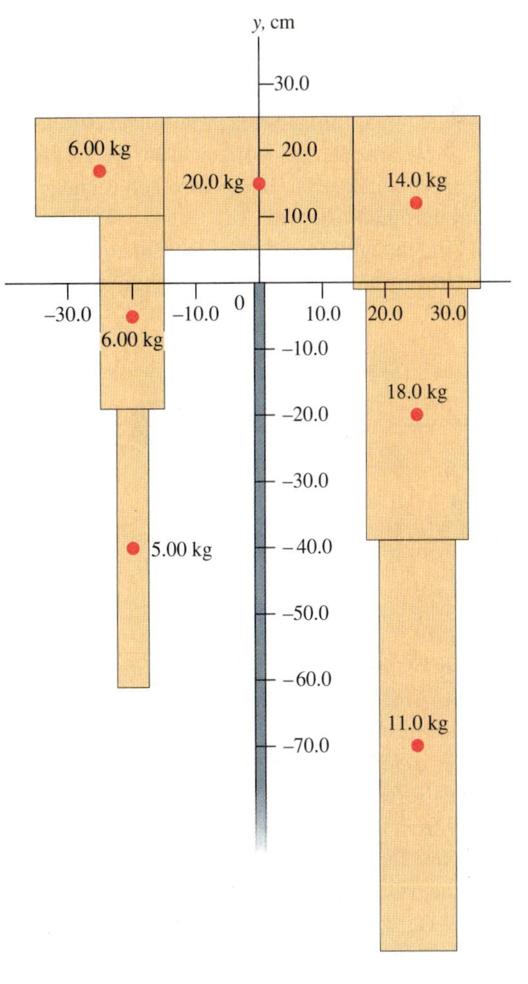

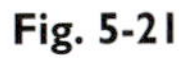

Fig. 5-21

★★ **33** Each side of an equilateral triangular plate has a length of 1.00 m. Find the distance of the plate's center of mass from the center of one side of the plate. (HINT: Use symmetry.)

34 The location of two particles at $t = 0$ is shown in Fig. 5-22. The particles are initially at rest. Each particle is subject to a *constant* force, as indicated in the figure.
(a) Find the position of the center of mass at $t = 0$.
(b) Find the resultant force acting on the system.
(c) Find the location of the center of mass at $t = 5.00$ s.

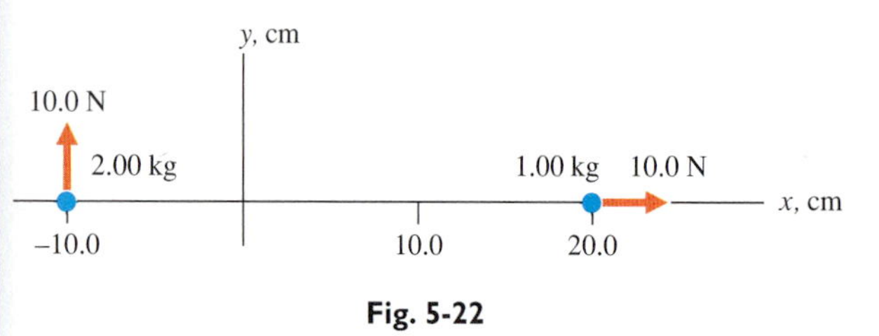

Fig. 5-22

35 Spheres A and B are attached to the ends of a rod of negligible mass (Fig. 5-23). The spheres are initially at rest on a frictionless, horizontal surface. A horizontal string is attached to B at point P and is under a constant tension of 6.00 N.
(a) Find the initial location of the center of mass.
(b) How far will the center of mass move in 2.00 s?

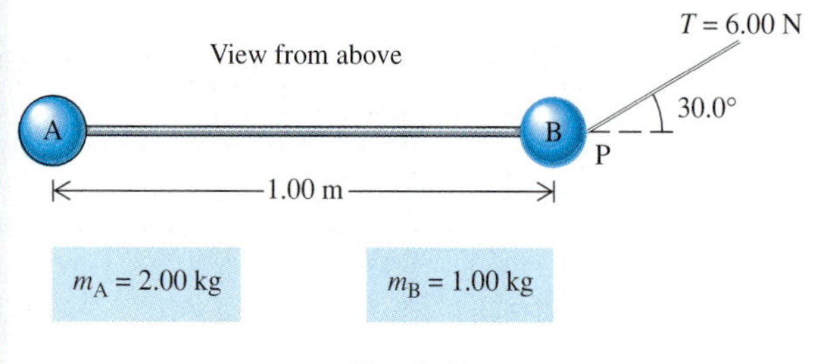

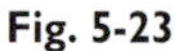

Fig. 5-23

36 A boy of mass 60 kg is about to disembark from a canoe of mass 40 kg. The canoe is initially at rest, with the bow just touching the dock (Fig. 5-24). The center of the canoe is 3.0 m behind the bow. As the boy moves forward 6.0 m to the bow, the canoe moves away from the dock.
(a) How far is the boy from the dock when he reaches the bow?
(b) Is the canoe moving at this point? Assume that there is no friction between the canoe and the water.

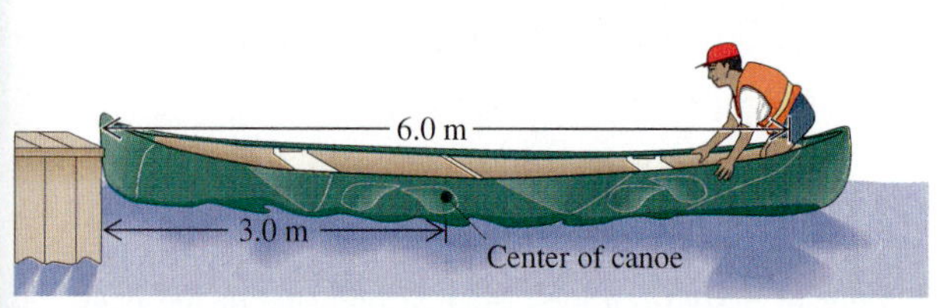

Fig. 5-24

37 An airplane of mass 1.0×10^4 kg falls vertically downward, crashing to the ground. The plane explodes just before impact. Investigators looking into the cause of the crash search for the wreckage. Large fragments of mass 4.0×10^3 kg, 3.0×10^3 kg, and 1.0×10^3 kg are found respectively at the point of impact, 30 m north, and 50 m east. Where should one look for the remainder of the plane; in other words, where is the center of mass of the remaining wreckage?

38 Olympic gold medalist Carl Lewis performs a long jump, starting at rest on a 12.0 m long flatbed railroad car of mass 450.0 kg. Lewis, whose mass is 90.0 kg, runs along the length of the car and jumps off the end. From the beginning of the run to the end of the jump, he travels a total distance of 15.0 m relative to the earth. How far has the railroad car moved by the time he hits the ground, if no frictional force acts between car and earth?

Additional Problems

39 The two blocks shown in Fig. 5-25 are attached to opposite ends of a string, which passes over a frictionless pulley of negligible mass. The pulley is attached to a wall on the right. The top block is pulled to the left with a force **P** of magnitude 10 N. As the top block moves to the left, the bottom block moves to the right.
(a) Draw the appropriate free-body diagrams necessary to solve for the acceleration of each block, but DO NOT COMPLETE the problem.
(b) What additional information must be provided in order to solve the problem?

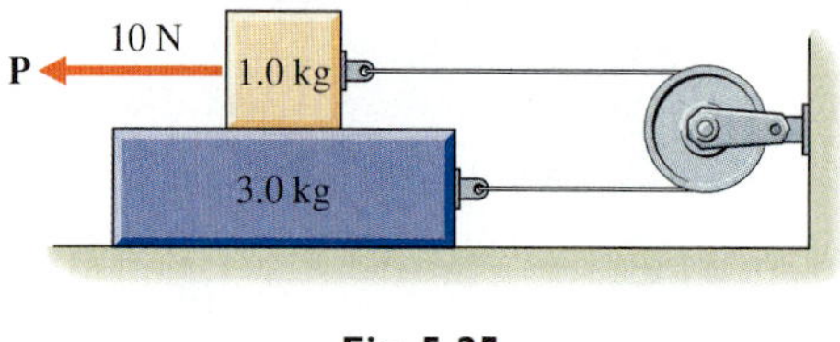

Fig. 5-25

40 Draw complete free-body diagrams for the block and the wedge shown in Fig. 5-26.

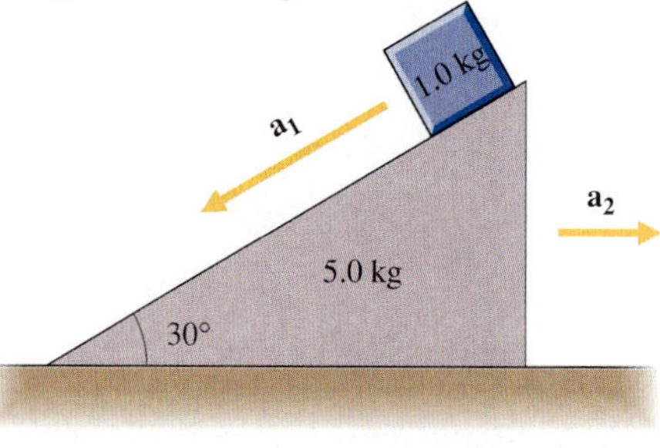

Fig. 5-26

★ **41** Find the magnitude of the force **P** necessary to drag the crate shown in Fig. 5-27 at constant velocity. The crate weighs 500 N, and $\mu_k = 1.0$.

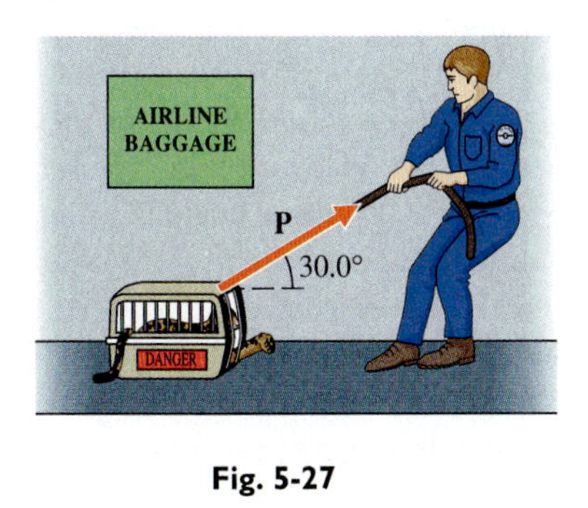

Fig. 5-27

★ **42** Find the acceleration of the two blocks sliding down the incline in Fig. 5-28.

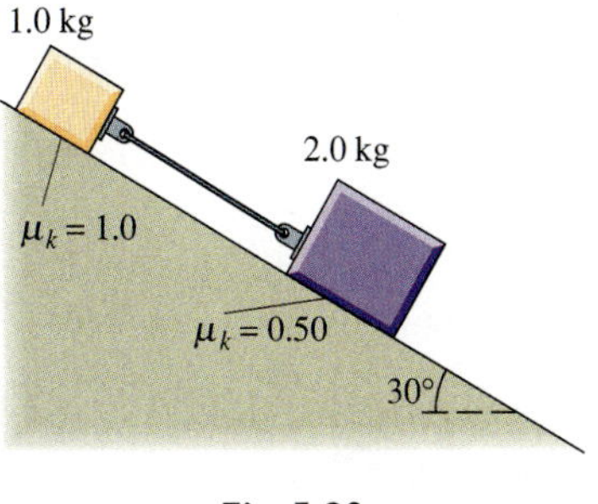

Fig. 5-28

★ **43** Find the acceleration of the 2.0 kg block in Fig. 5-29.

★ **44** The 1.0 kg block in Fig. 5-29 is replaced by a 4.0 kg block. Find the acceleration of the 2.0 kg block.

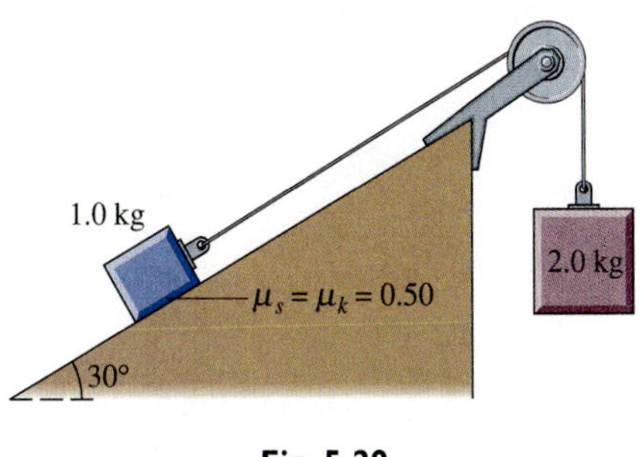

Fig. 5-29

★★ **45** Find the minimum mass of a block to replace the 1.0 kg block in Fig. 5-29, such that the 2.0 kg block moves upward.

★ **46** Find the acceleration of the 5.0 kg block in Fig. 5-30.

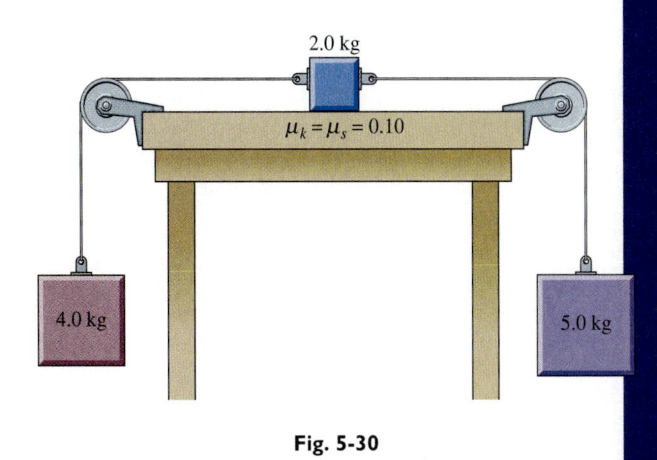

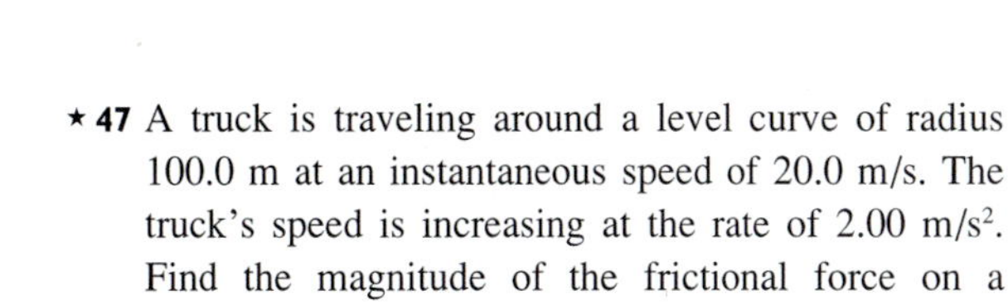

Fig. 5-30

★ **47** A truck is traveling around a level curve of radius 100.0 m at an instantaneous speed of 20.0 m/s. The truck's speed is increasing at the rate of 2.00 m/s^2. Find the magnitude of the frictional force on a 100.0 kg crate resting on the floor of the truck.

48 A certain roller coaster design uses a vertical loop of radius 8.00 m. Assuming that the roller coaster remains on the track, what is the minimum speed of a car at the top of the loop? (HINT: In general, the centripetal force is the vector sum of the car's weight **w** and the normal force **N** exerted by the tracks. The value of N varies with the speed. The minimum value of N is zero, which occurs when the roller coaster is just about to leave the track.)

49 Two books, each having a mass of 1.0 kg, are shown in Fig. 5-31. If $P = 30$ N and $\mu_s = \mu_k = 0.50$ for all surfaces, find the acceleration of each book.

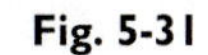

Fig. 5-31

★ **50** A truck and its contents have a total mass of 4.00×10^3 kg. The truck is initially moving at a constant speed of 20.0 m/s. Some of the truck's cargo, which is initially near the front end, shifts and moves toward the back. A mass of 4.00×10^2 kg moves 5.00 m in the backward direction during a 1.00 s interval. How far does the truck travel during this interval?

★ **51** The x coordinates of the centers of mass for two arbitrary bodies A and B may be written

$$x_A = \frac{\Sigma_A mx}{\Sigma_A m}$$

and

$$x_B = \frac{\Sigma_B mx}{\Sigma_B m}$$

where the subscripts A and B indicate that the sums are over the particles of the respective bodies. The system of the two bodies has a center of mass with an x coordinate that you can find by applying the definition

$$x_{cm} = \frac{\Sigma\, mx}{\Sigma\, m}$$

taking the sums over all the particles in both bodies. By separating the sums into two sums, each involving only particles in one of the bodies, show that x_{cm} may be expressed as

$$x_{cm} = \frac{M_A x_A + M_B x_B}{M_A + M_B}$$

where M_A and M_B are the respective masses of bodies A and B. Show that the y coordinate of the center of mass can be expressed in a similar way.

★ **52** Initially a wedge of mass 5.00 kg is at rest on a frictionless horizontal surface, and a 1.00 kg block is at rest near the top of the wedge (Fig. 5-26). The block begins to slide down the incline. After moving 1.00 m along the incline, the block again comes to rest. How far does the wedge move to the right?

★★ **53** An amusement park ride called the Rotor consists of a room in the shape of a vertical cylinder (2.00 m in radius) which, once the riders are inside, begins to rotate, forcing them to the wall. When the room reaches a speed of one rotation every 1.50 s, the floor suddenly drops out. What is the minimum coefficient of static friction between riders and wall necessary to prevent them from sliding down the wall?

★ **54** Suppose you suddenly discover that you are heading your car toward the edge of a cliff. Would it be better to try to stop while traveling straight ahead or to turn the car to one side? Assume that, if you turn the car, your path will be circular and that the coefficient of static friction is the same in either case. Explain your answer.

★★ **55** A board of uniform thickness is shown in Fig. 5-32. The board's center of mass is initially at point O. If a hole centered at P is cut out of the board, how far to the right of O is the new center of mass, if cutting the hole removes 10.0% of the board's mass? (HINT: Express the center of mass of the entire board in terms of the center of mass of the circular section and of the board with the hole.)

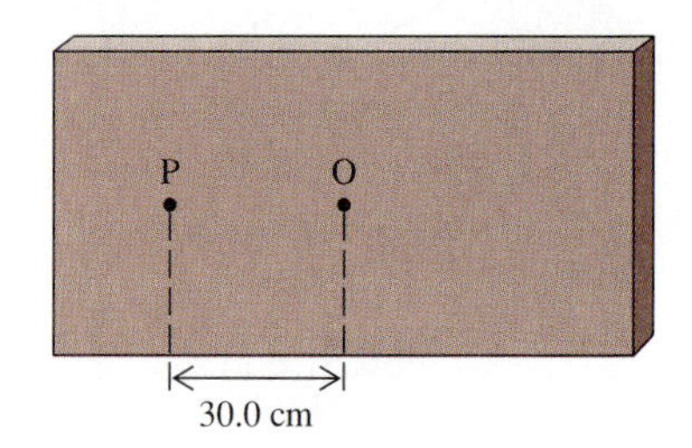

Fig. 5-32

CHAPTER 6 Gravitation

The paths of Mercury, Mars, Jupiter, and Saturn from the earth's reference frame, shown in the Munich planetarium.

We tend to think of gravity as that familiar downward pull of the earth we all feel. In this chapter we shall see that gravity is a much more universal phenomenon. Any two bodies exert a mutual attractive force. They "gravitate" toward each other. Of course, we don't normally see bodies drawn together as a result of the gravitational force between them. The gravitational force is much too weak to be observed in that way. It is nonetheless quite real, and under the right conditions we can observe the attractive gravitational force between two small objects. Their gravitation has nothing to do with the earth and would be the same anywhere in the universe.

6-1 Universal Gravitation

Newton's essential insight concerning gravity was that the earth's gravity is just one manifestation of a **universal tendency of matter to attract other matter.** This universal attraction is not obvious because the gravitational force between two bodies of ordinary dimensions is very small. Newton was never able to observe the effect of gravitational attraction between two terrestrial bodies. This was achieved more than 100 years later by Cavendish. Nevertheless Newton realized that the existence of a universal gravitational force was necessary to explain the motion of the moon and planets.

It is a popular myth that Newton's insight came as he observed an apple fall from a tree in the garden where he was absorbed in thought. The essential truth in this story is that Newton realized that the earth's gravity, which acts on apples and all other objects, is a universal phenomenon and that the earth's gravitational pull extends out to the orbit of the moon.

It must have been clear to Newton that some force is necessary to hold the moon to its circular orbit—a centripetal force. You can see that this force must exist by considering a rock attached to a string and swung overhead in a horizontal circle. The force on the rock, provided by the string, is necessary to maintain the circular motion. Break the string, and the rock flies off tangent to the circle. Similarly, there must be some force pulling on the moon and preventing it from flying off, and it was reasonable for Newton to assume that this force was the earth's gravity.

The next question was how this gravitational force depends on distance. Newton might well have answered this question, using a result Johannes Kepler had discovered from analysis of planetary orbits—**Kepler's third law** of planetary motion. (Kepler's work is described in detail in the end-of-chapter essay on the origins of the theory of universal gravitation.) Kepler's third law relates the mean distance of a planet from the sun to the planet's orbital period (the length of time for one complete orbit, or, in other words, the length of its year). The orbital period T increases with the mean distance r; T is proportional to the $\frac{3}{2}$ power of r:

$$T \propto r^{3/2} \qquad \text{(Kepler's third law)} \qquad (6\text{-}1)$$

(In the case of a circular orbit, r is the radius of the circle. The earth's orbit is approximately circular.)

We shall now show how Newton might have discovered a general expression for gravitational force by applying Kepler's third law to a circular planetary orbit. A planet moving around the sun along a circular orbit of radius r experiences a force **F**, whose magnitude is equal to the product of the planet's mass m and centripetal acceleration $a = v^2/r$:

$$F = ma = \frac{mv^2}{r}$$

We can express the planet's speed v as the circumference of the orbit ($2\pi r$) divided by the period T. Inserting $v = 2\pi r/T$ in the equation above, we obtain:

$$F = \frac{4\pi^2 mr}{T^2}$$

Using Kepler's third law to substitute for T, we find that:

$$F \propto \frac{m}{r^2}$$

The gravitational force exerted by the sun on a planet is proportional to the mass of the planet divided by the square of the distance from the sun.

It occurred to Newton that if the earth exerts a gravitational force on the moon and if the sun exerts a gravitational force on each planet then it is reasonable to suppose that there is a universal gravitational force between any two objects.

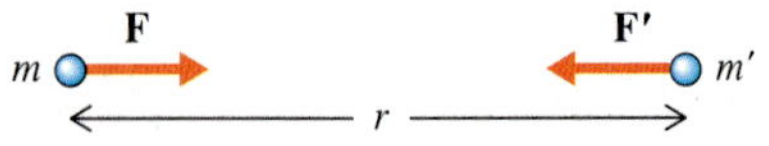

Consider the mutual attractive force between two particles of mass m and m', separated by a distance r (Fig. 6-1). We have found that the gravitational force on a planet is proportional to m/r^2. This suggests that the gravitational force on each particle in Fig. 6-1 should also be proportional to its mass divided by r^2. Thus F is proportional to m/r^2, and F' is proportional to m'/r^2. But according to Newton's third law, F and F' are equal. So F and F' are each proportional to the mass of either particle divided by r^2.

m F F' m' r

Fig. 6-1 Gravitational force between two particles.

If we denote the constant of proportionality by G, we may express this fundamental force law as

$$F = \frac{Gmm'}{r^2} \qquad (6\text{-}2)$$

The constant G is called the **universal gravitational constant.** The value of G was unknown to Newton, though he realized it was a very small number. By carefully measuring F for known masses, G has been found to be

$$G = 6.672 \times 10^{-11}\ \text{N-m}^2/\text{kg}^2$$

EXAMPLE 1 Attraction Between Two Small Masses

Find the gravitational force between two particles, of mass 1.00 kg each, separated by a distance of 1.00 m.

SOLUTION Fig. 6-1 shows the mutual attractive force between the two masses, m and m'. Applying Eq. 6-2, we find

$$F = \frac{Gmm'}{r^2}$$

$$= \left(6.67 \times 10^{-11}\ \frac{\text{N-m}^2}{\text{kg}^2}\right)\frac{(1.00\ \text{kg})(1.00\ \text{kg})}{(1.00\ \text{m})^2} = 6.67 \times 10^{-11}\ \text{N}$$

This is an extremely small force—about equal to the weight of a microscopic speck of dust of radius 0.01 mm.

Several Gravitating Particles

If several particles are present, each pair will experience a mutual gravitational attraction. **The total gravitational force on a given particle is the vector sum of the pair forces.**

EXAMPLE 2 Discovery of Neptune

The planet Neptune was discovered in 1846 on the basis of predictions made by astronomers who had noted that the orbit of the planet Uranus had certain irregularities. These irregularities were interpreted as arising from the gravitational effect of an unseen planet. The location of this unseen planet was predicted, and its existence was soon verified by direct observation. Compute the ratio of the gravitational force of Neptune on Uranus to the gravitational force of the sun on Uranus in 1822, when the sun, Uranus, and Neptune were aligned, with Uranus at a distance of 2.9×10^{12} m from the sun and Neptune at a distance of 4.5×10^{12} m from the sun. The mass of the sun is 2.0×10^{30} kg and the mass of Neptune is 1.0×10^{26} kg.

SOLUTION Fig. 6-2 shows the distances between the bodies and the forces $\mathbf{F}_S$ and $\mathbf{F}_N$ exerted on Uranus by the sun and Neptune respectively. Notice that we have drawn a free-body diagram only of Uranus. We do not draw the free-body diagrams of the other two bodies because we are interested here only in the forces acting on Uranus. We express the magnitude of both forces, using Eq. 6-1.

$$F_S = \frac{GM_SM_U}{r_1^2}$$

$$F_N = \frac{GM_NM_U}{r_2^2}$$

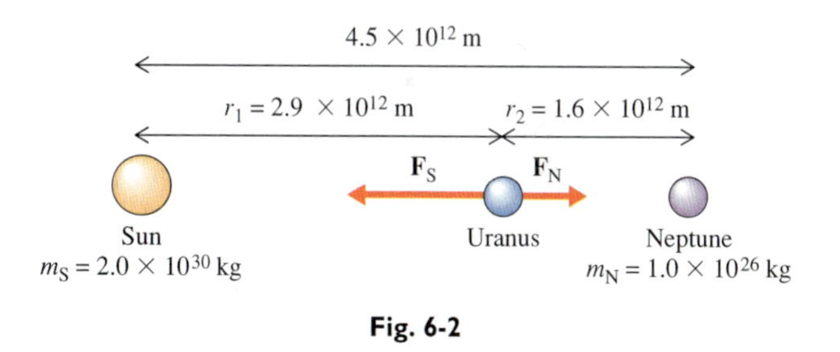

Fig. 6-2

We divide the second expression by the first, the factors G and M_U cancel, and using the data shown in Fig. 6-2, we find for the ratio of the forces:

$$\frac{F_N}{F_S} = \frac{M_N/r_2^2}{M_S/r_1^2} = \left(\frac{M_N}{M_S}\right)\frac{r_1^2}{r_2^2}$$

$$= \left(\frac{1.0 \times 10^{26}\ \text{kg}}{2.0 \times 10^{30}\ \text{kg}}\right)\frac{(2.9 \times 10^{12}\ \text{m})^2}{(1.6 \times 10^{12}\ \text{m})^2}$$

$$= 1.6 \times 10^{-4}$$

This calculation shows that the force exerted on Uranus by Neptune is much smaller than the force exerted by the sun. Yet the force of Neptune is big enough to perturb the orbit of Uranus, which allowed astronomers to predict the existence of this previously unseen planet.

EXAMPLE 3 Resultant of Three Gravitational Forces on a Particle

Four particles, each of mass 10.0 kg, are isolated from all other bodies, including the earth, and are located at the corners of a square, whose sides are 10.0 cm. Compute the net force on the particle in the upper left corner.

SOLUTION First we must use Eq. 6-1 to compute the magnitude of the individual pair forces $\mathbf{F}_1$, $\mathbf{F}_2$, and $\mathbf{F}_3$, shown in Fig. 6-3.

$$F_1 = F_3 = (6.67 \times 10^{-11}\ \text{N-m}^2/\text{kg}^2)\frac{(10.0\ \text{kg})^2}{(0.100\ \text{m})^2}$$

$$= 6.67 \times 10^{-7}\ \text{N}$$

To find F_2 we must first compute the distance to the particle exerting this force. This distance is the length of the square's diagonal, r, related to the square's edge length a by the Pythagorean theorem, $r = \sqrt{a^2 + a^2} = \sqrt{2}a$.

$$F_2 = (6.67 \times 10^{-11}\ \text{N-m}^2/\text{kg}^2)\frac{(10.0\ \text{kg})^2}{[\sqrt{2}(0.100\ \text{m})]^2}$$

$$= 3.33 \times 10^{-7}\ \text{N}$$

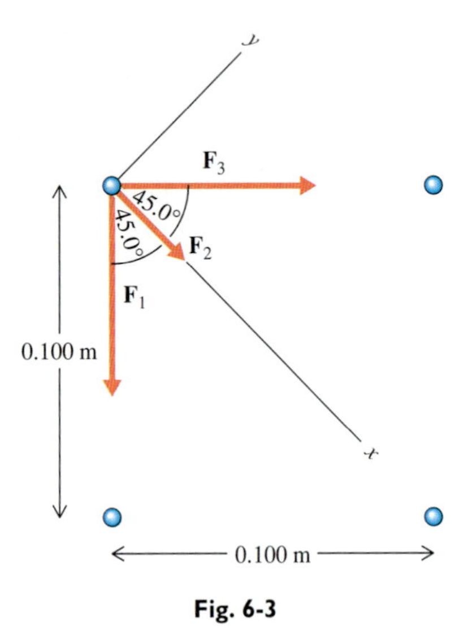

Fig. 6-3

From the symmetry of the forces in Fig. 6-3, we anticipate that the resultant force will be along the diagonal of the square, and so we choose the x-axis of our coordinate system along the diagonal. We then compute the components of the resultant force on the particle:

$$\begin{aligned}\Sigma F_y &= -F_1 \sin 45.0° + 0 + F_3 \sin 45.0° \\ &= 0\end{aligned}$$

$$\begin{aligned}\Sigma F_x &= F_1 \cos 45.0° + F_2 + F_3 \cos 45.0° \\ &= 2(6.67 \times 10^{-7}\ \text{N})(\cos 45.0°) + 3.33 \times 10^{-7}\ \text{N} \\ &= 1.28 \times 10^{-6}\ \text{N}\end{aligned}$$

The net force has magnitude 1.28×10^{-6} N and is directed along the positive x-axis, that is, toward the opposite corner of the square. The net force on each of the other particles has the same magnitude and could be calculated in a similar manner.

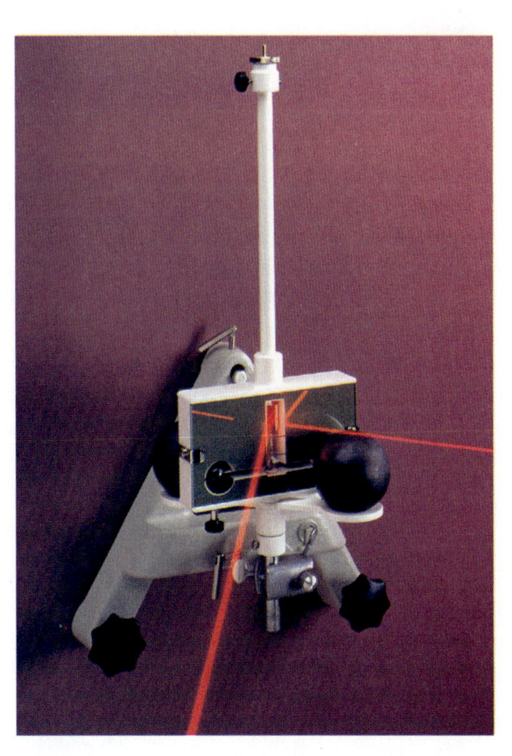

Fig. 6-4 A Cavendish balance, used to measure G.

Measurement of *G*

In 1798 Henry Cavendish reported an experiment in which he was able to measure for the first time the value of the gravitational constant. The determination of G allowed various other quantities to be calculated—the mass of the earth and the mass of the sun, for example. Aside from the importance of obtaining a precise value of G, the experiment provides a beautiful confirmation of Newton's prediction that there exists a mutual gravitational force between any two masses.

Fig. 6-4 shows a version of the Cavendish apparatus frequently used in student laboratories. Two small lead spheres are attached to the ends of a light, horizontal rod, which is suspended at its midpoint by a very thin bronze ribbon; a small mirror is attached to the center of the rod (Fig. 6-5). The fine suspension makes the rod and spheres very sensitive to horizontal forces. (This sensitivity makes it important to minimize vibrations. Air currents can even have an appreciable effect; therefore the rod is usually enclosed by glass.) When two larger lead spheres are brought close to the smaller spheres, the attractive forces on the smaller spheres produce a small but perceptible acceleration toward the larger spheres, and the rod begins to turn (Fig. 6-5b). The rotation is opposed by the suspending ribbon, which is twisted. Equilibrium is reached after the rod has rotated through a small angle, typically on the order of 1°. Measurement of this angle is facilitated by means of a beam of light reflected from the mirror. The rotation angle, the masses of the spheres, and their separation are all used in the computation of G. (See Problem 15 for a simplified hypothetical experiment for measuring G.)

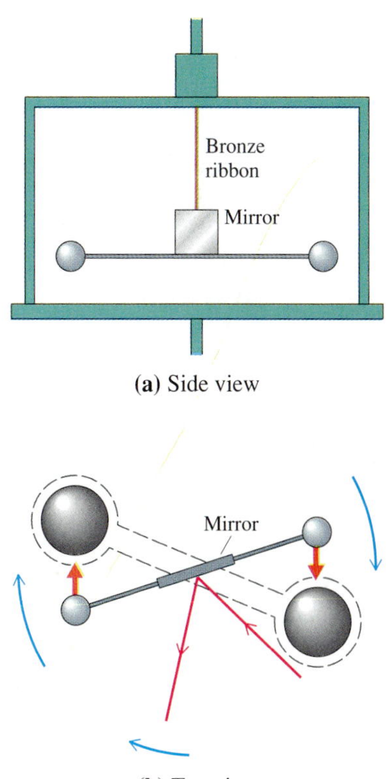

Fig. 6-5 **(a)** In a Cavendish balance two small lead spheres at the ends of a light rod are suspended by a very fine bronze ribbon. **(b)** When large lead spheres are brought near the small spheres, the gravitational attraction between the spheres causes the rod to rotate. A light ray reflected from an attached mirror rotates with the rod, allowing its rotation to be more easily measured.

6-2 Gravitational Attraction of the Earth

We can apply the law of universal gravitation to the gravitational interaction of the earth with bodies on or near its surface. Such bodies experience mutual gravitational attraction with each particle of the earth, as illustrated for an apple in Fig. 6-6a. Note that there is **no shielding of the gravitational force.** Each particle of the earth, no matter how deep below the surface, pulls on the apple as though there were no other earth particles between it and the apple. The apple's weight is the vector sum of all these forces. It is a remarkable fact that this resultant force on the apple is the same as it would be if the earth's mass were concentrated at its center* (Fig. 6-6b and c).

Thus we may obtain an expression for a body's weight w by applying the law of universal gravitation to the interaction between the body and the earth, treating them as particles of mass m and M, respectively, separated by a distance r (the distance of the body from the center of the earth):

$$F = w = \frac{GmM}{r^2} = m\left(\frac{GM}{r^2}\right)$$

where the factor GM/r^2 is a constant if the body is a fixed distance r from the center of the earth. Denoting this constant by the symbol g, we may express the body's weight as

$$w = mg \tag{6-3}$$

where

$$g = \frac{GM}{r^2} \tag{6-4}$$

The physical significance of the constant g is seen by considering the acceleration of a body subject only to the force of the earth's gravity, in other words, a body in free fall. According to Newton's second law, the body's acceleration a is

$$a = \frac{F}{m} = \frac{w}{m}$$

From Eq. 6-3, we find

$$a = g$$

Thus the law of universal gravitation applied to terrestrial gravity predicts that all bodies fall with the same constant acceleration g. This prediction is, of course, a well-known experimental fact recognized by Galileo. Here we see how this result follows from the law of universal gravitation. Furthermore, we have found that the value of g depends on the earth's mass M and on the distance r of the falling body from the center of the earth, as indicated by Eq. 6-4.

(a)

$\mathbf{F}_{\text{resultant}}$ m r $\mathbf{F}'_{\text{resultant}}$ M

(b)

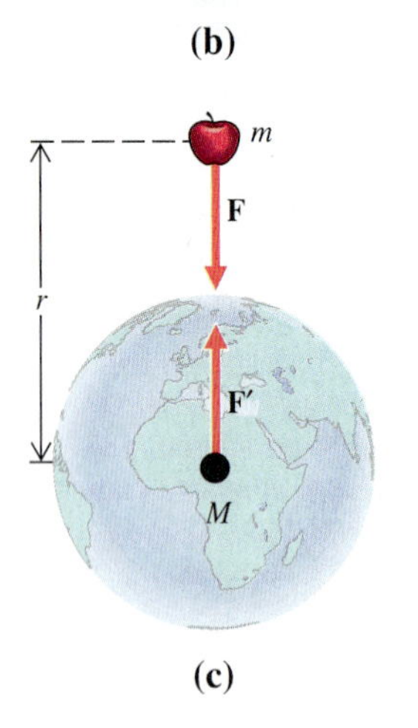

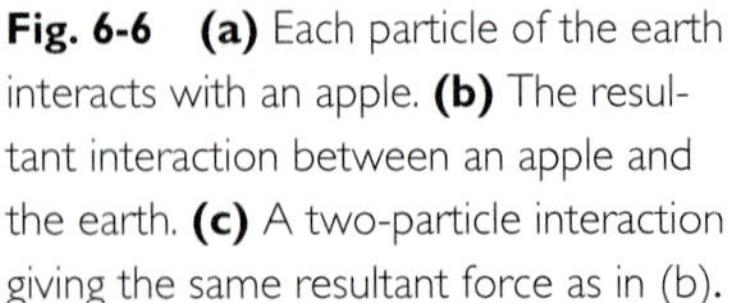

Fig. 6-6 **(a)** Each particle of the earth interacts with an apple. **(b)** The resultant interaction between an apple and the earth. **(c)** A two-particle interaction giving the same resultant force as in (b).

*This result is not obvious and not easy to prove. Its proof was one of the major factors delaying publication of the *Principia* for over 20 years. Newton developed integral calculus in order to solve problems of this kind.

For a body very near the surface of the earth, r can be accurately approximated by the earth's radius R, in which case g has the value

$$g = \frac{GM}{R^2} = 9.8 \text{ m/s}^2 \qquad \text{(at earth's surface)} \qquad (6\text{-}5)$$

As r increases, g decreases. For example, at a distance of two earth radii ($r = 2R$), $g = GM/(2R)^2 = \frac{1}{4}(9.8 \text{ m/s}^2) = 2.5 \text{ m/s}^2$.

Eqs. 6-3 and 6-4 may also be applied to the gravity produced on other planets or to the gravity produced by *any* spherically symmetric mass M.

EXAMPLE 4 Comparing the Motion of the Moon and a Falling Apple

One of Newton's earliest checks on his theory of gravitation involved a comparison of the moon's acceleration with that of an object on earth. The moon orbits the earth* in a nearly circular orbit of radius $r = 3.84 \times 10^8$ m (240,000 mi) in a period of 2.36×10^6 s (27.3 days) (Fig. 6-7). (a) Compute the centripetal acceleration of the moon, and compare it with the acceleration of a falling apple on earth. (b) Show that the law of universal gravitation accounts for the relative magnitude of these two accelerations, given that the earth's radius is 6.38×10^6 m.

Fig. 6-7

SOLUTION (a) The moon's centripetal acceleration a is

$$a = \frac{v^2}{r} = \frac{(2\pi r/T)^2}{r} = \frac{4\pi^2 r}{T^2}$$

$$= \frac{4\pi^2(3.84 \times 10^8 \text{ m})}{(2.36 \times 10^6 \text{ s})^2}$$

$$= 2.7 \times 10^{-3} \text{ m/s}^2$$

The apple's acceleration g is 9.8 m/s². Thus

$$\frac{g}{a} = \frac{9.8 \text{ m/s}^2}{2.7 \times 10^{-3} \text{ m/s}^2} = 3600$$

We know from our study of projectile motion in Ch. 3 that if an apple is either dropped from rest or thrown horizontally, it will drop a distance of $\frac{1}{2}gt^2$ in time t, or 4.9 m in 1 s. The moon can be thought of as falling from a straight-line path by $\frac{1}{3600}$ of this distance, or 1.4 mm in 1 s (Fig. 6-8).

(b) According to the law of gravitation, these accelerations are merely the two values of gravitational acceleration at two distances from the center of the earth. We apply Eq. 6-4, using g to denote the gravitational acceleration of an object at the earth's surface and g' to denote the gravitational acceleration of the moon, a distance r from the earth:

$$\frac{g}{a} = \frac{g}{g'} = \frac{GM/R^2}{GM/r^2} = \frac{r^2}{R^2} = \frac{(3.84 \times 10^8 \text{ m})^2}{(6.38 \times 10^6 \text{ m})^2} = 3600$$

in agreement with the measured value of this ratio.

*The moon does not exactly orbit the center of the earth. Both the earth and the moon orbit the center of mass of the earth-moon system. This point is much closer to the center of the earth than to the center of the moon because of the earth's much greater mass.

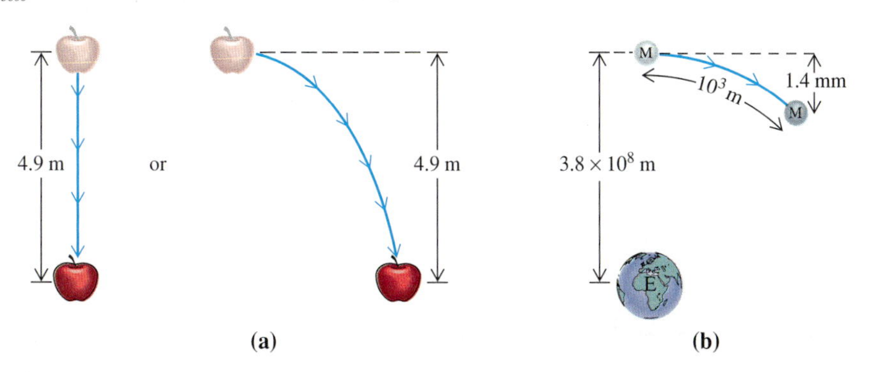

Fig. 6-8 **(a)** An apple falls 4.9 m during a 1 s interval. **(b)** The moon departs from a straight straight line path by 1.4 mm during a 1 s interval.

EXAMPLE 5 Speed of a Satellite in a Low Earth Orbit

Find an expression for the speed of an earth satellite in a circular orbit, as a function of the orbit's radius (Fig. 6-9). Evaluate the speed for an orbital radius of 6.50×10^6 m, given that the mass of the earth $M = 5.98 \times 10^{24}$ kg.

SOLUTION The satellite's centripetal acceleration is the gravitational acceleration g at the distance r from the center of the earth. Thus we can equate the expression for centripetal acceleration to the expression for g (Eq. 6-4):

$$\frac{v^2}{r} = \frac{GM}{r^2}$$

Solving for v, we obtain

$$v = \sqrt{\frac{GM}{r}} \qquad (6\text{-}6)$$

Fig. 6-9

In order for a satellite to have a circular orbit of radius r, it must have a speed given by this equation, which shows that the orbital speed decreases as the radius of the orbit increases. At a radius of 6.50×10^6 m (about 100 miles above the surface of the earth), we have

$$v = \sqrt{\frac{(6.67 \times 10^{-11}\ \text{N-m}^2/\text{kg}^2)(5.98 \times 10^{24}\ \text{kg})}{6.50 \times 10^6\ \text{m}}}$$
$$= 7.83 \times 10^3\ \text{m/s} \quad (\text{or } 17{,}500\ \text{mi/h})$$

Any object given this initial velocity, directed horizontally, at this elevation will maintain a circular orbit around the earth.

You may calculate the period of the orbit by dividing the circumference by the speed:

$$T = \frac{2\pi r}{v} = \frac{2\pi(6.50 \times 10^6\ \text{m})}{7.83 \times 10^3\ \text{m/s}}$$
$$= 5.22 \times 10^3\ \text{s} \quad (\text{or } 1.45\ \text{h})$$

EXAMPLE 6 Earth's Mass and Density

Use the relationship between G and g to find the mass and average density of the earth.

SOLUTION Eq. 6-5 relates gravitational acceleration g at the surface of the earth to the universal constant G and to the mass and radius of the earth:

$$g = \frac{GM}{R^2}$$

Thus we may use this equation and the known values for g, G, and R to obtain the mass of the earth:

$$M = \frac{gR^2}{G} = \frac{(9.8\ \text{m/s}^2)(6.4 \times 10^6\ \text{m})^2}{6.67 \times 10^{-11}\ \text{N-m}^2/\text{kg}^2}$$

$$= 6.0 \times 10^{24}\ \text{kg}$$

The average density of the earth is found when the mass is divided by the volume of a sphere of radius R:

$$\text{Density} = \frac{M}{\frac{4}{3}\pi R^3} = \frac{6.0 \times 10^{24}\ \text{kg}}{\frac{4}{3}\pi(6.4 \times 10^6\ \text{m})^3}$$
$$= 5.5 \times 10^3\ \text{kg/m}^3$$

The density of water is 1 g/cm^3, or 10^3 kg/m^3. Thus the earth's average density is 5.5 times the density of water. This means that the interior of the earth has a density considerably greater than that of solid surface matter, which typically has a density two or three times that of water. Evidence such as this indicates that the earth's core is metallic.

Other useful geological information is provided by precise measurements of g, which sometimes reveal very slight local variations because of deposits below the surface. For example, oil is often discovered by drilling near salt domes within the earth's crust. These salt deposits have a relatively low density and therefore produce a somewhat reduced value of g directly above the deposit. Ore deposits have relatively high density and produce an enhanced value of g above the deposits.

Fig. 6-10 High and low tides in the Bay of Fundy.

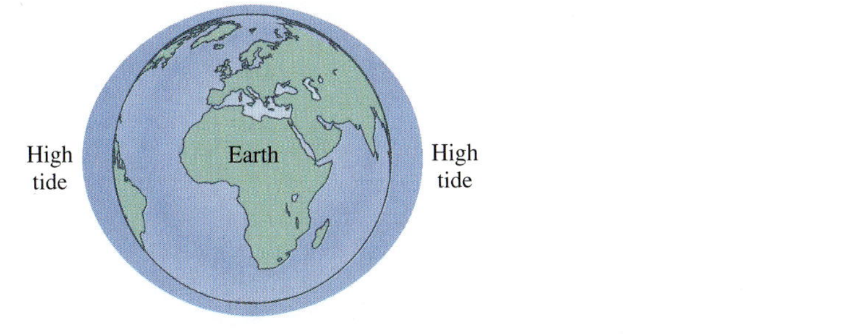

Fig. 6-11 High tides occur simultaneously on opposite sides of the earth.

Ocean tides result from the nonuniform pull of the moon on different parts of the earth, with the part of the earth closest to the moon experiencing the greatest force (Fig. 6-10). High tides occur at a given location whenever the moon is approximately directly overhead. Another high tide occurs simultaneously on the opposite side of the earth (Fig. 6-11), where the oceans experience the least gravitational pull. The combined result of the earth's rotation and the moon's orbital motion is that the moon is directly overhead a given location every 25 hours. Thus there is a high tide once every 12.5 hours.

*6-3 Noninertial Reference Frames

In Chapter 4 we formulated Newton's laws of motion and found that they, as well as all other laws of physics, are valid only in inertial reference frames (Section 4-3). Therefore an inertial reference frame should normally be used in the solution of physics problems. There are times, however, when an observer views events from a noninertial reference frame, and it is useful to understand how such an observer will interpret these events.

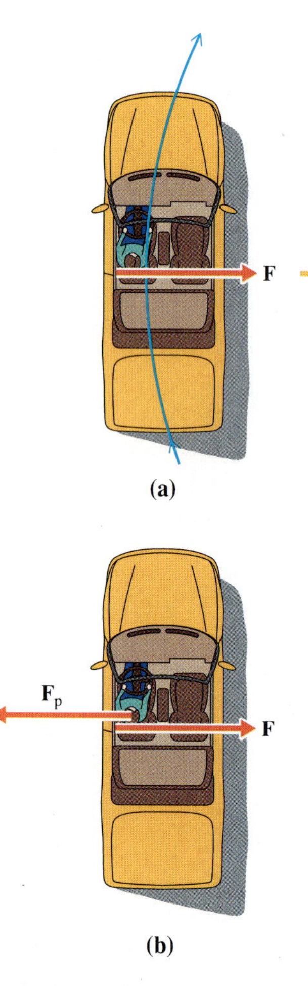

Fig. 6-12 **(a)** As a car rounds a curve, the door exerts a force **F** on the driver, producing centripetal acceleration directed toward the inside of the curve. **(b)** In the noninertial reference frame of the car, there is a pseudoforce $\mathbf{F}_p$ pushing the driver outward and a cancelling force **F** holding him in.

For example, suppose you drive a car fast around a sharp curve and find yourself pushed outward against the car door. From the earth's reference frame, a description of your motion is simple. You experience centripetal acceleration as a result of the force exerted by the door. The centripetal force is directed toward the inside of the curve, as indicated in Fig. 6-12a.

And yet you feel as though something is pushing you outward and as though only the inward force of the door prevents you from flying out, as illustrated in Fig. 6-12b. The outward pull certainly feels like a force, but it is not a force in the strict meaning of the term because there is no physical object producing the effect and therefore no force law to calculate it. The outward push appears only in the noninertial reference frame from which you describe your motion. This is an example of a **pseudoforce.** This particular kind of pseudoforce is called "centrifugal force." **It has the same effect as a true force, but it is not a true force because there is no physical object producing it.**

We shall now derive a general expression for a pseudoforce as a function of the acceleration of a uniformly accelerated reference frame. This discussion is limited to reference frames that are uniformly accelerated with respect to an inertial frame. Thus rotating reference frames are excluded.

Suppose a body of mass m is observed to have an acceleration $\mathbf{a}'$ relative to a noninertial reference frame. Viewed from an inertial reference frame, the body has a different acceleration $\mathbf{a}$, which is related to $\mathbf{a}'$ by the equation

$$\mathbf{a} = \mathbf{a}' + \mathbf{A} \tag{6-7}$$

where $\mathbf{A}$ is the acceleration of the noninertial reference frame relative to the inertial frame.* Applying Newton's second law in the inertial frame, we have

$$\Sigma\,\mathbf{F} = m\mathbf{a}$$

Substituting Eq. 6-7 into this equation, we obtain

$$\Sigma\,\mathbf{F} = m(\mathbf{a}' + \mathbf{A})$$

or

$$\Sigma\,\mathbf{F} - m\mathbf{A} = m\mathbf{a}' \tag{6-8}$$

The term $-m\mathbf{A}$ appears in this equation in the same way a force does. It is a vector quantity that, when added to the true forces acting on the body, equals the product of the body's mass and its observed acceleration $\mathbf{a}'$. In other words, in the noninertial frame the term $-m\mathbf{A}$ has the same effect as a force. Therefore we define it to be a pseudoforce, denoted by $\mathbf{F}_p$:

$$\mathbf{F}_p = -m\mathbf{A} \tag{6-9}$$

Using this definition, we may express Eq. 6-8 as

$$\Sigma\,\mathbf{F} + \mathbf{F}_p = m\mathbf{a}' \tag{6-10}$$

This equation may be applied in noninertial reference frames in the same way that Newton's second law is applied in inertial reference frames.

---

*Eq. 6-7 expresses a relationship between relative accelerations, which is like the relationship between relative velocities. The acceleration of the mass relative to the inertial frame equals its acceleration relative to the noninertial reference frame plus the acceleration of the noninertial frame relative to the inertial frame. This result follows from the corresponding equation for relative velocities.

EXAMPLE 7 Pseudoforce on a Driver Rounding a Curve

A car rounds a curve of radius 100 m at a speed of 20.0 m/s. As viewed from the car, what are the real forces and pseudoforces acting on the 80.0 kg driver, who is stationary relative to the car?

SOLUTION First we must compute the acceleration of the car (our accelerated reference frame) relative to the earth (our inertial frame). The car's centripetal acceleration **A** is directed toward the inside of the curve, as shown in Fig. 6-12a, and has magnitude

$$A = \frac{v^2}{r} = \frac{(20.0\ \text{m/s})^2}{100\ \text{m}} = 4.00\ \text{m/s}^2$$

So the driver experiences an outwardly directed pseudoforce $\mathbf{F}_p$ (shown in Fig. 6-12b) of magnitude

$$F_p = mA = (80.0\ \text{kg})(4.00\ \text{m/s}^2) = 320\ \text{N}$$

Since the driver does not move relative to the car, his acceleration relative to this noninertial reference frame is zero:

$$\mathbf{a}' = \mathbf{0}$$

According to Eq. 6-10, the pseudoforce $\mathbf{F}_p$ must be balanced by an opposing force **F**, such that

$$\mathbf{F} + \mathbf{F}_p = m\mathbf{a}' = \mathbf{0}$$

The force **F** is an inwardly directed, real force of magnitude 320 N. Either the seat or the car door provides this force.

Notice that the pseudoforce on an object is directly proportional to its mass. In this respect, a pseudoforce is like a gravitational force, that is, a weight. We can describe the vector sum of the true weight **w** and the pseudoforce $\mathbf{F}_p$ as the apparent weight, denoted by $\mathbf{w}_a$:

$$\mathbf{w}_a = \mathbf{w} + \mathbf{F}_p = m\mathbf{g} - m\mathbf{A} = m(\mathbf{g} - \mathbf{A})$$

In a noninertial frame, the vector quantity $\mathbf{g} - \mathbf{A}$ appears in the force law for apparent weight in the same way **g** appears in the force law for true weight in an inertial frame. Therefore we define $\mathbf{g} - \mathbf{A}$ to be the apparent gravitational acceleration and denote it by $\mathbf{g}_a$:

$$\mathbf{g}_a = \mathbf{g} - \mathbf{A} \qquad (6\text{-}11)$$

Then we may express the apparent weight as

$$\mathbf{w}_a = m\mathbf{g}_a \qquad \text{(apparent weight)} \qquad (6\text{-}12)$$

When no force other than gravity acts on a body, its observed acceleration in a noninertial reference frame will be $\mathbf{g}_a$. You can see this by applying the equation of dynamics, Eq. 6-10:

$$\Sigma\,\mathbf{F} + \mathbf{F}_p = m\mathbf{a}'$$

Since there is no true force other than gravity in this case and since we denote the combination of the true gravitational force and the pseudoforce by $\mathbf{w}_a$, we have

$$\mathbf{w}_a = m\mathbf{a}'$$

Applying Eq. 6-12, we obtain

$$m\mathbf{g}_a = m\mathbf{a}'$$

or

$$\mathbf{a}' = \mathbf{g}_a$$

All bodies in free fall are observed to have the same acceleration $\mathbf{g}_a$ in a noninertial frame, independent of their mass.

EXAMPLE 8 Apparent Gravity in a Centrifuge

The sedimentation rate of particles suspended in a liquid is proportional to g. This rate can be greatly increased when a centrifuge is used to create an apparent gravitational acceleration many times greater than g. An ultracentrifuge rotates at the rate of 1.0×10^3 rev/s in a horizontal plane. What is the apparent gravitational acceleration g_a in the centrifuge at a point 5.0 cm from the axis of rotation?

SOLUTION For a point P in the centrifuge, the acceleration **A** relative to the earth is indicated in Fig. 6-13. We compute the magnitude of this centripetal acceleration, using the rotational period $T = 1.0 \times 10^{-3}$ s:

$$A = \frac{v^2}{r} = \frac{(2\pi r/T)^2}{r} = \frac{4\pi^2 r}{T^2}$$

$$= \frac{4\pi^2(0.050 \text{ m})}{(1.0 \times 10^{-3} \text{ s})^2}$$

$$= 2.0 \times 10^6 \text{ m/s}^2$$

The apparent gravitational acceleration $\mathbf{g}_a$ can be computed using Eq. 6-11: $\mathbf{g}_a = \mathbf{g} - \mathbf{A}$. The vector subtraction is indicated in the figure. Since A is much greater than g in this example, we may ignore g, so that

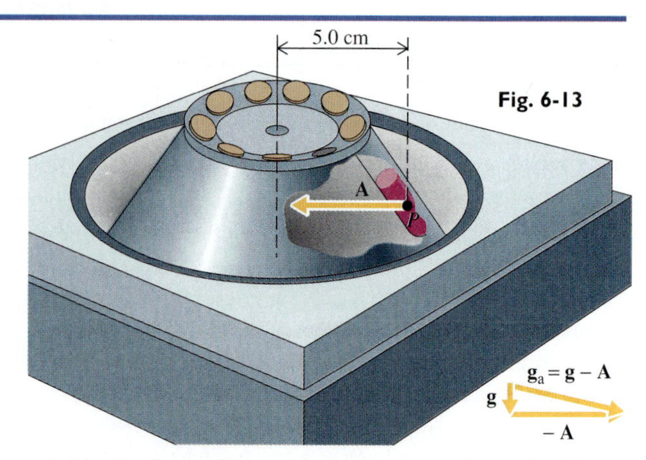

Fig. 6-13

$$\mathbf{g}_a \approx -\mathbf{A}$$

The apparent gravitational acceleration vector is directed radially outward (opposite **A**) and has a magnitude of 2.0×10^6 m/s², or about 200,000 times g. Thus sedimentation of suspended particles should be greatly enhanced in such a centrifuge.

EXAMPLE 9 Apparent Weightlessness in Orbit

Compute the true weight and the apparent weight of an astronaut of mass 75 kg, orbiting the earth in a space station 500 km above the earth's surface (Fig. 6-14).

SOLUTION We can easily calculate the astronaut's true weight by using the force law $w = mg$, but first we must find g. We apply Eq. 6-4, with a distance r equal to the sum of the earth's radius (6.4×10^6 m) plus the elevation of 5×10^5 m:

$$g = \frac{GM}{r^2}$$

$$= \frac{(6.67 \times 10^{-11} \text{ N-m}^2/\text{kg}^2)(6.0 \times 10^{24} \text{ kg})}{(6.4 \times 10^6 \text{ m} + 5 \times 10^5 \text{ m})^2}$$

$$= 8.4 \text{ m/s}^2$$

Applying the force law for weight, we obtain

$$w = mg = (75 \text{ kg})(8.4 \text{ m/s}^2)$$

$$= 630 \text{ N} \quad (\text{or } 140 \text{ lb})$$

This is somewhat smaller than the astronaut's true weight at the earth's surface, which is 75 kg × 9.8 m/s² = 740 N, or 170 lb.

The orbiting satellite is a noninertial reference frame that experiences an acceleration **g** relative to the earth. So the apparent gravitational acceleration inside the satellite is zero:

$$\mathbf{g}_a = \mathbf{g} - \mathbf{A}$$

$$= \mathbf{g} - \mathbf{g}$$

$$= 0$$

Fig. 6-14

Thus the apparent weight of the astronaut is zero:

$$\mathbf{w}_a = m\mathbf{g}_a = 0$$

All objects in the satellite float in apparent weightlessness. The effect within the satellite is exactly the same as though there were no gravitational force. For purposes of describing an experiment confined to the satellite, you can treat it as an inertial reference frame by simply ignoring the interaction with the earth.

Earth as an Inertial Reference Frame

A reference frame attached to the surface of the earth is only approximately inertial and this produces measurable effects. A truly inertial reference frame is one that is unaccelerated with respect to the distant matter in the universe. Our galaxy moves relative to this distant matter, and our sun moves relative to the galaxy. The magnitude of the accelerations associated with these motions is negligible, however (see Chapter 3, Problem 39). The orbital and rotational motions of the earth produce greater accelerations and should be considered.

The centripetal acceleration of the earth in its orbital motion about the sun is 6×10^{-3} m/s^2 (Chapter 3, Problem 40). Each particle on the earth has essentially this same acceleration with respect to the sun, produced by the sun's gravitational attraction. So, for purposes of describing phenomena on earth, we may ignore the orbital motion of the earth by completely ignoring its interaction with the sun. This is similar to the case of the orbiting space station in Example 9.

Consideration of the earth's rotational motion is more complicated. Rotating reference frames are beyond the scope of the present discussion. However, if we consider only objects that are either fixed or else moving slowly over a very small part of the earth's surface, we may treat a local reference frame attached to the earth's surface as being uniformly accelerated. The value of the centripetal acceleration is a function of latitude. At the poles it is zero, at 40° latitude it is 2.6×10^{-2} m/s^2 (Chapter 3, Problem 41), and at the equator it is 3.4×10^{-2} m/s^2.

The experimentally measured value of gravitational acceleration results from the combined effect of the earth's gravitational force and its centripetal acceleration $\mathbf{A}$. On the surface of the earth, we observe an apparent gravitational acceleration $\mathbf{g}_a = \mathbf{g} - \mathbf{A}$, where $\mathbf{g}$ is the gravitational acceleration relative to the earth's axis, the magnitude of which is computed as GM/R^2. At the equator, $\mathbf{g}$ and $\mathbf{A}$ are both in the same direction, directed toward the center of the earth. The vector subtraction in this case is very simple. Taking g to be the value of gravitational acceleration observed at the poles (9.83 m/s^2), we find at the equator

$$\begin{aligned} g_a = g - A &= 9.83 \text{ m/s}^2 - 3.4 \times 10^{-2} \text{ m/s}^2 \\ &= 9.80 \text{ m/s}^2 \end{aligned}$$

Actually, centripetal acceleration accounts for only about two thirds of the observed variation in gravitational acceleration over the surface of the earth. We have assumed that the earth is a perfect sphere, but in fact it has a slightly nonspherical shape, bulging at the equator because of the apparent force produced by the rotation on its axis. As a result, a person standing on the equator is farther away from the earth's center than a person standing at any other latitude, and this causes g to be smaller at the equator.

Origins of the Theory of Universal Gravitation

To appreciate Newton's work it is necessary to understand something of the theories preceding it. The story begins early in recorded history. Although many societies developed mythological descriptions of the universe (Fig. 6-A), the ancient Greeks observed the stars carefully and described their observations using various models. In the third century B.C., Aristarchus formulated a system in which the earth rotated on its axis and revolved about the sun. Unfortunately his model was promptly rejected and forgotten. No details of his system are known today; only fragmentary references to it by other writers survive.

In other models of the universe, the Greeks asserted what must have seemed a certainty—that the earth is stationary and that the sun, moon, and stars move. From the time of Plato, these models were based on the Platonic dictum that the motion of celestial bodies must be in perfect circles at uniform speed. The fact is that the paths of other planets, as seen by an observer on earth, are not even approximately circular. Greek astronomers later formulated complex systems in which the motion of each body was a combination of circular motions, or "epicycles." About A.D. 150 a very accurate system of this type was perfected by Ptolemy. However, Ptolemy had to sacrifice the ideal of uniform speed to achieve this accuracy.

The Ptolemaic system of the universe was forgotten during the Dark Ages but then rediscovered by Europe in the twelfth century and soon became firmly established.

In 1543, as Nicolaus Copernicus lay on his deathbed, the work he had kept to himself 30 years for fear of ridicule was at last published. Copernicus had formulated a system that coincides qualitatively with the modern view of the solar system: all the planets, including earth, orbit the sun, while the moon orbits the earth.

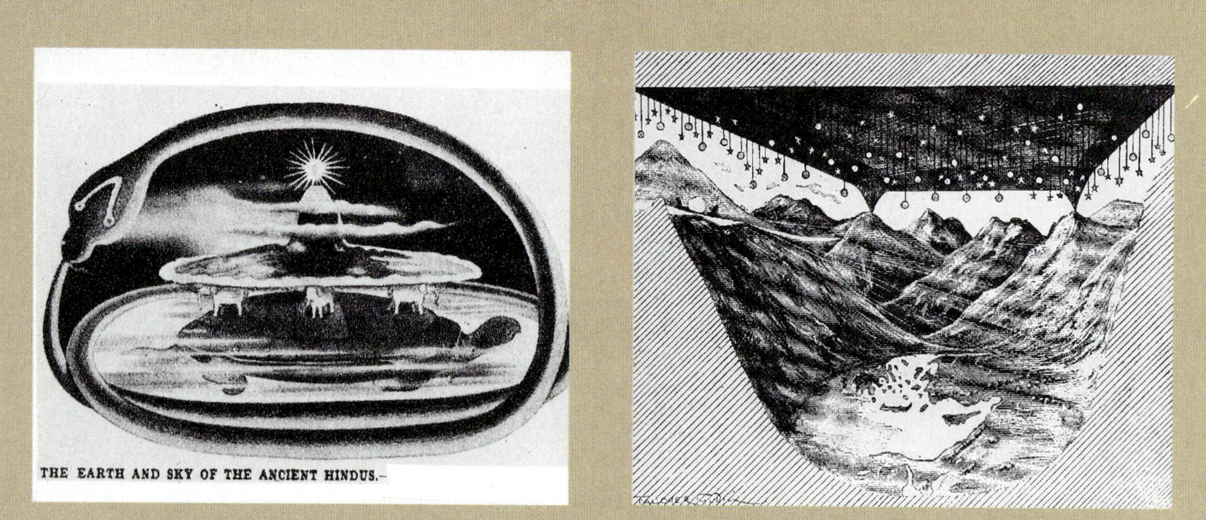

Fig. 6-A **(a)** The ancient Hindu universe was enclosed by a giant cobra. The tortoise, symbol of force and creative power, floated on a sea of milk. Upon its back stood four elephants supporting the earth at the east, west, north, and south points of the horizon. **(b)** In the ancient Egyptian universe, star lamps, suspended from the ceiling of the sky, were lighted and extinguished by the gods.

In recognizing the earth's motion, Copernicus made a great advance; however, he was still limited by the self-imposed constraint that all celestial motion should be circular. Like Ptolemy, he used epicycles. The Copernican system offered no greater simplicity or accuracy, and so there was no compelling reason to prefer it to the Ptolemaic system.*

During the last quarter of the sixteenth century, the Danish nobleman Tycho Brahe provided what was needed to resolve the problem—careful observation of the planets over an extended period of time. Brahe, whose island observatory was a gift from the king of Denmark, hoped to use his data to verify his own model of the solar system, in which the sun orbited the earth and all other planets orbited the sun. Brahe was a commanding, dictatorial figure—a rich nobleman who wore a silver nose because of a dueling injury and who had a passionate devotion to both astronomical observation and verification of his own system.

Kepler's Laws

Toward the end of his life, Brahe was joined in his work by Johannes Kepler. Kepler was young, brilliant, and fanatically committed to verifying the Copernican system. There was great personal strife between the two men, but each realized that he needed the other. Brahe possessed the best experimental data, and Kepler had the theoretical ability to interpret it. Kepler was given the difficult task of describing the orbit of Mars using Brahe's data. Of all the outer planets known at that time, Mars has the least circular orbit and was therefore the least able to fit into any circular scheme.

Kepler believed in the correctness of the Copernican system, though not in its details. From the beginning he rejected the Copernican epicycles and the assumption of constant speed; however, he clung to the idea of circular orbits. Finally, after 6 years of struggling through wrong guesses and calculational error, Kepler realized that the shape of the Martian orbit was elliptical. At last the battle was won. The earth and all the other planets orbit the sun, but the orbits are ellipses, not circles. This was the only system in close agreement with Brahe's precise astronomical data. In other words, the sun is the preferred reference frame for the description of planetary motion.

Kepler summarized his laborious study of planetary motion with the following three laws:

I The planets move in elliptical orbits with the sun at one focus.

II A line from the sun to any planet sweeps out equal areas in equal times.

III The square of a planet's period* is proportional to the cube of the planet's mean distance from the sun.

The primary importance of Kepler's laws of planetary motion is the basis they provided for Newton. Kepler regarded his laws as a first step toward his goal of describing the universe in a way that would show an interrelationship between astronomy, astrology, mathematics, and music.*

In *A New Astronomy,* the work in which Kepler announced the first two of his laws, there is also the first serious attempt to explain celestial motion in terms of physical causes. Kepler believed that there was some kind of force exerted on planets by the sun and that this force weakens with increasing distance from the sun, accounting for the outer planets' longer periods. His ideas were incomplete, however, and for more than 80 years after Kepler's discovery, no one was able to formulate a system of gravitation that explained his laws. Then in 1687 Isaac Newton published the *Principia,* containing the law of universal gravitation, which accounted for Kepler's laws and much more.

But the essence of Newton's discovery had been made 22 years earlier. In 1665 he was a 23-year-old student who had left Cambridge University and secluded himself at Woolsthorpe in the English countryside to escape the Great Plague that ravaged London. In that quiet setting he solved perhaps the most important scientific problem of all time: he discovered universal gravitation and convinced himself that it was responsible for planetary motion.

*Copernicus apparently considered his system an improvement because bodies moved along circular paths at *constant speed.*

*The period of a planet is the time for one complete orbit, or, in other words, the length of its year.

*Kepler had a mystical fascination with numbers. This, combined with his interest in astrology, led him to determine what he claimed to be the exact minute of his conception.

Isaac Newton (1642–1727)

Isaac Newton was born to an English farm family on Christmas day, 1642, the year of Galileo's death. Newton's father died before he was born, and when Isaac was 3 years old, his mother remarried and moved away, leaving him to be raised by his grandmother. Perhaps it was this early isolation that fostered Newton's introspective personality and began to develop his capacity for intense concentration.

As a teenager, Newton often amused himself by constructing mechanical gadgets and toys. One of these was a lantern-bearing kite that resembled a comet when he flew it at night, to the bewilderment of the local villagers. Another was a working model of a mill for grinding wheat, with a mouse as a source of power. Mechanical tinkering gradually gave way to scientific investigations. Newton was later to recount how his first scientific experiment occurred on his seventeenth birthday, during a storm. He decided to measure the strength of the wind by long jumping. He jumped first in the direction of the wind and then against it. The difference in the distances jumped was his measure of the wind's strength.

Newton rejected his mother's plans for him to become a farmer; instead he entered Cambridge at 18 years of age. Just after he completed his bachelor's degree, Cambridge was closed because of the bubonic plague in London, and he returned home. During the next 18 months he had time for quiet reflection. The result was an explosion of creativity. At this, the beginning of his scientific career, he conceived the ideas for most of his major discoveries—differential and integral calculus, a theory of color, centripetal force, and the inverse-square law of gravity. However, none of this work was published at that time.

In 1667 he returned to Cambridge, continued his study of light, and two years later was appointed professor of mathematics. In 1672 Newton published his *New Theory about Light and Colors*. The article was immediately attacked by Robert Hooke, a famous scientist of the day. Hooke claimed priority for most of the discoveries reported in Newton's paper. This was the beginning of a lifelong antagonism between the two men. Newton found the dispute so annoying that he decided to avoid further public scrutiny of his work by withholding it from publication.

One of the most important scientific problems of this period was the nature of the force responsible for the planetary motion described by Kepler's laws. The philosopher and mathematician René Descartes had developed a theory of attraction based on invisible whirlpools connecting the sun and the planets. A number of scientists, including Hooke, had speculated that the planetary force might be mathematically described by an inverse-square law. In 1684 Newton was approached by his friend Edmund Halley, who asked what the orbit of a planet would be if the planet were attracted to the sun by a force with an inverse square distance dependence. Newton replied that he had calculated such an orbit and found it to be an ellipse. Halley recognized the importance of this calculation and urged Newton to publish it. Through Halley's coaxing, Newton completed his work on gravitation, and the *Principia* was published in 1687 at Halley's expense.

Inspection of the thousands of pages of Newton's manuscript, full of multiple corrections and rewritings, reveals his attention to detail. His ability to concentrate the powers of his intellect for extended periods of time was one of his chief characteristics. He once described his method of discovery: "I keep the subject of my inquiry constantly before me, and wait till the first dawning opens gradually, by little and little, into a full and clear light."

The physical scope and mathematical virtuosity of the *Principia* immediately made an enormous impression on European science. However, it was not free from criticism. In England Hooke claimed priority for the inverse-square law. In France followers of Descartes defended the Cartesian theory. And in Holland the scientist Christian Huygens objected to the lack of an intuitive, mechanistic explanation of gravity, as opposed to a mathematical one: "I am by no means satisfied by . . . his Principle of Attraction, which to me seems absurd. . . . And I have often wondered how he could have given himself all the trouble of making such a number of investigations and difficult calculations that have no other foundation than this very principle."

There is no doubt that Newton believed that there was some sort of underlying mechanism for gravity. In a letter to a friend, he wrote:

> That gravity should be innate, inherent, and essential to matter, so that one body may act upon another at a distance through a vacuum, without the mediation of anything else, by and through which their action and force may be conveyed from one to another, is to me so great an absurdity, that I believe no man, who has in philosophical matters a competent faculty of thinking, can ever fall into it.

Despite repeated attempts, Newton was never able to arrive at a satisfactory mechanistic explanation of gravity. He stated in another letter: ". . . the cause of gravity is what I do not pretend to know, and therefore would take more time to consider of it." In the *Principia* Newton was careful to avoid speculative notions concerning the possible "causes" of gravity; his postulates were simple, and his deductions rigorously derived and amply supported by experiment, in sharp contrast to Descartes's writing on gravity. Followers of Newton were to accept the phenomenon of universal gravitational attraction as an inherent and fundamental property of matter, requiring no explanation.

Newton left Cambridge and accepted a government position as director of the Mint in 1699, at the age of 56. From this time until his death at age 85, he did little scientific work. But even in his later years, he possessed astounding mathematical abilities. On two separate occasions, he was given the challenge of solving mathematical problems which for months the greatest mathematicians of Europe had failed to solve. On both occasions he produced the solution in a single evening.

In 1712 the German mathematician Leibniz claimed that some of Newton's statements concerning the absolute nature of space led to atheistic beliefs. This must have deeply offended Newton, for he was a religious man and very much concerned with theological matters. Other theologians and philosophers had hailed the *Principia* as a kind of divine blueprint for the structure of the universe. In 1692 Newton had stated, "When I wrote my treatise about our system, I had an eye on such principles as might work with considering men for the belief of a Deity; and nothing can rejoice me more than to find it useful for that purpose." In the third edition of the *Principia*, which appeared in 1713, Newton responded to the charges of atheistic implications of his concepts with several pages devoted to a discussion of the nature of God and His relationship to space and time. One hundred years later, the French scientist Laplace was to base a deterministic philosophy on Newton's system. Laplace considered God an unnecessary "hypothesis." He wrote:

> Given for one instant an intelligence which could comprehend all the forces by which nature is animated and the respective positions of the beings which compose it, if moreover this intelligence were vast enough to submit these data to analysis, it would embrace in the same formula both the movements of the largest bodies in the universe and those of the lightest atom: to it nothing would be uncertain, and the future as the past would be present to its eyes.

On the basis of classical mechanics, Laplace's statement seems justified. However, the twentieth-century discovery of quantum mechanics has removed the scientific foundation for this deterministic philosophy and injected an inherent element of uncertainty. Although no longer regarded as our most fundamental picture of nature, Newtonian mechanics remains an intellectual triumph of immense importance in our continuing struggle to understand the physical universe.

CHAPTER 6 SUMMARY

The law of universal gravitation states that a mutually attractive force acts between any two particles a distance r apart. The magnitude of the force is

$$F = \frac{Gmm'}{r^2}$$

where the universal gravitational constant $G = 6.672 \times 10^{-11}$ N-m²/kg².

The earth, because of its spherically symmetric distribution of mass, attracts objects as though all of its mass were concentrated at its center. The weight of any object is the magnitude of this attractive force and may be expressed as a function of the object's mass m, the earth's mass M, and the distance r between the earth's center and the object:

$$w = mg$$

where

$$g = \frac{GM}{r^2}$$

At or near the earth's surface, r is approximately the radius of the earth and g is 9.8 m/s².

Pseudoforces arise in noninertial reference frames. In a noninertial frame that has a uniform acceleration **A** with respect to an inertial frame, any body of mass m will experience a pseudoforce $\mathbf{F}_p$, where

$$\mathbf{F}_p = -m\mathbf{A}$$

The combined effect of this pseudoforce and the true weight of an object is its apparent weight $\mathbf{w}_a$ in the noninertial frame. Apparent weight is proportional to apparent gravitational acceleration $\mathbf{g}_a$:

$$\mathbf{w}_a = m\mathbf{g}_a$$

where $\mathbf{g}_a$ is found by vector subtraction:

$$\mathbf{g}_a = \mathbf{g} - \mathbf{A}$$

Questions

1 The earth exerts a gravitational force of 2.0×10^{20} N on the moon. Is the force exerted by the moon on the earth (a) less; (b) greater; (c) the same?

2 Suppose the masses of two particles were both increased to three times their original values. By what factor would the gravitational force between them change: (a) $\frac{1}{3}$; (b) $\frac{1}{6}$; (c) $\frac{1}{9}$; (d) $\frac{1}{27}$?

3 Suppose the distance between two particles were decreased to half the original value. By what factor would the gravitational force between them change: (a) $\frac{1}{2}$; (b) $\frac{1}{4}$; (c) 2; (d) 4?

4 A free-fall experiment is performed on Mars. An object is released from rest and falls through a vertical distance of 2 m in 1 s. On earth the same object falls 5 m in 1 s. How much would a person who weighs 500 N on earth weigh on Mars: (a) 200 N; (b) 300 N; (c) 500 N?

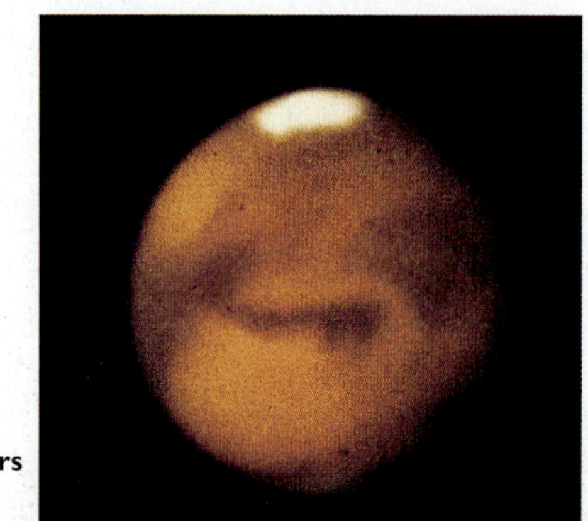

Mars

5 The average distance of the planet Saturn from the sun is roughly nine times the average distance of the earth from the sun. Approximately how long is a year on Saturn in units of earth years: (a) 1 y; (b) 3 y; (c) 9 y; (d) 27 y; (e) 81 y?

Saturn

6 Suppose a planet has a radius twice as large as the earth's radius but the same mass. How much would an object weighing 100 N on earth weigh on this planet: (a) 100 N; (b) 200 N; (c) 50 N; (d) 25 N?

7 Suppose a planet has a radius twice as large as the earth's radius but the same density. How much would an object weighing 100 N on earth weigh on this planet: (a) 100 N; (b) 200 N; (c) 800 N?

8 A man weighs 800 N at the surface of the earth. What would his weight be at the center of the earth: (a) 0; (b) 800 N; (c) 400 N?

9 Is it possible to shield a body from the gravitational force of another body by placing something between the two?

10 You are at a point midway between two stars of equal mass M in a binary star system. Suppose your weight on earth is 980 N. At that point what are your weight and mass?

11 Suppose you and a friend are orbiting the earth in separate satellites but in the same circular orbit. You are some distance ahead of your friend, and you want to send her a package. Assuming you could give the package any desired velocity, which of the following would work: (a) just release the package so that she could pick it up as she comes around; (b) throw it backward with a velocity such that it has zero velocity relative to the earth; (c) throw it backward with a velocity such that its velocity relative to earth is the negative of the satellite's velocity?

12 The sun as well as the moon contributes to ocean tides, but the moon's effect is greater than the sun's even though the sun exerts a larger gravitational force on the earth. Explain.

13 You are at the beach and see a full moon directly overhead. Would you expect the high tide to be typical, unusually high, or unusually low? Explain.

14 Imagine you are riding on a Ferris wheel, moving along a circular path in the vertical plane. What is the direction of the pseudoforce acting on you when you are at the highest point on the ride: (a) up; (b) down; or (c) horizontal?

15 During a jump off a diving board, is your apparent weight (a) equal to your true weight, (b) slightly less than your true weight, (c) slightly more than your true weight, or (d) zero?

Answers to Odd-Numbered Questions

1 c; **3** d; **5** d; **7** b; **9** no; **11** c; **13** unusually high; **15** d

Problems (listed by section)

6-1 Universal Gravitation

Kepler's Third Law

1 The average distance of the planet Mercury from the sun is 0.39 times the average distance of the earth from the sun. How long is a year on Mercury in units of earth years?

2 Mercury is the planet in our solar system closest to the sun, whereas Pluto is usually the planet farthest from the sun. The average distance of Pluto from the sun is approximately 100 times the average distance of Mercury from the sun. How long is a year on Pluto in units of "Mercury years"?

Mercury

Law of Universal Gravitation

3 Find the gravitational force exerted by a 0.500 kg mass on a 0.100 kg mass 2.00 m away.

4 How close would the two masses in the previous problem have to be in order for the gravitational force between them to have a magnitude of 1.00 N? Is this distance realistic?

5 Two particles each weigh 1.00 N. Find the magnitude of their mutual gravitational force when they are separated by a distance of 20.0 cm.

6 John weighs 710 N, and Marcia weighs 535 N. Estimate the gravitational force between them when they are 0.50 m apart.

7 A body of mass m is attracted toward a 10.0 kg mass, 30.0 cm away, with a force of magnitude 6.67×10^{-8} N. Find m.

8 Two particles A and B, with respective masses m_A and m_B, experience a mutual gravitational attraction. Show that the acceleration of A is independent of m_A but proportional to m_B and that the acceleration of B is independent of m_B but proportional to m_A.

9 Astrologers believe that the positions of the stars at the instant of a person's birth affect personality and that the mechanism for this supposed effect is gravitation. Suppose a baby is born under the constellation Scorpius, a group of 18 stars approximately 400 light-years (3.8×10^{18} m) away from the earth. The largest star in Scorpius is Antares, which has a mass of 2.0×10^{31} kg.

(a) Calculate the gravitational force exerted by Antares on the newborn "Scorpio" baby, whose mass is 4.0 kg.

(b) For comparison, calculate the gravitational force exerted on the baby by the doctor, who is 1.0 m away and has a mass of 60.0 kg.

10 Find the resultant force on (a) the 0.100 kg mass and (b) the 0.200 kg mass in Fig. 6-15 (the masses are isolated from the earth).

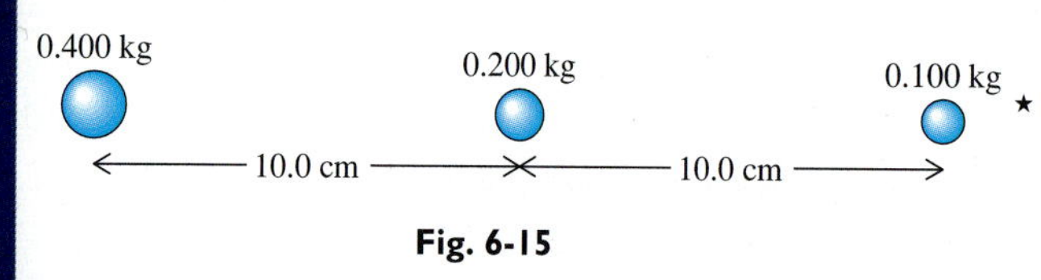

Fig. 6-15

★11 Three isolated particles, each having a mass of 2.00 kg, are at the vertices of an equilateral triangle with 1.00 m sides. Find the magnitude and direction of the resultant gravitational force on each particle.

★12 A spaceship of mass 1.0×10^6 kg is accelerated at a rate of 1.0 m/s² toward a binary star, which consists of two stars of equal mass m, as shown in Fig. 6-16. Find the mass m of each star.

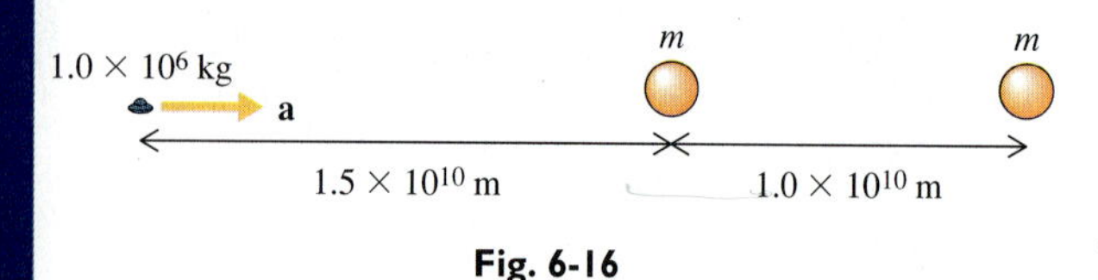

Fig. 6-16

★13 A particle of mass m is between a 1.00×10^2 kg mass and a 4.00×10^2 kg mass, which are 10.0 m apart. Find the distance of the particle from the 100 kg mass, such that the resultant force on the particle is zero.

14 Find the resultant gravitational force on the particle at the origin in Fig. 6-17. The particles are isolated.

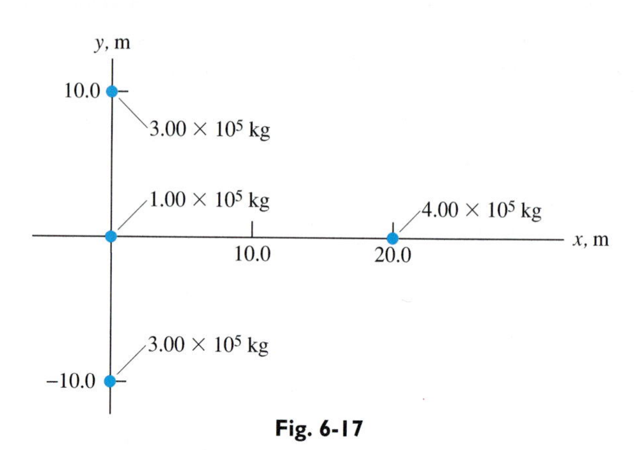

Fig. 6-17

Measurement of G

15 Two spheres of masses $m = 1.00$ g and $m' = 1.00 \times 10^2$ kg are isolated from all other bodies and are initially at rest, with their centers a distance $r = 15.0$ cm apart. One minute later, the smaller sphere has moved 0.534 mm toward the larger sphere. Compute the acceleration and the value of G.

★16 The centers of two 1.00×10^3 kg spheres are initially at rest, isolated from all other bodies and separated by a distance of 1.00 m. The spheres move toward each other because of gravitational attraction. After 7.00 minutes, they are 1.20 cm closer. Determine G from these data.

6-2 Gravitational Attraction of the Earth

Spherical Distribution of Mass

17 Jupiter has a radius of 7.14×10^7 m and a mass of 1.90×10^{27} kg.

(a) Find the force on a 1.00 kg mass at the surface of Jupiter.

(b) Find the magnitude of gravitational acceleration on Jupiter.

(c) How far would an object released from rest fall during a 1.00 s interval?

★18 How far above the surface of the earth would you have to be before your weight is reduced by 10%?

★19 Compute the gravitational acceleration on the moon and on Mars, given the following data:

$$m_{\text{moon}} = 0.0123 m_{\text{earth}}$$
$$m_{\text{Mars}} = 0.107 m_{\text{earth}}$$
$$R_{\text{moon}} = 0.272 R_{\text{earth}}$$
$$R_{\text{Mars}} = 0.530 R_{\text{earth}}$$

20 (a) A 1.00 kg mass is at the center of a uniform spherical shell of mass 1.00×10^{20} kg and radius 1.00×10^{6} m. Find the resultant gravitational force on the mass.

(b) Find the resultant gravitational force on the mass if it is placed just outside the shell.

(c) Find the value of g just outside the shell.

21 A boy of mass 50.0 kg stands on a plain near a mountain range that rises steeply from the plain. Estimate the magnitude of the gravitational force **F** exerted on the boy by a nearby mountain. Treat the mountain as a uniform sphere of mass 1.00×10^{14} kg and radius 2.00×10^{3} m (Fig. 6-18).

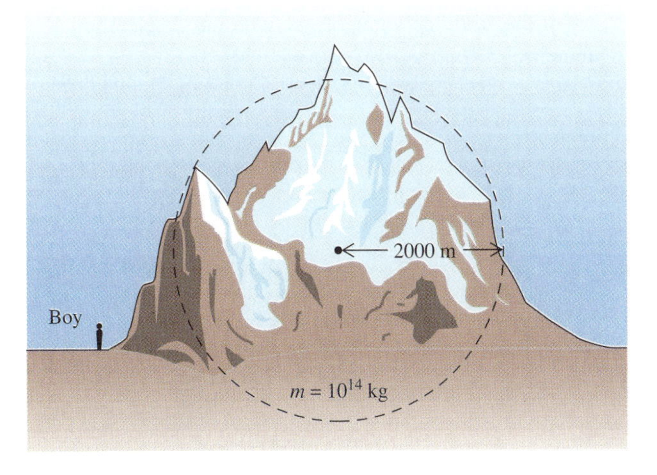

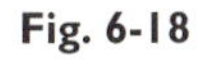
Fig. 6-18

Satellites

22 (a) Find the speed of a satellite moving around the earth in a circular orbit that has a radius equal to twice the earth's radius of 6.38×10^{6} m.

(b) Find the satellite's orbital period.

23 (a) Find the speed of a satellite moving in a circular orbit just above the surface of the moon. The moon has a radius of 1.74×10^{6} m and a mass of 7.36×10^{22} kg.

(b) Find the satellite's orbital period.

★ **24** Compute the orbital radius of an earth satellite that has an equatorial orbit and always remains above a fixed point P on the earth's surface. Communication satellites have such "geosynchronous" orbits (Fig. 6-19). These satellites are used to relay radio and television signals around the world.

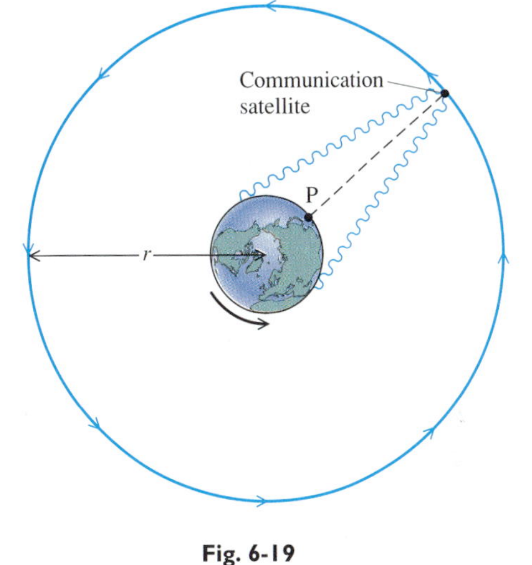

Fig. 6-19

25 The earth orbits the sun in an approximately circular orbit of radius 1.496×10^{11} m (about 93 million miles) in a period of $365\frac{1}{4}$ days. Use these data to determine the mass of the sun.

26 The sun orbits the center of the Milky Way galaxy (Fig. 6-20) in a period of 2.5×10^{8} years.

(a) Assuming a circular orbit of radius 2.4×10^{20} m, compute the mass of the Milky Way.

(b) Estimate the number of stars in the Milky Way, assuming the average star mass equals the mass of the sun—2.0×10^{30} kg.

(a)

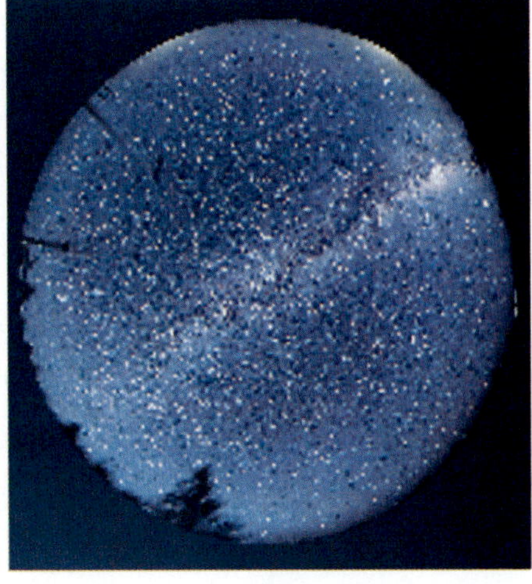
(b)

Fig. 6-20 The Milky Way as seen from **(a)** space and **(b)** earth (an extreme wide angle, 360° view).

*6-3 Noninertial Reference Frames

27 The driver of a car rounds a level, circular curve of radius 10.0 m at a speed of 8.00 m/s.
(a) Find the magnitude and direction of the pseudoforce acting on the driver, whose mass is 50.0 kg.
(b) What is the magnitude of the driver's apparent weight?

28 As a car is rounding the top of a certain hill, its motion is approximately circular (in a vertical plane) with a radius of curvature of 25.0 m. Find the apparent weight of the 575 N driver if the speed of the car is 50.0 km/h.

29 An elevator is moving upward. Find the apparent weight of a person inside, whose true weight is 7.00×10^2 N, if the elevator is (a) accelerating at 2.00 m/s^2; (b) decelerating at 5.00 m/s^2.

⋆ **30** An object is dropped from the ceiling of a train to the floor 2.00 m below. At the instant the object is released, the train's speed is 30.0 m/s. During the fall the train decelerates at a rate of 4.00 m/s^2. Describe the motion of the object from (a) the earth's reference frame and (b) the reference frame of the train.

⋆ **31** Find the apparent weight of a passenger of mass 85.0 kg on a roller coaster when (a) the track is straight and sloping downward at an angle of 45.0° with the horizontal; (b) the track is level but curving inward with a radius of 30.0 m and the coaster moves at 25.0 m/s; (c) the track is concave upward with a radius of curvature of 30.0 m and the coaster moves at 25.0 m/s; (d) the track is concave downward with a radius of curvature of 30.0 m and the coaster moves at 25.0 m/s.

32 In the film *2001: A Space Odyssey,* a space station in the shape of a large wheel rotates at a rate sufficient to produce an apparent gravitational acceleration equal to 9.80 m/s^2 (Fig. 6-21). Find the period of rotation.

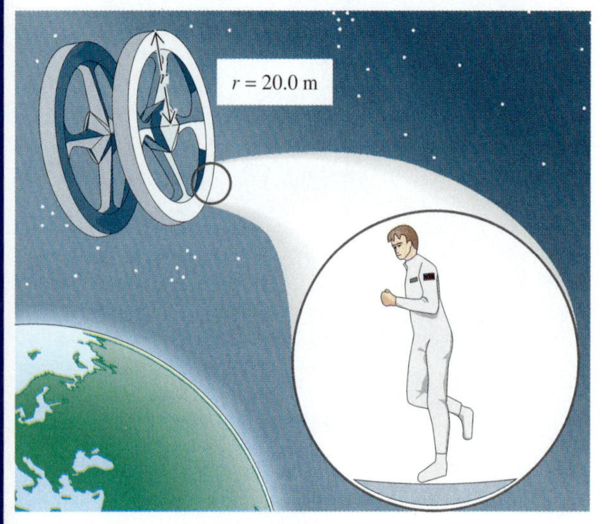

Fig. 6-21

Additional problems

⋆ **33** The gravitational constant would be easy to measure if ordinary matter were more dense. What would the density of a sphere of radius 1.00 m have to be in order to produce a gravitational acceleration of 1.00 m/s^2 at the surface of the sphere?

⋆ **34** Two 1.00 kg spheres are in orbit around the earth. Initially the spheres are 1.00 m apart (center to center) and are not moving with respect to each other. How long will it take for the spheres to move 1 cm closer to each other?

⋆⋆ **35** Calculate the apparent value of gravitational acceleration at 40° N latitude assuming that relative to the center of the earth the value of gravitational acceleration there is 9.83 m/s^2.

36 Find the resultant gravitational force on the dumbbell shown in Fig. 6-22. Ignore the mass of the connecting bar.

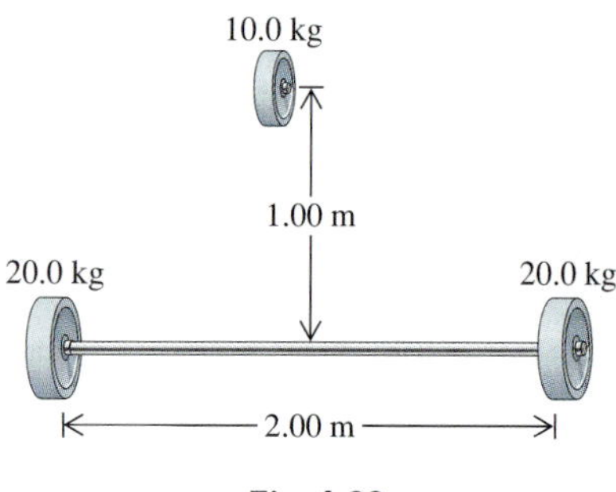

Fig. 6-22

⋆⋆ **37** On a certain amusement park ride, people stand against the walls of a cylindrical room of radius 3.00 m. The cylinder begins to rotate. When the rate of rotation becomes sufficiently great, the floor is removed and the riders remain pinned to the walls. If the coefficient of friction between riders and wall is 0.400, how fast must the cylinder rotate before the floor is removed? What is the apparent value of g inside?

⋆ **38** Compute the resultant gravitational force exerted on the moon by the sun and the earth, when they are aligned as shown in Fig. 6-23. Why can we ignore the force of the sun on the moon when considering the moon's orbit around the earth?

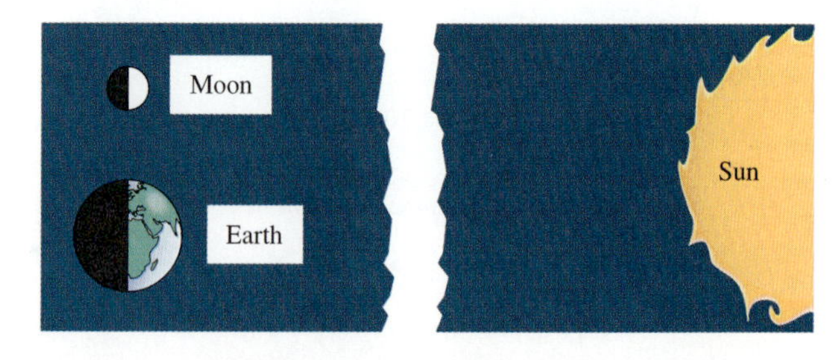

Fig. 6-23

★ **39** Find the radius of a planet made of solid lead, such that gravitational acceleration at the surface of the planet is 1.00 m/s^2. The density of lead is 1.13×10^4 kg/m^3.

★ **40** What is the vector sum of all the gravitational forces acting on the system of four masses shown in Fig. 6-24?

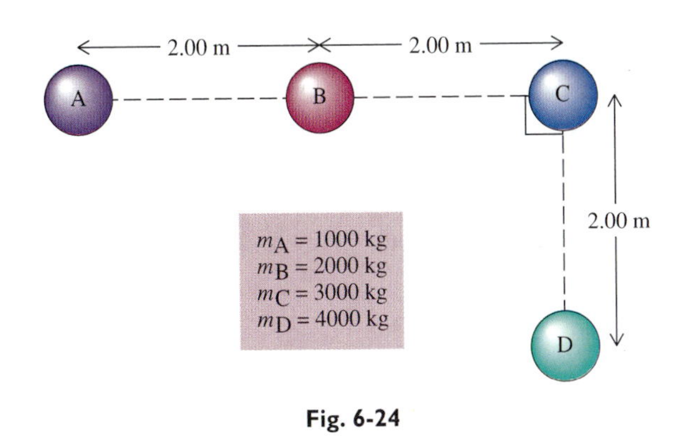

Fig. 6-24

★ **41** Mass is a measure of a body's inertia, and the gravitational force on a body is proportional to its mass. Thus there are two conceptually different properties of matter to which we give the name "mass." To distinguish these properties, mass is sometimes referred to as either "inertial mass, m_I," or "gravitational mass, m_G." Conceivably, the two masses could have slightly different values. Show that this would lead to a free-fall acceleration a that would depend on the ratio of a body's gravitational mass to its inertial mass:

$$a = \frac{m_G}{m_I} g \quad \text{where } g = \frac{GM}{R^2}$$

As the result of numerous experiments, we know that m_G and m_I are equal to within one part in 10^{12}.

42 Suppose the earth rotated on its axis so fast that at the equator objects had an apparent weight of zero. What would the length of a day be?

★★ **43** Show that for a spherical planet of uniform density, the orbital period of a satellite in a low, circular orbit is independent of the planet's radius but is inversely proportional to the square root of the planet's density.

44 Europa, one of the moons of Jupiter, was discovered by Galileo in 1610. Europa has a circular orbit of radius 6.708×10^5 km and period 3.551 days. Find the mass of Jupiter.

Jupiter

45 A Cavendish apparatus is shown in Fig. 6-25. You can estimate the angle of rotation of the rod in the following way. Assume that the rod is massless, that there is no resistance to rotation, and that the same force acts on both masses. Then the acceleration of each mass may be computed independently, using Newton's second law. Furthermore, if the distance each mass moves is small compared with the separation between m and M, then the acceleration can be treated as constant.

(a) Find the acceleration of each mass.

(b) How long a time is required for each mass to move a distance of 1 mm?

(c) Through what angle does the rod rotate during this interval?

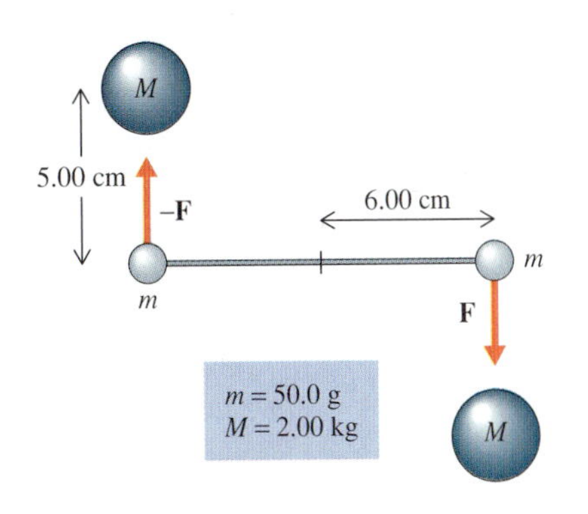

Fig. 6-25

CHAPTER 7 Energy

A drawn bow stores energy, which is transferred to the arrow as it is shot. Some bows store enough energy to shoot an arrow half a mile. *(McEwen E et al: Sci Am 264:76 [cover], June 1991.)*

The Concept of Energy

How fast must a spacecraft move to escape the earth? How much electric power can be generated using water from a certain waterfall? How many calories do you burn riding a bicycle uphill? To answer questions such as these we shall use a fundamental law of nature—**the law of conservation of energy.** This law states that **there exists a numerical quantity called "energy" that remains fixed in any process that occurs in nature.** We express the law more concisely by saying that **"energy is conserved."** The law of energy conservation applies without exception to all systems. If a certain isolated system has, say, 50 units of energy initially, that system will continue to have 50 units of energy, no matter what changes the system undergoes. It is possible for a system to lose energy only if that system is not isolated. Then the energy lost shows up as energy gained by some other system with which the first system has interacted.

Energy comes in many forms. Electrical energy, chemical energy, nuclear energy, and thermal energy are some forms of energy we shall study in later chapters. In this chapter we shall study only **mechanical energy,** which consists of two distinct types: (1) **kinetic energy,** associated with the motion of a body, and (2) **potential energy,** associated with the position of a body and a particular kind of mechanical force.

In general, the law of conservation of energy applies to the numerical sum of all forms of energy. If we add up mechanical energy, electrical energy, chemical energy, and so forth, the total energy of an isolated system is always constant. In this chapter we shall see that, under certain special circumstances, a system's mechanical energy alone is conserved. We shall show how this principle of conservation of mechanical energy follows from Newton's laws of motion. Rather than use Newton's laws directly to analyze the forces acting on a system, it is often easier to apply energy principles. For example, we shall use conservation of energy to calculate how fast a spacecraft must move to escape the earth (Ex. 7), to find the electric power generated in a certain Bavarian home using water from a small waterfall (Problem 59), and to estimate the energy needed to ride a bicycle uphill (Ex. 15).

7-1 Work and Kinetic Energy

Definition of Work Done by a Constant Force in One Dimension

In this section we shall define work and kinetic energy and then show how they are related through the work-energy theorem. The full significance of work and kinetic energy can be appreciated only after you see how they are connected through this important theorem.

When a force acts through a distance, we say, "The force does work." More precisely, the **work** W done by a constant force **F** acting on a body moving in a straight line (Fig. 7-1) is defined to be the product of the force component F_x in the direction of motion times the distance Δx the body moves:

$$W = F_x \, \Delta x \qquad \text{(constant force; linear motion)} \qquad (7\text{-}1)$$

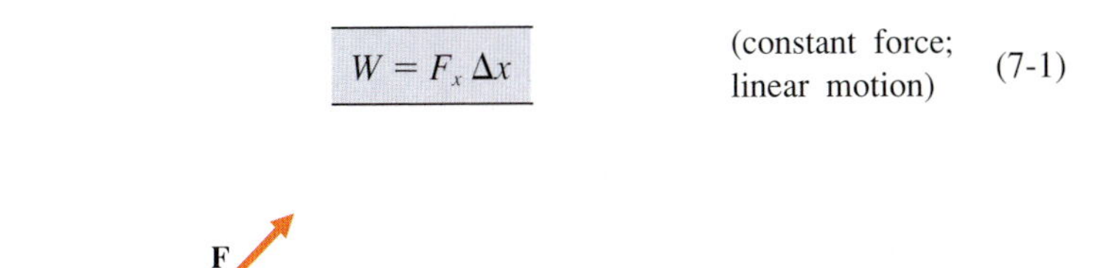

Fig. 7-1 As a block moves a distance Δx, the force **F** does work $F_x \, \Delta x$.

If a body does not move, $\Delta x = 0$, and so, even though forces may act on the body, no work is done by those forces (Fig. 7-2a), and no work is done on a moving body by any force that is perpendicular to the direction of the body's motion (Fig.7-2b), since such a force has a zero component in the direction of motion.

"Work" is a word commonly used to mean human effort. No such meaning is implied by the definition of work used in physics. For example, as you sit studying physics, you may be making an enormous effort, but there is no work being done, according to our definition of work as force acting through a distance. On the other hand, little or no effort is required to fall onto your bed. And yet work is done by the force you exert on your mattress and springs as they are being compressed. It is important not to confuse the physical concept of work with effort or with any other meaning attached to the word "work" in everyday language.

Fig. 7-2 **(a)** No work is done on a stationary barbell either by the barbell's weight **w** or by the force **F** exerted by the weight lifter. **(b)** No work is done by forces **N** and **w** acting on a skater, since neither force has a component in the direction of motion.

Units

The unit of work is the unit of force times the unit of distance—the N-m in SI. This unit is given the name "joule" (abbreviated J), in honor of James Joule, who demonstrated by numerous experiments in the nineteenth century that heat is a form of energy:

$$1 \text{ joule} = 1 \text{ N-m} = 1 \text{ kg-m}^2/\text{s}^2 \qquad (7\text{-}2)$$

In the cgs system the unit of work is the *erg*, defined as a dyne-cm. Since 1 N = 10^5 dyne and 1 m = 10^2 cm, 1 N-m = 10^7 dyne-cm or

$$1 \text{ J} = 10^7 \text{ erg} \qquad (7\text{-}3)$$

In the British system the unit of work is not given a separate name; it is simply called a "foot-pound," ft-lb. From the relationships between N and lb and between m and ft, it is easy to relate J to ft-lb:

$$1 \text{ J} = 0.738 \text{ ft-lb} \qquad (7\text{-}4)$$

EXAMPLE 1 Pulling a Suitcase

An airline passenger pulls his suitcase a horizontal distance of 40.0 m, exerting a force **F** of magnitude 25.0 N, directed 30.0° above the horizontal (Fig. 7-3). Find the work done by the force **F**.

SOLUTION To find the work we apply the definition (Eq. 7-1), using the component of force in the forward direction, the direction of motion.

$$W = F_x \Delta x = F \cos 30.0° \, \Delta x$$
$$= (25.0 \text{ N})(\cos 30.0°)(40.0 \text{ m})$$
$$= 866 \text{ J}$$

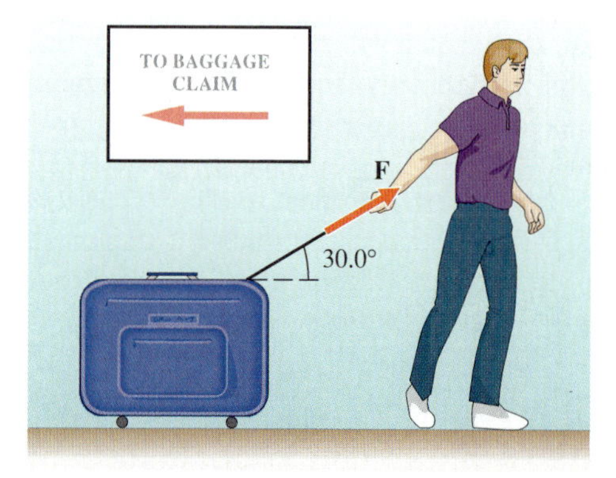

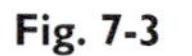
Fig. 7-3

Net Work

We define the **net work** on a body, W_{net}, to be the sum of the work done by all the forces acting on the body:

$$W_{net} = \Sigma W \tag{7-5}$$

Net work is important because the effect of work on the energy of a body depends only on the net work, as we shall see when we discuss the work-energy theorem.

EXAMPLE 2 Lifting a Box

A woman slowly lifts a box weighing 40.0 N from the floor to a shelf 1.50 m above (Fig. 7-4).
(a) Find the work done by the force **F** the woman exerts on the box.
(b) Find the work done on the box by its weight **w**.
(c) Find the net work done on the box.

SOLUTION **(a)** Since the box is lifted slowly, we assume that acceleration is negligible and therefore no net force acts on the box. This means that the woman exerts an upward force **F** of magnitude 40.0 N, balancing the box's weight. The force **F** acts in the direction of motion, and so the force component used in calculating the work W_F done by **F** is the full force of 40.0 N.

$$W_F = F_x \Delta x = F \Delta x = (40.0 \text{ N})(1.50 \text{ m})$$
$$= 60.0 \text{ J}$$

(b) The box's weight **w** acts opposite the direction of motion, and so its component in the direction of motion is negative ($-w$). Thus the work W_w done by **w** is negative.

$$W_w = w_x \Delta x = -w \Delta x = -(40.0 \text{ N})(1.50 \text{ m})$$
$$= -60.0 \text{ J}$$

(c) The net work done on the box is the sum of the work done by each of the forces acting on the box. Net work equals zero:

$$W_{net} = \Sigma W = W_F + W_w = +60 \text{ J} - 60 \text{ J} = 0$$

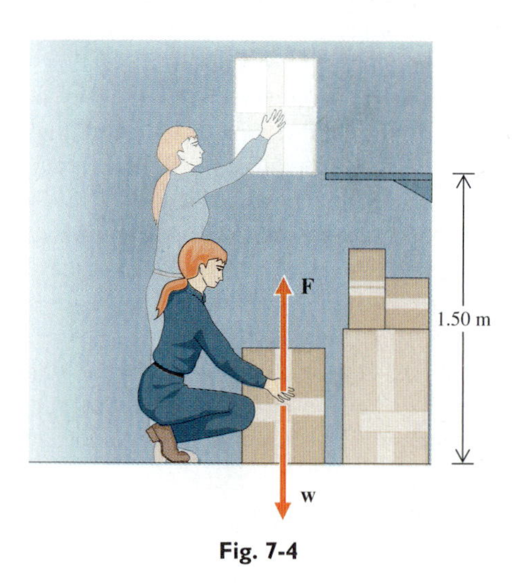

Fig. 7-4

Another way to find net work is to calculate the work done by the net force. Here the net force is zero, and so the work done by the net force must also be zero. Thus we get the same answer for W_{net} as we found by adding the work done by each force. It is easy to show that net work always equals the work done by the net force:

$$W_{net} = \Sigma W = \Sigma (F_x \Delta x) = (\Sigma F_x)(\Delta x) = F_{net} \Delta x$$

Kinetic Energy

A body's **kinetic energy** K is defined to be half its mass m times the square of its speed v.

$$K = \tfrac{1}{2}mv^2 \tag{7-6}$$

From its definition, kinetic energy must have units equal to mass units times velocity units squared—SI units of kg-(m/s)^2. Since 1 N = 1 kg-m/s^2, the SI unit of kinetic energy is N-m, or J, the same as the unit of work.

Sometimes kinetic energy is a conserved quantity. The simplest case of this is when a body moves at constant speed. Since both mass m and speed v are constant, the body's kinetic energy $\frac{1}{2}mv^2$ is also constant. Kinetic energy is conserved. A more interesting example of conservation of kinetic energy occurs in the game of pool. Suppose a cue ball is shot into a rack of balls (Fig. 7-5). If the cue ball has a mass of 0.2 kg and is initially moving at 10 m/s, its initial kinetic energy

$$K = \tfrac{1}{2}mv^2 = \tfrac{1}{2}(0.2 \text{ kg})(10 \text{ m/s})^2 = 10 \text{ J}$$

The other balls are initially at rest and so have no kinetic energy. Just after the collision, the kinetic energy of 10 J is shared among all balls. That is, if we add up the kinetic energies of all the balls just after the collision, the total is approximately 10 J. Kinetic energy is approximately* conserved in the collision of pool balls.

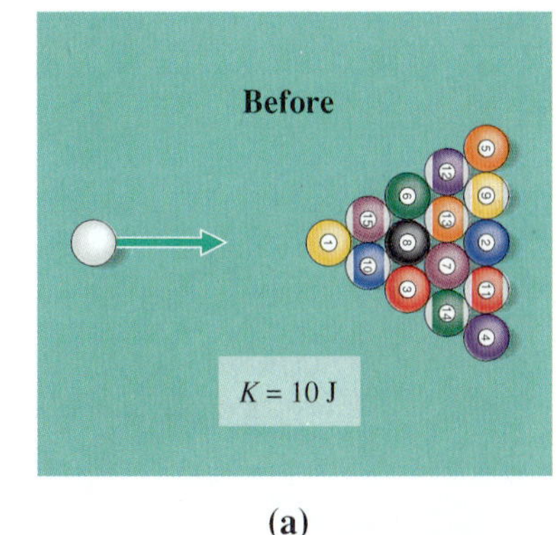

(a)

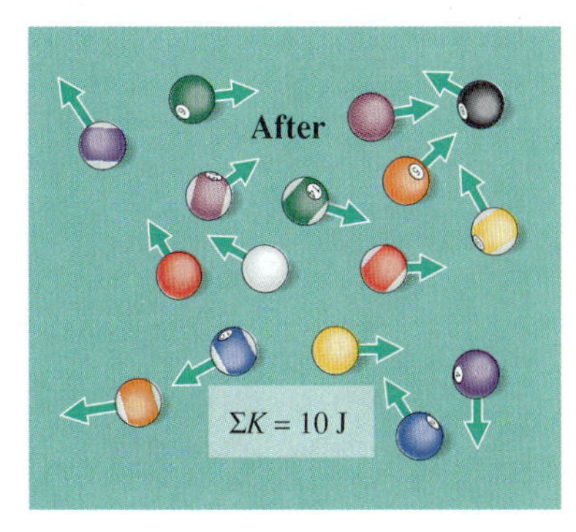

(b)

Fig. 7-5 The total kinetic energy of pool balls just after the "break" equals the kinetic energy of the cue ball before the break. Kinetic energy is conserved.

Work-Energy Theorem

Suppose a body moves along the x-axis and is subject to a number of constant forces $\mathbf{F}_1, \mathbf{F}_2, \mathbf{F}_3, \ldots$, whose resultant $\Sigma\, \mathbf{F}$ is a constant force directed along the x-axis (Fig. 7-6). According to Newton's second law, the body experiences an acceleration a_x given by

$$a_x = \frac{\Sigma\, F_x}{m} \tag{7-7}$$

where m is the body's mass. The body is accelerated, and so, as it moves through the distance $\Delta x = x - x_0$, its velocity changes. The final velocity v_x is related to the initial velocity v_{x0}, to the distance Δx, and to the acceleration a_x by the kinematic equation (from Chapter 2) $v_x^2 = v_{x0}^2 + 2a_x\, \Delta x$. There is only one component of velocity, $v^2 = v_x^2$, and so we may write the kinematic equation as

$$v^2 = v_0^2 + 2a_x\, \Delta x$$

Substituting for a_x from Newton's second law (Eq. 7-7), we obtain

$$v^2 = v_0^2 + 2\left(\frac{\Sigma\, F_x}{m}\right)(\Delta x)$$

Multiplying this equation by $m/2$ and rearranging, we can express this result

$$\tfrac{1}{2}mv^2 - \tfrac{1}{2}mv_0^2 = \Sigma\, (F_x\, \Delta x) \tag{7-8}$$

Thus we equate the change in kinetic energy ($\Delta K = K - K_0 = \frac{1}{2}mv^2 - \frac{1}{2}mv_0^2$) to the net work [$W_{\text{net}} = \Sigma\, W = \Sigma\, (F_x\, \Delta x)$].

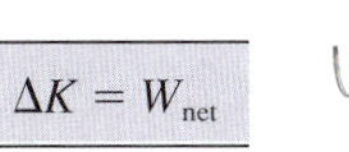

$$\Delta K = W_{\text{net}} \tag{7-9}$$

This result is known as the **work-energy theorem.**

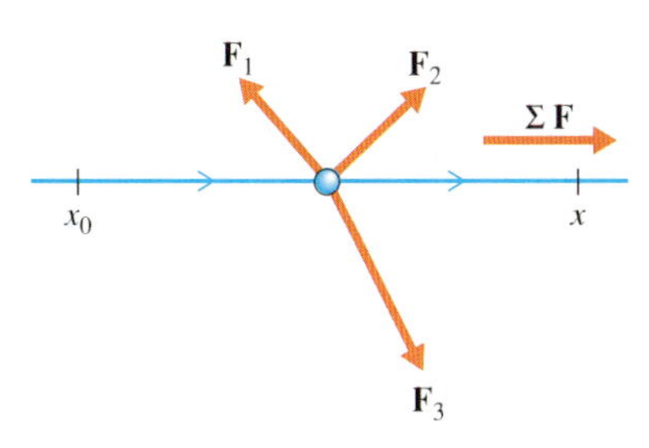

Fig. 7-6 A body moves from x_0 to x, and the forces acting on the body do work.

*A small part of the initial kinetic energy is converted to thermal energy and sound energy during the collision, and so kinetic energy is not exactly conserved.

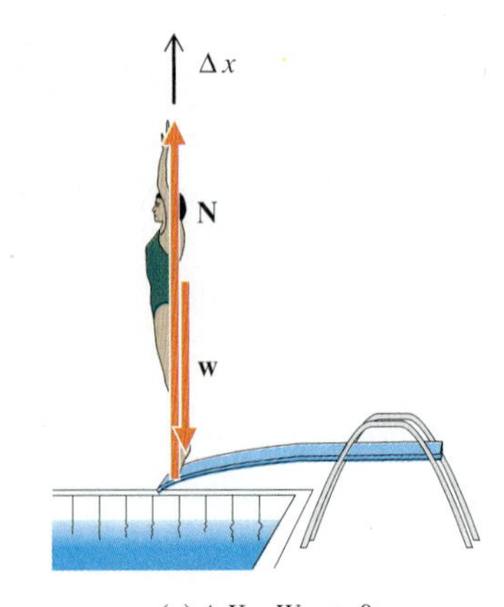

(a) $\Delta K = W_{net} > 0$

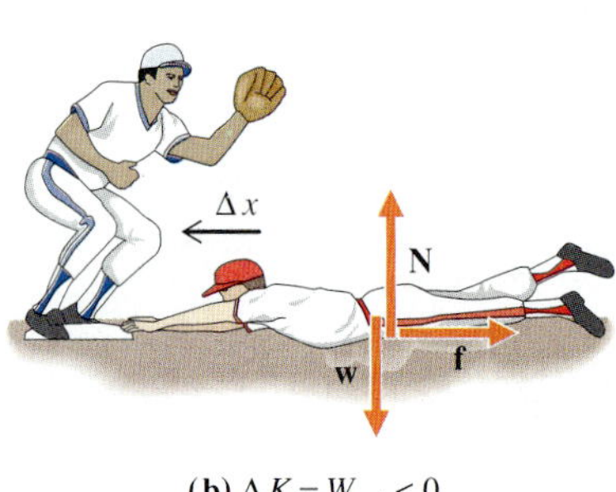

(b) $\Delta K = W_{net} < 0$

According to the work-energy theorem, when there is no net work done on an object, the object's change in kinetic energy is zero, or, in other words, kinetic energy is conserved. In Fig. 7-2a no work is done on the barbell, and so the barbell's kinetic energy remains constant—equal to zero. The skater in Fig. 7-2b has nonzero kinetic energy that remains constant, assuming that forces **F** and **w** are the only forces, since neither of these forces does work.

Fig. 7-7 shows two examples in which kinetic energy is not conserved. In Fig. 7-7a positive work is done by the normal force on a diver and negative work is done by the diver's weight, as the diver springs upward. Since the normal force is greater than the weight, the net work is positive. So, according to the work-energy theorem, the diver's kinetic energy increases ($\Delta K > 0$); in other words, the diver's speed increases. We can also see this from Newton's second law: the resultant force produces an upward acceleration.

In Fig. 7-7b only the force of friction does work on a baseball player sliding into second base. The other two forces have no component in the direction of motion and therefore do no work. The work done by friction is negative, since this force has a negative component along the line of motion. Thus the net work on the sliding player is negative, and, from the work-energy theorem, the player loses kinetic energy ($\Delta K < 0$); in other words, the player slows down. We can also predict this by applying Newton's second law: the resultant force is the frictional force, which produces an acceleration opposite the direction of motion and therefore slows the player.

Fig. 7-7 **(a)** A diver's kinetic energy increases as she springs upward because positive net work is done on her, since $N > w$. **(b)** A baseball player's kinetic energy decreases as he slides into second because friction does negative work on him.

EXAMPLE 3 Final Speed of a Sled

A child and sled having a combined weight of 335 N start from rest and slide 25.0 m down a 15.0° slope. Find the speed of the sled at the bottom of the slope, assuming negligible air resistance and a constant force of kinetic friction of 20.0 N.

SOLUTION The speed at the bottom of the slope may be calculated once the kinetic energy at that point is found from the work-energy theorem. Since the sled is initially at rest, $K_i = 0$. So the change in kinetic energy, which according to the work-energy theorem equals the net work done by the forces acting on the sled (shown in Fig. 7-8), is

$$\Delta K = K_f - 0 = W_{net} = W_N + W_w + W_f$$

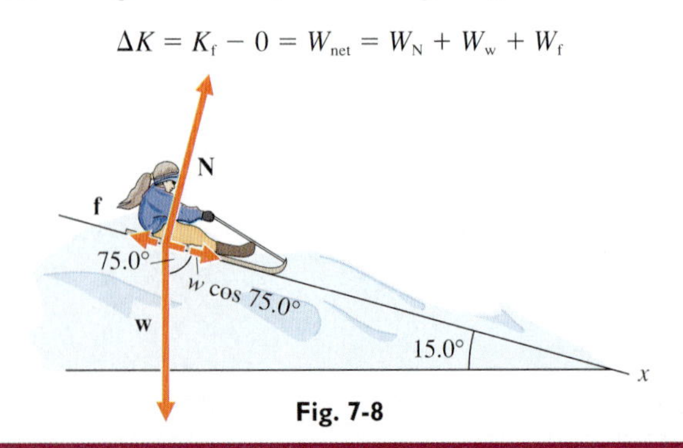

Fig. 7-8

The normal force does no work, since it is perpendicular to the motion:

$$W_N = 0$$

The weight does positive work, since there is a positive component w_x in the direction of motion:

$$W_w = w_x\,\Delta x = (335\text{ N})(\cos 75.0°)(25.0\text{ m}) = 2170\text{ J}$$

The work done by friction is negative since **f** opposes the motion:

$$W_f = f_x\,\Delta x = (-20.0\text{ N})(25.0\text{ m}) = -500\text{ J}$$

Adding the various work terms, we obtain

$$K_f = 0 + 2170\text{ J} - 500\text{ J} = 1670\text{ J}$$

Since $K_f = \frac{1}{2}mv_f^2$, v_f may be expressed

$$v_f = \sqrt{\frac{2K_f}{m}}$$

Since the mass $m = w/g = (335\text{ N})/(9.80\text{ m/s}^2) = 34.2$ kg, we find

$$v_f = \sqrt{\frac{2(1670\text{ J})}{34.2\text{ kg}}} = 9.88\text{ m/s}$$

Variable Force in Three Dimensions

We could have solved the previous example by first calculating the resultant force on the sled, then using Newton's second law to find the sled's acceleration, and finally applying the kinematic equation relating the velocity to the acceleration and distance. So the work-energy theorem has merely provided an alternative method for solving this kind of problem. However, the work-energy theorem may be generalized to deal with problems in which the forces are not constant and for which the path may not be linear. Direct solution of such problems from Newton's second law is much more difficult, since acceleration is not constant and the kinematic equations derived in Chapter 2 are not valid. The energy method then offers a significant advantage.

Consider a particle moving along a curved path and subject to a single variable force **F**, as shown in Fig. 7-9a. Let the path be divided into small intervals of length Δs, each of which is approximately linear and over each of which **F** is approximately constant in magnitude and direction, with a component F_s along the path (Fig. 7-9b). For each small interval, the change in kinetic energy is approximately equal to $F_s\,\Delta s$, and therefore the total change in kinetic energy from i to f is approximately equal to the sum of the $F_s\,\Delta s$ terms:

$$\Delta K = K_{\mathrm{f}} - K_{\mathrm{i}} \approx \Sigma\,(F_s\,\Delta s) \tag{7-10}$$

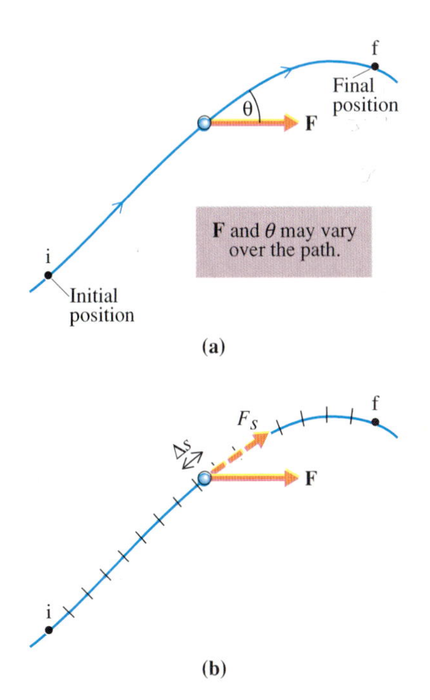

Fig. 7-9 **(a)** Force **F** acts on a particle as it moves along a curved path from i to f. **(b)** The total work done by **F** on the particle is the sum of the work done over small subintervals of length Δs.

The smaller the intervals, the better the approximation becomes, since then the intervals are more nearly linear and the force more nearly constant over each interval. Eq. 7-10 leads us to generalize our definition of work as follows. **The total work done by a force acting on a particle as the particle moves from position i to position f is a sum of terms $F_s\,\Delta s$:**

$$W = \Sigma\,(F_s\,\Delta s) \tag{7-11}$$

The intervals of length Δs used in this definition of work must be small enough that F_s is nearly constant over each interval.

Combining the two preceding equations, we may write

$$\Delta K = W$$

When two or more forces act on a particle moving along a curved path, it is the sum of the work done by all the forces that equals the change in kinetic energy. In other words, the change in kinetic energy equals the net work:

$$\Delta K = W_{\mathrm{net}} \qquad \text{(work-energy theorem)} \tag{7-12}$$

Graphical Interpretation of Work

Next we shall interpret our definition of work graphically, a technique that is useful in evaluating the work done by a varying force. Suppose we graph F_s as a function of s—in other words, graph the component of force acting on a particle as a function of the distance the particle moves over some interval i to f. Such a graph is shown in Fig. 7-10 for an arbitrary force. The interval i to f is divided into small subintervals of length Δs over which F_s is nearly constant. The product $F_s\,\Delta s$ is the area of a single rectangle. According to our definition of work (Eq. 7-11), the work is equal to a sum of terms $F_s\,\Delta s$ for very short intervals. Since $F_s\,\Delta s$ is the area of a rectangle, the work is equal to the sum of the areas of all the rectangles. But this is very nearly just the area under the curve, shaded blue.* Thus we see that the **work done by a force is the area under the F_s versus s curve between the initial and final points.**

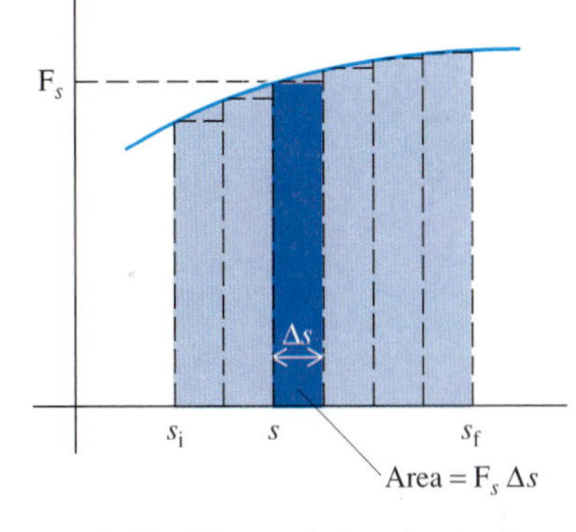

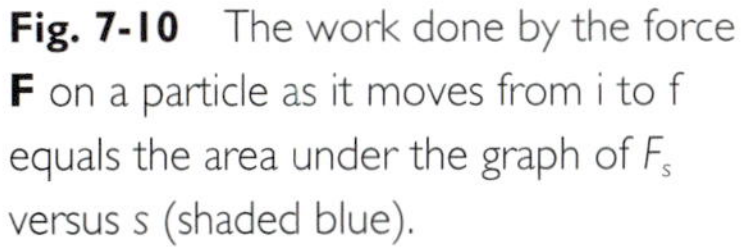
Fig. 7-10 The work done by the force **F** on a particle as it moves from i to f equals the area under the graph of F_s versus s (shaded blue).

*The difference between the area of the rectangles and the area under the curve disappears as we make the rectangles narrower.

EXAMPLE 4 Speed of an Arrow as it Leaves a Bowstring

The force exerted by a certain bow on an arrow decreases linearly after the arrow is released by the archer, starting at a value $F_s = 275$ N when the bow is fully drawn and decreasing to $F_s = 0$ as the arrow leaves the bowstring. The tail of the arrow moves from $s = 0$ to $s = 0.500$ m as the arrow is shot (Fig. 7-11a). Find the final speed of the arrow, which has a mass of 3.00×10^{-2} kg.

SOLUTION After we find the net work, we may use the work-energy theorem to find the final kinetic energy and the final velocity. Only the force **F** does work. This work equals the shaded area under the curve in Fig. 7-11b—the area of the triangle of base 0.500 m and height 275 N:

$$W_{\text{net}} = \tfrac{1}{2}(275\text{ N})(0.500\text{ m}) = 68.8\text{ J}$$

We apply the work-energy theorem, setting the initial kinetic energy equal to zero, since the arrow is initially at rest:

$$\Delta K = K_f - 0 = W_{\text{net}} = 68.8\text{ J}$$

$$K_f = \tfrac{1}{2}mv_f^2 = 68.8\text{ J}$$

Solving for v_f, we find

$$v_f = \sqrt{\frac{2K_f}{m}} = \sqrt{\frac{2(68.8\text{ J})}{3.00 \times 10^{-2}\text{ kg}}} = 67.7\text{ m/s}$$

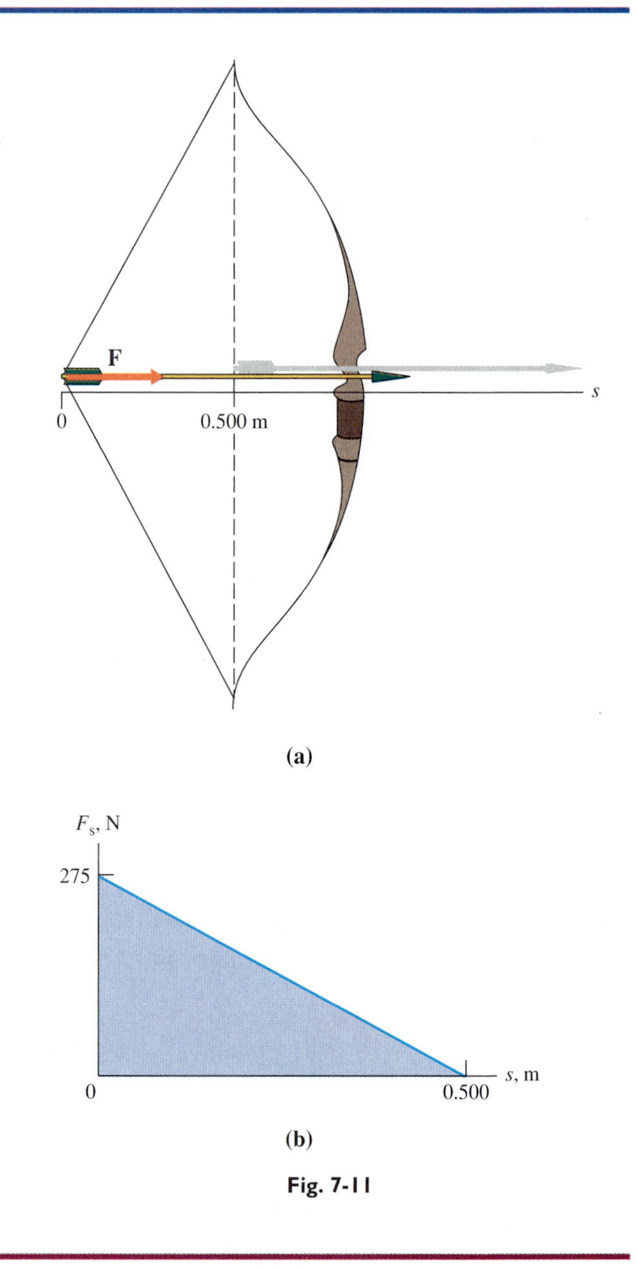

Fig. 7-11

In the next two sections we shall see how work done by certain forces is related to another form of energy, called "potential energy." Under certain conditions the sum of a system's kinetic energy plus potential energy is conserved.

7-2 Gravitational Potential Energy; Constant Gravitational Force

In this section we shall find a simple general expression for the work done on a body on or near the earth's surface by the constant force of gravity. We shall find that this work always equals the decrease in a quantity called "gravitational potential energy," which depends on the body's elevation. We shall see that when gravity is the only force doing work on a body, the sum of the body's kinetic energy plus its gravitational potential energy is conserved.

Work Done by a Constant Gravitational Force

Suppose a roller coaster starts from rest and accelerates down a curving track, falling through a vertical distance $y_i - y_f$, as it moves from point i to point f (Fig. 7-12). We shall obtain an expression for the work done on the roller coaster by its weight. Rather than use the actual path from i to f, we use an alternative path (path I in Fig. 7-12) to derive an expression for work, since it is much easier to derive the work for this alternative path than for the actual path. The expression we obtain, however, will apply to *any* path between points i and f, as shown at the end of this section.

Path I consists of a vertical displacement followed by a horizontal displacement. Work is done by the gravitational force only along the vertical part, for which there is a constant force $m\mathbf{g}$ along the direction of motion (Fig. 7-12). The work W_G equals the product of this force and the distance $y_i - y_f$:

$$W_G = mg(y_i - y_f)$$

$$W_G = mgy_i - mgy_f \tag{7-13}$$

There is no work done along the horizontal part of path I because the force $m\mathbf{g}$ has no component along the direction of motion.

We have derived Eq. 7-13 by considering the work done by gravity on a roller coaster for a specific path. However, this equation applies to the work done on *any* body of mass m by its weight $m\mathbf{g}$, as the body moves from initial elevation y_i to final elevation y_f along *any* path. According to Eq. 7-13, the work equals the difference in the values of the quantity mgy, which we call **gravitational potential energy** and denote by U_G:

$$U_G = mgy \tag{7-14}$$

Thus the work equals the decrease in gravitational potential energy—the initial value $U_{G,i}$ minus the final value $U_{G,f}$:

$$W_G = U_{G,i} - U_{G,f} \tag{7-15}$$

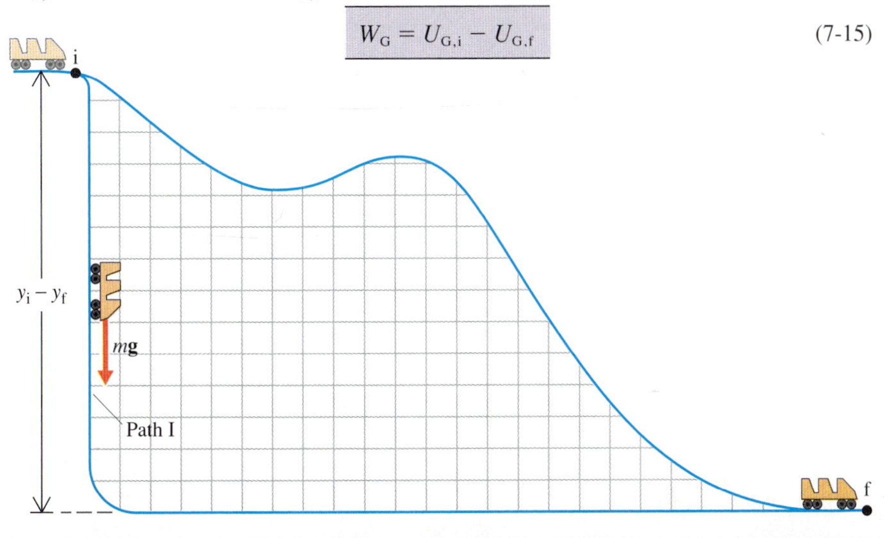

Fig. 7-12 A roller coaster moves along a curved track from point i to point f. Path I is an alternate path between the same points.

For example, suppose a roller coaster weighing 10^4 N starts at an elevation of 40 m, where its potential energy $mgy = 4 \times 10^5$ J, and falls to an elevation of 10 m, where its potential energy $mgy = 10^5$ J. No matter what path the roller coaster follows, the gravitational force does work on it equal to its decrease in potential energy of 3×10^5 J.

Conservation of Energy

Suppose the gravitational force alone does work. Then

$$W_{net} = W_G = U_{G,i} - U_{G,f}$$

From the work-energy theorem, however, we also know that

$$W_{net} = \Delta K = K_f - K_i$$

Equating these two expressions for the net work, we obtain

$$K_f - K_i = U_{G,i} - U_{G,f}$$

Thus, for example, in the case of the roller coaster, if there is negligible work done by friction or any other force except gravity, the roller coaster will gain kinetic energy equal to its lost potential energy of 3×10^5 J.

Rearranging terms in the equation above, we can express our result:

$$K_f + U_{G,f} = K_i + U_{G,i} \tag{7-16}$$

We define the total mechanical energy E to be the sum of the kinetic and gravitational potential energies:

$$E = K + U_G \tag{7-17}$$

Then Eq. 7-16 may be written

$$E_f = E_i \tag{7-18}$$

When gravity is the only force doing work on a body, the sum of the body's kinetic energy plus its gravitational potential energy—the total mechanical energy—is conserved.

As a simple example of conservation of mechanical energy, consider a body in free fall. As a body falls, its speed increases. Its kinetic energy increases while its potential energy decreases, so that the sum of the two—the total mechanical energy—remains constant. This is illustrated in Fig. 7-13 for a 1 kg body falling from rest through a distance of 1 m.

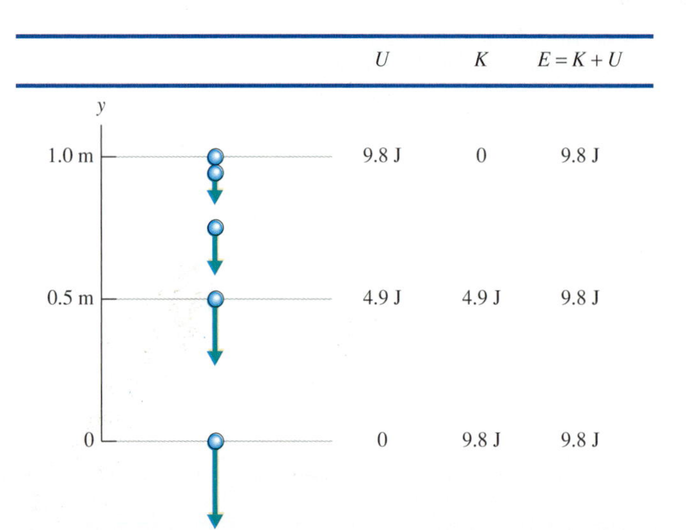

Fig. 7-13 The sum of this falling body's kinetic energy and its gravitational potential energy is a constant 9.8 J. The body's total mechanical energy is conserved.

EXAMPLE 5 Energy of a Thrown Ball

A ball of mass 0.200 kg is thrown vertically upward with an initial velocity of 10.0 m/s. Find (a) the total mechanical energy of the ball, (b) its maximum height, and (c) its speed as it returns to its original level. Neglect air resistance.

SOLUTION (a) The total mechanical energy is the sum of the kinetic energy plus the gravitational potential energy:

$$E = K + U_G = \tfrac{1}{2}mv^2 + mgy$$

We take the origin of the y-axis to be the initial position; then the initial energy E_i is purely kinetic:

$$E_i = \tfrac{1}{2}mv^2 + 0 = \tfrac{1}{2}(0.200\text{ kg})(10.0\text{ m/s})^2 = 10.0\text{ J}$$

Since the gravitational force is the only force doing work on the ball (with air resistance being neglected), E will remain equal to 10.0 J throughout the motion of the ball.

(b) When the height is maximum, the ball is momentarily at rest and $K = 0$. The total mechanical energy E is purely potential energy at this point:

$$E = mgy$$

Solving for y, we obtain

$$y = \frac{E}{mg} = \frac{10.0\text{ J}}{(0.200\text{ kg})(9.80\text{ m/s}^2)} = 5.10\text{ m}$$

(c) When the ball returns to its initial height, y again equals 0, and the potential energy is zero. Then the total energy is again purely kinetic, and the kinetic energy therefore equals 10.0 J. This is the same as the initial value of kinetic energy, and so the speed of the ball must also be the same.

$$E = K = 10.0\text{ J}$$

$$v = 10.0\text{ m/s}$$

It is a general characteristic of projectile motion that, in the absence of air resistance, the projectile has the same speed for points at the same elevation (Fig. 7-14). This follows from the fact that the potential energy will be the same at such points, and conservation of total mechanical energy then implies that the kinetic energy will also be the same.

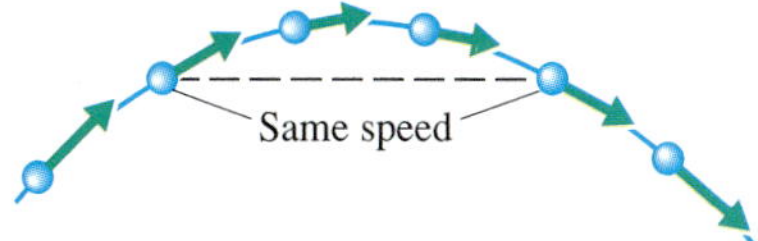

Fig. 7-14 A projectile has the same speed at points with the same elevation, if air resistance is negligible.

EXAMPLE 6 Speed of a Skier at the Bottom of a Hill

A skier starts from rest at the top of a ski slope and skis downhill (Fig. 7-15a). Find the skier's speed after her elevation decreases by 10.0 m, assuming no work is done by friction or air resistance.

Continued on next page.

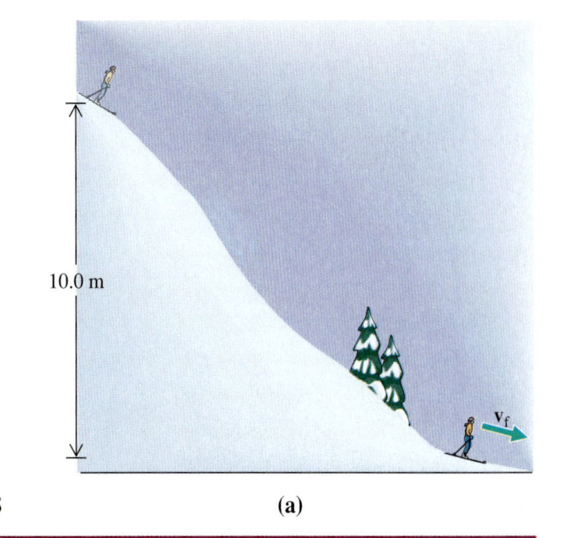

Fig. 7-15 (a)

EXAMPLE 6—cont'd

SOLUTION The forces acting on the skier are shown in Fig 7-15b. Since the normal force is perpendicular to the motion, it does no work. Only the weight does work. Therefore the total mechanical energy is conserved.

$$E_f = E_i$$

$$K_f + U_{G,f} = K_i + U_{G,i}$$

$$\tfrac{1}{2}mv_f^2 + 0 = 0 + mgy_i$$

Notice that we have arbitrarily chosen the origin so that $y_f = 0$. Solving for v_f, we find

$$v_f = \sqrt{2gy_i} = \sqrt{2(9.80\ \text{m/s}^2)(10.0\ \text{m})} = 14.0\ \text{m/s}$$

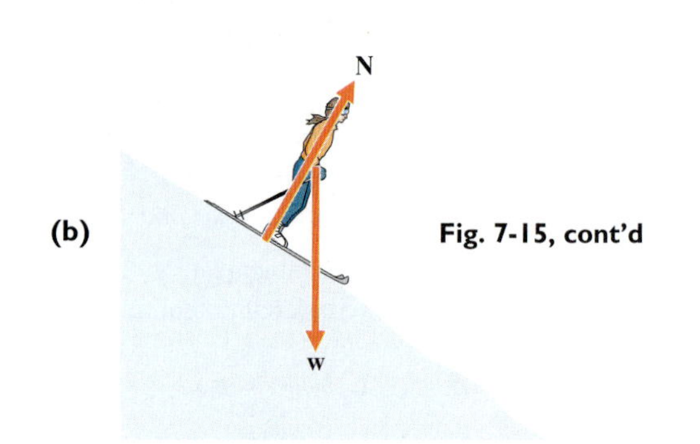

Fig. 7-15, cont'd

The skier would have this same final speed if she had fallen straight down through a vertical distance of 10.0 m, since her decrease in gravitational potential energy is determined solely by her vertical drop.

Proof That Work Done by Gravity Is Path-Independent

We shall now show that the work done by the gravitational force on a body is independent of the path the body travels from its initial position to its final position. Fig. 7-16 shows an arbitrary path between points i and f at respective elevations y_i and y_f. (To be specific you can think of a roller coaster moving along *any* path between points i and f.) Since the motion is not linear, we must use the general expression for work (Eq. 7-11):

$$W_G = \Sigma\,(F_s\,\Delta s)$$

Fig. 7-16 shows a blowup of a short, approximately linear, segment of the path. The component of the force $m\mathbf{g}$ along the path is $mg\cos\theta$, and so the work done by gravity over the interval Δs is

$$F_s\,\Delta s = (mg\cos\theta)(\Delta s) = mg\,(\Delta s\cos\theta)$$

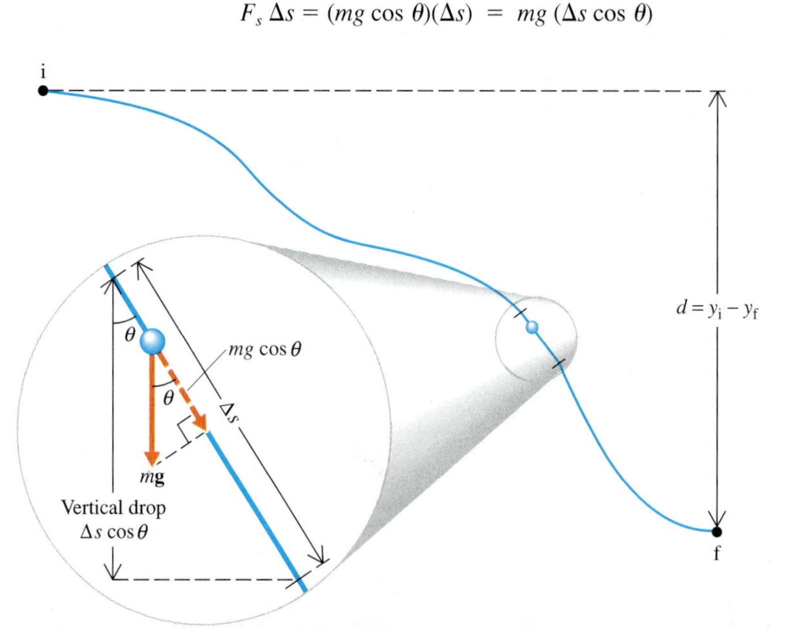

Fig. 7-16 Finding the work done by gravity as a body moves over an arbitrary path from i to f. The body's elevation decreases by $y_i - y_f$.

But $\Delta s \cos \theta$ equals the vertical drop, as indicated in the figure; thus

$$F_s \, \Delta s = mg(\text{vertical drop})$$

We obtain the total work over the entire path by adding the contributions arising from all vertical drops in the interval from i to f:

$$\begin{aligned} W_G &= mg[\Sigma \text{ (vertical drops)}] = mg(\text{net vertical drop}) \\ &= mg(y_i - y_f) \\ &= mgy_i - mgy_f \end{aligned}$$

This is the same result we obtained in Eq. 7-13, thus completing our proof that the work done by gravity is independent of path.

7-3 Gravitational Potential Energy; Variable Gravitational Force

In the last section we obtained an expression for the gravitational potential energy of a body on or near the earth's surface, where the gravitational force is constant. In this section we shall consider problems in which the gravitational force varies. Suppose a mass M (a planet, for example) exerts a gravitational force **F** on a smaller mass m (such as an approaching spacecraft). This force does work as m moves from an initial position i to a final position f (Fig. 7-17). If m moves over a significant distance compared to the separation of the two masses, the gravitational force **F** is not constant. Then one finds the work done by this force by breaking the path up into short intervals over which the force is nearly constant and calculating the sum:

$$W = \Sigma \, (F_s \, \Delta s)$$

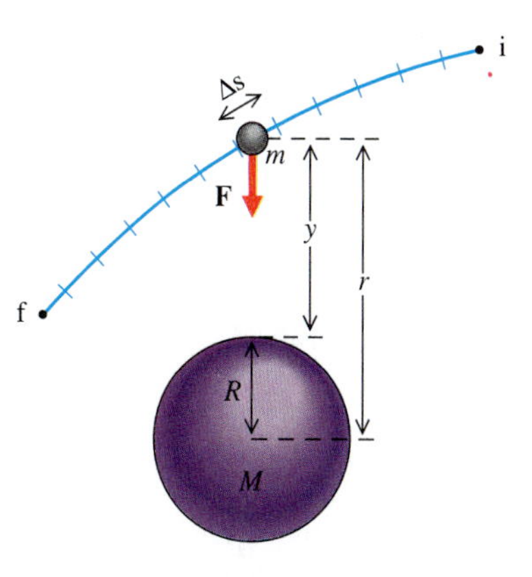

Fig. 7-17 Finding the work done by the gravitational force exerted by a mass M (a planet, for example) on a smaller mass m, as m moves from i to f.

Evaluation of this sum requires the use of integral calculus, and so we will not evaluate it here. But the result turns out to be quite simple. The work done by gravity equals the decrease in the gravitational potential energy, U_G:

$$W_G = U_{G,i} - U_{G,f}$$

where the potential energy depends on the distance r between the centers of the two masses, as given by the equation

$$U_G = -\frac{GmM}{r} \tag{7-19}$$

According to this equation, gravitational potential energy is always a negative quantity for all finite values of r. At $r = \infty$ the potential energy equals zero, whereas at $r = R$ (the radius of the larger body), $U_G = -\dfrac{GmM}{R}$. Suppose, for example, a spacecraft of mass m approaches the earth from a very great distance, so that its potential energy starts out equal to zero. The spacecraft's potential energy steadily decreases to a minimum value of $-\dfrac{GmM}{R}$ at the surface of the earth. The potential energy decreases by $\dfrac{GmM}{R}$.

Since the work done by gravity equals the difference in two values of the potential energy, we can always add an arbitrary constant to the potential energy at every point to provide a more convenient reference level of zero potential energy. Since the same constant is added to both the initial and final values, the difference in potential energy is unchanged. For example, if we add the constant $\frac{GmM}{R}$ to the expression for potential energy given in Eq. 7-19, we obtain a second equally valid potential energy function U'_G:

$$U'_G = -\frac{GmM}{r} + \frac{GmM}{R} \tag{7-20}$$

The zero of potential energy in this case occurs when $r = R$:

$$U'_G = -\frac{GmM}{R} + \frac{GmM}{R} = 0$$

whereas at $r = \infty$, we have

$$U'_G = \frac{GmM}{R}$$

The decrease in potential energy of a spacecraft of mass m, approaching the earth from a great distance, equals

$$U'_{G,i} - U'_{G,f} = \frac{GmM}{R} - 0 = \frac{GmM}{R}$$

the same decrease we calculated using the function U_G.

Most terrestrial bodies are always at nearly the same distance from the center of the earth. For example, when a batter hits a home run, the distance of the baseball from the center of the earth varies little over its trajectory. For such bodies the difference in gravitational potential energy between any two points can be calculated using one of the expressions just given (Eq. 7-19 or 7-20), or using Eq. 7-14 ($U_G = mgy$), which is valid when the gravitational force is constant. Problem 72 outlines a proof that these different equations for gravitational potential energy are consistent.

EXAMPLE 7 Speed of a Meteoroid Entering the Atmosphere

Find the speed of a meteoroid (Fig. 7-18) as it first enters the earth's atmosphere, if, when it is very far from the earth, it is moving relatively slowly, so that its initial kinetic energy is negligible.

SOLUTION Since the earth's gravitational force is the only force acting on the body, its mechanical energy is conserved:

$$E_f = E_i$$

or

$$K_f + U_{G,f} = K_i + U_{G,i}$$

Since the body is initially very far from the earth, its initial potential energy is approximately zero.

$$U_{G,i} \approx 0$$

And we are given that its initial kinetic energy is approximately zero.

$$K_i \approx 0$$

Substituting into the energy conservation equation, we obtain

$$K_f + U_{G,f} = 0$$

or

$$\tfrac{1}{2}mv_f^2 - \frac{GmM}{r_f} = 0$$

where m is the meteoroid's mass, M is the earth's mass, and r_f is approximately the earth's radius. Solving for v_f, we find

$$v_f = \sqrt{\frac{2GM}{r_f}}$$

$$= \sqrt{\frac{2\left(6.67 \times 10^{-11}\,\frac{\text{N-m}^2}{\text{kg}^2}\right)(5.98 \times 10^{24}\text{ kg})}{6.38 \times 10^6\text{ m}}}$$

$$= 1.12 \times 10^4\text{ m/s} = 11.2\text{ km/s}$$

Most meteoroids have higher speeds as they enter the atmosphere (typically 13 to 70 km/s), since they usually have a significant kinetic energy when they are far from the earth.

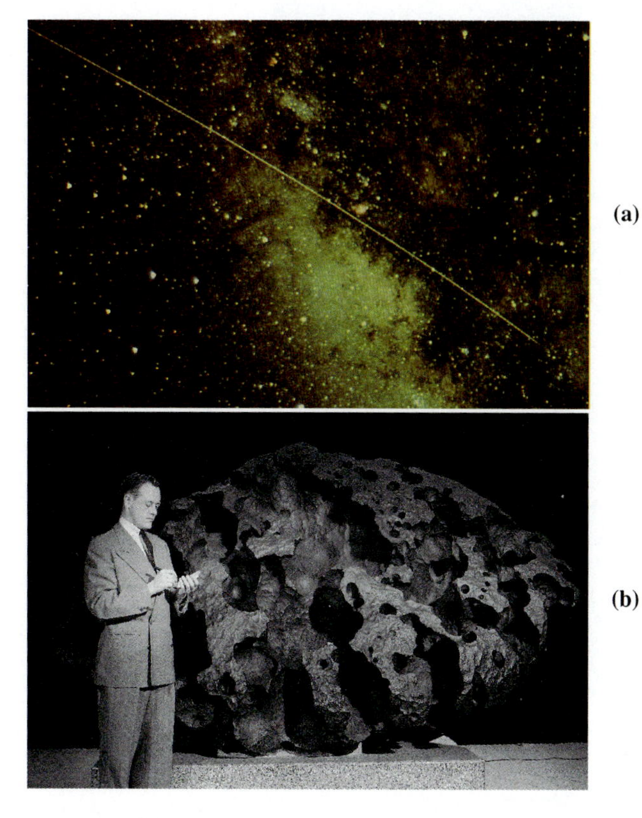

Fig. 7-18 **(a)** A meteor. **(b)** A 15-ton meteorite. A meteor, or "shooting star," is the bright streak of light that occurs when a solid particle (a "meteoroid") from space enters the earth's atmosphere and is heated by friction. Billions of meteoroids hit the earth each day. You can usually see several meteors per hour on a clear, moonless night. Fortunately, most meteoroids are no larger than a small pebble and vaporize before they reach the ground. Occasionally, a very large meteoroid strikes the earth (the fallen body is called a "meteorite"). For example, one such body formed the great meteor crater in Arizona over 5000 years ago (see p. 87). Another weighing about 10^5 tons destroyed hundreds of square miles of forest in Siberia in 1908. Most meteoroids originate from bodies that are already within the solar system.

Escape Velocity

What goes up must come down. If you throw a ball vertically upward, it always comes down again. But suppose you could give a ball an extremely large initial velocity, say, 50,000 km/h. The ball would escape the earth, never to return (if we assume negligible air resistance*). Initially, as the ball rose, it would decelerate at the rate of 9.80 m/s^2. However, because of the large initial velocity, it would rise to great heights, and the gravitational force and gravitational acceleration (which vary as $1/r^2$) would decrease as it rose. The ball could then rise still higher with less gravitational acceleration. Eventually, when the ball was far from the earth, the earth's gravity would no longer produce a significant effect. The ball would then continue with nearly constant velocity (Fig. 7-19).

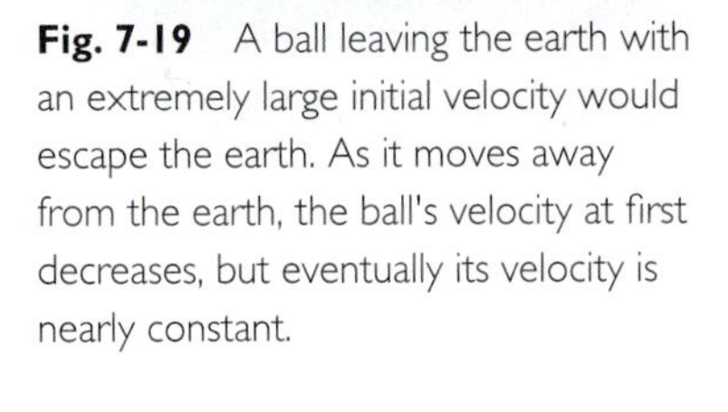

Fig. 7-19 A ball leaving the earth with an extremely large initial velocity would escape the earth. As it moves away from the earth, the ball's velocity at first decreases, but eventually its velocity is nearly constant.

Of course it is not possible to simply throw a ball with such a large initial velocity. However, rocket engines have given spacecraft large enough velocities to leave the earth's surface and explore the solar system. Some spacecraft are even able to escape the solar system.

Suppose that a spacecraft blasts off from a planet and reaches a large velocity while still close to the planet's surface. The rocket engines are then turned off. From that point on, the planet's gravitational force is the only force acting on the spacecraft, and its mechanical energy is therefore conserved.

$$E_i = E_f$$

$$\tfrac{1}{2}mv_i^2 - \frac{GmM}{r_i} = \tfrac{1}{2}mv_f^2 - \frac{GmM}{r_f}$$

The minimum initial velocity necessary to escape the planet is called the **escape velocity,** denoted by v_E. This is the value of the initial velocity that results in a final velocity v_f approaching zero as the distance r_f approaches infinity. We insert $v_f = 0$, $r_f = \infty$, $v_i = v_E$ and set r_i equal to the planet's radius R, and the energy conservation equation becomes

$$\tfrac{1}{2}mv_E^2 - \frac{GmM}{R} = 0$$

Solving for v_E, we obtain

$$v_E = \sqrt{\frac{2GM}{R}}$$

In deriving this equation, we did not need to assume any particular direction for the spacecraft's initial velocity vector, since kinetic energy is a scalar quantity, involving only the magnitude of velocity ($K = \tfrac{1}{2}mv^2$). So our conclusion is valid for a spacecraft moving away from the earth in any direction. If the initial speed exceeds v_E, the spacecraft escapes, never to return (unless acted upon by some other force) (Fig. 7-20).

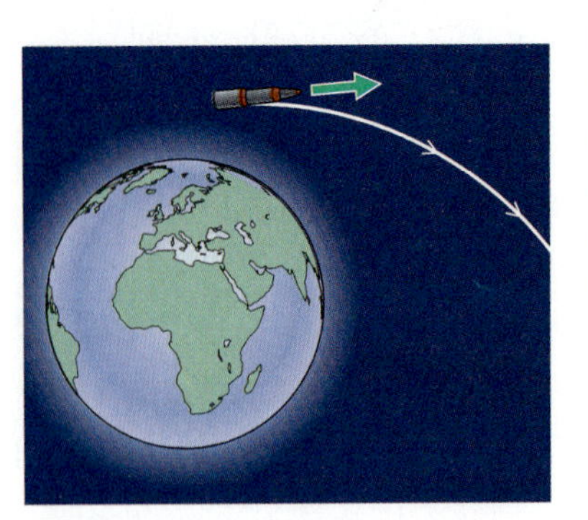

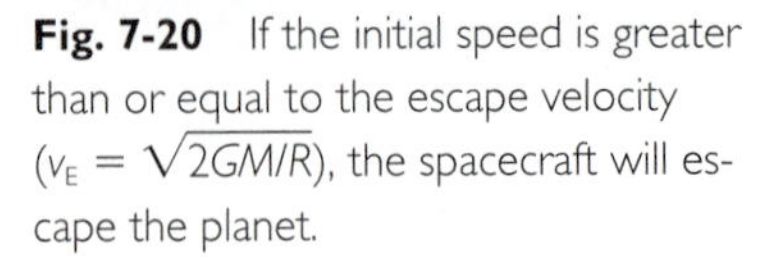

Fig. 7-20 If the initial speed is greater than or equal to the escape velocity ($v_E = \sqrt{2GM/R}$), the spacecraft will escape the planet.

*Air resistance would actually be a very large force on the ball at such a large velocity. However, our assumption of negligible air resistance is correct if we take the initial velocity to be the velocity of the ball at an elevation of about 100 km, above which the atmosphere is very thin.

EXAMPLE 8 Earth's Escape Velocity

Find the value of the escape velocity on earth.

SOLUTION Inserting values for the earth's mass and radius into Eq. 7-21, we find

$$v_E = \sqrt{\frac{2GM}{R}} = \sqrt{\frac{2(6.67 \times 10^{-11}\ \text{N-m}^2/\text{kg}^2)(5.98 \times 10^{24}\ \text{kg})}{6.38 \times 10^6\ \text{m}}}$$

$$= 1.12 \times 10^4\ \text{m/s} = 11.2\ \text{km/s}$$

Any object on earth with a velocity of at least 11.2 km/s in any direction away from the earth will leave the earth and never return, unless it is acted upon by forces other than just the earth's gravitational force.

7-4 Spring Potential Energy; Conservation of Energy

There are other kinds of potential energy besides gravitational; that is, there are other forces for which the work can be expressed as a decrease in some kind of potential energy. One of these is spring potential energy. For example, the compressed spring that launches the ball in a pinball machine stores spring potential energy (Fig. 7-21). We shall see that under certain circumstances spring potential energy can be converted into kinetic energy (for example, the kinetic energy of the ball in a pinball machine).

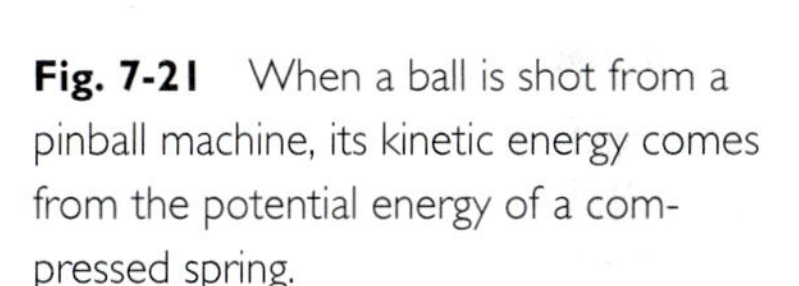

Fig. 7-21 When a ball is shot from a pinball machine, its kinetic energy comes from the potential energy of a compressed spring.

Work Done by a Spring Force

Suppose a body is attached to a spring that can be either stretched or compressed (Fig. 7-22). As discussed in Chapter 4, the force that the spring exerts on the body has an x component given by

$$F_x = -kx \tag{7-22}$$

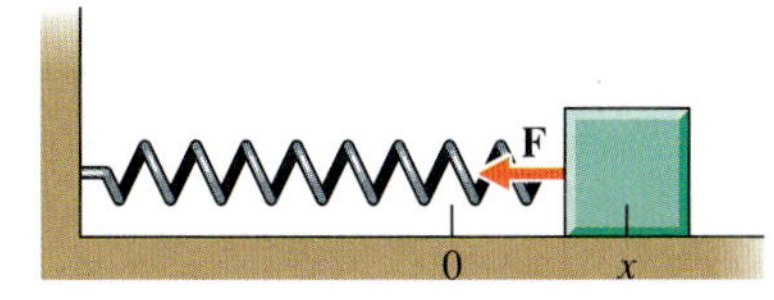

Fig. 7-22 The force **F** exerted on a body by a spring does work on the body as it moves.

where x is the displacement from the equilibrium position. Since this force varies with position, we compute the work by applying the general definition of work (Eq. 7-11), expressing the work as a sum:

$$W = \Sigma\ (F_x\ \Delta x)$$

As shown in Section 7-1, the work equals the area under the force versus displacement curve, between the initial and final points. The shaded area in Fig. 7-23 gives the work done by the spring force on a body in contact with the spring, as the body moves from x_i to x_f. This area is counted as negative because the force F_x is negative over the interval, the displacement is positive, and so each of the products in the sum ($F_x\ \Delta x$) is negative. The shaded area can be computed as the difference in the areas of two triangles. The larger triangle has a base extending from 0 to x_f and an area of $\frac{1}{2}(-kx_f)(x_f) = -\frac{1}{2}kx_f^2$. The smaller triangle has a base extending from 0 to x_i, and an area of $\frac{1}{2}(-kx_i)(x_i) = -\frac{1}{2}kx_i^2$. The work W_S done by the spring is the difference in these two areas:

$$W_S = \text{Area of larger triangle} - \text{Area of smaller triangle}$$

$$= -\tfrac{1}{2}kx_f^2 - (-\tfrac{1}{2}kx_i^2)$$

$$= \tfrac{1}{2}kx_i^2 - \tfrac{1}{2}kx_f^2$$

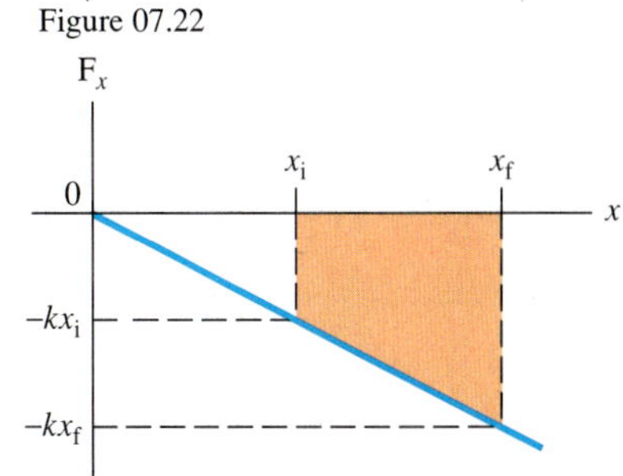

Fig. 7-23 Finding the work done by the spring force.

We see that the work is expressed as the difference in the values of the function $\frac{1}{2}kx^2$, evaluated at the two points x_i and x_f. This function we call the **spring potential energy** and denote by U_S.

$$U_S = \frac{1}{2}kx^2 \tag{7-23}$$

Now we can express the work W_S done by the spring as the decrease in spring potential energy.

$$W_S = U_{S,i} - U_{S,f} \tag{7-24}$$

For example, the work done by a spring of force constant 10^3 N/m on an attached mass moving from $x = 0$ to $x = 0.1$ m is

$$\begin{aligned} W_S &= \tfrac{1}{2}kx_i^2 - \tfrac{1}{2}kx_f^2 = 0 - \tfrac{1}{2}(10^3\text{ N/m})(0.1\text{ m})^2 \\ &= -5\text{ J} \end{aligned}$$

The spring does -5 J of work on the attached mass, meaning that the kinetic energy of the mass will decrease by 5 J, if no other forces act on it.

Conservation of Energy

If the spring force is the only force that does work on the body, then $W_{net} = W_S = U_{S,i} - U_{S,f}$, and applying the work-energy theorem, we find

$$\Delta K = W_{net}$$

$$K_f - K_i = U_{S,i} - U_{S,f}$$

or

$$K_f + U_{S,f} = K_i + U_{S,i} \tag{7-25}$$

In this case we define the total mechanical energy E to be the sum of the kinetic energy and the spring potential energy:

$$E = K + U_S \quad \text{(when only the spring force does work)} \tag{7-26}$$

and Eq. 7-25 may be written

$$E_f = E_i \tag{7-27}$$

Total mechanical energy is conserved when the spring force is the only force that does work. This is the same result obtained for the gravitational force, except that a different kind of potential energy is used here in defining the mechanical energy. Both the spring force and the gravitational force are called **conservative forces.**

EXAMPLE 9 Energy of a Bow and Arrow

When an archer pulls an arrow back in a bow, potential energy is stored in the stretched bow. Suppose the force required to draw the bowstring back in a certain bow varies linearly with the displacement of the center of the string, so that the bow behaves as a stretched spring. A force of 275 N is required to draw the string back 50.0 cm. (a) Find the potential energy stored in the bow when fully drawn. (b) Find the speed of an arrow of mass 3.00×10^{-2} kg as it leaves the bow, assuming that the arrow receives all the mechanical energy initially stored in the bow.

SOLUTION (a) First we find the force constant, using Eq. 7-22 ($F_x = -kx$):

$$k = \frac{-F_x}{x}$$

A force of 275 N is exerted by the string on the arrow in the forward direction when the string is displaced 50.0 cm backwards. Thus

$$k = \frac{-275\ \text{N}}{-0.500\ \text{m}} = 550\ \text{N/m}$$

Now we can apply Eq. 7-23 to find the potential energy.

$$U_S = \tfrac{1}{2}kx^2 = \tfrac{1}{2}(550\ \text{N/m})(-0.500\ \text{m})^2$$

$$= 68.8\ \text{J}$$

(b) Here we assume that no other forces do work, and so mechanical energy is conserved. As the bow leaves the string, the system's energy is the kinetic energy of the arrow.

$$E_f = E_i$$

$$K_f + U_{S,f} = K_i + U_{S,i}$$

$$\tfrac{1}{2}mv_f^2 + 0 = 0 + U_{S,i}$$

$$v_f = \sqrt{\frac{2U_{S,i}}{m}} = \sqrt{\frac{2(68.8\ \text{J})}{3.00 \times 10^{-2}\ \text{kg}}}$$

$$= 67.7\ \text{m/s}$$

Work Done by Both a Spring Force and a Gravitational Force

Suppose that both a spring force and a gravitational force do work on a body and that these are the only forces doing work. The net work is then the sum of the work done by the two forces.

$$\Sigma W = W_G + W_S$$

The work done by each force can still be expressed as a decrease in potential energy of the respective type.

$$\Sigma W = U_{G,i} - U_{G,f} + U_{S,i} - U_{S,f}$$

or

$$\Sigma W = (U_{G,i} + U_{S,i}) - (U_{G,f} + U_{S,f}) \tag{7-28}$$

We shall find that a generalization of our definition of mechanical energy will allow us to maintain the principle of conservation of mechanical energy. We first define the total potential energy to be the sum of the gravitational potential energy and the spring potential energy.

$$U = U_G + U_S \tag{7-29}$$

Then Eq. 7-28 may be written

$$\Sigma W = U_i - U_f$$

But according to the work-energy theorem, $\Sigma W = K_f - K_i$. Therefore

$$K_f - K_i = U_i - U_f$$

or

$$K_f + U_f = K_i + U_i$$

If we define the total mechanical energy E to be the sum of the kinetic energy and the *total* potential energy, we find once again that the total mechanical energy is conserved.

$$E_f = E_i \tag{7-30}$$

$$E = K + U \tag{7-31}$$

EXAMPLE 10 Maximum Height of an Arrow

Suppose the arrow described in the preceding example is shot vertically upward. Find the maximum height the arrow rises before falling back to the ground. Neglect air resistance.

SOLUTION The forces acting on the arrow are the spring-like force of the bowstring (as the arrow is shot) and the gravitational force. There are no other forces that do work on the arrow (neglecting friction and air resistance). Therefore the total mechanical energy (the sum of kinetic energy, spring potential energy, and gravitational potential energy) is conserved. To find the maximum height the arrow rises, we equate the initial energy (when the bow is drawn and the arrow is at rest) to the final energy (at the top of the flight).

$$E_f = E_i$$

$$K_f + U_{G,f} + U_{S,f} = K_i + U_{G,i} + U_{S,i}$$

Both the initial and final kinetic energies equal zero. We choose the origin of our y-axis at the arrow's starting point, so that the initial gravitational potential energy is zero. The final spring potential energy is zero because the bow is no longer stretched. Thus the conservation of energy equation becomes

$$U_{G,f} = U_{S,i}$$

$$mgy_f = U_{S,i}$$

Using the potential energy found in Ex. 9, we solve for y_f.

$$y_f = \frac{U_{S,i}}{mg} = \frac{68.8 \text{ J}}{(3.00 \times 10^{-2} \text{ kg})(9.80 \text{ m/s}^2)}$$

$$= 234 \text{ m}$$

In solving this problem we did not need to calculate the arrow's speed as it left the bow. Since mechanical energy is conserved throughout, we simply equated the energy when the bow was drawn to the energy when the arrow reached its highest point.

Fig. 7-24 The work done by friction on a skier moving from A to B depends on the skier's path between these points. Friction is a nonconservative force.

7-5 Conservative and Nonconservative Forces

Conservation of Mechanical Energy

The principle of conservation of total mechanical energy is satisfied when any number of forces act, so long as the work done by each of the forces can be expressed as a decrease in some kind of potential energy. When only such forces, called **conservative forces,** act on a body, the body's mechanical energy is conserved—the mechanical energy being defined as kinetic energy plus the sum of the potential energies corresponding to each of the conservative forces. As we have seen, both the gravitational force and the spring force are conservative. Another example of a conservative force is the electric force. In Chapter 18 we introduce electrical potential energy. Friction is an example of a force that is not conservative. There is no potential energy associated with friction, and so the mechanical energy of a body is not conserved when friction does work on it.

For example, suppose a skier skis down a slope from point A to point B (Fig. 7-24). The work done by friction depends on the path the skier chooses between points A and B. Little work is done by friction if the path is direct. But if the skier turns back and forth, there is considerable negative work done by friction, which tends to cancel the positive work done by gravity and to keep the skier's kinetic energy more or less constant.

Nonconservation of Mechanical Energy

In general both conservative and nonconservative forces act on a body. If we label the sum of the work done by all the conservative forces ΣW_c and the sum of the work done by all the nonconservative forces ΣW_{nc}, then the net work is the sum of these two terms.

$$W_{net} = \Sigma W_c + \Sigma W_{nc}$$

But ΣW_c equals the decrease in the total potential energy, $U_i - U_f$, and W_{net} equals the increase in kinetic energy, $K_f - K_i$, and so we find

$$K_f - K_i = U_i - U_f + \Sigma W_{nc}$$

or

$$K_f + U_f = K_i + U_i + \Sigma W_{nc}$$

Using the definition of the total mechanical energy ($E = K + U$), we may write this as

$$E_f = E_i + \Sigma W_{nc} \quad (7\text{-}32)$$

Using ΔE to denote the change in mechanical energy, $E_f - E_i$, we may express this result as

$$\Sigma W_{nc} = \Delta E \quad (7\text{-}33)$$

The nonconservative work may be either positive or negative. If it is positive, E increases, and if it is negative, E decreases. For example, friction is a nonconservative force that, since it always opposes the motion of a body, always does negative work on the body. Therefore, when friction is the only nonconservative force acting on a body, the body's mechanical energy decreases: $\Delta E = W_f < 0$.

EXAMPLE 11 Increasing the Energy of a Barbell by Lifting It

A weight lifter lifts a 1.00×10^3 N (225 lb) weight a vertical distance of 2.00 m (Fig. 7-25). (a) Find the increase in the total mechanical energy of the weight, assuming that there is little or no increase in the weight's kinetic energy. (b) Find the work done by the force **F** exerted on the weight by the weight lifter.

SOLUTION (a) The weight's change in mechanical energy, ΔE, equals its increase in gravitational potential energy.

$$\Delta E = \Delta U_G = mgy_f - mgy_i$$
$$= mg(y_f - y_i) = (1.00 \times 10^3 \text{ N})(2.00 \text{ m})$$
$$= 2.00 \times 10^3 \text{ J}$$

(b) The contact force **F** exerted on the weight by the weight lifter is a nonconservative force. The work W_F done by this force equals the total nonconservative work done on the weight and, according to Eq. 7-33, equals the weight's increase in mechanical energy.

$$W_F = \Sigma W_{nc} = \Delta E = 2.00 \times 10^3 \text{ J}$$

Fig. 7-25 A weight lifter provides a nonconservative force **F**, doing positive work and increasing the weight's mechanical energy.

It is easy to verify this result by directly computing the work done by **F** as the product of the force times the distance. Since the force **F** balances the weight, it has a magnitude of 1.00×10^3 N. We can compute the work as

$$F_x \Delta x = (1.00 \times 10^3 \text{ N})(2.00 \text{ m}) = 2.00 \times 10^3 \text{ J}$$

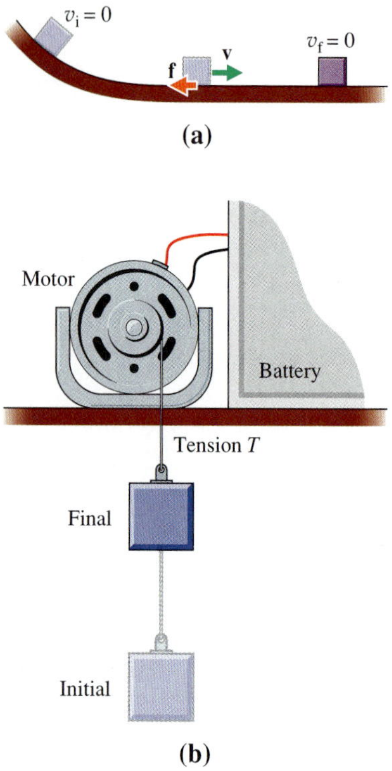

Fig. 7-26 **(a)** A block slides down an incline and then comes to rest. The mechanical energy of the block decreases. **(b)** An electric motor, powered by a battery, raises a weight. The mechanical energy of the weight increases.

Other Forms of Energy

Eq. 7-32 ($E_f = E_i + W_{nc}$) might *seem* to imply that the principle of conservation of energy is not always valid. However, this equation implies only that *mechanical* energy is not always conserved. It is always possible to identify a change in energy of some system that exactly balances a change in the mechanical energy of a body, though this compensating energy may be some other form of energy. To have a universal law of conservation of energy, we must enlarge the definition of energy to include more than just mechanical energy.

Fig. 7-26 shows examples of nonconservation of mechanical energy. In Fig. 7-26a the friction force on a block sliding down an incline causes the block to come to rest. Friction does negative work on the block, which accounts for the block's loss of mechanical energy ($\Delta E = W_f < 0$). But there is an increase in the temperature and the "thermal energy" of both the block and the surface. It turns out that the block's loss of mechanical energy is exactly balanced by the increase of thermal energy (to be studied in Chapter 13). In Fig. 7-26b an electric motor, powered by a battery, raises a weight. The mechanical energy of the weight increases as the result of the positive work done by the nonconservative force exerted on the weight by the tension **T** in the line ($\Delta E = W_T > 0$). It turns out that the weight's gain in mechanical energy is balanced by a loss in the battery's chemical energy (if we assume there is negligible friction and electrical resistance). The chemical energy of batteries is discussed in Chapter 19.

7-6 Power

The rate at which work is performed by a force is defined to be the power output of the force. The average power, denoted by $\overline{P}$, is the work divided by the time Δt over which the work is performed.

$$\overline{P} = \frac{W}{\Delta t} \qquad \text{(average power)} \qquad (7\text{-}34)$$

The instantaneous power, P, is the limiting value of this ratio, for a time interval approaching zero.

$$P = \lim_{\Delta t \to 0} \frac{W}{\Delta t} \qquad \text{(instantaneous power)} \qquad (7\text{-}35)$$

In our previous discussion of work and energy, we did not consider the time during which work is performed, or the rate at which work is performed. But this is an important consideration in many applications. For example, suppose you carry a heavy piece of furniture up a flight of stairs. The work required is the same, whether you walk or run up the stairs. However, you may find it difficult or impossible to run with such a heavy load. The key difference is that if you run, the rate at which work is done is much greater; that is, the power required is much greater. And you may not be capable of producing that much power.

Units

The SI unit of power is the J/s, which is called the "watt" (abbreviated W), in honor of James Watt, the inventor of the steam engine.

$$1 \text{ W} = 1 \text{ J/s} \tag{7-36}$$

In the British system, the unit of power is the ft-lb/s. The horsepower, abbreviated hp, is a larger, more commonly used unit.

$$1 \text{ hp} = 550 \text{ ft-lb/s} \tag{7-37}$$

This definition was introduced by Watt, based on his estimate of the maximum average power that could be delivered by a typical horse, over a period of a work day. One horsepower equals approximately three fourths of a kilowatt, or more precisely

$$1 \text{ hp} = 746 \text{ W} \tag{7-38}$$

A convenient unit of work or energy is the kilowatt-hour, abbreviated kWh. It is defined as the work or energy delivered at the rate of 1 kilowatt for a period of 1 hour. Since $W = \overline{P}\,\Delta t$,

$$1 \text{ kWh} = (1 \text{ kW})(1 \text{ h}) = (10^3 \text{ W})(3.60 \times 10^3 \text{ s})$$

$$1 \text{ kWh} = 3.60 \times 10^6 \text{ J} \tag{7-39}$$

The kilowatt-hour is commonly used by utility companies to measure the use of electrical energy. For example, if you are using electrical energy at the rate of 2 kW for a period of 10 hours, your energy consumption is 20 kWh.

EXAMPLE 12 The Power Required to Lift a Chamber From the Ocean Floor

A deep sea, underwater observation chamber is raised from the bottom of the ocean, 1 mile below the surface, by means of a steel cable. The chamber moves upward at constant velocity, reaching the surface in 5.00 minutes. The cable is under a constant tension of 2.00×10^3 lb. Find the power output required of the electric motor that pulls the cable in.

SOLUTION The power output of the motor is the rate of production of work by the tension force it supplies.

$$\overline{P} = \frac{W}{\Delta t} = \frac{F_x\,\Delta x}{\Delta t} = \frac{(2.00 \times 10^3 \text{ lb})(5280 \text{ ft})}{(5.00 \text{ min})(60 \text{ s/min})}$$

$$= (3.52 \times 10^4 \text{ ft-lb/s})\left(\frac{1 \text{ hp}}{550 \text{ ft-lb/s}}\right)$$

$$= 64.0 \text{ hp}$$

It is sometimes convenient to express the instantaneous power in terms of the instantaneous speed of the body on which work is performed. Since the work done by force **F** during a displacement Δs is $F_s\,\Delta s$, the power P may be written

$$P = \lim_{\Delta t \to 0} \frac{W}{\Delta t} = \lim_{\Delta t \to 0} \frac{F_s\,\Delta s}{\Delta t}$$

or, since speed $v = \lim_{\Delta t \to 0} \dfrac{\Delta s}{\Delta t}$,

$$\boxed{P = F_s v} \tag{7-40}$$

EXAMPLE 13 The Power Required to Drive Uphill or to Accelerate

(a) A car weighing 1.00×10^4 N travels at a constant velocity of 20.0 m/s up a 5.00° incline. Find the power that must be supplied by the force moving the car up the hill, assuming negligible friction and air resistance. (b) Find the power that must be supplied to accelerate the car at $0.100g$ on level ground when its speed is 10.0 m/s.

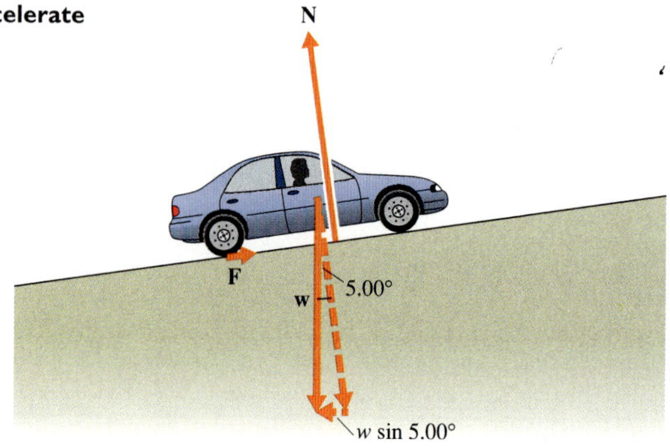

SOLUTION (**a**) Since the car's velocity is constant, the road must exert a force **F** that balances the component of weight down the incline (Fig. 7-27).

$$F = w \sin 5.00°$$

The power delivered by this force is found by applying Eq. 7-40.

$$\begin{aligned} P &= F_s v = (w \sin 5.00°)v \\ &= (1.00 \times 10^4 \text{ N})(\sin 5.00°)(20.0 \text{ m/s}) \\ &= 1.74 \times 10^4 \text{ W} \end{aligned}$$

or, since 1 hp = 746 W,

$$P = (1.74 \times 10^4 \text{ W})\left(\frac{1 \text{ hp}}{746 \text{ W}}\right) = 23.4 \text{ hp}$$

Fig. 7-27

(**b**) From Newton's second law, we know that the force **F** accelerating the car has magnitude

$$F = ma = m(0.100\text{g}) = 0.100w$$

The power supplied by this force is

$$\begin{aligned} P &= F_s v = (0.100w)v \\ &= (0.100)(1.00 \times 10^4 \text{ N})(10.0 \text{ m/s}) \\ &= 1.00 \times 10^4 \text{ W} \quad \text{(or 13.4 hp)} \end{aligned}$$

A Closer Look

The Energy to Run

Why is it so much harder to run than to ride a bicycle at the same speed? When you ride a bicycle, it is after all your own body that produces your motion, just as when you run. And yet cycling requires much less effort than running. After 30 minutes or an hour of running along a level road at a moderate pace, even a well-conditioned runner may tire, whereas a cyclist can keep the same pace with little effort (Fig. 7-A). In everyday language, we say that "running burns calories" or that "running uses a lot of energy." To understand the physical basis of such expressions, to see why running requires so much energy and is so much less energy efficient than bicycle riding, we shall apply concepts of work and energy to the human body. In the next (optional) section, we shall show in detail how to extend concepts of work and energy to systems of particles such as human bodies and machines. Here we shall simply describe in a general way how energy is used by the body when muscles contract and specifically how that energy is used in running and cycling.

The following are some general properties of work and energy associated with muscular exertion:

Work Done by Muscles

Muscles consist of bundles of muscle fibers. Under tension, these fibers can shorten, or "contract," as protein filaments within the fibers slide over each other. (See Chapter 10, Fig. 10-4 for a detailed description of the mechanism of muscle contraction.) Contraction of a muscle fiber means that a force (the tension in the muscle fiber) acts through a distance (the distance the fiber contracts). Hence work is done by contracting muscle fibers. The direct effect of a muscle's contraction may be to move one of the body's limbs. The moving limb, in turn, may exert a force on the surroundings and do work on the surroundings. For example, if you hold a weight in your hand and contract the biceps muscle in your arm, your hand and forearm swing upward, raising the weight. The work done by your biceps muscle is approximately equal to the work done by the force your hand exerts on the weight. The effect of this work is to increase the weight's gravitational potential energy. (See Chapter 10, Example 3 and Problem 36 for details.)

Heat Generated by the Body When Muscles Contract

Heat, a disordered form of energy, is generated whenever muscles do work. Typically the quantity of heat generated when muscles contract is about three times as great as the work done by the muscles. When your muscles do very much work, you can usually feel the heat generated by your body. You may begin to sweat, which is a way the body gets rid of excess heat.

Internal Energy of the Body

The body's internal energy is the total energy of all the particles within the body. Chemical reactions within the body provide the energy necessary to produce muscle contraction. The energy released by these chemical reactions produces the work and heat associated with muscle contraction. The body thereby loses some of its internal energy. Conservation of energy implies that the body's loss of internal energy equals the sum of the work and heat generated.

$$\text{Loss of internal energy} = \text{Work done by muscles} + \text{Heat generated}$$

When your body loses much internal energy in a short time interval, you tend to feel tired. Your body's internal energy is replenished by the consumption of food.

Now we can use these basic concepts of work and energy to understand why cycling requires less energy than running. A good bicycle is an exceptionally efficient means of using the body's internal energy to produce motion. Suppose you ride a high-tech bicycle with thin, well-inflated tires and very little friction in its moving parts. Riding such a bike over flat, level pavement at, say, 10 km/h, requires little effort. Once moving, both the kinetic energy and the gravitational poten-

Fig. 7-A An exhausted runner and a still fresh cyclist have traveled the same distance at the same speed.

Continued.

tial energy of the bicycle and your body stay constant with just a little pedaling required. At such a low bike speed, there is not much air resistance. Consequently, only a little work needs to be done by your legs as they push against the pedals and your body loses little internal energy in producing this small amount of work. The work that is done by your legs is needed to compensate for the small negative work done by friction and air resistance. If you did not pedal at all, your bike would gradually slow down.

If you were to ride a bike uphill or at a much higher speed, or if your tires were not well inflated, or if there were much friction in your wheel bearings, you would have to do considerably more work.

In contrast to riding a bike, when you run on a flat, level surface, your kinetic energy and gravitational potential energy can never be exactly constant. Watch a runner and you will see that the runner's head moves up and down somewhat, an indication of some change in elevation of the runner's center of mass. This means that the runner's gravitational potential energy is not constant. Some of that energy is lost each time the runner's body moves downward, and this energy must then be supplied as the body moves upward again. More efficient runners, especially champion marathoners, bob up and down less than average runners do and thereby use less energy.

A runner's center-of-mass kinetic energy also necessarily varies somewhat, again in contrast to that of a cyclist. Although this effect is more difficult to see, a runner's center of mass continually alternates between speeding up and slowing down with each stride. Although the variation in center-of-mass speed is slight, it does require a significant amount of work for the legs to increase the center-of-mass kinetic energy from the minimum value to the maximum value during each stride.

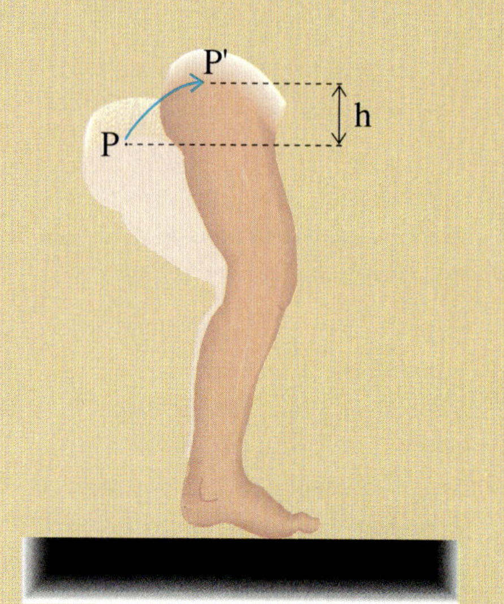

Fig. 7-B The runner's center of mass moves from point P to point P′, increasing elevation by a distance *h*.

Fig. 7-C The road exerts a backward force on a runner's foot as it hits the surface. This force tends to reduce the runner's speed.

To obtain a more detailed understanding of just how runners use energy, experimental studies have been performed using force platforms and high-speed photography (see Chapter 4, Example 11). Such studies* indicate that the body's internal work is used in four ways:

1. To raise the body's center of mass a few centimeters each step, increasing **gravitational potential energy.** Fig. 7-B indicates how running tends to produce an up-and-down motion. Gravitational potential energy is lost as the center of mass falls to its original level, with the energy being converted first into kinetic energy and then lost as the foot strikes the ground. So even though the ground you run on may be level, you are in a sense always going uphill. Of course, if you do actually run up a hill, your leg muscles must do more work to provide extra gravitational potential energy. Running downhill can reduce the work your legs do.
2. To increase slightly the body's center-of-mass speed and hence its **center-of-mass kinetic energy** at the beginning of each stride. This is necessary because as each stride is completed a backward force must be exerted on the forward foot by the road to stop the foot's forward motion and begin its backward motion. The entire body then experiences a backward external force (Fig. 7-C). This force slightly decelerates the runner's center of mass, which then must be accelerated again to its maximum speed. This is accomplished by the leg muscles doing work, pushing the foot backward against the road, so that the road now pushes the runner forward.

*See Alexander R McN: *Biomechanics,* 1975, Halstead Press, New York, pp 28-30.

A Closer Look

3. To provide the kinetic energy of the legs as they swing back and forth. This energy, called **rotational kinetic energy,** is lost as the foot strikes the ground and must be provided by the leg muscles that do work as they push the legs back and forth relative to the body's center of mass. (Rotational kinetic energy will be discussed in Chapter 9, Section 9-4.)
4. To compensate for the negative work done by **air resistance.** If you run at a slow or moderate pace with no wind, air resistance is not a very significant factor—certainly less significant than the other three. However, if you run directly into a strong wind, air resistance can become very significant, sometimes requiring more work than any of the other forms of energy used in running.

In Section 9-6, we shall use a rough mechanical model of running to estimate the power that must be provided by a runner's legs for each of the four types of energy use we have just described. There, for a running speed of 3.0 m/s (9 minutes per mile), we obtain the following estimates:

1. Center-of-mass potential energy, 94 W
2. Center-of-mass kinetic energy, 77 W
3. Rotational kinetic energy of legs, 58 W
4. Air resistance, 11 W

This gives a total power estimate of 240 W, or about one third horsepower! Although this is just a crude estimate,* it does give some idea of the considerable energy we use when we run.

*Studies of the power output of muscular forces have been made, using situations where this power is easily measured directly. For example, in the case of cycling, the power output of the muscular forces is very nearly the power delivered to the bike's pedals, and this power can be measured. The maximum power output that the human body can produce depends very much on the strength and conditioning of the individual. Exceptional athletes can maintain a power output of $\frac{1}{3}$ hp for about 1 hour, or 1 hp for about 1 minute, or 2 hp for about 6 seconds.

*7-7 Energy of a System of Particles

When a gymnast performs the "iron cross" (Fig. 7-28), his body is stationary and therefore no work is done by any force he exerts on his surroundings. And yet the gymnast's muscles tire after a few seconds of performing this difficult feat. Energy is used because the muscle fibers are under tension and are continually contracting and relaxing. Work is done by these internal tension forces, and energy must be supplied by the gymnast's body to do this work. In an example such as this, energy does not belong to a single particle. Instead, we have to regard the total energy as being distributed over the system of particles in the gymnast's body. We need to extend our treatment of energy to systems of particles so that we will be able to introduce some of the most interesting applications of energy concepts, such as work done by internal forces within a human body or a machine.

Fig. 7-28 A gymnast performs the iron cross.

Eq. 7-33 ($\Sigma W_{nc} = \Delta E$) applies to each particle in a system of n particles; that is, a particle's increase in mechanical energy equals the net work done by internal or external forces acting on that particle. If we write out this equation for each particle of the system and add the equations, we obtain a useful equation, applicable to the system as a whole.

$$\Sigma W_{nc,1} = \Delta E_1 \qquad \text{(for particle 1)}$$

$$\Sigma W_{nc,2} = \Delta E_2 \qquad \text{(for particle 2)}$$

$$\vdots \qquad \vdots$$

$$\Sigma W_{nc,1} + \Sigma W_{nc,2} + \cdots = \Delta E_1 + \Delta E_2 + \cdots$$

$$= \Delta(E_1 + E_2 + \cdots) \qquad \text{(7-41)}$$

We define the system's energy E to be the **sum of the single particle energies.**

$$E = E_1 + E_2 + \cdots \tag{7-42}$$

If we now let $\Sigma\, W_{nc}$ denote the **sum of the work done on all particles** of the system, we can express Eq. 7-41 as

$$\Sigma\, W_{nc} = \Delta E \tag{7-43}$$

This equation looks identical to Eq. 7-33 for a single particle. Here, however, we interpret E as the energy of a system and $\Sigma\, W_{nc}$ as the total work done on the system.

Sometimes internal forces do work, for example, forces within the muscles of the gymnast in Fig. 7-28, or forces on the pistons in the cylinders of an automobile engine. However, internal work is not present in a rigid body, a body in which there is no relative motion of the body's particles. This follows from the fact that the internal forces occur in oppositely directed, action-reaction pairs. Interacting particles within a rigid body experience the same displacement. Therefore the work done on one particle by an internal force **F** is the negative of the work done on the other interacting particle by the reaction force −**F**. The net work is then zero.

The nonconservative forces acting on a body deliver average power, $\Sigma\, \overline{P}_{nc}$, which is the net work per unit time performed by these forces.

$$\Sigma\, \overline{P}_{nc} = \Sigma\, \frac{W_{nc}}{\Delta t} = \frac{\Sigma\, W_{nc}}{\Delta t}$$

Using Eq. 7-43 ($\Sigma\, W_{nc} = \Delta E$), this may be expressed

$$\Sigma\, \overline{P}_{nc} = \frac{\Delta E}{\Delta t} \tag{7-44}$$

EXAMPLE 14 The Power Required to Run Up a Flight of Stairs

Compute the mechanical power provided by internal forces within the body of a person of mass 80.0 kg who runs up a flight of stairs, rising a vertical distance of 3.00 m in 3.00 s.

SOLUTION The only nonconservative forces doing work on the body are internal forces within the body. We find the average power output of these forces, $\overline{P}_{int}$, by applying Eq. 7-44.

$$\overline{P}_{int} = \Sigma\, \overline{P}_{nc} = \frac{\Delta E}{\Delta t}$$

If the body's kinetic energy is approximately constant during the climb, the only change in energy is the increase in gravitational potential energy.

$$\Delta E = \Delta U_G = mgy_f - mgy_i = mg(y_f - y_i)$$

Inserting this into the expression for power, we obtain

$$\overline{P}_{int} = \frac{mg(y_f - y_i)}{\Delta t} = \frac{(80.0\text{ kg})(9.80\text{ m/s}^2)(3.00\text{ m})}{3.00\text{ s}}$$

$$= 784\text{ W} \quad (\text{or } 1.05\text{ hp})$$

EXAMPLE 15 Energy and Power Required for Hill Climbing on a Bicycle

A cyclist rides up Latigo Canyon, rising from sea level to a final elevation of 869 m (2850 ft) (Fig. 7-29). The combined weight of the cyclist and the bicycle is 825 N (185 lb).
(a) Find the increase in mechanical energy in the system of the cyclist and the bicycle.
(b) What is the minimum work done by internal forces in the cyclist's body?
(c) How much power is supplied by the cyclist if he reaches the top in 1 hour?

SOLUTION (a) Kinetic energy is approximately constant, and so the increase in mechanical energy equals the increase in the cyclist's gravitational potential energy, which is quite large.

$$\Delta E = \Delta U_G = mg(\Delta y) = (825\text{ N})(869\text{ m})$$
$$= 7.17 \times 10^5\text{ J}$$

(b) There are several nonconservative forces acting on and within the system. The external forces of air resistance and road friction both do negative work on the system. In addition there is some internal friction in the bicycle wheel bearings and crank; these forces also do negative work. It is the positive work done by tension forces in the cyclist's muscles that is responsible for the increase in mechanical energy. In the most ideal case of negligible friction and air resistance, the work W_{int} done by the cyclist's muscles equals the net nonconservative work. According to Eq. 7-43, this equals the system's increase in mechanical energy:

$$W_{int} = \Sigma W_{nc} = \Delta E = 7.17 \times 10^5\text{ J}$$

More realistically, the internal work of the muscles must be somewhat greater, since part of it is used to balance the negative work done by friction and air resistance.

Fig. 7-29 The cyclist is ascending Latigo Canyon in Malibu, California. Beginning at sea level the road rises 2850 ft over a distance of 7 miles.

The source of energy here is stored "internal energy" in the cyclist's body. We shall study internal energy in Chapter 14 on thermodynamics. Here we simply note that the human body is at best only about 25% efficient; that is, the work performed by the muscles equals about 25% of the internal energy used by the body (the other 75% is converted to heat). Thus the loss in internal energy equals $4(7.17 \times 10^5\text{ J}) = 2.87 \times 10^6\text{ J}$, or 685 Calories (about the number of Calories provided by a half-pint of Häagen-Dazs ice cream).
(c) The power is found when we divide the work by the time interval.

$$P = \frac{W_{int}}{\Delta t} = \frac{7.17 \times 10^5\text{ J}}{3600\text{ s}} = 199\text{ W} \quad (\text{or } 0.27\text{ hp})$$

CHAPTER 7 SUMMARY

The kinetic energy of a particle of mass m moving at speed v is given by

$$K = \frac{1}{2}mv^2$$

The work done on a particle is given by

$$W = \Sigma\,(F_s\,\Delta s)$$

where the sum is over short intervals of length Δs, and F_s is the component of the force along the interval. If the path is linear and the force constant,

$$W = F_x\,\Delta x$$

Work and kinetic energy are related through the work-energy theorem, which states that a particle's increase in kinetic energy equals the net work done by the force acting on the particle.

$$\Delta K = W_{\text{net}}$$

A force is called conservative if the work done by it can be expressed as a decrease in some kind of potential energy. The gravitational force and the spring force are both conservative. Their respective potential energies are:

$$U_G = -\frac{GmM}{r}$$

or

$$U_G = mgy \quad \text{(if the distance from the center of the earth is essentially constant)}$$

and

$$U_S = \frac{1}{2}kx^2$$

Friction is a nonconservative force.

The total mechanical energy is defined to be the sum of the kinetic energy and the various potential energies.

$$E = K + U = K + U_G + U_S + \cdots$$

If only conservative forces do work on a particle, its total mechanical energy is conserved.

$$E_f = E_i \quad \text{(conservative forces only)}$$

More generally, a particle's mechanical energy may increase or decrease, if positive or negative work is done by nonconservative forces.

$$\Sigma\,W_{nc} = \Delta E = E_f - E_i$$

This equation may also be applied to a system of particles if the energy E is interpreted as the sum of single particle energies.

$$E = E_1 + E_2 + \cdots + E_n$$

Power is the rate at which work is done.

Average power $$\bar{P} = \frac{W}{\Delta t}$$

Instantaneous power $$P = \lim_{\Delta t \to 0} \frac{W}{\Delta t}$$

The instantaneous power provided by a force **F** to a body moving at speed v may be expressed

$$P = F_s v$$

The net power provided by nonconservative forces may be expressed

$$\Sigma\,\bar{P}_{nc} = \frac{\Delta E}{\Delta t}$$

Questions

1 Can the kinetic energy of a body ever be negative?

2 You drive your car along a curving road at constant speed.
(a) Does your kinetic energy change?
(b) Is the work done by the force accelerating your car positive, negative, or zero?

3 (a) Does the kinetic energy of a body depend on the reference frame of the observer?
(b) Does the work done on a body depend on the reference frame of the observer?

4 A child on a skateboard grabs the rear bumper of a car and is towed up a hill. The speed of the car is 20 mph at the bottom and 10 mph at the top of the hill. Is the work done on the child and skateboard by the following forces positive, negative, or zero: (a) force of the bumper on the child; (b) force of gravity on the child; (c) resultant force on the child?

5 Fig. 7-15 shows a skier skiing down a hill. Could the skier have chosen another path between the same two points such that (a) the work done by gravity was greater; (b) the work done by friction was greater?

6 Two objects are simultaneously released from the same height. One falls straight down, and the other slides without friction down a long inclined plane.
(a) Do both have the same acceleration?
(b) Do both have the same final speed?
(c) Do both take the same time to descend?

7 A satellite goes from a low circular earth orbit to a higher circular earth orbit.
(a) Does the gravitational force on the satellite increase, decrease, or remain the same?
(b) Does the satellite's gravitational potential energy increase, decrease, or remain the same?

8 Can the work done by a force always be expressed as a decrease in potential energy?

9 A boy rides a bicycle along level ground at *approximately* constant velocity, without pedaling. Is mechanical energy approximately conserved?

10 A boy rides a bicycle along level ground at constant velocity, pedaling at a steady rate.
(a) Is mechanical energy conserved?
(b) Is there any work done by individual nonconservative forces?
(c) Is there any net work done by nonconservative forces?

11 Suppose you run up a hill at constant speed.
(a) Is there net work done on your body?
(b) Is there a net nonconservative work done on your body?
(c) Is your mechanical energy conserved?

12 (a) As you drive a car from the top of a mountain to its base, is the car's mechanical energy conserved? Explain.
(b) Would you expect to get better gas mileage than on a flat road?

13 A fountain of water shoots high in the air. The water then falls back into a surrounding pool.
(a) Describe the transformation of energy the water undergoes, beginning with the water shooting upward at the base of the fountain.
(b) Is the mechanical energy of the water conserved as the water completes its cycle?
(c) What provides the nonconservative force that does positive work on the water?

14 A worker raises a load of bricks from the ground to a platform. The worker can lift one brick at a time or all the bricks together.
(a) In which case is the work greater, or is it the same in either case? Neglect the work done in raising the body each time he bends over.
(b) Take into account the work done in raising the body. In which case is the work greater?
(c) In which case is the required power greater?

15 Given that an automobile can develop only a limited amount of power, does this put a limit on the maximum slope of a mountain road that a given automobile can drive up at a given speed?

16 Weight lifters find that the greatest gains in strength (and the greatest muscle soreness) occur as a result of doing "negatives," that is, doing negative work on very large weights. Is negative work done by raising or lowering a weight?

Answers to Odd-Numbered Questions

1 no; **3** (a) yes; (b) yes; **5** (a) no; (b) yes; **7** (a) decrease; (b) increase; **9** yes; **11** (a) no; (b) yes; (c) no; **13** (a) kinetic energy to gravitational potential energy to kinetic energy to heat; (b) yes; (c) the water pump; **15** yes

Problems (listed by section)

7-1 Work and Kinetic Energy

1 Find the kinetic energy of (a) a bullet of mass 5.00 g traveling at a speed of 300 m/s; (b) a woman of mass 50.0 kg running at a speed of 9.00 m/s; (c) a car of mass 1.00×10^3 kg moving at a speed of 20.0 m/s.

2 The moon has a mass of 7.36×10^{22} kg and moves about the earth in a circular orbit of radius 3.80×10^8 m with a period of 27.3 days.
(a) Find the moon's kinetic energy as observed on earth.
(b) Would the moon's kinetic energy be the same from the sun's reference frame?

3 A man of mass 80.0 kg walks down the aisle of an airplane at a speed of 1.00 m/s in the forward direction while the plane moves at a speed of 300 m/s relative to the earth. Find the man's kinetic energy relative to (a) the plane; (b) the earth.

4 Suppose you carry a bag of groceries weighing 125 N from your car to your kitchen, a distance of 50 m, without raising or lowering the bag.
(a) What is the work done by the force you exert on the bag?
(b) Would the work be different if your kitchen were in an upstairs apartment?

5 Tarzan, who weighs 875 N, swings from a vine through the jungle. How much work is done by the tension in the vine as he drops through a vertical distance of 4.00 m?

6 One boat tows another boat by means of a tow line, which is under a constant tension of 500 N. The boats move at a constant speed of 5.00 m/s. How much work is done by the tension in 1.00 min?

7 You lift a box weighing 200 N from the floor to a shelf 1.50 m above.
(a) What is the minimum work done by the force you exert on the box?
(b) When would the work be greater than this minimum?

8 A weight lifter raises a 900 N weight a vertical distance of 2.00 m. Compute the work done by the force exerted on the weight by the weight lifter.

9 You are loading a refrigerator weighing 2250 N onto a truck, using a wheeled cart. The refrigerator is raised 1.00 m to the truck bed when it is rolled up a ramp. Calculate the minimum work that must be done by the force you apply and the magnitude of the force if the ramp is at an angle with the horizontal of (a) 45.0°; (b) 10.0°.

10 A man drags a table 4.00 m across the floor, exerting a constant force of 50.0 N, directed 30.0° above the horizontal.
(a) Find the work done by the applied force.
(b) How much work is done by friction? Assume the table's velocity is constant.

11 The driver of a 1500 kg car, initially traveling at 10.0 m/s, applies the brakes, bringing the car to rest in a distance of 20.0 m.
(a) Find the net work done on the car.
(b) Find the magnitude and direction of the force that does this work. (Assume this force is constant.)

12 A child on a sled is initially at rest on an icy horizontal surface. The sled is pushed until it reaches a final velocity of 6.00 m/s in a distance of 15.0 m. The coefficient of friction between the ice and runners of the sled is 0.200, and the weight of the child and the sled is 350 N. Find the work done by the force pushing the sled.

★**13** Air bags are used in cars to decelerate the occupants slowly when a car is suddenly decelerated in a crash.
(a) Compute the work done by the decelerating force acting on a 55.0 kg driver if the car is brought to rest from an initial speed of 20.0 m/s.
(b) Find the minimum thickness of the air bag if the average decelerating force is not to exceed 8900 N (2000 lb), and the center of the car moves forward 0.800 m during impact.

★**14** A particle moves along the x-axis from $x = 0$ to $x = 4$ m while acted upon by a force whose x component is given in Fig. 7-30. Estimate the work done by the force.

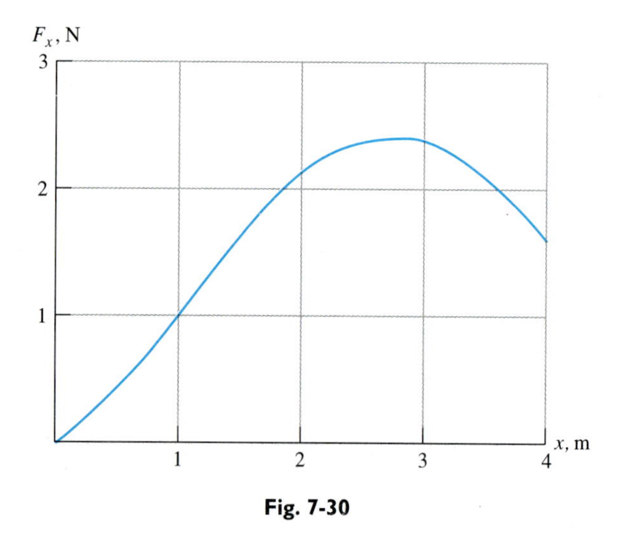

Fig. 7-30

7-2 Gravitational Potential Energy; Constant Gravitational Force

15 Suppose you are driving in the High Sierras from Mammoth Mountain to the Owens Valley 1500 m below. If the mechanical energy of your car were conserved, what would be your approximate final speed?

16 A small weight is suspended from a string of negligible weight and given an initial horizontal velocity of 2.00 m/s, with the string initially vertical. Find the maximum angle θ that the string makes with the vertical if the string is 1.00 m long.

17 You ski straight down a 45.0° slope, starting from rest and traveling a distance of 10.0 m along the slope. Find your final velocity, assuming negligible air resistance and friction.

18 A skier of mass 70.0 kg rides a ski lift to the top, which is 500 m higher than the base of the lift.
(a) Find the increase in the skier's gravitational potential energy.
(b) Find the minimum work done by the force exerted on the skier by the lift.
(c) When the skier skis down the run, what would her final velocity be if no force other than gravity did work? What other forces do work?

19 A cyclist coasts up a 10.0° slope, traveling 20.0 m along the road to the top of the hill. If the cyclist's initial speed is 9.00 m/s, what is the final speed? Ignore friction and air resistance.

★ **20** On a ski jump a skier accelerates down a ramp that curves upward at the end, so that the skier is launched through the air like a projectile. The objective is to attain the maximum distance down the hill. Prove that the vertical drop h must equal at least half the horizontal range R (Fig. 7-31).

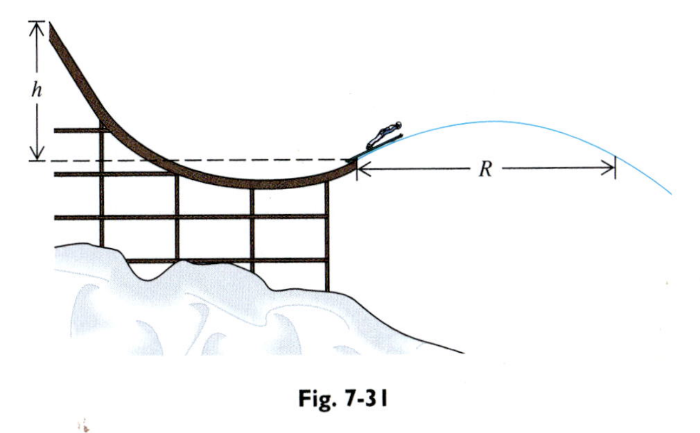

Fig. 7-31

21 Find the minimum work required to carry a truckload of furniture weighing 2.00×10^4 N to a third-story apartment, 20.0 m above the truck.

22 A ladder 2.50 m long, weighing 225 N, initially lies flat on the ground. The ladder is raised to a vertical position. Compute the work done by the force lifting the ladder. The ladder's center of gravity is at its geometric center.

23 What is the average force exerted on the diver in Fig. 7-7a by the diving board, if she weighs 700 N and accelerates from rest to a speed of 4.00 m/s while moving 0.300 m upward.

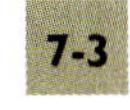

7-3 Gravitational Potential Energy; Variable Gravitational Force

24 Compute the escape velocity on Jupiter, which has a radius of 7.14×10^7 m and mass 318 times the earth's mass.

25 Find the minimum initial speed of a projectile in order for it to reach a height of 2000 km above the surface of the earth.

26 Suppose a rocket is at an elevation of 100 km and has an initial velocity of 1.00×10^4 m/s, directed vertically upward. If the rocket engines do not burn and no force other than the earth's gravity acts on the rocket, how far does it go?

27 If a space probe has a speed of 2.00×10^4 m/s as it leaves the earth's atmosphere, what is its speed when it is far from the earth?

★ **28** The Little Prince is a fictional character who lives on a very small planet (Fig. 7-32). Suppose that the planet has a mass of 2.00×10^{13} kg and a radius of 1.00×10^3 m.
(a) How long would it take for an object to fall from rest a vertical distance of 1.00 m?
(b) Suppose the Little Prince throws a ball vertically upward, giving it an initial velocity of 1.00 m/s. What would be the maximum height reached by the ball? (HINT: Don't assume g to be constant.)
(c) At what speed could a ball be thrown horizontally so that it would travel in a circular orbit just above the surface of the planet?

Fig. 7-32 The Little Prince.

★ **29** A satellite of mass m is in a circular earth orbit of radius r.
(a) Find an expression for the satellite's mechanical energy.
(b) Calculate the satellite's energy and speed if $m = 1.00 \times 10^4$ kg and $r = 1.00 \times 10^7$ m.

★ **30** Halley's comet is in an elongated elliptical orbit around the sun and has a period of about 76 years. Last seen in 1986, it will again be close to the sun and the earth in 2061. The comet's maximum distance from the sun is 5.3×10^{12} m, at which point (called "aphelion") its speed is 910 m/s.
(a) Find its speed when it is at its point of closest approach (perihelion), 8.8×10^{10} m from the sun, which has a mass of 2.0×10^{30} kg.
(b) The radius of the earth's orbit is 1.5×10^{11} m, a distance defined as an astronomical unit. Estimate the time required for Halley's comet to travel a distance equal to one astronomical unit, when it is near perihelion.

7-4 Spring Potential Energy; Conservation of Energy

31 A spring with a force constant of 1500 N/m is compressed 10.0 cm. Find the work done by the force compressing the spring.

32 When an archer pulls an arrow back in his bow, he is storing potential energy in the stretched bow.
(a) Compute the potential energy stored in the bow, if the arrow of mass 5.00×10^{-2} kg leaves the bow with a speed of 40.0 m/s. Assume that mechanical energy is conserved.
(b) What average force must the archer exert in stretching the bow if he pulls the string back a distance of 30.0 cm?

33 A toy consists of a plastic head attached to a spring of negligible mass. The spring is compressed a distance of 2.00 cm against the floor, and then the toy is released. The toy has a mass of 100 g and rises to a height of 60.0 cm above the floor. What is the spring constant?

34 An elevator car of mass 800 kg falls from rest 3.00 m, hits a buffer spring, and then travels an additional 0.400 m, as it compresses the spring by a maximum of 0.400 m. What is the force constant of the spring?

35 A 4.00 kg block starts from rest and slides down a frictionless incline, dropping a vertical distance of 3.00 m, before compressing a spring of force constant 2.40×10^4 N/m. Find the maximum compression of the spring.

36 A mass of 0.250 kg is attached to the end of a massless spring of unknown spring constant. The mass is dropped from rest at point A, with the spring initially unstretched. As the mass falls, the spring stretches. At point B the mass is as shown in Fig. 7-33.
(a) Find the force constant of the spring.
(b) Find the magnitude of the acceleration of the mass at point B.

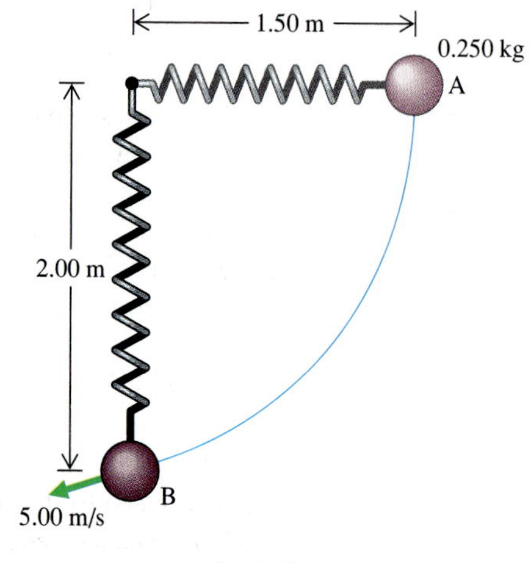

Fig. 7-33

37 A pole-vaulter begins a jump with a running start. He then plants one end of the pole and rotates his body about the other end, thereby rising upward (Fig. 7-34). At the beginning of the vault the pole bends, storing potential energy somewhat in the manner of a compressed spring. Near the top of the arc, the pole unbends, releasing its potential energy and pushing the pole-vaulter higher. With what speed must the pole-vaulter approach the bar if he is to raise his center of mass 5.00 m? Assume that mechanical energy is conserved. The world record in the pole vault is 6.0 m, and the fastest speed achieved by a runner is about 10 m/s.

Fig. 7-34

7-5 Conservative and Nonconservative Forces

38 A 1.00 kg block starts from rest at the top of a 20.0 m long 30.0° incline. Its kinetic energy at the bottom of the incline is 98.0 J. How much work is done by friction?

39 A 1.00 kg block slides down a 20.0 m long 30.0° incline at constant velocity. How much work is done by friction?

40 A ball of mass 0.300 kg is thrown upward, rising 10.0 m above the point at which it was released. Compute the average force exerted on the ball by the hand, if the hand moves through a distance of 20.0 cm as the ball is accelerated.

41 A person jumps from a burning building onto a fireman's net 15.0 m below. If the average force exerted by the net on the person is not to exceed 20 times the body weight, by how much must the center of the net drop as the person comes to rest?

★ **42** A skier skis down a steep slope, maintaining a constant speed by making turns back and forth across the slope as indicated in Fig. 7-35. The side edge of the skis cuts into the snow so that there is no chance of sliding directly down the slope; that is, there is a large static frictional force perpendicular to the length of the skis. There is a much smaller kinetic friction along the length of the skis. If the coefficient of kinetic friction is 0.10, what must be the total length of the skier's tracks as he drops a vertical distance of 100 m down a 40° slope at constant speed?

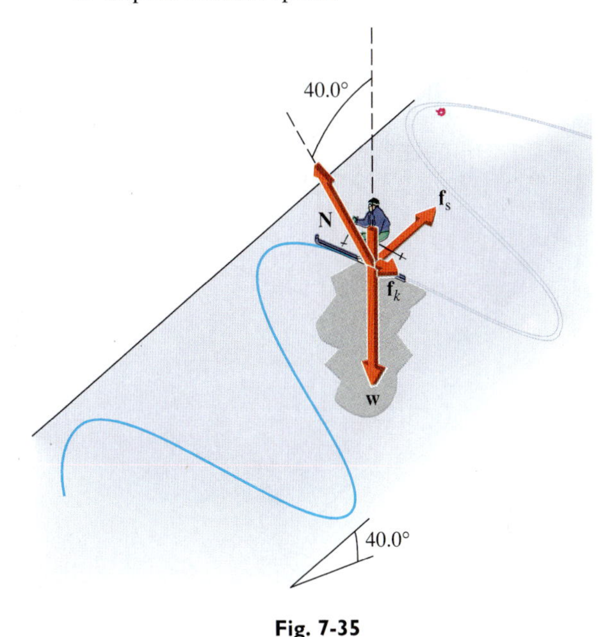

Fig. 7-35

43 A cyclist competing in the Tour de France coasts down a hill, dropping through a vertical distance of 30.0 m. The cyclist has an initial speed of 8.00 m/s and a final speed of 20.0 m/s. What fraction of the cyclist's initial mechanical energy is lost? What nonconservative forces cause this?

7-6 Power

44 How much work could be performed by a 746 W (1 hp) motor in 1 hour?

45 How much mechanical power must be supplied by a car to pull a boat on a trailer at a speed of 20.0 m/s if the force exerted by the car on the trailer is 2000 N?

46 A weight lifter raises a 1000 N weight a vertical distance of 2.00 m in a time interval of 2.00 s. Compute the power provided by the weight lifter's force.

47 Find the weight that could be lifted vertically at the constant rate of 10.0 ft/s, using the mechanical power provided by a 3.00 hp motor.

48 Ten boxes, each 20.0 cm high and weighing 200 N, initially are all side by side on the floor. The boxes are lifted and placed in a vertical stack 2.00 m high in a time interval of 5.00 s. Compute the power necessary to stack the boxes.

★ **49** Compute the minimum power necessary to operate a ski lift that carries skiers along a 45.0° slope. The lift carries 100 skiers of average weight 700 N at any one time, at a constant speed of 5.00 m/s.

50 (a) Compute the electrical energy used by a household whose monthly electrical bill is $30, computed at the rate of $0.06 per kWh.
(b) How long could this energy be used to burn ten 100 W light bulbs?
(c) If this energy were used to raise a car of mass 2000 kg, to what height would the car be raised?

51 The total annual use of energy in the United States is approximately 10^{19} J. Solar energy provides 1400 W per square meter of area in direct sunlight if this area is perpendicular to the sun's rays. Suppose that solar energy could be used with 100% efficiency. Find the total area of solar energy collectors needed to provide the nation's energy needs if on an average day these collectors could be used for 8.0 hours.

*7-7 Energy of a System of Particles

52 A hiker weighing 575 N carries a 175 N pack up Mt. Whitney (elevation, 4420 m), increasing her elevation by 3000 m.

(a) Find the minimum internal work done by the hiker's muscles.

(b) If she is capable of producing up to 746 W (1.0 hp) for an extended time, what is the minimum time for her to ascend?

53 Find the minimum internal work that must be done by the muscles of a shot-putter to impart an initial velocity to a 16.0 lb shot sufficient to give it a horizontal range of 60.0 ft.

54 A pole-vaulter of mass 80.0 kg is initially at rest before beginning his approach to the bar. He is instantaneously at rest at the high point of his jump, having raised his center of mass 5.00 m. Find the minimum internal work done by the vaulter's muscles.

55 Two blocks of mass 100 g each are initially at rest on a frictionless horizontal surface. The blocks are in contact with opposite ends of a spring of force constant 500 N/m, which is compressed 20.0 cm. Find the final speed of each block, after the spring is allowed to expand.

★**56** A car's engine develops mechanical power at the rate of 30.0 hp while moving along a level road at a speed of 60.0 mi/h. Half the mechanical energy developed by the engine is delivered to the wheels, with the remainder being wasted because of internal friction. Find the mechanical power developed by the engine in order for the 2500 lb car to travel at the same speed up a 6.00% grade ($\sin\theta = 6.00 \times 10^{-2}$).

Fig. 7-36 This house in the Bavarian forest near Munich derives its electric power from a small generator, powered by water from a canal.

★**57** A small canal diverts water from the Perlbach, a river in Bavaria. Water flowing through the canal drops through a small distance and turns a waterwheel, which powers an electric generator, providing electricity for the house shown in Fig. 7-36. Calculate the maximum electric power that can be generated if water moves through the canal at the rate of 1.00×10^3 kg/s and drops through a vertical distance of 2.00 m. (Only about 1 kW is used by the household, the remainder being sold to the local power company.)

Additional Problems

★**58** A pendulum swings through an arc of 90.0° (45.0° on either side of the vertical). The mass of the bob is 3.00 kg and the length of the suspending cord is 2.00 m. Find (a) the tension in the cord at the end points of the swing; (b) the velocity of the bob as it passes its lowest point and the tension in the cord at this point.

★**59** Tarzan grabs a vine, which is initially horizontal, and attempts to swing to the ground (Fig. 7-37). Tarzan weighs 890 N, and the breaking strength of the vine he knows to be 1780 N. As Tarzan is swinging, he is surprised to find that the vine breaks at a certain angle θ. Find θ.

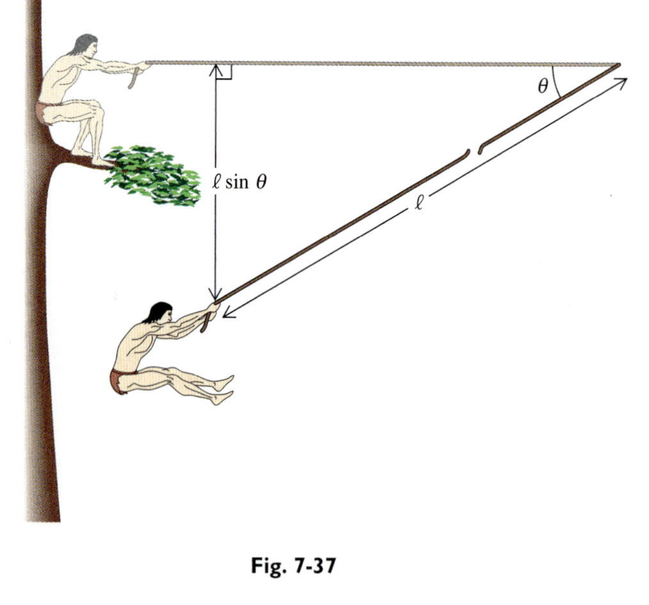

Fig. 7-37

★ **60** Find the minimum initial height h of the roller coaster in Fig. 7-38 if the roller coaster is to complete the 20.0 m diameter loop. Neglect friction.

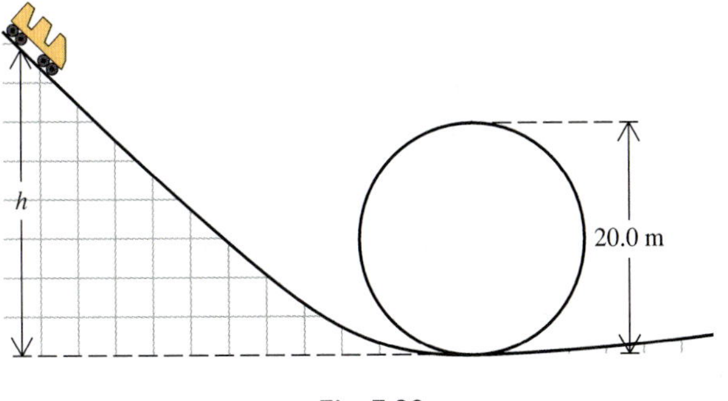

Fig. 7-38

61 The roller coaster in Fig. 7-39 has an initial speed of 7.00 m/s at point A. Find the apparent weight of a 450 N (100 lb) passenger at points B and C. Neglect friction.

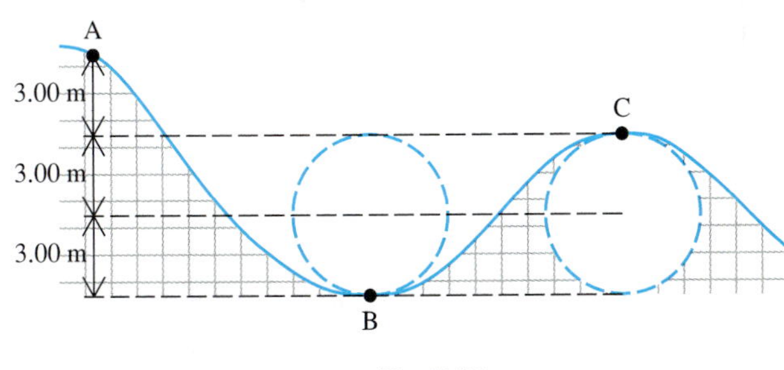

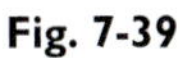

Fig. 7-39

★ **62** In his novel *From the Earth to the Moon,* published in 1865, Jules Verne first suggested that it might be possible to travel to the moon by firing a very high velocity projectile at the moon. Find the minimum initial velocity of such a projectile as it leaves the earth's atmosphere. Take into account the moon's gravitational force.

★ **63** In the novel *The Moon Is a Harsh Mistress,* a colony on the moon threatens the earth with bombardment by heavy stones. These stones are relatively easy to propel from the moon, with its low gravity, and yet reach a very high velocity as they strike the earth.

(a) Compute the minimum initial velocity necessary for a projectile at the surface of the moon in order for it to reach the earth.

(b) Find the velocity of the projectile as it enters the earth's atmosphere.

(c) Calculate the ratio of the projectile's final kinetic energy to its initial kinetic energy.

★ **64** It is estimated that artificial earth satellites have produced approximately 40,000 pieces of debris larger than a pea. Suppose that one of the larger pieces of debris of mass 100 kg is in a circular orbit at two earth radii from the center of the earth and that the mass strikes a satellite in the same orbit, traveling in the opposite direction. Calculate the kinetic energy of the mass relative to the satellite. For comparison, the energy released by 1 million tons of TNT (or a 1 megaton nuclear bomb) equals 4.18×10^9 J.

65 A person weighing 170 lb produces mechanical power of 0.10 hp in walking on a horizontal surface. Suppose that the person can provide a maximum mechanical power of 0.20 hp for 5 hours. What is the maximum height the person could climb up a mountain in this length of time? Assume that the extra 0.10 hp is used to provide gravitational potential energy.

★★ **66** Two blocks are attached to opposite ends of a string that passes over a massless, frictionless pulley (Fig. 7-40). Block A of mass 10.0 kg lies on a 60.0° incline with a coefficient of friction of 0.500, and block B of mass 1.00 kg is attached to a vertical spring of force constant 200 N/m. The blocks are initially at rest with the spring at equilibrium. Find the maximum height that block B rises.

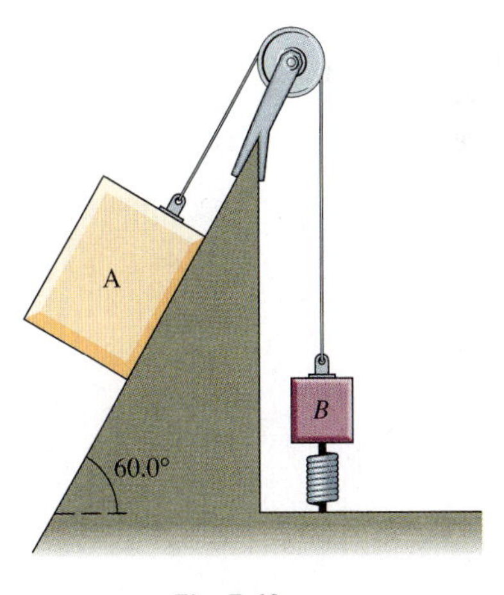

Fig. 7-40

★★ **67** A football of mass 0.500 kg is thrown by a quarterback, who accelerates the ball over a path of length 40.0 cm, releasing the ball with an initial velocity at an angle of 45.0° above the horizontal. The horizontal range of the football is 55.0 m. Find the average force exerted on the ball by the quarterback's hand. Ignore air resistance.

★ **68** An athlete who weighs 800 N is able to raise his center of mass 0.500 m in a vertical jump.
(a) Compute the internal work done by the athlete's leg muscles as he pushes off from the ground.
(b) Find the athlete's speed as the feet leave the ground.
(c) Find the time during which the feet are in contact with the ground and the body accelerates upward, assuming that the center of mass moves through a distance of 0.400 m at constant acceleration.
(d) Calculate the average mechanical power produced.

★ **69** Only a few hundred comets have been observed. (Halley's comet is the most spectacular of these.) However, it is now believed that there are perhaps 10^{12} comets, composing what is called the Oort cloud, with orbits much larger than the planetary orbits (Fig. 7-41). It may be perturbation of these comets' orbits by other bodies (a nearby star, for example) that occasionally sends one of them into a new orbit much closer to the sun, so that then, like Halley's comet, it becomes visible on earth. Some of these comets would very likely strike the earth, with devastating effects.* Suppose that a comet from the Oort cloud is slowed by a passing star, so that it falls toward the sun ($m = 1.99 \times 10^{30}$ kg) and strikes the earth. Find the comet's speed when it reaches earth, 1 astronomical unit (Au), or 1.49×10^{11} m, from the sun, if initially the comet is 50,000 Au from the sun and is moving at a speed of only a few m/s. Ignore the earth's gravitational effect, which is relatively small.

★★ **70** (a) When the comet in the last problem collides with the earth, an enormous cloud of dust is thrown into the atmosphere. Estimate the mass of the dust, assuming that the comet's mass is 1.0×10^{14} kg (the approximate mass of Halley's comet), that half the comet's energy is converted to gravitational potential energy of the dust cloud, and that the dust is uniformly spread in a layer 20 km thick (where most of the earth's atmosphere is concentrated).
(b) Calculate the density of the dust and compare with the density of air at the surface of the earth (1.2 kg/m^3). The dust would likely remain for months or years, would cut off solar radiation, shrouding the earth in darkness, and could well result in extinction of many species, including the human species. A similar outcome has been predicted for a nuclear war; the scenario in this case is referred to as "nuclear winter."

★★ **71** Repeat Problem 69, this time taking into account the earth's gravity.

★ **72** Use Eq. 7-20 to express the gravitational potential energy U'_G of a body of mass m at a distance r from the center of the earth, where r is the sum of the earth's radius R and the distance y of the mass above the earth's surface. Show that when $y << R$, this expression reduces to $U'_G \approx mgy$. (The symbol $<<$ means 'much less than'.)

Collision of a comet with Jupiter, July 16-23, 1994, as seen in ultraviolet light.

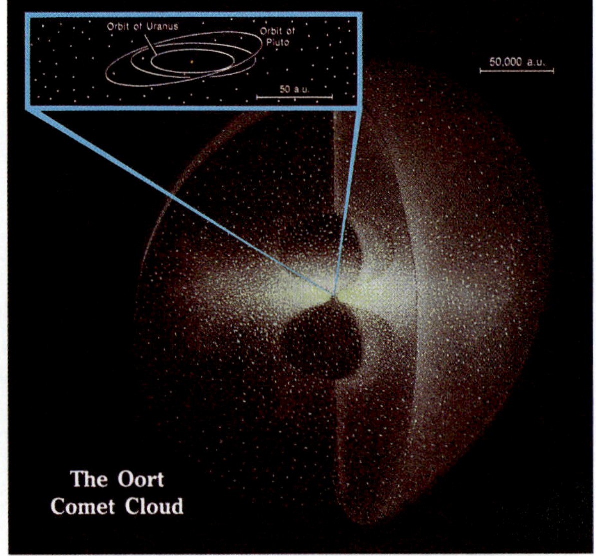

Fig. 7-41 (Diagram by Steven Simpson: *Sky and Telescope* 73:239, March 1987.)

*It has been proposed that a shower of new comets might rain down on the planets every 30 million years or so, as the solar system passes through a heavily populated part of our galaxy. This would account for increased geological activity on earth about every 30 million years. It might even account for the sudden mass extinction of dinosaurs and many other species, which occurred about 65 million years ago, about the same time that there was deposited on the earth's crust a thin sedimentary layer rich in iridium, which is otherwise rare on earth.

CHAPTER 8 Momentum

A tennis player uses his racket to impart momentum to a tennis ball.

In the last chapter we developed the concept of energy and applied the principle of conservation of energy. In this chapter we introduce the concept of linear momentum. Newton's laws of motion can be used to show that, like energy, linear momentum is a conserved quantity for an isolated system. We shall find that the analysis of certain kinds of problems is greatly simplified when this conservation law is applied.

8-1 Impulse and Linear Momentum

Suppose you bounce a tennis ball off a wall. The ball is initially moving to the left, hits the wall, and rebounds with its speed nearly unchanged but with the direction of its velocity vector reversed (Fig. 8-1). By applying Newton's second law, we can relate the average force on the ball to its change in velocity.

Fig. 8-1 The velocity of a tennis ball is reversed as it bounces off a wall.

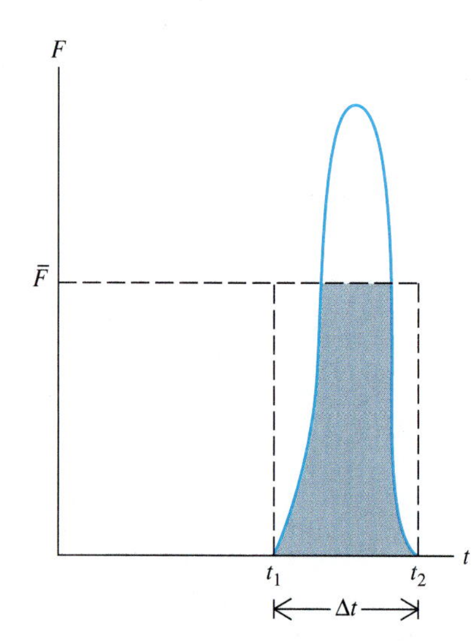

Fig. 8-2 The magnitude of an impulsive force as a function of time. The average value $\bar{F}$ and the collision time Δt determine the effect of the force in changing the velocity of a body.

Fig. 8-2 indicates the general nature of a contact force of this type. The force is quite large but acts over a short time interval. This kind of force is referred to as an "impulsive force." As indicated in the figure, such a force is typically not constant. However, the effect on the ball is independent of the detailed time dependence of the force. The effect is determined solely by the force's average value $\overline{\mathbf{F}}$ and its duration Δt. According to Newton's second law, the resultant force acting on a body equals its mass times its acceleration. Since the contact force during a collision is usually much greater than any other force acting on a body (such as its weight), we can ignore these other forces during the collision and equate the average collision force $\overline{\mathbf{F}}$ to mass times average acceleration $\bar{\mathbf{a}} = \Delta\mathbf{v}/\Delta t$:

$$\overline{\mathbf{F}} = m\bar{\mathbf{a}} = m\frac{\Delta\mathbf{v}}{\Delta t} = m\frac{\mathbf{v}_\mathrm{f} - \mathbf{v}_\mathrm{i}}{\Delta t} = \frac{m\mathbf{v}_\mathrm{f} - m\mathbf{v}_\mathrm{i}}{\Delta t} \tag{8-1}$$

We define the **linear momentum** of a body of mass m moving at velocity $\mathbf{v}$ to be the product $m\mathbf{v}$ and denote the linear momentum by the symbol $\mathbf{p}$:

$$\mathbf{p} = m\mathbf{v} \qquad \text{(linear momentum)} \tag{8-2}$$

The SI unit for momentum is kg-m/s.

Using the definition of linear momentum, we may express Eq. 8-1 in the form

$$\overline{\mathbf{F}} = \frac{\mathbf{p}_\mathrm{f} - \mathbf{p}_\mathrm{i}}{\Delta t}$$

or

$$\overline{\mathbf{F}} = \frac{\Delta\mathbf{p}}{\Delta t} \tag{8-3}$$

Multiplying this expression by Δt, we can put this result in the form

$$\overline{\mathbf{F}}\,\Delta t = \Delta\mathbf{p} \tag{8-4}$$

The product $\overline{\mathbf{F}}\,\Delta t$ is called the **impulse.** Thus Eq. 8-4 states that the impulse acting on a body equals its change in momentum. This result is called the **impulse-momentum theorem.**

Fig. 8-3 Many sports involve the striking of a ball. The application of a force by the hand, foot, bat, club, racket, or stick results in a change in the momentum of the ball. In each case there is a rather large contact force of short duration.

From the definition of impulse as the product of force and time, it follows that the SI unit for impulse is the N-s. Eq. 8-4 equates impulse and momentum, and the units for these two quantities must therefore be equal, as we can readily verify: N-s = (kg-m/s^2)(s) = kg-m/s.

According to the definition $\mathbf{p} = m\mathbf{v}$, momentum is a vector in the same direction as velocity and has a magnitude p equal to the product of a body's mass m and its speed v. Thus if two bodies move at the same speed, the one with more mass has more momentum. For example, a large truck has more momentum than a car moving at the same speed. However, even a large mass such as a truck has no momentum when it is at rest, whereas a very small mass can have considerable momentum if it is moving at high speed (for example, a speeding bullet).

Changing the momentum of a body requires an impulse equal to the change in momentum, according to the impulse-momentum theorem. Impulse is defined as the product of the average force and the time interval over which it acts. This means that a given change in momentum can be accomplished either by a large force acting over a short time or by a smaller force acting over a longer time. For example, a large truck moving at high speed has a very large momentum. Stopping the truck (that is, reducing its momentum to zero) requires a large impulse. Normally the truck is stopped when the brakes are applied, so that a moderately large force (friction) acts over a time interval of many seconds. However, the truck may be stopped much more rapidly in a collision; in this case the force must be enormously large, since it acts over a much shorter time interval. In either case the impulse is the same.

EXAMPLE 1 Force on a Tennis Ball Being Served

(a) Compute the impulse that must be imparted to a tennis ball ($m = 6.00 \times 10^{-2}$ kg) to serve it at a speed of 40.0 m/s. (b) If the time of contact is 5.00×10^{-3} s, what is the average force? The ball is struck when it is instantaneously at rest, after being tossed vertically upward (Fig. 8-4).

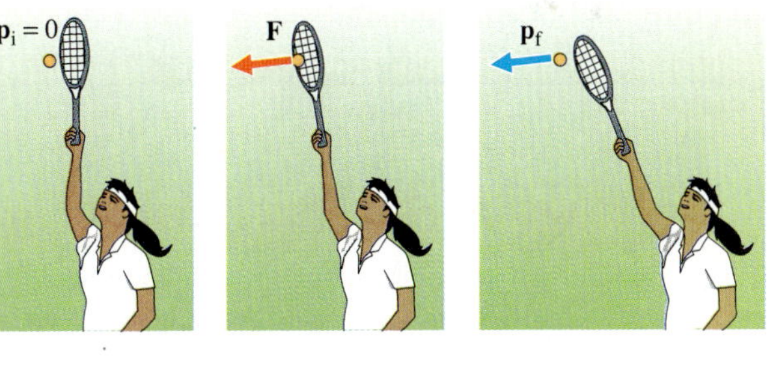

Fig. 8-4

SOLUTION (a) The tennis ball is initially at rest, and so its initial momentum is zero. The impulse momentum theorem then implies that the impulse applied to the ball is equal to its final momentum:

$$\overline{\mathbf{F}}\, \Delta t = \Delta \mathbf{p} = \mathbf{p}_f$$

The impulse is in the direction of the final momentum and has magnitude

$$\overline{F}\, \Delta t = p_f = mv_f = (6.00 \times 10^{-2}\ \text{kg})(40.0\ \text{m/s}) = 2.40\ \text{kg-m/s}$$

(b) The magnitude of the average force can be found by solving the preceding equation for $\overline{F}$:

$$\overline{F} = \frac{p_f}{\Delta t} = \frac{2.40\ \text{kg-m/s}}{5.00 \times 10^{-3}\ \text{s}} = 480\ \text{N}$$

EXAMPLE 2 Force on a Batted Ball

A baseball ($m = 0.150$ kg) thrown by a pitcher reaches a bat with a velocity of 36.0 m/s in a horizontal direction. (a) Compute the impulse that must be applied to the ball if it is to leave the bat with an initial velocity of 45.0 m/s, directed 45.0° above the horizontal (Fig. 8-5). (b) Find the average force acting on the baseball in the 1.00×10^{-3} s interval during which it is in contact with the bat.

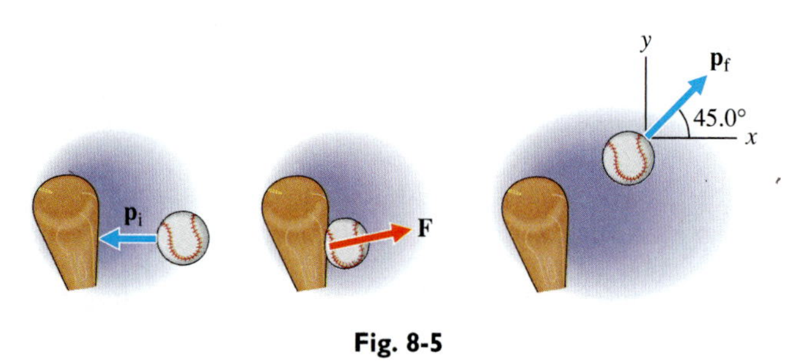

Fig. 8-5

SOLUTION **(a)** According to the impulse-momentum theorem, the impulse applied to the ball equals its change in momentum:

$$\overline{\mathbf{F}}\,\Delta t = \Delta\mathbf{p} = \mathbf{p}_f - \mathbf{p}_i$$

We can express this vector equation in component form and solve for the components of the impulse, where the coordinate axes are as indicated in Fig. 8-5:

$$\begin{aligned}\overline{F}_x\,\Delta t &= p_{fx} - p_{ix}\\ &= (0.150\text{ kg})(45.0\text{ m/s})(\cos 45.0°) - (0.150\text{ kg})(-36.0\text{ m/s})\\ &= 10.2\text{ N-s}\end{aligned}$$

$$\begin{aligned}\overline{F}_y\,\Delta t &= p_{fy} - p_{iy} = mv_{fy} - 0\\ &= (0.150\text{ kg})(45.0\text{ m/s})(\sin 45.0°)\\ &= 4.77\text{ N-s}\end{aligned}$$

This is an impulse of magnitude

$$\sqrt{(10.2\text{ N-s})^2 + (4.77\text{ N-s})^2} = 11.3\text{ N-s}$$

directed above the horizontal at an angle

$$\theta = \arctan(4.77/10.2) = 25.1°$$

(b) The average force acting on the ball is in the direction of the impulse (25.1° above the horizontal) and has magnitude

$$\overline{F} = \frac{\overline{F}\,\Delta t}{\Delta t} = \frac{11.3\text{ N-s}}{1.00 \times 10^{-3}\text{ s}}$$

$$= 11300\text{ N}$$

8-2 Momentum of a System of Particles; Conservation of Linear Momentum

The concept of momentum is particularly useful if we have a system consisting of a number of interacting particles—colliding billiard balls, say, or subatomic particles produced by a particle accelerator. We define the **total momentum P** to be the vector sum of the momenta of the particles in the system:

$$\mathbf{P} = \Sigma\,\mathbf{p} = \mathbf{p}_1 + \mathbf{p}_2 + \mathbf{p}_3 + \cdots \qquad (8\text{-}5)$$

When particles interact, their momenta change: $\mathbf{p}_1$ becomes $\mathbf{p}_1 + \Delta\mathbf{p}_1$, $\mathbf{p}_2$ becomes $\mathbf{p}_2 + \Delta\mathbf{p}_2$, and so forth. Thus the total momentum of the system changes by

$$\Delta\mathbf{P} = \Delta\mathbf{p}_1 + \Delta\mathbf{p}_2 + \Delta\mathbf{p}_3 + \cdots$$

Dividing this expression by the time interval Δt of the interaction, we find

$$\frac{\Delta\mathbf{P}}{\Delta t} = \frac{\Delta\mathbf{p}_1}{\Delta t} + \frac{\Delta\mathbf{p}_2}{\Delta t} + \frac{\Delta\mathbf{p}_3}{\Delta t} + \cdots$$

But, according to Eq. 8-3 ($\overline{\mathbf{F}} = \Delta\mathbf{p}/\Delta t$), each term on the right side of the preceding equation equals the average force acting on the corresponding particle. Thus we may express our result

$$\frac{\Delta\mathbf{P}}{\Delta t} = \overline{\mathbf{F}}_1 + \overline{\mathbf{F}}_2 + \overline{\mathbf{F}}_3 + \cdots$$

or

$$\frac{\Delta\mathbf{P}}{\Delta t} = \Sigma\,\overline{\mathbf{F}}$$

Since we are assuming our system to be isolated, there are no external forces acting on the particles; there are only internal forces acting between particles of the system. The sum of these internal forces equals zero, since according to Newton's third law such forces occur in action-reaction pairs. It follows from the equation above that there is no change in the total linear momentum of an isolated system. **The total linear momentum of an isolated system is conserved:**

$$\Delta\mathbf{P} = 0$$

or

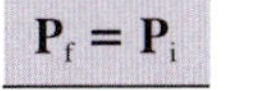

$$\mathbf{P}_f = \mathbf{P}_i \qquad \text{(isolated system)} \qquad (8\text{-}6)$$

Using the definition of total linear momentum as the vector sum of the particles' momenta ($\mathbf{P} = \Sigma\,\mathbf{p}$), we may express this result

$$\mathbf{p}_{1f} + \mathbf{p}_{2f} + \mathbf{p}_{3f} + \cdots = \mathbf{p}_{1i} + \mathbf{p}_{2i} + \mathbf{p}_{3i} + \cdots$$

Another useful expression for the total linear momentum can be obtained from the expression for center of mass velocity (Eq. 5-10):

$$M\mathbf{v}_{cm} = m_1\mathbf{v}_1 + m_2\mathbf{v}_2 + \cdots$$
$$= \mathbf{p}_1 + \mathbf{p}_2 + \cdots = \Sigma\,\mathbf{p}$$

Thus

$$\mathbf{P} = M\mathbf{v}_{cm} \qquad (8\text{-}7)$$

The law of conservation of linear momentum is quite general. The linear momentum of an isolated system is always conserved. Although we have derived this law from Newton's second law, it turns out to be valid even in the domain of subatomic physics, where Newtonian mechanics fails. For example, the principle of momentum conservation is routinely used in the analysis of bubble-chamber photographs, which show the paths of elementary particles (Fig. 8-6). The interactions analyzed in these photographs are complex, and it is useful to know that, whatever the process, the total linear momentum is conserved.

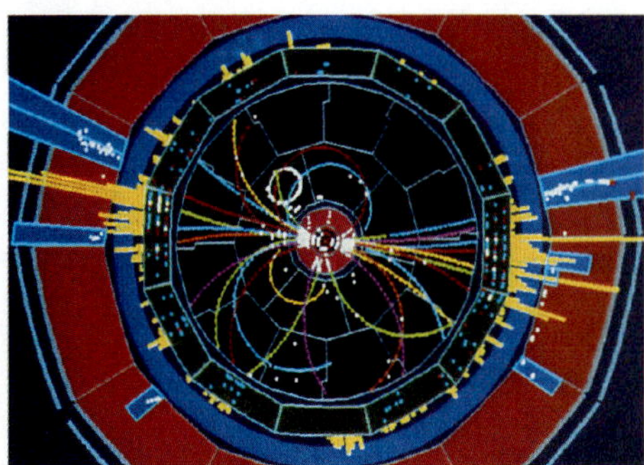

Fig. 8-6 A bubble chamber photograph of the paths of elementary particles.

EXAMPLE 3 Recoil of a Shotgun

A shotgun of mass 3.00 kg fires shot having a total mass of 5.00×10^{-2} kg. The shot leaves the barrel of the gun with a "muzzle velocity" of 525 m/s. (a) Find the recoil velocity of the shotgun, assuming it is free to move. (b) What impulse is required to stop the recoiling gun?

SOLUTION **(a)** Before the gun is fired, the total linear momentum of the gun and shot is zero, since both are at rest. The only significant force acting on the system during the firing is an internal force. Therefore total linear momentum is conserved, and so the final momentum must also be zero:

$$\mathbf{P}_f = \mathbf{P}_i = \mathbf{0}$$

This means that the final momentum of the gun $\mathbf{p}_g$ and that of the shot $\mathbf{p}_s$ must cancel:

$$\mathbf{P}_f = \mathbf{p}_g + \mathbf{p}_s = \mathbf{0}$$

or

$$\mathbf{p}_g = -\mathbf{p}_s$$

that is, the gun and shot have momenta of equal magnitude in opposite directions. Expressing this result in terms of the horizontal velocity components, we obtain

$$m_g v_{gx} = -m_s v_{sx}$$

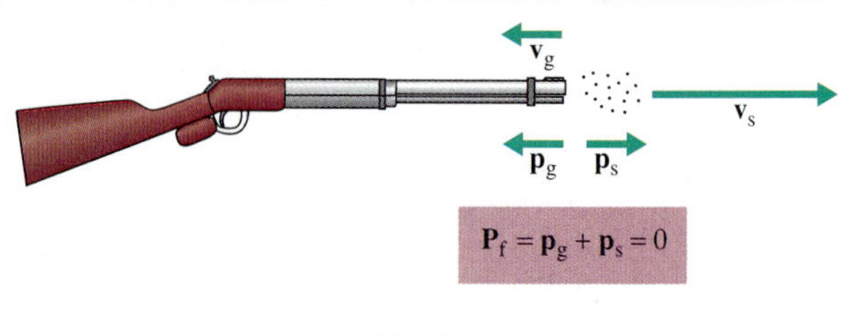

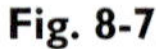

Fig. 8-7

Since the gun has a much larger mass than the shot, its speed is much less than the speed of the shot (Fig. 8-7). Solving for v_{gx}, we obtain:

$$v_{gx} = -\frac{m_s v_{sx}}{m_g} = -\frac{(5.00 \times 10^{-2}\text{ kg})(525\text{ m/s})}{3.00\text{ kg}}$$

$$= -8.75\text{ m/s}$$

(b) Bringing the shotgun to rest requires an impulse equal to the change in the gun's momentum (Eq. 8-4):

$$\overline{\mathbf{F}}\,\Delta t = \Delta\mathbf{p} = \mathbf{0} - \mathbf{p}_g = -\mathbf{p}_g$$

The impulse is directed opposite the gun's momentum and has magnitude

$$\overline{F}\,\Delta t = p_g = m_g v_g = (3.00\text{ kg})(8.75\text{ m/s})$$

$$= 26.3\text{ N-s}$$

8-3 Collisions and Kinetic Energy

Fig. 8-8a shows two bodies of equal mass about to collide head on. One of the masses is initially at rest. Suppose the bodies are on a frictionless horizontal surface so that no external horizontal forces act on the two-body system. It follows that the total linear momentum of the system is conserved. This condition does not, however, *uniquely* determine the final state of the system. Momentum could be conserved with any of the final states shown in Fig. 8-8b, as you can easily verify by adding the momenta of the two bodies for each final state. In all cases the sum equals $m\mathbf{v}$, the system's original momentum.

We can distinguish between the various final states shown in Fig. 8-8b on the basis of the system's final kinetic energy, which is different for each state. A collision in which the total kinetic energy of the system is conserved is called an **elastic collision.** (The idea that kinetic energy is sometimes conserved was discovered by Christian Huygens, a contemporary of Newton, who made a systematic study of collisions.)

For the collision in Fig. 8-8, the system's initial kinetic energy is just the kinetic energy of the incoming body, that is, $\frac{1}{2}mv^2$, since the other body is at rest. For the first final state in Fig. 8-8b, the system's final kinetic energy is again $\frac{1}{2}mv^2$, and so kinetic energy is conserved and the collision is, by definition, elastic. Such a collision is well approximated by the head-on collision of billiard balls.

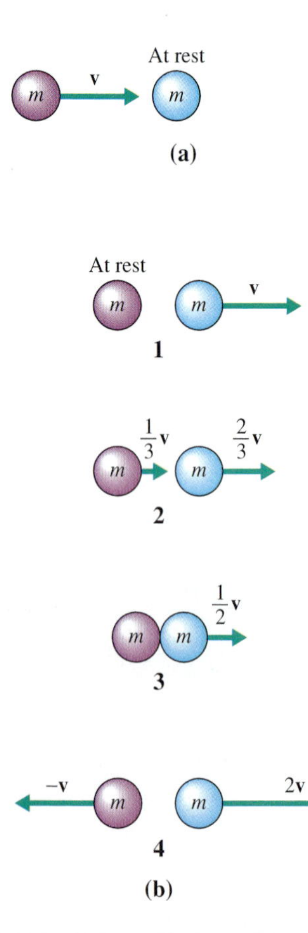

Fig. 8-8 **(a)** Two bodies of equal mass about to undergo a head-on collision. **(b)** Final states that conserve momentum.

For the second final state, the total kinetic energy is $\frac{1}{2}m(v/3)^2 + \frac{1}{2}m(2v/3)^2 = \frac{5}{9}(\frac{1}{2}mv^2)$. The total kinetic energy has decreased to $\frac{5}{9}$ of its initial value. When, as in this case, kinetic energy is not conserved, the collision is said to be **inelastic.** The second final state might occur, for example, when two soft rubber balls collide.

For the third final state, the two bodies have the same final speed, and so they move as a single unit. In other words, the two bodies stick together. This is called a **completely inelastic** collision. The final kinetic energy here is $\frac{1}{2}(2m)(v/2)^2 = \frac{1}{4}mv^2$. The total kinetic energy has decreased to half its initial value. When kinetic energy is lost during a collision, as in both the second and third final states, there is an accompanying temperature rise of the colliding bodies. This is an indication of another (microscopic) form of energy, thermal energy or heat, into which some of the kinetic energy is converted during collision.

For the fourth final state the total kinetic energy exceeds the initial kinetic energy. Such a collision does not ordinarily occur. The total kinetic energy cannot increase during a collision unless there is some other source of energy (for example, if the collision sets off an explosion or releases a compressed spring).

Elastic Collisions in One Dimension

Fig. 8-9 shows an elastic, head-on collision between two particles of *arbitrary mass.* We shall obtain general expressions for the final velocities $\mathbf{v}_1'$ and $\mathbf{v}_2'$ as a function of the initial velocity $\mathbf{v}_1$ and the particle masses m_1 and m_2. A unique solution for $\mathbf{v}_1'$ and $\mathbf{v}_2'$ is possible only when the collision is head on, so that all velocities, before and after, are along the same straight line, which we shall identify as the x-axis. Then we have two unknowns, v_{1x}' and v_{2x}', which we may find by using the two equations expressing conservation of momentum and conservation of kinetic energy. (In the more general two-dimensional case, we would have to know something about the force acting during collision.)

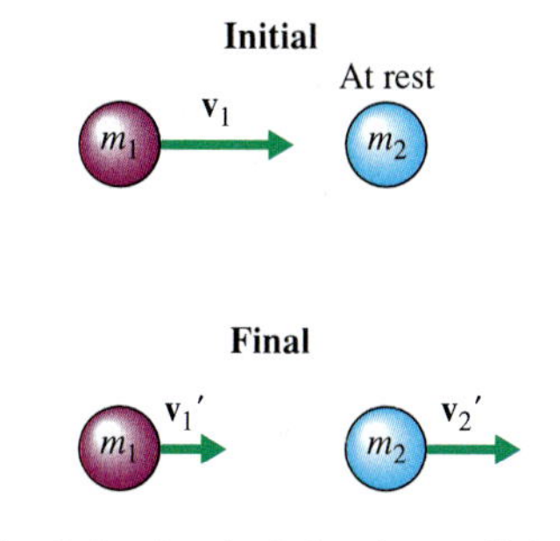

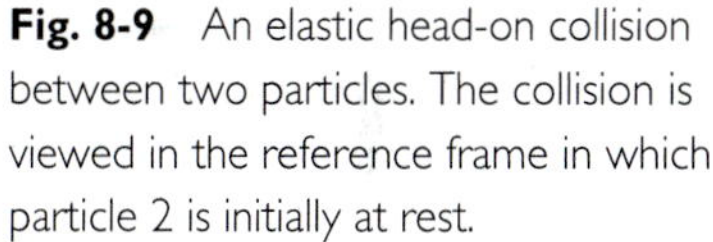
Fig. 8-9 An elastic head-on collision between two particles. The collision is viewed in the reference frame in which particle 2 is initially at rest.

The total linear momentum of the system is conserved:

$$P_{fx} = P_{ix}$$

or

$$m_1 v_{1x}' + m_2 v_{2x}' = m_1 v_{1x} \qquad \text{(8-8)}$$

The total kinetic energy is also conserved:

$$K_f = K_i$$

or

$$\tfrac{1}{2}m_1(v_{1x}')^2 + \tfrac{1}{2}m_2(v_{2x}')^2 = \tfrac{1}{2}m_1 v_{1x}{}^2 \qquad \text{(8-9)}$$

Now we can find v_{2x}' by solving Eq. 8-8 for v_{1x}' and substituting into Eq. 8-9. After performing the required algebra, we find

$$v_{2x}' = \frac{2m_1}{m_1 + m_2} v_{1x} \qquad \text{(elastic collision)} \qquad \text{(8-10)}$$

This result can be substituted into Eq. 8-8, and the resulting equation can be solved for v_{1x}'.

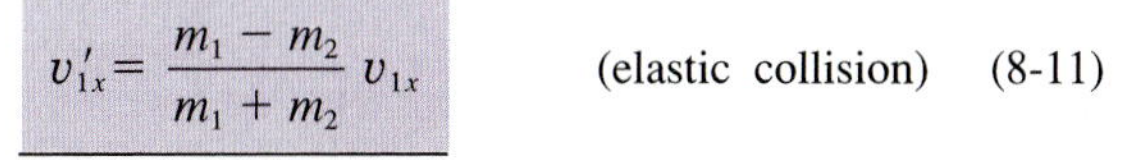

$$v_{1x}' = \frac{m_1 - m_2}{m_1 + m_2} v_{1x} \qquad \text{(elastic collision)} \qquad \text{(8-11)}$$

Notice that when $m_1 = m_2$, these equations reduce to $v'_{2x} = v_{1x}$ and $v'_{1x} = 0$. After collision, the first mass stops and the second mass begins to move with a velocity equal to the initial velocity of the first mass. You can see this demonstrated in the head-on collision of two pool balls.

In deriving Eqs. 8-10 and 8-11, we have assumed that particle 2 is initially at rest. We may apply these equations to problems in which both bodies are initially moving if, in performing the calculation, we simply use the reference frame in which the second body is at rest. This method is illustrated in Ex. 6.

EXAMPLE 4 Velocity of a Golf Ball as It Leaves a Club

A golf club of mass m_1 strikes a golf ball that is initially at rest and has a mass m_2 (Fig. 8-10). Treat the collision as an elastic, head-on collision, with m_1 much greater than m_2. (a) Show that the final velocity v'_{2x} of the ball is approximately twice the initial velocity v'_{1x} of the club. (b) What is the final velocity of the club?

SOLUTION **(a)** Since $m_1 >> m_2$, we can ignore the m_2 term in the denominator of Eq. 8-10, which is then approximated by

$$v'_{2x} \approx \frac{2m_1}{m_1} v_{1x} = 2v_{1x}$$

(b) Using Eq. 8-11 and again assuming m_2 is negligible compared with m_1, we find that the final velocity of the club is approximately unchanged:

$$v'_{1x} \approx \frac{m_1}{m_1} v_{1x} = v_{1x}$$

Careful inspection of the images in Fig. 8-10 shows that the velocity of the club head is unchanged by the impact and that the ball is moving approximately twice as fast as the club head.

Fig. 8-10

EXAMPLE 5 Rebound of a Racquetball

A racquetball is initially moving to the right at a speed of 30.0 m/s. The ball collides with a solid wall whose surface is perpendicular to the ball's initial velocity. The ball rebounds from the wall, reversing the direction of its velocity. Find the ball's final velocity. Assume the collision is elastic.

SOLUTION The wall's mass m_2 is obviously much greater than the ball's mass m_1. So when we use Eq. 8-11 for the final velocity of the ball, the equation can be approximated by

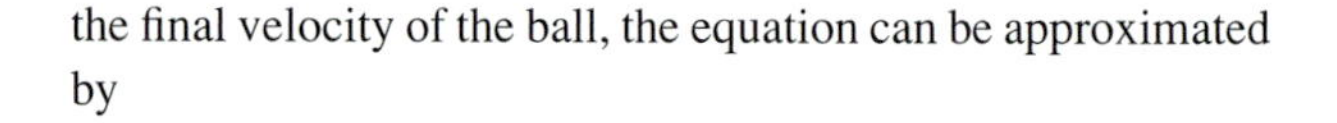

$$v'_{1x} \approx -\frac{m_2}{m_2} v_{1x} = -v_{1x} = -30.0 \text{ m/s}$$

The ball's speed as it leaves the wall is the same as its speed just before striking the wall. Not surprisingly, Eq. 8-10 predicts that the final velocity of the wall is zero.

EXAMPLE 6 An Elastic, Head-on Collision

Fig. 8-11 shows two masses about to undergo an elastic, head-on collision. Find their final velocities.

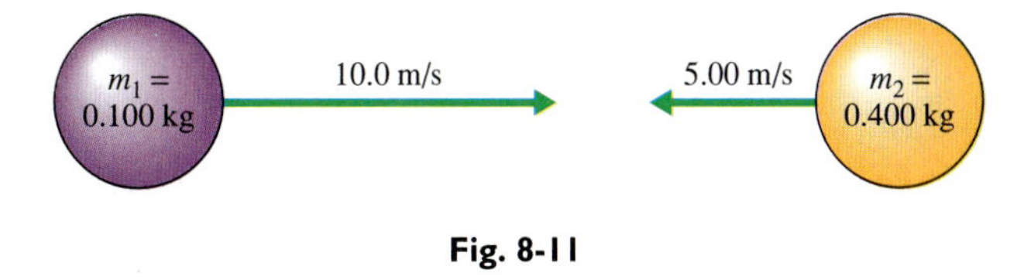

Fig. 8-11

SOLUTION First we must find the velocity of the 100 g mass, m_1, as seen in the reference frame in which the 400 g mass, m_2, is at rest. Relative to m_2, m_1 moves at a velocity of 15.0 m/s to the right. Thus we use $v_{1x} = 15.0$ m/s in applying Eqs. 8-10 and 8-11, and we find

$$v'_{2x} = \frac{2m_1}{m_1 + m_2} v_{1x} = \frac{2(0.100 \text{ kg})}{0.500 \text{ kg}}(15.0 \text{ m/s})$$
$$= 6.00 \text{ m/s}$$

and

$$v'_{1x} = \frac{m_1 - m_2}{m_1 + m_2} v_{1x} = \frac{0.100 \text{ kg} - 0.400 \text{ kg}}{0.500 \text{ kg}}(15.0 \text{ m/s})$$
$$= -9.00 \text{ m/s}$$

These values are relative to a reference frame that is moving to the left at 5.00 m/s relative to the original reference frame. So if we want the velocities observed in the original reference frame, we must add −5.00 m/s to each. Denoting the velocities in the original reference frame by u'_{1x} and u'_{2x}, we have

$$u'_{2x} = 6.00 \text{ m/s} - 5.00 \text{ m/s} = 1.00 \text{ m/s}$$

and

$$u'_{1x} = -9.00 \text{ m/s} - 5.00 \text{ m/s} = -14.0 \text{ m/s}$$

Elastic Collisions in Two Dimensions

Next we consider briefly the general case of an elastic collision between two particles, viewed in the rest frame of the second particle. Let m_1 have initial velocity $\mathbf{v}_1$ along the positive x-axis. When m_1 and m_2 collide, if the collision force is not along the x-axis, m_2 will experience a component of acceleration perpendicular to the x-axis and will have a final velocity $\mathbf{v}'_2$ at some angle θ_2 relative to the x-axis (Fig. 8-12). Let the xy plane be the plane of vectors $\mathbf{v}_1$ and $\mathbf{v}'_2$. Then conservation of momentum requires that velocity $\mathbf{v}'_1$ also lie in the xy plane. (Otherwise there would be a final component of total momentum in the z direction without an initial momentum in the z direction, violating the principle of momentum conservation).

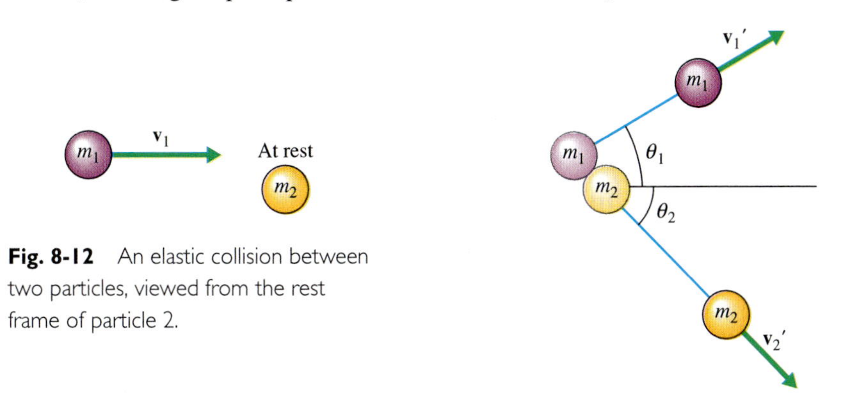

Fig. 8-12 An elastic collision between two particles, viewed from the rest frame of particle 2.

Now suppose we try to find the final velocities $\mathbf{v}'_1$ and $\mathbf{v}'_2$ as a function of the particle masses and the initial velocity $\mathbf{v}_1$. There are four unknowns: v'_{1x}, v'_{1y}, v'_{2x}, v'_{2y}. But there are only three equations available to solve for these unknowns: conservation of momentum provides two (one each for the x and y components of momentum) and conservation of kinetic energy gives one more. Thus the problem cannot be solved without some information about the force acting during the collision. It is possible to show that if $m_1 = m_2$ then the directions of the final velocity vectors differ by an angle of 90° ($\theta_1 + \theta_2 = 90°$), as illustrated in Fig. 8-13. (See Problem 30.)

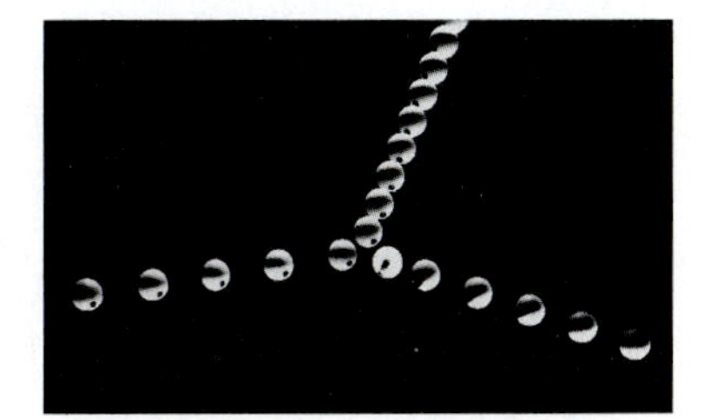

Fig. 8-13 Collision of billiard balls.

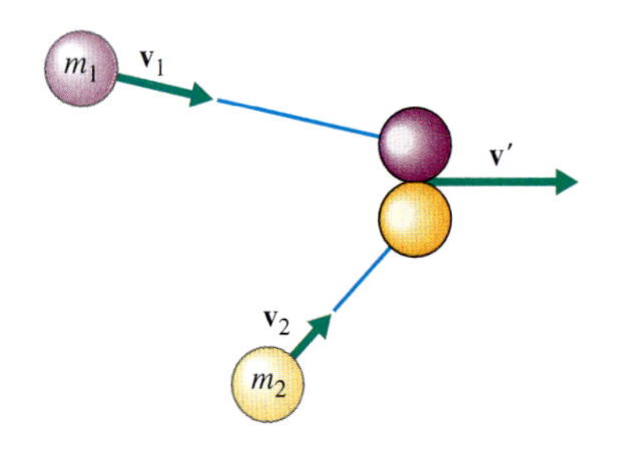

Fig. 8-14 A completely inelastic collision.

Completely Inelastic Collisions

Finally we consider the case of a completely inelastic collision. Two isolated particles with velocities $\mathbf{v}_1$ and $\mathbf{v}_2$ collide, stick together, and move off with final velocity $\mathbf{v}'$, as shown in Fig. 8-14.

Conservation of momentum gives an equation that can immediately be solved for $\mathbf{v}'$:

$$\mathbf{P}_f = \mathbf{P}_i$$

or

$$(m_1 + m_2)\mathbf{v}' = m_1\mathbf{v}_1 + m_2\mathbf{v}_2$$

Solving for $\mathbf{v}'$, we find

$$\mathbf{v}' = \frac{m_1\mathbf{v}_1 + m_2\mathbf{v}_2}{m_1 + m_2} \tag{8-12}$$

Notice that $\mathbf{v}'$ is also the system's center-of-mass velocity, which remains constant both before and after collision, since center of mass velocity is proportional to the conserved total momentum (Eq. 8-7: $\mathbf{P} = M\mathbf{v}_{\text{cm}}$).

EXAMPLE 7 A Completely Inelastic Automobile Collision

Two cars approach an icy intersection. One car has mass 1.00×10^3 kg and travels east at 20.0 m/s, and the other has mass 2.00×10^3 kg and travels north at 15.0 m/s. The cars collide and become coupled together, as shown in Fig. 8-15. Find their common final velocity, assuming no significant external force acts on them during the collision.

SOLUTION Conservation of momentum requires that $\mathbf{v}'$ be related to $\mathbf{v}_1$ and $\mathbf{v}_2$ by Eq. 8-12:

$$\mathbf{v}' = \frac{m_1\mathbf{v}_1 + m_2\mathbf{v}_2}{m_1 + m_2}$$

In component form this becomes

$$v'_x = \frac{m_1 v_{1x} + m_2 v_{2x}}{m_1 + m_2} = \frac{(1.00 \times 10^3 \text{ kg})(20.0 \text{ m/s}) + 0}{1.00 \times 10^3 \text{ kg} + 2.00 \times 10^3 \text{ kg}}$$
$$= 6.67 \text{ m/s}$$

and
$$v'_y = \frac{m_1 v_{1y} + m_2 v_{2y}}{m_1 + m_2} = \frac{0 + (2.00 \times 10^3 \text{ kg})(15.0 \text{ m/s})}{1.00 \times 10^3 \text{ kg} + 2.00 \times 10^3 \text{ kg}}$$
$$= 10.0 \text{ m/s}$$

The final velocity vector has magnitude

$$\sqrt{(6.67 \text{ m/s})^2 + (10.0 \text{ m/s})^2} = 12.0 \text{ m/s}$$

and is directed at an angle $\theta = \arctan(10.0/6.67) = 56.3°$ above the positive x-axis.

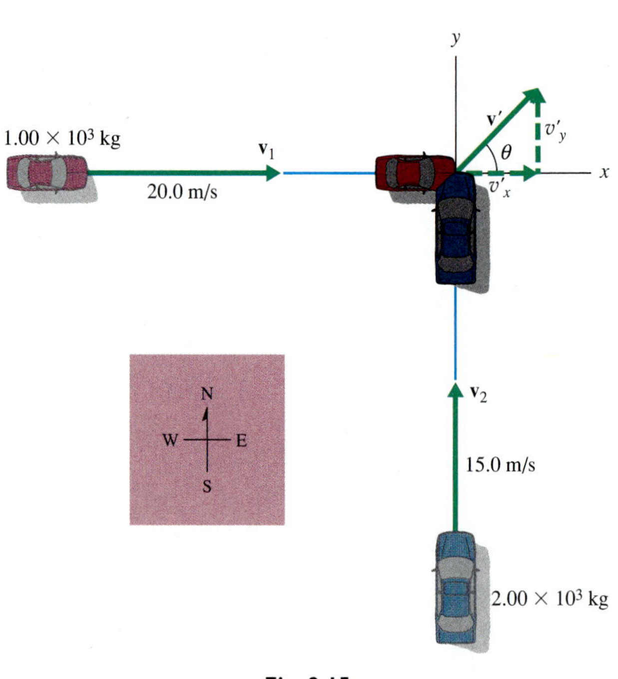

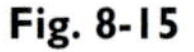

Fig. 8-15

CHAPTER 8 SUMMARY

A particle's linear momentum $\mathbf{p}$ is defined as the product of its mass m and its velocity $\mathbf{v}$:

$$\mathbf{p} = m\mathbf{v}$$

The total linear momentum $\mathbf{P}$ of a system of particles is defined as the vector sum of the single particle momenta:

$$\mathbf{P} = \Sigma\,\mathbf{p} = \mathbf{p}_1 + \mathbf{p}_2 + \mathbf{p}_3 + \cdots$$

The momentum of a system may also be expressed as the product of its total mass M and its center-of-mass velocity $\mathbf{v}_{cm}$:

$$\mathbf{P} = M\mathbf{v}_{cm}$$

The total linear momentum of an isolated system is conserved:

$$\mathbf{P}_f = \mathbf{P}_i \qquad \text{(isolated system)}$$

or

$$\mathbf{p}_{1f} + \mathbf{p}_{2f} + \cdots = \mathbf{p}_{1i} + \mathbf{p}_{2i} + \cdots$$

The rate of change of a particle's momentum equals the average force acting on the particle:

$$\overline{\mathbf{F}} = \frac{\Delta\mathbf{p}}{\Delta t}$$

or

$$\overline{\mathbf{F}}\,\Delta t = \Delta\mathbf{p}$$

The product of the average force $\overline{\mathbf{F}}$ acting on a particle and the duration of the force Δt is called the "impulse." The preceding expression equates the impulse acting on a particle to its change in momentum. This result is known as the impulse-momentum theorem.

An elastic collision is defined as a collision in which the system's total kinetic energy is conserved. A completely inelastic collision is one in which the two bodies stick together after the collision so that there is only a single final velocity.

In a head-on collision, all forces and velocities are along a single line. For a head-on, elastic collision between two masses, viewed in the reference frame in which m_2 is initially at rest, the respective final velocities of m_1 and m_2 are

$$v'_{1x} = \frac{m_1 - m_2}{m_1 + m_2}\,v_{1x}$$

$$v'_{2x} = \frac{2m_1}{m_1 + m_2}\,v_{1x}$$

For a completely inelastic collision between masses m_1 and m_2, with respective initial velocities $\mathbf{v}_1$ and $\mathbf{v}_2$, (which are not necessarily along the same axis), the masses have a common final velocity $\mathbf{v}'$, given by

$$\mathbf{v}' = \frac{m_1\mathbf{v}_1 + m_2\mathbf{v}_2}{m_1 + m_2}$$

Questions

1 If the kinetic energy of a particle is zero, is its linear momentum necessarily zero?

2 If the linear momentum of a particle is zero, is its kinetic energy necessarily zero? If the linear momentum of a system of particles is zero, is the kinetic energy of the system necessarily zero?

3 A boxer naturally tends to pull his head away from an opponent's punch as the opponent's hand makes contact. Does this pulling away tend to reduce or increase (a) the impulse of the blow; (b) the average force; (c) the collision time?

4 Suppose a Ping-Pong ball collides in midair with a bowling ball. Does the bowling ball undergo the same change in momentum as the Ping-Pong ball?

5 If you drop a drinking glass onto a concrete floor, the glass will very likely break. But if you dropped the same glass from the same height onto a carpeted floor, it might not break. Is the change in momentum of the glass the same in both cases? Explain. Compare with a high-wire circus performer falling onto a safety net as opposed to falling onto the floor.

6 Air bags are used as a safety device in some autos. The bags automatically inflate in front of passengers during a collision. Which of the following are not affected by the air bags during a collision: change in linear momentum, average force, collision time?

7 An inventor claims to have made a super lightweight rifle having less mass than the mass of the bullets it fires. Would such a rifle be a good idea?

8 A rocket's engines are fired in interplanetary space, and the rocket accelerates forward. Since there are no external forces acting on the rocket, its linear momentum should be conserved. What balances the increasing forward momentum of the rocket?

9 A pendulum swings back and forth. Is its linear momentum conserved?

10 Suppose you are standing and toss a ball vertically upward and then catch it. Is the normal force on your feet constant, or does it increase or decrease as you throw the ball and as you catch it?

Answers to Odd-Numbered Questions

1 yes; **3** (a) neither; (b) reduce; (c) increase; **5** Yes, the change in momentum and the impulse are the same, but compression of the carpet lengthens the collision time and reduces the average force; a safety net serves the same function for the circus performer, lengthening the time of contact; **7** No, the rifle's recoil velocity would be greater than the bullet's velocity; **9** no, there is an external gravitational force acting on it.

Problems (listed by section)

8-1 Impulse and Linear Momentum

1 Find the linear momentum and kinetic energy of (a) a bullet of mass 5.00×10^{-2} kg moving at a speed of 325 m/s; (b) a football player of mass 112 kg moving at a speed of 10.0 m/s; (c) a truck of mass 1.00×10^{4} kg moving at a speed of 20.0 m/s.

2 A bowling ball weighing 70.0 N initially moves at a speed of 5.00 m/s. How long must a force of 45.0 N be applied to the ball to stop it?

3 Find the average horizontal force exerted by the ground on the feet of a sprinter of mass 50.0 kg who accelerates from rest to a speed of 10.0 m/s in a time interval of 1.00 s.

★ **4** A rubber ball of mass 30.0 g is dropped from a height of 2.00 m onto a floor. The velocity of the ball is reversed by the collision with the floor, and the ball rebounds to a height of 1.50 m. What impulse was applied to the ball during the collision?

5 As you bounce a tennis ball of mass 6.00×10^{-2} kg off a wall, its velocity changes from $v_x = +10.0$ m/s to $v_x = -8.00$ m/s. Find the impulse applied to the ball.

6 (a) Compute the impulse necessary to bring to rest the 50.0 kg driver of an automobile initially moving at 25.0 m/s.

(b) Find the average force on the driver if the deceleration is accomplished gradually by the application of brakes over a 10.0 s interval.

(c) Find the average force if the car is stopped in a collision that brings the driver's seat to rest in 0.100 s, assuming the driver is restrained by a seat belt. What are the functions of seat belts and air bags in a collision?

★ **7** (a) What minimum impulse must be imparted to a football of mass 0.420 kg to have a horizontal range of 40.0 m?

(b) Assuming that contact with the football is maintained as it moves 20.0 cm, estimate the time of contact and the average force on the football.

8 (a) What impulse must be imparted to a golf ball of mass 0.100 kg to give it a velocity of 70.0 m/s?

(b) If the golf club is in contact with the ball as it is accelerated over an interval of 1.00 cm, estimate the average force exerted on the ball by the club.

9 A runner of mass 60.0 kg initially moves at a speed of 9.00 m/s. How long must an average external force of 2.00×10^{2} N act to bring the runner to rest?

8-2 Momentum of a System of Particles; Conservation of Linear Momentum

10 A 90.0 kg fullback attempts to score a touchdown by diving over the goal line. When he is at the goal line and moving horizontally at 6.00 m/s, he is met in midair by an opposing 110 kg (240 lb) linebacker, who is initially moving at 4.00 m/s in the opposite direction. The two meet in a head-on collision at the goal line. Does the fullback cross the goal line?

11 A boy of mass 50.0 kg is initially on a skateboard of mass 2.00 kg, moving at a speed of 10.0 m/s. The boy falls off the skateboard, and his center of mass moves forward at a speed of 11.0 m/s. Find the final velocity of the skateboard.

12 A skater of mass 80.0 kg initially moves in a straight line at a speed of 5.00 m/s. The skater approaches a child of mass 40.0 kg, whom he lifts on his shoulders. Assuming there are no external horizontal forces, what is the skater's final velocity?

13 A missile of mass 1.00×10^2 kg is fired from a plane of mass 5.00×10^3 kg initially moving at a speed of 3.00×10^2 m/s. If the speed of the missile relative to the plane is 1.00×10^3 m/s, what is the final velocity of the plane?

14 A person of mass 50.0 kg stands at rest. How much will the average force exerted on the feet by the floor differ from the person's weight during a 0.200 s interval in which the blood's upward component of momentum increases by 4.00×10^{-2} kg-m/s?

★ **15** A girl weighing 500 N jumps from a tree, and her center of mass falls a vertical distance of 2.00 m. Find the impulse necessary to bring her to rest. Over what time interval must the deceleration last in order that the average force does not exceed 200 N?

★ **16** A person of mass 80.0 kg is initially at rest on the edge of a large stationary platform of mass 200 kg, supported by frictionless wheels on a horizontal surface. The person jumps off the platform, traveling a horizontal distance of 1.00 m while falling a vertical distance of 0.500 m to the ground. What is the final speed of the platform?

8-3 Collisions and Kinetic Energy

★ **17** A tennis ball initially moves with a velocity of 30.0 m/s horizontally to the right. A tennis racket strikes the ball, giving it a velocity of 30.0 m/s to the left. Treat the collision as elastic and the ball and racket as particles of mass 0.0500 kg and 0.500 kg respectively.

(a) Does the speed of the racket change as a result of the collision?

(b) What is the value of the total momentum of the system?

(c) Find the initial velocity of the racket.

18 Two billiard balls are initially traveling toward each other at speeds of 2.00 m/s and 4.00 m/s. The balls undergo an elastic, head-on collision. Find their final velocities.

19 An object of mass 1.00 kg is released from rest and drops 2.00 m to the floor. The collision is completely inelastic. How much kinetic energy is lost during the collision?

★ **20** As shown in Fig. 8-16, two billiard balls are initially near a railing, which can be thought of as an infinite mass. Find the final velocities of the balls, assuming head-on, elastic collisions.

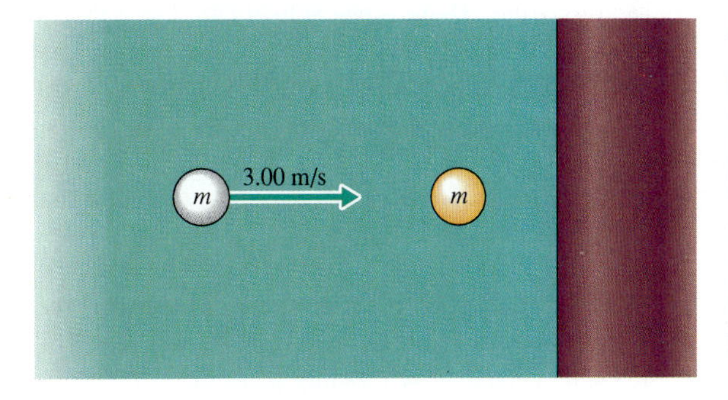

Fig. 8-16

21 Two blocks move along a linear path on a nearly frictionless air track. One block, of mass 0.100 kg, initially moves to the right at a speed of 5.00 m/s, while the second block, of mass 0.200 kg, is initially to the left of the first block and moving to the right at 7.00 m/s. Find the final velocities of the blocks, assuming the collision is elastic.

22 Two cars approach an ice-covered intersection. One car, of mass 1.20×10^3 kg, is initially traveling north at 10.0 m/s. The other car, of mass 1.60×10^3 kg, is initially traveling east at 10.0 m/s. The cars reach the intersection at the same instant, collide, and move off coupled together. Find the velocity of the center of mass of the two-car system just after the collision.

23 A basketball of mass 0.400 kg is passed to a player of mass 80.0 kg who jumps off the floor while catching the ball. The player's center of mass is instantaneously at rest just before he catches the ball, which is initially moving at a velocity of 10.0 m/s, directed 30.0° below the horizontal. Find the horizontal component of the center of mass velocity of the system of player and ball just after the catch.

24 A bullet of mass 10.0 g is fired into an initially stationary block and comes to rest in the block. The block, of mass 1.00 kg, is subject to no horizontal external forces during the collision with the bullet. After the collision, the block is observed to move at a speed of 5.00 m/s.

(a) Find the initial speed of the bullet.

(b) How much kinetic energy is lost?

25 A 1.00 kg block and a 2.00 kg block are initially at rest on a frictionless, horizontal surface, with a compressed spring of negligible mass between them. When the blocks are released, they move off in opposite directions. If the final velocity of the 1.00 kg block is 3.00 m/s to the left, what is the final velocity of the 2.00 kg block? How much does the kinetic energy of the system increase?

Additional Problems

26 In a game of pool, a cue ball with an initial velocity of 2.00 m/s directed to the right collides with the five ball, which has equal mass and is initially at rest. The five ball then moves at 1.73 m/s at an angle of 30.0° to the direction of the cue ball's initial velocity.
(a) Find the cue ball's final velocity.
(b) Is kinetic energy conserved?

27 A particle with an initial linear momentum of 2.00 kg-m/s directed along the positive x-axis collides with a second particle, which has an initial linear momentum of 4.00 kg-m/s, directed along the positive y-axis. The final momentum of the first particle is 3.00 kg-m/s, directed 45.0° above the positive x-axis. Find the final momentum of the second particle.

★ **28** A baseball catcher moves his glove away from a 45.0 m/s fastball as he catches it. If the average force he exerts on the ball is 1.00×10^3 N, how far does the glove move? Assume constant deceleration of the ball, which has a mass of 0.150 kg.

★ **29** Ballistocardiography is a branch of medical technology used in the diagnosis of heart disease. The patient lies on a platform that is supported on a nearly frictionless surface. The beating of the heart and the flow of blood produce small but measurable oscillations in the position of the platform. A record of the displacement of the platform as a function of time is related to the patient's blood flow and can be used to detect cardiovascular disorders. The total linear momentum of the blood is a function of the speed and quantity of blood moving through the body.
(a) Show that if the system of the patient and the platform is completely isolated from horizontal external forces any change in the x component of the blood's total linear momentum is accompanied by an opposite change in the momentum of the remainder of the system (that is, the platform and all the patient's body except the blood).
(b) Find the change in linear momentum of the blood during an interval of 1.0×10^{-2} s in which the platform moves 3.0×10^{-4} cm to the right at an approximately constant acceleration of 2.0×10^{-2} m/s^2. The mass of the entire system is 100 kg, of which 5.0 kg is the mass of the blood.
(c) What is the change in the x component of the blood's center-of-mass velocity during the interval described in part b?

★★ **30** Consider the collision of two billiard balls as the elastic collision of two isolated particles of equal mass. Ignore friction and rotation. The first ball has an initial velocity $\mathbf{v}_1$, and the second ball is initially at rest. Denote the mass of either by m. Show that if the collision is not head on, the final velocities of the two balls are perpendicular to each other (Fig. 8-17). HINT: First use linear momentum conservation to show that

$$v_1^2 = (v_1')^2 + (v_2')^2 + 2v_1'v_2' \cos(\theta_1 + \theta_2)$$

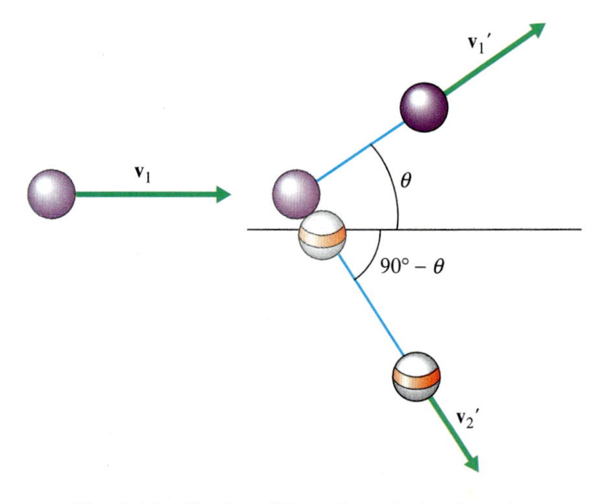

Fig. 8-17 Elastic collision of two isolated particles of equal mass.

★31 A pool player is attempting to shoot the eight ball into the side pocket by hitting a glancing blow with the cue ball (Fig. 8-18). The final velocity of the eight ball is in the direction of the force **F** that acts on it during the collision. It is characteristic of the collision between smooth, hard spheres that the collision force is a repulsive force acting at the point of contact along the line between the centers of the two spheres.

(a) Find the distance b between the initial path of the cue ball and a parallel line through the center of the eight ball. Both have the same mass m and radius r.

(b) What angle will the cue ball's final velocity make with the x-axis?

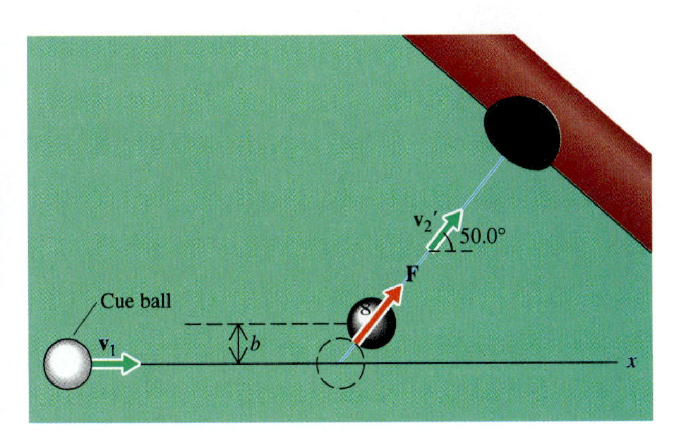

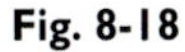
Fig. 8-18

★★32 A body of mass m_1 undergoes an elastic collision with a second body of mass m_2 that is initially at rest.

(a) Show that if $m_1 << m_2$ the speed of m_1 is unchanged by the collision: $v_1' = v_1$. [HINT: Use conservation of linear momentum and of kinetic energy to show that $1 - (v_1')^2/v_1{}^2 = (m_1/m_2)|\mathbf{v}_1/v_1 - \mathbf{v}_1'/v_1|^2$, which approaches zero as m_1/m_2 goes to zero.]

(b) Apply the result of part (a) to the collision of a ball with a stationary wall (Fig. 8-19). Assume that the wall is smooth and can exert only a force perpendicular to its surface (along the x-axis). Show that v_{1y} is unchanged by the collision whereas v_{1x} is reversed ($v_{1x}' = -v_{1x}$).

★★33 (a) Consider the collision of a ball with a smooth wall moving at velocity **V** away from the ball. Assuming an elastic collision, show that the component of the ball's velocity perpendicular to the wall changes from v_x to v_x', where $v_x' = -v_x + 2V$. (HINT: First solve the problem in the rest frame of the wall, using the results of Problem 32.) If we assume that $V << v_x$, this result shows that the speed and kinetic energy of the ball are reduced by the collision, and so the kinetic energy of the wall must be increased.

(b) Show that if the wall is moving toward the ball at velocity **V**, $v_x' = -v_x - 2V$. In this case the kinetic energy of the ball increases while that of the wall decreases.

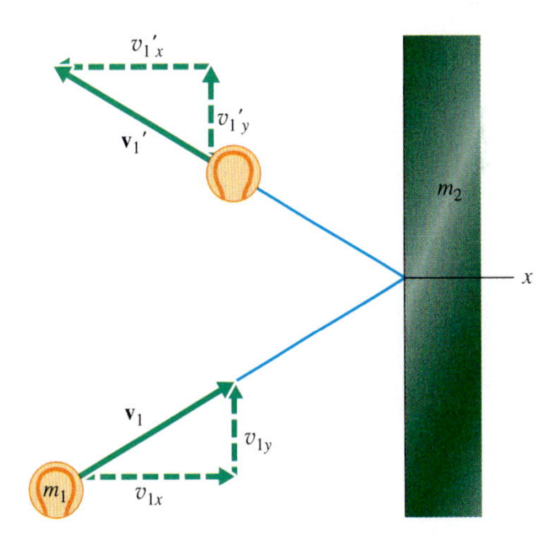

Fig. 8-19 Collision of a ball with a smooth wall.

CHAPTER 9 Rotation

The planet Jupiter and its Great Red Spot, a remarkably stable pattern about the size of Earth, formed by circulating atmospheric gases where opposing high speed jet streams meet. Jupiter rotates once every 10 hours, producing an average surface speed of 10^4 m/s.

A car's wheels rotate as the car moves, a gymnast's body rotates as she dismounts the parallel bars, a football rotates as it is thrown through the air, a door rotates on its hinges, a compact disk rotates on a CD player, the earth rotates on its axis, ... In this chapter, we shall use Newton's laws of motion to analyze rotating bodies such as these. But first we need to develop a description of rotational motion analogous to the description of linear motion in Chapters 1 and 2.

9-1 Description of Rotational Motion

Suppose a rigid body rotates about an axis fixed in direction; for example, a door turns on its hinges or a wheel turns on a stationary axle. The axis of rotation is a line about which the body rotates. Any point P on the rotating body moves in a circle about some point O on this line (Fig. 9-1a).

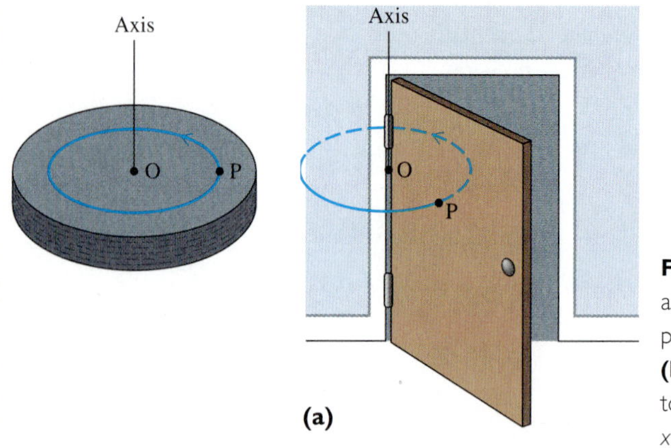

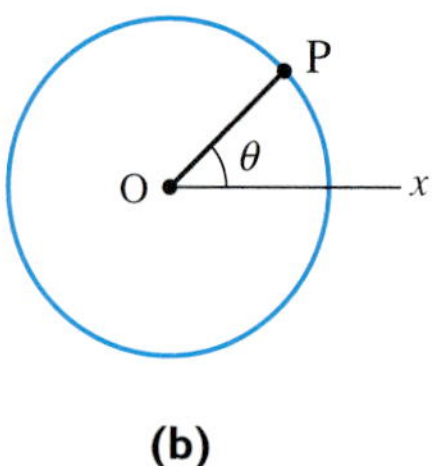

Fig. 9-1 **(a)** An arbitrary point P on a rotating body moves along a circular path centered on the axis of rotation. **(b)** A line from point O on the axis to point P makes an angle θ with the x-axis.

Angular Velocity

Fig. 9-1b shows the plane of rotation of a point P on the body. The line OP forms an angle θ with the x-axis of a fixed coordinate system. The angle θ changes as the body rotates. The orientation of the entire rigid body is determined by the single variable θ. This is analogous to the case of linear motion, where the position of a particle is specified by a single Cartesian coordinate x. Recall from Section 1-6 that motion along the x-axis is indicated by the velocity component v_x—the rate of change of the x coordinate

$$\bar{v}_x = \frac{\Delta x}{\Delta t} \qquad \text{or} \qquad v_x = \lim_{\Delta t \to 0} \frac{\Delta x}{\Delta t}$$

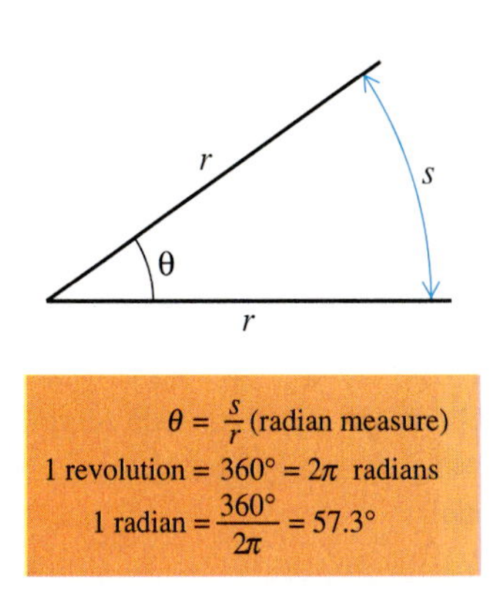

Fig. 9-2 The radian measure of an angle.

Similarly, rotational motion is indicated by the rate of change of the angle θ; we define this rate of change of θ to be the angular velocity and denote it by the Greek letter ω (omega):

$$\bar{\omega} = \frac{\Delta \theta}{\Delta t} \qquad \text{(average angular velocity)} \qquad (9\text{-}1)$$

$$\omega = \lim_{\Delta t \to 0} \frac{\Delta \theta}{\Delta t} \qquad \text{(instantaneous angular velocity)} \qquad (9\text{-}2)$$

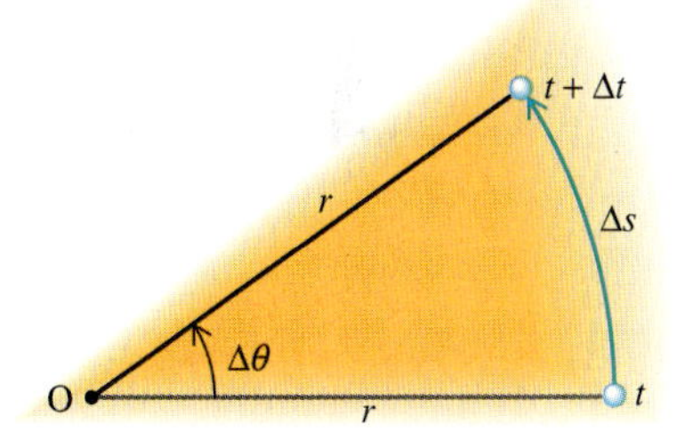

Fig. 9-3 During a time interval Δt, a body rotates through an angle $\Delta\theta$ and a point on the body moves a distance Δs.

When ω is constant, the rate of rotation is uniform and ω may be calculated as $\Delta\theta/\Delta t$.

Angular velocity indicates both the rate of rotation and the sense of rotation, that is, whether the rotation is clockwise or counterclockwise. A positive value of ω means that θ is increasing; in other words, the rotation is counterclockwise. A negative value of ω means that θ is decreasing; that is, the rotation is clockwise.

Angular velocity may be expressed in various units—degrees per second, revolutions per minute (rpm), radians per second (rad/s). (The radian measure of an angle is defined in Fig. 9-2.) If a body is rotating at a uniform rate of, say, 10 degrees per second, a line drawn in the plane of rotation from any point on the body to the axis of rotation will move through an angle of 10° in 1 s.

Velocity of a Point on a Rotating Rigid Body

Radian measure is useful because of the simple relationship it provides between linear and angular quantities. In Fig. 9-3, a body rotates through an angle $\Delta\theta$, and a particle at a distance r from the axis of rotation moves through a linear distance Δs during a time interval Δt. Using the radian definition of an angle ($\Delta\theta = \Delta s/r$), we may relate the linear distance Δs to the angle $\Delta\theta$:

$$\Delta s = r\,\Delta\theta$$

Fig. 9-4 This device was designed by Leonardo da Vinci for the purpose of measuring linear distances along a curved path. Each revolution of the wheel causes a pebble to drop from its place in a horizontal wheel into a box. (*Physics Today*, Nov. 1978.)

Dividing this equation by the time interval Δt, we obtain a relationship between the average linear speed $\bar{v}$ and the average angular velocity $\bar{\omega}$:

$$\bar{v} = \frac{\Delta s}{\Delta t} = \frac{r\,\Delta\theta}{\Delta t} = r\bar{\omega}$$

As Δt approaches zero, both $\bar{v}$ and $\bar{\omega}$ approach their instantaneous values v and ω, and the previous equation becomes

$$v = r\omega \qquad (\omega \text{ in rad/s}) \qquad (9\text{-}3)$$

EXAMPLE 1 Speed on a Merry-Go-Round

A merry-go-round rotates at a steady rate and completes one revolution in 15.0 s. Find the linear speed of a child standing on the merry-go-round (a) 5.00 m from the center; (b) 10.0 m from the center.

SOLUTION The constant angular velocity of the merry-go-round is easily calculated in units of revolutions per second:

$$\omega = \overline{\omega} = \frac{\Delta\theta}{\Delta t} = \frac{1.00 \text{ rev}}{15.0 \text{ s}}$$

$$= 6.67 \times 10^{-2} \text{ rev/s}$$

However, to calculate the child's linear speed, we must express ω in units of radians per second so that we may validly apply Eq. 9-3.

(a) At a radial distance $r = 5.00$ m,

$$v = r\omega = (5.00 \text{ m})(6.67 \times 10^{-2} \text{ rev/s})\left(\frac{2\pi \text{ rad}}{1 \text{ rev}}\right)$$

$$= 2.09 \text{ m/s}$$

(b) The child is now standing twice as far from the axis of rotation as in part a. Since speed is proportional to this distance, the child moves twice as fast:

$$v = 2(2.09 \text{ m/s}) = 4.18 \text{ m/s}$$

Angular Acceleration

The rate of change of angular velocity is defined to be the **angular acceleration** $\overline{\alpha}$:

$$\overline{\alpha} = \frac{\Delta\omega}{\Delta t} \quad \text{(average angular acceleration)} \quad (9\text{-}4)$$

$$\alpha = \lim_{\Delta t \to 0} \frac{\Delta\omega}{\Delta t} \quad \text{(instantaneous angular acceleration)} \quad (9\text{-}5)$$

The definition of angular acceleration parallels the definition of linear acceleration a_x for linear motion, where a_x is defined as the rate of change of linear velocity v_x:

$$\overline{a}_x = \frac{\Delta v_x}{\Delta t} \quad \text{or} \quad a_x = \lim_{\Delta t \to 0} \frac{\Delta v_x}{\Delta t}$$

EXAMPLE 2 Angular Acceleration of a CD

A compact disk starts from rest and accelerates to its final angular velocity of 3.50 rev/s in 1.50 s. Find the disk's average angular acceleration in revolutions per second squared and in radians per second squared.

SOLUTION Applying Eq. 9-4, we obtain

$$\overline{\alpha} = \frac{\Delta\omega}{\Delta t} = \frac{3.50 \text{ rev/s}}{1.50 \text{ s}}$$

$$= 2.33 \text{ rev/s}^2$$

or

$$\overline{\alpha} = (2.33 \text{ rev/s}^2)\left(\frac{2\pi \text{ rad}}{1 \text{ rev}}\right) = 14.7 \text{ rad/s}^2$$

Acceleration of a Point on a Rotating Body

Each particle of a rotating body moves along a circular path and therefore experiences a centripetal acceleration $\mathbf{a}_R$ in the radial direction, toward the axis. The magnitude of this centripetal or **radial acceleration** is given by Eq. 3-13:

$$a_R = \frac{v^2}{r}$$

Using Eq. 9-3 ($v = r\omega$), we may express a_R in terms of ω:

$$a_R = \frac{(r\omega)^2}{r}$$

$$a_R = r\omega^2 \qquad (9\text{-}6)$$

Since the linear speed of a particle in a rotating body is proportional to angular velocity (Eq. 9-3: $v = r\omega$), whenever the body experiences a change in angular velocity $\Delta\omega$, each particle of the body experiences a proportional change in linear speed Δv:

$$\Delta v = r\,\Delta\omega$$

Dividing this equation by the time interval Δt during which the change occurs, we obtain an expression for the rate of change of the particle's linear speed:

$$\frac{\Delta v}{\Delta t} = r\frac{\Delta\omega}{\Delta t} = r\alpha$$

Taking the limit as Δt approaches zero, we obtain an expression for the instantaneous rate of change of linear speed:

$$\lim_{\Delta t \to 0} \frac{\Delta v}{\Delta t} = r\alpha$$

This is the particle's instantaneous acceleration along the instantaneous direction of motion, that is, tangent to the particle's trajectory, as indicated in Fig. 9-5. This instantaneous acceleration is referred to as the **tangential acceleration** and denoted by $\mathbf{a}_T$:

$$a_T = r\alpha \qquad (\alpha \text{ in rad/s}^2) \qquad (9\text{-}7)$$

Fig. 9-6 shows the radial acceleration and the tangential acceleration of a point P on a rotating body that has an increasing angular velocity.

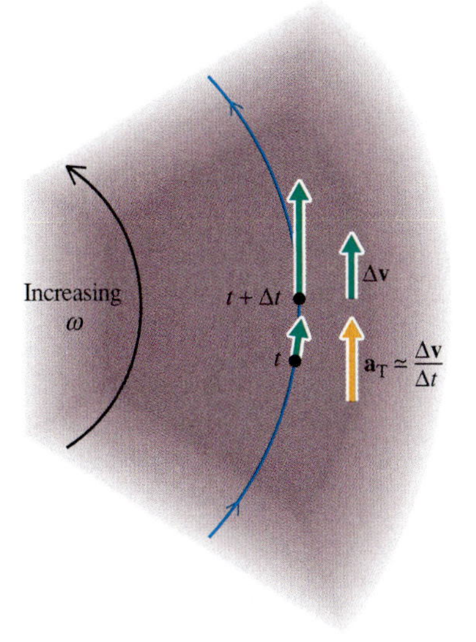

Fig. 9-5 When a rotating body experiences angular acceleration, a point on the body has a tangential linear acceleration $\mathbf{a}_T$, tangent to the circular path.

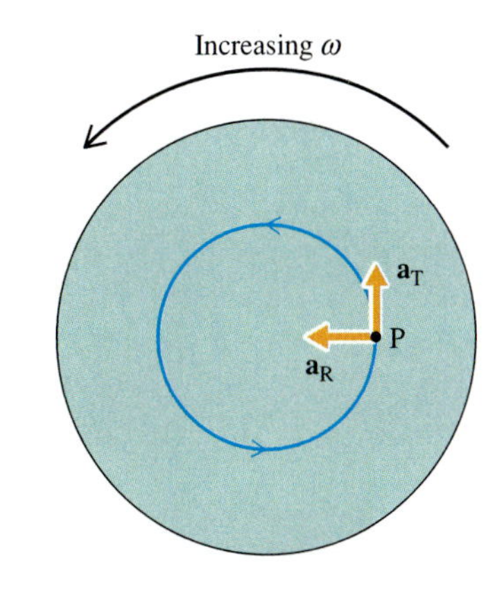

Fig. 9-6 The angular velocity of a rotating body increases in the counterclockwise direction. A point P on the body has both radial acceleration $\mathbf{a}_R$ and tangential acceleration $\mathbf{a}_T$.

EXAMPLE 3 Crack the Whip

In the game "crack the whip," a group of people hold hands and form a line. Then the person at one end begins to rotate the line about a fixed axis close to her or his own position. A person near the axis has small velocity and acceleration, but someone at the other end has large velocity and acceleration. Assuming that the line has a fixed shape (not necessarily a straight line), find the speed and acceleration of the person at the end of the line, 10.0 m from the axis of rotation, if a point 1.00 m from the axis moves at an instantaneous speed of 0.500 m/s and has a tangential acceleration of 1.00 m/s².

SOLUTION First we shall calculate the angular speed and angular acceleration of the line from the linear speed and tangential acceleration at the 1.00 m point, using Eqs. 9-3 ($v = r\omega$) and 9-7 ($a_T = r\alpha$):

$$\omega = \frac{v}{r} = \frac{0.500 \text{ m/s}}{1.00 \text{ m}} = 0.500 \text{ rad/s}$$

$$\alpha = \frac{a_T}{r} = \frac{1.00 \text{ m/s}^2}{1.00 \text{ m}} = 1.00 \text{ rad/s}^2$$

Now we may use these angular rates, which represent the motion of the entire line, to find the linear speed and acceleration of the person 10.0 m from the center. Applying Eq. 9-3, we find

$$v = r\omega = (10.0 \text{ m})(0.500 \text{ rad/s}) = 5.00 \text{ m/s}$$

The linear acceleration has both radial and tangential components, found from Eqs. 9-6 and 9-7:

$$a_R = r\omega^2 = (10.0 \text{ m})(0.500 \text{ rad/s})^2 = 2.50 \text{ m/s}^2$$

$$a_T = r\alpha = (10.0 \text{ m})(1.00 \text{ rad/s}^2) = 10.0 \text{ m/s}^2$$

Thus the person at the end of the line experiences a large acceleration and will find it difficult to hold on.

Since the definitions of angular velocity [$\omega = \underset{\Delta t \to 0}{\text{limit}} (\Delta\theta/\Delta t)$] and angular acceleration [$\alpha = \underset{\Delta t \to 0}{\text{limit}} (\Delta\omega/\Delta t)$] parallel the definitions of linear velocity [$v_x = \underset{\Delta t \to 0}{\text{limit}} (\Delta x/\Delta t)$] and linear acceleration [$a_x = \underset{\Delta t \to 0}{\text{limit}} (\Delta v_x/\Delta t)$], we can easily obtain the kinematic equations describing angular motion at **constant angular acceleration** by exploiting this correspondence. We merely make the appropriate substitutions in the linear kinematic equations derived in Chapter 2.

$$x \to \theta$$

$$v_x \to \omega$$

$$a_x \to \alpha$$

$$v_x = v_{x0} + a_x t \quad \to \quad \boxed{\omega = \omega_0 + \alpha t} \tag{9-8}$$

$$x = x_0 + v_{x0}t + \tfrac{1}{2}a_x t^2 \quad \to \quad \boxed{\theta = \theta_0 + \omega_0 t + \tfrac{1}{2}\alpha t^2} \tag{9-9}$$

$$x = x_0 + \tfrac{1}{2}(v_{x0} + v_x)t \quad \to \quad \boxed{\theta = \theta_0 + \tfrac{1}{2}(\omega_0 + \omega)t} \tag{9-10}$$

$$v_x^2 = v_{x0}^2 + 2a_x(x - x_0) \quad \to \quad \boxed{\omega^2 = \omega_0^2 + 2\alpha(\theta - \theta_0)} \tag{9-11}$$

9-2 Torque

Now that we know how to describe the motion of a body rotating about some axis, let us look at how forces cause rotation. Suppose a bicycle wheel is mounted on an axle and a force **F** is applied to the wheel, causing it to rotate counterclockwise, as indicated in Fig. 9-7. We say that "**F** produces a torque about O." **Torque** is a measure of the tendency of a force to rotate a body. The torque associated with a force depends not only on the force, but also on where the force is applied. For example, if the same force **F** were applied at point Q in Fig. 9-7, rather than at point P, it would pull directly against the axle and therefore could produce no rotation about O; in this case, there would be no torque about O.

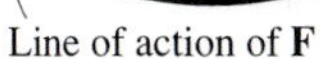

Fig. 9-7 Force **F** causes this wheel to rotate counterclockwise.

Next we shall give a precise definition of torque. Consider first the line shown in Fig. 9-7 extending out from the point P along the line of the force vector **F**. This line is called the **line of action** of the force. The perpendicular distance from the rotational axis O to the line of action is called the **moment arm** of the force and is denoted by $r_\perp$. **Torque is defined to be the product of the magnitude of a force and its moment arm.** Denoting torque by the Greek letter τ (tau), we write this

$$\tau = Fr_\perp$$

From this equation we see that applying the force **F** at point Q in Fig. 9-7 produces no torque about O because the force then has a moment arm equal to zero (the line of action passes through O). On the other hand, the same nonzero torque is produced by **F** applied either at P or at P′ because the line of action is the same in either case and the moment arm is therefore the same.

To complete our definition of torque, we must consider the rotation direction. An object rotating about a fixed axis has two possible directions of rotation: clockwise and counterclockwise. We define torque to be positive if it tends to produce counterclockwise rotation and negative if it tends to produce clockwise rotation. Thus the complete equation defining torque is

$$\tau = \pm Fr_\perp \quad (+\text{ counterclockwise; } -\text{ clockwise}) \tag{9-12}$$

The force **F** in Fig. 9-7 tends to rotate the wheel counterclockwise; hence the torque is positive. If the same force were applied at point R, the torque would be negative because the wheel would then tend to rotate clockwise. Fig. 9-8a shows forces, all of which tend to produce counterclockwise rotation about O and which therefore produce positive torque about O. Fig. 9-8b shows forces, all of which tend to produce clockwise rotation about O and which therefore produce negative torque about O. Notice that in either of these figures the sign of the torque is not determined solely by the signs of force components F_x and F_y. Thus, for example, a force with a positive F_x can produce either positive torque or negative torque, depending on where it is applied relative to the axis of rotation. Torques, like force components, are signed quantities, whose signs are determined by the direction in which they tend to rotate a body about an axis.

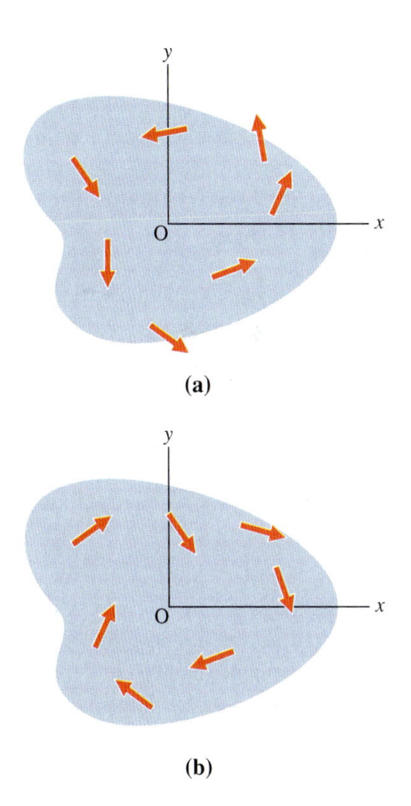

Fig. 9-8 **(a)** The forces acting on this rigid body all tend to produce counterclockwise rotation about O. The torques are therefore positive. **(b)** The forces acting on this rigid body all tend to produce clockwise rotation about O. The torques are therefore negative.

From Eq. 9-12, we see that the units of torque are a force unit multiplied by a distance unit. The standard SI unit for torque is therefore the newton-meter.

EXAMPLE 4 Applying Torque with a Wrench

Find the torque applied to the wrench in Fig. 9-9, with the center of the nut as the axis of rotation.

SOLUTION First notice that the torque tends to rotate the wrench counterclockwise about O; therefore τ is positive.

Applying Eq. 9-12, we find

$$\tau = +Fr_{\perp} = Fr \sin 60.0°$$
$$= (70.0 \text{ N})(0.200 \text{ m})(\sin 60.0°)$$
$$= 12.1 \text{ N-m}$$

If the 70.0 N force were perpendicular to the wrench handle, the torque would be greater: (70.0 N)(0.200 m) = 14.0 N-m.

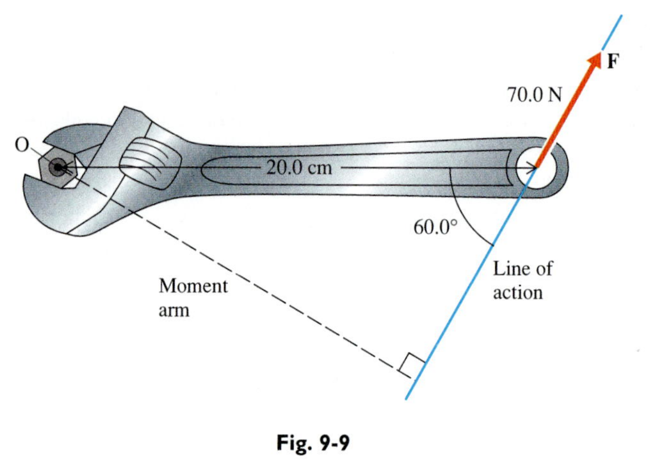

Fig. 9-9

EXAMPLE 5 Computing Torques for Various Forces

Find the value of τ with respect to O for each of the forces shown in Fig. 9-10a, all of which have a magnitude of 1.00 N.

SOLUTION Force $\mathbf{F}_1$ tends to produce clockwise rotation about O. Hence the corresponding torque τ_1 is negative. Fig. 9-10b shows the line of action of $\mathbf{F}_1$. We can read the moment arm directly: 2.00 m. The torque is thus

$$\tau_1 = -F_1 r_{1\perp} = -(1.00 \text{ N})(2.00 \text{ m}) = -2.00 \text{ N-m}$$

Force $\mathbf{F}_2$ tends to produce counterclockwise rotation about O and has a moment arm of 1.50 m. Hence it produces a torque

$$\tau_2 = +F_2 r_{2\perp} = +(1.00 \text{ N})(1.50 \text{ m}) = +1.50 \text{ N-m}$$

The line of action of $\mathbf{F}_3$ passes through O. Hence its moment arm is zero and

$$\tau_3 = 0$$

It is easiest to compute the torque produced by $\mathbf{F}_4$ by resolving the force into horizontal and vertical components and summing the torques produced by the components. When using this method, you must be careful to draw the components at the point where the force is applied, so that you calculate moment arms correctly (Fig. 9-10c). The horizontal component, $F_4 \cos 60.0°$, tends to produce counterclockwise rotation and therefore contributes a positive torque. The vertical component, $F_4 \sin 60.0°$, tends to produce clockwise rotation and therefore contributes a negative torque. Thus we find

$$\tau_4 = +(1.00 \text{ N})(\cos 60.0°)(2.00 \text{ m}) - (1.00 \text{ N})(\sin 60.0°)(2.00 \text{ m})$$
$$= -0.732 \text{ N-m}$$

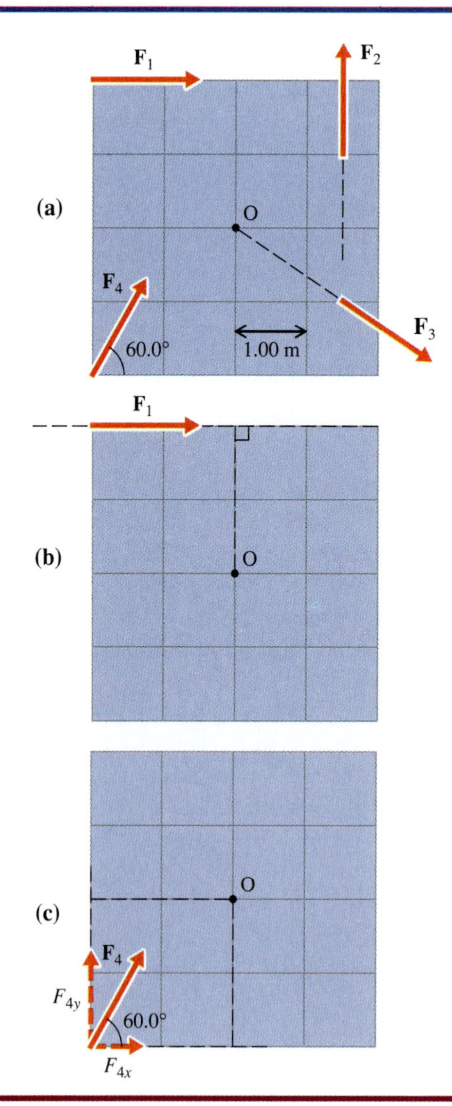

Fig. 9-10

9-3 Dynamics of Rotation about a Fixed Axis

A door is pushed open by a force applied to the edge opposite the hinges (Fig. 9-11a). A wheel mounted on an axle begins to turn because of a force applied tangent to its edge (Fig. 9-11b). Both of these situations are examples of rigid bodies experiencing angular acceleration because of an unbalanced torque. In this section, we shall find the relationship between the angular acceleration of a body rotating about a fixed axis and the net torque acting on the body. We shall see that this relationship follows from Newton's second law.

Consider a rigid body free to rotate about a fixed axis, which we choose to lie along the z-axis (Fig. 9-12a). Each particle in the body is free to move only in a plane perpendicular to the z-axis and can experience an acceleration only in this plane. Thus the net force acting on any particle must also lie in this plane. Fig. 9-12b shows a view looking down on the body, so that the z-axis (the axis of rotation) is perpendicular to the page. There is a resultant force **F** acting on a particle of mass m located a distance r from the z-axis.

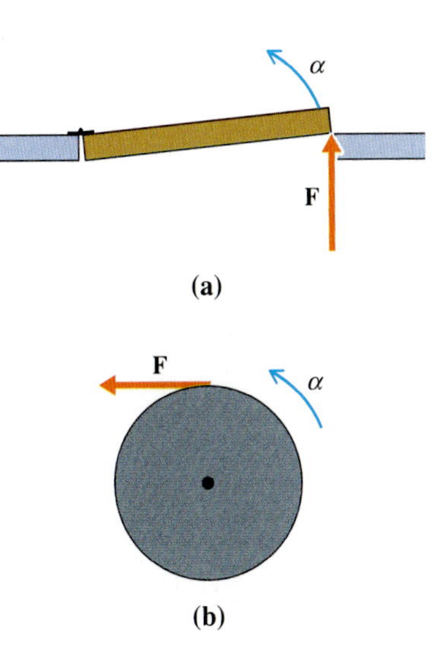

Fig. 9-11 **(a)** A door opens when a torque is applied to it. **(b)** A wheel begins to turn when a torque is applied to it.

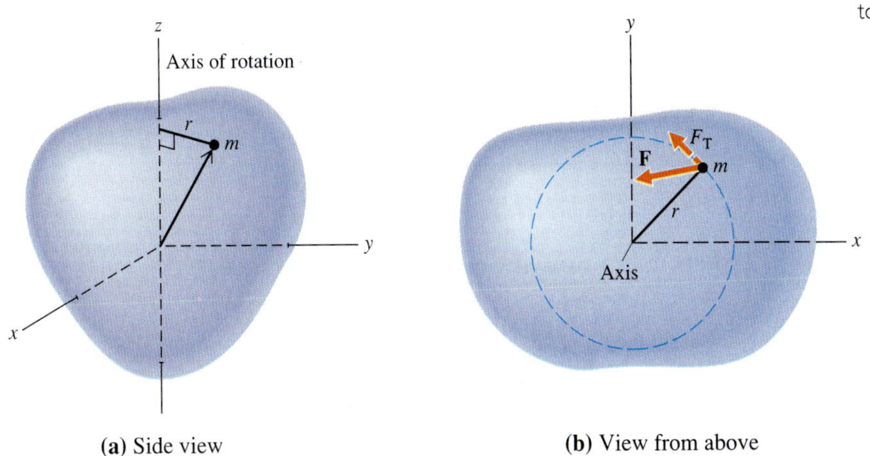

Fig. 9-12 **(a)** A body is free to rotate only about the z-axis. Within the body any particle of mass m can move only in a plane parallel to the xy plane. **(b)** A cross-sectional view of the body showing the xy plane.

The component of **F** tangent to the particle's path of rotation, labeled F_T, produces a torque

$$\tau = F_T r$$

According to Newton's second law, there is a tangential acceleration a_T produced by F_T ($F_T = ma_T$). Thus

$$\tau = ma_T r$$

We can now use Eq. 9-7 ($a_T = r\alpha$) to relate the tangential acceleration of m to the angular acceleration of the entire body. Substituting the expression for a_T given by Eq. 9-7 into the preceding equation, we obtain

$$\tau = mr^2\alpha$$

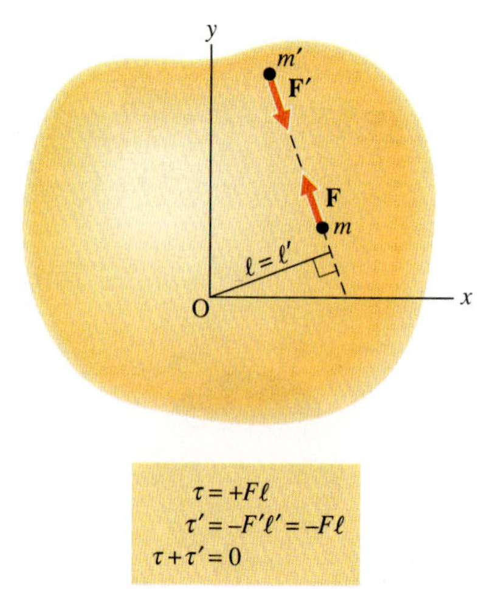

Fig. 9-13 The internal forces **F** and **F′** acting between any pair of particles m and m' produce no net torque. This follows from Newton's third law, which states that the two forces have the same magnitude and the same line of action but are oppositely directed. Since all internal forces occur in such action-reaction pairs, all internal torques cancel.

By itself, this equation is not particularly useful, but this same relationship may be applied to each and every particle of the system. When the resulting single particle equations are summed, a useful dynamical equation is obtained, one that describes the dynamics of rigid-body rotation. We label the particles 1, 2, 3, . . . , n, and write the corresponding equations:

$$\tau_1 = m_1 r_1^{\,2}\alpha$$

$$\tau_2 = m_2 r_2^{\,2}\alpha$$

$$\vdots$$

$$\tau_n = m_n r_n^{\,2}\alpha$$

Summing these equations, we obtain

$$\begin{aligned} \tau_1 + \tau_2 + \cdots + \tau_n &= m_1 r_1^{\,2}\alpha + m_2 r_2^{\,2}\alpha + \cdots + m_n r_n^{\,2}\alpha \\ &= (m_1 r_1^{\,2} + m_2 r_2^{\,2} + \cdots + m_n r_n^{\,2})\alpha \\ \Sigma\, \tau &= (\Sigma\, mr^2)\alpha \end{aligned} \tag{9-13}$$

Now, the sum of the torques would seem to be a very difficult thing to compute, since there are an enormous number of internal forces and internal torques within a rigid body. Fortunately, all the internal torques sum to zero, as shown in Fig. 9-13. Thus only the external torques must be considered when one is evaluating the sum. It is this fact that makes torque a useful quantity.

Moment of Inertia

The bracketed quantity in Eq. 9-13 is called the **moment of inertia** and denoted by the symbol I:

$$I = \Sigma\, mr^2 \tag{9-14}$$

This equation indicates that a body's moment of inertia depends on its particles' masses m and distances r from the axis of rotation. This means that the way a body's mass is distributed about the axis of rotation determines the body's moment of inertia with respect to that axis.

You can see from Eq. 9-14 that moment of inertia has units of mass times length squared. Thus the SI unit for moment of inertia is kg-m^2.

Using the definition of moment of inertia given in Eq. 9-14, we can write Eq. 9-13 in the form

$$\Sigma\, \tau = I\alpha \tag{9-15}$$

This is the fundamental dynamical equation for describing the rotation of a rigid body about a fixed axis. It is the rotational analog of Newton's second law ($\Sigma\, F_x = Ma_x$), with the resultant external force $\Sigma\, F_x$ replaced by the resultant external torque $\Sigma\, \tau$, linear acceleration a_x replaced by angular acceleration α, and mass M replaced by moment of inertia I.

$$\Sigma\, F_x \rightarrow \Sigma\, \tau$$
$$a_x \rightarrow \alpha$$
$$M \rightarrow I$$

Mass is a measure of resistance to linear acceleration: for a given resultant force, the greater the mass of a body, the less its acceleration. Similarly, Eq. 9-15 tells us that **moment of inertia is a measure of resistance to rotational acceleration:** for a given resultant torque, the greater the moment of inertia of a body, the less its angular acceleration.

When we can regard a body as having all its mass localized at only a few points, it is quite easy to calculate its moment of inertia, as illustrated in Fig. 9-14 for a baton. Think of the baton as a rod of negligible mass with "particles" of mass at the ends. We can then find the baton's moment of inertia about an axis through its midpoint by using Eq. 9-14, as shown in the figure. If a baton twirler wishes to rotate the baton about its midpoint, she must apply a torque. The magnitude of this torque depends on the value of I. For example, if she wishes to give the baton shown in the figure an angular acceleration of 5.0 rad/s², Eq. 9-15 tells us that she must apply a torque of $I\alpha = (0.080 \text{ kg-m}^2)(5.0 \text{ rad/s}^2) = 0.40$ N-m.

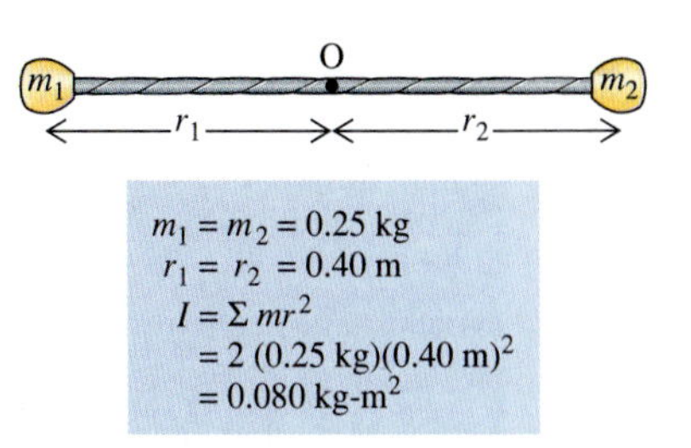

Fig. 9-14 The moment of inertia of a baton.

The mass of most bodies is distributed continuously over a region of space, and therefore the calculation of moments of inertia is generally more complicated than in our baton example. We shall not be concerned with such calculations because they require the use of integral calculus. We shall simply apply the results of calculus derivations, using simple general expressions that have been obtained for the moments of inertia of bodies of various shapes. However, we can derive without calculus one such expression—the moment of inertia of a thin-walled hollow cylinder. Since all the cylinder's mass M is at the same distance R from the axis, the moment of inertia is $I = MR^2$, as shown in Fig. 9-15. Expressions for I for several simple shapes are given in Fig. 9-16. Notice that the moment of inertia depends on both the body and the axis. So, for example, a thin rod of mass M and length ℓ rotating about its midpoint, has a moment of inertia $\frac{1}{12}M\ell^2$, but the same rod rotating about an axis through one end has a moment of inertia $\frac{1}{3}M\ell^2$. It makes sense that the moment of inertia is greater in the second case, since parts of the rod are farther from the axis.

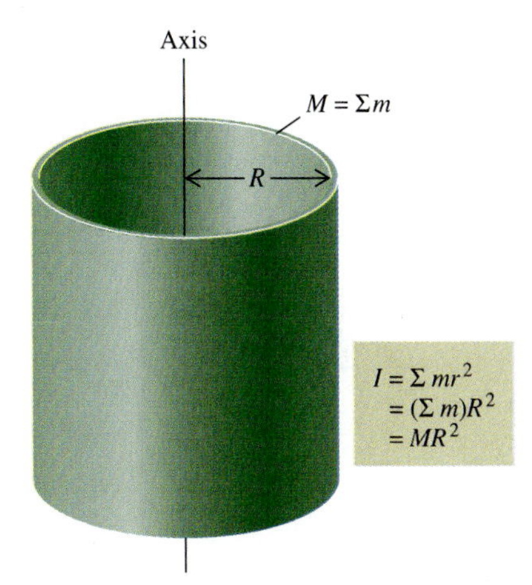

Fig. 9-15 The moment of inertia of a thin-walled hollow cylinder.

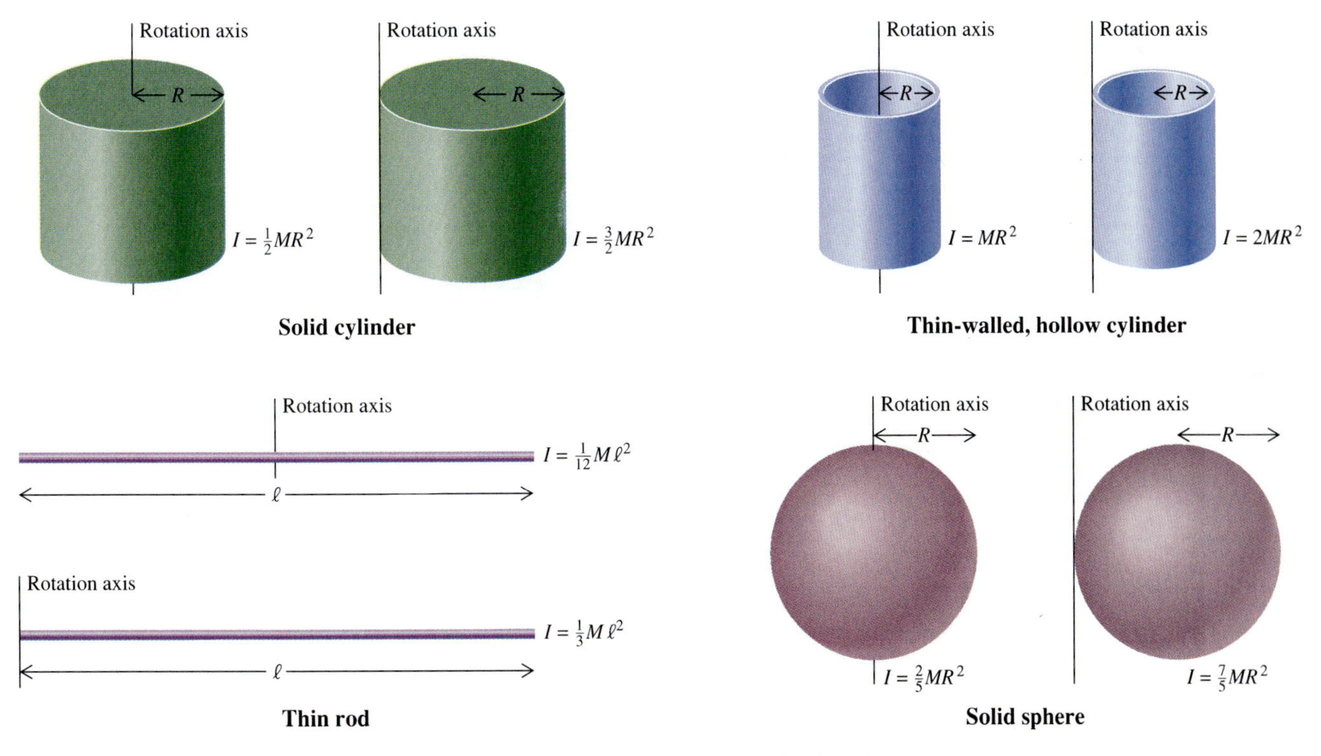

Fig. 9-16 Moments of inertia for bodies of various shapes.

EXAMPLE 6 Spinning a Bicycle Wheel

A bicycle wheel is rotating counterclockwise about a stationary axis (Fig. 9-17). A constant torque τ is applied to the wheel, which starts from rest. When the wheel has rotated through an angle of 30.0° (0.524 rad), the torque is removed. Find τ if the wheel's final angular velocity ω is 10.5 rad/s, assuming negligible frictional torque in the bearings. The wheel's moment of inertia I is 0.200 kg-m^2.

SOLUTION The single applied torque can be found from Eq. 9-15 ($\tau = I\alpha$), but first we must determine the angular acceleration α. It may be found from Eq. 9-11:

$$\omega^2 = \omega_0^2 + 2\alpha(\theta - \theta_0)$$

Substituting $\omega_0 = 0$ and solving for α, we obtain

$$\alpha = \frac{\omega^2}{2(\theta - \theta_0)} = \frac{(10.5 \text{ rad/s})^2}{2(0.524 \text{ rad})}$$
$$= 105 \text{ rad/s}^2$$

Fig. 9-17

Now we may evaluate τ:

$$\tau = I\alpha = (0.200 \text{ kg-m}^2)(105 \text{ rad/s}^2)$$
$$= 21.0 \text{ N-m}$$

If the radius of the wheel is 0.333 m, this torque could be produced by a constant tangential force of magnitude 63.0 N (14 lb) applied to the edge of the wheel.

9-4 Rotational Kinetic Energy

In this section we shall investigate the kinetic energy associated with rotation. We begin by considering a body rotating about a fixed axis. Each particle of mass m has kinetic energy

$$K = \tfrac{1}{2}mv^2$$

where the particle's speed v is related to the body's angular velocity ω and the distance r to the axis of rotation by Eq. 9-3: $v = r\omega$. Substituting into the previous equation, we obtain

$$K = \tfrac{1}{2}mr^2\omega^2$$

Now the total kinetic energy of the rotating body is the sum of the kinetic energies of its particles:

$$K = \Sigma\,(\tfrac{1}{2}mr^2\omega^2)$$
$$= \tfrac{1}{2}(\Sigma\, mr^2)\omega^2$$

The quantity $\Sigma\, mr^2$ is the body's moment of inertia I. Thus we may express this result.

$$K = \tfrac{1}{2}I\omega^2 \qquad (9\text{-}16)$$

EXAMPLE 7 Earth's Rotational Energy

Compute the rotational kinetic energy of the earth.

SOLUTION The earth is approximately spherical in shape, has a mass of 5.98×10^{24} kg and a radius of 6.38×10^6 m, and rotates on its axis with a period of 24 hours. Since the moment of inertia of a sphere is $\frac{2}{5}MR^2$, and the earth rotates through an angle of 2π radians during its 24-hour period, the earth's kinetic energy may be written

$$K = \tfrac{1}{2}I\omega^2 = \tfrac{1}{2}(\tfrac{2}{5}MR^2)(2\pi/T)^2 = \frac{\frac{4}{5}\pi^2MR^2}{T^2}$$
$$= \frac{\frac{4}{5}\pi^2(5.98 \times 10^{24} \text{ kg})(6.38 \times 10^6 \text{ m})^2}{[24(3600 \text{ s})]^2}$$
$$= 2.57 \times 10^{29} \text{ J}$$

To put this figure into perspective, the annual use of energy in the United States is approximately 10^{19} J.

Rotational and Translational Kinetic Energy

The total kinetic energy of a symmetrical body of mass M, rotating with angular velocity ω about a center of mass moving at velocity $\mathbf{V}$, can always be expressed as a sum of two terms: the translational kinetic energy $\frac{1}{2}MV^2$ and the kinetic energy of rotation about the center of mass $\frac{1}{2}I\omega^2$, where I is the moment of inertia about the center of mass.

$$K = \frac{1}{2}MV^2 + \frac{1}{2}I\omega^2 \tag{9-17}$$

EXAMPLE 8 How to Win a Pinewood Derby Race

A boy wishes to build a model car to race in a "Pinewood Derby" contest (Fig. 9-18). The car is to start from rest and roll down an inclined plane, dropping a vertical distance of 2.00 m. The boy is supplied with a block of wood, plastic wheels, and nails to serve as axles. No motor or other internal source of energy is allowed. Both the car's body and wheels can be shaped, and lead weights can be inserted, but the car's total mass cannot exceed 500 grams. Find the final speed of the car if the mass m of each wheel is (a) 5.00% of the total mass M; (b) 1.00% of the total mass. Assume that air resistance is negligible and that internal work done by friction is reduced to a negligible amount through lubrication.

Fig. 9-18

SOLUTION With our assumptions, no nonconservative forces do work, and thus the car's mechanical energy is conserved.

$$E_f = E_i$$

The car's initial energy is gravitational potential energy, with the car's center of mass at a height $y = 2.00$ m above its final position.

$$E_i = Mgy$$

The car's final energy is in the form of kinetic energy—translational kinetic energy $\frac{1}{2}MV^2$, where V is the final speed, and rotational kinetic energy $\frac{1}{2}I\omega^2$ for each of the four wheels.

$$E_f = \frac{1}{2}MV^2 + 4\left(\frac{1}{2}I\omega^2\right)$$

A wheel's angular velocity ω is related to the car's linear speed V and the wheel's radius r. In time t the wheel rotates through an angle $\theta = s/r$, where s is the distance traveled both by the edge of the wheel and by the car's body. Thus $s = Vt$ and $\theta = s/r = Vt/r$, and so $\omega = \theta/t = V/r$. The moment of inertia of each cylindrical wheel is given by $I = \frac{1}{2}mr^2$. Substituting these expressions for I and ω into the equation for E_f and equating it to E_i, we obtain

$$\frac{1}{2}MV^2 + mV^2 = Mgy$$

Solving for V, we find

$$V = \sqrt{\frac{2gy}{1 + 2m/M}}$$

(a) Using $m/M = 0.0500$, we find

$$V = \sqrt{\frac{2(9.80 \text{ m/s}^2)(2.00 \text{ m})}{1 + 2(0.0500)}} = 5.97 \text{ m/s}$$

(b) If $m/M = 0.0100$, we get

$$V = \sqrt{\frac{2(9.80 \text{ m/s}^2)(2.00 \text{ m})}{1 + 2(0.0100)}} = 6.20 \text{ m/s}$$

We see that reduced wheel mass produces a significantly greater final velocity. A good strategy in building such a car is to reduce as much as possible the mass of each wheel, or its moment of inertia.

9-5 Angular Momentum

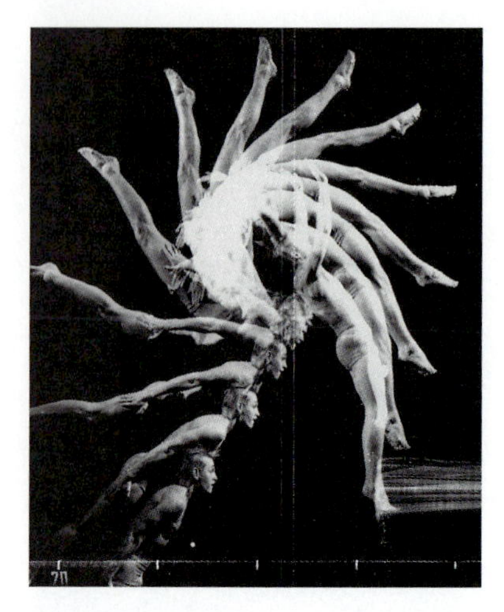

Fig. 9-19 A diver's angular momentum is conserved.

We have seen that velocity, acceleration, mass, and force each have rotational analogs, namely, angular velocity, angular acceleration, moment of inertia, and torque. Linear momentum also has a rotational counterpart, called **angular momentum.** Since momentum is defined as the product of mass and velocity, it is perhaps not surprising that angular momentum, denoted by L, is defined as the product of the corresponding angular quantities—moment of inertia I and angular velocity ω:

$$L = I\omega \tag{9-18}$$

Like linear momentum, angular momentum is sometimes a conserved quantity. One can show, by applying Newton's second law to the particles of a rotating body, that when there is no resultant external torque acting on a body, its angular momentum is conserved*:

$$L_f = L_i \qquad (\text{when } \Sigma\, \tau = 0) \tag{9-19}$$

Some sports activities involve the spin angular momentum of the human body. A diver who initiates a dive with the body outstretched and a small angular velocity can greatly increase that angular velocity by tucking the arms and legs in close to the body. The moment of inertia of the body about the axis of rotation is thereby reduced. And since no external torque acts on the diver while in the air, his total angular momentum ($L = I\omega$) is conserved. Thus a smaller moment of inertia I results in a larger angular velocity ω. The same principle applies to an ice skater who initiates a slow rotation about a vertical axis and then reduces her moment of inertia by bringing her arms and legs in close to the axis and thereby increases her angular velocity ω (Fig. 9-20).

Fig. 9-20 An ice skater's angular momentum is approximately conserved, if there is negligible frictional torque. Therefore her angular velocity increases as her moment of inertia decreases.

*A proof requires the use of vector calculus.

EXAMPLE 9 Spinning on a Swivel Seat

A student sits on a swivel seat that is mounted on good bearings and so is free to turn about a vertical axis with very little frictional resistance. Thus the student and the seat are not subject to any external torques. Initially the student rotates at an angular velocity ω_i, with arms outstretched and a 2.0 kg mass in each hand. Then the arms are brought in close to the rotational axis, and the student rotates faster, with angular velocity ω_f (Fig. 9-21). Initially the weights are held 0.75 m from the axis, and finally they are a negligible distance from the axis. The change in position of the arms contributes relatively little to the change in the moment of inertia and therefore can be ignored in an approximate calculation. The mass distribution of the rest of the body and the seat can be approximated by a cylinder of mass 72 kg and radius 25 cm. If $\omega_i = 1.0$ rev/s, what is the value of ω_f?

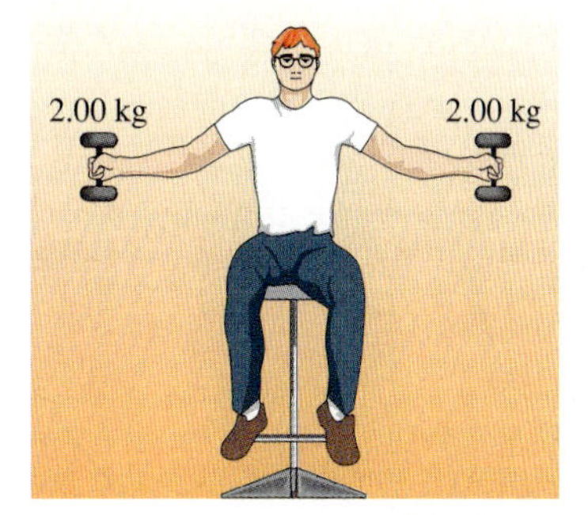

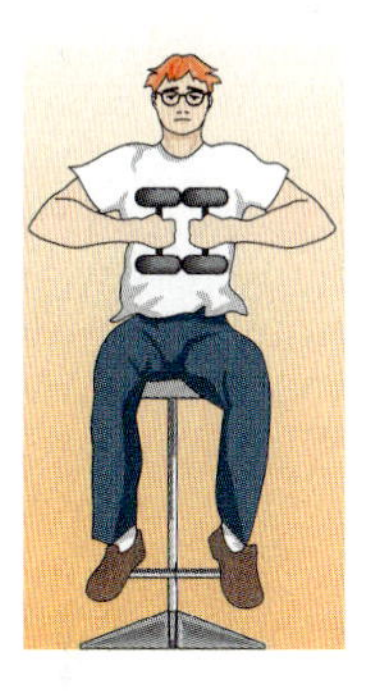

Fig. 9-21

SOLUTION Since there are no external torques on the system, its total angular momentum is conserved. The angular velocity changes as the moment of inertia is changed, so that the product $L = I\omega$ is constant.

First we must calculate the contributions to the system's moment of inertia. The fixed moment of inertia I_B of the body and seat is approximated using the expression for the moment of inertia of a cylinder given in Fig. 9-16:

$$I_B = \tfrac{1}{2}MR^2 = \tfrac{1}{2}(72 \text{ kg})(0.25 \text{ m})^2 = 2.3 \text{ kg-m}^2$$

The moment of inertia I_W of the outstretched weights is given by

$$I_W = 2mr^2 = 2(2.0 \text{ kg})(0.75\text{m})^2 = 2.3 \text{ kg-m}^2$$

The total moment of inertia of the system is initially given by

$$I_i = I_B + I_W = 2.3 \text{ kg-m}^2 + 2.3 \text{ kg-m}^2$$
$$= 4.6 \text{ kg-m}^2$$

After the weights are pulled in, they no longer contribute significantly to the system's moment of inertia, which is reduced to its final value:

$$I_f = I_B = 2.3 \text{ kg-m}^2$$

By conservation of angular momentum,

$$L_f = L_i$$

or

$$I_f\omega_f = I_i\omega_i$$

Therefore

$$\omega_f = \frac{I_i}{I_f}\omega_i = \frac{4.6 \text{ kg-m}^2}{2.3 \text{ kg-m}^2}(1.0 \text{ rev/s})$$
$$= 2.0 \text{ rev/s}$$

*9-6 Energy Analysis of Running

An essay in Chapter 7 described how energy is used by a runner. In this section we shall use a simplified model of running to estimate the power output required for each of the four functions described in the Chapter 7 essay. We assume the runner moves along level ground at an average speed of 3.0 m/s (about 9 minutes per mile). Fig. 9-22 gives data used in this analysis.

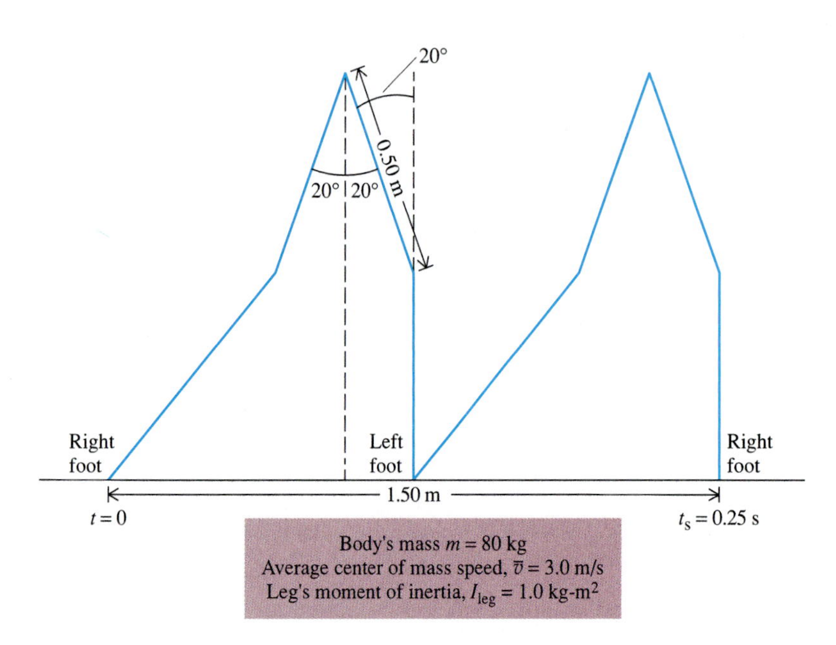

Fig. 9-22 When each foot has moved through its 1.5 m stride, one cycle is completed and the center of mass has moved forward 1.5 m in a time interval of 1.5 m/(3.0 m/s) = 0.50 s. This means that the time t_s for each stride equals 0.25 s.

Rotational Kinetic Energy

The maximum rotational kinetic energy of the leg occurs approximately when the leg is vertical, having accelerated from rest through an angle of 20° or 0.35 rad in about 0.13 s, half the time for the entire stride. Assuming constant angular acceleration, the final angular velocity ω_{max} is twice the average angular velocity:

$$\omega_{max} = 2\overline{\omega} = 2\frac{\Delta\theta}{\Delta t} = \frac{2(0.35 \text{ rad})}{0.13 \text{ s}} = 5.4 \text{ rad/s}$$

The maximum rotational kinetic energy is given by $\frac{1}{2}I\omega_{max}^2$, and the average power provided for rotation is this energy divided by the time for one stride.

$$\overline{P}_R = \frac{\Delta E}{\Delta t} = \frac{\frac{1}{2}I\omega_{max}^2}{t_s} = \frac{\frac{1}{2}(1.0 \text{ kg-m}^2)(5.4 \text{ rad/s})^2}{0.25 \text{ s}}$$
$$= 58 \text{ W} \quad \text{(or 0.078 hp)}$$

Gravitational Potential Energy

As a simple model for determining the increase in height of the center of mass, we assume that the forward upper leg rotates about the knee, thereby raising both the upper end of the leg and the center of mass a distance $\Delta y = 3.0$ cm, as shown in Fig. 9-23. This produces an increased gravitational potential energy $Mg\,\Delta y$, and thus the average power provided for gravitational potential energy is given by

$$\bar{P}_G = \frac{\Delta E}{\Delta t} = \frac{Mg\,\Delta y}{t_s} = \frac{(80\text{ kg})(9.8\text{ m/s}^2)(0.030\text{ m})}{0.25\text{ s}}$$
$$= 94\text{ W} \quad (\text{or } 0.13\text{ hp})$$

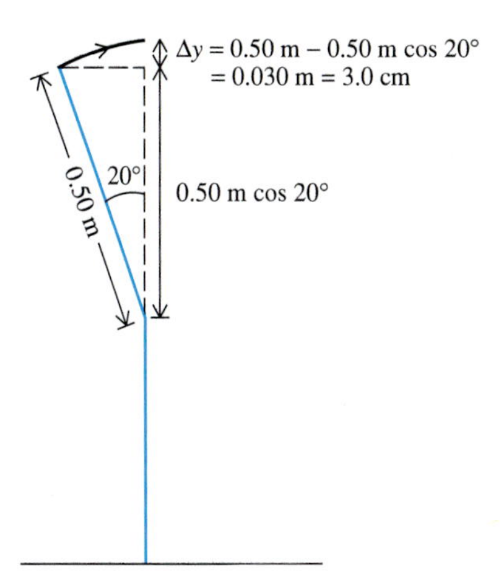

Fig. 9-23 The forward upper leg rotates about the knee, raising the body's center of mass 3.0 cm.

Translational Kinetic Energy

First we must determine the change in the body's center of mass speed v. Applying Newton's second law, we relate the average force $\bar{F}$ exerted by the ground on the forward foot striking the ground to the average acceleration of the center of mass, $\Delta v/\Delta t$.

$$\bar{F} = M\frac{\Delta v}{\Delta t}$$

or

$$\Delta v = \frac{\bar{F}\,\Delta t}{M}$$

The force $\bar{F}$ produces a torque τ, which serves to stop the forward rotation of the leg about the hip, reducing the angular velocity from ω_{max} to 0. The torque τ is the product of the force $\bar{F}$ and the moment arm d, shown in Fig. 9-24. Thus

$$\tau = I\bar{\alpha}$$

$$\bar{F}d = I\frac{\Delta\omega}{\Delta t} = I\frac{0 - \omega_{max}}{\Delta t}$$

Solving for $\bar{F}$ and inserting into the expression for Δv, we obtain

$$\Delta v = -\frac{I\omega_{max}}{Md} = -\frac{(1.0\text{ kg-m}^2)(5.4\text{ rad/s})}{(80\text{ kg})(0.90\text{ m})} = -0.075\text{ m/s}$$

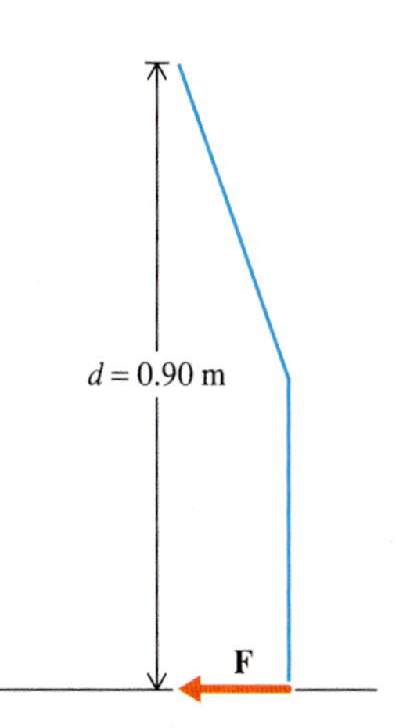

Fig. 9-24 The force of the ground on the forward foot stops the rotation of the leg about the hip and reduces the body's center of mass speed.

Thus the center of mass slows by 0.075 m/s each time a foot strikes the ground. It then is accelerated when the opposite leg pushes against the ground. The average power used to produce translational kinetic energy is given by

$$\bar{P}_T = \frac{\Delta E}{\Delta t} = \frac{\Delta K}{t_s} = \frac{\frac{1}{2}M(v_f^2 - v_i^2)}{t_s}$$

Since the average speed is 3.0 m/s and v_f and v_i differ by 0.075 m/s, $v_i = 3.0\text{ m/s} - \frac{1}{2}(0.075\text{ m/s}) = 2.96$ m/s and $v_f = 3.0\text{ m/s} + \frac{1}{2}(0.075\text{ m/s}) = 3.04$ m/s. Thus

$$\bar{P}_T = \frac{\frac{1}{2}(80\text{ kg})[(3.04\text{ m/s})^2 - (2.96\text{ m/s})^2]}{0.25\text{ s}}$$
$$= 77\text{ W} \quad (\text{or } 0.10\text{ hp})$$

Air Resistance

The magnitude of the force of air resistance is given by

$$F = CAv^2$$

where C is a constant and A is the cross-sectional area. (See Problem 54, Chapter 4.) This force does (negative) work at the rate of $-Fv = -CAv^3$. Thus the power used to compensate for air resistance is given by

$$P_A = +CAv^3$$

Assuming $C = 0.80$ N-s²/m⁴ and $A = 0.50$ m², we obtain

$$\begin{aligned} P_A &= (0.80 \text{ N-s}^2/\text{m}^4)(0.50 \text{ m}^2)(3.0 \text{ m/s})^3 \\ &= 11 \text{ W} \quad (\text{or } 0.01 \text{ hp}) \end{aligned}$$

The total average power that must be provided by muscular force is the sum of these four terms:

$$\begin{aligned} \bar{P} &= \bar{P}_R + \bar{P}_G + \bar{P}_T + P_A \\ &= 58 \text{ W} + 94 \text{ W} + 77 \text{ W} + 11 \text{ W} \\ &= 240 \text{ W} \quad (\text{or } 0.32 \text{ hp}) \end{aligned}$$

A running person is an exceedingly complex mechanical system, and the model we have used here is only a crude approximation. Therefore the numbers calculated above should not be taken too seriously. For example, we have overestimated the term $\bar{P}_G$. The lower leg bends forward as the upper leg rises, so that the increase in center of mass height is less than the 3.0 cm calculated in Fig. 9-24.

CHAPTER 9 SUMMARY

The equations describing rigid body rotation are analogous to the equations describing linear motion:

$$x \leftrightarrow \theta$$

$$v_x \leftrightarrow \omega$$

$$a_x \leftrightarrow \alpha$$

Constant a_x		Constant α
$v_x = v_{x0} + a_x t$	$\leftrightarrow$	$\omega = \omega_0 + \alpha t$
$x = x_0 + v_{x0}t + \frac{1}{2}a_x t^2$	$\leftrightarrow$	$\theta = \theta_0 + \omega_0 t + \frac{1}{2}\alpha t^2$
$x = x_0 + \frac{1}{2}(v_{x0} + v_x)t$	$\leftrightarrow$	$\theta = \theta_0 + \frac{1}{2}(\omega_0 + \omega)t$
$v_x^2 = v_{x0}^2 + 2a_x(x - x_0)$	$\leftrightarrow$	$\omega^2 = \omega_0^2 + 2\alpha(\theta - \theta_0)$

A point on a rotating rigid body a distance r from the axis of rotation has a linear speed v related to the body's angular speed ω by the equation

$$v = r\omega$$

The point has radial and tangential components of acceleration a_R and a_T, related to the body's angular speed ω and angular acceleration α by the equations

$$a_R = r\omega^2$$

$$a_T = r\alpha$$

Torque is the measure of the tendency of a force to produce rotation. The magnitude of the torque produced by a force is the product of the force's magnitude and its moment arm. The sign of the torque is determined by the direction in which the force tends to rotate the body, positive for counterclockwise rotation and negative for clockwise rotation:

$$\tau = \pm F r_\perp \qquad (+\text{, counterclockwise; } -\text{, clockwise})$$

There is a one-to-one correspondence between the variables of rotational dynamics and those of linear dynamics:

$$M \leftrightarrow I = \Sigma\, mr^2$$

$$F_x \leftrightarrow \tau$$

$$\Sigma\, F_x = Ma_x \leftrightarrow \Sigma\, \tau = I\alpha$$

The angular momentum L of a rotating body is defined to be the product of the body's moment of inertia I and angular velocity ω:

$$L = I\omega$$

When there is no resultant external torque acting on the body, its angular momentum is conserved:

$$L_f = L_i \qquad (\text{when } \Sigma\, \tau = 0)$$

Questions

1 A meter stick is mounted on an axle through its midpoint and is subject to the forces shown in Fig. 9-25. In which direction will it rotate?

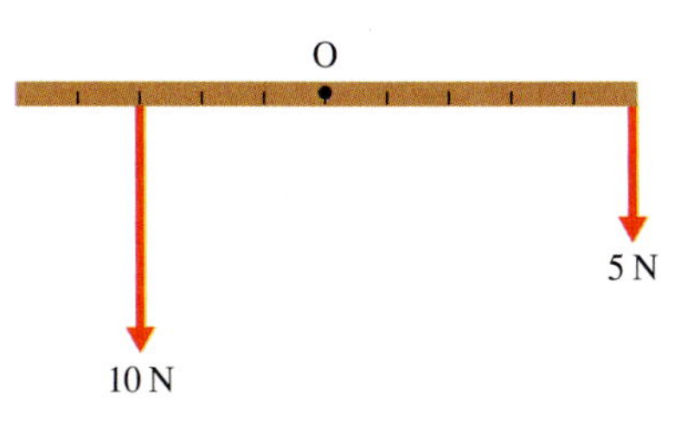

Fig. 9-25

2 What property of a tennis racket determines the ease with which it can be rapidly rotated?

3 What appears to be a massive door made of solid marble can be opened easily and quickly with fingertip pressure. It is explained that the door is easy to open because it is perfectly balanced on nearly frictionless hinges. What can you conclude about the door?

4 When a car stops quickly, the front end dips down.
- (a) What force produces the torque responsible for this slight rotation?
- (b) What force produces a countertorque that prevents the car from rotating farther?
- (c) What special feature of a car could cause it to flip over?

5 What feature of a Jeep or a trailer makes them both more likely than a car to turn over on an incline?

6 When you flip a coin, what force could change its angular velocity while it is in the air?

Answers to Odd-Numbered Questions

1 counterclockwise; **3** It's not solid marble; **5** The relatively high center of gravity means that on an incline the line of action of the weight is farther from the midpoint between the wheels than in a car. Thus the vehicle can be turned over by a much smaller torque (provided by the force of the wind, for example).

Problems (listed by section)

9-1 Description of Rotational Motion

1 A compact disk on a CD player rotates clockwise. The disk completes one revolution in 0.286 s.
(a) Find the angular velocity of the disk.
(b) Find the speed of a point on the disk 6.00 cm from the axis of rotation.

2 A helicopter propeller blade starts from rest and experiences a constant angular acceleration of 10.0 rad/s^2. Find the speed of a point on the blade 3.00 m from the axis of rotation when the blade has completed 10 revolutions.

3 A centrifuge is used to increase the sedimentation rate of particles suspended in a liquid, over the rate produced by gravity alone. Find the angular velocity of a centrifuge in which a particle 10.0 cm from the axis of rotation experiences a radial acceleration of 1.00×10^3 times gravitational acceleration g.

4 Find the angular velocity of a clock's second hand and the linear speed and linear acceleration of a point on the second hand 5.00 cm from the rotation axis.

5 An ice skater rotates about a vertical axis through the center of his body. Find his angular velocity, if the radial acceleration at a point on the body 20.0 cm from the axis of rotation is not to exceed 10.0 times gravitational acceleration g.

6 A merry-go-round starts from rest and accelerates uniformly over 20.0s to a final angular velocity of 6.00 rev/min. Find (a) the maximum linear speed of a person sitting on the merry-go-round 5.00 m from the center; (b) the person's maximum radial acceleration; (c) the angular acceleration of the merry-go-round; (d) the person's tangential acceleration.

★ 7 Calculate the angular velocity of the earth about its rotation axis. Find the linear acceleration of a point on the earth's surface at 40.0° N latitude.

★ 8 A compact disk player has just been turned off. A disk decelerates uniformly from an initial angular velocity of 8.00 rev/s. The disk rotates through 3.00 revolutions before coming to rest. What is the tangential acceleration of a point on the disk 5.00 cm from the axis?

9-2 Torque

9 A heavy concrete panel is being lifted into position in a building by means of a crane (Fig. 9-26). The tension of 2.00×10^4 N in the supporting cable produces a torque with respect to point O.
(a) Does this torque tend to rotate the panel clockwise or counterclockwise about an axis through point O?
(b) Find the torque.

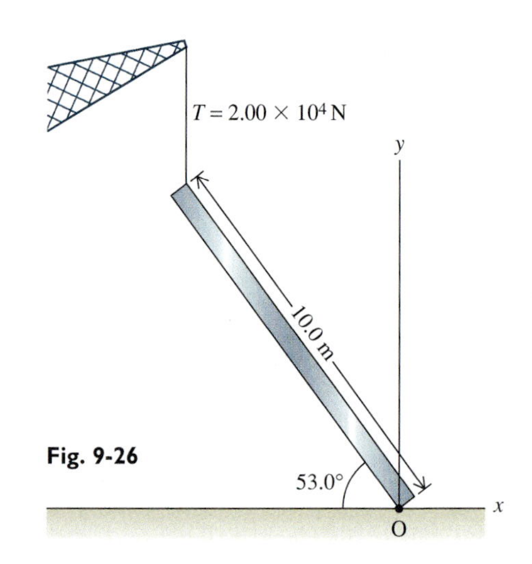

Fig. 9-26

10 In opening a door, a 40.0 N force is applied as shown in Fig. 9-27. Find the torque produced by this force with respect to point O.

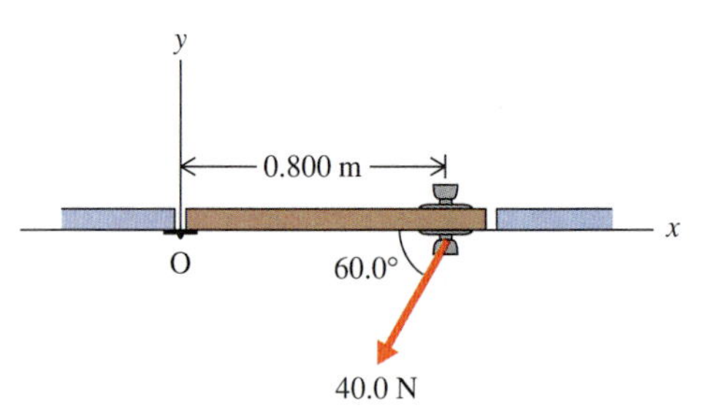

Fig. 9-27 View of a door from above.

11 A cyclist exerts a downward force on a bicycle pedal, as shown in Fig. 9-28. Find the torque about O.

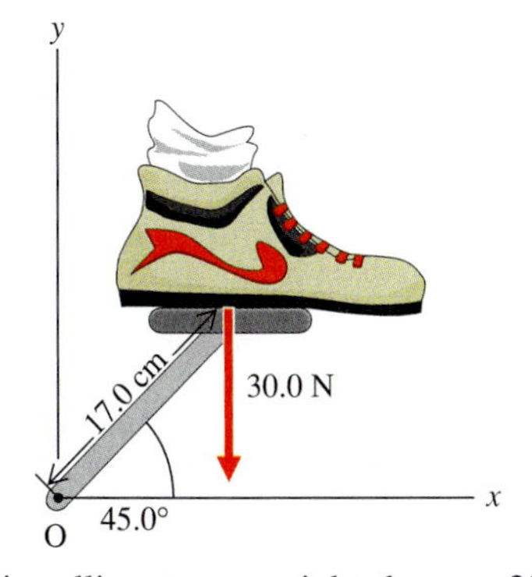

Fig. 9-28

12 A car is rolling to your right down a 30.0° incline. The front axle exerts a downward vertical force of 2.00×10^3 N on the front passenger-side wheel, which has a radius of 35.0 cm.

(a) Does this force tend to rotate the wheel clockwise or counterclockwise about an axis through the point of contact of the wheel with the ground?

(b) Find the torque on the wheel with respect to this axis.

13 Find the value of τ for forces $\mathbf{F}_1$, $\mathbf{F}_2$, and $\mathbf{F}_4$ in Fig. 9-10a with respect to an origin at the upper right-hand corner of the plate. All three forces have a magnitude $F = 1.00$ N.

14 As a bowling ball of radius 15.0 cm rolls down the alley to the right, a frictional force of magnitude 1.00 N, directed to the left, acts on it. Find the frictional torque with respect to an axis through the center of the ball.

★**15** The gate in Fig. 9-29, which is 2.00 m wide and 1.00 m tall, is supported by hinges and a guy wire. For $F_1 = 66.4$ N, $F_2 = 87.8$ N, $F_3 = 175$ N, $F_4 = 109$ N, find the torque resulting from each force about an axis through point P.

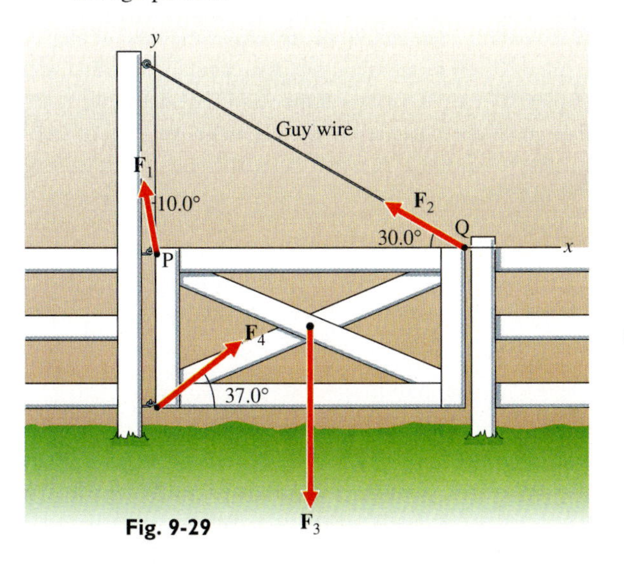

Fig. 9-29

★**16** Repeat the preceding problem using an axis through point Q.

9-3 Dynamics of Rotation About a Fixed Axis

17 A baseball pitcher pivots his extended arm about his shoulder joint, applying a constant torque of 180 N-m for 0.10 s to his arm, which has a moment of inertia of 0.50 kg-m^2.

(a) Find his arm's angular acceleration and final angular velocity, assuming it starts from rest.

(b) Find the final speed of the ball, relative to his shoulder, which is 0.80 m from the ball.

18 A child sits down on one end of a horizontal seesaw of negligible weight, 2.00 m from the pivot point. No one balances on the other side. Find the instantaneous angular acceleration of the see-saw and the tangential linear acceleration of the child.

19 A large fish strikes bait cast by a fisherman with a rod and reel. The fish creates tension in the line of 65 N. How many revolutions of the reel occur in 1.0 s if the reel has a moment of inertia of 4.0×10^{-3} kg-m^2 and the radial distance of the line from the reel's axis is 4.0 cm? Assume the reel has negligible friction.

20 A boxer receives a horizontal blow to the head that topples him over. The force, of magnitude 1.00×10^3 N, is applied for 1.0×10^{-2} s at a point 1.70 m above the floor. The boxer has a moment of inertia of 80.0 kg-m^2 for rotation about an axis at his feet.

(a) Find the boxer's final angular velocity.

(b) How long would it take him to hit the floor if he continued with the same angular velocity? Why does his angular velocity increase as he falls?

★**21** A dart player holds a dart in her hand with her forearm initially in a vertical position. She quickly accelerates the forearm forward and releases the dart, holding the upper arm fixed, as good dart players usually do. Find the torque that must be applied to the forearm by muscles connecting it to the upper arm in order for the dart to leave the hand with a speed of 12.0 m/s after the forearm has rotated through an angle of 0.300 rad. The radial distance from the dart to the pivot point at the elbow is 0.400 m. Treat the forearm as a thin rod of mass 2.00 kg and length 0.400 m.

★**22** A 1.00 kg mass is attached to a light cord that is wrapped around a pulley of radius 5.00 cm, which turns with negligible friction. The mass falls at a constant acceleration of 3.00 m/s^2. Find the moment of inertia of the pulley.

★ **23** Two masses, $m_1 = 1.00$ kg and $m_2 = 2.00$ kg, are attached to the ends of a light cord, which passes over a frictionless pulley in the shape of a uniform disk of mass 3.00 kg. How long does it take the 2.00 kg mass to fall a vertical distance of 1.00 m? What is the tension on either side of the pulley?

★ **24** A 20.0 kg door that is 1.00 m wide opens slowly as a pull of constant magnitude 50.0 N is exerted on the doorknob. This applied force, which is always perpendicular to the door, is necessary to balance the frictional torque in the hinges. How large must an applied force of constant magnitude be in order to open the door through an angle of 90.0° in 0.500 s?

25 A hollow cylinder and a solid cylinder have the same diameter. Determine which has the greater moment of inertia with respect to an axis of rotation along the axis of the cylinder: (a) when the cylinders have equal mass; (b) when they have equal density.

9-4 Rotational Kinetic Energy

26 A rotating flywheel has been proposed as a means of temporarily storing mechanical energy in an automobile, providing an energy source for the car. The energy that can be stored in this way is limited by the size and weight of the flywheel and by the maximum angular velocity it can attain without flying apart. Suppose a solid cylindrical flywheel of radius 90.0 cm and mass 85.0 kg rotates at a maximum angular velocity of 225 rad/s. Find the maximum rotational energy of the flywheel. How many liters of gasoline would provide the same mechanical energy given that each liter provides 8.50×10^6 J of mechanical energy?

27 A sphere of radius r starts from rest and rolls without slipping along a curved surface, dropping through a vertical distance of 0.500 m. Find the final speed v of the sphere's center of mass.

28 A sphere, a solid cylinder, and a thin-walled hollow cylinder all have the same mass and radius. The three are initially held side by side on an incline. They are all released simultaneously and roll down the incline.
(a) Which one reaches the bottom of the incline first?
(b) Which one reaches the bottom of the incline last?

29 A car with a total mass of 1.75×10^3 kg has wheels of radius 32.0 cm and moment of inertia 0.800 kg-m^2. Find the car's (a) translational kinetic energy and (b) rotational energy when the car is moving at a speed of 25.0 m/s.

★ **30** Suppose that in the previous problem the wheels were replaced by oversized wheels of the same shape and made of the same materials, but 50% wider and with a 50% larger radius. By what factor would the rotational energy increase?

31 Masses A (1.00 kg) and B (2.00 kg) are attached to opposite ends of a light rod 1.00 m long that is mounted on a frictionless, horizontal axle through its midpoint. The rod is initially horizontal. Find the speed of the masses as the rod swings to a vertical position.

9-5 Angular Momentum

32 Compute the linear momentum and angular momentum of a Frisbee of mass 0.160 kg if it has a linear speed of 2.00 m/s and an angular velocity of 50.0 rad/s. Treat the Frisbee as a uniform disk of radius 15.0 cm.

33 Compute the angular momentum of the earth arising from (a) its orbital motion around the sun; (b) its rotation on its axis.

★ **34** A woman doing aerobics jumps vertically upward 0.500 m and, while in the air, rotates her body about a vertical axis. If the rotation is done with the legs extended out from the body, her body rotates through 180°. If, however, she brings her legs in close to the axis of rotation, her body rotates through 360°.
(a) Find the average angular speed for each case.
(b) Find the ratio of her body's moment of inertia with her legs extended to her moment of inertia with her legs in.

35 A weather vane initially at rest has a moment of inertia of 0.100 kg-m^2 about its axis of rotation. A 50.0 g piece of clay is thrown at the vane and sticks to it at a point 20.0 cm from the axis. The initial velocity of the clay is 15.0 m/s, directed perpendicular to the vane. Find the angular velocity of the weather vane just after it is struck.

★ **36** A student sits on a swivel seat, as in Example 9. Initially the student is at rest, holding a spinning bicycle wheel that has an angular velocity of 10.0 rad/s (Fig. 9-30). The student turns the spinning wheel upside down, reversing the direction of its angular velocity. This causes the student to begin to rotate. Find the student's final angular velocity. The moments of inertia of the wheel and of the student are 0.300 kg-m^2 and 2.50 kg-m^2 respectively.

Fig. 9-30

37 A diver jumps off a diving board and begins a slow forward rotation about a horizontal axis, so that in dropping a vertical distance of 5.00 m her body undergoes half a rotation. She then repeats the dive, but this time tucks her body so that it undergoes 2.50 rotations. If her initial moment of inertia is 20.0 kg-m^2, what is her final moment of inertia after tucking her body?

*9-6 Energy Analysis of Running

38 Suppose that the runner described in Section 9-6 gains weight. Estimate the increase in power required to run at the same speed if the runner's mass increases from 80 to 90 kg. Assume no change in the leg's moment of inertia.

★ **39** The runner described in this section now wears ankle weights, which increase the leg's moment of inertia from 1.0 kg-m^2 to 2.0 kg-m^2 without adding significantly to the overall mass. Estimate the increase in power required to run at the same speed.

★ **40** Estimate the power required for the 80 kg runner described in this section to run up a 5.0° slope at the same speed of 3.0 m/s.

★ **41** Calculate the power expended against air resistance if the runner described in this section runs at 3.0 m/s directly into a 5.0 m/s (11 mi/h) wind.

Additional Problems

42 The letter A is printed on a flat metal plate, which rotates at a constant angular velocity of 60.0 rpm. When the letter is illuminated only by a flashing stroboscopic light, it may appear to be stationary. Find the lowest flash rate necessary to see (a) one stationary image; (b) n stationary images.

43 It is common in machinery for a belt to connect the shaft of an electric motor to the part of the machine to be rotated (Fig. 9-31). Find the relationship between the two angular velocities.

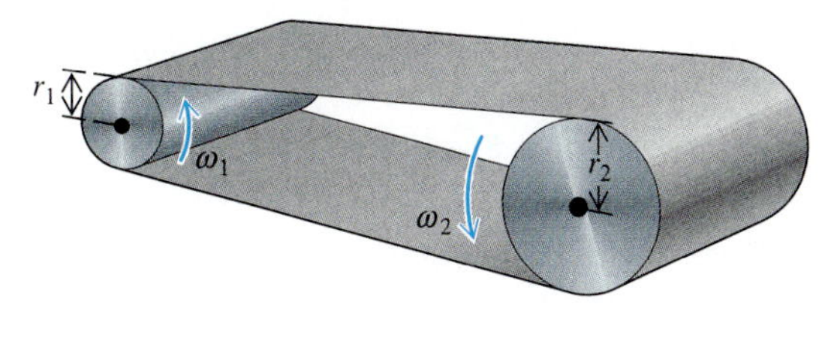

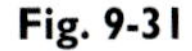

Fig. 9-31

44 A runner whose legs are 1.00 m long moves at a constant speed of 8.00 m/s. Each stride covers 1.20 m. Treat the leg as a rigid body.
(a) Find the average angular speed of the legs with respect to the runner's center of mass.
(b) Find the average speed of the runner's foot with respect to the center of mass. (Remember that speed is the magnitude of velocity, and so these averages are not zero.)

45 Fig. 9-32 shows a rope passing over a pulley on a stationary axle. Prove that the two tensions have equal magnitude if the pulley has negligible moment of inertia and negligible friction in its bearings.

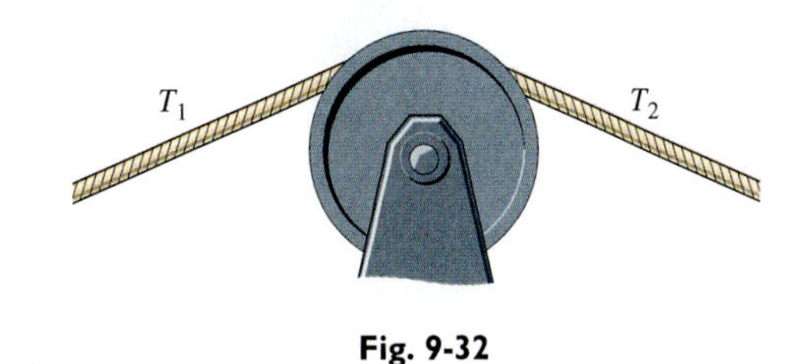

Fig. 9-32

46 Express the pseudoforce acting on a particle in the centrifuge described in Problem 3 as a function of the particle's mass.

47 Two blocks are attached to the ends of a light cord that passes over a frictionless pulley. Compute the acceleration of the blocks (Fig. 9-33).

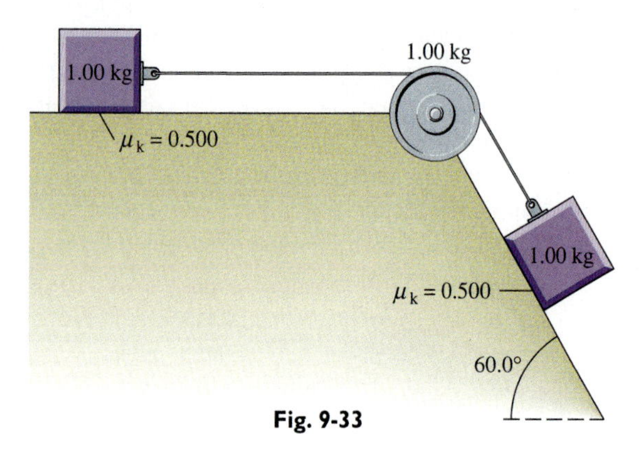

Fig. 9-33

★★ **48** A Ping-Pong ball of radius R is struck by a paddle that exerts a constant-direction force **F**. This force may be resolved into vector components **N** and **f**, as shown in Fig. 9-34. Show (a) that as a result of the collision the ball acquires a change in both spin angular momentum ΔL and linear momentum $\Delta \mathbf{p}$ and (b) that the ratio $\Delta L/\Delta p$ has a maximum value of $\mu_s R/\sqrt{\mu_s^2 + 1}$, where μ_s is the coefficient of static friction. Find the angle θ corresponding to this maximum value if $\mu_s = 2$.

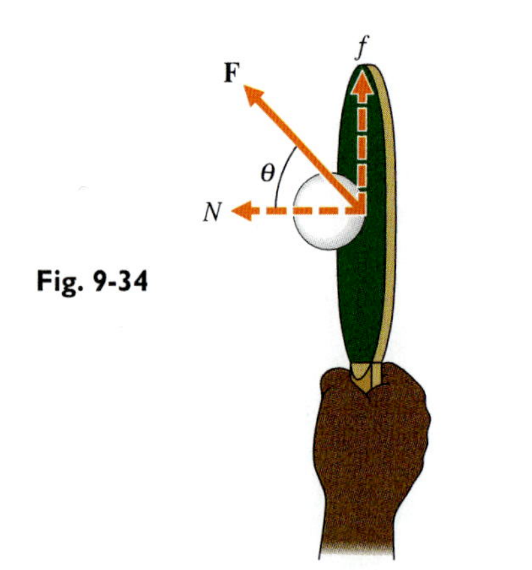

Fig. 9-34

★ **49** Gases are ejected from a spacecraft through exhaust jets, as shown in Fig. 9-35. The ejection of these gases imparts spin angular momentum to the spacecraft. In this way its orientation can be controlled. The spacecraft has a moment of inertia of 1.00×10^3 kg-m^2. Find the mass of gas that must be ejected to bring the spacecraft to rest if it has an initial angular velocity of 3.00 rpm in the clockwise direction as viewed from behind and gas is ejected at a speed of 100 m/s..

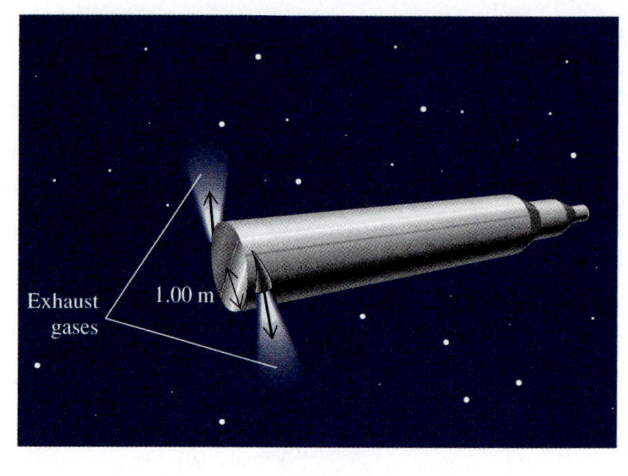

Fig. 9-35

★ **50** The mechanical power provided by a cyclist's legs is used to: (i) compensate for rolling friction and internal friction; (ii) compensate for air resistance; (iii) increase gravitational potential energy when hill climbing. Assuming a drag coefficient $C = 0.800$ N-s^2/m^4, estimate the mechanical power expended against air resistance at a speed of (a) 7.00 m/s (15 mi/h); (b) 14.0 m/s (30 mi/h). The technique of "drafting" other cyclists is used to reduce air resistance.

★ **51** (a) Compute the power dissipated by air resistance for a car moving at a speed of 55.0 mi/h, with a drag coefficient of 1.50×10^{-3} lb-s^2/ft^4 and cross-sectional area of 20.0 ft^2.

(b) Repeat the calculation for a speed 75.0 mi/h.

CHAPTER 10 Static Equilibrium

Giant sequoias up to 100 m high stand in static equilibrium.

In the last chapter we saw how rotation can result from external forces producing torque. In this chapter we shall see how external forces can be balanced to produce no rotation of an extended body—for example, how the forces on a ladder must be balanced to prevent it from tipping over. We shall also see how excessive internal forces in a body can result in the body breaking, for example, the breaking of a bone.

10-1 Conditions for Static Equilibrium

The external forces acting on a body that stays at rest in an inertial reference frame always satisfy certain conditions. For a single particle, there is only one condition, and it is quite simple: the vector sum of the forces must equal zero. For an extended body, things are more complicated. A resultant force of zero ensures only that the center of mass of the body remains at rest, as in the following example.

Suppose a meter stick is initially at rest on a table. If two equal-magnitude, horizontal forces **F** and −**F** are applied as shown in Fig. 10-1a, the stick will begin to rotate. It will not remain at rest, despite the fact that the resultant force acting on it is zero. Only its center of mass does not move. This one point remains at rest, and the meter stick is said to be in **translational equilibrium.**

The meter stick remains at rest when **F** and −**F** are applied as shown in Fig. 10-1b. In this case, the stick neither rotates nor translates. It is in **rotational equilibrium** as well as translational equilibrium.

The meter stick rotates when the forces do not act along the same line (Fig. 10-1a) because in this case there is a net torque. Both forces produce negative (clockwise) torque. Therefore the stick will rotate with negative angular acceleration about its center of mass. On the other hand, when the two opposing forces have the same line of action (Fig. 10-1b), the stick does not rotate because there is no net torque. (The two forces **F** and −**F** have equal magnitudes and equal moment arms, and they produce torques of opposite sign, and so these torques cancel.)

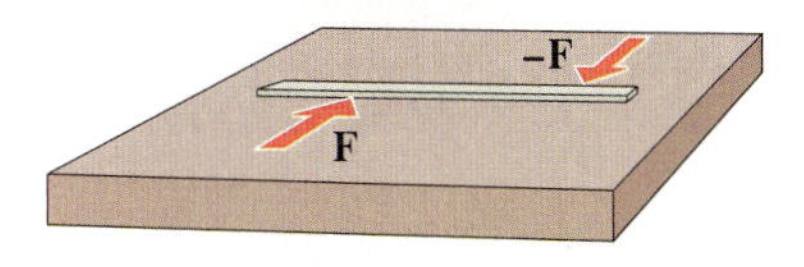

(a)

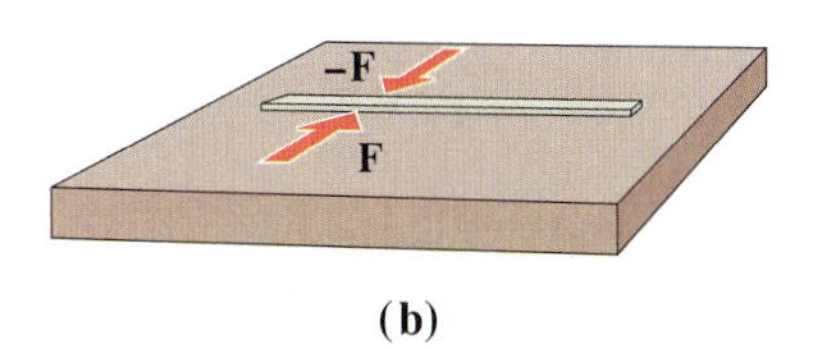

(b)

Fig. 10-1 **(a)** Forces **F** and **-F** cause a meter stick to rotate. **(b)** Forces **F** and **-F**, applied now at the same place on the meter stick, cause no rotation.

The general **condition for rotational equilibrium** follows as a special case of Eq. 7-15 ($\Sigma\, \tau = I\alpha$). When **the sum of the torques produced by external forces equals zero** ($\Sigma\, \tau = 0$), a body will experience no angular acceleration ($\alpha = 0$). If the body is initially not rotating ($\omega_0 = 0$), it will continue in a state of nonrotation ($\omega = 0$):

$$\Sigma\, \tau = 0 \qquad \text{(rotational equilibrium)} \qquad (10\text{-}1)$$

The **condition for translational equilibrium** is that **the sum of the external forces must equal zero.** For forces acting in the xy plane, we express this condition:

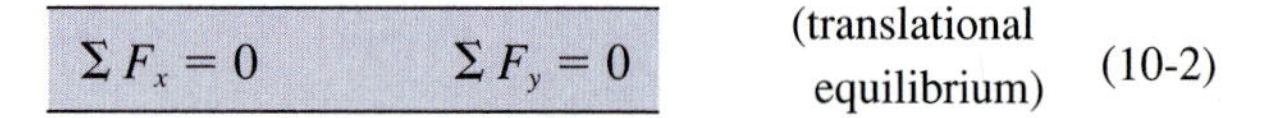

$$\Sigma\, F_x = 0 \qquad \Sigma\, F_y = 0 \qquad \text{(translational equilibrium)} \qquad (10\text{-}2)$$

A body in both rotational and translational equilibrium is said to be in **static equilibrium.** Eqs. 10-1 and 10-2 give the conditions that must be satisfied by the external forces acting on a body in order for that body to be in static equilibrium. These equations are very useful in solving for unknown forces acting on a stationary body.

Choice of Rotation Axis

In applying the condition for rotational static equilibrium ($\Sigma\, \tau = 0$), you may choose *any* rotation axis to compute the torques. Although the values of the individual torques will depend on the axis chosen, the *resultant* torque will always be zero when the body is in rotational equilibrium, since there is no rotation about *any* axis. You should choose an axis that makes the calculation as simple as possible. As we shall see in the examples that follow, you can often choose the axis so that some of the unknown forces have zero moment arms and therefore zero torques with respect to that axis. Then these unknown forces will not appear in the torque equation, which reduces to a simple equation you can solve for a single unknown force.

EXAMPLE 1 Balancing on a Seesaw

Two children sit balanced on a seesaw (Fig. 10-2a). Find the weight of the smaller child, who sits 2.00 m from the center of the board, if the larger child, who sits 1.00 m from the center, weighs 648 N.

SOLUTION Fig. 10-2b shows the free-body diagram of the board, which is subjected to three contact forces—two downward forces equal to the weights of the children, $\mathbf{w}_1$ and $\mathbf{w}_2$, and one upward force $\mathbf{N}$ provided by the support. We assume that the board has negligible weight.

We choose the pivot point O as the origin. Since the unknown force $\mathbf{N}$ acts directly at O, this force produces no torque about O. Therefore $\mathbf{N}$ will not appear in the torque equation. We apply the torque equation (Eq. 10-1: $\Sigma\, \tau = 0$), expressing the magnitude of each torque as the product of its force and its moment arm. The torque τ_1 produced by force $\mathbf{w}_1$ is positive because this torque tends to rotate the board counterclockwise. The torque τ_2 produced by force $\mathbf{w}_2$ is negative because this torque tends to rotate the board clockwise:

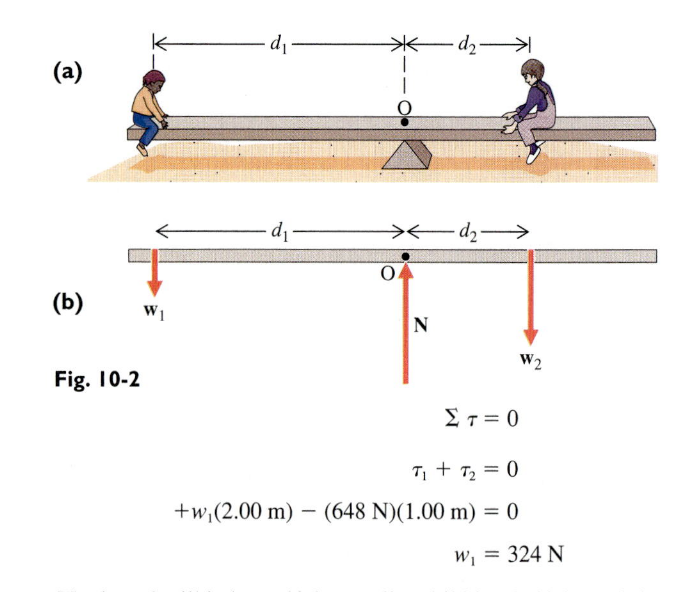

Fig. 10-2

$$\Sigma\, \tau = 0$$

$$\tau_1 + \tau_2 = 0$$

$$+w_1(2.00\text{ m}) - (648\text{ N})(1.00\text{ m}) = 0$$

$$w_1 = 324\text{ N}$$

The board will balance if the smaller child has half the weight of the larger child and sits twice as far from the center.

EXAMPLE 2 Forces on a Leaning Ladder

An aluminum ladder of negligible weight and length ℓ leans against a frictionless wall (Fig. 10-3a). A man weighing 756 N (170 lb) stands halfway up the ladder. The coefficient of friction between the ground and the ladder is 0.500. (a) Find the forces exerted on the ladder by the ground and the wall. (b) How high can the man climb before the ladder begins to slip?

SOLUTION **(a)** First we draw a free-body diagram of the ladder and man (Fig. 10-3b). We choose point O as our origin so that both **N** and **N′** produce no torque.* Then these two unknowns do not appear in the torque equation, allowing us to easily solve for f:

$$\Sigma \tau = 0$$

$$f(\ell \sin 60.0°) - w(\tfrac{1}{2}\ell \cos 60.0°) = 0$$

Solving for f, we find

$$f = \frac{w \cos 60.0°}{2 \sin 60.0°} = \frac{756 \text{ N}}{2 \tan 60.0°} = 218 \text{ N}$$

Cancellation of the x components of force implies that this is also the value of N', the force exerted by the wall on the ladder:

$$\Sigma F_x = 0$$

$$f - N' = 0$$

$$N' = f = 218 \text{ N}$$

Finally, we apply the condition for equilibrium in the y direction to calculate N, the normal force exerted by the ground on the ladder:

$$\Sigma F_y = 0$$

$$N - w = 0$$

$$N = w = 756 \text{ N}$$

To be sure that the ladder does not slip we need to check that the frictional force exerted by the ground on the ladder does not exceed the maximum possible value:

$$f_{max} = \mu_s N = (0.500)(756 \text{ N}) = 378 \text{ N}$$

The actual frictional force is less than this maximum allowable value.

(b) If the man climbs higher on the ladder, the torque associated with his weight increases. If the ladder is not to slip, this increased torque must be balanced by an increased frictional torque. Thus, as the man climbs, the frictional force increases so that equilibrium is maintained. However, f cannot exceed the maximum value of 378 N. We shall solve for the maximum distance x the man may safely climb, corresponding to f = 378 N. Applying the condition for rotational equilibrium to the free-body diagram of Fig. 10-3c, we obtain

$$f\ell \sin 60.0° - wx \cos 60.0° = 0$$

Next we solve for x:

$$x = \left(\frac{f}{w}\right)(\tan 60.0°)\ell = \frac{378\text{N}}{756\text{N}}(\tan 60.0°)\ell = 0.866\ell$$

Thus with the ladder leaning at this angle, the man can climb only about 90% of the total length. How could the ladder's position be changed so that the man can climb to the top?

*The base of the ladder is another reasonable choice for the origin. In this case, N and f are eliminated from the torque equation, and N' can be easily found.

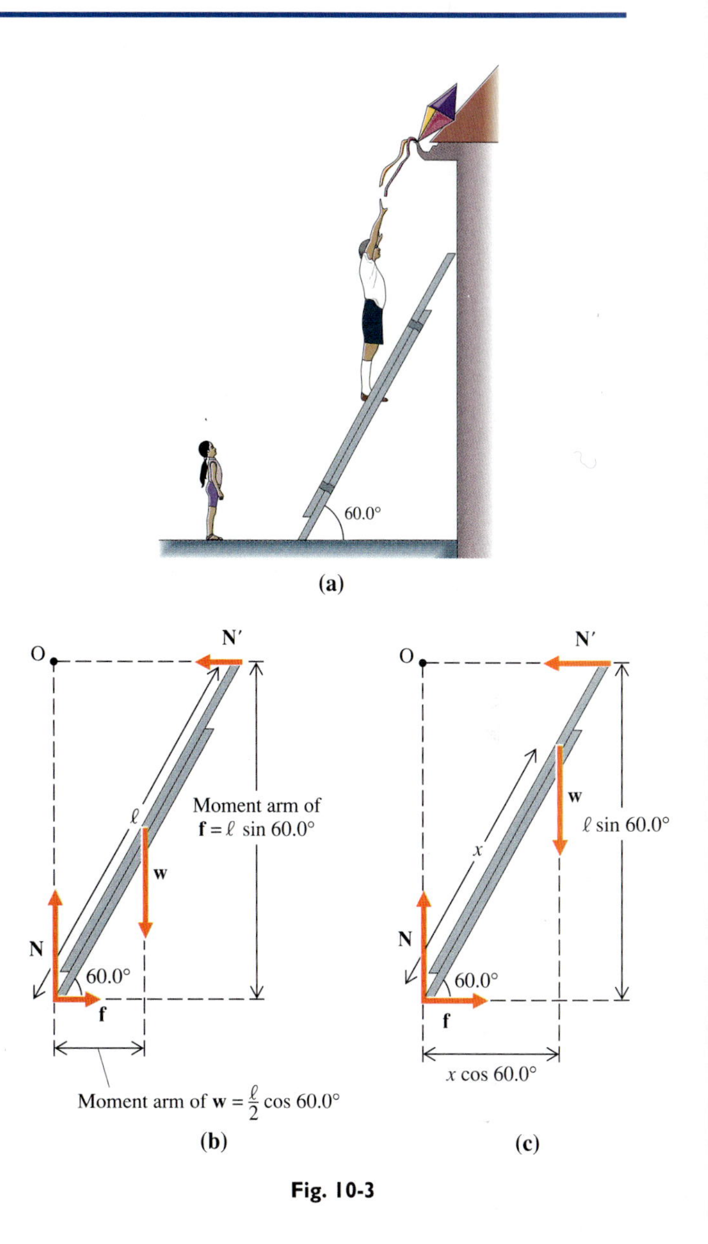

Fig. 10-3

Muscles

The human body moves by means of muscles and bones. A muscle consists of a bundle of fibers (Fig. 10-4), the ends of which are attached by tendons to adjacent bones. When muscle fibers receive an electrical stimulus from nerve endings connected to the brain, they shorten, or contract. Contracting muscles exert tension forces on the bones to which they are attached. If the force exerted by the muscle on a bone is not balanced by other forces, the bone will rotate about a pivot point located between itself and the bone adjacent to it.

The primary muscle responsible for raising the forearm is the biceps, shown in Fig. 10-5a, where a heavy weight is held in the hand. A simplified model of Fig. 10-5a is shown in Fig. 10-5b, where a string under tension represents the biceps. The angle θ_1 that the upper arm makes with the vertical is determined by the shoulder, and the angle θ_2 that the forearm makes with the horizontal is determined by the contraction of the biceps.

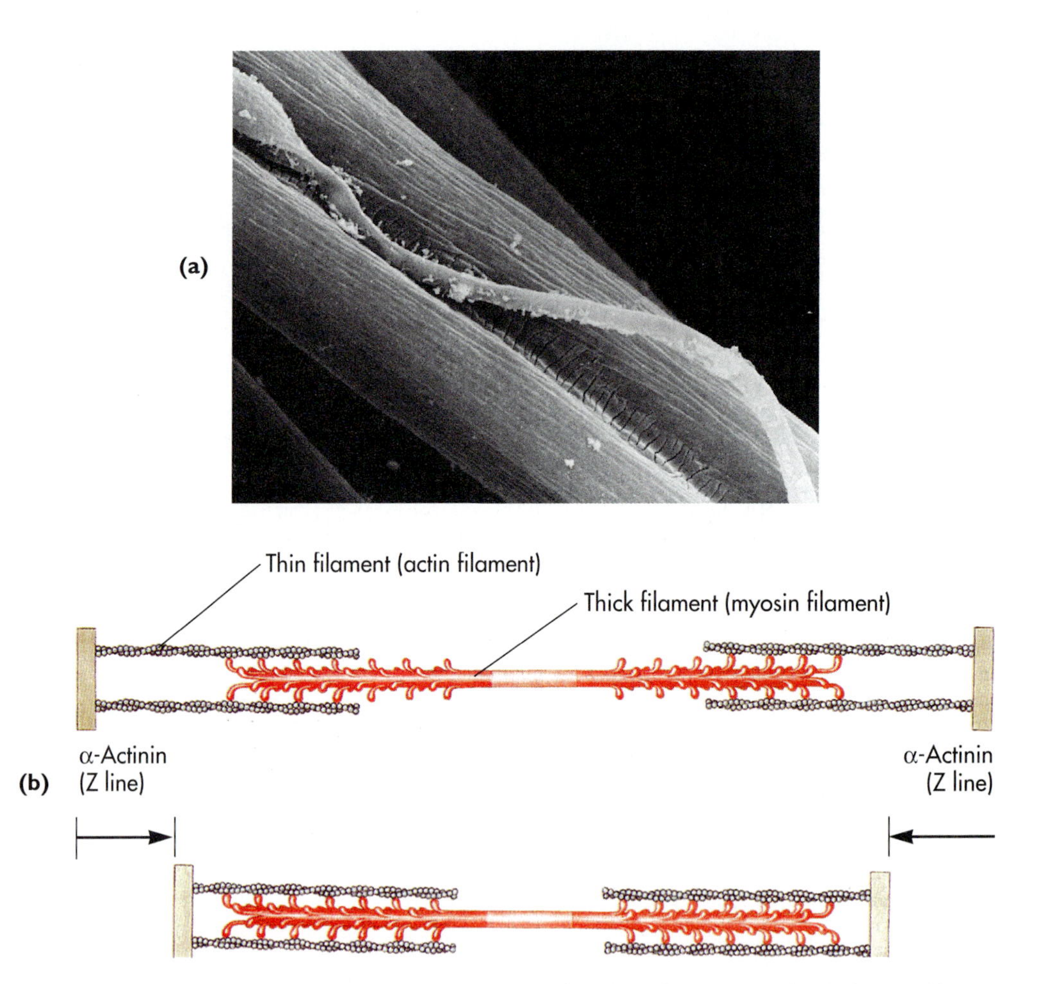

Fig. 10-4 **(a)** Scanning electron micrograph of a muscle fiber and the (smaller) nerve fiber that activates it. (From Hubel DH: *Sci Am* 241[3]:53, Sept. 1979.) **(b)** The mechanism of muscle contraction. Muscle fibers consist of overlapping protein filaments—myosin and actin. When the myosin is activated by a nerve impulse, the projections (myosin "heads") pull it into the space between the actin filaments, shortening the muscle fiber. (From Raven PH, Johnson GB: *Understanding Biology*, ed 3, St. Louis, 1992, Mosby.)

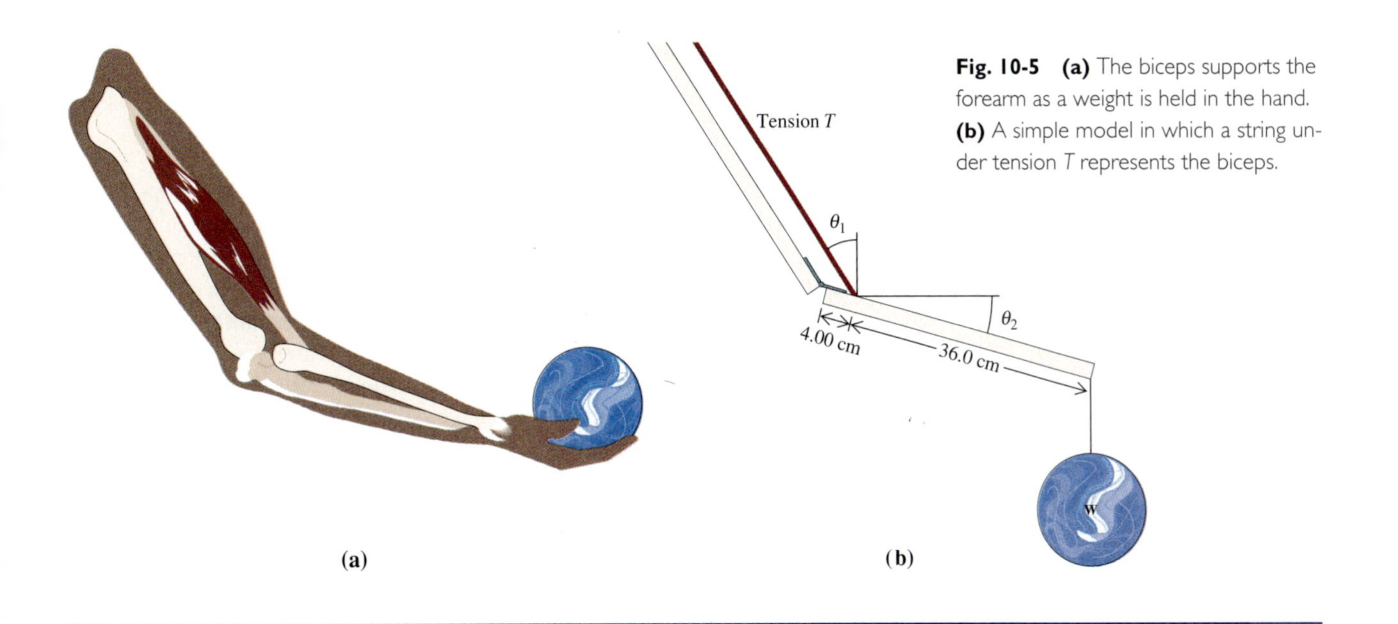

Fig. 10-5 **(a)** The biceps supports the forearm as a weight is held in the hand. **(b)** A simple model in which a string under tension T represents the biceps.

EXAMPLE 3 Biceps Tension Required to Lift a Weight

(a) Find the tension in the biceps, using the model in Fig. 10-5b, when $\theta_1 = 0°$ and $w = 134$ N (30 lb). (b) Muscles can produce a maximum tension of about 70.0 N per square centimeter of the muscle's largest cross section. What is the minimum diameter of the biceps for this problem? (Assume a circular cross-sectional area.)

SOLUTION (a) We choose as a free body the forearm and draw a free-body diagram in Fig. 10-6. We neglect the weight of the forearm because it is small compared with the other forces acting on the forearm. Setting $\theta_1 = 0°$ in Fig. 10-5b means that the biceps exerts a tension **T** directed vertically upward. Since both **T** and **w** are vertical, the force **E** on the elbow must also be vertical. Otherwise there would be a resultant force along the x-axis. And since **w** produces a negative torque about the point P where **T** is applied to the forearm, **E** must be directed downward so that it produces an opposing positive torque about P. (If the direction of **E** had been indicated incorrectly in the free-body diagram, this could have been discovered by subsequent application of the conditions for static equilibrium. Here we shall choose the point O as our origin, so that there is no torque produced by **E** about O. The direction of **E** will therefore not affect the torque equation.)

We apply the condition for rotational equilibrium. The moment arm of **T** is (4.00 cm)(cos θ_2) and the moment arm of **w** is (40.0 cm)(cos θ_2):

$$\Sigma \tau = 0$$

$$+T(4.00 \text{ cm})(\cos \theta_2) - w(40.0 \text{ cm})(\cos \theta_2) = 0$$

Solving for T, we obtain

$$T = 10.0w = 10.0(134 \text{ N}) = 1340 \text{ N} \quad \text{(or 300 lb)}$$

Notice that this result is independent of θ_2.

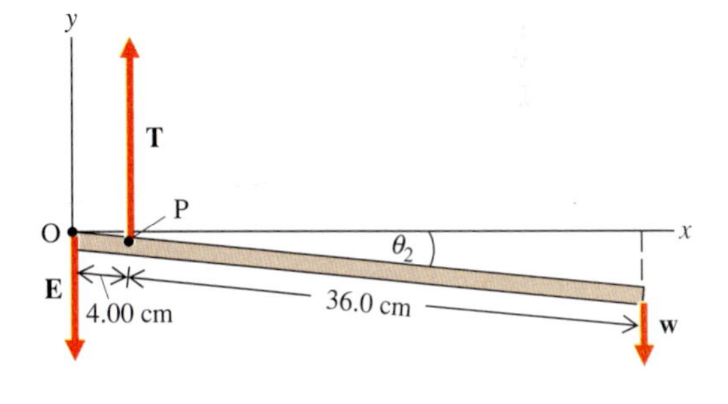

Fig. 10-6

Because the biceps is connected to the forearm so close to the elbow, a relatively large tension (1340 N) is produced by a small applied force (134 N). If the muscle were attached, say, 8 cm from the elbow, twice as much weight could be supported with the same tension of 1340 N. Having the attachment point close to the elbow does have an advantage, however. It permits the forearm to rotate through a large angle when the biceps contracts only slightly, thereby allowing the forearm to move rapidly in the short time necessary for this small contraction. Evolution has favored a human anatomy designed for rapid limb movement rather than weight lifting.

(b) The minimum cross-sectional area is

$$A = \frac{1340 \text{ N}}{70.0 \text{ N/cm}^2} = 19.1 \text{ cm}^2$$

We are assuming a circular cross section, and so this area is related to the diameter d by the formula

$$A = \pi r^2 = \frac{\pi d^2}{4}$$

Therefore $\qquad d = \sqrt{4A/\pi} = 4.93$ cm

This is a rather large biceps.

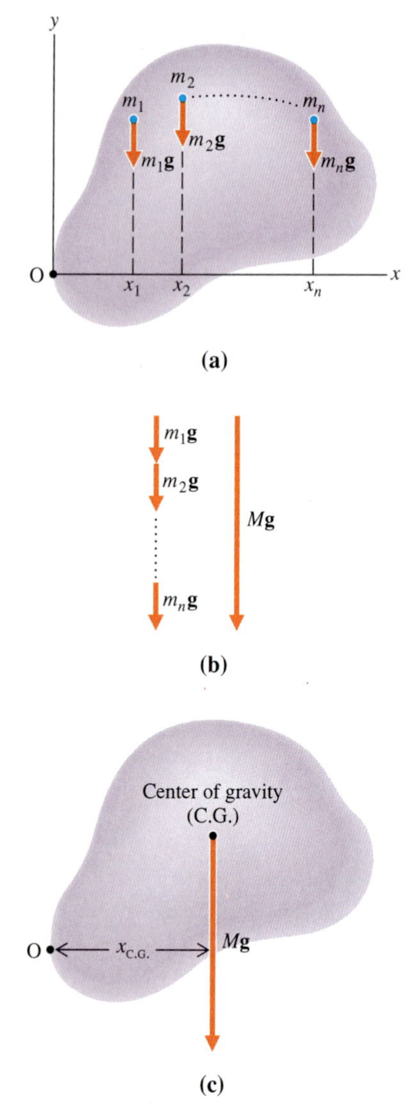

Fig. 10-7 **(a)** A body composed of *n* particles, each with its own weight and moment arm. **(b)** The body's total weight, the vector sum of the particles' weights, equals *M***g**, where *M* is the body's total mass. **(c)** We can compute the total gravitational torque on the body by treating the weight as though it acts at a single point, the center of gravity.

10-2 Center of Gravity

The torque produced by the gravitational force on an extended body is due to the weight of each particle of the body. Fortunately, we do not have to compute and sum the torques for each of the large number of particles that compose any extended body. Instead the total gravitational torque on an extended body may be computed as though the weight were concentrated at a single point, called the **center of gravity.**

The body shown in Fig. 10-7a consists of n particles of masses $m_1, m_2, \ldots, m_n$ and weights $m_1\mathbf{g}, m_2\mathbf{g}, \ldots, m_n\mathbf{g}$. As shown in Fig. 10-7b, the resultant gravitational force on the body is the vector sum of the particle weights $m_1\mathbf{g} + m_2\mathbf{g} + \cdots + m_n\mathbf{g}$, or simply $M\mathbf{g}$, where $M = m_1 + m_2 + \cdots + m_n$. We shall show that, for purposes of calculating torque, the total weight $M\mathbf{g}$ can be thought of as acting at a point, the center of gravity, as illustrated in Fig. 10-7c.

Referring to Fig. 10-7a, we see that the total torque produced by the weights of the n particles can be expressed in terms of the x coordinates of the particles, $x_1, x_2, \ldots, x_n$:

$$\begin{aligned}\tau &= -m_1 g x_1 - m_2 g x_2 - \cdots - m_n g x_n \\ &= -g(m_1 x_1 + m_2 x_2 + \cdots + m_n x_n)\end{aligned} \tag{10-3}$$

Recalling that the body's center of mass has an x coordinate given by Eq. 5-5:

$$x_{\text{cm}} = \frac{m_1 x_1 + m_2 x_2 + \cdots + m_n x_n}{M}$$

we may write

$$m_1 x_1 + m_2 x_2 + \cdots + m_n x_n = M x_{\text{cm}}$$

Substituting this equation into Eq. 10-3, we obtain

$$\tau = -Mg x_{\text{cm}} \tag{10-4}$$

This equation shows that the torque may be computed as though the body's entire weight $M\mathbf{g}$ is concentrated at its center of mass. This point serves as the "center of gravity." Thus the center of gravity illustrated in Fig. 10-7c is the same as the center of mass.* As we saw in Chapter 5, the center of mass of a symmetric body is at its geometric center. So the geometric center is also the center of gravity.

*The center of mass and the center of gravity are logically distinct and in fact will not be the same point if the body is large enough so that g is not uniform over the entire body and could not therefore be factored in Eq. 10-3.

EXAMPLE 4 Forces on a Book Balanced on a Table

The edge of a 9.00 N book extends over the edge of a table. Find the force exerted by the table on the book and the effective line of action of this force.

SOLUTION The book is acted upon by upward forces along its surface of contact with the table. The vector sum of these forces is the normal force **N**. Because the book is in translational equilibrium, this normal force must balance the book's weight:

$$\mathbf{N} = -\mathbf{w}$$

$$N = w = 9.00 \text{ N}$$

Although **N** is distributed over the whole surface of contact, we can think of it as having a single effective line of action, the same as for the book's weight. Because the book is in rotational equilibrium, this line of action must pass through the center of gravity (Fig. 10-8). You can think of **N** as acting at a single point directly beneath the book's center of gravity. Of course, this is physically possible only so long as part of the table is beneath the center of gravity.

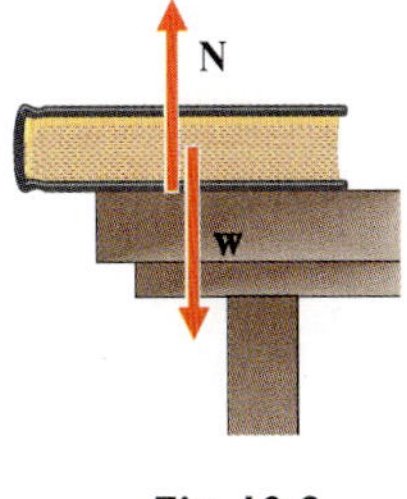

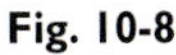
Fig. 10-8

EXAMPLE 5 Forces Supporting a Car's Front and Rear Wheels

A car weighing 1.80×10^4 N is parked on level ground. Its center of gravity is located 1.30 m in front of the rear axle and 2.00 m behind the front axle. Find the force exerted by the ground (a) on the front wheels and (b) on the back wheels.

SOLUTION The free-body diagram of the car is given in Fig. 10-9; $\mathbf{N}_R$ is the net force on the rear wheels, and $\mathbf{N}_F$ is the net force on the front wheels.

(a) Applying the condition for rotational equilibrium with the origin at the base of $\mathbf{N}_R$, we can eliminate N_R from the torque equation and solve for N_F:

$$\Sigma \tau = 0$$

$$+N_F(3.30 \text{ m}) - w(1.30 \text{ m}) = 0$$

$$N_F = (1.30/3.30)w$$

$$= (1.30/3.30)(1.80 \times 10^4 \text{ N})$$

$$= 7.09 \times 10^3 \text{ N}$$

(b) We apply the condition for translational equilibrium to find N_R:

$$\Sigma F_y = 0$$

$$N_R + N_F - w = 0$$

$$N_R = w - N_F = 1.80 \times 10^4 \text{ N} - 7.09 \times 10^3 \text{ N}$$

$$= 1.09 \times 10^4 \text{ N}$$

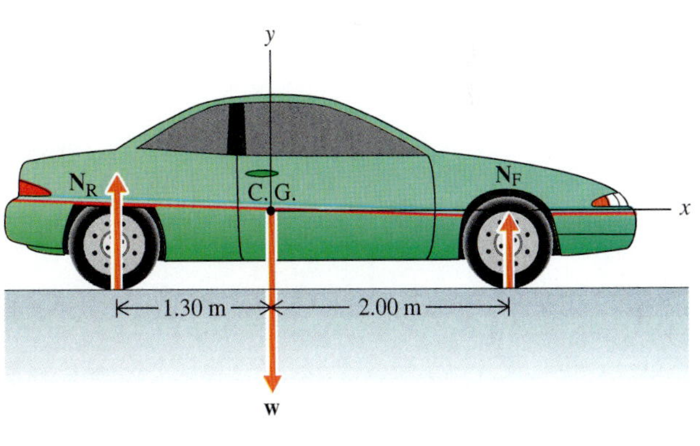

Fig. 10-9

*10-3 Stress and Strain

When external forces are applied to a rigid body, its particles shift their relative positions, thereby changing the internal forces between these particles. When a body is in static equilibrium, the vector sum of the internal and external forces on each particle equals zero.

To gain some idea of the nature of these internal forces, we shall analyze the forces acting within a board being used as a child's seesaw (Fig. 10-10a). We choose a section of the board to the left of point O as our free body. The external forces acting on this section are the smaller child's weight $\mathbf{w}_1$ and the forces exerted by the adjacent section of board to the right. The forces exerted on the free body along the surface of contact between the free body and the adjacent section must balance the force $\mathbf{w}_1$ and the torque produced by $\mathbf{w}_1$. It is clear that the net force acting on this surface of contact is $-\mathbf{w}_1$; otherwise our free body would not be in translational equilibrium. But the force $-\mathbf{w}_1$ alone would produce a net positive torque and counterclockwise rotation. So there must be a balancing clockwise torque produced by other forces.

A simple combination that would produce equilibrium is shown in Fig. 10-10b. The large opposing horizontal forces can be accounted for if we understand that the board bends under its load. In a thin plank, the bending is large enough to be obvious (Fig. 10-10c), whereas in a thicker one it is not. But in either case, there is bending. This bending means that the particles along the top of the board are stretched apart, producing a tension (as in a stretched rope) and that the particles along the bottom of the board are pushed together, producing a compression (as in a compressed spring).

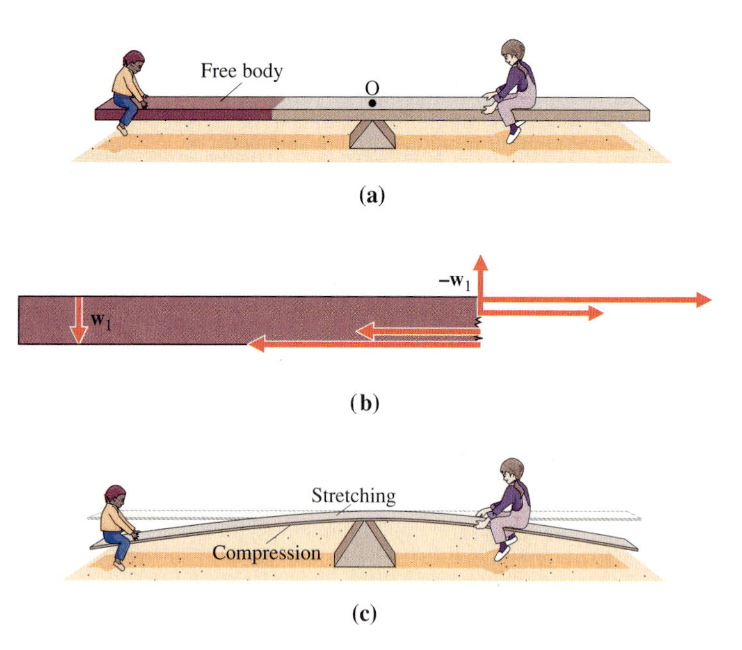

Fig. 10-10 **(a)** Two children balance on a seesaw. **(b)** The forces exerted on the section of the board on the left by the adjacent section must balance both the force $\mathbf{w}_1$ and the torque produced by $\mathbf{w}_1$. **(c)** The board's upper surface is stretched and is under tension. The board's lower surface is compressed.

Stress

Internal forces acting on surfaces within a body are said to produce **stress** in the body. Stress, denoted by the Greek letter σ (sigma), is defined as the force per unit area on the surface on which it acts:

$$\sigma = \frac{F}{A} \quad (10\text{-}5)$$

When the force is perpendicular to the surface, it is called **normal stress.** When the force is tangent to the surface, it is called **shear stress.** A normal stress can be either tensile or compressive. In Fig. 10-10c the normal stress along the upper part of the board is tensile stress and the normal stress along the lower part is compressive stress. The stress produced by the vertical force $-\mathbf{w}_1$ is shear, since this force is tangent to the vertical cross section on which it acts (Fig. 10-10c).

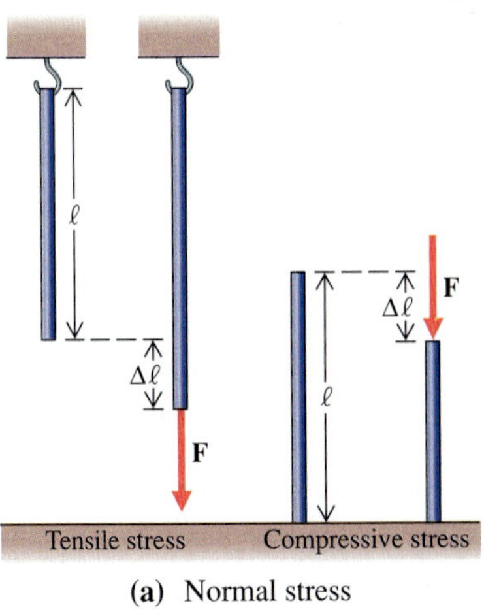

(a) Normal stress

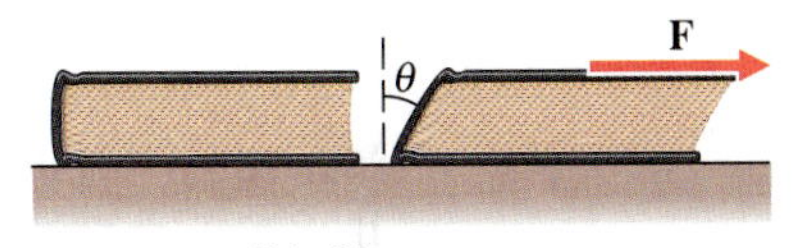

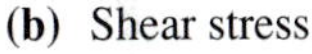

(b) Shear stress

Fig. 10-11 **(a)** A normal stress either stretches or compresses a rod. **(b)** A shear stress causes an angular displacement of a book.

Strain

The change in shape of a body subjected to stress is called **strain.** The strain in the weight-bearing board of Fig. 10-10 is rather complex, since both shear and normal stresses are present. In Fig. 10-11, however, we see bodies subject to pure normal stress and pure shear stress.

The Greek letter ϵ (epsilon) is used to denote strain. Normal strain is denoted by ϵ_N and defined to be the fractional change in length of the body that results when a normal stress is applied:

$$\epsilon_N = \frac{\Delta \ell}{\ell} \quad (10\text{-}6)$$

Shear strain is denoted by ϵ_S and is defined to be the tangent of the angular shift θ that results when a shear stress is applied:

$$\epsilon_S = \tan \theta \quad (10\text{-}7)$$

Stress is a function of strain. In Fig. 10-12 a typical stress-strain curve is shown for a rigid material, such as metal, wood, or bone, under either compression or tension. For small strains the relationship is approximately linear. The slope of the linear section is called Young's modulus, denoted by Y:

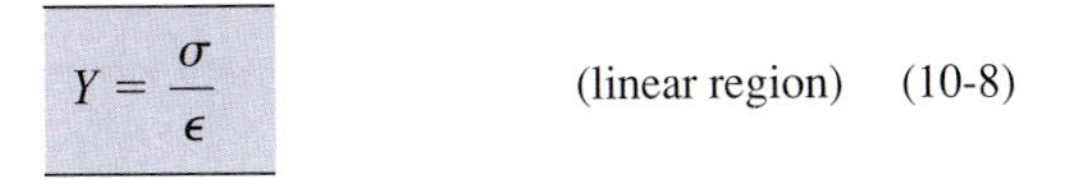

$$Y = \frac{\sigma}{\epsilon} \quad \text{(linear region)} \quad (10\text{-}8)$$

Young's modulus may be used to compute the stress required to produce a given strain. For example, it follows from Eq. 10-8 that to produce a strain of 0.010, the stress required is $\sigma = \epsilon Y = 0.010Y$. For steel $Y = 2.0 \times 10^{11}$ N/m^2, and so a stress of 2.0×10^9 N/m^2 must be applied to produce this 1% strain. For rubber $Y = 1.0 \times 10^6$ N/m^2, and so a stress of only 1.0×10^4 N/m^2 must be applied to produce the same strain. Rubber is much easier to stretch or compress than steel, and this is reflected in rubber's lower value of Y.

The point where the stress-strain curve ends corresponds to fracture of the material. The value of the stress at this point is typically on the order of 10^7 to 10^8 N/m^2.

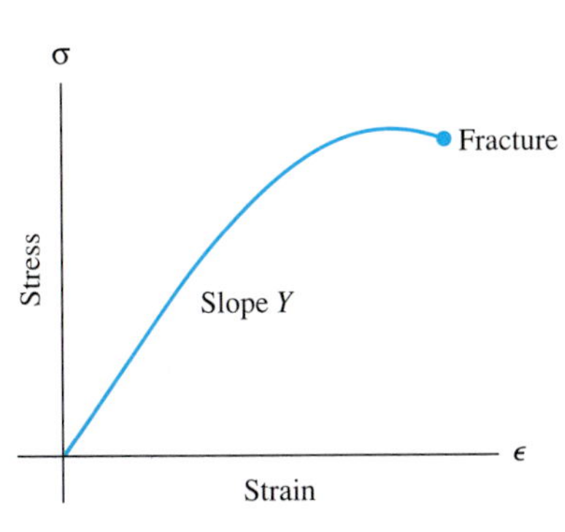

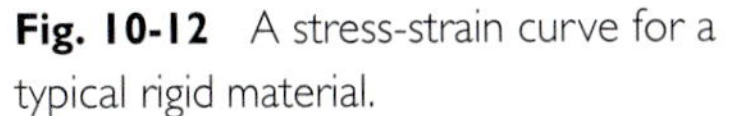

Fig. 10-12 A stress-strain curve for a typical rigid material.

EXAMPLE 6 Strained Spinal Disks

A weight lifter holds 1.0×10^3 N (230 lb) directly overhead. In the midsection of his body this weight is supported by his spine. Spinal vertebrae are separated by easily compressible disks, each disk having an average cross-sectional area of 1.0×10^{-3} m². Young's modulus for a disk is 7.0×10^6 N/m². The total thickness of all the intervertebral disks is 15 cm. Compute the weight lifter's decrease in height resulting from the compression of the disks. For simplicity, ignore the effect of the upper body's weight on the spine.

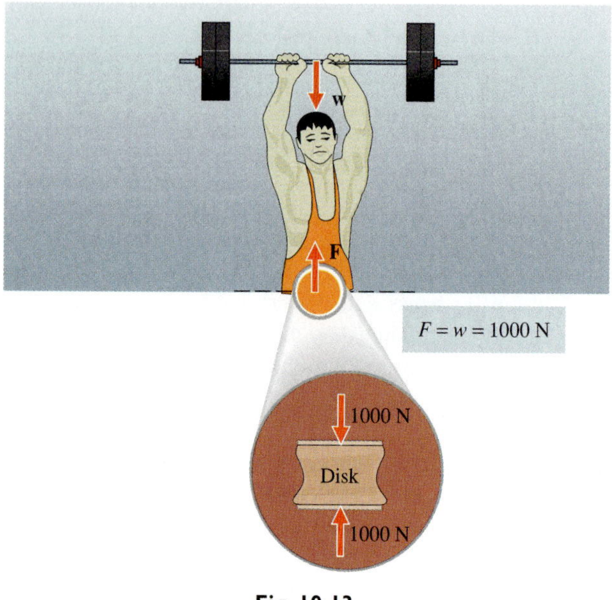

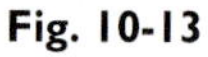
Fig. 10-13

SOLUTION Consider a free-body diagram of the 1000 N weight and the part of the weight lifter's body directly above one of the disks (Fig. 10-13). The force **F** exerted by the disk on the upper body must balance the 1000 N weight. The upper body exerts a downward reaction force of 1000 N on the disk. The disk also experiences a balancing upward force of 1000 N, exerted by contact with the body below. Each disk experiences the same opposing forces of 1000 N, produced by contact with the vertebrae above and below it. The stress on each disk is this force divided by the area:

$$\sigma = \frac{F}{A}$$

$$= \frac{1.0 \times 10^3 \text{ N}}{1.0 \times 10^{-3} \text{ m}^2} = 1.0 \times 10^6 \text{ N/m}^2$$

The strain in each disk may be computed by use of Young's modulus. From Eq. 10-8, we find

$$\epsilon = \frac{\sigma}{Y}$$

$$= \frac{1.0 \times 10^6 \text{ N/m}^2}{7.0 \times 10^6 \text{ N/m}^2} = 0.14$$

Each disk will be compressed by 14%, and hence the total thickness of the disks is reduced by 14%. From the definition of strain ($\epsilon_N = \Delta\ell/\ell$), it follows that the change in thickness $\Delta\ell$ is the product of the strain and the total length ℓ of the disks:

$$\Delta\ell = \epsilon\ell = (0.14)(15 \text{ cm}) = 2.1 \text{ cm}$$

This reduction in thickness should be approximately equal to the weight lifter's decrease in height.

When a rigid body is subject to a small external torque, large stress may be produced by the body's internal forces. This can be seen in Fig. 10-10b, where the horizontal internal forces acting on the left section of the board are much larger than the weight $\mathbf{w}_1$—an external force. The large magnitude of these internal forces is necessary to produce sufficient torque to balance the positive torque produced by $\mathbf{w}_1$ and $-\mathbf{w}_1$, since the horizontal forces have much smaller moment arms than $\mathbf{w}_1$ and $-\mathbf{w}_1$.

A detailed analysis of the internal forces in a rigid rod shows that the rod will break when the torque acting on it reaches a certain maximum value τ_{max}. This torque is related to the maximum stress at fracture σ_{max} by the equation

$$\tau_{max} = \frac{\pi}{4} r^3 \sigma_{max} \tag{10-9}$$

where r is the rod's radius.

EXAMPLE 7 Breaking a Leg

(a) Compute the torque required to break a tibia (leg bone) that has a radius of 1.00 cm at its thinnest point, if the maximum stress at fracture is 2.00×10^8 N/m^2. (b) Suppose a football player is being tackled by one opponent who holds his ankle in a fixed position while another opponent applies a horizontal force to his knee, 0.500 m above the ankle. What is the maximum value of the force that can be applied before the leg will break?

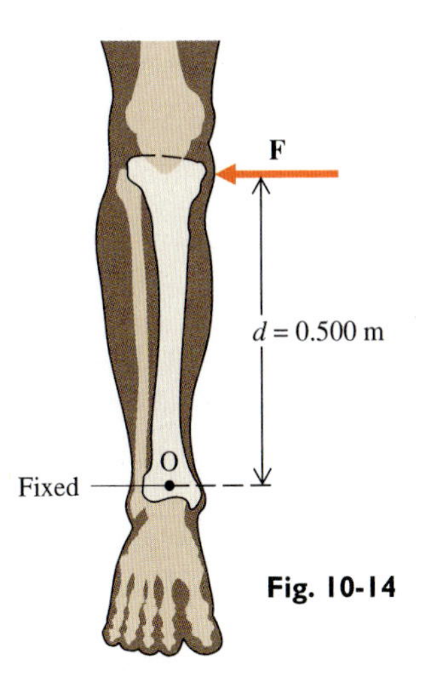

Fig. 10-14

SOLUTION (a) Applying Eq. 10-9, we find

$$\tau_{max} = \frac{\pi}{4} r^3 \sigma_{max} = \frac{\pi}{4}(1.00 \times 10^{-2}\text{ m})^3(2.00 \times 10^8\text{ N/m}^2)$$

$$= 157\text{ N-m}$$

(b) From Fig. 10-14, we see that the external torque applied to the leg is related to the applied force's magnitude F by the equation

$$\tau = Fd$$

Breaking occurs when $\tau = \tau_{max}$, or

$$F = \frac{\tau_{max}}{d} = \frac{157\text{ N-m}}{0.500\text{ m}}$$

$$= 314\text{ N} \quad (\text{or } 71\text{ lb})$$

This force is surprisingly small. If the leg were free to move, a much larger force could be applied without serious injury. It is the fact that the ankle is pinned in place that causes the problem. To hold it in place, other external forces must be acting on the ankle. It is the combination of all the forces that produces the large internal stress within the leg.

Scaling

Structures cannot be increased arbitrarily in size while maintaining the same proportions. The weight of a body is proportional to its volume, but the strength of the supporting members is proportional to their cross-sectional area ($F = \sigma A$). If all dimensions of a body are proportionately increased, the ratio of strength to weight decreases. Thus a structure that is strong at one size may not be greatly scaled up in size without greatly diminishing its strength. To maintain sufficient strength, larger structures must be proportionately thicker (Fig. 10-15).

Galileo recognized this effect. In *Two New Sciences* he wrote:

> . . . you can plainly see the impossibility of increasing the size of structures to vast dimensions either in art or in nature; likewise the impossibility of building ships, palaces, or temples of enormous size in such a way that their oars, yards, beams, iron-bolts, and, in short, all their other parts will hold together; nor can nature produce trees of extraordinary size because the branches would break down under their own weight; so also it would be impossible to build up the bony structures of men, horses, or other animals so as to hold together and perform their normal functions if these animals were to be increased enormously in height; for this increase in height can be accomplished only by employing a material which is harder and stronger than usual, or by enlarging the size of the bones, thus changing their shape until the form and appearance of the animals suggest a monstrosity. . . . If one wishes to maintain in a great giant the same proportion of limb as that found in an ordinary man he must either find a harder and stronger material for making bones, or he must admit a diminution of strength in comparison with men of medium stature; for if his height be increased inordinately he will fall and be crushed under his own weight.

Fig. 10-15 Giant redwood trees have disproportionately large trunks and small branches.

HAPTER 10 SUMMARY

For a rigid body to be in static equilibrium, the following equations must be satisfied by the external forces acting on it:

$$\Sigma F_x = 0$$

$$\Sigma F_y = 0$$

$$\Sigma \tau = 0$$

Although the torque produced by a given force depends on the choice of rotation axis, the resultant torque with respect to any axis equals zero if the body is in static equilibrium. An axis should be chosen so as to minimize computation.

The weight of any body is distributed over the entire body. When calculating torque, however, that weight can be treated as though it acts at a single point, called the "center of gravity." When **g** is uniform over a body, the center of gravity is the same as the center of mass.

The stress σ in a body is the force per unit area acting on a surface within the body:

$$\sigma = \frac{F}{A}$$

Normal stress is perpendicular to the surface along which it acts, and shear stress is tangent to the surface. The strain ϵ in a body is a measure of the change in body shape that results from stress. Normal strain ϵ_N is defined as the fractional change in length of a body:

$$\epsilon_N = \frac{\Delta \ell}{\ell}$$

Shear strain ϵ_S is defined as

$$\epsilon_S = \tan \theta$$

where θ is the angular shift resulting from a shear stress.

Stress is a function of strain. For many materials, stress is proportional to strain when the strain is small. The constant of proportionality is called Young's modulus:

$$Y = \frac{\sigma}{\epsilon}$$

Fracture of a material generally occurs when σ is increased to about 10^7 or 10^8 N/m^2.

If a rigid rod is subject to a torque that tends to make it bend, stress is produced within the rod. If a torque and stress are great enough, the rod breaks. The maximum applied torque τ_{max} is related to the maximum stress σ_{max} by the equation

$$\tau_{max} = \frac{\pi}{4} r^3 \sigma_{max}$$

where r is the rod's radius.

Questions

1 A tightrope walker carries a long pole, which helps her maintain her balance (Fig. 10-16). Suppose that at some instant her center of gravity is displaced to the side so that there is a torque tending to rotate her body off the rope.
(a) What quantity does the pole serve to increase?
(b) How does this affect the time it takes her body to rotate appreciably?

Fig. 10-16

2 The wheels in Fig. 10-17 are subject to various forces, all of the same magnitude. Which of the wheels undergoes: (a) translational acceleration but not rotational acceleration; (b) rotational acceleration but not translational acceleration; (c) both translational and rotational acceleration?

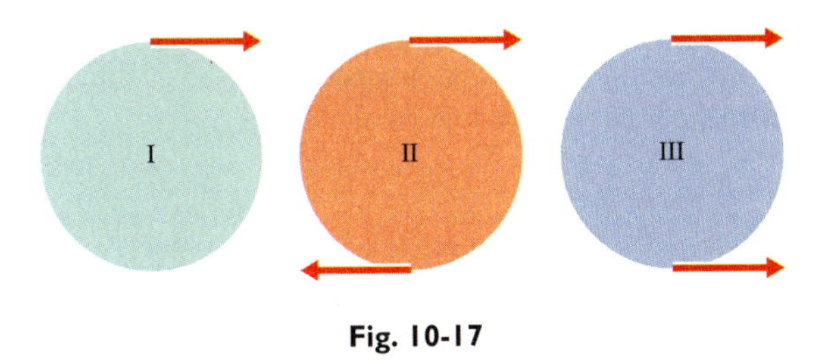

Fig. 10-17

3 Suppose the force **F** applied to the bicycle wheel in Fig. 9-7 has a magnitude of 100 N. What two forces applied at R and Q could produce static equilibrium of the wheel?

4 Forces $\mathbf{F}_1$ and $\mathbf{F}_2$ act on a flat disk, as shown in Fig. 10-18. The vector sum of the two forces equals 10 N, directed east. A third force, of magnitude 10 N and directed west, is applied to the perimeter of the disk. To produce static equilibrium, this force must be applied at (a) P; (b) Q; (c) R; (d) S; (e) P or R; (f) Q or S.

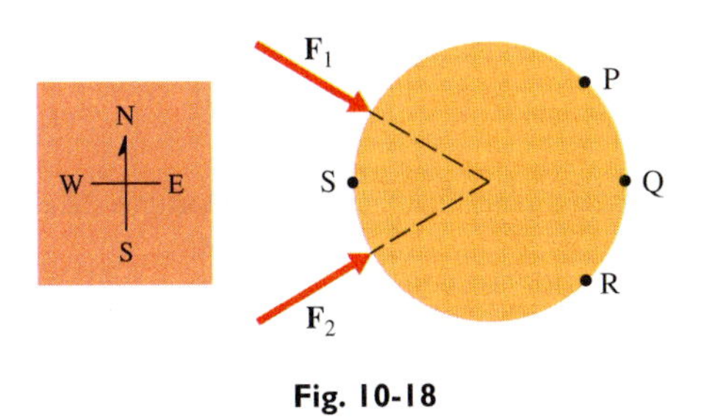

Fig. 10-18

5 Is the free body in Fig. 10-19 in static equilibrium?

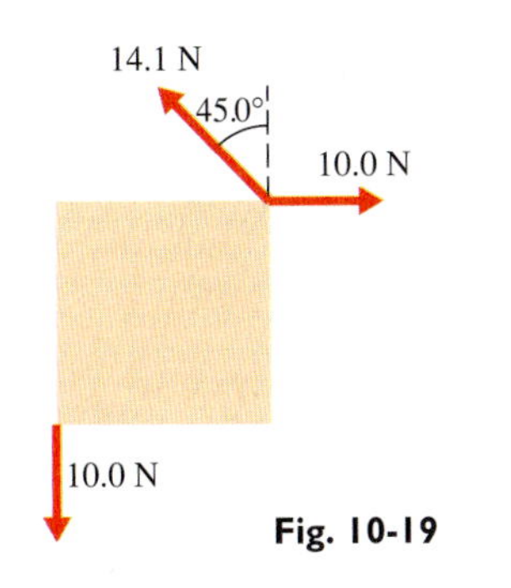

Fig. 10-19

6 A square block is subject to the three forces shown in Fig. 10-20. A fourth force is applied to the block so as to produce static equilibrium.

(a) Find the magnitude and direction of the fourth force.

(b) Where is the fourth force applied to the block?

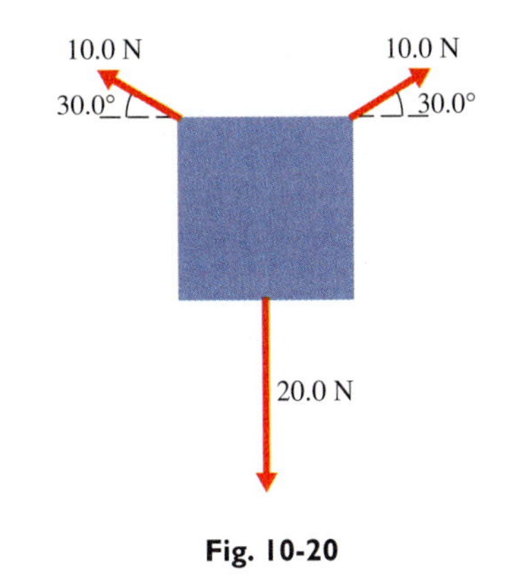

Fig. 10-20

7 If the car in Fig. 10-9 is parked headed downhill rather than on level ground, would the following forces increase, decrease, or remain the same: (a) Force of the road on the rear wheels; (b) Force of the road on the front wheels; (c) Total force of the road on all the wheels?

8 A long board is held horizontally with two hands. If the board is held with both hands near one end, the magnitude of the force exerted by each hand will be: (a) greater than the weight of the board; (b) less than the weight of the board.

9 (a) Is it possible for the center of gravity of a body to be at a point where there is no matter?

(b) Where is the center of gravity of a doughnut?

10 The center of gravity of a standing, stationary person must be (a) directly above a point halfway between the two feet; (b) directly above a point of contact between the feet and the floor; (c) either directly above a point of contact between the feet and the floor or between two points of contact.

11 A waiter carries a heavy tray of food in front of him.

(a) In what direction will he have to lean to maintain his balance?

(b) Will he lean less if he holds the tray close to his body or at arm's length?

12 A physical therapist treating a patient with a damaged disk in the lower back emphasizes the importance of correct posture in reducing irritation of the disk. In trying to explain to the patient the physical effect of good posture, the therapist claims that if the patient stands and sits correctly, the upper part of the spinal column will absorb more of the body's weight and less will be transmitted to the lower spinal column.
(a) Is the therapist correct?
(b) When your body is in a vertical position, how can you prevent some of its weight from being supported by the lower back?
(c) How is a very curved spine likely to cause greater irritation to a disk?

13 Ski bindings rigidly attach a boot to the ski and are designed to automatically release the boot when the force on the binding exceeds a certain value—usually about 90 N (20 lb). Why is this an important safety feature?

Answers to Odd-Numbered Questions

1 (a) moment of inertia; (b) reduces angular acceleration and therefore increases the time needed to rotate through a given angle; **3** At R, a 100 N force in the same direction as **F**; at Q, a 200 N force in the direction opposite **F**; **5** no; **7** (a) decrease; (b) increase; (c) same; **9** (a) yes; (b) in the hole; **11** (a) backward; (b) close; **13** The release feature prevents the foot from being trapped in a fixed position by the ski. If the foot were trapped in a fixed position, a small sideways force could break the leg.

Problems (listed by section)

10-1 Conditions for Static Equilibrium

1 In Fig. 10-21 a crowbar is used to lift a weight. If a downward force of 225 N is applied to the end of the bar, how much weight does the other end bear? The crowbar itself has negligible weight.

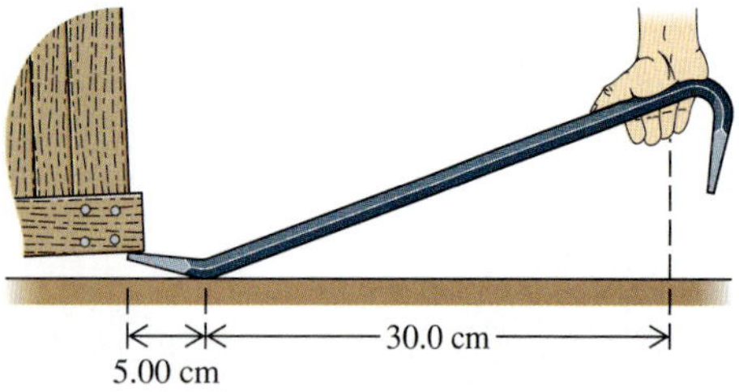

Fig. 10-21

2 An artist designs a mobile of light horizontal rods connected by vertical strings and supporting various shaped weights (Fig. 10-22). Find the magnitudes w_2, w_3, and w_4 if $w_1 = 1.00$ unit of weight. The numerical values given in the figure all have units of length.

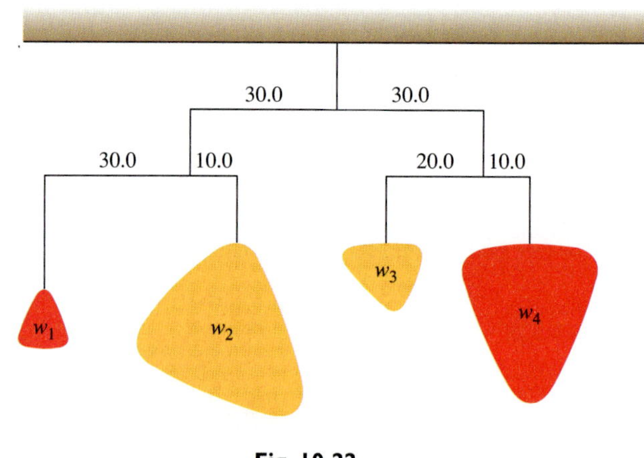

Fig. 10-22

3 A newly planted tree is supported against the wind by a rope tied to a stake in the ground, as shown in Fig. 10-23. The force of the wind, though distributed over the tree, is equivalent to a single force **F**, as shown in the figure. Find the tension in the rope. Assume that neither the tree's weight nor the force of the ground on the tree's roots produces any significant torque about point O.

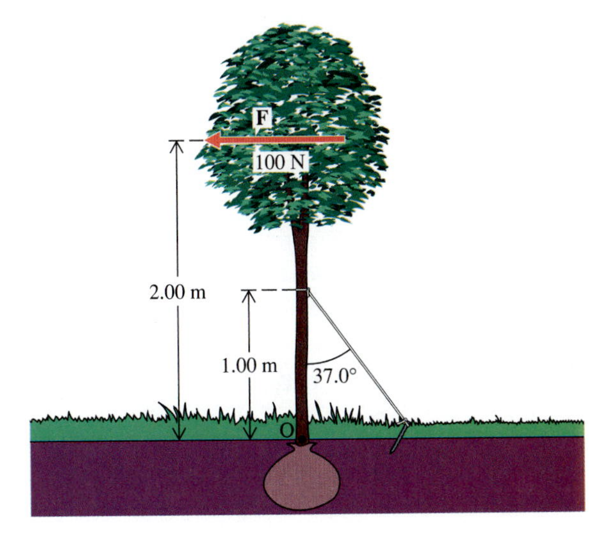

Fig. 10-23

4 A deep-sea fisherman struggles to bring in a big salmon (Fig. 10-24). Find the force each hand must exert on the pole if the tension in the line is 175 N. Neglect the weight of the pole and assume that the force exerted by the left hand is directed along the horizontal.

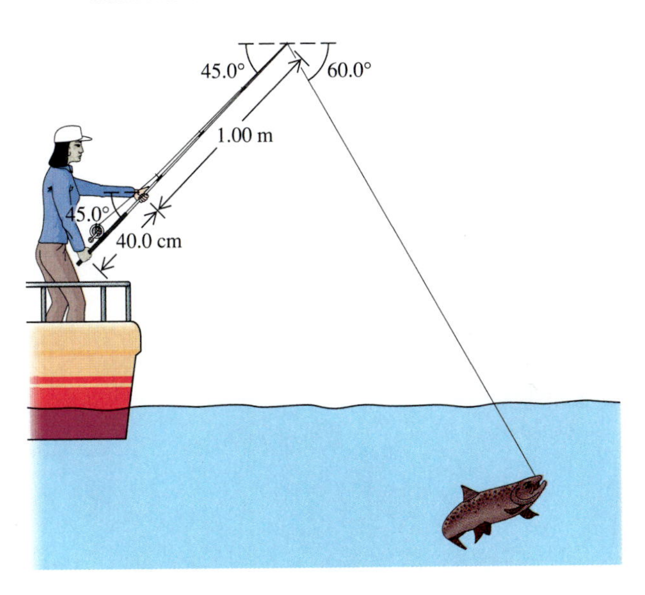

Fig. 10-24

5 A 90.0 N weight is held in the hand. The upper arm makes an angle of 30.0° with the vertical, and the lower arm is 15.0° above the horizontal. Find the tension in the biceps tendon. Refer to Fig. 10-5.

6 The swimmer in Fig. 10-25 propels himself forward by exerting a 125 N backward force on the water with his hand. He experiences a forward reaction force **F** of magnitude 125 N. Find the magnitude and direction of the torque exerted on the arm by **F** about an axis through point O.

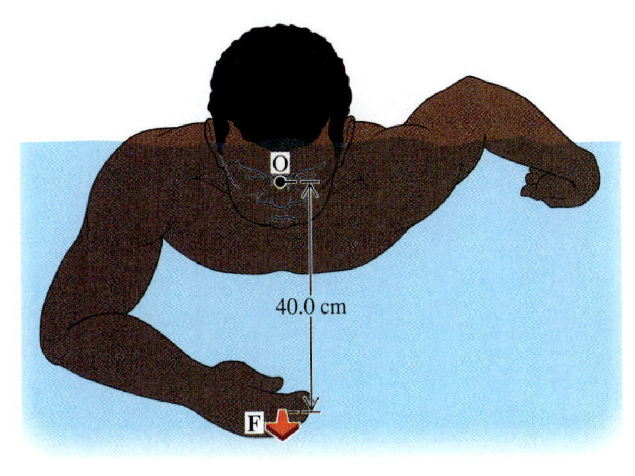

Fig. 10-25

7 As a swimmer pulls his arm through the water, various muscles exert forces on the upper arm. Fig. 10-26 shows a force **F** exerted on the humerus (upper arm bone) by the pectoral muscle. The muscle is connected to the bone 8.0 cm from the center point O of the shoulder joint. Find the magnitude of **F**, if this force's torque on the arm provides half of the total torque balancing the torque produced by the water pushing against the hand.

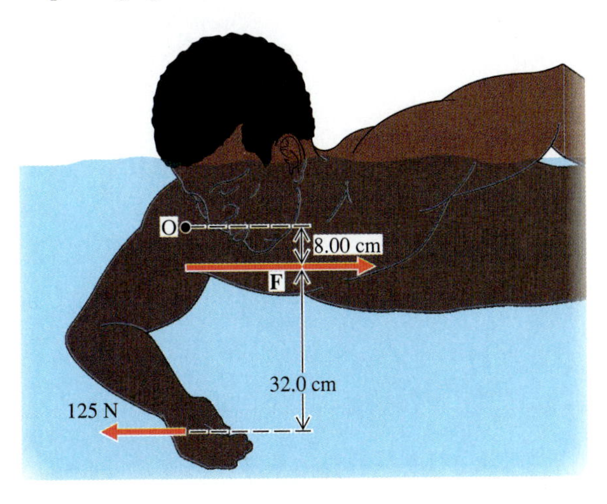

Fig. 10-26

★ **8** An athlete strengthens her leg and knee by doing "leg curls." While sitting on a bench with the leg initially bent at the knee, she supports a weight with the foot. She then slowly raises the foot and lower leg, with the upper leg held stationary. Find the tension in the quadriceps tendon for the position shown in Fig. 10-27.

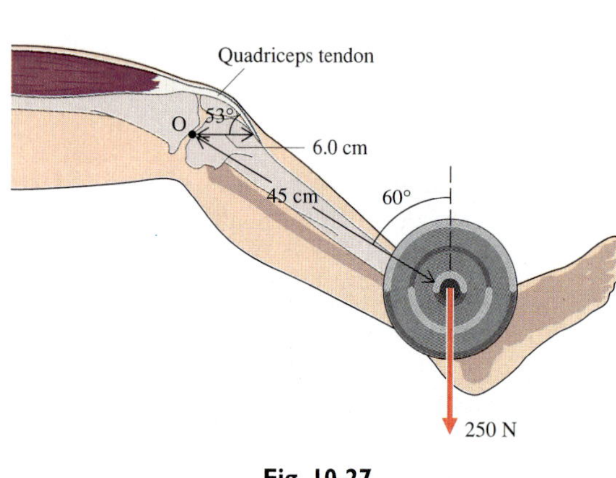

Fig. 10-27

★ **9** In Fig. 10-28 the foot of a walking person is momentarily in contact with the ground and approximately in static equilibrium. The foot bears the person's full weight of 535 N ($N = w = 535$ N). Compute (a) the tension T in the Achilles' tendon; (b) the horizontal and vertical components of the force **P** at the ankle.

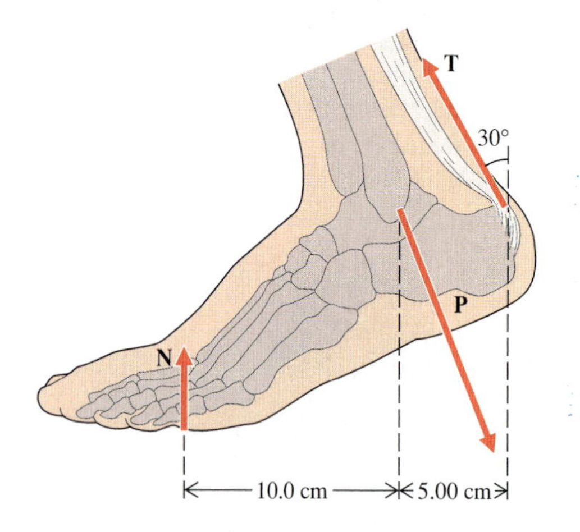

Fig. 10-28

★ **10** The weightless strut in Fig. 10-29 is *not* attached to the wall; it is prevented from falling only by friction. Find (a) the magnitude of the force of friction between the wall and the strut; (b) the normal force exerted by the wall on the strut; (c) the minimum coefficient of static friction.

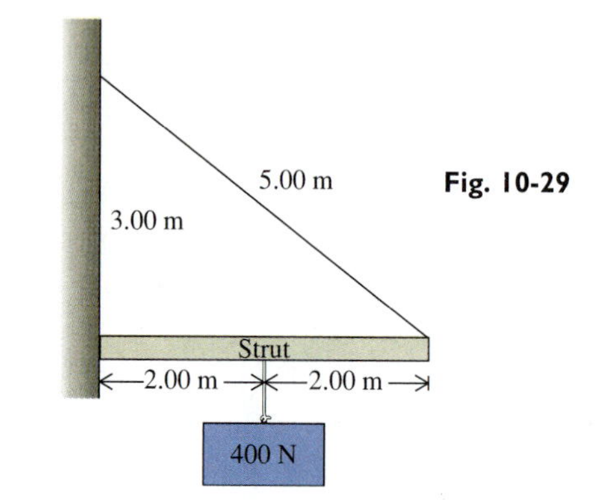

Fig. 10-29

10-2 Center of Gravity

11 A man doing a slow push-up is approximately in static equilibrium. His body is horizontal, with his weight of 756 N supported by his hands and feet, which are 1.35 m apart. One hand rests on a spring scale, which reads 267 N. If each hand bears an equal weight, how far from the shoulders is the man's center of gravity?

12 A man and a boy carry a canoe 6.00 m long and weighing 375 N. The canoe is held overhead in a horizontal position. The boy holds one end of the canoe. How far from the opposite end should the man stand so that the boy exerts a force of only 125 N?

★ **13** A piece of furniture weighing 1.00×10^3 N is carried up stairs, as shown in Fig. 10-30. Find the vertical forces exerted by the two people at P and Q.

Fig. 10-30

★ **14** The two planks in Fig. 10-31 have a combined weight of 200.0 N. Assuming that the floor is frictionless, find the forces the floor exerts at the two points of contact.

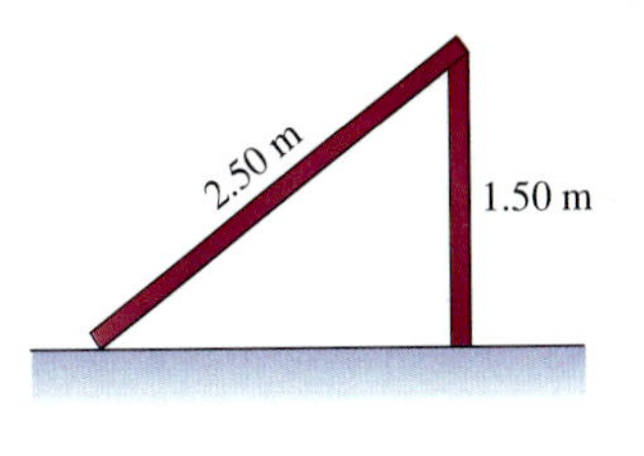

Fig. 10-31

★ **15** A raccoon wants to tip over a garbage can by pulling on its top edge with a force **P** acting 60.0° below the horizontal. The can is a cylinder of height 80.0 cm and diameter 50.0 cm, weighing 40.0 N. The can's center of gravity is directly beneath its geometric center. What is the minimum value of P that will cause the can to tip?

★ **16** A bench seat is attached to a wall by hinges and supported from above by two ropes, as shown in Fig. 10-32. The seat weighs 135 N and is designed to support a person weighing up to 1350 N, sitting on the edge. How much tension must each rope be able to withstand?

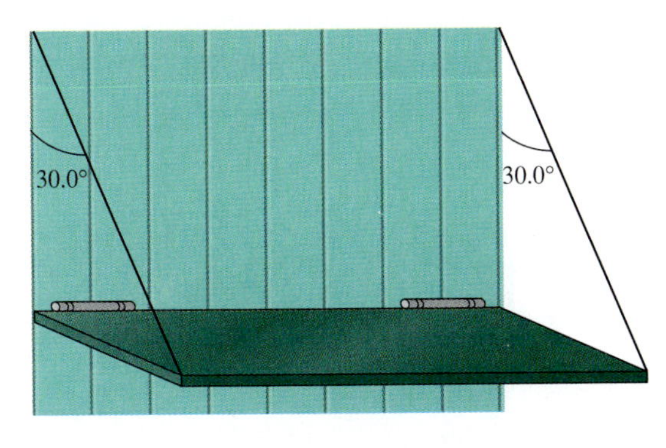

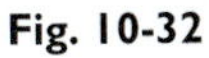

Fig. 10-32

★ **17** A horizontal board weighing 200.0 N is supported at points A and B in Fig. 10-33. The plank serves as a platform for a painter, who weighs 600.0 N. Find the maximum distance D, such that the plank will not tip, no matter where the painter stands.

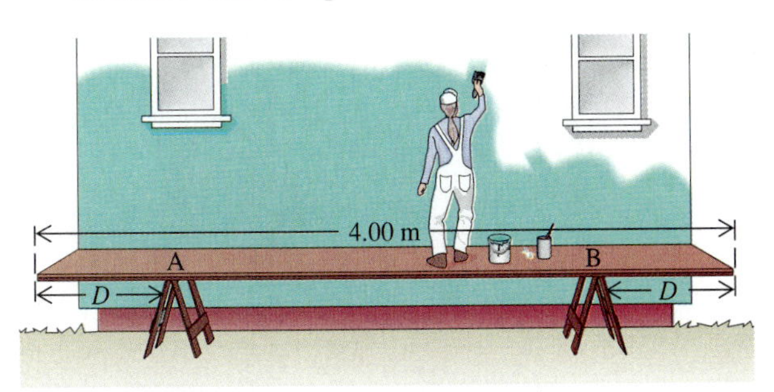

Fig. 10-33

18 What is the minimum horizontal force **P** necessary to tip over the 5.00 kg block shown in Fig. 10-34?

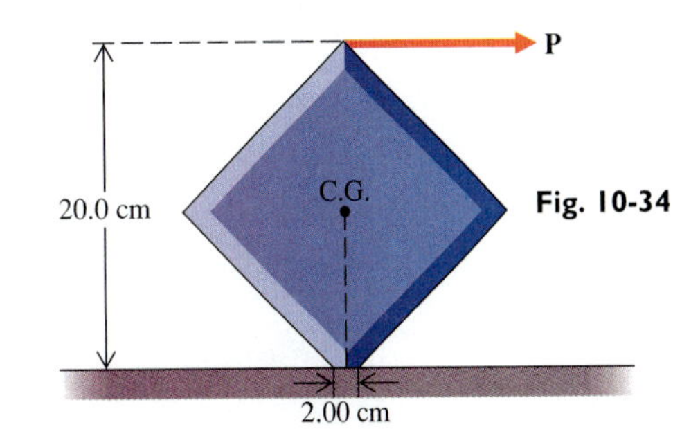

Fig. 10-34

★ **19** The deltoid muscle connects the shoulder blade to the upper arm (humerus), and so contraction of the deltoid raises the arm (Fig. 10-35). The deltoid tendon connects the humerus 15.0 cm from the shoulder joint at an angle of 20.0° with the horizontal. Suppose the arm weighs 35.0 N and its center of gravity is 38.0 cm from the shoulder joint. A weight W is held in the outstretched hand 90.0 cm from the shoulder joint. Find the tension in the deltoid tendon in terms of W. Find the force the shoulder joint exerts on the humerus if $W = 40.0$ N. Why is this force so large even for such a small weight?

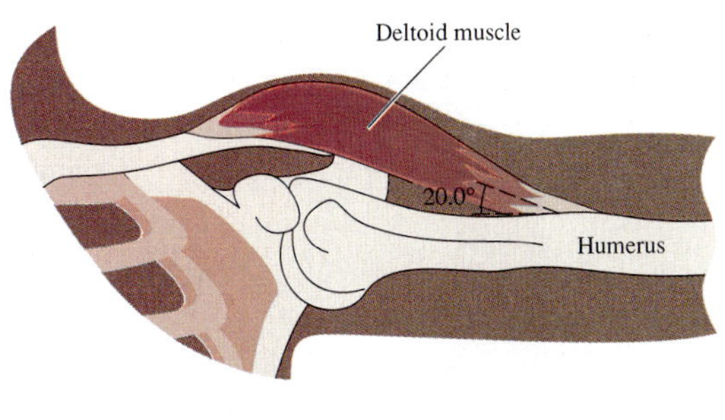

Fig. 10-35

20 Find the magnitude of the vertical force **F** necessary to lift the end of a loaded wheelbarrow of weight 1400 N, shown in Fig. 10-36.

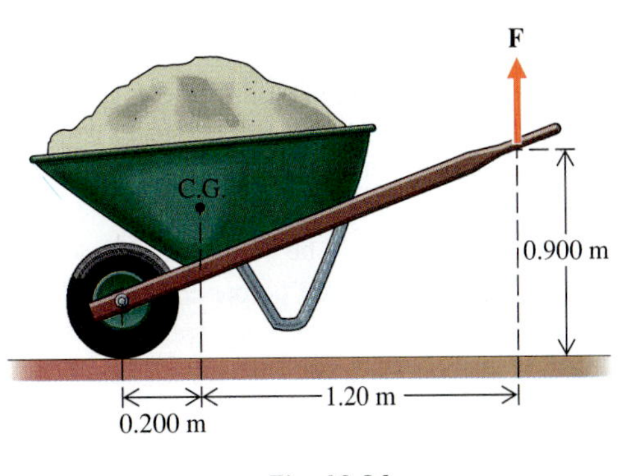

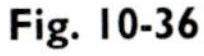

Fig. 10-36

★**21** Find the minimum horizontal force that must be applied to the handles of the wheelbarrow in the last problem, in order that the wheel of diameter 20.0 cm will slowly go over a brick 6.00 cm high, as shown in Fig. 10-37. (No vertical component of force is applied to the handles.)

Fig. 10-37

★**22** A rigid lean-to shelter of weight 90.0 N is supported by a horizontal rope, as shown in Fig. 10-38. The center of gravity of the lean-to is at its center. What is the tension in the rope? The answer is independent of *d*.

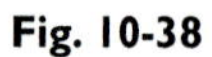

Fig. 10-38

*10-3 Stress and Strain

★**23** An old camping trick for breaking thick sticks for firewood is to wedge one end of a stick between two small trees that are very close together, and then apply a force to the other end. Determine the largest radius a stick may have if it is to be broken in this way by a 4.00×10^2 N force applied perpendicular to the stick 1.00 m from where the stick is wedged. The breaking stress is 2.00×10^8 N/m^2.

24 (a) Compute the maximum longitudinal force that may be supported by a bone before breaking, given that the compressive stress at which bone breaks is 2.00×10^8 N/m^2. Treat the bone as a solid cylinder of radius 1.00 cm.
(b) Young's modulus for bone is 2.00×10^{10} N/m^2. Estimate the strain in the bone just before breaking.

25 A certain concrete will support a maximum compressive stress of 1.20×10^7 N/m^2 before being crushed. A column of uniform width is to be made with this concrete. How tall can the column be if the stress in it at any point is not to exceed one fourth the maximum value? The weight density of the concrete is 2.40×10^4 N/m^3.

★**26** A girl accidentally steps into a gopher hole, causing her foot and ankle to be stuck. If all her weight of 525 N is borne by the one stuck foot, what is the maximum distance her center of gravity can be displaced horizontally from the point of support before her leg will break?

Additional Problems

★**27** An irregularly shaped, flat object is suspended from a string, first at point P and then at point Q (Fig. 10-39). In each case, lines are marked on the object, extending downward from the point of support. Prove that the object's center of gravity lies at the point of intersection of the lines.

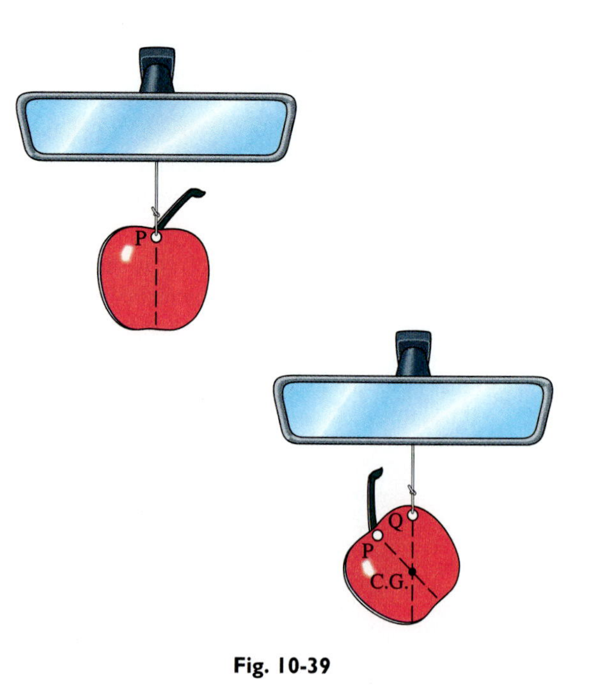

Fig. 10-39

★ **28** Three identical blocks are stacked with each extending as far as possible over the block beneath it (Fig. 10-40). Find the distance D that the top block overhangs the bottom block if the length of each block is 20.0 cm.

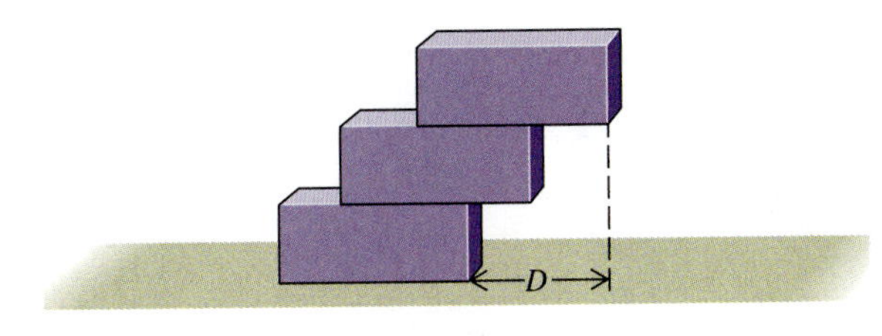

Fig. 10-40

★ **29** A block of height 30.0 cm and width 10.0 cm is initially at rest on a table. One end of the table is then slowly raised so that there is an increasingly large angle θ between the table's surface and the horizontal. If the coefficient of friction between block and table is 0.500, will the block tip over or slide down the incline?

★ **30** A symmetrical table of height 0.800 m, length 1.50 m, and weight 445 N is dragged across the floor by a force applied to its front edge. The force is directed to the right and upward and makes an angle of 30.0° with the horizontal.

(a) Find the minimum force necessary to drag the table across the floor. The coefficient of sliding friction between table and floor is 0.400.

(b) Calculate the normal force and the frictional force on each leg, and show that the table will not tip.

★ **31** A sprinter of mass 80.0 kg has one foot on the ground, as shown in Fig. 10-41. The force **F** exerted by the ground at point P is directed so that the torque about the body's center of mass is approximately zero.

(a) Find the angle θ this force makes with the horizontal.

(b) Find the instantaneous acceleration of the center of mass, assuming that it is directed horizontally.

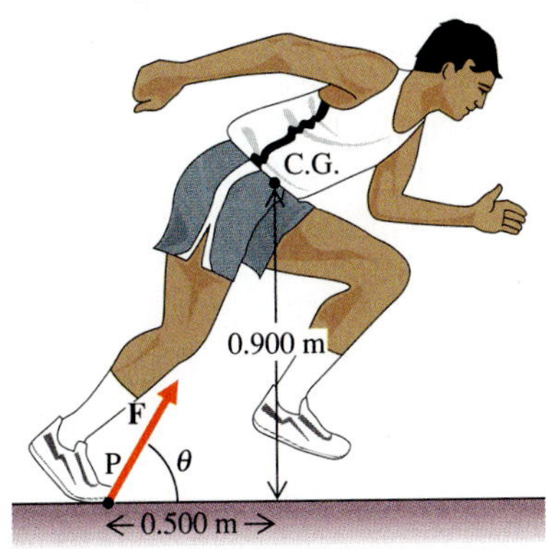

Fig. 10-41

★ **32** A car weighing 1.00×10^4 N is at rest on a 30.0° incline. Each wheel bears an equal part of the car's weight. Thus the axles apply a vertical downward force of 2.50×10^3 N to the center of each wheel.

(a) Find the frictional force exerted by the road on each wheel.

(b) Find the sum of the magnitudes of the frictional forces $\mathbf{f}_B$ applied by the brakes to the wheel, as shown in Fig. 10-42.

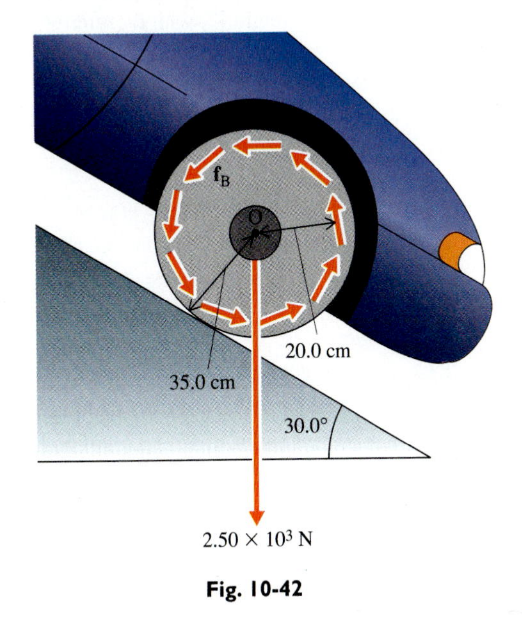

Fig. 10-42

★★ **33** An aluminum ladder of negligible weight leans against a vertical wall at an angle of 60.0° with the horizontal. The coefficient of static friction between both ladder and wall and ladder and ground is 0.300. A sculptor climbs the ladder and reaches a point at which the ladder begins to slip. What fraction of the length of the ladder has the sculptor climbed?

★★**34** When a person bends the body at an angle θ while lifting a weight w, large forces are produced near the base of the spine. The back is supported by a system of ligaments and muscles connecting the spine to the pelvis (Fig. 10-43a). The effect of these muscles is equivalent to having a cord connected at a point P two thirds up from the lower end of the spine and making an angle of 12° with the spine (Fig. 10-43b). The weight of the upper body supported by the spine is typically about 60% of body weight W, with a center of gravity at P. Additional weight w supported by the arms and shoulders acts at the top end of the spine.

(a) Find the tension in the "cord" and the force on the base of the spine if $\theta = 30°$, $W = 800$ N, and $w = 0$.

(b) Repeat the calculation with $w = 400$ N.

(c) The disk at the base of the spine is subject to a compressive force. If the disk has an area of 1.0×10^{-3} m^2, what stress does it experience for the situation described in (b)? By how much does the disk compress if its original thickness is 0.80 cm? Although the relationship between stress and strain in this region is not linear, we may use an effective value of Young's modulus equal to 2.0×10^7 N/m^2.

(d) How great would the weight w have to be to cause a stress of 1.5×10^7 N/m^2—enough to rupture the disk?

This problem illustrates why most people are subject to lower back pain at some time in their lives. It is rather easy to produce a large stress on the lower disks. A heavily strained disk will bulge and press on the spinal nerve, causing pain in the lower back and the legs.

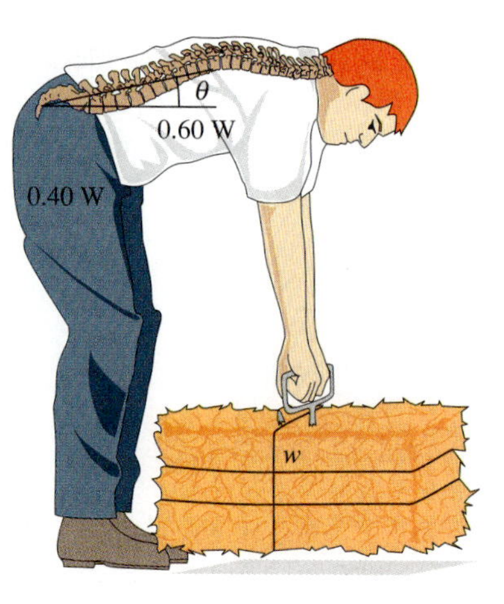

(a)

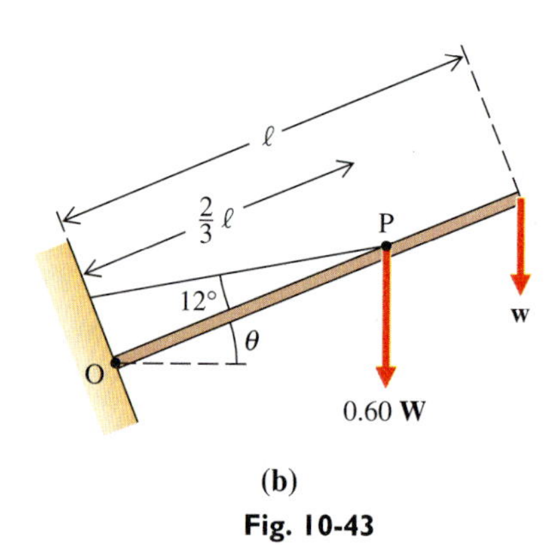

(b)

Fig. 10-43

35 The weight in Fig. 10-44 hangs from a long board of negligible weight. Find the magnitude of the tension T in the horizontal rope and the horizontal and vertical components of force exerted on the board at P.

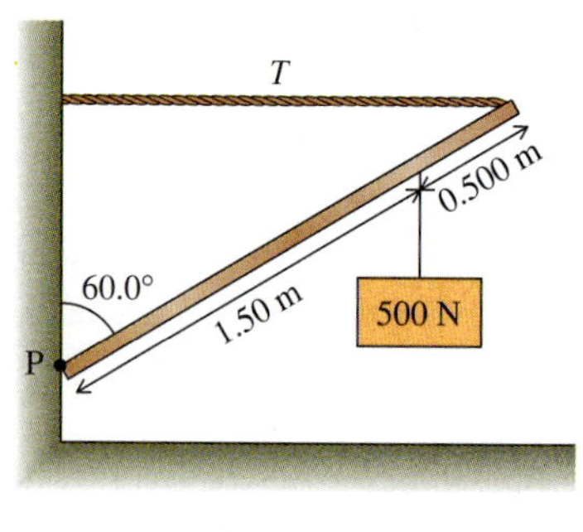

Fig. 10-44

36 Suppose the hand-held weight in Ex. 3 is raised because the biceps muscle contracts. Show that, as the forearm is raised to a horizontal position, the work done by the tension force in the muscle equals the work done on the weight by the force the hand exerts on the weight.

CHAPTER 11 Fluids

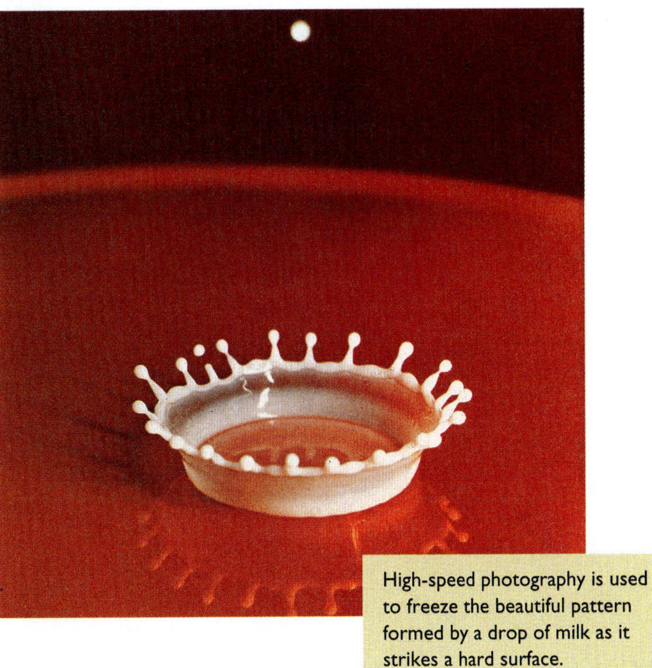

High-speed photography is used to freeze the beautiful pattern formed by a drop of milk as it strikes a hard surface.

Have you ever wondered why some insects can walk on water, or why a huge metal boat floats, but a small metal coin tossed into a pool sinks to the bottom? Did you know that the tremendous forces generated in a hurricane or tornado result from an imbalance in air pressure—pressure that is always present in our atmosphere but that we normally do not feel only because it is perfectly balanced on all sides of our bodies? These phenomena are examples of forces exerted by fluids. **A fluid is a substance that can flow.** Its shape is dependent on the forces exerted on it. Since both liquids and gases can flow, both are fluids. Solids cannot flow; they maintain a rigid shape, which is changed only slightly under the application of large forces, as we saw in Section 7-5.

In this chapter we shall first study fluids at rest and then, beginning in Section 11-5, fluids in motion. In Section 11-7 we shall apply the laws of fluid flow to the flow of blood in the human circulatory system.

11-1 Properties of Fluids

Density

An important property of any substance is its density. The **density** of a body is denoted by the Greek letter ρ (rho) and is defined as the ratio of the body's mass m to its volume V:

$$\rho = \frac{m}{V} \tag{11-1}$$

Table 11-1 Densities

Solids	Density (kg/m^3)	Liquids	Density (kg/m^3)
Aluminum	2.70×10^3	Alcohol, ethyl	0.791×10^3
Bone	$1.7–2.0 \times 10^3$	Blood, plasma	1.03×10^3
Brick	$1.4–2.2 \times 10^3$	Blood, whole	1.05×10^3
Copper	8.9×10^3	Gasoline	$0.66–0.69 \times 10^3$
Gold	19.3×10^3	Mercury	13.6×10^3
Granite	2.7×10^3	Milk	1.03×10^3
Ice	0.917×10^3	Olive oil	0.918×10^3
Iron	7.8×10^3	Water (0° to 30° C)	1.00×10^3
Lead	11.3×10^3	Seawater	1.025×10^3
Marble	$2.6–2.8 \times 10^3$		
Paper	$0.7–1.2 \times 10^3$	Gases*	Density (kg/m^3)
Quartz	2.65×10^3		
Rubber	$1.1–1.2 \times 10^3$	Air	1.20
Salt	2.18×10^3	Carbon dioxide	1.84
Styrofoam	0.1×10^3	Helium	0.17
Wood		Hydrogen	0.084
Balsa	$0.11–0.14 \times 10^3$	Methane	0.67
Ebony	$1.11–1.33 \times 10^3$	Nitrogen	1.16
Maple	$0.62–0.75 \times 10^3$	Oxygen	1.33
Oak	$0.60–0.90 \times 10^3$	Ozone	1.99
Pine	$0.35–0.60 \times 10^3$	Steam (100° C)	0.60

*Except for steam, all gas densities are at 20° C and 1 atmosphere.

Density is a characteristic property of a substance. For example, a piece of aluminum may have any volume or mass, but because it is aluminum, the ratio of mass to volume, that is, its density, will always be the same. Density is measured in units of kg/m^3. Densities of various substances are given in Table 11-1. Notice that solids and liquids have densities that are on the order of a thousand times larger than typical gas densities.

The **specific gravity** of a material is defined to be the ratio of its density to the density of water. Water has a density of 1.00×10^3 kg/m^3. Thus aluminum, which has a density of 2.70×10^3 kg/m^3, has a specific gravity of 2.70.

EXAMPLE 1 The Weight of Air in a Room

Find the mass of air in an empty room with dimensions 5.00 m by 4.00 m and a ceiling 3.00 m high. What is the weight of this mass of air?

SOLUTION It follows from the definition of density (Eq. 11-1: $\rho = m/V$) that the mass of air is directly proportional to the volume of the room containing it:

$$m = \rho V$$

$$= (1.20 \text{ kg/m}^3)(5.00 \text{ m} \times 4.00 \text{ m} \times 3.00 \text{ m}) = 72.0 \text{ kg}$$

The weight of this volume of air is

$$w = mg = (72.0 \text{ kg})(9.80 \text{ m/s}^2) = 706 \text{ N} \quad \text{(or 159 lb)}$$

The air in a room this size weighs about as much as an average person.

Stress in a Fluid

A swimmer can pull a hand through the water slowly with the slightest force, though a very large force is required to pull a hand through the water rapidly. The water can exert no static frictional force on the hand; it can exert only a kinetic frictional force. This is a general property of fluids and applies to internal forces within the fluid as well as to external forces. A force applied tangent to any fluid surface, internal or external, will produce motion of that surface. **A fluid is not able to support a static shear stress.** This is in contrast to a solid, which can support large static shear stresses, the only effect being a slight deformation of the solid (Section 10-3). Both fluids and solids can support a static normal stress (Fig. 11-1).

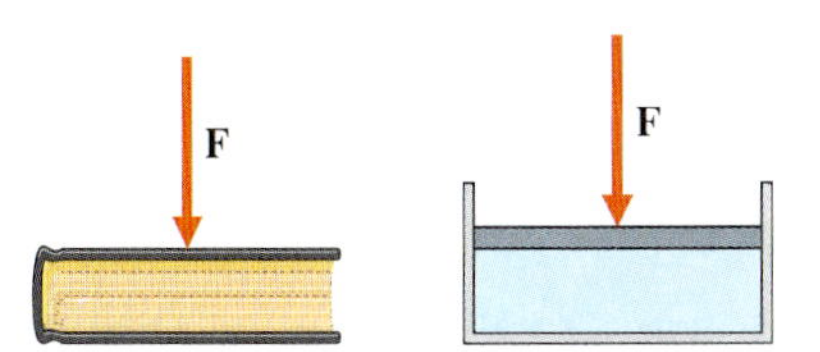

Fig. 11-1 A book and a fluid, each subject to a static normal force that produces normal stress within the material.

11-2 Pressure in a Fluid at Rest

The normal stress within a fluid at rest* leads to the definition of fluid pressure. A fluid exerts a force on a body with which it is in contact and that force is ordinarily directed toward the body on which it acts. That is, it is usually a push, not a pull. (Exceptions are discussed in Section 11-4.) The pressure P at any point in a fluid is defined with reference to a small surface area A centered at that point. This surface could be, for example, a small section of the wall of a container holding a fluid (Fig. 11-2a and 11-2b), or it could be a small surface area of an object submerged in a fluid (Fig. 11-2c). **The magnitude of the force F exerted by the fluid on the surface, divided by the surface area A, is defined to be the pressure at that point:**

$$P = \frac{F}{A} \tag{11-2}$$

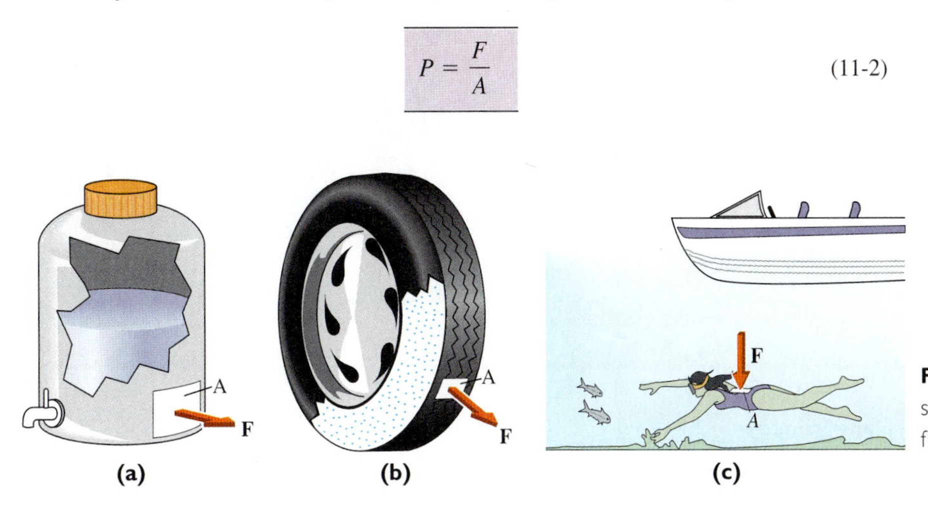

Fig. 11-2 A fluid exerts a force **F** on a surface of area A in contact with the fluid.

The SI unit of pressure is the N/m², which is called a "pascal" (abbreviated Pa). Other commonly used units are lb/in², torr, and atmospheres:

$$1 \text{ Pa} = 1 \text{ N/m}^2 \tag{11-3}$$

$$1 \text{ torr} = 133 \text{ Pa} \tag{11-4}$$

$$1 \text{ atm} = 1.01 \times 10^5 \text{ Pa} = 14.7 \text{ lb/in}^2 \tag{11-5}$$

We could use the device shown in Fig. 11-3 to measure fluid pressure. A piston of negligible weight is sealed in a cylinder, in which the piston moves without friction. The piston and base of the cylinder are connected by a spring. Fluid pressure on the outer surface of the piston compresses the spring; the distance the spring is compressed is directly proportional to the pressure. With such a device we would find that the pressure at a given point in a fluid is the same no matter what the orientation of the surface experiencing that pressure. This property of a fluid can be shown to follow from the fact that only a normal stress can exist in a fluid at rest (Problem 17).

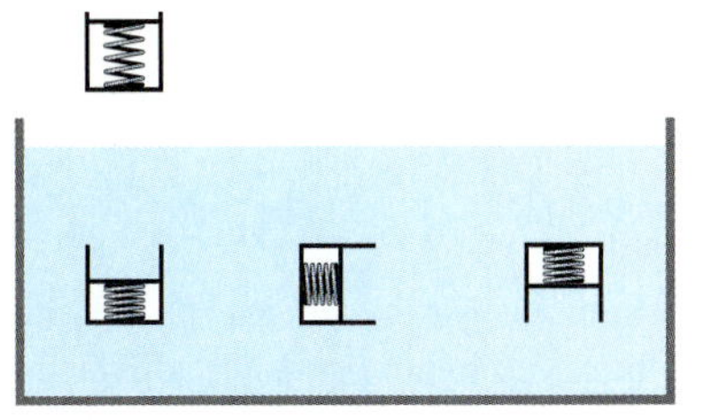

Fig. 11-3 Fluid pressure compresses the spring. The pressure is the same in all directions in a fluid at rest.

*When a gas is in static equilibrium, its molecules will be moving rapidly, perhaps over large distances. Yet there is no net transport of a significant part of the gas, and so it may be treated on a macroscopic level as though it were at rest.

A form of fluid pressure that we are constantly exposed to is atmospheric pressure. An impressive demonstration of the magnitude of atmospheric pressure was performed in 1654 by Otto von Guericke, the inventor of the vacuum pump. He pumped the air out of a spherical cavity formed by two hollow hemispheres placed together. Two teams of horses then tried to pull the hemispheres apart (Fig. 11-4). The external air pressure held the hemispheres together, and the horses were unable to separate them. Once a valve was opened, allowing air to fill the cavity, the hemispheres were easily separated.

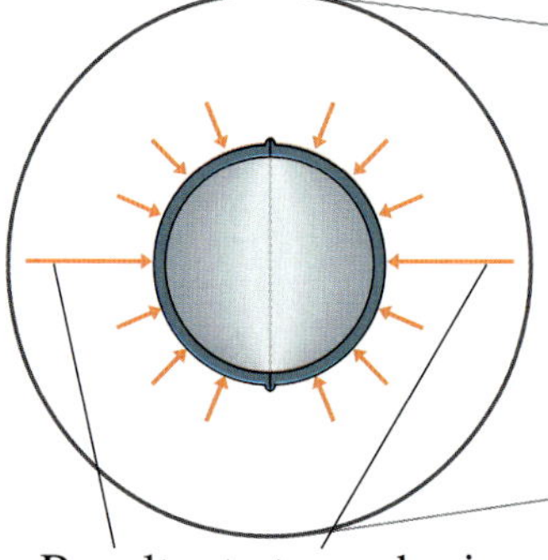

Fig. 11-4 A dramatic demonstration of atmospheric pressure. Teams of horses could not pull apart two evacuated hollow hemispheres that were held together only by the air pressure outside the hemispheres.

EXAMPLE 2 Air Pressure on an Airplane Window

An airplane window has an area of 825 cm². Cabin pressure is 1.000 atm, and the outside pressure is 0.300 atm. Find the net force produced by air pressure on the window.

SOLUTION The forces on the window produced by air pressure are shown in Fig. 11-5. This combination of forces results in a net fluid force $\mathbf{F}_f$, directed outward, of magnitude

$$F_f = P_iA - P_oA$$
$$= (P_i - P_o)A$$
$$= (1.000 \text{ atm} - 0.300 \text{ atm})\left(\frac{1.01 \times 10^5 \text{ Pa}}{1 \text{ atm}}\right) \times (825 \text{ cm}^2)\left(\frac{10^{-2} \text{ m}}{1 \text{ cm}}\right)^2$$
$$= 5.83 \times 10^3 \text{ N}$$

In order that the window be held in place, this force must be balanced by an oppositely directed force of equal magnitude (provided by the window frame).

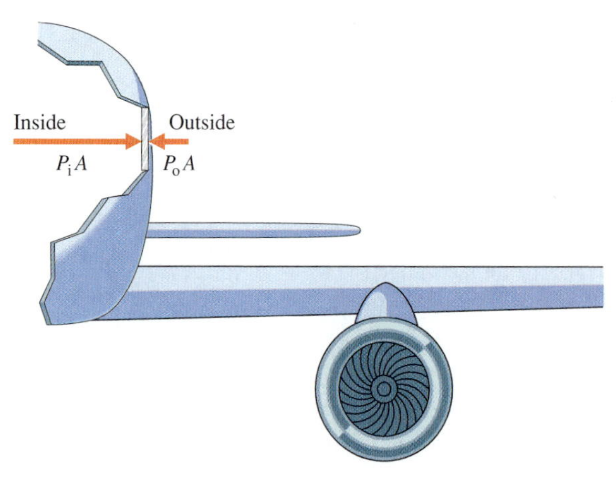

Fig. 11-5

Variation of Pressure with Depth in an Incompressible Fluid

Although fluid pressure does not vary with orientation, it does vary with elevation or depth in the fluid. For example, the water pressure at the bottom of a lake is considerably greater than the water pressure just below the surface. A swimmer who dives a few meters below the surface will feel pressure on the eardrums, and a deep-sea diver far below the water's surface must use compressed air to balance the increased external pressure on the chest. We shall now find an expression for the pressure in a fluid at some distance below a reference level (which may be, for example, the surface of a liquid).

Consider a thin cylinder of fluid extending from a reference point at a pressure P_0 to a point a distance h below, where the pressure is P (Fig. 11-6). The external forces exerted on this cylinder of fluid are shown in the figure. They include the weight $m\mathbf{g}$ and the forces exerted by the surrounding fluid. This cylinder of fluid is in static equilibrium, and so the resultant force on it must vanish, according to Newton's second law. The horizontal forces exerted on the cylinder by the surrounding fluid balance each other. However, the vertical fluid forces do not balance. The upward fluid force on the bottom surface must be greater than the downward fluid force on the top surface, so as to balance the weight of the column of fluid. We set the sum of the vertical forces equal to zero to obtain an expression for P:

$$\Sigma F_y = 0$$

$$PA - P_0A - mg = 0$$

The mass m of the cylinder of fluid is the product of the fluid's density ρ and its volume Ah. Substituting $m = \rho Ah$ into the equation above and solving for P, we obtain

$$P = P_0 + \rho gh \tag{11-6}$$

We have assumed here that the fluid has a fixed density ρ, independent of variation in pressure. This assumption of incompressibility is a good approximation for most liquids, including water, whose density increases only slightly when the pressure increases by many atmospheres. Gas density, however, is strongly dependent on pressure. This means that our equation for fluid pressure will not be valid for gases when there is a large change in pressure. For example, it is a poor approximation to apply Eq. 11-6 to the variations in atmospheric pressure for elevation differences of more than about 1000 m.

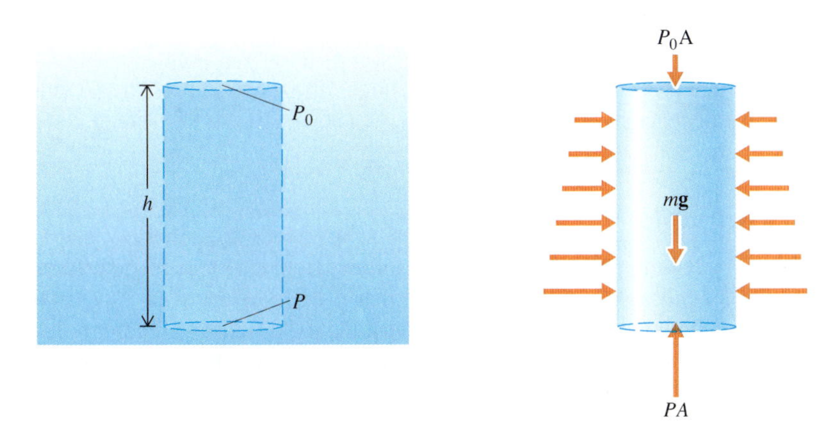

Fig. 11-6 A column of fluid of height h is chosen as a free body so that we can derive an expression for the fluid pressure P a distance h below a point where the pressure is P_0.

EXAMPLE 3 Pressure Beneath the Surface of the Sea

Find the increase in water pressure as depth increases by 10 m in seawater.

SOLUTION Eq. 11-6 relates the pressures at two points a vertical distance h apart.

$$P = P_0 + \rho g h$$

Using the density of seawater from Table 11-1, we find the pressure increase $P - P_0$ when $h = 10$ m.

$$\begin{aligned} P - P_0 &= \rho g h \\ &= (1.025 \times 10^3 \text{ kg/m}^3)(9.80 \text{ m/s}^2)(10 \text{ m}) \\ &= 1.0 \times 10^5 \text{ Pa} = 1.0 \text{ atm} \end{aligned}$$

Pressure increases by 1.0 atm for every 10 m increase in depth (Fig. 11-7). A scuba diver breathes air at a pressure regulated to match the pressure of the surrounding water so that the pressure inside the diver's body equals the pressure outside.

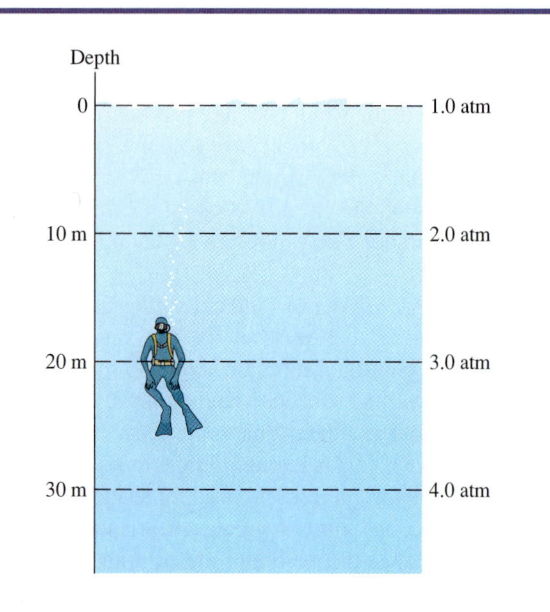

Fig. 11-7 Pressure at various depths in seawater.

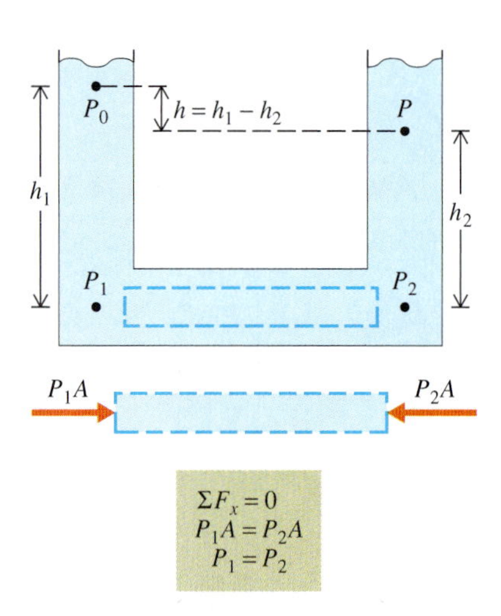

Fig. 11-8 The equation $P = P_0 + \rho g h$ applies to any two points connected by a path through the liquid, where h is the difference in the depth of the two points.

Although Eq. 11-6 was derived for the situation in which the point at pressure P is directly beneath the point at pressure P_0, it can be applied to any two points, so long as there is a path through the liquid connecting them. The quantity h is the difference in elevation of the two points, as illustrated in Fig. 11-8. We can obviously apply Eq. 11-6 to the points at pressures P_0 and P_1 and again to the points at pressures P and P_2:

$$P_1 = P_0 + \rho g h_1$$

$$P_2 = P + \rho g h_2$$

But pressures P_1 and P_2 are equal, as shown in the figure. Thus we may equate the two expressions above:

$$P + \rho g h_2 = P_0 + \rho g h_1$$

or

$$P = P_0 + \rho g (h_1 - h_2)$$

The distance $h_1 - h_2$ is the difference in elevation of the points at pressures P_0 and P. Denoting this distance by h, we obtain

$$P = P_0 + \rho g h$$

Pascal's Principle

Pascal's principle is an interesting result that follows from the preceding equation for the variation of pressure with depth. Notice that if the pressure P_0 is increased by some arbitrary amount the equation predicts that the pressure P at all other points in the fluid increases by the same amount. For example, suppose that the pressure P_0 increases by 5 atm. Then, since the value of $\rho g h$ is unchanged at each point, the pressure P at each point must also increase by 5 atm. This is the essence of **Pascal's principle,** which is usually stated: **Pressure applied to an enclosed fluid is transmitted undiminished to every portion of the fluid and to the walls of the containing vessel.**

Pascal's principle implies that fluid pressure can be used in force-multiplying machines. When an external force is applied to a fluid at one point, the pressure at that point is increased, and, according to Pascal's principle, the pressure at all other points in the fluid is increased by the same amount. Thus, if a small force $\mathbf{F}_1$ is applied to the fluid by means of a small piston of area A_1, the pressure throughout the fluid increases by F_1/A_1. Automobile brakes are based on this principle (Fig. 11-9a). If there is another piston in contact with the fluid, one having a larger area A_2, the force on this piston will increase by an amount F_2, which is larger than F_1. Equating the increases in pressure, we find

$$\frac{F_2}{A_2} = \frac{F_1}{A_1}$$

This equation shows that the force that can be supported by the second piston is greater than the force supported by the first, as indicated in Fig. 11-9b. This effect is used in various hydraulic devices, including the hydraulic lifts used to raise cars for repairs (Fig. 11-9c). For example, a car weighing 9000 N can be supported on a hydraulic lift with a column of fluid having a cross section of 300 cm^2, when one applies a force of 300 N to a connecting column of fluid of cross-sectional area 10 cm^2. In either column the pressure is 30 N/cm^2.

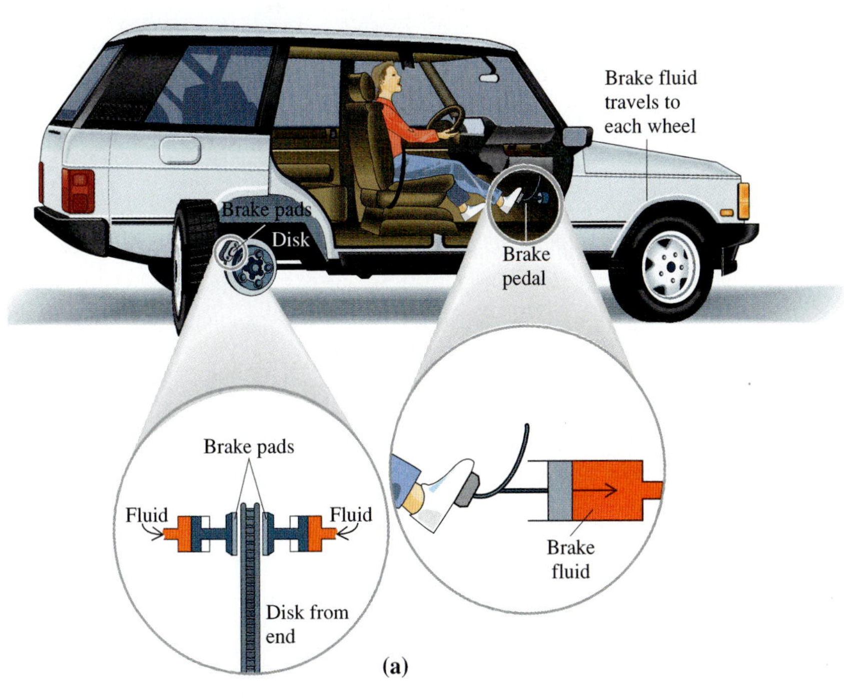

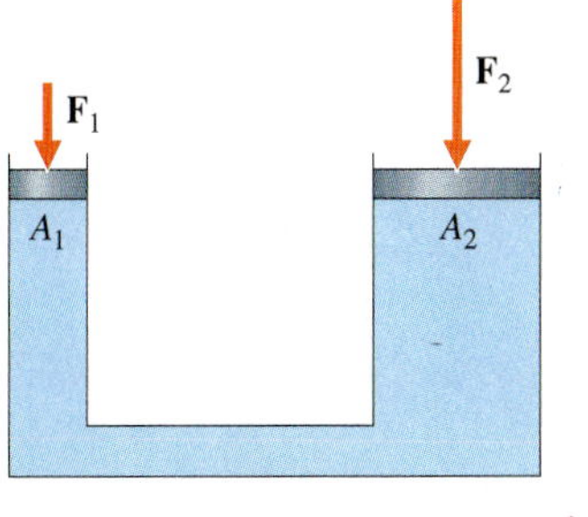

Fig. 11-9 **(a)** Pressure applied to a car's brake pedal is transmitted by brake fluid to the car's wheels. **(b)** The same force per unit area is supported by different-sized pistons that are at the same height and are in contact with a static fluid, because the fluid pressure on each piston is the same. Thus a small force applied to a small piston balances a large force on a large piston. **(c)** The force-multiplying effect shown in (b) is applied in this hydraulic lift.

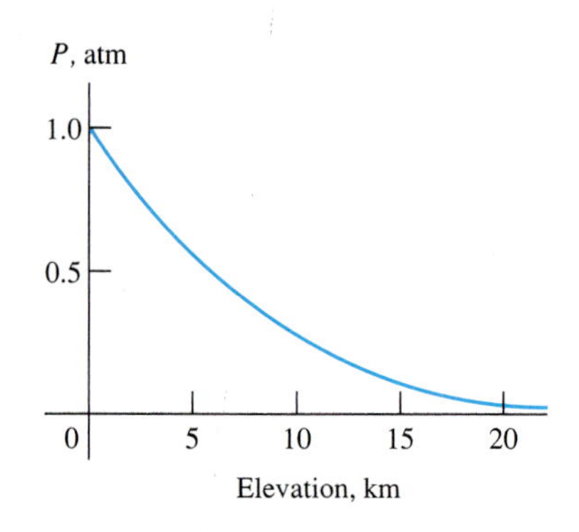

Fig. 11-10 Variation in normal atmospheric pressure as a function of elevation.

Variation in Atmospheric Pressure with Elevation

In liquids, fluid pressure varies linearly with elevation (or depth), but this is not the case with gases. Fig. 11-10 shows the nonlinear variation in atmospheric pressure with elevation. Gas pressure varies with height in a nonlinear way because a gas, unlike a liquid, is easily compressed. The density of a gas increases with increasing pressure. This causes the air at the surface of the earth to be far more dense and at a far higher pressure than the air at an elevation of a few kilometers. For example, in Denver, Colorado (elevation 1 mile), atmospheric pressure is only about 80% of atmospheric pressure at sea level. The seventeenth-century scientists who first measured atmospheric pressure were amazed to find such large drops in pressure at high elevations.

Decreased atmospheric pressure can have practical consequences. For example, performance in some athletic events, such as swimming, tends to be poorer at high altitudes, because of the decreased supply of oxygen. And it is difficult to cook foods by boiling because water boils at a lower temperature when atmospheric pressure is reduced. Pressure cookers solve this problem.

Fig. 11-11 Nearly all the earth's atmosphere is contained in a thin shell only about 100 km thick.

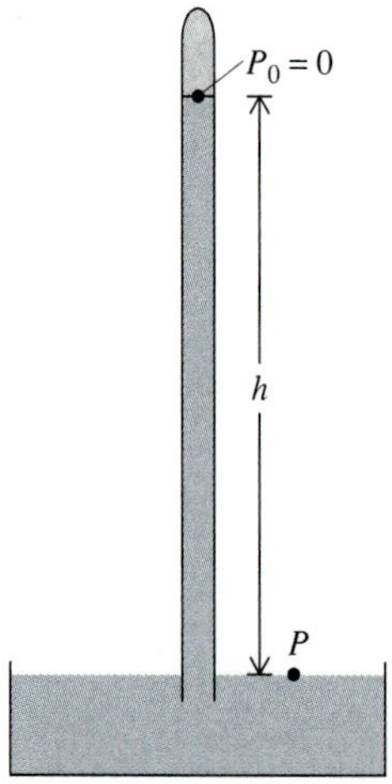

The Barometer

The barometer, a device invented by Torricelli in 1643, uses liquid pressure to measure atmospheric pressure. One can construct a barometer by taking a glass tube that is closed at one end and about 80 cm long, filling it with mercury, and inverting it in a container of mercury open to the atmosphere.* The level of liquid in the tube will drop a few centimeters but will maintain a level higher than in the container (Fig. 11-12). The space above the mercury in the tube is a vacuum (except for a very small amount of mercury vapor), and so the pressure at this point is practically zero. Atmospheric pressure P is the pressure at a distance h below the top of the mercury column. Since P is related to h by Eq. 11-6 ($P = P_0 + \rho g h$), the height of the mercury column indicates the value of atmospheric pressure—as atmospheric pressure increases, the column's height h increases.

Fig. 11-12 A barometer. Atmospheric pressure P equals the pressure at the bottom of a column of mercury of height h. Normal atmospheric pressure corresponds to $h = 76.0$ cm.

*Pascal constructed a barometer using wine in a glass tube over 10 m long.

EXAMPLE 4 Measuring Atmospheric Pressure With a Mercury Barometer

Find the value of atmospheric pressure when the height of the mercury column in a barometer is 76.0 cm.

SOLUTION Atmospheric pressure P is the pressure at the surface of the mercury pool, while the pressure P_0 at the top of the column equals zero. From Table 11-1, we find that the density of mercury is 13.6×10^3 kg/m^3. Applying Eq. 11-6, we obtain

$$\begin{aligned} P &= P_0 + \rho g h \\ &= 0 + (13.6 \times 10^3\ \text{kg/m}^3)(9.80\ \text{m/s}^2)(0.760\ \text{m}) \\ &= 1.01 \times 10^5\ \text{Pa} \end{aligned}$$

This is the standard value of atmospheric pressure at sea level. Depending on weather conditions, pressures will vary by a few percent.

Atmospheric pressure is often quoted—for example, in weather forecasts—in terms of the height of the mercury column in a barometer. Since the atmospheric pressure is directly proportional to this height, it is easy to convert to standard units. One simply takes the ratio of h in centimeters to its standard value, 76.0 cm, to find the pressure in atm:

$$P = (1\ \text{atm})\frac{h\ (\text{in cm})}{76.0\ \text{cm}}$$

Measurement of blood pressure is normally expressed in terms of millimeters of mercury. Since 76.0 cm or 760 mm corresponds to 1 atm of pressure, each mm of mercury in a barometer corresponds to a pressure of 1 atm/760, which is defined as 1 torr:

$$1\ \text{torr} = \frac{1\ \text{atm}}{760} = \frac{1.01 \times 10^5\ \text{Pa}}{760} = 133\ \text{Pa}$$

Gauge Pressure

A tire-pressure gauge (Fig. 11-13) is another device used to measure air pressure, in this case the air pressure inside a tire. When the end of the gauge is inserted into the valve of an inflated tire, the gauge reads the excess pressure in the tire over the air outside. This pressure difference is called **gauge pressure.** Normal gauge pressure for an automobile tire is about 2 atmospheres (2×10^5 Pa, or 30 lb/in^2), meaning that the pressure inside the tire is about 3 atmospheres.

When air at a pressure greater than 1 atmosphere enters a tire-pressure gauge, the air pushes a piston upward, compressing a spring attached to the piston. The compression of the spring is proportional to the force exerted by the air, and this force is proportional to the gauge pressure ΔP. The gauge pressure is read on a scale that is pushed upward by the piston. Because the scale is not attached to the piston, it remains extended after the gauge is removed from the tire and the piston moves down.

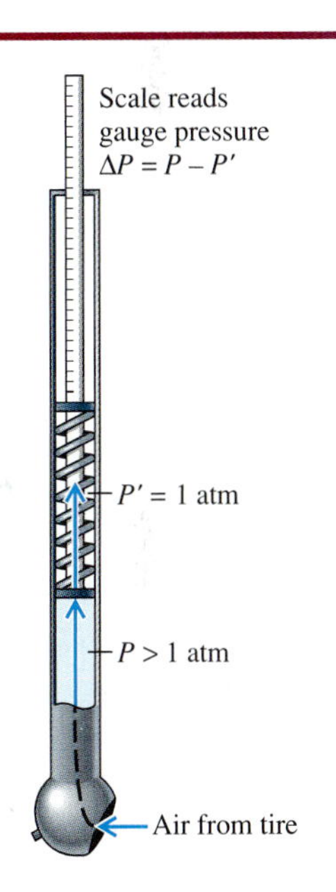

Fig. 11-13 A tire-pressure gauge.

11-3 Archimedes' Principle

When an object is either partially or totally submerged in a fluid, the object is buoyed up by the fluid. For example, a helium-filled balloon is lifted upward by the surrounding air (Fig. 11-14a), and a person floating in water is supported by the water in contact with the body (Fig. 11-14b). Even a dense object, such as a rock, experiences an upward force when submerged in a fluid, because of the pressure of the surrounding fluid. Evidence of this is that you may be able to lift a heavy stone from the bottom of a river bed, though the same stone would be too heavy to be lifted on dry land. The net effect of the water's pressure, acting on all sides of the stone, is to provide an upward force called the "buoyant force."

(a) 

(b) 

Fig. 11-14 **(a)** A helium balloon is lifted by the buoyant force exerted on it by the surrounding air. **(b)** A person floats because the body is supported by the surrounding water.

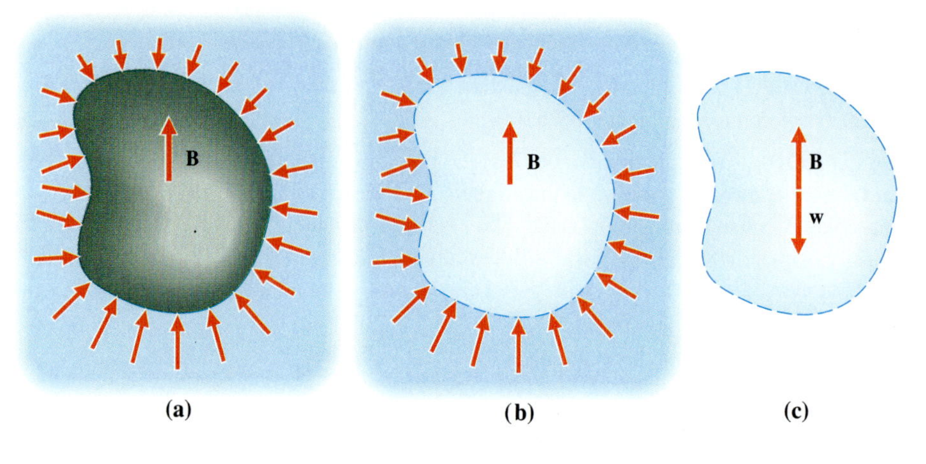

Fig. 11-15 **(a)** The buoyant force **B** on a submerged object is the resultant of the forces exerted along the surface of the object by the surrounding fluid. **(b)** The fluid displaced by a submerged object is a body of fluid having the same size and shape as the object. The forces exerted on this body of fluid by the surrounding fluid are exactly the same as those exerted on the submerged object, and thus the buoyant force **B** is the same in both cases. **(c)** The body of fluid is in static equilibrium. Therefore the buoyant force and the weight of the body of fluid are equal in magnitude but opposite in direction and have the same line of action.

The phenomenon of buoyancy is a direct consequence of the increase in fluid pressure with depth. As seen in Fig. 11-15a, the average fluid pressure on the bottom of a submerged object is necessarily greater than the average fluid pressure on the top. The sum of the fluid forces acting on all sides is an upward force **B**, the **buoyant force.**

It is a simple matter to find a useful expression for the magnitude of the buoyant force. Suppose the submerged object is removed and replaced by fluid identical to the surrounding fluid (Fig. 11-15b). We call this **the fluid displaced by the object.** This body of fluid has the same size and shape as the submerged object and will therefore be subject to the same forces exerted by the surrounding fluid along its surface. Thus the buoyant force **B** on this quantity of fluid is the same as the buoyant force on the submerged object. But the fluid is in equilibrium and is subject only to the buoyant force and to its own weight (Fig. 11-15c). Therefore these forces must cancel and must have the same line of action (so that the net torque is zero). **The buoyant force on an object submerged in a fluid is equal in magnitude to the weight of the fluid displaced by the object and effectively acts at the displaced fluid's center of gravity.**

This result was discovered in the third century B.C. by the great scientist and mathematician Archimedes* and is known as **Archimedes' principle.** Letting m denote the mass of the displaced fluid, ρ the fluid's density, and V the volume displaced, Archimedes' principle may be expressed

$$B = mg = (\rho V)g$$

or

$$B = \rho g V \tag{11-7}$$

*Archimedes discovered his principle of buoyancy while working on a very practical problem for the king of Syracuse in Sicily. The king asked Archimedes to determine whether a certain crown was pure gold or was partially made of some other cheaper metal. Archimedes knew that if the crown were entirely gold, its density would be that of gold. However, determining its density was a problem, since its volume could not be measured directly because of its irregular shape. It occurred to Archimedes that he could submerge the crown in a filled container of water; the crown's volume would equal the volume of water displaced by the crown. The volume of displaced water could easily be measured and compared with the volume of water displaced by an equal weight of pure gold. Unfortunately for the crownmaker, Archimedes found that the crown displaced significantly more water than the gold. The crown's density was less than that of gold, meaning that a less dense metal had been added. The crownmaker was executed.

EXAMPLE 5 Why a Helium Balloon Rises

Calculate the buoyant force on a helium-filled balloon having a volume of 14,000 cm³.

SOLUTION We apply Eq. 11-7, where ρ is the density of air found in Table 11-1 and V is the displaced air's volume (equal to the balloon's volume):

$$B = \rho g V$$

$$= (1.20 \text{ kg/m}^3)(9.80 \text{ m/s}^2)(14{,}000 \text{ cm}^3)\left(\frac{10^{-6} \text{ m}^3}{1 \text{ cm}^3}\right)$$

$$= 0.165 \text{ N}$$

Helium has a much lower density than air, and so the weight of helium inside the balloon is considerably less than the weight of the displaced air. The total weight of the balloon, including both the rubber and the helium inside, will normally be less than the buoyant force, and the balloon will therefore rise.

EXAMPLE 6 Determining Whether a Crown is Solid Gold

An object is suspended from a scale, first with the object in air (Fig. 11-16a) and then with it submerged in water (Fig. 11-16b).

(a) Show that the object's specific gravity equals the ratio of the object's weight to the change in the scale readings.

(b) Suppose that the object is a crown and that the scale reads 9.80 N in air and 9.00 N in water. Is the crown solid gold?

SOLUTION **(a)** Newton's second law is applied to the object in air (Fig. 11-16c) and in water (Fig. 11-16d) in order to obtain expressions for the respective tensions (equal to the scale readings).

$$T = w$$

$$T' = w - B$$

Subtracting these expressions, we obtain

$$T - T' = B$$

An object's specific gravity is its density divided by the density of water. Since weight is proportional to density, specific gravity may be expressed as the weight of an object divided by the weight of an equal volume of water. But, according to Archimedes' principle, the weight of an equal volume of water (the volume displaced by a submerged object) equals the buoyant force B:

$$\text{Specific gravity} = \frac{\text{Weight of object}}{\text{Weight of an equal volume of water}}$$

$$= \frac{w}{B}$$

Inserting the expression relating B to the scale readings, we obtain

$$\text{Specific gravity} = \frac{w}{T - T'}$$

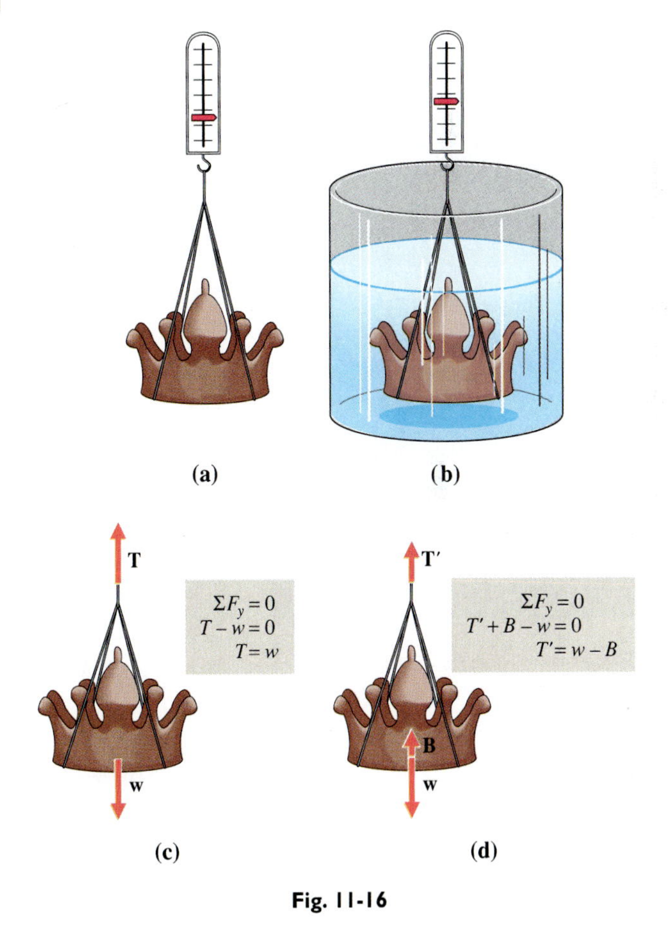

Fig. 11-16

(b) Applying the preceding equation, we find

$$\text{Specific gravity} = \frac{9.80 \text{ N}}{9.80 \text{ N} - 9.00 \text{ N}} = 12.3$$

The crown's specific gravity is much less than that of gold (19.3, Table 11-1). Therefore it is not solid gold.

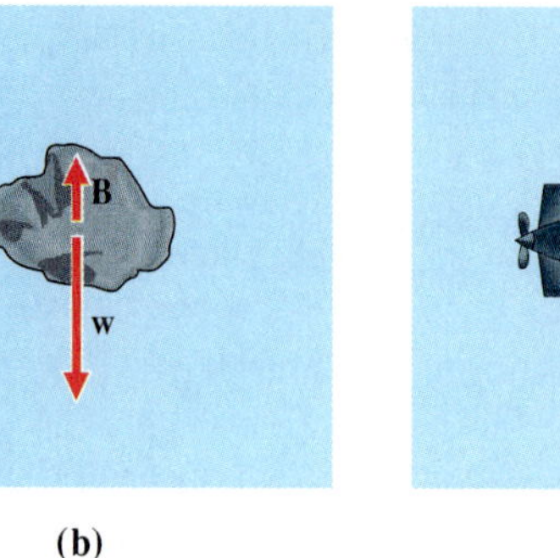

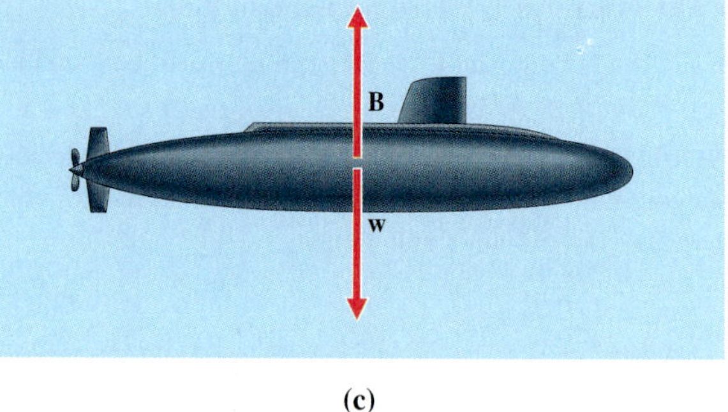

(a) (b) (c)

Fig. 11-17 **(a)** A beach ball submerged in water is less dense than the water and is accelerated upward. **(b)** A rock submerged in water is more dense than the water and is accelerated downward. **(c)** When its density equals that of the surrounding water, a submerged submarine maintains equilibrium.

Fig. 11-18 A scuba diver adjusts buoyancy so that she can either float on the surface, descend to the bottom, or maintain a fixed position between the bottom and the surface, as shown here. The diver's vest is inflated with air supplied by the same compressed air source used for breathing. By inflating or deflating the vest, the diver becomes more or less buoyant.

If a submerged body is less dense than the surrounding fluid, the weight of the fluid displaced exceeds the weight of the body. There will then be an upward resultant force, and the body will rise (Fig. 11-17a). If a submerged body is more dense than the surrounding fluid, the weight of the fluid displaced is less than the weight of the body, the resultant force is downward, and the body will sink (Fig. 11-17b). Finally, if a submerged body has the same density as the surrounding fluid, the weight of the fluid displaced equals the weight of the body, the resultant force is zero, and the body may remain stationary (Fig. 11-17c and Fig. 11-18).

A body that floats on the surface of a liquid is in static equilibrium. Thus the resultant force on the body must be zero—the buoyant force must balance the weight of the body. According to Archimedes' principle the floating body must then displace a quantity of liquid having a weight equal to its own weight. When calculating the buoyant force on an object that is not totally submerged, be careful to use the volume of the fluid displaced, not the entire volume of the object (Fig. 11-19).

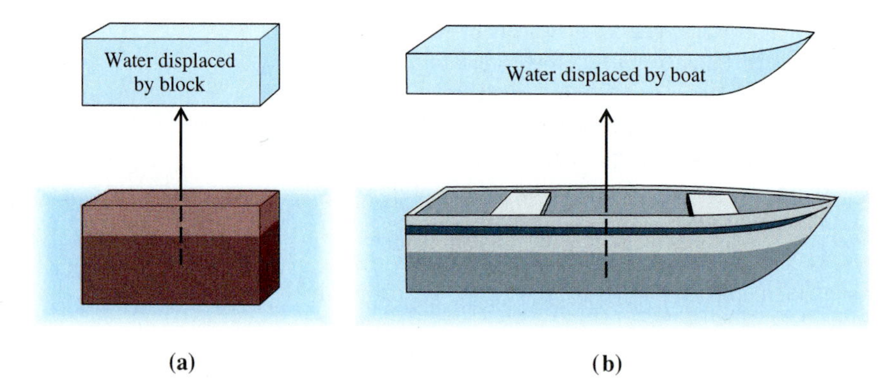

Fig. 11-19 **(a)** A block of wood, less dense than water, floats with its top above the water's surface. The block displaces a volume of water less than the block's volume, with the same weight as the block, so that the buoyant force on the block balances the weight. **(b)** A boat may be constructed of metal, which has a higher density than water. However, because of its hollowed-out shape, the boat is able to displace a volume of water weighing as much as the boat but having a volume less than that of the boat's entire volume. Thus the boat floats.

EXAMPLE 7 How Much Weight Can a Canoe Carry?

A hollow log in the shape of a half cylinder of length 3.00 m and radius 0.350 m is used as a canoe. If the canoe weighs 1.00×10^3 N, what is the maximum weight it can hold without sinking?

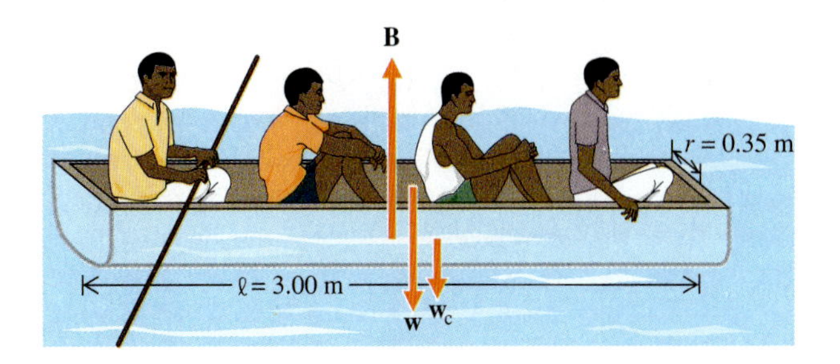

Fig. 11-20

SOLUTION As the weight it carries increases, the canoe will sink lower in the water. It carries the maximum weight w when its top is at the level of the water (Fig. 11-20). The canoe is subject to the upward bouyant force **B** and to two downward forces—the weight of the canoe $\mathbf{w}_c$, and the passengers' weight **w**. Since the canoe is in static equilibrium, it follows from Newton's second law that

$$\Sigma F_y = 0$$

$$B - w_c - w = 0$$

$$w = B - w_c$$

According to Archimedes' principle the magnitude of the buoyant force equals the weight of the displaced water, which we find by applying Eq. 11-7, using as the volume of the displaced water the volume of the canoe—the volume of a half cylinder of radius $r = 0.350$ m and length $\ell = 3.00$ m.

$$B = \rho g V = \rho g(\tfrac{1}{2}\pi r^2 \ell)$$

$$= (1.00 \times 10^3 \text{ kg/m}^3)(9.80 \text{ m/s}^2)(\tfrac{1}{2}\pi)(0.350 \text{ m})^2(3.00 \text{ m})$$

$$= 5.66 \times 10^3 \text{ N}$$

Substituting into the equation for w, we obtain

$$w = 5.66 \times 10^3 \text{ N} - 1.00 \times 10^3 \text{ N}$$

$$= 4.66 \times 10^3 \text{ N} \quad \text{(or 1050 lb)}$$

EXAMPLE 8 The Hidden Danger of an Iceberg

Icebergs are often a hidden danger to ships, since much of an iceberg's volume lies beneath the level of the water. One sees only the tip of the iceberg. Find the fraction of an iceberg's volume that is below the surface (Fig. 11-21).

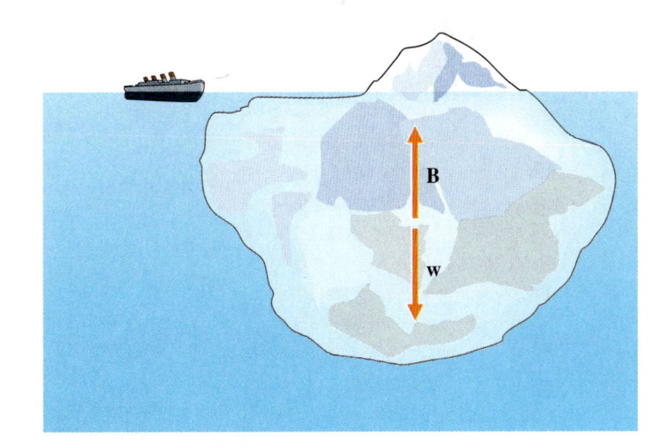

Fig. 11-21

SOLUTION We apply Newton's second law and Archimedes' principle:

$$\Sigma F_y = 0$$

$$B = w$$

$$\rho_w g V_w = \rho_i g V_i$$

The volume of the displaced water V_w equals the part of the iceberg's volume V_i that is below the surface. Thus the fraction beneath the surface is V_w/V_i. Solving the equation above for this ratio and inserting the respective densities of ice and seawater from Table 11-1, we find

$$\frac{V_w}{V_i} = \frac{\rho_i}{\rho_w} = \frac{0.917}{1.025} = 0.895$$

The iceberg is 89.5% submerged.

*11-4 Surface Tension and Capillarity

Fig. 11-22 Small drops of water form spherical beads on a waxed surface.

When the flow of water from a faucet is very slow, the water falls not as a continuous stream but as a series of drops. Water on the waxed hood of a car forms spherical beads (Fig. 11-22). Water forms drops or beads because of a tendency of water's surface to contract. The surface of the water behaves somewhat like a stretched membrane. This effect is called "surface tension." For very small quantities of water, the force of surface tension is much greater than the water's weight. If no other significant forces act on the water, its surface will contract to give the smallest possible surface area, which means a spherical shape. For example, water beads on the waxy surface of a leaf or a car's hood because there is little attractive force between the water and the adjacent surface. Small water drops in mist or fog are also spherical. (The shape of larger drops falling in air is affected by the force of air resistance.) Around the time of the American Revolution, round musket balls were produced when molten lead was dropped from a tower into a tank of water. The surface tension of the lead caused it to form spheres as it fell through the air.

The phenomenon of surface tension is common to all liquids. At any liquid-gas interface, the surface of the liquid is in a state of tension; that is, molecules on the surface exert an attractive force on adjacent surface molecules. The effect is somewhat like tension in a stretched string; surface tension is a pull on any section of the surface toward adjacent sections of the surface and is always directed tangent to the surface (Fig. 11-23). However, unlike tension in a string, surface tension is *not* produced by an external force "stretching the surface." Indeed, if the shape of the liquid is changed in such a way as to change its surface area, the magnitude of the surface tension force is unchanged.

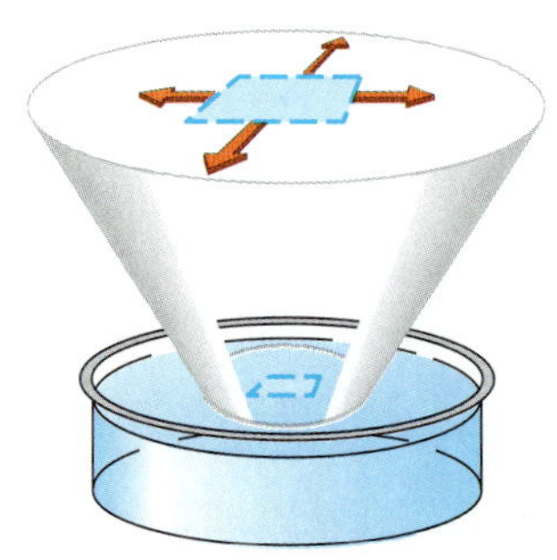

Fig. 11-23 The force of surface tension on a section of a liquid surface.

The nature of the surface tension force can be better understood when one sees how it arises from molecular forces. When the surface of a liquid is open to the atmosphere, some of the liquid's molecules escape the surface and form a gas just above the surface. (In this context a gas is often referred to as a vapor.) Molecules at the surface of the liquid are in a very dynamic state. Some of them are leaving the liquid to become part of the vapor, while equal numbers are going from the vapor to the liquid. The result is that, on the average, there are fewer molecules in the surface layer of the liquid than in an equal volume of liquid below the surface. The spacing between these surface molecules is somewhat greater than the molecular spacing within the liquid, and the attractive forces between the surface molecules tend to pull them together (Fig. 11-24). Within the liquid, the spacing of the molecules is such that the average intermolecular force is zero.* If a liquid's surface area is increased, molecules are simply drawn from the bulk liquid up to the surface, and both the average molecular spacing and the force of surface tension remain constant.

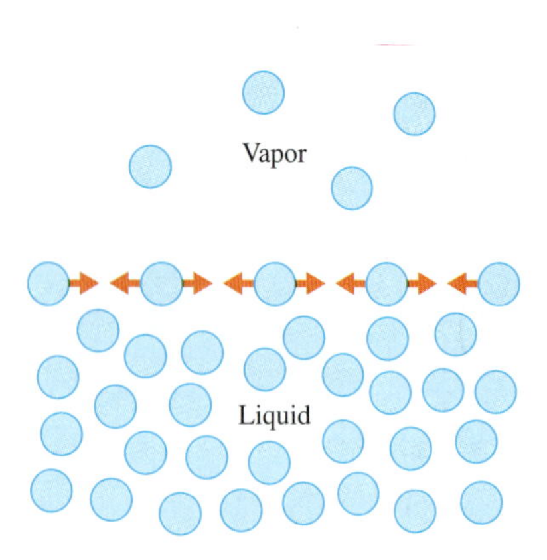

Fig. 11-24 The spacing of molecules in the surface boundary layer between a liquid and its vapor results in an attractive force between these molecules, tending to pull them together, producing the liquid's surface tension.

Surface tension does, however, depend on temperature. As temperature increases, the rate of exchange between liquid and vapor increases. The mean separation of molecules in the surface layer then increases, and the mean intermolecular force decreases; that is, the surface tension decreases.

*For simplicity, we ignore here the effects of pressure on the fluid, which would be absent if there were no gravitational field. Also, molecular motion within the liquid gives rise to frequent collisions and large *instantaneous* forces, but these forces are randomly directed and average to zero.

To define surface tension quantitatively, we consider an imaginary line of length ℓ separating adjacent sections of the surface of a liquid, as indicated in Fig. 11-25. The surface tension, denoted by the Greek letter γ (gamma), is defined to be the tension per unit of length along which it acts:

$$\gamma = \frac{T}{\ell} \tag{11-8}$$

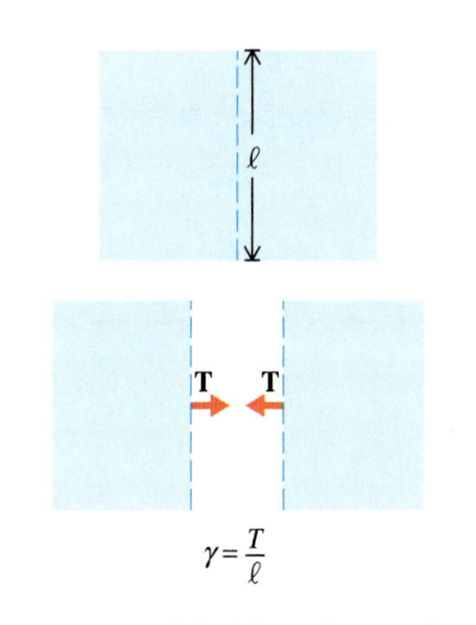

Fig. 11-25 A liquid's surface tension γ is defined to be the tension per unit length (T/ℓ) between adjacent sections of the surface.

Except for mercury, water has the highest surface tension of all common liquids: 0.073 N/m at 20° C, 0.059 N/m at 100° C. The surface tension of mercury at 20° C is 0.47 N/m.

Surface tension makes it possible for small objects having a density greater than the density of water to float on the surface of water (Fig. 11-26). When the object pushes the surface downward, there is a component of surface tension in the upward direction, supporting the weight, as illustrated in the following example.

Fig. 11-26 A paper clip supported on the surface of water by surface tension.

EXAMPLE 9 How Surface Tension Allows an Insect to Walk on Water

Calculate the weight supported by an insect's leg as the insect stands on water. The end of the leg has a spherical shape, with a radius of 0.50 mm. Suppose the water's surface makes an angle of 50° with the vertical at the highest point where the water makes contact with the leg (Fig. 11-27a).

SOLUTION Fig. 11-27b shows a free-body diagram of the bottom of the insect's leg and a thin layer of water in contact with it. The external forces are the weight **w** borne by the leg and the force of surface tension exerted on the layer of water by the surrounding water. From Eq. 11-8, it follows that the magnitude of the surface tension force is the product of the surface tension γ and the length of the line along which it acts. Since the tension acts along the circumference of a circle of radius r at an angle of 50° with the vertical, the total upward component of this force is given by $\gamma(2\pi r)(\cos 50°)$. Newton's second law predicts that this force must balance the weight:

$$\Sigma F_y = 0$$

$$\gamma(2\pi r)(\cos 50°) - w = 0$$

$$w = \gamma(2\pi r)(\cos 50°)$$

$$= (0.073 \text{ N/m})(2\pi)(5.0 \times 10^{-4} \text{ m})(\cos 50°)$$

$$= 1.5 \times 10^{-4} \text{ N}$$

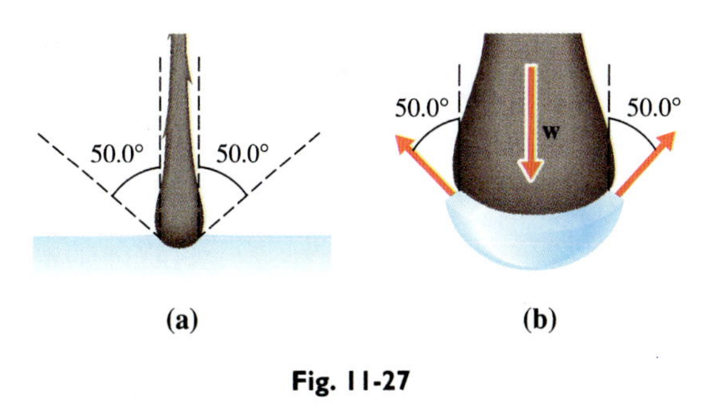

Fig. 11-27

If each of the insect's six legs bears an equal weight, the insect's weight is

$$6(1.5 \times 10^{-4} \text{ N}) = 9.0 \times 10^{-4} \text{ N}$$

The insect's mass is

$$m = \frac{w}{g} = \frac{9.0 \times 10^{-4} \text{ N}}{9.8 \text{ m/s}^2} = 9.2 \times 10^{-5} \text{ kg} = 0.092 \text{ g}$$

Large insects cannot be supported by surface tension because the ratio of weight to the length of the line along which surface tension acts increases rapidly with increasing size.

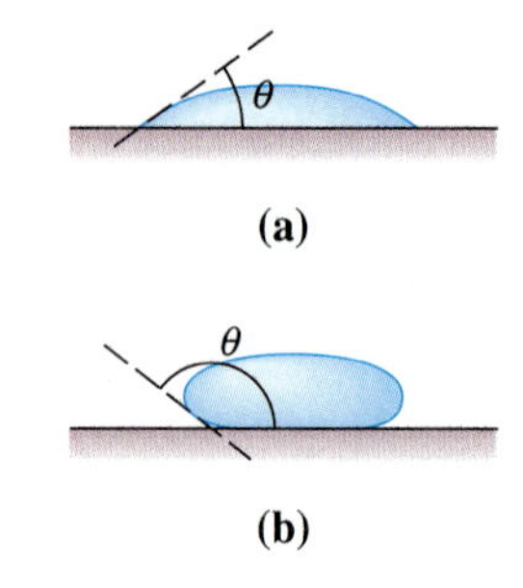

Fig. 11-28 **(a)** Adhesion of this liquid to the surface of this solid is stronger than the liquid's surface tension, and the liquid tends to spread over the surface. **(b)** Adhesion of this liquid to the surface of this solid is weaker than the liquid's surface tension, and the liquid tends to contract or bead.

Contact Angle

When a liquid is in contact with a solid, the liquid's surface molecules are attracted to the surface of the solid. If this force of adhesion to the solid is stronger than the surface-tension force, the liquid will tend to spread along the surface, or "wet the surface," as illustrated in Fig. 11-28a. If the adhesion to the solid is weaker than the force of surface tension, the liquid tends to contract, or bead, as illustrated in Fig. 11-28b. A quantitative measure of this phenomenon is the **contact angle** between the liquid-vapor surface and the liquid-solid surface. As seen in Fig. 11-28, wetting occurs when $\theta < 90°$ and beading occurs when $\theta > 90°$. For water on *clean* glass, $\theta \approx 0$, meaning that water will spread over a horizontal glass surface. On the other hand, water on wax has a contact angle $\theta > 90°$, meaning that water beads on a waxed surface.

The surface of water in contact with the vertical walls of a glass container will also tend to spread over the surface of the glass. The water's surface curves upward at the sides of the container (Fig. 11-29). In this case the strong adhesion of the water to the glass is opposed by the weight of the water drawn upward. The strength of the adhesive force is sufficient to support only the small volume of water pulled up around the sides of the container.

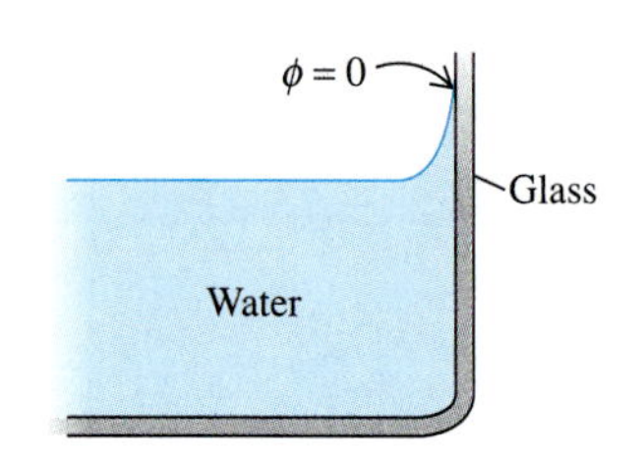

Fig. 11-29 The surface of water in a glass container curves upward at the sides, making an angle of 0° with the glass surface at the top.

Capillarity

If, however, a very thin glass tube, called a **capillary,** is inserted in a container of water, the water in the tube will rise to a considerable height above the level of the surrounding water. This phenomenon, called **capillarity,** may be explained by applying Newton's second law to a column of water whose radius is slightly smaller than the radius of the tube opening (Fig. 11-30). This free body is chosen so that the only contact forces are those exerted by the fluid surrounding the column; the adhesive forces do not act on this free body and therefore do not enter the calculation.

Notice that the pressure at both the top and the bottom of the column equals 1 atm, and so these forces balance. Likewise, the horizontal pressure forces along the sides of the column balance. According to Newton's second law, the resultant force equals zero. Thus the remaining vertical forces must cancel. The weight mg of the column of liquid is supported by the upward force of surface tension, which acts along a circle of radius r:

$$\Sigma F_y = 0$$

$$\gamma(2\pi r) - mg = 0$$

The water's mass m equals the product of its density ρ and its approximately cylindrical volume $V = \pi r^2 h$:

$$\gamma(2\pi r) - \rho(\pi r^2 h)g = 0$$

Solving for h, we find

$$h = \frac{2\gamma}{\rho g r} \tag{11-9}$$

Notice that h is inversely proportional to the tube's radius. Only for a very thin tube will the height of the column of water be large.

When the contact angle is greater than 90°, as for mercury on glass, the liquid in a capillary tube is depressed, rather than elevated (see Problem 36).

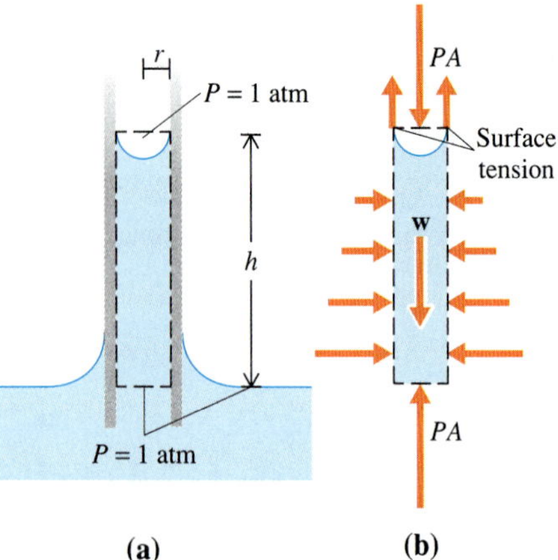

Fig. 11-30 **(a)** Water rises in a thin glass tube. **(b)** The free-body diagram of a cylinder of water.

EXAMPLE 10 Water in a Capillary

Calculate the height of water in a capillary of radius 0.10 mm.

SOLUTION Applying Eq. 11-9, using the values for the surface tension and density of water, we find

$$h = \frac{2\gamma}{\rho g r}$$

$$= \frac{2(0.073 \text{ N/m})}{(1.0 \times 10^3 \text{ kg/m}^3)(9.8 \text{ m/s}^2)(1.0 \times 10^{-4} \text{ m})}$$

$$= 0.15 \text{ m} = 15 \text{ cm}$$

The pressure within the column of water in a capillary increases with depth, as predicted by Eq. 11-6 ($P = P_0 + \rho gh$). Since the pressure at the base of the column is 1 atm, the pressure within the liquid near the top of the column equals 1 atm $- \rho gh$. The pressure just above the liquid's surface is 1 atm. This means that there is a pressure difference $\Delta P = \rho gh$ across the surface; the pressure in the liquid is less than the pressure in the air above the liquid, as illustrated in Fig. 11-31. We can use Eq. 11-9 for h to express the pressure difference ΔP in terms of the surface tension and the radius r of the capillary:

$$\Delta P = \rho gh$$

$$\Delta P = \frac{2\gamma}{r} \tag{11-10}$$

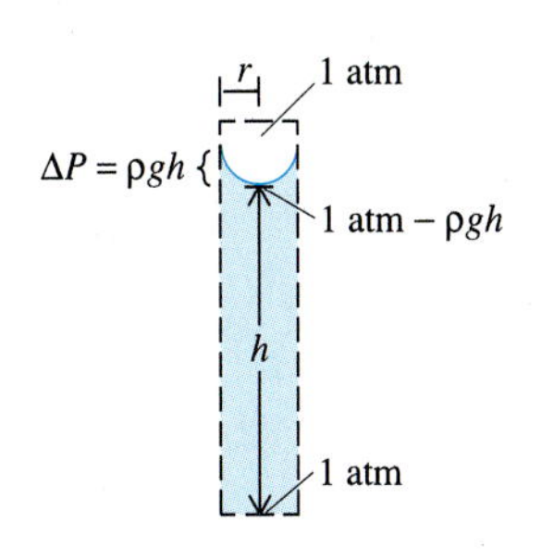

Fig. 11-31 There is a pressure difference ΔP across the hemispherical surface of water at the top of a capillary.

It is possible to prove that this equation describes the pressure drop from vapor to liquid across any spherical shaped liquid-vapor surface of radius r (see Problem 35). Indeed, the shape of the liquid's surface at the top of a capillary is hemispherical.

Negative Pressure

If r is small enough, Eq. 11-10 predicts that ΔP can exceed 1 atm. This would mean that the liquid pressure at the top of the column is less than zero. **Negative pressure** means that the bulk liquid is in a state of tension rather than compression. Adhesion to the walls of the capillary stretches the liquid upward. In water, negative pressures of -300 atm have been obtained.

Water and nutrients are transported upward to the tops of trees through xylem, a system of capillaries formed from dead cells. In tall trees the column of liquid is high enough to produce negative pressures within the xylem. Pressures of -20 atm have been observed near the tops of very tall trees. This is a capillary effect. However, there is not a direct liquid-vapor surface at the top of the xylem; therefore we cannot apply Eq. 11-9. Instead, there is a network of small openings leading from the sides of the xylem to the leaves (Fig. 11-32). It is the liquid-vapor surfaces in these very small openings that produce the large negative pressures predicted by Eq. 11-10.

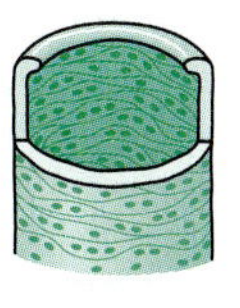

Fig. 11-32 Tiny holes in the sides of xylem contain liquid-vapor surfaces of sufficient area to support the liquid in the xylem.

11-5 Fluid Dynamics; Bernoulli's Equation

Steady Flow

"But today he saw only one of the river's secrets, one that gripped his soul. He saw that the water continually flowed and flowed and yet it was always there; it was always the same and yet every moment it was new." In this passage from the novel *Siddhartha,* Herman Hesse expresses a metaphor for the continuity of life. In so doing, he notes the fact that in a river, although everything appears the same from one mo-

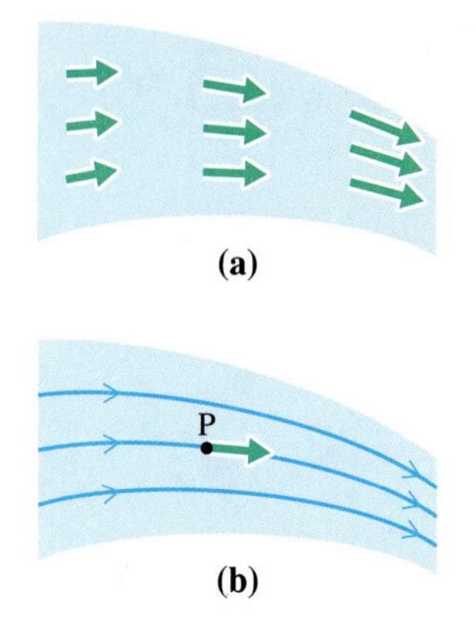

Fig. 11-33 The velocity of a fluid may be indicated by **(a)** drawing velocity vectors at selected points or **(b)** drawing flow lines whose tangent indicates the direction of the fluid's velocity at a point.

ment to the next at a given point, it is new water passing that point. We can either focus on a particle of water and follow it or focus on the river at a fixed point, where everything appears unchanging. In our study of fluid dynamics, we shall usually adopt the second point of view. We shall consider the velocity of a moving fluid at a fixed point in the fluid rather than follow a particular quantity of the fluid as it moves. If each particle of fluid has the same velocity when it passes this fixed point and if the fluid density remains constant at the point, we say there is a **steady flow.** We can then represent the velocity of the fluid by drawing vectors at selected points in the fluid (Fig. 11-33a). An alternative way of representing fluid velocity is to draw continuous flow lines (Fig. 11-33b). The tangent to a flow line at a given point indicates the direction of the fluid velocity at that point.

We shall now obtain an expression for the rate at which a volume of fluid moves through a surface perpendicular to the direction of flow. Consider the surface of area A shown in Fig. 11-34. All the fluid initially contained in the volume element ΔV will move a distance $v\,\Delta t$ during the time interval Δt. Thus ΔV is the volume of fluid passing through the surface in this time interval and may be expressed as the product of the length $v\,\Delta t$ and the area A:

$$\Delta V = (v\,\Delta t)A$$

The **flow rate** Q is the rate at which fluid flows through the surface—the volume per unit time passing through the surface:

$$Q = \frac{\Delta V}{\Delta t} \tag{11-11}$$

Substituting the expression for ΔV into this equation, we obtain

$$Q = vA \tag{11-12}$$

Flow rate is measured in m^3/s.

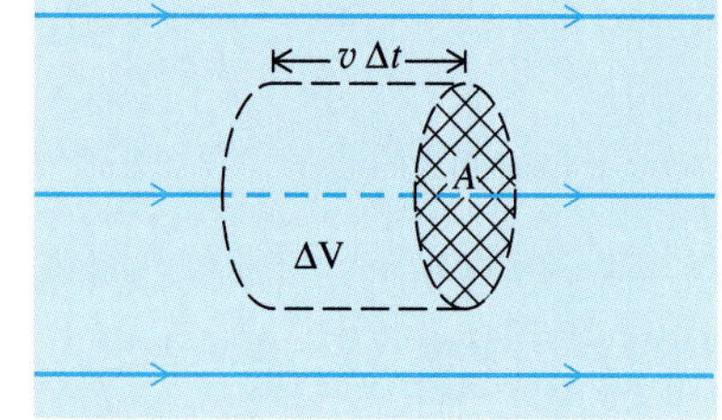

Fig. 11-34 The cylindrical volume of fluid ΔV will pass through the surface of area A during the time interval Δt.

EXAMPLE 11 Filling a Bathtub

A bathtub is being filled with water flowing at a speed of 1.00 m/s through a faucet with an opening having a cross-sectional area of 8.00 cm^2. How long will it take to fill the tub with 0.200 m^3 of water?

SOLUTION First we calculate the flow rate Q, using Eq. 11-12:

$$Q = vA = (1.00\text{ m/s})(8.00\text{ cm}^2)\left(\frac{10^{-2}\text{ m}}{1\text{ cm}}\right)^2$$
$$= 8.00 \times 10^{-4}\text{ m}^3/\text{s}$$

Since $Q = \Delta V/\Delta t$, it follows that the time Δt necessary to fill the tub is given by

$$\Delta t = \frac{\Delta V}{Q} = \frac{0.200\text{ m}^3}{8.00 \times 10^{-4}\text{ m}^3/\text{s}}$$
$$= 250\text{ s} \approx 4\text{ min}$$

Continuity Equation

For steady flow, the volume of fluid in a particular region is fixed. Fluid flowing out of the region is continuously replaced by an equal volume of fluid flowing into the region. This fact implies a simple relationship between the velocity at different points in the fluid. Consider a region of the fluid of volume V bounded on the sides by flow lines and at the ends by two surfaces perpendicular to the flow lines, with areas A_1 and A_2, as shown in Fig. 11-35. Such a region is called a "flow tube." Fluid enters this flow tube only through surface 1 and leaves only through surface 2. The condition that must be satisfied for steady flow is that the flow rates through the two surfaces be equal, so that there is as much fluid* entering the volume V as there is leaving it:

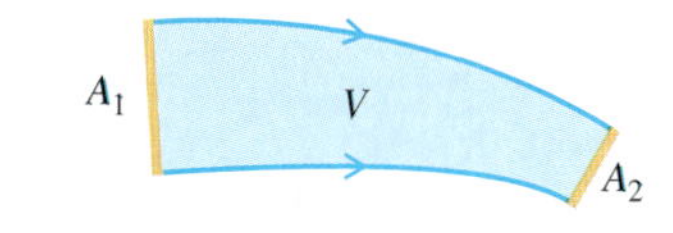

Fig. 11-35 A flow tube.

$$Q_1 = Q_2$$

Using Eq. 11-12 for the flow rate, we may express this result

$$v_1 A_1 = v_2 A_2 \tag{11-13}$$

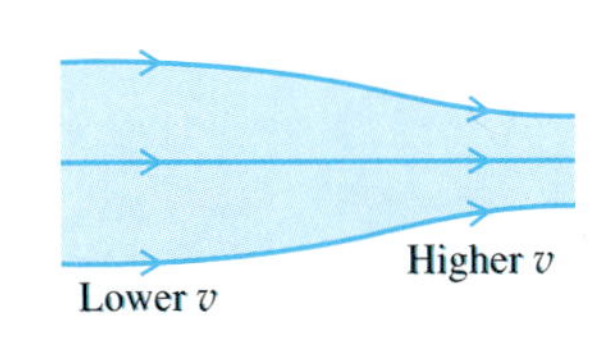

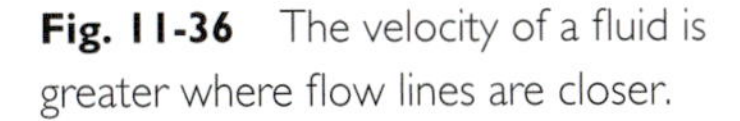

Fig. 11-36 The velocity of a fluid is greater where flow lines are closer.

This equation is called the **continuity equation.** According to it, when the cross-sectional area of a flow tube decreases, the speed of the fluid increases and vice versa. The cross-sectional area of a flow tube decreases as the flow lines come closer together. Thus the velocity of fluid is greater where the flow lines are closer together (Fig. 11-36). Consider, for example, the flow of water in a river or stream. The water moves most swiftly at points where the channel is narrow (Fig. 11-37a). Another example is water flowing from a faucet. The velocity of the water increases as it falls, and so the cross-sectional area of the flow tube decreases (Fig. 11-37b).

(a)

(b)

Fig. 11-37 **(a)** A stream flows more swiftly where the channel narrows. **(b)** This stream of water narrows as it flows downward from the faucet and the water's velocity increases.

*We assume here that the fluid is incompressible, and so its density is the same at both points.

EXAMPLE 12 Branching of Blood Vessels

An artery of radius 0.500 cm carries blood at a flow rate $Q = 10.0\ \text{cm}^3/\text{s}$. This artery branches into smaller and smaller blood vessels until finally the blood travels in capillaries of radius 5.00×10^{-4} cm at a speed of 0.100 cm/s. What is the flow rate in a single capillary? How many capillaries does the artery branch into?

SOLUTION The flow rate Q_1 in a single capillary is the product of the blood's speed v_1 and the capillary's cross-sectional area A_1 (Eq. 11-12):

$$Q_1 = v_1 A_1 = v_1 \pi r^2 = (0.100\ \text{cm/s})\pi(5.00 \times 10^{-4}\ \text{cm})^2$$
$$= 7.85 \times 10^{-8}\ \text{cm}^3/\text{s}$$

The flow rate Q through the artery must equal the combined flow rates of all the capillaries into which the artery branches. Thus Q equals the number n of such capillaries times the flow rate Q_1 in each:

$$Q = nQ_1$$

$$n = \frac{Q}{Q_1} = \frac{10.0\ \text{cm}^3/\text{s}}{7.85 \times 10^{-8}\ \text{cm}^3/\text{s}}$$
$$= 1.27 \times 10^8$$

This one artery branches into 127 million capillaries!

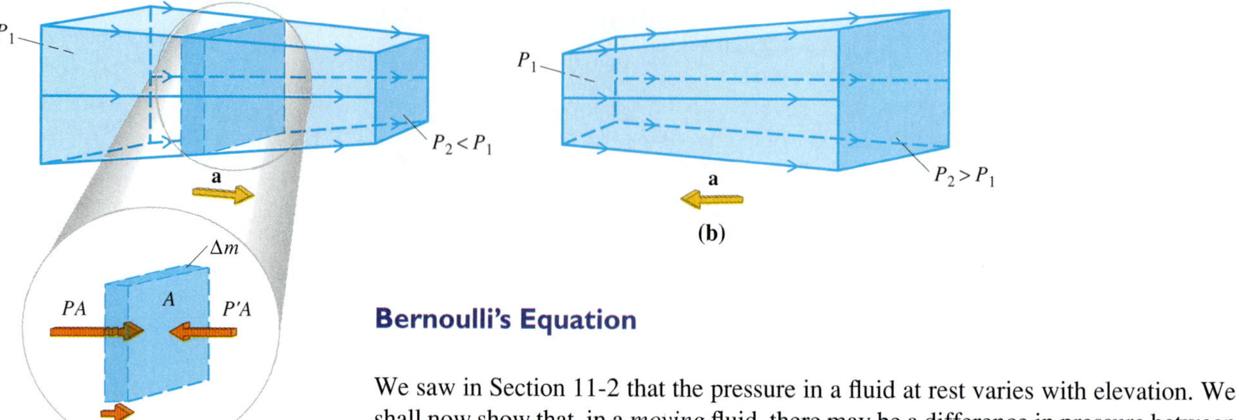

Fig. 11-38 **(a)** The fluid in a flow tube with converging flow lines is accelerated; therefore the pressure on the right is less than the pressure on the left.. **(b)** Fluid decelerates as it moves from left to right. Therefore the pressure on the right is greater than the pressure on the left.

Bernoulli's Equation

We saw in Section 11-2 that the pressure in a fluid at rest varies with elevation. We shall now show that, in a *moving* fluid, there may be a difference in pressure between points at the same elevation. Consider a flow tube in which the flow lines are converging (Fig. 11-38a). As the fluid moves from left to right, its velocity increases. Thus the fluid is accelerated to the right, and, according to Newton's second law, there must be a net force to the right acting on each element of the fluid. This force is provided by the pressure of the surrounding fluid. Thus for any small section of fluid in this region, the pressure P on the left side is greater than the pressure P' on the right side, as indicated in Fig. 11-38a. If the pressure on both sides were the same, there could be no resultant force, and one is obviously needed to provide the observed acceleration. Thus there is a gradual decrease in the fluid's pressure as it moves from left to right. The pressure in a fluid decreases as its velocity increases (Fig. 11-38a, $P_2 < P_1$). Conversely, the pressure increases as the velocity decreases (Fig. 11-38b, $P_2 > P_1$).

We can obtain a general expression relating the pressures at two points in a fluid by analyzing the change in the total mechanical energy of the quantity of fluid initially contained within the boundaries of a flow tube, shown in Fig. 11-39a. The total mechanical energy of the fluid is not conserved because the fluid is subject to the forces exerted by fluid adjacent to the system at surfaces S_1 and S_2, and these forces are nonconservative. According to Eq. 7-43, the net work done by the nonconservative forces equals the change in the system's mechanical energy:

$$\Sigma W_{nc} = \Delta E = E_f - E_i \tag{11-14}$$

We apply this equation to the body of fluid initially in the flow tube to obtain a relationship between the pressure P, velocity v, and elevation y at S_1 and S_2.

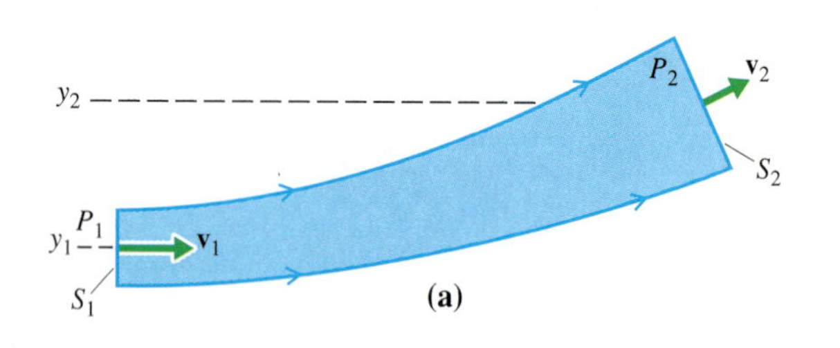

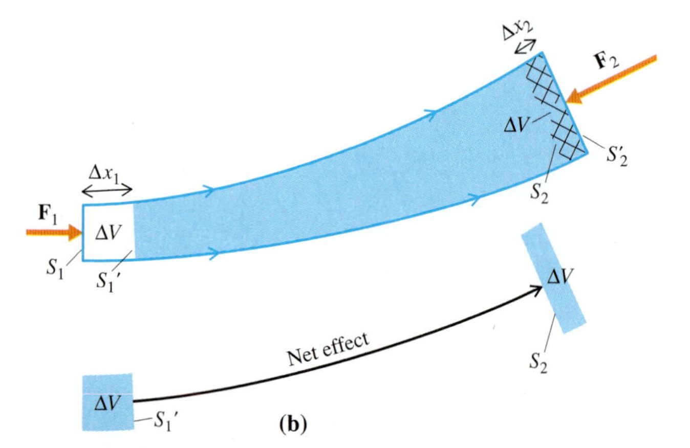

Fig. 11-39 **(a)** Fluid is initially between surfaces S_1 and S_2. **(b)** Fluid shown in (a) has moved so that it is now between S_1' and S_2'. The net effect of the flow is to transfer a volume of fluid ΔV from one end of the flow tube to the other.

Consider the same body of fluid after a short time interval, during which the fluid at the first surface moves a distance Δx_1 and the fluid at the second surface moves a distance Δx_2, so that the boundary surfaces of the fluid are changed to S_1' and S_2', as shown in Fig. 11-39b. Positive work W_1 is done on the system by the force $\mathbf{F}_1$ exerted by fluid to the left of the first surface:

$$W_1 = F_{1x}\,\Delta x_1 = P_1A_1\,\Delta x_1 = P_1\,\Delta V$$

Negative work W_2 is done on the system by the force $\mathbf{F}_2$ exerted by fluid to the right of the second surface:

$$W_2 = F_{2x}\,\Delta x_2 = -P_2A_2\,\Delta x_2 = -P_2\,\Delta V$$

The net nonconservative work is the sum of W_1 and W_2:

$$\Sigma\, W_{\text{nc}} = (P_1 - P_2)\Delta V \qquad (11\text{-}15)$$

Next we find an expression for the change in the total mechanical energy of the system. In examining Fig. 11-39b, we see that the quantity of fluid between surfaces S_1' and S_2 is unchanged, and the net effect of the flow is to shift a volume of fluid ΔV from one end of the flow tube to the other. This volume has mass

$$m = \rho\,\Delta V \qquad (11\text{-}16)$$

and mechanical energy equal to the sum of its kinetic energy $K = \frac{1}{2}mv^2$ and its gravitational potential energy $U = mgy$. The change in the mechanical energy of the system equals the change in energy of this mass:

$$\begin{aligned}\Delta E &= E_f - E_i = K_f + U_f - (K_i + U_i)\\ &= \tfrac{1}{2}mv_2^2 + mgy_2 - (\tfrac{1}{2}mv_1^2 + mgy_1)\end{aligned}$$

Inserting the expression for mass (Eq. 11-16) and rearranging terms, we obtain

$$\Delta E = \tfrac{1}{2}(\rho\,\Delta V)(v_2^2 - v_1^2) + (\rho\,\Delta V)g(y_2 - y_1)$$

Finally we equate the expression for the nonconservative work (Eq. 11-15) to the expression above for the change in mechanical energy and obtain

$$P_1 - P_2 = \tfrac{1}{2}\rho(v_2^2 - v_1^2) + \rho g(y_2 - y_1) \qquad (11\text{-}17)$$

This equation, first derived by Daniel Bernoulli in 1738, is known as **Bernoulli's equation.** Notice that in the special case of static fluid ($v_1 = v_2 = 0$) Bernoulli's equation reduces to the result we found before for a fluid at rest (Eq. 11-6):

$$P_1 - P_2 = \rho g(y_2 - y_1) = \rho gh$$

An alternative form of Bernoulli's equation is obtained if we bring all the variables with subscript 1 to the left side of the equation and all those with subscript 2 to the right side:

$$P_1 + \tfrac{1}{2}\rho v_1^2 + \rho g y_1 = P_2 + \tfrac{1}{2}\rho v_2^2 + \rho g y_2 \qquad (11\text{-}18)$$

The sum of the pressure P, the kinetic energy term $\frac{1}{2}\rho v^2$, and the potential energy term ρgy has the same value at any two points in the fluid connected by a flow tube. For points in a flow tube for which the potential energy terms are approximately equal, Bernoulli's equation predicts that the **pressure increases as the speed decreases.** This prediction is called **Bernoulli's principle** and explains many commonly observed phenomena, such as those shown in Fig. 11-40.

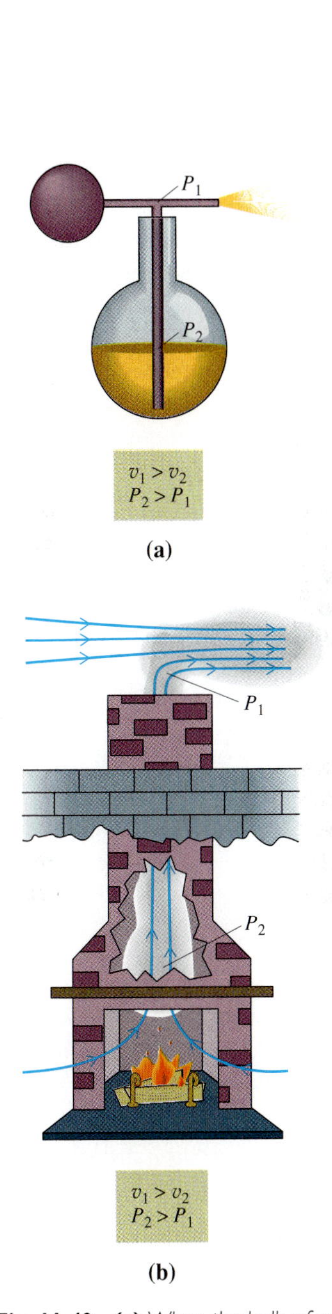

Fig. 11-40 **(a)** When the bulb of an aspirator is squeezed, air rushes out through the tube, reducing the pressure there. This causes liquid to flow up through the stem and out the tube. **(b)** A chimney works best when it is tall and exposed to air currents, which reduce the pressure at the top and enhance the upward flow of smoke.

EXAMPLE 13 Water Pressure Required for a Sprinkler

Water flows through a garden hose and out through a sprinkler. The hose has an inner radius of 1.00 cm, and the sprinkler head has 20 holes, each with a radius of 0.500 mm. The water moves through the hose at a speed of 0.500 m/s. (a) Find the speed at which the water leaves the sprinkler head. (b) Calculate the pressure in the hose required to produce this flow.

SOLUTION (a) The water in the hose can be divided into 20 flow tubes, each of which narrows to enter a hole in the sprinkler head. One such flow tube is shown in Fig. 11-41. We apply the continuity equation and relate the speed v_2 of water in the hole to its speed v_1 in the hose. The cross-sectional area of a single flow tube is $\frac{1}{20}$ of the cross-sectional area of the hose.

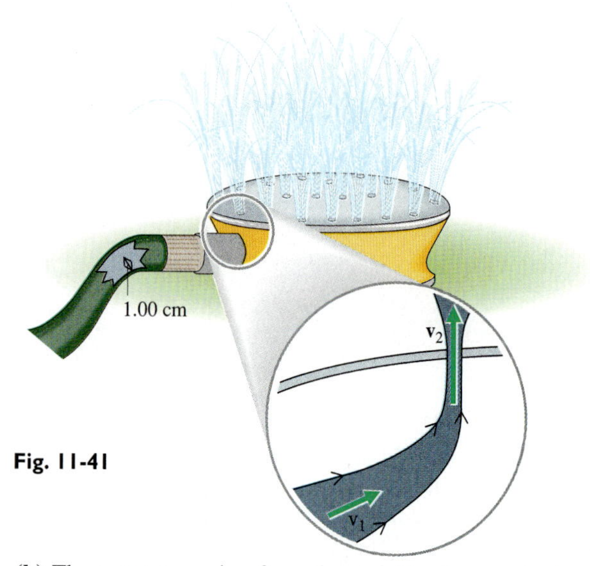

Fig. 11-41

$$v_2 A_2 = v_1 A_1$$

$$v_2 = v_1 \frac{A_1}{A_2} = (0.500 \text{ m/s})\frac{\pi(1.00 \text{ cm})^2/20}{\pi(0.0500 \text{ cm})^2} = 10.0 \text{ m/s}$$

The stream of water that leaves the hole behaves much like a simple projectile. We can calculate the horizontal range of this stream using Eq. 3-10 for projectile motion:

$$R = v_0^2 \sin(2\theta_0)/g$$

Supposing that the stream is directed at an angle of 45.0° above the horizontal, we find

$$R = \frac{(10.0 \text{ m/s})^2(\sin 90.0°)}{9.80 \text{ m/s}^2} = 10.2 \text{ m}$$

(b) The water emerging from the sprinkler is open to the atmosphere, and so it is at atmospheric pressure ($P_2 = 1$ atm). The hose and sprinkler are assumed to be at the same elevation ($y_1 = y_2$). We apply Bernoulli's equation (Eq. 11-17) to find the pressure P_1 in the hose:

$$\begin{aligned} P_1 - P_2 &= \tfrac{1}{2}\rho(v_2^2 - v_1^2) + \rho g(y_2 - y_1) \\ &= \tfrac{1}{2}(1.00 \times 10^3 \text{ kg/m}^3)[(10.0 \text{ m/s})^2 - (0.500 \text{ m/s})^2] + 0 \\ &= 4.99 \times 10^4 \text{ Pa}\left(\frac{1 \text{ atm}}{1.01 \times 10^5 \text{ Pa}}\right) \\ &= 0.49 \text{ atm} \end{aligned}$$

or

$$P_1 = P_2 + 0.49 \text{ atm} = 1.49 \text{ atm}$$

EXAMPLE 14 Flow Rate of Water in a Cooler

A container of water is open to the atmosphere. Water flows out of the container through a small opening at its base (Fig. 11-42).

(a) Find the speed v_2 of the water at point 2, where it passes out through the opening, as a function of the height h of the water's upper surface above the opening. Assume that the cross-sectional area of the container is much greater than the cross-sectional area of the opening. (b) Calculate v_2 and the flow rate if h = 20.0 cm and the cross-sectional area of the opening is 0.500 cm². (c) At what rate does the upper surface of the water fall if the cross-sectional area of the container is 400 cm²?

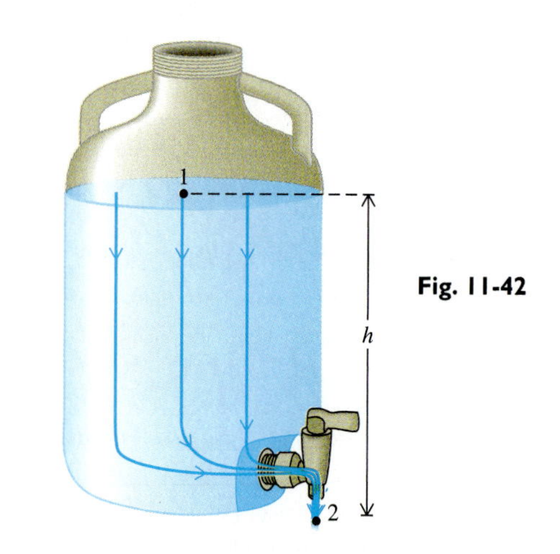

Fig. 11-42

EXAMPLE 14—cont'd

SOLUTION (a) We apply Bernoulli's equation at the two points. Both points are open to the atmosphere, and so the pressure at both equals atmospheric pressure: $P_1 = P_2 =$ 1 atm. Using this result in Bernoulli's equation and solving for v_2^2, we obtain

$$v_2^2 = v_1^2 + 2g(y_1 - y_2)$$

From the continuity equation we know that $v_2 = v_1A_1/A_2$. But $A_1 >> A_2$. Therefore $v_2 >> v_1$, and we introduce no significant error in setting v_1 equal to zero in the equation above:

$$v_2^2 \approx 2g(y_1 - y_2) = 2gh$$

$$v_2 = \sqrt{2gh}$$

Notice that this speed is the same as the speed of an object that has fallen from rest through a distance h. This result is known as "Torricelli's law." In deriving it, we have assumed nothing about the direction of the fluid velocity $\mathbf{v}_2$. Thus Torricelli's law applies to a stream of fluid leaving the container in any direction.

(b) If h = 20.0 cm, we find

$$v_2 = \sqrt{2gh} = \sqrt{2(9.80 \text{ m/s}^2)(0.200 \text{ m})}$$

$$= 1.98 \text{ m/s} = 198 \text{ cm/s}$$

The flow rate Q is the product of the speed and the cross-sectional area of the opening:

$$Q = v_2A_2 = (198 \text{ cm/s})(0.500 \text{ cm}^2)$$

$$= 99.0 \text{ cm}^3\text{/s}$$

(c) The rate at which the upper surface of the fluid falls is just the speed v_1 of the fluid at point 1. And since, according to the continuity equation, the flow rate is the same here, we find

$$Q = v_1A_1$$

or

$$v_1 = \frac{Q}{A_1} = \frac{99.0 \text{ cm}^3\text{/s}}{400 \text{ cm}^2}$$

$$= 0.248 \text{ cm/s}$$

Thus $v_1 << v_2$, as assumed in part a.

*11-6 Viscosity

In the last section we considered only applications of fluid flow for which the shear stress in the moving fluid could be ignored. In this section we shall see the effects of such stress.

Consider a cylindrical container, inside which is mounted a second, smaller cylinder that is free to rotate without friction about its axis. Let the volume between the two cylinders contain water (Fig. 11-43). If a horizontal force **F** is applied tangent to the surface of the inner cylinder, the cylinder will begin to rotate, no matter how small the force. Of course, the larger the force, the faster the inner cylinder will rotate. The water undergoes a shear stress that is velocity dependent. When the velocity of the inner cylinder approaches zero, so does the shear stress.

If the water were replaced by another fluid, the force required for a given rate of rotation would change. For example, a much larger force would be required if the fluid were honey, whereas only a very small force would be required if the fluid were air. The property of a fluid that is observed in such an experiment is called "viscosity," and the device described above is called a "viscosimeter." Honey is a highly viscous fluid, whereas water has a much lower viscosity, and the viscosity of air is much lower still.

Viscosity is strongly dependent on temperature. The viscosity of a fluid increases as its temperature is reduced. Thus a cold viscous fluid ("molasses in January") flows very slowly.

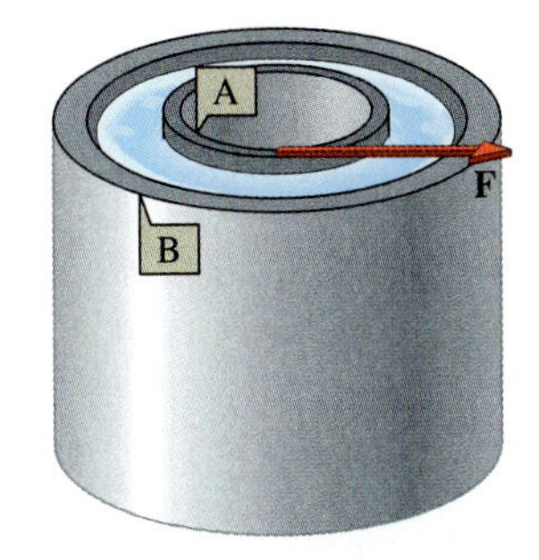

Fig. 11-43 Water fills the space between a stationary outer cylinder and an inner cylinder that is free to rotate. A force **F** applied to the inner cylinder causes it to rotate.

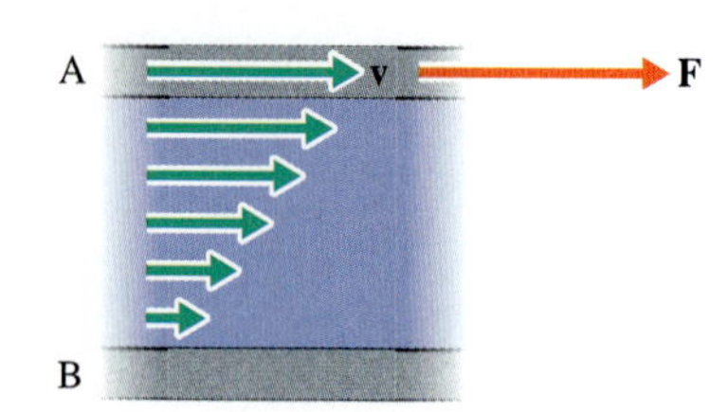

Fig. 11-44 Fluid between two solid surfaces in relative motion. The fluid's velocity gradually changes from one surface to another. The fluid in contact with either solid surface does not move relative to that surface.

Laminar and Turbulent Flow

When a fluid flows past a stationary solid surface, the fluid adjacent to the surface adheres to it and so is at rest. The velocity of the fluid gradually increases as the distance y from the surface increases (Fig. 11-44). (This figure could represent, for example, a top view of a small rectangular section of water in the viscosimeter in Fig. 11-43). One can think of the fluid as being divided into layers that slide over each other. Elements of the fluid in adjacent layers do not mix with each other, and the flow lines form a regular pattern. This kind of flow is called **laminar flow.**

There is another kind of flow, known as **turbulent flow,** in which the motion is not so orderly. For turbulent flow a random mixing of adjacent layers occurs, and the flow lines cross. The existence of turbulence does not necessarily imply that the flow is not steady. Even for turbulent flow there still may be a steady average flow velocity at each point in the fluid.

Viscosity

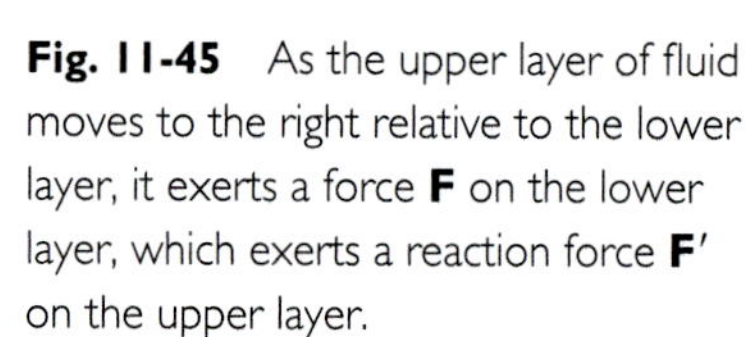

Fig. 11-45 As the upper layer of fluid moves to the right relative to the lower layer, it exerts a force **F** on the lower layer, which exerts a reaction force **F′** on the upper layer.

A shear stress must exist within a fluid in laminar flow. This is illustrated in Fig. 11-45, where two adjacent layers of fluid are shown. There is kinetic friction between the two layers. The upper layer exerts a force **F** on the lower layer as it slides over it, pulling the lower layer to the right. The reaction force **F′** is exerted on the upper layer by the lower layer. The force per unit surface area, F/A, is the shear stress. This stress is found to be proportional to the ratio $\Delta v/\Delta y$:

$$\frac{F}{A} \propto \frac{\Delta v}{\Delta y}$$

The constant of proportionality is called the **viscosity** and is denoted by the Greek letter η (eta):

$$\frac{F}{A} = \eta \frac{\Delta v}{\Delta y}$$

or

$$F = \eta A \frac{\Delta v}{\Delta y} \tag{11-19}$$

This equation implies that $\eta = \dfrac{F}{A\,\Delta v/\Delta y}$, and so the SI unit of viscosity is

$$\frac{\text{N}}{(\text{m}^2)\left(\dfrac{\text{m/s}}{\text{m}}\right)} = \text{N-s/m}^2 = \text{Pa-s}$$

Other commonly used units are poise (P) and centipoise (cP), defined as

$$1\text{P} = 0.1 \text{ Pa-s}$$

$$1 \text{ cP} = 10^{-2} \text{ P} = 10^{-3} \text{ Pa-s}$$

Some representative values of η are given in Table 11-2.

Table 11-2 Viscosities

Fluid	Temperature (°C)	Viscosity (cP = 10^{-3} Pa-s)
Air	20	1.8×10^{-2}
	40	1.9×10^{-2}
	100	2.2×10^{-2}
Water	0	1.79
	20	1.00
	40	0.66
	100	0.28
Blood, whole	37 (body temp.)	4.0
Blood, plasma	37	1.5
Glucose	22	9.1×10^{15}
	100	2.5×10^{4}
Glass	575	1.1×10^{15}

Reynolds Number

Laminar and turbulent flow of water through tubes was studied by Osborne Reynolds toward the end of the nineteenth century. He used a thin capillary tube to insert a fine stream of dye along a flow line of water flowing through a glass tube. The dye made the flow line visible. When the flow was relatively slow, the dye was confined to a thin stream, parallel to the axis of the tube. In this case the flow was obviously laminar. For greater flow velocities, however, the flow was turbulent; the dye was quickly dispersed throughout the water, indicating the crossing of flow lines and mixing of fluid levels (Fig. 11-46).

Reynolds was able to show that the flow of fluid in a tube is laminar or turbulent depending on the values of four physical parameters: the diameter d of the tube, the average flow speed $\bar{v}$, the fluid's density ρ, and its viscosity η. The condition that is satisfied by these variables when turbulence begins is surprisingly simple. Reynolds discovered that the onset of turbulence could in all cases be determined by the value of a single dimensionless function R:

$$R = \frac{\rho \bar{v} d}{\eta} \tag{11-20}$$

The value of this function is called the "Reynolds number." When $R < 2000$, flow is laminar, and when $R > 3000$, flow is turbulent. For values between 2000 and 3000, the flow is unstable and may be either laminar or turbulent. Only if the flow is laminar will Bernoulli's equation be valid.

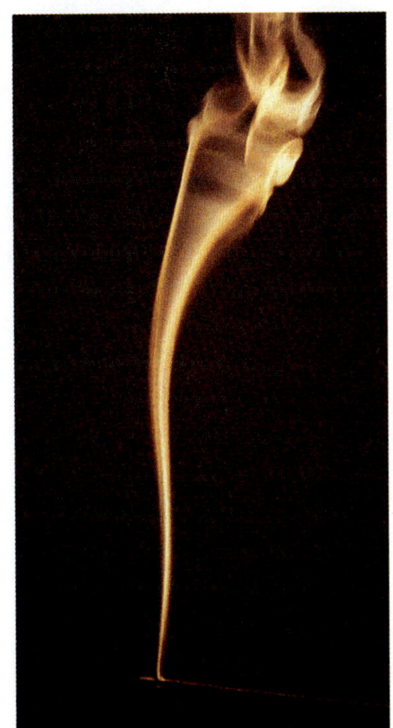

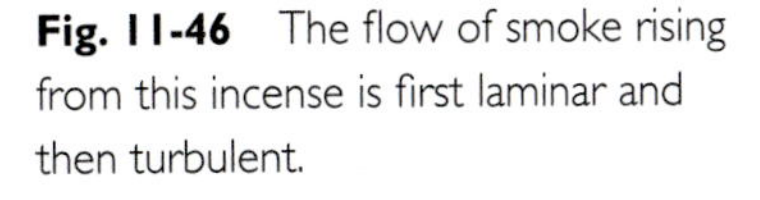

Fig. 11-46 The flow of smoke rising from this incense is first laminar and then turbulent.

EXAMPLE 15 Laminar Flow of Water and Blood

Find the maximum value of the product $\bar{v}d$ to be certain of laminar flow, for water at 20° C and for blood at 37° C.

SOLUTION In order to be certain that the flow is laminar, the Reynolds number must be less than 2000:

$$R = \frac{\rho \bar{v} d}{\eta} < 2000$$

Therefore the condition on $\bar{v}d$ is that

$$\bar{v}d < \frac{2000\eta}{\rho}$$

Using the values given in Tables 11-1 and 11-2 for ρ and η of water and blood, we obtain the following results. For water at 20° C,

$$\bar{v}d < \frac{2000(1.00 \times 10^{-3}\ \text{Pa-s})}{1.00 \times 10^{3}\ \text{kg/m}^3}$$

or

$$\bar{v}d < 2.00 \times 10^{-3}\ \text{m}^2/\text{s}$$

Thus, if $d = 1.00$ cm, the water can have at most a speed of 0.200 m/s in order for one to be certain that the flow will be laminar. But if the diameter is reduced to 1.00 mm, the flow will be laminar at speeds up to 2.00 m/s. The flow of water in household water pipes is normally turbulent.

For blood,

$$\bar{v}d < \frac{2000(4.0 \times 10^{-3}\ \text{Pa-s})}{1.05 \times 10^{3}\ \text{kg/m}^3} = 7.6 \times 10^{-3}\ \text{m}^2/\text{s}$$

If $d = 1.0$ cm, the flow of blood will be laminar for flow speeds less than 0.76 m/s. And if $d = 1.0$ mm, the speed may be as great as 7.6 m/s and the flow will still be laminar. Blood flow in the circulatory system is generally laminar. In the smaller blood vessels, the flow speed is quite low, and so the product $\bar{v}d$ is largest in the veins and arteries. Only in the aorta, which has a diameter of approximately 2.5 cm, is the flow occasionally turbulent. Normally the average flow speed in the aorta is about 20 cm/s, and so $\bar{v}d \approx 5.0 \times 10^{-3}$ m²/s, and the flow is laminar. However, during vigorous activity $\bar{v}$ may be as great as 1.0 m/s, so that $\bar{v}d = 2.5 \times 10^{-2}$ m²/s, and the flow is definitely turbulent. When this occurs, the heart must do more work to maintain blood flow.

*11-7 Poiseuille's Law

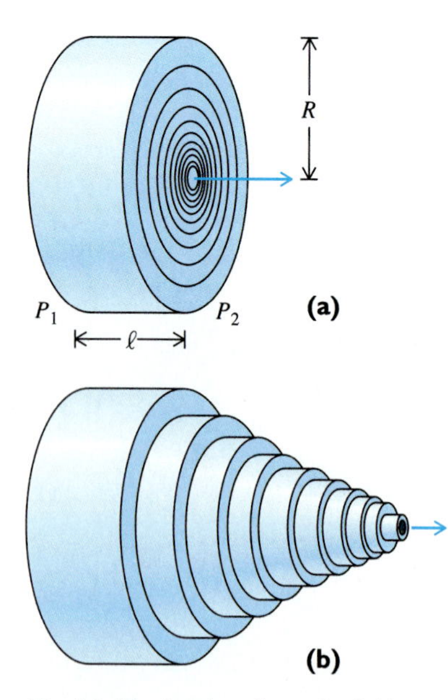

Fig. 11-47 Laminar flow of a fluid through a tube or pipe. **(a)** Layers of fluid initially in a cylindrical section of the tube. **(b)** A short time later the faster-moving inner layers have moved farther than the outer layers.

Consider the laminar flow of a viscous fluid through a tube or pipe. Fig. 11-47 shows cylindrical layers of fluid sliding past each other as the fluid flows to the right through a cylindrical tube of length ℓ and radius R.

Each layer of fluid is acted on by viscous forces that oppose its motion. Thus the fluid must be pushed from left to right. This push is provided by the fluid pressure, which must be greater on the left side than on the right ($P_1 > P_2$). The fluid pressure must gradually decrease as one moves downstream. This is an important aspect of blood flow in the smaller blood vessels of the circulatory system. This result is not predicted by Bernoulli's equation. However, as we shall see, if the fluid's viscosity and velocity are not too large and if the tube's radius is not too small, the variation in pressure caused by viscosity will be negligible, and Bernoulli's equation will be valid. For example, in the human circulatory system, blood flow in the larger veins and arteries is not greatly affected by viscous forces; in the smaller blood vessels, however, viscosity causes a drop in pressure.

It is possible to apply Newton's second law to an element of the fluid and derive an expression relating the flow rate Q to the tube's length ℓ and radius R, to the pressure difference $P_1 - P_2$, and to the fluid's viscosity η. This derivation gives the result

$$Q = \frac{\pi(P_1 - P_2)}{8\eta\ell} R^4 \tag{11-21}$$

which is known as **Poiseuille's law,** after the French physiologist Jean-Louie-Marie Poiseuille (pronounced /pwä-zœ́y/), who discovered it in 1840.

Poiseuille's Law and the Circulatory System

Poiseuille's law has important applications to blood flow in the human circulatory system. The fact that Q is proportional to R^4 means that a small change in the size of a blood vessel will produce a large change in the flow rate. For example, if R increases by a factor of 2, Q increases by a factor of $2^4 = 16$. Blood flow is controlled by changes in the size of the arterioles in various parts of the body. The arterioles are tiny blood vessels (inner radius $\approx$ 0.01 mm), through which blood flows after leaving the arteries—the major blood vessels carrying blood from the heart. Contraction of muscle fibers in the walls of the arterioles in one part of the body reduces the blood flow through them and thereby increases blood flow through arterioles in other parts of the body. In this way, the body can adjust the flow of blood in response to changes in conditions. For example, when the body is cold, blood flow near the body's surface is reduced so as to reduce the heat lost through the skin. This is accomplished by stimulation of nerve impulses, which produces contraction of arterioles supplying blood near the skin.

The dependence of the flow rate on R has very serious negative implications for a person whose arteries are obstructed by the buildup of plaque on the arterial walls. The decrease in the effective inner radius of the blood vessel means that either the flow rate Q must decrease or the pressure gradient $(P_1 - P_2)/\ell$ must increase. This means that the heart must do more work to maintain the flow. The heart is a pump that takes in the blood from the veins at approximately zero gauge pressure and, after oxygenating the blood in the lungs, expels it through the aorta and into the major arteries. The normal average gauge pressure in the large arteries is 100 torr, ranging between a maximum of 120 torr (systolic pressure) and a minimum of 80 torr (diastolic pressure) during each heartbeat.

The overall average pressure drop of 100 torr is necessary to overcome forces of fluid friction and pump the blood through the complex network of blood vessels that permeates the human body (Fig. 11-48). The heart must do work on the blood, exerting a force on it as it is expelled at the higher pressure. When the small arteries are partially blocked, a larger pressure difference across the arteries is required to maintain the same flow rate, and the average blood pressure in the large arteries will then be greater than 100 torr. Thus the heart must do more work during each heartbeat to maintain this higher blood pressure.

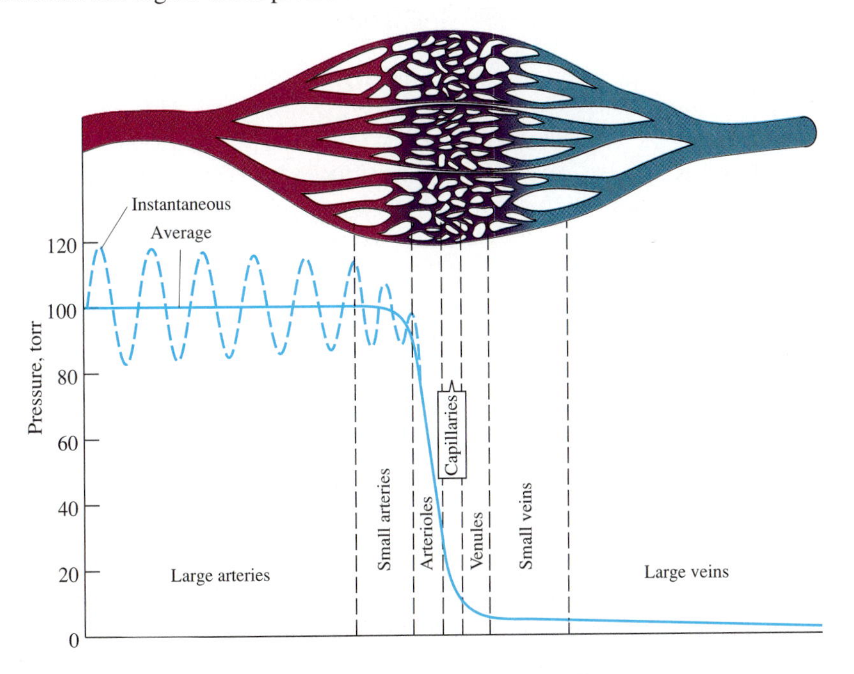

Fig. 11-48 Normal values of instantaneous and average pressure for various blood vessels in the circulatory system. Although the arterioles, capillaries, and venules are very short, most of the pressure drop is across them, because of their very small diameters.

Poiseuille or Bernoulli

Eq. 11-21 [$Q = \pi(P_1 - P_2)R^4/(8\eta\ell)$] can be solved for $P_1 - P_2$ as a function of the other variables, so that Poiseuille's law takes the form

$$P_1 - P_2 = \frac{8\eta\ell Q}{\pi R^4} \tag{11-22}$$

Written in this way, Poiseuille's law may be easily compared with Bernoulli's equation, Eq. 11-17:

$$P_1 - P_2 = \tfrac{1}{2}\rho(v_2^2 - v_1^2) + \rho g(y_2 - y_1)$$

These are, of course, two very different expressions for the same quantity—the pressure difference between two points in a moving fluid. This can be confusing. However, it should not be too difficult to decide which equation to use if the assumptions made in the derivations are kept in mind. In deriving Poiseuille's law, we assumed that both the elevation and the cross-sectional area of the pipe were constant, and so it may be applied only when these conditions are satisfied. In deriving Bernoulli's equation, we assumed that the fluid had negligible viscosity, or, equivalently, that the effects of viscosity on $P_1 - P_2$ were negligible. We shall see in the following examples that both Bernoulli's equation and Poiseuille's equation are useful in understanding various aspects of blood pressure in the human circulatory system.

EXAMPLE 16 Blood Pressure in the Major Arteries

(a) Compute the pressure drop per centimeter for a typical major artery of radius 0.50 cm, carrying blood at a flow rate of 1.0×10^{-5} m^3/s (10 cm^3/s), as predicted by Poiseuille's law.
(b) Show that the decrease of pressure with increasing fluid velocity predicted by Bernoulli's equation is normally not a significant effect in the circulatory system. The maximum flow speed occurs along the center of the aorta and is normally about 0.40 m/s.
(c) Compute the average blood pressure in the major arteries of the head and legs of a standing person, at points that are respectively 0.30 m above and 1.20 m below the heart. Assume that the average blood pressure at the level of the heart is 100 torr.

SOLUTION **(a)** For a 1.0 cm length of a typical major artery, Poiseuille's law predicts a pressure drop

$$P_1 - P_2 = \frac{8\eta\ell Q}{\pi R^4}$$

$$= \frac{8(4.0 \times 10^{-3}\ \text{Pa-s})(1.0 \times 10^{-2}\ \text{m})(1.0 \times 10^{-5}\ \text{m}^3/\text{s})}{\pi(5.0 \times 10^{-3}\ \text{m})^4}$$

$$= 1.6\ \text{Pa}\left(\frac{1\ \text{torr}}{133\ \text{Pa}}\right) = 0.012\ \text{torr}$$

Thus, over a distance of 1 m, viscous effects should produce a pressure drop of only about 1 torr. This is almost negligible compared to typical arterial pressures of about 100 torr. Since viscous effects are so small, it is valid to apply Bernoulli's equation in the major arteries.

(b) For the pressure difference between two points connected by major arteries, Bernoulli's equation predicts

$$P_1 - P_2 = \tfrac{1}{2}\rho(v_2^2 - v_1^2) + \rho g(y_2 - y_1)$$

We wish to show that the kinetic energy terms, $\frac{1}{2}\rho v^2$, do not significantly affect the pressure. The quantitiy $\frac{1}{2}\rho v^2$ is largest in the aorta, where it has the value

$$(\tfrac{1}{2}\rho v^2)_{\text{max}} = \tfrac{1}{2}(1.0 \times 10^3\ \text{kg/m}^3)(0.40\ \text{m/s})^2$$

$$= 80\ \text{Pa}\left(\frac{1\ \text{torr}}{133\ \text{Pa}}\right) = 0.60\ \text{torr}$$

The difference in the values of $\frac{1}{2}\rho v^2$ at two points will be less than 0.60 torr, and so changes in the blood's velocity do not produce significant changes in blood pressure. In applying Bernoulli's equation to the arteries, it is a good approximation to ignore the kinetic energy terms.

EXAMPLE 16—cont'd

(c) Parts a and b show that hydrostatic effects produce the only significant variations in the average blood pressure. We shall apply the hydrostatic equation to find the pressure at points 1 and 3 in Fig. 11-49a, given that the gauge pressure at point 2 is 100 torr. At point 1, we find

$$P_1 = P_2 + \rho g(y_2 - y_1)$$

$$= 100 \text{ torr} + \frac{(1.05 \times 10^3 \text{ kg/m}^3)(9.8 \text{ m/s}^2)(-0.30 \text{ m})}{133 \text{ Pa/torr}}$$

$$= 77 \text{ torr}$$

And at point 3, we obtain

$$P_3 = P_2 + \rho g(y_2 - y_3)$$

$$= 100 \text{ torr} + \frac{(1.05 \times 10^3 \text{ kg/m}^3)(9.8 \text{ m/s}^2)(1.20 \text{ m})}{133 \text{ Pa/torr}}$$

$$= 193 \text{ torr}$$

The pressure in torr is the height in millimeters of the column of mercury that would be supported by such a pressure. The density of blood is $\frac{1}{13}$ the density of mercury, and hence a column of blood 13 times as high as the column of mercury could be supported. Thus, if a catheter connected to a long thin tube were inserted into the arteries at points 1, 2, and 3, blood would rise in the tubes to heights h_1, h_2, and h_3, where

$$h_1 = 13(77 \text{ mm}) = 1.0 \text{ m}$$

$$h_2 = 13(100 \text{ mm}) = 1.3 \text{ m}$$

$$h_3 = 13(193 \text{ mm}) = 2.5 \text{ m}$$

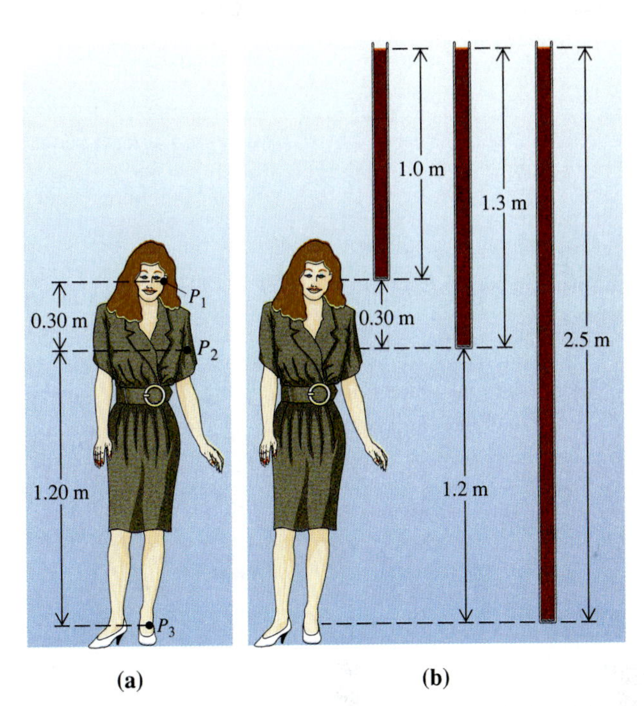

Fig. 11-49 **(a)** Blood pressure in major arteries in the head, heart, and ankle are respectively P_1, P_2, and P_3. **(b)** If catheters were inserted in arteries at each of these points and connected to vertical tubes, blood would rise to the same level in each—1.0 m above the head, 1.3 m above the heart, and 2.5 m above the ankle.

This is illustrated in Fig. 11-49b. Of course, if the body is horizontal, there will be no significant variation in ρgy and the average pressure in all major arteries will be 100 torr.

EXAMPLE 17 Blood Pressure in the Arterioles

Blood enters the arterioles at an average pressure of about 90 torr and leaves at an average pressure of about 30 torr, and so most of the 100 torr pressure drop in the circulatory system occurs across the arterioles. Arterioles average 1.0 cm in length, have an inner radius of 0.010 mm, and carry blood at an average speed of 0.50 cm/s. Show that if one assumes the viscosity of blood is 4.0×10^{-3} Pa-s the pressure drop across an arteriole is double the measured value of 60 torr.

SOLUTION The pressure drop predicted by Poiseuille's law depends on the flow rate Q:

$$P_1 - P_2 = \frac{8\eta\ell Q}{\pi R^4}$$

The flow rate is the product of the average speed $\bar{v}$ and the cross-sectional area A:

$$Q = \bar{v}A = v\pi R^2$$

Substituting this expression into Poiseuille's equation, we obtain

$$P_1 - P_2 = \frac{8\eta\ell\bar{v}}{R^2}$$

$$= \frac{8(4.0 \times 10^{-3} \text{ Pa-s})(1.0 \times 10^{-2} \text{ m})(5.0 \times 10^{-3} \text{ m/s})}{(1.0 \times 10^{-5} \text{ m})^2(133 \text{ Pa/torr})}$$

$$= 120 \text{ torr}$$

Clearly this exceeds the entire pressure drop in the circulatory system. If a value for η of 2.0×10^{-3} Pa-s is used, one obtains the measured value $P_1 - P_2 = 60$ torr. Thus the apparent viscosity of blood in the arterioles is half the normal value of 4.0×10^{-3} Pa-s. (This value was obtained by measurement of the blood flow through glass tubes with a radius much greater than 0.01 m.) This result is known as the Fahreus-Lindquist effect (see Problem 64).

CHAPTER 11 SUMMARY

Density is denoted by ρ and is defined as mass per unit volume:

$$\rho = \frac{m}{V}$$

Pressure P is defined as the force per unit area on a surface:

$$P = \frac{F}{A}$$

A fluid exerts pressure on any surface with which it is in contact. This pressure is a function of position in the fluid but is independent of the orientation of the surface, if the surface is small.

In an incompressible liquid at rest, the pressure P a distance h below a point where the pressure is P_0 is

$$P = P_0 + \rho gh$$

Archimedes' principle states that if an object is either partially or totally submerged in a fluid the fluid exerts on the object an upward force, called the "buoyant force," which is equal in magnitude to the weight of the fluid displaced and acts effectively at the center of gravity of the displaced fluid. The buoyant force may be expressed as the product of the displaced fluid's weight density and volume:

$$B = \rho gV$$

The surface of a liquid is in a state of tension. The tension T per unit of length ℓ along which it acts is defined to be the surface tension γ:

$$\gamma = \frac{T}{\ell}$$

The flow rate Q of a fluid through a surface is defined to be the volume of fluid crossing the surface per unit time. For steady flow, at any point in the fluid, Q has the same value, equal to the product of the fluid's speed v and the surface area A:

$$Q = \frac{\Delta V}{\Delta t} = vA$$

Bernoulli's equation relates the pressures P_1 and P_2 at two points along a flow tube:

$$P_1 + \tfrac{1}{2}\rho v_1^2 + \rho g y_1 = P_2 + \tfrac{1}{2}\rho v_2^2 + \rho g y_2$$

When a fluid flows, a shear stress exists within it. This stress is proportional to $\Delta v/\Delta y$, the variation of the flow speed v with distance y perpendicular to the flow. The constant of proportionality is the viscosity η:

$$\frac{F}{A} = \eta \frac{\Delta v}{\Delta y}$$

The flow of a fluid may be either laminar or turbulent. The Reynolds number R is used to predict which kind of flow will occur:

$$R = \frac{\rho \bar{v} d}{\eta}$$

where ρ is the fluid density, $\bar{v}$ the average speed, d the diameter of the tube through which the fluid flows, and η the fluid's viscosity. For $R < 2000$, flow is laminar; for $R > 3000$, flow is turbulent.

Poiseuille's law predicts the flow rate of a viscous fluid through a tube of radius R as a result of a pressure gradient $(P_1 - P_2)/\ell$:

$$Q = \frac{\pi(P_1 - P_2)}{8\eta\ell} R^4$$

Questions

1 You are standing on rocks, first in shallow water and then in deeper water. At which place will the pressure of the rocks on your feet be greater?

2 What happens to a soft, slightly underinflated football when it is taken from sea level to a high elevation in the mountains?

3 The heated air inside a hot-air balloon weighs 5000 N. If the balloon supports a weight of 2000 N, what is the weight of the air displaced by the balloon?

4 An ice cube floats in a glass of water. When the ice melts, will the water level rise, fall, or remain the same?

5 A mixed drink consists of a mixture of alcohol and water. If the alcohol content of the drink exceeds a certain fraction, ice cubes will sink to the bottom of the glass rather than float. Estimate this fraction, using for the specific gravities of alcohol and ice the approximate values of 0.8 and 0.9, respectively.

6 The center of gravity of the boat shown in Fig. 11-50 is at point P, and the center of gravity of the water it displaces is at point Q. What happens to the boat when it tilts to one side?

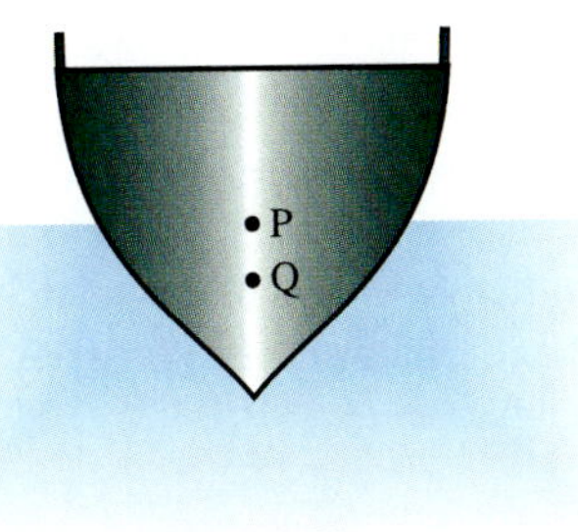

Fig. 11-50

7 A certain liquid has the same contact angle on glass as water, but the liquid has a higher density and a lower surface tension than water. Compared with water, will a column of this liquid rise to a greater or lesser height in a glass capillary tube?

8 In a gravity-free region of space, would you expect the rise of water in a glass capillary tube to increase, decrease, or remain the same?

9 A thin drinking straw made of waxed paper is placed in a glass of water. Is the water level inside the straw slightly higher or slightly lower than the water level in the glass?

10 The left end of a sheet of paper is held horizontally so that the right end hangs, curving downward. What happens to the right end when you blow a horizontal stream of air over the top of the paper?

11 A crowd tries to leave an auditorium through a small passageway (Fig. 11-51). When people push against each other, the "flow" is much like that of an incompressible fluid. Is the pressure greater in the auditorium or in the passageway?

12 For which position will the maximum blood pressure in the body have the smallest value: (a) standing upright; (b) sitting; (c) lying horizontally; (d) standing on one's head?

13 The pressure drop along a certain small artery initially equals 1 torr. Suppose that buildup of plaque on the artery's wall reduces its effective radius to one third its original value. What will be the pressure drop along the artery, if the flow through the artery remains the same?

14 In Ex. 6, when the object is submerged in water, does the pressure of the water on the bottom of the container increase, decrease, or remain constant?

15 (a) In an orbiting space station, would the blood pressure in major arteries in the leg ever be greater than the blood pressure in major arteries in the neck?
(b) Would you expect the blood pressure drop across an arteriole to be less than it would be on earth?

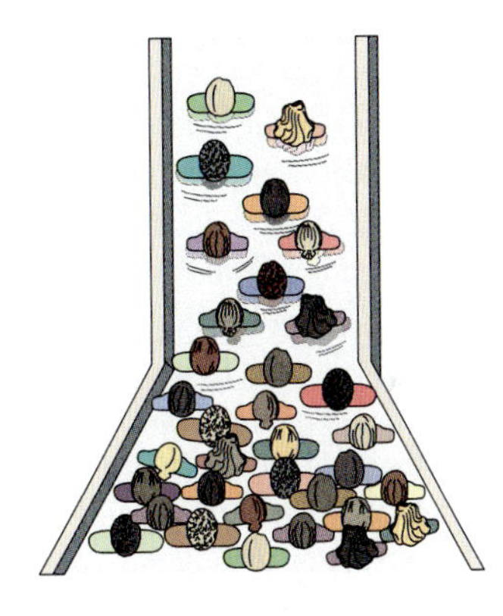

Fig. 11-51 A crowd flowing like an incompressible fluid.

Answers to Odd-Numbered Questions

1 In shallow water; **3** 7000 N; **5** $\frac{1}{2}$; **7** lesser; **9** slightly lower; **11** in the auditorium; **13** 81 torr; **15** (a) no; (b) no

Problems (listed by sections)

11-2 Pressure in a Fluid at Rest

1 (a) A man weighing 765 N lies on a bed of nails. Find the average pressure exerted by a nail on his skin if there are 1000 nails and the flattened point on each nail has a radius of 0.500 mm.

(b) Calculate the pressure exerted on the skin by an acupuncture needle of radius 0.100 mm applied with a force of 5.00 N.

2 A sheet of paper with dimensions 20.0 cm by 30.0 cm lies on a desk. Calculate the downward force exerted on the paper by the atmosphere. Considering the magnitude of this force, why is it so easy to lift the paper?

3 Calculate the force required to remove a suction cup of area 4.00 cm^2 from the surface to which it is stuck.

4 The air pressure on the bottom of an airplane's wings is greater than the pressure on the top, providing an upward lift. Calculate the surface area of the wings required to lift a plane weighing 4.00×10^5 N if the pressure difference is 0.100 atm.

★ **5** An automobile weighing 1.00×10^4 N has weight equally distributed over its wheels. The weight of the car causes each tire to be slightly flattened on the bottom so that there is a surface area in contact with the road. The gauge pressure of the air in the tire is 2.00×10^5 Pa (30 lb/in^2). Find the area of contact, assuming that the only forces on the section of tire in contact with the road are the air pressure and the normal force exerted by the surface of the road.

★ **6** In order for a person to breathe, the chest must be expanded so that the air pressure inside the lungs is slightly less than the air pressure outside. A partial vacuum is created, and air flows into the lungs. When you breathe underwater through a snorkel (a tube open to air at the surface), you must expand your chest against the water's pressure. Suppose that 20,000 N/m^2 is the maximum pressure the chest can support. What is the maximum depth at which you could breathe through a snorkel?

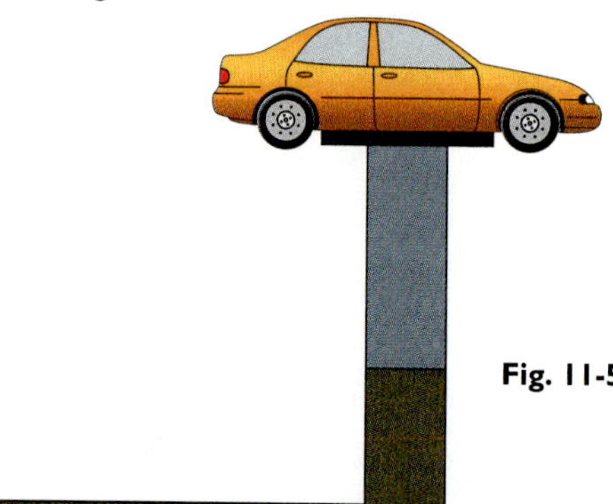

Fig. 11-53

7 When you lift an inverted glass from a basin of water, the glass remains filled with water, with the water in the glass above the level of the water in the basin (Fig. 11-52). Calculate the maximum height of the water column that can be supported in this way, and compare with the length of the mercury column in a barometer. This problem illustrates why mercury is used in a barometer.

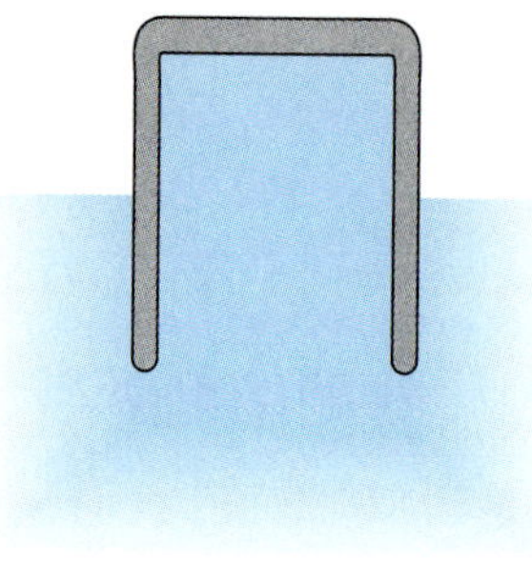

Fig. 11-52

8 Air pressure inside the lungs can be reduced to a minimum of about −80 torr. How high can water be sucked up a straw?

9 A vacuum pump is used to raise water from a well a distance h below the pump. What is the maximum value of h (a) at sea level; (b) on Mt. Everest, 10 km above sea level, where atmospheric pressure is 0.30 atm?

★ **10** A hollow metal cylinder of radius 10.0 cm, closed at both ends, is cut in two. The halves are heated and sealed together while the air inside is hot. When the cylinder and air cool, the pressure inside the cylinder drops to 0.700 atm. How great a force must be applied to each half of the cylinder to separate them?

★ **11** (a) Find the force **P** that must be applied to a piston of area 10.0 cm^2 to produce sufficient fluid pressure to support a car weighing 10,000 N by means of a column of fluid of cross-sectional area 300 cm^2 (Fig. 11-53).

(b) Find the increase in the car's gravitational potential energy when it is raised 1.00 m.

(c) How far must the smaller piston move in order for the larger one to move 1.00 m?

(d) Calculate the work done by **P** in moving the smaller piston, and compare with the answer to part b.

★12 A bowl of water has a flat base of area 20.0 cm² and rests on a table. The height of the water is 10.0 cm, the weight of the water is 5.00 N, and the weight of the bowl is negligible (Fig. 11-54).

(a) Find the force exerted on the base of the bowl by the water in contact with it.

(b) Find the force exerted on the base of the bowl by the surface of the table.

(c) If the base of the bowl is in equilibrium, why are the answers to a and b not the same?

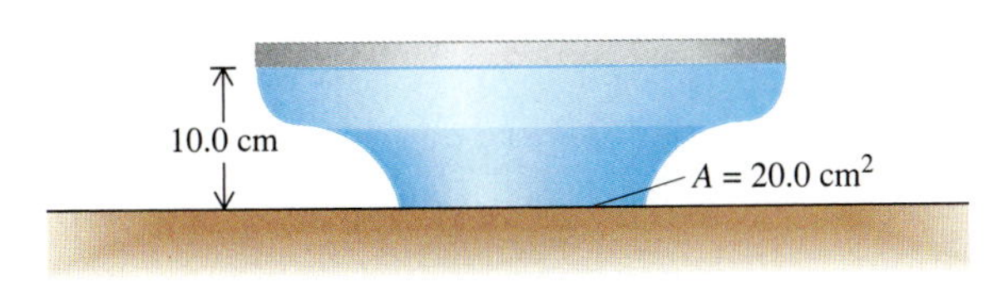

Fig. 11-54

13 Scuba divers breathe compressed air stored in metal tanks. What must the pressure of the air supplied to a diver at a depth of 10.0 m in salt water be if breathing is to require the same effort as at the surface?

14 A pressure cooker is a covered pot sealed everywhere except for a small circular opening in the top. A small weight placed over this opening determines the pressure at which steam is released from the pot. Suppose that the radius of the opening is 0.500 mm and the weight is 0.800 N. Find the maximum pressure inside the pot.

★15 A long, thin metal tube of cross-sectional area 1.00 cm² is sealed into the top of a hollow cube of edge length 0.100 m. Water fills the cube and extends upward a distance h into the tube. Find h if the water exerts a force of 1.00×10^3 N on the base of the cube. What is the weight of the water?

16 A watertight car is submerged under water. What force is required to open a door of area 1.00 m² if the average depth of water at the door is 2.00 m?

★17 Fig. 11-55 shows the free-body diagram of a wedge of fluid within a larger body of fluid. The forces exerted by the surrounding fluid are normal to the surface of the wedge. Assuming the wedge to be small and its weight negligible, prove that the pressure on the three sides is the same.

11-3 Archimedes' Principle

18 Find the displacement in cubic meters of a boat weighing 4450 N.

19 What is the weight of the water displaced by a boat weighing 9000 N?

20 (a) Calculate the pressure on a man's feet if his weight is 1000 N and the soles of his feet have a total surface area of 400 cm².

(b) Find the pressure on his feet when his body is 70% submerged in water if his density is assumed to be 1.00 g/cm³.

21 An irregularly shaped object weighs 10.0 N in air. When immersed in water, the object has an apparent weight of 4.00 N. Find its density.

★22 A fishing line has attached to it a hollow plastic float 2.50 cm in diameter and having a mass of 2.00 g. Find the mass of a lead weight that, when attached to the bottom of the float, will cause the float to be half submerged.

23 A man weighing 675 N is floating in water with 2% of his body's volume above the surface of the water. Find the volume of the man's body.

★24 A 133 lb woman in a swimming pool expels air from her lungs and then has a density just barely greater than that of water. If she makes no effort to swim or tread water, she will slowly sink to the bottom of the pool. By taking a deep breath of air she is able to float with her nose above water. What volume of air must be inhaled if 3% of her body's volume is above the surface of the water?

★25 If a man is capable of lifting a 100 lb stone on land, how heavy a stone could he lift under water? The specific gravity of the stone is 2.50.

★26 A cylindrical buoy having a diameter of 0.700 m floats in seawater with its top 0.700 m above the surface of the water. Find the distance that the buoy sinks when a person weighing 625 N stands on it.

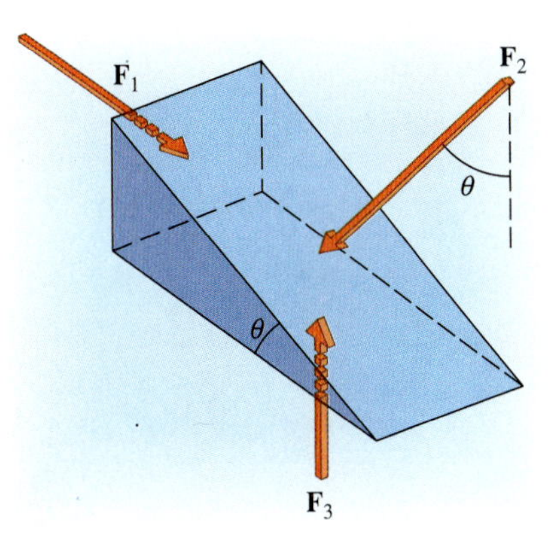

Fig. 11-55

★ **27** In the movie *The Great Race*, the characters find themselves floating in the ocean with their car on a large slab of ice (Fig. 11-56). Suppose that the ice is 95% submerged, the car weighs 9000 N, and the slab's surface area is 12.5 m^2. Find the thickness of the ice.

Fig. 11-56

28 Hydrotherapy is a medical technique used to allow patients to exercise weakened limbs. The limb is submerged in water so that its apparent weight is reduced. Find the apparent weight of an arm of weight 50.0 N and volume 3.50×10^3 cm^3 when it is under water.

29 A solid piece of iron will obviously not float in water. It will, however, float in mercury. Find the fraction of the iron submerged in the mercury.

★ **30** A life preserver made of Styrofoam is designed to support a person weighing 575 N, with 80% of the body and 90% of the preserver submerged. Find the volume of Styrofoam required if the body's volume is 0.0500 m^3.

★ **31** A hollow sphere made of iron floats half submerged in water. Find the ratio of the thickness of the iron to the outer radius of the sphere.

*11-4 Surface Tension and Capillarity

32 One can measure surface tension by measuring the force required to lift a wire ring from the surface of a liquid (Fig. 11-57). Find the force required to lift a ring with negligible mass and a radius of 10.0 cm from the surface of water.

33 At 20° C, how high will capillarity cause water to rise in a glass tube of radius 1.00 cm?

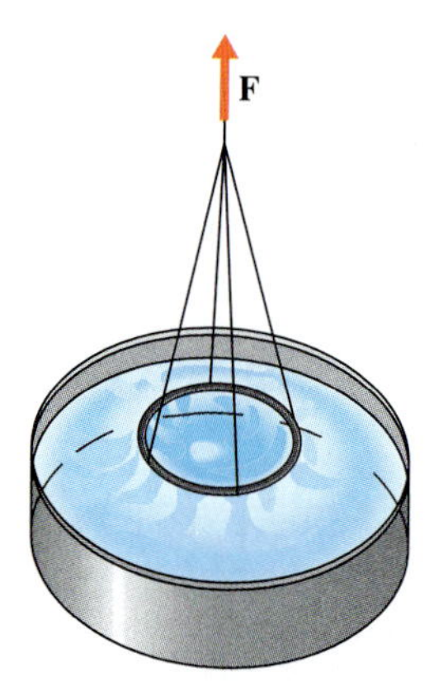

Fig. 11-57

★ **34** Prove that the vapor pressure inside a soap bubble exceeds atmospheric pressure by $4\gamma/r$, where r is the radius of the bubble and γ is the surface tension of the soap film. Calculate the pressure inside a soap bubble of radius 1.00 cm if $\gamma = 0.0500$ N/m.

★ **35** Fig. 11-58 shows a cylindrical free body consisting of a liquid and vapor separated by a hemispherical surface. Show that the vapor pressure exceeds the liquid pressure by $2\gamma/r$.

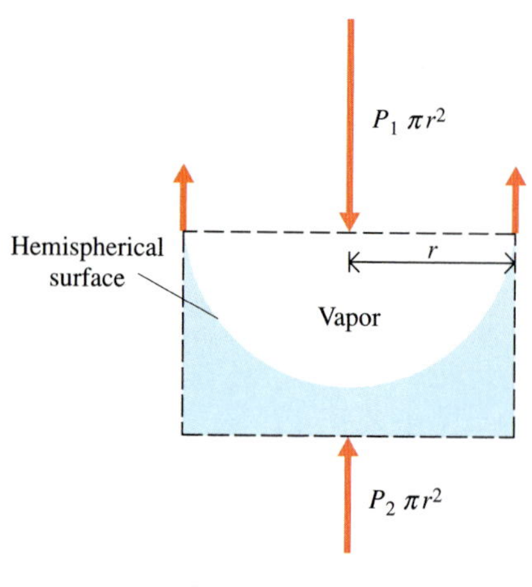

Fig. 11-58

★ **36** A glass capillary is inserted into a pool of mercury (Fig. 11-59). The mercury inside the capillary is depressed a distance h below the level outside.
(a) Derive an expression for h in terms of the capillary's radius r, mercury's surface tension γ, and the contact angle θ.
(b) Calculate h for a radius of 1.00 mm.

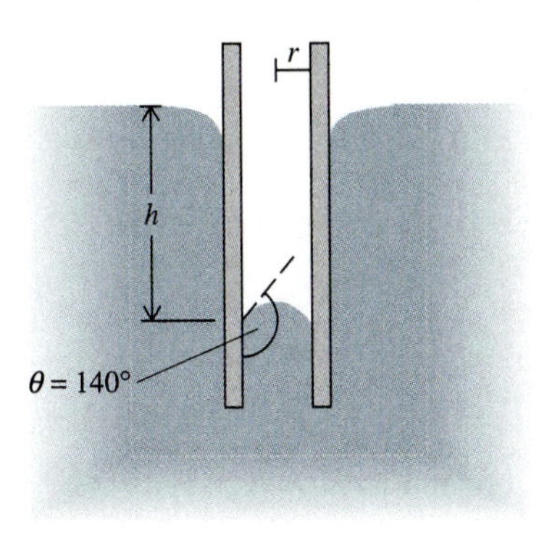

Fig. 11-59

37 The pressure in the xylem near the base of a tall tree is 1.00 atm.
(a) What is the pressure in the xylem 50.0 m above, near the top of the tree? Assume that the fluid in the xylem has the same density as water.
(b) Suppose that a hole is drilled into the xylem at the top of the tree, exposing the fluid to the atmosphere. Would there be a flow of fluid outward or air inward?

★ **38** Two plates of glass, parallel to each other, are held perpendicular to the surface of water in a tub. If the plates are 1.00 mm apart and the bottom edges are lowered halfway into the water, how high will the water rise between the plates?

39 The pressure in a column of water in a glass tube is −300 atm. Calculate the magnitude and direction of the force on a small, approximately flat section of the glass of area 1.00 cm^2.

11-5 Fluid Dynamics; Bernoulli's Equation

40 Water is ejected from a garden hose of radius 1.00 cm at a speed of 5.00 m/s. Find the flow rate through the hose.

41 A container of cross-sectional area 100 cm^2 contains liquid of density 2.00 g/cm^3 to a height of 6.00 cm above a small hole of area 1.00 cm^2 in the side.
(a) Find the speed of the fluid as it leaves the hole.
(b) Find the flow rate through the hole.

★ **42** Water is supplied to an office building 30.0 m high through a pipe at ground level.
(a) Find the minimum pressure in the supply pipe so that water can reach the top floor of the building.
(b) Suppose the pressure in the supply pipe is designed so that water can flow out through a much smaller pipe on the top floor at a speed of 5.00 m/s. Find the pressure in the larger pipe on the ground floor.

★ **43** A fountain consists of a large pool of water and a column of water projected upward by means of a pump. The column has a cross-sectional area of 100 cm^2 at its base and reaches a height of 3.00 m.
(a) Compute the speed of water as it leaves the pool and the flow rate.
(b) Find the increase in mechanical energy per unit mass $\Delta E/\Delta m$, as water starts from rest in the pool and then is expelled by the pump.
(c) Find the power supplied by the pump by computing the increase in mechanical energy per unit time.

★ **44** The heart takes in blood at a pressure P_1 and expels it at a higher pressure P_2. Thus there is an increase of pressure across the heart, $\Delta P = P_2 - P_1$, which typically averages about 100 torr.
(a) Show that the power expended by the heart in pumping blood at the rate Q is given by $\Delta P\, Q$.
(b) Calculate the power expended by the heart in pumping blood at the rate of 100 cm^3/s (corresponding to a state of rest) and at 500 cm^3/s (corresponding to vigorous activity).

★ **45** Fig. 11-60 shows a Venturi meter, which is a device used to measure the flow speed of a fluid.
(a) Show that

$$v_1{}^2 = \frac{2(P_1 - P_2)}{\rho[(A_1/A_2)^2 - 1]}$$

(b) Let the fluid flowing in the pipe be water. Find v_1 if h = 10.0 cm and $A_1 = 4A_2$. The silver fluid is mercury.

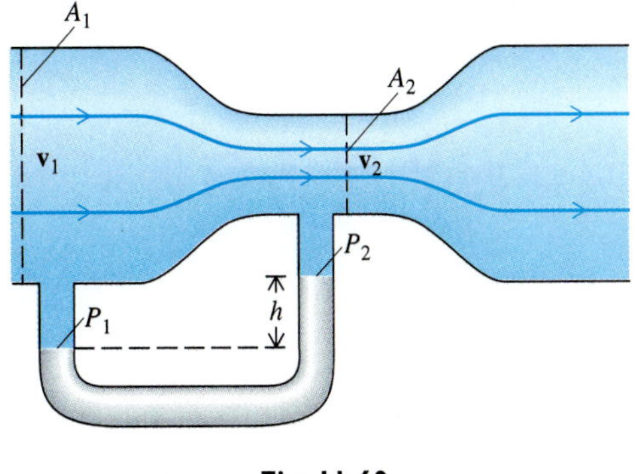

Fig. 11-60

★ **46** The piston of a small pump for a fountain is 4.00 cm in diameter, and the discharge nozzle is 2.00 cm in diameter (Fig. 11-61). The pump ejects water into the atmosphere at a rate of 40.0π cm^3/s. The lower face of the piston is open to the atmosphere. What force **F** must be exerted on the piston when it is 2.00 cm below the nozzle? Neglect viscosity, friction, and the weight of the piston. The velocity of the piston is constant.

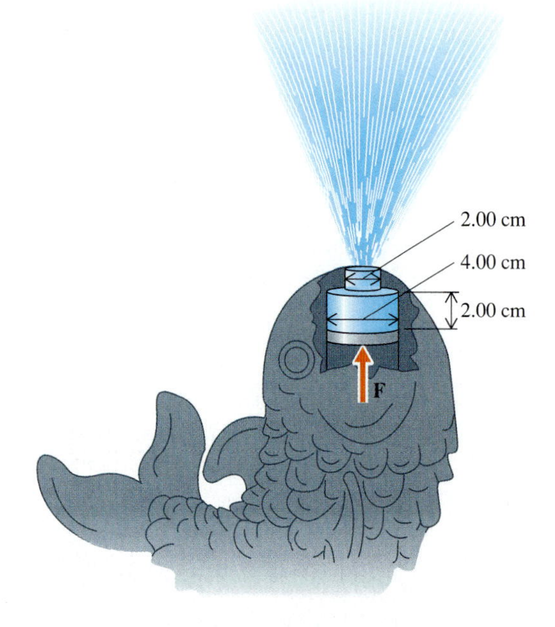

Fig. 11-61

★ **47** A water pipe is inclined 30.0° below the horizontal. The radius of the pipe at the upper end is 2.00 cm. If the gauge pressure at a point at the upper end is 0.100 atm, what is the gauge pressure at a point 3.00 m downstream, where the pipe has narrowed to a 1.00 cm radius? The flow rate is 20.0π cm^3/s.

48 For an airplane in flight, the average speed of the air relative to the plane is 300 m/s below the wings and 330 m/s above the wings (Fig. 11-62). The bottom of each wing has a surface area of 30.0 m^2. Calculate the lift, which is the net upward force of air on the plane.

*11-6 Viscosity

49 Find the maximum diameter of a tube that air can flow through with laminar flow at a flow speed of 1.00 m/s. Would you expect the earth's winds to typically be laminar or turbulent?

50 The human body contains about 5500 cm^3 of blood. How long does the blood take to make a complete circuit through the body: (a) at rest, when the flow rate through the aorta is 100 cm^3/s; (b) during vigorous exercise, when the flow rate through the aorta is 500 cm^3/s?

51 The venae cavae are two large veins that carry blood to the heart. Each has a radius of 1.50 cm. Find the average speed of the blood through the venae cavae when blood is moving from the heart through the aorta (radius 1.25 cm) at an average speed of 20.0 cm/s.

*11-7 Poiseuille's Law

52 When a major artery is cut, the blood squirts out because of the high arterial pressure. Compute the vertical rise of a stream of blood from a small cut in a major artery in the leg, 1.00 m below the heart. Use the average arterial pressure.

53 In 1732 Stephen Hales made the first measurement of blood pressure when he inserted a brass tube into a major artery of a horse. The brass tube was connected to a vertical glass tube 3 m high. The horse's average blood pressure was 194 torr. How high above the artery did the column of blood rise in the glass tube?

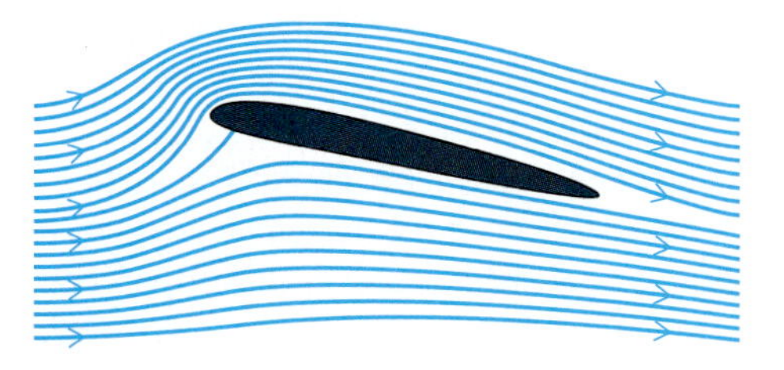

Fig. 11-62

54 Blood pressure is normally measured in the following way. A cuff is placed around the arm just above the elbow and inflated with air by means of a hand pump (Fig. 11-63). Initially the pressure of the cuff is sufficient to stop the flow of blood in the arm. The air pressure is then slowly reduced until blood begins to flow through the artery during part of each heartbeat. At first blood flows only briefly during each beat, at the time when the blood pressure reaches its maximum value, called *systolic pressure*. At this point the artery is constricted and the flow of blood is turbulent. The turbulence produces characteristic sounds, which can be detected with a stethoscope placed just below the cuff. When the sounds are first detected, the air pressure is read from a pressure gauge and is recorded as the systolic pressure. The pressure in the cuff is then reduced until there are no sounds of turbulence. This means that the artery is not constricted even during the minimum pressure, called *diastolic pressure*. The air pressure corresponding to the disappearance of turbulent flow is recorded as the diastolic pressure. Normal values of systolic and diastolic pressures are 120 torr and 80 torr, which is stated "120 over 80." Suppose that an inexperienced person measured blood pressure with the arm raised so that the cuff was 20.0 cm above the level of the heart. If the blood pressure is actually normal, what readings would be obtained from this measurement?

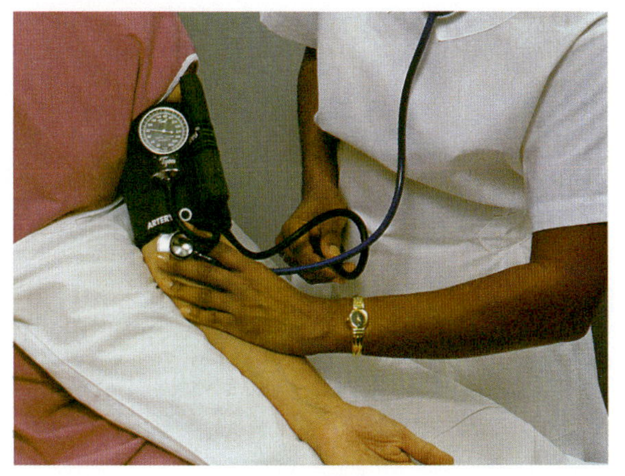

Fig. 11-63

55 A woman hangs by her feet from "gravity boots" that support her weight from above. The average blood pressure in her aorta is 100 torr. What is the average blood pressure in a major artery in her ankle, at a point 1.10 m above the aorta?

56 A blood vessel of inner radius 1.00 mm carries blood at an average speed of 3.00 cm/s. Find the pressure drop along a 1.00 cm length of the blood vessel.

57 As the skin becomes cold, a capillary near the skin's surface contracts. If the pressure drop per centimeter along the capillary is constant, what ratio of final capillary radius to initial radius is necessary to reduce the flow rate by one half?

Additional Problems

★★ **58** In the movie *Once Upon a Time in America,* set in the Prohibition era, a clever ploy is used by bootleggers transporting liquor by boat. Attached to each case of liquor is a balloon of negligible weight and a cloth bag filled with salt (Fig. 11-64). When police approach, the liquor is dumped overboard. Initially the cases sink, but after a time the salt dissolves and the balloons float to the surface. Suppose that a liquor case weighs 180 N and displaces 130 N of water and that, when the balloon floats, it is half submerged. Find the volume of the balloon and the required weight of salt.

Fig. 11-64

★ **59** A balloon filled with helium has a weight of 2.00 N and a volume of 0.500 m^3. It is secured to the ground by means of two strings, as shown in Fig. 11-65. Find the tension in the strings.

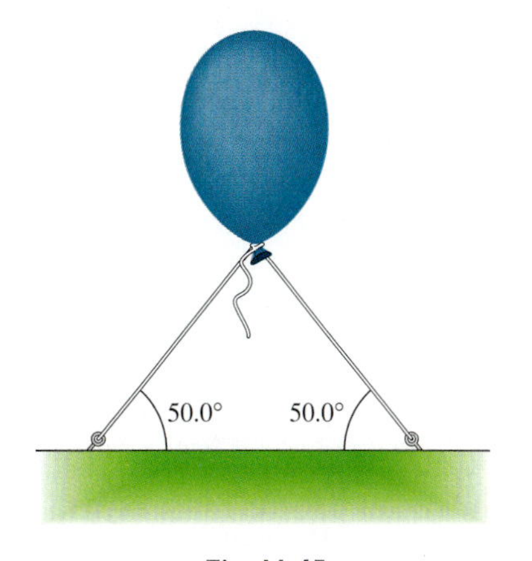

Fig. 11-65

★★ **60** Cyclists racing in a velodrome sometimes fill their tires with helium rather than air, thereby reducing the weight of the tire and (more important) the wheel's moment of inertia. A certain bicycle has a total mass of 8.00 kg, and each wheel has a diameter of 70.0 cm and a moment of inertia of 7.00×10^{-2} kg-m^2. The cross-sectional area of the tire's cavity is 4.00 cm^2. Find the percentage reduction in mass and moment of inertia.

★ **61** A mountain climber of mass 80.0 kg wishes to reduce by 20% the work required to climb a mountain. Therefore he attaches to his body a large helium-filled balloon. Find the volume of the balloon required.

62 A person with high blood pressure has an average pressure of 150 torr in the aorta. What is the pressure in a large artery in the ankle, 1.20 m below the aorta?

★ **63** Water flows into a cylindrical, open-topped container at the steady rate of 100 cm^3/s. Water flows out through a small hole of radius 4.00 mm in the bottom of the can, which is initially empty. Find the rate at which the volume of water in the can is increasing (a) initially and (b) when the water level is 10.0 cm high. (c) What is the maximum height of water in the can?

★★ **64** Fig. 11-66 shows a flow tube of blood with one end in a large blood vessel, where the flow lines are equally spaced, and the other end in an arteriole, where the flow lines are concentrated toward the center.

(a) Use Bernoulli's equation to show that the pressure P_1 at the center of the arteriole is less than the pressure P_2 at the walls.

(b) Blood consists of solid cells suspended in a plasma. The viscosity of the plasma is 1.5 cP and that of whole blood is 4.0 cP. Apply the results of part (a) to account for the apparent decrease in viscosity in the arterioles.

★ **65** A paper clip formed by iron wire of radius 0.40 mm floats on the surface of water. Calculate the maximum radius of wire that will float, assuming that the surface tension force is vertical.

★★ **66** To support a considerable weight on the surface of water by surface tension, a square grid, consisting of n thin aluminum wires of radius 0.500 mm, is constructed with a 1.00 cm spacing. Calculate the edge length of each square if it supports a weight of 500 N. Assume that the force of surface tension is directed vertically upward.

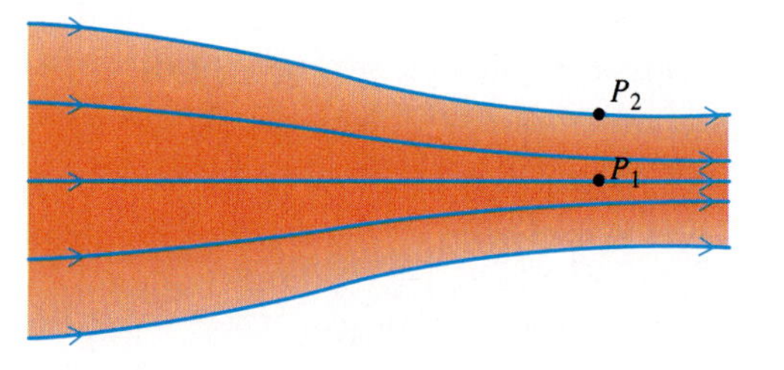

Fig. 11-66

CHAPTER 12 Temperature and Kinetic Theory

We can feel the blazing heat of the summer sun or the biting cold of a winter blizzard. Our bodies are sensitive even to small changes in the temperature of our surroundings. We respond to these changes with several adaptive mechanisms, like sweating or shivering, to maintain a nearly constant internal body temperature. This sensitivity to the thermal environment is the basis for our concepts of hot and cold, out of which the scientific definition of temperature evolved. Temperature is a quantitative measure of how hot or cold something is. In this chapter we shall see how various kinds of thermometers are used to measure temperature and how temperature can be interpreted as a measure of molecular kinetic energy.

12-1 Temperature Measurement

Thermometers

Temperature is measured with a thermometer. Galileo invented the first thermometer, which made use of air's property of expanding as it is heated. The air's volume indicated the temperature. Today there are various kinds of thermometers, each appropriate for the range of temperatures and the system to be measured. For example, in addition to the common mercury thermometer, there are thermocouple thermometers, thermistor thermometers, and optical pyrometers (Fig. 12-1).

Each thermometer depends on the existence of some **thermometric property** of matter. For example, the expansion or contraction of mercury in a fever thermometer correlates with the body's sensation of hot and cold. A person who has a high fever both is hot to the touch and will register a higher than normal temperature on a mercury thermometer. The length of the mercury column in the glass stem of the thermometer gives us a quantitative measure of temperature.

The constant volume gas thermometer maintains a quantity of gas in a fixed volume (Fig. 12-1e). As the gas is heated, its pressure increases, as indicated by a pressure gauge. Pressure is the thermometric property in this thermometer. Pressure readings are used to indicate temperature.

Suppose a mercury thermometer is used to measure the temperature of a hot liquid. When the thermometer is first immersed in the liquid, the length of the mercury column increases. A short time later it reaches an equilibrium level and remains constant. This final equilibrium level indicates the common temperature of the liquid and the thermometer. The interaction between the thermometer and the system, during which the mercury expands, is called "thermal interaction." The final state, in which the length of the mercury column no longer changes, is called a state of **thermal equilibrium.** The length of the mercury column then indicates the temperature of both the thermometer and the liquid. When two bodies are in thermal equilibrium, they are at the same temperature.

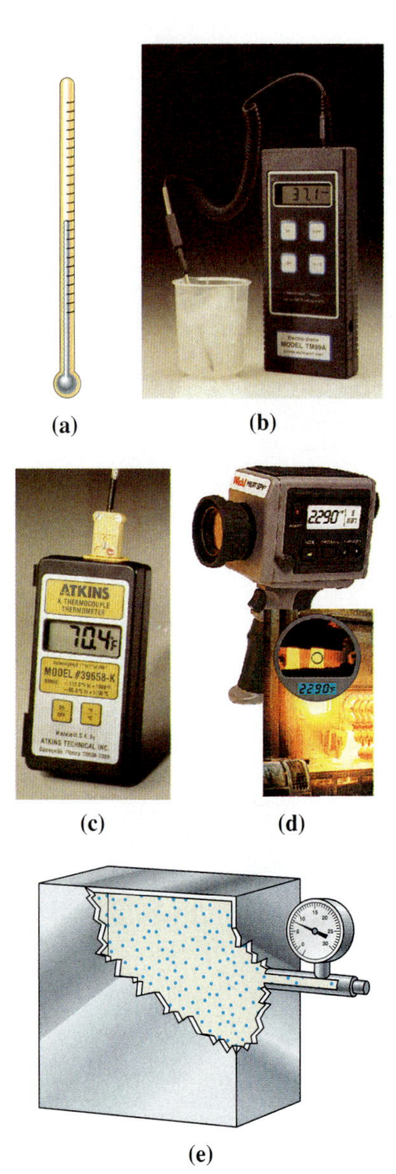

Fig. 12-1 Thermometers. The length of a column of mercury indicates temperature in a mercury thermometer **(a)**. Electrical properties of a probe are used to measure temperature in a thermistor thermometer **(b)** and a thermocouple thermometer **(c)**. In an optical pyrometer **(d)**, used to measure very high temperatures, light emitted by the hot body is compared with light produced by an electrically heated filament inside the pyrometer. **(e)** Gas pressure indicates temperature in a constant-volume gas thermometer.

Temperature Scales

To assign a numerical value to the temperature of a body, we need a temperature scale. The two most common scales in everyday use are the Fahrenheit scale and the Celsius scale (formerly known as the centigrade scale). In establishing a temperature scale, one could use any kind of thermometer and any thermometric property. For example, the Celsius scale was based on the expansion of a column of liquid such as mercury in a thin glass tube. One could just as well base a temperature scale on the volume of a fixed quantity of air at atmospheric pressure, as in Galileo's first thermometer. There is no point in having many independently defined temperature scales. The **absolute**, or **Kelvin**, scale is now the standard in terms of which all other scales, such as Celsius and Fahrenheit, are defined.

The Kelvin scale is chosen as the standard for important reasons. First, various laws of physics are most simply expressed in terms of this scale. We shall see an example of this in the ideal gas law, described in the next section. Second, zero on the absolute scale has fundamental significance. It is the lowest possible temperature a body can approach.

The definition of the Kelvin scale is based on a constant-volume gas thermometer and uses the thermal properties of water to establish a reference temperature. Liquid water can coexist with both water vapor and ice in thermal equilibrium at only one temperature. This state, called the "triple point," is achieved when a vacuum-sealed container of liquid water is cooled until it is in equilibrium with both water vapor and ice. The triple-point temperature T_{tr} is defined to be 273.16 kelvin (abbreviated K):

$$T_{tr} = 273.16\text{ K} \tag{12-1}$$

The triple point is very close to the normal freezing point of water at atmospheric pressure (273.15 K). The reason for assigning the particular value 273.16 to the triple point is so that the Kelvin scale will be related to the older Celsius scale in a simple way, as we shall see.

Temperature on the Kelvin scale is defined according to the following prescription. A container of low-density gas with a fixed volume (a constant-volume gas thermometer) is allowed to reach thermal equilibrium with water at the triple point. The pressure of the gas at the triple point, P_{tr}, is measured. Then the gas thermometer is placed in thermal contact with the body whose temperature is to be determined, and when thermal equilibrium is reached, the gas pressure P is again measured. The ratio P/P_{tr} *defines* the temperature of the body according to the equation

$$T = (273.16 \text{ K})\frac{P}{P_{tr}} \tag{12-2}$$

For example, when the pressure of the gas in the thermometer is exactly double its pressure at the triple point, the temperature of the body by definition is 273.16 K $\times$ 2 = 546.32 K. The gas used in this thermometer can be virtually any gas, so long as the density of the gas is low enough. All gases give the same value for the temperature of a given system, in the limit as the gas density approaches zero.

Historically, when the **Celsius** scale was established, the normal freezing and boiling points of water were assigned respective values of 0° C (zero degrees Celsius) and 100° C. A mercury thermometer was brought to thermal equilibrium with water at each of these temperatures, and the level of the mercury column was marked as 0° C for ice water and 100° C for boiling water. The mercury column between these two marks was then divided into 100 equal intervals, corresponding to temperature intervals of 1 Celsius degree (1 C°). Today the Celsius temperature scale is defined in terms of the Kelvin scale. Celsius temperature T_C is now defined by the equation

$$T_C = T - 273.15 \tag{12-3}$$

On the Celsius scale, the triple point of water is 273.16 − 273.15, or 0.01° C. The normal freezing point of water open to the air at one atmosphere of pressure is 273.15 K, or 0.00° C. The normal boiling point of water is 373.15 K, or 100.00° C. This definition of the Celsius scale conforms to the earlier definition based on the freezing and boiling points of water. Temperature intervals on the Celsius and Kelvin scales are the same. For example, the difference in temperature between the boiling point of water and its freezing point is 100 Celsius degrees (C°), or 100 kelvins.

Table 12-1 The Kelvin, Celsius, and Fahrenheit temperature scales

	Kelvin scale		Celsius scale		Fahrenheit scale	
Sun's surface temperature	6000 K		6000° C		10,000° F	
Gold melts	1336 K		1063° C		1945° F	
Water boils	373 K	100 K	100° C	100 C°	212° F	180 F°
Human body temperature	310 K		37° C		99° F	
Typical room temperature	293 K		20° C		68° F	
Water freezes	273 K		0° C		32° F	
Mercury freezes	234 K		−39° C		−38° F	
Ethyl alcohol freezes	143 K		−130° C		−202° F	
Nitrogen liquifies	77 K		−196° C		−321° F	
Helium liquifies	4.2 K		−269° C		−452° F	
Absolute zero	0 K		−273° C		−460° F	

Fahrenheit temperature T_F, measured in degrees Fahrenheit (°F), is defined relative to the Celsius temperature T_C by the equation

$$T_F = 32 + \tfrac{9}{5}T_C \quad (12\text{-}4)$$

The normal freezing and boiling points of water on the Fahrenheit scale are 32° F and 212° F respectively. The interval between these points is 180° F. There are only 100 C° between these same points. Thus Celsius degrees are bigger than Fahrenheit degrees: 1 C° is $\frac{180}{100}$, or $\frac{9}{5}$, times 1 F°.

EXAMPLE 1 Measuring a Fever on the Celsius Scale

Normal internal body temperature is 98.6° F. A temperature of 106° F is considered a high fever. Find the corresponding temperatures on the Celsius scale.

SOLUTION Solving Eq. 12-4 for T_C, we obtain

$$T_C = \tfrac{5}{9}(T_F - 32.0)$$

A temperature of 98.6° F corresponds to

$$T_C = \tfrac{5}{9}(98.6 - 32.0) = 37.0° \text{ C}$$

and a temperature of 106° F corresponds to

$$T_C = \tfrac{5}{9}(106 - 32.0) = 41.1° \text{ C}$$

12-2 Ideal Gas Law

The pressure of a gas can be changed in several ways. One way to increase the pressure of a gas confined to a fixed volume is to increase the number of gas molecules in the volume. You do this, for example, when you pump air into a bicycle tire or an automobile tire. Another way to change the pressure of a gas is to change its temperature. For example, when the air in an automobile tire heats up, its pressure increases significantly. A third way to change gas pressure is to change the volume containing the gas; decreasing volume causes an increase in pressure (Fig. 12-2).

For low-density gases, there is a simple, universal relationship between the gas pressure P, volume V, Kelvin temperature T, and number of gas molecules N. The product of P and V is proportional to the product of N and T:

$$PV \propto NT$$

or

$$PV = NkT \quad (12\text{-}5)$$

This equation is called the **ideal gas law.** The constant k is known as "Boltzmann's constant" and is found from experiment to have the value

$$k = 1.380 \times 10^{-23} \text{ J/K} \quad (12\text{-}6)$$

The ideal gas law is most accurate in describing noble gases like neon or helium at low densities. But the ideal gas law provides a good approximate description of the behavior of other gases, so long as they are not close to the liquid state. In applying the ideal gas law, temperature must be expressed in kelvins, not in °C or °F.

Fig. 12-2 If you squeeze one end of a balloon, you force the air inside the balloon into a smaller volume, and the pressure of the air increases.

Special cases of the gas law are found when one considers the variation of two of the variables P, V, N, and T, while the other two variables are held constant. For example, if N and T are fixed, the ideal gas law implies that the product PV is constant:

$$PV = \text{constant} \qquad \text{(for constant } N \text{ and } T\text{)} \qquad (12\text{-}7)$$

This result is known as Boyle's law, in honor of Robert Boyle, who discovered it in 1660. Boyle's law implies that if the volume of a gas is reduced to half its original value the pressure of the gas is doubled.

If P and N are fixed, the ideal gas law implies that the volume of the gas is directly proportional to its temperature:

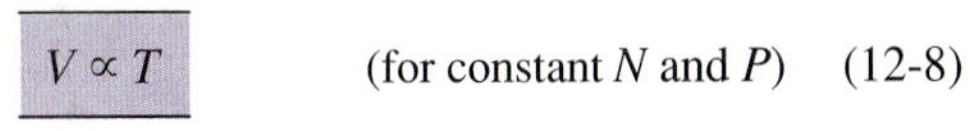

$$V \propto T \qquad \text{(for constant } N \text{ and } P\text{)} \qquad (12\text{-}8)$$

This result was discovered by Joseph Gay-Lussac in 1802.

If V and N are fixed, the ideal gas law implies that

$$P \propto T \qquad \text{(for constant } N \text{ and } V\text{)} \qquad (12\text{-}9)$$

The very definition of temperature on the Kelvin scale requires that this relationship be satisfied, at least in the limit of a very low-density gas.

EXAMPLE 2 The Temperature of an Ideal Gas After Compression

An ideal gas initially has a volume of 1.0 liter (L), a pressure of 1.0 atmosphere (atm), and a temperature of 27° C. The pressure is raised to 2.0 atm, compressing the volume of the gas to 0.60 L. Find the final temperature of the gas.

SOLUTION We are given the following initial and final values of P, V, and T:

$$P_i = 1.0 \text{ atm} \qquad P_f = 2.0 \text{ atm}$$
$$V_i = 1.0 \text{ L} \qquad V_f = 0.60 \text{ L}$$
$$T_{Ci} = 27° \text{ C} \qquad T_{Cf} = ?$$

The number of molecules, N, is constant. The problem is to find the final temperature T_{Cf}. We can do this simply by first writing the ideal gas law for the initial state of the gas and again for the final state and then taking the ratio of the two expressions:

$$P_iV_i = NkT_i$$
$$P_fV_f = NkT_f$$
$$\frac{T_f}{T_i} = \frac{P_fV_f}{P_iV_i}$$

Since this equation involves ratios of pressures and volumes, we may insert these quantities in the units in which they are given, that is, atmospheres and liters, rather than converting to standard units of Pa and m^3. The conversion to standard units would simply introduce identical factors for both initial and final values, and these factors would cancel. We must be careful, however, to convert temperature from degrees Celsius to kelvins, even when a ratio is used, as it is here, since this change in units involves an additive term, rather than a multiplicative factor. Thus we must use $T_i = 27 + 273 = 300$ K. Substituting values into the preceding equation, we obtain

$$\frac{T_f}{300 \text{ K}} = \frac{(2.0 \text{ atm})(0.60 \text{ L})}{(1.0 \text{ atm})(1.0 \text{ L})} = 1.2$$

or

$$T_f = (1.2)(300 \text{ K}) = 360 \text{ K}$$

This corresponds to a final Celsius temperature T_{Cf} of

$$T_{Cf} = T_f - 273 = 360 - 273$$
$$= 87° \text{ C}$$

Dalton's Law of Partial Pressures

The ideal gas law may also be applied to a mixture of noninteracting ideal gases. Suppose that N_1 molecules of a single ideal gas contained in a volume V produce a pressure P_1 at temperature T when this is the only gas in the volume, and suppose that N_2 molecules of a second ideal gas contained in a volume V produce a pressure P_2 at temperature T when the second gas is the only one contained in the volume. According to **Dalton's law of partial pressures,** a mixture of these two ideal gases produces a pressure P that is the sum of P_1 and P_2 (if there is no chemical interaction between the two gases):

$$P = P_1 + P_2$$

Using the ideal gas law to express P_1 and P_2 in terms of T, V, and the respective numbers of molecules, N_1 and N_2, we find

$$P = \frac{N_1 kT}{V} + \frac{N_2 kT}{V} = \frac{(N_1 + N_2)kT}{V}$$

Letting N denote the total number of molecules of both gases, that is, $N = N_1 + N_2$, we obtain

$$P = \frac{NkT}{V}$$

Dalton's law may be generalized to apply to a mixture of any number of noninteracting ideal gases. Thus the ideal gas law may be applied to air, with N representing the total number of molecules of any type—nitrogen, oxygen, and so forth.

EXAMPLE 3 The Number of Air Molecules in a Hot-air Balloon

The air inside a hot-air balloon (Fig. 12-3) is at a temperature of 100.0° C, while the temperature of the surrounding air in the atmosphere is 20.0° C. Find the ratio of the number of air molecules inside the balloon to the number of air molecules contained in an equal volume of air outside the balloon. Assume that the air pressure is the same inside and outside.

SOLUTION We first use the ideal gas law (Eq. 12-5) to obtain an expression for the number of air molecules in a volume V.

$$N = \frac{PV}{kT}$$

Using subscripts i and o to denote air inside and outside the balloon, we obtain the ratio of the number of air molecules inside to the number outside:

$$\frac{N_i}{N_o} = \frac{P_i V_i / kT_i}{P_o V_o / kT_o}$$

Since pressures P_i and P_o are equal and volumes V_i and V_o are equal, this reduces to

$$\frac{N_i}{N_o} = \frac{T_o}{T_i}$$

Fig. 12-3 Hot-air balloon.

Converting temperatures to kelvins, we find

$$\frac{N_i}{N_o} = \frac{273 + 20}{273 + 100} = \frac{293\text{ K}}{373\text{ K}} = 0.786$$

According to Archimedes' principle, the hot-air balloon will experience a buoyant force equal to the weight of the displaced air. Since there is less air inside the balloon than in the volume of atmosphere displaced, the buoyant force is greater than the weight of the hot air, and the balloon will rise if the weight carried by the balloon is not too great. (Problem 50 asks you to calculate the volume of a hot-air balloon required to support 5000 N.)

EXAMPLE 4 A Deep Dive Must be Short

A Scuba diver breathes air stored in a tank carried on the back. At the beginning of a dive air pressure in the tank is about 200 atm, or 3000 lb/in^2. Air pressure inside the tank gradually decreases as the diver uses the air. A pressure regulator adjusts the pressure of the air coming out of the tank so that when the diver breathes it the air is at the same pressure as the surrounding water. A certain diver has sufficient air to stay 60 min at a depth of 10 m below the ocean's surface, where the pressure is 2.0 atm (Fig. 12-4). How long would this diver's air supply last at a depth of 30 m, where the pressure is 4.0 atm? The diver breathes the same volume of air per unit time at any depth. Assume a constant temperature.

Fig. 12-4 Scuba diver.

SOLUTION From the ideal gas law, we know that the number of air molecules in a volume V at pressure P and temperature T is given by

$$N = \frac{PV}{kT}$$

This equation shows that the number of air molecules in a given volume of air breathed by the diver is directly proportional to the air's pressure. Thus, as the diver descends to greater depths where the air pressure must be greater, a given volume of inhaled air contains more air molecules, and the air supply is therefore consumed more quickly. Since the volume of air consumed per unit time is constant, the volume consumed in time t is proportional to t.

$$V \propto t$$

Since N is proportional to PV, it follows that

$$N \propto Pt$$

A fixed number of air molecules are available to the diver. Thus the product of the air pressure and the time to consume the air has the same value at any depth.

$$P't' = Pt$$

Inserting $P = 2.0$ atm and $t = 60$ min for a depth of 10 m and $P' = 4.0$ atm for a depth of 30 m, we find that the air supply at this depth lasts a time t', where

$$(4.0 \text{ atm})t' = (2.0 \text{ atm})(60 \text{ min})$$
$$t' = 30 \text{ min}$$

Because the pressure is doubled, the air lasts only half as long.

Atomic Mass

It is often convenient to express the ideal gas law in a slightly different form, known as the "molar form." To accomplish this, we first define atomic mass and the mole. **Atomic mass is the mass of an atom relative to other atoms, using a scale in which the most common type of carbon atom is defined to have a mass of exactly 12.** A hydrogen atom has about $\frac{1}{12}$ the mass of a carbon atom and so has an atomic mass of approximately 1. A helium atom has about $\frac{4}{12}$ the mass of a carbon atom and so has an atomic mass of approximately 4. The periodic table of the elements shows the atomic masses of all the elements. The atomic masses listed there are actually averages over the different types of atoms naturally occurring for each element. For example, the atomic mass of carbon is given as 12.01, rather than exactly 12, because roughly 1% of all naturally occurring carbon atoms have a mass of 13.

The molecular mass of a molecule is the sum of the atomic masses of the atoms making up the molecule. For example, the molecular mass of the H_2O molecule equals the atomic mass of oxygen plus twice the atomic mass of hydrogen, that is, approximately $16 + 2(1) = 18$.

The unit of mass on the atomic mass scale is called the **atomic mass unit,** denoted by u. We can relate this unit to the gram. Experiment shows that

$$1 \text{ u} = 1.6606 \times 10^{-24} \text{ g}$$

Mole

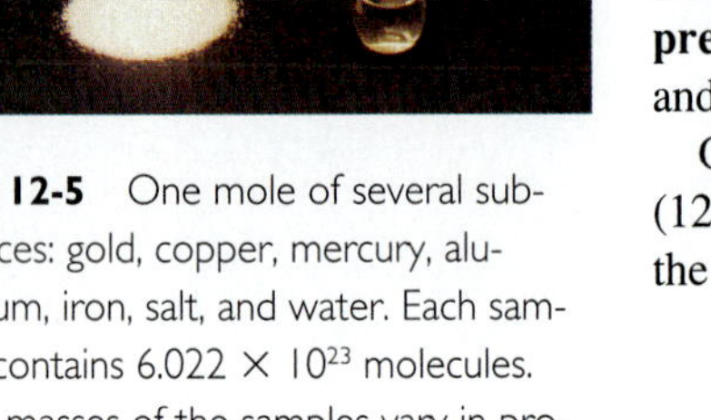

Fig. 12-5 One mole of several substances: gold, copper, mercury, aluminum, iron, salt, and water. Each sample contains 6.022×10^{23} molecules. The masses of the samples vary in proportion to the molecular masses: from 18g of water up to 201g of mercury.

Even very small quantities of matter consist of an enormously large number of molecules, and so it is convenient to express the quantity of matter in terms of a large unit, called the "mole." **A mole is defined as a certain number of atoms or molecules, called Avogadro's number,** denoted by N_A. The value of Avogadro's number is such that **one mole** of a substance, consisting of any kind of atom or molecule, **has a mass numerically equal to the atomic or molecular mass** of that substance **expressed in grams.** For example, one mole of carbon-12 atoms has a mass of 12 g, and one mole of H_2O molecules has a mass of 18 g (Fig. 12-5).

One can compute Avogadro's number by dividing the mass of 1 mole of carbon 12 (12 g) by the mass of a single carbon-12 atom, equal to 12 atomic mass units, where the atomic mass unit is related to the gram by the preceding equation.

$$N_A = \frac{12 \text{ g}}{12 \text{ u}} = \frac{12 \text{ g}}{12(1.6606 \times 10^{-24} \text{ g})}$$

$$\boxed{N_A = 6.022 \times 10^{23}} \tag{12-10}$$

We may express the number of molecules, N, of a substance as the product of Avogadro's number, N_A, and the number of moles, denoted by n:

$$N = nN_A \tag{12-11}$$

EXAMPLE 5 Number of Atoms in a Nail

Find the number of atoms in an iron nail of mass 5.00 g.

SOLUTION First we inspect the periodic table (shown on the inside back cover) and find that the atomic mass of iron (Fe) is 55.847. This means that 1 mole of naturally occurring iron has a mass of 55.847 g. We can now calculate the number of moles of iron in the nail, which we denote by n:

$$n = (5.00 \text{ g})\left(\frac{1 \text{ mole}}{55.847 \text{ g}}\right) = 8.95 \times 10^{-2} \text{ mole}$$

Since 1 mole contains Avogadro's number of atoms, the nail contains a number of atoms equal to the number of moles times N_A:

$$N = nN_A = (8.95 \times 10^{-2} \text{ mole})\left(\frac{6.022 \times 10^{23} \text{ atoms}}{1 \text{ mole}}\right)$$

$$= 5.39 \times 10^{22} \text{ atoms}$$

Molar Form of the Ideal Gas Law

To obtain the molar form of the ideal gas law, we substitute $N = nN_A$ (Eq. 12-11) into our original form of the gas law (Eq. 12-5):

$$PV = NkT = nN_AkT$$

The product N_Ak is called the ideal gas constant, denoted by R.

$$R = N_Ak$$

$$= (6.022 \times 10^{23})(1.380 \times 10^{-23} \text{ J/K})$$

$$\boxed{R = 8.31 \text{ J/K}} \tag{12-12}$$

Substituting R for N_Ak in the ideal gas law, we obtain the molar form of the gas law.

$$\boxed{PV = nRT} \tag{12-13}$$

EXAMPLE 6 Finding the Mass of a Volume of Air

Find the mass of air in a room with dimensions 5.00 m $\times$ 4.00 m $\times$ 3.00 m, if the air pressure is 1.00 atm and the temperature is 27.0° C.

SOLUTION First we apply the ideal gas law (Eq. 12-13) to find the number of moles, using the Kelvin temperature ($T = 273 + 27.0 = 300$ K) and expressing pressure in Pa (Eq. 11-5: 1.00 atm = 1.01×10^5 Pa).

$$n = \frac{PV}{RT}$$

$$= \frac{(1.00\,\text{atm})(1.01 \times 10^5\,\text{Pa/atm})(5.00\,\text{m} \times 4.00\,\text{m} \times 3.00\,\text{m})}{(8.31\,\text{J/K})(300\,\text{K})}$$

$$= 2.43 \times 10^3 \text{ moles}$$

Since 1 mole has a mass equal to the molecular mass in grams, the mass of air equals the product of the number of moles times the molecular mass in grams. A nitrogen molecule N_2 has a molecular mass of 2(14) = 28, and an oxygen molecule O_2 has a molecular mass of 2(16) = 32. Air consists of approximately 80% nitrogen and 20% oxygen, and so the average molecular mass is $0.8 \times 28 + 0.2 \times 32 = 28.8$ Thus

$$m = (2.43 \times 10^3\text{ moles})(28.8\text{ g/mole}) = 7.00 \times 10^4\text{ g}$$

$$= 70.0\text{ kg}$$

12-3 Kinetic Theory; Model of an Ideal Gas

Kinetic Theory

Kinetic theory is an area of physics that was developed in the late nineteenth century by Rudolph Clausius, James Clerk Maxwell, Ludwig Boltzmann, and others. Kinetic theory provides an explanation for the behavior of a macroscopic system in terms of its microscopic components—atoms or molecules, which obey dynamical laws. In this section we shall use kinetic theory to provide an explanation for the pressure of an ideal gas in terms of a molecular model.

A gas contained in a volume of macroscopic dimensions consists of an enormously large number of molecules. These molecules move in a random, chaotic way throughout the volume of the container (Fig. 12-6). When a molecule strikes a surface, it bounces off, exerting a small force on the surface (Fig. 12-7a). At any instant there will be many molecules colliding with the surface. The effect of these collisions is to produce a resultant force, which may be quite large (Fig. 12-7b).

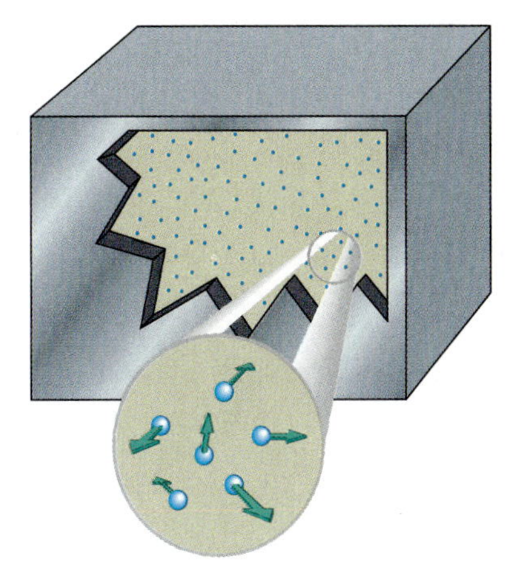

Fig. 12-6 A container of gas consists of a large number of molecules moving randomly and colliding with the walls of the container.

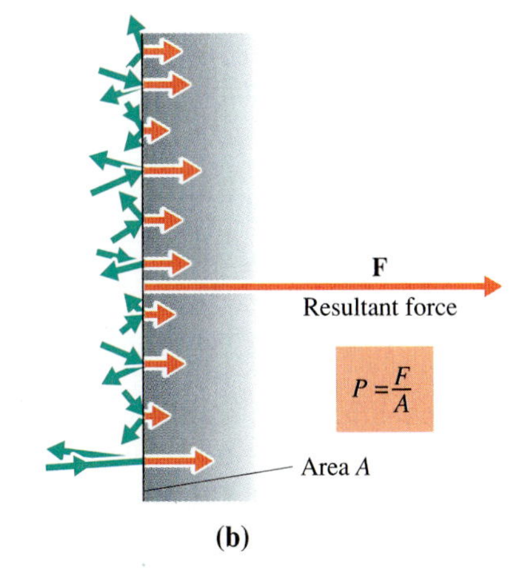

Fig. 12-7 **(a)** A single molecule collides with a wall of the container; during the collision the molecule exerts a small force perpendicular to the surface. **(b)** At any instant a large number of molecules collide with the surface. The effect of all these collisions is a resultant force **F** and pressure P, which are effectively constant.

The resultant force is not steady but fluctuates rapidly, depending on the number of molecules striking the surface at any instant. But, for a surface of macroscopic size, the number of molecules involved is so large that fluctuations in the net force are negligibly small. Thus the molecules exert a pressure on the container walls that is effectively constant.

By way of analogy, consider the pressure that a strong, fine spray of water from a shower head produces on your back. There are many individual collisions with the back, resulting in a fairly steady pressure. In the case of gas pressure, each molecule exerts an extremely small force on the surface with which it collides. And yet there is such an enormously large number of collisions that the net effect is a pressure that is much greater and much more steady than that produced by a stream of water from a shower. Indeed atmospheric pressure is 1.01×10^5 Pa, or 14.7 lb/in^2, an indication that one's back is constantly subject to a total force amounting to several hundred pounds because of air pressure. Of course in normal circumstances atmospheric pressure is the same on all surfaces of the body, and so it gives rise to no resultant force.

Kinetic Interpretation of Temperature

Rudolf Clausius first derived the ideal gas law from kinetic theory in 1857. He used a very oversimplified model of a gas, in which the molecules were assumed to move in three orderly lines, perpendicular to the sides of the box. In the next section we shall derive the ideal gas law, using the more realistic model of molecules moving randomly in all directions. This derivation, first presented by James Clerk Maxwell in 1859, shows that **the average kinetic energy of an ideal gas molecule equals $\frac{3}{2}kT$.** A molecule of mass m traveling at speed v has kinetic energy $K = \frac{1}{2}mv^2$. Denoting the average values of K and v^2 by $\overline{K}$ and $\overline{v^2}$, we may express Maxwell's result as:

$$\overline{K} = \frac{1}{2}m\overline{v^2} = \frac{3}{2}kT \tag{12-14}$$

This equation shows that **the temperature of an ideal gas is a measure of the average kinetic energy of its molecules.** In light of this result, the phenomenon of thermal equilibrium is easy to understand. Two systems of gas that are initially at different temperatures have different values of average kinetic energy per molecule. When the systems are placed in thermal contact, the system with the higher temperature will lose energy as the system with the lower temperature gains energy. This process continues until the average molecular kinetic energies and hence also the temperatures of the two systems are the same. Thus thermal equilibrium is simply a consequence of the equal sharing of kinetic energy among the molecules of both systems.

Although Eq. 12-14 is easily derived only for an ideal gas, it applies to any system, including liquids and solids. The average translational kinetic energy of the molecules in a body at absolute temperature T equals $\frac{3}{2}kT$.

Maxwell Distribution

Maxwell carried his analysis a step farther and derived an expression for the distribution of molecular speeds. Maxwell's theoretical prediction is shown graphically in Fig. 12-8 for oxygen at a temperature of 300 K.

Fig. 12-9 indicates how the distribution changes with temperature. As T increases, the distribution becomes broader and shifts to the right, toward higher values of v. The Maxwell distribution has been amply verified by experiments.

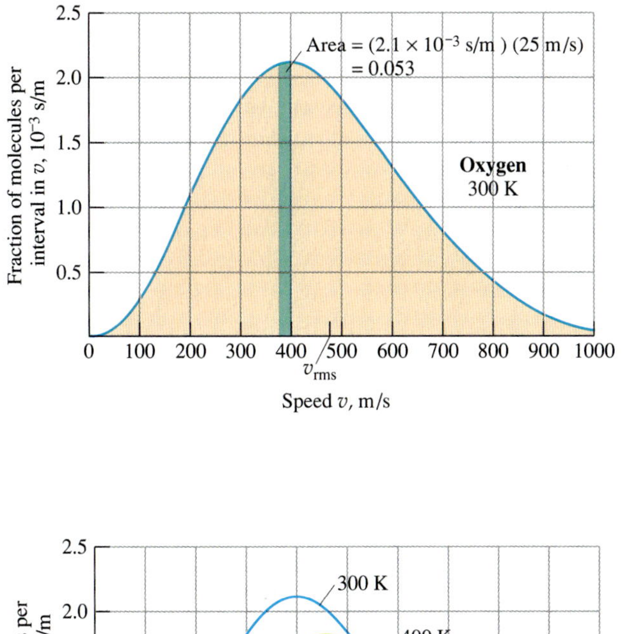

Fig. 12-8 Distribution of molecular speeds for oxygen at 300 K. The area under the curve between any two values of v gives the fraction of the molecules having speed somewhere in the interval between the two values of v. For example, the cross-hatched area is 0.053, meaning that 5.3% of the molecules have values of v between 375 m/s and 400 m/s.

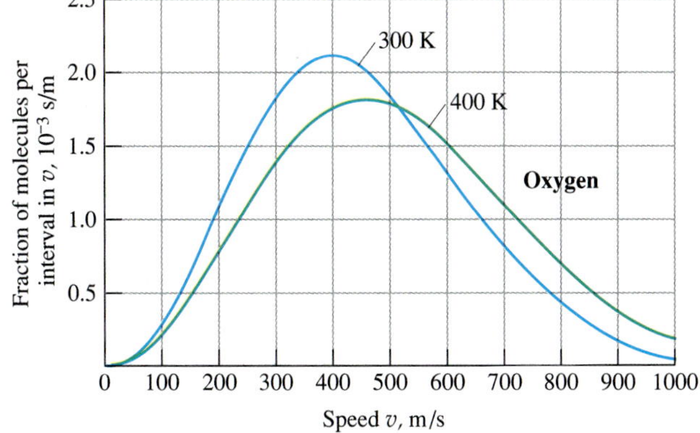

Fig. 12-9 Distribution of molecular speeds for oxygen at 300 K and at 400 K.

Root Mean Square Speed

It is convenient to be able to characterize the Maxwell distribution by a single number, typical of molecular speeds at a given temperature. For this purpose one may use either the average speed or the root mean square (rms) speed, defined as the square root of the average squared speed and denoted by v_{rms}.

$$v_{rms} = \sqrt{\overline{v^2}} \tag{12-15}$$

For the Maxwell distribution v_{rms} is approximately equal to the average speed.

Since we have already derived an expression for the average molecular kinetic energy (Eq. 12-14), it is easy to obtain an expression for v_{rms}.

$$\tfrac{1}{2}m\overline{v^2} = \tfrac{3}{2}kT$$

Thus

$$\overline{v^2} = \frac{3kT}{m}$$

and

$$v_{rms} = \sqrt{\frac{3kT}{m}} \tag{12-16}$$

EXAMPLE 7 RMS Speed of Oxygen Molecules at Room Temperature

Calculate v_{rms} for oxygen molecules at a temperature of 300 K.

SOLUTION The molecular mass of oxygen is 32, which means that an oxygen molecule has a mass of 32 atomic mass units (32 u), where $\mathrm{u} = 1.66 \times 10^{-27}$ kg. Applying Eq. 12-16, we find

$$v_{\mathrm{rms}} = \sqrt{\frac{3kT}{m}} = \sqrt{\frac{3kT}{32\ \mathrm{u}}} = \sqrt{\frac{3(1.38 \times 10^{-23}\ \mathrm{J/K})(300\ \mathrm{K})}{32(1.66 \times 10^{-27}\ \mathrm{kg})}} = 484\ \mathrm{m/s}$$

This speed, faster than a speeding bullet, is near the center of the molecular speed distribution shown in Fig. 12-8.

*12-4 Derivation of the Ideal Gas Law

We shall now derive the ideal gas law, using the kinetic theory model of an ideal gas described in the last section. We shall do this by first obtaining an expression for the pressure exerted by the molecules on a surface. This expression turns out to be a function of the number of molecules per unit volume and the molecules' average kinetic energy. By relating temperature to molecular kinetic energy, we shall obtain the ideal gas law. This derivation serves two purposes: (1) to understand how the empirical ideal gas law is the result of molecular dynamics and (2) to give a mechanical interpretation of temperature as a measure of molecular kinetic energy.

To derive the ideal gas law, we shall find it necessary to make certain assumptions:

1. **It is valid to apply Newton's laws to gas molecules,** just as we would to billiard balls or other macroscopic bodies.*

2. **Collisions of molecules with the walls of the container are elastic, and the walls are smooth.** It follows that when a molecule strikes a wall, the molecule rebounds, with its component of velocity perpendicular to the wall reversed, as shown in Problem 32, Chapter 8.

3. **The number of molecules is large, and their motion is random.**

4. **The time during which molecules are in contact either with the walls or with each other is negligible** compared to the time during which they are moving freely. This requires of course that the molecules not be too tightly packed, that is, that the density of the gas be sufficiently low.

Let the container for the gas be a rectangular box with sides of length ℓ_x, ℓ_y, and ℓ_z, aligned with the x, y, z coordinate axes (Fig. 12-10). First we shall derive an expression for the pressure on the right side of the container, based on the assumption of no collisions between molecules. After deriving this result, we shall consider the effect of intermolecular collisions.

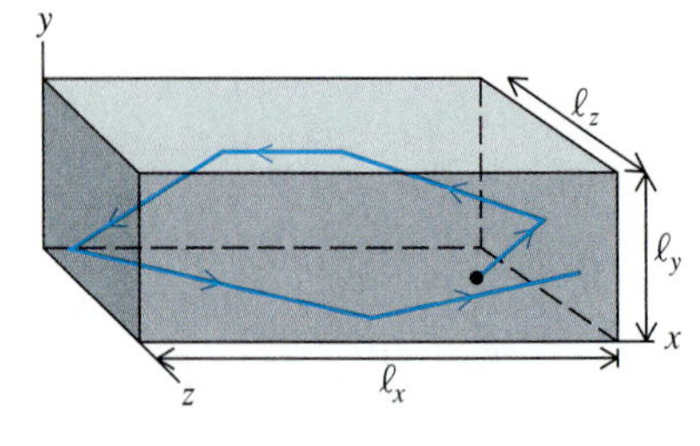

Fig. 12-10 The path of a single gas molecule in a box.

*Many features of atoms and molecules can be described only by use of the more general (and difficult) methods of quantum theory, rather than Newton's laws, as we shall see in Chapter 29. It is correct to use classical mechanics whenever the results obtained by classical mechanics are a good approximation to quantum predictions, as is the case here.

Consider the path of a single molecule, beginning at the instant just before it strikes the right side of the box. The effect of the collision with the right side is to reverse the x component of the molecule's velocity: v_x changes to $-v_x$. If this molecule does not collide with any other molecule, it will strike the left side of the box a time ℓ_x/v_x later, bounce off the left wall with its x component of the velocity reversed again ($-v_x$ changes to $+v_x$), and return to the right side a time ℓ_x/v_x after this collision. Thus the complete time Δt for the molecule to traverse the length of the box and return to the right side is given by

$$\Delta t = \frac{2\ell_x}{v_x}$$

Of course, the molecule may strike other sides of the box between collisions with the right and left ends, as shown in Fig. 12-10. These collisions do not change v_x, and so they do not affect the time interval between successive collisions with the right side of the box.

Each time the molecule collides with the right surface, the molecule exerts a force F_x to the right. This is just the reaction to the force F_x' the surface exerts on the molecule, and by Newton's third law $F_x = -F_x'$. According to Newton's second law, the average value of the force on the molecule, $\overline{F}_x'$, is the product of the molecule's mass and its average x component of acceleration, $\overline{a}_x$. Thus

$$\overline{F}_x = -\overline{F}_x' = -m\overline{a}_x$$

$$= -m\,\frac{\Delta v_x}{\Delta t}$$

For each collision with the wall, the molecule undergoes a change in v_x given by $\Delta v_x = (-v_x) - v_x = -2v_x$. And since the time interval Δt between collisions is $2\ell_x/v_x$, the average force may be expressed

$$\overline{F}_x = \frac{-m(-2v_x)}{2\ell_x/v_x} = \frac{mv_x^2}{\ell_x}$$

Within the box there are N molecules, each having its own value of v_x^2. The resultant force $\Sigma\,\overline{F}_x$ is just the sum of the forces exerted by the N molecules.

$$\Sigma\,\overline{F}_x = \Sigma\,\frac{mv_x^2}{\ell_x} = \frac{m}{\ell_x}\,\Sigma\,v_x^2$$

We may express $\Sigma\,v_x^2$ in terms of the average value of v_x^2, denoted by $\overline{v_x^2}$, and defined as $\overline{v_x^2} = \frac{1}{N}\,\Sigma\,v_x^2$.

$$\Sigma\,\overline{F}_x = \frac{mN\overline{v_x^2}}{\ell_x}$$

Dividing $\Sigma\,\overline{F}_x$ by the area of the surface $\ell_y\ell_z$, we obtain the pressure P.

$$P = \frac{\Sigma\,\overline{F}_x}{\ell_y\ell_z} = \frac{Nm\overline{v_x^2}}{\ell_x\ell_y\ell_z}$$

Since the volume $V = \ell_x\ell_y\ell_z$, we have

$$P = \frac{Nm\overline{v_x^2}}{V}$$

or

$$PV = Nm\overline{v_x^2} \qquad (12\text{-}17)$$

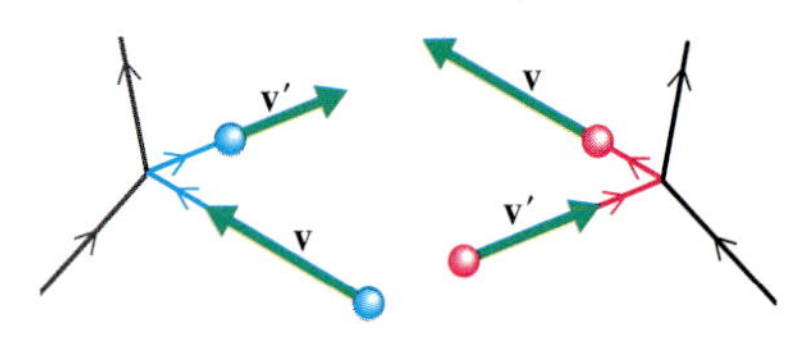

Fig. 12-11 A molecule changes its velocity from **v** to **v′** because of a collision. A second molecule, at approximately the same location, undergoes a collision that changes its velocity from **v′** to **v**.

The model we have given so far is a reasonable model for a very low-density gas. However, it is a terribly unrealistic picture of a gas such as air at normal density, since we have neglected collisions between molecules. An air molecule may undergo something on the order of 10^5 collisions for every cm it travels. These collisions will certainly destroy the regular path we have assumed for a single molecule moving across the length of the box. We shall see, however, that the very large number of molecules and the randomness of their motion cause the result to be the same as though there were no collisions. At this point assumption 3 must be used in a very specific way: for every molecule that has its velocity changed from **v** to **v′** by a collision, we assume that there is another molecule at approximately the same location that undergoes a collision that changes its velocity from **v′** to **v**. See Fig. 12-11. Whenever a molecule undergoes a collision, it is replaced by another molecule at essentially the same position and moving with the same velocity, and so the preceding calculation of pressure is unaffected.

To show that this assumption is plausible, we note that at typical gas densities of 10^{25} molecules per cubic meter a tiny cube of edge length 10^{-2} mm, or 10^{-5} m, and volume 10^{-15} m^3 contains on the order of 10^{10} molecules. Although only a small fraction of these molecules undergo collisions at any one instant (perhaps one out of 10^3 or 10^4), still the total number in the cube undergoing simultaneous collisions is large (about 10^6 or 10^7), and the collisions are completely random. So it is very reasonable to assume that with this large number of random collisions the distribution of molecular velocities remains unchanged.

Finally, if the preceding calculation of pressure is to remain unchanged by the effects of intermolecular collisions, we must assume that the time of intermolecular contact is small compared to the time between collisions (assumption 4), and so our calculation of Δt is unaffected.

It is convenient to express Eq. 12-17 in terms of the average value of v^2, rather than the average value of the component v_x^2. Since $v^2 = v_x^2 + v_y^2 + v_z^2$, the average value of v^2 is the sum of the averages of v_x^2, v_y^2, and v_z^2.

$$\overline{v^2} = \overline{v_x^2} + \overline{v_y^2} + \overline{v_z^2}$$

But there is no reason for the average squared components of velocity along the x, y, and z axes to differ, and so $\overline{v_x^2} = \overline{v_y^2} = \overline{v_z^2}$, and we can express the equation above as

$$\overline{v^2} = 3\overline{v_x^2}$$

or

$$\overline{v_x^2} = \tfrac{1}{3}\overline{v^2}$$

Substituting this result into Eq. 12-17, we obtain

$$PV = Nm\left(\tfrac{1}{3}\overline{v^2}\right) = N\left(\tfrac{1}{3}m\overline{v^2}\right)$$

We can complete our derivation of the ideal gas law ($PV = NkT$) if we identify $\frac{1}{3}m\overline{v^2}$ in the expression above as kT. Thus in the process of deriving the gas law, we arrive at a mechanical interpretation of the temperature of an ideal gas.

$$\tfrac{1}{3}m\overline{v^2} = kT$$

or

$$\tfrac{1}{2}m\overline{v^2} = \tfrac{3}{2}kT \tag{12-18}$$

We have used an atomic model of a gas to derive the ideal gas law. Although today this seems a very natural way to explain gas pressure, it was not always so. In the early 1900s the very existence of atoms was not universally accepted, since there was no direct evidence of them. Ludwig Boltzmann was one of the primary advocates of the atomic theory and used the atomic hypothesis in his analysis of macroscopic systems. But Boltzmann's work was severely criticized by other famous physicists, including Ernst Mach, on the grounds that physical theory should not be based on hypothetical entities such as atoms. Boltzmann committed suicide in 1906, just 2 years before direct confirmation of the atomic theory and universal acceptance of his methods.

*12-5 Vapor Pressure and Humidity

If you fill a glass with water and leave it, within a day or two the water level in the glass will drop noticeably, unless the surrounding air is very damp. The water "evaporates." This means that water molecules leave the liquid and form a vapor or gas that mixes with the air.

We can understand evaporation on a molecular level, using concepts of kinetic theory. The molecules in a liquid have a distribution of velocities, similar to the Maxwell-Boltzmann distribution of molecular velocities in a gas. Although intermolecular forces bind most of the molecules close together in the liquid, some of the molecules move fast enough to leave the surface of the liquid, like a rocket with a velocity greater than escape velocity leaving the earth. Since the molecules that evaporate are those with the greatest velocity and kinetic energy, the average kinetic energy of the molecules remaining in the liquid decreases, and so the temperature of the liquid decreases. You can feel the cooling effect of evaporation when you step out of a shower and water evaporates from your skin. The effect is more dramatic when the air is very dry. If you step out of a swimming pool in the desert, even though the air may be quite hot, you can be chilled by water evaporating from your skin.

To better understand the process of evaporation, consider a liquid in a closed container with a piston (Fig. 12-12a). If the piston is raised, evaporation begins. Vapor fills the space above the liquid (Fig. 12-12b) and creates a pressure, called **vapor pressure.** As more and more molecules enter the vapor, some molecules begin to go from the vapor back into the liquid. Initially more molecules leave the liquid than enter it, and both the density and pressure of the vapor increases. There is soon reached an equilibrium state, in which as many molecules enter the liquid as leave it. We say the vapor is then "saturated," since there can be no further increase in the number of molecules in the vapor (Fig. 12-12c). In this equilibrium state, vapor pressure reaches its maximum value, called **saturated vapor pressure.** If the piston is raised higher, more molecules enter the vapor phase until the vapor pressure again reaches the same saturated vapor pressure.

As the temperature of a liquid increases, more of its molecules have sufficient kinetic energy to escape the liquid. Thus, as temperature increases, the rate at which molecules leave the surface of the liquid increases; that is, the rate of evaporation increases. An equilibrium state is not reached until the density of the vapor phase increases enough that the rate at which molecules reenter the liquid matches the new higher rate at which molecules leave the liquid (Fig. 12-12d). The higher-density equilibrium state is one of higher pressure. Thus, as temperature increases, saturated vapor pressure increases.

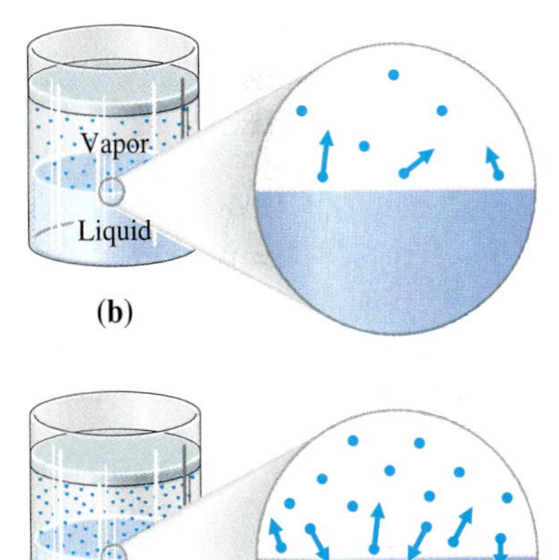

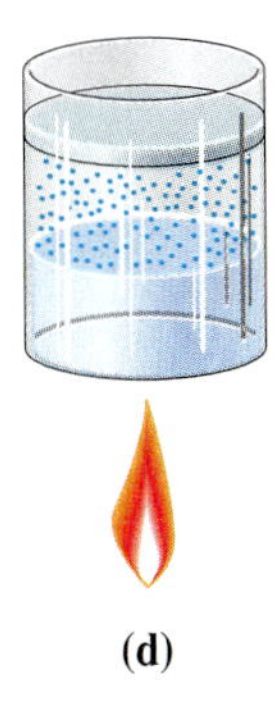

Fig. 12-12 **(a)** A liquid in a closed container with a movable piston. **(b)** Molecules begin to leave the liquid and form a vapor when the piston is raised. **(c)** Saturated vapor in equilibrium with the liquid. Equal numbers of molecules enter and leave the liquid. **(d)** Saturated vapor at a higher temperature has greater density and pressure.

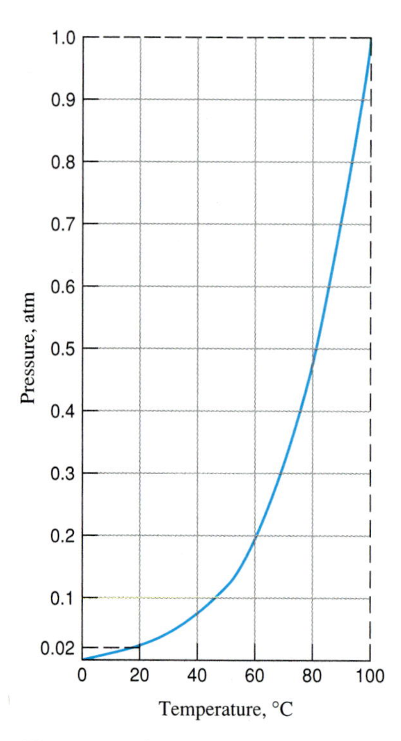

Fig. 12-13 Saturated vapor pressure of water as a function of temperature.

Fig. 12-13 shows a graph of saturated vapor pressure versus temperature for water. At 20° C water's saturated vapor pressure is only 0.02 atm, at 30° C it doubles to 0.04 atm, and at 100° C it increases to 1.0 atm—50 times greater than at 20° C.

When water is exposed to the open air, evaporation proceeds in the same way as when no air is present. According to Dalton's law of partial pressures, a mixture of gases produces a pressure that is the sum of the pressures produced by each of the separate gases. Thus nitrogen, oxygen, and water vapor each contribute their own partial pressures to the total pressure of our atmosphere. At a temperature of 20° C and a pressure of 1 atm, the partial pressure of water vapor in the air can be no greater than 2% of the total pressure, or 0.02 atm, since the saturated vapor pressure of water is 0.02 atm at 20° C.

We refer to the water content of air as its **humidity.** Humidity is at a maximum at a given temperature when the partial pressure of water in the air equals the saturated vapor pressure of water at that temperature. Since saturated vapor pressure increases rapidly with increasing temperature, the maximum humidity of warm air is much greater than the maximum humidity of cooler air. Typically air contains water vapor at a partial pressure below its saturated value. The **relative humidity** of air is defined as the ratio of the partial pressure of the water vapor in the air to the saturated vapor pressure of water at that temperature.

$$\text{Relative humidity} = \frac{\text{Partial pressure of water vapor}}{\text{Saturated vapor pressure of water}} \times 100 \qquad (12\text{-}19)$$

For example, if the partial pressure of water vapor in air at 20° C is 0.01 atm, since the saturated vapor pressure of water is 0.02 atm at 20° C, the relative humidity of the air is $\frac{0.01\ \text{atm}}{0.02\ \text{atm}} \times 100 = 50\%$.

Dew, Fog, and Rain

When air containing water vapor is cooled sufficiently, some of the vapor condenses to the liquid state. This condensation begins when the air drops below the temperature at which the water vapor's partial pressure equals the saturated vapor pressure, in other words, when it drops below the temperature at which relative humidity is 100%. This temperature is called the **dew point.** Such condensation often occurs at night when dew forms on the ground, as the temperature of the ground falls and cools the surrounding air. For example, suppose air temperature during the day is 30° C with a relative humidity of 75%, meaning that the partial pressure of water vapor in the air is 75% of the saturated vapor pressure of 0.04 atm at 30° C, or 0.03 atm. Then, if the temperature of the air near the ground falls to 20° C at night, since the saturated vapor pressure of water at that temperature is only 0.02 atm, the partial pressure of water vapor in the air must drop by one third from 0.03 atm to 0.02 atm, meaning that one third of the water vapor must condense.

Fog results when a humid warm air mass mixes with cooler air and tiny droplets of water form as the temperature of the warm air drops below the dew point. Clouds form when air rises and is cooled below the dew point. When a cloud cools suddenly, condensation is more rapid, water droplets increase in size, and then come together to form larger drops, which fall as rain.

Boiling

If the temperature of a liquid is raised enough so that the liquid's vapor pressure equals the pressure of the surrounding air, the liquid begins to boil. That is, bubbles of vapor form within the bulk liquid. These vapor bubbles push outward against the liquid, which is at approximately the same pressure as the air. Water boils at a temperature of 100° C when the surrounding air is at a pressure of 1.00 atm because the saturated vapor pressure of air at 100° C is 1.00 atm. If the surrounding air is at a lower pressure, water will boil at a lower temperature. For example, on a mountain at an elevation of 3000 m, where atmospheric pressure is only 0.7 atm, water boils at a temperature of 90° C, since its saturated vapor pressure at that temperature equals 0.7 atm.

12-6 Thermal Expansion

Nearly all solids and liquids expand as they are heated. The fractional increase in volume, $\Delta V/V$, is often found to be directly proportional to the increase in temperature, ΔT. The constant of proportionality is called the "volume coefficient of expansion," denoted by β. Thus

$$\frac{\Delta V}{V} = \beta\,\Delta T$$

or

$$\Delta V = \beta V\,\Delta T \tag{12-20}$$

The change in volume ΔV is proportional to the original volume V, as well as to the temperature change ΔT. Thus, for example, if 1 liter (1000 cm^3) of water is heated from 20° C to 25° C, its volume increases by only about 1 cm^3. But if the water in a swimming pool of volume 1000 m^3 is heated over the same temperature interval, the water increases in volume by 1 m^3, or 10^6 cm^3. In both cases the ratio $\Delta V/V$ is 10^{-3}.

In the case of solids, the volume expansion is accomplished by an increase in all linear dimensions. As a solid is heated, the distance between any two points in the solid increases. The fractional increase in length is normally the same in all directions. Thus, if a block of marble expands thermally by 0.1% in length, the block's height and width will each also increase by 0.1%. The increase in size is like a photographic enlargement (Fig. 12-14).

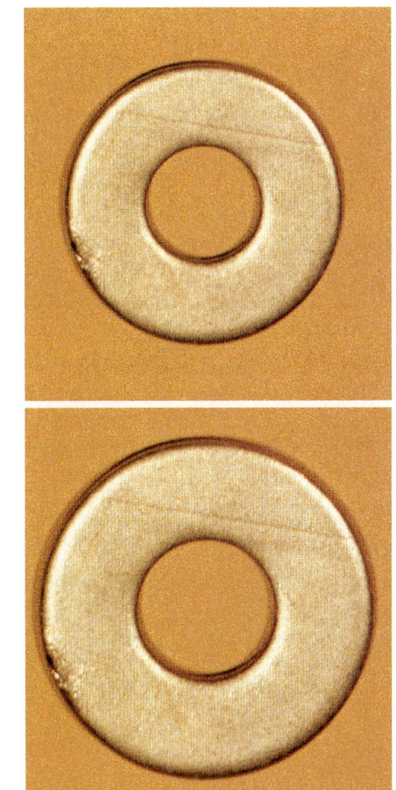

Fig. 12-14 As a washer is heated from temperature T to temperature $T + \Delta T$, all its linear dimensions get bigger. Even the hole gets bigger. The actual expansion, however, is much smaller than indicated here. The expansion shown here is 20%, which is approximately 100 times greater than the expansion of aluminum heated 100 C°.

Instead of using a volume coefficient of expansion for solids, we normally use a linear expansion coefficient α, which is a measure of the fractional change in the linear dimensions of the solid. For a temperature change ΔT, a length ℓ changes by $\Delta \ell$, where

$$\frac{\Delta \ell}{\ell} = \alpha\,\Delta T$$

or

$$\Delta \ell = \alpha \ell\,\Delta T \tag{12-21}$$

It is possible to show that the volume coefficient of expansion for a solid equals 3 times its linear coefficient:

$$\beta = 3\alpha \qquad \text{(for solids)} \tag{12-22}$$

A proof of this result is outlined in Problem 42. Coefficients of expansion for various materials are given in Table 12-2. For liquids there is no measure of linear expansion, since liquids must conform to the shapes of their containers.

Table 12-2 Coefficients of thermal expansion

Solids	α, $(C°)^{-1}$	Liquids and gases	β, $(C°)^{-1}$
Aluminum	2.4×10^{-5}	Ethyl alcohol	1.1×10^{-3}
Brass	1.9×10^{-5}	Mercury	1.8×10^{-4}
Brick	1×10^{-5}	Olive oil	7.2×10^{-4}
Concrete	1.2×10^{-5}	Water	2.1×10^{-4}
Copper	1.7×10^{-5}	Air	3.67×10^{-3}
Diamond	1.2×10^{-6}	Hydrogen	3.66×10^{-3}
Glass	4×10^{-6} to 1×10^{-5}	Nitrogen	3.67×10^{-3}
Gold	1.4×10^{-5}		
Graphite	2×10^{-6}		
Ice	5.1×10^{-5}		
Paraffin	1×10^{-4}		
Steel	1.2×10^{-5}		
Wood, parallel to fiber	5×10^{-6} to 1×10^{-5}		
Wood, across fiber	3×10^{-5} to 6×10^{-5}		

EXAMPLE 8 Expansion of the Golden Gate Bridge

Find the change in the total length of the 2700 m long Golden Gate Bridge, as the temperature of the bridge increases from 5° C to 25° C. The bridge is constructed of steel and concrete.

SOLUTION From Table 12-1 we find that steel and concrete both have the same coefficient of thermal expansion: 1.2×10^{-5} $(C°)^{-1}$. Thus both materials expand equally. We apply Eq. 12-21 to find the change in the bridge's length for the 20.0 C° increase in temperature.

$$\Delta\ell = \alpha\ell\,\Delta T = [1.2 \times 10^{-5}\,(C°)^{-1}](2700\text{ m})(20.0\text{ C°})$$
$$= 0.65\text{ m} = 65\text{ cm}$$

Bridges must be constructed with expansion joints to allow for such thermal expansion (Fig. 12-15); otherwise they might buckle. Sidewalks can also buckle. If a concrete sidewalk is poured at a cool temperature with no allowance for thermal expansion, the sidewalk can buckle when it gets hot (Fig. 12-16).

Fig. 12-15 An expansion joint on a bridge.

Fig. 12-16 This sidewalk buckled because of thermal expansion.

EXAMPLE 9 Overflow of an Expanding Liquid

A glass container that has a volume of 1.0 liter and is filled with alcohol is initially refrigerated at a temperature of 7° C. How much of the alcohol will overflow if the container is placed in a warm room where the temperature is 27° C? The linear coefficient of expansion for the glass is $1.0 \times 10^{-5}\ (\mathrm{C}^\circ)^{-1}$.

SOLUTION The volume coefficient of expansion for the glass is 3 times its linear coefficient, or $3.0 \times 10^{-5}\ (\mathrm{C}^\circ)^{-1}$. Thus

$$\begin{aligned}\Delta V_{\text{glass}} &= \beta_{\text{glass}} V\, \Delta T \\ &= [3.0 \times 10^{-5}\ (\mathrm{C}^\circ)^{-1}](1.0 \times 10^3\ \mathrm{cm}^3)(20\ \mathrm{C}^\circ) \\ &= 0.60\ \mathrm{cm}^3\end{aligned}$$

From Table 12-1 we find that the volume coefficient of expansion for alcohol is 1.1×10^{-3}—much larger than the coefficient for glass. Therefore the alcohol will expand more than the glass.

$$\begin{aligned}\Delta V_{\text{alc}} &= \beta_{\text{alc}} V\, \Delta T \\ &= [1.1 \times 10^{-3}\ (\mathrm{C}^\circ)^{-1}](1.0 \times 10^3\ \mathrm{cm}^3)(20\ \mathrm{C}^\circ) \\ &= 22\ \mathrm{cm}^3\end{aligned}$$

Since the volume of the glass container increases by only about 1 cm^3, about 21 cm^3 of the alcohol will overflow.

EXAMPLE 10 Thermal Stress

A 2.0 m long aluminum window frame is mounted snugly in a brick wall at a temperature of 10° C. Find the stress in the frame when the temperature rises to 30° C. Young's modulus for aluminum is $6.9 \times 10^{10}\ \mathrm{N/m}^2$.

SOLUTION From Table 12-1 we find that the coefficients of thermal expansion for aluminum and brick are $2.4 \times 10^{-5}\ (\mathrm{C}^\circ)^{-1}$ and $1 \times 10^{-5}\ (\mathrm{C}^\circ)^{-1}$ respectively. Thus, if both were free to expand as the temperature increases, the aluminum would expand more. However, the expansion of the aluminum is limited by the surrounding brick. The window can expand no more than the opening, which expands in the same way as the brick itself. The brick walls exert stress on the aluminum, in effect compressing it from the dimensions it would have if it were free to expand to the dimensions of the expanded opening.

First we apply Eq. 12-21 to compute the increase in the length of the opening, $\Delta\ell_{\text{opening}}$, for a 20 C° temperature increase, using the thermal expansion coefficient for brick.

$$\begin{aligned}\Delta\ell_{\text{opening}} &= \alpha\ell\, \Delta T = [1 \times 10^{-5}\ (\mathrm{C}^\circ)^{-1}](2.0\ \mathrm{m})(20\ \mathrm{C}^\circ) \\ &= 4 \times 10^{-4}\ \mathrm{m} = 0.4\ \mathrm{mm}\end{aligned}$$

Next we compute the increase in length the aluminum *would* have if it were free to expand thermally.

$$\begin{aligned}\Delta\ell_{\text{Al}} &= \alpha\ell\, \Delta T = [2.4 \times 10^{-5}\ (\mathrm{C}^\circ)^{-1}]\,(2.0\ \mathrm{m})(20\ \mathrm{C}^\circ) \\ &= 9.6 \times 10^{-4}\ \mathrm{m} = 0.96\ \mathrm{mm}\end{aligned}$$

The brick's effect on the window frame is to reduce its expansion from 0.96 mm to 0.4 mm. That is, the brick in effect compresses the aluminum frame by 0.96 mm − 0.4 mm = 0.6 mm. This means that the brick exerts a stress σ, which we can compute using Eqs. 10-6 ($\epsilon = \Delta\ell/\ell$) and 10-8 ($Y = \sigma/\epsilon$)

$$\begin{aligned}\sigma &= Y\epsilon = Y\frac{\Delta\ell}{\ell} \\ &= (6.9 \times 10^{10}\ \mathrm{N/m}^2)\frac{0.6 \times 10^{-3}\ \mathrm{m}}{2.0\ \mathrm{m}} \\ &= 2 \times 10^7\ \mathrm{N/m}^2\end{aligned}$$

This means that every 1 cm^2 section of the window frame adjacent to the brick is subjected to a force

$$F = \sigma A = (2 \times 10^7\ \mathrm{N/m}^2)(10^{-2}\ \mathrm{m})^2 = 2000\ \mathrm{N}$$

Such a large force is likely to bend the frame and break the window glass.

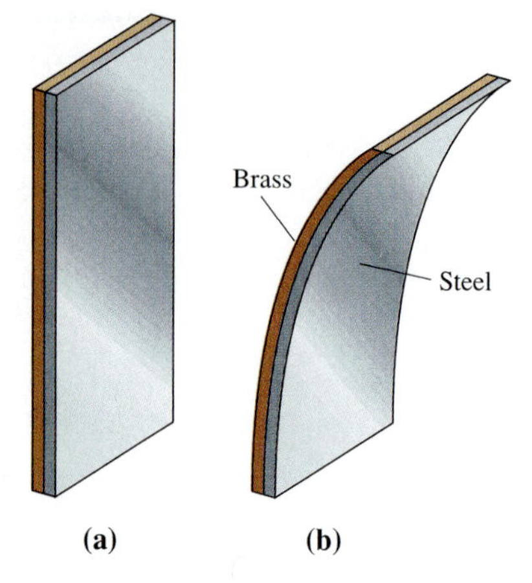

Fig. 12-17 **(a)** A bimetallic strip. Both metals initially have the same length. **(b)** When heated, the brass expands more than the steel.

Bimetallic Strip

Thermal expansion is applied in a bimetallic strip, which can be used to make a thermometer or a thermostat. The bimetallic strip is formed by welding or riveting together two thin strips of metal with different α's (Fig. 12-17). For example, a bimetallic strip may consist of brass and steel, with thermal expansion coefficients of $1.9 \times 10^{-5}\,(\mathrm{C}^\circ)^{-1}$ and $1.2 \times 10^{-5}\,(\mathrm{C}^\circ)^{-1}$ respectively. When heated, the brass tends to expand more than the steel, since α is greater for brass than for steel. But along the surface where the two metals are bonded together, their expansion must be the same. The result is that the two strips bend toward the steel, and so the average separation between molecules in the brass is greater than the average separation between molecules in the steel. Fig. 12-18 shows how a bimetallic strip can be used in a thermostat, which controls a heating system. The thermostat switch is on when the temperature is low, and the switch turns off when the temperature increases.

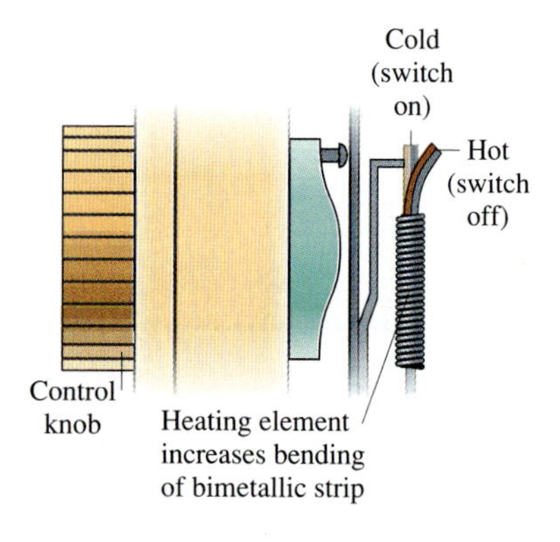

Fig. 12-18 The thermostat switch opens when the temperature increases enough to bend the bimetallic strip to the right.

Thermal Expansion of Water

In many cases the simple linear dependence of ΔV on ΔT expressed by Eq. 12-20 is valid over all temperature ranges commonly encountered, with a constant value for β. However, for some substances the variation of volume with temperature is more complicated. Water is such a substance. Fig. 12-19 shows the density of water as a function of temperature. Notice that at most temperatures the density of water decreases as its temperature increases; that is, water expands as it is heated. But in the temperature range from 0° C to 4° C water contracts as it is heated. Water is one of the few materials that have this property. This has an important effect on the rate at which lakes freeze. As air temperatures drop, the temperature of the water in a lake drops also, with the cooling occurring first at the surface of the lake. For temperatures above 4° C this cooling proceeds very efficiently. As water cools, it becomes more dense and sinks to a lower level in the lake, as warmer, less dense water rises to take its place. Thus there is a natural mixing of warmer and colder water, causing rapid cooling of water beneath the surface. However, when the water reaches a uniform temperature of 4° C, the process changes. Cooling of the surface water below 4° C decreases its density. Thus it stays at the surface, and further cooling of the water beneath proceeds more slowly. The surface of the lake may freeze. But in even the coldest weather, large lakes do not freeze solid. The water at the bottom of the lake remains at 4° C, enabling the marine life there to survive.

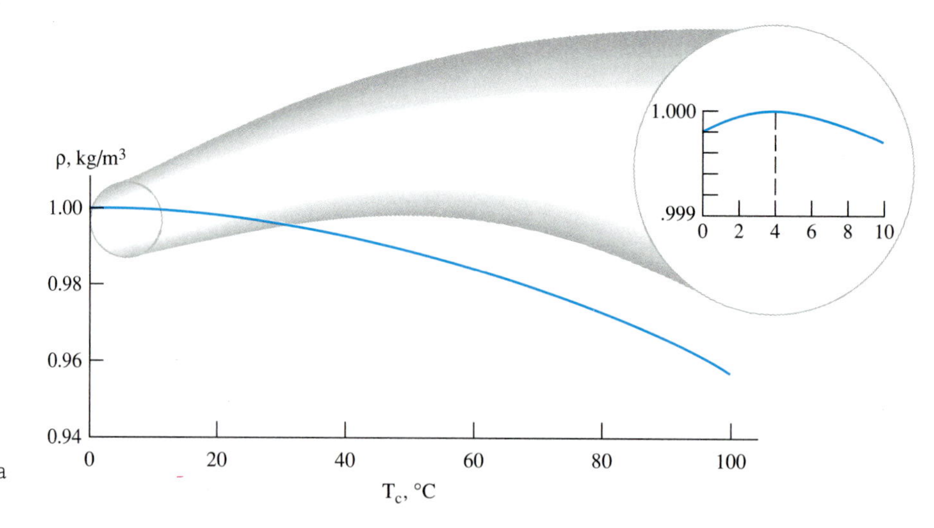

Fig. 12-19 The density of water as a function of temperature.

CHAPTER 12 SUMMARY

The Kelvin temperature scale is based on measurements of pressure for a low-density gas at constant volume. The temperature 273.16 K is assigned to the triple point of water (the point at which water can coexist in the solid, liquid, and vapor states). The ratio of the gas pressure P to its value at the triple point, P_{tr}, defines the temperature.

$$T = (273.16\ \text{K})\frac{P}{P_{tr}}$$

Celsius and Fahrenheit temperature scales are defined relative to the Kelvin scale.

$$T_C = T - 273.15$$

$$T_F = 32 + \tfrac{9}{5}T_C$$

The ideal gas law relates the pressure P, volume V, number of molecules N, and temperature T for an ideal gas:

$$PV = NkT$$

where k is Boltzmann's constant.

$$k = 1.380 \times 10^{-23}\ \text{J/K}$$

One mole of a substance contains Avogadro's number of molecules, N_A, and has a mass in grams numerically equal to the molecular mass of the substance.

$$N_A = 6.022 \times 10^{23}$$

The ideal gas law may also be expressed in terms of the number of moles, n.

$$PV = nRT$$

where $$R = N_A k = 8.31\ \text{J/K}$$

Dalton's law of partial pressures states that a mixture of noninteracting ideal gases produces a pressure that is the sum of the pressures that would be produced by each of the gases alone.

Kinetic theory uses a molecular model to derive results such as the ideal gas law.

At a given temperature, the distribution of molecular speeds is given by the Maxwell distribution. A typical molecular speed at any given temperature T is the rms speed:

$$v_{rms} = \sqrt{\overline{v^2}} = \sqrt{\frac{3kT}{m}}$$

where m is the molecular mass.

For many materials, heating produces an increase in volume that is proportional to the increase in temperature, ΔT.

$$\Delta V = \beta V\, \Delta T$$

where V is the original volume and β is the coefficient of expansion.

For solids there is a linear coefficient of expansion, α, that determines the change in length $\Delta\ell$, corresponding to a temperature change ΔT.

$$\Delta\ell = \alpha\ell\, \Delta T$$

For a solid the volume coefficient is 3 times the linear coefficient:

$$\beta = 3\alpha$$

Questions

1 Two bodies, A and B, are initially at the same temperature. If the temperature of A increases by 1 C° and the temperature of B increases by 1 F°, which has the higher final temperature?

2 Will a thermometer in direct sunlight accurately measure the temperature of the surrounding air?

3 Air bubbles are expelled by a scuba diver on the ocean floor (Fig. 12-20). Does the size of a bubble change as it rises to the surface?

Fig. 12-20

4 What is the average velocity of gas molecules?

5 A container is filled with a mixture of hydrogen and helium gases. Compared to helium molecules, do the hydrogen molecules on the average have greater, lesser, or the same: (a) speed; (b) kinetic energy?

6 If you increase the temperature of a gas, while holding its volume fixed, will the time interval between a molecule's collisions with one wall of the container increase, decrease, or remain constant?

7 Is a molecule of oxygen gas at 300 K more likely to have a speed of 100 m/s or 600 m/s?

8 Two drinking glasses are stuck together, one inside the other. It is possible to separate the glasses by immersing the outer glass in water and filling the inner glass with water of a different temperature. Hot tap water and ice water work well. Where should the ice water be?

9 Water fills a large industrial cooking pot of height 1 meter. The pot is heated from beneath, and when the water at the bottom of the pot reaches a certain temperature, it begins to boil. Will this temperature be the normal boiling point of 100° C, or will it be at a slightly higher temperature or at a slightly lower temperature?

Answers to Odd-Numbered Questions

1 A; **3** Yes, volume increases because of a decrease in pressure; **5** (a) greater; (b) same; **7** 600 m/s; **9** Slightly higher temperature

Problems (listed by section)

12-1 Temperature Measurement

1 Find the temperature in °F corresponding to the following temperatures on the Celsius scale: 0.00° C, 10.0° C, 20.0° C, 30.0° C, 40.0° C.

2 Find the temperature in °C corresponding to the following temperatures on the Fahrenheit scale: 0.00° F, 10.0° F, 20.0° F, 40.0° F, 60.0° F, 80.0° F.

3 Find the pressure of a constant-volume gas thermometer at a temperature of 300 K if its pressure at the triple point is 0.100 atm.

4 Find the triple point pressure of a constant-volume gas thermometer if its pressure is 0.400 atm at a temperature of 100° C.

5 At what temperature are the Celsius and Fahrenheit temperatures the same?

12-2 Ideal Gas Law

6 Find the mass in grams of a carbon dioxide molecule.

7 Find the number of atoms in a gold ring of mass 4.00 g.

8 Find the number of H_2O molecules in 1.00 liter (1000 cm^3) of water.

9 Find the number of molecules in 1.00 cm^3 of air at a pressure of 1.00 atm and a temperature of 300 K.

10 Find the volume occupied by 1.00 mole of ideal gas at standard conditions of pressure and temperature ($P = 1.00$ atm, $T = 273$ K).

11 A gas consists of 5.00 grams of oxygen and 5.00 grams of nitrogen and occupies a volume of 1.00 liter at a temperature of 350 K.
(a) Find the number of gas molecules.
(b) Find the partial pressure of the oxygen.
(c) Find the partial pressure of the nitrogen.
(d) Find the total pressure of the gas.

12 In deflating an air mattress you press the valve release so that air begins to escape. After some of the air has escaped, the pressure inside the air mattress decreases to atmospheric pressure, and no more air flows out. By squeezing all the air out from the end of the air mattress opposite the valve, you reduce the volume and increase the pressure in the remaining air, so that it will flow out. The air initially occupies a volume of 0.120 m^3. By how much do you need to reduce its volume in order to increase the pressure to 1.20 atm?

13 By how much does the air pressure in a house increase if the house is sealed and the air temperature increases from 10.0° C to 20.0° C? The initial air pressure is 1.00 atmosphere.

14 Air is pumped into a bicycle tire. The air initially in the tire has a volume of 2500 cm^3, a temperature of 20.0° C, and a *gauge* pressure of 2.00 atm. How many molecules of air must be pumped into the tire in order to raise the gauge pressure to 5.00 atm? Assume that the volume and temperature of the air inside the tire are approximately constant.

15 A driver measures her "cold" tire gauge pressure to be 1.93×10^5 Pa. The measurement is made before driving, when the tires are at the temperature of the surroundings, 20.0° C. After driving several miles, she checks the gauge pressure again and finds that it has increased to 2.20×10^5 Pa. Find the final temperature of the air in the tires, assuming negligible change in volume.

16 A balloon is filled with 5.00 liters of helium at a pressure of 1.20 atm and a temperature of 27.0° C.
(a) Find the mass of the helium.
(b) Find the buoyant force on the balloon if it is placed in the atmosphere with air pressure at 1.00 atm and temperature at 23.0° C.
(c) Find the acceleration of the balloon if the rubber has a mass of 4.00 g.

17 A 1.00 liter flask contains a certain quantity of ideal gas at 300 K. Then an equal quantity of the same gas is added to the flask, after which the absolute pressure is 1.50 times its original value. What is the final temperature?

18 A 1.00 liter cylinder contains helium and oxygen, with partial pressures of 0.100 atm and 0.200 atm respectively, at a temperature of 300 K. Find the mass density of each gas.

12-3 Kinetic Theory; Model of an Ideal Gas

19 Suppose that a 1.00 cm^3 box in the shape of a cube is perfectly evacuated, except for a single particle of mass 1.00×10^{-3} g. The particle is initially moving perpendicular to one of the walls of the box at a speed of 400 m/s. Assume that the collisions of the particle with the walls are elastic.
(a) Find the mass density inside the box.
(b) Find the average pressure on the walls perpendicular to the particle's path.
(c) Find the average pressure on the other walls.
(d) Find the temperature inside the box. Discuss the assumption of elastic collisions, in light of this result.

20 Find the rms speed of H_2O molecules in atmospheric water vapor at a temperature of 20.0° C.

21 A volume of nitrogen gas has a density of 1.00 kg/m^3 and a pressure of 1.00 atm. Find the temperature of the gas and the rms speed of its molecules.

22 Find the rms speed of (a) a helium molecule at 300 K; (b) an iron atom of mass 55.9 u at 300 K; (c) a pollen particle of mass 5.00×10^{-13} g suspended in water at a temperature of 300 K.

23 (a) Find the ratio of the average energy of a neon gas molecule to the average energy of a helium gas molecule if both gases are at the same temperature.
(b) Is either kind of molecule more likely than the other to lose energy during molecular collisions?

24 Estimate the fraction of oxygen molecules with speeds between 200 m/s and 600 m/s at 300 K. Use Fig. 12-8.

*12-5 Vapor Pressure and Humidity

25 What air pressure is required for water to boil at a temperature of 60° C?

26 What is the maximum partial pressure of water vapor in air at a temperature of 80° C?

27 The water vapor in air at 20° C has a partial pressure of 0.0050 atm. What is the relative humidity?

28 When the relative humidity is 60% and the temperature is 30° C, what is the partial pressure of water vapor in the air?

29 What is the dew point for air that has a relative humidity of 50% and a temperature of 30° C?

12-6 Thermal Expansion

30 A brick is initially 10.00 cm high and an aluminum can is initially 10.01 cm high. By how much must their temperatures be raised in order for the brick and the can to have exactly the same height?

31 Thin copper wire of length 100 m is wound on a cylindrical copper spool of length 10 cm. What is the change in length of the wire when the temperature is raised from 10° C to 40° C?

32 A solid wood door 0.90 m wide is hung in a metal doorway at 0.0° C with a clearance of 2.0 mm on the side. If the linear coefficient of expansion of the wood is 6.0×10^{-5} $(C°)^{-1}$, at what temperature would the door begin to touch the side of the doorway? Assume that the metal doorway does not expand significantly.

★ **33** An aluminum canteen is initially filled with 600 cm^3 of water at 0.0° C. If the canteen is not tightly closed, how much of the water will overflow when its temperature rises to 30° C?

34 The exterior brick surface of a house has a surface area of 200 m^2 in the winter when the temperature is 0.0° C. How much does the surface area increase in the summer when the temperature rises to 30° C?

★ **35** Two drinking glasses of diameter 8.0 cm are stuck together, one inside the other. The glasses can be separated without force if the outer glass is thermally expanded while the inner glass is thermally contracted. Initially both glasses are at 20° C. How much clearance can one produce by cooling the inner glass to 10° C by filling it with cold water while heating the outer glass to 30° C by running hot water over it? The linear expansion coefficient for the glass is $1.0 \times 10^{-5}\ (C°)^{-1}$.

★★ **36** Steel wire of radius 1 mm is cooled from 20° C to 0° C. How much tension would you have to apply to the wire to stretch it back to its original length? Young's modulus for steel is $2.0 \times 10^{11}\ N/m^2$.

37 By how much will the volume of coolant in an automobile radiator increase when its temperature increases by 100 C° if the initial volume of coolant is 12 liters and its thermal coefficient of expansion is $5.0 \times 10^{-4}\ (C°)^{-1}$?

Additional Problems

★ **38** An air bubble is expelled by an underwater diver at a depth of 10.0 m. As the bubble rises to the surface, the pressure decreases and the volume of the bubble therefore increases. The temperature of the water is constant. Find the final volume of the bubble if its initial volume is 10.0 cm^3.

★ **39** A cylindrical chamber, open at the bottom, is to be submerged in the ocean and used as a diving chamber (Fig. 12-21). Initially air at a pressure of 1.00 atm and a temperature of 300 K fills the cylinder of height 3.00 m. What will be the height h of the air in the cylinder if it is at a depth of 10.0 m where the temperature is 290 K?

★★ **40** An ideal gas is heated from T to $T + \Delta T$. Show that the volume coefficient of expansion is $1/T$.

★ **41** Almost the entire volume of mercury in a thermometer is contained in the bulb. The remainder extends up into a capillary tube. Suppose that the bulb has a volume of 1.00 cm^3. Find the inner radius of the capillary tube if the level of the mercury rises 2.00 cm per Celsius degree increase in temperature. Thermometer glass has a volume coefficient of expansion much less than that of mercury. Therefore you can ignore the expansion of the glass.

★★ **42** A cube of edge length ℓ has a volume $V = \ell^3$. When the cube's temperature is increased by ΔT, the edge length increases to $\ell + \Delta\ell$ and the volume increases to $V + \Delta V = (\ell + \Delta\ell)^3$. Obtain an expression for $\Delta V/V$ and show that the volume coefficient of expansion β is related to the linear expansion coefficient α by the equation $\beta = 3\alpha$, assuming $\alpha\,\Delta T$ is small.

43 Two boxes, each having a volume of 1.00 liter, contain helium gas. The gas in one box has a mass of 4.00 g and is initially at a temperature of 350 K, and the gas in the other box has a mass of 8.00 g and is initially at a temperature of 300 K. The two gases are brought into thermal contact and reach a common final temperature T.
(a) Find the initial pressure and rms speed for each gas.
(b) Find the final rms speed.
(c) Find the final temperature T.
(d) Find the final pressure for each gas.

★★ **44** A sealed glass jar contains air initially at a pressure of 1.00 atm and a temperature of 0° C. Find the air pressure when the jar is heated to 400° C. The linear expansion coefficient of the glass is $1.00 \times 10^{-5}\ (C°)^{-1}$.

45 At a temperature of 0.0° C the steel rectangle shown in Fig. 12-22 has a gap of 1.0 mm. What is the change in the gap if the steel is heated to 1000° C?

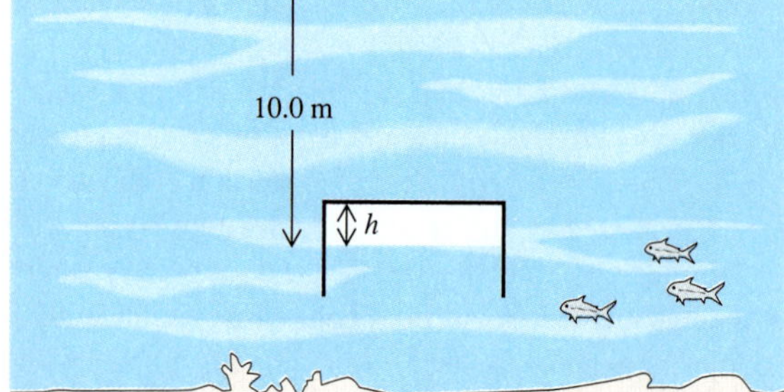

Fig. 12-21

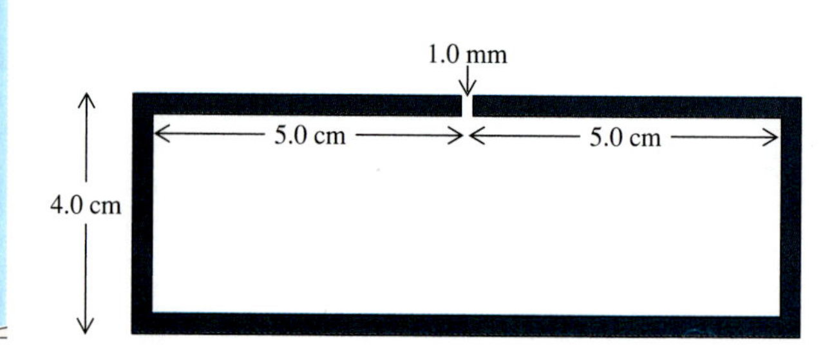

Fig. 12-22

★ **46** Suppose that a steel band encircled the earth. By how much would you have to raise the temperature of the metal to expand it enough so that it would be lifted 100 m away from the surface?

★ **47** Global warming causes the sea level to rise due to both the melting of glaciers and the thermal expansion of the sea. Estimate the rise in sea level from thermal expansion as the earth's mean temperature increased by 0.5 C° from 1860 to the present.

★★ **48** Estimate the thickness of the layer of water that would cover the earth if all the water vapor in the earth's atmosphere were to condense. Treat the air as a uniform layer 10 km thick, with a temperature of 20° C and a relative humidity of 50%.

★★ **49** Wood paneling with an unusually high thermal coefficient of expansion of 1.0×10^{-4} $(C°)^{-1}$ is installed at a temperature of 5° C, with no space between boards to allow for thermal expansion. The temperature later increases to 30° C, and the paneling buckles because it expands considerably while the surface to which it is nailed does not. Find the average angle at which each piece of paneling protrudes from the wall.

50 Find the volume of air inside a hot-air balloon required to support a weight of 5000 N if the temperature of the air inside the balloon is 100° C and the temperature of the surrounding air is 20° C. Use the result of Ex. 3, where we found that the density of the hot air is 78.6% of the cooler air's density (1.2 kg/m^3).

CHAPTER 13 Heat

Standing on the beach, feeling the summer sun, a cool breeze, and the hot sand beneath your feet, you experience three forms of heat: radiation from the sun, convection of heat from your body to the air, and conduction of heat from the sand to your feet. In this chapter we shall study radiation, convection, and conduction. We shall also see how heat can affect a body that absorbs or releases it, for example, by a change in the body's temperature.

13-1 Definition of Heat

Caloric Theory

The modern concept of heat evolved from an earlier theory known as the "caloric theory." This theory, which was generally accepted in the eighteenth century, held that a body is heated by absorbing a substance called "caloric" and that a body is cooled by the release of caloric. When two bodies at different temperatures were placed in contact, the warmer body was believed to transfer caloric to the cooler body until thermal equilibrium was reached. Caloric was believed to be a conserved quantity; that is, caloric could not be created or destroyed; the total amount of caloric in the universe was supposed to be constant. Around 1800 a few scientists began to recognize that the caloric theory could not explain situations in which mechanical energy caused the temperature of a body to increase. One of the first to be aware of this problem was Benjamin Thompson, who later became Count Rumford. Thompson's insight came when he observed the boring of cannon barrels. Water used to cool a cannon boiled away and had to be continually replenished. Thompson noted that there seemed to be an infinite capacity for the boring tool to create caloric. This was inconsistent with the idea that caloric was a conserved substance and suggested to Thompson that the motion of the boring tool itself was responsible for the increased temperature, that heat was nothing more than invisible motion of the particles within a body, and that the boring tool was transforming some of its visible motion into this invisible form, thereby heating the cannon and water.

Thompson's work was too qualitative to have much of an impact on the well-established caloric theory. But gradually, through the work of Hermann Helmholtz, Robert Mayer, and James Joule, the caloric theory was completely discredited. Joule, the son of an English brewery owner, devised a series of experiments, published in 1843, which succeeded in establishing heat as a form of energy. The energy unit "joule" is named in honor of his achievement.

Heat and Internal Energy

Today we are left with a trace of the caloric theory through its influence on the development of language. We still tend to speak of heat as though it were a substance that resides in a body. For example, we might say of a slice of apple pie, just out of an oven, that "it has a lot of heat in it." This is an incorrect use of the term "heat," as we define it in physics. By **heat** we mean energy in the process of being *transferred* from one system to another by **conduction, convection,** or **radiation.** We define the **internal energy** of a body to be the total energy of the particles *within* the body. A body contains internal energy; it does not contain heat. The internal energy of a body increases when heat flows into the body. For example, when a pie is in the oven, it is heated by the oven and its internal energy increases. As the pie cools, its internal energy decreases, as it heats the surroundings. If you bite into the pie too soon, you will be aware of heat conducted from the pie to your mouth.

Heat can cause a change in a body's temperature by changing the internal energy of the body. The temperature of a body is a measure of its molecules' average kinetic energy, as we saw in the previous chapter (Eq. 12-15: $\frac{1}{2}m\overline{v^2} = \frac{3}{2}kT$). The internal energy of a body is the total energy (kinetic and potential) of its molecules. As the internal energy of a body increases, its average molecular kinetic energy usually increases as well. Its temperature therefore also increases. For example, as an oven supplies heat to a pie, both the internal energy and the temperature of the pie increase.

We distinguish so carefully between heat and internal energy because heat is only one of the ways that the internal energy of a body can change. In the next chapter, we shall see that a body's internal energy can also change when work is done on the body. For example, we can raise the temperature and internal energy of a liquid by stirring it—exerting a force through a distance. In this case there is work, but no heat. In an automobile engine, the temperature of the fuel-air mixture rises as the mixture is compressed. Work is done on the gases by the piston compressing them, and the internal energy and temperature rise without any heat being supplied. So it is meaningless to speak of the "heat content of a body." A body contains internal energy, and heat is merely one of the ways in which that internal energy can change.

13-2 Calorimetry

A **calorie** (abbreviated cal) is a unit of heat, which was originally defined as the quantity of heat necessary to raise the temperature of 1 gram of water by 1 Celsius degree. Since heat is a form of energy, the calorie, like the joule, is an energy unit. The relationship between these two independently defined units was determined ex-

Fig. 13-1 Each of these foods contains about 100 Calories, or 100 × 10^3 calories.

perimentally. The same rise in the temperature of a body of water was produced both when heat, measured in calories, was added and when work, measured in joules, was done. This was the kind of experiment performed by Joule. It was found that 1 cal = 4.186 J. This relationship was referred to as the "mechanical equivalent of heat." Maintaining two independent energy units is unnecessary, and so we now *define* the calorie as 4.186 J.

$$1 \text{ cal} = 4.186 \text{ J} \tag{13-1}$$

The energy value of food in dietary studies is measure in "Calories," with a capital C (Fig. 13-1). This is a confusing unit because it equals 10^3 calories, or 1 kilocalorie.

$$1 \text{ Cal} = 1 \text{ kcal} = 10^3 \text{ cal} \tag{13-2}$$

Calorimetry deals with processes in which the only energy exchange is in the form of heat, that is, processes in which there is no work. Since the total energy of an isolated system must be conserved and since there is no work being done, it follows that the heat absorbed by one body equals the heat released by another.

Let the symbol Q denote the heat absorbed by a body, so that ***Q* has a positive value if heat is absorbed and a negative value if heat is released.** Then we can express conservation of energy for the special case of calorimetry problems (that is, when no work is done) by the equation

$$\Sigma Q = 0 \qquad \text{(when no work is done)} \tag{13-3}$$

Specific Heat

The heat absorbed by a body may result in a phase transition (melting or boiling) or in a temperature change. In the latter case the heat Q absorbed is proportional to both the body's mass and its change in temperature ΔT. The constant of proportionality is called the **specific heat** and is denoted by c. Thus

$$Q = mc\,\Delta T \tag{13-4}$$

Specific heat is a property of the kind of material. For example, iron has a specific heat of 0.11 cal/g-C°. Thus, to heat a 10 g piece of iron from 20° C to 40° C, one must supply to the iron a quantity of heat

$$Q = mc\,\Delta T = (10 \text{ g})(0.11 \text{ cal/g-C°})(20 \text{ C°}) = 22 \text{ cal}$$

The specific heat of a substance is in general a function of temperature; that is, the amount of energy required to change the temperature by 1 C° may vary, depending on the particular temperature interval. For example, the value for the interval from 20° C to 21° C may be quite different from the value for the interval from 40° C to 41° C. But for many materials c is nearly constant over a broad range of temperatures. The specific heat of water varies by less than 1% from 0° C to 100° C. Values of specific heat for a number of substances are given in Table 13-1.

Table 13-1 Specific heats

Values are for a constant pressure and a temperature of 20° C unless otherwise noted.

Substance	Specific heat, c	
	(cal/g-C°)	(J/kg-C°)
Air	0.24	1010
Aluminum	0.214	896
Copper	0.092	390
Ethyl alcohol	0.58	2400
Glass	0.20	840
Human body	0.83	3500
Ice (−10° C)	0.50	2100
Iron	0.11	460
Lead	0.03	130
Steam	0.48	2000
Steel	0.11	2100
Water	1.00	4190
Wood	0.42	1800

Phase Transitions

In the case of a phase transition (melting or boiling), heat is absorbed or released without a change in temperature. Different phases of the same substance at the same temperature contain different amounts of internal energy. For example, 1 g of liquid water at 0° C contains more internal energy than 1 g of ice at 0° C, and 1 g of steam at 100° C contains more internal energy than 1 g of liquid water at 100° C. Changing the phase of a substance requires a transfer of energy to or from the system. Here we assume that this energy is in the form of heat. It is found that the heat required is proportional to the mass undergoing the phase transition. The heat per unit mass is called the "latent heat" and is denoted by L. So, for a mass m to be transformed from the low-energy phase to the high-energy phase, the quantity of heat Q that must be supplied is

$$Q = mL \tag{13-5}$$

The latent heat of melting for ice is 79.7 cal/g, or 3.35 × 10⁵ J/kg. Thus, if 10.0 g of ice at 0° C are to be converted to water at 0° C, the quantity of heat that must be supplied is $Q = mL = (10\text{ g})(79.7\text{ cal/g}) = 797$ cal. **The latent heat of vaporization for water is 539 cal/g, or 2.26 × 10⁶ J/kg.** This is the heat per unit mass required to convert water at 100° C to steam at 100° C.

Evaporation

At temperatures below 100° C, water can be transformed from the liquid to the vapor state through evaporation if the water's surface is exposed to an environment that is not saturated with water vapor. The rate of evaporation depends on the relative humidity of the air. Evaporation of water is utilized in cooling the human body through sweating. The heat of vaporization of water at body temperature (37° C) is 580 cal/g. Thus each gram of sweat that evaporates carries away from the skin 580 calories of heat. This can be a very effective way of cooling the body if the surrounding air is dry, so that the rate of evaporation is rapid. For example, if you step out of a swimming pool in the desert, you will initially feel chilly, even though the dry desert air is quite hot.

Heat of Combustion and Food

When a substance burns, it loses internal energy in the form of heat. The heat produced per gram of combustible material is called the **heat of combustion.** For gasoline the heat of combustion is about 11 kcal/g. When the human body uses food, the process is much more complex than simple combustion. But the final products are just the same as though the food were burned in oxygen. For example, when glucose burns, it reacts with oxygen and produces carbon dioxide, water, and about 4 kcal/g of heat. When glucose is utilized by the body, various enzymes are involved, but the final products are the same, including the same amount of energy released, though not all in the form of heat. It is for this reason that calorimetry experiments that measure the heats of combustion of various foods are meaningful for understanding human metabolism. It is found that carbohydrates provide approximately 4 kcal/g, alcohol provides 7 kcal/g, and fats provide about 9 kcal/g. The internal energy in proteins is not completely utilized by the body. The energy used is about 25% less than that released in a calorimeter experiment, with the remainder being lost in excretion. The energy per gram provided to the body by protein is the same as for carbohydrates, 4 kcal/g. For example, a 100 g potato contains 22 g of carbohydrates, 2 g of protein, no fat, and 76 g of water, which provides no energy. Therefore the potato contains 96 Calories (24 g × 4 kcal/g). This means that eating the potato will provide 96 kcal of energy to the body.

EXAMPLE 1 Pouring Boiling Water Into a Cup

An aluminum cup of mass 0.300 kg is initially at room temperature, 20.0° C. If 0.200 kg of boiling water is poured into the cup, what will be the final temperature of water and cup just after pouring? Neglect heat loss to the surroundings.

SOLUTION Applying Eqs. 13-3 and 13-4 to the system of water and aluminum we have

$$\Sigma Q = 0$$

$$m_{\mathrm{w}} c_{\mathrm{w}}\, \Delta T_{\mathrm{w}} + m_{\mathrm{Al}} c_{\mathrm{Al}}\, \Delta T_{\mathrm{Al}} = 0$$

Substituting values for the masses and specific heats and expressing the temperature changes in terms of the common final temperature T and the initial temperatures of water and aluminum, we obtain

$$(0.200\ \mathrm{kg})(4190\ \mathrm{J/kg\text{-}C^\circ})(T - 100^\circ\ \mathrm{C}) + (0.300\ \mathrm{kg})(896\ \mathrm{J/kg\text{-}C^\circ})(T - 20.0^\circ\ \mathrm{C}) = 0$$

Solving for T, we find

$$T = 80.6^\circ\ \mathrm{C}$$

The temperature change of the cup is 80.6° C − 20° C = 60.6° C, whereas the temperature change of the water is 80.6° C − 100° C = −19.4° C. Why is the water's temperature change so much smaller than the cup's temperature change?

EXAMPLE 2 Adding Ice Cubes to a Pot of Water

Ice cubes of mass 0.100 kg at 0.0° C are added to 0.600 kg of water contained in a copper pot of mass 0.400 kg. The copper and water are initially at 30.0° C. Find the final temperature of the system, assuming negligible heat loss to the surroundings.

SOLUTION This example is more complicated than the previous one because it is not initially obvious how much of the ice will melt. It is possible that only part of the ice will melt and the final temperature of the ice-water mixture will be 0° C. Or it is possible that all the ice will melt and the final temperature might be greater than 0° C. First we must determine the answer to this question by calculating the heat that must be absorbed by the ice to melt it and the heat that must be lost by the water and copper to reduce their temperature to 0° C. Comparison of these two numbers will indicate whether all the ice melts.

In order for all the ice to melt, it must absorb heat

$$Q_1 = mL = (0.100 \text{ kg})(3.35 \times 10^5 \text{ J/kg}) = 3.35 \times 10^4 \text{ J}$$

To reduce the temperature of the water and pot to 0° C, they must lose heat or absorb a negative heat Q_2, where

$$Q_2 = m_w c_w \Delta T_w + m_{Cu} c_{Cu} \Delta T_{Cu}$$

The temperature change $\Delta T_w = \Delta T_{Cu} = -30.0$ C°. Thus

$$\begin{aligned} Q_2 &= (m_w c_w + m_{Cu} c_{Cu}) \Delta T_{Cu} \\ &= [(0.600 \text{ kg})(4190 \text{ J/kg-C°}) + (0.400 \text{ kg})(390 \text{ J/kg-C°})] \times (-30 \text{ C°}) \\ &= -8.01 \times 10^4 \text{ J} \end{aligned}$$

The minus sign tells us that we have a heat loss of 8.01×10^4 J. Since this number is greater than 3.35×10^4 J, all the ice melts. Now we find the final temperature of the system by applying Eqs. 13-3, 13-4, and 13-5.

$$\Sigma Q = 0$$

$$m_i L + m_i c_w \Delta T_i + (m_w c_w + m_{Cu} c_{Cu}) \Delta T_{Cu} = 0$$

Notice that there are two heats for the ice: the term "$m_i L$" for the phase transition (the 3.35×10^4 J already calculated) and the term "$m_i c_w \Delta T_i$" for the heat absorbed by the melted ice in order to raise its temperature from 0° C to the system's final temperature T. Substituting numerical values into the heat equation, we obtain a linear equation in T.

$$\begin{aligned} &3.35 \times 10^4 \text{ J} + (0.100 \text{ kg})(4190 \text{ J/kg-C°})(T - 0.0° \text{ C}) + \\ &[(0.600 \text{ kg})(4190 \text{ J/kg-C°}) + \\ &(0.400 \text{ kg})(390 \text{ J/kg-C°})](T - 30.0° \text{ C}) = 0 \end{aligned}$$

Solving for T, we find

$$T = 15.1° \text{ C}$$

13-3 Radiation

We are all aware of the warming effect of the sun's rays. Solar radiation is but one example of electromagnetic radiation. We will discuss this electrical phenomenon in more detail in a later chapter. Our brief discussion here will be just enough to understand some of the thermal effects of radiation.

Electromagnetic radiation consists of waves and is therefore characterized by a wavelength. The human eye is sensitive to electromagnetic radiation with a wavelength in a certain small range. This visible electromagnetic radiation we call "light." There are, however, many other wavelengths of electromagnetic radiation to which the eye is not sensitive. These include radio waves, microwaves, and ultraviolet and infrared radiation (Fig. 13-2). Only about 45% of solar radiation is in the visible part of the spectrum; another 45% is infrared, and 10% is ultraviolet.

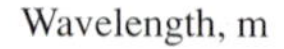

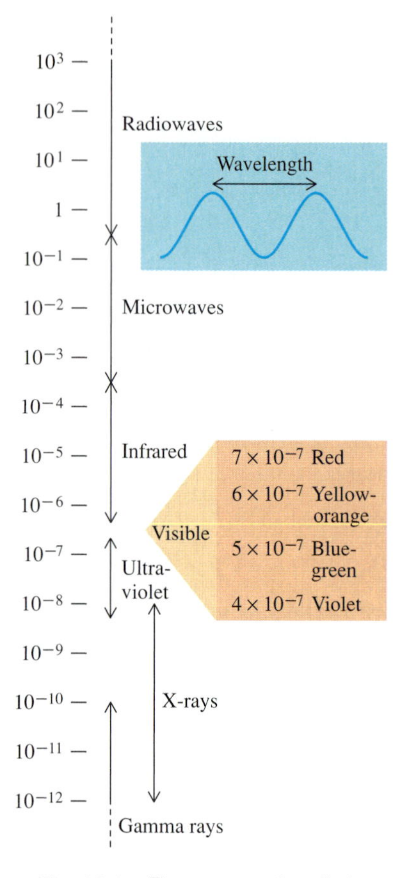

Fig. 13-2 Electromagnetic radiation.

Emission of Radiation

Hot objects emit radiation of various wavelengths. The distribution of wavelengths depends on the temperature of the object. At higher temperatures, shorter wavelengths are dominant. For example, an object at room temperature emits only invisible infrared radiation, whereas the heating element in an oven, when it reaches a temperature of about 800° C, begins to glow red, emitting, in addition to infrared radiation, the shorter-wavelength red light (Fig. 13-3). The still higher temperature of the filament in an incandescent light bulb (about 2000° C) results in the emission of white light—a mixture of red light as well as shorter wavelengths—yellow, green, and so forth. Thermographs record infrared radiation emitted by a body (Fig. 13-4).

It is found that the total energy radiated per unit time by an object increases rapidly as the object is heated. Let H denote the rate of emission, the heat per unit time emitted.

$$H = \frac{Q}{t} \tag{13-6}$$

Fig. 13-3 An artist's kiln.

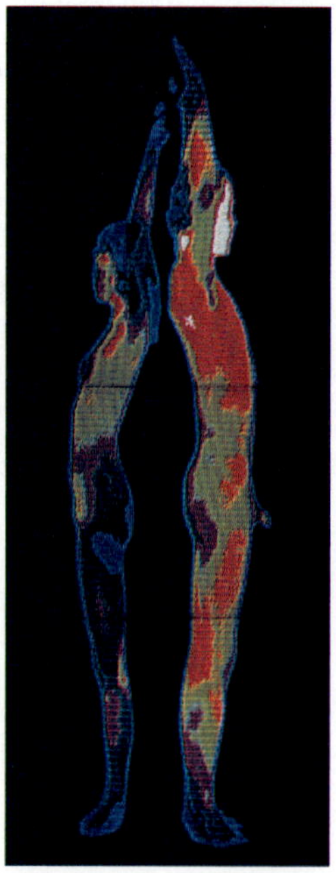

Fig. 13-4 The thermograph shown here records infrared radiation from the human body. The brightest areas correspond to the highest temperatures. Notice that the air is cool. An unusual temperature often corresponds to some kind of disorder (such as a tumor), which can be revealed by a thermograph.

Since H is an energy per unit time, it has units of watts (W). **Stefan's law** states that, for a body of surface area A, H equals the product of the following four quantities: A, a constant e called the **emissivity** of the surface, a universal constant σ called the Stefan-Boltzmann constant, and the fourth power of the *absolute* temperature T.

$$H = Ae\sigma T^4 \qquad \text{(emission of radiation)} \qquad (13\text{-}7)$$

where $\sigma = 5.67 \times 10^{-8}\ \text{W-m}^{-2}\text{-K}^{-4}$

and e is a number between 0 and 1, depending on the surface.

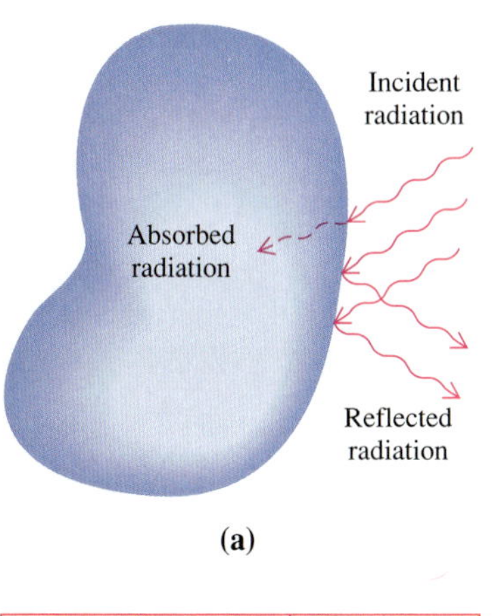

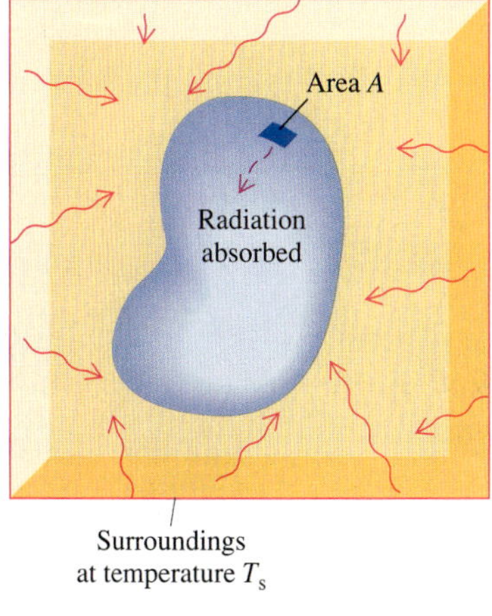

Fig. 13-5 **(a)** Radiation incident on a body is partly absorbed and partly reflected. **(b)** A body enclosed by walls absorbs some of the energy emitted by the walls.

Absorption of Radiation

In addition to emitting electromagnetic radiation, bodies can also absorb it. For example, if you stand in front of a fireplace, you can feel the warmth as your body absorbs radiation from the glowing logs. All the radiation incident on an opaque body is either absorbed or reflected, as indicated in Fig. 13-5a.

The rate of absorption depends on several factors, including the amount of incident radiation. Suppose the body is completely enclosed by surroundings at a uniform temperature T_s. For example, the body could be inside a box whose walls are at a uniform temperature (Fig. 13-5b). The amount of radiant energy absorbed is proportional to the amount of radiation incident on the surface. But this incident radiation in turn is proportional to the radiation emitted by the walls of the box, which, according to Eq. 13-7, varies as T_s^4. Thus the heat absorbed is also proportional to T_s^4. The complete expression for the rate at which radiation is absorbed by a section of the body of area A is

$$H = Ae\sigma T_s^4 \qquad \text{(absorption of radiation)} \qquad (13\text{-}8)$$

Notice that this expression has the same form as the expression for the rate of emission (Eq. 13-7), except that the temperature T_s is the temperature of the surroundings, not the temperature of the body itself. For a surface of area A, the rates of both emission and absorption are proportional to $Ae\sigma$, where e is the emissivity of the surface. So for a surface with a small value of e (close to zero) the surface absorbs little radiation and emits little radiation. A shiny metallic surface is a good example. It reflects most of the radiation incident on it and absorbs relatively little. Such a surface is also a poor emitter of radiation. In contrast, a very black surface, charcoal, for example, is both a good absorber and a good emitter, with a value of e close to 1. The lining of a Thermos bottle is silvered to minimize radiation to the outer walls of the bottle.

That both the rates of emission and absorption are proportional to $Ae\sigma$ is no accident, as we shall now show. Suppose that a body of surface area A and emissivity e is in a container and that both the body and the container are at the same temperature T. The body emits energy at the rate $Ae\sigma T^4$, as predicted by Stefan's law. The body's temperature T must remain constant, since the walls are at the same temperature. But the only way that the temperature can remain constant is if the energy lost by emission of radiation is balanced by an equal rate of absorption of radiation. Since this holds true for a body of any surface area or emissivity, the dependence of H on both these factors must be the same for absorption as for emission.

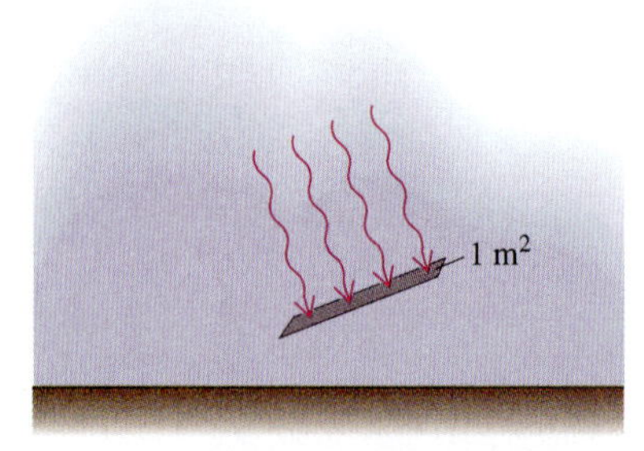

Fig. 13-6 On a cloudless day solar radiation on earth delivers energy of 1400 J each second to a 1 m² surface perpendicular to the radiation.

For a body at temperature T, enclosed by surroundings at a different temperature T_s, the net radiation absorbed is given by the rate of absorption minus the rate of emission.

$$H_{net} = Ae\sigma T^4 - Ae\sigma T_s^4$$

or

$$H_{net} = Ae\sigma(T^4 - T_s^4) \quad \text{(net radiation)} \quad (13\text{-}9)$$

Emissivity is a function of both the surface and the wavelength of radiation. For example, the emissivity of human skin varies from about 0.6 for very light skin to about 0.8 for very dark skin for visible light. For infrared radiation, however, all human skin has an emissivity close to 1. So all human skin is very black, that is, nonreflecting, in the infrared region.

Solar Radiation

On a clear day **solar radiation delivers energy to the surface of the earth at a rate of approximately 1400 watts per square meter** of surface area $A_\perp$, perpendicular to the radiation (Fig. 13-6). Fig. 13-7 shows a home with solar panels, which utilize solar radiation to assist in the heating needs of the home.

Fig. 13-7 Homes heated by solar panels.

EXAMPLE 3 Cooling the Human Body by Radiation

Most home heating systems heat air, rather than the interior walls. If the air temperature is kept moderate, say, 20° C (68° F), in winter the walls will usually be significantly colder. Radiative heat losses from the human body to the walls may then be significant, making the room uncomfortable, unless heavy clothing is worn. Find the rate at which energy is lost by the body through radiation for a standing person who is naked with 1.5 m² of exposed skin at a temperature of 32° C (90° F) if the walls are at a temperature of 10° C (50° F).

SOLUTION Applying Eq. 13-13 and setting $e = 1.0$ for infrared radiation absorbed or emitted by human skin, we find

$$H = Ae\sigma(T^4 - T_s^4)$$

$$= (1.5 \text{ m}^2)(1.0)(5.67 \times 10^{-8} \text{ W-m}^{-2}\text{-K}^{-4})[(305 \text{ K})^4 - (283 \text{ K})^4]$$

$$= 190 \text{ W}$$

In an hour the total heat lost through such radiation would amount to

$$Q = Ht = (190 \text{ W})(3600 \text{ s})$$

$$= 6.8 \times 10^5 \text{ J}$$

or

$$Q = (6.8 \times 10^5 \text{ J})\left(\frac{1 \text{ cal}}{4.186 \text{ J}}\right)\left(\frac{1 \text{ kcal}}{10^3 \text{ cal}}\right)$$

$$= 160 \text{ kcal}$$

Thus a large amount of energy would be lost to the walls, and unless the person were generating quite a lot of energy through physical activity, the skin would feel quite cold, even though the air temperature is moderate. The same air temperature on a sunny day at the beach could be very comfortable.

13-4 Convection

Convection is the heating that occurs through the motion of a fluid. An example is the cooling of your body by contact with air. Since your skin is normally at a higher temperature than the surrounding air, when air molecules collide with skin molecules, the air molecules rebound with an increased kinetic energy on the average. So they carry energy from your body. This process is much more efficient when a breeze blows. In this case there is a more rapid replacement of the warmer air molecules near the skin with cooler molecules from farther away. This is called "forced convection." If you sweat, the cooling is much faster yet because, in addition to the air cooling the skin, water vapor is also leaving the skin as the sweat evaporates. The more energetic water molecules leave first, a process that decreases the average molecular energy of the remaining molecules, thereby lowering the skin's temperature.

An automobile engine is cooled by forced convection of radiator fluid that flows around the engine's cylinders, and a computer is cooled by forced convection of air that is blown through it by a fan (Fig. 13-8).

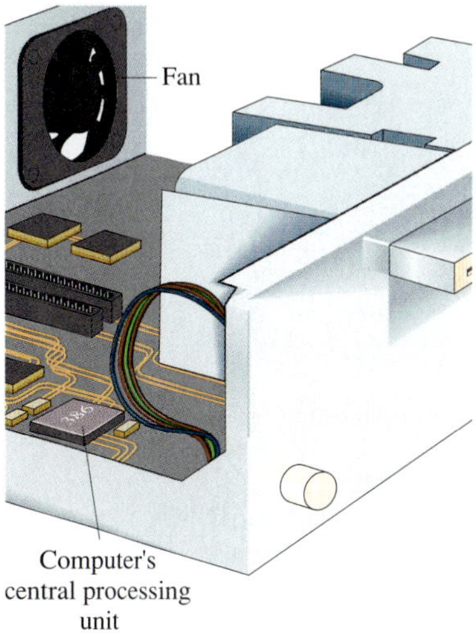

Convection is a complex process, and so a complete mathematical description of convection is difficult to obtain. Convective heating depends on several factors—the temperature of the body, the temperature and velocity of the fluid surroundings, and other factors, including even the detailed geometry of the body's surface. The rate of convective heating is found to be approximately proportional to the product of the body's surface area A and the temperature difference ΔT between the body and the surrounding fluid. The constant of proportionality, h, is called the "convection coefficient."

$$H = hA\,\Delta T \qquad \text{(convection)} \qquad (13\text{-}10)$$

The value of the convection coefficient h depends on many variables and has been found by experimental measurement for various applications.

Fig. 13-8 A computer's central processing unit is cooled by forced convection. The unit is isolated from other parts so that air from the cooling fan circulates freely around the unit.

EXAMPLE 4 Cooling the Human Body by Convection

Calculate the rate of heat loss by convection for the situation described in Example 3—exposed skin of area 1.5 m² at 32° C and surrounding air at 20° C: (a) in still air, for which $h = 7.0\ \text{W/m}^2\text{-C}°$; (b) in air moving at a speed of 2.0 m/s, for which $h = 20\ \text{W/m}^2\text{-C}°$.

SOLUTION (a) Applying Eq. 13-10, we find

$$\begin{aligned} H &= hA\,\Delta T \\ &= (7.0\ \text{W/m}^2\text{-C}°)(1.5\ \text{m}^2)(32°\ \text{C} - 20°\ \text{C}) \\ &= 130\ \text{W} \end{aligned}$$

In an hour the total heat lost through convection would be

$$\begin{aligned} Q &= Ht = (130\ \text{W})(3600\ \text{s}) \\ &= 4.7 \times 10^5\ \text{J} \end{aligned}$$

or

$$\begin{aligned} Q &= (4.7 \times 10^5\ \text{J})\left(\frac{1\ \text{kcal}}{4186\ \text{J}}\right) \\ &= 110\ \text{kcal} \end{aligned}$$

Notice that this is considerably less than the 160 kcal lost through radiation, found in Example 3.

(b) Repeating the calculation for the larger convection coefficient, corresponding to moving air, we find

$$\begin{aligned} H &= hA\,\Delta T \\ &= (20\ \text{W/m}^2\text{-C}°)(1.5\ \text{m}^2)(32°\ \text{C} - 20°\ \text{C}) \\ &= 360\ \text{W} \end{aligned}$$

and

$$\begin{aligned} Q &= Ht = (360\ \text{W})(3600\ \text{s})\left(\frac{1\ \text{kcal}}{4186\ \text{J}}\right) \\ &= 310\ \text{kcal} \end{aligned}$$

We find that the heat loss in 1 hour through convection is now much greater than the loss through radiation.

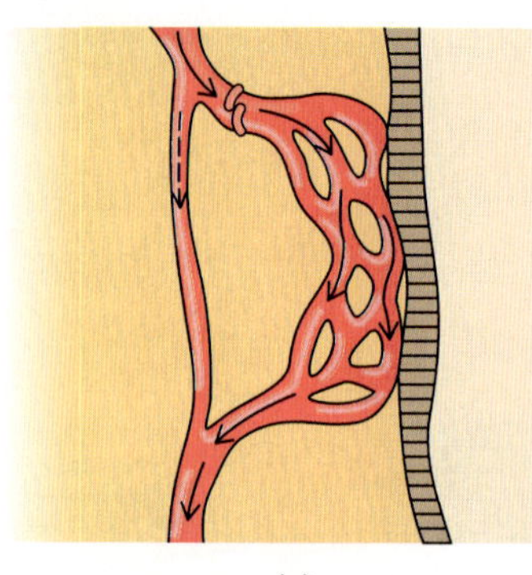

(a)

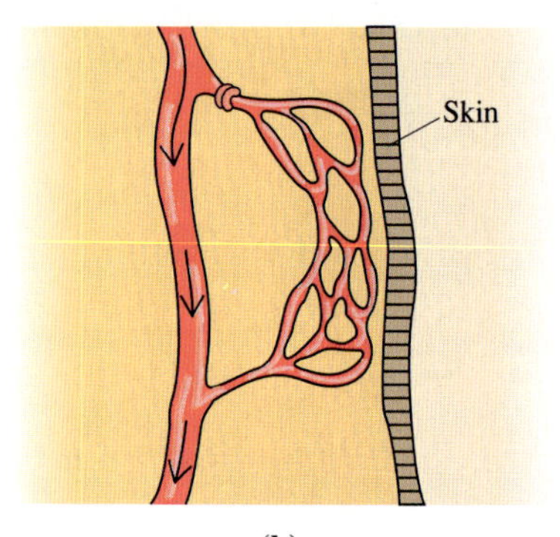

(b)

Fig. 13-9 **(a)** When the skin is warm, expanded capillaries enhance the flow of blood and heat near the skin. **(b)** When the skin is cold, constriction of capillaries reduces the flow of blood and heat.

The human body is partially protected against heat loss in a cold environment by the autonomic nervous system, which responds in various ways, especially by (1) increasing production of heat within the muscles and internal organs (shivering is one way this is accomplished) and (2) redistributing blood flow (Fig. 13-9). The transfer of heat from the body's interior to the surface is achieved primarily through the flow of blood. In cold surroundings blood flow near the skin is reduced by constriction of capillaries. This allows the skin and subcutaneous fat to provide better thermal insulation. Skin temperature can drop to as low as 15° C, reducing heat loss from the skin to the environment, while the body maintains its normal internal temperature of about 37° C. This mechanism is significantly more effective in women than in men, and consequently women are better able to tolerate cold.

13-5 Conduction

If you place one end of a metal poker into a fire while holding the other end in your hand, you will feel the end you are holding become warmer. This is an example of conduction. The metal poker is conducting heat from the fire to your hand. The end of the rod in contact with the flame is at a higher temperature than other sections of the rod, and its atoms and electrons therefore have more kinetic energy. Through collisions with neighboring atoms and electrons, this excess energy is shared, raising the temperature of adjacent parts of the rod. Gradually energy is conducted along the length of the rod, and you feel the warmth in your hand.

The process is similar to convection, except that since the atoms conducting the heat are part of a solid they are locked in position. They are not free to move over large distances but can only vibrate about a fixed point. In a metal, however, electrons are free to move over considerable distances. Only metals have these free electrons. It is this property of metals that makes them such good conductors of heat, as well as good conductors of electricity.

Next we shall describe quantitatively the process of heat conduction. Consider two bodies of very large heat capacity at temperatures T_1 and T_2, where $T_1 > T_2$. If the amount of heat absorbed or released by these bodies is not enough to significantly change their temperatures, we call the bodies "thermal reservoirs" and consider their temperatures fixed. Suppose that heat is transferred from reservoir 1 at temperature T_1 to reservoir 2 at temperature T_2 by means of heat conduction through a body of length ℓ and constant cross-sectional area A. See Fig. 13-10.

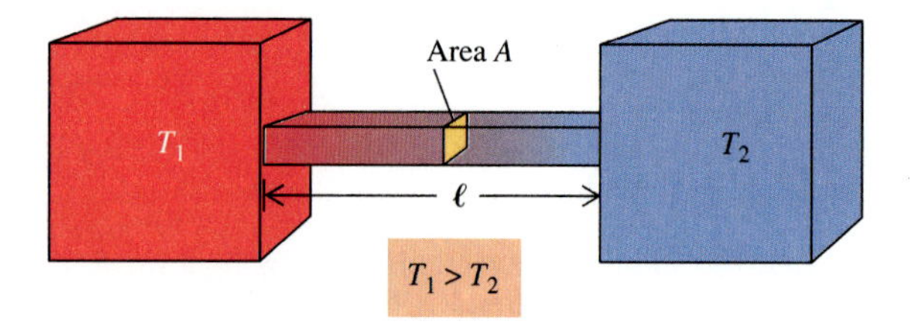

Fig. 13-10 A rod conducts heat between two thermal reservoirs.

When the two reservoirs are first connected by the conductor, the temperature at either end of the conductor may be quite different from that of the reservoir with which it is in contact. After some time interval thermal equilibrium is established between each end of the conductor and its respective thermal reservoir. And at each intermediate point along the conductor an equilibrium temperature is reached. The temperature is found to drop linearly from T_1 to T_2; that is, the drop in temperature per distance traveled along the conductor is constant. Once this equilibrium is established, no net heat is delivered to any section of the conductor (otherwise its temperature would change). Thus the heat flowing across any cross section of the conductor is the same. This process, called "steady heat flow," is analogous to steady fluid flow, in which the same mass of fluid flows across each cross section of a flow tube. We shall consider only the case of steady heat flow, for which there are many important applications.

It is found that the rate of heat flow H is proportional to the product of the cross-sectional area A and the temperature drop ΔT per interval of length $\Delta \ell$ (Fig. 13-11). The constant of proportionality is called the **thermal conductivity,** denoted by k.

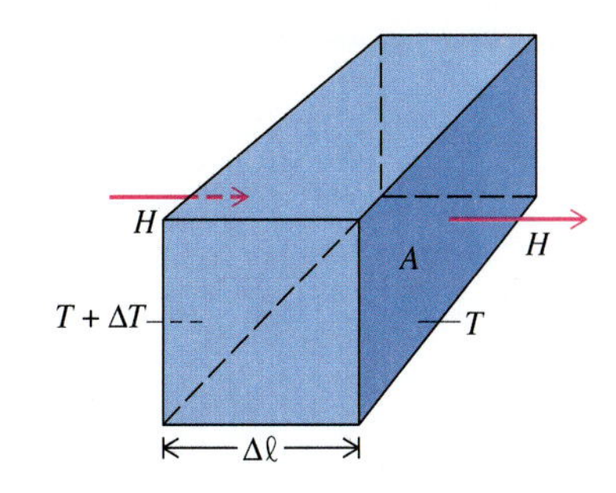

Fig. 13-11 Heat conduction through a body depends on its cross-sectional area A, and the temperature drop per interval of length $\Delta T/\Delta \ell$.

$$H = kA \frac{\Delta T}{\Delta \ell} \qquad (13\text{-}11)$$

This equation applies to a section of *any* material of uniform cross section perpendicular to the direction of steady heat flow so long as there is no motion of the material itself. Thus it applies to fluids so long as the fluids are static. Some representative values of thermal conductivity are given in Table 13-2. Materials with large values of thermal conductivity are good heat conductors. Materials with small values of thermal conductivity are poor heat conductors or good thermal insulators. The values in the table indicate that metals are good heat conductors and materials like asbestos, wood, cork, and air are poor conductors or good thermal insulators. The range in values has effects that are familiar to you. For example, if you touch both metal and wood at a common temperature greater than skin temperature, the metal will feel hotter than the wood. On the other hand, if the temperature of the metal and wood is less than skin temperature, the metal will feel colder than the wood. The reason for this is that the metal conducts heat to or from the area of contact with your skin more rapidly than wood. So in either case the heat transfer is greater with the metal than with the wood.

The thermal insulating property of air is important in clothing. Clothes that trap a relatively thick layer of air next to the skin are most effective in reducing heat loss to the environment. Wool clothing has this property. Coats filled with a thick layer of down are both lightweight and very effective in providing thermal insulation because of the thick layer of trapped air, which is not able to flow and is therefore prevented from transferring heat by convection.

Table 13-2 Thermal conductivity

Material	k (W/m-C°)
Silver	410
Copper	390
Aluminum	200
Steel	50
Marble	~2
Concrete	0.8
Glass	0.8
Animal tissue	0.2
Asbestos	0.08
Wood	~0.08
Cork	0.04
Air	0.024

EXAMPLE 5 Cooling the Human Body by Conduction

Clothing traps a layer of air of average thickness 2.0 mm next to the skin of a person whose surface area is 1.5 m². Compute the heat lost by conduction through the clothing if the skin temperature is 29° C and the outer surface of the clothing is at a temperature of 22° C.

SOLUTION Applying Eq. 13-11 and using the value for the thermal conductivity given in Table 13-2, we find

$$H = kA \frac{\Delta T}{\Delta \ell}$$

$$= (0.024 \text{ W/m-C°})(1.5 \text{ m}^2) \frac{29° \text{ C} - 22° \text{ C}}{0.0020 \text{ m}}$$

$$= 130 \text{ W}$$

EXAMPLE 6 R-Value of Insulation

The effectiveness of various home insulators is indicated by their "*R*-value," which is a measure of their resistance to conducting heat and is defined by the equation

$$H = A\frac{\Delta T}{R} \quad (13\text{-}12)$$

By comparing Eq. 13-12 with Eq. 13-11 $\left(H = kA\frac{\Delta T}{\Delta \ell}\right)$ we see that

$$R = \frac{\Delta \ell}{k} \quad (13\text{-}13)$$

For a given temperature difference, a high value of R means that a relatively small amount of heat will flow through a unit area of insulation. Find the rate of conduction through a 4-inch thick, 1 square-foot layer of fiberglass insulation, with $R = 13$ ft^2-h-F°/Btu, if the temperature difference across the sides of the insulation is 20 F°. The engineering unit Btu (British thermal unit) equals 1055 J.

SOLUTION Applying Eq. 13-12, we find

$$H = A\frac{\Delta T}{R} = \frac{(1.0\text{ ft}^2)(20\text{ F}°)}{13\text{ ft}^2\text{-h-F}°/\text{Btu}}$$

$$= 1.5\text{ Btu/h}$$

or converting to standard units, we find

$$H = (1.5\text{ Btu/h})\left(\frac{1\text{ h}}{3600\text{ s}}\right)\left(\frac{1055\text{ J}}{1\text{ Btu}}\right)$$

$$= 0.45\text{ W}$$

This result is for 1 ft^2 of surface area. Suppose this rate of heat conduction represented the average heat flow through the ceiling, floors, and outer walls of a home in which the combined area of these surfaces was 10^4 ft^2. Then the total heat conducted out through these surfaces would be 10^4 times the value calculated above, that is, 15,400 Btu/h, or 4500 W.

CHAPTER 13 SUMMARY

Heat Q is energy being transferred between systems by either conduction, convection, or radiation. The rate of heat transfer H is defined by

$$H = \frac{Q}{t}$$

For conduction, H is proportional to the cross-sectional area A of the conducting body and to the rate of change of temperature with distance, $\Delta T/\Delta \ell$.

$$H = kA \frac{\Delta T}{\Delta \ell} \qquad \text{(conduction)}$$

The constant k is the thermal conductivity. Convective heating occurs through the motion of a fluid in contact with a body at a different temperature; H is approximately proportional to the surface area A of the body and to the temperature difference ΔT between the body and the fluid.

$$H \approx hA\,\Delta T \qquad \text{(convection)}$$

The convection coefficient h is not a constant. It is determined experimentally for various fluids as a function of fluid velocity. For radiation, H is proportional to the surface area A of the radiating body, to the emissivity e of the surface, and to the *absolute* temperature T raised to the fourth power.

$$H = Ae\sigma T^4 \qquad \text{(Stefan's law)}$$

The Stefan-Boltzmann constant $\sigma = 5.67 \times 10^{-8}$ W-m^{-2}-K^{-4}. If a body at temperature T is in surroundings at a uniform temperature T_S, the net radiative heat loss is

$$H = Ae\sigma(T^4 - T_S^4)$$

If systems exchange energy only in the form of heat, the total heat absorbed by all systems must equal zero.

$$\Sigma\, Q = 0 \quad \text{(when no work is done)}$$

A positive value of Q means heat is absorbed by the system, and a negative value means heat is released by the system. Heat is measured in units of joules, calories, and Calories, where

$$1 \text{ cal} = 4.186 \text{ J}$$

$$1 \text{ Cal} = 1 \text{ kcal} = 10^3 \text{ cal} = 4186 \text{ J}$$

Absorption of heat Q may cause a mass m of a substance to undergo a phase change, in which case

$$Q = mL$$

where L is called the "latent heat" of the substance. For water the latent heat of melting is 79.7 cal/g and the latent heat of vaporization is 539 cal/g. Heat Q that causes a temperature change ΔT of a substance of mass m is given by

$$Q = mc\,\Delta T$$

where c is the specific heat of the substance, expressed in cal/g-C° or J/kg-C°.

Questions

1 A 1 kg iron pan contains 1 kg of water. In heating the pan and water equally, what percentage of the total heat supplied goes to the pan?

2 Glass is transparent to visible light but opaque to infrared radiation. Explain why the glass walls of a greenhouse are effective in raising the temperature of its interior.

3 The interior of a closed car parked in direct sunlight on a summer day can become quite hot, much hotter than the air temperature outside. The car's interior is like a greenhouse. After the windows are opened, the air inside the car quickly cools down, but the seatcovers remain hot for a while. To minimize this time, is it better to have white or black seatcovers?

4 Very hot coffee is poured into a glass cup and into an aluminum cup.
(a) Which cup feels hotter when you touch it?
(b) Suppose you drink the coffee 5 minutes after it is poured. In which cup would the coffee be hotter?

5 A tiled bathroom floor usually feels colder to bare feet than a carpeted floor. Which of the following statements explains this?
(a) The tile is at a lower temperature than the carpet.
(b) The tile has more heat capacity.
(c) The tile has lower heat capacity.
(d) The tile has higher thermal conductivity.
(e) The tile has lower thermal conductivity.

6 Normally a breeze feels cooler than still air. But in very hot desert air, a breeze can be hotter than still air. What is the minimum temperature at which this enhanced heating begins to take effect? Explain in terms of molecular collisions.

7 Compared to an equal mass of cold water, is it correct to say that a cup of hot tea contains: (a) more heat; (b) more internal energy; (c) more caloric?

8 You are driving in the winter over wet roads at nightfall as the temperature begins to drop below freezing. You notice that ice is beginning to form on a section of the road passing over a bridge but not elsewhere. Explain.

9 A potato will bake faster if you stick a nail in it. What physical property of the nail are you using to rapidly raise the temperature of the potato?

10 A chair is made of wood and aluminum. Normally the aluminum part feels cooler than the wood, though they are at the same temperature. For what temperature would the aluminum feel neither cooler nor hotter than the wood?

11 Is it possible to cool a closed room with a fan that circulates air within that room?

12 Is it possible to cool a person with a fan that circulates air in a closed room?

Answers to Odd-Numbered Questions

1 10%; **3** black; **5** (d); **7** (a) no; (b) yes; (c) no; **9** conductivity; **11** no

Problems (listed by section)

13-2 Calorimetry

1 How much heat is required to raise by 10 C° the temperature of 1.0 kg of (a) water; (b) iron?

2 How much energy expressed in units of food Calories (that is, kilocalories) is required to operate a 100 W light bulb for 10.0 hours?

3 The heat of combustion of a 25 g sample of bread (1 slice) is to be measured in a "bomb" calorimeter. Enclosed in a 2.0 kg steel capsule is the dried bread, a thin electric wire used to ignite the bread when an electric current passes through the wire, and a sufficient supply of oxygen to ensure complete combustion. Surrounding the capsule is 4.0 kg of water. The system has an initial temperature of 20° C and a final temperature of 48° C. Find the heat of combustion of the bread. Neglect the heat generated in the wire.

4 A 50.0 g ice cube, initially at 0° C, is dropped into a Styrofoam cup containing 350 g of water, initially at 20.0° C. What is the final temperature of the water, if no heat is transferred to the Styrofoam or the surroundings?

5 A pot containing 1.00 liter of water, initially at 20.0° C, is placed on a 1000 W electric heating element.

(a) How much heat must be supplied to the water to bring it to a boil?

(b) How much heat is necessary to boil all the water away? What is the minimum time required to boil all the water away?

6 A porcelain cup of mass 300 g and specific heat 0.260 cal/g-C° contains 150 cm^3 of coffee, which has a specific heat of 1.00 cal/g-C°. If the coffee and cup are initially at 70° C, how much ice at 0° C must be added to lower the temperature to 50° C?

7 An object of mass 50.0 g is initially placed in boiling water so that it is at 100° C. Then the object is dropped into a calorimeter, consisting of 100 g of water in an aluminum cup of mass 150 g, thermally insulated from the surroundings. The initial temperature of the calorimeter is 20.0° C and the final temperature is 40.0° C. Find the specific heat of the object.

8 Ice of mass 50.0 g at −10.0° C is added to 200 g of water at 15.0° C in a 100 g glass container of specific heat 0.200 cal/g-C° at an initial temperature of 25.0° C. Find the final temperature of the system.

★ **9** A certain car requires 2.00×10^6 J of thermal energy to operate its engine while traveling a distance of 1.00 km. Gasoline has a density of 740 kg/m^3 and a heat of combustion of 4.60×10^7 J/kg. Find the volume of gasoline required to operate the car over this distance.

10 How much heat is carried away from the body of a sweating person in 1 hour by the evaporation of 0.500 liter of water from the skin?

13-3 Radiation

11 Find the rate of radiation per unit area by a body of emissivity 0.500 at a temperature of 20.0° C.

12 A person sitting quietly in a room on a winter day may lose heat primarily by radiation from the outer surface of her clothes to the walls of the room. If the walls are at 10.0° C and the rate of radiation is 80.0 kcal/h (a typical basal metabolic rate), what is the temperature of the outer surface of the clothes, which have an emissivity of 0.800?

13 At what temperature would a body of emissivity 0.500 emit radiation at the same rate as a body of equal surface area and emissivity 1.00 at 0.00° C?

★ **14** A black body has a mass of 4.00 kg, specific heat of 2000 J/kg-C°, surface area of 0.200 m^2, emissivity 1.00, and initial temperature 100° C. The surroundings are at a temperature of 20.0° C. Find the drop in the black body's temperature during a 5.00-minute interval that would result from heat loss by radiation only.

★ **15** An aluminum tray used for making ice cubes is filled with water and placed in a freezer at −20.0° C. The emissivity of the aluminum is close to zero, but the emissivity of water or ice is 0.950. The exposed top surface area of each ice cube is 20.0 cm^2.

(a) Find the rate at which heat is removed from an ice cube by radiation just as the water begins to freeze.

(b) How long would it take for radiation alone to remove enough heat to freeze the ice cubes if each has a mass of 80.0 g?

16 The outer surface of a skier's clothes of emissivity 0.700 is at a temperature of 5.00° C. Find the rate of radiation if the skier has a surface area of 1.50 m^2 and the surroundings are at −15.0° C.

17 The sun is 1.50×10^{11} m from the earth.

(a) Over what surface area is solar radiation spread at this distance?

(b) What is the total power radiated by the sun?

(c) Treating the sun as a body of emissivity 1.00, estimate its temperature. The sun's radius is 6.96×10^8 m.

★ **18** Find the rate at which solar energy is delivered to a garden plot of dimensions 5.00 m × 10.0 m at noon on (a) a summer day when the sun's rays are at an angle of 10.0° from vertical; (b) a winter day when the sun's rays are at an angle of 57.0° from vertical.

★ **19** Suppose you are lying on the beach with 0.750 m^2 of skin exposed to solar radiation at noon on a hot summer day when the sun is at an angle of 10.0° from the vertical. Find the radiation absorbed, assuming that 80.0% of the incident radiation is absorbed.

13-4 Convection

20 In Problem 19 suppose that your skin temperature is 35° C.

(a) Find the rate at which energy is radiated by the body, assuming an emissivity of 1.0.

(b) Find the rate at which heat is lost by the body through convection if the surrounding air is at 20° C and the convection coefficient is 20 W/m^2-C°.

(c) If 95 W of heat are generated in the body internally and transmitted to the skin, how much heat must be carried away by processes other than radiation and convection if the skin's total incoming heat and outgoing heat balance? What are these other processes?

21 The exterior walls of a house have a total area of 200 m^2 and are at 12.0° C and the surrounding air is at 7.0° C. Find the rate of convective cooling of the walls, assuming a convection coefficient of 3.0 W/m^2-C°.

★ **22** Suppose that you are outdoors in air at −18° C (0° F) and the outer surface of your clothing is at 10° C. Convective cooling of your body is enhanced by a wind, in which the convection coefficient is double its value for still air. Find the temperature of still air that would produce the same rate of convective cooling, that is, the "wind-chill factor."

23 Food is cooked in a conventional oven primarily by convection. The average temperature of a 9.0 kg turkey rises from 20° C to 24° C in an oven at a temperature of 180° C over a 20-minute interval. Assuming the specific heat of the turkey is the same as that of the human body and using 0.20 m^2 for the surface area, find the value of the convection coefficient.

24 Find the emissivity of a surface that radiates heat at the same rate that it loses heat through convection when its temperature is 100° C and the surroundings are at 20° C if the convection coefficient is 5.0 W/m^2-C°.

★ **25** Find the radius of an aluminum spherical shell of mass 2.0 kg such that its temperature drops by 0.10 C° per second by convective cooling when the surrounding air is 10 C° cooler than the aluminum. The convection coefficient is 4.0 W/m^2-C°.

13-5 Conduction

26 Aluminum and wood rods of the same thickness and length conduct heat as a result of the same temperature difference between the ends of each rod. Find the ratio of heat conducted by the aluminum to the heat conducted by the wood.

27 A down jacket has a surface area of 0.80 m^2 and is filled with a 4.0 cm thick layer of down, which has a thermal conductivity of 6.0×10^{-6} kcal/s-m-C°. The coat is worn by a person whose skin temperature is maintained at 30° C. The outer surface of the coat is at −20° C. At what rate is heat conducted through the coat?

★ **28** A copper rod of cross-sectional area 10 cm^2 has one end immersed in boiling water and the other in an ice-water mixture, which is thermally well insulated except for its contact with the copper. The length of the rod between the containers is 20 cm, and the rod is covered with a thermal insulator to prevent heat loss from the sides. How many grams of ice melt each second?

29 Ice of mass 10.0 kg at 0.00° C is placed in an ice chest. The ice chest has 2.00 cm thick walls of thermal conductivity 1.00×10^{-5} kcal/s-m-C° and a surface area of 1.30 m^2.
(a) How much heat must be absorbed by the ice before it melts?
(b) If the outer surface of the ice chest is at 30.0° C, how long will it take for the ice to melt?

30 On a cold night the temperatures of the two surfaces of a glass windowpane in a house are 7.8° C on the outer surface and 8.0° C on the inner surface when the air temperature is 0° C outside and 20° C inside. The glass is 3.0 mm thick.
(a) Find the rate at which heat is conducted through the glass per square meter of window area.
(b) Find the value of the convection coefficients for the heating of the inner surface of the glass and for the cooling of the outer surface.

31 When your skin temperature is 29° C, what is the effective average thickness of the insulating layer of tissue separating the skin's surface at 29° C from the body's interior at 37° C? Use the value of thermal conductivity provided in Table 13-2. It is appropriate to use this value only when the skin is relatively cold and little heat is carried to the surface by blood.

32 Find the average *R*-value of the insulation in a home with a total exterior surface area of 6000 ft^2 that conducts heat outward at the rate of 4000 Btu/h when there is a temperature difference of 10 F° between the inner and outer walls.

★ **33** An exterior wall of area 150 ft^2, exposed to direct sunlight, has a surface temperature of 120° F. The wall is well insulated with insulation having an *R*-value of 13 ft^2-h-F°/Btu. If the temperature of the interior wall is 70° F, how long will it take to conduct enough heat through the wall to raise the temperature of the 50 kg of air inside the room by 1.0 F°? Assume that all of the heat is absorbed by the air.

★ **34** Find the diameter of an aluminum rod that would conduct heat at the same rate as a 1.0 cm diameter copper rod of the same length, with the same temperature difference between the ends of each rod.

★ **35** The ends of a copper rod and an aluminum rod of the same length and radius are joined together so that they form one long rod, which conducts heat. Find the temperature of the point where the rods are joined if the opposite end of the copper is at 100° C and the opposite end of the aluminum is at 0° C.

Additional Problems

★ **36** At the earth's surface, radiation from the sun has an intensity of 1400 W per square meter perpendicular to the radiation. This radiation is partly reflected and partly absorbed. The energy that is absorbed is reemitted in the form of infrared radiation from the entire surface of the earth. Determine a constant, uniform temperature for the earth's surface that would achieve this balance.

★ **37** Suppose a lake has a layer of ice 10 cm thick on its surface. The top of the ice layer is cooled by the atmosphere to −5.00° C. The lower surface is in contact with water at 0.0° C. The thermal conductivity of ice is 0.004 cal/s-cm-C°.
(a) At what rate is heat transferred through a 1.0 cm^2 area of the ice?
(b) How long will it take to freeze an additional 1.0 mm of ice onto the lower surface of the ice?

★ **38** Steam of mass 10.0 g at 100° C is added to 100 g of ice at 0.00° C. Find the final temperature of the system.

★ **39** Steam of mass 20.0 g at 100° C is added to 100 g of ice at 0.00° C. Find the final temperature of the system.

★ **40** Water heated by the sun during the day is the source of heat for a solar house at night. Suppose that the house loses heat to the outdoors at a rate of 5.00 kcal/s during a 12.0 h period. The water is heated during the day to 40.0° C and cools to 30.0° C overnight as it supplies heat to the house. Find the volume of water required.

★ **41** The roof of a house is heated by solar radiation. The heat absorbed is 700 W/m^2, the temperature of the roof is 50° C, and its emissivity is 1.0 in the infrared region. The temperature of the surrounding air is 35° C, with a convection coefficient of 5.0 W/m^2-C°. The roof is insulated with a 6.0 cm thick layer of insulation with thermal conductivity 0.050 W/m-C°.

(a) For a 1.0 m^2 area, find the heat lost by the roof through radiation, convection, and conduction. Assume that the temperature of the roof is stable so that there is a balance between heat gained and heat lost.

(b) Find the temperature of the inner surface of the insulation.

★ **42** The filament in an incandescent light bulb is a very thin tungsten wire. The wire has high electrical resistance, which results in a high temperature and the emission of visible radiation (along with infrared radiation). The wire is in the shape of a double coil (Fig. 13-12) so that in a small space there is a relatively large surface area that emits radiation.

(a) Find the surface area of tungsten required for a 100 W light bulb. The emissivity of tungsten is 0.26, and the temperature of the filament is 2000° C.

(b) If the radius of the tungsten wire is 0.020 mm, what is the length of the wire?

Fig. 13-12 A micrograph of a light bulb filament shows that it is a coil within a coil.

CHAPTER 14 Thermodynamics

A builder's fantasy: constructing a home by dynamiting a pile of lumber.

Watching a film run backwards can be amusing. We see things that we could never see in real life. For example, if we watch a backwards film of someone spilling milk, we see the splattered milk springing upward from the floor and back very neatly into the glass. Or, if we watch a backwards film of a building being demolished, we see a pile of rubble come together to form the building. In either case a state of disorder spontaneously changes to a state of order. Because we know such changes cannot be real, we immediately know that the film is being played backwards. And yet nothing in either of these backward events violates the laws of physics we have studied so far. In terms of an analysis based solely on forces and energy, spontaneous transformations of disorder into order are possible. In this chapter we shall study the law that describes the practical impossibility of such events—the second law of thermodynamics.

Thermodynamics is a branch of physics dealing with thermal processes involving energy exchange in the form of heat and work. Thermodynamics began in the early nineteenth century as a result of efforts to understand and improve on the design of heat engines, devices that use heat to produce work (Fig. 14-1). The analysis of heat engines led to an understanding of the limitations on our ability to transform disordered thermal energy into ordered energy in the form of work, or, more generally, to transform any kind of disorder into order.

Fig. 14-1 An early steam engine. Steam escaping from the two jets caused the sphere to rotate. Disordered thermal energy is partially transformed into the ordered kinetic energy of the rotating sphere.

14-1 Thermodynamic Systems

In the study of mechanics we often focused attention on a particular system and considered the forces exerted on that system by the surroundings. In thermodynamics, too, we consider the interaction between a particular system and its surroundings. However, a **thermodynamic system** always consists of a large number of particles, for example, the molecules in a gas or the atoms in a magnet. Of course, in mechanics we often choose as the system of interest a body consisting of a large number of particles; for example, we may study the external forces acting on a 1 kg block containing 10^{25} atoms. But the description in mechanics is not concerned with the fact that there is a large number of particles. For example, we may be describing the center-of-mass motion of the 1 kg block. In thermodynamics the fact that there is a large number of particles is essential to the description.

Suppose we attempted to describe a system consisting of only 10 gas molecules, using the laws of mechanics. A complete mechanical description, which would involve the positions and velocities of the molecules, and the intermolecular forces, would be very difficult. For a macroscopic system, consisting of, say, 10^{23} particles, a complete mechanical description would be impossible. But this is neither necessary nor desirable. The very size of a macroscopic system permits a different kind of description, a thermodynamic description. We have seen an example of this in our discussion of the ideal gas law in Chapter 12.We don't need to think about the position and velocity of individual gas molecules to understand their macroscopic effect of producing pressure on the walls of a container. The enormous number of molecules in a gas actually simplifies the description because it makes the pressure constant, rather than fluctuating.

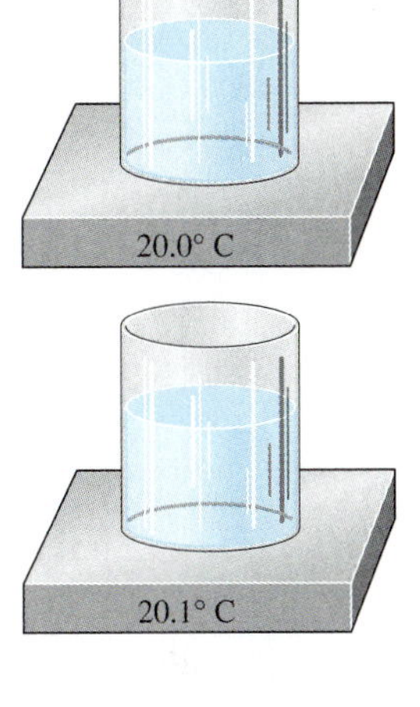

Thermodynamic Variables

In describing a thermodynamic system we need only a few variables, called **thermodynamic variables.** For example, the thermodynamic variables for an ideal gas are **pressure** P, **volume** V, **temperature** T, **number of molecules** N, **internal energy** U, and **entropy** S. (Internal energy and entropy will be defined later.) Not all thermodynamic variables are independent. For example, ideal gas variables P, V, N, and T are related by the ideal gas law: $PV = NkT$ (Eq. 12-5).

An **equilibrium state** of a thermodynamic system is defined to be one for which the thermodynamic variables have constant values if the system does not interact with the surroundings. And, for most of our applications, an equilibrium state will also imply that values of thermodynamic variables such as pressure and temperature are the same everywhere within the system. For example, a gas in an equilibrium state will have a pressure and temperature that are the same in all parts of the gas and that remain constant if the system is left alone.

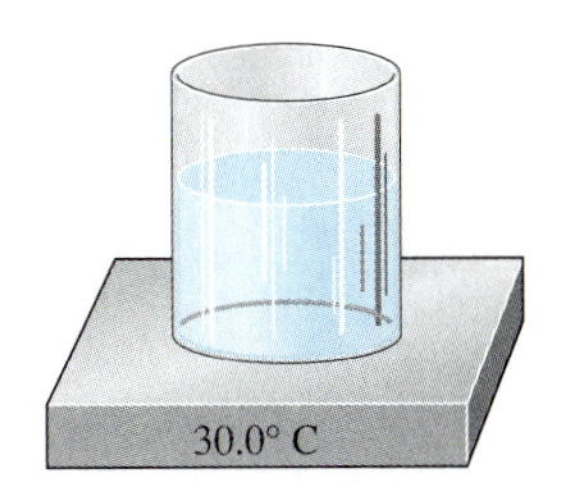

Fig. 14-2 A container of water is heated very slowly from 20.0° C to 30.0° C by bringing it into contact with a series of metal blocks at progressively higher temperatures. During each contact a small amount of heat is conducted from the block to the water. In this way the water is always close to thermal equilibrium. This process approximates a quasi-static process.

Quasi-static or Reversible Process

A **quasi-static process** is any process that takes a system through a continuous succession of equilibrium states. Although quasi-static processes are very important in the analysis of thermodynamic systems, they are never realized in practice; they are only approximated by real processes. For example, suppose that heat is slowly added to water in a container by gradually warming the bottom of the container. As heat is added, the temperature of the water near the bottom is somewhat greater than the temperature near the top. The more slowly heat is added, the more uniform the temperature within the cylinder will be at any instant. If heat is added very slowly, the temperature would be nearly uniform. We would then have an approximately quasi-static process (Fig. 14-2).

A quasi-static process is sometimes referred to as a **reversible process** because, if the system is in equilibrium at all times, reversing the external conditions will take the system through the same set of states in reverse. For example, if after gradually warming a container of gas, you slowly cool a container of gas, the system goes through the same succession of equilibrium states but in reverse order.

Internal Energy

In the Chapter 7 essay "The energy to run," we applied the concept of "internal energy" to the human body. There we saw that a runner uses considerably more internal energy than a cyclist moving at the same speed as the runner. The **internal energy** of a thermodynamic system, denoted by U, **is defined to be the sum of the kinetic and potential energies of all the particles in the system, as observed in a reference frame in which the center of mass of the system is at rest.** For a system whose center of mass is moving relative to the observer, the internal energy of the system equals the total energy less the center-of-mass kinetic energy.

In Chapter 12 we found an expression for the average translational kinetic energy of an ideal gas molecule (Eq. 12-15: $\frac{1}{2}m\overline{v^2} = \frac{3}{2}kT$). The translational kinetic energy of an ideal gas, consisting of N molecules, equals N times the average molecular kinetic energy, $\frac{3}{2}kT$, for a total energy of $\frac{3}{2}NkT$. Since $NkT = PV = nRT$, we can express the total energy in molar form as $\frac{3}{2}nRT$.

$$\text{Translational kinetic energy of an ideal gas} = \tfrac{3}{2}nRT$$

In formulating our model of an ideal gas, we assumed that the time during which a molecule is in contact with any other molecule is negligible compared to the time it is moving freely. This means that at any instant the great majority of the gas molecules are free from contact with other molecules. Intermolecular forces and the associated potential energy equal zero, unless molecules are in contact or are very close. So potential energy arising from intermolecular forces is negligible in an ideal gas, and the internal energy of the gas is therefore equal to its kinetic energy. For an ideal gas that consists of monatomic (one-atom) molecules the only kind of kinetic energy possible is *translational* kinetic energy, and so the internal energy equals the translational kinetic energy.

$$U = \tfrac{3}{2}nRT \quad \text{(for a monatomic ideal gas)} \quad (14\text{-}1)$$

For an ideal gas of diatomic (two-atom) molecules there is an additional contribution to the internal energy, arising from the rotation of molecules. There are two possible axes of rotation, perpendicular to the line between the two atoms in the molecule. The molecule can rotate about either of these axes. It is possible to show that the kinetic energy associated with each of these rotations shares in the system's total energy equally with the translational kinetic energy associated with each direction of motion. There are 3 directions of translational motion and 2 axes of rotation, and so the rotational kinetic energy equals two thirds the translational kinetic energy of $\frac{3}{2}nRT$; that is, rotational kinetic energy equals nRT. Adding this to the translational kinetic energy, we find that the internal energy equals $\frac{5}{2}nRT$.

$$U = \tfrac{5}{2}nRT \quad \text{(for a diatomic ideal gas)} \quad (14\text{-}2)$$

We can not always express the internal energy of every system so simply. For example, there is no simple expression for the internal energy of water. However, even when there is no equation that can be used to compute internal energy, *changes* in internal energy can always be found by use of the first law of thermodynamics, as we shall see in the next section.

EXAMPLE 1 The Internal Energy of Helium in a Balloon

Calculate the internal energy of 17.0 g of helium inside a helium balloon at a temperature of 300 K.

SOLUTION First we calculate the number of moles of helium, using its atomic mass of 4.00:

$$n = (17.0 \text{ g})\left(\frac{1 \text{ mole}}{4.00 \text{ g}}\right) = 4.25$$

Since helium is a monatomic gas and since at moderate temperatures and densities we can treat helium as an ideal gas, we can apply Eq. 14-1 for the internal energy of a monatomic ideal gas.

$$U = \tfrac{3}{2}nRT = \tfrac{3}{2}(4.25)(8.31 \text{ J/K})(300 \text{ K})$$
$$= 1.59 \times 10^4 \text{ J}$$

This is a considerable amount of energy when compared, for example, with the center-of-mass kinetic energy the helium would have if the balloon were moving at, say, 2.0 m/s:

$$K = \tfrac{1}{2}mv^2 = \tfrac{1}{2}(17.0 \times 10^{-3} \text{ kg})(2.0 \text{ m/s})^2 = 3.4 \times 10^{-2} \text{ J}$$

14-2 First Law of Thermodynamics

Isolated System

The **first law of thermodynamics** is essentially just a statement that energy is conserved. **For an isolated system** this means that **the sum of the internal energy and the center-of-mass energy is constant.**

EXAMPLE 2 A Temperature Change Caused by Free Fall

A piece of clay falls to the floor from a height of 2.00 m. Compute the increase in temperature of the clay if it has a specific heat of 920 J/kg-C°.

SOLUTION If none of the clay's energy is transferred to the floor during the collision, the clay can be considered an isolated system. Therefore the sum of its internal energy and center-of-mass energy is conserved. Initially the center-of-mass energy is in the form of gravitational potential energy. As the clay falls, its potential energy is converted to kinetic energy. As the clay strikes the floor, the center of mass is brought to rest. The ordered center-of-mass kinetic energy is converted to internal energy. Thus the clay's increase in internal energy, ΔU, equals the original gravitational potential energy of the center of mass, mgy.

$$\Delta U = mgy$$

The increased internal energy has the same effect as if internal energy were added in the form of heat. So the increase in internal energy may be expressed

$$\Delta U = mc\,\Delta T$$

Equating the two expressions for ΔU, we obtain the clay's change in temperature, ΔT.

$$mc\,\Delta T = mgy$$

or

$$\Delta T = \frac{gy}{c} = \frac{(9.80 \text{ m/s}^2)(2.00 \text{ m})}{920 \text{ J/kg-C°}} = 2.13 \times 10^{-2} \text{ C°}$$

The internal energy of the clay would have to increase much more to produce a rise in temperature of even 1 C°.

Interacting System; Thermodynamic Work

The most important applications of the principle of energy conservation in thermodynamics are for systems that are not isolated but rather are free to exchange energy with their surroundings. Energy may be transferred between a system and its surroundings in the form of either heat or work. In the last chapter we found that heating a system can cause a temperature increase or a phase change, either of which indicates an increase in the system's internal energy. Work is the other way in which energy can be transferred between a system and its surroundings. For example, in an automobile engine expanding gases in the cylinders push pistons and do work on them (Fig. 14-3). Utimately that work is used to turn the wheels of a car—to move

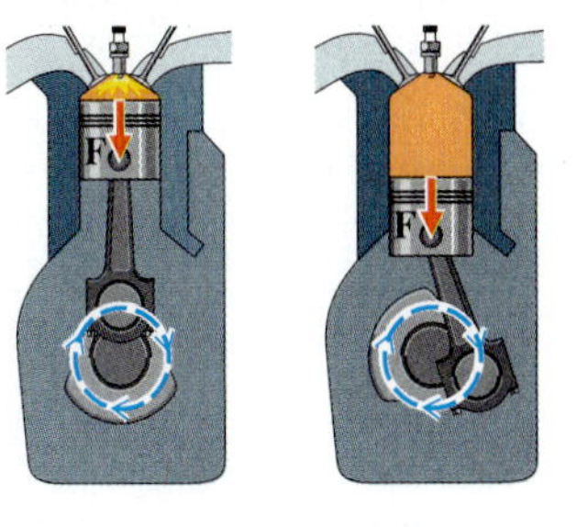

Fig. 14-3 In the cylinder of an automobile engine expanding gases exert a force on a piston, moving the piston and doing work on it.

the car, to give it kinetic energy. The energy supplied to the car comes from the internal energy initially stored in the gasoline. Burning the gasoline liberates that internal energy, using it to do work on the pistons, providing kinetic energy to the pistons and, through them, to the entire car.

When a thermodynamic system does work on its surroundings, if there is no heat exchange and no change in the system's center-of-mass energy, the system's change in internal energy, ΔU, is the negative of the work W done by the system:

$$\Delta U = -W$$

This result follows from Eq. 7-33 ($\Delta E = W_{nc}$), where ΣW_{nc} is the net nonconservative work done *on the system* — the negative of the work W done *by the system* ($\Sigma W_{nc} = -W$) — and the system's energy change is its change in internal energy ($\Delta E = \Delta U$), since by assumption there is no change in the system's center-of-mass energy.

If a system does work on the surroundings and also absorbs heat, the resulting change in the internal energy ΔU equals the heat Q absorbed minus the work W done by the system.

$$\Delta U = Q - W \quad \text{(first law of thermodynamics)} \quad (14\text{-}5)$$

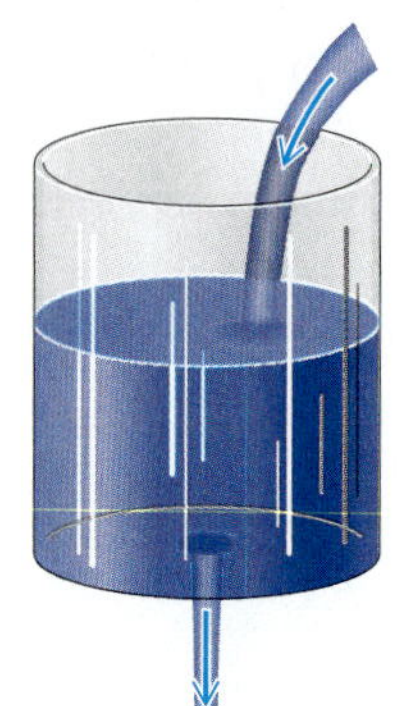

Fig. 14-4 Water is being poured into a container while some of the water flows out through a hole in the bottom. The container of water is analogous to a thermodynamic system interacting with its surroundings. The water inside the container represents the system's internal energy *U*, which will tend to increase because of heat absorbed (water flowing in) and tend to decrease because of work performed by the system (water flowing out).

This equation expresses the **first law of thermodynamics.** It is simply a statement of energy conservation for a system that interacts with its surroundings. Again we can derive this equation from Eq. 7-33 ($\Delta E = \Sigma W_{nc}$), where the net nonconservative work is now allowed to take its two possible forms, called heat and thermodynamic work. The heat Q absorbed by a system is simply the work done on the system by means other than macroscopic, thermodynamic work—for example, heat conduction in which work is done on the system on the microscopic level through molecular collisions. The system's internal energy change ΔU is the difference between the incoming heat Q (the microscopic work done *on* the system) and the outgoing work W (the macroscopic work done *by* the system)(Fig. 14-4).

Eq. 14-5 is valid for both positive and negative values of Q and W. It is important to remember the meaning of the signs of Q and W. **A positive value of Q means that heat is absorbed by the system, whereas a negative value of Q means that heat is released by the system. A positive value of W means that positive work is done by the system, and a negative value of W means that negative work is done by the system** (that is, positive work is done *on* the system).

EXAMPLE 3 Work Done on a Piston by Steam

Water is contained in a cylinder of cross-sectional area 1.00×10^{-2} m^2 (100 cm^2) with a movable, frictionless piston at one end (Fig. 14-5). The area above the piston is a vacuum. The piston weighs 1.00×10^3 N, and the water exerts a force of 1.00×10^3 N on the piston, and so the system is in equilibrium at a pressure of

$$\frac{1.00 \times 10^3 \text{ N}}{1.00 \times 10^{-2} \text{ m}^2} = 1.00 \times 10^5 \text{ N/m}^2 \quad \text{(about 1 atm)}$$

The water, initially at 100° C, is supplied with 2260 J of heat, which results in 1.00 g of water being converted to steam. Steam occupies a volume of 1670 cm^3 at this pressure.

(a) Find the work done by the system (the water) on the piston.

(b) Find the change in the internal energy of the system.

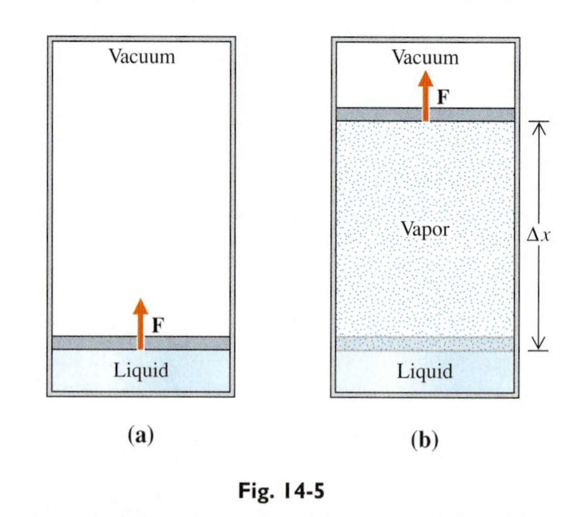

Fig. 14-5

EXAMPLE 3—cont'd

SOLUTION **(a)** As the 1.00 g of water is converted from liquid to vapor, the volume of the system increases by about 1670 cm^3, and so the piston rises a distance $\frac{1670\ \text{cm}^3}{100\ \text{cm}^2} =$ 16.7 cm. The work done by the system on the piston is

$$W = F_x\,\Delta x = (1.00 \times 10^3\ \text{N})(0.167\ \text{m}) = 167\ \text{J}$$

This is also the increase in potential energy of the piston.

(b) The first law of thermodynamics can be used to find the system's increase in internal energy. As 167 J of work is done by the system, 2260 J of heat is added to the system, and so

$$\Delta U = Q - W = 2260\ \text{J} - 167\ \text{J} = 2093\ \text{J}$$

It is this increase in internal energy that is responsible for the 1.00 g of water being converted from the liquid to the vapor state. The vapor state is a state of higher internal energy.

Expressions for Work

Next we shall consider ways of determining the work resulting from a change in the volume of a thermodynamic system. For the special case of a system undergoing a volume change ΔV at constant pressure P, the work is given by $P\,\Delta V$, as shown in Fig. 14-6.

$$W = P\,\Delta V \qquad \text{(at constant } P\text{)} \qquad (14\text{-}6)$$

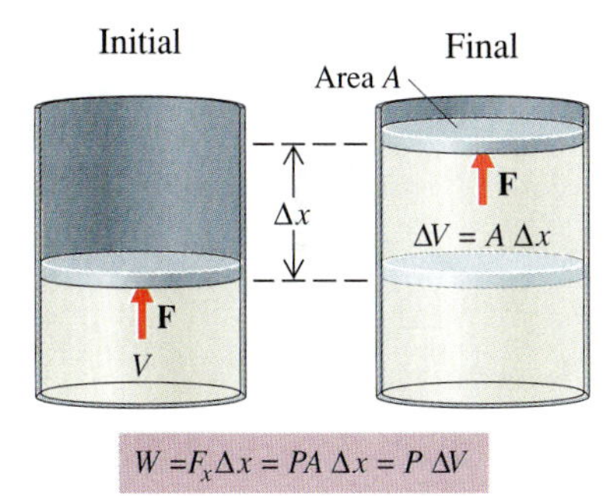

Fig. 14-6 Finding the work done by a system on the surroundings as the volume of the system increases at constant pressure.

This equation applies to both expansion and compression. Expansion implies a positive ΔV and positive work (as shown in the figure); compression implies negative ΔV and negative work.

For a system whose pressure is not constant during a change in volume, one can find the work by considering the process as consisting of a series of small changes in volume ΔV, for each of which the pressure P is nearly constant. The work done during each part is then $P\,\Delta V$, and the total work for the entire volume change is given by the sum of $P\,\Delta V$ terms.*

$$W = \Sigma\,(P\,\Delta V) \qquad (14\text{-}7)$$

We can obtain a graphical interpretation of work by plotting P versus V, as in Fig. 14-7. As shown in this figure, **the work equals the area under the curve** between the initial and final points. For the simple case of a constant-pressure process, the area under the curve is the area of a rectangle (Fig. 14-8).

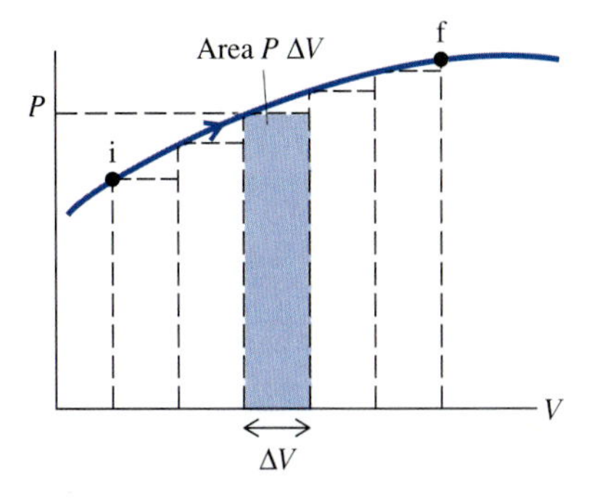

Fig. 14-7 $W = \Sigma\,(P\Delta V)$ = Area of rectangles ≈ Area under the curve. (The equality becomes exact if we let the width of the rectangles approach zero.)

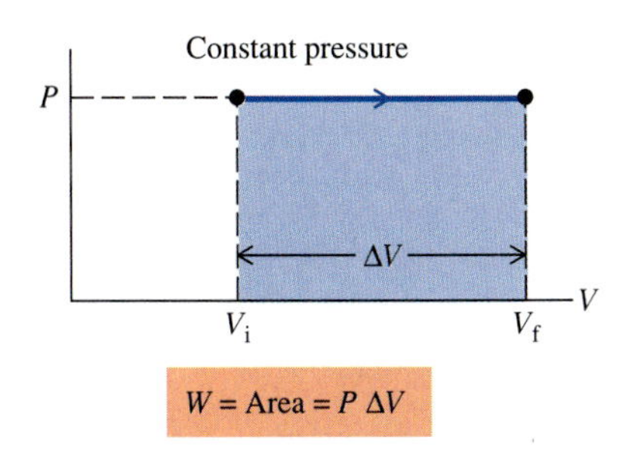

Fig. 14-8 Expansion at constant pressure.

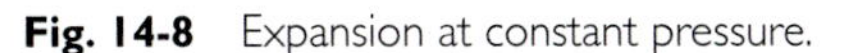

*More precisely the work is the limiting value of this sum as the volume increments ΔV approach zero; that is, $W = \lim_{\Delta V \to 0} [\Sigma\,(P\,\Delta V)]$.

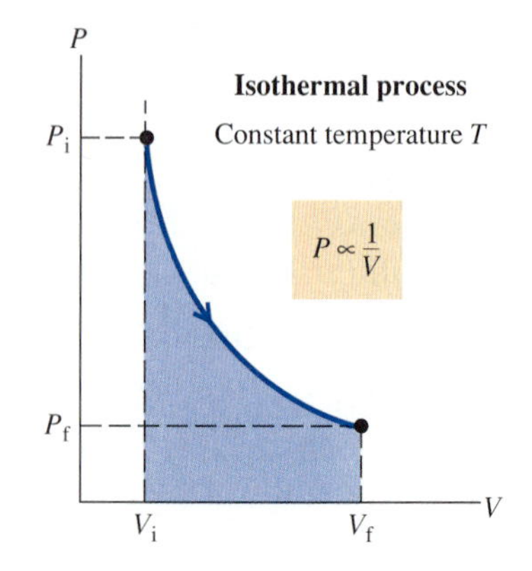

Fig. 14-9 Expansion of an ideal gas at constant temperature, an "isothermal" process.

An **isothermal process** is one in which a system expands or is compressed at constant temperature. If the system undergoing an isothermal process is an ideal gas, it follows from the ideal gas law ($PV = nRT$) that the pressure is inversely proportional to the volume ($P = nRT/V = \text{Constant}/V$). The PV diagram for an ideal gas undergoing an isothermal process is shown in Fig. 14-9. It is possible to obtain a general expression for the work in an isothermal process as a function of the temperature T and the initial and final volumes V_i and V_f. Since the derivation requires the use of calculus, we state the result without proof:

$$W = nRT \ln \frac{V_f}{V_i} \qquad \text{(along an isotherm)} \qquad (14\text{-}8)$$

where "ln" means the logarithm to the base e (2.718 . . .), called the "natural log."

An **adiabatic process** is one in which no heat is absorbed ($Q = 0$). For such a process, the system's change in internal energy equals minus the work done by the system:

$$\Delta U = Q - W = -W$$

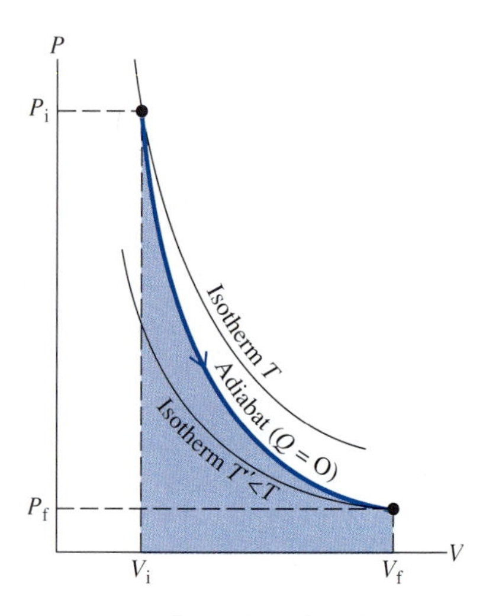

Fig. 14-10 Expansion of an ideal gas with no heat ($Q = 0$), an "adiabatic" process. Temperature decreases during this process.

When positive work is done by the system in an adiabatic process, the system's internal energy must decrease. This generally means that the temperature of the system decreases, as illustrated in Fig. 14-10 for an ideal gas undergoing an adiabatic expansion.

Negative work is done by a system that is compressed, that is, for a process in which the system's final volume is less than its initial volume. This corresponds to a PV diagram in which the process proceeds from right to left, as illustrated in Fig. 14-11. The work is then the negative of the area under the curve.

A **cycle** is a process in which the system completes a closed path. Over one complete cycle the system returns to its initial state. Positive work is done during that part of the process in which the system is expanding, whereas negative work is done during the part in which the system is compressed. Fig. 14-12 shows a green-shaded region, which is beneath both the expansion and compression parts of the cycle, and a blue-shaded region, which is beneath only the expansion part. The net work done during a cycle equals the positive area under the expansion phase minus the area under the compression phase. Thus the net work equals the area enclosed by the curve (the area of the blue-shaded region).

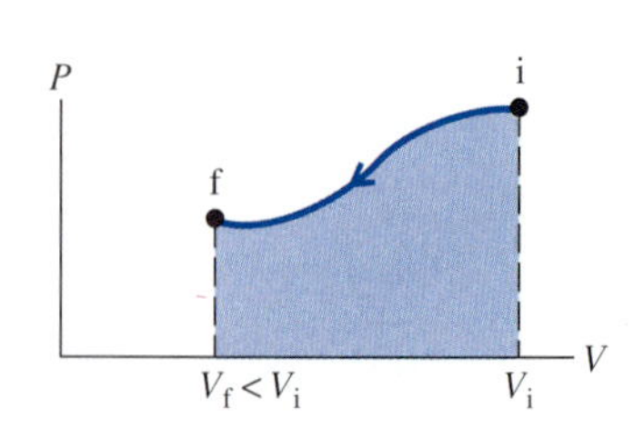

Fig. 14-11 Negative work is done by a system being compressed.

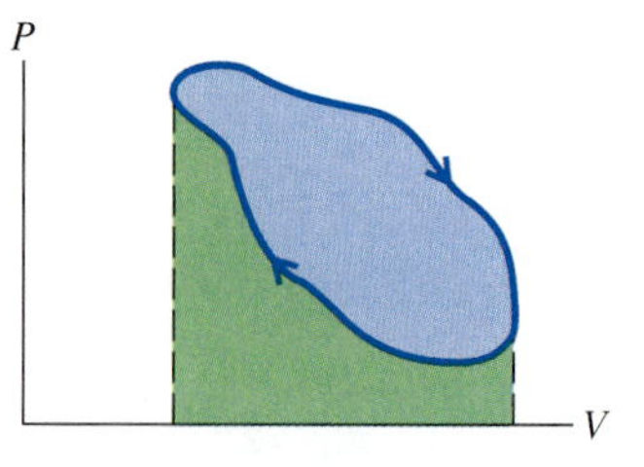

Fig. 14-12 The net work done by a system over a cycle equals the area enclosed by the curve.

EXAMPLE 4 Work Done By an Expanding Ideal Gas

(a) Compute the work done by an ideal gas as it expands from the initial state i to the final state f, as indicated in Fig. 14-13 for each of the three quasi-static processes, I, II, and III.
(b) What is the change in internal energy of the gas?

SOLUTION **(a) I** This process consists of two parts: an expansion from i to A at constant pressure and a constant-volume decrease in pressure from A to f. During the constant-pressure expansion the system does work:

$$W_{iA} = P\,\Delta V = (1.00 \text{ atm})(2.50 \text{ L} - 2.00 \text{ L}) = 0.500 \text{ L-atm}$$

(The energy unit L-atm $= 1.00 \times 10^{-3} \text{ m}^3 \times 1.01 \times 10^5 \text{ N/m}^2 = 101$ J.) As the system goes from state A to state f, the volume is constant. This means that there is no work done by the system. The total work done by the system for process I is

$$W_I = W_{iA} = 0.500 \text{ L-atm}$$

II In this process the system goes from i to B to f, with work being done only from B to f. Again this work is at constant pressure, but since the pressure is lower than in the process from i to A, there is less work done.

$$W_{II} = W_{Bf} = P\,\Delta V = (0.800 \text{ atm})(2.50 \text{ L} - 2.00 \text{ L}) = 0.400 \text{ L-atm}$$

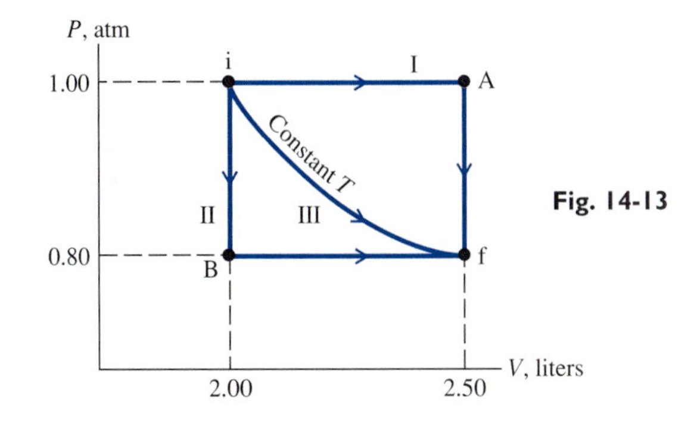

Fig. 14-13

III We can find the work done during this isothermal process by applying Eq. 14-8:

$$W_{III} = nRT \ln \frac{V_f}{V_i}$$

Using the ideal gas law, we may substitute P_iV_i for nRT. Thus

$$W_{III} = P_iV_i \ln \frac{V_f}{V_i} = (1.00 \text{ atm})(2.00 \text{ L})\left(\ln \frac{2.50}{2.00}\right) = 0.446 \text{ L-atm}$$

(b) The change in internal energy of the system from i to f is zero, since there is no change in temperature, and the internal energy of an ideal gas is a function only of temperature. Notice that, although W depends on the process connecting initial and final states, the change in internal energy ΔU does not. Internal energy, unlike work or heat, depends only on the thermodynamic state of the system.

EXAMPLE 5 Heat Absorbed by an Expanding Ideal Gas

For each of the processes described in the previous example, compute the heat absorbed by the system.

SOLUTION We can calculate the heat absorbed by applying the first law of thermodynamics:

$$\Delta U = Q - W$$

In the last example we found that there was no change in internal energy from state i to state f. So $\Delta U = 0$ for any of the processes connecting these states, and so for each process the heat absorbed equals the work done by the system.

$$0 = Q - W$$
$$Q = W$$

Using the results for the work computed in the last example, we find

$$Q_I = W_I = 0.500 \text{ L-atm}$$
$$Q_{II} = W_{II} = 0.400 \text{ L-atm}$$
$$Q_{III} = W_{III} = 0.446 \text{ L-atm}$$

Notice that Q, like W, depends on the process connecting the initial and final states.

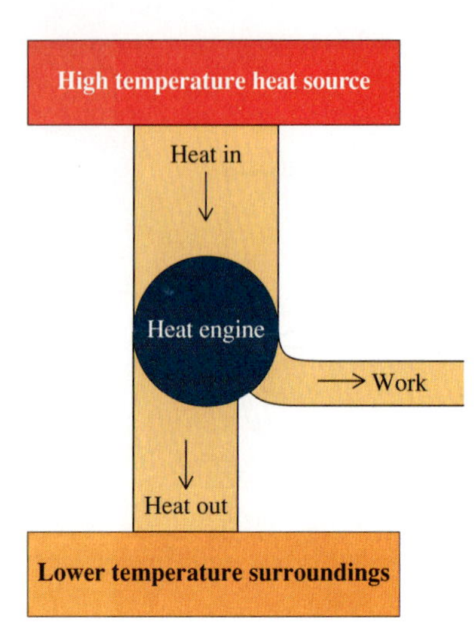

Fig. 14-14 A heat engine absorbs heat at a high temperature and releases heat at a lower temperature.

14-3 Heat Engines and Refrigerators

A heat engine is a device that extracts heat from a source and uses some of that energy to do work. The first practical heat engine was the steam engine, whose invention in the eighteenth century ushered in the industrial revolution. The automobile's internal combustion engine can also be treated as a heat engine, though the heat actually comes from combustion of gasoline, rather than from a source outside the engine.

In an automobile engine, burning fuel provides heat that results in the expansion of gases in the engine's cylinders. The expanding gases do work on the pistons. This work is transmitted by the crankshaft to the wheels. The wheels exert a backward force on the pavement, and the pavement exerts a forward force on the car. If frictional losses are neglected, work done on the car by the force of the pavement is equal to the work done by the expanding gases in the engine's cylinders. As we shall see, however, this work is much less than the internal energy lost by the gas; most of this energy is necessarily lost in the form of waste heat.

A schematic representation of a heat engine is shown in Fig. 14-14. It is a general characteristic of heat engines that some of the heat absorbed is wasted; heat is absorbed at a high temperature, some of that heat is used to do work, and the remaining "waste heat" is expelled at a lower temperature. The efficiency e of a heat engine is defined as the work performed divided by the heat absorbed:

$$e = \frac{\text{Work performed}}{\text{Heat absorbed}} \tag{14-9}$$

For example, a heat engine that does 200 J of work for every 1000 J of absorbed heat has an efficiency of $\frac{200 \text{ J}}{1000 \text{ J}} = 0.200$, or 20.0%. This example is fairly typical of the efficiencies of actual heat engines.

A refrigerator is in essence the opposite of a heat engine. It uses mechanical work to transfer heat from the low-temperature interior of the refrigerator to higher-temperature surroundings. Work must be done to operate a refrigerator, indicated schematically in Fig. 14-15. The work is used to run the compressor, which compresses a refrigeration fluid that circulates through pipes inside and outside the refrigerator, as shown in Fig. 14-16. The fluid carries heat from the refrigeration compartment to the condenser. The heat expelled by a refrigerator can be felt in the hot air blowing from beneath it. (The air is blown over the condenser in order to transfer heat from the fluid inside.)

In a refrigerator the heat and work are just the reverse of these quantities in a heat engine, as is evident from Figs. 14-14 and 14-15. An air conditioner is essentially the same as a refrigerator. A house or car is refrigerated by the air-conditioning unit, as heat is pumped outside.

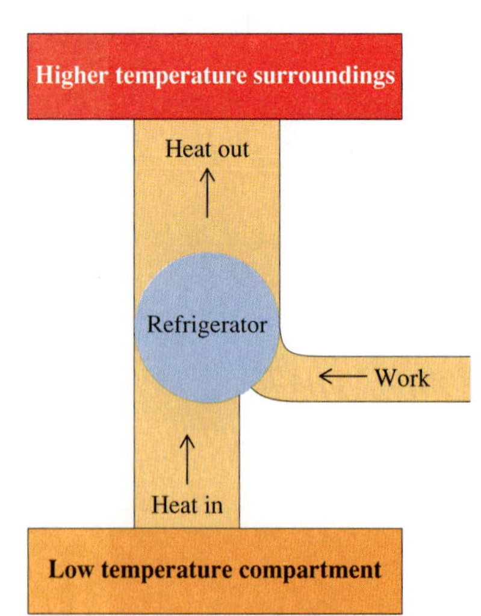

Fig. 14-15 A refrigerator transfers heat from a low-temperature compartment to higher-temperature surroundings.

In both heat engines and refrigerators there is a thermodynamic system that repeatedly goes through a cycle of states. In an automobile engine this system is a mixture of gasoline vapor and air, whereas in a refrigerator it is a refrigeration fluid such as Freon.

A simple heat engine is shown in Fig. 14-17. A monatomic ideal gas is the system that is taken through a sequence of states, with the result that the net heat absorbed is used to do work by raising a weight. A complete analysis of this heat engine shows that it has low efficiency (12%) and produces very little work per cycle (60 J).

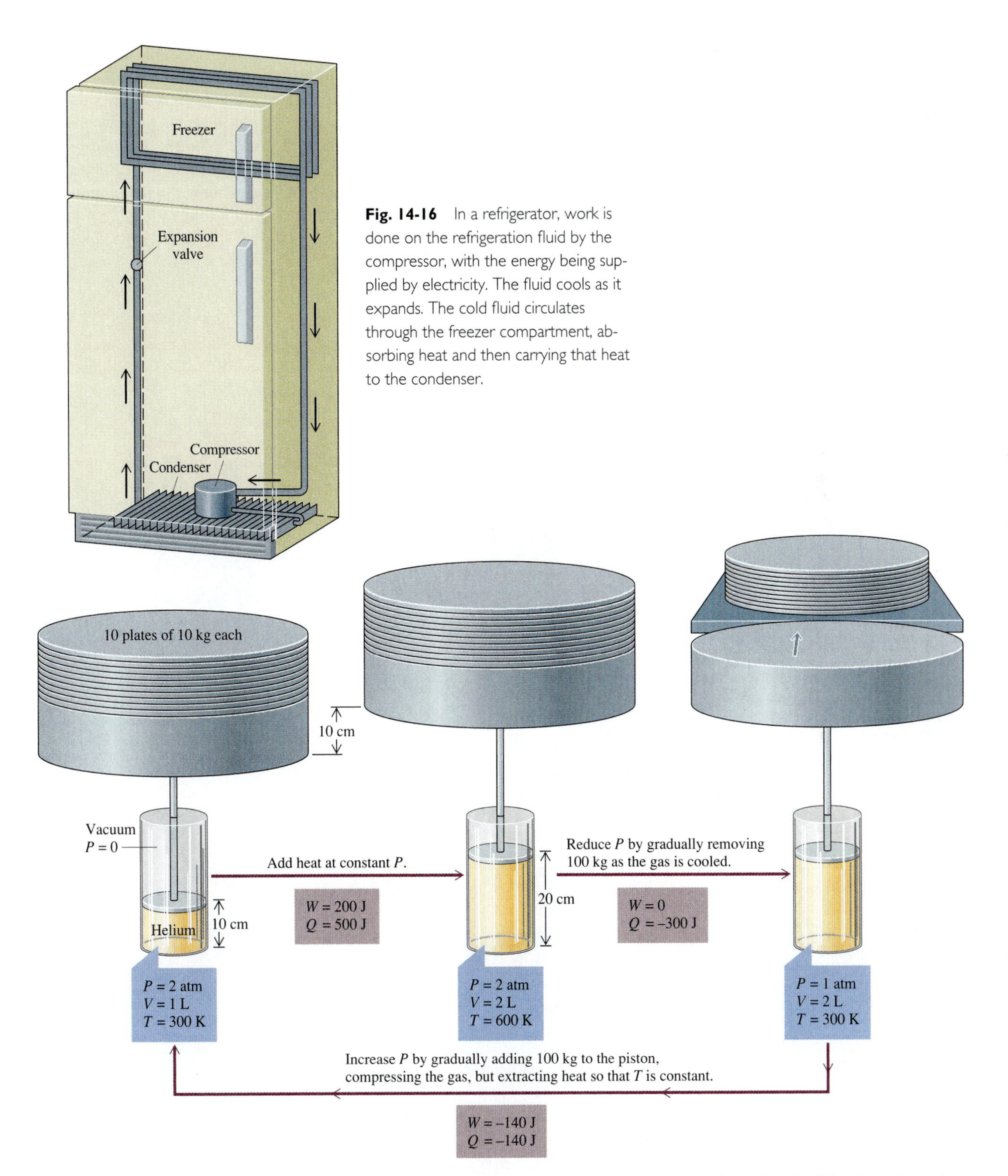

Fig. 14-16 In a refrigerator, work is done on the refrigeration fluid by the compressor, with the energy being supplied by electricity. The fluid cools as it expands. The cold fluid circulates through the freezer compartment, absorbing heat and then carrying that heat to the condenser.

Fig. 14-17 A simple heat engine. Heating the helium in the cylinder increases its temperature and volume, thereby lifting the weights that maintain the external pressure on the gas. Weight is removed as the gas is cooled, reducing pressure and temperature. Finally, the gas is returned to its initial state by gradually adding weights, increasing pressure, reducing volume, and extracting heat. Weights are removed at points higher than the points where they are added. Thus the work done by the helium in a cycle is used to raise the weights. For the cycle, net work = net heat = 60 J.

A more practical engine is the gasoline engine (Fig. 14-18). The system, which does positive net work, is a mixture of air and gasoline vapor. A spark from a spark plug causes the gasoline to ignite. The heat absorbed in this case is just the heat of combustion of the gasoline. The heat released by the system is the heat carried away by the exhaust gases. Once each cycle, the thermodynamic system (gasoline and air) is replaced by an identical new system in the initial state. The *PV* diagram for the gasoline engine, corresponding to an idealized quasi-static process, with an ideal gas as the working substance, is called the "Otto cycle" and is shown in the figure.

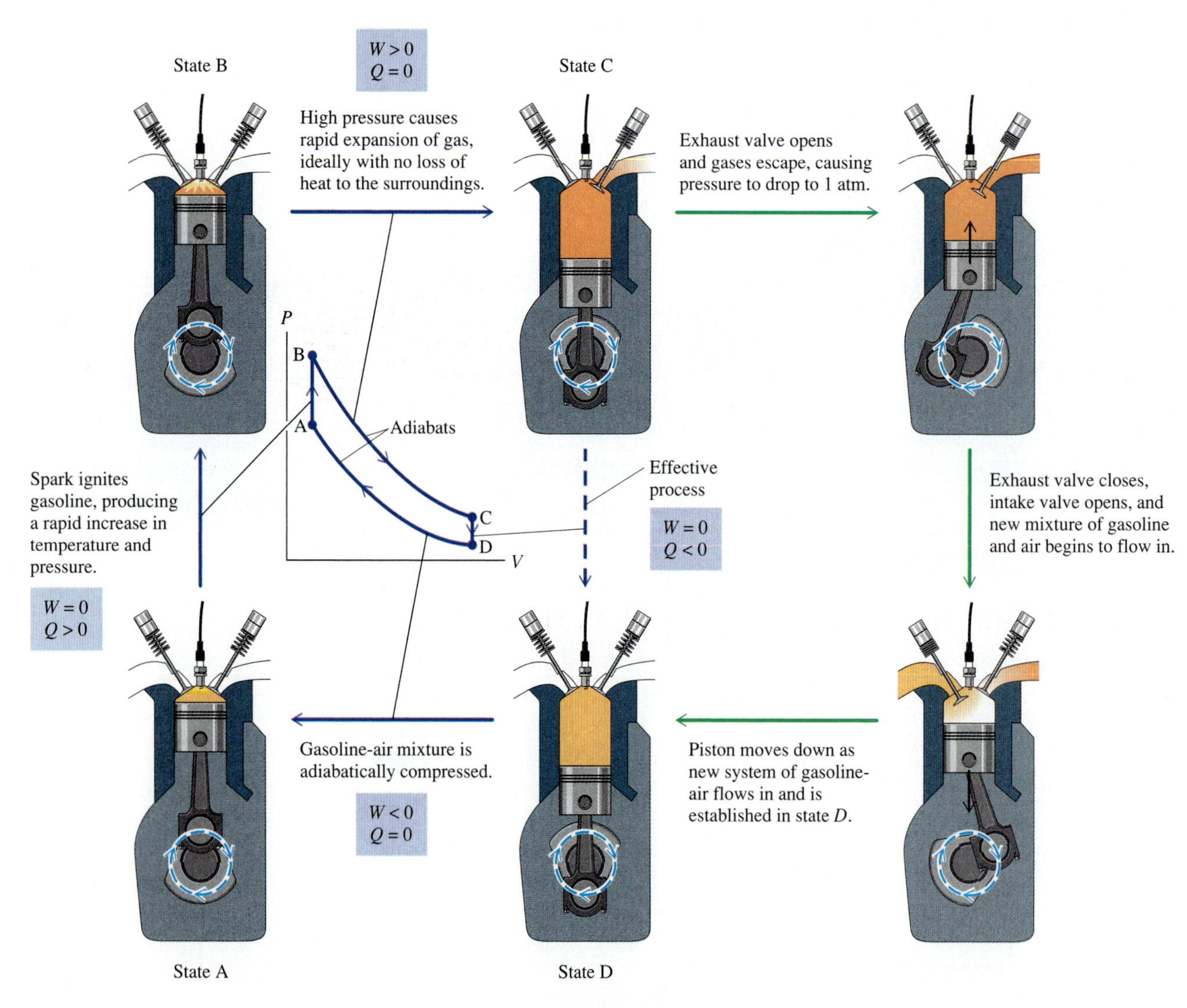

Fig. 14-18 The Otto cycle of a gasoline engine.

14-4 Second Law of Thermodynamics

Suppose that after thoroughly shuffling a deck of cards, you found that they were arranged in the sequence

A,A,A,A, 2,2,2,2, 3,3,3,3, . . . ,Q,Q,Q,Q, K,K,K,K

You would certainly be very surprised. For although such a sequence is a possibility, it is extremely unlikely. It would be hopeless to predict *any* sequence resulting from a shuffle of the cards. But almost all the sequences of cards that can occur are very unremarkable—that is, they do not represent any easily describable state of order. If one were to begin with the deck arranged in the previously mentioned ordered sequence (4 aces, 4 deuces, . . .), and then shuffled, one could predict with near certainty that the deck would become disordered. This would happen simply because there are so many more possible disordered sequences than there are ordered sequences.

Fig. 14-19 There is a strong element of luck in most card games. This sequence of cards, a royal flush, is one of the luckiest poker hands.

All mechanical interactions involving the random interactions of many particles demonstrate this same tendency for an ordered state to change to a disordered state. There is nothing in the laws of mechanics that requires this to be so. It is simply that there are many more disordered states of a many-particle system than there are ordered states, and so it is highly unlikely that a state of greater order will result from a random process. Consider a cue ball smashing into an orderly rack of balls on a pool table (Fig. 14-20a). The collision will almost certainly result in the balls moving off in various directions, with the energy shared by many balls. According to the laws of mechanics, it is perfectly possible for the reversed situation to occur, that is, for a group of balls to be initially moving in different directions and undergo collisions that result in all but the cue ball coming to rest in an orderly arrangement while the cue ball carries away all the energy (Fig. 14-20b). All that is required for this to happen is that the initial velocities of the balls in Fig. 14-20b be the negatives of the final velocities in Fig. 14-20a. Of course the spontaneous occurrence of such an event is extremely unlikely.

Before collision

After collision

(a) Ordinary collision

Before collision

After collision

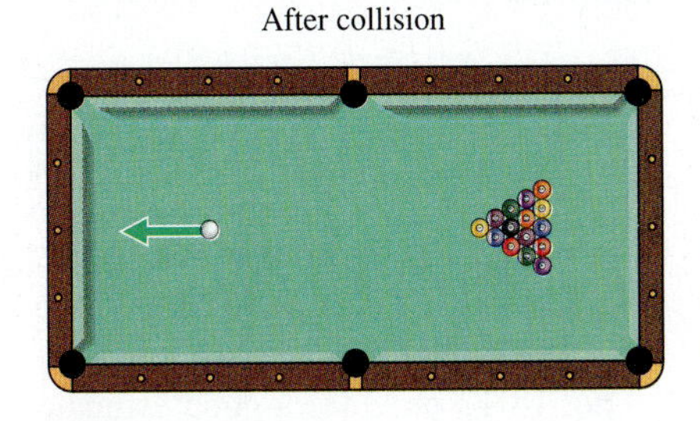

(b) Reversed collision

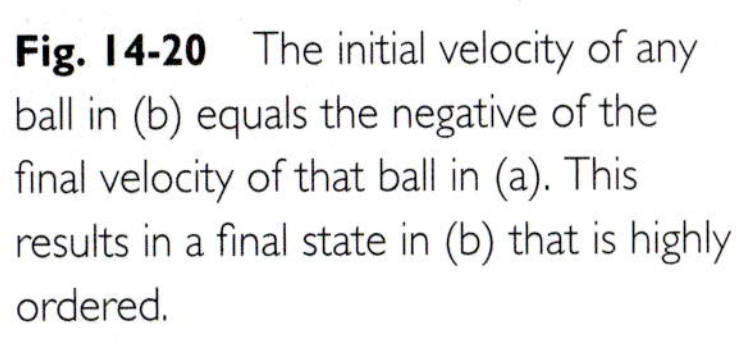

Fig. 14-20 The initial velocity of any ball in (b) equals the negative of the final velocity of that ball in (a). This results in a final state in (b) that is highly ordered.

Definition of Entropy

The second law of thermodynamics describes the tendency of systems containing large numbers of randomly interacting particles to go from states of relative order to states of relative disorder. To precisely formulate the second law, we must first introduce the concept of entropy, a thermodynamic variable that provides a relationship between mechanical states and thermodynamic states. As we already noted at the beginning of Section 14-1, a complete mechanical description of a macroscopic system is much more involved than a thermodynamic description of the same system. For example, the mechanical state of a gas is specified when the position and velocity of every molecule of the gas are given, whereas the thermodynamic state is specified when only the number of molecules N, the volume V, and the internal energy U are given. It is not surprising then that there is an enormously large number of mechanical states corresponding to any one thermodynamic state. For example, a gas with a given amount of internal energy can have a huge number of arrangements of molecular positions and velocities. The entropy of a system is a measure of the number of possible mechanical states of the system for a given thermodynamic state. More precisely, **the entropy S is defined to be Boltzmann's constant k times the natural log of the number Ω of possible mechanical states.**

$$S = k \ln \Omega \qquad (14\text{-}10)$$

The units for entropy are the same as for k: J/K.

The entropy of a system is a function of the system's independent variables (N, V, and U for a gas). It is possible to show that for a monatomic ideal gas the entropy is given by

$$S = Nk \ln (VU^{3/2}) + C \quad \text{(monatomic ideal gas)} \qquad (14\text{-}11)$$

where C is a constant. (A partial derivation of this result is outlined in Problem 51.)

We can use the preceding expression for the entropy of an ideal gas to discuss some general properties of entropy. For example, this expression shows that if any of the quantities N, V, or U increases then the entropy increases. (A logarithm always increases as its argument increases.) Thus, if you increase the volume from V to $2V$, the entropy becomes

$$S = Nk(\ln 2VU^{3/2}) + C = Nk \ln 2 + Nk(\ln VU^{3/2}) + C$$

which is an increase in entropy of $Nk \ln 2$.

For most systems there is no simple expression for the entropy in terms of the independent variables. However, there is a simple expression that allows us to calculate changes in the entropy of *any system* when the initial and final states of the system are at the same temperature. **The entropy change equals the heat absorbed in a quasi-static, isothermal process connecting the states, divided by the absolute temperature.**

$$\Delta S = \frac{Q}{T} \quad \text{(at constant } T) \qquad (14\text{-}12)$$

It is possible to show that this result follows from the definition of entropy (Eq. 14-10: $S = k \ln \Omega$). A derivation for the special case of a monatomic ideal gas is outlined in Problem 58.

Eq. 14-12 provides a good estimate for the entropy change of a system even if there is some small change in its absolute temperature.

Entropy Statement of the Second Law; Isolated System

The second law of thermodynamics states that **the entropy of an isolated system never decreases.** The change in entropy is either positive or zero.

$$\Delta S \geq 0 \quad \text{(for an isolated system)} \tag{14-13}$$

Since entropy is defined as $k \ln \Omega$, the second law of thermodynamics requires that an isolated system never undergo a process for which the final number of states is less than the initial number of states. If the final number Ω_f were less than the initial number Ω_i, there would be a negative entropy change ($\Delta S = k \ln \Omega_f - k \ln \Omega_i$), in violation of the second law.

We can express the entropy change in terms of the ratio of the numbers of final and initial states.

$$\Delta S = S_f - S_i = k \ln \Omega_f - k \ln \Omega_i$$

$$\Delta S = k \ln \frac{\Omega_f}{\Omega_i} \tag{14-14}$$

We can also obtain the inverse relationship:

$$\frac{\Delta S}{k} = \ln \frac{\Omega_f}{\Omega_i}$$

or

$$\frac{\Omega_f}{\Omega_i} = e^{\Delta S/k} \tag{14-15}$$

In the examples that follow we shall see that a small change in entropy corresponds to an enormous change in the number of available states. From this we see the essential meaning of the second law. A system tends to go to a final state for which there is an enormous increase in the number of corresponding mechanical states. This happens simply because there are so many ways for it to happen. The second law of thermodynamics is a very unusual kind of fundamental physical law in that it is purely statistical, and so one could conceive of its being violated by a random event. One could imagine, for example, all the molecules of a gas spontaneously crowding into one corner of a container, thereby reducing the volume and the value of the entropy. Although in principle this is possible, in practice it is extremely unlikely—much less likely than finding a randomly shuffled deck of cards in the ordered sequence A,A,A,A, . . . , K,K,K,K.

EXAMPLE 6 Change in Entropy of an Expanding Gas

1.00 mole of a monatomic ideal gas is initially confined to an insulated container of volume V. The gas is then allowed to escape through an opening into an initially evacuated region of equal volume (Fig. 14-21), so that the final volume of the gas is $2V$.

(a) Find the change in entropy of the gas.

(b) Find the ratio of the number of final states to the number of initial states.

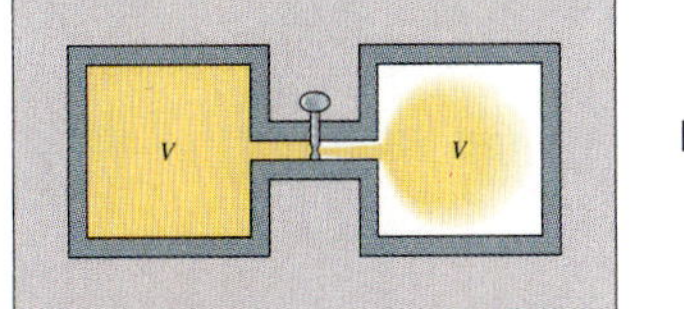

Fig. 14-21

Solution on next page.

EXAMPLE 6—cont'd

SOLUTION **(a)** We apply Eq. 14-12, setting N equal to Avogadro's number, since there is 1 mole.

$$\Delta S = S_f - S_i = Nk(\ln 2VU^{3/2}) + C - Nk(\ln VU^{3/2}) - C$$
$$= Nk \ln 2 = (6.022 \times 10^{23})(1.38 \times 10^{-23}\text{ J/K})(\ln 2)$$
$$= 5.76\text{ J/K}$$

(b) To find the ratio of the number of final states to the number of initial states, we apply Eq. 14-15, using the value for ΔS found in part a.

$$\frac{\Omega_f}{\Omega_i} = e^{\Delta S/k} = e^{(5.76\text{ J/K})/(1.38\times10^{-23}\text{J/K})}$$
$$= e^{4.17\times10^{23}}$$

Since $e = 10^{0.43}$, we can express our result as

$$\frac{\Omega_f}{\Omega_i} = (10^{0.43})^{4.17\times10^{23}} = 10^{1.8\times10^{23}} = 10^{\overbrace{180\ldots0}^{22\text{ zeroes}}}$$

This is an extremely large number. When there are so many more ways for the gas to occupy the entire volume $2V$, it is a practical certainty that the molecules will never again crowd into one half of the container, though there is nothing in the laws of mechanics that would prevent this from happening.

Entropy Statement of the Second Law; Interacting System

When a system interacts with its surroundings, we may consider the combination of the system and the surroundings as forming an "isolated system" and apply the second law of thermodynamics to the combined system. Suppose that when both the system and the surroundings are in given thermodynamic states, there are Ω_{sys} mechanical states possible for the system and Ω_{surr} mechanical states possible for the surroundings. Since for each of the Ω_{sys} states of the system, there are Ω_{surr} states of the surroundings, the total number Ω of unique states of the combined system equals the product of Ω_{sys} and Ω_{surr}.*

$$\Omega = \Omega_{sys}\Omega_{surr} \tag{14-16}$$

The entropy of the combined system is given by

$$S = k \ln \Omega = k \ln (\Omega_{sys}\Omega_{surr}) = k \ln \Omega_{sys} + k \ln \Omega_{surr}$$

or

$$S = S_{sys} + S_{surr}$$

Thus the total entropy change of the combined system is the sum of the changes in entropy of each part.

$$\Delta S = \Delta S_{sys} + \Delta S_{surr}$$

Applying the second law of thermodynamics ($\Delta S \geq 0$) to the combined system, we find

$$\boxed{\Delta S_{sys} + \Delta S_{surr} \geq 0} \tag{14-17}$$

From this expression we see that it is possible for the entropy of an interacting system to decrease ($\Delta S_{sys} < 0$) but only if the entropy of the surroundings increases at least enough to balance the system's entropy loss, so that the sum of the two entropy changes is positive (or zero). Since entropy is a measure of disorder, this means that it is possible for a system to become more ordered but only at the expense of the surroundings' becoming less ordered.

*This is analogous to the calculation of the number of unique ways of drawing a card from each of two decks. Since there are 52 cards possible from the first deck and for each of these 52 you can draw any one of the 52 cards from the second deck, the number of possible combinations is 52 × 52.

EXAMPLE 7 Change in Entropy of Boiling Water

(a) Find the change in entropy of 100 g of water as it boils at atmospheric pressure. (b) Find the ratio of the number of final states to the number of initial states. (c) Find the change in the entropy of the surroundings if the heat is supplied to the system (the water) by the thermal reservoir that is only slightly hotter than the water, that is, about 100° C.

SOLUTION (a) Since this process occurs at a constant temperature ($T = 373$ K), we can apply Eq. 14-12 to find ΔS. Setting the heat absorbed by the water equal to its mass times the latent heat of boiling (2.26×10^6 J/kg), we find that

$$\Delta S = \frac{Q}{T} = \frac{mL}{T} = \frac{(0.100 \text{ kg})(2.26 \times 10^6 \text{ J/kg})}{373 \text{ K}}$$

$$= 606 \text{ J/K}$$

(b) Applying Eq. 14-15, we obtain

$$\frac{\Omega_f}{\Omega_i} = e^{\Delta S/k} = e^{(606 \text{ J/K})/(1.38\times 10^{-23}\text{J/K})} = e^{4.4\times 10^{25}}$$

Since $e = (10^{0.43})$, we can express our result as

$$\frac{\Omega_f}{\Omega_i} = (10^{0.43})^{4.4\times 10^{25}} = 10^{1.9\times 10^{25}} = 10^{190\overbrace{\ldots}^{\text{24 zeroes}}0}$$

(c) The heat released by the reservoir equals the heat absorbed by the water, and both the reservoir and the water are at nearly the same temperature as the heat is transferred. Applying the equation $\Delta S = \dfrac{Q}{T}$ to the reservoir, we see that, except for the sign of the heat absorbed (negative for the reservoir), the calculation of the entropy change is identical to that in part (a).

$$\Delta S_{\text{surr}} = \frac{Q}{T} = -606 \text{ J/K}$$

Thus the total entropy change of the system and the surroundings in this case equals zero. If the surroundings had been at a higher temperature, the negative entropy change of the surroundings would not have balanced the positive entropy change of the system, and the total entropy change of the combined system would be positive. Physical processes are normally not quasi-static and therefore result in a net increase in entropy.

Kelvin Statement of the Second Law

Historically the second law of thermodynamics has been expressed in different ways. Although these statements of the second law appear to be quite different from each other, they can be shown to be equivalent. The second law was first discovered by Sadi Carnot in 1824 through his analysis of the efficiencies of heat engines. The clearest formulation of his ideas was given later by Lord Kelvin. The **Kelvin statement of the second law** says that: **It is impossible to have a transformation whose only result is to extract heat from a source that is at the same temperature throughout and transform it completely into work.** Thus the Kelvin statement says that the kind of heat engine indicated in Fig. 14-22 is an impossibility.

Fig. 14-22 According to the Kelvin statement of the second law of thermodynamics, the process pictured here is impossible. Heat from a source at a single temperature cannot be converted entirely into work.

It is easy to derive the Kelvin statement of the second law from the entropy statement of the second law. Suppose that a heat engine absorbs positive heat from a source at a single temperature T and converts this heat entirely into work. Since the heat engine goes through a cycle, its thermodynamic state, including its entropy, is unchanged. The work done by the engine may be used to increase the mechanical energy of some system (lift a weight or turn a wheel, for example), leaving the thermodynamic state of that system unchanged. The only change in thermodynamic state is that of the body supplying the heat. Since it releases heat, that is, absorbs negative heat, it would undergo a negative change in entropy ($\Delta S = Q/T < 0$). Then the total entropy change would be negative, in violation of the entropy statement of the second law. Therefore such a process is impossible, as predicted by the Kelvin statement.

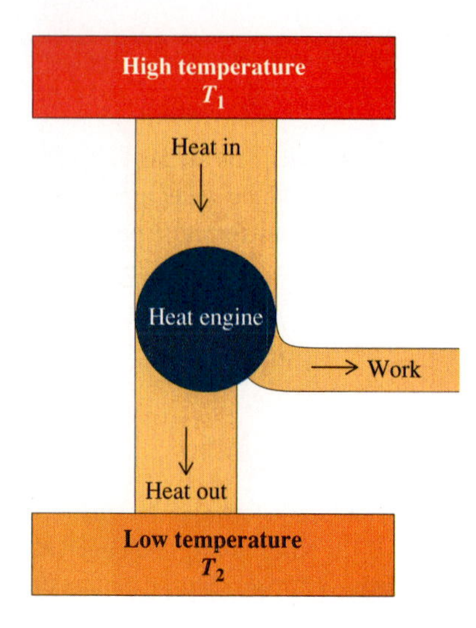

Fig. 14-23 A Carnot engine. Heat is absorbed only at temperature T_1, heat is rejected only at temperature T_2, and the process is quasi-static.

Carnot Engine

The simplest possible kind of heat engine consistent with the second law of thermodynamics is one that absorbs heat at a temperature T_1 and releases heat at a lower temperature T_2, as illustrated in Fig. 14-23. The net work done during each cycle equals the heat absorbed minus the heat released. If the cycle is quasi-static, the engine is said to be a **Carnot engine,** and the cycle is called a **Carnot cycle.** Heat is absorbed only at temperature T_1 and rejected only at temperature T_2 in a Carnot engine. So a Carnot cycle consists of two isothermal processes in which heat is exchanged and two adiabatic processes taking the system back and forth between temperatures T_1 and T_2.

Since no real thermodynamic process is ever quasi-static, the Carnot engine is an idealized heat engine. However, the Carnot engine is of great theoretical importance in the design of real heat engines. It can be shown that **a Carnot engine operating between temperatures T_1 and T_2 has the highest attainable efficiency e_C of any engine operating between these same two temperature extremes:**

$$e \leq e_C \tag{14-18}$$

where the Carnot efficiency e_C is given by

$$e_C = 1 - \frac{T_2}{T_1} \tag{14-19}$$

A derivation of this result, called **Carnot's theorem,** is outlined in Problem 52.

The efficiency of all Carnot engines operating between the same two temperatures is the same, independent of the thermodynamic system. Two very different kinds of Carnot cycles are shown in Fig. 14-24, one using an ideal gas as the system and the other using steam as the system.

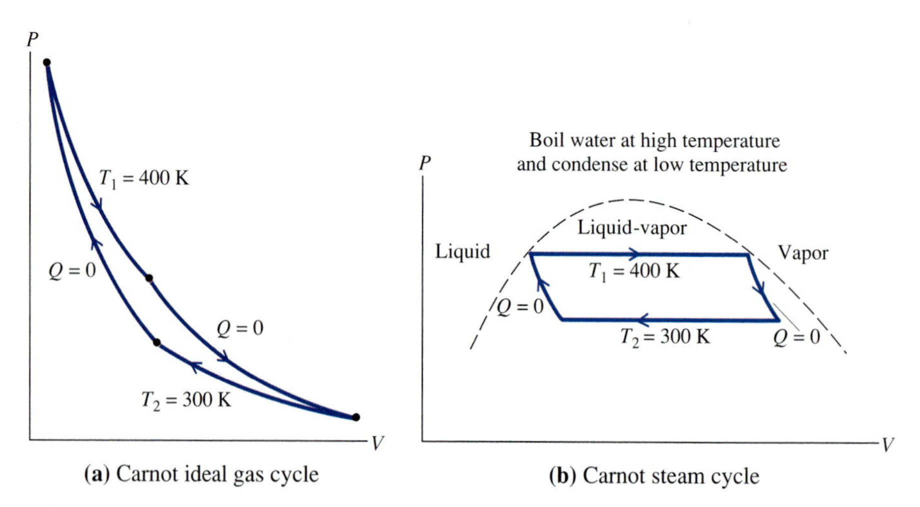

Fig. 14-24 Each of the cycles shown is a Carnot cycle, since each consists of two adiabatic, quasi-static processes connecting two isothermal, quasi-static processes. And since they operate between the same temperatures, $T_1 = 400$ K and $T_2 = 300$ K, they have the same efficiency $e_C = 1 - \frac{300}{400} = 0.25$.

EXAMPLE 8 The Minimum Heat Wasted by an Engine

A certain heat engine operates between 27° C and 327° C and absorbs 1000 J of heat during each cycle. What is the minimum amount of heat expelled during each cycle?

SOLUTION A Carnot engine operating between the same temperatures will have the highest efficiency and therefore will expel the least heat. Any real engine will have lower efficiency and will expel more of the heat absorbed. Converting temperatures to the Kelvin scale, we find the Carnot efficiency

$$e_C = 1 - \frac{T_2}{T_1} = 1 - \frac{300 \text{ K}}{600 \text{ K}}$$

$$= 0.500$$

In a Carnot engine operating between these temperatures, 50% of the heat absorbed is converted to work and the remaining 50% is expelled as waste heat. Thus the minimum heat expelled in a real heat engine operating between these temperatures is 50% of the absorbed heat.

$$\text{Heat expelled} \geq (0.500)(1000 \text{ J}) = 500 \text{ J}$$

Clausius Statement of the Second Law

Another expression of the second law of thermodynamics is the **Clausius statement,** which says that **heat will not spontaneously flow from a colder body to a hotter body.** In other words, heat does not flow "uphill" (Fig. 14-25). The Clausius statement is in accord with everyday experience. As an example violating the Clausius statement, suppose you drop an ice cube into a glass of water at room temperature and positive heat flows to the water from the ice cube so that the ice becomes colder and the water becomes warmer. Of course, this would never happen. You will not observe this or any other violation of the Clausius statement of the second law.

The entropy statement of the second law can be used to derive the Clausius statement (Problem 59).

Fig. 14-25 The Clausius statement of the second law says that heat does not spontaneously flow from a colder body to a hotter body.

*14-5 Human Metabolism

Basal Metabolic Rate

In our discussion of calorimetry in Chapter 13 we noted the energy per gram provided by various foods (4 kcal per gram of protein or carbohydrate, 7 kcal per gram of alcohol, and 9 kcal per gram of fat). Another measure of the body's utilization of the internal energy in food is the energy released per liter of oxygen consumed. This turns out to be very nearly the same for all foods—approximately 5 kcal/liter. This means that measurement of a person's instantaneous rate of oxygen consumption can be used as a means of determining the instantaneous rate at which the body is utilizing the internal energy stored in food. A person of mass 75 kg consumes about 16 liters of oxygen per hour* when in a state of complete rest. This means that for one just to stay alive the minimum energy required is (5 kcal/L)(16 L/h) = 80 kcal/h, or 1900 kcal/day. This is known as the **basal metabolic rate,** abbreviated BMR.

It is found experimentally that the BMR is approximately proportional to body mass raised to the three-fourths power.

$$\text{BMR} \approx M^{3/4}$$

This result is known as **Kleiber's law** and is found to hold not only for the human species, but also for other mammals, from mice to elephants. Physiologists some-

*About 40 liters per hour is inhaled in the resting state, but 24 liters is exhaled.

times measure the BMR per unit of body surface area, the assumption being that BMR is approximately proportional to the surface area. It turns out that both this assumed relationship and Kleiber's law adequately describe the variation in BMR of different-sized humans. On the average BMR is approximately proportional to surface area and to $M^{3/4}$. These two results are in approximate agreement over the small range of body sizes considered. But Kleiber's law turns out to be applicable to various species over an enormous range of masses, whereas the prediction that BMR is proportional to surface area is badly in error over such a range.

In addition to its dependence on mass, the BMR is a function of age and sex. It is found to be slightly lower for women than for men of the same mass. The BMR is also found to decrease gradually with age. The BMR is very roughly a constant for different individuals of the same age and sex who have the same mass. However, there are significant variations among individuals. An abnormally high or low BMR can be an indication of disease (such as hyperthyroidism or hypothyroidism), but variations of as much as 20% have been found in healthy, otherwise normal individuals. People with relatively low BMR's require less food than those with higher BMR's and consequently must restrict their food intake to something less than the average amount to avoid storing the extra food in the form of excess fat. If the appetite is not correspondingly reduced, this condition can make it very difficult for the person to avoid becoming obese. Reduced BMR, however, is apparently not the cause of most cases of obesity.

Work and Metabolism

Of course, the rates of oxygen consumption and energy utilization increase as the level of physical activity increases. For a 75 kg person, light activity requires 125 to 250 kcal/h. Heavy activity, such as heavy manual work or strenuous sport, requires 500 to 750 kcal/h. In cases of extreme exertion, the rate of energy utilization can be as high as 1500 kcal/h, but such levels are attained only by individuals in excellent physical condition and then only for a limited period of time. The maximum attainable rate of oxygen consumption is one measure of physical conditioning.

Suppose that a person's day consists of 10 hours sleeping or resting, requiring 80 kcal/h, 13 hours of light activity, averaging 150 kcal/h, and 1 hour of very strenuous activity, requiring 750 kcal/h; then the total energy required is $10 \times 80 + 13 \times 150 + 1 \times 750 = 3500$ kcal. The energy requirements of such a person will be met if food provides 3500 kcal. If the caloric intake were reduced to 2500 kcal, the extra energy would be obtained from energy stored in the body's tissues. Body fat is used first. Since fat provides 9 kcal/g, the extra 1000 kcal of energy could be obtained from 110 g, or $\frac{1}{4}$ lb of fat. The person would lose $\frac{1}{4}$ lb of weight in 1 day. On the other hand, if the caloric intake were increased to 4500 kcal, 1000 kcal more than required, the extra energy would be stored in the form of body tissue—$\frac{1}{4}$ lb of extra fat or perhaps a somewhat greater weight of muscle.

The body uses the internal energy provided by food. Consider the body as a thermodynamic system over a period of time during which no food enters the body and no waste is excreted. The only changes in the body's internal energy will be produced by the transfer of heat to or from the surroundings or by the performance of work. According to the first law of thermodynamics, the change in the body's internal energy, ΔU, is given by

$$\Delta U = Q - W$$

Ordinarily the body transfers heat to the environment. Thus the heat Q absorbed by the body is negative. The work W done by the body on the surroundings is positive. Thus the first law implies that both the heat generated by the body and the work done by the body contribute to a negative value for ΔU, that is, a decrease in the body's internal energy. Most of the body's energy loss to the environment is in the form of heat.

The biochemical processes utilizing the internal energy in food do not directly produce heat, and yet most of this energy ultimately ends up as heat. For example, measurement of the oxygen used by the heart indicates that the beating of the heart requires 130 kcal per day of mechanical energy for a 75 kg person. This is about 7% of the BMR. This energy is used to pump blood through the circulatory system, where the energy ends up as heat—heat that is eventually carried away from the body.

The work done by the human body on its environment is often very difficult to measure because of the complexity of the motion involved in many activities (see, for example, the energy analysis of running in Chapter 9). However, the work done in pedaling a bicycle is fairly easy to measure. One need simply measure the force on the pedals and the distance they move. Such measurements indicate that the mechanical work done by the body is never more than about 25% of the internal energy being used (as observed by measurement of the oxygen consumed). The remaining 75% (or more) is lost as heat. And yet the bicycle is one of the more efficient means of converting human internal energy into work.* Thus the efficiency of a human being as a machine is no more than about 25%. One might be tempted to treat the body as a heat engine and compare its actual efficiency to the theoretical efficiency. Certainly the second law of thermodynamics applies to the human body. However, since the body does not initially convert food energy into heat before it is utilized, it cannot be considered a heat engine, and the second law cannot be applied in an obvious way.

*The most important advantage of the bicycle, however, is the fact that very little work is required to ride a bicycle on a flat, level surface compared to the work required to walk or run the same distance.

EXAMPLE 9 Burning Calories on a Lifecycle

In 1.00 h, a 50.0 kg woman does work of 75 kcal in pushing on the pedals of her bicycle. During this interval she also consumes 60 liters of air. (a) Find her power output in watts. (b) Find the decrease in her internal energy. (c) How much heat does she transfer to her environment?

Fig. 14-26

SOLUTION (a) $P = \frac{W}{t} = \frac{(75 \times 10^3 \text{ cal})(4.20 \text{ J/cal})}{3600 \text{ s}}$

$$= 88 \text{ W}$$

(b) Since 5.0 kcal of energy is used for each liter of oxygen consumed, we have

$$\Delta U = -(5.0 \text{ kcal/L})(60.0 \text{ L}) = -300 \text{ kcal}$$

(c) The heat absorbed by her body is found by application of the first law:

$$Q = \Delta U + W$$

$$= -300 \text{ kcal} + 75 \text{ kcal} = -225 \text{ kcal}$$

Thus 225 kcal of heat is lost by her body to the environment. She "burns" 300 kcal—75 kcal by doing work and another 225 kcal by generating heat.

CHAPTER 14 SUMMARY

A **thermodynamic system** consists of a large number of particles and is described by means of a few **thermodynamic variables,** the values of which specify a thermodynamic state. Some thermodynamic variables are pressure P, volume V, number of particles N, temperature T, internal energy U, and entropy S.

The **internal energy** is defined to be the total energy of all the particles of the system minus the center-of-mass energy.

The **entropy** is defined by

$$S = k \ln \Omega$$

where k is Boltzmann's constant ($k = 1.38 \times 10^{-23}$ J/K) and Ω is the number of possible mechanical states corresponding to a particular thermodynamic state of the system.

Thermodynamic variables are not all independent; they are related by equations of state. The ideal gas law relates P, V, N, and T for an ideal gas: $PV = NkT$. The entropy of a monatomic ideal gas is related to its volume, internal energy, and number of molecules by the equation

$$S = Nk(\ln VU^{3/2}) + C \quad \text{(for a monatomic ideal gas)}$$

For certain ideal systems there is a simple relationship between internal energy and temperature:

$$U = \tfrac{3}{2}nRT \quad \text{(for a monatomic ideal gas)}$$

$$U = \tfrac{5}{2}nRT \quad \text{(for a diatomic ideal gas)}$$

Equilibrium states of a thermodynamic system are characterized by values of thermodynamic variables, which remain constant if the system is not disturbed. A **quasi-static process** consists of a continuous succession of equilibrium states. Although quasi-static processes are never realized in practice, they are approximated by real processes, for example, when heat is added very slowly to a system. An **isothermal process** is one for which T is constant, and an **adiabatic process** is one for which $Q = 0$.

The **first law of thermodynamics** states that, for an **isolated system,** the total internal energy plus center-of-mass energy is constant and, for an **interacting system,** the internal energy may change either by the absorption of heat, Q, or by the performance of work, W.

$$\Delta U = Q - W$$

When the volume of a system changes from V_i to V_f in a quasi-static process, the system does work W, which is equal to the area under a P versus V curve. The work is positive if V increases and negative if V decreases. Important special cases are the constant-pressure and isothermal processes.

$$W = P\,\Delta V \quad \text{(at constant pressure)}$$

$$W = nRT \ln \frac{V_f}{V_i} \quad \text{(along an isotherm)}$$

The **second law of thermodynamics** states that the total change in entropy of a system and its surroundings is never negative.

$$\Delta S_{sys} + \Delta S_{surr} \geq 0$$

Since $\Delta S = k \ln \frac{\Omega_f}{\Omega_i}$, when S increases, the system goes to a thermodynamic state with a larger number of corresponding mechanical states—a state of greater disorder. If the initial and final states of a system are at the same temperature, the entropy change may be computed by use of the equation

$$\Delta S = \frac{Q}{T} \quad \text{(at constant } T\text{)}$$

where Q is computed for any quasi-static process connecting the initial and final states.

The **Clausius statement** of the second law says that heat will not spontaneously flow from a colder to a hotter body. The **Kelvin statement** of the second law states that a transformation whose only result is to transform into work heat extracted from a source that is at the same temperature throughout is impossible.

A heat engine uses heat to produce work. A **Carnot engine** absorbs all heat at a single temperature T_1 and expels all heat at a single (lower) temperature T_2 and goes through a quasi-static cycle, called a **Carnot cycle.** The efficiency e of a heat engine is defined as the ratio:

$$e = \frac{\text{Net work}}{\text{Heat absorbed}}$$

The efficiency e_C of a Carnot engine is given by $e_C = 1 - \frac{T_2}{T_1}$. No heat engine operating between the temperature extremes T_1 and T_2 can have a greater efficiency than a Carnot engine operating between these same temperatures

$$e \leq e_C$$

Questions

1 Is it possible to determine whether an increase in a system's temperature was caused by heat added to the system or by work done on the system if you examine only the final state?

2 When you vigorously pump air with a bicycle pump, the pump becomes warm.
(a) Has heat been absorbed by the air?
(b) Has work been done on the air?

3 A gas initially confined to an insulated container of volume V is allowed to escape through an opening into an initially evacuated region of equal volume V, as shown in Fig. 14-21.
(a) Does the gas do work?
(b) Does the gas absorb heat?
(c) Does the internal energy of the gas change?
(d) If the gas is ideal, does its temperature change?
(e) If the gas is not ideal and initially has a negative potential energy contributing to its internal energy, what happens to its temperature after expansion when the gas molecules are farther apart?

4 Suppose an ideal gas goes from an initial state of volume V to a final state of volume $2V$ either by an isothermal or by an adiabatic process.
(a) For the adiabatic process does the temperature increase, decrease, or remain constant?
(b) Is more work done by the gas in the isothermal process or is it the same in either case?

5 If you add 100 J of heat to a container of gas, will the rise in temperature be greater if you hold the volume fixed or if you hold the pressure fixed?

6 Which has more entropy, 50 g of ice at 0° C or 50 g of water at 0° C? Which is in a state of greater order?

7 Is it possible to cool a kitchen by leaving a refrigerator door open?

8 Suppose an air-conditioner unit is operated inside a room so that the exhaust goes into the room. What effect will the air conditioner have on the temperature of the room?

9 In Example 2 a piece of clay falls to the floor from a height of 2 m, with the result that the temperature of the clay increases slightly.
(a) Has heat been absorbed by the clay?
(b) Has the entropy of the clay increased?
(c) Is it possible to apply the formula $\Delta S = \frac{Q}{T}$ to the clay?

10 Would you expect the efficiency of a heat engine to increase as a result of increasing the maximum temperature at which the engine absorbs heat?

11 Would you expect the efficiency of a heat engine to increase as a result of increasing the minimum temperature at which the engine releases heat?

Answers to Odd-Numbered Questions

1 no; **3** (a) no; (b) no; (c) no; (d) no; (e) decreases; **5** volume; **7** no; **9** (a) no; (b) yes; (c) yes; **11** no

Problems (listed by sections)

14-1 Thermodynamic Systems

1 Calculate the internal energy of 32.0 g of low-density oxygen gas (O_2) at a temperature of 300 K.

2 Calculate the internal energy of 2.00 g of low-density helium gas at a temperature of 300 K.

3 The cockpit of a supersonic jet traveling at 680 m/s (twice the speed of sound) contains 2.00 kg of oxygen at 10.0° C.
(a) Find the kinetic energy of the center of mass of the oxygen molecules relative to the earth.
(b) Find the total kinetic energy of the oxygen molecules relative to the jet.
(c) Find the internal energy of the oxygen.

4 A tank contains 1 mole of low-density helium gas at 20.0° C. How fast would the tank have to move in order that the center-of-mass kinetic energy of the gas equaled its internal energy?

14-2 First Law of Thermodynamics

5 A tennis ball of mass 6.00×10^{-2} kg, initially moving at a speed of 30.0 m/s, hits a wall and rebounds with a speed of 20.0 m/s. Assuming no energy loss to the wall, find the increase in the internal energy of the ball.

6 James Joule once attempted to measure the increase in temperature of the water in a waterfall resulting from its decrease in gravitational potential energy (Fig. 14-27). How high would the waterfall have to be for the temperature at the bottom to be 1 C° higher than at the top?

Fig. 14-27

7 A car of mass 1.00×10^3 kg is initially moving on a level road at a speed of 20.0 m/s. Compute the increase in temperature of the brakes, assuming that all the mechanical energy ends up as internal energy in the brake system. Assume a total heat capacity of 10,000 J/C°.

8 A steel paper clip is bent by a force of 10.0 N exerted on it through a distance of 5.00 cm. By how much does the temperature of the midsection of the paper clip increase if all the energy goes into the internal energy of this section, which has a mass of 0.100 g?

★ **9** Water of mass 1.00 kg is heated when it is stirred with a paddle that exerts a constant force of 5.00 N on the water and moves through it at a speed of 2.00 m/s. How long will it take for the temperature of the water to increase by 1 C°, if all the mechanical energy goes into increasing the internal energy of the water?

10 An electric motor powered by a battery lifts 1.00 kg through a vertical distance of 0.500 m. Find the decrease in internal energy of the battery.

11 When 1.00 g of water is converted to steam at 1.00 atm of pressure, the increase in internal energy is 2093 J, as calculated in Example 3. Find the increase in the average molecular energy of steam over water.

12 How much does the internal energy of a gas increase when the gas absorbs 800 J of heat while doing 200 J of work?

13 How much heat must be absorbed by a system to do 4000 J of work while its internal energy increases by 5000 J?

14 Find the change in temperature of 2.00 moles of an ideal diatomic gas that absorbs 900 J of heat and does 400 J of work.

★ **15** Heat is added to 1.00×10^3 cm^3 of water at atmospheric pressure. The water's temperature increases by 50.0 C°.
(a) Find the work done by the water as it expands.
(b) Find the increase in the water's internal energy.

★ **16** A 1.00 m^3 container of air absorbs 6000 J of heat at a constant pressure of 1.00 atm. Find (a) the increase in the temperature of the air; (b) the work done by the air; (c) the increase in the air's internal energy. Air has a density of 1.20 kg/m^3, a constant-pressure specific heat of 1010 J/kg-C°, and a thermal coefficient of expansion of 3.67×10^{-3} (C°)$^{-1}$.

17 (a) Find the work done by 2.00 moles of an ideal gas if its volume doubles at a constant temperature of 23.0° C.
(b) Find the heat absorbed by the gas.

★ **18** A nonideal gas goes through the cycle ABCA (Fig. 14-28). If 7000 J of heat is added to the gas in the process AB and 1000 J of heat is removed in BC, how much heat is added in the process CA?

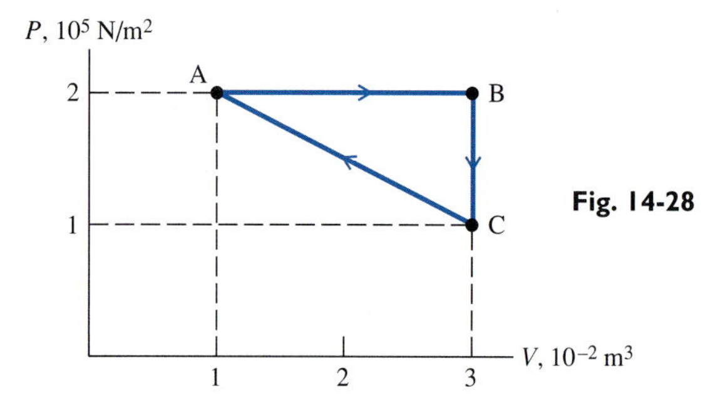

Fig. 14-28

19 Fig. 14-29 is a PV diagram for an ideal gas. Processes I, II, and III take place at constant volume, temperature, and pressure respectively. In process III the heat added to the gas is −61 cal and the work done by the gas is −24 cal.
(a) Find the change in internal energy for process III.
(b) Find Q, ΔU, and W for process I.

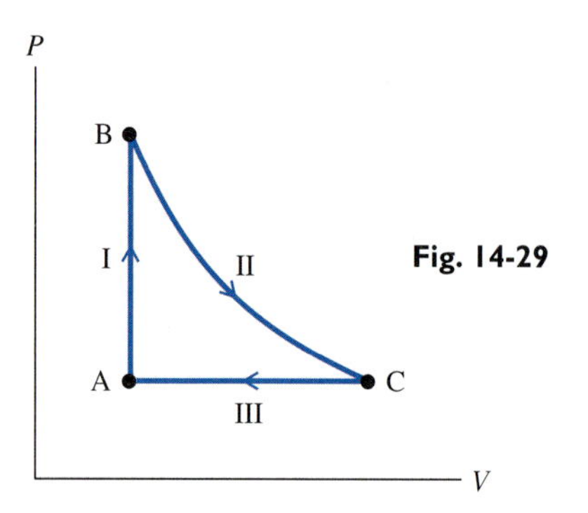

Fig. 14-29

20 Compute the net work done by the system for the cycle indicated in Fig. 14-30.

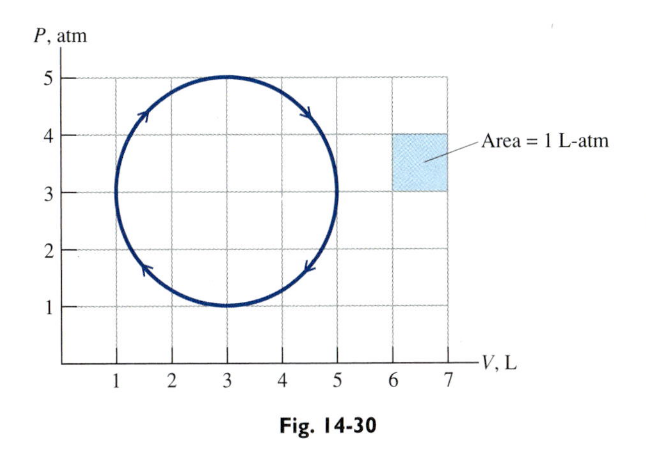

Fig. 14-30

14-3 Heat Engines and Refrigerators

21 In a certain automobile engine with an efficiency of 20.0%, combustion of gasoline in one cylinder releases 500 J of heat.

(a) Find the work done by the system of gasoline vapor and air on the surroundings during one cycle.

(b) Find the total mechanical power produced by all four cylinders, operating at 50 cycles/s, or 3000 rpm.

22 A certain automobile engine is performing work at the rate of 30,000 W (40 hp) with an efficiency of 25%. At what rate is heat being expelled by the engine?

23 An automobile uses 1.5×10^4 W (20 hp) of mechanical power against air resistance and rolling friction while traveling at a constant speed of 25 m/s.

(a) The heat of combustion of gasoline is 11 kcal/g and its density is 0.74 g/cm^3. Find the energy content of gasoline in units of J/L.

(b) If the efficiency of the engine is 25%, find the energy used by the car per distance traveled in units of J/km.

(c) Find the distance traveled by the car per volume of gasoline used in units of km/L and mi/gal (1 mi = 1.61 km, and 1 gal = 3.79 L).

24 The coefficient of performance of a refrigerator is defined as the ratio of heat extracted from the refrigeration compartment to the work required to operate the refrigerator. A certain refrigerator has a coefficient of performance of 3 and requires 500 W of electrical power to operate its compressor. At what rate will this refrigerator deliver heat to the room in which it is operating when it is running?

25 It can be shown, using calculus, that the efficiency of an engine undergoing the Otto cycle is

$$e = \left[1 - \left(\frac{V_B}{V_C}\right)^{2/5}\right]$$

which is determined by the compression ratio V_C/V_B. Evaluate the efficiency for a compression ratio of 8.0.

14-4 Second Law of Thermodynamics

26 (a) Find the entropy change of the piece of clay described in Example 2.

(b) Find the ratio of the number of final mechanical states to the number of initial mechanical states of the clay.

27 A car of mass 1.00×10^3 kg is initially traveling at a speed of 20.0 m/s. The brakes are applied and the car comes to rest. Assuming that no energy is transferred between the car and its surroundings, find the car's increase in entropy and the ratio of the car's number of final mechanical states to the number of initial states. Assume the car's temperature stays at about 300 K.

28 Two copper blocks, each of mass 1.00 kg, initially have different temperatures, 20° C and 28° C. The blocks are placed in contact with each other and come to thermal equilibrium. No heat is lost to the surroundings.

(a) Find the final temperature of the blocks and the heat transferred between them.

(b) Find the entropy change of each block during the time interval in which the first joule of heat flows.

(c) Estimate the entropy change of each block after it has reached thermal equilibrium. Use each block's average temperature during the process in calculating the estimated values of ΔS.

29 An ice chest initially contains 3.00 kg of ice at 0° C.

(a) Find the change in the entropy of the system of the ice chest and its contents just after the ice has melted.

(b) Find the change in the entropy of the surroundings, assuming the air outside the ice chest is at 20° C.

(c) Find the total entropy change of the universe as a result of this process.

(d) Find the ratio Ω_f/Ω_i for the system, for the surroundings, and for the combination of the system and its surroundings.

30 A heat conducting rod is connected between two insulated containers, one containing water and steam at 100° C, the other containing ice water at 0° C. There is a steady heat flow at the rate of 10.0 W along the rod.
(a) During a 1 s interval, calculate the entropy changes of (1) the water and steam, (2) the rod, (3) the ice water, (4) the universe
(b) Suppose that there are initially $10^{10^{23}}$ possible mechanical states for the combined system of the two containers and rod. How many final states are available to the combined system?

31 Suppose that by stirring 100 g of water you are able to increase its temperature from 19.0° C to 21.0° C. Find the increase in the water's entropy.

32 An object of heat capacity 100 cal/C° is initially at 30.0° C.
(a) Estimate the entropy change of the object after it is dropped into a swimming pool at 26.0° C.
(b) Estimate the resulting entropy change of the universe.

33 The temperature of the air inside a house is 20° C, and the air temperature outside is 0° C. Heat is conducted through the walls and roof at the rate of 5000 W. Find the total entropy change of the house and its surroundings resulting from the heat conduction over a 20.0-minute interval.

34 A 1.00 m² black surface is exposed to direct sunlight and completely absorbs 1400 J of radiation during a 1.00 s interval.
(a) Calculate the change in entropy of the surface.
(b) Calculate the change in the sun's entropy when this radiation was emitted from the surface of the sun at a temperature of 5500 K.

★ **35** Exactly 1.00 mole of an ideal gas undergoes an isothermal expansion (T = 300 K) from state A to state B and then returns to state A by another process. The volume of the gas in state B is three times its initial volume.
(a) For the process AB, find the work done by the gas and its change in entropy.
(b) Find the gas's change in entropy for the process BA.

★ **36** A gas undergoes the cycle shown in Fig. 14-31, consisting of an isotherm (AB), a constant-volume process (BC), and an adiabatic process (CA). During the isothermal process 100 J of heat is absorbed by the gas. Find the system's change in entropy for each of the processes.

37 (a) What is the probability of throwing a 1 on a single roll of a die?
(b) If two dice are rolled, what is the probability of throwing a 1 on the first die and a 2 on the second die?
(c) If two dice are rolled, what is the probability of throwing a 1 and a 2?
(d) If two dice are rolled, what is the probability of throwing a sum of 7?

38 A heat engine absorbs heat from a source at 500 K and exhausts heat at 300 K. If 100 J of work is done in one cycle, what is the minimum heat exhausted?

39 A heat engine operates surrounded by air at 20° C. If the heat engine has an efficiency of 40.0%, the highest temperature inside the heat engine during the cycle must be greater than what value?

40 Find the efficiency of a Carnot engine operating between 0° C and 300° C.

41 During one cycle a Carnot engine does net work of 6000 J and expels 12,000 J of heat. If the engine's high temperature is 500 K, what is its low temperature?

42 Find the maximum power that can be delivered by a heat engine operating at 3000 rpm between 20° C and 2000° C and absorbing 1000 J of heat per cycle.

*14-5 Human Metabolism

43 How many peanuts should be consumed by the woman in Example 10 to replenish the energy lost during the hour of exercise? An average peanut supplies 10 kcal.

44 A 75.0 kg man, who is engaged in light activity and who is not sweating, uses internal energy at a rate of 200 kcal/h. The man then begins strenuous exercise, requiring 700 kcal/h. Assuming that the extra heat generated is dissipated through sweating and that 25% of the sweat evaporates, find the man's rate of sweat production in L/h.

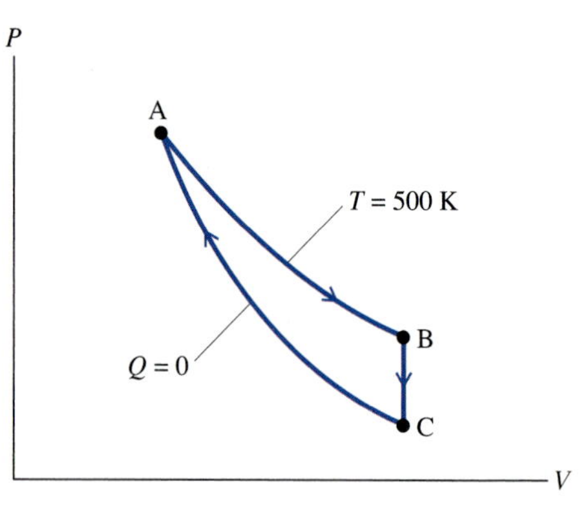

Fig. 14-31

45 Calculate the length of time required to lose 20 lb on the following dietary plans. The weight reduction comes from utilization of body fat, which provides 9000 kcal of internal energy per kg of fat, or 4100 kcal per lb.

(a) A complete starvation diet, with no physical activity. Assume a BMR of 1900 kcal/day.

(b) An extreme diet of 1200 kcal/day, with only light activity requiring 150 kcal/h for 14 h and sleep requiring 80 kcal/h for 10 h.

(c) A moderate diet of 2000 kcal/day, with the same energy expenditure as in diet b, except for 1 hour of exercise, requiring an additional 500 kcal.

46 Assuming that the total energy consumption scales in the same way as BMR, find the proper ratio of caloric intake for two average individuals of the same age and sex who weigh 150 lb and 100 lb respectively. If the person weighing 100 lb requires 2000 Calories per day, how many Calories are needed by the person weighing 150 lb?

47 How many additional Calories are necessary for a mountain climber of mass 60.0 kg to climb 1000 m if the body does this additional work with 10.0% efficiency?

48 How many extra pounds of food must one consume to gain 1 pound of fat if the food consumed is pure (a) protein; (b) carbohydrates; (c) alcohol; (d) fat?

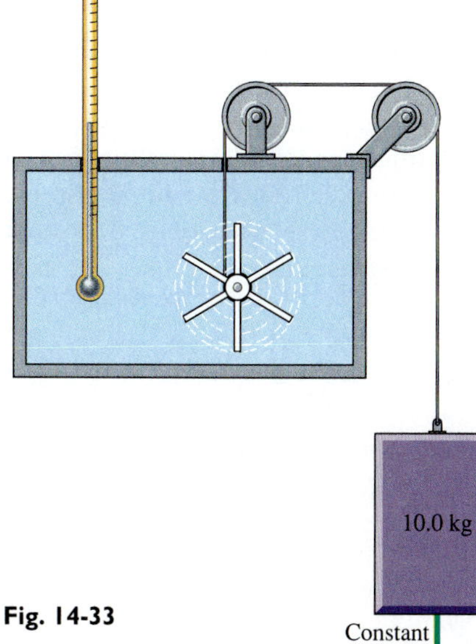

Fig. 14-33

Additional Problems

★ **49** One gram of water undergoes the following Carnot cycle (Fig. 14-32):

AB The liquid water is heated at constant temperature (100° C) and constant pressure (1.0 atm) until it is completely converted to steam.

BC The vapor is adiabatically expanded until the temperature drops to 50° C. Part of the vapor is converted to liquid in this process.

CD The liquid and vapor are compressed at constant temperature and pressure.

DA The liquid and vapor are adiabatically compressed to the original state in which the water is entirely liquid. The heat of vaporization of water is 540 cal/g at 100° C and 569 cal/g at 50° C.

(a) Find the net work done in one cycle.

(b) Find the mass of vapor condensed into liquid in the process CD.

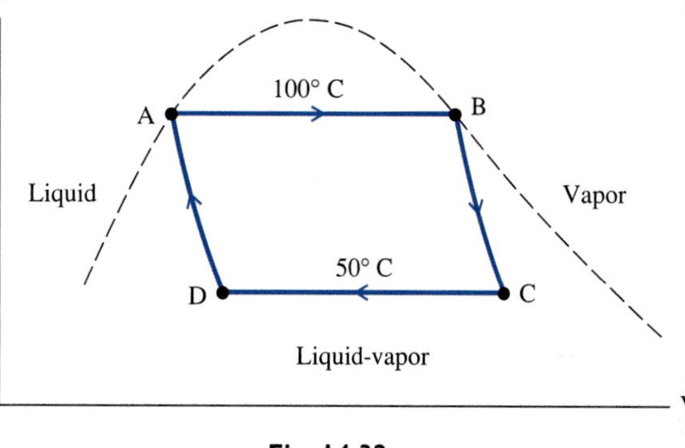

Fig. 14-32

50 A 10.0 kg mass is attached by a rope to a paddlewheel immersed in 1.00 liter of water in a thermally insulated container (Fig. 14-33). The paddlewheel turns as the mass falls through a distance of 3.00 m at constant velocity. This results in an increase in the water's temperature of 7.00×10^{-2} C°. The water could have been taken from the same initial thermodynamic state to the same final state by the addition of heat. Determine the "mechanical equivalent of heat" by equating the mechanical work to the heat necessary to produce this same temperature increase; that is, find the relationship between the calorie and the joule *as measured in this experiment*. The experiment described here is essentially the same as one performed by Joule.

★ **51** The mechanical state of a gas is specified when the position and velocity of each gas molecule are given. In this problem we shall indicate how to find the dependence of the number of states Ω on the volume V of the gas. The analysis is somewhat oversimplified, but it describes the essential elements of an exact analysis while avoiding mathematical complexity. Let the volume V be divided into a number of small regions of equal volume V_0. Specification of a particle's position will mean indicating which subvolume the particle is in. How many states are available to (a) one particle; (b) two particles; (c) N particles? (d) Show that $\Omega \propto V^N$. (The dependence of Ω on U can be shown by a similar analysis of velocity states.) (e) Show that $S = Nk(\ln V) + f$, where f is independent of V. Show that this result is consistent with Eq. 14-11.

★★ **52** (a) Show that the entropy statement of the second law implies that for *any* engine operating between two temperatures T_1 and T_2, the ratio of the heat released at T_2 to the heat absorbed at T_1 is greater than or equal to the ratio T_2/T_1.

(b) Show that the efficiency $e \leq 1 - \dfrac{T_2}{T_1}$.

(c) Show that if the engine undergoes a quasi-static process and is thus a Carnot engine, $e = 1 - \dfrac{T_2}{T_1}$. Thus a Carnot engine has the maximum possible efficiency.

★★ **53** A heat engine absorbs heat $Q_1, Q_2, \ldots, Q_N$ from thermal reservoirs at the respective temperatures $T_1, T_2, \ldots, T_N$ and rejects heat $Q_{N+1}, Q_{N+2}, \ldots, Q_M$ to thermal reservoirs at temperatures $T_{N+1}, T_{N+2}, \ldots, T_M$. Let $Q_{in} = Q_1 + Q_2 + \cdots + Q_N$ and $Q_{out} = Q_{N+1} + Q_{N+2} + \cdots + Q_M$. Show that the entropy statement of the second law of thermodynamics requires that $\dfrac{Q_{out}}{Q_{in}} > \dfrac{T_M}{T_1}$, and therefore the efficiency of the heat engine is less than that of a Carnot engine operating between temperatures T_1 and T_M:

$$e < e_C = 1 - \frac{T_M}{T_1}.$$

54 A heat pump is a device used to heat a house. The heat pump extracts heat from the outdoors at temperature T_2, does work on the system, and delivers heat to the indoors, which is at the higher temperature T_1. Thus it operates in the same way as a refrigerator, except that the refrigerated area is the outdoors. Find the minimum power that must be supplied to operate a heat pump that delivers heat to the interior of a house at the rate 10.0 kcal/s when the temperature inside is 20° C and the temperature outside is 0° C.

★ **55** A Carnot refrigerator goes through a cycle that is the reverse of a Carnot heat engine (Fig. 14-34).

(a) Show that for a Carnot refrigerator the coefficient of performance (see Problem 24) equals $T_2/(T_1 - T_2)$.

(b) Evaluate for $T_1 = 293$ K and $T_2 = 263$ K.

(c) Find the average power required to operate this Carnot refrigerator if heat leaks into the refrigeration chamber at an average rate of 3.00 kcal/min.

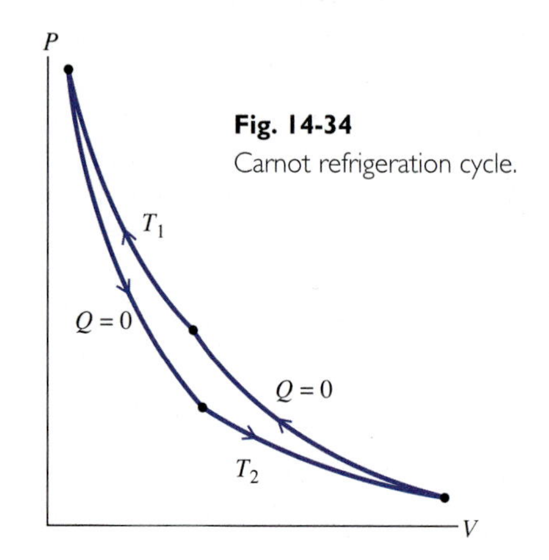

Fig. 14-34 Carnot refrigeration cycle.

★★ **56** The earth's atmosphere is heated mainly by convective heat transfer from the earth's surface, rather than by direct solar radiation. Air near the earth's surface is heated, and this air mixes with air at higher elevations and transfers energy to it. The temperature of the atmosphere varies with climate and weather conditions, but there is a fairly constant rate of decrease of temperature with increasing altitude. A simple model that gives fair agreement with this observed effect is the adiabatic model. We assume that an equilibrium thermal gradient exists, such that a volume of air that rises or falls in the atmosphere and does not exchange heat with the surrounding air will be at the same temperature as the surrounding air. Since the pressure drops as elevation increases, the air expands, does work, and drops in temperature. Using calculus, one can show that

$$\frac{\Delta T}{\Delta P} = \frac{2}{7}\left(\frac{T}{P}\right)$$

Show that temperature varies with elevation z according to the equation

$$\frac{\Delta T}{\Delta z} = -\frac{2}{7}\left(\frac{Mg}{R}\right)$$

where M is the molecular mass of air. Evaluate this expression.

★ **57** Find the minimum time required to transform 1.0 kg of water at 0° C to ice cubes at 0° C in the freezer compartment of a refrigerator that requires 100 W of electrical power and has a value of 2.0 for its coefficient of performance (defined in Problem 24).

★ **58** Show that N molecules of a monatomic ideal gas, undergoing an isothermal expansion from an initial volume V_i to a final volume V_f, absorb heat $Q = NkT \ln \frac{V_f}{V_i}$. Show that the gas undergoes an entropy change of $\Delta S = Nk \ln \frac{V_f}{V_i}$, or $\frac{Q}{T}$.

★ **59** Use the entropy statement of the second law of thermodynamics to derive the Clausius statement. First, assume that the Clausius statement is violated: suppose a system at a temperature T_1 absorbs positive heat Q_1 released by another system at a temperature $T_2 < T_1$. Show that this assumption would mean that the total entropy of the two systems would decrease, in violation of the entropy statement. Thus you will show that the Clausius statement cannot be violated without violating the entropy statement. Since we assume that the entropy statement is true, it follows that the Clausius statement is also true.

CHAPTER 15 Harmonic Motion

The regular swinging of a pendulum is the time-keeping mechanism in a pendulum clock.

Harmonic motion is the name given to any motion that repeats over the same path in a fixed period of time. Such motion is also called periodic motion. The back-and-forth swinging of a pendulum, the bobbing up and down of a cork floating in a lake, and the sound-producing vibrations of a guitar string, tuning fork, or vocal cords are just a few examples of harmonic (periodic) motion. Since vibrations are the source of waves, our study of harmonic motion in this chapter will lead to the study of wave motion in the following chapter.

15-1 Simple Harmonic Motion

Suppose you suspend a mass from a spring and then give the mass an upward push so that it oscillates up and down; that is, it moves repeatedly up and down. You can record the way the mass's y coordinate changes with time if you attach a pen to the mass, with the pen's tip touching a strip of paper that moves laterally at constant speed (Fig. 15-1). Another way to show the motion is to draw images of the system at equal time intervals (Fig. 15-2). The motion of a mass oscillating on a spring is a particularly simple kind of harmonic motion, which is appropriately termed **simple harmonic motion,** abbreviated SHM. As illustrated in the figure, the y coordinate of the mass varies with time according to the equation

$$y = A \sin\left(2\pi \frac{t}{T}\right) \tag{15-1}$$

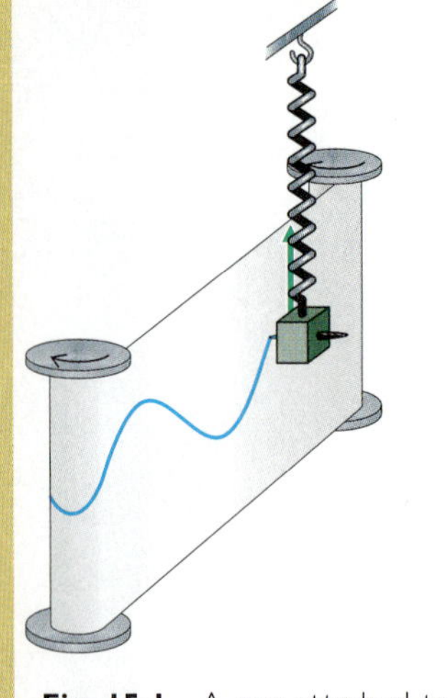

Fig. 15-1 A pen attached to an oscillating mass on a spring marks a curve on a moving strip of paper.

where A is the maximum value of y, called the **amplitude,** t is time, and T is the **period** of the motion, that is, the time needed for the mass-spring system to complete one

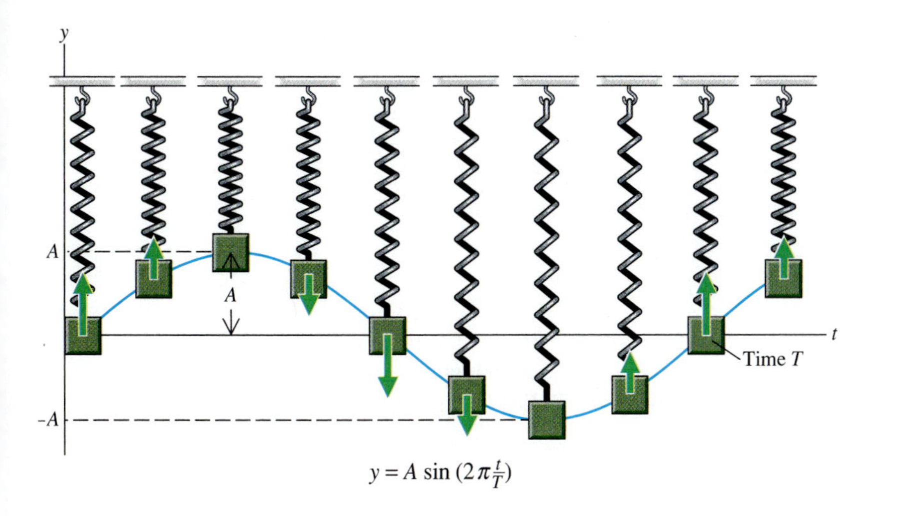

Fig. 15-2 A mass oscillating on a spring is shown at equal time intervals. After eight such time intervals, the motion repeats.

cycle (one period) of its harmonic motion. The quantity $2\pi\frac{t}{T}$ is called the **phase angle** and is measured in radians. The value of this phase angle at a given time t tells us the phase of the motion (the part of the cycle) that the mass is experiencing at that time. To better understand this, we shall now evaluate y for certain values of t.

At time $t = 0$, Eq. 15-1 gives

$$y = A \sin 0 = 0$$

indicating that the mass starts at the origin, as Fig. 15-2 shows. As t increases, y begins to increase. When $t = \frac{T}{4}$, we find that

$$y = A \sin\left(2\pi\frac{T/4}{T}\right) = A \sin\left(\frac{\pi}{2}\right) = A$$

At this point, the sine has its maximum value of 1, and so the displacement of the mass reaches its maximum value A, the amplitude. The mass then begins to move downward. At $t = \frac{T}{2}$, we find the mass is back at the origin:

$$y = A \sin\left(2\pi\frac{T/2}{T}\right) = A \sin \pi = 0$$

The mass reaches the origin with a downward velocity and so overshoots it. The y coordinate becomes negative, reaching a minimum value $y = -A$ when $t = \frac{3T}{4}$:

$$y = A \sin\left(2\pi\frac{3T/4}{T}\right) = A \sin\left(\tfrac{3}{2}\pi\right) = -A$$

The mass then begins to move upward again, and when $t = T$, it is back to the origin and moving upward:

$$y = A \sin\left(2\pi\frac{T}{T}\right) = A \sin(2\pi) = 0$$

One full cycle of the motion has been completed in the time interval from $t = 0$ to $t = T$. The cycle then repeats. Every time a period T elapses, the system goes through one cycle.

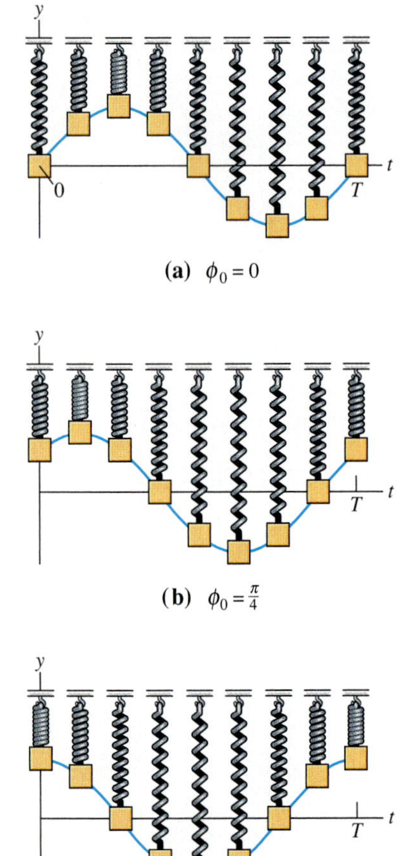

Fig. 15-3 SHM of a mass on a spring with different initial conditions.

A more general expression for the SHM of a mass oscillating on a spring is

$$y = A \sin\left(2\pi\frac{t}{T} + \phi_0\right) \tag{15-2}$$

In this more general expression the phase angle is

$$\phi = 2\pi\frac{t}{T} + \phi_0 \tag{15-3}$$

where ϕ_0 is the initial value of the phase angle, that is, the value of ϕ when $t = 0$. The initial phase angle ϕ_0 determines the initial value of y. If $\phi_0 = 0$, as for the motion shown in Fig. 15-2, then when $t = 0$, Eq. 15-2 gives $y = A \sin \phi_0 = A \sin 0 = 0$. If, on the other hand, $\phi_0 = \frac{\pi}{2}$, then at $t = 0$, $y = A \sin \phi_0 = A \sin\left(\frac{\pi}{2}\right) = A$, meaning that the mass starts its motion at $t = 0$ at its maximum upward displacement. Fig. 15-3 shows the SHM of a mass on a spring for several values of ϕ_0.

The **frequency** f of any periodic motion is defined as the **number of cycles per second.** Since one cycle is completed in time T, f is the inverse of T:

$$f = \frac{1}{T} \tag{15-4}$$

For example, if $T = \frac{1}{2}$ s, $f = \frac{1}{\frac{1}{2}\text{s}} = \frac{2}{\text{s}}$, which is read as "2 cycles per second." The frequency unit $\frac{1}{\text{s}}$ ("cycles per second") is given the name **hertz,** abbreviated Hz. Thus for our example $f = 2$ Hz.

As illustrated in Fig. 15-4, definitions of amplitude, period, and frequency can be applied to any periodic motion, not just SHM.

We shall study two important examples of SHM in this chapter: the motion of a pendulum and the motion of a mass attached to a spring. In both cases we shall analyze forces and show how **a net force opposing displacement and proportional to the magnitude of the displacement leads to SHM.** To accomplish this, however, we must first relate SHM to uniform circular motion.

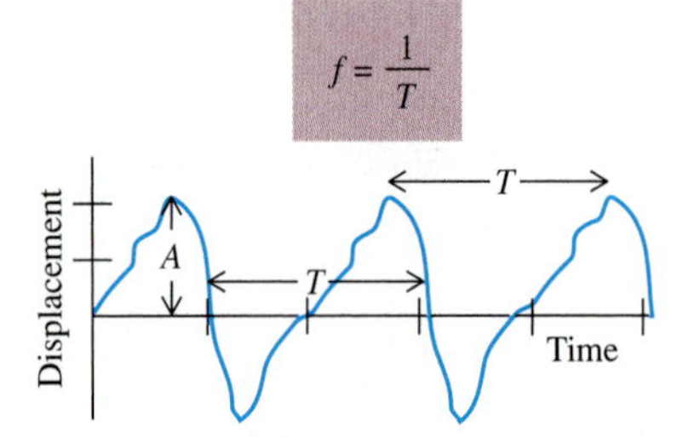

Fig. 15-4 Periodic motion that is not SHM.

15-2 Relationship between SHM and Circular Motion

Fig. 15-5a shows a peg on a wheel that is rotating in a vertical plane; the wheel is alongside an oscillating mass on a spring. The uniform circular motion of the peg keeps it at the same height as the mass, which undergoes SHM. This means that the peg's vertical coordinate y must be described by the equation for SHM, just as the y coordinate of the mass is.

To prove this relationship we shall show that the y coordinate of a particle, moving like the peg in Fig. 15-5b along a circular path, can be described by an equation of the simple harmonic form. Fig. 15-6 shows a particle on the edge of a wheel of radius A, rotating at constant angular velocity ω. At time t, a line from the origin to the particle makes an angle ϕ with the x-axis. The y coordinate is related to ϕ by the equation

$$y = A \sin \phi$$

We can apply the definition of angular velocity (Eq. 9-1) to express ϕ in terms of t:

$$\omega = \frac{\Delta\phi}{\Delta t} = \frac{\phi - \phi_0}{t - 0}$$

Solving for ϕ, we find

$$\phi = \omega t + \phi_0$$

Substituting this expression for ϕ into the equation above for y, we obtain

$$y = A \sin (\omega t + \phi_0) \qquad (15\text{-}5)$$

Angular velocity ω is measured in radians per second, and since there are 2π radians in one complete circle and the time interval corresponding to one complete circle is the period T of the motion, we can express the angular velocity as

$$\omega = \frac{2\pi}{T} \qquad (15\text{-}6)$$

or, since $f = \frac{1}{T}$,

$$\omega = 2\pi f \qquad (15\text{-}7)$$

If we substitute Eq. 15-6 in Eq. 15-5, we obtain an expression identical to Eq. 15-2:

$$y = A \sin\left(2\pi\frac{t}{T} + \phi_0\right)$$

Thus we see that the y coordinate of a particle in uniform circular motion is described by the equation describing SHM. Any body whose motion is described by a *single* equation of this form is said to be in SHM. Of course, a particle in uniform circular motion moves in a plane and therefore a second coordinate x is needed to describe its motion. Motion in the x direction can also be described by an equation of the simple harmonic form. Circular motion therefore is a combination of two simple harmonic motions in perpendicular directions.

In circular motion, ω represents angular velocity. However, in simple harmonic motion, ω does not represent angular velocity. For example, there is no angular motion or angular velocity for a mass oscillating on a spring. In SHM ω represents **angular frequency,** which is the rate of change of the phase angle ϕ. The physical significance of ω in SHM is given by its relationship to either the period or the frequency $\left(\omega = 2\pi f = \frac{2\pi}{T}\right)$.

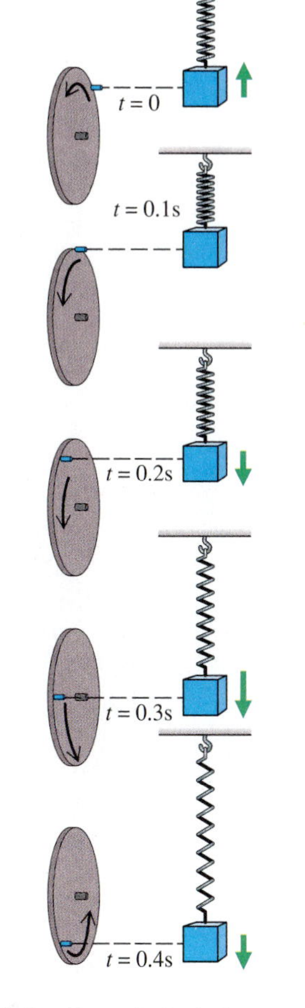

Fig. 15-5 At any instant, the peg on a wheel uniformly rotating in a vertical plane has the same elevation as a mass oscillating on a spring.

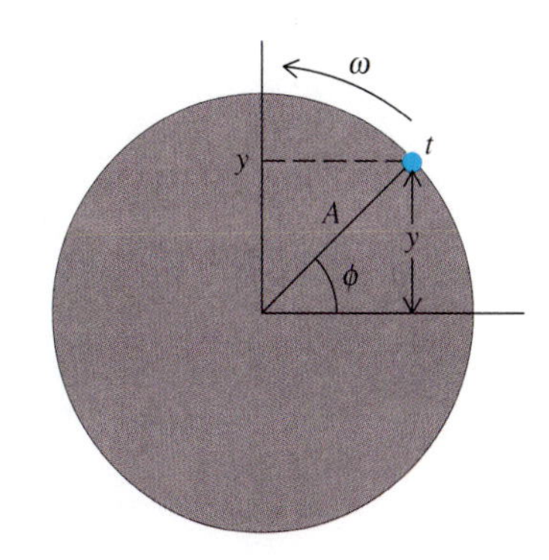

Fig. 15-6 A particle on the edge of a uniformly rotating wheel has a y coordinate that varies with time according to the equation for SHM: $y = A \sin \phi = A \sin (\omega t + \phi_0)$.

Next we shall use the connection between circular motion and SHM to show how acceleration is related to displacement for SHM. For a particle moving at constant speed about a circle of radius A, we know that the particle's acceleration is directed toward the center of the circle and has magnitude given by Eq. 3-12:

$$a = \frac{v^2}{r} = \frac{v^2}{A}$$

Since the speed v is related to the angular velocity ω by Eq. 9-3 ($v = r\omega = A\omega$), we may write the equation above as

$$a = \frac{(A\omega)^2}{A} = A\omega^2$$

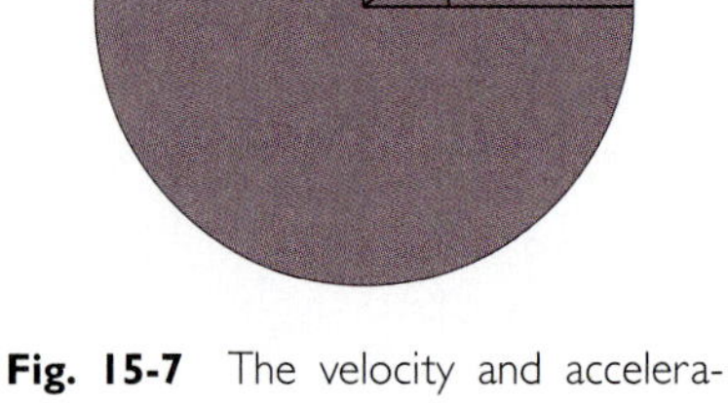

Fig. 15-7 The velocity and acceleration of a particle on the edge of a uniformly rotating wheel.

From Fig. 15-7 we see that $a_y = -a \sin \phi$; using the preceding expression for a, we obtain

$$a_y = -A\omega^2 \sin \phi$$

But since $y = A \sin \phi$, we may express our result

$$a_y = -\omega^2 y$$

To simplify the form of this equation and to emphasize that ω may have any constant value, we define a constant

$$C = \omega^2 \tag{15-8}$$

and then rewrite our equation for a_y in the form

$$a_y = -Cy \tag{15-9}$$

Since motion in the y direction is the same for circular motion and SHM, it follows that this equation must apply to SHM. Whenever we can describe the acceleration of a body by a single equation of this form (acceleration proportional to displacement but in the opposite direction), we know that the system undergoes SHM and therefore that its displacement varies with time according to Eq. 15-2:

$$y = A \sin\left(2\pi\frac{t}{T} + \phi_0\right) \tag{15-10}$$

where A and ϕ_0 are determined by initial conditions and where the period T is related to the constant C by Eqs. 15-8 and 15-6:

$$C = \omega^2 = \left(\frac{2\pi}{T}\right)^2$$

or

$$T = \frac{2\pi}{\sqrt{C}} \tag{15-11}$$

EXAMPLE 1 SHM of a System With a Given Value of C

Suppose $a_y = (-100\ \text{s}^{-2})y$. Find (a) the period and frequency of the motion; (b) the position y at any time t.

SOLUTION (a) Noting that the equation we are given for a_y is of the SHM form given by Eq. 15-9, with $C = 100\ \text{s}^{-2}$, we apply Eq. 15-11 to find the period:

$$T = \frac{2\pi}{\sqrt{C}} = \frac{2\pi}{\sqrt{100\ \text{s}^{-2}}} = 0.628\ \text{s}$$

We find the frequency by taking the inverse of T (Eq. 15-4):

$$f = \frac{1}{T} = \frac{1}{0.628\ \text{s}} = 1.59\ \text{s}^{-1} = 1.59\ \text{Hz}$$

(b) We find the y coordinate by applying Eq. 15-10:

$$y = A\sin\left(2\pi\frac{t}{T} + \phi_0\right) = A\sin\left(2\pi\frac{t}{0.628\ \text{s}} + \phi_0\right)$$

$$= A\sin\left[(10\ \text{s}^{-1})t + \phi_0\right]$$

Of course y is not completely specified unless A and ϕ_0 are given. These values can be determined only if further information is given about the physical system. In particular, the initial conditions at $t = 0$ determine A and ϕ_0.

15-3 Mass and Spring

Period of the Motion

Consider a mass m that is on a frictionless horizontal surface and is attached to a spring of force constant k, as shown in Fig. 15-8. The force the spring exerts on the mass depends on how much the spring is stretched or compressed, and so this force is a function of the mass's position. As we discussed in Chapter 4, the relationship between the force F_x exerted on the mass and its position x is given by Hooke's law (Eq. 4-6):

$$F_x = -kx \tag{15-12}$$

This force tends to restore the mass to its equilibrium position. We shall show how this force leads to SHM. We apply Newton's second law to obtain an expression for the acceleration of the mass:

$$a_x = \frac{\Sigma F_x}{m} = -\frac{kx}{m}$$

If we let C stand for the constant ratio k/m, we may express this result

$$a_x = -Cx \tag{15-13}$$

where

$$C = \frac{k}{m} \tag{15-14}$$

Notice that Eq. 15-13 is of the SHM form—Eq. 15-9 (only here we use x instead of y). Thus the solution must be of the SHM form. Substituting Eq. 15-14 into Eq. 15-11, we find for the period

$$T = \frac{2\pi}{\sqrt{C}} = \frac{2\pi}{\sqrt{k/m}}$$

or

$$T = 2\pi\sqrt{\frac{m}{k}} \tag{15-15}$$

This expression tells us that the period increases as m increases, meaning that oscillations are slower for greater masses than for smaller ones, as one might intuitively expect. Our expression for the period also agrees with our intuition in saying that the

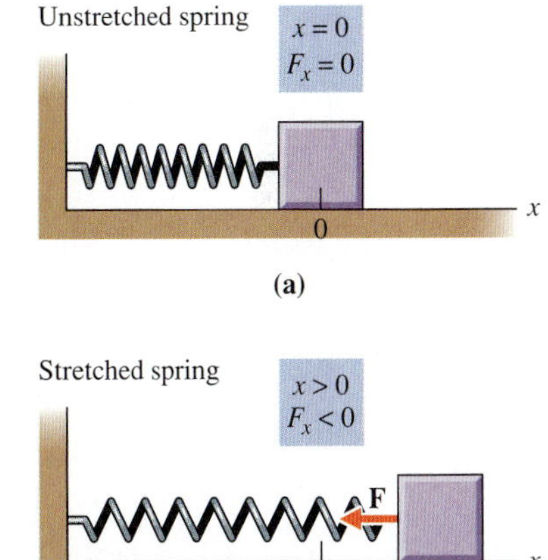

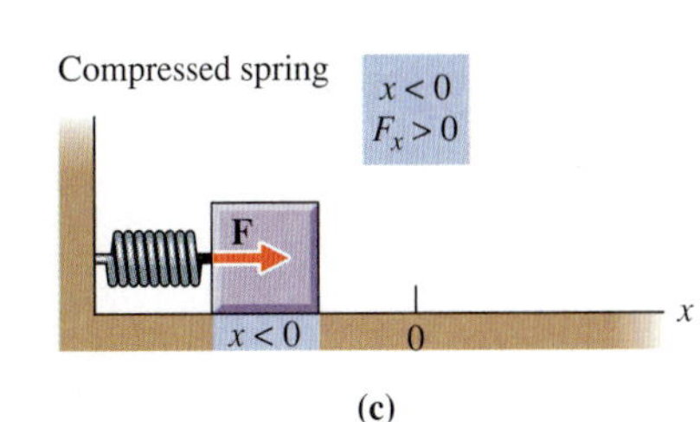

Fig. 15-8 The force exerted by a spring on a mass attached to the spring.

period decreases as k increases, meaning that oscillations are faster for a stiffer spring with its larger k than for a weaker spring with its smaller k. However, as for all other SHM, the period derived here is independent of amplitude. If we double the amplitude by stretching the spring out twice as far as before, the mass must cover twice the distance per cycle but must do so in exactly the same time T. The mass moves faster when the amplitude is increased because it experiences a stronger maximum accelerating force.

EXAMPLE 2 Period of the Motion of a Mass Attached to a Spring

A 200-g mass is on a frictionless horizontal surface and is attached to a spring that exerts a force of 5.0 N to the left when the spring is stretched to the right by 0.10 m. Find the period and frequency of the resulting motion.

SOLUTION Applying Hooke's law (Eq. 15-12: $F_x = -kx$), we first find the spring's force constant:

$$\text{k} = -\frac{F_x}{x} = -\frac{-5.0\text{ N}}{0.10\text{ m}} = 50\text{ N/m}$$

Applying Eq. 15-15, we find the period:

$$T = 2\pi\sqrt{\frac{m}{k}} = 2\pi\sqrt{\frac{0.20\text{ kg}}{50\text{ N/m}}} = 0.40\text{ s}$$

The frequency f is the inverse of T:

$$f = \frac{1}{T} = \frac{1}{0.40\text{ s}} = 2.5\text{ Hz}$$

Energy

When we studied the spring force in Chapter 7, we found that it is a conservative force, with potential energy given by Eq. 7-23: $U_s = \frac{1}{2}kx^2$. Thus the total mechanical energy of a mass m on a spring, when no other forces do work, is given by the sum of the kinetic and potential energies:

$$E = K + U_s$$

$$E = \tfrac{1}{2}mv^2 + \tfrac{1}{2}kx^2$$

It is easy to evaluate E in terms of the amplitude A of the motion; since $v = 0$ when x is at its maximum value of A, we have

$$E = \tfrac{1}{2}kA^2 \qquad (15\text{-}16)$$

Substituting for E in the preceding equation and solving for v, we obtain

$$v = \sqrt{\frac{k}{m}(A^2 - x^2)} = \sqrt{\frac{k}{m}}\sqrt{A^2 - x^2}$$

or, since $T = 2\pi\sqrt{\frac{m}{k}}$, we can substitute $\sqrt{\frac{k}{m}} = \frac{2\pi}{T}$:

$$v = \frac{2\pi}{T}\sqrt{A^2 - x^2} \qquad (15\text{-}17)$$

This equation may be applied to find v for any value of x. The maximum value of v occurs when $x = 0$:

$$v_{max} = \frac{2\pi A}{T} \qquad (15\text{-}18)$$

Although Eq. 15-18 has been derived for the special case of a mass oscillating on a spring, it turns out to be valid for any kind of SHM.

EXAMPLE 3 Speed of a Mass Attched to a Spring

A mass attached to a horizontal spring oscillates with an amplitude of 10.0 cm at a frequency of 5.00 Hz. (a) Find the maximum speed of the mass. (b) Find the speed when $x = 5.00$ cm.

SOLUTION (a) Applying Eq. 15-18, we find

$$v_{max} = \frac{2\pi A}{T} = 2\pi f A = 2\pi(5.00 \text{ Hz})(0.100 \text{ m}) = 3.14 \text{ m/s}$$

(b) Applying Eq. 15-17, we find

$$v = \frac{2\pi}{T}\sqrt{A^2 - x^2} = \frac{2\pi}{0.200 \text{ s}}\sqrt{(0.100 \text{ m})^2 - (0.050 \text{ m})^2}$$

$$= 2.72 \text{ m/s}$$

Vertical Motion

Although our discussion of a mass on a spring in this section has been limited to a mass sliding on a horizontal frictionless surface, a mass attached to a vertical spring also oscillates in SHM, as we saw in Fig. 15-1. The analysis in the vertical case is a bit more complicated because the motion results from the combination of the spring force and the weight of the mass, but the expression for the period turns out to be the same (Eq. 15-15).

EXAMPLE 4 Finding the Mass of a Car With Weak Shock Absorbers

When a driver and four passengers having a total mass of 340 kg get into a car, the car's springs are compressed and the car is lowered by 3.00 cm. The car, which has weak shock absorbers, hits a bump and oscillates vertically with a period of 0.750 s. Find the combined mass of the car and its occupants.

SOLUTION The system of the car and passengers may be treated as a mass m attached to a single spring, as indicated in Fig. 15-9. First we must calculate the force constant of the spring. Since, according to Hooke's law, $F_x = -kx$, a change in position Δx is related to a change in force ΔF_x by

$$\Delta F_x = -k\,\Delta x$$

Fig. 15-9 After hitting a bump, a car oscillates vertically. The car and passengers are like a mass oscillating on a spring.

When the people get into the car, the force applied to the spring changes by w, their combined weight ($\Delta F_x = w$). As a result, the spring is compressed a distance $\Delta x = -3.00$ cm. Thus

$$k = -\frac{\Delta F_x}{\Delta x} = -\frac{w}{\Delta x} = -\frac{(340 \text{ kg})(9.80 \text{ m/s}^2)}{-3.00 \times 10^{-2} \text{ m}}$$

$$= 1.11 \times 10^5 \text{ N/m}$$

The period is related to k and m (the total mass of the car and its occupants) by Eq. 15-15:

$$T = 2\pi\sqrt{\frac{m}{k}}$$

Solving for m, we find

$$m = \frac{kT^2}{4\pi^2} = \frac{(1.11 \times 10^5 \text{ N/m})(0.750 \text{ s})^2}{4\pi^2}$$

$$= 1580 \text{ kg}$$

Fig. 15-10 As you slow down on a swing, your decreasing speed is related to your decreasing displacement in such a way that the period of your motion remains constant. The period is also independent of your mass.

15-4 The Pendulum

The pendulum was the basic time-keeping mechanism of the first precision clocks, developed in the seventeenth century after Galileo's study of pendulums. Galileo's work was inspired by his observation of swaying chandeliers in a cathedral. He discovered that the periodic motion of a pendulum is independent of both the mass of the pendulum and the amplitude of the motion. The period depends only on the length of the pendulum. Thus the swings of a pendulum can be used to mark time. Even as the pendulum gradually slows down, its amplitude decreasing, its period does not change (Fig. 15-10). This is a general characteristic of SHM, and we shall see that pendulum motion is approximately SHM for small-amplitude oscillations.

Simple Pendulum

A **simple pendulum** consists of a small mass m suspended from a string of length ℓ with negligible mass (Fig. 15-11). The only forces acting on the mass are its weight $m\mathbf{g}$ and the tension $\mathbf{T}$ in the string. When the string makes an angle θ with the vertical, the component of weight $mg \cos \theta$ is balanced by the tension. There is, however, an unbalanced component of the weight $mg \sin \theta$ that acts along the arc length s in the negative s direction; this component pushes the pendulum to the left, tending to restore it to the equilibrium state, in which the string is vertical and m is at point O. However, when the mass reaches the equilibrium position, it has a velocity, and so it then moves to the left of the equilibrium point where it experiences a component of weight directed to the right, tending again to restore it to the equilibrium position. The back-and-forth swinging of a pendulum is caused by the weight continually pushing the mass toward its equilibrium position and the mass overshooting that position.

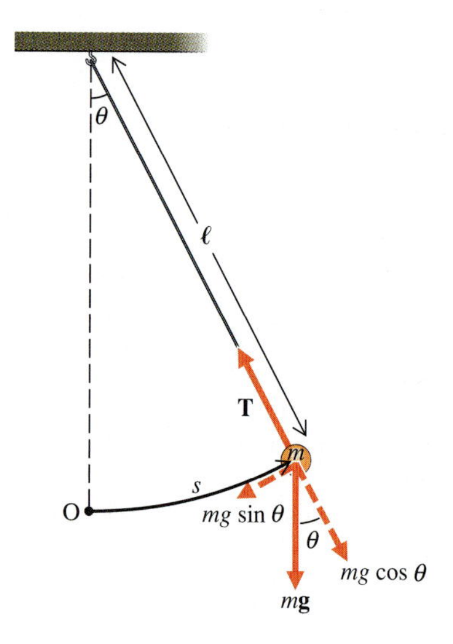

Fig. 15-11 A simple pendulum.

Applying Newton's second law along the s direction, we find

$$\Sigma F_s = ma_s$$

$$-mg \sin \theta = ma_s$$

or

$$a_s = -g \sin \theta$$

As it stands, this equation does not have a simple solution. However, if we consider only small values of θ, we can simplify the solution by applying the **small angle approximation** $\sin \theta \approx \theta$, where θ is measured in radians (Fig. 15-12). For angles up to about 0.3 rad (17°) the small angle approximation is quite accurate. Using it, we find that our equation simplifies to

$$a_s \approx -g\theta$$

Expressing θ as s/ℓ, we obtain

$$a_s \approx -\frac{g}{\ell}s$$

or

$$a_s \approx -Cs \tag{15-19}$$

where

$$C = \frac{g}{\ell}$$

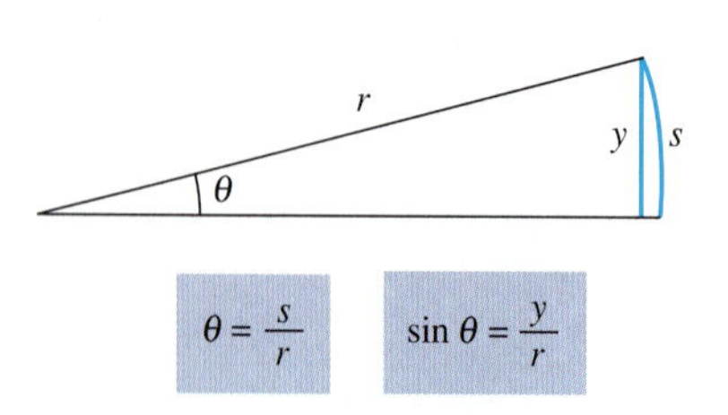

Fig. 15-12 For small angles, $y \approx s$ and $\sin \theta \approx \theta$, with θ measured in radians:
sin 0.100 = 0.0998
sin 0.200 = 0.199
sin 0.300 = 0.296
sin 0.400 = 0.389

Notice that Eq. 15-19 is of the SHM form (Eq. 15-9: $a_y = -Cy$). Thus the solution must be of the SHM form. Substituting the preceding expression for C into Eq. 15-11 ($T = \frac{2\pi}{\sqrt{C}}$), we find

$$T = \frac{2\pi}{\sqrt{g/\ell}}$$

or

$$T = 2\pi\sqrt{\frac{\ell}{g}} \quad \text{(for small amplitudes)} \tag{15-20}$$

This result shows that the period of a simple pendulum undergoing small oscillations is independent of both mass and amplitude.* The period does depend on g. Since it is quite easy to measure accurately both the period and the length of a pendulum, the relationship between T and g means that pendulums can be used to measure g with ease and accuracy. Pendulums are used by geologists to perform very precise measurements of g at different points on earth, allowing detection of small, local variations caused by massive ore deposits.

Fig. 15-13 Measurement of the period of a pendulum can be used to measure g.

EXAMPLE 5 The Period of a Swing

A small child weighing 200 N sits on a swing of length 2.2 m. Find the period for small-amplitude oscillations, treating the swing as a simple pendulum.

SOLUTION Applying Eq. 15-20, we find the period of a simple pendulum of length 2.2 m:

$$T = 2\pi\sqrt{\frac{\ell}{g}} = 2\pi\sqrt{\frac{2.2 \text{ m}}{9.8 \text{ m/s}^2}} = 3.0 \text{ s}$$

This result is completely independent of the child's weight.

Physical Pendulum

A **physical pendulum** is a rigid body of any shape, supported at a single point O, some distance from the body's center of gravity, and free to oscillate about O (Fig. 15-14a). A simple pendulum is a special case of a physical pendulum in which all the mass is concentrated at a single point.

We can analyze the motion of the physical pendulum, using the dynamical equation describing rotation about a fixed axis—Eq. 9-15:

$$\Sigma \tau = I\alpha$$

where $\Sigma \tau$ is the net external torque, I is the moment of inertia, and α is the angular acceleration.

Consider the physical pendulum shown in Fig. 15-14b. The line of length d from the point of support to the center of gravity of the pendulum would be oriented exactly vertically if the pendulum were in its equilibrium position. Because the pendulum is displaced from equilibrium, however, the line makes an angle θ with the vertical. If there is no frictional torque at O, the only external torque acting on the pendulum is that produced by its weight $m\mathbf{g}$. This torque tends to rotate the pendulum clockwise when θ is positive, as in Fig. 15-14b. This means that the torque is negative when θ is positive. When θ is negative (that is, when the center of gravity is to the left of O), the torque tends to rotate the pendulum counterclockwise, and so the torque is positive. Thus the sign of the torque is opposite the sign of θ, and since the moment arm is $d \sin \theta$, we may express the net torque as

$$\Sigma \tau = -mgd \sin \theta$$

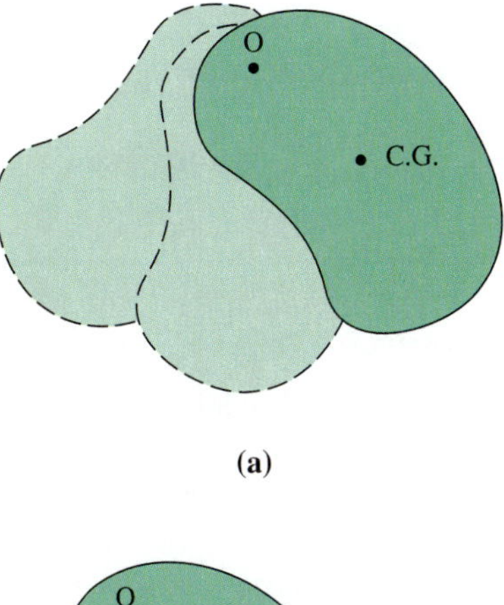

(a)

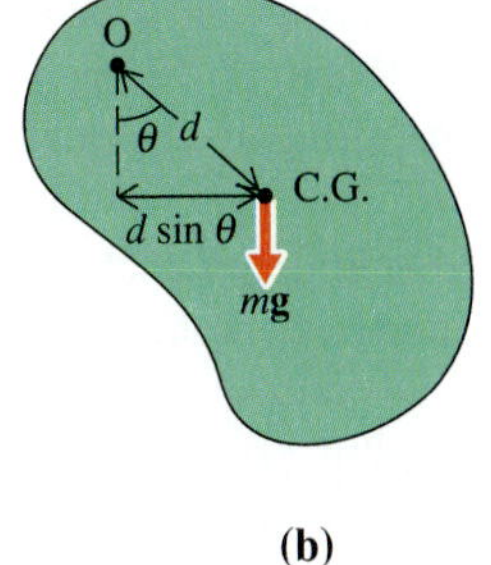

(b)

Fig. 15-14 **(a)** A physical pendulum oscillates about point O. **(b)** The only torque on the pendulum is produced by the pendulum's weight $m\mathbf{g}$, acting at the center of gravity.

*This is true only to the extent that the amplitude is small, less than about 15°. For moderately large amplitudes, up to about 60°, the period is greater than the result predicted by Eq. 15-20 by a factor of $(1 + A^2/16)$.

Substituting this value for $\Sigma\,\tau$ into the preceding equation and solving for α, we obtain

$$\alpha = -\frac{mgd}{I}(\sin\theta)$$

which for small angles may be expressed

$$\alpha \approx -C\theta$$

where

$$C = \frac{mgd}{I}$$

Since the angular acceleration α of the pendulum equals a negative constant times θ, we see that once again the motion is simple harmonic. Substituting this expression for C into Eq. 15-11 $\left(T = \frac{2\pi}{\sqrt{C}}\right)$, we have

$$T = 2\pi\sqrt{\frac{I}{mgd}} \tag{15-22}$$

EXAMPLE 6 The Period of Oscillation of a Meter Stick

A meter stick is mounted on an axle through one end and is free to swing without friction. Find the period of small-amplitude oscillations.

SOLUTION The moment of inertia of a thin rod of mass M and length ℓ is $\frac{1}{3}M\ell^2$ (Fig. 9-16c), and its center of gravity is at a distance $d = \frac{\ell}{2}$ from the axis of rotation. Applying Eq. 15-22, we find

$$T = 2\pi\sqrt{\frac{I}{mgd}} = 2\pi\sqrt{\frac{M\ell^2/3}{Mg\ell/2}} = 2\pi\sqrt{\frac{2}{3}\left(\frac{\ell}{g}\right)}$$

$$= 2\pi\sqrt{\frac{2}{3}\left(\frac{1.00\text{ m}}{9.80\text{ m/s}^2}\right)}$$

$$= 1.64\text{ s}$$

EXAMPLE 7 The Natural Period of Walking

The period of walking is the time for one complete cycle, consisting of two steps. During this cycle each leg completes approximately a half cycle of pendulum-like motion: first one leg swings forward, and then the other leg swings forward. When a person walks in a natural, relaxed way, the motion of the lower half of the leg is approximately that of a free-swinging physical pendulum through most of the forward swing. Find the natural period of walking for a person whose lower leg is 0.50 m long if the amplitude is small. Assume that the moment of inertia of the lower leg is the same as that of a thin rod.

SOLUTION Since we are treating the lower leg as a thin rod mounted at one end, we can apply the result found in the previous example:

$$T = 2\pi\sqrt{\frac{2}{3}\left(\frac{\ell}{g}\right)} = 2\pi\sqrt{\frac{2}{3}\left(\frac{0.50\text{ m}}{9.80\text{ m/s}^2}\right)} = 1.2\text{ s}$$

Fig. 15-15

We can test our answer by asking subjects to walk in a relaxed fashion, timing 20 steps, or 10 cycles, and dividing the time by 10. The result is usually fairly close to our calculated period of 1.2 s for a person whose lower leg is about 50 cm long. For very tall or very short individuals, the predicted period varies only slightly, usually no more than 5%.

Energy

Since the only force doing work on a frictionless pendulum is the weight of the pendulum $m\mathbf{g}$, the pendulum's total mechanical energy E is conserved. A pendulum's gravitational potential energy U_G is given by $U_G = mgy$, where y is the vertical coordinate of the center of gravity, and its kinetic energy is given by either $K = \frac{1}{2}mv^2$ for a simple pendulum or $K = \frac{1}{2}I\omega^2$ for a physical pendulum. Thus we may write for the two cases

$$E = \tfrac{1}{2}mv^2 + mgy \quad \text{(simple pendulum)} \qquad (15\text{-}23)$$

$$E = \tfrac{1}{2}I\omega^2 + mgy \quad \text{(physical pendulum)} \qquad (15\text{-}24)$$

EXAMPLE 8 The Energy of a Child On a Swing

Suppose that the swing in Example 5 has an amplitude of 15°. Find the total mechanical energy of the system.

SOLUTION As shown in Fig. 15-16, when the swing is at its maximum angle, the y coordinate of the child's center of gravity is 7.5 cm. At this point, the child is instantaneously at rest, and hence has no kinetic energy. As stated in Example 5, the child has weight $mg = 200$ N.

Thus

$$E = K + U_G$$
$$= 0 + mgy = (200\ \text{N})(0.075\ \text{m})$$
$$= 15\ \text{J}$$

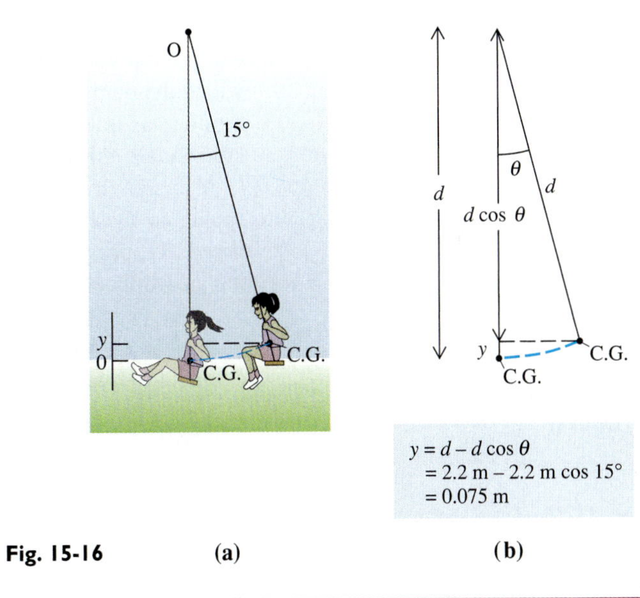

Fig. 15-16 (a) (b)

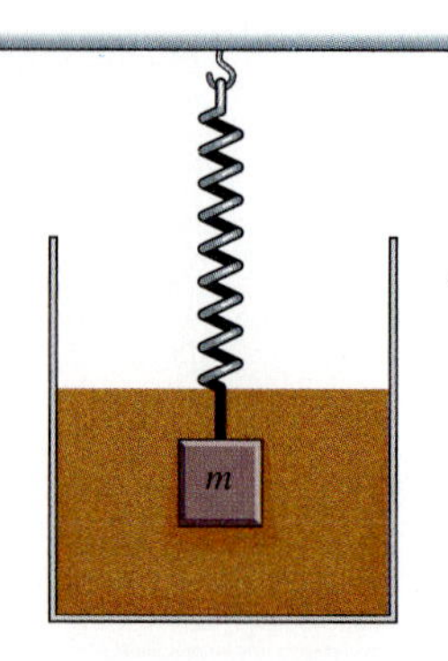

Fig. 15-17 A mass oscillating on a spring is submerged in a viscous fluid.

15-5 Damped and Forced Oscillations

Damped Oscillations

The only kind of harmonic motion we have considered so far is simple harmonic motion. The equations describing SHM imply that it continues indefinitely. For example, Eq. 15-1, describing the oscillations of a mass on a spring, implies that the mass continues to oscillate up and down forever, with a constant amplitude. And our analysis of pendulum motion implied that the back-and-forth swinging of a pendulum never ends. Both systems are idealized, ignoring such factors as friction and air resistance. The total mechanical energy of either ideal system is conserved. Of course, in any real physical system there are factors that eventually cause the motion to stop if the system is left alone. In such systems, mechanical energy is dissipated through some kind of friction and converted to internal energy. This effect is called **damping.** For example, the oscillations of a mass on a spring are rather quickly damped out when the mass is submerged in a viscous fluid (Fig. 15-17). A system for which the damping is slight is called **underdamped.** In this case oscillation slowly dies out, as indicated in Fig. 15-18. If damping is very great, there may be no oscillation at all; this **overdamped** motion is indicated in Fig. 15-19.

Also shown in Fig. 15-19 is the special case of **critical damping,** which corresponds to the fastest approach to the final equilibrium position. When a car hits a bump, its springs tend to make it oscillate up and down, as in Example 4. If the car has good shock absorbers, however, they will provide sufficient damping so that when the car is pushed away from its equilibrium position, it quickly returns to it, with little or no oscillation. The shock absorbers are designed with just enough damping to make the system close to being critically damped.

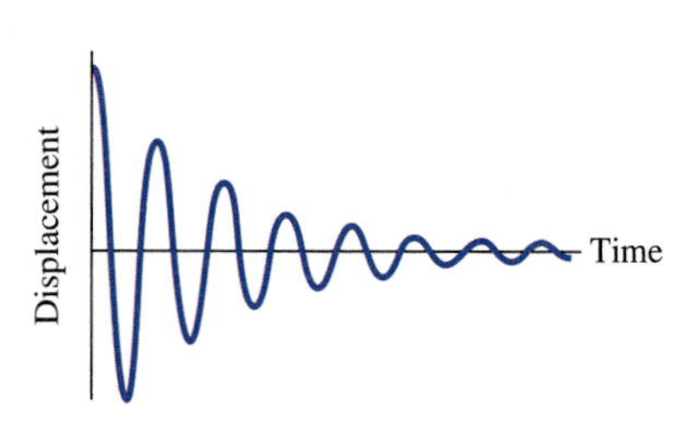

Fig. 15-18 Oscillations of an underdamped harmonic oscillator. A car with bad shock absorbers could undergo such oscillations after hitting a bump. Underdamped motion quickly reaches the equilibrium position but overshoots it and oscillates about the equilibrium position; the oscillations eventually die out and equilibrium is reached.

Forced Oscillations

The motion of an isolated damped oscillator eventually ends. It is possible to maintain oscillations indefinitely, however, if energy is supplied to the oscillator. Consider, for example, a pendulum clock. The oscillations of the pendulum are continuous because of the energy provided by a spring mechanism inside the clock. The spring stores potential energy, which was supplied to it when the clock was wound. So long as this mechanical energy can be transferred to the pendulum, the energy loss caused by damping can be compensated for, and the oscillations can continue at constant amplitude. The spring exerts a periodic applied force that is synchronized with the natural oscillations of the pendulum. In the same way, the motion of a child on a swing may continue indefinitely if the swing is pushed periodically with an applied force that is synchronized with the natural frequency of the swing. The pendulum clock and the swing are both examples of forced harmonic motion.

Fig. 15-19 Overdamped motion is slow to return to the final equilibrium position. Critically damped motion returns quickly to the equilibrium position without overshooting it and would be the most desirable motion for a spring-shock absorber system on a car.

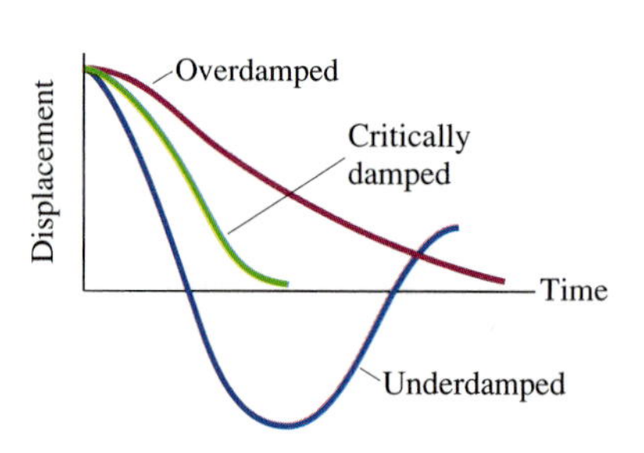

Resonance

When a periodic force is applied to an oscillatory system, the system responds by oscillating at the frequency of the applied force. However, the amplitude of the oscillations is strongly dependent on the frequency of the applied force. The amplitude is largest when the frequency of the force is close to the natural frequency of oscillation of the system. This is indicated in Fig. 15-20 for both a lightly damped and a heavily damped system. This phenomenon is called **resonance.** For the lightly damped case, the amplitude at resonance is 10 times greater than the oscillator displacement that would result if a constant force equal to the amplitude of the periodic force were to be applied to the oscillator.

Suppose, for example, that a playground swing has a natural period of 3 s. If we put a child on that swing and then push at intervals of 2 s, we will be interrupting the natural cycle of the swing, and as a result the amplitude of the motion will be small. If we push at 3 s intervals, however, we enhance the natural cycle and get a much larger amplitude. For a periodic force with a maximum magnitude of, say, 20 N, applied to the swing at its natural period of 3 s, the swing's amplitude might well be 40° or more, perhaps 10 times greater than the angular displacement produced by a constant 20 N force.

The phenomenon of resonance is easily demonstrated with a simple pendulum. Let the pendulum be supported by a hand, which moves back and forth over a distance of a few centimeters. The system experiences an oscillating apparent force at a frequency equal to the frequency of the hand. When this frequency equals the resonant frequency of the system, large oscillations are observed (Fig. 15-21). The pendulum oscillates with an amplitude many times greater than the amplitude of oscillation of the hand. If, however, the frequency of the hand's motion is significantly different from the resonant frequency of the system, only small-amplitude oscillations are observed.

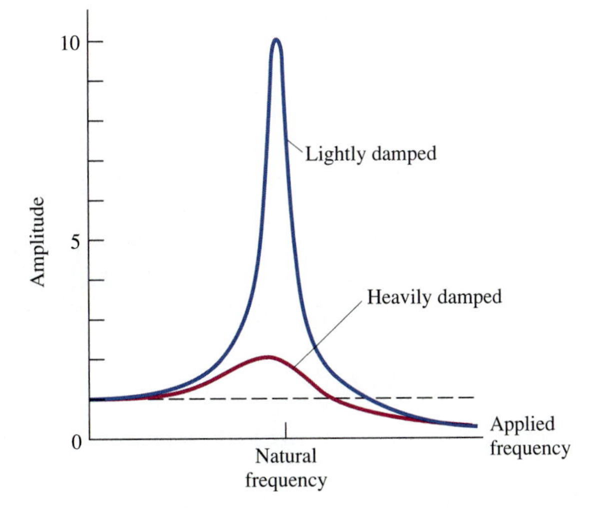

Fig. 15-20 Resonance in two oscillating systems, one heavily damped and the other lightly damped. If you were to apply a constant force to either system, the oscillator would be displaced from equilibrium by the amount indicated by the dashed line. Apply a *periodic* force of the same amplitude as the constant force, however, and if the frequency of the applied force is close to the system's natural frequency, you get a dramatic increase in the amplitude of the oscillation, especially in the lightly damped system.

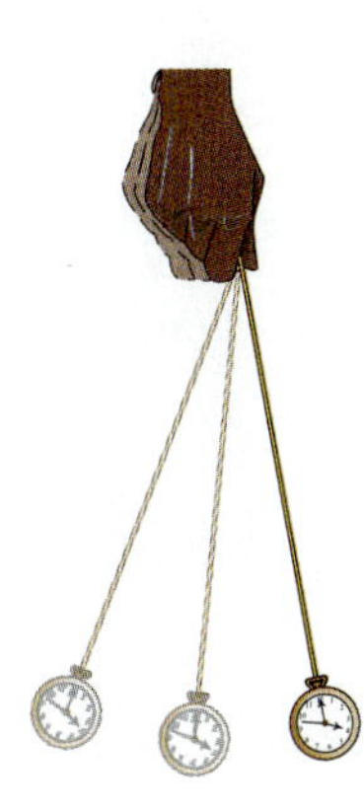

Fig. 15-21 Moving the point of support of a simple pendulum back and forth produces large-amplitude oscillations of the pendulum if the frequency of the motion is the pendulum's resonant frequency.

Resonance is common to many physical systems other than the simple mechanical ones described here. For example, as we shall see in Chapter 22, an electric circuit may have a resonant frequency. A radio is tuned by turning a knob that adjusts a circuit so that the resonant frequency of the circuit is the frequency of the desired station. Then only radio waves of that particular frequency are highly amplified.

Rigid bodies also have resonant frequencies. Unlike the mass on a spring or the simple pendulum, they usually have more than one resonant frequency. When a body is forced to vibrate at a resonant frequency, the effect can be dramatic. For example, if a singer produces a note of just the right frequency and the sound is sufficiently amplified* and directed at a thin glass, the glass may break as a result of the resonant vibrations induced in it (Fig. 15-22). The frequency of the sound must be equal to one of the resonant frequencies of the glass. The values of these resonant frequencies depend on the thickness and shape of the glass.

Structures must be designed so that resonant vibrations are not accidentally produced. The Tacoma Narrows bridge was destroyed in 1940, a few months after it was completed, as a result of resonant vibrations produced by gusting winds (Fig. 15-23). When a column of soldiers or a marching band crosses a small bridge, the marchers are often instructed to break step to avoid producing vibration at a resonant frequency of the bridge.

*Contrary to popular myth, there is no evidence that anyone has ever accomplished this feat without amplifying the voice.

Fig. 15-22 An amplified human voice causes a glass to break when the glass vibrates at a resonant frequency.

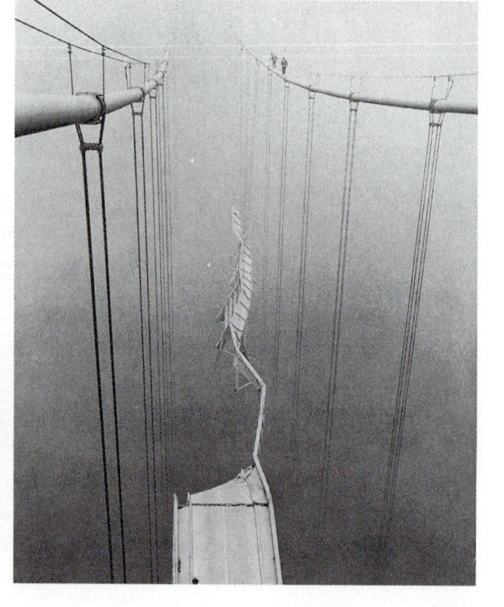

Fig. 15-23 Collapse of the Tacoma Narrows Bridge.

CHAPTER 15 SUMMARY

For periodic motion, we denote the period, or time for one cycle, by T. The frequency, measured in cycles/s, or Hz, is given by

$$f = \frac{1}{T}$$

and the angular frequency by

$$\omega = 2\pi f = \frac{2\pi}{T}$$

For any body undergoing simple harmonic motion (SHM), acceleration a_y is related to displacement y by the equation

$$a_y = -Cy$$

where C is a constant. This equation has as its general solution

$$y = A \sin\left(2\pi\frac{t}{T} + \phi_0\right)$$

where $$T = \frac{2\pi}{\sqrt{C}}$$

The amplitude A (the maximum value of displacement y) and initial phase angle ϕ_0 are determined by initial conditions.

A mass m attached to a spring of force constant k oscillates with a period

$$T = 2\pi\sqrt{\frac{m}{k}}$$

A physical pendulum is a body of mass m that rotates about some axis located a distance d from the body's center of gravity and has a moment of inertia I about the axis of rotation. For small-amplitude oscillations (up to about 0.3 rad), the angular displacement θ of the pendulum from its equilibrium position is simple harmonic with a period

$$T = 2\pi\sqrt{\frac{I}{mgd}}$$

A simple pendulum is a special case of the physical pendulum in which all of the mass is concentrated at a point a distance ℓ from the rotation axis. For small amplitudes, the period is given by

$$T = 2\pi\sqrt{\frac{\ell}{g}}$$

For SHM, both the amplitude and the total mechanical energy are constant. For a damped harmonic oscillator, mechanical energy is dissipated at a rate determined by the amount of damping. In forced harmonic motion, the resonant frequency is the value of the applied frequency that maximizes the amplitude.

Questions

1 An electrocardiogram (ECG or EKG) is a graph showing electrical oscillations resulting from the beating of a human heart (Fig. 15-24). Is such motion approximately (a) harmonic? (b) simple harmonic? (See Chapter 18 for further discussion of ECGs.)

2 Suppose you increase the amplitude of a child on a swing from 10° to 20°. How does the child's average speed change?

3 By what percent does the period of a simple pendulum change if you double its (a) amplitude; (b) mass; (c) length?

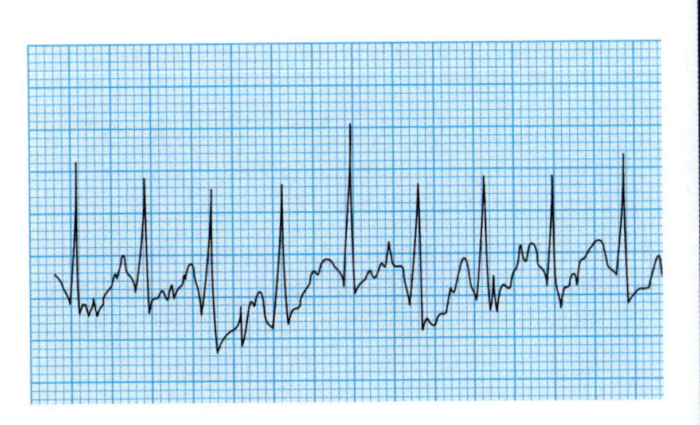

Fig. 15-24

4 How could you change either the mass or the force constant of a mass oscillating on a spring to double its period?

5 Suppose a pendulum clock is placed on an elevator. Would the clock be faster or slower if the elevator (a) moves downward at constant speed; (b) is accelerated upward?

6 Would the period on the moon be longer than, shorter than, or the same as on earth for (a) a mass on a spring; (b) a simple pendulum?

7 When we describe the angular frequency of a simple pendulum undergoing small-amplitude oscillations as $\omega = 2\pi$ rad/s, does this mean that the pendulum moves through an angle of 2π rad each second?

8 Would you tend to walk faster or slower on the moon than on earth if you walk in a relaxed fashion?

9 When you "pump" a swing to increase the amplitude of swinging, you are applying a periodic force at a certain frequency. Should this be close to the swing's resonant frequency?

10 Suppose your car is vibrating as you travel along a road because at regular intervals you hit small spaces between concrete segments in the road. The force of the road on the car is periodic and near a resonant frequency of the car. What can you do to reduce the vibration?

Answers to Odd-Numbered Questions

1 (a) yes (b) no; **3** (a) 0 (for small amplitudes); (b) 0; (c) increases by 41%; **5** (a) no change (b) faster; **7** no; **9** yes

Problems (listed by sections)

15-1 Simple Harmonic Motion

1 A tuning fork oscillates at a frequency of 512 Hz. Find the period of the motion.

2 A certain radio station broadcasts at 102.7 megahertz. This is the frequency of oscillation of electrons in the transmitting antenna producing the radio wave. Find the period of the electrons' motion.

3 Find the amplitude, period, and frequency of the motion described by Fig. 15-25.

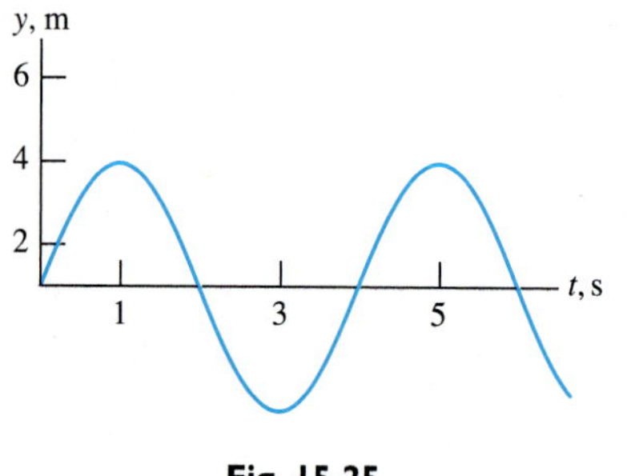

Fig. 15-25

4 For the motion described by Fig. 15-25, find the displacement at (a) $t = 12.0$ s; (b) $t = 13.0$ s; (c) $t = 13.5$ s.

5 For the ECG shown in Fig. 15-24, the time scale is 0.0400 s per small division. What is the heartbeat rate in units of (a) Hz; (b) beats per minute? (The ECG was recorded when the subject was running on a treadmill set at an incline.)

6 Suppose that, for the graph shown in Fig. 15-4, each division on the time scale represents 1.00×10^{-3} s and divisions on the displacement scale are in cm. Find the following: (a) amplitude; (b) period; (c) frequency; (d) displacement at $t = 0.020$ s.

7 The motion of a body is described by the equation $y = 3.00 \sin (0.200\pi t)$, where t is in s and y is in m. Find (a) the amplitude; (b) the period; (c) the frequency; (d) the displacement at $t = 4.00$ s; (e) the displacement at $t = 24.0$ s.

8 Functions of the form $A \sin\left(2\pi\frac{t}{T} + \phi_0\right)$ are shown in Fig. 15-26. Find the initial phase angle for each.

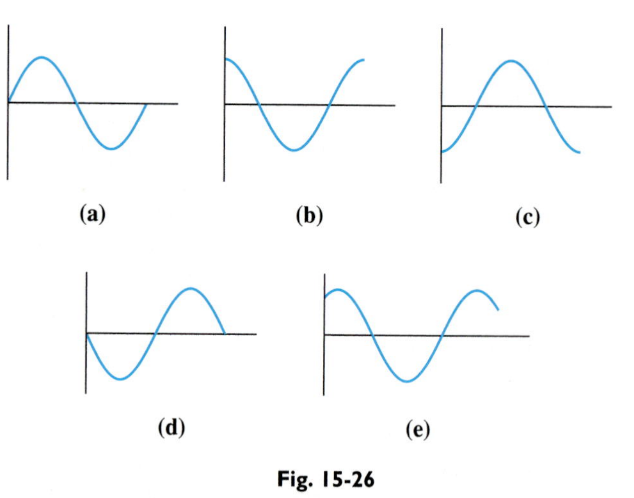

Fig. 15-26

9 Find the amplitude, angular frequency, and initial phase angle for the motion described by the graph shown in Fig. 15-27.

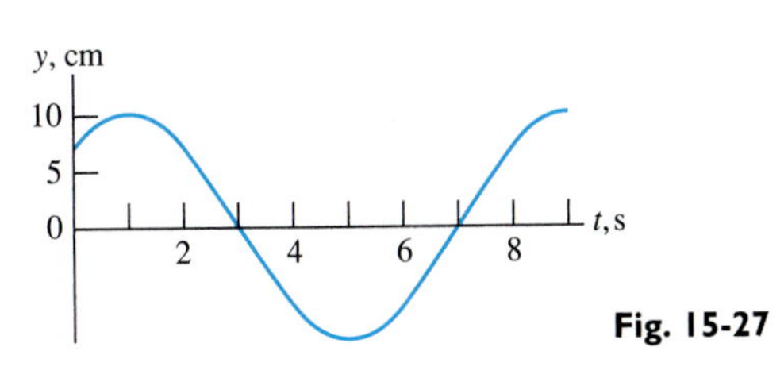

Fig. 15-27

10 Two bodies oscillate in SHM with the same amplitude (4.00 cm) and frequency but out of phase so that body A always has a lead in phase angle of 30.0° over body B. Find the displacement of A when the displacement of B is 4.00 cm.

11 Functions y_1 and y_2 are shown in Fig. 15-28. Find the difference in the phase angles, $\phi_1 - \phi_2$, at $t = t_0$.

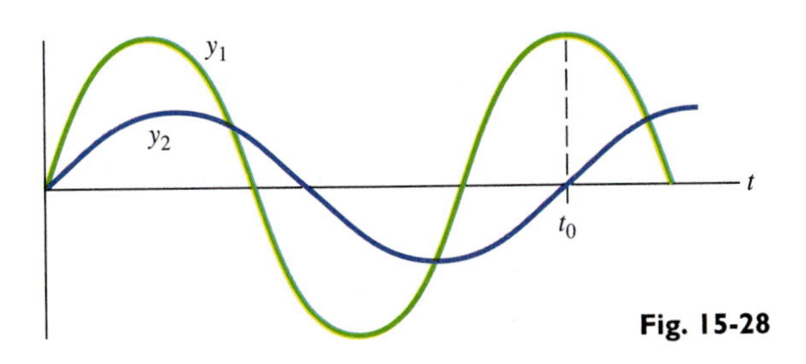

Fig. 15-28

15-2 SHM and Circular Motion

12 For a record playing at 33⅓ rpm, find the frequency in Hz and the angular velocity in rad/s.

13 A child repeatedly opens and closes a door at regular intervals of 2 s. The door swings through a 50° arc. Find the angular frequency of the motion.

14 The motion of a body is described by the equation $a_y = -10y$. Find (a) the angular frequency; (b) the frequency; (c) the period.

★**15** A fan blade rotates at a constant angular velocity of 10.0 rev/s. A point on the blade at some instant has coordinates $x = 10.0$ cm, $y = 20.0$ cm. Find the components of acceleration a_x and a_y of that point.

16 An ant is at a fixed point on a 33⅓ rpm record 10.0 cm from the center. The record moves at a constant angular velocity (33⅓ rpm). Write expressions for the x and y coordinates of the ant as a function of time, assuming that (a) $x = 0$ when $t = 0$; (b) $x = 10.0$ cm when $t = 0$.

★**17** A body of mass 3.00 kg moves along the x-axis, and its motion is described by the equation $x = 4 \sin (18t + 2)$, where x is in m and t is in s. Find the resultant force on the body at (a) $t = 0.100$ s; (b) $t = 0.200$ s.

15-3 Mass and Spring

18 A 10.0 kg mass attached to a spring oscillates with a period of 3.14 s. Find the force constant of the spring.

19 A patio swing is suspended by two springs, each of which has a force constant of 500 N/m, equivalent to a single spring force constant of 1000 N/m. Find the mass of a person sitting on the swing oscillating up and down at a frequency of 0.500 Hz.

20 A boy of mass 50.0 kg standing on the end of a diving board depresses it vertically downward a distance of 20.0 cm. By pushing down on the board with a force a little greater than his weight, the boy can depress the end of the board a bit farther. The boy and the board then oscillate up and down. Estimate the period of oscillation, assuming that the force the board exerts is approximately like that of a compressed spring, in other words, that it obeys Hooke's law.

21 A small mass m is supported by a vertical spring. When an additional 100 g is attached to the original mass, the system begins to oscillate at a frequency of 0.500 Hz. When the oscillations die out, the spring is found to have increased in length by 10.0 cm. Find m.

22 A 2.00 kg mass attached to a spring oscillates with a period of 0.400 s and an amplitude of 20.0 cm. Find (a) the total mechanical energy of the system; (b) the maximum speed of the mass.

23 A 50.0 kg mass attached to a spring oscillates with an amplitude of 2.00 m. The total mechanical energy of the system is 400 J. Find the position of the mass when its speed is 2.00 m/s.

★**24** A block rests on a horizontal surface, and the coefficient of static friction between the block and the surface is 0.400. The surface begins to move back and forth in the horizontal direction, vibrating in SHM with a constant amplitude of 10.0 cm but a gradually increasing frequency. Find the frequency at which the block begins to slide on the surface.

25 A point on the A string of a guitar oscillates in approximately simple harmonic motion with a frequency of 220 Hz and an amplitude of 1.00 mm. Find the maximum speed and acceleration of this point.

26 A spring scale being used to measure the weight of an object reads 12.0 N when it is used on earth. The spring stretches 5.00 cm under the load. The same object is weighed on the moon, where gravitational acceleration is $\frac{1}{6}g$. Find the reading of the spring scale on the moon and the period for vertical oscillations of the spring.

15-4 The Pendulum

27 Find the period of a simple pendulum of length 9.80 m.

28 Find the length of a simple pendulum that completes 10.0 oscillations in 20.0 s.

29 Find g at a point on earth where $T = 2.01$ s for a simple pendulum of length 1.00 m, undergoing small-amplitude oscillations.

★ 30 A pendulum clock is accurate on earth, where the period of its pendulum is 1.00 s.

(a) Find its period on the moon, where gravitational acceleration is 1.6 m/s^2.

(b) During a 1 h interval, what would be the elapsed time shown on the clock?

★ 31 A uniform rectangular sign of width 50.0 cm, height 30.0 cm, and negligible thickness hangs vertically from supporting hinges attached at its upper edge. Find the period of small-amplitude oscillations of the sign.

32 An irregularly shaped body of mass 2.00 kg is mounted on an axle located 25.0 cm from the center of gravity of the body. The body oscillates as a pendulum with a period of 3.00 s and an amplitude of 0.100 rad. Find the body's moment of inertia about this axis.

33 Two small spheres of masses 1.00 kg and 2.00 kg are attached to opposite ends of a thin rod of length 1.00 m and of negligible mass. A hole is drilled through the center of the rod, so that it can hang from a nail and oscillate in a vertical plane. Find the period of small-amplitude oscillations.

15-5 Damped and Forced Oscillations

34 The amplitude of oscillation of a mass attached to a spring decreases by 10% with each oscillation. By what percent does the energy of the system decrease with each oscillation?

35 Suppose that the child-swing system in Example 5 gradually slows down in such a way that on each successive swing the maximum y coordinate of the child's center of gravity is 3.00 mm less than on the previous swing.

(a) How much energy would have to be supplied by a parent pushing the child to maintain a constant amplitude?

(b) At what rate should this energy be supplied?

Additional Problems

★ 36 The period of a physical pendulum is measured first at sea level, where the pendulum completes 100 cycles in 1550 s, and then on a mountaintop, where it completes 100 cycles in 1551 s. Assuming that the change in period is caused only by the variation in g with altitude, find the height of the mountain.

★ 37 Find the natural walking speed v as a function of leg length ℓ. Assume that the upper and lower legs have the same length $\ell/2$, that the lower leg swings like a physical pendulum (a thin rod), and that the leg makes a maximum angle of 20° with the vertical (Fig. 15-29). Show that $v = \frac{2}{\pi}\sqrt{3\ell g}\sin 20°$. Evaluate for $\ell = 1.0$ m and $\ell = 0.80$ m.

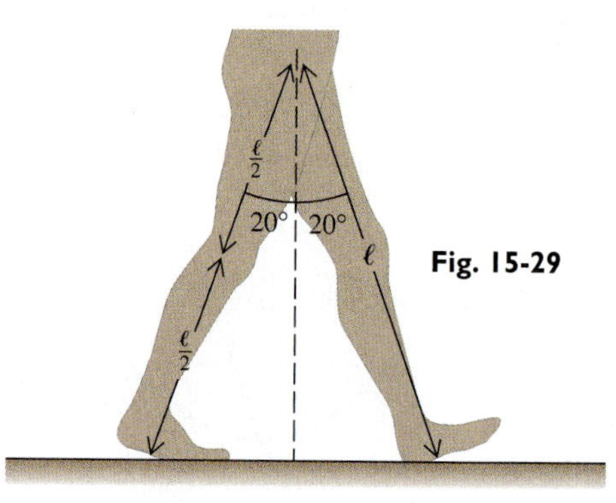

Fig. 15-29

★★ 38 Suppose a tunnel were dug through the earth, as shown in Fig. 15-30.

(a) Show that a body in the tunnel a distance x from the center of the tunnel will undergo an acceleration $a_x = -g_r \cos\theta$, where g_r is the gravitational acceleration at a distance r from the center of the earth.

(b) Assuming that the earth has uniform density ρ, show that $g_r = -\frac{4}{3}\pi G\rho r$.

(c) Show that, in the absence of friction, a body in the tunnel will experience SHM.

(d) Find the angular frequency and period. Show that both these quantities are independent of the length of the tunnel.

(e) Assuming that the body's maximum speed is 70.0 m/s, what is the length of the tunnel? What is the average speed?

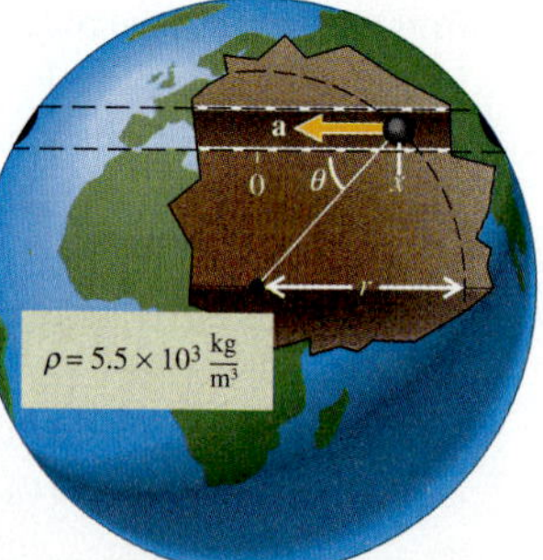

Fig. 15-30

★ **39** A baseball bat of mass 1.8 kg has a hole drilled through the center of the grip and is mounted on a stationary axle about which the bat is free to swing. The bat's center of gravity is at a distance $d = 50$ cm from the axis of rotation O, as shown in Fig. 15-31. The period of small-amplitude oscillations is 1.7 s.

(a) Find the bat's moment of inertia I.

(b) A force $\mathbf{F}_1$ of magnitude 15000 N is applied at a distance r from O, as shown in the figure. (A maximum instantaneous force of this order of magnitude is typical in striking a baseball.) Show that the instantaneous angular acceleration of the bat is given by F_1r/I and that the acceleration of the center of mass is F_1rd/I.

(c) Show that there is a force $\mathbf{F}_2$ exerted by the axle at point O, where $F_{2x} = F_1\left(\frac{Mrd}{I} - 1\right)$, and that F_{2x} vanishes if $r = I/Md$. Thus if $\mathbf{F}_1$ is applied at this point, a distance I/Md from the axis of rotation, no additional force is required to produce pure rotational motion about O. This point is called the **center of percussion.** In any sport where a ball is struck, it is important that the ball strikes near the center of percussion of the bat, racket, or club. When this happens, there is no additional force that must be supplied by the hands; the bat, club, or racket does not sting the hands.

(d) Find the distance r to the center of percussion for the bat.

(e) Find F_{2x} if $r = 80$ cm.

(f) Show that if all the bat's mass were concentrated at the center of percussion the period would be unchanged. For this reason the center of percussion is also called the **center of oscillation.**

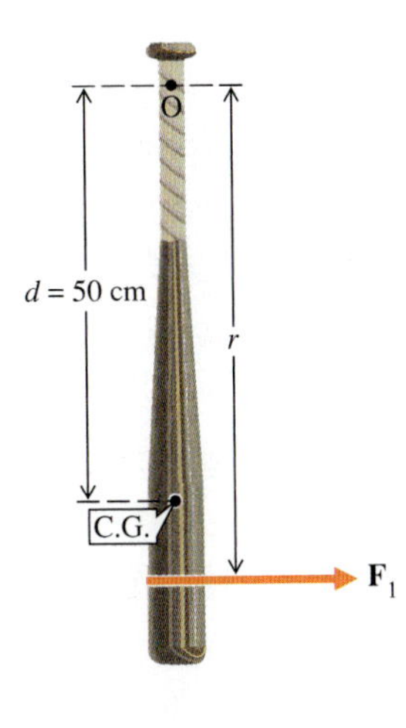

Fig. 15-31

★ **40** A hoop of radius 0.500 m hangs from a nail in a wall so that the hoop can swing about its point of support. Find the period of small-amplitude oscillations of the hoop.

★ **41** A small mass slides on a frictionless hemispherical surface (Fig. 15-32). Show that this system is similar to a simple pendulum. Find the period of small-amplitude oscillations.

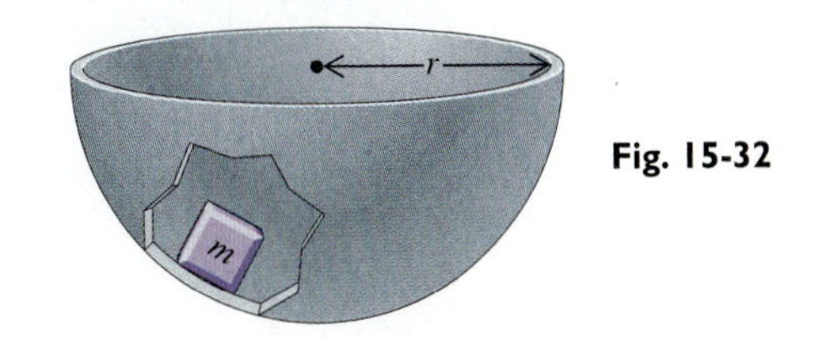

Fig. 15-32

★★ **42** A cylindrical buoy of mass 20.0 kg and cross-sectional area 400 cm^2 floats at equilibrium in calm water. Show that if the buoy is pushed a distance d below its equilibrium position it will experience an upward restoring force proportional to d and thus will oscillate in SHM. Find the period.

★ **43** The pendulum in an antique clock can be approximated by a thin rod with a coefficient of thermal expansion of 2.00×10^{-5} (C°)$^{-1}$. At 20.0° C, the period of the pendulum is 1.00 s. If the temperature increases to 30.0° C, how many seconds per day will the clock gain or lose?

CHAPTER 16 Mechanical Waves; Sound

Ocean waves, generated by the wind, sometimes travel many kilometers before breaking as they enter shallow water. Although the wave and its energy move over great distances, the water experiences only small-amplitude, periodic motion.

The world around us is filled with waves—sound waves, radio waves, microwaves, X rays, light waves, water waves, earthquake waves, and many others. Some of these waves require a material medium for their transmission. These are called **mechanical waves.** Water waves, sound, and earthquake waves are all mechanical waves, since each requires a medium through which to propagate. Water waves travel through water, and earthquake waves travel through the earth. Sound travels through air or some other medium. None of these waves can be transmitted through a vacuum.

In this chapter we shall study only mechanical waves, in particular, water waves, sound, and waves on a string. In later chapters we shall study radio waves, microwaves, X rays, and visible light, all of which are examples of electromagnetic waves. Electromagnetic waves are not mechanical waves. No physical medium is necessary to transmit these waves; they can travel through a vacuum. For example, when we look at a starry sky, we see light that has traveled through the vacuum of interplanetary space.

Although we consider only mechanical waves in this chapter, we shall find that many of the wave concepts learned here are applicable to electromagnetic waves as well. For example, all kinds of waves can transmit energy.

16-1 Description of Waves

Wave Pulses; Dominoes, Strings, and Springs

The phenomenon of a **wave pulse** is easily demonstrated with a line of dominoes, standing on end and closely spaced (Fig. 16-1). If the domino at the left end is pushed over to the right, the effect is transmitted down the line. All the dominoes fall in turn.

Each domino has kinetic energy only for the brief interval during which it is falling. A pulse of energy is transmitted from one end of the line to the other, though each domino moves only a very short distance. This is a general characteristic of mechanical wave motion: **energy is transported through matter without the transport of the matter itself.** A mechanical wave transmits energy from one place to another while the matter through which it is transmitted remains in place.

Fig. 16-1 When the domino at the left end of this line is pushed over to the right, a wave pulse passes through the line.

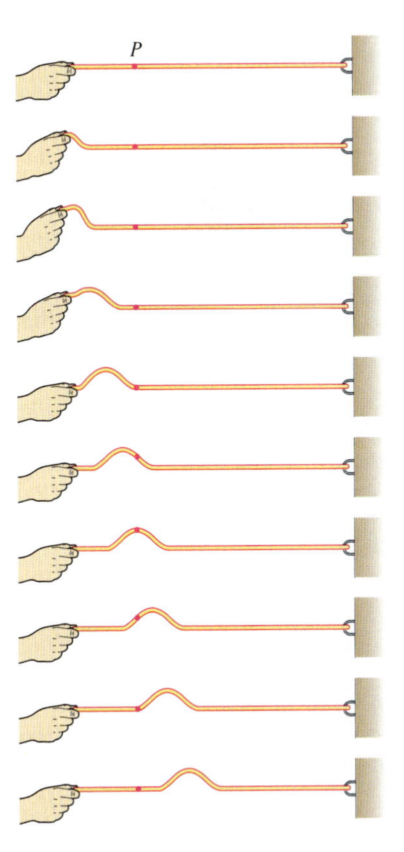

Fig. 16-2 A quick up-and-down motion of the left end of this string produces a wave pulse that travels horizontally along the string to the right. As the wave pulse passes each point on the string, that point undergoes the same up-and-down motion as the left end.

A wave pulse can also be generated on a string under tension. If the left end of a taut string is moved up and down once with a quick flip of the wrist, a wave pulse of fixed shape moves to the right at constant velocity (Fig. 16-2). *This motion of the wave pulse is not the same as the motion of the string.* As illustrated in the figure, a point P on the string moves vertically while the wave moves horizontally and transmits energy along the string. A wave such as this, in which the motion of the medium is perpendicular to the wave motion, is called a **transverse wave.**

Waves for which the motion of the medium is parallel to the direction of wave propagation are called **longitudinal waves.** Longitudinal waves on a spring are illustrated in Fig. 16-3. If the left end of the stretched, horizontal spring is pushed to the right and then pulled back to the left, the adjacent section of the spring first compresses and then stretches. This motion is transmitted to the right all along the spring: as the wave pulse passes, each section moves first to the right and then to the left. This passage of the pulse results in first a compression and then a stretching, or "rarefaction," of the spring.

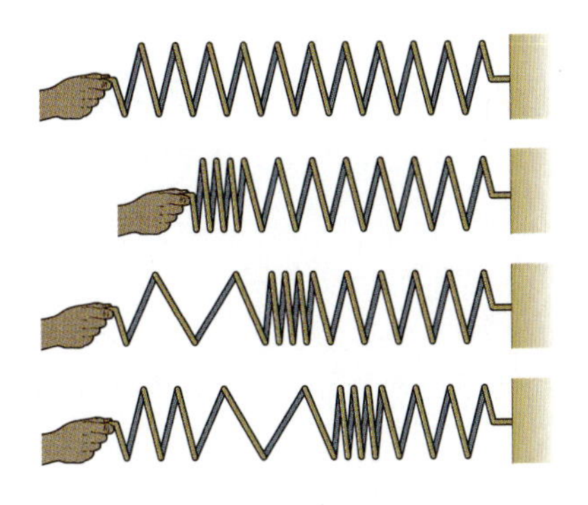

Fig. 16-3 A quick back-and-forth horizontal motion of the left end of this spring produces a wave pulse that travels along the spring to the right. As the wave pulse passes each segment of the spring, that segment undergoes the same back-and-forth motion as the left end.

Water Waves: Two-Dimensional Waves

Fig. 16-4 A wave produced by a raindrop falling on the still surface of a lake.

The preceding examples were one dimensional waves; that is, the wave energy moved along a line. A water wave is an example of wave motion in two dimensions. When the surface of a body of water is disturbed, a wave propagates radially outward (in two dimensions) from the disturbance along the surface of the water. For example, a water wave is produced by a raindrop striking the surface of a lake, as shown in Figs. 16-4 and 16-5. As the wave moves outward from the point of disturbance, the energy being transmitted by the wave spreads out and the height of the wave gradually diminishes. This is true for all two-dimensional waves as well as for all three-dimensional waves, which we consider next.

Sound Waves: Three-Dimensional Waves

A sound wave in air is an example of a three-dimensional, longitudinal wave. It consists of compressions and rarefactions of air molecules spreading out in all directions. The click you hear when two billiard balls collide is a sound-wave pulse. Air molecules are compressed between the colliding balls. This compression is followed by a rarefaction, a volume in which the density of the air molecules is much lower than in undisturbed air. The effect is transmitted outward to the surrounding air, and the wave pulse propagates in all directions.

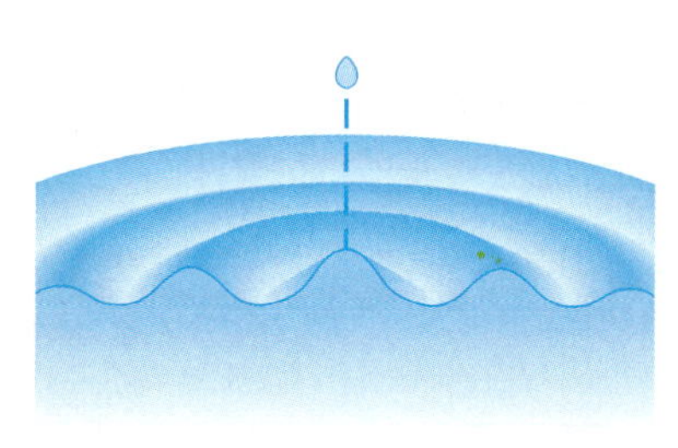

Fig. 16-5 Side view of the wave produced by a raindrop on the surface of a lake.

If a sound wave originates at a point and propagates outward equally in all directions, the wave disturbance takes the shape of a spherical surface centered on the source, as indicated in Fig. 16-6a. This is called a **spherical wave.**

If a wave disturbance is the same everywhere over the surface of a plane, the wave is said to be a **plane wave.** If a small part of a spherical wave is viewed at a large distance from the source, this part of the spherical surface is approximately a plane, and the wave may be represented by a plane wave in this region (Fig. 16-6b). A plane wave varies only in one direction—its direction of motion, which is perpendicular to the plane. Since variation in a plane wave depends on only one spatial variable, it may be described in the same way as the variation in a one-dimensional wave.

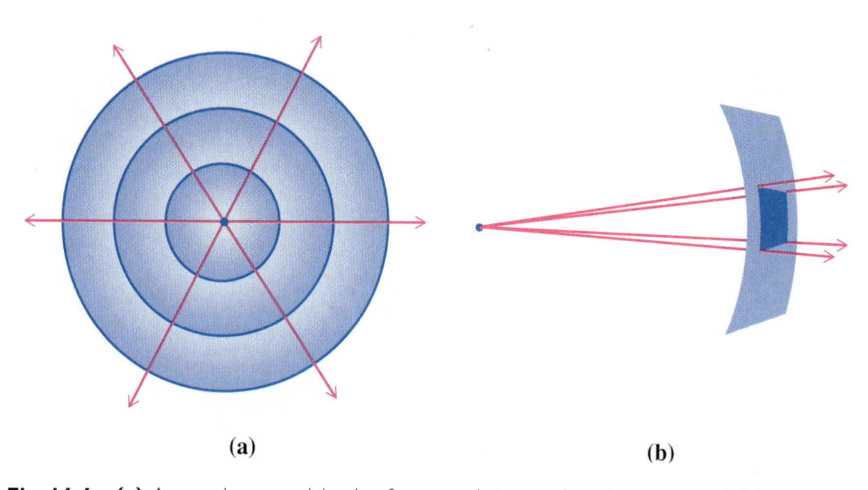

Fig. 16-6 **(a)** A sound wave originating from a point spreads out spherically. **(b)** At a great distance from the source of a spherical wave, small sections of the spherical surfaces are approximately planes.

Periodic Waves; Wavelength, Frequency, and Speed

Many waves are repetitive, or periodic. Periodic motion of the wave's source results in a wave whose form is periodic in space. For example, you can generate a periodic wave on a string by repeatedly flicking the string up and down. Some periodic wave forms are shown in Fig. 16-7. A wave form may represent the observed shape of a wave on a string, or it may represent the spatial variation of any other kind of wave disturbance. For example, the wave form may represent the spatial variation in the density of air molecules in a sound wave.

The **wavelength** of a wave, denoted by the Greek letter λ (lambda), is the distance between any two successive identical points on the wave—from one crest to the next, say, or from one trough to the next (Fig. 16-7).

If the motion of the source is SHM, the wave form that results has a sine wave shape (Fig. 16-7a) and is called a **harmonic wave.** A harmonic wave on a string is generated when one end of the string is repeatedly moved up and down in SHM (Fig. 16-8a). A vibrating tuning fork generates a harmonic sound wave (Fig. 16-8b). Harmonic waves are of particular importance in the study of waves, as we shall see in Section 16-6 when we study superposition of waves.

As a periodic wave passes a given particle in the medium, that particle undergoes periodic motion. The **frequency** f of a periodic wave is the frequency of the periodic motion experienced by each particle of the medium. For example, each air molecule in the path of a 512 Hz sound wave vibrates at a frequency of 512 Hz.

A wave imparts the motion of the source to each particle of the medium, but for each particle this motion is delayed by the time interval required for the wave to travel to that particle. For example, as a harmonic wave travels along a string, all particles of the string undergo SHM. In Fig. 16-8a particles P and Q undergo the same SHM, but Q is $\frac{1}{2}$ cycle behind P because a time interval of $\frac{1}{2}$ period is required for the wave to travel from P to Q.

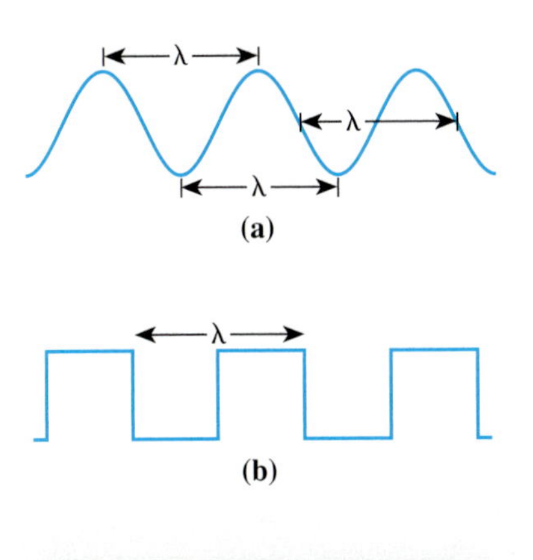

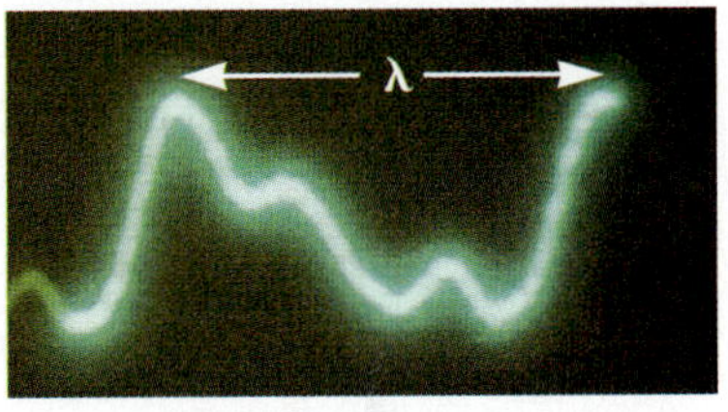

Fig. 16-7 Periodic waves: **(a)** harmonic wave; **(b)** square wave; **(c)** wave generated by a vibrating guitar string.

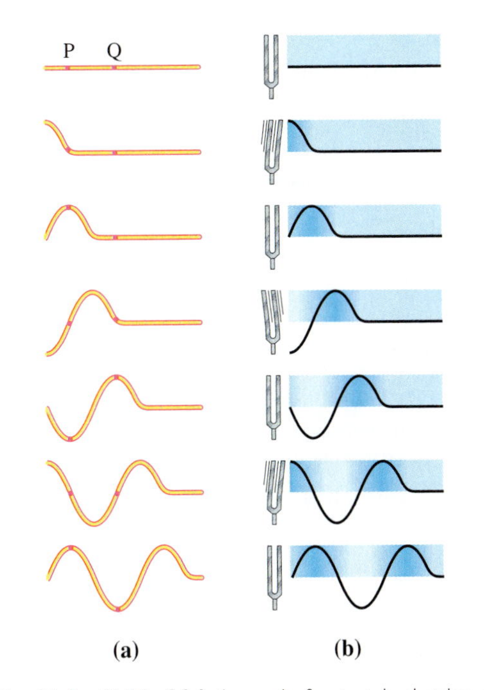

Fig. 16-8 SHM of **(a)** the end of a stretched string or **(b)** a tuning fork generates a harmonic wave.

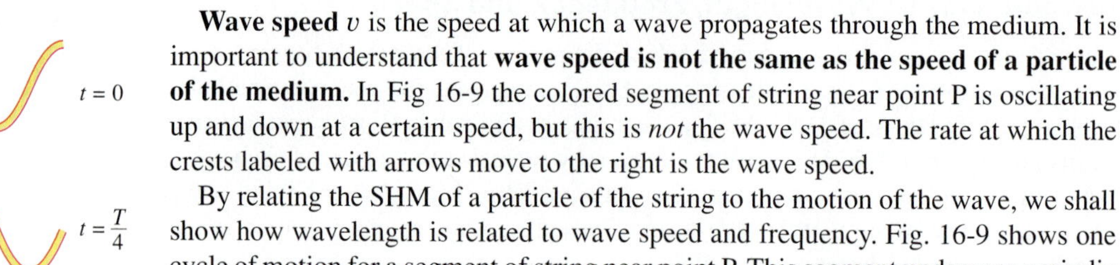

Fig. 16-9 Each segment of a string moves up and down with a period T as a wave of wavelength λ moves along the string to the right. In time T one wavelength has passed a fixed point P.

Wave speed v is the speed at which a wave propagates through the medium. It is important to understand that **wave speed is not the same as the speed of a particle of the medium.** In Fig 16-9 the colored segment of string near point P is oscillating up and down at a certain speed, but this is *not* the wave speed. The rate at which the crests labeled with arrows move to the right is the wave speed.

By relating the SHM of a particle of the string to the motion of the wave, we shall show how wavelength is related to wave speed and frequency. Fig. 16-9 shows one cycle of motion for a segment of string near point P. This segment undergoes periodic motion (up and down in the y direction) with period T. During the same time T, each crest of the wave moves a distance λ to the right. The wave crest moves to the right at the wave speed v, where

$$v = \frac{\lambda}{T}$$

Indeed, each part of the wave form moves at this speed. Since the frequency f equals $1/T$, we may express the equation above in the form

$$v = \lambda f \tag{16-1}$$

This relationship is valid for waves of any kind (water waves, sound, light, and so forth), and although we have illustrated a harmonic wave in Fig. 16-9, the relationship is valid for any wave form.

The frequency of a wave is always determined solely by the wave source. Thus, once a wave is formed, its frequency doesn't change, even when the wave passes from one medium to another. In contrast, **the speed of a wave is determined by the medium through which the wave travels.** For example, the speed of sound in water (1480 m/s) is quite different from the speed of sound in air (340 m/s). Wave speed in a given medium *may* depend on the frequency of the wave; that is, waves of some frequencies may travel faster than waves of other frequencies. This phenomenon is called **dispersion.***

If both the speed and the frequency of a wave are known, we can find its wavelength by using Eq. 16-1 ($v = \lambda f$) to solve for λ. Thus **wavelength depends on both the source and the medium.**

EXAMPLE 1 Wavelength of Sound for a Musical Note

Frequencies of sound produced by a piano range from about 30 Hz for the lowest notes to about 4000 Hz for the highest notes. Find the wavelength in air of a 262 Hz sound wave produced by striking middle C on a piano. What would the wavelength of this sound be under water? The speed of sound is 340 m/s in air and 1480 m/s in water.

SOLUTION We apply Eq. 16-1:

$$v = \lambda f$$

or

$$\lambda = \frac{v}{f}$$

In air, $\lambda = \dfrac{340 \text{ m/s}}{262 \text{ Hz}} = 1.30 \text{ m}$

In water, $\lambda = \dfrac{1480 \text{ m/s}}{262 \text{ Hz}} = 5.65 \text{ m}$

Notice that the frequency remains unchanged, even when the wave leaves the air and enters water. Since wave speed increases as the wave goes from air to water, the relationship $\lambda = \dfrac{v}{f}$ shows that the wavelength must increase as v increases.

*Electromagnetic waves traveling through matter also exhibit dispersion. For example, red light travels through glass faster than blue light; this is responsible for the colors one sees in looking through a prism (see Section 23-2).

Reflection

Any abrupt change in the medium a wave travels through will result in the wave's being reflected. An echo is a sound wave that reflects as its medium changes from air to some solid. If you yell in a mountainous area, you can often hear your echo a few seconds later. The sound wave produced by your voice has been reflected by the surrounding mountains. Similarly, when a wave on a string reaches the end of the string, the wave is reflected. Fig. 16-10 shows the reflection of wave pulses. If the end of the string is held fixed, as in Fig. 16-10a, the wave pulse is inverted when it is reflected. If the end is free to move, as in Fig. 16-10b, the reflected wave pulse is not inverted.

Fig. 16-11 shows a wave pulse in a changing medium, going from a lighter string to a heavier one. In this case the pulse is partially transmitted into the second medium and partially reflected back into the first medium. The reflected pulse is inverted when the second medium is denser than the first one. When the first medium is the denser one, the reflected wave is not inverted.

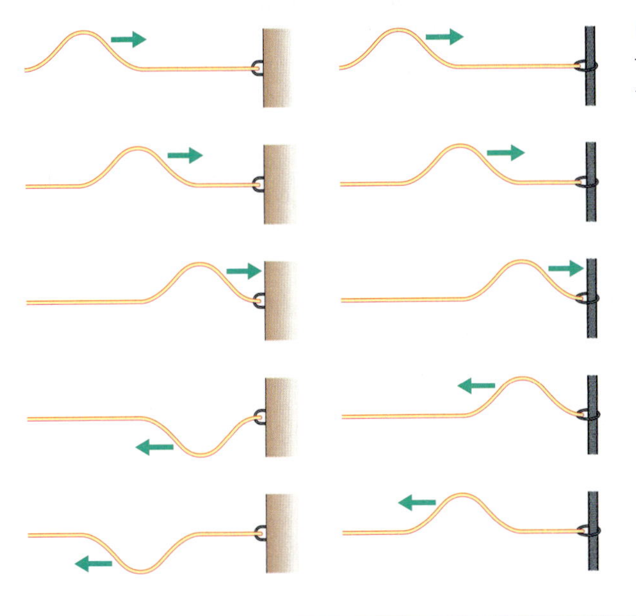

Fig. 16-10 When a wave pulse on a string reaches the string's end, the wave is **(a)** inverted if the end is fixed; **(b)** not inverted if the end is free.

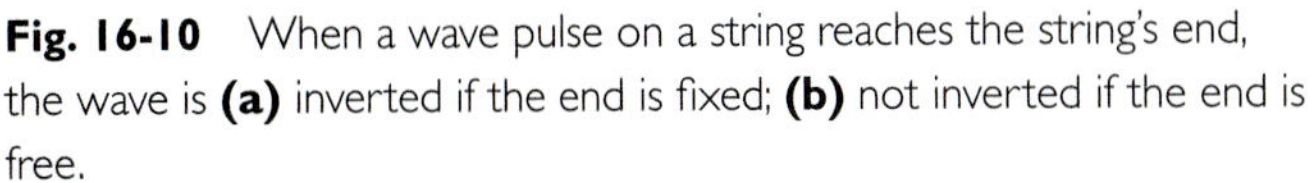

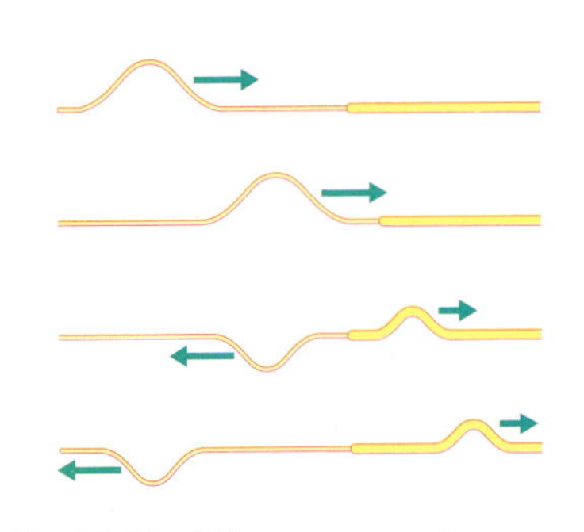

Fig. 16-11 When a wave pulse goes from one string to another of different density, the pulse is partially transmitted and partially reflected.

EXAMPLE 2 Ultrasound Imaging

Only sounds in the frequency range from about 20 Hz to about 20,000 Hz are audible to humans. Ultrasound is the name given to sound at frequencies above 20,000 Hz. Ultrasound can be used to produce images inside the human body (Fig. 16-12). Ultrasound waves penetrate the body, traveling at a speed of 1500 m/s, and are reflected from surfaces inside.

For a good ultrasonic picture having sufficient detail, the wavelength should be no greater than about 1.0 mm. Find the frequency of such an ultrasonic wave.

SOLUTION Solving Eq. 16-1 ($v = \lambda f$) for f, we obtain

$$f = \frac{v}{\lambda} = \frac{1500 \text{ m/s}}{1.0 \times 10^{-3} \text{ m}}$$

$$= 1.5 \times 10^{6} \text{ Hz} \quad (\text{or } 1.5 \text{ MHz})$$

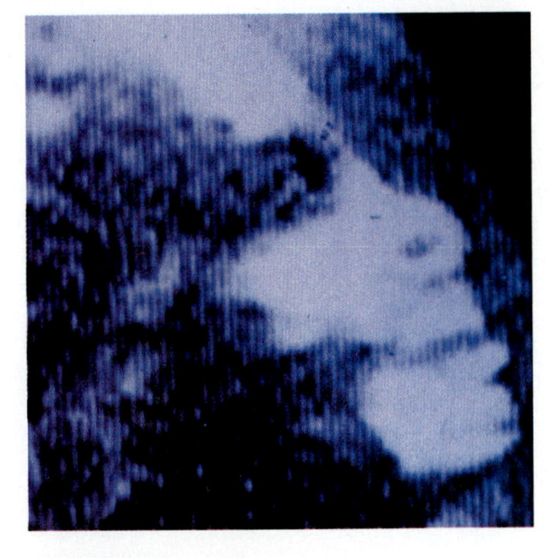

Fig. 16-12 Ultrasound image of a fetus.

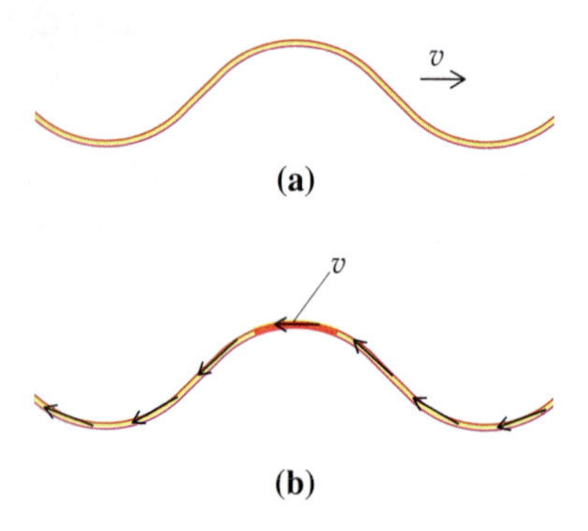

Fig. 16-13 **(a)** Viewed from the laboratory reference frame, a wave pulse moves to the right at constant speed v, while the string moves in the vertical direction as the pulse passes. **(b)** Viewed from a reference frame moving to the right at speed v with respect to the laboratory, the wave pulse appears stationary and the string has a horizontal component of velocity v to the left.

16-2 Wave Speed

In this section we shall discuss the speed of propagation of waves of various types and see which characteristics of the media determine the wave speed. Again, as stressed in the preceding section, it is important to understand that what we are considering here is not the speed of individual particles of the medium but rather the speed at which the wave form moves through the medium.

Speed of a Wave on a String

Consider a small-amplitude pulse transmitted along a horizontal string under a tension **F**. A stationary observer sees the pulse moving horizontally at a constant speed v and the string moving vertically but not horizontally (Fig. 16-13a).

If, however, the observer moves with the pulse at speed v, the pulse will appear to be stationary but the string will be moving horizontally as well as vertically. For the moving observer, the string has a horizontal velocity component v to the left (Fig. 16-13b). A point on the string that is instantaneously at the top of the wave form has *only* a horizontal component of velocity v. A small segment of the string centered on this point (shaded in the figure) follows an approximately circular path at speed v.

We can obtain an expression for the wave speed by analyzing the forces acting on this string segment of length ℓ and mass M, shown in an expanded form in Fig. 16-14. The two components of tension $F \sin \theta$ produce a resultant force in the radial direction, which, according to Newton's second law, produces centripetal acceleration v^2/R:

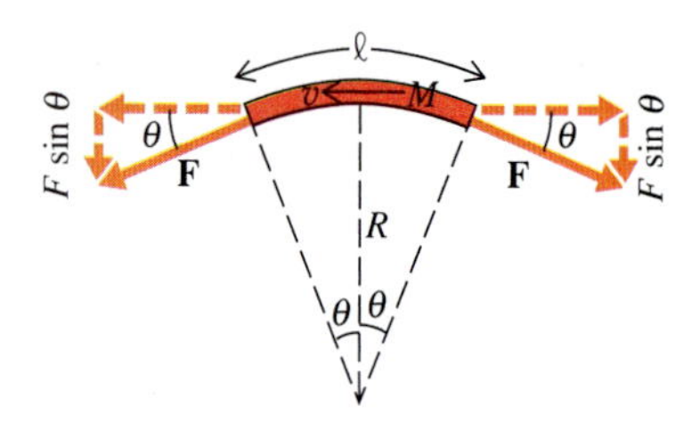

Fig. 16-14 An expanded view of the segment of string near the top of the pulse in Fig. 16-13.

$$2(F \sin \theta) = Ma = \frac{Mv^2}{R}$$

The angle θ is small, and so $\sin \theta$ may be approximated by θ (as in Chapter 15, Fig. 15-11). We relate θ to ℓ and R, using Fig. 16-14:

$$\sin \theta \approx \theta = \frac{\ell/2}{R}$$

Substituting this expression for $\sin \theta$ into the equation above, we obtain

$$2F\frac{\ell/2}{R} = M\frac{v^2}{R}$$

Solving for v, we find

$$v = \sqrt{\frac{F\ell}{M}}$$

This expression indicates that the wave speed depends only on tension and on the mass per unit length, which we shall denote by the Greek letter μ:

$$\mu = \frac{M}{\ell} \tag{16-2}$$

With this definition we may write v more concisely:

$$v = \sqrt{\frac{F}{\mu}} \tag{16-3}$$

This equation predicts that if we increase the tension in a string the wave speed increases and if we replace the string by one having greater mass the wave speed decreases.

EXAMPLE 3 Transmitting a Wave Pulse on a String

Two people hold opposite ends of a 10.0 m long rope having a mass of 1.00 kg. The person at one end gives the rope a small upward jerk. How long is it before the person at the other end feels the jerk, if the rope is held with a tension of (a) 40.0 N; (b) 10.0 N?

SOLUTION (a) We first apply Eq. 16-3 to find the wave speed on the rope, which has a mass density of (1.00 kg)/(10.0 m), or 0.100 kg/m:

$$v = \sqrt{\frac{F}{\mu}} = \sqrt{\frac{40.0 \text{ N}}{0.100 \text{ kg/m}}}$$
$$= 20.0 \text{ m/s}$$

At this speed the wave will travel the 10.0 m length of the rope in a time interval

$$\Delta t = \frac{x}{v} = \frac{10.0 \text{ m}}{20.0 \text{ m/s}}$$
$$= 0.500 \text{ s}$$

(b) When the tension is reduced to 10.0 N, the speed decreases and the time interval increases:

$$v = \sqrt{\frac{F}{\mu}} = \sqrt{\frac{10.0 \text{ N}}{0.100 \text{ kg/m}}}$$
$$= 10.0 \text{ m/s}$$

The pulse is now slower and so it takes longer to transmit:

$$\Delta t = \frac{x}{v} = \frac{10.0 \text{ m}}{10.0 \text{ m/s}} = 1.00 \text{ s}$$

Speed of Sound

Table 16-1 gives the speed of sound in various media. Notice that sound travels considerably faster in solids and liquids than in gases. Unlike gas molecules, the molecules of solids and liquids are in constant contact with their neighbors. Consequently, these molecules respond more quickly to a wave pulse than do gas molecules, which interact only through occasional collisions.

In Chapter 12 we found that the rms speed of molecules in an ideal gas is given by Eq. 12-16:

$$v_{rms} = \sqrt{\frac{3kT}{m}}$$

where k is Boltzmann's constant, T is the absolute temperature, and m is the mass of a molecule. It is possible to use the laws of mechanics to prove that the speed of sound in an ideal gas is proportional to v_{rms}. The exact result of this derivation is

$$v = \sqrt{\frac{\gamma}{3}}\, v_{rms}$$

where γ equals 1.40 for diatomic gases like nitrogen and oxygen. Thus

$$v = \sqrt{\frac{1.40kT}{m}} \quad \text{(For a diatomic ideal gas)} \quad (16\text{-}4)$$

The fact that wave speed is proportional to rms molecular speed in an ideal gas is certainly plausible, since the wave propagates as a result of the interaction of the gas molecules during collisions and the average time between collisions depends on the rms speed.

Table 16-1 Speed of sound

Medium	Speed (m/s)
Air (20° C)	344
Air (0° C)	332
Hydrogen (0° C)	1270
Water (20° C)	1480
Average body tissue (37° C)	1570
Aluminum	5100
Copper	3560
Iron	5130

EXAMPLE 4 The Wavelength of Sound Produced by a 512 Hz Tuning Fork

Suppose you strike a tuning fork that has a resonant frequency of 512 Hz. Find the speed and wavelength of the wave that propagates through the air if the air temperature is (a) 0° C; (b) 20.0° C.

SOLUTION **(a)** Air consists of approximately 80% nitrogen and 20% oxygen, and so the molecular mass of air is approximately $0.8(28) + 0.2(32) = 28.8$. This means that an average air molecule has a mass of 28.8 atomic mass units (u), where $1 \text{ u} = 1.66 \times 10^{-27}$ kg. Since both oxygen and nitrogen are diatomic molecules, we can apply Eq. 16-4:

$$v = \sqrt{\frac{1.40kT}{m}} = \sqrt{\frac{(1.40)(1.38 \times 10^{-23} \text{ J/K})(273 \text{ K})}{(28.8)(1.66 \times 10^{-27} \text{ kg})}}$$

$$= 332 \text{ m/s}$$

The wavelength is found when Eq. 16-1 ($v = \lambda f$) is applied:

$$\lambda = \frac{v}{f} = \frac{332 \text{ m/s}}{512 \text{ Hz}} = 0.648 \text{ m} = 64.8 \text{ cm}$$

(b) At 20.0° C, or 293 K, we find

$$v = \sqrt{\frac{1.40\,kT}{m}} = \sqrt{\frac{(1.40)(1.38 \times 10^{-23} \text{ J/K})(293 \text{ K})}{(28.8)(1.66 \times 10^{-27} \text{ kg})}}$$

$$= 344 \text{ m/s}$$

Sound travels somewhat faster through warmer air. Since f remains the same regardless of temperature, the wavelength increases:

$$\lambda = \frac{v}{f} = \frac{344 \text{ m/s}}{512 \text{ Hz}} = 0.672 \text{ m} = 67.2 \text{ cm}$$

Speed of Deep-Water Waves

Although water waves are easy to observe, they are relatively complex waves that are difficult to analyze in general. So we shall discuss only the special case of long-wavelength waves in deep water.

A wave is considered a **deep-water wave** if the depth of the water is much greater than the wavelength of the wave. In a deep-water wave each particle of water at the surface moves along a path in the vertical plane. If the wave is harmonic, a water particle's path is circular and a particle's speed is constant, as indicated in Fig. 16-15. As the wave propagates in a horizontal direction, the water undergoes both horizontal and vertical motion. Thus a water wave is neither transverse nor longitudinal. A body floating in the water will experience the same circular motion as the water. Suppose you are swimming in the ocean a good distance from shore. As a wave passes, you are first lifted, then pushed forward, then let down, and finally pushed back to your starting place, as indicated in the sequence of drawings in Fig. 16-15.

This circular motion is not limited to water at the surface; it extends to the water below, but the amplitude of the motion diminishes rapidly with depth. At a distance of 0.73λ below the surface, the amplitude is only 1% of the amplitude at the surface.

The speed of deep-water waves having wavelengths greater than about 10 cm is given by the following approximate expression:

$$v = \sqrt{\frac{g\lambda}{2\pi}} \quad \text{(deep-water waves; } \lambda \geq 10 \text{ cm)} \qquad \text{(16-5)}$$

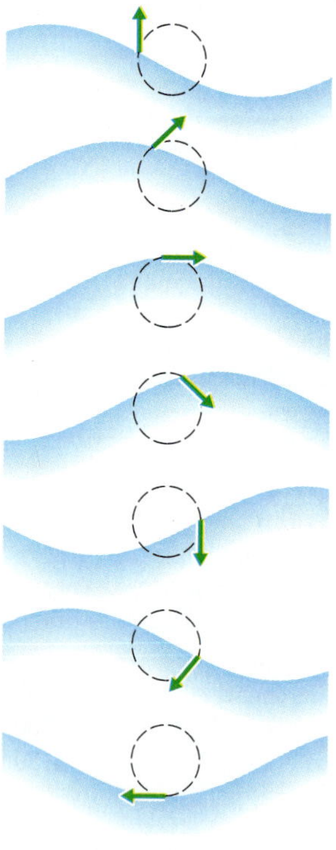

Fig. 16-15 As a deep-water wave passes, the water at the surface moves along a circular path whose radius is the amplitude of the wave.

One can derive this result by applying Bernoulli's equation (Problem 74). Notice that Eq. 16-5 implies that dispersion occurs for water waves; waves of different frequency (and therefore different wavelength) travel at different speeds.

EXAMPLE 5 Deep-Water Wave Speed

Find the speed of deep-water waves of wavelength (a) 2.00 m; (b) 5.00 m.

SOLUTION Applying Eq. 16-5, we find

(a)
$$v = \sqrt{\frac{g\lambda}{2\pi}} = \sqrt{\frac{(9.80 \text{ m/s}^2)(2.00 \text{ m})}{2\pi}} = 1.77 \text{ m/s}$$

(b)
$$v = \sqrt{\frac{(9.80 \text{ m/s}^2)(5.00 \text{ m})}{2\pi}} = 2.79 \text{ m/s}$$

16-3 Moving Sources and Observers: The Doppler Effect

Suppose you are standing beside a highway while cars pass by at high speed. As each car approaches, it produces a sound with a high frequency, or pitch. Just as the car passes, the pitch you hear drops significantly. Although the sound produced by the engine is unchanged, the frequency of the sound you hear is higher while the car is approaching than while it is moving away. This phenomenon is known as the **Doppler effect.** It occurs when either a source of sound or an observer of sound is in motion. Although the source produces a sound wave of a certain frequency f_S, the observed frequency f_O may be quite different.

Stationary Source, Moving Observer

Consider first the situation in which the source is stationary and the observer is moving. (All motion is measured relative to the medium of the sound wave.) The wavelength of the sound is determined by the frequency f_S of the source and the speed v of the sound through the medium:

$$\lambda = \frac{v}{f_S}$$

If the observer O is moving toward the source S at speed v_O as shown in Fig. 16-16, the observer passes more wavefronts per second than were emitted per second; that is, the observer hears a frequency that is greater than f_S. Since the wavefronts move *relative* to the observer at speed $v + v_O$, the observed frequency f_O will be this relative speed divided by λ:

$$f_O = \frac{v + v_O}{\lambda}$$

Substituting in this equation the value of λ from the preceding equation, we find

$$f_O = \frac{v + v_O}{v/f_S}$$

or

$$f_O = f_S\left(1 + \frac{v_O}{v}\right)$$

Next suppose that an observer is moving away from the source, as is observer O′ in Fig. 16-16. Now fewer wavefronts per second pass the observer; that is, a frequency lower than f_S is observed. The wavefronts move relative to the observer at a speed $v - v_O$. So the observed frequency is

$$f_O = \frac{v - v_O}{\lambda} = \frac{v - v_O}{v/f_S}$$

or

$$f_O = f_S\left(1 - \frac{v_O}{v}\right)$$

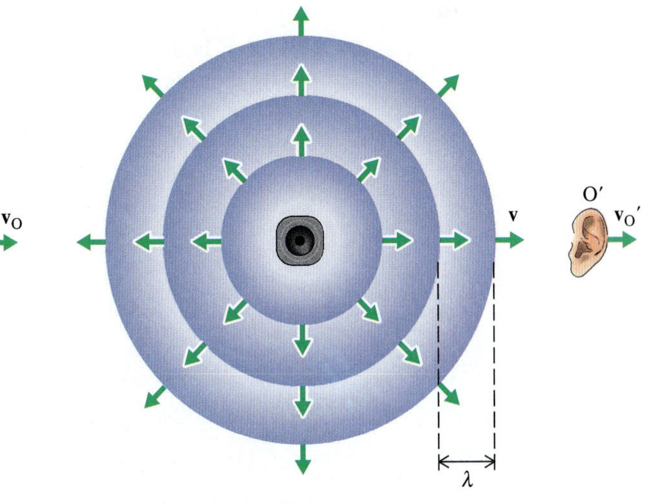

Fig. 16-16 Observers O and O′ hear sound produced by a stationary source. Observer O is moving toward the source at speed v_O, and observer O′ is moving away from the source at speed v_O'.

Summarizing the results for the observed frequency when the observer is moving either toward or away from the source, we have

$$f_O = f_S\left(1 \pm \frac{v_O}{v}\right) \quad \begin{array}{l}(+,\text{ observer moving toward source;}\\ -,\text{ observer moving away from source)}\end{array} \quad (16\text{-}6)$$

Moving Source, Stationary Observer

Next we consider a moving source and a stationary observer. Fig. 16-17 shows two wavefronts produced by a source that is moving to the right at constant velocity v_S. At $t = 0$, the source was at point S, at which instant it emitted the larger wavefront shown in the figure—a spherical surface centered at S. At $t = T = 1/f_S$, the source emitted from point S′ the smaller wavefront shown in the figure—a spherical surface centered at S′, a distance $v_S T$ to the right of S. Adjacent wavefronts in front of the moving source are squeezed together, whereas adjacent wavefronts behind the moving source are spread out. The observed wavelengths are respectively $\lambda - v_S T$ and $\lambda + v_S T$. Since the observer is at rest with respect to the medium, the relative speed of the wavefronts is just the speed v of waves through the medium. The observed frequency in front of the source is

$$f_O = \frac{v}{\lambda - v_S T}$$

where λ may be expressed as v/f_S:

$$f_O = \frac{v}{v/f_S - v_S/f_S}$$

or

$$f_O = \frac{f_S}{1 - v_S/v}$$

Behind the moving source, the observed frequency is $v/(\lambda + v_S T)$, and

$$f_O = \frac{f_S}{1 + v_S/v}$$

Summarizing these results for a moving source, we have

$$f_O = \frac{f_S}{1 \mp v_S/v} \quad \text{(}-\text{, source moving toward observer; } +\text{, source moving away from observer)} \qquad (16\text{-}7)$$

If both the observer and the source are moving through the medium, factors from both Eq. 16-6 and Eq. 16-7 are simultaneously present. We find in general

$$f_O = f_S\left(\frac{1 \pm v_O/v}{1 \mp v_S/v}\right) \quad \text{(upper signs if toward; lower signs if away)} \qquad (16\text{-}8)$$

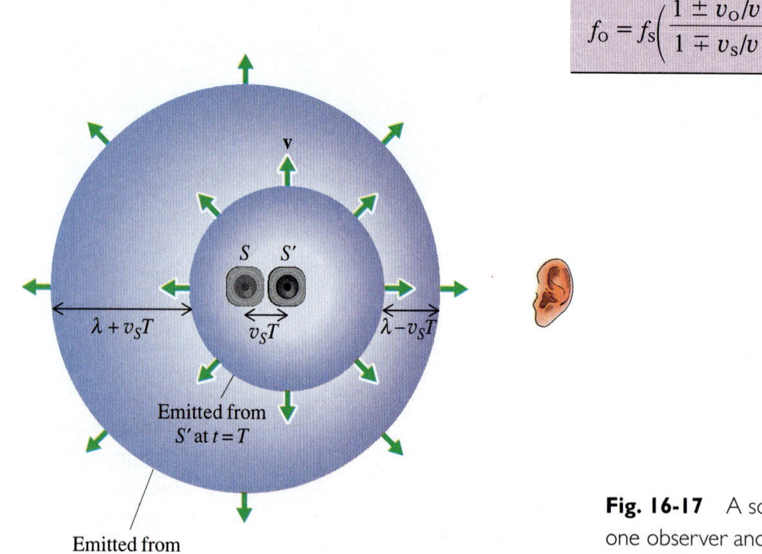

Fig. 16-17 A source of sound moves at speed v_S away from one observer and toward another observer. Wavefronts move at the speed of sound v.

EXAMPLE 6 Listening to the Whistle of an Approaching Train

A train travels parallel to a highway at a speed of 35.0 m/s. A car traveling on the highway at 30.0 m/s in the opposite direction is approaching the train. The driver of the car hears the train's whistle at a frequency of 650 Hz. (Fig. 16-18).
(a) What is the frequency of the whistle as heard on the train?
(b) After the train passes, what is the frequency of the whistle as heard by the car's driver? Use 344 m/s as the speed of sound.

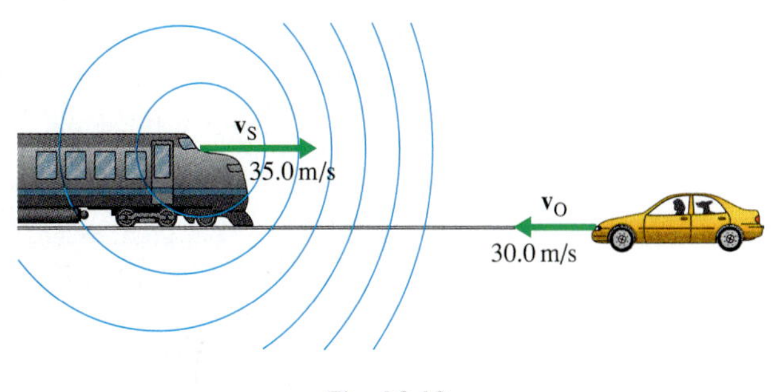

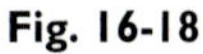

Fig. 16-18

SOLUTION (a) We apply Eq. 16-8 with the two upper signs, since the observer and the source are moving toward each other:

$$f_O = f_S\left(\frac{1 + v_O/v}{1 - v_S/v}\right)$$

$$650\text{ Hz} = f_S\left(\frac{1 + (30.0\text{ m/s})/(344\text{ m/s})}{1 - (35.0\text{ m/s})/(344\text{ m/s})}\right)$$

$$650\text{ Hz} = 1.21 f_S$$

or

$$f_S = \frac{650\text{ Hz}}{1.21} = 537\text{ Hz}$$

(b) After the source passes, observer and source are moving away from each other. So we use the lower signs in Eq. 16-8 to find the frequency heard by the driver:

$$f_O = f_S\left(\frac{1 - v_O/v}{1 + v_S/v}\right)$$

$$= (537\text{ Hz})\left(\frac{1 - (30.0\text{ m/s})/(344\text{ m/s})}{1 + (35.0\text{ m/s})/(344\text{ m/s})}\right)$$

$$= 445\text{ Hz}$$

As the train passes, the driver hears a dramatic drop in frequency, from 650 Hz to 445 Hz.

The Electromagnetic Doppler Effect

The Doppler effect is not limited to sound waves. All waves experience a similar effect. In particular, all electromagnetic waves, including visible light, undergo a Doppler shift when there is relative motion of observer and source. However, because of the unique nature of electromagnetic waves, the analysis of their Doppler shift is different from the analysis we used for sound waves.

Here we simply state without proof the relationship between the source frequency f_S and the observed frequency f_O. For electromagnetic waves, it is only the *relative* velocity of observer and source that counts. Denoting the relative speed by v and the speed of light by c, the following equation provides a good approximation to the observed frequency, so long as v is much less than c, where $c = 3.00 \times 10^8$ m/s:

$$f_O \approx f_S\left(1 \pm \frac{v}{c}\right) \quad (+,\text{ toward; } -,\text{ away}) \quad (v << c) \qquad (16\text{-}9)$$

The Doppler effect is important in astronomy, where it is used to determine the speed of a star emitting light that is observed at a frequency f_O shifted somewhat from the frequency f_S that would be emitted by the same kind of source if it were stationary.

The electromagnetic Doppler effect is also utilized in police radar units. Electromagnetic radiation of radar frequency is reflected from a moving car back to the radar unit. The speed of the car is determined when the Doppler shift of the radar frequency is observed.

Supersonic Speeds

So far we have considered only sources of sound moving at speeds less than the speed of sound, or at "subsonic" speeds. If the source moves at a "supersonic" speed, that is, faster than the speed of sound, a region of intense sound is created. This is known as a **sonic boom.** The effect is caused by the concentration of sound energy in a relatively small region of space. Wherever wavefronts are very close together, there is a concentration of energy. As these closely spaced wavefronts sweep past an observer, the observed intensity of the sound is quite large.

Fig. 16-19a shows a source moving to the right at a subsonic speed. Notice that there is some concentration of energy in front of the source. Fig. 16-19b shows a source moving slightly faster than the speed of sound, and Fig. 16-19c shows a source moving much faster than the speed of sound. In all three figures, the region of concentrated sound energy is inside the dashed lines. In Fig. 16-19c the concentrated region is along the surface of a cone.

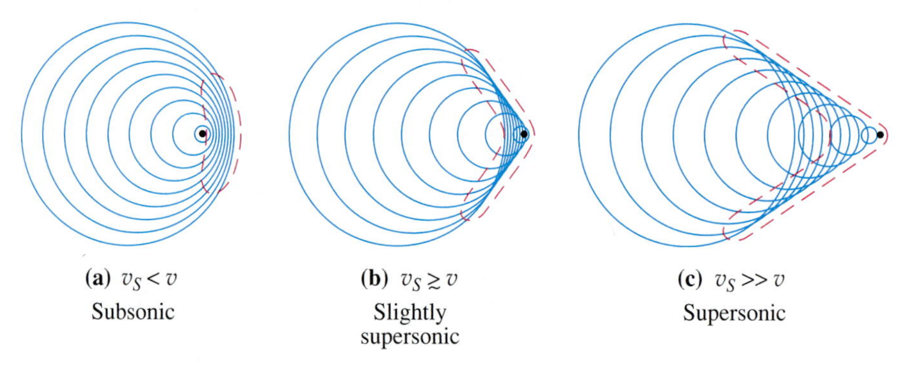

Fig. 16-19 A source of sound travels at a speed v_S **(a)** less than the speed of sound; **(b)** slightly greater than the speed of sound; **(c)** much greater than the speed of sound. Dashed lines show regions where energy is most concentrated.

Fig. 16-20 shows that the surface of this cone makes an angle θ with the direction of motion, where

$$\sin \theta = \frac{v}{v_S} \tag{16-10}$$

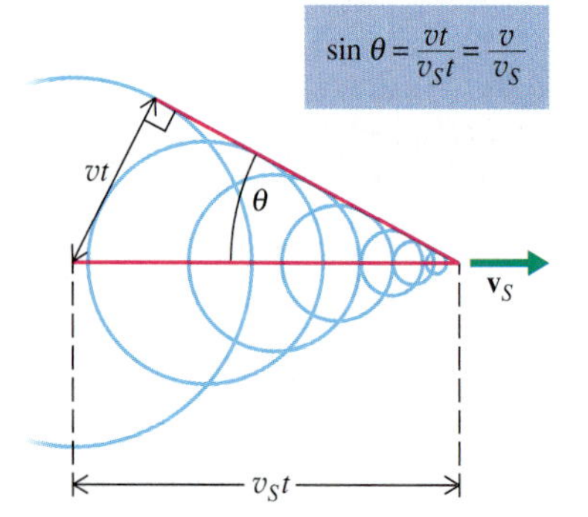

Fig. 16-20 When a source moves at a speed v_S greater than the speed of sound v, sound energy is concentrated along the surface of a cone. The surface makes an angle θ with the direction of motion.

The inverse of this ratio, v_S/v, is called the "mach number." For example, if an object moves at mach 2, its speed is twice the speed of sound, or about 680 m/s.

As the surface of the cone passes an observer, a loud sound of short duration is heard—a sonic boom. Notice that this sonic boom occurs not just as the "sound barrier" is broken, as many people erroneously believe. It is produced for as long as the source moves at supersonic speed, and anyone in the path of the conical edge of the wavefront will hear it. Fig. 16-21 shows a supersonic plane at equal time intervals. The surface of the cone sweeps along the ground at the same speed as the plane.

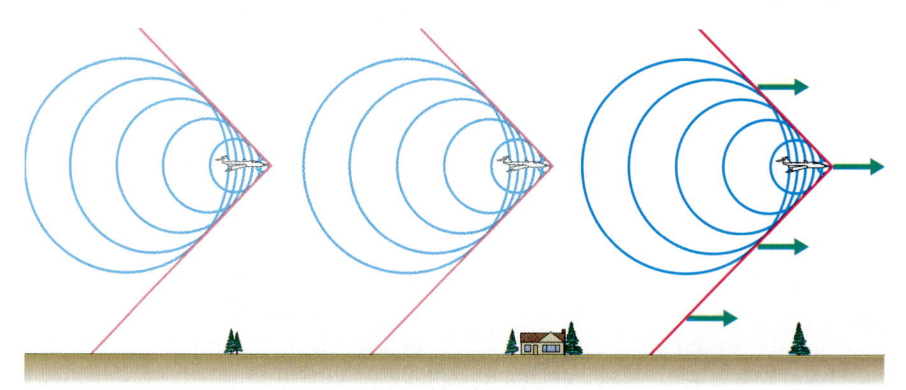

Fig. 16-21 A conical surface of concentrated sound energy sweeps over the ground as a supersonic plane passes overhead.

The crack of a whip is the shock wave, or sonic boom, produced by the tip of the whip moving at supersonic speed. The sound of a bullet is a similar effect. When a boat moves through water at a speed greater than the speed of wave propagation, the pattern of waves in its wake is a two-dimensional version of the wave pattern in a sonic boom (Fig. 16-22). The front edge of the wake is the region with the greatest concentration of energy.

Fig. 16-22 The passing of the leading edge of a boat's wake is the water wave equivalent of a sonic boom and is created when a boat moves through water at a speed greater than that of the waves it produces.

16-4 Power and Intensity; the Decibel Scale

Waves sometimes transmit large amounts of energy in short intervals of time. For example, when Hurricane Iniki hit Hawaii in 1992, the waves spawned by the hurricane carried enough energy to cause great destruction.

In this section we shall study the rate at which energy is transmitted in various waves, that is, the power carried by a wave (Fig. 16-23).

Fig. 16-23 Water waves sometimes carry large amounts of energy.

Harmonic Waves on a String

Consider first the power transmitted by a harmonic wave on a string. The power P is defined as the energy per unit time transmitted past a given particle on the string. In Fig. 16-24 a particle at point O is about to experience the passage of one cycle of a harmonic wave. At the end of one cycle, the wave form has passed O. So during a time interval equal to the period of the motion, all the energy contained in a segment of the string of length λ has been transmitted past O. The power transmitted equals the energy E contained in one wavelength divided by the period T, the time during which the energy passes:

$$P = \frac{E}{T}$$

The energy E in one wavelength can be calculated from the fact that the motion of each particle of the string is SHM, like the motion of a mass m on a spring of force constant k. Using Eq. 15-15 $\left(T = 2\pi\sqrt{\frac{m}{k}}\right)$, we can solve for k in terms of m and the period T:

$$k = \frac{4\pi^2 m}{T^2}$$

The mass m is the mass of one wavelength of the string. Using Eq. 16-2 $\left(\mu = \frac{m}{L} = \frac{m}{\lambda}\right)$, we find

$$m = \mu\lambda$$

We insert the two preceding equations into the expression for the total energy of a mass on a spring oscillating with amplitude A (Eq. 15-16: $E = \frac{1}{2}kA^2$) to obtain

$$E = \frac{2\pi^2\mu\lambda A^2}{T^2}$$

To find the power, we divide E by T:

$$P = \frac{E}{T} = \frac{2\pi^2\mu\lambda A^2}{T^3}$$

Substituting $T = \frac{1}{f}$ and $\lambda = \frac{v}{f}$, we obtain

$$P = 2\pi^2\mu v f^2 A^2 \qquad (16\text{-}11)$$

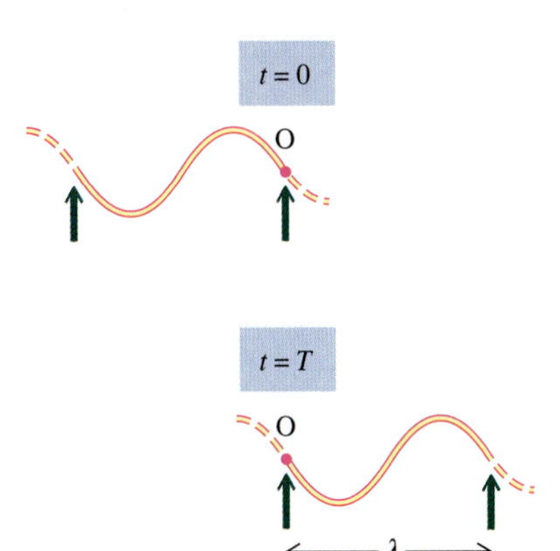

Fig. 16-24 The kinetic and potential energy contained in one wavelength of string at $t = 0$ is transmitted through the particle O in the time interval T.

EXAMPLE 7 Power Transmitted by a Wave on a String

A 10 Hz harmonic wave of amplitude 5.0 cm travels at 30 m/s along a string having a mass density of 0.020 kg/m. Find the power transmitted.

SOLUTION Applying Eq. 16-11, we find

$$P = 2\pi^2 \mu v f^2 A^2$$
$$= 2\pi^2 (0.020 \text{ kg/m})(30 \text{ m/s})(10 \text{ Hz})^2 (0.050 \text{ m})^2$$
$$= 3.0 \text{ W}$$

Harmonic Sound Waves

In contrast to a wave on a string, which transmits energy along a line, a sound wave spreads energy over a volume of space. We shall find an expression for the rate at which sound energy is transported from one region of space to another, passing through a cross-sectional area perpendicular to the direction of motion.

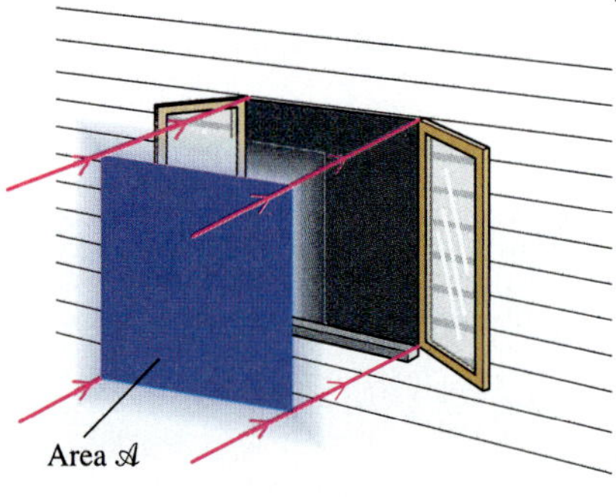

(a)

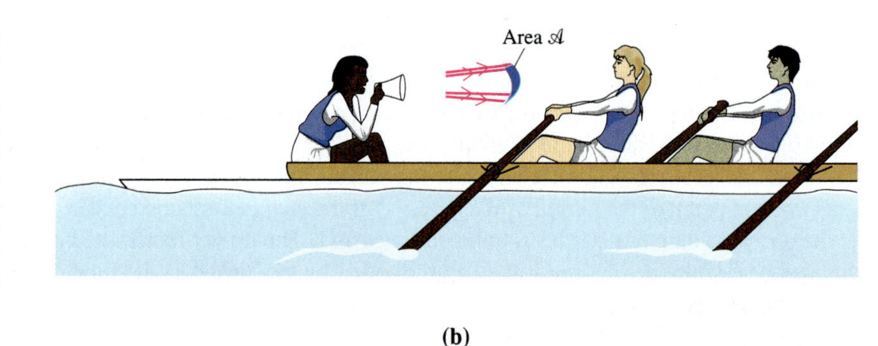

(b)

Fig. 16-25 **(a)** A plane wave passes through an open window that is oriented perpendicular to the direction of motion. Sound energy passes into the room through the window of area $\mathcal{A}$. **(b)** A spherical wave passes through a cross-sectional area $\mathcal{A}$, which is part of a spherical surface.

Motion of a sound wave through a surface area is illustrated in Fig. 16-25 for both plane and spherical waves. If we apply to a harmonic sound wave the same reasoning used to obtain an expression for the power carried by a harmonic wave on a string, we obtain the same result, Eq. 16-11:

$$P = 2\pi^2 \mu v f^2 A^2$$

where μ, the mass per unit length of the medium, is the product of the density ρ of the medium and the cross-sectional area $\mathcal{A}$, as shown in Fig. 16-26 for a plane wave:

$$\mu = \rho \mathcal{A}$$

Substituting this value for μ into the preceding equation, we obtain

$$P = 2\pi^2 \rho \mathcal{A} v f^2 A^2 \tag{16-12}$$

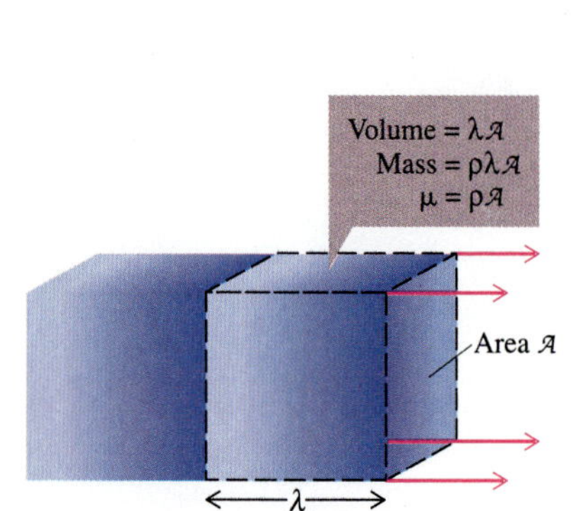

Fig. 16-26 Finding the mass per unit length of the medium for a sound wave

Often the concentration of power in a sound wave is of more importance than the total power; that is, we may wish to know how much power is transmitted per unit area of the wave. We define the power divided by the cross-sectional area perpendicular to the direction of wave motion to be the **intensity** I of the sound wave:

$$I = \frac{P}{\mathcal{A}} \tag{16-13}$$

Combining Eqs. 16-12 and 16-13, we obtain an expression for the intensity of a harmonic sound wave of amplitude A and frequency f, moving at speed v through a fluid of density ρ:

$$I = 2\pi^2 \rho v f^2 A^2 \quad (16\text{-}14)$$

The Decibel Scale

The loudness of a sound is related to its intensity. Normally the human ear is capable of hearing sounds of even very low intensity over a wide range of frequencies. The greatest sensitivity of the ear is at frequencies of a few thousand hertz. The most sensitive individuals can hear sounds having intensities as low as about 10^{-12} W/m^2 at a frequency of about 4000 Hz. We shall take this intensity as our reference level—the lowest sound intensity perceptible—and denote it by I_0:

$$I_0 = 1.00 \times 10^{-12}\ \text{W/m}^2 \quad (16\text{-}15)$$

The sounds we commonly hear are somewhere in the range I_0 to $10^{12}I_0$. When sound intensity reaches about 1 W/m^2, or $10^{12}I_0$, the sound becomes painful. This level of intensity is typical of the sound near the loudspeakers at a rock concert.

Two sounds that differ significantly in loudness will have intensities whose ratio will be some power of 10. For example, the rustling of leaves produces sound nearby of intensity roughly $10I_0$. A whisper, which is perceived as just a little louder than the leaves, might have an intensity of about 10^2I_0. It is therefore useful to define a logarithmic scale for measuring sounds. The **intensity level,** denoted by the Greek letter β (beta), is defined as

$$\beta = 10 \log\left(\frac{I}{I_0}\right) \quad (16\text{-}16)$$

Intensity level is a measure of the intensity of sound relative to the reference intensity I_0. The number β is expressed as decibels, abbreviated dB. The lowest audible sound has an intensity I_0 and an intensity level

$$\beta = 10 \log\left(\frac{I_0}{I_0}\right) = 10 \log 1 = 0$$

As mentioned above, the threshold of pain occurs at an intensity of 1 W/m^2, which corresponds to an intensity level

$$\beta = 10 \log\left(\frac{I}{I_0}\right) = 10 \log\left(\frac{1\ \text{W/m}^2}{10^{-12}\ \text{W/m}^2}\right)$$

$$= 10 \log 10^{12} = (10)(12)$$

$$= 120\ \text{dB}$$

Audibility curves in Fig. 16-27show the minimum audible intensity levels for individuals with very sensitive, average, and severely handicapped hearing. Notice that 1% of the population cannot hear sounds below about 70 dB.

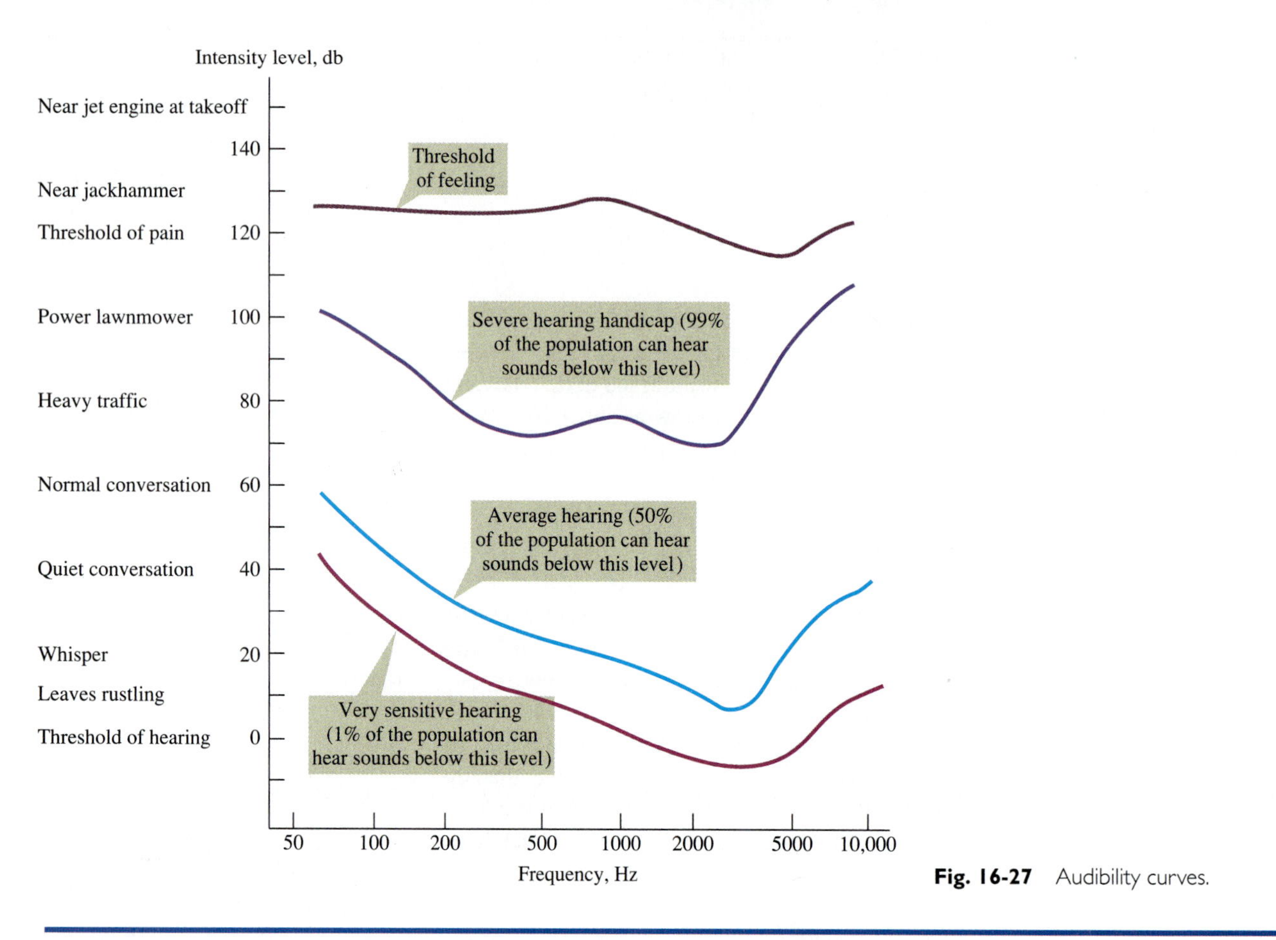

Fig. 16-27 Audibility curves.

EXAMPLE 8 Decibels Decrease as Distance from a Source of Sound Increases

A person 2.00 m away from you and talking in a normal voice produces 60.0 dB sound at your ear. (a) Assuming that the sound propagates equally in all directions, how much power is delivered by the person talking? (b) What is the intensity level at a distance of 6.00 m?

SOLUTION (a) Solving for I in Eq. 16-16:

$$\beta = 10 \log\left(\frac{I}{I_0}\right),$$

we find

$$I = I_0 10^{\beta/10}$$
$$= (1.00 \times 10^{-12}\ \text{W/m}^2)(10^{60.0/10})$$
$$= 1.00 \times 10^{-6}\ \text{W/m}^2$$

If we assume that the sound propagates equally in all directions, the intensity of sound is the same everywhere on a spherical surface of radius $r = 2.00$ m. The total power is spread over this surface and is equal to the product of the intensity and the surface area $\mathscr{A} = 4\pi r^2$:

$$P = I\mathscr{A} = I4\pi r^2$$
$$= (1.00 \times 10^{-6}\ \text{W/m}^2)(4\pi)(2.00\ \text{m})^2 = 5.03 \times 10^{-5}\ \text{W}$$

(b) At $r = 6.00$ m, the distance has tripled, and since $\mathscr{A} \propto r^2$, the surface area is 9 times as great as before. Thus the intensity is reduced by a factor of 9:

$$I' = \frac{I}{9} = \frac{1.00 \times 10^{-6}\ \text{W/m}^2}{9}$$
$$= 1.11 \times 10^{-7}\ \text{W/m}^2$$

And the new intensity level β' is found from the definition:

$$\beta' = 10 \log\left(\frac{I}{I_0}\right)$$
$$= 10 \log\left(\frac{1.11 \times 10^{-7}\ \text{W/m}^2}{1.00 \times 10^{-12}\ \text{W/m}^2}\right)$$
$$= 50.5\ \text{dB}$$

EXAMPLE 9 Small Amplitude and Large Amplitude Sound Waves

(a) Find the amplitude of vibration of air molecules at 20.0° C at a frequency of 1000 Hz for sounds of intensity level 0 dB and 160 dB. (b) Repeat the calculation for a frequency of 100 Hz.

SOLUTION (a) Solving for A in Eq. 16-14 ($I = 2\pi^2\rho v f^2 A^2$), we find

$$A = \sqrt{\frac{I}{2\pi^2\rho v f^2}}$$

A 0 dB sound corresponds to an intensity $I_0 = 1.00 \times 10^{-12}$ W/m². We find from Table 10-1 that the density of air at 20.0° C is 1.20 kg/m³ and from Table 16-1 that the speed of sound at this temperature is 344 m/s. Inserting these values into the equation above, we find

$$A = \sqrt{\frac{1.00 \times 10^{-12}\ \text{W/m}^2}{2\pi^2(1.20\ \text{kg/m}^3)(344\ \text{m/s})(1.00 \times 10^3\ \text{Hz})^2}}$$

$$= 1.11 \times 10^{-11}\ \text{m}$$

This is an extremely small displacement, about $\frac{1}{10}$ the diameter of the smallest atom, and yet the human ear is capable of hearing such sounds.

A 160 dB sound is 10^{16} times as intense as a 0 dB sound. Since $A \propto \sqrt{I}$, this means that the new amplitude A' is $\sqrt{10^{16}}$ times A:

$$A' = \sqrt{10^{16}}\,A = (10^8)(1.11 \times 10^{-11}\ \text{m})$$

$$= 1.11 \times 10^{-3}\ \text{m}$$

(b) Notice that the amplitude is proportional to $1/f$. Thus, when the frequency is reduced to 100 Hz, or one tenth of its former value, the amplitude is 10 times as great. At 0 dB, we have

$$A = (10.0)(1.11 \times 10^{-11}\ \text{m})$$

$$= 1.11 \times 10^{-10}\ \text{m}$$

And at 160 dB, we find

$$A = (10.0)(1.11 \times 10^{-3}\ \text{m})$$

$$= 1.11 \times 10^{-2}\ \text{m}$$

$$= 1.11\ \text{cm}$$

So for very low-frequency, high-intensity sounds, the displacement of air molecules is quite large. This effect can be dramatically demonstrated. A candle flame held near a loudspeaker will flicker whenever the loudspeaker emits very loud, low-frequency sounds (Fig. 16-28).

Fig. 16-28

16-5 Time Dependence of the Displacement of a Particle of the Medium

Next we shall obtain an expression for the time dependence of a wave's displacement at a given point. We shall consider only harmonic one-dimensional waves and harmonic plane waves, so that spatial variations occur only in one direction, which we take to be the x-axis. Suppose the wave source is located at the origin ($x = 0$) and oscillates in SHM. If we assume zero displacement when $t = 0$, the displacement y at the origin is given by Eq. 15-1:

$$y = A \sin\left(2\pi \frac{t}{T}\right)$$

At any point along the wave, the motion of any particle of the medium is of the same type, that is, SHM of period T and amplitude A. However, the particle motion at any point is delayed by the time it takes for the wave to travel to that point. If the wave travels in the positive x direction at speed v, the time delay $\Delta t = \frac{x}{v}$. So the particle motion at point x at time t is the same as the motion at the origin at time $t - \frac{x}{v}$, as given by the equation

$$y = A \sin\left(2\pi \frac{t - x/v}{T}\right) = A \sin\left[2\pi\left(\frac{t}{T} - \frac{x}{vT}\right)\right]$$

or, since $vT = \frac{v}{f} = \lambda$, we may express this result

$$y = A \sin\left[2\pi\left(\frac{t}{T} - \frac{x}{\lambda}\right)\right] \qquad (+x \text{ direction}) \qquad (16\text{-}17)$$

The following example illustrates how this expression represents wave motion along the positive x-axis. For motion along the negative x-axis, we replace v by $-v$ in the expression for the time delay ($t - x/v \rightarrow t - x/{-v} = t + x/v$), which leads to the expression

$$y = A \sin\left[2\pi\left(\frac{t}{T} + \frac{x}{\lambda}\right)\right] \qquad (-x \text{ direction}) \qquad (16\text{-}18)$$

EXAMPLE 10 Snapshots of a Waveform

A wave source at the origin oscillates in SHM at a frequency of 5.0 Hz and with an amplitude A. A plane wave travels in the $+x$ direction at a speed of 10 m/s. Find an expression for the displacement y at any point x at $t = 0$, $t = 0.025$ s, and $t = 0.050$ s. Graph y versus x for each of these times. (These three graphs correspond to three "snapshots" of the waveform at equal time intervals.)

SOLUTION First we must calculate T and λ and then apply Eq. 16-17:

$$T = \frac{1}{f} = \frac{1}{5.0\ \text{Hz}} = 0.20\ \text{s}$$

$$\lambda = \frac{v}{f} = \frac{10\ \text{m/s}}{5.0\ \text{Hz}} = 2.0\ \text{m}$$

$$y = A \sin\left[2\pi\left(\frac{t}{0.20} - \frac{x}{2}\right)\right]$$

where t is in s and x is in m.

At $t = 0$: $y = A \sin(-\pi x)$

At $t = 0.025$ s: $y = A \sin\left(\frac{\pi}{4} - \pi x\right)$

At $t = 0.050$ s: $y = A \sin\left(\frac{\pi}{2} - \pi x\right)$

The graphs of y as a function of x are given in Fig. 16-29. These graphs show a waveform moving to the right at a speed of 10 m/s.

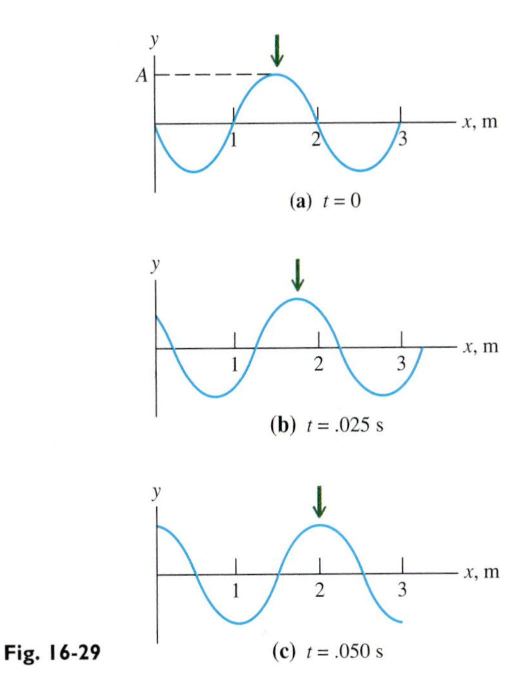

Fig. 16-29

16-6 Superposition of Waves; Beats; Standing Waves

Superposition Principle

When two (or more) waves move through a medium, the net effect on the medium is a wave whose displacement at any point is found by adding the two (or more) separate wave displacements. This principle, called the **superposition principle,** is illustrated in Fig. 16-30 by two pulses traveling in opposite directions along a spring. When they cross, what is seen is a displacement of the medium that equals the sum of the displacements of the two pulses.

The superposition principle applies to sound waves. For example, suppose that 10 people are in a room. If only one person is talking, the sound wave will have—at some time and at a given point in the room—a certain displacement y_1. If that person stops talking and a second person begins talking, there will be a displacement y_2 at that same point; a third person speaking alone would produce a displacement y_3 at the point, and so forth. When all 10 speak at once, the displacement y is the sum of the individual displacements: $y = y_1 + y_2 + \ldots + y_{10}$. (Despite the presence of all these waves producing an additive net effect, it is a remarkable fact that the human ear is often able to pick out individual voices in such a situation—at a party, for example.)

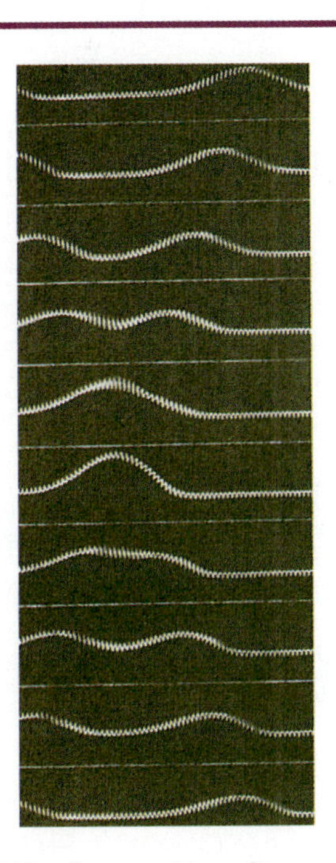

Fig. 16-30 Superposition of two wave pulses.

Constructive and Destructive Interference

Suppose that two harmonic waves of the same frequency both travel in the positive x direction through some medium. The waves may differ in amplitude and in initial phase angle. The two displacements are

$$y_1 = A_1 \sin\left[2\pi\left(\frac{t}{T} - \frac{x}{\lambda}\right)\right] \tag{16-19}$$

$$y_2 = A_2 \sin\left[2\pi\left(\frac{t}{T} - \frac{x}{\lambda}\right) + \phi\right] \tag{16-20}$$

There are two important special cases: $\phi = 0$ and $\phi = \pi$.

When $\phi = 0$, the two waves are said to be "in phase." We find the resulting wave displacement either graphically, as indicated in Fig. 16-31, or algebraically, by adding the two preceding expressions, with $\phi = 0$:

Fig. 16-31 Constructive interference.

$$\begin{aligned} y &= y_1 + y_2 \\ &= A_1 \sin\left[2\pi\left(\frac{t}{T} - \frac{x}{\lambda}\right)\right] + A_2 \sin\left[2\pi\left(\frac{t}{T} - \frac{x}{\lambda}\right)\right] \\ &= (A_1 + A_2) \sin\left[2\pi\left(\frac{t}{T} - \frac{x}{\lambda}\right)\right] \end{aligned}$$

The total displacement y has the same dependence on x as the two individual waves have. The total amplitude is the sum of the two amplitudes A_1 and A_2. This effect is called **constructive interference.** When $\phi = \pi$, that is, when the two waves are 180° out of phase, they tend to cancel each other. The effect, called **destructive interference,** is seen graphically in Fig. 16-32, or we can find it by adding Eqs. 16-19 and 16-20 after setting $\phi = \pi$ in Eq. 16-20:

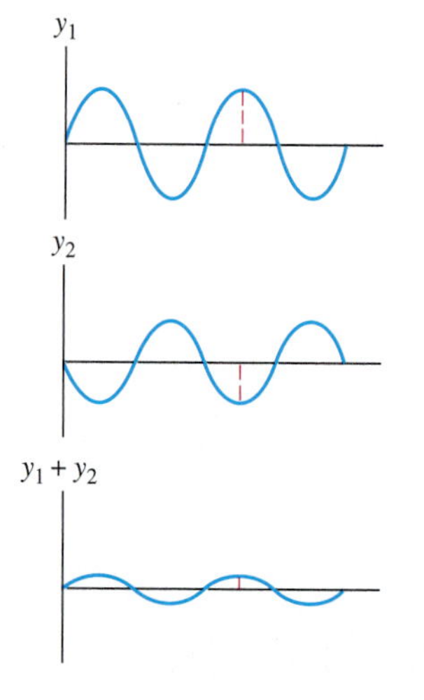

Fig. 16-32 Destructive interference.

$$y = y_1 + y_2 = A_1 \sin\left[2\pi\left(\frac{t}{T} - \frac{x}{\lambda}\right)\right] + A_2 \sin\left[2\pi\left(\frac{t}{T} - \frac{x}{\lambda}\right) + \pi\right]$$

But since $\sin(\theta + \pi) = -\sin\theta$, we may express this as

$$y = A_1 \sin\left[2\pi\left(\frac{t}{T} - \frac{x}{\lambda}\right)\right] - A_2 \sin\left[2\pi\left(\frac{t}{T} - \frac{x}{\lambda}\right)\right]$$

or

$$y = (A_1 - A_2) \sin\left[2\pi\left(\frac{t}{T} - \frac{x}{\lambda}\right)\right]$$

Here the amplitude is the difference in the two amplitudes A_1 and A_2. If A_1 and A_2 are equal, the destructive interference of these waves results in an amplitude of zero. In this case, the two waves cancel. The presence of these two waves results in no wave at all!

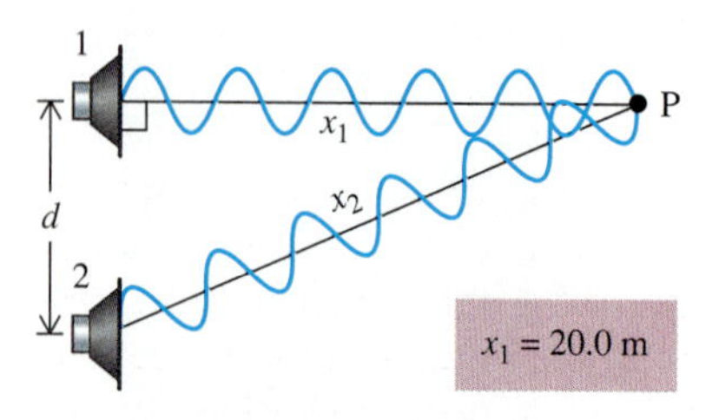

Fig. 16-33 Sound waves from two stereo speakers interfere destructively at P because sound from speaker 2 travels ½ wavelength farther than sound from speaker 1.

Difference in Path Length

Interference effects often arise when a phase difference results from different distances to the sources of the waves. Consider a point P located a distance x_1 from source 1 and a distance x_2 from source 2 (Fig. 16-33). We assume that each wave has the same amplitude A, the same frequency f, and the same initial phase angle $\phi_0 = 0$. Sources that either are in phase or have a constant phase difference are said to be **coherent.** Since the waves move through the same medium at the same speed and with the same frequency, it follows that they also have the same wavelength $\lambda = \frac{v}{f}$. The respective displacements at point P are

$$y_1 = A \sin\left[2\pi\left(\frac{t}{T} - \frac{x_1}{\lambda}\right)\right]$$

$$y_2 = A \sin\left[2\pi\left(\frac{t}{T} - \frac{x_2}{\lambda}\right)\right]$$

The difference in the phase angles is

$$\Delta\phi = 2\pi\frac{x_2}{\lambda} - 2\pi\frac{x_1}{\lambda}$$

or, letting $\Delta x = x_2 - x_1$,

$$\Delta\phi = 2\pi\frac{\Delta x}{\lambda} \qquad (16\text{-}21)$$

If there is no difference in the path length ($\Delta x = 0$), then $\Delta\phi = 0$ and we have constructive interference. Constructive interference can also occur if Δx equals any integer multiple of λ, for then the phase difference is a multiple of 2π, and this is equivalent to no phase difference because it leaves the sine unchanged [$\sin(\theta + 2\pi n) = \sin\theta$]. So constructive interference occurs when the path-length difference is an integer multiple of λ:

$$\Delta x = 0, \lambda, 2\lambda, 3\lambda, \ldots \qquad \text{(constructive interference)} \qquad (16\text{-}22)$$

Destructive interference occurs whenever $\Delta\phi = \pi$, or 3π, or 5π, ... From Eq. 16-21, we see that such phase differences result when Δx takes on the following values:

$$\Delta x = \frac{\lambda}{2}, \frac{3\lambda}{2}, \frac{5\lambda}{2}, \ldots \qquad \text{(destructive interference)} \qquad (16\text{-}23)$$

EXAMPLE 11 Stereo Speakers Producing Sounds That Cancel Each Other Out

Two stereo speakers are connected to an oscillator that causes the speakers to produce identical harmonic sound waves of wavelength 20.0 cm and frequency 1720 Hz. The two sources are coherent; that is, they oscillate in phase. Let P be a point 20.0 m from the first speaker, as shown in Fig. 16-33. How far from speaker 1 must speaker 2 be placed for there to be destructive interference at P?

SOLUTION For destructive interference to occur, the minimum spacing between the speakers must be such that there is a difference in the path lengths x_1 and x_2 equal to half a wavelength, as indicated in the figure:

$$x_2 - x_1 = \frac{\lambda}{2}$$

$$x_2 = x_1 + \frac{\lambda}{2} = 20.0\text{ m} + \frac{0.200\text{ m}}{2} = 20.1\text{ m}$$

Applying the Pythagorean theorem in Fig. 16-33, we find

$$d = \sqrt{x_2^2 - x_1^2} = \sqrt{(20.1\text{ m})^2 - (20.0\text{ m})^2} = 2.00\text{ m}$$

If the speakers are placed 2.00 m apart, there is no sound at P. A listener at P hears nothing. If the listener moves 1.00 m laterally from P so that she is equidistant from the two speakers, sound is heard. In this case the two waves interfere constructively, producing sound with an amplitude equal to twice the amplitude produced by one speaker alone. And since intensity is proportional to the square of the amplitude, the intensity of the sound is four times as great as the intensity produced by one speaker. A pattern of alternating constructive and destructive interference is produced in the entire region of space around the speakers. This two-source interference pattern has an analog in optics, which we shall study in Chapter 26.

Beats

When two waves have slightly different frequencies, interference at any point in space can alternate between constructive and destructive. Suppose, for example, you are tuning a guitar by comparing the frequencies of the sounds produced by two of its strings. When the frequencies are close but not equal, you hear beats—a sound that varies periodically in intensity—a throbbing or pulsing variation from loud to weak to loud. Fig. 16-34 shows the displacement of two harmonic waves of slightly different frequencies and periods (T_1 and T_2) as a function of time. The bottom of the figure shows the sum of the two waves, which represents the wave disturbance when both waves are simultaneously present. This resultant wave shows a periodic variation in amplitude.

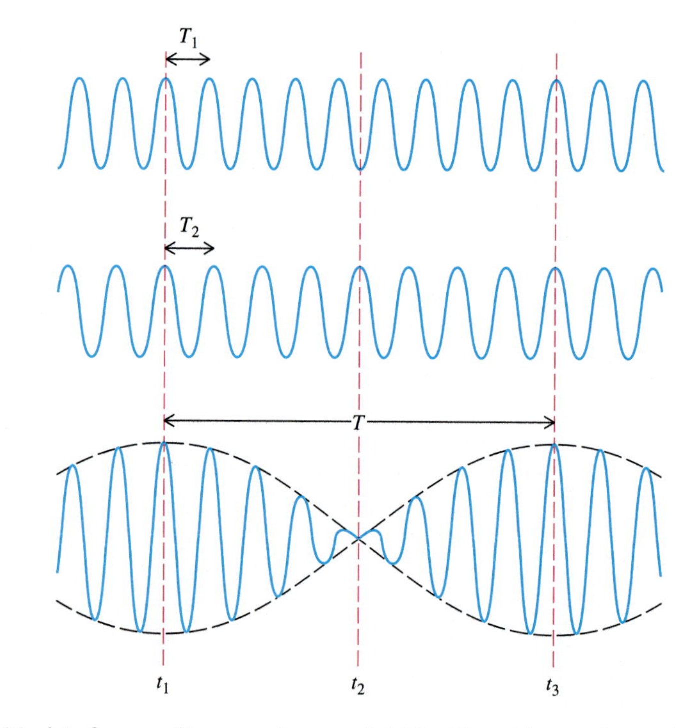

Fig. 16-34 Interference of two sound waves of slightly different frequencies produces beats—a sound that varies in intensity.

Notice that at time t_1 the waves interfere constructively but at time t_2 they interfere destructively. At time t_3 they interfere constructively again. The resultant wave's period $T = t_3 - t_1$ can be expressed in terms of the periods T_1 and T_2 of the interfering waves. Constructive interference occurs when both waves are at a peak. Each wave must undergo a number of complete cycles before constructive interference can occur again. The wave with a shorter period (T_1) must undergo one more cycle than the wave with the longer period (T_2). Thus

$$T = nT_2$$

and

$$T = (n + 1)T_1$$

Equating these expressions and solving for n, we find

$$n = \frac{T_1}{T_2 - T_1}$$

and, since $T = nT_2$,

$$T = \frac{T_1 T_2}{T_2 - T_1}$$

This result is better expressed in terms of the beat frequency f_B, the inverse of the beat period T. Taking the inverse of the equation above, we find

$$f_B = \frac{1}{T} = \frac{T_2 - T_1}{T_1 T_2} = \frac{1}{T_1} - \frac{1}{T_2}$$

or

$$f_B = f_1 - f_2 \qquad (16\text{-}24)$$

For example, if two guitar strings have frequencies of 440 Hz and 445 Hz, when both strings are plucked together, you hear beats at a beat frequency of 5 Hz; that is, the amplitude and intensity of the sound vary from loud to soft to loud with a period of $\frac{1}{5}$ second.

Standing Waves on a String

Suppose a harmonic wave is generated by an oscillator at the left end of a stretched string, with the right end held fixed. Fig. 16-35 shows the string at equal time intervals. The wave travels to the right, reaches the right end, and is reflected back. As soon as the reflected wave begins to propagate from the right end, it interferes with the incident wave. As indicated in Fig. 16-35, the superposition of the incident wave (dotted line) and the reflected wave (dashed line) produces, near the right end, a wave that is different from the original traveling wave (shaded part of figure).

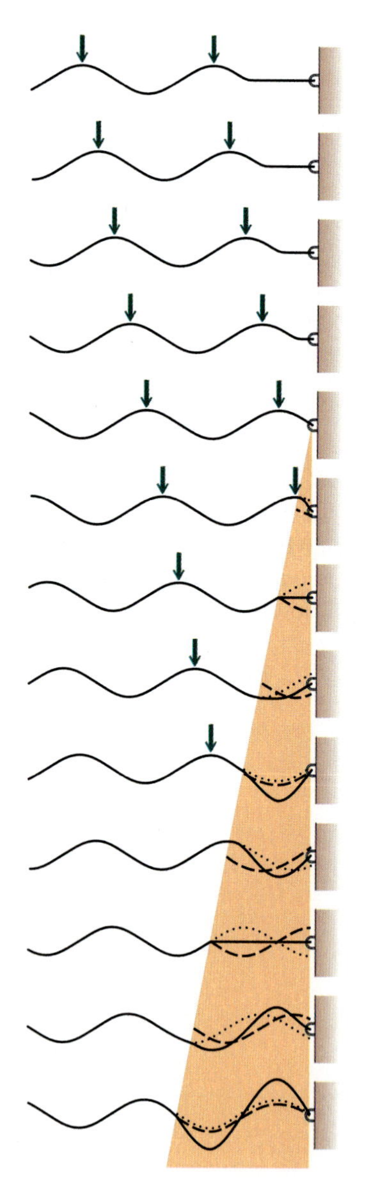

Fig. 16-35 A wave traveling to the right is reflected, and a standing wave begins to be formed at the right end of the string, indicated by shading. The string's displacement in this shaded region is a superposition of the original wave traveling to the right, *dotted line,* and a reflected wave traveling to the left, *dashed line.*

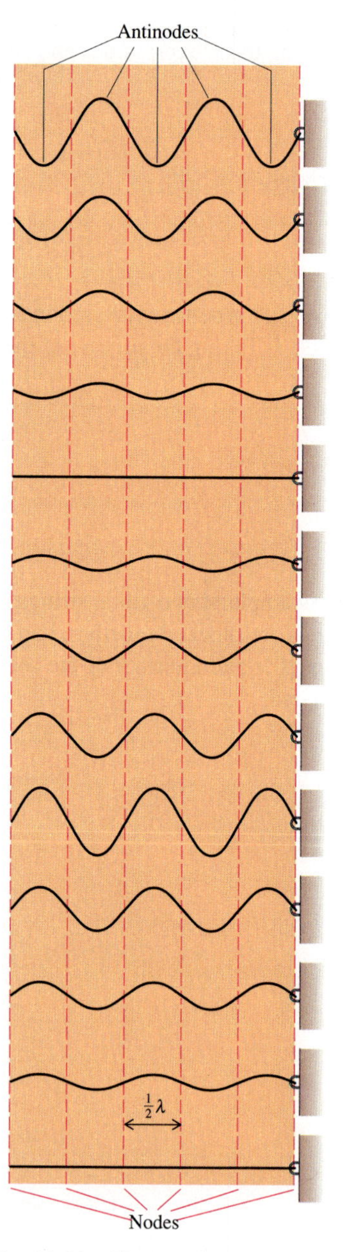

Fig. 16-36 The standing wave now extends over a larger region, indicated by the shading.

As the reflected wave travels farther to the left, the region of wave superposition grows (Fig. 16-36). At some points in the superposition region, the displacement is at times greater than the amplitude of the incident wave; at other points, *the displacement is always zero*. The points of **zero displacement** are called **nodes,** and the points of **maximum displacement** are called **antinodes.** The location of the nodes and antinodes is fixed, and the wave is therefore called a **standing wave.** The distance between adjacent nodes is $\frac{1}{2}$ wavelength, as indicated in Fig. 16-36:

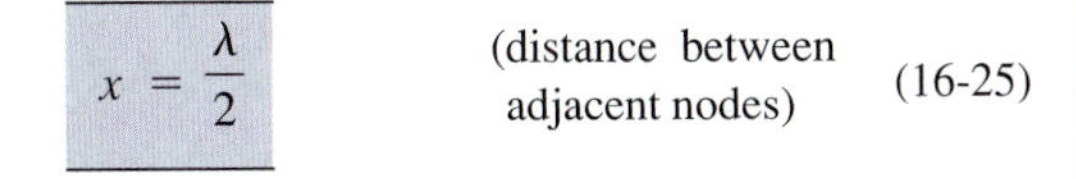

$$x = \frac{\lambda}{2} \qquad \text{(distance between adjacent nodes)} \qquad (16\text{-}25)$$

This is also the distance between adjacent antinodes, which are midway between the nodes.

The fact that standing waves result from the superposition of waves traveling in opposite directions can be shown algebraically. This is accomplished when we add the expressions $y_1 = A \sin\left[2\pi\left(\frac{t}{T} - \frac{x}{\lambda}\right)\right]$ and $y_2 = A \sin\left[2\pi\left(\frac{t}{T} + \frac{x}{\lambda}\right)\right]$, representing the displacements of the incident and reflected waves, respectively. See Problem 64.

Eventually the standing wave reaches the left end of the string (in the time it takes for the reflected wave to travel from the right end). When this happens, the wave is again reflected—this time to the right. If this second reflected wave happens to be in phase with the original wave, the standing wave pattern is reinforced—its amplitude increases (Fig. 16-37). If the second reflected wave is not in phase with the original wave, the standing-wave pattern is destroyed. The ends of the string are fixed.* So, to have a standing wave, the ends of the string must be at nodes of the standing wave. Since there is a node at every half wavelength, the string's length ℓ must be equal to an integral number n of half wavelengths:

$$\ell = \frac{n\lambda}{2}$$

This relationship between ℓ and λ may be stated as a condition on the values of the wavelength of standing waves that can be produced on a string of length ℓ. Solving the equation above for λ, we find

$$\lambda = \frac{2\ell}{n} \qquad (\text{for } n = 1, 2, 3, \ldots) \qquad (16\text{-}26)$$

that is,

$$\lambda = 2\ell, \ell, \frac{2\ell}{3}, \ldots$$

In order for a standing wave to be produced on a string of length ℓ, the oscillator must have a frequency f_n, found by inserting Eq. 16-26 into the equation $f = \frac{v}{\lambda}$:

$$f_n = n\frac{v}{2\ell} \qquad (\text{for } n = 1, 2, 3, \ldots) \qquad (16\text{-}27)$$

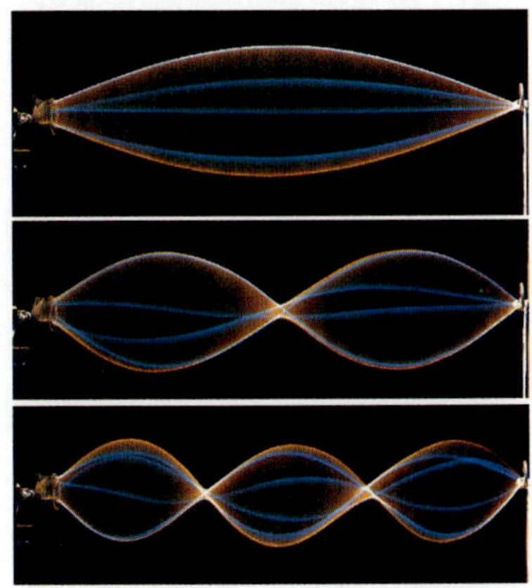

Fig. 16-37 A standing-wave pattern is formed when the length of the string is an integral number of half wavelengths; otherwise no standing wave is formed.

*The end of the string to which the oscillator is attached is not exactly fixed but oscillates with an amplitude that is small compared with the oscillation amplitude at an antinode. So the oscillation end of the string is close to being a node of the standing wave.

The oscillator is the source of energy. In an ideal system the amplitude would increase without limit as the waves are reflected at each end of the string. In practice, the amplitude does increase to the extent that the amplitude of oscillation at the antinodes is much greater than the amplitude of the oscillator. Hence the oscillator is close to a node of the system, and the oscillator end of the string can be considered a fixed end. If the oscillator is stopped, ideally the standing wave would continue indefinitely. In actuality, the wave quickly dies out. The energy delivered by the oscillator is necessary to compensate the energy loss during the reflection at the ends and the energy loss caused by air resistance.

If the frequency of the oscillator is not one of the standing-wave frequencies, the reflected waves will be out of phase at each reflection, and there will be destructive interference as often as constructive interference. The resulting wave will have an irregular shape and a small amplitude.

Resonant Frequencies; Harmonics

In Chapter 15 we described the phenomenon of resonance and introduced the concept of resonant frequency. When a body is forced to oscillate at one of its resonant frequencies, its amplitude of oscillation is large. The resonant frequencies of a vibrating string of length ℓ are the standing-wave frequencies given by Eq. 16-27 $\left(f_n = n\frac{v}{2\ell}, \text{ where } n = 1, 2, 3, \ldots\right)$. The lowest resonant frequency, called either the "fundamental frequency" or the "first harmonic," corresponds to setting $n = 1$ in this equation:

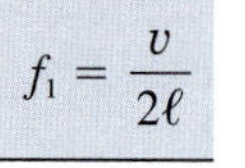

$$f_1 = \frac{v}{2\ell} \qquad \textbf{(fundamental frequency, or first harmonic)} \qquad (16\text{-}28)$$

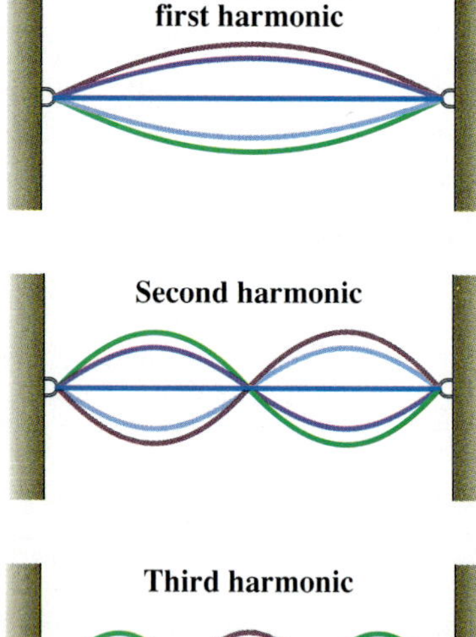

The frequency corresponding to $n = 2$ in Eq. 16-27 has twice the value of the first harmonic and is called the second harmonic:

$$f_2 = 2f_1 \qquad \textbf{(second harmonic)}$$

The third harmonic ($n = 3$) equals three times the fundamental f_1:

$$f_3 = 3f_1 \qquad \textbf{(third harmonic)}$$

The nth harmonic equals n times the fundamental frequency:

$$f_n = nf_1 \qquad \textbf{(\textit{n}th harmonic)} \qquad (16\text{-}29)$$

Fig. 16-38 shows the first three harmonics of a stretched string.

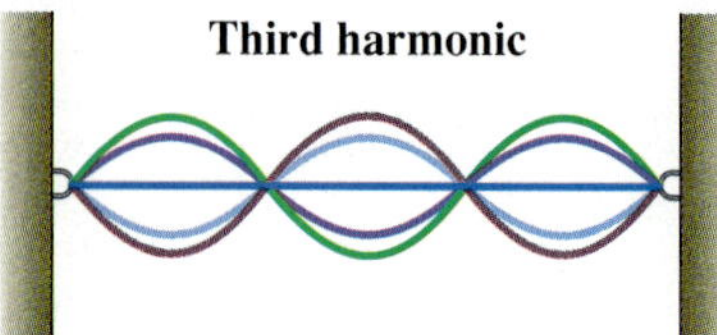

Fig. 16-38 Harmonics of a vibrating string.

EXAMPLE 12 The First Three Harmonics of a Vibrating String

A string of mass 2.00 g and length 1.00 m is fixed at one end and attached at the other end to an oscillator of variable frequency. The string is under a tension of 51.0 N. Find the three lowest oscillator frequencies for which standing waves will be formed.

SOLUTION Waves propagate along the string at a speed that may be found by applying Eq. 16-3:

$$v = \sqrt{\frac{F}{\mu}}$$

The mass density μ is 2.00 g/m, or 2.00×10^{-3} kg/m. Thus

$$v = \sqrt{\frac{51.0\ \text{N}}{2.00 \times 10^{-3}\ \text{kg/m}}} = 160\ \text{m/s}$$

Applying Eq. 16-28, we find the minimum oscillator frequency f_1 for the formation of a standing wave:

$$f_1 = \frac{v}{2\ell} = \frac{160\ \text{m/s}}{2(1.00\ \text{m})} = 80.0\ \text{Hz}$$

This is the lowest resonant frequency of the string, that is, its fundamental frequency, or first harmonic. The corresponding wavelength $\lambda_1 = \dfrac{v}{f_1} = \dfrac{160\ \text{m/s}}{80.0\ \text{Hz}} = 2.00$ m. This wavelength equals twice the length of the string ($\lambda = 2\ell$), as illustrated in the sketch of first harmonic vibrations in Fig. 16-38.

The two next-lowest standing-wave, or resonant, frequencies are the second harmonic f_2 (two times the fundamental) and the third harmonic f_3 (three times the fundamental):

$$f_2 = 2f_1 = 2(80.0\ \text{Hz}) = 160\ \text{Hz}$$

$$f_3 = 3f_1 = 3(80.0\ \text{Hz}) = 240\ \text{Hz}$$

Corresponding to the second harmonic is the wavelength $\lambda_2 = \dfrac{v}{f_2} = \dfrac{160\ \text{m/s}}{160\ \text{Hz}} = 1.00$ m, which equals the length of the string. Corresponding to the third harmonic is the wavelength $\lambda_3 = \dfrac{v}{f_3} = \dfrac{160\ \text{m/s}}{240\ \text{Hz}} = 0.667$ m, two thirds the length of the string. See Fig. 16-38.

Harmonic Analysis

Consider a wave displacement y that is a sum of two displacements y_1 and y_2, each representing a different resonant frequency of a string fixed at both ends. According to the superposition principle, y represents a possible wave on the string. For example, it is possible for the string to vibrate simultaneously at both its first and second harmonics, as illustrated in Fig. 16-39.

Conversely, one can consider any vibration of the string to be a superposition of its resonant frequencies. For example, if a stretched string is plucked at its center, the wave motion that results can be resolved into a sum of harmonics, each having its own amplitude.

A vibrating string will cause the surrounding air to vibrate at the same frequency, producing a sound wave. (If the string is part of a musical instrument, an air cavity, the "sound box," in the instrument enhances and amplifies the sound, forming standing sound waves, which we shall examine next.)

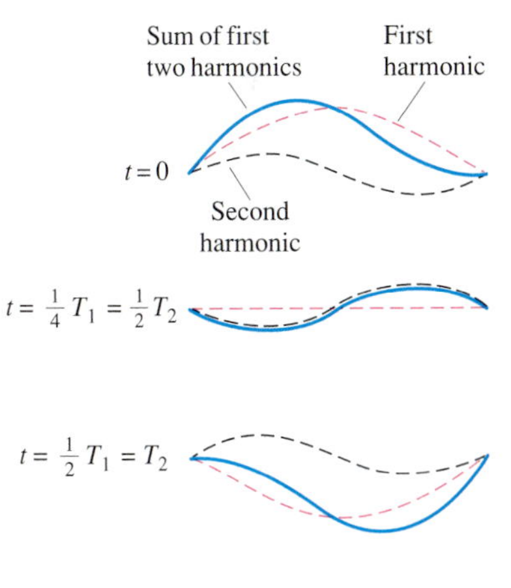

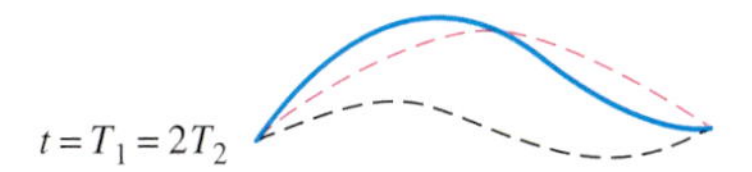

Fig. 16-39 A string vibrates at both its first and second harmonics.

EXAMPLE 13 The Right Tension for a Guitar's Thickest String

The thickest string on a guitar has a mass per unit length of 5.60×10^{-3} kg/m (Fig. 16-40). The string is stretched along the neck of the guitar and is free to vibrate between two fixed points 0.660 m apart. When plucked, this string vibrates at a fundamental frequency of 165 Hz and produces a sound wave of the same frequency, which corresponds to the musical note E below middle C. In order for the string to have this fundamental frequency, the tension in it must be adjusted to the right value. Find that value.

Fig. 16-40

SOLUTION We apply Eq. 16-28 $\left(f_1 = \dfrac{v}{2\ell}\right)$ for the fundamental frequency and Eq. 16-3 $\left(v = \sqrt{\dfrac{F}{\mu}}\right)$ for the speed of the wave on a string of mass per unit length μ, under tension F.

$$f_1 = \frac{v}{2\ell} = \frac{1}{2\ell}\sqrt{\frac{F}{\mu}}$$

Solving for F, we find

$$F = (2\ell f_1)^2\mu = [2(0.660\text{ m})(165\text{ Hz})]^2(5.60 \times 10^{-3}\text{ kg/m})$$
$$= 266\text{ N}$$

The same kind of analysis we have applied to a vibrating string can be applied to any other vibrating body. The vibrations can always be resolved into a linear sum of vibrations at certain resonant frequencies. The values of the resonant frequencies will depend on the particular body considered and will not in general be a simple multiple of the fundamental frequency. Consider, for example, the vibration of a drum. When struck, the drum's flexible membrane vibrates at various resonant frequencies. The fundamental frequency f_1 depends on the size of the drum—the larger the drum, the lower the fundamental frequency is. The next few resonant frequencies are approximately $1.59f_1$, $2.14f_1$, $2.30f_1$, $2.65f_1$. Because these are not integral multiples of f_1, they are not harmonics. This fact is responsible for the nonmelodious nature of the sound produced by a drum, in contrast to that produced by string or wind instruments. These latter instruments have resonant frequencies that are integral multiples of their fundamental frequencies, and so the sounds produced are pleasing to the ear, hence the name "harmonics."

A tuning fork is a device used to tune musical instruments. It is designed so that, when struck, it vibrates at essentially only one frequency; in other words, the amplitude of vibration of its fundamental frequency is much greater than the amplitude of any of its other resonant frequencies. The size of the tuning fork determines its frequency.

Standing Sound Waves

Suppose that a vibrating tuning fork is placed near the left end of a pipe that is open at both ends. A sound wave is generated, and sound can be heard at the right end of the pipe. The amplitude and loudness of the sound will depend on the frequency of the tuning fork and on the length of the pipe. The air in the pipe has certain resonant frequencies. If the tuning fork frequency is close to one of the pipe's resonant frequencies, the sound heard at the right end will have a relatively large amplitude. This situation is analogous to the formation of a standing wave on a string. The sound wave that originates at the left end of the pipe travels to the other end and is partially reflected; the part of the wave energy not reflected is transmitted out through the opening. The reflected sound wave then travels back to the left and is again partly reflected. This process is repeated. Unless the various reflected waves are all in phase, there is much destructive interference and only a small-amplitude wave results. This will be the case unless the frequency of the tuning fork is at a resonant frequency of the pipe. If the tuning fork is at a resonant frequency, the reflected waves are all in phase and a standing wave is formed.

Each of the pipe's open ends is a displacement antinode* since the air is free to move there. A standing wave will be formed only if the wavelength of the sound is related to the length of the pipe in such a way that antinodes are located at the ends (Fig. 16-41). Since the antinodes are a half wavelength apart, the pipe's length ℓ must be an integral number n of half wavelengths:

$$\ell = n\frac{\lambda}{2} \qquad (\text{for } n = 1, 2, 3, \ldots)$$

or

$$\lambda = \frac{2\ell}{n}$$

The resonant frequencies of the pipe are found when these values of λ are inserted into the equation $f = \dfrac{v}{\lambda}$:

$$f_n = n\frac{v}{2\ell} \qquad (\text{for } n = 1, 2, 3, \ldots) \qquad (16\text{-}30)$$

The lowest resonant frequency, that is, the **fundamental frequency,** or **first harmonic,** corresponds to setting $n = 1$ in the preceding equation:

$$f_1 = \frac{v}{2\ell} \qquad \text{(first harmonic of an open-ended pipe)} \qquad (16\text{-}31)$$

Second-, third-, and higher-order harmonics correspond to setting $n = 2$, $n = 3$, and so on in Eq. 16-30. These harmonics are multiples of the first harmonic:

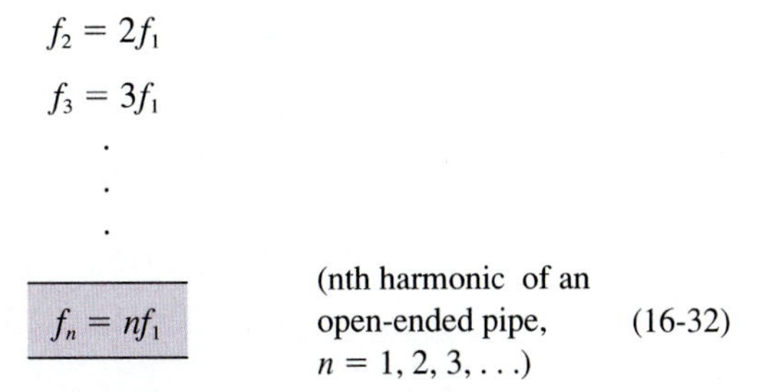

$$f_2 = 2f_1$$

$$f_3 = 3f_1$$

$$\vdots$$

$$f_n = nf_1 \qquad (\text{nth harmonic of an open-ended pipe, } n = 1, 2, 3, \ldots) \qquad (16\text{-}32)$$

The first three harmonics of an open-ended pipe are shown in Fig. 16-41.

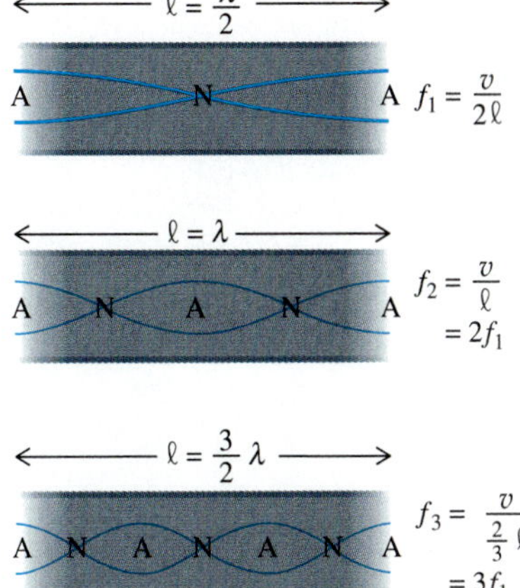

Fig. 16-41 The first three harmonics of sound in a pipe with both ends open.

*The antinode does not occur exactly at the pipe's end. But if the pipe's diameter is small compared with the wavelength of the sound (as it is in most musical instruments), an antinode is located close to an open end.

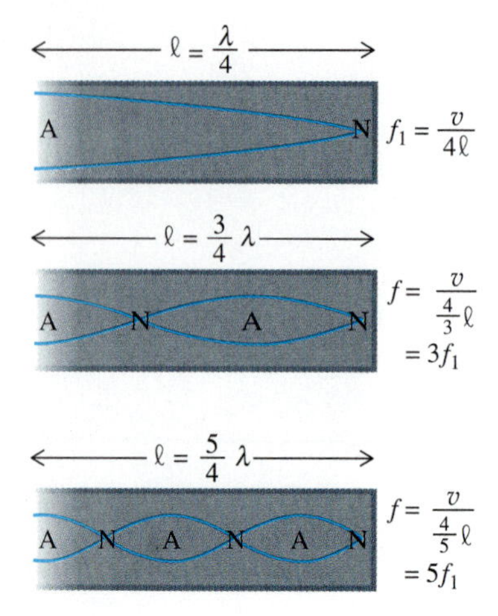

Fig. 16-42 The lowest three harmonics of sound in a pipe with one end closed. Notice that only odd harmonics are present.

A standing wave can also be formed in a pipe that is closed at one end. The closed end is a displacement node, since the air is not free to move there. The open end is a displacement antinode. For a standing wave to be produced, the sound's wavelength must be such that an antinode is formed at one end and a node at the opposite end. The longest wavelength that can satisfy this condition is one for which the pipe's length ℓ equals $\lambda/4$, or $\lambda = 4\ell$ (Fig. 16-42). The corresponding frequency, the pipe's lowest resonant frequency, the fundamental frequency, or first harmonic, is

$$f_1 = \frac{v}{\lambda}$$

or

$$f_1 = \frac{v}{4\ell} \quad \text{(first harmonic of a pipe closed at one end)} \qquad (16\text{-}33)$$

The second longest wave that can satisfy the condition of a node at one end and an antinode at the other end corresponds to $\ell = \frac{3}{4}\lambda$, or $\lambda = \frac{4}{3}\ell$ (Fig. 16-42). The corresponding frequency $\frac{v}{\lambda} = \frac{3}{4}\frac{v}{\ell}$, which is three times the fundamental frequency:

$$f_3 = 3f_1$$

The third longest wave corresponds to $\ell = \frac{4}{5}\lambda$, and a frequency $\frac{v}{\lambda} = \frac{5}{4}\frac{v}{\ell}$, which is five times the fundamental frequency f_1:

$$f_5 = 5f_1$$

In general, we have as our resonant frequencies all odd multiples of the fundamental frequency f_1:

$$f_n = nf_1 \quad \text{(nth harmonic of a pipe closed at one end, } n = 1, 3, 5, \ldots) \qquad (16\text{-}34)$$

No even harmonics are present for a pipe closed at one end.

Standing waves can also be produced in a pipe when a turbulent airflow is directed near one end (Fig. 16-43). This method is utilized in musical wind instruments, such as the organ and flute. In this case, the various resonant frequencies are produced simultaneously.

Fig. 16-43 A standing wave is produced by a turbulent flow of air through one end of a wind instrument such as an organ pipe.

EXAMPLE 14 Adjusting the Length of a Column of Air to Produce a Standing Wave

A tube is completely filled with water, and a tuning fork vibrating at 512 Hz is placed above it. The level of water in the tube is gradually reduced as water is drained from the bottom until a condition of resonance is reached, at which point the sound is loudest. Find the length ℓ of the air-filled cavity (Fig. 16-44).

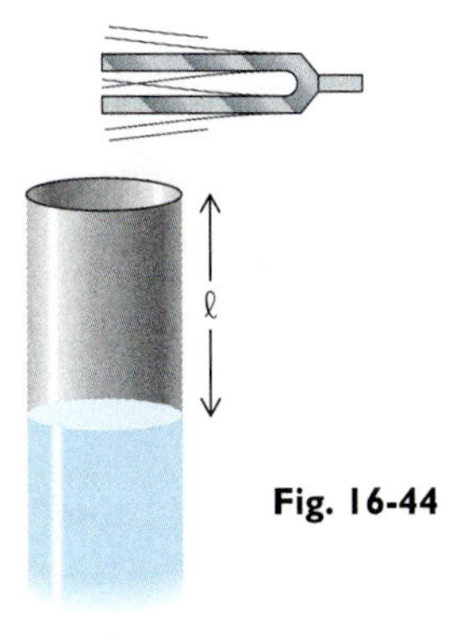

Fig. 16-44

SOLUTION What we are interested in here is the empty part of the tube. As we drain water out, we are in essence changing the length ℓ of a pipe closed at one end (the top of the water column causes the "pipe" to be closed at one end). Resonance occurs when the frequency of the tuning fork equals a resonant frequency of this "pipe" of length ℓ. The wavelength of the sound produced by the tuning fork is

$$\lambda = \frac{v}{f} = \frac{340\text{ m/s}}{512\text{ Hz}} = 0.67\text{ m} = 67.2\text{ cm}$$

From Fig. 16-41, we see that the shortest value of ℓ for resonance is $\frac{1}{4}$ wavelength. Thus $\ell = \frac{1}{4}\lambda = 17$ cm.

EXAMPLE 15 The Frequency of Sound From a Flute

All the holes of a flute of length 65.6 cm are covered, and a musical note is produced when the flute is blown into. The flute acts as a pipe open at both ends. Find the frequency of the sound produced.

SOLUTION We apply Eq. 16-31 to find the fundamental frequency of the flute, using the speed of sound at 20.0° C (344 m/s; Table 16-1):

$$f_1 = \frac{v}{2\ell} = \frac{344\text{ m/s}}{2(0.656\text{ m})} = 262\text{ Hz}$$

This is the frequency of the musical note middle C.

A string instrument, such as a guitar, violin, or piano, utilizes an air cavity to amplify the sound produced. Each instrument, of course, has its own characteristic sound. Two instruments can produce the same note and yet sound quite different from each other. The musical note is determined by the fundamental frequency of the sound; the differences are in the harmonic structure of the sounds. Different instruments produce different relative amplitudes of the harmonics when the same musical note is played. The human ear detects the harmonic structure of the sounds and identifies them with the respective instruments. See Fig. 16-45. This harmonic structure is determined by the size and shape of the air cavity, the structure of the sounding board, the characteristics of the string, and the interaction of these three factors, together with where and how the string is sounded—whether it is plucked, bowed, or hammered.

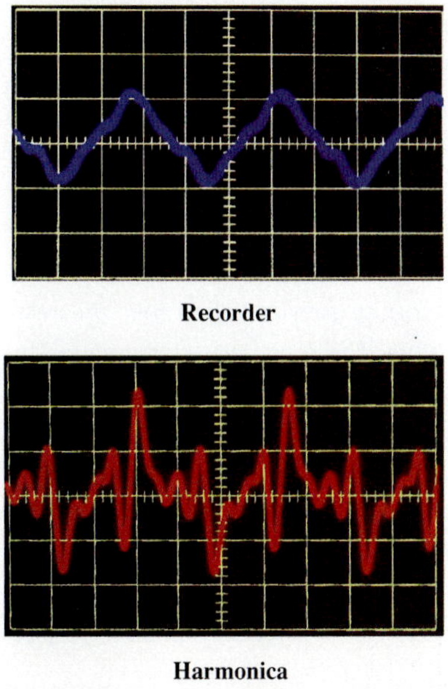

Recorder

Harmonica

Fig. 16-45 The same musical note A, $f_1 = 440$ Hz, is produced on two instruments. The sounds produced are quite different because of their different harmonic structures.

The Ear

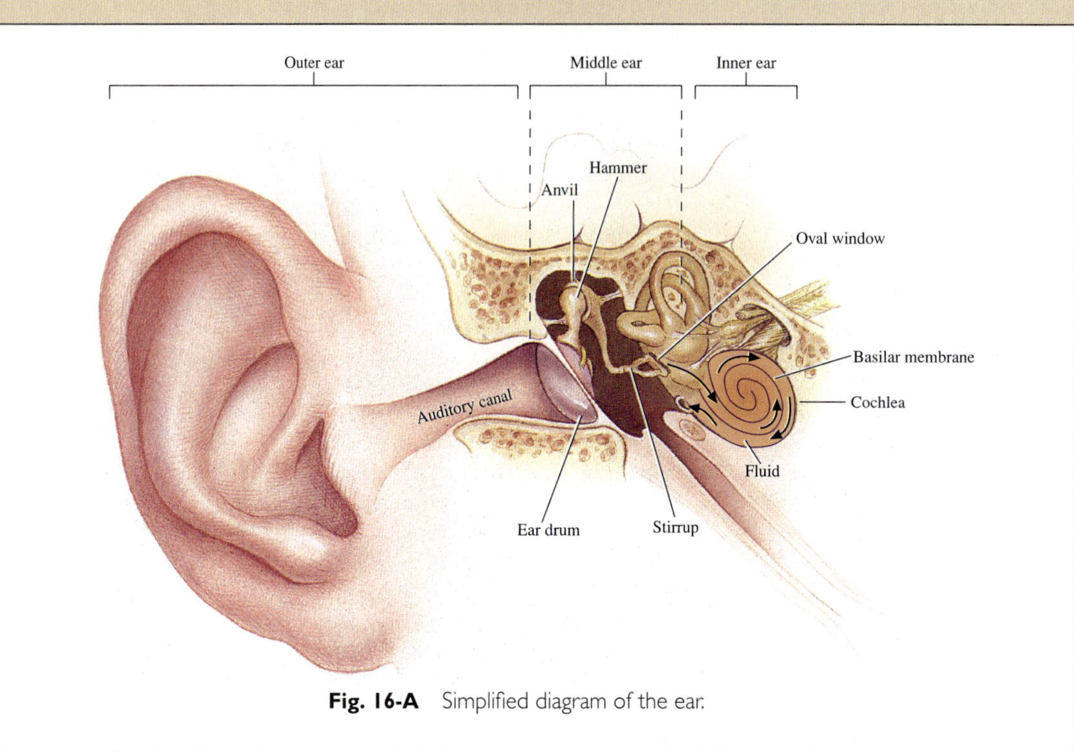

Fig. 16-A Simplified diagram of the ear.

The human ear is a remarkably sensitive detector of sounds. It is able to detect faint sounds with displacements smaller than an atomic diameter, as seen in Example 9. Furthermore, the ear is capable of harmonic analysis of the sounds it hears. For example, you can immediately recognize the difference between a guitar and a piano playing the same musical note. Somehow your ear senses the harmonic components of the two sounds, and, utilizing your memory of the harmonics of each instrument, you identify one source as a guitar and the other as a piano. Similarly, you are usually able to recognize familiar voices over the telephone. The sound waves transmitted by the telephone receiver are completely determined by their harmonic components. Thus recognition of voices consists of some kind of harmonic analysis by the ear and brain.

The 3 cm long auditory canal in the outer ear is essentially an air-filled pipe closed at one end. Its fundamental frequency corresponds to a wavelength $\lambda = 4\ell = 4(3 \text{ cm}) = 12$ cm, or a frequency $f = \frac{v}{\lambda} = \frac{340 \text{ m/s}}{0.12 \text{ m}} = 2800$ Hz. The maximum sensitivity of human hearing occurs at about this frequency (roughly that of a baby's cry). Thus the length of the auditory canal explains why the ear should be particularly sensitive to frequencies close to 2800 Hz. However, this factor alone cannot explain the broad range of frequencies to which the ear can be highly sensitive (roughly 20 Hz to 20,000 Hz for 1% of the population). The most important frequency-dependent effects occur in the cochlea of the inner ear, shown in Fig. 16-A.

Connecting the cochlea's oval window to the eardrum at the end of the auditory canal are the three bones of the middle ear (hammer, anvil, and stirrup). These bones serve as a system of levers that roughly doubles the force of vibrations transmitted from the eardrum to the oval window. Since the area of the oval window is much less than the area of the eardrum, the pressure of sound waves is greatly increased as the wave passes into the viscous fluid within the cochlea. Sound waves propagate through that fluid from the oval window to the tip of the cochlea, returning along the other side of the basilar membrane, as indicated by arrows in the figure. Pressure differences between the fluids on either side of the basilar membrane displace the membrane laterally. Tens of thousands of nerve endings along the membrane sense this displacement. These nerve endings then transmit electrical pulses to the brain, and we hear. Information about the harmonics present in the sound wave are provided by the location of the stimulated nerve endings along the basilar membrane and by the rate of transmission of pulses.

CHAPTER 16 SUMMARY

Mechanical waves, such as sound, water waves, and waves on a string, require a material medium through which the wave travels. Electromagnetic waves, including light, require no medium; they can travel through a vacuum.

For a continuous, periodic wave (as opposed to a wave pulse) of frequency f, the motion of the medium at each point in the wave is a periodic disturbance that has the same frequency f. This frequency is determined by the wave source.

The distance between two successive identical points on a wave is called the "wavelength," denoted by λ. The speed v at which the wave energy and wave form move is determined by the medium. The speed, wavelength, and frequency of any wave are related by the equation

$$v = \lambda f$$

A wave on a string of mass per unit length μ, under a tension F, travels at a speed

$$v = \sqrt{\frac{F}{\mu}}$$

Sound travels through a diatomic, ideal gas at a speed

$$v = \sqrt{\frac{1.40kT}{m}}$$

where k is Boltzmann's constant, T is absolute temperature, and m is the molecular mass.

Water waves having wavelengths that are greater than 10 cm but much less than the depth of the water travel at a speed

$$v = \sqrt{\frac{g\lambda}{2\pi}}$$

The Doppler effect is the phenomenon that occurs when either the source of a wave or an observer moves relative to the medium through which the wave travels. The result is that the observed frequency f_O is different from the source frequency f_S. For mechanical waves, the two frequencies are related by the equation

$$f_O = f_S\left(\frac{1 \pm v_O/v}{1 \mp v_S/v}\right) \quad \text{(upper signs if toward; lower signs if away)}$$

where the upper sign in both numerator and denominator applies when the motion of either the observer or the source is toward the other and the lower sign applies when either the observer or the source moves away from the other.

The power transmitted by a harmonic wave on a string is

$$P = 2\pi^2 \mu v f^2 A^2$$

where μ is the string's mass per unit length, v is the wave speed, f is the frequency, and A is the amplitude. The intensity of a harmonic sound wave is given by a similar expression (with the mass density ρ replacing μ):

$$I = 2\pi^2 \rho v f^2 A^2$$

where intensity is defined as the power per cross-sectional area $\mathscr{A}$:

$$I = \frac{P}{\mathscr{A}}$$

The intensity level β of a sound wave, measured in decibels, is defined as

$$\beta = 10 \log\left(\frac{I}{I_0}\right)$$

where the reference level $I_0 = 10^{-12}$ W/m^2 is the lowest sound intensity perceptible by individuals with normal hearing.

The wave displacement y in a harmonic wave traveling in the positive x direction is expressed in terms of position x and time t as

$$y = A \sin\left[2\pi\left(\frac{t}{T} - \frac{x}{\lambda}\right)\right]$$

where A is the amplitude of the wave, T is the period, and λ is the wavelength. For a wave traveling in the negative x direction, the corresponding expression is

$$y = A \sin\left[2\pi\left(\frac{t}{T} + \frac{x}{\lambda}\right)\right]$$

The superposition principle states that when two or more wave disturbances are present in a medium the displacement at any point is the sum of the individual displacements.

When two sources produce waves that either are in phase or have a constant phase difference, the sources are said to be coherent. Coherent wave sources can produce observable interference effects at various points in space.

When the waves at a given point are in phase, they interfere constructively, and the resulting amplitude is maximum—the sum of the two amplitudes. When waves are 180° out of phase, they interfere destructively, and the amplitude is the difference of the two amplitudes—equal to zero if the amplitudes are equal.

In general, the phase difference $\Delta\phi$ between two waves, arising from a path-length difference Δx, is

$$\Delta\phi = 2\pi \frac{\Delta x}{\lambda}$$

Constructive interference occurs when

$$\Delta x = 0, \lambda, 2\lambda, 3\lambda, \ldots$$

Destructive interference occurs when

$$\Delta x = \frac{\lambda}{2}, \frac{3\lambda}{2}, \frac{5\lambda}{2}, \ldots$$

Two waves of slightly different frequencies f_1 and f_2 produce beats with a beat frequency $f_B = f_1 - f_2$.

Standing waves are waves in which, at certain points, called "nodes," the displacement is always zero, whereas at other points, called "antinodes," the displacement is maximum. Standing waves are formed when waves are reflected from the boundaries of the wave medium. The distance Δx between adjacent nodes in any standing wave equals $\frac{1}{2}$ wavelength:

$$\Delta x = \frac{\lambda}{2}$$

For a standing wave on a string having both ends fixed, the two ends are located at nodes of the standing wave, and so the length is a multiple of $\frac{1}{2}$ wavelength:

$$\ell = \frac{n\lambda}{2}$$

or

$$\lambda = \frac{2\ell}{n} \qquad (\text{for } n = 1, 2, 3, \ldots)$$

A wave having such a wavelength will be produced if the frequency of the source is appropriate for the wave's speed v and length ℓ:

$$f = \frac{v}{\lambda}$$

with λ given above, or

$$f_n = n\frac{v}{2\ell} \qquad (\text{for } n = 1, 2, 3, \ldots)$$

These frequencies are the resonant frequencies of the string. The lowest resonant frequency f_1 is called the "fundamental frequency," or the "first harmonic," f_2 is called the "second harmonic," f_3 is the "third harmonic," and so forth.

For a standing sound wave in a pipe with both ends open, each end is an antinode, and the harmonic frequencies are the same as for a string:

$$f_n = n\frac{v}{2\ell} \qquad (\text{for } n = 1, 2, 3, \ldots)$$

In a pipe open at one end and closed at the other, there is a node at the closed end and an antinode at the open end; the harmonics, which are lower than for an open pipe of the same length, are given by

$$f_n = n\frac{v}{4\ell} \qquad (\text{for } n = 1, 3, 5, \ldots, \text{ odd only})$$

Any standing wave can be regarded as a superposition of various harmonics.

Questions

1 Suppose you throw someone a ball, thereby transporting kinetic energy. Could you consider the motion of the ball to be a mechanical wave pulse?

2 Which quantities determine the speed of a particle in a medium through which a wave propagates: (a) wave speed; (b) wavelength and frequency; (c) amplitude and frequency; (d) wave speed and period?

3 Would it be possible to detect on earth sounds produced on another planet if you had a detector sensitive enough to very low-intensity sounds?

4 Suppose you were standing on the moon as a nearby lunar landing module approached the surface, firing its engines to slow its descent. Could you hear the engine?

5 When cars begin to move in a long line of stalled traffic, the motion passes through the line as a wave pulse.
(a) What is the direction of motion of the pulse relative to the motion of the cars?
(b) Is the wave speed affected by the drivers' reaction times?

6 Creating a "wave" is a popular pastime at football games and other sporting events. The wave begins when everyone in one section of the stadium quickly stands up and then sits down. Then people sitting in an adjacent section, say, to the right, respond with the same motion but delayed a bit. Next the section to their right follows. The result is a wave pulse through the spectators. Does human reaction time affect the wave's (a) amplitude; (b) period; (c) wave speed; (d) wavelength?

7 When lightning strikes at some distance, will you see the lightning first or hear the accompanying thunder first? Explain.

8 Suppose you are in a rowboat on a quiet lake as a fast motorboat passes at a considerable distance. The motorboat produces waves, which are a superposition of harmonic waves of various wavelengths. Do the waves of longer or shorter wavelengths reach you first?

9 A source of sound and an observer both move through the air along the same line at the same velocity so that there is zero velocity of the source relative to the observer.
(a) Will the observer hear a Doppler-shifted frequency?
(b) Will the time it takes for the sound to travel from the source to the observer be different from what it would be if both were at rest?

10 Large supersonic transport planes (SST's) produce a sonic boom, which many people find very objectionable. Suppose an SST is to travel coast to coast. Would it be reasonable for the plane to head out over the ocean immediately after takeoff, break the sound barrier there and then head back over land to avoid creating a sonic boom over land?

11 By what factor must the amplitude of a sound wave be increased in order to increase the intensity level by 10 dB?

12 If a 100 Hz harmonic sound wave produces a 60 dB sound at a given point, what is the intensity level of a 1000 Hz harmonic sound wave having the same amplitude at the same point?

13 If a jet engine produces at takeoff a 100 dB sound 100 m from the aircraft, at what distance should the sound's intensity level be 80 dB? Assume that the sound is not reflected or absorbed by the surroundings.

14 Some of the strings on a classical guitar are single strands of plastic of varying diameters, whereas others are plastic wrapped with wire. Thus the mass of the strings varies considerably. Will the sound produced by the heaviest string have a higher or lower frequency than the sound produced by the other strings?

15 A guitar is tuned when the tension in its strings is adjusted. Should you increase or decrease the tension in a string to produce a higher-frequency sound?

16 Will standing sound waves have a lower fundamental frequency in pipe A, which is open at both ends, or in pipe B, of equal length, open at one end and closed at the other?

17 Suppose you replace the air in a pipe by helium.
(a) How would this affect the resonant frequencies of sound waves in the pipe?
(b) Suppose you inhale some helium from a helium-filled balloon. What happens to your voice? Explain.

Answers to Odd-Numbered Questions

1 no; **3** no; **5** (a) opposite; (b) yes; **7** see it first; **9** (a) no; (b) yes; **11** $\sqrt{10}$; **13** 1 km; **15** increase; **17** (a) increases them; (b) much higher pitch

Problems (listed by sections)

16-1 Description of Waves

1 A cork floating on a lake bobs up and down as a small wave passes. The cork completes 4.00 cycles in 1.00 s. The wave peaks are 10.0 cm apart. Find the speed of the wave.

2 Tsunamis (or tidal waves) are very-long-wavelength water waves generated by earthquakes. A tsunami originating in Japan has a wavelength of 10^5 m and a period of 10 minutes. How long does it take to travel across the Pacific Ocean to California, 8000 km away?

3 Radio waves are electromagnetic waves that travel at a speed of 3.00×10^8 m/s, the speed of light. An AM radio station has an assigned frequency of 890 kHz, which means that the radio waves broadcast by the station are at this frequency. Find the wavelength of these radio waves.

4 Suppose you are listening to a song on your radio. You change to another station playing the same song. If the second station broadcasts at a higher radio frequency than the first, will that affect the wavelength of the sound you hear?

5 You see lightning strike a mountaintop, and 3.00 s later you hear the accompanying thunder. How far away is the mountaintop? The speed of sound is 340 m/s, and the speed of light is 3.00×10^8 m/s.

★ **6** Earthquakes generate two types of waves, which travel through the earth: (1) *primary,* or P, waves are longitudinal *pressure* waves that have the greater speed (about 5 km/s in the crust) and are therefore the first to be felt at some distance from the center of the earthquake, and (2) *secondary,* or S, waves are transverse, *shear* waves that are somewhat slower ($v \approx 3$ km/s). The location of the earthquake can be determined by recording the arrival times for these waves on seismographs at various locations. Suppose an S wave is recorded 2 min after a P wave. About how far away from the seismograph is the earthquake's center?

★ **7** Suppose that there is a seismic P wave traveling at 5.0×10^3 m/s with a wavelength of 2.0×10^3 m.
(a) Find the wave's frequency.
(b) Find the average speed of a particle of the earth's surface at a point where the wave amplitude is 2.0 cm.

★ **8** When a transverse wave of amplitude 2.00 cm and wavelength 20.0 cm passes through a medium, the average speed of a particle of the medium is 4.00 m/s. Find the speed of the wave.

9 A nerve impulse is a wave pulse that travels along a nerve, typically at a speed of 50 m/s. If the pulse sweeps past one point in the nerve from $t = 0$ to $t = 2$ ms, during what time interval will it pass a point in the nerve 1 m away?

16-2 Wave Speed

10 By what percent must one increase the tension in a guitar string to change the speed of waves on the string from 300 m/s to 330 m/s?

11 A 1.00 m long string of mass 0.0100 kg, under a tension of 100 N, transmits a wave of amplitude 2.00 cm and wavelength 10.0 cm.
(a) Find the speed of the wave form.
(b) How far does a particle of the string travel during one cycle?
(c) What is the average speed of a particle of the string?

★ **12** A 2.00 m long string of mass 10.0 g is attached to a 3.00 m long string of mass 30.0 g (Fig. 16-46). The strings are under a tension of 100 N.
(a) How long will it take for a wave pulse to travel from point A to point C?
(b) How long will it take for reflected pulses to return to point A? Indicate the orientation of the reflected pulses relative to the original pulse.

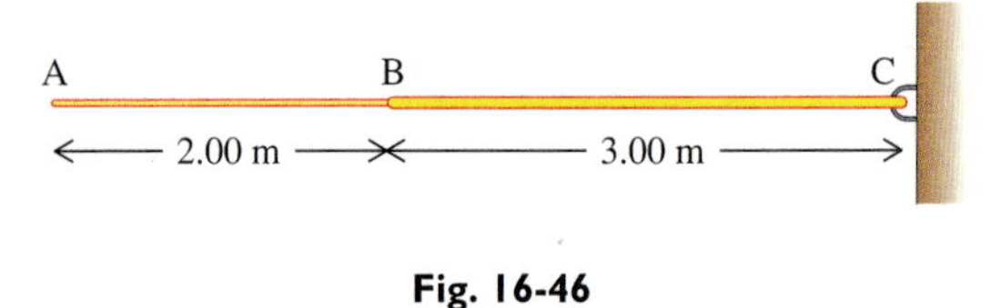

Fig. 16-46

13 The normal human ear is sensitive to sounds with frequencies from about 20 Hz to about 20,000 Hz. What is the corresponding range of wavelengths (a) in air; (b) in water?

14 If you hear your echo 3.00 s after you shout, how far away is the mountain reflecting the sound?

15 Suppose you are in your dorm listening on your radio to a football game being played on campus, 500 m away. As a touchdown is scored, you hear on the radio the roar of the crowd. When do you hear the crowd's cheers directly from the stadium?

16 Find the frequency of a deep-water wave of wavelength 30.0 cm.

17 You are floating in the ocean in deep water as a 5.00 m wavelength wave passes. How long do you wait between wave peaks?

18 Find the speed of deep-water waves having a frequency of 1.00 Hz.

★ **19** A fisherman sits in a stationary boat in the middle of a lake as a high-powered motorboat passes at some distance. The fisherman notices that the wavelengths of the waves from the wake gradually decrease, with 10.0 cm waves arriving 60.0 s after 30.0 cm waves. How far away was the motorboat?

★ **20** A boat floating at rest encounters an unusually big deep-water ocean wave with an amplitude of 15.0 m and a wavelength of 200 m. Find the boat's speed and the magnitude of its acceleration.

21 Sonar depth finders are used on boats to determine the depth of water by reflecting a pulse of sound from the bottom. What is the depth of water if there is a 0.10 s delay between emission of the pulse and detection of the reflected pulse?

22 At what temperature would sound travel through air at 400 m/s?

23 How much longer would it take for sound to travel 1.0 km through arctic air at −50° C than to travel 1.0 km through desert air at +50° C?

24 A bat emits ultrasonic pulses and uses them to navigate and to locate flying insects. If these pulses are sent at a rate of 4 per second, what is the maximum distance a reflecting object can be if the reflected pulse is to be received by the bat before the next pulse is emitted?

25 What is the wavelength of a 5.00×10^4 Hz sound wave pulse emitted by a bat?

16-3 The Doppler Effect

★ **26** Suppose you are in a high-speed boat moving at 20.0 m/s directly into approaching 1.00 m wavelength waves. What is the time interval between wave peaks hitting the boat?

27 A surfer rides a wave, moving at the same speed as the wave. What is the frequency of the wave, as observed by the surfer?

★ **28** Suppose you want to demonstrate the Doppler effect for deep-water waves, using a 6.00 Hz source moving toward a stationary observer. How fast would the source have to move through the water if the frequency of the waves seen by the observer is to be twice the frequency of the source?

29 A sound source moves through air toward a stationary observer. The frequency of the sound the observer hears is 20.0% higher than the source frequency. How fast is the source moving?

30 Suppose that a 1024 Hz tuning fork moves at 10.0 m/s through (a) air and (b) water. For each medium, find the observed frequency of the sound at a point directly in front of the source.

★★ **31** The driver of a car hears the hum of its engine at a frequency of 200 Hz.

(a) Find the frequency of the sound heard by a pedestrian standing beside the road, first as the car approaches at a speed of 20.0 m/s and then after it passes.

(b) Now suppose that the wind is blowing at a speed of 20.0 m/s in the direction of the car's motion. What would be the frequencies of the sound heard by the pedestrian as the car approaches and passes?

★ **32** Bats use the Doppler effect as a directional guide and to detect insects. As it flies at a speed of 5.00 m/s, a bat emits a brief pulse of 60.0 kHz ultrasound in the forward direction. A nearly stationary insect in front of the bat reflects sound back to the bat. Find the frequency of the sound detected by the bat. (Bats are particularly sensitive to 61.8-kHz sound.)

★ **33** As you stand beside the German Autobahn (where there are no speed limits), a Porsche passes while sounding its horn. The frequency of the sound you hear drops from 500 Hz as the car approaches to 300 Hz after the car passes. How fast was it moving?

34 Sound from a foghorn in a lighthouse has a frequency of 100 Hz. Suppose that during a storm 40.0 m/s winds blow by the lighthouse. What is the frequency of sound heard by a stationary observer (a) downwind and (b) upwind from the lighthouse?

★ **35** A driver sounds his horn as he drives at 25.0 m/s toward a tunnel in the side of a mountain. The mountainside reflects the sound, and the driver hears an echo. Find the frequency of the echo heard by the driver if the frequency emitted by the horn is 500 Hz.

★★ **36** Suppose that, as you are driving, a stationary police radar unit in front of you detects a radar signal reflected from your car. The ratio of the reflected signal frequency to the emitted frequency differs from unity by 2×10^{-7}. How fast are you going?

★ **37** A police car moving at 20.0 m/s follows a speeding car moving at 30.0 m/s. A radar unit in the police car emits a 1.00×10^{10} Hz signal, which is reflected by the other car. What is the difference between this emitted frequency and the frequency of the reflected signal detected by the radar unit?

★ **38** A star is moving away from the earth at 50.0 km/s. By how much will the H_α line ($f = 4.571 \times 10^{14}$ Hz) be shifted and in what direction?

★ **39** A supersonic jet passes directly overhead at an altitude of 12,000 m, traveling at 680 m/s, or mach 2. How much time elapses before you hear the sonic boom?

16-4 Power and Intensity; the Decibel Scale

40 Two people talk simultaneously. If the intensity level is 60 dB when either one speaks alone, what is the intensity level when both speak at once?

41 One hundred people at a party are talking at once. On the average, each person alone produces 65 dB sound near the center of the room. Find the intensity level produced by all 100.

42 When a 100 dB sound wave comes through an open window of area 0.500 m², how much acoustic energy passes through the window in a 10.0 min interval?

43 At a football game, 100,000 spectators produce sound that is heard 1.00 km away, where the intensity level is 60.0 dB. Assuming that no sound is reflected or absorbed, we can treat the sound as radiating equally in all directions, so that the intensity is constant over a hemispherical surface. How much acoustic power is generated by the fans?

44 The driver of a car honks her horn as she enters a narrow tunnel. If the intensity level is 80.0 dB 20.0 m in front of the car, what is the intensity level 100 m in front of the car? Assume that no sound is absorbed by the tunnel walls.

45 Find the intensity level of a 50.0 Hz sound wave in which the amplitude of vibration is 1.00 mm.

46 Only half the population can hear a 60 dB, 60 Hz sound, but nearly everyone can hear a 100 dB, 60 Hz sound. What is the ratio of the amplitudes of these waves?

47 Find the intensity of a 53.0 dB sound.

48 What is the minimum power required for a loudspeaker to produce 105 dB sound at a distance of 10.0 m in any direction?

16-5 Time Dependence of the Displacement of a Particle of the Medium

49 A wave on a string is described by the equation $y = 5 \sin(4\pi t - 0.1\pi x)$, where x and y are in cm and t is in s. What are the values of the amplitude, frequency, wavelength, and speed of this wave? In what direction does it travel?

50 For the wave described in Problem 49, graph y versus x at $t = 0$, $t = 0.125$ s, and $t = 0.250$ s.

51 A distant 512 Hz tuning fork produces a plane sound wave that moves along the positive x-axis and has an amplitude of 1.00×10^{-8} m. Write an expression for the displacement as a function of x and t.

52 A wave is described by the equation $y = 3 \sin(2x + 10t)$, where x is in cm and t is in s. Find the wave's amplitude, frequency, wavelength, speed, and direction of motion.

★ **53** A wave of wavelength 40.0 cm and amplitude 4.00 cm propagates along a string at 10.0 m/s, in the positive x direction. At $t = 0$, the displacement at the origin is zero and the string is moving upward. Find the displacement at $x = 3.00$ m at $t = 0.500$ s.

★★ **54** The wave form shown in Fig. 16-47 travels along the positive x-axis at 10.0 m/s. Find the displacement y at $x = 25.0$ m at $t = 20.0$ s.

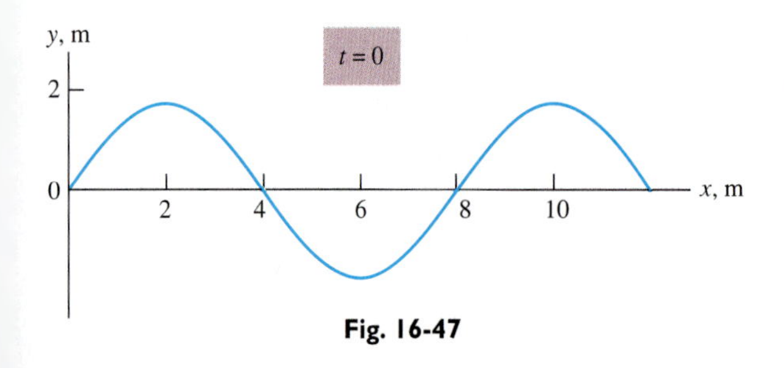

Fig. 16-47

16-6 Superposition of Waves; Beats; Standing Waves

55 Find an expression for the displacement y produced when there are two waves present with displacements $y_1 = 4 \sin(2t - 5x)$ and $y_2 = 2 \sin(2t - 5x)$.

56 Two upward wave pulses are generated at opposite ends of a string and travel toward each other. One has an amplitude of 5.00 cm and the other has an amplitude of 3.00 cm. What is the maximum displacement of any particle of the string, and when does this displacement occur?

57 A positive (upward) wave pulse and a negative (downward) wave pulse are generated simultaneously at opposite ends of a string. The maximum displacements of the positive and negative pulses are 5.00 cm and 3.00 cm respectively. What is the maximum displacement at the midpoint of the string?

58 Two small loudspeakers are connected to a single source—an 860 Hz sine-wave generator—so that they are coherent. Suppose that initially the speakers are side by side, a few centimeters apart, equidistant from a listener 2.00 m in front of the speakers.

(a) One of the speakers is slowly moved straight back away from the listener until at some point she hears almost no sound. How far is the speaker moved?

(b) How much farther back should the speaker be moved for the sound to be about as loud as it was initially?

★ **59** Two coherent harmonic sound sources produce destructive interference at a point that is 1.23 m from one source and 1.26 m from the other. What are the three lowest possible frequencies of the waves?

60 Find the beat frequency produced by an 800 Hz source and an 804 Hz source.

61 A piano tuner simultaneously strikes a 262 Hz tuning fork and the middle C key on a piano and hears beats with a beat period of 0.500 s. What are the possible frequencies of the slightly out-of-tune key? Should the tension in the piano wire be changed in such a way as to shorten or to lengthen the period of the beats?

62 Find the wavelengths of the three longest standing waves that can be formed on a 1.00 m long string fixed at both ends.

63 Find the first three harmonics of a 80.0 cm guitar string if the speed of waves on the string is 704 m/s.

★ **64** Use the trigonometric identity $\sin(\alpha \pm \beta) = \sin\alpha\cos\beta \pm \sin\beta\cos\alpha$ to prove that the superposition of two harmonic waves traveling in opposite directions produces a standing wave; i.e., prove that if

$$y_1 = A\sin\left[2\pi\left(\frac{t}{T} - \frac{x}{\lambda}\right)\right]$$

and $$y_2 = A\sin\left[2\pi\left(\frac{t}{T} + \frac{x}{\lambda}\right)\right]$$

then $$y = y_1 + y_2 = 2A\sin\left(\frac{2\pi t}{T}\right)\cos\left(\frac{2\pi x}{\lambda}\right)$$

Sketch y at various points in a cycle.

65 A musician playing a violin, guitar, or other string instrument changes the fundamental frequency of one of the strings by "fingering," that is, by pressing the string against the neck of the instrument with a finger so that the length of the string is effectively shortened. To change a string's fundamental frequency from 440 Hz (A) to 512 Hz (C), how much should one shorten the string if its original length is 70 cm?

66 The E string on a guitar has a length of 66.0 cm. The string's fundamental frequency is 165 Hz. Pressing the string against one of the frets along the neck of the guitar effectively shortens the length of the string. What length will give the E string a frequency of 262 Hz (middle C)? Assume the tension is constant.

67 The longest pipe in a certain organ is 4.00 m long. What is the lowest frequency the organ will produce if the pipe is (a) open at both ends; (b) closed at one end?

68 Given that most people cannot hear sounds outside the frequency range 50 Hz to 10,000 Hz, what are reasonable minimum and maximum lengths for musical wind instruments, which are open at both ends?

★ **69** If the temperature of the air in an organ pipe drops from 300 K to 250 K, what will the fundamental frequency of the pipe be if it originally was 440 Hz?

70 At a certain point in space, two waves have displacements y_1 and y_2, which are graphed as functions of time in Fig. 16-48. Sketch the resultant displacement $y = y_1 + y_2$.

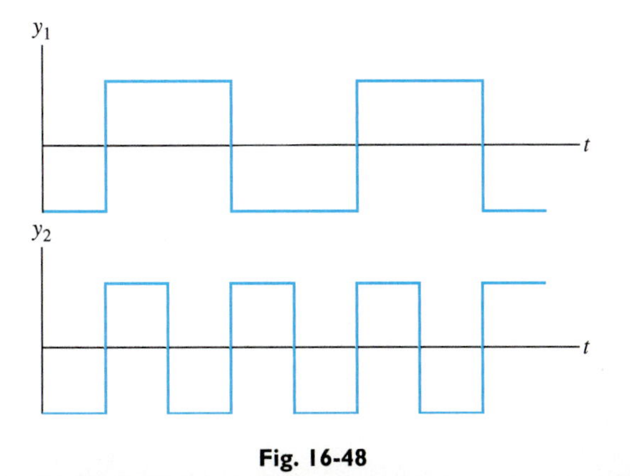

Fig. 16-48

Additional Problems

71 The Richter scale is a logarithmic scale for measuring the total energy released in an earthquake. The Richter magnitude M is defined by the equation $M = \frac{2}{3}\log\left(\frac{E}{E_0}\right)$, where E_0 is a reference energy equal to 25,000 J.

(a) How much energy was released in the 1994 Los Angeles earthquake, which registered 6.8 on the Richter scale?

(b) By what factor would the energy of this quake have to be increased in order to have a magnitude of 7.8, or of 8.8?

Fig. 16-49 Aftermath of the 1906 San Francisco earthquake.

★ **72** The 1906 San Francisco earthquake lasted 1 min and registered 8.2 on the Richter scale (defined in problem 71).

(a) What was the average power generated during this earthquake?

(b) Estimate the intensity 10 km from the center of the quake.

73 An earthquake on the opposite side of the earth from you generates waves. How long do you wait for the arrival of P waves that travel through the earth at a speed of 5.00 km/s?

★★ 74 In a harmonic water wave, the water moves along a circular path of radius A.

(a) Prove that the speed of the water v_w is related to the wave speed v by the equation $v_w = \frac{2\pi A}{\lambda} v$.

(b) By applying Bernoulli's equation in the rest frame of the wave, show that the wave speed is $v = \sqrt{\frac{g\lambda}{2\pi}}$.

75 An autofocus camera sends out an ultrasonic pulse that is reflected by an object and then detected by the camera's rangefinder. If the time delay between emission and detection is 0.0100 s, how far is the object from the camera?

★★ 76 Show that the acceleration of the water at the peak of a harmonic water wave equals g when the amplitude of the wave equals $\frac{\lambda}{2\pi}$. Would you expect the amplitude of a water wave to ever exceed this value?

★ 77 Find the maximum speed and acceleration of a particle of a string that transmits a wave of amplitude 3.00 cm and frequency 20.0 Hz.

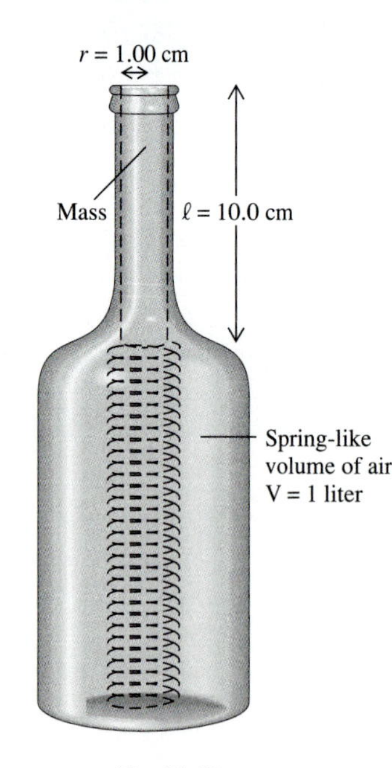

Fig. 16-50

★ 78 You can easily produce a standing wave by blowing across the top of an empty bottle. The fundamental frequency of the standing wave is surprisingly low, however. The air in the neck of the bottle oscillates in SHM as the air lower down in the bottle is alternately compressed and expanded. The air in the neck behaves like a mass attached to a spring. The springlike force is provided by the lower air. The air in the neck must move a considerable distance before there is a significant opposing force from the compressed lower air. Thus the force constant is small, and the resonant frequency $\frac{1}{2\pi}\sqrt{\frac{k}{m}}$ is low. A detailed mechanical analysis yields the following approximate formula for the resonant frequency: $f = \frac{1}{2}\left(\frac{vr}{\sqrt{\pi \ell V}}\right)$, where v is the speed of sound and r, ℓ, and V are as indicated in Fig. 16-50.

(a) Find the fundamental frequency of the bottle shown.

(b) Find the minimum length of an organ pipe having the same fundamental frequency.

★ 79 We perceive the direction of sound sources through slight time differences between the arrival of waves at each ear. The waves reach the ears simultaneously when the sound source is directly in front of the listener.

(a) Find the time difference when sound comes from a distant source at an angle of 30.0° relative to the forward direction, as indicated in Fig. 16-51.

(b) What would the time difference be if the source were directly behind the listener? How can the listener distinguish between sources directly in front and those directly behind? Will the effect of rotating the head a few degrees give the same result for sources in front and those behind?

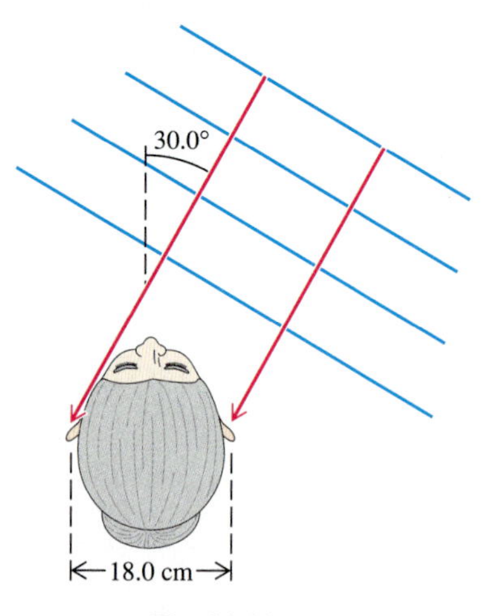

Fig. 16-51

CHAPTER 17 The Electric Field

This electrical discharge was produced at the turn of the century by inventor Nikola Tesla, shown in the photo.

You can observe "static electricity" in various ways. Walk across a wool rug with rubber-soled shoes on a dry day and your body becomes "charged," and then, when you touch a metal doorknob, there is a sudden, painful spark. Remove clothes from a dryer, and they cling together because electric charges have been transferred between the clothes. Brush your hair on a very dry day, and both hair and brush become charged. The brush then attracts dust or small bits of paper and your hair stands on end. All these phenomena result from forces acting between electric charges at rest.

This chapter and the next deal with **electrostatics,** the study of electric charges at rest. This will serve as a foundation for understanding electric current (charges in motion), with its many applications to modern technology. After studying current, we shall study magnetism and show its connection with electricity. The culmination of all this will be our study of the laws of electromagnetism, in which electric and magnetic phenomena are shown to be intimately connected.

17-1 Electric Charge

Historical Origins

One of the earliest observations of electric attraction was that amber that has been rubbed with a piece of cloth will attract light objects, such as feathers or straw. This was known to the Greeks as early as 600 B.C. In 1600 A.D. William Gilbert, personal physician to Queen Elizabeth I, published the first systematic study of electricity and magnetism. Gilbert showed that, contrary to popular opinion, amber was not unique in its attractive property; many other substances could produce similar attractive effects after frictional contact. This attractive force was obviously a very general property of nature, and so he gave it a special name—"electric force." (*Elektron* is the Greek word for 'amber.')

No further significant progress was made in the science of electricity until 1734, when Charles du Fay attempted to account for his observation that the electric force could be repulsive as well as attractive. Du Fay believed that there must be two kinds of "electrical fluid" that could flow into a body and "electrify" it. Two bodies charged with the same kind of fluid would repel each other, whereas two bodies charged with different kinds of fluid would attract each other.

In 1750 Benjamin Franklin, on the basis of extensive experiments, sought to replace the two-fluid theory by a one-fluid theory. He believed that an excess of this single fluid was responsible for one type of electrical charging and a deficiency of it was responsible for the other type of charging. The body containing the excess fluid he described as "positively charged," and the body with the deficiency he called "negatively charged." For example, Franklin believed that when someone rubs a piece of glass with the bare hand both glass and skin become charged because some of the electric fluid in the skin is transferred to the glass. This movement of the fluid gives the skin a deficiency of the fluid and the glass an excess. Thus he assigned a negative charge to the skin and a positive charge to the glass. Franklin proposed that when two oppositely charged bodies are brought together the body with the excess fluid gives it up to the body with the deficiency; both bodies become uncharged and no longer produce electrical force.

Today we know that matter normally consists of equal quantities of positive and negative electric charges—protons and electrons—and therefore has no net charge. The protons are tightly bound in the nuclei of atoms, whereas the electrons, which are far less massive than the protons, are more weakly bound to atoms and therefore are more easily removed. The masses of the proton and electron are

$$m_p = 1.67 \times 10^{-27} \text{ kg}$$

$$m_e = 9.11 \times 10^{-31} \text{ kg}$$

When a body does not contain an equal number of protons and electrons, it has a net positive or negative charge. Transfer of electrons to or from a body gives that body a net charge.* If we think of electrons as Franklin's electrical fluid, we see that his ideas concerning electrical charging were essentially correct. Franklin's analysis of the charging of skin and glass was incorrect in one aspect, however. Frictional contact between skin and glass results in the flow of the "fluid" (electrons) from glass to skin, rather than from skin to glass as supposed by Franklin; hence the skin, to which Franklin assigned a negative charge, has an excess of electrons. It follows that the electrons must have negative charge. This is just a convention, arising from Franklin's arbitrary assignment of negative charge to the skin.

Fig. 17-1 Many of the early experiments with static electricity involved human subjects who were used to detect the presence of electric charge by the electric shock it produced.

*Sometimes a net charge is the result not of electron transfer but of the transfer of either positively charged "ions" (atoms missing one or more electrons) or negatively charged ions (atoms with one or more extra electrons).

Demonstration of Electrostatic Forces

The following simple experiment demonstrates several important properties of electric charge. First charge a piece of aluminum foil by rubbing it with plastic wrap. Then let the aluminum touch two Ping-Pong balls, each suspended from a string. When the balls are brought close together, they experience a mutual repulsive force (Fig. 17-2a). Next charge a piece of rubber by rubbing it with wool, and then let the rubber touch two other Ping-Pong balls. Like the first two, these balls experience a mutual repulsive force (Fig. 17-2b).

If one of the balls charged by rubber is now brought close to one of the balls charged by aluminum, a mutual attractive force is observed (Fig. 17-2c), but if these two balls are allowed to touch each other, the electrical attraction quickly disappears (Fig. 17-2d). (We assume here that the balls are equally charged.)

This behavior can be described in terms of excess and deficiency of electrons. Electrons are taken away from the aluminum when it is rubbed, leaving it with a deficiency of electrons. There are then more protons than electrons and therefore a net positive charge. When the aluminum touches two balls, they become positively charged as some of their electrons flow to the electron-deficient foil. The two positively charged balls (A) repel each other. Electrons from the wool are added to the rubber when it is rubbed, leaving the rubber and the balls it touches negatively charged. The negatively charged balls (R) repel each other. When a positively charged ball (A) is close to a negatively charged ball (R), they attract. When the two are brought into contact, electrons flow from R to A, leaving each electrically neutral, and they no longer experience any force.

This experiment demonstrates an important general property of electric charges: **like charges repel,** and **unlike charges attract.**

An electroscope is a device that uses the repulsion of like charges to show the presence of charge. It consists of a metallic rod with two small pieces of gold leaf at the bottom. When charged, the leaves repel each other (Fig. 17-3).

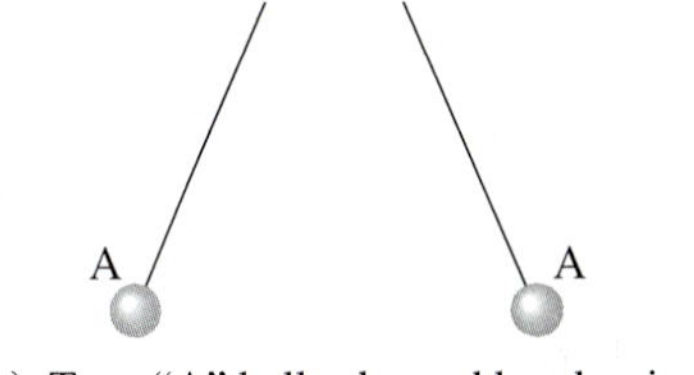

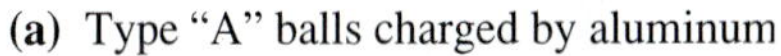
(a) Type "A" balls charged by aluminum

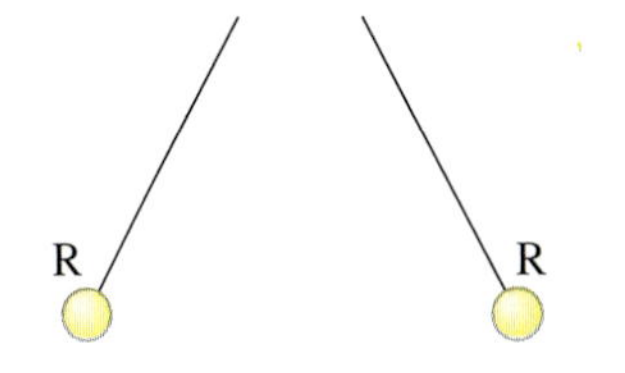

(b) Type "R" balls charged by rubber

(c) Types A and R before contact

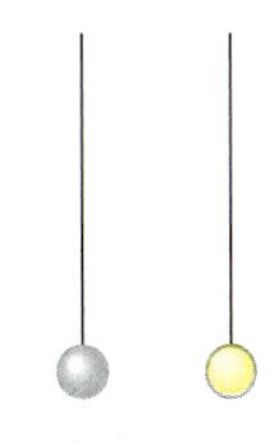

(d) Types A and R after contact

Fig. 17-2 Forces between charged Ping-Pong balls.

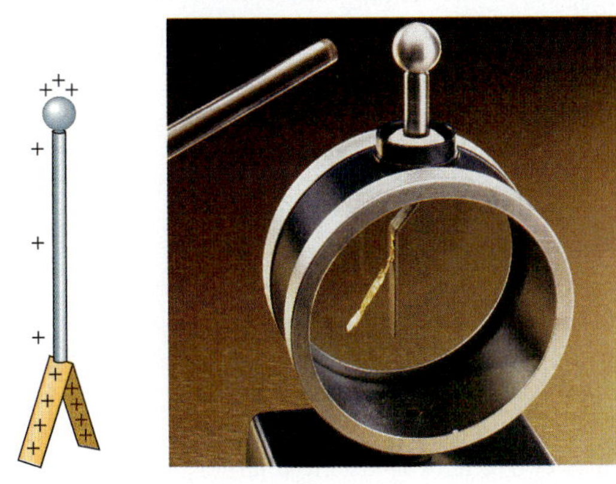

Fig. 17-3 When charged, the leaves of an electroscope stand out.

Insulators and Conductors

When some materials are charged by contact with a charged object, the excess or deficiency of electrons very quickly leaves the point of contact and distributes itself over the entire surface of the newly charged material. Such materials are called **conductors.** They readily conduct electric charge from one point to another. Metals are good conductors, as are the human body and the earth. For other materials, called **insulators** or **dielectrics,** charge placed on one part of the surface remains localized. Most nonmetals are insulators. For example, wood, rubber, glass, and plastic are all insulators.

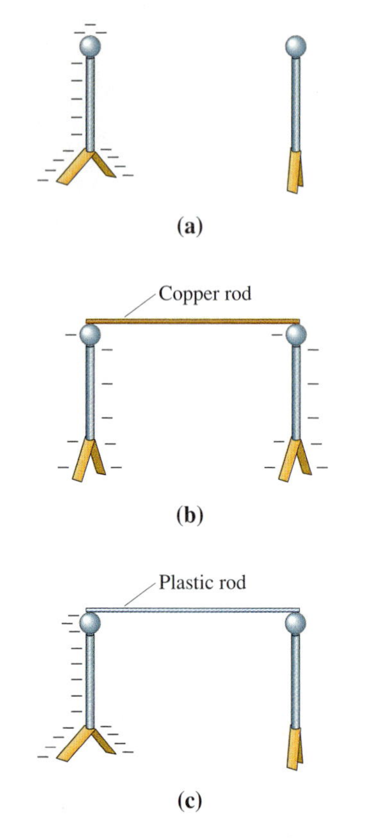

Fig. 17-4 A copper rod transfers charge from one body to another **(b)**, but a plastic rod does not **(c)**.

Metals are effective in transporting electrons from one body to another. Suppose you have a negatively charged body and want to transfer some of its charge to a second, initially uncharged body (Fig. 17-4a). If you connect a copper rod between the two, some of the excess electrons in the first body will pass through the rod and onto the second body (Fig. 17-4b). If, on the other hand, you connect the two bodies by an insulator—a plastic rod, for example—there will be no transfer of charge, and the second body will remain uncharged (Fig. 17-4c).

It is the microscopic structure of insulators and conductors that is responsible for their very different conducting properties. All the electrons in an insulator are bound to individual atoms, but in a conductor each atom gives up some small number of electrons so that these electrons are free to move through the conductor. The remaining electrons and the positively charged nucleus of each atom are bound together to form a positively charged ion. The positive ions are bound together in the conductor in a fixed structure (Fig. 17-5).

Although a conductor may have no net charge, it still contains an enormous reservoir of free electrons that can quickly respond to an electric force imposed by a nearby charged body. These free electrons then move through the conductor.

Conservation of Charge

Although charge moves from place to place, net charge is never created or destroyed at any point in space. This principle is known as "conservation of charge" and is a general law of nature. If electrons and protons were indestructible particles, charge conservation would be obvious. Experiments indicate, however, that these particles are not indestructible. They, along with many other particles, may be created and destroyed. However, in all such experiments there is never creation or destruction of *net* charge. For example, an electron may interact with a "positron" with the result that the two annihilate each other, leaving no charge behind. (The positron is a particle that has exactly the same mass as the electron but carries a positive charge.) Since the electron and positron have opposite charges of equal magnitude, the net charge of the system is zero both before and after the annihilation, and so the principle of charge conservation is satisfied.

Units of Charge

An obvious and natural choice for the basic unit of charge is the magnitude of the electron's charge, which we denote by e. Thus we express the electron's charge as $-e$ and the proton's charge as $+e$:

$$\text{electron charge} = -e$$

$$\text{proton charge} = +e$$

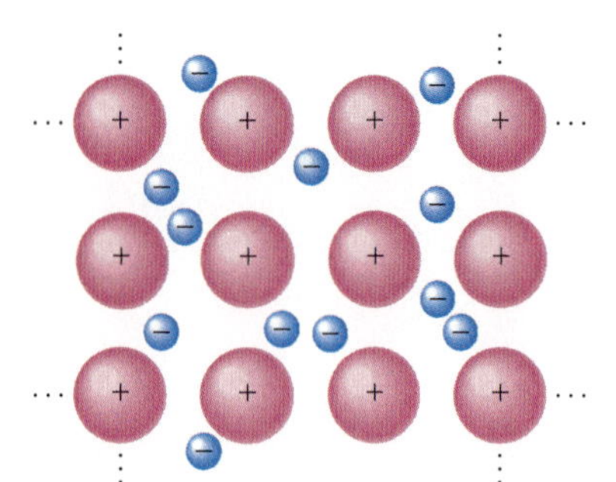

Fig. 17-5 The charge distribution in an uncharged conductor. The positively charged ions are bound to fixed positions, but some electrons are free to move throughout the conductor.

A charged macroscopic body usually has an excess or deficiency of an enormously large number of electrons, and so it is convenient to define a larger unit of charge. The coulomb (abbreviation C) is such a unit. It is an experimentally determined unit* that may be expressed as

$$1 \text{ C} = (6.24 \times 10^{18})e \tag{17-1}$$

Alternatively, e may be expressed as $1 \text{ C}/6.24 \times 10^{18}$, or

$$e = 1.60 \times 10^{-19} \text{ C} \tag{17-2}$$

*Eq. 17-1 is not the definition of the coulomb. Instead, the coulomb is defined in terms of the ampere, the SI unit of electric current, which we shall define in Chapter 20.

The symbol q is used to denote charge. The charge on a body can be expressed as an integral multiple of e:

$$q = \pm ne \tag{17-3}$$

where n is the number of electrons that have been either taken from or added to the body.

EXAMPLE 1 Finding the Number of Excess Electrons on a Charged Body

A body has a net charge of -1.00×10^{-9} C. How many excess electrons are contained in the body?

SOLUTION Solving Eq. 17-3 for n, we have

$$n = \frac{-q}{\mathrm{e}} = \frac{-(-1.00 \times 10^{-9}\ \mathrm{C})}{1.60 \times 10^{-19}\ \mathrm{C}}$$

$$= 6.25 \times 10^{9}$$

17-2 Coulomb's Law

Two Point Charges

In 1785 Charles Coulomb established the fundamental force law for two static point charges*:

1. Each of the two charges experiences a force that is directed along the line between the two charges; the force is repulsive for charges of like sign and attractive for charges of opposite sign (Fig. 17-6).
2. The magnitude of the force is proportional to the product of the magnitudes of the two charges and inversely proportional to the square of the distance r between them:

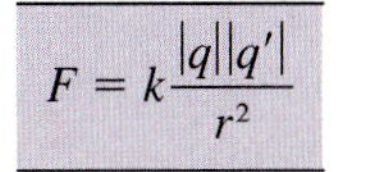

$$F = k\frac{|q||q'|}{r^2} \tag{17-4}$$

where k is a force constant that experiment shows to have the value

$$k = 8.99 \times 10^{9}\ \mathrm{N\text{-}m^2/C^2}$$

In problem solving, we shall sometimes round this off to $k = 9.0 \times 10^9\ \mathrm{N\text{-}m^2/C^2}$.

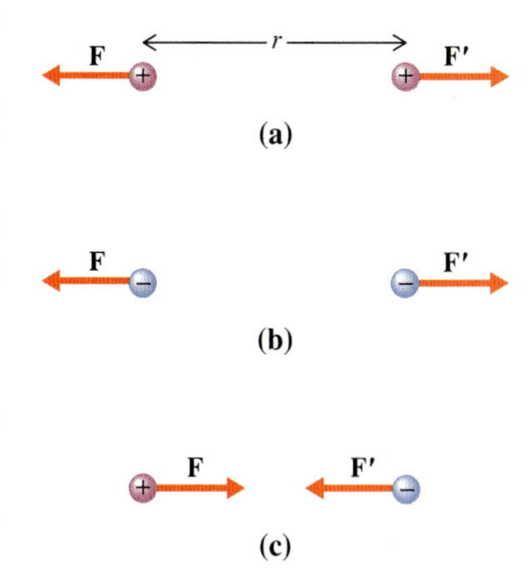

Fig. 17-6 Forces between charges.

*Coulomb used a torsion balance to perform his experiments. This device was imitated by Cavendish years later to study gravitational force. The Cavendish torsion balance was described in Chapter 6.

EXAMPLE 2 Force Exerted by One Charge on Another

Two point charges, $q = +1.00$ C and $q' = -1.00$ C, are located 2.00 m apart (Fig. 17-7a). What are the magnitude and direction of the force that q exerts on q'?

SOLUTION Since the charges are of opposite sign, the force between them is attractive, and so the force **F** that q exerts on q' is directed toward q, as shown in Fig. 17-7b. (Of course there is also an oppositely directed reaction force acting on q, but we are not asked to compute it, and so we have not shown it in the figure.)

The magnitude of **F** is found from Eq. 17-4:

$$F = k\frac{|q||q'|}{r^2} = (8.99 \times 10^9 \text{ N-m}^2/\text{C}^2)\frac{(1.00 \text{ C})(1.00 \text{ C})}{(2.00 \text{ m})^2}$$

$$= 2.25 \times 10^9 \text{ N}$$

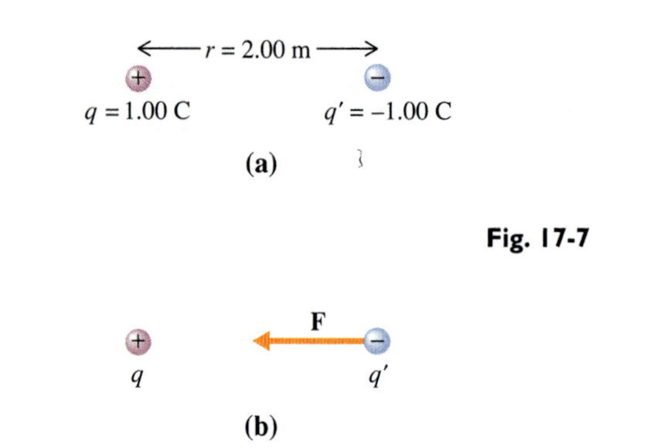

Fig. 17-7

This is an enormously large force, and thus we would normally never observe such large concentrations of charge. In ordinary static charging of bodies through friction, the charge might typically be something like 10^{-10} C.

Coulomb's Law and Universal Gravitation

Coulomb's law is the second fundamental force law we have encountered in our study of physics. It is similar in form to the fundamental force law introduced in Chapter 6, Newton's law of universal gravitation. Recall that the gravitational force between two particles is proportional to the product of their masses and inversely proportional to the square of the distance between them (Eq. 6-1):

$$F = G\frac{mm'}{r^2} \tag{17-5}$$

where $G = 6.67 \times 10^{-11} \text{ N-m}^2/\text{kg}^2$

Thus the Coulomb force depends on charge in the same way that the gravitational force depends on mass, and both forces have the same $1/r^2$ dependence.

There are, of course, important differences between Coulomb's law and the gravitational force law. Although there is only one kind of mass, there are two kinds of charge, positive and negative, and although the gravitational force is always attractive, the electrostatic force may be attractive or repulsive, depending on the signs of the charges.

Another obvious difference between the two force laws is that the electrical force constant k is much larger than the gravitational constant G. Thus, as we saw in Example 2, two bodies 2.00 m apart, each carrying a charge of magnitude 1.00 C, experience a force of 2.25×10^9 N, or about half a billion pounds. On the other hand, two 1.00 kg bodies 2.00 m apart experience a mutual gravitational force of only 1.67×10^{-11} N. The large value of the electric force constant means that the interaction of even relatively small charges can produce significant forces.

Superposition Principle

When several point charges are present in a region of space, **each charge experiences forces exerted by the other charges. The resultant force acting on each charge is the *vector* sum of all the Coulomb forces acting on that charge.** This principle is called the **superposition principle** and is illustrated for three positive charges in Fig. 17-8.

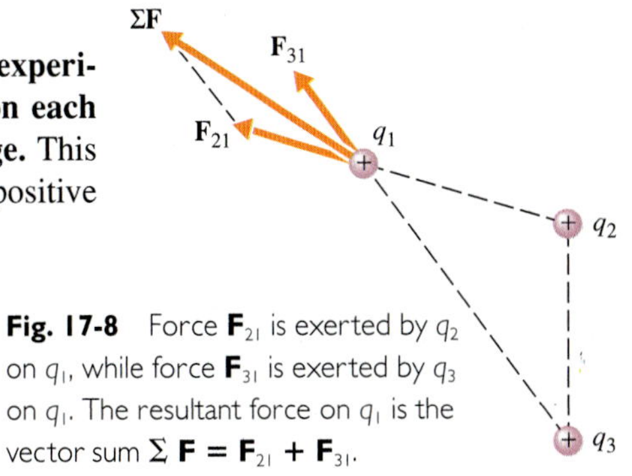

Fig. 17-8 Force $\mathbf{F}_{21}$ is exerted by q_2 on q_1, while force $\mathbf{F}_{31}$ is exerted by q_3 on q_1. The resultant force on q_1 is the vector sum $\Sigma\,\mathbf{F} = \mathbf{F}_{21} + \mathbf{F}_{31}$.

EXAMPLE 3 The Resultant of Two Electrostatic Forces on a Charge

Find the resultant force on q_3 in Fig. 17-9a.

SOLUTION The first step is to indicate in a figure the forces $\mathbf{F}_{13}$ and $\mathbf{F}_{23}$ exerted on q_3 by q_1 and q_2 respectively (Fig. 17-9b). Notice that $\mathbf{F}_{13}$ is drawn with the tail of the arrow on q_3, since that is the charge this force acts on, and is directed away from q_1, since both q_1 and q_3 are positive and charges of like sign repel. Force $\mathbf{F}_{23}$ is directed toward q_2, since q_2 and q_3 have opposite signs and q_2 therefore exerts an attractive force on q_3. We have also indicated in the figure the distances from q_3 to the other two charges.

The next step is to calculate the magnitudes of the forces. Applying Coulomb's law (Eq. 17-4), we find

$$F_{13} = k\frac{|q_1||q_3|}{r^2}$$

$$= (9.0 \times 10^9\ \text{N-m}^2/\text{C}^2)\frac{(1.0 \times 10^{-6}\ \text{C})(5.0 \times 10^{-6}\ \text{C})}{(3.0 \times 10^{-2}\ \text{m})^2}$$

$$= 50\ \text{N}$$

$$F_{23} = k\frac{|q_2||q_3|}{r^2}$$

$$= (9.0 \times 10^9\ \text{N-m}^2/\text{C}^2)\frac{(4.0 \times 10^{-6}\ \text{C})(5.0 \times 10^{-6}\ \text{C})}{(3.0\sqrt{2} \times 10^{-2}\ \text{m})^2}$$

$$= 100\ \text{N}$$

The final step is to find the resultant force on q_3 by taking the sum of vectors $\mathbf{F}_{13}$ and $\mathbf{F}_{23}$, using the usual method of vector addition by components. (It may be helpful to review Section 1-5.)

$$\Sigma\ F_x = +F_{13} - F_{23}\cos 45° = 50\ \text{N} - (100\ \text{N})(\cos 45°)$$

$$= -21\ \text{N}$$

$$\Sigma\ F_y = +F_{23}\sin 45° = (100\ \text{N})(\sin 45°)$$

$$= 71\ \text{N}$$

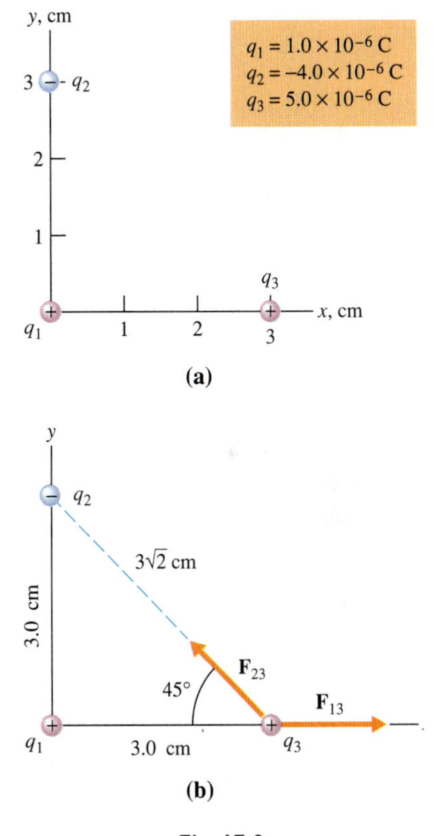

Fig. 17-9

The resultant force has a magnitude given by

$$|\Sigma\ \mathbf{F}| = \sqrt{(\Sigma F_x)^2 + (\Sigma F_y)^2} = \sqrt{(-21\ \text{N})^2 + (71\ \text{N})^2}$$

$$= 74\ \text{N}$$

This force is directed at an angle θ above the negative x-axis, where

$$\theta = \arctan\frac{71\ \text{N}}{21\ \text{N}} = 74°$$

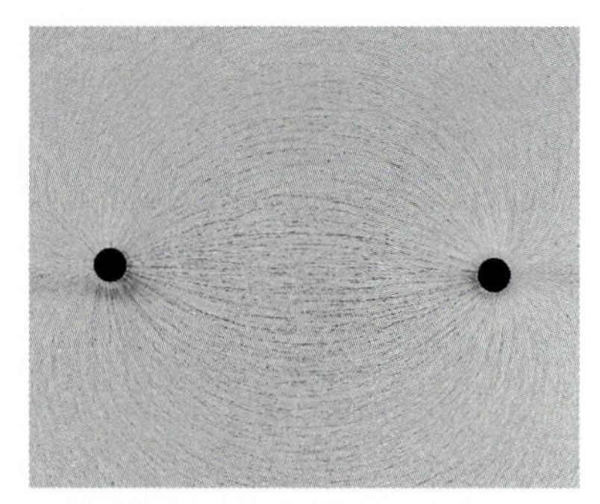

Fig. 17-10 One can show the electric field by suspending bits of thread in oil around electric charges. The threads align with the electric field.

Fig. 17-11 Electrons in a radio receiving antenna respond to the electric field produced by electrons in a distant transmitting antenna.

17-3 The Electric Field

The Concept of a Field

There is another way to view the interaction between charges. Rather than considering directly the forces acting between charges, we can instead introduce the concept of an **electric field.** A charge creates an electric field in the region of space around the charge (Fig. 17-10). The electric field may then act on other charges in that region. Gravitational force may be regarded in a similar way. The earth sets up a gravitational field in space. When a mass is placed near the earth, the earth's gravitational field acts on the mass, and the mass experiences a force.

If you have seen the "Star Trek" television series, you may recall an often repeated scene that can help you understand the concept of an electric field. When the starship Enterprise is in danger of attack, Captain Kirk activates an invisible "force field" around the ship. If any incoming missiles enter that force field, they experience a force, causing them to explode before they reach the ship. Like the Enterprise's force field, an electric field is an invisible force field—one that exerts force on electric charges entering the field.

The electric field concept is an alternative to Coulomb's law for viewing the interaction between charges. For example, in the case of two interacting charges, you can either think of the Coulomb force between the two charges, or think of one charge as the source of an electric field that exerts a force on the second charge. As long as all charges are at rest, the two approaches are equivalent.

Introducing the electric field may seem like an unnecessary complication, since we replace something fairly simple and direct, the Coulomb force between charges, by a less direct approach. However, the field approach turns out to be absolutely essential in later chapters, when we need to describe moving charges. When charges move, their fields change, but the effect is not immediately communicated to other charges. Instead, changes in the field propagate through space at the speed of light. The field affecting a given charge at a certain instant is created by another charge that produced the field at an earlier time. For example, the motion of electrons in a television or radio transmitting antenna creates an electric field. The effect of this electric field can be experienced for many miles around the transmitting antenna. When a television or radio in the vicinity is tuned to that particular station, the electrons in the receiving antenna experience a force that, with the help of the receiver's circuits, results in the images and/or sounds produced (Fig. 17-11). The electric field to which the receiving antenna's electrons are responding at any instant was produced earlier by the motion of the transmitting antenna's electrons.

Definition of the Electric Field

We define the electric field E at a point in space to be the force per unit charge that a test charge would experience if placed at that point. Denoting the test charge by q', we have

$$\mathbf{E} = \frac{\mathbf{F}}{q'} \tag{17-6}$$

From this definition it follows that the electric field is a vector quantity and that the SI unit for the electric field is the force unit divided by the charge unit: N/C.

It is important to understand that although the test charge q' enters into the formal definition of the electric field the presence of an electric field at a given point in space does *not* depend on the presence of a test charge at that point. The electric field is the force per unit charge a test charge experiences *if* placed in the field. The electric field is present whether or not there is a test charge to experience its effect.

The source of the electric field may be a single charge or any number of charges (Fig. 17-10). In this section we develop methods for calculating electric fields.

If the electric field at a given point in space is known, we can use the equation defining **E** to find the force **F** on *any* charge q' placed at that point.* Solving Eq. 17-6 for **F**, we obtain

$$\mathbf{F} = q'\mathbf{E} \quad (17\text{-}7)$$

Fig. 17-12 illustrates how an electric field is produced by source charges and how this field produces a force on other charges placed in it.

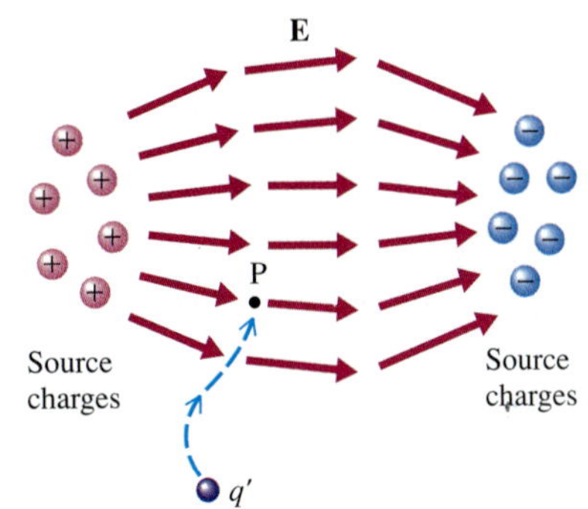

(a) Place charge q' at point P and it experiences a force $\mathbf{F} = q'\mathbf{E}$

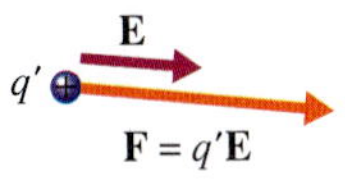

(b) Force on q' when q' is placed at P, assuming q' is positive

Fig. 17-12 Source charges produce an electric field **E** that exerts a force **F** on a charge q' placed in the field.

Single Source Charge

We shall begin by showing how to determine the strength and direction of the electric field produced by a single point charge q. We imagine placing a positive test charge q' at any point P near the source charge q (Fig. 17-13a). The point P where we place the test charge is called the **field point,** defined as that point at which we wish to determine the electric field

If the source charge q is positive, q' is repelled by it. The electric field is defined as the force per unit charge on q' ($\mathbf{E} = \mathbf{F}/q'$), and so the electric field is in the same direction as the force on q'; that is, **the electric field produced by a positive source charge is directed away from the source charge** (Fig. 17-13b).

If the source charge is negative, q' is attracted to it. It follows that **the electric field produced by a negative source charge is directed toward the source charge** (Fig. 17-13c).

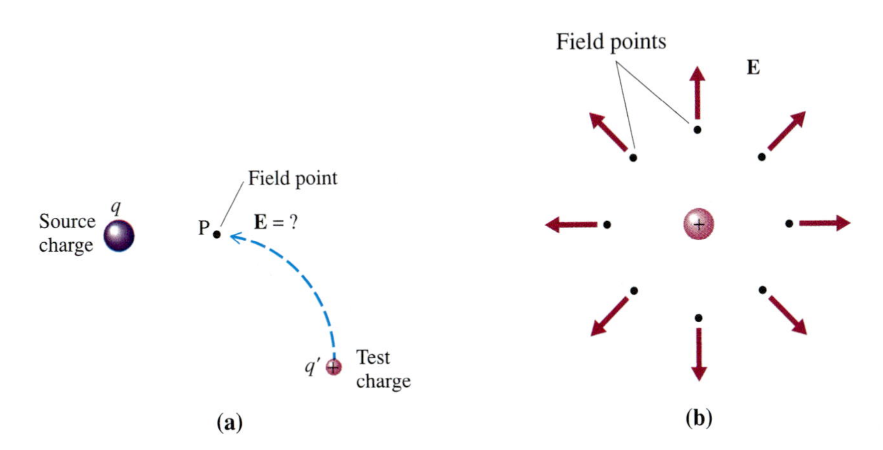

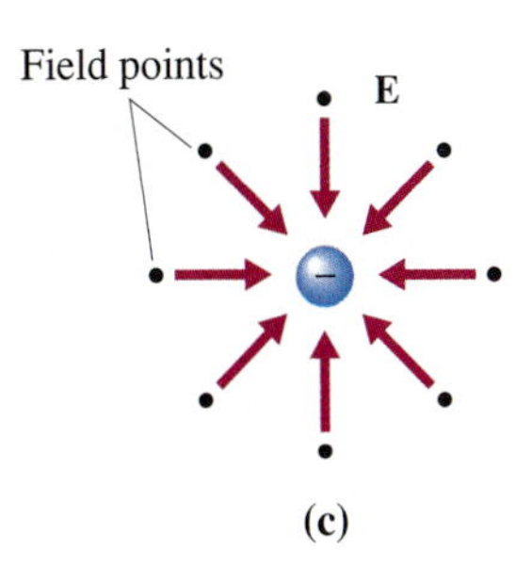

Fig. 17-13 The electric field produced by a single point charge at various field points. **(a)** A test charge at P responds to the field at that point. **(b)** The electric field of a positive charge is directed away from the charge. **(c)** The electric field of a negative charge is toward the charge.

---

*We assume that the presence of q' does not affect the distribution of charge that is the source of the electric field; either q' is so small that it exerts a negligibly small force on the source charges, or the source charges are somehow held in place.

The magnitude of the force q exerts on q' is given by Coulomb's law (Eq. 17-4):

$$F = k\frac{|q||q'|}{r^2}$$

The electric field is defined as the force per unit charge exerted on a test charge q', and so we find the magnitude of the electric field produced by q by dividing both sides of the preceding equation by $|q'|$:

$$E = k\frac{|q|}{r^2} \tag{17-8}$$

This equation gives the magnitude of the electric field produced by a point charge q at any distance r from the charge.

EXAMPLE 4 The Electric Field of a Negative Charge At Various Field Points

Find the magnitude and direction of the electric field produced by a source charge $q = -1.00 \times 10^{-9}$ C (a) at field point a, located 1.00 m to the right of q, and (b) at field point b, located 2.00 m to the left of q. (c) Find the force on either a $+1.00 \times 10^{-12}$ C charge or a -1.00×10^{-11} C charge placed at field point a.

SOLUTION (a) Applying Eq. 17-8 we find

$$E = \frac{k|q|}{r^2} = (8.99 \times 10^9 \text{ N-m}^2/\text{C}^2)\frac{1.00 \times 10^{-9}\text{ C}}{(1.00\text{ m})^2}$$

$$= 8.99 \text{ N/C}$$

Since the source charge is negative, the field is directed toward it, that is, to the left (Fig. 17-14).

(b) Again applying Eq. 17-8, we find

$$E = \frac{k|q|}{r^2} = (8.99 \times 10^9 \text{ N-m}^2/\text{C}^2)\frac{1.00 \times 10^{-9}\text{ C}}{(2.00\text{ m})^2}$$

$$= 2.25 \text{ N/C}$$

At this field point, the field is directed to the right, which is again toward the negative source charge.

The electric field has the same magnitude at all points that are the same distance r from the source charge. The field is thus spherically symmetric. Fig. 17-14 shows the electric field at a and b, as well as at other selected field points at distances of 1 m and 2 m from q. A positive source charge would produce a similar electric field pattern, except that the direction of **E** would be reversed.

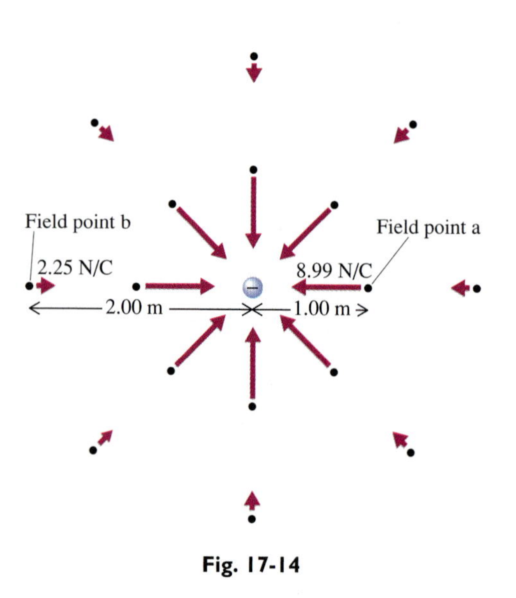

Fig. 17-14

(c) If we place a charge q' in the electric field, according to Eq. 17-7, the charge experiences a force $\mathbf{F} = q'\mathbf{E}$. Thus a charge of $+1.00 \times 10^{-12}$ C placed at a, where $\mathbf{E} = 8.99$ N/C to the left, would experience a force to the left of magnitude 8.99×10^{-12} N. A charge of -1.00×10^{-11} C placed at a would experience a force to the right (in the $-\mathbf{E}$ direction) of magnitude 8.99×10^{-11} N.

EXAMPLE 5 The Gravitational Field of the Earth

The gravitational field is defined as the gravitational force per unit mass that would be experienced by a test mass placed at the point where the field is to be evaluated. Find the magnitude and direction of the gravitational field of the earth at field points that are at distances R and $2R$ from the center of the earth, where R is the earth's radius. Use 6.0×10^{24} kg as the mass of the earth and 6.4×10^6 m as its radius.

SOLUTION The earth is a spherically symmetric distribution of mass M that exerts an attractive force **F** on any other mass m. As seen in Chapter 6, this force is directed toward the center of the earth and has magnitude

$$F = G\frac{mM}{r^2}$$

where G is the gravitational constant (6.67×10^{-11} N-m^2/kg^2) and r is the distance from m to the center of the earth. Applying the definition of the gravitational field, which we denote by **g**, we find

$$g = \frac{F}{m} = \frac{GM}{r^2}$$

The gravitational field equals gravitational acceleration. The units are either N/kg or, equivalently, m/s^2. At $r = R$, we find

$$g = \frac{GM}{R^2} = \frac{(6.67 \times 10^{-11}\ \text{N-m}^2/\text{kg}^2)(6.0 \times 10^{24}\ \text{kg})}{(6.4 \times 10^6\ \text{m})^2}$$

$$= 9.8\ \text{N/kg}$$

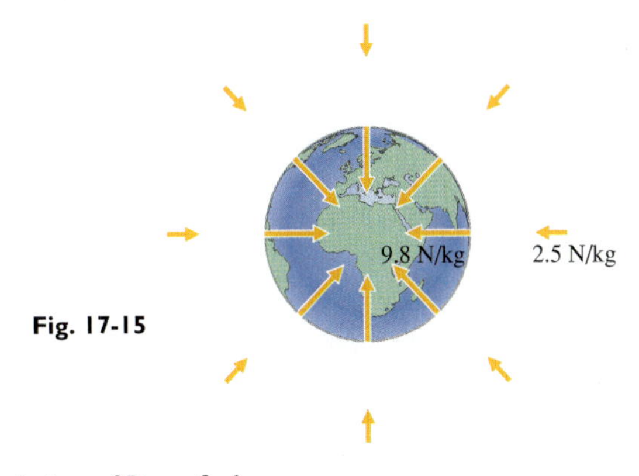

Fig. 17-15

And at $r = 2R$, we find

$$g = \frac{GM}{(2R)^2} = \frac{1}{4}\left(\frac{GM}{R^2}\right) = \tfrac{1}{4}(9.8\ \text{N/kg})$$

$$= 2.5\ \text{N/kg}$$

Fig. 17-15 shows the earth's gravitational field evaluated at selected field points at distances R and $2R$ from the center. The field pattern is the same as for a negative charge (Fig. 17-14).

It follows from the definition of the gravitational field as force per unit mass that if we place a mass m in a gravitational field **g** the mass experiences a force $\mathbf{F} = m\mathbf{g}$. Thus a 10 kg mass at the surface of the earth, where $\mathbf{g} = 9.8$ N/kg directed toward the earth's center, will experience a force of magnitude 98 N in the direction of **g**. At a distance $r = 2R$ from the center of the earth, where $\mathbf{g} = 2.5$ N/kg directed inward, the same 10 kg mass would experience a force of magnitude 25 N in the direction of **g**.

Group of Point Source Charges

When the source of an electric field is a group of point charges ($q_1, q_2, \ldots, q_n$), the electric field is the **resultant** electric force per unit charge on a test charge q' placed at a field point; that is,

$$\mathbf{E} = \frac{\mathbf{F}_1 + \mathbf{F}_2 + \cdots + \mathbf{F}_n}{q'}$$

$$= \frac{\mathbf{F}_1}{q'} + \frac{\mathbf{F}_2}{q'} + \cdots + \frac{\mathbf{F}_n}{q'}$$

Each term in this equation is the field that would be produced by one charge alone. **So the total electric field E is the vector sum of the fields produced by the individual charges:**

$$\mathbf{E} = \mathbf{E}_1 + \mathbf{E}_2 + \cdots + \mathbf{E}_n \qquad (17\text{-}9)$$

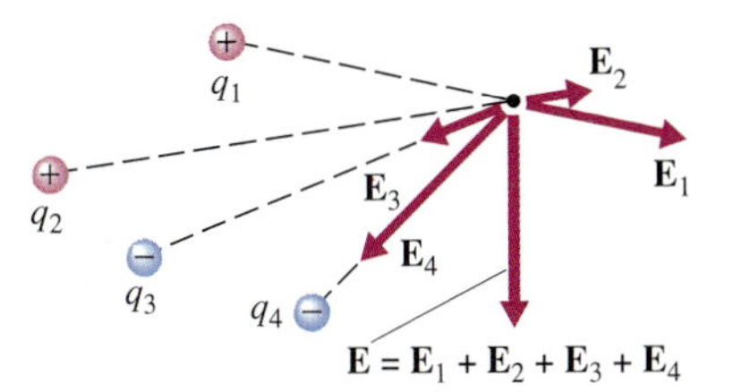

Fig. 17-16 The field produced by a group of point charges is the vector sum of the single charge fields.

Fig. 17-16 illustrates the calculation of the electric field. You first calculate the electric field of individual charges as before and then compute the **vector** sum of these fields.

EXAMPLE 6 The Electric Field of Two Charges

Point charges $q_1 = +5.00 \times 10^{-9}$ C and $q_2 = -5.00 \times 10^{-9}$ C are on the x-axis with respective coordinates $x = -5.00$ cm and $x = +5.00$ cm. Find the electric field at three points: point a at the origin; point b on the x-axis at $x = -10.0$ cm; and point c on the y-axis, 10.0 cm from each charge.

SOLUTION

At field point a, charges q_1 and q_2 produce fields $\mathbf{E}_1$ and $\mathbf{E}_2$, both directed to the right, away from q_1 and toward q_2 (Fig. 17-17). The fields have equal magnitudes, since the source charges have equal magnitudes and are equal distances away from a:

$$E_2 = E_1 = k\frac{|q_1|}{r_1^2} = (8.99 \times 10^9 \text{ N/C})\frac{5.00 \times 10^{-9} \text{ C}}{(5.00 \times 10^{-2} \text{ m})^2}$$
$$= 1.80 \times 10^4 \text{ N/C}$$

The total field $\mathbf{E}$ at point a is directed to the right and has magnitude

$$E = E_1 + E_2 = 2(1.80 \times 10^4 \text{ N/C})$$
$$= 3.60 \times 10^4 \text{ N/C}$$

At field point b, q_1 produces a field $\mathbf{E}_1$ to the left (away from q_1) and q_2 produces a field $\mathbf{E}_2$ to the right (toward q_2):

$$E_1 = k\frac{|q_1|}{r_1^2} = (8.99 \times 10^9 \text{ N/C})\frac{5.00 \times 10^{-9} \text{ C}}{(5.00 \times 10^{-2} \text{ m})^2} = 1.80 \times 10^4 \text{ N/C}$$

$$E_2 = k\frac{|q_2|}{r_2^2} = (8.99 \times 10^9 \text{ N/C})\frac{5.00 \times 10^{-9} \text{ C}}{(15.0 \times 10^{-2} \text{ m})^2} = 0.200 \times 10^4 \text{ N/C}$$

The total electric field is directed to the left and has magnitude

$$E = E_1 - E_2 = 1.60 \times 10^4 \text{ N/C}$$

At point c, $\mathbf{E}_1$ and $\mathbf{E}_2$ are directed as indicated in Fig. 17-17. The fields have equal magnitudes because the two source charges are equidistant from c:

$$E_2 = E_1 = k\frac{|q_1|}{r_1^2} = (8.99 \times 10^9 \text{ N/C})\frac{5.00 \times 10^{-9} \text{ C}}{(0.100 \text{ m})^2}$$
$$= 4.50 \times 10^3 \text{ N/C}$$

The total field is found by vector addition of $\mathbf{E}_1$ and $\mathbf{E}_2$:

$$\mathbf{E} = \mathbf{E}_1 + \mathbf{E}_2$$

$$E_x = E_{1x} + E_{2x} = (4500 \text{ N/C})(\cos 60.0°) + (4500 \text{ N/C})(\cos 60.0°)$$
$$= 4.50 \times 10^3 \text{ N/C}$$

The vertical components of $\mathbf{E}_1$ and $\mathbf{E}_2$ cancel, and so the y component of $\mathbf{E}$ is 0. Thus the field is directed to the right and has a magnitude of 4.50×10^3 N/C.

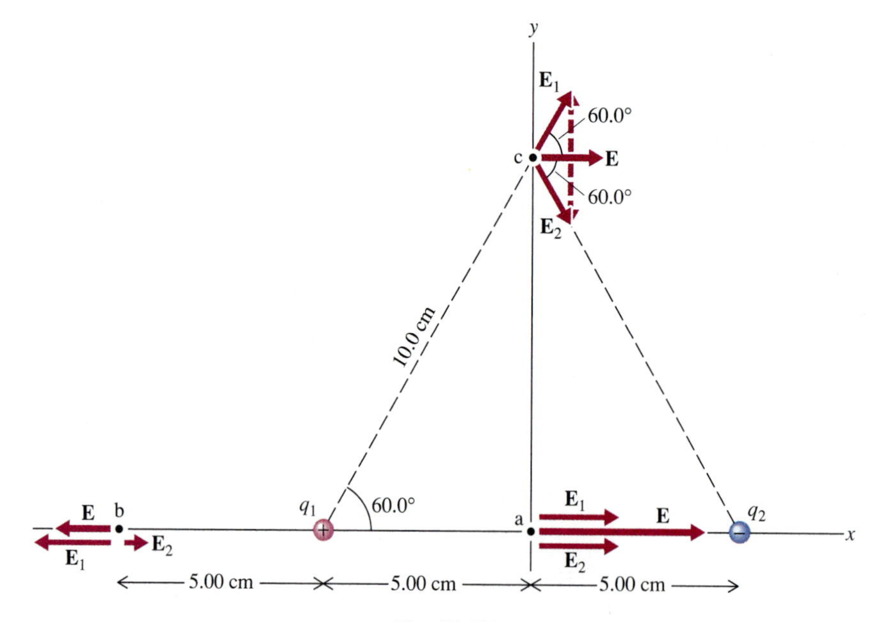

Fig. 17-17

EXAMPLE 7 Acceleration of an Electron in an Electric Field

Find the instantaneous acceleration of an electron placed at field point c in the previous example.

SOLUTION First we calculate the force on the electron, using Eq. 17-7 ($\mathbf{F} = q'\mathbf{E} = -e\mathbf{E}$). Since the charge is negative, the direction of the force is opposite the direction of the electric field. The field at c is directed to the right, and so the force on the electron is directed to the left. We find the magnitude of the force by multiplying the magnitude of the electron's charge times the magnitude of the field at c, found in the previous example to be 4.50×10^3 N/C:

$$F = |q'|E = eE$$
$$= (1.60 \times 10^{-19}\ \text{C})(4.50 \times 10^3\ \text{N/C})$$
$$= 7.20 \times 10^{-16}\ \text{N}$$

Since this is the only force acting on the electron, the acceleration of the electron, according to Newton's second law, is also directed to the left and has magnitude

$$a = \frac{F}{m} = \frac{7.20 \times 10^{-16}\ \text{N}}{9.11 \times 10^{-31}\ \text{kg}}$$
$$= 7.90 \times 10^{14}\ \text{m/s}^2$$

EXAMPLE 8 The Electric Field of Three Charges

Charges $q_1 = 1.0 \times 10^{-6}$ C, $q_2 = -1.0 \times 10^{-6}$ C, and $q_3 = 2.0 \times 10^{-6}$ C are located on the y-axis, with respective y coordinates 10 cm, 0, and −10 cm. Find the electric field at the field point having coordinates $x = 10$ cm, $y = 0$.

SOLUTION First we indicate in Fig. 17-18 the directions of $\mathbf{E}_1$, $\mathbf{E}_2$, and $\mathbf{E}_3$ produced by q_1, q_2, and q_3. Next we calculate the magnitude of the fields:

$$E_1 = k\frac{|q_1|}{r_1^2} = (9.0 \times 10^9\ \text{N-m}^2/\text{C}^2)\frac{1.0 \times 10^{-6}\ \text{C}}{(0.10\sqrt{2}\ \text{m})^2} = 4.5 \times 10^5\ \text{N/C}$$

$$E_2 = k\frac{|q_2|}{r_2^2} = (9.0 \times 10^9\ \text{N-m}^2/\text{C}^2)\frac{1.0 \times 10^{-6}\ \text{C}}{(0.10\ \text{m})^2} = 9.0 \times 10^5\ \text{N/C}$$

$$E_3 = k\frac{|q_3|}{r_3^2} = (9.0 \times 10^9\ \text{N-m}^2/\text{C}^2)\frac{2.0 \times 10^{-6}\ \text{C}}{(0.10\sqrt{2}\ \text{m})^2} = 9.0 \times 10^5\ \text{N/C}$$

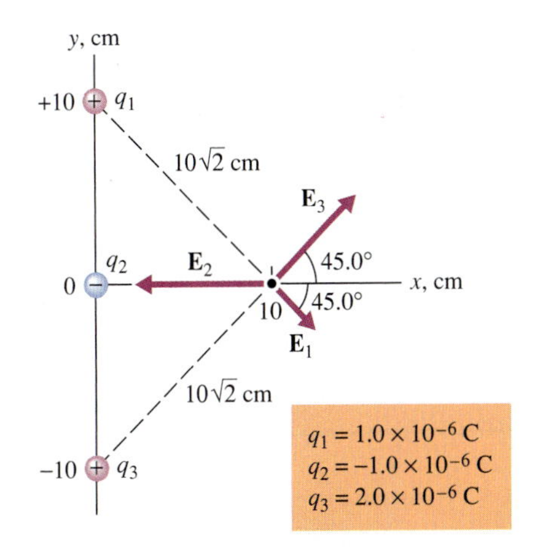

Fig. 17-18

Finally we calculate the vector sum, using the calculated magnitudes and the directions indicated in Fig. 17-18:

$$E_x = E_{1x} + E_{2x} + E_{3x}$$
$$= (4.5 \times 10^5\ \text{N/C})(\cos 45°) - 9.0 \times 10^5\ \text{N/C} + (9.0 \times 10^5\ \text{N/C})(\cos 45°)$$
$$= 5.5 \times 10^4\ \text{N/C}$$

$$E_y = E_{1y} + E_{2y} + E_{3y}$$
$$= -(4.5 \times 10^5\ \text{N/C})(\sin 45°) + 0 + (9.0 \times 10^5\ \text{N/C})(\sin 45°)$$
$$= 3.2 \times 10^5\ \text{N/C}$$

This is a vector of magnitude 3.2×10^5 N/C, directed 80° above the positive x-axis.

In the examples above we identified certain charges as sources of an electric field and then calculated the magnitude and direction of that field. You should realize that these source charges can also respond to a field. Thus each charge both is the source of an electric field and also experiences a force caused by the electric field of all other charges.

17-4 Fields Produced by Continuous Distributions of Charge

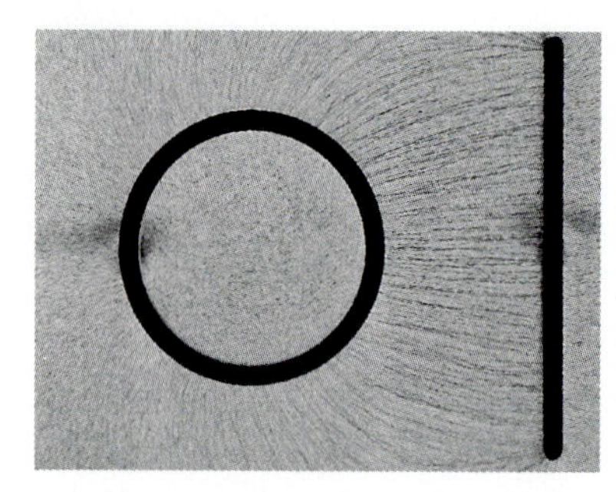

Fig. 17-19 The electric field produced by two charged conductors is made visible by a suspension in oil of bits of thread, which align with the field.

Often the source of an electric field is more than just a few point charges, and so the methods of the last section are not sufficient to compute the field. For example, a charged metal plate might have perhaps 10^{10} excess electrons on its surface. In computing the electric field produced by the plate, we certainly can't compute the fields of 10^{10} electrons, point by point. Instead we can think of the excess electrons as forming a continuous distribution of charge on the plate's surface and in so doing simplify the computation of the field. In this section we describe the electric fields produced by such charge distributions (Fig. 17-19).

The first step in evaluating the electric field produced at a field point P by a continuous distribution of charge is to divide the charge into tiny elements of charge Δq and then consider the field $\Delta\mathbf{E}$ produced by each such element at P (Fig. 17-20). Since Δq is so small, we can use the formula for a point charge to express the magnitude of the field $\Delta\mathbf{E}$ produced by Δq. Applying Eq. 17-8 $\left(E = k\frac{|q|}{r^2}\right)$ to our "point" charge Δq, we may write

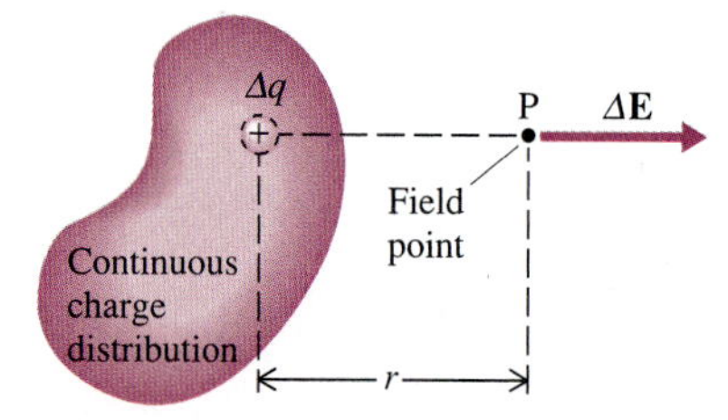

Fig. 17-20 Field $\Delta\mathbf{E}$ produced at a field point P by a small charge element Δq (assumed positive), part of a continuous charge distribution.

$$|\Delta\mathbf{E}| = k\frac{|\Delta q|}{r^2} \tag{17-10}$$

The field $\Delta\mathbf{E}$ is directed away from Δq if Δq is positive and toward Δq if Δq is negative. We find the total electric field at P by applying the superposition principle, that is, by summing the $\Delta\mathbf{E}$'s that were calculated using Eq. 17-10:

$$\mathbf{E} = \Sigma\,(\Delta\mathbf{E}) \tag{17-11}$$

EXAMPLE 9 Electric Field of a Charged Ring

A thin ring is uniformly charged with a positive charge Q. A field point P is located on the axis of the ring at a distance r from the edge of the ring (Fig. 17-21). Derive an expression for the electric field at P.

SOLUTION Charge Q is uniformly spread over the ring, and since each element of charge Δq is the same distance r from the field point, each contributes a field of the same magnitude ΔE. Fig. 17-21 shows a pair of charge elements, Δq and $\Delta q'$, located on opposite sides of the ring, along with their respective fields, $\Delta\mathbf{E}$ and $\Delta\mathbf{E}'$, at field point P. The components perpendicular to the axis, $\Delta E_\perp$ and $\Delta E'_\perp$, cancel. This means that the only components contributing to the total field are those along the x-axis, ΔE_x and $\Delta E'_x$. Since the entire ring can be divided into other pairs of charges comparable to Δq and $\Delta q'$, all field components perpendicular to the x-axis cancel. So only components along the x-axis need to be summed to obtain the total electric field E_x:

$$E_x = \Sigma\ (\Delta E_x)$$

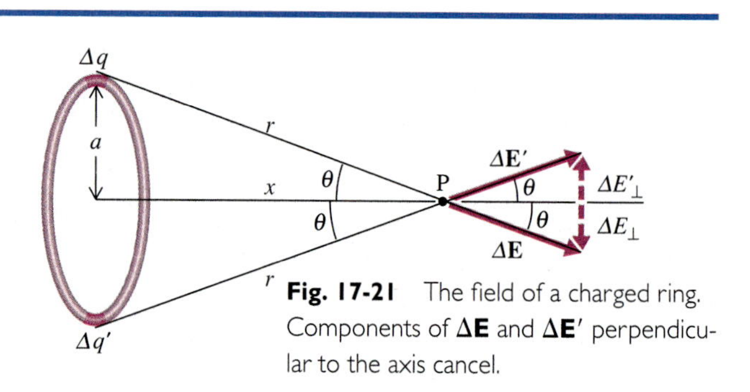

Fig. 17-21 The field of a charged ring. Components of $\Delta\mathbf{E}$ and $\Delta\mathbf{E}'$ perpendicular to the axis cancel.

Each of the n charge elements Δq contributes an equal component $\Delta E_x = \Delta E \cos\theta$, where ΔE is given by Eq. 17-10 $\left(\Delta E = k\frac{|\Delta q|}{r^2}\right)$.

Thus

$$E_x = \Sigma\ \left[k\frac{|\Delta q|}{r^2}(\cos\theta)\right] = nk\frac{\Delta q}{r^2}(\cos\theta)$$

Since $n\,\Delta q$ equals the total charge Q on the ring, we may express this result as

$$E_x = k\frac{Q}{r^2}(\cos\theta) \tag{17-12}$$

Charge Density

In the preceding example, charge was distributed over a line. More often, however, we encounter problems in which charge is distributed over either a surface area or a volume. Such distributions lead to the definitions of surface charge density and volume charge density. **Surface charge density,** denoted by the Greek letter σ (sigma), is defined as **charge per unit area. Volume charge density,** denoted by the Greek letter ρ (rho), is defined as **charge per unit volume.**

$$\sigma = \frac{Q}{A} \tag{17-13}$$

$$\rho = \frac{Q}{V} \tag{17-14}$$

The units for σ are C/m^2, and the units for ρ are C/m^3.

EXAMPLE 10 Charge Density in a Uranium Nucleus

The uranium nucleus has a radius of 7.4×10^{-15} m and carries a charge of +92e. Find the charge density within the nucleus.

SOLUTION Since the nuclear charge is spread over the volume of the nucleus, we apply the definition of volume charge density (Eq. 17-14) and use the formula for the volume of a sphere of radius r ($V = \frac{4}{3}\pi r^3$).

$$\rho = \frac{Q}{V} = \frac{+92e}{\frac{4}{3}\pi r^3} = \frac{+92(1.6 \times 10^{-19}\text{ C})}{\frac{4}{3}\pi(7.4 \times 10^{-15}\text{ m})^3}$$

$$= 8.7 \times 10^{24}\text{ C/m}^3$$

Such a large charge density can occur within a nucleus because the large forces of electrical repulsion between the protons are balanced by attractive nuclear forces.

Uniformly Charged Infinite Plane

An important special case of a continuous charge distribution is the infinite, uniformly charged plane (Fig. 17-22a). The solution of a problem involving an infinite distribution of charge might seem to be an exercise of no physical significance. However, the solution to this problem is a good approximation to the field a *finite* plane of charge produces at field points that are close enough to the surface of the plane and far enough from the edges to make the plane "look" infinite (Fig. 17-22b). For such field points, the missing part of the infinite plane (from the edges of the actual plane out to infinity) will contribute a negligible amount to the total electric field. Thus the realistic problem of a finite plane of charge can be approximated by the problem of an infinite plane, which has a simple solution.

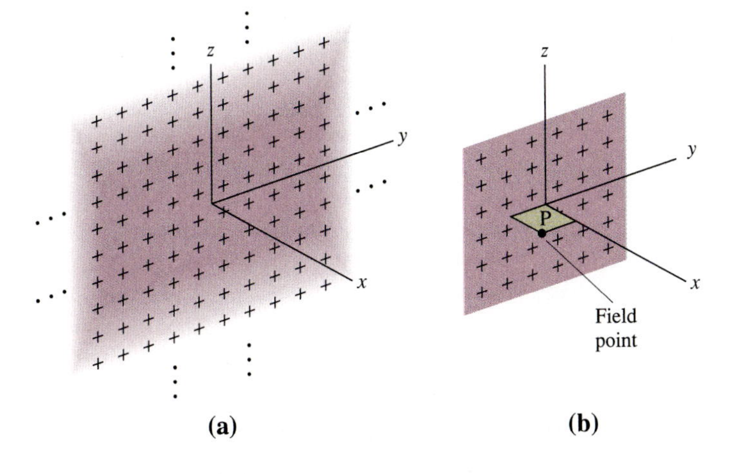

Fig. 17-22 **(a)** An infinite, uniformly charged plane. **(b)** A finite, uniformly charged plane. Point P is close enough to the plane and far enough from the edges that the plane looks infinite.

It is possible to derive the infinite plane's electric field by applying Eq. 17-11, but since this requires the use of integral calculus, we shall state the result here without proof. (However, a derivation based on Gauss's law is provided in Appendix B.)

The electric field of an infinite plane having a uniform surface charge density σ is uniform, is directed perpendicular to the plane, and has magnitude

$$E = 2\pi k|\sigma| \quad \text{(infinite plane; constant } \sigma\text{)} \qquad (17\text{-}15)$$

The field of a positively charged plane is directed away from the plane (Fig. 17-23a). The field of a negatively charged plane is directed toward the plane (Fig. 17-23b).

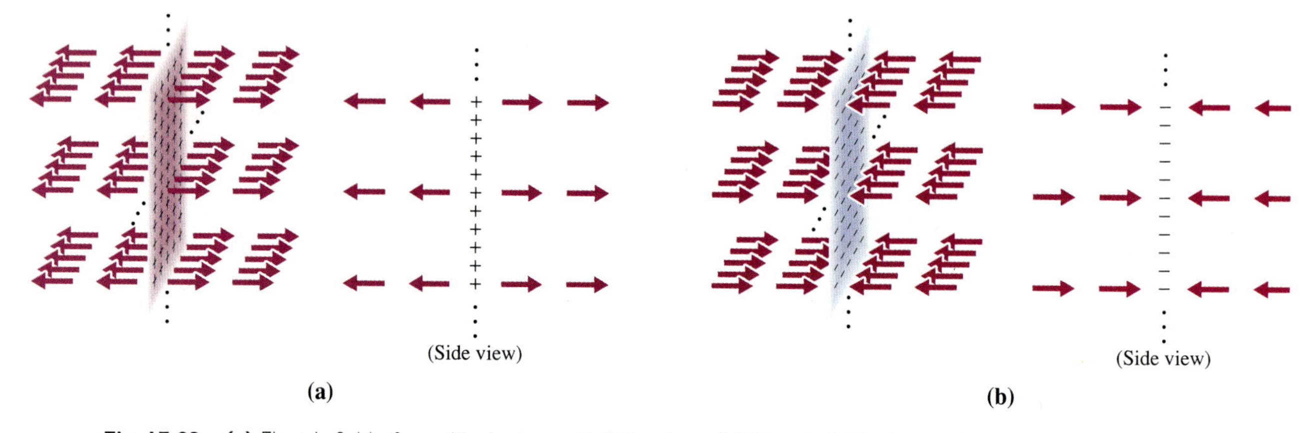

Fig. 17-23 **(a)** Electric field of a positively charged infinite plane. **(b)** Electric field of a negatively charged infinite plane.

EXAMPLE 11 Electric Field of a Charged Sheet of Photocopy Paper

Inside a photocopier, a sheet of copy paper with dimensions of 20 cm by 30 cm has 1.0×10^{11} electrons removed from one side, producing a uniform positive surface charge. The purpose of this positive charge is to attract the negatively charged black ink, or "toner," from the photocopier drum, where the image of the original is first formed; when the toner is attracted off the drum, the image is transferred onto the copy paper. Find the electric field at a point 1.0 cm from the surface of the paper and not too close to the edge.

SOLUTION The uniformly charged sheet has a surface charge density

$$\sigma = \frac{Q}{A} = \frac{(1.0 \times 10^{11})e}{A} = \frac{(1.0 \times 10^{11})(1.60 \times 10^{-19}\text{ C})}{(0.20\text{ m})(0.30\text{ m})}$$
$$= +2.7 \times 10^{-7}\text{ C/m}^2$$

This charge produces an electric field that is directed away from the sheet and has a magnitude given by Eq. 17-15:

$$E = 2\pi k|\sigma| = 2\pi(9.0 \times 10^9\text{ N-m}^2/\text{C}^2)(2.7 \times 10^{-7}\text{ C/m}^2)$$
$$= 1.5 \times 10^4\text{ N/C}$$

Isolated Conducting Plate

Suppose a net positive charge is placed on a large, isolated metal plate, say, by removal of some electrons through static charging. The remaining free electrons in the metal will arrange themselves in such a way that the net positive charge will be distributed equally over the surface of the plate, with equal charge on the two sides (Fig. 17-24). The electric field at any point is the vector sum of the fields produced by the sides, each of which has a positive charge density σ. Each side produces a field that is approximately the field of an infinite plane, that is, a field of magnitude $2\pi k\sigma$, directed away from the surface. To either side of the plate, the fields $\mathbf{E}_1$ and $\mathbf{E}_2$ produced by the two surface charges are in the same direction, and thus the total field to either side is directed away from the plate and has a magnitude twice that produced by one surface:

$$E = 2(2\pi k\sigma) = 4\pi k\sigma \qquad (17\text{-}16)$$

Inside the metal, the direction of $\mathbf{E}_1$ is opposite the direction of $\mathbf{E}_2$. Thus these fields cancel and the resultant field inside is zero:

$$\mathbf{E} = \mathbf{E}_1 + \mathbf{E}_2 = \mathbf{0} \qquad \text{(inside conductor)} \qquad (17\text{-}17)$$

Indeed, the static charge arranges itself on the surface of the plate in such a way as to produce zero field inside. If a nonzero field $\mathbf{E}$ were present inside, the free electrons in the metal would experience a force $\mathbf{F} = -e\mathbf{E}$ and would keep moving until the arrangement of charge produced zero field inside.

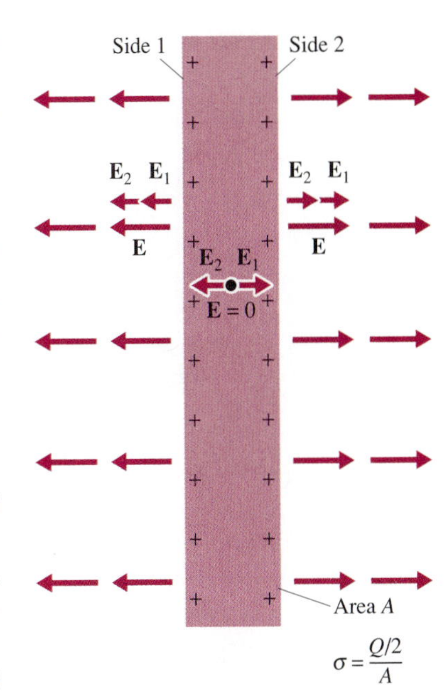

Fig. 17-24 Side view of an isolated conducting plate, with charge equally divided between the two sides. The two charged sides produce fields ($\mathbf{E}_1$ and $\mathbf{E}_2$) that cancel inside the plate but not outside.

Properties of Conductors

Several features of the charged metal plate are properties of charged conductors of *any* shape. The most important general properties of charged conductors are:

1. **The electric field inside a statically charged conductor always equals zero.** The free electrons in the conductor keep moving until the charge distribution produces zero electric field inside.
2. **Any net charge on a conductor is on the surface.** It is possible to prove this property by applying Gauss's law (Appendix B). Gauss's law shows why a net charge inside a conductor must always produce a nonzero field inside, and since there is no field inside, there can be no charge.
3. **The electric field within a cavity in a conductor is always zero,** if there is no charge inside the cavity. This property can be proved by use of Gauss's law and the fact that the electric force is a conservative force. This feature of conductors can be utilized to shield a region of space from electric fields produced by charges outside the region. All that is required is a closed conducting surface surrounding the region of space to be shielded. An automobile's metal body almost completely surrounds the automobile's interior with a conductor and hence provides good protection against the car being penetrated by electric fields, produced, for example, by an electrical storm (Fig. 17-25).
4. **The electric field just outside the surface of a conductor is perpendicular to the surface and has magnitude**

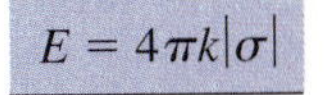

$$E = 4\pi k|\sigma| \qquad (17\text{-}18)$$

where $|\sigma|$ is the magnitude of the surface charge density on the part of the conductor closest to the field point (Fig. 17-26). This property is a generalization of the result for a conducting plate and, like properties 2 and 3, can be derived from Gauss's law (Appendix B).

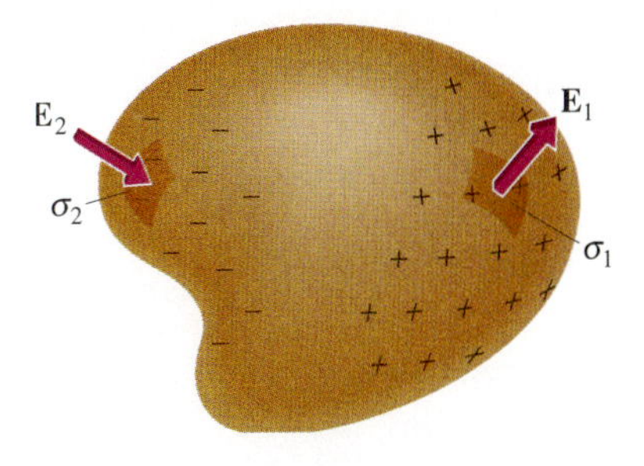

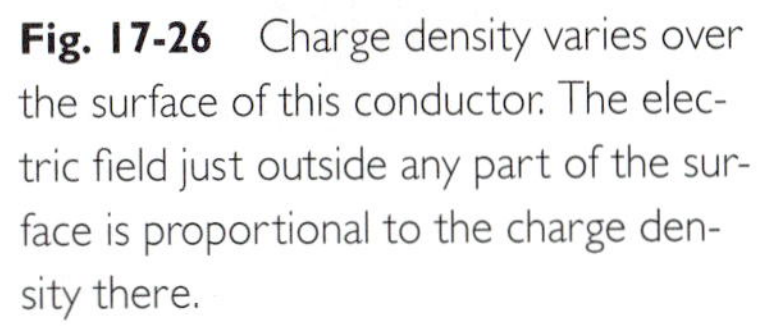

Fig. 17-26 Charge density varies over the surface of this conductor. The electric field just outside any part of the surface is proportional to the charge density there.

Fig. 17-25 The metal body of a car shields the interior from electric fields.

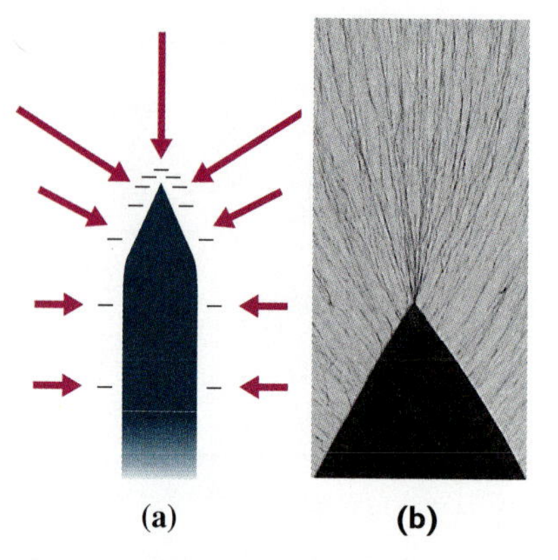

Fig. 17-27 **(a)** The electric field around a charged conductor is strongest near a sharp point. **(b)** Bits of thread suspended in oil, surround a sharply pointed, charged conductor. Notice that the threads are concentrated near the point, indicating that the field is strongest there.

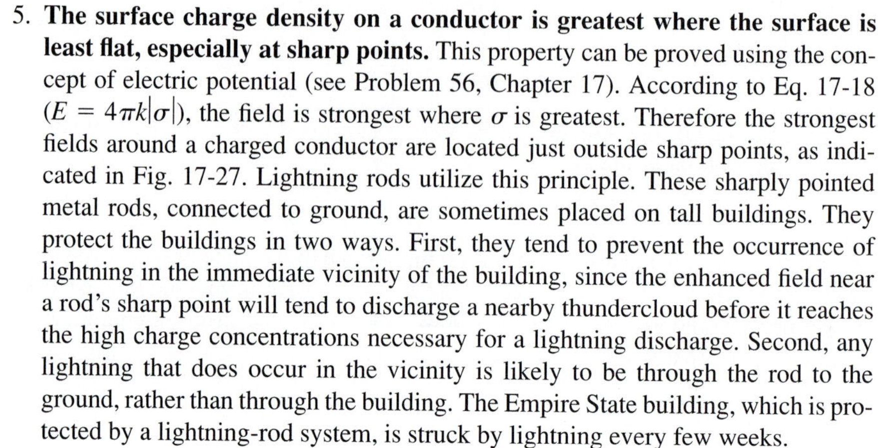

5. **The surface charge density on a conductor is greatest where the surface is least flat, especially at sharp points.** This property can be proved using the concept of electric potential (see Problem 56, Chapter 17). According to Eq. 17-18 ($E = 4\pi k|\sigma|$), the field is strongest where σ is greatest. Therefore the strongest fields around a charged conductor are located just outside sharp points, as indicated in Fig. 17-27. Lightning rods utilize this principle. These sharply pointed metal rods, connected to ground, are sometimes placed on tall buildings. They protect the buildings in two ways. First, they tend to prevent the occurrence of lightning in the immediate vicinity of the building, since the enhanced field near a rod's sharp point will tend to discharge a nearby thundercloud before it reaches the high charge concentrations necessary for a lightning discharge. Second, any lightning that does occur in the vicinity is likely to be through the rod to the ground, rather than through the building. The Empire State building, which is protected by a lightning-rod system, is struck by lightning every few weeks.

17-5 Field Lines

The direction of an electric field can be graphically represented by continuous lines called **field lines. The direction of the field at any point is the direction of the tangent to the field line at that point** (Fig. 17-28).

The magnitude of the electric field can also be indicated by field lines but not by their length, since the lines are continuous. Instead, **the spacing of the lines indicates the strength of the field.** Consider any surface perpendicular to the field lines (Fig. 17-29). The lines are drawn so that the magnitude of **E** is proportional to the number n of field lines per unit area through a surface of area $A_\perp$, perpendicular to the field lines:

$$E \propto \frac{n}{A_\perp} \tag{17-19}$$

Direction of the electric field at P

E

P

Q

Direction of the electric field at Q

Fig. 17-28 Field lines.

Where the field lines are closer together, the number of lines per unit area is greater and so the electric field is stronger. Where the field lines are farther apart, the electric field is weaker (Fig. 17-29).

The field line representation is useful because it is often possible to draw field lines that are continuous through most regions of space. **Field lines begin and end only at points where there is electric charge.** This general property of field lines is proved in Appendix B on Gauss's law. However, it is easy to verify the continuity of field lines for two special cases: the uniformly charged, infinite plane and the point charge.

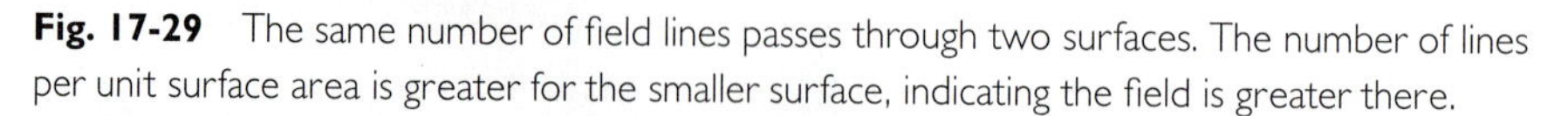

Fig. 17-29 The same number of field lines passes through two surfaces. The number of lines per unit surface area is greater for the smaller surface, indicating the field is greater there.

Field Lines for a Uniformly Charged, Infinite Plane

As described in the previous section, the field of a uniformly charged, infinite plane is uniform and perpendicular to the plane. The field line representation of such a field consists of continuous, equally spaced lines extending out in either direction from the plane and perpendicular to it (Fig. 17-30).

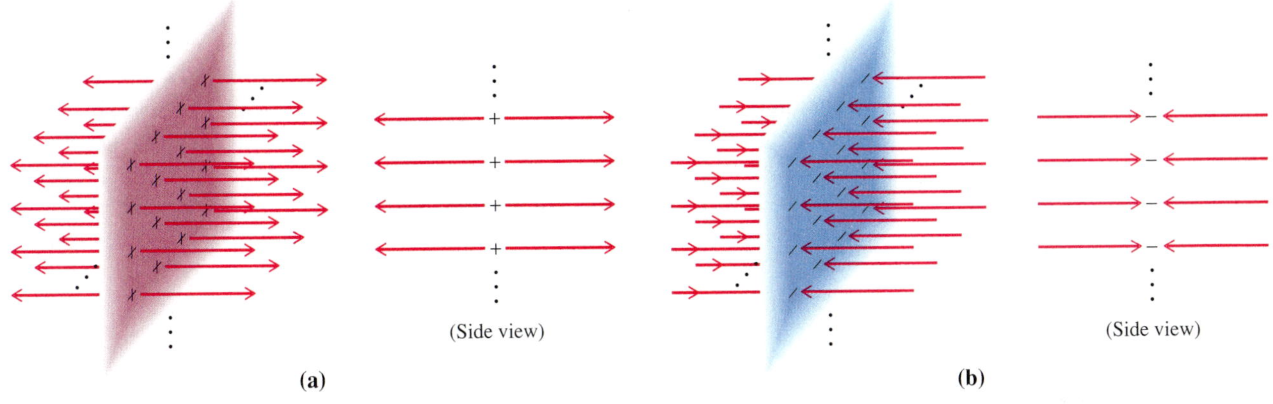

Fig. 17-30 Field lines for a uniformly charged infinite plane. **(a)** Positively charged plane. **(b)** Negatively charged plane.

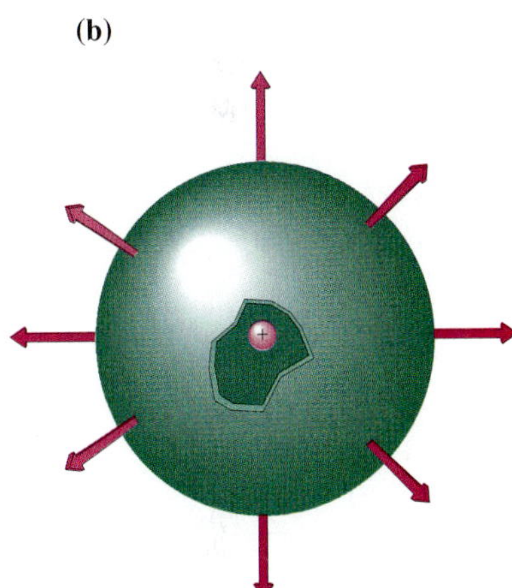

Fig. 17-31 This spherical surface is perpendicular to the electric field produced by a point charge at the center of the sphere.

According to Eq. 17-19, electric field strength is proportional to the number of field lines n per unit area $A_\perp$. Thus the number of field lines through a perpendicular surface is proportional to the product of field strength and surface area:

$$n \propto EA_\perp$$

Or we may introduce a constant of proportionality c and express this result as

$$n = cEA_\perp \tag{17-20}$$

The constant c is a scale factor, chosen so as to obtain the desired number of lines in a drawing.

Field Lines for a Single Point Charge

We shall apply Eq. 17-20 to the problem of representing by field lines the electric field of a positive point charge q. Since the field is directed radially outward from the charge, we choose as a perpendicular surface a spherical surface that has radius r and is centered on the charge (Fig. 17-31). We apply Eq. 17-20, using the expression for the field of a point charge q at a distance r and the equation for the surface area of a sphere of radius r:

$$n = cEA_\perp = c\left(k\frac{q}{r^2}\right)(4\pi r^2)$$

or

$$n = 4\pi kqc \tag{17-21}$$

Notice that this expression for n involves only constants. This means that the number of field lines passing through a sphere of *any* radius r is the same, independent of r. Thus we can represent the field by continuous field lines (Fig. 17-32). All the lines that originate at q pass through any spherical surface centered on q. The number of lines drawn from q will depend on the value chosen for the scale factor c.

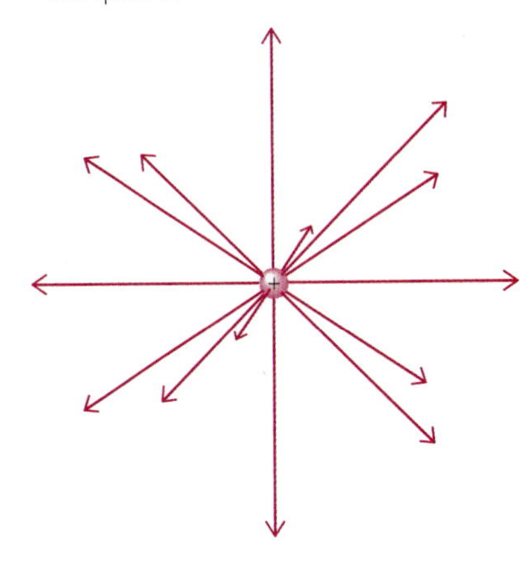

Fig. 17-32 Field lines of a point charge.

Field of a Uniformly Charged Sphere

A charged sphere in which charge is uniformly distributed throughout the volume of the sphere produces an electric field that, for field points outside the sphere, is the same as though the charge were concentrated at the center of the sphere; that is, the field outside the sphere is the field of a point charge.* This result, proved in Appendix B on Gauss's Law, means that the field lines outside a uniform sphere of charge look the same as for a point charge (Fig. 17-32).

Two-Dimensional Drawings of Field Lines

It is difficult to indicate a three-dimensional field line pattern by a perspective drawing. Therefore, we shall often use two-dimensional drawings of field lines. For example, Fig. 17-33 shows in two dimensions the field lines of an electric dipole, which consists of two point charges of opposite sign but equal magnitude. When viewing such figures, you should keep in mind that the actual pattern of field lines is three-dimensional.

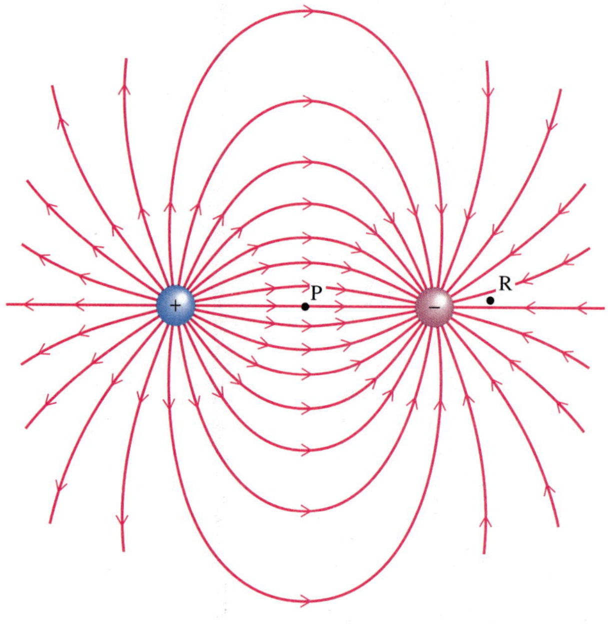

Fig. 17-33 Field lines of a dipole.

EXAMPLE 12 Finding the Strength of a Field From the Number of Field Lines

Suppose that two small surfaces of equal area are drawn in Fig. 17-33—one a plane surface passing through point P and perpendicular to field lines there and the other a spherical surface centered on the negative charge and passing through point R. Suppose that 9 field lines pass through the plane surface and 72 pass through the spherical surface. If the strength of the electric field at P is 200 N/C, what is it at R?

SOLUTION Since field strength is proportional to the number of lines per unit area and the two surface areas are equal, the field strength is proportional to the number of lines. Thus the field strength at R is

$$E_R = \tfrac{72}{9}E_P = 8(200 \text{ N/C})$$

$$= 1600 \text{ N/C}$$

*Similarly, the earth's gravitational field is the same for points outside the earth as though the earth's mass were concentrated at its center. (See Chapter 6, Fig. 6-6.)

CHAPTER 17 SUMMARY

Electric charge q, measured in coulombs (C), may be either positive or negative. Charges of like sign repel one another; charges of opposite sign attract one another. Uncharged matter consists of equal numbers of protons ($q = +e$) and electrons ($q = -e$), where

$$e = 1.60 \times 10^{-19}\text{ C}$$

Electrons may be transferred to or from a body, giving it a net charge proportional to the number n of electrons transferred:

$$q = \pm ne$$

The magnitude of the mutual attractive or repulsive force between point charges is proportional to the product of the magnitude of the charges and inversely proportional to the square of the distance r between them, according to Coulomb's law:

$$F = k\frac{|q||q'|}{r^2}$$

where $k = 8.99 \times 10^9\text{ N-m}^2/\text{C}^2$

The superposition principle states that the resultant force produced by a number of charges on a charge q is found by first calculating the magnitudes and directions of the Coulomb forces each charge exerts on q and then finding the vector sum of these forces.

The electric field $\mathbf{E}$ is a vector quantity produced by a distribution of charge in the region of space surrounding that charge. A point in space where the field is evaluated is called a "field point."

A charge q' in an electric field experiences a force:

$$\mathbf{F} = q'\mathbf{E}$$

where $\mathbf{E}$ is the field at the location of q'; $\mathbf{E}$ is produced by other charges.

The field produced by a single point charge q is directed away from q if q is positive and toward q if q is negative. The magnitude of the field of a point charge at a distance r away from the charge is

$$E = k\frac{|q|}{r^2}$$

The field produced by a distribution of charge is the vector sum of the fields produced by each individual charge:

$$\mathbf{E} = \mathbf{E}_1 + \mathbf{E}_2 + \cdots + \mathbf{E}_n$$

For a continuous distribution of charge, we find the electric field by dividing the charge into tiny elements of charge Δq, calculating the field $\Delta\mathbf{E}$ of each element Δq as though it were a point charge, and then applying the superposition principle to find the total field:

$$\mathbf{E} = \Sigma\,(\Delta\mathbf{E})$$

where $|\Delta\mathbf{E}| = k\dfrac{|\Delta q|}{r^2}$

Surface charge density σ is defined as charge per unit area, and volume charge density ρ is defined as charge per unit volume:

$$\sigma = \frac{Q}{A}$$

$$\rho = \frac{Q}{V}$$

An infinite plane having a uniform charge density σ produces a uniform field, directed perpendicular to the plane—away from the plane if $\sigma > 0$ and toward the plane if $\sigma < 0$. The magnitude of the field is

$$E = 2\pi k|\sigma|$$

Both the electric field and the net charge inside a statically charged conductor equal zero. Any net charge on a conductor is on its surface, producing a surface charge density σ. The field at a field point just outside the conductor is perpendicular to the conductor's surface and has a magnitude determined by σ on the part of the surface near the field point, according to the equation

$$E = 4\pi k|\sigma|$$

Field lines are directed lines used to represent an electric field. They are continuous at most points in space and are often curved. Field lines terminate only at points in space where there is electric charge, beginning only on positive charges and ending only on negative charges. The tangent to a field line at any point indicates the direction of the field at that point. The magnitude of the field is indicated by the spacing of the lines—the number of lines per cross-sectional area is proportional to the strength of the field; thus the field is strongest where the lines are closest.

Questions

1 Is the electric force between an electron and a proton attractive or repulsive?

2 A positron is the antiparticle of an electron, which means that the positron has the same mass as the electron but it has a positive charge +e. Will a proton attract or repel a positron?

3 Physicists now know that protons and other particles previously thought to be indivisible are in fact made up of smaller units called "quarks." Protons consist of three quarks. There are various kinds of quarks, but all have electric charge of either $\pm\frac{2}{3}$e or $\pm\frac{1}{3}$e.

(a) If one of the quarks in a proton has a charge of $+\frac{2}{3}$e, what are the charges of the other two quarks?

(b) A neutron is an uncharged particle that, together with the proton, makes up most of the mass of an atom. Like the proton, the neutron consists of three quarks, one of which has an electric charge of $+\frac{2}{3}$e. What are the charges of the other two quarks?

4 The magnitude of the electric force between two point charges is initially 180 N. What will the force be if the distance between the two charges is (a) doubled; (b) tripled; (c) halved?

5 The water molecule has an electric dipole moment. This means that one end of the molecule is positively charged and the other end is negatively charged. These dipole moments result in electric forces between the molecules. Which of the pairs of dipoles shown in Fig. 17-34 experiences a net attractive force?

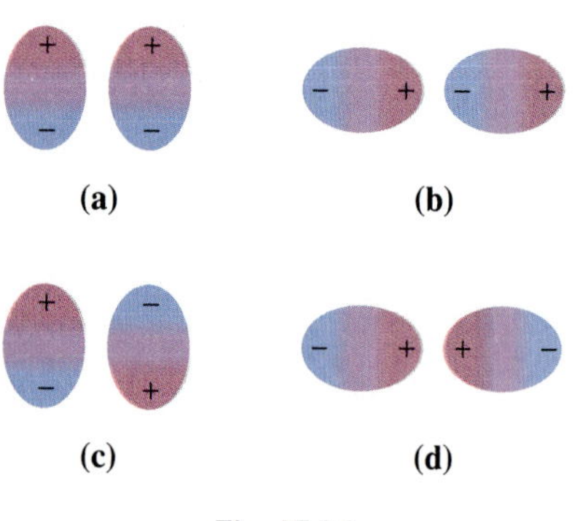

Fig. 17-34

6 A small positively charged object is brought close to one end of a long metal rod that is electrically insulated and initially uncharged. The object does not touch the rod. Does the rod exert a force on the object? Explain.

7 Two initially uncharged metal spheres are connected by a copper wire. A positively charged object is placed near one of the spheres but not touching it. What can you do to cause the two spheres to retain a charge even after the object is moved away? This process of charging without contact is called "charging by induction."

8 Fig. 17-35 shows two protons and one electron. What is the direction of the electric field at R?

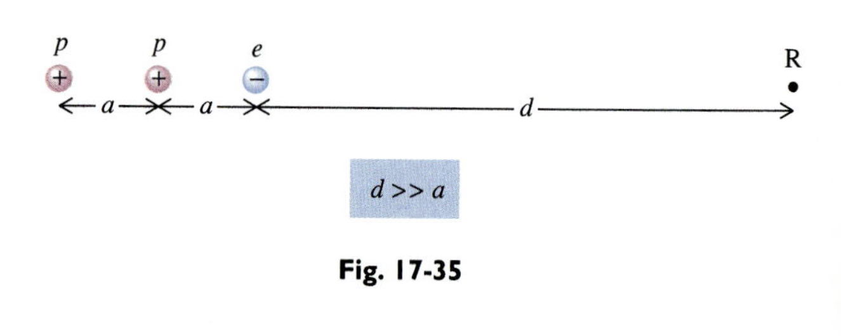

Fig. 17-35

9 All four charges in Fig. 17-36 have the same magnitude. At point P the electric field produced by these charges is directed toward the left. Charge q_1 is negative. What are the signs of q_2, q_3, and q_4?

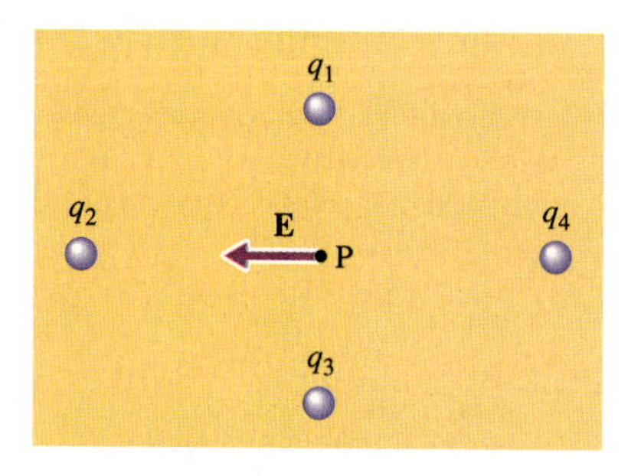

Fig. 17-36

10 Small pieces of paper are near a comb that carries a static charge. Although each piece of paper has no net charge, the comb's field causes the paper to become polarized, as shown in Fig. 17-37. All the positive charge in a piece of paper is pushed down in the direction of **E**, and all the negative charge is pushed up. By considering the magnitude and direction of the forces on the charges at the two ends of each piece of paper, determine whether this nonuniform field exerts a net force on the paper and, if so, in what direction.

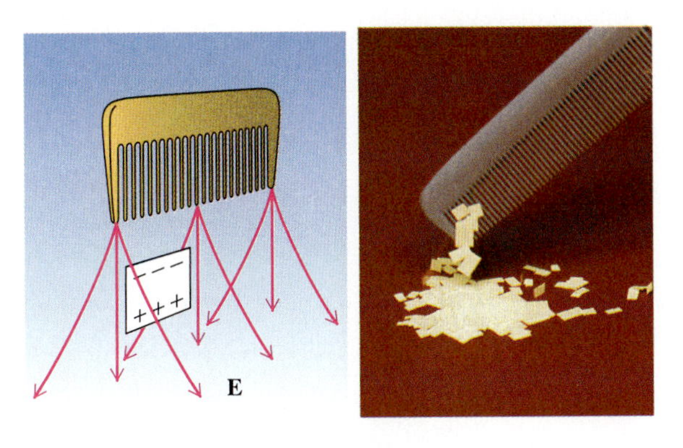

Fig. 17-37

11 In the "Star Trek" television series, prisoners on the starship *Enterprise* were confined to their cabins by means of an invisible force field in an open doorway (Fig. 17-38). When they attempted to pass through the doorway, they received a painful shock. Suppose the force field is just a strong electric field. A prisoner reasons that since his body is uncharged an electric field should not bother him. What's wrong with that reasoning?

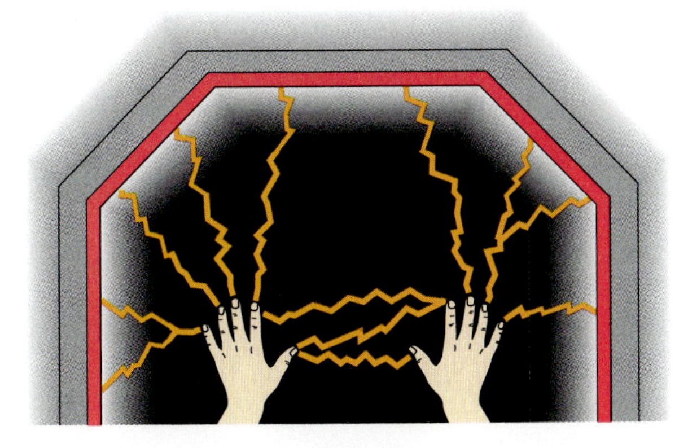

Fig. 17-38

12 Which of the sketches in Fig. 17-39, if any, can represent electric field lines?

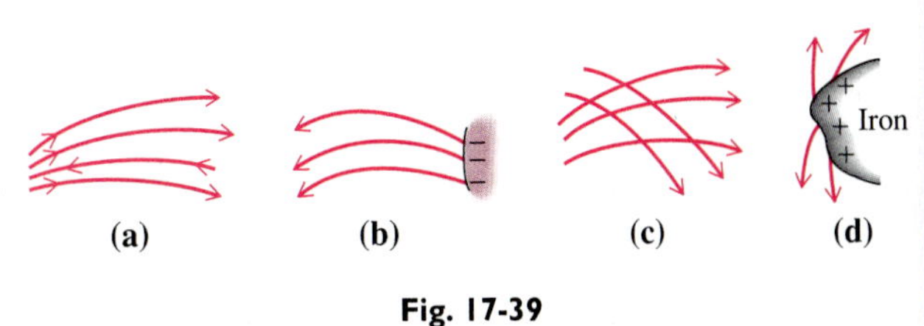

Fig. 17-39

13 Two uniformly charged infinite planes are shown in Fig. 17-40. If a positively charged balloon of negligible weight is attached by a string to point P, what will the direction of the string be when the balloon comes to rest?

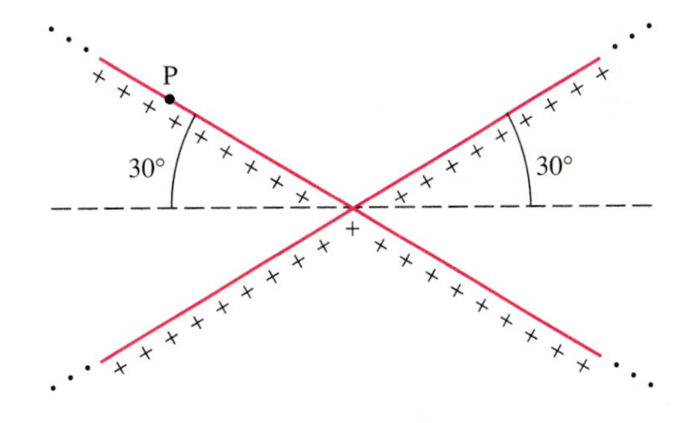

Fig. 17-40

14 If no other charge distribution is nearby, is it possible for the right side of a flat conducting plate to be charged while the left side remains uncharged? Explain.

15 Fig. 17-41 shows electric field lines around a conductor. (a) At which of the four points is the field strongest? (b) At which point is there negative charge?

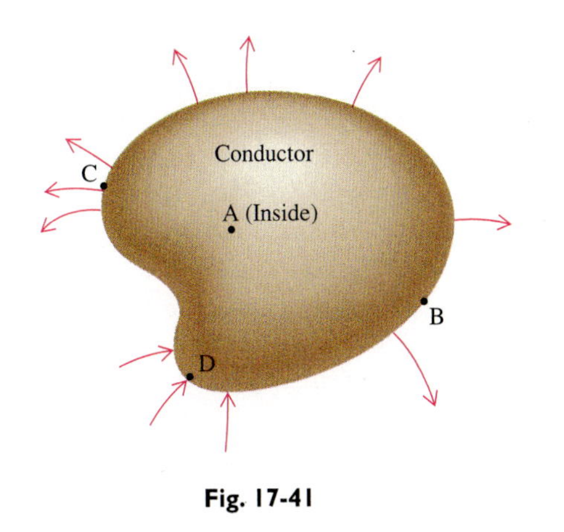

Fig. 17-41

16 Suppose you are driving through a thunderstorm. Would it be safer to be (a) inside a car with a metal body; (b) inside a fiberglass car; (c) outside your car?

17 Suppose you are in a cave, deep within the earth. Are you safe from electrical storms?

18 Suppose a positive charge of +20e is moved through a small opening in a hollow metal sphere and placed at a point P, which is connected to the inside of the sphere by means of a metal wire (Fig. 17-42).

(a) What is the final charge distribution, and how is it achieved?

(b) Is there any limit to the amount of charge that can be transferred in this way, if the sphere is in a vacuum, so that there is no way for the charge to leak off?

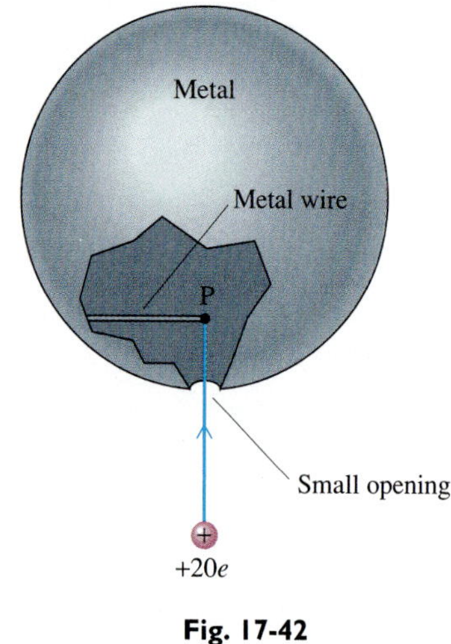

Fig. 17-42

19 A Van de Graaff generator operates on the principle described in the preceding question. Positive charges are supplied to the bottom end of a conveyor belt through frictional contact (Fig. 17-43a). The belt then carries the positive charge to the inside of a metal sphere, where the charge is transferred to the sphere's outer surface.

(a) Why must an upward force be applied to the positively charged side of the belt to lift it?

(b) When the girl in Fig. 17-43b touches the metallic surface, her body becomes an extension of the conductor, causing her skin to become positively charged and her hair to stand on end because of repulsion of like charges. Which of the following forces acts on the end of a hair, balancing the electric force of repulsion: attractive electric force, normal force, tension in the hair, or weight of the hair?

20 Determine the magnitude and sign of q_2 in Fig. 17-44.

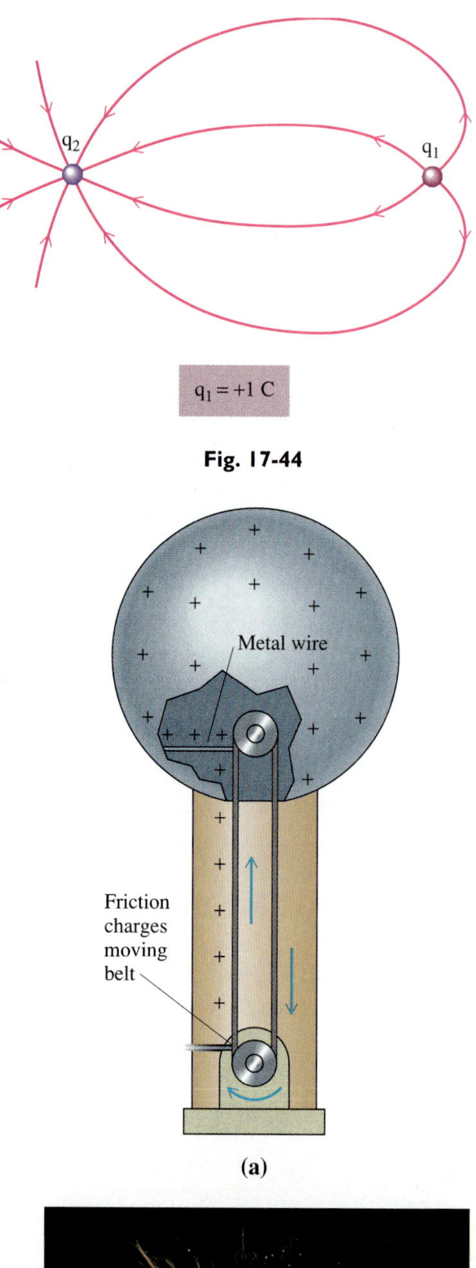

Fig. 17-44

(a)

(b)

Fig. 17-43

21 St. Elmo's fire is a glowing of the air that occurs under certain conditions around a charged object discharging into the atmosphere. This kind of discharge, also called a "corona discharge," is sometimes observed around the wings of an airplane in flight. Are you more likely to see St. Elmo's fire near the rounded front edge of the airplane's wing or near the sharp back edge?

Answers to Odd-Numbered Questions

1 attractive; **3** (a) $+\frac{2}{3}$e, $-\frac{1}{3}$e; (b) $-\frac{1}{3}$e, $-\frac{1}{3}$e; **5** b, c; **7** disconnect the wire; **9** −, −, +; **11** Although the body has no net charge, there are certainly charges within it; **13** vertical; **15** (a) C; (b) D; **17** yes; **19** (a) to balance the force of the downward-directed field acting on the belt's charges as a result of the charges already on the sphere (b) tension in the hair; **21** the sharp back edge

Problems (listed by sections)

17-2 Coulomb's Law

1 A copper wire 90.0 cm long and 1.00 mm in diameter has a mass of 6.35 g.
(a) Find the number of electrons in the wire. (Copper has an atomic number of 29; that is, there are 29 protons in the copper atom. Copper's atomic mass is 63.5.)
(b) There is one free electron per atom in copper. Find the number of free electrons in the wire.

2 Donna and John have masses of 50.0 kg and 80.0 kg respectively.
(a) How many protons are there in each person? (Protons make up roughly 55% of the mass of the human body.)
(b) How many electrons are in each person?
(c) Suppose John and Donna stand 5.00 m apart. Calculate the force exerted on John's protons by (1) Donna's protons and (2) her electrons.
(d) What is the resultant force on his protons?

★ **3** The structure of a sodium chloride (table salt) crystal is shown in Fig. 17-45. Each sodium ion Na^+ has a charge +e and is adjacent to a chloride ion Cl^-, which has a charge −e. The electric force of attraction between sodium ions and chlorine ions holds the crystal together.
(a) What is the magnitude of the force between adjacent sodium and chlorine ions, 2.82×10^{-10} m apart?
(b) What is the resultant force on any ion in the crystal?
(c) Suppose you attempt to break a cubic salt crystal, 1.00 mm on a side, by applying forces **F** and −**F** perpendicular to opposite sides of the cube, trying to pull it apart. How great would F have to be to overcome the attractive forces of all the ions in a 1.00 mm^2 plane of the crystal?

★ **4** There are extremely large electrical forces of repulsion between the protons in the nucleus of an atom. However, these forces are normally not as great as the "strong force," which is the force that binds all the protons and neutrons in the nucleus together. The strong force has a very short range—on the order of 2×10^{-15} m. When protons or neutrons are separated by a distance greater than this, the strong force does not act. Thus, if for some reason a nucleus splits in two, or "fissions," each fragment can experience an electrical repulsive force without any other force to balance it. Suppose a uranium nucleus (92 protons) splits into two nuclei having 46 protons each.
(a) Calculate the repulsive force between these nuclei just after the split, when they are 10^{-14} m apart.
(b) Suppose that all the nuclei in one cross section of a 1 mm^3 cube (about 1.3×10^{11} nuclei) simultaneously split and experience a force perpendicular to the cross section. What would be the total force splitting the cube apart?

5 Find the ratio of the magnitudes of the electrical and the gravitational forces acting between a proton and an electron separated by an arbitrary distance d.

6 Find the force on a negative charge that is placed midway between two equal positive charges. All charges have the same magnitude.

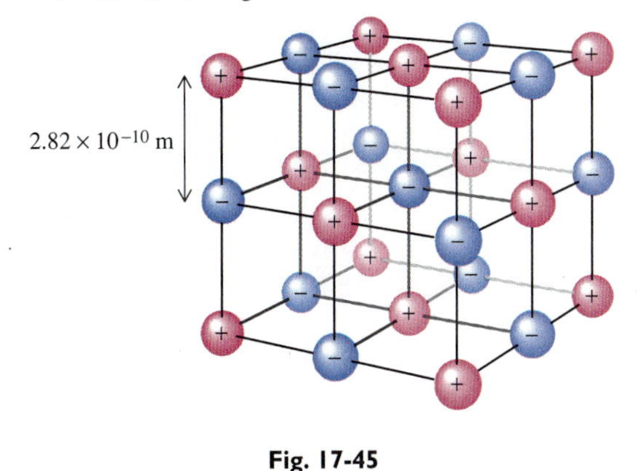

Fig. 17-45

★7 Hanging from threads are two charged balls made of pith (a very light, spongy material that comes from inside the stems of certain plants). The balls each have a mass of 0.100 g and a charge of the same magnitude q and are attracted toward each other, as shown in Fig. 17-46.

(a) What is the magnitude of the electric force?

(b) What is the magnitude of q?

(c) How many electrons have been transferred to or from each ball?

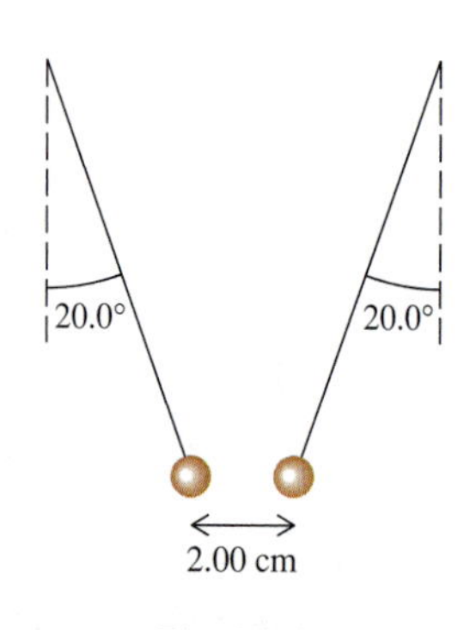

Fig. 17-46

8 Three charges, $q_1 = +2.00 \times 10^{-9}$ C, $q_2 = -3.00 \times 10^{-9}$C, and $q_3 = +1.00 \times 10^{-9}$C, are located on the x-axis at $x_1 = 0$, $x_2 = 10.0$ cm, and $x_3 = 20.0$ cm. Find the resultant force on q_3.

9 An electron is near a positive ion of charge +9e and a negative ion of charge −8e (Fig. 17-47).

(a) Find the magnitude and direction of the resultant force on the electron.

(b) Find the magnitude and direction of the electron's instantaneous acceleration.

Fig. 17-47

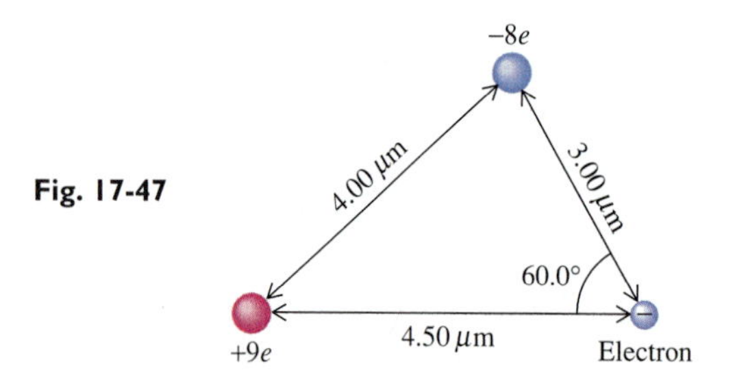

★10 (a) In Fig. 17-48, what are the magnitude and direction of the resultant force on q_1?

(b) What is the resultant force on the center of mass of the four charges?

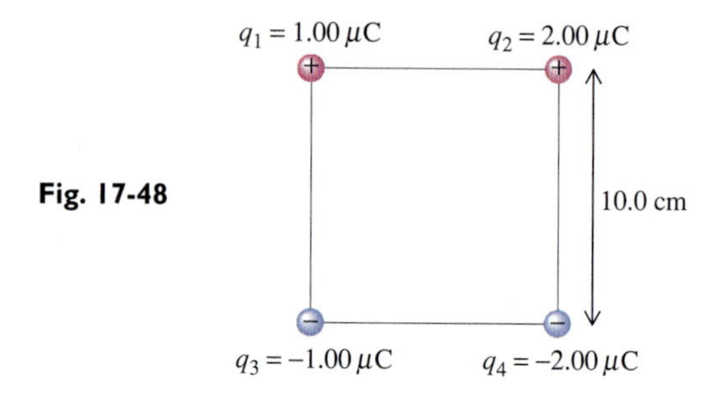

Fig. 17-48

11 In a Cartesian coordinate system, the charge $q_1 = -2.00 \times 10^{-4}$ C is at the origin, the charge $q_2 = 1.00 \times 10^{-3}$ C has coordinates $x = 10.0$ m, $y = 0$, and the charge $q_3 = -1.00 \times 10^{-4}$ C has coordinates $x = 0$, $y = -5.00$ m. Find the magnitude and direction of the resultant force on q_1.

★12 Given two point charges q_1 and $q_2 = 4q_1$, find the position of a third charge q_3 relative to the other two charges, such that the resultant force on q_3 is zero.

17-3 The Electric Field

13 A charge q_1 on the y-axis produces a field of magnitude 3 N/C at the origin, and a charge q_2 on the x-axis produces a field of magnitude 4 N/C at the origin. What is the magnitude of the total field at the origin?

14 Fig. 17-49 shows fields $\mathbf{E}_1$, $\mathbf{E}_2$, $\mathbf{E}_3$, and $\mathbf{E}_4$ at point P, produced respectively by charges q_1, q_2, q_3, and q_4. What is the sign of each charge?

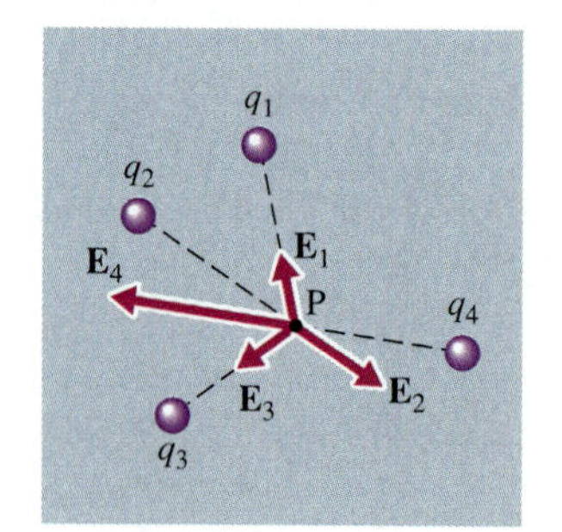

Fig. 17-49

15 Fig. 17-50 shows positive charges q_1 and q_2. At which of the five points, A, B, C, D, or G, could you place a negative charge q_3 of the right magnitude so that the field at point P is the field produced by q_1 alone?

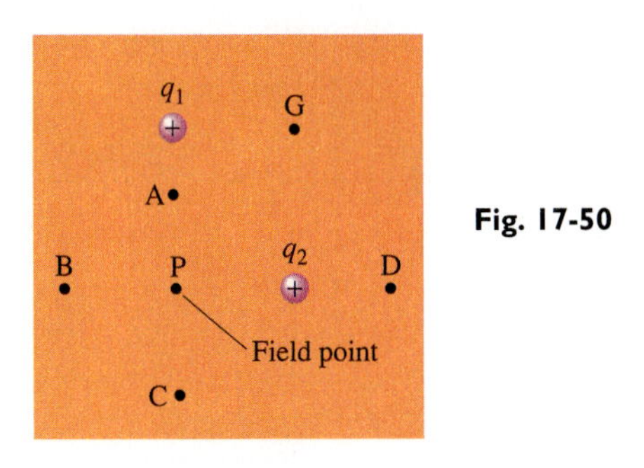

Fig. 17-50

16 In Fig. 17-51 two electrons are the same distance from a field point P. At which of the points A, B, C, or D could a proton be placed so that the electric field at P is zero?

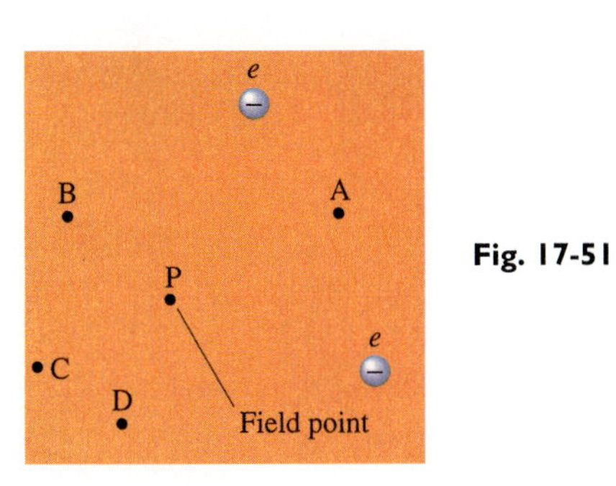

Fig. 17-51

17 Find the electric field at P in Fig. 17-52.

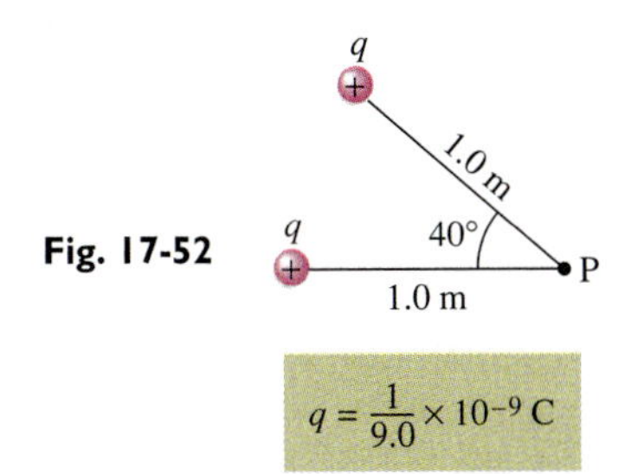

Fig. 17-52

$q = \frac{1}{9.0} \times 10^{-9}$ C

18 At what distance from a proton does its electric field have a magnitude of 1 N/C?

19 Points A, B, and C are at the vertices of an equilateral triangle. A certain positive charge q placed at A produces an electric field of magnitude 100 N/C at C. Suppose a second, identical charge is placed at B. What is the magnitude of the new electric field at C?

★ **20** Two point charges q_1 and q_2 are separated by 20.0 cm. The electric field at their midpoint is 600 N/C, directed away from q_1, which is $+1.00 \times 10^{-9}$ C. Find q_2.

21 A 2.00×10^{-9} C charge has coordinates $x = 0$, $y = -2.00$; a 3.00×10^{-9} C charge has coordinates $x = 3.00$, $y = 0$; and a -5.00×10^{-9} C charge has coordinates $x = 3.00$, $y = 4.00$, where all distances are in cm. Determine magnitude and direction for (a) the electric field at the origin and (b) the instantaneous acceleration of a proton placed at the origin.

22 Point charges of $+1.00 \times 10^{-9}$ C, $+1.00 \times 10^{-9}$ C, and -2.00×10^{-9} C are placed at the vertices of an equilateral triangle. Find the magnitude of the electric field at the center of the triangle, which is 10.0 cm from each vertex.

23 Find the x and y components of the electric field produced by q_1 and q_2 in Fig. 17-53 at (a) point A; (b) point B.

Fig. 17-53

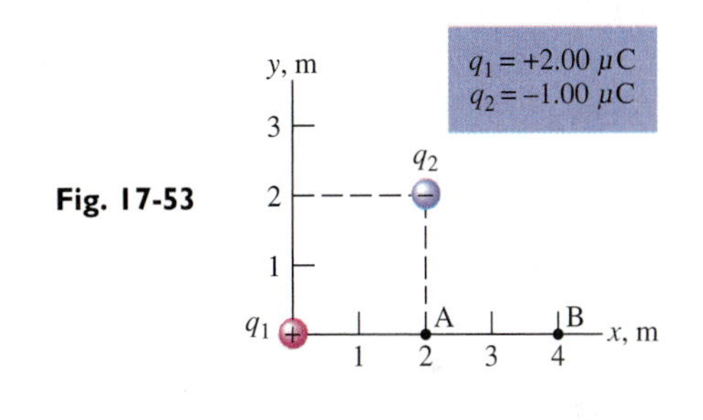

24 A Ping-Pong ball that has a mass of 2.40 g is charged when 1.00×10^{11} electrons are added. The ball is stationary in an electric field. Find the magnitude and direction of the field.

★ **25** A small Styrofoam ball weighing 1.00×10^{-2} N is supported by a thread in a horizontal electric field of magnitude 1.00×10^{6} N/C. The thread makes an angle of 45.0° with the vertical. Find the magnitude of the charge on the ball.

26 Millikan measured the electron's charge by observing tiny charged oil drops in an electric field. Each drop had a charge imbalance of only a few electrons. The strength of the electric field was adjusted so that the electric and gravitational forces on a drop would balance and the drop would be suspended in air. In this way the charge on the drop could be calculated. The charge was always found to be a small multiple of 1.60×10^{-19} C. Find the charge on an oil drop weighing 1.00×10^{-14} N and suspended in a downward field of magnitude 2.08×10^{4} N/C.

★ **27** A beam of electrons is shot into a uniform downward electric field of magnitude 1.00×10^3 N/C. The electrons have an initial velocity of 1.00×10^7 m/s, directed horizontally. The field acts over a small region, 5.00 cm in the horizontal direction.
(a) Find the magnitude and direction of the electric force exerted on each electron.
(b) How does the gravitational force on an electron compare with the electric force?
(c) How far has each electron moved in the vertical direction by the time it has emerged from the field?
(d) What is the electron's vertical component of velocity as it emerges from the field?
(e) The electrons move an additional 20.0 cm after leaving the field. Find the total vertical distance that they have been deflected by the field.

28 During a thunderstorm the electric field at a certain point in the earth's atmosphere is 1.00×10^5 N/C, directed upward. Find the acceleration of a small piece of ice of mass 1.00×10^{-4} g, carrying a charge of 1.00×10^{-11} C.

17-4 Continuous Charge Distributions

29 The earth's surface has a charge density of about -1.0×10^{-9} C/m². Find the total charge on the surface.

★ **30** The earth's surface charge density (-1.0×10^{-9} C/m²) is roughly balanced by a net positive charge in the lower 10 km of the earth's atmosphere. What is the average volume charge density of this atmospheric charge?

31 A thin gold ring of radius 1.00 cm carries a uniform charge per unit length of 1.00×10^{-14} C/m. Find the electric field on the axis of the ring 1.00 cm from the center.

★ **32** A 20.0 cm by 30.0 cm sheet of paper has a uniform surface charge density of 5.00×10^{-8} C/m². Find the electric field at a distance from the paper's surface of (a) 1.00 mm; (b) 1.00 cm; (c) 3.00 cm; (d) 5.00 m. None of the field points is close to the paper's edge.

★ **33** A rectangular slab of dimensions 1.00 m × 1.00 m × 10.0 cm has a uniform charge distribution of 1.00×10^{-7} C spread throughout its volume. Find the electric field at a point just outside the charge distribution, close to the center of one of the large faces of the slab.

★ **34** The earth is a good electrical conductor and, during periods of clear weather, has a downward-directed electric field of about 100 N/C at low altitudes. Find the charge density on the earth's surface when its field has this value.

35 Two large parallel planes each carry a uniform distribution of charge of the same magnitude σ but of opposite signs (Fig. 17-54). Find the electric field at points a, b, and c.

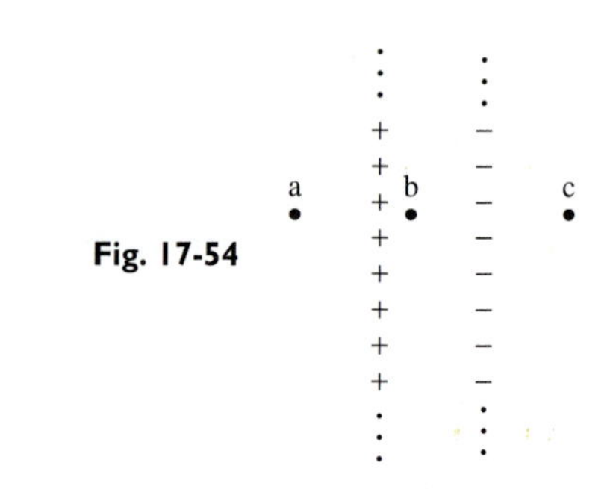

Fig. 17-54

36 Two large flat dielectric surfaces are parallel to each other and carry uniform charge densities (Fig. 17-55). Find the electric field at points a, b, and c.

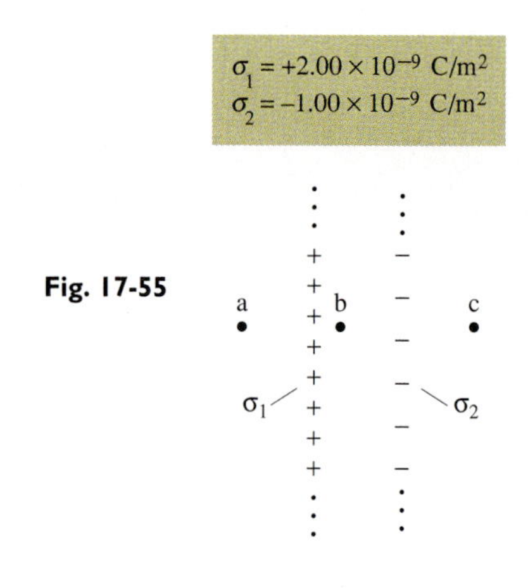

Fig. 17-55

37 Fig. 17-56 shows two large flat surfaces that have uniform charge densities. Find the electric field at points a and b.

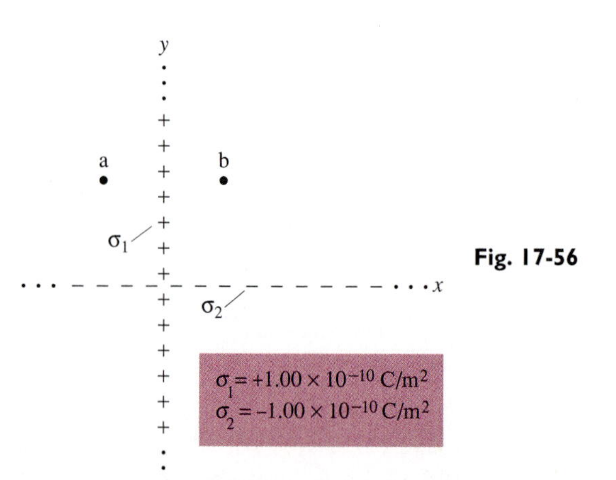

Fig. 17-56

38 A point charge q is near a uniformly charged, large flat surface of a dielectric (Fig. 17-57). Find the electric field at P.

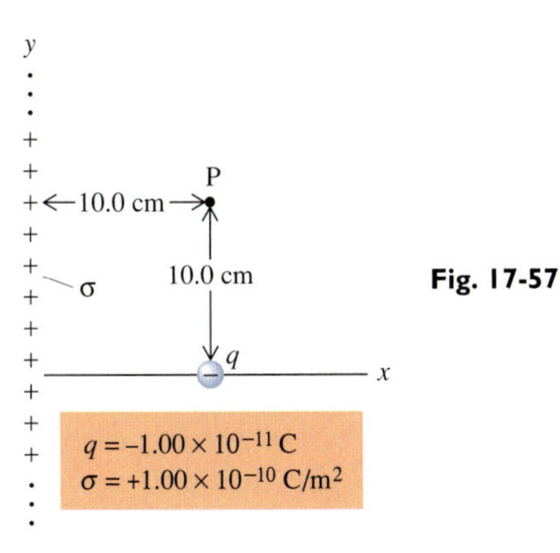

Fig. 17-57

17-4 Field Lines

39 Fig. 17-58 shows the field that results when a conducting sphere is placed in a uniform external field.

(a) Find the field at P, inside the conductor.

(b) At which of the points R, S, or T is the field weakest, and at which of these points is it strongest?

(c) Find the charge density on the surface of the conductor near R if the field at that point has magnitude 1.00×10^5 N/C.

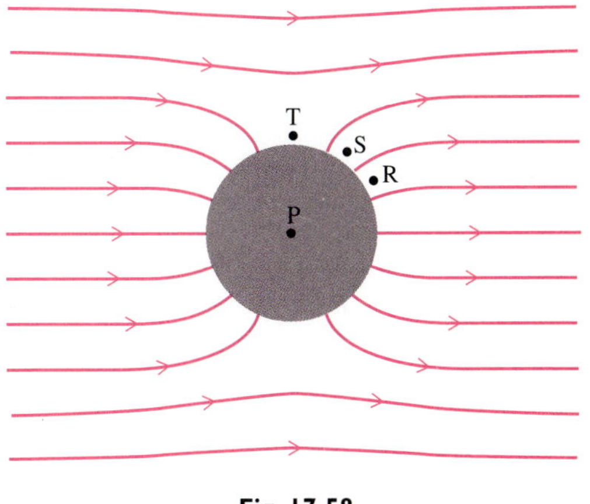

Fig. 17-58

40 A coaxial cable consists of an inner cylindrical conductor of radius 1.00 mm inside a hollow cylindrical conductor of radius 3.00 mm. Charges on the surfaces of these conductors produce the field lines shown in Fig. 17-59. The field lines are equally spaced along the axis of the cable. If the magnitude of the field just outside the inner conductor is 600 N/C, what is the magnitude of the field just inside the outer conductor?

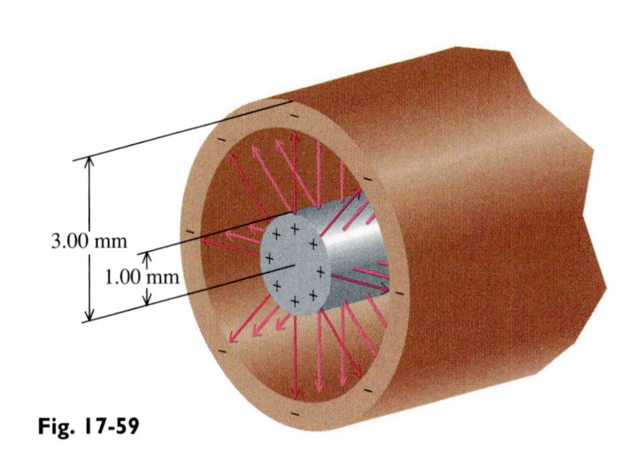

Fig. 17-59

Additional Problems

★ **41** Charge $q_1 = 1.00 \times 10^{-9}$ C is at the origin and charge $q_2 = 9.00 \times 10^{-9}$ C has coordinates $x = 40.0$ cm, $y = 0$. Find the coordinates of a field point where $\mathbf{E} = \mathbf{0}$.

★ **42** An aluminum sphere of radius 1.00 m carries a charge of -1.00×10^{-4} C. The sphere is isolated except for a particle of mass 1.00×10^{-15} kg and charge 1.00×10^{-17} C, which orbits the sphere in a circular orbit of radius 3.00 m. Find the period of the orbit.

★★ **43** Electric charge on the earth's surface and in the earth's atmosphere produces an electric field, which on a clear day (no thunderstorms) is directed vertically downward and has a magnitude of about 100 N/C just above the earth's surface. Very small negative particles can be supported by the field. For larger particles, the charge that must be carried may be so great that the field around the charge causes the surrounding air to conduct the charge away. This occurs at fields of about 3.0×10^6 N/C in dry air. Find the radius of the largest water drop that could be supported in the earth's electrostatic field. Assume that the drop is spherical.

Fig. 17-60 Above this field of wheat is an invisible atmospheric electric field. Even though you can't see it, the electric field is just as real as the wheat field.

★★44 Find the radius of the largest (spherical) water drop whose weight could be supported by any electrostatic field in dry air. The air becomes conducting when the *total* field at any point exceeds 3.0×10^6 N/C.

45 An electron is initially at rest just outside a large copper surface that has a charge density of 1.00×10^{-6} C/m^2. How far has the electron moved in 1.00×10^{-9} s if the field is uniform over this region?

46 A thundercloud contains a large concentration of charged particles: ionized molecules, charged drops of water, bits of ice, and specks of dust. There is a concentration of positive charge in the upper part of the cloud and of negative charge in the lower part.* Suppose that the charge distribution in a certain cloud can be approximated by two uniform spheres of charge +100 C and −100 C, centered at points P and Q (Fig. 17-61). Find the magnitude and direction of the electric field (a) at P and (b) at the location of an airplane 1.00 km directly above P.

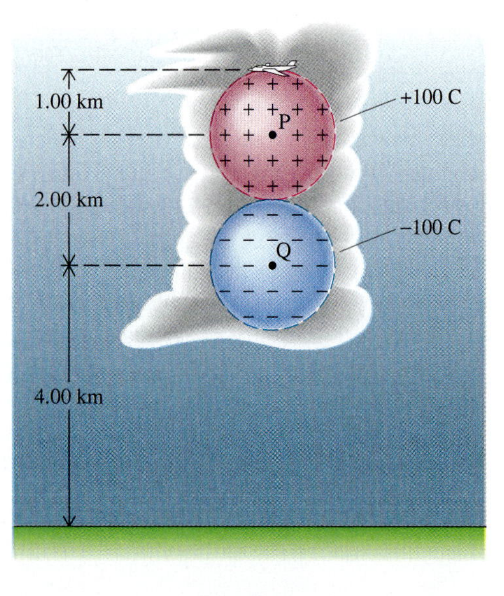

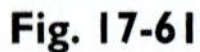
Fig. 17-61

*Negative charges are periodically transferred to the ground in the form of lightning strokes. Worldwide electrical storms throughout the day maintain the negative charge density on the earth's surface. This sporadic downward flow is balanced by a slow upward flow of negative charge during clear weather.

★47 A point charge q is placed midway between two identical positive point charges of magnitude 4.00×10^{-9} C. The resultant force on *each* of the three charges is zero. Find q.

48 The field just outside a point on the surface of a copper wire of radius 1.00 mm has magnitude 1.00×10^4 N/C.

(a) Find the magnitude of the surface charge density at that point on the wire.

(b) If the surface charge density is uniform around the circumference of the wire, how much charge is on a 1.00 m length of the wire?

★★49 A ring of radius 20.0 cm has uniform charge density of 1.00×10^{-6} C/cm. A 1.00 cm section of the ring on the right side is removed. What are the magnitude and direction of the electric field at the center of the ring?

★★50 Two point charges 0.600 m apart experience a repulsive force of 0.400 N. The sum of the two charges equals 1.00×10^{-5} C. Find the values of the two charges.

CHAPTER 18 Electric Potential

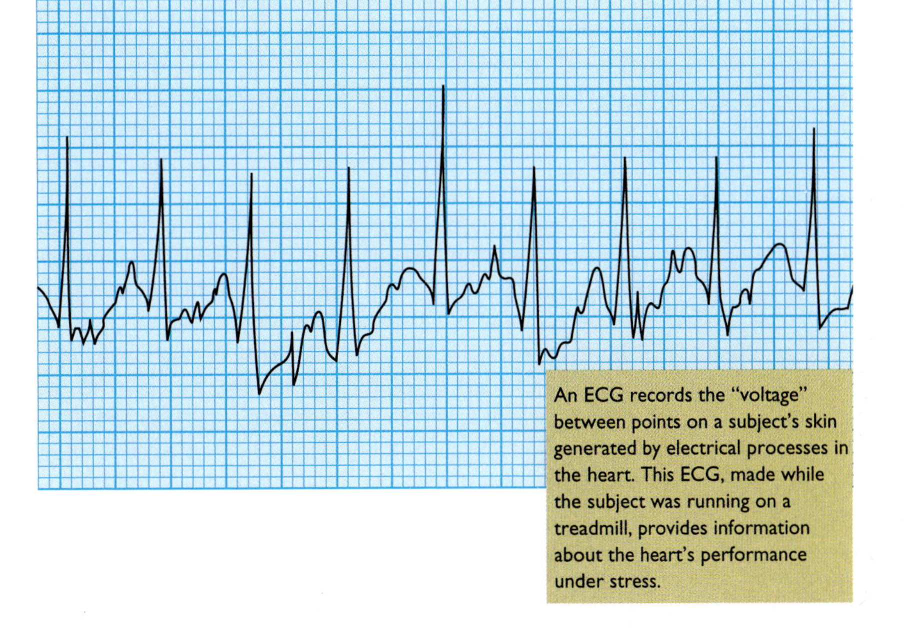

An ECG records the "voltage" between points on a subject's skin generated by electrical processes in the heart. This ECG, made while the subject was running on a treadmill, provides information about the heart's performance under stress.

Both nature and technology utilize electric energy. A lightning stroke unleashes an enormous amount of electrical potential energy, converting it to heat and light. On a much smaller scale, pulses of electric energy within the human nervous system regulate body functions. Power plants generate electrical energy, which is then distributed through vast electrical networks and used for a variety of purposes (Fig. 18-1). An electrical outlet in your home is part of such a network and provides you with electrical potential energy, which you can use to operate electric lights, appliances, and home electronics. In this chapter we shall begin to use energy concepts in our study of electricity. Before reading this chapter, it is a good idea to review the energy concepts introduced in Chapter 7.

Fig. 18-1 Electrical energy for the lights of Las Vegas is generated many kilometers away at Hoover Dam. The electrical energy comes from the gravitational potential energy of the water stored by the dam.

18-1 Electrical Potential Energy and Electric Potential

In Chapter 7 we found that the work done by certain forces is independent of path and can be expressed as a decrease in some kind of potential energy. Such forces are called "conservative forces." In particular we found that the gravitational force is conservative. When only conservative forces do work on a body, the sum of the body's kinetic energy and potential energy is conserved. For example, a book of mass 1.0 kg held 1.0 m above the floor has a gravitational potential energy $U_G = mgy = (1.0\text{ kg})(9.8\text{ m/s}^2)(1.0\text{ m}) = 9.8\text{ J}$ and a kinetic energy $K = 0$. If the book is dropped to the floor, its total mechanical energy $K + U_G$ is conserved because only gravity does work on it and the gravitational force is conservative. Just before it hits the floor, the book's potential energy $U_G = 0$, but its kinetic energy equals 9.8 J, since the sum $K + U_G$ is a constant 9.8 J.

As noted in the preceding chapter, Coulomb's law has the same mathematical form as the gravitational force law. Thus the Coulomb force, like the gravitational force, must be conservative. And since any electrostatic force can be considered a sum of Coulomb forces, **the total electrostatic force is always conservative.** This means that any charge in an electrostatic field has "electrical potential energy," U_E. This electrical potential energy, like mechanical potential energy, can be converted to other forms of energy, for example, heat and light in the case of a lightning stroke.

Definition of Electric Potential

The concept of electric potential is closely related to the concept of electrical potential energy. **We define the electric potential V at a point in space to be the electrical potential energy per unit charge that a test charge q' would have if placed at that point:**

$$V = \frac{U_E}{q'} \tag{18-1}$$

It follows from this definition that the SI unit for electric potential is joules per coulomb, which is defined to be a volt (abbreviated V) in honor of Alessandro Volta, who invented the electric battery in 1800:

$$1\text{ V} = 1\text{ J/C} \tag{18-2}$$

Since electric potential is measured in volts, it is often loosely referred to as "voltage." Keep in mind that we use nearly the same symbol for potential, V, and for the unit of potential, the volt, V, except that the symbol for the unit is not italicized.

The definition of electric potential as electrical potential energy per unit charge is analogous to the definition of electric field as the force per unit charge on a test charge. Like the electric field, the electric potential is present whether or not there is a test charge present to experience it. The source of the electric potential is the same charge that is the source of the electric field. However, unlike the electric field, the electric potential is a scalar quantity. It is therefore typically easier to work with the electric potential than with the electric field.

When the electric potential at a field point is known, we can easily calculate the electrical potential energy of a charge q' placed at that point. It follows from the definition $V = \dfrac{U_E}{q'}$ that the charge q' will have an electrical potential energy U_E:

$$U_E = q'V \tag{18-3}$$

EXAMPLE 1 Using a Test Charge to Find the Potential

(a) A test charge $q' = 10^{-6}$ C is placed at a point P in space where q' has electrical potential energy of 10^{-4} J (Fig. 18-2). What is the value of the electric potential at that point? (b) Suppose that the original test charge is replaced by a second test charge $q' = 2 \times 10^{-5}$ C at P. Find the electrical potential energy of the new test charge.

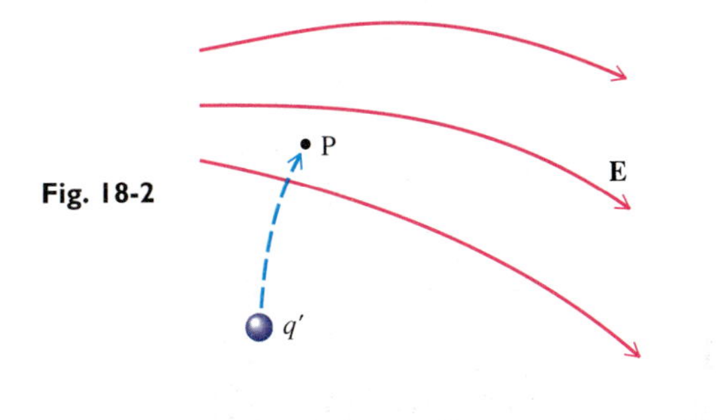

Fig. 18-2

SOLUTION **(a)** Applying Eq. 18-1, we find

$$V = \frac{U_E}{q'} = \frac{10^{-4}\text{ J}}{10^{-6}\text{ C}} = 100\text{ V}$$

Remember that P is at an electric potential of 100 V regardless of the presence or absence of the test charge. Later in this section we shall see how the electric potential at a point can be calculated from a knowledge of either the electric field or the source charge producing the electric field.

(b) Since we know that the electric potential at this point is 100 V, we can apply Eq. 18-3 to find the electrical potential energy of the second test charge placed there.

$$U_E = q'V = (2 \times 10^{-5}\text{ C})(100\text{ V}) = 2 \times 10^{-3}\text{ J}$$

Potential Difference

In Chapter 7 we found that the reference level, or zero point, of gravitational potential energy is arbitrary. Consider, for example, the gravitational potential energy of a book of mass 1.0 kg falling 1.0 m from a tabletop to the floor. The book's gravitational potential energy U_G is mgy (Eq. 7-14). If we set $y = 0$ at the floor, the book has zero potential energy on the floor and potential energy $U_G = (1.0\text{ kg})(9.8\text{ m/s}^2) \times (1.0\text{ m}) = 9.8$ J on the table. If we set $y = 0$ at the table, the book has zero potential energy there and potential energy -9.8 J on the floor at $y = -1.0$ m. With either choice of zero point, the physically significant fact remains unchanged: the book's potential energy decreases by 9.8 J as it falls from the table to the floor.

The reference level for electrical potential energy is also arbitrary. Only *differences* in electrical potential energy have physical significance. And since electric potential is defined as electrical potential energy per unit charge, it follows that **only differences in electric potential** (sometimes called the "voltage drop") **have physical significance**. We can attach a value to the electric potential at a point only after we have defined a reference level.

A voltmeter is a device used to measure the difference in electric potential between two points in an electric circuit (Fig. 18-3). Voltmeters will be discussed in Chapter 20.

Fig. 18-3 This voltmeter measures a potential difference of 7.55 V.

EXAMPLE 2 Charge Travelling From One Battery Terminal to Another

The terminals of a 12 V battery differ in electric potential by 12 V, with the positive terminal being at the higher potential, as indicated in Fig. 18-4. Suppose that a 3.0 C charge travels from A to B. How much electrical potential energy does it lose?

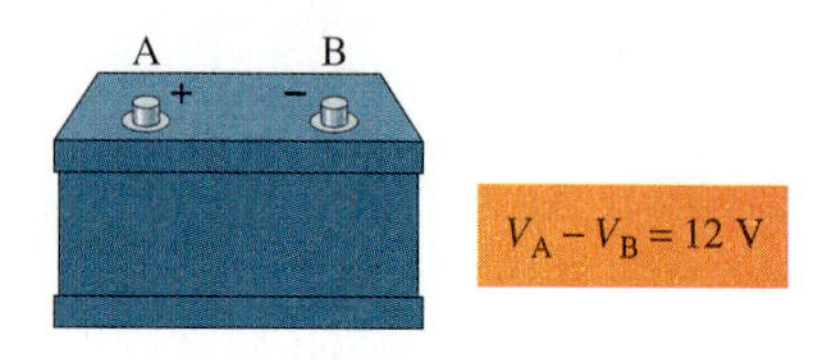

Fig. 18-4

SOLUTION We calculate the loss in potential energy by relating the potential energy to the potential at points A and B, using Eq. 18-3:

$$U_{E,A} - U_{E,B} = q'V_A - q'V_B = q'(V_A - V_B)$$
$$= (3.0\text{ C})(12\text{ V}) = 36\text{ J}$$

Notice that we have not assigned values to the potential at A and B. The result is independent of these values, so long as V_A is 12 V higher than V_B. For example, we could have $V_B = 0$, $V_A = 12$ V, or $V_B = 100$ V, $V_A = 112$ V, or $V_B = -12$ V, $V_A = 0$.

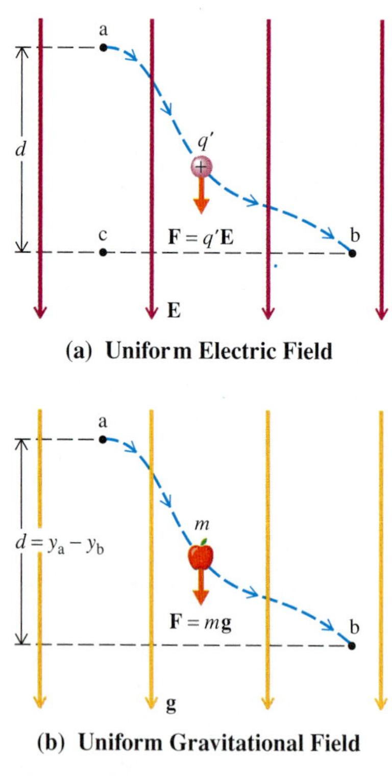

Fig. 18-5 We can find **(a)** the work done by a uniform electric field on a charge q' by considering **(b)** the work done by gravity on a mass m.

Potential Difference in a Uniform Electric Field

Calculating electric potential is easiest when the electric field is uniform (Fig. 18-5a). Imagine a positive test charge q' traveling from a to b in a uniform field **E** and subject to a constant force $\mathbf{F} = q'\mathbf{E}$. By applying the definition of work, it is possible to show that the work done by the electric field on q' can be expressed as a decrease in the electrical potential energy of q'. However, we shall not use this direct but somewhat tedious approach. Instead we shall obtain an expression for electrical potential energy by noting how this situation is mathematically identical to a mass m moving in a uniform gravitational field (Fig. 18-5b). The charge q' corresponds to the mass m, and the constant electric field **E** corresponds to the constant gravitational field **g**. We solved the gravitational problem in Chapter 7, showing that the work W_G done by gravity is independent of path and can be expressed as the decrease in gravitational potential energy from a to b:

$$W_G = U_{G,a} - U_{G,b} = mgy_a - mgy_b = mg(y_a - y_b)$$

or

$$U_{G,a} - U_{G,b} = mgd$$

where $d = y_a - y_b$. Substituting q' for m and E for g, we obtain the corresponding expression for the decrease in the electrical potential energy U_E of charge q' moving from a to b:

$$U_{E,a} - U_{E,b} = q'Ed$$

Applying the definition of electric potential as electrical potential energy per unit charge, we divide the equation above by q' to obtain an expression for the drop in electric potential from a to b:

$$V_a - V_b = Ed \quad \text{(in a uniform field)} \quad (18\text{-}4)$$

We can also obtain this result directly by calculating the work W_E the electric field does on q' along a particular path from a to b. Let the charge move from a to b by first moving parallel to the field from a to point c (shown in Fig. 18-5a) and then moving perpendicular to the field from c to b. Work is done on q' only for the part of the path from a to c. Since there is a constant component of force along this path, the work is the product of this component ($F_s = q'E$) and the path length d:

$$W_{E,a \to b} = W_{E,a \to c} = F_s \, \Delta s = q'Ed$$

The work equals the decrease in the charge's electrical potential energy from a to b. If we divide by q', we again find that the drop in electric potential from a to b is Ed, as we found in Eq. 18-4 for an arbitrary path connecting a and b.

The difference in electric potential at two points in an electrostatic field is not always as simply expressed as in Eq. 18-4, which is valid only for a uniform field. However, it is always true that the difference in electric potential depends only on the two points and is the same no matter which path connects the points.

We have drawn the electric field in Fig. 18-5a directed vertically downward to emphasize the correspondence with the gravitational field. However, our expression for the drop in electric potential (Eq. 18-4) is applicable to a uniform electric field in *any* direction, as long as the distance d is measured parallel to field lines and as long as the higher-potential point is "upstream" in the field from the lower-potential point.*

*The term "upstream" comes from visualizing the electric field as a flowing river, with the direction of the water current in the direction of the electric field.

EXAMPLE 3 Finding the Potential Difference Between Two Points in a Uniform Field

A uniform electric field of magnitude 2000 N/C is directed 37.0° below the horizontal (Fig. 18-6). (a) Find the potential difference between P and R. (b) If we define the reference level of potential so that the potential at R is 500 V, what is the potential at P?

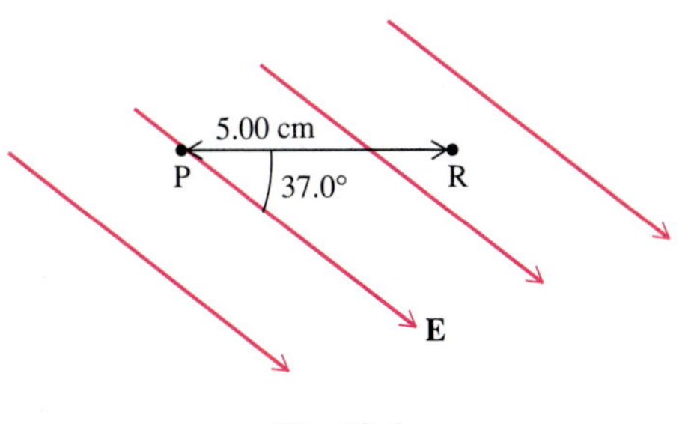

Fig. 18-6

SOLUTION **(a)** Point P is upstream and is therefore at the higher potential. We apply Eq. 18-4 to compute the drop in potential from P to R.

$$V_P - V_R = Ed$$

We must be careful not to identify d as the distance between P and R. Rather, d is the distance along the direction of **E** and is equal to (5.00 cm)(cos 37.0°) = 4.00 cm. Substituting into the equation above, we find

$$V_P - V_R = (2000 \text{ N/C})(4.00 \times 10^{-2} \text{ m})$$
$$= 80.0 \text{ V}$$

(b) If the potential at R is 500 V, then solving the equation above for V_P, we find

$$V_P = V_R + 80 \text{ V} = 500 \text{ V} + 80 \text{ V} = 580 \text{ V}$$

EXAMPLE 4 Potential at Points in the Earth's Atmosphere

Electric charge on the earth's surface and in the earth's atmosphere produces an electric field that, on a clear day (no thunderstorms), is directed vertically downward and has an approximately constant magnitude of about 100 N/C just above the earth's surface (Fig. 18-7). (a) Find the value of the potential V at an elevation of 10 m and at an elevation of 20 m. Set the potential of the ground equal to zero. (b) If a small positive ion of negligible weight is initially at rest in the atmosphere, which way will the earth's electric field cause it to move, up or down? Which way will the field cause a small negative ion initially at rest to move, up or down?

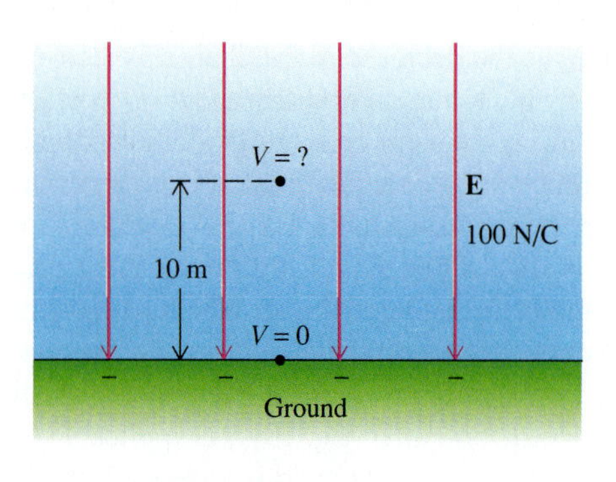

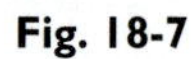

Fig. 18-7

SOLUTION (a) We apply Eq. 18-4 for the potential difference between a point G on the ground and a point P at an elevation of 10 m, setting the ground potential equal to zero:

$$V_P - V_G = Ed$$

$$V_P - 0 = (100\ \text{N/C})(10\ \text{m})$$

$$V_P = 1000\ \text{V}$$

We calculate the potential at a point R at an elevation of 20 m in the same way:

$$V_R - V_G = Ed$$

$$V_R - 0 = (100\ \text{N/C})(20\ \text{m})$$

$$V_R = 2000\ \text{V}$$

The potential is the same anywhere on a horizontal surface. Any point G on the ground is at a potential of 0, any point P at an elevation of 10 m is at a potential of 1000 V, and any point R at an elevation of 20 m is at a potential of 2000 V.

(b) As indicated in Fig. 18-8, a positive ion experiences downward force in the direction of the electric field and, in the absence of other forces, will be accelerated downward. A negative ion experiences an upward force, opposite the electric field, and, in the absence of other forces, will be accelerated upward.* **A positive charge tends to move from higher to lower potential; a negative charge tends to move from lower to higher potential.**

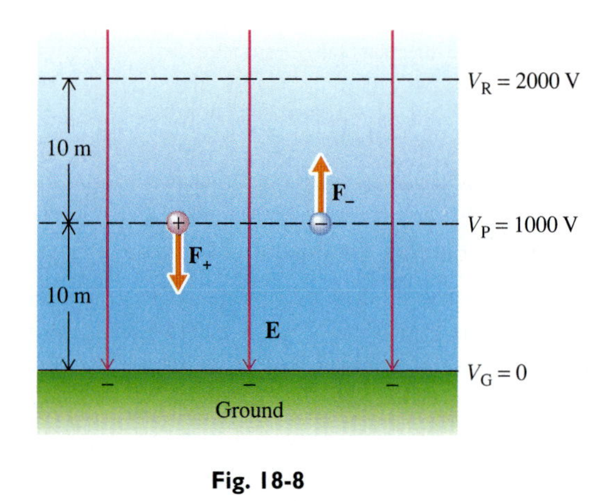

Fig. 18-8

*There are thousands of positive and negative ions in a cubic centimeter of air on a clear day. The ions move in response to the earth's electric field. These ions are continuously produced by sources of radioactivity on the earth's surface and by cosmic rays entering the earth's atmosphere. The steady flow of positive charges downward and negative charges upward would soon eliminate the negative charge on the earth's surface, were it not for the periodic transfer of negative charges to the ground in lightning strokes. (There are always electrical storms somewhere on earth. See Chapter 19, Problem 33, for calculation of the average number of lightning strokes worldwide at any instant and of your chance of being struck by one.)

Since we often encounter situations in which electrons, protons, or other small charges are placed in an electric field, it is convenient to define a new unit of energy that is related to e, the magnitude of the electron or proton charge. This unit, called the **electron volt, abbreviated eV, is defined as the electrical potential energy of a charge +e at a point where the potential is 1 V.** In terms of joules, we express this as

$$1\ \text{eV} = (1.60 \times 10^{-19}\ \text{C})(1\ \text{J/C})$$

or

$$1\ \text{eV} = 1.60 \times 10^{-19}\ \text{J} \qquad (18\text{-}5)$$

EXAMPLE 5 An Ion Accelerated in the Atmosphere

Suppose a small positive ion of negligible weight and with a charge $q' = 10e$ is initially at rest in the earth's atmosphere at an elevation of 10 m on a clear day. Use the results of the last example to find the ion's kinetic energy when it hits the ground, assuming (unrealistically) that the ion does not collide with other particles in the atmosphere.

SOLUTION In Example 4 we found that the potential at an elevation of 10 m is 1000 V and the potential at ground level is zero. Thus the ion moves from a point at a potential of 1000 V to a point at zero potential.

We are given that the ion's weight is negligible; that is, the gravitational force on the ion is negligible compared with the electric force. So the ion's gravitational potential energy is negligible compared with its electrical potential energy.

If we assume that the only significant force acting on the ion is the electric force, the ion's total mechanical energy is conserved; that is , the sum of its kinetic energy and electrical potential energy is constant:

$$E_f = E_i$$

$$K_f + U_{E,f} = K_i + U_{E,i}$$

Setting the initial kinetic energy equal to zero and applying Eq. 18-3 to relate the ion's electrical potential energy to the potential, we find

$$K_f + q'V_f = 0 + q'V_i$$

$$K_f = q'(V_i - V_f) = (10e)(1000 \text{ V} - 0)$$

$$= 10{,}000 \text{ eV}$$

or, converting to joules,

$$K_f = (1.0 \times 10^4 \text{ eV})\left(\frac{1.6 \times 10^{-19} \text{ J}}{1 \text{ eV}}\right) = 1.6 \times 10^{-15} \text{ J}$$

EXAMPLE 6 Acceleration of Electrons in an Electron Microscope

In an electron microscope electrons are accelerated across a high voltage to give a high-energy electron beam that is then scattered off the object being examined. Magnetic fields focus the scattered electrons, forming a highly magnified image on a fluorescent screen. Fig. 18-9a shows an electron microscope and the image it produced of individual uranium atoms. The electron beam is accelerated across a potential difference of 100,000 V (Fig. 18-9b). Each electron starts from rest at a potential of 0 and moves to a point where the potential is 1.00×10^5 V. Find the change (a) in an electron's electrical potential energy and (b) in its kinetic energy.

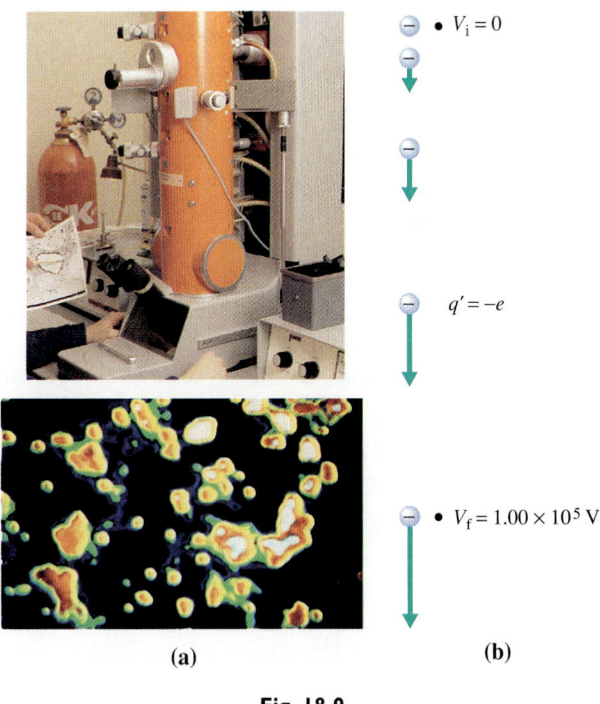

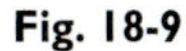

(a) (b)

Fig. 18-9

SOLUTION (a) It follows from Eq. 18-3 ($U_E = q'V$) that

$$\Delta U_E = q'\Delta V = (-e)(1.00 \times 10^5 \text{ V} - 0)$$

$$= -1.00 \times 10^5 \text{ eV}$$

Notice that the electron, a negative charge, moves from lower to higher potential. This movement corresponds to a loss of potential energy.

(b) The electron's weight is a negligible force compared with the electric force. Thus only the electric force does significant work on the electron, and therefore the electron's total mechanical energy is conserved; that is, the sum of the electron's kinetic energy and electrical potential energy is constant. The loss of 1.00×10^5 eV in potential energy must be compensated by a gain of 1.00×10^5 eV in kinetic energy:

$$\Delta K = -\Delta U_E = -(-1.00 \times 10^5 \text{ eV})$$

$$= 1.00 \times 10^5 \text{ eV}$$

Equipotential Surfaces

Suppose a test charge q' moves through an electric field in such a way that the path of the charge is always perpendicular to the field (Fig. 18-10). Then the electric force has no component along the path ($F_s = 0$), and it follows from the definition of work [Eq. 7-11: $W = \Sigma\,(F_s\,\Delta s)$] that the electric force does no work on the test charge. Because no work is done on q', there is no change in the electrical potential energy of q'. Thus this is a path of constant potential. Any path along a surface perpendicular to the field lines is a path of constant potential (Fig. 18-11). The potential is constant over the entire surface, which is called an **equipotential surface.**

An example of equipotential surfaces was seen in Example 4, where we found that, in the earth's vertical electric field, horizontal surfaces are surfaces of constant potential (Fig. 18-8).

Since the electric field at the surface of a conductor is everywhere perpendicular to the surface, **the surface of any conductor is an equipotential surface** (Fig. 18-12). The following example illustrates how equipotential surfaces can be used to determine the magnitude and direction of an electric field.

$W_E = \Sigma F_s\,\Delta s = 0$

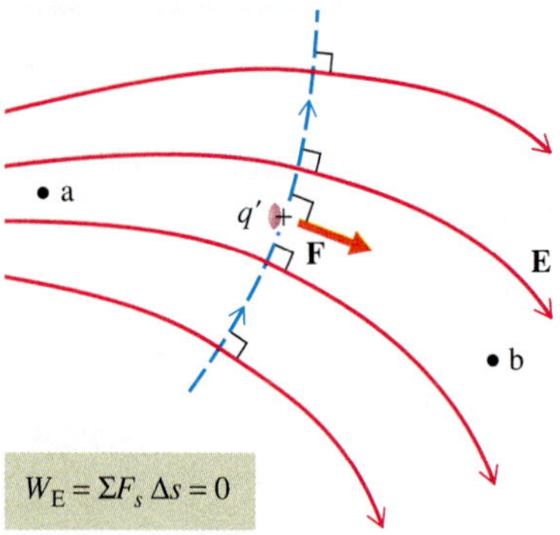

Fig. 18-10 The electric field does no work on a charge q' that moves along a path perpendicular to the field, and so the charge's potential energy is constant.

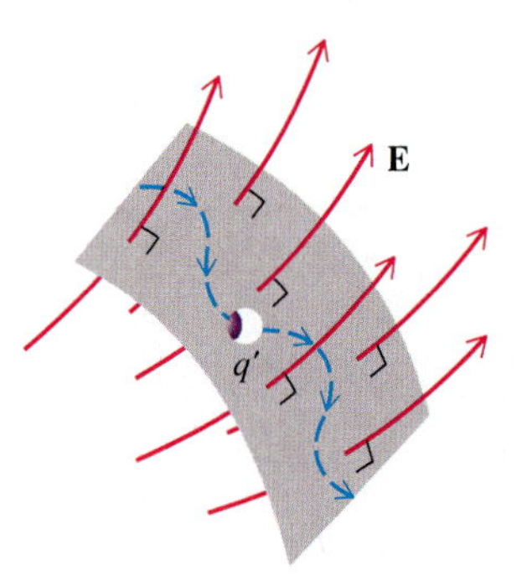

Fig. 18-11 A test charge q' has the same potential energy everywhere on a surface perpendicular to the field lines.

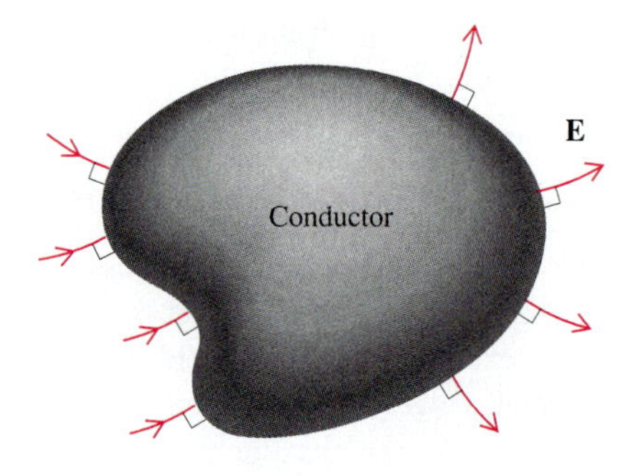

Fig. 18-12 A conductor's surface is an equipotential surface because it is perpendicular to the field lines.

EXAMPLE 7 Using Equipotential Surfaces to Find an Electric Field

Three equipotential surfaces are shown in Fig. 18-13. Draw the corresponding field lines and estimate the field strength at point a, where the distance between the surfaces is 4 cm.

SOLUTION The field lines are perpendicular to the equipotential surfaces, as shown in Fig. 18-14. In the vicinity of point a, the surfaces are nearly flat, and so, in order to estimate the field strength, it is a reasonable approximation to use Eq. 18-4 for the potential difference in a uniform field:

$$V_a - V_b \approx Ed$$

$$E \approx \frac{V_a - V_b}{d} = \frac{8\text{ V} - 6\text{ V}}{4 \times 10^{-2}\text{ m}}$$

$$\approx 50\text{ V/m}$$

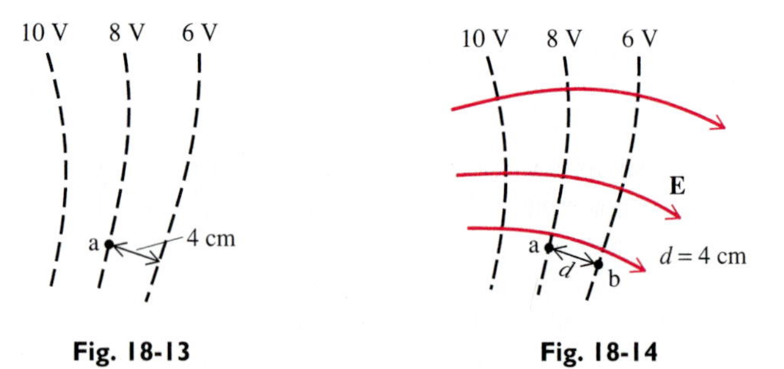

Fig. 18-13 **Fig. 18-14**

The strength of the electric field is often expressed in volts per meter. This is the same unit as newtons per coulomb, as we can see using the definition of a volt:

$$\frac{\text{V}}{\text{m}} = \frac{\text{J/C}}{\text{m}} = \frac{\text{N-m/C}}{\text{m}} = \text{N/C}$$

Potential of a Single Point Charge

Next we obtain an expression for potential in the field of a point charge q. Since the electric field produced by q is directed radially, the equipotential surfaces are spherical, as illustrated in Fig. 18-15 for a positive charge q. We could find an expression for the difference in potential energy of a test charge q' moving between arbitrary points a and b by computing the work the electric field does on q'. Since this calculation requires the use of integral calculus, however, we shall simply state the result without proof. The potential energy of the test charge q' at a distance r from the source charge q is

$$U_E = \frac{kqq'}{r} \tag{18-6}$$

Applying the definition of the electric potential ($V = U_E/q'$), we obtain an expression for the potential at a distance r from a point charge q:

$$V = \frac{kq}{r} \qquad \text{(single point charge)} \tag{18-7}$$

This equation applies to both positive and negative charges.

Remember that only potential differences have physical significance, and so when we write an expression for the potential at a point, as we have done here, we are implying that a choice for the reference level of potential has been made. Eq. 18-7 implies that the reference potential is at infinity; that is, when $r = \infty$, the equation gives $V = 0$.

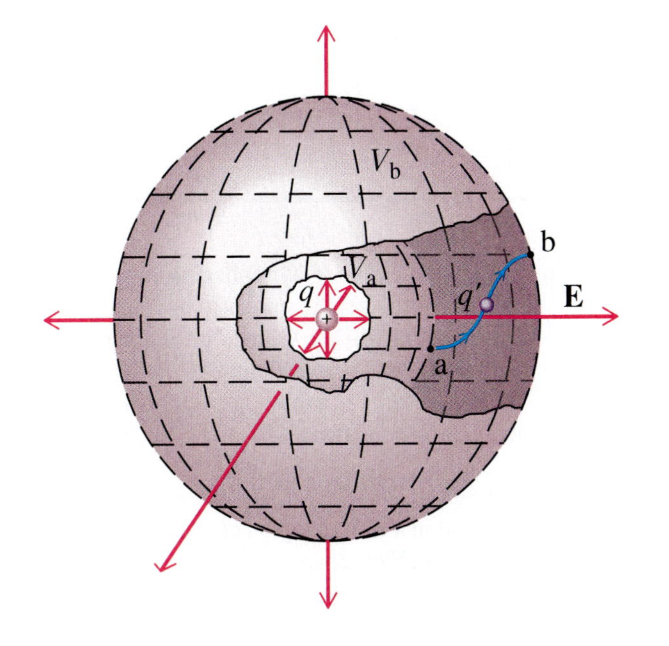

Fig. 18-15 Equipotential surfaces of a positive point charge.

EXAMPLE 8 Equipotential Surfaces for a Point Charge

Plot equipotential surfaces and field lines for (a) a point charge $q = 1.0 \times 10^{-9}$ C; (b) a point charge $q = -1.0 \times 10^{-9}$ C.

SOLUTION (a) Applying Eq. 18-7 for any distance r from q, we find

$$V = \frac{kq}{r} = \frac{(9.0 \times 10^9 \text{ N-m}^2/\text{C}^2)(1.0 \times 10^{-9} \text{ C})}{r} = \frac{9.0}{r} \text{ V-m}$$

Since V depends only on r, the equipotential surfaces are spheres. Equipotential surfaces are usually drawn so that adjacent surfaces represent equal voltage intervals. Using 3 V intervals, we find that

$$\begin{aligned} &\text{at } r = 1.0 \text{ m}, && V = 9.0 \text{ V} \\ &\text{at } r = 1.5 \text{ m}, && V = 6.0 \text{ V} \\ &\text{at } r = 3.0 \text{ m}, && V = 3.0 \text{ V} \\ &\text{at } r = \infty, && V = 0 \end{aligned}$$

The first three equipotential surfaces in this list are shown in Fig. 18-16a. The fourth surface corresponds to $V = 0$ and is at infinity.

(b) For a negative point charge of the same magnitude, only the sign of the potential changes. The equipotential surfaces for this case are shown in Fig. 18-16b. Notice that for both the positive and negative charges the field lines are perpendicular to the equipotential surfaces and are directed from regions of higher potential to regions of lower potential.

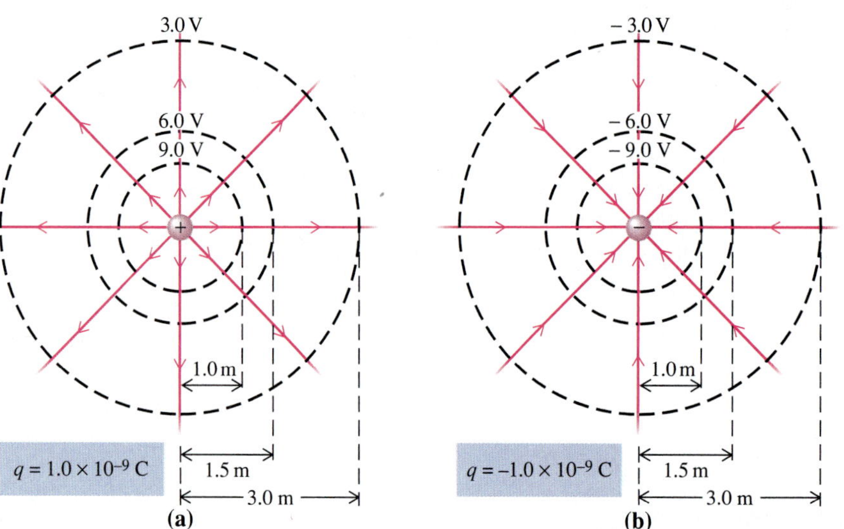

Fig. 18-16 Cross-sectional views of equipotential surfaces for **(a)** a positive point charge and **(b)** a negative point charge.

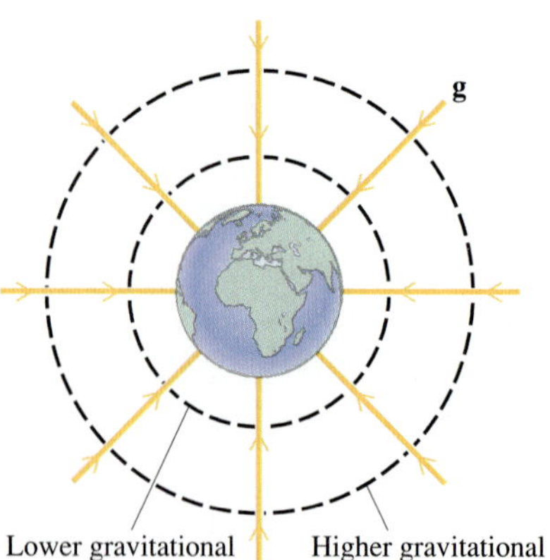

Fig. 18-17 Spherical surfaces of constant gravitational potential energy.

The relationship between the earth's gravitational field and the gravitational potential energy of a 1 kg mass in the field is similar to the relationship between V and **E** for a negative point charge (Fig. 18-17). Gravitational potential energy is constant along spherical surfaces, which are perpendicular to the gravitational field **g**. The field lines are directed from surfaces of higher gravitational potential energy to surfaces of lower gravitational potential energy.

Potential of Several Point Charges

Suppose that the electric field in a certain region of space is produced by a collection of n point charges $q_1, q_2, \ldots, q_n$. Each charge q_i produces a field $\mathbf{E}_i$, and the total field **E** is the vector sum of the single-particle fields. The work **E** does on a test charge q' is therefore the sum of the work done by the single-particle fields. The electrical potential energy of q' at any point is a sum of potential energy terms, each of which corresponds to one of the charges $q_1, q_2, \ldots, q_n$. And the potential (the electrical potential energy per unit charge) is the sum of the single-particle potentials:

$$V = \frac{kq_1}{r_1} + \frac{kq_2}{r_2} + \cdots + \frac{kq_n}{r_n}$$

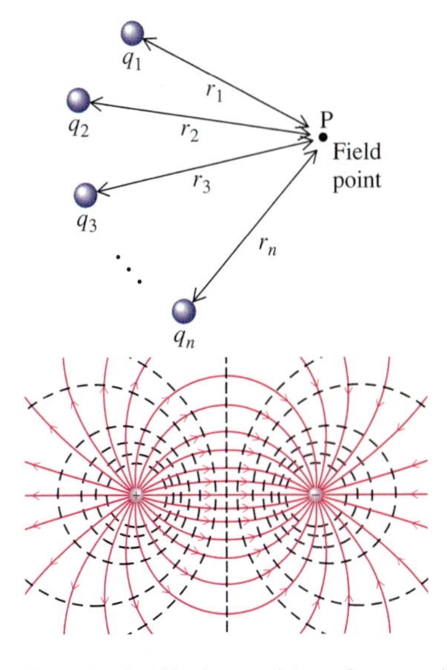

Fig. 18-18 The field point where V is to be evaluated is at various distances r_1, r_2, . . . , r_n from the charges q_1, q_2, . . . , q_n that are the source of V.

We can factor k from this equation and express the result more concisely using summation notation:

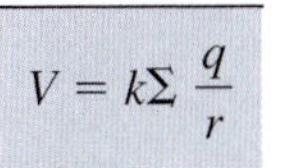

$$V = k\Sigma \frac{q}{r} \qquad \text{(point charges)} \qquad (18\text{-}8)$$

As illustrated in Fig. 18-18, the field point P at which V is to be evaluated is at a distance r_1 from source charge q_1, a distance r_2 from source charge q_2, and so on. Unlike the calculation of the electric field, the calculation of V involves a simple scalar sum.

One of the simplest charge configurations is a set of two point charges of equal magnitude but opposite sign separated by some distance. Two such charges are said to form an **electric dipole.** Equipotential surfaces and field lines are sketched for an electric dipole in Fig. 18-19.

Fig. 18-19 Equipotential surfaces and field lines of an electric dipole.

EXAMPLE 9 Dipole Potential

An electric dipole consists of charges $q_1 = \frac{20}{9.0} \times 10^{-9}$ C and $q_2 = -\frac{20}{9.0} \times 10^{-9}$ C separated by a distance of 6.0 m. Find the potential produced by the dipole for the following field points, all of which are on the line joining the two charges: (a) $r_1 = 1.0$ m, $r_2 = 5.0$ m; (b) $r_1 = 2.0$ m, $r_2 = 4.0$ m; (c) $r_1 = 3.0$ m, $r_2 = 3.0$ m; (d) $r_1 = 4.0$ m, $r_2 = 2.0$ m; (e) $r_1 = 5.0$ m, $r_2 = 1.0$ m. A proton is released from rest at point a; find its speed when it reaches point d.

SOLUTION Applying Eq. 18-8 for arbitrary values of r_1 and r_2, we find

$$V = k\Sigma \frac{q}{r}$$
$$= (9.0 \times 10^9 \text{ N-m}^2/\text{C}^2)\left(\frac{\frac{20}{9.0} \times 10^{-9}\text{ C}}{r_1} + \frac{-\frac{20}{9.0} \times 10^{-9}\text{ C}}{r_2}\right)$$
$$= (20 \text{ V-m})\left(\frac{1}{r_1} - \frac{1}{r_2}\right)$$

Evaluating this expression at field points a to e, we find

$$V_a = (20 \text{ V-m})\left(\frac{1}{1.0\text{ m}} - \frac{1}{5.0\text{ m}}\right) = 16 \text{ V}$$
$$V_b = (20 \text{ V-m})\left(\frac{1}{2.0\text{ m}} - \frac{1}{4.0\text{ m}}\right) = 5.0 \text{ V}$$
$$V_c = (20 \text{ V-m})\left(\frac{1}{3.0\text{ m}} - \frac{1}{3.0\text{ m}}\right) = 0 \text{ V}$$
$$V_d = (20 \text{ V-m})\left(\frac{1}{4.0\text{ m}} - \frac{1}{2.0\text{ m}}\right) = -5.0 \text{ V}$$
$$V_e = (20 \text{ V-m})\left(\frac{1}{5.0\text{ m}} - \frac{1}{1.0\text{ m}}\right) = -16 \text{ V}$$

The values of V at the respective field points are indicated in Fig. 18-20.

A proton released from rest at point a will be accelerated to the right in the direction of the electric field. It will go from a higher potential to a lower potential. Its change in potential energy from a to d is found when Eq. 18-3 ($U_E = q'V$) is applied:

$$\Delta U_E = q'\,\Delta V = (+e)(V_d - V_a)$$
$$= e(-5.0 \text{ V} - 16 \text{ V}) = -21 \text{ eV}$$

Since the proton's total mechanical energy must be conserved, its kinetic energy must increase by 21 eV:

$$\Delta K = K_f - K_i = 21 \text{ eV}$$

And since the proton was initially at rest, $K_i = 0$. Thus

$$K_f = 21 \text{ eV}$$

or $$K_f = (21 \text{ eV})(1.6 \times 10^{-19} \text{ J/eV}) = 34 \times 10^{-19} \text{ J}$$

Using the definition of kinetic energy and the proton's mass, we find

$$K_f = \tfrac{1}{2}mv_f^2$$
$$v_f = \sqrt{\frac{2K_f}{m}} = \sqrt{\frac{2(34 \times 10^{-19}\text{ J})}{1.67 \times 10^{-27}\text{ kg}}}$$
$$= 6.4 \times 10^4 \text{ m/s}$$

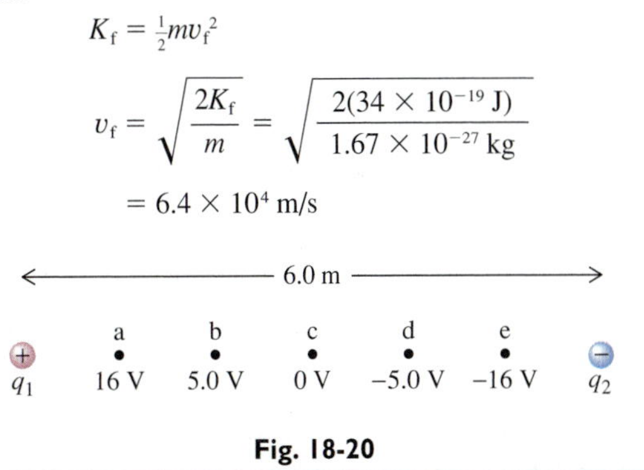

Fig. 18-20

EXAMPLE 10 Potential Near Three Point Charges

Point charges are located at three corners of a square (Fig. 18-21). (a) Find the potential at P and R. (b) An electron initially at rest at R moves to P and is acted upon only by the electrostatic force. Find the speed of the electron at P.

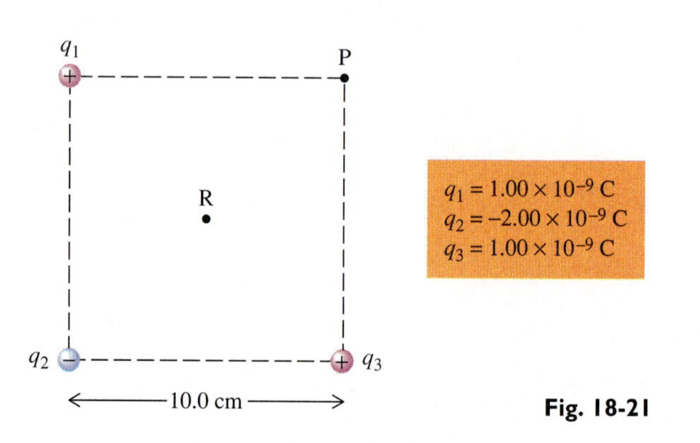

Fig. 18-21

SOLUTION (a) At P, we find

$$V_P = k\Sigma \frac{q}{r}$$

$$= (8.99 \times 10^9 \text{ N-m}^2/\text{C}^2)\left(\frac{1.00 \times 10^{-9}\text{ C}}{0.100\text{ m}} + \frac{-2.00 \times 10^{-9}\text{ C}}{0.100\sqrt{2}\text{ m}} + \frac{1.00 \times 10^{-9}\text{ C}}{0.100\text{ m}}\right)$$

$$= 52.7 \text{ V}$$

At R we find

$$V_R = (8.99 \times 10^9 \text{ N-m}^2/\text{C}^2)\left(\frac{1.00 \times 10^{-9}\text{ C}}{0.0500\sqrt{2}\text{ m}} + \frac{-2.00 \times 10^{-9}\text{ C}}{0.0500\sqrt{2}\text{ m}} + \frac{1.00 \times 10^{-9}\text{ C}}{0.0500\sqrt{2}\text{ m}}\right)$$

$$= 0$$

(b) The potential energy of the electron at a point where the potential is V is $U_E = q'V = -eV$. Thus the change in the electron's potential energy is

$$\Delta U_E = -e\Delta V$$

If no other forces act on the electron, its mechanical energy is conserved and hence its change in kinetic energy is

$$\Delta K = -\Delta U_E$$
$$= e\Delta V$$

Expressing kinetic energy in terms of the electron's mass and speed, and using the fact that the electron's initial speed is zero, we find

$$\tfrac{1}{2}mv^2 = \Delta K = e\Delta V$$

or

$$v = \sqrt{\frac{2e\Delta V}{m}} = \sqrt{\frac{2(1.60 \times 10^{-19}\text{ C})(52.7\text{ V} - 0)}{9.11 \times 10^{-31}\text{ kg}}}$$
$$= 4.30 \times 10^6 \text{ m/s}$$

18-2 Capacitance

A capacitor is a simple device for storing charge. It consists of two conductors separated by a small space. Each conductor carries a net charge. The charges are equal in magnitude but opposite in sign (Fig. 18-22). Thus the net charge on the entire capacitor is zero.

Capacitors of all shapes and sizes are important elements in electric circuits (Fig. 18-23a). Some capacitors are used to tune radio circuits; others are used to store energy that can be quickly discharged and used as the source of energy for a laser (Fig. 18-23b). Whatever their purpose, all capacitors have the common function of storing charge.

If the charge shown on the conductors in Fig. 18-22 is increased, the electric field will increase, and so the potential difference between the two conductors will also increase. The relationship between the charge Q and the potential difference is particularly simple: These two quantities are directly proportional. To have more concise notation we now **adopt the convention of using V to denote a potential difference** rather than the potential at a single point:

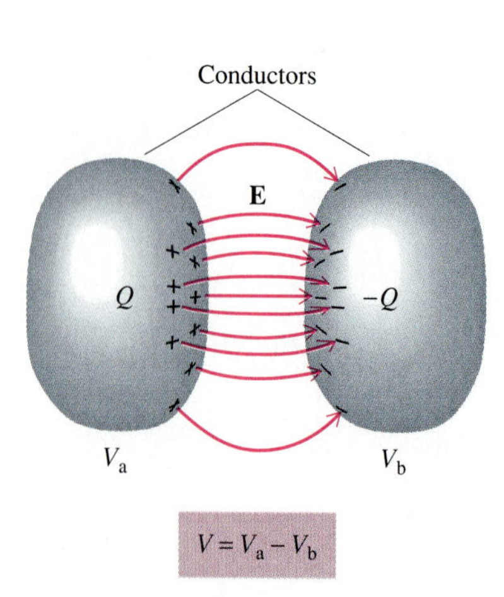

Fig. 18-22 A capacitor consists of two conductors carrying opposite charges of equal magnitude.

$$V = V_a - V_b \tag{18-9}$$

Using this convention, we can express very concisely the relationship between charge and potential difference:

$$Q \propto V \tag{18-10}$$

This result is proved in general in advanced texts on electricity. We shall verify it shortly in one important special case.

The constant of proportionality in the relationship above is called **capacitance** and is denoted by C:

$$Q = CV \tag{18-11}$$

It follows from this equation that capacitance must have units of coulombs per volt. We call this unit a farad (abbreviation F), in honor of Michael Faraday:

$$1\text{ F} = 1\text{ C/V} \tag{18-12}$$

Capacitance is a measure of how much charge a capacitor can store for a given potential difference V. For example, a 2 F capacitor stores twice as much charge as a 1 F capacitor, if both have the same potential difference between their conductors.

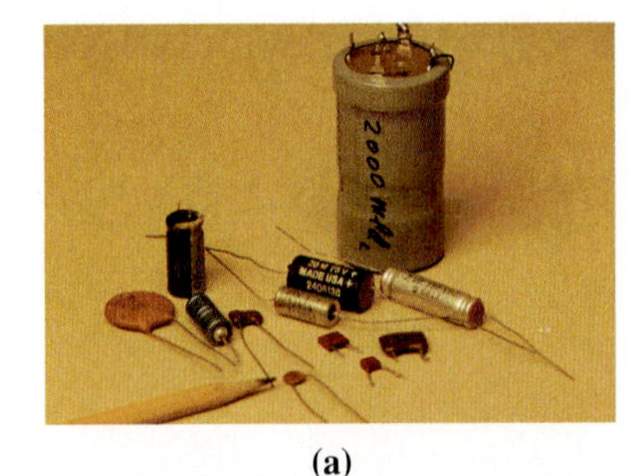

(a)

(b)

Fig. 18-23 **(a)** Capacitors found in electric circuits. **(b)** Capacitors used to energize a very powerful laser.

Parallel-Plate Capacitor

The size and shape of a capacitor determine the value of its capacitance. A common and particularly simple geometry is the parallel-plate capacitor (Fig. 18-24). We assume here that the plates are separated by a vacuum. Since the electric field between the plates is uniform, we may apply Eq. 18-4 and express the potential difference as the product of the field strength E and the plate separation d:

$$V = Ed$$

In Chapter 17 we found that the field strength just outside a conductor is a function of the surface charge density (Eq. 17-21):

$$E = 4\pi k\sigma = 4\pi k\frac{Q}{A}$$

Inserting this expression for E into the equation above, we obtain

$$V = \left(4\pi k\frac{Q}{A}\right)d$$

Solving for Q, we find

$$Q = \left(\frac{A}{4\pi kd}\right)V$$

This verifies the linear relationship between Q and V for the parallel-plate geometry (Eq. 18-11: $Q = CV$) and gives an expression for the capacitance, which is just the factor in parentheses in the equation above:

$$C = \frac{A}{4\pi kd}$$

The capacitance of any capacitor is a constant the value of which depends on the geometry of the capacitor. For the parallel-plate capacitor, capacitance depends only on the area A of the plates and the distance d between them. One can increase capacitance by either increasing A or decreasing d.

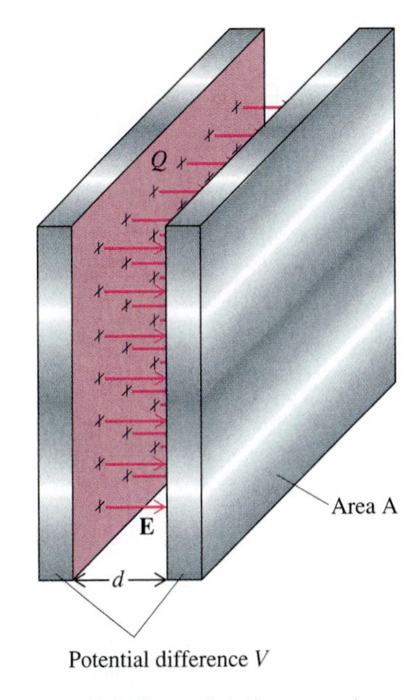

Fig. 18-24 A parallel-plate capacitor.

It is convenient to define a new quantity ϵ_0, which will simplify the preceding equation as well as other equations in later chapters. Defining

$$\epsilon_0 = \frac{1}{4\pi k} \tag{18-13}$$

we may express the capacitance

$$C = \frac{\epsilon_0 A}{d} \quad \text{(parallel-plate capacitor)} \tag{18-14}$$

We find the value of ϵ_0 by substituting the value of k into Eq. 18-13:

$$\epsilon_0 = \frac{1}{4\pi(8.99 \times 10^9 \text{ N-m}^2/\text{C}^2)} = 8.85 \times 10^{-12} \text{ C}^2/\text{N-m}^2$$

The units for ϵ_0 can be simplified: $\text{C}^2/\text{N-m}^2 = \text{C/V-m} = \text{F/m}$. Thus

$$\epsilon_0 = 8.85 \times 10^{-12} \text{ F/m} \tag{18-15}$$

EXAMPLE 11 Plate Area of a 1F Capacitor

Calculate the area of one plate of a parallel-plate capacitor having a capacitance of 1.00 F if the plates are separated by a distance of 1.00 mm.

SOLUTION Solving Eq. 18-14 for A, we find

$$A = \frac{Cd}{\epsilon_0}$$

$$= \frac{(1.00 \text{ F})(1.00 \times 10^{-3} \text{ m})}{8.85 \times 10^{-12} \text{ F/m}}$$

$$= 1.13 \times 10^8 \text{ m}^2$$

Such a capacitor would be enormously large, which means that 1 F is a huge amount of capacitance. Units of microfarads (μF) and picofarads (pF) are commonly used, where

$$1 \ \mu\text{F} = 10^{-6} \text{ F}$$

$$1 \text{ pF} = 10^{-12} \text{ F}$$

Typically we encounter capacitors with capacitance on the order of a few μF or less.

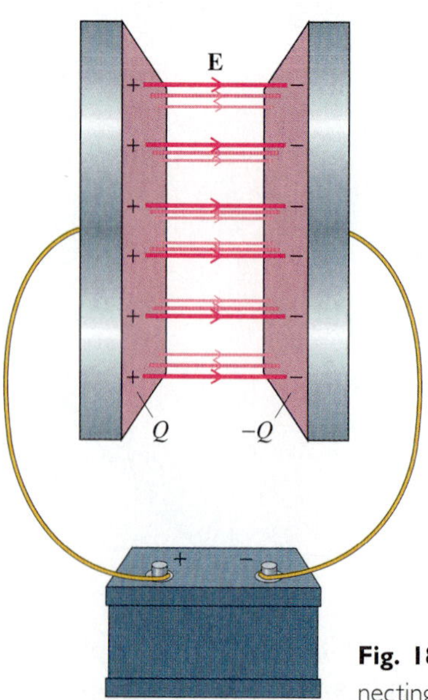

Fig. 18-25 We can charge a capacitor by connecting its plates to a battery.

Charging a Capacitor

We can charge a capacitor by connecting its plates to the terminals of a battery or power supply (Fig. 18-25). (A power supply is a device used like a battery but which receives its energy from the electrical outlet it plugs into.) The battery or power supply is a source of electrical energy and produces a fixed potential difference across its terminals. Each terminal is connected to one capacitor plate by a metal wire. The terminal, connecting wire, and capacitor plate form one continuous conductor. And since a conductor is at a constant potential, each capacitor plate is at the same potential as the terminal to which it is connected. So the battery or power supply maintains a fixed potential difference across the capacitor, and this means there must be charge on the capacitor.

The plates of a capacitor always have charges of opposite sign and equal magnitude. Whenever a charge is established on one capacitor plate, an opposite charge is quickly drawn toward it from the other plate. The opposing charges are attracted to the inner surfaces of the two plates. When the opposite charges have equal magnitude, the electric fields produced by the two charged surfaces cancel everywhere except between the plates. If the charge magnitudes were unequal, a net electric field would exist inside the metal of the plates. Charge would then flow until the field disappeared, that is, until the opposite charges were equal in magnitude.

If a potential difference is established across a capacitor and then the energy source is disconnected, the capacitor will continue to store charge on the inner surfaces of its plates. If you then provide a conducting path across the plates, they will quickly discharge through the conducting path. If the stored charge is large enough, the discharge is accompanied by a bright spark and a loud sound like a firecracker (Fig. 18-26). It can be dangerous to handle large capacitors. If your body happens to provide a conducting path through which such a capacitor discharges, the result could be painful or even fatal.

Fig. 18-26 A discharging capacitor.

Parallel Capacitors

In electric circuits we often find a combination, or "network," of many capacitors. To simplify the analysis of a circuit, it is convenient to be able to represent the effect of all the capacitors in the network by a single **equivalent capacitor** that produces the same effect for circuit points outside the network. We shall consider two important ways of combining capacitors: series and parallel combinations.

When two or more capacitors are connected in such a way that each must have the **same potential difference** across its plates, the capacitors are said to be connected in **parallel.** A parallel combination is accomplished when the plates are connected as shown in Fig. 18-27. All the plates and wires connected to each other must be at the same potential because they are all part of one continuous conductor. Thus in Fig. 18-27 the two plates on the left side are both at the same potential V_a, and the two plates on the right side are both at potential V_b. The potential difference across the two capacitors must then be the same: $V_a - V_b$.

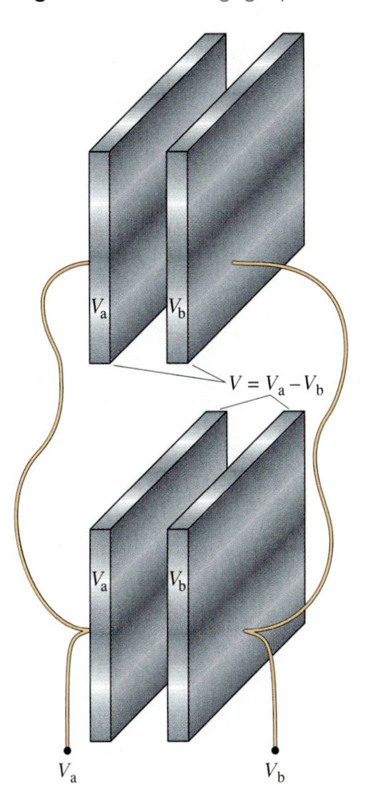

Fig. 18-27 Two capacitors connected in parallel.

It is convenient to represent a capacitor by the symbol ⊣⊢ and to represent ideal conducting wires by straight lines. Then the parallel combination shown in Fig. 18-27 can be represented as shown in Fig. 18-28. (Although the symbol for a capacitor suggests parallel plates, it can represent a capacitor with any kind of geometry.)

The total charge Q stored on a parallel combination of capacitors is the sum of the charges on the individual capacitors:

$$Q = Q_1 + Q_2$$

where the charge on each capacitor equals the product of its capacitance and the same potential difference V:

$$Q = C_1V + C_2V = (C_1 + C_2)V$$

A single capacitor having a capacitance $C = C_1 + C_2$ stores the same charge Q when the same potential V is applied across it. Thus C is the equivalent capacitance for this combination:

$$C = C_1 + C_2$$

The derivation above is easily generalized to three or more capacitors connected in parallel, with the result that the equivalent capacitance is just the sum of the individual capacitances:

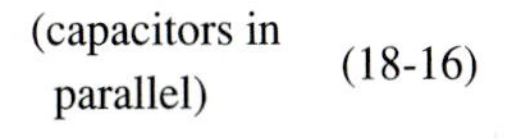

$$C = C_1 + C_2 + C_3 + \cdots \quad \text{(capacitors in parallel)} \qquad (18\text{-}16)$$

For example, if 5 μF, 10 μF, and 20 μF capacitors are connected in parallel, this is equivalent to a single 35 μF capacitor.

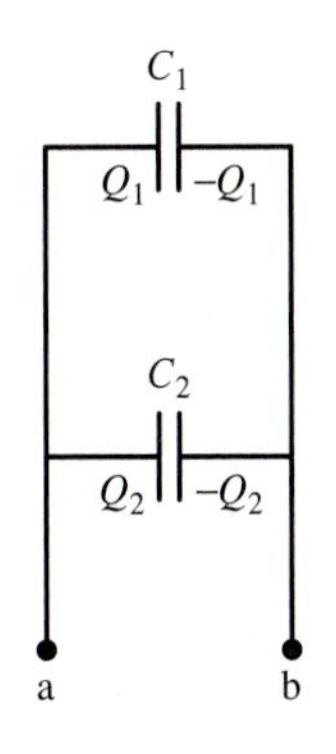

Fig. 18-28 Circuit diagram of two capacitors connected in parallel.

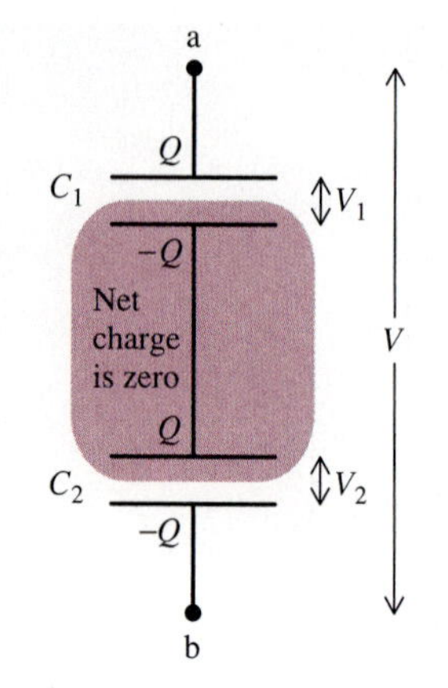

Fig. 18-29 Circuit diagram of two capacitors connected in series.

Series Capacitors

When two capacitors are connected as in Fig. 18-29, they are said to be in **series.** With this kind of connection, the **same charge** Q is stored on each capacitor. We can understand this by realizing that before the capacitors are connected to some power source they are uncharged. When the capacitors are charged, the negative charge on the bottom plate of C_1 must come from electrons drawn from the upper plate of C_2, since these two plates are not connected to anything else. So these two isolated plates must have opposite charges, Q and $-Q$ (Fig. 18-29).

The total potential difference V between points a and b in Fig. 18-29 must equal the sum of the potential differences V_1 and V_2 across the two capacitors because it is only between the capacitors' plates that there is any change in potential (remember that there is no change in potential along the connecting wires):

$$V = V_1 + V_2$$

When we express the potential difference across each capacitor in terms of its capacitance, the equation above becomes

$$V = \frac{Q}{C_1} + \frac{Q}{C_2} = Q\left(\frac{1}{C_1} + \frac{1}{C_2}\right)$$

We wish to find the equivalent capacitance for this series combination; that is, we want the value of capacitance C for the single equivalent capacitor that will store the same charge Q for the same potential difference V. The potential difference across this single equivalent capacitor may be expressed

$$V = \frac{Q}{C}$$

Equating the two expressions above for V, we obtain an equation for the equivalent capacitance C:

$$\frac{Q}{C} = Q\left(\frac{1}{C_1} + \frac{1}{C_2}\right)$$

or

$$\frac{1}{C} = \frac{1}{C_1} + \frac{1}{C_2}$$

The derivation above is easily generalized to three or more capacitors connected in series, with the result that the inverse of the equivalent capacitance is just the sum of the inverses of all the individual capacitances:

$$\frac{1}{C} = \frac{1}{C_1} + \frac{1}{C_2} + \frac{1}{C_3} + \cdots \quad \text{(capacitors in series)} \quad (18\text{-}17)$$

Thus, for example, if three 5 μF capacitors are connected in series, the sum of the inverses is $\frac{3}{5}$ $(\mu\text{F})^{-1}$, and the equivalent capacitance is $\frac{5}{3}$ μF.

Complex networks of capacitors can often be reduced to a single equivalent capacitor when the rules for series and parallel combinations are applied.

EXAMPLE 12 Equivalent Capacitance of a Network

(a) Find the equivalent capacitance of the network shown in Fig. 18-30. (b) Find the charge stored on the 12 μF capacitor when the potential difference between a and b is 100 V.

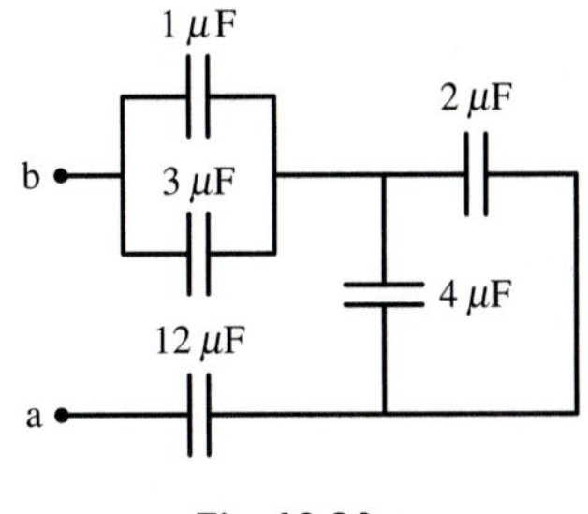

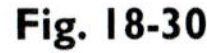

Fig. 18-30

SOLUTION (a) We reduce the network in steps, looking for either series or parallel combinations. The 1 μF and 3 μF capacitors are in parallel, and the 2 μF and 4 μF capacitors are in parallel. Thus these combinations can be replaced by 4 μF and 6 μF equivalent capacitors, as indicated in Fig. 18-31a. Now we have three capacitors in series. Adding inverses, we find the inverse of the equivalent capacitance (Fig. 18-31b).

$$\frac{1}{C} = \frac{1}{4\ \mu\text{F}} + \frac{1}{6\ \mu\text{F}} + \frac{1}{12\ \mu\text{F}}$$

$$= \frac{6}{12\ \mu\text{F}} = \tfrac{1}{2}\,(\mu\text{F})^{-1}$$

Thus $$C = 2\ \mu\text{F}$$

(b) For a 100 V potential difference applied between a and b, the equivalent capacitor stores charge

$$Q = CV = (2\ \mu\text{F})(100\ \text{V}) = 200\ \mu\text{C}$$

The original network stores the same charge. Since the 12 μF capacitor is connected directly to point a, 200 μC must be stored on this capacitor.

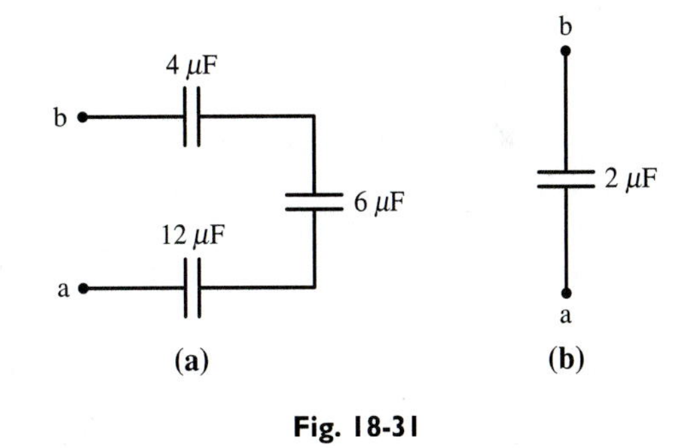

Fig. 18-31

Energy Stored by a Capacitor

Since a capacitor maintains charge at a potential difference, it stores electrical potential energy. We can calculate the amount of energy stored by recognizing that the effect of charging a capacitor is to transfer a quantity of charge from one plate to the other (Fig. 18-32). Suppose that the charge is transferred bit by bit. As a small quantity of positive charge Δq moves from the negative plate to the positive plate, it moves to a point at a higher potential and its electrical potential energy therefore increases. The increase in electrical potential energy equals the product of the charge Δq and the potential difference between the plates. But the value of the capacitor's potential difference depends on how much charge has already been transferred. Just before the charging begins, this potential difference is zero; at the end of the process, the potential difference has reached a final value V. The *average* potential during the charging is half this final value:

$$\text{Average potential difference} = \tfrac{1}{2}V$$

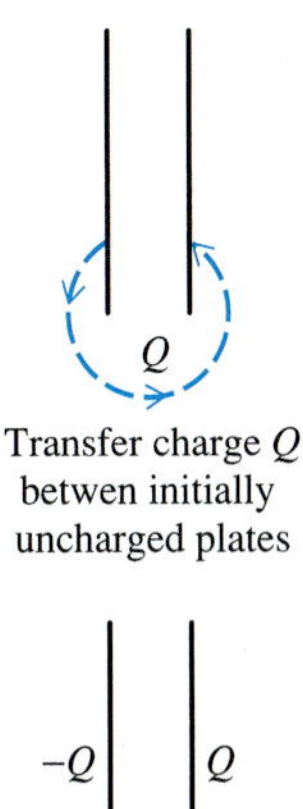

Fig. 18-32 Transferring charge Q from a capacitor's left plate to its right plate gives the left plate a charge −Q and the right plate a charge Q.

The total potential energy U_E stored by the charged capacitor equals the sum of the increases in potential energy of all the charge increments Δq, that is, the increase in potential energy of all the transferred charge Q. This increase in potential energy equals the product of the charge times the average value of the potential difference:

$$U_E = Q(\tfrac{1}{2}V)$$

$$U_E = \tfrac{1}{2}QV \tag{18-18}$$

Since $Q = CV$, we may substitute for either Q or V in Eq. 18-18 and express U_E in two alternative forms:

$$U_E = \tfrac{1}{2}CV^2 \tag{18-19}$$

$$U_E = \frac{1}{2}\frac{Q^2}{C} \tag{18-20}$$

EXAMPLE 13 Energy Stored by Capacitors Before and After Connecting Them

A 4.00 μF capacitor is initially isolated and has a potential difference of 10.0 V across its plates. The plates of this capacitor are then connected in parallel to the plates of an initially uncharged 12.0 μF capacitor. Find the electrical potential energy stored before and after connection.

SOLUTION Applying Eq. 18-19, we compute the initial electrical potential energy U_E:

$$U_E = \tfrac{1}{2}CV^2$$

$$= \tfrac{1}{2}(4.00 \times 10^{-6}\ \text{F})(10.0\ \text{V})^2$$

$$= 2.00 \times 10^{-4}\ \text{J}$$

After the capacitors are connected in parallel, we have an equivalent capacitance of 16.0 μF. The total charge stored on the two capacitors equals the charge initially stored on the 4.00 μF capacitor:

$$Q = CV = (4.00 \times 10^{-6}\ \text{F})(10.0\ \text{V})$$

$$= 4.00 \times 10^{-5}\ \text{C} = 40.0\ \mu\text{C}$$

Applying Eq. 18-20, we compute the final potential energy U'_E:

$$U'_E = \frac{1}{2}\frac{Q^2}{C}$$

$$= \frac{1}{2}\frac{(4.00 \times 10^{-5}\ \text{C})^2}{16.0 \times 10^{-6}\ \text{F}}$$

$$= 5.00 \times 10^{-5}\ \text{J}$$

This energy is considerably less than the 2.00×10^{-4} J originally stored by the 4.00 μF capacitor. Thus most of the initial electrical potential energy has been lost. We can understand this loss by considering what happens when the two capacitors are connected. They are connected in parallel and so must have the same potential difference between their plates. Since $Q = CV$, the final charge stored on each capacitor is proportional to its capacitance. Thus the 12.0 μF capacitor stores three times as much charge as the 4.00 μF capacitor, that is, three fourths of the available 40.0 μC.

Therefore 30.0 μC of charge must flow between each plate of the 4.00 μF capacitor and the plate of the 12.0 μF capacitor to which it is connected, as indicated in Fig. 18-33. During this flow of charge through the connecting wires, most of the initial electrical potential energy is converted to thermal energy.

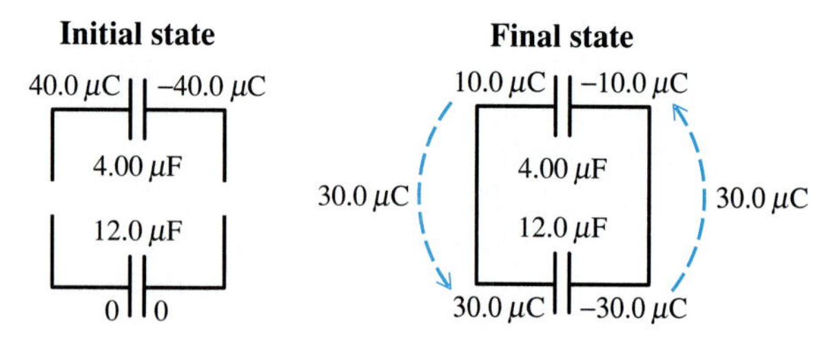

Fig. 18-33 In order for the system to go from the initial state to the final state, 30.0 μC of charge must flow through the connecting wires.

Field Energy

At this point, you might ask, "Exactly *where* is the potential energy of a charged capacitor stored?" To answer this question, we shall first obtain an expression for the electrical potential energy as a function of the electric field between the capacitor plates for the special case of a parallel-plate capacitor. We shall then think of the energy as being stored in the electric field. Such an interpretation will be essential when we consider time-varying fields in later chapters.

According to Eq. 18-19, the potential energy stored in a capacitor may be expressed

$$U_E = \tfrac{1}{2}CV^2$$

Using Eq. 18-14 for the capacitance of a parallel-plate capacitor and Eq. 18-4 for the potential difference in a uniform field, we obtain

$$\begin{aligned} U_E &= \frac{1}{2}\left(\frac{\epsilon_0 A}{d}\right)(Ed)^2 \\ &= \tfrac{1}{2}\epsilon_0 A d E^2 \end{aligned}$$

The volume of space between the plates of the capacitor is filled with a uniform field **E**. This volume is the product of the area A of each plate and the plate separation d. Thus the energy per unit volume, called the **energy density,** which we shall denote by u, is found when the expression above is divided by Ad:

$$u = \tfrac{1}{2}\epsilon_0 E^2 \qquad (18\text{-}21)$$

We can think of this energy as being stored in the field between the capacitor plates.

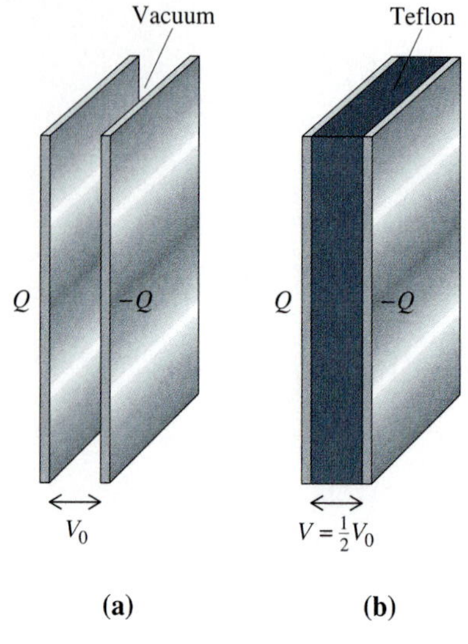

Fig. 18-34 Inserting Teflon between the plates of a capacitor reduces the potential difference from V_0 to $\frac{1}{2}V_0$.

18-3 Dielectrics

Consider an isolated capacitor having charge Q, potential difference V_0 across its plates, and a vacuum between the plates (Fig. 18-34a). Its capacitance C_0 is then

$$C_0 = \frac{Q}{V_0}$$

Now suppose that a dielectric material, for example, a layer of Teflon, is inserted between the plates (Fig. 18-34b). If we now measure the potential difference across the plates, we find that V is about half the original value V_0. Since there is no way for Q to change in this experiment, this decrease in V means that the ratio of charge to voltage has approximately doubled; there is a new value of the capacitance C that is about twice the vacuum value C_0:

$$C = \frac{Q}{V} = \frac{Q}{\frac{1}{2}V_0} = 2\frac{Q}{V_0}$$

or

$$C = 2C_0$$

When any dielectric is placed between the plates of an isolated capacitor, there is a reduction in potential from the vacuum value and hence an increase in the capacitance. The size of the increase depends on the kind of dielectric. The ratio C/C_0 is called the **dielectric constant** of the dielectric placed between the plates and is denoted by the Greek letter κ (kappa):

$$\kappa = \frac{C}{C_0} \tag{18-22}$$

Teflon has a dielectric constant of about 2. Constants for some other common dielectrics are given in Table 18-1.

Also given in Table 18-1 are **dielectric strengths,** defined as the largest electric fields the dielectrics can withstand before breaking down and becoming conductors. For example, air has a dielectric strength of 3×10^6 V/m. When the electric field in air exceeds this value, air molecules become ionized and are accelerated by the field, and so the air becomes conducting. This happens, for example, in an electrical storm or when the electric field around a high-voltage transmission line becomes too great.

Table 18-1 Dielectric Constants and Dielectric Strengths

Material	Dielectric constant	Dielectric strength (V/m)
Vacuum	1 (exactly)	∞
Air	1.0006	3×10^6
Teflon	2	6×10^7
Paper	4	2×10^7
Glass (Pyrex)	5	1×10^7
Neoprene rubber	7	1×10^7
Porcelain	7	2×10^7
Water	80	
Titanium dioxide	100	2×10^8

If we apply Eq. 18-22 to a parallel-plate capacitor, which we found has a vacuum capacitance $C_0 = \frac{\epsilon_0 A}{d}$, we obtain an expression for its capacitance when a dielectric is placed between the plates:

$$C = \kappa C_0$$

$$C = \frac{\kappa \epsilon_0 A}{d} \quad \text{(parallel-plate capacitor)} \quad (18\text{-}23)$$

Solid dielectrics are used in most capacitors. They offer several advantages over either vacuum or air capacitors: (1) capacitance is increased because of the κ factor; (2) the dielectric strength is larger than that of air, and so the plates can be closer (smaller d); since C is inversely proportional to d, capacitance is again increased; and (3) the small separation of the plates is easier to maintain when a solid layer of dielectric is inserted between them.

EXAMPLE 14 Capacitor with Titanium Dioxide Between the Plates

The plates of a parallel-plate capacitor containing a titanium dioxide dielectric have an area of $1.0 \times 10^3\ \text{cm}^2$ and are separated by a distance of 0.10 mm. (a) Find the capacitance. (b) What is the maximum voltage that can be appled to this capacitor?

SOLUTION (a) Applying Eq. 18-23 and using the dielectric constant of titanium dioxide given in Table 18-1 ($\kappa = 100$), we find

$$C = \frac{\kappa \epsilon_0 A}{d} = \frac{(100)(8.85\times10^{-12}\text{F/m})(1.0\times10^3\text{cm}^2)(10^{-2}\text{m}/1\,\text{cm})^2}{1.0\times10^{-4}\text{m}}$$

$$= 8.9 \times 10^{-7}\ \text{F} = 0.89\ \mu\text{F}$$

(b) The maximum voltage is the product of the maximum electric field magnitude E (the dielectric strength of titanium dioxide given in Table 18-1) and the plate separation d.

$$V = Ed = (2 \times 10^8\ \text{V/m})(1.0 \times 10^{-4}\ \text{m})$$

$$= 2 \times 10^4\ \text{V}$$

Physical Origin of the Dielectric Constant

When a dielectric is placed between the plates of a capacitor, the capacitance increases. We can understand this phenomenon in terms of how a dielectric's molecules respond to an electric field. If the charge distribution in a molecule is not symmetric, so that it has positive and negative sides, or "poles," it is called a **polar molecule.** In the absence of an external electric field, polar molecules are randomly oriented (Fig. 18-35a). When an external field is present, polar molecules tend to align with the field so that the positive poles are more likely to point in the direction of the field (Fig. 18-35b). In nonpolar molecules, the charge distribution inside each molecule is electrically symmetric in the absence of a field. However, nonpolar mol-

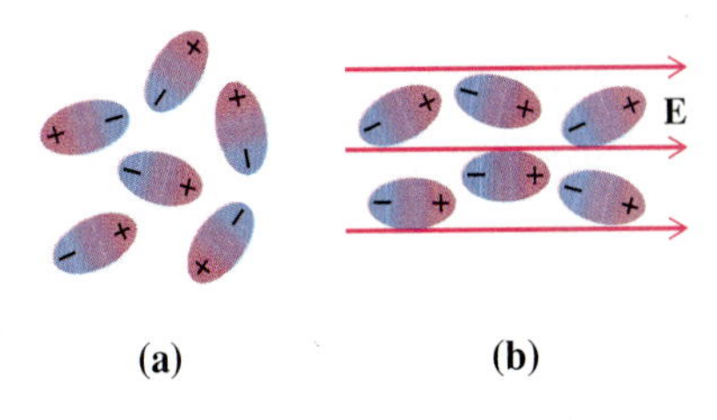

Fig. 18-35 **(a)** Polar molecules are randomly oriented in the absence of an external field. **(b)** Polar molecules are partially aligned with an external electric field.

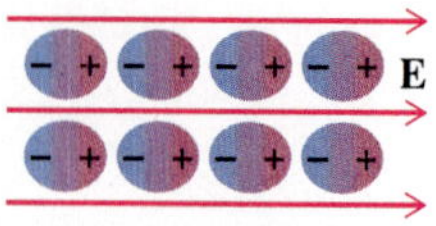

E

Fig. 18-36 Nonpolar molecules polarized by an external electric field.

Charge cancels in this region.

E

Negative surface charge

Positive surface charge

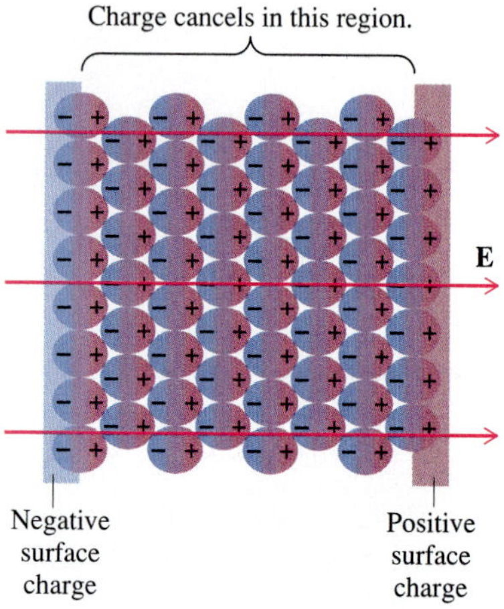

Fig. 18-37 The net effect of a dielectric in a uniform electric field is to produce positive and negative surface charges on opposite sides of the dielectric.

ecules develop poles—or a "dipole moment" as it is called—when an electric field is present (Fig. 18-36). The amount of polarization developed by nonpolar molecules, however, is not as great as it is in polar molecules. For both polar and nonpolar dielectrics, the net effect of the field is to give the dielectric positive and negative surface charges, as illustrated in Fig. 18-37 for a nonpolar dielectric.

When a dielectric is inserted between the plates of a capacitor, dielectric surface charge partially cancels the charge on the surfaces of the capacitor plates. Thus the field between the plates is reduced (Fig. 18-38), and this reduction in field strength reduces the potential difference. The lower potential difference for the same amount of charge stored on the plates means that there is a larger value of capacitance. The effect is greater for polar molecules than for nonpolar ones. For example, water molecules are polar and water has a high dielectric constant ($\kappa = 80$), whereas the dielectric constant for paper, a nonpolar material, is only 4.

The argument used to derive the electrical potential energy of a vacuum-filled capacitor also applies when a dielectric is present, and so in either case we may use Eqs. 18-18 to 18-20:

$$U_E = \tfrac{1}{2}QV = \tfrac{1}{2}CV^2 = \tfrac{1}{2}\frac{Q^2}{C}$$

Eq. 18-21 for the energy density of the electric field must, however, be modified. That equation was derived from the expression $U_E = \frac{1}{2}CV^2$, from which we found the energy density $u = \frac{1}{2}\epsilon_0 E^2$. If we substitute $C = \kappa C_0$ into the expression for U_E, we may repeat the derivation, which is changed only by the factor κ. Thus the energy density in a dielectric is

$$u = \tfrac{1}{2}\kappa\epsilon_0 E^2 \qquad (18\text{-}24)$$

This equation does not imply that the energy density is enhanced over the vacuum value by the factor κ because E is reduced when a dielectric is placed near free charge.

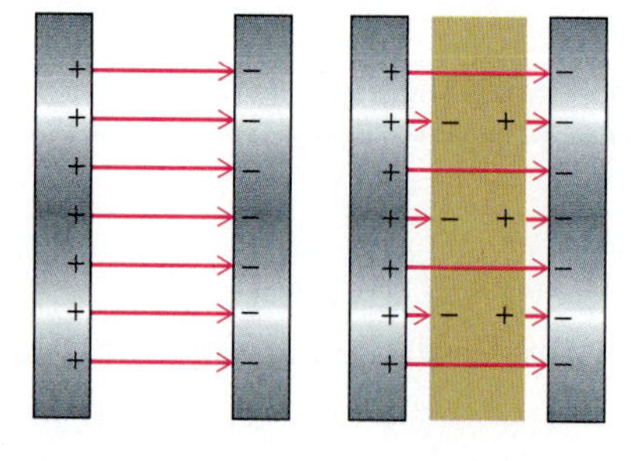

Fig. 18-38 Insertion of a dielectric in a capacitor cancels some of the charge on the plates and therefore reduces the electric field and the potential difference between the plates.

*18-4 The Oscilloscope

In this section we shall study the motion of an electron in a uniform electric field and the application of this motion to the oscilloscope, an instrument used to display and measure electrical signals (Fig. 18-39). The oscilloscope allows us to see the time dependence of a time-varying potential difference, in effect, to graph the potential difference versus time.

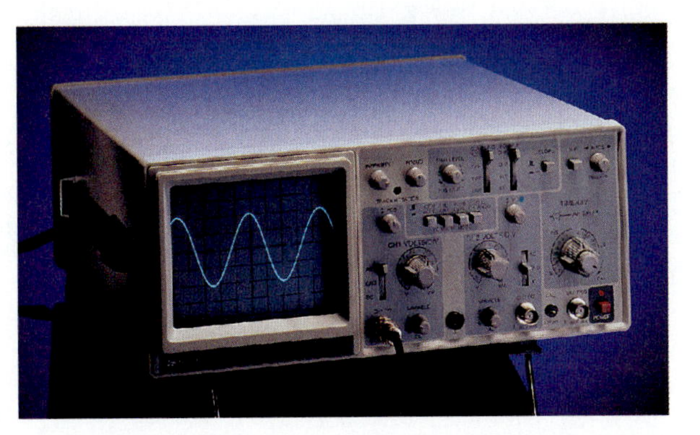

Fig. 18-39 An oscilloscope.

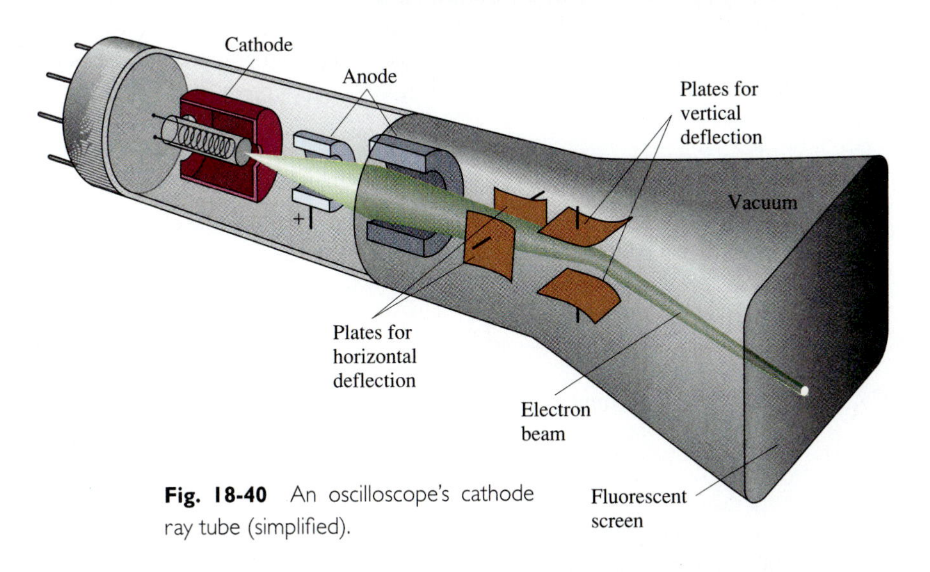

Fig. 18-40 An oscilloscope's cathode ray tube (simplified).

The oscilloscope utilizes a cathode ray tube (CRT) (Fig. 18-40), which is a vacuum-sealed, glass tube that is also used as the picture tube in a television. When heated, a metal plate called the cathode emits a beam of electrons (originally called cathode rays). Because there is a potential difference between the cathode and the anode, another metal plate that is maintained at a positive potential relative to the cathode, the electrons are accelerated toward the anode and pass through the hole in its center. The electrons then strike a point on the screen, and the fluorescent material on the screen produces a bright spot of light at that point. The location of the spot can be varied by bending the path of the electron beam. This is accomplished with electric fields generated by the horizontal and vertical deflection plates. We shall show how the deflection produced by each set of plates depends on the potential difference applied to them and on the cathode-to-anode potential difference, which accelerates the electrons.

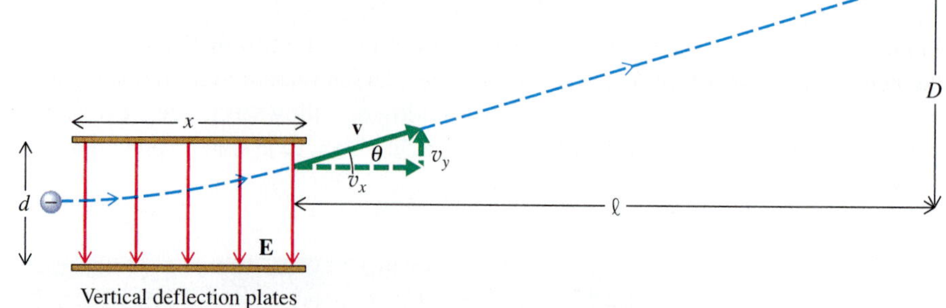

Fig. 18-41 When the electric field between the vertical deflection plates of a CRT is directed downward, the electron beam is deflected upward.

Consider the effect on the electrons of the electric field produced by the vertical deflection plates (Fig. 18-41). Suppose that at some instant there is a potential difference V between the plates, with the upper plate at the higher potential. Then the field is directed downward and has magnitude

$$E = \frac{V}{d}$$

where d is the spacing between the plates. Because the electrons have a negative charge, this field produces a constant upward force $\mathbf{F} = q'\mathbf{E} = -e\mathbf{E}$, and hence a constant upward acceleration a_y. The electrons enter the electric field with only a horizontal velocity v_x. They leave the field having moved upward only slightly, as shown in the figure, but with a newly acquired vertical velocity component v_y, given by

$$v_y = a_y t = \frac{F_y t}{m} = \frac{eE}{m} t$$

where m is the mass of an electron and t is the time during which the electrons are in the field. This time t is related to the length x of the plates and to the horizontal component of velocity v_x by the equation

$$t = \frac{x}{v_x}$$

Combining the three preceding equations, we obtain

$$v_y = \frac{exV}{dmv_x}$$

The deflection D of the electron beam is determined by the angle θ at which the beam leaves the field. From the figure we see that

$$\frac{D}{\ell} = \tan\theta = \frac{v_y}{v_x}$$

Inserting the preceding expression for v_y, we find

$$D = \frac{\ell exV}{dmv_x^2}$$

The velocity v_x at which the electrons enter the field is determined by the cathode-to-anode potential difference that accelerated them as they left the cathode. We denote this potential difference by V' and note that the electron's gain in kinetic energy $\frac{1}{2}mv_x^2$ equals its loss in potential energy eV', or $mv_x^2 = 2eV'$. Using this in the equation above, we obtain

$$D = \frac{\ell xV}{2dV'} \tag{18-25}$$

Thus we see that the deflection is directly proportional to the potential difference V across the deflecting plates and inversely proportional to the potential difference V' between anode and cathode. The same equation applies to horizontal deflections produced by applying a potential difference to the horizontal deflection plates.

EXAMPLE 15 Electron Beam Deflection When 1V is Applied to an Oscilloscope's Plates

Suppose that an oscilloscope's CRT has the following dimensions: $\ell = 40$ cm, $x = 5$ cm, and $d = 1$ cm. Let the accelerating potential V' between cathode and anode be 1000 V and the deflection voltage V be 1 V across either set of deflection plates. Find the deflection of the electron beam.

SOLUTION Applying Eq. 18-25, we find

$$D = \frac{\ell x V}{2dV'} = \frac{(0.4 \text{ m})(0.05 \text{ m})(1 \text{ V})}{2(0.01 \text{ m})(1000 \text{ V})} = 10^{-3} \text{ m} = 0.1 \text{ cm}$$

The deflection is directly proportional to V. Thus a 10 V potential difference applied to the deflection plates produces a 1 cm deflection, 20 V produces a 2-cm deflection, and so on. The deflection of the electron beam for a given deflection-plate voltage is determined by a setting of the oscilloscope's controls, which determine the cathode-to-anode accelerating voltage. This example corresponds to a setting of 10 V/cm.

Eq. 18-25 shows that the deflection D of an oscilloscope's electron beam is proportional to the potential difference V applied across its deflection plates, as illustrated in the preceding example. Thus an oscilloscope can be used to measure the potential difference across its plates. With no voltage applied to either set of plates, a dot of light is seen at the center of the screen (Fig. 18-42a). When the beam is deflected 3 cm vertically (Fig. 18-42b), with the control set at 10 V/cm, we know that there is a potential difference of 30 V across the vertical deflection plates. The oscilloscope is equipped with metal probes that connect the internal deflection plates with points outside in order to measure the potential difference between those points (Fig. 18-43). Since they are connected to the external points by good conductors, the plates are at the same potential as those points.

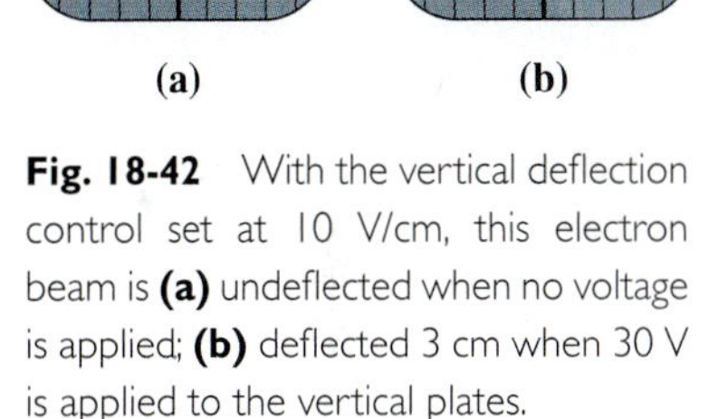

Fig. 18-42 With the vertical deflection control set at 10 V/cm, this electron beam is **(a)** undeflected when no voltage is applied; **(b)** deflected 3 cm when 30 V is applied to the vertical plates.

As mentioned in Section 18-1, voltmeters are used to measure a constant potential difference. The oscilloscope, however, may also be used to observe and measure potential differences that vary with time, especially periodic potential differences. This is one of the most important functions of an oscilloscope, and so we describe it here, although it involves time-varying fields, which we shall not discuss in detail until Chapter 21.

The oscilloscope may be set to "internal sweep." With this setting a circuit inside the oscilloscope applies a time-varying voltage to the horizontal deflection plates, causing the electron beam to sweep horizontally across the face of the screen at a constant speed from left to right and, when it reaches the right side, to jump back almost instantly to the left and repeat the motion. If the sweep rate is slow, one sees a spot of light moving at constant speed across the screen. The image remains on the screen for a brief time after the electron beam sweeps by, however, and so if the sweep rate is rapid enough, one sees a continuous horizontal line.

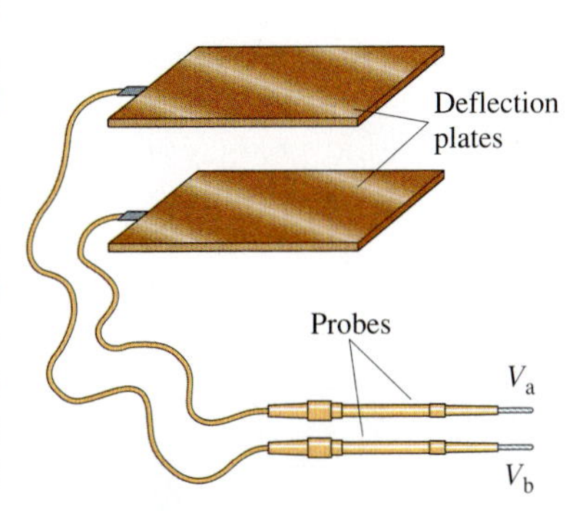

Fig. 18-43 Metal probes connected to an oscilloscope's deflection plates allow one to measure the potential difference between two points.

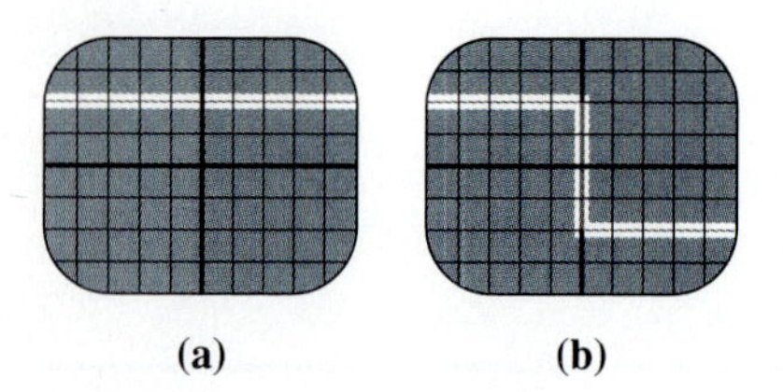

Fig. 18-44 Oscilloscope image when **(a)** a constant voltage is applied to the vertical plates; **(b)** alternating positive and negative voltages are applied to the vertical plates.

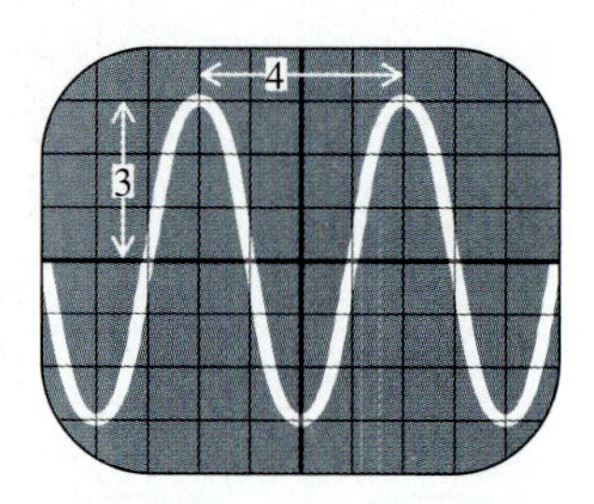

Fig. 18-45 The voltage measured by this oscilloscope is a sine function of time.

Suppose that, with a fast horizontal sweep, a constant voltage is applied to the vertical plates. Then the beam is deflected as shown in Fig. 18-44a. Next suppose that a time-varying voltage is applied to the vertical plates. Then the time variation will be graphically displayed on the screen. For example, if the voltage alternates between equal positive and negative values, such as $+10$ V for 10^{-2} s, then -10 V for 10^{-2} s, then $+10$ V again, and if the sweep rate is adjusted so that the beam travels across every 2×10^{-2} s, the display will be as shown in Fig. 18-44b.

The constant horizontal velocity of the electron beam makes the horizontal axis effectively a time axis. Adjustable controls on the oscilloscope indicate the number of seconds or milliseconds per cm on the horizontal axis, that is, the time required for the beam to move past one horizontal division. Thus the oscilloscope allows us to obtain detailed information about the voltage applied to its vertical plates and how that voltage varies with time. For example, if the display shown in Fig. 18-45 is seen on the screen of an oscilloscope that has its vertical deflection set at 1 mV/cm and its sweep rate set at 2 ms/cm, we know that the voltage being measured is time-dependent—a sine function of time having an amplitude of 3 mV (3 cm $\times$ 1 mV/cm) and a period of 8 ms (4 cm $\times$ 2 ms/cm), or a frequency of $\dfrac{1}{8 \times 10^{-3}\ \text{s}} = 125$ Hz.

The oscilloscope can be used to analyze any kind of time-dependent phenomenon that can be converted to a time-dependent voltage. For example, a sound produced by the human voice or by a musical instrument can be converted by means of a microphone to a voltage that has the same time dependence as the sound wave and can be displayed on the screen of an oscilloscope (Fig. 18-46a). One can also use the oscilloscope to monitor electrical voltages generated by the human body. For example, electrodes connected from the oscilloscope to points on the head can pick up tiny voltages, on the order of microvolts, produced by processes within the brain. (The signals must be amplified first.) The subject's mental state (alert, relaxed, drowsy) is reflected in the kind of pattern that is observed. The record of potential versus time is called an "electroencephalogram," abbreviated EEG (Fig. 18-46b). Similarly, the heart's activity can be monitored by electrodes connected to various parts of the body, and the resulting record of potential versus time is called an "electrocardiogram," abbreviated ECG, or EKG. An example of an ECG is shown at the beginning of this chapter. Irregularities in the functioning of the brain or heart are reflected in abnormal EEG's or ECG's respectively.

A television picture tube is similar to an oscilloscope tube except that in the television tube the deflection is accomplished by means of magnetic fields (discussed in Chapter 21) rather than electric fields. In this case the electron beam sweeps horizontally first across the top of the screen, varying in intensity according to the image being displayed, sweeps across again but displaced a bit lower on the screen, and repeats this process until the screen is covered with 525 lines in a total time of $\frac{1}{30}$ of a second. The eye sees a complete image covering the entire screen.

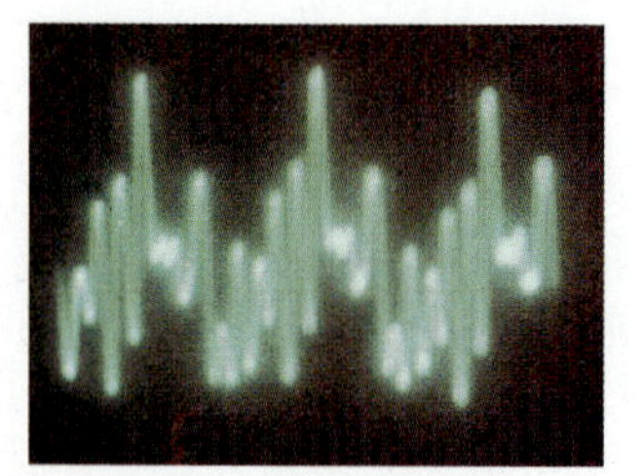

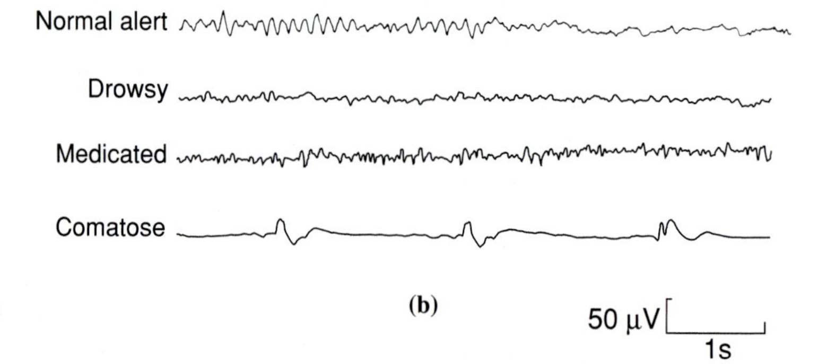

Fig. 18-46 An oscilloscope can be used to observe **(a)** sound waves and **(b)** brain waves.

CHAPTER 18 SUMMARY

The electrostatic force is conservative. The work the electric field does on a charge moving from a to b is independent of the path from a to b. The work equals the decrease in the charge's electrical potential energy U_E.

The electric potential is defined as electrical potential energy per unit charge and is measured in volts:

$$V = \frac{U_E}{q'}$$

$$1 \text{ V} = 1 \text{ J/C}$$

The electron volt is a unit of energy equal to the potential energy of a proton at a point where the potential is 1 V:

$$1 \text{ eV} = (e)(1 \text{ V}) = 1.60 \times 10^{-19} \text{ J}$$

Only differences in potential, or "voltage drops," have physical meaning. The voltage drop between points a and b is sometimes denoted simply by V:

$$V = V_a - V_b$$

Equipotential surfaces are surfaces on which the potential has a constant value. The surface of a conductor is an equipotential surface. Electric field lines are perpendicular to equipotential surfaces and are directed from higher to lower potential.

In a uniform electric field, equipotential surfaces are planes, perpendicular to the field. The potential difference between two such planes is

$$V = Ed \qquad \text{(in a uniform field)}$$

where d is the distance between the planes.

For a collection of point charges, we may find the potential at any point by applying the equation

$$V = k\Sigma \frac{q}{r}$$

where r is the distance between charge q and the point where V is evaluated. This equation is based on the convention of setting the potential at infinity equal to zero.

A capacitor is a device designed to store charge. It consists of two conductors separated by a dielectric or by a vacuum, carrying charges Q and $-Q$. The charge Q on a capacitor is proportional to the potential difference between the conductors. The constant of proportionality is the capacitance C:

$$Q = CV$$

Capacitance is measured in farads, where

$$1 \text{ F} = 1 \text{ C/V}$$

Capacitance depends only on the geometry of the capacitor's two conductors. For a parallel-plate capacitor with a vacuum between the plates:

$$C = \frac{\epsilon_0 A}{d}$$

where A is the area of each plate, d is the separation between the plates, and

$$\epsilon_0 = \frac{1}{4\pi k} = 8.85 \times 10^{-12} \text{ F/m}$$

The effect of a network of capacitors, for points outside the network, is represented by a single equivalent capacitor. When two or more capacitors are connected so that they have the same potential difference across their plates, they are said to be in parallel. The equivalent capacitance of parallel capacitors is the sum of the C's:

$$C = C_1 + C_2 + C_3 + \cdots$$

Capacitors connected in series each store the same charge. We find the equivalent capacitance C of several series capacitors by applying the equation

$$\frac{1}{C} = \frac{1}{C_1} + \frac{1}{C_2} + \frac{1}{C_3} + \cdots$$

When a dielectric is inserted between the conductors of a capacitor, the value of the capacitance C is greater than the value C_0 for a vacuum. The ratio C/C_0, a constant for any particular kind of dielectric material, is called the "dielectric constant" and is denoted by κ:

$$\kappa = \frac{C}{C_0}$$

The dielectric strength is the maximum electric field the dielectric can withstand before breaking down and becoming a conductor.

The electrical potential energy stored by a capacitor may be expressed

$$U_E = \frac{1}{2}\frac{Q^2}{C} = \frac{1}{2}CV^2 = \frac{1}{2}QV$$

The energy density of the electrostatic field is

$$u = \frac{1}{2}\kappa\epsilon_0 E^2$$

which in vacuum reduces to

$$u = \frac{1}{2}\epsilon_0 E^2$$

An oscilloscope is often used to observe the time variation of a potential difference applied to the oscilloscope's vertical plates. The oscilloscope's display screen becomes a graph of potential difference as a function of time.

Questions

1 (a) Would a positively charged object have more electrical potential energy if placed on the positive terminal of a battery or on the negative terminal?
(b) On which terminal would a negatively charged object have greater electrical potential energy?

2 Equipotential surfaces are shown in Fig. 18-47.
(a) If a proton were placed at rest at point P, in which direction would the proton move?
(b) If an electron were placed at rest at point R, in which direction would the electron move?

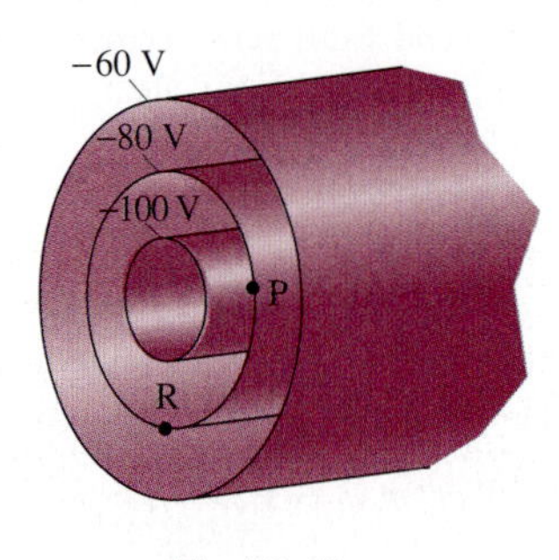

Fig. 18-47

3 A positive point charge is placed at the center of a hollow, metallic sphere that carries no net charge (Fig. 18-48).
(a) What is the sign of the charge density (+, −, or 0) at P, R, and S?
(b) What is the potential difference between P and S?

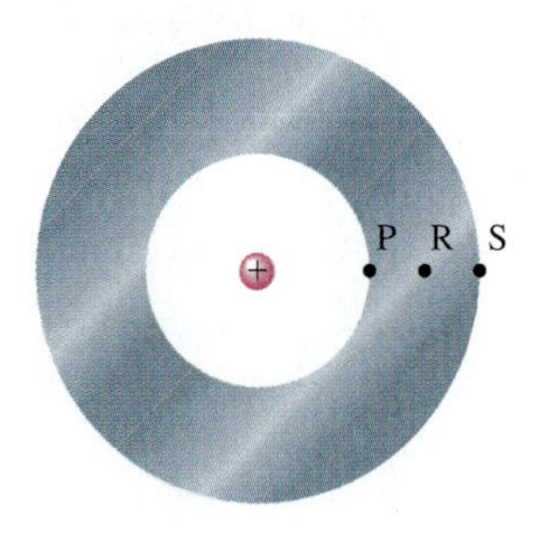

Fig. 18-48

4 Fig. 18-49 shows a cross-sectional view of equipotential surfaces that are perpendicular to the page.
(a) At which of the points A, B, C, D is the electric field directed toward the right?
(b) At which of the points is the electric field greatest?

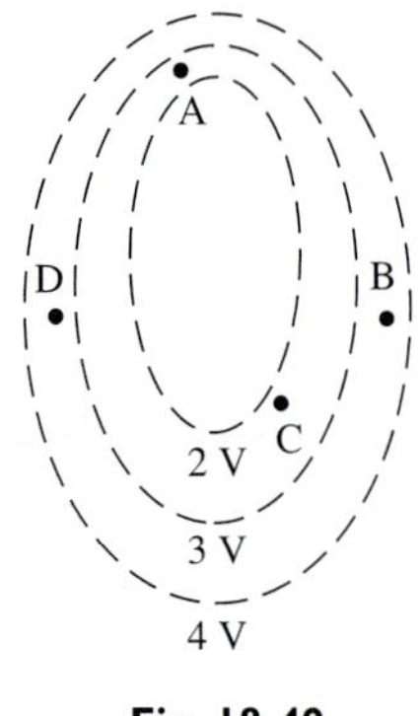

Fig. 18-49

5 (a) A proton is initially at rest at P in Fig. 18-50. Does it move? If so, in which direction?
(b) Repeat part (a) for an electron at P.

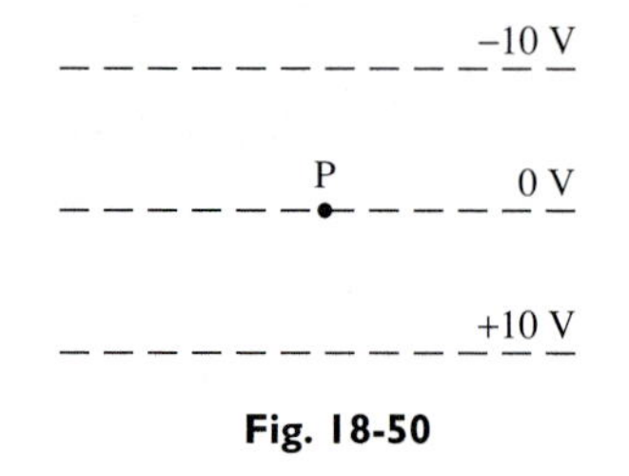

Fig. 18-50

6 At which of the points A, B, C, D in Fig. 18-51 is the potential greatest? The lines are electric field lines.

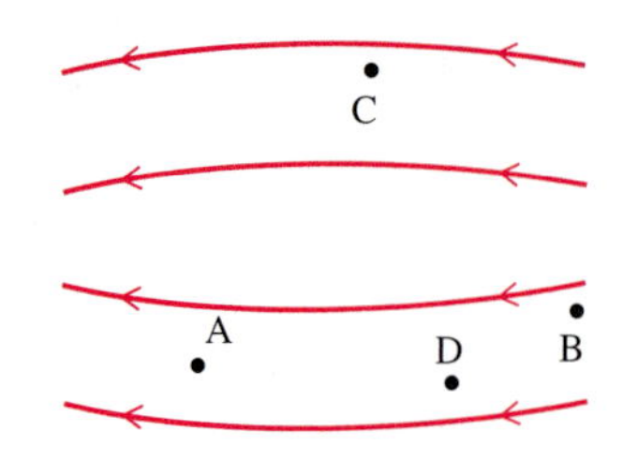

Fig. 18-51

7 Fig. 18-52 shows equipotential surfaces.
(a) What is the direction of **E** at P and R?
(b) Is the magnitude of the electric field greater at P or at R?

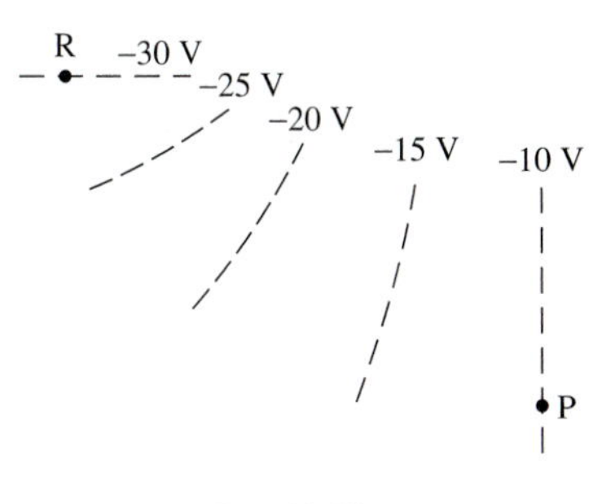

Fig. 18-52

8 Charges q_1 and q_2 produce a potential V at a certain point P. If q_1 alone were present, the value of the potential at P would be $V_1 = 3$ V; if q_2 alone were present, the potential would be $V_2 = 4$ V. With both charges present, what is the potential at P?

9 Fig. 18-53 shows two protons and one electron. At which of the points A, B, C, or D at ∞ is the potential least?

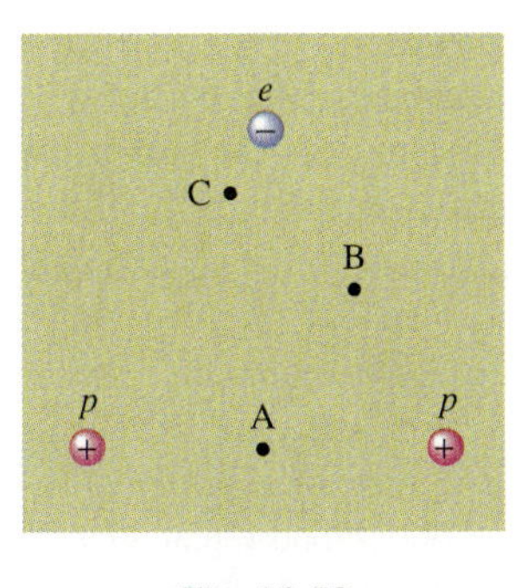

Fig. 18-53

10 An alpha particle has twice the charge and 4 times the mass of the proton. If an alpha particle and a proton both start from rest and are accelerated through the same potential difference, which, if either, has the greater final speed?

11 For the field shown in Fig. 18-10, are adjacent equipotential surfaces closer to each other near point a or point b?

12 (a) If $V = 0$ at a certain point, is the electric field at that point necessarily zero?
(b) If $V = 0$ everywhere within a certain volume of space, is the electric field necessarily zero within that volume?

13 Suppose that the electric field is zero everywhere in a certain volume of space.
(a) Can the potential vary over this volume?
(b) Is it necessarily true that $V = 0$ within the volume?

14 Points a and b are connected by a path along which there is an electric field perpendicular to the path and having a magnitude of 100 N/C. If $V = 10$ V at a, what is the potential at b?

15 Do equipotential surfaces ever intersect?

16 In solving problems involving electric circuits, we usually adopt the convention that the earth is at a potential of zero. Using this convention, can we apply the formula $V = k\Sigma \frac{q}{r}$ to a problem involving a collection of point changes near the surface of the earth?

17 Does a capacitor's capacitance depend on (a) the voltage across the plates; (b) the charge on the plates; (c) the size of the plates; (d) the separation between the plates?

18 A capacitor initially stores 1 J of electrical potential energy. If the charge on each plate were tripled, what would be the new value of the stored potential energy?

19 A capacitor disconnected from a battery or any other source carries a constant charge on its plates. A dielectric is inserted between the plates.
(a) Does the capacitor's electrical potential energy increase, decrease, or remain the same?
(b) Does the electric field between the plates increase, decrease, or remain the same?

20 A capacitor connected to a battery maintains a constant potential difference across its plates. A dielectric is inserted between the plates.
(a) Does the charge on the plates increase, decrease, or remain the same?
(b) Does the electric field between the plates increase, decrease, or remain the same?
(c) Does the electrical potential energy increase, decrease, or remain the same?

Answers to Odd-Numbered Questions

1 (a) positive; (b) negative; **3** (a) −, 0, +; (b) 0; **5** (a) up; (b) down; **7** (a) to the left; upward; (b) R; **9** C; **11** a; **13** (a) no; (b) no; **15** no; **17** (a) no; (b) no; (c) yes; (d) yes; **19** (a) decrease; (b) decrease

Problems (listed by sections)

18-1 Electrical Potential Energy and Electric Potential

1 The potential at point P in an electric field is 100 V.
(a) What is the electrical potential energy of a proton placed at P, in eV and in joules?
(b) What is the electrical potential energy of an electron at P, in eV and in joules?

2 A proton is initially at rest at a point where the potential is 500 V. Some time later the proton is at a point where the potential is 200 V. If no force other than the electrostatic force acts on the proton, what is its final kinetic energy?

3 Lightning strikes a point on earth, delivering −20.0 C of charge from cloud to earth across a potential difference of 1.00×10^8 V.
(a) Is the cloud or the earth at the higher potential?
(b) How much energy is delivered to the ground?
(c) If this energy could be used to light a 100 W light bulb, how long would it last?

4 There is a 10 V drop in potential from point A to point B and a 60 V rise in potential from point A to point C. If the potential at C is 100 V, what is the potential at B?

5 On a clear day the electric field in the atmosphere near the earth's surface is 100 N/C, directed vertically downward.
(a) If we adopt the convention that the potential at the earth's surface is 0, what is the potential 100 m above the surface?
(b) What is the potential at the top of Pike's Peak, 14,110 ft above sea level?
(c) What is the potential at the top of a man's head if the man is 2 m tall and is standing?

6 Two identical charges $q = +\dfrac{1}{9.0} \times 10^{-9}$ C are shown in Fig. 18-54. Find the potential at P.

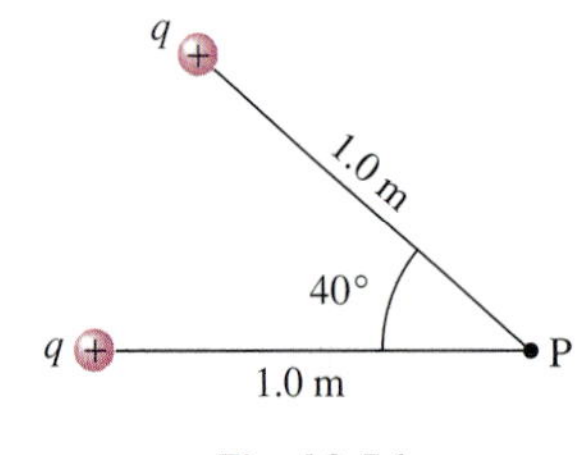

Fig. 18-54

7 At what distance from an isolated proton does the potential equal 1.00 V?

8 Fig. 18-55 shows two point charges $q_1 = +5.0 \times 10^{-8}$ C and $q_2 = -6.0 \times 10^{-8}$ C.
(a) Find the potential at A.
(b) Find the potential at B.
(c) Find the potential difference $V_A - V_B$.

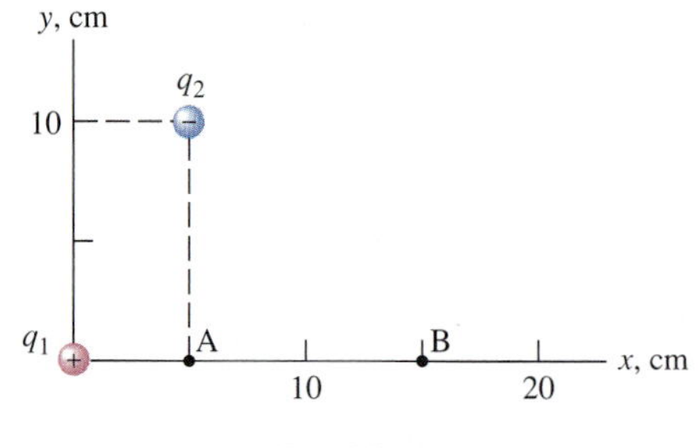

Fig. 18-55

9 Find the potential at the center of an equilateral triangle, 10 cm from each of the following charges, which are located at the vertices of the triangle: +1 μC, +1 μC, −2 μC.

10 In 1910 Rutherford performed a classic experiment in which he directed a beam of alpha particles at a thin gold foil. He unexpectedly observed a few of the particles scattered almost directly backward. This result was not consistent with then current models of atomic structure and led Rutherford to propose the existence of a very dense concentration of positive charge at the center of an atom—the atomic nucleus. The alpha particle has a charge of +2e and the gold nucleus a charge of +79e. Suppose that an alpha particle is initially a great distance from the gold, has a kinetic energy of 4.00 MeV (4.00×10^6 eV), and is headed directly at a gold nucleus. How close will the particle come to the center of the nucleus? Treat the nucleus and the alpha particle as point charges.

11 Find the voltage drop from the origin to the point $x = 0$, $y = 2.00$ m resulting from the following charges: +1.00 μC at $x = -1.00$ m, $y = 0$; +2.00 μC at $x = +2.00$ m, $y = 0$; and −2.00 μC at $x = 0$, $y = -2.00$ m.

12 The potential equals 8.00 V at the midpoint between two point charges that are 2.00 m apart. One of the charges is 1.00×10^{-9} C. Find the value of the other charge.

★ **13** The circular arc shown in Fig. 18-56 has a uniform charge per unit length of 1.00×10^{-8} C/m. Find the potential at P, the center of the circle.

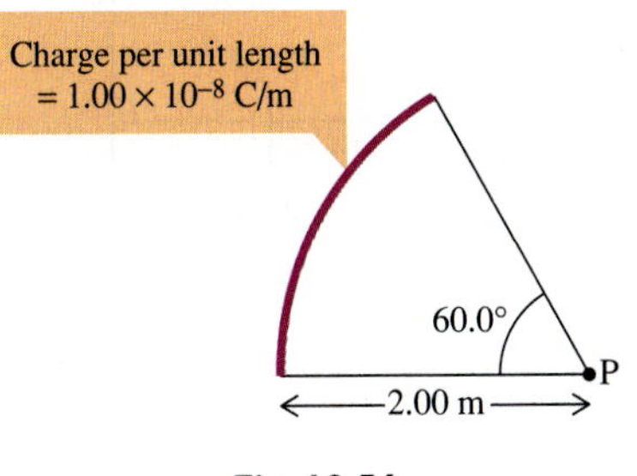

Fig. 18-56

★ **14** A uniformly charged thin ring has a radius a and carries a total charge Q. Derive an expression that gives the potential at a point that is on the axis of the ring and a distance x from the ring's center.

15 A uniform electric field of magnitude 200 N/C is directed along the negative y-axis. Find the potential at the point $x = 30.0$ cm, $y = -40.0$ cm if the potential at the origin equals zero.

★ **16** A proton initially at rest is accelerated by a uniform electric field. The proton moves 5.00 cm in 1.00×10^{-6} s. Find the voltage drop through which the proton moves.

★ **17** The potential equals 400 V at 1.00 m from a very large, flat dielectric sheet that carries a surface charge density of -1.00×10^{-8} C/m^2. Find the potential 3.00 m from the sheet.

18 Points A and B in Fig. 18-57 are in a *uniform, vertical* electric field. What are the magnitude and direction (up or down) of the electric field?

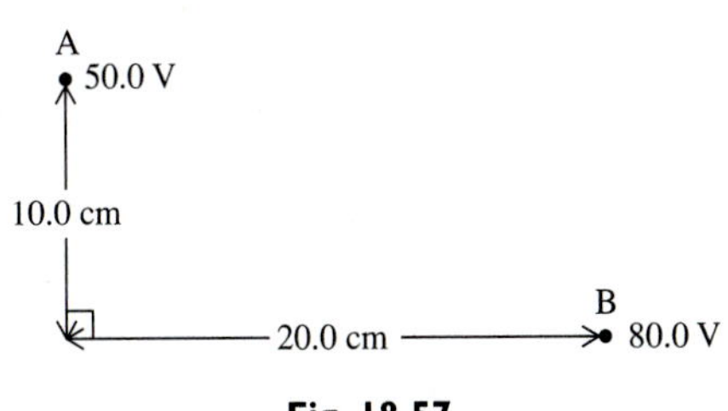

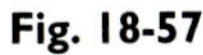
Fig. 18-57

18-2 Capacitance

19 A capacitor has parallel plates that have an area of 1.00 cm^2 and are 1.00 mm apart. There is a vacuum between the plates, and the electric field magnitude is 1.00×10^6 N/C. Find (a) the potential across the plates; (b) the charge.

20 A 1.00×10^{-10} F capacitor has parallel plates with dimensions 10.0 cm × 20.0 cm and a vacuum between the plates.
(a) Find the distance between the plates.
(b) A potential difference of 100 V is applied to the plates. Find the charge and the magnitude of the field between the plates.

21 Find the equivalent capacitance when two 1 μF capacitors are connected (a) in parallel; (b) in series.

22 A single isolated conductor can store charge. We can define its capacitance as the ratio of the charge to the potential at the surface. (Alternatively, one can think of a second conductor at infinity storing a negative charge of equal magnitude. This gives the same result, since $V = 0$ at infinity and the potential difference is therefore just the potential on the first conductor's surface.)
(a) Find an expression for the capacitance of an isolated spherical conductor of radius R. (HINT: The field outside the sphere is the same as though all of the charge were concentrated at the center. Gauss's law can be used to prove this result.)
(b) Find the capacitance of a copper sphere of radius 1.00 cm.
(c) Find the capacitance of an isolated spherical conductor the size of the earth. (Although the earth stores charge, it is not a single isolated conductor. Negative charge is stored on the ground and a roughly equal quantity of positive charge is spread through the lower atmosphere.)

23 A 1.00 μF capacitor is connected in parallel with a 2.00 μF capacitor. The 1.00 μF capacitor carries a charge of +10.0 μC on one plate, which is at a potential of 50.0 V. Find (a) the potential on the negative plate of the 1.00 μF capacitor; (b) the equivalent capacitance of the two capacitors; (c) the charge on the negative plate of the 2.00 μF capacitor.

24 A 1.00 μF capacitor is connected in series with a 2.00 μF capacitor. The 1.00 μF capacitor carries a charge of +10.0 μC on one plate, which is at a potential of 50.0 V.
(a) Find the potential on the negative plate of the 1.00 μF capacitor.
(b) Find the equivalent capacitance of the two capacitors.

★ **25** A 1 μF capacitor is charged when it is connected across a 9 V battery. The capacitor is then disconnected from the battery and connected in parallel with an initially uncharged 2 μF capacitor. Find the final charge and the voltage across the plates for (a) the 1 μF capacitor; (b) the 2 μF capacitor.

★ **26** A 1 μF capacitor is charged when it is connected across a 9 V battery. The capacitor is then disconnected from the battery and connected in parallel with an initially uncharged 1 μF capacitor. Find the final charge and the voltage across the plates for each capacitor.

★ **27** The potential difference across the 1.00 μF capacitor in Fig. 18-58 is 10.0 V. Find (a) the equivalent capacitance between A and B; (b) the potential difference between C and D.

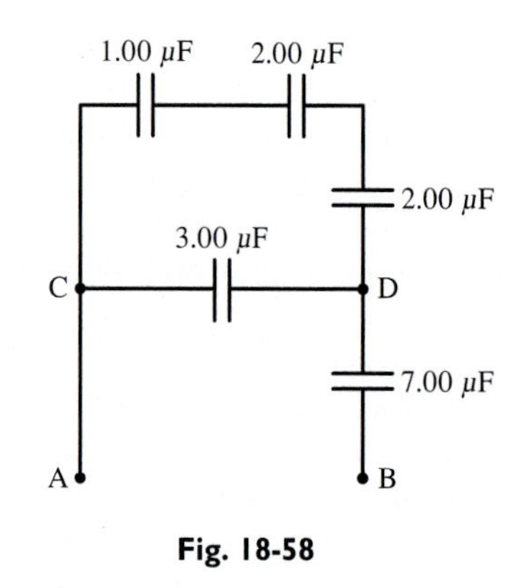

Fig. 18-58

★ **28** Three identical isolated capacitors, each with capacitance C, are connected in parallel, with a 10.0 V potential difference across their plates. Each capacitor carries a charge of 5.00 μC.

(a) Find C.

(b) Find the equivalent capacitance.

(c) Suppose that one of the capacitors is first disconnected and then reconnected with its terminals reversed. Find the final voltage drop across the capacitors.

★ **29** A 12.0 V battery is connected across points A and B in Fig. 18-59. Find the potential difference across the 3.00 μF capacitor.

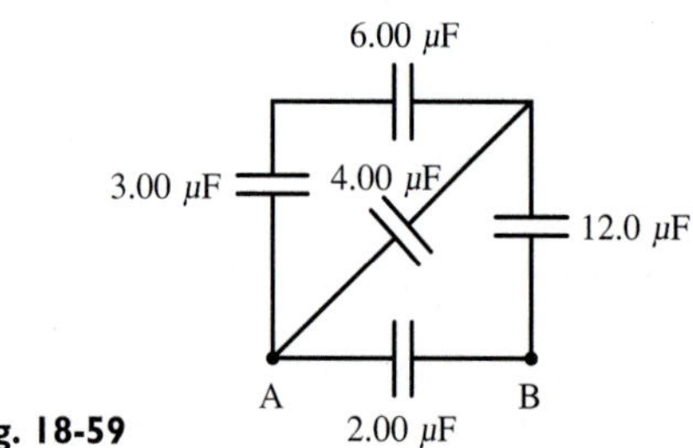

Fig. 18-59

★ **30** In Fig. 18-60 the potential difference between A and B is 40.0 V. Find the charge on the 12.0 μF capacitor.

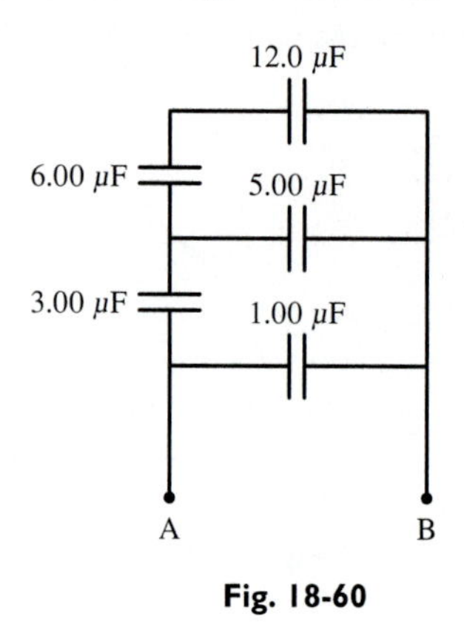

Fig. 18-60

31 A capacitor is discharged when a wire is connected across its terminals. The electrical energy stored in the capacitor is quickly released in the form of heat in the wire and a spark at the point of contact. How much energy is released for a 1.00 μF capacitor with 10.0 V across its plates?

★ **32** A 60.0 V power supply is connected between A and B in Fig. 18-61. Find the charge on the 10.0 μF capacitor with the switch (a) open and (b) closed. (c) How much charge flows through the switch when it is closed? (The open switch can be thought of as a capacitor of zero capacitance.)

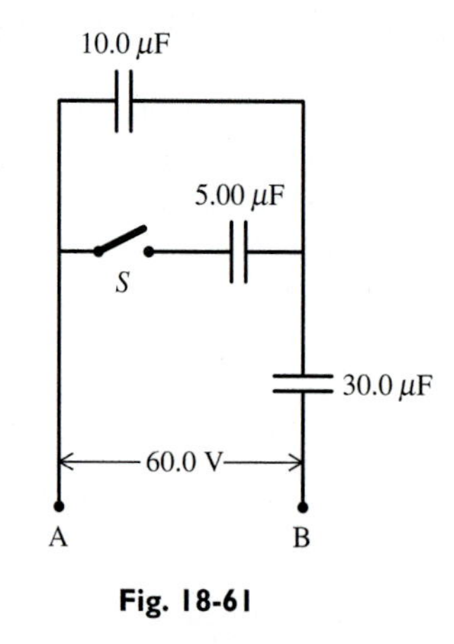

Fig. 18-61

33 A pulsed carbon dioxide laser has an energy output of 1.00 J. This energy is released in the form of a 10^9 W pulse of infrared radiation lasting 10^{-9} s. The laser needs an energy source capable of rapidly releasing 3.00 J of electrical energy. An array of parallel capacitors is used for this purpose. The potential difference across the network is 1.00×10^4 V. Find the equivalent capacitance of the network.

34 A capacitor is not the most efficient device for storing energy. Batteries can store more energy in much less space. For example, a typical 12 V automobile battery stores on the order of 10^6 J.

(a) Find the capacitance necessary to store 10^6 J with a potential difference of 10^4 V across the capacitor's terminals.

(b) Suppose that such a capacitor was made in the form of a parallel-plate capacitor with a vacuum between the plates and an electric field no greater than 10^6 V/m. What is the minimum area of the plates?

★ **35** The plates of a capacitor are oppositely charged and therefore experience a mutual attractive force.

(a) Show that the force on each plate of a parallel-plate capacitor has a magnitude of $F = \dfrac{Q^2}{2Cd}$, where Q is the charge stored, C is the capacitance, and d is the plate separation.

(b) Evaluate the force for a 1.00×10^{-2} μF capacitor with $d = 1.00$ mm and $Q = 10.0$ μC.

(c) Suppose that one plate of the capacitor is held fixed while the other is allowed to move a distance d, so that the two plates meet. Prove that the work done by the electric force equals the initial electrical potential energy stored by the capacitor.

★ **36** Initially a 1.00 μF capacitor and a 3.00 μF capacitor are connected in parallel, with 10.0 V across their plates. The two capacitors are not connected to anything else. The connecting wires are then reversed on the terminals of one capacitor, so that the positive plate of each capacitor is connected to the negative plate of the other capacitor.

(a) Find the initial total charge stored.

(b) Find the final total charge stored.

(c) Find the loss of electrical potential energy.

18-3 Dielectrics

★ **37** The fluids inside and outside a cell are good conductors separated by the cell wall, which is a dielectric. Thus the cell has capacitance; charge may be stored on its inner and outer surfaces (Fig. 18-62). It is a good approximation to treat the thin charged layer as a parallel-plate capacitor. Typically the wall is 1.00×10^{-8} m thick and has a dielectric constant of 5.00.

(a) Find the capacitance per cm^2 of cell membrane .

(b) Suppose the potential difference across the cell wall is 0.100 V. Find the magnitude of the charge stored on either side of the cell wall per cm^2.

(c) Find the field strength across the membrane.

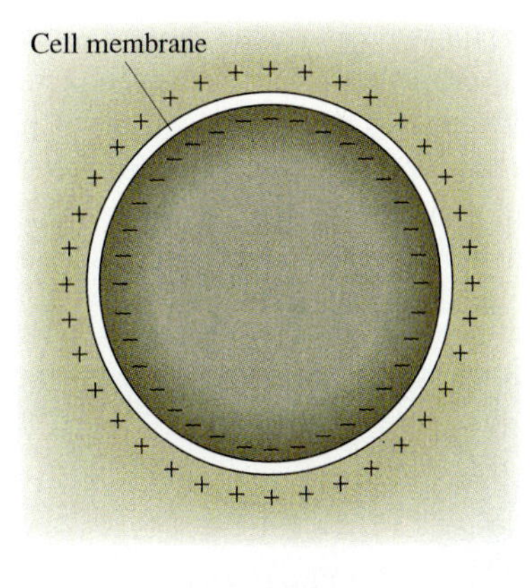

Fig. 18-62

38 The plates of a parallel-plate capacitor each have an area of 0.1 m^2 and are separated by a 0.5 mm thick layer of rubber. The capacitor is connected to a 12 V battery. Find (a) the capacitance; (b) the charge stored; (c) the electric field between the plates.

★ **39** A person who has become electrostatically charged is about to touch a large metal plate with his fingertip. The positive charge on his finger induces an opposite charge of equal magnitude on the surface of the plate (Fig. 18-63). There is a spark between the fingertip and the plate when the separation is 2 mm. Approximate the system by treating it as a parallel-plate capacitor with plate area of 1 cm^2. Find (a) the potential difference between the finger and the metal plate; (b) the charge on the fingertip; (c) the force exerted on the finger by the charge on the metal plate. (d) Why is the large potential difference not lethal?

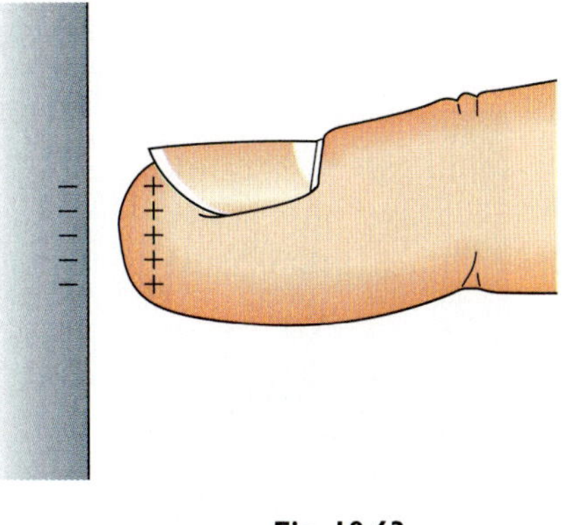

Fig. 18-63

40 A spark plug initiates combustion of gasoline in an automobile engine by producing a spark across a small gap between two conductors whenever the electric field in the gap exceeds the dielectric strength of air (3×10^6 V/m) (Fig. 18-64). For a gap of 0.8 mm, what is the minimum potential difference that must be applied to produce a spark?

Fig. 18-64

★41 Find the minimum volume of a capacitor that will store electrical potential energy of (a) 1.00 J (typical energy used to energize a high-power laser); (b) 1.00×10^6 J (typical energy stored in an automobile battery). Assume a dielectric constant of 10.0 and a dielectric strength of 1.00×10^7 V/m.

42 The Leyden jar, invented by chance in 1746, was one of the earliest types of capacitors (Fig. 18-65), made by covering the inner and outer surfaces of a glass jar with conducting layers (for example, aluminum foil). It is a good approximation to treat the Leyden jar as a parallel-plate capacitor.

(a) Find the capacitance for a cylindrical jar of height 20 cm and radius 5 cm, made of 2 mm thick Pyrex glass.

(b) What is the maximum charge that could be stored on it?

Fig. 18-65

18-4 The Oscilloscope

43 Fig. 18-66 shows a wave form displayed on an oscilloscope set at a sweep rate of 1.00 ms/cm and a vertical deflection of 0.100 V/cm. Find the (a) amplitude; (b) period; and (c) frequency of the wave form.

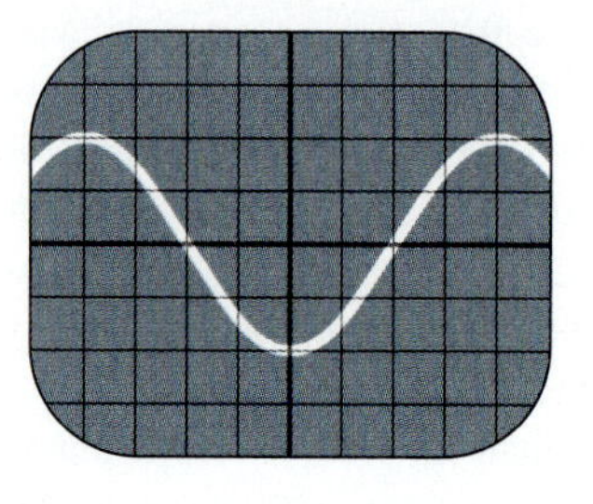

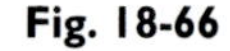

Fig. 18-66

44 An electron in an oscilloscope is accelerated as a result of the potential difference between the cathode and anode. The electron has negligible initial speed and a final speed of 1.00×10^7 m/s. Find the voltage drop from cathode to anode.

45 The horizontal scale on the ECG shown at the beginning of this chapter is 0.0400 s per division. How fast was the heartbeat?

46 Sketch the display on an oscilloscope screen when the signal is a sine wave of frequency 2000 Hz and amplitude 20.0 V, the sweep rate is 50.0 μs/cm, and the vertical gain is 5.00 V/cm.

Additional Problems

★47 A 1.0×10^{-9} C charge is at the origin, and a -1.0×10^{-9} C charge is 2.0 m to the right on the positive x-axis. The electric potential is -1.2 V at a point on the x-axis that is a distance d to the right of the negative charge. Find d.

48 A positively charged copper sphere of radius 10.0 cm is placed close to a negatively charged metal plate (Fig. 18-67). The charge density is 1.00×10^{-7} C/m^2 at a point on the sphere closest to the plate, and the potential at this point is 50.0 V. What is the potential at a point on the opposite side of the copper sphere, where the charge density is zero?

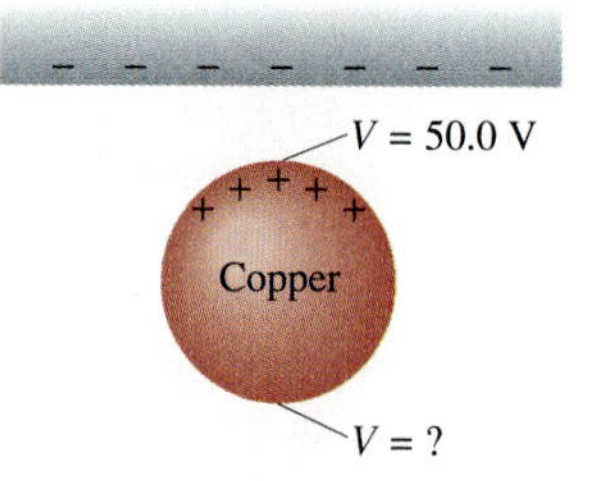

Fig. 18-67

★★**49** A thin ring of radius 10.0 cm is uniformly charged with 2.00×10^{-8} C. A particle of mass 1.00 g and charge 1.00×10^{-10} C is initially at the center of the ring, moving along the axis at a speed of 1.00 mm/s. Find the speed of the particle after it has moved 20.0 cm.

★**50** An electron having an initial kinetic energy of 1.00×10^3 eV is directed at a large metal plate that carries a surface charge density of -1.00×10^{-7} C/m². The electron stops just before reaching the plate. What was the original distance between the electron and the plate?

★★**51** A particle of mass 100 g and charge 1.00×10^{-6} C is attached to a string and allowed to swing as a pendulum in a uniform electric field of magnitude 1.00×10^6 N/C (Fig. 18-68). Find the speed of the particle at B if it starts from rest at A.

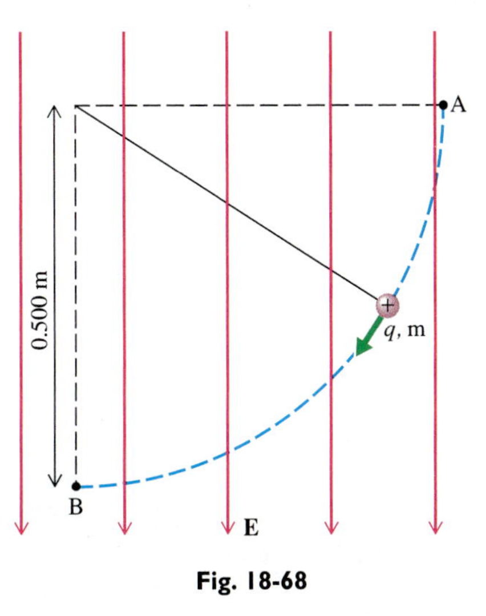

Fig. 18-68

★★**52** A particle of mass 1.00×10^{-5} kg and charge 1.00×10^{-6} C is initially at point P in an industrial chimney, moving upward at a speed of 6.00 m/s. As a result of the combined effect of gravity and an electric field doing negative work on it, the particle is brought to rest at point R, 2.00 m directly above P. The potential at R is 400 V higher than the potential at P. An upward force is exerted on the particle by the upward air current. Find the work done by the air.

★★**53** A proton moving at 1.00×10^7 m/s is projected into a region where there is a radial electric field produced by coaxial metal cylinders (Fig. 18-69). The proton moves along a circular path of radius 20.0 cm. Find the potential difference between the cylinders. (The magnitude of the field is approximately what it would be if the cylinders were flat parallel plates.)

54 Charge Q is uniformly distributed over a sphere of radius R. Using calculus, it is possible to show that the total electrical potential energy of the charge distribution is $\frac{3}{5}\left(\frac{kQ^2}{R}\right)$.

(a) Evaluate the electrical potential energy of the uranium nucleus, which has a charge of +92e and a radius of 7.40×10^{-15} m.

(b) How much energy is stored in 1.00 mole of uranium?

★★**55** The earth is a good electrical conductor with an approximately uniform distribution of charge on its surface. In clear weather, the field just above the surface of the earth has a magnitude of 100 N/C and is directed vertically downward.

(a) How much charge is stored on the earth's surface?

(b) What is the potential of the earth, if we let $V = 0$ at infinity?

(c) How much electrical potential energy would be produced by this charge distribution alone? (Actually, charge in the upper atmosphere also contributes to the electrical potential energy of the earth.)

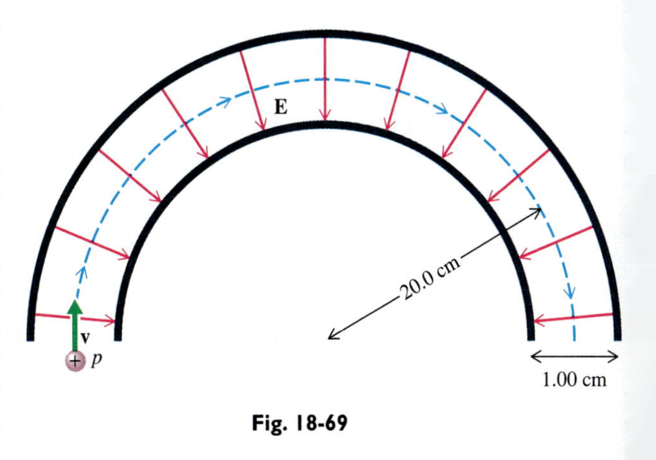

Fig. 18-69

★ **56** Fig. 18-70 shows two conducting spheres of radii r_A and r_B connected by a long, thin conducting wire.

(a) Show that the electric field magnitudes at A and B are related by the equation $\frac{E_A}{E_B} = \frac{r_B}{r_A}$. This relationship shows that the electric field is greater at the surface of the smaller sphere and, more generally, illustrates how the electric field is greatest at places on a conductor's surface where the radius of curvature is least. The electric field will therefore be much stronger at sharp points on the surface. This can produce dielectric breakdown of the surrounding air and discharge of the conductor.

(b) Find the radius r_A of the smaller sphere at which dielectric breakdown will occur if $E_B = 10^3$ N/C and $r_B = 1$ m.

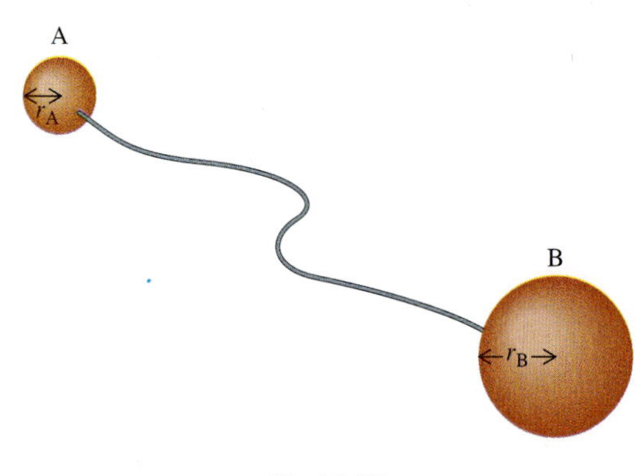

Fig. 18-70

57 Electrolytic capacitors give relatively high capacitance in a small volume. The dielectric consists of a very thin metal oxide covering a thin metal foil. The foil is immersed in an electrolyte (a conducting fluid). The foil and the electrolyte are the two conductors.

(a) A certain electrolytic capacitor can withstand at most 10 V. Assuming a dielectric strength of 2.0×10^7 V/m for the oxide layer, what is its thickness?

(b) Estimate the volume of a 0.10 F capacitor. Assume the metal foil is 0.010 mm thick and the dielectric constant is 50.

★★ **58** A spherical conductor of radius r_1 is centered inside a hollow spherical conductor of radius r_2. Prove that the conductors have a capacitance equal to $\frac{r_1 r_2}{k(r_2 - r_1)}$.

59 In the movie *Back to the Future,* Doc tells Marty that, to power his automotive time machine, he needs at least 1.2 gigawatts (1.2×10^9 W) and that he can obtain this from a stroke of lightning. Suppose that a certain lightning stroke lasts 0.20 s and transfers 20 C of charge across a potential difference of 5.0×10^7 V.

(a) How much energy is delivered, and what is the power of the stroke?

(b) If the time machine uses energy at the rate of 1.2 gigawatts, how long will the energy obtained from the lightning last?

(c) The time machine uses a so-called "flux" capacitor. Assuming that it has characteristics similar to an ordinary capacitor, with a dielectric constant of 100 and a dielectric strength of 1.0×10^8 V/m, find the minimum volume it can have to store the energy from the lightning stroke.

(d) A typical automobile battery stores about 10^6 J in a volume of 10^{-2} m^3. How large would an automobile battery have to be to store the energy from the lightning?

★ **60** Suppose you have a supply of 1.00 μF capacitors capable of withstanding 1000 V before dielectric breakdown occurs. By considering series and parallel combinations of these capacitors, design a network capable of withstanding 2000 V and having an equivalent capacitance of 2.00 μF.

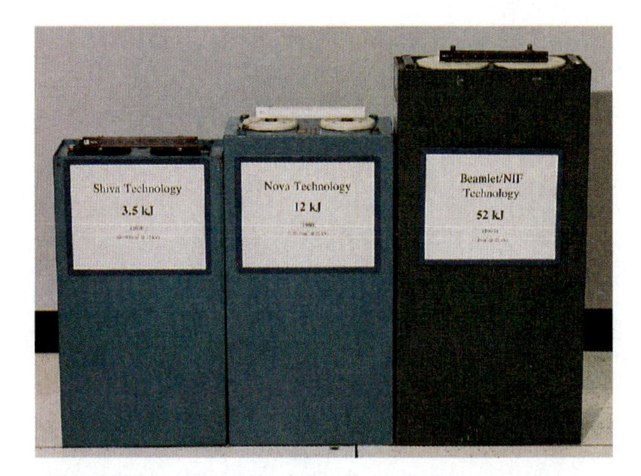

Fig. 18-71 Capacitors capable of storing ever greater quantities of energy have been developed in recent years for use in high power lasers.

CHAPTER 19 Electric Current

During an electrical storm huge amounts of electric charge are transferred between the earth's surface and the atmosphere.

Have you ever wondered exactly what happens when you turn on an electric light—what causes the bulb to glow? Or have you wondered what lightning is and what causes it? Unfortunately, electricity remains a great mystery even to most educated people today. In this chapter we shall remove some of that mystery. We shall see what forces act on the electrons in a light-bulb filament, causing it to light up. We shall describe lightning, see how it balances other charge-flow processes on earth (Problem 32), and even estimate our chances of being struck by it (Problem 33). We shall also describe how batteries supply energy to an electric circuit and how high-temperature superconductors may radically change electrical technology in the future.

In the two preceding chapters we studied electrostatics—electric charges at rest. In this chapter we study charges in motion. We begin by introducing the concept of electric current, a concept used to describe the motion of charges.

19-1 Electric Current

If a conducting wire is connected to the terminals of a battery, an electric field is produced inside the wire, directed along its length (Fig. 19-1). The free electrons in the wire experience a force and move in the direction opposite the field.

Fig. 19-1 An electric field inside a conducting wire.

The **rate of flow of charge** through a cross section of the wire is called **electric current.** This term applies to the motion of any charge through any cross-sectional area. We define the electric current I to be the charge ΔQ per unit time passing through an area.

$$I = \frac{\Delta Q}{\Delta t} \quad \text{(average current)} \quad (19\text{-}1)$$

If the rate of flow of charge is not constant, the definition above represents only the average flow rate. It is then useful to define an instantaneous current as the limiting value of the expression above as Δt approaches zero:

$$I = \lim_{\Delta t \to 0} \frac{\Delta Q}{\Delta t} \quad \text{(instantaneous current)} \quad (19\text{-}2)$$

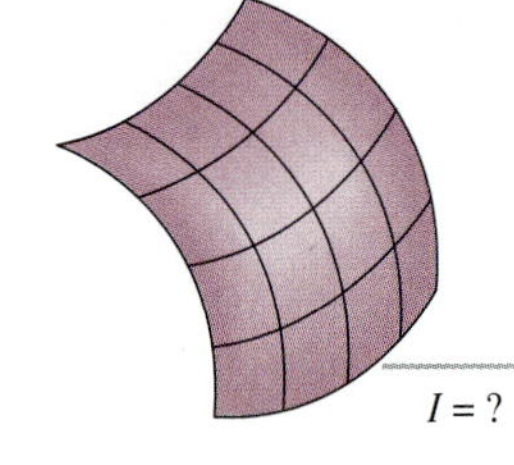

Fig. 19-2 Finding the current through a surface.

The SI unit for electric current is the coulomb per second, which is called an "ampere" (amp for short) and denoted by A.

$$1 \text{ A} = 1 \text{ C/s} \quad (19\text{-}3)$$

In a lightning stroke, typically 20 C of charge may pass a point in 10^{-3} s. This means that the current through a cross section of the atmosphere perpendicular to the direction of charge flow at that point is $\dfrac{20 \text{ C}}{10^{-3} \text{ s}} = 20{,}000$ A.

Since the electric charge can be either positive or negative, electric current can also be positive or negative. Before we determine the sign of a current, we must specify the direction in which we want to find the current. Suppose we want to find the current to the right through a certain surface (Fig. 19-2). If positive charge flows to the right, ΔQ is positive on the right side of the surface, and so the current to the right is positive. If negative charge flows to the right, ΔQ is negative on the right side and so the current to the right is negative. For example, if -10 C of charge flows to the right through the surface in 1 s, the current to the right is -10 A.

A negative current in one direction is equivalent to a positive current in the opposite direction. For example, if -10 C of charge flows to the right through the surface in Fig. 19-2, the current to the right is negative, but the current to the left is positive because the flow of *negative* charge from left to right means ΔQ on the left side is positive. So a current of -10 A to the right is equivalent to a current of $+10$ A to the left.

Fig. 19-3 shows a proton moving to the right through a surface and an electron moving to the left through the same surface. In either case there is a positive current to the right (or, equivalently, a negative current to the left). In most of our applications moving electrons will produce the electric current. The direction of positive current will then be opposite the direction of the electrons' motion. However, even though electrons produce a current, we shall often find it easier to **think of the current as the flow of positive charge in the direction in which the current is positive.**

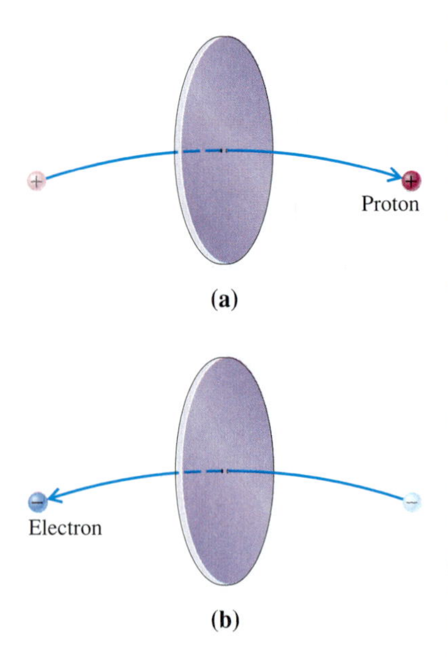

Fig. 19-3 **(a)** A proton moves through a surface to the right and produces a positive current to the right. **(b)** An electron moves through a surface to the left and produces a positive current to the right.

EXAMPLE 1 Current Charging a Capacitor

A wire is connected to one plate of a capacitor (Fig. 19-4). Find the current to the right through the wire if 1.0×10^{19} electrons per second move from the plate through the wire to the left.

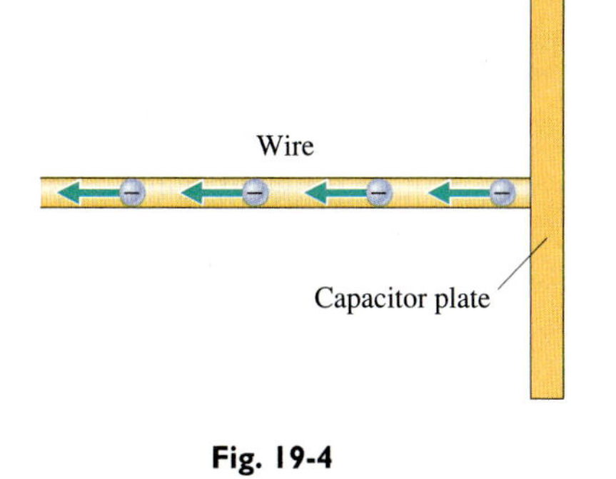

Fig. 19-4

SOLUTION The charge removed from the plate in 1.0 s is $(1.0 \times 10^{19})(-1.6 \times 10^{-19}\text{ C}) = -1.6$ C. This charge passes through any cross section of the wire from right to left, meaning that there is a negative current to the left of −1.6 C per second, or −1.6 A. Equivalently, there is a positive current to the right of +1.6 A. You can think of this current as a flow of 1.6 C of positive charge to the right through any cross section of the wire during a 1.0 s interval. Such a positive charge flow would increase the charge on the capacitor plate by 1.6 C each second—the same effect produced by the actual process of removing −1.6 C from the plate each second as the electrons are removed.

19-2 Ohm's Law

When you turn on a car's headlights, electric current flows through the bulbs and connecting wires (Fig. 19-5). When you turn on an electric heater or hairdryer, current flows through a heating element. Suppose you want to know just how much current flows in these situations. To solve such problems we shall often apply Ohm's law—a law relating the electric current I through some conducting medium to the potential difference V applied across the ends of the conductor. **Ohm's law** states that **current is proportional to potential difference.**

$$I \propto V$$

Many materials (metals, for example) are found to obey this simple relationship. Such materials are called "ohmic." For other, "non-ohmic" materials (transistors, for example), no such simple relationship between current and voltage exists.

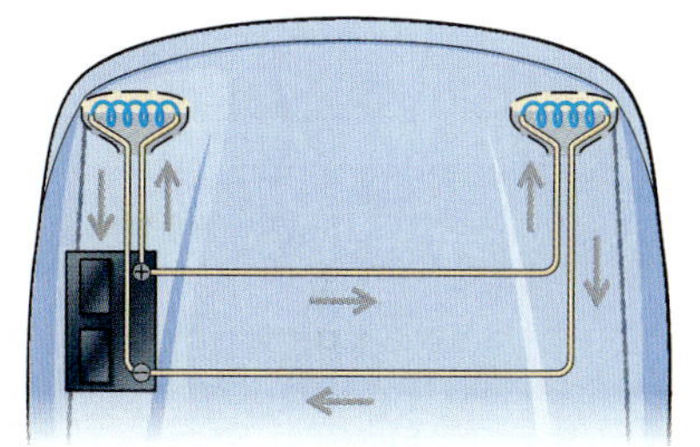

Fig. 19-5 Electric current flows through a car's headlights when the lights are on.

Resistance

Since voltage and current are proportional, the ratio of voltage to current is a constant for a particular piece of ohmic material. Thus another way to express Ohm's law is to say that the ratio of V to I is a constant. We call this ratio the **electrical resistance** of the piece of material and denote it by R:

$$R = \frac{V}{I} \qquad (19\text{-}4)$$

Units of resistance are volts per amp, which we call **ohms** and abbreviate by the Greek letter Ω:

$$1\ \Omega = 1\text{ V/A} \qquad (19\text{-}5)$$

We can solve Eq. 19-4 for either V or I and express Ohm's law either as

$$V = IR \tag{19-6}$$

or

$$I = \frac{V}{R} \tag{19-7}$$

This last form of Ohm's law can be used to find the current I through a piece of material when we know both its resistance R and the voltage V applied across it. For a given applied voltage, the greater the resistance of a piece of material, the less current will flow through it.

EXAMPLE 2 Current Through a Headlight

Find the current through an automobile headlight of resistance 2.5 Ω, when the potential difference across the light is 12 V.

SOLUTION Applying Eq. 19-7, we find the current

$$I = \frac{V}{R} = \frac{12\text{ V}}{2.5\ \Omega} = 4.8\text{ A}$$

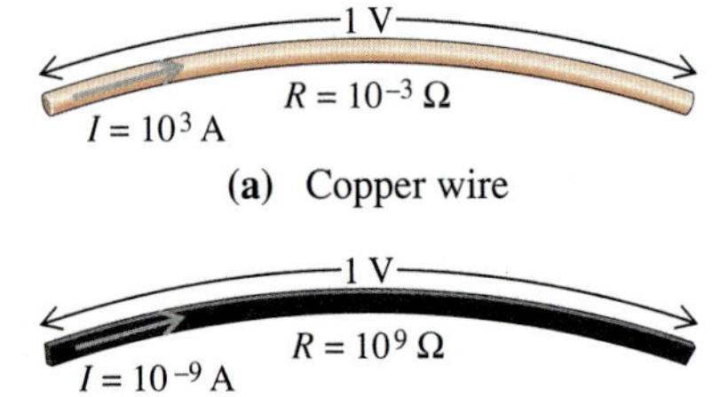

Fig. 19-6 The current through **(a)** a copper wire is much greater than the current through **(b)** a strip of rubber when the same voltage is applied to each.

Resistivity

Different kinds of materials can give radically different values of resistance. For example, you may apply a 1 V potential difference both across the ends of a copper wire and across the ends of a strip of rubber (Fig. 19-6). The wire might have a resistance of only $10^{-3}\ \Omega$ and therefore carry a current

$$I = \frac{V}{R} = \frac{1\text{ V}}{10^{-3}\ \Omega} = 10^{3}\text{ A}$$

The rubber, however, will have enormous resistance—perhaps $10^{9}\ \Omega$, which would mean that the current through it would be

$$I = \frac{V}{R} = \frac{1\text{ V}}{10^{9}\ \Omega} = 10^{-9}\text{ A}$$

Copper is a much better conductor than rubber, and so the copper wire has much less resistance than the strip of rubber.

Resistivity is the physical property that indicates copper is a good conductor and rubber is a poor conductor. Experiment shows that the resistance R of a particular piece of material is given by

$$R = \frac{\rho \ell}{A} \tag{19-8}$$

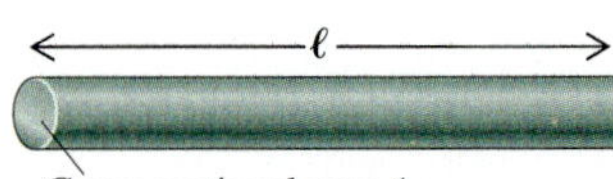

Fig. 19-7 A piece of material of length ℓ and cross-sectional area A has resistance $R = \rho\ell/A$, where ρ is the resistivity of the material.

where ρ is the resistivity of the material, ℓ is the length of the piece, and A is its cross-sectional area (Fig. 19-7).

According to Eq. 19-8, for a given kind of material of resistivity ρ, resistance will increase if either ℓ increases or A decreases. Thus a long, thin copper wire has greater resistance than a short, thick copper wire. Thicker wires are used when large currents must be carried, for example, to connect appliances, such as electric stoves, that require large currents (Fig. 19-8a). Thinner wires are used when the intended current is smaller, for example, to connect a light bulb (Fig. 19-8b).

The unit of resistivity is most easily expressed in terms of ohms. Solving Eq. 19-8 for ρ, we see that $\rho = \frac{RA}{\ell}$, and so ρ has units of $\frac{\Omega\text{-m}^2}{\text{m}} = \Omega\text{-m}$.

Resistivity varies with temperature. For most metals ρ increases linearly with temperature over a fairly broad temperature range. Thus we may express ρ at temperature T as

$$\rho = \rho_0[1 + \alpha(T - T_0)] \tag{19-9}$$

where ρ_0 is the value of ρ at the reference temperature T_0 (usually 20° C) and α is a constant called the **temperature coefficient of resistance.** Since resistance is directly proportional to resistivity, the resistance of a particular sample is related to its temperature by the same relationship:

$$R = R_0[1 + \alpha(T - T_0)] \tag{19-10}$$

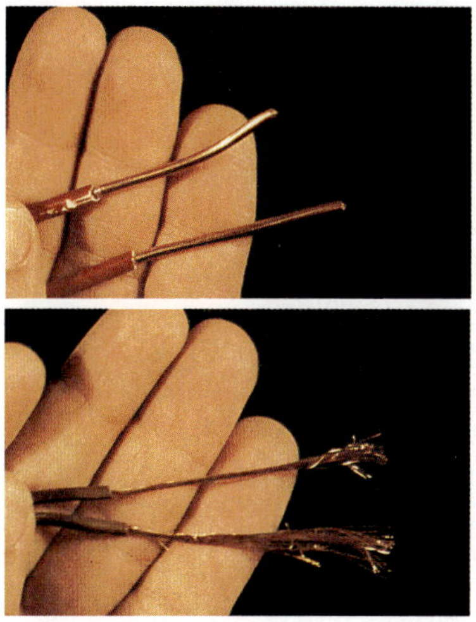

Table 19-1 gives resistivities and temperature coefficients for various materials. The table shows a tremendous range of values for ρ, from about 10^{-8} Ω-m for good conductors like silver, copper, and gold up to about 10^8 to 10^{16} Ω-m for glass, wood, and rubber.

Fig. 19-8 **(a)** Heavy copper wire designed to carry a large current. **(b)** Light copper wire designed to carry a small current.

Table 19-1 **Resistivities and temperature coefficients**

Material	ρ_0 (Ω-m) at 20° C	α (°C^{-1})
Silver	1.59×10^{-8}	3.8×10^{-3}
Copper	1.72×10^{-8}	3.9×10^{-3}
Gold	2.44×10^{-8}	3.4×10^{-3}
Aluminum	2.82×10^{-8}	3.9×10^{-3}
Tungsten	5.51×10^{-8}	4.5×10^{-3}
Iron	10×10^{-8}	5.0×10^{-3}
Lead	22×10^{-8}	3.9×10^{-3}
Mercury	96×10^{-8}	0.9×10^{-3}
Nichrome (an alloy used in heating elements)	100×10^{-8}	0.4×10^{-3}
Carbon	3.5×10^{-5}	-0.5×10^{-3}
Body fluids	~ 0.1	
Germanium*	~ 0.5	-50×10^{-3}
Silicon*	$\sim 10^3$	
Wood	10^8–10^{12}	
Polyethylene	2×10^{11}	
Glass	10^{10}–10^{14}	
Hard rubber	10^{13}–10^{16}	

*Resistivities of these semiconductors are strongly dependent on the presence and concentration of impurities. For example, doping silicon with aluminum can reduce the resistivity of the silicon by a factor of 10^{-6}.

Germanium and silicon are typical examples of materials known as **semiconductors,** which have much higher resistivity than metallic conductors but much lower resistivity than insulators. Semiconductors are the materials used to construct various electronic components, such as transistors, which we shall discuss in Chapter 29.

At very low temperatures certain materials have zero resistance. Such materials are called **superconductors.** Mercury is a superconductor below 4.2 K, and lead is a superconductor below 7.2 K. Experiments have shown that once an electric current is produced in a superconducting lead ring, the current will continue for more than a year without any electric field or source of energy needed to produce the current, as long as the lead's temperature is kept below 7.2 K. Applications of superconductivity and recent discoveries are discussed in an essay at the end of this section.

EXAMPLE 3 Current Through a Copper Wire with 1.50 V across it

A potential difference of 1.50 V is applied to a 1.00 m long copper wire of radius 0.500 mm (Fig. 19-9). Find the resistance of the wire and the current through it at 20.0° C.

SOLUTION We apply Eq. 19-8 using the value of ρ for copper at 20° C given in Table 19-1 and using $A = \pi r^2$ for the circular cross section of the wire of radius $r = 0.500$ mm:

$$R = \frac{\rho \ell}{A} = \frac{\rho \ell}{\pi r^2} = \frac{(1.72 \times 10^{-8}\ \Omega\text{-m})(1.00\ \text{m})}{\pi (0.500 \times 10^{-3}\ \text{m})^2}$$

$$= 2.19 \times 10^{-2}\ \Omega$$

We find the current by applying Ohm's law:

$$I = \frac{V}{R} = \frac{1.50\ \text{V}}{2.19 \times 10^{-2}\ \Omega} = 68.5\ \text{A}$$

Because of copper's low resistivity, the resistance of the wire is quite small and the current through it is large.

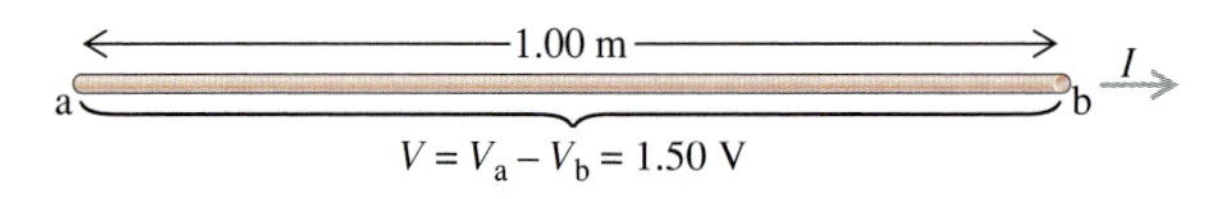

Fig. 19-9

Placing the wire directly across battery terminals the way we have indicated here is not advisable, since it would soon waste the battery's energy by converting it to heat. Placing any good conductor directly across the terminals of a larger battery can be dangerous because of the very large current produced.

The wire described here is relatively thin. A copper wire with a larger radius would have even less resistance and so would carry even more current. If, however, the 0.500 mm radius wire were made of Nichrome, it would have considerably more resistance (about 1.3 Ω) and carry much less current for the same applied voltage $\left(\frac{1.5\ \text{V}}{1.3\ \Omega} = 1.2\ \text{A}\right)$.

EXAMPLE 4 Resistance of a Hot Copper Wire

Find the resistance of the copper wire described in the last example at 100° C.

SOLUTION The resistance at 100° C is found when Eq. 19-10 is applied:

$$R = R_0[1 + \alpha(T - T_0)]$$

The reference resistance R_0 is the value of $2.19 \times 10^{-2}\ \Omega$ found in the previous example for 20° C. Using the value of α for copper found in Table 19-1, we find that at 100° C,

$$R = (2.19 \times 10^{-2}\ \Omega)[1 + (3.9 \times 10^{-3}\ (^\circ\text{C})^{-1})(100^\circ\text{C} - 20^\circ\ \text{C})]$$

$$= 2.9 \times 10^{-2}\ \Omega$$

The preceding examples show that copper wire with a diameter of 1 mm will have resistance much less than 1 Ω per meter of length. Larger-diameter copper wire will have even less resistance. Such small resistance means that copper wire can carry relatively large currents with only a very small voltage drop along its length. This property makes copper useful in connecting elements of an electric circuit. For example, we use copper wire to connect an electric light bulb to a power supply. The same current passes through the wire and the light-bulb filament (an extremely thin tungsten wire), but the filament has much higher resistance than the wire. Therefore the voltage drop across the filament is much greater than the voltage drop along the copper wire. Nearly the entire voltage drop applied to the connecting wires appears across the filament, almost none of it across the wires (Fig. 19-10).

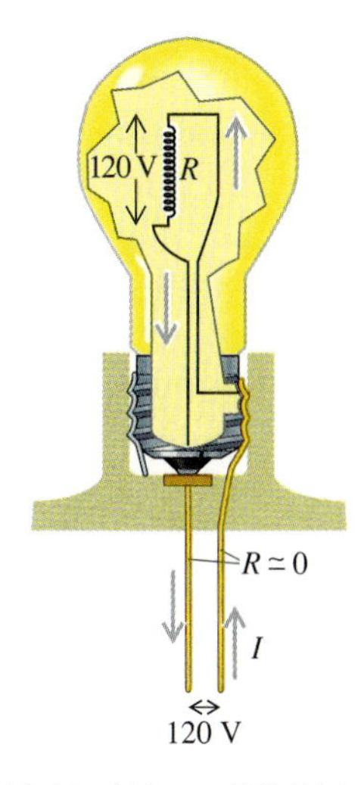

Fig. 19-10 When 120 V is applied across the ends of copper wire connected to a light bulb, there is a negligible voltage drop along the connecting wires because of their negligible resistance. The entire 120 V acts across the light-bulb filament, which has significant resistance.

Resistors

All kinds of electric devices—light bulbs, electric heaters, electric motors, and so on—have resistance, represented in circuit diagrams by the symbol -WW-. The same symbol is used to represent the resistance of **resistors,** small devices designed to provide electrical resistance. When inserted in an electric circuit, resistors are used to control the amount of current in various parts of the circuit. Some resistors consist of a very long coil of fine wire. Others are made of a composition material containing carbon. The resistance of the resistors shown in Fig. 19-11 varies from 0.5 Ω to 10^6 Ω.

Connecting wires having negligible resistance are represented in a circuit diagram by straight lines. Thus two resistors connected by wire of negligible resistance are represented as shown in Fig. 19-12.

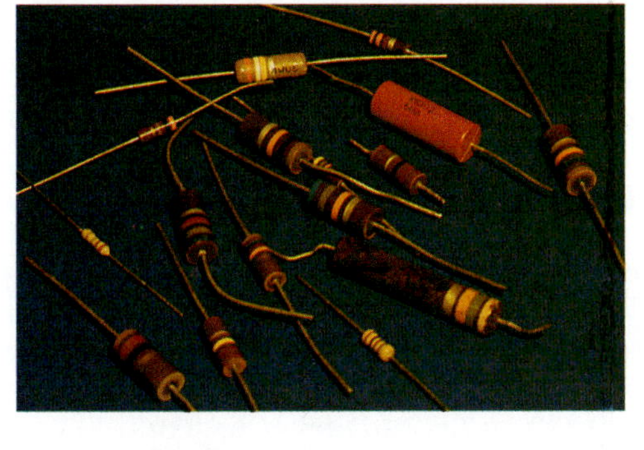

Fig. 19-11 Various resistors.

Fig. 19-12 Two resistors connected by a wire of negligible resistance.

EXAMPLE 5 Finding the Current Through a Resistor

Find the current through the resistor in Fig. 19-13.

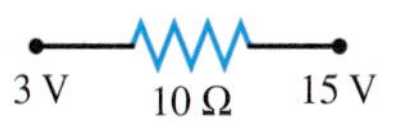

Fig. 19-13

SOLUTION Between the two ends of the resistor there is a potential difference $V = 15\text{ V} - 3\text{ V} = 12\text{ V}$. We find the current by applying Ohm's law:

$$I = \frac{V}{R} = \frac{12\text{ V}}{10\ \Omega} = 1.2\text{ A}$$

Positive current is directed to the left, from the higher potential to the lower potential—in other words, in the direction of the electric field.

EXAMPLE 6 Finding the Voltage Drop Across a Resistor

A positive current of 2 A is directed from a to b through the 4 Ω resistor in Fig. 19-14. Find the voltage drop from a to b.

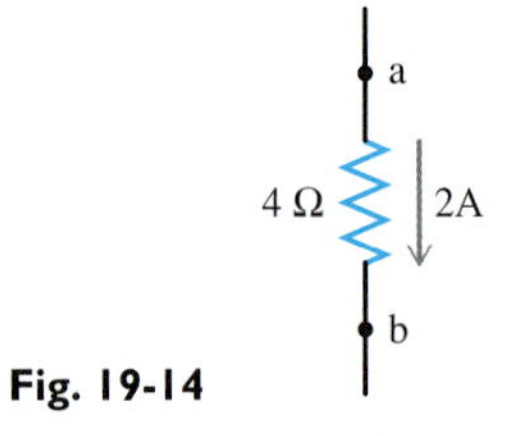

Fig. 19-14

SOLUTION Positive current is in the direction of the electric field and hence is directed from the end of the resistor at the higher potential to the end at the lower potential. Thus the potential at a is greater than the potential at b, and there is a positive voltage drop from a to b:

$$V_{ab} = V_a - V_b = +IR = +(2\text{ A})(4\ \Omega) = +8\text{ V}$$

On the other hand, the voltage drop from b to a, $V_{ba} = V_b - V_a$, is −8 V.

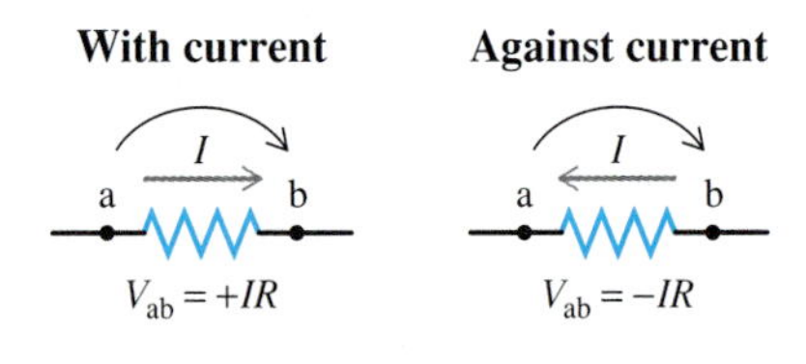

Fig. 19-15 The voltage drop across a resistor.

Voltage Drop Across a Resistor

The preceding example illustrates an important rule: When a resistor is crossed from a to b, the voltage drop V_{ab} is either $+IR$ or $-IR$, depending on whether you are moving in the indicated current direction or opposite that direction—$+IR$ with the current, $-IR$ against it (Fig. 19-15):

$$V_{ab} = \pm IR \quad \begin{pmatrix} +\text{, with current;} \\ -\text{, against current} \end{pmatrix} \tag{19-14}$$

This rule, which works for both positive and negative values of I, will prove useful in solving circuit problems in the next chapter.

19-3 Electric Power; Batteries and AC Sources

Electrical energy is used for a variety of purposes, for example, to provide mechanical energy to an electric motor or to provide thermal energy to an electric heater, an electric stove, or a toaster (Fig. 19-16). In this section we shall see how to calculate the electrical energy provided to such devices, and we shall also describe sources of electrical energy.

Fig. 19-16 Electrical energy is being converted to thermal energy in the heating element of this toaster.

Consider the process that occurs as current flows through a resistor. We can think of this current as being produced by positive charges flowing from the higher potential end to the lower potential end. This means that the charges lose electrical potential energy as they pass through the resistor. Conservation of energy requires that this loss be compensated by a gain in some other form of energy. The energy gained is not in the form of kinetic energy of the charges because the current is the same at both ends of the resistor, and so the charges must have the same average speed at both ends. The constant average speed is caused by continual collisions of charges with the lattice formed by the atoms of the resistor. These collisions cause the average kinetic energy of the *lattice* to increase, but the average velocity and the average kinetic energy of the *current producing charges* are unchanged. The increase in kinetic energy of the lattice means that the thermal energy of the resistor increases; that is, the resistor heats up. The rate of production of thermal energy is just the rate at which electrical energy is used.

Next we shall obtain an expression for the rate at which electrical potential energy is used, that is, for electric power consumption by devices such as resistors and electric motors.

A Closer Look

Superconductivity

The phenomenon of superconductivity has been known since 1911, when Kammerlingh Onnes discovered that the resistance of mercury is exactly zero at temperatures below 4.2 K. Recent developments in superconductivity and its application have generated great worldwide interest in superconducting materials.

Superconducting wires used as coils in electromagnets produce extremely strong magnetic fields without the usual heating produced by current in conventional coils. Superconducting electromagnets are used in physics high-energy accelerator labs and for magnetic resonance imaging, a medical diagnostic technique (Fig. 19-A). In Japan and Germany new high-speed experimental trains are both levitated and propelled by magnetic forces between the train's superconducting electromagnets and magnetic fields induced in the track below (Fig. 19-B).

Superconductors are now being used in electronics. An invention used to measure magnetic fields with great accuracy is the "superconducting quantum interference device" (called "squid" for short), which contains very thin layers of superconducting material.

Until the last few years, no known superconductors exhibited their superconducting behavior above 20 K. At higher temperatures these superconductors have the properties of ordinary conductors; that is, they have nonzero resistance. Maintaining the superconducting state required cooling with expensive liquid helium to 4.2 K. Therefore the practical applications of superconductivity have so far been limited to very specialized and expensive devices.

However, in 1986 a new class of ceramic "high-temperature" superconductors was discovered. The new materials are superconducting at much higher temperatures than any previous materials. One compound (TlBaCaCuO) becomes superconducting below 125 K, and some others become superconducting below about 90 K. Of course these are still very low temperatures, −150° C or less! However, such temperatures are not too difficult to attain using liquid nitrogen at 77 K. Liquid nitrogen is much more plentiful and much less expensive than liquid helium, costing less than some bottled water. Fig. 19-C shows a small permanent magnet suspended by magnetic forces over a ceramic superconductor cooled with liquid nitrogen. (The magnetic effects involved in this demonstration are explained in Chapter 22.)

The discovery of "high-temperature" superconductors immediately gave rise to speculation about wonderful new inventions—smaller and faster computers, powerful miniature electric motors, practical electric cars. However, there are two fundamental problems with the new ceramic superconductors: (1) they are brittle and difficult to shape into wires; (2) the superconducting state is maintained only for very low currents in bulk superconductors, and practical applications require high currents. There is hope that these problems can be overcome. Perhaps one day even room-temperature superconductors will be developed.

Progress is difficult to predict, since the theoretical understanding of superconductivity in these new materials is very incomplete. There was no satisfactory explanation of superconductivity at all until Bardeen, Cooper, and Schrieffer published the "BCS" theory in 1957, 46 years after Kammerlingh Onnes's discovery of superconductivity. The BCS theory is not able to explain the mechanism of superconductivity in the new high-temperature superconductors, however. Perhaps a better theoretical understanding will provide the insight to come up with new and better superconducting compounds. Or perhaps better superconductors will be discovered without an understanding of the mechanism. The history of scientific discovery indicates that it could go either way.

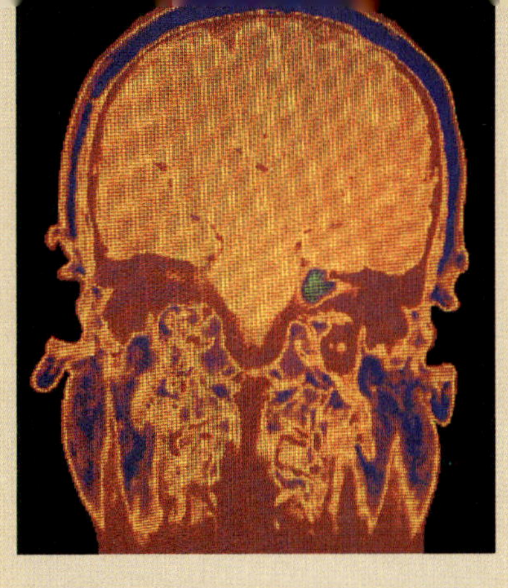

Fig. 19-A Magnetic resonance imaging.

Fig. 19-B This experimental train is capable of speeds up to 500 km/h.

Fig. 19-C A permanent magnet is levitated over a disk of $YBa_2Cu_3O_7$, cooled by liquid nitrogen.

Electric Power Loss

Fig. 19-17 Current through an electrical device with a potential difference $V = V_a - V_b$ across it.

Consider a device with terminals a and b maintained at constant potential V_a and V_b respectively and through which passes a constant positive current I from a to b (Fig. 19-17). During a short time interval Δt, a quantity of charge $\Delta Q = I\,\Delta t$ enters the device at a and an equal quantity of charge leaves the device at b. If we assume that $V_a > V_b$, the effect of the device during this time interval is to lower the electrical potential energy of the charge passing through it:

$$\text{Electrical potential energy loss} = \Delta QV_a - \Delta QV_b = \Delta Q(V_a - V_b) = (I\,\Delta t)V$$

where we use V to denote the voltage drop $V_a - V_b$.

The **electric power loss** P is the **rate of loss* of electrical potential energy** and is obtained when the expression above is divided by the time interval Δt:

$$P = IV \tag{19-15}$$

For example, if a current of 2 A passes through an electric motor that has a potential difference of 10 V across its terminals, the motor uses electric power at the rate $P = IV = (2\text{ A})(10\text{ V}) = 20$ W. This is the rate at which the electric motor converts electrical potential energy to mechanical work.

The electric power loss in a resistor carrying a current I is found by applying Eq. 19-15. In this case, we can utilize Ohm's law to express the power loss in alternative forms; substituting either $V = IR$ or $I = V/R$ in the expression $P = IV$, we get

$$P = IV = I^2R = \frac{V^2}{R} \qquad \text{(for resistors)} \tag{19-16}$$

*In some devices—for example, a battery—positive current enters the low-potential terminal and leaves the high-potential terminal. There is then an increase in electrical potential energy as charge passes through the device. The rate of increase of this energy, the electric power gain, is also found by use of Eq. 19-15.

EXAMPLE 7 Melting Ice with an Electric Window Defroster

The heating element in the rear-window defroster of a Mazda RX-7 has a resistance of 3.00 Ω. The element is connected directly across the car's 12.0 V battery. How much heat is produced in the element in 10.0 min and how much ice will melt?

SOLUTION First we find the electric power used by the resistor. Since we know V and R, we use the third form of Eq. 19-16.

$$P = \frac{V^2}{R} = \frac{(12.0\text{ V})^2}{3.00\ \Omega} = 48.0\text{ W}$$

Using the definition of power as energy per unit time and the definition of a watt as a J/s, we find that in a time interval of 10.0 min, the electrical energy used, or the thermal energy produced, is

$$P\,\Delta t = (48.0\text{ J/s})(10.0\text{ min})\left(\frac{60\text{ s}}{1\text{ min}}\right) = 2.88 \times 10^4\text{ J}$$

Since 335 J/g is required to melt ice, if all the heat produced by the defroster were absorbed by ice at 0° C, the mass of ice that would melt is $\dfrac{2.88 \times 10^4\text{ J}}{335\text{ J/g}} = 86.0$ g.

Batteries

A current-carrying resistor continuously uses electrical energy and so must be connected to a source that continuously produces electrical energy. A battery is a common source of electrical energy.

The development of the modern electric battery began with a chance discovery by Luigi Galvani in 1780. While dissecting a frog, Galvani found that he could sometimes make the muscles in the frog's leg twitch by touching the leg simultaneously with his dissecting knife and some other metal instrument (Fig. 19-18). Galvani believed that the effect was electrical and that the source was within the frog. He referred to the phenomenon as "animal electricity." Volta, a contemporary of Galvani, performed his own experiments and was the first to realize that the effect Galvani had observed was produced not by the frog but by the proximity of different metals in a conducting fluid (the frog's body fluid).

Volta found that he could create a source of electricity by placing paper moistened in a salt solution between two metallic disks, one made of zinc and the other of silver. When the ends of a metallic object were touched to the zinc and silver simultaneously, a small spark was observed. The effect was enhanced when he connected a series of cells, each consisting of zinc, paper, and silver (Fig. 19-19). Thus Volta created the first electric battery.

The batteries shown in Fig. 19-20 are similar to Volta's original battery. The dry-cell flashlight battery consists of an inner carbon rod and an outer zinc cylinder and is filled with an acid absorbed in a solid material that separates the carbon from the zinc (Fig. 19-21).

The chemical reactions within the cell are complex, but the following simplified description gives the essentials of what goes on. The acid slowly dissolves the zinc; that is, negative ions in the acid pull positive zinc ions from the metallic lattice structure of the zinc cylinder. The removal of the zinc ions leaves the cylinder with an excess of free electrons and therefore a net negative charge. An opposite kind of reaction occurs on the carbon rod. Free electrons from the carbon are attracted to positive ions in the acid and are pulled away from the rod, leaving it positively charged. The reactions continue until the potential difference across the terminals is 1.5 V, at which point there is a sufficiently strong electrostatic field within the battery to oppose the further interaction of ions, and so the reaction stops. Within the battery, there is a delicate equilibrium between the interionic forces favoring the chemical reaction and the electrostatic force opposing the reaction. When the battery is connected in a circuit, electrons are removed from the negative terminal and travel around the circuit. The chemical reaction then proceeds at a rate sufficient to replace the electrons removed and maintain the 1.5 V potential difference.*

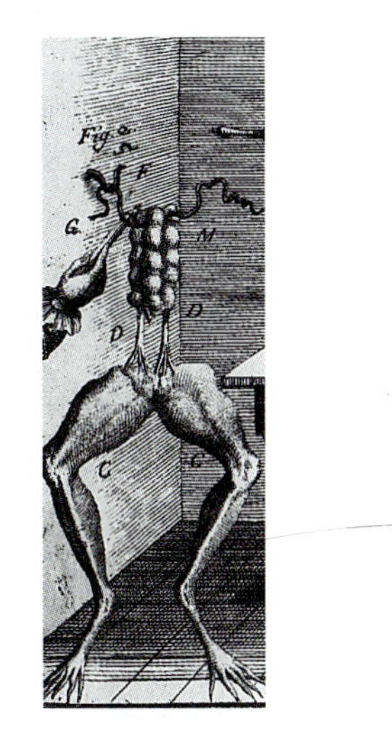

Fig. 19-18 Galvani's experiment.

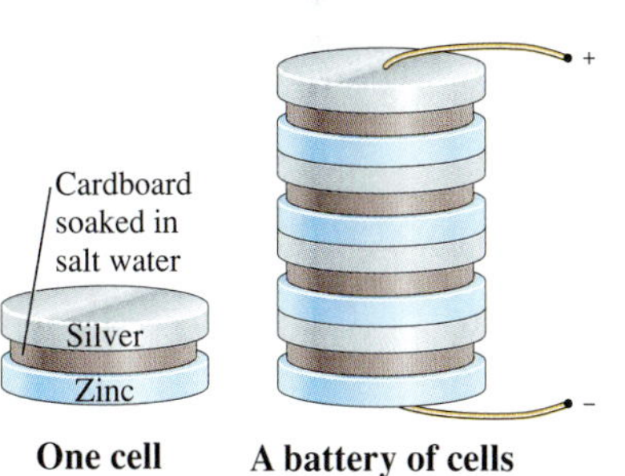

Fig. 19-19 Volta's first battery.

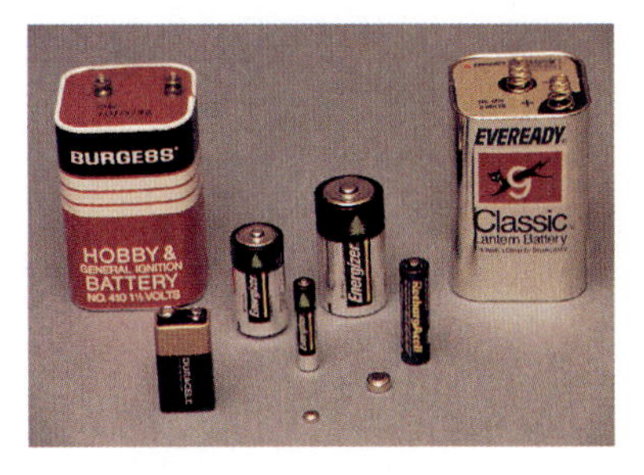

Fig. 19-20 Common batteries.

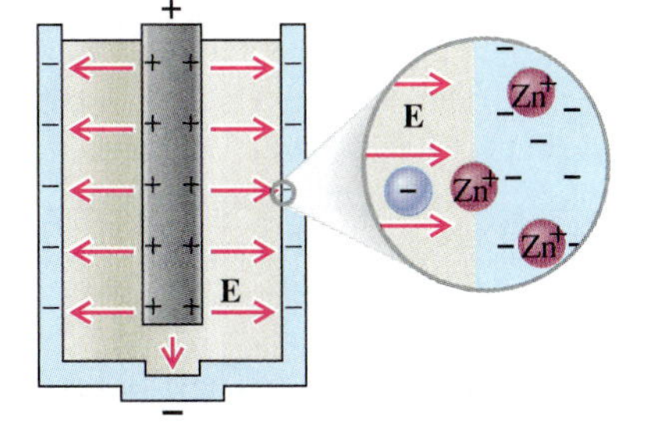

Fig. 19-21 Schematic of a dry cell battery. The attraction between positive zinc ions and negative ions in the acid is opposed by the electric field within the cell.

*However, if the electrons are drawn away at a fast enough rate—that is, if the current through the battery is large enough—there is a significant resistance to the motion of the ions. This means that the equilibrium electric field and the terminal potential difference opposing the flow will be less. We shall return to a discussion of internal resistance at the end of this section.

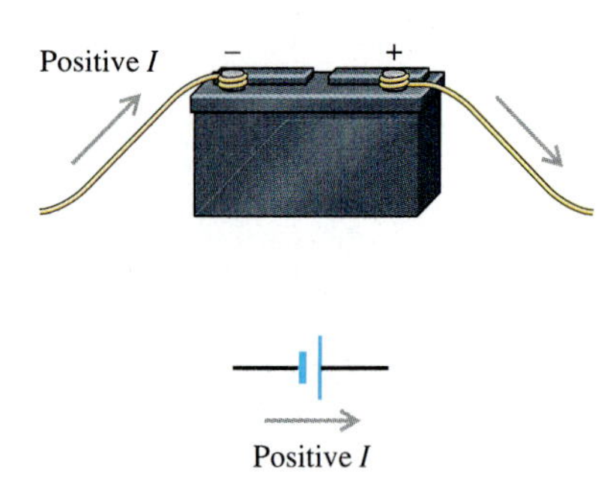

Fig. 19-22 Chemical energy is converted to electrical potential energy inside a battery used as a source of energy.

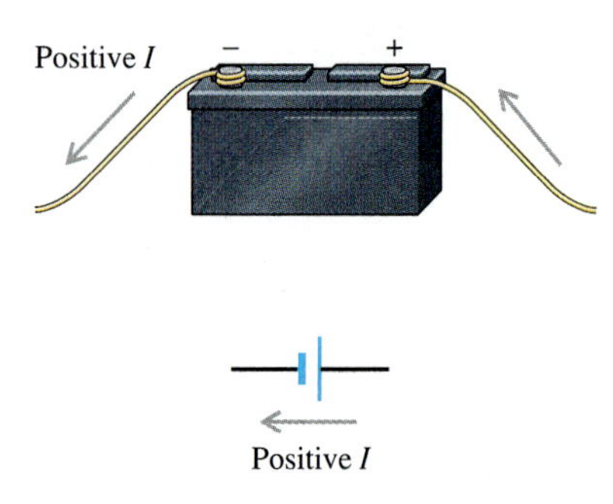

Fig. 19-23 Electrical potential energy is being converted to chemical energy; in other words, the battery is being energized, or "charged."

As charge passes through the battery, the charge gains electrical potential energy. This energy is supplied by the loss of internal energy, or "chemical energy," of the system of electrodes and acid within the battery. Eventually the battery's terminals are dissolved, and the chemicals are used up; the battery is dead, no longer capable of converting internal energy to electrical energy.

The voltage across the terminals of different batteries varies, depending on the chemical reactions and on the strength of interaction between terminals and acid. An automobile battery contains six cells, each consisting of a lead electrode and a lead dioxide electrode separated by sulfuric acid. The chemical reaction of sulfuric acid, lead, and lead dioxide results in a potential difference of 2 V across each cell. Therefore the entire battery of six cells maintains a total potential difference of 12 V between its terminals.

A battery is represented in a circuit diagram by the symbol $-|\!\vdash$, with the longer line representing the higher potential terminal, designated by a + on the battery. Fig. 19-22 shows a battery and its circuit representation. The battery is being used as a source of electrical energy. Thus, inside the battery, positive current flows from the negative terminal to the positive terminal as the battery transforms chemical energy to electrical energy. Outside the battery positive current flows through the connecting wires away from the positive terminal and toward the negative terminal.

For some batteries, such as the automobile battery, the conversion of chemical energy to electrical energy can be reversed. Reversal occurs if the chemical reaction is reversible and if there is another, stronger source that can be connected to the battery in such a way that the current through the battery is reversed (Fig. 19-23). In passing through the battery, positive charge loses electrical potential energy, and chemical energy is stored. The usual chemical process is reversed, and the original supply of chemicals (electrodes and acid) is replenished. When an automobile engine is running, the electric generator or alternator is the source that maintains the chemical energy of the battery. This process is loosely referred to as "charging" the battery. In the common flashlight battery, however, irreversible processes occur, and they prevent the battery from being "recharged."

Emf

A battery is quantitatively described by its "emf" value.* The **emf, denoted by $\mathcal{E}$, is defined to be the energy per unit charge provided by a source as charge crosses the source's terminals.** Let ΔU denote the energy provided (chemical energy in the case of a battery) to charge ΔQ crossing the terminals. Then we can express the definition of emf as

$$\mathcal{E} = \frac{\Delta U}{\Delta Q} \tag{19-17}$$

Since emf is an energy per unit charge, the SI unit of emf is joules per coulomb, or volts, which is the same as the unit of electric potential. The rating of a battery or other source is its emf, measured in volts. An automobile battery has an emf of 12 V, whereas a flashlight battery has an emf of 1.5 V. These values represent the energy supplied to charge as the charge crosses the terminals. For example, when 1 C of charge crosses the terminals of a discharging 12 V battery, the battery supplies to the charge energy $\Delta U = (\Delta Q)\mathcal{E} = (1 \text{ C})(12 \text{ V}) = 12 \text{ J}$. Thus the battery loses 12 J of its chemical energy.

The definition of emf applies not only to batteries and chemical energy but to other energy sources as well. For example, electric generators, solar cells, thermocouples, and even electric eels are all sources of emf (Fig. 19-24). In each, electrical potential energy is created from some other form of energy. For example, a solar cell converts

*The name emf originated as an abbreviation for "electromotive force." Since emf is *not* a force, we shall avoid using the term electromotive force.

Fig. 19-24 A generator, solar cells, and an electric eel.

sunlight to electrical energy, and a thermocouple converts thermal energy to electrical energy.

Let P denote the **power provided by a source of emf,** the rate at which chemical energy or some other form of energy is used:

$$P = \frac{\Delta U}{\Delta t}$$

Applying Eq. 19-17, we find

$$P = \frac{(\Delta Q)\mathcal{E}}{\Delta t}$$

or

$$P = I\mathcal{E} \tag{19-18}$$

EXAMPLE 8 Energy Conversion in a Flashlight Battery

A flashlight is powered by a 1.5 V battery that delivers power at the rate of 3.0 W. Find (a) the current passing through the battery; (b) the charge crossing the battery terminals in 20 min; (c) the battery's loss of chemical energy during this interval.

SOLUTION **(a)** Applying Eq. 19-18, we find

$$I = \frac{P}{\mathcal{E}} = \frac{3.0\text{ W}}{1.5\text{ V}} = 2.0\text{ A}$$

(b) We use the definition of current $\left(I = \frac{\Delta Q}{\Delta t}\right)$ to find the charge ΔQ passing through the battery during a time interval Δt:

$$\Delta Q = I\,\Delta t = (2.0\text{ A})(20\text{ min})\left(\frac{60\text{ s}}{1\text{ min}}\right) = 2400\text{ C}$$

(c) Solving Eq. 19-17 for ΔU, we obtain

$$\Delta U = (\Delta Q)\mathcal{E} = (2400\text{ C})(1.5\text{ V}) = 3600\text{ J}$$

We could also obtain this result by applying the definition of power $\left(P = \frac{\Delta U}{\Delta t}\right)$:

$$\Delta U = P\,\Delta t = (3.0\text{ W})(20\text{ min})\left(\frac{60\text{ s}}{1\text{ min}}\right) = 3600\text{ J}$$

This is the total energy delivered by the battery. This energy is converted partly to radiant energy in the flashlight bulb and partly to heat, both in the bulb and inside the battery.

In an **ideal source** of emf, some form of energy (chemical, solar, or other) is completely converted to electrical energy. If positive charge ΔQ crosses a potential difference V from the negative to the positive terminal of a source, the charge gains electrical potential energy $(\Delta Q)V$ while the source loses energy $(\Delta Q)\mathcal{E}$. If the source is ideal, the energy it loses equals the electrical energy gained by the charge, that is

$$(\Delta Q)\mathcal{E} = (\Delta Q)V$$

or

$$V = \mathcal{E}$$

Thus in an ideal source of emf the voltage across the terminals equals the emf. For example, a normal automobile battery is a nearly ideal source of emf and so has a terminal voltage very close to its emf of 12 V.

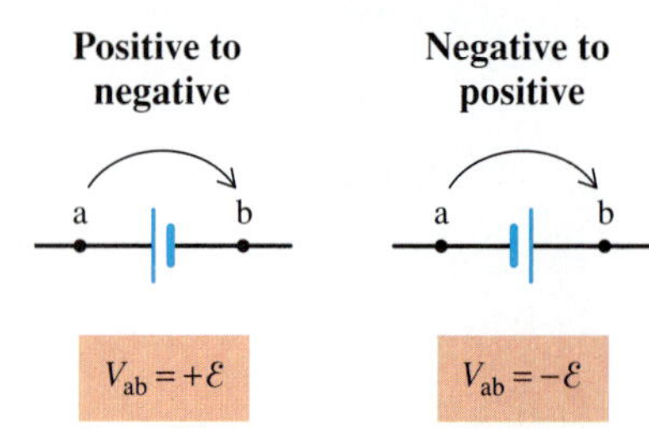

Fig. 19-25 The voltage drop across an ideal source of emf.

Voltage Drop Across a Source of emf

In describing electric circuits, we often need to calculate the voltage drop across the terminals of a source. If one goes from a positive terminal a to a negative terminal b of an ideal source, there is a positive drop in potential $V_{ab} = V_a - V_b = +\mathcal{E}$. If one goes from a negative terminal a to a positive terminal b, there is a rise in potential, which is the same as a negative voltage drop, $V_{ab} = V_a - V_b = -\mathcal{E}$ (Fig. 19-25).

ideal source $$V_{ab} = \pm\mathcal{E} \quad \begin{pmatrix} +, \text{ from } + \text{ to } -; \\ -, \text{ from } - \text{ to } + \end{pmatrix} \tag{19-19}$$

Fig. 19-26 shows some examples applying this result.

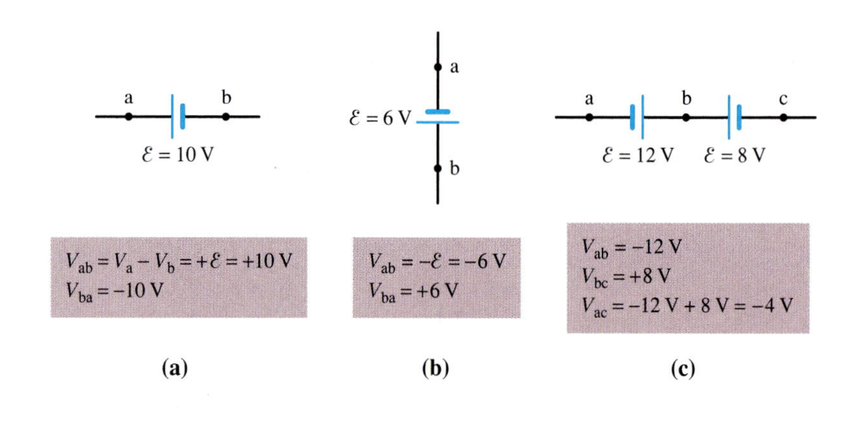

Fig. 19-26 Voltage drops across ideal sources of emf.

EXAMPLE 9 Energy Conversion in a Simple Circuit

A simple circuit consisting of an ideal battery, a resistor, and connecting wires is shown in Fig. 19-27, along with its circuit diagram. Describe the energy conversion that occurs as charge flows through the circuit.

SOLUTION The battery maintains a certain potential difference between its terminals, with a at the higher potential. There is no change in potential along the connecting wires, and hence points a and c are at the same potential and points b and d are at the same potential. Thus the battery maintains the same potential difference across the resistor as it does across its own terminals. Ohm's law ($V = IR$) then predicts that the voltage drop across the resistor produces a current through it from c to d. This same steady current exists throughout the circuit. (If this were not so, charge would pile up at some point and create a large electric field. Such an electric field would immediately cause the charge to disperse.)

As indicated in the figure, there is a positive counterclockwise current. We can think of this current as positive charge moving counterclockwise around the circuit. As the charge passes from point b through the battery to point a, its electrical potential energy increases while the chemical energy stored in the battery decreases. Then the charge moves around the remainder of the circuit, flowing through the resistor from c to d and losing electrical potential energy while producing thermal energy in the resistor. Completing the circuit, the charge returns to point b, where its electrical potential energy is the same as when it began at b.

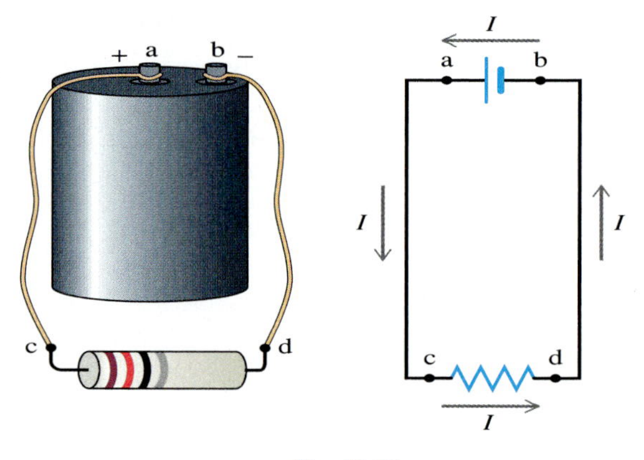

Fig. 19-27

The charge then goes around the circuit again. The process is continuous. The current is the same at all points in the circuit at any instant. As some charge is passing through the battery, an equal amount of charge is moving through the resistor. Inside the battery, chemical energy is continuously being used to produce electrical potential energy, while inside the resistor, electrical potential energy is continuously converted to thermal energy.

Often in a battery the production of electrical potential energy is accompanied by some internal heating of the battery. The battery can then be treated as an ideal source $\mathcal{E}$ connected in series with a resistor r, where r represents the battery's internal resistance (Fig. 19-28). When current flows through the battery, there will be either a positive or a negative voltage drop ($\pm Ir$) across this internal resistance, the sign depending on the direction of current. The terminal voltage of the battery will then be either greater than or less than $\mathcal{E}$. If there is no current through the battery, then $Ir = 0$ and the terminal voltage equals the emf. Thus you can measure the emf of a battery by measuring the potential difference across its terminals when it is disconnected from a circuit.

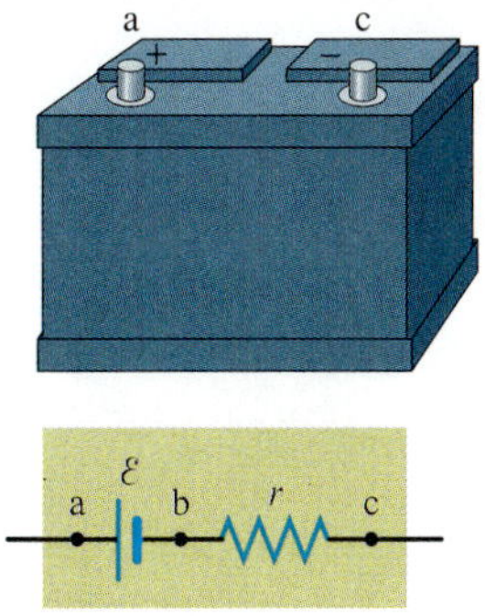

Fig. 19-28 A real battery can be treated as an ideal source $\mathcal{E}$ in series with a resistor r. This resistor represents the internal resistance of the battery.

EXAMPLE 10 Terminal Voltage for a Non-Ideal Battery

Suppose that the battery in Fig. 19-28 has an emf $\mathcal{E} = 1.5$ V, an internal resistance $r = 1\ \Omega$, and a positive current $I = 0.2$ A directed toward the left from c to a. Find the terminal voltage drop $V_{ac} = V_a - V_c$.

SOLUTION We find the total voltage drop from a to c by summing the voltage drops V_{ab} and V_{bc}:

$$V_a - V_c = V_a - V_b + V_b - V_c$$

or

$$V_{ac} = V_{ab} + V_{bc}$$

Applying Eq. 19-19, we find $V_{ab} = +\mathcal{E}$. And since in crossing the resistor from b to c we are going against the current, application of Eq. 19-14 gives $V_{bc} = -Ir$. Thus

$$V_{ac} = +\mathcal{E} - Ir = +1.5\text{ V} - (0.2\text{ A})(1\ \Omega)$$
$$= +1.3\text{ V}$$

The physical significance of this result can be understood when we consider what happens to a charge of $+1.0$ C as it crosses the battery terminals in the direction of the current from c to a. There is a loss of chemical energy of $(1.0\text{ C})(1.5\text{ V}) = 1.5$ J, a gain in electrical potential energy of $(1.0\text{ C})(1.3\text{ V}) = 1.3$ J, and a gain in thermal energy of $(1.0\text{ C})(0.2\text{ V}) = 0.2$ J. Thus energy is conserved. Chemical energy is converted to both electrical potential energy and thermal energy inside the battery.

Alternating Current Source

Batteries provide an energy source for devices such as flashlights, portable radios, and calculators. The current through a battery flows in one direction. We call this **direct current,** or "DC." The electrical energy provided by electrical outlets in your home produces **alternating current,** or "AC," which means that the current alternates in direction back and forth. In Chapter 20 we shall discuss household electricity in more detail, and in Chapter 22 we shall see how such electricity is generated at power plants. For now, we note that the conducting wires connected to an electrical outlet are maintained at a potential difference that varies with time. The same time-dependent voltage is applied to any device plugged into an electrical outlet, an electric light, for example (Fig. 19-29).

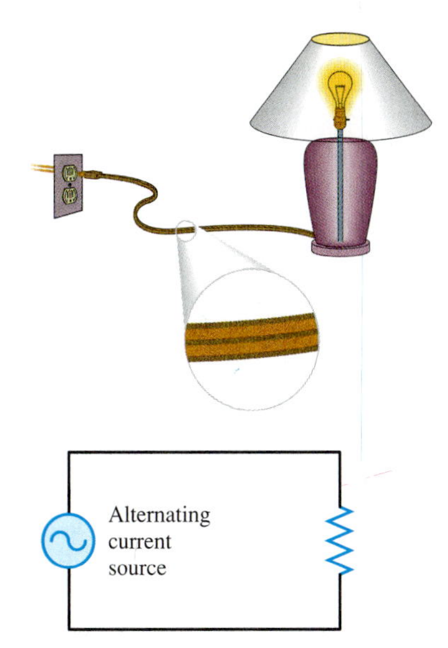

Fig. 19-29 A light bulb connected to a wall outlet and its circuit representation.

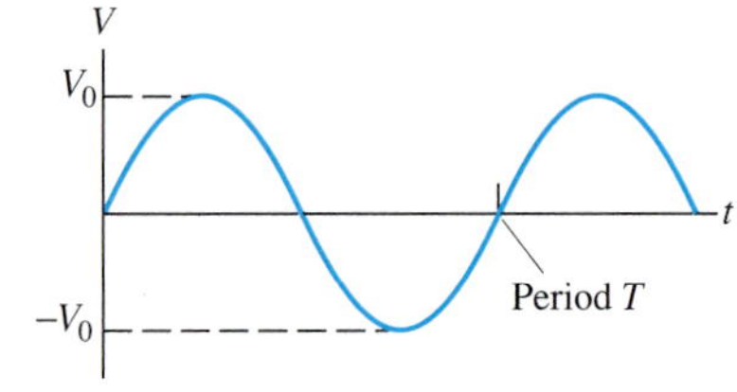

Fig. 19-30 The voltage provided by a source of alternating current.

The time dependence of the voltage is harmonic—a sine wave of frequency f and amplitude V_0 (Fig. 19-30):

$$V = V_0 \sin(2\pi f t) \tag{19-20}$$

Any harmonic voltage source provides a voltage of this form. For a standard household outlet in the United States, $V_0 = 170$ V and $f = 60$ Hz. A resistor connected to an alternating voltage source will carry an alternating current, found by applying Ohm's law.

$$I = \frac{V}{R} = \frac{V_0}{R}\sin(2\pi f t)$$

or

$$I = I_0 \sin(2\pi f t) \tag{19-21}$$

where $I_0 = \dfrac{V_0}{R}$ is the current amplitude.

We can find the instantaneous power dissipated by a resistor by using a form of Eq. 19-16:

$$P = I^2 R$$

or

$$P = \frac{V^2}{R}$$

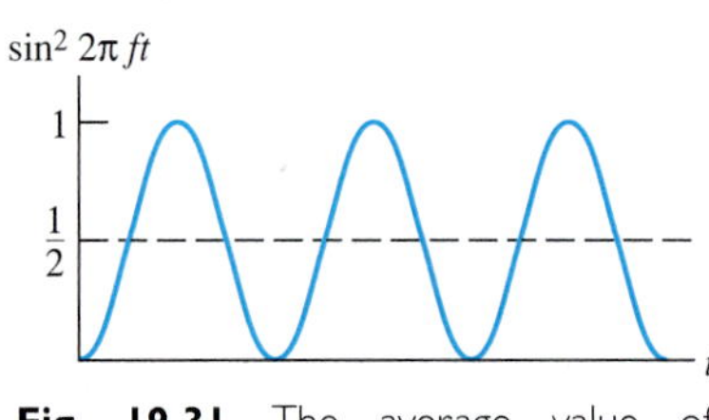

Fig. 19-31 The average value of $\sin^2(2\pi f t)$ equals $\frac{1}{2}$.

We are usually more interested in the time-averaged power dissipated by a resistor, rather than the instantaneous power. This average power $\overline{P}$ is determined by the average values of I^2 or V^2—that is, $\overline{I^2}$ or $\overline{V^2}$. Both I^2 and V^2 are proportional to $\sin^2(2\pi f t)$, the average value of which is $\frac{1}{2}$ (Fig. 19-31). Thus from Eqs. 19-20 and 19-21 we find

$$\overline{I^2} = I_0^2\,\overline{\sin^2(2\pi f t)} = \tfrac{1}{2}I_0^2$$

and

$$\overline{V^2} = V_0^2\,\overline{\sin^2(2\pi f t)} = \tfrac{1}{2}V_0^2$$

It is convenient to express average power in terms of the **"root-mean-square"** of either current or voltage—defined as the square root of the mean current or voltage squared. From the preceding equation we see that the rms value of either I or V equals $1/\sqrt{2}$ times the respective amplitude.

$$I_{\text{rms}} = \sqrt{\overline{I^2}} = \frac{1}{\sqrt{2}} I_0 \tag{19-22}$$

$$V_{\text{rms}} = \sqrt{\overline{V^2}} = \frac{1}{\sqrt{2}} V_0 \tag{19-23}$$

For a standard electrical outlet in the United States, $V_{\text{rms}} = \dfrac{1}{\sqrt{2}}(170\text{ V}) = 120$ V. We usually refer to this simply as a 120 V outlet. In most of Europe, the rms voltage is 240 V.

We can use rms values of I and V to express average power. First we take the average of the expressions for instantaneous power:

$$\overline{P} = \overline{I^2}R = \frac{\overline{V^2}}{R}$$

Then we use the definitions of I_{rms} and V_{rms} to substitute $\overline{I^2} = I_{\text{rms}}^2$ and $\overline{V^2} = V_{\text{rms}}^2$ and obtain

$$\overline{P} = I_{\text{rms}}^2 R = \frac{V_{\text{rms}}^2}{R} \tag{19-24}$$

An rms current or voltage produces the same average power as a constant current or voltage of the same value.

EXAMPLE 11 Rate at Which Heat is Produced by a Toaster

A Nichrome wire having a resistance of 10.0 Ω is the heating element in an electric toaster. Find the power used by the toaster when it is connected to a standard electrical outlet.

SOLUTION The outlet provides 120 V, rms. From Eq. 19-24 we find

$$P = \frac{V_{\text{rms}}^2}{R} = \frac{(120 \text{ V})^2}{10.0\ \Omega} = 1440 \text{ W}$$

This is the rate at which electrical energy is being converted to thermal energy in the toaster.

*19-4 Electric Current and Ohm's Law on the Microscopic Level

In this section we shall describe electric current in a conductor on the microscopic level and obtain a more fundamental understanding of Ohm's law.

Electric Current and Moving Charges

First we shall consider the various factors that determine the value of electric current when charges flow through a region of space. For simplicity, suppose that identical positive charges q are uniformly spread over some region of space and are all moving with the same velocity **v** to the right (Fig. 19-32). We want to find an expression for the current I through a surface of area A. We shall see that the current depends on q, v, and A and also on the number of charges per unit volume, denoted by n.

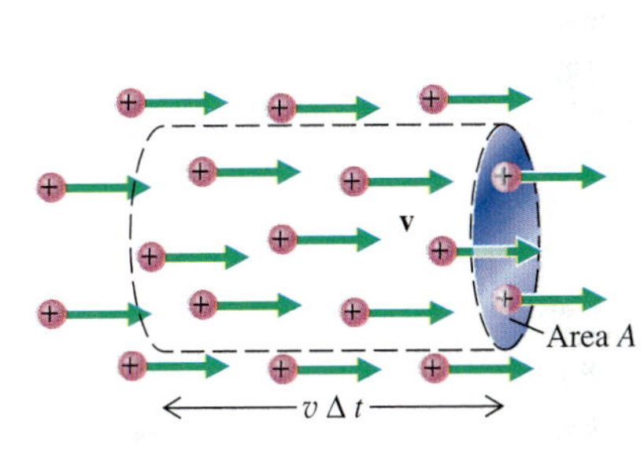

Fig. 19-32 Moving charges produce a current through surface area A.

What is the quantity of charge passing through the surface area A in time Δt? Since all the charges move a distance $v\,\Delta t$ to the right during the time interval Δt, the quantity of charge passing through A is all of the charge initially contained in the cylindrical region of length $v\,\Delta t$, to the left of A, shown in Fig. 19-32. The number of charges in this region equals the product of n, the number of charges per unit volume, and the volume of the cylinder, $(v\,\Delta t)A$:

$$\text{Number of charges passing through } A \text{ in time } \Delta t = nvA\,\Delta t$$

The quantity of charge ΔQ passing through A during Δt equals the number of charges times the value q of each individual charge:

$$\Delta Q = qnvA\,\Delta t$$

Dividing by Δt, we obtain an expression for the current through the surface:

$$I = \frac{\Delta Q}{\Delta t} = qnvA$$

If, instead of all charges having the same speed v, there is some distribution of speeds with an average value $\overline{v}$, the preceding equation is modified by use of this average value $\overline{v}$ instead of v. Furthermore, if we want to apply our result to negative charge carriers as well as positive, we can do so by replacing q by its absolute value.

$$I = |q|n\overline{v}A \qquad (19\text{-}25)$$

EXAMPLE 12 How Long for a Light Bulb to Glow?

A copper wire with a cross-sectional area of 1.0 mm² carries a positive current of 20 A to the right. The density of free electrons in copper is $8.5 \times 10^{28}/\text{m}^3$. Find the average speed of the free electrons. Is this related to the time it takes for a light to turn on?

SOLUTION Recall from Chapter 17 that in metals it is electrons that are free to move. Thus the positive current to the right is produced by the flow of negatively charged free electrons to the left. We apply Eq. 19-25 to find the average speed of the free electrons, whose motion produces the current:

$$I = |q|n\bar{v}A$$

$$\bar{v} = \frac{I}{|q|nA} = \frac{20\text{ A}}{(1.6 \times 10^{-19}\text{ C})(8.5 \times 10^{28}/\text{m}^3)(1.0 \times 10^{-6}\text{ m}^2)}$$

$$= 1.5 \times 10^{-3}\text{ m/s} = 1.5\text{ mm/s}$$

The electrons move to the left through the wire at an average speed of 1.5 mm/s. Thus on the average it requires 1 s for an electron to move 1.5 mm along the wire, or 1000 s (17 min) to move 1.5 m. This average electron speed should not be confused with the speed at which electron motion is communicated from one end of a current-carrying wire to the other. When a wire is connected as part of a circuit, the current begins almost simultaneously at all points in the wire. An electric field is very quickly established along the length of the wire. This is an electromagnetic wave phenomenon, and the field propagates at the speed of light even though electrons move very slowly in response to the field. Thus when you turn on a light switch, there is not a delay of minutes before the light bulb glows. Instead, an electric field and electric current are established in the filament of the light bulb and in the connecting wires almost instantly, and so the glow of the light bulb is practically immediate.

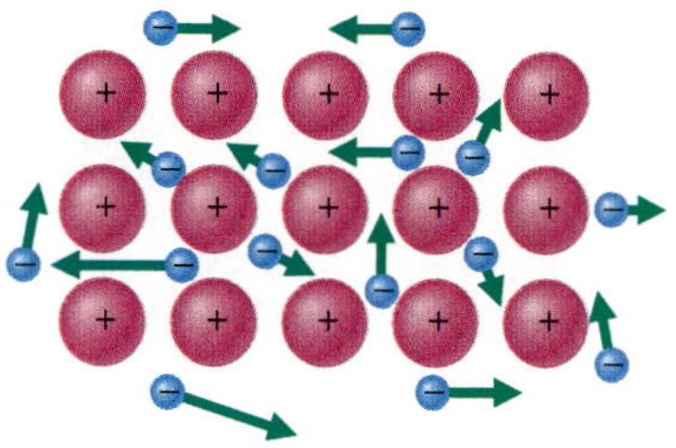

Microscopic Form of Ohm's Law

To get a better idea of the physical meaning of Ohm's law, let us picture what happens when a potential difference is applied across the ends of a metallic conductor. A small section of a metal consists of atoms that form a regular array, called a "lattice." In forming the lattice, each atom loses one or two electrons, and so the lattice consists of positive ions rigidly bound in place, whereas the electrons given up are free to move throughout the lattice (Fig. 19-33). These free electrons behave much like gas molecules, moving in random directions and colliding with positive ions.

Fig. 19-33 In a metallic lattice, positive ions form a fixed, periodic pattern, whereas conduction electrons are free to move over large distances.

An electric field may be set up within the conductor by applying a potential difference across it, for example, by connecting it to the terminals of a battery. The electric field then accelerates the free electrons in the direction opposite the field. The electrons continue to collide frequently with the positive ions, and so the acceleration of the electrons by the field is interrupted frequently.

An analogous effect is the motion of marbles rolling down an incline and colliding with a series of nails (Fig. 19-34). If there were no nails, a marble would experience a constant acceleration down the incline, and its velocity would increase linearly with time. Instead, a marble accelerates briefly, collides with a nail and stops or bounces back and then begins to accelerate down the incline again. The continual alternation between acceleration and collision for each marble gives rise to an average downward flow of marbles that is fairly constant. The steeper the incline, the greater will be the average speed of the marbles.

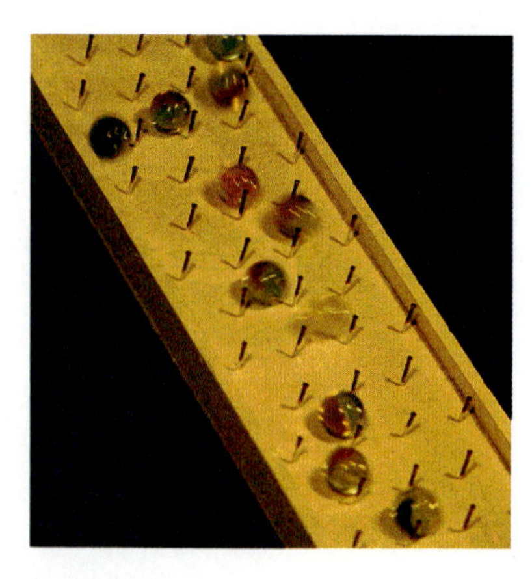

Fig. 19-34 A mechanical model illustrating the motion of conduction electrons through a metal.

Similarly, conduction electrons are accelerated by an electric field for the brief time interval between collisions. This intermittent acceleration gives a low average electron velocity directed opposite the electric field.

The average speed of marbles down an incline will increase as the steepness of the incline increases. So too does the average speed of conduction electrons increase as the strength of the electric field increases. The electrons' average speed $\bar{v}$ is directly proportional to the field strength E:

$$\bar{v} \propto E$$

From Eq. 19-25 ($I = |q|n\bar{v}A$), we see that

$$\bar{v} \propto \frac{I}{A}$$

Since $\bar{v}$ is proportional to both E and I/A, it follows that E and I/A are proportional to each other.

$$E \propto \frac{I}{A}$$

The constant of proportionality is the resistivity ρ of the metal, and so we may express this result as

$$E = \rho\frac{I}{A} \tag{19-26}$$

This equation is the microscopic form of Ohm's law, from which we shall now derive the more familiar macroscopic form of Ohm's law, $V = IR$.

Macroscopic Form of Ohm's Law

Consider a straight wire of length ℓ and cross-sectional area A carrying a uniform current I to the right (Fig. 19-35).

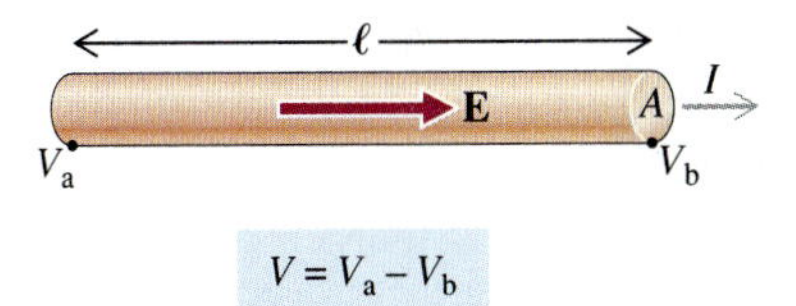

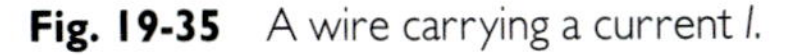
Fig. 19-35 A wire carrying a current *I*.

According to the microscopic form of Ohm's law, there must also be a uniform electric field to the right of strength

$$E = \rho\frac{I}{A}$$

This uniform electric field produces a potential difference $V = V_a - V_b$, which, according to Eq. 18-4 ($V = Ed$), is the product of E and the length of the wire:

$$V = E\ell$$

Inserting the preceding expression for E into this equation gives

$$V = I\frac{\rho\ell}{A}$$

Defining the wire's electrical resistance R as

$$R = \frac{\rho\ell}{A}$$

and inserting this into the preceding equation, we obtain the macroscopic form of Ohm's law:

$$V = IR$$

The picture of metallic conduction we have presented here is greatly oversimplified. A treatment that is more complete and quantitative is provided in Appendix C.

CHAPTER 19 SUMMARY

Electric current is defined as the rate of flow of charge and is measured in amperes (A), coulombs per second:

$$I = \frac{\Delta Q}{\Delta t}$$

$$1 \text{ A} = 1 \text{ C/s}$$

Ohm's law relates the voltage drop V across a particular piece of material to the current I through it:

$$V = IR$$

where the resistance R is measured in ohms (Ω), or volts per ampere:

$$1 \ \Omega = 1 \text{ V/A}$$

The resistance of a cylindrical conductor of length ℓ and cross-sectional area A is related to the resistivity ρ of the medium by the equation

$$R = \frac{\rho \ell}{A}$$

Resistivity depends on temperature and on the medium's temperature coefficient of resistance α:

$$\rho = \rho_0[1 + \alpha(T - T_0)]$$

In crossing a resistor R carrying a current I, the voltage drop V_{ab} from point a to point b is given by either $+IR$ or $-IR$, depending on the direction of the current relative to the path from a to b:

$$V_{ab} = \pm IR \quad (+\text{, with current; } -\text{, against current})$$

A battery is a source of energy that maintains an approximately constant potential difference across its terminals. A source's emf $\mathcal{E}$ is the energy per unit charge provided by the source to the charge crossing the terminals of the source.

$$\mathcal{E} = \frac{\Delta U}{\Delta Q}$$

In an ideal battery, the terminal voltage equals the emf. The voltage drop V_{ab} across an ideal source from a to b is either $+\mathcal{E}$ or $-\mathcal{E}$, depending on direction:

$$V_{ab} = \pm \mathcal{E} \quad (+\text{, from } + \text{ to } -; \ -\text{, from } - \text{ to } +)$$

A battery with internal resistance can be represented in a circuit by an ideal battery in series with a resistor. The rules for calculating voltage drops across resistors and ideal sources may then be applied.

Electrical outlets maintain a potential difference between the wires connected to the outlet. The potential difference

$$V = V_0 \sin(2\pi f t)$$

produces an alternating current I in a resistor connected to the outlet, where

$$I = I_0 \sin(2\pi f t)$$

The rms value of current or voltage is defined as the square root of the mean current or voltage squared.

$$I_{rms} = \sqrt{\overline{I^2}} = \frac{1}{\sqrt{2}} I_0$$

$$V_{rms} = \sqrt{\overline{V^2}} = \frac{1}{\sqrt{2}} V_0$$

Standard electrical outlets in the United States provide 120 V, rms, at a frequency $f = 60$ Hz.

Electric power is the rate of conversion of electrical energy to some other form. For a device with a voltage drop V across its terminals and current I through it, the electric power is

$$P = IV$$

For a resistor, the power may also be expressed

$$P = I^2 R$$

or

$$P = \frac{V^2}{R}$$

These same expressions may be used for the average power provided by an alternating current source to a resistor if we use rms values of I and V:

$$\overline{P} = I_{rms}^2 R = \frac{V_{rms}^2}{R}$$

In a source of emf $\mathcal{E}$, the rate at which chemical energy (or some other form of energy) is used is

$$P = I\mathcal{E}$$

where I is the current through the source.

Questions

1 Because of the earth's atmospheric electric field, which is directed downward during clear weather, positive ions flow downward in the atmosphere and negative ions flow upward. What is the sign of the current in the downward direction, resulting from the flow of the (a) positive charges; (b) negative charges? What is the sign of the current in the upward direction resulting from the flow of the (c) positive charges; (d) negative charges?

2 Fig. 19-36 shows a glowing circle caused by a beam of electrons moving along a circular path in a magnetic field. (See Chapter 21 for a further discussion of such phenomena.) The electrons move around the circle clockwise. Is the current positive in the clockwise or counterclockwise direction?

Fig. 19-36

3 A single strand of wire of resistance R is cut in half, and the two halves are placed side by side to form parallel strands. What is the resistance of the two-strand combination?

4 Two nails, one made of aluminum and the other made of iron, have exactly the same size and shape. Suppose the same potential difference is applied across the ends of each nail.

(a) In which nail will the current be greater?

(b) In which nail will the thermal energy increase faster?

5 Does a 60 W light bulb or a 100 W light bulb have greater electrical resistance?

6 Electrical potential energy is lost as an electric current passes through a resistor. Does this loss mean that the current leaving the resistor is less than the current entering?

7 Conduction electrons moving through a copper wire produce an electric current. Does the electric field within the wire do positive, negative, or zero work on the electrons?

8 A lemon with a copper rod and a steel paperclip stuck in the sides acts as a battery (Fig. 19-37).

(a) What is the emf?

(b) This battery is not at all practical as a source of energy because of its very high internal resistance. What would happen to the voltmeter's reading if the lemon were connected to a 1 Ω resistor to form a complete circuit?

Fig. 19-37

9 Suppose your car's battery is dead. You wish to "charge" it by connecting it to a good battery with jumper cables—heavy copper cables used to connect the terminals of the two batteries. To energize your dead battery and avoid a dangerously large current, should the cables connect the terminals: (a) + to + and − to − or (b) + to − and − to +?

10 A lightweight electrical extension cord is intended to be used to connect low-power electrical devices, such as radios and lights. Suppose you use such a cord to connect a heavyweight power tool to an electrical outlet. What happens and what should you do?

11 In a certain region of space, moving electrons produce a current through a surface. Suppose the electrons are replaced by protons having the same number density n and the same average velocity. Would the current change? If so, how?

12 When you turn an electric heater on, which of the following factors account for the time delay before the heating element begins to glow: (a) the time for electrons to move from one end of the element to the other; (b) the time for electrons to move from the power supply to the heating element; (c) the time for electromagnetic waves to move through the wires; (d) the time for enough charge to pass through the element to provide sufficient thermal energy to increase the temperature?

13 Two copper wires of the same length but different diameters are connected in series in a circuit. Indicate for each of the following quantities whether it is greater in the larger or in the smaller wire, or whether it is the same in both: (a) resistance; (b) current; (c) temperature.

14 Two copper wires of the same length but different diameters are connected in parallel in a circuit. Indicate for each of the following quantities whether it is greater in the larger or in the smaller diameter wire, or whether it is the same in both: (a) resistance; (b) current; (c) temperature.

15 A copper wire and an aluminum wire, both having the same length and diameter, are connected in series. Indicate for each of the following quantities whether it is greater in the copper or in the aluminum wire, or whether it is the same in both: (a) resistance; (b) current; (c) temperature.

Answers to Odd-Numbered Questions

1 (a) +; (b) +; (c) −; (d) −; **3** $\frac{1}{4}R$; **5** 60 W; **7** positive; **9** a; **11** Yes. It would have the same magnitude but opposite direction; **13** (a) smaller; (b) same; (c) smaller; **15** (a) Al; (b) same; (c) Al

Problems (listed by sections)

19-1 Electric Current

1 A copper wire carries a positive current I to the right. Each second 1.00×10^{20} electrons pass through a cross section of the wire.
(a) In what direction are the electrons moving?
(b) Calculate I.

2 A current of 2.50 A flows through a 1.00 mm diameter copper wire that is connected to a capacitor plate. The current is directed toward the plate. Find the charge transferred to the plate in 1.00×10^{-3} s.

3 A copper wire carries a current of 10.0 A. How many electrons pass a point in the wire in 1 hour?

4 In what appears to be a single lightning bolt, there are typically several distinct strokes over the same path, each lasting on the order of 10^{-4} s, with an interval of about 0.05 s between them. The duration of the several strokes is about 0.20 s. Thus, although the peak current may reach a maximum of 20,000 A, the average current over the 0.20 s is much less—typically about 120 A upward. Find the charge transferred to one point on the earth's surface during a typical 0.20 s discharge.

19-2 Ohm's Law

5 A certain resistor is made with a 50.0 m length of fine copper wire, 5.00×10^{-2} mm in diameter, wound onto a cylindrical form and having a fiber insulator separating the coils. Calculate the resistance.

6 There is a potential difference of 1.0 V between the ends of a 10 cm long graphite rod that has a cross-sectional area of 1.0 mm^2. The resistivity of graphite is 7.5×10^{-6} Ω-m. Find (a) the resistance of the rod; (b) the current; (c) the electric field inside the rod.

7 A certain electric extension cord has a resistance of 2.00 Ω. Suppose it is replaced by an extension cord made of the same material. The replacement cord is twice the diameter of the original and twice as long. What is the resistance of the new cord?

8 A lead wire of cross-sectional area 1.00 mm^2 is at a temperature of 2.00 K and carries a current of 0.500 A.
(a) Find the voltage drop along a 10.0 m length of the wire.
(b) Will the temperature of the lead rise as a result of the current?

9 When an electric heater is turned on, the Nichrome heating element is at 0° C and draws a current of 16 A, with a potential difference of 120 V between the ends of the element. The element heats up to 800° C, with the potential difference constant. Find the final current.

10 Find the potential at b in Fig. 19-38.

★ **11** In order to double its resistance, by how much would you have to increase the temperature of a piece of (a) copper; (b) Nichrome?

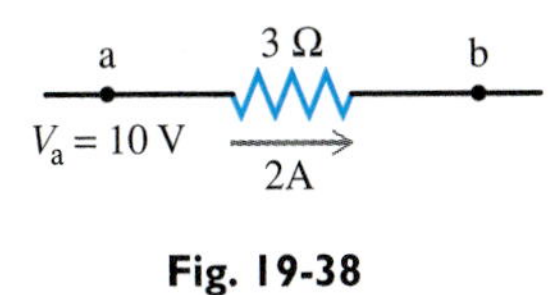

Fig. 19-38

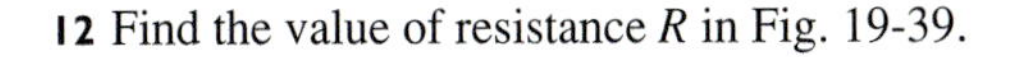

12 Find the value of resistance R in Fig. 19-39.

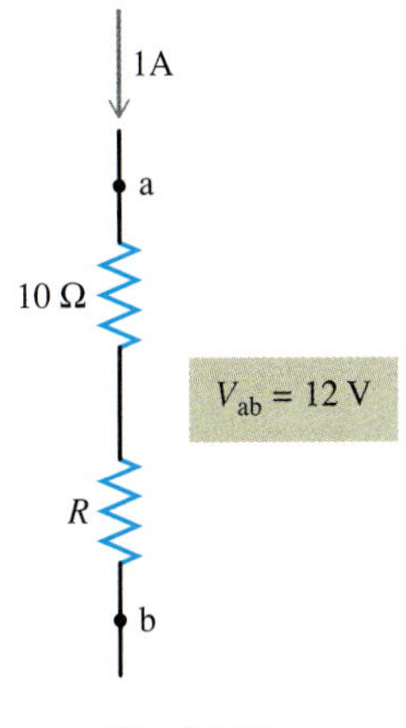

Fig. 19-39

13 (a) Find the current through the resistor in Fig. 19-40.
(b) If the resistor is made of carbon, what will its resistance be when its temperature is increased by 100 C°?

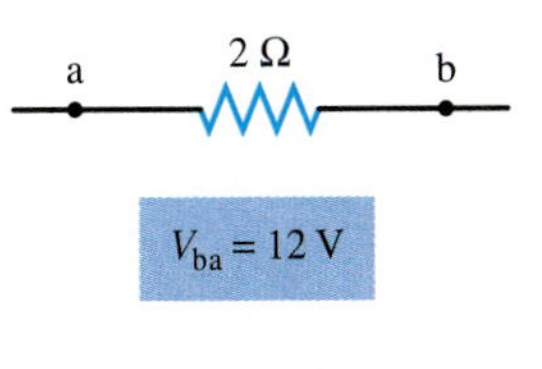

Fig. 19-40

19-3 Electric Power; Batteries and AC Sources

14 A battery with an emf of 10.0 V and internal resistance of 1.00 Ω carries a positive current of 0.500 A from its negative terminal to its positive terminal.
(a) How long is required for 1.00 C of charge to pass through the battery?
(b) By how much will the electrical potential energy of 1.00 C of charge increase in crossing the battery terminals?
(c) By how much will the chemical energy decrease?
(d) How much electrical power is supplied by the battery?
(e) At what rate is chemical energy used?

15 A positive current of 2.00 A flows from the negative to the positive terminal inside a battery that has an emf of 10.0 V. How long does it take for this battery to lose 100 J of chemical energy?

16 The terminal voltage of a certain battery is measured with the battery disconnected and is found to be 6.00 V. The internal resistance of the battery is 1.00 Ω. Find the voltage across the battery terminals when there is a current of 2.00 A from the negative to the positive terminal.

17 (a) Find the potential at B in Fig. 19-41.
(b) Find the emf of the battery.
(c) Is the battery being "charged" or "discharged"; that is, is electrical energy being converted to chemical energy or vice versa?

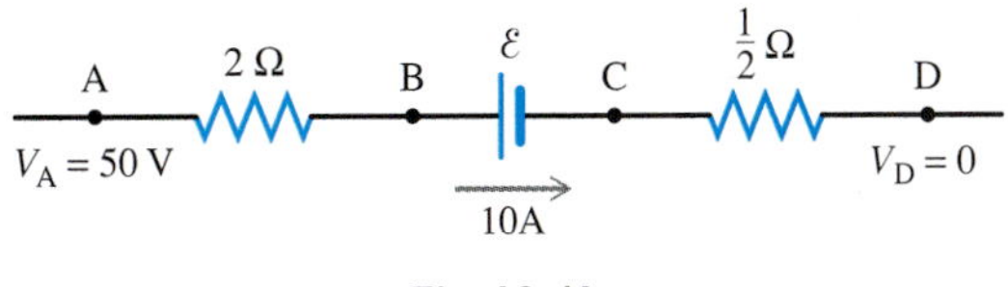

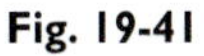

Fig. 19-41

★ **18** When a car's starter is in use, it draws a large current. The car's lights draw much less current. As a certain car is starting, the current through the battery is 60.0 A and the potential difference across the battery terminals is 9.00 V. When only the car's lights are used, the current through the battery is 2.00 A and the terminal potential difference is 11.9 V. Find the battery's emf and internal resistance.

19 Car batteries are often rated in amp-hours. This rating indicates the quantity of charge that can pass through the battery before the battery's chemical energy must be restored by the car's generator or some other source. The amp-hour unit is convenient for simple calculations. For example, a 60-amp-hour battery can supply a current of 1.0 A for 60 h (or 60 A for 1.0 h) before going dead. Find the energy stored in (a) a 60 A-h, 12 V car battery; (b) a 1.0 A-h, 1.5 V flashlight battery.

20 A 2000 W air conditioner and a 500 W refrigerator each operate on 120 V, rms. Find the current through (a) the air conditioner; (b) the refrigerator.

21 Find the resistance of (a) a 100 W light bulb; (b) a 2000 W heating element in a stove. Power ratings are based on the assumption of a potential difference of 120 V, rms across the terminals.

22 The total power input to all the electrical appliances in a certain household averages 500 W throughout the day. What will the electric bill be for a 30-day period if the rate is 6 cents per kWh?

23 An electrical outlet for an electric stove provides a potential difference of 240 V, rms. Find the amplitude of the potential difference.

24 The instantaneous potential difference V_{ab} across an electric light bulb has a maximum value of +170 V and a minimum value of −170 V. What is the rms voltage?

25 If the instantaneous current through a heating element in an electric heater reaches a maximum value of 10 A to the right at some instant, what is the current through the element $\frac{1}{120}$ s later?

26 The rms voltage across a 5.00 Ω resistor is 35.0 V. Find the maximum instantaneous current through the resistor.

★27 (a) What is the minimum length of time necessary to bring to a boil 1.00 L (1.00×10^3 cm^3) of water, initially at 20.0° C, using as a source of heat the 2000 W heating element on a stove?

(b) What additional minimum time is necessary to boil the water away?

28 Two ordinary 100 W, 120 V light bulbs are connected in series, with 120 V, rms across the combination. Find the total power used by the two bulbs.

*19-4 Electric Current and Ohm's Law on the Microscopic Level

29 Find the average speed of the electrons in a 1.0 cm diameter, copper power line, when it carries a current of 20 A.

30 The peak instantaneous current in a certain lightning stroke is 20,000 A vertically upward. The current is approximately constant at this value for 1.0×10^{-4} s.

(a) Find the charge transferred to the ground during this time interval.

(b) The current at one point in the lightning bolt is produced by electrons moving at a speed of 1.0×10^7 m/s through a circular cross section of radius 1.0 cm. What is the number of electrons per m^3 in the discharge?

(c) Assuming this same electron density throughout, what is the total charge contained in a 1.0 km long lightning bolt that has an average cross-sectional area of 1.0 cm^2?

★31 An automobile starter motor is an electric motor used to start the gasoline engine. To operate the starter motor, the ignition switch is turned on, and a current of 50 A is supplied through a 2.0 mm radius copper wire that connects the motor to the automobile battery.

(a) If electric charge had to travel the 0.50 m distance from the battery to the motor before the motor would operate, how long would it take?

(b) How far do the electrons move in the 2.0 s required to start the engine?

★32 There is a net charge of -5.9×10^5 C on the surface of the earth and a net positive charge spread throughout the earth's atmosphere. This charge distribution produces an electric field that, just above the earth's surface, has an average value of 130 N/C, downward, during clear weather. (During thunderstorms the field is much stronger and oppositely directed.) Ions in the atmosphere move in response to the field, producing an average downward current of 3.5×10^{-12} A per square meter of cross-sectional area.

(a) Find the total ionic current into the earth's surface.

(b) If the ionic current remained constant and there were no other currents at the earth's surface, how long would it take to remove the earth's surface charge?

(c) Find the potential difference between the ground and a point 1.0 km above it, assuming the field remains constant.

(d) Find the resistance of a 1.0 km thick layer of the earth's atmosphere just above the surface.

(e) Find the resistivity of the earth's lower atmosphere.

(f) What solid materials have roughly the same resistivity?

★33 Problem 32 shows that, at the earth's surface, there is a global, downward, ionic current of 1800 A. Lightning provides an equal average upward current and thereby maintains the earth's charge distribution. A lightning discharge at one point lasts about 0.20 s, with an average upward current of 120 A.

(a) Find the average number of simultaneous lightning discharges occurring somewhere on the earth.
(b) If there were equal likelihood of lightning striking anywhere on the earth, what would be the chance of lightning striking at any instant the 0.10 m^2 area within which you stand (Fig. 19-42)?
(c) What would be the chance of lightning striking such an area one time in a 1-year interval?*

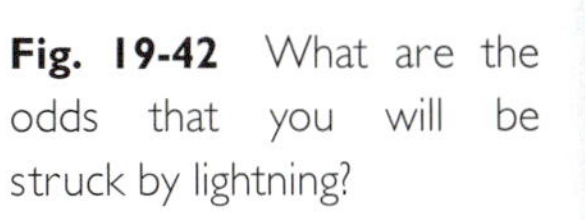

Fig. 19-42 What are the odds that you will be struck by lightning?

Additional Problems

34 A "dead" but rechargeable car battery has an emf of 11.8 V and internal resistance of 1.00 Ω. To start the car, a good battery ($\mathcal{E}$ = 12.0 V, $r \approx 0$) is connected to the dead battery by jumper cables of negligible resistance (Fig. 19-43). The battery terminals are connected + to + and − to −. The weak battery is charged by the strong battery when connected in this way.
(a) How long does it take for the weak battery to receive a 2.00 amp-hour charge? Assume that the emfs and resistances remain constant.
(b) How long would it take to acquire this same 2.00 amp-hour charge if the engine were running and the generator were charging the battery with a current of 4.00 A?

Fig. 19-43

*Each year deaths from lightning in the United States number about 100 (about 1 out of every 2 million Americans). There are 10 times this number of deaths caused by household electrocution and 500 times this number in auto accidents.

★ **35** Fig. 19-44 shows copper and aluminum wires connected together and carrying a current of 5.00 A to the right. The right end of the aluminum is at a potential of zero. Find (a) the resistance of the copper wire; (b) the resistance of the aluminum wire; (c) the potential at the left end of the copper wire. (d) In which wire is the electric field greater?

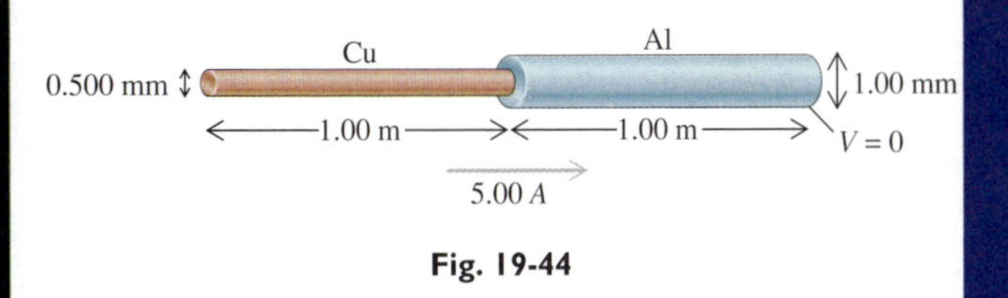

Fig. 19-44

★ **36** When the tungsten filament in a 100 W, 120 V light bulb reaches 2000° C, 100 W of electric power is used to produce light and heat.
(a) Find the current through the filament under these conditions, which are established very shortly after the light is turned on.
(b) Find the filament current and the power used when the light is first turned on, assuming an initial filament temperature of 20° C.
(c) Estimate the time needed to heat the filament to 2000° C. The filament has a mass of 0.040 g and a specific heat of 0.032 cal/g-C°.

★★ **37** The filament in an incandescent light bulb is a very thin, coiled tungsten wire. To radiate 100 W at 2000° C, the surface area of the tungsten must equal 2.54 cm^2 (see Problem 42 in Chapter 13). Find the radius and the length of the wire.

★ **38** An aluminum wire has the same resistance as a copper wire of the same length. Find the ratio of the aluminum wire's diameter to the copper wire's diameter.

★ **39** The electric field equals 1.00×10^6 N/C at the surface of the 20.0 cm radius dome of a Van de Graaf generator, described in Question 19 in Chapter 17. What is the current carried by the belt if the power required to operate the generator is 10.0 W? Assume negligible friction.

40 An electric heater draws an rms current of 10 A when there is a potential difference of 120 V, rms across the heating element. What does it cost to operate the heater constantly for 30 days if the power company charges 6.0 cents per kWh?

41 Find the rms current in a 1.00 hp, 120 V electric motor, assuming that the mechanical power output of 1.00 hp equals the electrical power input.

CHAPTER 20 Direct Current Circuits

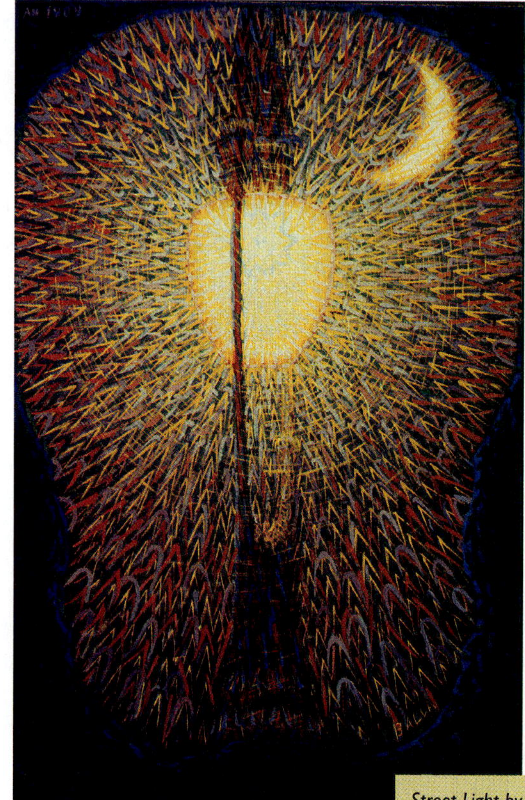

Street Light by Giacomo Balla, 1909. (Museum of Modern Art, New York.)

Fig. 20-1 Replica of the first incandescent lamp, which used sewing thread as its filament.

Near the turn of the century, Thomas Edison built, in New York City, the world's first permanent electric power plant. Edison generated and distributed electricity as direct current (DC). He had also invented the first practical incandescent light bulb. So for those with access to Edison's system, nights began to be much brighter, both indoors and out. Electric power consumption has steadily increased throughout the twentieth century, though power companies now supply electricity as alternating current (AC). In this chapter we shall study mostly DC circuits, though we shall discuss some aspects of household alternating current at the end of the chapter. In Chapter 22, we shall return to a detailed discussion of AC circuits because we can understand such circuits thoroughly only after we have learned about magnetism.

In this chapter we shall also learn how electrical currents within the human body are essential to life and why the body is therefore so susceptible to electric shock. Finally, we shall see how household electricity works and how to avoid electric shock.

20-1 Description of Circuits

DC circuits consist of circuit elements such as resistors and batteries (or other sources of constant emf) connected in such a way as to form one or more closed paths, or **loops.** Fig. 20-2 shows a one-loop DC circuit and a two-loop DC circuit.

When a DC circuit is first completed by the closing of a switch, there are time-dependent, or "transient," effects. The current initially is a function of time. Typically the current soon reaches a steady-state, or constant, value at each point in the circuit. In Section 20-6 we shall consider some transient effects. Here we shall be concerned only with **steady-state current.**

If our circuit consists of a single loop, the steady-state current is the same in all elements of the loop. This follows from the fact that charge does not accumulate at any point in the circuit. (If any charge did accumulate, an electric field would be created; the field would then cause the charge to dissipate.) For example, in Fig. 20-2a the current is the same at points P and R. If this were not true, charge would accumulate in the resistor between these two points. Such accumulation of charge does not occur.

In any multiloop circuit there are points at which three or more conductors are joined. These points are called **branch points.** In Fig. 20-2b, points C and D are branch points. The current divides at these points, and there can be different currents through the branches of the circuit connecting the points.

An **open circuit** is an incomplete loop—a path that is not closed. Examples of open circuits are shown in Fig. 20-3. In each open circuit there is no direct conducting path between points a and b. In other words, the resistance between these points is infinite. Thus the current between the points must be zero, and therefore the steady-state current throughout the circuit is zero.

A **short circuit** is an ideal conducting path connected across a circuit element or across a combination of elements (Fig. 20-4). When a resistor is short-circuited, it is effectively eliminated from the circuit; all the current that would otherwise pass through the resistor goes through the ideal conducting path (the "short") instead. Short circuits often occur unintentionally when wires accidentally touch (Fig. 20-5). Such shorts can cause a dangerously large increase in the current through the circuit.

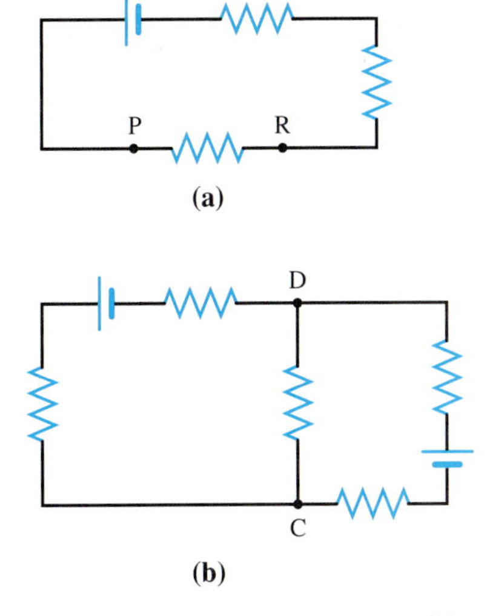

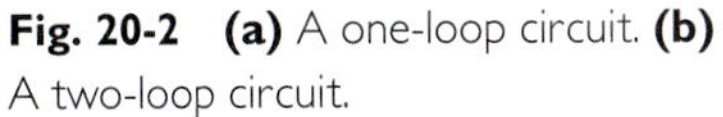

Fig. 20-2 **(a)** A one-loop circuit. **(b)** A two-loop circuit.

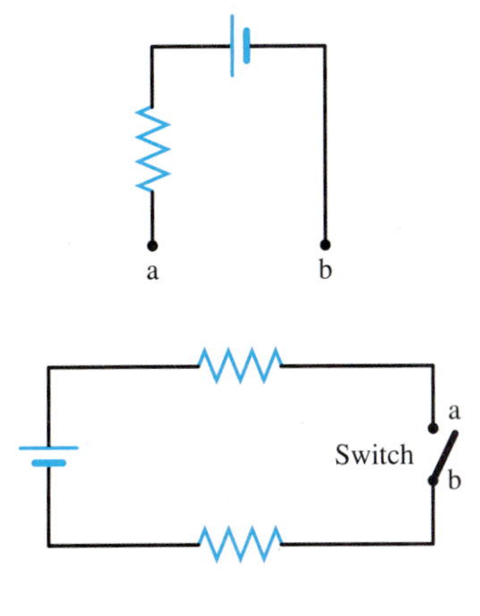

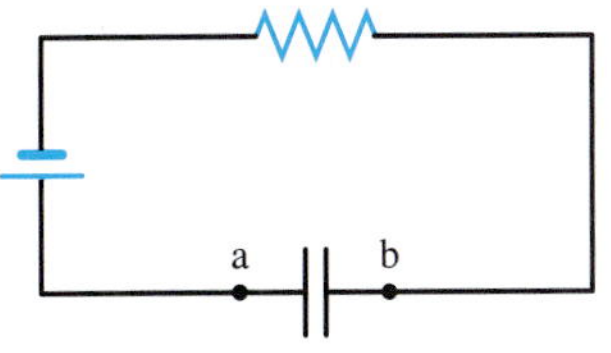

Fig. 20-3 Open circuits. The steady-state current is zero.

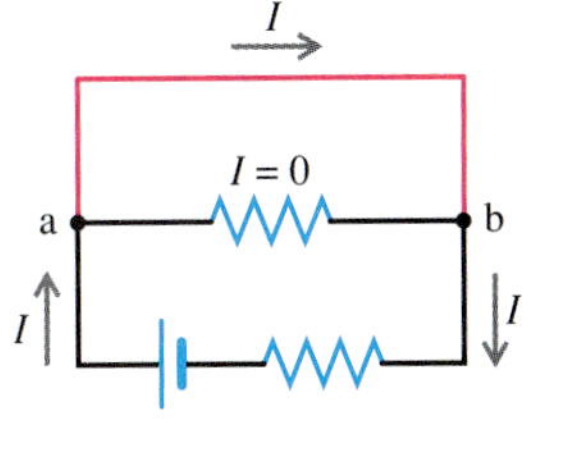

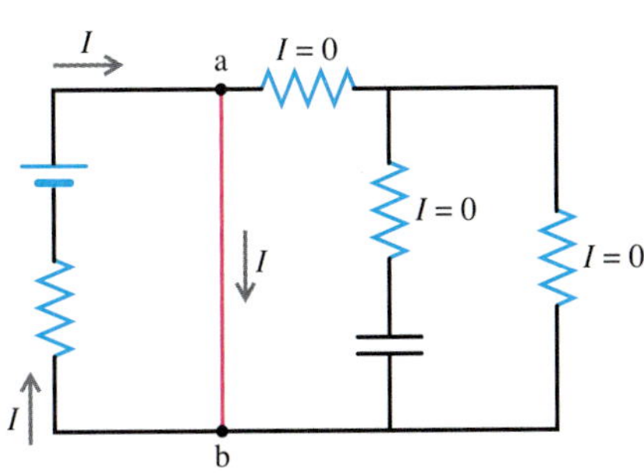

Fig. 20-4 The red wire in each of these circuits creates a short circuit between points a and b.

(a)

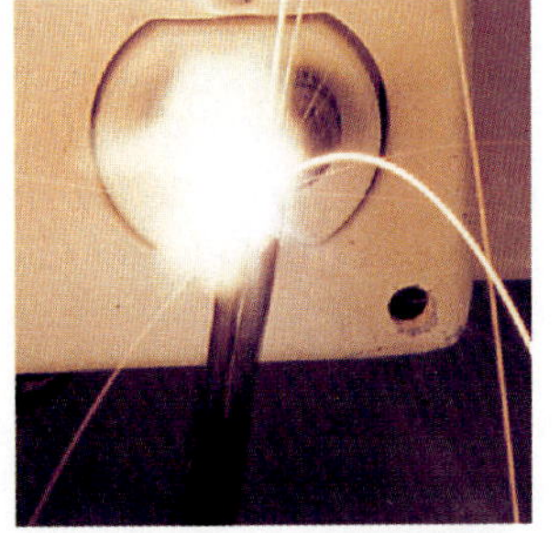

(b)

Fig. 20-5 **(a)** The insulation has worn away between these wires. **(b)** The wires touch when plugged into a power supply, creating a short circuit directly across the power source and therefore a large current.

20-2 Kirchhoff's Rules

Often it is important to determine the current through a circuit element or the potential difference across it. Such circuit problems are solved in general by application of Kirchhoff's rules. These two rules are simply applications of the principles of charge conservation and energy conservation. They may be stated as follows:

Kirchhoff's first rule

The sum of the currents entering any branch point is zero:

$$\Sigma I = 0 \tag{20-1}$$

Kirchhoff's second rule

The sum of the voltage drops around any closed path is zero:

$$\Sigma V = 0 \tag{20-2}$$

According to Kirchhoff's first rule, the positive and negative currents into a branch point sum to zero. Since a negative current inward is the same as a positive current outward, the first rule means that positive inward current is balanced by equal positive outward current. In other words, equal quantities of charge flow into and away from a branch point. This follows from the fact that charge is conserved and does not accumulate at the branch point.

EXAMPLE 1 Balancing Currents at a Branch Point

Find the current I_3 into branch point P in Fig. 20-6.

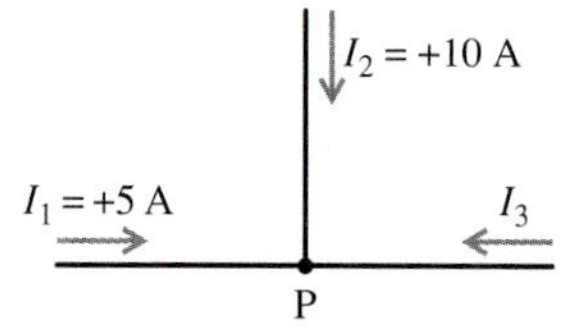

$I_2 = +10$ A

$I_1 = +5$ A $\quad I_3$

P

Fig. 20-6

SOLUTION We apply Eq. 20-1 and solve for the unknown current I_3:

$$\Sigma I = 0$$

$$I_1 + I_2 + I_3 = 0$$

$$I_3 = -I_1 - I_2 = -5 \text{ A} - 10 \text{ A} = -15 \text{ A}$$

This negative current into the branch point is equivalent to a positive current of 15 A away from the branch point. Thus there is a total inward current of 15 A balanced by an outward current of 15 A.

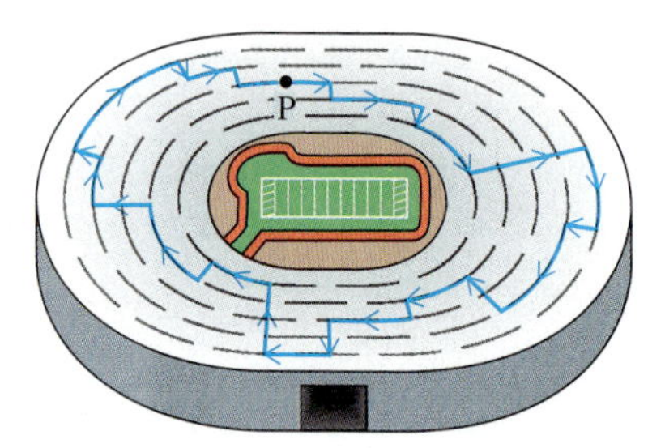

Fig. 20-7 A food vendor takes as many steps up as down in completing a loop around the stadium.

To understand Kirchhoff's second rule, we first note that the potential has a definite value at each point in a circuit. The value changes from point to point, but if you follow a closed path (that is, start at a point and follow a path that ends at the starting point), the final value of the potential must be the same as the initial value. Thus all the positive voltage drops (decreases in potential) must be canceled by negative voltage drops (increases in potential); the sum of the voltage drops must equal zero.

By way of analogy, consider the path of a food vendor in a football stadium. Suppose her path is a complete loop around the stadium, so that she ends at the same point P at which she began (Fig. 20-7). Obviously there has been no change in her elevation once she returns to P, and therefore the number of downward steps must have been exactly equal to the number of upward steps.

In the lithograph by M.C. Escher shown in Fig. 20-8, the steps seem to be going constantly upward all the way around a clockwise loop. This is obviously not possible; it is an illusion created by the artist. Similarly, it is not possible to have all positive (or all negative) voltage drops around a circuit. The sum of all voltage drops must equal zero.

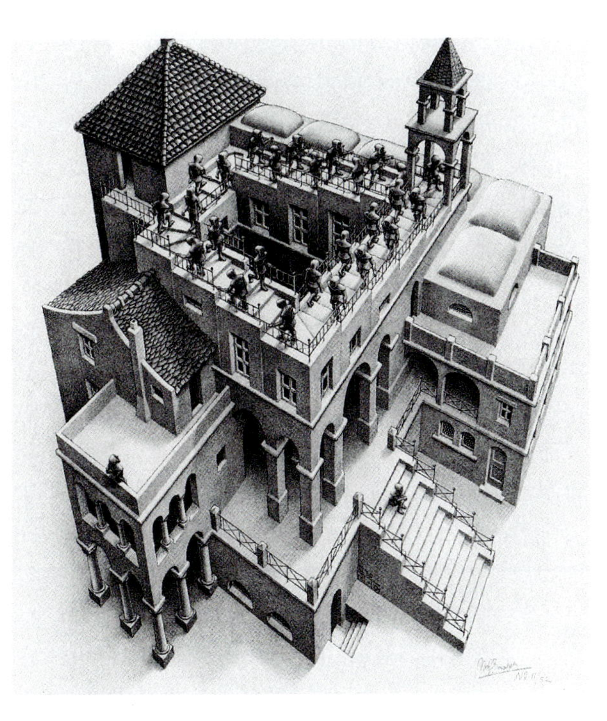

Fig. 20-8 In this lithograph by M.C. Escher the steps seem to be going up all the way around in the clockwise direction.

EXAMPLE 2 Verifying Kirchhoff's Second Rule For a Circuit

For the circuit in Fig. 20-9, verify that the sum of the voltage drops is zero for a counterclockwise loop starting at point a.

SOLUTION We calculate a voltage drop across a resistor as $\pm IR$ (Eq. 19-14), and we calculate the voltage drop across an ideal source of emf as $\pm\mathcal{E}$ (Eq. 19-19). Thus, starting at a and proceeding counterclockwise, we have

$$V_{ab} = +IR = +(2\text{ A})(6\ \Omega) = +12\text{ V}$$

$$V_{bc} = +IR = +(2\text{ A})(3\ \Omega) = +6\text{ V}$$

$$V_{cd} = +\mathcal{E} = +10\text{ V}$$

$$V_{de} = +IR = +(2\text{ A})(1\ \Omega) = +2\text{V}$$

$$V_{ea} = -\mathcal{E} = -30\text{ V}$$

$$\Sigma V = +12\text{ V} + 6\text{ V} + 10\text{ V} + 2\text{ V} - 30\text{ V}$$

$$= 0$$

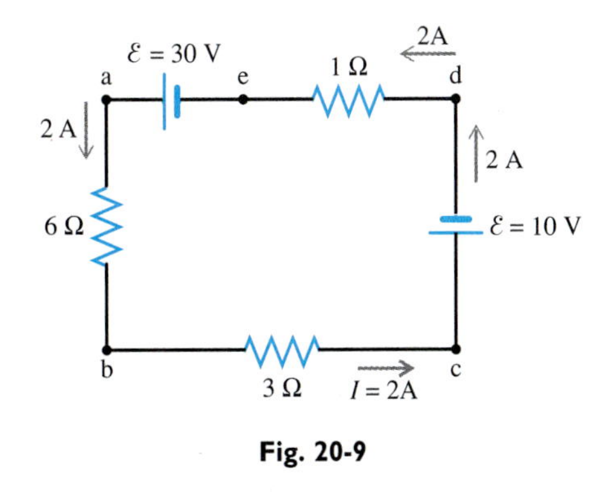

Fig. 20-9

We could just as well have started computing voltage drops at point b or at any other point. The same terms are summed and thus the sum is still zero. Or we could go around the loop clockwise. In that case the voltage drop across each circuit element would be the negative of the value computed above and the sum would again be zero.

Single-Loop Circuits

In a single-loop DC circuit, if the values of resistors and emfs are known, it is a simple matter to find the current in the loop by applying Kirchhoff's second rule.

EXAMPLE 3 A Single Loop Circuit With One Battery

(a) Find the current in the circuit shown in Fig. 20-10. (b) Find the voltage drop across the 7.0 Ω resistor.

SOLUTION **(a)** Positive current flows away from the positive battery terminal, that is, counterclockwise around the circuit. Starting at point a and proceeding counterclockwise (in other words, moving with the current), we set the sum of the voltage drops equal to zero:

$$\Sigma V = 0$$

$$+I(7.0\ \Omega) + I(4.0\ \Omega) + I(1.0\ \Omega) - 6.0\ \text{V} = 0$$

$$I = \frac{6.0\ \text{V}}{12.0\ \Omega} = 0.50\ \text{A}$$

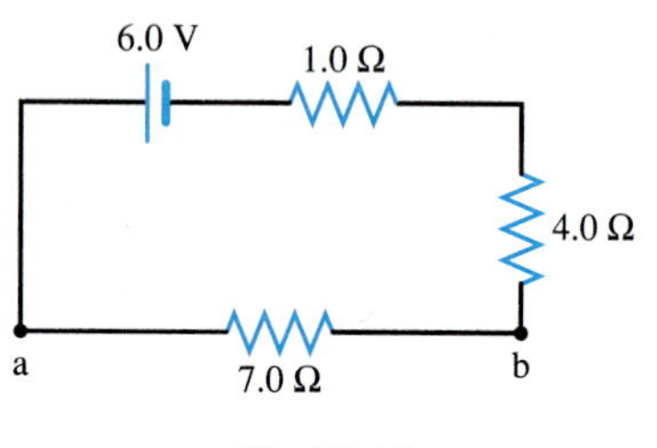

Fig. 20-10

(b) Now we can compute the voltage drop across the 7.0 Ω resistor:

$$V_{ab} = +IR = (0.50\ \text{A})(7.0\ \Omega) = 3.5\ \text{V}$$

EXAMPLE 4 A Single Loop Circuit With Two Batteries

(a) Find the current in the circuit shown in Fig. 20-11.
(b) Find the voltage drop V_{ad}.

SOLUTION **(a)** We first assume a direction for the current. If we obtain a negative value for I, this means that the current is positive in the opposite direction. Here we assume a clockwise current, as indicated in the figure. Starting at point g and proceeding clockwise around the circuit, we sum the voltage drops and solve for I:

$$\Sigma V = 0$$

$$+12.0\ \text{V} + I(0.5\ \Omega) + I(2.0\ \Omega) + I(0.5\ \Omega) - 4.0\ \text{V} + I(3.0\ \Omega) + I(4.0\ \Omega) = 0$$

$$8.0\ \text{V} + I(10.0\ \Omega) = 0$$

$$I = -\frac{8.0\ \text{V}}{10.0\ \Omega} = -0.80\ \text{A}$$

The negative value of I means that current is positive in the opposite direction—counterclockwise. We could have guessed this by observing that the 12 V battery tends to produce positive counterclockwise current and the 4 V battery tends to produce positive clockwise current. The 12 V battery is larger and therefore determines the direction of the positive current.

Positive current flows through the 12 V battery from the negative to the positive terminal, and so electrical potential energy is being supplied by the 12 V battery; in other words, this battery is discharging. Positive current through the 4 V battery goes from the positive to the negative terminal, and so electrical potential energy is being used by the 4 V battery; in other words, the 4 V battery is being charged.

Fig. 20-11

(b) We use the path abd to calculate V_{ad}. Remember positive current is counterclockwise. So, in following this path, we are moving with the current:

$$V_{ad} = +(0.80\ \text{A})(3.0\ \Omega) + 4.0\ \text{V} + (0.80\ \text{A})(0.5\ \Omega) = 6.8\ \text{V}$$

We can also obtain this result by computing voltage drops along the upper path connecting a and d, in other words, by using path agfed.

20-3 Equivalent Resistance

Many circuits contain segments consisting of a combination, or "network," of resistors (Fig. 20-12a). When such a network is connected between points a and b as part of a complete circuit containing a source of emf, there will be a voltage drop V_{ab} across the network and a current I through it, as indicated in the figure. We define the **equivalent resistance** of the network as **the resistance of the single resistor that would produce, in the remainder of the circuit, the same effect as that produced by the network.** That is, the equivalent resistance R_{eq} would carry the same current I for the same potential difference V_{ab} between points a and b (Fig. 20-12b):

$$V_{ab} = IR_{eq}$$

or

$$R_{eq} = \frac{V_{ab}}{I} \qquad (20\text{-}3)$$

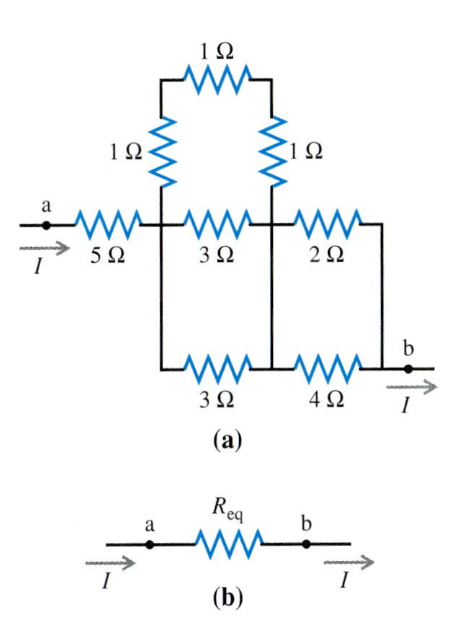

Fig. 20-12 **(a)** A network of resistors. **(b)** The equivalent resistance of the network shown in (a).

Series Resistors

The equivalent resistance is particularly easy to compute when resistors are connected in series. Resistors are said to be in **series** when they are **connected end to end, with no branch points in between.** Thus for resistors connected in series, **the current is the same in all the resistors.** Fig. 20-13 shows three resistors connected in series; these three resistors have a common current I from a to b and a voltage drop V_{ab} across them. This voltage drop may be expressed as the sum of the voltage drops across the individual resistors, V_1, V_2, and V_3:

$$V_{ab} = V_1 + V_2 + V_3$$

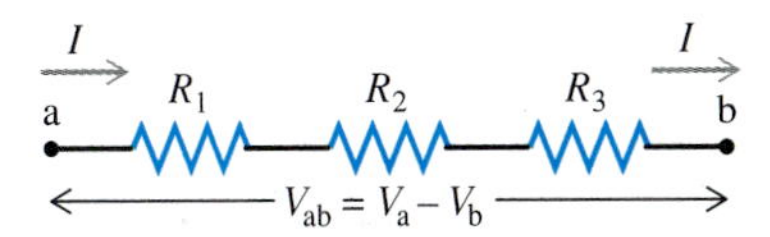

Fig. 20-13 Three resistors in series.

Now Ohm's law may be used to express the voltage drop across each resistor in terms of its resistance and the current I:

$$V_1 = IR_1$$
$$V_2 = IR_2$$
$$V_3 = IR_3$$

Substituting into the equation for V_{ab}, we obtain

$$V_{ab} = IR_1 + IR_2 + IR_3$$

Applying Eq. 20-3 ($R_{eq} = V_{ab}/I$), we divide the preceding equation by I to obtain an expression for the equivalent resistance of three resistors connected in series:

$$R_{eq} = R_1 + R_2 + R_3 \qquad \text{(series)} \qquad (20\text{-}4)$$

Thus to find the equivalent resistance we simply **add the resistances of the series resistors.** This same rule applies to any number of resistors connected in series. For example, if a 5 Ω resistor is connected in series with a 10 Ω resistor, the equivalent resistance is 15 Ω. The equivalent resistance of the four series resistors shown in Fig. 20-14 is 20 Ω.

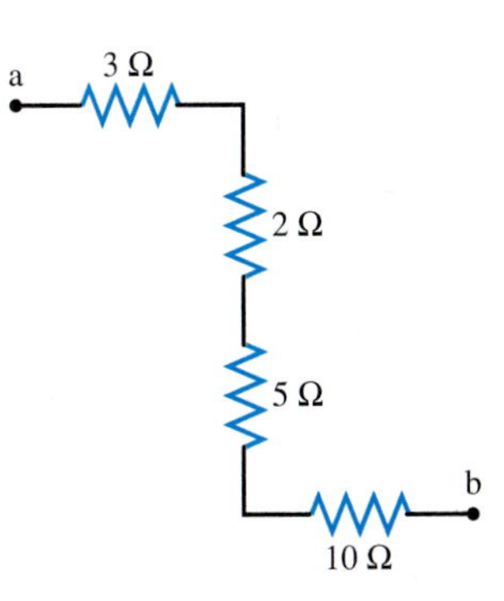

Fig. 20-14 Series resistors.

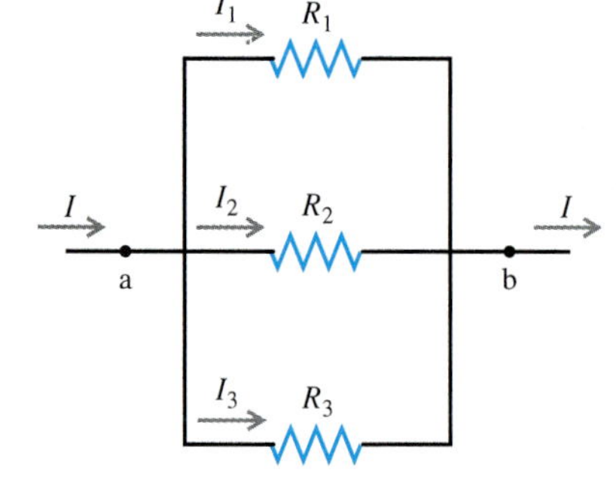

Fig. 20-15 Three resistors in parallel.

Parallel Resistors

Resistors are said to be in **parallel** when **both ends of each resistor are connected to the ends of the other resistors by ideal conducting paths.** Since the connected ends must be at the same potential, **the potential difference across each of the parallel resistors is the same.** Fig. 20-15 shows three resistors connected in parallel. The left ends of all three resistors are connected and the right ends of all three are connected; thus there is a common potential drop V_{ab} across each resistor. The total current I entering the network branches into three currents: I_1 through R_1, I_2 through R_2, and I_3 through R_3. Applying Kirchhoff's first rule either at branch point a or at branch point b, we find that the current I is the sum of the currents through the three branches:

$$I = I_1 + I_2 + I_3 \tag{20-5}$$

It follows from the definition of equivalent resistance (Eq. 20-3: $R_{eq} = V_{ab}/I$) that I can be expressed

$$I = \frac{V_{ab}}{R_{eq}} \tag{20-6}$$

And Ohm's law can be applied to each resistor to express the current through it in terms of its resistance and V_{ab}:

$$I_1 = \frac{V_{ab}}{R_1} \qquad I_2 = \frac{V_{ab}}{R_2} \qquad I_3 = \frac{V_{ab}}{R_3} \tag{20-7}$$

Substituting Eqs. 20-6 and 20-7 into Eq. 20-5, we obtain

$$\frac{V_{ab}}{R_{eq}} = \frac{V_{ab}}{R_1} + \frac{V_{ab}}{R_2} + \frac{V_{ab}}{R_3}$$

or

$$\frac{1}{R_{eq}} = \frac{1}{R_1} + \frac{1}{R_2} + \frac{1}{R_3} \qquad \text{(parallel)} \tag{20-8}$$

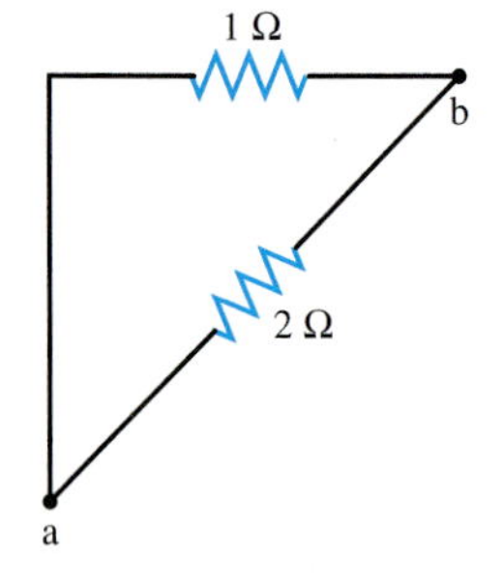

Fig. 20-16 Parallel resistors.

To find the **inverse** of the **equivalent resistance** we **add the inverse resistances of the parallel resistors.** This rule applies to any number of resistors connected in parallel. For the two parallel resistors shown in Fig. 20-16, we find

$$\frac{1}{R_{eq}} = \frac{1}{1\ \Omega} + \frac{1}{2\ \Omega} = \frac{3}{2}\ \Omega^{-1}$$

$$R_{eq} = \frac{2}{3}\ \Omega$$

Notice that this equivalent resistance ($\frac{2}{3}$ Ω) is less than the resistance of either of the two parallel resistors (1 Ω and 2 Ω). This means that, for a given voltage drop from a to b, the two paths provided by the parallel combination of resistors allow more current to flow from a to b than would flow if either resistor were removed. The two paths for current produce less resistance than one path alone. In general, the equivalent resistance of a parallel network is always less than the resistance of any resistor in the network (Problem 16).

Series-Parallel Combinations

We can often reduce a complex network of resistors to a single equivalent resistor by employing the rules for calculating the equivalent resistance of series and parallel resistors.

EXAMPLE 5 Equivalent Resistance of a Network
(a) Find the equivalent resistance of the network in Fig. 20-17.
(b) Find the current through the 4.0 Ω resistor if $V_{ab} = 30$ V.

SOLUTION (a) The 4.0 Ω resistor and the two 8.0 Ω resistors are connected in parallel. We can find their equivalent resistance by applying Eq. 20-8:

$$\frac{1}{R_{eq}} = \frac{1}{R_1} + \frac{1}{R_2} + \frac{1}{R_3} = \frac{1}{4.0\ \Omega} + \frac{1}{8.0\ \Omega} + \frac{1}{8.0\ \Omega} = 0.50\ \Omega^{-1}$$

$$R_{eq} = 2.0\ \Omega$$

Thus we could replace the three parallel resistors by a single 2.0 Ω resistor, as shown in Fig. 20-18a. The network is further simplified when we recognize that the 2.0 Ω and 1.0 Ω resistors are in series. Thus the final equivalent resistance of the network is

$$R_{eq} = 2.0\ \Omega + 1.0\ \Omega = 3.0\ \Omega$$

(b) If 30 V is applied between points a and b in Fig. 20-17, the effect is the same (for points outside a and b) as applying 30 V to the 3.0 Ω equivalent resistor in Fig. 20-18b. The total current is found by applying Ohm's law:

$$I = \frac{V_{ab}}{R_{eq}} = \frac{30\ \text{V}}{3.0\ \Omega} = 10\ \text{A}$$

Thus the current through the network is 10 A. This 10 A current enters the network at a and branches there into three currents through the three parallel resistors (Fig. 20-18c). Applying Kirchhoff's first rule at a, we find

$$I_1 + I_2 + I_3 = 10\ \text{A} \qquad (20\text{-}9)$$

Using the fact that the voltage drop is the same across all the parallel resistors, we can obtain equations relating the respective currents. The voltage drops across the 4.0 Ω and 8.0 Ω resistors are the same. Thus

$$I_1R_1 = I_2R_2$$

$$I_1 = I_2\frac{R_2}{R_1} = I_2\left(\frac{8.0\ \Omega}{4.0\ \Omega}\right)$$

$$I_1 = 2I_2 \qquad (20\text{-}10)$$

The currents I_2 and I_3 through the parallel 8.0 Ω resistors are obviously the same:

$$I_3 = I_2 \qquad (20\text{-}11)$$

Substituting Eqs. 20-10 and 20-11 into Eq. 20-9, we obtain

$$4I_2 = 10\ \text{A}$$

$$I_2 = 2.5\ \text{A}$$

Then Eq. 20-10 gives

$$I_1 = 2I_2 = 5.0\ \text{A}$$

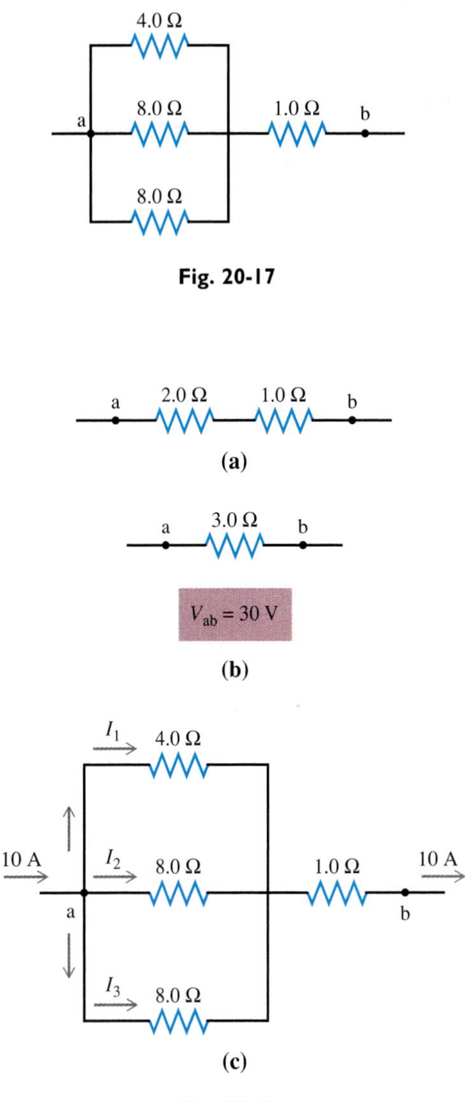

Fig. 20-17

Fig. 20-18

An alternative method for finding the current through any of the parallel resistors is to apply Ohm's law ($I = V/R$), after first determining the voltage across the parallel resistors; this voltage is the same as the voltage across the 2.0 Ω equivalent resistor in Fig. 20-18a: $V = IR = (10\text{A})(2.0\ \Omega) = 20$ V. Thus the current through the 4.0 Ω resistor is $I_1 = V/R = 20\ \text{V}/4.0\ \Omega = 5.0$A.

EXAMPLE 6 Simplifying a Circuit by Means of Equivalent Resistance

Find the current through the 12 V battery in Fig. 20-19a.

SOLUTION Although this circuit appears to be a multiloop circuit, it can be reduced to a single loop when we apply the rules for calculating equivalent resistance. This is done in a series of steps, and the results are indicated in the progressively simpler circuit diagrams of Fig. 20-19b and 20-19c. The final figure (Fig. 20-19c) is a single-loop circuit, with a 12 V emf tending to produce positive clockwise current and a 6 V emf tending to produce positive counterclockwise current. Thus in Fig. 20-19c we indicate the current direction as clockwise so that the current will be positive. We apply Kirchhoff's second rule and solve for the current I:

$$\Sigma V = 0$$

$$+I(6\ \Omega) + 6\ \text{V} + I(4\ \Omega) + I(8\ \Omega) - 12\ \text{V} = 0$$

$$I(18\ \Omega) = 6\ \text{V}$$

$$I = \tfrac{1}{3}\ \text{A}$$

A shortcut for finding the current in a single loop circuit like the one in Fig. 20-19c is to divide the net emf by the total resistance in the loop:

$$I = \frac{\mathcal{E}\ \text{net}}{\Sigma R}$$

Taking clockwise emf as positive and counterclockwise emf as negative, we find the clockwise current:

$$I = \frac{12\ \text{V} - 6\ \text{V}}{18\ \Omega} = \tfrac{1}{3}\ \text{A}$$

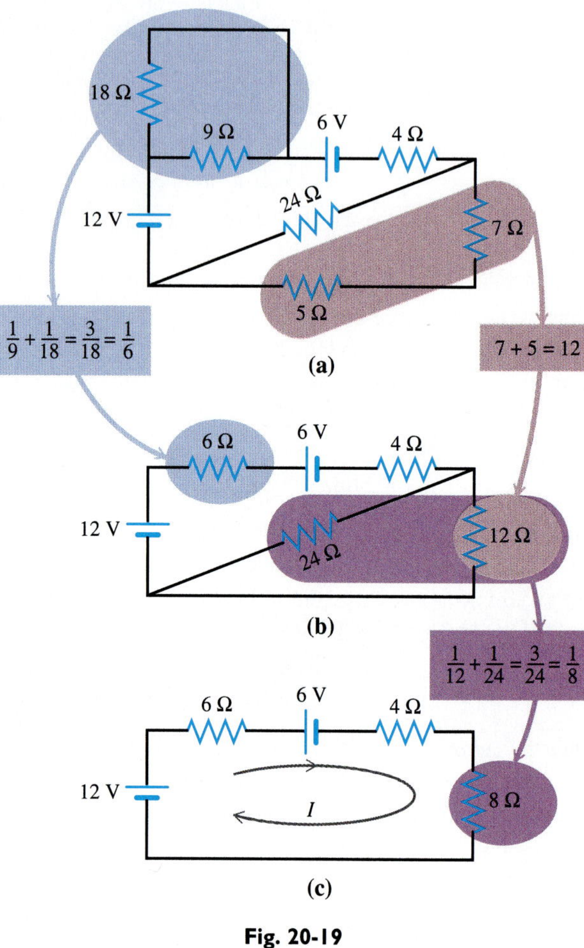

Fig. 20-19

*20-4 Multiloop Circuits

It is not always possible to reduce a circuit to a single loop. Fig. 20-20 shows two circuits that cannot be reduced. If the emf's and resistances are known, the currents can be found by the following steps:

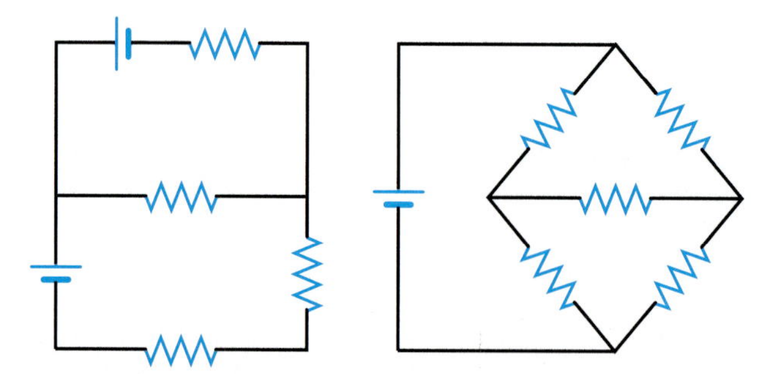

Fig. 20-20 Multiloop circuits that cannot be reduced to single-loop circuits.

1 Assume a direction for each current, applying Kirchhoff's first rule so as to minimize the number of unknowns. For example, in Fig. 20-21a we have labeled currents I_1 and I_2. According to Kirchhoff's first rule, the remaining current is $I_1 + I_2$ in the direction indicated. It is often not possible to correctly guess the direction of positive current, but this is not important. If positive current is actually opposite the assumed direction, the solution will give a negative sign. For example, if the currents in Fig. 20-21a are $I_1 = -5$ A, $I_2 = 2$ A, then the positive currents are as indicated in Fig. 20-21b.

2 Apply Kirchhoff's second rule around as many loops as you have unknown currents. Doing so will give a number of independent linear equations involving an equal number of unknown currents.

3 Solve for the currents.

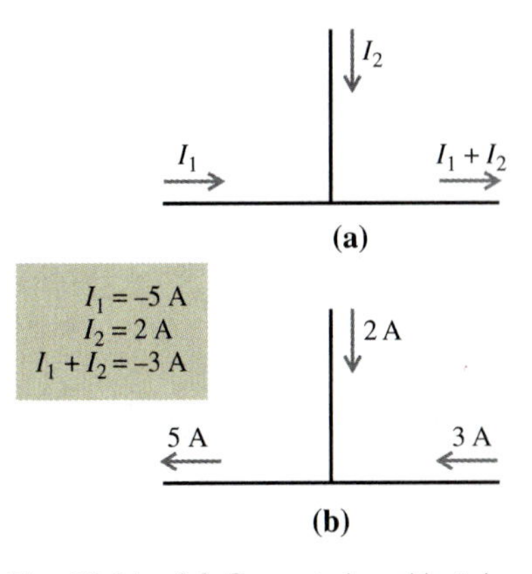

Fig. 20-21 **(a)** Currents I_1 and $I_1 + I_2$ are negative. **(b)** Equivalent positive currents.

EXAMPLE 7 Currents and Voltages in a Multiloop Circuit

(a) Find the currents in Fig. 20-22a. (b) Find the voltage across the terminals of the 15 V battery, V_{ba}. (The 1.0 Ω resistor is the battery's internal resistance.)

SOLUTION (a) First we label the currents in such a way that Kirchhoff's first rule is satisfied at each branch point. This is indicated in Fig. 20-22b. Next we apply Kirchhoff's second rule to the two loops, starting in the upper left-hand corner of each loop and proceeding clockwise. For loop 1, we have

$$\Sigma V = 0$$

$$-15.0 \text{ V} + I_1(1.0\ \Omega) + I_1(8.0\ \Omega) - I_2(20.0\ \Omega) - I_2(4.0\ \Omega) + 5.0 \text{ V} = 0$$

We may simplify this equation and drop the units, remembering that the current is in amps:

$$9.0I_1 - 24I_2 = 10.0 \qquad (20\text{-}12)$$

For loop 2, we get

$$\Sigma V = 0$$

$$-5.0 + 24I_2 + 18(I_1 + I_2) = 0$$

or

$$18I_1 + 42I_2 = 5.0 \qquad (20\text{-}13)$$

Finally, we solve Eqs. 20-12 and 20-13 for I_1 and I_2. Multiplying Eq. 20-12 by 2 and subtracting from Eq. 20-13, we obtain

$$90I_2 = -15.0$$

$$I_2 = -0.17 \text{ A}$$

Returning to Eq. 20-12, we find

$$9.0I_1 - 24(-0.17) = 10.0$$

$$9.0I_1 = 6.0$$

$$I_1 = 0.67 \text{ A}$$

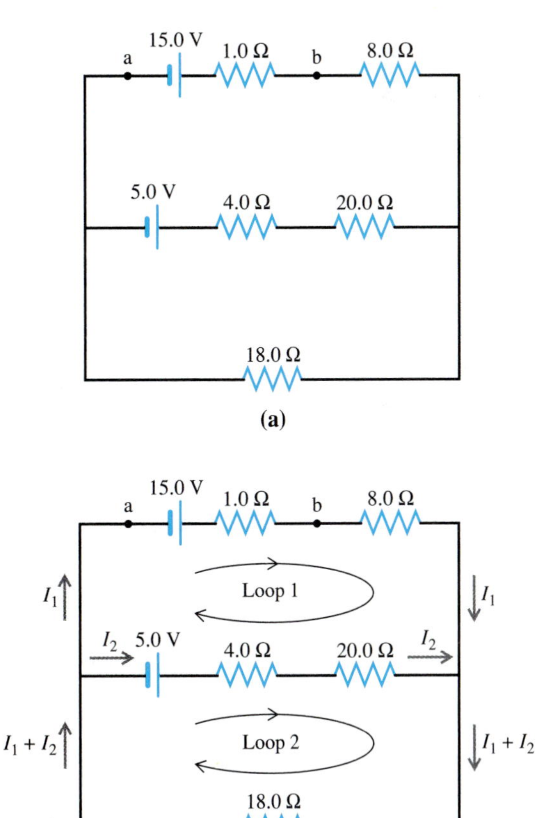

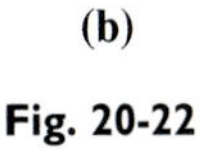

Fig. 20-22

(b) Now we can calculate the voltage drop across the 15 V battery:

$$V_{ba} = -(0.67 \text{ A})(1.0\ \Omega) + 15.0 \text{ V}$$

$$= 14.3 \text{ V}$$

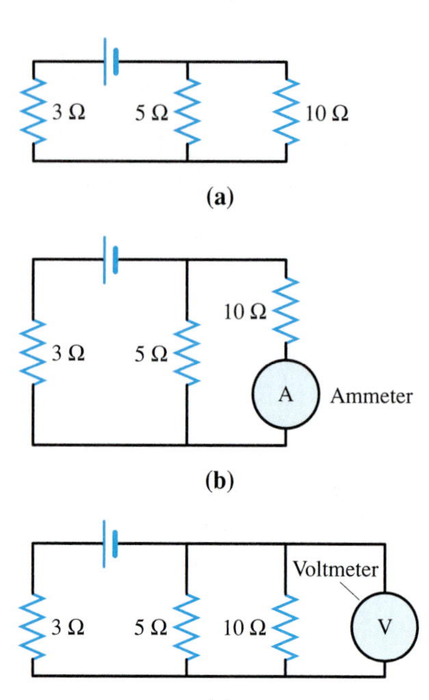

Fig. 20-23 **(a)** Original circuit. **(b)** Circuit as modified by insertion of an ammeter to measure the current through the 10 Ω resistor. **(c)** Circuit as modified by connection of a voltmeter to measure the potential difference across the 10 Ω resistor.

20-5 Measurement of Current, Potential Difference, and Resistance

Ammeter

An **ammeter** is a meter used to measure electric current. The meter reads the current passing through it. We measure the current at a particular point in a circuit by inserting an ammeter at that point, so that it is **connected in series** with the adjacent circuit elements. Fig. 20-23b shows how an ammeter must be connected to the circuit of Figure 20-23a in order to measure the current through the 10 Ω resistor. Ideally the ammeter should have zero resistance so that the current through the circuit without the ammeter is unchanged when the ammeter is inserted. In practice, ammeters always have some resistance. However, the effect of the ammeter will be negligible if its resistance is much less than the total resistance of other resistors connected in series with it. The resistance of an ammeter is often much less than 1 Ω.

Voltmeter

A **voltmeter** is a meter used to measure potential difference. The reading of the meter is the potential difference, or voltage drop, between its two terminals. We find the potential difference between two points in a circuit by connecting the voltmeter's terminals to these points, so that the voltmeter is **connected in parallel** with other circuit elements connected between the two points. Fig. 20-23c shows the connection of a voltmeter to measure the voltage drop across the 10 Ω resistor.

Ideally, a voltmeter should have infinite resistance so that the equivalent resistance between points of connection is unchanged by the voltmeter. In practice, voltmeters have finite resistance. Indeed, some current must pass through the voltmeter to produce a reading. This means that the original circuit is modified by the addition of a parallel resistor. If the voltmeter's resistance is much greater than the resistance of other circuit elements between the two points, however, only a small amount of current will pass through the voltmeter, and the effect of the voltmeter on the circuit will be negligible.

EXAMPLE 8 Finding An Unknown Resistance by Current and Voltage Measurements

Fig. 20-24 shows a circuit containing a resistor of unknown resistance R. Suppose that you have a voltmeter and an ammeter at your disposal. The ammeter has a resistance of 1 Ω and reads current in milliamperes (10^{-3} A), abbreviated mA. The voltmeter has a resistance of 10^6 Ω. Show how the meters can be used to measure the current through R, the voltage drop across R, and the value of R.

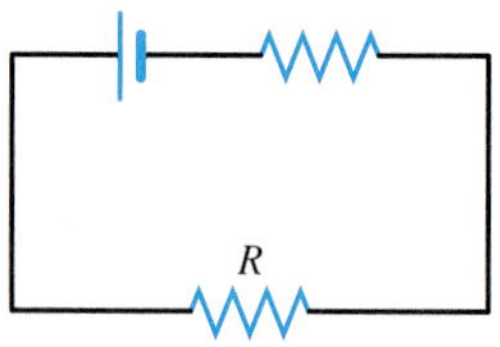

SOLUTION Fig. 20-25 shows the connection of the voltmeter and ammeter. The readings of these meters are respectively the voltage drop V across R and the current I through R. The value of R may then be calculated from Ohm's law. Suppose, for example, the voltmeter reads 11 V and the ammeter reads 50 mA. Then

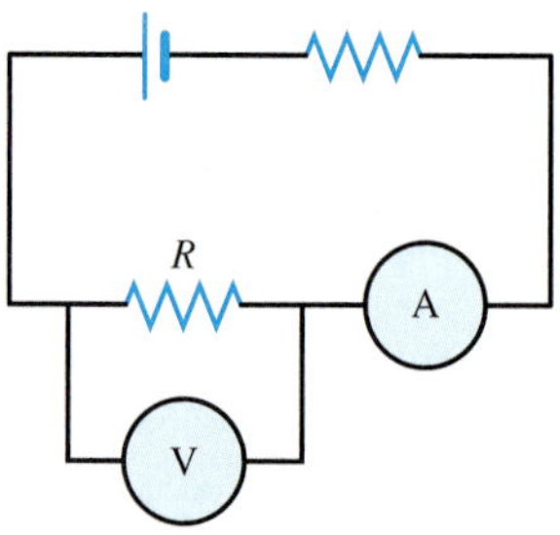

$$R = \frac{V}{I} = \frac{11\text{ V}}{0.050\text{ A}} = 220\ \Omega$$

Fig. 20-24

Fig. 20-25

Since the ammeter's resistance (1 Ω) is much less than 220 Ω and the voltmeter's resistance (10^6 Ω) is much greater than 220 Ω, the values found for R, V, and I should be very nearly the values in the original circuit.

Ohmmeter

Resistance can be measured either indirectly through measurements of voltage and current, as in the preceding example, or directly with an **ohmmeter.** An ohmmeter is connected across the ends of the resistor to be measured (Fig. 20-26). The instrument shown in the photos is a **multimeter,** which can serve as an ammeter, a voltmeter, or an ohmmeter, depending on its dial setting.

Resistance can be measured with great precision using a **Wheatstone bridge,** described in Problem 22. Problems 31 and 32 illustrate how ammeters and voltmeters can be constructed with a galvanometer, a meter that responds to the magnetic force acting on its coil when current passes through it.

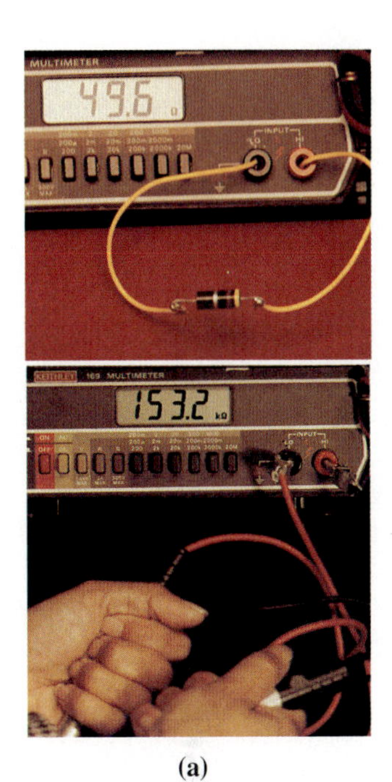

(a)

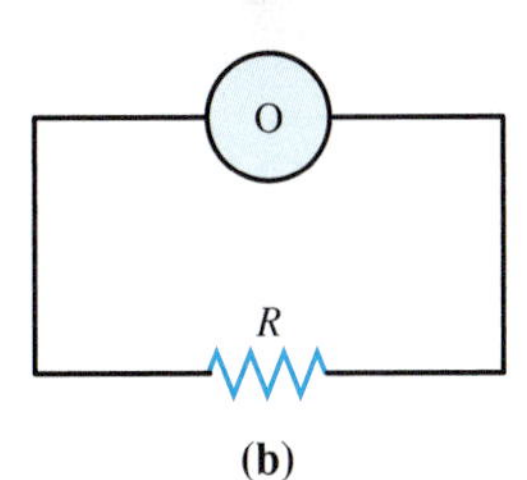

(b)

Fig. 20-26 **(a)** An ohmmeter can be used to measure the resistance of a resistor or of the human body. **(b)** Circuit representation of an ohmmeter used to measure resistance *R*.

20-6 RC Circuits

Charging a Capacitor

Up to now we have considered only steady-state current. In some circuits, however, time-dependent effects are important. A particularly simple example of this is the circuit shown in Fig. 20-27, which consists of a capacitor C, a resistor R, and a source of emf $\mathcal{E}$ connected in series with each other, and a switch S, which completes the loop when it is closed. This is called an "RC circuit."

Suppose that the capacitor is initially uncharged and the switch is open. Then at $t = 0$ let the switch be closed. Obviously no steady-state current can exist in this circuit because the capacitor makes it an open circuit. As soon as the switch is closed, however, there is a temporary, or "transient," current. At any instant this current is the same at all points in the circuit except for the region between the capacitor plates, where there is no current. We find the instantaneous current by applying Kirchhoff's second rule:

$$\Sigma V = 0$$

Starting at a and going around the loop counterclockwise (in the direction of the current I), we have

$$V_{\text{ab}} + V_{\text{bc}} + V_{\text{ca}} = 0$$

$$+IR + \frac{Q}{C} - \mathcal{E} = 0 \tag{20-14}$$

At $t = 0$, there is no charge on the capacitor and hence the voltage drop $\frac{Q}{C} = 0$. Thus the initial current, which we shall denote by I_0, is easily found from the equation above:

$$I_0 R - \mathcal{E} = 0$$

$$I_0 = \frac{\mathcal{E}}{R} \tag{20-15}$$

This expression gives the value of the current at the instant the switch is closed; positive charge flows at this rate counterclockwise through the battery, resistor, and connecting wires. A positive current in this direction means that positive charge Q begins to accumulate on the capacitor's left plate and negative charge $-Q$ begins to accumulate on the right plate.

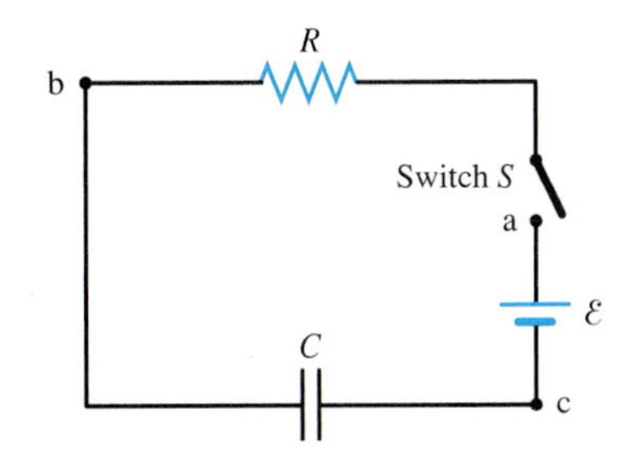

Fig. 20-27 An RC circuit. The switch is closed at $t = 0$.

Eq. 20-14 relates the charge Q on the plates at any instant to the instantaneous current I, which equals the rate of change of Q. This equation can be solved by use of calculus. The result is

$$I = I_0 e^{-t/\tau} \tag{20-16}$$

and

$$Q = C\mathcal{E}(1 - e^{-t/\tau}) \tag{20-17}$$

where e is the base of the natural logarithm ($e \approx 2.718$) and τ is a quantity called the **time constant,** defined as

$$\tau = RC \tag{20-18}$$

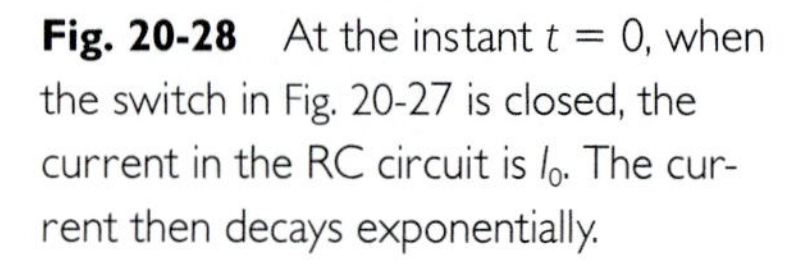

Fig. 20-28 At the instant $t = 0$, when the switch in Fig. 20-27 is closed, the current in the RC circuit is I_0. The current then decays exponentially.

According to Eqs. 20-15 and 20-16, the current starts at the value $I_0 = \dfrac{\mathcal{E}}{R}$ at $t = 0$ and then decreases as t increases because of the $e^{-t/\tau}$ factor. The current is said to "decay exponentially." Such decay is most easily described in terms of the time constant. From Eq. 20-16 we can find the value of I as a fraction of I_0 at times that are multiples of the time constant:

$$t = 0: \qquad I = I_0$$

$$t = \tau: \qquad I = I_0 e^{-1} = 0.37 I_0$$

$$t = 2\tau: \qquad I = I_0 e^{-2} = 0.14 I_0$$

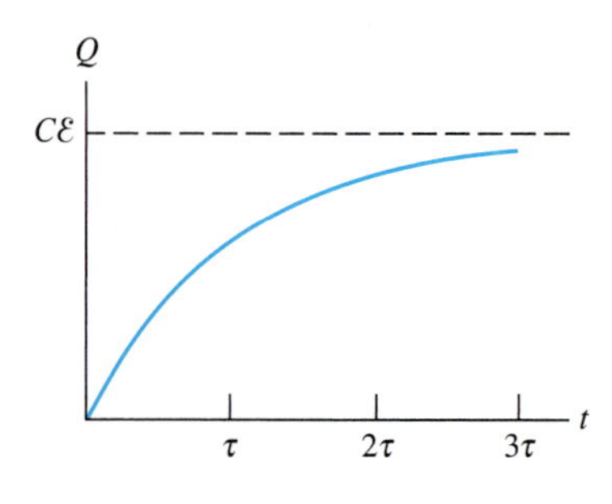

Fig. 20-29 The charge on the capacitor in Fig. 20-27 increases from $Q = 0$ at $t = 0$ to $Q \approx C\mathcal{E}$ for $t >> \tau$.

Although the value of $e^{-t/\tau}$ is never identically zero, after a period of time much greater than τ, the value is nearly zero. For example, at $t = 7\tau$, $I \approx 10^{-3} I_0$, and at $t = 14\tau$, $I \approx 10^{-6} I_0$. Fig. 20-28 shows a graph of I as a function of t.

As the current decreases, the charge on the capacitor increases. Evaluating Eq. 20-17 for Q, we find that, at $t = 0$, $Q = 0$ and, at $t >> \tau$, $e^{-t/\tau} \approx 0$ so that $Q \approx C\mathcal{E}$. After the current through the resistor has effectively stopped, the total charge stored on the capacitor is $C\mathcal{E}$; the full voltage drop $\mathcal{E}$ provided by the battery then appears across the capacitor plates. A graph of Q as a function of t is shown in Fig. 20-29.

Discharging a Capacitor

We can discharge a charged capacitor by connecting a resistor directly across its plates (Fig. 20-30). In this case, application of Kirchhoff's second law and the methods of calculus shows that both the current I through the resistor and the capacitor's charge Q are exponentially decaying functions of time:

$$I = I_0 e^{-t/\tau} \tag{20-19}$$

$$Q = Q_0 e^{-t/\tau} \tag{20-20}$$

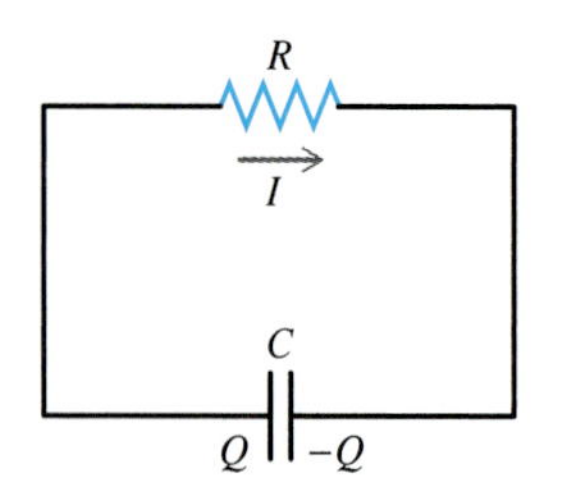

Fig. 20-30 An RC circuit in which an initially charged capacitor discharges through a resistor.

These functions are graphed in Fig. 20-31.

Circuits in which a capacitor is alternately charged and discharged have important practical applications, including pacemakers for the heart and stroboscopic lights. See Problems 38 and 39.

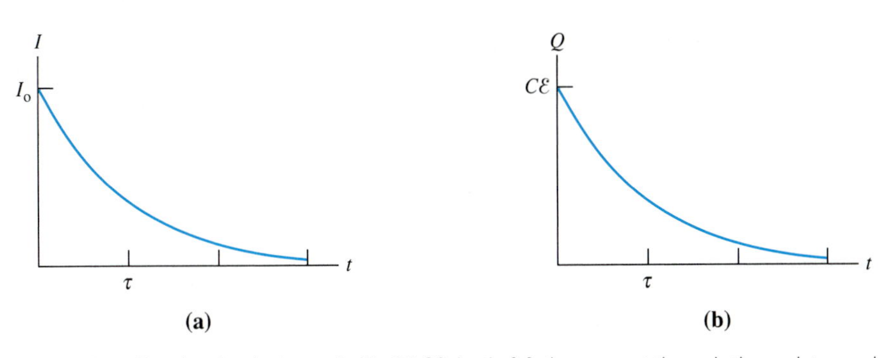

Fig. 20-31 For the circuit shown in Fig. 20-30, both **(a)** the current through the resistor and **(b)** the charge on the capacitor decay exponentially.

EXAMPLE 9 Charging a Capacitor in an RC Circuit

Let the RC circuit in Fig. 20-27 have $R = 50\ \Omega$, $C = 100\ \mu\text{F}$, and $\mathcal{E} = 10$ V. Find the current when the switch is first closed, the final charge stored on the capacitor, and the circuit's time constant.

SOLUTION Applying Eq. 20-15, we find the initial current:

$$I_0 = \frac{\mathcal{E}}{R} = \frac{10\ \text{V}}{50\ \Omega} = 0.20\ \text{A}$$

The final charge on the capacitor is

$$Q = C\mathcal{E} = (100 \times 10^{-6}\ \text{F})(10\ \text{V}) = 1.0 \times 10^{-3}\ \text{C}$$

From Eq. 20-18 we calculate the time constant:

$$\tau = RC = (50\ \Omega)(100 \times 10^{-6}\ \text{F})$$
$$= 5.0 \times 10^{-3}\ \text{s}$$

Since the time constant is so small, the current has all but disappeared in much less than 1 s. An increase in the resistance by a factor of 10^3 (to 50,000 Ω) would increase the time constant by the same factor, to 5.0 s. In this case, we would have to wait much longer for the current effectively to stop or for the 10^{-3} C to be deposited on the capacitor. The longer wait is due to the much smaller initial current; increasing R to 50,000 Ω causes $I_0 = \frac{\mathcal{E}}{R}$ to decrease to $\frac{10\ \text{V}}{50{,}000\ \Omega} = 2.0 \times 10^{-4}$ A.

Electrical Effects in the Human Body

Nerves

Electrical processes are essential to the functioning of the human body. For example, the transmission of information in the body is an electrical phenomenon. This information is in the form of electrical pulses carried by nerves. Some nerves transmit sensory information from the skin to the brain or to the spinal column. Other nerves transmit signals within the brain. And still other nerves transmit signals from the brain or spinal column to muscles, causing them to contract. A nerve consists of a bundle of nerve cells; a nerve cell is called a "neuron." Each neuron has an input end, a long, thin body called the "axon," and an output end (Fig. 20-A). The input end either converts a stimulus such as heat or pressure to electrical form or receives an electrical stimulus from the output end of another neuron. The axon transmits the electrical signal from the input end to the output end. The output end either stimulates a muscle cell or transmits an electrical signal across a gap (called a "synapse") to another neuron.

The length of axons varies considerably, depending on their function. An axon extending from the spinal column to the foot may be over 1 meter long. The diameter of axons in the human body varies somewhat, but a typical value is 0.01 mm. (Squid have axons up to 0.5 mm in diameter; their much larger diameters makes these axons easier to use in experimental work on nerve conduction, and so they are commonly used for this purpose.)

In the undisturbed state, the concentrations of potassium, sodium, and chlorine ions inside and outside the axon produce a static potential difference across the cell membrane—with the inside of the cell at a lower potential than the outside. The potential difference is approximately 70 mV.

When a sufficiently strong stimulus is produced at the receptor end of a neuron, the neuron responds by transmitting a signal along the axon to the output end. For example, heat at the receptor end of a sensory neuron produces a change in the electrical state of that end, increasing the potential inside the cell. If the potential increases by as much as 20 mV, there is a sudden surge of ion conduction across the cell membrane at that point, and the potential inside the cell rapidly increases further until it is about 40 mV higher than the potential outside. Thus the total change in potential is 70 mV + 40 mV = 110 mV, or about 0.1 V. Then the potential at that point rapidly decreases to its original value (Fig. 20-B). This change in potential, which originates at the receptor end of the neuron, propagates along the axon as a wave pulse at a speed that depends on the type of axon. A typical speed is 50 m/s. The duration of the pulse at any point on the axon is about 2×10^{-3} s.

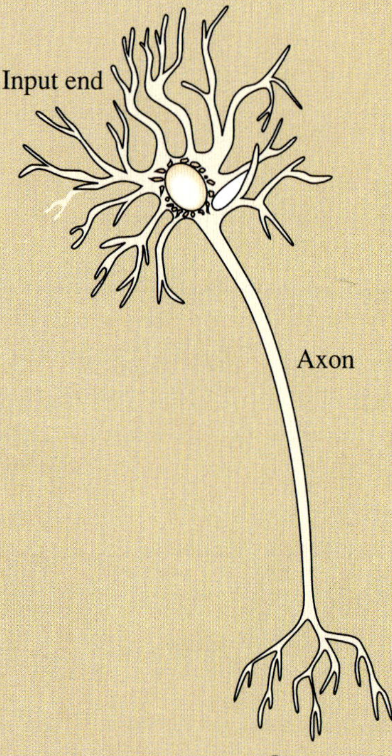

Fig. 20-A A neuron.

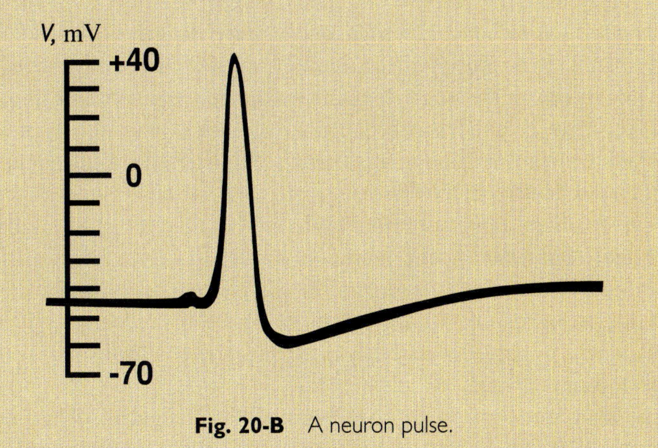

Fig. 20-B A neuron pulse.

The Heart

The beating of the human heart also has important electrical aspects. The contraction of the muscle cells in the heart is spontaneous and coordinated. The spontaneity of the heartbeat is demonstrated by the fact that animal hearts that have been removed and placed in a suitable solution continue to beat. The coordinated contraction of the heart is accomplished by electrical pulses that travel between the cells. These voltage pulses are similar in magnitude to the pulses transmitted by nerve cells—on the order of 0.1 V.

Pacemaker

The rate at which the heart beats is of course variable and is regulated by a set of nerves called the "pacemaker nerves." These nerves stimulate contraction of heart cells at one point, and then a wave of contraction passes through the heart. Failure of the pacemaker nerves to function properly is a very common heart problem. In such cases, an artificial pacemaker in the form of a tiny electric circuit may be implanted in the body. The artificial pacemaker serves the same function as the body's own pacemaker nerves—to initiate and regulate the beating of the heart by electrical stimulation.

Electrocardiogram

During the cycle of a heartbeat, variations in the electric potential within the heart result in small but measurable variations in the potential difference between various points on the skin. The potential differences there are on the order of 10 μV to 100 μV. If metallic leads are connected to the skin at various points, they will have small, variable potential differences between them. The potential differences may then be amplified and displayed on a monitor or recorded on a graph. As noted in Chapter 18, the graphical record of the variation in potential is called an "electrocardiogram" (ECG). Normally several measurements are made using different contact points on the skin. These graphs are related to the heartbeat and can be read by a skilled diagnostician for characteristic signs of heart disease.

Also mentioned in Chapter 18, an electroencephalogram (EEG) is a graph of variations in potential at points on the head. The shape of the EEG is related to states of awareness and may also indicate brain disorders.

20-7 Electric Shock and Household Electricity

Electric Shock

Because electrical processes are essential to the human body, the body is very sensitive to externally imposed electric currents. You can feel a current of only 1 mA through any part of your body. Electrical currents are dangerous principally because they interfere with the normal functioning of either the heart or the nerves governing breathing. **The effect on the body depends on the current's magnitude, duration, and path.** The path is important because it determines how much of the current passes through the heart or through vital nerves. For example, a current between two fingers of the same hand is much less dangerous than the same current from one hand to the other or from any limb to another limb; little of the current through the fingers reaches the heart, but a significant fraction of a hand-to-hand current passes directly through the heart. Thus, when you are working with potentially dangerous electrical equipment, it is wise to keep your feet well insulated from contact with good conductors and, where possible, to use only one hand, thereby protecting yourself from a large limb-to-limb current.

A current from one limb to another begins to be dangerous when it exceeds a few milliamps. More than about **10 to 20 mA will cause uncontrollable muscular contractions** that may prevent the victim from letting go of the current source. Currents of this magnitude will also stop respiration, which is the reason they are fatal if they persist for a few minutes. Current in the approximate range from **100 mA to 1 A will cause a condition known as "ventricular fibrillation,"** in which the contraction of cardiac cells is uncoordinated and as a result the heart is unable to pump blood.* Once they begin, these irregular contractions do not stop spontaneously, even if the current is stopped, and the victim dies in a few minutes. Fibrillation can be stopped by a series of intense bursts of current, each on the order of a few amps and lasting for a few milliseconds. This momentarily stops contraction of all cardiac cells, so that coordinated contractions can then resume.

Currents through the body result when a potential difference exists between two points on the body. Ohm's law predicts that the magnitude of that current depends on two factors—the potential difference V and the resistance R of the path connecting the points: $I = V/R$. The concentration of ions in body fluids makes the interior of the body a relatively good conductor. Skin, on the other hand, is a poor conductor. Thus, when a potential difference is applied between two points on the skin, most of the resistance of the path connecting the two points is through the skin. The magnitude of this resistance depends on whether the skin is wet or dry. For dry skin, the resistance from one hand to the other is usually about 10,000 Ω. If the skin is wet, the hand-to-hand resistance is only about 1000 Ω. This means that with a potential difference across the hands of 10 V the current through the body may be about 1 mA ($10\text{ V}/10^4\ \Omega = 10^{-3}$ A) if the skin is dry or 10 mA ($10\text{ V}/10^3\ \Omega = 10^{-2}$ A) if the skin is wet. Thus **a 10 V potential difference can be quite dangerous.**

Ground

Differences in potential produce current through conducting paths. The earth is a good conductor, and it carries a net negative charge on its surface of about -10^{-9} C/m^2. When an object touches the earth or is connected to the earth by a good conductor, charge will flow between the object and the earth if there is a potential dif-

---

*A current of 100 mA from limb to limb may result in only about 0.1 mA through the heart. Thus internal currents less than 1 mA can be dangerous. For example, if a hospital patient has a catheter inserted near the heart, a very small leakage of current through the catheter can produce a lethal shock.

ference between them. Since only *differences* in potential are important, we may arbitrarily define the potential at some point to be zero. In most applications it is convenient to define the potential of the earth to be zero. Whenever a point is connected to the earth by a conductor of negligible resistance, that point too is at a potential of 0 V and is said to be **grounded.** Grounding is usually accomplished by connecting the point with a thick copper wire to a metal water pipe, which makes good contact with the ground. Grounding in an electric circuit is denoted by the symbol ⏚.

Power Lines

Electric power lines are conducting wires that are maintained at some potential relative to ground. One line, the "ground wire," is maintained at zero potential and is connected to ground at the house. Other, so-called "hot wires" are maintained at a potential that varies with time. The time dependence is harmonic (that is, a sine function) at a frequency of 60 Hz. The amplitude of the potential on a hot wire is 170 V, and the rms voltage, according to Eq. 19-23 ($V_{rms} = \frac{1}{\sqrt{2}}V_0$), is $\frac{1}{\sqrt{2}}(170\text{ V}) =$ 120 V. Most households are supplied with two hot wires. The potential on these lines oscillates 180° out of phase (Fig. 20-32). The potential of one wire relative to the other, $V_1 - V_2$, is a time-varying potential difference of amplitude 2(170 V) = 340 V, which means the rms potential difference is $\frac{1}{\sqrt{2}}(340\text{ V}) = 240$ V. This 240 V potential difference is used to operate appliances requiring large power output. Since $P = IV$, a potential difference of 240 V allows the appliance to draw only half the current of a 120 V connection. Consequently, there is less heating of the connecting wires.

(a)

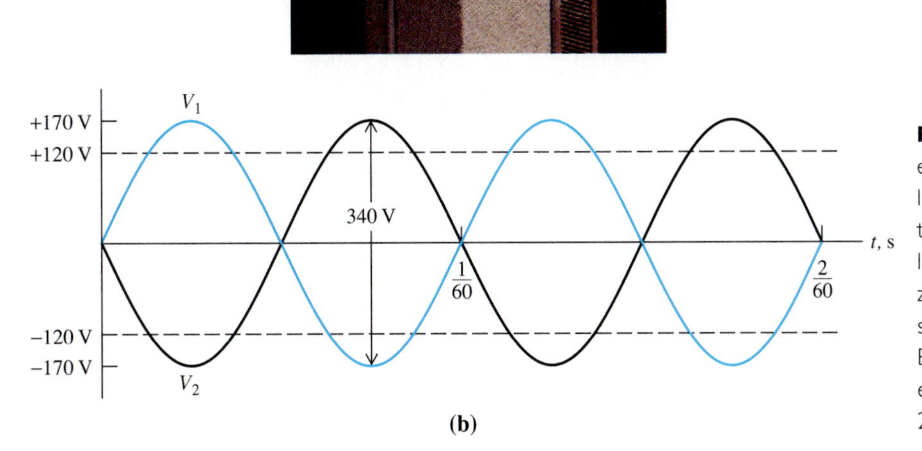

Fig. 20-32 **(a)** Three lines provide electric power to this house. Two of the lines, "hot wires," are maintained at potentials that vary with time. The third line, the "ground wire," is maintained at zero potential. **(b)** V_1 and V_2 are the respective potentials of the two hot wires. Each is 120 V, rms. The potential difference between the wires, $V_1 - V_2$, is 240 V, rms.

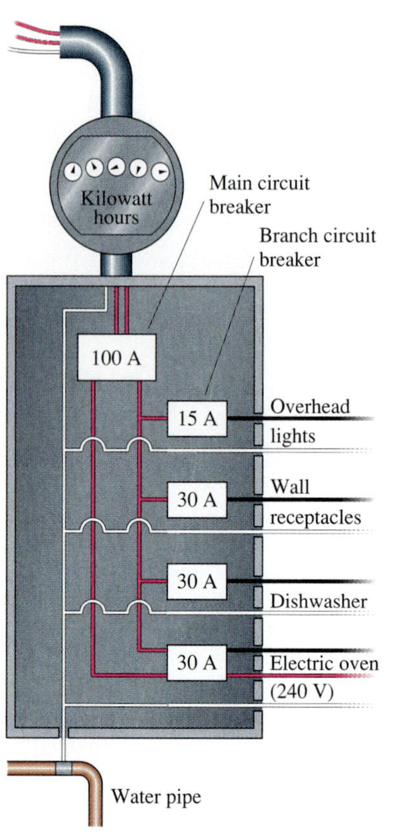

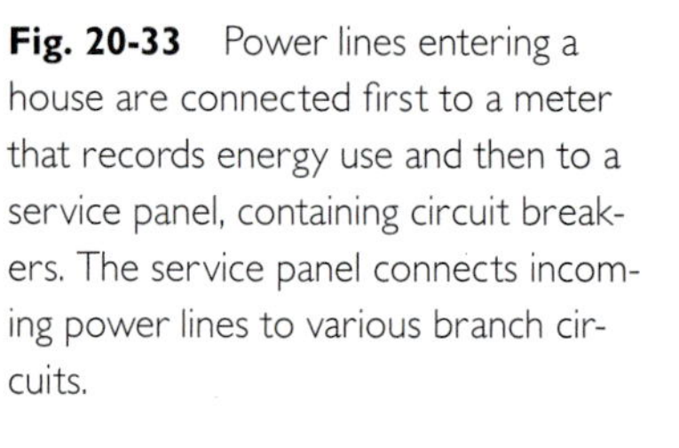
Fig. 20-33 Power lines entering a house are connected first to a meter that records energy use and then to a service panel, containing circuit breakers. The service panel connects incoming power lines to various branch circuits.

Household Circuits

The power lines entering a house are connected first to a meter that records the electrical energy used (Fig. 20-33) and then to a main circuit breaker or, in an older house, to a fuse. The circuit breaker or fuse is used as a safety device that breaks the circuit (in other words, opens it) whenever current exceeds a certain designated amount, usually about 100 A. If it were not for such safety devices, very large currents could be produced, and wiring could overheat and start a fire. For example, if the insulated wires in a frayed electrical cord touch each other, a short circuit and a large instantaneous current are created. The circuit breaker will then quickly open the circuit.

Leading from the main circuit breaker are several branch circuits, each with its own circuit breaker, usually from 15 A to 30 A. These branch circuits are connected in parallel, as indicated in Fig. 20-33. According to standard electrical code, hot wires should be black or red and ground wires should be white.

A branch circuit may lead to a single appliance, such as a dishwasher or electric oven, or it may be connected to a number of other devices connected in parallel with each other. For example, a branch circuit may lead to wall receptacles connected in parallel so that they can be used independently (Fig. 20-34a). Or a branch circuit may lead to overhead lights connected in parallel, each with its own series switch (Fig. 20-34b).

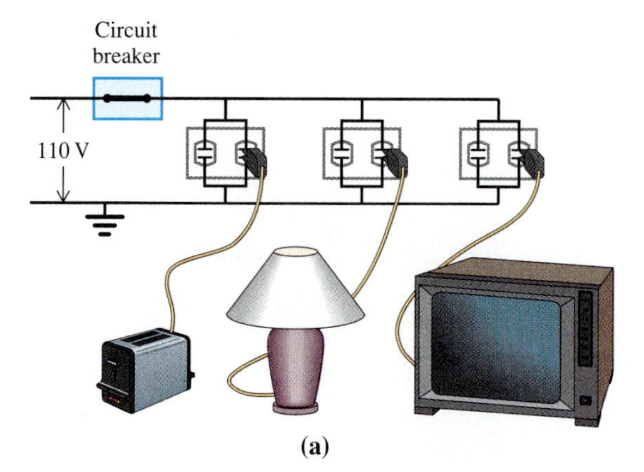

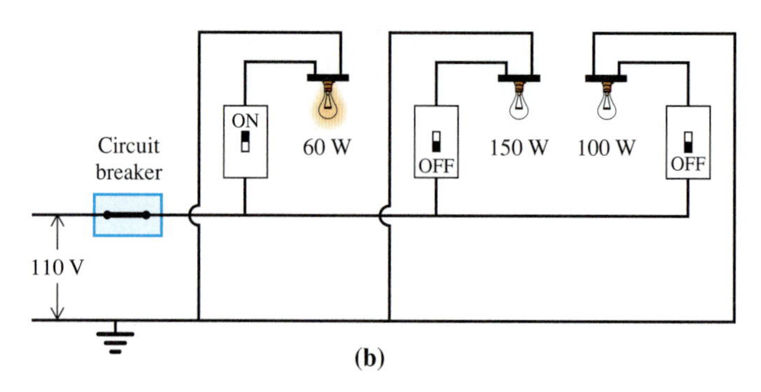

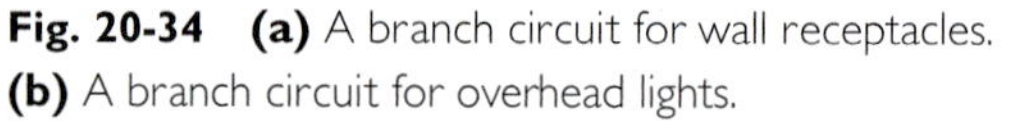
Fig. 20-34 **(a)** A branch circuit for wall receptacles. **(b)** A branch circuit for overhead lights.

Most modern 120 V appliances are equipped with three-prong plugs, designed for an electrical receptacle that is connected to a hot wire, a ground wire, and a "grounding" wire (Fig. 20-35). The potential difference between the hot wire and the ground wire provides the energy to operate the appliance, and it is these two wires that carry current. The grounding wire connects the metal casing of the appliance to ground, thereby protecting the user from electrical shock. Fig. 20-36a shows a hot wire accidentally touching the metal casing of an ungrounded appliance, so that the casing is at a potential of 120 V. A person touching the casing may provide a relatively low resistance path to ground, with the result that a possibly lethal current passes through the body. This current would not normally be sufficient to cause the circuit breaker to open the circuit. Fig. 20-36b shows the difference that a grounded casing can make. In this case, a very low resistance path from P to ground through the grounding wire will draw a current sufficient to activate the circuit breaker.

It is possible that a short to the metal casing will not be sufficient to trip a 20 A or 30 A circuit breaker. This might happen if the path to ground through the casing has too much resistance or if the potential of the casing is only at 10 V or 20 V rather than 120 V. This can be very dangerous because a person touching the casing can still receive a lethal shock. For this reason, grounding wires are more effective when they are connected to a device called a "ground fault interrupter," which opens circuits whenever there is a current of more than a few milliamps through a grounding wire.

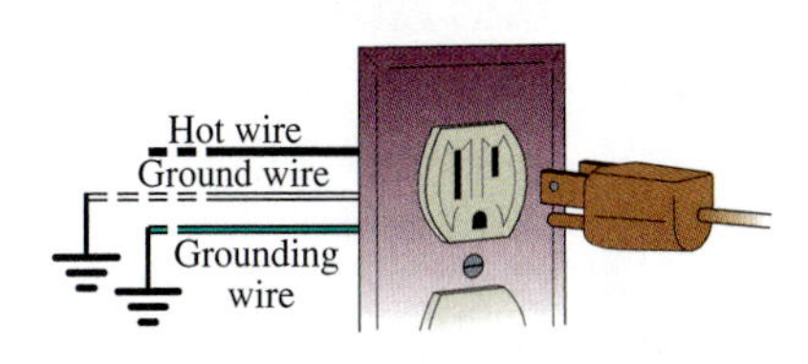

Fig. 20-35 A receptacle for a 120 V, three-pronged plug is connected to two current carrying wires, the hot wire and the ground wire, and to a third wire, the grounding wire, which grounds the device plugged into the receptacle.

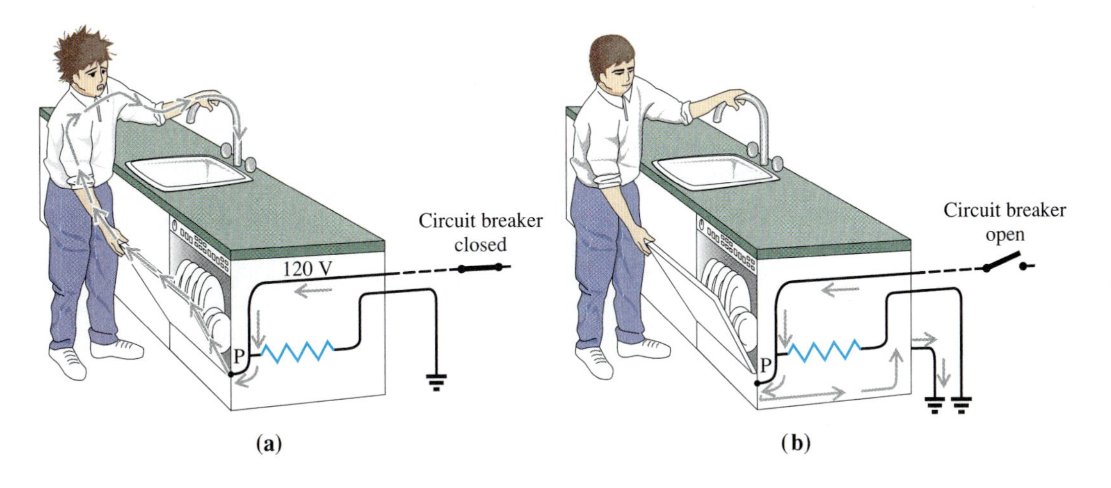

Fig. 20-36 **(a)** An ungrounded electric appliance can cause serious electrical shock. **(b)** A grounded appliance can sometimes prevent electric shock by causing a circuit breaker to open when a "hot" wire touches the appliance's metal casing.

CHAPTER 20 SUMMARY

In order to calculate currents and potential differences in circuits, we apply Kirchhoff's rules:

1 The sum of the currents entering any branch point is zero:

$$\Sigma I = 0$$

2 The sum of the voltage drops around any closed path is zero:

$$\Sigma V = 0$$

We analyze multiloop circuits by first applying Kirchhoff's first rule at branch points to minimize the number of unknown currents, then applying Kirchhoff's second rule around as many loops as there are unknown currents in order to obtain equations for these unknowns, and finally solving these equations.

The equivalent resistance of a network of resistors is the single resistor that carries the same current as the network when the potential difference across it is the same as that across the network. Series resistors are connected end to end, with no branch points in between, so that the current is the same in each. The equivalent resistance of resistors in series is the sum of the resistances:

$$R_{eq} = R_1 + R_2 + R_3 + \cdots \qquad \text{(resistors in series)}$$

Parallel resistors are connected so that the potential difference is the same across all of them. We find the inverse of the equivalent resistance for parallel resistors by adding the inverses of the resistances:

$$\frac{1}{R_{eq}} = \frac{1}{R_1} + \frac{1}{R_2} + \frac{1}{R_3} + \cdots \qquad \text{(resistors in parallel)}$$

An ammeter measures electric current through a circuit element. An ammeter is connected in series with the element. A voltmeter measures the potential difference between the two points to which its terminals are connected. Thus the voltmeter is in parallel with the circuit elements connected between the two points.

An RC circuit consists of a battery $\mathcal{E}$, a resistor R, a capacitor C, and a switch. If initially the capacitor is uncharged and the switch is open, closing the switch produces an initial current

$$I_0 = \frac{\mathcal{E}}{R}$$

At time t the current reduces to

$$I = I_0 e^{-t/\tau}$$

where τ, called the time constant, is the product of the circuit's resistance and capacitance:

$$\tau = RC$$

The value of the time constant is the time during which the current falls to $1/e$ of its original value. As current flows through the resistor, charge accumulates on the plates of the capacitor. At time t the charge Q is

$$Q = C\mathcal{E}(1 - e^{-t/\tau})$$

Nerve cells transmit electrical signals with potential differences on the order of 0.1 V. The human body is therefore very sensitive to externally imposed voltages and currents. Electric currents from limb to limb begin to be dangerous when they exceed about 1 mA. Even smaller internal currents, for example, within the heart, can be lethal. The body's electrical resistance depends on whether the skin is wet or dry but is usually in the range of $10^3\ \Omega$ to $10^4\ \Omega$. A potentially fatal current of 10 mA can sometimes result from a potential difference as small as 10 V.

Questions

1 Which of the resistors in Fig. 20-37 carries (a) the greatest current; (b) the least current?

2 If the current through the 4 Ω resistor in Fig. 20-37 is 3 A, what is the current through (a) the 1 Ω resistor; (b) the 2 Ω resistor?

3 Suppose that the circuit of Fig. 20-37 is broken at point P; that is, the wire is cut at this point. Indicate whether the current increases, decreases, or remains the same through (a) the 2 Ω resistor; (b) the 5 Ω resistor.

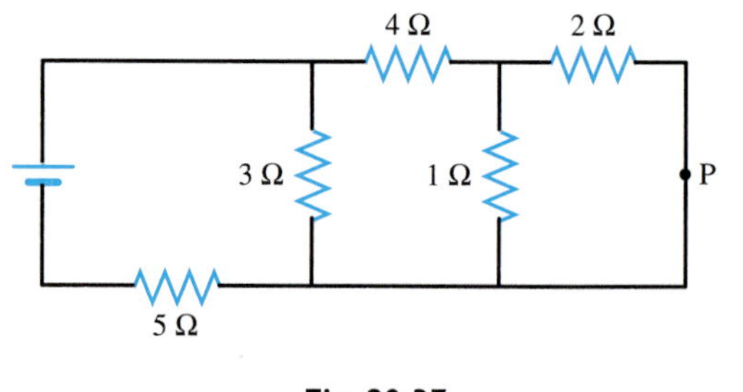

Fig. 20-37

4 Fig. 20-38 shows currents in some of the wires connected at branch points a and c. Find the current at b.

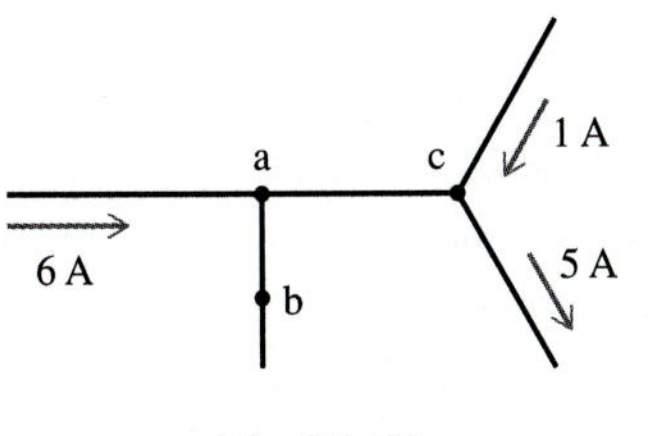

Fig. 20-38

5 Indicate which of the points a, b, c, d in Fig. 20-39 could be connected with copper wire in order that the current in each 10 Ω resistor equals 1 A.

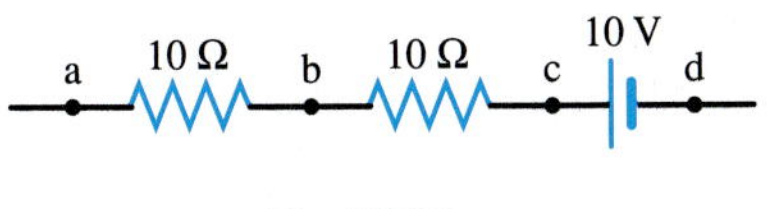

Fig. 20-39

6 The resistors R_1, R_2, and R_3 in Fig. 20-40 represent light bulbs 1, 2, and 3.

(a) Under what conditions of the three switches S_1, S_2, and S_3 will you have an *open* circuit?

(b) Under what conditions of the switches will you have a *short* circuit?

(c) Is it possible to light bulb 2 without lighting bulb 3? If so, indicate how.

(d) Is it possible to light bulb 1 without lighting either bulb 2 or bulb 3? If so, indicate how.

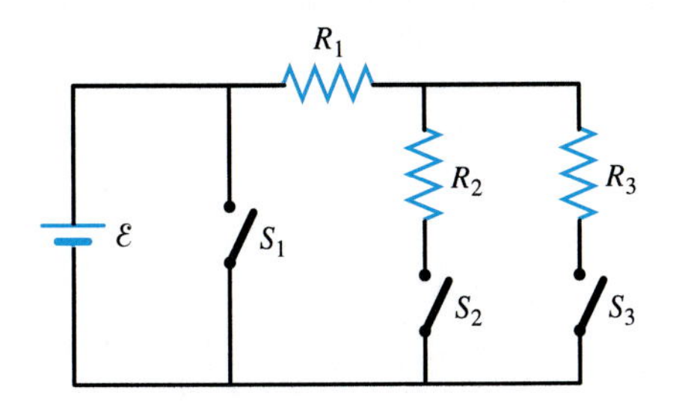

Fig. 20-40

7 A flashlight is initially powered by a single 1.5 V battery of negligible internal resistance. Suppose that a second 1.5 V battery is connected in series with the first battery. How would the current through the flashlight bulb change when the batteries are connected (a) + to −; (b) + to +?

Suppose next that the two batteries are connected in parallel (+ to +).

(c) How will the current through the flashlight bulb change from the initial single-battery value?

(d) How will the current through the first battery change?

(e) How will the operating life of the first battery change?

8 (a) In which circuit in Fig. 20-41 will the ammeter A measure the current in the resistor?

(b) What effect will the ammeter have in the other circuit?

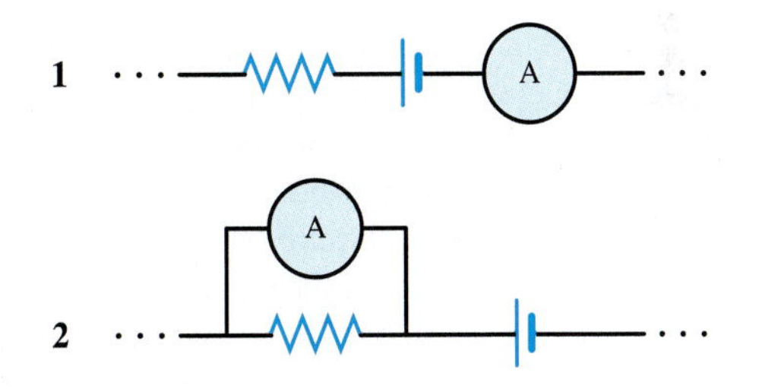

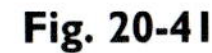
Fig. 20-41

9 (a) In which circuit in Fig. 20-42 will the voltmeter V measure the potential difference between a and b?

(b) What effect will the voltmeter have in the other circuit?

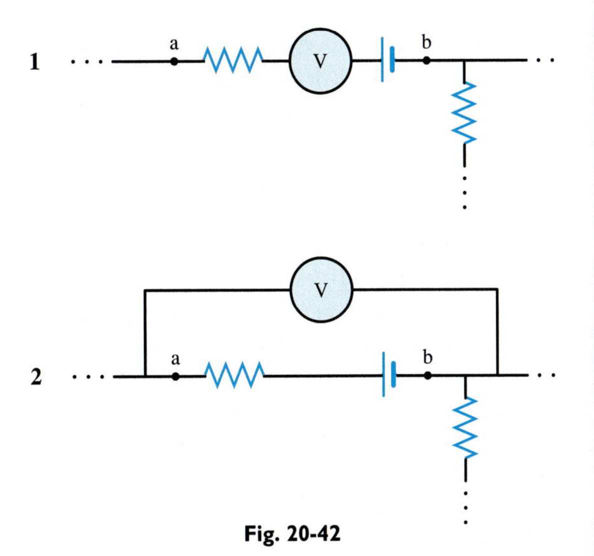

Fig. 20-42

10 Fig. 20-43 shows an RC circuit.

(a) Initially switch S is open and the capacitor is uncharged. What are the initial voltage drops across $\mathcal{E}$, R, C, and S?

(b) Now suppose switch S is closed. What are the voltage drops across $\mathcal{E}$, R, C, and S 1 min after the switch is closed?

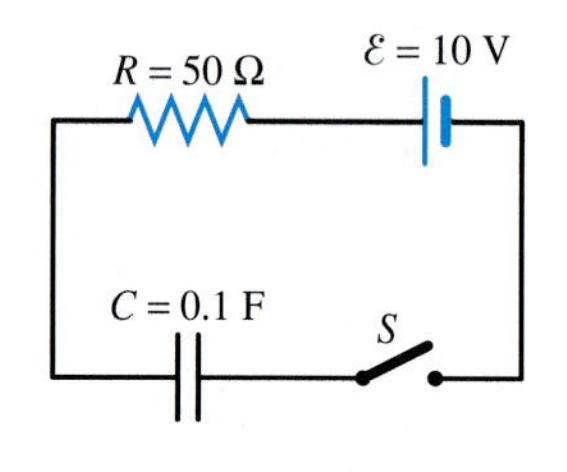

Fig. 20-43

11 You are trying to change a light bulb in an overhead fixture while standing barefooted on a counter top next to your kitchen sink. One foot touches the metal water faucet while your hand accidentally touches an exposed hot wire in the fixture. The wire is at a potential of 120 V. The wiring system is equipped with circuit breakers and a ground fault interrupter. Will you get a serious electrical shock? Explain.

12 An electric hair dryer plugged into a wall receptacle accidentally falls into a bathtub full of water. Is it possible that a person in the bathtub would receive a lethal shock? Explain.

13 Does a ground fault interrupter guarantee that current through household wiring will stop when a significant current passes through (a) a human body; (b) a grounding wire?

14 A small insulated metal sphere carries a charge of 10^{-7} C and is at a potential of 10,000 V. Your left hand is grounded. Would it be more dangerous for your right hand to touch the sphere or to touch a power line at 120 V? Explain.

15 When you blow a fuse, is it a good idea to replace the fuse with a penny, allowing the circuit to function again? Why or why not?

16 A 100 W table lamp, a 200 W television, and a 1500 W electric heater are all plugged into wall receptacles and turned on. When you plug in a power floor sander to a wall receptacle on the same branch circuit, the circuit breaker is tripped, turning off all these appliances.

(a) Will the overhead lights on a separate branch circuit go out? Assume that the current was not sufficient to trip the main circuit breaker.

(b) Before resetting the circuit breaker and again trying to operate the floor sander, turning off which one of the other appliances would be most likely to help?

Answers to Odd-Numbered Questions

1 (a) 5 Ω; (b) 2 Ω; **3** (a) decreases to 0; (b) decreases; **5** a to c and b to d; **7** (a) current doubles; (b) current goes to 0; (c) same current; (d) current is halved; (e) battery life is doubled; **9** (a) 2; (b) increases resistance between a and b; **11** Yes; the current from the wire will travel through your body to ground through the water pipes. No current will pass through the ground fault interrupter, and the circuit breaker will not open; **13** (a) no; (b) yes; **15** No; the fuse is a safety device, and its blowing is a signal that there is too much current in the circuit; a current too large for the wiring can cause the wires to overheat, and a fire could start. What you should do is (1) remove one or more of the electrical devices from the circuit and (2) replace the blown fuse with a fresh one.

Problems

20-2 Kirchhoff's Rules

1 Three wires are connected at a branch point. One wire carries a positive current of 6 A into the branch point, and a second wire carries a positive current of 4 A away from the branch point. Find the current carried by the third wire *into* the branch point.

2 Find the current through the battery in Fig. 20-44.

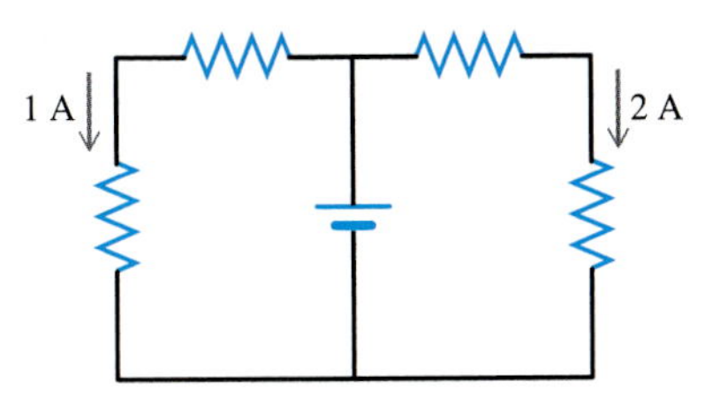

Fig. 20-44

3 Find the current at point a in Fig. 20-45.

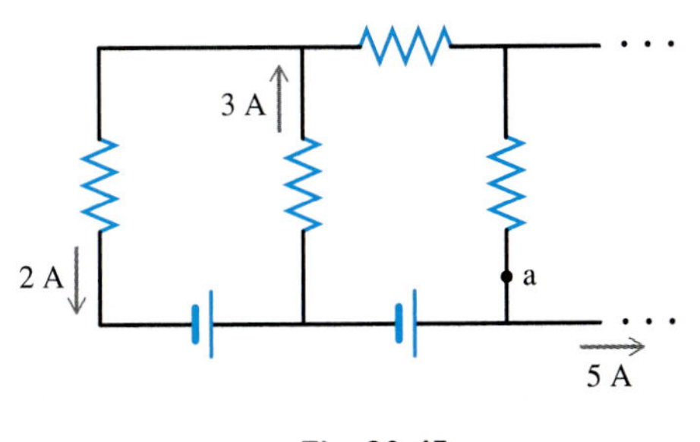

Fig. 20-45

4 (a) For the circuit in Fig. 20-46, what is the voltage drop across the 5 Ω resistor when switch S is open?

(b) What is the voltage drop across the switch when it is open?

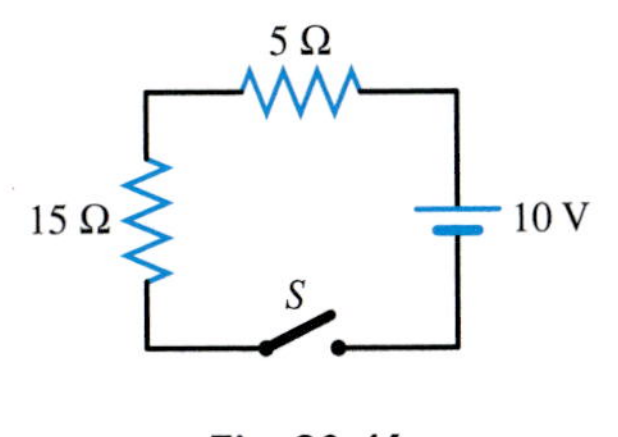

Fig. 20-46

★**5** Find the voltage drop from b to c in Fig. 20-47.

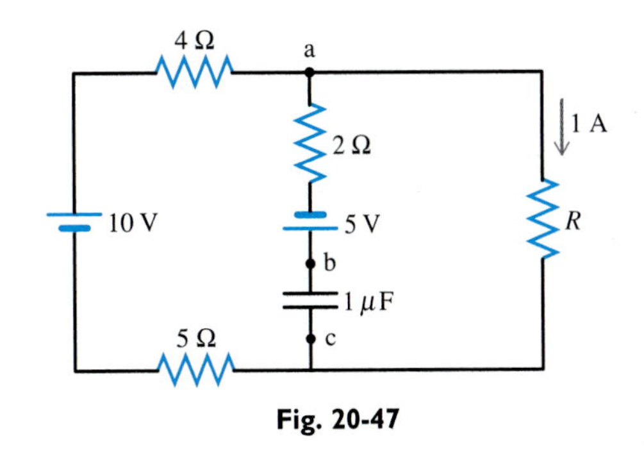

Fig. 20-47

6 For the circuit in Fig. 20-48, find (a) the magnitude of the current and the direction of positive current; (b) the potential difference across the battery terminals a and b; (c) the rate at which heat is being produced in the 5 Ω resistor; (d) the rate at which heat is being produced inside the battery; (e) the rate at which the battery's chemical energy is being used.

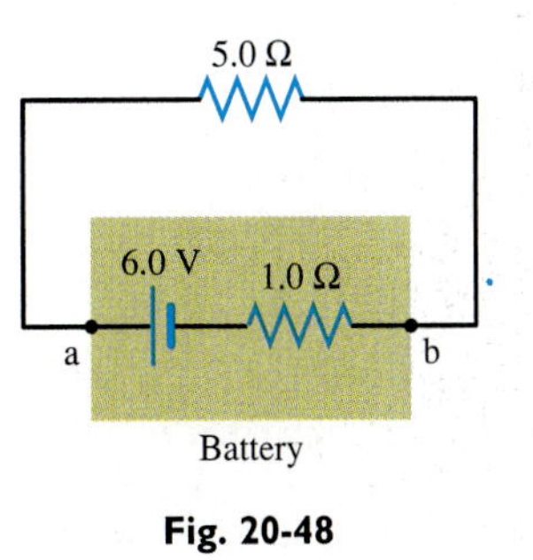

Fig. 20-48

7 The current in Fig. 20-49 is 0.50 A. Find (a) the voltage drop V_{ab}; (b) the power dissipated in R; (c) the electric power supplied by the 9.0 V battery; (d) the rate at which chemical energy is stored in the 6.0 V battery.

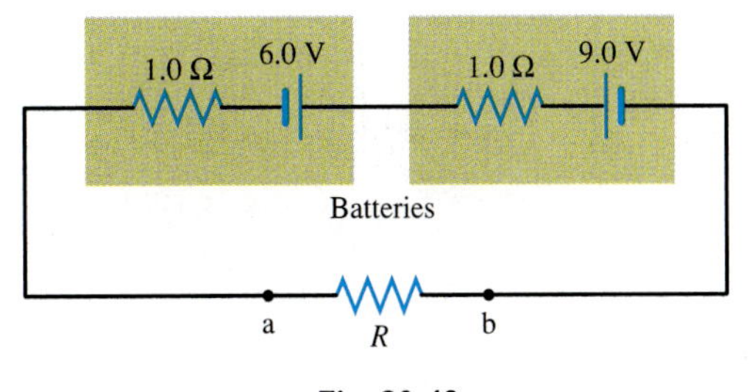

Fig. 20-49

8 In Fig. 20-50, find the potential difference across the terminals of the battery in which chemical energy is being converted to electrical potential energy.

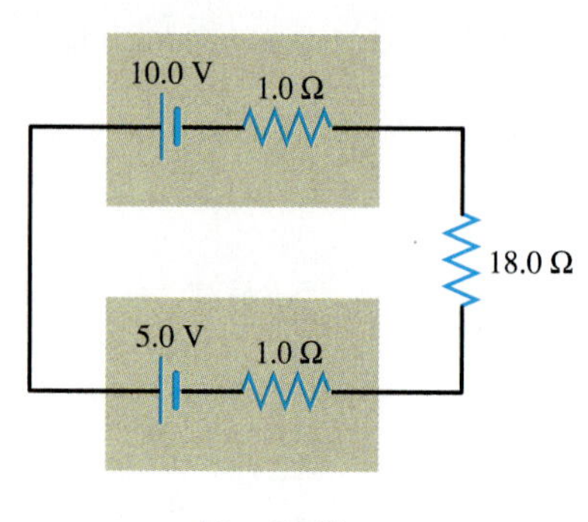

Fig. 20-50

9 (a) Find the magnitude and direction of the current in Fig. 20-51.
(b) Find the voltage drop V_{ab}.

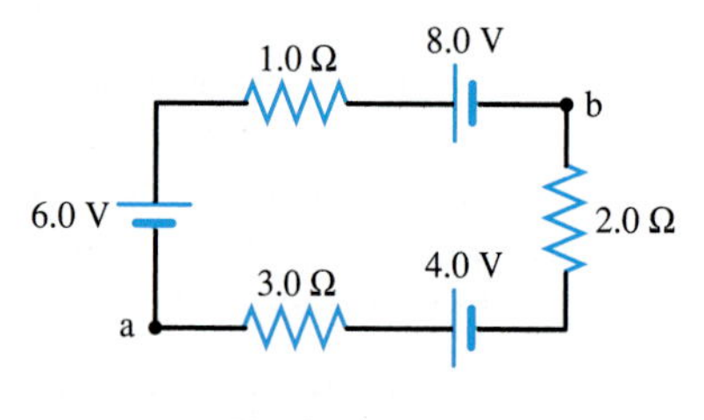

Fig. 20-51

20-3 Equivalent Resistance

10 Draw a circuit diagram of the circuit in Fig. 20-52.

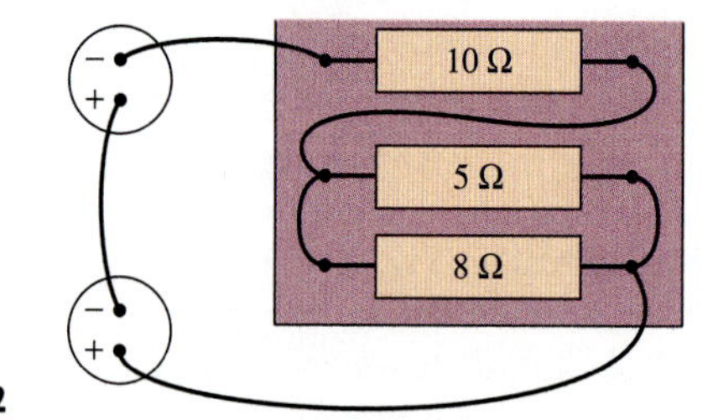

Fig. 20-52

11 Draw a circuit diagram of the circuit in Fig. 20-53.

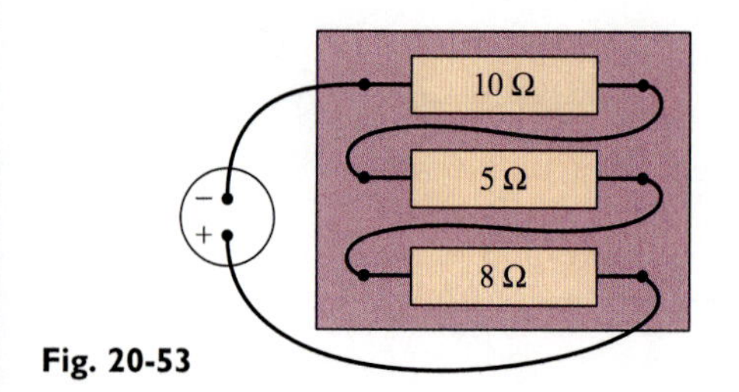

Fig. 20-53

12 For each network in Fig. 20-54, find the equivalent resistance between a and b.

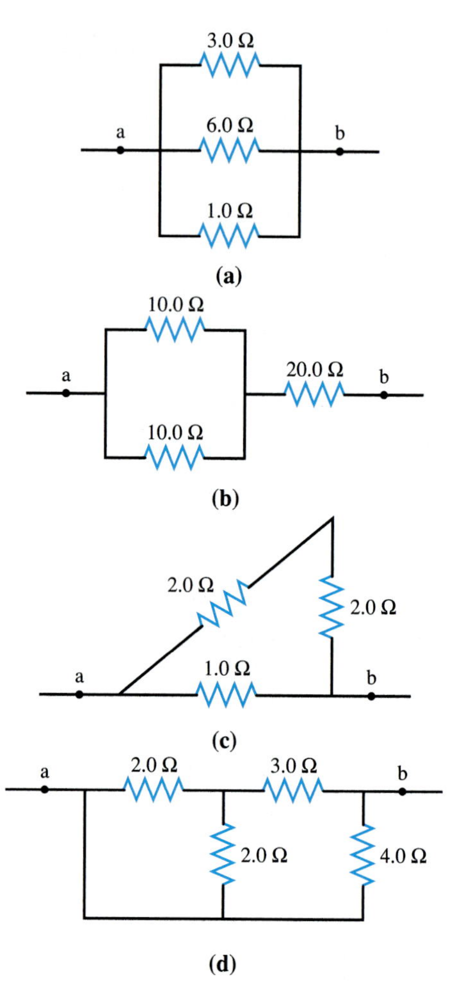

Fig. 20-54

13 The current through the 4.0 Ω resistor in Fig. 20-55 is 0.30 A. What is the current through the 5.0 Ω resistor?

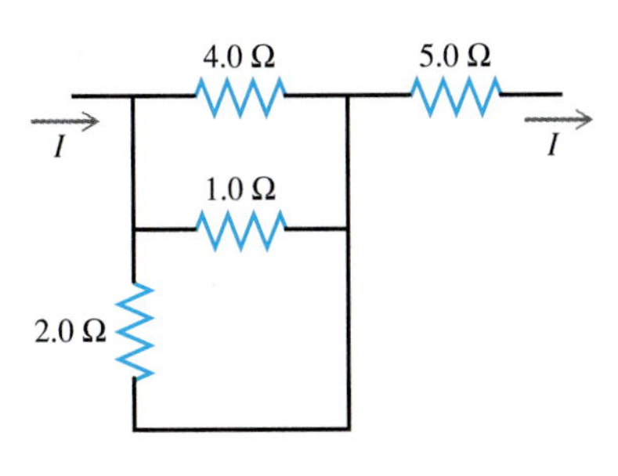

Fig. 20-55

14 (a) Find the equivalent resistance between points a and b in Fig. 20-56.
(b) Find the potential difference V_{ab} that will produce a current of 1.0 A in the 2.0 Ω resistor.

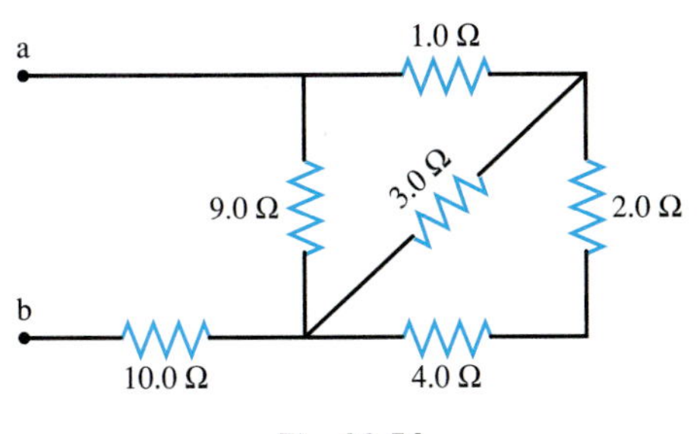

Fig. 20-56

★**15** (a) The 6.0 Ω resistor in Fig. 20-57 carries a current of 1.0 A. Find the current through the 5.0 Ω resistor. (b) If the 3.0 Ω resistor were replaced by a different resistor, which, if any, of the other resistors would carry the same current as before?

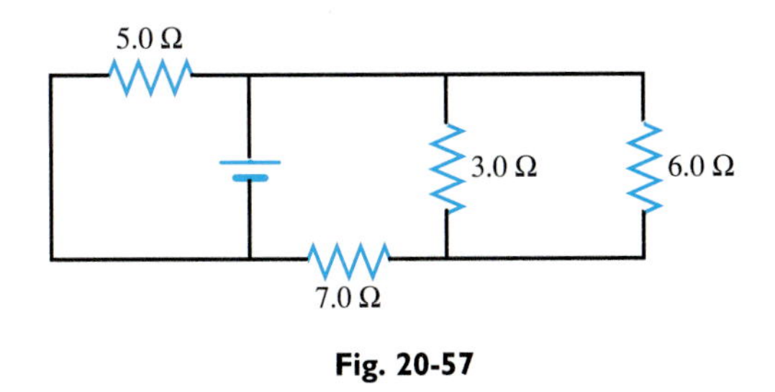

Fig. 20-57

★**16** For resistors connected in series, the equivalent resistance is obviously greater than the resistance of any of the individual resistors. Prove that, when resistors are connected in parallel, the equivalent resistance is less than the resistance of any of the individual resistors.

★**17** Four resistors—a 1 Ω, a 10 Ω, and two 100 Ω—are connected as shown in Fig. 20-58. To give minimum equivalent resistance between a and b, should the 1 Ω resistor be in parallel with the 10 Ω resistor or with one of the 100 Ω resistors?

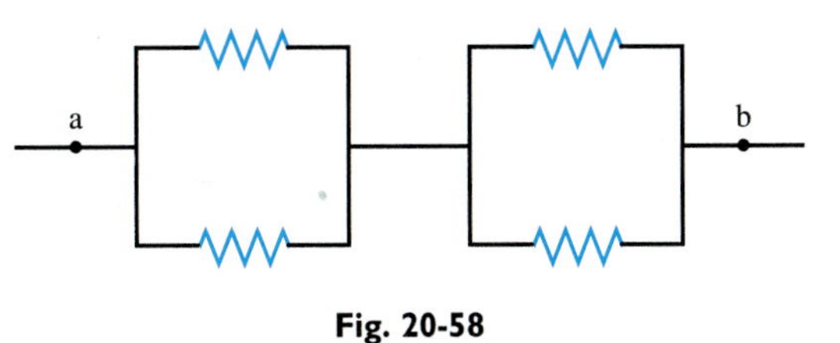

Fig. 20-58

18 Find .the current through the 6.0 Ω resistor in Fig. 20-59.

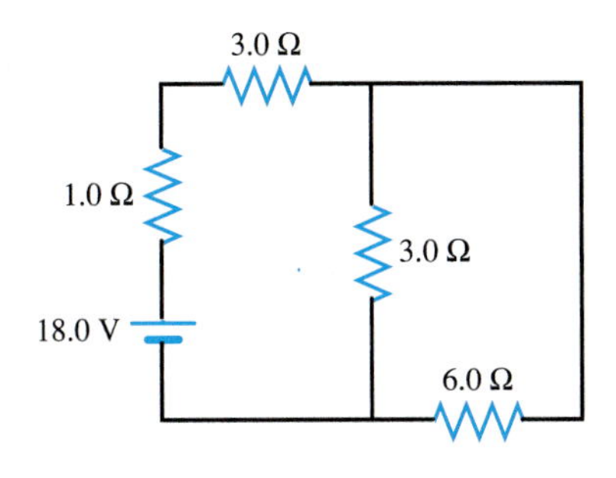

Fig. 20-59

19 Find the current through the 2.0 Ω resistor in Fig. 20-60.

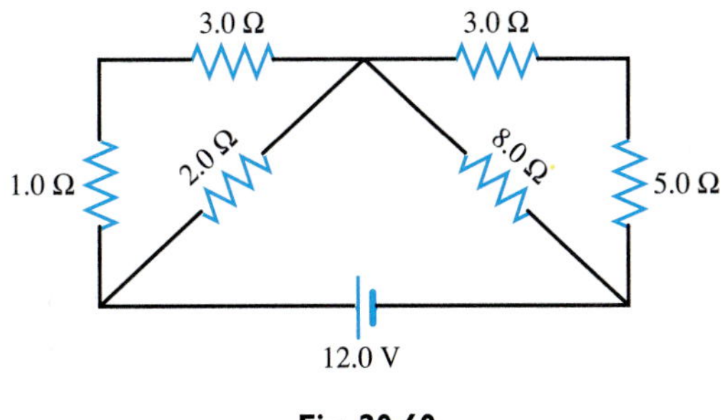

Fig. 20-60

20 In Fig. 20-61, find the voltage drop V_{ab} and the current through the 1.0 Ω resistor when the switch is (a) open and (b) closed.

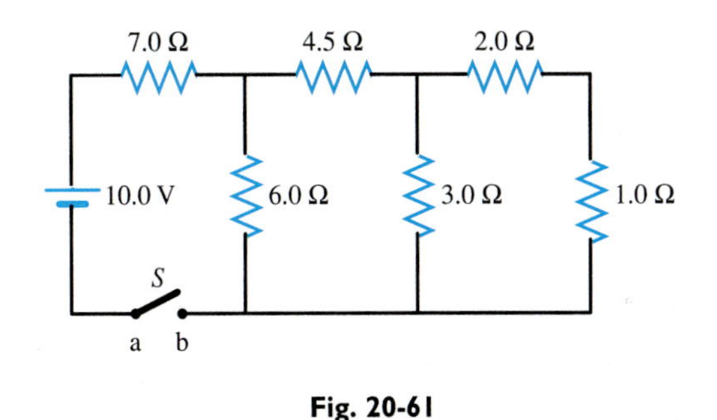

Fig. 20-61

21 Find the voltage drop V_{ab} in Fig. 20-62.

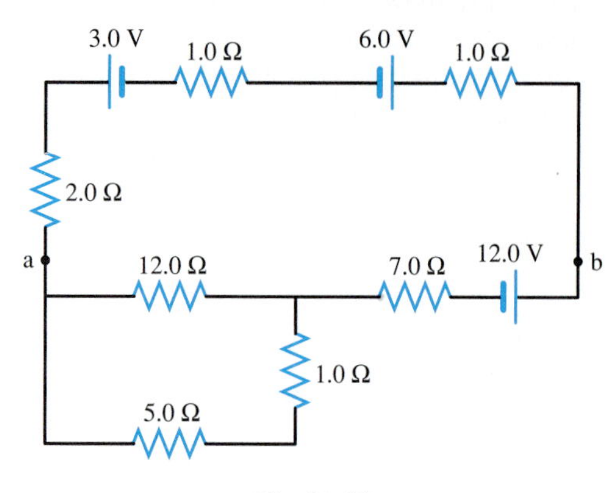

Fig. 20-62

★★**22** A Wheatstone bridge, shown in Fig. 20-63, is a network used to measure an unknown resistance R. The values R_1, R_2, and R_3 are known. The resistance R_3 is adjusted so that the potential at a is the same as at b. One can achieve this by varying R_3 until no current flows through a meter connected between a and b. Show that $R = \dfrac{R_2R_3}{R_1}$.

Fig. 20-63

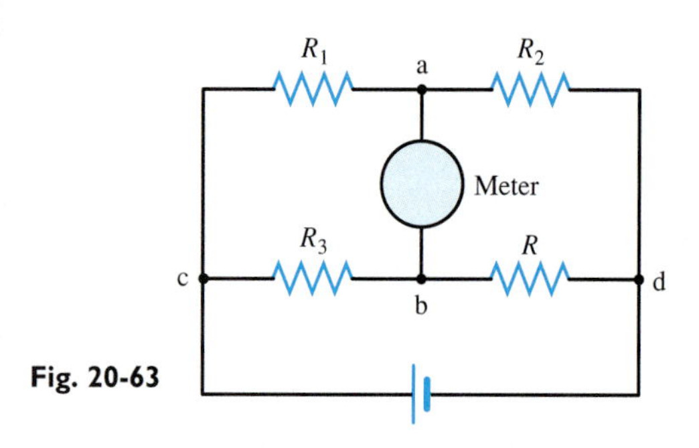

*20-4 Multiloop Circuits

23 For the circuit in Fig. 20-64, find (a) the current through the battery; (b) the voltage drop V_{ab}.

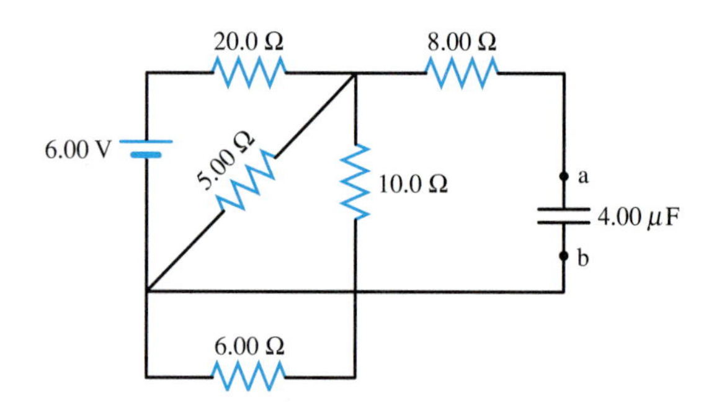

Fig. 20-64

★**24** For the circuit in Fig. 20-65, find the current through (a) the 10.00 V battery; (b) the 6.00 V battery.

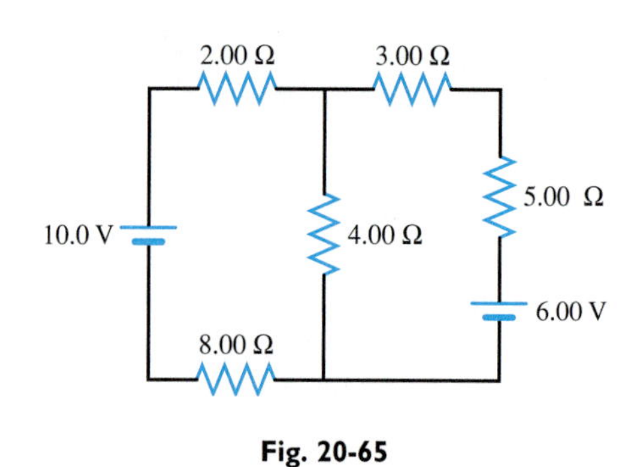

Fig. 20-65

★**25** For the circuit in Fig. 20-66, find the current (a) at a; (b) at b.

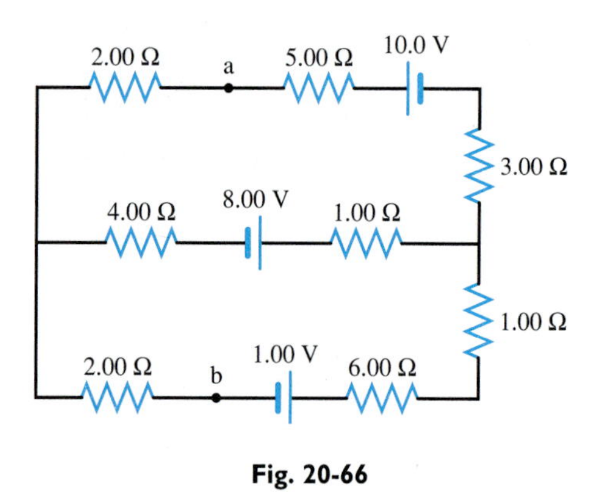

Fig. 20-66

★**26** For the circuit in Fig. 20-67, find the current I_1.

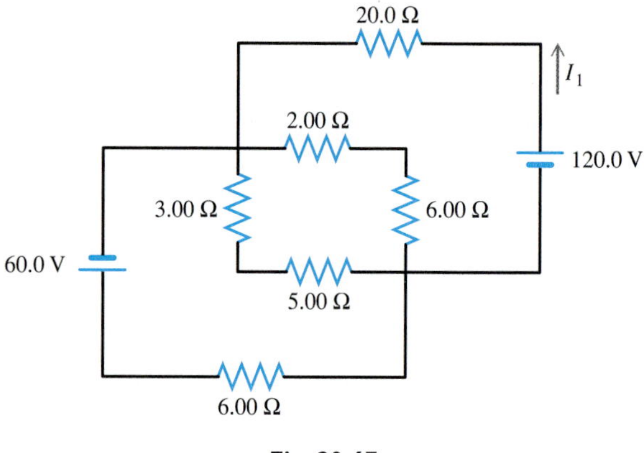

Fig. 20-67

★★**27** Find I_1, I_2, and I_3 in Fig. 20-68.

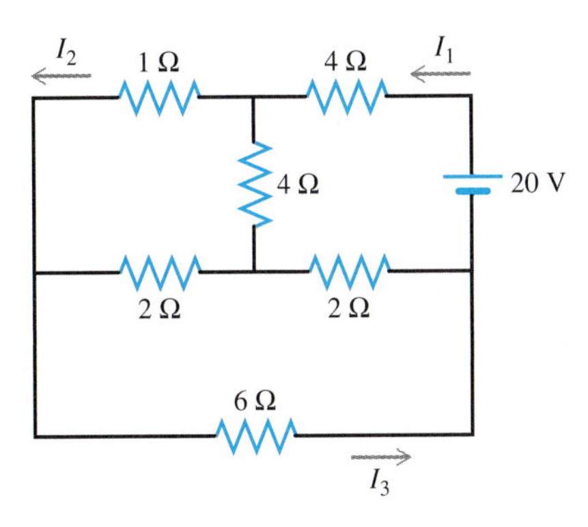

Fig. 20-68

20-5 Measurement of Current, Potential Difference, and Resistance

28 Find the readings of ammeters A_1 and A_2 in Fig. 20-69. All resistors are 2.00 Ω.

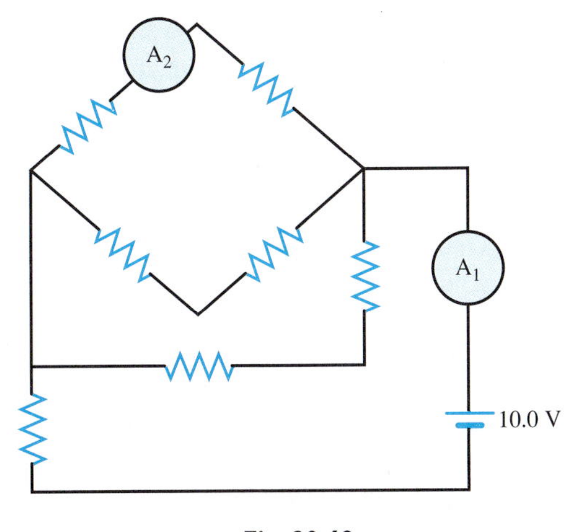

Fig. 20-69

29 Ammeter A in Fig. 20-70 reads 1.0 A. Find the battery's emf $\mathcal{E}$.

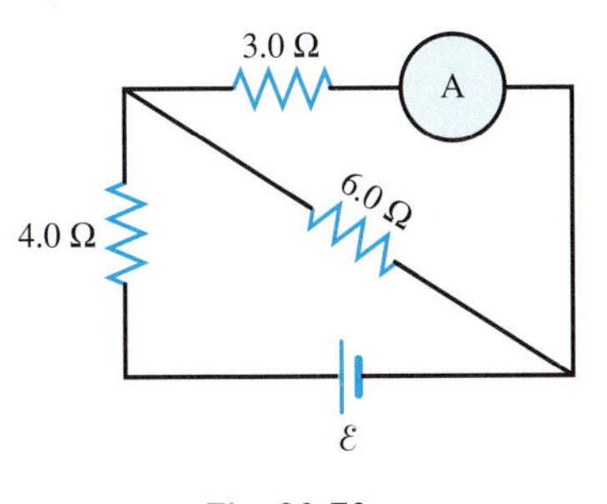

Fig. 20-70

30 The ammeter in Fig. 20-71 reads 300 mA. Find the battery's internal resistance r and the reading of the voltmeter V.

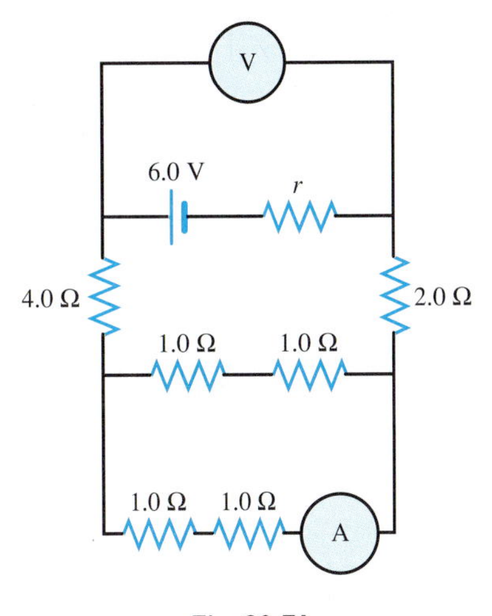

Fig. 20-71

31 A galvanometer, such as that shown in Fig. 20-72, has an internal resistance R_G = 300 Ω and full-scale deflection for a current of 20 μA through its coil. A parallel resistor R_P is connected to the galvanometer as indicated in Fig. 20-73 to form an ammeter.

(a) Find the value of R_P and the ammeter's equivalent resistance if the ammeter is to measure currents in the range 0 to 2 A.

(b) When the meter shows half-scale deflection, what are the currents through the galvanometer and through the ammeter?

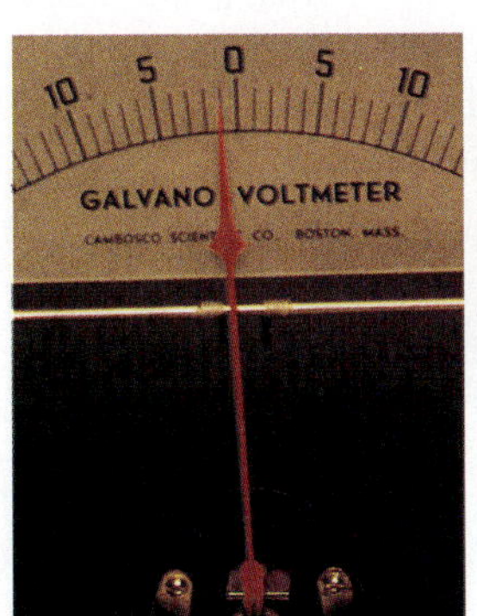

Fig. 20-72 The demonstration galvanometer shown here measures the current passing through its coil. Deflection of the needle is proportional to the current.

An enlarged view of the galvanometer's copper coil at the bottom of the galvanometer needle.

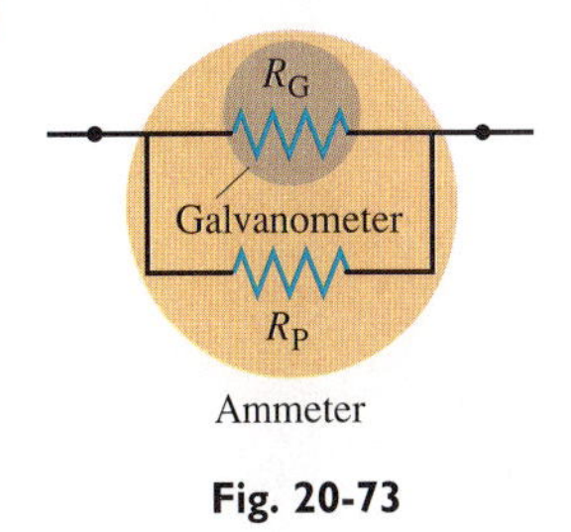

Fig. 20-73

32 The galvanometer described in problem 31 is connected to a series resistor R_S as indicated in Fig. 20-74 to form a voltmeter.

(a) Find the value of R_S and the voltmeter's equivalent resistance if the voltmeter is to measure voltages in the range 0 to 20 V.

(b) When the voltmeter reads 10 V, what is the current through the galvanometer, and what are the voltage drops across R_G and R_S?

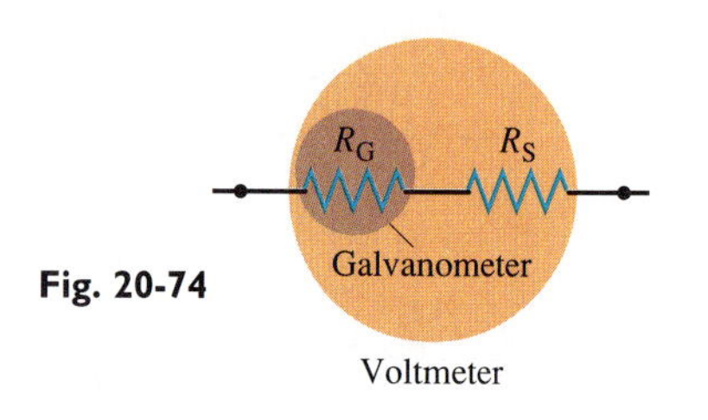

Fig. 20-74

20-6 RC Circuits

33 For the circuit shown in Fig. 20-27, let $\mathcal{E} = 10$ V. Initially the switch is open and the capacitor is uncharged. When the switch is closed, there is an initial instantaneous current of 1.00 mA in the resistor. Two seconds later the current through the resistor has fallen to 0.37 mA. What are the values of R and C?

34 A battery, a resistor, and a capacitor are connected in series. Does the final charge on the capacitor depend on the value of the resistance?

★35 (a) For the circuit shown in Fig. 20-27, let $\mathcal{E} = 10$ V, $R = 5.0\ \Omega$, and $C = 0.1$ F. What is the current at b just after the switch is closed?

(b) How much charge will have passed b by the time the current goes to zero?

(c) Find the current at the instant the capacitor has a charge of 0.20 C.

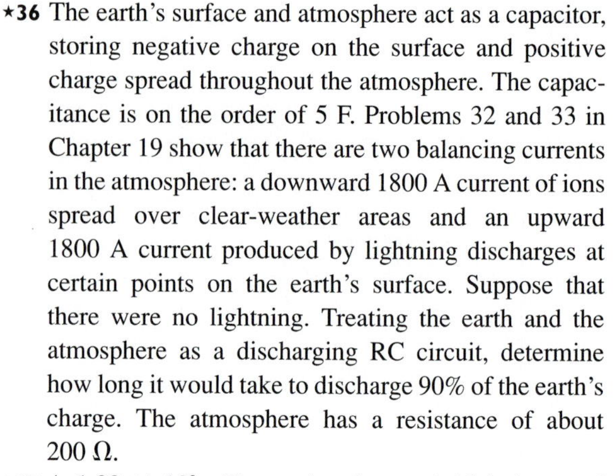

★36 The earth's surface and atmosphere act as a capacitor, storing negative charge on the surface and positive charge spread throughout the atmosphere. The capacitance is on the order of 5 F. Problems 32 and 33 in Chapter 19 show that there are two balancing currents in the atmosphere: a downward 1800 A current of ions spread over clear-weather areas and an upward 1800 A current produced by lightning discharges at certain points on the earth's surface. Suppose that there were no lightning. Treating the earth and the atmosphere as a discharging RC circuit, determine how long it would take to discharge 90% of the earth's charge. The atmosphere has a resistance of about 200 Ω.

★37 A 1.00×10^3 μF capacitor has an initial charge of 0.100 C. When a resistor is connected across the capacitor plates, there is an initial current through the resistor of 1.00 A. What is the current 1.00 s later?

★★38 A neon tube* connected to an RC circuit as indicated in Fig. 20-75a can be used to produce an oscillating voltage V across the tube, which then emits a periodic flashing light. (A strobe light works this way.) Discharge begins when $V = 60$ V and continues until V falls below 40 V, at which point the tube becomes an open circuit again. During discharge, the tube's resistance R' is $1.0 \times 10^3\ \Omega$. Thus there is an alternation between (1) a charging process during which the capacitor voltage V rises to 60 V and no current flows through the tube and (2) a discharging process during which V falls from 60 V to 40 V and which, to a good approximation, involves only the capacitor and the tube (since $R' << R$). This charge-discharge cycle produces the periodic voltage shown in Fig. 20-75b, with times t_1, t_2, t_3, and period T related to the time constants of the two RC circuits. Find (a) the time constant for the charging circuit; (b) the time constant for the discharging circuit; (c) t_1; (d) t_2; (e) t_3; and (f) T.

*The neon tube is an example of a gas-discharge tube, a glass tube from which the air is evacuated and which is filled with a vapor. Sufficient voltage applied to electrodes within the tube causes an electrical discharge through the gas, which then emits light.

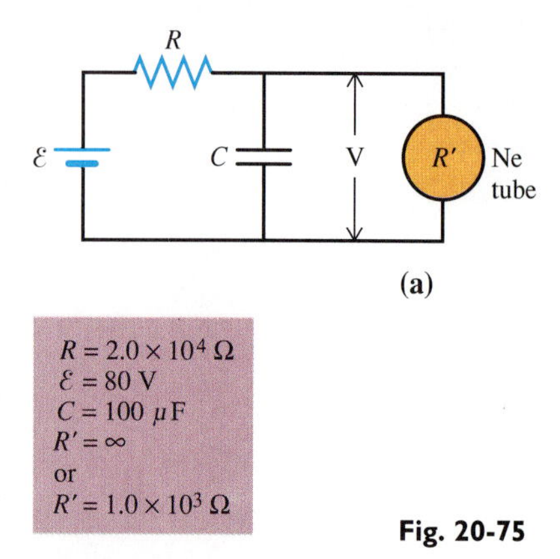

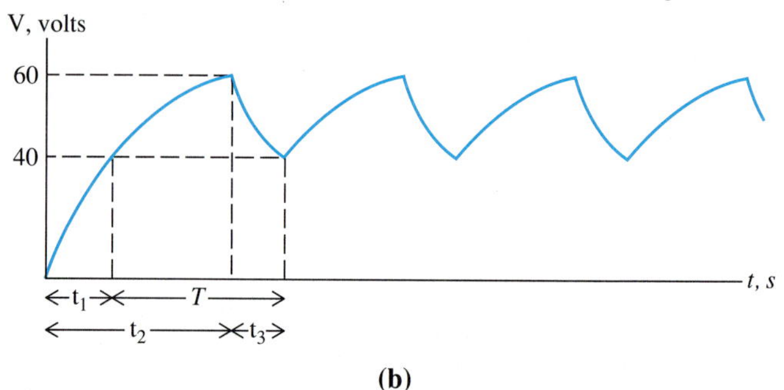

Fig. 20-75

★★**39** An electronic pacemaker circuit often utilizes an RC oscillator circuit similar to the one described in Problem 38. When the charge on the capacitor reaches a certain value, the capacitor discharges, providing an electrical pulse into the heart and stimulating contraction. The period of the pulse determines the period of the heartbeat and is related to the time constants of the charging and discharging circuits. Suppose that a certain pacemaker's charging circuit uses a 6.0 V battery, storing 1.0×10^5 J of energy, has a time constant and a period both equal to 1.0 s, and delivers 1.0×10^{-3} J of energy to the capacitor during each cycle.

(a) Find the charging circuit's capacitance, resistance, and peak current, assuming the capacitor loses half of its stored energy during the discharging part of its cycle and that the peak voltage across the capacitor is 63% of 6.0 V.

(b) How long will the batteries last?

20-7 Electric Shock and Household Electricity

40 Estimate the hand-to-hand resistance of the human body, excluding the resistance of the skin. Use 2 Ω-m for the resistivity of body fluids and treat the path between the hands as a uniform cylinder of radius 4 cm and length 1.5 m.

★**41** Suppose that each of a person's hands makes contact with a different conductor over a surface area of 10.0 cm^2. We measure the hand-to-hand resistance of the body by connecting the conductors to an ohmmeter that reads 1.00×10^4 Ω when the hands are dry. From problem 40 we know that internal body resistance is only about 600 Ω.

(a) Find the resistance across a 10.0 cm^2 section of the skin.

(b) Find the resistivity of the skin, assuming it is 1.00 mm thick.

(c) Suppose the area of contact of each hand with a conductor is reduced to 1.00 cm^2. Find the resistance between the hands.

42 At some instant, three of the circuits in a house are connected to 120 V devices with the following power requirements: 2000 W, 600 W, 400 W, 200 W, 150 W, 60 W, 40 W.

(a) Find the total current carried by the power lines leading into the house.

(b) If the main circuit breaker limits current to 100 A, what is the maximum total electrical power that can be utilized by all 120 V devices in the house?

43 What is the average electrical power used in a house in which 1200 kWh of electrical energy is used in a 30-day period?

★**44** Heavy-gauge, high-voltage transmission lines are used to transmit electrical power with little loss of energy along the line. Suppose that the lines supplying power to one part of a city have a total resistance of only 2.00 Ω over a distance of several kilometers. The potential difference between the wires at the input end is 24,000 V.

(a) Find the current in the line when power consumption is 1.00×10^3 kW.

(b) How much of this power is lost by heating of the wires?

(c) Find the potential difference between the wires at the end of the line.

(d) Now suppose the power were transmitted with an input potential difference of 2400 V, rather than 24,000 V. Compute the power loss and the potential difference at the opposite end.

45 A 24,000 V transmission line has resistance of 3.0×10^{-4} Ω per meter of length and carries a current of 40 A.

(a) Find the potential difference of points 1.0 m apart on one wire.

(b) What is the potential difference between the two adjacent wires?

(c) Would a person receive an electric shock if hanging from a high-voltage line and making contact with *only* one wire?

★★**46** A two-way switch connects two conductors in either of its two positions. Two such switches are sometimes used in a household lighting circuit in a way that allows lights to be switched on or off from two locations. Design a circuit of this type.

★**47** Because they make contact with the ground and have low electrical resistance, water pipes are often used for grounding.

(a) Find the resistance of a 10.0 m long iron water pipe that has an inner diameter of 2.00 cm and an outer diameter of 3.00 cm.

(b) A typical branch circuit will carry up to 30.0 A before a circuit breaker opens. If 30.0 A is the maximum current that a circuit breaker could permit to pass through the pipe in part (a), what is the maximum potential at any point on the pipe?

Additional Problems

48 Draw a circuit diagram corresponding to the circuit shown in Fig. 20-34a.

49 Draw a circuit diagram corresponding to the circuit shown in Fig. 20-34b.

50 (a) Fig. 20-76a shows a typical flashlight circuit, consisting of two 1.5 V batteries, a bulb with a resistance of 9.0 Ω, and a switch. How much electrical power is supplied to the bulb when the switch is closed?

(b) Now suppose that two additional pairs of 1.5 V batteries are connected as shown in Fig. 20-76b. How much electrical power is supplied to the bulb when the switch is closed?

(c) When a single pair of batteries is used (Fig. 20-76a), the flashlight will operate for 3.0 h with little change in power output. How long will it operate using the six batteries shown in Fig. 20-76b?

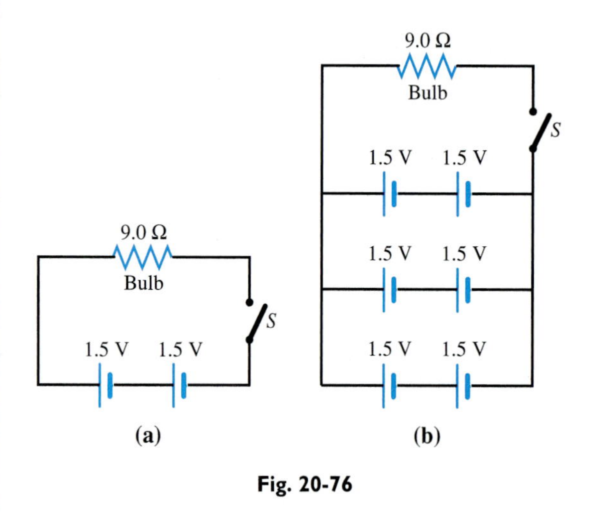

Fig. 20-76

★51 How large a resistor should be placed in series with a 288 W, 120 V heating element so that the power used is reduced to 100 W when 120 V is applied across the combination?

★★52 To avoid overheating a resistor, its maximum power input rating should not be exceeded. Suppose you have a number of 1 Ω, 4 W resistors. By considering series and parallel combinations, design a network with an equivalent resistance of 2 Ω for which the power input can be 20 W.

★53 The dielectric in a real capacitor will always have finite resistivity, and therefore an isolated capacitor will gradually lose its charge by conduction through the dielectric. Suppose that the dielectric in a certain 100 μF capacitor has a dielectric constant of 10.0. When a 1.00 MΩ voltmeter is connected across the capacitor plates, the circuit is as indicated in Fig. 20-77, where R is the capacitor's resistance. The voltmeter initially reads 100 V. After 75.0 s, the voltmeter reads 37.0 V.

(a) Find R.

(b) How long would it take the capacitor to lose 90% of its charge if the voltmeter were not connected?

(c) Find the dielectric's resistivity.

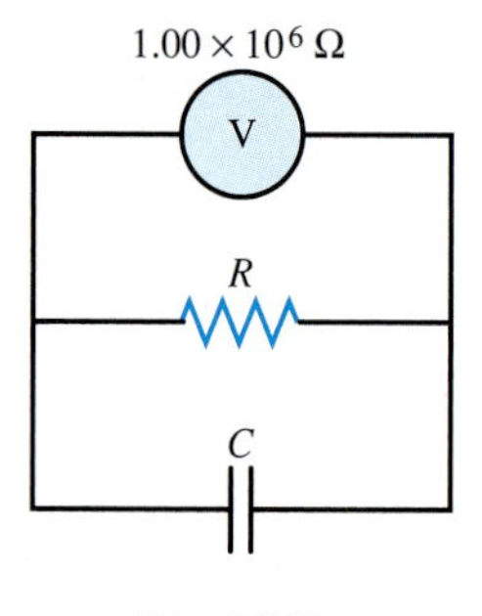

Fig. 20-77

★54 In a discharging RC circuit with a time constant of 10.0 s, how long does it take for half of the capacitor's energy to be lost?

★55 In Fig. 20-78 the capacitor is initially uncharged and the switch is open.

(a) Find the current through the 1000 Ω resistor just after the switch is closed.

(b) Find the current through the 1000 Ω resistor 1.00 h later.

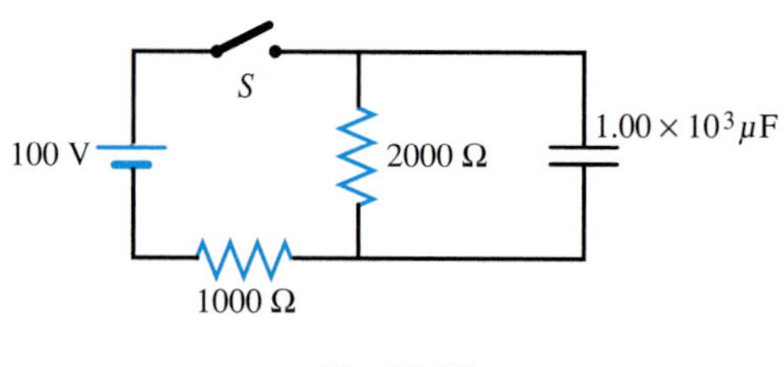

Fig. 20-78

56 In Fig. 20-79, find the current through each resistor and the charge on each capacitor long after the switch is (a) opened; (b) closed.

★57 The axoplasm fluid inside the axon of a nerve cell has a resistivity of 2.00 Ω-m, and the membrane surrounding the cell has resistivity of 5.00×10^{7} Ω-m. For an axon of length 1.00 m, diameter 1.00×10^{-5} m, and membrane thickness 4.00×10^{-9} m, find the resistance (a) of the entire length of the axoplasm; (b) across the membrane.

★★58 The terms "sodium pump" and "potassium pump" are used to describe the (poorly understood) mechanism responsible for maintaining the ion concentrations and potential difference across an axon membrane. The membrane acts as a capacitor that leaks some current across its terminals. The pumps can be likened to batteries that maintain the potential difference across the membrane. The passage of a nerve pulse involves a rapid change in membrane conductivity and therefore in the charge stored on the membrane, followed by restoration of the original charge distribution. It is not surprising that the duration of the pulse is roughly equal to the time constant of an RC circuit having the resistance and capacitance of the membrane. An axon membrane is 4×10^{-9} m thick and has a resistivity of 5×10^{7} Ω-m and a dielectric constant of 5. Show that a length ℓ of axon has membrane resistance R, capacitance C, and a time constant $\tau = RC$ that is independent of ℓ and also is independent of the axon's radius and membrane thickness. Find the value of τ.

★★59 The "ladder" circuit in Fig. 20-80 can be used to represent the resistance of an axon. Let R_A represent the resistance of the axoplasm inside the axon to flow along the axon, and let R_M represent resistance to flow across the membrane. The fluid outside the membrane has negligible resistance.

(a) Find the equivalent resistance of the network between a and b. Assume an infinite number of resistors with finite equivalent resistance R. HINT: Compare R with the equivalent resistance of the network to the right of a′ and b′.

(b) Show that if the length of each axoplasm resistor between successive membrane resistors approaches zero then $R_M >> R_A$ and $R = \sqrt{R_A R_M}$. Evaluate R for an axon of radius 1.00×10^{-5} m, membrane thickness 4.00×10^{-9} m, membrane resistivity of 5.00×10^{7} Ω-m, and axon resistivity of 2.00 Ω-m.

★60 (a) Find the current through the battery and the charge on the capacitor in Fig. 20-81, long after the switch is closed.

(b) If the switch is opened, how long will it take for the charge on the capacitor to decrease to 400 μC?

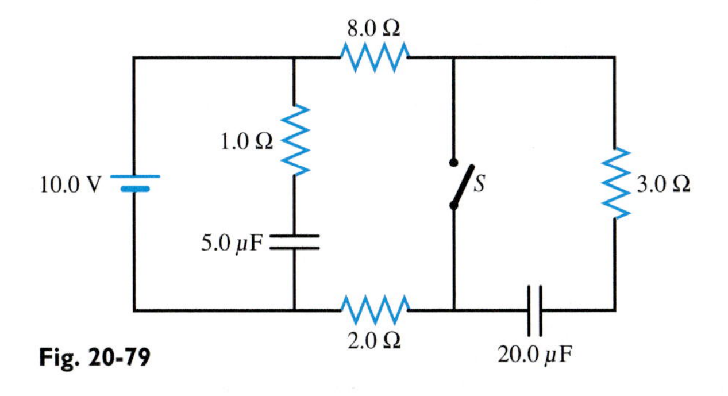

Fig. 20-79

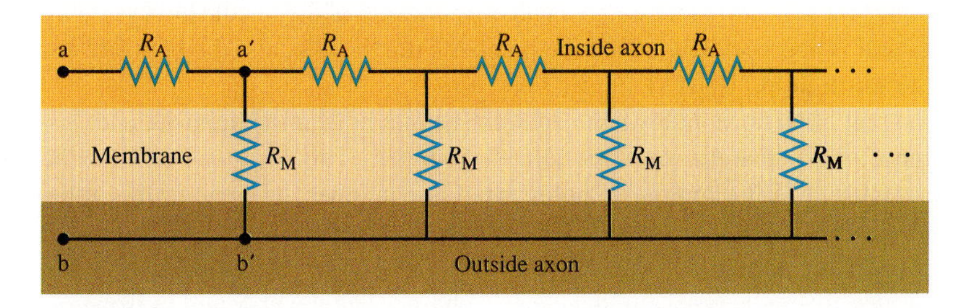

Fig. 20-80

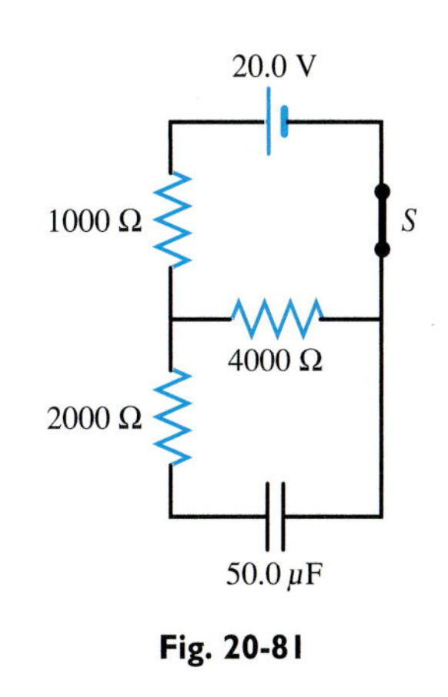

Fig. 20-81

CHAPTER 21 Magnetism

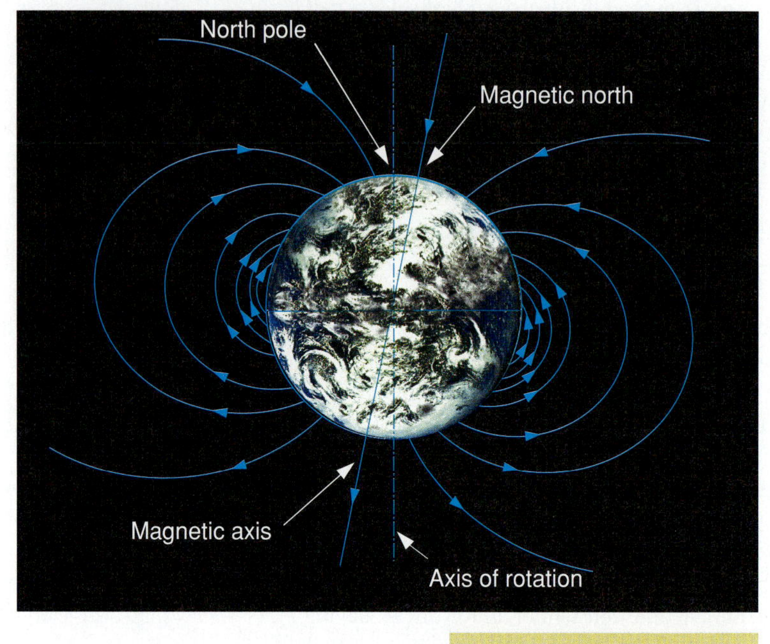

The earth's magnetic field.

Children are fascinated by the way a magnet exerts force on an object without touching it. This phenomenon has been known for thousands of years. The Greeks were familiar with magnetite, a natural magnetic rock found in abundance near a city in Turkey known then as Magnesia, now called Manisa.

In this chapter we shall see how study of the curious properties of magnets gradually led to an understanding of the "magnetic field" produced by magnets and other sources and also led to such practical applications as magnetic resonance imaging (Example 2), stereo speakers (Example 4), and electric motors (Example 5).

21-1 The Magnetic Field

Historical Background

By the eleventh century the Chinese were using a magnetic compass as a navigational aid. A compass needle is a long, thin magnet that is supported at its center so that it is free to pivot. If there are no other magnets nearby, one end of the needle always points north. This end is appropriately called the magnet's **north pole,** and the opposite end is called the **south pole.**

Every magnet has a north pole at one end and a south pole at the opposite end. It is near a magnet's poles that the magnetic force is strongest. This force can be either attractive or repulsive. **Like magnetic poles repel** one another, whereas **unlike magnetic poles attract** one another (Fig. 21-1). Note that the interaction of magnetic poles is similar to the interaction of electric charges (like charges repel; unlike charges attract). However, magnetic poles, unlike electric charges, are never found in isolation; they always occur in pairs.

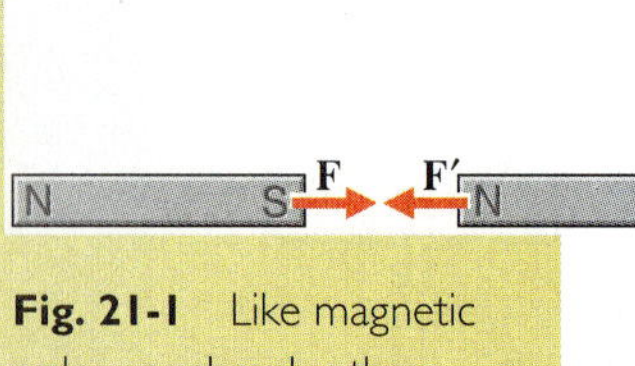

Fig. 21-1 Like magnetic poles repel each other; opposite poles attract each other.

In the early nineteenth century electricity and magnetism were regarded as unrelated phenomena. However, Hans Christian Oersted, a professor at the University of Copenhagen, suspected a connection. In 1820, while performing a lecture demonstration, he discovered that when a compass is placed close to a wire carrying an electric current, the compass needle turns (Fig. 21-2). Thus Oersted showed that electric current could be the source of a magnetic force. Other discoveries soon followed. Magnets were found to exert a force on current-carrying wires (the electric motor is an application of this principle). Within a few weeks of Oersted's discovery the French scientist André Ampère showed that two current-carrying wires exert a mutual magnetic force on each other. Ampère and two other French physicists, Biot and Savart, were able to give a complete quantitative description of the magnetic interaction of electric currents. Once he realized that electric current could exert a magnetic force, Ampère suggested that the magnetism of permanent magnets was also due to electric current, specifically, to permanent, microscopic current loops within the magnetic material. A more complete understanding of the nature of magnets had to await the twentieth-century development of quantum physics, but Ampère was essentially correct in supposing the origin of magnetism to be the interaction of moving charges.

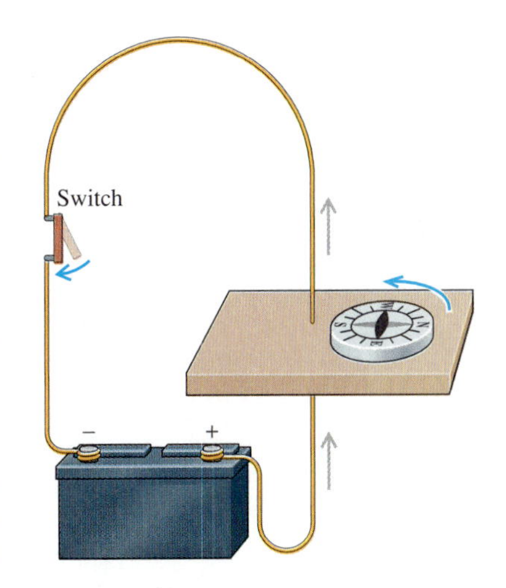

Fig. 21-2 When the switch is closed, the direction of the compass needle changes from north to west.

We have seen that an electric charge produces an electric field in the region of space around the charge and that this field exerts a force on other electric charges placed in the field. Similarly, both permanent magnets and electric currents produce another kind of field. This "magnetic field" exerts a force on other magnets or currents. In more fundamental terms, **the source of a magnetic field is moving charge, and the effect of a magnetic field is to exert a force on other moving charge placed in the field.**

The parallel between electric and magnetic fields is not accidental. There is a fundamental relationship between these fields. Under the most general conditions, electric charge produces both electric and magnetic fields. These general, time-varying "electromagnetic fields" will be discussed in Chapter 22. In certain special cases, electromagnetic fields reduce either to purely electric fields (produced by charges at rest, as studied in Chapter 17) or to purely magnetic fields (to be discussed in this chapter).

Definition of the Magnetic Field

In Sections 21-4 and 21-5 we shall see how the magnetic field can be calculated for various sources. For now, however, we can simply assume the existence of a magnetic field (produced perhaps by a permanent magnet) and explain the meaning of the field in terms of the effect it produces on a test charge.

The magnetic field, like the electric field, is a vector quantity. **We denote the magnetic field by the symbol B and represent it graphically by field lines,*** as we did for the electric field. Figs. 21-3 and 21-4 show magnetic field lines produced by various magnets.

The electric field was defined as the force per unit charge experienced by a test charge. For the magnetic field the definition is a bit more complicated. If we place a stationary test charge in a magnetic field (for example, near a permanent magnet), the charge experiences no force.

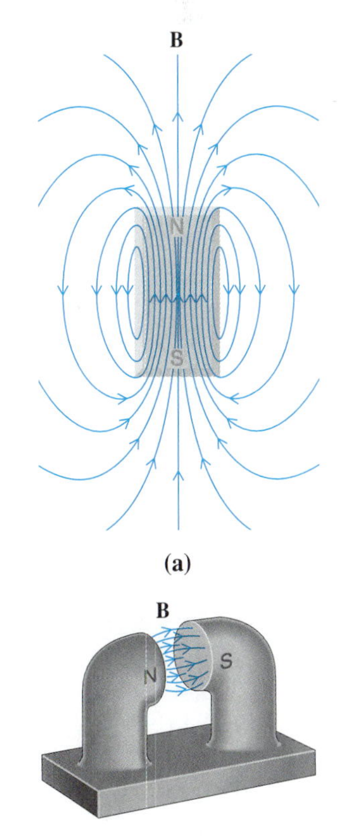

Fig. 21-3 Magnetic field lines produced by permanent magnets: **(a)** a bar magnet; **(b)** a magnet shaped to produce a uniform field in a gap—analogous to the electric field between capacitor plates.

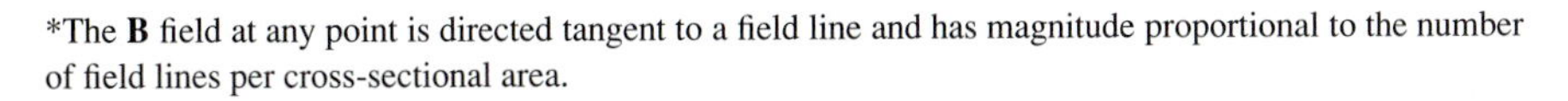
*The **B** field at any point is directed tangent to a field line and has magnitude proportional to the number of field lines per cross-sectional area.

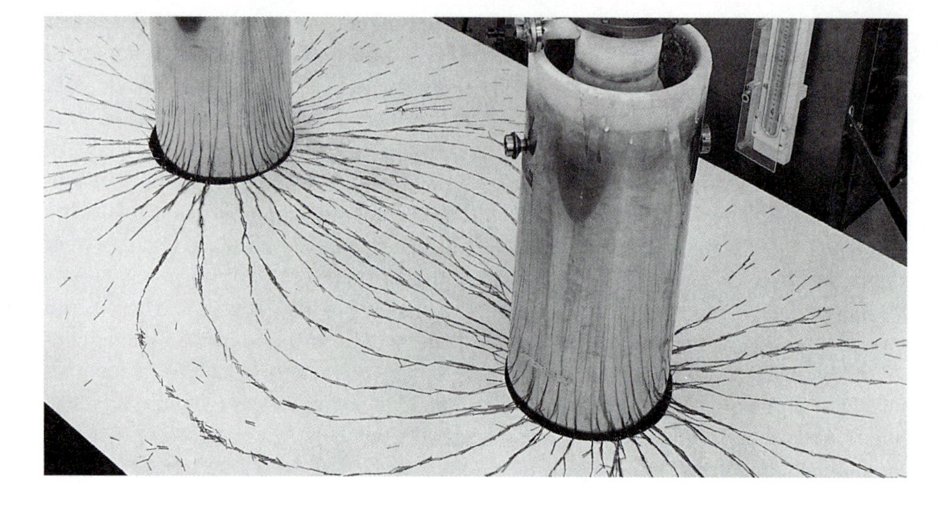

Fig. 21-4 Iron nails line up along field lines in the field of a strong superconducting electromagnet.

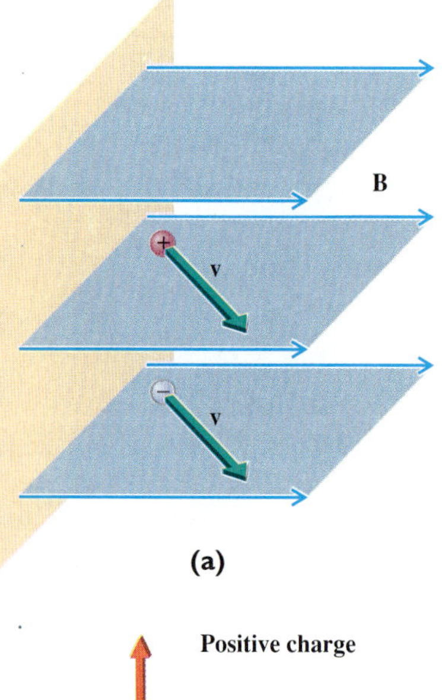

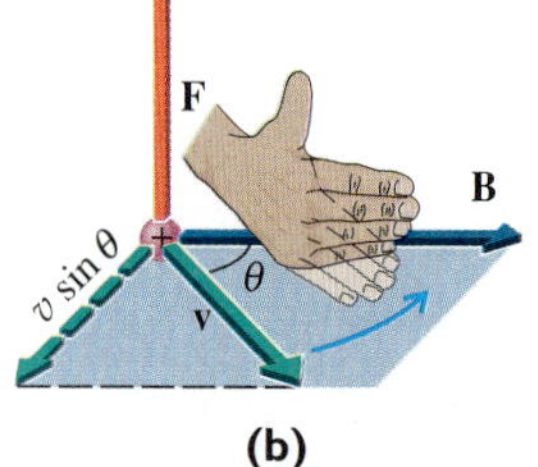

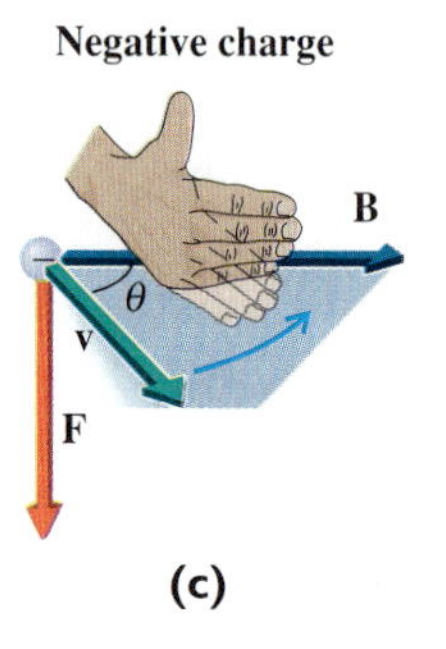

If, however, we give our test charge a velocity **v** having a component perpendicular to the magnetic field, the charge experiences a force. **This force is perpendicular to the plane formed by vectors v and B in a direction given by the right-hand rule** described in Fig. 21-5. This rule is used to determine the direction of the magnetic force relative to the directions of **v** and **B**.

The magnitude of the force **F** acting on a test charge in a magnetic field depends on the charge q, its speed v, the field strength B, and the angle θ between **v** and **B**:

$$F = |q|vB \sin \theta \tag{21-1}$$

Solving this equation for B, we obtain

$$B = \frac{F}{|q|v \sin \theta} \tag{21-2}$$

This expression defines the magnetic field in terms of the force **F** it exerts on a test charge q moving through it.

The SI unit of the magnetic field is the tesla (abbreviated T). From Eq. 21-2 we see that this unit may be defined in terms of the units of force, charge, and velocity:

$$1 \text{ T} = 1 \frac{\text{N}}{\text{C-m/s}} = 1 \frac{\text{N}}{\text{A-m}} \tag{21-3}$$

Another common unit of the magnetic field is the gauss (abbreviated G), where

$$1 \text{ G} = 10^{-4} \text{ T} \tag{21-4}$$

The earth's magnetic field is about $0.5 \text{ G} = 5 \times 10^{-5}$ T. Permanent magnets produce fields of up to about $10^4 \text{ G} = 1$ T. Electromagnets can produce even stronger fields. Superconducting electromagnets can produce fields of up to about 20 or 30 T, and even somewhat stronger fields in brief pulses.

Fig. 21-5 **(a)** Positive and negative charges moving in a magnetic field. **(b)** The direction of the instantaneous force acting on a positive charge is found by the right-hand rule: place the right hand along the **v** vector in such a way that the fingertips point in the direction of **v** and the fingers can rotate from **v** to **B**. When the hand is in this position, the extended right thumb points in the direction of the force. **(c)** The force on a negative charge is in the direction opposite that indicated by the right-hand rule.

Suppose we propel an object of mass 10 g and charge 10^{-6} C through a 1 T magnetic field at a velocity of 100 m/s, directed perpendicular to the field. From Eq. 21-1, we find $F = |q|vB \sin 90° = (10^{-6}\text{ C})(10^2\text{ m/s})(1\text{ T})(1) = 10^{-4}$ N. If no other forces act on the object, it will experience an acceleration of magnitude $a = F/m = 10^{-4}\text{ N}/10^{-2}\text{ kg} = 10^{-2}\text{ m/s}^2$. This small acceleration shows that the object will experience a negligible change in velocity in the brief time during which it shoots through the magnetic field (assumed to be limited to a small region of space). In order for a magnetic field to have a significant effect on a single charged object, the charge-to-mass ratio of the object must be much larger than the ratio in this example. Since it is difficult to place a sufficiently large charge on a macroscopic body, applications of magnetic force are ordinarily limited to very small particles—electrons, protons, ionized atoms, and so forth. In the next section we shall see how the magnetic forces acting on electrons in current-carrying wires can give rise to large macroscopic forces and practical applications, such as the electric motor.

EXAMPLE 1 Acceleration of an Electron in a Magnetic Field

Find the instantaneous acceleration of an electron that is moving at 1.0×10^7 m/s in the xy plane, at an angle of 30° with the y-axis. A uniform magnetic field of magnitude 10 T is in the positive y direction (Fig. 21-6a).

SOLUTION First we determine the instantaneous force on the electron. Vectors **v** and **B** both lie in the xy plane. The force vector **F**, which is always perpendicular to the plane of **v** and **B**, must therefore be in either the positive or negative z direction. Applying the right-hand rule, as indicated in Fig. 21-6b, we align the right hand along **v** so that the fingertips point in the direction of **v** and the fingertips can rotate from **v** to **B**; the extended right thumb points in the positive z direction. Since the electron is a negative charge, the force on it is in the opposite direction, that is, along the negative z-axis.

The magnitude of the magnetic force is given by Eq. 21-1:

$$F = |q|vB \sin \theta$$
$$= (1.6 \times 10^{-19}\text{ C})(1.0 \times 10^7\text{ m/s})(10\text{ T})(\sin 30°)$$
$$= 8.0 \times 10^{-12}\text{ N}$$

Applying Newton's second law, we find the electron's instantaneous acceleration **a**, which is directed along the negative z-axis and has magnitude

$$a = \frac{F}{m} = \frac{8.0 \times 10^{-12}\text{ N}}{9.1 \times 10^{-31}\text{ kg}} = 8.8 \times 10^{18}\text{ m/s}^2$$

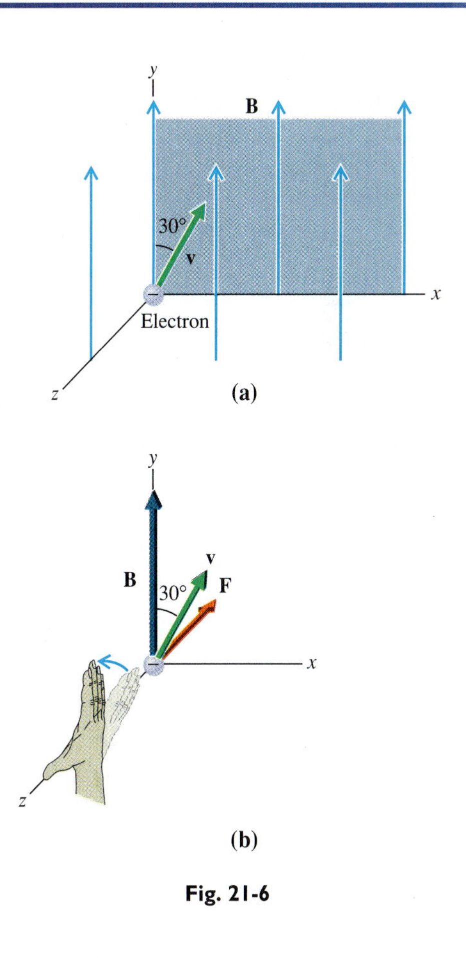

Fig. 21-6

EXAMPLE 2 Magnetic Resonance Imaging

Magnetic resonance imaging (MRI) is a medical diagnostic technique used to obtain high-resolution pictures of the body's internal soft tissue, which cannot be seen well using other imaging methods, such as X rays. The patient is exposed to a very strong magnetic field, on the order of a few teslas. (Watches or any other objects containing iron must be removed before one enters the field, to avoid having them pulled off by the magnet.) Some patients are concerned about possible harmful effects of such a strong field. Assuming there are no iron particles in the body, the most obvious possible effect of a static magnetic field* would be on ions moving through the bloodstream. These ions have their highest concentration and move most rapidly in the aorta, the large artery in which the blood's velocity is greatest, and so the magnetic force should be greatest there.

(a) Find the maximum magnetic force on a sodium ion of charge +e moving at 0.50 m/s through a 2.0 T magnetic field.

(b) Compare this magnetic force to the electric force exerted on the ion by a second sodium ion at a distance of 1.0×10^{-8} m, which is roughly the average distance between these ions in the blood.

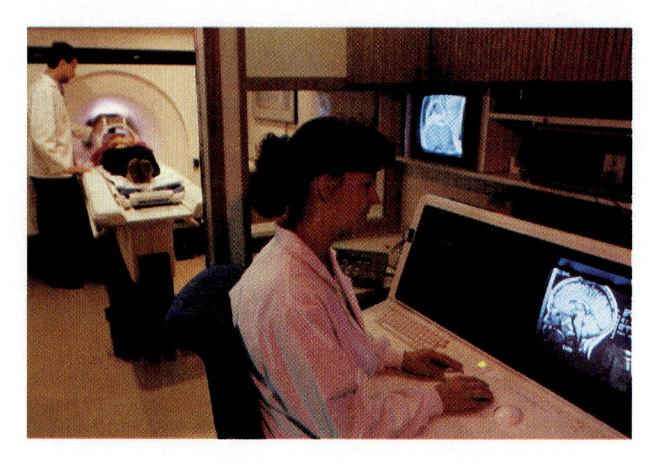

Fig. 21-7

SOLUTION **(a)** Applying Eq. 21-1 with $\sin\theta = 1$ so that the force is as large as possible, we find

$$F = |q|vB\sin\theta = (1.6 \times 10^{-19}\text{ C})(0.50\text{ m/s})(2.0\text{ T})(1)$$
$$= 1.6 \times 10^{-19}\text{ N}$$

(b) Applying Coulomb's law (Eq. 17-4), we calculate the interionic force:

$$F = k\frac{|q_1||q_2|}{r^2} = \frac{(9.0 \times 10^9\text{ N-m}^2/\text{C}^2)(1.6 \times 10^{-19}\text{ C})^2}{(1.0 \times 10^{-8}\text{ m})^2}$$
$$= 2.3 \times 10^{-12}\text{ N}$$

This value is more than 10^7 times greater than the magnetic force. So it appears that the magnetic force on ions in the bloodstream is negligible compared with other forces and should not pose a health hazard.

*There may also be cause for concern about the induction effects of the weaker, time-varying fields.

21-2 Magnetic Forces on Current-Carrying Conductors

When a current-carrying wire is placed in a magnetic field, each of the electrons contributing to the current experiences a magnetic force. Fig. 21-8a shows such a wire in a **B** field, along with an enlarged view of the interior of a section of the wire. The X's in the enlargement indicate that the $B_\perp$ component of **B** is going into the plane of the page. (Think of the X's as the end view of tail feathers on arrows representing **B** vectors.) The resultant external force **F** on the section of the wire is the vector sum of the forces on all the electrons contained in that section. To obtain the direction of the force $\mathbf{F}_e$ on each electron, we apply the right-hand rule: when we align the fingers so that they point in the direction of **v** and are able to rotate from **v** to **B**, the right thumb points downward. Since the electron has a negative charge, the force $\mathbf{F}_e$ is in the opposite direction, that is, upward, as indicated in the figure. All these parallel forces $\mathbf{F}_e$ on the electrons produce a resultant force **F** in the same direction.

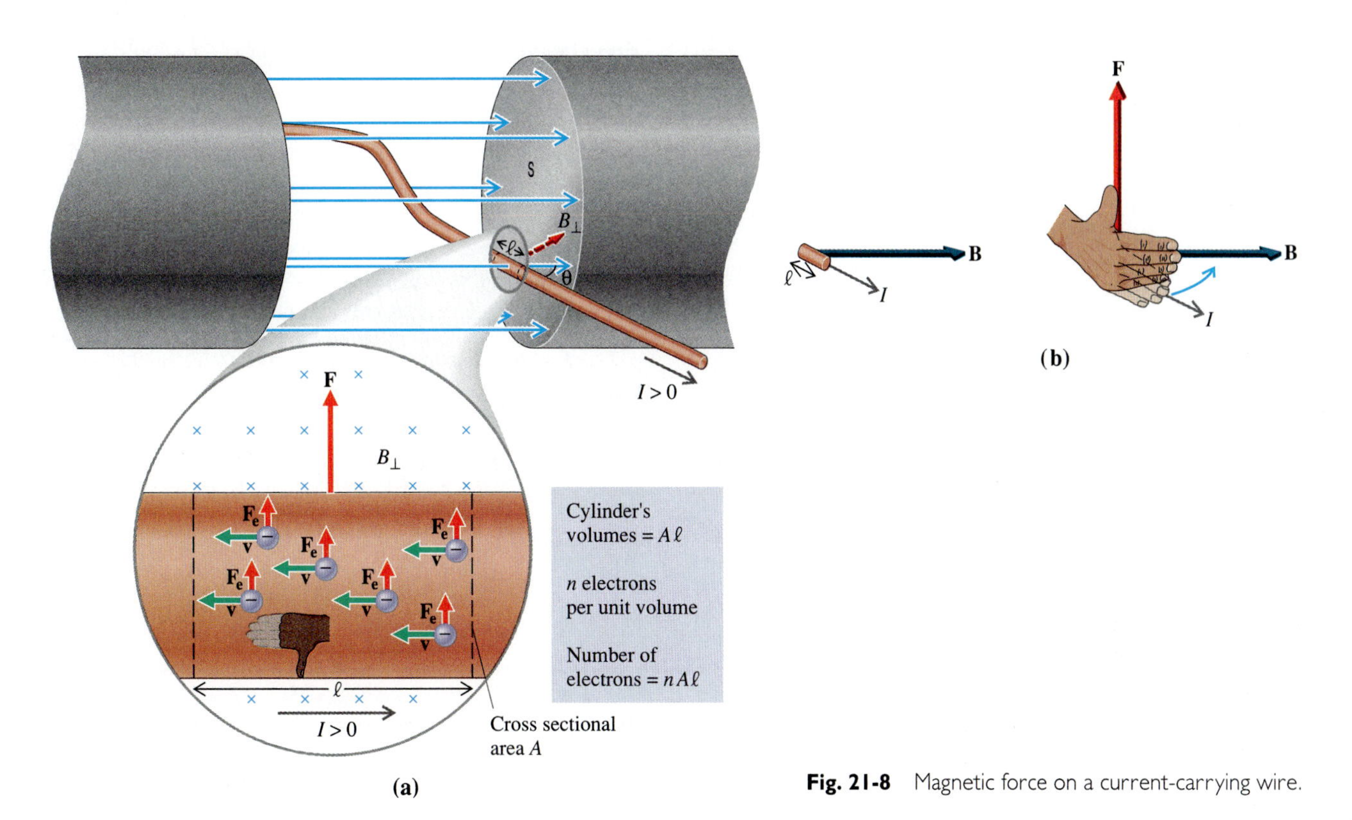

Fig. 21-8 Magnetic force on a current-carrying wire.

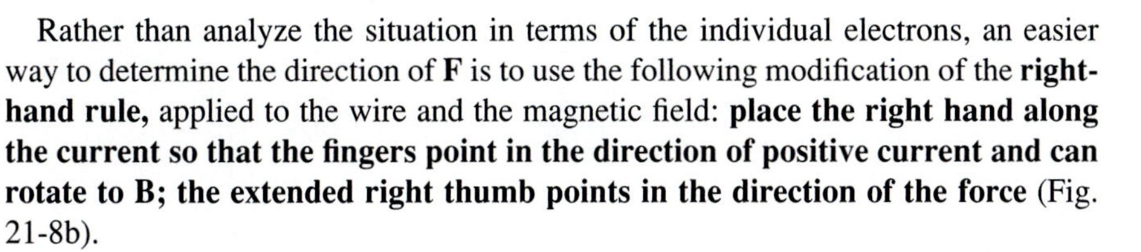

Rather than analyze the situation in terms of the individual electrons, an easier way to determine the direction of **F** is to use the following modification of the **right-hand rule,** applied to the wire and the magnetic field: **place the right hand along the current so that the fingers point in the direction of positive current and can rotate to B; the extended right thumb points in the direction of the force** (Fig. 21-8b).

The only component of the magnetic field **B** that contributes to the magnetic force on the wire is the component perpendicular to the electrons' velocity **v**; that is, only the component of **B** perpendicular to the wire contributes. This component $B_\perp = B \sin\theta$.

The magnitude of the force on a single electron moving at the average velocity **v** is given by Eq. 21-1:

$$F_e = |q|vB \sin\theta$$

As indicated in the figure, the number of electrons contained in a length ℓ of wire equals $nA\ell$, where n is the number of electrons per unit volume and A is the wire's cross sectional area. The magnitude of the total force F on the section equals the number of electrons times the force F_e on an average electron:

$$F = (nA\ell)|q|vB \sin\theta$$

Rearranging factors, we have

$$F = (n|q|vA)\ell B \sin\theta$$

The quantity in parentheses in the equation above is the current I in the wire, according to Eq. 19-25. Thus

$$F = I\ell B \sin\theta \tag{21-5}$$

EXAMPLE 3 Magnetic Force on a Wire

Fig. 21-9 shows a wire in the magnetic field of a permanent magnet. The field has a nearly uniform magnitude of 0.50 T over a cylindrical region (shaded in the drawing) of diameter 6.0 cm and is approximately zero elsewhere. Find the magnetic force on the wire, which carries a current of 40 A. The wire is perpendicular to the field and passes through the center of the cylindrical region.

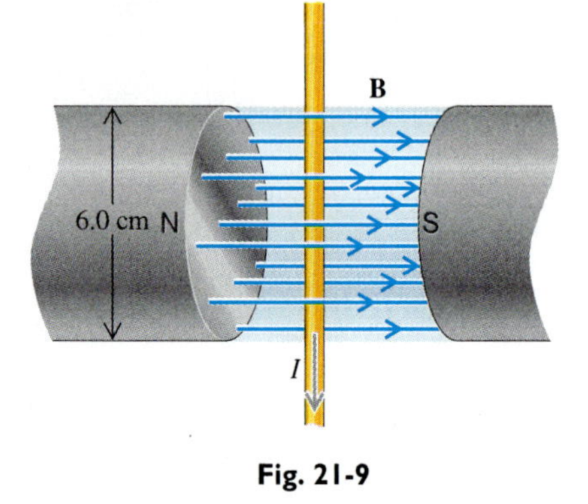

Fig. 21-9

SOLUTION First we apply the right-hand rule to find the direction of the magnetic force. When we place the right hand such that the fingertips point in the direction of I and the fingers can rotate toward the direction of **B**, the extended right thumb points out of the page, indicating that this is the direction of the magnetic force.

To find the magnitude of this force, we apply Eq. 21-5 to the section of the wire in the field—a section of length 6.0 cm. The current is perpendicular to the field, and so $\theta = 90°$:

$$F = I\ell B \sin\theta = (40\text{ A})(0.060\text{ m})(0.50\text{ T})(\sin 90°)$$
$$= 1.2\text{ N}$$

EXAMPLE 4 Magnetic Force Makes Music

Fig. 21-10 shows a stereo speaker, which produces sound waves when its cone, made of a stiff, lightweight paper, vibrates. The vibration is caused by a time-varying magnetic force applied to the cone. A coil of wire is wrapped around a hollow cylinder at the end of the cone. This cylinder slides back and forth inside a narrow cylindrical slot in a magnet that produces a radially directed magnetic field. The coil is connected to the stereo's amplifier, which provides a signal in the form of a time-varying current. The magnetic field exerts a force on the coil, and this force varies in magnitude and direction (left or right) according to the magnitude and direction of the current in the coil.

(a) Find the resultant magnetic force on the coil when the current is directed as shown if the magnetic field in the slot equals 0.20 T, the total length of wire in the coil is 4.0 m, and the current in the coil is 3.0 A.

(b) Estimate the instantaneous acceleration of the cone and coil if their combined mass is 40 g.

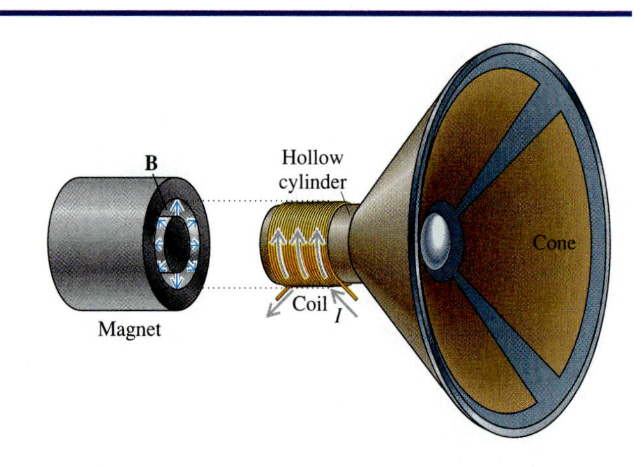

Fig. 21-10 View of a stereo speaker's cone, removed from slot in the speaker's magnet.

SOLUTION (a) Fig. 21-11 shows the forces acting on segments of one loop of the coil. When the current is directed clockwise as shown, the force on each loop segment $\Delta\ell$ is directed out of the page. (This direction is indicated by the symbol $\odot$. Think of this as an end view of a point on an arrow representing the force vector.) The force on each segment has magnitude ΔF given by Eq. 21-5, with $\theta = 90°$:

$$\Delta F = I\,\Delta\ell\,B \sin\theta = I\,\Delta\ell\,B \sin 90°$$
$$= I\,\Delta\ell\,B$$

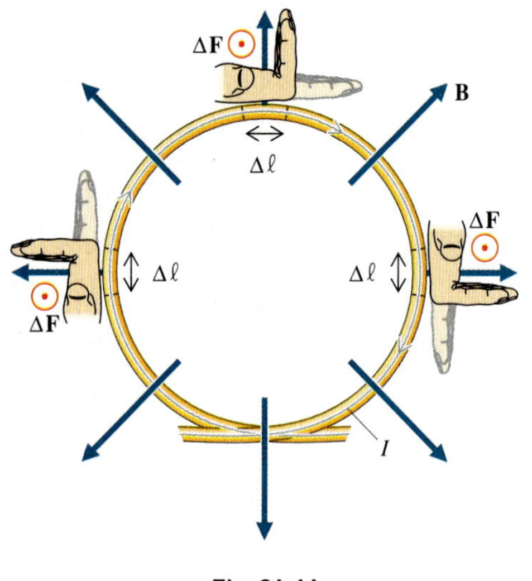

Fig. 21-11

EXAMPLE 4—cont'd

Each segment of each turn in the coil produces a force in the same direction, and so the total magnetic force on the coil has a magnitude F equal to the sum of the ΔF magnitudes. Summing the ΔF's is equivalent to replacing $\Delta \ell$ in the preceding equation by the sum of the $\Delta \ell$'s, the coil's total length ℓ:

$$F = I\ell B = (3.0 \text{ A})(4.0 \text{ m})(0.20 \text{ T})$$
$$= 2.4 \text{ N}$$

The direction of the magnetic force on the coil is perpendicular to the plane of the loops, in other words, to the right in Fig. 21-10. When the current in the coil changes direction, the force is to the left. If the current in the coil varies as a sine wave, the force varies accordingly. This means that the cone oscillates in SHM, which generates a harmonic sound wave.

(**b**) We ignore the force exerted on the cone by the material that holds the cone in its frame and estimate the acceleration using the magnetic force as the net force acting on the cone and coil:

$$a = \frac{F}{m} = \frac{2.4 \text{ N}}{40 \times 10^{-3} \text{ kg}} = 60 \text{ m/s}^2$$

Torque on a Current Loop in an External Field

Electric motors are an important application of the magnetic force on a current-carrying wire. Inside an electric motor, copper wire is wrapped around a rotor, a metal cylinder to which the motor's shaft is attached (Fig. 21-12). The rotor is mounted on bearings so that it is free to turn. When current passes through the wire, a magnetic field exerts forces on sections of the wire. The resultant magnetic force on the wire is zero, but the forces do produce a nonzero resultant torque on the wire. The rotor therefore turns and is able to do mechanical work.

The detailed design of practical motors is rather involved, but we can understand the basic mechanism by considering the effect of an external magnetic field on a rigid current-carrying coil (Fig. 21-13). The coil, which encloses a rectangular area A, consists of N parallel turns of wire, carrying current I. We define the **magnetic moment vector m** to be a vector perpendicular to the plane of the coil in the direction indicated in the figure, with a magnitude given by

$$m = NIA \qquad (21\text{-}6)$$

(This definition applies to a coil or current loop of any shape.)

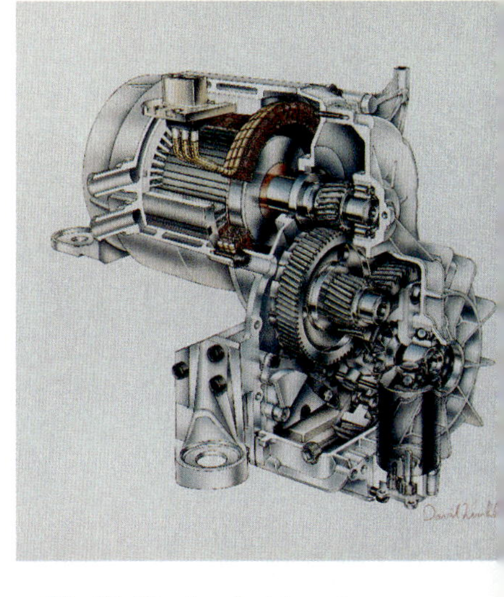

Fig. 21-12 An electric motor.

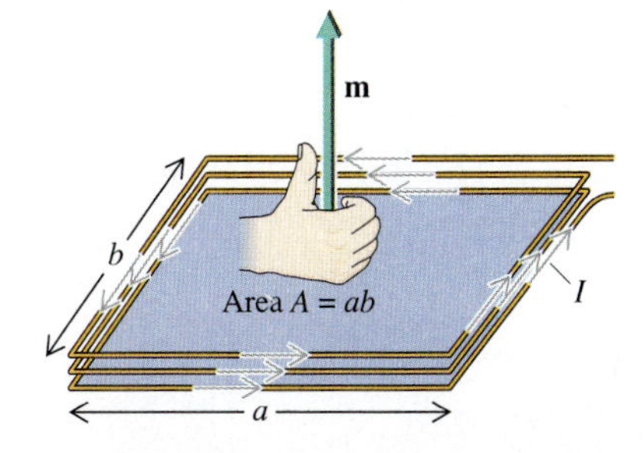

Fig. 21-13 The magnetic moment vector is perpendicular to the plane of the coil, in the direction of the right thumb when the fingers of the right hand curve around the coil in the direction of the current.

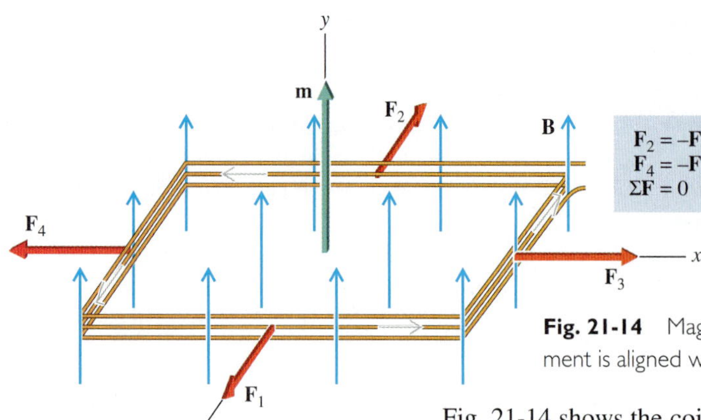

Fig. 21-14 Magnetic forces on a current-carrying coil when its magnetic moment is aligned with an external magnetic field.

Fig. 21-14 shows the coil in a uniform, external magnetic field. The **B** field is directed along the positive y-axis, perpendicular to the plane of the coil, so that **m** and **B** are in the same direction. The magnetic forces on the coil are found when we apply the right-hand rule and Eq. 21-5 to each side of the rectangle. Opposite sides carry opposite currents of equal magnitude. Consequently the forces acting on opposite sides are equal in magnitude and opposite in direction. Thus the net external force on the coil equals zero:

$$\Sigma \mathbf{F} = 0$$

For the orientation of the coil shown in Fig. 21-14, each pair of opposing forces acts along the same line and therefore produces no torque. However, for other orientations of the coil, the magnetic forces will produce a resultant torque. This is illustrated in Fig. 21-15a, where the coil is now shown rotated through an angle α about the z-axis. Forces on opposite sides still cancel, and so the resultant magnetic force on the coil is still zero. However, forces $\mathbf{F}_3$ and $\mathbf{F}_4$ now have different lines of action, and each force produces a torque that tends to rotate the coil about the z-axis. We can calculate the net torque using the side view of the coil shown in Fig. 21-15b. Each force tends to rotate the coil around the z-axis in the counterclockwise direction, and each force therefore gives a positive torque. Notice that **the torque tends to align the magnetic moment vector with B.**

The magnitude of either $\mathbf{F}_3$ or $\mathbf{F}_4$ is found by applying Eq. 21-5 ($F = I\ell B \sin\theta$) to each wire segment of length b, perpendicular to **B.** Since there are N turns, we have N segments on each side. Each segment experiences a force of magnitude $IbB \sin 90^\circ = IbB$. Thus the magnitude of the total force on either side is

$$F_3 = F_4 = NIbB$$

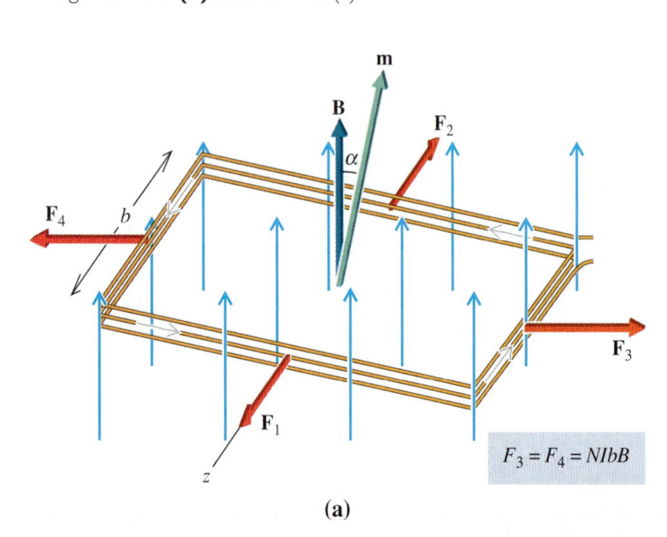

(a)

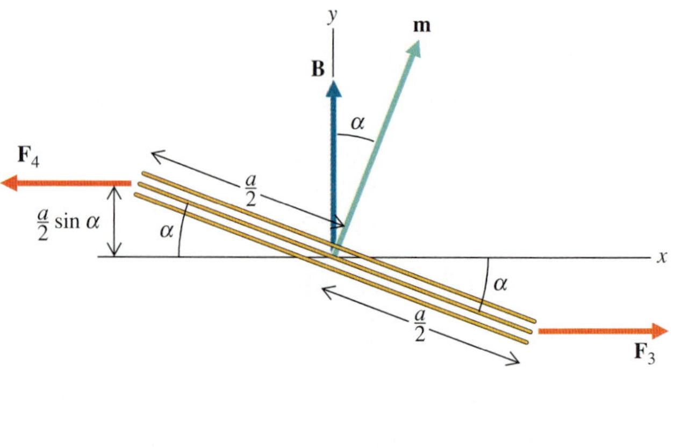

(b) Side view

Fig. 21-15 **(a)** Magnetic forces on a current-carrying coil when its magnetic moment is not aligned with an external magnetic field. **(b)** Side view of (a).

From Fig. 21-15b we see that the moment arm of each force equals $\frac{a}{2}\sin\alpha$. So each force contributes an equal torque, the product of force times moment arm, and thus the total magnetic torque is

$$\tau = 2\left(NIbB\frac{a}{2}\sin\alpha\right)$$

$$= NIabB\sin\alpha \qquad (21\text{-}7)$$

The coil's rectangular area $A = ab$, and so in the preceding equation the factor $NIab = NIA = m$, the magnitude of the coil's magnetic moment, according to Eq. 21-6. Thus we can express the torque more concisely as

$$\boxed{\tau = mB\sin\alpha} \qquad (21\text{-}8)$$

This result turns out to be valid for a coil or current loop of any shape.

EXAMPLE 5 Torque Produced by a Motor

Fig. 21-16 shows a diagram of an electric motor. Coils of wire are wrapped around the rotor in several planes. The commutator and brushes allow the current to pass through any coil only when that coil is in position to experience maximum magnetic torque; in other words, current passes through a coil only when the plane of the coil is approximately parallel to **B** or, equivalently, when its magnetic moment vector is approximately perpendicular to **B**. Each coil consists of 100 turns of wire, enclosing an area of 2.00×10^{-2} m^2. The current through the wire is 2.00 A, and the magnitude of the field is 0.500 T. Find the magnetic torque, assuming **m** and **B** are perpendicular to each other.

SOLUTION The magnitude of the coil's magnetic moment vector is found by applying Eq. 21-6:

$$m = NIA = (100)(2.00\text{ A})(2.00 \times 10^{-2}\text{ m}^2) = 4.00\text{ A-m}^2$$

The magnetic torque tends to rotate the rotor counterclockwise as viewed from the left side of the motor. The magnitude of the magnetic torque is given by Eq. 21-8:

$$\tau = mB\sin\alpha = (4.00\text{ A-m}^2)(0.500\text{ T})(\sin 90.0°) = 2.00\text{ N-m}$$

If this torque is not balanced by an opposing torque, provided by an external load, the rotor will experience an angular acceleration. If there is an opposing torque of equal magnitude, the net torque on the rotor equals zero and the angular velocity of the rotor is constant. Mechanical work is then provided by the motor. The rate at which this work is done is the mechanical power P, equal to the product of the torque τ and the angular velocity ω:*

$$P = \tau\omega \qquad (21\text{-}9)$$

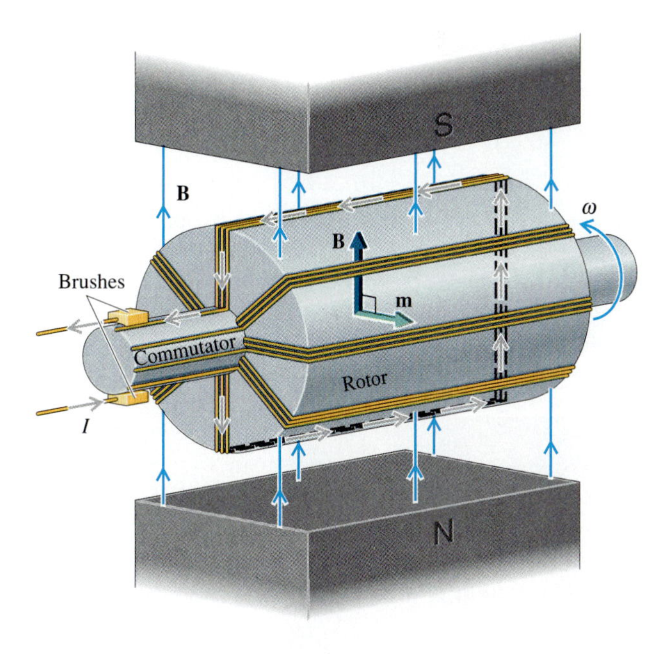

Fig. 21-16 An electric motor.

At an angular velocity of 1000 rpm, the electric motor described here would provide power

$$P = (2.00\text{ N-m})(1000\text{ rpm})\left(\frac{2\pi\text{ rad}}{1\text{ rev}}\right)\left(\frac{1\text{ min}}{60\text{ s}}\right)$$

$$= 209\text{ W}$$

or $$P = 209\text{ W}\left(\frac{1\text{ hp}}{746\text{ W}}\right) = 0.281\text{ hp}$$

*This equation is the rotational analog of Eq. 7-40: $P = Fv$.

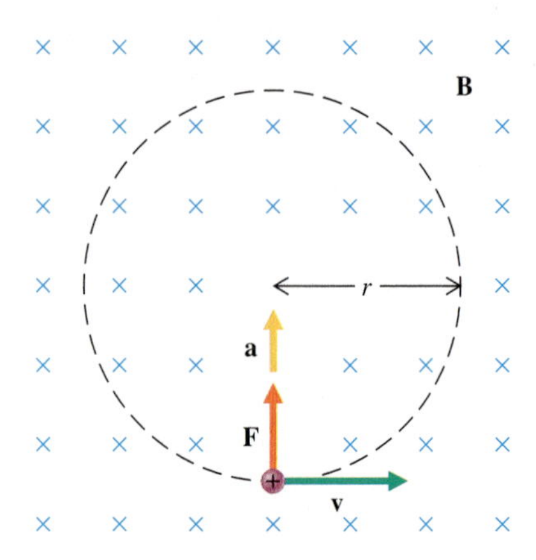

Fig. 21-17 A charge has an initial velocity perpendicular to a uniform magnetic field. The charge experiences a force perpendicular to its velocity, and so the charge moves along a circular path.

21-3 Motion of a Point Charge in a Magnetic Field

Uniform Magnetic Field

When projected into a magnetic field, a point charge will follow a trajectory that in some cases is quite simple. Since the magnetic force is always perpendicular to the charge's velocity, there can be no component of this force along the trajectory, and hence **the magnetic force does no work on the charge.** From the work-energy theorem, it follows that if the magnetic force is the only force acting on the charge then its kinetic energy is conserved. This means that **the charge's speed is constant;** the magnetic field can affect only the *direction* of motion.

Consider first the special case of a charge with initial velocity **v** perpendicular to a uniform magnetic field (Fig. 21-17). The charge experiences a force **F** of magnitude $F = |q|vB \sin 90° = |q|vB$. As we have seen, the speed v is constant, and so this equation shows that the magnitude of the force is therefore also constant. Thus at any instant the charge experiences an acceleration that is directed perpendicular to its instantaneous velocity and has a constant magnitude a, where

$$a = \frac{F}{m} = \frac{|q|vB}{m}$$

In Chapter 3 we found that acceleration of constant magnitude directed perpendicular to velocity is associated with circular motion. Thus the charge travels along a circular trajectory, and the equation for centripetal acceleration may be applied (Eq. 3-13):

$$a = \frac{v^2}{r}$$

Equating the two preceding expressions for a and solving for r, the radius of the circular path, we obtain

$$r = \frac{mv}{|q|B} \qquad (21\text{-}10)$$

Fig. 21-18 shows some paths of charges in magnetic fields.

The motion of the charge along its circular path is periodic. The period T equals the circumference of the circle divided by the speed of the charge:

$$T = \frac{2\pi r}{v}$$

Substituting for r from Eq. 21-10, we obtain

$$T = \frac{2\pi m}{|q|B} \qquad (21\text{-}11)$$

Notice that the period is independent of speed. Ernest Lawrence made use of this result in his design of the first **cyclotron** in 1932, for which he received the Nobel prize.

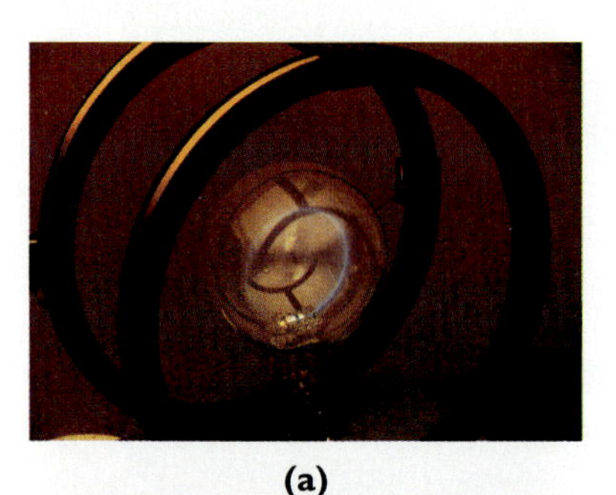

(a)

(b)

Fig. 21-18 (a) The circular path of an electron beam in a magnetic field is made visible by collisions of electrons with atoms of mercury vapor in an evacuated glass tube. After collisions, the atoms radiate light. (See Problem 28 for a discussion of this device, used to measure the electron's charge-to-mass ratio.) (b) A bubble chamber, the invention of which earned the Nobel Prize for Donald Glaser in 1960, shows the paths of various elementary particles created in high energy accelerators. By analyzing the motion of particles in a magnetic field, physicists have discovered many new particles and learned much about fundamental forces. The electrons shown here spiral inward as they lose energy because of their interaction with the bubble chamber; that is, as v decreases, r decreases, as predicted by Eq. 21-10.

EXAMPLE 6 Circular Electron Beam

A beam of electrons moving at 1.00×10^7 m/s describes a circular path in a uniform magnetic field of magnitude 1.00×10^{-3} T (Fig. 21-19a). Find (a) the direction of motion (clockwise or counterclockwise); (b) the period; (c) the radius of the circle.

SOLUTION **(a)** Both acceleration and magnetic force are directed toward the center of the circle. Knowing the direction of the force **F**, we can use the right-hand rule to determine the direction of the electron's motion. Since the electron has negative charge, the right thumb must be directed away from the center of the circle (that is, pointing in the direction of the force any *positive* charge would experience). The only way the thumb can point outward when the fingers are rotated from **v** to **B** is if the electron moves clockwise (Fig. 21-19b).

(b) The period is found when we apply Eq. 21-11:

$$T = \frac{2\pi m}{|q|B} = \frac{2\pi(9.11 \times 10^{-31}\ \text{kg})}{(1.60 \times 10^{-19}\ \text{C})(1.00 \times 10^{-3}\ \text{T})}$$

$$= 3.58 \times 10^{-8}\ \text{s}$$

(c) The radius of the path is found from Eq. 21-10:

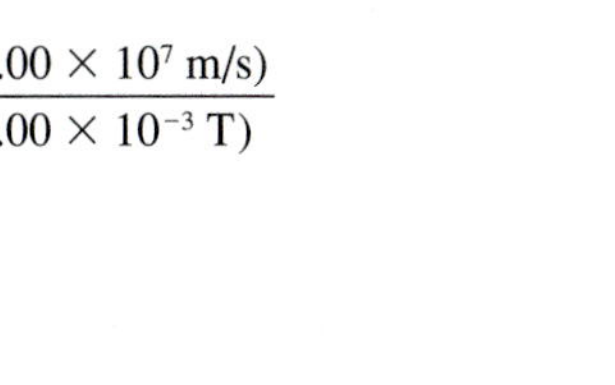

$$r = \frac{mv}{|q|B} = \frac{(9.11 \times 10^{-31}\ \text{kg})(1.00 \times 10^7\ \text{m/s})}{(1.60 \times 10^{-19}\ \text{C})(1.00 \times 10^{-3}\ \text{T})}$$

$$= 5.69 \times 10^{-2}\ \text{m} = 5.69\ \text{cm}$$

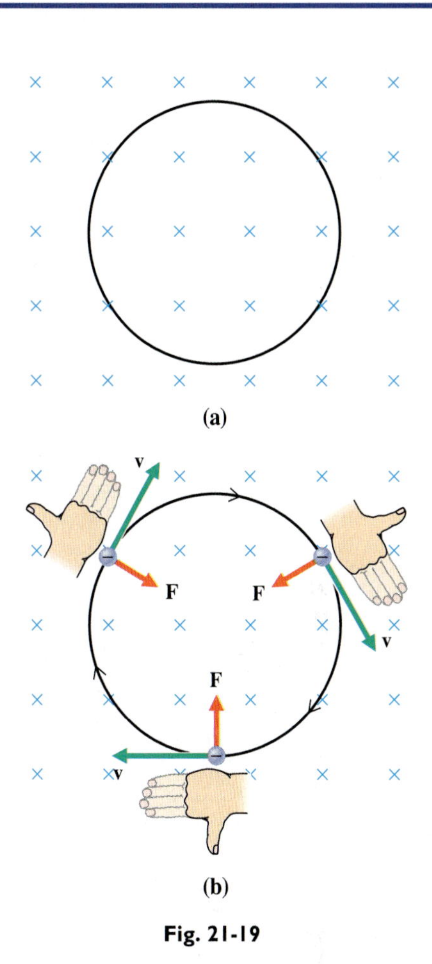

Fig. 21-19

EXAMPLE 7 Separating Atoms With Different Masses

The mass spectrometer is a device used to separate atoms or molecules according to mass. Chemists use mass spectrometers to help identify unknown molecules. Fig. 21-20 shows a simplified diagram of a mass spectrometer. The source molecules are first ionized and accelerated through a potential difference. They are then projected into a uniform magnetic field, which causes them to follow a circular trajectory with a radius r that depends on mass. Find the ratio of the masses m_1 and m_2 of two ions, corresponding to radii $r_1 = 21.0$ cm and $r_2 = 20.0$ cm. Assume that both are singly ionized atoms with the same initial velocity.

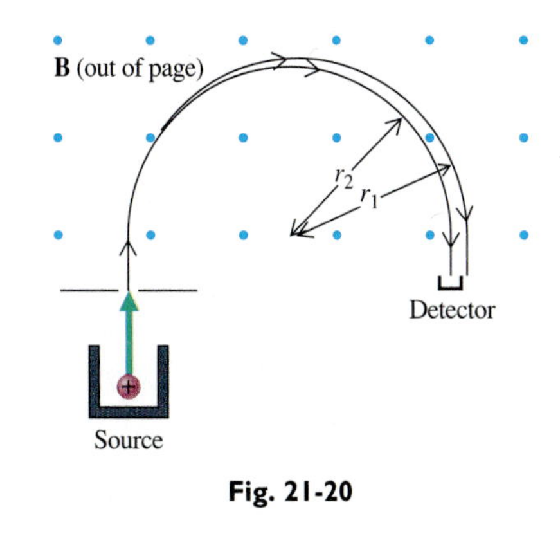

Fig. 21-20

SOLUTION From Eq. 21-10, we see that, since v, $|q|$, and B are constant, r is proportional to m. Thus

$$\frac{m_1}{m_2} = \frac{r_1}{r_2} = \frac{21.0\ \text{cm}}{20.0\ \text{cm}} = 1.05$$

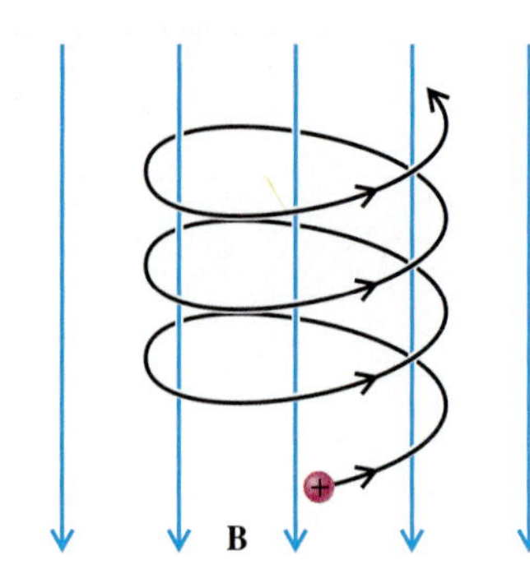

Fig. 21-21 A positive charge moving in a uniform magnetic field.

If a charge is projected into a uniform magnetic field with a component of velocity parallel to the field lines, the trajectory that results is a combination of circular motion in the plane perpendicular to the field and uniform motion along the field direction. This means that the path is a helix, as illustrated in Fig. 21-21 for a positive charge.

Nonuniform Magnetic Field

In a nonuniform magnetic field, the trajectory of a charged particle is more complicated. For example, in the field shown in Fig. 21-22a, positive charge first spirals upward with a decreasing radius and an upward component of velocity that decreases to zero and then spirals downward with an increasing radius and an increasing downward component of velocity. The nonuniform field reflects the particle downward and is referred to as a "magnetic mirror." A field configuration that combines two magnetic mirrors is called a "magnetic bottle" (Fig. 21-22b) and is used to confine charged particles. However, particles that have too large a component of velocity along the axis of the bottle will not be confined.

When a gas of high-energy, ionized particles, called a "plasma," is confined in a magnetic bottle, collisions between the particles give some particles enough velocity to escape. The confinement problem is important in the design of a fusion reactor, which utilizes sophisticated versions of the magnetic-bottle idea. The objective in a fusion reactor is to confine a high-density, high-temperature plasma for a sufficiently long time that nuclei will fuse, releasing energy. (The fusion process is the source of the sun's energy and will be discussed in Chapter 30.) Since fuel for such a reaction is abundant on earth, a successful fusion reactor could provide a cheap, virtually inexhaustible source of energy.

The confinement of charged particles in a magnetic field also occurs naturally in two regions outside the earth's atmosphere. These regions, called the "Van Allen belts" (Fig. 21-23), were discovered by Van Allen in 1958, using data from the first U.S. satellite. The belts are composed of "cosmic rays" (mostly protons and electrons), which are expelled by the sun or by distant stars and become trapped in the earth's magnetic field. Particles sometimes escape the belts and enter the earth's atmosphere near the poles, where they ionize the air, which then radiates a beautiful glowing light, known in northern latitudes as the "aurora borealis," or "northern lights" (Fig. 21-24).

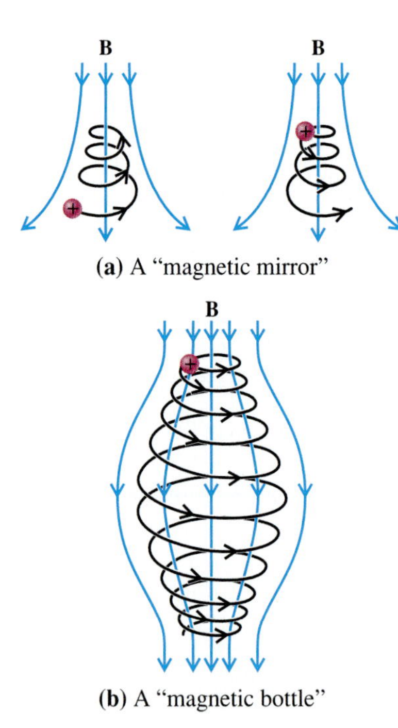

Fig. 21-22 Charges moving in nonuniform magnetic fields.

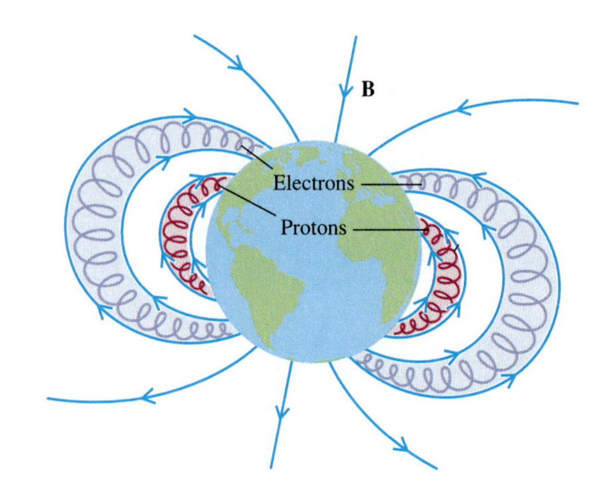

Fig. 21-23 Van Allen belts.

Fig. 21-24 Aurora borealis, as seen from space and from earth.

21-4 Magnetic Fields Produced by Electric Currents

In Chapter 17 we saw that the source of any electric field is electric charge. The electric field $\Delta\mathbf{E}$ produced by a small charge element Δq (assumed positive) is directed as indicated in Fig. 21-25 and has magnitude

$$|\Delta\mathbf{E}| = k\frac{|\Delta q|}{r^2}$$

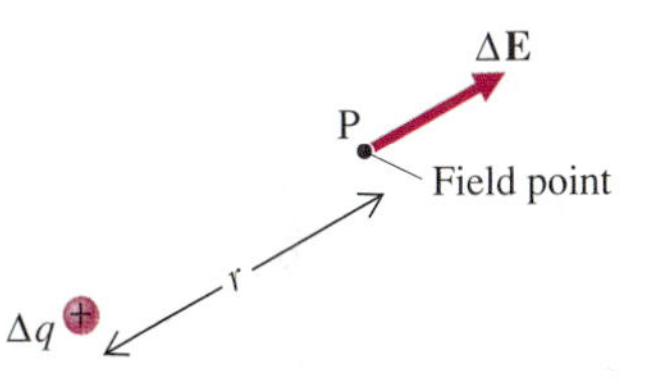

Fig. 21-25 The electric field at point P produced by a small, positive charge element Δq.

We can find the total electric field $\mathbf{E}$ produced by any distribution of charge by taking the vector sum of the fields $\Delta\mathbf{E}$ produced by charge elements Δq:

$$\mathbf{E} = \Sigma\,(\Delta\mathbf{E})$$

Biot-Savart Law

The source of any magnetic field is moving charge. In 1820 Biot and Savart discovered a law for finding the magnetic field $\Delta\mathbf{B}$ produced by the electric current in a small element of wire.* This result is analogous to the rule for finding the electric field of a charge element. The **Biot-Savart law** states that a small element of a wire of length $\Delta\ell$ carrying a current I produces at any field point a magnetic field $\Delta\mathbf{B}$. The field is perpendicular to the plane formed by the wire segment and the vector $\mathbf{r}$, a displacement vector from the wire segment to the field point. The direction of $\Delta\mathbf{B}$ is determined by the **right-hand rule** indicated in Fig. 21-26. The magnitude of the field is

$$|\Delta\mathbf{B}| = \frac{\mu_0}{4\pi}\left(\frac{I\,\Delta\ell\,\sin\theta}{r^2}\right) \tag{21-12}$$

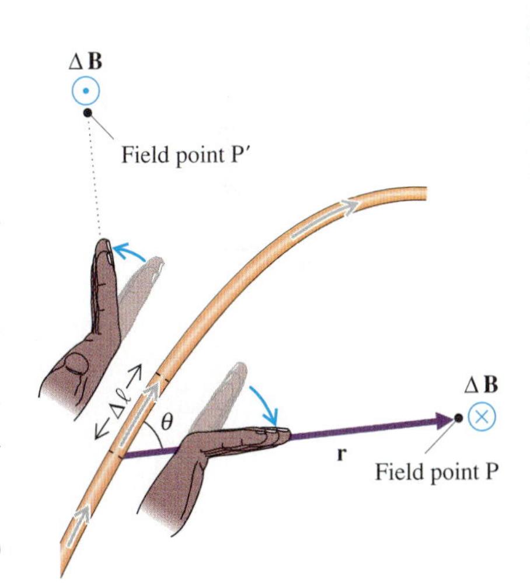

Fig. 21-26 Place the right hand along the wire in such a way that the fingertips point in the direction of the positive current and the fingers can rotate from the wire to the vector **r** locating the field point. When the hand is in this position, the extended right thumb points in the direction of $\Delta\mathbf{B}$. At point P the field is into the page, and at point P′ the field is out of the page.

where r (the magnitude of vector $\mathbf{r}$) is the distance from the current element to the field point, θ is the angle between the current element and $\mathbf{r}$, and the constant μ_0 (called the "permeability of free space") has the value

$$\mu_0 = 4\pi \times 10^{-7}\ \text{T-m/A} \tag{21-13}$$

*We could formulate an expression for the magnetic field of a single charge moving at velocity **v**. However, electric currents are a more common, practical source of magnetic fields. Therefore we shall take current elements as the basic elements for calculating magnetic fields.

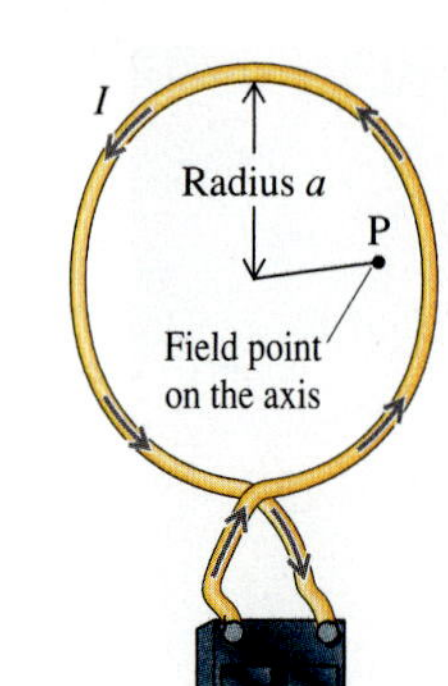

(a)

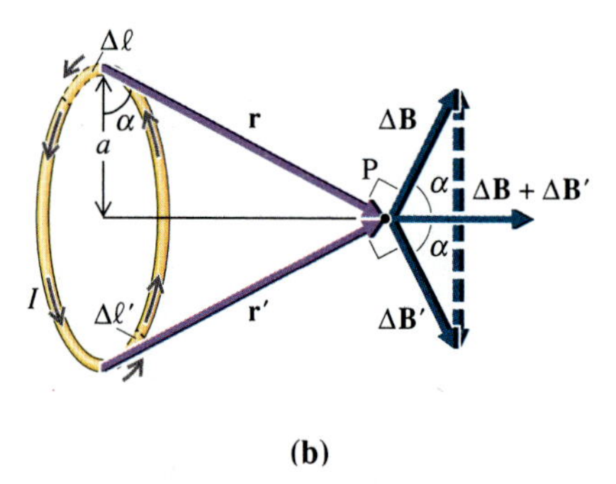

Fig. 21-27 The magnetic field of a circular current loop.

Just as the electric field produced by a charge distribution is the vector sum of the fields produced by charge elements within the distribution, the magnetic field of any distribution of electric currents is the vector sum of the fields produced by small elements of that current distribution:

$$\mathbf{B} = \Sigma\,(\Delta\mathbf{B}) \tag{21-14}$$

Circular Current Loop

We shall apply the Biot-Savart law to find an expression for the magnetic field produced by a circular current loop of radius a at any point P on the loop axis (Fig. 21-27a). Each current element $\Delta\ell$ is the same distance r from the field point and contributes a field $\Delta\mathbf{B}$ of the same magnitude and making the same angle α with the axis (Fig. 21-27b). The components of the $\Delta\mathbf{B}$'s perpendicular to the axis cancel in pairs, leaving only components $|\Delta\mathbf{B}|\cos\alpha$ along the axis to be summed. Thus the total magnetic field is directed along the axis and has magnitude

$$B = \Sigma\,(|\Delta\mathbf{B}|\cos\alpha)$$

where we see from the figure that $\cos\alpha = \dfrac{a}{r}$. The field $|\Delta\mathbf{B}|$ produced by an element $\Delta\ell$ is given by Eq. 21-12:

$$|\Delta\mathbf{B}| = \frac{\mu_0}{4\pi}\left(\frac{I\,\Delta\ell\,\sin\theta}{r^2}\right)$$

From the figure we see that the current in $\Delta\ell$ is perpendicular to $\mathbf{r}$, and so $\theta = 90°$. Using the value $\sin 90° = 1$ in the equation above and substituting for $|\Delta\mathbf{B}|$ and $\cos\alpha$ in the preceding expression for B, we obtain

$$B = \sum \frac{\mu_0}{4\pi}\left(\frac{Ia\,\Delta\ell}{r^3}\right) = \frac{\mu_0}{4\pi}\left(\frac{Ia}{r^3}\right)\Sigma\,(\Delta\ell)$$

When we sum the lengths $\Delta\ell$ of the current elements, we obtain the total length of the loop, which is the circle's circumference, equal to $2\pi a$. Inserting this value for $\Sigma\,(\Delta\ell)$ in the equation above, we find

$$B = \frac{\mu_0 I a^2}{2r^3} \tag{21-15}$$

With the current flowing around the loop in the direction indicated in Fig. 21-27, the field is directed along the axis *away* from the center of the loop. If the current were in the opposite direction, the direction of $\mathbf{B}$ would be *toward* the center of the circle.

A convenient way to remember the direction of the magnetic field relative to the current is to let the fingers of the right hand curve around the loop in the direction of the current; the extended right thumb then points in the direction of $\mathbf{B}$ (Fig. 21-28).

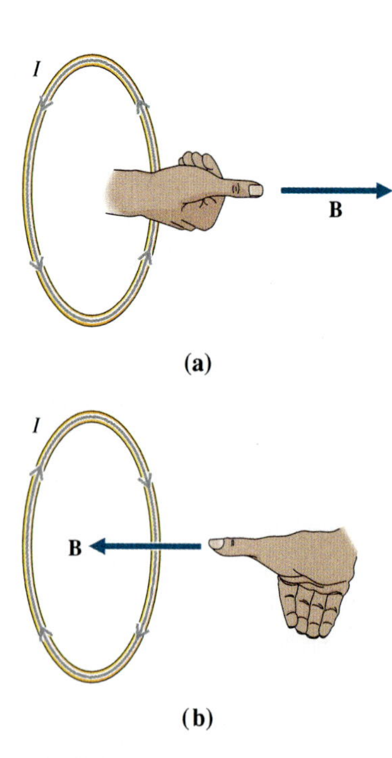

Fig. 21-28 The direction of the magnetic field produced by a circular current loop.

The magnetic field produced by a current loop may be expressed in an alternative form. Notice that the loop's magnetic moment $\mathbf{m}$ is in the direction of the magnetic field and has magnitude equal to the product of the current I and the area πa^2:

$$m = I(\pi a^2)$$

or

$$Ia^2 = \frac{m}{\pi}$$

Substituting into Eq. 21-15, we obtain

$$B = \frac{\mu_0}{2\pi}\left(\frac{m}{r^3}\right)$$

or, since **B** is in the same direction as **m**, we can express this as a vector equation:

$$\mathbf{B} = \frac{\mu_0}{2\pi}\left(\frac{\mathbf{m}}{r^3}\right) \qquad (21\text{-}16)$$

It turns out that this equation gives the magnetic field of a current loop of any shape, for a distant field point in the direction of **m**.

Iron filings show the magnetic field of a circular current loop.

EXAMPLE 8 The Magnetic Field of a Circular Current Loop

Find the magnetic field at the center of a circular loop of radius 2.00 cm, carrying a current of 10.0 A clockwise.

SOLUTION When we let the fingers of the right hand curve around the loop in the direction of the current, the extended right thumb points inward, indicating the direction of the magnetic field (Fig. 21-29). We find the magnitude of the field by applying Eq. 21-15:

$$B = \frac{\mu_0 I a^2}{2r^3}$$

Since the field point we are interested in is at the center of the circle, the distance r to the wire equals the circle's radius a. Thus

$$B = \frac{\mu_0 I}{2a} \qquad (21\text{-}17)$$

$$= \frac{(4\pi \times 10^{-7}\ \text{T-m/A})(10.0\ \text{A})}{2(2.00 \times 10^{-2}\ \text{m})} = \pi \times 10^{-4}\ \text{T} = \pi\ \text{G}$$

This field is only a little stronger than the earth's magnetic field and is typical of a magnetic field near a current-carrying wire.

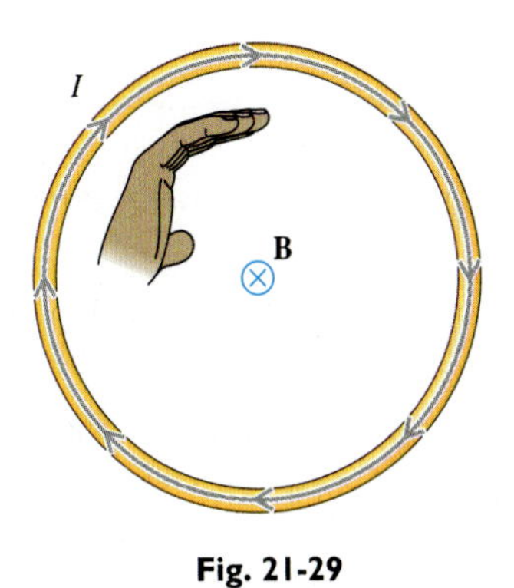

Fig. 21-29

Long Straight Wire

The magnetic field produced by a long straight wire can be determined by applying the Biot-Savart law for short current elements and summing, using Eqs. 21-12 and 21-14, as we did for the circular loop. Unfortunately, in the straight-wire case, it is more difficult to sum the $\Delta \mathbf{B}$'s, and the methods of integral calculus must be used. Therefore we shall simply state the result of the calculation without proof. The magnetic field lines are circles centered on the wire, with the field direction found by using the right hand as shown in Fig. 21-30: When the right thumb points along the direction of positive current, the fingers of the right hand curve in the direction of the field lines.

Notice that the field lines are complete circles, with no beginning and no end. It is a general characteristic of magnetic field lines that they always form closed paths. **Magnetic field lines never begin or end at any point in space.** This is in contrast to electric field lines, which begin on positive charge and end on negative charge.

The magnetic field's magnitude at a distance r from the wire is

$$B = \frac{\mu_0}{2\pi}\left(\frac{I}{r}\right) \qquad \text{(long, straight wire)} \qquad (21\text{-}18)$$

I

r P

(a) Side view

Inward current

Outward current

(b) End views

Fig. 21-30 The magnetic field of a long, straight, current-carrying wire.

Iron filings show the magnetic field of a long, straight wire.

EXAMPLE 9 Magnetic Field of a Long Wire

A long straight wire is directed perpendicular to the page and carries a current of 100 A into the page. Find the magnetic field at a field point 4.00 cm to the right of the wire.

SOLUTION Using the right hand, we find that the field lines are clockwise (Fig. 21-31). This means that at point P, the tangent to the field line is downward, indicating that the direction of $\mathbf{B}$ is downward at that point. Applying Eq. 21-18, we find the field's magnitude:

$$B = \frac{\mu_0}{2\pi}\frac{I}{r} = \frac{4\pi \times 10^{-7}\ \text{T-m/A}}{2\pi}\left(\frac{100\ \text{A}}{4.00 \times 10^{-2}\ \text{m}}\right)$$
$$= 5.00 \times 10^{-4}\ \text{T} = 5.00\ \text{G}$$

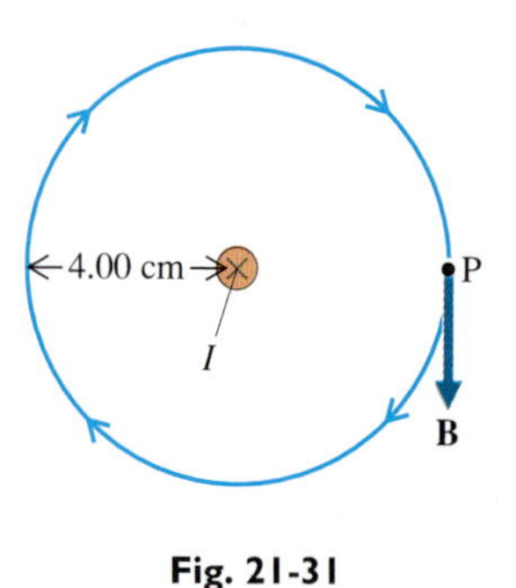

Fig. 21-31

Magnetic Force between Parallel Wires

When two current-carrying wires are placed close to each other, each wire may experience a force caused by the other wire's magnetic field. The simplest case is that of two long, straight, parallel wires. Suppose the wires carry current in the same direction, as shown in Fig. 21-32. The field $\mathbf{B}_1$, produced by current I_1 flowing through wire 1, is directed into the page at the location of wire 2. The magnitude of $\mathbf{B}_1$ at the location of wire 2, a distance r away from wire 1, is given by Eq. 21-18:

$$B_1 = \frac{\mu_0}{2\pi}\left(\frac{I_1}{r}\right)$$

The field $\mathbf{B}_1$ exerts a force on wire 2. We find the direction of this force by applying the right-hand rule: When we rotate the fingers of the right hand from the direction of I_2 (upward) to the direction of $\mathbf{B}_1$ (inward), the extended right thumb points to the left, indicating the direction of the force $\mathbf{F}_{12}$ that wire 1 exerts on wire 2. The magnitude of this force is given by Eq. 21-5:

$$F_{12} = I_2 \ell B_1 \sin\theta$$

Since the current is perpendicular to the magnetic field, $\theta = 90°$. Using $\sin 90° = 1$ in the equation above and inserting the expression for B_1, we obtain

$$F_{12} = \frac{\mu_0}{2\pi}\left(\frac{I_1 I_2 \ell}{r}\right) \qquad (21\text{-}19)$$

It is easy to show that the force $\mathbf{F}_{21}$ exerted by wire 2 on an equal length ℓ of wire 1 has the same magnitude and is directed toward the right, as indicated in the figure. In other words, there is a mutual attractive force between the two wires.

If the current in one of the wires were reversed, the force on each wire would be reversed (Problem 36), meaning that there would then be a mutual repulsive force. **Parallel wires carrying current in the same direction attract each other. Parallel wires carrying opposing currents repel each other.**

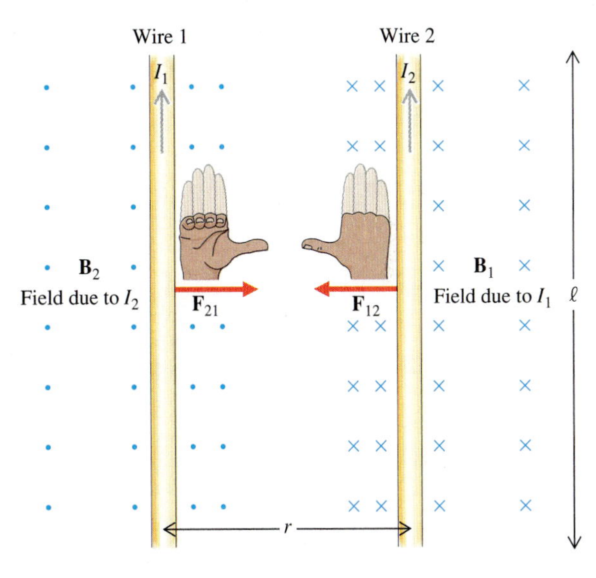

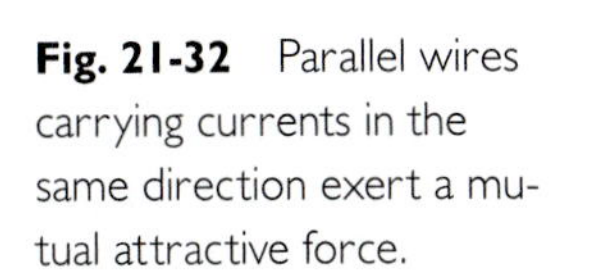

Fig. 21-32 Parallel wires carrying currents in the same direction exert a mutual attractive force.

EXAMPLE 10 Magnetic Force Between Parallel Wires

Two long, straight, parallel wires 1.00 cm apart each carry a current of 1.00 A in the same direction. Find the magnitude of the force on a 1.00 m length of either wire.

SOLUTION Applying Eq. 21-19, we find

$$F_{12} = \frac{\mu_0}{2\pi}\left(\frac{I_1 I_2 \ell}{r}\right) = \frac{(4\pi \times 10^{-7}\ \text{T-m/A})(1.00\ \text{A})(1.00\ \text{A})(1.00\ \text{m})}{2\pi(1.00 \times 10^{-2}\ \text{m})}$$

$$= 2.00 \times 10^{-5}\ \text{N}$$

Definition of the Ampere and the Coulomb

The definitions of the SI units of current and charge are based on measurement of the force between two parallel, current-carrying wires. The wires are connected in series, so that they carry the same current. The ampere can be defined as the current that produces on each wire a force of exactly 2×10^{-5} N per m of wire length when the separation between the wires is exactly 1 cm. The coloumb is defined as the quantity of charge that passes a point in a wire carrying a current of exactly 1 A in exactly 1 s.

Fig. 21-33 A solenoid.

Solenoids

A solenoid is a coil of wire wound around a cylindrical form, in the shape of a helix (Fig. 21-33). The purpose of the solenoid shape is to produce a strong magnetic field in a limited region of space. Solenoids are frequently used in electrical circuits. For example, a solenoid is used as part of an automobile's electrical starter system.

Fig. 21-34 shows the magnetic field produced by a loosely wound solenoid carrying a current I. Field lines are shown only in a single plane through the solenoid's axis. Notice that close to the wire the field is approximately that of a single straight wire; that is, the field lines are circles centered on the wire. Inside the solenoid the fields produced by all the current elements tend to be in the same direction, which produces a strong magnetic field along the solenoid's axis. Outside the solenoid the fields produced by the various current elements tend to cancel, and so the field outside is relatively weak.

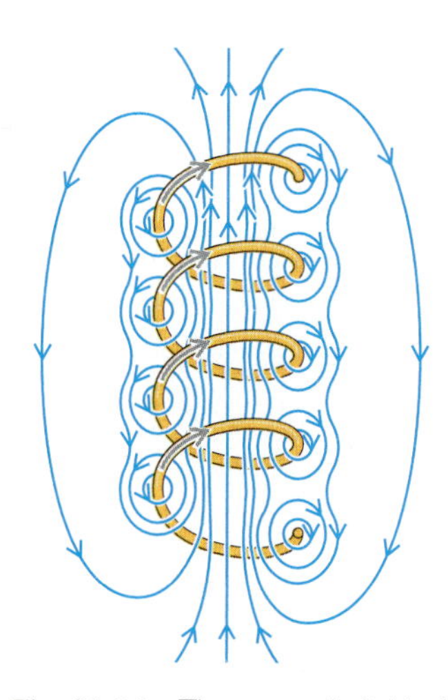

Fig. 21-34 The magnetic field of a loosely wound solenoid.

A tightly wound solenoid is nearly equivalent to a stack of identical circular current loops. We can define an **ideal solenoid** as an infinitely long stack of such loops. It is possible to derive an exact expression for the magnetic field of such a solenoid. However, a complete derivation of this result is quite difficult, and so we shall simply state the result without proof. The field outside the ideal solenoid is exactly equal to zero:

$$\mathbf{B} = 0 \quad \text{(outside an ideal solenoid)} \quad (21\text{-}20)$$

The field inside the ideal solenoid is directed along the solenoid's axis, in the direction indicated by the right thumb, when the fingers curve around the solenoid in the direction of positive current (the same rule as for a circular current loop). For a solenoid carrying current I, with n turns per unit of length along the axis, the magnetic field inside is

$$B = \mu_0 n I \quad \text{(inside an ideal solenoid)} \quad (21\text{-}21)$$

Fig. 21-35a shows the magnetic field of an ideal solenoid, and Fig. 21-35b shows the magnetic field of a tightly wound, real solenoid of finite length. The field inside the ideal solenoid is a good approximation to the field inside the real solenoid, except at field points very close to the wires or close to either end. Near the ends of the real solenoid, the field is approximately half the ideal solenoid field.

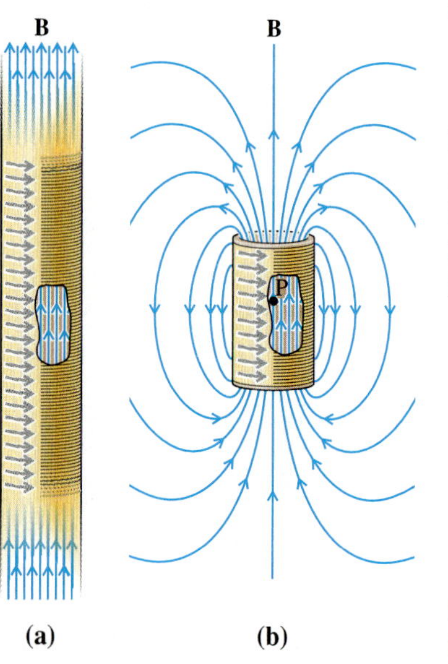

Fig. 21-35 The magnetic field produced by **(a)** an ideal solenoid; **(b)** a real solenoid.

Iron filings show the magnetic field of a solenoid.

EXAMPLE 11 Magnetic Field Inside a Solenoid

Find the approximate magnetic field at point P inside the real solenoid shown in Fig. 21-35b if the solenoid carries a current of 20.0 A and there are 10.0 turns for each cm of the solenoid's length.

SOLUTION For such a field point we can apply the ideal solenoid result, Eq. 21-21, with n = 10.0 per cm, or 1.00×10^3 per m:

$$B = \mu_0 nI = (4\pi \times 10^{-7}\ \text{T-m/A})(1.00 \times 10^3\ \text{m}^{-1})(20.0\ \text{A})$$

$$= 2.51 \times 10^{-2}\ \text{T} = 251\ \text{G}$$

21-5 Magnetic Fields Produced by Permanent Magnets

Electron Spin

The electron has a magnetic moment that manifests itself in various ways, including the magnetism of permanent magnets. We can think of the electron as a spinning sphere of negative charge. This spin is an intrinsic property of the electron; that is, the electron never stops spinning. Since the electron is charged, its spin produces a current and a magnetic moment.

EXAMPLE 12 Magnetic Field Near an Electron

Find the magnetic field produced by an electron's magnetic moment at a point 1.00×10^{-10} m from the electron in the direction of its magnetic moment vector, which has a magnitude of 9.27×10^{-24} A-m^2.

SOLUTION Since the field point is in the direction of **m**, we can apply Eq. 21-16:

$$\mathbf{B} = \frac{\mu_0}{2\pi}\left(\frac{\mathbf{m}}{r^3}\right)$$

Thus the electron's magnetic field at this point is in the direction of the magnetic moment vector **m** and has magnitude

$$B = \frac{\mu_0}{2\pi}\left(\frac{m}{r^3}\right) = \frac{(4\pi \times 10^{-7}\ \text{T-m/A})(9.27 \times 10^{-24}\ \text{A-m}^2)}{2\pi(1.00 \times 10^{-10}\ \text{m})^3}$$

$$= 1.85\ \text{T}$$

Thus an electron produces a strong magnetic field at a distance of 10^{-10} m, roughly equal to the diameter of an atom. This is an indication of the strong magnetic fields at much greater distances produced by the collective effect of all the electrons in a permanent magnet.

Magnetization

Suppose that we place a piece of iron in a very strong external magnetic field. We could do this, for example, by placing it inside a solenoid carrying a large current. Electron magnetic moments within the iron will align with the external field (Fig. 21-36a). If we then turn off the external field (by turning off the current in the solenoid, for example), the alignment of magnetic moments will remain (Fig. 21-36b).* The iron is said to be magnetized. The aligned magnetic moments are equivalent to currents circulating in the plane perpendicular to the magnetic moment vectors. These currents will of course produce their own magnetic field.

Consider a very small volume within the magnetized iron (Fig. 21-37). Since the electrons are all aligned, with their magnetic moment vectors in the same direction, the magnetic moment of the volume equals the number N of electrons within the volume times the magnetic moment $\mathbf{m}_e$ of a single electron:

$$\mathbf{m} = N\mathbf{m}_e \tag{21-22}$$

Since N is proportional to the volume, this equation shows that the magnetic moment vector is likewise proportional to the volume. It is therefore useful to define a quantity called the **magnetization M** as the **magnetic moment per unit volume:**

$$\mathbf{M} = \frac{\mathbf{m}}{V} \tag{21-23}$$

Combining the two preceding equations, we find

$$\mathbf{M} = \frac{N}{V}\mathbf{m}_e \tag{21-24}$$

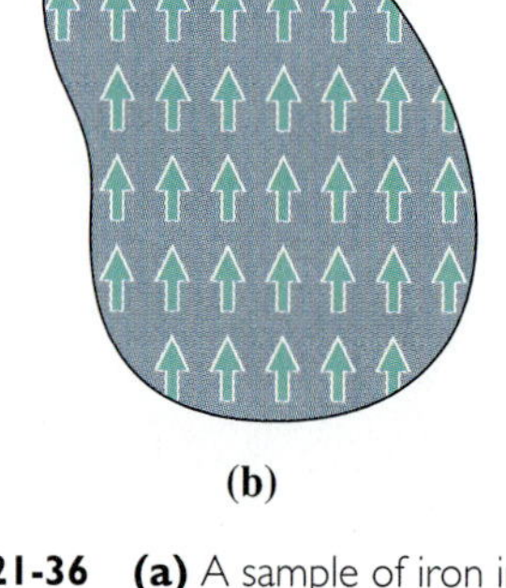

Fig. 21-36 **(a)** A sample of iron in a strong external magnetic field. Electrons' magnetic moments are aligned with the field. **(b)** The same sample after the external field is turned off. Electron magnetic moments remain aligned.

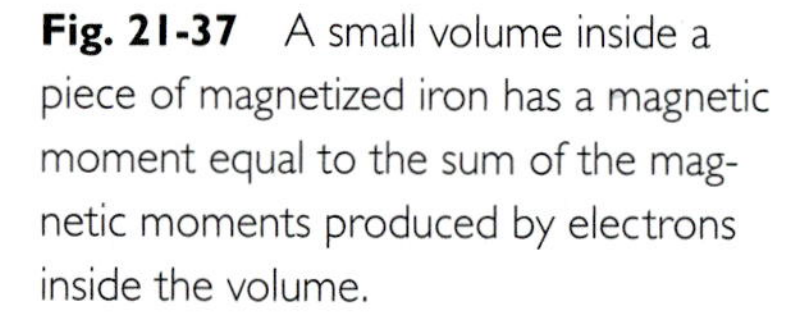

Fig. 21-37 A small volume inside a piece of magnetized iron has a magnetic moment equal to the sum of the magnetic moments produced by electrons inside the volume.

EXAMPLE 13 How Many Atomic Electrons Contribute to Iron's Magnetization?

The maximum magnetization of iron is 1.7×10^6 A/m. Calculate the number of aligned electron magnetic moments per unit volume of the iron.

SOLUTION Applying Eq. 21-24 and using the value for m_e given in the last example, we find

$$\frac{N}{V} = \frac{M}{m_e} = \frac{1.7 \times 10^6 \text{ A/m}}{9.27 \times 10^{-24} \text{ A-m}^2} = 1.8 \times 10^{29}/\text{m}^3$$

There are approximately 9×10^{28} iron atoms per m³ of volume. Our calculation therefore indicates that two electron spins per atom are aligned in fully magnetized iron.

*The picture we present here is somewhat oversimplified. After the external field is turned off, the alignment of all electron spins is no longer complete. There will be "domains" within the sample in which spins are aligned in a different direction, as discussed in the next section.

Magnets and Solenoids

A cylindrical magnet having uniform magnetization **M** along its axis produces a magnetic field that is identical to that of a solenoid with the same dimensions, if the product of the solenoid's number of turns per unit length n and current I equals the magnet's magnetization M:

$$M = nI \tag{21-25}$$

For example, the magnetic field of the solenoid shown in Fig. 21-35b is identical to the field of the magnet shown in Fig. 21-3a. The proof of this correspondence between magnets and solenoids is outlined in Problem 50, but we can qualitatively understand the similarity between the two by using Fig. 21-38. A thin cross section of the magnet can be divided into small volume elements, each with its own magnetic moment. Although the magnetic moment of a volume element is produced by electron spins, the same magnetic moment could be produced by current circulating around the sides of the volume element, as indicated in the figure. Since the magnetic field of the magnet is determined solely by the magnetic moments of its volume elements, such a current would produce the same magnetic field as the electron spins.

The current around the side of any volume element is exactly cancelled by current around the side of an adjacent element. The exceptions are the sides that are at the surface of the cylinder. The currents around these sides are not cancelled, and they give a current around the surface of the cylinder. Such a surface current is exactly like the current produced by a tightly wound solenoid.

We can apply the correspondence between a cylindrical magnet and a solenoid to obtain an expression for the magnetic field produced by a long, cylindrical magnet at field points far from the magnet's ends. Using the expressions for the field of an ideal solenoid (Eqs. 21-20 and 21-21) and Eq. 21-25 to substitute $nI = M$, we find

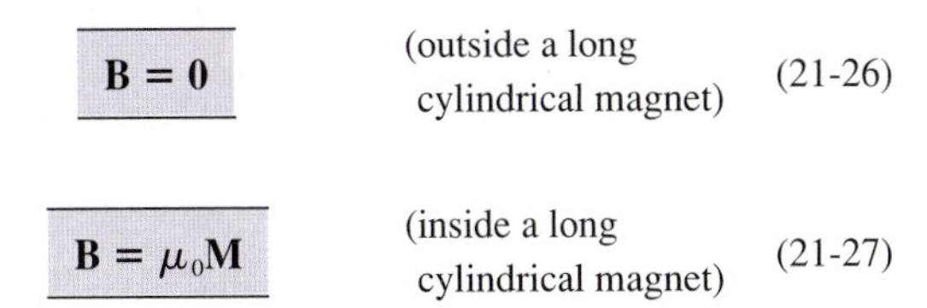

$$\mathbf{B} = 0 \quad \text{(outside a long cylindrical magnet)} \tag{21-26}$$

$$\mathbf{B} = \mu_0 \mathbf{M} \quad \text{(inside a long cylindrical magnet)} \tag{21-27}$$

It should be emphasized that these equations are only approximate, and the approximation is good only for a long magnet and only at field points that are far from the magnet's poles.

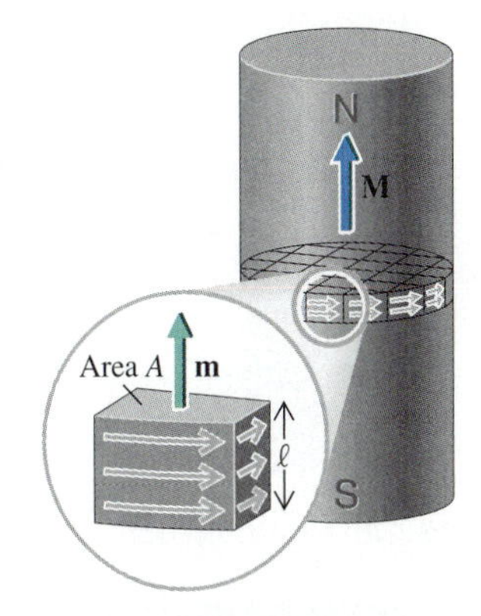

(a) Uniformly magnetized cylindrical magnet

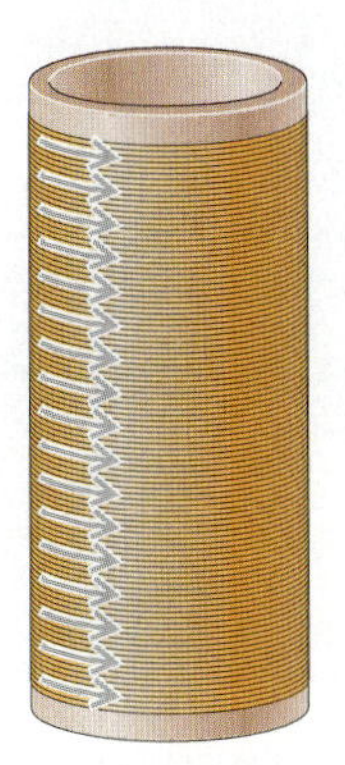

(b) Solenoid

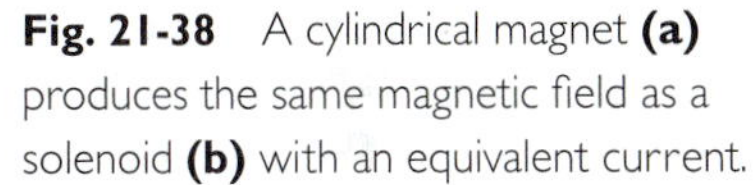

Fig. 21-38 A cylindrical magnet **(a)** produces the same magnetic field as a solenoid **(b)** with an equivalent current.

EXAMPLE 14 Magnetic Field Inside a Long, Cylindrical Magnet

Find the magnetic field within a long, cylindrical, iron magnet, which has a uniform magnetization along its axis of 1.7×10^6 A/m, assuming the field point is far from the ends of the magnet.

SOLUTION Applying Eq. 21-27, we find that the magnetic field within the magnet is in the same direction as the magnetization, that is, along the axis, and has magnitude

$$B = \mu_0 M = (4\pi \times 10^{-7}\ \text{T-m/A})(1.7 \times 10^6\ \text{A/m})$$
$$= 2.1\ \text{T}$$

This magnetic field is much stronger than any of the others we have calculated in preceding examples. The magnetic field produced by strongly magnetized matter is always much greater than the magnetic field produced by current in a wire, unless the wire is a superconductor carrying a very large current.

Fig. 21-39 An electromagnet.

An iron core is often placed inside a solenoid to enhance its magnetic field. Such an arrangement is called an **electromagnet** (Fig. 21-39). The total magnetic field is the sum of the fields produced by the solenoid current and by the iron; thus the core produces a strong resultant field* (Fig. 21-40). The field is especially strong if the coil of wire is superconducting and can therefore carry a very large current without overheating.

Fig. 21-40 This electromagnet exerts a magnetic force strong enough to lift a car.

*21-6 Magnetic Materials

Iron is not the only material that can be magnetized. Indeed, *any material* at all when placed in an external magnetic field will have within it some magnetization. In most materials, however, the magnetization is very small and is directed *opposite* the external field. These materials are called **diamagnetic.** Almost all other materials, when placed in an external magnetic field, develop a small magnetization in the direction of the external field. Such materials are called **paramagnetic.** Graphite, quartz, and glass are all diamagnetic. Aluminum, platinum, and sodium are paramagnetic. The magnetization in a typical paramagnet or diamagnet is less than one millionth the magnetization in iron placed in the same external field. In both paramagnets and diamagnets, the magnetization disappears as soon as the external field is turned off.

In addition to iron, there are a few other special materials that can have very large values of magnetization and retain their magnetization even after removal from an external field. These materials are called **ferromagnetic,** after the word *ferrum*, which is Latin for 'iron'. Iron, cobalt, nickel, gadolinium, and dysprosium are the only ferromagnetic elements. Various alloys and compounds are also ferromagnetic, including of course magnetite, the original magnetic material.

Any macroscopic piece of matter contains an enormous number of electrons, each of which has an intrinsic spin. We might wonder, then, why the phenomenon of ferromagnetism is relatively rare. The answer is related to the fact that most of the electrons in any atom tend to be paired with electrons of opposite spin. (We shall discuss this aspect of atomic structure at greater length in Chapter 29 on atomic physics.) Each such pair of electrons has two opposing magnetic moment vectors and therefore contributes no net magnetic moment to the atom. In most materials the atoms have no magnetic moment, and the materials are then diamagnetic. In some materials, however, each atom has a net magnetic moment from electrons whose magnetic moments are not cancelled out by other electrons. There is then the possibility that the material will be either paramagnetic or ferromagnetic.

In the paramagnet, the atomic magnetic moments are independent of each other. In the presence of an external magnetic field, each magnetic moment experiences a torque that tends to align it with the field. But this tendency to align is opposed by a tendency toward randomness in orientation because of thermal motion. The net effect is a small overall magnetization in the direction of the external field (Fig. 21-41a). The magnetization disappears when the field is removed (Fig. 21-41b).

In a ferromagnet, atomic magnetic moments are not independent of each other.

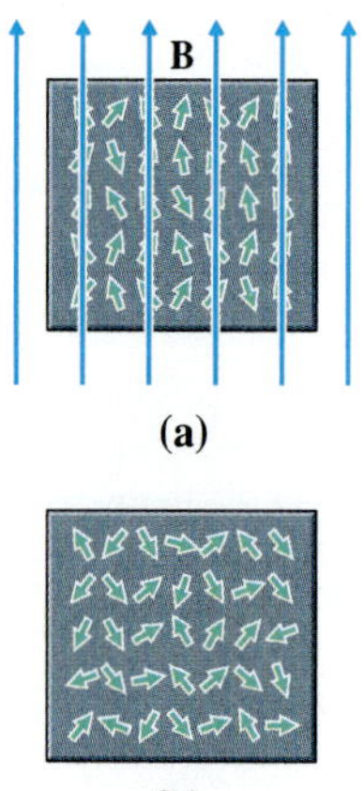

Fig. 21-41 **(a)** Atomic magnetic moments within a paramagnet in a strong, external magnetic field, directed upward. The atoms produce a small upward magnetization. **(b)** With no external magnetic field, the paramagnet's atomic magnetic moments are oriented completely randomly, and there is therefore no magnetization.

*The main function of the coil in an electromagnet is to keep the magnetization within the iron maximized in one direction, in other words, to align domains within the iron. See the next section for a discussion of domains.

The electrons that give an atom its net magnetic moment interact with the corresponding electrons in neighboring atoms in a way that tends to align the magnetic moments. These electrons experience a mutual force, called the "exchange force," the nature of which is impossible to explain without a knowledge of quantum theory. Here we simply note in passing that the exchange force is related to the electrostatic repulsive force between the interacting electrons. There are only a few materials in which the electrons that give the atom its magnetic moment are favorably located within the atom in such a way that the exchange force becomes a significant effect. Thus only these few materials are ferromagnetic.

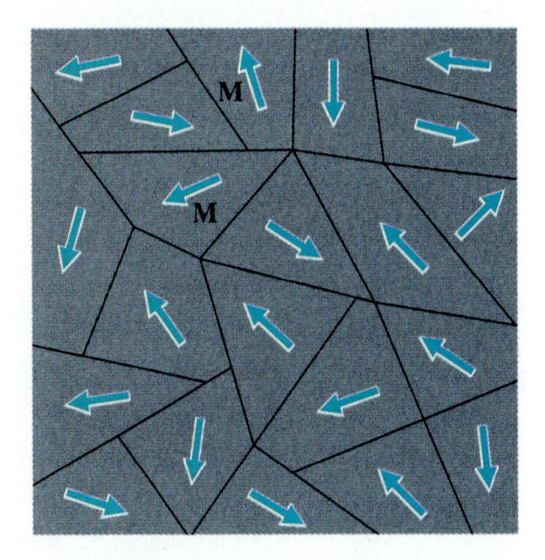

Fig. 21-42 Magnetic domains within an unmagnetized ferromagnet.

Ferromagnetism is the cooperative, collective effect of an enormous number of interacting electrons. When a typical ferromagnetic material at room temperature is placed in a strong external magnetic field, the torque exerted on the electrons' magnetic moments by the external field is aided by the internal exchange force between the electrons. The result may be a nearly perfect alignment of all the magnetic moments and a large magnetization, as indicated for iron in Fig. 21-36 and Example 13. When the ferromagnet is removed from the field, the electron magnetic moments continue to interact and a significant magnetization remains, as does the strong magnetic field resulting from that magnetization.

When the temperature of a ferromagnet is raised beyond a certain point, called the **critical temperature,** thermal effects dominate the exchange force. At temperatures above the critical temperature, a ferromagnet behaves like a paramagnet, with a small magnetization appearing only in the presence of an external field. The critical temperature for iron is 770° C.

Ordinarily, a piece of iron (an iron nail at room temperature, for example) gives no evidence of producing a magnetic field. Indeed, a typical piece of iron does not have a significant magnetization at room temperature, in the absence of an external field. Instead the sample is divided into a number of tiny **domains,** with dimensions of millimeters or less. Within each domain, the alignment of electron magnetic moments is nearly perfect. Thus each domain has a large magnetization. But the various domains have magnetization vectors that are in different directions (Fig. 21-42). Neighboring domains tend to have opposite magnetizations (Fig. 21-43a). This gives the entire sample zero average magnetization and produces no magnetic field. This tendency for adjacent domains to have opposite magnetizations is a result of the magnetic interaction between the two domains. The domains can be thought of as magnets whose opposite poles are attracted.

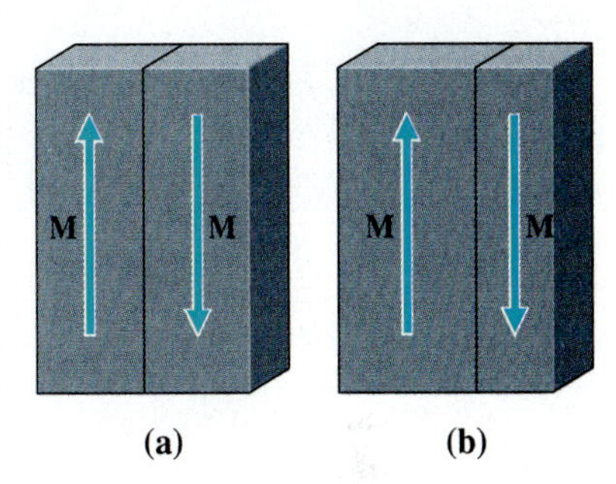

Fig. 21-43 **(a)** Adjacent domains in an unmagnetized ferromagnet. **(b)** Adjacent domains in a ferromagnet, magnetized by an upwardly directed, external magnetic field.

An external magnetic field causes either rotation of domain magnetization toward the direction of the external field or growth of domains having magnetizations in the field direction (Fig. 21-43b). The degree of alignment with an external magnetic field depends on the strength of the field. If it is strong enough, all domains will be aligned, and the sample will have the maximum possible magnetization—a condition called "saturation."

When a magnetized object is removed from an external magnetic field, there is always some realignment of domains. However, there usually remains a significant average magnetization in the direction of the external field. The size of this "remanent" magnetization and the ease with which the remanent magnetization is destroyed depend on the material. Iron is a "soft" magnetic material, meaning that its remanent magnetization is easily removed. A tap with a hammer may be enough to realign the domains within a piece of iron, in such a way as to give it zero average magnetization. Thus iron makes a poor permanent magnet. There are other magnetic materials in which the remanent magnetization is not easily removed. These "hard" magnetic materials make good permanent magnets. One such material is Alnico V, a special alloy developed for this property, consisting of 51% iron, 8% aluminum, 14% nickel, 24% cobalt, and 3% copper.

A Closer Look

Biomagnetism

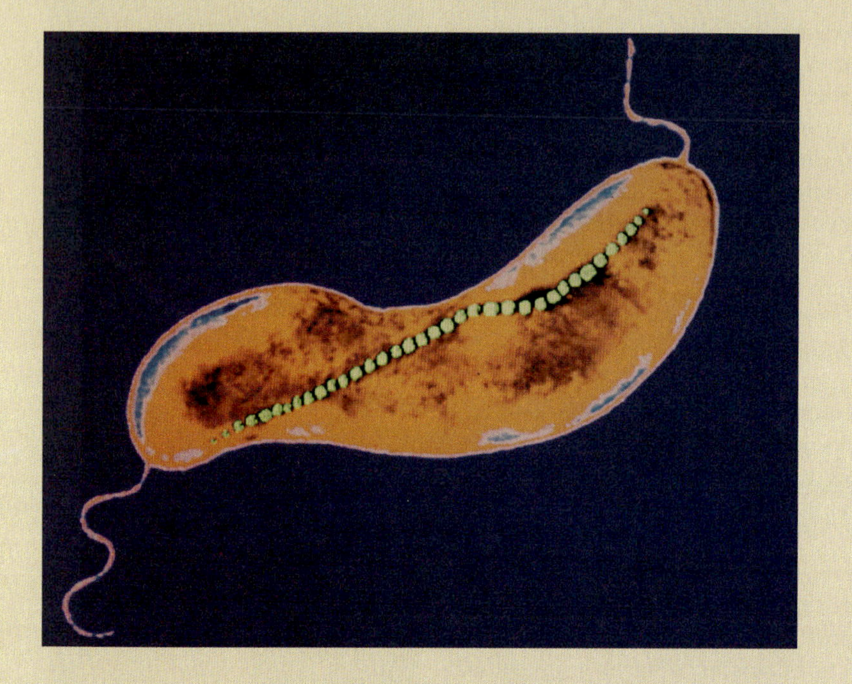

All organisms on earth live in the earth's magnetic field, and so it is reasonable to expect that some organisms might be sensitive to magnetic forces. Generally, detection of such effects, isolated from other factors, has been very difficult. Although biologists have speculated that bees, homing pigeons, and migratory birds navigate by means of the earth's magnetic field, no one has ever succeeded in proving this. However, in the early 1970s, a graduate student discovered a clear biomagnetic effect, through a combination of luck and careful observation.

Richard Blakemore was studying sulfide-rich mud samples from Eel Pond, a marshy area near Woods Hole, Massachusetts. When Blakemore put a drop of mixed mud sediments and bacterial culture under a microscope, he noticed that the bacteria always swam to one edge of the little pool. When he turned the microscope around, or moved it to another place in the lab, or covered it with a cardboard box to see if the bacteria were swimming toward or away from the light, the microbes always swam in the same direction—to the north. Upon bringing a magnet near the microscope, he reported: "To my astonishment, the hundreds of swimming cells instantly turned and rushed away from the end of the magnet! They were always attracted by the end that also attracted the north-seeking end of a compass needle and they were repelled by its opposite end. Their swimming speed was very fast, on the order of 100 micrometers per second, and the entire population consisting of hundreds of freely and independently swimming cells swerved in unison as the magnet was moved about nearby."

Bacteria that show this effect are called **magnetotactic.** Such bacteria, which have now been found worldwide, thrive in sediments with low oxygen content. Apparently they evolved early in the earth's history before the atmosphere was rich in oxygen. Thus they are best suited to the low oxygen conditions within the mud.

Magnetotactic bacteria are rod shaped and about 10^{-3} mm long and contain crystals of the magnetic mineral magnetite. The tiny internal magnets cause the bacteria to align in the direction of an external magnetic field. As they move forward, they move in the direction of the magnetic field. Normally they are exposed to the earth's magnetic field, which means they move to the north.

What biological advantage results from motion of these bacteria to the north? None, as it turns out. But remember the earth's magnetic field dips into the earth in the Northern Hemisphere. Motion along the direction of the field also means motion downward into the earth—and into the oxygen-scarce mud these bacteria love. So when cells are disturbed, there is a mechanism for getting them back down.

Blakemore tested this reasoning by going to New Zealand in the Southern Hemisphere. The earth's magnetic field there is directed outward from the earth (and northward). This means that in order for the magnetotactic bacteria to move downward there, they would need to have a reversed magnetic polarity so that they would move opposite to the earth's magnetic field—southward and down. Blakemore was happy to find his hypothesis confirmed; the magnetotactic bacteria in the Southern Hemisphere do indeed have reversed polarity.

When bacterial cells divide, each offspring cell usually receives a magnetic chromosome, which gives the offspring its parent's magnetic orientation. Occasionally, an offspring cell receives no such chromosome. As the crystals form in these cells, they may have the same or opposite polarity to the parent (depending on their location at the time). Thus, although the majority of the population is north seeking in the Northern Hemisphere, and the majority is south seeking in the Southern Hemisphere, in both cases there is a small population that moves in the opposite direction. This minority population is not well adapted for survival, unless of course the earth's magnetic field reverses, which geologists believe that it sometimes does. Then the minority population would thrive and become the dominant population. So this group provides some insurance for the survival of the species.

Magnetotactic bacteria could play a role in determining how the earth's magnetic field might have changed through the ages. Researchers are looking for deposits of dead bacteria within sedimentary rocks to indicate the orientation of the earth's magnetic field at the time of death. These deposits could confirm the hypothesis that the poles have migrated and reversed at least once, as tectonic plates have carried land masses closer to or farther from the equator. For example, in Pre-Cambrian times, 500 million years ago, the magnetic north pole may have been located near Hawaii. Bacteria that lived that long ago could provide the answer.

CHAPTER 21 SUMMARY

The source of any magnetic field **B** is moving charge, and the effect of a magnetic field is to exert a force on moving charge placed in the field. The moving charge may be an isolated point charge or an electron moving through a current-carrying wire or a spinning electron in a magnet. The magnetic field produced by a small wire segment of length $\Delta\ell$ carrying current I is in a direction determined by rotating the fingers of the right hand from the direction of the current to the direction of the vector locating the field point. The direction of the extended right thumb indicates the direction of the field $\Delta\mathbf{B}$, whose magnitude is given by

$$|\Delta\mathbf{B}| = \frac{\mu_0}{4\pi}\left(\frac{I\,\Delta\ell \sin\theta}{r^2}\right)$$

where r is the distance to the field point, θ is the angle between the wire segment and the line to the field point, and the constant μ_0 (called the "permeability of free space") has the value

$$\mu_0 = 4\pi \times 10^{-7} \text{ T-m/A}$$

The magnetic field is measured in units of teslas (T), the SI unit, or gauss (G), where

$$1 \text{ T} = 1 \text{ N/A-m}$$

and

$$1 \text{ G} = 10^{-4} \text{ T}$$

One can find the total magnetic field of any distribution of current by dividing the current up into small linear elements $\Delta\ell$, calculating the field $\Delta\mathbf{B}$ produced by each element at a given field point, and then taking the vector sum of the $\Delta\mathbf{B}$'s:

$$\mathbf{B} = \Sigma\,(\Delta\mathbf{B})$$

A long straight wire carrying a current I produces a magnetic field having circular field lines centered on the wire. The direction of the field is such that, if the thumb of the right hand is in the direction of positive current, the fingers of the right hand curve around the wire in the direction of the field lines. The magnitude of the field is determined by the distance r to the wire:

$$B = \frac{\mu_0}{2\pi}\left(\frac{I}{r}\right)$$

The magnetic moment **m** of a current-carrying coil is a vector perpendicular to the plane of the coil, in a direction determined by the direction of the extended right thumb when the right fingers curve around the coil in the direction of positive current. The magnitude of the magnetic moment is defined as the product of the number of turns N, the current I, and the enclosed area A:

$$m = NIA$$

The magnetic field produced by a circular current loop at a point on the axis of the loop is in the direction of the extended right thumb when the fingers of the right hand curve around the loop in the direction of positive current. The magnitude of the loop's magnetic field **B** at a point on the axis is

$$B = \frac{\mu_0}{2}\left(\frac{Ia^2}{r^3}\right)$$

where a is the loop's radius and r is the distance from the field point to an edge of the loop. Alternatively, the field of a circular current loop may be expressed in terms of the loop's magnetic moment **m**:

$$\mathbf{B} = \frac{\mu_0}{2\pi}\left(\frac{\mathbf{m}}{r^3}\right)$$

This equation gives the approximate magnetic field produced by a current loop of any shape if **m** of the loop is directed toward the field point and if the distance from the loop is large compared with the loop's dimensions.

The magnetic field of a long, tightly wound solenoid, at field points far from either end of the solenoid, is approximately zero outside the solenoid. Inside the solenoid (far from either end), the magnetic field is uniform, with a magnitude given by

$$B = \mu_0 n I$$

where n is the number of turns per unit length. The direction of the field inside the solenoid is along the axis, in the direction indicated by an extended right thumb when the fingers of the right hand curve around the coil in the direction of the positive current.

The magnetization **M** of a permanent magnet is defined as the magnet's magnetic moment per unit volume:

$$\mathbf{M} = \frac{\mathbf{m}}{V}$$

The magnetic field of a cylindrical permanent magnet with a uniform magnetization **M** along its axis is the same as that of a tightly wound solenoid with the same dimensions as the magnet if the product of the solenoid's turns per unit length n and current I equals M. If such a magnet is long, then for field points far from either pole the magnet's magnetic field is approximately given by

$$\mathbf{B} = \mu_0\mathbf{M} \quad \text{(inside a long cylindrical magnet)}$$

$$\mathbf{B} = 0 \quad \text{(outside a long cylindrical magnet)}$$

At a point where there is a magnetic field **B**, a point charge q moving with a velocity **v** experiences a magnetic force **F** of magnitude

$$F = |q|vB \sin \theta$$

where θ is the angle between vectors **v** and **B**. The force **F** is perpendicular to the plane of vectors **v** and **B**, in the direction of the extended right thumb when the fingers of the right hand rotate from **v** to **B**.

A wire of length ℓ carrying a current I in a uniform, external magnetic field **B** experiences a magnetic force **F** of magnitude

$$F = I\ell B \sin \theta$$

where θ is the angle between the wire and the magnetic field. The force **F** is perpendicular to the plane formed by the wire and vector **B**, in the direction of the extended right thumb when the right fingers rotate from the direction of positive current to the direction of **B**.

Parallel current-carrying wires exert a mutual magnetic force that is attractive if the currents are in the same direction and repulsive if the currents are in opposite directions. The magnitude of the force is determined by the currents I_1 and I_2, the length ℓ of the segment experiencing the force, and the distance r between the wires:

$$F_{12} = \frac{\mu_0}{2\pi}\left(\frac{I_1 I_2 \ell}{r}\right)$$

A current loop in a uniform, external magnetic field experiences zero net force. However, the loop does experience a torque that tends to align its magnetic moment **m** with the direction of the **B** field. The torque is given by

$$\tau = mB \sin \alpha$$

where α is the angle between **m** and **B**.

When a particle of charge q and mass m is projected into a uniform magnetic field **B** with a velocity vector **v** perpendicular to the field, the charge will move along a circular path of radius r given by

$$r = \frac{mv}{|q|B}$$

with a period T given by

$$T = \frac{2\pi m}{|q|B}$$

Questions

1 Positive and negative charges are located at the points shown in Fig. 21-44. A uniform magnetic field is directed along the positive y-axis. The direction of each charge's velocity vector is indicated in the figure. Find the direction of the instantaneous force on each charge.

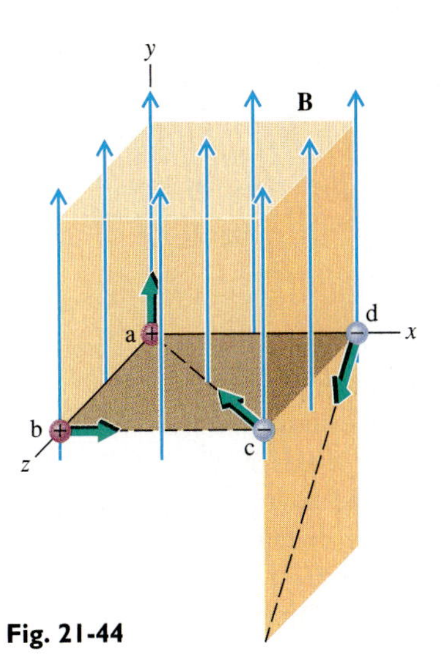

Fig. 21-44

2 Because they are moving through a uniform magnetic field, the electrons at points a and b in Fig. 21-45 experience the instantaneous forces indicated in the figure.

(a) Find the possible directions of the magnetic field.

(b) In what plane does each electron's velocity vector lie?

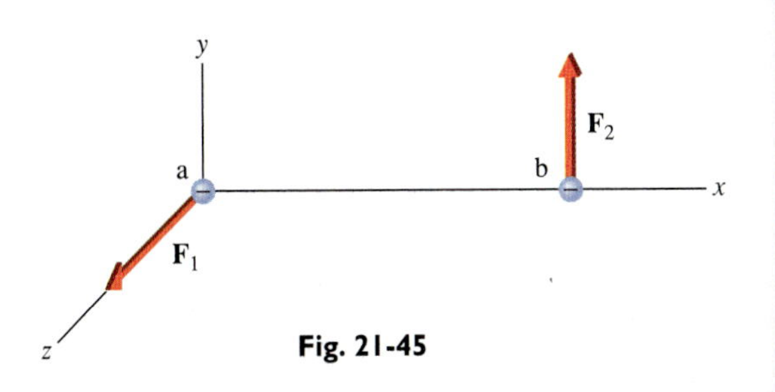

Fig. 21-45

3 Is it possible for a magnetic field to do work on a charged particle moving through the field?

4 A segment of wire carries a positive current I along the positive y-axis (Fig. 21-46). Find the direction of the magnetic field produced by the current at each of the points a through f.

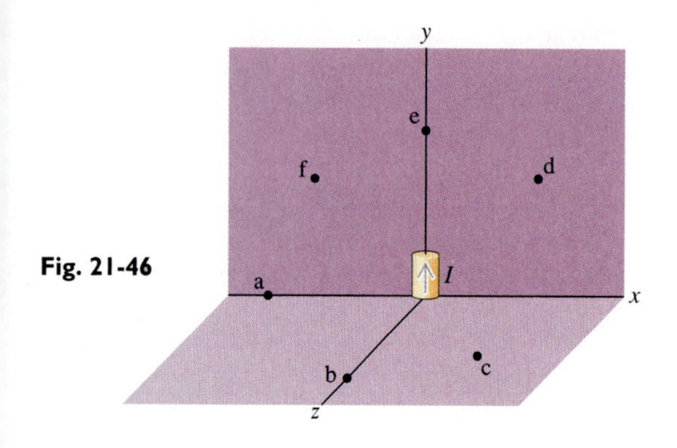

Fig. 21-46

5 Points a to f in Fig. 21-46 are all the same distance from the origin. At which points is the magnetic field strongest?

6 The triangular wire in Fig. 21-47 carries a positive current in the direction indicated. A uniform magnetic field is directed along the positive z-axis. Find the directions of the magnetic forces acting on segments a, b, and c.

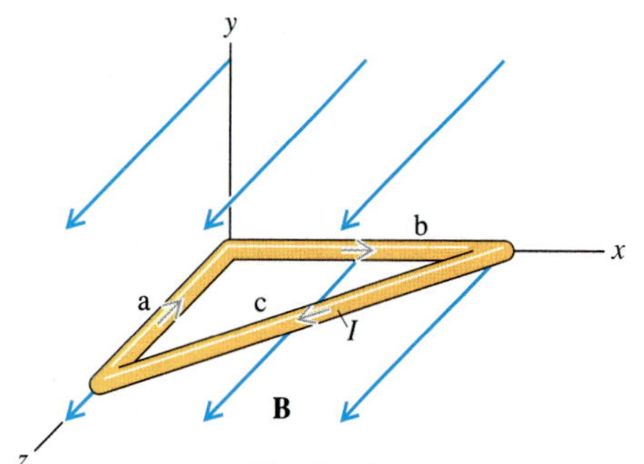

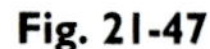

Fig. 21-47

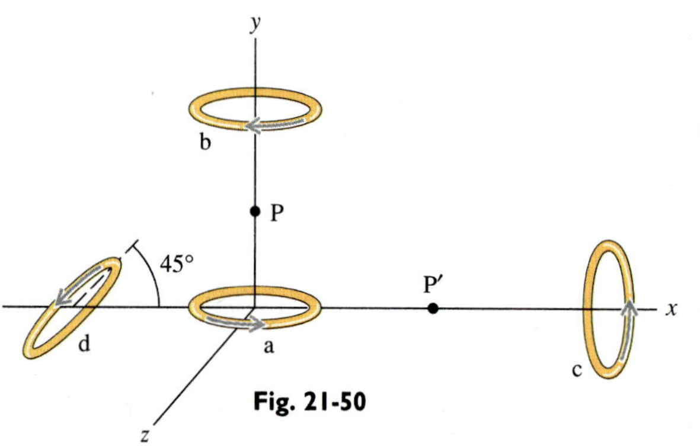

Fig. 21-50

7 A loop of wire carries a positive clockwise current in a plane perpendicular to an external magnetic field (Fig. 21-48). Will the magnetic force on the wire produce tension or compression in the wire?

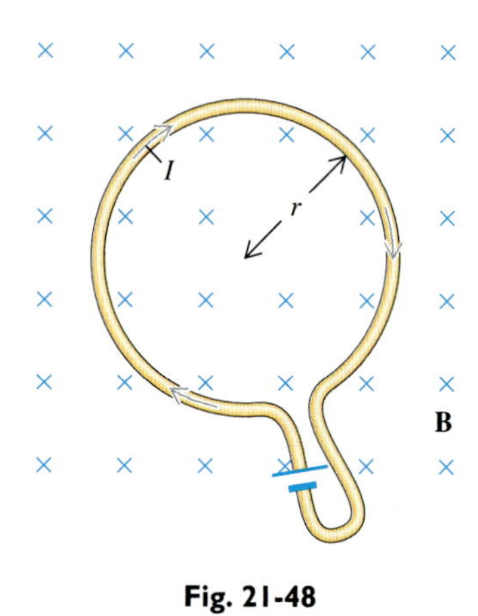

Fig. 21-48

8 Electrons move along a counterclockwise circular path in the xy plane. Find the direction of the magnetic field that produces this motion.

9 An electron beam moves in a horizontal, straight line through an *electric* field directed into the page (Fig. 21-49). Find a possible direction of the *magnetic* field that is also present in this region.

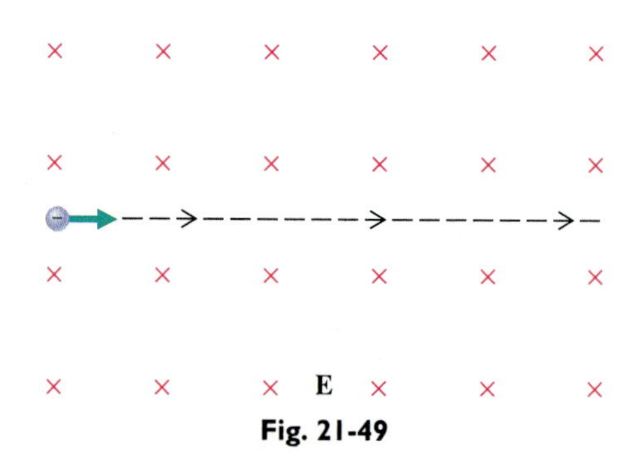

Fig. 21-49

10 Find the direction of the magnetic moment vector for each current loop in Fig. 21-50.

11 A uniform, external magnetic field is directed along the positive x-axis in Fig. 21-50. For each current loop, does the magnetic torque tend to rotate the loop clockwise or counterclockwise as viewed from the positive z-axis?

12 At point P in Fig. 21-50, find the directions of the magnetic fields produced by current loops a and b. At point P′, find the direction of the magnetic field produced by current loop c.

13 A long, straight wire carries a positive current I into the page (Fig. 21-51). Find the direction of the magnetic field at each point indicated in the figure.

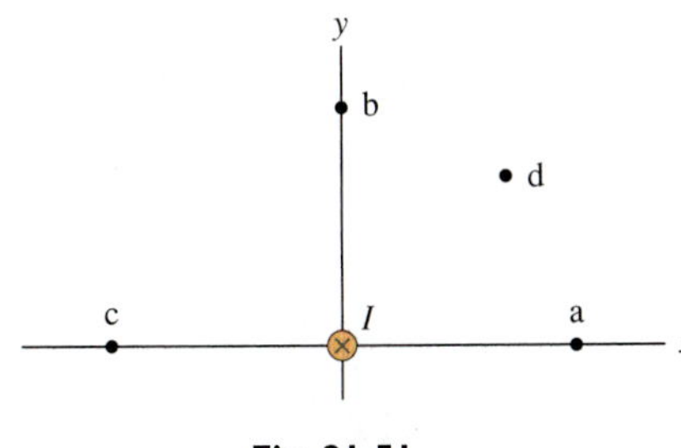

Fig. 21-51

14 Two long, straight wires carry currents of equal magnitude directed perpendicular to the page, one out and one in (Fig. 21-52). Find the direction of the magnetic field at each point shown in the figure.

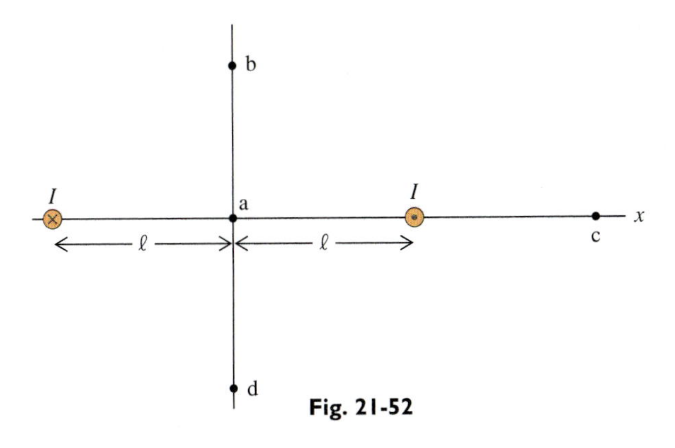

Fig. 21-52

15 Two long, straight wires carry the same current I in perpendicular directions (Fig. 21-53). Find the direction of the magnetic field at each point shown in the figure. Each point is the same distance from both wires.

Fig. 21-53

16 Four parallel wires each carry current I perpendicular to the page in the directions indicated in Fig. 21-54. Find the direction of the magnetic field at point P, at the center of the square formed by the wires.

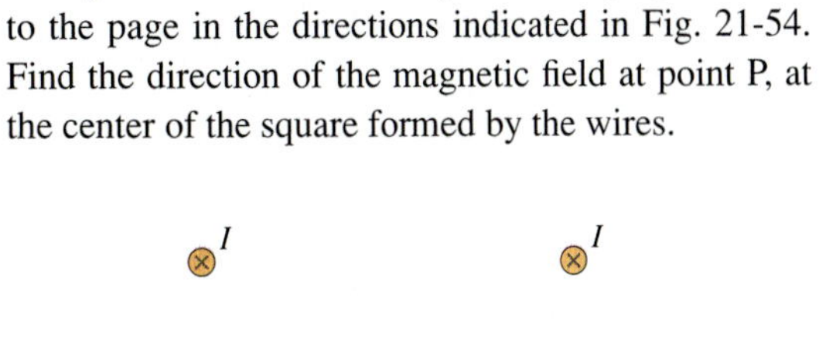

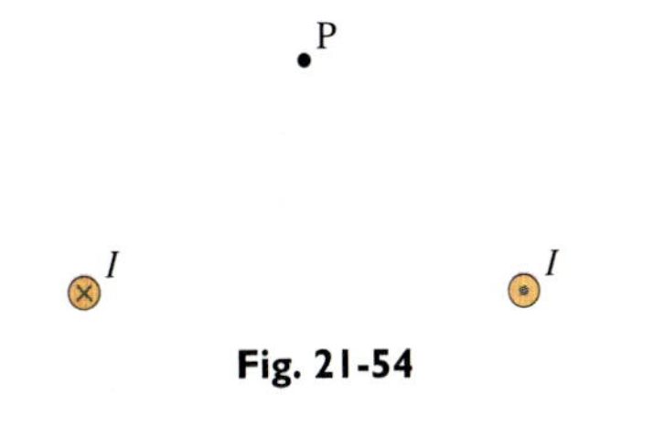

Fig. 21-54

17 Find the direction of the resultant magnetic force on the wire in the upper left corner of Fig. 21-54.

18 A square current loop is close to a long, straight, current-carrying wire (Fig. 21-55).

(a) Is there a resultant magnetic force on the loop? If so, where is it directed?

(b) Does the loop exert a net force on the wire?

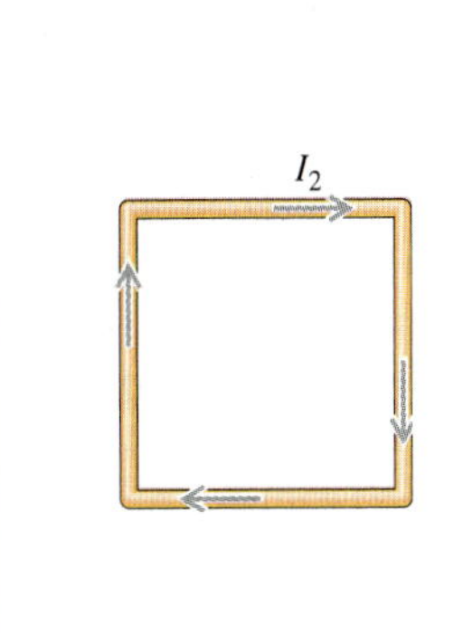

Fig. 21-55

19 The wires in household electrical cord are parallel and only a few millimeters apart. Describe the magnetic field produced by electric current in the cord.

20 Observer O sees a stationary point charge, which produces an electric field but no magnetic field. Is it possible for a second observer O′, moving with respect to O, to observe the same point charge producing a magnetic field?

21 Which of the lines in Fig. 21-56 could be field lines representing a magnetic field?

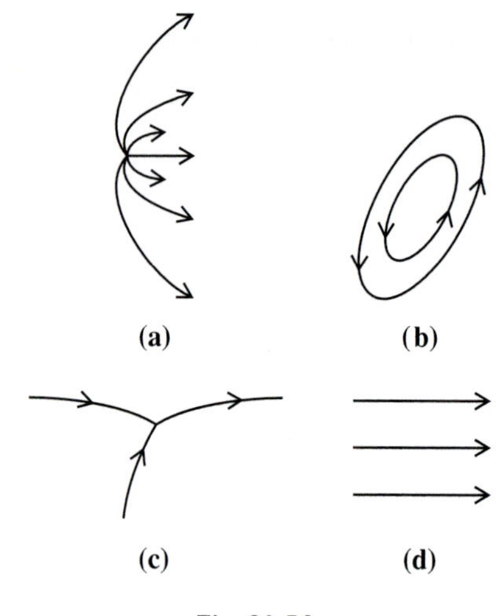

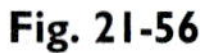

Fig. 21-56

22 Which of the lines in Fig. 21-56 could be field lines representing an electrostatic field?

23 Is the earth's magnetic north pole more like the north pole of a magnet or the south pole of a magnet?

24 Which way would a compass needle point at the earth's magnetic north pole?

25 An unmagnetized piece of iron is placed directly above the north pole of a strong permanent magnet (Fig. 21-57).

(a) What is the direction of the iron's magnetization?

(b) Is the force on the iron attractive or repulsive?

(c) Now suppose the iron is turned upside down. What is the direction of the iron's magnetization?

(d) Is the force on the iron attractive or repulsive?

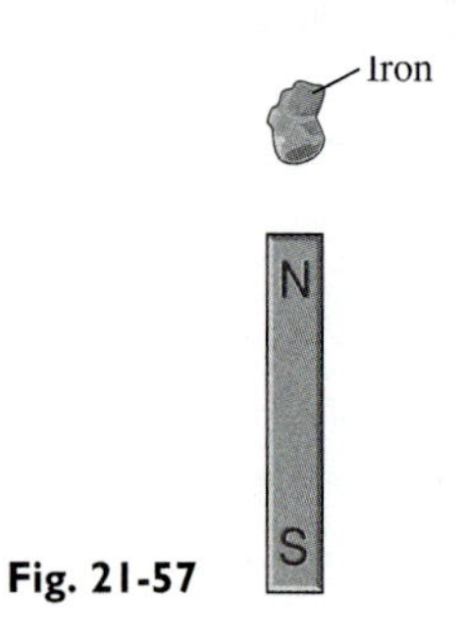

Fig. 21-57

26 In Fig. 21-16 does the magnet producing the torque on the motor's coil experience a torque itself?

27 A hollow, cylindrical magnet has a uniform magnetization **M** directed along the cylinder's axis (Fig. 21-58). The field at point P, inside the hollow area, equals approximately (a) $\mu_0\mathbf{M}$; (b) $-\mu_0\mathbf{M}$; (c) 0; (d) $2\mu_0\mathbf{M}$.

Fig. 21-58

Answers to Odd-Numbered Questions

1 charge at a: experiences no force; charge at b: along the $+z$-axis; charge at c: in the x-z plane, 45° to the left of the $+z$-axis; charge at d: along the $+x$-axis; **3** no; **5** a, b, and c; **7** tension; **9** in the plane of the page, upward, perpendicular to **v** and **E**; **11** loop a: clockwise; loop b: counterclockwise; loop c: no rotation; loop d: counterclockwise; **13** at a: $-y$ direction; at b: $+x$ direction; at c: $+y$ direction; at d: in the x-y plane, directed 45° below the $+x$ axis; **15** at a: no magnetic field; at b: out of the page; at c: no magnetic field; at d: into the page; **17** toward the lower right corner of the square; **19** The fields of the two wires tend to cancel, since the wires carry equal currents in opposite directions. A small resultant magnetic field is produced at a point closer to one wire than to the other. The direction of this resultant field oscillates because of the alternating direction of the current in the wires; **21** b and d; **23** south pole, since a magnet's north pole is directed toward it; **25** (a) upward; (b) attractive; (c) upward; (d) attractive; **27** c

Problems (listed by sections)

21-1 The Magnetic Field

1 What are the signs of charges 1, 2, and 3 in Fig. 21-59?

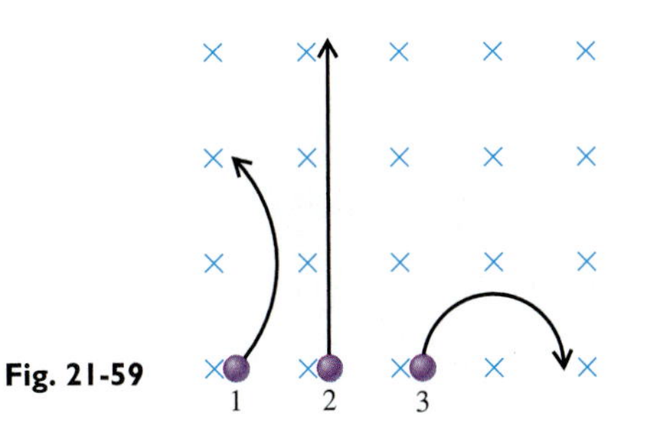

Fig. 21-59

2 Charges 1 and 3 in Fig. 21-59 have the same mass and the same speed. Which has the greater magnitude?

3 The earth is constantly bombarded by cosmic rays, which are charged particles from outer space. Find the force on one such cosmic ray—a proton that is instantaneously moving vertically downward at 1.00×10^7 m/s at a point outside the atmosphere over New York, where the magnetic field dips 74.0° below the horizontal and has a horizontal component of 1.70×10^{-5} T (Fig. 21-60).

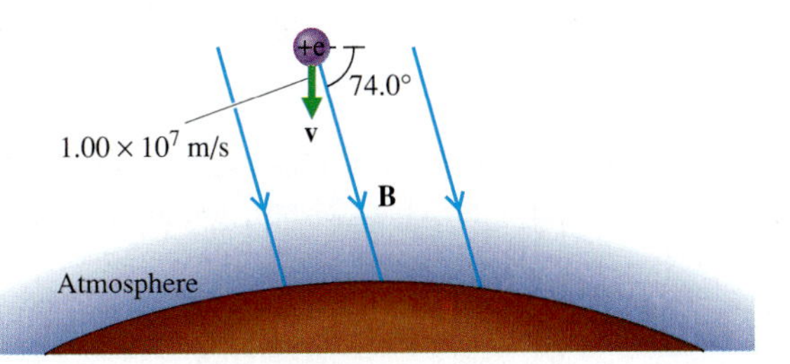

Fig. 21-60 View of a cosmic ray, as one looks west above New York.

4 Find the possible directions of a uniform magnetic field that exerts the forces $\mathbf{F}_1$ and $\mathbf{F}_2$ respectively on charges q_1 and q_2 in Fig. 21-61.

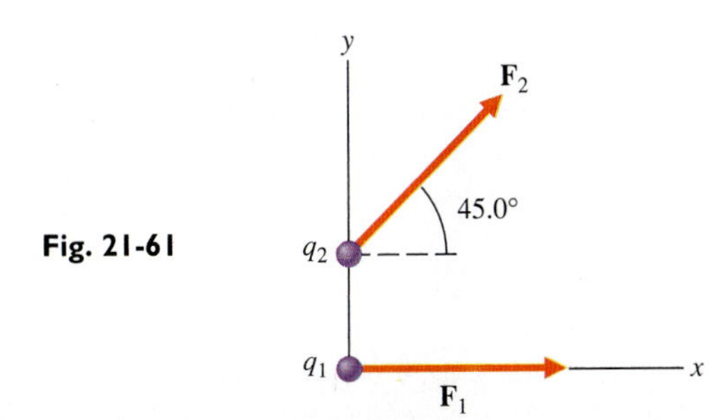

Fig. 21-61

5 Find the direction and magnitude of the minimum velocity of charge $q = +2e$ that will produce the magnetic force $\mathbf{F}$ shown in Fig. 21-62.

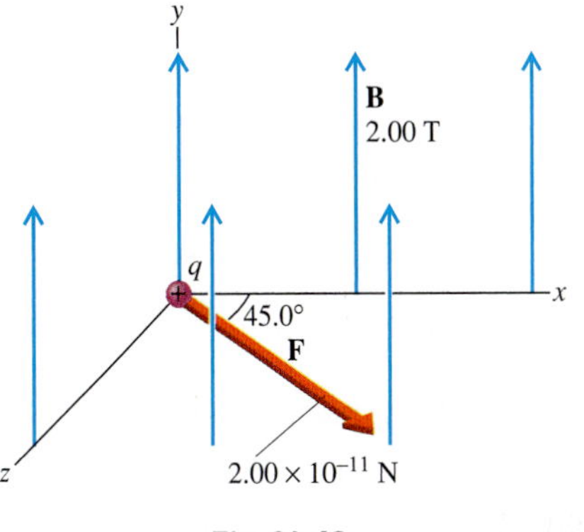

Fig. 21-62

6 Charging by frictional contact typically produces static charge of about 10^{-9} C on small macroscopic objects. Suppose a small object of mass 2.00 g has a charge of 1.00×10^{-9} C. How fast would the object have to move through a 10.0 T magnetic field to experience a magnetic force equal to its weight?

★ **7** In an attempt to observe a large magnetic effect on a macroscopic charged body, suppose you let a charged capacitor free-fall through a magnetic field (Fig. 21-63). Using the data shown in the figure, find (a) the force on each plate of the capacitor; (b) the resultant magnetic force on the capacitor; (c) the resultant magnetic torque on the capacitor.

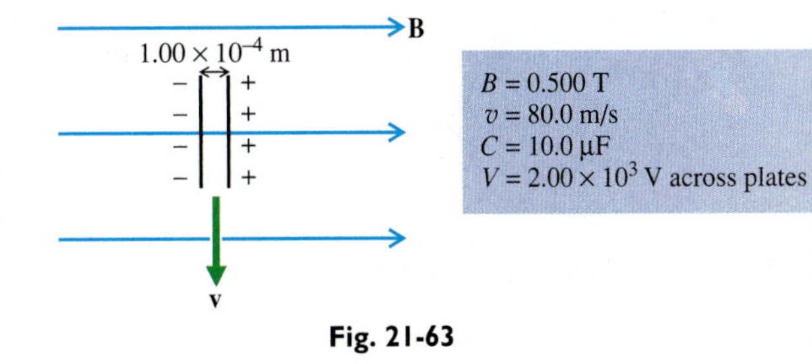

Fig. 21-63

8 The magnetic field near the surface of a neutron star has a magnitude of 1.00×10^8 T. Find the acceleration of an electron near the star moving at 1.00×10^7 m/s if both **v** and **B** are in the same plane, with **B** directed to the right and **v** directed 30.0° above **B**.

★9 The earth's magnetic field can affect the electron beam in an oscilloscope or a television tube. An electron that has been accelerated through a potential difference of 1.00×10^4 V has a horizontal initial velocity, directed north. Find the magnetic force on the electron when the earth's field has a horizontal component of 0.250 G and dips 60.0° below the horizontal (values typical in California).

★10 Estimate the deflection of the electron beam produced by the earth's magnetic field in the preceding problem.

21-2 Magnetic Forces on Current-Carrying Conductors

11 Each wire segment in Fig. 21-64 has a length of 20.0 cm and carries a current of 3.00 A in the directions indicated. Find the magnitude and direction of the magnetic force on each segment if the magnetic field has a magnitude of 0.400 T.

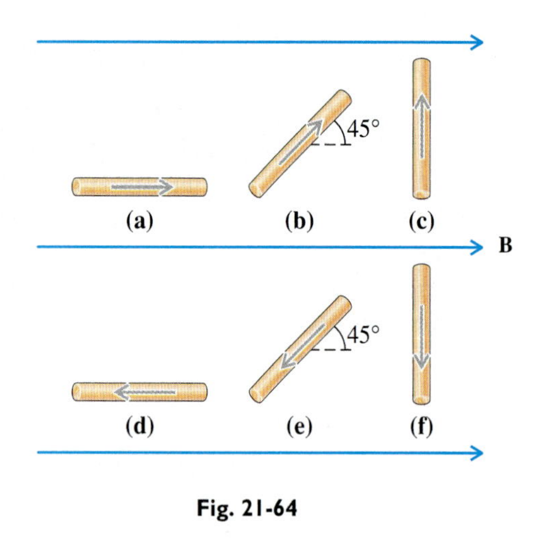

Fig. 21-64

12 Find the magnitude and direction of the weakest magnetic field required to levitate a wire of length 5.00 cm and mass 2.00 g, carrying a current of 8.00 A.

13 Stainless-steel poles were arranged by an artist for the purpose of producing an enhanced display of lightning in a field of nearly one square mile (Fig. 21-65). Calculate the magnetic force on one of these poles, 6.30 m long, when it transmits an upward current of 2.00×10^4 A during a lightning discharge. The magnetic field in New Mexico, where the rods are located, dips 64.0° below the horizontal and has a horizontal component of magnitude 2.50×10^{-5} T.

Fig. 21-65 A field of lightning rods as a work of art.

14 (a) Show that, to magnetically levitate a wire in a magnetic field of magnitude B, the wire must carry a minimum current $I = \frac{\rho g A}{B}$, where ρ is the mass density of the wire and A is its cross-sectional area. (b) Find the current required to levitate a copper wire of radius 1 mm in the earth's magnetic field at a point where the field has a horizontal component of 2×10^{-5} T.

15 The end of a screwdriver is used to discharge a 10.0 μF capacitor (Fig. 21-66), producing a spark and a large instantaneous current through the screwdriver. The potential difference across the capacitor is initially 400 V, and the resistance is 1.00×10^{-4} Ω along the path of the discharge. Find the instantaneous magnetic force on the screwdriver produced by (a) the earth's magnetic field, indicated in the figure; (b) a magnetic field of magnitude 2.00 T directed downward.

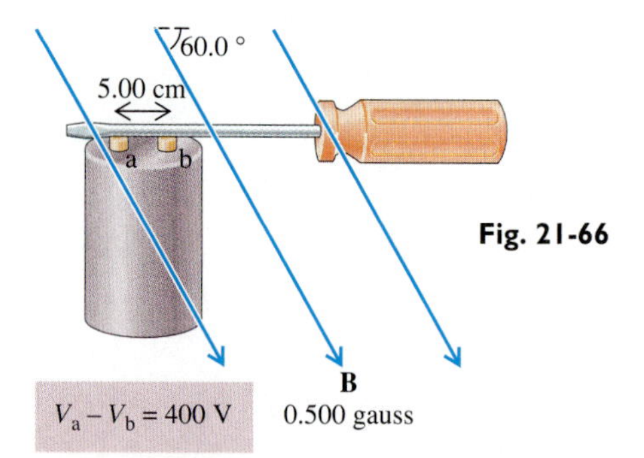

Fig. 21-66

★ **16** The triangular loop of wire in Fig. 21-67 carries a current of 2.00 A in the direction indicated. There is a uniform magnetic field of magnitude 0.500 T in the positive y direction.

(a) Find the magnetic force on each segment of the wire.

(b) By considering the torque produced by each force, calculate the net torque about the z-axis.

(c) Calculate the loop's magnetic moment.

(d) Use the magnetic moment to calculate the magnetic torque on the loop.

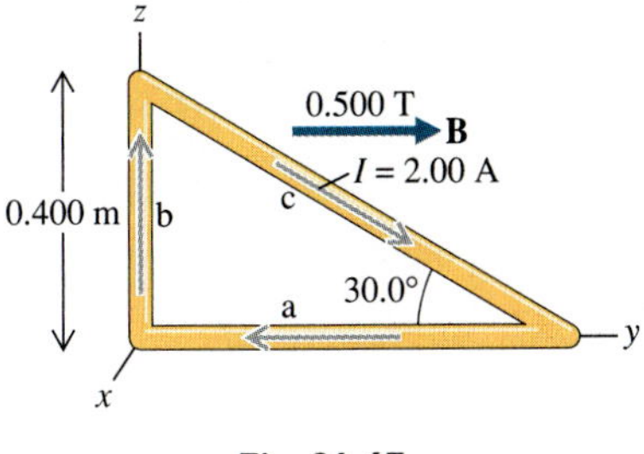

Fig. 21-67

17 A rigid wire with resistance of 0.100 Ω forms a nearly complete circular loop (Fig. 21-68). A 9.00 V battery is connected between points a and b. Find the torque exerted on the wire by a magnetic field of magnitude 0.400 T, directed to the right, in the plane of the loop.

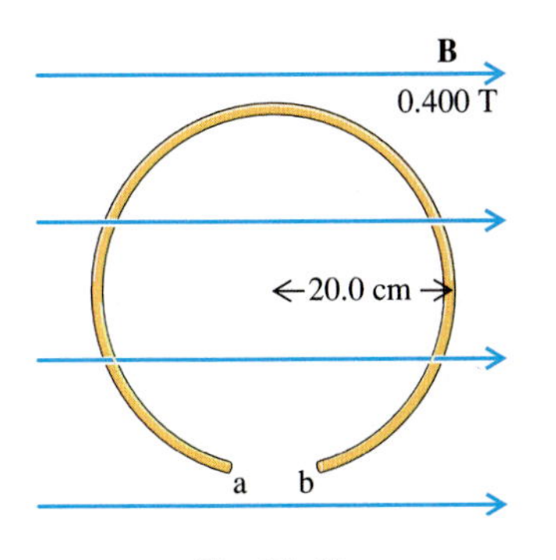

Fig. 21-68

18 The mechanism of a galvanometer is illustrated in Fig. 21-69. A coil of wire is mounted so that it is free to rotate in a magnetic field when current passes through the coil. Since the field is radially directed, it is always parallel to the plane of the coil, no matter what the coil's orientation. As current passes through the coil, the coil rotates until the magnetic torque is balanced by a torque exerted by an attached spring. The spring torque is proportional to the angle of rotation. Since the balancing magnetic torque is proportional to the current, the current through the coil is also proportional to the angle of rotation, indicated by the galvanometer needle. Let the coil consist of 25 turns enclosing an area of 4.00 cm^2. The magnetic field strength is 5.00×10^{-3} T. Find the current in the coil when the needle rotates through 20.0° if the spring torque at that point is 1.00×10^{-6} N-m.

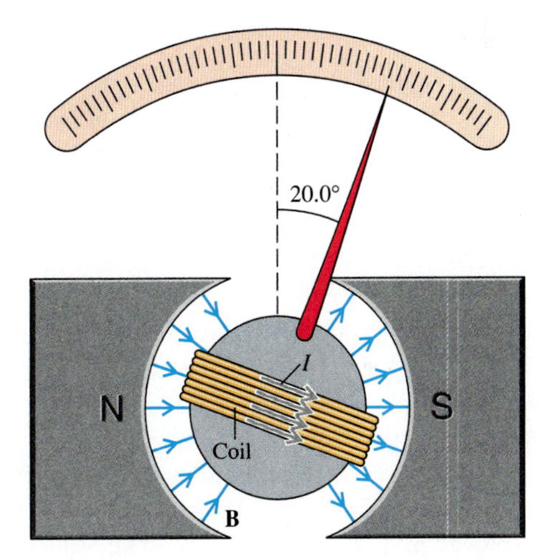

Fig. 21-69 A galvanometer.

19 Find the minimum current that must pass through the coil of an electric motor to produce a torque of 20.0 N-m if the coil consists of 200 turns enclosing an area of 300 cm^2 and the magnetic field strength is 0.400 T.

20 Suppose you wanted to design an automatic door opener that operated by means of magnetic torque applied to a coil of wire wrapped around the edge of the door (Fig. 21-70). A magnetic field is directed to the right, parallel to the door.

(a) What direction of current in the coil would exert a torque tending to open the door outward (toward the reader).

(b) The coil carries a current of 10.0 A, the door has dimensions 1.00 m by 2.00 m, and the magnetic field strength is 100 G. How many turns of wire in the coil are required to produce a torque of 40.0 N-m?

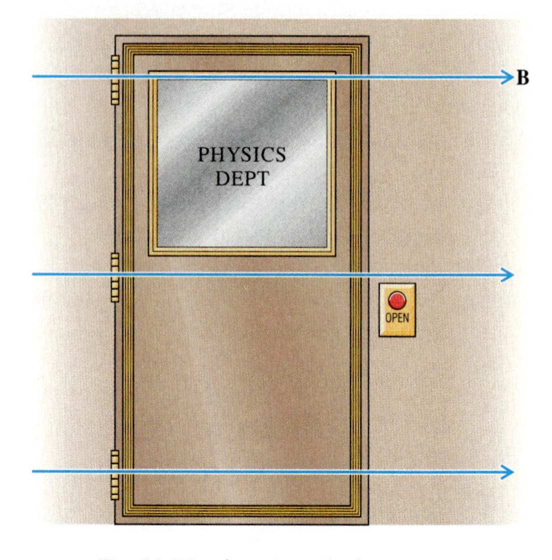

Fig. 21-70 An automatic door opener.

21-3 Motion of a Point Charge in a Magnetic Field

21 A proton is initially moving west at a speed of 1.00×10^6 m/s in a uniform magnetic field of magnitude 0.300 T directed vertically upward. Describe in detail the proton's trajectory, including its shape and orientation.

22 An electron with a kinetic energy of 1.00×10^4 eV is initially moving south in a uniform magnetic field of magnitude 500 G, directed east. Describe in detail the electron's trajectory, including its shape and orientation.

23 (a) Derive an expression for the speed v of a beam of charged particles if, when moving through perpendicular electric and magnetic fields, the beam moves in a straight line at constant velocity directed to the east. The magnetic field is directed north, and the electric field is directed vertically downward (in other words, toward the center of the earth).

(b) Evaluate the expression for v if $E = 3.00 \times 10^4$ N/C and $B = 2.00 \times 10^{-2}$ T.

(c) What would happen to charges moving faster than v?

(d) What would happen to charges moving slower than v? (Crossed electric and magnetic fields serve as a velocity selector and are used to produce a beam that has a well-defined velocity as the beam enters a mass spectrometer.)

24 A singly charged ^{7}Li ion has a mass of 1.16×10^{-26} kg. It is accelerated through a potential difference of 500 V and subsequently enters a uniform magnetic field of magnitude 0.400 T perpendicular to the ion's velocity. Find the radius of its path.

25 A chemist uses a mass spectrometer to determine the amounts of specific kinds of air pollution in an air sample. Each molecule is ionized and follows a circular trajectory in the spectrometer's magnetic field. The heaviest concentration of molecules corresponds to a circle of diameter 28 cm. What is the diameter of the circular path of NO (nitrous oxide) molecules?

★ 26 In a region where there is a uniform 0.200 T magnetic field perpendicular to the xy plane, a stream of protons is to be shot from the origin at a fixed initial speed v_0 (Fig. 21-71).

(a) At what angle θ_0 must the velocity vector $\mathbf{v}_0$ be directed in order for the protons to reach as great a height y as possible?

(b) If the maximum height is 1.00 m, what is the initial speed of the protons?

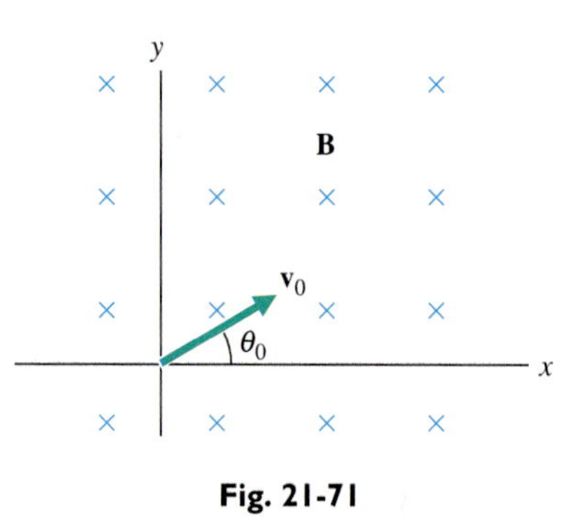

Fig. 21-71

27 An electron moving in the positive x direction at a speed of 9.00×10^7 m/s enters a uniform magnetic field of magnitude 0.0500 T directed into the page and perpendicular to the xy plane.
(a) At what time after entering the field will the electron's velocity vector be parallel to the y-axis?
(b) What are the x and y coordinates of the electron at that time if the electron starts at the origin?

★ **28** The electron's charge-to-mass ratio e/m can be measured with the device shown in Fig. 21-18a. The electrons in the beam have been accelerated through a potential difference V and follow a circular path of radius r; the path is perpendicular to a uniform magnetic field of magnitude B. Prove that $\frac{e}{m} = \frac{2V}{B^2r^2}$. Evaluate this expression for $V = 40.0$ V, $B = 5.00 \times 10^{-4}$ T, and $r = 4.33$ cm.

21-4 Magnetic Fields Produced by Electric Currents

29 Each of the wires in Fig. 21-72 carries a current of 3.00 A in the directions indicated. For each wire, find the magnitude and direction of the magnetic field at a point P that is 1.00 cm from the wire.

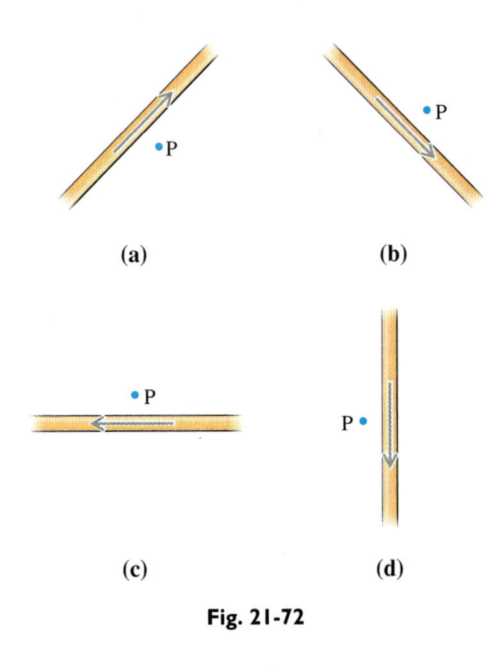

Fig. 21-72

30 A long, straight wire perpendicular to this page carries a current of 5.00 A directed into the page. Find the magnetic field at the following points, all 10.0 cm from the wire: (a) to the left of the wire; (b) directly below the wire; (c) to the right of the wire; (d) directly above the wire.

31 Find the strength of the magnetic field 1.00 cm from a power line carrying a current of 125 A.

32 Estimate the magnitude of the magnetic field 1.00 cm from the screwdriver in Fig. 21-66 when the instantaneous current through it is 4.00×10^4 A during the discharge of the capacitor.

33 Find the magnetic field at P and P′ on the axis of a circular coil of radius 20.0 cm (Fig. 21-73). The coil consists of 10 turns of wire carrying a current of 4.00 A in the direction indicated.

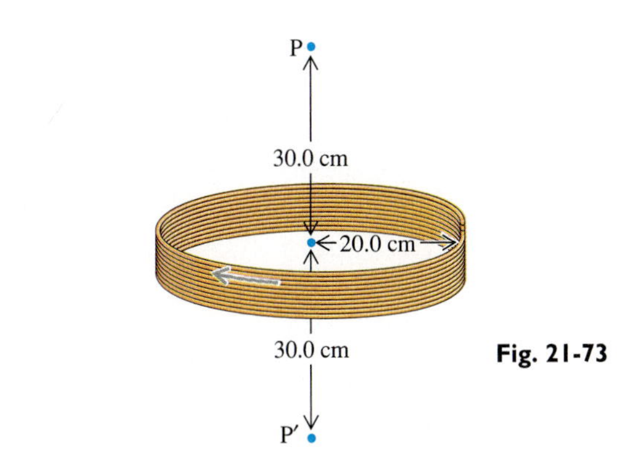

Fig. 21-73

34 Fig. 21-74 shows a Helmholtz coil, which consists of two identical circular coils having the same axis and separated by a distance equal to the radius of each coil. This arrangement produces a magnetic field that is fairly uniform over an extended region near the midpoint P between the coils. Therefore the Helmholtz coil is frequently used in experimental research requiring a moderately strong, nearly uniform, magnetic field over a fairly large, accessible region. Find the magnetic field at P if $r = 33.0$ cm and each coil consists of 36 turns of wire carrying a current of 4.00 A.

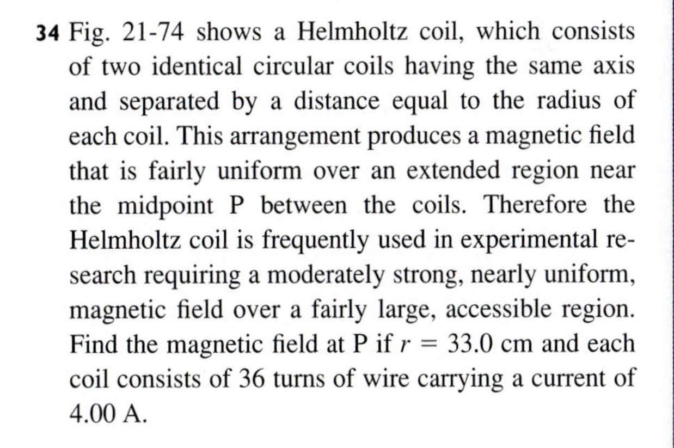

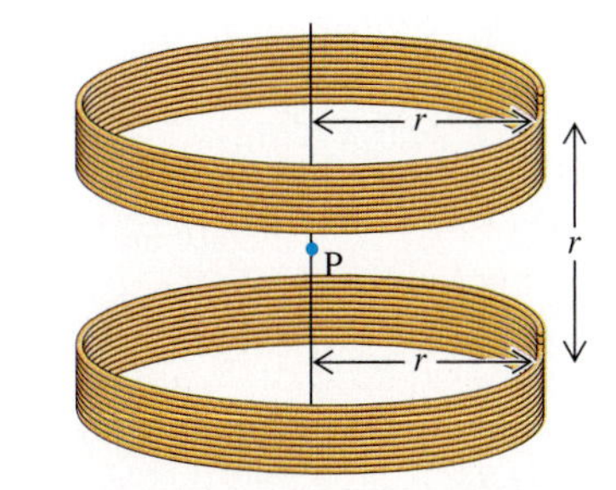

Fig. 21-74 Helmhotz coil.

35 The earth's magnetic field at the magnetic north pole is directed vertically downward and has a magnitude of 0.62 G. The earth's field is approximately that of a current loop located at the center of the earth.

(a) Find the magnitude and direction of the earth's magnetic moment.

(b) It is believed that the earth's magnetic field is produced by electric current in the earth's liquid metallic outer core. Suppose that the magnetic moment were produced by a current through a simple circular path of radius equal to the average radius of the outer core (2300 km). Find the direction and magnitude of this current.

36 Derive Eq. 21-19 for the case of parallel wires carrying oppositely directed currents. Show that the force is repulsive in this case.

★ **37** Estimate the force between the two circular current loops in Fig. 21-75.

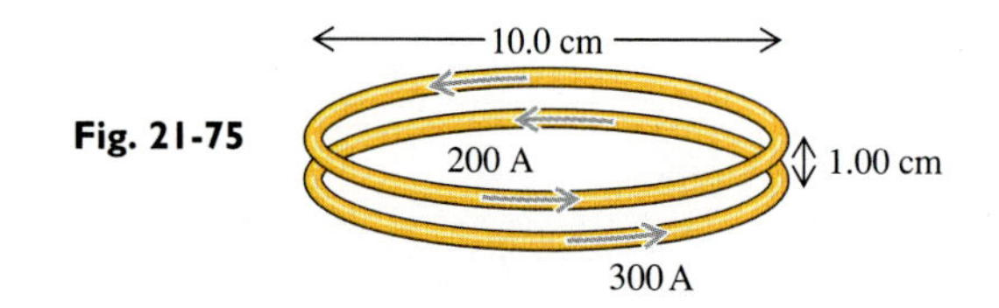

Fig. 21-75

38 Two vertical, long, straight wires each carry a current of 5.00 A in the upward direction and are separated by 20.0 cm. Find the resultant magnetic field at a point P midway between the wires.

★ **39** The sides of the square formed by the four wires in Fig. 21-54 are 8.00 cm long. Each wire carries a current of 10.0 A in the direction shown.

(a) Find the magnetic field at point P at the center of the square.

(b) Find the resultant magnetic force on the wire at the lower right.

40 Two long, straight wires lie in the same plane and cross at right angles (Fig. 21-76). Find the resultant magnetic field at point P, which lies in the plane of the wires and is 40.0 cm from each.

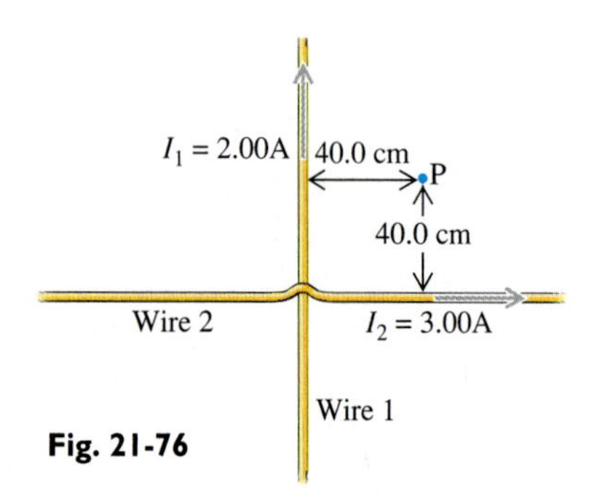

Fig. 21-76

★ **41** Two long, straight wires lie along the x- and z-axes, and each carries a steady current of 4.00 A (Fig. 21-77). Find the resultant magnetic field at point P on the y-axis at y = 25.0 cm.

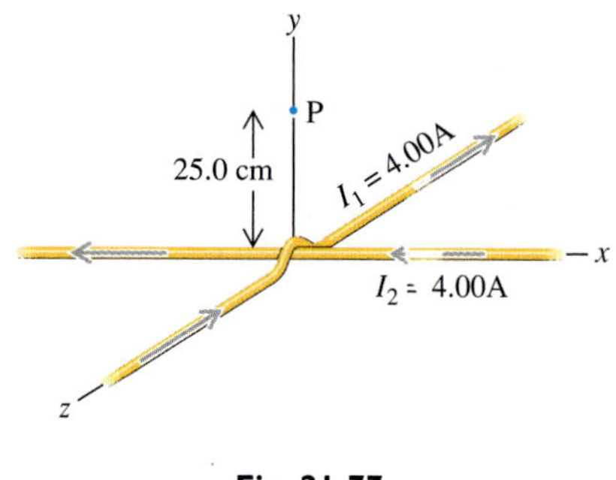

Fig. 21-77

★ **42** Two long, straight wires lie in the same plane and cross at right angles (Fig. 21-76). What is the resultant force exerted by wire 1 on wire 2? Neglect any effect of the small loop where wire 1 passes over wire 2.

★★ **43** Suppose you are designing a long solenoid capable of producing a magnetic field of 1.00 T inside.

(a) What would be the value of the product of the current and the number of turns per unit length?

(b) Show that the electric power dissipated by the solenoid's wire per unit area of the wire's surface is $\dfrac{P}{A} = \dfrac{2(nI)^2\rho}{\pi^2 r}$, where ρ is the wire's resistivity and r is its radius.

(c) Evaluate P/A for copper wires of radii 1.00 mm and 1.00 cm. Are these realistic numbers? For comparison, the filament of a light bulb dissipates about 10^5 W/m^2.

44 At what point on the axis of a circular current loop is the magnetic field greatest?

★ **45** The magnetic field of a long, straight wire is strongest at the surface of the wire. Copper wire of varying thickness can safely carry a current of up to about 1.0×10^6 A per m^2 of cross-sectional area without overheating. Show that a larger-radius wire can safely produce a stronger magnetic field than a smaller-radius wire. Evaluate the magnetic field at the surface of a wire of radius 0.50 cm when the current in the wire divided by its cross-sectional area is 1.0×10^6 A/m^2.

★ **46** (a) Find the resultant magnetic force exerted by the long wire on the square current loop in Fig. 21-55 if I_1 = 8.00 A and I_2 = 4.00 A. The edge length of the square is 20.0 cm, and the distance from the long wire to the closest edge of the loop is 5.00 cm.

(b) Find the magnetic torque on the current loop.

21-5 Magnetic Fields Produced by Permanent Magnets

47 A cylindrical permanent magnet has a volume of 200 cm^3 and a uniform magnetization of 1.0×10^6 A/m along its axis. The magnet is placed in a uniform external magnetic field of magnitude 0.40 T. The magnet's magnetization vector is initially at an angle of 30° with the direction of the external field. Find the magnetic torque exerted on the magnet.

★ 48 A cylindrical electromagnet contains a uniformly magnetized core having a magnetization of 1.7×10^6 A/m, surrounded by a coil 20 cm long with 3000 turns of wire carrying a current of 4.0 A. Find the magnetic field in the core.

★★ 49 Estimate the magnetic force between the two identical disk magnets in Fig. 21-78. The magnets have a uniform magnetization of 1.0×10^6 A/m along their axes. (HINT: Consider each magnet's equivalent current.)

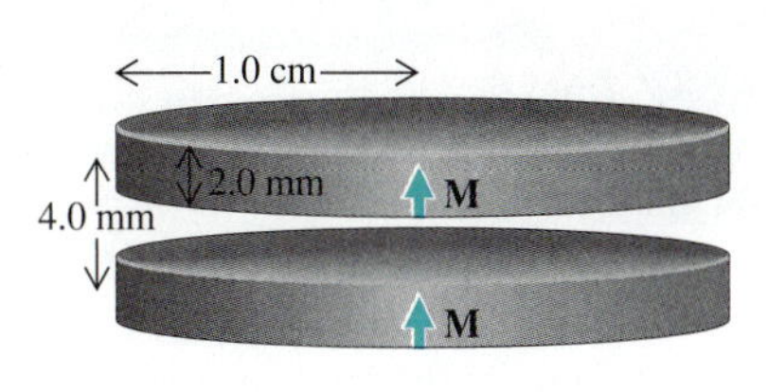

Fig. 21-78

★ 50 Derive Eq. 21-25. The first step is to use Fig. 21-38 to show that the current around the edge of a small volume element of thickness ℓ equals the product of ℓ and the magnetization M.

Additional Problems

★ 51 A long, straight wire lies along the y-axis and carries a current of +10.0 A in the positive y direction. A second long, straight wire, parallel to the first and 10.0 cm to the right, carries a current of +5.00 A in the negative y direction. At what value of the x coordinate does the resultant magnetic field vanish?

★★ 52 The circular loop of wire of radius r in Fig. 21-48 carries a current I in an external magnetic field **B** perpendicular to the plane of the loop. Derive an expression for the tension in the wire and evaluate the tension for $r = 50.0$ cm, $I = 40.0$ A, $B = 2.00$ T.

★ 53 A metal rod of length 20.0 cm and mass 100 g is free to slide over two parallel, horizontal metallic rails connected to a 10.0 V battery (Fig. 21-79). The battery has internal resistance of 0.100 Ω, and the rod and rails have negligible resistance. A 2.00 T magnetic field is directed perpendicular to the plane of the circuit.

(a) Find the magnetic force on the rod.

(b) Find the rod's speed after it has moved 30.0 cm, assuming negligible friction.

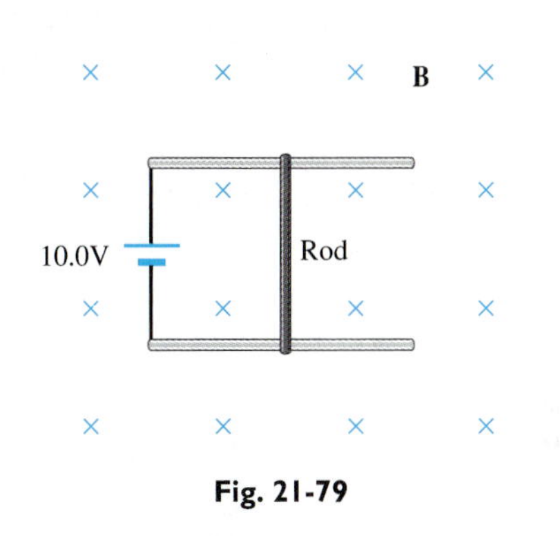

Fig. 21-79

★ 54 The electrical power input to a stereo speaker is 100 W when the volume is turned up all the way. The acoustic efficiency of the speaker is 1.00%, meaning that there is only 1.00 W of power in the sound wave produced by the speaker. The remainder of the power is dissipated as heat in the speaker's coil or used to overcome friction where the cone is mounted in the frame. The front of the speaker's cone has a cross-sectional area of 0.500 m^2.

(a) Find the intensity and intensity level of the sound as it leaves the speaker.

(b) Find the amplitude of the air at this point if the sound wave is harmonic and has a frequency of 1000 Hz.

(c) The cone has a mass of 100 g. Assuming that the cone undergoes harmonic motion of the same amplitude as the air, find the maximum force on the cone.

(d) The speaker's coil consists of 60.0 m of copper wire of radius 0.200 mm. Find the coil's resistance.

(e) Find the peak current in the coil. Assume that the average power dissipated as heat is 99 W, meaning that the peak instantaneous rate of heat production is 198 W.

(f) Calculate the maximum magnetic force on the coil if the magnetic field strength is 0.210 T.

CHAPTER 22 Electromagnetic Induction and AC Circuits

This electric generator, shown being inspected, will be used to provide electric power. Inside the generator a changing magnetic field "induces" an electric current in huge coils of copper wire.

Modern society relies on the electric generator for the abundant electrical power it produces. Fig. 22-1 shows a sketch of the first generator, invented by Michael Faraday. Today's large and elaborate generators operate on the same physical principle as Faraday's generator. Although realizing he had discovered a device with great practical potential, Faraday was not interested in developing it himself. When asked by a British government official of what use his invention was, he replied, "I know not, but I wager that one day your government will tax it."

In this chapter we shall study the electric generator and the alternating current it produces. In preceding chapters we have described electric fields produced by stationary charge distributions (electrostatics) and magnetic fields produced by constant currents (magnetostatics). Electricity and magnetism were studied under conditions in which it was possible to have one field without the other—either an electric field and no magnetic field or a magnetic field and no electric field. In this chapter and the next we shall consider the more general situation of charge distributions and currents that vary with time. As these sources vary, they produce time-varying electric and magnetic fields. We shall find that there are relationships between electric and magnetic fields that make it impossible to have a time-varying electric field without an associated magnetic field and impossible to have a time-varying magnetic field without an associated electric field. We shall often refer to these associated electric and magnetic fields as a single entity—the "electromagnetic" field.

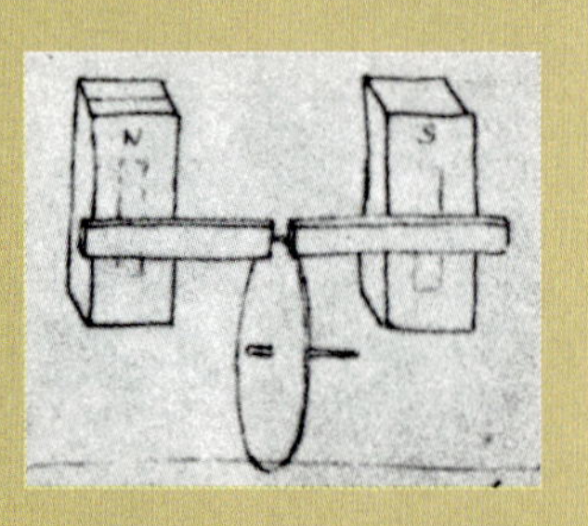

Fig. 22-1 Drawing from Faraday's diary of the first generator.

22-1 Faraday's Law

Induction Experiments

In 1831 Michael Faraday observed an effect that even today, over 160 years later, is rather amazing. Faraday produced electric current in a coil of wire that was not connected to a battery and had no other obvious source of current. The current was "induced" in the coil when there was a changing electric current in a nearby circuit (Fig. 22-2). The induced current lasted only very briefly, as the switch was being closed. When there was a steady current in the circuit containing the battery, Faraday detected no current in the coil.

Faraday also found that he could induce an electric current in a coil of wire simply by moving a permanent magnet through the coil (Fig. 22-3). This effect is the basis for electric generators.

Faraday's discoveries followed years of experimentation, based on his intuitive belief that it should be possible to induce electric currents with other electric currents. (See the essay at the end of this chapter for a discussion of Faraday's life and work.)

A particularly simple induction experiment is illustrated in Fig. 22-4. A copper ring is in a uniform time-varying magnetic field, directed perpendicular to the plane of the ring. When the magnetic field changes, current is induced in the ring. This current can be detected in various ways, for example, by the heat it produces in the ring.

The presence of an electric current in a conductor means that there must be a source of energy; that is, there must be a source of emf. A battery is a common source of emf. However, in the case of the ring in Fig. 22-4, there is no battery. The source of emf must be related to the changing magnetic field. The time-varying magnetic field "induces" an emf in the ring as though there were a battery inside the ring.

We can get a picture of what is happening within the ring on the microscopic level by noting that the presence of a current in the ring means that there must be an electric field within the ring (Eq. 19-26: $E = \rho I/A$). The changing magnetic field produces this electric field. From symmetry, the electric field is the same throughout the ring. The induced emf is the work per unit charge done by the induced electric field.

The electric field is not limited to the metal ring. Electric field lines circulate throughout the region of space where the magnetic field is varying.

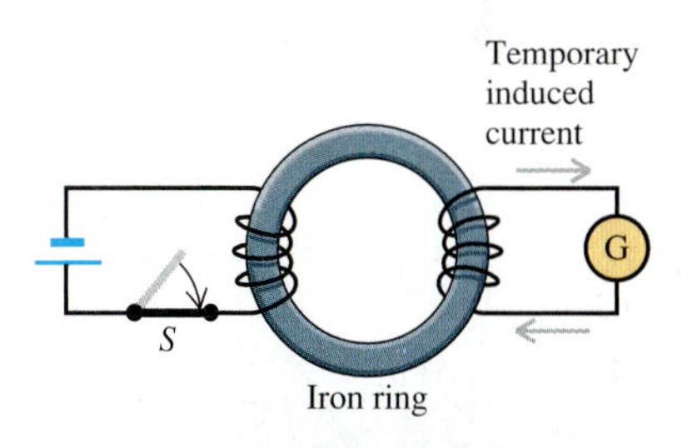

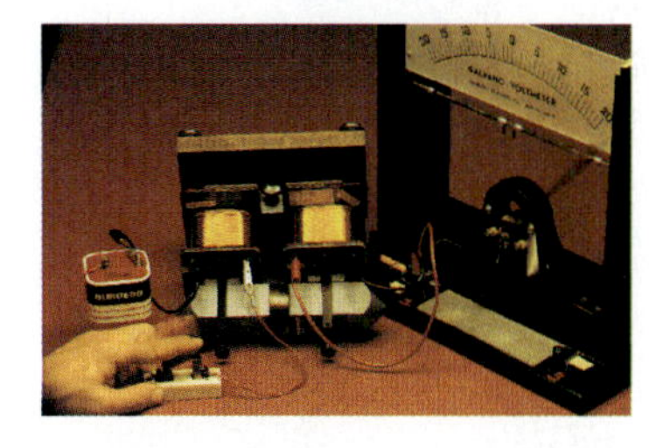

Fig. 22-2 As switch S is closed, there is a momentary deflection of the galvanometer needle, indicating a temporary current induced in the second circuit. An iron ring enhances the effect, which would otherwise be too small to be seen. The circuits are electrically insulated from the ring. Opening the switch also produces a brief current.

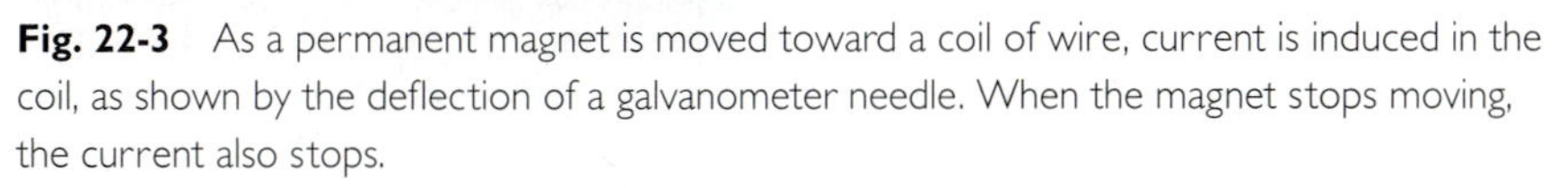

Fig. 22-3 As a permanent magnet is moved toward a coil of wire, current is induced in the coil, as shown by the deflection of a galvanometer needle. When the magnet stops moving, the current also stops.

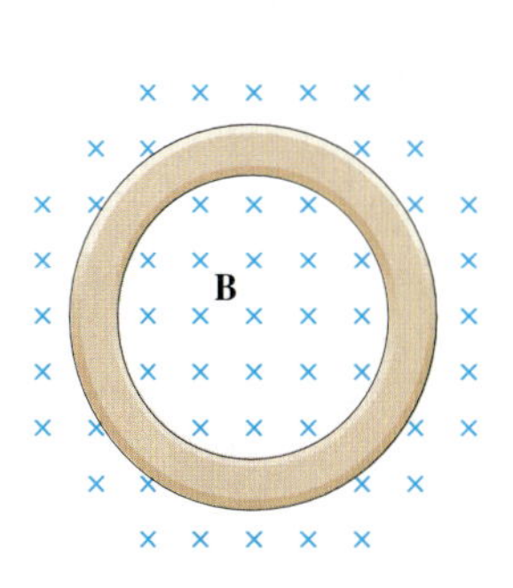

Fig. 22-4 A time-varying magnetic field induces a current in a copper ring.

Magnetic Flux

To discuss induced emf quantitatively, we must first introduce the concept of **magnetic flux.** Fig. 22-5 shows a magnetic field that is uniform over a surface of area A. For a uniform field such as this, the magnetic flux Φ is defined as the product of the area A and the magnetic field component $B_\perp$, perpendicular to the surface.*

$$\Phi = B_\perp A = BA\cos\theta \qquad \text{(for a uniform field)} \qquad (22\text{-}1)$$

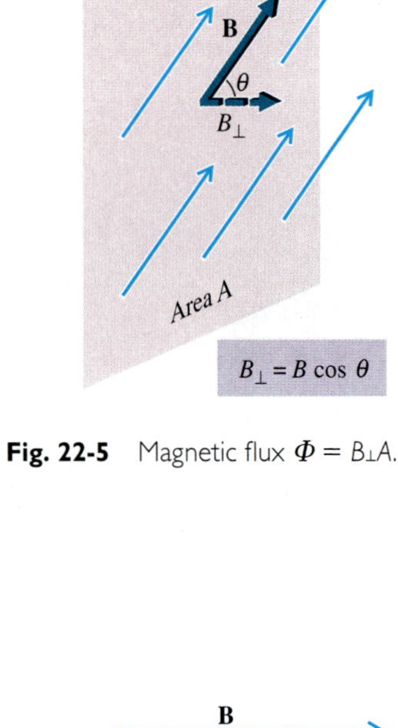

Fig. 22-5 Magnetic flux $\Phi = B_\perp A$.

This definition implies that the magnetic flux through a surface is proportional to the number of magnetic field lines passing through that surface. For example, when a surface is oriented so that field lines are parallel to the surface ($B_\perp = 0$), the flux equals zero and the number of lines passing through the surface also equals zero (Fig. 22-6a). Both magnetic flux and the number of field lines through a surface are maximized if the surface is oriented perpendicular to the field lines (Fig. 22-6b).

Faraday's Law; Constant $\mathcal{E}$

Faraday's law states that the emf induced around any loop equals the negative rate of change of outward magnetic flux through the surface bounded by the loop. If the flux changes at a constant rate $\Delta\Phi/\Delta t$, Faraday's law is expressed

$$\mathcal{E} = -\frac{\Delta\Phi}{\Delta t} \qquad \text{(Faraday's law; constant } \mathcal{E}) \qquad (22\text{-}2)$$

Positive emf means that the direction of the induced electric field in the loop will produce **counterclockwise current,** whereas **negative emf**† means an induced field that produces **clockwise current** (Fig. 22-7).

Rather than use Faraday's law to determine the direction of an induced emf, it is usually easier to use Faraday's law to determine only the magnitude of the induced emf. Taking the magnitude of both sides of Eq. 22-2, we obtain

$$|\mathcal{E}| = \left|\frac{\Delta\Phi}{\Delta t}\right| \qquad \text{(constant } \mathcal{E}) \qquad (22\text{-}3)$$

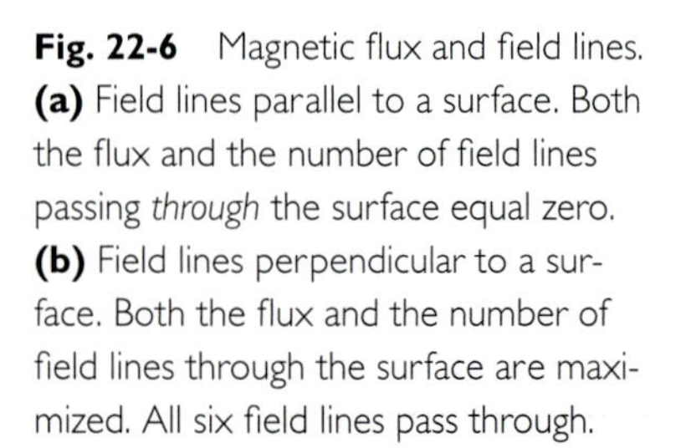

Fig. 22-6 Magnetic flux and field lines. **(a)** Field lines parallel to a surface. Both the flux and the number of field lines passing *through* the surface equal zero. **(b)** Field lines perpendicular to a surface. Both the flux and the number of field lines through the surface are maximized. All six field lines pass through.

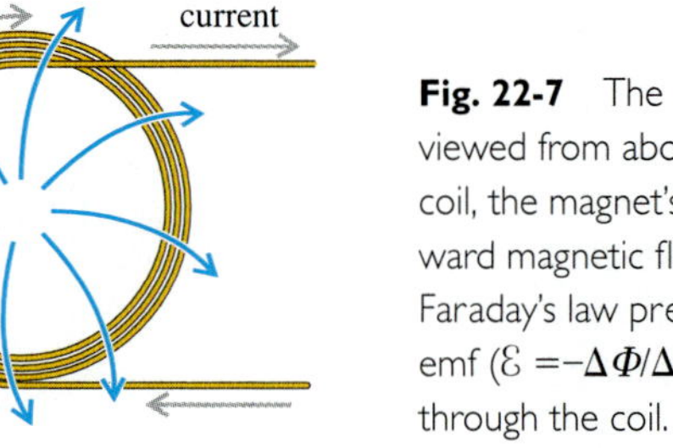

Fig. 22-7 The same coil shown in Fig. 22-3 is viewed from above. As the magnet moves toward the coil, the magnet's **B** field produces an increasing outward magnetic flux through the coil ($\Delta\Phi/\Delta t > 0$). Faraday's law predicts a negative (clockwise) induced emf ($\mathcal{E} = -\Delta\Phi/\Delta t < 0$), resulting in clockwise current through the coil.

*If the field is not uniform over a surface, we find the flux by dividing the surface into small elements of area ΔA, computing the flux $B_\perp\,\Delta A$ for each element, and summing to compute the total flux $[\Phi = \Sigma\,(B_\perp\,\Delta A)]$.

†For a constant source of emf, like a battery, the emf was defined as a positive quantity—the energy per unit charge supplied by a source, or, equivalently, the work per unit charge done by a source. Here we need to define a sign for emf in order to deal with situations in which the source is an induced electric field, which may be changing in direction. It is as though we were using a battery with the orientation of the positive and negative terminals being continually reversed.

Lenz's Law

The direction of the induced current is easiest to find from **Lenz's law,** which may be derived from Faraday's law and which states that **an induced current produces a magnetic field that opposes the change in magnetic flux that produced it.** For example, the induced current in the coil shown in Fig. 22-7 produces its own magnetic field. If the right fingers curve around the coil in the direction of the current, the extended right thumb points in the direction of the magnetic field produced by the current, according to the rule formulated in Chapter 21 (see Fig. 21-28). Applying this rule to the coil in Fig. 22-7 we see that the magnetic field of the induced current shown in the figure is *inward* and therefore opposes the increase in the outward magnetic field of the approaching magnet, as predicted by Lenz's law.

The magnetic field produced by the induced current is much weaker than the magnetic field of the permanent magnet. So there is still an increasing magnetic flux. It is simply reduced slightly by the field of the induced current.

To find the direction and magnitude of an induced current, we use both Lenz's law and Faraday's law, as seen in the following example.

EXAMPLE 1 Current Induced in a Copper Ring

The ring shown in Fig. 22-4 has a radius of 4.00 cm and a resistance of $1.00 \times 10^{-3}\ \Omega$. The magnetic field is increasing at a constant rate from 0.200 T to 0.400 T in a time interval of 1.00×10^{-2} s. Find the current in the ring.

SOLUTION We apply Lenz's law to find the direction of the current. The current must be counterclockwise so that the magnetic field of this induced current is directed outward, as shown in Fig. 22-8, opposing the increase in the inward flux.

To find the magnitude of the induced current, we must first calculate the change in magnetic flux and apply Faraday's law to determine the induced emf. The external magnetic field is uniform over the area of the ring and perpendicular to this surface area. We apply Eq. 22-1:

$$\Phi = B_\perp A = BA$$

Since the ring has a fixed area ($A = \pi r^2$), the change in magnetic flux, $\Delta\Phi$, may be expressed in terms of the change in magnetic field, ΔB, as

$$\Delta\Phi = \Delta B(A) = \Delta B(\pi r^2)$$

$$= (0.400\text{ T} - 0.200\text{ T})(\pi)(4.00 \times 10^{-2}\text{ m})^2 = 1.01 \times 10^{-3}\text{ T-m}^2$$

Applying Faraday's law (Eq. 22-3), we find

$$|\mathcal{E}| = \left|\frac{\Delta\Phi}{\Delta t}\right| = \frac{1.01 \times 10^{-3}\text{ T-m}^2}{1.00 \times 10^{-2}\text{ s}} = 0.101\text{ V}$$

The magnitude of the current resulting from this emf may be calculated in the usual way for a single loop circuit consisting of a constant counterclockwise emf $\mathcal{E} = 0.101$ V and a resistance $R = 1.00 \times 10^{-3}\ \Omega$ (Fig. 22-9).

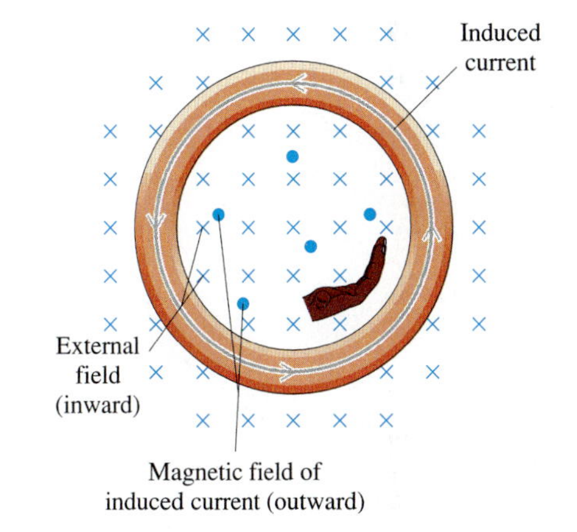

Fig. 22-8 An increasing external magnetic field (directed inward) induces a counterclockwise current, which produces an outward magnetic field, opposing the change that produced it.

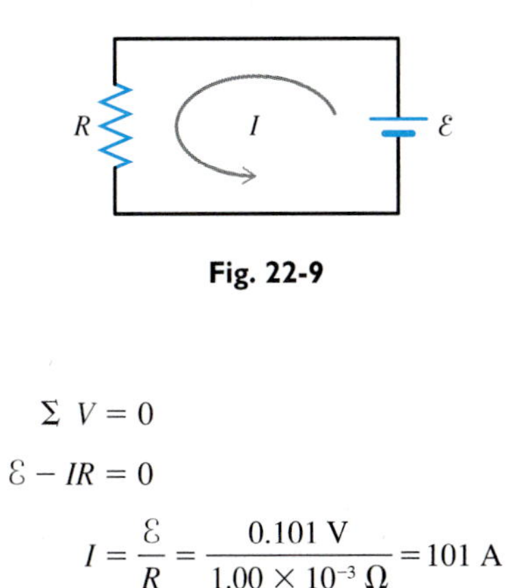

Fig. 22-9

$$\Sigma V = 0$$

$$\mathcal{E} - IR = 0$$

$$I = \frac{\mathcal{E}}{R} = \frac{0.101\text{ V}}{1.00 \times 10^{-3}\ \Omega} = 101\text{ A}$$

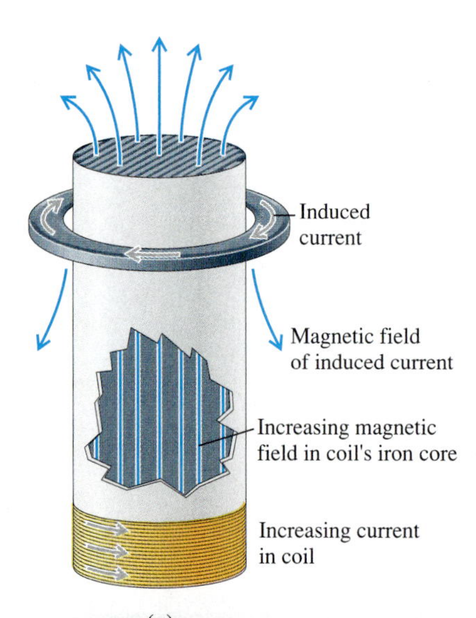

Fig. 22-10 Magnetic propulsion.

Fig. 22-10 shows a rather dramatic demonstration of induction. A long iron bar inserted in a coil with a time-varying electric current forms an electromagnet, with a time-varying magnetic field in the iron. An emf is induced in a conducting ring placed over the iron bar. The ring's induced current produces a magnetic field opposing the changing magnetic field in the iron (Fig. 22-10a). The induced current in the ring is opposite the current in the coil. Therefore the ring is repelled by the coil and flies upward* (Fig. 22-10b).

Motional emf

Faraday's law and Lenz's law may also be applied to a moving circuit, for example, to a flexible loop of wire in a uniform, constant magnetic field (Fig. 22-11). As the sides of the loop are pulled outward, the area enclosed by the loop decreases. According to Lenz's law, there must be a clockwise current induced in the loop so that the induced current produces a magnetic field directed inward, in the same direction as the external field, and so the induced field opposes the decrease in flux through the loop. The emf induced by the motion of a conductor through a magnetic field is called **motional emf.**

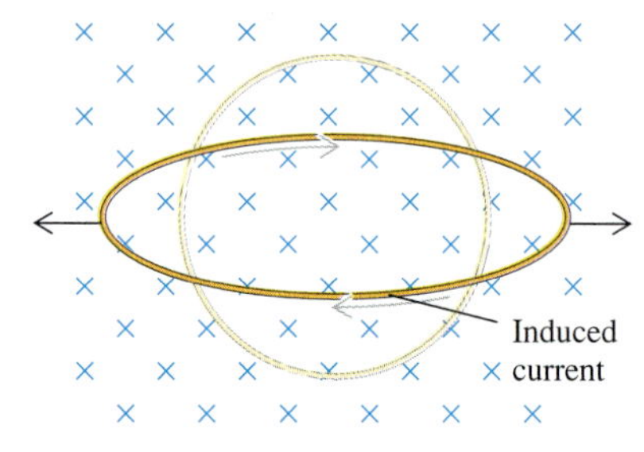

Fig. 22-11 Reducing the area enclosed by a flexible loop of wire in a magnetic field induces a current in the loop.

Another example of motional emf is illustrated in Fig. 22-12. A metal rod slides to the right at constant velocity **v** while in contact with a stationary, U-shaped conductor. A complete conducting loop is formed by the rod and the part of the U-shaped conductor to the left of the rod. A uniform, constant magnetic field **B** is directed downward, perpendicular to the plane of the loop (Fig. 22-12a). As the rod slides, the area enclosed by the loop increases, and so the flux through the loop increases. Thus there will be an induced emf and an induced current in the loop. From Lenz's law we see that the induced current is counterclockwise, producing inside the loop a magnetic field directed opposite the external field (Fig. 22-12b). Since **B** is uniform and perpendicular to the plane of the loop, the magnitude of the flux through the loop at any instant is the product of the field magnitude B and the area A enclosed by the loop at that instant.

$$|\Phi| = BA = B\ell x$$

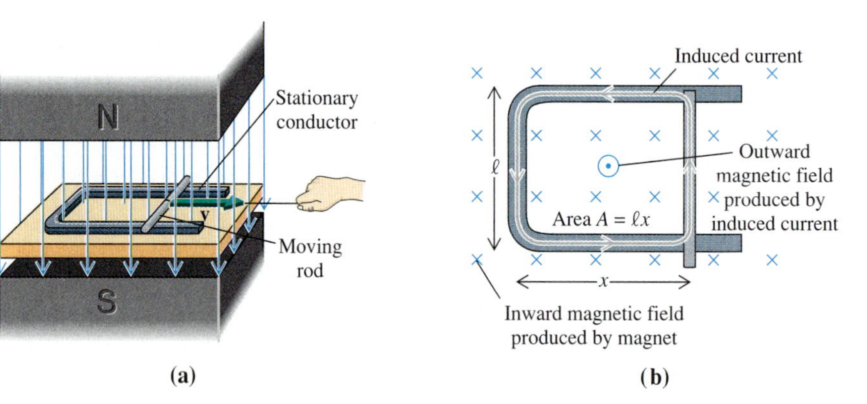

Fig. 22-12 **(a)** A metal rod slides over a stationary U-shaped conductor in a uniform, constant magnetic field. **(b)** Because the conducting path encloses an increasing area, a current is induced along the path. The magnetic field of the induced current is outward, opposing the increasing inward magnetic flux.

*This explanation accounts only for the magnetic force on the ring during that part of the AC cycle during which there is increasing current in the coil in the direction indicated. The repulsive force on the ring during other parts of the cycle is more difficult to explain.

The rod moves a distance $\Delta x = v\ \Delta t$ during the time interval Δt. The magnitude of the change in flux during this time interval is

$$|\Delta \Phi| = B\ell\ \Delta x = B\ell v\ \Delta t$$

Dividing by Δt and applying Faraday's law (Eq. 22-3: $|\mathcal{E}| = |\Delta \Phi / \Delta t|$), we find

$$|\mathcal{E}| = B\ell v \qquad (22\text{-}4)$$

EXAMPLE 2 Current Induced by Pulling a Metal Rod Through a Magnetic Field

Suppose that the left side of the U-shaped conductor in Fig. 22-12 has resistance of 0.10 Ω and that there is negligible resistance in each of the other three sides of the loop, including the rod, which has a length of 10 cm and moves at a constant speed of 4.0 m/s through a magnetic field of magnitude 0.50 T. Calculate the current induced in the loop.

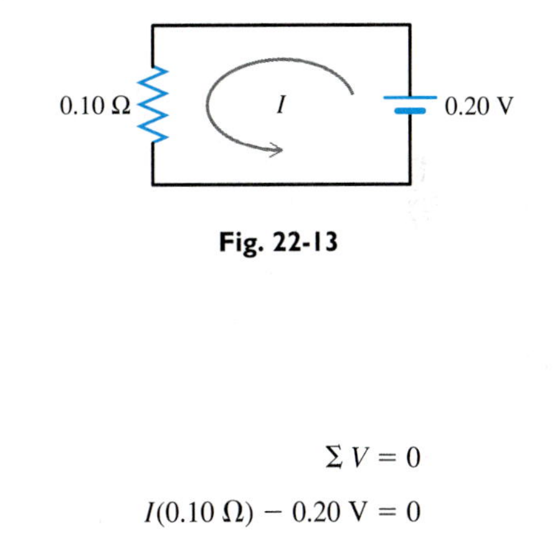

Fig. 22-13

SOLUTION Applying Eq. 22-4, we find for the magnitude of the emf

$$|\mathcal{E}| = B\ell v = (0.50\ \text{T})(0.10\ \text{m})(4.0\ \text{m/s})$$
$$= 0.20\ \text{V}$$

The loop is equivalent to the circuit shown in Fig. 22-13, with counterclockwise induced current I, which we find by applying Kirchhoff's second rule.

$$\Sigma V = 0$$
$$I(0.10\ \Omega) - 0.20\ \text{V} = 0$$
$$I = \frac{0.20\ \text{V}}{0.10\ \Omega} = 2.0\ \text{A}$$

Faraday's Law; Variable $\mathcal{E}$

In general, the rate of change of magnetic flux through a surface is not constant. Then Faraday's law must be expressed in its more general form, in which the emf equals the negative of the *instantaneous* flux rate:

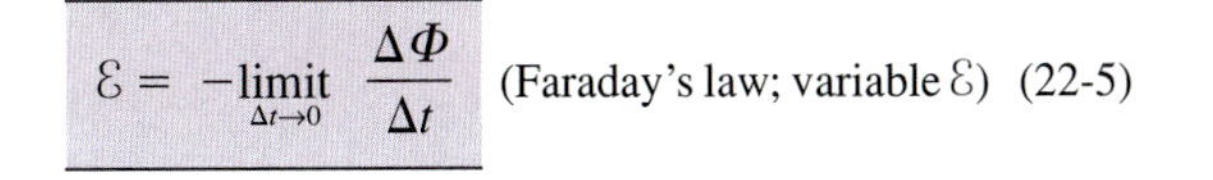

$$\mathcal{E} = -\lim_{\Delta t \to 0} \frac{\Delta \Phi}{\Delta t} \qquad \text{(Faraday's law; variable } \mathcal{E}\text{)} \quad (22\text{-}5)$$

Generators

The electric generator is an important practical application of motional emf. A simple alternating-current generator is shown in Fig. 22-14. A coil of wire is rotated in an external magnetic field. The magnetic flux through the coil varies as the coil rotates. When the plane of the coil is parallel to the magnetic field, the flux is zero. The flux through the coil is greatest when the plane of the coil is perpendicular to the field, so that the plane's normal **n** is aligned with the field. In the figure, the generator shaft is shown rotating counterclockwise at a point in the motion where **n** is aligning with the magnetic field. The magnetic flux through the surface is increasing; therefore, according to Faraday's law, there must be an emf induced in each loop of the coil. The induced current is directed as indicated in the figure (so that it produces a magnetic field that opposes the increase in flux), in accord with Lenz's law. As the shaft continues to turn, the normal **n** will point above the horizontal, and the flux through the surface will begin to decrease. The induced current will then be in the opposite direction. Thus the generator produces current that alternates in direction—alternating current (AC).

We can obtain an expression for the alternating induced emf by applying Faraday's law. If the shaft is rotated at constant angular velocity ω (that is, a fixed number of radians per second), the angle θ between **B** and **n** may be expressed

$$\theta = \omega t$$

The flux through the surface of area A bounded by the coil is given by Eq. 22-1:

$$\Phi = BA \cos \theta = BA \cos \omega t$$

In this case flux changes at a rate that is not constant. Thus, in applying Faraday's law to find the induced emf, we must use Eq. 22-5:

$$\mathcal{E} = -\lim_{\Delta t \to 0} \frac{\Delta \Phi}{\Delta t}$$

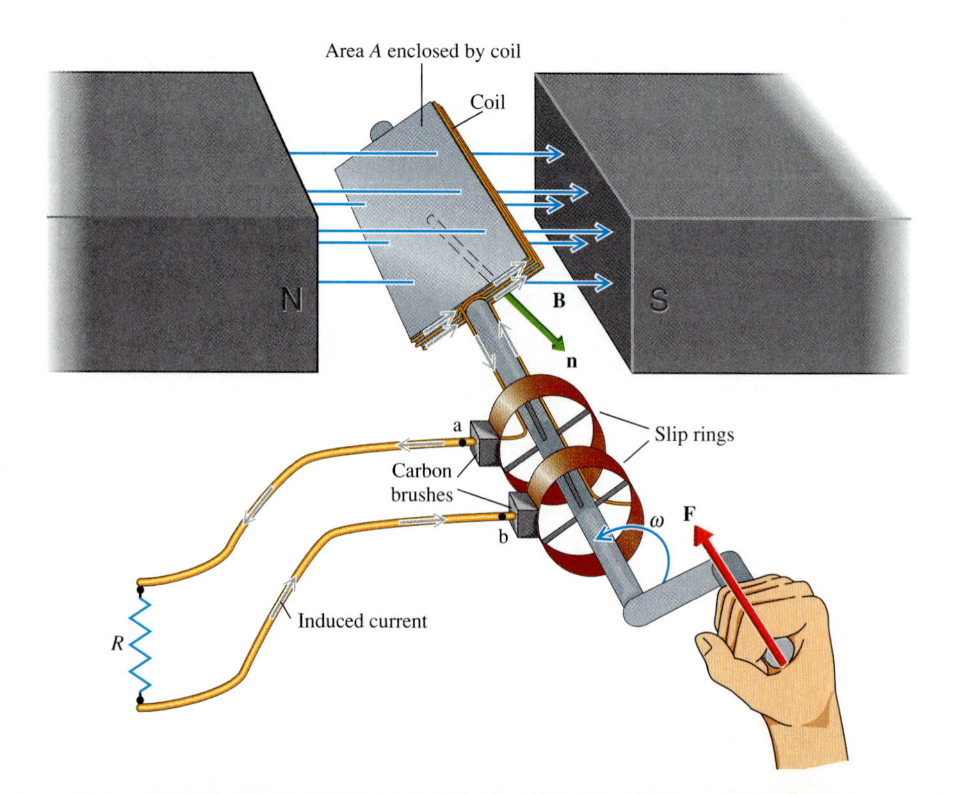

Fig. 22-14 An alternating-current generator connected to a resistor R.

The methods of calculus must be used to calculate the limit. We present the result without proof:*

$$\mathcal{E} = BA\omega \sin \omega t$$

This is the emf induced in each turn of the coil. For a coil with N turns, the total emf is N times the emf in a single turn.

$$\mathcal{E} = NBA\omega \sin \omega t \tag{22-6}$$

The maximum induced emf occurs when $\sin \omega t = 1$, that is, when $\omega t = 90°$, so that **n** is perpendicular to **B**. It is at this point that the flux is changing most rapidly. Denoting the maximum value of induced emf by $\mathcal{E}_0$, we may express the induced emf as

$$\mathcal{E} = \mathcal{E}_0 \sin \omega t \tag{22-7}$$

where

$$\mathcal{E}_0 = NBA\omega \tag{22-8}$$

The generator's angular velocity ω is related to the induced emf's period T and frequency f by Eqs. 15-6 and 15-7:

$$\omega = 2\pi f = 2\pi \frac{t}{T}$$

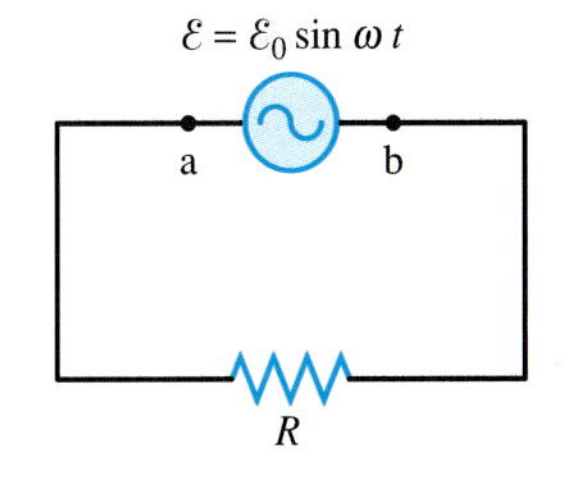

Fig. 22-15 The circuit representation of the generator and resistor in Fig. 22-14.

Commercial generators are designed somewhat differently from the generator we have described here. These generators utilize an electromagnet that rotates inside a stationary coil, in which the current is induced. The principle is the same: changing magnetic flux through the coil induces the emf.

The symbol ⊝ is used to represent an AC generator. Using this symbol, the generator and resistor shown in Fig. 22-14 are represented by the circuit diagram in Fig. 22-15. Fig. 22-16 shows a graph of the generator's emf as a function of time. Also shown in this figure, for selected values of t, are DC circuits with the same emf as the instantaneous emf in the AC circuit.

To find the voltage drop V_{ab} across the generator's terminals, we follow the rule developed for batteries in Chapter 19, which gives

$$V_{ab} = \mathcal{E} = \mathcal{E}_0 \sin \omega t \tag{22-9}$$

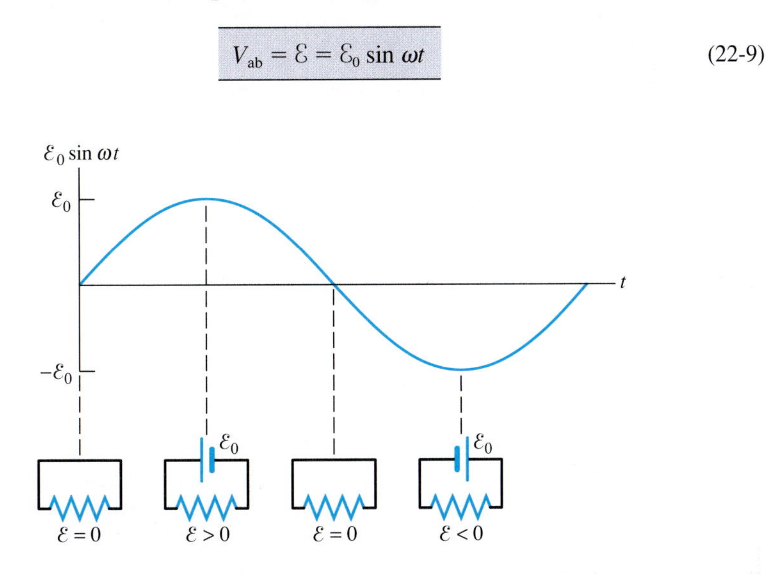

Fig. 22-16 The emf produced by an AC generator.

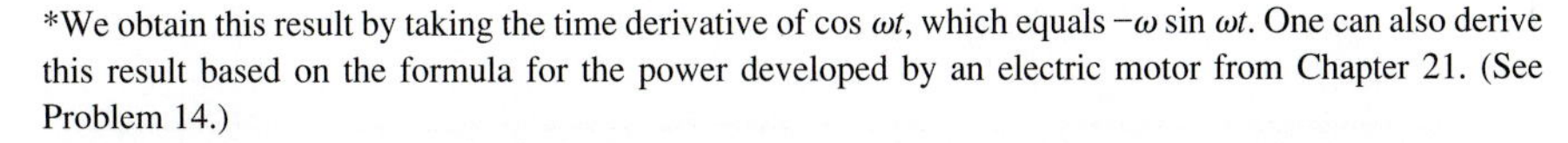

*We obtain this result by taking the time derivative of $\cos \omega t$, which equals $-\omega \sin \omega t$. One can also derive this result based on the formula for the power developed by an electric motor from Chapter 21. (See Problem 14.)

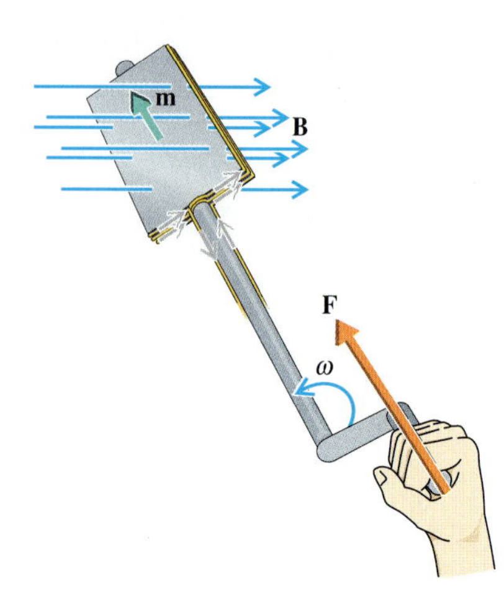

Fig. 22-17 Counterclockwise rotation of the generator coil induces a current and a magnetic moment **m**. The coil experiences a clockwise magnetic torque, which tends to align **m** with the external field. A counterclockwise torque must be applied to the shaft to cancel the magnetic torque and maintain the counterclockwise rotation.

Chemical energy is the source of the electrical potential energy provided by a battery. What then is the source of the electrical potential energy provided by a generator? The key to answering this question is to realize that the generator coil will not turn spontaneously. An applied external torque is required to balance the torque exerted on the coil by the magnetic field, as seen in Fig. 22-17. Thus the generator's energy source can be anything that provides a torque to turn the generator's shaft. For example, in a hydroelectric plant (Fig. 22-18a) the kinetic energy of water is used to turn turbine blades connected to generator coils. In other power plants coal is often used as the energy source (Fig. 22-18b). The coal is burned and used to produce steam, which turns turbine blades and generator coils. Nuclear fuel is used as the heat source in place of coal in nuclear power plants (Fig. 22-18c). In the desert outside Palm Springs, California, windmills utilize the kinetic energy of the wind to power electric generators (Fig. 22-18d). Human energy can even be used to generate electrical energy, for example, to power a small bicycle light (Problem 16).

Fig. 22-18 Various ways of generating electricity. **(a)** A hydroelectric power plant, outside and inside. **(b)** A coal-powered plant. **(c)** A nuclear powered plant. **(d)** Wind-powered generators.

EXAMPLE 3 Electric Power Supplied by a Generator

An electric generator operating at 60.0 Hz has a coil consisting of 100 turns enclosing an area of 5.00 m^2 in a magnetic field of 0.100 T. (a) Calculate the maximum value of the emf generated. (b) Find the maximum instantaneous power supplied by the generator when the current is 50.0 A.

SOLUTION **(a)** Applying Eq. 22-8, we find

$$\mathcal{E}_0 = NBA\omega = (100)(0.100\text{ T})(5.00\text{ m}^2)(2\pi \times 60.0\text{ Hz})$$
$$= 18{,}800\text{ V}$$

(b) Applying Eq. 19-18, we obtain

$$P = I\mathcal{E}_0 = (50.0\text{ A})(18{,}800\text{ V}) = 940{,}000\text{ W}$$
$$= 940\text{ kW} = 0.940\text{ MW}$$

The greater the current drawn from a generator, the more electric power is supplied by the generator, and of course the more mechanical power must be delivered to the generator to turn it. The current through the generator is determined by the load connected to it. Every time someone switches on an electric device, a parallel branch is added to the vast electric network connected to a generator, and more current is drawn from it. For example, when you turn on your air conditioner, more current must pass through the generator providing your electricity, and some energy source such as coal must be used to provide additional energy to the generator.

22-2 Inductance

Self-induction

In previous examples of induction, the source of the changing magnetic field was external to the conductor in which an emf was induced. However, when a circuit carries a time-varying current, this creates a time-varying magnetic field in the space around the circuit and therefore induces an emf in the circuit. So, in this case, the circuit is the source of its own induced emf. This particular kind of induction is appropriately called **self-induction.**

Lenz's law tells us that any induced emf opposes the change in magnetic flux that produces it. In the case of self-induced emf, the changing magnetic flux is created by a changing current in the circuit. **This self-induced emf always opposes a change in current.** Self-induced emf is sometimes referred to as **back emf.**

Consider, for example, a simple DC circuit consisting of a 3 V battery, a 1 Ω resistor, and a switch S (Fig. 22-19). Initially the switch is open, and the current $I = 0$. At $t = 0$ the switch is closed. The current does not instantly change from zero to its steady-state value of 3 A, for this would imply an infinite rate of change in magnetic flux through the surface bounded by the circuit, which would produce an infinite back emf in the circuit. Instead the current grows gradually over a finite time interval from zero to its final value of 3 A. The time interval over which current changes is actually very short—on the order of 10^{-7} s for this circuit. However, if a large solenoid were inserted into the circuit, it would take much longer to reach a steady-state current. We shall return to a detailed discussion of the time variation of current in such circuits, but first we must develop the precise relationship between back emf and the rate of change of current.

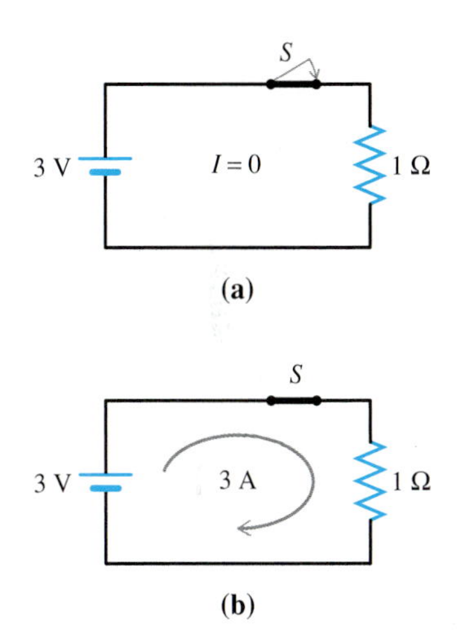

Fig. 22-19 **(a)** When the switch S is first closed, there is no current in the circuit. **(b)** A short time after the switch is closed there is a steady current of 3 A.

The magnetic field produced by a circuit at any field point is proportional to the current I through the circuit, according to the Biot-Savart law (Eq. 21-12). The magnetic flux through the surface bounded by the circuit is proportional to the average magnetic field, and so the flux itself is also proportional to the current.

$$\Phi \propto I$$

We can express the relationship between Φ and I as an equality by introducing a constant of proportionality L.

$$\Phi = LI \tag{22-10}$$

The constant L, called the **self-inductance,** depends only on the size and shape of the circuit. We can see the significance of the self-inductance by applying Faraday's law to the equation above.*

$$\mathcal{E} = -\frac{\Delta \Phi}{\Delta t}$$

$$\mathcal{E} = -L\,\frac{\Delta I}{\Delta t} \tag{22-11}$$

The value of the self-inductance determines the size of the back emf $\mathcal{E}$ for a given rate of change of current $\Delta I/\Delta t$. The back emf opposes the change in current. So the

*For the remainder of this chapter we will omit the limit notation when referring to a varying emf; that is, we will take it to be understood that if $\Delta\Phi/\Delta t$ is not constant what is really meant is the limiting value of this ratio.

greater a circuit's self-inductance, the more it will resist a change in current. Thus self-inductance can be thought of as a measure of a circuit's "electrical inertia," analogous to mass in mechanics.

The SI unit of self-inductance is the **henry,** abbreviated H, after the nineteenth century American physicist Joseph Henry* We can relate this unit to other electrical units, using Eq. 22-11. Since the units for $\mathcal{E}$ and $\Delta I/\Delta t$ are respectively V and A/s, we have

$$1\ \text{V} = 1\ \text{H-A/s}$$

or

$$1\ \text{H} = 1\ \text{V-s/A} \tag{22-12}$$

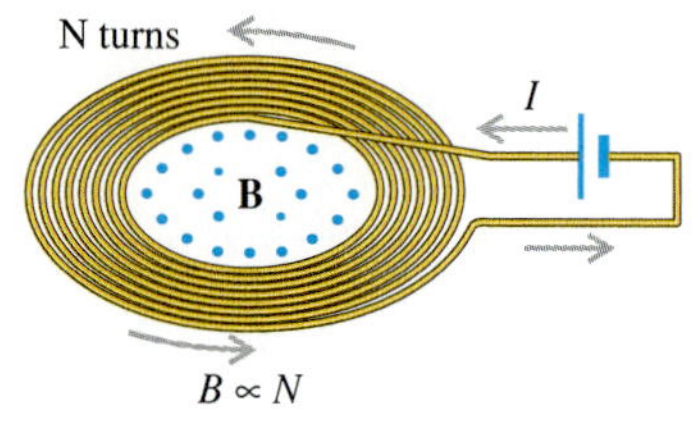

Fig. 22-20 The magnetic field produced by a coil is proportional to N, the number of turns in the coil. The coil's self-inductance is proportional to N^2.

Unless a circuit contains some kind of coil, such as a solenoid, its self-inductance will be extremely small and, for most purposes, can be ignored. For example, the self-inductance of a circular loop of wire of radius 10 cm is on the order of 10^{-7} H. However, when a circuit does contain a coil, the self-inductance can be quite significant.

The magnetic field is much stronger inside a coil, such as that shown in Fig. 22-20, than it is near a single conducting wire; the field is enhanced by a factor of N, where N is the number of turns in the coil. This means that when current changes at a certain rate $\Delta I/\Delta t$ the changing magnetic flux inside the coil is enhanced by a factor of N and so also is the induced electric field circulating around the changing magnetic field lines. Each turn of wire in the coil experiences the induced electric field, and so the total emf induced in the N turns is N times the emf around a single turn, or N^2 times the emf induced by a single turn on itself. Thus the self-inductance of a coil with N turns is proportional to N^2. Coils with thousands of turns obviously have much greater self-inductance than that of a single loop.

Inductors

An **inductor** is a coil whose primary function is to provide self-inductance to a circuit in which it is inserted (Fig. 22-21). Whenever there is a changing current through the inductor, it provides the circuit with back emf. Inductors are often used in electric circuits; for example, an inductor is used as part of a tuning circuit in a radio.

Since an inductor consists of many turns of wire, it usually has a significant resistance. However, we can think of a real inductor as a resistor in series with an "ideal inductor," which has no resistance. An **ideal inductor** is represented in circuits by the symbol  .

We shall obtain an expression for the self-inductance of a long, thin solenoid of length ℓ and cross-sectional area A with N turns. The magnetic field inside the solenoid is, to a good approximation, given by the equation found in Chapter 21 for an ideal (infinite) solenoid with $n = N/\ell$ turns per unit length. The magnetic field is directed along the solenoid's axis and has magnitude given by Eq. 21-21:

$$B = \mu_0 n I = \mu_0 \frac{NI}{\ell}$$

The flux through any cross section of the solenoid is the product of the field magnitude B and the cross-sectional area A (Eq. 22-1):

$$\Phi = BA$$

or

$$\Phi = \mu_0 \frac{NA}{\ell} I$$

Fig. 22-21 An inductor.

*Henry built the first large electromagnets, including one that could lift two tons of iron. He also apparently discovered induction independently of Faraday, though he published his results somewhat later.

According to Faraday's law, there is an emf $\mathcal{E} = -\dfrac{\Delta\Phi}{\Delta t}$ in each turn of the solenoid. For all N turns, the total emf is N times the value for one turn:

$$\mathcal{E} = -N\frac{\Delta\Phi}{\Delta t}$$

Using the preceding expression for Φ to compute $\Delta\Phi$, we find

$$\Delta\Phi = \mu_0\frac{NA}{\ell}\,\Delta I$$

and

$$\mathcal{E} = -\mu_0\frac{N^2A}{\ell}\frac{\Delta I}{\Delta t}$$

Comparing this expression with Eq. 22-11, we see that the solenoid's self-inductance is given by

$$L = \frac{\mu_0 N^2 A}{\ell} \qquad (22\text{-}13)$$

EXAMPLE 4 A Solenoid's Self Inductance

Find the self-inductance of a solenoid of length 10 cm, with 1000 turns and a cross-sectional area of 2.0 cm².

SOLUTION Applying Eq. 22-13, we find

$$L = \frac{\mu_0 N^2 A}{\ell} = \frac{(4\pi \times 10^{-7}\ \text{T-m/A})(1000)^2(2.0 \times 10^{-4}\ \text{m}^2)}{0.10\ \text{m}}$$

$$= 2.5 \times 10^{-3}\ \text{H} = 2.5\ \text{mH}$$

The fact that an inductor produces an emf in a circuit means that the voltage drop across the terminals of an (ideal) inductor must have a magnitude equal to the magnitude of the emf, just as for a battery or a generator.

$$|V_{ab}| = L\left|\frac{\Delta I}{\Delta t}\right|$$

To determine the sign of the voltage drop across an inductor, we note that its back emf means that the inductor acts like a battery, with terminals that are always oriented so that they oppose a change in the current, as indicated in Fig. 22-22. As shown in the figure, when one crosses the inductor terminals in the direction of positive current, the voltage drop has the same sign as $\Delta I/\Delta t$: positive if I is increasing and negative if I is decreasing. Using this result and the preceding expression for the magnitude of the voltage drop, we see that the voltage drop across an inductor can be expressed

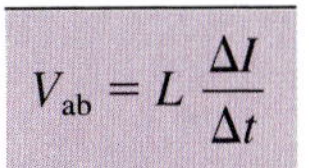

$$V_{ab} = L\frac{\Delta I}{\Delta t} \qquad \text{(for positive current directed from a to b)} \qquad (22\text{-}14)$$

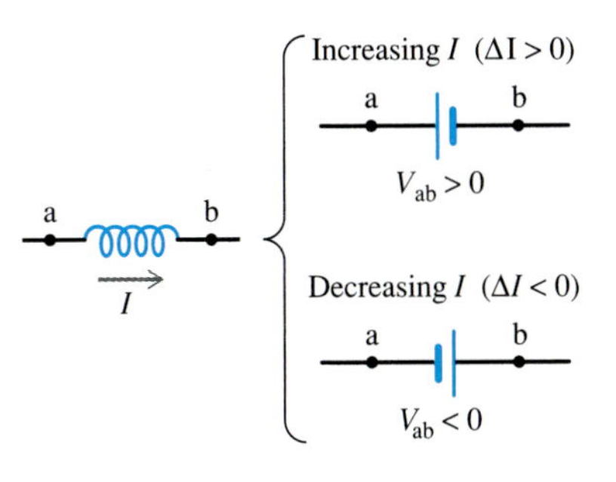

Fig. 22-22 Back emf of an inductor.

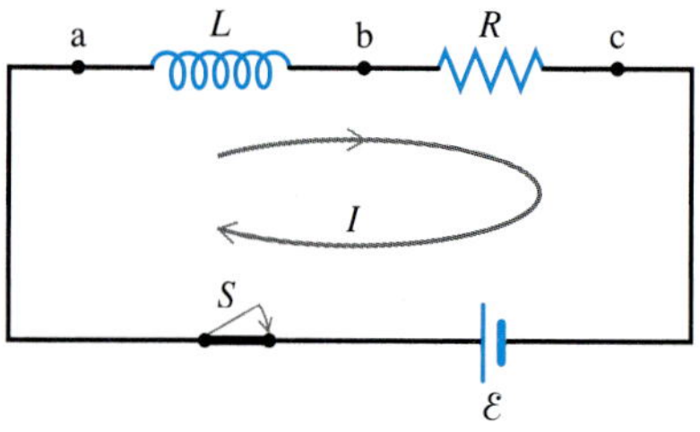

RL Circuits

Fig. 22-23 An RL circuit.

Next we shall describe the time variation of current in a simple circuit containing a resistor R and an ideal inductor L—an "RL" circuit. Fig. 22-23 shows a resistor and an inductor connected to a battery through a switch, which is closed at $t = 0$. We apply Kirchhoff's second rule around the circuit in the clockwise direction (starting at point a), and use the rules for the voltage drops across an inductor, a resistor, and a source of emf.

$$\Sigma V = 0$$

$$+L\frac{\Delta I}{\Delta t} + IR - \mathcal{E} = 0 \qquad (22\text{-}15)$$

This equation has the same form as the equation describing the charge on a capacitor being charged by a battery in an RC circuit (discussed in Section 20-6). Like the charge on the capacitor, the current in the RL circuit is an exponential function:

$$I = \frac{\mathcal{E}}{R}(1 - e^{-t/\tau}) \qquad (22\text{-}16)$$

where τ is the **time constant** of the RL circuit and is given by

$$\tau = \frac{L}{R} \qquad (22\text{-}17)$$

The methods of calculus are used to verify that Eq. 22-16 is the solution to Eq. 22-15.

When $t = 0$, Eq. 22-16 predicts that $I = \frac{\mathcal{E}}{R}(1 - 1) = 0$. This is as we expected. The current cannot change instantaneously from zero before the switch is closed to a nonzero value at the instant the switch is closed. At $t = \infty$, $I = \frac{\mathcal{E}}{R}(1 - e^{-\infty}) = \frac{\mathcal{E}}{R}$. This is the final constant value of the current.

The time constant is a measure of the time required to approach the final current. It is easy to find I as a fraction of the final current $\mathcal{E}/R$ at times that are multiples of τ.

$$t = \tau: \quad I = \frac{\mathcal{E}}{R}(1 - e^{-1}) = 0.63\frac{\mathcal{E}}{R}$$

$$t = 2\tau: \quad I = \frac{\mathcal{E}}{R}(1 - e^{-2}) = 0.86\frac{\mathcal{E}}{R}$$

$$t = 5\tau: \quad I = \frac{\mathcal{E}}{R}(1 - e^{-5}) = 0.99\frac{\mathcal{E}}{R}$$

$$t = 7\tau: \quad I = \frac{\mathcal{E}}{R}(1 - e^{-7}) = 0.999\frac{\mathcal{E}}{R}$$

Thus when $t >> \tau$, the current is very nearly equal to $\mathcal{E}/R$.

EXAMPLE 5 Current in an RL Circuit

A very large coil has self-inductance of 2.0 H and resistance of only 10 Ω.* The coil is connected across a 12 V battery at $t = 0$.

(a) Find the steady-state current.

(b) Find the current through the coil at $t = 0.50$ s, and indicate this point on a graph of I versus t.

SOLUTION The coil can be treated as an ideal 2.0 H inductor in series with a 10 Ω resistor. The circuit is that shown in Fig. 22-23.

(a) The steady-state current is simply $\mathcal{E}/R$, as for a circuit containing only a battery and a resistor.

$$I = \frac{\mathcal{E}}{R} = \frac{12\text{ V}}{10\ \Omega} = 1.2\text{ A}$$

(b) Applying Eqs. 22-16 and 22-17, we find

$$\tau = \frac{L}{R} = \frac{2.0\text{ H}}{10\ \Omega} = 0.20\text{ s}$$

$$I = \frac{\mathcal{E}}{R}(1 - e^{-t/\tau}) = (1.2\text{ A})(1 - e^{-0.50\text{ s}/0.20\text{ s}})$$

$$= 1.1\text{ A}$$

Fig. 22-24 shows a graph of I versus t.

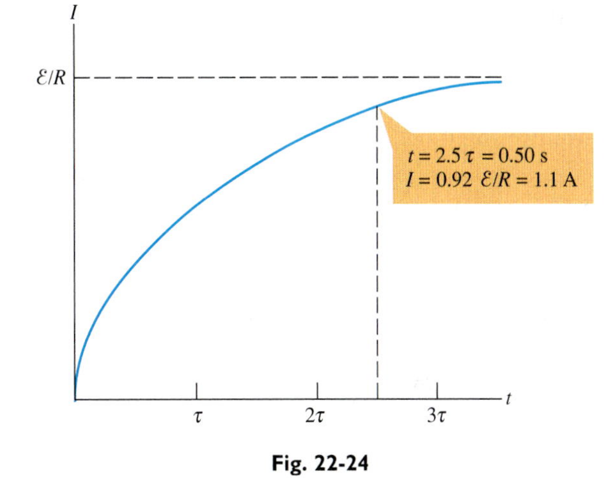

Fig. 22-24

*The ratio L/R for a coil usually has a value of 10^{-3} s or less.

Suppose that in an inductive circuit, in which a steady current has been established, a switch is suddenly opened. The current will not instantly change from its initial value to zero. The circuit's back emf tends to maintain the current. The continuing current causes charge to accumulate on either side of the switch until the potential difference across the switch equals the applied emf. If the voltage across the switch is great enough, there may be a spark resulting from dielectric breakdown of the air. For example, when you turn off a toaster by unplugging it, you can see a spark between the plug and the electrical outlet.

Magnetic Energy

When electric current passes through a resistor, electrical potential energy is converted to heat within the resistor. The loss of electrical potential energy equals the product of the charge passing through the resistor and the voltage drop across the resistor. Similarly, there is a loss of electrical potential energy when there is an increasing electric current I through an inductor ($\Delta I/\Delta t > 0$). This follows from the fact that positive current flows through a positive voltage drop $V_{ab} = L\,(\Delta I/\Delta t)$. Thus the charge Δq passing through the inductor loses electrical potential energy, where

$$\text{Electrical energy lost} = \Delta q\, V_{ab} = \Delta q\, L\, \frac{\Delta I}{\Delta t} \qquad (22\text{-}18)$$

This loss of electrical potential energy differs, however, from the loss in a resistor in two respects:

1 The energy is not converted to heat, since no heat is generated in an ideal inductor.

2 The energy can be recovered when the current decreases; when $\Delta I/\Delta t < 0$, the equation above predicts a negative loss of electrical potential energy, meaning that there is a gain in electrical potential energy.

Conservation of energy requires that the loss (or gain) of electrical potential energy of charge passing through an inductor must be accompanied by an equal gain (or loss) in some other form of energy. We shall find that this other energy is associated with the magnetic field inside an inductor. It is therefore called **magnetic energy,** denoted by U_m. Equating the gain in magnetic energy, ΔU_m, to the loss in electrical potential energy expressed in Eq. 22-18, we obtain an expression for ΔU_m.

$$\Delta U_m = \Delta q\, L\, \frac{\Delta I}{\Delta t} = L\, \frac{\Delta q}{\Delta t}\, \Delta I$$

As the current increases from zero to a final value I, more and more magnetic energy is stored in the inductor. The total magnetic energy U_m is found from the equation above by inserting for the change in current, $\Delta I = I$, and for $\Delta q/\Delta t$ the average current, $\frac{1}{2}I$.

$$U_m = \tfrac{1}{2}LI^2 \qquad (22\text{-}19)$$

This equation shows that magnetic energy stored in the inductor at any instant depends on the current in the inductor at that instant. Notice that this expression has the same form as the equation for the electrostatic energy stored by a capacitor C with a voltage V across its plates (Eq. 18-19: $U_e = \frac{1}{2}CV^2$).

Next we shall show how the magnetic energy may be written in terms of the magnetic field for the special case of a long, thin solenoid. The expression we shall obtain, though derived only for this simple special case, turns out to be valid in general. Inserting the expression for the self-inductance of the solenoid (Eq. 22-13) into the equation above, we obtain

$$U_m = \tfrac{1}{2}\mu_0 \frac{N^2 A I^2}{\ell}$$

We can think of this energy as being stored inside the solenoid of length ℓ and cross-sectional area A. The **magnetic energy per unit volume,** which we shall denote by u_m, is found when we divide the expression above by the volume of the solenoid $A\ell$, which gives

$$u_m = \tfrac{1}{2}\mu_0 \frac{N^2 I^2}{\ell^2}$$

Since the magnitude of the uniform magnetic field inside the solenoid is given by $B = \mu_0 NI/\ell$ (Eq. 21-21), we may substitute B/μ_0 for NI/ℓ in the equation above to obtain the general expression for the magnetic energy per unit volume at any point in space:

$$u_m = \frac{1}{2}\left(\frac{B^2}{\mu_0}\right) \tag{22-20}$$

Notice that this expression for magnetic energy density in a magnetic field is similar in form to the expression for electric energy density in an electric field, obtained in Chapter 18 (Eq. 18-21: $u_e = \tfrac{1}{2}\epsilon_0 E^2$).

The magnetic energy in an inductor may be compared with the electrostatic energy in a capacitor. When a capacitor is charged, electric potential energy is stored in the electric field between the capacitor plates; the energy is released as the capacitor discharges and the electric field is reduced. When current is established in an inductor, electrical potential energy is converted to magnetic energy, which is stored in the magnetic field inside the inductor. This energy is released when the inductor's current and its associated magnetic field decrease.

EXAMPLE 6 Large Scale Energy Storage in a Magnetic Field

Energy storage is important for electrical power companies because of fluctuations in the rate of electrical energy consumption. Superconducting coils have been used for large-scale energy storage.

(a) Find the magnetic energy density inside a superconducting coil that produces a magnetic field of 10.0 T.

(b) Find the volume of space necessary to store 1.00×10^5 kWh of energy at the density found in part (a). (This is enough energy to supply the electrical energy for roughly 5000 households for one day.)

SOLUTION **(a)** Applying Eq. 22-20, we find

$$u_m = \frac{1}{2}\left(\frac{B^2}{\mu_0}\right) = \frac{1}{2}\left(\frac{(10.0\text{ T})^2}{4\pi \times 10^{-7}\text{ T-m/A}}\right)$$
$$= 3.98 \times 10^7 \text{ J/m}^3$$

This is a very high energy density compared to the energy densities attainable with other methods of storing energy. See Problem 29.

(b) Applying the definition of magnetic energy density (magnetic energy per unit volume), we solve for the volume V required to store 1.00×10^5 kWh of energy.

$$u_m = \frac{U_m}{V}$$

$$V = \frac{U_m}{u_m} = \frac{(1.00 \times 10^5\text{ kWh})(3.60 \times 10^6\text{ J/kWh})}{3.98 \times 10^7\text{ J/m}^3}$$
$$= 9050 \text{ m}^3$$

This volume corresponds to a cube with edge length $V^{1/3} = 20.8$ m.

*22-3 Alternating Current Circuits

Although Edison's first power company supplied electrical energy in the form of direct current, today electricity is supplied by power companies as alternating current. We will discuss the practical advantage of alternating current, or AC, when we explain transformers at the end of this section. But first we shall describe a few simple AC circuits, containing an alternating source of emf, such as a generator, and either a resistor, a capacitor, an inductor, or all three together.

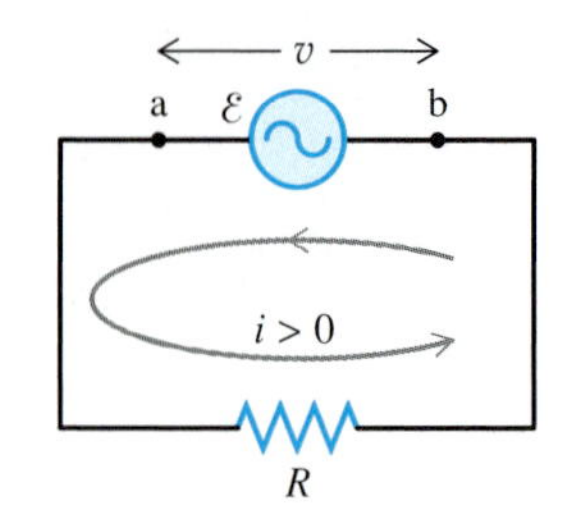

Fig. 22-25 A resistor connected across an AC source.

In this section we shall adopt the notation that lower-cased letters i and v will be used to denote instantaneous, time-varying currents and voltages respectively. Capital letters I and V will be used to denote constant values of current and voltage.

Resistor and an AC Source

Fig. 22-25 shows an AC circuit consisting of a resistor R connected across the terminals of a source of alternating emf $\mathcal{E}$. For example, this circuit could represent a lamp plugged into an electrical outlet.

The voltage drop v across the source's terminals is given by Eq. 22-9:

$$v = V \sin \omega t \tag{22-21}$$

where we have used V instead of $\mathcal{E}_0$ to denote the amplitude of the voltage drop. (Household line voltage has an amplitude $V = 170$ V, a frequency $f = 60$ Hz, and angular frequency $\omega = 2\pi f = 120\pi$ rad/s.) The alternating voltage will produce in the circuit an instantaneous current, which alternates in direction, but which at any instant is the same throughout the loop. We wish to find the current i in the counterclockwise direction, as indicated in the figure. The voltage drop across the resistor equals the voltage drop v across the source. So we simply apply Ohm's law to find the current through the resistor.

$$i = \frac{v}{R} \tag{22-22}$$

Inserting the preceding expression for v, we obtain

$$i = \frac{V}{R} \sin \omega t$$

or

$$i = I \sin \omega t \tag{22-23}$$

where

$$I = \frac{V}{R} \tag{22-24}$$

Fig. 22-26a shows graphs of i and v versus time. Notice that both i and v are proportional to $\sin \omega t$. The current i through the resistor and the voltage v across the resistor both oscillate at the same frequency. Furthermore, both i and v go to zero at the same time, and both reach their peak values at the same time. **The current and voltage are in phase.** This is a feature of the current through a resistor and the voltage across a resistor that is satisfied in any circuit.

Another way of representing current and voltage is by means of a **phasor diagram** (Fig. 22-26b). Current and voltage are represented by phasors **I** and **V** respectively. Phasors are like vectors, in that they have both a direction and a magnitude. The magnitude of a phasor is the amplitude of the time-varying quantity it is used to represent. The respective magnitudes of **I** and **V** are amplitudes I and V. The direction of a phasor varies with time. Phasors rotate at angular velocity ω, starting at $t = 0$ in the x direction. At any time t the phasors make an angle ωt with the x-axis. The instantaneous values of either current or voltage are found when the respective phasor is projected onto the y-axis, as indicated in the figure. Phasor diagrams are useful in the analysis of more complex circuits.

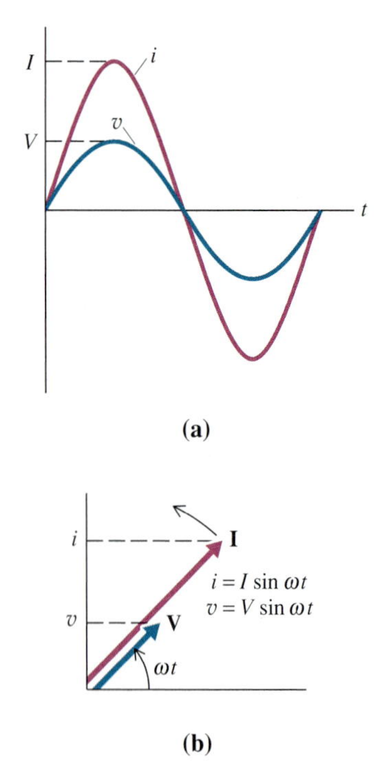

Fig. 22-26 **(a)** Time variation of i and v for the circuit in Fig. 22-25. **(b)** The phasor representation for the current and voltage in (a). Rotating phasors **V** and **I** represent current and voltage.

Notice that Fig. 22-26a shows the counterclockwise current i oscillating between positive and negative values, with an average value equal to zero. The current changes direction every half-cycle. Electrons oscillate back and forth, and on the average go nowhere! This does *not* mean that there is no electrical energy supplied to the resistor. As the electrons move back and forth, they collide with the lattice and produce thermal energy. The rate at which this thermal energy is produced is given by Eq. 19-16 for the power loss in a resistor:

$$P = i^2 R \tag{22-25}$$

Since P is proportional to i^2, the instantaneous power is always positive. Substituting Eq. 22-23 for the current into Eq. 22-25, we obtain

$$P = I^2R \sin^2 \omega t$$

Fig. 22-27 shows a graph of $\sin^2 \omega t$ versus ωt. The function $\sin^2 \omega t$ is always positive and varies from 0 to 1. The average value of $\sin^2 \omega t$ is $\frac{1}{2}$, as indicated in the figure. This means that we can find an expression for the average power dissipated by the resistor, P_{av}, by replacing $\sin^2 \omega t$ by $\frac{1}{2}$ in the equation above.

Fig. 22-27 The value of $\sin^2 \omega t$ varies between 0 and 1, oscillating above and below its average value of $\frac{1}{2}$.

$$P_{av} = \tfrac{1}{2}I^2R \qquad (22\text{-}26)$$

Another way to express this result is in terms of a quantity called the **root-mean-square current,** abbreviated rms current. The rms current is defined to be the square root of the mean (or average) current squared. Applying this definition to Eq. 22-23 and using the fact that the average value of $\sin^2 \omega t$ equals $\frac{1}{2}$, we find

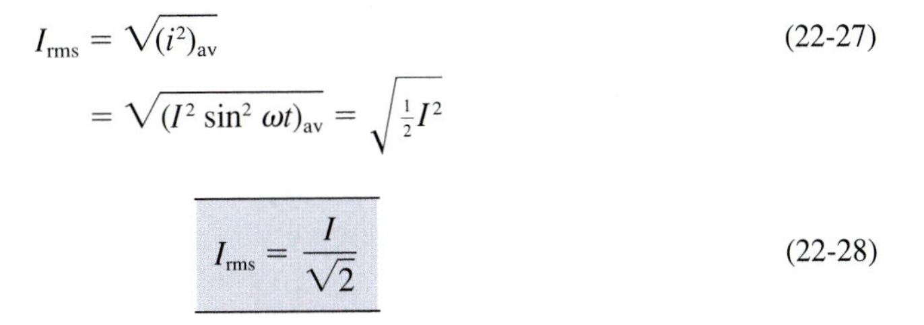

$$I_{rms} = \sqrt{(i^2)_{av}} \qquad (22\text{-}27)$$

$$= \sqrt{(I^2 \sin^2 \omega t)_{av}} = \sqrt{\tfrac{1}{2}I^2}$$

or

$$I_{rms} = \frac{I}{\sqrt{2}} \qquad (22\text{-}28)$$

Substituting $I_{rms}^2 = \frac{1}{2}I^2$ into Eq. 22-26, we can express the average power loss in the resistor as

$$P_{av} = I_{rms}^2 R \qquad (22\text{-}29)$$

The rms voltage is defined in the same way, and since v has the same time dependence as i, the rms voltage V_{rms} is related to the voltage amplitude V in the same way that I_{rms} is related to I; that is,

$$V_{rms} = \frac{V}{\sqrt{2}} \qquad (22\text{-}30)$$

AC ammeters and voltmeters read rms values. Thus, if a voltmeter measuring household line voltage reads 120 V, this means that the root mean square of the instantaneous voltage across the lines is 120 V. Then, according to Eq. 22-30, the amplitude of the line voltage is $\sqrt{2}V_{rms} = \sqrt{2}(120 \text{ V}) = 170$ V.

We have seen that Ohm's law can be expressed either in terms of instantaneous values of current and voltage (Eq. 22-22: $i = v/R$) or in terms of amplitudes (Eq. 22-24: $I = V/R$). It is easy to show that Ohm's law can also be expressed in terms of rms values. Substituting $I = \sqrt{2}I_{rms}$ and $V = \sqrt{2}V_{rms}$ into Eq. 22-24, we find

$$I_{rms} = \frac{V_{rms}}{R}$$

or

$$V_{rms} = I_{rms}R \qquad (22\text{-}31)$$

EXAMPLE 7 An AC Voltmeter's Reading

Find the reading of an AC voltmeter placed across a 50.0 Ω resistor, which carries a current $i = 5.00 \sin(120\pi t)$, where i is in amps.

SOLUTION The current amplitude $I = 5.00$ A. Applying Eq. 22-28, we find the rms current.

$$I_{rms} = \frac{I}{\sqrt{2}} = \frac{5.00 \text{ A}}{\sqrt{2}} = 3.54 \text{ A}$$

We apply Ohm's law (Eq. 22-31) to find the rms voltage, which gives the voltmeter's reading.

$$V_{rms} = I_{rms}R = (3.54 \text{ A})(50.0 \ \Omega) = 177 \text{ V}$$

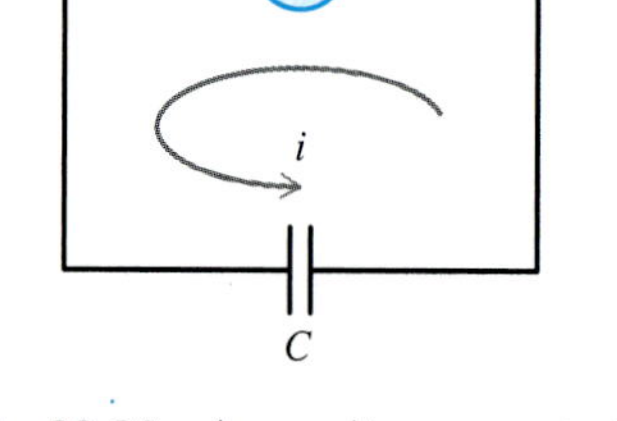

Fig. 22-28 A capacitor connected to an AC source.

Capacitor and an AC Source

Fig. 22-28 shows an AC circuit consisting of a capacitor connected across the terminals of a source of alternating emf, producing across the capacitor a voltage drop

$$v = V \sin \omega t \tag{22-32}$$

The figure shows positive counterclockwise current, which results in positive charge on the capacitor's left plate. The charge q stored on this plate at any instant is the product of the capacitance C and the instantaneous voltage v.

$$q = Cv = CV \sin \omega t$$

The current i flowing through the connecting wires equals the rate of change of q.

$$i = \frac{\Delta q}{\Delta t}$$

The methods of calculus can be used to find an expression for i from the preceding equation for q. The result is

$$i = I \cos \omega t \tag{22-33}$$

where

$$I = \omega CV \tag{22-34}$$

Fig. 22-29a shows graphs of v and i versus ωt. Notice that although the current and voltage both oscillate at the same frequency they are not in phase. **The current leads the voltage by $\frac{1}{4}$ cycle, or 90°.** This is also shown in the corresponding phasor diagram (Fig. 22-29b).

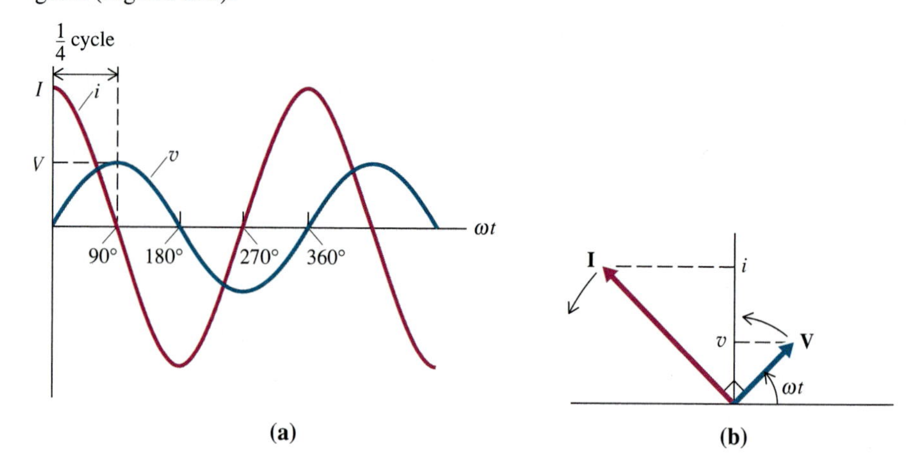

Fig. 22-29 **(a)** Time variation of i and v for the circuit in Fig. 22-28. **(b)** The phasor representation for the current and voltage in (a).

Although no steady-state direct current can exist in a circuit containing a capacitor, our solution indicates that an alternating current can continue indefinitely. Of course no current actually crosses the capacitor's plates. The current exists in the wires connected to the capacitor, as the charge on its plates alternates in sign, in response to the alternating applied voltage. Initially there is a positive current in the counterclockwise direction, meaning that positive charge is being deposited on the left plate of the capacitor. But when the voltage reaches its peak value and starts to reduce (when $\omega t = 90°$), the capacitor begins to discharge. This means that the current is positive in the clockwise direction (or negative in the counterclockwise direction, as the negative value of i in Fig. 22-29a indicates).

Next we shall express the relationship between V and I given in Eq. 22-34 in a different form. First we define the capacitor's **capacitive reactance,** denoted by X_C, as

$$X_C = \frac{1}{\omega C} \tag{22-35}$$

Substituting $\omega C = \dfrac{1}{X_C}$ in Eq. 22-34, we obtain

$$I = \frac{V}{X_C} \tag{22-36}$$

This equation has the same form as Ohm's law for a resistor. The reactance of a capacitor is related to the capacitor's current and voltage in the same way that the resistance of a resistor is related to its current and voltage. Thus reactance must have the same units as resistance, units of ohms.

The same relationship exists between rms values of current and voltage, since each equals $\sqrt{2}$ times the respective amplitude.

$$I_{rms} = \frac{V_{rms}}{X_C} \tag{22-37}$$

We can understand the physical reason for the current's dependence on reactance by considering why increasing the value of either capacitance or frequency should tend to increase the current, as predicted by Eqs. 22-35 and 22-36. During a quarter cycle, from $\omega t = 0$ to $\omega t = 90°$, charge CV is deposited on the left plate of the capacitor, as a result of the positive current. If C increases, both the charge and the current must increase. And if ω increases, the charge is deposited in a shorter time interval, meaning that again the current increases.

The electrical power supplied to the capacitor at any instant is given by Eq. 19-15:

$$P = iv$$

Substituting for i and v from Eqs. 22-32 and 22-33, we obtain

$$P = IV \sin \omega t \cos \omega t \tag{22-38}$$

or, using the trigonometric identity $\sin 2\omega t = 2 \sin \omega t \cos \omega t$, we get

$$P = \tfrac{1}{2} IV \sin 2\omega t$$

The average value of $\sin 2\omega t$ is zero, and so the average electrical power supplied to the capacitor is zero.

$$P_{av} = 0 \qquad \text{(capacitor)} \tag{22-39}$$

A charged capacitor stores energy in the field between its plates. As the capacitor is being charged, the source of emf is supplying energy to the capacitor. But when the capacitor discharges, the energy that was stored in the electric field is delivered back to the source. So the result is like a battery that is alternately discharging (giving up energy) and being charged (being reenergized).

EXAMPLE 8 The Frequency at which a Capacitor's Reactance is 50 Ω.

At what frequency would a 5.00 μF capacitor produce the same current as the 50.0 Ω resistor in Example 6, when connected to the same voltage source as the resistor?

SOLUTION If the current and voltage amplitudes are the same as for the resistor, it follows that the ratio V/I must be the same:

$$X_C = \frac{V}{I} = 50.0\ \Omega$$

Applying Eq. 22-35, we solve for f.

$$X_C = \frac{1}{\omega C} = \frac{1}{2\pi f C}$$

$$f = \frac{1}{2\pi C X_C} = \frac{1}{2\pi(5.00 \times 10^{-6}\ \text{F})(50.0\ \Omega)}$$

$$= 637\ \text{Hz}$$

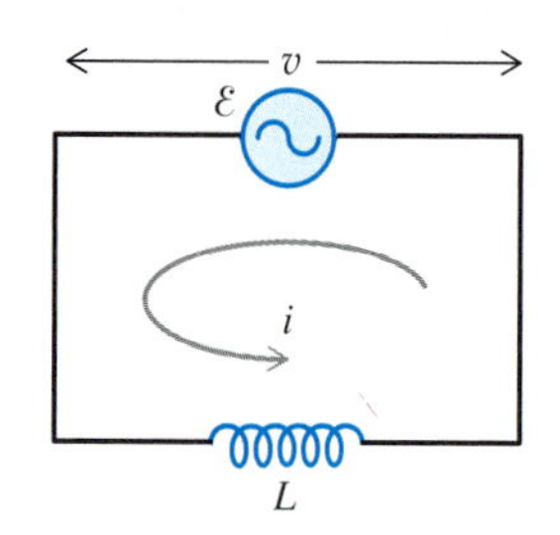

Fig. 22-30 An inductor connected to an AC source.

Inductor and an AC Source

Fig. 22-30 shows an AC circuit consisting of an inductor connected across the terminals of a source of alternating emf, producing across the inductor a voltage drop

$$v = V \sin \omega t \tag{22-40}$$

This voltage drop is related to the rate of change of current through the inductor by Eq. 22-14:

$$v = L\frac{\Delta i}{\Delta t}$$

Equating the two expressions for v and solving for $\Delta i/\Delta t$, we obtain

$$\frac{\Delta i}{\Delta t} = \frac{V}{L} \sin \omega t$$

The methods of calculus can be used to solve this equation for i. The result is

$$i = -I \cos \omega t \tag{22-41}$$

where

$$I = \frac{V}{\omega L}$$

We define **inductive reactance,** denoted by X_L, as

$$\boxed{X_L = \omega L} \tag{22-42}$$

The preceding equation can then be expressed in the form

$$\boxed{I = \frac{V}{X_L}} \tag{22-43}$$

As usual, the same relationship exists between rms values of current and voltage.

$$I_{rms} = \frac{V_{rms}}{X_L} \qquad (22\text{-}44)$$

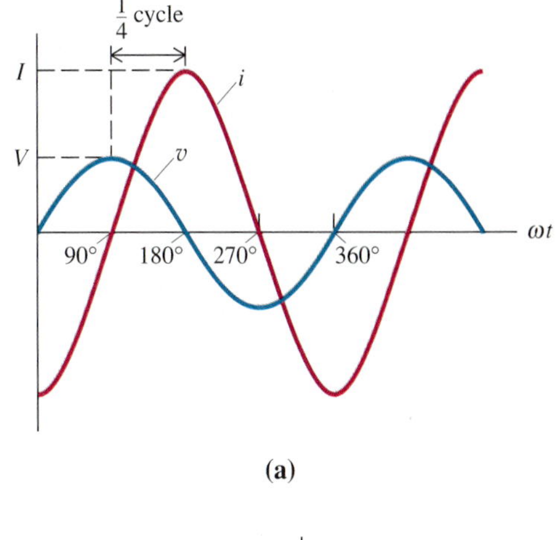

Like capacitive reactance, inductive reactance has units of ohms.

Fig. 22-31a shows graphs of v and i versus ωt for the inductor circuit. As in the case of the capacitor circuit, though the current and voltage both oscillate at the same frequency, they are not in phase. But here **the current lags the voltage by $\frac{1}{4}$ cycle, or 90°.** This is also shown in the corresponding phasor diagram (Fig. 22-31b). The electrical inertia of the inductor causes this current lag.

The electrical power supplied to the inductor at any instant is found by applying Eq. 19-15:

$$P = iv$$

Substituting for i and v from Eqs. 22-40 and 22-41, we obtain

$$P = -IV \sin \omega t \cos \omega t$$

Except for the minus sign, this is the same expression as Eq. 22-38, which gives the instantaneous power supplied to a capacitor. As we found for the capacitor, the average power supplied is zero.

$$P_{av} = 0 \qquad \text{(inductor)} \quad (22\text{-}45)$$

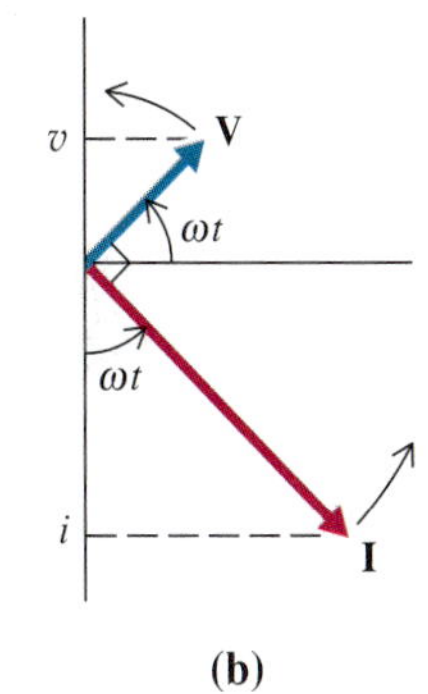

As the current increases, the magnetic field and magnetic energy inside the inductor increase, with the energy being supplied by the source. As the current decreases, the magnetic energy decreases, with the energy being returned to the source.

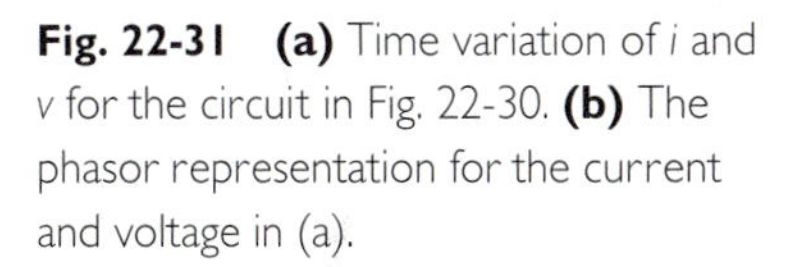

Fig. 22-31 **(a)** Time variation of *i* and *v* for the circuit in Fig. 22-30. **(b)** The phasor representation for the current and voltage in (a).

EXAMPLE 9 Finding the Current Through an Inductor

A 20.0 mH inductor is connected across a 60.0 Hz source with an rms voltage of 120 V. Find the rms current.

SOLUTION First we apply Eq. 22-42 to find the inductive reactance

$$X_L = \omega L = 2\pi f L = 2\pi(60.0 \text{ Hz})(2.00 \times 10^{-2} \text{ H})$$
$$= 7.54\ \Omega$$

The rms current is found by applying Eq. 22-44.

$$I_{rms} = \frac{V_{rms}}{X_L} = \frac{120 \text{ V}}{7.54\ \Omega} = 15.9 \text{ A}$$

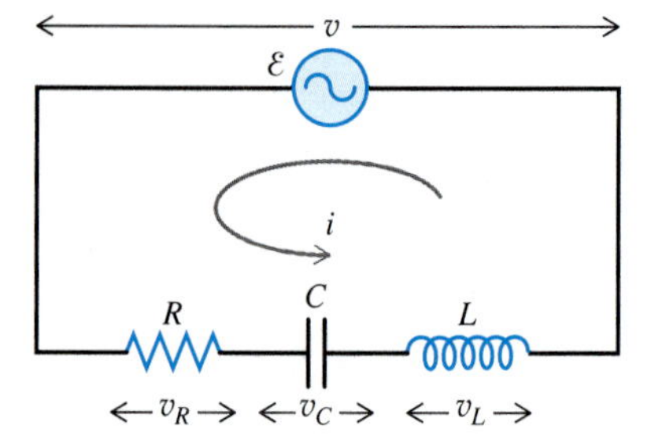

Fig. 22-32 A resistor, a capacitor, and an inductor connected in series with an AC source.

RCL Series Circuit

Fig. 22-32 shows an alternating current source connected across a series combination of a resistor, a capacitor, and an inductor. The source produces a voltage drop v across the combination, where

$$v = V \sin \omega t \tag{22-46}$$

This voltage drop equals the sum of the voltage drops across the circuit elements R, C, and L.

$$v = v_R + v_C + v_L \tag{22-47}$$

Inserting into this expression the preceding equation for v and the expressions for the voltage drops across R, C, and L, we obtain

$$V \sin \omega t = iR + \frac{q}{C} + L\frac{\Delta i}{\Delta t} \tag{22-48}$$

This equation may be solved for i, using methods of calculus. The result is a current that oscillates at the same frequency as the source but not necessarily in phase with it.

$$i = I \sin(\omega t - \phi) \tag{22-49}$$

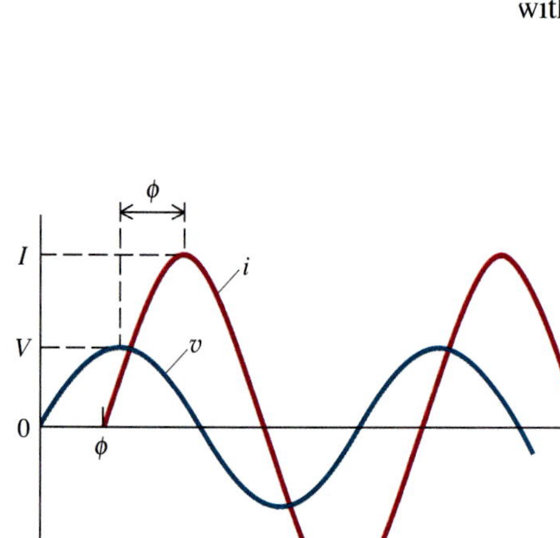

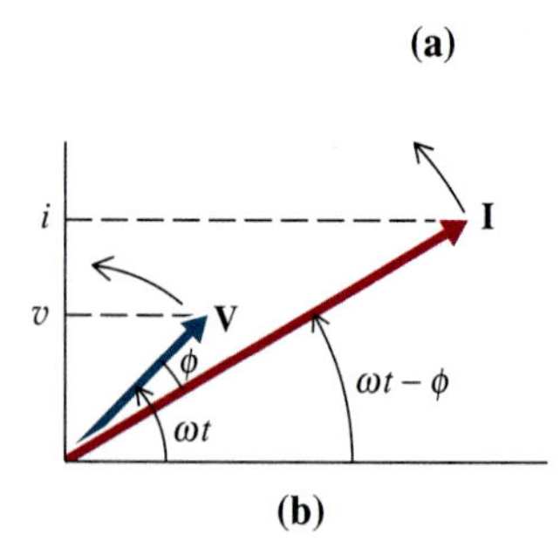

Fig. 22-33 **(a)** Time variation of the voltage v across the source and the current i for the circuit in Fig. 22-32. **(b)** The phasor representation for the voltage and current in (a).

The amplitude I and phase angle ϕ are determined by V, R, C, L, and ω.

Fig. 22-33a shows graphs of v and i versus ωt, and Fig. 22-33b shows the corresponding phasor diagram. We can find the phase angle ϕ between the current and voltage, as well as the relationship between the amplitudes V and I, by inserting into our phasor diagram phasors representing voltages across each of the elements R, C, and L. This expanded phasor diagram, shown in Fig. 22-34, makes use of the following phase relationships (the same as in the simpler circuits containing only one element: R, C, or L):

1. The voltage v_R across the resistor is in phase with the current.
2. The current leads the capacitor voltage v_C by 90°.
3. The current lags the inductor voltage v_L by 90°.

In order for the projections of the phasors on the y-axis to satisfy Eq. 22-47, the phasors themselves must satisfy the corresponding vector relationship:

$$\mathbf{V} = \mathbf{V}_R + \mathbf{V}_C + \mathbf{V}_L \tag{22-50}$$

Since $(\mathbf{V}_L + \mathbf{V}_C)$ has magnitude $(V_L - V_C)$ and is perpendicular to $\mathbf{V}_R$, as indicated in Fig. 22-34, the magnitude of phasor $\mathbf{V}$ is related to the magnitudes of the other phasors by the equation

$$V^2 = V_R^2 + (V_L - V_C)^2$$

The voltage amplitudes are related to the current amplitude by the equations

$$V_R = IR$$

$$V_L = IX_L$$

$$V_C = IX_C$$

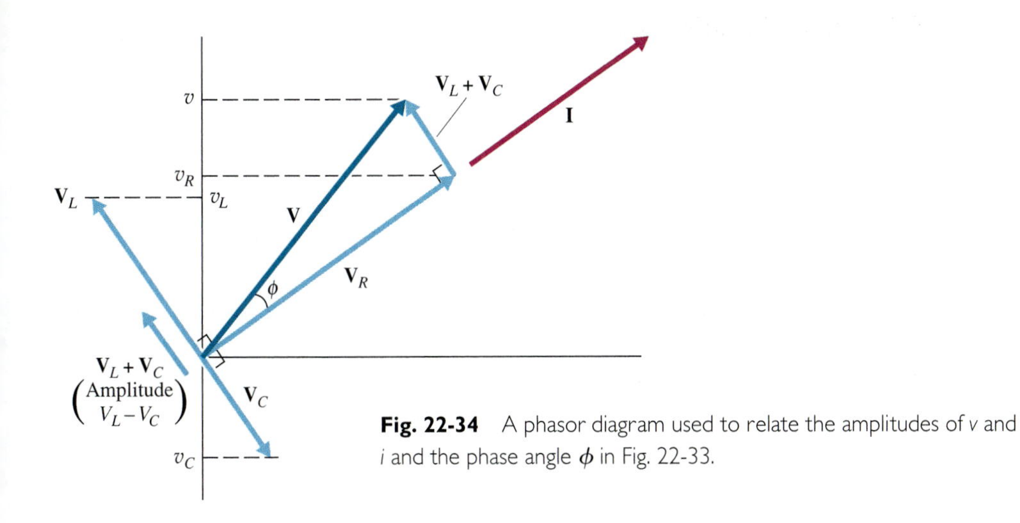

Fig. 22-34 A phasor diagram used to relate the amplitudes of v and i and the phase angle ϕ in Fig. 22-33.

Substituting these three equations into the preceding equation, we find

$$V^2 = I^2R^2 + I^2(X_L - X_C)^2$$

or

$$V = I\sqrt{R^2 + (X_L - X_C)^2}$$

We may express this result as

$$V = IZ \tag{22-51}$$

where Z, called the **impedance,** is defined as

$$Z = \sqrt{R^2 + (X_L - X_C)^2} \tag{22-52}$$

Once again the rms values of current and voltage satisfy the same equation as the amplitudes; that is:

$$V_{rms} = I_{rms}Z \tag{22-53}$$

We can think of impedance as a generalization of resistance. The impedance appears in Eq. 22-51 the same way that resistance appears in Ohm's law. And, according to Eq. 22-52, Z reduces to R when X_L and X_C both equal zero.

EXAMPLE 10 Voltmeter Readings Across RCL Circuit Elements

Suppose the RCL circuit shown in Fig. 22-32 has the following values for the circuit elements: $R = 300\ \Omega$, $C = 5.60\ \mu\text{F}$, and $L = 200$ mH. Let the rms voltage across the 60.0 Hz source be 120 V.

(a) Find the rms current.

(b) Find the rms voltage drop across each of the circuit elements.

SOLUTION **(a)** First we calculate the capacitive and inductive reactances, using Eqs. 22-35 and 22-42:

$$X_C = \frac{1}{\omega C} = \frac{1}{2\pi f C} = \frac{1}{2\pi(60.0\text{ Hz})(5.60 \times 10^{-6}\text{ F})}$$
$$= 474\ \Omega$$

$$X_L = \omega L = 2\pi f L = 2\pi(60.0\text{ Hz})(200 \times 10^{-3}\text{ H})$$
$$= 75.4\ \Omega$$

Next we insert these values, along with the resistance, into Eq. 22-52 to find the impedance.

$$Z = \sqrt{R^2 + (X_L - X_C)^2} = \sqrt{(300\ \Omega)^2 + (75.4\ \Omega - 474\ \Omega)^2}$$
$$= 499\ \Omega$$

Finally, we apply Eq. 22-51 to find the rms current.

$$I_{rms} = \frac{V_{rms}}{Z} = \frac{120\text{ V}}{499\Omega}$$
$$= 0.240\text{ A}$$

(b) We find the rms voltage drop across each circuit element by multiplying the rms current times that element's impedance, either R, X_C, or X_L.

$$V_{R,\,rms} = I_{rms}R = (0.240\text{ A})(300\ \Omega) = 72.0\text{ V}$$
$$V_{C,\,rms} = I_{rms}X_C = (0.240\text{ A})(474\ \Omega) = 114\text{ V}$$
$$V_{L,\,rms} = I_{rms}X_L = (0.240)(75.4\ \Omega) = 18.1\text{ V}$$

These values correspond to the readings of an AC voltmeter placed across each of the circuit elements. If we place the voltmeter across all three elements, it will not read the sum of these voltage drops (204 V). Instead it will read the rms voltage of the source, 120 V. The rms voltage drops are not additive because the voltages are not in phase.

Resonance

Consider an RCL circuit with fixed values of resistance, capacitance, and inductance, but with a source frequency f, which can be varied. Since both inductive and capacitive reactance depend on frequency, varying the frequency causes the circuit's impedance and current to vary. According to Eq. 22-51, current is maximized when impedance is minimized. From Eq. 22-52, we see that, at a frequency such that $X_L - X_C = 0$, the impedance will have its smallest value, $Z = R$. Setting the two expressions for reactance equal to each other, and solving for the corresponding frequency f_0, we find

$$X_L = X_C$$

$$2\pi f_0 L = \frac{1}{2\pi f_0 C}$$

$$f_0 = \frac{1}{2\pi\sqrt{LC}} \tag{22-54}$$

This frequency at which the current is maximized is the circuit's **resonant frequency.** See Chapter 15, Section 15-5, for a discussion of resonance.

Resonance is used in tuning a radio or television. Tuning adjusts the receiving circuit so that it is in resonance at a resonant frequency equal to the broadcast frequency, so that only the signal from that one source is amplified.

Transformers

Electrical power companies provide electrical energy by means of the alternating potential difference maintained between power lines. The potential difference between the lines entering a residence must be great enough to operate the typical electrical devices to which they are connected but not so great as to be extremely dangerous. (For example, you would not want to plug an appliance into a 100,000 V outlet.) So residences are supplied with lines at 120 V and 240 V, rms. See Chapter 20, Section 20-7.

On the other hand, long transmission lines leading from a distant power generator need to be at very high voltage to minimize energy loss in the lines. Some of the electrical energy supplied by a generator is wasted as heat in the power lines. These lines are thick wires, but since the lines are very long, their total resistance becomes significant. (The resistance might be on the order of 1 Ω per mile.) The average rate of production of heat in lines having resistance R and carrying rms current I_{rms} is given by

$$P_{av,\ waste} = I_{rms}^2 R \tag{22-55}$$

The average rate at which energy is supplied to lines, carrying rms current I_{rms} and with an rms potential difference V_{rms} between the lines, is given by

$$P_{av,\ supply} = I_{rms} V_{rms} \tag{22-56}$$

To minimize the production of waste heat while supplying the required power, it is best to maintain the lines at as high a voltage as possible, so that the current is as small as possible. For example, from Eq. 22-56, we see that lines transmitting 100,000 W of power could carry 100 A with a potential difference of 1000 V, or the lines could transmit the same power by carrying 1 A with a potential difference of 100,000 V. According to Eq. 22-55, in the first case the rate of production of waste heat equals $(100\ \text{A})^2R$, or 10^4 W per Ω of line resistance, whereas in the second case the rate of heat production is $(1\ \text{A})^2 R$, or 1 W per Ω of line resistance. Clearly the higher the voltage, the more efficient will be the transmission of power. But this voltage requirement differs from the requirements for energy utilization. So what is needed is a device to change voltages, so that power can be transmitted over long distances at high voltage and then, close to the point of energy use, transmitted at much lower voltages. The transformer is such a device.

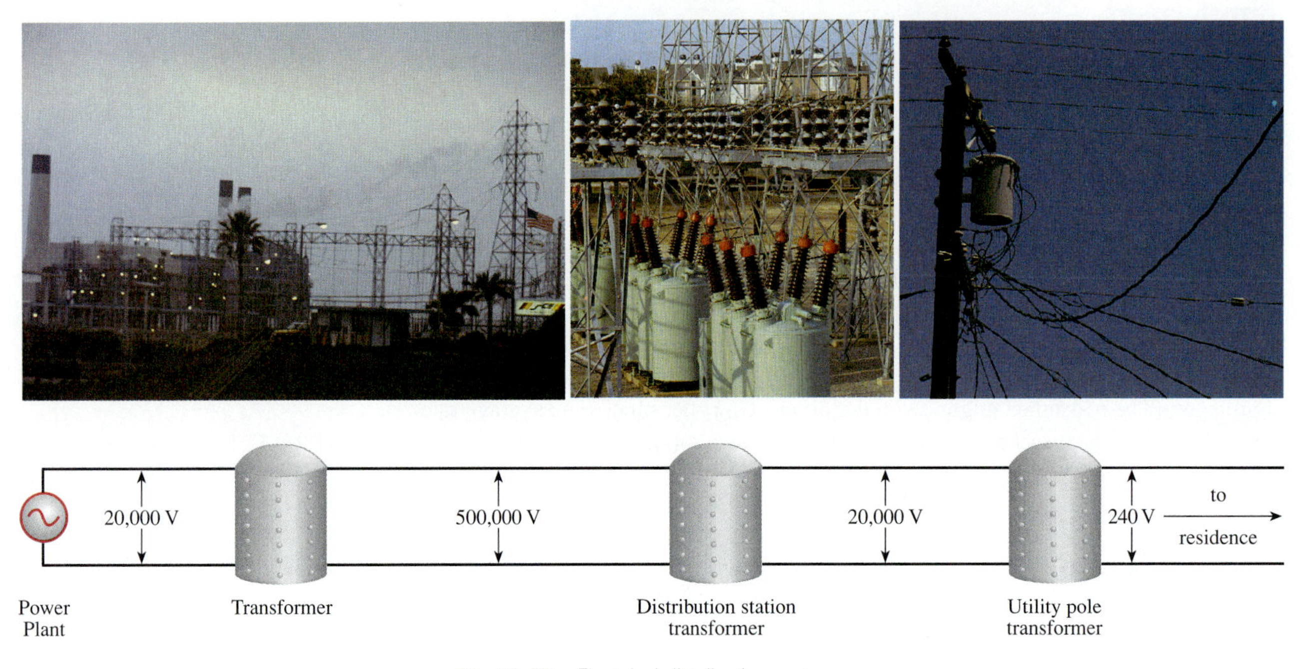

Fig. 22-35 Electrical distribution system.

Fig. 22-35 shows a schematic diagram indicating how voltages are first increased and then decreased in lines leading from a power plant to points of energy consumption.

Fig. 22-36 shows a schematic diagram of a transformer consisting of two electrically insulated coils wound around a common iron core. The primary coil, consisting of N_1 turns, is shown connected to a generator that maintains an rms voltage $V_{1,\text{rms}}$ across it. The alternating current in the primary coil produces in the iron core a time-varying magnetic field, which induces an emf in the secondary coil, and a voltage $V_{2,\text{rms}}$ across the secondary, which is different from $V_{1,\text{rms}}$. Each coil of the transformer acts like an inductor. The magnetic flux is the same through each turn of either coil. Therefore, according to Faraday's law, any one turn of either coil must have the same emf, which we shall designate $\mathcal{E}_{\text{turn}}$. The primary coil, with N_1 turns, has a total emf $\mathcal{E}_1$, where

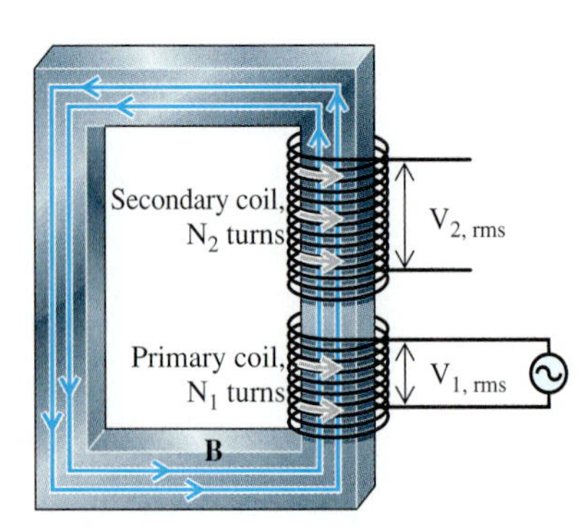

Fig. 22-36 Schematic diagram of a transformer.

$$\mathcal{E}_1 = N_1\mathcal{E}_{\text{turn}} \tag{22-57}$$

and the secondary coil, with N_2 turns, has a total emf $\mathcal{E}_2$, where

$$\mathcal{E}_2 = N_2\mathcal{E}_{\text{turn}} \tag{22-58}$$

Equating two expressions for $\mathcal{E}_{\text{turn}}$ obtained from Eqs. 22-57 and 22-58, we obtain

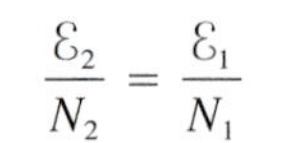

$$\frac{\mathcal{E}_2}{N_2} = \frac{\mathcal{E}_1}{N_1}$$

As for an ideal battery, inductor, or generator, the voltage drop across either coil of an ideal transformer (one with negligible energy loss) equals the emf induced in the coil. Thus we can substitute voltages for emf's in the equation above, which yields

$$\frac{V_{2,\text{rms}}}{N_2} = \frac{V_{1,\text{rms}}}{N_1}$$

or

$$V_{2,\text{rms}} = \frac{N_2}{N_1} V_{1,\text{rms}} \tag{22-59}$$

If the number of turns in the secondary is greater than that in the primary, $\frac{N_2}{N_1} > 1$, and so the output voltage is greater than the input voltage. Such a transformer steps up voltage. If, however, there are fewer turns in the secondary than in the primary, $\frac{N_2}{N_1} < 1$, the output voltage is less than the input voltage; the transformer steps down voltage.

A transformer depends on time-varying voltage for its operation, since it is based on the principle of induction. There is no comparable device for changing the voltage of a direct-current source. Hence alternating current, rather than direct current, is used today to supply society's electrical needs.

The fact that a step-up transformer can actually put out a higher voltage than is input to the transformer might lead you to believe that you can get more energy out of a transformer than you put in. This would be true if the currents in the primary and secondary coils were equal. However, they are not. In an ideal transformer (one with negligible energy loss from heat in the windings and so forth), conservation of energy requires that the power input and power output are the same. Since the average power is the product of rms current and voltage, we have the following relationship between primary and secondary currents and voltages:

$$I_{1,\text{rms}} V_{1,\text{rms}} = I_{2,\text{rms}} V_{2,\text{rms}} \tag{22-60}$$

EXAMPLE 11 Power Loss in Transmission Lines at Different Voltages

A power station transmits electrical power a long distance through lines having a total resistance of 200 Ω. Calculate the power lost through heating of the lines, if 4.00×10^2 kW of power is supplied by the generator to lines at a potential difference of (a) 2.00×10^4 V; (b) 5.00×10^5 V.

SOLUTION (a) First we calculate the current carried by the lines, applying Eq. 22-56.

$$P_{\text{av,supply}} = I_{\text{rms}} V_{\text{rms}}$$

$$I_{\text{rms}} = \frac{P_{\text{av,supply}}}{V_{\text{rms}}} = \frac{4.00 \times 10^5 \text{ W}}{2.00 \times 10^4 \text{ V}} = 20.0 \text{ A}$$

Applying Eq. 22-55, we find the power wasted as heat in the lines.

$$P_{\text{av,waste}} = I_{\text{rms}}^2 R = (20.0 \text{ A})^2 (200\ \Omega) = 8.00 \times 10^4 \text{ W} = 80.0 \text{ kW}$$

This is one fifth of the 400 kW of power generated—a significant loss.

(b) Repeating our calculations at the higher line voltage, we find

$$I_{\text{rms}} = \frac{P_{\text{av,supply}}}{V_{\text{rms}}} = \frac{4.00 \times 10^5 \text{ W}}{5.00 \times 10^5 \text{ V}} = 0.800 \text{ A}$$

$$P_{\text{av,waste}} = I_{\text{rms}}^2 R = (0.800 \text{ A})^2 (200\ \Omega) = 128 \text{ W}$$

This is a very small fraction of the power generated. Stepping up the voltage from 20,000 V to 500,000 V can be accomplished by a transformer in which the ratio of secondary turns to primary turns, N_2/N_1, equals the ratio of the voltages, V_2/V_1, according to Eq. 22-59. Thus

$$\frac{N_2}{N_1} = \frac{500{,}000}{20{,}000} = 25$$

Michael Faraday

Michael Faraday (1791-1867).

Fig. 22- A Faraday lecturing at the Royal Institution.

In 1805 Michael Faraday, the 14-year-old son of a London blacksmith, began work as an apprentice bookbinder. During the next several years, young Faraday read many of the books that passed through the shop where he worked. *Encyclopædia Britannica* articles on electricity and chemistry stimulated his interest, and he began to perform simple experiments and to attend scientific lectures. At 21 years of age Faraday was hired as a laboratory assistant at the Royal Institution, an important center of scientific research. Having only the most elementary formal education, Faraday had gained access to the best laboratory and library facilities, as well as to the institute's director, the chemist Sir Humphrey Davy, who tutored him. Faraday was given increasing responsibility, and over a period of years gradually progressed from assisting others in their work to performing his own original research. By age 30 his discoveries had made him world famous. Eventually Faraday became the director of the institute. In that capacity, he often lectured to the public on a broad range of scientific subjects. Faraday was a skillful and popular lecturer. At his annual Christmas lecture for children, he loved to inspire and entertain his young audience with dramatic demonstrations.

Faraday's research was directed by his belief in unifying principles in nature and by his search for symmetries in natural phenomena. He believed, for example, that gravitational, electric, and magnetic forces were all related.* Faraday's vision led him to perform important experiments in chemistry, electricity, magnetism, optics, and sound.

Faraday studied and repeated other scientists' experiments in electricity and magnetism. Oersted's demonstration of the force exerted on a magnet by a conducting wire had been the first step toward the unification of electricity and magnetism. Ampère, Biot, and Savart then studied the interaction of electric currents. Faraday's attention was particularly drawn to the experiments and theories of Ampère, who proposed that the magnetism displayed by permanent magnets is due to hidden electrical currents within the magnet, as discussed in Chapter 21. This suggested to Faraday that the magnetism induced in iron by a permanent magnet† might be an indication of a

*The unification of the gravitational force with the other fundamental forces was a goal of Einstein; physicists today still search for a "grand unified field theory."

†This induced magnetism is actually caused by alignment of magnetized, randomly oriented domains with an external magnetic field, as discussed in Section 21-6.

general property of "induction," which all electric currents should possess. Thus Faraday thought that an electric current in one conductor might induce a current in another nearby conductor, and in 1825 he tried to observe such an effect. One current carrying wire was placed close to a second wire, which was a part of a complete circuit that included a sensitive galvanometer G, used to detect any current that might be induced in the second circuit (Fig. 22-B). No current was detected.

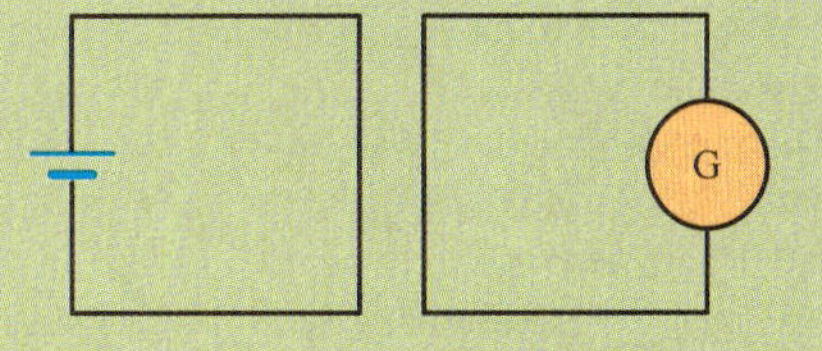

Fig. 22-B The constant current in the circuit on the left induced no current in the circuit on the right, contrary to Faraday's expectations.

However, Faraday was never satisfied with one experiment. He was an extremely careful, thorough, and open-minded experimentalist, who constantly tried to challenge his own theories and to discover any faulty preconceptions. Over the next several years, he performed many experiments in the search for induced current. Finally in 1831 Faraday observed induced electric current using the apparatus illustrated in Fig. 22-2. Two coils of electrically insulated wire were wound around an iron ring. A battery provided the current in the primary circuit on the left, and a galvanometer was used to detect any current induced in the secondary circuit on the right. Faraday correctly supposed that the iron would enhance any induced current that was produced, making it easier to detect. While the primary circuit containing the battery was being connected, Faraday noticed a momentary deflection of the galvanometer. With a steady current through the primary circuit, no deflection of the galvanometer was observed. But when the primary circuit was disconnected, there was again a momentary deflection of the galvanometer needle, this time in the opposite direction. Thus Faraday discovered that although a steady current produces no induction effect a changing current in one circuit does induce a current in a nearby circuit. He then began a series of other experiments in which he showed that electric currents were induced in conductors moving near permanent magnets, as illustrated in Fig. 22-3. Using this principle, Faraday built the first primitive electric generator; by rotating a conducting disk in a magnetic field, he produced a continuous current in wires connected to the disk (Fig. 22-1). Eventually the electric generator would make possible the production of cheap, abundant electric current, which batteries could never do.

One of Faraday's most important contributions was the conceptual framework he used in observing electric and magnetic phenomena. In the early nineteenth century, electric and magnetic forces were described mathematically, but, lacking an understanding of atomic structure, the conceptual basis of the description was confused, with various theories competing to "explain" the phenomena. Faraday did not understand all of the mathematics that was being used, but he had a great intuitive sense. He was the first to use the concept of a field and to picture field lines, which he referred to as "lines of force." This approach was to flourish in the hands of James Maxwell, who possessed the mathematical education that Faraday lacked. Maxwell read Faraday's work, as well as the more mathematical work of others. His synthesis resulted in the unification of electricity and magnetism, using the first field theory. Field theory has been an important part of theoretical physics ever since. In the introduction to his great work *A Treatise on Electricity and Magnetism,* in 1873 Maxwell wrote:

> As I proceeded with the study of Faraday, I perceived that his method of conceiving the phenomena was also a mathematical one, though not exhibited in the conventional form of mathematical symbols. . . . For instance, Faraday, in his mind's eye, saw lines of force traversing all space where the mathematicians saw centres of force attracting at a distance: Faraday saw a medium where they saw nothing but distance: Faraday sought the seat of the phenomena in real actions going on in the medium, they were satisfied that they had found it in a power of action at a distance. . . .

HAPTER 22 SUMMARY

The magnetic flux Φ through a surface of area A is defined as the product of the area and the field's component $B_\perp$, perpendicular to the surface.

$$\Phi = B_\perp A$$

The flux is proportional to the number of field lines passing through the surface.

When the magnetic flux changes, an electric field is created or induced in the surrounding space. The result of this induced field can be an emf and a current induced in a conducting loop. According to **Faraday's law,** the magnitude of the emf induced around any loop equals the magnitude of the rate of change of magnetic flux through the surface bounded by the loop.

$$|\mathcal{E}| = \left|\frac{\Delta\Phi}{\Delta t}\right| \quad \text{(constant } \mathcal{E}\text{)}$$

or

$$|\mathcal{E}| = \left|\lim_{\Delta t \to 0} \frac{\Delta\Phi}{\Delta t}\right| \quad \text{(variable } \mathcal{E}\text{)}$$

The direction of the induced current resulting from the emf is found from **Lenz's law,** which states that the induced current produces a magnetic field that opposes the change in magnetic flux that produced it.

Rotating a coil of wire in an external magnetic field at an angular velocity ω results in an induced emf $\mathcal{E}$ and a terminal potential difference V_{ab}, where

$$V_{ab} = \mathcal{E} = \mathcal{E}_0 \sin \omega t$$

The amplitude $\mathcal{E}_0$ is the product of the number of turns N, the field strength B, and the coil's area A.

$$\mathcal{E}_0 = NBA$$

A time-varying current in a coil produces a self-induced emf, opposing the emf $\mathcal{E}$ in the coil. The magnitude of the induced emf depends on a property of the coil called its self-inductance, denoted by L and defined by the relationship between the current I through the coil and the magnetic flux through the coil resulting from that current:

$$\Phi = LI$$

A coil designed to have large self-inductance is called an inductor. The self-inductance of a long, tightly wound solenoid of length ℓ and cross-sectional area A, with N turns, is given by

$$L = \frac{\mu_0 N^2 A}{\ell}$$

The voltage drop V_{ab} across an inductor is given by

$$V_{ab} = L\frac{\Delta I}{\Delta t}$$

if positive current flows through the inductor in the direction from point a to point b.

When a resistor R and an inductor L are initially connected in series across a battery of emf $\mathcal{E}$, current is not immediately established in the circuit because of the electrical inertia of the inductor. Instead the current grows with time according to the equation

$$I = \frac{\mathcal{E}}{R}(1 - e^{-t/\tau})$$

where the time constant τ is given by

$$\tau = \frac{L}{R}$$

An inductor L carrying current I stores magnetic energy U_m, where

$$U_m = \tfrac{1}{2}LI^2$$

The magnetic energy per unit volume, u_m, inside the inductor (or anywhere there is a magnetic field) may be expressed in terms of the field strength as

$$u_m = \frac{1}{2}\left(\frac{B^2}{\mu_0}\right)$$

An RCL series circuit is formed when a resistor R, a capacitor C, and an inductor L are connected in series across an AC source supplying a voltage $V \sin \omega t$ across the combination. The current through the circuit is given by

$$i = I \sin(\omega t - \phi)$$

The amplitude I is determined by the voltage amplitude V and the impedance Z from the equation

$$V = IZ$$

where Z is related to the resistance R and to the capacitive reactance

$$X_C = \frac{1}{\omega C}$$

and to the inductive reactance

$$X_L = \omega L$$

by the equation

$$Z = \sqrt{R^2 + (X_L - X_C)^2}$$

It is often convenient to express the current by its rms (root-mean-square) value, I_{rms}, which is related to the current amplitude I by the equation

$$I_{rms} = \frac{I}{\sqrt{2}}$$

Similarly, rms voltage V_{rms} is related to the voltage amplitude V by the equation

$$V_{rms} = \frac{V}{\sqrt{2}}$$

The rms current and voltage are related in the same way as the current and voltage amplitudes:

$$V_{rms} = I_{rms}Z$$

A resistor converts electrical energy to heat at an average rate

$$P_{av} = I_{rms}^2 R \qquad \text{(for a resistor)}$$

whereas neither an inductor nor a capacitor uses electrical power:

$$P_{av} = 0 \qquad \text{(for a capacitor or inductor)}$$

The current in the circuit is maximized at a frequency f_0, called the "resonant frequency," which is given by

$$f_0 = \frac{1}{2\pi\sqrt{LC}}$$

Transformers transform voltages from an input voltage $V_{1,rms}$ to an output voltage $V_{2,rms}$, where the voltages are related to the number of turns of wire in the primary and secondary, N_1 and N_2, by the equation

$$V_{2,rms} = \frac{N_2}{N_1} V_{1,rms}$$

Questions

1 Do more of the earth's magnetic field lines pass outward through the earth's surface or inward, or is there an equal number in each direction?

2 Can there ever be more magnetic field lines passing outward through a spherical surface than passing inward?

3 What is the direction of the current through resistor R in Fig. 22-37 as switch S is (a) closed; (b) opened?

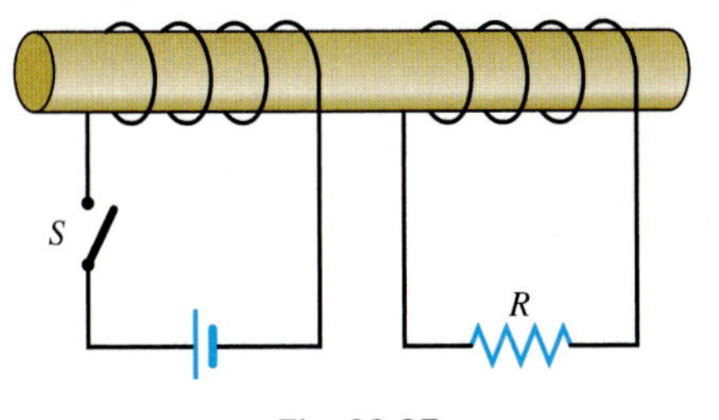

Fig. 22-37

4 Is the direction of the induced current in the circular loop in Fig. 22-38 clockwise or counterclockwise as switch S is (a) closed; (b) opened?

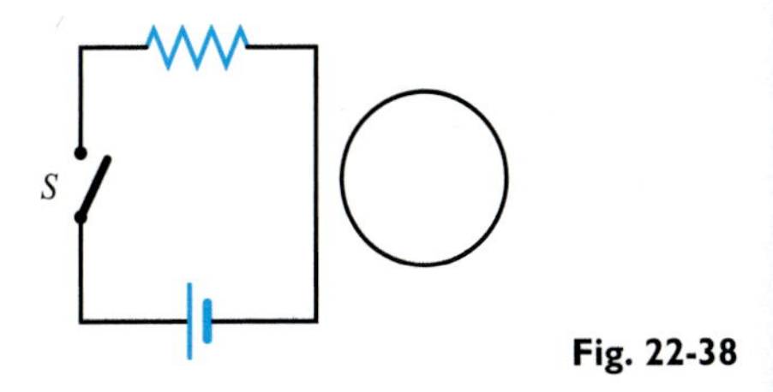

Fig. 22-38

5 Fig. 22-39 shows a metal ring moving upward toward the south pole of a permanent magnet. Is the current induced in the ring in the clockwise or counterclockwise direction, when viewed from above?

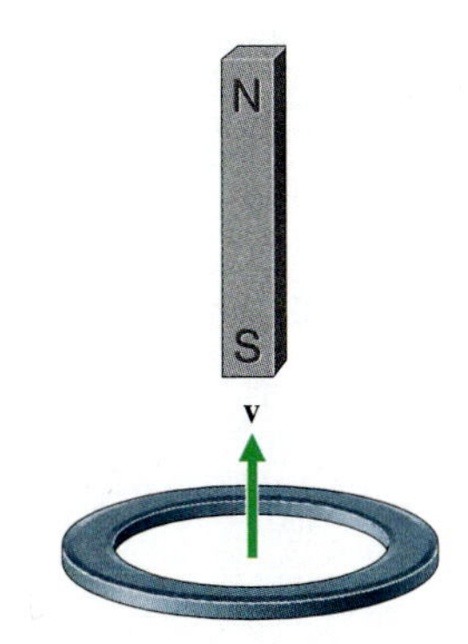

Fig. 22-39

6 Fig. 22-40 shows a square coil of wire falling through a magnetic field. Indicate for each of the three positions shown in the figure the direction of the current (if any) induced in the coil and the direction of the net magnetic force (if any) acting on the coil. This illustrates how induced currents can tend to decelerate a conductor moving through a magnetic field.

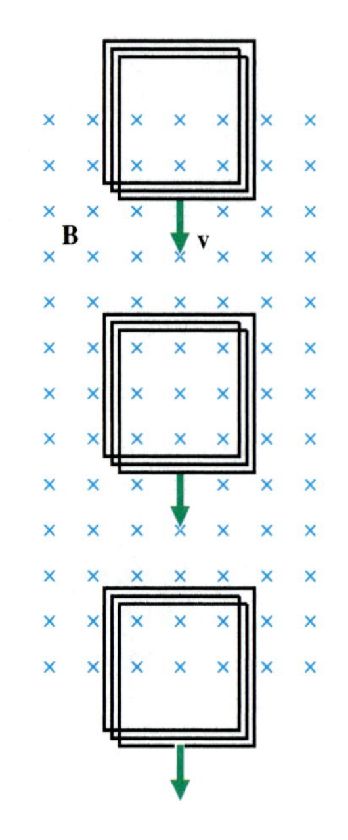

Fig. 22-40

7 Fig. 22-41 shows a coil of wire in the xz plane, with a magnetic field, directed along the x-axis. Around which of the three coordinate axes should the coil be rotated in order to generate an emf and a current in the coil?

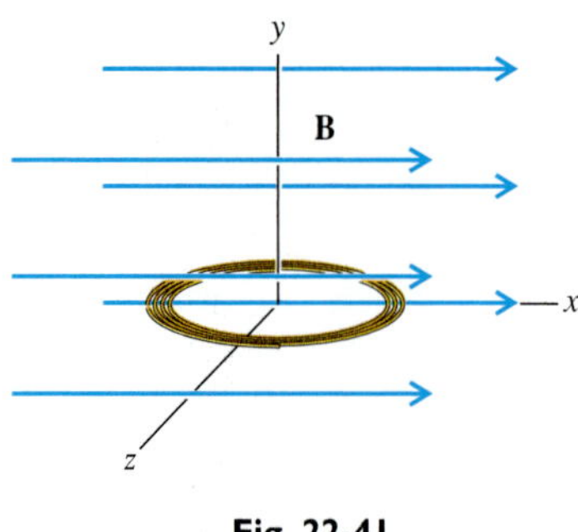

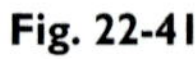

Fig. 22-41

8 A coil of wire is connected to the pedals of a bicycle, so that the coil turns as the bicycle is pedaled (Fig. 22-42). Pedaling will generate an emf in the coil if a magnetic field is present, unless the magnetic field is in which of the following directions: (a) vertical; (b) horizontal, from the front to the back of the bicycle; (c) horizontal, from one side of the bicycle to the other? For which of these directions of the magnetic field will the force required to pedal the bicycle be greater than when there is no magnetic field? For which directions will the force be less than when there is no magnetic field?

Fig. 22-42

9 When an iron bar is inserted into a solenoid, the self-inductance of the solenoid: (a) increases; (b) decreases; (c) remains the same.

10 Switch S in Fig. 22-43 is initially closed.

(a) Find the sign of the voltage drop V_{ab} across the inductor just after the switch is opened.

(b) Estimate the current through the resistor 10^{-2} s after the switch is opened.

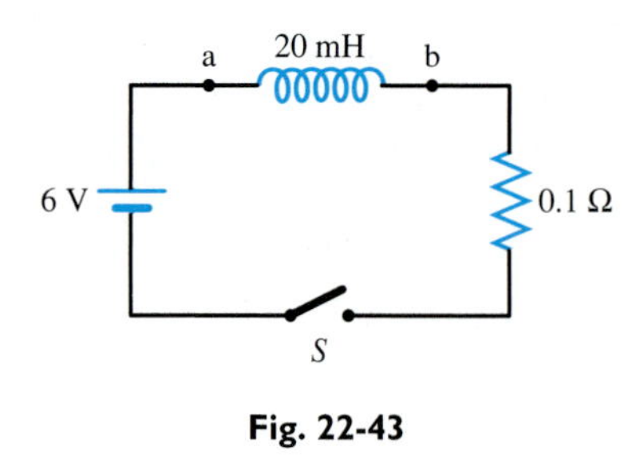

Fig. 22-43

11 Suppose you insert an ideal inductor in series with a resistor in a DC circuit. What effect will the inductor have on the steady-state current through the resistor?

12 Does a very high-frequency, single-loop AC circuit containing an inductor behave like (a) an open circuit or (b) a short circuit?

13 Does a very high-frequency, single-loop AC circuit containing a capacitor behave like (a) an open circuit or (b) a short circuit?

14 The primary of a transformer is connected to a 12 V battery. What is the voltage across the transformer's secondary if it has 10 times as many turns as the primary?

15 The current through the coil of a stereo speaker determines the intensity of the sound it produces, as discussed in Chapter 21 (Example 4 and Problem 55). A typical large speaker box contains two speakers, as indicated in Fig. 22-44. The larger one, called a "woofer," is better at producing high-quality low-frequency, long-wavelength sound. The smaller one, called a "tweeter," is better at producing high-frequency, short-wavelength sound. The electric current from the stereo amplifier is a time-varying signal, with both high-frequency and low-frequency components. The two speakers are connected in parallel as shown in the figure.

(a) To reduce the amplitude of the low-frequency component of the current through the tweeter, relative to the high-frequency component, should a capacitor or an inductor be inserted at point P, in series with the tweeter's coil?

(b) Should a capacitor or an inductor be inserted at point Q in series with the woofer's coil?

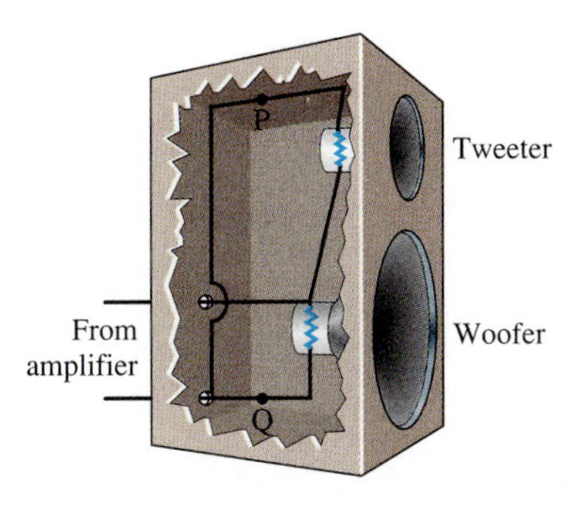

Fig. 22-44 Stereo speakers.

16 Consider plugging each of the following directly into an electrical outlet in a home: (a) a 100 Ω resistor; (b) a 100 μF capacitor; (c) a 100 μH ideal inductor. Which would be more likely to blow a fuse or trip a circuit breaker?

17 Which of the following 120 V AC devices costs more to operate: (a) one that draws an rms current of 10 A, or (b) one that draws a peak current of 12 A?

18 Which of the following 120 V AC devices, each of which has an impedance of 100 Ω, requires more electrical power: (a) one with 50 Ω of inductance, or (b) one with 10 Ω of inductance?

19 In a certain RCL series circuit, $L = 30\ \mu\text{H}$ and $C = 10\ \mu\text{F}$. The rms current is 10 A and the rms voltage across the capacitor is 20 V. Is the average energy stored by the inductor or the capacitor greater?

20 An airport metal detector consists of a large coil connected as part of an RCL series circuit. When one walks through the coil with a significant amount of metal, an alarm is set off (Fig. 22-45). The detector is sensitive to small changes because the circuit is initially at resonance, and a small change in the circuit produces a large drop in current, which is used to activate the alarm. Which circuit parameter does one change by bringing metal through the detector: (a) resistance; (b) capacitance; (c) inductance; or (d) frequency?

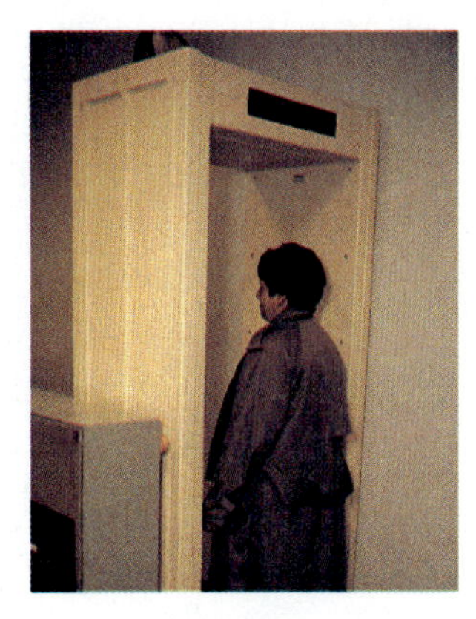

Fig. 22-45

Answers to Odd-Numbered Questions

1 equal number in each direction; **3** (a) to the right; (b) to the left; **5** clockwise; **7** *z*-axis only; **9** a; **11** none; **13** b; **15** (a) capacitor; (b) inductor; **17** a; **19** capacitor

Problems (listed by sections)

22-1 Faraday's Law

1 A magnetic field of magnitude 0.40 T is directed vertically upward. Find the magnetic flux through a flat surface of area 20 cm^2, if the surface is (a) vertical; (b) horizontal; (c) at an angle of 60° with the horizontal.

2 Find the flux of the earth's magnetic field through a horizontal section of the earth's surface of area 100 m^2 in a region where the magnetic field has a magnitude of 0.400 gauss and a horizontal component of 0.200 gauss.

3 You stand in the earth's magnetic field, directly facing its horizontal component (in other words, facing magnetic north). The field has a magnitude of 0.500 gauss and is directed 60.0° below the horizontal.
(a) Calculate the inward magnetic flux through your body's front surface, of area 0.500 m^2.
(b) Compute the outward flux through your body's back surface.
(c) What is the net inward magnetic flux through your body's entire surface?

4 The Houston Astrodome encloses a ground area of 37,000 m^2. Find the flux through the dome of the earth's magnetic field, which has a magnitude there of 0.50 gauss and is directed 60° below the horizontal.

★ **5** A magnetic field of magnitude 0.10 T is directed along the positive x-axis and passes through a wedge-shaped object, as indicated in Fig. 22-46. Find the *outward* directed magnetic flux through (a) the bottom; (b) the front (triangular) side; (c) the left side; (d) the right (inclined) side.

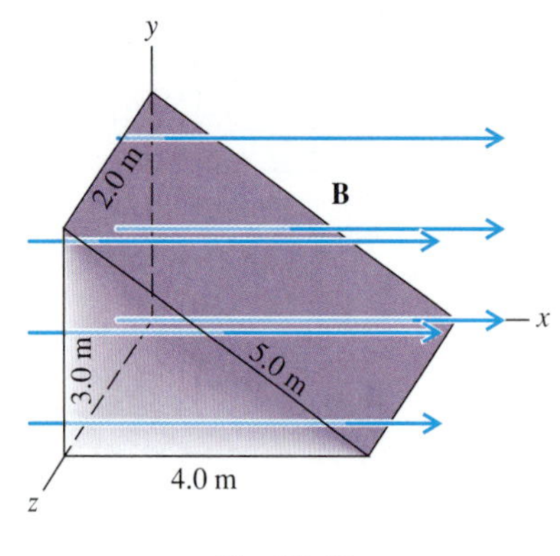

Fig. 22-46

6 A circular loop of wire of radius 10 cm moves at a speed of 5.0 m/s in the direction of a 0.10 T uniform magnetic field, perpendicular to the plane of the loop. What is the emf induced in the loop?

7 In a time interval of 1.0×10^{-7} s, the magnetic flux through the circular loop shown in Fig. 22-38 increases from zero to 3.0×10^{-6} T-m^2, as switch S is closed. What is the average emf induced in the loop?

★ **8** A small rectangular loop of wire is located close to a long straight wire initially carrying a current of 100 A (Fig. 22-47). Find the average emf induced in the loop if the current in the wire reverses direction in a time interval of $\frac{1}{120}$ s.

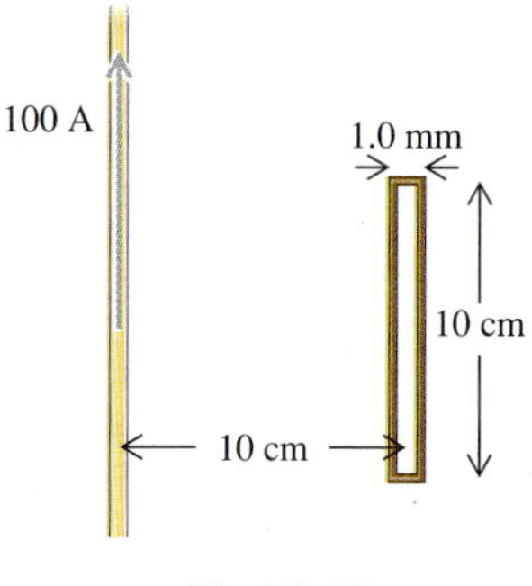

Fig. 22-47

★ **9** A small circular loop of wire of radius 5.0 cm and resistance $1.0 \times 10^{-3}\ \Omega$ is centered inside a large circular loop of wire of radius 50 cm (Fig. 22-48). The larger loop, which initially carries a current of 8.0 A, is cut and its current is reduced to zero over a time interval of 1.0×10^{-6} s. Find the average current in the smaller loop during this time interval. (The magnetic field of the larger loop is approximately constant over the smaller loop.)

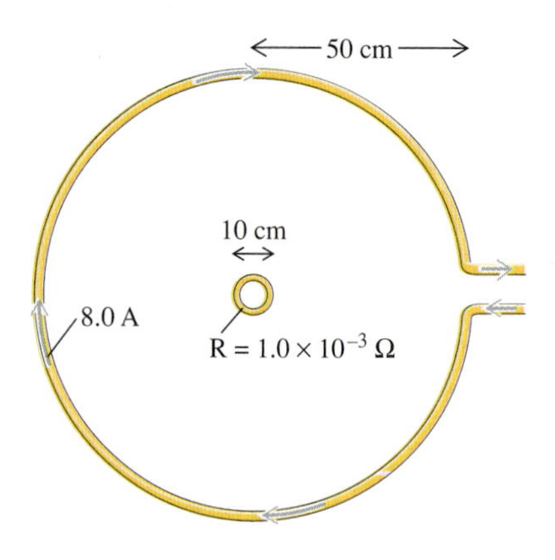

Fig. 22-48

10 The area enclosed by the wire loop in Fig. 22-11 is reduced from 30 cm^2 to 10 cm^2 in a time interval of 0.10 s. The magnitude of the magnetic field is 0.50 T. Find the emf induced in the wire.

11 Suppose that the rod in Fig. 22-12 moves to the left, so that the area enclosed by the conducting path decreases at the rate of 1.0×10^3 cm^2/s. Find the magnitude and direction of the current if the resistance of the conducting path is 0.10 Ω and the magnetic field strength is 0.50 T.

★★12 A horizontal copper rod of mass 5.0 g and length 10 cm slides downward with negligible friction as it maintains contact with vertical copper wires, which connect it to a 1.0×10^{-2} Ω resistor (Fig. 22-49). The rod and wires have negligible resistance. A uniform magnetic field of 0.60 T is directed horizontally, perpendicular to the plane of the circuit. Initially the rod accelerates downward but soon approaches a constant terminal speed v. Find v.

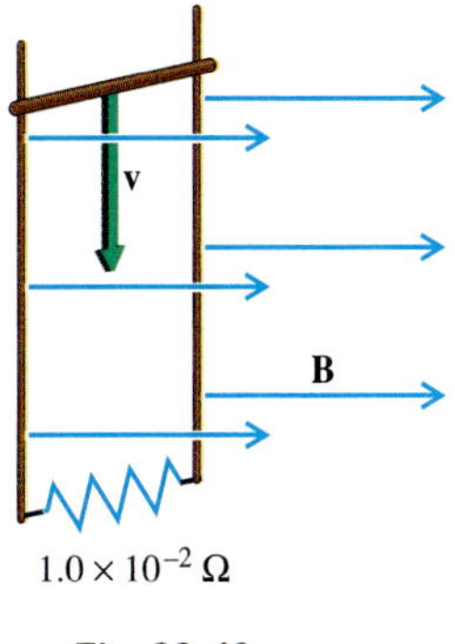

Fig. 22-49

★★13 Fig. 22-40 shows a square metallic loop of edge length 10.0 cm, mass 4.00 g, and resistance 1.00×10^{-3} Ω moving downward at a speed o· 4.00 cm/s, and entering a uniform magnetic field **B**.

(a) What is the strength of the field if the loop experiences no acceleration initially?

(b) When will its speed begin to change?

★14 When a torque τ turns the shaft of an electric generator at an angular velocity ω, the mechanical power input is given by $P = \tau\omega$ (the rotational analog of Eq. 7-40: $P = Fv$). Equating the mechanical torque to the magnetic torque on the current loop and equating the mechanical power input to the electrical power supplied by the generator, derive Eq. 22-7.

★15 Suppose that the electric motor discussed in Example 5, Chapter 21, is operated in reverse, so that it acts as an electric generator (Fig. 22-50). At any time during the rotation of the coils, the single coil that is connected through the commutator to the external circuit is approximately perpendicular to the magnetic field. Thus, instead of an emf that varies as $\mathcal{E}_0 \sin \omega t$, we have an emf $\mathcal{E} \approx \mathcal{E}_0 \sin 90° = \mathcal{E}_0$, with $\mathcal{E}_0 = NBA\omega$. This means that the generator acts as a nearly constant source of emf, a DC generator. Suppose that the generator supplies 210 W of electric power while connected to a 50.0 Ω resistor. The coil has negligible resistance. Find the generator's emf and the current in the circuit.

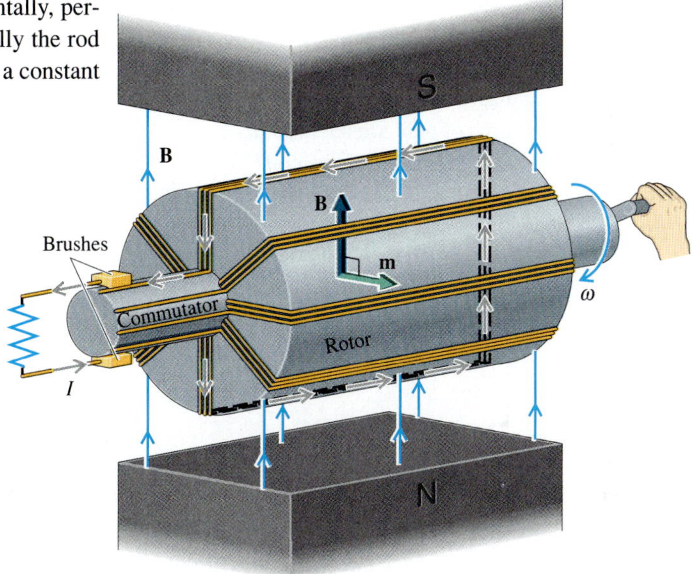

Fig. 22-50 A DC generator.

★16 Suppose that the electric generator shown in Fig. 22-42 consists of 1000 turns of wire enclosing an area of 100 cm^2 in a magnetic field of 0.30 T. The generator is connected to an 18 Ω resistor. By pedaling the bicycle you are able to supply power of 75 W (~0.1 hp) to the generator.

(a) Find the current and emf generated, and the rate of pedaling in rev/s.

(b) Suppose you decrease your rate of pedaling to half the original rate. How much power is supplied to the generator?

17 In a region where there is a vertical magnetic field of 2.00×10^{-2} T, a ring of radius 1.00 cm is flipped in the air as one would flip a coin, so that it begins rotating about a horizontal axis at the rate of 50.0 rev/s.

(a) Find the maximum instantaneous emf induced in the ring and the maximum current if the ring has a resistance of 1.00×10^{-4} Ω.

(b) What effect does the magnetic field have on the rate of rotation?

★ **18** Suppose you are stranded on a desert island and have no magnets, but you do have plenty of 0.50 mm radius copper wire and flowing water as a source of power. You also have various electrical devices that require for their operation an electric generator producing 170 V peak emf (or an rms emf of 120 V) at 60 Hz. You hope to use the earth's magnetic field to generate the electricity you need, by turning a 1 meter square coil of wire at 60 Hz about an axis perpendicular to the earth's field of magnitude 0.50 gauss.

(a) How many turns of wire are required?

(b) What is the total resistance of the coil?

(c) What is the maximum (short-circuit) peak current provided by the generator, and what peak instantaneous power is required to turn the generator in this case?

(d) Could your generator be used to operate a device requiring 50 W of electrical power?

22-2 Inductance

19 The self-inductance of the circuit in Fig. 22-17 is 1.00×10^{-7} H. Find the back emf when the current is increasing at the rate of 2.00×10^{7} A/s.

20 Find the cross-sectional area of a 5.0 cm long solenoid with 500 turns and a self-inductance of 40 μH.

21 A solenoid with 200 turns and a cross-sectional area of 1.00 cm² is 3.00 cm long. How much current must the solenoid carry in order that the flux produced by its own magnetic field is equal to the flux produced by earth's magnetic field of magnitude 0.500 G, directed along the solenoid's axis?

22 (a) Find a possible set of values for the number of turns, length, and cross-sectional area of a solenoid with a self-inductance of 1.00 H.

(b) What is the back emf produced by such an inductor when the current through it is increasing at the rate of 100 A/s?

23 Find the voltage drop V_{ab} across each of the inductors in Fig. 22-51.

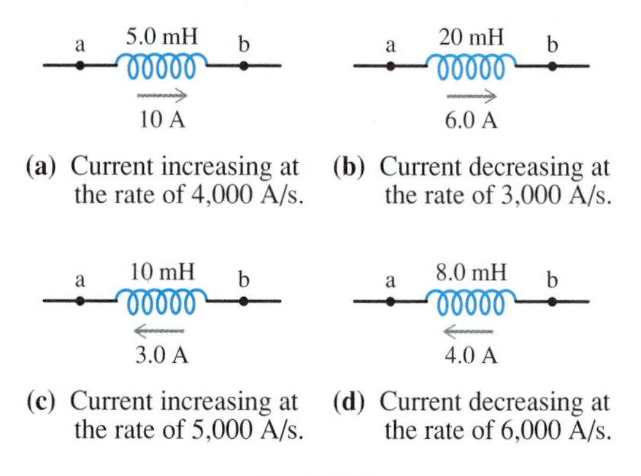

Fig. 22-51

★ **24** An extension cord consists of two long, thin wires separated by a small distance d, carrying current of equal magnitude but in opposite directions. When the current I changes, there is a changing magnetic flux through the surface area between the wires. Show that if one approximates the magnetic field by its value at the midpoint between the wires the cord's self-inductance is approximately $2\mu_0\ell/\pi$, where ℓ is the length of the cord. Evaluate for a cord of length 10 m.

★ **25** Find the rate of change of current in the circuit shown in Fig. 22-23 at $t = 0$ just as the switch is closed, if $L = 2.00$ mH and $\mathcal{E} = 6.00$ V.

★ **26** At $t = 0$ an inductor is connected to a 9.0 V battery. At $t = 2.0$ s, the current through the inductor is 0.37 A, and after about 10 s, the current reaches a stable value of 1.0 A. Find the resistance and the inductance of the inductor.

27 Find the magnetic energy stored by a 5.00 mH inductor when the current through it is 10.0 A.

28 Estimate roughly the magnetic energy stored in the earth's magnetic field by calculating the magnetic energy in a sphere with twice the earth's radius, within which is a uniform magnetic field of 0.5 G. For comparison, the earth's rotational kinetic energy is 2.6×10^{29} J (Example 7, Chapter 9.)

29 Calculate the gravitational potential energy stored in 1.00 m³ of water raised a vertical distance of 100 m. Find the ratio of this gravitational energy density to the magnetic energy density found in Example 6.

*22-3 AC Circuits

30 The rms current through a light bulb is 0.833 A when the bulb is connected to a standard 120 V rms source. Find (a) the bulb's resistance; (b) the peak instantaneous current; (c) the peak instantaneous voltage; (d) the average electrical power used by the bulb.

31 An electric heater is rated at 1000 W, 120 V rms. Find the heater's: (a) resistance; (b) rms current; (c) peak current; (d) peak instantaneous power; (e) minimum instantaneous power.

32 Find the reactance of a 2.00 μF capacitor at a frequency of (a) 60.0 Hz; (b) 600 Hz; (c) 6000 Hz.

33 Find the rms current that results from connecting a 2.00 μF capacitor directly across a 120 V rms source at a frequency of (a) 60.0 Hz; (b) 600 Hz; (c) 6000 Hz.

★34 A capacitor draws a 10.0 A rms current from a 20.0 V rms source. Find (a) the maximum instantaneous electrical power supplied to the capacitor; (b) the minimum instantaneous electrical power supplied to the capacitor.

35 Find the reactance of a 2.00 mH inductor at a frequency of (a) 60.0 Hz; (b) 600 Hz; (c) 6000 Hz.

36 Find the rms current that results from connecting a 2.00 mH inductor directly across a 120 V rms source, at a frequency of (a) 60.0 Hz; (b) 600 Hz; (c) 6000 Hz.

37 For the circuit in Fig. 22-52, the AC voltmeter V_1 reads 50.0 V rms. Find the readings of the AC ammeter A and of voltmeters V_2 and V_3. Assume that the ammeter and voltmeters operate ideally; that is, the ammeter has zero impedance, and the voltmeters have infinite impedance.

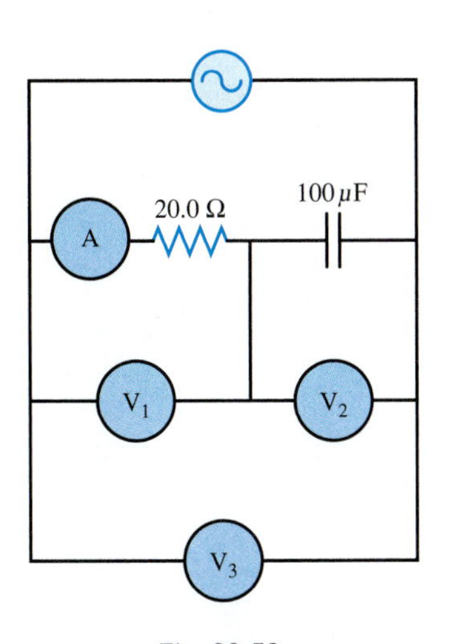

Fig. 22-52

38 For the circuit shown in Fig. 22-53, the AC ammeter A reads 4.00 A rms. Find the readings of the four AC voltmeters. Assume that the ammeter has zero impedance and the voltmeters have infinite impedance.

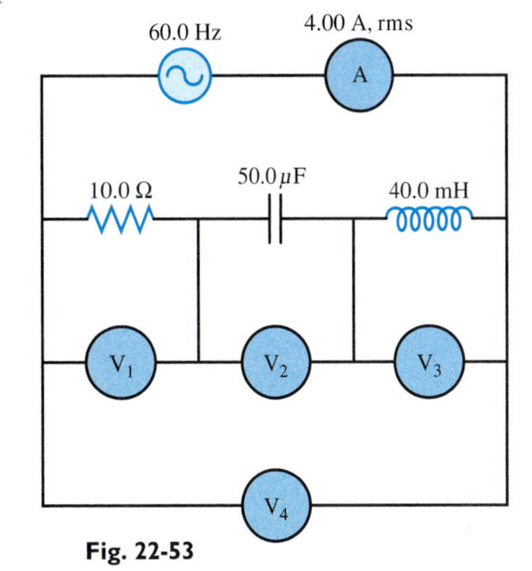

Fig. 22-53

★39 An RCL series circuit has the values $R = 8.00\ \Omega$, $C = 4.00\ \mu\text{F}$, and $L = 80.0\ \mu\text{H}$. Find the average power dissipated when the rms voltage across the three circuit elements is 15.0 V, and the frequency is 1.00×10^4 Hz.

40 A 10.0 Ω resistor, a 10.0 μF capacitor, and a 30.0 mH inductor are connected in series with a 120 V (rms) source with variable frequency.

(a) At what frequency will the current be maximum?

(b) Find the maximum rms current.

(c) Suppose you now replace the 10.0 μF capacitor by a 20.0 μF capacitor. What is the new value of the rms current?

41 Find the resonant frequency for the circuit in Fig. 22-53.

42 (a) Sketch the phasor diagram corresponding to the RCL series circuit in Fig. 22-32, with a voltage amplitude of 100 V, a frequency of 60 Hz, a resistance of 100 Ω, a capacitance of 25 μF, and an inductance of 200 mH. Draw the phasors at $t = 0$, and again at $t = \frac{1}{240}$ s.

(b) What are the values of the instantaneous current at $t = 0$ and at $t = \frac{1}{240}$ s?

★43 (a) Find the current in the circuit shown in Fig. 22-32 if $R = 6.00\ \Omega$, $C = 0.500\ \mu\text{F}$, and $L = 0$, and the source supplies an emf of 40.0 V (rms) at a frequency of 3.00 kHz.

(b) What circuit element (other than a source) can we insert into this series circuit to increase the current as much as possible?

(c) Find this maximum rms current.

44 A transformer is used to operate a model train. The transformer, which is plugged into a standard electrical outlet, produces an output voltage of 20.0 V (rms), applied to the train's tracks.
(a) What is the ratio of the number of turns in the primary to the number of turns in the secondary?
(b) If the train uses an average power of 10.0 W, what are the rms currents in the secondary and in the primary? Assume an ideal transformer, with no energy loss.

45 Power lines at 20,000 V (rms) are connected to a utility-pole transformer. The output of the transformer's secondary, which leads to a residence, is at 240 V (rms). Find the ratio of the number of turns in the primary to the number of turns in the secondary.

★ **46** An ideal step-up transformer has 20 times as many turns in its secondary as in its primary. An rms current of 10 A passes through the primary when the secondary is connected to a circuit with an impedance of 8000 Ω. What is the rms voltage across the primary?

47 Neon lights are an example of light produced as a result of an electrical discharge in a tube filled with a gas, which emits a characteristic color (Fig. 22-54). The light emitted by discharge tubes is important in the study of atomic physics. (See Sections 23-2 and 29-1.) To produce the electrical discharge, a high voltage must be applied to electrodes sealed in the ends of the tube. The resistance of a tube is 1.52×10^6 Ω and the rms current through it is 3.00 mA. The transformer used to produce the high voltage is plugged into a standard electrical outlet. Find the ratio of the number of turns in the secondary to the number of turns in the primary.

Fig. 22-54 Neon lights and other gas discharge tubes require a high-voltage source.

48 A 120 V (rms) power line is connected to the primary of a step-down transformer, with a turn ratio of 12:1. The transformer's secondary is connected to a doorbell, which requires 10.0 W of electric power for its operation. What is the peak instantaneous current in the primary?

Additional Problems

★ **49** A square loop 10 cm on a side is placed in a uniform, time-varying magnetic field perpendicular to the plane of the loop, as indicated in Fig. 22-55. The field has an instantaneous magnitude of 0.50 T and is increasing at the rate of 0.20 T/ms. The loop has a resistance of 0.010 Ω.
(a) Find the loop's magnetic moment.
(b) Find the torque on the loop.

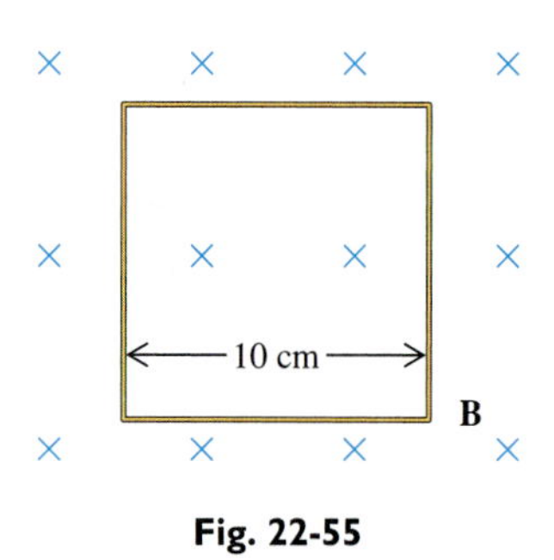

Fig. 22-55

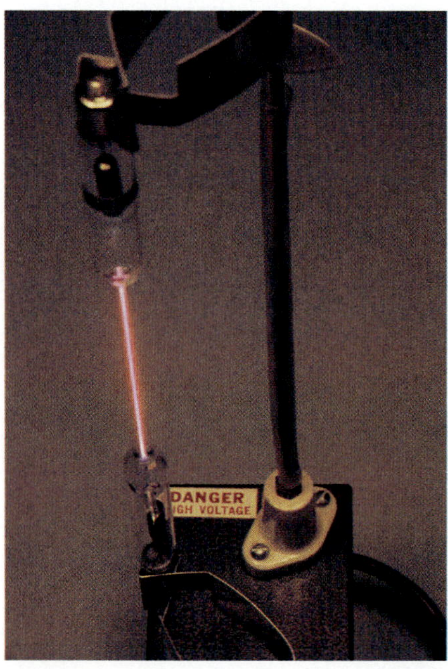

★★**50** A plane loop of wire enclosing an area of 5.00 cm^2 and having a resistance of 1.0×10^{-3} Ω is placed inside a long solenoid. The plane of the loop makes an angle of 60° with the solenoid's axis. The solenoid has 100 turns per cm of length, and at some instant carries a current of 5.0 A. The current is increasing at the rate of 3.0×10^3 A/s. Find the torque on the loop.

★**51** A solenoid of length 1.00 m and cross-sectional area 0.200 m^2 has a self-inductance of 1.00 H. Find (a) the number of turns; (b) the radius of the copper wire used to form the solenoid (assuming one layer of wire); (c) the solenoid's resistance; (d) the time constant.

★**52** The edges of an inclined plane are made of rigid conducting rods separated by 20.0 cm and connected at the base by a 1.00×10^{-2} Ω resistor (Fig. 22-56). The plane makes an angle of 20.0° with the horizontal. There is a vertical downward magnetic field of magnitude 0.400 T. A rigid rod slides down the plane at a constant speed of 60.0 cm/s. Neglect friction and the resistance of the rods.

(a) What current flows in the resistor?

(b) What is the mass of the rod?

53 A small canal diverts water from the Perlbach, a river in Bavaria. Water flowing through the canal drops through a small distance and turns a waterwheel, which powers an electric generator, providing electricity for the house shown in Fig. 22-57.

(a) Calculate the maximum electrical power that can be generated if water moves through the canal at the rate of 1.00×10^3 kg/s and drops through a vertical distance of 2.00 m. (Only about 1 kW is used by the household, the remainder being sold to the local power company.)

(b) The generator's coil consists of 100 turns enclosing an area of 400 cm^2. The magnetic field strength is 0.225 T, and the generator operates at 60.0 Hz. Find the rms value of the emf generated.

(c) Find the maximum rms current through the generator.

★**54** Suppose that the resistance of the woofer and tweeter coils shown in Fig. 22-44 are each 8.00 Ω. Suppose that a 10.0 μF capacitor is inserted at point P, and a 1.00 mH inductor is inserted at point Q in the figure.

(a) Calculate the rms current through the woofer and tweeter coils respectively when a 20.0 V (rms), 500 Hz voltage is input to the speaker.

(b) Repeat the calculation for a frequency of 5000 Hz.

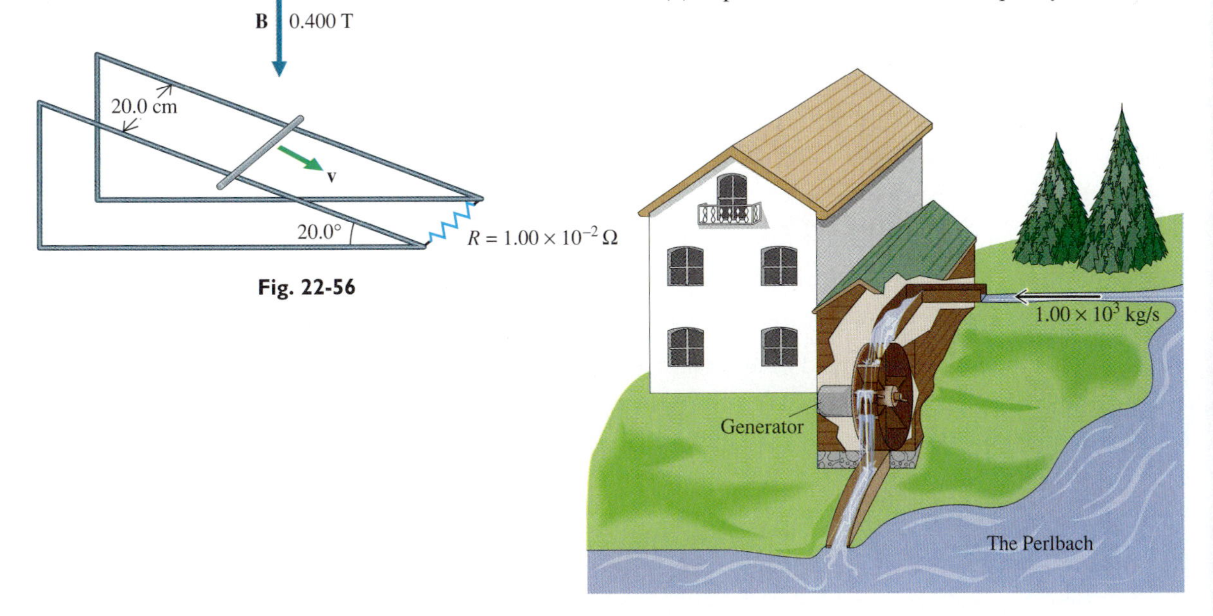

Fig. 22-56

Fig. 22-57 Water flowing through the canal provides the energy necessary to operate a small generator and generate electricity.

CHAPTER 23 Light

The Crab nebula is believed to be the remnant of a supernova explosion recorded in the year 1054. The giant star that appeared in that year was visible even in daylight for several weeks. This photograph records the radiation from accelerated charged particles in the debris of that ancient explosion.

In 1865 a young Englishman named James Maxwell made a remarkable discovery. Based on his mathematical study of Faraday's law and other laws governing the behavior of electric and magnetic fields, Maxwell predicted that light is an electromagnetic wave. In other words, Maxwell claimed that a beam of light consists of electric and magnetic fields,* moving and vibrating as waves, in much the same way that air molecules vibrate in a sound wave. Maxwell further predicted that it would be possible to produce other invisible electromagnetic waves.

It is hard to overstate the impact that Maxwell's discovery has had on the modern world. Our environment is permeated by artificially produced electromagnetic radiation—radio and television waves used for communication, X rays used for medical diagnosis, radar used to detect aircraft or speeding cars, even microwaves used to cook our food in microwave ovens. Nobel laureate Richard Feynman has remarked:

> From a long view of the history of mankind—seen from say, ten thousand years from now—there can be little doubt that the most significant event of the 19th century will be judged as Maxwell's discovery of the laws of electrodynamics. The American Civil War will pale into provincial insignificance in comparison with this important scientific event of the same decade.

*Faraday had speculated in 1845 that light is related to oscillating electric fields but was unable to develop this insight.

In this chapter we shall describe some elements of Maxwell's theory of electromagnetic waves. Then, in the remainder of this chapter and the next three chapters, we shall study visible light and its applications, a subject known as optics.

23-1 Electromagnetic Waves

Maxwell's Theory

James Maxwell was a mathematical prodigy as a child and became during his short lifetime one of the most prolific physicists of all time. In contrast to Michael Faraday, Maxwell was from a wealthy family and received the best formal education possible. He had varied interests in the arts as well as in science and was able to speak with authority on almost any subject. Maxwell made important contributions in several areas of physics. For example, he discovered the Maxwell distribution of molecular velocities, discussed in Chapter 12 (Section 12-3). But Maxwell's greatest achievement was his insight into electromagnetic phenomena. He carefully read the work of Faraday, as well as more specialized, mathematical papers written by others. Because he had the mathematical training and aptitude that Faraday lacked, Maxwell was able to put the results of Faraday's experimental studies into precise, mathematical language and to refine the concept of electric and magnetic fields.

Fig. 23-1 James Clerk Maxwell (1831–1879).

Maxwell developed four equations that summarized what was previously known about electric and magnetic fields and added an extra term to one equation to make the four equations logically consistent. **Maxwell's equations show how electric and magnetic fields are related to each other and also how the fields are related to electric charges and currents.** Unfortunately, a quantitative understanding of Maxwell's equations (displayed in Fig. 23-2) requires a knowledge of mathematics beyond the level of this book. Here we shall state without proof some of the most important predictions arising from Maxwell's equations.

Maxwell's theory shows that **accelerated charge produces waves,** consisting of electric and magnetic fields. Furthermore, the theory predicts that **electromagnetic waves always travel through empty space at a fixed speed** c, which is related to the electric constant ϵ_0 and the magnetic constant μ_0 by the equation

$$c = \frac{1}{\sqrt{\epsilon_0 \mu_0}} \tag{23-1}$$

Substituting values of ϵ_0 and μ_0 into this equation, we find

$$c = \frac{1}{\sqrt{(8.85 \times 10^{-12}\ \text{F/m})(4\pi \times 10^{-7}\ \text{T-m/A})}}$$

$$= 3.00 \times 10^{8}\ \text{m/s}$$

This is precisely the **speed of light**! When Maxwell discovered this implication of his theory, it convinced him that **light is an electromagnetic wave.** In Maxwell's own words,

> that the velocity of propagation is the velocity . . . so nearly that of light, that it seems we have strong reason to conclude that light itself (including radiant heat and other radiations if any) is an electromagnetic disturbance in the form of waves propagated through the electromagnetic field according to electromagnetic laws.

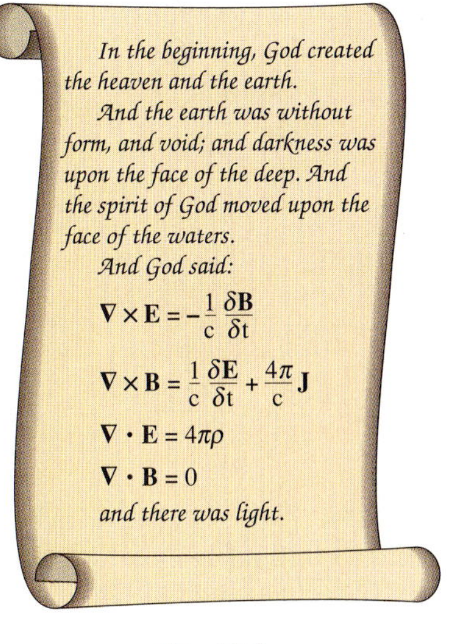

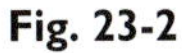

Fig. 23-2

We learned in Chapter 15 that when an oscillating body produces a wave the frequency of that wave is the body's frequency of oscillation. Thus, **when oscillating charge produces an electromagnetic wave, the frequency of the wave is the frequency of oscillation of the electric charge that is the source of the wave.** The frequency of visible light is very high, and consequently the wavelength is very short (on the order of 10^{-3} mm). So perhaps it is not surprising that one cannot easily recognize how sources of visible light involve accelerated charges.* However, according to Maxwell's theory, it should be possible to produce electromagnetic radiation of much lower frequency and longer wavelength by causing electric charges to oscillate at an appropriate frequency. Thus Maxwell set the stage for the discovery of radio waves, microwaves, and so forth.

Today we recognize that visible light is only one small part of the electromagnetic spectrum, shown in Fig. 23-3. But Maxwell's ideas were not immediately accepted by other scientists. It was not until 1887, 22 years after Maxwell's work, that Heinrich Hertz began to publish the results of his experimental studies of electromagnetic waves, providing convincing experimental proof of the correctness of Maxwell's predictions. Hertz was the first to generate electromagnetic waves by means that were obviously electrical. Hertz produced a spark across a gap between two conductors connected to a high-voltage, high-frequency source of emf. This served as his transmitter of electromagnetic waves (Fig. 23-4). The waves were detected by means of a loop of wire with a small gap at one end, where sparks were detected.

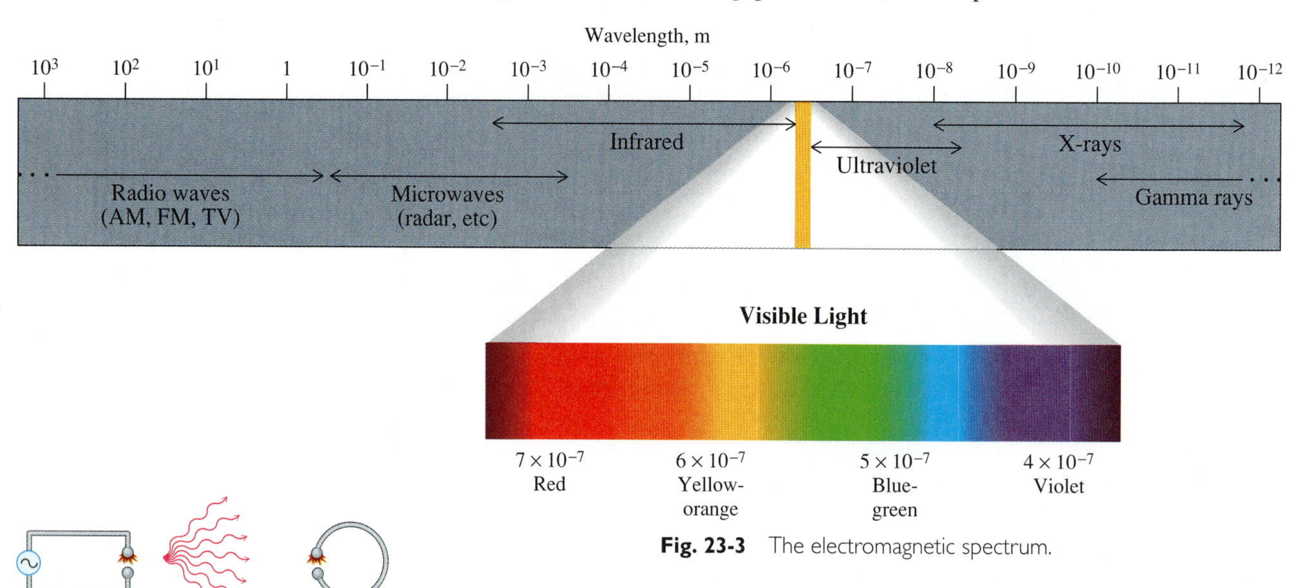

Fig. 23-3 The electromagnetic spectrum.

Fig. 23-4 Hertz's production and detection of electromagnetic waves.

The waves that Hertz generated had wavelengths ranging from 0.5 m to several meters. He showed that these waves had many of the properties of light. For example, they were reflected and refracted like light. He even constructed a 1.5 m tall wooden frame in the triangular shape of a prism and then filled it with asphalt to study its effect on his radio waves. He compared the bending of the radio waves by his asphalt prism to the way a glass prism bends visible light. Hertz concluded:

> The experiments described appear to me, at any rate, eminently adapted to remove any doubt as to the identity of light, radiant heat, and electromagnetic wave motion.

---

*It was not until the development of quantum theory in the twentieth century that it was quantitatively understood how visible light is produced. Maxwell's theory was not sufficient to explain sources of visible light. But Maxwell's conclusion that light propagates as an electromagnetic wave remains valid.

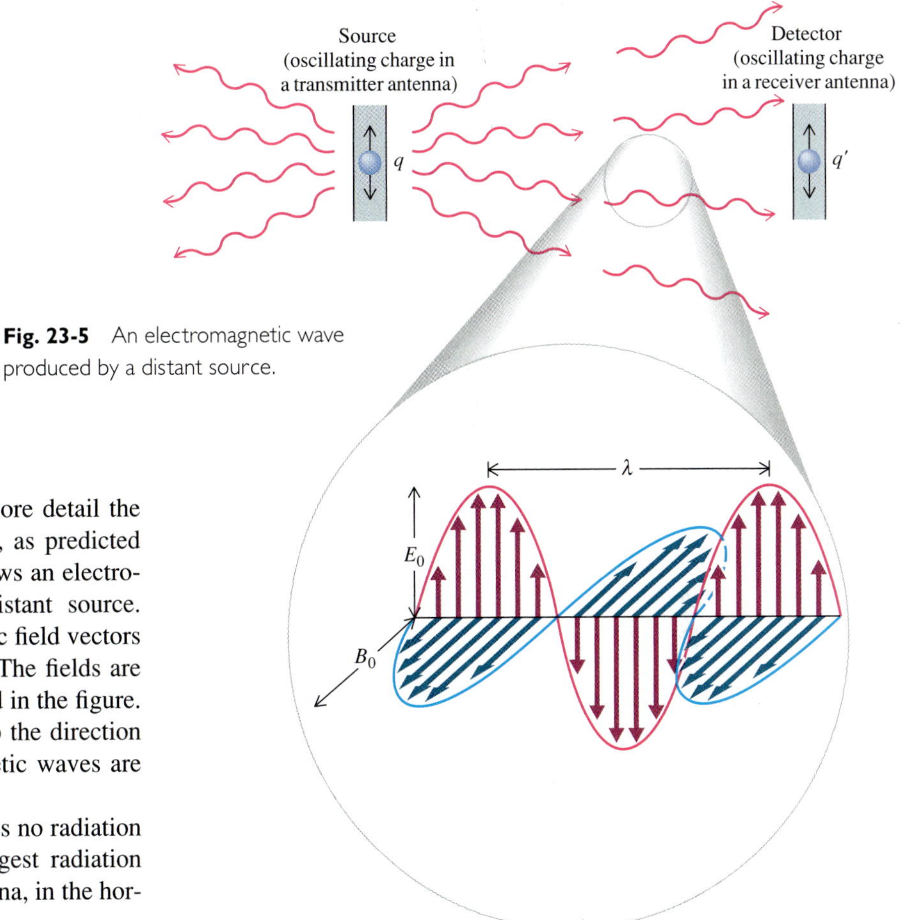

Fig. 23-5 An electromagnetic wave produced by a distant source.

Next we shall discuss in a bit more detail the nature of an electromagnetic wave, as predicted by Maxwell's theory. Fig. 23-5 shows an electromagnetic wave produced by a distant source. Notice that the electric and magnetic field vectors oscillate in phase with each other. The fields are mutually perpendicular, as indicated in the figure. The fields are also perpendicular to the direction of wave propagation; electromagnetic waves are transverse waves.

A linear vertical antenna produces no radiation in the vertical direction; the strongest radiation fields are perpendicular to the antenna, in the horizontal direction.

The amplitude of the magnetic field B_0 is related to the electric field amplitude E_0 by the equation

$$B_0 = \frac{E_0}{c} \tag{23-2}$$

In previous chapters we have seen that both static electric fields and magnetic fields contain energy. (See the discussion of field energy, Section 18-2, and of magnetic energy, Section 22-2.) It is not surprising, then, that electromagnetic fields also contain energy. The time average of an electromagnetic wave's energy per unit volume, denoted by u_{av}, is related to the electric field amplitude E_0 by the equation

$$u_{av} = \tfrac{1}{2}\epsilon_0 {E_0}^2 \tag{23-3}$$

The energy in an electromagnetic wave is not stationary. The energy moves as the wave moves, at the speed of light. The **intensity** of a light wave is defined in the same way as the intensity of a sound wave was defined in Chapter 16 (Eq. 16-13); that is, the intensity I is the power P (energy per unit time) passing through a small surface perpendicular to the direction of propagation of the wave, divided by the area A (Fig. 23-6).

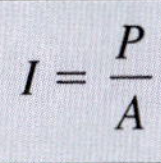

$$I = \frac{P}{A} \tag{23-4}$$

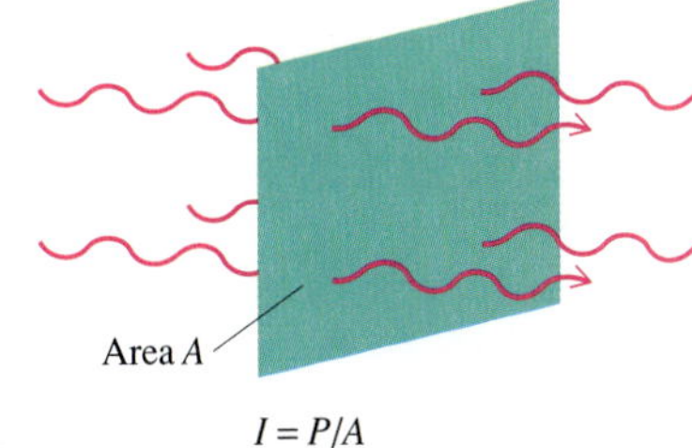

Fig. 23-6 The intensity of light passing through a surface.

Intensity is measured in units of watts/m^2.

It is possible to show that an electromagnetic wave's average intensity I_{av} is the product of the wave's average energy per unit volume (u_{av}) and the wave's speed c. (See Problem 54.)

$$I_{av} = u_{av}c$$

Using Eq. 23-3 to substitute for u_{av}, we obtain an expression for the average intensity of an electromagnetic wave, expressed in terms of the electric field amplitude E_0.

$$I_{av} = \tfrac{1}{2}\epsilon_0 c E_0^2 \qquad (23\text{-}5)$$

EXAMPLE 1 Electric and Magnetic Fields in Sunlight

By measuring the heating of a material as it absorbs light from the sun, one finds that the intensity of sunlight at the surface of the earth is 1300 W/m^2. Find the amplitudes of the electric and magnetic fields in sunlight.

SOLUTION We apply Eq. 23-5 to find the electric field amplitude.

$$I_{av} = \tfrac{1}{2}\epsilon_0 c E_0^2$$

or

$$E_0 = \sqrt{\frac{2I_{av}}{\epsilon_0 c}} = \sqrt{\frac{2(1300 \text{ W/m}^2)}{(8.85 \times 10^{-12} \text{ F/m})(3.00 \times 10^8 \text{ m/s})}}$$
$$= 990 \text{ N/C}$$

This is a strong electric field! When you get a sunburn, it is the result of prolonged exposure to this intense, high-frequency electric field.

Applying Eq. 23-2, we find the magnetic field amplitude.

$$B_0 = \frac{E_0}{c} = \frac{990 \text{ N/C}}{3.00 \times 10^8 \text{ m/s}} = 3.30 \times 10^{-6} \text{ T}$$

In contrast to the electric field, the magnetic field in sunlight is extremely weak.

Electromagnetic Waves and Photons

Although Maxwell and Hertz showed that light is an electromagnetic wave, experiments in modern physics show that light is something more than a wave phenomenon. When light is either emitted or absorbed, it does so in discrete amounts of energy—localized particle-like bits of energy called **photons.** The dual nature of light—both particle-like and wavelike—is at the heart of quantum physics and will be discussed more fully in Chapter 28. Our study of light in this and the next three chapters will be concerned mainly with the wave properties of light.

23-2 The Nature of Light

The Visible Spectrum

Visible light is that small part of the electromagnetic spectrum to which the human eye is sensitive. Electromagnetic radiation is visible in the frequency range from about 4×10^{14} Hz to about 8×10^{14} Hz. We perceive light of any particular frequency in the visible range as having a definite color. For example, we perceive light of frequency 6.5×10^{14} Hz as blue and light of frequency 4.5×10^{14} Hz as red. Table 23-1 shows the frequency range of the various colors. Also given in the table are the corresponding wavelengths of these waves in a vacuum. White light is actually a mixture of colored light—red, yellow green, violet, and so on. Newton demonstrated this fact by using a prism to separate sunlight into a spectrum of colors. Newton also showed that these colors could be recombined to produce white light again.

Table 23-1 The Visible Spectrum

Spectral colors and the approximate frequencies and vacuum wavelengths corresponding to each color.

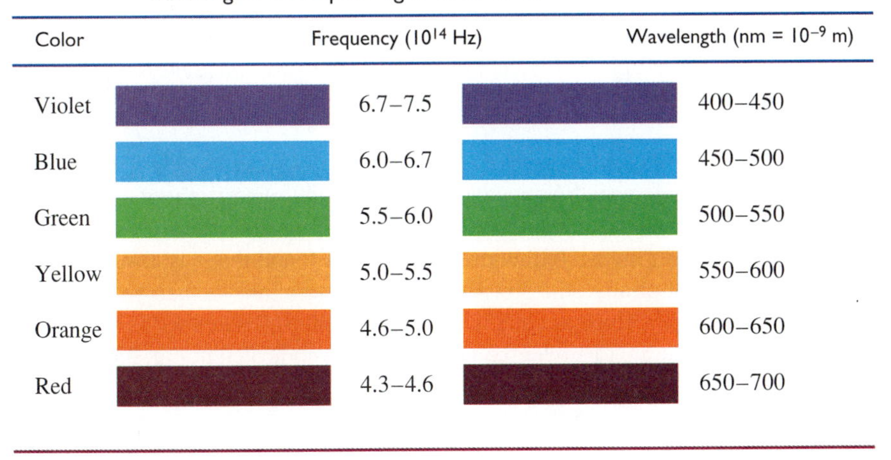

Color	Frequency (10^{14} Hz)	Wavelength (nm = 10^{-9} m)
Violet	6.7–7.5	400–450
Blue	6.0–6.7	450–500
Green	5.5–6.0	500–550
Yellow	5.0–5.5	550–600
Orange	4.6–5.0	600–650
Red	4.3–4.6	650–700

Although objects in our surroundings are illuminated with white light, they have a multitude of colors. These colors are produced by an object's selective reflection or absorption of parts of the visible spectrum.

Human perception of color is a complex phenomenon. For example, the right combination of red light and green light can be perceived by the human eye to be the same color as a pure yellow light; more specifically, red-orange light of frequency 4.6×10^{14} Hz combined with green light of frequency 5.7×10^{14} Hz looks exactly the same as yellow light of frequency 5.2×10^{14} Hz. Color perception is discussed in an essay in Chapter 25.

Various sensors can detect electromagnetic radiation beyond the range of human vision. For example, some photographic film is sensitive to **infrared radiation** (abbreviated IR). Infrared radiation is beyond the red end of the visible spectrum; that is, it has somewhat lower frequency than visible light. The range of infrared frequencies is approximately 3×10^{11} Hz to 4×10^{14} Hz. Infrared film is useful in night photography, since all bodies constantly emit infrared radiation as a result of their thermal energy. Such thermal radiation increases with temperature. Thermography is a medical diagnostic technique in which infrared film is used to detect tumors and other disorders, which result in higher than normal temperatures in affected areas of the body and therefore produce greater radiation (Fig. 23-7).

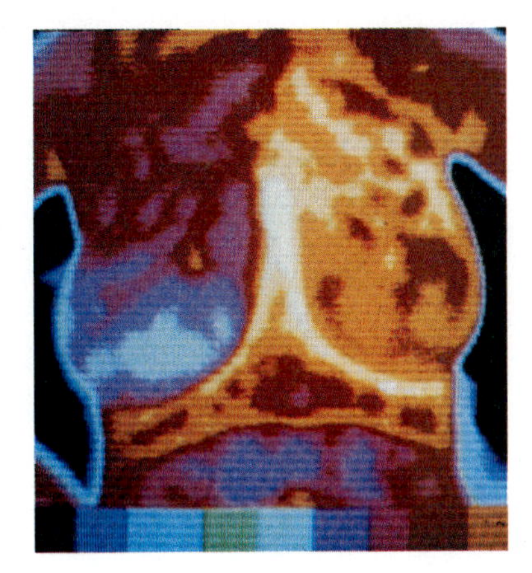

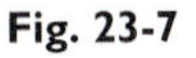

Fig. 23-7

Ultraviolet radiation (abbreviated UV) is electromagnetic radiation that is beyond the violet end of the visible spectrum; that is, its frequency is somewhat higher than that of visible light—from about 8×10^{14} Hz to about 3×10^{17} Hz. Although ultraviolet light is invisible to human eyes, bees have eyes that are sensitive to it. Flowers that appear white to the human eye actually have patterns of color over their surfaces, but these colors are visible only to the bee because they are "ultraviolet colors," resulting from different amounts of ultraviolet light of each frequency reflected from different areas of the flowers' surfaces.

Visual Sensitivity

The brightness of light as perceived by the human eye depends both on the intensity of the light and its frequency. Of course, if the frequency is outside the visible range, the radiation is invisible, no matter how intense. Within the visible range, the eye is more sensitive to light near the center of the visible spectrum, with maximum sensitivity to yellow-green light of frequency 5.4×10^{14} Hz and a vacuum wavelength of 550 nm. The eye is much less sensitive to light near the ends of the visible range—red or violet light.

We can measure the sensitivity of the eye to light of various wavelengths by having an observer compare such light with a 550 nm yellow-green light of arbitrary intensity I_0. The intensity I of the other light is increased until the light appears as bright to the observer as the yellow-green light. It is found that I is always greater than I_0; that is, light of any other color needs to be more intense than yellow-green light in order to be perceived as equally bright. The ratio I_0/I is defined as sensitivity. Sensitivity varies from 0 to 1 over the visual range, as indicated in Fig. 23-8a.

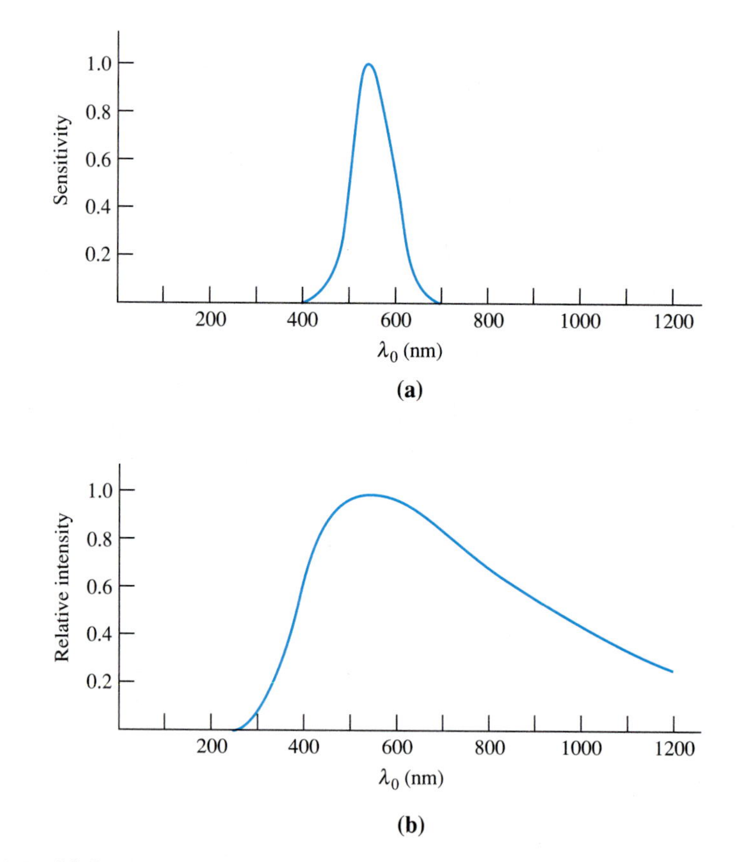

Fig. 23-8 **(a)** Sensitivity of the eye to light of various vacuum wavelengths. **(b)** Relative intensity of solar radiation at the earth's surface as a function of vacuum wavelength.

Sources of Light

Sources of light can be described in terms of the frequencies of light they produce. Our most common source of light is, of course, the sun. A graph of the intensity of solar radiation at the surface of the earth (Fig. 23-8b) shows a broad maximum from about 500 nm to 600 nm. Fig. 23-8 indicates that the human eye is most sensitive precisely to the part of the spectrum for which solar radiation is maximum. Apparently the eye has evolved in such a way as to make the best use of the sun's light.

The most common types of artificial illumination are incandescent lights and fluorescent lights, both of which produce a continuous spectrum of light somewhat like the solar spectrum. Gas discharge tubes, on the other hand, are sources that produce narrow bands of color. Neon lights and sodium or mercury vapor lamps are examples of gas discharge tubes (Fig. 23-9). These light sources are made by first evacuating the air from a glass tube and then filling the tube with a particular kind of vapor—neon, sodium, mercury, or hydrogen, for example. When a potential difference is applied across the tube, it radiates light (Fig. 23-10).

Fig. 23-9 Neon lights and sodium street lights are both examples of gas discharge tubes.

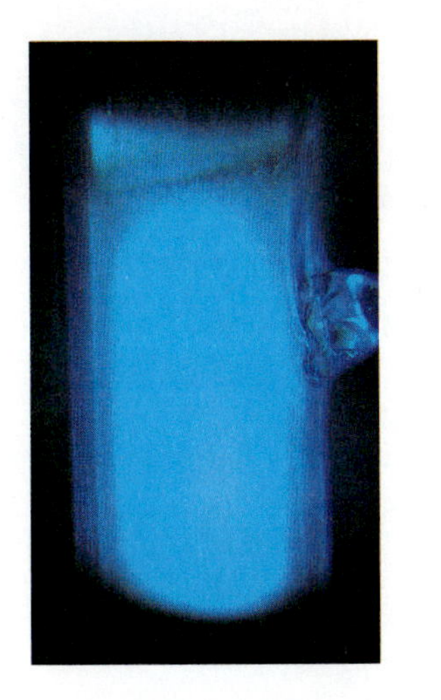

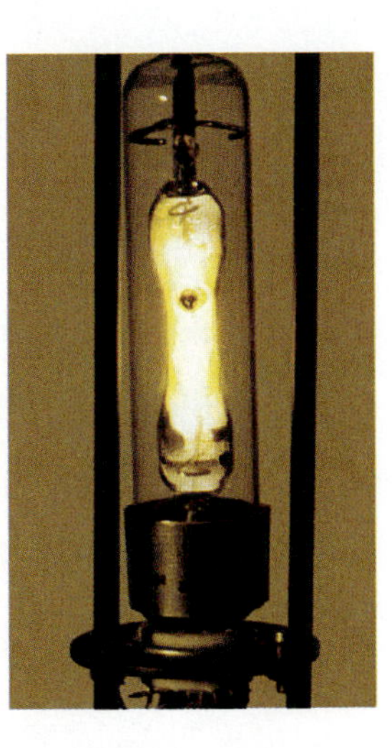

Fig. 23-10 Mercury and sodium discharge tubes.

The light emitted by a gas discharge tube consists of a large number of separate, narrow bands of frequency. The colors produced are characteristic of the kind of vapor in the tube (Fig. 23-11). Each frequency band is typically very narrow, and so one can think of the light within a band as having only one frequency or color. We call light of a single frequency **monochromatic light.** Real light sources always emit some range of frequencies; so, although the frequency band may be narrow, it is never truly monochromatic.

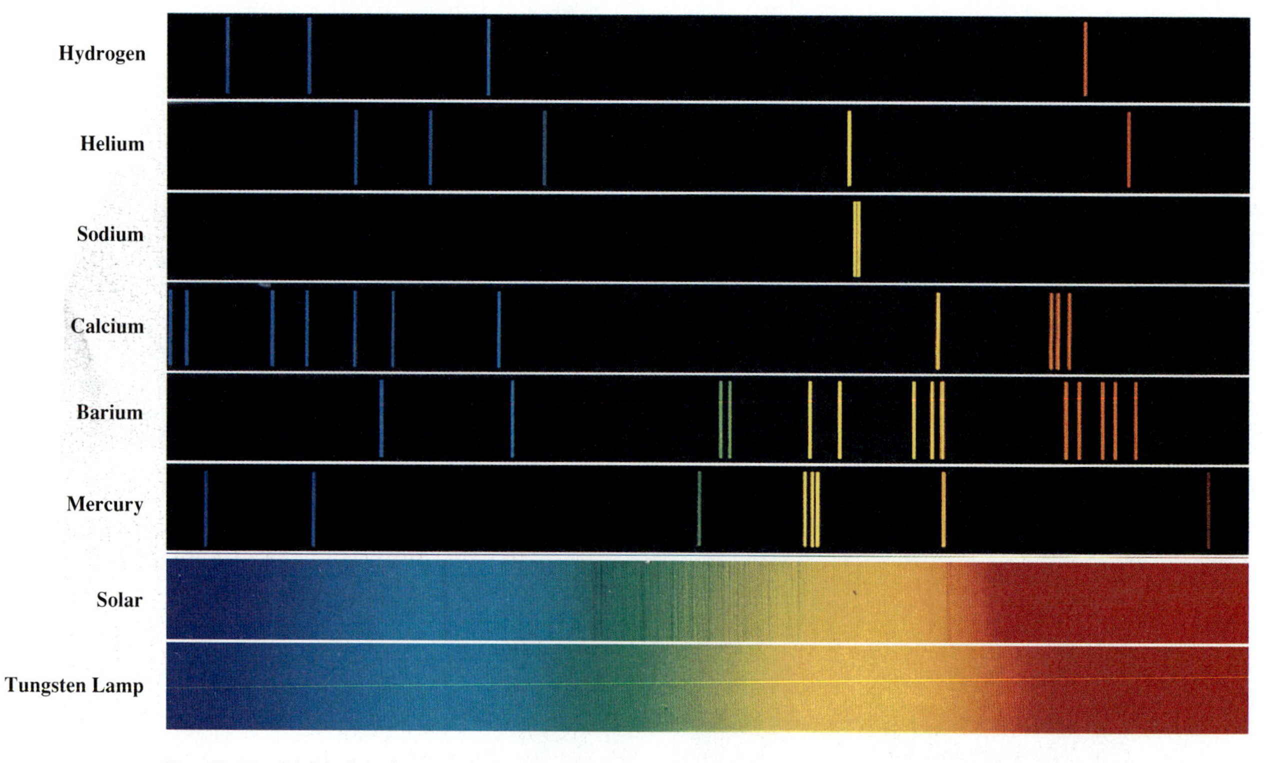

Fig. 23-11 Visible-light frequencies produced by the sun, a tungsten incandescent lamp, and by various gas discharge tubes. These sources also produce frequencies outside the visible range.

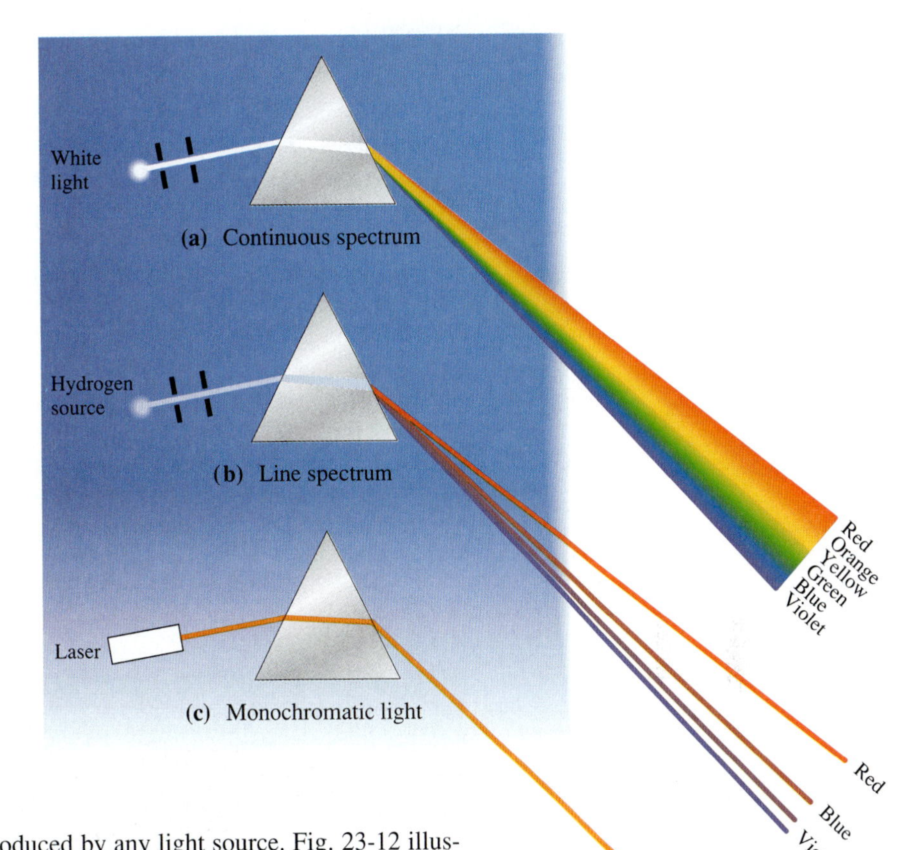

Fig. 23-12 A prism separates the colors within an incident beam of light. **(a)** White light. **(b)** Light from a hydrogen-gas discharge tube. **(c)** Laser light.

A glass prism can separate colors produced by any light source. Fig. 23-12 illustrates the effect of a prism on light from a continuous source, a gas discharge tube, and a laser. Lasers produce light that is more nearly monochromatic than light from other sources. For example, the helium-neon (He-Ne) laser produces orange light of frequency 4.741×10^{14} Hz, with a frequency range, or "bandwidth," on the order of only 10^8 Hz.

Another obvious characteristic of laser light is that it is very concentrated. Lasers produce narrow intense beams of light. Even a relatively low-power laser like the common and inexpensive He-Ne laser produces rather high-intensity light. A 1 mW laser producing a beam of cross-sectional area 1 mm^2 has an intensity of $\dfrac{10^{-3}\text{ W}}{(10^{-3}\text{ m})^2} = 10^3\text{ W/m}^2$, which is comparable to the intensity of sunlight. The laser light actually appears brighter than sunlight because, unlike sunlight, all the radiation is in the visible range. On the other hand, a 50 W incandescent light bulb produces light and heat that spread out spherically, and at a distance of 1 m from the bulb the energy is spread over a spherical surface of area 4π m^2, and so its intensity is $\dfrac{50\text{ W}}{4\pi\text{ m}^2} = 4\text{ W/m}^2$, much less than the intensity of a He-Ne laser. There are other lasers that produce light that is much more intense than that of the He-Ne laser. For example, the invisible infrared light from a carbon dioxide laser may have an intensity of 10^6 W/m^2 or more. Lasers in which the radiation is focused and delivered in pulses can produce a pinpoint burst of radiant energy with an intensity of over 10^{15} W/m^2. High-intensity lasers are used in many industries for cutting, drilling, or welding materials, and laser scalpels are becoming routine in many types of surgery.

23-3 The Propagation of Light

Measurement of the Speed of Light

Nothing in our ordinary experience indicates that the propagation of light is anything but instantaneous. Actually light travels at a finite speed, though the speed is very great. Galileo made the first recorded attempt to measure this speed. He carried a lantern to the top of a mountain one night while an assistant, also equipped with a lantern, was stationed on another nearby mountain. Both lanterns were initially covered. Galileo first uncovered his lantern, and when the assistant saw the light from Galileo's lantern, he uncovered the second lantern. Thus a short time after uncovering the first lantern, Galileo saw the second lantern. By measuring the time interval and the distance between the two lanterns, Galileo hoped to determine the speed of light. The time interval he actually observed, of course, was the reaction time of his assistant. The time required for light to travel between two points on earth is extremely short and so could not be measured in this way. Galileo correctly concluded that the speed of light is either infinite or extraordinarily great.

Fig. 23-13 Galileo's attempt to measure the speed of light.

From our late twentieth-century perspective, it is easy to dismiss Galileo's attempt to measure the speed of light as incredibly naïve. However, if we give some thought to the historical context in which he performed his experiment, we begin to appreciate his insight. Galileo was the first to recognize that light might have a finite speed—a speed that might be measurable. Later measurements by others often relied on methods similar to Galileo's, though using more sophisticated ways of measuring very short time intervals.

The earliest estimates of the speed of light were made in the seventeenth century from astronomical observations. The first very precise measurements were accomplished in the nineteenth century. From 1880 until 1926 the American physicist Albert Michelson performed a series of experiments, using a rotating mirror technique invented by Foucault (a variation on a method originally developed by Fizeau in 1849). Michelson's later experiments were performed in California, using mirrors located on nearby mountains, Mount Wilson and Mount San Antonio. The basic concept of the measurement was the same as Galileo's: to measure the time required for light to travel back and forth between the two mountains.

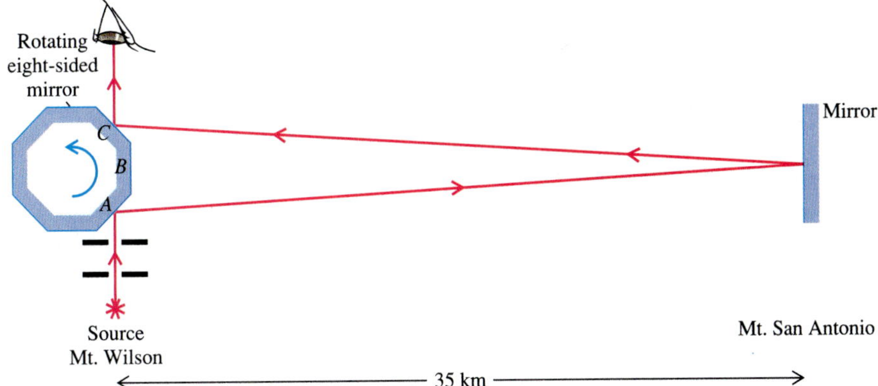

Fig. 23-14 Michelson's apparatus used to measure the speed of light.

A diagram of Michelson's apparatus is shown in Fig. 23-14. A light source, a rotating eight-sided mirrored wheel, and the observer were all located on Mt. Wilson, and a stationary mirror was placed on Mt. San Antonio 35 km away. With the eight-sided mirror stationary at first, light from the source was directed so that it reflected from one of the eight sides (A), traveled 35 km to the mirror M where it was again reflected, returned 35 km, and finally reflected from another side of the mirrored wheel (C) toward an observer. Then the wheel was rotated at a gradually increasing speed. At first the reflected light previously seen by the observer disappeared because the position of side C changed as the light traveled between the two mountains. However, when the rotation speed was great enough, the wheel completed one eighth of a revolution during the light's round trip, and side B replaced side C, so that the light was reflected from side B toward the observer and could again be seen. The wheel's rotation rate for this condition was found to be 540 revolutions per second. Thus the time required for the wheel to complete one eighth of a revolution was $\frac{1}{8}(\frac{1}{540})$ s. The speed of light, denoted by c, was computed to be the round trip distance of 70 km, divided by this time interval.

$$c = \frac{70 \times 10^3 \text{ m}}{\frac{1}{8}(\frac{1}{540}) \text{ s}}$$
$$= 3.0 \times 10^8 \text{ m/s}$$

By 1983 the most accurate measurements had established the speed of light (in a vacuum) as 2.99792458×10^8 m/s, with an estimated error of only ± 1 m/s. At that time precision in these measurements was limited by the precision of the standard meter, which was then based on the wavelength of light from a krypton source. The speed of light in a vacuum is a fundamental constant of nature. Since time could be measured on an atomic clock with much greater precision than distance could be measured, it made sense to turn the definition of the unit of length around, defining it in terms of speed and time. And so the meter is now defined to be the distance traveled by light in a vacuum during a time interval of 1/299,792,458 s. By definition, the speed of light is 299,792,458 m/s.

To three significant figures, in either air or a vacuum,

$$c = 3.00 \times 10^8 \text{ m/s} \quad (23\text{-}6)$$

Light in a Medium

Although no medium is necessary for the propagation of light, light can travel through certain media; that is, some objects are transparent (glass, water, and so on), whereas others are opaque. Opaque objects either reflect or absorb all the light incident on them. A medium may be opaque to some frequencies while transmitting other frequencies. For example, when sunlight shines through a piece of blue stained

Fig. 23-15 The color of this fish is affected by the water, which selectively absorbs red from the sunlight that illuminates the fish.

glass, the glass appears blue because only the blue part of the spectrum is transmitted; other frequencies are absorbed in the glass. Water has this same kind of property, though to a lesser degree. Light passing through several meters of water has a blue-green tint because the water weakly absorbs red light (Fig. 23-15).

Some media that are completely opaque to visible light are transparent to other parts of the electromagnetic spectrum; for other media the opposite is true. For example, radio waves can travel through wood, which is opaque to light, whereas glass transmits visible light but strongly absorbs infrared radiation. A 1 cm thick piece of glass transmits only 2% of incident infrared radiation of frequency 10^{14} Hz. This property of glass has an important function in a greenhouse. Sunlight passes through glass into the greenhouse, and plants and other objects inside absorb much of the radiant energy. These objects reemit infrared electromagnetic radiation, which would ordinarily be radiated away. But because glass is opaque to the infrared, this radiation is trapped inside, and as a result it is considerably warmer inside the greenhouse than outside. The same effect occurs when a car, with its windows up, is exposed to direct sunlight. The temperature inside the car can be much higher than the temperature outside.

Water vapor in the atmosphere is also rather opaque to infrared radiation. Thus cloudy nights tend to be warmer than clear nights because the clouds tend to prevent the infrared radiation emitted by the earth's surface from being radiated away into space. Carbon dioxide is another component of the atmosphere that contributes to its warming by trapping infrared radiation. This phenomenon is known as the "greenhouse effect." Some scientists believe that the earth is beginning a potentially disastrous global warming trend, through the greenhouse effect, as a result of increasing atmospheric concentrations of carbon dioxide—the result of burning fossil fuels, deforestation, and other phenomena, not all well understood.

The air in our atmosphere, especially the ozone, absorbs some of the higher frequency ultraviolet (UV) light from the sun. This is fortunate in that the absorbed frequencies would be very damaging to human skin. Sunburn results mainly from UV light that penetrates our atmosphere. The amount of UV light removed from sunlight is dependent on the thickness of the layer of air through which the light passes. Thus there is much more UV light at very high altitudes than at sea level, and one needs more protection from the sun at high elevations (Fig. 23-16). The thickness of the air layer through which sunlight passes also depends on the angle of the sun in the sky. For this reason in northern latitudes there is very little UV light at sea level during the winter, and in the summer there is little UV light before 10:00 A.M. and after 2:00 P.M.

Because ozone is important in blocking UV light, the recent discovery of a worldwide depletion of the ozone layer is cause for great concern, not merely because of the potential damage to human skin, but more importantly the damage this might cause to the entire planetary food chain.

Fig. 23-16 Sunburned skier.

Index of Refraction

The speed of light through matter is always somewhat less than the speed of light in vacuum. A measure of light's speed in a medium is the **index of refraction.** We define the index of refraction, denoted by n, to be the ratio of the speed c in vacuum to the speed v in the medium.

$$n = \frac{c}{v} \tag{23-7}$$

Since v is always less than c, n is always greater than unity. For example, the speed of light in water is approximately $\frac{3}{4}c$, and so the index of refraction of water is approximately $\frac{4}{3}$, or 1.33. Values of n for various media are given in Table 23-2.

Table 23-2 Refractive Indices

Gases (1 atm)	n	Solids	n
Helium	1.000036	Ice	1.310
Hydrogen	1.000132	*Plastics:*	
Air	1.000293	Lucite	1.500
Carbon dioxide	1.000451	Polyethylene	1.500
Liquids		Polystyrene	1.600
Methyl alcohol	1.329	*Glasses:*	
Water	1.333	Ordinary crown	1.517
Ethyl alcohol	1.361	Light flint	1.580
Carbon tetrachloride	1.461	Dense flint	1.655
Benzene	1.501	Densest flint	1.890
The eye		Rock salt	1.544
Aqueous humor	1.336	Quartz	1.544
Vitreous humor	1.337	Zircon	1.923
Cornea	1.376	Diamond	2.409
Crystalline lens	1.413		

The refractive indices in this table are for yellow light of frequency 5.09×10^{14} Hz, or vacuum wavelength 589 nm, corresponding to the yellow light of the sodium spectrum. The refractive index varies somewhat with temperature. For example, n for water ranges from 1.333 at 20° C to 1.318 at 100° C. Except for ice, all values in the table are for a temperature of 20° C.

Dispersion

The index of refraction of a medium usually varies somewhat with the frequency of the light. For example, n varies over the visible spectrum from 1.51 to 1.53 for ordinary crown glass and from 1.87 to 1.95 for the densest flint glass. This phenomenon is called **dispersion.** In the next section we shall see how this effect results in the spreading of spectral colors when white light is viewed through a prism.

The speed of light v is related to its wavelength λ and its frequency f by the usual wave relationship (Eq. 16-1):

$$v = \lambda f \tag{23-8}$$

Although the speed v depends on the medium through which light travels, the frequency of a light wave is unaffected by the medium. The frequency of a light wave is determined by the light source. The equation above then implies that the wavelength λ will vary in proportion to v. For the special case of light in a vacuum, $v = c$ and $\lambda = \lambda_0$, where λ_0 denotes the vacuum wavelength. The wave relationship is then expressed

$$c = \lambda_0 f \tag{23-9}$$

Dividing Eq. 23-9 by Eq. 23-8, we obtain an equation that shows the dependence of the wavelength on the index of refraction.

$$n = \frac{c}{v} = \frac{\lambda_0}{\lambda}$$

or

$$\lambda = \frac{\lambda_0}{n} \tag{23-10}$$

Wavelengths of light are often measured in units of nanometers, abbreviated nm, where

$$1 \text{ nm} = 10^{-9} \text{ m}$$

Another unit commonly used is the angstrom, denoted by Å, where

$$1 \text{ Å} = 10^{-10} \text{ m}$$

Vacuum wavelength rather than frequency is often used to describe light. For example, yellow light from the sodium spectrum is usually described by its vacuum wavelength of 589 nm, or 5890 Å, rather than by its frequency, 5.09×10^{14} Hz.

Fig. 23-17 shows a typical dispersion curve for a type of glass; the index of refraction is shown as a function of the vacuum wavelength λ_0.

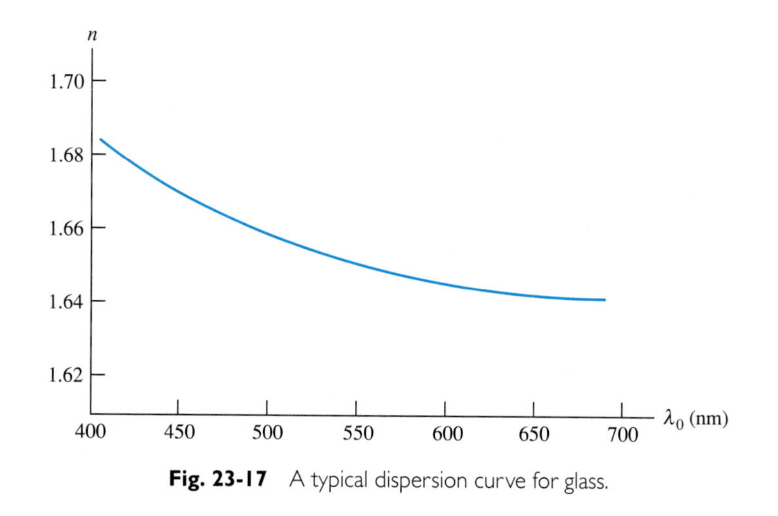

Fig. 23-17 A typical dispersion curve for glass.

EXAMPLE 2 Laser Light in Air and in Glass

The light emitted by a He-Ne laser has a frequency of 4.74×10^{14} Hz. Find the wavelength of the light in a vacuum and in ordinary crown glass.

SOLUTION Applying Eq. 23-9 ($c = \lambda_0 f$), we find the vacuum wavelength λ_0.

$$\lambda_0 = \frac{c}{f} = \frac{3.00 \times 10^8 \text{ m/s}}{4.74 \times 10^{14} \text{ Hz}}$$

$$= 6.33 \times 10^{-7} \text{ m}$$

$$= 633 \text{ nm}$$

In glass, where the speed of light is less, the wavelength is also less. From Table 23-2 we find the index of refraction for glass, and apply Eq. 23-10.

$$\lambda = \frac{\lambda_0}{n} = \frac{633 \text{ nm}}{1.52}$$

$$= 416 \text{ nm}$$

Light Rays

A point source of light (for example, a tiny light bulb) radiates light equally in all directions. Such a source produces a spherical wave. The wavefronts are spherical surfaces over which the wave disturbance (the electric field) has a maximum value. The distance between adjacent wavefronts is the wavelength λ. Wavefronts propagate outward from the source. **Rays** are arrows drawn perpendicular to the wavefronts (Fig. 23-18). The rays indicate the direction in which the wavefronts move—radially outward from a point source.

Fig. 23-19 shows a plane wave. The wavefronts are parallel planes, and the rays, which are perpendicular to the wavefronts, are parallel to each other.

A small section of a spherical wave produced by a distant point source is approximately a plane wave (Fig. 23-20). A laser beam is also approximately a plane wave.

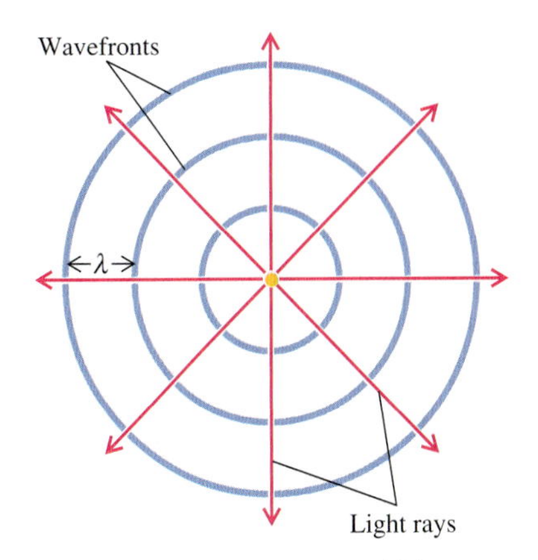

Fig. 23-18 A point source of light produces spherical wavelengths, shown in cross-section. Light rays are perpendicular to wavefronts.

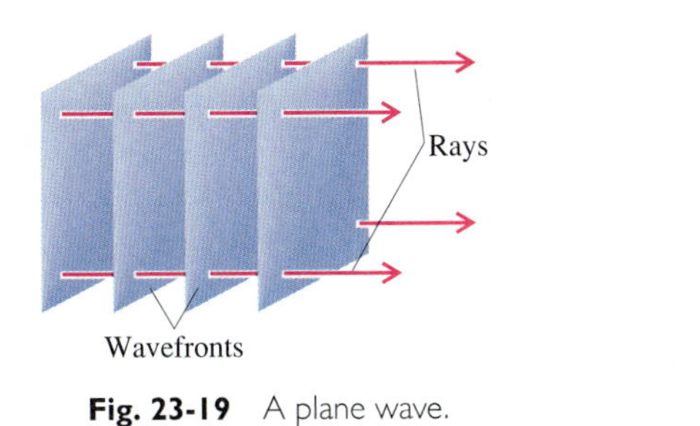

Fig. 23-19 A plane wave.

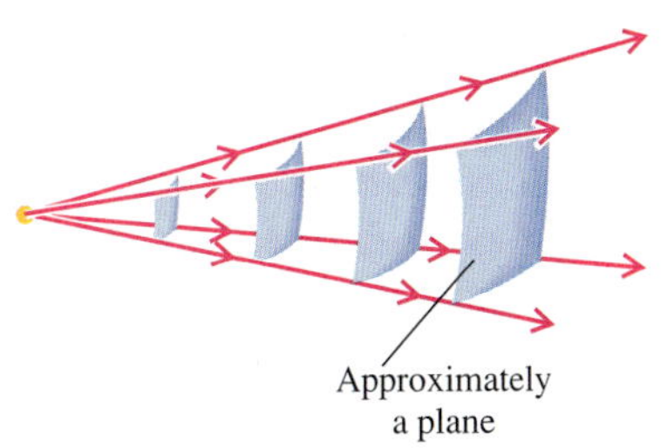

Fig. 23-20 Small segments of large spherical wavefronts approximate a plane wave.

We see when light rays enter our eyes (Fig. 23-21). The light rays are always interpreted by the brain as having traveled in a straight line. We will discuss details of the vision mechanism in Chapter 25.

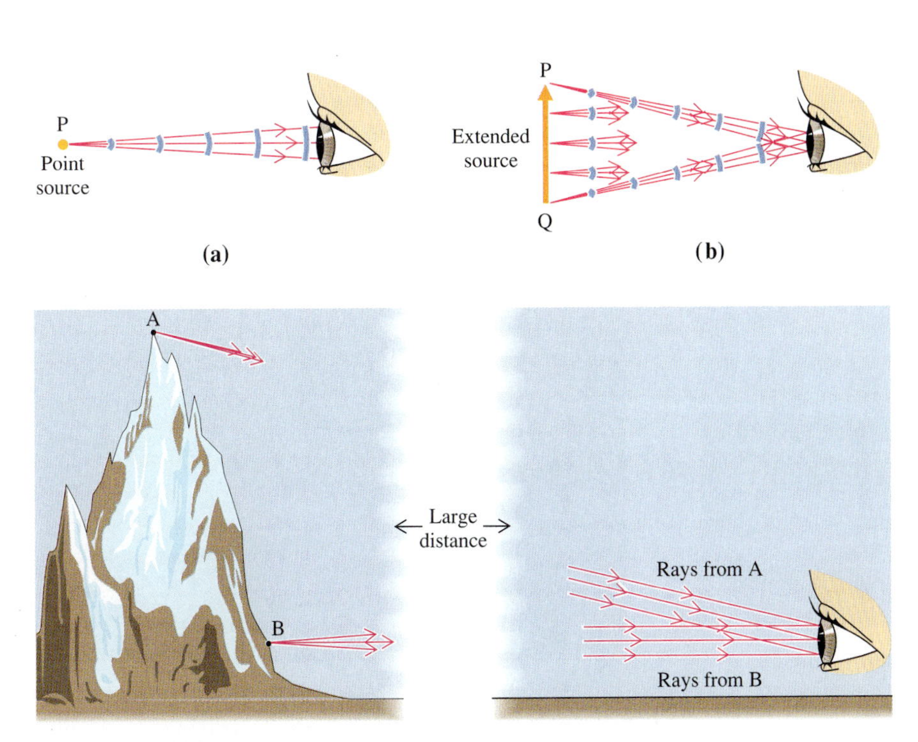

Fig. 23-21 **(a)** A point source. **(b)** An extended source. Each point of the source produces its own spherical wave and associated rays. **(c)** Parallel light rays from points at the top and bottom of a large distant source.

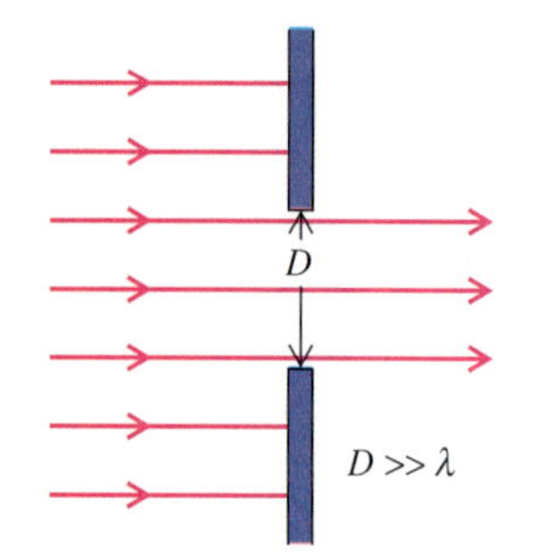

Fig. 23-22 When the width of an opening is much greater than the wavelength of the light passing through it, light rays pass straight through; that is, bending of light around the edges of the opening is negligible.

Diffraction; Geometrical and Wave Optics

All waves bend around objects in their paths. This phenomenon is known as **diffraction.** For sound waves, diffraction effects are usually pronounced. For example, you can hear street noise throughout the interior of a room that has an open window facing the street. The sound wave passes through the window opening and spreads in all directions to all parts of the room.

For certain situations diffraction of light waves is significant too. The study of light for such cases is called **wave optics,** or **physical optics.** But light waves have much shorter wavelengths than sound waves. Consequently, as we shall see, diffraction effects for light are typically much less than for sound. Many phenomena are much more easily explained if we ignore the wave nature of light and its associated diffraction effects and think of light rays as traveling in a straight line as they pass through openings or pass by objects. This approximation is known as **geometrical optics.** The validity of geometrical optics will depend on the size of the opening in comparison with the wavelength of the light. If the opening is much larger than the wavelength, diffraction effects will be small (Fig. 23-22). As the opening becomes smaller, diffraction effects become more pronounced and the approximation of geometrical optics becomes less valid. In this chapter and the next we shall study geometrical optics. Then in Chapter 26 we shall study wave optics.

23-4 Reflection and Refraction

Suppose that light traveling through one medium strikes the surface of a second medium. Light may be both transmitted into the second medium and reflected back into the first medium. For example, when sunlight strikes the surface of a lake, some light is reflected from the surface and some is transmitted into the water (Fig. 23-23). Or when light strikes a glass windowpane, some of the light is reflected from the surface of the glass, while most of the light is transmitted through the window. It is also possible in some cases for light to be reflected from a surface, with no light transmitted. Shiny metallic surfaces and mirrors produce this effect (Fig. 23-24).

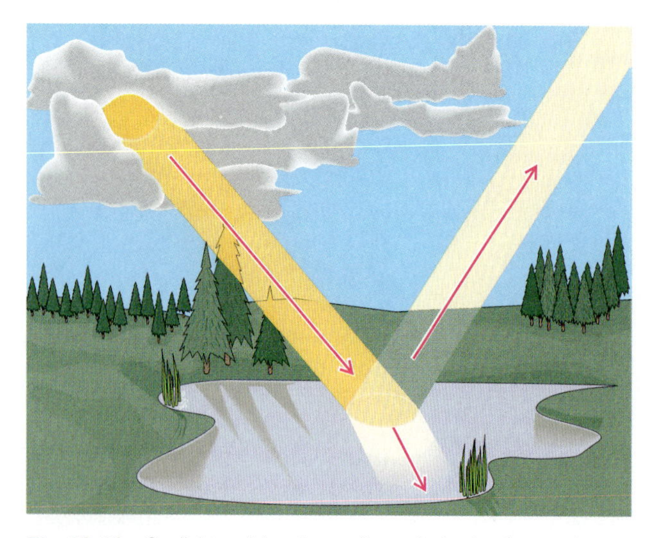

Fig. 23-23 Sunlight striking the surface of a body of water is partially reflected and partially transmitted into the water.

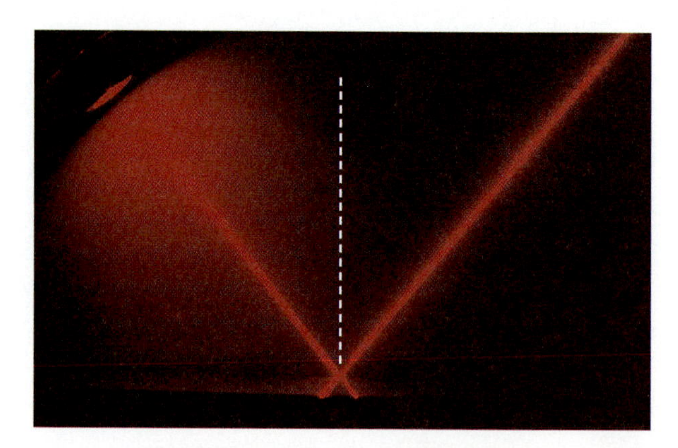

Fig. 23-24 Reflection of a laser beam.

Most objects in our surroundings have surfaces that are rougher than a shiny metal surface; consequently, we see the light reflected from them as though the objects themselves were the source of that light (Fig. 23-25).

Fig. 23-26 shows a light ray incident on a surface between two media. The first medium has an index n, and the medium into which light is transmitted has an index n'. The direction of all rays is measured by the angles made by the rays with the **normal,** a line perpendicular to the surface. The reflected ray leaves the surface at the same angle as the incident ray, as indicated in the figure. This result is known as the **law of reflection.**

$$\text{Angle of incidence} = \text{Angle of reflection} \tag{23-11}$$

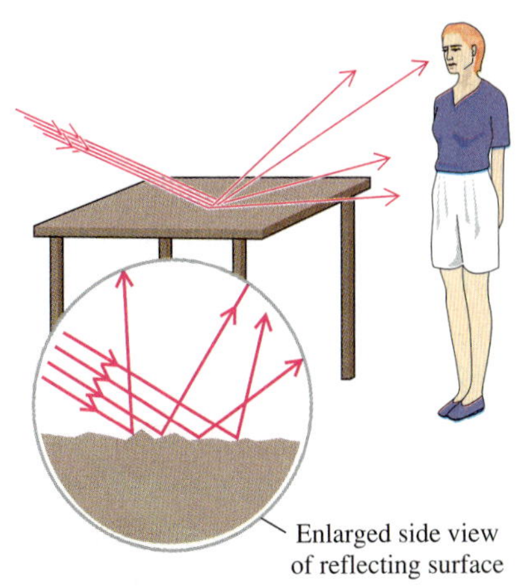

Fig. 23-25 Light rays incident on a small section of an object's surface are typically reflected in all directions because of the irregularity of the surface. Thus each section of the source acts as a point source of light, and the object as a whole acts as an extended source of light.

The laser beam shown in Fig. 23-24 satisfies the law of reflection. Both the incident and reflected beams make the same angle with the normal (indicated as a dashed line in the figure).

A transmitted ray is bent, or "refracted," as it enters the second medium. The amount of refraction depends on the angle of incidence and the refractive indices of the two media. The incident and refracted angles and the indices n and n' are related by the **law of refraction,** also called **Snell's law:**

$$n \sin \phi = n' \sin \phi' \tag{23-12}$$

Given the indices n and n' and the angle of incidence ϕ, one can solve for the angle ϕ'. In the next chapter we will apply Snell's law to explain how images are produced by lenses (Section 24-3).

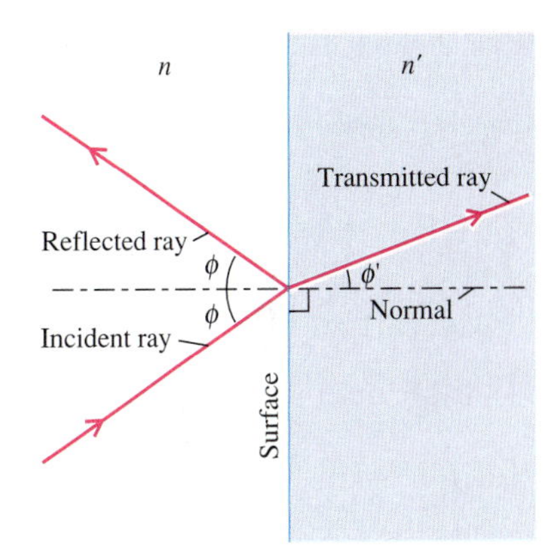

Fig. 23-26 A light ray incident on the surface between two media with different refractive indices results in a reflected ray and a transmitted ray.

Snell's law predicts that the transmitted ray will be bent toward the normal if the index increases (Fig. 23-27a), or away from it if the index decreases (Fig. 23-27b). For example, if light goes from air ($n = 1$) to glass ($n' = 1.5$), Snell's law predicts that the angle of refraction ϕ' satisfies the equation

$$1.5 \sin \phi' = 1 \sin \phi$$

$$\sin \phi' = \left(\frac{1}{1.5}\right)(\sin \phi)$$

For any angle ϕ, this equation predicts that ϕ' will be smaller than ϕ. Thus, when light goes from air to glass, rays are bent toward the normal (Fig. 23-27a). When light goes from a higher-index medium to a lower-index medium (for example, glass to air), rays are bent away from the normal (Fig. 23-27b).

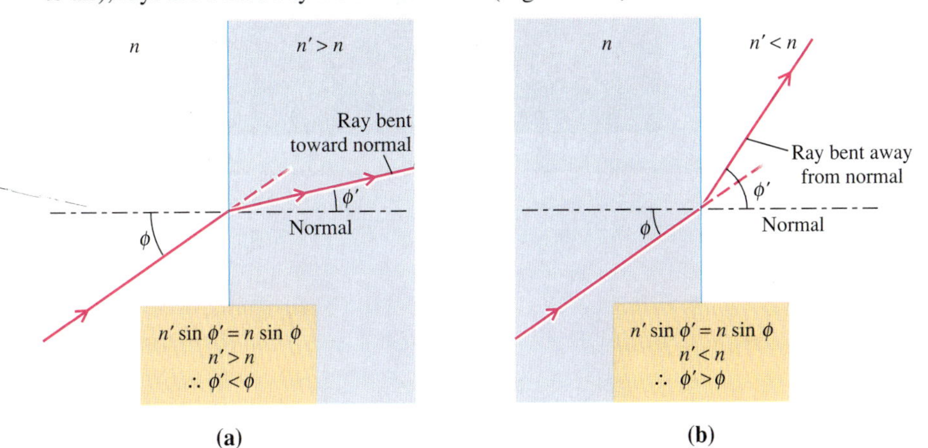

Fig. 23-27 **(a)** A light ray bends toward the normal when it passes into a higher index medium. **(b)** A light ray bends away from the normal when it passes into a lower index medium.

EXAMPLE 3 **Water Is Deeper Than It Appears**

An observer looking across a swimming pool views a light ray coming from a point P on the opposite side of the pool at the bottom (Fig. 23-28). The ray leaves the water at a point 1.50 m from the opposite side of the pool and makes an angle of 45.0° with the vertical. Thus the pool appears to be only 1.50 m deep. Find the pool's true depth d.

SOLUTION First we apply Snell's law and solve for the angle of incidence ϕ.

$$n \sin \phi = n' \sin \phi'$$

$$1.33 \sin \phi = 1.00 \sin 45.0°$$

$$\sin \phi = 0.532$$

$$\phi = 32.1°$$

From the geometry of the figure, we see that

$$\frac{1.50 \text{ m}}{d} = \tan \phi = \tan 32.1°$$

$$d = 2.39 \text{ m}$$

The pool is considerably deeper than it appears. The apparent shortening and bending of objects partially submerged in water are related effects (Fig. 23-29).

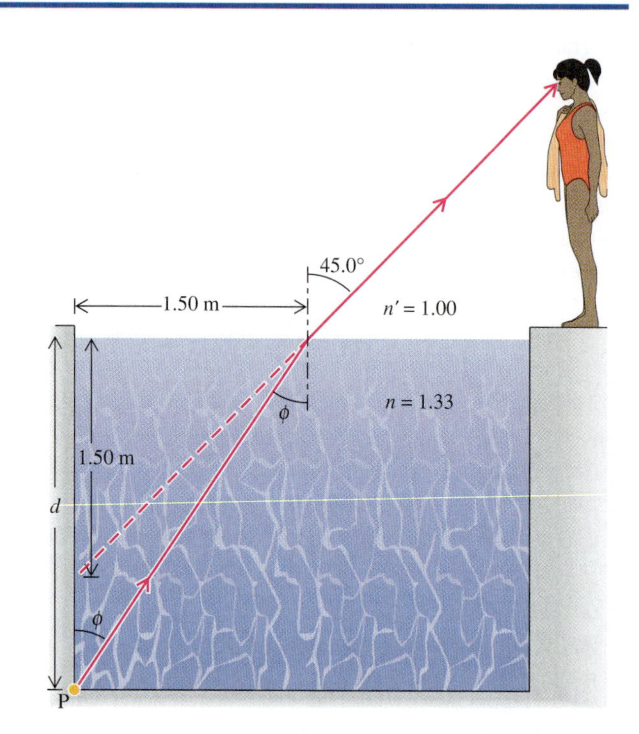

Fig. 23-28

Fig. 23-29 Note how both legs appear shortened and how the left leg appears unnaturally bent sideways at the knee.

EXAMPLE 4 Light Through a Windowpane

A ray of light is incident at an angle ϕ on a glass windowpane (Fig. 23-30). Find the angle at which the ray leaves the glass.

SOLUTION Here refraction occurs twice. First, in going from air to glass, the ray is bent toward the normal. Then, in going from glass to air, the ray is bent away from the normal (Fig. 23-31). We apply Snell's law at the first surface to obtain

$$1 \sin \phi = n \sin \phi'$$

We apply Snell's law at the second surface, using the fact that the two normals are parallel and the angle of incidence therefore equals ϕ', as indicated in the figure.

$$n \sin \phi' = 1 \sin \phi''$$

Comparing the two preceding equations, we find

$$\sin \phi'' = \sin \phi$$

or

$$\phi'' = \phi$$

This means that the ray emerges from the glass at the same angle that it entered, for *any* incident angle ϕ. The sheet of glass has not changed the direction of the light. If the two surfaces of the glass were not parallel, the direction of the light would be changed, as seen in the following example of a prism.

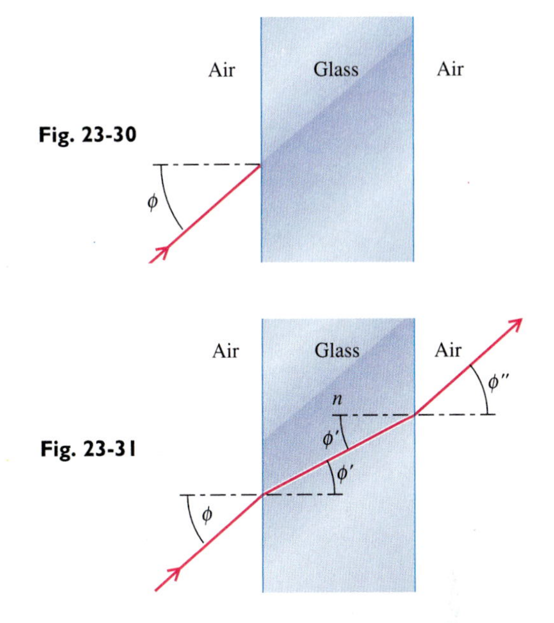

Fig. 23-30

Fig. 23-31

EXAMPLE 5 Separating Colors With a Prism

The refractive index of a certain glass has a value of 1.67 for blue light of vacuum wavelength 450 nm and a value of 1.64 for red light of vacuum wavelength 700 nm. An equilateral prism is made from this glass, and a beam of white light is incident on a face of the prism at an angle of 45.0°. Find the respective angles at which blue light and red light emerge from the opposite side of the prism.

SOLUTION Fig. 23-32 indicates the bending of a light ray toward the normal as it enters the prism and away from the normal as it leaves the prism. Inspection of the figure indicates that the angle of incidence at the second surface is $60.0° - \phi$, where ϕ is the angle of refraction at the first surface. We apply Snell's law at each surface for both the red and blue light, using the appropriate index for each.

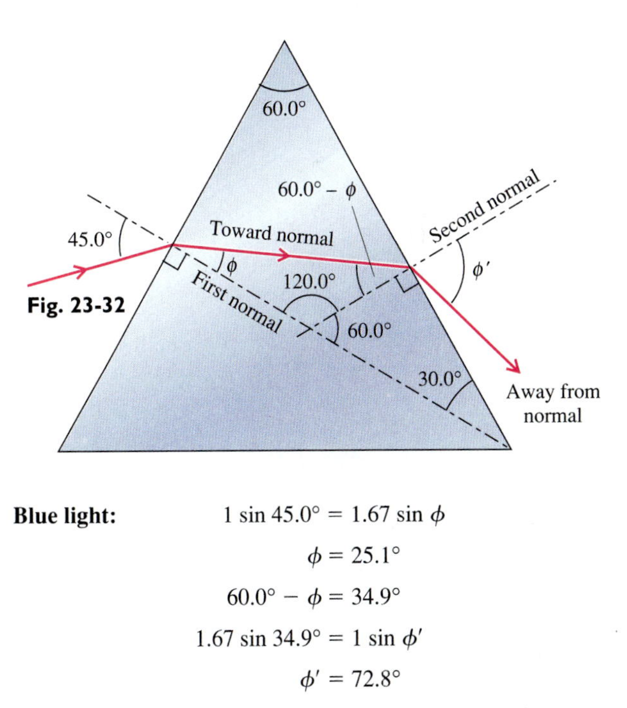

Fig. 23-32

Red light:

$$1 \sin 45.0° = 1.64 \sin \phi$$
$$\phi = 25.5°$$
$$60.0° - \phi = 34.5°$$
$$1.64 \sin 34.5° = 1 \sin \phi'$$
$$\phi' = 68.3°$$

Blue light:

$$1 \sin 45.0° = 1.67 \sin \phi$$
$$\phi = 25.1°$$
$$60.0° - \phi = 34.9°$$
$$1.67 \sin 34.9° = 1 \sin \phi'$$
$$\phi' = 72.8°$$

Thus blue light emerges from the prism at an angle 4.5° greater than that of the red light. Other colors (orange, yellow, green) have indices between 1.64 and 1.67 and would emerge at angles between 68.3° and 72.8°. Thus one sees a continuous spectrum of colors in the diverging beam that emerges from the prism, as illustrated in Fig. 23-12a.

Reflectance and Transmittance

At a surface between two transparent media, part of the incident light is transmitted and part is reflected. We define the **reflectance** R as the ratio of the intensity of the reflected light I_r to the intensity of incident light I_i.

$$R = \frac{I_r}{I_i} \qquad (23\text{-}13)$$

Transmittance is similarly defined as the ratio of transmitted intensity I_t to incident intensity.

$$T = \frac{I_t}{I_i} \qquad (23\text{-}14)$$

Adding the expressions above for R and T, we obtain

$$R + T = \frac{I_r}{I_i} + \frac{I_t}{I_i} = \frac{I_r + I_t}{I_i}$$

If there is no absorption of light at the surface, conservation of energy requires that $I_r + I_t = I_i$, and this expression reduces to

$$R + T = 1 \qquad (23\text{-}15)$$

Values of reflectance and transmittance depend on the angle of incidence and the refractive indices of the two media. Expressions for R and T may be obtained from Maxwell's equations, which can also be used to derive the laws of reflection and refraction. In the special case of a ray incident normal to the surface, R is found to be a simple function of the indices n and n':

$$R = \left(\frac{n' - n}{n' + n}\right)^2 \quad \text{(for normal incidence)} \qquad (23\text{-}16)$$

Once R is found, we may find the value of T by applying Eq. 23-15 ($R + T = 1$). From Eq. 23-16, we see that if $n' = n$, $R = 0$. No light is reflected if there is no difference in the refractive indices of the two media. When a glass rod has the same index as a liquid in which it is immersed, the rod becomes invisible because the glass-liquid surface reflects no light (Fig. 23-33).

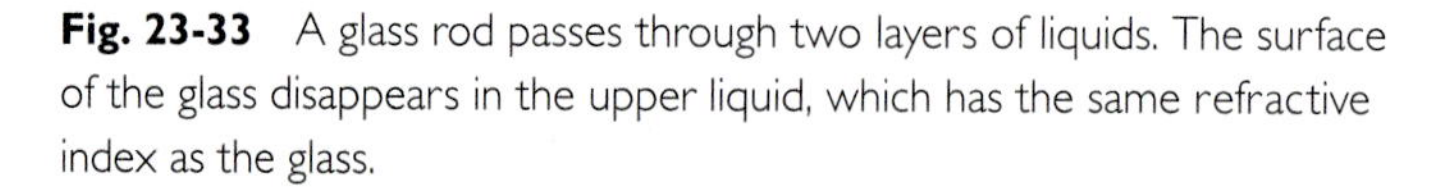

Fig. 23-33 A glass rod passes through two layers of liquids. The surface of the glass disappears in the upper liquid, which has the same refractive index as the glass.

EXAMPLE 6 Reflectance of Water, Glass, and Diamond

Find the reflectance of light normally incident from air onto surfaces of water, crown glass, and diamond.

SOLUTION Applying Eq. 23-16 and using the values given in Table 23-2 for the refractive indices of the respective media, we find:

$$R = \left(\frac{n' - n}{n' + n}\right)^2$$

Air to water: $$R = \left(\frac{1.33 - 1}{1.33 + 1}\right)^2 = 0.020$$

Air to glass: $$R = \left(\frac{1.52 - 1}{1.52 + 1}\right)^2 = 0.043$$

Air to diamond: $$R = \left(\frac{2.41 - 1}{2.41 + 1}\right)^2 = 0.171$$

Values of transmittance are easily found when we apply Eq. 23-15 ($R + T = 1$). For example, for air to water:

$$T = 1 - R = 1 - 0.020 = 0.980$$

Total Internal Reflection

Light is sometimes reflected from a glass surface, as though it were a perfect mirror (Fig. 23-34). This remarkable phenomenon, called **total internal reflection,** occurs only when the light is incident on the surface from the glass side and only if the angle of incidence is greater than a certain angle, 41.1° for crown glass. Total internal reflection occurs whenever light is incident from a higher-index medium onto a lower-index medium at an angle of incidence greater than a certain **critical angle** ϕ_c, the value of which depends on the media.

$$R = 1 \qquad (\text{when } n' < n \text{ and } \phi > \phi_c) \qquad (23\text{-}17)$$

Maxwell's equations predict the phenomenon of total internal reflection. It occurs only in those cases for which Snell's law fails to give a physically meaningful answer for the angle of refraction. Suppose, for example, in going from glass to air the angle of incidence is 60°. Then Snell's law predicts an angle of refraction given by

$$n' \sin \phi' = n \sin \phi$$

$$1 \sin \phi' = 1.5 \sin 60^\circ$$

$$\sin \phi' = 1.3$$

Obviously there is no angle whose sine is greater than 1; Snell's law clearly fails here. Physically what happens in this case is that *there is no refracted ray.* The light is totally reflected. Any time Snell's law gives $\sin \phi' > 1$, there is total internal reflection.

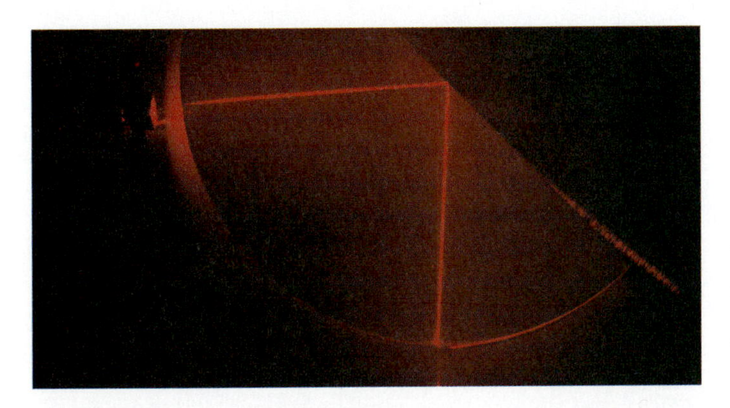

Fig. 23-34 A laser beam is incident from the left onto a semi-circular glass plate. The beam travels to the right through the glass and is reflected downward from the flat side of the plate.

The critical angle at which total internal reflection begins corresponds to an angle of refraction whose sine equals 1, that is, $\phi' = 90°$. Fig. 23-35 shows the angle of incidence approaching the critical angle. The intensity of the refracted ray approaches zero as the angle of refraction approaches 90°. Fig. 23-36 shows quantitatively how the reflectance of light increases as the critical angle is approached, and how $R = 1$ for all angles greater than the critical angle.

To obtain an expression for the critical angle, we use Snell's law, setting the angle of refraction equal to 90°.

$$n \sin \phi_c = n' \sin 90°$$
$$= n'$$

or

$$\sin \phi_c = \frac{n'}{n} \tag{23-18}$$

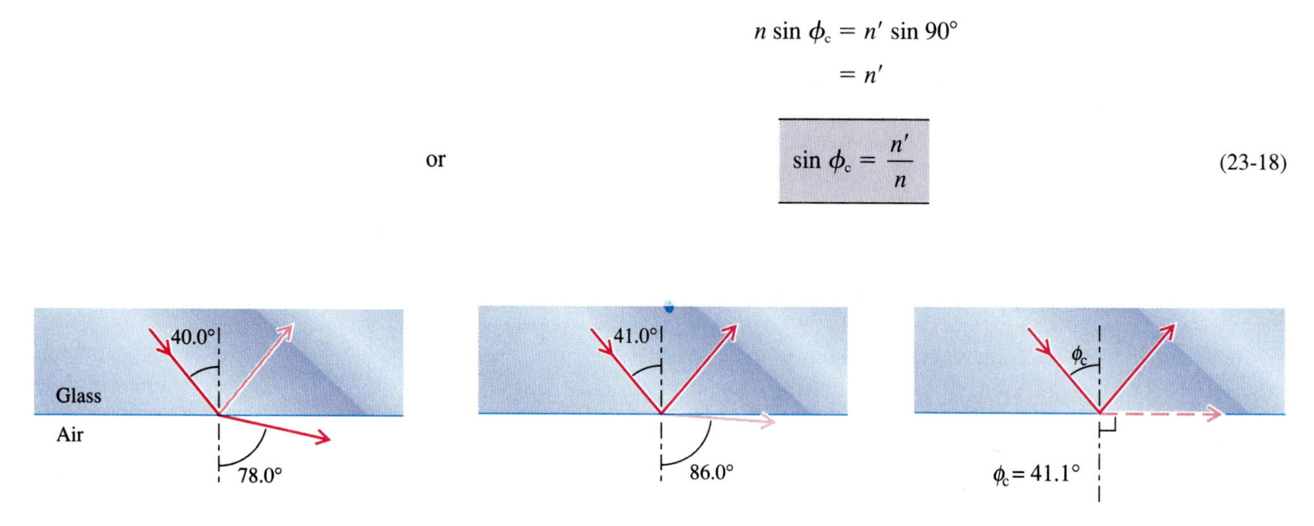

Fig. 23-35 The angle of incidence approaches the critical angle, with light incident from crown glass ($n = 1.52$) onto air ($n' = 1.00$).

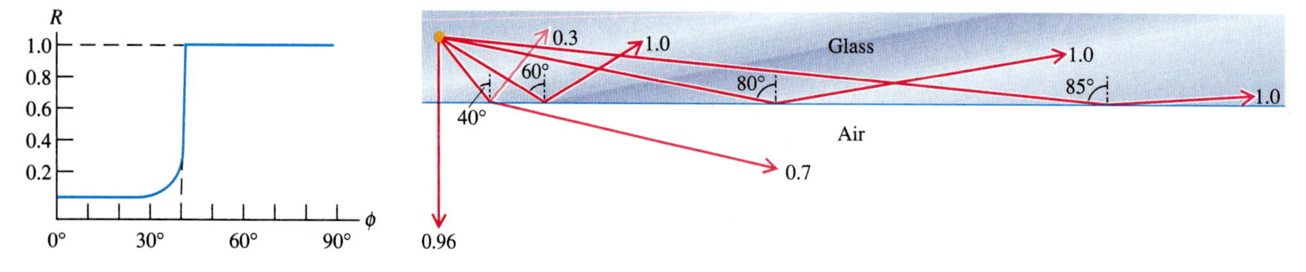

Fig. 23-36 Reflectance of light from glass to air for various angles of incidence.

EXAMPLE 7 Why a Diamond Sparkles

Compute the critical angle for light incident onto air from water, crown glass, and diamond. How does diamond's small critical angle relate to the appearance of a diamond gemstone?

SOLUTION Applying Eq. 23-18, we find

$$\sin \phi_c = \frac{n'}{n}$$

Water to air: $$\sin \phi_c = \frac{1}{1.33} = 0.752$$

$$\phi_c = 48.8°$$

Glass to air: $$\sin \phi_c = \frac{1}{1.52} = 0.658$$

$$\phi_c = 41.1°$$

Diamond to air: $$\sin \phi_c = \frac{1}{2.41} = 0.415$$

$$\phi_c = 24.5°$$

The unusually high refractive index of diamond leads to its small critical angle in air of 24.5°. Diamond gemstones are cut with many facets in such a way that much of the incident light undergoes multiple total internal reflections within the diamond before passing out again into the air (Fig. 23-37a).

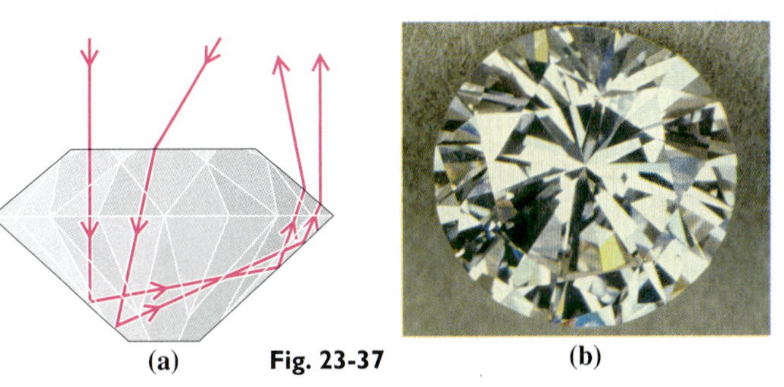

(a) **Fig. 23-37** (b)

Many of the rays incident on the diamond from above are reflected back upward. Like a prism, the diamond is a dispersive medium (that is, n varies somewhat with λ), and so the various colors composing white light travel somewhat different paths and emerge in different directions. Hence as a diamond is turned, different colors are seen, giving it its characteristic sparkling quality (Fig. 23-37b).

To keep a diamond ring sparkling you must keep the bottom of the diamond clean so that the sharp change in refractive index from diamond to air is maintained, and the critical angle is kept low. Dirt or moisture on the bottom of the diamond increases the critical angle, allowing much of the light incident on the diamond to escape through the bottom.

Fiber Optics

Light rays can be transmitted along a glass or plastic fiber, by being repeatedly internally reflected at the surface of the fiber (Fig. 23-38).

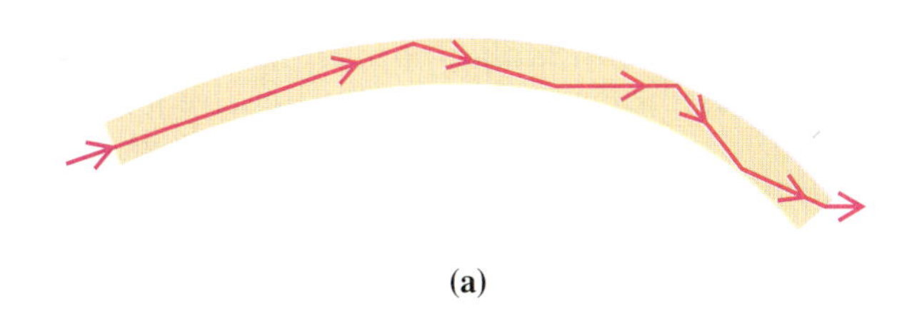

(a)

(b)

Fig. 23-38 **(a)** Total internal reflection in an optical fiber allows light to travel along a curved path. **(b)** A decorative display using optical fibers.

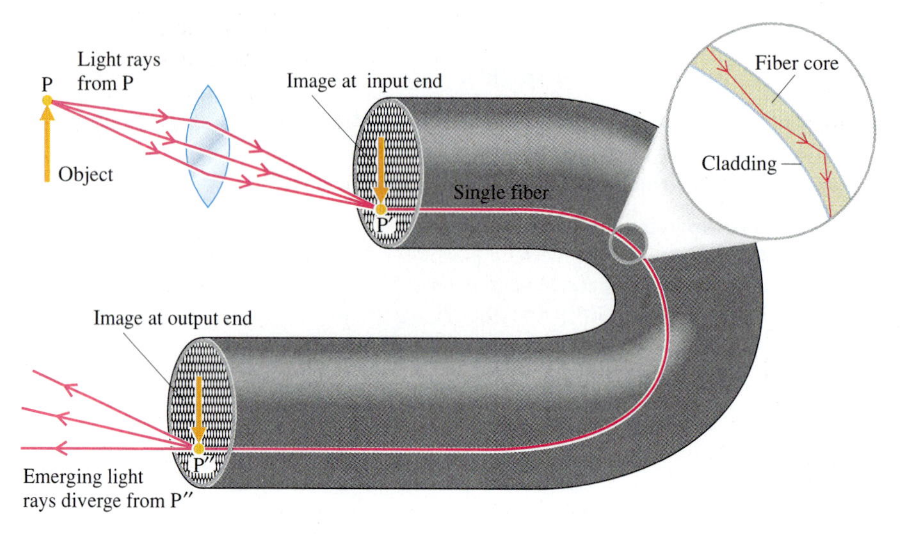

Fig. 23-39 An image is transmitted by a bundle of optical fibers. The relative positions of the fibers must be the same at the input and output ends. Light travels through one of the fibers from P′ to P″. That single fiber is shown, greatly exaggerated in size. There may be 10^4 fibers in a bundle, each as fine as a human hair.

Bundles of fibers can be used to transmit images (Figs. 23-39 and 23-40). The fibers each have a diameter that is typically about 50 μm or 0.05 mm (about the thickness of a human hair). Such thin fibers are required to allow transmission of a sharp image. Each fiber is clad with a material of lower index, so that light rays traveling within the fiber are internally reflected each time they strike the surface between the fiber and the cladding material, so long as their angle of incidence exceeds the critical angle. If the fiber has an index of 1.6 and is sheathed in a material of index 1.5, the critical angle is 70° ($\sin \phi_c = n'/n = 1.5/1.6 = 0.94$; $\phi_c = 70°$).

The reflection is more perfect than that of a high-quality mirror, for which R is about 0.97. Even after thousands of reflections there is no reduction in intensity. Indeed the main limitation to use of very long optical fibers is the absorption of light within the medium. Optical fibers up to several meters in length are routinely used in medical diagnosis to observe internal organs. Some remarkable internal pictures of the human body have been obtained by the photographer Lennart Nilsson, using fiber optics (Fig. 23-41).

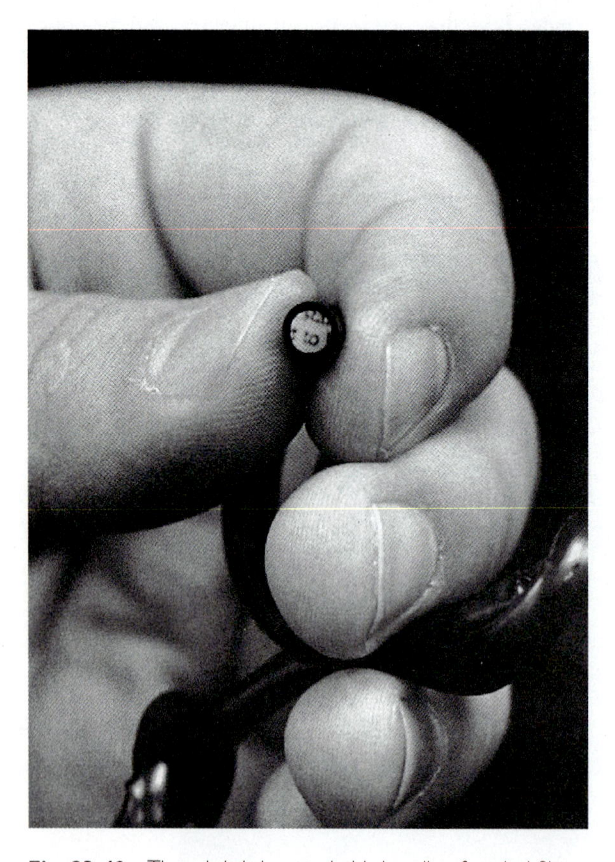

Fig. 23-40 Though it is knotted, this bundle of optical fibers transmits an image of print from a page over which the other end of the bundle is placed.

In arthroscopic surgery, a fiber-optic cable, with a surgical instrument mounted at the end, is inserted through a tiny incision into the area of the body where the surgery is to be performed. Damage to the surrounding tissue is much less than for traditional surgery. Arthroscopic surgery is often used, for example, to remove torn cartilage in the knee. Athletes often undergo such surgery and usually recover very quickly.

Fiber optics is used for telephone communication. By varying the intensity of the transmitted light, information is encoded in the light wave and transmitted along with the light, in much the same way information is transmitted in radio waves for radio communication. Fiber-optic cables offer significant advantages over traditional telephone communication via microwave transmission and telephone lines. Optical fibers are able to transmit more information in much less space than copper telephone lines. Optical fibers are also immune to electrical interference, unlike both microwave telephone signals and telephone lines.

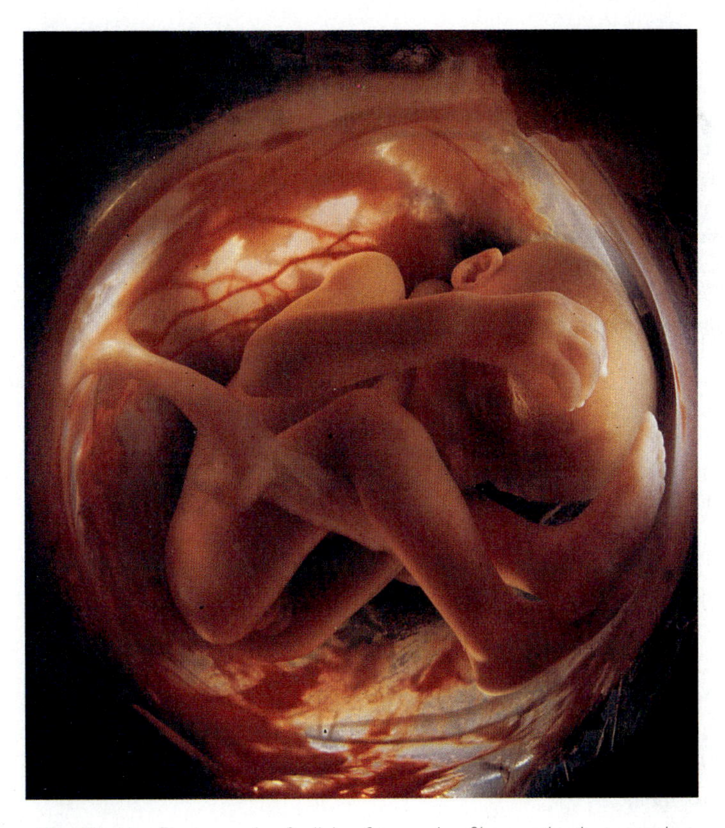

Fig. 23-41 Photograph of a living fetus, using fiber-optic photography.

CHAPTER 23 SUMMARY

Visible light is that part of the electromagnetic spectrum to which the eye is sensitive. It consists of wavelengths (in a vacuum) ranging from 400 nm to 700 nm, with corresponding colors from violet to red.

Light travels through a vacuum at a speed $c = 3.00 \times 10^8$ m/s. Light travels through a material medium at a speed $v < c$. A material's index of refraction n is determined by the speed of light in the medium:

$$n = \frac{c}{v}$$

Although all frequencies of electromagnetic radiation travel at the same speed c in a vacuum, in matter v and n vary with frequency. This phenomenon is called "dispersion."

The wavelength of light is related to its speed and frequency by the usual wave relationship

$$v = \lambda f$$

Rays are arrows pointing in the direction in which light travels. Rays are perpendicular to wavefronts. In the approximation known as geometrical optics, rays travel in a straight line through any homogeneous medium. Geometrical optics is valid and the wave nature of light can be ignored so long as any opening through which light passes has dimensions that are much greater than the wavelength of light.

When light travels from one medium to another medium with a different index, light will usually be both reflected and transmitted. The directions of the incident, reflected, and transmitted rays are measured by the angles these rays make with the normal to the surface separating the media. The angles of reflected and refracted rays are determined respectively by the laws of reflection and refraction.

Law of reflection:

$$\text{Angle of incidence} = \text{Angle of reflection}$$

Law of refraction or Snell's law:

$$n \sin \phi = n' \sin \phi'$$

When the index n of the first medium is greater than the index n' of the second medium, Snell's law applies only for angles of incidence less than or equal to the critical angle ϕ_c, where

$$\sin \phi_c = \frac{n'}{n}$$

For angles of incidence greater than the critical angle, no light is transmitted; the incident light is totally reflected.

Light intensity I is defined as the power P carried by light per cross-sectional area A perpendicular to the direction of propagation.

$$I = \frac{P}{A}$$

Reflectance R and transmittance T measure the respective intensities of reflected and transmitted light, I_r and I_t, relative to the intensity of incident light I_i.

$$R = \frac{I_r}{I_i}$$

$$T = \frac{I_t}{I_i}$$

Conservation of energy requires that

$$R + T = 1$$

For normal incidence,

$$R = \left(\frac{n' - n}{n' + n}\right)^2$$

If $n > n'$ and $\phi > \phi_c$, $R = 1$.

Questions

1 An electromagnetic wave from a distant radio transmitting antenna travels eastward, with the magnetic field oscillating along a north-south line.

(a) One kind of radio-receiving antenna is a long conductor, which, for best reception, is aligned so that the electric field vector can accelerate electrons along its length (Fig. 23-42a). What should be the orientation of this antenna?

(b) Another kind of receiving antenna is a circular loop of wire, in which an emf is induced by the electromagnetic wave (Fig. 23-42b). What should be the direction of the loop's axis?

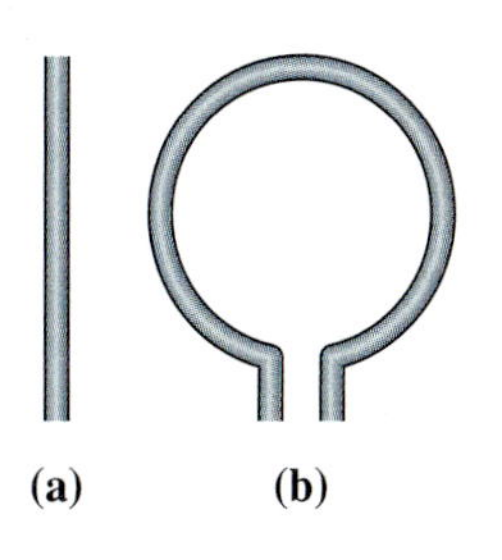

Fig. 23-42 Receiving antennas.

2 The microwaves reflected from the walls of a microwave oven can form standing waves. As a result, food that is not turned may not cook at points corresponding to the nodes in the standing wave. If the closest cold spots in a frozen dinner are 6.1 cm apart, what is the wavelength of the microwaves?

3 Monochromatic light enters the eye of a swimmer underwater. The light has a wavelength of 600 nm in water.

(a) Is the light visible underwater?

(b) Is this light visible if it passes through air before entering the eye?

4 Certain monochromatic light has a wavelength of 300 nm inside glass with a refractive index of 1.5. What color is the light?

5 Light travels through a 2 m long optical fiber, which has an index of 1.5. How long does it take for light to travel the length of the fiber?

6 Which of the points A through E could be the source for the light ray shown in Fig. 23-43?

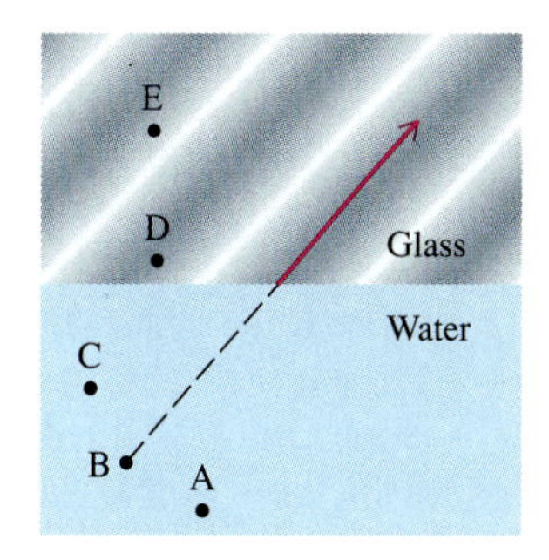

Fig. 23-43

7 With the line of sight directed at point P, which of the points in Fig. 23-44 could the observer see?

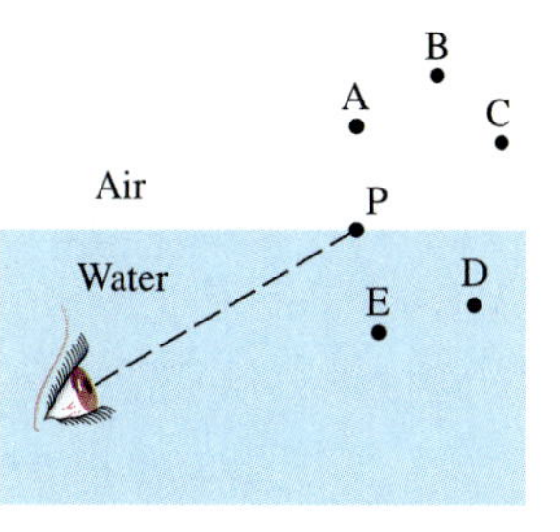

Fig. 23-44

8 Which reflects a greater percentage of incident light: the cornea or the surface of a lake? Assume normal incidence.

9 A greater percentage of incident light will be transmitted into the eye if (a) contact lenses are worn; (b) no contact lenses are worn. The surface of the eye (the cornea) has a refractive index of 1.38, and the contact lens has an index of 1.5.

10 A vertically directed laser beam passes through a glass of water. Will more light be transmitted if the light is directed: (a) downward, going from air to water to glass to air, or (b) upward, going from air to glass to water to air; or (c) will it be the same in either direction?

11 Is the critical angle of glass less in air or in water?

12 Certain optical fibers have an index of 1.5. The fibers are each clad with a material having a different index, so that the fibers are separated. Which would be the best value for the index of the cladding material: 1.2, 1.4, or 1.6?

Answers to Odd-Numbered Questions

1 (a) vertical; (b) north-south; **3** (a) no; (b) no; **5** 10^{-8} s; **7** D; **9** b; **11** air

Problems (listed by sections)

23-1 Electromagnetic Waves

1 What kind of electromagnetic waves have a frequency of (a) 10^8 Hz; (b) 10^{10} Hz; (c) 10^{13} Hz; (d) 10^{16} Hz; (e) 10^{18} Hz?

2 TV channel 2 broadcasts in the frequency range 54 to 60 MHz. What is the corresponding range of wavelengths?

3 FM radio station KRTH in Los Angeles broadcasts on an assigned frequency of 101 MHz with a power of 50,000 W.

(a) What is the wavelength of the radio waves produced by this station?

(b) Estimate the average intensity of the wave at a distance of 30.0 km from the radio transmitting antenna. Assume for the purpose of this estimate that the antenna radiates equally in all directions, so that the intensity is constant over a hemisphere centered on the antenna.

(c) Estimate the amplitude of the electric field at this distance.

4 The human eye can detect light with an intensity as low as about 10^{-9} W/m^2. A typical large galaxy, such as our own Milky Way, radiates energy at the rate of about 4×10^{36} W. What is the maximum distance from earth of a typical galaxy that can be seen with the unaided eye? (There are only three galaxies that are visible without instruments. One of these, the Andromeda galaxy, shown in Fig. 23-45, is about 2×10^6 light-years from earth—where 1 light-year, the distance traveled by light in 1 year, is 9.46×10^{15} m.)

Fig. 23-45 The Andromeda galaxy.

5 The intensity of sunlight at the earth's surface is 1300 W/m^2. What is the total power radiated by the sun? The distance from the earth to the sun is 1.49×10^{11} m.

6 A certain microwave oven produces microwaves of frequency 2450 MHz with a power of 500 W. How long should it take to heat 200 g of water, initially at room temperature, to the boiling point? Assume that all the microwave energy is absorbed by the water. (The microwave frequency is significant because the water molecule has a rotational resonant frequency at 2450 MHz. This means that water very effectively absorbs radiant energy at this frequency, thereby raising its rotational kinetic energy and its temperature.)

★★ **7** "Ghost" images are a common sight on a TV screen (Fig. 23-46). These weak duplicate images, shifted by as much as 1 cm from the primary image, can result from a weak secondary TV signal reaching the receiving antenna after being reflected from a building or hillside. Calculate the additional path length traveled by a signal producing a ghost image displaced 1.00 cm from the primary image on a 26-inch diagonal screen. The electron beam sweeps through the 525 lines on the screen 30 times per second.

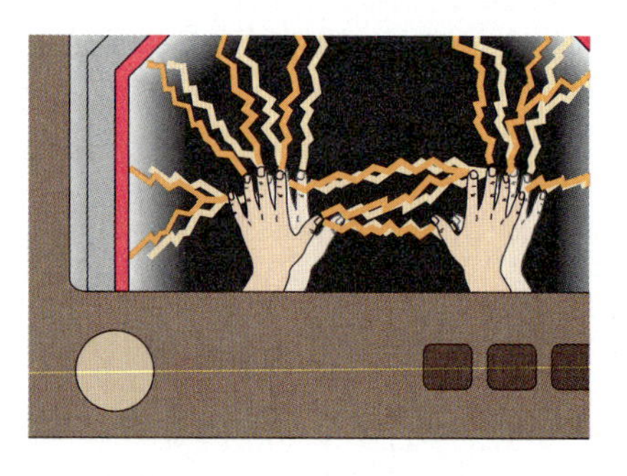

Fig. 23-46 Ghost images.

★ **8** The reception on a portable TV with an attached antenna can sometimes be improved by simply moving the TV. If the improvement is due to going from a node to an antinode in a standing electromagnetic wave, what is the maximum distance you should have to move a TV for improved reception of a VHF station (channels 2 to 13, f = 54 to 216 MHz)?

23-3 The Propagation of Light

9 Long-distance telephone communication utilizes transmission of electromagnetic signals in the microwave part of the spectrum. How long does it take to transmit a signal cross country, a distance of 5000 km?

10 A light-year is the distance light travels in 1 year. How many meters are there in a light-year?

11 The star Sirius is a distance of 8 light-years from the earth. If Sirius suddenly stopped shining, how much longer would it be seen on earth?

12 (a) What is the speed of light in glass with a refractive index of 1.50?
(b) How long does it take for light to travel through 1.00 cm of glass?

13 Find the index of refraction in a medium in which the speed of light is 1.50×10^8 m/s.

14 Find the frequency of light that has a vacuum wavelength of 420 nm.

15 Find the vacuum wavelength of light of frequency 4.00×10^{14} Hz.

16 Find the speed of light in a medium where light of frequency 5.00×10^{14} Hz has a wavelength of 300 nm.

17 Light of frequency 6.00×10^{14} Hz passes through Lucite plastic.
(a) Find the vacuum wavelength of the light.
(b) Find the wavelength inside the Lucite.

18 What color is light that has a wavelength of 700 nm inside glass of index 1.6?

19 Orange light of frequency 5.00×10^{14} Hz is incident normally on a glass plate of thickness 2.00 cm and index of refraction 1.50. Find the number of waves in the plate at any instant.

20 Find the wavelength in ice of yellow-green monochromatic light.

21 Find the intensity of light in a flashlight beam of diameter 10.0 cm with 2.00 W of power.

★ **22** Sunlight enters a skylight with dimensions 2.00 m by 3.00 m. The intensity of sunlight is 1300 W/m². Find the total solar energy incident on the skylight during a 1-hour interval if (a) the sun's rays are perpendicular to the surface of the skylight; (b) the sun's rays are at an angle of 30.0° with the normal to the surface.

★ **23** The sun's rays are focused, using a lens as a "burning glass." All the visible sunlight entering a lens of radius 3.00 cm is concentrated in a circular spot of radius 1.00 mm. Find the intensity of visible light in the spot if the intensity of visible light striking the lens is 700 W/m².

23-4 Reflection and Refraction

24 A He-Ne laser beam with an intensity of 1.00×10^3 W/m² is incident from air to water, perpendicular to the surface of the water. Find the intensity of the beam transmitted into the water.

★ **25** Light of intensity 100 W/m² is incident on a glass windowpane (n = 1.5), with rays approximately perpendicular to the surface, as shown in Fig. 23-47. Find the intensities I_1, I_2, and I_3.

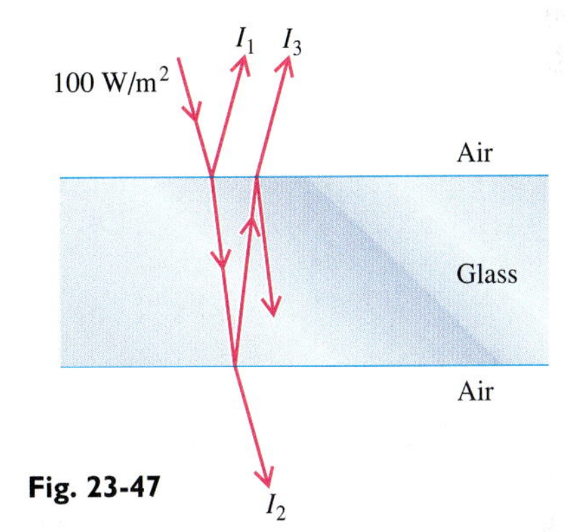

Fig. 23-47

★ **26** Light reflected from the front surface of a glass windowpane (n = 1.5) has an intensity I_1. Light entering the glass undergoes multiple reflections. Let I_2 be the intensity of light reflected after only one reflection at the back surface of the glass. Find the ratio I_2/I_1. Assume normal incidence.

27 A narrow beam of light is incident on the center of a human eye, perpendicular to the surface of the cornea. What percentage of the incident light is transmitted into the cornea?

28 A light ray is incident on the outer surface of the cornea at an angle of 40.0° with the normal. Find the angle the transmitted ray makes with the normal.

★ **29** Light passes from air into another medium at point A (Fig. 23-48). How long does it take the light ray to travel from A to B?

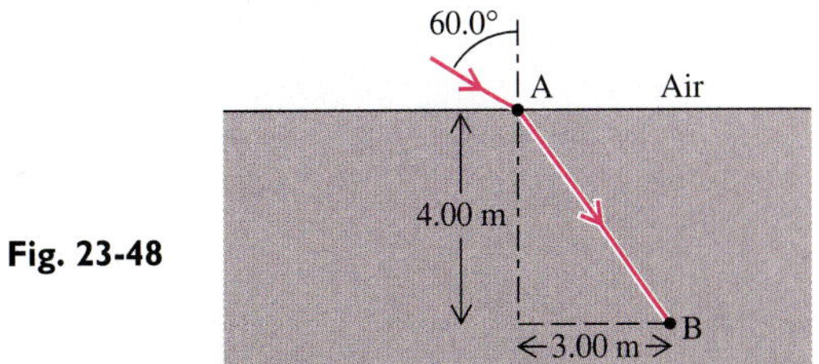

Fig. 23-48

30 A light ray passes from water into ice. If the transmitted ray makes an angle of 20.0° with the normal, what is the angle of incidence?

31 Light travels from air through a flat sheet of glass ($n = 1.5$) into water. If the angle of incidence in air is 30°, what is the angle of the refracted ray in water?

32 Find the refractive index of the glass in Fig. 23-49.

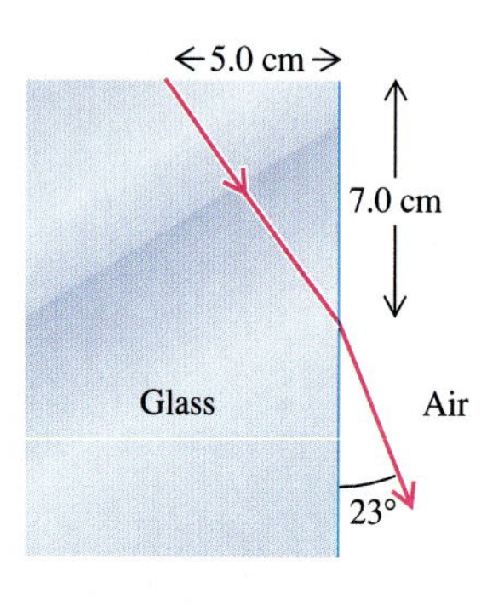

Fig. 23-49

33 A light ray passes through a glass lens, shown in cross section in Fig. 23-50. Find the angle of incidence ϕ.

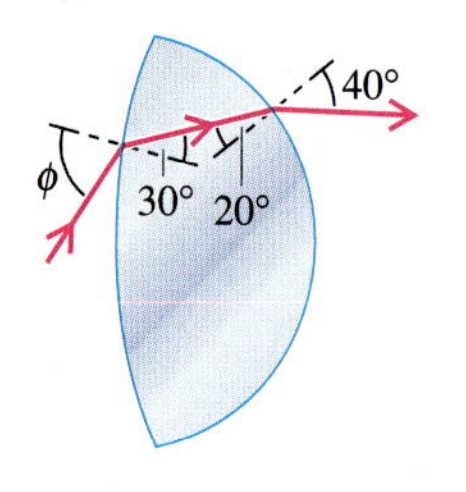

Fig. 23-50

34 A light ray travels from air to water and then is reflected by a mirror underwater (Fig. 23-51).

(a) Find the angle of incidence as the ray enters the water.

(b) Find the angle of refraction as the ray leaves the water.

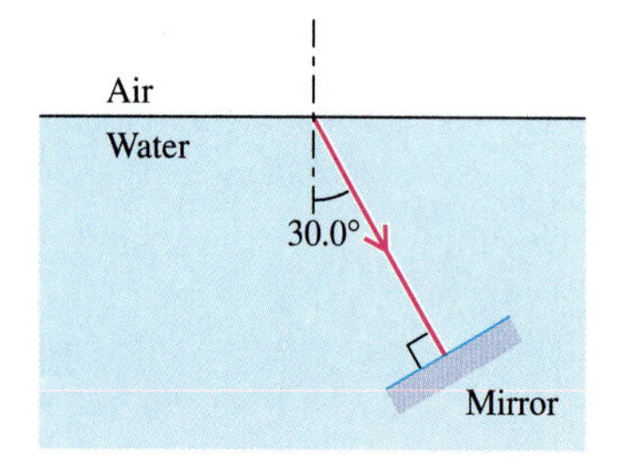

Fig. 23-51

★ **35** A light ray is incident on a glass plate at an angle of 45° (Fig. 23-52).

(a) Find angles α, β, and γ.

(b) Will any of the light striking the side of the glass plate at point P pass out through the side?

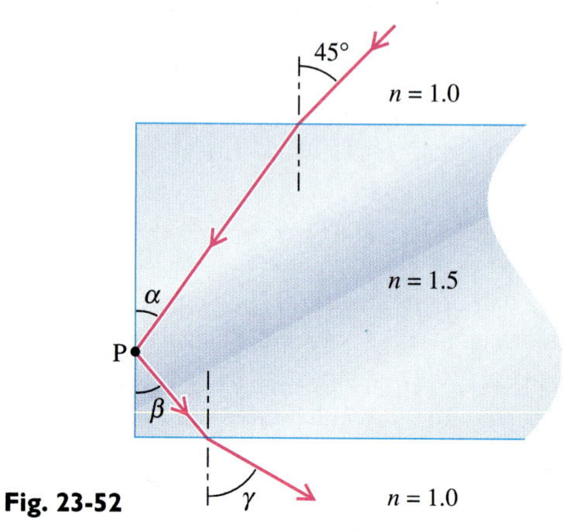

Fig. 23-52

★ **36** Blue light of frequency 6.30×10^{14} Hz is incident at an angle of 48.6° on an equilateral glass prism, which has an index of 1.50 for light of this frequency. An incident ray strikes the center of one side of the prism.

(a) Find the angle at which the ray emerges from the prism.

(b) Find the wavelength of the light in the glass.

(c) How many wavelengths of light are contained in the glass at any instant if the length of a side is 2.50 cm?

37 A light ray is incident along a radius of a semicircular glass plate (Fig. 23-53). For what range of values of θ will no light emerge from the bottom of the plate?

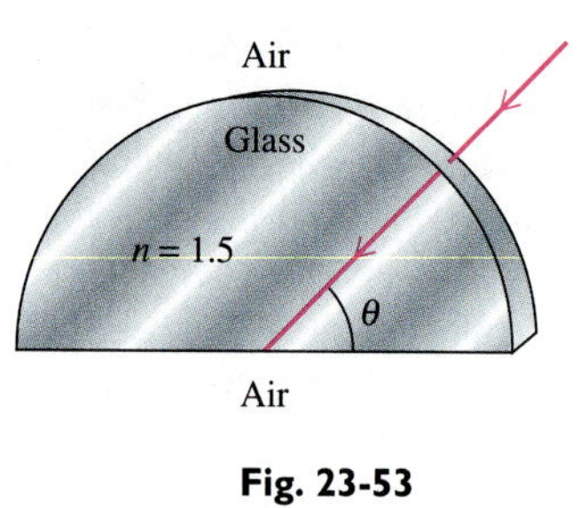

Fig. 23-53

38 Light is incident normally on a prism (Fig. 23-54). Find the path of the light ray through the prism, which has a refractive index of 1.5.

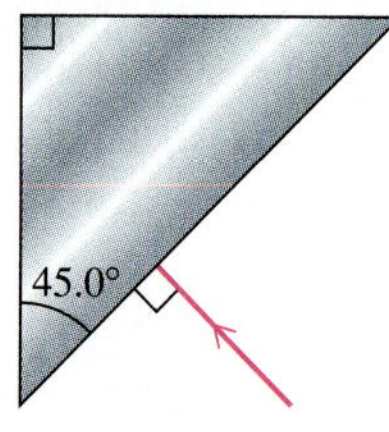

Fig. 23-54

★ **39** A light ray is incident normally on a prism (Fig. 23-55). Find the angle (relative to the normal) of the light ray transmitted back into air at the point where light first emerges from the prism.

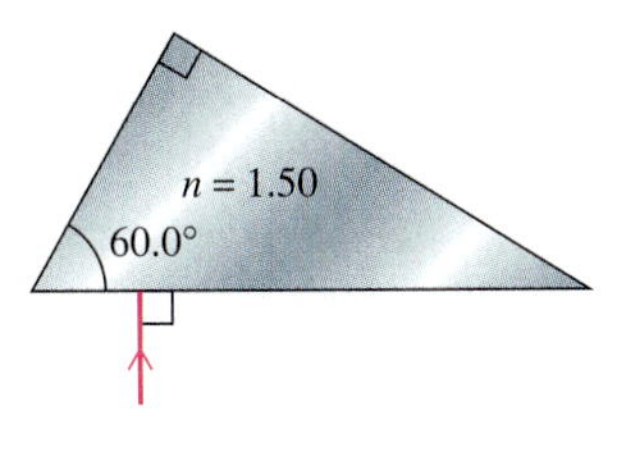

Fig. 23-55

★ **40** When a submerged swimmer looks up at the smooth surface of the water above, she sees a mirrorlike surface with a circular hole, through which she sees objects outside the water (Fig. 23-56). Explain. Find the radius of the hole if the swimmer's eye is 1.00 m below the surface.

Fig. 23-56

Additional Problems

★ **41** A solar eclipse occurs when the moon is between the earth and the sun. The earth is then in the shadow of the moon. From the moon, the moon's shadow could be seen to cover a small part of the earth's surface. The shadow just begins to cover a given point on the earth (that is, the complete eclipse begins) at a certain instant. Is this seen first from the moon or from the earth, or would it be seen at the same time in either place?

Fig. 23-57

★ **42** This problem shows why the shape of the sun appears to be somewhat flattened at sunset (Fig. 23-57). The sun subtends an angle of approximately 0.5° at the earth. Suppose that a ray from the bottom of the sun's disk is incident on the earth's atmosphere at an angle of 89.0°, and a ray from the top of the sun's disk is incident on the atmosphere at an angle of 88.5°. Find the angle between the two rays as they strike the earth's surface. Treat the atmosphere as a flat layer with a uniform refractive index of 1.0003.

★ **43** When light is incident from air perpendicular to the surface of a certain medium, 88.9% of the light is transmitted into the medium. Find the speed of light in the medium.

★★ **44** Find the total intensity of light transmitted by a glass windowpane when light of intensity 1.00×10^3 W/m^2 is incident on the glass, normal to the surface. Take multiple reflections into account. Take the index of the glass to be 1.50.

★ **45** Two mirrors are placed 1.00 m apart, with their surfaces parallel. Suppose that light of intensity I_0 is initially incident on one mirror, normal to its surface. Each mirror reflects 98.0% of the incident light at each reflection. After how many reflections is the intensity of reflected light reduced to $(1.00 \times 10^{-6})I_0$? How long does this take?

★ **46** Can a light ray enter a circular glass disk at such an angle that it becomes trapped inside because of total internal reflection?

★ **47** The light ray in Fig. 23-58 strikes the hypotenuse of the prism at an angle of incidence greater than the critical angle. Find the angle ϕ of the ray as it leaves the base of the prism. The incident ray is perpendicular to the left side of the prism.

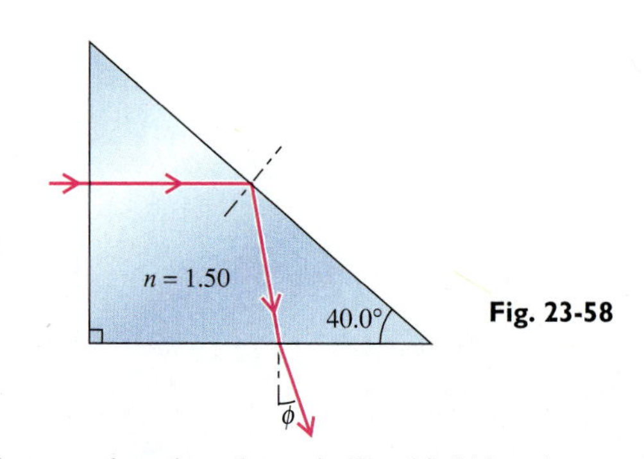

Fig. 23-58

48 Suppose the prism shown in Fig. 23-54 is submerged in water, and the incident light ray shown in the figure has its direction reversed by two total internal reflections inside the prism. Find the minimum refractive index of the prism.

49 Most mirrors are made by silvering the back surface of a plate of glass so that light entering the glass is almost perfectly reflected from the back surface. Find the index of the glass in Fig. 23-59.

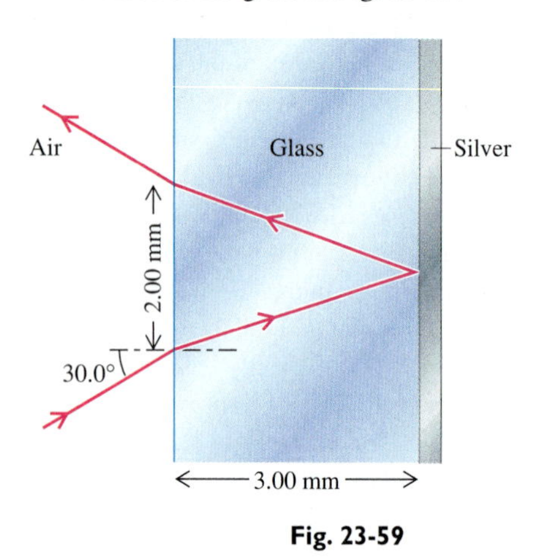

Fig. 23-59

★ **50** An observer, sitting some distance from the edge of a swimming pool 2.00 m deep, sees an object at the bottom of the pool. The observer's line of sight is directed a few degrees below horizontal at a point on the water's surface just beyond the edge of the pool. How far is the object from the side of the pool?

★★ **51** A fisherman sees rays of light apparently diverging from a point P′ on a fish (Fig. 23-60). The fish appears to be 1.0 m below the surface. What is its actual depth?

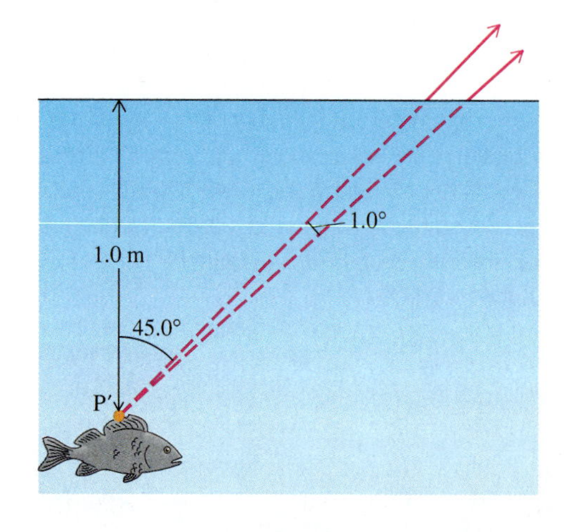

Fig. 23-60

★ **52** A thin tree rises 10.0 m vertically from the edge of a lake. The sun casts a shadow of the tree on the bottom of the lake. The sun's rays make an angle of 45.0° with the vertical. A fisherman in a boat 11.0 m from the base of the tree sees the tip of the shadow directly beneath him. How deep is the lake at the point directly beneath the fisherman?

★★**53** Beer mugs are often made with thick sides (near the bottom), so that they appear to hold more than they really do (Fig. 23-61a). As seen in the photo, the beer appears to extend all the way to the *outer* surface of the mug. Fig. 23-61b shows a view from above of the path of a ray from a point P in the beer at the inner surface of the mug. The figure is drawn so that the ratio of the wall thickness t to the inner radius r is as large as possible. In this case the light from point P appears to be coming from P′ at the edge. (For thicker walls, light from point P would appear to be coming from inside the edge. For thinner walls, light from point P could not be seen from this direction, and light apparently coming from P′ would be originating at another point in the beer.)

(a) Find t/r, assuming beer has a refractive index of 1.33. (HINT: First calculate angles ϕ_c, ϕ_c', and θ.)

(b) Find the ratio of the apparent volume of beer to the actual volume of beer in the mug.

★**54** Radiation is incident perpendicular to a surface area A. The radiation that will pass through this surface during time Δt is initially in a rectangular volume of space, extending back a distance $c\,\Delta t$ from the surface. Show that the rate at which electromagnetic energy flows through the surface can be expressed in terms of the electromagnetic energy per unit volume as $u_{av}cA$ and that the average intensity of radiation is given by $I_{av} = u_{av}c$.

★**55** An emf is induced in a VHF TV antenna, a circular loop of radius 10.0 cm, by a 60.0 MHz TV signal of intensity 1.00×10^{-4} W/m^2 coming from a westerly direction.

(a) To maximize the induced emf, what should be the direction of the antenna's axis? Assume that the electric field is oscillating in a vertical plane.

(b) For this orientation find the average emf induced over a quarter cycle, during which the magnetic field increases from zero to its maximum value.

(a)

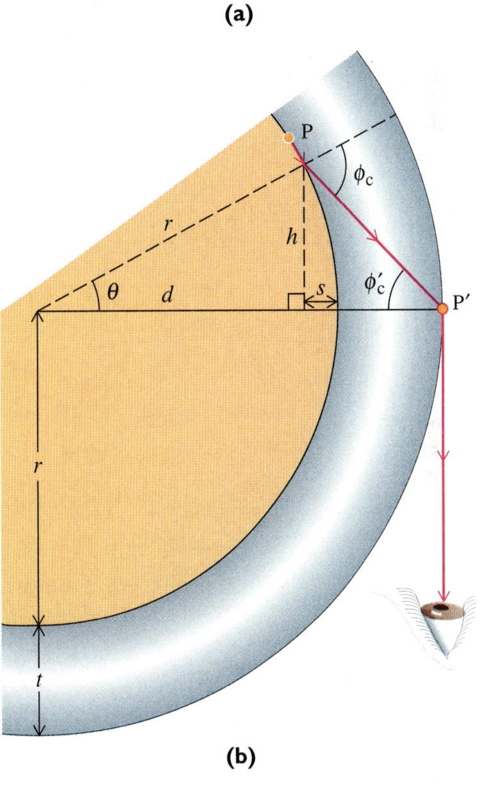

(b)

Fig. 23-61

CHAPTER 24 Geometrical Optics

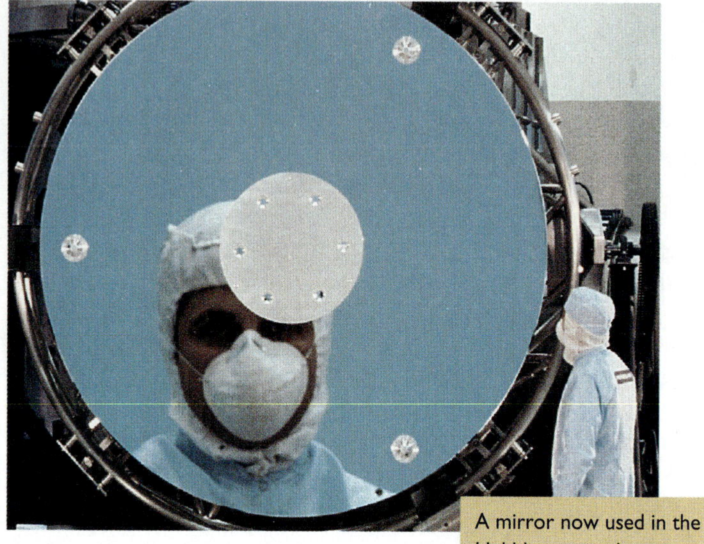

A mirror now used in the Hubble space telescope

Have you ever entered an unfamiliar room in which one wall was covered with a mirror and thought at first that the room extended beyond the mirrored wall—that the image you saw in the mirror was real? Or have you seen, at an amusement park or novelty store, an object appear to be floating in air, when in fact there is really nothing there (Fig. 24-2)? You may know that "it's all done with mirrors." But have you wondered just how mirrors can be used to create such convincing illusions? In this chapter we will answer such questions, developing concepts and equations to describe the behavior of mirrors as an application of the law of reflection, presented in the last chapter. We shall see, for example, that a plane mirror can create a perfect optical image. There is no way not to be fooled by a plane mirror covering a wall if the mirror is very clean and very flat.

In this chapter we shall also apply the law of refraction to gain an understanding of lenses and their uses. For example, we shall see why focusing a camera requires changing the distance between the camera lens and the film.

In the following chapter we will apply our knowledge of mirrors and lenses, developed in this chapter, to explain optical systems: magnifiers, microscopes, telescopes, eyeglasses, and even the human eye itself.

Fig. 24-1 Which is the mirror image? How do you know?

Fig. 24-2 Illusions created with mirrors. Part of what appears to be real in each of these pictures is actually just an image formed by light. Can you tell which parts? None of the coins you see are really where they appear to be, and where you see a light bulb, no real light bulb is present. The meter stick passes through the mirror image of the bulb. For explanations, see Example 4 and Problem 22.

24-1 Plane Mirrors

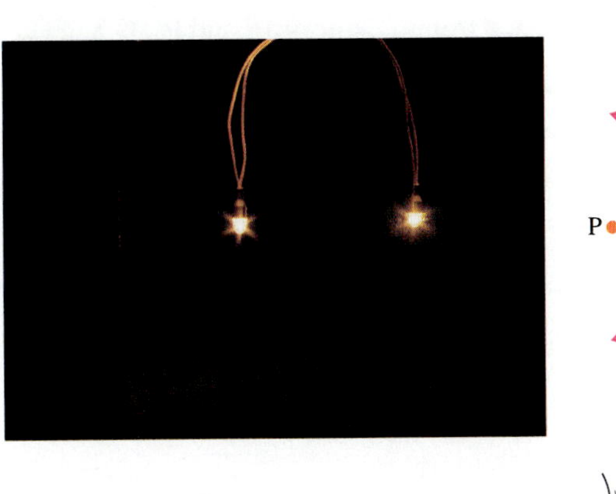

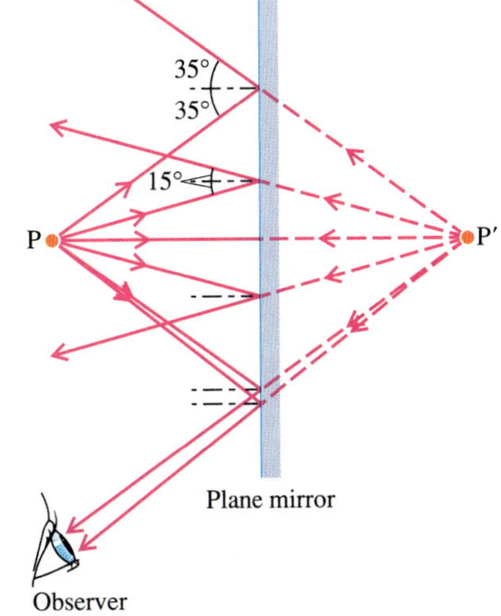

Fig. 24-3 A point source of light P and its image P′, in a plane mirror. Each light ray from P that strikes the mirror satisfies the law of reflection (angle of incidence = angle of reflection). Extending the reflected rays back behind the mirror *(dashed lines)*, we see that the reflected rays are diverging from P′, the image point.

Fig. 24-3 indicates how a plane mirror forms an image of a point source of light P. (The point source could be a tiny light bulb, for example.) Rays of light from P are reflected from the mirror, each ray satisfying the law of reflection. If the reflected rays are extended back behind the mirror, these extensions all intersect at a point P′, as indicated in the figure. The reflected rays diverge from this point. The reflected light appears to an observer to be coming from P′ because the eye interprets light rays as having traveled in a straight line. In other words, the eye sees an image at P′. A plane mirror is a perfect optical instrument if its surface is a perfect plane. All rays coming from P are imaged precisely at P′.

The drawing of reflected rays in Fig. 24-3 indicates **an image point directly behind the mirror** and **at the same distance from the mirror as the source point.**

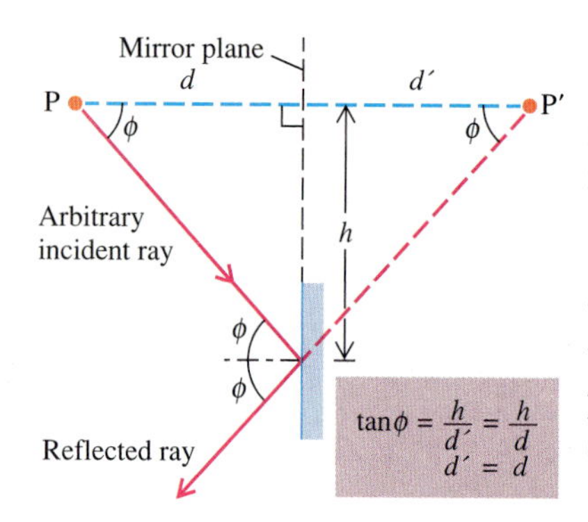

A geometric proof of this result is given in Fig. 24-4. Notice that P need not be directly in front of part of the mirror. A ray incident on the mirror at *any* angle will be reflected along a line from P′, where P′ is a point directly behind the plane of the mirror, the same distance d from the plane as the point P. The image at point P′ is called a "virtual" image point. In general, images are referred to as either real or virtual. A **real image point** is defined as **a point at which converging light rays intersect.** A **virtual image point** is defined as **a point from which rays of light appear to be diverging,** though the rays do not actually intersect at the image point. Although a plane mirror produces a virtual image, we shall see that curved mirrors and lenses sometimes produce real images.

Fig. 24-4 The blue dashed line through object point P is perpendicular to the plane of a mirror. The extension of any reflected light ray behind the mirror intersects this line at a pont P′. Because the angle of reflection equals the angle of incidence, P′ is the same distance d from the mirror plane as the object point. This conclusion is valid for any light ray at any angle of incidence. Thus *all* reflected rays appear to be coming from P′.

For an observer to see an image, the mirror must cross the line between the observer's eye and the image point (Fig. 24-5).

When a mirror forms an image of an extended object, each point on the object serves as a point source of light, from which light rays are emitted. The image formed by the mirror consists of image points corresponding to the object points (Fig. 24-6). Image and object points are an equal distance from the mirror (P and P′, Q and Q′, and so forth). Reflected light appears to be coming from the image points. Notice that the image has the same size as the object.

P
P′
O
O′

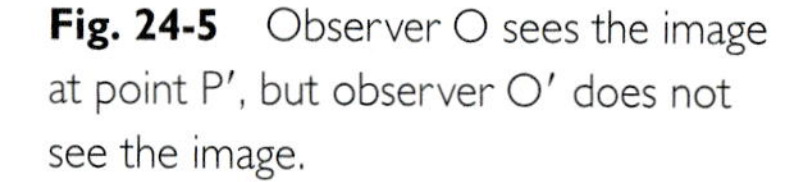

Fig. 24-5 Observer O sees the image at point P′, but observer O′ does not see the image.

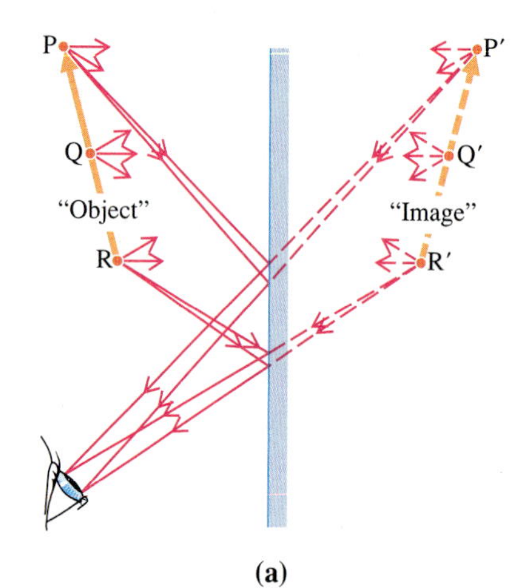

(a)

(b)

Fig. 24-6 The image of an extended object in a plane mirror.

EXAMPLE 1 Multiple Images In Perpendicular Mirrors

Locate all the images of a point source P located near two perpendicular plane mirrors M_1 and M_2 (Fig. 24-7).

SOLUTION Each mirror will produce an image of the object: image points P_1' in M_1 and P_2' in M_2 (Fig. 24-8a). These images correspond to reflections from either M_1 or M_2. It is also possible to have rays reflected by both mirrors. As indicated in Fig. 24-8b, if light strikes M_2 after being reflected from M_1, the ray incident on M_2 appears to be coming from P_1'. (Remember all light reflected from M_1 appears to be coming from P_1'.) Thus the point P_1' serves as an object for the second mirror.

Applying our rule for locating a mirror image, we find that the image of P_1' in mirror M_2 is formed at P_3' (Fig. 24-8c). To an observer, the ray shown in this figure appears to be coming from P_3'. In following this simple method of locating images, the rays we have constructed automatically satisfy the law of reflection at each mirror.

In this example we have found the image produced by one optical element (mirror M_1) and then used that as an object for a second optical element (mirror M_2) to find an image (P_3'). We shall use this general procedure in solving many optical problems involving combinations of mirrors or lenses.

Fig. 24-7

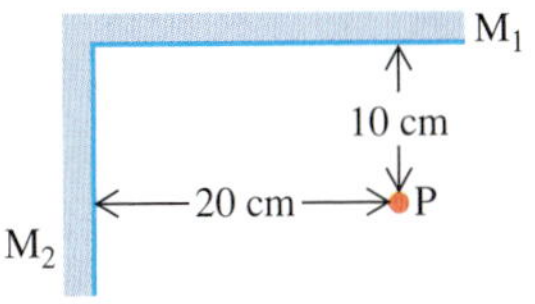

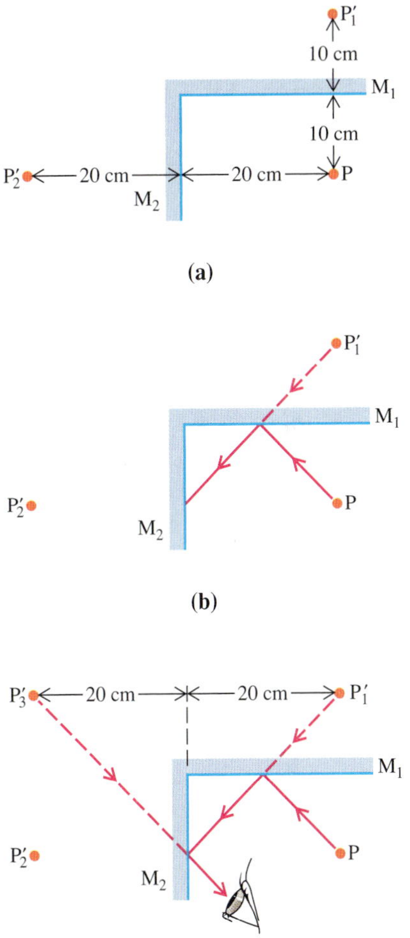

Fig. 24-8 (a) (b) (c)

Fig. 24-9 Three virtual images of a candle flame are seen in two perpendicular mirrors.

24-2 Spherical Mirrors

Many optical devices use curved mirrors, either concave or convex (Fig. 24-10). A concave mirror is used, for example, as part of a reflecting telescope. Convex mirrors are often seen in stores, where they are used to give a wide-angle view. The reflecting surface of a concave or convex spherical mirror is part of a spherical surface.

Fig. 24-10 **(a)** A concave mirror. **(b)** A convex mirror.

Concave Spherical Mirrors

First we shall describe qualitatively how a concave spherical mirror forms images and then obtain a formula for locating images. Fig. 24-11 shows image formation by a concave mirror in a series of ray diagrams. A point object is infinitely far away in the first drawing and is brought closer in each successive drawing.

(a) Parallel rays from a distant point source are reflected by a concave mirror and converge to form a real image. The distant object point is on the **optical axis,** a line perpendicular to the center of the reflecting surface. The image point F is called the **focal point.** The distance f from the mirror to the focal point is called the **focal length.** The **center of curvature** C is the center of the spherical reflecting surface.

(b) A point object at P a distance s from the mirror produces a real image at P′ a distance s′ from the mirror. In this drawing the incident and reflected rays at the top of the mirror make a smaller angle ϕ than in the preceding drawing. The object is closer to the mirror and the image is farther away than in (a).

(c) The object P is at the same point occupied by the image in (b). Therefore the incident rays are just the reverse of the reflected rays in (b). The law of reflection then predicts that the reflected rays here will be the reverse of the former incident rays. The paths of the rays in both figures are the same, but the directions of the rays are opposite. This means that object and image points are interchanged. This result is an example of a general principle, the **principle of optical reversibility: One may always interchange an object and an image, reversing the directions of all rays.** This principle applies to both reflection and refraction and follows from the fact that the laws of reflection and refraction do not involve the direction of light rays.

(d) The object is placed at the focal point F. Applying the principle of optical reversibility to (a), where F was an image point, we find that the image here is at infinity, the location of the object point in (a).

(e) When the object is placed inside the focal point, the incident and reflected rays form larger angles, and the reflected rays no longer intersect. Instead the reflected rays diverge from a point P′ behind the mirror, forming a virtual image as in the case of a plane mirror.

Fig. 24-11 Image formation by a concave mirror. Reflected rays obey the law of reflection (angle of incidence equals angle of reflection). Depending on the location of the object, the mirror forms either a real image (where light rays actually intersect) or a virtual image (the same kind of image formed by a plane mirror). Notice that in all cases a concave spherical mirror tends to converge rays; that is, either the reflected rays converge to form a real image, or, as in Fig. 24-11e, the reflected rays diverge from the optical axis less than the incident rays—that is, reflected rays make a smaller angle with the optical axis than the incident rays do. We sometimes refer to a concave mirror as a **converging mirror.**

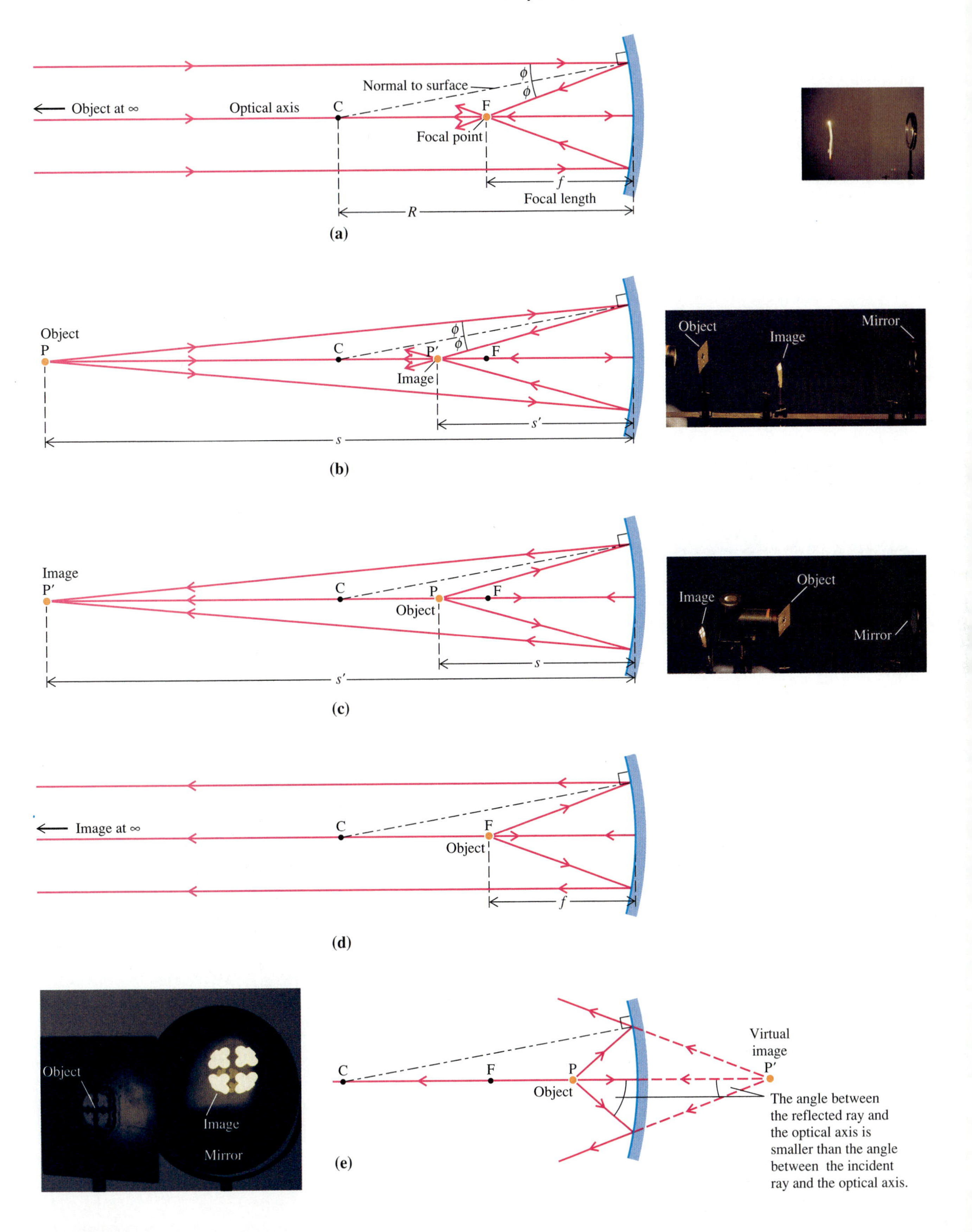
Normal to surface
Object at ∞
Optical axis
C
F
Focal point
f
Focal length
R
(a)
Object
P
C
P′
Image
F
s′
s
Mirror
(b)
Image
P′
C
P
Object
F
s
s′
Mirror
(c)
Image at ∞
C
F
Object
f
(d)
Object
Image
Mirror
C
F
P
Object
Virtual
image
P′
The angle between
the reflected ray and
the optical axis is
smaller than the angle
between the incident
ray and the optical axis.
(e)

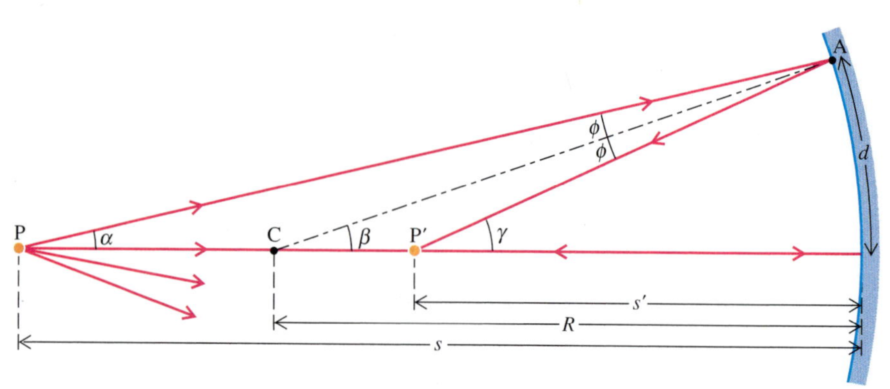

Fig. 24-12 Light rays diverging from object point P are reflected from a concave mirror and converge at image point P′. The image distance s′ depends on the object distance s and on the mirror's radius of curvature.

Next we shall derive an equation relating object and image distances. Fig. 24-12 shows a light ray originating at object point P, reflecting from a concave mirror, and crossing the optical axis at P′, the image point. We shall see that all rays originating at P cross the axis at the same point P′ if the angles α, β, and γ are all fairly small, say, less than 10°.

We shall apply the geometric theorem that states that an exterior angle of a triangle equals the sum of the two opposite interior angles (Fig. 24-13). Applying this first to triangle PAC and then to triangle PAP′, we find

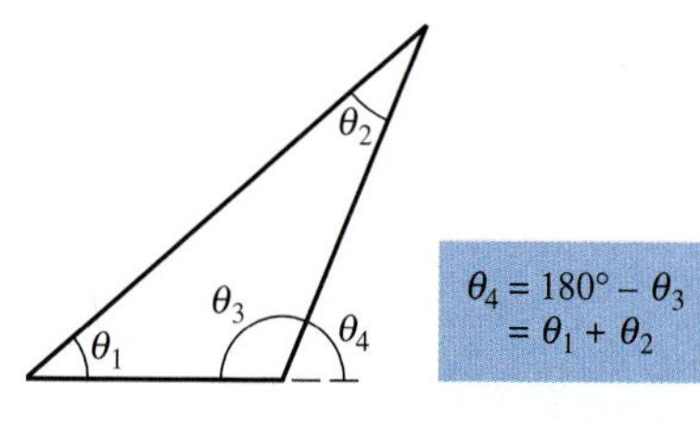

Fig. 24-13 Exterior angle θ_4 equals the sum of the two opposite interior angles, θ_1 and θ_2.

$$\beta = \alpha + \phi$$

$$\gamma = \alpha + 2\phi$$

Eliminating ϕ between these two equations gives

$$\alpha + \gamma = 2\beta \tag{24-1}$$

From inspection of Fig. 24-12, we can relate angles α, β, and γ to the respective linear distances s, R, and s', using the radian definition of an angle (the arc length of a circular segment divided by its radius). We assume that the angles α, β, and γ are all small angles (less than about 10°), and so the circular arcs corresponding to these angles all have approximately the same length d. Thus

$$\alpha \approx \frac{d}{s} \qquad \beta = \frac{d}{R} \qquad \gamma \approx \frac{d}{s'}$$

Substitution into Eq. 24-1 yields

$$\frac{d}{s} + \frac{d}{s'} = \frac{2d}{R}$$

or

$$\frac{1}{s} + \frac{1}{s'} = \frac{2}{R} \tag{24-2}$$

This result is independent of the particular angle α of the incident ray; that is, it applies to all incident rays that make small angles with the optical axis. All such rays, called **paraxial rays,** converge to the image point P′ located a distance s' from the vertex.

For a distant object point P, we can set s equal to ∞. The image then is at the focal point ($s' = f$), and the equation above gives

$$\frac{1}{\infty} + \frac{1}{f} = \frac{2}{R}$$

or

$$f = \frac{R}{2} \qquad \text{(24-3)}$$

Substituting this expression for f into Eq. 24-2 gives:

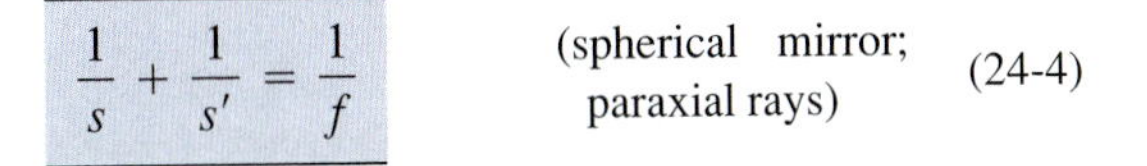

$$\frac{1}{s} + \frac{1}{s'} = \frac{1}{f} \qquad \text{(spherical mirror; paraxial rays)} \qquad \text{(24-4)}$$

This mirror equation may be used to locate an image point, given the position of the object point.

If we interchange s and s' in the mirror equation, the equation is unchanged. This implies the principle of optical reversibility, since interchanging s and s' in this equation corresponds to physically interchanging object and image. For example, we have seen that $s = \infty$, $s' = f$ satisfies the mirror equation. If we interchange these values, we obtain $s = f$, $s' = \infty$, again satisfying the mirror equation. The two meanings of the focal point, indicated in Fig. 24-11a and 24-11d, correspond to these two choices of s and s'.

EXAMPLE 2 Image of a Distant Point in a Concave Mirror

A concave spherical mirror has a radius of curvature of 20.0 cm. A small source of light is directly in front of the mirror, 10 m away. Find the position of the real image produced by the mirror.

SOLUTION We first apply the equation for the focal length (Eq. 24-3).

$$f = \frac{R}{2} = 10.0 \text{ cm} = 0.100 \text{ m}$$

Then we apply the mirror equation, setting the object distance $s = 10$ m, and solve for the image distance s'.

$$\frac{1}{s} + \frac{1}{s'} = \frac{1}{f}$$

$$\frac{1}{s'} = \frac{1}{f} - \frac{1}{s} = \frac{1}{0.100 \text{ m}} - \frac{1}{10 \text{ m}} = 9.90 \text{ m}^{-1}$$

$$s' = 0.101 \text{ m} = 10.1 \text{ cm}$$

The image is located on the optical axis, 10.1 cm in front of the mirror, very close to the focal point. Whenever the object distance is much greater than the focal length, the image will be very close to the focal point.

If either incident or reflected rays make relatively large angles with the optical axis, these rays will not all converge at exactly the same image point. This means that the image produced will be somewhat fuzzy or blurred. This is illustrated in Fig. 24-14a for a distant point source on the optical axis. This effect, known as **spherical aberration**, is not present in parabolic mirrors (Fig. 24-14b), which are commonly used in astronomical telescopes.

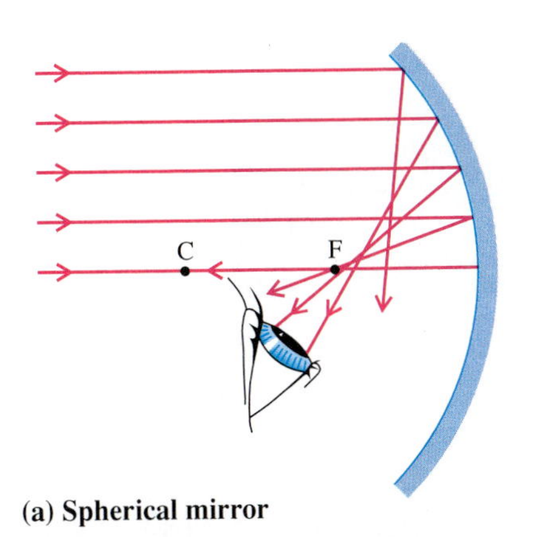

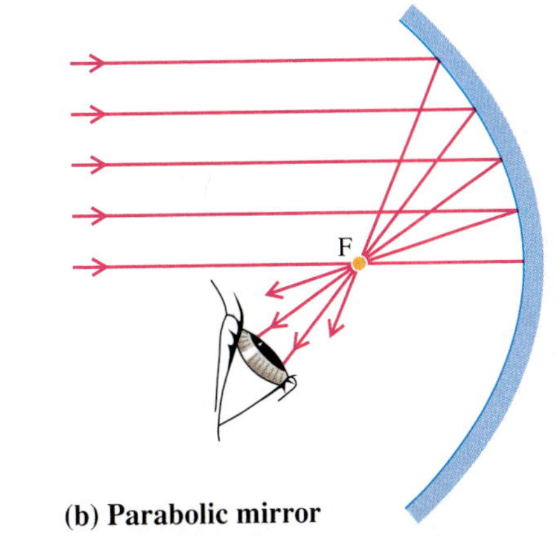

Fig. 24-14 **(a)** Rays from a distant point source on the optical axis are incident on a spherical mirror. Only paraxial rays converge at the focal point. **(b)** For a parabolic mirror all rays from a distant point source on the optical axis converge at the focal point.

Virtual Images

The mirror equation may also be applied when a virtual image is produced, but in this case we must *treat the image distance s' as a negative quantity,* as shown in Fig. 24-15. Here we introduce the first of several sign conventions that are essential to successful problem solving in geometrical optics:

Image distance			
	$s' > 0$	Reflected rays converge toward a **real image in front of the mirror.**	(24-5)
	$s' < 0$	Reflected rays diverge from a **virtual image behind the mirror.**	(24-6)

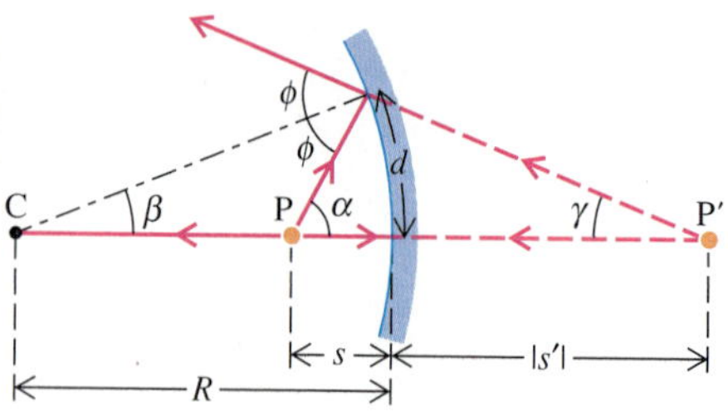

$$\alpha = \beta + \phi$$
$$2\phi = \alpha + \gamma$$

Eliminating ϕ, we obtain

$$\alpha - \gamma = 2\beta$$

or, for small angles,

$$\frac{d}{s} - \frac{d}{|s'|} = \frac{2d}{R}$$
$$\frac{1}{s} - \frac{1}{|s'|} = \frac{2}{R} = \frac{1}{f}$$

Taking s' to be a negative quantity, $|s'| = -s$, giving

$$\frac{1}{s} + \frac{1}{s'} = \frac{1}{f}$$

Fig. 24-15 Derivation of an equation to locate the virtual image formed by a concave mirror when an object is inside the mirror's focal point. If we take the image distance s' to be negative, we obtain the same equation used for a real image.

With this sign convention for s' and others for s and R, we shall be able to apply the mirror equation to any problem involving spherical mirrors.

Virtual Objects

When incident light rays converge toward a point P behind a mirror, as in Fig. 24-16, we refer to the point of convergence as a virtual object point. **Virtual objects** occur when a mirror interrupts converging light rays before a real image can be formed. These converging rays must be produced by some other converging mirror or lens (Fig. 24-17).

Once again the mirror equation is applicable (Problem 12 outlines a proof), but here we must treat the object distance s as a negative quantity. The sign convention for the object distance therefore is:

Object distance			
	$s > 0$	Incident rays diverge from a **real object in front of the mirror.**	(24-7)
	$s < 0$	Incident rays converge toward a **virtual object behind the mirror.**	(24-8)

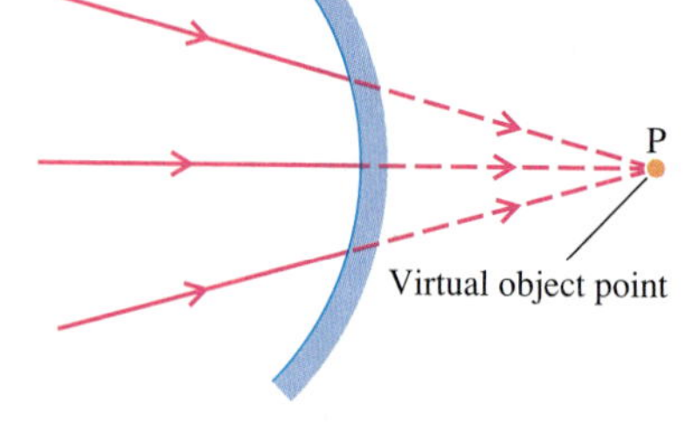

Fig. 24-16 A virtual object point P. Light rays converging toward P are interrupted by a mirror before reaching that point.

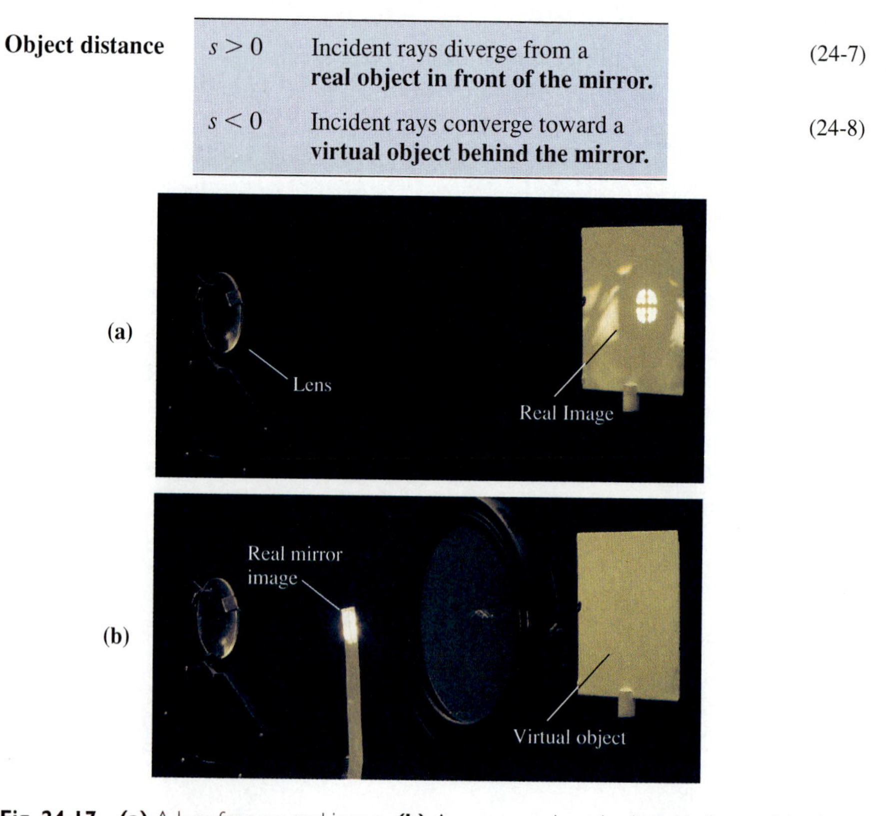

Fig. 24-17 **(a)** A lens forms a real image. **(b)** A concave mirror is placed in front of the image, which then serves as a "virtual" object for the mirror. The mirror image of this object is formed in front of the mirror.

Convex Spherical Mirrors

Both the mirror equation and the focal length equation ($f = R/2$), which we have derived for concave mirrors, turn out to be applicable to convex mirrors as well (Problem 13 outlines a proof) if we *interpret the radius of curvature R as a signed quantity* and use the following sign convention:

Radius of curvature

$R > 0$	**concave mirror**	(24-9)
$R < 0$	**convex mirror**	(24-10)

The focal length then is positive for concave mirrors and negative for convex mirrors.

$$f = \frac{R}{2} > 0 \quad \text{for a concave mirror} \qquad (24\text{-}11)$$

$$f = \frac{R}{2} < 0 \quad \text{for a convex mirror} \qquad (24\text{-}12)$$

Fig. 24-18 shows how images are formed by a convex mirror. Notice that in all cases the convex mirror tends to diverge rays; that is, the reflected rays either diverge more than the incident rays (Fig. 24-18a and 24-18b), or they converge less (Fig. 24-18c). A convex mirror is a **diverging mirror.**

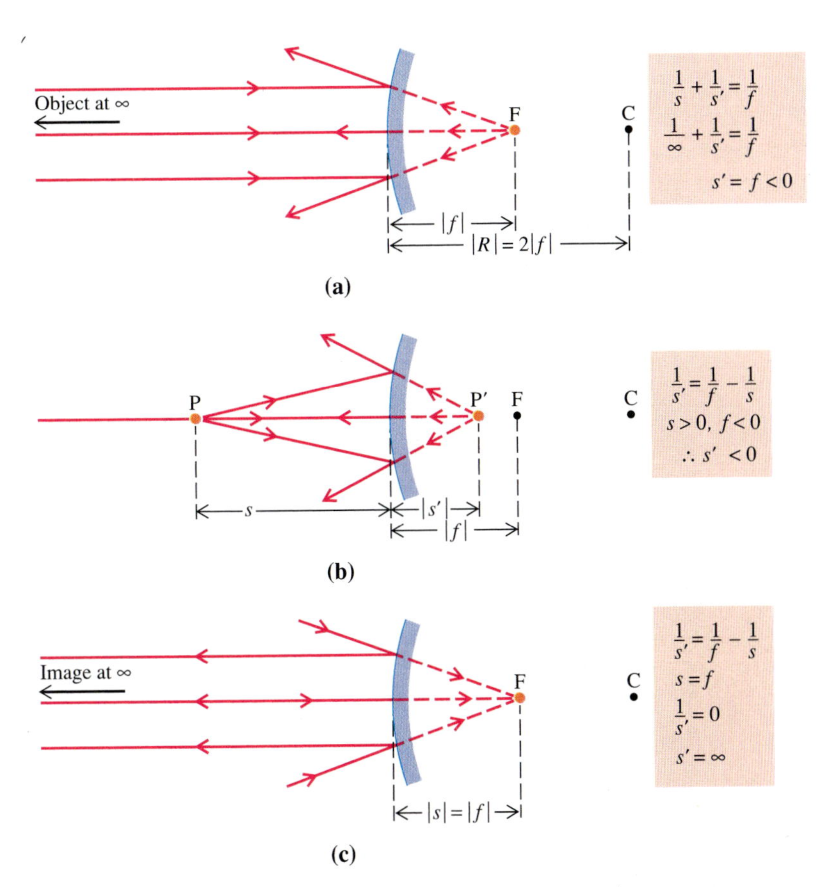

Fig. 24-18 Image formation by a convex mirror.

Linear Magnification

A spherical mirror can be used to "magnify" an object, that is, to produce an image larger than the object (Fig. 24-19). We shall obtain an equation that will allow us to determine the size of a mirror image from its location.

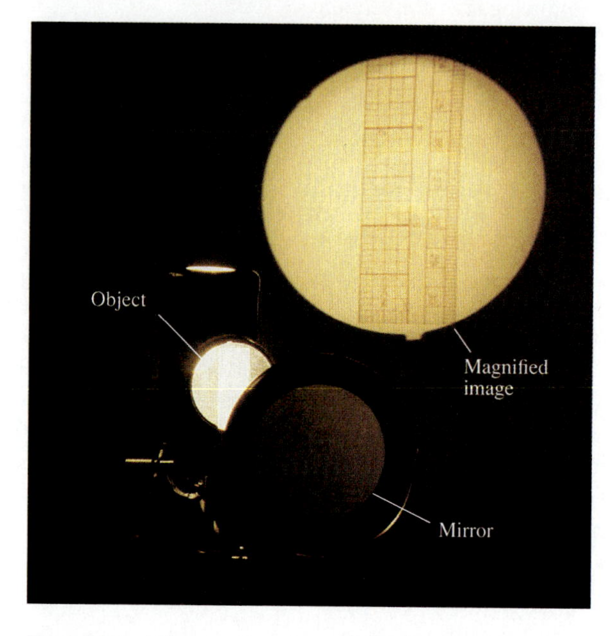

Fig. 24-19 When a mirror image is farther from the mirror than the object, it is also larger than the object.

Each point on an extended object has a corresponding image point. Fig. 24-20 shows an object arrow with points P and Q at the ends; rays from P are imaged at P′ and rays from Q are imaged at Q′. Two rays are shown coming from Q. One of these passes through the center of curvature, and since this ray is perpendicular to the mirror's surface, it is reflected back on itself. The second incident ray strikes the mirror at an angle ϕ above the optical axis and is therefore reflected at the same angle below the axis. The intersection of these two rays locates the image point Q′.

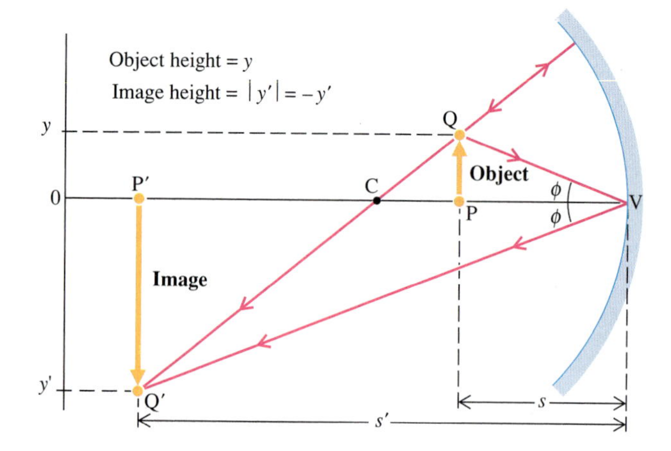

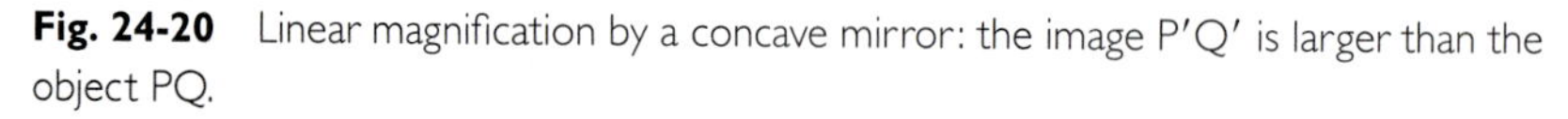

Fig. 24-20 Linear magnification by a concave mirror: the image P′Q′ is larger than the object PQ.

We can obtain a relationship between object and image heights by inspecting triangles PQV and P′Q′V, from which we see that

$$\tan \phi = \frac{y}{s} = -\frac{y'}{s'}$$

The minus sign in this equation is due to the fact that the image is inverted, and therefore image point Q′ has a negative y coordinate. We define the **linear magnification** m to be the ratio y'/y of the coordinates of image point Q′ and object point Q. Solving the equation above for y'/y, we find

$$m = \frac{y'}{y} = -\frac{s'}{s} \tag{24-13}$$

This equation allows us to calculate both the size and the orientation of the image relative to the object. When $|s'| > |s|$, that is, when the image distance is greater than the object distance, $|m| > 1$, meaning that the image is larger than the object, as in Fig. 24-20. When the image is nearer the mirror than the object ($|s'| < |s|$), $|m| < 1$, meaning that the image is smaller than the object. The sign of m indicates the relative orientation of image and object. If m is positive, this means that the y coordinates of image and object have the same sign—the image is upright. If m is negative, the y coordinates have opposite signs—the image is inverted, as in Fig. 24-20. Fortunately Eq. 24-13 may be applied to all situations for which the mirror equation is applicable: concave or convex mirrors, real or virtual objects and images.

EXAMPLE 3 A Woman's Image in a Convex Mirror

A woman of height 1.7 m (5′6″) stands directly in front of a convex spherical mirror 2.0 m away. The mirror has a radius of curvature $R = -50$ cm. Find the location and size of the woman's image.

SOLUTION First we apply the equation for the focal length (Eq. 24-3).

$$f = \frac{R}{2} = \frac{-50 \text{ cm}}{2} = -25 \text{ cm}$$

The convex mirror has a negative focal length. Next we apply the mirror equation (Eq. 24-4), and solve for the image distance s'.

$$\frac{1}{s} + \frac{1}{s'} = \frac{1}{f}$$

$$\frac{1}{200 \text{ cm}} + \frac{1}{s'} = \frac{1}{-25 \text{ cm}}$$

$$s' = -22 \text{ cm}$$

The plane of the woman's image is 22 cm behind the vertex, or about 3 cm in front of the focal point.

Applying the formula for linear magnification (Eq. 24-13), we find

$$m = -\frac{s'}{s} = -\frac{-22 \text{ cm}}{200 \text{ cm}} = 0.11$$

or

$$\frac{y'}{y} = 0.11$$

Solving for y', we obtain

$$y' = 0.11y = (0.11)(1.7 \text{ m})$$

$$= 0.19 \text{ m} = 19 \text{ cm}$$

The woman's upright image is just 19 cm high. The mirror produces a miniaturized image of distant objects, as does the mirror in Fig. 24-10b. You can also see this effect if you look at your reflection in the convex side of a shiny spoon.

Graphical Method

The position and size of an image may be found without any calculation, using a graphical method. Several special rays are drawn from the tip of the object arrow. The rays, called **principal rays,** make use of properties of the focal point and center of curvature so that the paths of these rays can be drawn. The rays locate the image point at the tip of the image arrow. The graphical method is illustrated in Fig. 24-21, where it is applied to several different situations. There are three kinds of principal rays, which are used in all cases. The incident principal rays are:

1 A ray **parallel to the optical axis,** as though coming from an object at ∞;

2 A ray in a direction as though coming from an object at the **focal point** (or, in the case of a convex mirror, directed toward an object at the focal point);

3 A ray along a line through the **center of curvature.**

It is easy to draw the complete paths of these rays. For example, for a concave mirror: the first ray is reflected in a direction such that it passes through the focal point; the second ray is reflected parallel to the optical axis; the third ray is reflected back through the center of curvature. Any two of the reflected rays are sufficient to locate the image. The third ray serves as a check. Once the image is found, any other ray from the object may be drawn.

The graphical method is a helpful complement to the use of equations in problem solving. However, in applying this method, you must be careful to draw to scale the various distances along the optical axis: the object distance, the radius of curvature, and the focal length.

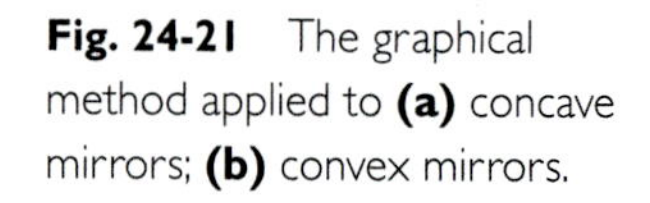

Fig. 24-21 The graphical method applied to **(a)** concave mirrors; **(b)** convex mirrors.

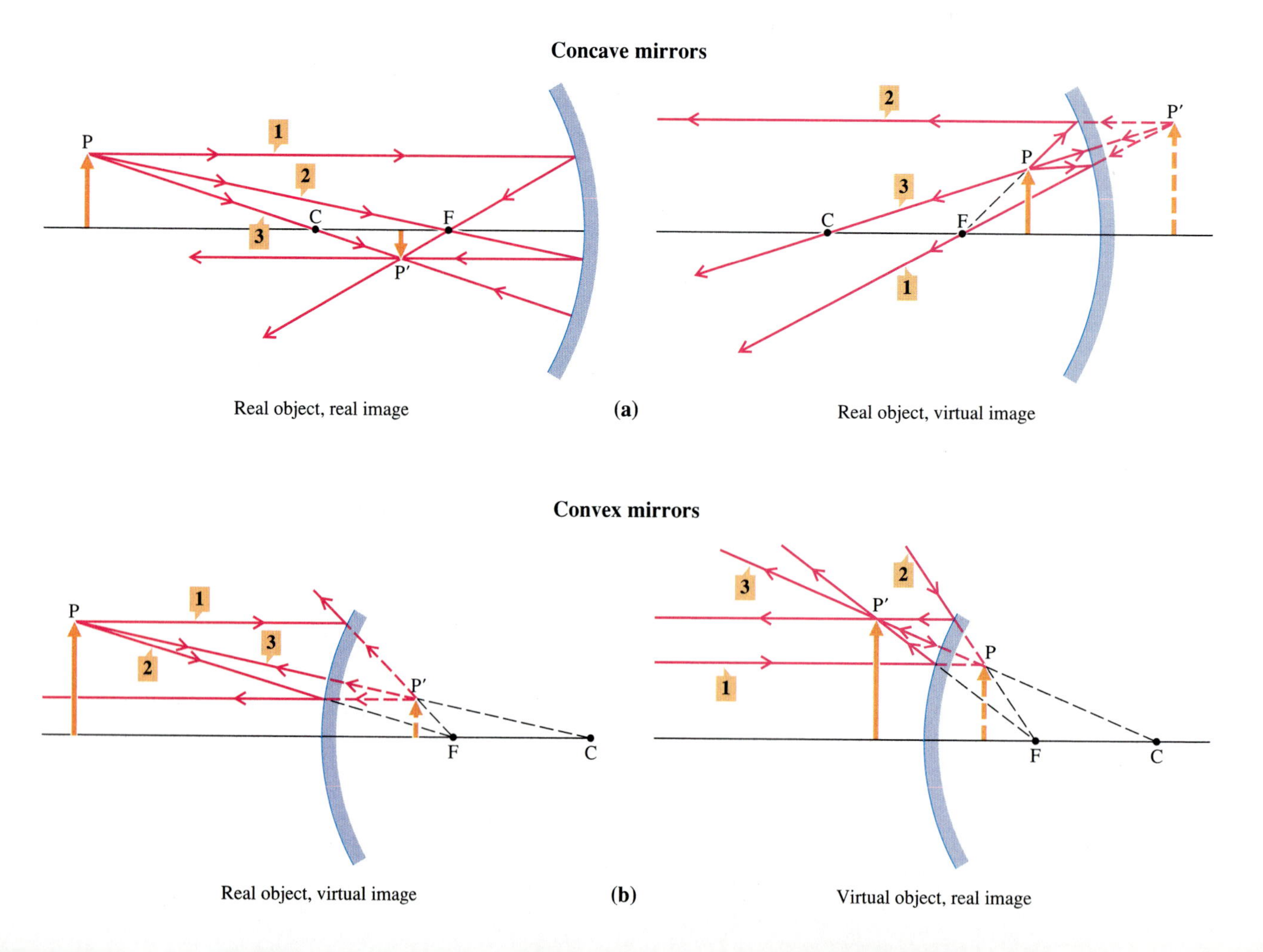

EXAMPLE 4 The Phantom Light Bulb

This example shows how the image of a light bulb in Fig. 24-2 is created. Let the base of an inverted light bulb be placed at the center of curvature of a concave spherical mirror of radius 80 cm. Find the location, size, and orientation of the image.

SOLUTION We apply the mirror equation, setting the object distance equal to the radius of curvature and using the fact that $f = \frac{R}{2}$.

$$\frac{1}{s} + \frac{1}{s'} = \frac{1}{f}$$

$$\frac{1}{s'} = \frac{1}{f} - \frac{1}{s} = \frac{1}{R/2} - \frac{1}{R} = \frac{1}{R}$$

$$s' = R = 80 \text{ cm}$$

Thus the image distance is the same as the object distance. The image lies in the same plane as the object. Next we calculate the magnification.

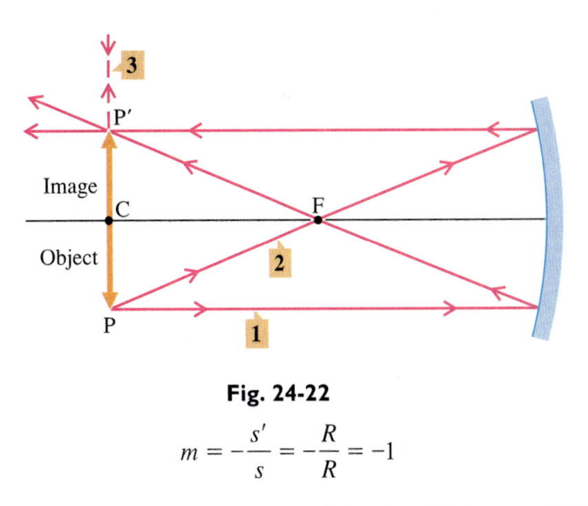

Fig. 24-22

$$m = -\frac{s'}{s} = -\frac{R}{R} = -1$$

The image is the same size as the object, but it is inverted (Fig. 24-22). The "object" bulb is directly beneath the image in the photograph in Fig. 24-2.

Application of the graphical method, shown in Fig. 24-22, helps in understanding how the image is formed.

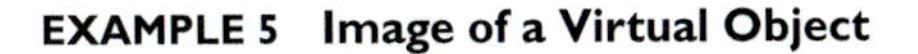

EXAMPLE 5 Image of a Virtual Object

A real image of height 2 cm is formed by a concave mirror. Then a second mirror of radius 20 cm is placed 10 cm in front of the image (Fig. 24-23a), so that the converging rays strike the second mirror before the image is formed. Find the image formed by the second mirror.

SOLUTION The image formed by the first mirror serves as a virtual object for the second mirror, with an object distance $s = -10$ cm. We apply the mirror equation, using the focal length of $\frac{20 \text{ cm}}{2} = 10$ cm.

$$\frac{1}{s} + \frac{1}{s'} = \frac{1}{f}$$

$$\frac{1}{-10 \text{ cm}} + \frac{1}{s'} = \frac{1}{10 \text{ cm}}$$

$$\frac{1}{s'} = \frac{2}{10 \text{ cm}} = \frac{1}{5 \text{ cm}}$$

$$s' = 5 \text{ cm}$$

A real image is formed 5 cm in front of the mirror. To find the size and orientation of the image, we apply the equation for linear magnification.

$$m = \frac{y'}{y} = -\frac{s'}{s}$$

$$y' = -\frac{s'}{s}y = -\frac{5 \text{ cm}}{-10 \text{ cm}}(-2 \text{ cm}) = -1 \text{ cm}$$

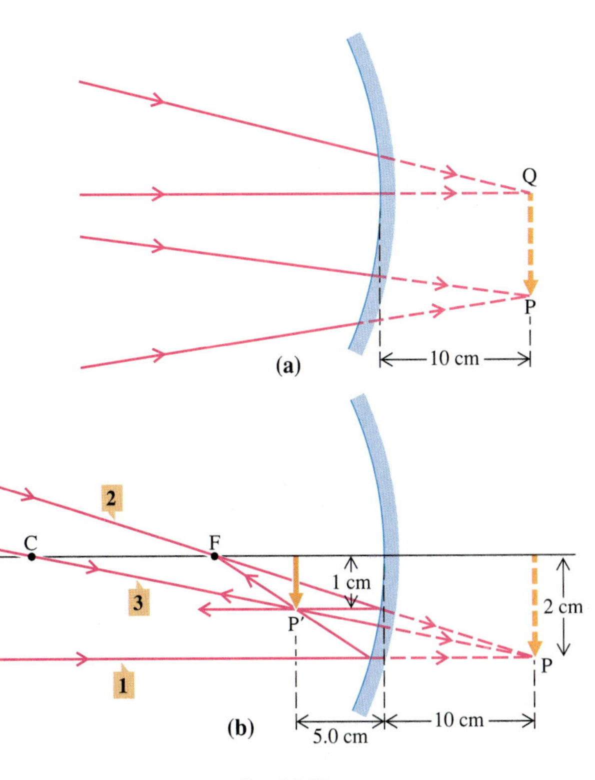

Fig. 24-23

Thus the tip of the image arrow is 1 cm below the optical axis. Application of the graphical method is shown in Fig. 24-23b. Notice that the vertical and horizontal scales are different.

Fig. 24-24 Lens grinding.

24-3 Lenses

One of the oldest and most important applications of optical principles is the design of lenses for use in various instruments—eyeglasses, cameras, microscopes, and so forth. Most lenses are made by forming spherical surfaces on both sides of a circular glass disk. Spherical surfaces are most common because they are the easiest to make (Fig. 24-24). Refraction at a spherical surface is also relatively easy to describe mathematically. Thus we begin our study of lenses with a description of refraction at various spherical surfaces.

Fig. 24-25 shows parallel rays of light incident on spherical surfaces, separating two media with different refractive indices n and n' (for example, air and glass). As seen in the figure, transmitted rays either converge toward the optical axis, forming a real image, or diverge from the axis, forming a virtual image. Which they do is dependent both on the kind of surface (concave or convex) and whether the second medium has a larger or smaller index than the first.

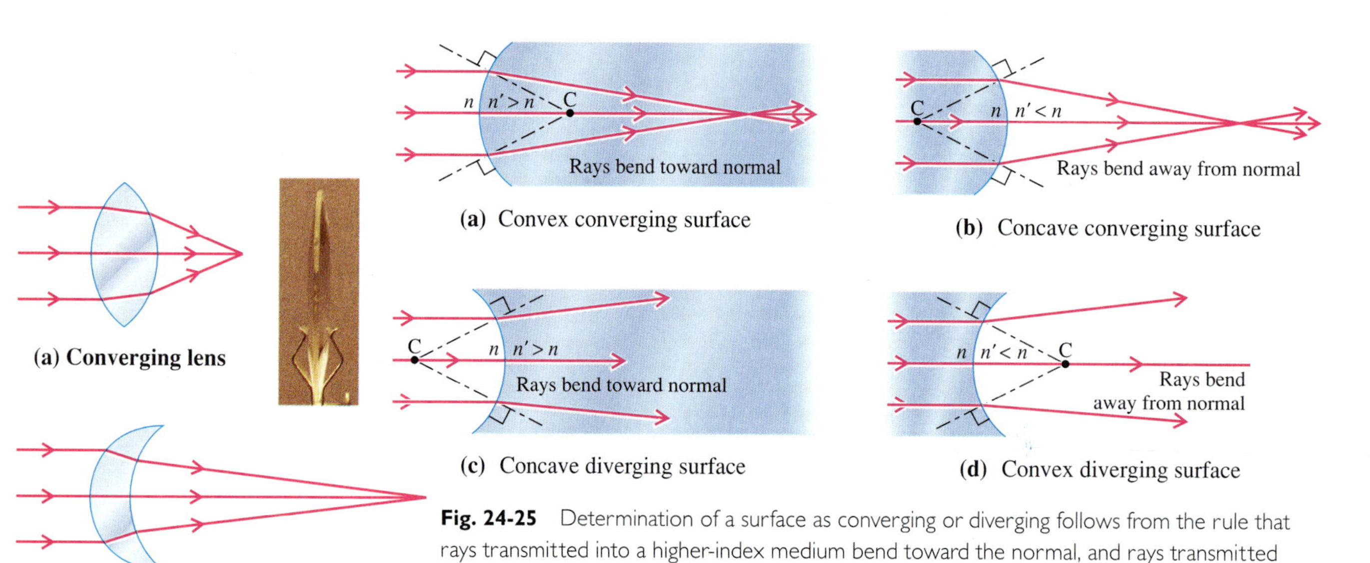

Fig. 24-25 Determination of a surface as converging or diverging follows from the rule that rays transmitted into a higher-index medium bend toward the normal, and rays transmitted into a lower-index medium bend away from the normal.

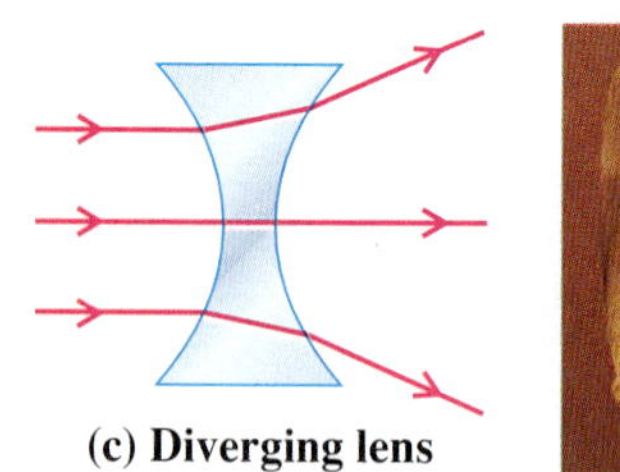

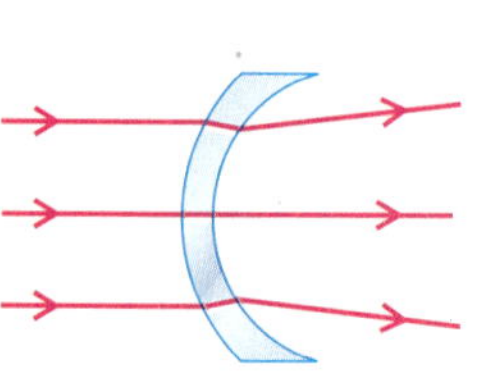

Fig. 24-26 Spherical lenses.

One must be careful in the designation of surfaces as concave or convex because there are two sides from which to view the surface, and a surface that is convex from one side is concave from the other. Here we adopt the convention of always viewing the surface from the side of the incident light. This convention is used in labeling surfaces in Fig. 24-25.

The amount of bending at a surface depends on the magnitude of the radius of curvature. The more curved a surface (that is, the shorter the radius of curvature), the greater is the refraction.

Normally a lens is made of glass, a relatively high-index medium, and is surrounded by air, a lower-index medium. There are several general shapes, illustrated in Fig. 24-26. In each case, two surfaces from Fig. 24-25 are combined to produce a lens. For example, Fig. 24-26a shows a lens formed when the surfaces in Figs. 24-25a and 24-25b are combined.

All the lenses in Fig. 24-26 illustrate the following simple but general rule: **converging lenses are thicker in the middle than at the edge, and diverging lenses are thinner in the middle than at the edge.**

Refraction at a Spherical Surface

First we shall show how to locate the image formed by refraction at a single spherical surface, and then we shall describe image formation by a thin lens with two such surfaces.

Paraxial rays from an object point P are imaged by a spherical refracting surface at an image point P′ (Fig. 24-27). Object and image distances, s and s', are related by the following equation (for which a derivation is provided at the end of this section):

$$\frac{n}{s} + \frac{n'}{s'} = \frac{n' - n}{R} \tag{24-14}$$

where n and n' are the refractive indices of the two media and R is the radius of curvature of the surface.

We may use Eq. 24-14 to locate images in much the same way we used the mirror equation. If we adopt the following sign conventions for s, s', and R, this one equation may be applied to both concave and convex surfaces and to both real and virtual objects and images:

Object distance	$s > 0$	Incident rays diverge from a **real object in front of the surface** (incident side).	(24-15)
	$s < 0$	Incident rays converge toward a **virtual object behind the surface** (transmitted side).	(24-16)
Image distance	$s' > 0$	Transmitted rays converge toward a **real image behind the surface** (transmitted side).	(24-17)
	$s' < 0$	Transmitted rays diverge from a **virtual image in front of the surface** (incident side).	(24-18)
Radius of curvature	$R > 0$	**convex** surface	(24-19)
	$R < 0$	**concave** surface	(24-20)

Notice that the sign conventions for s and s' are the same as for mirrors, except that images are formed by transmitted light rather than by reflected light. Real images are formed by transmitted light behind the refracting surface, whereas for mirrors real images are formed by reflected light in front of the mirrored surface. Virtual images are formed in front of a refracting surface, in contrast to mirrors, for which virtual images are behind the surface. The sign convention for the radius of curvature is the opposite of that for mirrors.

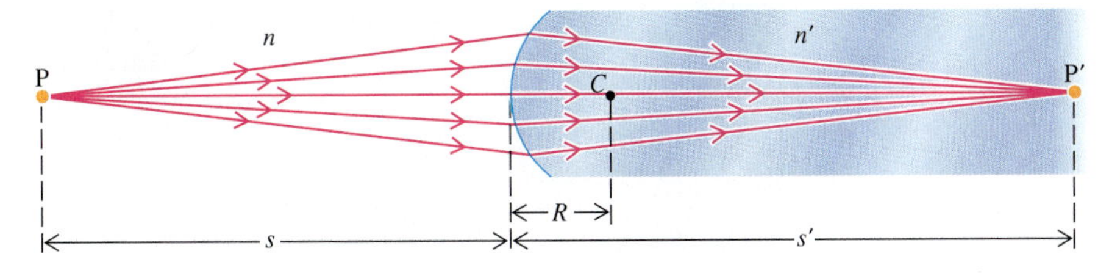

Fig. 24-27 Refraction at a spherical surface produces an image at P′ of an object at P.

A flat surface can be thought of as the surface of an infinite sphere. Thus Eq. 24-14 may be applied to refraction of paraxial (small-angle) rays at a flat surface if we let $R = \infty$ so that the right side of this equation equals zero.

EXAMPLE 6 The Apparent Depth of a Fish

Find the apparent depth of a small fish 1.00 m below the surface of a lake when viewed from directly overhead.

SOLUTION Light rays diverging from a point on the fish are bent away from the normal as they pass from water to air (Fig. 24-28a). Since the fish is viewed from directly above, the light rays entering the observer's eye make a small angle with the normal; that is, they are paraxial. Thus we can apply Eq. 24-14 to find the image distance s'.

$$\frac{n}{s} + \frac{n'}{s'} = \frac{n' - n}{R}$$

Light travels a distance $s = 1.00$ m from water of refractive index 1.33 to air of index 1.00. The flat refracting surface corresponds to $R = \infty$. Thus

$$\frac{1.33}{1.00 \text{ m}} + \frac{1.00}{s'} = \frac{1.33 - 1.00}{\infty} = 0$$

$$\frac{1}{s'} = -1.33 \text{ m}^{-1}$$

$$s' = -0.75 \text{ m} = -75 \text{ cm}$$

The water forms a virtual image of the fish 75 cm below the surface. The fish, like all underwater objects, appears to be closer to the surface than it really is.

If we view a submerged object from the side, rather than from directly above the object, the effect is much more dramatic (Fig. 24-28b). The object can be more than twice as deep as it appears. See Problem 51, Chapter 23. (Eq. 24-14 does not apply unless the object is viewed from directly above because only then is the assumption of paraxial rays satisfied.)

To a Scuba diver looking through a diving mask, underwater objects also appear closer than they are (Fig. 24-29). The mask traps a layer of air in front of the diver's eye. The thin flat layer of plastic at the front of the mask has little effect on the incident light (like the windowpane in Example 4, Chapter 23). Thus the light effectively passes from water to air, and everything the diver sees appears closer than it really is, just as for an observer looking down into a body of water.

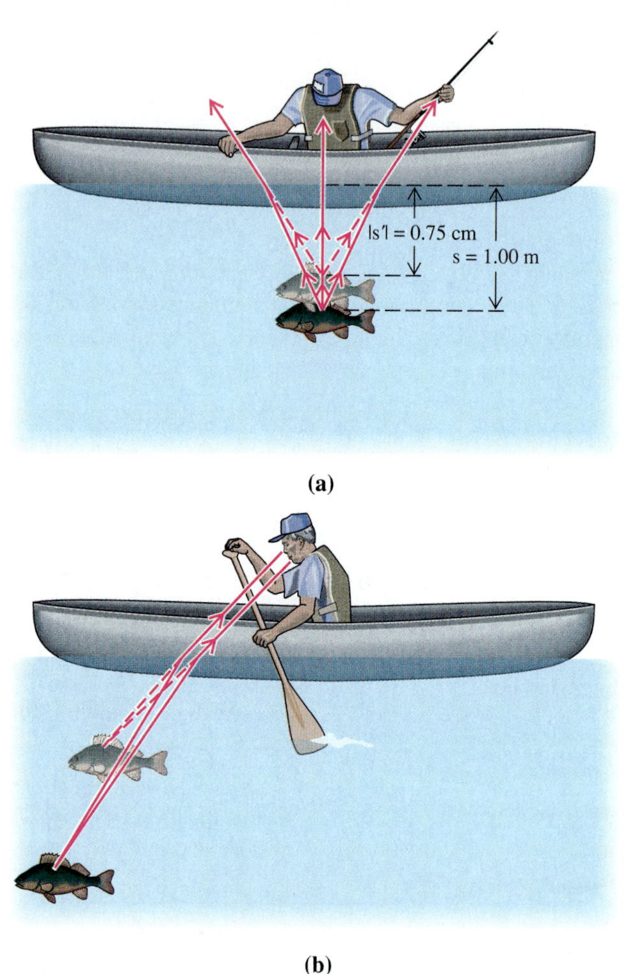

Fig. 24-28

Fig. 24-29

Thin Lenses

We shall now derive an equation describing image formation by a thin lens with spherical surfaces. We do this by twice applying the equation for refraction at a spherical surface (Eq. 24-14). At the first surface, light travels from air with index 1 to glass with index n. Thus we have

$$\frac{1}{s_1} + \frac{n}{s_1'} = \frac{n-1}{R_1} \tag{24-21}$$

At the second surface, light travels from glass to air, and we obtain

$$\frac{n}{s_2} + \frac{1}{s_2'} = \frac{1-n}{R_2} \tag{24-22}$$

The object for the second surface is the image produced by the first. Since the lens is assumed to be thin, the image produced by the first surface is approximately the same distance from either surface. Thus

$$|s_2| \approx |s_1'|$$

If the first image is real, the second object is virtual (Fig. 24-30), and if the first image is virtual, the second object is real. Therefore in all cases s_1' and s_2 have opposite signs.

$$s_2 = -s_1'$$

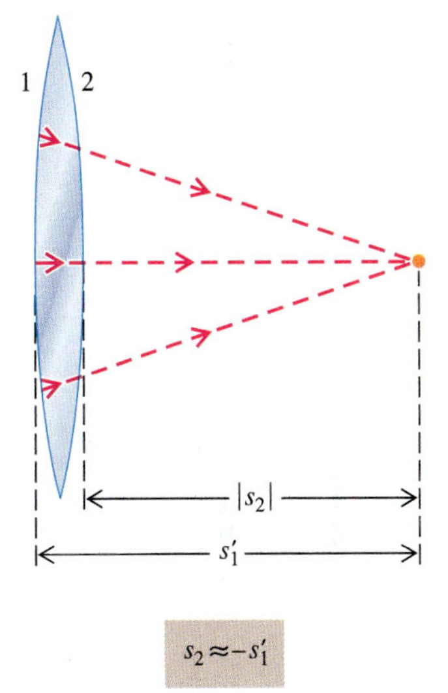

Fig. 24-30 The image formed by a lens's first surface serves as an object for its second surface.

Substituting this result into Eq. 24-22 and adding Eq. 24-21, we obtain

$$\frac{1}{s_1} + \frac{1}{s_2'} = (n-1)\left(\frac{1}{R_1} - \frac{1}{R_2}\right)$$

Here s_1 is the distance from the object to the first surface and s_2' is the distance from the second surface to the final image. We now drop the subscripts on object and image distances, and, since the lens is assumed to be thin, we measure these distances from the center of the lens.

$$\frac{1}{s} + \frac{1}{s'} = (n-1)\left(\frac{1}{R_1} - \frac{1}{R_2}\right) \tag{24-23}$$

For distant objects, $s \to \infty$ and $1/s \to 0$. We identify the corresponding image point as a "focal point"; the focal length f is the value of the image distance, the distance from the lens to the focal point.

$$s = \infty \qquad s' = f$$

Inserting these values into Eq. 24-23, we obtain an expression for the focal length, known as the "lensmaker's equation":

$$\frac{1}{f} = (n-1)\left(\frac{1}{R_1} - \frac{1}{R_2}\right) \quad \text{(thin lens in air)} \tag{24-24}$$

Substitution of this expression into Eq. 24-23 gives an equation relating object and image distances, which is identical to the mirror equation. Here we shall call it the thin lens equation:

$$\frac{1}{s} + \frac{1}{s'} = \frac{1}{f} \quad \text{(thin lens; paraxial rays)} \tag{24-25}$$

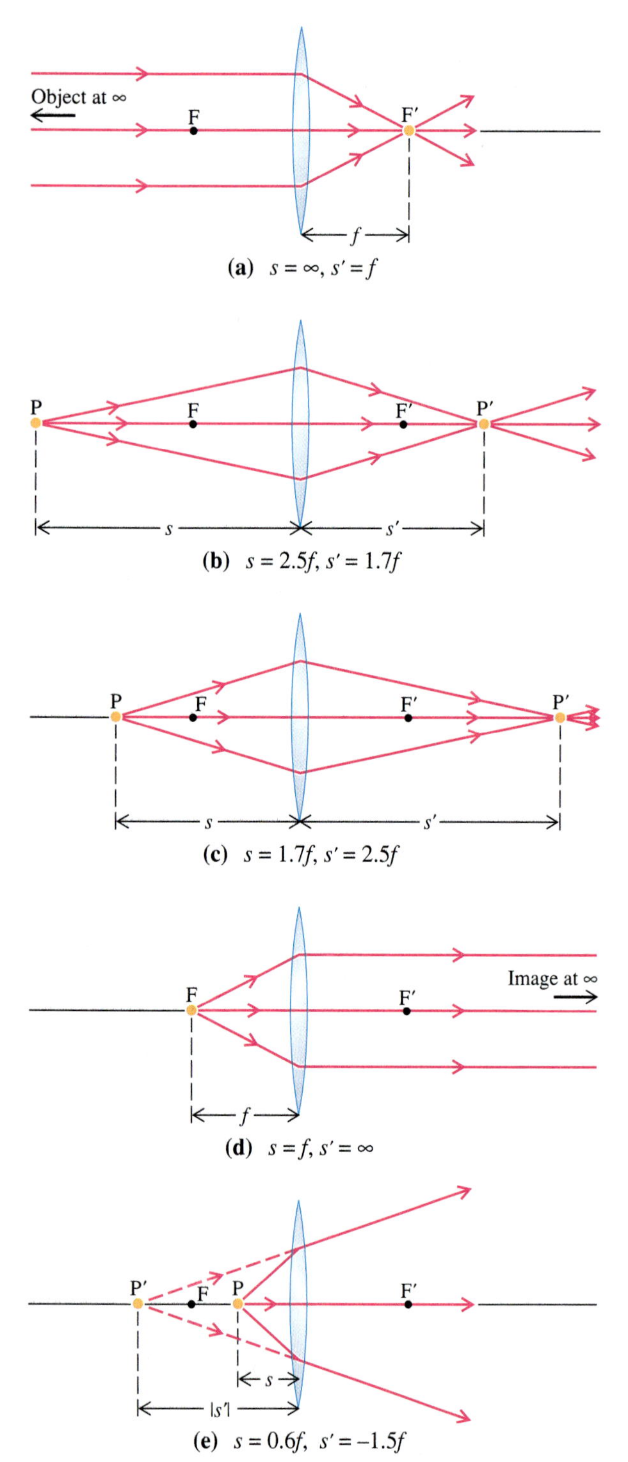

Fig. 24-31 Image formation by a positive lens. A point object on the optical axis is at a great distance in **(a)** and is closer to the lens in each successive illustration. As the object comes closer, the image moves away. The principle of optical reversibility is illustrated by the interchange of objects and images in **(a)** and **(d)** and again in **(b)** and **(c).**

The thin lens equation is applied to lenses in the same way that it is applied to mirrors, except, of course, that lens images are formed by transmitted light, rather than by reflected light.

For a lens in a medium other than air, one can show (Problem 40) that the derivation of the lensmaker's equation is modified in such a way that the lens index n is replaced by the ratio of the lens index to the index of the surrounding medium; that is, $n \to n/n_o$, where n_o is the index of the outer medium.

$$\frac{1}{f} = \left(\frac{n}{n_o} - 1\right)\left(\frac{1}{R_1} - \frac{1}{R_2}\right) \quad \text{(thin lens; outer medium } n_o\text{)} \tag{24-26}$$

Positive Lenses

Fig. 24-31 shows effects predicted by the thin lens equation for a positive-focal-length lens, with a point object at various positions on the optical axis. A distant object produces an image at the focal point ($s = \infty$, $s' = f$) in Fig. 24-31a. Fig. 24-31d shows another interpretation of the focal point. In this figure rays diverge from a point F located a distance f to the left of the lens. The values $s = f$, $s' = \infty$ also satisfy the thin lens equation, meaning that an object at the focal point produces an image at ∞. Notice that in all cases the lens tends to converge rays; that is, the transmitted rays either converge, or they diverge less than the incident rays (as in Fig. 24-31e). See Fig. 24-11 for comparison with a converging mirror.

When an object is just outside the focal point of a positive lens, the distance from lens to image is much greater than the focal length.

Negative Lenses

A negative-focal-length lens is a diverging lens, as illustrated in Fig. 24-32. Fig. 24-32a shows parallel incident rays from a distant object producing a virtual image at the focal point ($s = \infty$, $s' = f < 0$). A negative lens forms a virtual image of any real object. A real image may be formed, however, if the object is virtual. In particular, a virtual object at the focal point results in an image at ∞ (Fig. 24-32c). See Fig. 24-18 for comparison with a diverging mirror.

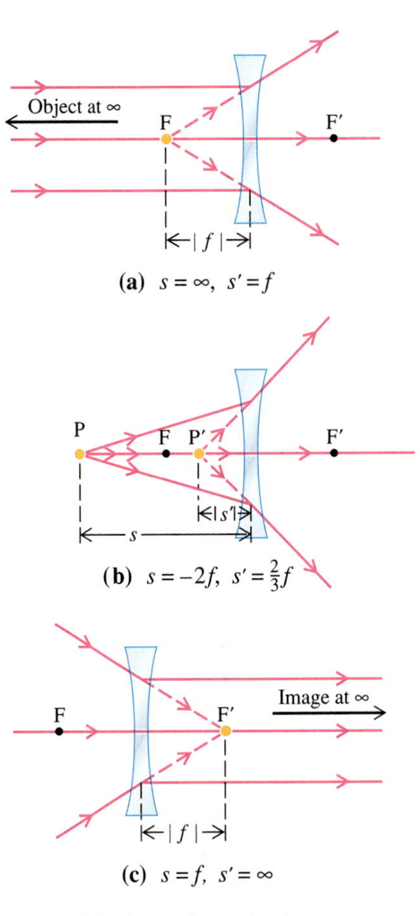

Fig. 24-32 Image formation by a negative lens.

Optical Power

The lensmaker's equation (Eq. 24-24) indicates how the focal length of a lens is determined by the index of the glass and the curvature of the two surfaces. The lensmaker's equation predicts that flatter surfaces with larger radii of curvature will have longer focal lengths. Fig. 24-33 shows three lenses with progressively shorter focal lengths. The focal lengths decrease as the surfaces become more curved (smaller magnitude R's). For very short focal lengths, the lens must be small.

The **optical power** P_o of a lens is defined as the inverse of its focal length.

$$P_o = \frac{1}{f} \tag{24-27}$$

The standard unit of optical power is the diopter, abbreviated d, defined to be an inverse meter.

$$1\text{ d} = 1\text{ m}^{-1}$$

For example, a 2 d lens has a focal length of $\frac{1}{2}$m, and a $-\frac{1}{3}$d lens has a focal length of −3 m. Optical power is a measure of the ability of a lens to refract light; it measures the "bending power" of the lens. A short-focal-length, positive lens converges incident parallel rays in a short distance. Hence it is a high-power lens. High-power lenses are small (Fig. 24-33c). Similarly, a short-focal-length, negative lens is a small, high power lens.

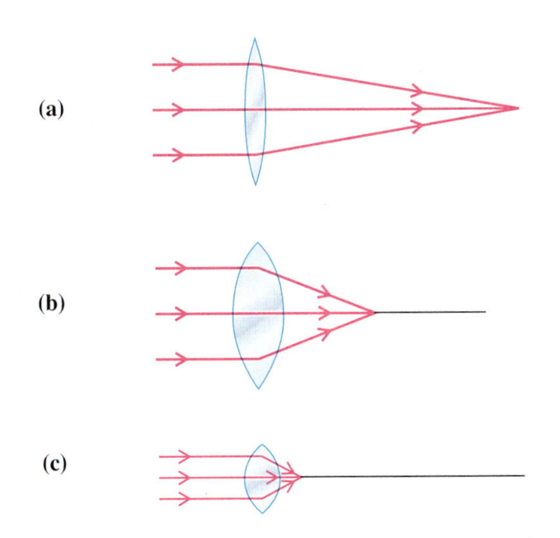

Fig. 24-33 Three lenses with progressively shorter focal lengths. A very short focal length, or high optical power, requires very small radii of curvature, which means a high power lens must be small.

EXAMPLE 7 Focusing the Sun's Rays With a Lens

The first surface of a certain lens is convex, and the second surface is concave. Each surface has a radius of curvature of magnitude 10 cm. (a) Sketch the lens. (b) Find the optical power and focal length of the lens. (c) The lens is used as a "burning glass." The sun's rays are focused by the lens on a dry leaf, causing it to ignite. How far is the leaf from the lens? (d) Now suppose the leaf is viewed through the lens, with the leaf placed 1.0 cm from the lens. Where does the leaf appear to be?

SOLUTION **(a)** Fig. 24-34a shows a sketch of the lens, with the surfaces labeled in reference to their curvature as seen from the left side, assumed to be the side of the incident light.

(b) To find the focal length, we apply the lensmaker's equation (Eq. 24-24):

$$\frac{1}{f} = (n - 1)\left(\frac{1}{R_1} - \frac{1}{R_2}\right)$$

Following the sign convention for spherical refracting surfaces, we have

$$\text{convex: } R_1 = +10 \text{ cm}$$

$$\text{concave: } R_2 = -10 \text{ cm}$$

Using these values for the radii and $n = 1.5$ for the index, the lensmaker's equation gives

$$\frac{1}{f} = (1.5 - 1.0)\left(\frac{1}{10 \text{ cm}} - \frac{1}{-10 \text{ cm}}\right) = \frac{1}{10 \text{ cm}}$$

Thus $$f = 10 \text{ cm} = 0.10 \text{ m}$$

and $$P = \frac{1}{f} = \frac{1}{0.10 \text{ m}} = 10 \text{ d}$$

Notice that $f = R_1$. One can show that this simple relationship holds for any symmetric lens made of glass with an index of 1.5 (Problem 37).

(c) To concentrate solar energy on the leaf, we should place it at the focal point of the lens, so that parallel rays from a point on the sun converge to a point on the leaf (Fig. 24-34b).

(d) We apply the thin lens equation (Eq. 24-25), and solve for the image distance, using an object distance of 1.0 cm.

$$\frac{1}{s} + \frac{1}{s'} = \frac{1}{f}$$

$$\frac{1}{s'} = \frac{1}{f} - \frac{1}{s} = \frac{1}{10 \text{ cm}} - \frac{1}{1.0 \text{ cm}} = -\frac{9}{10 \text{ cm}}$$

$$s' = -\frac{10}{9} \text{ cm} = -1.1 \text{ cm}$$

This means the virtual image of the leaf is very close to the object, as shown in Fig. 24-34c. The cone of rays seen by the observer passes through the center section of the thin lens, which is nearly flat and which therefore produces almost no bending of rays (see Example 4, Chapter 23). In general, as an object is brought very close to a lens, so that s is much smaller than f, the virtual image formed by the lens is at nearly the same position as the object.

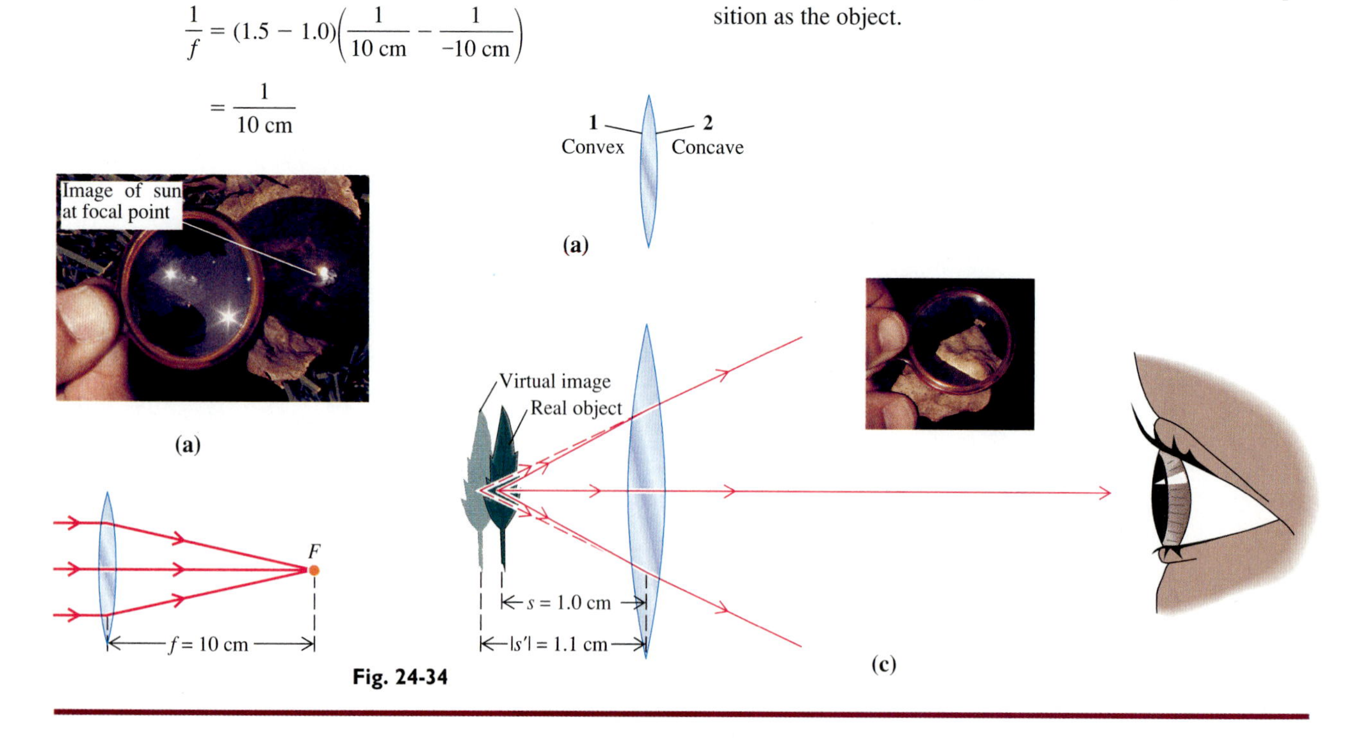

Fig. 24-34

EXAMPLE 8 Finding the Focal Length of a Lens

A certain thin lens has surfaces with radii $R_1 = +20$ cm and $R_2 = +10$ cm. Sketch the lens. Find the focal length and the image distance for an object distance of 20 cm.

SOLUTION Since both radii are positive, both surfaces are convex. The first surface has the larger radius, meaning that it is flatter (Fig. 24-35). Applying the lensmaker's equation, we find

$$\frac{1}{f} = (n - 1)\left(\frac{1}{R_1} - \frac{1}{R_2}\right)$$

$$= (1.5 - 1.0)\left(\frac{1}{20 \text{ cm}} - \frac{1}{10 \text{ cm}}\right) = -\frac{1}{40 \text{ cm}}$$

$$f = -40 \text{ cm}$$

Next we apply the thin lens equation to find s'.

$$\frac{1}{s} + \frac{1}{s'} = \frac{1}{f}$$

$$\frac{1}{s'} = \frac{1}{f} - \frac{1}{s} = \frac{1}{-40 \text{ cm}} - \frac{1}{20 \text{ cm}}$$

$$s' = -13 \text{ cm}$$

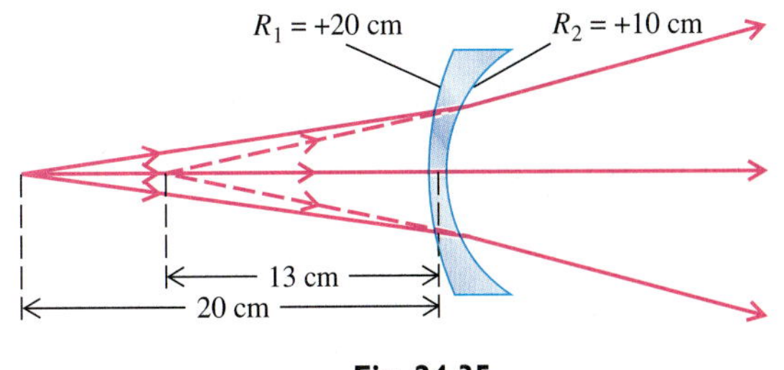

Fig. 24-35

This image is virtual and is located 13 cm in front of the lens. The effect of the lens is to diverge the incident rays, so that the object appears closer to the lens than it really is (Fig. 24-35).

Linear Magnification

Each point on an extended object will have a corresponding image point produced by a lens. As shown in Fig. 24-36, the linear magnification of such images is described by the same equation used to describe spherical mirrors:

$$m = \frac{y'}{y} = -\frac{s'}{s} \tag{24-28}$$

The derivation in Fig. 24-36 is for a real image of a real object, but the result is valid for all cases described by the thin-lens equation.

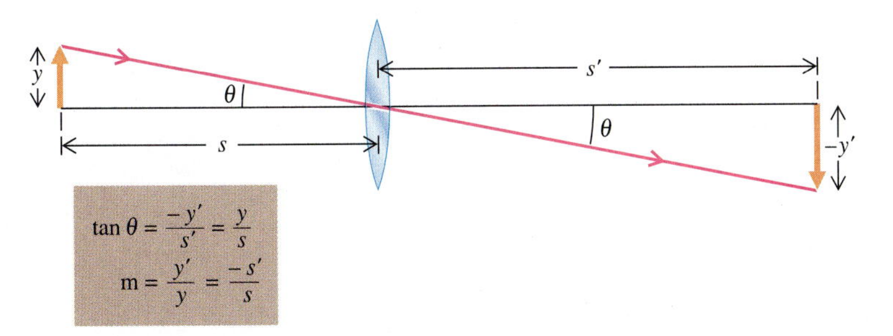

Fig. 24-36 Derivation of the magnification equation for a real image of a real object. The ray striking the center of the lens passes straight through. This section of the lens is approximately like a flat, thin sheet of glass and, as seen in Example 4, Chapter 23, will not change the direction of the incident ray.

EXAMPLE 9 Photographing Someone at a Distance and Close-up

A certain 35 mm camera uses a lens with a focal length of 50.0 mm (Fig. 24-37). Each frame of photographic film is 24 mm wide and 35 mm long. (a) If you photograph someone 10.0 m away from the camera, what must the distance between the lens and the film be? Treat the lens as a thin lens. (b) If the subject is 2.0 m tall, how high is his image on the film? (c) How much will the distance between the film and the lens need to be changed so that you can shoot a close-up of the subject at a distance of 40 cm from the lens?
(d) What are the maximum dimensions of the subject that will be seen in the close-up photo?

Fig. 24-37

SOLUTION (a) We apply the thin lens equation (Eq. 24-25) to find the image distance.

$$\frac{1}{s'} = \frac{1}{f} - \frac{1}{s} = \frac{1}{50.0 \times 10^{-3}\text{ m}} - \frac{1}{10.0\text{ m}}$$

$$s' = 5.03 \times 10^{-2}\text{ m} = 50.3\text{ mm}$$

The image is formed 50.3 mm behind the lens, and this should be the location of the photographic film for a clear image to be formed on it. Notice that this distance is nearly equal to the focal length. Whenever the object distance is much greater than the focal length, as it is here, the image lies approximately in the focal plane. All object distances from 10 m to infinity have corresponding image distances, which are all very nearly equal to 50 mm.

(b) We apply the magnification equation (Eq. 24-28) to find the height of the image on the film (Fig. 24-38).

$$m = \frac{y'}{y} = -\frac{s'}{s}$$

$$y' = -\frac{ys'}{s} = -\frac{(2.0\text{ m})(5.03 \times 10^{-2}\text{ m})}{10\text{ m}}$$

$$= -1.0 \times 10^{-2}\text{ m} = -10\text{ mm}$$

(c) We again apply the thin lens equation.

$$\frac{1}{s'} = \frac{1}{f} - \frac{1}{s} = \frac{1}{50.0 \times 10^{-3}\text{ m}} - \frac{1}{40 \times 10^{-2}\text{ m}}$$

$$s' = 5.7 \times 10^{-2}\text{ m} = 57\text{ mm}$$

Thus the distance between the film and the lens must be increased by 7 mm so that you can focus on the closer object.

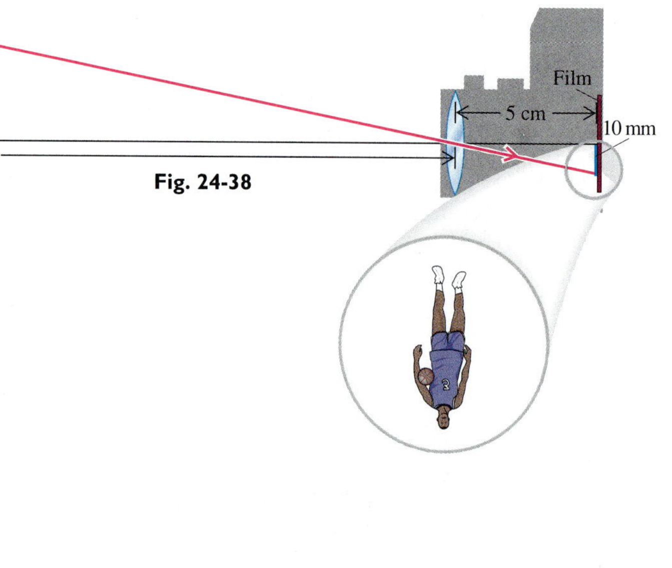

Fig. 24-38

(d) We apply the magnification equations, solving for the object height y:

$$m = \frac{y'}{y} = -\frac{s'}{s}$$

$$y = -\frac{y's}{s'}$$

We can apply this equation for both horizontal and vertical directions. Corresponding to the film width ($y' = 24$ mm), we have an object distance

$$y = -\frac{(24\text{ mm})(40\text{ cm})}{57\text{ mm}} = -17\text{ cm}$$

and corresponding to the film length ($y' = 35$ mm), we have an object distance

$$y = -\frac{(35\text{ mm})(40\text{ cm})}{57\text{ mm}} = -25\text{ cm}$$

The close-up photo covers a 17 cm by 25 cm field of view, a head shot.

Graphical Method

A graphical method may be used to obtain the position and size of an image, without calculation. The method is essentially the same as that used for spherical mirrors. The following principal rays are drawn from the tip of the object arrow:

1 A ray **parallel to the optical axis,** as though coming from an object at ∞;
2 A ray in a direction as though coming from an object at the **focal point** (or, in the case of a negative lens, directed toward an object at the focal point);
3 A ray through the **center of the lens.**

The only difference from the principal rays of mirrors is the third principal ray. The third ray for mirrors passes through the center of curvature and therefore is reflected back along the same path. The third principal ray for a lens passes straight through the center of the lens, without being bent. The graphical method is illustrated in Fig. 24-39.

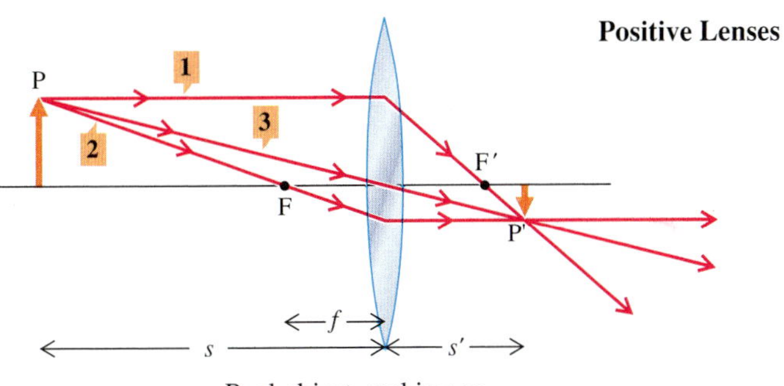

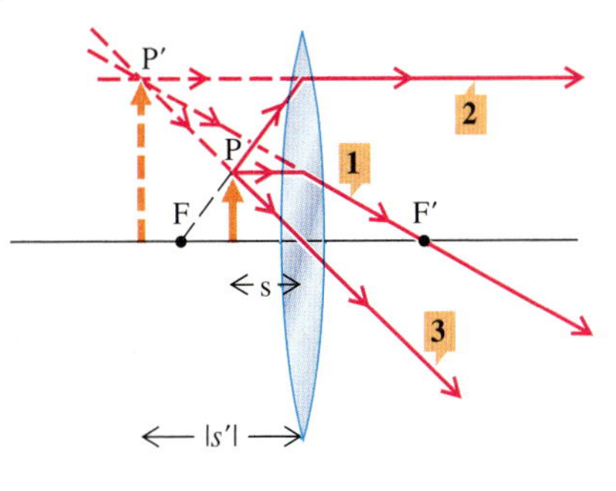

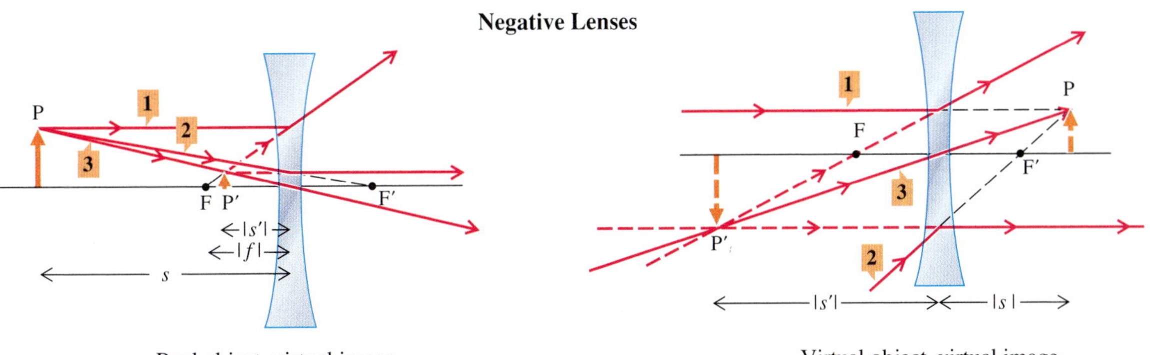

Fig. 24-39 The graphical method applied to positive and negative lenses.

EXAMPLE 10 Finding the Final Image Produced by Two Lenses

(a) An object 1 cm high is placed 15 cm in front of a converging lens with a focal length of 10 cm. Find the position, size, and orientation of the image, both by calculation and by using the graphical method. (b) A second lens of focal length −7.5 cm is placed 25 cm to the right of the first lens. Find the final image.

SOLUTION **(a)** The graphical solution is shown in Fig. 24-40a. (Notice that horizontal and vertical distances are not drawn to the same scale.) The image is 30 cm to the right of the lens, is inverted, and is twice the size of the object. These results are verified when you apply the thin lens equation and the magnification equation:

$$\frac{1}{s'} = \frac{1}{f} - \frac{1}{s} = \frac{1}{10 \text{ cm}} - \frac{1}{15 \text{ cm}} = \frac{1}{30 \text{ cm}}$$

$$s' = 30 \text{ cm}$$

$$m = -\frac{s'}{s} = -\frac{30 \text{ cm}}{15 \text{ cm}} = -2$$

(b) The real image formed by the first lens serves as a virtual object for the second lens (Fig. 24-40b). Notice that the three principal rays are all directed toward the virtual object. The final image is 15 cm to the right of the second lens and is three times the size of the previous image, or 6 cm high.

Again the results are verified by calculation:

$$\frac{1}{s'} = \frac{1}{f} - \frac{1}{s}$$

$$= \frac{1}{-7.5 \text{ cm}} - \frac{1}{-5 \text{ cm}} = \frac{1}{15 \text{ cm}}$$

$$s' = 15 \text{ cm}$$

$$m = -\frac{s'}{s} = -\frac{15 \text{ cm}}{-5 \text{ cm}} = 3$$

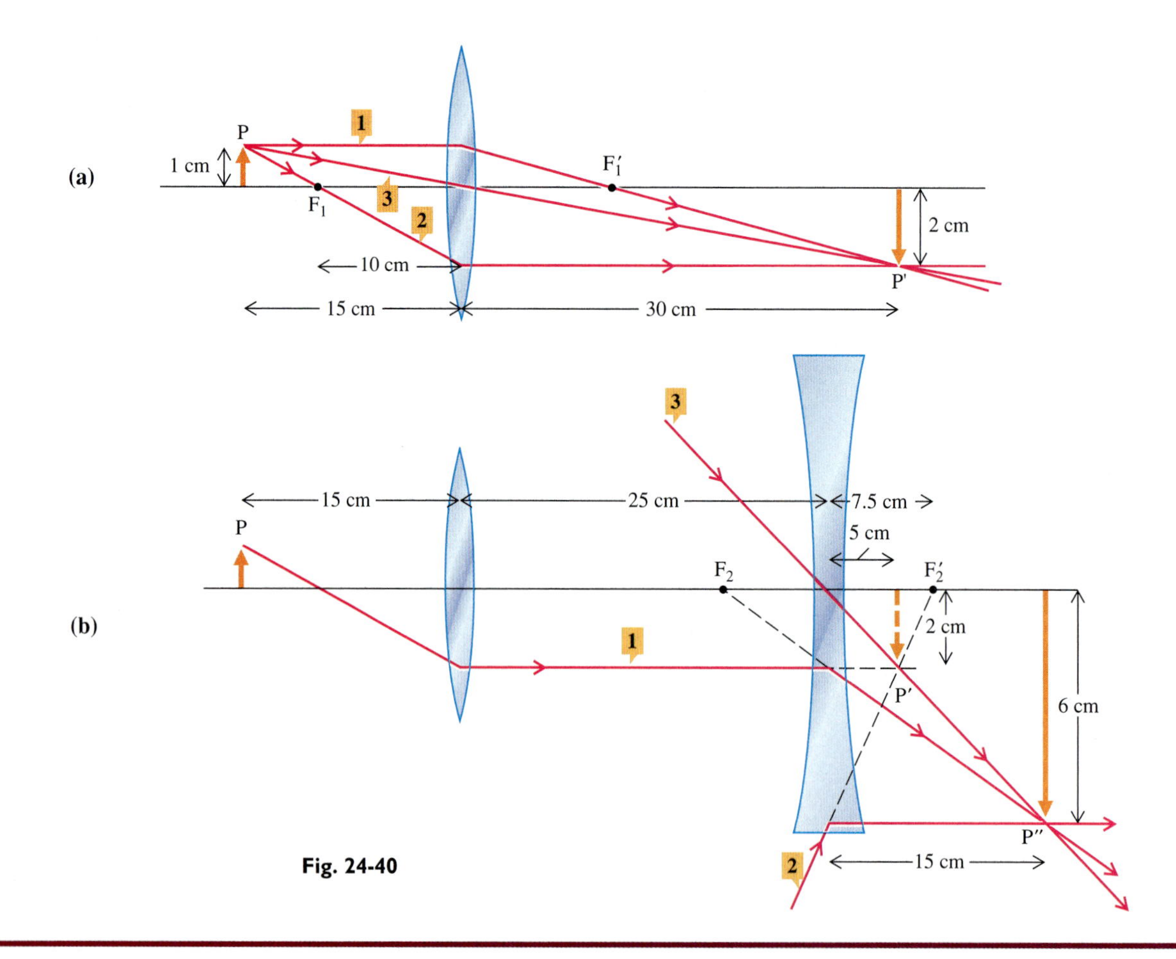

Fig. 24-40

Aberrations

Various approximations, both explicit and implicit, have gone into our analysis of spherical thin lenses: (1) the lens thickness is negligible compared to other distances; (2) the rays are paraxial; (3) the refractive index of the lens glass is a constant n for light of any wavelength, so that dispersion is negligible; (4) the objects are relatively small and close to the optical axis, so that the image is undistorted and lies approximately in a plane. These approximations are often not well satisfied in actual applications of a single thin lens. Variations from our ideal description, resulting from the failure of various approximations, are called **aberrations;** some of these are illustrated in Figs. 24-41 to 24-43. Fortunately, in any particular application it is possible to design lens combinations to correct for the aberrations that are of most significance. For example, two glass lenses with different dispersion curves can be combined to eliminate **chromatic aberration,** as shown in Fig. 24-41.

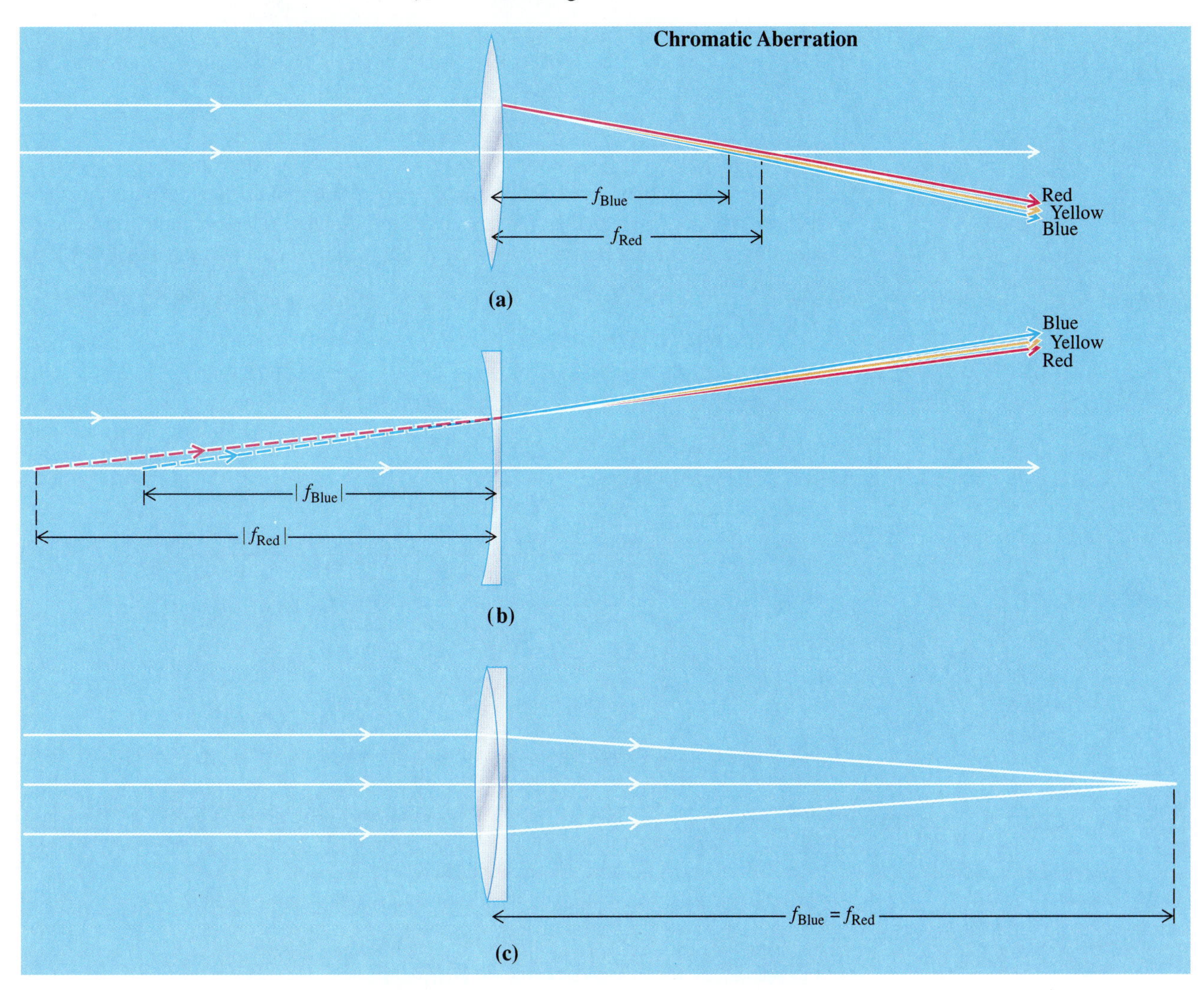

Fig. 24-41 Chromatic aberration, illustrated in **(a)** and **(b),** is caused by dispersion of light in glass, a prismlike effect. The amount of dispersion depends on the kind of glass used for the lens. The negative lens in (b) uses a glass with greater dispersion (a larger variation in n) than the positive lens in (a), but because the shape of the negative lens results in less refraction (a longer focal length), the angular spread of the spectrum for the two lenses is the same. Thus combining the lenses, as shown in **(c),** results in a positive-focal-length lens, which displays no dispersion. Such a combination is called an "achromatic doublet." Illustrations (a) and (b) show an exaggerated amount of dispersion, about 20 times that of typical glass.

Fig. 24-42 The distorted, multi-colored image of graph paper seen through the lens is the result of chromatic and other aberrations.

Correction of chromatic aberration is very important in the design of camera lenses. When uncorrected lenses are used, the images on color photographs often have bands of color at the edges because of slight displacement of the images by light of each color (Fig. 24-42).

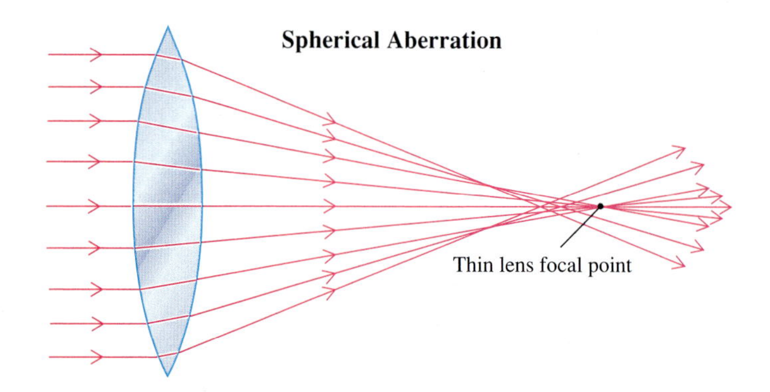

Fig. 24-43 Spherical aberration. Rays striking the lens near its edge have relatively large angles of incidence and refraction. Hence the small-angle approximation, used to obtain the thin-lens focal length, fails. These nonparaxial rays cross the optical axis inside the thin-lens focal point. Spherical aberration can be reduced by placing a screen with a small aperture in front of the lens so as to block rays near the edge of the lens.

Derivation of Single-Surface Refraction

We shall now derive the equation describing image formation by refraction at a single spherical surface (Eq. 24-14). We shall derive the equation only for the special case of a real image formed by a convex surface, with the second medium having the higher index. (Problems 23 and 24 outline derivations of the equation for a concave surface and for a virtual image.)

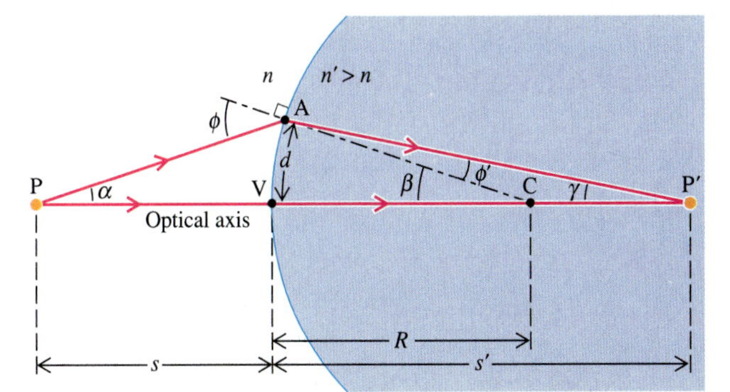

Fig. 24-44 Light rays diverging from object point P are refracted by a convex spherical surface and converge at image point P′. The image distance s' depends on the object distance s, on the radius of curvature R, and on the refractive indices n and n'.

We draw a single ray from an arbitrary object point P on the optical axis (Fig. 24-44). The incident ray makes a small (but otherwise arbitrary) angle α with the axis; that is, we consider only paraxial rays. The angles of incidence and refraction, ϕ and ϕ', are related by Snell's law:

$$n \sin \phi = n' \sin \phi'$$

If angles α and β are small, angles ϕ and ϕ' will also be small, and we can apply the small-angle approximation, in which the sine of an angle is replaced by the angle itself, measured in radians (see Chapter 15, Fig. 15-12). With this approximation the equation above becomes

$$n\phi = n'\phi' \tag{24-29}$$

Next we apply the exterior-angle theorem (used in the mirror derivation—see Fig. 24-13) to the triangles PAC and CAP′, shown in the figure:

$$\phi = \alpha + \beta \tag{24-30}$$

$$\beta = \phi' + \gamma$$

Solving Eq. 24-29 for ϕ' and substituting into the equation above, we obtain

$$\beta = \frac{n}{n'}\phi + \gamma \tag{24-31}$$

Now Eqs. 24-30 and 24-31 may be used to eliminate ϕ. Combining these equations and rearranging terms, we find

$$n\alpha + n'\gamma = (n' - n)\beta \tag{24-32}$$

Again assuming small angles, we may express angles α, β, and γ by the arc length d divided by the respective lengths s, R, and s'.

$$\alpha \approx \frac{d}{s} \qquad \beta = \frac{d}{R} \qquad \gamma \approx \frac{d}{s'}$$

Using these expressions in Eq. 24-32 and cancelling the d's, we obtain Eq. 24-14:

$$\frac{n}{s} + \frac{n'}{s'} = \frac{n' - n}{R}$$

CHAPTER 24 SUMMARY

Light rays striking a lens or mirror either diverge from a real object or converge toward a virtual object. Rays leaving the lens or mirror either converge toward a real image or diverge from a virtual image.

A plane mirror forms a virtual image of a real object. The image is the same size as the object. Each image point is the same distance from the plane of the mirror as the corresponding object point.

The location of images formed with paraxial rays, either by spherical mirrors or by spherical thin lenses, may be found by use of the equation

$$\frac{1}{s} + \frac{1}{s'} = \frac{1}{f}$$

where s is the object distance, s' is the image distance, and f is the focal length of the mirror or lens. For a real object or a real image, the corresponding distance s or s' is positive. For a virtual object or a virtual image, the corresponding distance s or s' is negative. A positive value of the focal length f means that the mirror or lens is converging, whereas a negative value of f means that it is diverging.

For spherical mirrors, the focal length is half the radius of curvature R.

$$f = \frac{R}{2}$$

where $R > 0$ for a concave mirror

and $R < 0$ for a convex mirror

The sign convention for spherical refracting surfaces is the opposite of the convention for mirrors:

$R > 0$ for a convex refracting surface

$R < 0$ for a concave refracting surface

To find the focal length in air of a thin lens of index n, with surfaces having radii R_1 and R_2, one applies the lensmaker's equation:

$$\frac{1}{f} = (n - 1)\left(\frac{1}{R_1} - \frac{1}{R_2}\right)$$

For a thin lens of index n immersed in a medium of index n_o, we use

$$\frac{1}{f} = \left(\frac{n}{n_o} - 1\right)\left(\frac{1}{R_1} - \frac{1}{R_2}\right)$$

For image formation by a single refracting surface of radius R, object and image distances are related by the equation

$$\frac{n}{s} + \frac{n'}{s'} = \frac{n' - n}{R}$$

where n is the index of the medium in which the light is incident and n' is the index of the transmitting medium.

For both spherical thin lenses and spherical mirrors, the size and orientation of an image relative to the object is found by use of the equation for linear magnification

$$m = \frac{y'}{y} = -\frac{s'}{s}$$

Aberration is the failure of all rays incident on a mirror or lens from a single object point to form a single image point. There are various causes and types of aberrations. Chromatic aberration is caused by dispersion within a lens, and spherical aberration is caused by rays at the edge of a spherical lens or mirror being bent too much. Optical instruments often use lens systems designed to correct for aberrations.

The optical power of a lens or mirror is defined as the inverse of its focal length and is measured in units of inverse meters, or diopters.

$$P_o = \frac{1}{f}$$

The graphical method of finding an image consists in using three principal rays drawn from the tip of an object arrow to locate the tip of the image arrow.

Questions

1 (a) Does the size of a mirror image depend on the size of the mirror?
(b) Does the part of a mirror image that is seen by an observer depend on the size of the mirror?

2 You see the image of a person in a mirror (Fig. 24-45). Will the person necessarily be able to see you by looking in the mirror if there is sufficient illumination? Explain.

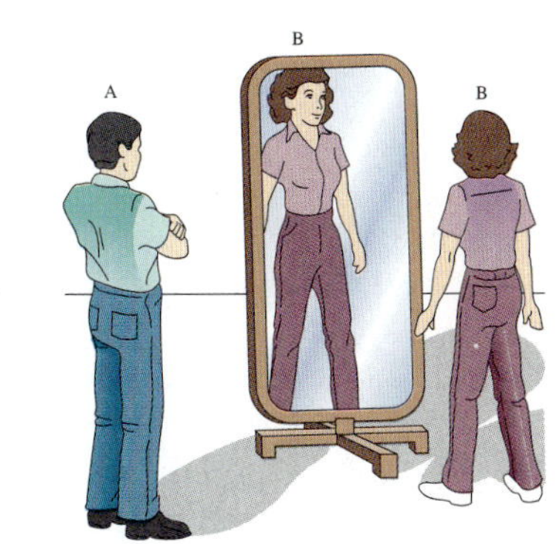

Fig. 24-45

3 (a) Does a vertical plane mirror reverse right and left?
(b) Does it reverse top and bottom?
(c) Does it appear to reverse either? Explain.

4 Words appear reversed when viewed in a mirror, though the mirror itself does not reverse them. Whether the reversal appears to be from right to left or from up to down depends on how objects are rotated for viewing in the mirror. Suppose that you view in a vertical plane mirror the image of the letter "p" written on a piece of paper. The paper is rotated 180° so that it faces the mirror and you can observe the image.
(a) If the paper is rotated about a vertical axis, what do you see in the mirror (Fig. 24-46)?
(b) If the paper is rotated about a horizontal axis, what do you see?

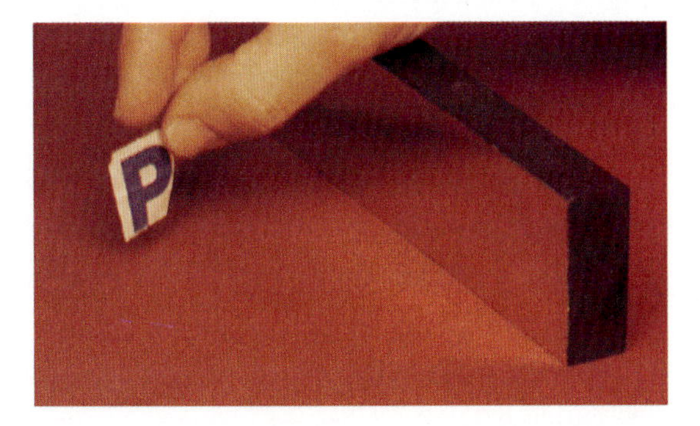

Fig. 24-46

5 Is it possible that light rays reflected from a plane mirror could form a real image? If so, how?

6 You see your image in a hemispherical silver bowl, and then you turn the bowl over and see your image in the other side. In one side the image is upright, but in the other it is inverted.
(a) For which side is it inverted—the inside or the outside of the bowl?
(b) For which side is the image closer to your eye?
(c) For which side is the image closer to the bottom of the bowl?
(d) For which side is the linear magnification greater?

7 A spherical mirror is placed underwater. Will its focal length increase, decrease, or remain the same?

8 Are the images seen in Fig. 24-2 real or virtual?

9 Is it possible to obtain a photographic image by placing a piece of photographic film at the location of (a) a real image; (b) a virtual image?

10 Is it possible to use a camera to photograph (a) a real image; (b) a virtual image?

11 A lens is to be used as a burning lens, as in Example 7. Which of the following properties affect its ability to concentrate the sun's energy: magnitude of the focal length, sign of the focal length, index of the glass, diameter of the lens?

12 A certain glass lens has a positive focal length in air. The lens is placed underwater.
(a) Will its focal length be positive or negative in water?
(b) Will its focal length be greater or less than in air?
(c) Will its optical power be greater or less than in air?

13 Will a lens that has a positive focal length when surrounded by one medium necessarily have a positive focal length when surrounded by any other medium?

14 A clear plastic bag shaped like a positive lens is filled with water. It acts as a converging "water lens" in air. (The plastic simply serves to contain and shape the water and has no significant optical effect itself.) Now suppose the water lens is immersed in benzene ($n = 1.50$).
(a) Will the lens be a converging or a diverging lens?
(b) Will the magnitude of the focal length be greater or smaller than in air?

15 A spherical air bubble in water acts as a lens.
(a) Is it a converging or a diverging lens?
(b) Does it have a positive or a negative focal length?
(c) Would the magnitude of its focal length increase or decrease if the diameter of the bubble were increased?

16 An "air lens" for use as a positive lens underwater is made when a plastic bag is filled with air under pressure. The sides of the bag have spherical shapes. Should the bag be thicker in the middle or at the edge?

17 Is the spherical aberration of a spherical mirror affected if the mirror is submerged in water?

18 Does chromatic aberration occur for spherical mirrors (a) in air; (b) in water?

19 Could you start a fire with a "burning lens" made of ice?

20 The dining room on the top floor of the Fairmont Hotel in San Francisco offers a magnificent view of the city at night. A tourist looking out through a window takes a flash photo. Of what advantage or disadvantage is the flash?

Answers to Odd-Numbered Questions

1 (a) no; (b) yes; **3** (a) no; (b) no; (c) yes, right and left *appear* reversed in the sense that if you look at the mirror image of a person raising her right hand you would interpret the raised image hand to be a left hand. Suppose you view someone facing you with her right hand raised; that hand would be on your left side. Although the mirror does not reverse right and left, rotating an object about a vertical axis for viewing does reverse right and left, and so the mirror appears to reverse right and left. See question 4; **5** yes, if the object is virtual; **7** remain the same; **9** (a) yes; (b) no; **11** all of the properties; **13** no; **15** (a) diverging; (b) negative; (c) increase; **17** no; **19** yes

Problems (listed by sections)

24-1 Plane Mirrors

1 In Fig. 24-47 which of the object arrows, A, B, or C, can be seen in its entirety by observer O in mirror M?

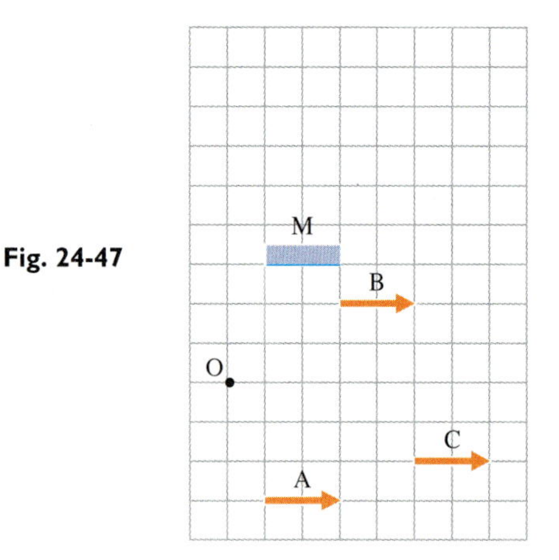

Fig. 24-47

2 In Fig. 24-48 construct rays from point P reflected from the mirror at points A and B. These rays should satisfy the law of reflection. Extend the line of these rays behind the mirror to find the image point P′.

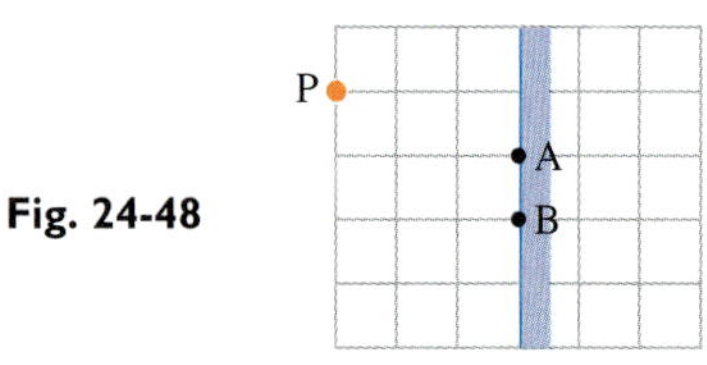

Fig. 24-48

★**3** A cowboy, 7 feet tall with his hat, views his full image in a plane mirror.

(a) What is the height of the image?

(b) What is the minimum length of the mirror? Does this result depend on the distance of the cowboy from the mirror?

4 The driver of a car sees the image of another car in the rear-view mirror, a plane mirror 50 cm from the driver's eyes. The image covers 2 cm of the mirror. Estimate the distance to the other car.

★**5** The ceiling and two adjacent walls in a room are mirrored so that there are three adjacent plane mirrors, each perpendicular to the others. How many images of yourself could you see in the mirrors?

6 Two parallel, vertical, plane mirrors, 20 cm apart, face each other. A light source at point P is 15 cm from the mirror on the left and 5 cm from the mirror on the right.

(a) How many images of point P are formed by the mirrors?

(b) Find the distance from the mirror on the right to the two nearest images behind the mirror.

(c) Find the number of reflections of light rays for each of these images.

7 You are walking toward a plane mirror at a speed of 2 m/s.

(a) How fast is your mirror image moving relative to the mirror?

(b) How fast is your image moving relative to you?

★ **8** A laser beam is incident at an angle ϕ on a plane mirror. The mirror is then rotated through an angle θ, so that the angle of incidence increases by θ. Through what angle is the reflected beam rotated as a result of the rotation of the mirror?

24-2 Spherical Mirrors

9 Fig. 24-49 shows a concave spherical mirror with a focal point F. A small object is located on the mirror's optical axis at a distance much greater than the mirror's focal length. At which point, A, B, or C, would an observer be able to see the image of this object?

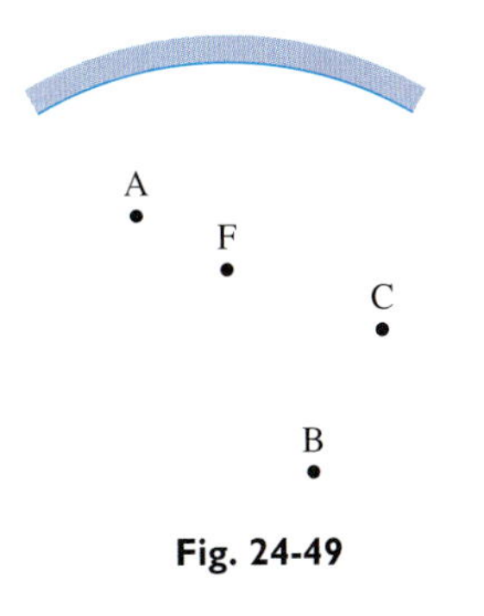

Fig. 24-49

10 For each of the incident light rays in Fig. 24-50 construct the reflected ray by drawing it parallel to the appropriate line in the figure. Show that the image distance $s' \approx R/2$.

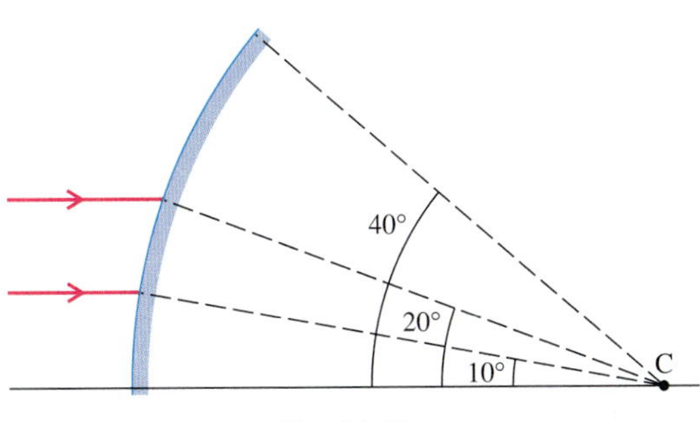

Fig. 24-50

11 A boy looks at his real image in a concave spherical mirror.

(a) Is the image upright or inverted?

(b) Does the mirror reverse right and left?

(c) If the boy raises his right hand, which hand does the image appear to raise?

★ **12** Use Fig. 24-51 to derive the mirror equation for a concave spherical mirror with a virtual object. Show that the object distance s must be treated as a negative quantity.

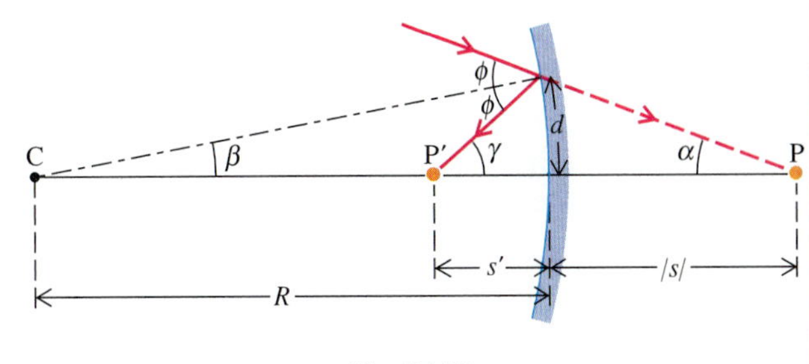

Fig. 24-51

★ **13** Use Fig. 24-52 to derive the mirror equation for a convex mirror. Show that the radius of curvature R must be treated as a negative quantity.

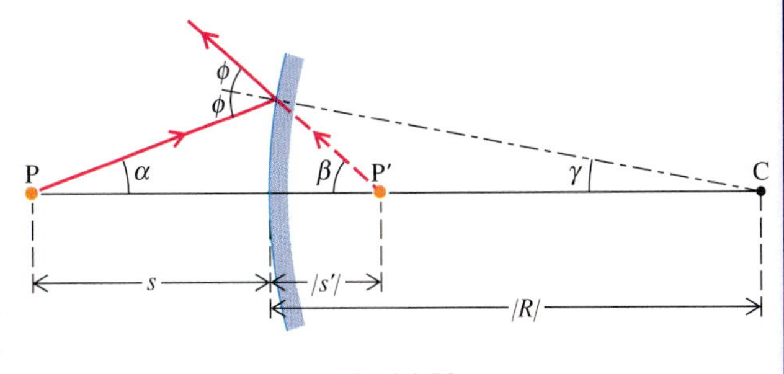

Fig. 24-52

★ **14** Use the spherical mirror equation to prove that (a) the image of a real object in a convex mirror is always virtual; that is, prove that if $s > 0$ and $f < 0$, $s' < 0$; (b) the image of a real object in a concave mirror is real only if the object is outside the focal point of the mirror; that is, if $s > 0$ and $f > 0$, $s' > 0$ only if $s > f$.

15 A spherical silver Christmas tree ornament, 10 cm in diameter, acts as a mirror.

(a) Are the images one sees in it real or virtual, upright or inverted?

(b) Find the image height of a man 2.0 m tall, standing 3.0 m from the ornament.

★ **16** A hiker is lost somewhere near Mt. Whitney, which she knows has an elevation of 4600 m. To determine her distance from the mountain, she uses her compact mirror, a small concave mirror with a focal length of 20 cm. She views an image of the mountain in her mirror, and notes that when the mirror is held about 50 cm from her eyes, the image covers a section of the mirror about 5 cm high. How far is Mt. Whitney?

★★ **17** Michelle looks into David's eyes and sees her image reflected (Fig. 24-53). The image of her head (20 cm high) just fills his pupil (4 mm in diameter) when it is 20 cm from his eye. Calculate the radius of curvature of his cornea. To simplify the solution, make reasonable approximations about the size and location of the image.

Fig. 24-53

18 Draw the reflected light rays for each of the principal rays shown in Fig. 24-54. Draw a fourth ray reflected from the mirror at an arbitrary point.

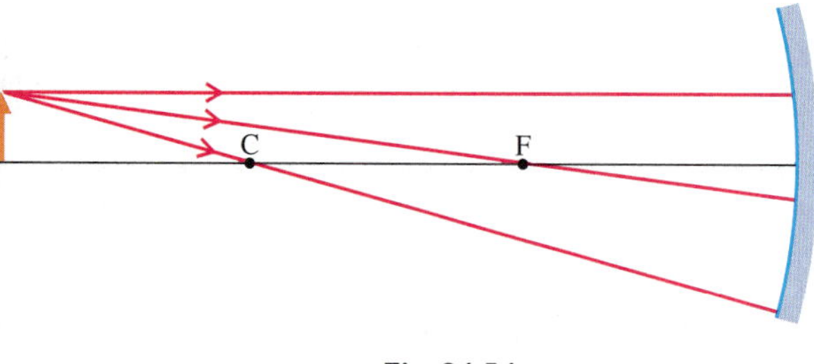

Fig. 24-54

19 An object 2.0 cm high is placed 45 cm from a concave mirror of radius 10 cm and radius of curvature 60 cm. Find the location, size, and orientation of the image. Solve both by calculation and by drawing principal rays.

20 An object 2.0 cm high is placed 10 cm from the surface of a polished silver sphere of radius 10 cm. Find the location, size, and orientation of the image. Solve both by calculation and by drawing principal rays.

★ **21** A concave spherical mirror with a focal length of 10 cm faces a plane mirror with the optical axis of the spherical mirror perpendicular to the plane mirror. A small object is placed at point P on the optical axis, 10 cm from the plane mirror and 30 cm from the vertex of the spherical mirror. Find the distance from the plane mirror to the three nearest images.

★ **22** In Fig. 24-2 coins appear to be on top of a shiny convex surface, but the coins are not really there. The illusion is created by two opposing concave mirrors, each with a radius of curvature of 20 cm with vertices 10 cm apart, and the coins placed at point P at the vertex of the lower mirror (Fig. 24-55). There is a small hole in the center of the upper mirror.

(a) Locate the image of P in the upper mirror.

(b) Use the first image as the object for the lower mirror and find its image.

(c) Draw rays from P reflected by both mirrors and emerging through the hole.

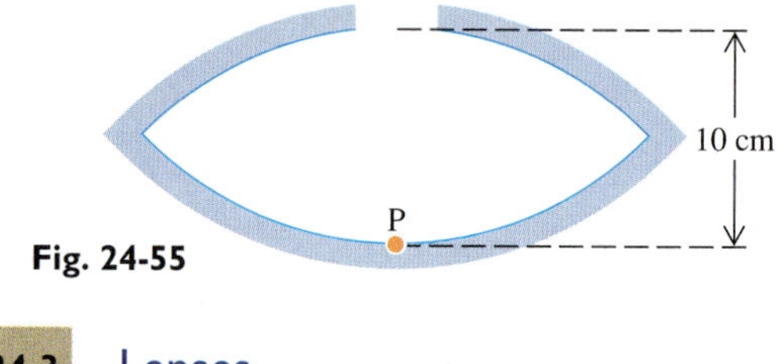

Fig. 24-55

24-3 Lenses

Refraction at a Spherical Surface.

★ **23** Use Fig. 24-56 to derive Eq. 24-14 for refraction at a concave spherical surface. Show that the radius of curvature of the concave surface must be treated as a negative quantity.

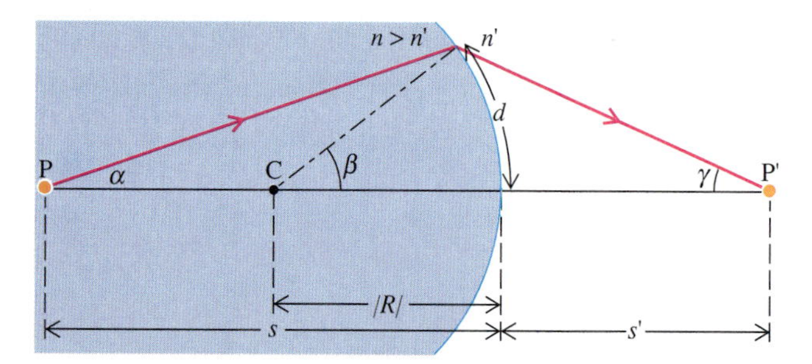

Fig. 24-56

★ **24** Use Fig. 24-57 to derive Eq. 24-14 for a spherical refracting surface producing a virtual image. Show that the image distance s' must be treated as a negative quantity.

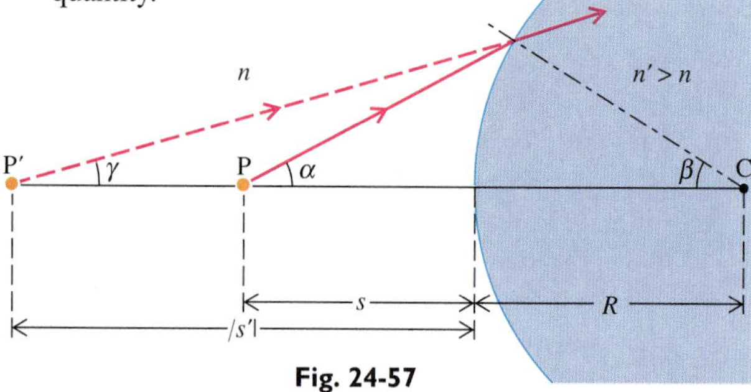

Fig. 24-57

25 Light rays from a distant point source are incident perpendicular to the flat side of a semicircular glass plate of radius 10 cm and index 1.5. Find the distance of the image from the center of the curved side.

26 A fish underwater views a fisherman directly above. The fisherman's head is 1.00 m above the surface of the water. How far above the surface does it appear to the fish to be?

27 A swimming pool appears to be 1.50 m deep, when you view the bottom of the pool from directly above the water's surface. What is the actual depth?

28 The bottom of a wine glass has a spherical shape with a 4.0 cm radius of curvature. The glass is filled with transparent white wine of index 1.36. The top surface of the wine is flat, and the bottom surface is spherical. A light fixture is directly over the wine and far above it. How far below the bottom surface of the wine will an image be formed? Ignore refraction by the glass.

Thin Lenses

★ **29** A laser beam 1.0 mm in diameter is directed along the optical axis of a thin lens of focal length +5.0 cm (Fig. 24-58).
(a) How far from the lens will the beam be focused?
(b) A second positive lens is placed to the right of the first. Light emerges from the second lens in a parallel beam of diameter 4.0 mm. Thus the combination of lenses acts as a beam expander. Find the focal length of the second lens and the distance between the lenses.

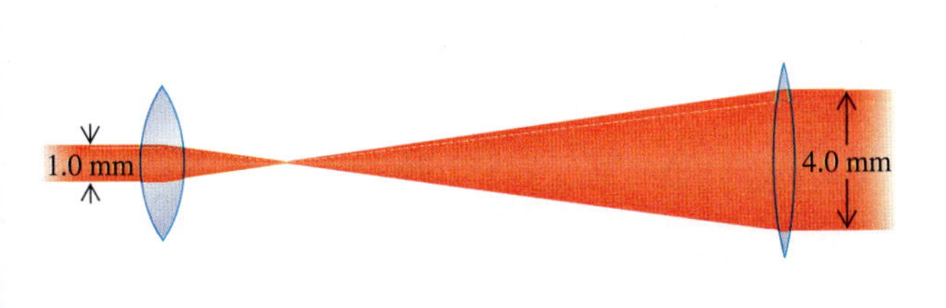

Fig. 24-58

30 An object 1.0 cm high is placed on the optical axis of a +10 cm focal length lens, 15 cm from the lens. Find the location, size, and orientation of the image.

31 A child of height 1.0 m is photographed with a camera 10 m away. The camera has a "zoom" lens with a focal length that varies from 28 mm for wide-angle shots to 150 mm for telephoto shots. Find the minimum and maximum lengths of the child's image on the film.

32 Find the image distance and linear magnification for a small object on the optical axis, 20 cm from a lens of focal length −10 cm.

33 Find the focal length of a lens that produces an inverted image four times the size of a small object placed on the optical axis, 20 cm from the lens.

★ **34** Prove that the image of a real object produced by a negative lens is always smaller than the object.

★★ **35** Prove that the optical power of two thin lenses in contact with each other equals the sum of the powers of the individual lenses.

36 The lenses in Fig. 24-59 are all made of glass of index 1.50 and have surfaces with radii of curvature of magnitude either 10.0 cm or 20.0 cm. Find the radii of curvature and the focal length of each lens.

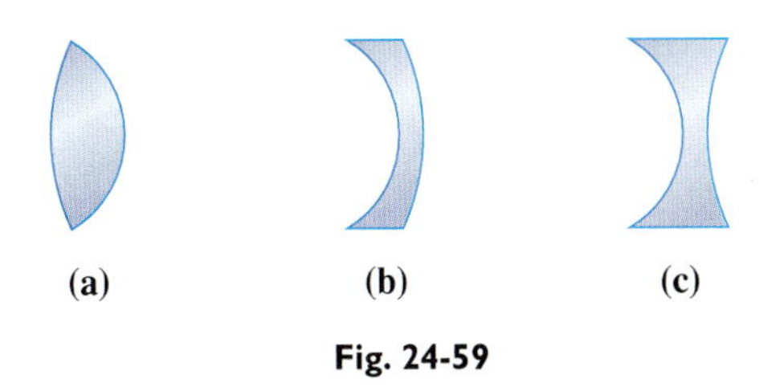

Fig. 24-59

★ **37** Prove that for a symmetric lens of index 1.5 the focal length equals the radius of curvature of the first surface: $f = R_1$.

★ **38** Prove that if a thin lens is turned around, so that convex surfaces become concave and vice versa, the focal length remains the same.

39 A thin lens with index 1.50 has two convex surfaces with radii $R_1 = +5.0$ cm, $R_2 = +15$ cm. Find the image distance for an object 10 cm from the lens.

★ **40** Derive Eq. 24-26; that is, show that for a thin lens of index n, surrounded by a medium of index n_o, object and image distances are related by the equation

$$\frac{1}{s} + \frac{1}{s'} = \left(\frac{n}{n_o} - 1\right)\left(\frac{1}{R_1} - \frac{1}{R_2}\right)$$

★ **41** A crown glass lens ($n = 1.52$) has an optical power of 4.0 d in air. What is its power in benzene ($n = 1.50$)?

42 A symmetric glass lens of index 1.50 has a focal length of −10 cm in air. A plastic bag having the same shape as the glass lens is filled with pressurized air, forming an "air lens" to be used underwater. Find its focal length.

43 Complete the paths of the principal rays in Fig. 24-60, and draw the images.

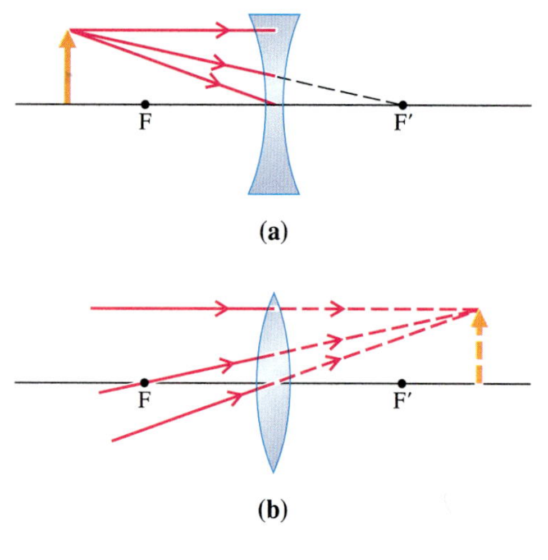

Fig. 24-60

44 Draw principal rays for the objects in Fig. 24-61, and find the images.

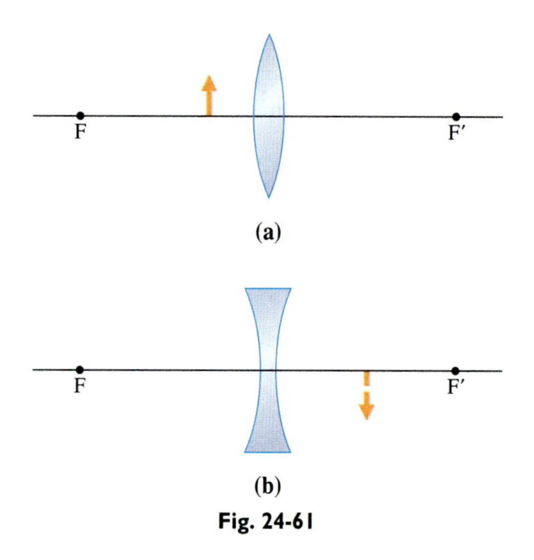

Fig. 24-61

45 A small object is on the optical axis, 25 cm from a +20 cm lens. Find the image distance and magnification, both by calculation and by the graphical method.

46 A small object is 10 cm to the left of a converging lens ($f = +5.0$ cm), and a second diverging lens ($f = -3.0$ cm) is 4.0 cm to the right of the first. Find the position and magnification of the final image, both by calculation and by the graphical method.

47 An object 5.0 mm high is on the optical axis of two lenses with focal lengths $f_1 = +5.0$ cm, $f_2 = +10$ cm. The object is 6.0 cm to the left of the first lens, and the second lens is 35 cm to the right of the first lens. Find the position (relative to the second lens) and the size of the final image, both by calculation and by the graphical method.

Additional Problems

48 What are the minimum dimensions of a rectangular plane mirror used by a woman to see her full image if the woman's height is 170 cm and her maximum width is 40 cm?

49 In the surface of a calm lake you see the image of a tree of height 25 m (Fig. 24-62). The tree is on the opposite shore, 65 m away. How far is the image of the treetop from your eye?

Fig. 24-62

50 A plane mirror is inclined at an angle of 45° with the vertical. A boy standing in front of the mirror tosses a ball vertically upward. In what direction does the ball's image move?

★ **51** A certain camera is able to focus on objects from 1.00 m to ∞. Suppose you want to take an extreme close-up shot, with the camera 20 cm from the object. Find the focal length of a thin lens that could be placed directly in front of the camera lens such that the image of the close object would be in focus (Fig. 24-63).

Fig. 24-63 Improvising a close-up shot.

★ **52** Find the image of the letter "L" in Fig. 24-64.

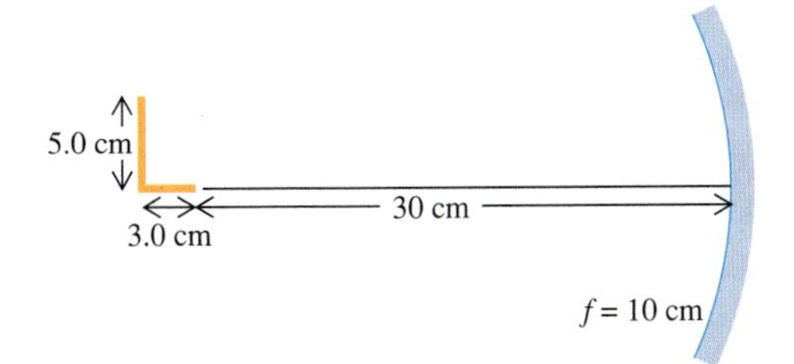

Fig. 24-64

★ **53** Light from a point source P passes through a lens, is reflected by a plane mirror, and passes through the lens again (Fig. 24-65). Find the position of the final image relative to the lens.

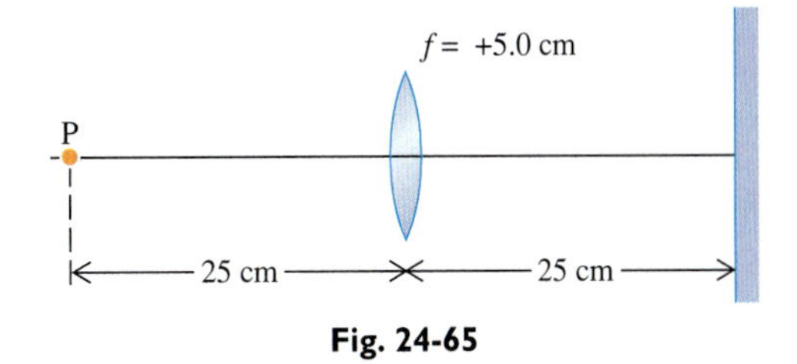

Fig. 24-65

★★ **54** A positive lens forms a real image at point P′ of an object at point P. Where could you position a mirror relative to P′ so that reflected light would pass through the lens again, forming an image at the original object point P? Solve and sketch for (a) a plane mirror; (b) a convex mirror; (c) a concave mirror.

★★ **55** Light from a distant point source on the optical axis is incident on a symmetric negative lens. Some of the light is reflected from the front surface, and some is transmitted. Thus the surface of the lens acts as a concave spherical mirror. Find the index of the glass if the image formed by reflected and transmitted light are at the same point.

CHAPTER 25 The Eye and Optical Instruments*

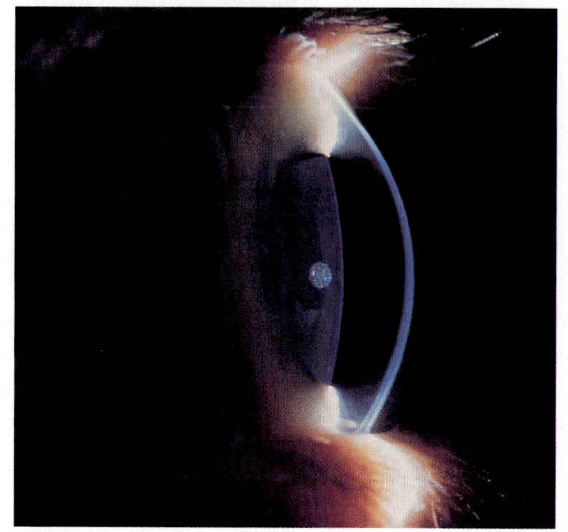

The human eye produces an exquisitely detailed and accurate image of the external world. What we see, however, is certainly not as simple as a photographic image. Vision is a dynamic, selective process in which the eye and brain work together. Some objects "seen" by the eye are hardly sensed at all, but other minor details or fleeting impressions are not only sensed, but also may be consciously or unconsciously stored in memory. The eye not only produces an image, but also processes the image, partially analyzing it. Thus the eye may be regarded as an extension of the brain.

Look at your hand and consider what is happening. Light reflected from your hand enters your eye and is refracted first at the eye's outer surface and again at surfaces within the eye. A real image of your hand is formed on your retina, at the back of your eye. Photosensitive receptors in your retina absorb light where the image is formed; these receptors then undergo a chemical change and generate an electrical potential difference that propagates along a receptor cell in a manner similar to nerve transmission (described in the Chapter 20 essay on electrical effects in the human body). However, nerve impulses are not transmitted directly from the image receptors to your brain. Electrical pulses from various receptors combine and interact in the retina in a complex process, from which emerge electrical signals in the form of a series of pulses. The signals contain coded information about features of the image. The optic nerve carries these signals from the retina to the visual center of your brain. The result of this entire process is the sensation of sight.

*This chapter may be covered after Chapter 26.

Although there is much that is not yet understood about image reception and interpretation, image formation by the eye is easily understood on the basis of simple optical principles. In this chapter we shall describe in detail the structure of the eye and the process of image formation. We shall also explain how various optical instruments, such as the microscope and the telescope, can magnify images. Color sensitivity of the eye is discussed in an essay at the end of the chapter.

25-1 The Human Eye

Structure of the Eye

Fig. 25-1 shows a horizontal cross section of a human eye with typical dimensions. As light enters the eye, it passes first through the **cornea,** a thin transparent membrane of approximately spherical shape, with a refractive index of **1.38.** Most of the refraction of light occurs at the air-cornea surface, since the change in the index is much greater there than at any of the other surfaces inside the eye.

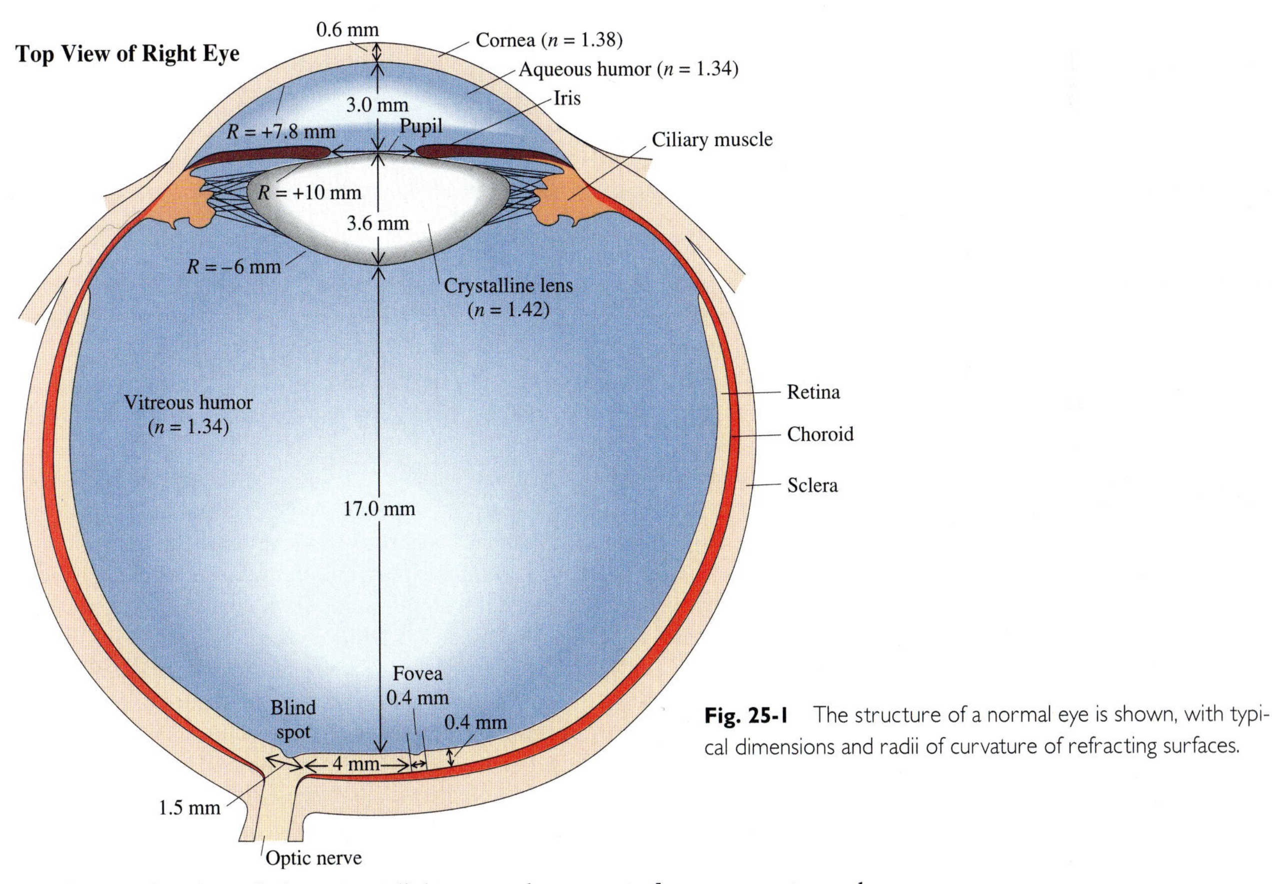

Fig. 25-1 The structure of a normal eye is shown, with typical dimensions and radii of curvature of refracting surfaces.

After passing through the cornea, light enters the **aqueous humor,** a watery substance with a refractive index of **1.34.** The incoming light then encounters the **iris,** the colored curtain that opens and closes to regulate the amount of light entering the eye. The opening itself is called the **pupil.** Light entering the pupil passes through the **crystalline lens.** The lens is a small, pliable, transparent body, consisting of layers with different refractive indices. The effective overall index of the lens is **1.42.** The **ciliary muscles** attach the lens to the sides of the eye.

Light next passes through the **vitreous humor,** which is a watery substance similar to the aqueous humor and having the same index of **1.34.**

Finally light passes into the **retina,** forming an image on its back surface, where the photosensitive receptors are located. The most sensitive part of the retina is the **fovea,** a small section approximately 0.4 mm in diameter, near the center of the retina. It is the part of an image falling on the fovea that is most clearly seen. Approximately 4 mm from the fovea there is a blind spot, where there are no photosensitive receptors and where the nerves from the various parts of the retina come together and leave the eye on their way to the brain. The blind spot is approximately 1.5 mm in diameter. Its effect is described in Problem 1.

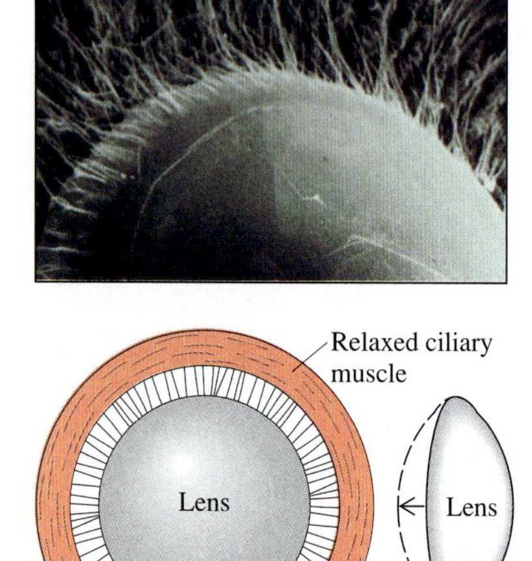
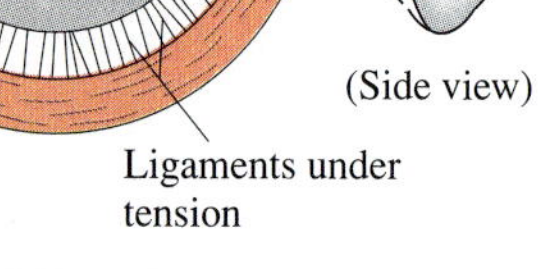

Fig. 25-2 When the circular ciliary muscle fibers contract, tension in the suspensory ligaments is reduced, and the front of the lens bulges outward.

Accommodation

Most of the refraction of light rays entering the eye occurs as the light travels from air through the cornea to the aqueous humor. However, essential additional converging power is provided by the crystalline lens. (That it acts as a converging lens is apparent from its shape and the fact that it is immersed in a medium of lower refractive index.) The curvature of the crystalline lens is quite variable, especially in a young eye. It is this property of the lens that allows the eye to focus on objects at different distances. If the lens were rigid, there could be only one object distance for which a perfect image is formed on the retina. But the normal eye can, of course, focus on objects over a wide range of distances, from very close objects to very distant ones. The closest object point is called the **near point** and the farthest object point is called the **far point.** The eye can focus on objects anywhere between these points. For a young adult with normal vision, the near point is typically 25 cm from the eye and the far point is at infinity.

This process of adjusting the lens for varying object distances, called "accommodation," occurs without conscious effort, through the action of ciliary muscles. When the ciliary muscles are relaxed, the lens is relatively flat, and distant objects form a clear image on the retina. When the ciliary muscles contract, the front surface of the lens becomes more curved, giving the lens a shorter focal length or higher power, which is necessary to form a sharp image on the retina when the object is close to the eye (Fig. 25-2).

Fig. 25-3 shows light rays from a point source imaged by the eye under various conditions. The drawings show the actual size and shape of a young, normal eye, with typical dimensions and curvatures, capable of focusing from 10 cm to ∞.

If a very close object is viewed with the ciliary muscles relaxed, as in Fig. 25-3b, the optical power of the eye is not sufficient to produce an image on the retina. Instead the image point is behind the retina. The light rays converging toward the image point are detected by the retina, but the receptors are spread over an extended circular area, called a "circle of confusion." This means that the detected image of an extended object consists of overlapping circles of confusion (Fig. 25-4). Points on the object can be seen as distinct and separate only for those object points that are sufficiently far apart that their image circles do not overlap. Thus the finer details of the image are lost, and the visual sensation is that the image is blurred or fuzzy (Fig. 25-5).

Fig. 25-3c shows the effect of exerting the ciliary muscles. The lens becomes more curved and the refracting power of the eye is therefore increased so that the image point of a close object point is on the retina. The eye has focused on the close object. For any particular object distance, the eye must accommodate. By adjusting the curvature of the lens between the extremes shown in Figs. 25-3a and 25-3c the eye can focus on objects located anywhere from 10 cm to ∞.

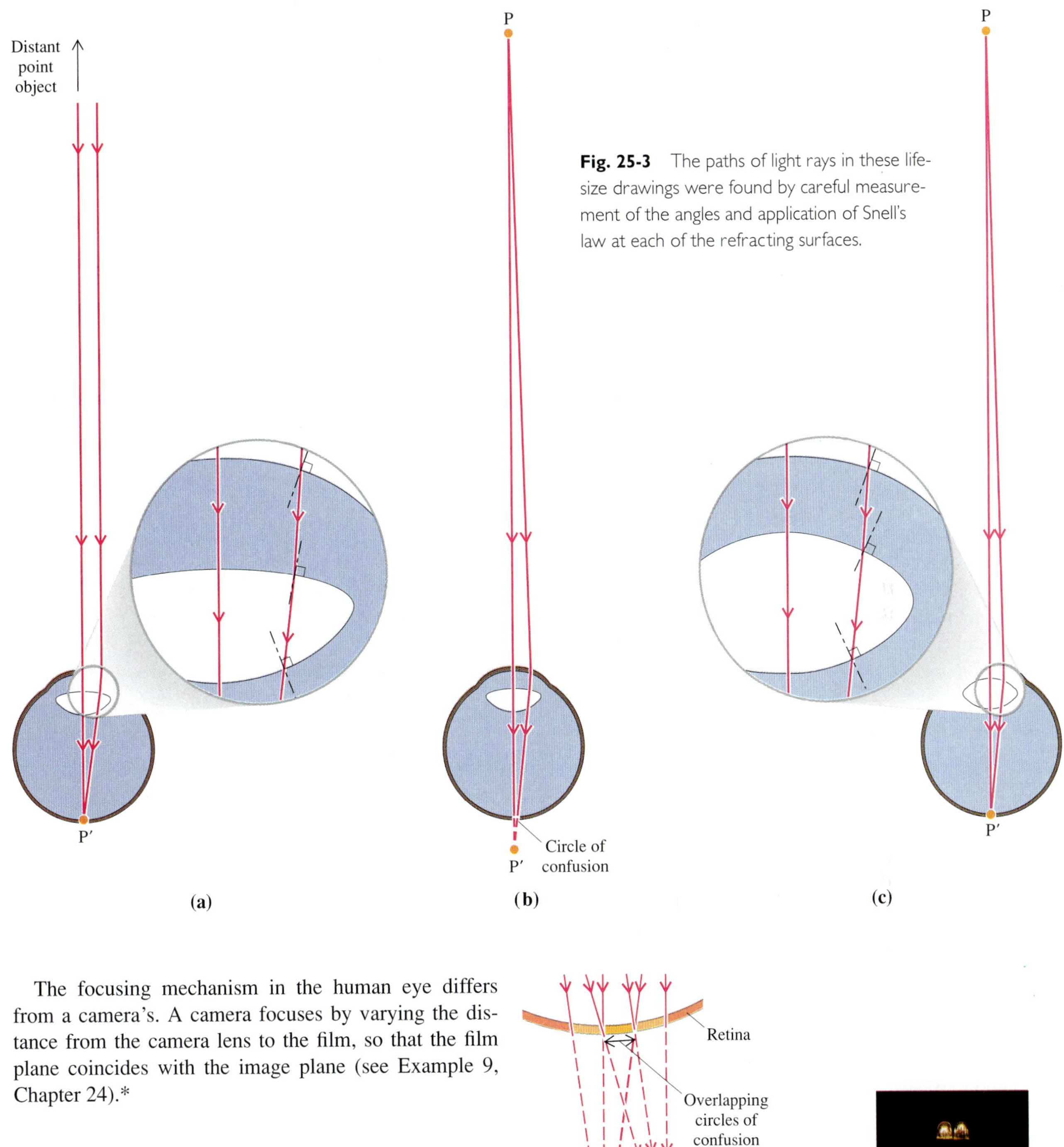

Fig. 25-3 The paths of light rays in these life-size drawings were found by careful measurement of the angles and application of Snell's law at each of the refracting surfaces.

The focusing mechanism in the human eye differs from a camera's. A camera focuses by varying the distance from the camera lens to the film, so that the film plane coincides with the image plane (see Example 9, Chapter 24).*

Fig. 25-4 An out-of-focus eye forms an image behind the retina, not on it. Rays converging toward each image point form a circle of confusion where they strike the retina.

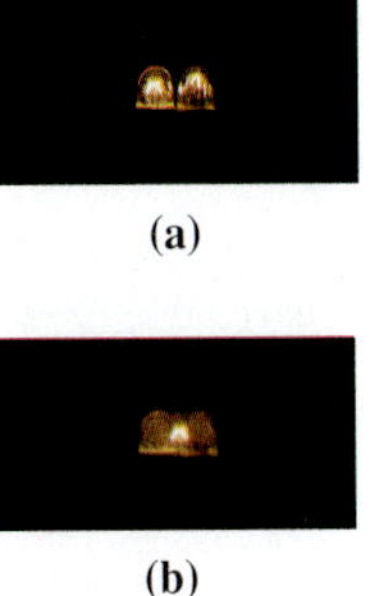

Fig. 25-5 Two small lights, as seen with the eyes **(a)** in focus; **(b)** out of focus.

*In certain fish the focusing mechanism is similar to that used in a camera. The distance between the lens and the retina is adjusted for viewing objects at different distances. When the eyes are in the relaxed state, these fish can see only objects that are very close, which is exactly opposite the case for human vision. Still another focusing mechanism is used by owls and hawks. These birds focus by changing the curvature of the cornea.

EXAMPLE 1 How The Eye Produces an Image

Show that light from a distant object point on the eye's optical axis forms an image on the retina when the eye is relaxed and has the dimensions and curvatures given in Fig. 25-1.

SOLUTION We can locate the final image by repeatedly applying the equation for image formation by a spherical refracting surface; the image produced by each surface is used as the object for the next surface. We can simplify the calculation somewhat by recognizing that the cornea is thin and has approximately parallel sides so that the refraction that occurs as light travels from air to cornea to aqueous humor is essentially the same as though the aqueous humor were directly in contact with the air. (We saw a similar effect when light passed straight through a windowpane in Example 4, Chapter 23.) Thus there are only three surfaces to be considered: (1) air to aqueous humor; (2) aqueous humor to lens; and (3) lens to vitreous humor.

(1) We apply Eq. 24-14, using an infinite object distance at the first surface, a convex surface from air ($n = 1.00$) to aqueous humor ($n' = 1.34$), with a radius of curvature $R = 7.8$ mm.

$$\frac{n}{s} + \frac{n'}{s'} = \frac{n' - n}{R}$$

$$\frac{1}{\infty} + \frac{1.34}{s'} = \frac{1.34 - 1}{7.8 \text{ mm}}$$

$$s' = 31 \text{ mm}$$

Fig. 25-6 shows the first image point P_1', 31 mm behind the cornea.

(2) The point P_1' serves as a virtual object 27 mm behind the second surface—a convex surface from aqueous humor ($n = 1.34$) to lens ($n' = 1.42$), with $R = 10$ mm. We treat the object distance as negative and again apply Eq. 24-14.

$$\frac{n}{s} + \frac{n'}{s'} = \frac{n' - n}{R}$$

$$\frac{1.34}{-27 \text{ mm}} + \frac{1.42}{s'} = \frac{1.42 - 1.34}{10 \text{ mm}}$$

$$s' = 25 \text{ mm}$$

The second image point P_2' is 25 mm behind the front surface of the lens, as shown in the figure.

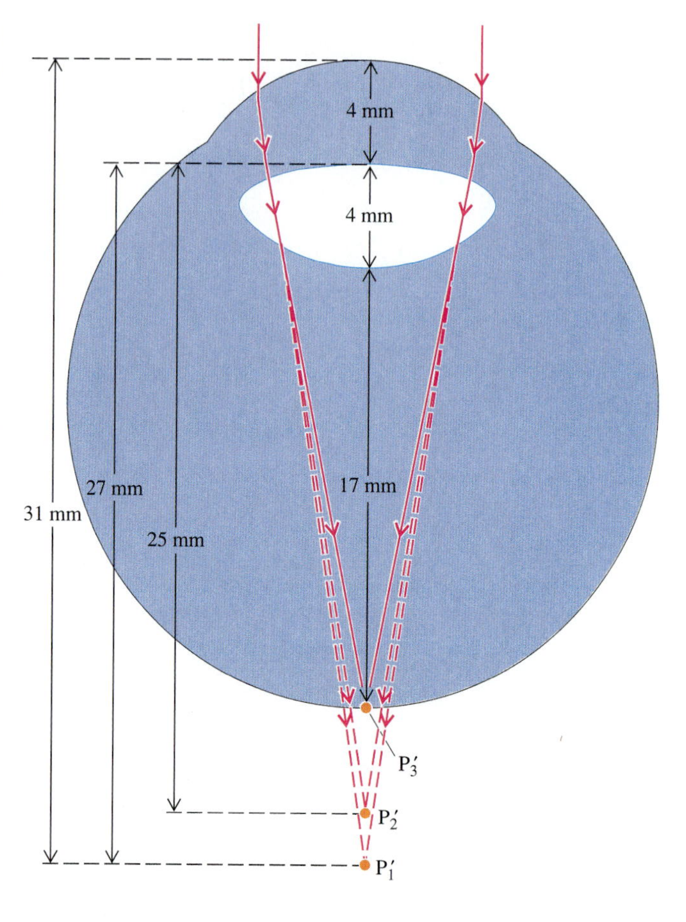

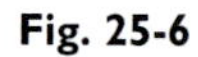

Fig. 25-6

(3) Point P_2' serves as a virtual object 21 mm behind the third surface—a concave surface from lens ($n = 1.42$) to vitreous humor ($n' = 1.34$), with $R = -6.0$ mm. Applying Eq. 24-14 once again, we find

$$\frac{n}{s} + \frac{n'}{s'} = \frac{n' - n}{R}$$

$$\frac{1.42}{-21 \text{ mm}} + \frac{1.34}{s'} = \frac{1.34 - 1.42}{-6.0 \text{ mm}}$$

$$s' = 17 \text{ mm}$$

This is exactly the distance from the lens to the retina, showing that the final image is located on the retina.

Defects in Image Formation; Correction of Vision

Myopia, or **nearsightedness,** is a defect of the eye in which the eye is unable to clearly see distant objects. Myopia occurs when the image formed by the relaxed eye is in front of the retina. For each point source, the light striking the retina is spread over a circle of confusion (Fig. 25-7). The sensation is blurred vision (Fig. 25-8), similar to the case of an unfocused normal eye.

Myopia results when the combined optical power of the cornea and lens is too great for the length of the eyeball. The eye can see a clear image only if the object is brought close to the eye. The far point (the farthest point at which an object can be seen) is often less than 1 m, compared to a far point at ∞ for a normal eye.

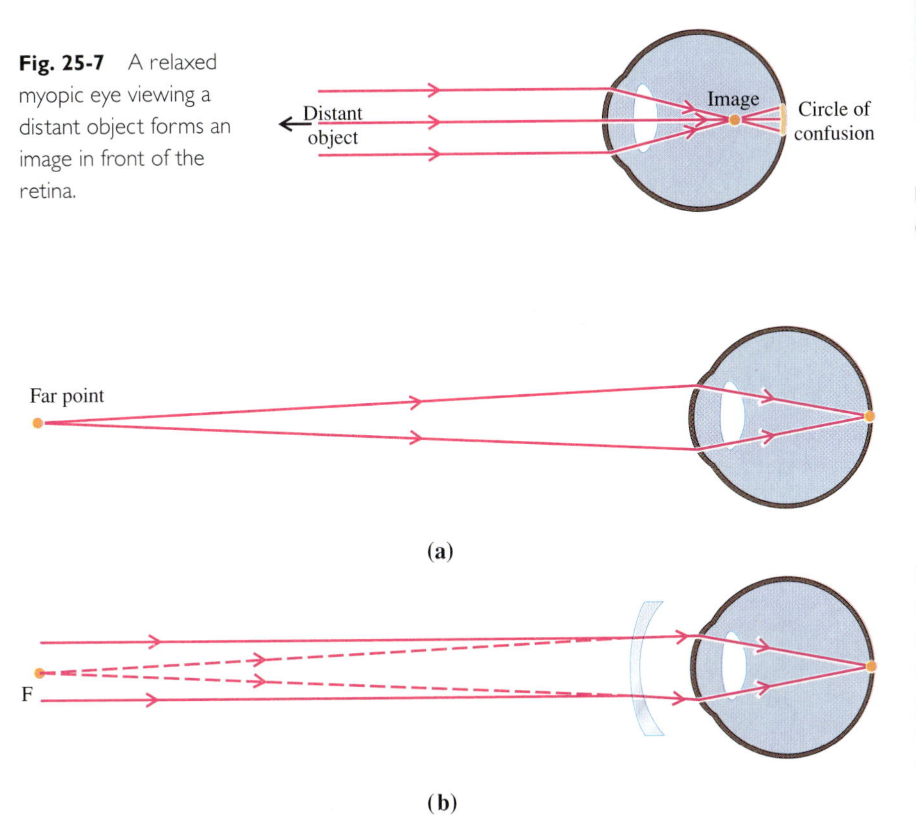

Fig. 25-7 A relaxed myopic eye viewing a distant object forms an image in front of the retina.

(a)

(b)

Fig. 25-8 A distant view as seen by **(a)** a normal eye; **(b)** a myopic eye.

Fig. 25-9 **(a)** A relaxed myopic eye views an object at the eye's far point. **(b)** Myopia is corrected by placing in front of the eye a negative lens, the focal point of which is the eye's far point. The relaxed eye then sees rays from a distant object apparently diverging from the eye's far point.

Nearsightedness can be corrected with eyeglasses or contact lenses.* An appropriate negative lens is placed in front of each eye, so that light from a distant object diverges as though it were coming from the eye's far point (Fig. 25-9). The lens produces an image of a distant object at the eye's far point. Thus **the far point is the focal point of the lens.**

A near-sighted person usually has a normal range of optical power—that is, a normal ability to vary the shape of the crystalline lens. This results in the near point of the myopic eye being closer than that of a normal eye. A lens designed to correct the far point of a myopic eye will also affect the near point, as in the following example.

*Surgery is also sometimes used in extreme cases to change the shape of the cornea.

EXAMPLE 2 Correcting Myopia

(a) Find the optical power of a lens necessary to correct an eye with a far point of 50 cm. Neglect the distance from the lens to the cornea.
(b) If the eye's near point is 10 cm, what is the eye's corrected near point?

SOLUTION **(a)** A diverging or negative lens with a focal length of magnitude 50 cm is required.

$$f = -50 \text{ cm} = -0.50 \text{ m}$$

Thus the power of the lens is

$$P_o = \frac{1}{f} = \frac{1}{-0.50 \text{ m}} = -2.0 \text{ d}$$

(b) An object will be at the "corrected near point" when the image produced by the lens is at the eye's actual near point. Here the eye's near point is 10 cm in front of the lens. Thus we must find the object distance s for which the lens produces a virtual image 10 cm in front of the lens, that is, for which $s' = -10$ cm. Applying the thin lens equation, we find

$$\frac{1}{s} = \frac{1}{f} - \frac{1}{s'} = P_o - \frac{1}{s'}$$

$$= -2.0 \text{ d} - \frac{1}{-0.10 \text{ m}} = +8.0 \text{ m}^{-1}$$

or

$$s = \frac{1}{8.0 \text{ m}} = 13 \text{ cm}$$

The original range of vision for this myopic eye was 10 cm to 50 cm. After correction, the range becomes 13 cm to ∞.

Hyperopia, or **farsightedness,** is a defect of the eye that occurs when the combined optical power of the cornea and lens is too little for the length of the eyeball. This results in the inability of the eye to see close objects clearly. When objects are placed at a normal near point, the hyperopic eye is unable to converge the rays, even with maximum accommodation. Instead, the image point is behind the retina, resulting in a blurred image (Fig. 25-10). To be clearly seen, an object must be moved away from the eye—beyond the eye's near point, which is often more than 50 cm away for a young adult with hyperopia.

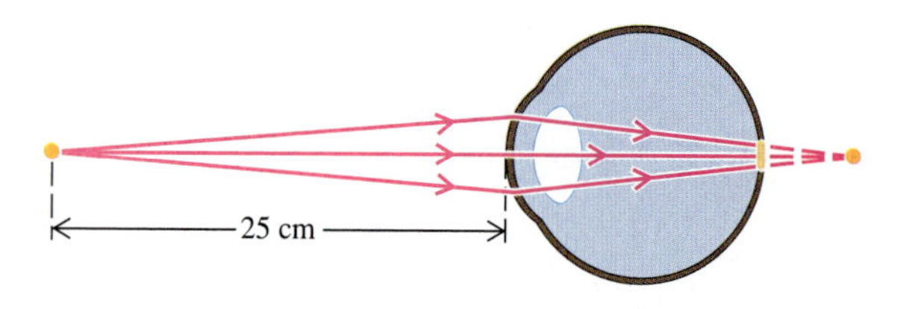

Fig. 25-10 A hyperopic eye, viewing a close object, forms an image behind the retina, even though the eye is focused as close as possible through maximum exertion of ciliary muscles and maximum power of the crystalline lens.

Farsightedness can be corrected by the appropriate positive lens. The lens must bend the rays diverging from an object so that the light entering the eye appears to be coming from a point farther away from the eye than the actual object. For example, when correcting hyperopia for close work, such as reading, the rays must appear to come from the eye's near point (Fig. 25-11).

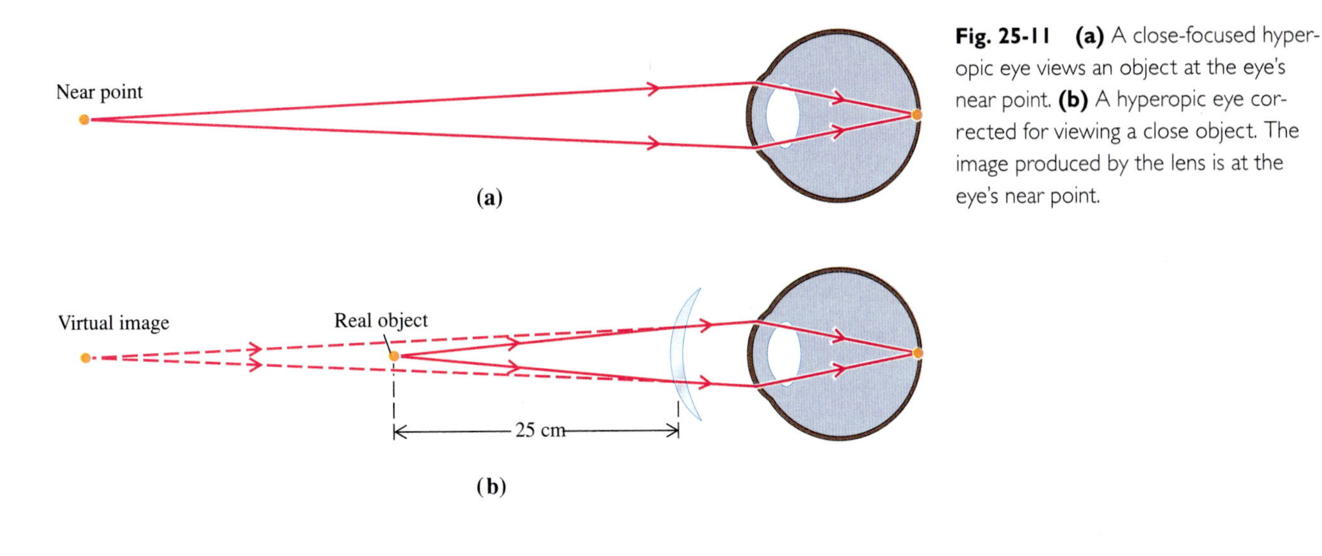

Fig. 25-11 **(a)** A close-focused hyperopic eye views an object at the eye's near point. **(b)** A hyperopic eye corrected for viewing a close object. The image produced by the lens is at the eye's near point.

EXAMPLE 3 Correcting Hyperopia

Find the power of a lens designed for reading purposes to correct an eye with a near point of 75 cm. Assume a standard reading distance of 25 cm. Neglect the distance from the corrective lens to the cornea.

SOLUTION A converging or positive lens is required, with a power such that when an object is placed 25 cm in front of the lens, the lens forms a virtual image as shown in Fig. 25-11b. The image distance $s' = -75$ cm, so that this image serves as an object for the eye at the eye's near point. Applying the thin lens equation, we find

$$P_o = \frac{1}{f} = \frac{1}{s} + \frac{1}{s'} = \frac{1}{0.25 \text{ m}} + \frac{1}{-0.75 \text{ m}}$$
$$= 2.7 \text{ d}$$

Fig. 25-12 The image of a printed page is brought into sharp focus for a hyperopic eye when reading glasses are used.

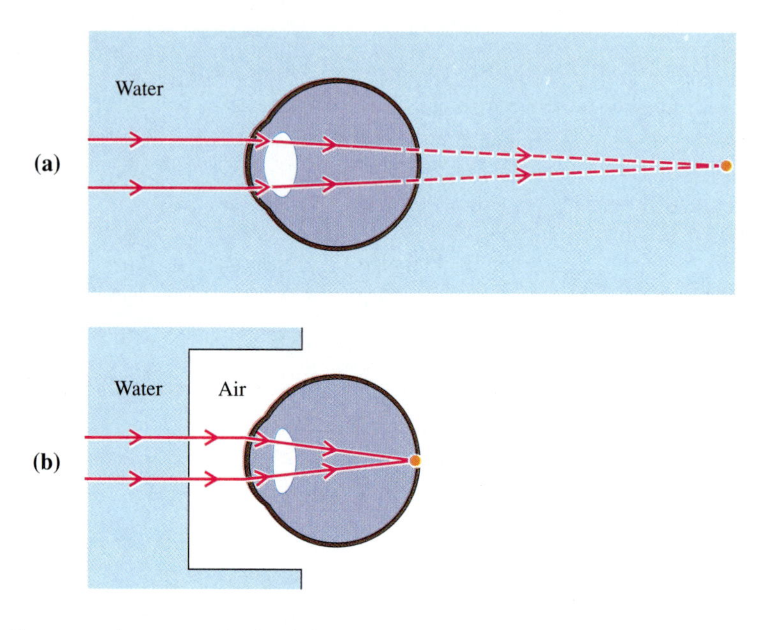

Fig. 25-13 Underwater vision **(a)** without goggles; **(b)** with goggles.

Everyone is extremely farsighted under water. When the eye is in direct contact with water, it is impossible to form a clear image of an object at any distance. Even for very distant objects, the optical power of the human eye under water is not sufficient to form an image on the retina (Fig. 25-13a). The reason for this is quite simple. There is very little bending of light as it enters the eye, because the refractive index of water (1.33) is too close to the indices of the eye's media. (In particular, the aqueous humor index of 1.34 differs from the index of water by less than 1%.) However, if swimming goggles or a diving mask is worn, the normal eye can form a clear image on the retina (Fig. 25-13b). The layer of air next to the eye establishes the essential air-cornea refracting surface.

Astigmatism is a vision defect resulting from a lack of symmetry of the eye with respect to the optical axis. This is usually caused by irregular curvature of the cornea. For example, the cornea's refractive power for rays along the horizontal may be different from its power for rays along the vertical. Astigmatism is corrected with cylindrical rather than spherical lenses, as shown in Question 9.

As one grows older, the crystalline lens gradually loses its flexibility, thereby reducing the eye's range of accommodation, so that the near point gradually recedes with age. This condition is called "presbyopia." Fig. 25-14 shows a graph of the variation of the near point with age for those with normal vision. The graph indicates that after about 50 years of age the normal near point is beyond 30 cm, which is a typical reading distance. Thus reading glasses are usually necessary at about that age for one who has normal vision. For one who is nearsighted, the aging of the lens often means that glasses worn to correct myopia must be removed when reading, or bifocals must be worn. See Problem 15.

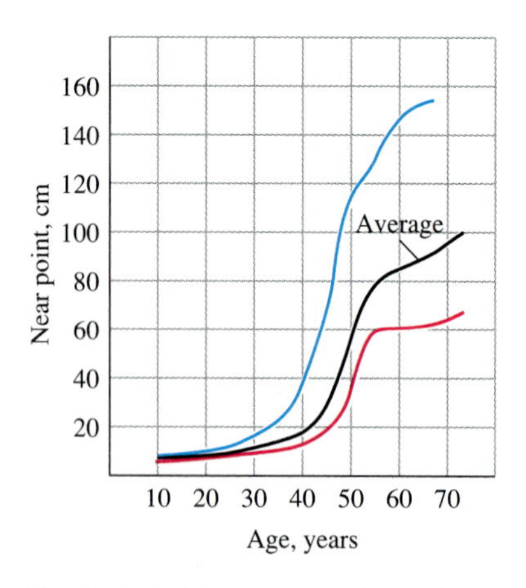

Fig. 25-14 The average value of the near point is shown as a function of age, for one with normal vision. The curves to either side of the average curve indicate the range of variation in the near point among normal individuals.

Fig. 25-1 shows "typical" dimensions of the normal eye. However, there is considerable variation of these dimensions among individuals. The length of the eyeball varies from 22 mm to 26 mm, and the shape of the cornea and lens vary in such a way that corneal power may vary by as much as 8 d, and the power of the relaxed lens may vary by as much as 10 d. If these parameters were combined in a random fashion, the occurrence of myopia and hyperopia would be far greater than it actually is. Apparently, in the early years of life, while the eye is still growing, there is a tendency for the length of the eye to adjust to corneal and lens shape, in such a way that normal vision results. Genetic factors seem to play a role in determining whether the developing eye ends up normal, myopic, or hyperopic.

Extended Objects; The Reduced Eye

In one sense, we see everything upside down; that is, our eyes form real, inverted images of the external world on our retinas, just as a positive lens forms a real, inverted image. However, seeing is much more than just image formation, and our eye-brain perception of the world is right side up.

The idea that the eye forms an inverted image was hard to accept when first proposed by Kepler in the seventeenth century. But the phenomenon was demonstrated by the German astronomer Christoph Scheiner, who removed the outer layers from an animal's eye and, looking through the transparent retina, observed on the retina a small, inverted image of objects in front of the eye.

One can sometimes observe the image formation in a living human eye. If the choroid layer behind the retina is very light, as is often the case for a person of light complexion, the image of a bright object actually shines through the eyeball. If a suitable subject is placed in a dark room, with the right eye turned toward a lighted candle far to the right side, one can see through the left side of the right eye, a small, inverted image of the candle flame shining through.

By drawing a ray diagram for an extended object, we can show that the image formed on the retina is inverted, and we can determine its size. Fig. 25-15 shows rays from a point P at the tip of a distant object arrow. As these nearly parallel rays enter the eye, they are refracted at the cornea and lens surfaces and converge to an image point P′ on the retina, below the optical axis, as shown in the figure.

Notice that one of the rays focused at P′ follows a nearly straight line from P to P′. We can locate the image point P′ simply by drawing a single straight ray from the object point through a point O on the optical axis to the retina. Thus point O is like the center of a thin lens, and **the ray drawn through O is like the principal ray drawn straight through the center of a thin lens.**

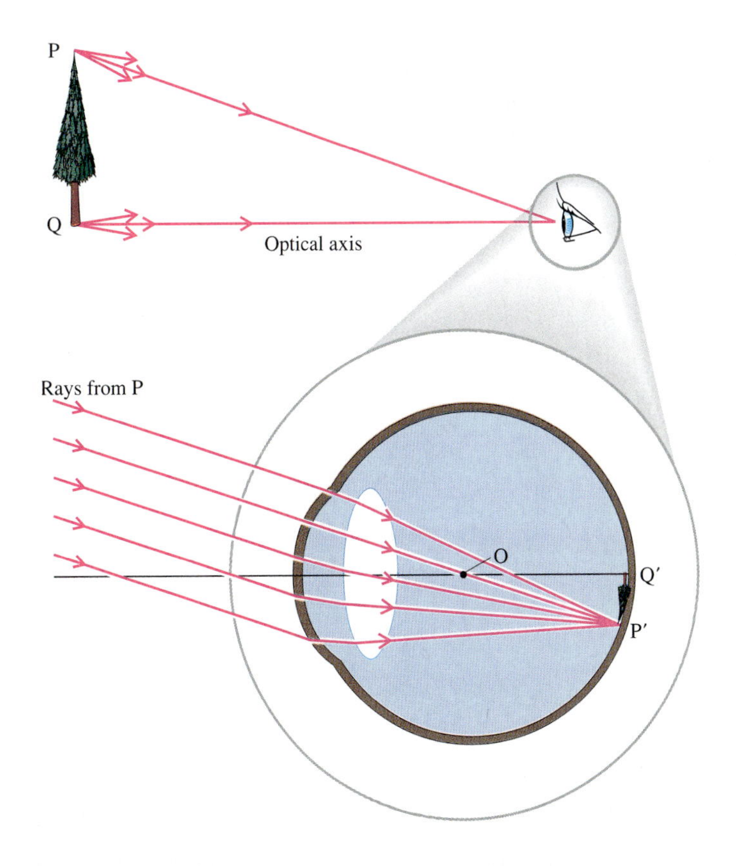

Fig. 25-15 Rays from an object point P converge to an image point P′ on the retina. One ray travels in a straight line from P to P′, crossing the optical axis at O.

The point O is always located 16 mm in front of the retina, independent of object distance and object size. When the eye accommodates for vision at the near point, point O moves only about $\frac{1}{2}$ mm. The ray through O is very useful in determining the size of an image, without the need for drawing a detailed ray diagram. We simply use a 24 mm diameter circle to represent the eye, draw an optical axis along the line of sight, and place point O on the axis 16 mm in front of the retina. This drawing, shown in Fig. 25-16, is called the **reduced eye.** From the figure, we see that if an object subtends an angle θ at the eye, the height h of the image is given by

$$h = (16 \text{ mm})(\tan \theta) \tag{25-1}$$

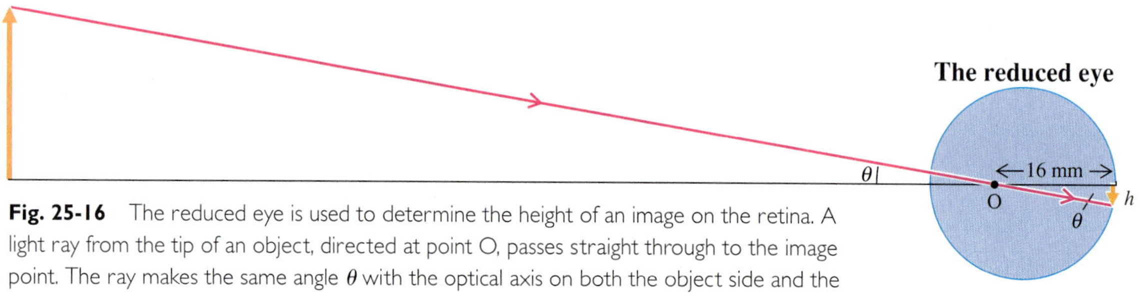

The reduced eye

Fig. 25-16 The reduced eye is used to determine the height of an image on the retina. A light ray from the tip of an object, directed at point O, passes straight through to the image point. The ray makes the same angle θ with the optical axis on both the object side and the image side. It is this angle that determines the image height h.

EXAMPLE 4 Image of a Mountain on a Retina

Find the size of the image of a mountain 3 miles high, in the eye of an observer located 12 miles from the mountain.

SOLUTION From Fig. 25-17, we see that the mountain subtends an angle θ, where $\tan \theta = \frac{3 \text{ mi}}{12 \text{ mi}} = \frac{1}{4}$. Applying Eq. 25-1, we obtain the size of the image on the observer's retina.

$$h = (16 \text{ mm})(\tan \theta) = (16 \text{ mm})(\tfrac{1}{4}) = 4 \text{ mm}$$

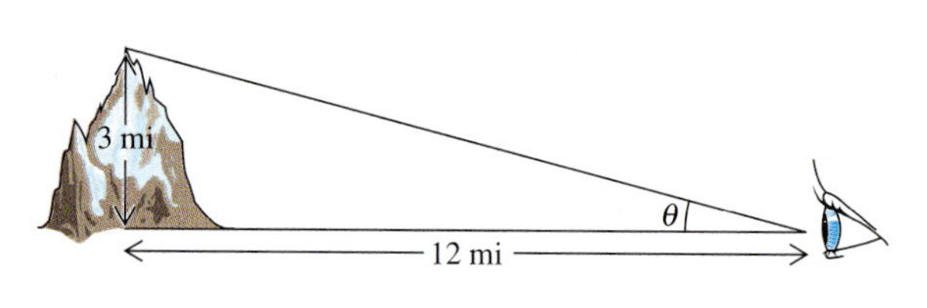

Fig. 25-17

This example shows how even a very large object, filling a large part of the field of view, produces a very small image on the retina.

Visual Acuity

Visual acuity is the eye's ability to see fine detail. A measure of acuity is the minimum angle subtended by two points that are barely able to be resolved by the eye, that is, seen as separate points (Fig. 25-18). This minimum angle, which depends on the individual, is easy to measure. A subject with normal or corrected vision views two dots, 1 mm apart, from a distance of about 5 m. The subject will be able to see the dots but will be unable to distinguish them; they will appear to merge into a single mark. If the subject then approaches the dots, at some distance (usually about 2 or 3 m) the dots begin to appear separate. The dots then subtend the smallest angle resolvable by the subject's eyes. That angle, θ_{min}, measured in radians, is the ratio of the distance between the dots (1 mm) to the distance d of the dots from the eye (Fig. 25-19).

$$\theta_{min} = \frac{1 \text{ mm}}{d} \tag{25-2}$$

The value of θ_{min} depends not only on the individual tested, but also on conditions of illumination. Acuity is somewhat better (θ_{min} is smaller) for monochromatic light than for white light. Acuity also improves as the brightness of the light is increased to moderately high intensity. The eye's limited acuity results from several factors, including aberrations and diffraction, which are discussed in the optional Section 25-5.

Under conditions of good illumination with white light, the distance d might be 2 m, for example. Then, applying Eq. 25-2, we find

$$\theta_{min} = \frac{10^{-3}\text{ m}}{2\text{ m}} = 5 \times 10^{-4}\text{ rad} \tag{25-3}$$

For most subjects the minimum angle is between 4×10^{-4} rad and 5×10^{-4} rad, but for some individuals with exceptional acuity, the minimum angle may be as small as 2×10^{-4} rad.

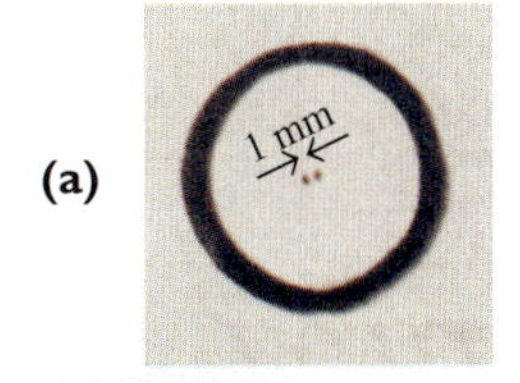

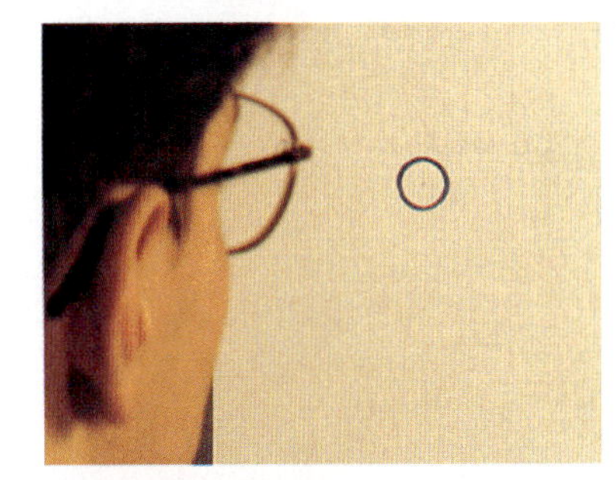

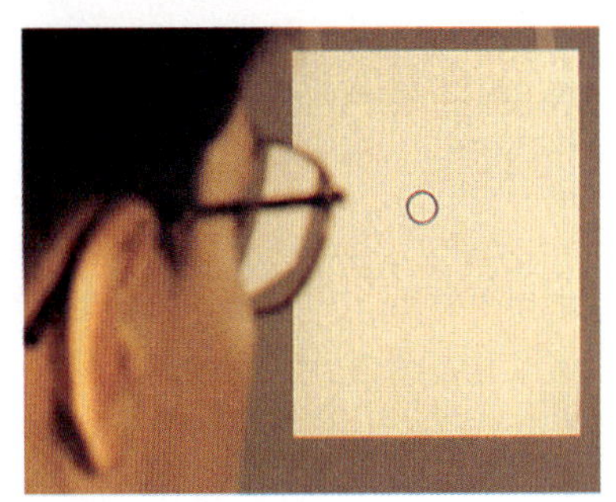

Fig. 25-18 **(a)** Two pencil dots 1 mm apart. **(b)** The viewer is just close enough to the dots to be able to distinguish them. **(c)** The viewer is farther from the dots than he is in (b). Consequently, though he can see a mark inside the circle, he can't see two separate dots.

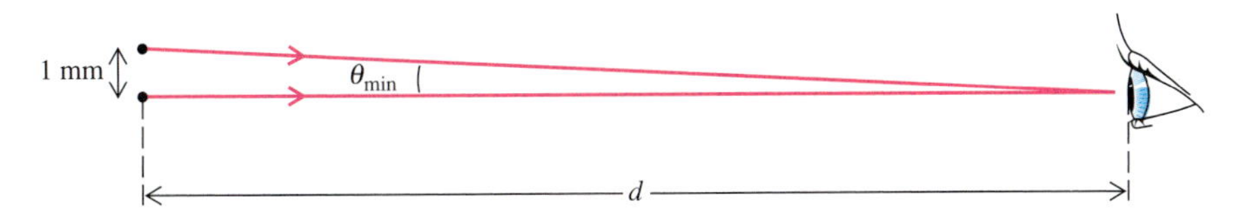

Fig. 25-19 Two dots 1 mm apart are viewed from a distance d. The angle θ_{min} subtended by the points decreases as the distance increases.

EXAMPLE 5 Identifying a Face at a Distance

What is the minimum distance between two points that you are able to resolve from a distance of 100 m, if you have normal visual acuity? Could you recognize a familiar face 100 m away?

SOLUTION The angle θ_{min} subtended by these points is 5×10^{-4} radians (Eq. 25-3). Using Fig. 25-20, we can express the separation s of the points in terms of θ_{min}.

$$\theta_{min} = \frac{s}{r}$$

$$s = r\theta_{min} = (100\text{ m})(5 \times 10^{-4}\text{ rad})$$
$$= 5 \times 10^{-2}\text{ m} = 5\text{ cm}$$

At a distance of 100 m you will not be able to distinguish details smaller than 5 cm. Thus you may not be able to identify a familiar face at this distance (Fig. 25-21).

Fig. 25-20

θ_{min} — s — $r = 100$ m

Fig. 25-21

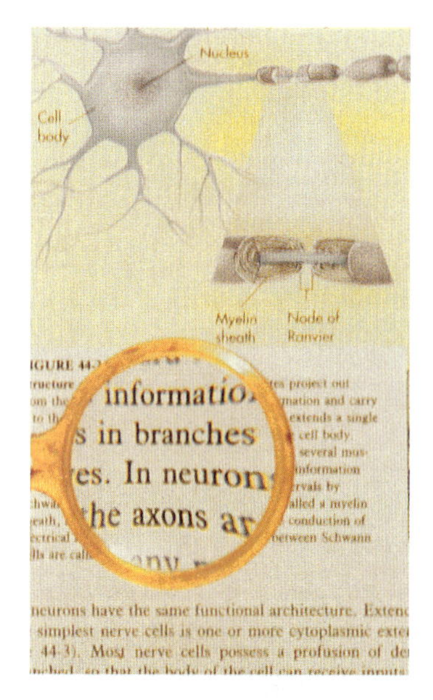

Fig. 25-22 Fine print seen under a magnifier.

25-2 The Magnifier

The magnifying glass, the microscope, and the telescope are optical instruments, all of which have the same general purpose: to increase the size of the retina's image of an object viewed through them (Fig. 25-22). The particular instrument used depends on whether the object is close or distant and how much magnification is required.

Suppose you want to examine a small object that can be brought close to your eye. If you view the object without an instrument, it will appear largest and will be seen in greatest detail if you bring it as close as possible to your eye—to your near point, so that the size of the image on your retina is as large as possible. (If the object were brought inside the near point, the retinal image would be even larger, but the image would then be blurred.)

When placed at the near point, the object subtends some angle θ at the eye, as illustrated in Fig. 25-23a for a typical near point of 25 cm. The figure shows the central ray drawn from the tip of the object arrow straight through point O in the reduced eye to the tip of the image arrow on the retina. The retinal image subtends the same angle θ as the object does. The length of the image is proportional to this angle.

A **magnifying glass,** or **magnifier,** is a converging lens or lens system used to increase the size of the image on the retina when an object is viewed through it. The magnifier produces an image, which serves as an object for the eye. This object subtends a larger angle at the eye than the original object. This is illustrated in Fig. 25-23b for a typical magnifier ($f = +10$ cm), with the object placed at the first focal point of the lens, so that rays from a single point on the object are parallel as they emerge from the lens. The parallel light rays entering the eye appear to be coming from a point at ∞. The object subtends an angle θ' at the lens and the image on the retina subtends the same angle θ'. The focal length of the lens is less than the near-point distance for the eye. Thus the object is closer to the lens than it was previously to the eye, and the angle θ' is greater than θ. This means that the image on the retina is bigger and the object therefore appears bigger.

A measure of the magnifying power of an optical instrument is the angular magnification M, which we define to be the ratio of the respective angles subtended by the retinal images with and without the instrument:

$$M = \frac{\theta'}{\theta} \tag{25-4}$$

We can use Fig. 25-23 to obtain an expression for the angular magnification of a magnifier used under conditions of relaxed viewing by a normal eye. We assume that angles θ and θ' are small. From Figs. 25-23a and 25-23b we see that $\theta \approx \frac{y}{d}$ and $\theta' \approx \frac{y}{f}$. Thus

$$M = \frac{\theta'}{\theta} \approx \frac{y/f}{y/d}$$

or

$$M = \frac{d}{f} \qquad \text{(for relaxed viewing)} \tag{25-5}$$

For the special case illustrated in the figure, $d = 25$ cm and $f = 10$ cm, which gives

$$M = \frac{25 \text{ cm}}{10 \text{ cm}} = 2.5$$

The image on the retina is 2.5 times larger with the magnifying glass than without it.

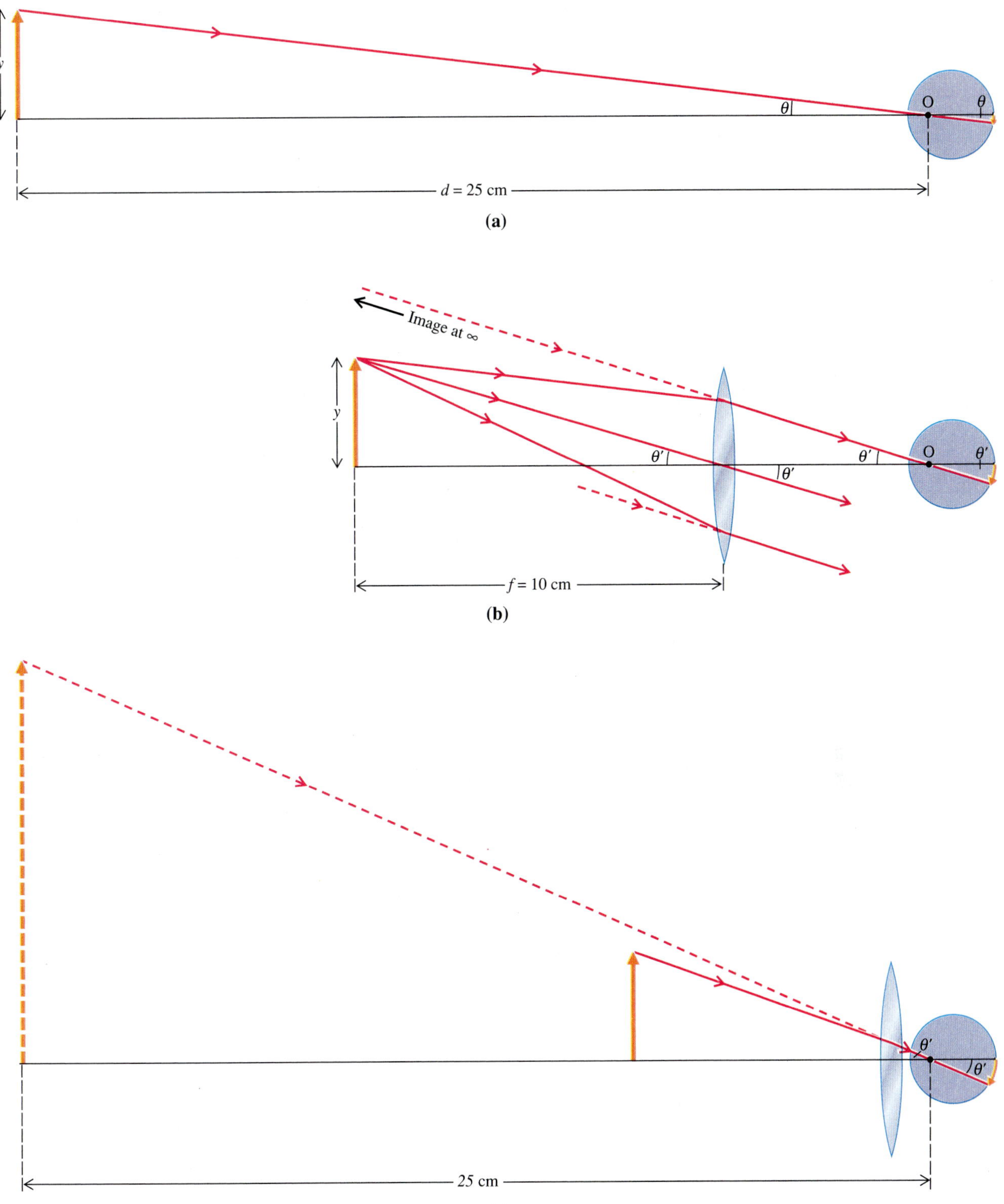

Fig. 25-23 Scale drawing (about two-thirds life size) of the eye's view of an object when seen **(a)** by the unaided eye at the eye's near point; **(b)** through a magnifying glass with the object placed at the focal point of the lens; **(c)** with the lens and object as close as possible to the eye, so that the image in the lens is at the eye's near point.

We have obtained Eq. 25-5 as an expression for the angular magnification under relaxed viewing conditions by a normal eye. The derivation did not depend on the distance from the magnifier to the eye. Thus the magnification is independent of this distance. The object, as seen through the magnifier, appears just as big, independent of the lens-to-eye distance, so long as the distance between the object and the lens is fixed, with the object at the first focal point of the lens.

If the object is moved inside the focal point of a magnifier, the angular magnification can be greater than predicted by Eq. 25-5. Then, instead of viewing with the eye relaxed, the eye is focused on a virtual image formed by the lens. In this case the distance from lens to eye is important. For maximum magnification the magnifier should be close to the eye, and the image produced by the lens should be at the eye's near point (Fig. 25-23c). The angular magnification then increases by 1 over the value d/f for relaxed viewing (Problem 26).

$$M = \frac{d}{f} + 1 \quad \text{(with image at near point and lens close to eye)} \qquad (25\text{-}6)$$

For example, using a 10 cm focal length lens, one who has a 25 cm near point will be able to obtain an angular magnification ranging from 2.5 to 3.5, depending on how the object and lens are positioned.

When the magnifier is used for maximum magnification its function is simply to allow the object to be as close as possible to the eye and therefore to subtend as large an angle as possible.

When a magnifier is used by one who has an abnormal range of vision (for example, one who is nearsighted), the magnification will be different from that for a person with normal vision. See Problem 27.

Fig. 25-24 A glass bead provides a highly magnified but poor-quality image.

It is possible to produce relatively high angular magnification with a lens of very short focal length. For example, if $f = 2.5$ cm, Eq. 25-5 gives $M = 25 \text{ cm}/2.5 \text{ cm} = 10$. However, when one attempts to actually use such a short focal length lens, one finds that, although the image is large, the quality of the image is very poor because of aberrations—spherical, chromatic, distortion, field curvature, and so on, which become very large effects for a small, high-power lens (Fig. 25-24). For this reason single lens magnifiers are usually limited to longer-focal-length lenses, with an angular magnification of no more than about 2 or 3. We can make magnifiers producing higher magnification, with good quality images, by designing a system of lenses that reduces aberrations. The achromatic doublet shown in Fig. 24-41c is one such design. Lens systems can achieve angular magnifications of up to about 10 or 20. High- power magnifiers, whether made of a single lens or a combination of lenses, must have short focal lengths. They are therefore small, and the field of view through them is limited.*

*In the seventeenth century the biologist Anton Leeuwenhoek used tiny glass beads with focal lengths less than 1 cm as magnifiers.

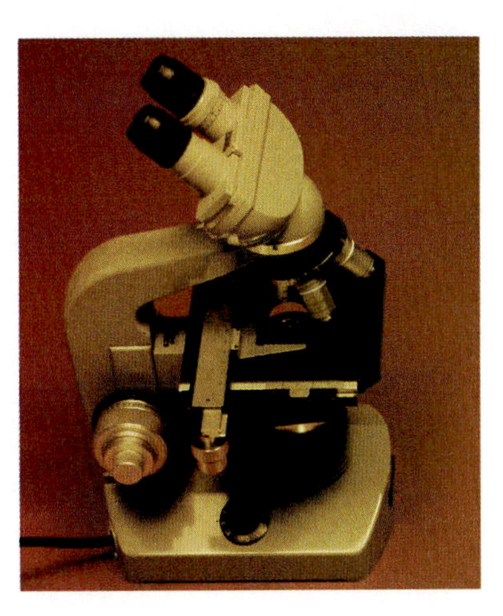

Fig. 25-25 A compound microscope.

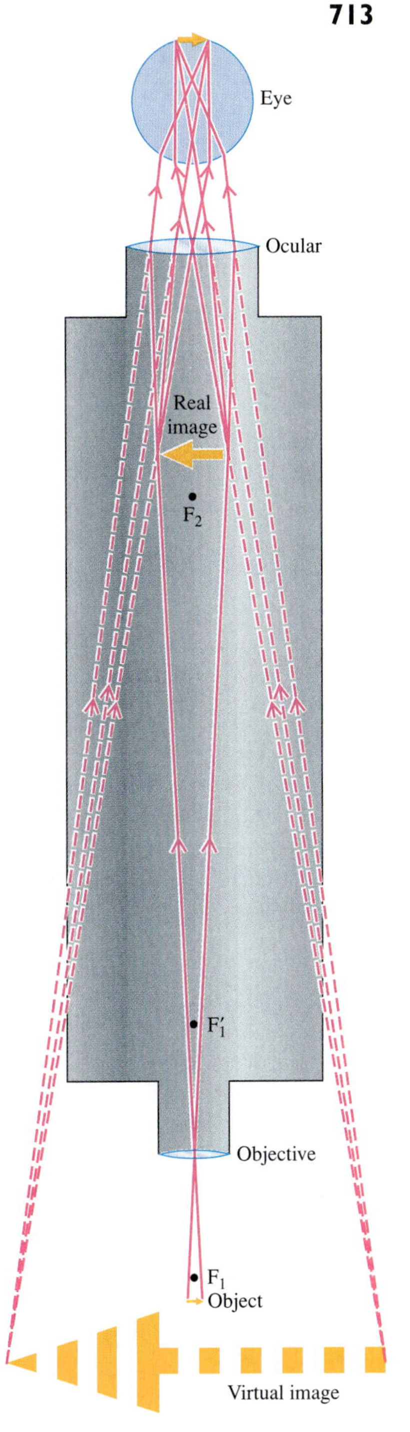

Fig. 25-26 A real image is produced by the objective lens. This image serves as an object for the ocular lens, which produces a virtual image to be viewed by the eye. The figure is drawn to scale, about two-thirds life size. The objective and ocular each have an angular magnification of 5, giving an overall angular magnification of 25.

25-3 The Microscope

The **compound microscope** (Figs. 25-25 and 25-26) is a lens system that produces a magnification much higher than that of a simple magnifying glass. The microscope utilizes two elements, an **objective** and an **eyepiece,** or **ocular,** each of which is represented in Fig. 25-26 by a single lens. In practice the objective and ocular are both multielement lens systems, carefully designed to correct for aberrations. The objective has a short focal length. The object is placed just outside the first focal point of this lens, so that a real image is formed at an image distance that is much greater than the object distance. The objective therefore produces a highly magnified real image of the object being examined. This real image serves as an object for the ocular. The ocular serves the same function as a magnifier. It allows the eye to come very close to the real image formed by the objective, so that the real image, as seen through the ocular, subtends a large angle at the eye.

Without the ocular, the eye could examine at its near point the real image formed by the objective. The effect of the objective would then be to replace the object at the near point by a magnified real image. The resulting angular magnification M_1 achieved by the objective alone is therefore simply the linear magnification m_1 of the objective, determined by the ratio of image-to-object distances ($m = -s'/s$).

$$M_1 = m_1 \quad \text{(objective)} \quad (25\text{-}7)$$

The use of the ocular allows this magnified image to be magnified again. The real image viewed through the ocular is like an object viewed through a magnifying glass. If M_2 is the angular magnification of the ocular, this means that the real image viewed through the ocular is magnified by this factor. Since the real image is already magnified by the factor M_1, the microscope produces an overall magnification M, which is the product of M_1 and M_2.

$$M = M_1 M_2 \quad (25\text{-}8)$$

Values of angular magnification marked on objectives and oculars correspond to an assumed near point of 25 cm. In Fig. 25-26 both the objective and the ocular have angular magnifications of 5, which gives an overall angular magnification of 25. This means that the image formed on the retina is 25 times larger with the microscope than without it.

EXAMPLE 6 Distinguishing Close Points with a Microscope

(a) Find the angular magnification obtained with a microscope, using an objective with an angular magnification of 50 and an ocular with an angular magnification of 10.
(b) Find the minimum separation of two points that can be resolved under the microscope; that is, find the distance between two points that the eye is barely able to distinguish as separate points. Assume that the microscope produces a perfect image, so that the minimum distance resolvable under the microscope is limited only by the eye's visual acuity. Assume that the minimum angle between points resolvable by the eye is 5.0×10^{-4} rad, a typical value.

SOLUTION (a) Applying Eq. 25-8 we find

$$M = M_1 M_2 = (50)(10) = 500$$

(b) The eye can resolve two points with an angular separation of 5.0×10^{-4} rad at the eye, whether seen with or without the microscope. Let θ'_{min} denote this minimum angle of resolution for two object points viewed through the microscope, and let θ_{min} denote the angular separation of the two points, when placed at the eye's near point and viewed without the microscope. Then, applying the definition of angular magnification, we can solve for θ_{min}.

$$M = \frac{\theta'_{\text{min}}}{\theta_{\text{min}}}$$

or

$$\theta_{\text{min}} = \frac{\theta'_{\text{min}}}{M} = \frac{5.0 \times 10^{-4}\ \text{rad}}{500} = 1.0 \times 10^{-6}\ \text{rad}$$

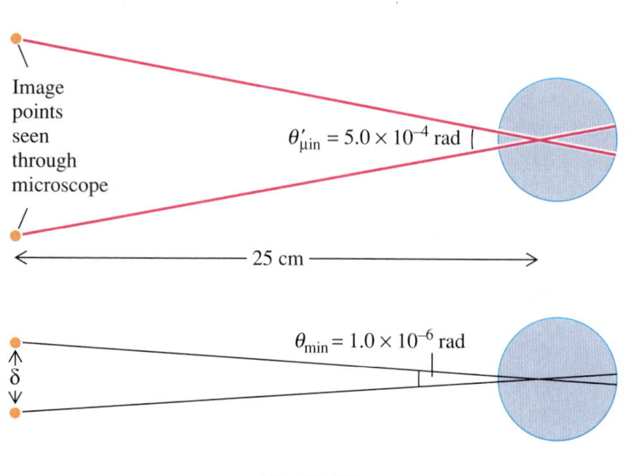

Fig. 25-27

Let δ denote the separation between the object points that are barely resolved by the eye when they are viewed through the microscope. From Fig. 25-27 we see that

$$\frac{\delta}{25\ \text{cm}} = 1.0 \times 10^{-6}\ \text{rad}$$

or

$$\delta = (25\ \text{cm})(1.0 \times 10^{-6}\ \text{rad}) = 25 \times 10^{-8}\ \text{m} = 0.25\ \mu\text{m} = 250\ \text{nm}$$

With this microscope the eye is able to resolve points that are separated by a distance roughly equal to half the wavelength of light. For such small distances, however, our assumption of a perfect microscopic image begins to break down.

In the last example we found that a microscope with a magnification of 500 allows the eye to resolve points separated by a distance of roughly half the wavelength of light. One might wonder whether even higher resolution can be achieved with a higher magnification microscope. This turns out to be impossible. The resolution of a microscope can be limited both by aberrations and by diffraction (the bending of light waves through narrow openings, mentioned in Chapter 23). By use of highly corrected, multielement lenses for both the objective and the ocular, aberrations are virtually eliminated in a high-quality microscope. Diffraction, however, places a fundamental limit on the magnification that can be achieved. We shall see in the next chapter that **the wave nature of light prevents us from using light to resolve points closer than about half the wavelength of the light.** From the last example, we see that this limitation on resolution is reached at an angular magnification of about 500. Although higher magnifications can be achieved, this would do nothing to improve the resolution of an image. Increasing magnification would produce a bigger image, but one that was fuzzy, lacking in fine detail.*

An electron microscope can achieve much higher resolution than an optical microscope. Although the process of image formation is quite different for an electron microscope, utilizing magnetic fields instead of lenses, resolution is limited by the same basic factors as in an optical telescope—aberrations and diffraction. Modern physics reveals that, like light, electrons have wave properties. The wavelengths of electrons used in an electron microscope may be less than 10^{-6} times the wavelength of light, which indicates that it might be possible to achieve magnifications 10^6 times greater than that of an optical microscope, or about 5×10^8. Although aberrations preclude magnifications this great, angular magnifications of up to about 10^6 have been achieved.

---

*Even for a magnification of 500 one begins to see diffraction effects. These can be reduced somewhat by placing a drop of oil between the object and the objective. The oil may have a refractive index of 1.5, which means that the wavelength λ is smaller than the vacuum wavelength λ_0 (Eq. 23-10: $\lambda = \lambda_0/n = \lambda_0/1.5$).

Fig. 25-28 An electron microscope.

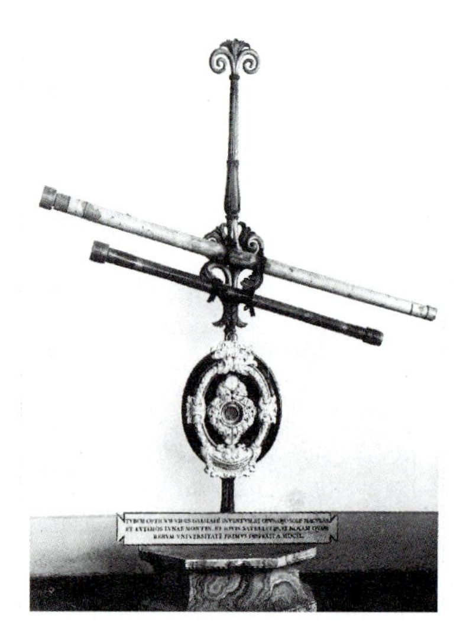

Fig. 25-29 Galileo's telescope.

25-4 The Telescope

Galileo was one of the first to build a telescope (Fig. 25-29). With it he studied the stars and planets, marking the beginning of astronomical observation, which has given us our modern picture of the universe, with its galaxies, quasars, neutron stars, black holes, and big-bang background radiation.

The telescope is used to produce an enlarged retinal image of a distant object. A telescope, like a microscope, has an objective and an eyepiece. Fig. 25-30a shows a distant object viewed by the unaided eye. Fig. 25-30b shows the same object viewed through a refracting astronomical telescope. The same angle θ is subtended by the object at the eye in Fig. 25-30a and at the telescope objective in Fig. 25-30b. The objective forms a real image of the object. Since the object is distant, the image is at the objective's focal point. The eyepiece is used as a magnifier to view the image formed by the objective, with the eye relaxed, and so the objective image must be at the focal point of the eyepiece. Thus the distances of the objective and eyepiece from the image are equal to the respective focal lengths f_1 and f_2.

As we saw in the last section, both the objective and the eyepiece of a microscope have short focal lengths. A telescopic eyepiece, which serves the same function as a microscope's eyepiece, should also have a short focal length. But the focal length of a telescopic objective should not be short. The magnification of a telescope is determined by the size of the image produced by the objective at its focal point. The image height increases in direct proportion to the focal length f_1. Therefore, for large angular magnification, f_1 should be large.

From Fig. 25-30b, we see that for small angles,

$$\theta \approx \frac{h}{f_1} \qquad \text{and} \qquad \theta' \approx \frac{h}{f_2}$$

The angles θ and θ' are the respective angles subtended by the retinal image with and without the telescope. Therefore the ratio θ'/θ is the angular magnification.

$$M = \frac{\theta'}{\theta} = \frac{h/f_2}{h/f_1}$$

or

$$M = \frac{f_1}{f_2} \tag{25-9}$$

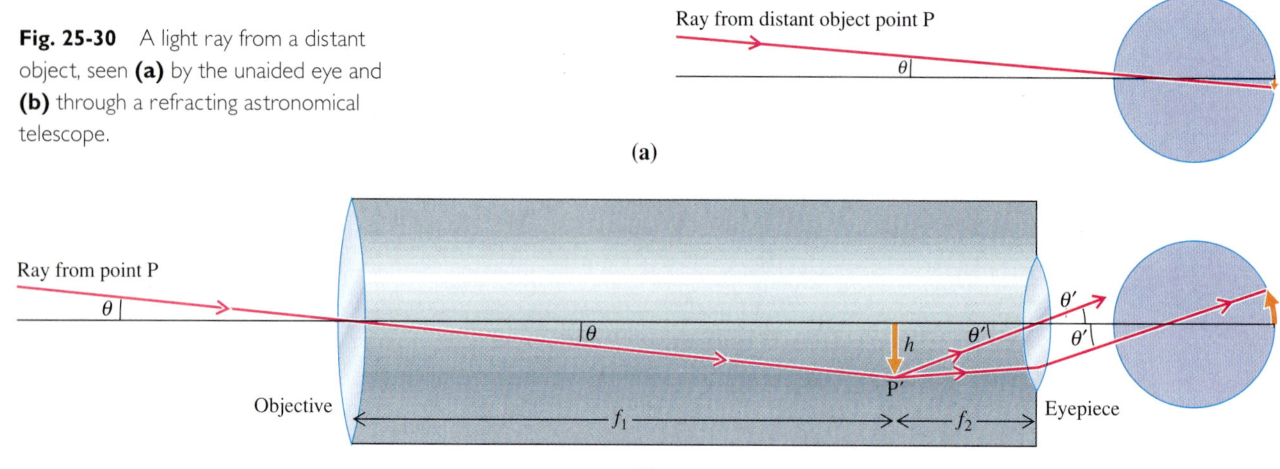

Fig. 25-30 A light ray from a distant object, seen **(a)** by the unaided eye and **(b)** through a refracting astronomical telescope.

For example, a telescope with an 80 cm focal-length objective and a 2 cm focal-length eyepiece will produce an angular magnification of 40. This means that if a distant object is viewed through the telescope the image on the retina is 40 times larger than if the same object were viewed without the telescope.

Diffraction does not place a fundamental limit on the maximum angular magnification for a telescope, as it does for a microscope. In principle, one can keep increasing magnification and improving resolution by increasing both the size of the objective and its focal length. Increasing the focal length gives higher magnification. Increasing the diameter of the objective improves resolution by decreasing diffraction. Resolution can also be limited, however, by aberrations. Aberrations are easier to reduce for a reflecting telescope, for which the objective is a converging mirror, rather than a converging lens (Fig. 25-31). There are no chromatic aberrations for a converging mirror, and one can eliminate spherical aberration by giving the mirror a parabolic shape, rather than a spherical shape (see Fig. 24-14).

Astronomical telescopes are sometimes very large. The Hale telescope at the Mount Palomar observatory in California has an objective with a 5.1 m diameter (200 inches) and a focal length of 17 m; Russia has an even larger telescope, 6 m in diameter; but presently it is the Keck I telescope in Hawaii that is the largest—with a 9.82 m diameter collecting surface of 36 segmented hexagonal mirrors. The resolution of these large telescopes is limited by atmospheric conditions, rather than by diffraction effects. Their great size is actually for another purpose—to detect very faint sources of light. As indicated in Fig. 25-31, parallel rays from a distant point source are concentrated by the telescope. The light energy entering the telescope per unit time is directly proportional to the cross-sectional area of the telescope. Hence very large telescopes are better able to detect a faint source of light.

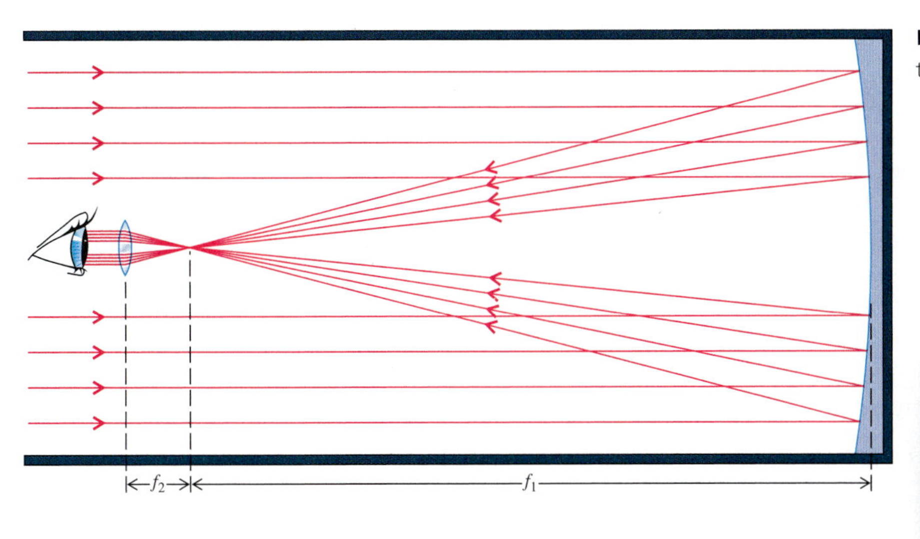

Fig. 25-31 A reflecting astronomical telescope.

Fig. 25-32 The Hubble space telescope.

The Hubble space telescope, launched on April 25, 1990 and operating above the earth's atmosphere, was expected to have a much higher resolution than any previous telescope (Fig. 25-32). Unfortunately, a flaw of about 1 mm in the curvature of the telescope's huge primary mirror produced so much spherical aberration that its resolution was not much better than telescopes on the ground. However, in 1994 an optical system that corrects for the mirror's flaw was installed, and the Hubble began to realize its full potential, producing exciting new pictures of great clarity and beauty (Fig 25-33).

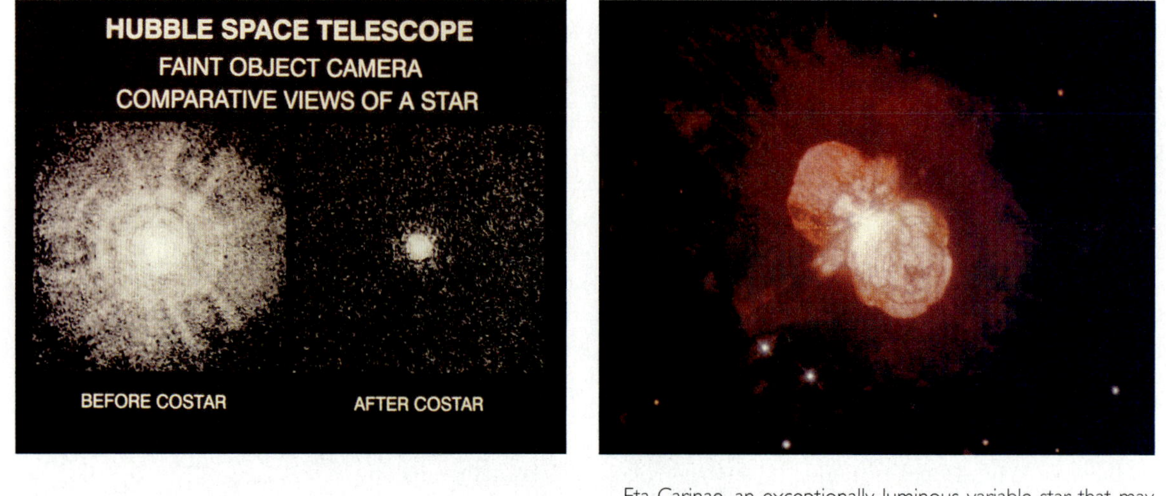

Eta Carinae, an exceptionally luminous variable star that may one day explode as a supernova

Fig. 25-33

EXAMPLE 7 The Moon Seen Through a Telescope

The Mt. Palomar telescope is used to observe the moon, 3.8×10^8 m away. The objective has a focal length of 17 m and the eyepiece has a focal length of 17 cm. Find the minimum distance between object points on the moon that are just barely resolved by an eye looking through the telescope. Assume that resolution is limited by the eye's acuity and that the minimum angle of resolution is 5.0×10^{-4} rad.

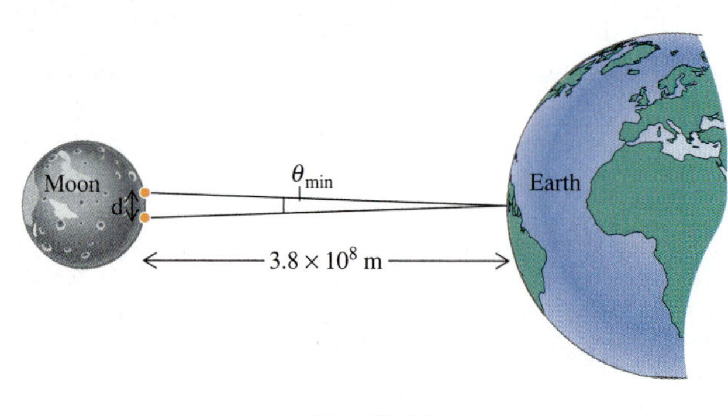

Fig. 25-34

SOLUTION Applying Eq. 25-9, we find

$$M = \frac{f_1}{f_2} = \frac{17 \text{ m}}{0.17 \text{ m}} = 100$$

The image on the retina subtends an angle

$$\theta'_{\text{min}} = 5.0 \times 10^{-4} \text{ rad}$$

From the definition of angular magnification, we find the angle subtended by the object.

$$M = \frac{\theta'_{\text{min}}}{\theta_{\text{min}}}$$

$$\theta_{\text{min}} = \frac{\theta'_{\text{min}}}{M} = \frac{5.0 \times 10^{-4} \text{ rad}}{100} = 5.0 \times 10^{-6} \text{ rad}$$

From Fig. 25-34 we see that the distance d between the two points on the moon is given by

$$d = (3.8 \times 10^8 \text{ m})\theta_{\text{min}} = (3.8 \times 10^8 \text{ m})(5.0 \times 10^{-6} \text{ rad}) = 1.9 \times 10^3 \text{ m}$$

Points closer than 1.9 km cannot be resolved.

The Retina and Color Sensitivity

Structure of the Retina

Light entering the retina first passes through outer layers, consisting of nerve endings and connecting cells, before reaching the receptor cells, where an image is formed and detected by photosensitive molecules. There are two types of receptor cells, called "rods" and "cones" because of the rod or cone shapes of the ends of these cells (Fig. 25-A).

Receptor cells are not connected directly to nerve fibers. Between the receptor cells and the nerve fibers is a layer of "bipolar" cells. One end of these cells interacts electrically with receptor cells across small gaps, or "synapses"; the other end interacts across synapses with nerve fibers (Fig. 25-B). It is the complexity of this electrical network that gives the retina its brainlike capacity to analyze visual images.

Fig. 25-A A scanning electron micrograph of the receptor ends of rods and cones.

Intensity Range: Scotopic and Photopic Vision

The eye is able to adapt to an enormously wide range of light intensities. For example, you can read a book in bright sunlight or look at the stars at night. The intensity of starlight reaching your eyes is only about 10^{-6} W/m^2, but light reflected from a book in direct sunlight has an intensity at the eye on the order of 10 W/m^2. So given time to adapt to different lighting conditions, the eye is effective over a range of intensities varying by a factor of at least 10^7. Actually the entire range of sensitivity is much greater than this. Experiments on the threshold of vision for the dark-adapted eye have shown a sensitivity to light much weaker than starlight. Light can be detected when as few as 100 photons* enter the eye in an interval of 0.001 s. When the eye is exposed to a scene in bright sunlight, in an interval of 0.001 s the eye receives on the order of 10^{32} photons—roughly 10^{30} times the number at the threshold of vision.

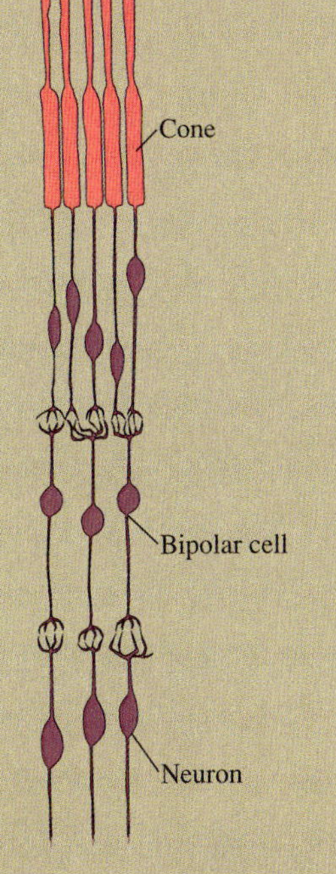

Fig. 25-B Typical connection of foveal cones with nerve fibers through bipolar cells.

Part of the eye's adaptation between light and dark involves the size of the pupil, but the pupil diameter can vary only from about 2 mm to about 6 mm. This means that the area of the pupil can vary only by a factor of $3^2 = 9$. This accounts for only a small part of the eye's ability to adapt. Most of it comes from the way the retina responds to vastly different intensities.

The eye possesses a dual system for detecting light over two intensity ranges. Light-adapted vision, called photopic vision, utilizes only cones. Dark-adapted vision, called scotopic vision, utilizes only rods. Pure scotopic vision occurs in light of intensity less than about 10^{-3} W/m^2 (a little less bright than a moonlit night). The cones are then completely inoperative, and light is detected solely through absorption of photons by the rhodopsin molecules, located at the ends of the rod cells. With scotopic vision there is no color sense; although the eye is sensitive to various frequencies, these are sensed not as colors but as shades of gray, as in a

*Only about 5 of these 100 photons are actually absorbed by receptors in the retina. Absorption of a single photon by a photosensitive molecule is enough to produce an electrical response in the receptor cell, but the circuitry of the bipolar layer needs about five photons to produce a nerve impulse. This mechanism apparently prevents the retina from responding to the few thermal photons that are present.

Continued.

black and white photograph. Visual acuity, the ability to see fine details, is very limited under scotopic conditions. For example, it is impossible to read fine print under these conditions.

The Visible Spectrum

Although "visible light" constitutes only that part of the electromagnetic spectrum with vacuum wavelengths between about 400 nm to 700 nm, the retina is actually somewhat sensitive to ultraviolet light, with wavelengths shorter than 400 nm. These shorter wavelengths are not normally sensed because most such light is absorbed by other parts of the eye before reaching the retina. The cornea absorbs wavelengths below 300 nm, and the lens absorbs almost all light of wavelength below 400 nm. The lens thus protects the retina from the potentially damaging UV light from 300 nm to 400 nm.*

Spectral Sensitivity

For scoptic vision, the eye's sensitivity to various frequencies is not the same as for photopic vision (shown at the beginning of Chapter 23). Fig. 25-C shows both scotopic and photopic sensitivities in a single graph. Notice that under photopic conditions the eye has maximum sensitivity to yellow-green light, but under scotopic conditions the eye has maximum sensitivity to the blue-green part of the spectrum (though of course it is not sensed as blue-green) and is completely blind to the red end of the spectrum. This is known as the Purkinje effect and can be observed in a garden at night, where red flowers may appear black and blue flowers appear white or gray.

Rhodopsin molecules have been chemically extracted from rod cells. When exposed to light of various wavelengths, these molecules absorb some wavelengths more readily than others. The "absorption spectrum," which shows the relative amount of each wavelength absorbed, is nearly identical to the eye's scotopic sensitivity curve. Thus it is the rhodopsin molecule that is responsible for the eye's spectral sensitivity under scotopic conditions.

After absorbing light in the retina, rhodopsin is regenerated in a complex chemical process involving vitamin A and requiring about 5 minutes for half of a large sample of molecules to be regenerated. After about 30 minutes the regeneration is nearly complete. Apparently the time required for the eye's adaptation to the dark, after exposure to bright light, is related to this regeneration time, and a severe deficiency in vitamin A will prevent it and result in night blindness.

It is much more difficult to extract the photosensitive molecules from cone cells than it is to extract the photosensitive rhodopsin molecule from rod cells. However, beginning in the 1960s delicate experiments were performed, using a technique in which a narrow light beam was directed onto individual cone cells in retinal segments taken from either human, monkey, or goldfish retinas. The frequency of the light was varied, and the fraction of light absorbed was measured as a function of frequency. In all these experiments three distinct kinds of cells were identified, each with its own absorption spectrum. Approximate absorption curves for human cones are shown in Fig. 25-D. Each of these curves covers a broad spectral range, with maximum absorption at 450 nm, 540 nm, and

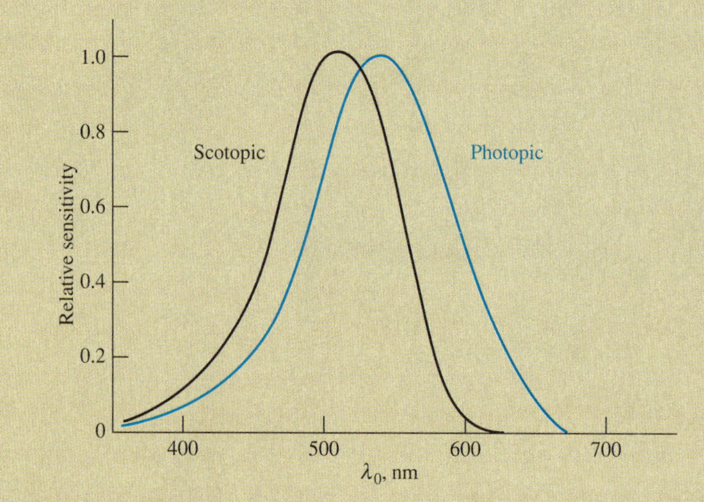

Fig. 25-C Spectral sensitivity of the eye for photopic (light-adapted) and scotopic (dark-adapted) vision.

*In time the lens itself may be damaged by the absorption of UV light. (This often occurs in older individuals who have spent a lifetime exposed to sunlight, especially in tropical areas where sunlight is most intense, or in the Arctic zone where much sunlight is reflected from snow and ice.) As the damage increases, the lens gradually becomes opaque. This is called a "cataract." A person with a cataract may have the lens surgically removed. It is then found that the eye has considerable sensitivity to ultraviolet light. Some people have even reported seeing X rays.

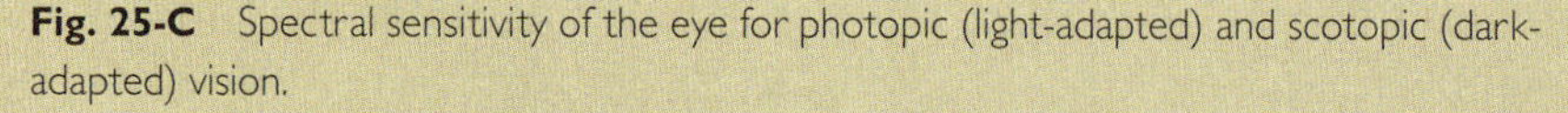

580 nm. These maxima correspond to blue, green, and yellow light respectively.

This spectrometry work verifies a theory proposed by Thomas Young in 1802. Young believed that human perception of color comes from three distinct color receptors in the human eye, sensitive respectively to blue, green, and red light,* and that all color sensations were combinations of excitations of these receptors. Maxwell and Helmholtz further developed this "trichromatic" theory, which is very successful in accounting for color-mixing phenomena. Any spectral color can be matched when the right combination of blue, green, and red light—the "primary colors"—are mixed (Fig. 25-E). For example, you see yellow when your red and green cones are equally stimulated and your blue cones are not stimulated. You can accomplish this either by looking at monochromatic yellow light or by looking at a combination of green light and red light. In either case, the perception is yellow.

You should not confuse the primary colors of light with the primary paint colors used by an artist who mixes paints. The primary paint colors are yellow, cyan (a brilliant shade of blue), and magenta (a purplish red). Paint pigments create color by absorbing or subtracting out other colors in the light that illuminates them. For example, when blue paint is illuminated by white light, the pigment absorbs the red end of the spectrum and reflects the blue end. Yellow paint illuminated by white light absorbs the blue part of the light and reflects red and green light equally. If you mix yellow and blue paints, you get green paint, since the yellow pigment absorbs the blue light and the blue pigment absorbs the red light. On the other hand, if you mix yellow light and blue light, you get white light!

*Although the third pigment peaks in the yellow, rather than red, it does extend far enough into the red to account for detection of this color.

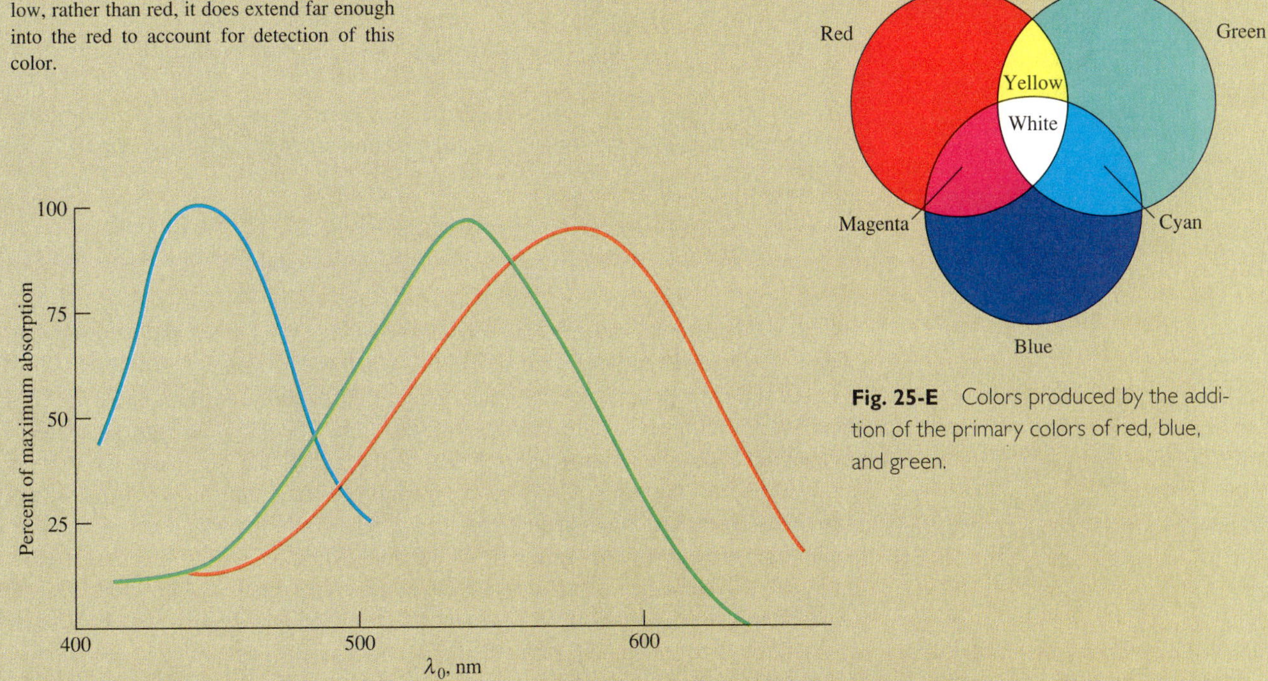

Fig. 25-E Colors produced by the addition of the primary colors of red, blue, and green.

Fig. 25-D Absorption of light by the three types of cones in the retina of the human eye.

Continued.

Color blindness occurs when one or more of the cones are missing, or, more commonly, when the absorption spectra of the red and green cones are somewhat different from those shown in Fig. 25-D. "Red-green" color blindness is about 10 times more common in men than in women, occurring in about 6% of all males. Other kinds of color blindness are more rare. Only about one person in 30,000 is completely blind to colors. Those who are red-green color blind can distinguish bright primary colors but have difficulty with certain shades of color.

Although the idea of three distinct kinds of color receptors now seems to be well established, it seems certain that color perception does not consist merely of receptor cells sending one of three messages (blue, green, or red) along the optic nerve directly to the brain. There is an abundance of color phenomena that is impossible to explain in this way. For example, Land has performed a series of experiments that dramatically illustrate how the human eye is able to maintain a constant determination of the color of an object, independent of the frequency of the illuminating light and also therefore independent of the frequency of the light reflected by the object. Apparently the eye is able to compare the light reflected from various objects in the visual field and to somehow determine from these data the "true" color of each object. This is just one indication of the processing of visual information that takes place both within the retina in the network of receptors, bipolar cells, and nerve fibers and in the visual cortex of the brain.

*25-5 Factors Limiting Visual Acuity

Visual acuity could conceivably be limited by several factors: scattering of light by small particles in the vitreous humor, spherical aberration, chromatic aberration, the finite size of the foveal cones, and diffraction. We will consider the effect of each of these factors.

We can apply Eq. 25-1 to find the distance h between two image points on the retina corresponding to two object points that subtend angle θ_{min}. Assuming $\theta_{min} = 5 \times 10^{-4}$ rad for a typical observer, we find

$$h = (16 \text{ mm})(\tan \theta_{min}) \approx (16 \text{ mm})\theta_{min} = (16 \times 10^{-3} \text{ m})(5 \times 10^{-4} \text{ rad})$$
$$= 8 \times 10^{-6} \text{ m} = 8 \ \mu\text{m}$$

This result is illustrated in Fig. 25-35.

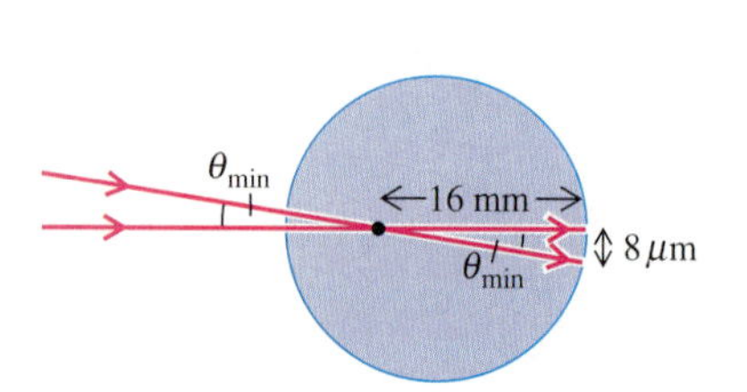

Fig. 25-35 Image points on the retina that are barely resolvable by the eye are 8 μm apart.

Small particles in the aqueous humor scatter some of the light passing through. If enough of this scattered light were actually detected by the retina, it would add optical noise to the image and thereby reduce acuity. Fortunately, scattering is not a significant factor in limiting acuity because of the retina's directional sensitivity, which prevents light from being detected if it enters at a large angle relative to the optical axis. The shape of the cones is responsible for this phenomenon, known as the **Stiles-Crawford effect.** The cone-shaped end of the cell channels image-forming rays to the photosensitive molecules at the tip of the cone while tending to reflect scattered light rays, which enter the cone at relatively large angles relative to the cone axis.

Spherical aberration is a defect present in spherical lenses, in which light rays passing through the outer edge of the lens are refracted more and imaged closer to the lens than rays passing through the center of the lens (see Fig. 24-43). There is no significant spherical aberration in the eye because of its structure. The iris blocks the rays that would contribute most to spherical aberration, allowing only rays close to the optical axis to enter. The inhomogeneous structure of the crystalline lens tends to further reduce spherical aberration. The lens index of refraction varies from 1.41 at its relatively rigid core to 1.39 at its softer outer layers.* This means that the tendency of the rays passing through the outer part of the lens to be refracted more, as a result of spherical aberration, is offset by a reduction in refraction because of a lower average index experienced by those rays.

Chromatic aberration in the eye, caused by the variation of the eye's index over the visible spectrum, is a significant factor limiting visual acuity. The situation is qualitatively the same as for a converging lens (Fig. 24-41a). The refractive index is greater for blue light than for red light, and therefore the optical power of the eye is greater for blue light than for red light. This means that for a distant white-light source, blue light forms an image about 0.5 mm in front of the image formed by red light. This could seriously limit acuity. Fortunately, however, chromatic aberration in the fovea is greatly reduced by the presence of a pigment layer, called the **macula lutea,** which covers the fovea and absorbs the blue part of the visible spectrum, allowing through only those wavelengths from about 500 nm to 700 nm (green to red). (Blue-sensitive cones are in fact completely absent from the very center of the fovea.) It is the blue part of the spectrum for which dispersion is greatest. The refractive index is nearly constant for green through red, and as a result the focal points for green and red light are separated by only about 0.1 mm. The image on the cone layer will be sharpest if this layer is between the green and red images. As shown in Fig. 25-36, this arrangement produces the smallest circle of confusion on the cones. The diameter of the circle of least confusion can be estimated, using this figure (Problem 42). For pupil diameters of 2 mm to 4 mm, the circle of least confusion varies in diameter from about 5 μm to 10 μm. Two points, whose (ideal) point images span a distance smaller than the diameters of the chromatic circles, will not be resolvable. Thus chromatic aberration sets a limit to visual acuity close to the typical experimental value of 8 μm, previously calculated.

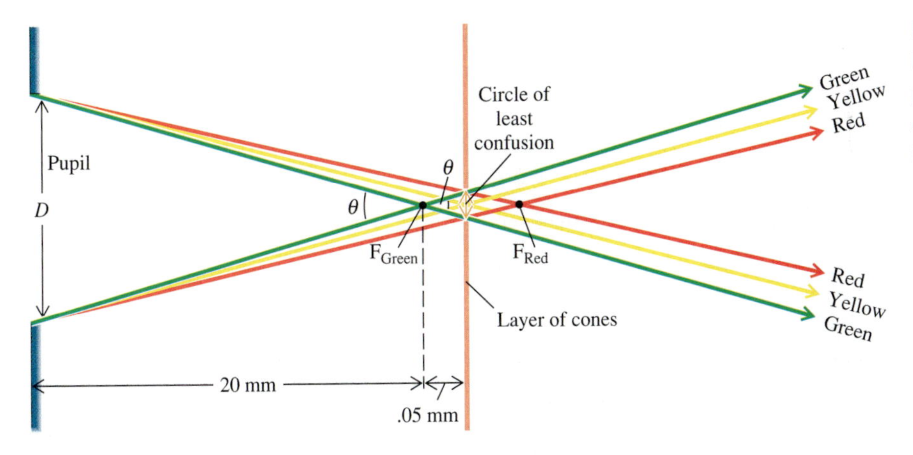

Fig. 25-36 Green, yellow, and red light are imaged at slightly different points within the eye, and so circles of confusion are formed on the retina, limiting visual acuity.

The spacing of foveal cones is about 3 μm, and this certainly sets a fundamental limit to the eye's acuity. Image points closer than 3 μm would fall on the same cone. To be resolved as separate image points, these points have to fall on two cones, 6 μm apart, separated by one unstimulated cone (Fig. 25-37). Again this is close to our typical experimental value of 8 μm.

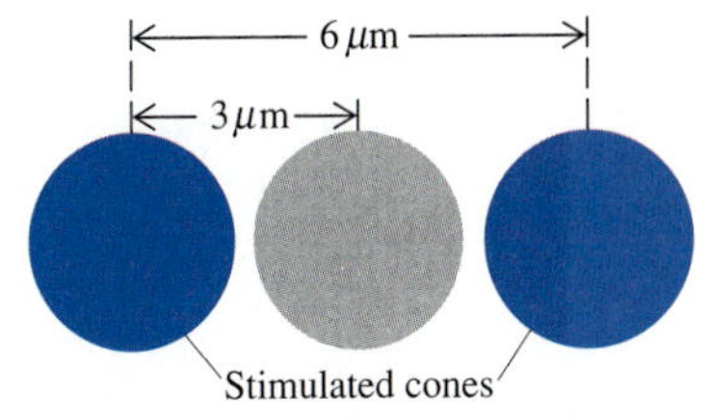

Fig. 25-37 The finite size of cones limits the eye's acuity.

*The effective overall refractive index of the lens is 1.42; that is, if the lens were replaced by a homogeneous body of the same shape, producing the same refraction, the replacement would have an index of 1.42.

Another fundamental limitation to visual acuity is the phenomenon of diffraction, to be discussed in Chapter 26. When light from a distant point source passes through a circular opening or a circular lens, the light tends to bend outward, and so, when an image is formed, the image is not a point but rather a circle. The size of this circle increases as the diameter of the opening is reduced. For the eye, the size of the pupil determines the size of the diffraction circles on the retina. A 2 mm diameter pupil results in a 7 μm diameter diffraction circle, and a 4 mm diameter pupil gives a 5 μm circle. Again we have an acuity-limiting factor close to the typical experimental value of 8 μm.

The evolution of the human eye has produced a finely balanced system. Chromatic aberration, cone size, and diffraction all place roughly the same limits on acuity. Increasing the size of the pupil would reduce diffraction but would increase chromatic aberration. Decreasing pupil size would have the opposite effect. See Problem 41. Only by having both a larger pupil and a larger eye could both be reduced.

Smaller cones would reduce one limitation to acuity, but diffraction and chromatic effects would make this change useless, unless the eye were increased in size. And a larger eye would make rapid movement of the eye more difficult.

We have so far ignored eye movement. The image on the retina is normally not static. Muscles attached to the outer surface of the eye keep it in constant motion, even when looking at a stationary object, as seen in Fig. 25-38. The large scale scanning motion of the eye is needed so that various parts of the image are focused on the fovea and each part can be examined in detail. Scanning motion terminates in small dots, called "fixation points." The system of lines and fixation points in Fig. 25-38 resembles a connect-the-dots drawing. When these dots are enlarged and carefully examined, it is found that they are actually very small areas over which the eye rapidly moves back and forth in a complex way. Experimental studies indicate that the eye apparently needs to see a dynamic image in order to see at all. When an optical device is used to produce a constant image on the retina, that image gradually fades away.

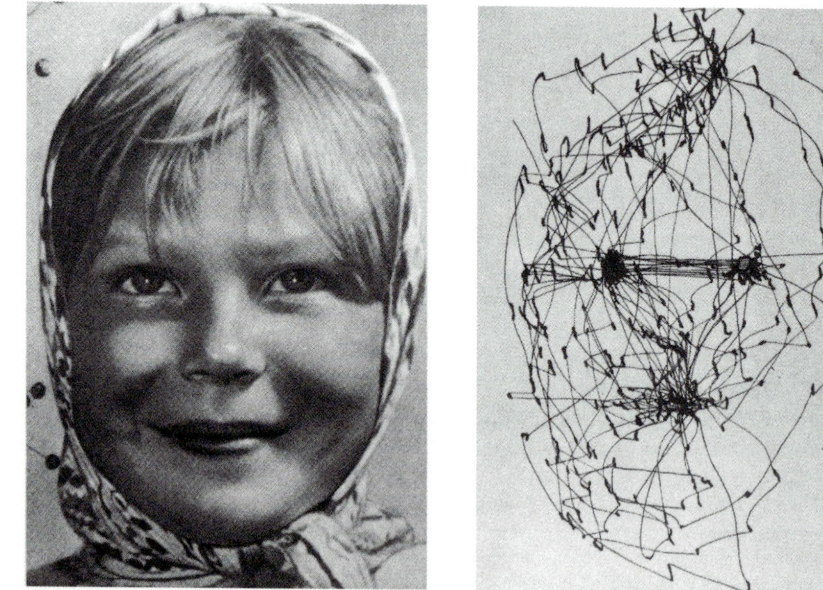

Fig. 25-38 Examination of the photograph on the left by an observer resulted in a pattern of movements by the observer's eye, shown in the figure on the right. The eye's motion was measured by recording the position of light rays reflected from a tiny mirror attached to the surface of the eye.

CHAPTER 25 SUMMARY

The human eye refracts incoming light rays at the cornea and again at the crystalline lens, forming a real inverted image on the retina. The eye adjusts for varying object distances by accommodation—varying the shape of the crystalline lens and hence the total optical power of the eye. This allows the eye to have clear vision over a range of object distances from a near point to a far point. For a person with normal vision the far point is ∞, and for a normal young adult the near point is about 25 cm.

Myopia, or nearsightedness, is the condition of the eye in which there is too much converging power for the length of the eye. Consequently, a distant object is imaged in front of the retina and appears blurred. Myopia may be corrected by placement of an appropriate negative lens in front of the eye.

Hyperopia, or farsightedness, is the condition of the eye in which there is too little converging power for the length of the eye. Consequently, a close object is imaged behind the retina and appears blurred. Hyperopia may be corrected by placement of an appropriate positive lens in front of the eye.

Astigmatism usually results from an asymmetrically shaped cornea and is corrected by an asymmetric lens.

The location of an image point on the retina can be found when a ray is drawn from the object point straight through a point O on the eye's optical axis, 16 mm in front of the retina. The reduced eye is a drawing in which the eye is represented by a circle, with the one ray drawn through O to determine image size.

Visual acuity is the eye's ability to see fine detail. A measure of acuity is the minimum angle between points that can be resolved by the eye. This angle depends on the individual. Typically

$$\theta_{\text{min}} = 5 \times 10^{-4} \text{ rad}$$

The angular magnification M of an optical instrument is defined by the equation

$$M = \frac{\theta'}{\theta}$$

where θ is the angle subtended at point O in the reduced eye by the retinal image without the instrument and θ' is the angle subtended by the retinal image with the instrument.

A magnifier is a positive lens or lens system. Used by a relaxed normal eye, the magnifier gives an angular magnification

$$M = \frac{d}{f}$$

where d is the near-point distance and f is the focal length of the lens. For maximum magnification the magnifier is held close to the eye, and the object is placed so that an image is formed at the eye's near point. This gives a magnification

$$M = \frac{d}{f} + 1$$

The compound microscope has an objective and an ocular. The objective is a short-focal-length, positive lens system, which forms a magnified real image of a small object placed just outside its focal point. The ocular serves as a magnifier, allowing the eye to come very close to the image produced by the objective. The overall angular magnification of the microscope is the product of objective and ocular magnifications.

$$M = M_1 M_2$$

The resolution of a microscope may be limited by aberrations or diffraction.

A telescope has an objective and an ocular. The objective is a long-focal-length, positive lens system or mirror, which is used to form a real image of a distant object. The ocular serves the same function as in the microscope—to allow the eye to come close to the image formed by the objective. The angular magnification of the astronomical telescope equals the ratio of the respective focal lengths of the objective and ocular.

$$M = \frac{f_1}{f_2}$$

Questions

1 A contact lens is worn in direct contact with the cornea. Suppose that a contact were displaced outward a short distance from the eye, while maintaining its same shape.

(a) Would its effect on the eye be the same, or would there be more refractive power or less refractive power?

(b) If the same contact lens were surgically implanted inside the eye, would its effect be the same, or would there be more or less refractive power?

2 Which of the lens shapes shown in Fig. 24-26 would be most appropriate for contact lenses correcting farsightedness?

3 Soft contact lenses are nearly invisible when they are immersed in water for storage. This indicates that the refractive index of these lenses is (a) much greater than 1.33; (b) close to 1.33; (c) much less than 1.33.

4 Squinting can allow a nearsighted person who is not wearing glasses or contact lenses to see distant objects more clearly. The reason for this is that (a) less light enters the eye; (b) narrower cones of light rays enter the eye, forming smaller circles of confusion on the retina; (c) pressure is exerted on the lens, causing light rays to be better focused on the retina; (d) there is more diffraction; (e) there is less diffraction.

5 Suppose you are nearsighted and are trying to read distant road signs without glasses or contact lenses to correct your vision. Is this (a) easier in bright daylight; (b) easier at night; or (c) the same day or night?

6 In order to produce the same correction of vision as glasses, should the power in diopters of contact lenses be greater than, less than, or equal to the power of glasses?

7 Fig. 25-39 shows a plastic lens used to replace surgically a damaged lens in the human eye. Does the focal length of the lens change from its value in air after the lens is implanted?

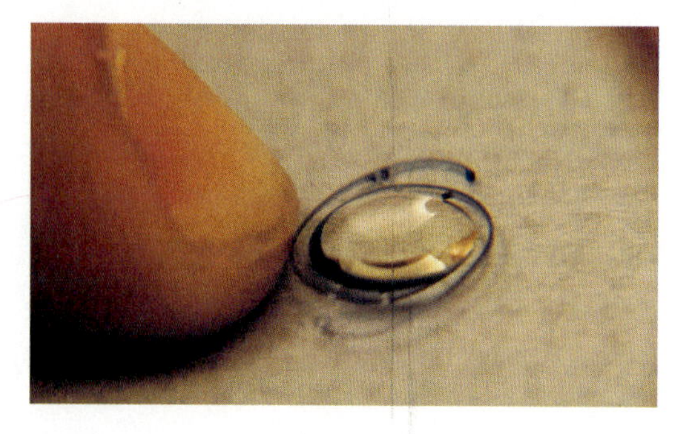

Fig. 25-39

8 In the 1970s Nicholas Brown discovered that as the eye ages, the crystalline lens becomes more curved in its relaxed state, which is used for distant vision. Assuming that the dimensions of the eyeball do not change and the normal eye still produces a sharp image on the retina, this increased curvature must be compensating for some other progressive change in the eye. Which of the following would account for it: (a) increased refractive index of the lens; (b) decreased refractive index of the lens; (c) decreased refractive index of the vitreous humor? (See J.F. Koretz and G.H. Handelman: "How the Human Eye Focuses," *Sci Am* 259:92, July 1988.)

9 Fig. 25-40 shows the focusing of rays from a point source by an asymmetric cornea—that is, one with astigmatism. All rays in the vertical plane are imaged at point P, whereas all rays in the horizontal plane are imaged at point Q. As a result there are line images at P and Q. Suppose we wish to place a negative cylindrical lens in front of the cornea to focus rays in the vertical plane at Q, rather than P, so that a single point image is formed at Q. Should the lens have orientation (a) or (b)?

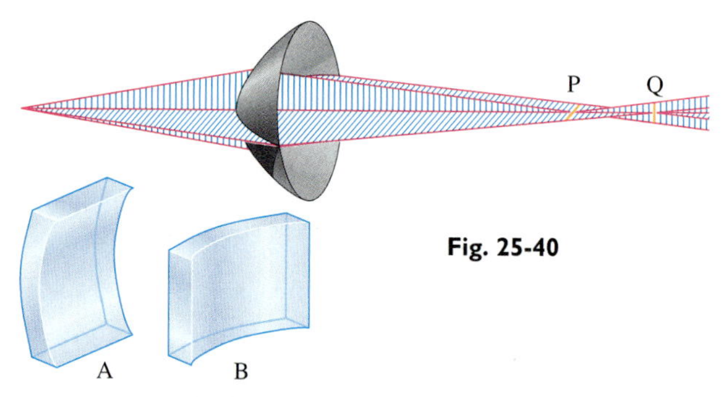

Fig. 25-40

10 Which would be easier to resolve with the normal, unaided eye under favorable viewing conditions: two people standing side by side 1 m apart at a distance of 10^3 m, or two stars, 0.1 light years apart, each of which is exactly 10 light-years from earth?

11 Suppose that a magnifying glass was made of diamond in the same shape as a certain glass magnifier. Would the glass or the diamond magnifier produce greater magnification, or would it be the same for each?

12 Would a nearsighted person using a magnifier achieve higher magnification with or without wearing glasses? That is, which way would the image on the retina be as large as possible?

13 Which of the following focal lengths would not be appropriate for the ocular of a compound microscope or an astronomical telescope: 2.5 cm, −2.5 cm, 5 cm, −10 cm, 40 cm?

14 Which of the following focal lengths would not be appropriate for a telescope's objective: 10 cm, 20 cm, 50 cm, 2 m, 100 m?

Answers to Odd-Numbered Questions

1 (a) same; (b) less; **3** b; **5** a; **7** yes, increases; **9** a; **11** diamond; **13** −2.5 cm, −10 cm, 40 cm

Problems (listed by sections)

25-1 The Human Eye

1 To demonstrate your blind spot, position your eyes directly above Fig. 25-41, close your left eye and focus your right eye on the word "blind." Then vary the distance of your eye from the words until at some distance the word "spot" disappears. The image of this word will then lie on the blind spot in your right eye. Approximately how far is the page from your eyes when this happens?

2 A relaxed crystalline lens has a refractive index n = 1.42 and radii of curvature R_1 = +10.0 mm, R_2 = −6.00 mm. The lens is surrounded by two media of index 1.34. Calculate the focal length and optical power of the lens, treating it as a thin lens.

3 In Example 1 we found that a cornea with R = 7.8 mm forms an image of a distant object 31 mm from the cornea. This image serves as an object for the crystalline lens. Treat the lens as a thin lens, 5 mm behind the cornea. From problem 2 the relaxed lens has an optical power of 15.9 d. Use the thin lens equation to estimate the location of the final image.

4 An object is placed 10 cm in front of the cornea.
(a) What is the image distance for the image formed by the cornea alone?
(b) The image formed by the cornea serves as an object for the lens. Treat the lens as a thin lens 5 mm behind the cornea. Find the optical power of the lens necessary to form an image on the retina, 19 mm from the center of the lens.

5 Find the power of the lens necessary to correct an eye with a far point of (a) 25 cm; (b) 50 cm.

6 Find the far point of an eye for which a prescribed lens has an optical power of (a) −0.50 d; (b) −3.0 d.

7 What is the minimum power lens prescription that would allow an eye with a near point of 60 cm to see a clear image of a newspaper held 25 cm from the eye?

8 A contact lens has an optical power of −1.5 d. What is the uncorrected far point of a person for whom a pair of such lenses is prescribed?

9 A certain eye, when relaxed, needs a +0.50 d lens in front of the eye to form a clear image on the retina of a distant object. Where is the (virtual) object seen by the eye?

10 A nearsighted person, wearing identical contact lenses, has a corrected range of clear vision from 20 cm to ∞. The person's uncorrected near point is 10 cm.
(a) What is the power of each lens?
(b) What is the person's uncorrected range of clear vision?

11 A nearsighted man has a near point of 10 cm and a far point of 50 cm. He shaves without using his glasses. In order to see a clear image, what are the minimum and maximum distances of the mirror from his face?

★★**12** When a certain myopic eye looks through a corrective lens, it sees an image of a tree 30 cm from the eye when the tree itself is 50 m away. Does this affect depth perception? Explain.

Fig. 25-41

13 Corrective swimming goggles are designed for a nearsighted person. Water is the medium on the outside of each lens, and air is the medium on the inside, between the lens and the eye. Will the lenses be thicker in the middle or at the edge?

★**14** Fig. 25-42 shows a model eye 10 times the actual size of the human eye. The model is filled with water. A thin glass wall at the front acts like a cornea, forming a spherical surface for the water. The crystalline lens is represented by a thin glass lens immersed in the water.

(a) Find the focal length of the lens necessary to form an image of a distant object on the retina.

(b) Find the focal length of the lens necessary to form a retinal image of an object 25 cm in front of the cornea.

(c) With no lens inside the model eye, the lens from part (a) is placed in front of the cornea. Is the image of a distant object now formed on the retina, behind it, or in front of it?

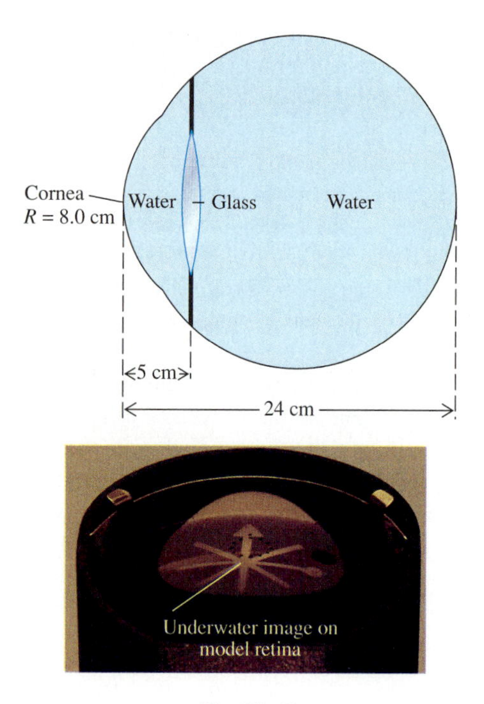

Fig. 25-42

15 Find the power of the upper and lower halves of bifocals* designed to correct the vision of a 50-year-old nearsighted man who has a near point of 25 cm and a far point of 50 cm.

★**16** Suppose a certain eye has a near point of 50 cm and, when completely relaxed, forms an image of a distant object 30 cm behind the cornea. Find the power of the upper and lower halves of bifocals* designed to correct both close and distant vision. Hint: for the upper half, treat the lens and the eye as two thin lenses in contact, for which the combined power is the sum of the two powers

17 Find the height of the retinal image of a person 1.5 m tall, standing 4.0 m away.

18 (a) What is the angle subtended by an object whose image just covers the fovea, 0.4 mm in diameter?

(b) At what distance would the image of a person's face, of height 20 cm, just fill the fovea?

Visual acuity

19 Two laser beams fall on a screen and illuminate small circular areas. What is the minimum distance between the centers of the circles in order for an observer with normal visual acuity to be able to distinguish them from a distance of 10 m?

20 Which of the following are impossible to resolve with the normal, unaided eye? (a) two retinal cones, 2.5 μm apart, held at the eye's near point of 25 cm; (b) two grains of sand, 1 mm apart, at a distance of 1 m; (c) two pine needles, 2 mm apart on a tree 2 km away; (d) car headlights 1000 m away; (e) two moon craters, 2 miles apart, 240,000 miles away; (f) two stars, 2 light-years apart, in the Andromeda galaxy, 2×10^6 light-years away from earth.

21 A marksman with exceptional visual acuity is able to resolve points subtending an angle as small as 2×10^{-4} rad. What is the maximum distance at which he can resolve points 1 cm apart?

25-2 The Magnifier

22 A person with a range of vision from 15 cm to ∞ uses a 10 cm focal length magnifier. Find the angular magnification when the magnifier is used with the eye (a) relaxed; (b) focused at its near point.

23 An object 1.0 cm high is viewed under a magnifier of focal length 10 cm by a person with a near point of 25 cm and a far point of ∞. Find the angle subtended by the image formed by the magnifier when the eye is (a) relaxed; (b) focused at its near point.

*Bifocals have lenses that are split into an upper section, through which the wearer looks for vision at a distance, and a lower section, for viewing close objects. Each half has a curvature appropriate for that kind of vision.

24 A certain object subtends an angle of 0.010 rad when at the near point of a normal eye, 18 cm from the eye. When the eye views this object under a certain magnifier, the image subtends an angle that can be as much as 0.040 rad. Find the magnifier's focal length.

25 A 60-year-old man, who has normal vision, has a near point of 80 cm. Instead of using reading glasses, he chooses to use a magnifier of focal length 16 cm for reading. He reads with his eyes relaxed.
(a) Find the magnification M.
(b) What is the distance from the page to the magnifier?
(c) What is the maximum angular magnification that can be obtained with this magnifier by a child with a near point of 8 cm?

26 Derive Eq. 25-6 ($M = d/f + 1$), which gives the angular magnification of a magnifier when the magnifier is next to the eye and forms an image at the near point.

27 Example 2 describes correction of vision for a nearsighted person with a range of vision from 10 cm to 50 cm. As calculated in the example, after correction the range is 12.5 cm to ∞. An object 1.0 cm high is viewed by this person under various circumstances. Find the angle subtended by the object seen by the eye (a) without glasses, object at the near point; (b) with glasses, object at the corrected near point; (c) under a magnifier ($f = +10$ cm), without glasses, image at the near point; (d) under a magnifier ($f = +10$ cm), with glasses, image at the corrected near point.

25-3 The Microscope

28 What is the angular magnification of a microscope using an objective with a magnification of 25 and an ocular with a magnification of 10?

29 A microscope objective with a magnification of 10 is used in a microscope with a magnification of 50. Find the focal length of the ocular if the ocular's image is formed at the eye's near point, 25 cm from the eye. Neglect the distance between the eye and the ocular.

30 The objective of a certain microscope provides an angular magnification of 50. The image is 16 cm from the objective. How far is the object from the objective?

31 Can two points 10^{-4} mm apart be resolved with a light microscope?

32 (a) Can two points 10^{-2} mm apart be resolved with a good-quality microscope with an angular magnification of 100?
(b) What is the angle subtended by the microscope image?

33 The resolution of a microscope is diffraction limited. A drop of oil is placed between the object and the objective. Would the image be better if the oil drop has a refractive index of 1.4 or 1.6?

★**34** An observer views a small insect at the near point of her eye, at which point it subtends an angle of 2° (Fig. 25-43a). Fig. 25-43b, when completed, will illustrate the effect of viewing the insect through a low-power microscope. For clarity, the figure shows the eye drawn disproportionately large. Draw principal rays to find first the image formed by the objective and then to find the image on the retina. What is the angle subtended by the retinal image?

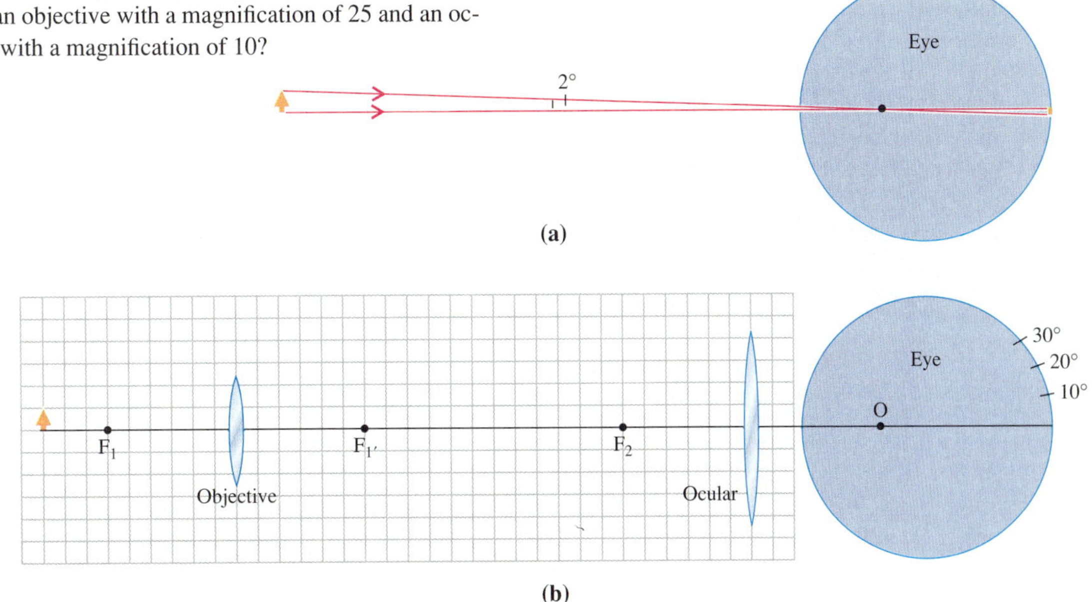

Fig. 25-43

25-4 The Telescope

35 An astronomical telescope has an objective with a focal length of 1.0 m and an ocular with a focal length of 2.0 cm. What is its angular magnification?

36 The objective and ocular of an astronomical telescope are 1.55 m apart. The ocular has a focal length of 5.00 cm. What is the telescope's angular magnification?

37 The objective mirror of the Mt. Palomar telescope has a 200-inch diameter. If an ocular with a diameter of 2 inches is used, what is the ratio of the intensity of light emerging from the ocular to the intensity of incident light?

38 What is the angular separation between two stars that are barely resolved by the average eye, when viewed through a good-quality astronomical telescope with an angular magnification of 50?

39 Suppose that you have a microscope and a telescope, each capable of producing an angular magnification of 100. Which of the pairs of objects listed in Problem 20 could not be resolved even when either the microscope or telescope is used?

*25-5 Factors Limiting Visual Acuity

40 A penny has a diameter of 2 cm and is 1 m from the eye. How many cones are covered by the image on the retina?

41 The size of diffraction circles on the retina, formed by the images of point objects, have diameters inversely proportional to the pupil's diameter. If the diameter of the pupil could be doubled, diffraction effects in the eye would be reduced by half. But from Fig. 25-36 it is apparent that this change alone would double the size of the circles of confusion produced by chromatic aberration. How much bigger would the eye have to be to reduce the size of chromatic circles by half, if the pupil's diameter is doubled?

42 Use Fig. 25-36 to estimate the diameter of the circle of confusion on the retina produced by chromatic aberration when the pupil diameter is (a) 2 mm; (b) 4 mm.

Additional Problems

★★ **43** A nearsighted man has a near point of 10.0 cm and a far point of 50.0 cm. He looks at images in a concave spherical mirror of objects at distances ranging from 1.00 m to ∞. He is able to see all these images clearly without glasses by focusing his eyes over their entire range of accommodation. Find the mirror's focal length.

★ **44** Consider a glass sphere of index 1.5 that, when placed in front of the normal eye, will correct underwater vision. Both the eye and the sphere are immersed in water. Distant underwater objects viewed through the sphere are imaged on the retina. Let the back surface of the sphere be 1.0 cm from the cornea. Find the radius of the sphere. Ignore refraction at the cornea and within the eye, except for the crystalline lens. From Example 1, the lens needs a virtual object 31 mm behind the cornea to form an image on the retina. This virtual object is the image that the glass sphere must produce.

★ **45** Fig. 25-44 shows a concave mirror acting as a magnifier. A small object is placed in the focal plane of the mirror.

(a) Prove that the angular magnification of the mirror is the same as for a glass magnifier: $M = d/f$, where d is the near-point distance and f is the focal length.

(b) Find the radius of curvature if the angular magnification is 2.0 for a person with a near point of 20 cm.

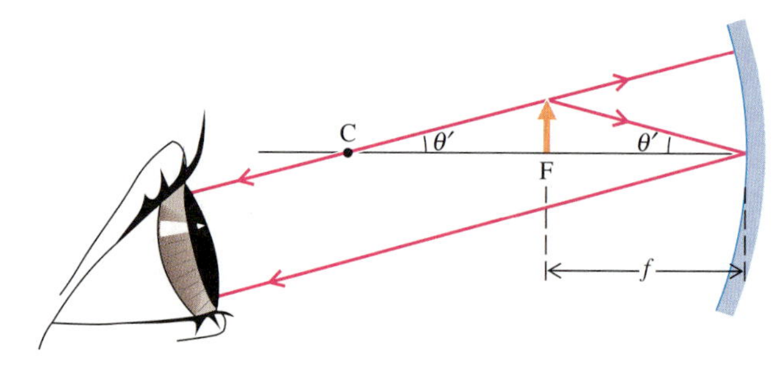

Fig. 25-44

★★ **46** (a) Show that in general the angular magnification of a magnifier is given by $M = \dfrac{d}{s + x(1 - s/f)}$, where d is the eye's near point distance, s is the distance from the object to the lens, and x is the distance from the lens to the eye. Notice that this expression reduces to $M = d/s$, when either $x = 0$ or $s = f$.

(b) Show that when the image formed by a magnifier is at the eye's near point the angular magnification is given by $M = \dfrac{d - x}{s}$.

★★ **47** You start walking from a great distance toward a large concave mirror. You can see a clear image of yourself in the mirror until you are 1.00 m from the mirror, whereupon the image you see becomes blurred because it is at your near point. Assume the near point of your eye is 25.0 cm. Find the radius of curvature of the mirror.

CHAPTER 26 Wave Optics

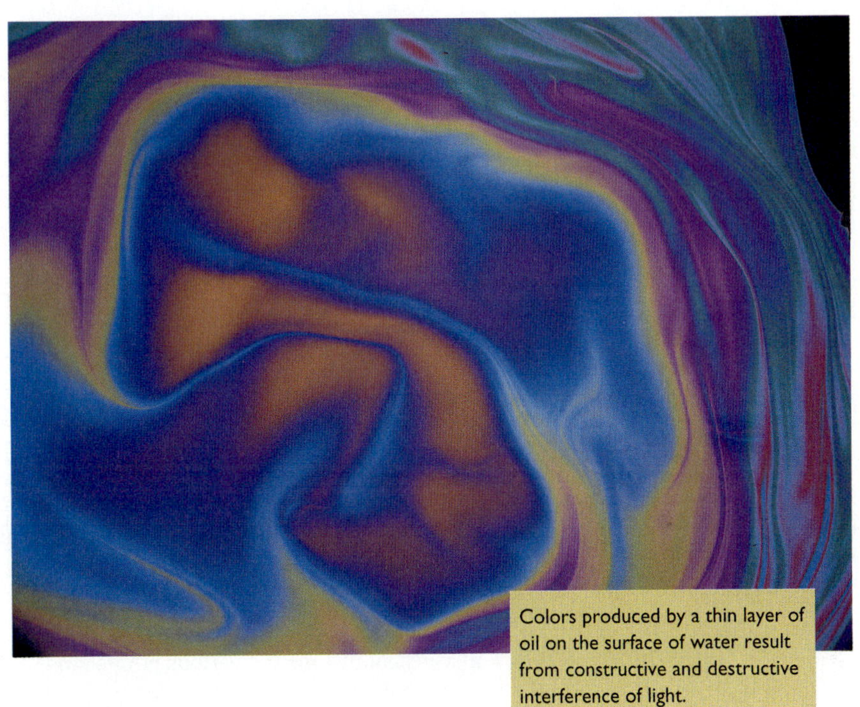

Colors produced by a thin layer of oil on the surface of water result from constructive and destructive interference of light.

Why is the sky blue? What causes the beautiful colors in a soap bubble or an oil film? Why are clouds and ocean surf white, though both are formed by tiny drops of clear, colorless water? Why do Polaroid sunglasses reduce reflected glare? In this chapter we shall answer these questions. Additionally, we shall see how two light beams can combine to produce darkness, we shall show how to measure the wavelength of light using a meter stick, and we shall see why the magnification of any optical microscope is limited by the wave properties of light.

We begin by giving a brief qualitative introduction to the wave phenomena of polarization, diffraction, and interference before returning to a more quantitative discussion of each. In our description of various experiments we shall often use a laser as our light source because of its wonderfully simple properties.

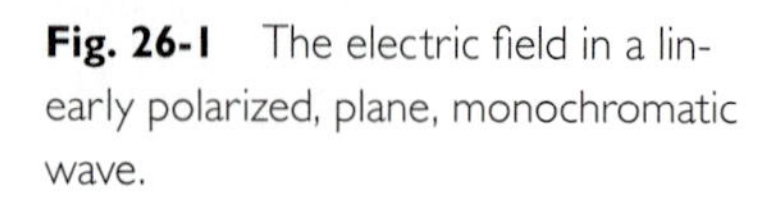
Fig. 26-1 The electric field in a linearly polarized, plane, monochromatic wave.

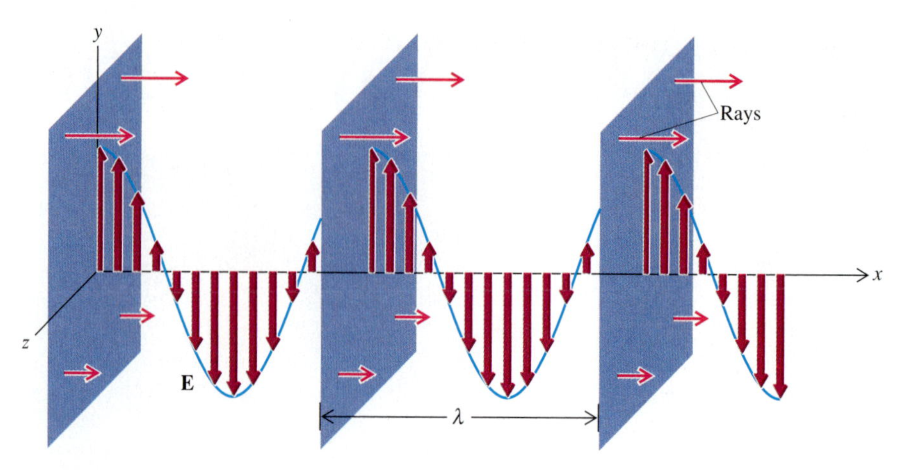

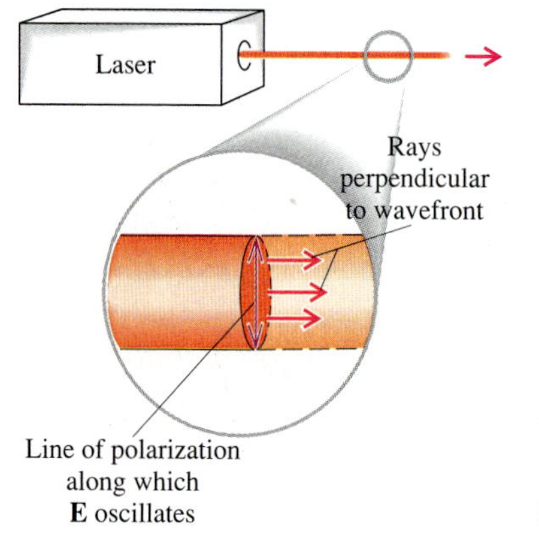

Fig. 26-2 A polarized laser beam.

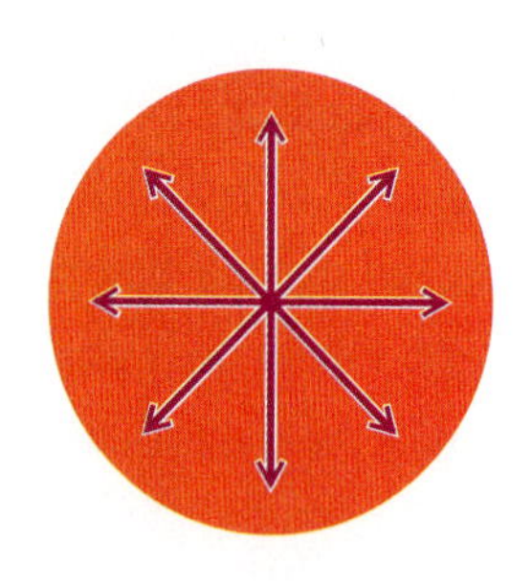

Fig. 26-3 Cross section of an unpolarized light beam.

26-1 Wave Properties of Light

Polarization

The simplest kind of light wave is a plane, monochromatic wave, which is linearly polarized (Fig. 26-1). **Linear polarization means that the electric field vector is always directed parallel to a single line** (the y-axis in the figure). Fig. 26-1 shows only the electric field. An associated magnetic field is parallel to the z-axis and oscillates in phase with the electric field. Values of the electric field are shown at a particular instant of time for various points along the x-axis. In a plane wave the value of the electric field is the same along any plane perpendicular to the direction of the wave's motion. The figure shows plane wavefronts, along which the electric field is maximum. Rays show the direction of motion of the wave—along the x-axis.

The plane wave described in Fig. 26-1 is approximated by a section of a polarized laser beam (Fig. 26-2). Within the beam, rays are approximately parallel, and wavefronts are approximately cross sections of the beam.* By turning the laser on its side, rotating it 90°, we can produce a wave linearly polarized in the horizontal direction, rather than in the vertical direction. By rotating the laser through some other angle we can get polarization in any direction perpendicular to the beam.

Most natural light sources and many lasers have random polarization. This means that at a given instant the electric field at any point in the wave is just as likely to be directed along any line perpendicular to the direction of motion. The light is then said to be unpolarized, and we indicate this state as shown in Fig. 26-3.

Frequency Bandwidths

Any real source of light is not exactly monochromatic; that is, there is never just one precise value of frequency. Instead there is a range or band of frequencies, which may be wide or narrow. The narrower the band, the more nearly the wave approximates a monochromatic wave. Laser light is nearly monochromatic. A common helium-neon laser emits light at a frequency of 4.74×10^{14} Hz with a bandwidth of about 10^8 Hz. This means that the frequency range is less than 1 part in 10^6. More ex-

*There is some slight spreading of the beam, and hence the beam is not exactly a plane wave. A plane wave is also approximated by a small section of a spherical wave from a distant point source.

pensive, frequency-stabilized lasers have bandwidths as low as 10^4 Hz. Some lasers have even achieved a stabilized frequency range of less than 100 Hz. By way of comparison, spectral lines emitted by various gas discharge tubes typically have bandwidths of roughly 10^9 Hz, and white light, ranging in frequency from 4×10^{14} Hz to 7×10^{14} Hz, has a bandwidth of 3×10^{14} Hz. The order of magnitude of these bandwidths is summarized in Table 26-1.

Table 26-1 Frequency bandwidths typical of various kinds of light

Light	Δf (Hz)
Stabilized He-Ne laser	10^4
Common He-Ne laser	10^8
Spectral line	10^9
White light	10^{14}

Coherence

It is often important to be able to predict the relationship between the phase of a light wave at two different times at the same point in space. For example, suppose that at some instant t_0, at one point in a laser beam, the electric field vector has its maximum value; that is, you are at a peak in the wave. At a time Δt later, will the wave again have its peak value or will it have some other value (Fig. 26-4)? If the laser light were truly monochromatic, the solution would be easy. We could simply determine the exact number of cycles elapsed in a given time interval Δt by multiplying the frequency (the number of cycles per second) by the time interval Δt. (If the result were a whole number, the wave would again be at a peak. Or if the result were a whole number plus $\frac{1}{4}$, the electric field would be zero at that instant.) However, even laser light is not exactly monochromatic. There is always some frequency range f to $f + \Delta f$. The number of cycles per second during any particular time interval can be anywhere in this range. So the number of cycles completed during a time interval Δt is somewhere in the range $f\,\Delta t$ to $(f + \Delta f)\,\Delta t$. If the time interval Δt is small enough, the product $\Delta f\,\Delta t$ will be much less than 1 cycle, and there will be little uncertainty in the number of cycles completed, or in the final phase of the cycle. We can then predict the final phase from the initial phase, and we say that the wave is coherent over the time interval Δt. This means that there is a definite, predictable phase relationship. The condition for coherence then is that the time interval be small enough that

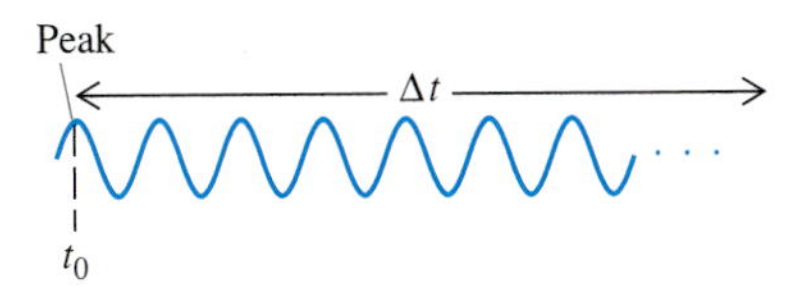

Fig. 26-4 What phase of the cycle occurs at a time Δt after t_0?

$$\Delta f\,\Delta t << 1$$

or that

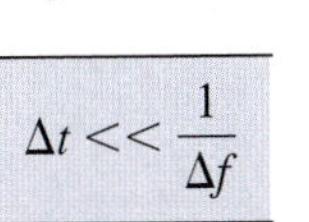

$$\Delta t << \frac{1}{\Delta f} \qquad \text{(condition for coherence)} \qquad (26\text{-}1)$$

EXAMPLE 1 Coherence of Light Sources

Will you have coherence over a time interval of 10^{-6} s for light from (a) a gas discharge tube; (b) a stabilized He-Ne laser?

SOLUTION (a) For a single line from a gas discharge tube, we see from Table 26-1 that even for a single spectral line $\Delta f = 10^9$ Hz. Thus $\frac{1}{\Delta f} = 10^{-9}$ s, and the time interval $\Delta t = 10^{-6}$ s is much too long to satisfy Eq. 26-1, since 10^{-6} s is certainly not less than 10^{-9} s. Thus this kind of light is not coherent over such a time interval. The number of cycles completed is uncertain by $\Delta f\,\Delta t = (10^9\ \text{s}^{-1})(10^{-6}\ \text{s}) = 10^3$ cycles, and so there is no ability to predict the phase over such a time interval.

(b) For light from a stabilized He-Ne laser, $\Delta f = 10^4$ Hz. Thus the condition that must be satisfied for such light is

$$\Delta t << \frac{1}{\Delta f}$$

or

$$\Delta t << 10^{-4}\ \text{s}$$

A time interval of 10^{-6} s satisfies this condition. Thus the laser light is coherent over this time interval. Notice that $\Delta f \Delta t = (10^4\ \text{s}^{-1})(10^{-6}\ \text{s}) = 10^{-2}$, and so we know the number of elapsed cycles with an uncertainty of only one hundredth of a cycle.

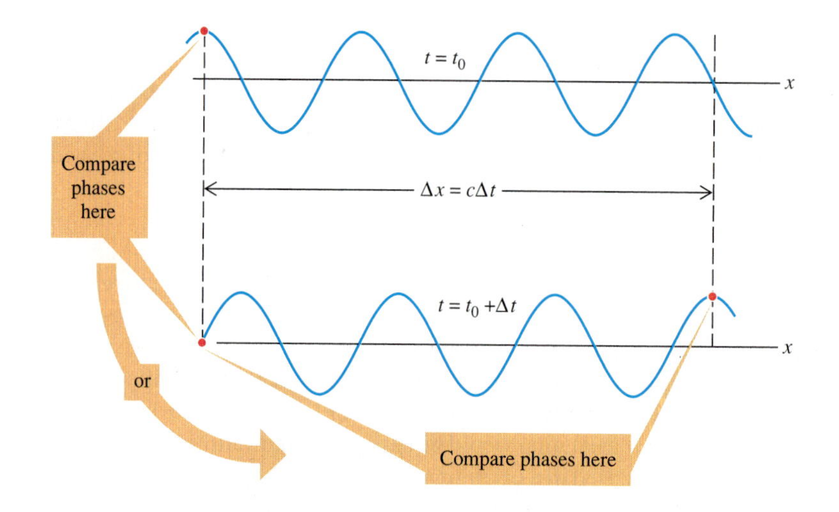

Fig. 26-5 During a time interval Δt a wave peak advances a distance $\Delta x = c\,\Delta t$. Thus comparing the phases at two points a distance Δx apart at a fixed time is the same as comparing the phases at the same point in space over a time interval Δt.

Each wavefront in a light wave advances at the speed of light. Therefore, we can relate the coherence of light at a fixed point in a plane wave at two different times to coherence at two different points in a plane wave at the same time. As illustrated in Fig. 26-5, during a time interval Δt a wavefront advances a distance $\Delta x = c\,\Delta t$, and so comparison of phases at two points Δx apart is equivalent to comparing the phases at a fixed point over a time interval Δt. Since $\Delta t << \frac{1}{\Delta f}$ is the condition for coherence, two points in a wavefront will be coherent if

$$\Delta x << \frac{c}{\Delta f}$$

The distance $\frac{c}{\Delta f}$ is called the **coherence length,** denoted by x_c.

$$x_c = \frac{c}{\Delta f} \qquad (26\text{-}2)$$

The condition for coherence may be expressed in terms of x_c:

$$\Delta x << x_c \qquad \text{(coherence condition)} \qquad (26\text{-}3)$$

For a stabilized He-Ne laser with a frequency range of 10^4 Hz, we find

$$x_c = \frac{c}{\Delta f} = \frac{3 \times 10^8 \text{ m/s}}{10^4 \text{ Hz}} = 3 \times 10^4 \text{ m}$$

Two points in the laser beam have a predictable phase relationship, as long as they are much less than 30,000 m apart! For a common laboratory He-Ne laser, the bandwidth is of the order of 10^8 Hz. Thus

$$x_c = \frac{c}{\Delta f} = \frac{3 \times 10^8 \text{ m/s}}{10^8 \text{ Hz}} = 3 \text{ m}$$

The two points must be much closer than 3 m. Certainly two points a few cm apart are coherent. For white light,

$$x_c = \frac{c}{\Delta f} = \frac{3 \times 10^8 \text{ m/s}}{10^{14} \text{ Hz}} = 3 \times 10^{-6} \text{ m}$$

Thus points in a beam of white light must be considerably less than a thousandth of a millimeter apart to be coherent.

Diffraction

Suppose you are standing behind an open doorway, listening to a conversation in the next room. You can easily hear the voices from the room because the sound waves bend around the doorway. This phenomenon is called **diffraction.** It is a property common to all waves to bend or diffract around an obstacle. However, the amount of bending depends on the wavelength of the wave and the dimensions of the obstacle. In general, the longer the wavelength, the greater is the diffraction. Light, with its relatively short wavelength, bends or diffracts very little around an open doorway, but sound waves with their much longer wavelengths, diffract a great deal. Thus you can hear the conversation though you cannot see those who are talking.

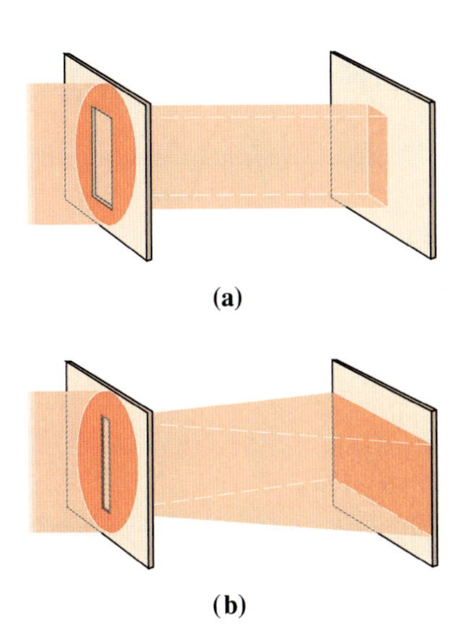

Fig. 26-6 **(a)** Little diffraction is produced by a slit wider than 1 mm. **(b)** Considerable diffraction is produced by a slit less than 0.1 mm wide.

Seeing diffraction of light requires careful observation. Suppose we pass an intense beam of light through a narrow slit in an opaque screen and project it onto a white screen (Fig. 26-6). If the slit is relatively wide, (say, at least a millimeter), we get an image of the slit on the screen (Fig. 26-6a). As predicted by geometrical optics, the rays passing through the slit travel straight to the screen. But if we make the slit very narrow (say, less than about 0.1 mm), the image on the screen actually gets wider (Fig. 26-6b), violating the prediction of geometrical optics. We find that the narrower we make the slit, the more the light bends outward. Of course if we make the slit much less than 0.1 mm, there will be too little light to be seen, even if the light source illuminating the slit is very intense. But if we could make a slit with a width much less than the wavelength of light and still have enough light intensity to see the small amount of light passing through, we would see the light spread out in all directions, forming a cylindrical wave. Or if we replaced the slit by a tiny circular hole, with a diameter much less than the wavelength of light, we would produce a spherical wave. This result, illustrated in Fig. 26-7, is the essence of the **Huygens-Fresnel principle,** according to which **each section of wavefront in the diffracting aperture is the source of a spherical wave.** Fig. 26-8 shows a photograph, illustrating this principle for water waves. The waves in Fig. 26-8 are incident on an aperture much smaller than the wavelength.

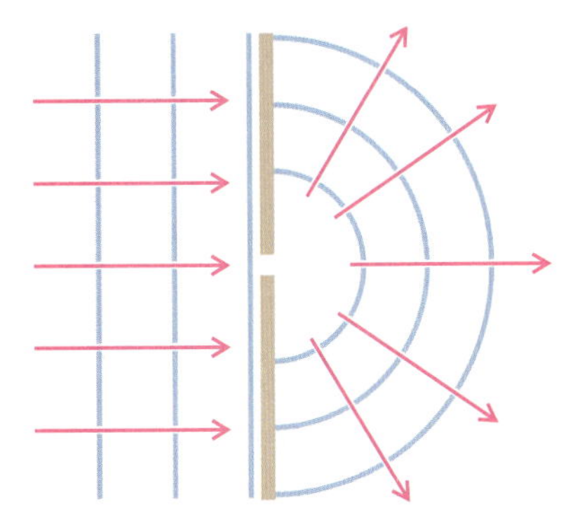

Fig. 26-7 Cross section of a spherical wave, resulting from diffraction of light by a circular hole with a diameter much smaller than the wavelength of the light.

Diffraction of light was first observed and recorded by the Jesuit priest Francesco Grimaldi, a contemporary of Newton. Grimaldi observed the spreading of a narrow beam of sunlight entering a darkened room. Geometrical optics, which was then based on a picture of light consisting of particles,* could not account for this phenomenon, and so Grimaldi proposed that light is a wave. Grimaldi's idea was rejected by Newton, who believed that if light were a wave, the diffraction effect would be much greater than observed. It must have seemed unlikely to Newton that light could have the incredibly small wavelength necessary to explain such a small amount of diffraction. Newton's authority was so great that the wave theory was not accepted for another 200 years.

Interference

Like sound waves or waves on a string, light waves can interfere constructively or destructively. Constructive interference occurs when two light waves are in phase, and destructive interference occurs when two light waves are 180° out of phase. The colors in the photograph of the oil slick at the beginning of this chapter result from interference of the light reflected from the top and bottom surfaces of the thin film of oil. The thickness of the film varies, and, as a result, different colors of light interfere

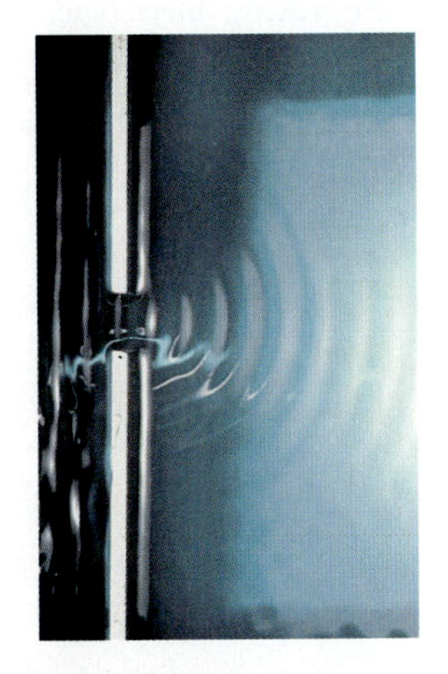

Fig. 26-8 Diffraction of water waves.

*In the twentieth century it was discovered that light does indeed have some particle-like properties. However this modern idea of a photon as a "particle" of light refers to emission or absorption of light, not to the way it propagates. Light travels as a wave, not as a bunch of particles, contrary to Newton's belief. The dual character of light as wavelike (in transmission) and particle-like (in absorption and emission) is at the heart of modern quantum physics, to be discussed in Chapter 28.

constructively. For example, where the film appears blue the thickness is such that blue light reflected from the two surfaces interferes constructively, while red light interferes destructively. In the sections that follow we shall investigate interference and diffraction phenomena quantitatively.

26-2 Interference

Fig. 26-9a illustrates **constructive interference** of light, which occurs when two light waves are in phase at a certain point in space over a period of time. Fig. 26-9b shows **destructive interference,** which occurs at a point in space where the waves are 180° out of phase over a period of time. If the two waves are of equal amplitude and 180° out of phase, the presence of two sources of light actually produces darkness! Fig. 26-10 illustrates the intensity of light that is seen at a given point in space where two waves of equal amplitude interfere either constructively or destructively. Since the intensity of a wave is proportional to the square of its amplitude, constructive interference of two equal-amplitude waves produces in the resultant wave 2 times the amplitude or 4 times the intensity of the individual waves. Whether the interference is constructive or destructive at a given point in space depends on the position of the point relative to the sources of light. Interference of light typically produces a pattern of light and dark areas (Fig. 26-11).

In order for the eye to perceive interference of light, there must be a definite phase relationship between the two waves over a time interval that the eye can detect. The eye has a response time on the order of $\frac{1}{20}$ of a second. Thus interference effects must be stable for at least this long to be visible. This is longer than the coherence time of even the most monochromatic sources available today. The relative phases of two independent light sources (say, two different lasers) will vary randomly over time intervals greater than the coherence time. Thus, if we illuminate an area with two different sources, the interference of their light waves at a given point will rapidly oscillate from constructive to destructive, and so no interference pattern is visible. All one sees is a uniform illumination equal to the sum of the two intensities. For example, two equal-amplitude waves from separate sources produce instantaneous intensities rapidly oscillating between 0 and 4 times each wave's intensity I, at each point in space. Thus one sees only the time average of the instantaneous intensity, which at all points is the same: twice the intensity of each source's wave, since the average of 0 and $4I$ equals $2I$. With the present state of technology, it is impossible to see interference of light from two independent light sources.*

*It is possible to detect electronically interference of two independent sources, as demonstrated by Brown and Twiss in 1952.

E_{1y}

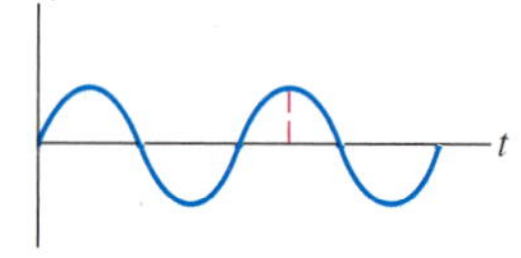

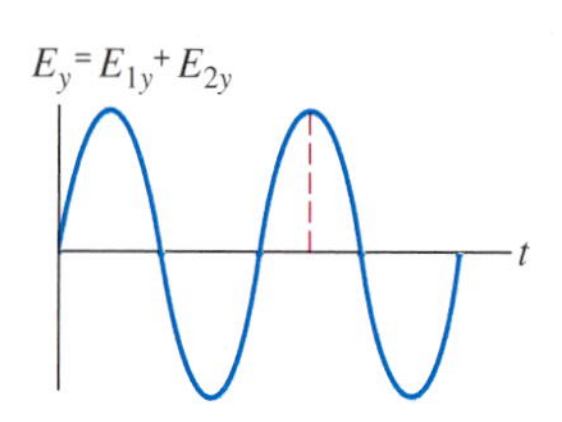

(a) Constructive interference

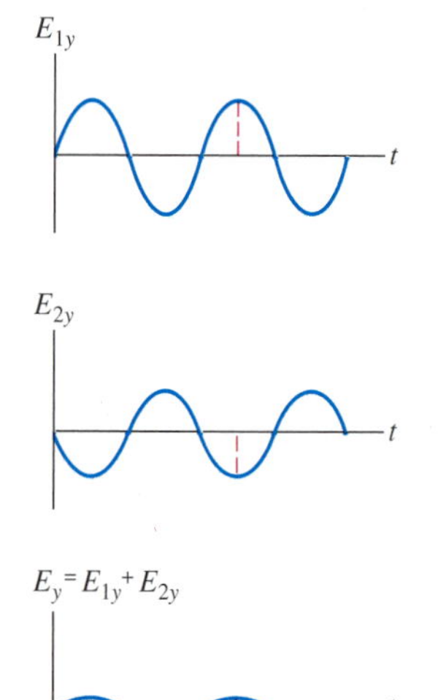

(b) Destructive interference

Fig. 26-9 Two light waves can interfere either **(a)** constructively or **(b)** destructively.

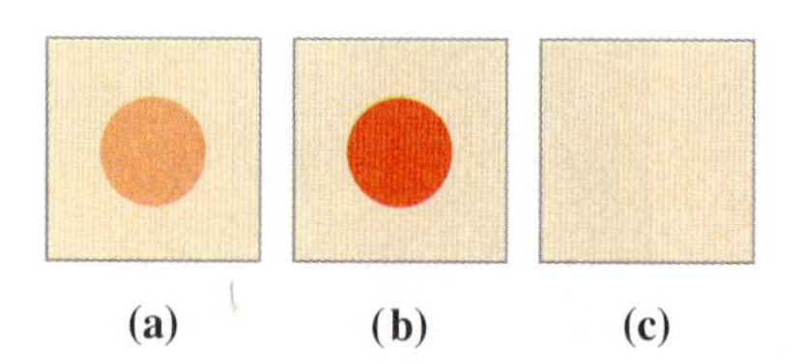

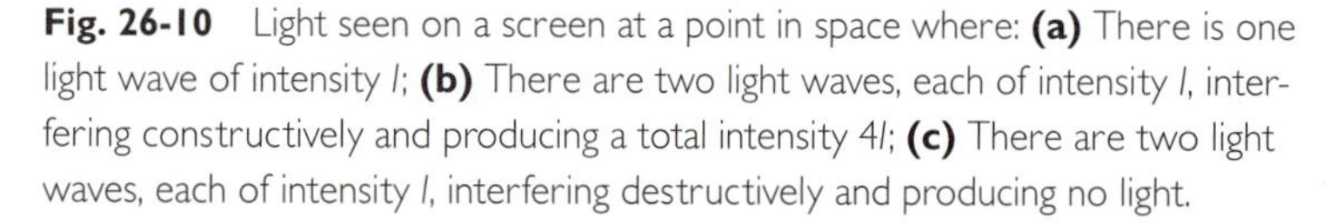

Fig. 26-10 Light seen on a screen at a point in space where: **(a)** There is one light wave of intensity I; **(b)** There are two light waves, each of intensity I, interfering constructively and producing a total intensity $4I$; **(c)** There are two light waves, each of intensity I, interfering destructively and producing no light.

Fig. 26-11 An interference pattern.

Interference effects are easily observable when a single wavefront is divided into two separate parts, which then follow separate paths to the point where interference is observed. Since the two waves arise from a common wavefront, changes in phase are common to each, and the interference pattern remains constant.

Young's Double Slit

One of the simplest ways to produce interfering light is to use a double-slit arrangement like the one studied by the English physician Thomas Young in 1801.* If a monochromatic plane wave is incident on a pair of thin, closely spaced slits, the two slits serve as sources of coherent light. The slits must be narrow enough and close enough that there is a significant amount of diffraction and overlap of the two wavefronts. As illustrated in Fig. 26-12, what is seen on a screen in front of the slits is a pattern of alternating light and dark fringes. Fig. 26-13 shows how the location of the fringes relates to the distance to each of the slits. Point P is equidistant from the two slits, and so the two waves are exactly in phase at this point, and therefore point P is at the center of an interference maximum—a bright fringe. The first dark fringe above the central bright fringe is at a point Q, which is one-half wavelength farther from slit 2 than from slit 1. The next bright fringe is at point R, which is 1 wavelength farther from slit 2 than from slit 1.

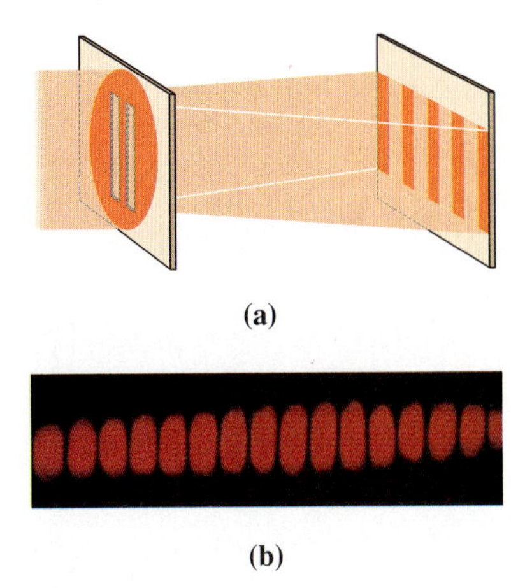

Fig. 26-12 **(a)** Overlapping wavefronts interfere constructively at certain points and destructively at others. (Figure is not drawn to scale.) **(b)** Photograph of interference fringes from double slits illuminated with a He-Ne laser.

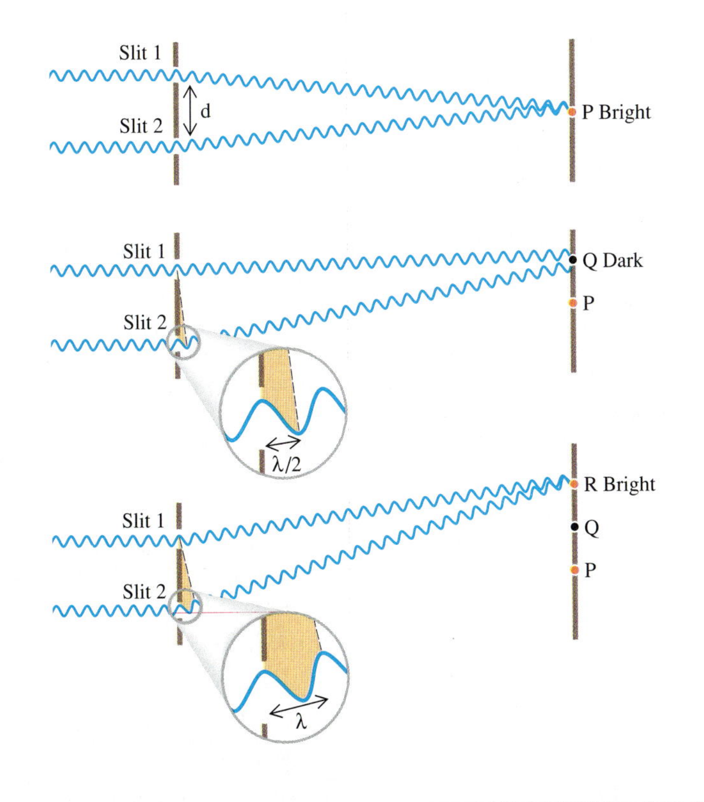

Fig. 26-13 Constructive interference occurs at P and R. Destructive interference occurs at Q. (Figures are not drawn to scale.)

*In 1801 Young used his double-slit experiment to measure the wavelength of light and to provide support for his belief that light is a wave. His ideas gained wide acceptance only after many years.

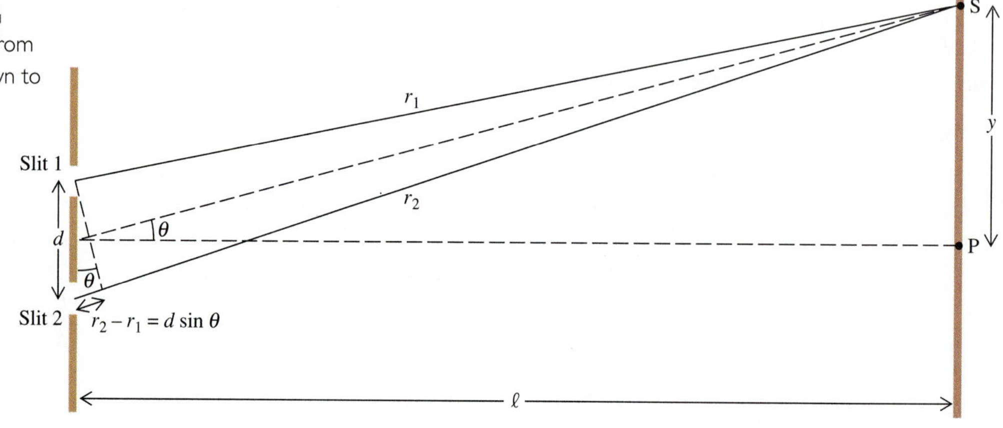

Fig. 26-14 A point S on a screen is located a distance r_1 from slit 1 and a distance r_2 from slit 2. (This figure is not drawn to scale.)

The location of the fringes can be determined with the aid of Fig. 26-14. The figure shows an arbitrary point S some distance y from the center of the interference pattern at point P. The angular displacement of point S is measured by the angle θ. The difference in the path lengths from S to each of the slits is related to the same angle θ. As shown in the figure, this path-length difference is $d \sin \theta$. Constructive interference occurs when this distance equals a whole number of wavelengths:

$$d \sin \theta = m\lambda \qquad m = 0, 1, 2, \ldots \qquad \text{(constructive interference)} \qquad (26\text{-}4)$$

Destructive interference occurs when the difference in path lengths equals a whole number of wavelengths plus $\frac{1}{2}$ wavelength:

$$d \sin \theta = (m + \tfrac{1}{2})\lambda \qquad m = 0, 1, 2, \ldots \qquad \text{(destructive interference)} \qquad (26\text{-}5)$$

EXAMPLE 2 Measuring Light's Wavelength With a Meter Stick

Light from a He-Ne laser illuminates two narrow slits, 0.20 mm apart, producing interference fringes on a wall 6.67 m from the slits (Fig. 26-12). The centers of the bright fringes are 2.1 cm apart. (a) Determine the wavelength of the laser light. (b) What would the fringe separation be if the slits were illuminated with violet light of wavelength 400 nm?

SOLUTION (a) From Fig. 26-14, we see that the distance y from the center of the interference pattern to any point S is related to the angle θ and the distance ℓ from slits to screen:

$$\tan \theta = \frac{y}{\ell}$$

Any point in the interference pattern is at a very small angle θ, for which $\sin \theta$ and $\tan \theta$ are very nearly identical. Thus

$$\sin \theta \approx \frac{y}{\ell}$$

Inserting this equation into Eq. 26-4 yields an expression for the distance y_m to the mth fringe

$$d\frac{y_m}{\ell} = m\lambda \qquad m = 0, 1, 2, \ldots$$

or

$$y_m = \frac{m\lambda\ell}{d}$$

EXAMPLE 2—cont'd

Thus $y_0 = 0, \quad y_1 = \frac{\lambda \ell}{d}, \quad y_2 = \frac{2\lambda \ell}{d}, \ldots$

The fringes are equally spaced, separated by a distance

$$\Delta y = \frac{\lambda \ell}{d} \qquad (26\text{-}6)$$

Solving for λ, we find

$$\lambda = \frac{d\,\Delta y}{\ell} = \frac{(2.0 \times 10^{-4}\text{ m})(2.1 \times 10^{-2}\text{ m})}{6.67\text{ m}}$$

$$= 6.3 \times 10^{-7}\text{ m}$$

$$= 630\text{ nm}$$

Thus, using measurements in cm and m, we indirectly measure the wavelength of the laser light, a length less than a thousandth of a mm.

(b) Applying Eq. 26-6, we find that for violet light of wavelength 400 nm, the fringe spacing changes to

$$\Delta y = \frac{\lambda \ell}{d} = \frac{(400 \times 10^{-9}\text{ m})(6.67\text{ m})}{2.0 \times 10^{-4}\text{ m}} = 1.3 \times 10^{-2}\text{ m}$$

$$= 1.3\text{ cm}$$

Because of its shorter wavelength, violet light produces interference fringes that are closer together.

Thin Films

There is another simple way to split a single light wave into separate, coherent waves, which can then interfere. When light is incident on a partially reflecting surface (for example, a glass surface), part of the incident light is reflected and part is transmitted into the second medium. If the two light waves again come together, after having followed paths of somewhat different length, they will interfere. The difference in path lengths, however, must be less than the coherence length of the light, or the two waves will be totally incoherent. The Michelson interferometer, described in Problem 21, uses a half-silvered mirror to equally divide the incident wavefront from a monochromatic source.

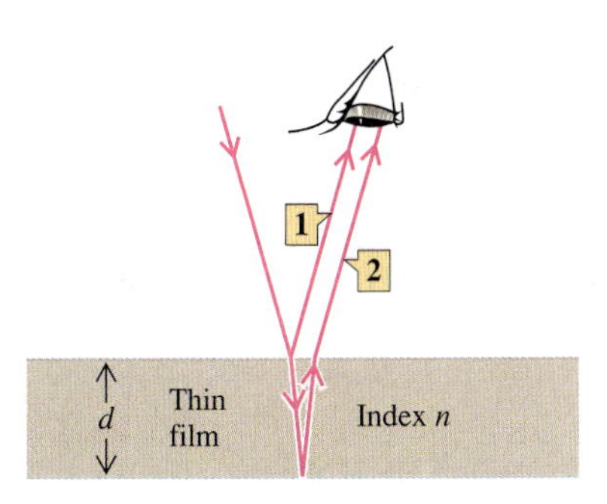

Fig. 26-15 The top and bottom surfaces of a thin film reflect light upward.

A thin transparent film can also serve to produce two coherent light waves, one reflected from the top and one from the bottom of the film (Fig. 26-15). If the film is very thin, you can see the interference, even with a white light source, as indicated by the photo of the colored oil film at the beginning of this chapter.

In Fig. 26-15 the second light wave travels a distance greater than the first. The difference in the path lengths may cause a phase difference between the two waves. In addition, there may also be phase changes produced by the reflections. **When light is incident on the surface of a medium with a higher refractive index than that of the incident medium, reflected light experiences a 180° phase change.** If the second medium has a lower index than the first, reflection causes no phase change. The situation is the same as that of a wave pulse on a rope, partially reflected because of a change in density of the rope. As illustrated in Chapter 16, Fig. 16-11, the reflected pulse is inverted if the second section of rope has higher density than the first. If the second section is of lower density than the first, there is no inversion of the reflected pulse, that is, no change in phase.

The microscope slides in Fig. 26-16 show a pattern of interference fringes produced by light reflected on either side of the thin film of air trapped between the slides. The air film varies in thickness, and hence the interference alternates between constructive (bright) and destructive (dark).

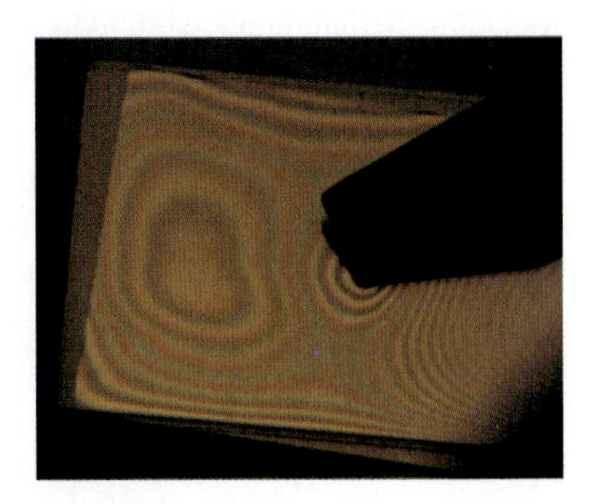

Fig. 26-16 An interference pattern is formed by reflection from the thin film of air between two microscope slides.

EXAMPLE 3 Colors On An Oil Slick

What color will be brightest when white light is reflected at normal incidence from a film of oil 250 nm thick on the surface of a puddle of water? The oil has a refractive index of 1.4.

SOLUTION The light reflected from the upper surface of the oil film undergoes a 180° phase change, since oil's refractive index is higher than air's (Fig. 26-17). The light reflected from the lower surface of the oil experiences no phase change, since water's index is lower than oil's. Thus, if light reflected from the lower surface is to arrive at the upper surface in phase with the light reflected from the top surface, the extra distance traveled must give it a net 180° phase change. This means that the additional path length, equal to twice the film's thickness t for normal incidence, must equal a whole number of wavelengths plus $\frac{1}{2}$ wavelength.

$$2t = (m + \tfrac{1}{2})\lambda_n \qquad \text{where } m = 0, 1, 2, \ldots \qquad (26\text{-}7)$$

The wavelength λ_n is of course the wavelength in the oil. This is related to the vacuum wavelength λ_0 by Eq. 23-10:

$$\lambda_n = \frac{\lambda_0}{n}$$

where n is the refractive index of the oil. Inserting this expression into Eq. 26-7 and solving for λ_0, we find

$$\lambda_0 = \frac{2tn}{m + \frac{1}{2}} \qquad \text{for } m = 0, 1, 2, \ldots \qquad (26\text{-}8)$$

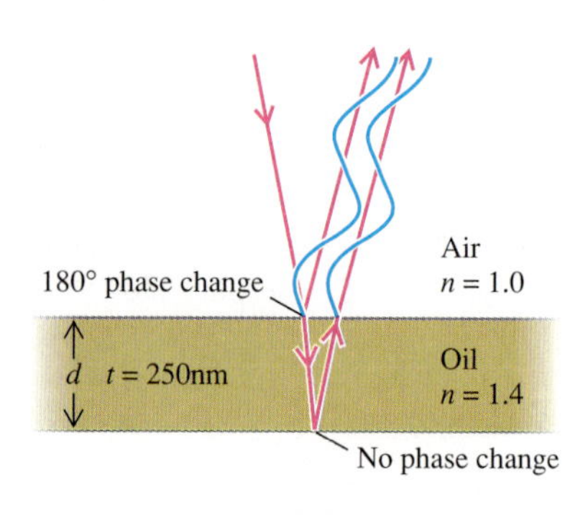

Fig. 26-17

Trying possible values of m, we find

$$m = 0: \qquad \lambda_0 = \frac{2tn}{\frac{1}{2}} = \frac{2(250\text{ nm})(1.4)}{\frac{1}{2}} = 1400\text{ nm}$$

$$m = 1: \qquad \lambda_0 = \frac{2tn}{\frac{3}{2}} = \frac{1400\text{ nm}}{3} = 470\text{ nm}$$

$$m = 2: \qquad \lambda_0 = \frac{2tn}{\frac{5}{2}} = \frac{1400\text{ nm}}{5} = 280\text{ nm}$$

Only for $m = 1$ do we find a wavelength in the visible range (400 to 700 nm). Blue light of wavelength 470 nm will interfere constructively, and therefore blue will be the color most strongly reflected by the film. The film will appear blue. One can show that red light will experience destructive interference (Problem 14).

EXAMPLE 4 Nonreflective Glass Coating

Uncoated glass reflects 4% of the light incident on its surface at normal incidence. Sometimes glass is coated with a thin layer of a transparent material so that the intensity of the reflected light is reduced. Find the minimum thickness of a coating of magnesium fluoride, MgF_2 ($n = 1.38$), which will produce destructive interference at a wavelength in the middle of the visible spectrum (550 nm).

SOLUTION Both reflected waves experience a 180° phase change, since both are reflected from a medium with a higher index than that of the incident medium (Fig. 26-18). The only relative change in phase results from a difference in path length. Destructive interference occurs for a minimum path difference of $\frac{1}{2}$ wavelength.

$$2t = \tfrac{1}{2}\lambda_n = \frac{\lambda_0}{2n}$$

or

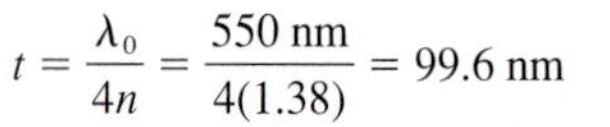

$$t = \frac{\lambda_0}{4n} = \frac{550\text{ nm}}{4(1.38)} = 99.6\text{ nm}$$

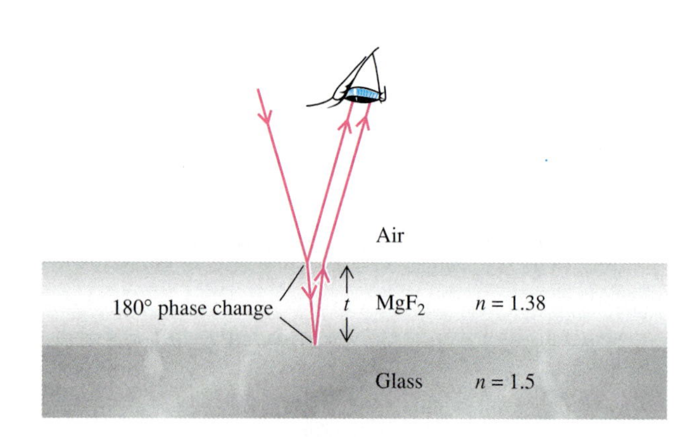

Fig. 26-18

EXAMPLE 4—cont'd

This is the minimum thickness of MgF_2 that will produce destructive interference for $\lambda_0 = 550$ nm. Such a layer will not completely eliminate reflection at this wavelength, since the amplitudes of the two interfering waves are not equal, and so there is only partial cancellation of the waves, as in Fig. 26-9b.

One might wonder whether this coating will produce a reduction in reflected intensity at other wavelengths or whether it could perhaps enhance reflection by constructive interference at some wavelengths. The condition for constructive interference here is that the path-length difference equals a whole number of wavelengths.

$$2t = m\lambda_n = m\frac{\lambda_0}{n}$$

or

$$\lambda_0 = \frac{2tn}{m} \qquad \text{where } m = 1, 2, 3, \ldots$$

$$m = 1: \quad \lambda_0 = 2tn = 2(99.6 \text{ nm})(1.38) = 275 \text{ nm}$$

$$m = 2: \quad \lambda_0 = \frac{2tn}{2} = 138 \text{ nm}$$

Other values of m yield smaller wavelengths. Thus no value of m gives a wavelength in the visible range. No visible light interferes constructively. A more detailed analysis, which takes into account the intensities of the interfering waves, shows that the effect of the coating is to reduce reflection fairly uniformly across the visible spectrum to an average of about 1% of incident intensity, compared to a 4% reflection for uncoated glass. However, there is a slight enhancement of reflected intensity in the blue part of the spectrum. Coating glass with several thin layers of different materials, carefully selected for index and thickness, can provide a further reduction in intensity. One of the most important applications of such coatings is for lenses in optical instruments, which use a large number of lenses that would otherwise produce much unwanted reflected light. Fig. 26-19 shows eyeglasses that have a nonreflective coating on one lens.

Fig. 26-19

26-3 Diffraction

Historical Background

The quantitative study of diffraction was of great historical importance in establishing the wave nature of light. Although Young's double-slit experiment supported the wave theory, many nineteenth-century scientists clung to Newton's particle theory of light. Full acceptance of the wave theory followed careful quantitative studies of diffraction by various scientists, especially Fresnel and Arago. In evaluating the wave theory of diffraction proposed by Fresnel, Poisson objected that it led to a rather strange and, to Poisson, an obviously false prediction: that at the center of the shadow of a round object would be a bright spot. Poisson argued that waves diffracted around the edges would travel an equal distance to the center and therefore interfere constructively there, if the wave theory were correct. Poisson presented this argument as a proof that the wave theory was wrong. Arago promptly performed the crucial experiment and found the predicted bright spot at the center of the shadow (Fig. 26-20). Based on such results, the wave theory of light was strongly established by 1820.

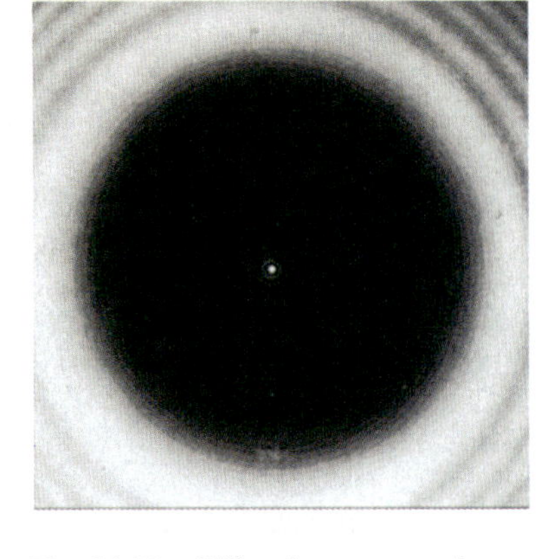

Fig. 26-20 Diffraction pattern of a penny. Constructive interference of light diffracted around the edge of the penny produces a bright spot at the center of the shadow.

Fraunhofer and Fresnel Diffraction

When an object producing diffraction is illuminated by a plane wave and the resulting diffraction pattern is viewed on a screen at a large enough distance from the object, detailed analysis of the diffraction pattern is greatly simplified. This case is referred to as "Fraunhofer diffraction." When the illuminating source is not a plane wave or the screen is not far enough away, the diffraction is called "Fresnel diffraction." Analysis of Fresnel diffraction is more complicated than analysis of Fraunhofer diffraction, and the Fresnel diffraction pattern itself looks quite different from the Fraunhofer diffraction pattern of the same object. Figs. 26-20 and 26-21 are examples of Fresnel diffraction. We shall analyze only Fraunhofer diffraction because of its relative simplicity.

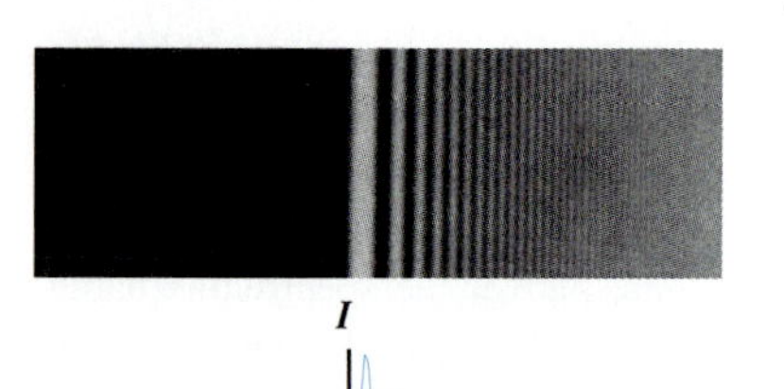

Fig. 26-21 Diffraction by a straight edge and the corresponding graph of intensity versus position. The dashed line indicates the intensity predicted by geometrical optics, implying a sharp-edged shadow.

Single Slit

A particularly simple case, Fraunhofer diffraction by a single slit, is shown in Fig. 26-22. In Section 26-1 we described diffraction of light around the edges of the slit (Fig. 26-6b). Actually the phenomenon is somewhat more complicated because light coming from various parts of the slit interferes to form a series of light and dark bands, as shown in Fig. 26-22.

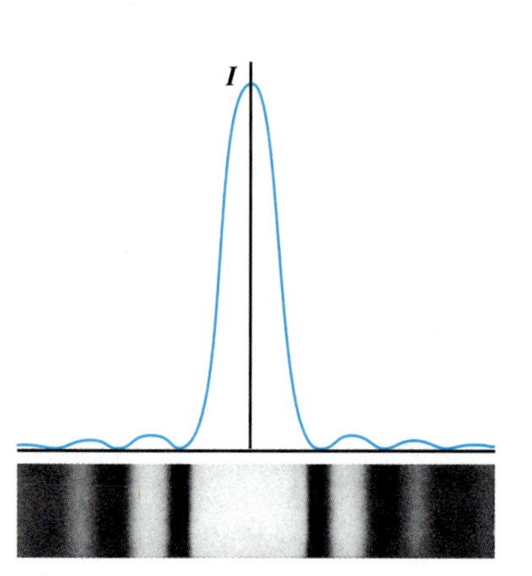

Fig. 26-22 Single-slit diffraction. Light diffracted by the slit forms a series of light and dark bands on a distant screen.

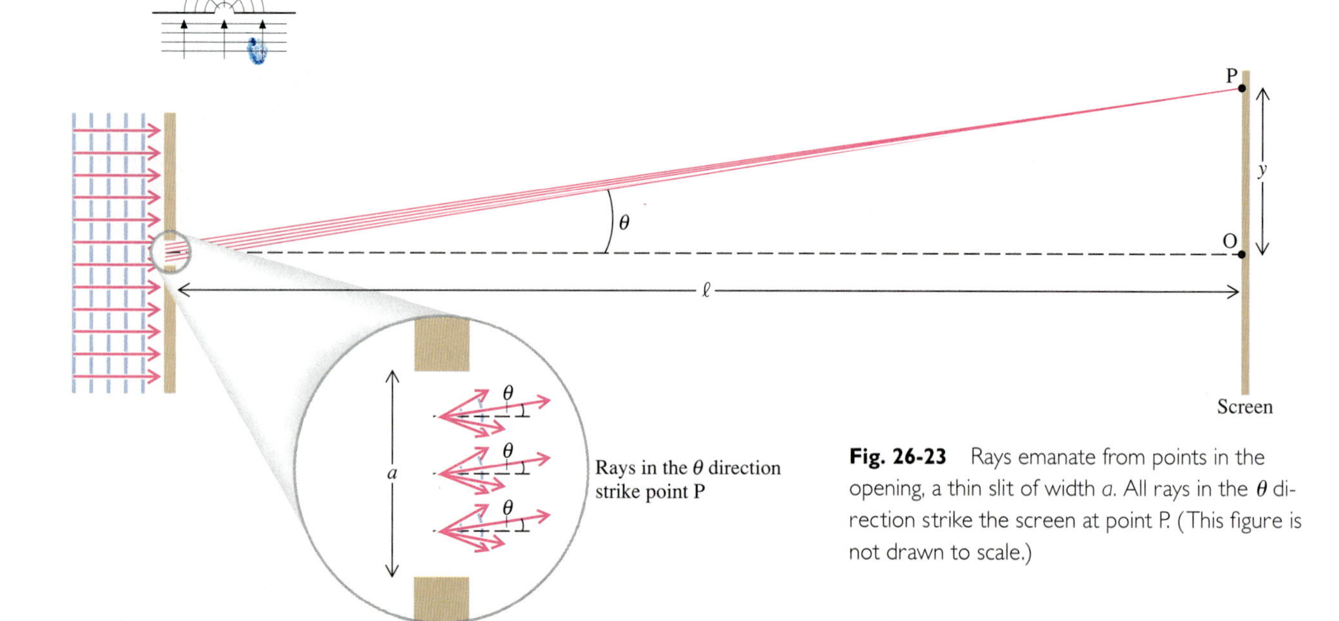

Fig. 26-23 Rays emanate from points in the opening, a thin slit of width a. All rays in the θ direction strike the screen at point P. (This figure is not drawn to scale.)

We can use the **Huygens-Fresnel principle** to analyze single-slit diffraction. According to this principle, **each section of a wavefront in the diffracting aperture is the source of a spherical wavelet. The amplitude of the light wave at any point beyond the aperture is the superposition of all these wavelets.** Fig. 26-23 shows a plane wave incident on a narrow slit and a distant screen for viewing the resulting diffraction pattern. An enlarged section of the figure shows the spherical wavelets and associated rays emanating from several points in the opening. Rays striking the screen at point P interfere constructively or destructively, depending on the relative phases of the waves. Since P is at a great distance, the rays reaching P are nearly parallel. These rays form an angle θ with a line drawn to point O, which is directly opposite the slit.

Fig. 26-24 shows rays directed at the center of the diffraction pattern ($\theta = 0$). Since these parallel rays travel equal distances to the screen, interference is constructive, and the center of the pattern is therefore a diffraction maximum.

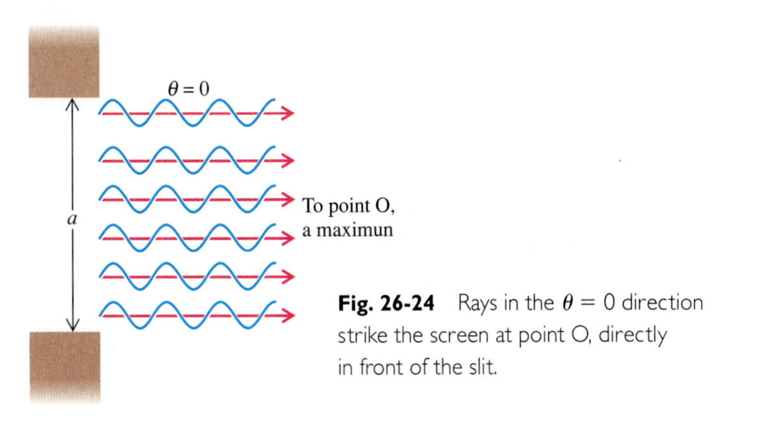

Fig. 26-24 Rays in the $\theta = 0$ direction strike the screen at point O, directly in front of the slit.

Next we shall locate the diffraction minima produced by a single slit. We can use Fig. 26-25 to find the values of θ corresponding to diffraction minima. The first minimum is indicated in Fig. 26-25a, where the slit of width a has been divided into two halves. We compare *any* ray from the top half with a ray a distance $a/2$ below it. Let the difference in path length of two such rays be $\lambda/2$, so that they interfere destructively. As shown in the figure this occurs for rays at an angle θ, where

$$a \sin \theta = \lambda$$

Since *all* rays from the top half of the slit can be paired with a cancelling ray from the bottom half, this angle corresponds to a diffraction minimum. A similar argument can be applied to find other diffraction minima. In Fig. 26-25b the slit is divided into 4 segments, and it is shown that another diffraction minimum occurs at an angle θ, such that

$$a \sin \theta = 2\lambda$$

In Fig. 26-25c, the slit is divided into 6 segments, and a diffraction minimum is found at an angle θ, where

$$a \sin \theta = 3\lambda$$

In general, single-slit diffraction minima occur at any angle θ satisfying the equation

$$a \sin \theta = m\lambda \quad m = 1, 2, 3, \ldots \quad \text{(diffraction minima)} \quad (26\text{-}9)$$

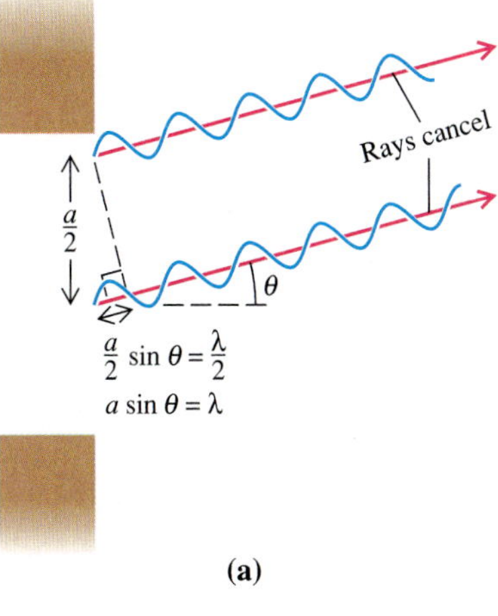

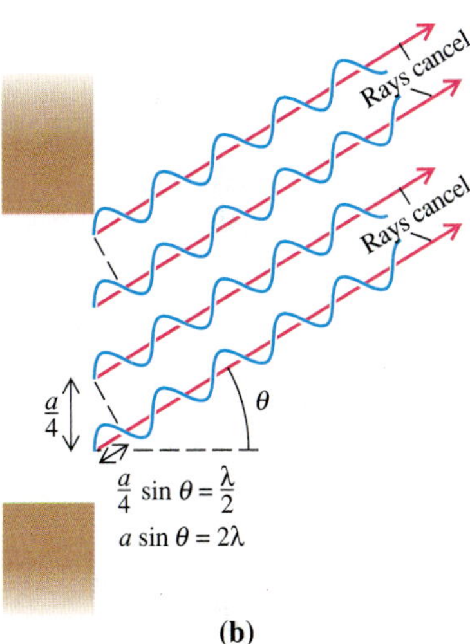

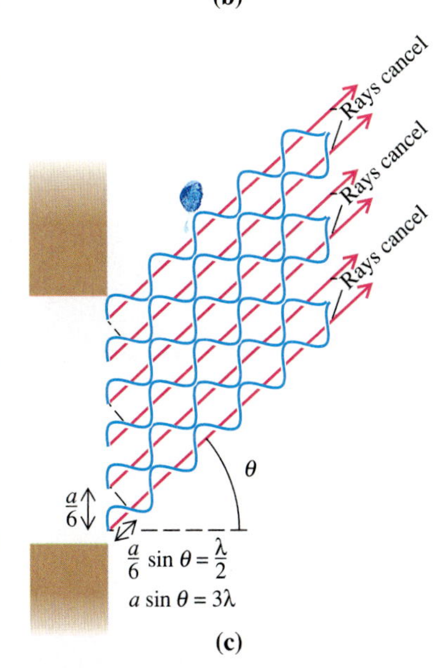

Fig. 26-25 Locating the **(a)** first, **(b)** second, and **(c)** third diffraction minima for a single slit of width a.

EXAMPLE 5 A Wide Image of a Thin Slit

Find the width of the central diffraction maximum of a slit of width 0.100 mm, as seen on a screen 2.00 m from the slit. The slit is illuminated by a He-Ne laser beam ($\lambda = 633$ nm).

SOLUTION The edge of the central diffraction maximum will correspond to the first minimum ($m = 1$). Applying Eq. 26-9, we find the angle θ corresponding to this point.

$$a \sin \theta = m\lambda = \lambda$$

or

$$\sin \theta = \frac{\lambda}{a} = \frac{633 \text{ nm}}{1.00 \times 10^{-4} \text{ m}} = 6.33 \times 10^{-3}$$

$$\theta = 6.33 \times 10^{-3} \text{ rad} \quad (\text{or } 0.363^\circ)$$

Next we find the linear distance y from the center of the pattern to the point on the screen corresponding to this angle.

Since the angle is quite small, we may approximate $\sin\theta$ by y/ℓ, where y and ℓ are shown in Fig. 26-23. Thus

$$\frac{y}{\ell} \approx \sin \theta = 6.33 \times 10^{-3}$$

or

$$y = (6.33 \times 10^{-3})\ell = (6.33 \times 10^{-3})(2.00 \text{ m})$$

$$= 1.27 \times 10^{-2} \text{ m} = 1.27 \text{ cm}$$

Since the central maximum extends an equal distance below the midpoint, the width of the maximum is double this value, or 2.54 cm. Thus the central maximum has a width 254 times the slit width (0.0100 cm). In other words, the width of the slit's central image is 254 times greater than the width of the image predicted by geometrical optics.

Circular Aperture

Fraunhofer diffraction by a circular aperture is of particular importance because of its application to the eye and to optical instruments, which generally have circular apertures. However, quantitative analysis of the circular aperture is considerably more complicated than analysis of the slit. Therefore we shall simply state without proof the one most important result of that analysis: the first diffraction minimum of an aperture of diameter D is at an angle θ, where

$$D \sin \theta = 1.22\lambda$$

Notice that this equation is similar to the equation for the first minimum of a slit of width a (Eq. 26-9 with $m = 1$: $a \sin \theta = \lambda$). Since $\sin \theta$ is very nearly equal to θ in radians, for the angles we normally encounter, we can express this result

$$\theta \approx 1.22\frac{\lambda}{D} \quad \text{(first minimum; circular aperture)} \quad (26\text{-}10)$$

Fig. 26-26 shows the Fraunhofer diffraction pattern of a circular aperture and the corresponding graph of intensity versus θ. The circular aperture was illuminated by a plane wave. One can think of this wave as originating from a distant point source. Since most of the light energy is concentrated in the central disk, called the "Airy disk," for simplicity we can regard this disk as the image of the point.

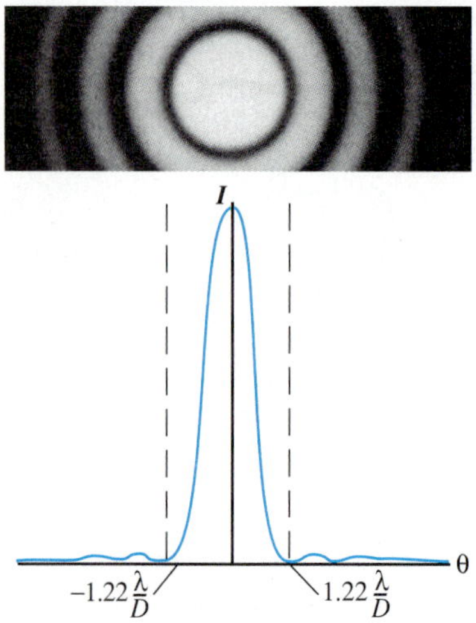

Fig. 26-26 Fraunhofer diffraction by a circular aperture. The central maximum is called the "Airy disk," which contains 84% of the light in the pattern. Since the diameter of the Airy disk is inversely proportional to the aperture diameter D, as the diameter of the aperture decreases, the disk gets bigger.

Rayleigh Criterion for Resolution

When a point source of light is imaged by an optical system with a circular aperture, the image is an Airy disk. For example, the image of a star formed by a telescope is such a disk. If two points are very close, their Airy disks will overlap, and you may not be able to distinguish separate images. Fig. 26-27 shows the image of two points that are (a) clearly resolved, (b) barely resolved, or (c) unresolved. As a quantitative measure of the resolution of two points, Lord Rayleigh proposed the following criterion—**the Rayleigh criterion: two points are barely resolved when the center of one's Airy disk is at the edge of the other's Airy disk.**

Fig. 26-28 illustrates the formation of the image of two points that are barely resolved according to Rayleigh's criterion. Notice that the angular separation θ_{min} of the two points P and Q is the angle from the center of an Airy disk to the first minimum, expressed by Eq. 26-10. Thus two points are resolved only if they subtend an angle at least as big as this minimum value θ_{min}, where

$$\theta_{min} = 1.22\frac{\lambda}{D} \quad \text{(Rayleigh's criterion for resolution)} \qquad (26\text{-}11)$$

Any image formed by an optical system consists of a set of Airy disks, each of which is the image of a single point on the object. The size of these disks determines the resolution.

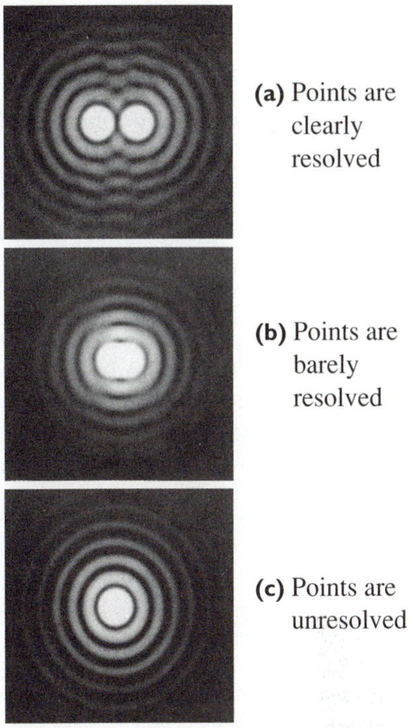

(a) Points are clearly resolved

(b) Points are barely resolved

(c) Points are unresolved

Fig. 26-27 Resolution of two point sources of light diffracted by a circular aperture.

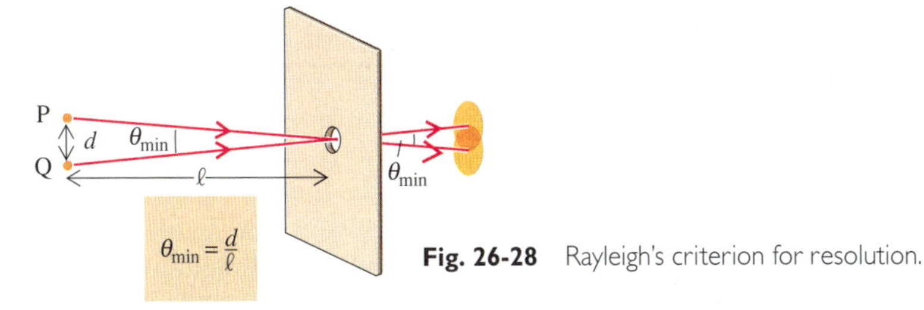

Fig. 26-28 Rayleigh's criterion for resolution.

EXAMPLE 6 Diffraction of Light by the Eye.

(a) Find the prediction of diffraction theory for the minimum angle subtended by two points that are barely resolved by the eye. Assume a pupil diameter of 2.0 mm, and use a wavelength at the center of the visible spectrum. (b) Find the distance between the two points if they are 25 cm from the eye, at its near point.

SOLUTION (a) We apply Eq. 26-11, using for λ the wavelength inside the eye, where the refractive index $n = 1.34$. At the center of the visible spectrum the vacuum wavelength $\lambda_0 = 550$ nm, and

$$\lambda = \frac{\lambda_0}{n}$$

Thus $$\theta_{min} = 1.22\frac{\lambda}{D} = 1.22\frac{\lambda_0}{nD} = 1.22\frac{5.5 \times 10^{-7}\text{ m}}{(1.34)(2.0 \times 10^{-3}\text{ m})}$$

$$= 2.5 \times 10^{-4}\text{ rad}$$

(b) From Fig. 26-28 we see that this angle equals the separation d between the points divided by the distance of 25 cm.

$$\frac{d}{25\text{ cm}} = 2.5 \times 10^{-4}\text{ rad}$$

$$d = (25\text{ cm})(2.5 \times 10^{-4}\text{ rad}) = 6.3 \times 10^{-3}\text{ cm}$$

The eye should be unable to resolve points closer than about 0.06 mm.

The minimum angle we have calculated is fairly close to the measured minimum angle between points barely resolved by the normal eye. The factors other than diffraction that affect this minimum angle are discussed in Chapter 25, Section 25-5 (Factors Limiting Visual Acuity).

EXAMPLE 7 Diffraction Limit of a Microscope.

Find an expression for the minimum separation between two points that are barely resolved by a microscope with an objective of diameter D and focal length f.

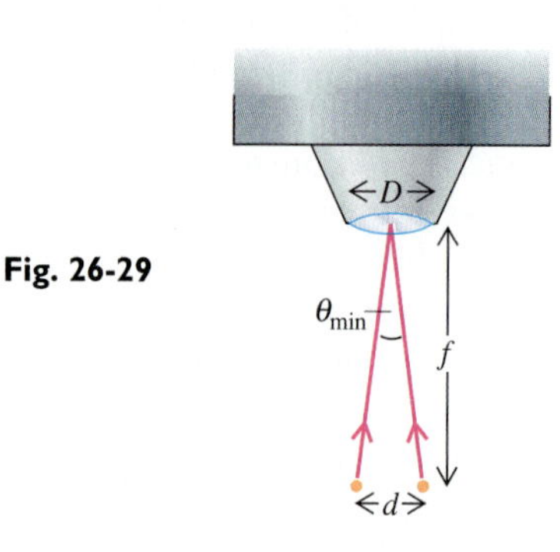

Fig. 26-29

SOLUTION Using Fig. 26-29 and applying Eq. 26-11, we find

$$d = f\theta_{min} = 1.22\frac{f\lambda}{D} \quad (26\text{-}12)$$

To make d as small as possible we need to minimize the ratio f/D. But it is not possible to make f less than about $D/2$, the radius of the lens.* Setting $f = D/2$ in the expression above, we obtain

$$d = 0.61\lambda$$

Thus **the minimum distance between two points that can be resolved by any microscope equals roughly half the wavelength of the light used to illuminate the points.** For example, using the value $\lambda = 550$ nm for the center of the visible spectrum, we find

$$d = (0.61)(550 \times 10^{-9}\text{ m})$$
$$= 3.4 \times 10^{-7}\text{ m}$$

*For a simple symmetrical lens made of glass with a refractive index of 1.5, the focal length equals the radius of curvature of the first surface, as shown in Problem 24-37. But the radius of curvature can be no smaller than the radius of the lens itself. Therefore the focal length is always less than the radius $D/2$.

Diffraction Gratings

A diffraction grating consists of thousands of very narrow, closely spaced slits, made by etching precisely spaced grooves on a glass plate (Fig. 26-30). The slits are the transparent spaces between the grooves. Typically a grating consists of thousands or tens of thousands of transparent lines per cm. It follows from our discussion of single-slit diffraction that each line, because of its extremely narrow width, produces diffracted light spread over a considerable angle—perhaps 20° or 30°. Of course, the diffraction pattern of one such line alone would not produce enough intensity to be seen by itself. But when diffracted light from thousands of lines interfere, bright, sharp diffraction maxima are produced, (Figs. 26-31 and 26-32).

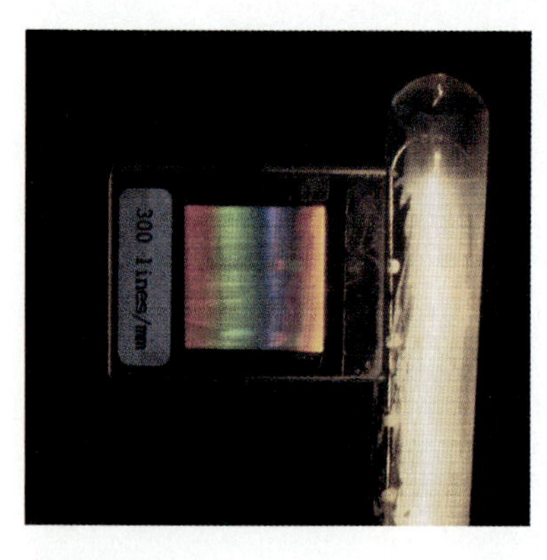

Fig. 26-30 A diffraction grating consists of thousands of narrow, closely spaced slits.

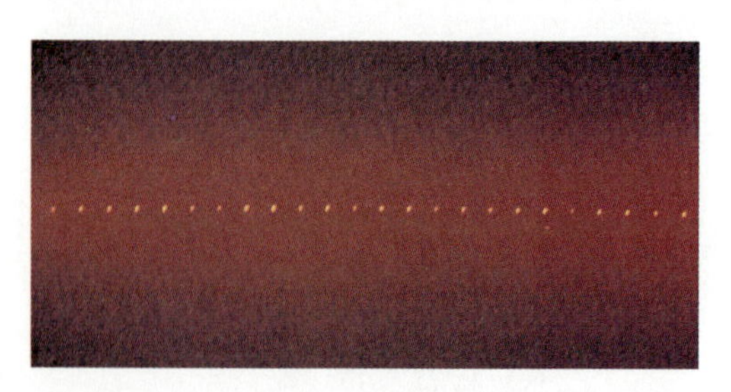

Fig. 26-31 The diffraction pattern produced by a grating illuminated by a He-Ne laser.

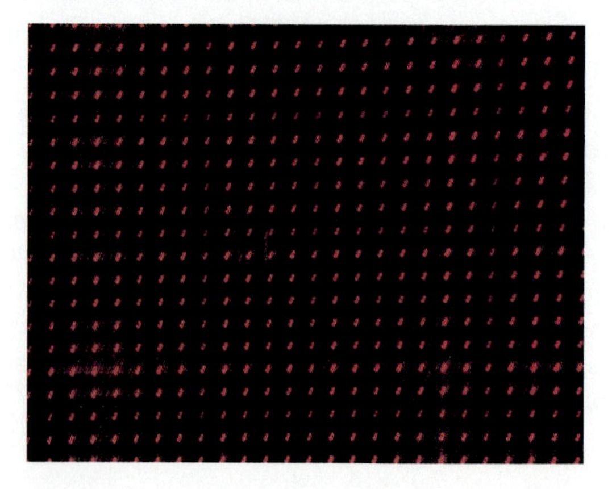

Fig. 26-32 The diffraction pattern produced by two perpendicular diffraction gratings illuminated by a He-Ne laser.

Fig. 26-33 can be used to locate the diffraction maxima of a grating with spacing d between adjacent slits. As indicated in the figure, for light in the θ direction, the difference in path length of rays from adjacent slits is $d \sin \theta$. These adjacent rays will interfere constructively if the difference in path length is an integral multiple of λ, that is, 0, λ, 2λ, 3λ, and so forth. Indeed rays from all the slits will interfere constructively if they are directed at an angle θ, such that

$$d \sin \theta = m\lambda \qquad m = 0, 1, 2, \ldots \qquad (26\text{-}13)$$

Unlike the maxima produced by a double slit, diffraction grating maxima are very narrow and sharp, as shown in Fig. 26-31. This can be understood when we consider how a slight change in the angle θ away from a value satisfying Eq. 26-13 will affect the intensity. Suppose that the change in angle is so slight that light from adjacent slits is still nearly in phase. If there were only two slits interfering, such an angle would still give an intensity close to the maximum value. However, with the thousands of slits in a grating, there can be destructive interference in many ways. For example, light from slits 100 spacings apart might interfere destructively. If all pairs of slits 100 spacings apart interfere destructively, there will be no light in that particular direction.

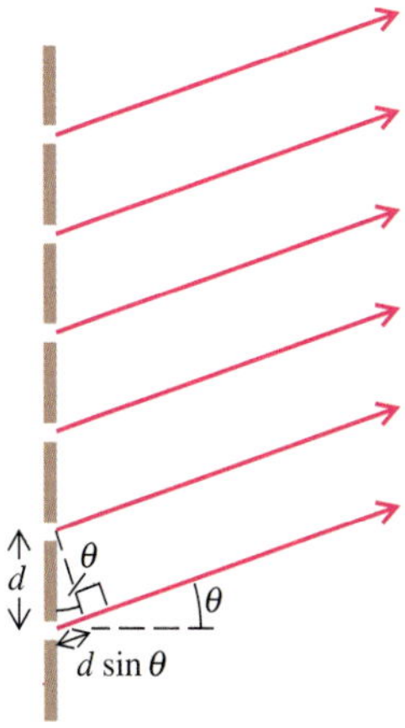

Fig. 26-33 Rays from a diffraction grating.

Diffraction gratings can be used to measure the wavelength of light, as illustrated in the following example.

EXAMPLE 8 Separating the Sodium Doublet

Find the first-order ($m = 1$) diffraction angles for the sodium doublet, using a grating with 10^6 lines/m. The sodium doublet consists of two yellow lines in the spectrum of sodium, with nearly identical wavelengths: 589.00 nm and 589.59 nm.

SOLUTION Applying Eq. 26-13 to each of the wavelengths, with $d = 10^{-6}$ m and $m = 1$, we find

$$\sin \theta = \frac{m\lambda}{d}$$

$$\sin \theta_1 = \frac{(1)(589.00 \times 10^{-9}\text{ m})}{10^{-6}\text{ m}} = 0.58900$$

$$\theta_1 = 36.09°$$

Fig. 26-34

$$\sin \theta_2 = \frac{(1)(589.59 \times 10^{-9}\text{ m})}{10^{-6}\text{ m}} = 0.58959$$

$$\theta_2 = 36.13°$$

Thus the angular separation between the lines is 0.04°, or 7×10^{-4} rad. Viewed at a distance of 1 m, the lines are 0.7 mm apart.

X-ray diffraction is a technique that utilizes the small spacing between the atoms in a crystal as a three-dimensional diffraction grating. The atomic spacing is on the same order as wavelengths in the X-ray portion of the electromagnetic spectrum. So X rays, rather than visible light, are diffracted by a crystal. And the resulting diffraction pattern can be used to discover the crystal structure. X-ray diffraction of DNA was used by Watson and Crick in discovering the structure of DNA in 1951 (Fig. 26-35).

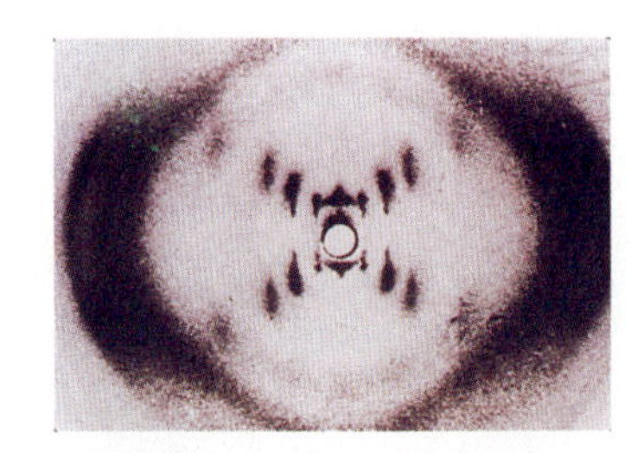

Fig. 26-35 X-ray diffraction pattern of DNA. The double helix structure of DNA was revealed by this historic photograph taken by Rosalind Franklin.

26-4 Polarization

Polarization by Absorption

Most light sources produce unpolarized light (Fig. 26-3), as opposed to the polarized light produced by some lasers (Figs. 26-1 and 26-2). However, there are ways to polarize light that is initially unpolarized, or to change the direction of polarization of polarized light. One way is to pass the light through a Polaroid sheet, a synthetic material first produced by Edwin Land in 1928 when he was an undergraduate.

There is a direction along each Polaroid sheet called its "transmission axis." Light linearly polarized along this axis passes through the sheet (Fig. 26-36a), whereas light polarized in the perpendicular direction is completely absorbed (Fig. 26-36b). If the incident light is linearly polarized at some angle θ relative to the transmission axis, the light will be partially absorbed and partially transmitted. As illustrated in Fig. 26-36c, the component of the electric field parallel to the axis is transmitted. The light that emerges is thus polarized along the direction of the transmission axis and has an amplitude E related to the incident amplitude E_0 by the equation

$$E = E_0 \cos \theta$$

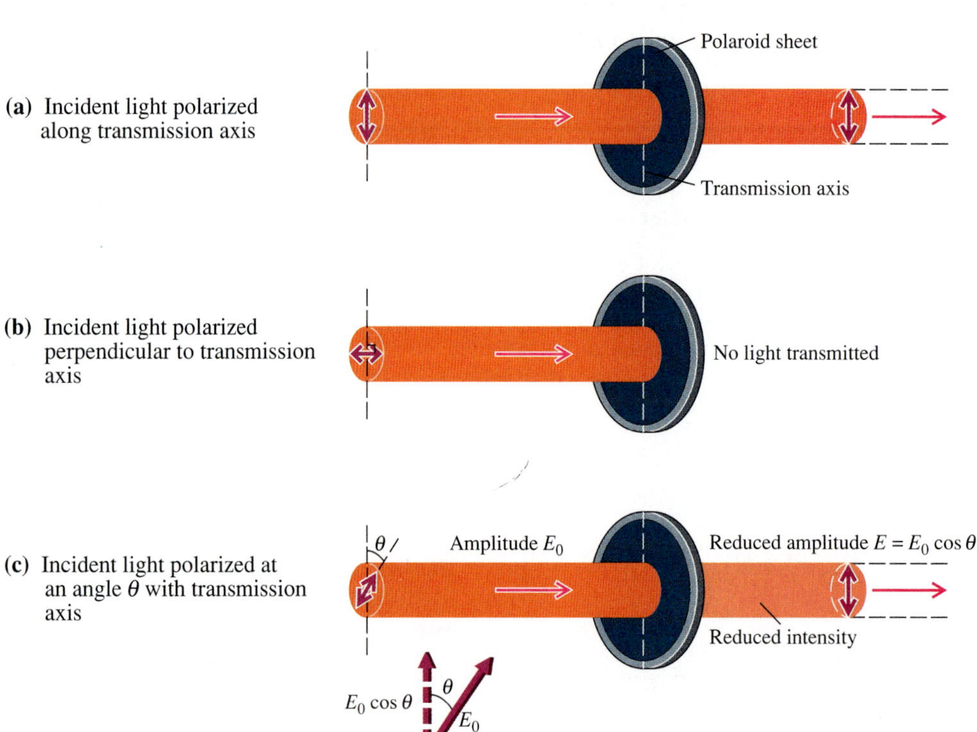

Fig. 26-36 The effect of a Polaroid sheet on initially polarized light depends on the direction of initial polarization relative to the direction of the sheet's transmission axis.

The intensity of light is proportional to the square of its amplitude (Eq. 23-5: $I_{av} = \frac{1}{2}\epsilon_0 c E_0^2$). Squaring the equation above, we obtain

$$E^2 = E_0^2 \cos^2 \theta$$

Multiplying both sides of this equation by the appropriate constant ($\frac{1}{2}\epsilon_0 c$), we obtain a relationship between the average transmitted intensity I and the average incident intensity I_0:

$$I = I_0 \cos^2 \theta \tag{26-14}$$

This result is known as the **law of Malus.** The intensity of the transmitted light has its maximum value, $I = I_0$, when $\theta = 0$, and has its minimum value, $I = 0$, when $\theta = 90°$.

Unpolarized light consists of a superposition of linearly polarized waves, with varying directions of polarization, as illustrated in Fig. 26-37a. The electric field associated with each of these waves can be resolved into x and y components, relative to an arbitrary coordinate system. Since the direction of polarization is random, the resultant x and y components are equal. We can replace the many randomly directed linearly polarized waves by just two linearly polarized waves of equal intensity, with mutually perpendicular polarization directions, as illustrated in Figs. 26-37b and 26-37c.

When unpolarized light is incident on a Polaroid sheet, only the component along the transmission axis is transmitted. Since the two components have equal intensity in unpolarized light, this means that the intensity of the transmitted light is half the intensity of the incident light (Fig. 26-38).

$$I = \tfrac{1}{2}I_0 \quad \text{(initially unpolarized light)} \quad (26\text{-}15)$$

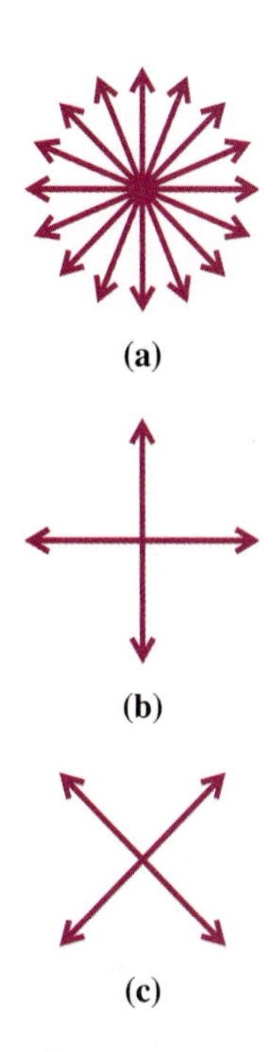

Fig. 26-37 Equivalent representations of unpolarized light.

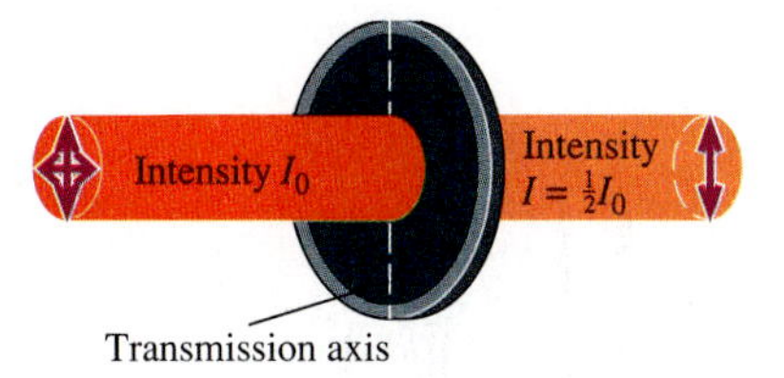

Fig. 26-38 A Polaroid sheet polarizes initially unpolarized light.

EXAMPLE 9 Light Passing Through Two Polarizers

An unpolarized laser beam of intensity 1000 W/m² is incident on a Polaroid sheet with a vertical transmission axis. The light passing through this sheet strikes a second Polaroid sheet, with a transmission axis at an angle of 30.0° from the vertical (Fig. 26-39). Find the polarization and the intensity of the light emerging from the second sheet.

SOLUTION The light transmitted by the first Polaroid sheet is vertically polarized and, according to Eq. 26-15, has an intensity equal to half the incident intensity.

$$I = \tfrac{1}{2}I_0 = \tfrac{1}{2}(1000\ \text{W/m}^2) = 500\ \text{W/m}^2$$

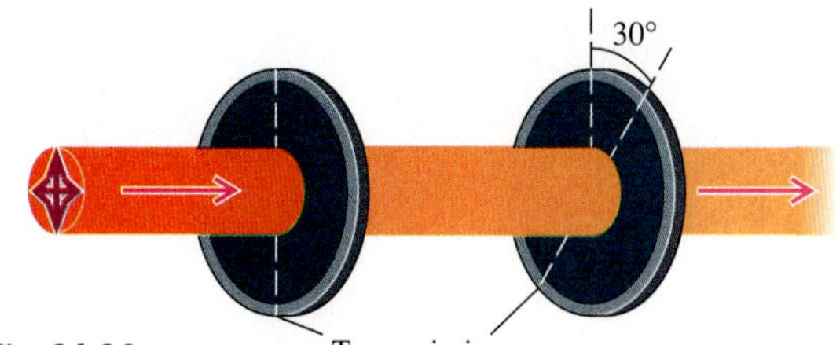

Fig. 26-39

The light incident on the second sheet is polarized at an angle of 30° relative to this sheet's transmission axis and, according to the law of Malus (Eq. 26-14), has intensity

$$I' = I_0' \cos^2\theta = (500\ \text{W/m}^2)(\cos^2 30.0°) = 375\ \text{W/m}^2$$

Polarization by Reflection

When light is reflected from the surface of a dielectric, such as water or glass, the intensity of the reflected light depends on the angle of incidence and on the polarization of the incident light. Light polarized parallel to the reflecting surface (Fig. 26-40a) is always more strongly reflected than light polarized in a perpendicular direction (Fig. 26-40b). Unpolarized light can be thought of as consisting of two equal-intensity polarized waves—one polarized parallel to the surface and a second polarized perpendicular to the first. After reflection, the component polarized parallel to the surface is more intense than the other component. Fig. 26-41 shows the intensities of the two components in a beam of initially unpolarized light reflected by water, for several angles of incidence. The intensities are predicted by Maxwell's equations. Notice that both components are more strongly reflected for very large angles of incidence. Any smooth dielectric surface becomes mirror-like as the angle of incidence approaches 90°. You can observe this effect by holding up a smooth sheet of paper so that rays from a light source are reflected at glancing incidence.

The angle of incidence at which the reflected light is 100% polarized is known as "Brewster's angle," denoted by θ_B. Brewster's angle has a value of 53° for reflection by water, as indicated in Fig. 26-41c. Maxwell's equations can be used to derive an expression for Brewster's angle, relating it to the refractive index n of the incident medium and the index n' of the reflecting medium. We present the result here without proof:

$$\tan \theta_B = \frac{n'}{n} \quad \text{(Brewster's angle)} \quad (26\text{-}16)$$

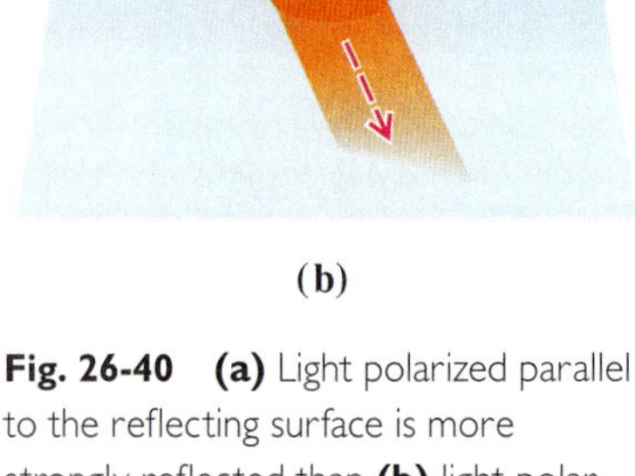

(a)

(b)

Fig. 26-40 **(a)** Light polarized parallel to the reflecting surface is more strongly reflected than **(b)** light polarized in a perpendicular direction.

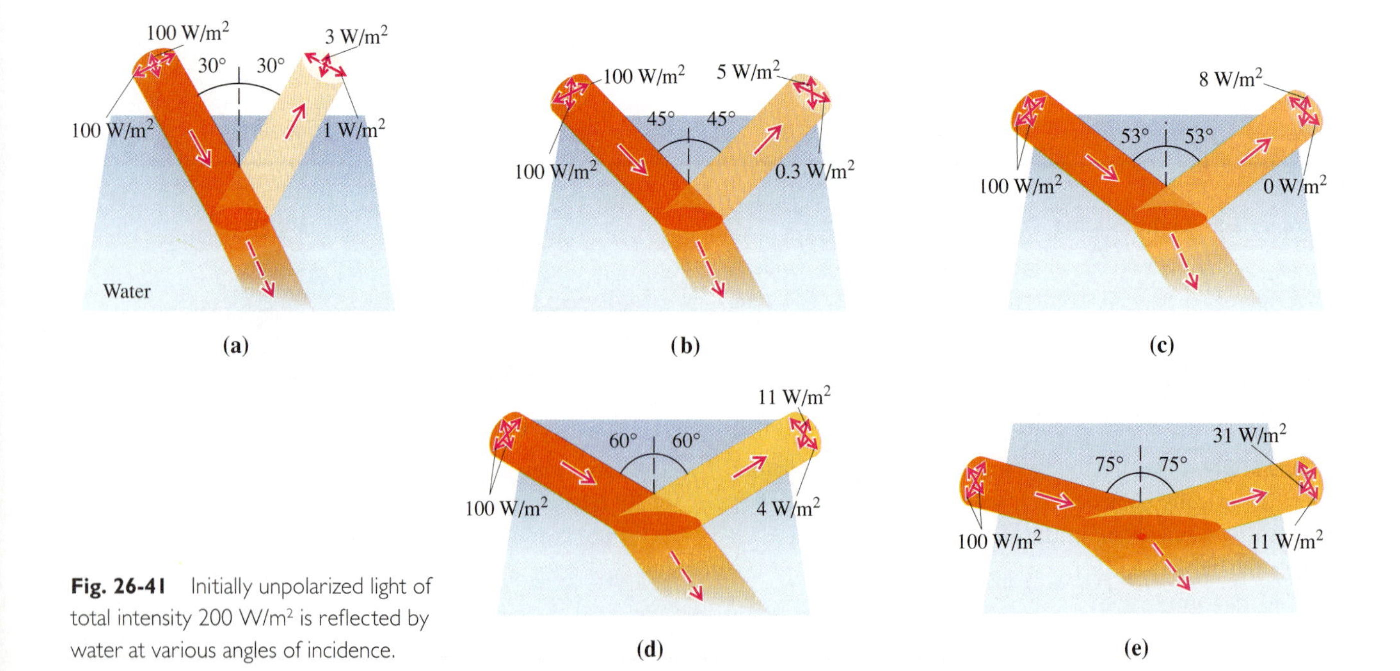

Fig. 26-41 Initially unpolarized light of total intensity 200 W/m² is reflected by water at various angles of incidence.

EXAMPLE 10 Brewster's Angle for Glass

Calculate Brewster's angle for light incident from air onto a glass surface if the glass has a refractive index of 1.5.

SOLUTION Applying Eq. 26-16, using the refractive indices for glass and air, we find

$$\tan \theta_B = \frac{n'}{n} = \frac{1.5}{1.0} = 1.5$$

$$\theta_B = 56°$$

Polaroid sunglasses are effective at reducing reflected glare from the surface of a body of water or from other surfaces (Fig. 26-42). The lenses are made of Polaroid sheets with vertical transmission axes. The reflected light consists mainly of horizontally polarized light, and such light is completely absorbed by the lenses.

Fig. 26-42 An early ad for Polaroid sunglasses.

EXAMPLE 11 A Sunset Seen Through Polaroid Sunglasses

The setting sun is reflected from the surface of a lake at an angle of incidence of 75°. The intensity of the incident light is 200 W/m², as in Fig. 26-41e. Find the intensity of the reflected light reaching the eye of an observer wearing Polaroid sunglasses (Fig. 26-43).

SOLUTION From Fig. 26-41e we see that the intensity of the reflected light having a polarization along the transmission axis of the sunglasses is 11 W/m². Therefore this is the intensity of the reflected light reaching the eye.

Without sunglasses, the observer would see reflected light of both polarizations. From Fig. 26-41e, the intensity of reflected light seen by the observer would be 11 W/m² + 31 W/m² = 42 W/m².

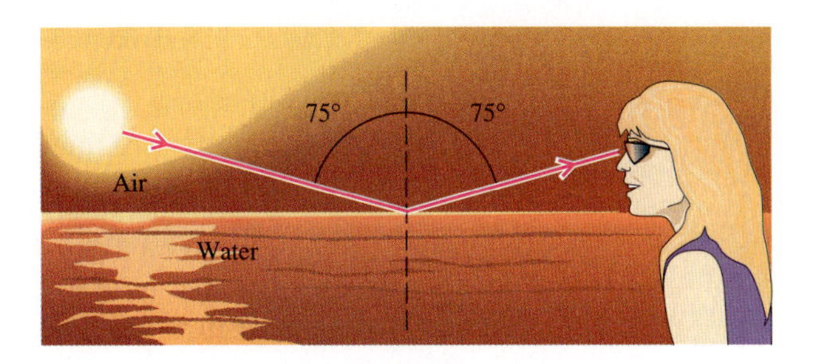

Fig. 26-43

Non-Polaroid sunglasses that produce the same darkening as these Polaroid sunglasses would absorb 50% of *all* incident light, and so they would transmit to the observer reflected light of intensity 21 W/m².

Polarization by Scattering

When an electromagnetic wave is incident on an atom, the atom's electrons oscillate in response to the oscillating electric field. The electrons behave like tiny antennas; they emit their own radiation with the same frequency as the incident electromagnetic wave, but scatter the radiation in various directions. The intensity of this scattered radiation depends on the light's frequency. Blue light is scattered much more effectively than red light. Fig. 26-44 shows how scattering of sunlight by the earth's atmosphere gives us blue skies and red sunsets. Observers A and B both see blue sky as a result of the blue part of the sun's spectrum being scattered toward their eyes by the atmosphere. Meanwhile, observer C sees an orange or red sun as a result of the blue part of the spectrum's having been scattered out of the beam of direct sunlight. This kind of scattering, called "Rayleigh scattering," is the result of independent, incoherent radiation by many atoms.

Fig. 26-44 Scattering of sunlight by the earth's atmosphere results in blue skies (seen by A and B) and red sunsets (seen by C).

Scattering is also the basic mechanism at the heart of reflection and refraction by a solid or a liquid. But the higher density and relative immobility of the atoms in the liquid or solid mean that light scattered by neighboring atoms is coherent and can therefore interfere constructively or destructively. The result of this interference is remarkably simple. The scattered waves interfere destructively in all directions, except those corresponding to the reflected and refracted waves, for which the interference is constructive. So we see only a reflected wave and a refracted wave.

The particles of water in a cloud or in ocean surf also "scatter" sunlight, but in this case the scattered light is white, in contrast to the blue sky. The tiny water droplets simply reflect and refract incident light. The result of multiple reflections and refractions by a very large number of droplets is to redirect or scatter the incident white light in all directions.

Scattering of sunlight by the atmosphere tends to polarize the light. Fig. 26-45 shows how this polarization arises. A beam of unpolarized sunlight, incident on the atmosphere, travels along the x-axis. This transverse electromagnetic wave has an electric field that oscillates in the yz plane; there is no component along the x-axis, the direction of propagation of the wave. Electrons within atoms in the atmosphere oscillate in the yz plane, in response to the incident wave, and scatter light in various directions. The nature of radiation produced by any source of radiation is that there can be a component of the electric field in a given direction only if there is a component of motion of the radiating source parallel to that direction. Therefore there can be no x-component of the electric field in scattered radiation, since there is none in the incident wave. This implies that **radiation scattered** along the yz plane, **perpendicular to the incident beam, must be polarized.** For radiation scattered in any such direction there is only a single line perpendicular to the direction of propagation, along which the electric field vector can oscillate, as indicated in Fig. 26-45. Light scattered in other directions is partially polarized.

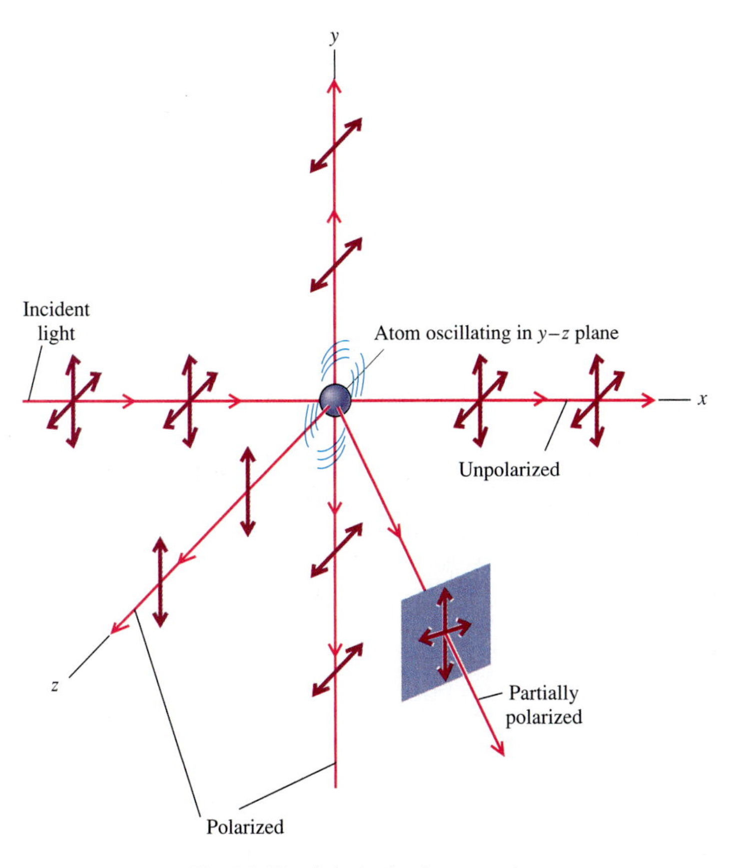

Fig. 26-45 Polarization by scattering.

A Closer Look

Magic in the Sky

Mirage! Looming! Mountain specter! What do you think of when you hear these terms? Illusions or ghosts, perhaps. The names are suggestive of the fear and bewilderment that these phenomena have aroused over the centuries. It may surprise you to learn, however, that the names describe real optical effects in the atmosphere. You already know enough about the basic principles of optics to understand such phenomena.

Mirages

You have heard of people lost in the desert who imagined that they saw tempting pools of blue water just beyond the next sand dune and who dragged themselves forward in hopeless pursuit of them. You may think that such mirages are simply hallucinations caused by heat and thirst. Actually, they are almost always real phenomena—real light is behaving in a way that creates an illusion in which anyone, hot and thirsty or not, can share (Fig. 26-A).

Mirages are refraction phenomena. Light rays are bent by layers of air at different temperatures. Warm air has lower density than cool air and has a lower refractive index. As a ray coming from a cooler layer enters a warmer layer—which is what happens when the ray is moving downward toward a searingly hot desert surface—the ray is refracted away from the normal. The ray can be bent so much that it curves back upward (Fig. 26-B). When it reaches the viewer's eye, it is automatically traced back by the brain as if it came from a source directly in line with its final segment. That is, it is seen as if it were ahead and below, rather than ahead and above. Light coming from a clear blue sky produces the illusion of a bright blue pool on the ground ahead.

Fig. 26-A A fairly common kind of mirage: the dry surface of a road appears to be wet.

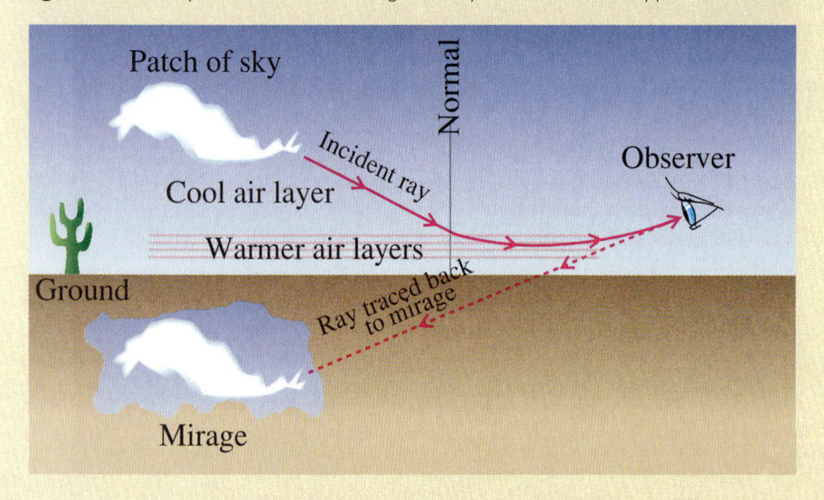

Fig. 26-B Formation of an inferior mirage.

A mirage of the sort seen in a hot desert is called an "inferior mirage" because it appears below the light source (the sky). There is another kind of mirage, called a "superior mirage," which appears above the source of light (Fig. 26-C). A superior mirage is typically caused when light moves upward from a layer of cool, dense air into warmer, less dense layers. The rays are bent away from the normal, as before, but this time they turn down rather than upward (Fig. 26-D). When they strike the eye, they are traced up to a mirage seen at their apparent point of origin. Under such conditions, a ship moving on the water below the horizon can look like a ghost ship sailing through the sky! The appearance of a superior mirage is sometimes called **looming,** for obvious reasons: the mirage looms above its source.

Under special conditions, looming can produce truly uncanny effects. One particular type of looming is called **fata morgana,** after Morgan le Fay, the fairy-enchantress of the King Arthur legends who lived on a magical island. Fata morgana is most often seen in the Strait of Messina, a waterway that separates Sicily

Fig. 26-C A superior mirage of a ferry, which appears to be vertically elongated.

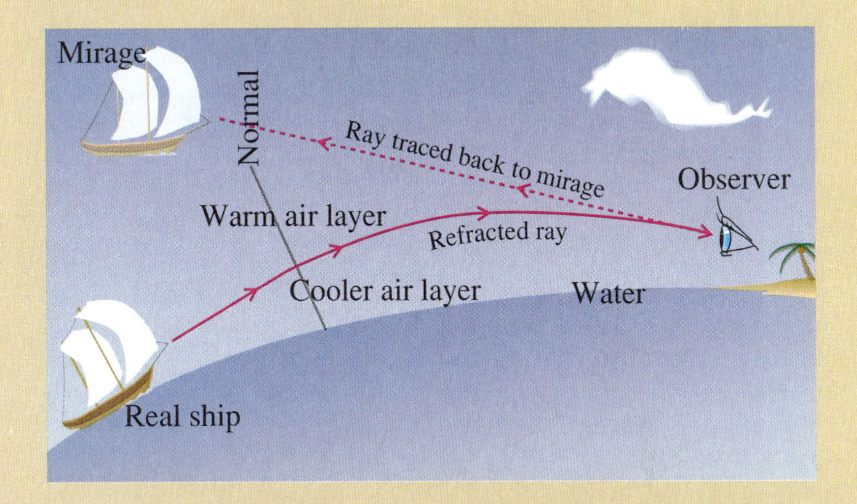

Fig. 26-D Formation of a superior mirage.

from Italy and that was long dreaded for its deadly currents, rocks, and whirlpools. The mirage is caused by irregular layerings of air of various densities, which produce multiple refractions and multiple overlapping images. The result is an apparent vertical elongation of the source object, sometimes to enormous proportions. For example, when seen from a ship in the strait, objects such as trees or hills on the shore can look like huge, weirdly shaped figures that can be disorienting and dangerous to unsuspecting mariners (Fig. 26-E).

Coronas and Anticoronas

Interference or diffraction effects can occur when light rays pass near the edges of tiny objects in the atmosphere. If the diffraction appears around the light source, the effect is usually a ring, called a **corona.**

Coronas can appear around the sun or moon when light rays pass near the edges of water droplets in the atmosphere. Rings of different colors can be seen because different wavelengths are diffracted to varying degrees.

Fig. 26-E What appears in this picture to be icy castles in the sky is actually an extremely rare, mirage-like effect known as fata morgana.

Continued.

A Closer Look

A related phenomenon caused by the presence of hexagonal ice crystals is called a **halo** (Fig. 26-F). If the ice crystals line up in just the right way on a sunny day, they can produce two separate bright spots, one to either side of the sun. These ghost suns are called **parhelia** or more popularly, **sundogs.**

When diffraction occurs around a shadow area, the effect is called an **anticorona.** Anticoronas are also known by the names **glory** (Fig. 26-G) and **mountain specter.** These are rare and awesome phenomena.

The glory is seen as a halo surrounding the shadow of the observer's head. Be cause of the extreme directional dependence of this effect, no one else sees the halo around the observer's head. Two observers might each see a halo around his or her own head, but each observer will not be able to see the other's halo.

Fig. 26-F A street light blocks the sun's direct rays, allowing a halo around the sun to be clearly seen.

Fig. 26-G The bright circle around the plane's shadow is called a glory.

One particularly famous example of an anticorona occurs occasionally near the Brocken, a mountain in the Harz range of central Germany. Because it is known in German legend as the site of the Walpurgis Night witchcraft rituals, the Brocken is an appropriate location for the anticorona, which is there given the name "Brocken specter." The specter can be seen at twilight on sunny days by observers who stand near the foot of the mountain when there are misty banks of fog or cloud just above them that do not reach as high as the mountain top. The low-lying sun then casts a huge shadow of the mountain peak onto the upper surface of the mist. This silhouette of the peak appears surrounded by rings of colored light, as rays passing around the edge of the peak are bent and separated out according to their wavelength (Fig. 26-H).

Fig. 26-H Viewing the Brocken Specter.

CHAPTER 26 SUMMARY

Light propagates as an electromagnetic wave. An ideal monochromatic source produces a wave with a definite frequency and wavelength. If the wave is also linearly polarized, the electric field oscillates parallel to a line perpendicular to the direction of wave propagation. If the wave is unpolarized, it can be thought of as a superposition of two linearly polarized waves, with perpendicular lines of polarization. Any real source is not perfectly monochromatic. There is always some uncertainty in frequency, Δf.

Light is said to be coherent over a time interval Δt if there is a predictable phase relationship between the phases of the wave at the beginning and end of the interval. The condition for coherence is that the time interval be much less than the inverse of the uncertainty in frequency:

$$\Delta t << \frac{1}{\Delta f} \qquad \text{(coherence condition)}$$

Two points in a wave separated by a distance Δx are coherent, that is, have a predictable phase relationship, if they are much closer than the wave's coherence length x_c, where

$$x_c = \frac{c}{\Delta f}$$

$$\Delta x << x_c \qquad \text{(coherence condition)}$$

When two light waves are in phase at a given point in space, they are said to interfere constructively. The intensity of the light is maximum at such a point. When two light waves are 180° out of phase at a point, they are said to interfere destructively. At such a point the intensity of the light is minimum. If the two interfering waves have equal intensity, this minimum intensity equals zero; that is, the superposition of two light waves produces darkness. An interference pattern is a pattern of light and dark areas, corresponding to constructive and destructive interference by overlapping waves. In order for interference patterns to be visible, the phase relationship between the interfering waves must remain stable; that is, the waves must be coherent over a sufficiently long time interval. In practice this means that the two waves must be produced by splitting of a single wave into two separate parts, which travel over different paths. Then, if the difference in path lengths of the two waves is less than the coherence length, an interference pattern is visible.

If a plane light wave of wavelength λ is incident on two narrow slits separated by a distance d, a pattern of equal-width light and dark fringes is seen on a distant screen. An angle θ locating the center of a bright fringe satisfies the equation

$$d \sin \theta = m\lambda \qquad \text{where } m = 0, 1, 2, \ldots$$

and an angle θ locating the center of a dark fringe satisfies the equation

$$d \sin \theta = (m + \tfrac{1}{2})\lambda \qquad \text{where } m = 0, 1, 2, \ldots$$

Light reflected from the top and bottom surfaces of a thin film will interfere if the film's thickness is much less than the coherence length of the light. The phase difference in the two interfering waves arises from a difference in path length and may also arise from a 180° phase change of light reflected from a medium with a higher refractive index than that of the incident medium. When there is *no relative phase change* between the two waves caused by reflection, constructive interference by a film of thickness t and index n occurs for light with vacuum wavelengths

$$\lambda_0 = \frac{2tn}{m} \qquad \text{where } m = 1, 2, 3, \ldots$$

When there is a 180° relative phase change caused by reflection, constructive interference occurs for wavelengths

$$\lambda_0 = \frac{2tn}{m + \frac{1}{2}} \qquad \text{where } m = 1, 2, 3, \ldots$$

Light bends or diffracts around obstacles placed in its path. The Huygens-Fresnel principle states that each section of a wavefront in a diffracting aperture is the source of a spherical wave. The superposition of these waves produces a diffraction pattern—a pattern of light and dark areas, corresponding to constructive and destructive interference of the waves. The simplest kind of diffraction, known as "Fraunhofer diffraction," occurs when the diffracting object is illuminated by a plane wave and the diffraction pattern is viewed on a distant screen. Fraunhofer diffraction by a single slit of width a produces a broad central maximum and weaker successive maxima. The diffracting minima occur at angles θ satisfying the equation

$$a \sin \theta = m\lambda \qquad \text{where } m = 1, 2, 3, \ldots$$

Fraunhofer diffraction by a circular aperture of diameter D produces a circular diffraction maximum, called an "Airy disk." The first diffraction minimum occurs at an angle given by the following approximate expression:

$$\theta \approx 1.22\frac{\lambda}{D}$$

The image of a point source of light produced by an optical system with a circular aperture is an Airy disk. Two object points may not be resolvable as separate points if they are too close, because their Airy disks overlap. According to the Rayleigh criterion, two points are barely resolved if the center of one's Airy disk is at the edge of the other's Airy disk. The minimum angle subtended by points that are barely resolved by a circular aperture of diameter D is given by

$$\theta_{\min} = 1.22\frac{\lambda}{D}$$

A diffraction grating, which consists of thousands of very narrow, closely spaced slits, produces bright, sharp diffraction maxima at angles θ satisfying the equation

$$d \sin \theta = m\lambda \qquad m = 0, 1, 2, \ldots$$

where d is the slit spacing.

Light that is initially unpolarized becomes polarized as it passes through a Polaroid sheet. The light that emerges is polarized along the transmission axis of the sheet and has an intensity I equal to half the intensity I_0 of the unpolarized light.

$$I = \tfrac{1}{2}I_0$$

If polarized light is incident on a Polaroid sheet, the light that emerges is polarized along the sheet's transmission axis and has an intensity I related to the initial intensity I_0 by the law of Malus:

$$I = I_0 \cos^2 \theta$$

where θ is the angle between the incident light's polarization direction and the transmission axis.

Light reflected from the surface of a dielectric is at least partially polarized, with the component of the electric field parallel to the surface more strongly reflected than the component in the perpendicular direction. For an angle of incidence known as "Brewster's angle," the polarization is complete; that is, the reflected light is polarized parallel to the reflecting surface. Brewster's angle θ_B is related to the refracting indices n and n' of the incident medium and the reflecting medium by the equation

$$\tan \theta_B = \frac{n'}{n}$$

Scattering of sunlight by atoms in the earth's atmosphere produces complete polarization of the scattered light at a scattering angle of 90°, or partial polarization at other scattering angles, as indicated in Fig. 26-45.

Questions

1 (a) If visible light had a much longer wavelength—on the order of meters—would it be possible to see around a corner?

(b) Would you be able to see much detail?

2 Linearly polarized light travels along the y-axis. Is it possible for there to be a component of the electric field (a) in the x direction; (b) in the y direction; (c) in the z direction; (d) in both the x and the y directions; (e) in any direction perpendicular to the y-axis?

3 Would you expect to be able to see an interference pattern produced by sunlight reflected from the surfaces of a glass windowpane? Explain.

4 Suppose you have just invented a new laser with an incredible bandwidth of only 1 Hz. Would it be possible to produce a visible interference pattern by allowing light from two such lasers to overlap?

5 When two equally intense, coherent light waves are in phase at a point in space, the total intensity is four times the intensity of each of the waves alone—more than the sum of the intensities of the two waves. Explain why this does not violate the principle of conservation of energy.

6 Suppose that Young's double-slit experiment is performed underwater. Would the interference fringes be closer, farther apart, or spaced the same as for the experiment performed in air?

7 A high-quality, diffraction-limited microscope will have better resolution if illuminated with (a) white light; (b) blue light; (c) red light; (d) orange laser light.

8 The soap bubble in Fig. 26-46 is colored because of interference of light reflected from its surfaces. Just before a bubble bursts, its thickness is much less than the wavelength of light. What color will the bubble then appear? Explain.

Fig. 26-46

9 A diffraction pattern is produced by a rectangular aperture with a width of 0.1 mm and a height of 0.2 mm. Will the central maximum have greater height or width?

10 What happens to the width of the central diffraction maximum produced by a thin slit as the slit width is decreased?

11 When unpolarized light is normally incident on a plane surface, is there any possibility that the reflected light will be polarized? Explain.

12 Two Polaroid sheets are oriented with perpendicular transmission axes, as shown in Fig. 26-47. Unpolarized light is incident from the left side.

(a) Will any light be transmitted through the sheet on the right?

(b) Suppose a third Polaroid sheet is placed between the two sheets shown in the figure. Could any light possibly be transmitted through the sheets?

(c) What are the orientations of the middle sheet's transmission axis that will result in no light being transmitted through all three sheets?

13 Suppose you are swimming underwater. Is the light that reaches your eyes partially polarized? Explain.

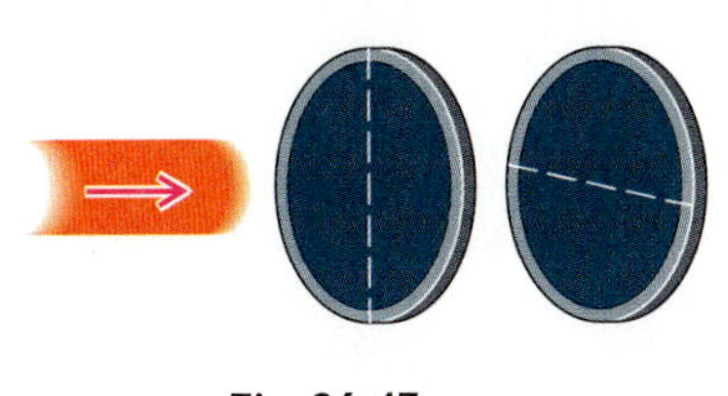

Fig. 26-47

14 A beam of sunlight is reflected by a store window. The angle of incidence is 56°. Will polarized sunglasses reduce the glare?

15 Suppose you are driving a car in the late afternoon, looking directly into the sun. Would Polaroid sunglasses be more effective than ordinary sunglasses in reducing the intensity of direct sunlight?

16 An intense unpolarized laser beam is incident on a horizontal glass surface at Brewster's angle. The reflected beam then strikes a Polaroid sheet. The sheet will heat up most if the transmission axis is (a) horizontal; (b) vertical; (c) at an angle of 45° relative to the vertical.

17 (a) What is the color of the sky on the moon?

(b) What is the color of the setting sun seen from the moon?

18 Is light scattered by a cloud polarized?

Answers to Odd-Numbered Questions

1 (a) yes; (b) no; **3** No. The coherence length of white light is much less than the thickness of the glass; **5** The total energy equals the sum of the energy in the separate waves, but energy is redistributed, greater where interference is constructive and less where it is destructive; **7** b; **9** width; **11** No. All directions perpendicular to the direction of wave motion are physically equivalent. There is nothing to distinguish a polarization direction; **13** Yes, slightly. The component of polarization parallel to the surface is somewhat weaker, since this component is reflected more strongly; **15** no; **17** (a) black; (b) white

Problems (listed by sections)

26-1 Wave Properties of Light

1 TV channel 2 broadcasts over a frequency range from 54 MHz to 60 MHz. Suppose that at $t = 0$ at a certain point in space the electric field of the station's signal is 0.1 N/C, directed upward, its peak value. Can you estimate the electric field at that point at (a) $t = 1\ \mu s$; (b) $t = 1$ ns?

2 VHF TV channels have a bandwidth of 6 MHz.

(a) What is the coherence length of a VHF signal?

(b) Would it be possible for TV signals to form an interference pattern by superposition of a direct signal and one reflected from a hill, with the reflected wave traveling an additional 1000 m?

★ **3** A CO_2 laser produces infrared radiation with a wavelength of 10.6 μm and a bandwidth of 10^{-5} nm. What is the coherence length of the radiation?

4 The coherence length of an ordinary white light source can be increased if we place a color filter in front of the source, so that the light that passes through the filter is somewhat monochromatic. The minimum wavelength of the emerging light is 600 nm. What is the maximum wavelength in order for the coherence length to be 0.100 mm?

5 A laser beam of intensity 1.00×10^3 W/m² with a wavelength of 600 nm is incident on an opaque screen with a tiny circular aperture of radius 1.00×10^{-7} m. Apply the Huygens-Fresnel principle to find the intensity of the diffracted light at a distance of 1.00 cm from the center of the aperture.

26-2 Interference

6 Two thin slits separated by 0.100 mm are illuminated by light from a He-Ne laser (λ = 633 nm), producing interference fringes on a distant screen. Find the angle between the centers of the central bright fringe and the next bright fringe.

7 Two thin slits separated by 0.15 mm are illuminated by a monochromatic plane wave, producing interference fringes on a distant screen. If the angle between adjacent fringes is 3.5×10^{-3} rad, what is the color of the fringes?

8 Blue interference fringes are formed on a screen 2.0 m away from a double slit illuminated by monochromatic light of wavelength 480 nm. The distance between the centers of adjacent fringes is 4.0 mm. Find the separation between the two slits.

9 Two narrow slits separated by 0.40 mm are illuminated by monochromatic light of wavelength 500 nm. How many bright fringes can be seen on a screen 1.0 cm wide placed 1.0 m in front of the slits?

10 How far is the second dark fringe to the right of the central bright fringe in the double-slit interference pattern seen on a screen 3 m from the slits? The slits are separated by 0.3 mm and are illuminated by monochromatic light of wavelength 600 nm.

★11 The experiment described in Example 2a is performed underwater. Find the separation between the centers of the bright fringes. What color are the fringes?

★12 Two narrow slits, separated by 0.10 mm, are illuminated by white light. Is it possible to see interference fringes? Explain.

13 Two narrow slits, illuminated by light consisting of two distinct wavelengths, produce two overlapping colored interference patterns on a distant screen. The center of the fourth bright fringe in one pattern coincides with the center of the third bright fringe in the other pattern. What is the ratio of the two wavelengths?

14 White light is reflected at normal incidence from a 250 nm thick oil film of index 1.40 (Fig. 26-17). Light of which vacuum wavelength will experience destructive interference?

15 A layer of oil, with a refractive index of 1.40798, exactly 1 mm thick, floats on water.
(a) Light of vacuum wavelength 632.800 nm, emitted by a stabilized He-Ne laser, is reflected at normal incidence by the oil. How many wavelengths are contained in the light wave passing back and forth through the oil?
(b) Does the light reflected from the two surfaces of the oil interfere constructively or destructively?

★16 White light is reflected at normal incidence by a thin film of water on glass. Both 500 nm and 700 nm light reflected from the two surfaces interferes constructively. Find the minimum thickness of the film.

17 White light is normally incident on two glass microscope slides with a thin layer of air of variable thickness between the slides, as seen in Fig. 26-16. Which wavelengths interfere constructively when reflected at a point where the air layer is 800 nm thick?

★★18 An interference pattern, consisting of alternating bright and dark rings and known as Newton's rings, is formed by interference of monochromatic light reflected from the two air-glass surfaces between a convex lens and a glass plate (Fig. 26-48).
(a) Show that the radius of the mth bright ring is given by

$$r = \sqrt{R(m + \tfrac{1}{2})\lambda - \tfrac{1}{4}(m + \tfrac{1}{2})^2\lambda^2}$$

where R is the radius of curvature of the convex surface.
(b) Find the radius of the tenth ring for light of wavelength 550 nm if R = 200 cm.

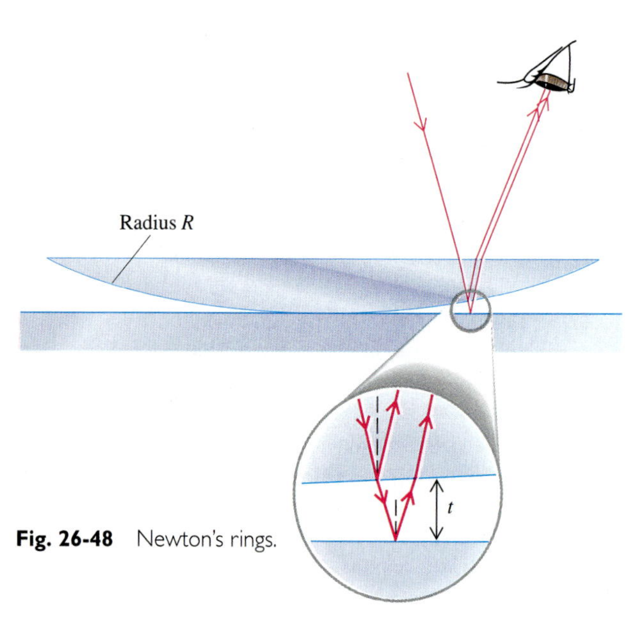

Fig. 26-48 Newton's rings.

19 A human hair is placed between the ends of two microscope slides, forming a thin wedge of air between the slides, as indicated in Fig. 26-49. An expanded He-Ne laser beam of wavelength 633 nm is incident on the slides from above. An observer viewing the slides from above sees a series of thin, straight, alternating light and dark fringes, with 120 bright fringes spread over the 5.00 cm length of the slides. Find the diameter of the hair.

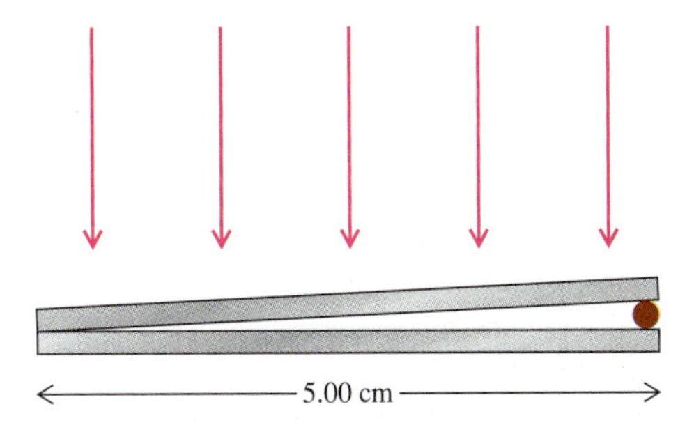

Fig. 26-49

Fig. 26-50 Reflection of white light by a thin soap film.

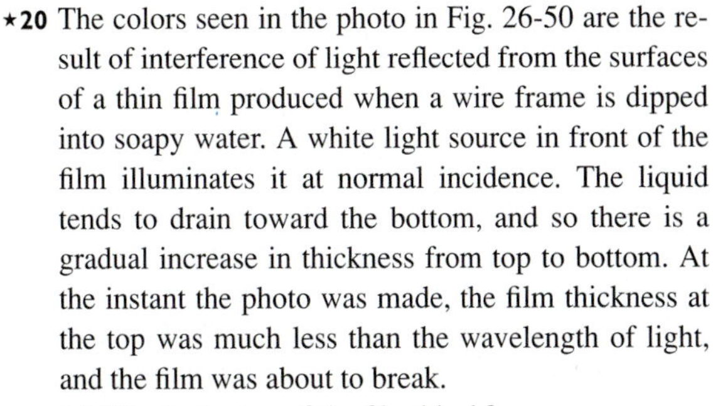

★20 The colors seen in the photo in Fig. 26-50 are the result of interference of light reflected from the surfaces of a thin film produced when a wire frame is dipped into soapy water. A white light source in front of the film illuminates it at normal incidence. The liquid tends to drain toward the bottom, and so there is a gradual increase in thickness from top to bottom. At the instant the photo was made, the film thickness at the top was much less than the wavelength of light, and the film was about to break.

(a) Why is the top of the film black?

(b) Just beneath the dark section of the film is a section that appears white, indicating constructive interference for all wavelengths. Show that if the film is 100 nm thick the difference in path lengths of the two reflected waves is such that all visible wavelengths come closer to satisfying the condition for constructive interference than the condition for destructive interference.

(c) Find a film thickness such that yellow light of wavelength 575 nm interferes constructively and light near either end of the spectrum interferes destructively.

★21 A Michelson interferometer is shown in Fig. 26-51a. Fig. 26-51b shows the main elements of the interferometer. A light wave incident from the left is split into two separate waves by a half-silvered mirror M_h inclined at an angle of 45° relative to the incident beam. Half of the incident light is transmitted, while the other half is reflected by M_h, producing two perpendicular beams of equal intensity. Mirrors M_1 and M_2 reflect the respective beams back to M_h, which then reflects half and transmits half of each beam. Half of the light reflected from M_1 is transmitted through M_h and falls on the screen, while half the light reflected from M_2 is reflected by M_h and falls on the screen.

(a)

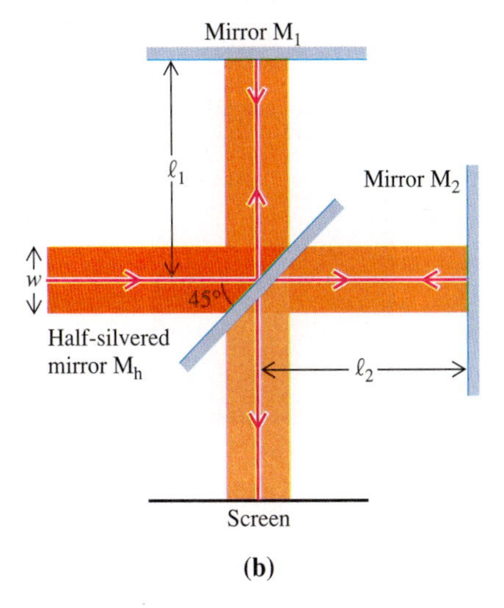

(b)

Fig. 26-51 Michelson interferometer.

Thus the two beams come together again, after having followed different paths. Mirror M_2 is fixed, but one may move mirror M_1 by turning a micrometer screw, so that the path length ℓ_1 can be varied precisely. Do the light waves interfere constructively or destructively if the value of $\ell_1 - \ell_2$ is (a) 275 nm (using white light); (b) 1.0000 mm (using white light); (c) 1.0000 mm (using He-Ne laser light)?

★★ **22** In the Michelson interferometer, described in the last problem, suppose that mirror M_1 is rotated by a small angle θ, as indicated in Fig. 26-52a. This increases the path length traveled by the left side of the beam reflected from M_1. Assume that the right side of the beam interferes constructively, so that a bright fringe appears at point P on the screen.
(a) Find an equation expressing the condition for constructive interference at point Q at the opposite edge of the beam.
(b) Find the number of fringes appearing on the screen, using an expanded He-Ne laser beam of width 10.0 cm, when $\theta = 1.00 \times 10^{-4}$ rad (Fig. 26-52b).

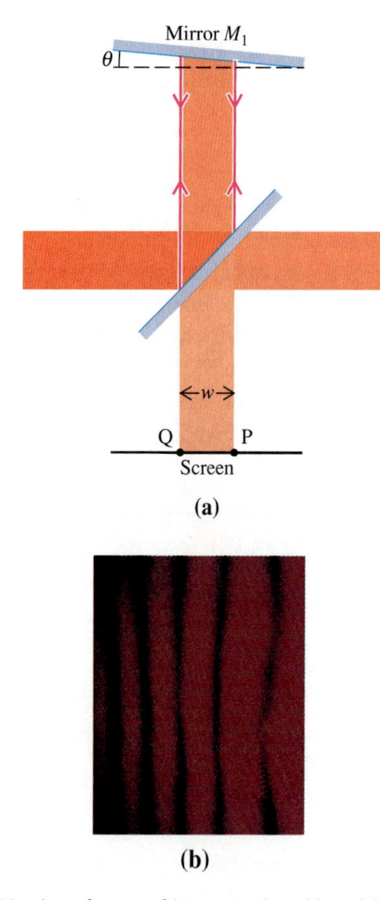

Fig. 26-52 Interference fringes produced by a Michelson interferometer.

26-3 Diffraction

23 (a) Find the angle θ locating the first minimum in the Fraunhofer diffraction pattern of a single slit of width 0.200 mm, using light of wavelength 400 nm.
(b) Find the angle locating the second minimum.

24 How many diffraction maxima are contained in a region of the Fraunhofer single-slit pattern, subtending an angle of 2.00°, for a slit width of 0.100 mm, using light of wavelength 580 nm?

25 A Fraunhofer diffraction pattern of a single slit of width 0.300 mm is produced using light of wavelength 600 nm. Find the angle subtended by the central diffraction maximum.

26 At what distance from a single slit of width 0.150 mm should a screen be placed so that the width of the Fraunhofer central diffraction maximum is 15.0 mm (100 times the slit width), using light of wavelength 500 nm?

27 A tall narrow doorway of width 0.8 m opens into a large empty room. Plane sound waves of wavelength λ are normally incident on the doorway from the outside. In some directions within the room you may hear no sound. Find the angles specifying these directions (measured relative to the perpendicular to the doorway) for a wavelength of (a) 20.0 cm (f = 1720 Hz); (b) 78.0 cm (f = 440 Hz); (c) 1.00 m (f = 344 Hz)? (Assume that you can ignore the effect of sound reflected by the floor of the room and that you are far enough inside the room that the conditions of Fraunhofer diffraction apply.)

28 Water waves of wavelength 10.0 m are normally incident on a 30.0 m wide entrance to a harbor channel. Find the angle, relative to the direction of motion of the incident waves, corresponding to the first diffraction minimum.

★ **29** Light consisting of two wavelengths (400 nm and 600 nm) is used to produce overlapping Fraunhofer diffraction patterns of a single slit of width 2.00 mm. Find an angle θ at which diffraction minima for the two wavelengths coincide.

30 Find the angle subtended by the Airy disk in the Fraunhofer diffraction pattern of a circular aperture of diameter 0.200 mm illuminated by light from a He-Ne laser (λ = 633 nm).

31 A monochromatic plane wave of wavelength 500 nm is incident on an opaque screen with a pinhole of diameter 0.100 mm. Find the diameter of the Airy disk on a wall 5.00 m in front of the pinhole.

★ **32** The planet Venus has a diameter of 12,100 km and its distance from the earth is 4.2×10^7 km at its closest approach to the earth. On the basis of diffraction theory, should the human eye, with the pupil fully dilated to a diameter of 6.0 mm, be able to distinguish Venus from a star because of the larger angle subtended by Venus? Use 550 nm for the vacuum wavelength of light and 1.34 for the refractive index of the eye. (Actually there are other limits to the visual acuity of the eye, which normally cannot resolve points with an angular separation of less than about 5.0×10^{-4} rad. See Section 25-5.)

33 You are looking at starlight of wavelength 600 nm with your pupils fully dilated to a diameter of 6.00 mm. Find the diameter of the corresponding Airy disk on each retina. The distance from pupil to retina is approximately 21.0 mm, and the refractive index within the eye is 1.34.

34 You are driving at night on a long, straight highway in the desert as another car approaches. What is the maximum distance at which you can tell that the car is not a motorcycle by seeing its two headlights, which are 1.5 m apart?

(a) Assume that your visual acuity is limited only by diffraction. Use 550 nm for the vacuum wavelength, and assume that your pupils are dilated to a 6.0 mm diameter.

(b) Assume a more realistic, typical visual acuity with $\theta_{min} = 5.0 \times 10^{-4}$ rad.

35 On the basis of diffraction theory, calculate the minimum angle between objects barely resolvable by the human eye for the following values of the pupil diameter D and vacuum wavelength λ_0: (a) $D = 2.00$ mm, $\lambda_0 = 400$ nm; (b) $D = 2.00$ mm, $\lambda_0 = 700$ nm; (c) $D = 6.00$ mm, $\lambda_0 = 400$ nm; (d) $D = 6.00$ mm, $\lambda_0 = 700$ nm.

36 The diffraction pattern shown in Fig. 26-31 was produced by a grating with 500 lines/cm illuminated by a He-Ne laser ($\lambda = 633$ nm). Find the angle between adjacent maxima.

37 A monochromatic plane wave is normally incident on a diffraction grating with 1.00×10^5 lines/m. For what visible wavelengths would this grating produce a maximum at $\theta = 10.0°$?

38 Looking at a white light source through a diffraction grating with 2.00×10^5 lines/m, you see a spectrum of colors (Fig. 26-30). Find the wavelength of the light seen in the first-order spectrum at an angle of 5°.

39 The surface of a compact disk (CD) behaves as a kind of diffraction grating. When light is reflected from the surface, each of the closely spaced ridges on a CD behaves as the source of a spherical light wave. Except for the fact that light is reflected rather than transmitted, the effect is the same as for the transmission gratings described in the text; that is, one sees a spectrum of colors (Fig. 26-30). Suppose a beam of sunlight is normally incident on a CD. You see yellow light of wavelength 575 nm reflected at an angle of 30.0° relative to the normal. Find the number of ridges per mm on the CD.

26-4 Polarization

40 You look through a Polaroid sheet at a 100 W light bulb 3.00 m away.

(a) Find the intensity of the light you see.

(b) Next you place a second Polaroid sheet in front of the first sheet with an angle of 45.0° between the two transmission axes. What intensity do you observe?

41 Unpolarized light is incident on a series of two Polaroid sheets. The intensity of the light emerging from the second sheet is 10.0% of the intensity of the light incident on the first sheet. Find the angle between the transmission axes of the two sheets.

★★ **42** Initially light is incident on a series of two Polaroid sheets, with perpendicular transmission axes, so that no light emerges from the second sheet. Without changing these sheets, a third Polaroid sheet is inserted between the other two. Is it possible that light will now emerge from the series of three sheets? If so, what should be the angle between the transmission axes of the inserted sheet and the first sheet in order that the final intensity of the emerging light will be maximized?

43 Suppose you are underwater, wearing Polaroid sunglasses. Looking upward toward the water's surface, you view the setting sun by seeing light rays from the sun that strike the water at an angle of incidence of 75°. The intensity of the incident sunlight is 200 W/m^2. What is the intensity of the light you see?

44 Suppose you are photographing a quiet lake that reflects a mirrorlike image of trees on the opposite shore. You place a Polaroid sheet in front of your camera. What orientation of its transmission axis will increase the relative intensity of the reflected light compared to direct light from objects in the background, thereby making the reflected image appear brighter in the photograph?

45 Calculate Brewster's angle for light incident from glass into air if the glass has an index of 1.5.

★★ **46** Let θ denote Brewster's angle for light reflected from a surface when the light is incident from medium A onto medium B. Let θ' denote Brewster's angle for light reflected from the same surface but with the light incident from medium B onto medium A. Prove that $\theta + \theta' = 90°$.

47 Calculate Brewster's angle for light incident from water onto glass of refractive index 1.5.

★ **48** An unpolarized laser beam enters a container of water. The beam is partially reflected from the water-glass surface, as indicated in Fig. 26-53. For what angle of incidence ϕ will this reflected beam be completely polarized?

Fig. 26-53

Additional Problems

49 A Michelson interferometer (described in Problem 21) can be used to accurately measure the wavelength of its light source. Mirror M_1 is moved a distance of 0.148 mm. What is seen at a certain point on the screen changes from bright to dark to bright 542 times as the mirror is being moved. Find the wavelength of the source.

★ **50** A certain diffraction grating, illuminated by a gas discharge tube, produces a diffraction pattern in which the higher-order spectra overlap. Both a violet line and an orange line appear side by side at an angle of approximately 40°. How many lines per meter does the grating have?

★★ **51** In the double-slit interference pattern, what is seen is the interference of the diffracted waves produced by each of the slits. The brightest fringes are those that correspond to the central diffraction maximum of each slit. Suppose that each of the slits is 0.200 mm wide, the slits' centers are 0.900 mm apart, and they are illuminated by a source of wavelength 600 nm. How many bright fringes will be seen in the central maximum?

★★ **52** The width of each opening in a diffraction grating is 2.00×10^{-3} mm, and the distance between adjacent openings is 2.00×10^{-2} mm.

(a) Find the angle subtended by the central diffraction maximum of each opening when illuminated by a He-Ne laser ($\lambda = 633$ nm).

(b) How many bright spots are seen in the central maximum?

★★ **53** Suppose you are trying to demonstrate diffraction by a single aperture in such a way that maximizes the angle subtended by the central diffraction maximum so that the effect is as dramatic as possible. Thus you want as small an aperture as possible. If, however, you make the aperture too small, there will be too little light to be seen. So there is a practical limit to how small you can make it. Suppose a circular aperture is illuminated by a He-Ne laser beam ($\lambda = 633$nm), with an intensity of 1.00×10^3 W/m^2. The Airy disk, seen on a screen 10.0 cm in front of the aperture, contains 84.0% of the radiant energy passing through the aperture. In order to be visible, the average intensity of the light forming the Airy disk should be at least 1.00×10^{-2} W/m^2. Find the minimum diameter of the aperture and the diameter of the corresponding Airy disk.

★ **54** Find the minimum angular magnification of a microscope, such that an eye with normal acuity ($\theta_{min} = 5.0 \times 10^{-4}$ rad) and a normal near point of 25 cm can distinguish points as close as possible with a light microscope. Assume $\lambda = 550$ nm.

55 Estimate the maximum distance at which you should be able to recognize a familiar face, looking through binoculars that provide an angular magnification of 7.0. Suppose that to accomplish this you need to be able to resolve points 1.0 cm apart. Assume that in bright sunlight resolution is limited only by diffraction of light by your eye, which then has a pupil diameter of 2.0 mm. Use $\lambda = 550$ nm for the vacuum wavelength of light.

★★ **56** (a) Estimate the minimum diameter for an aperture of a camera used to detect ground-based missiles from a spy satellite.

(b) Estimate the minimum focal length of the camera's lens if the 9 μm diameter of the grains in the photographic film are not to limit the resolution.

CHAPTER 27 Relativity

A scene from the movie *The Time Machine*, based on the 1895 novel by H.G. Wells

In H.G. Wells's classic novel *The Time Machine*, the hero invents a device that transports him hundreds of years into the future. As fantastic as this sounds, a time machine is possible. As we shall see in this chapter, all that is required is a spaceship capable of moving at a speed close to the speed of light. If you traveled for a few years in such a spaceship and then returned to earth, you would find that many more years had elapsed on earth. If you were traveling fast enough, perhaps hundreds or even thousands of years would have passed. Although extremely high-speed, long-distance space travel is not yet practical, someday it may be. An astronaut might blast off from the earth in a spaceship that takes her to distant regions of space, reaching speeds approaching the speed of light. The astronaut, still young, might return to earth hundreds of years later. However, unlike the Wells story, she would never be able to go back to the past or to those she left behind. Time travel is apparently limited to a one-way trip—into the future, never into the past.

We can predict today the possibility of time travel, based on work first published in 1905 by a man who was then an obscure Swiss patent clerk. The man was Albert Einstein (Fig. 27-1), and his work was the theory of relativity—a theory that was to make him the most famous scientist of all time. As much as any athlete or film star, he was a great celebrity of his day. His name became synonymous with genius. Einstein's life is described in an essay at the end of this chapter.

Although Einstein's insights were brilliant, his ideas did not arise in a vacuum. In a sense they were the natural result of the kinds of questions that were being asked at the turn of the century. However, we shall not discuss this historical background of relativity theory here. In this way we can present without interruption the essence of Einstein's ideas, so that you can better appreciate their logical simplicity and beauty.

Fig. 27-1 Albert Einstein in 1905.

Fig. 27-2 Various clocks: a simple pendulum being used to mark time, a pendulum clock in Berne, Switzerland, a quartz wristwatch, earth clock.

27-1 Measurement of Time; Einstein's Postulates

The Nature of Time

What is time? Physicists, philosophers, and poets have all struggled with this question. Simply defining time as "duration" or "period" does no good, since such a definition merely substitutes for time another undefined quantity. Comedian Woody Allen jokingly defined time as "nature's way of keeping everything from happening at once." Of course, we all have a sense of what time is, though we may not be able to express it in words. In the *Principia,* Isaac Newton wrote: "I do not define time, space, place, and motion, since they are well known to all." However it is important that we examine more carefully the meaning of time, so that we do not harbor false or misleading concepts. Newton himself believed that "absolute, true, and mathematical time, of itself, and from its own nature, flows equably without relation to anything external. . . ." We shall see that such a concept of absolute time is unfounded. Einstein showed that time is relative, not absolute.

The concept of time as it is used in physics always relates to measurement with a clock. Suppose, for example, a certain plane arrives at an airport at 4:00 P.M. Saying that the time of arrival is 4:00 P.M. means simply that the arrival of the plane is simultaneous* with a clock reading 4:00. **The time of any event is given by the reading of a nearby clock, simultaneous with the event.**

Clocks are generally based on some kind of repetitive or periodic motion (Fig. 27-2). Old-fashioned pendulum clocks count oscillations of a pendulum. A modern quartz wristwatch measures time by counting the oscillations of a quartz crystal. The quartz watch is more accurate than a pendulum clock because the period of oscillation of the quartz is more regular than the period of the pendulum. The most accurate clocks are atomic clocks, which are based on the period of the electromagnetic radiation emitted by an atom. Motion of the earth and moon provides a kind of natural clock, with time units of days, months, or years. And the human body is also a natural clock, with various natural cycles or intervals: the heartbeat, circadian rhythms (cycles of about 1 day, such as the sleep cycle), monthly menstrual cycles, and even lifetimes.

Measuring the time of a very distant event is not as direct as for a nearby event, since observation of the event requires light, which travels at a finite speed.† When you look up at the night sky, you see light from stars at various distances. For example, you may see the stars Sirius and Betelgeuse simultaneously. This does not mean that the light from these stars is emitted simultaneously. Astronomers have determined that the distances to these stars are approximately 9 light-years and 490 light-years respectively; that is, light from Sirius takes 9 years to reach earth, and light from Betelgeuse takes 490 years to reach earth. The light you see coming from Betelgeuse was emitted 481 years earlier than the light from Sirius, though you see the light from each at the same instant!

We compute the time of distant events based on the known speed of light and the place where the event occurred. We can use that computation to coordinate clocks at

*When we say that the plane's arrival and the clock's reading are simultaneous, you know what that means. The idea of simultaneous events at the same place is a basic undefinable concept; that is, we cannot define this concept in terms of anything more basic.

†The earth is small enough that we can witness any event on earth with almost no delay time. Light can travel half way around the world in less than 0.1 s. For example, we can witness on television a political uprising in Beijing, China, as it is occurring, by means of electromagnetic waves bounced off of communication satellites.

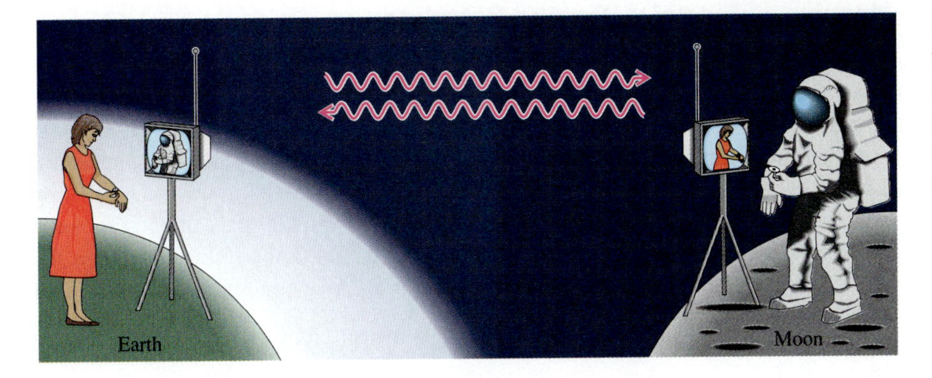

Fig. 27-3 Observers on the earth and the moon synchronize their watches by means of electromagnetic signals. Compared to the speed of light, the observers are nearly at rest with respect to one another.

very distant locations. For example, an astronaut on the moon can synchronize her clock by communicating with earth via visible light or other electromagnetic radiation (Fig. 27-3). In principle, one could set up a system of synchronized clocks throughout the universe. If these clocks are at rest relative to one another at known locations, they form a reference frame for observing events throughout the universe. **The time of any event, relative to this reference frame, is defined as the reading of one of the reference frame's clocks close to the event.**

The next question that arises is whether systems of clocks in relative motion at very high speed can be synchronized so that passing clocks always agree. The answer is no, as we shall see in the next section. But to arrive at this conclusion we must first introduce Einstein's postulates.

Einstein's Postulates

Einstein based his theory of relativity on the following two postulates:

I The principle of relativity: All laws of physics are valid in any inertial reference frame.

II Light always travels through a vacuum at a fixed speed c, relative to any inertial reference frame, independent of the motion of the light source.

Meaning of Einstein's First Postulate

An inertial reference frame is any reference frame in which the principle of inertia, Newton's first law, is valid. As discussed in Chapter 4 (Section 4-3), the earth's surface is approximately an inertial reference frame. Any reference frame moving at constant velocity with respect to the earth is also inertial. Suppose you are in a plane moving at a constant velocity of 1000 km/h westward. According to the principle of relativity, all the laws of physics work for you, just as they would on the ground. It follows that if the plane's windows were covered there would be no way to discover your motion by means of any experiment confined to the plane. For example, if you were to perform a free-fall experiment, you would measure the same gravitational acceleration as though you were at rest with respect to the earth. Other experiments in mechanics, electricity, optics, and so forth would all give the same results as though performed at rest on the ground, if in all these experiments there were no interactions with anything outside the plane. The principle of relativity implies that absolute motion is meaningless. Only relative motion has meaning. Thus, for example, it is just as valid to use the reference frame of the plane and describe the plane as being at rest and the earth as moving at 1000 km/h eastward.

Accelerated reference frames are not inertial. The laws of physics do not work in such reference frames. For example, you can easily detect a plane's motion at takeoff and landing, when it is accelerating. You can feel the force on your body producing the acceleration.

Meaning of Einstein's Second Postulate

Einstein's second postulate is at first hard to accept. According to this postulate, if we observe light from *any* inertial reference frame, the light travels at a fixed speed $c = 3.00 \times 10^8$ m/s no matter how fast the source of the light may be moving. For example, suppose that a spaceship (A) is passing the earth at a speed of 2.00×10^8 m/s and emits a laser pulse directed at the moon (Fig. 27-4). According to Einstein's second postulate, the laser light travels from the spaceship to the moon at a speed of 3.00×10^8 m/s—not at a speed of 5.00×10^8 m/s! This means that the pulse is received on the moon, 3.84×10^8 m away, after a time delay of $\dfrac{3.84 \times 10^8 \text{ m}}{3.00 \times 10^8 \text{ m/s}} = 1.28$ s. This time interval is measured by synchronized clocks on the earth and the moon.

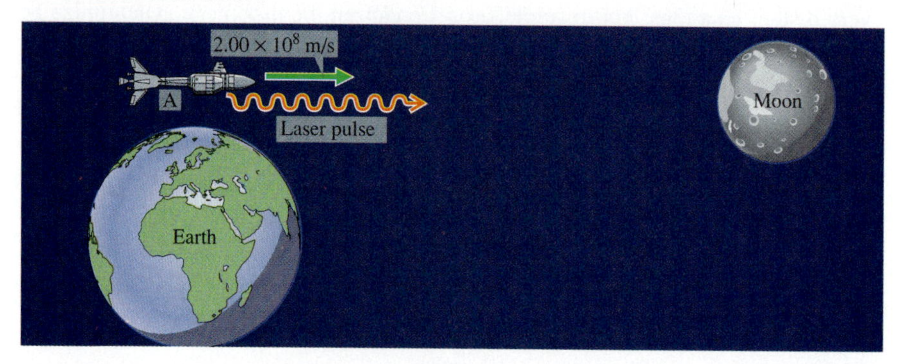

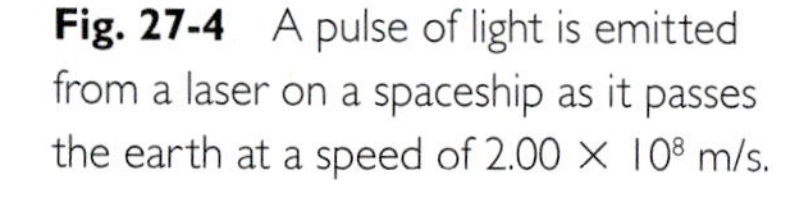
Fig. 27-4 A pulse of light is emitted from a laser on a spaceship as it passes the earth at a speed of 2.00×10^8 m/s.

The really amazing thing about the motion of light is that it moves at the same speed relative to *any* inertial reference frame. Suppose, for example, the light pulse is observed by an astronaut who also happens to be passing the earth but moving in the opposite direction at a speed of 1.00×10^8 m/s in spaceship B, as shown in Fig. 27-5. According to Einstein's second postulate, the laser pulse travels at a speed of 3.00×10^8 m/s, as seen by the astronaut on ship B. The laser pulse moves at the same speed (3.00×10^8 m/s) relative to the earth and relative to each of the spaceships, despite the fact that the spaceships are moving relative to the earth. Obviously the equations for determining relative velocities, presented in Chapter 3 (Section 3-4), are not valid for light. (And these equations are invalid for any bodies moving at speeds comparable to the speed of light, as we shall see in Section 27-4.)

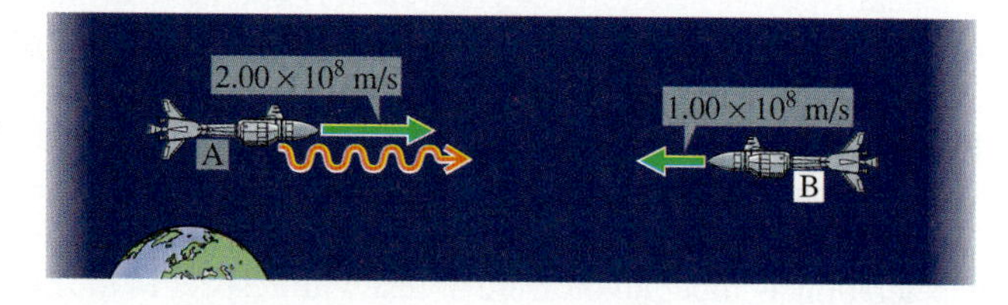

Fig. 27-5 Spaceship B, relative to the earth, travels at a velocity of 1.00×10^8 m/s directed toward the left. This ship is an inertial reference frame. Therefore the laser pulse emitted by ship A travels at a speed of 3.00×10^8 m/s, relative to B.

By way of contrast with the behavior of light, consider how sound waves travel when the source of sound is moving. The speed of wave propagation is independent of the motion of the source of the sound. However, the speed is not the same in all reference frames. Sound waves move through a medium at a fixed speed *relative to the medium.* Sound travels through air at a speed of 340 m/s, relative to the air, even if the sound source is moving through the air. For example, as illustrated in Fig. 27-6, sound from a race car engine moves at 340 m/s relative to the air (and relative to a stationary observer), even though the car (A) is moving at 70 m/s. However, the speed of sound is different when measured by an observer moving relative to the air. If car B approaches car A at a speed of 50 m/s (relative to the ground and the air), the sound wave moves relative to B at a speed of 390 m/s (Fig. 27-7).

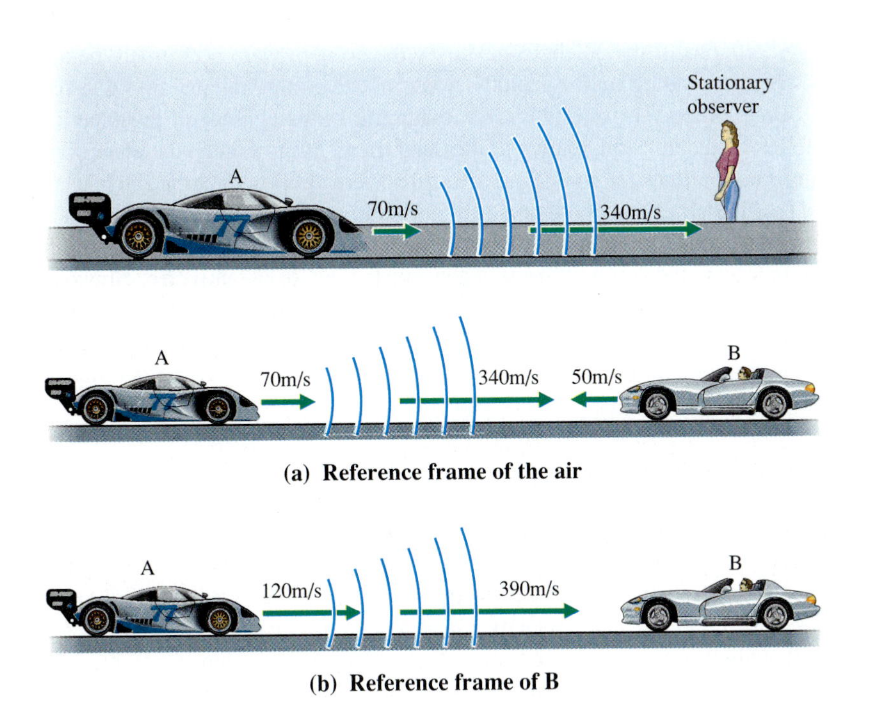

Fig. 27-6 Relative to a stationary observer, sound waves emitted by a moving race car (A) travel at a speed of 340 m/s, independent of the speed of the car.

Fig. 27-7 Sound emitted by A, as observed in two different reference frames.

Although car B is an inertial reference frame and therefore the laws of physics are just as good in B's reference frame as in any other, the law of sound propagation must take account of the relative motion of the air—the medium through which the sound propagates. And so the observed speed of the sound wave is different for B from that observed by one who is at rest with respect to the air. For light propagation, no medium is required, and observers in all inertial reference frames observe the same velocity of light.

Einstein's second postulate is actually implied by his first postulate. Remember that the laws of electromagnetism predict the existence of electromagnetic waves that travel at a speed c, which can be calculated in terms of the electric constant ϵ_0 and the magnetic constant μ_0:

$$c = \frac{1}{\sqrt{\epsilon_0 \mu_0}} = 3.00 \times 10^8 \text{ m/s}$$

If the laws of electromagnetism are correct (and they are), it follows from the principle of relativity that the speed of light should be the same in all inertial reference frames.

Experimental Support for Einstein's Postulates

Einstein's postulates are supported by various experiments, some performed before and some after Einstein first published his theory. The Michelson-Morley experiment, described in detail in Problem 4, was the most famous such experiment. Michelson and Morley, along with other nineteenth-century physicists, assumed that light would behave like sound waves, in the sense that there would be a privileged reference frame for the propagation of light. This reference frame would be like a material medium for light waves, but it would be present even in a vacuum, that is, when no matter was present. This strange and elusive medium was called **"the ether"** (not to be confused with ether gas, an anesthetic). Michelson and Morley believed that the earth, at least some time during the year, would move relative to the ether. They were determined to detect this relative motion, using sensitive optical interference methods. (The instrument they used, a Michelson interferometer, was described in Chapter 26, Problem 21).

Many versions of the Michelson-Morley experiment were performed, beginning in 1887 and extending into the 1900s. Some of these experiments used a terrestrial light source; others used sunlight—a moving light source. None of the experiments was able to detect motion relative to the ether. In an effort to keep the ether concept, various explanations for the negative results were proposed. Some physicists believed that the ether somehow clung to the earth, so that no matter how the earth's motion changed throughout the year, it was always in the privileged ether reference frame. In a way this was a return to a geocentric view of the universe. However, all such explanations were shown to be inconsistent with various experimental observations. Eventually physicists became convinced that no one would ever find a way of detecting motion relative to the ether and that indeed there was no ether, that is, no privileged reference frame. The principle of relativity prevailed.

27-2 Time Dilation

In this section we will use Einstein's postulates to show how observers in relative motion measure different values for the duration of any sequence of events. Measurement of a time interval depends on the observer's reference frame. For example, suppose that a football game in Denver lasts 3.00 hours as measured on a clock at the football stadium (or on any other good clock in the earth's reference frame). As we shall see in Example 1, if the game is viewed by observers in a reference frame moving at a speed of 2.70×10^8 m/s relative to the earth, the game lasts 6.88 hours.

We can derive an equation relating time intervals measured in different reference frames by considering the following experiment. Let a laser emit a pulse of light that travels a distance D and is then reflected by a mirror back toward the laser (Fig. 27-8). The light pulse is absorbed by a detector adjacent to the laser. Since the light travels a distance $2D$ at a speed c, the time elapsed between emission and absorption is given by

$$\Delta t_0 = \frac{2D}{c} \tag{27-1}$$

A time interval such as this, **measured on a single stationary clock,** is referred to as a **proper time interval** and is denoted by the zero subscript on the Δt. We shall refer to the arrangement of laser, mirror, and detector in Fig. 27-8 as a "light clock."

The emission and absorption of light by a light clock can also be viewed by an observer in a reference frame relative to which the light clock is moving to the right at velocity $\mathbf{v}$. Fig. 27-9 shows the emission, reflection, and absorption of a pulse of light, as seen by such an observer. The observer's reference frame is an inertial frame. Therefore, according to Einstein's postulates, (1) the laws of physics apply (in particular the law of reflection), and (2) light travels at speed c. Relative to this reference frame, the light must travel a longer, diagonal path, and so the time interval between emission and absorption is greater than that in the other reference frame, in which the light travels vertically back and forth. The time elapsed between emission and absorption, during which light travels the distance 2ℓ, is given by

$$\Delta t = \frac{2\ell}{c}$$

Using the figure and applying the Pythagorean theorem, we see that

$$\ell = \sqrt{D^2 + (v\,\Delta t/2)^2}$$

Inserting this expression for ℓ into the preceding equation, we obtain

$$\Delta t = \frac{2\sqrt{D^2 + (v\,\Delta t/2)^2}}{c}$$

Fig. 27-8 A pulse of light is emitted by a laser, reflected by a mirror a distance D from the laser, and absorbed by a detector next to the laser. The pulse travels from laser to detector in time $2D/c$. We call this arrangement a "light clock."

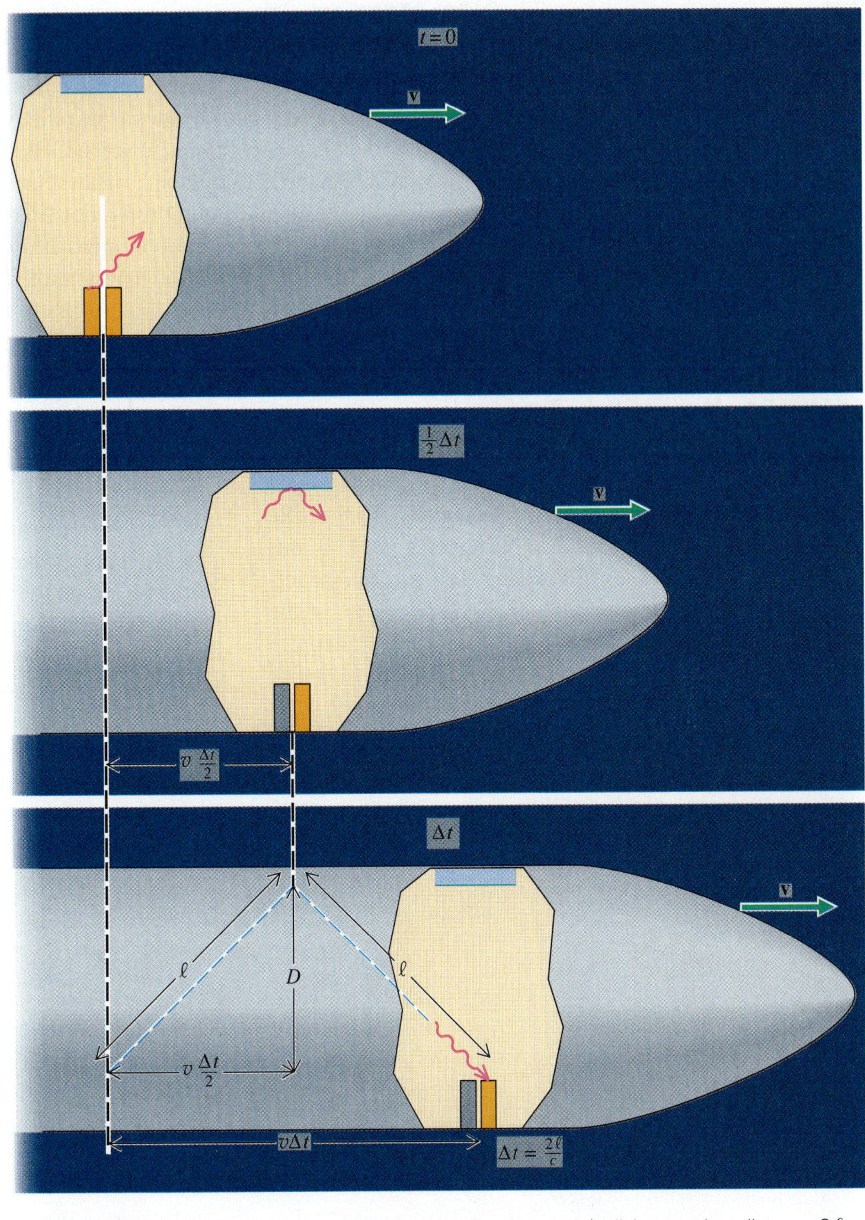

Fig. 27-9 A moving light clock. During the time interval Δt, light travels a distance 2ℓ while the laser, mirror, and detector travel a distance $v\ \Delta t$.

The quantity Δt appears on both sides of this equation. Squaring both sides and solving for Δt, we find

$$\Delta t = \frac{2D/c}{\sqrt{1 - (v/c)^2}}$$

Substituting $2D/c = \Delta t_0$ (Eq. 27-1), we obtain

$$\Delta t = \frac{\Delta t_0}{\sqrt{1 - (v/c)^2}} \tag{27-2}$$

The time interval Δt is not a proper time interval; Δt is measured on a system of two synchronized clocks—one at the point where the pulse is emitted and a second at the point where the pulse is absorbed. Eq. 27-2 expresses the relationship between the two different measurements of the elapsed time of any sequence of events, as seen by observers in different reference frames. It is important to remember that Δt_0 represents a proper time interval measured on a single clock and Δt represents the corresponding time interval on a system of clocks.

EXAMPLE 1 Moving Clocks Run Slow

(a) A football game in Denver lasts 3.00 hours. The game is viewed by space travelers in a convoy of spaceships, which happen to be passing the earth as the game is in progress. The spaceships all move at a constant velocity of 2.70×10^8 m/s relative to the earth's surface. How long is the game in the reference frame of the spaceships? (b) A clock on one of the spaceships is compared with clocks in the earth's reference frame. How much time has elapsed in the earth's reference frame while 3.00 hours elapses on the spaceship clock?

SOLUTION (a) We apply Eq. 27-2 using $\Delta t_0 = 3.00$ hours, since this is a proper time interval measured on a single clock at the game. This clock and the earth move at a speed $v = 2.70 \times 10^8$ m/s relative to the space travelers, who measure the game's duration to be

$$\Delta t = \frac{\Delta t_0}{\sqrt{1-(v/c)^2}} = \frac{3.00 \text{ h}}{\sqrt{1-\left(\frac{2.70 \times 10^8 \text{ m/s}}{3.00 \times 10^8 \text{ m/s}}\right)^2}}$$

$$= 6.88 \text{ h}$$

Fig. 27-10 shows how the beginning and end of the game are seen (close up) by different members of the space convoy.

(b) Now the proper time interval $\Delta t_0 = 3.00$ hours is measured on the clock on one of the spaceships while the time interval Δt is measured in the earth's reference frame. The spaceship clock moves at a speed of 2.70×10^8 m/s relative to earth. The calculation is identical to that in part (a), and so we conclude that 6.88 hours elapses in the earth's reference frame. Fig. 27-11 shows the spaceship clock compared with two clocks in the earth's reference frame.

Note the symmetry of the situation. Relative to either inertial reference frame, **a moving clock runs slow!**

EXAMPLE 1 —cont'd

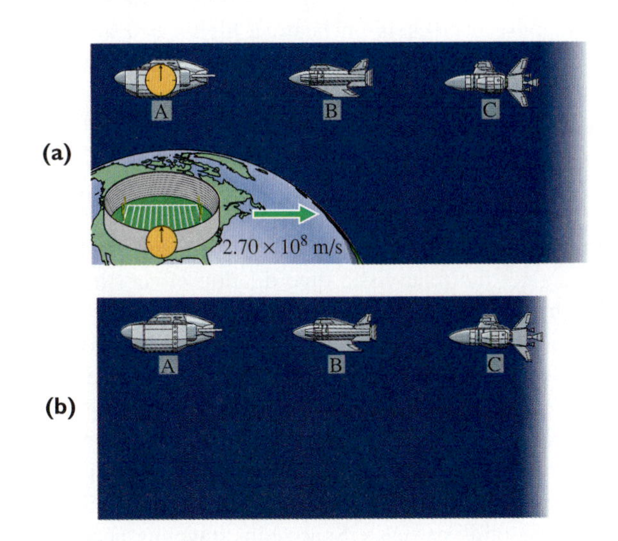

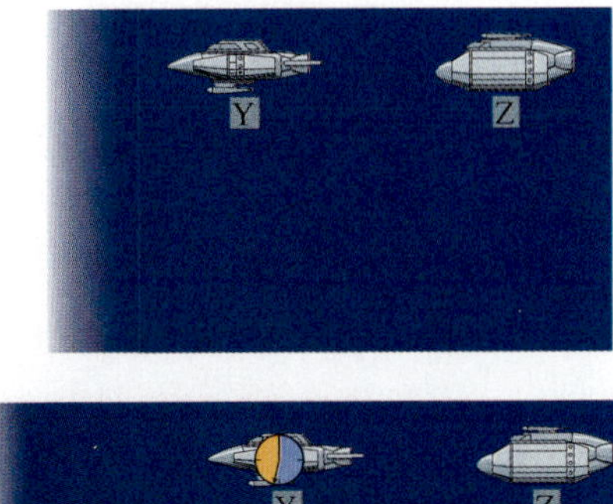

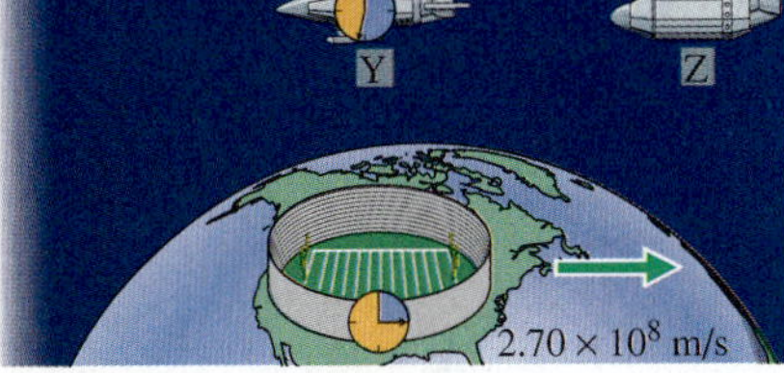

Fig. 27-10 **(a)** The beginning of a football game is witnessed by an observer on board a passing spaceship (A), part of a convoy of spaceships. Relative to the convoy, "spaceship *Earth*" moves at a constant velocity of 2.70×10^8 m/s to the right. **(b)** The end of the football game is witnessed by an observer on board spaceship Y.

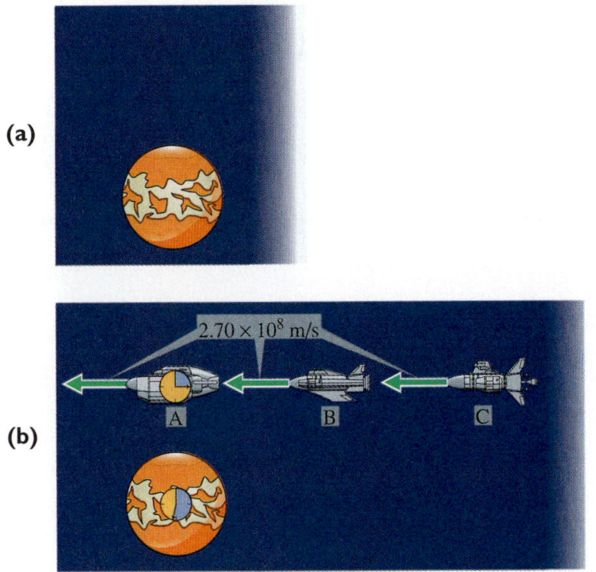

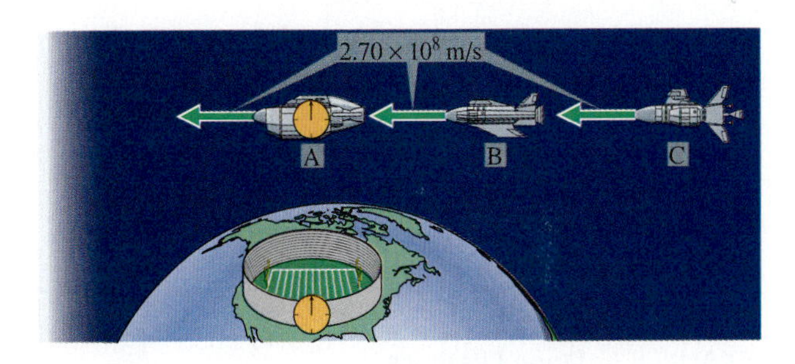

Fig. 27-11 **(a)** Spaceship A compares its clock with an earth-bound clock. **(b)** Spaceship A compares its clock with a clock located on a distant planet colonized by earth. The planet is assumed to be at rest with respect to earth, and the planetary clock is synchronized with the earth clock.

The time dilation formula applies to clocks moving at any speed whatsoever. However our experience with the relativistic phenomenon of time dilation is very limited because we never see clocks or any material bodies moving at speeds approaching the speed of light. Suppose a clock is on board a high-speed aircraft moving at 900 m/s (2010 mi/hr). The time dilation factor for this clock has the value

$$\frac{1}{\sqrt{1-(v/c)^2}} = \frac{1}{\sqrt{1-\left(\dfrac{900\text{ m/s}}{3.00\times 10^8\text{ m/s}}\right)^2}} = 1 + 4.50\times 10^{-12}$$

The time dilation factor differs from one by only 4.50×10^{-12}. This means that for all practical purposes you can neglect the relativistic time dilation effect; that is, $\Delta t \approx \Delta t_0$. Measurements of time by the clock on the aircraft and measurement by a system of clocks on earth give the same value. If the pilot compares his watch with an earthbound clock before and after his flight, he would not be able to observe a difference in the readings.

From the preceding examples you might suppose that meaningful applications of relativity are limited to futuristic or fantasy situations. Such is not the case. Physicists in research labs all over the world routinely use the theory of relativity in describing the motion of electrons or other subatomic particles. Spaceships traveling near the speed of light are only fantasy today. But subatomic particles do move at relativistic speeds, as in the following example.

EXAMPLE 2 Muon Lifetimes

Muons are elementary particles with the same charge as an electron and a mass 207 times the mass of an electron. Muons are unstable, with a mean lifetime of 2.2×10^{-6} s, as measured in their own rest frame. This means that when muons are created in the laboratory by the decay of some other particle, if the muons are either at rest or moving at much less than the speed of light, the muons will decay into other particles, on the average 2.2×10^{-6} s after they are created. Muons created in a high-energy accelerator move at a speed 99.9% of the speed of light, relative to the lab. How long (on the average) after the muons are created do they decay, as measured in the lab?

SOLUTION Applying Eq. 27-2, we find for the mean lifetime, measured in the laboratory:

$$\Delta t = \frac{\Delta t_0}{\sqrt{1-(v/c)^2}} = \frac{2.2\times 10^{-6}\text{ s}}{\sqrt{1-(0.999)^2}}$$

$$= 49 \times 10^{-6}\text{ s}$$

Experimental measurement of high-energy muon lifetimes are in complete agreement with values calculated using the time dilation formula.

Muons are created naturally by cosmic rays in the earth's upper atmosphere. These muons move at very high speed and therefore experience considerable time dilation. Only because the time dilation effect is so large is it possible for muons to live long enough to reach the surface of the earth, where they may be detected.

EXAMPLE 3 The Twin Paradox

Twins part on their twentieth birthday; one remains at home on earth, and the other leaves the earth on a long, high-speed space journey. The spaceship quickly accelerates* to a speed of $0.95c$, maintains this speed for time $\Delta t_0/2$ (ship time), quickly turns around, and travels home at a speed of $0.95c$, arriving at time Δt_0, as measured on the ship and by the astronaut's biological clock. While time Δt_0 has elapsed for the astronaut, the time elapsed for his stationary brother is given by Eq. 27-2:

$$\Delta t = \frac{\Delta t_0}{\sqrt{1-(v/c)^2}} = \frac{\Delta t_0}{\sqrt{1-(0.95)^2}} = 3.2\ \Delta t_0$$

Thus 3.2 years have elapsed on earth for every year on the spaceship. If the astronaut arrives home at age 40, after 20 years of travel, the earthbound twin is 84 years old, having aged (3.2)(20) = 64 years during his brother's absence (Fig. 27-12).

A paradox arises if one attempts to describe the space flight from the reference frame of the spaceship. Relative to the spaceship the earthbound twin is traveling first away from the ship at 0.95c and later toward the ship at 0.95c. The astronaut applies Eq. 27-2, computing his elapsed time corresponding to the time elapsed for his moving earthbound brother: $\Delta t = \frac{\Delta t_0}{\sqrt{1-(v/c)^2}} = 3.2\ \Delta t_0$. The astronaut predicts that 3.2 years elapses on the spaceship for every year elapsed on earth. So, if the astronaut returns to earth after a journey of 20 years (ship time), he expects to find that his twin has aged only $\frac{20 \text{ years}}{3.2} = 6.3$ years; that is, the astronaut predicts that when he is 40, his earthbound twin will be only 26, not 84! These contradictory predictions cannot both be correct. Resolve the paradox.

*To reach relativistic speeds without having the astronaut experience extreme acceleration, the spaceship would have to accelerate over a period of a few years. See Problem 33.

Fig. 27-12 Twins part and years later reunite.

SOLUTION The difficulty lies in the asymmetry between the twins, introduced during the periods of acceleration of the spaceship. It is these accelerations that prevent the spaceship from being inertial. Only the earthbound twin remains in an inertial reference frame continously, and therefore only his calculations are to be trusted. The astronaut does arrive home much younger than his twin brother. The acceleration of the spaceship is somehow responsible for the slower aging of the astronaut.

Although the twin experiment described in this example has never been performed, in 1971 a rather direct experimental verification of the effect was accomplished by Hafele and Keating. Time intervals were measured on two identical cesium atomic clocks. One of the clocks was flown around the world on commercial airlines. The time dilation formula predicted that the traveling clock should run slow by $(184 \pm 23) \times 10^{-9}$ s. When compared to the stationary clock, the traveling clock was found to have lost $(203 \pm 10) \times 10^{-9}$ s, in complete agreement with the prediction of time dilation.

27-3 Length Contraction

In this section we shall show how distances traveled at very high speed appear to the traveler to be shortened or contracted. Such length contraction could theoretically make possible travel to stars hundreds of light years away in a voyage lasting only a few decades. For example, suppose that, in an attempt to reach an extraterrestrial civilization, a space expedition is launched from earth to the star Antares, 424 light-years away. Even if a spaceship were to travel at nearly the speed of light, the trip would take about 424 years, as measured in the earth's reference frame. It might then seem impossible for anyone to live long enough to make such a journey. But remember, moving clocks run slow. So, although approximately 424 years elapses on earth, the duration of the trip could be much shorter, as measured on the spaceship, if it is moving fast enough. Suppose, for example, the average speed is $0.999c$. The trip's duration Δt_0 observed on the spaceship is related to the earth-measured time interval $\Delta t = 424$ years by the time dilation formula (Eq. 27-2):

$$\Delta t = \frac{\Delta t_0}{\sqrt{1 - (v/c)^2}}$$

Solving for Δt_0, we find

$$\Delta t_0 = \sqrt{1 - (v/c)^2}\,\Delta t = \sqrt{1 - (0.999)^2}\,(424 \text{ years})$$
$$= 19.0 \text{ years}$$

From the reference frame of the spaceship, the behavior of clocks on board is entirely normal. An observer on the spaceship sees the earth and Antares* moving relative to the spaceship at nearly the speed of light. And yet the trip takes only 19 years. The observer therefore concludes that the distance from earth to Antares is approximately 19 light-years, not 424 light-years. From the reference frame of the spaceship, the distance from earth to Antares contracted from 424 light-years to 19 light-years after the spaceship left the earth and accelerated to a very high velocity.

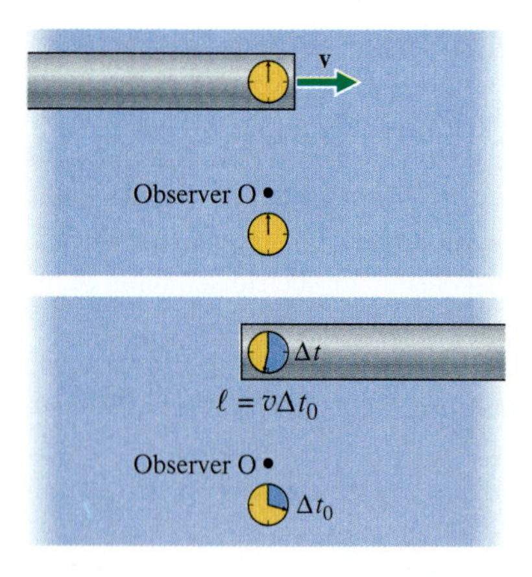

Fig. 27-13 A rod moves to the right at speed v relative to observer O.

The preceding example illustrates that measurement of length, like measurement of time, depends on the reference frame of the observer. We shall now obtain a general formula for length contraction. When the length of a body (or a system of bodies) is measured in the reference frame in which the body is at rest, the measurement can proceed in the usual way—by comparison of the length with a standard, say, a meter stick. We refer to this length, measured in the usual way in the body's rest frame, as a **proper length,** and denote it by ℓ_0.

Now suppose the body is moving parallel to its length, relative to an observer who attempts to measure its length. To be specific, we can think of the body as a rod moving to the right at speed v, relative to an observer O (Fig. 27-13). A very simple way for the observer to measure the rod's length is to first measure the proper time interval Δt_0 elapsed on the observer's clock as the rod passes. The rod's length ℓ must equal the product of the rod's speed and the elapsed time.

$$\ell = v\,\Delta t_0 \tag{27-3}$$

The corresponding time interval Δt, measured on a system of clocks attached to the moving rod, is related to Δt_0 by Eq. 27-2:

$$\Delta t = \frac{\Delta t_0}{\sqrt{1 - (v/c)^2}}$$

or

$$\Delta t_0 = \Delta t\sqrt{1 - (v/c)^2}$$

*Although Antares is not at rest with respect to earth, its velocity is much less than the speed of light, only a few thousand meters per second.

Substituting this expression for Δt_0 into Eq. 27-3, we obtain

$$\ell = v\,\Delta t\,\sqrt{1 - (v/c)^2}$$

From the reference frame of the rod, the observer travels the length ℓ_0 of the rod at speed v in the time interval Δt (Fig. 27-14). Thus

$$\ell_0 = v\,\Delta t$$

Substituting ℓ_0 for $v\,\Delta t$ in the preceding equation, we obtain the length-contraction formula:

$$\ell = \ell_0\sqrt{1 - (v/c)^2} \tag{27-4}$$

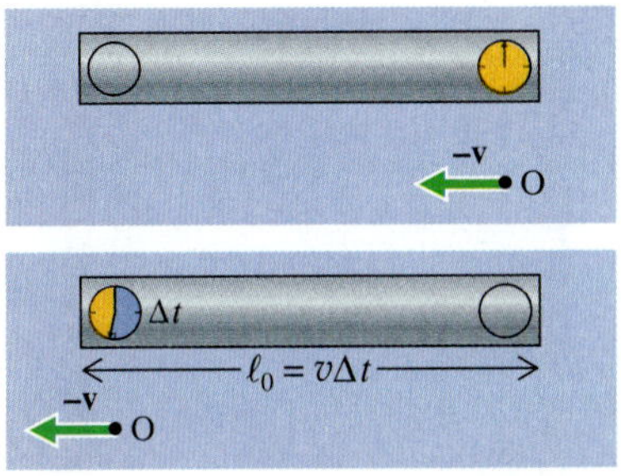

Fig. 27-14 Relative to the rod, observer O moves to the left at speed v.

EXAMPLE 4 Relativistic Contraction of a Meter Stick

Find the length of a meter stick, as measured by an observer, relative to whom the meter stick is moving parallel to its length at a speed of $0.95c$.

SOLUTION Applying Eq. 27-4, we find

$$\ell = \ell_0\sqrt{1 - (v/c)^2} = (1\text{ m})\sqrt{1 - (0.95)^2} = 0.31\text{ m}$$
$$= 31\text{ cm}$$

The length-contraction formula applies not only to the length of a single rigid body, but also to the distance between two bodies at rest (or nearly at rest) in a single reference frame. Although most stars in our galaxy are moving relative to earth at speeds that are quite large by terrestrial standards (on the order of thousands of meters per second), these speeds are quite small compared to the speed of light. Thus the length-contraction formula can be applied to the distance from earth to these stars, as in the following example.

EXAMPLE 5 Relativistic Contraction of an Astronomical Distance

The proper distance between earth and Antares is 424 light-years. Find the distance between the two as seen by an observer traveling between the two bodies at a speed of either $0.900c$ or $0.999c$.

SOLUTION Applying Eq. 27-4, at a speed $v = 0.900c$, we find

$$\ell = \ell_0\sqrt{1 - (v/c)^2} = (424\text{ LY})\sqrt{1 - (0.900)^2}$$
$$= 185\text{ LY}$$

At a speed $v = 0.999c$, we find

$$\ell = (424\text{ LY})\sqrt{1 - (0.999)^2}$$
$$= 19.0\text{ LY}$$

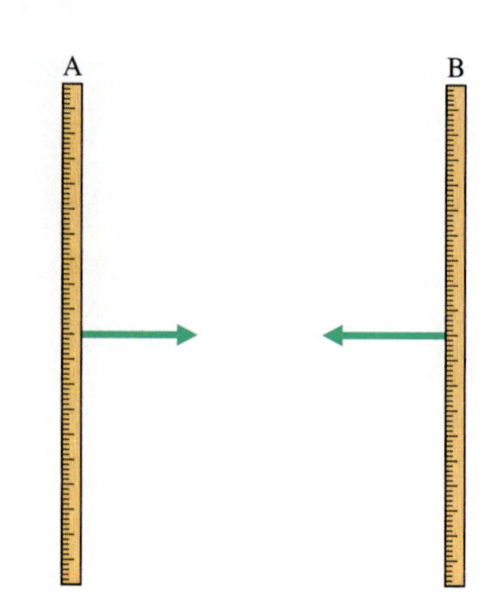

Fig. 27-15 Meter sticks in relative motion.

It is important to note that only lengths parallel to the direction of motion are contracted. Lengths perpendicular to the direction of motion are unaffected by the motion. To see that this must be so, consider two parallel meter sticks, A and B, in relative motion, with a relative velocity directed perpendicular to their lengths (Fig. 27-15). The position of the ends of the meter sticks can be compared as they pass. Observers in the reference frame of either meter stick must agree if the two lengths differ; that is, if A is shorter than B, all would see this contraction of A relative to B. But such contraction would violate the principle of relativity. If relativity predicted that A contracted relative to B, then the same law of contraction, applied in the reference frame of A, would mean that B should be contracted relative to A. Since all observers must agree on which meter stick is shorter, the only possible answer is that neither is shorter. The lengths are the same.

EXAMPLE 6 Observers Moving in Perpendicular Directions

Find the distance from the earth to the moon, as measured by observers O and O′, each of whom are in the spacecraft moving relative to the earth at a speed of $0.80c$, as indicated in Fig. 27-16.

SOLUTION For observer O, the earth-moon system is moving at a velocity of $0.80c$, directed along the line from the earth to the moon, that is, parallel to the length to be measured. We apply Eq. 27-4, and find that observer O measures a length

$$\ell = \ell_0\sqrt{1-(v/c)^2} = (3.8 \times 10^8 \text{ m})\sqrt{1-(0.80)^2}$$
$$= 2.3 \times 10^8 \text{ m}$$

For observer O′ the motion of the earth and moon is to the left, perpendicular to the length to be measured. Therefore O′ measures the same length as in the earth's reference frame:

$$\ell = \ell_0 = 3.8 \times 10^8 \text{ m}$$

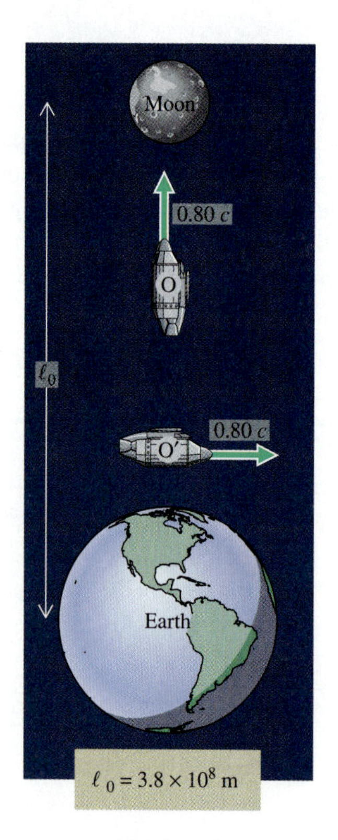

Fig. 27-16

27-4 Relative Velocity

Suppose that an object on board a spaceship moves at a velocity **u**′ relative to the spaceship, which in turn is moving at velocity **v** away from the earth (Fig. 27-17). Relative to the earth, the object in the spaceship is moving away from the earth at a velocity **u**, which depends on the values of **u**′ and **v**. But contrary to what one would intuitively expect, the value of u_x is not equal to $u_x' + v_x$. The rule for determining relative velocities, learned in Chapter 3, is not correct when speeds approach the speed of light. Einstein showed that, in general, u_x is given by

$$u_x = \frac{u'_x + v_x}{1 + \frac{u_x' v_x}{c^2}} \qquad (27\text{-}5)$$

Problem 37 outlines a derivation of this equation. We shall see in the following example that if either u' or v is much less than c, this equation reduces to the usual nonrelativistic result:

$$u_x \approx u_x' + v_x \qquad (\text{if } u' << c \text{ or } v << c)$$

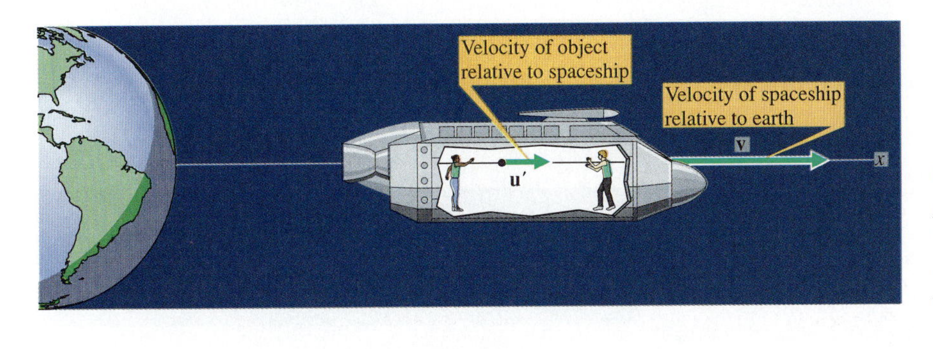

Fig. 27-17 The velocity **u** of the object relative to earth depends on its velocity **u**′ relative to the spaceship and on the velocity **v** of the spaceship relative to earth.

EXAMPLE 7 Addition of Nonrelativistic Velocities

Suppose that a rocket is fired from a spacecraft at a velocity $u_x' = 2.00 \times 10^3$ m/s as the spacecraft moves away from the earth at a velocity $v_x = 3.00 \times 10^3$ m/s. Find the velocity of the rocket, relative to the earth.

SOLUTION Applying Eq. 27-5, we find

$$u_x = \frac{u_x' + v_x}{1 + \frac{u_x' v_x}{c^2}} = \frac{2.00 \times 10^3 \text{ m/s} + 3.00 \times 10^3 \text{ m/s}}{1 + \frac{(2.00 \times 10^3 \text{ m/s})(3.00 \times 10^3 \text{ m/s})}{(3.00 \times 10^8 \text{ m/s})^2}}$$

$$= 5.00 \times 10^3 \text{ m/s}$$

To an excellent approximation, u_x equals the sum of velocities u_x' and v_x because each of these velocities is much less than the speed of light.

The relative velocity formula (Eq. 27-5) applies to light as well as to material bodies, as illustrated in the following example.

EXAMPLE 8 Light From a Moving Spaceship

As a spaceship moves away from the earth at velocity **v**, it emits a pulse of laser light in the forward direction. Show that the velocity of the light, relative to earth, equals c, independent of the velocity of the spaceship, in accord with Einstein's second postulate.

SOLUTION Applying Eq. 27-5, we find

$$u_x = \frac{u_x' + v_x}{1 + \dfrac{u_x' v_x}{c^2}} = \frac{c + v_x}{1 + \dfrac{c v_x}{c^2}} = \frac{c + v_x}{(c + v_x)/c}$$

$$= c$$

EXAMPLE 9 Relative Motion of Two Spacecraft

Spacecraft A and B both approach a planet at half the speed of light, as shown in Fig. 27-18. Find the velocity of B relative to A.

Fig. 27-18

SOLUTION Relative to A, the planet is approaching at half the speed of light (Fig. 27-19). Since the velocity is along the negative x direction, $v_x = \frac{-c}{2}$. To find the velocity of B relative to A, denoted by u_x, we apply Eq. 27-5, using $u_x' = \frac{-c}{2}$ for the velocity of B relative to the planet.

$$u_x = \frac{u_x' + v_x}{1 + \dfrac{u_x' v_x}{c^2}} = \frac{-c/2 - c/2}{1 + \dfrac{(-c/2)^2}{c^2}}$$

$$= \frac{-c}{1 + \frac{1}{4}} = -\tfrac{4}{5}c$$

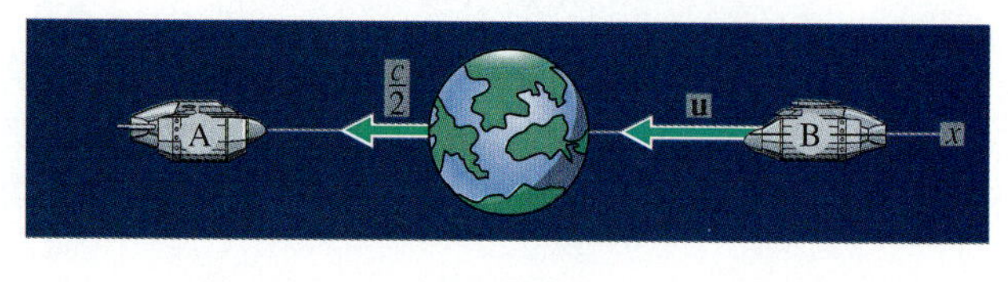

Fig. 27-19

27-5 Relativistic Mass and Energy

We have found that both time and length are relative, not absolute, quantities. It should therefore not be too surprising to find that mass and energy are also relative quantities—that measured values of mass and energy depend on the reference frame of the observer. Einstein used the theory of relativity to derive formulas for the mass and energy of moving bodies. Einstein showed that, if a body has mass m_0 when it is at rest, then when the body moves at a speed v, its mass m is given by

$$m = \frac{m_0}{\sqrt{1 - (v/c)^2}} \tag{27-6}$$

Fig. 27-20 shows how the relativistic mass of a body approaches infinity as its speed approaches the speed of light. Since a body's mass is a measure of its resistance to being accelerated, infinite mass means infinite resistance to acceleration. Thus the closer a body comes to the speed of light, the harder it is to accelerate further. The speed of light therefore is a fundamental speed limit, imposed by nature. A body can come very close to the speed of light but can never reach or exceed it.

From Fig. 27-20 we see that only for very high speeds will a body have a mass that differs significantly from its rest mass m_0. Thus in ordinary applications of mechanics we don't have to be concerned about changing mass. For example, the mass of a car does not change measurably, as it is accelerated from 0 to 50 km/h.

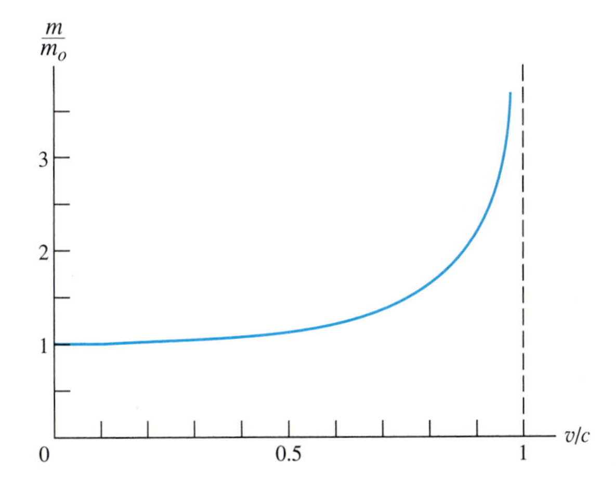

Fig. 27-20 The dependence of the mass of a body on its speed.

EXAMPLE 10 The Mass of a Moving Electron

Find the mass of an electron moving at a speed of $0.999c$.

SOLUTION Applying Eq. 27-6, we find that the electron's mass increases from its rest mass $m_0 = 9.11 \times 10^{-31}$ kg to

$$m = \frac{m_0}{\sqrt{1 - (v/c)^2}} = \frac{m_0}{\sqrt{1 - (0.999)^2}}$$

$$= 22.4m_0 = 22.4(9.11 \times 10^{-31}\ \text{kg})$$

$$= 2.04 \times 10^{-29}\ \text{kg}$$

One of the most interesting predictions of the theory of relativity is that **mass is a form of energy.** This result is expressed by Einstein's famous equation

$$E = mc^2 \tag{27-7}$$

This equation relates a body's total energy E to its relativistic mass m. The two quantities are proportional. The energy of a body equals its mass times a constant (c^2). So it is fair to say that mass and energy are equivalent, or that mass is a form of energy.

In deriving this equation, Einstein united two fundamental principles. Before Einstein's discovery, conservation of mass and conservation of energy were believed to be two unrelated laws of nature. But Einstein showed that mass and energy are proportional, and so conservation of mass implies conservation of energy.

We can express a body's energy in terms of its rest mass m_0 and velocity v by substituting into the preceding equation for energy the expression for relativistic mass (Eq. 27-6).

$$E = \frac{m_0 c^2}{\sqrt{1 - (v/c)^2}} \tag{27-8}$$

When a body is at rest, $m \rightarrow m_0$ and the energy reduces to $m_0 c^2$, which we refer to as **rest energy** and denote by E_0.

$$E_0 = m_0 c^2 \tag{27-9}$$

This equation predicts that a body of moderate mass has an enormous amount of energy. For example, a 1 kg mass at rest has energy $E_0 = (1 \text{ kg})(3 \times 10^8 \text{ m/s})^2 = 9 \times 10^{16}$ J—thousands of times greater than the energy released in an atomic bomb. How is it then that we don't notice the tremendous rest energy stored in ordinary objects? As long as nothing changes, there is nothing to notice. Only when a significant amount of rest energy is converted to some other form is the energy observed. For example, when matter and antimatter meet, both the matter and antimatter are annihilated and rest energy is completely converted to radiant energy. However, this has never been observed except on the level of subatomic particles. For example, an electron and its antimatter particle, a positron, can mutually annihilate, producing two photons with about 10^{-13} J of radiant energy.

The most dramatic large-scale conversion of mass into radiant energy occurs in transformations of nuclei in bombs, as described in Chapter 30. In the explosion of an atomic bomb about one thousandth of the rest energy is converted to radiant energy and heat. Even in ordinary combustion processes, such as the combustion of gasoline, the total rest mass of the products of combustion is slightly less than the original rest mass. The difference in rest mass is used to produce heat. However, the heat released is small enough that the reduction in rest mass is not noticeable, as we shall see in the following example.

EXAMPLE 11 Loss of Rest Mass by Burning Gasoline

How much rest mass is lost during combustion of 1 liter of gasoline? Gasoline has a density of 0.74 kg/L and a heat of combustion of 3.4×10^7 J/L.

SOLUTION The reduction in the gasoline's rest energy equals the heat of combustion. Applying Eq. 27-9, we find an extremely small loss of mass.

$$m_0 = \frac{E_0}{c^2}$$

$$\Delta m_0 = \frac{\Delta E_0}{c^2} = \frac{-3.4 \times 10^7 \text{ J}}{(3.0 \times 10^8 \text{ m/s})^2}$$

$$= -3.8 \times 10^{-10} \text{ kg}$$

The difference between the energy of a moving body and the energy that body would have at rest is defined to be **kinetic energy,** K.

$$K = E - E_0 \qquad (27\text{-}10)$$

Thus

$$K = mc^2 - m_0c^2$$

or

$$K = m_0c^2\left[\frac{1}{\sqrt{1-(v/c)^2}} - 1\right] \qquad (27\text{-}11)$$

This expression for kinetic energy reduces to our previous definition of kinetic energy, $K = \frac{1}{2}mv^2$ (Eq. 7-6), when $v << c$. See Problem 30.

$$K \approx \tfrac{1}{2}m_0v^2 \qquad (\text{when } v << c) \qquad (27\text{-}12)$$

EXAMPLE 12 A Small Relativistic Correction to Kinetic Energy

Find the kinetic energy of a body having a rest mass of 2.0 kg moving at a speed of $0.10c$.

SOLUTION Applying Eq. 27-11, we find

$$K = m_0c^2\left[\frac{1}{\sqrt{1-(v/c)^2}} - 1\right]$$

$$= (2.0\text{ kg})(3.0 \times 10^8\text{ m/s})^2\left[\frac{1}{\sqrt{1-(0.10)^2}} - 1\right]$$

$$= 9.1 \times 10^{14}\text{ J}$$

If we wish to calculate the kinetic energy by applying the approximate classical expression (Eq. 27-12), we obtain

$$K \approx \tfrac{1}{2}m_0v^2 = \tfrac{1}{2}(2.0\text{ kg})(3.0 \times 10^7\text{ m/s})^2$$

$$= 9.0 \times 10^{14}\text{ J}$$

The error in this approximate calculation is only about 1%. Generally when v is no more than $0.1c$, there is little error in using classical formulas.

EXAMPLE 13 Energy of an Accelerated Electron

An electron is accelerated from rest through a potential difference of 1.00×10^6 V, so that it has a kinetic energy of 1.00×10^6 eV, or 1.00 MeV. Find the electron's rest energy, final total energy, and final speed.

SOLUTION Applying Eq. 27-9, we find the rest energy.

$$E_0 = m_0c^2 = (9.11 \times 10^{-31}\text{ kg})(3.00 \times 10^8\text{ m/s})^2$$

$$= 8.20 \times 10^{-14}\text{ J}$$

We can express this result in units of MeV, using the conversion 1 MeV = 10^6 eV = 1.60×10^{-13} J.

$$E_0 = (8.20 \times 10^{-14}\text{ J})\left(\frac{1\text{ MeV}}{1.60 \times 10^{-13}\text{ J}}\right)$$

$$= 0.51\text{ MeV}$$

Next we calculate the electron's final total energy, using Eq. 27-10:

$$K = E - E_0$$

or

$$E = K + E_0 = 1.00\text{ MeV} + 0.51\text{ MeV}$$

$$= 1.51\text{ MeV}$$

Finally we relate energy to speed (Eqs. 27-7 to 27-9).

$$E = mc^2 = \frac{m_0c^2}{\sqrt{1-(v/c)^2}} = \frac{E_0}{\sqrt{1-(v/c)^2}}$$

or

$$\sqrt{1-(v/c)^2} = \frac{E_0}{E} = \frac{0.51\text{ MeV}}{1.51\text{ MeV}} = 0.338$$

Solving for v/c, we find

$$v/c = 0.941$$

or

$$v = 0.941c = 0.941(3.00 \times 10^8\text{ m/s})$$

$$= 2.82 \times 10^8\text{ m/s}$$

General Relativity

Soon after Einstein developed his "special theory of relativity," he became convinced that the principle of relativity should extend beyond the special case of inertial reference frames to any reference frame whatsoever. He then set out to develop a more general theory in which the limitation to inertial reference frames could be removed. Roughly 10 years later he succeeded in formulating such a theory—the general theory of relativity.

Einstein was driven by his belief that the laws of physics should apply to all observers in all reference frames. There seemed to be no obvious reason why those reference frames that are inertial should be so. The singling out of special reference frames seemed to endow space itself with an absolute quality that Einstein did not believe it possessed.

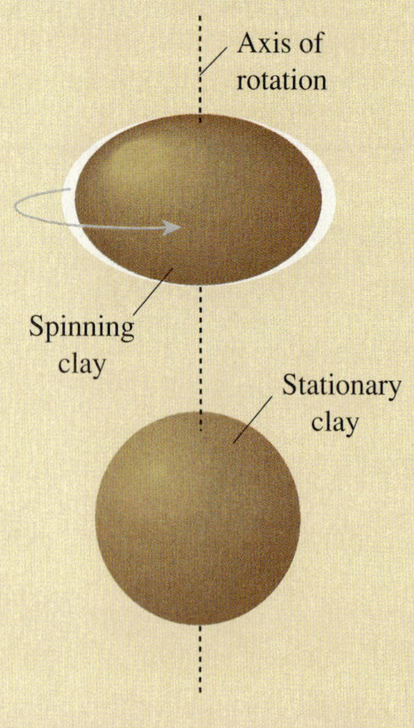

Fig. 27-A Two balls of clay in intergalactic space, viewed from an inertial reference frame. The ball on top spins and bulges outward.

Consider, for example, the following experiment. Two balls of potter's clay are located in intergalactic space, far from all matter, so that nothing exerts force on the balls, which we view from an inertial reference frame. Suppose that one of the balls is spinning about an axis passing through the centers of both balls (Fig. 27-A). The clay in the spinning ball pushes outward, just as it would on a spinning potter's wheel on earth. Thus the spinning ball of clay bulges, while the stationary ball is spherical. (The same effect is seen in spinning planets: the earth bulges slightly at the equator).

Consider the *relative* motion of the two balls. Each is rotating with respect to the other. With this symmetry of motion, why is it that one ball bulges and the other does not? We say that the spherical ball is at rest in an inertial reference frame, while the bulging ball, relative to that inertial reference frame, is spinning and its parts are therefore accelerated. But what is it that makes the reference frame in which the one sphere is at rest an inertial frame? Is it space itself?

Ernst Mach was the first to suggest an answer to this kind of question—an answer that was helpful in guiding Einstein to the general theory. According to Mach, we must look to the distant matter in the universe that we had assumed to have no influence on the balls of clay. Only one of the balls bulges because only one is accelerated with respect to that distant matter. The other ball of clay is at rest (or moving at constant velocity) with respect to the distant stars. Mach's ideas suggested to Einstein that it really is only relative motion that counts after all.

One of Einstein's early insights on the road to discovering the general theory

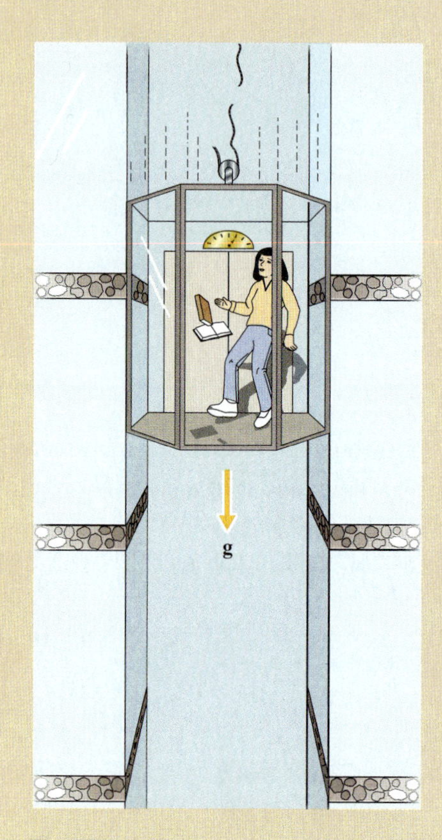

Fig. 27-B To an observer outside a freely falling elevator, a passenger inside is accelerated by the force of gravity. But so long as the elevator continues to fall, it is impossible for the passenger to detect the earth's gravity by any experiment confined to the elevator.

was that the observed effects of gravity depend on the reference frame of the observer. For example, suppose you are in a glass-walled elevator and the cable suddenly snaps, so that both you and the elevator are in free fall (Fig. 27-B). An observer outside the elevator sees you falling freely because there is nothing to support you—no opposing force to balance your weight. But within the elevator (your freely falling reference frame), you see things differently. You do not see

Fig. 27-C Astronauts in training experience weightlessness.

yourself falling relative to the elevator, and you do not "feel" your own weight.* In a freely falling elevator you feel just as weightless as you would in intergalactic space. Astronauts in an earth satellite such as Skylab (Fig. 27-C) experience weightlessness for precisely this reason: they are continuously falling as they orbit the earth. Not only do you feel no weight during free fall, but also it is impossible to detect gravity by *any* experiment confined to a freely falling reference frame. For example, if you drop an object, it falls with you. Relative to the reference frame, it does not move.

Einstein postulated that in any reference frame freely falling in a uniform gravitational field, all the laws of physics would be the same as in an inertial reference frame with no gravitational field present. Einstein claimed that the two reference frames would be completely equivalent. He called this principle the **equivalence principle.**

Einstein claimed further that in an isolated region of space, a reference frame accelerating with respect to an inertial reference frame is completely equivalent to an inertial reference frame with a gravitational field present (Fig. 27-D). In other words, not only can we effectively eliminate a gravitational field by falling in it, we can also effectively create a gravitational field by accelerating in a field-free region of space.

Fig. 27-D Two equivalent reference frames: **(a)** A spaceship accelerating at 9.8 m/s^2; **(b)** A reference frame at rest on earth. Any experiment confined to one or the other of these two reference frames gives exactly the same results, according to the equivalence principle.

*You can experience the feeling of partial weightlessness very briefly even in a functioning elevator, for example, when it is moving up and quickly comes to a stop. The feeling of weightlessness is actually just the absence of the normal feeling of weight we are accustomed to. This feeling of weight is caused by the compression of our bodies' tissue and the pressing of our internal organs against each other. The usual contact and compression are caused by the opposing forces of earth's gravity (your weight) and whatever solid matter such as a chair or a floor that prevents you from falling. In free fall there is no such compression.

Continued.

A Closer Look

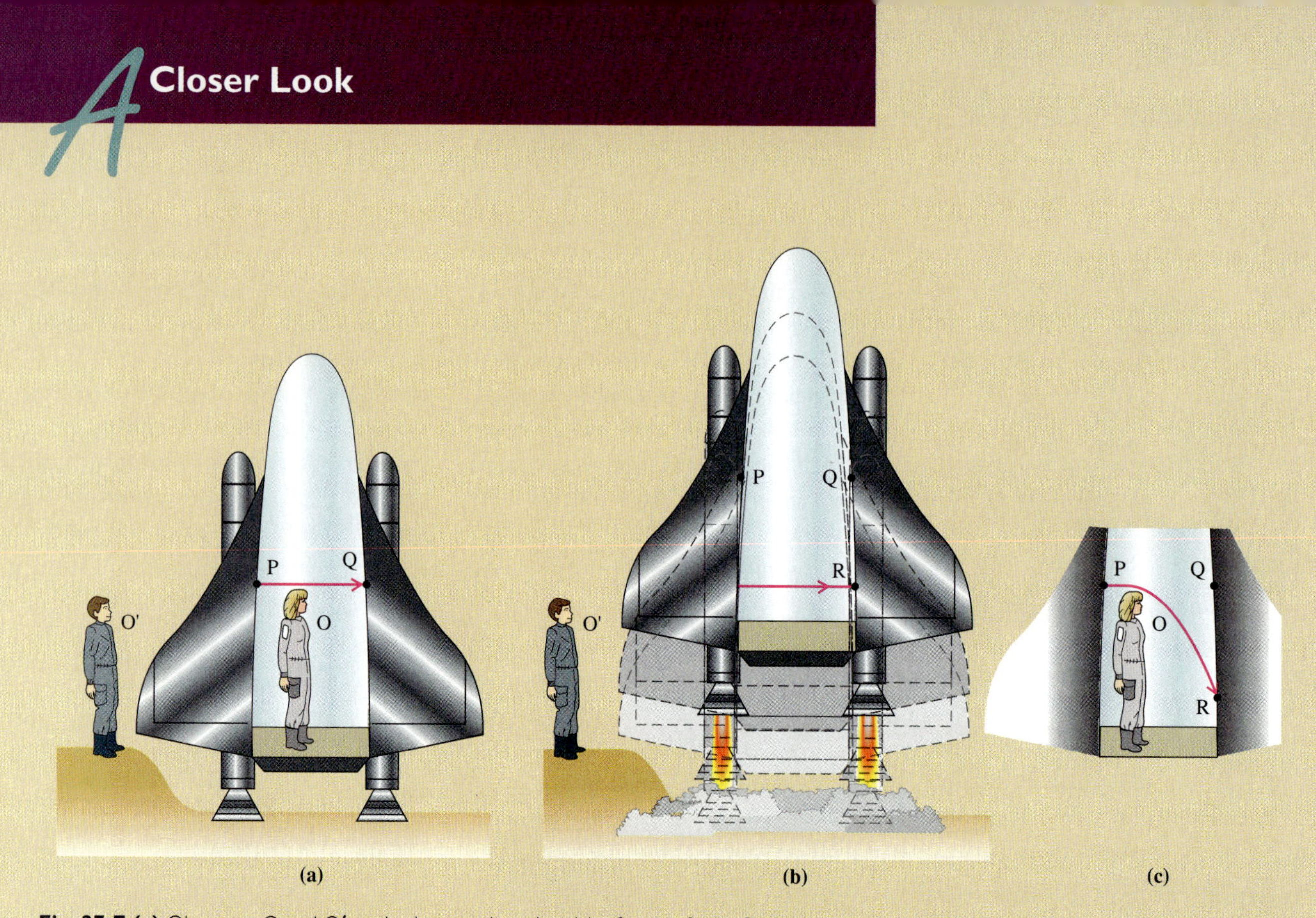

Fig. 27-E (a) Observers O and O′ are both at rest in an inertial reference frame. Both observe a pulse of light travel across the stationary spaceship from P to Q. **(b)** The spaceship accelerates as a laser pulse travels across. The pulse travels in a straight line, as seen by O′, and so strikes a point R on the wall lower than P, since the wall moves up as the pulse travels across. **(c)** As seen by O, the path of the second laser pulse bends downward as it travels across the ship.

Einstein used his equivalence principle as a first step in developing his general theory of relativity. Unforunately, the theory is too mathematically complex to describe in detail. However, we can use the equivalence principle to understand one of the key experimental predictions of the general theory: the effect of a gravitational field on the path of a light ray. Consider the following experiment. A spaceship is initially at rest in an inertial reference frame in a field-free region of space. An observer O inside the ship directs a laser pulse across the ship from point P to point Q (Fig. 27-Ea). A second observer at rest in the same reference frame, but outside the ship, also observes the experiment through the ship's glass walls. Now suppose the experiment is repeated, but this time the ship accelerates upward (Fig. 27-Eb). The path of the light ray, as seen by O′, who remains at rest in the inertial reference frame, must follow the same straight path as before, and must therefore strike the wall of the upward accelerating ship at a point R lower than P, as indicated in the figure. This means that, as seen by observer O, the light bends downward (Fig. 27-Ec), a conclusion that, according to the equivalence principle, is the same whether O is in an accelerating space ship or is at rest in an inertial frame with a downward-directed gravitational field. So we must conclude that the path of a light ray is bent downward by a gravitational field. The effect, however, is much too small to be seen in the earth's gravitational field. Light travels far too fast for it to drop appreciably even after traveling great distances along the earth's surface.

A Closer Look

Einstein predicted that the effect could be seen if you could view a star when the light from it passed very close to the surface of the sun on its way to the earth. The bending of the light toward the sun would cause a change in the apparent position of the star. When the earth is at a point in its orbit around the sun where the earth, sun, and a certain star are approximately aligned so that light from the star passes very close to the sun's surface on its way to the earth, the position of the star relative to other stars appears to change (Fig. 27-F). Of course, under ordinary conditions the sun's light is much too bright for the star's light to be seen then. However, during a total solar eclipse, the sun's light is blocked, permitting the star to be seen. A solar eclipse is a relatively rare and exciting natural phenomenon (Fig. 27-G). Thus great excitement surrounded an expedition of British scientists who set out to observe the solar eclipse on May 29, 1919, to find out if Einstein was right. The deflection predicted by Einstein is small, even though the sun has a gravitational field hundreds of times larger than earth's. Einstein predicted a change in the angular position of the star of only 1.7 seconds of arc, or about 0.0005°. Careful measurement of data and analysis of results were completed months after the expedition. The scientists then announced that light is indeed bent by gravity, as predicted by Einstein, and by precisely the predicted 1.7 seconds. Einstein became an instant celebrity.

Over the years solutions to equations of general relativity have contributed to our fundamental understanding of the universe. For example, based on certain relativistic solutions and some astronomical observations, we now believe that our universe began about 10 billion years ago as a point in space-time, an event referred to as the Big Bang. Another prediction of the general theory is the existence of black holes—stars that collapse under their own gravity and generate fields so intense that nothing can escape, not even light (Fig. 27-H). In May, 1994, improved photographs from the Hubble Space Telescope gave the most convincing evidence yet for the existence of black holes.

Fig. 27-F Light from a distant star bends in the sun's gravitational field on its way to earth, thereby changing its apparent position. An eclipse of the sun by the moon allows this starlight to be seen.

Fig. 27-G An annular solar eclipse, as seen in Eolia, Missouri on May 10, 1994. The earth is somewhat closer to the sun during an annular eclipse than during a total eclipse, and so the sun's disk is not entirely blocked from view.

Fig. 27-H Artist's concept of a black hole.

Albert Einstein

The early life of Albert Einstein, the man who was to change forever our concepts of space and time, gave no hint of genius. Einstein was born in the town of Ulm, in southern Germany, on March 14, 1879, to a family of small-businessmen, not known for great learning or revolutionary outlook. As a child, Albert did not seem gifted. Indeed he was very slow in learning to talk. School was a boring and depressing experience for him. He was a daydreamer who did not accept the attitudes his strict, convention-bound teachers attempted to impart.

Surprisingly, it may have been his reluctance to part with his seemingly childish ways of looking at the world that made possible the eventual blossoming of his genius. Later in life, he was to remark that the main requirement for creativity in science is not knowledge but rather a freshness of vision, a childlike openness in seeing the patterns of nature unfettered by the false and narrow traditional assumptions that hinder most thinking. As he wrote in his autobiography, *The World As I See It:* "We have forgotten what features in the world of experience caused us to frame concepts, and we have great difficulty in representing the world of experience to ourselves without the spectacles

of the old-established conceptual interpretation."

In his teens, Einstein fantasized about questions that would have seemed foolish to most adults—questions such as: "What would the world look like if I rode on a beam of light?" Thanks to an uncle, who aroused the boy's interest in mathematics, Albert began to give his speculations mathematical form. He later continued his education in Switzerland and enrolled at the Swiss Federal Polytechnic School in Zürich. Once again, however, he found formal schooling deadening, and he turned his attention to the original writings of earlier scientists, especially Maxwell, whom he admired. Academically he managed to do just enough to satisfy his instructors as to his basic competence, and he received his diploma in 1900.

After searching without success for a teaching post, he finally took a rather routine job at a patent office in Berne, the Swiss capital. Fortunately, this position left him enough time to begin to work out some of the profound relationships in nature he had glimpsed early on.

In the year 1905, he wrote and published four scientific articles that were to change the science of physics forever. In one of these articles, he established on a sure footing the atomic theory of matter through his explanation of Brownian motion. In another, he carried the earlier insights of Max Planck to new heights with the first formulation of a photon theory of light. As if these two breakthroughs were not enough, in two other papers he dazzled the scientific world with a wholly original theory of mass, energy, and motion—his special theory of relativity—an attempt, as it were, to tell the world about his imaginary boyhood ride on a light beam.

The catalogue of things changed by his theory is impressive. Suddenly, mass and energy were no longer unalterably different kinds of things, but the same thing in different forms, united by the famous equation $E = mc^2$. Suddenly, the familiar world of Galilean mechanics, with its simple additivity of motions, was overthrown, and with it the possibility of absolute frames of reference. The length and mass of an object now depended on the way in which the object moved, especially as its speed approached that of light, which became an upper limit on all speeds. Not even time could be saved from this new universe of relativity. In short, not a single one of the fundamental concepts of physics would ever again mean what it had meant before. Einstein had changed these concepts forever. Scientists who once would have laughed at his imaginary ride on a light beam now had to take that ride along with him—a ride that would leave the familiar world unreachably far behind.

In the decade that followed, Einstein went on to become a professor at the University of Berlin. There, he succeeded in formulating a general theory of relativity. In that theory, he resolved the long-standing puzzle of the equivalence of inertial and gravitational mass. Einstein had formulated a new theory of gravitation that replaced the old established Newtonian theory. Now gravity could be seen in terms of the curvature of space-time.

Throughout the 1920s, Einstein continued his attempts to unify and simplify the laws of physics and in 1921 received the Nobel Prize in physics. He undertook a valiant search, one that would continue throughout the remainder of his life, for a unified field theory that would bring together the fundamental forces of nature—an effort that others continue. Einstein also offered an ongoing critique of quantum mechanics as a complete description of reality. He objected to the Heisenberg uncertainty principle as creating a fuzzy area of indeterminacy within the atom, saying "God does not play dice with the world." Although it resulted in some philosophical refinement of quantum theory, his critique of quantum theory did not fundamentally change it, and is now generally regarded as a misguided effort.

The rise of the Nazis to power in 1933 led Einstein to resign his position in Berlin and to emigrate to the United States. He soon settled at Princeton, New Jersey, and joined the Institute for Advanced Study there. He later became a citizen of the United States and spent the remainder of his life at Princeton, diligently continuing his research and also working toward the cause of peace and human justice until his death in 1955.

Throughout his life, Einstein had been a modest man, unconcerned with fame. He dressed casually and comfortably, even when giving lectures, often wearing an old sweater, house slippers, and no socks. He was loved as much for his simplicity, kindness, and concern for social justice as he was for his monumental intelligence.

Einstein in his garden.

It can be argued that Einstein's scientific contributions might not have been possible without his very human personal qualities—among them, his sense of the oneness of nature, his boyish wonder, his feeling of awe before the beauty of the universe, and his faith in our ability to grasp its workings. He believed that "the eternal mystery of the world is its comprehensibility."

CHAPTER 27 SUMMARY

The theory of relativity is based on two postulates:

I The principle of relativity: All the laws of physics are valid in any inertial reference frame.

II Light always travels through a vacuum at a fixed speed c, relative to any inertial reference frame, independent of the motion of the light source.

A **proper time interval** is a time interval measured on a single stationary clock and is denoted by Δt_0. A time interval Δt can also be measured on a system of synchronized clocks at rest in any inertial reference frame. When a clock is observed moving at high speed v relative to a system of clocks, **the moving clock is found to run slow;** that is, a longer time interval Δt elapses on the system of clocks than the time interval Δt_0 on the moving clock. The two time intervals are related by the time dilation formula:

$$\Delta t = \frac{\Delta t_0}{\sqrt{1 - (v/c)^2}}$$

When the length of a body (or the distance between two bodies) is measured by an observer in the rest frame of the body (or bodies), this is defined to be the **proper length** and is denoted by ℓ_0. If a body (or system of bodies) moves at a very high speed v relative to some reference frame, observers in that reference frame will measure lengths parallel to the direction of motion contracted to a value ℓ related to the proper length ℓ_0 by the equation

$$\ell = \ell_0\sqrt{1 - (v/c)^2}$$

Measurement of distances perpendicular to the direction of motion is not affected by the motion.

The velocity u_x of an object relative to an inertial reference frame is related to the velocity u_x' of the object relative to a second reference frame and the velocity v_x of the second frame relative to the first by the equation

$$u_x = \frac{u_x' + v_x}{1 + \dfrac{u_x' v_x}{c^2}}$$

A body's total energy E is related to its mass m by the equation

$$E = mc^2$$

where the mass varies with speed v from its rest value m_0, as given by

$$m = \frac{m_0}{\sqrt{1 - (v/c)^2}}$$

When a body is at rest ($v = 0$), its mass reduces to m_0 and its energy reduces to the rest energy E_0:

$$E_0 = m_0c^2$$

A body's kinetic energy K is the difference between its total energy E when moving and its rest energy E_0.

$$K = E - E_0 = mc^2 - m_0c^2$$

or

$$K = m_0c^2\left[\frac{1}{\sqrt{1 - (v/c)^2}} - 1\right]$$

Unless v is on the order of $0.1c$ or greater, the preceding expression for K is well approximated by the classical result:

$$K \approx \tfrac{1}{2}m_0v^2 \qquad \text{(when } v << c\text{)}$$

Questions

1 As a child Einstein wondered what he would see in a mirror if he were able to run at the speed of light, holding the mirror at arm's length in front of his face. Would Einstein have seen his reflection in the mirror if he could have run that fast? (We can suppose that he runs at a speed very close to but less than the speed of light, to make this thought experiment more realistic.)

2 Is there any way for an astronaut in a spacecraft moving at constant velocity relative to the earth to determine his velocity, or whether he is moving at all, by means of observations or measurements that are completely confined to the spacecraft?

3 The explosion of a supernova is a rare and spectacular astronomical event. Suppose that an observer on earth witnesses, at nearly the same time, the explosion of two supernovas, A and B, and that supernova A is closer to the earth than B. If A is seen first, can the observer conclude that A exploded before B? If B is seen first, can the observer conclude that B exploded before A?

4 Stationary observers at points A, B, and C witness events that occur simultaneously at points P and P′ (Fig. 27-21). At which of the observation points will the events be seen simultaneously?

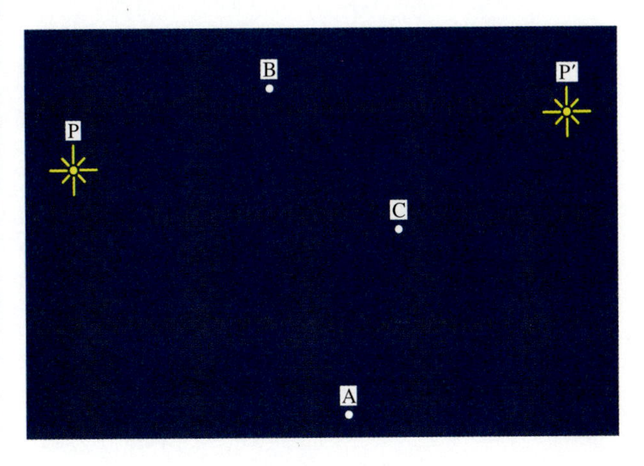

Fig. 27-21

5 Aristotle claimed that motion of the earth was impossible, for if the earth moved it would produce tremendous winds and other effects that could be felt. At a latitude of 40° the surface of the earth moves at a speed of 356 m/s relative to the center of the earth. Why don't we feel a 356 m/s wind?

6 Galileo described the following interesting thought experiment:

> Shut yourself up with some friend in the main cabin below decks on some large ship, and have with you there some flies, butterflies, and other small flying animals. Have a large bowl of water with some fish in it; hang up a bottle that empties drop by drop into a wide vessel beneath it. With the ship standing still, observe carefully how the little animals fly with equal speed to all sides of the cabin. The fish swim indifferently in all directions; the drops fall into the vessel beneath; and, in throwing something to your friend, you need to throw it no more strongly in one direction than another, the distances being equal; jumping with your feet together, you pass equal spaces in every direction. When you have observed all these things carefully (though there is no doubt that when the ship is standing still everything must happen in this way), have the ship proceed with any speed you like, so long as the motion is uniform and not fluctuating this way and that.

Which, if any, of these experiments would be affected by motion on the ship?

7 A stone is dropped from the masthead of a ship moving to the right at constant velocity (Fig. 27-22).
(a) Where will the stone strike the deck of the ship—to the right, to the left, or at the base of the mast?
(b) Describe the path of the stone viewed from the ship's reference frame.
(c) Describe the path of the stone viewed from land.

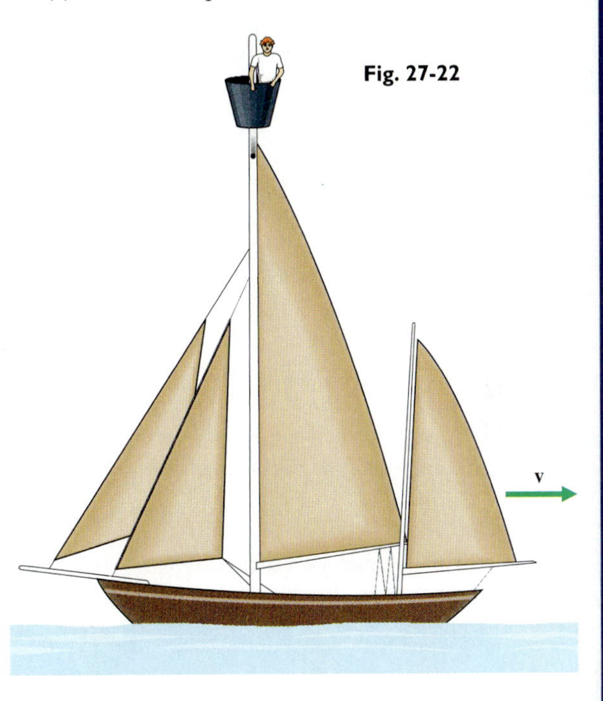

Fig. 27-22

8 Suppose that lightning strikes each end of a train, which is moving in a straight line at a speed comparable to the speed of light (Fig. 27-23). An observer O, riding on the train and located at the midpoint, sees the flashes of light simultaneously and knows that observers at each end of the train would record precisely the same instant for the lightning strokes, using synchronized clocks at those points (A and B). But what about the lightning strokes as seen by observers on the ground? Are they simultaneous, or does one occur before the other?

Fig. 27-23 Lightning strokes hit both ends of a train and are simultaneously seen by observer O at the midpoint of the train.

9 Suppose that the situation described in the preceding question is changed so that lightning strikes B before it strikes A, as seen on the train. Is it possible that, viewed from the ground, the lightning might be seen to strike A before B? In other words is it possible that the order of two events could depend on the reference frame of the observer?

10 Suppose that an alien spacecraft passes over the earth one night, just as lights in two adjacent houses are turned on.

(a) Is it possible that the alien could see one light before the other?

(b) Are simultaneous events occurring at the same point, or nearly the same point, simultaneous in all reference frames?

11 A farmer wishes to store a ladder in a barn that is, unfortunately, shorter than the ladder (Fig. 27-24a). The farmer decides that if he runs with the ladder fast enough it will contract enough so that it will fit inside the barn (Fig. 27-24b). A neighbor says that running will only make the fit worse because, viewed from the ladder's reference frame, it is the barn that contracts (Fig. 27-24c).

(a) Who is right?

(b) Will the farmer be able to store the ladder by quickly closing the barn door while the ladder is contracted inside?

12 Which of the following quantities are conserved in any physical process: mass, total energy, kinetic energy, time, velocity, acceleration?

13 Which of the following quantities have the same value in any inertial reference frame: mass, total energy, kinetic energy, time, velocity, acceleration, speed of light?

14 Two identical blobs of putty of rest mass m_0 collide (Fig. 27-25). After they stick together, is the total rest mass greater than $2m_0$, less than $2m_0$, or equal to $2m_0$?

15 The hydrogen atom consists of a proton and an electron bound together by the attractive electrical force between the opposite charges. To ionize the atom, that is, to remove the electron, one must supply 13.6 eV of energy. Is the rest mass of a hydrogen atom greater than, equal to, or less than the sum of the masses of a proton and an electron?

16 Is there a finite limit on the energy of a moving body?

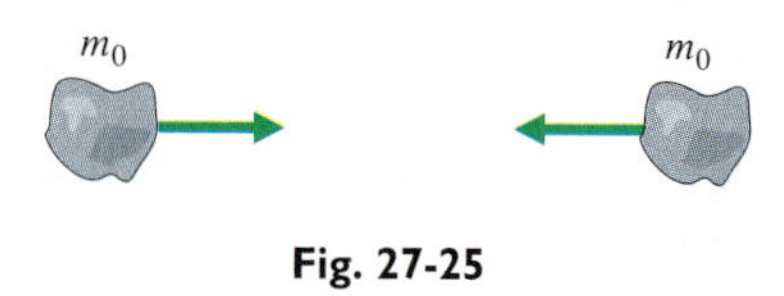

Fig. 27-25

Answers to Odd-Numbered Questions

1 yes; **3** no; yes; **5** the air moves with the earth; **7** (a) at the base; (b) straight down; (c) parabolic trajectory; **9** yes; **11** (a) both; (b) not without breaking it; **13** speed of light; **15** less than

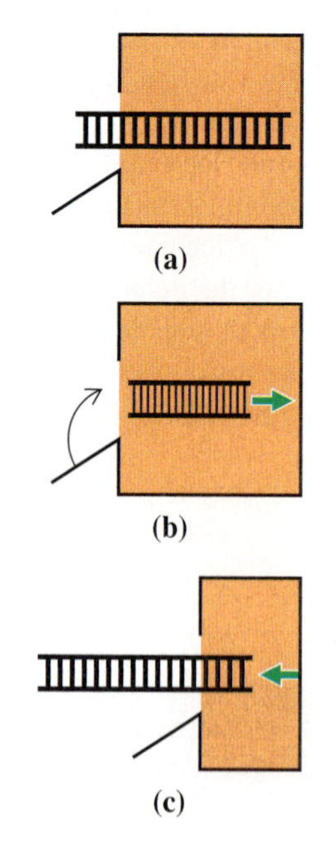

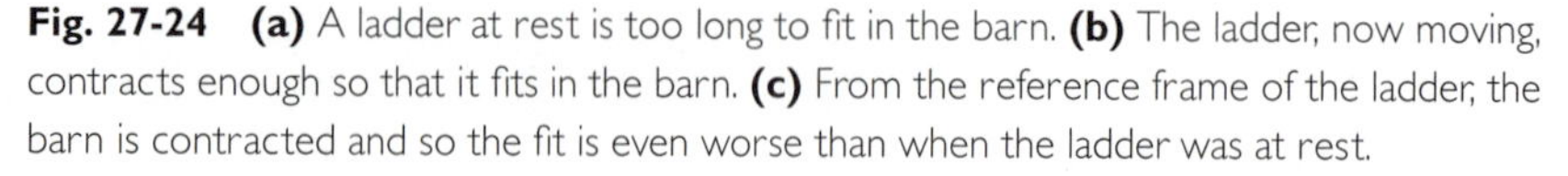

Fig. 27-24 **(a)** A ladder at rest is too long to fit in the barn. **(b)** The ladder, now moving, contracts enough so that it fits in the barn. **(c)** From the reference frame of the ladder, the barn is contracted and so the fit is even worse than when the ladder was at rest.

Problems (listed by sections)

27-1 Measurement of Time; Einstein's Postulates

1 (a) In communicating with an astronaut on the moon, 3.8×10^8 m from earth, what is the minimum time delay in getting a response to a question?
(b) What would the minimum time delay be for communication with an astronaut at a distance of 5 light-years from earth?

2 Suppose you are on Mars when it is 9.00×10^{10} m from earth. You call earth for the correct time and receive a recorded message indicating that the time at the tone is exactly 5:00 P.M. What time should you set on your watch?

★3 Two spaceships are traveling, one in front of the other, in the same direction, at the same speed of 1.50×10^8 m/s relative to the earth. A laser pulse is emitted by one of the ships and is reflected back by a mirror on the other ship. The total round trip time for the laser pulse is 4.00×10^{-3} s. How long would it take for a small landing module to travel between the ships if its speed, relative to the ships, is 1.00×10^3 m/s?

★★4 The Michelson interferometer, described in Chapter 26 (Problems 21 and 22), is a device in which an incident light beam is split by a half-silvered mirror into two beams, which then travel over perpendicular paths before coming together again to form an interference pattern. The interferometer was first used by Michelson and Morley in the hope that it would detect motion of the earth relative to the ether. They assumed that the earth travels relative to the ether at a speed v comparable to the earth's orbital speed of 3.0×10^4 m/s, at least sometime during the year. They also assumed that light travels at a fixed speed c relative to the ether.

(a) Show that it follows from these assumptions that the times required for light to travel along the perpendicular paths shown in Fig. 27-26 are given by

$$t_{\parallel} = \frac{2\ell_{\parallel}/c}{1-(v/c)^2}$$

and

$$t_{\perp} = \frac{2\ell_{\perp}/c}{\sqrt{1-(v/c)^2}}$$

(b) Show that the phase difference between the two waves is given by

$$\Delta\phi = \frac{4\pi}{\lambda}\left[\frac{\ell_{\parallel}}{1-(v/c)^2} - \frac{\ell_{\perp}}{\sqrt{1-(v/c)^2}}\right]$$

(c) Since phase differences could be caused by slight path-length differences, as well as motion through the ether, the apparatus was rotated through 90° so that the roles of the two paths were reversed. Show that this should give a change in phase

$$\Delta\phi' - \Delta\phi = \frac{4\pi(\ell_{\parallel} + \ell_{\perp})}{\lambda}\left[\frac{1}{\sqrt{1-(v/c)^2}} - \frac{1}{1-(v/c)^2}\right]$$

Evaluate the phase change using $\ell_{\parallel} = \ell_{\perp} = 11$ m, $\lambda = 550$ nm, and $v = 3.0 \times 10^4$ m/s. No phase change was observed.

★5 Two race cars are traveling in the same direction, one in front of the other, 340 m apart, at a speed of 70.0 m/s. The driver of the car in front sounds his horn. How much time elapses before the other driver hears the horn?

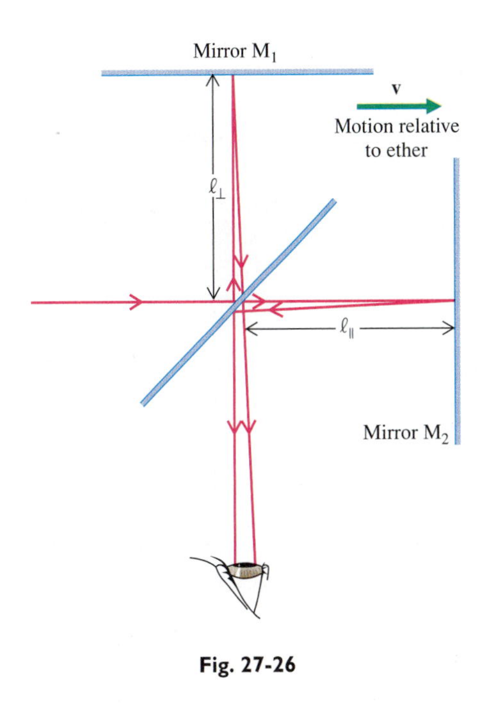

Fig. 27-26

27-2 Time Dilation

6 Clock A travels to the right at a speed of $\sqrt{\frac{3}{4}}c$ relative to clocks B and C. As A passes B, all clocks read 12:00 (Fig. 27-27). These readings could be seen simultaneously by a stationary observer O, midway between B and C, though O would make the observation at some later time.

(a) Clock C reads 6:00 as A passes C. What does clock A read?

(b) At what time does observer O see the three clocks read 12:00?

7 A UFO is observed directly above the earth moving at a constant velocity. An astronaut on the moon (3.8×10^8 m from earth) observes the UFO passing overhead 2.0 s later. From the UFO's reference frame, how long did it take to travel from the earth to the moon?

8 An astronaut with a steady heartbeat of 60 beats per minute travels at a constant speed of $0.90c$, relative to earth. During the voyage his heart beats 5.0×10^6 times. From the earth's reference frame, how many times did the astronaut's heart beat, and what was the heart rate?

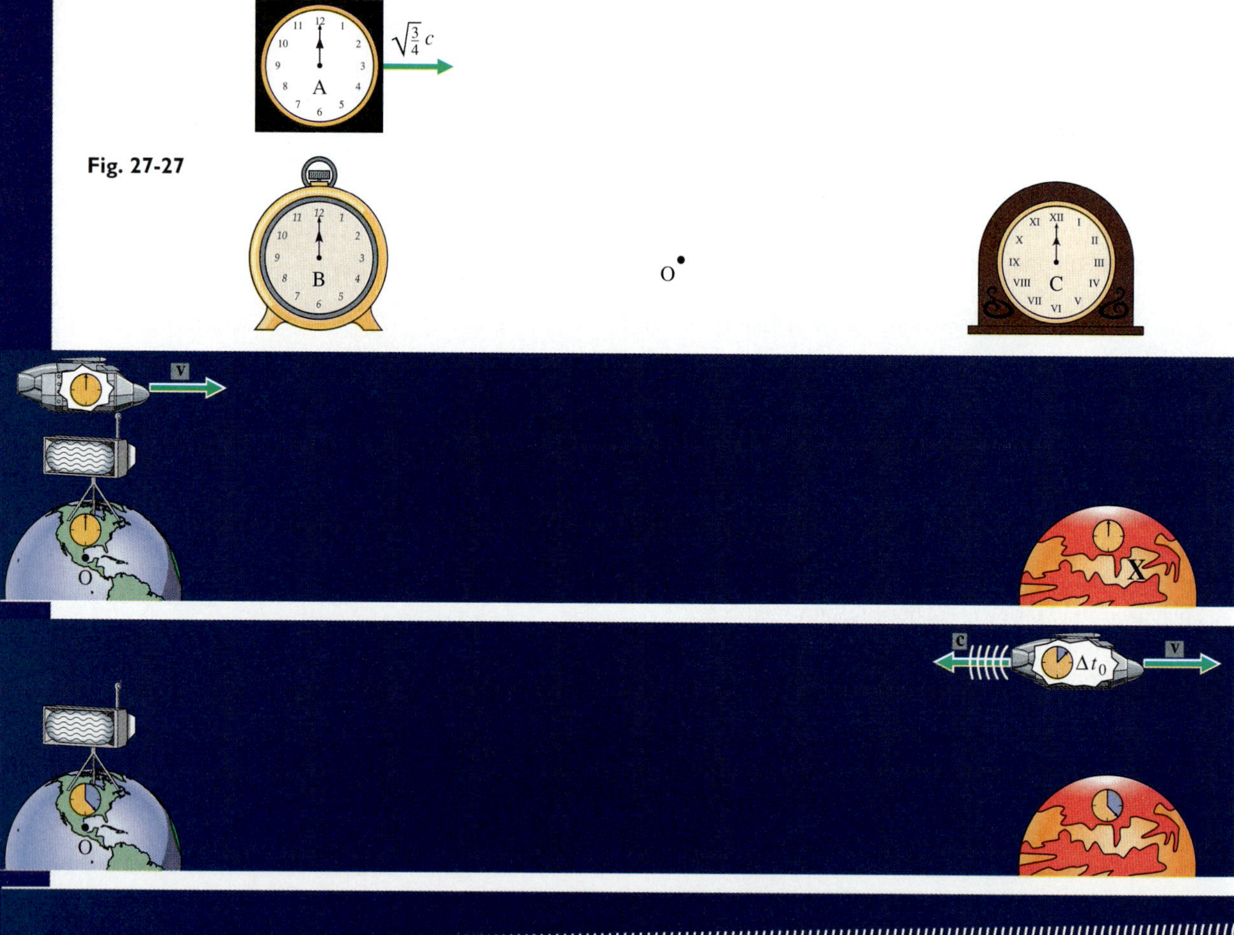

Fig. 27-27

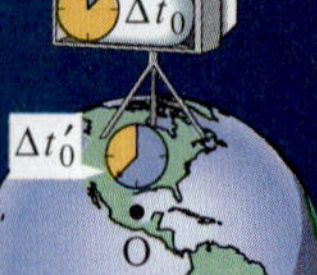

Fig. 27-28 **(a)** A clock on board a spaceship reads the same time time as an earth clock at the instant the ship passes the earth. **(b)** At time Δt_0 later, as measured on the ship, the ship passes planet X, which is at rest with respect to earth and has a clock synchronized with the earth clock. At the instant the ship passes planet X, the ship emits a video signal showing its elapsed time Δt_0. **(c)** At time $\Delta t_0'$ after the ship passes earth, as measured on the earth clock, the video signal from the ship is received on earth and a television monitor shows the ship's time as Δt_0. All 3 images could be seen by an observer in the earth's reference frame, located midway between earth and planet X.

9 After being created in a high-energy particle accelerator, a pi meson at rest has an average lifetime of 2.60×10^{-8} s. Traveling at a speed very close to the speed of light, a pi meson travels a distance of 120 m before decaying. How fast is it moving?

★ **10** An atomic clock is placed on board a high-speed aircraft that travels at a constant speed of 900 m/s for exactly 1 hour, as measured by the clock. How much time is the clock found to have gained or lost when it is compared with a second atomic clock after the flight? (HINT: $(1 - x)^{-1/2} \approx 1 + \frac{1}{2}x$, when $x << 1$.)

★★ **11 Relativistic Doppler effect**

(a) Using Fig. 27-28, show that when a single stationary observer views a clock moving away the time interval Δt_0 that is seen to elapse on the moving clock is related to the time interval $\Delta t_0'$ elapsed on the observer's own clock by the equation

$$\Delta t_0' = \Delta t_0 \sqrt{\frac{1 + v/c}{1 - v/c}}$$

(b) Use this result to show that the observed frequency f_o of a radiation source, moving away from the observer at speed v, is shifted toward the red from the source frequency f_s, according to the equation

$$f_o = f_s \sqrt{\frac{1 - v/c}{1 + v/c}}$$

(c) Astronomers observe that the light from a certain quasar (3C 273) is shifted down in frequency by 15.0%. How fast is this quasar moving?

★★ **12** Pulses of laser light are emitted at 5.00 s intervals from a laser on earth. The pulses are directed at a spaceship, which is moving away from the earth at a speed of $0.850c$. The pulses are reflected from the ship back to earth. Find the time interval measured on earth between reception of the reflected pulses. (HINT: Use the result of the previous problem.)

27-3 Length Contraction

★ **13** The starship *Enterprise* leaves a distant earth colony, which is nearly at rest with respect to the earth. The ship travels at nearly constant velocity toward earth. When the *Enterprise* reaches earth 1.00 y has elapsed, as measured on the ship. Clocks in the earth's reference frame show that the voyage lasted 3.00 y. How far from earth is the colony, as measured in the reference frame of (a) the earth; (b) the ship?

14 How far from the earth would you be able to travel in a lifetime of 80 years if you were able to travel at a constant speed of $0.99c$?

15 How much would you age in traveling a distance of 5.00 light years from earth if you were able to travel at a speed of $0.990c$?

16 A spaceship with a proper length of 100 m passes an observer at a speed of $0.99999c$. What is the length of the ship as seen by the observer?

17 How fast must a meter stick move relative to an observer so that she would measure its length to be 1 cm?

★ **18** Two parallel meter sticks, moving in opposite directions, pass each other. The speed of each meter stick, relative to the other, is close to the speed of light. The entire length of one meter stick passes a single point on the other stick in a time interval of 1.00×10^{-9} s according to a clock fixed at that point. How fast is each meter stick moving relative to the other?

★★ **19** Two parallel meter sticks, moving in opposite directions, pass each other at very high speed. The time interval during which they pass is 1.0×10^{-8} s according to a system of clocks in the reference frame of either stick; that is, during this time interval there is some overlap of the two lengths. How fast is each meter stick moving relative to the other?

27-4 Relative Velocity

20 As spaceship A moves away from the earth at $0.500c$, the captain observes a second spaceship (B) moving in the same direction at a speed of $0.500c$ relative to A. Find the speed of B relative to the earth.

21 Spaceships A and B move in opposite directions at the same speed of $0.5c$ relative to earth, with A moving away from earth and B moving toward earth. Find the velocity of B relative to A.

★ **22** Two spaceships, initially side by side, move away from earth at speeds, relative to earth, of $0.900c$ and $0.990c$ respectively. What is the distance between the ships 10.0 s later, as seen by a passenger on the slower ship?

23 The starship *Enterprise* is fleeing from a pursuing Klingon spaceship at a speed of $0.800c$ relative to the other ship. The *Enterprise* fires a laser pulse as well as a rocket back at the pursuing Klingons. The speed of the rocket, relative to the *Enterprise*, is $0.200c$. Find the velocities of the laser pulse and of the rocket, as seen by the Klingons.

24 An observer sees two spaceships moving in opposite directions, each at a speed of $0.900c$, relative to the observer. Find the speed of one spaceship relative to the other.

★★ **25** A spaceship moving at a speed of $0.500c$ relative to an observer fires a missile. The velocity of the missile relative to the observer is the negative of the missile's velocity relative to the ship. Find the missile's speed relative to the spaceship.

27-5 Relativistic Mass and Energy

26 Electrical energy of approximately 1.00×10^{19} J is used annually in the United States. How much rest mass must be used to generate this much energy if no energy is wasted?

27 Quasars are faint, distant sources of radio waves. (*Quasar* is short for "quasi-stellar source." They are so named because, like a star, they appear to the astronomer to be pointlike.) From the shift in the frequency of their emitted light toward the red, called the "redshift," we know that quasars are moving very fast (see Problem 11). Astronomers observe that the more distant an object is from the earth, the faster it moves. In this way they determine that quasars are billions of light years from earth. To be visible at this great distance, quasars must have enormous luminosity. Typically a quasar radiates energy at a rate on the order of 10^{40} W, roughly 10^{14} times greater than the sun or 40 times greater than the most luminous galaxy. At what rate is rest mass being consumed to produce this much radiation?

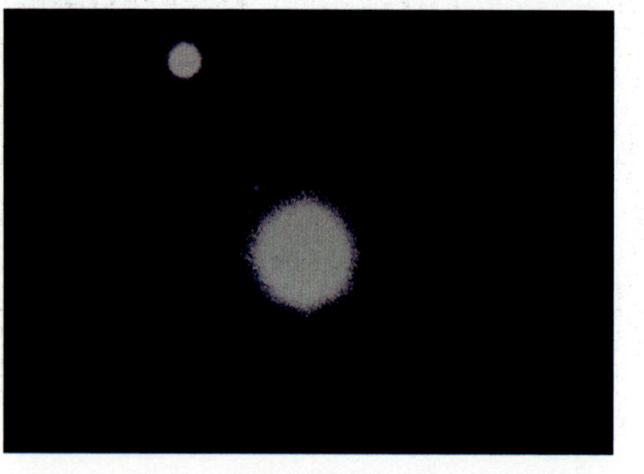

Quasar 3C-273

★ **28** Find the potential difference through which a proton must be accelerated to give it a final speed of $0.990c$, starting from rest.

29 How much does the mass of 1.00 kg of water increase as its temperature is raised from 0° C to 100° C?

★ **30** Show that the relativistic expression for kinetic energy $\left(\text{Eq. 27-11: } K = m_0c^2\left[\dfrac{1}{\sqrt{1-(v/c)^2}} - 1\right]\right)$ reduces to the classical expression (Eq. 7-6: $K = \frac{1}{2}m_0v^2$) when $v << c$. HINT: According to the binomial expansion, $(1 + x)^\alpha \approx 1 + \alpha x$, when $x << 1$.

31 In Example 11 we found that burning 1 liter, or 0.74 kg, of gasoline reduces its rest mass by 3.8×10^{-10} kg. In an atomic bomb, 8×10^{13} J of energy are released for every kg of nuclear fuel undergoing fission—the same energy released in the explosion of roughly 20,000 tons of TNT. Find the ratio of the fractional decrease in mass of the nuclear fuel to the fractional decrease in mass of gasoline when each is burned.

32 Suppose you had as your personal energy source a quantity of matter and antimatter, which, when brought together, annihilated itself, that is, completely converted to radiant energy. What mass of fuel would be required in a lifetime of 80.0 years to provide power at a steady rate of 1.00 kW?

Additional Problems

★ **33** (a) As seen on earth, how long would it take to accelerate a spaceship to a speed of $0.900c$ if the ship could be given a constant acceleration of 9.80 m/s^2 relative to the earth?

(b) The value of an observed acceleration, like velocity, depends on the reference frame of the observer. The spaceship's acceleration a_x', measured relative to a reference frame with respect to which it is instantaneously at rest, is related to its acceleration a_x, measured relative to a frame in which it is moving at speed v, by the equation $a_x' = \dfrac{a_x}{[1-(v/c)^2]^{3/2}}$. Find the maximum acceleration of the spaceship described in (a) as measured in its own reference frame. What does this imply about the length of time (observed on earth) to accelerate a spaceship to $0.900c$?

★34 How far from the earth could you travel in a lifetime of 80.0 years if you were moving at a constant speed of (a) $0.900c$; (b) $0.990c$; (c) $0.999c$?

★★35 A certain spaceship travels so close to the speed of light that while 1.00 s elapses on the spaceship 1.00 y elapses in the earth's reference frame. Find the difference between the speed of light and the speed of the spacecraft, expressed as a fraction of c.

★36 Twin astronauts, Donna and Michelle, leave the earth at the same time, traveling in separate spaceships moving at different (constant) velocities. They reach their common destination at different times. Donna measures the distance she travels to be twice as great as the distance Michelle measures for the length of her own journey. Find the ratio of the time elapsed for Donna to the time elapsed for Michelle, who travels at a speed of $0.900c$.

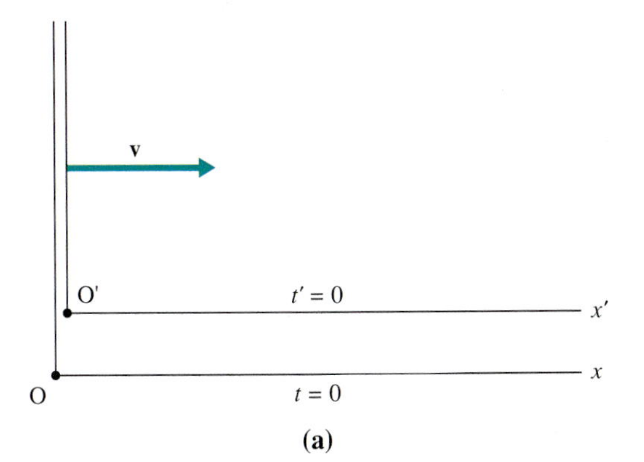

(a)

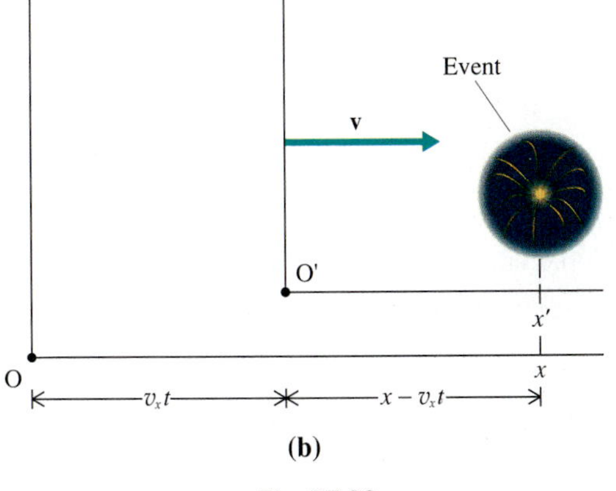

(b)

Fig. 27-29

★★37 This problem outlines the derivation of Eq. 27-5, which relates the values of a body's x component of velocity as measured in two different inertial reference frames. We will refer to one of the reference frames as the "stationary frame." The other reference frame, which we will refer to as the "moving frame," moves relative to the stationary frame along the positive x-axis at speed v. We denote coordinates in the moving frame by primes (x', t', . . .) and denote coordinates in the stationary frame by unprimed letters (x, t, . . .).

Observers at the two frames' origins (O and O′) pass at some instant and agree to define that instant as the zero point of their time measurement; that is, the origins pass at $t = t' = 0$ (Fig. 27-29a). At any later time t, as observed in the stationary frame, the point O′ will have moved a distance $v_x t$ along the x-axis (Fig. 27-29b).

(a) An observer in the stationary frame observes an event at time t at a certain point x. That same event has coordinates x' and t' in the moving frame. Show that the coordinates x, t, and x' are related by the equation

$$x' = \frac{x - v_x t}{\sqrt{1 - (v_x/c)^2}}$$

HINT: The interval from O′ to x' is a proper length in the moving frame, and the measurement of this interval in the stationary frame is the difference in the coordinates of the two end points, $x - v_x t$. See Fig. 27-29b.

(b) Use symmetry to show that

$$x = \frac{x + v_x t}{\sqrt{1 - (v_x/c)^2}}$$

(c) Use the equations found in parts (a) and (b) to show that

$$t = \frac{t' + x' v_x/c^2}{\sqrt{1 - (v_x/c)^2}}$$

(d) Use the definition of velocity, applied in the two reference frames ($u_x = \Delta x/\Delta t$ and $u_x' = \Delta x'/\Delta t'$), and results found in parts (b) and (c), to show that

$$u_x = \frac{u_x' + v_x}{1 + u_x' v_x/c^2}$$

CHAPTER 28 Quantum Concepts

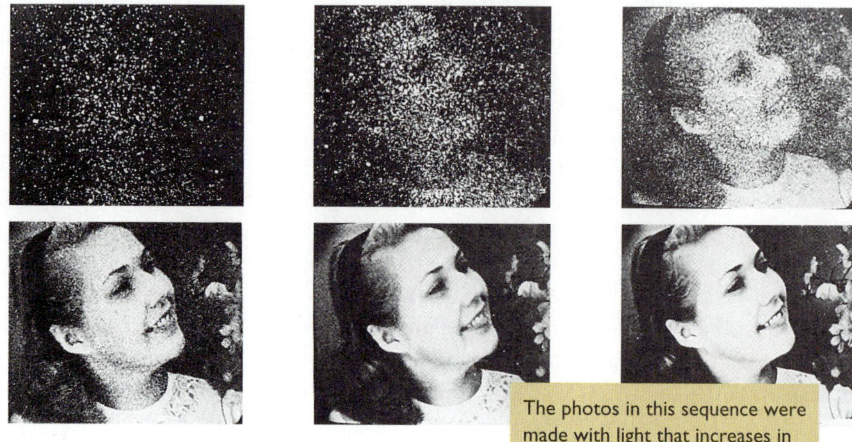

The photos in this sequence were made with light that increases in intensity from one photo to the next. Each spot corresponds to a single photon.

What is light? At the beginning of the twentieth century the answer seemed clear: light is a wave. In the nineteenth century Young and Fresnel had established the wave theory of light as the basis for explaining interference and diffraction, and Maxwell had further developed the description of light as a wave, explaining that it is an electromagnetic wave. Maxwell believed in the ether as the medium through which light waves propagate. In the last chapter we saw how Einstein's theory of relativity eliminated the concept of an ether. Light waves can propagate in a vacuum; no medium is necessary.

Another radical revision in our understanding of the nature of light occurred at the beginning of the twentieth century. Einstein was one of many who contributed to the development of this new theory of light and matter, which was called "quantum theory." It was found that when the energy in a "light wave" is absorbed by a detector (as in the photos above) the energy is localized in space and given up in fixed increments, or "quanta," of energy. A quantum of light is called a "photon." Initially this seemed to be in conflict with the wave theory of light. But eventually it was recognized that light has a dual character—wavelike in propagation (accounting for interference phenomena) but particle-like in interaction or detection. The quantum concept was also applied to subatomic particles, such as electrons. They too were found to have the dual nature of waves and particles. For example, an electron beam directed at a double slit can exhibit the same kind of interference pattern as light. This means that the electrons propagate as waves, not as tiny particles of matter! The study of the quantum theory of light and matter is fascinating, involving concepts that are often strange and counter to our intuition. Perhaps this should not be surprising, since these concepts are used to describe phenomena observed on the subatomic scale—a scale much smaller than that accessible to normal human experience.

Quantum physics has transformed both our understanding of nature and modern technology. Quantum theory explains the periodic table of the elements and provides the physical basis for predicting chemical properties of the elements. Principles of quantum physics led to the invention of the laser, with its striking ability to produce a very narrow beam of monochromatic, high-intensity light. Quantum physics is being used to develop semiconductor devices, which are the heart of powerful microcomputers and other electronic technology. In this chapter we shall introduce basic quantum concepts, and in the next chapter we shall apply quantum physics to atoms, lasers, and semiconductors.

28-1 Photons

Blackbody Radiation; Planck's Constant

An important problem in physics at the end of the nineteenth century was to explain how radiation is emitted by an ideal blackbody radiator. As you learned in Chapter 13, a hot object emits radiation of various wavelengths, and the distribution of those wavelengths depends on the temperature of the body. For example, an object at room temperature emits only invisible infrared radiation, whereas the heating element in an oven, when it reaches a temperature of about 800° C, begins to glow red, emitting, in addition to infrared radiation, shorter-wavelength red light. The still higher temperature of the filament in an incandescent light bulb (about 2000° C) results in the emission of white light—a mixture of red light and shorter wavelengths—yellow, green, and so forth.

Some objects radiate energy more rapidly than others at the same temperature. We found in Chapter 13 that the most effective radiator is an ideal blackbody—a body that absorbs all radiation incident on it. A good approximation to an ideal blackbody is a cavity with a small opening (Fig. 28-1). Radiation entering the cavity is very unlikely to be reflected out.

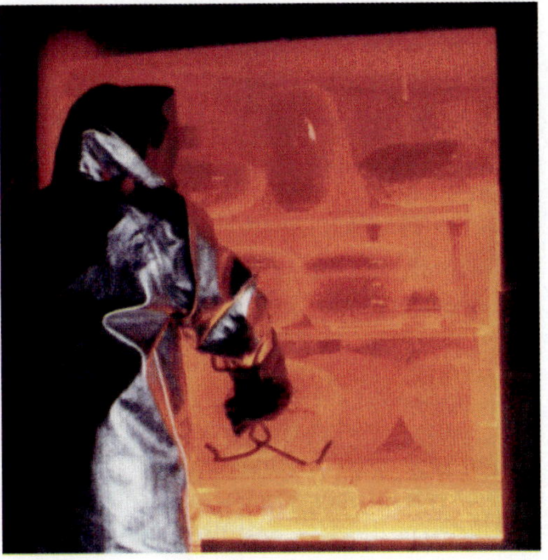

Fig. 28-1 The cavity in this pottery kiln approximates a blackbody radiator. When the kiln is hot, it emits visible light.

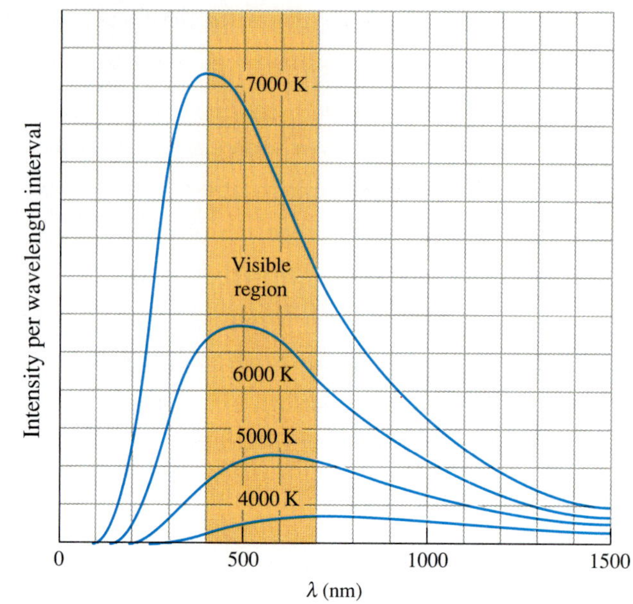

Fig. 28-2 Blackbody radiation.

Nineteenth century physicists carefully measured blackbody radiation. They found that the intensity of the radiation emitted by any blackbody depends only on wavelength and temperature. Fig. 28-2 shows blackbody radiation curves. Attempts to use classical theory to explain these curves were unsuccessful.

In 1900, after years of effort, the German physicist Max Planck was able to derive an equation that fit perfectly the experimental curves. However, Planck found it necessary to modify classical concepts to obtain his results. Classically it was understood that the radiation emitted by a hot body results from oscillations of the atoms in the walls. (Remember, according to Maxwell, the source of all radiation is accelerated charge.) Classical theory predicts that the blackbody radiation of any frequency f arises from atoms oscillating at that frequency. The energy of the atom was supposed to vary continuously with the amplitude of oscillation, and this would determine how much energy was radiated. However, in order for Planck to derive his result, he had to assume that atoms oscillating at any frequency f could only have an energy that was some multiple of a small constant h times the frequency; that is, the possible energies of atomic oscillators were given by $E = 0, hf, 2hf, 3hf, \ldots$, or

$$E = nhf \qquad \text{(for } n = 0, 1, 2, \ldots) \tag{28-1}$$

This means, for example, that it is impossible for an atom oscillating at frequency f to have energy less than hf or to have any value of energy between hf and $2hf$. Atoms can have only discrete values of energy, integer multiples of hf. The quantity h, now known as **Planck's constant,** has the value

$$h = 6.626 \times 10^{-34} \text{ J-s} \tag{28-2}$$

Photoelectric Effect

Five years after Planck's work, Albert Einstein proposed that in blackbody radiation the radiation itself is emitted in discrete packets of energy $E = hf$. Later these "particles" of radiant energy came to be called **photons.**

$$E = hf \qquad \text{(energy of a photon)} \tag{28-3}$$

Since h is a very small number, light of any significant intensity must consist of many photons.

Einstein used the photon concept to explain a phenomenon known as the **photoelectric effect,*** which could not be explained using classical concepts. In the photoelectric effect light incident on the surface of a metal causes electrons to be emitted by the surface (Fig. 28-3). For most metals the light must have a frequency at least in the ultraviolet range. But for some metals, such as sodium, visible light can produce the effect.

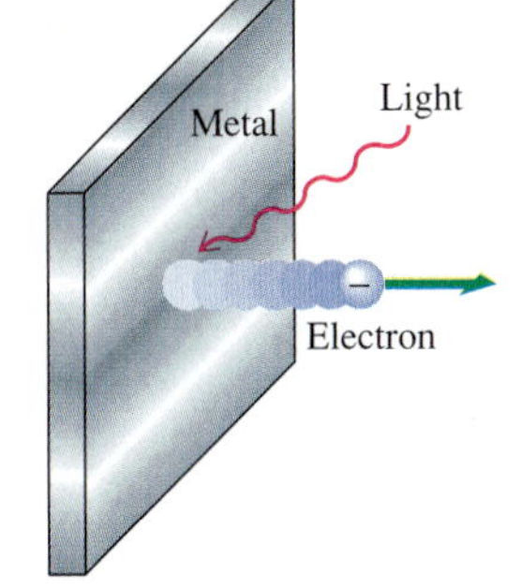

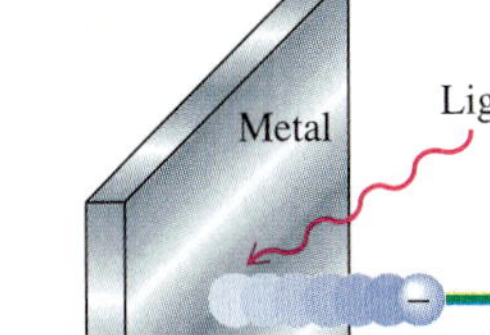

Fig. 28-3 The photoelectric effect.

A photocell is a device that makes use of the photoelectric effect. An evacuated glass tube contains the photosensitive metal, which is connected to one terminal of the tube, and a second metal plate is connected to the other terminal (Fig. 28-4). When light of an appropriate frequency shines on the photosensitive surface, electrons are ejected. If the photocell is connected as part of a complete circuit, current will flow around the circuit only as long as electrons are emitted by the surface. Thus the photocell can act as a light-sensitive switch. For example, a photocell can be used as a motion detector. When someone passes through a beam of light directed at the photocell, current through the cell turns off, activating a magnetic relay, used to turn on an automatic door opener, or perhaps a burglar alarm. A photocell can also be used to turn on electric lights automatically at nightfall.

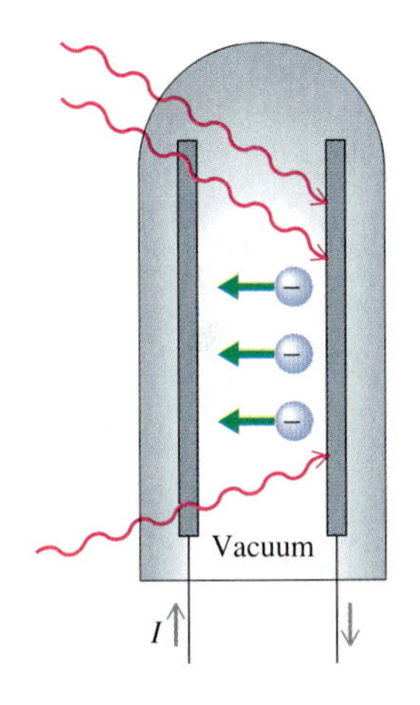

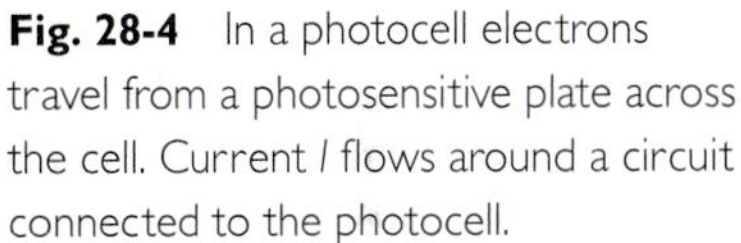

Fig. 28-4 In a photocell electrons travel from a photosensitive plate across the cell. Current I flows around a circuit connected to the photocell.

Some electrons are emitted from a photosensitive surface with a considerable initial velocity. Even if a reverse potential is applied to the terminals, so as to oppose the electrons' motion (Fig. 28-5), some electrons may traverse the tube and a current may still be detected. One can measure the maximum kinetic energy of the photoelectrons by measuring the potential difference V_s required to stop the current. An electron's decrease in kinetic energy from its maximum value K_{max} to zero must equal its increase in potential energy eV_s.

$$K_{max} = eV_s \tag{28-4}$$

On the basis of classical theory, one would suppose that this maximum kinetic energy would depend on the intensity of the light ejecting the photons but not on the frequency of that light. Instead, one finds experimentally that the maximum kinetic energy depends on frequency but not on intensity.

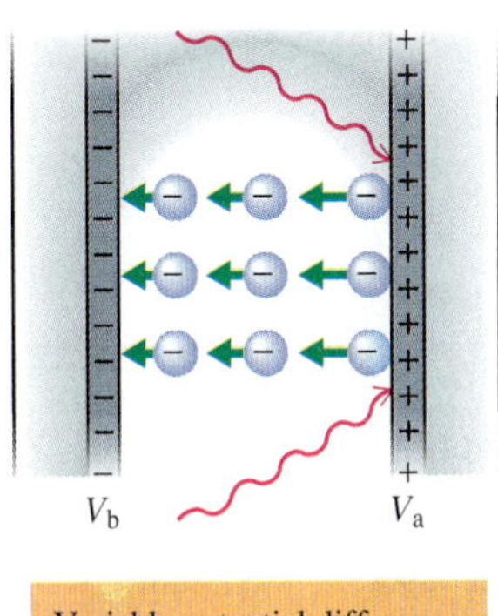

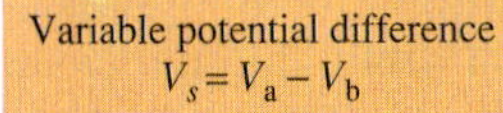

Fig. 28-5 A potential difference that is reversed in sign from the tube's normal operating voltage acts to oppose motion of photoelectrons across the tube. If the voltage is not too great, some current may still be observed.

*Einstein published his papers on the photoelectric effect and the theory of relativity, as well as an important paper explaining Brownian motion, in a single volume of the journal *Annalen der Physik* in 1905. He was later awarded a Nobel prize for his work on the photoelectric effect.

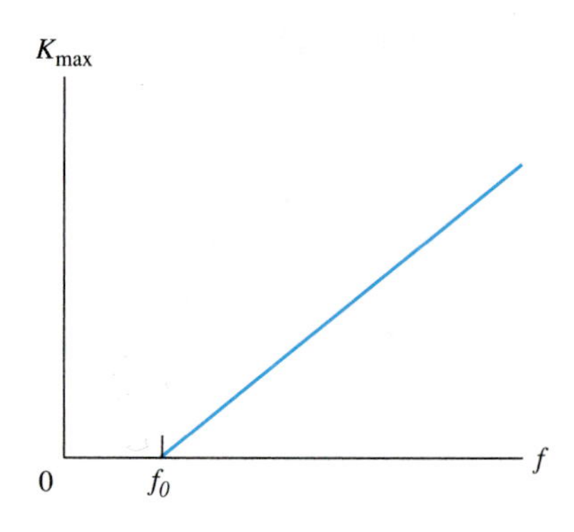

Fig. 28-6 The maximum kinetic energy of photoelectrons varies linearly with the frequency of the incident light.

If the frequency of the incident light is reduced, the maximum kinetic energy of photoelectrons is found to be reduced proportionately, as indicated in Fig. 28-6. Below a certain frequency f_0, the maximum kinetic energy of photoelectrons is reduced to zero—meaning that no photoelectrons are emitted at all.

Einstein was able to explain the photoelectric effect by assuming that the light energy was absorbed by the photoelectrons in discrete quanta, or photons. Each electron absorbs one photon of energy $E = hf$. Some of the energy acquired from the photon must be used to do work against the attractive electrical forces, exerted by the metallic surface; these forces tend to keep electrons on the surface. The minimum work that must be done to liberate the most weakly bound of the surface electrons is called the **work function** and is denoted by W. The maximum kinetic energy a photoelectron can have is the difference between the absorbed photon's energy hf and the work function W.

$$K_{max} = hf - W \tag{28-5}$$

This equation is of a form consistent with the experimental data in Fig. 28-6. Numerous experiments performed by the American physicist Robert Millikan from 1905 to 1915 generated data in complete agreement with Einstein's theory, providing strong support for belief in the reality of light quanta. Different metals were found to have different values for the work function, but for each the graph of K_{max} versus f gave a straight line with the same slope, equal to h, Planck's constant.

EXAMPLE 1 Photoelectric Effect for Sodium

The work function for sodium is 2.27 eV. What is the minimum frequency of light incident on a sodium surface in order that electrons will be emitted?

SOLUTION The minimum light frequency f_0 necessary to emit electrons corresponds to electrons emitted with zero kinetic energy. We set $K_{max} = 0$ in Eq. 28-5:

$$K_{max} = hf_0 - W = 0$$

or

$$hf_0 = W$$

In this case a photon provides barely enough energy to liberate an electron. Solving for f_0 and converting electron volts to joules, we find

$$f_0 = \frac{W}{h} = \frac{(2.27 \text{ eV})(1.60 \times 10^{-19} \text{ J/eV})}{(6.63 \times 10^{-34} \text{ J-s})}$$

$$= 5.48 \times 10^{14} \text{ Hz}$$

This corresponds to yellow-green light of wavelength

$$\lambda = \frac{c}{f_0} = \frac{3.00 \times 10^8 \text{ m/s}}{5.48 \times 10^{14} \text{ Hz}} = 547 \text{ nm}$$

EXAMPLE 2 Laser Photons

How many photons per second are emitted by a 1.0-milliwatt He-Ne laser ($\lambda = 633$ nm)?

SOLUTION The laser produces radiant energy at the rate of 1.0×10^{-3} W, or 1.0×10^{-3} J/s. In one second the laser emits radiant energy $E = 1.0 \times 10^{-3}$ J. The radiation consists of n photons, each with energy hf, where f is the frequency of the laser light.

$$E = nE_{photon} = nhf$$

Solving for the number of photons emitted, we find

$$n = \frac{E}{hf} = \frac{E}{h(c/\lambda)} = \frac{1.0 \times 10^{-3} \text{ J}}{(6.63 \times 10^{-34} \text{ J-s})\left(\dfrac{3.0 \times 10^8 \text{ m/s}}{633 \times 10^{-9} \text{ m}}\right)}$$

$$= 3.2 \times 10^{15}$$

Even this weak, 1 mW source emits an enormous number of photons per second.

Compton Effect

In 1922 the American physicist Arthur Compton performed an experiment that provided further support for Einstein's photon theory. Compton directed a beam of monochromatic X rays* at a graphite target. The atomic electrons in the target absorbed some of the energy of the incident beam, scattering the X rays at various angles (Fig. 28-7). Compton found that the scattered radiation was shifted somewhat in wavelength. This was yet another experiment that classical theory was unable to explain. Classically there should have been no shift in wavelength. Compton used Einstein's photon theory to successfully explain his data. The electrons in the target are struck by a stream of particle-like photons, each of which has a definite energy and momentum. A photon's momentum p is the product of its relativistic mass m and its velocity c.

$$p = mc$$

According to the theory of relativity, $E = mc^2$. Substituting $m = \frac{E}{c^2}$ in the equation above, we obtain

$$p = \frac{E}{c}$$

Expressing the photon's energy E as hf, we find

$$p = \frac{hf}{c}$$

The light's frequency f and wavelength λ are related by the equation $c = \lambda f$, which we can use to substitute for c in the equation above and obtain

$$p = \frac{h}{\lambda} \qquad \text{(momentum of a photon)} \qquad (28\text{-}6)$$

Compton applied the principles of conservation of energy and momentum to the collision of a photon and an electron. An electron receives energy and momentum from a photon, leaving the photon with less energy and momentum, meaning a lower frequency, or longer wavelength. The change in wavelength depends on the scattering angle, the radiation scattered through larger angles corresponding to a greater change in wavelength. As shown in Problem 20, the maximum shift in wavelength, at a scattering angle of 180°, is given by

$$\Delta\lambda = \frac{2h}{mc} \qquad (28\text{-}7)$$

where m is the electron mass (Fig. 28-8).

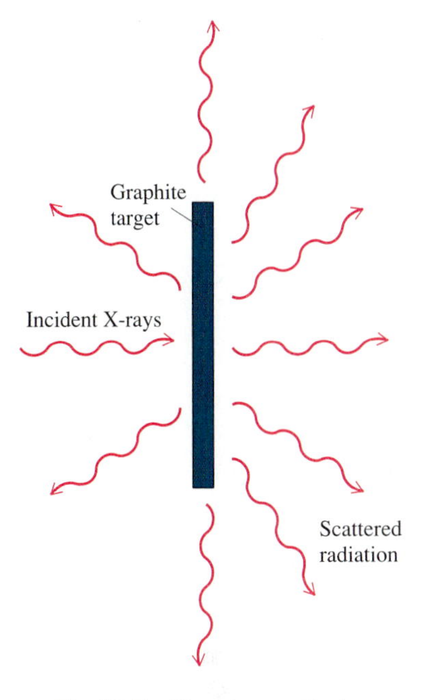

Fig. 28-7 Compton scattering.

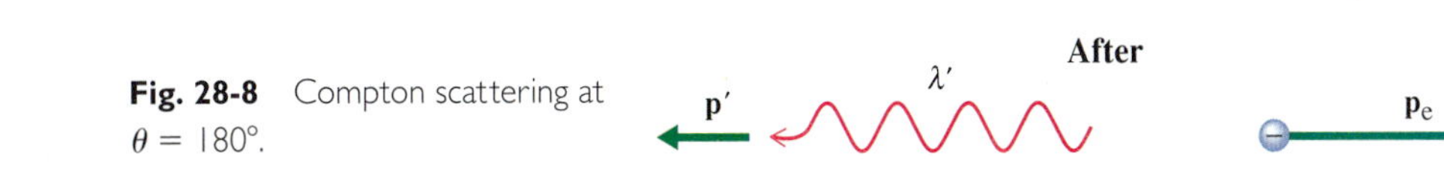

Fig. 28-8 Compton scattering at $\theta = 180°$.

*X rays are high-frequency electromagnetic radiation, which will be discussed in greater detail at the end of Chapter 29.

EXAMPLE 3 Energy and Momentum of an Electron After Compton Scattering

Find the energy and momentum absorbed by an electron in Compton's experiment when the incident X rays ($\lambda = 0.0710$ nm, $f = 4.23 \times 10^{18}$ Hz) are scattered through 180°.

SOLUTION Using Eq. 28-7, we find a wavelength shift

$$\Delta\lambda = \frac{2h}{mc} = \frac{2(6.63 \times 10^{-34} \text{ J-s})}{(9.11 \times 10^{-31} \text{ kg})(3.00 \times 10^{8} \text{ m/s})}$$

$$= 4.85 \times 10^{-12} \text{ m}$$

This gives a final wavelength

$$\lambda' = \lambda + \Delta\lambda = 7.10 \times 10^{-11} \text{ m} + 4.86 \times 10^{-12} \text{ m}$$

$$= 7.59 \times 10^{-11} \text{ m}$$

which corresponds to a final frequency

$$f' = \frac{c}{\lambda'} = \frac{3.00 \times 10^{8} \text{ m/s}}{7.59 \times 10^{-11} \text{ m}} = 3.95 \times 10^{18} \text{ Hz}$$

The scattered photon has changed its energy by

$$\Delta E = hf' - hf = h\,\Delta f$$

$$= (6.63 \times 10^{-34} \text{ J-s})(3.95 \times 10^{18} \text{ Hz} - 4.23 \times 10^{18} \text{ Hz})$$

$$= -1.86 \times 10^{-16} \text{ J}$$

This means that the electron has absorbed energy of 1.86×10^{-16} J. We can find the electron's momentum by applying the principle of conservation of momentum and equating the initial and final total momentum of the system. Using Fig. 28-8, with the notation indicated there, we have

$$\mathbf{p}_e + \mathbf{p}' = \mathbf{p}$$

or

$$p_e - p' = p$$

$$p_e = p + p' = \frac{h}{\lambda} + \frac{h}{\lambda'} = h\left(\frac{1}{\lambda} + \frac{1}{\lambda'}\right)$$

$$= (6.63 \times 10^{-34} \text{ J-s})\left(\frac{1}{7.10 \times 10^{-11} \text{ m}} + \frac{1}{7.59 \times 10^{-11} \text{ m}}\right)$$

$$= 1.81 \times 10^{-23} \text{ kg-m/s}$$

28-2 Wave-Particle Duality

When Einstein proposed that light was emitted or absorbed as particle-like quanta, or photons, he was fully aware of the phenomena that the wave theory of light explains. The explanation of interference and diffraction phenomena given in Chapter 26 was based on the wave theory of light. For example, in Section 26-2 we discussed Young's double-slit experiment, in which light waves passing through two slits interfere constructively or destructively, producing a pattern of light and dark areas on the screen (Fig. 28-9). Light was described as an electromagnetic wave. The intensity of the wave at any point was found to be proportional to the square of the amplitude of the electric field at that point.

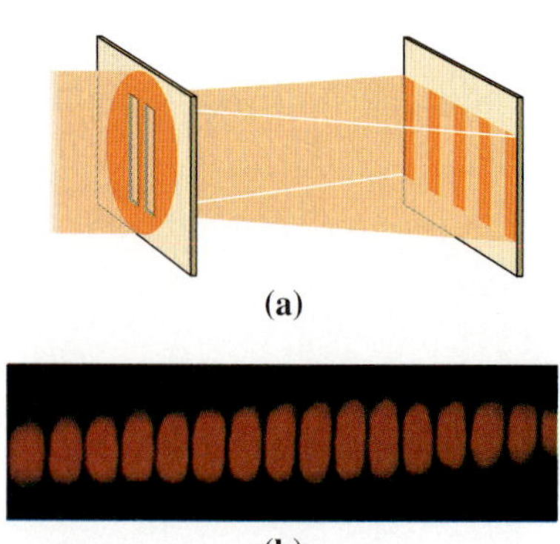

Fig. 28-9 **(a)** Overlapping wavefronts interfere constructively at certain points and destructively at others. (Figure is not drawn to scale.) **(b)** Photograph of interference fringes from double slits illuminated with a He-Ne laser.

How can we reconcile the wave theory of light with Einstein's concept of photons? A resolution of this problem took many years and involved the efforts of many scientists. One of the problems that had to be overcome was the natural human tendency to cling to mental pictures derived from ordinary experience, for example, to think of light as either little bullets shot out by a light source, or to picture it as similar to a water wave spreading out from the source. Light is like neither, though it has aspects of both. Let us examine the double-slit experiment in greater detail to arrive at a better understanding of the nature of light.

Imagine performing the double-slit experiment with light of such a low intensity that only one photon at a time passes through the slit and strikes the screen. Suppose that a very fine-grained photographic film is placed on the screen so that a photon is absorbed by a grain of the film at a very localized spot on the screen, essentially a point. The detection of a photon at a single point (Fig. 28-10a) bears no resemblance to the interference pattern that we get with light of greater intensity. If we continue the experiment with very low-intensity light, we see more photons absorbed on the screen. Their locations seem at first to be random (Fig. 28-10b). But gradually the dots of light begin to resemble the familiar interference pattern (Fig. 28-10d).

This experiment reveals the relationship between photons and the smooth, continuous interference pattern predicted by wave theory. **The probability of detecting a photon at any point is proportional to the intensity of the light wave at that point.** Photons are most likely to be detected near the center of the bright fringes, where the light wave has greatest intensity. No photons are detected in the dark areas, where the intensity of the light wave is zero.

This experiment would seem to indicate that we should regard light as consisting of a large number of very small particles (called "photons") and that a "light wave" is nothing more than a way of describing how the photons are distributed as they move through space. This is certainly a tempting interpretation of the experiment we have just described. However, it is wrong, as we shall now show.

Suppose we insist on thinking of light as really consisting of tiny particles. Now suppose we perform the double-slit experiment with light of low enough intensity that only a single photon passes through the slits in a given time interval. Then we would picture the light as a single particle (photon) following some trajectory on its way to the screen (Fig. 28-11). This means that the particle must travel through one slit or the other, though we don't know which one if we simply look at the spot of light on the screen.

On the basis of this picture, we should be able to get the double-slit interference pattern by first closing slit 2 and letting photons go only through slit 1 (one at a time) and then closing slit 1 and letting the photons go only through 2. After all, we are assuming that each photon goes through only one of the slits. If we do the experiment this way, however, we no longer have the same results. We no longer see interference fringes. Instead we see two overlapping single-slit diffraction patterns—a broad bright band of light in the central area (Fig. 28-12). There are no dark fringes within this band because it is no longer possible for two light waves to pass simultaneously through the two slits and cancel each other out on the screen. We have changed the experiment, and so we should not be too surprised that the results are different.

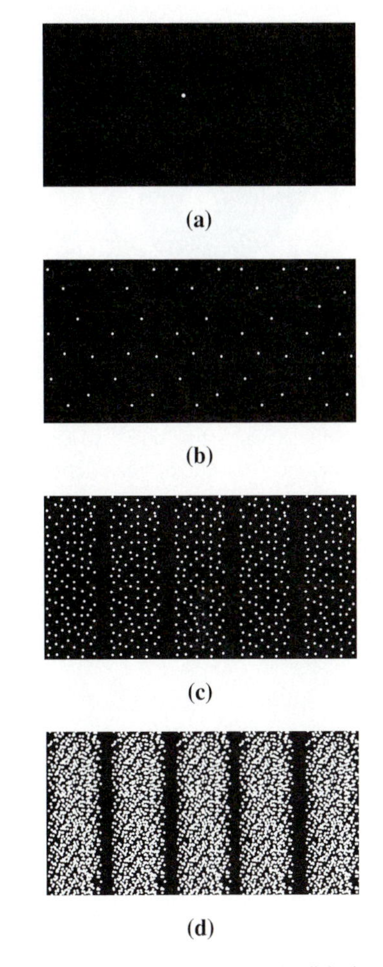

Fig. 28-10 Very low intensity light is used in the double-slit experiment so that the absorption of individual photons can be detected on the film. This series of simulated photos shows first one photon, then 50, then 500, and finally 5000 photons.

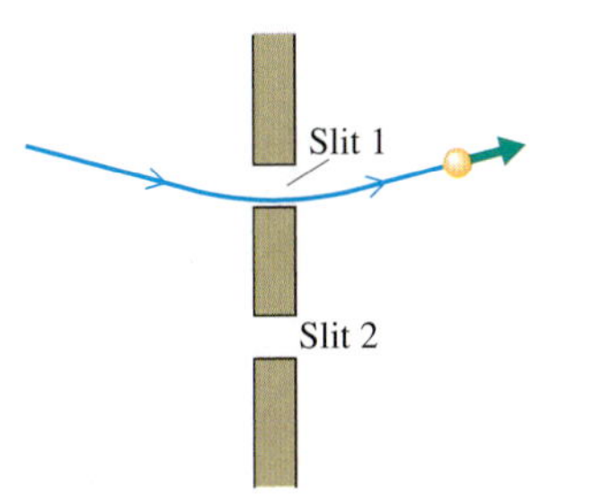

Fig. 28-11 Supposed trajectory of a hypothetical light particle. There is no such trajectory or particle.

Fig. 28-12 Central part of diffraction pattern.

Fig. 28-13 The Chinese symbol for *yin* and *yang* expresses the unity of opposites in nature, for example, male and female. This symbol can be used to express the wave-particle duality of light and was incorporated by Niels Bohr, one of the founders of quantum theory, into his family coat of arms.

Thus we are led to the conclusion that even when only one photon strikes the screen in the double-slit experiment there is a light wave going through both slits simultaneously. The light behaves as a wave as it propagates through space. It is only when it is emitted or absorbed that the light behaves as a particle. Light has characteristics of both waves and particles. **Light propagates as an electromagnetic wave. The intensity of the wave at any point gives the probability of detecting the quantum particle, or photon, at that point, if one performs an experiment designed to detect the particle.**

Wave properties and particle properties of light are related in the equations for the energy and momentum of a photon (Eqs. 28-3 and 28-6):

$$E = hf$$

$$p = \frac{h}{\lambda}$$

Although f and λ are wave properties, E and p are respectively the energy and momentum of a single, particle-like photon.

Fig. 28-14 A fantasy "quantum skier."

EXAMPLE 4 Number of Photons in a Diffraction Pattern

Fig. 28-15 shows the graph of intensity versus position on a screen displaying the full diffraction pattern of a single slit. (Only the central part of the pattern was shown in Fig. 28-12.) Find the ratio of the number of photons found in a 1 mm interval at the center of the central maximum (red section) to the number found in an equal interval in the secondary maximum (green section).

SOLUTION The probability of detecting a photon at a point is proportional to the value of the intensity at that point. Thus the ratio of the numbers of photons in the two intervals equals the ratio of the corresponding intensities.

$$\frac{n_1}{n_2} = \frac{I_1}{I_2} = \frac{2.10 \text{ W/m}^2}{0.10 \text{ W/m}^2} = 21$$

For every photon detected in the secondary maximum, 21 photons are found in the central maximum. We assume that a large enough number of photons have been detected, so that the ratio of the measured numbers accurately reflects the ratio calculated above.

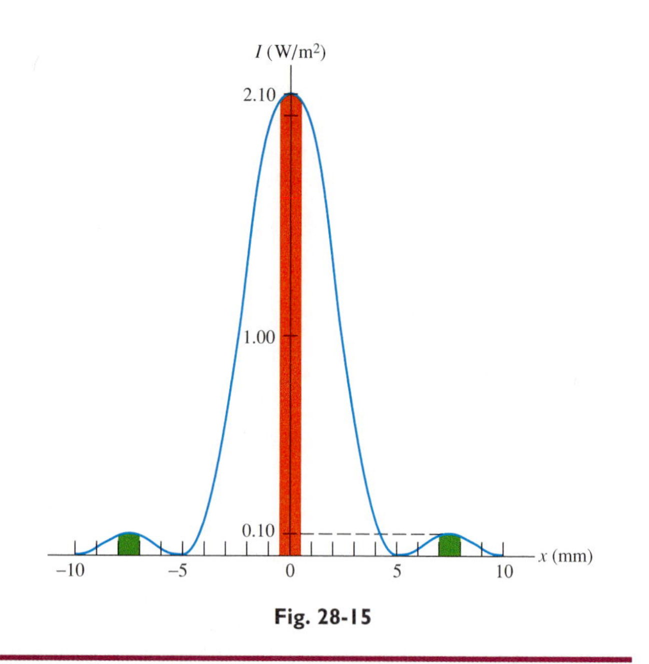

Fig. 28-15

28-3 The Uncertainty Principle

In any light wave there is always a range of frequencies and wavelengths present, a "bandwidth," as discussed in Chapter 26. Even laser light is not perfectly monochromatic. The uncertainty in frequency and wavelength implies an uncertainty in the energy and momentum of photons associated with the wave, since $E = hf$ and $p = \frac{h}{\lambda}$.

We shall show that the uncertainty in photon momentum is related to the uncertainty in the location at which a photon is emitted or absorbed. We shall also see that the uncertainty in photon energy is related to the uncertainty in the time at which a photon is emitted or absorbed.

Wave Packets

Suppose a single photon is emitted by a light source. The light wave associated with this photon typically has a very narrow bandwidth Δf. Although we cannot supply the proof here, it is possible to show that a wave consisting of a continuous range of frequencies gives a localized wave, called a "wave packet,"* illustrated in Fig. 28-16. The various frequency components of the wave cancel each other out, except over a limited interval Δx. This interval represents the uncertainty in the photon's position. If we try to measure the position of the photon, it is sure to be found somewhere in the interval Δx.

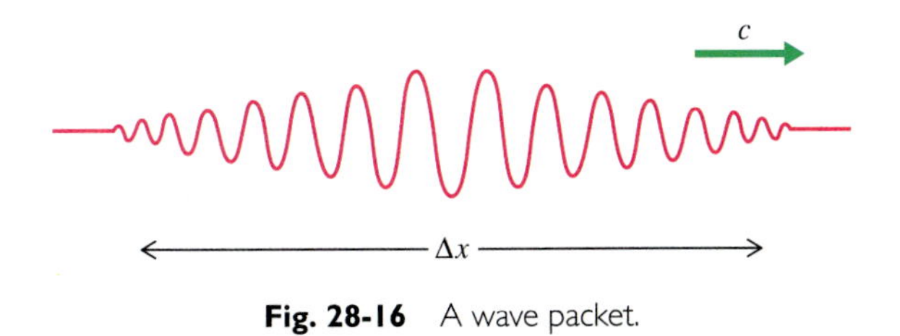

Fig. 28-16 A wave packet.

Uncertainty in Position and Momentum

There is a predictable relationship between the phases at two points in the wave packet if they are within the wave's coherence length $x_c = \frac{c}{\Delta f}$ (Eq. 26-2). Since there can be no phase relationship between a point in the wave packet and a second point outside the packet, the coherence length must certainly be less than the length of the wave packet, Δx:

$$x_c = \frac{c}{\Delta f} < \Delta x$$

or

$$\Delta x > \frac{c}{\Delta f}$$

*To get some idea of how superposition of waves of different frequency can give rise to wave packets, inspect Fig. 16-34, which shows the addition of two frequencies in the formation of beats.

The uncertainty in frequency means there is also uncertainty in wavelength and momentum. Since $p = \frac{h}{\lambda} = \frac{hf}{c}$,

$$\Delta p = \frac{h\,\Delta f}{c}$$

Multiplying the preceding expressions for Δx and Δp, we find

$$\Delta x\,\Delta p > h$$

The product of the uncertainty in the photon's position and the uncertainty in the photon's momentum is always greater than Planck's constant. Werner Heisenberg, one of the founders of quantum theory, was able to show with a more detailed analysis that the product of uncertainties in measurements of the x coordinate and corresponding component of momentum is always greater than or equal to $h/2\pi$:

$$\Delta x\,\Delta p_x \geq \frac{h}{2\pi} \tag{28-8}$$

This result is known as the **Heisenberg uncertainty principle.** We shall see in the next chapter that it is applicable not only to light, but also to matter, for example, electrons. Indeed it was for matter waves that Heisenberg first derived it.

EXAMPLE 5 Uncertainty in Position of a Photon

The momentum of an X-ray photon scattered at a certain angle in a Compton scattering experiment is found to have an x component of momentum of 9.00×10^{-24} kg-m/s, with an uncertainty of 1.00%. What is the minimum uncertainty in the photon's x coordinate?

SOLUTION Applying the Heisenberg uncertainty principle, we find

$$\Delta x\,\Delta p_x \geq \frac{h}{2\pi}$$

$$\Delta x \geq \frac{h}{2\pi\,\Delta p_x} = \frac{h}{2\pi(0.0100)p_x}$$

$$= \frac{6.63 \times 10^{-34}\text{ J-s}}{2\pi(0.0100)(9.00 \times 10^{-24}\text{ kg-m/s})}$$

$$\Delta x \geq 1.17 \times 10^{-9}\text{ m} = 1.17\text{ nm}$$

The minimum uncertainty in this photon's x coordinate is 1.17 nm.

Uncertainty in Energy and Time

A wave packet of length Δx moves at the speed of light, c. So at a given point in space there is a wave present only for some time interval $\Delta t = \frac{\Delta x}{c}$. If a photon is to be detected, its time of detection is uncertain by Δt.

There is a predictable relationship between the phases of the wave at any two times if these times are within the wave's coherence time $t_c = \frac{1}{\Delta f}$ (Eq. 26-1). This coherence time must certainly be less than or equal to the wave packet's time of passage, Δt.

$$t_c = \frac{1}{\Delta f} < \Delta t$$

or

$$\Delta t > \frac{1}{\Delta f}$$

The uncertainty in frequency implies an uncertainty in photon energy, ΔE. Since $E = hf$,

$$\Delta E = h\,\Delta f$$

Multiplying the two preceding expressions, we find

$$\Delta E\,\Delta t > h$$

The product of the uncertainty in the photon's energy and the uncertainty in the photon's time of detection is always greater than h. Heisenberg showed with his more careful analysis that

$$\Delta E\,\Delta t \geq \frac{h}{2\pi} \tag{28-9}$$

This is the second form of Heisenberg's uncertainty principle. It applies to any kind of measurement of a photon's energy . For example, if we know that a photon is emitted during a certain time interval Δt, the energy of that photon will be uncertain by at least $\frac{h}{2\pi\,\Delta t}$. Eq. 28-9 also applies to the uncertainties in E and t for electrons and other matter.

Richard Feynman (1918-1988)

Richard Feynman was born in New York on May 11, 1918. His father was a clothing sales manager with a deep interest in science. The young Feynman would sit on his father's lap and listen to extracts from the Encyclopedia Britannica, "translated" he later recalled, into images that captured his imagination. Feynman has said that his father "taught me how to notice things... with no pressure, just lovely, interesting discussions." This relationship led Feynman to develop such an interest in science that as he later declared, "when I was younger, I concentrated all my effort on it." This concentration of effort included performing simple experiments at home, developing magic tricks using basic chemistry, earning money by repairing schoolmates' radios, and developing an impressive ability to solve mathematical problems in a very personal intuitive way.

Feynman left High School and his New York home in 1935 to study at the Massachusetts Institute of Technology. He graduated with a B.S. in physics in 1939. By that time he had already decided that the quantum theory of electricity and magnetism presented the major challenge in theoretical physics. He realized that the understanding of this field, known as quantum electrodynamics, "was not completely satisfactory." He would eventually earn a Nobel prize for helping to make it much more satisfactory.

Quantum electrodynamics is the branch of quantum mechanics that deals with the interactions between electrically charged particles and the electromagnetic field. It therefore describes the interactions between photons of electromagnetic radiation and matter, and also describes the interactions between charged particles of matter such as electrons. The foundations of quantum electrodynamics were laid by Paul Dirac, Werner Heisenberg, and Wolfgang Pauli in the 1920s and 1930s.

Feynman began to explore his concerns and ideas in quantum electrodynamics during his graduate studies at Princeton University. His Ph.D. thesis, entitled *The Principle of Least Action in Quantum Mechanics*, was published in 1942.

By the time his thesis was published Feynman was working as a group leader on the Manhattan project, which led to the development of the atomic bomb. His group was involved in separating the isotopes of uranium needed to construct the bomb. Even during this challenging work, Feynman's mind was often applied to more fundamental theory. He later recalled that he found the time to further develop his ideas on quantum electrodynamics, working "on buses and so forth, with little pieces of paper."

Feynman was present when the first atomic bomb was tested at Alamogordo, New Mexico. In later years he was troubled by the contrast between the horror of Hiroshima and the wild celebratory parties he and his colleagues were enjoying after the bomb was dropped. In the months after Hiroshima and Nagasaki he developed a deep pessimism for the future, convinced that civilization would soon be destroyed in a worldwide nuclear conflagration.

In these first few months after the war Feynman spent some time working for Hans A. Bethe, his Manhattan project supervisor, at the General Electric Company in New York. He then accepted a post as associate professor of theoretical physics at Cornell University, where he was able to return to his interest in quantum electrodynamics. He made no immediate progress, however, and found his interest in the subject waning.

He later enjoyed telling the story of a trivial incident in the Cornell cafeteria that rekindled his enthusiasm and set him on the road to success. He recalled that "some kid threw up a plate" and he noticed that the blue Cornell emblem seemed to rotate at a different rate from the plate's wobble, Feynman began to play around with equations that might allow him to understand the motion of the plate. Solving these equations led him into the problem of the rotation of the spin of electrons. This, in turn, led to his groundbreaking work in quantum electrodynamics. As he later remarked, "the whole business that I got the Nobel prize for came from that piddling around with the wobbling plate."

As originally conceived, the theory of quantum electrodynamics presented seemingly insurmountable mathematical absurdities. In particular, it gave rise to situations in which the value of the electron's charge and mass apparently became infinite. Richard Feynman, Julian Schwinger (also in America) and Sin-itiro Tomonaga in Japan, independently found mathematical techniques to bypass these infinities. Their techniques involve the process of "renormalization," which basically allows all the positive infinities to cancel out all the negative infinities. The mass and charge of the electron can then simply be assigned to be the values that are experimentally determined for them. The sophisticated mathematics lying behind that simple summary allowed a modified version of quantum electrodynamics to be prepared, yielding predictions that agree very precisely with observation in the real world. So in essence, Feynman discovered a way to do calculations in quantum electrodynamics that yielded results that made sense, whereas previous methods had yielded only nonsense.

Feynman's particular approach involved calculating the probability of any event by taking into account the probabilities of all the possible ways in which the event might happen. This is known as the *path integral* or *sum over histories* approach to quantum mechanics. Much of Feynman's reasoning can be summarized using simple drawings that have become known as *Feynman diagrams*. These diagrams represent the interactions between particles indicating, for example, the exchange of photons between the particles (Fig. 28-A). Feynman diagrams greatly simplify the consideration of many quantum mechanical interactions, and they can be found scattered throughout all modern quantum mechanics texts.

The work of Feynman, Schwinger, and Tomonaga made quantum electrodynamics the most accurate physical theory in existence, allowing, for example, calculation of the energy levels of hydrogen atoms to an accuracy greater than 1 part in 10^{11}. Quantum electrodynamics has become a model to guide the development of similar theories aimed at describing other fields and interactions in physics. For their contributions toward making quantum electrodynamics such a success, Feynman, Schwinger, and Tomonaga were jointly awarded the Nobel prize in physics in 1965. Feynman was initially reluctant to accept the prize because of a deep-rooted mistrust of honors, which he attributed to his father's influence. He said, "I've already got the prize. The prize is the pleasure of finding the thing out." In the end, however, he was persuaded to accept his Nobel award, and thereafter he seemed to enjoy the glamour that came with it.

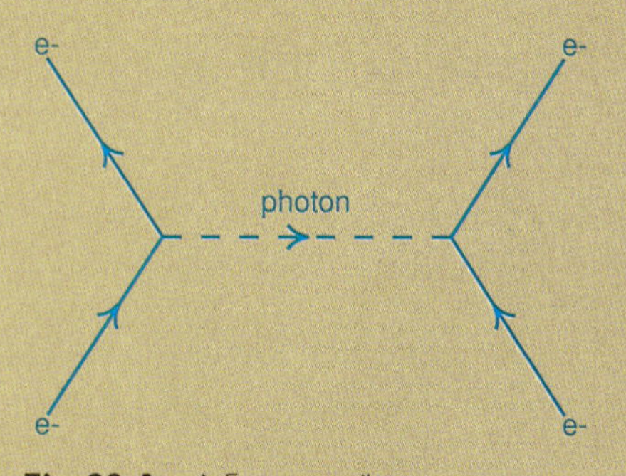

Fig. 28-A A Feynman diagram representing the interaction between two electrons. The electron on the left emits a photon, which is then absorbed by the electron on the right.

Although Feynman, Schwinger, and Tomonaga were able to make some mathematical sense of quantum electrodynamics, it is significant that they did it by finding ways to *avoid,* rather than truly *solve* some key mathematical difficulties. The challenge remains to describe quantum electrodynamics with mathematics that does not require what Feynman has called the "hocus pocus" of renormalization.

In 1950 Feynman became professor of theoretical physics at The California Institute of Technology. He continued to perform very significant work. In 1955 he found an explanation for the structure and behavior of liquid helium. When cooled to below 2.19 K, liquid helium changes into the strange "superfluid" state. In this state the liquid becomes frictionless as its viscosity effectively disappears. Feynman was able to explain this transition in terms of quantum mechanics.

Together with Murray Gell-Mann, Feynman also used his diagrammatic approach to quantum mechanics to develop a theory of the so-called "weak" interactions in physics, responsible for the emission of beta particles from certain radioactive atoms. He also explored the extent to which the basic framework of quantum electrodynamics might be applied to the "strong" interaction (or "strong force") responsible for binding protons together within atomic nuclei.

In his later years as much attention was given to Feynman's abilities as a teacher and his flamboyant personality as to his continuing work in physics. His famous textbook, *The Feynman Lectures in Physics,* was published in 1963 and soon inspired countless students of physics with its wonderful insights and its informal, stimulating style.

In the 1980s the publication of a collection of anecdotes in his book *Surely You're Joking, Mr. Feynman* and a series of TV appearances and programs devoted to his life gave Feynman a high celebrity profile.

Further publicity came in 1986 when he was invited to serve on the Presidential Commission set up to investigate the *Challenger* space shuttle disaster. Helped by some of the project's engineers, Feynman soon realized that seals known as

Continued.

O-rings had failed due to a loss of elasticity when exposed to low temperatures on the morning of the shuttle's launch. In a celebrated act of showmanship, he demonstrated the problem during a televised meeting of the Commission. As the meeting proceeded he clamped an O-ring in a glass of iced water. A few minutes later he removed the O-ring, which by that time had clearly lost the resilience and flexibility required for it to function properly. The Commission only published Feynman's findings in full after he had threatened to resign. His experience of the Commission left him disillusioned by the way in which Washington politics could hinder attempts to discover the truth.

The Challenger Commission was Feynman's last major public undertaking. In 1986 he was found to have a cancer of the bone marrow, and in 1987 he developed an abdominal cancer similar to one that had been removed 9 years earlier. He died in February 1988 at the age of 69. Some students at Caltech responded by hanging a large banner across the front of the Millikan library that simply declared: "We love you, Dick." That final tribute encapsulated the feelings of many thousands of students, including those who knew Feynman only through his books, who found his work and teaching a major source of inspiration.

Stephen Hawking

Stephen W. Hawking is widely considered the most brilliant theoretical physicist since Einstein. He is the Lucasian Professor of Mathematics at Cambridge University, a post once occupied by Isaac Newton. He is also perhaps the most famous disabled person in the world, having been confined to a wheelchair for the last two decades by Lou Gehrig's disease (amyotrophic lateral sclerosis). His fortitude and humor in the face of his illness are legendary. His scientific goal could not be more ambitious: he is working toward a grand unified theory of physics, a theory that would reconcile quantum mechanics and general relativity. He once called physics "playing the game of universe." His contributions to that game may well earn him the Nobel Prize.

Hawking has said that he was led to theoretical physics and cosmology by the "big questions": "Where did the universe come from? How and why did it begin? Will it come to an end, and if so, how?" Those questions caused him, as an Oxford undergraduate, to switch from mathematics to physics. During his first 2 years of graduate school at Cambridge, he worked very little: his illness had been diagnosed and he was not expected to live long. As he put it, "There didn't seem to be much point in completing my Ph.D." Then he met Jane Wilde, a student of modern languages. Hawking has said that his marriage to Jane in 1965 was the turning point in his life, that it galvanized him both to fight his disease and to apply himself to his work. "It made me determined to live," he said. "I started working hard for the first time in my life. To my surprise, I found I liked it."

His doctoral thesis and his work for several years after addressed a "big question": What did the beginning of the universe look like? He was led by this question into his work on black holes, for which he is best known. If Einstein's general relativity theory is correct, a black hole contains as its center a singularity, a single infinitely dense point containing all the black hole's mass. Anything that comes within the black hole's gravitational influence is crushed into the point. With his colleague and friend Roger Penrose, Hawking published in 1970 a paper in which they reversed this process on a grand scale. They showed mathematically that if general relativity is correct then our entire universe must have begun from a singularity. That is, they showed that general relativity theory calls for a Big Bang.

Many physicists disliked the concept of singularities, which are impossible under Newtonian physics, and the paper received, in Hawking's laconic words, "a lot of opposition. . . . However, one cannot really argue with a mathematical theorem." In fact Hawking has argued with many theorems, his own and others. He clearly enjoys intellectual combat, and he retracts his own work without mincing words if he decides he was wrong: "I think it much better and less confusing if you admit in print that you were wrong."

One argues with a mathematical theorem by replacing it with a better theorem. According to general relativity, nothing escapes from a black hole, not even light. After his 1970 paper Hawking began using quantum mechanics to examine that belief and was startled to find that as matter falls into a black hole quantum interactions should occur and particles should be emitted. This idea called for major changes in general relativity theory, and Hawking himself was reluctant to believe his results at first, but he had confidence in the physics he had used. The 1974 paper describing his work caused enormous controversy; the chairman of the conference where it was delivered called it "rubbish." The concept of "Hawking radiation" has gained considerable ground since then, though the radiation has not yet been detected. More importantly, Hawking's paper made use of both quantum mechanics and general relativity—a step toward a grand unification.

In recent years he has moved away from singularities in his theories; the laws of physics break down completely at a singularity and Hawking finds the implications for the Big Bang disturbing. "If the laws of science broke down at the beginning of the universe," he has said, "they could break down anywhere." For that reason, with his students and his colleague James Hartle, he has developed a theoretical model of the universe in which singularities are not required.

Such a model, if accepted by the physics community, would supersede not only his own earlier work but also, to a great extent, general relativity theory. In recent years Hawking has described himself as "over the hill," saying that physicists, as they age, lose the mental agility and the courage needed to propose radical ideas. Clearly that is not true of Hawking. He lives almost completely in his mind, but he does not find that life confining. "My goal is simple," he said in 1981. "It is complete understanding of the universe, why it is as it is and why it exists at all."

HAPTER 28 SUMMARY

Light displays properties of both particles and waves. Light propagates as an electromagnetic wave. But light behaves as a particle when it is emitted or absorbed; that is, the interaction is localized, and energy is transferred in discrete units, or quanta, called "photons."

A photon's energy E is related to the light's frequency f by the equation

$$E = hf$$

where h is Planck's constant, which has the value

$$h = 6.63 \times 10^{-34} \text{ J-s}$$

In the photoelectric effect light of an appropriate frequency shines on a photosensitive surface, and photons are absorbed by electrons near the surface. Electrons are then emitted from the surface with kinetic energy, which can have a maximum value given by

$$K_{max} = hf - W$$

where hf is the energy given up by a photon and the work function W is the work done to liberate the most weakly bound electrons.

A photon has momentum p related to the light's wavelength λ by the equation

$$p = \frac{h}{\lambda}$$

This relationship accounts for the Compton effect, in which X rays scattering off atomic electrons undergo a shift in wavelength as a result of momentum transferred to the electrons.

The probability of detecting a photon at any point is proportional to the intensity of the light wave at that point.

There is uncertainty not only in the position of a photon (Δx), but also in its momentum (Δp_x). These uncertainties are related by the Heisenberg uncertainty principle:

$$\Delta x \, \Delta p_x \geq \frac{h}{2\pi}$$

Another form of the uncertainty principle relates the uncertainty in a photon's energy (ΔE) and its time of detection (Δt):

$$\Delta E \, \Delta t \geq \frac{h}{2\pi}$$

Questions

1 Black-and-white photographic film is sensitive to light across the visible spectrum. The film must be developed in total darkness to avoid unwanted exposure. However, when the negative image on the film is converted to a positive image on photographic print paper, a red darkroom light can be used. The red light does not affect the print paper. A white light is used to project the image from the negative onto the paper, exposing the paper and producing the final positive image. Is the minimum photon energy required to initiate this chemical process on the print paper less than or greater than the minimum photon energy required to expose the film?

2 Assuming that the power rating in watts is the same for the green, yellow, and red lights of a traffic light, which light emits more photons during one second of operation?

3 X-ray machines are used to view the contents of luggage moving along conveyor belts at airport security stations. Although X rays can expose photographic film, these machines do not expose photographic film that happens to be in the luggage. The most likely explanation is that (a) the wavelength is too short; (b) the frequency is too low; (c) the number of photons absorbed by the film is too low; (d) the X rays are absorbed by the luggage.

4 Some photographers refuse to put their film through X-ray security machines, described in the preceding question. They claim that their film is sometimes slightly exposed, or "fogged," affecting the quality of their pictures. Which of the following would be factors that could cause too much exposure: (a) stopping the conveyor belt to carefully inspect the luggage; (b) too high an intensity of X rays; (c) too short an X-ray wavelength; (d) overcrowding the luggage?

5 If you increase the intensity of light shining on a photosensitive surface in a photocell while keeping the frequency of the light constant, which of the following will not increase: (a) current through the photocell; (b) the number of electrons emitted; (c) the momentum of each electron emitted; (d) the kinetic energy of each electron emitted?

6 Suppose the photosensitive surface in a photocell is sodium. Would this photocell produce a current if it were illuminated with (a) blue light; (b) red light? See Example 1.

7 A CO_2 laser emits infrared radiation. Would a visible light laser of the same intensity as a CO_2 laser emit more or less photons per second than the CO_2 laser?

8 Suppose that we perform Young's double-slit interference experiment with low enough intensity light so that one photon at a time passes through the slits. We detect which slit each photon passes through by observing the scattering of a beam of electrons in front of the slits. Will the diffraction pattern we see on the screen look more like a single-slit diffraction pattern or a double-slit diffraction pattern? Discuss how reducing either the number of electrons in the beam or the energy of each electron might affect the experiment.

Answers to Odd-Numbered Questions

1 greater than; **3** c; **5** c and d; **7** less

Problems (listed by sections)

28-1 Photons

1 The wavelength of visible light ranges from 400 nm to 700 nm. What is the corresponding range of photon energies?

2 Human skin is most sensitive to ultraviolet light of approximate wavelength 300 nm.
(a) Find the energy of a single photon of such light.
(b) The threshold for sunburn occurs when 1 cm^2 of skin absorbs about 10^{15} of these UV photons over an interval of several hours or less. How much energy is absorbed? What is the intensity of the UV light if it delivers this much energy in 10 minutes? (By way of comparison, the ultraviolet component of sunlight normally incident on the earth's surface at sea level ranges in wavelength from 310 nm to 400 nm, with a total intensity of about 40 W/m^2.)

3 TV channel 2 broadcasts in the frequency range 54 to 60 MHz.
(a) Find the energy of the average photon emitted.
(b) If the power of the broadcast is 50,000 W, how many photons are emitted per second?

★ 4 A He-Ne laser produces light of wavelength 633 nm with intensity 1.00×10^3 W/m^2, over a beam of cross-sectional area 1.00 mm^2. How many photons per second are emitted by the laser?

★ 5 Estimate how many photons of yellow light (λ = 575 nm) must be absorbed by 1.0 cm^2 of human skin to raise the skin temperature 1.0° C. Assume that the photon energy is absorbed by 1.0 cm^3 of tissue. As rough estimates of the tissue's density and specific heat, use the values for water.

★ 6 Sunlight of average wavelength 550 nm enters a car's windows and is absorbed by interior surfaces, raising the temperature inside the car by 10.0° C. If the heat capacity of the car's interior is 1.00×10^6 J/C°, how many photons were absorbed?

★ 7 (a) How many photons per second enter one eye if you look directly at a 100 W light bulb 2.00 m away? Assume a pupil diameter of 4.00 mm and a wavelength of 600 nm.
(b) How many photons per second enter your eye if a 1.00 mW laser beam is directed into your eye? (λ = 633 nm)

★★ 8 Light can be detected when as few as 100 photons enter the eye in an interval of 1.00×10^{-3} s. (Only about 5 of these photons are absorbed by receptors in the retina.) What is the maximum distance at which you could possibly see a 100 W sodium lamp emitting light of wavelength 589 nm if your pupil is dilated to a diameter of 6.00 mm?

9 To expose photographic film, a minimum energy of 1.0 eV must be absorbed from a photon by a silver bromide molecule on the film's surface. What is the maximum wavelength of light that will expose the film?

10 Find the work function of gold, which emits photoelectrons only if illuminated by light of frequency 1.16×10^{15} Hz or greater. Will gold emit photoelectrons when it is illuminated by visible light?

11 Blue light of frequency 6.3×10^{14} Hz falls on the surface of sodium, which has a work function of 2.27 eV. Find the maximum kinetic energy of emitted electrons.

12 Green light of wavelength 525 nm falls on the surface of potassium, which has a work function of 2.26 eV. Find the maximum speed of photoelectrons emitted from the surface.

13 The work functions of sodium, uranium, aluminum, and gold are respectively 2.27 eV, 3.63 eV, 4.08 eV, and 4.82 eV. Which of these elements will emit photoelectrons when illuminated by (a) red light (λ = 700 nm); (b) violet light (λ = 400 nm); (c) ultraviolet light (λ = 250 nm)?

14 A potential difference of 1.00 V must be applied to a photocell to stop the current when the photosensitive surface is illuminated with ultraviolet light of frequency 1.00×10^{15} Hz. Find the work function.

15 UV light of frequency 1.00×10^{15} Hz illuminates a sodium surface ($W = 2.27$ eV) in a photocell.
(a) Find the magnitude of the potential difference that must be applied to the photocell to stop the current.
(b) Is the sodium surface or the other terminal of the photocell at the higher potential?

16 Find the momentum of a photon of the following types of radiation: (a) visible light of wavelength 633 nm from a He-Ne laser; (b) UV light of wavelength 300 nm; (c) X rays of wavelength 0.100 nm.

17 A light source emits photons, each of which has a momentum of 1.40×10^{-27} kg-m/s.
(a) What color is the light?
(b) How much energy does each photon have?

18 An X-ray beam of initial wavelength 0.0500 nm is Compton scattered from an electron. Find the wavelength of the X rays scattered at 180°.

★★ **19** An X-ray photon that is Compton scattered from an electron at an angle of 180° imparts to the electron a speed of 1.00×10^7 m/s. Find the wavelength of the incident X rays.

★★ **20** In the Compton-scattering experiment, the wavelength of scattered X rays differs only slightly from the wavelength of incident X rays. For a scattering angle of 180°, (a) use conservation of momentum to show that an electron gains momentum of approximately $2h/\lambda$; (b) use conservation of energy to show that an electron gains kinetic energy of approximately $hc\,\Delta\lambda/\lambda^2$; (c) show that $\Delta\lambda \approx \frac{2h}{mc}$.

★★ **21** Find an expression for the shift in wavelength of X-ray photons scattered at an angle of 90°.

28-2 Wave-Particle Duality

22 In one small section of a diffraction pattern centered at point A, 1000 photons are detected during a certain time interval. During the same interval, in another equal-sized section of the diffraction pattern centered at point B, 3000 photons are detected. If the intensity of the light at point A is 5.00 W/m^2, what is the intensity at B?

23 The amplitude of the electric field at point A in an interference pattern is 10 times as great as the electric field amplitude at point B. Find the ratio of the number of photons per s in a small section of the pattern centered at A to the number per s in an equal-sized section centered at B.

24 In a certain X-ray diffraction pattern, 1.00×10^{12} X rays per s are absorbed by a detector at a point in the pattern where the intensity of the X rays is 100 W/m^2. How many X rays per s would the detector absorb at a point in the pattern where the X-ray intensity is 300 W/m^2?

28-3 The Uncertainty Principle

25 A stabilized He-Ne laser emits light of frequency 4.74×10^{14} Hz with a bandwidth of 1.00×10^4 Hz.
(a) Find the momentum of a photon emitted by the laser and the uncertainty in the photon's momentum.
(b) Find the minimum uncertainty in the position of a photon of this light.

26 Suppose a photon is equally likely to be found anywhere in an interval of 10.0 cm along the x-axis. What is the minimum uncertainty in the photon's x component of momentum?

27 A beam of light directed along the z-axis passes through a slit parallel to the y-axis. The slit's width is 0.100 mm.
(a) What is the uncertainty in the x coordinate of a photon in the light just emerging from the slit?
(b) What is the minimum uncertainty in the photon's x component of momentum?

★ **28** Laser light of wavelength 633 nm passes through a slit of width 0.200 mm. When the light emerges from the slit, the momentum of a photon may be directed over a range of angles, measured relative to the forward direction. Find the minimum value of that range.

29 A common He-Ne laser emits light of wavelength 633 nm with a bandwidth of 1.00×10^8 Hz. Find the uncertainty in energy of a photon emitted by the laser.

30 A light source emits a photon during a time interval of 1.00×10^{-8} s. Find (a) the minimum uncertainty in the photon's energy; (b) the minimum frequency bandwidth of the light.

Additional Problems

★ **31** The intensity of sunlight at the earth's surface is 1400 W/m^2, and the average wavelength is 550 nm.

(a) How many photons per second are absorbed by a perfect blackbody surface of area 1.00 m^2 perpendicular to the sun's rays?

(b) How much momentum is transferred to the plate in a 1-hour interval?

★ **32** Find an expression for the number of photons per unit time absorbed by a section of a blackbody's surface of area A from a light beam of frequency f and electric field amplitude E_0, normally incident on the surface.

★★ **33** Suppose you are designing a sail for an interplanetary sail plane operating on the momentum imparted by sunlight. Assume the intensity of sunlight is the intensity at the earth's surface, 1400 W/m^2.

(a) Find the maximum acceleration that could be achieved by a reflecting aluminum sail of thickness 0.100 mm.

(b) How much would its speed increase in 1 year?

CHAPTER 29 The Atom

A circle of 48 iron atoms on a copper surface. The circle's diameter is 14 nm. This remarkable image was produced by a scanning tunneling electron microscope.

For thousands of years atoms were unseen hypothetical entities dreamed up by philosophers. These atoms of antiquity were supposed to be indivisible—the smallest possible units of matter. Over the past 200 years, chemists and physicists developed the concept of an atom. In the nineteenth century, atoms, though still unseen, were used to explain an enormous variety of natural phenomena. Eventually atoms were found to have structure and to be divisible, though not by means of ordinary physical or chemical processes. Today we can produce images of individual atoms, as in the photo above. Atoms, which were once only a philosopher's dream, have become a physical reality. That all matter is composed of atoms is perhaps our single most important piece of scientific knowledge.

To really understand what an atom is, it is helpful to see how the concept of atoms developed historically. In this chapter we shall sketch this history—from the indestructible atoms of past centuries to the Bohr model of the atom in the early 1900s to the atom as we understand it today. The modern concept of an atom is based on quantum mechanics, a bizarre but powerful system for describing interacting particles. This "quantum" mechanics is radically different from Newton's classical mechanics. We shall introduce some of the principles of quantum mechanics and use them to explain atoms, as well as lasers and semiconductors.

29-1 Atomic Spectra and the Bohr Model of the Atom

Atomic Spectra

Various sources of light were described in Chapter 23 (Sect. 23-2). There we saw that the light emitted by a gas discharge tube consists of narrow bands of frequency. When this light is dispersed by a prism, it forms a spectrum consisting of narrow colored bands, or "spectral lines" (Fig. 29-1a). This is in contrast to the continuous range of colors produced by many other sources of light, for example, the sun, incandescent or fluorescent bulbs, or a blackbody(Fig. 29-1b).

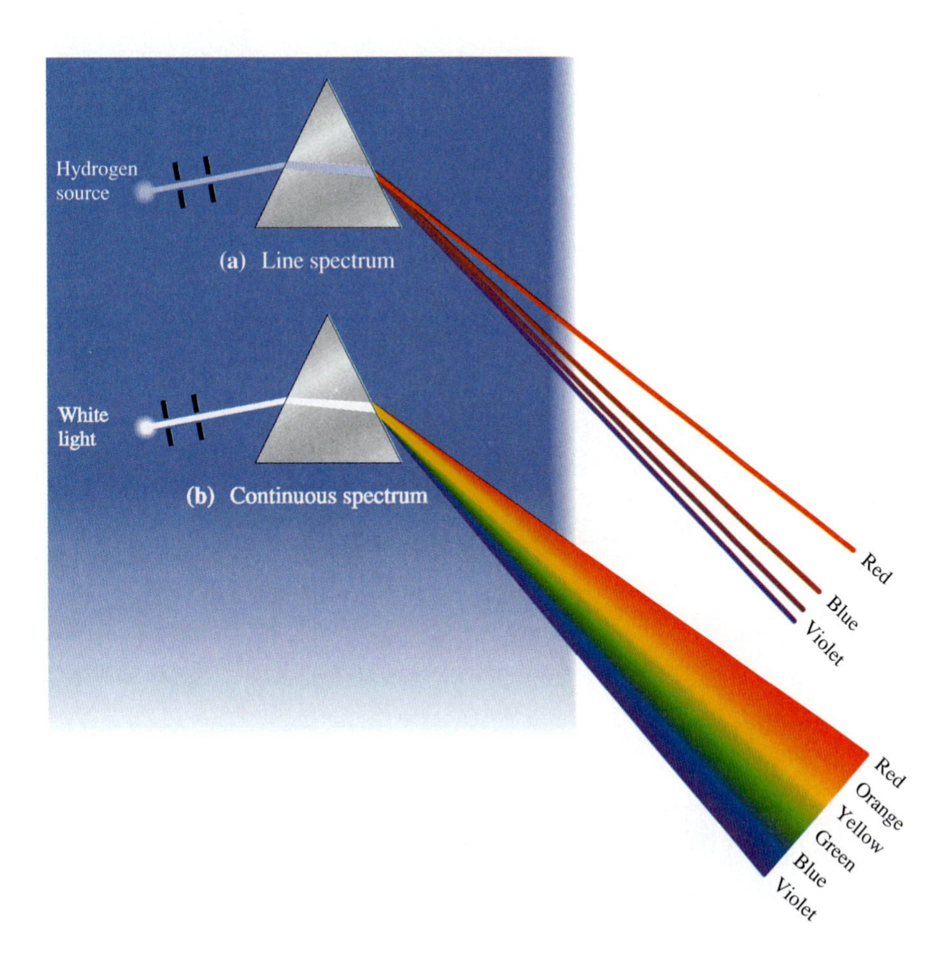

Fig. 29-1 **(a)** A gas discharge tube, containing a gas such as sodium or hydrogen (shown here) produces a line spectrum. **(b)** White light produces a continuous spectrum.

The gas in one of these discharge tubes is of low enough density that the gas atoms act independently as they radiate light. So what we see when we look at the spectral lines is the radiation produced by a single atom, only more intense. The specific pattern of lines that we see depends on the kind of atom: hydrogen, neon, sodium, and so forth. Each creates its own characteristic spectrum (Fig. 29-2). Such spectra are the principal tool used by astronomers to determine the composition of stars and through Doppler analysis to determine the motion of the stars.

We see the signature of an atom in the spectral lines it produces. Analysis of this signature was the key to understanding atomic structure for physicists at the turn of the century.

The hydrogen spectrum is the simplest. Hydrogen produces fewer lines than other gases, and the spacing between the lines varies in a systematic way. This is especially evident when we view some of the ultraviolet hydrogen lines, in addition to the visible lines (Fig. 29-3). There appears to be a systematic decrease in wavelengths from the visible into the ultraviolet.

Fig. 29-2 Atomic spectra. (The solar spectrum is included for reference.)

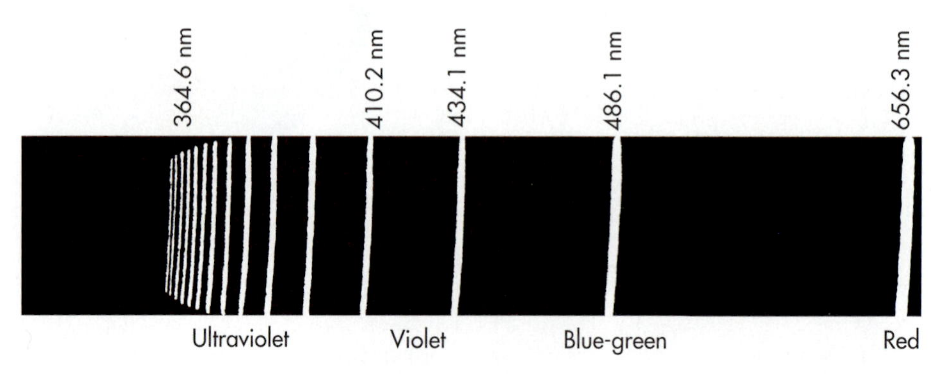

Fig. 29-3 The hydrogen spectrum, showing both visible and ultraviolet lines.

In 1885 Johann Balmer, a Swiss schoolteacher, played with the numbers corresponding to the wavelengths of the four visible hydrogen lines and discovered that they could be described by a single, simple formula:

Balmer series

$$\frac{1}{\lambda} = R\left(\frac{1}{2^2} - \frac{1}{n^2}\right) \qquad (n = 3, 4, 5, \text{ or } 6) \qquad \text{(29-1)}$$

where R, known as the **Rydberg constant,** has the value

$$R = 1.097 \times 10^7 \text{ m}^{-1} \qquad \text{(29-2)}$$

Substituting a particular value of n in Eq. 29-1 gives one of the hydrogen wavelengths. For example, $n = 3$ gives

$$\frac{1}{\lambda} = (1.097 \times 10^7 \text{ m}^{-1})\left(\frac{1}{2^2} - \frac{1}{3^2}\right) = 1.52 \times 10^6 \text{ m}^{-1}$$

or

$$\lambda = 6.56 \times 10^{-7} \text{ m} = 656 \text{ nm}$$

This wavelength corresponds to the red line in the hydrogen spectrum. The blue-green line ($\lambda = 486$ nm) corresponds to $n = 4$ and so forth.

The Balmer formula also describes the wavelengths of ultraviolet hydrogen lines if n is allowed to have an integer value greater than 6. For large values of n, the lines begin crowding together, approaching a limiting value corresponding to $n = \infty$:

$$\frac{1}{\lambda} = R\left(\frac{1}{2^2} - \frac{1}{\infty^2}\right) = \frac{R}{4}$$

or

$$\lambda = \frac{4}{R} = \frac{4}{1.097 \times 10^7 \text{m}} = 3.65 \times 10^{-7} \text{m} = 365 \text{ nm}$$

Later experiments resulted in the discovery of other hydrogen lines in the infrared and far ultraviolet. These lines too were described by simple formulas, identical to the Balmer formula, except that 2^2 was replaced by 1^2, 3^2, 4^2, and so forth, and n could be any integer giving a positive value for λ:

Lyman series (UV):

$$\frac{1}{\lambda} = R\left(\frac{1}{1^2} - \frac{1}{n^2}\right) \qquad (n = 2, 3, 4, \ldots)$$

Paschen series (IR):

$$\frac{1}{\lambda} = R\left(\frac{1}{3^2} - \frac{1}{n^2}\right) \qquad (n = 4, 5, 6, \ldots)$$

Brackett series (IR):

$$\frac{1}{\lambda} = R\left(\frac{1}{4^2} - \frac{1}{n^2}\right) \qquad (n = 5, 6, 7, \ldots)$$

Wavelengths in any of the hydrogen series can be expressed by a single formula:

$$\frac{1}{\lambda} = R\left(\frac{1}{(n')^2} - \frac{1}{n^2}\right) \qquad (n' = 1, 2, 3, \ldots \text{ and } n > n') \qquad \text{(29-3)}$$

It is the remarkably simple pattern of radiation by a hydrogen atom, summarized by this equation, that was to lead Niels Bohr to propose the first successful, detailed model of an atom and would ultimately lead to the modern quantum mechanical model of an atom.

Early Concepts of Atoms

The earliest concept of an atom was that of a small, indivisible unit of matter. Greek philosophers speculated about the existence of such atoms thousands of years ago. In the seventeenth century, Robert Hooke proposed that the symmetrical structure of crystals resulted from the regular stacking of atoms. In 1803 John Dalton proposed that each chemical element is composed of a different kind of atom, each with its own characteristic mass. Later in the nineteenth century Maxwell and Boltzmann derived gas laws and explained properties of gases, using the concept of atoms. They used a mechanical model of a gas, composed of atoms—tiny, billiard ball–like objects, subject to the laws of mechanics. But the existence of atoms was not universally accepted even as recently as the early 1900s. Ernst Mach was an influential physicist who argued against explaining macroscopic phenomena on the basis of these hypothetical entities that could not be seen. One of the final steps in winning universal acceptance of the existence of atoms was Einstein's analysis of "Brownian motion." Einstein analyzed the motion of a suspension of pollen particles in liquid. He showed how their erratic motion resulted from fluctuations in the number of collisions with molecules of the liquid.

Discovery of Electrons

In 1750 Benjamin Franklin introduced the concept of an electrical fluid flowing between bodies, resulting in their "charging," because of either an excess or deficiency of the fluid. In the nineteenth century it was recognized that the "electrical fluid" consisted of charged particles, which came to be called "electrons." These particles were much smaller than atoms and had a definite charge and mass. J.J. Thomson is credited with discovering the electron in 1897. He performed critical experiments on electron beams in cathode ray tubes* by deflecting the beams with electric and magnetic fields and determining the ratio of the electron's charge to its mass.

Thomson's Model of an Atom

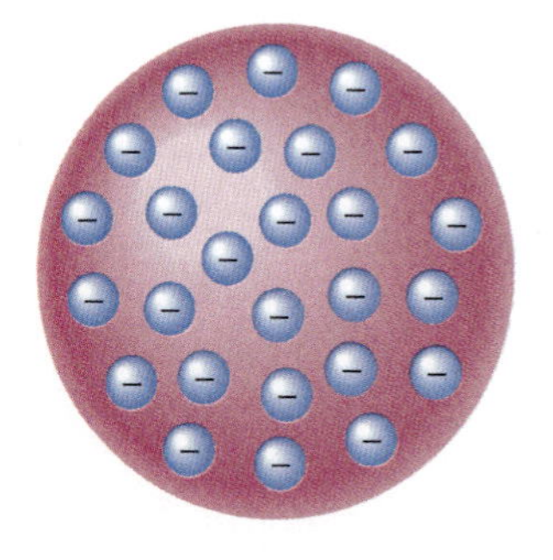

Fig. 29-4 The Thomson plum-pudding model of an atom.

After his work with electron beams, it was quite natural for Thomson to speculate on where these electrons came from. Any metal electrode in a cathode ray tube was capable of emitting a seemingly endless stream of electrons. Thus it was reasonable to assume that all matter had electrons in it. Since matter consisted of atoms, this suggested that electrons were particles within an atom. By the early 1900s people began to think of atoms for the first time not as indestructible units but as something with an internal structure. The problem was to determine that structure. Obviously matter was normally electrically neutral. So an atom must contain positive charge sufficient to balance the negative charge of its electrons. The model investigated by Thomson and favored by most other physicists at the turn of the century is illustrated in Fig. 29-4. The electrons were believed to be embedded in a smooth sphere of positive charge, like plums in a pudding.

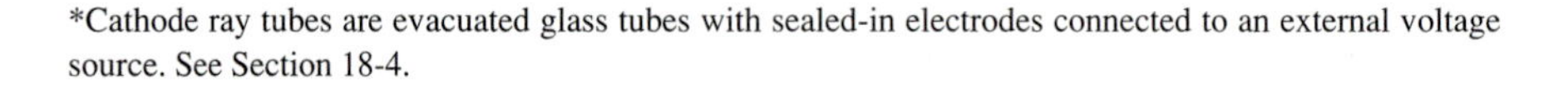

*Cathode ray tubes are evacuated glass tubes with sealed-in electrodes connected to an external voltage source. See Section 18-4.

Discovery of the Nucleus

Ernest Rutherford (Fig. 29-5), a former student of Thomson, began about 1909 to investigate the structure of matter, using as a probe positively charged rays, called "alpha (α) particles," produced by radium.* Rutherford placed a tiny radium sample in a lead tube, creating a beam of α particles, which he directed at a delicate gold foil, with a thickness of only 6×10^{-5} cm, or about 1000 atoms (Fig. 29-6). Since the α particles would not penetrate more than a few centimeters through air, the whole apparatus had to be placed in an evacuated container. The α particles were able to penetrate the gold foil and were scattered in various directions. Although the α particles themselves were invisible, their paths could be inferred from the tiny flash of light they produced as they struck a zinc sulfide screen.

Fig. 29-5 Ernest Rutherford (1871–1937). Rutherford received the Nobel Prize in chemistry in 1908 for his work on radioactive decay. It bothered him that his prize had been awarded in chemistry, rather than physics. In his acceptance speech, he remarked that he had observed many transformations, but none so rapid as his from physicist to chemist.

According to Thomson's model, the gold atoms in the target contained positive and negative charges, which could be expected to exert both attractive and repulsive forces on the α particles. A net force at any instant would result when the α particle was very close to an electron. As indicated in Fig. 29-7, an α particle would undergo only a very slight deflection from each close encounter with an electron, which has much less mass than the α particle. An α particle would experience only a small net deflection after thousands of such encounters. Detailed calculations based on this model indicated that α particles would typically be scattered through an angle of about 1°, with no α particles being scattered by more than a few degrees.

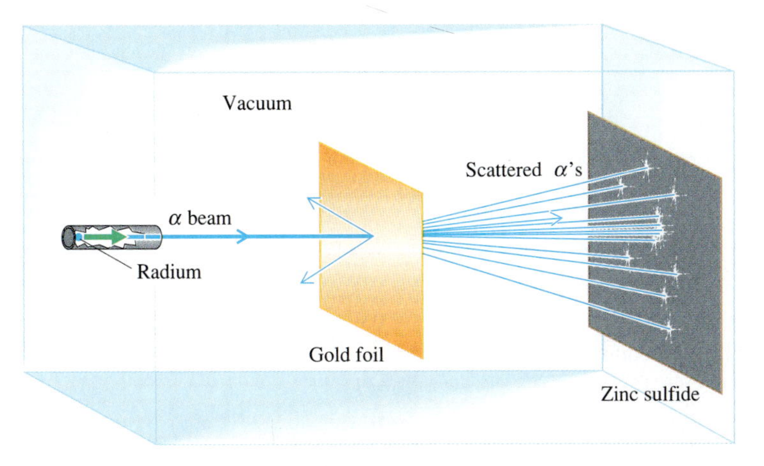

Fig. 29-6 Rutherford's scattering equipment.

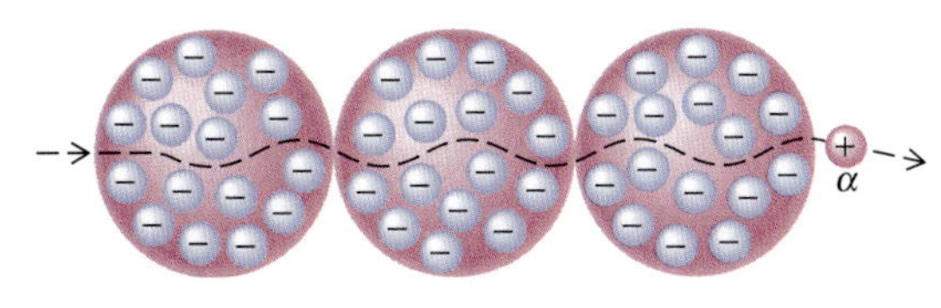

Fig. 29-7 Typical deflection of α particles, as predicted by the Thomson model.

*This highly radioactive element had been discovered and isolated by Marie Curie a few years earlier. The α particles emitted by radium were later shown to be helium nuclei, consisting of two protons and two neutrons. Rutherford's idea of smashing high-velocity subatomic particles into matter is still used today to explore matter on the subatomic scale, though the particle probes are now at much higher energies, produced by huge, elaborate particle accelerators.

When the Rutherford experiment was performed, the α particles seemed at first to scatter as expected, with 99% of the particles scattered at angles of 3° or less. To be thorough, Rutherford directed his student Hans Geiger to look for any α particles that might be scattered at very large angles. Rutherford expected that he would see nothing. To Rutherford's great surprise, Geiger found that roughly 1 in 10,000 α particles were scattered at angles greater than 90°; a few were even scattered backward, as though they were rebounding from an inpenetrable wall. As Rutherford later said, "It was almost as incredible as if you fired a 15-inch shell at a piece of tissue paper and it came back and hit you."

Two years later, in 1911, Rutherford was able to work out a mathematical theory accounting in detail for the α particles scattered in various directions. This theory was based on a radically altered model of the atom, in which all the positive atomic charge and nearly all the atomic mass are concentrated in a **nucleus.** The nuclear radius is less than one ten-thousandth the radius of the entire atom. Most α particles had experienced little or no deflection because they did not come close enough to any nucleus in the target. A particle passing through a layer of a few thousand atoms is unlikely to hit even one nucleus, or even come very close to one. But a few α particles will come close and will be scattered at large angles (Fig. 29-8).

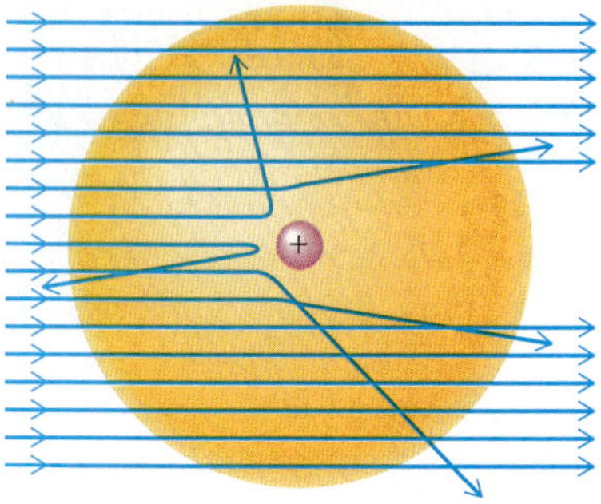

Fig. 29-8 Scattering of α particles by a gold atom. Only those α particles aimed at or close to the nucleus undergo much deflection.

EXAMPLE 1 An α Particle Colliding Head-on With a Nucleus

Rutherford determined that the α particles emitted by radium have a speed of about 1.5×10^7 m/s, or a kinetic energy of about 5 MeV. If such an α particle makes a head-on collision with a gold nucleus, how close will it come to the center of the nucleus before reversing its motion? The gold nucleus has a charge of +79e, and the α particle has a charge of +2e.

SOLUTION The distance of closest approach corresponds to the point where the α particle's velocity and kinetic energy are instantaneously equal to zero. We shall use the principle of conservation of energy to find that point, a distance r from the center of the nucleus. The final electrical potential energy of the α particle equals the product of its charge (+2e) and the potential $\left[V = \dfrac{k(79e)}{r}\right]$ produced by the gold nucleus at a distance r. The initial potential energy of the α particle is zero, since it starts at a great distance from the nucleus. Equating final and initial energies, we find

$$E_f = E_i$$
$$K_f + U_f = K_i + U_i$$
$$0 + q'V = K_i + 0$$
$$(2e)\frac{k(79e)}{r} = K_i$$
$$r = \frac{(2)(79)ke^2}{K_i}$$
$$= \frac{(2)(79)(9 \times 10^9 \text{ N-m}^2/\text{C}^2)(1.6 \times 10^{-19}\text{ C})^2}{(5 \times 10^6\text{ eV})(1.6 \times 10^{-19}\text{ J/eV})}$$
$$= 5 \times 10^{-14}\text{ m}$$

Even for a distance this small, the electric potential of the gold nucleus is like that of a point charge. This means that the nuclear radius must be less than this value.

There was an immediate problem with Rutherford's model. Certainly there must be electrons within an atom. They are necessary to cancel the positive charge of the nucleus and give the electrical neutrality normally displayed by matter. The electrons also have to be there to fill most of the space occupied by an atom. Atomic diameters are about 10^{-10} m, roughly 10^4 times nuclear diameters. The negatively charged electrons could not possibly be at rest because the electrical pull of the positive nucleus would quickly cause them to collapse inward. Even if, as some suggested, the electrons orbit the nucleus as the planets orbit the sun, it would not be long before such an orbit would have to decay. Electromagnetic theory predicts that any accelerated charge produces electromagnetic radiation. An orbiting electron would experience centripetal acceleration and so would radiate. The radiant energy would be supplied at the expense of the electron's mechanical energy. Calculation shows that an orbiting electron would radiate energy so fast that within about 10^{-9} s it would spiral into the nucleus.

Fig. 29-9 Niels Bohr (1885–1962). After receiving his Ph.D. at the age of 26, Bohr left Denmark, where he was a soccer star, and studied in England under Thomson and Rutherford. He returned to Copenhagen in 1912 as a professor at the university and proposed his atomic model shortly thereafter. Later he played an important role in developing the modern interpretation of quantum theory.

Bohr Model of the Atom

A bold answer to the dilemma was proposed in 1913 by Niels Bohr (Fig. 29-9). Bohr saw a way of applying to the hydrogen atom Einstein's concept of photons and Rutherford's concept of a nucleus, to derive the Balmer formula (Eq. 29-1), and thereby explain the hydrogen spectrum. In the process of doing this, Bohr provided the first successful, quantitative model of an atom.

Bohr's model was based on four assumptions:

1 **A single electron of charge −e orbits the nucleus of charge +e.** The attractive electric force exerted by the nucleus on the much less massive electron provides the centripetal acceleration of the electron in its circular orbit (Fig. 29-10). The electron's motion is analogous to the motion of a planet in its orbit around the sun. Like an orbiting planet or satellite, the greater the energy of the electron, the greater is the radius of its orbit.

2 **Only certain size orbits are allowed.** Only those orbits with radii r_n, corresponding to certain values of angular momentum L_n, are allowed. Angular momentum L is defined as the product of moment of inertia I and angular velocity ω: $L = I\omega$ (Eq. 9-18). An electron of mass m moving at speed v in a circular orbit of radius r has a moment of inertia $I = mr^2$ and angular velocity $\omega = \frac{v}{r}$. Taking the product of these two quantities, we find that the electron's angular momentum can be expressed as $L = mvr$. **The allowed values of the angular momentum L are assumed to be $h/2\pi$ (Plank's constant over 2π) times any positive integer n.**

$$L_n = mv_n r_n = n\frac{h}{2\pi} \qquad (n = 1, 2, 3, \ldots) \qquad (29\text{-}4)$$

Bohr was forced to make this arbitrary assumption to complete his derivation of the Balmer formula. There is nothing in classical mechanics that would lead you to expect that values of angular momentum should be restricted in this way. Bohr's efforts to explain the hydrogen spectrum led him to assume that angular momentum is "quantized," just as Planck's efforts to explain blackbody radiation led him to assume that energy is quantized.

3 **When in one of its allowed orbits, the electron emits no radiation.** This assumption says that the classical laws of electromagnetic radiation by an accelerated charge simply don't apply to an electron in its orbit.

4 **When an electron changes from a high-energy orbit to a lower-energy orbit, the excess energy is emitted as a photon.**

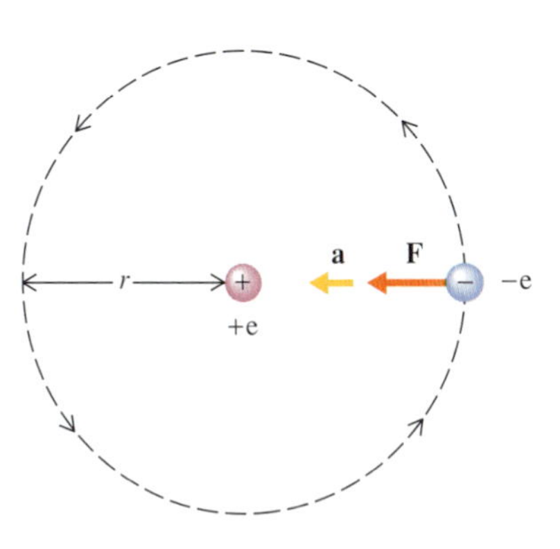

Fig. 29-10 An electron orbits a proton in the Bohr model of the hydrogen atom. The proton exerts an attractive electrical force on the electron, which gives the electron its centripetal acceleration.

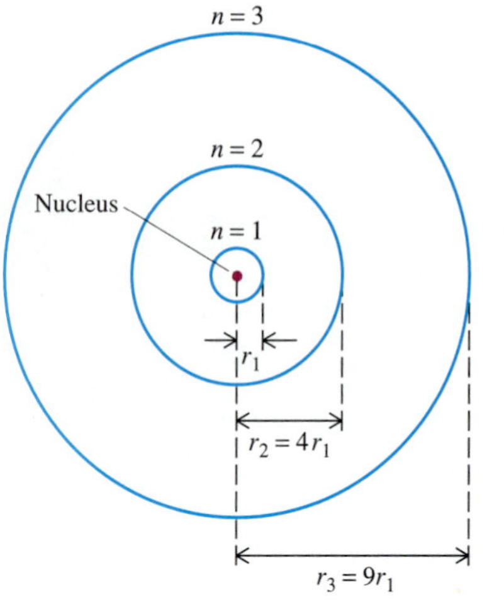

Fig. 29-11 Allowed electron orbits in the Bohr model of the hydrogen atom.

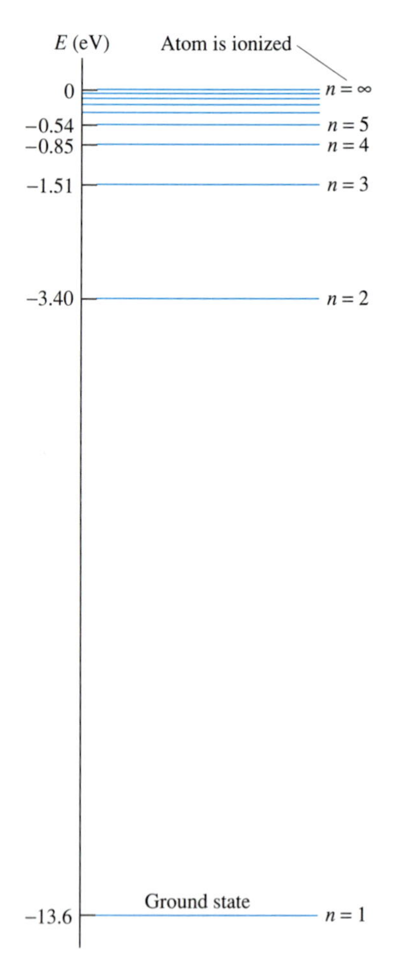

Fig. 29-12 Allowed electron energies in the Bohr model of the hydrogen atom.

As we shall show at the end of this section, Bohr was able to use his first two assumptions to derive the possible orbital radii and their corresponding energies:

$$r_n = n^2 r_1 \qquad (n = 1, 2, 3, \ldots) \qquad (29\text{-}5)$$

and

$$E_n = \frac{E_1}{n^2} \qquad (n = 1, 2, 3, \ldots) \qquad (29\text{-}6)$$

where

$$r_1 = \frac{h^2}{4\pi^2 mke^2} \qquad (29\text{-}7)$$

and

$$E_1 = -\frac{2\pi^2 mk^2 e^4}{h^2} \qquad (29\text{-}8)$$

Substituting numerical values into these expressions gives

$$r_1 = 5.29 \times 10^{-11}\ \text{m} \qquad (29\text{-}9)$$

$$E_1 = -13.6\ \text{eV} \qquad (29\text{-}10)$$

Atoms were known to have diameters on the order of 10^{-10} m. The Bohr model of the hydrogen atom is certainly consistent with this fact, since Eq. 29-9 predicts a minimum diameter of 1.06×10^{-10} m.

Some of the allowed Bohr orbits and corresponding electron **energy levels** are shown in Figs. 29-11 and 29-12 respectively. When the electron is in the first Bohr orbit ($n = 1$), it is then as close as it can be to the nucleus, and it is in the lowest possible energy state, called the **ground state.** Larger values of n correspond to larger orbits and higher energies.

From Eqs. 29-5 and 29-6, we see that when $n = \infty$, $r_n = \infty$ and $E_n = 0$. This means that when the electron has zero energy, it is far away from the nucleus; that is, when $E = 0$ the electron is no longer held by the nucleus. The atom has lost the electron and become ionized. Both the potential and kinetic energy equal zero in this state. In any of the other "bound states" the electron's energy is negative. The negative value of E results from the dominance of the negative potential energy term over the positive kinetic energy term. Potential energy is negative because of the way we defined the zero* of electrical potential energy in Chapter 18. Similarly, in Chapter 7 we found that a mass has negative gravitational potential energy when it is acted upon by a variable gravitational force $\left(\text{Eq. 7-19: } U_G = -G\frac{mM}{r}\right)$.

*If we wanted, we could redefine the zero of energy for the hydrogen atom by adding 13.6 eV to all states. The energy of the ground state would then be zero. Other bound states would have energies between 0 and 13.6 eV.

Although an electron can be in any of the allowed states, normally it is in the ground state. In an electrical discharge through a hydrogen discharge tube, collisions impart energy to electrons in some hydrogen atoms, putting them into higher-energy, **excited states.** An electron in an excited state spontaneously falls to a state of lower energy, emitting a photon to carry off the excess energy.

Transitions can occur between any of the energy levels. When an electron goes from energy level E_n to a lower level E_n', it emits a photon of energy

$$E_n - E_n' = hf \tag{29-11}$$

where f is the frequency of the emitted radiation. Substituting $f = \frac{c}{\lambda}$ and using Eq. 29-6 to relate the values of energy to the quantum numbers n and n', we find

$$\frac{E_1}{n^2} - \frac{E_1}{(n')^2} = h\frac{c}{\lambda}$$

or

$$\frac{1}{\lambda} = \frac{-E_1}{hc}\left(\frac{1}{(n')^2} - \frac{1}{n^2}\right) \tag{29-12}$$

If we identify the factor $-E_1/hc$ as Rydberg's constant, this equation is identical to Eq. 29-3, which describes the hydrogen spectral series. Setting R equal to this factor and using Eq. 29-8 to substitute for E_1, we find

$$R = \frac{2\pi^2 mk^2e^4}{h^3c} \tag{29-13}$$

Inserting numerical values into this expression, we find complete agreement with the experimental value of Rydberg's constant; that is, $R = 1.097 \times 10^7\ \text{m}^{-1}$. Thus Bohr's model is successful in accounting for the entire spectrum of radiation emitted by the hydrogen atom.

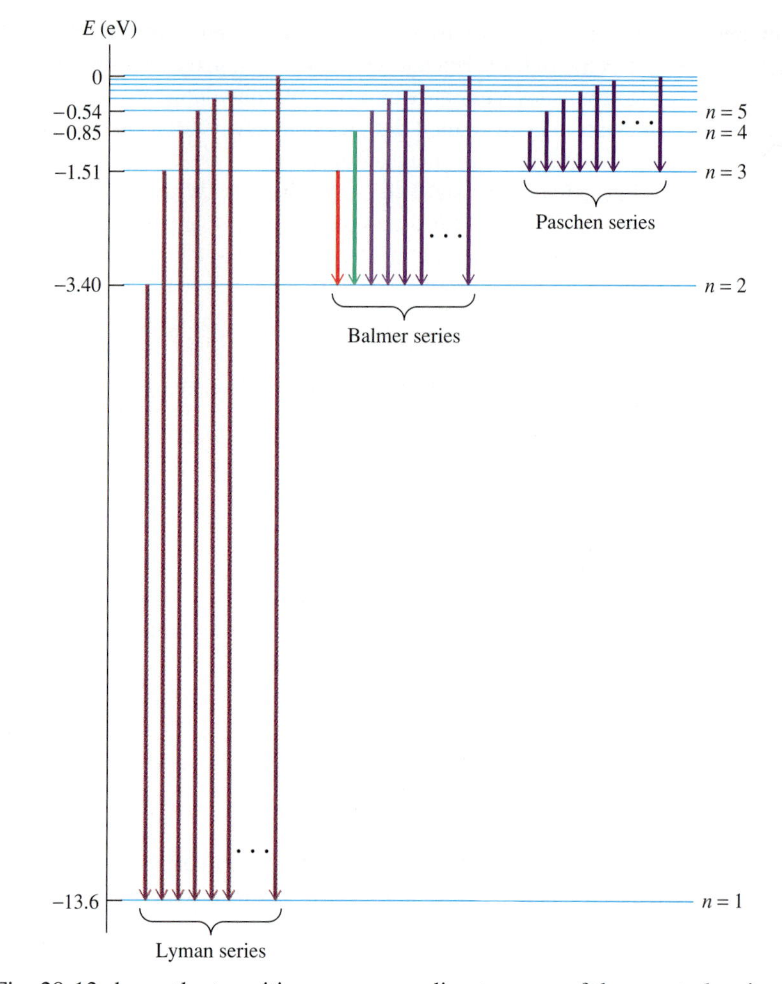

Fig. 29-13 Transitions between electron energy levels.

Fig. 29-13 shows the transitions corresponding to some of the spectral series. The Balmer series corresponds to transitions ending in the $n = 2$ state. For example the transition from the $n = 3$ state to the $n = 2$ state is accompanied by emission of a photon with energy equal to the energy difference between these two states; that is, $-1.51 \text{ eV} - (-3.40 \text{ eV}) = 1.89 \text{ eV}$. This corresponds to light of frequency

$$f = \frac{E}{h} = \frac{(1.89 \text{ eV})(1.602 \times 10^{-19} \text{ J/eV})}{6.626 \times 10^{-34} \text{ J-s}}$$

$$= 4.57 \times 10^{14} \text{ Hz}$$

and wavelength

$$\lambda = \frac{c}{f} = \frac{3.00 \times 10^{8} \text{ m/s}}{4.57 \times 10^{14} \text{ Hz}} = 6.56 \times 10^{-7} \text{ m}$$

$$= 656 \text{ nm}$$

This is the first (red) line in the Balmer series shown in Fig. 29-2 and 29-3.

The transition from the $n = 4$ state to the $n = 2$ state produces a photon of greater energy because of the greater energy difference between the states. This means that the light emitted in this transition has a higher frequency and a shorter wavelength, which turns out to be 486 nm, the blue-green line in the Balmer series.

As indicated in Fig. 29-13, the transitions in the Lyman series are between states with greater energy differences, so that higher-energy photons are emitted. Thus the Lyman series radiation is in the ultraviolet, at a higher frequency than the Balmer series. The much smaller energy differences associated with the Paschen series imply lower-frequency infrared radiation.

Although the Bohr model is successful in accounting for the hydrogen spectrum, it is not successful in describing multielectron atoms. Nor is it capable of predicting the intensity of spectral lines. In the next section we shall see how the Bohr model has been superseded by the quantum mechanical model of an atom. However, some elements of the Bohr model are a lasting legacy:

1 **angular momentum is quantized;**
2 **only certain discrete energies are allowed for atomic electrons;**
3 **electrons undergoing transitions between energy levels emit photons, which carry off the excess energy.**

Derivation of Energy Levels of the Bohr Atom

Next we shall show how Bohr's first two assumptions lead to the expressions for the radius and energy of the Bohr orbits (Eqs. 29-5 to 29-8). Equating the Coulomb force on the electron $\left(k\frac{e^2}{r^2}\right)$ to its mass m times its centripetal acceleration v^2/r, we obtain an expression for the electron's speed v in an orbit of radius r (shown in Fig. 29-10):

$$|\Sigma \mathbf{F}| = ma$$

$$k\frac{e^2}{r^2} = m\frac{v^2}{r} \tag{29-14}$$

Next we apply the postulate of quantized angular momentum (Eq. 29-4):

$$mv_n r_n = n\frac{h}{2\pi}$$

or

$$v_n = \frac{nh}{2\pi m r_n}$$

We use this expression to substitute for v in Eq. 29-14 and solve for the allowed radii r_n:

$$k\frac{e^2}{r_n^2} = m\frac{n^2h^2}{4\pi^2 m^2 r_n^3}$$

or

$$r_n = n^2\frac{h^2}{4\pi^2 mke^2} \tag{29-15}$$

This expression is equivalent to Eqs. 29-5 and 29-7.

The electron in its orbit is at a fixed distance r from the nuclear charge (+e), which, according to Eq. 18-7 $\left(V = k\frac{q}{r}\right)$, produces a potential

$$V = k\frac{+e}{r}$$

The electron's electrical potential energy is the product of its own charge (−e) and the nuclear potential (Eq. 18-3: $U_E = q'V$). Multiplying −e times V gives

$$U_E = -k\frac{e^2}{r}$$

This expression is analogous to the expression for the gravitational potential energy of a mass m at a distance r from a larger mass M $\left(\text{Eq. 7-19: } U_G = -G\frac{mM}{r}\right)$, for example, a planet of mass m orbiting the sun of mass M. Adding the expression for electrical potential energy to the electron's kinetic energy, we obtain the electron's total mechanical energy E.

$$E = K + U = \tfrac{1}{2}mv^2 - k\frac{e^2}{r}$$

From Eq. 29-14, we see that

$$mv^2 = k\frac{e^2}{r}$$

Substituting this expression into the preceding equation for E, we find

$$E = \tfrac{1}{2}k\frac{e^2}{r} - k\frac{e^2}{r}$$

$$= -\tfrac{1}{2}k\frac{e^2}{r}$$

Using Eq. 29-15 to substitute for r, we find the energy of the nth Bohr orbit.

$$E_n = -\frac{1}{n^2}\left(\frac{2\pi^2 mk^2e^4}{h^2}\right)$$

This expression is equivalent to Eqs. 29-6 and 29-8.

29-2 Wave Properties of Electrons; Quantum Mechanics

De Broglie Wavelength

In his 1923 doctoral thesis, the French physicist Louis de Broglie suggested that matter might exhibit the properties of both particles and waves, similar to the wave-and-particle properties of light. Before de Broglie's hypothesis, matter was believed to consist of particles. An electron, for example, was considered to be a particle, which always has a well-defined location in space. De Broglie imagined it would be a beautiful symmetry of nature if the electron turned out to exhibit wave properties, as light does, so that both matter and light would display wave-particle duality.

The momentum of a photon is related to the wavelength of light by Eq. 28-6:

$$p = \frac{h}{\lambda}$$

or

$$\lambda = \frac{h}{p}$$

De Broglie believed that this same equation might apply to a particle of matter. Associated with the particle would be a wave with a wavelength, called the **de Broglie wavelength,** found by substitution of the particle's momentum $p = mv$ into the preceding equation.

$$\lambda = \frac{h}{p} = \frac{h}{mv} \tag{29-16}$$

In 1927 the American physicists Clinton Davisson and Lester Germer performed an experiment that demonstrated the wave properties of electrons, thereby confirming de Broglie's hypothesis. A beam of electrons was directed at a nickel crystal. The crystal scattered the electrons, producing a diffraction pattern similar to that produced by X rays (Fig. 29-14). The diffraction pattern was used to measure the wavelength of the diffracted waves. The wavelength measured by Davisson and Germer was equal to the de Broglie wavelength, predicted by Eq. 29-16.

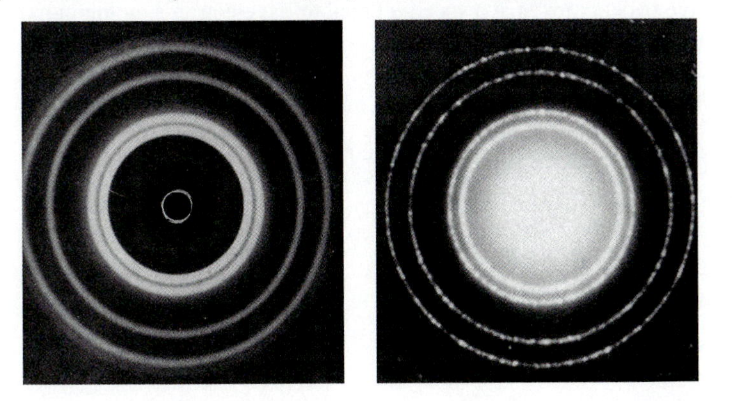

Fig. 29-14 Diffraction patterns formed when **(a)** X rays pass through an aluminum foil; **(b)** electrons pass through an aluminum foil.

EXAMPLE 2 Wavelengths of an Electron and a Particle of Dust

Find the de Broglie wavelength of (a) an electron moving at a speed of 1.0×10^5 m/s; (b) a dust particle of mass 1.0×10^{-10} kg, moving at a speed of 1.0×10^{-3} m/s.

SOLUTION (a) The electron has momentum

$$p = mv = (9.11 \times 10^{-31}\ \text{kg})(1.0 \times 10^5\ \text{m/s})$$
$$= 9.1 \times 10^{-26}\ \text{kg-m/s}$$

and a de Broglie wavelength given by Eq. 29-16:

$$\lambda = \frac{h}{p} = \frac{6.63 \times 10^{-34}\ \text{J-s}}{9.1 \times 10^{-26}\ \text{kg-m/s}} = 7.3 \times 10^{-9}\ \text{m} = 7.3\ \text{nm}$$

(b) The dust particle has momentum

$$p = mv = (1.0 \times 10^{-10}\ \text{kg})(1.0 \times 10^{-3}\ \text{m/s}) = 1.0 \times 10^{-13}\ \text{kg-m/s}$$

and a de Broglie wavelength

$$\lambda = \frac{h}{p} = \frac{6.63 \times 10^{-34}\ \text{J-s}}{1.0 \times 10^{-13}\ \text{kg-m/s}} = 6.6 \times 10^{-21}\ \text{m}$$

Although the electron's wavelength of 7.3 nm is small, it is on the order of 100 times the diameter of an atom. This wavelength is large enough that it can be measured under certain circumstances. For example, wave diffraction effects are evident in the Davisson-Germer experiment, where an electron beam is directed at a crystal with interatomic spacing much smaller than the electron's wavelength. (Remember that diffraction effects become important when the diffracting aperture is either smaller than or on the same order as the wavelength of the diffracted waves.)

In contrast to the electron, the dust particle has a de Broglie wavelength much too small to be measurable. Its wavelength of 6.6×10^{-21} m is much smaller than even the nucleus of an atom.

This example illustrates that de Broglie wavelengths, which are inversely proportional to mass, are too small to be measured for macroscopic bodies, even for dust particles, much less for larger objects. The wave nature of matter is significant only on the atomic or subatomic scale, which is the reason we are not normally aware of it!

Electron Wave Packets

From the beginning of our study of electricity in Chapter 17 we have treated electrons as tiny particles of matter. The de Broglie result indicates that this treatment is at best an oversimplification. We shall now describe more fully how this simple picture must be modified. Suppose we observe a single, isolated electron. The wave associated with the electron will not have a precise wavelength. There is always a range of wavelengths or frequencies present, as there is for light waves. But wavelength and momentum are related, according to the de Broglie relation. So there is a range of values of momentum, or an "uncertainty" in momentum Δp.

An electron with a fairly well-defined momentum or wavelength can be represented by a wave packet (Fig. 29-15), similar to a wave packet associated with a photon. The wave packet has a wavelength λ and a finite length Δx. If one tries to detect the electron, say by scattering an X ray off it, it will be found somewhere in the interval Δx. **The probability of finding the electron at any point is proportional to the intensity of the wave at that point.**

The wave packet moves at a speed v equal to the electron's momentum p divided by its mass m:

$$v = \frac{p}{m}$$

where p is related to λ by the de Broglie equation; that is:

$$p = \frac{h}{\lambda}$$

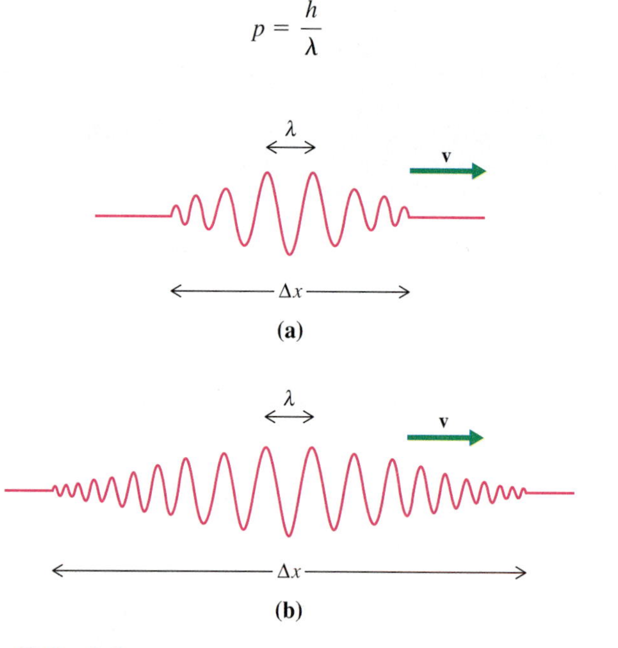

Fig. 29-15 Electron wave packets.

Uncertainty Principle

Heisenberg's uncertainty principle applies to electrons, as it does to photons. The product of the uncertainty in the electron's position Δx and the uncertainty in the electron's x component of momentum is greater than or equal to $h/2\pi$.

$$\Delta x \, \Delta p_x \geq \frac{h}{2\pi}$$

According to this relationship, if we have an electron wave packet that is fairly short (small Δx), it will have a fairly large uncertainty in momentum, Δp_x (Fig. 29-15a). On the other hand, if the uncertainty in momentum is smaller, that is, if the wave is more monochromatic, the uncertainty in position, Δx, will be larger (Fig. 29-15b).

Since the product $\Delta x\, \Delta p_x$ can be as small as $h/2\pi \approx 10^{-34}$ J-s, this means that it is possible for both the uncertainty in position and the uncertainty in momentum to be quite small, say, $\Delta x = 10^{-10}$ m, $\Delta p_x = 10^{-24}$ kg-m/s. In circumstances where dimensions of apertures and measurements of position and momentum are much coarser than the uncertainties in these quantities, wave properties of the electron will go unnoticed, and the electron will behave as a classical particle. We can always regard macroscopic bodies as classical particles, as we shall see in the following example.

EXAMPLE 3 Uncertainties in Position of an Electron and a Dust Particle

Find the width of wave packets associated with the electron and the dust particle in the last example, assuming a 1.0% uncertainty in the momentum of each. Let the width of each wave packet be the minimum allowed by the Heisenberg uncertainty principle.

SOLUTION Setting the product of uncertainties in position and momentum to the minimum value and solving for Δx, we find

$$\Delta x\, \Delta p = \frac{h}{2\pi}$$

$$\Delta x = \frac{h}{2\pi\, \Delta p}$$

For both the electron and the dust particle,

$$\Delta p = 0.010p$$

Inserting this in the preceding equation and substituting the values of momentum found in Example 2, we find for the electron

$$\Delta x = \frac{h}{2\pi(0.010p)} = \frac{6.63 \times 10^{-34}\text{ J-s}}{2\pi(0.010)(9.1 \times 10^{-26}\text{ kg-m/s})}$$

$$= 1.2 \times 10^{-7}\text{ m} = 120\text{ nm}$$

or roughly 16 times the electron's de Broglie wavelength of 7.3 nm. For the dust particle, we find

$$\Delta x = \frac{h}{2\pi(0.010p)} = \frac{6.63 \times 10^{-34}\text{ J-s}}{2\pi(0.010)(1.0 \times 10^{-13}\text{ kg-m/s})}$$

$$= 1.1 \times 10^{-19}\text{ m}$$

which is 17 times the dust particle's de Broglie wavelength of 6.6×10^{-21} m. The wave packets for both the electron and the dust particle are sketched in Figs. 29-16a and 29-16b respectively.

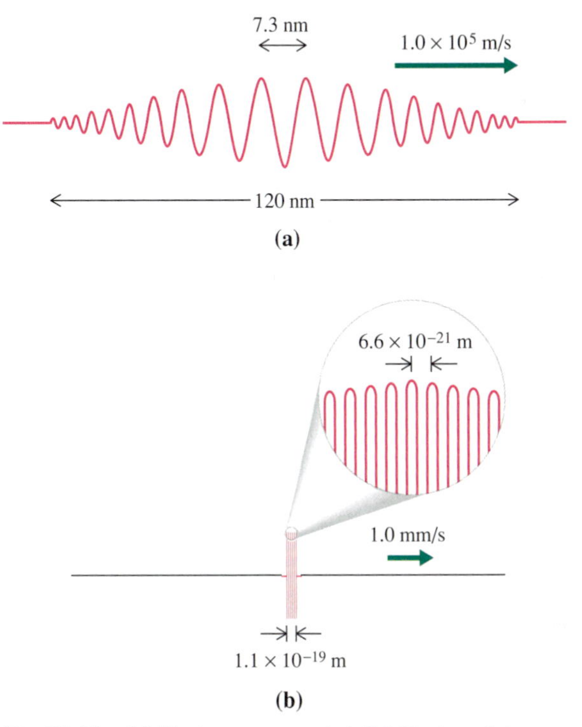

Fig. 29-16 **(a)** Electron wave packet. **(b)** Dust-particle wave packet.

Depending on the kind of observation of the electron that is attempted, it may or may not appear to behave as a classical particle. If we attempt to measure its location plus or minus a few hundred nm, we may regard the location of the electron as well defined. The electron can for the purpose of that kind of observation be regarded as a classical particle, moving at the speed of the wave packet (1.0×10^5 m/s).

But, if we attempt to measure the position more accurately, say, by scattering an X ray off it, we will find the electron localized at some one point within the wave packet. In this case the electron displays wave-particle duality. The electron propagates as a wave packet but interacts with the X ray as a particle.

The dust particle, on the other hand, behaves for all practical purposes as a classical particle, moving at the speed of its wave packet (1.0 mm/s). Both its wavelength and the length of its wave packet are so short that there is no way to detect wave properties.

Standing-Wave Explanation of Bohr's Angular Momentum Postulate

Shortly after proposing his matter wave hypothesis, de Broglie used it to provide some insight into the Bohr model of the hydrogen atom. There had been no physical basis for Bohr's postulate of quantized angular momentum. To derive Eq. 29-3 describing the hydrogen spectrum, Bohr had to postulate that the electron's angular momentum mvr was a multiple of $h/2\pi$:

$$mvr = n\frac{h}{2\pi} \qquad (n = 1, 2, 3, \ldots)$$

De Broglie realized that the factor mv is the electron's linear momentum, which should be related to the electron's de Broglie wavelength λ by Eq. 29-16 $\left(\lambda = \dfrac{h}{mv}\right)$, or

$$mv = \frac{h}{\lambda}$$

Inserting this expression into the preceding equation, we obtain

$$\frac{h}{\lambda}r = n\frac{h}{2\pi}$$

or

$$2\pi r = n\lambda \qquad (n = 1, 2, 3, \ldots) \qquad (29\text{-}17)$$

Since $2\pi r$ is the circumference of an orbit of radius r, this equation shows that Bohr orbits contain a whole number of electron waves: 1 wave in the first orbit, 2 waves in the second orbit, and so on. Orbits containing a fractional number of wavelengths do not occur. Fig. 29-17 illustrates de Broglie waves for the first three Bohr orbits.

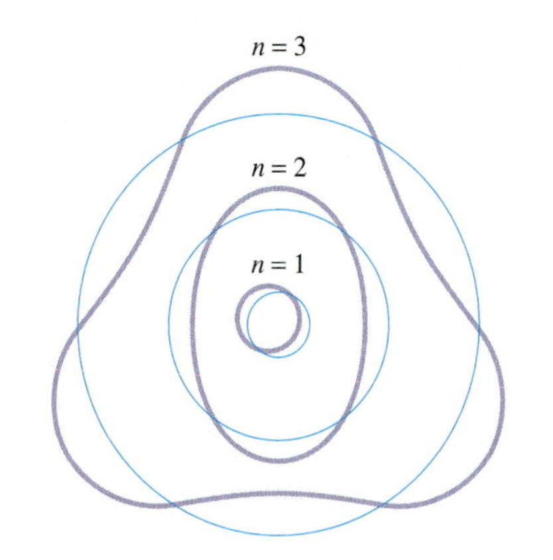

Fig. 29-17 De Broglie waves in the first three Bohr orbits.

Quantum Mechanics

Fig. 29-18 Erwin Schroedinger (1887–1961), one of the originators of quantum mechanics.

De Broglie's ideas concerning the wave properties of matter provided insight but did not give a complete theory for describing how matter waves would behave, in the way that Maxwell's equations provided a complete description of electromagnetic waves. A wave theory was developed a few years after de Broglie's work by Erwin Schroedinger (Fig. 29-18), who formulated a wave equation describing matter waves. About the same time Werner Heisenberg (Fig. 29-19) developed a different mathematical formalism, involving matrices for describing quantum phenomena. The Heisenberg and Schroedinger approaches were later shown to be equivalent. Many others contributed to the development of the new theory, which came to be called **quantum mechanics.** Unfortunately, the mathematics of quantum mechanics is far too complicated to explain in detail in an introductory text. We will therefore limit our discussion to a description of some of its more important predictions.

Quantum theory has been enormously successful in providing detailed understanding of interactions and processes on the molecular, atomic, and subatomic scales. Quantum theory is used to understand the electrical properties of conductors and semiconductors. Its success in this area has led to the development of an enormous technology of solid-state electronics. We shall discuss the basic elements of the quantum mechanical explanation of atomic structure in the next section. Essays on the laser and the quantum theory of solids are provided at the end of the chapter.

Mechanics, Classical and Quantum

Fig. 29-19 Werner Heisenberg (1901–1976) helped to originate quantum mechanics and discovered the uncertainty principle.

Quantum mechanics has replaced classical mechanics as our most fundamental way of describing physical processes. Quantum mechanics is successful at describing phenomena on the atomic and subatomic scale, where there is a complete failure of classical analysis in terms of forces and Newton's laws of motion. Quantum mechanics can also be used to explain everything that classical mechanics explains, that is, macroscopic phenomena such as blocks on inclined planes. However, because of its greater mathematical complexity, quantum mechanics is not applied to such problems. It is possible to show that for macroscopic phenomena, the predictions of classical mechanics are an excellent approximation to the predictions of quantum mechanics. Quantum mechanics provides the most fundamental picture of physical reality. It is a "covering" theory, a theory that covers everything that was known before in the classical domain as well as the new quantum phenomena. But as a practical matter, classical mechanics is still very useful.

29-3 Quantum Theory of Atomic Structure and Spectra; X Rays

The Hydrogen Atom

The Bohr theory of the atom was important historically in providing some ideas about atomic structure. But the Bohr model has been replaced by a description using quantum mechanics. The quantum mechanical explanation of atomic structure has succeeded in explaining many more phenomena in far greater detail than the Bohr model. For example, quantum mechanics successfully explains multielectron atoms and intensities of spectral radiation, which the Bohr model is unable to do.

Some elements of a quantum mechanical description of atoms are seen in the Bohr model. Like the Bohr theory, quantum mechanics describes states using **quantum numbers** to specify those states (numbers like Bohr's n used to label energy states E_n). Quantum mechanics also predicts **quantized values of energy and angular momentum,** which Bohr's theory assumes. Finally, like Bohr's theory, quantum mechanics predicts **emission of radiation by electrons undergoing transitions** between energy levels.

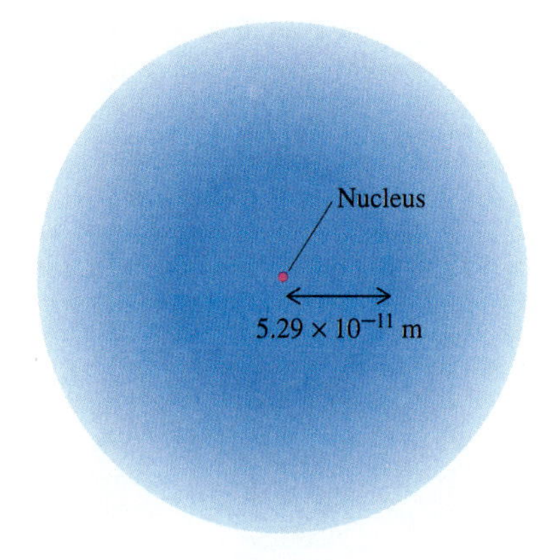

Fig. 29-20 Electron cloud for an electron in the ground state of the hydrogen atom.

The exact position of the electron within a hydrogen atom is not specified, as it was in the Bohr atom. No longer do we have the simple picture of a particle orbiting a nucleus. Instead we have a **wave function,** whose magnitude at any point indicates the probability of locating the electron at that point. The magnitude of an electron's wave function is constant in time but varies smoothly over some region of space within the atom. The magnitude of the wave function is often represented graphically by a cloud of varying density, as indicated in Fig. 29-20 for the ground state of the hydrogen atom. The density of the electron cloud at any point indicates the probability of finding the electron at that point.

Four quantum numbers are required to specify the wave function of an electron in a hydrogen atom. The **principal quantum number** n specifies the average distance of the electron from the nucleus and can have any positive integral value:

$$n = 1, 2, 3, \ldots \tag{29-18}$$

The **orbital quantum number** ℓ specifies the degree to which the electron cloud differs from a spherically symmetric distribution and can have any of the following values, related to the value of n:

$$\ell = 0, 1, \ldots, n - 1 \tag{29-19}$$

For example, if $n = 1$, the only possible value of ℓ is 0. For $n = 2$, ℓ can have either of the values 0 or 1. For $n = 3$, possible values of ℓ are 0, 1, and 2. The greater the value of ℓ, the less spherically symmetric is the electron cloud. Fig. 29-20 shows the $n = 1$, $\ell = 0$ state.

The **magnetic quantum number** m specifies the orientation in space of the electron cloud, with given values of n and ℓ. The magnetic quantum number is so named because only in an external magnetic field does the electron's energy depend on this quantum number, which can be any integer in the range $-\ell$ to $+\ell$.

$$m = -\ell, -\ell + 1, \ldots, \ell - 1, \ell \tag{29-20}$$

For example, if $\ell = 2$, possible values of m are $-2, -1, 0, +1$, and $+2$. If $\ell = 0$, then $m = 0$ is the only possibility; the distribution is spherically symmetric and there is therefore only one possible orientation.

The **spin quantum number** s specifies one of the two possible "spin" states of the electron:

$$s = +\tfrac{1}{2} \quad \text{(spin up)} \tag{29-21}$$

or

$$s = -\tfrac{1}{2} \quad \text{(spin down)} \tag{29-22}$$

When the spin quantum number was originally proposed, the electron was thought of as a small, spinning sphere of negative charge, and the two spin states were supposed to represent two possible directions of rotation—spin up meant counterclockwise rotation (viewed from above); spin down meant clockwise rotation. However, the simple picture of a spinning electron, like the Bohr picture of an electron as a particle orbiting the nucleus, has since proved untenable. Spin is simply an intrinsic property of electrons. Associated with this property is a quantum number, which can take two values, $+\frac{1}{2}$ or $-\frac{1}{2}$. We refer to these states as spin up or spin down, respectively.

Fig. 29-21 shows possible combinations of three of the quantum numbers specifying states of the hydrogen atom. The fourth quantum number s is omitted for clarity, since there are always just two spin states for any set of the other quantum numbers.

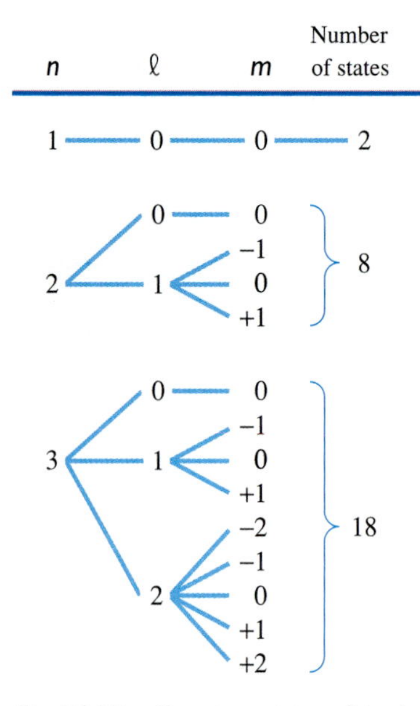

Fig. 29-21 Quantum states of the hydrogen atom. For each combination of quantum numbers n, ℓ, and m, there are two quantum states, corresponding to spin up or spin down.

Multielectron Atoms

A piece of lead looks quite different from a piece of gold because lead and gold are composed of different kinds of atoms. Each element has its own distinctive kind of atom, whose structure determines the element's physical and chemical properties. An element's **atomic number,** denoted by Z, specifies the number of protons in the nucleus. For a neutral atom, Z is also equal to the number of electrons. For example, $Z = 1$ for hydrogen, which means that a neutral hydrogen atom has one proton in the nucleus and one electron outside the nucleus. For lead, $Z = 82$, meaning that there are 82 protons in the lead nucleus and 82 electrons outside the nucleus. The quite different properties of gold are a consequence of its atomic number of 79.

The physical and chemical properties of an atom are determined by the arrangement of electrons in the volume of space surrounding the nucleus. The quantum states of individual electrons are specified by the same four quantum numbers used to specify the quantum states of the hydrogen atom, though the detailed descriptions of those states are different. For example, both the wave function and the energy of the electron closest to the nucleus are quite different for each atom, though the quantum numbers are the same: $n = 1,\ \ell = 0,\ m = 0,\ s = \pm\frac{1}{2}$.

In 1925 Austrian physicist Wolfgang Pauli discovered a simple principle that governs how electrons occupy various quantum states. This principle, known as the **Pauli exclusion principle,** states that **no two electrons can have the same set of quantum numbers.**

For an atom in a state of lowest possible total energy, called the **ground state,** the electrons will each have the lowest energy consistent with the Pauli principle. Those atomic electrons with less energy are closer to the nucleus than those with greater energy. (This is analogous to earth satellites, for which the energy is less in lower orbits than in higher ones.) The average distance of an electron from the nucleus increases with an increase in either n or ℓ. Therefore the energy of an electron also increases with an increase in either of these numbers. The lowest-energy states are those with the smallest values of n and ℓ. For helium ($Z = 2$), both electrons can be in a state with $n = 1$, $\ell = 0$, because there are two such states, one with spin up and the other with spin down. For lithium ($Z = 3$), two of the three electrons fill the two possible $n = 1$ states, and so the third electron must be in one of the $n = 2$ states. Groupings of some quantum states with the same principal quantum number are often referred to as "shells." Those states with $n = 1$ form the **K shell;** those with $n = 2$ form the **L shell;** and so forth. Fig. 29-22 shows the electron configurations of atoms with atomic numbers from 1 to 14.

The periodic table (on the inside back cover of this text) is a chart of the elements ordered according to atomic number and arranged in columns so that elements with similar chemical properties appear in the same column. The similarities can be understood on the basis of the atomic electron configurations. For example, He and Ne are both gases that tend not to react chemically with other elements. The reason is that in the atoms of each of these elements the number of electrons exactly completes shells. He has a complete K shell, and Ne has complete K and L shells. When shells are complete, electrons in them are tightly bound; therefore neither He nor Ne has much tendency to either gain or lose electrons. On the other hand, both Li and Na are highly chemically reactive because each has one electron beyond complete shells. These electrons, called "valence electrons," are easily given up in interactions with other atoms, which is the reason these elements are so reactive.

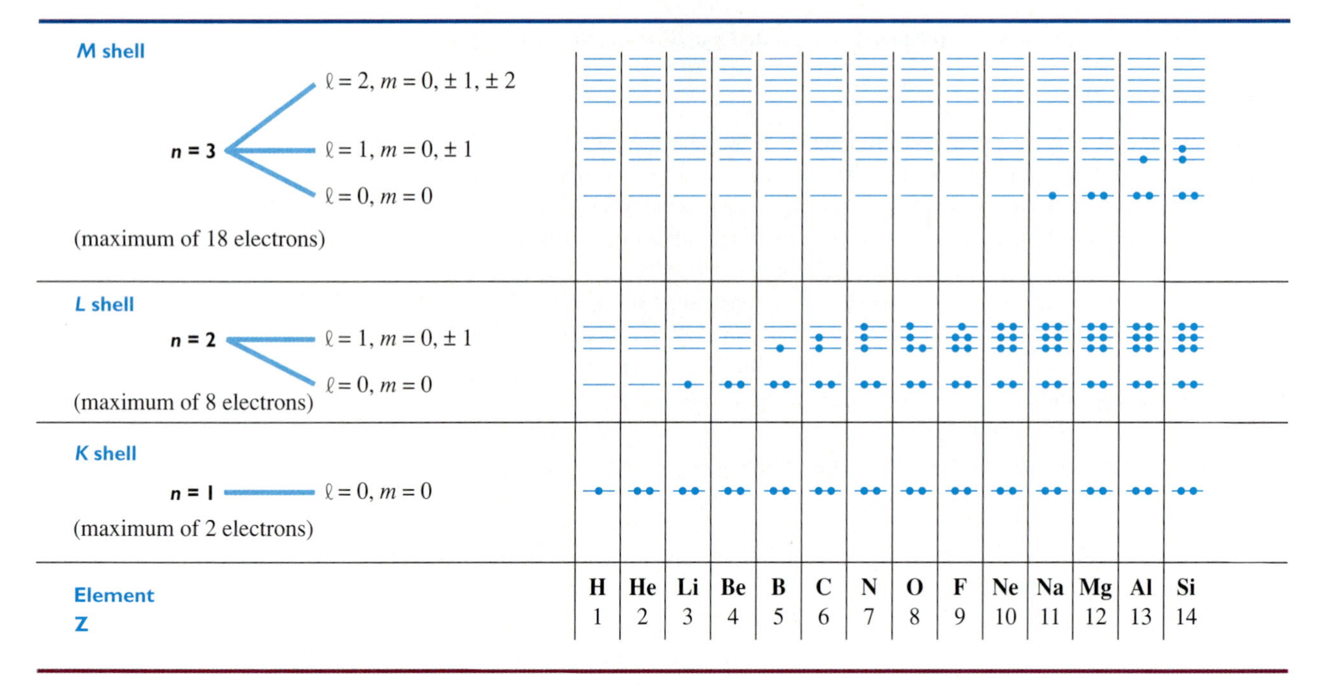

Fig. 29-22 Electron configurations of elements with atomic numbers from 1 to 14.

The Pauli principle governs structure of atoms and thereby determines the nature of chemical interactions. If the Pauli principle could be violated, an atom in its ground state would have all electrons in the lowest-energy state. Atoms would then be much smaller than they actually are. The myriad complex chemical interactions that occur among different kinds of atoms would be greatly changed and simplified. Indeed, life as we know it would be impossible. But the Pauli principle is obeyed, and atoms of increasing atomic number have increasing complexity.

Visible Atomic Spectra

Like hydrogen, each element produces its own characteristic line spectrum in a gas discharge tube (Fig. 29-2). Electrons and ions are accelerated through the tube as a result of the potential difference applied to the tube. Some of the kinetic energy supplied to these current carriers is given up through collisions with gas atoms in the tube. The energy given up in one of these collisions is on the order of a few eV, which is enough to excite an outer-shell electron to an energy level above its ground-state energy. The electron then falls back to the ground state, losing its excess energy in the form of a visible photon. By measuring the frequency of the emitted light, one can get information about the energy levels of atomic electrons.

EXAMPLE 4 Atomic Energy Levels of Sodium

Sodium vapor lamps emit yellow light (Fig. 29-2). The highest-intensity line in the sodium spectrum has a wavelength of 589 nm, which corresponds to a transition of the M-shell valence electron from an excited state ($n = 3$, $\ell = 1$) to the ground state ($n = 3$, $\ell = 0$). Find the energy difference between these states.

SOLUTION The energy difference equals the energy of the photon emitted:

$$\Delta E = hf = \frac{hc}{\lambda} = \frac{(6.63 \times 10^{-34}\ \text{J-s})(3.00 \times 10^{8}\ \text{m/s})}{(589 \times 10^{-9}\ \text{m})}$$

$$= 3.38 \times 10^{-19}\ \text{J}\,\frac{1\ \text{eV}}{1.60 \times 10^{-19}\ \text{J}}$$

$$= 2.11\ \text{eV}$$

Fluorescence and Phosphorescence

Fluorescence and phosphorescence are phenomena that occur in certain substances when they absorb ultraviolet light. Part of the energy absorbed from a UV photon gives the atoms a higher vibrational energy state (corresponding to a higher temperature). The rest of the energy results in a transition of an atomic electron to a higher energy level.

In the case of fluorescence, there is an immediate transition of the excited electron back to the ground state, accompanied by the emission of visible light. Fluorescent light bulbs are essentially gas discharge tubes containing mercury vapor emitting primarily ultraviolet light, which causes a powder on the inner surface of the bulbs to fluoresce. Day-Glo paints and laundry detergent whiteners contain fluorescent materials. The glowing appearance of white or light-colored clothes containing these materials is quite spectacular when illuminated by an ultraviolet "black" light on a dance floor.

The process is similar but more complicated in the case of phosphorescence, and so the deexcitation and emission of light is delayed somewhat. Depending on the phosphorescent material, the delay time can range from seconds to hours. You can observe this time-delayed emission in phosphorescent displays that glow in the dark, on watch dials, for example. The inner surface of television tubes are coated with phosphors, which absorb the energy of the television's electron beam, and emit light just long enough so that the entire image on the screen can be seen as the beam sweeps over it.

(a)

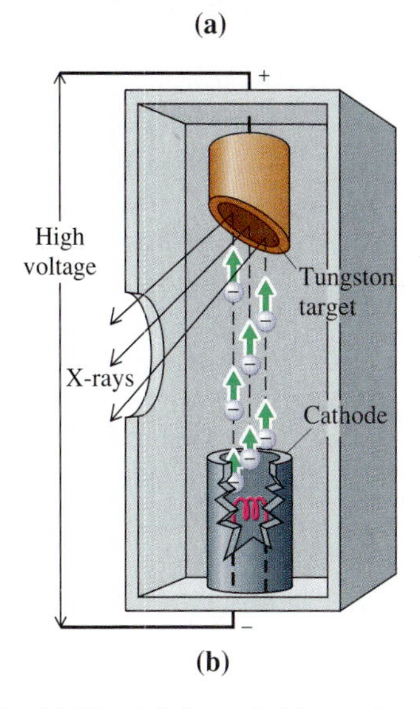

(b)

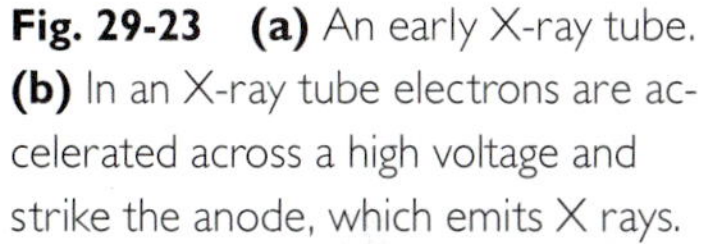

Fig. 29-23 **(a)** An early X-ray tube. **(b)** In an X-ray tube electrons are accelerated across a high voltage and strike the anode, which emits X rays.

Atomic X-Ray Spectra

It is possible to excite atomic electrons sufficiently to make them produce X rays as well as visible light. X rays were first observed quite by accident. In 1895 Wilhelm Roentgen noticed that when a very high voltage was applied to a cathode ray tube a fluorescent mineral some distance away began to glow. An invisible radiation was being produced by the tube (Fig. 29-23). The radiation emanating from the tube also exposed photographic film. Roentgen referred to the unknown rays he discovered as **X rays.** Eventually it was recognized that X rays are electromagnetic radiation of very short wavelength, on the order of about 0.1 nm.

Soon after his discovery, Roentgen began using X rays to produce photographs of the human body. Fig. 29-24 shows one of his first X-ray photographs. A beam of X rays directed at a human subject will be partially absorbed, with the amount of absorption depending on the density of tissue encountered. The X rays that pass through the body and strike a photographic plate provide an internal image of the body. X rays are commonly used in medical diagnosis—from dental X-rays to CAT scans.*

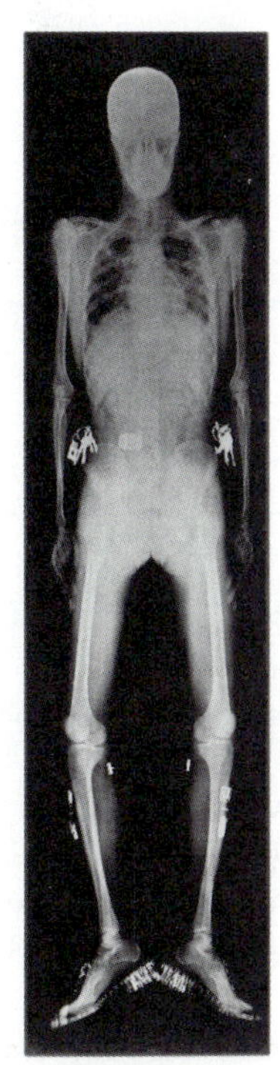

Fig. 29-24

*CAT stands for computerized axial tomography, where tomography refers to an image of layers of tissue. In a CAT scan, a beam of X rays successively scans layers within the body, and computerized analysis is used to enhance the image.

The intensity of radiation produced by a typical X-ray tube is indicated in Fig. 29-25. X-ray energy is provided by the kinetic energy of electrons accelerated across the X-ray tube. When these electrons strike the anode, they produce X rays through two mechanisms:

1. One of the anode's inner-shell atomic electrons can absorb sufficient energy from an accelerated electron to excite the atomic electron to a higher-energy state in one of the outer shells. When the electron falls back to its original state, it emits a photon with energy equal to the large energy difference between the states. The high energy of the photon means that the radiation is of high frequency, or short wavelength, that is, an X ray. This mechanism produces X rays of sharply defined frequency, or wavelength, since the photon energy corresponds to energy differences between precise atomic energy levels. These X rays show up as spikes in Fig. 29-25. Their spectrum is characteristic of the material of which the anode is made, and they are therefore called **characteristic X rays.**
2. Most of the electrons striking the anode do not transfer their energy to atomic electrons. Instead they directly emit X-ray photons as they lose kinetic energy, decelerating as a result of electrical forces acting on them as they penetrate the anode. This kind of radiation is called **Bremsstrahlung,** which is German for 'braking radiation'. A single electron with initial kinetic energy K can emit one photon or several photons. When one photon is emitted, its energy equals the maximum value K, and its wavelength is the minimum value (as indicated in Fig. 29-25). But when several photons are emitted, the energy of each can have any value, so long as the sum of the photon energies equals K.

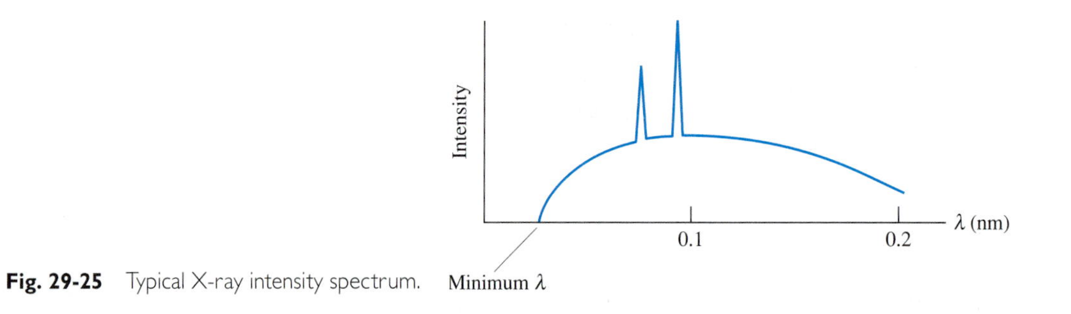

Fig. 29-25 Typical X-ray intensity spectrum.

EXAMPLE 5 X-ray Wavelength

Find the minimum wavelength produced in an X-ray tube, operating with a potential difference of 5.0×10^4 V across the tube.

SOLUTION The kinetic energy K of an electron accelerated from rest across a potential difference V equals the decrease in the electron's potential energy, that is, the product of the electron's charge magnitude e and V (Eq. 18-3: $U_E = q'V$). Thus each of the electrons accelerated through the potential difference of 5.0×10^4 V has 5.0×10^4 eV of kinetic energy. When all of an electron's kinetic energy goes into producing a single photon, that photon will have the maximum possible energy of 50,000 eV and the minimum possible wavelength. We relate the photon's energy E to its wavelength λ and solve for λ.

$$E = hf = \frac{hc}{\lambda}$$

or

$$\lambda = \frac{hc}{E} = \frac{(6.63 \times 10^{-34}\text{ J-s})(3.0 \times 10^8\text{ m/s})}{(5.0 \times 10^4\text{ eV})(1.6 \times 10^{-19}\text{ J/eV})}$$

$$= 2.5 \times 10^{-11}\text{ m}$$

$$= 0.025\text{ nm}$$

A Closer Look

Lasers

The invention of the first laser in 1960 gave the world a source of light that was like nothing seen before—intense, monochromatic light in a pencil-thin beam that does not spread out from the source like ordinary light. The unique properties of laser light were created by finding special electron energy levels and by preparing an environment that would produce intense light by stimulating transitions between those energy levels. But before lasers could be imagined, the process of "stimulated emission" had to be discovered. We shall first see how atoms emit light through this process and then show how we amplify such light in a laser. The term "laser" stands for **l**ight **a**mplification by **s**timulated **e**mission of **r**adiation.

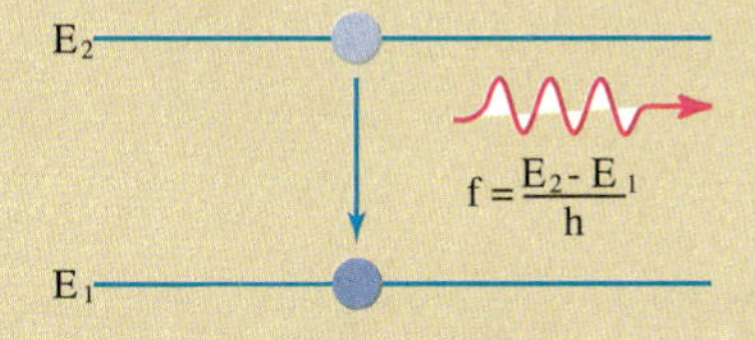

Fig. 29-A Spontaneous emission. An atomic electron spontaneously goes from a state of higher energy E_2 to a state of lower energy E_1. The emitted photon has energy equal to the energy lost by the electron.

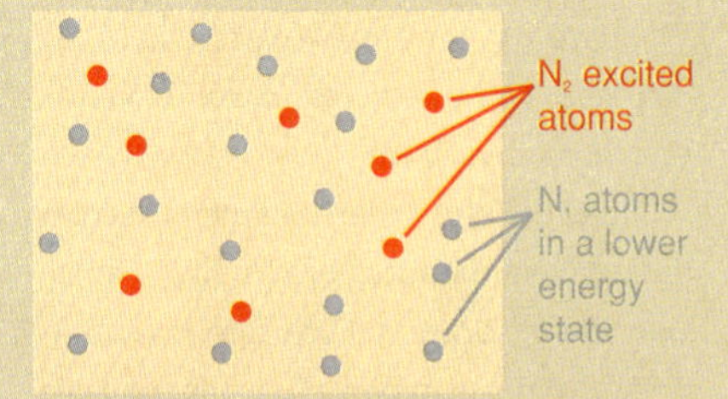

Fig. 29-B Atoms in excited and lower energy states. The excited atoms can spontaneously emit photons and go to lower energy states.

Spontaneous Emission

We have already discussed emission of radiation by atoms in gas discharge tubes, the process that produces atomic spectral lines. This process, called *spontaneous emission,* occurs when an electron in a higher energy ("excited") state* spontaneously goes to a lower energy state and emits a photon that carries off the excess energy (Fig. 29-A).

Suppose a certain system of radiating atoms consists of N_1 atoms in the lower energy state E_1 and N_2 atoms in the excited state E_2 (Fig. 29-B). The greater the number in the excited state, the more atoms there are to emit radiation. Thus it is not surprising that the rate of spontaneous emission by the system turns out to be proportional to the number N_2 in the excited state:

$$\text{Rate of spontaneous emission} \propto N_2$$

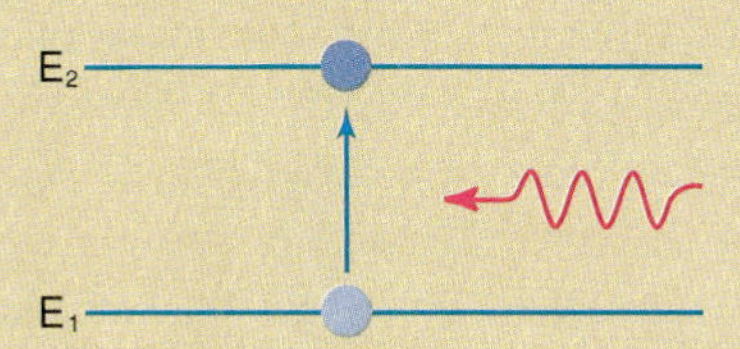

Fig. 29-C Stimulated absorption. Absorption of a photon by an atomic electron stimulates the electron to a higher energy state.

*Excited states can be created by collisions between electrons and atoms within a gas in a high-voltage gas discharge tube.

Stimulated Absorption

Consider the process that is the reverse of emission—the absorption of light by matter. In this process an atom goes from a lower energy state to a higher energy state by absorbing a photon (Fig. 29-C). An isolated atom cannot spontaneously undergo such a transition. A photon must first be present to be absorbed. Because the photon's presence is required to stimulate this process, we call it *stimulated absorption.* The rate of stimulated absorption of radiation by a system is proportional to both the number of photons (N_{ph}) and to the number of atoms available to absorb radiation, that is, the number N_1 in the lower energy state (Fig. 29-D).

$$\text{Rate of stimulated absorbtion} \propto N_1 \times N_{ph}$$

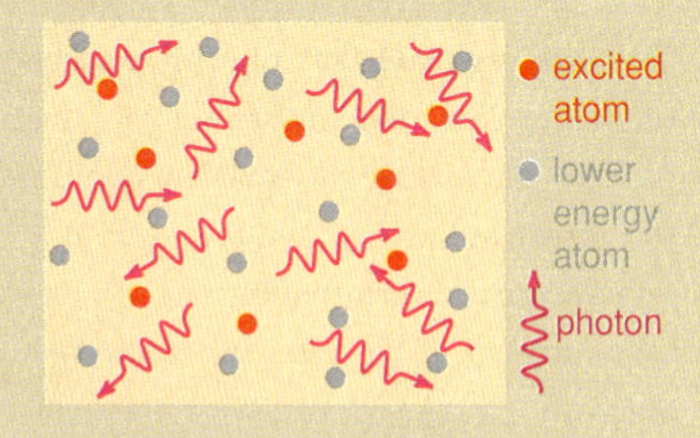

Fig. 29-D A system of photons and atoms. Photons are both emitted and absorbed.

Continued.

A Closer Look

Stimulated Emission

In 1917 Albert Einstein predicted the existence of a second kind of emission process, called *stimulated emission,* which, like stimulated absorption, requires an initial stimulating photon (Fig. 29-E). Einstein based his prediction on the following kind of thermodynamic argument.* When thermal equilibrium is established between a system of atoms and photons (for example, between the walls of a black body cavity and the radiation inside), the number of atoms in an excited state, the number of atoms in a lower enery state, and the number of photons are all constant. Thus the number of atoms absorbing radiation must exactly balance the number emitting radiation, so that the number of atoms in each state does not change. Spontaneous emission alone could not balance stimulated absorption because the absorption process is proportional to the number of photons present, whereas spontaneous emission is independent of the number of photons. For thermal equilibrium to be possible at different temperatures with different amounts of radiation present, the rates of both emission and absorption must increase as radiation increases. Thus we conclude that thermal equilibrium requires the existence of the process of stimulated emission, which, like stimulated absorption, proceeds at a rate proportional to the number of photons present. The rate of stimulated emission is also proportional to the number N_2 of atoms in an excited state—that is, to the number of atoms available for emission, just as the rate of stimulated absorption is proportional to the number of atoms in the lower energy state.

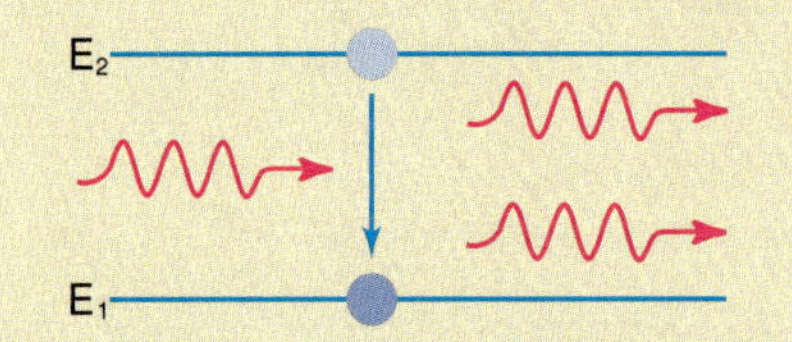

Fig. 29-E Stimulated emission. An incident photon stimulates an atomic electron in an excited state to emit a photon and go to a lower energy state. Thus two photons are produced where there was only one before.

$$\text{Rate of stimulated emission} \propto N_2 \times N_{ph}$$

Einstein showed that the emitted radiation would have the same frequency, phase, and direction as the incident radiation stimulating the transition.

Light Amplification

Once the process of stimulated emission is understood, it is easy to imagine how this process could be used to amplify light—to create a kind of chain reaction of transitions that produces more photons at each step (Fig. 29-F). Because the emitted radiation is always in phase with the incident radiation and also has the same frequency and wavelength, such amplification would produce intense, coherent, monochromatic, one-directional light, which is precisely what we see in a laser beam. There is, however, an obstacle. Remember, the rates of both stimulated absorption and stimulated emission are proportional to the number of photons present. Thus while radiation is being emitted by atoms in excited states, radiation is also being absorbed by atoms in lower energy states. And although increasing levels of radiation can stimulate more of the excited atoms to emit, that same radiation would also stimulate more of the lower energy atoms to absorb. Thus we cannot amplify light unless we can find some way to make emission dominate absorption.

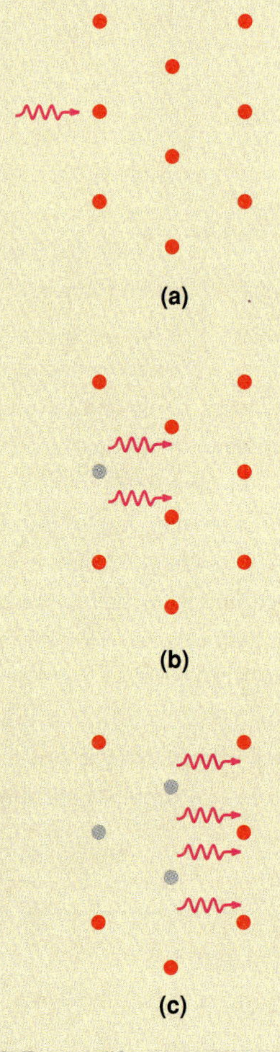

Fig. 29-F Amplification of light through repeated stimulated emission by excited atoms.

Consider the other important factors determining the rates of stimulated emission and absorption. Each rate is proportional to the number of atoms available for the respective process (that is, the rate of absorption is proportional to the number of atoms in the lower energy state, and the rate of emission is propor-

*The argument we present here is a simplified variation on Einstein's argument. Our reasoning is based on the assumption of approximately equal, temperature-independent populations of the two energy levels, a condition that is satisfied only for relatively high temperatures and a small difference between energy levels. Einstein's more detailed analysis applies to all temperatures and energies.

tional to the number of atoms in the excited state). Thus if we want the emission process to dominate, we must have more excited atoms than lower energy atoms. Now this presents a real problem because ordinarily in a many-particle system of atoms, many more of the atoms will be in the lower energy state than in the excited state, as illustrated for example, in Figs. 29-B and 29-D. So it would seem that we can absorb light much more efficiently than we can emit it.

The problem of creating an environment suitable for light amplification is to find a way to reverse the usual distribution of atoms among energy states. To create a laser we need to find a way to have more atoms in the higher energy state than in the lower energy state—a situation called *population inversion* (Fig. 29-G).

But how does one create a population inversion? Various methods have been used in different kinds of lasers. All involve some kind of "pumping" of electrons into a high energy state that has an unusually long lifetime. Such a state, called a *metastable* state, creates a kind of bottleneck for excited electrons trying to make transitions back to lower energy states. Electrons pile up in the metastable state and create a population inversion with respect to the lower energy state they decay into. Amplification can then take place.

Helium-Neon Laser

We shall describe in detail only one of the multitude of lasers available today: the helium-neon laser, which contains a mixture of helium and neon gases. Fig. 29-H shows the most important transitions in a helium-neon laser. The process begins with helium atoms being excited by a high-voltage discharge through the sealed tube containing the gases. Some of the helium atoms in this excited state transfer their excitation energy to neon atoms through atomic collisions. This transfer is possible because one of the neon levels has energy nearly identical to the excited helium state. The collision gives a neon electron enough energy to put it into the excited state of energy E_3, while the helium electron falls back to its ground state energy. Neon's E_3 state is a metastable state, whereas the lower energy states into which the neon eventually decays have much shorter, more normal lifetimes. Electrons pile up in the state E_3 but are rapidly removed from state E_2. Thus the desired population inversion is created between states E_3 and E_2, and amplification can take place.

Laser amplification is greatly enhanced by a feedback mechanism, a way to get the light produced by stimulated emission to continually stimulate further transitions. This is accomplished by placing parallel mirrors at opposite ends of the tube, so that a standing wave is set up between the mirrors (Fig. 29-I). The back mirror M_1 is perfectly reflecting, but the front mirror M_2 emits a small amount of the light incident on it. (Its reflectance is typically about 98%.) It is this small emitted fraction of light that is the laser's output.

The well-defined direction of the laser's beam is the result of the repeated reflections between the mirrors. Only waves traveling along the length of the tube stay within the tube, and only these waves are then strongly amplified by the gas medium. The typical angular spread in

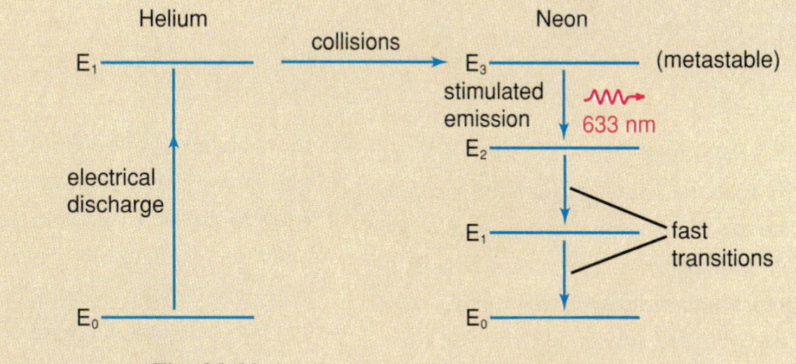

Fig. 29-H Important transitions in a helium-neon laser.

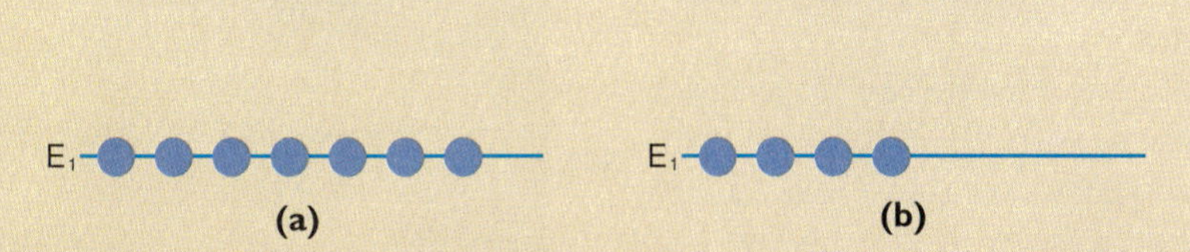

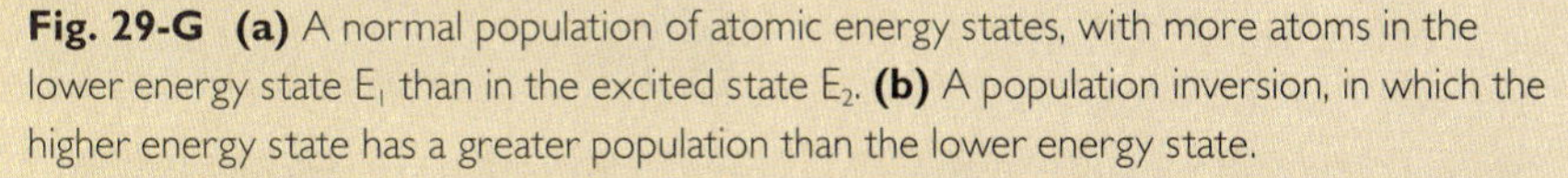

Fig. 29-G **(a)** A normal population of atomic energy states, with more atoms in the lower energy state E_1 than in the excited state E_2. **(b)** A population inversion, in which the higher energy state has a greater population than the lower energy state.

Continued.

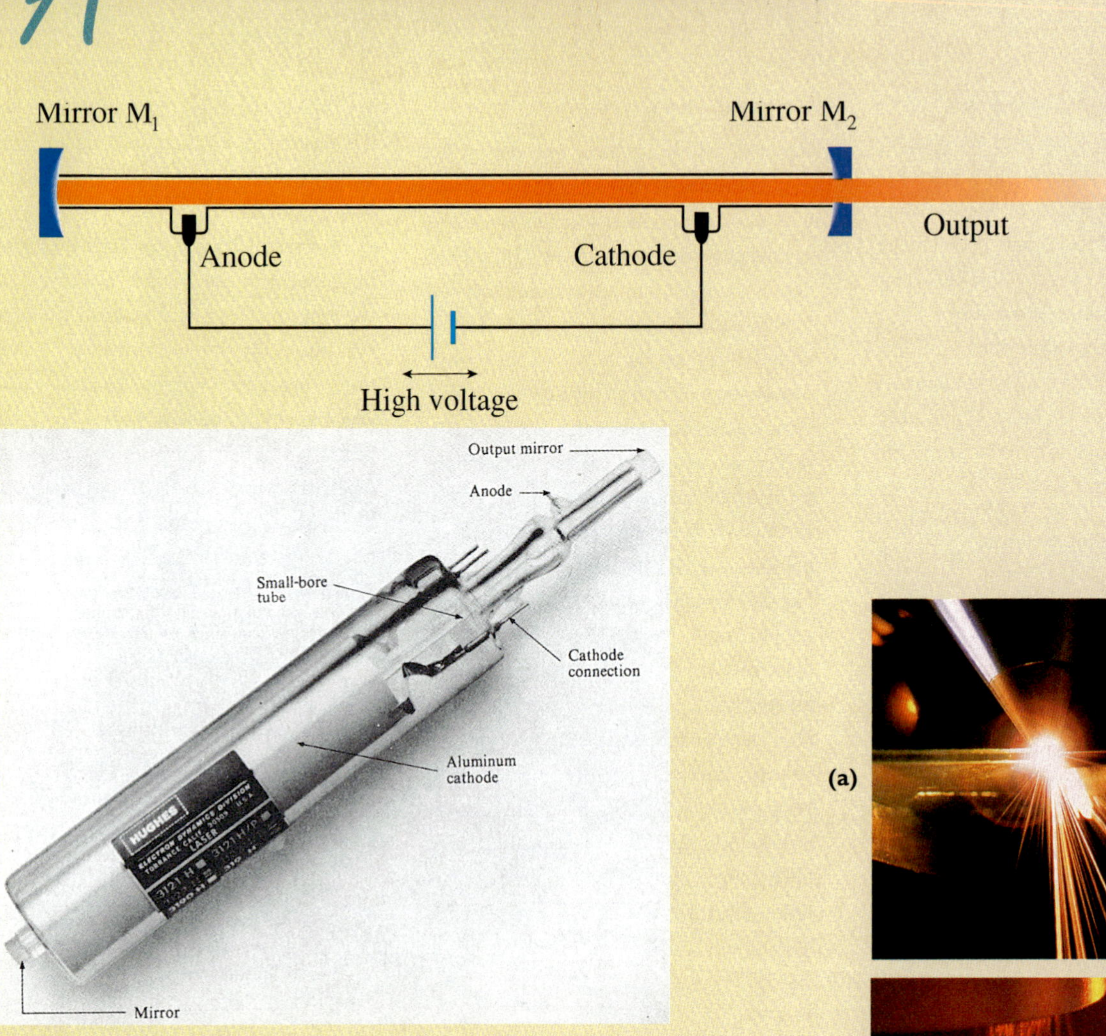

Fig. 29-I A helium-neon laser tube.

a helium-neon beam is about 10^{-3} rad; that is, the width of the beam increases by about 1 mm for every meter it travels.

Although the output of a laser is intense, monochromatic light, most lasers are not very energy efficient. For example, the efficiency of a helium-neon laser is only about 0.02%. This means that about 5 watts of electrical power must be supplied for every milliwatt of laser output. Carbon dioxide lasers, often used in high-power industrial applications, can have much higher efficiency—on the order of 30%.

Lasers are increasingly becoming a common part of technology. We have carbon dioxide lasers for welding and cutting metals, an assortment of semiconductor lasers for fiber-optic communication systems, laser printers and CD players, various kinds of lasers for eye and other surgery, and the common helium-neon laser for holography, for demonstrating principles and performing experiments in physics, and even for use in supermarket scanners that read product bar codes and automatically register sales.

Scientists and engineers have displayed great ingenuity in inventing lasers and applying them. The beauty of this amazing light and the story of its discovery and development are a testament to the seemingly endless wonders of nature that can be revealed through insight into nature's laws.

(a)

(b)

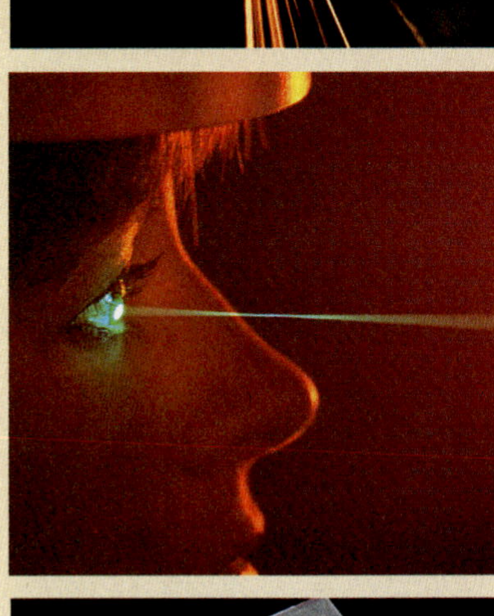

(c)

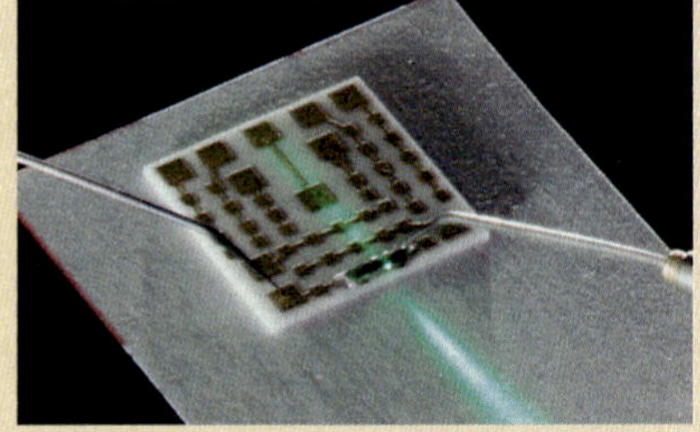

Fig. 29-J **(a)** Laser welding. **(b)** Laser surgery. **(c)** A solid state laser.

Semiconductor Electronics

Calculators, televisions, cameras, watches, dishwashers, automobiles—indeed nearly everything produced today that has electrical components—all make use of semiconductor electronics in the form of tiny "integrated circuits." These products all utilize a controlled flow of electrons in semiconductor materials such as silicon. An integrated circuit, or "IC," is a base chip of silicon, a single crystal roughly the size of a dime, with a huge number of microscopic electrical components (resistors, capacitors, and transistors) formed on the crystal's surface in a sequence of delicate steps, including depositing materials on the silicon, chemical etching, oxidation, and so on. The connected electrical components form a microscopic circuit, which is designed to achieve some specific purpose. For example, the purpose of a computer memory chip is to store information, and the purpose of an IC chip in a stereo amplifier is to amplify an input signal, converting a time-varying input voltage with an amplitude of millivolts or less to an output voltage of perhaps a few volts. One of the great advantages of ICs is that designers of electronic equipment have these chips, which are prepackaged integrated circuits that can be utilized for specific design objectives.

The number of components on a chip ranges from a few thousand on a stereo amplifier chip up to about a million on a computer memory chip. Much of the technological progress of the last 30 years is the direct result of IC applications. For example, a laptop personal computer can now have more computing power than the early room-sized, vacuum-tube computers of the 1950s. A personal portable radio (such as a Walkman), small enough to carry comfortably as you walk or run, may sound as good as an old-fashioned portable radio that had much bigger working parts, greater power requirements, and a much bigger battery. These technical advances began with the invention of the transistor in 1948. We shall see that the transistor, like the laser, exploits quantum energy levels—more specifically, quantum levels in semiconducting solids.

A computer memory chip smaller than a thimble.

Section of a computer memory chip magnified 300 times.

Energy Levels in Solids

The application of quantum mechanics to solids* shows that electrons in solids, like electrons in isolated atoms, occupy quantized energy levels. Once we see the structure of these energy levels, we can understand the process of electrical condution in solids.

*The solids that we describe here are "crystalline" solids, in which the atoms form a definite repeated pattern. Metals, semiconductors, and insulators such as diamond and quartz are all crystalline. Other solids, such as wood, glass, and rubber are not crystalline; their atoms are not arranged in a regular, orderly pattern.

To make this structure plausible, we consider first the structure of atomic energy levels. The simplicity of the hydrogen atom is reflected in the simplicity of its energy levels and its visible spectrum. Few levels of the right energy are available to produce visible photons, and so there are only a few visible lines (Fig. 29-2). Other atoms have more complex energy level structure and produce many more visible lines, but they still produce discrete *lines* (as opposed to a continuous spectrum), which means the atomic levels are discrete.

We can think of a solid as being formed by isolated atoms brought close enough together that atomic electron clouds overlap. The energy levels of the entire system of atoms that constitutes the solid is formed by the splitting of each atomic energy level into an enormous number of closely spaced levels with different energies. Thus solid state energy levels consist of broad bands, each of which contains many levels spaced so closely that they appear continuous. The effectively continuous range of energy differences between solid state energy levels means that when a solid is heated enough to emit light, the spectrum of colors it emits is continuous.

Continued.

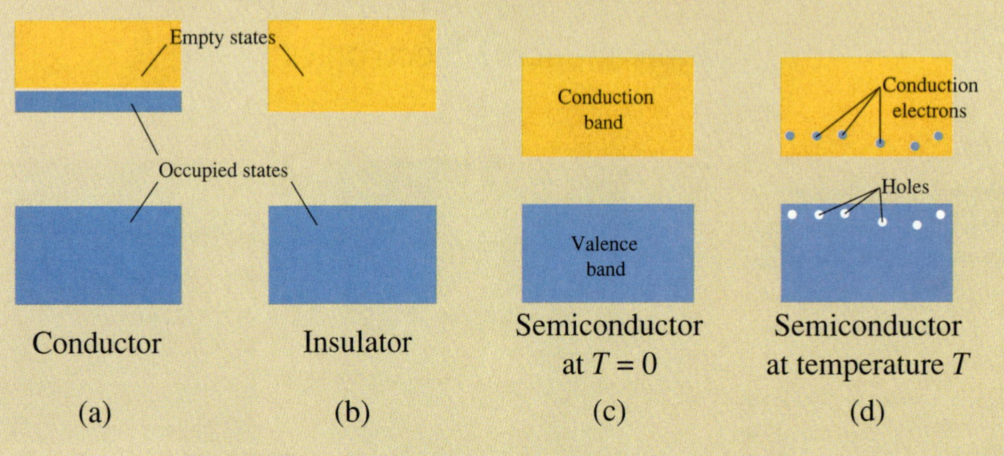

Fig. 29-K Band structure of conductors, insulators, and semiconductors.

Fig. 29-K shows typical energy bands for conductors, insulators, and semiconductors. According to the Pauli principle, each quantum state can be occupied by no more than one electron. Thus in the lowest energy state of a solid, each of the lowest energy, single electron states are filled from the lowest up to some state of maximum energy. In a conductor, the highest energy electrons do not completely fill the energy levels in a band (Fig. 29-Ka). Such partial filling of the band makes the material a conductor because an electric field can supply energy to electrons, since there are plenty of higher energy states available in this "conduction band."

In contrast, in an insulator the highest energy band containing electrons is completely filled. The band is called the "valence band." The energy band above it, the conduction band, contains no electrons (Fig. 29-Kb), and there is a large energy gap between the two bands. If an electric field is to cause conduction in a solid, it must supply energy continuously to the electrons; that is, it must take the electrons through a continuous sequence of higher energy states. But this is impossible in an insulator because there are no available vacant states. There is a significant energy gap between the highest filled state and the next available state.

The band structure of a semiconductor is similar to that of an insulator except that the energy gap between valence and conduction bands is much smaller in a semiconductor than in an insulator (Fig. 29-Kc). For example, the energy gap in the insulator diamond is 5.4 eV, whereas the gap in the semiconductor silicon is only 1.1 eV. The small energy gap in semiconductors means that it is possible for a few electrons to gain enough thermal energy to reach the conduction band.* This small number of electrons can then conduct. When thermally excited electrons leave the valence band, the energy levels they vacate can be filled by other electrons in the valence band. Thus there is a second form of conduction that can take place, caused by electrons that move to fill these vacancies, or "holes," as they are called. We can think of the holes as positive charges moving in a direction opposite the electrons. The relatively small number of electrons in the conduction band and holes in the valence band gives a semiconductor its conductivity. Though this conductivity is much less than that of a conductor, it is much greater than that of an insulator. For example, the order of magnitude of the resistivities (at room temperature) of the conductor copper, the semiconductor silicon, and the insulator diamond are 10^{-8} Ω-m, 10^{3} Ω-m, and 10^{11} Ω-m, respectively.

Doped Semiconductors

The conductivity of a semiconductor can be increased enormously by substituting another kind of atom at some locations throughout a semiconductor's crystal structure. This process is called "doping."

Suppose silicon, which has a valence of 4, is doped with phosphorus, which has a valence of 5; more specifically, suppose one in every million silicon atoms is replaced in the crystal by a phosphorus atom. Adding the phosphorus will not change the electrical neutrality of the semiconductor, but it will add significantly to the number of conduction electrons and therefore to the conductivity. Four of the five valence electrons from the phosphorus atoms will replace the silicon valence electrons; the fifth phosphorus va-

*Detailed calculation shows that at room temperature, where $kT = 0.03$ eV, a small but significant fraction of the electrons in silicon will have thermal energy greater than 1.1eV and will therefore reach the conduction band. On the other hand, there is practically zero probability of an electron at room temperature having thermal energy as great as the 5.4 eV energy gap in diamond, and so no electrons in diamond reach the conduction band.

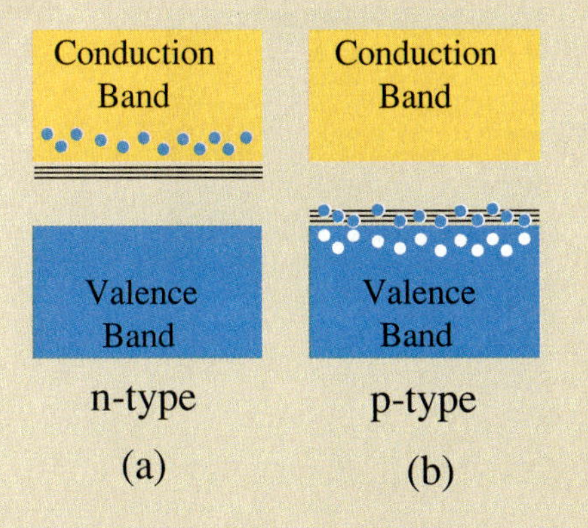

Fig. 29-L (a) Electrons in the conduction band of an n-type semiconductor.
(b) Holes in the valence band of a p-type semiconductor.

lence electron is not tightly bound to the atoms. It occupies an energy level just below the conduction band (about 0.04 eV below the conduction band, compared with 1.1 eV below for the electrons at the top of the valence band). A substantial fraction of these new electrons will have enough thermal energy to reach the conduction band (roughly 20% at room temperature) (Fig. 29-La). Although perhaps only one in a million silicon atoms is replaced by a phosphorus atom, so many more of the electrons associated with these new states reach the conduction band that the effect of doping is to dramatically increase conductivity. This kind of doping produces additional negative charge carriers, and so we call such a material an **n-type** semiconductor. In addition to the electrons added by doping, there will be a small number of electrons thermally excited to the conduction band from the valence band, which means that the valence band will have a small number of holes to contribute to the conductivity. Thus in an n-type semiconductor, although electric current comes mainly from electrons, the **majority carriers**, there is also a small contribution from holes, the **minority carriers**.

A second kind of doping can produce a great increase in the conductivity of a semiconductor. Suppose we replace some of the silicon atoms in a silicon crystal by aluminum atoms, which have a valence of 3. Again the crystal remains electrically neutral, but this time we have added more vacant states just above the valence band—0.06 eV above the valence band for aluminum-doped silicon. This means that a significant fraction of electrons near the top of the valence band will have sufficient energy to fill these states, leaving a large number of holes behind in the valence band (Fig. 29-Lb). These holes are effectively positive charge carriers, and so we call a silicon crystal doped with aluminum (or any other valence-3 atom) a **p-type** semiconductor. In addition to the holes arising from electrons excited to the states just above the valence band, we also have a small number of electrons thermally excited from the valence band to the conduction band. So in a p-type semiconductor we have holes as majority carriers and electrons as minority carriers.

Like an n-type semiconductor, a p-type semiconductor has a conductivity far greater than that of a pure semiconductor, though still far less than the conductivity of a metal.

P-N Junction

Transistors and other devices make use of the electrical properties of a **p-n junction,** formed by bringing into contact a p-type semiconductor and an n-type semiconductor (Fig. 29-Ma). When the junction is initially formed, electrons from the n side begin to diffuse into the p side. Likewise, holes begin to diffuse from the p side to the n side. Electrons and holes meet near the junction, the electrons fill the holes, and this leaves the central region depleted of both holes and electrons. Removing electrons from the n side leaves the positive doping ions there unbalanced and therefore gives the n side a static positive charge. Similarly, the depletion of holes from the p-side leaves its negative doping ions unbalanced and gives it a static negative charge. As the

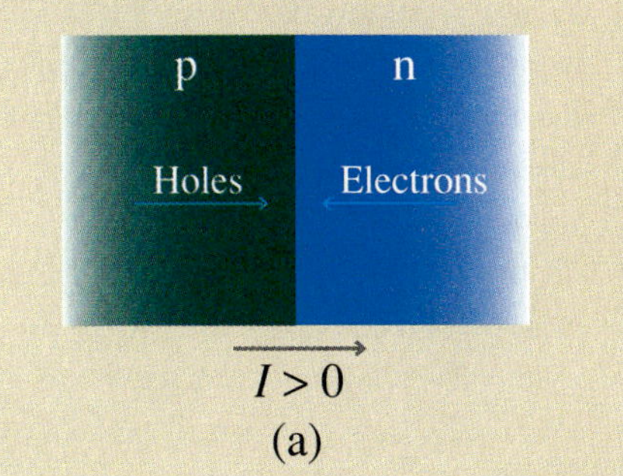

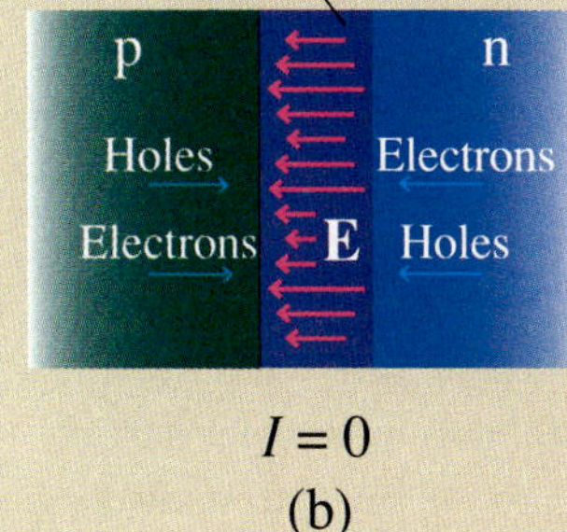

Fig. 29-M (a) When a p-n junction is first formed, a significant positive current flows from the p side to the n side as a result of the diffusion of majority carriers—electrons diffusing from n to p and holes diffusing from p to n. **(b)** Diffusion of majority carriers across the junction results in electrons filling holes near the junction, creating a depletion zone free of electrons and holes and therefore charged by the positive and negative doping ions that are no longer covered by electrons and holes, respectively. At equilibrium the diffusion of majority carriers is reduced to the point that its current just balances the opposing current of the minority carriers, and net current across the junction is zero.

Continued.

A Closer Look

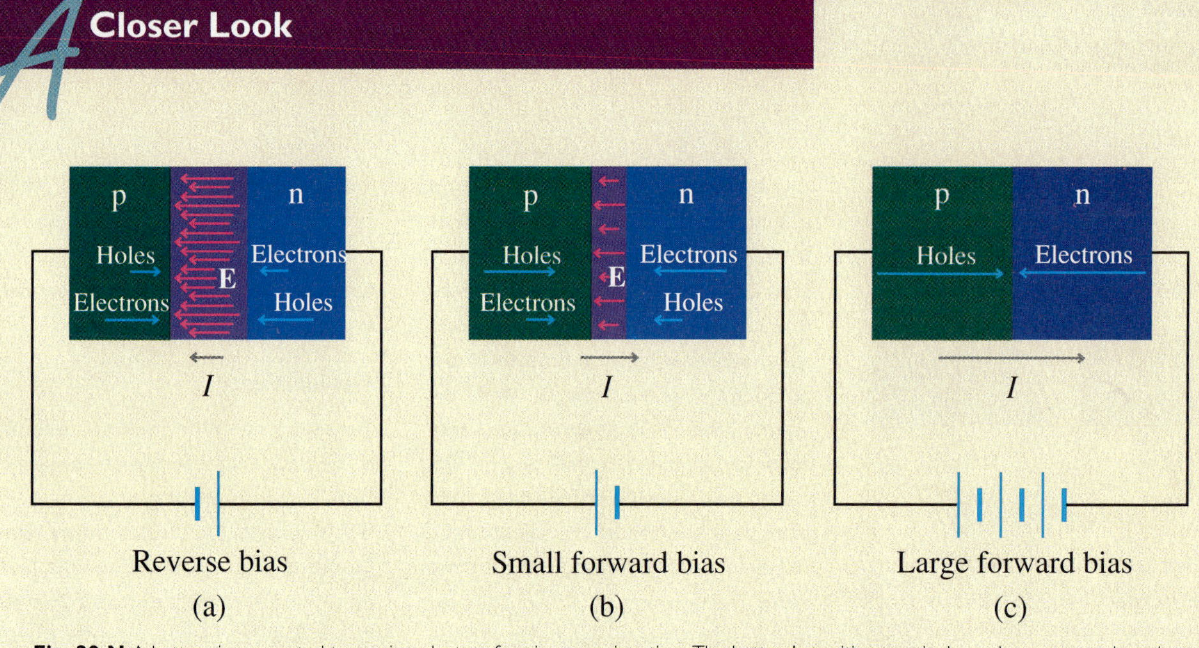

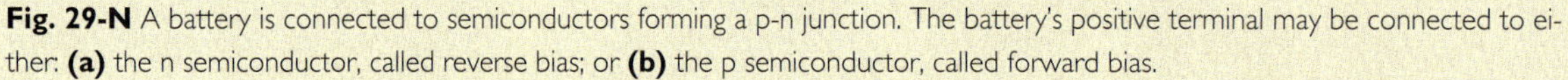

Fig. 29-N A battery is connected to semiconductors forming a p-n junction. The battery's positive terminal may be connected to either: **(a)** the n semiconductor, called reverse bias; or **(b)** the p semiconductor, called forward bias.

flow of the majority carriers continues, the charge density in the depletion zone near the junction increases along with an electric field that begins to oppose the further diffusion of majority carriers. However, this field has the opposite effect on the minority carriers on both sides. They are accelerated by the field. The majority carrier current quickly decreases until it just matches the small opposing current of the minority carriers. The net current is zero in this equilibrium situation (Fig. 29-Mb).

Suppose we apply an external voltage to a p-n junction by connecting a battery across it. If we connect the positive battery terminal to the n side and the negative terminal to the p side, we have what is called a **reverse bias.** The effect is to increase the already large potential difference from n to p—that is, to strengthen the electric field directed from n to p (Fig. 29-Na). This reduces the majority current, so that the very small minority current then exceeds the majority current. Thus reverse bias creates only a very small current.

Now suppose we reverse the battery terminals so that the p-n junction is **forward biased.** A small forward bias reduces the potential difference across the depletion zone, therefore creating a small net current as the majority current exceeds the minority current (Fig. 29-Nb). A larger forward bias can eliminate the depletion zone altogether, with the result that majority carriers diffuse freely across the junction and produce a much larger current (Fig. 29-Nc).

The typical current response of a p-n junction to a variable applied voltage is shown in Fig. 29-O, where we take the voltage as positive for forward bias and negative for reverse bias. Notice that as the forward bias voltage increases, the current increases at an increasing rate.

Junction Diode

Many electronic devices need to be supplied with DC voltage, and yet line voltage is AC. Thus we need to be able to convert AC to DC. This process, called **rectification,** can be accomplished with a **junction diode,** a simple device made by joining a p-type semiconductor to an n-type semiconductor, so that a p-n junction is formed between them. The current-voltage graph for a p-n junction (Fig. 29-O) shows that the diode transmits a significant current in only one direction and gives practically infinite resistance for current flow in the opposite direction.

Fig. 29-P shows the effect of a diode inserted into a simple circuit with a resistor and an alternating current source. The diode permits a significant positive current only in the direction indicated by the arrow in the diode circuit symbol. When the voltage V_{ab} is positive, a positive clockwise current flows around the circuit; the diode functions as a short circuit, and the full voltage V_{ab} appears across the resistor ($V_{cd} = V_{ab}$). But when the voltage V_{ab} becomes negative, because

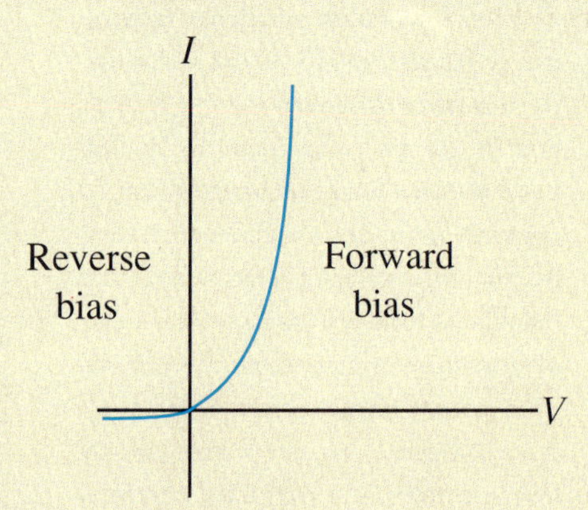

Fig. 29-O The current through a p-n junction (from p to n) as a function of the voltage applied to the junction.

the diode will permit no significant counterclockwise current, there is no significant current through the resistor and hence the voltage across the resistor is practically zero at such times ($V_{cd} = 0$). Thus the diode permits only the positive part of the time-varying voltage to appear across the resistor.

Two diodes are often used in combination with a "center-tapped" transformer to produce a more uniform positive output voltage (Fig. 29-Qa). During the part of the source's voltage cycle in which the voltage V_{ab} across the transformer is positive, the positive voltage drop V_{ad} from the transformer's top to its center tap causes current to flow through the upper diode shown in the circuit, but no current passes through the lower diode because the voltage drop V_{bd} from the bottom of the transformer to the center tap is negative in this part of the cycle. When V_{ab} is negative, V_{ad} is negative but V_{bd} is positive, and so current flows through the lower diode but not the upper diode. The two diodes are connected to the resistor in such a way that no matter what the sign of the voltage drop from a to b, positive current flows through the resistor in only one direction—**direct current** from c to d. The voltage drop V_{cd} across the resistor is always positive and therefore can be used as a direct current voltage source. The resistor's output voltage can be smoothed somewhat by inserting a capacitor in parallel with the resistor (Fig. 29-Qb). Other more elaborate "filter" circuits can produce an effectively constant DC voltage source.

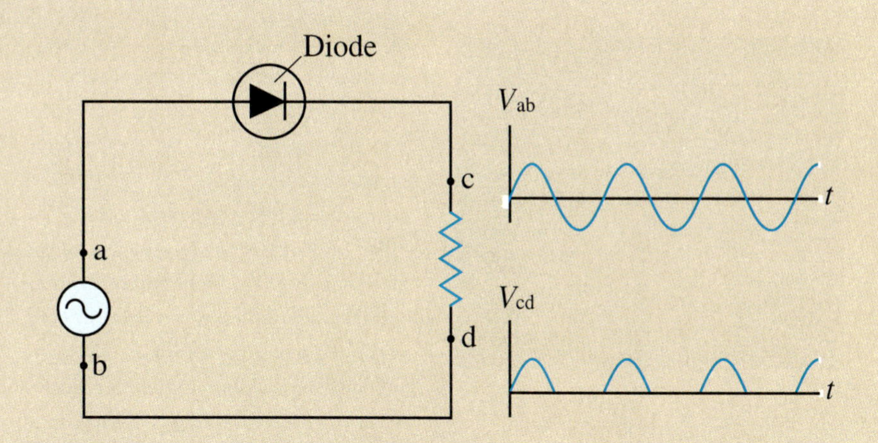

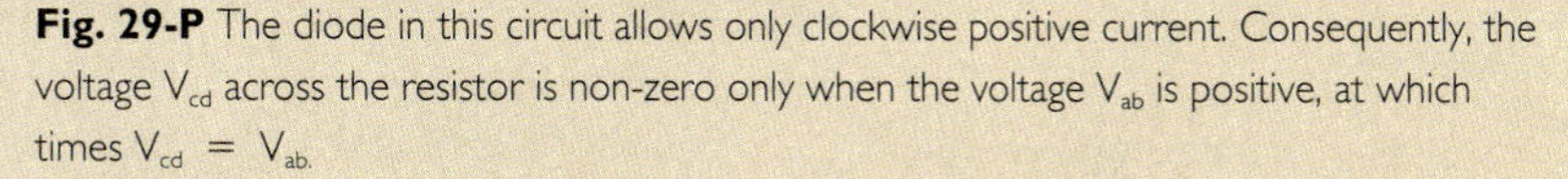

Fig. 29-P The diode in this circuit allows only clockwise positive current. Consequently, the voltage V_{cd} across the resistor is non-zero only when the voltage V_{ab} is positive, at which times $V_{cd} = V_{ab}$.

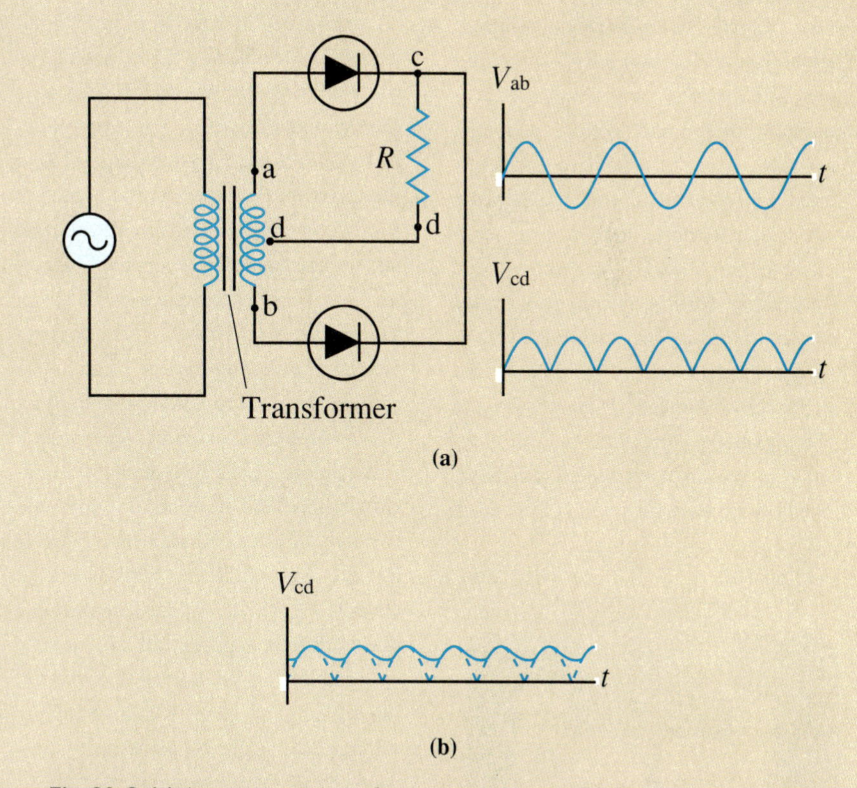

Fig. 29-Q (a) A center-tapped transformer and two diodes. **(b)** The voltage across the resistor is smoothed if a capacitor is placed in parallel with it.

Continued.

A Closer Look

Fig. 29-R The original transistor.

Transistors

Various types of transistors have been developed since the original transistor was invented in 1948 (Fig. 29-R). We shall describe only the original type and show how it can be used to amplify an electrical signal. Fig. 29-S shows such a transistor, which consists of two n-type semiconductors sandwiched around a p-type semiconductor, forming two p-n junctions. The n-type semiconductor on the left is much more heavily doped than either of the other two semiconductors. This heavily doped semiconductor, called the **emitter,** supplies practically all of the charge carriers that flow through the transistor. These charge carriers are the emitter's n-type majority carriers, electrons in the emitter's conduction band. The flow of these electrons is controlled by a small forward bias voltage between the emitter and the **base,** the center section of the npn transistor —a thin layer of lightly doped p-type semiconductor (about 10^{-6} m thick). This lightly doped material contains only a few holes, and consequently very few of the electrons that flow from emitter to base combine with holes in the base. Instead, these electrons pass through the base and into the third semiconductor material, called the **collector,** a lightly doped n-type semiconductor. The second p-n junction, the base-collector junction, is strongly reverse biased. This means that virtually no current arises from the action of the base and collector semiconductors. However, this reverse bias is actually an accelerating field for the electrons injected into the base. And so virtually all the electrons that enter the base pass through the collector and around the external part of the circuit. (Roughly 2% of the electrons flow through the branch of the circuit connected between the base and the biasing voltage sources.) The current is insensitive to the exact value of the reverse bias voltage on the base-collector junction. The current is controlled soley by the small forward bias voltage on the emitter-base junction, and even a small change in this voltage can dramatically change the current. (Notice how strongly current increases with increasing forward voltage on the current-voltage curve in Fig. 29-O). Fig. 29-T shows how this transistor, represented by the conventional circuit symbol for a transistor, can be used to amplify a time-varying voltage, for example, the voltage produced by a CD player or a microphone—a voltage with an amplitude of perhaps a few millivolts. The AC voltage source connected between points a and b is in series with the larger constant voltage source that produces the forward bias on the emitter-base junction and thus becomes a small part of the biasing voltage. Because the biasing voltage now varies with time, so also does the current, which is so dependent on the bias voltage. This current in turn determines the "output voltage."

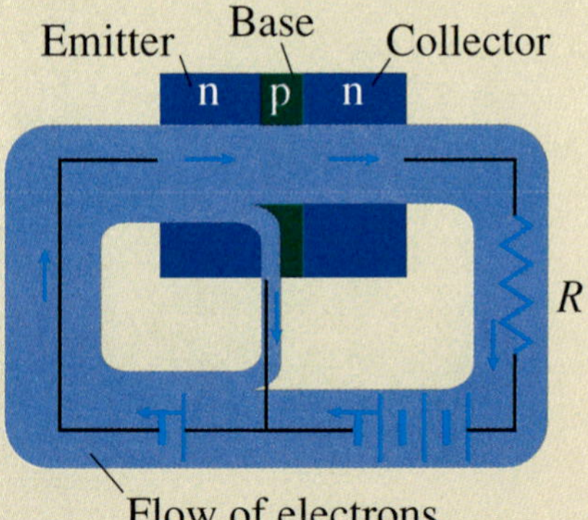

Fig. 29-S An npn transistor

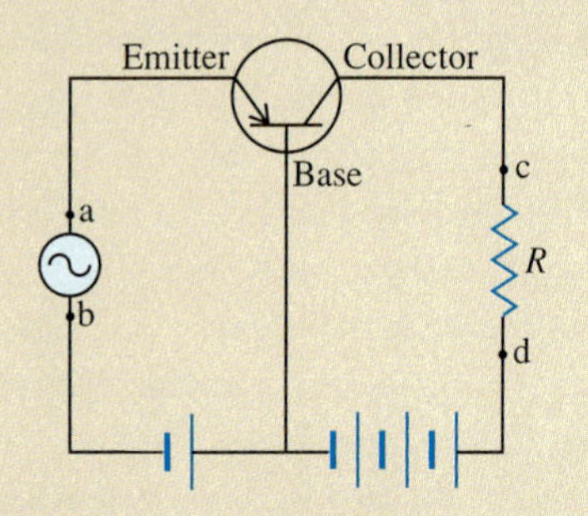

Fig. 29-T An npn transistor amplifies a small input voltage V_{ab}, producing a large output voltage V_{cd}.

The output voltage, which appears across the "load" resistor R in Fig. 29-T is directly proportioned to the current flowing through it, according to Ohm's law. And since this resistance is high (typically on the order of kΩ's), even small changes in the current produce relatively large voltage changes. For example, a 1 mA change in current produces a 1 V change in voltage across a 1 kΩ resistor ($V = IR = 10^{-3}A \times 10^3\Omega = 1V$). Thus the small time-varying input voltage gives an output voltage with the same time variation but with a much larger amplitude—perhaps 100 times as great. The transistor amplifies the signal.

The power provided by the output signal is also amplified since input power is the product of the input voltage and the current ($P = IV$) and output power is the product of the amplified output voltage and essentially the same current. The extra power is provided by the voltage source that provides the reverse bias to the collector junction.

CHAPTER 29 SUMMARY

The atoms of a gas in a discharge tube emit light, consisting of a set of wavelengths characteristic of the kind of atom. The hydrogen spectrum is particularly simple and consists of a number of series:

Lyman series (UV): $$\frac{1}{\lambda} = R\left(\frac{1}{1^2} - \frac{1}{n^2}\right) \quad (n = 2, 3, 4, \ldots)$$

Balmer series (visible): $$\frac{1}{\lambda} = R\left(\frac{1}{2^2} - \frac{1}{n^2}\right) \quad (n = 3, 4, 5, \ldots)$$

Paschen series (IR): $$\frac{1}{\lambda} = R\left(\frac{1}{3^2} - \frac{1}{n^2}\right) \quad (n = 4, 5, 6, \ldots)$$

Brackett series (IR): $$\frac{1}{\lambda} = R\left(\frac{1}{4^2} - \frac{1}{n^2}\right) \quad (n = 5, 6, 7, \ldots)$$

All of these series are expressed in the single equation

$$\frac{1}{\lambda} = R\left(\frac{1}{(n')^2} - \frac{1}{n^2}\right) \quad (n' = 1, 2, 3, \ldots \text{ and } n > n')$$

where the Rydberg constant R has the value

$$R = 1.097 \times 10^7 \text{ m}^{-1}$$

The Bohr model of the atom was successful in explaining the hydrogen spectrum. In the Bohr model of hydrogen, an electron is pictured as orbiting a proton in a circular orbit of radius r, where r is allowed to have only the values

$$r_n = n^2 r_1 \quad (n = 1, 2, 3, \ldots)$$

and $$r_1 = \frac{h^2}{4\pi^2 mke^2} = 5.29 \times 10^{-11} \text{ m}$$

The electron can have only the discrete values of energy E:

$$E_n = \frac{E_1}{n^2} \quad (n = 1, 2, 3, \ldots)$$

$$E_1 = -\frac{2\pi^2 mk^2e^4}{h^2} = -13.6 \text{ eV}$$

When an electron initially in an excited state of energy E_n goes to a state of lower energy E_n', it emits a photon of frequency f and energy hf equal to the difference in electron energy levels

$$E_n - E_n' = hf$$

The quantum theory of atoms, though it retained some features of the Bohr theory, such as discrete energy levels, eliminated the Bohr picture of a particle-like electron moving around a nucleus in a well-defined orbit. Quantum mechanics predicts that matter, like light, displays properties of both waves and particles. Associated with a particle of matter is a wave function, the intensity of which at any point is proportional to the probability of finding the particle at that point. A wave function has a de Broglie wavelength λ related to the particle's momentum $p = mv$ by the equation

$$\lambda = \frac{h}{p} = \frac{h}{mv}$$

Both forms of the Heisenberg uncertainty principle apply to particles of matter as well as to photons. The product of a particle's uncertainty in position Δx and its uncertainty in momentum Δp_x is greater than or equal to $h/2\pi$.

$$\Delta x \, \Delta p_x \geq \frac{h}{2\pi}$$

The uncertainty in a particle's energy ΔE and the time of detection Δt are related in the same way.

$$\Delta E \, \Delta t \geq \frac{h}{2\pi}$$

Quantum mechanics successfully explains the structure of multielectron atoms and also the frequencies and intensities of their spectral radiation. One atom is distinguished from another by its atomic number Z, which specifies the number of protons in the atom's nucleus or the equal number of electrons around the nucleus. The state of an atom is determined by the quantum states occupied by the atomic electrons. The quantum state of any one electron is specified by four quantum numbers, n, ℓ, m, and s, where

$$n = 1, 2, 3, \ldots$$

$$\ell = 0, 1, 2, \ldots, n - 1$$

$$m = -\ell, -\ell + 1, \ldots, \ell - 1, \ell$$

$$s = +\tfrac{1}{2}, -\tfrac{1}{2}$$

The way electrons occupy these quantum states is governed by the Pauli exclusion principle, which states that no two electrons can have the same set of quantum numbers. The electrons normally occupy states of minimum energy, consistent with this principle.

Atomic electrons can be excited by collisions with other atoms in a gas discharge tube. The electrons are excited to a higher-energy state and then fall back to a lower-energy state, emitting a photon to carry off the excess energy. Each atom emits a spectrum of wavelengths (visible, infrared, and ultraviolet) that are characteristic of that particular atom.

Questions

1 Suppose that in some remote part of the universe Planck's constant has the value of 1 J-s. Would atoms there be larger, smaller, or the same size as on earth?

2 Suppose you tried to fit $3\frac{1}{4}$ de Broglie wavelengths around a hypothetical Bohr orbit. Is it possible for the wave to have a single value at all points on the orbit?

3 After a hydrogen atom emits a photon of visible light, corresponding to one of the lines in the Balmer spectrum, what happens to that atom?

4 A beam of light with a continuous range of wavelengths ranging from IR through UV passes through a glass tube filled with a gas. In the light emerging from the tube certain wavelengths are missing. These wavelengths are absorbed by the gas; that is, atomic electrons absorb photons corresponding to these wavelengths and undergo transitions to higher energy levels. The positions of the dark absorption lines are exactly the same as in emission spectra of the same element, since they correspond to transitions between the same energy levels. For hydrogen gas at room temperature, the only lines observed are in the UV part of the spectrum—the Lyman series. The Balmer series and other series are not seen. Why not?

5 Will the electrical power input to a gas discharge tube affect (a) the distribution of spectral lines; (b) the intensities of spectral lines?

6 Suppose you have a 10 lb lead block, and you could somehow remove 3 protons from the nucleus of each atom in the block.
(a) What would happen to the block's appearance?
(b) What would happen to its value?
(c) What element would you have?

7 The element helium was discovered in 1868 on the sun, 27 years before it was found on earth. Which of helium's properties allowed for its discovery in this way: (a) chemical interaction with other elements; (b) wavelengths of spectral lines; (c) atomic weight; (d) speed of sound through helium?

8 An X-ray photon has an energy of at least 10^4 eV. In an atom of H, B, C, Au, or U, the energy of an electron in the lowest energy level is −13.6 eV, −188 eV, −285 eV, −81,000 eV, or −115,000 eV, respectively. Which of these elements could be used as a target in an X-ray tube and produce characteristic X rays?

9 If you increase the voltage across an X-ray tube, you will change (a) the minimum-wavelength X rays observed; (b) the minimum-frequency X rays observed; (c) the characteristic X rays observed.

10 Is it conceivable that a material could be found that fluoresces by absorbing IR photons, rather than UV photons, that is, a material that absorbs IR and emits visible light?

Answers to Odd-Numbered Questions

1 much larger; **3** Since it is still in the excited $n = 2$ state, it emits a UV photon, bringing it back to the ground state; **5** (a) no; (b) yes; **7** b; **9** a

Problems (listed by sections)

29-1 Atomic Spectra and the Bohr Model of the Atom

1 Calculate the second longest wavelength in the Balmer series. What color is this light?

2 Calculate the shortest wavelength of light emitted by the hydrogen atom.

3 What is the longest wavelength of light that can be absorbed by a hydrogen atom that is initially in the ground state?

4 What wavelength light will be emitted by a hydrogen atom shortly after emitting the 656 nm red line of the Balmer spectrum?

★ **5** A hydrogen atom emits light of wavelength 94.96 nm. Find the values of the quantum number n for the initial and final states of the atom.

6 An oil drop of radius 0.4 mm falls on a horizontal surface of water and spreads over an area of 0.1 m^2. Since the layer of oil must be at least one atom thick, the thickness of the layer gives an upper limit on the diameter of an atom. Calculate this maximum diameter.

7 The radius of the hydrogen nucleus is 1.2×10^{-15} m. Using as the radius of the hydrogen atom the radius of the first Bohr orbit, calculate the ratio of (a) the atom's radius to the nuclear radius; (b) the atom's volume to the nuclear volume.

★ **8** (a) Use the Bohr model to calculate the period and frequency of an electron in the first Bohr orbit of the hydrogen atom.
(b) What is the range of frequency of the light emitted by the hydrogen atom?

★ **9** (a) Use the Bohr model to calculate the frequency of an electron in the 100th Bohr orbit of the hydrogen atom.
(b) Find the frequency of light emitted in the transition from the 100th orbit to the 99th orbit.

10 Find the potential energy in MeV of an α particle at the edge of a uranium nucleus of radius 7.4×10^{-15} m.

★ **11** Each α particle in a beam of α particles has a kinetic energy of 5.0 MeV. Through what potential difference would you have to accelerate these α particles in order that they would have enough energy so that if one is fired head-on at a gold nucleus it could reach a point 1.0×10^{-14} m from the center of the nucleus?

29-2 Wave Properties of Electrons; Quantum Mechanics

12 Find the de Broglie wavelength of an electron accelerated across a 20,000 V potential difference in a TV tube. Is it reasonable to treat the electron beam as a beam of particles?

13 Find the de Broglie wavelength of a baseball of mass 0.15 kg moving at a speed of 40 m/s (90 mi/hr).

14 The uncertainty in position of a stationary 1 kg block is roughly an atomic diameter, or about 10^{-10} m.
(a) What is the minimum uncertainty in the block's velocity?
(b) How far might the block move in 1 year as a result of this velocity?

15 The uncertainty in position of an electron is 1.00×10^{-10} m. Find the minimum uncertainty in the electron's velocity.

★ **16** Find the de Broglie wavelength of an oxygen molecule of average kinetic energy in air at a temperature of 300 K.

★ **17** Electrons in an electron microscope have a kinetic energy of 1.00×10^5 eV.
(a) Find the de Broglie wavelength of the electrons.
(b) Find the ratio of this wavelength to the wavelength of light at the middle of the visible spectrum (550 nm).
(c) How many times greater magnification is *theoretically* possible with this microscope than with a light microscope?

29-3 Quantum Theory of Atomic Structure and Spectra; X rays

18 An atomic electron is in a state with principal quantum number $n = 3$. What are the possible combinations of quantum numbers ℓ and m?

19 How many quantum states are possible for an atomic electron for which $n = 5$ and $m = 0$?

20 How many quantum states are possible for an atomic electron for which $n = 5$?

21 Find the minimum wavelength produced in an X-ray tube operating with a potential difference of 40,000 V across the tube.

22 The minimum-wavelength X ray produced by a certain X-ray tube is 0.040 nm. Find the minimum voltage applied to the tube.

23 One of the brightest spectral lines produced by a neon light has a wavelength of 640 nm. Find the difference in energy levels of an atomic electron undergoing the transition necessary to produce this line.

Additional Problems

★ **24** Find the kinetic energy of an electron in the ground state of the hydrogen atom.

★ **25** Show that the maximum speed of an electron in a hydrogen atom, as calculated in the Bohr theory, is always much less than the speed of light, so that it is consistent to use nonrelativistic formulas.

★ **26** Quantum mechanics shows that the average value of the radial distance of an electron from the nucleus of a hydrogen atom is given by $n(n + \frac{1}{2})r_1$, where n is the principal quantum number and r_1 is the radius of the first Bohr orbit. Quantum theory also predicts the spread in radial distance to be the average value divided by $\sqrt{2n + 1}$. Show that for large quantum numbers the prediction gives a well-defined orbital radius, as predicted by the Bohr theory.

★★ **27** (a) Show that the orbital frequency of an electron in the nth Bohr orbit is given by $f = \dfrac{4\pi^2 mk^2e^4}{n^3h^3}$.
(b) Show that the frequency of the photon emitted in the transition from the $n + 1$ state to the n state is given by $f = \dfrac{2\pi^2 mk^2e^4}{h^3}\left[\dfrac{2n + 1}{n^2(n + 1)^2}\right]$. Show that for large quantum numbers this frequency is approximately equal to the electron's orbital frequency, in accord with the prediction of classical radiation theory.

CHAPTER 30 Nuclear Physics and Elementary Particles

The largest scientific instruments ever built are particle accelerators, used to study matter on the smallest possible scale. Shown here are an exterior view of the Fermilab accelerator and an interior view of the lab's underground tunnel.

In this final chapter we shall describe phenomena occurring on the incredibly small scale of atomic nuclei and elementary particles. We shall also describe biological effects of nuclear radiation and how radioactive decay of nuclei is used to determine the age of archeological specimens.

30-1 Nuclear Structure

In the last chapter we saw how the work of Rutherford, Bohr, Schroedinger, and others developed our understanding of atomic structure. We now know much about the structure of the atomic nucleus as well. In this section we shall briefly describe some of the most important features of that structure. Unfortunately, we will not be able to describe much of the fascinating history of discoveries in nuclear physics. Some sense of this history is provided at the end of this chapter in the essay on Marie Curie, who made some of the most important discoveries in nuclear physics.

Nucleons

Within an atom are negatively charged electrons, surrounding a nucleus, which contains all the atom's positive charge and nearly all its mass. The size of the nucleus is a small fraction of the size of the entire atom. The particles within the nucleus are called **nucleons** and are of two types: **protons,** positively charged particles, each with a charge equal in magnitude to the electron's charge, and **neutrons,** electrically neutral or uncharged particles, as the name implies.

Protons and neutrons have nearly the same mass. Denoting their respective masses by m_p and m_n, they are

$$m_p = 1.6726 \times 10^{-27}\ \text{kg}$$
$$m_n = 1.6750 \times 10^{-27}\ \text{kg}$$

Nucleons are much more massive than electrons, which have mass

$$m_e = 9.1094 \times 10^{-31}\ \text{kg}$$

The number of protons in an atom's nucleus is the atom's **atomic number,** denoted by Z. Normally atoms are electrically neutral because of the presence of electrons, equal in number to the protons. As described in the last chapter, an atom's atomic number gives its position on the periodic table and determines its chemical properties. The **neutron number** N specifies the number of neutrons in a given nucleus.

The **mass number** A specifies the total number of nucleons and, since these nucleons are either protons or neutrons, equals the sum of the atomic number Z and the mass number N.

$$A = Z + N \tag{30-1}$$

The mass number determines the approximate mass of a nucleus. A nuclear mass is roughly equal to the product of the number of nucleons (A) times the mass of a single nucleon (about 1.67×10^{-27} kg).

Nuclides

The different kinds of nuclei are called **nuclides.** We denote a specific nuclide by the symbol used to denote the chemical element, along with the nuclide's values of Z and A, in the following way:

$${}^{A}_{Z}\text{X}$$

where X is the symbol for the element. For example, the nuclide with two protons and two neutrons is a helium nucleus, and is denoted by

$${}^{4}_{2}\text{He}$$

and the nuclide consisting of six protons and six neutrons is a carbon nucleus and is denoted by

$${}^{12}_{6}\text{C}$$

Nuclides with the same atomic number but different mass numbers are called **isotopes.** There are two stable isotopes of carbon, ${}^{12}_{6}\text{C}$ and ${}^{13}_{6}\text{C}$. On earth 98.9% of carbon is the isotope ${}^{12}_{6}\text{C}$; 1.1% is the isotope ${}^{13}_{6}\text{C}$. Atomic masses given in the periodic table are averages that reflect the relative abundance of isotopes on earth. For example, the atomic mass of carbon is listed as 12.011, which means that on the average the mass of a carbon atom is 12.011 u. (The **atomic mass unit** u, introduced in Chapter 12, is defined as $\frac{1}{12}$ the mass of the ${}^{12}_{6}\text{C}$ atom.) One may compute the average of 12.011 u by adding 98.9% of the ${}^{12}_{6}\text{C}$ atomic mass of 12 u to 1.1% of the ${}^{13}_{6}\text{C}$ atomic mass of about 13 u:

$$0.989(12\ \text{u}) + 0.011(13\ \text{u}) = 12.011\ \text{u}$$

Table 30-1 gives the masses of selected atoms, as well as the masses of the electron, proton, and neutron in atomic mass units u, where

$$1\ \text{u} = 1.6606 \times 10^{-27}\ \text{kg}$$

Appendix D gives a more complete list of isotopes.

Table 30-1 Masses of selected particles and atoms

The atomic masses include the masses of atomic electrons.

Atom or subatomic particle		Mass, u	
electron		5.486×10^{-4}	
proton		1.007276	
neutron		1.008665	
hydrogen	$^{1}_{1}H$	1.007825	(99.99%)
	$^{2}_{1}H$	2.014102	(0.01%)
helium	$^{4}_{2}He$	4.002603	(~100%)
beryllium	$^{9}_{4}Be$	9.012182	(~100%)
carbon	$^{12}_{6}C$	12.000000	(98.9%)
	$^{13}_{6}C$	13.003355	(1.1%)
nitrogen	$^{14}_{7}N$	14.003074	(99.6%)
	$^{15}_{7}N$	15.000109	(0.4%)
oxygen	$^{16}_{8}O$	15.994915	(99.8%)
	$^{17}_{8}O$	16.999133	(0.04%)
	$^{18}_{8}O$	17.999159	(0.2%)
aluminum	$^{26}_{13}Al$	26.981541	(100%)
iron	$^{56}_{26}Fe$	55.934939	(100%)
lead	$^{206}_{82}Pb$	205.97446	(24.1%)
	$^{207}_{82}Pb$	206.97589	(22.1%)
	$^{208}_{82}Pb$	207.97664	(52.3%)
uranium	$^{235}_{92}U$	235.04393	(0.7%)
	$^{238}_{92}U$	238.05079	(99.3%)

Nuclear Size

The size and shape of nuclei have been determined from scattering experiments, similar to Rutherford's α-particle experiment, described in the last chapter. It is found that nuclei are roughly spherical, with an approximate radius* r, which depends on the mass number A, as given by the equation

$$r \approx (1.2 \times 10^{-15}\ \text{m})A^{1/3} \tag{30-2}$$

Since the volume of a sphere is proportional to r^3, this equation predicts that a nuclear volume is proportional to A. This means that the volume of a nucleus is directly proportional to the number of nucleons it contains. The nucleons are like hard, impenetrable spheres, tightly packed together (Fig. 30-1). All nuclei therefore have approximately the same density.

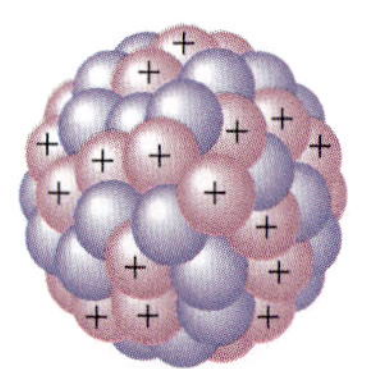

Fig. 30-1 The nucleus can be thought of as consisting of tightly packed, hard spheres—protons (+) and neutrons (0 charge).

*A nuclear radius is not a sharply defined quantity, any more than an atomic radius is sharply defined, since the wave properties of both atoms and nuclei imply an uncertainty in position.

EXAMPLE 1 Nuclear Density

Find the approximate density of any nucleus.

SOLUTION Density is defined as mass per unit volume, and so first we shall find expressions for the mass and volume of a nucleus with mass number A. Since A indicates the number of nucleons within the nucleus and since both protons and neutrons have the same approximate mass of 1.67×10^{-27} kg, we may express the nuclear mass m as

$$m \approx A(1.67 \times 10^{-27}\text{ kg})$$

An expression for the nuclear volume is found when Eq. 30-2 for the radius of a nucleus is inserted into the equation for the volume of a sphere of radius r.

$$V = \tfrac{4}{3}\pi r^3 \approx \tfrac{4}{3}\pi(1.2 \times 10^{-15}\text{ m})^3 A$$

When we divide the mass m by the volume V, the mass number A cancels and we obtain the approximate density ρ of any nucleus.

$$\rho = \frac{m}{V} \approx \frac{1.67 \times 10^{-27}\text{ kg}}{\tfrac{4}{3}\pi(1.2 \times 10^{-15}\text{ m})^3}$$

$$\approx 2.3 \times 10^{17}\text{ kg/m}^3$$

This density is enormously large, on the order of 10^{14} times larger than the density of ordinary solid or liquid matter, for which $\rho \approx 10^3$ kg/m^3.

Stable Nuclides

Stable nuclides are those nuclides that last forever if undisturbed. Unstable nuclides spontaneously decay by emitting particles. This process is called "radioactivity." All the nuclides in Table 30-1 are stable, except for the two isotopes of uranium. Both of these isotopes undergo radioactive decay but with a "half-life"* that is very long—on the order of 10^9 years.

Nearly all stable nuclides have at least as many neutrons as protons in the nucleus. The only exceptions are ${}^1_1$H (which has only a single proton) and ${}^3_2$He. For atomic numbers up to about 20, stable nuclides have very nearly equal numbers of protons and neutrons. For larger values of atomic number, stable nuclides have roughly 50% more neutrons than protons (Fig. 30-2).

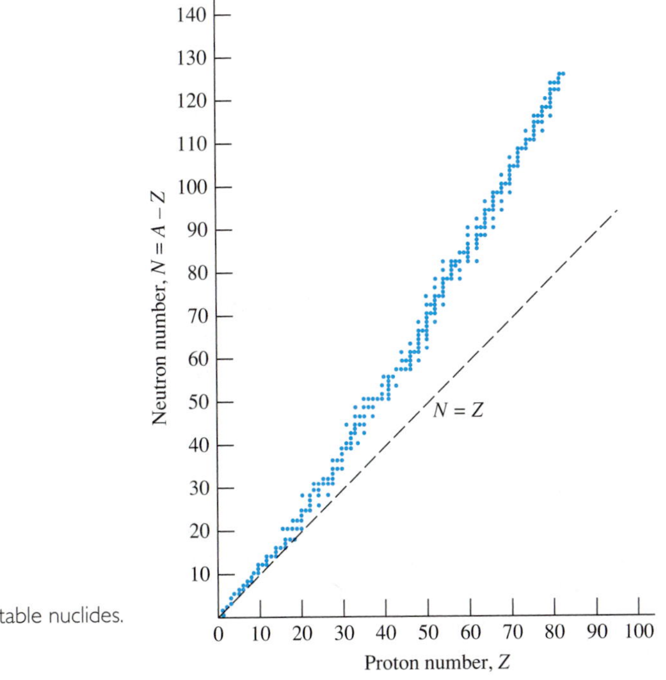

Fig. 30-2 Stable nuclides.

*Half-life, the time required for half of the nuclei in a radioactive sample to decay, is discussed in the following section on radioactivity.

Strong Nuclear Force

The protons within a nucleus are densely packed, positive charges. How is it possible that the protons stay together, in light of the repulsive Coulomb force that acts between like charges? Clearly there must be some other force that more than compensates for the electric force that tends to separate the protons. This other force, known as the **strong nuclear force,** acts between any two nucleons (proton and proton, neutron and neutron, or proton and neutron) and is responsible for holding the nucleons together in the tightly packed formation we find in any nucleus. This strong force is the third fundamental force we have encountered in our study of physics. The two we have already studied are the gravitational force and the electromagnetic force. A fourth force, the weak nuclear force, responsible for a type of radioactive decay (beta decay), is described in Section 30-5. Any force in nature is an example of one of these four kinds of force.

The strong force between adjacent nucleons is much stronger than the repulsive Coulomb force between adjacent protons. However, the strong force has a very short range, compared to the Coulomb force. For example, two protons on opposite sides of a uranium nucleus will experience a considerable Coulomb repulsive force. But at this distance the strong force between the protons is zero. So the nucleons are held together by the strong forces, which act directly between the nucleons that are closest together. The strong force pulls the nucleons together but does not allow the nucleons to collapse into each other. The nucleons are somewhat like a cluster of marbles glued together.

Because of the detailed nature of the strong force and some requirements of quantum mechanics, it turns out that stable arrangements of two or more nucleons always have either approximately equal numbers of protons and neutrons, or, for larger nuclei, more neutrons than protons, as seen in Fig. 30-2. The reason that there must be proportionately fewer protons in a large nucleus is due to the increased effect of the electrostatic force. Each proton in a large nucleus feels a significant repulsive force from each of the other protons, and so the total repulsive Coulomb force on each proton increases rapidly for increasing Z. To compensate for this, large stable nuclei have more neutrons than protons. Since neutrons are uncharged, they experience only the strong force. If the atomic number becomes too large, it is impossible to form a stable nucleus, no matter how large the number of neutrons. The largest stable nuclide is bismuth, ${}^{209}_{83}\text{Bi}$. Larger nuclei all undergo spontaneous radioactive decay.

Binding Energy; Mass Defect

Whenever particles are bound together by attractive forces, energy is required to pull them apart. For example, in Chapter 7 we saw that in order for a body at the surface of the earth to "escape" the earth's gravity the body must have a large initial kinetic energy, corresponding to an initial velocity at least equal to the escape velocity of 11,200 m/s. In Chapter 29 we saw that in order for an electron, initially in the ground state of a hydrogen atom, to leave the atom, 13.6 eV of energy must be supplied to the electron by absorption of a 13.6 eV photon. This energy of 13.6 eV is called the "binding energy" of the hydrogen atom. Unless at least this much energy is supplied to a hydrogen atom, it remains a bound system; the electron and proton remain bound together. More energy would have to be supplied to remove all the electrons from a multielectron atom; in other words, such atoms have higher binding energies.

The nucleons within a nucleus are bound together by attractive forces, and so energy must be supplied to the nucleons to separate them. The total energy required to break a nucleus apart into separated nucleons is called the **binding energy of the nucleus.**

Einstein's mass-energy formula (Eq. 27-9: $E_0 = m_0c^2$) gives the rest energy E_0 of a body of rest mass m_0. Since energy must be supplied to separate nucleons, it follows that the separated nucleons will have more energy and therefore more mass than a nucleus in which the same nucleons are bound together. The energy difference is large enough that such mass differences are significant and therefore easily measured, unlike the much smaller mass differences resulting from electron binding energies. The difference between the total mass of the separated nucleons and the mass of the bound nucleus is called the **mass defect** and is denoted by Δm. Since the masses of nucleons and nuclides are known with considerable accuracy, as indicated in Table 30-1, it is easy to calculate Δm. The energy equivalent of the mass defect Δm (the change in rest mass) is the binding energy ΔE, which, according to Einstein's mass energy formula, is given by

$$\Delta E = \Delta mc^2 \tag{30-3}$$

A common unit of energy for describing nuclear energies is the MeV (million electron volts), where an electron volt is 1.6022×10^{-19} J. Thus

$$1 \text{ MeV} = 1.6022 \times 10^{-13} \text{ J}$$

We shall often calculate mass-energy, with the mass appearing in atomic mass units, u. Therefore it is convenient to know the energy equivalent of 1 u in units of MeV. Multiplying the value of u expressed in kg by c^2, we obtain the energy equivalent in joules.

$$(1 \text{ u})c^2 = (1.6606 \times 10^{-27} \text{ kg})(2.9979 \times 10^8 \text{ m/s})^2$$

$$= 1.4924 \times 10^{-10} \text{ J}$$

Multiplying by the conversion factor from units of J to MeV, we have

$$(1 \text{ u})c^2 = (1.4924 \times 10^{-10} \text{ J})\left(\frac{1 \text{ MeV}}{1.6022 \times 10^{-13} \text{ J}}\right)$$

or

$$(1 \text{ u})c^2 = 931.5 \text{ MeV} \tag{30-4}$$

EXAMPLE 2 An Alpha Particle's Binding Energy

Calculate the binding energy of an alpha (α) particle (the helium nuclide ${}^4_2\text{He}$).

SOLUTION First we shall calculate the mass defect and then the binding energy. From Table 30-1 we find the mass of the ${}^4_2\text{He}$ atom, including its two electrons, and then subtract the mass of the two electrons to find the nuclear mass.

$$\text{Mass of nucleus} = 4.002603 \text{ u} - 2(5.486 \times 10^{-4} \text{ u})$$

$$= 4.001506 \text{ u}$$

Next we calculate the mass of 2 protons and 2 neutrons, again using data from Table 30-1.

$$\text{Mass of separated nucleons} = 2(1.007276 \text{ u}) + 2(1.008665 \text{ u})$$

$$= 4.031882 \text{ u}$$

The difference between the total mass of the separated nucleons and the mass of the nucleus is the mass defect, Δm.

$$\Delta m = 4.031882 \text{ u} - 4.001506 \text{ u}$$

$$= 0.030376 \text{ u}$$

Applying Eq. 30-3, we find the energy equivalent of this mass, that is, the binding energy.

$$\Delta E = \Delta mc^2 = (0.030376 \text{ u})c^2$$

Using Eq. 30-4, we can easily calculate ΔE in MeV.

$$\Delta E = (0.030376)(931.5 \text{ MeV})$$

$$= 28.3 \text{ MeV}$$

This is more than a million times greater than the binding energy of the hydrogen atom (13.6 eV).

EXAMPLE 2—cont'd

An alternative method of calculating Δm is to include the mass of the two electrons in both the initial and final calculations of mass, that is, to consider breaking a helium atom with its 2 electrons into 2 neutrons plus 2 hydrogen atoms, with 1 electron each.

$$\Delta m = \text{mass of (2 H atoms + 2 neutrons)} - \text{mass of He atom}$$
$$= 2(1.007825 \text{ u}) + 2(1.008665 \text{ u}) - 4.002603 \text{ u}$$
$$= 0.030377 \text{ u}$$

This result is in agreement with the result previously found.

In the last example, we found the binding energy of the helium nucleus to be 28.3 MeV. Since this nucleus contains 4 nucleons, the binding energy per nucleon is $\frac{1}{4}$(28.3 MeV) = 7.1 MeV. Fig. 30-3 shows a graph of the binding energy per nucleon for all stable nuclides, as well as for some radioactive nuclides. Notice that the data lie on a fairly smooth curve. The binding energy per nucleon is roughly the same for most nuclei—about 8 MeV. The curve peaks at about $A = 60$, which means that nuclei with atomic mass close to 60 are the most stable; nucleons within these nuclei are the most tightly bound.

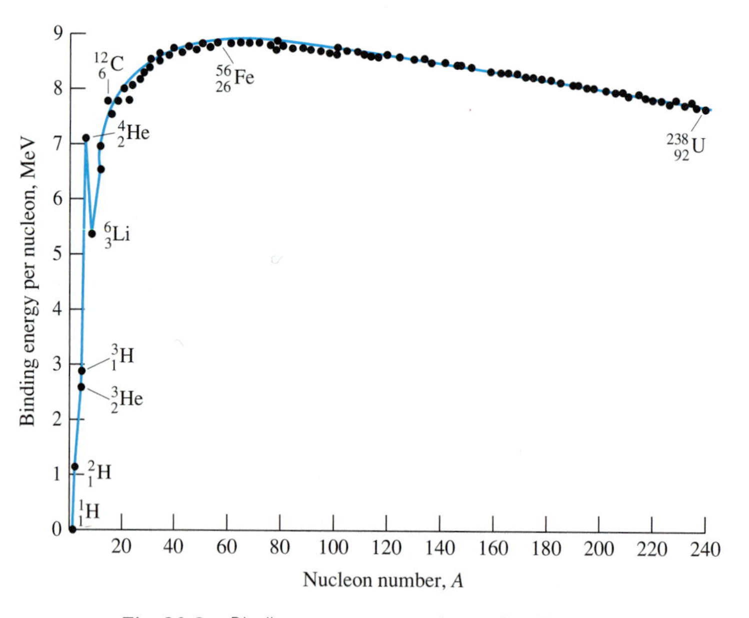

Fig. 30-3 Binding energy per nucleon of stable nuclei.

30-2 Radioactive Decay

Whenever a nucleus is in a state that is not the lowest-energy state for a nucleus with that number of nucleons, the nucleus is unstable. Sooner or later such a nucleus will spontaneously decay. Unstable nuclei decay through the emission of alpha (α) particles, beta (β) particles, or gamma (γ) rays. Although these particles and rays are invisible, their presence can be detected by a simple experiment, using a radioactive source such as radium, which emits all three. The source is placed at the bottom of a long narrow hole in a lead block. The block is enclosed in an evacuated chamber and placed in a horizontal, external magnetic field (Fig. 30-4).

The lead absorbs all radiation, except that which is directed straight out the hole. This provides a well-directed, vertical beam of radiation. When the beam enters the magnetic field, it separates into three beams, which expose photographic film in three spots, as indicated in the figure. Applying the right-hand rule, we see that the magnetic force $\mathbf{F}_+$ on a positive charge moving upward is directed to the left, and so the beam bent toward the left must consist of positively charged particles. They are called α particles. Further experiment shows these particles to be ${}^4_2\text{He}$ nuclei. The magnetic field exerts on negative charges a force $\mathbf{F}_-$, directed to the right. So the beam bent to the right consists of negatively charged particles, called β particles, now known to be electrons. The rays that are not deflected are called γ rays, which turn out to be electromagnetic radiation of very high frequency—even higher frequency than X rays.

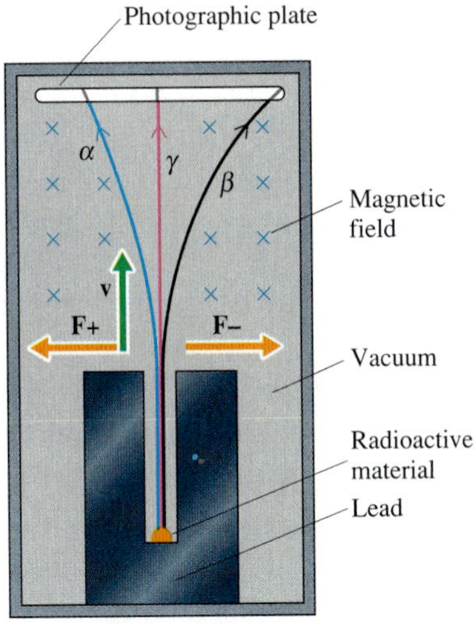

Fig. 30-4 A radioactive material emits α particles, β particles, and γ rays. A magnetic field bends the beam of α particles to the left and the beam of β particles to the right, indicating that the α particles are positively charged and the β particles are negatively charged.

Alpha Decay

Alpha decay commonly occurs in large nuclei, with atomic numbers greater than 82. The α particle consists of two protons and two neutrons, and so the decaying nucleus has its atomic number reduced by two and its mass number reduced by four. This means that the remaining nucleus is that of a different element. We refer to this process as the transmutation of the element. For example, uranium 238 decays to thorium 234 by emission of an α particle.

$$ {}^{238}_{92}\text{U} \rightarrow {}^{234}_{90}\text{Th} + {}^{4}_{2}\text{He} $$

A second example is the α decay of radium 226 to radon 222.

$$ {}^{226}_{88}\text{Ra} \rightarrow {}^{222}_{86}\text{Rn} + {}^{4}_{2}\text{He} $$

We see in these examples of α decay that both charge and nucleon number are conserved. The total value of Z is the same before and after the decay. Likewise, the total value of A is the same before and after decay.

We refer to a nucleus before undergoing radioactive decay as a "parent" nucleus, and we refer to a nucleus after decay as a "daughter" nucleus. With this terminology we can represent α decay in general as

$$ \text{parent } (Z, A) \rightarrow \text{daughter } (Z - 2, A - 4) + \alpha \qquad (\alpha \text{ decay}) $$

Quite often the daughter nucleus is itself radioactive, and so the daughter can become a parent.

After α decay the combined rest mass of the daughter nucleus and the α particle is less than the mass of the parent nucleus. The excess rest mass energy of the parent is carried off in the form of kinetic energy of the daughter nucleus and the α particle. Energy is conserved. Momentum is also conserved during the decay, and this fact can be used to find how the kinetic energy is distributed between the α particle and the daughter nucleus. Since the daughter nucleus is much more massive than the α particle, very little of the kinetic energy goes to the daughter nucleus (see Problem 20). The α particle carries away kinetic energy nearly equal to all the excess mass-energy provided by the parent nucleus.

EXAMPLE 3 Kinetic Energy of an α Particle From U-238

Calculate the kinetic energy of the decay products when uranium 238 decays to thorium 234 by emitting an α particle.

SOLUTION From Appendix D we find the respective masses of ${}^{238}_{92}\text{U}$, ${}^{234}_{90}\text{Th}$, and ${}^{4}_{2}\text{He}$ to be 238.05079 u, 234.04359 u, and 4.00260 u. (These masses include the masses of orbital electrons for the neutral atoms. Since the same number of electrons are present before and after decay, we do not need to subtract electron masses.) The parent nucleus has excess rest mass Δm, which equals its rest mass less that of the decay products.

$$\Delta m = 238.05079 \text{ u} - 234.04359 \text{ u} - 4.00260 \text{ u}$$
$$= 0.00460 \text{ u}$$

The energy equivalent of this mass ($\Delta E = \Delta mc^2$) equals the kinetic energy carried away

$$K = \Delta E = \Delta mc^2 = (0.00460 \text{ u})c^2$$

Applying Eq. 30-4 $[(1\text{u})c^2 = 931.5 \text{ MeV}]$, we can express the result in MeV.

$$K = (0.00460)(931.5 \text{ MeV}) = 4.28 \text{ MeV}$$

Most of this kinetic energy is given to the α particle. Because of its larger mass, only a small fraction is carried by the thorium nucleus.

The process of α decay is a consequence of the instability of a large nucleus because of the Coulomb repulsion between the protons within the nucleus. Protons alone are not emitted because the strong nuclear force binds each proton to other nucleons. There is a competition between the repulsive effect of the Coulomb force and the attractive effect of the strong nuclear force. The α particle is a very stable unit, and so α particles are emitted. The emission of the α particle removes two protons from the parent nucleus, in response to the Coulomb force, but allows the two protons to stay close to each other and to two neutrons, in response to the strong force.

Beta Decay

Fig. 30-4 shows negatively charged β particles emitted by a radioactive source. This process, called "β^- decay," occurs when one of the neutrons in an unstable nucleus transforms to a proton plus an electron and emits the electron. The result is that the parent nucleus is transformed to a daughter nucleus with the same number of nucleons but with an increase of one in atomic number.

$$\text{parent } (Z, A) \rightarrow \text{daughter } (Z + 1, A) + \text{electron} \qquad (\beta^- \text{ decay})$$

An example of β^- decay is the process

$${}^{14}_{6}\text{C} \rightarrow {}^{14}_{7}\text{N} + {}^{0}_{-1}\text{e}$$

where ${}^{0}_{-1}\text{e}$ represents the electron. Notice that both charge and nucleon number are conserved.

EXAMPLE 4 Energy Released in β Decay of Carbon 14

Find the energy released when $^{14}_{6}C$ undergoes β^- decay.

SOLUTION From Appendix D we find the mass of the $^{14}_{6}C$ atom, including its 6 orbital electrons, to be 14.003242 u. The daughter nucleus is $^{14}_{7}N$, and the mass of the neutral $^{14}_{7}N$ atom, including its 7 orbital electrons, is 14.003074 u. The daughter nucleus has only 6 orbital electrons just after decay, but we can use the mass of the neutral $^{14}_{7}N$ atom to account for the mass of the daughter nucleus and its 6 orbital electrons plus the mass of the one electron emitted by the parent nucleus. So the parent nucleus has excess rest mass equal to the difference in the masses of the $^{14}_{6}C$ atom and the $^{14}_{7}N$ atom.

$$\Delta m = 14.003242 \text{ u} - 14.003074 \text{ u}$$
$$= 0.000168 \text{ u}$$

The energy equivalent of this mass is

$$\Delta E = \Delta mc^2 = (0.000168 \text{ u})c^2$$

Since $(1 \text{ u})c^2 = 931.5$ MeV, we can express this as

$$\Delta E = (0.00168)(931.5 \text{ MeV}) = 0.156 \text{ MeV}$$

From the preceding example, one would expect the electron emitted in the β decay of $^{14}_{6}C$ to have a kinetic energy of nearly 0.156 MeV. (The much more massive daughter nucleus has negligible recoil velocity and kinetic energy.) However, experimental measurement of the emitted electron's kinetic energy indicates that it can have any value from zero up to a maximum value equal to the 0.156 MeV calculated value. This leaves unaccounted for energy of up to 0.156 MeV. When this discrepancy was first noticed, it caused serious concern because it seemed to indicate violation of one of the most fundamental physical principles—conservation of energy.

The solution to the dilemma was proposed in 1930 by Wolfgang Pauli. He suggested that an additional, unseen particle was emitted by a β decaying nucleus. The existence of this particle, called a **neutrino,** was verified in 1956. A neutrino is somewhat like a photon, in that it has zero charge and travels at the speed of light. However, the neutrino is much more difficult to detect because it interacts with matter only very weakly.

The Greek letter ν (nu) is used to denote a neutrino. The neutrino has an "antiparticle,"* the antineutrino, which is denoted by $\overline{\nu}$. An antineutrino is emitted in β^- decay. Thus the β^- decay of $^{14}_{6}C$ is more accurately shown as

$$^{14}_{6}C \rightarrow {}^{14}_{7}N + {}^{0}_{-1}e + \overline{\nu}$$

Some nuclei decay by a *second kind of β decay.* In these nuclei a proton is transformed into a neutron plus an emitted "positron" and a neutrino. The positron is the antiparticle to an electron. The positron has the same mass as the electron, but its charge of $+e$ is opposite the electron's charge. It is designated either as β^+, or as $^{0}_{1}e$. In β^+ decay, the parent nucleus is transformed to a daughter nucleus with the same number of nucleons but with a decrease of one in atomic number.

$$\text{parent } (Z, A) \rightarrow \text{daughter } (Z-1, A) + \text{positron} + \text{neutrino} \qquad (\beta^+ \text{ decay})$$

An example of β^+ decay is the process

$$^{15}_{8}O \rightarrow {}^{15}_{7}N + {}^{0}_{1}e + \nu$$

The third type of β decay that sometimes occurs is *electron capture*. This process is a kind of reverse β decay, in that an electron is absorbed rather than being emitted. An orbital electron is captured by the nucleus, and the electron and a proton within the nucleus are transformed to a neutron. An example of electron capture is

$$^{7}_{4}Be + {}^{0}_{-1}e \rightarrow {}^{7}_{3}Li + \nu$$

*The subject of antimatter and antiparticles is discussed in Section 30-5.

Gamma Decay

Gamma rays are electromagnetic radiation of very high frequency. The radiation is produced by nuclei undergoing a change in energy level. Like atoms, nuclei can exist in excited states, states of greater energy than the minimum energy, or ground, state. And like atoms, nuclei undergo transitions from higher to lower energy states by emitting photons. The difference in nuclear energy levels, typically on the order of MeV's, is considerably greater than the difference in atomic energy levels. Thus the emitted photons, γ-ray photons, are of high energy, corresponding to radiation of high frequency (Eq. 28-3: $E = hf$).

Nuclei are in initial, excited states as the result of some other process having just occurred. This process might be collision with another nucleus or an α or β decay. An example of β decay to an excited state is

$$^{60}_{27}\text{Co} \rightarrow {}^{60}_{28}\text{Ni}^* + {}^{0}_{-1}\text{e} + \overline{\nu}$$

where the * indicates an excited state of the nickel nuclide $^{60}_{28}$Ni. The β decay is followed by a transition of the nickel nucleus from the excited state to the ground state through the emission of a γ ray.

$$^{60}_{28}\text{Ni}^* \rightarrow {}^{60}_{28}\text{Ni} + \gamma$$

Notice that in γ decay there is no change in either the atomic number or the mass number. The same nucleus remains after the decay, only in a lower-energy state.

Decay Rates; Half-Life

A particular unstable nucleus may not decay for a very long time. The decay of such a nucleus is an uncertain event. The likely time interval before decay will vary greatly among different nuclides. There is some probability of decay of any unstable nucleus at any time. But this probability is often quite low. For example, the probability of radioactive decay of a single $^{226}_{88}$Ra nucleus during any 1-second interval is about one in 10^{11}. These are very small odds. Any particular radium-226 nucleus will almost certainly not decay during 1 second of observation. But if we examine a sample containing 10^{13} radium-226 nuclei, one out of 10^{11} nuclei will decay in a 1 s interval. Since there are 10^{13} nuclei present, this means that we can expect about $\left(\frac{1}{10^{11}}\right)(10^{13}) = 100$ nuclei† to undergo decay during any 1-second interval.

As parent nuclei decay, there are fewer and fewer parent nuclei remaining. Depending on the particular radioactive nuclide, the rate of decrease in the number of parent nuclei may be slow or fast. The fractional change in the number of parent nuclei, $\Delta N/N$, during a time interval Δt is proportional to Δt.

$$\frac{\Delta N}{N} \propto \Delta t$$

The constant of proportionality is called the **decay constant,** denoted by the Greek letter λ.

$$\frac{\Delta N}{N} = -\lambda\, \Delta t \qquad (30\text{-}5)$$

†More precisely, the expected number of decays is 137. This does not mean, however, that there are going to be exactly 137 decays in any 1-second interval. The process of radioactive decay is random and therefore subject to fluctuations. The number of observed decays in 1 second might be, say, 120. In the next second maybe 145 nuclei will decay. In the third second perhaps 132 will decay. The number of decays per second, averaged over many 1-second intervals, is 137.

The minus sign means that ΔN is negative; that is, the number of parent nuclei decreases. The decay constant indicates the rate of decay for a particular radioactive nuclide. For example, for ${}^{226}_{88}\text{Ra}$, $\lambda \approx 10^{-11}\ \text{s}^{-1}$, meaning that roughly one out of 10^{11} radium-226 nuclei decay in 1 second. For ${}^{92}_{36}\text{Kr}$, $\lambda \approx \frac{4}{10}\ \text{s}^{-1}$, meaning that roughly 4 out of 10 krypton-92 nuclei decay in 1 second.

We obtain an expression for the rate of change of parent nuclei, $\Delta N/\Delta t$, by multiplying Eq. 30-5 by $N/\Delta t$, which gives

$$\frac{\Delta N}{\Delta t} = -\lambda N \tag{30-6}$$

The **activity** $\mathcal{a}$ of a radioactive sample is defined as the **rate of decay,** which is the magnitude of $\Delta N/\Delta t$. Taking the magnitude of both sides of the preceding equation, we obtain

$$\mathcal{a} = \left|\frac{\Delta N}{\Delta t}\right| = \lambda N \tag{30-7}$$

The SI unit for activity is the **becquerel** (Bq), which is defined as one decay per second.

$$1\ \text{Bq} = 1\ \text{decay/s} \tag{30-8}$$

Another, more commonly used unit, is the **curie** (Ci), defined as exactly

$$1\ \text{Ci} = 3.7 \times 10^{10}\ \text{decays/s} \tag{30-9}$$

The original definition of the curie was based on the activity of 1 g of radium.

It is possible to obtain from Eq. 30-6 an expression describing how the number of parent nuclei decreases with time. Using calculus, one can show that the number N of parent nuclei present at time t is related to the initial number N_0 of parent nuclei at $t = 0$ by the equation

$$N = N_0 e^{-\lambda t} \tag{30-10}$$

where e is the base of the natural logarithm ($e \approx 2.718$).

Fig. 30-5 shows a graph of N versus t. This decay curve is described by the **half-life,** $T_{1/2}$, which is the **time interval during which half of the sample decays.** The

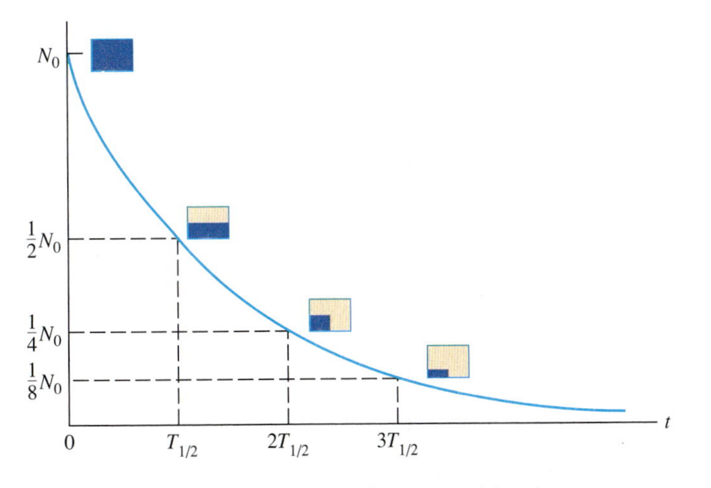

Fig. 30-5 The number of parent nuclei at time t.

exponential decay curve is such that **for each successive time interval of length $T_{1/2}$, the number N is reduced by half.** So if you have N_0 nuclei initially, after a time $T_{1/2}$, you will have $\frac{1}{2}N_0$ nuclei remaining. After a time $t = 2T_{1/2}$, the number present is reduced by half again to $\frac{1}{4}N_0$. After a time $t = 3T_{1/2}$, the number of nuclei present is reduced by half a third time to $\frac{1}{8}N_0$. In general, after a time interval equal to n half-lives, the number of nuclei remaining is reduced by a factor of $(\frac{1}{2})^n$.

$$N = (\tfrac{1}{2})^n N_0 \qquad (t = nT_{1/2}) \qquad (30\text{-}11)$$

For example, suppose a certain sample initially contains 6400 $^{212}_{83}Bi$ nuclei, with a half-life of about 1 hour. In 5 hours, which in this case is 5 half-lives, the number of parent nuclei remaining will be $(\frac{1}{2})^5(6400) = \frac{1}{32}(6400) = 200$ nuclei.

The half-life of a nuclide is related to its decay constant. If the decay constant is large, the nuclide decays rapidly and its half-life is short. If the decay constant is small, the nuclide decays slowly and its half-life is long. The two quantities can be related, using Eq. 30-10, setting $N = \frac{1}{2}N_0$ and $t = T_{1/2}$.

$$\tfrac{1}{2}N_0 = N_0 e^{-\lambda T_{1/2}}$$

Dividing this equation by N_0 and then taking the natural log, we find

$$\ln \tfrac{1}{2} = \ln e^{-\lambda T_{1/2}} = -\lambda T_{1/2}$$

Solving for λ and using $\ln \frac{1}{2} = -\ln 2$, we obtain

$$\lambda = \frac{\ln 2}{T_{1/2}} = \frac{0.693}{T_{1/2}} \qquad (30\text{-}12)$$

The activity $\mathcal{a}$ of a sample is proportional to the number N of radioactive nuclei, and so $\mathcal{a}$ and N have the same time dependence. Multiplying Eq. 30-10 ($N = N_0 e^{-\lambda t}$) by λ, we find

$$\lambda N = \lambda N_0 e^{-\lambda t}$$

At time t the activity has the value $\mathcal{a} = \lambda N$ (Eq. 30-7). At time t_0 the activity has its initial value $\mathcal{a}_0 = \lambda N_0$. Thus the preceding equation can be expressed

$$\mathcal{a} = \mathcal{a}_0 e^{-\lambda t} \qquad (30\text{-}13)$$

This is an equation of the same form as Eq. 30-10, and so the graph of activity versus time (Fig. 30-6) looks the same as the graph of number of parents versus time (Fig. 30-5).

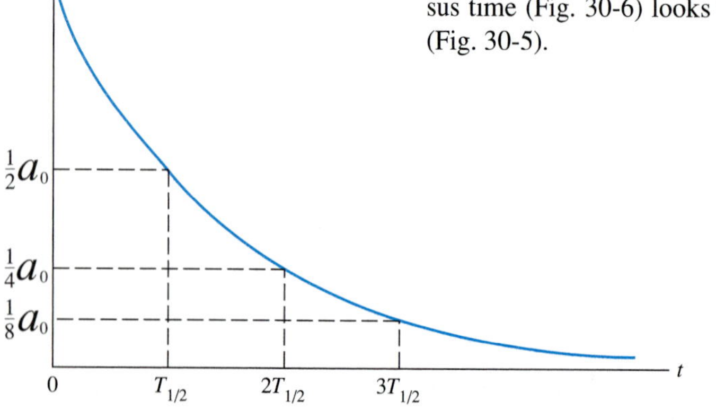

Fig. 30-6 The activity of a radioactive sample at time t.

EXAMPLE 5 How Much of a Sodium 24 Sample Decays in 2 Days?

Initially a certain radioactive sample contains 1.00×10^{10} atoms of the sodium isotope $^{24}_{11}Na$, which has a half-life of 15.0 h. (a) Find the number of parent nuclei present 48.0 h later. (b) Find the initial activity of the sample and the activity 48.0 h later.

SOLUTION (a) We can quickly estimate the number of parent nuclei after 48.0 h, by noting that this time interval is approximately 3 half-lives, and so the number is reduced by a factor of about $(\frac{1}{2})^3 = \frac{1}{8}$. So the number remaining is about $(\frac{1}{8})(10^{10})$, or roughly 10^9 parents. To calculate the number more precisely, we first calculate the number n of half-lives in 48.0 h.

$$n = \frac{48.0 \text{ h}}{15.0 \text{ h}} = 3.20$$

Next we find the final parent number N by applying Eq. 30-11.

$$N = (\tfrac{1}{2})^n N_0 = (\tfrac{1}{2})^{3.20}(1.00 \times 10^{10}) = 1.09 \times 10^9$$

An alternative method of solution is to apply Eq. 30-10:

$$N = N_0 e^{-\lambda t}$$

To use this equation, we must first calculate λ from Eq. 30-12:

$$\lambda = \frac{\ln 2}{T_{1/2}} = \frac{\ln 2}{15.0 \text{ h}} = 4.62 \times 10^{-2} \text{ h}^{-1}$$

Substituting this value of λ and the values of N_0 and t into the preceding equation, we find

$$N = (1.00 \times 10^{10})e^{-(4.62 \times 10^{-2} \text{h}^{-1})(48.0 \text{ h})}$$

$$= 1.09 \times 10^9$$

(b) We find the activity by applying Eq. 30-7.

At $t = 0$: $\quad \mathcal{A}_0 = \lambda N_0 = (4.62 \times 10^{-2} \text{ h}^{-1})(1.00 \times 10^{10})$

$$= 4.62 \times 10^8 \text{ decays/h}$$

or, converting to units of Bq (decays/s), we find

$$\mathcal{A}_0 = (4.62 \times 10^8 \text{ decays/h})\left(\frac{1 \text{ h}}{3600 \text{ s}}\right)$$

$$= 1.28 \times 10^5 \text{ Bq}$$

Using the definition of the curie unit (Eq. 30-9), we convert our answer to curies.

$$\mathcal{A}_0 = (1.28 \times 10^5 \text{ Bq})\left(\frac{1 \text{ Ci}}{3.70 \times 10^{10} \text{ Bq}}\right)$$

$$= 3.46 \times 10^{-6} \text{ Ci} = 3.46 \ \mu\text{Ci}$$

At $t = 48$ h: $\quad \mathcal{A} = \lambda N = (4.62 \times 10^{-2} \text{ h}^{-1})(1.09 \times 10^9)$

$$= (5.04 \times 10^7 \text{ decays/h})\left(\frac{1 \text{ h}}{3600 \text{ s}}\right) \times \left(\frac{1 \text{ Ci}}{3.70 \times 10^{10} \text{ Bq}}\right)$$

$$= 3.78 \times 10^{-7} \text{ Ci} = 0.378 \ \mu\text{Ci}$$

Alternatively, we could have found the final activity $\mathcal{A}$ from $\mathcal{A}_0$, using Eq. 30-13 ($\mathcal{A} = \mathcal{A}_0 e^{-\lambda t}$).

Carbon Dating

A technique known as **carbon dating** is used to determine the age of ancient archeological specimens. This method is based on a comparison of the small but measurable radioactivity of living matter with the reduced activity of a specimen that was once living—a piece of wood or an animal bone, for example. The activity begins to decline when the plant or animal dies, and so a measurement of activity indicates how long it has been dead.

All living matter contains carbon, a small fraction of which is the radioactive isotope $^{14}_{6}C$. The half-life of this isotope is 5700 years, which is a useful time interval for dating specimens that are anywhere from about 1000 to 30,000 years old. Of the carbon atoms in the carbon dioxide of the earth's atmosphere, a small fraction are $^{14}_{6}C$

atoms. This fraction has apparently been constant for tens of thousands of years.* Living plants take in carbon dioxide, and animals eat plants. Neither plants nor animals distinguish between isotopes. So in living plants and animals the ${}^{14}_{6}C$ is present in the same proportion as it is in the atmosphere—about one part in 8×10^{11}, which results in an activity of **0.23 Bq per gram of carbon in the living organism.** Once an organism dies, however, its ${}^{14}_{6}C$ nuclei decay and are not replaced. The organism's radioactivity decreases in proportion to the decrease in ${}^{14}_{6}C$.

EXAMPLE 6 Carbon Dating an Ancient Bone

A bone fragment found in an archeological dig contains 6.0 g of carbon and has an activity of 0.15 Bq. Find the age of the bone.

SOLUTION When the animal died, the initial activity $\mathcal{a}_0$ of the bone fragment was (6.0 g)(0.23 Bq/g) = 1.4 Bq. To find the time t since death, we relate the final activity $\mathcal{a}$ at time t to the initial activity $\mathcal{a}_0$, using Eq. 30-13.

$$\mathcal{a} = \mathcal{a}_0 e^{-\lambda t}$$

Dividing by $\mathcal{a}_0$ and taking the natural log, we find

$$\ln \frac{\mathcal{a}}{\mathcal{a}_0} = \ln e^{-\lambda t} = -\lambda t$$

or

$$t = -\frac{\ln(\mathcal{a}/\mathcal{a}_0)}{\lambda}$$

Using Eq. 30-12 $\left(\lambda = \frac{\ln 2}{T_{1/2}}\right)$ to substitute for λ, we find

$$t = -\frac{\ln(\mathcal{a}/\mathcal{a}_0)}{\ln 2} T_{1/2}$$

$$= \frac{\ln(1.4 \text{ Bq}/0.15 \text{ Bq})}{\ln 2}(5700 \text{ y})$$

$$= 18{,}000 \text{ y}$$

Decay Series

Quite often a parent nucleus decays into a daughter that is itself radioactive. In other words, the daughter becomes a parent, as it decays into some other nuclide. For example, after radium 226 α decays into radon 222, the radon in turn α decays into polonium 218. The polonium too is unstable and either α decays into lead 214 or β decays into astatine 218. Fig. 30-7 shows a radioactive decay series that begins with uranium 238 and ends with lead 206, which is stable.

Since uranium 238 has such a long half-life (4.5×10^9 y), there is still plenty of it around, even though it is believed to have been formed just before the formation of the solar system, about 5 billion years ago. Shorter half-life materials such as radon 222 ($T_{1/2}$ = 3.8 d) would soon disappear, were it not for the fact that they are part of the decay series that begins with long-lived uranium 238 and are therefore continually being replenished.

In addition to the radioactive series shown in Fig. 30-7, there are two other natural radioactive series, one beginning with uranium 235 and one beginning with thorium 232. Nearly all radioactive nuclides on earth are part of one of these three radioactive series. A fourth radioactive series begins with the nuclide neptunium 227, artificially produced in the laboratory through nuclear reactions, which we shall discuss in the next section.

*The ${}^{14}_{6}C$ in atmospheric carbon dioxide would continually decline as a result of radioactive decay were it not for another process at work in the atmosphere. Cosmic rays interact with the atmosphere and produce new ${}^{14}_{6}C$ nuclei. The rate of production of ${}^{14}_{6}C$ nuclei is about equal to the rate of decay, and so the supply of ${}^{14}_{6}C$ in the atmosphere is approximately constant.

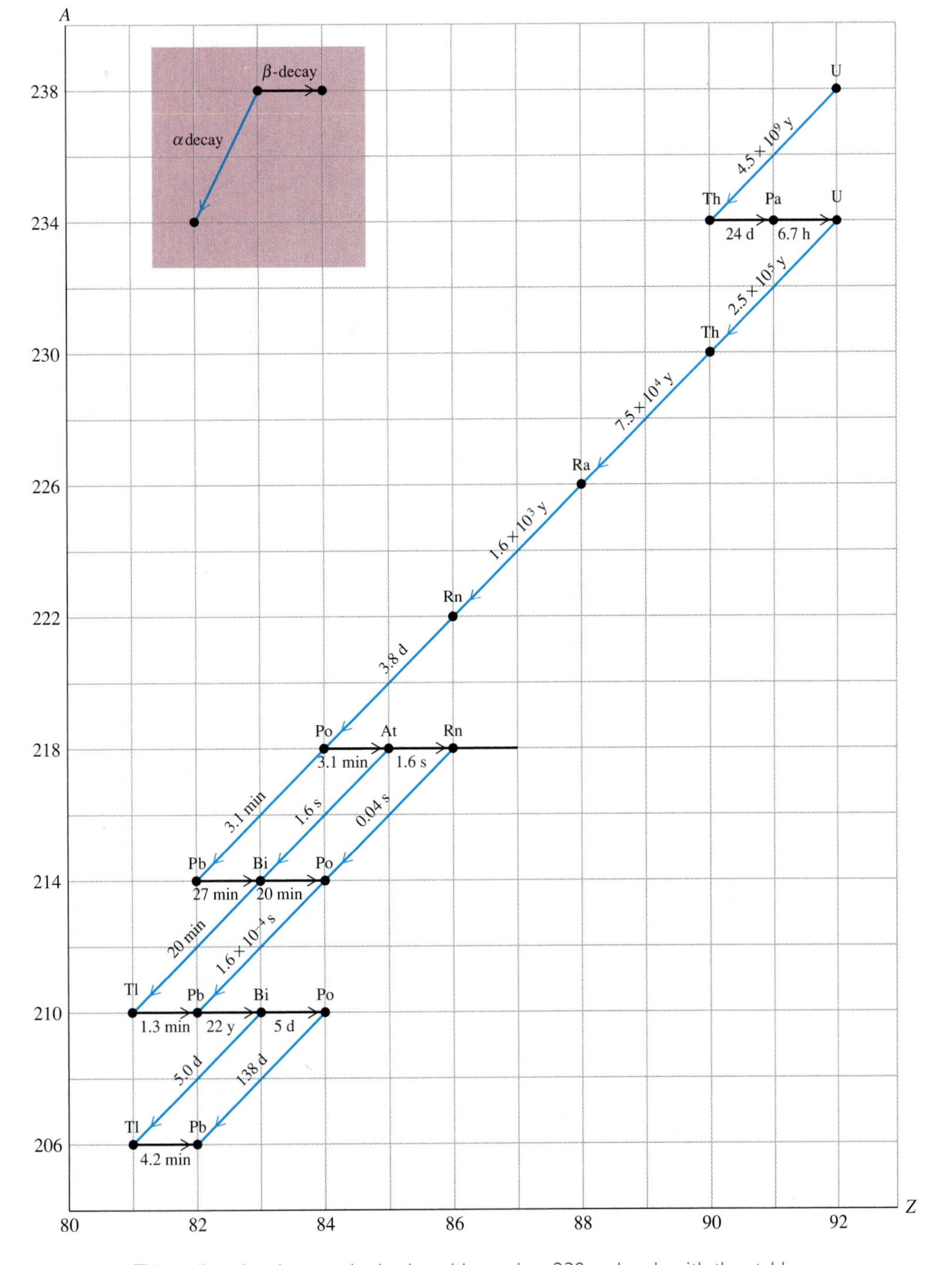

Fig. 30-7 This radioactive decay series begins with uranium 238 and ends with the stable nuclide lead 206. For each α or β decay, the figure shows the half-life in seconds (s), minutes (min), hours (h), or years (y).

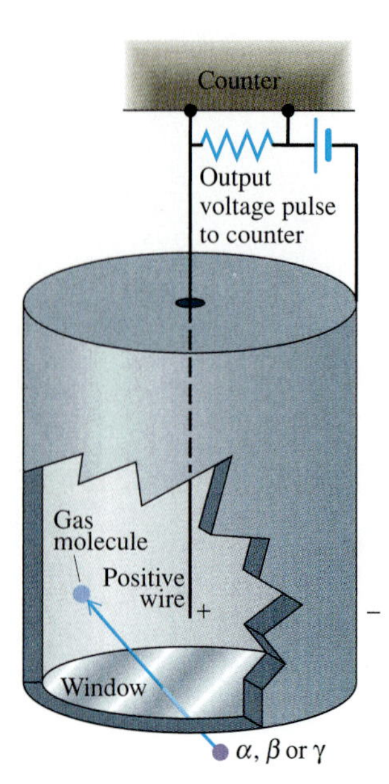

Fig. 30-8 A Geiger counter.

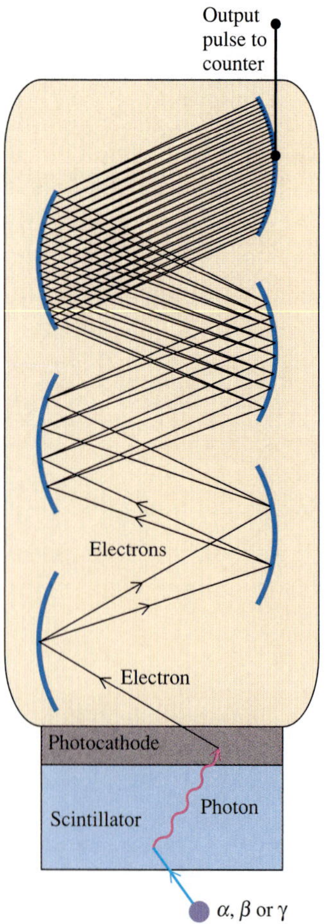

Fig. 30-9 A scintillation counter.

Radiation Detectors

We can observe and measure radioactive decay using various devices that detect the α, β, and γ radiation emitted by decaying nuclei. One of the most common radiation detectors is the **Geiger counter** (Fig. 30-8), which counts the number of α's, β's, or γ's absorbed. The radiation is detected in the Geiger tube, which is connected to a counting device. The Geiger tube is a metal tube with a thin glass or mica window at one end. A wire extending along the axis of the tube is maintained at a positive potential ($\sim 10^3$ V) relative to the cylinder. An α, β, or γ enters the tube through the window and ionizes a gas atom, liberating an atomic electron. This electron is then accelerated toward the positive wire at the center. As the electron moves toward the wire, it collides with and ionizes other gas atoms, and so a large number of electrons quickly move to the wire. This produces a short pulse of current both through the tube and through a resistor in series with the tube. The voltage across the resistor is amplified and used to activate a counter, which records the number of particles detected. If a loudspeaker is connected to the output, you can hear a click for each count.

A **scintillation counter** is another kind of detector. The scintillator is a material that emits a visible photon when an α, β, or γ is absorbed. The scintillator is connected to a **photomultiplier tube,** consisting of a cathode and a series of other electrodes at progressively higher potentials (Fig. 30-9). A photon emitted by the scintillator is absorbed by the cathode, which then emits an electron through the photoelectric effect. This electron is then accelerated toward the nearest positive electrode, which is at a potential of about 100 V above the cathode. The positive electrodes have the special property that, for every electron absorbed, several electrons are emitted. These electrons are then accelerated toward the next electrode, at a still higher potential. In a series of 10 or 15 such electrodes, there is a large multiplier effect, with perhaps 10^6 electrons at the last electrode, producing a current pulse big enough to be detected. Scintillation counters tend to be more sensitive detectors than Geiger counters.

Some radiation detectors show the paths of detected radiation. One of these is the **cloud chamber,** which contains a gas that has been cooled below its condensation temperature. An incident particle ionizes gas molecules, and the gas begins to condense around these ions, and so a visible trail of liquid drops is left by the particle as it passes through the chamber. A similar kind of detector is the **bubble chamber,** in which a liquid that is about to boil forms visible bubbles centered at points where a passing particle has ionized the liquid (Fig. 30-10).

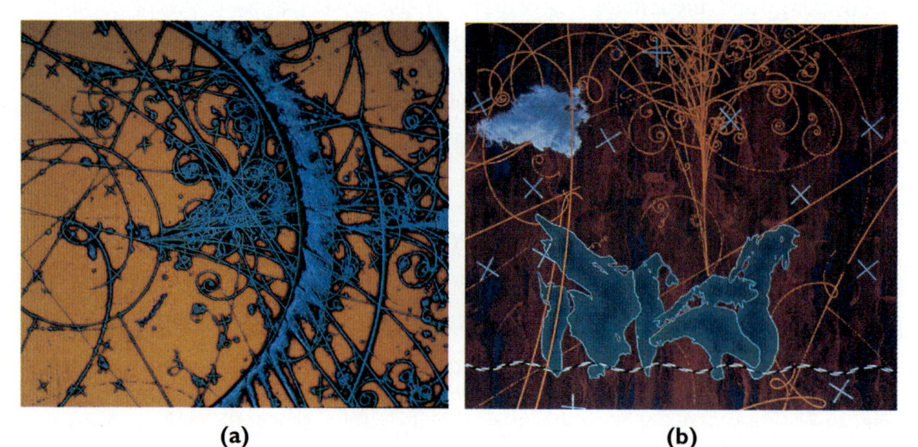

(a) (b)

Fig. 30-10 **(a)** Bubble-chamber photograph showing the paths of particles. **(b)** An artist's work, inspired by bubble chamber photographs.

30-3 Nuclear Reactions; Fission and Fusion

In the last section we described processes in which a nucleus spontaneously decays by α, β, or γ decay. It is also possible for a nucleus to be transformed as a result of a collision with either another nucleus or a particle such as a neutron. A collision that causes a change in a nucleus is called a **nuclear reaction.** In 1919 Rutherford first produced a nuclear reaction in the laboratory by directing α particles at nitrogen gas. Rutherford observed protons given off. He concluded that, for charge to be conserved, nitrogen nuclei must have been transformed to oxygen nuclei in the reaction

$$^{4}_{2}\text{He} + {}^{14}_{7}\text{N} \rightarrow {}^{17}_{8}\text{O} + {}^{1}_{1}\text{H}$$

Notice that this reaction conserves both charge and mass number; the total value of Z equals 9 before and after the reaction ($2 + 7 = 8 + 1$) and the total value of A equals 18 before and after the reaction ($4 + 14 = 17 + 1$). Conservation of charge and mass number are often used to determine the nuclide resulting from a nuclear reaction.

EXAMPLE 7 Transforming Beryllium by α Bombardment

A beryllium target is bombarded with α particles, and neutrons are given off. Determine the final nucleus resulting from this reaction.

SOLUTION From Table 30-1, we see that virtually all beryllium is the isotope $^{9}_{4}\text{Be}$, and so the nuclear reaction is

$$^{4}_{2}\text{He} + {}^{9}_{4}\text{Be} \rightarrow {}^{A}_{Z}\text{X} + {}^{1}_{0}\text{n}$$

where $^{A}_{Z}\text{X}$ is used to represent the unknown final nucleus. Balancing atomic number and mass number, we find $Z = 6$ and $A = 12$. This means that the final nucleus is carbon 12, and the reaction is

$$^{4}_{2}\text{He} + {}^{9}_{4}\text{Be} \rightarrow {}^{12}_{6}\text{C} + {}^{1}_{0}\text{n}$$

Q-Value of Reactions

Both energy and momentum are conserved in nuclear reactions. Mass energy mc^2 is often transformed to kinetic energy K or vice versa in such reactions. Since energy is conserved, a change in the system's total mass energy ($\Delta m\, c^2$) must be compensated by an opposite change in the system's kinetic energy (ΔK).

$$\Delta E = 0$$

$$\Delta K + \Delta m\, c^2 = 0$$

$$\Delta K = -(\Delta m)c^2$$

The energy Q transformed from mass energy to kinetic energy is called the "reaction energy," or the "Q-value" of the reaction.

$$Q = \Delta K = -(\Delta m)c^2 \tag{30-14}$$

If $Q > 0$, kinetic energy is gained ($\Delta K > 0$) and mass energy is lost ($\Delta m < 0$), and the reaction is called "exothermic." If $Q < 0$, kinetic energy is lost ($\Delta K < 0$) and mass energy is gained ($\Delta m > 0$), and the reaction is called "endothermic."

EXAMPLE 8 Calculating the Q-value of a Reaction

(a) Determine the Q-value of the reaction described in Example 7 ($^4_2\text{He} + {}^9_4\text{Be} \rightarrow {}^{12}_6\text{C} + {}^1_0\text{n}$).

(b) If the α particle has an initial kinetic energy of 10.0 MeV and the beryllium target nucleus is at rest, what is the total kinetic energy of the carbon nucleus and the neutron?

SOLUTION **(a)** We calculate the Q-value by applying Eq. 30-14 [$Q = -(\Delta m)c^2$], where $-(\Delta m)$ is the sum of the initial masses (^4_2He and ^9_4Be) minus the sum of the final masses ($^{12}_6\text{C}$ and ^1_0n). Using the values for the masses given in Table 30-1, we find

$$Q = -(\Delta m)c^2$$
$$= [(m_{\text{He}} + m_{\text{Be}}) - (m_{\text{C}} + m_{\text{n}})]c^2$$
$$= [(4.002603\text{ u} + 9.012183\text{ u}) - (12.000000\text{ u} + 1.008665\text{ u})]c^2$$
$$= (0.006121\text{ u})c^2$$

To express our answer in units of MeV, we use Eq. 30-4 [$(1\text{ u})c^2 = 931.5\text{ MeV}$] and obtain

$$Q = (0.006121)(931.5\text{ MeV}) = 5.702\text{ MeV}$$

(b) We find the total kinetic energy of the carbon nucleus and the neutron, the system's final kinetic energy K_f, by equating the change in kinetic energy to the Q-value of the reaction (Eq. 30-14).

$$K_\text{f} - K_\text{i} = \Delta K = Q$$
$$K_\text{f} = K_\text{i} + Q = 10.0\text{ MeV} + 5.702\text{ MeV}$$
$$= 15.7\text{ MeV}$$

EXAMPLE 9 Will 1-MeV α's Bombarding N-14 Produce a Reaction?

Would it have been possible for Rutherford to have used α particles with a kinetic energy of 1.0 MeV to produce the nuclear reaction $^4_2\text{He} + {}^{14}_7\text{N} \rightarrow {}^{17}_8\text{O} + {}^1_1\text{H}$?

SOLUTION First we calculate the Q-value of the reaction, applying Eq. 30-14 [$Q = -(\Delta m)c^2$] and using masses given in Table 30-1.

$$Q = -(\Delta m)c^2 = [(m_{\text{He}} + m_{\text{N}}) - (m_{\text{O}} + m_{\text{H}})]c^2$$
$$= [(4.002603\text{ u} + 14.003074\text{ u}) - (16.999133\text{ u} + 1.007825\text{ u})]c^2$$
$$= (-0.001281\text{ u})c^2 \left(\frac{931.5\text{ MeV}}{(1\text{ u})c^2}\right) = -1.193\text{ MeV}$$

This means that the system's change in kinetic energy is negative.

$$\Delta K = -1.193\text{ MeV}$$

The target nitrogen nuclei are at rest, and so the system's initial kinetic energy is that of the α particle, given to be 1.0 MeV. It is impossible for the system's kinetic energy to decrease by 1.193 MeV, since kinetic energy can never be negative. Thus this reaction is impossible with 1.0 MeV α particles. Higher energy α particles are required. You might think that 1.193 MeV would be sufficient kinetic energy for the α particles. Actually, somewhat more energy is required. To conserve momentum, the final particles must have momentum equal to the α particle's initial momentum. This means that these final particles must have some kinetic energy, and the α particle must therefore have kinetic energy 1.193 MeV greater than the system's final kinetic energy.

All the nuclear reactions described so far produce stable nuclides. Sometimes, however, nuclear reactions produce unstable nuclides. Nuclear reactions are often used to prepare radioactive samples used for various medical applications. For example, in the treatment of cancer patients, the radioactive isotope cobalt 60, produced through nuclear reactions in the laboratory, is used as a source of radiation to kill cancer cells. Artificially produced radioactive isotopes are also used as tracers to determine how well an organ is functioning. For example, a patient with a thyroid problem drinks a liquid containing a trace of the radioactive isotope iodine 131. Some time later, radiation detectors are used to measure the distribution of this isotope throughout the body, so that the physician can test the effectiveness of the thyroid gland in its function of distributing iodine.

Transuranic Elements

In the 1930s Enrico Fermi began experimenting with neutrons as projectiles fired at nuclei to produce nuclear reactions. Fermi reasoned that neutrons should be more effective than α particles in producing reactions because the neutrons are uncharged, unlike α particles, which are positively charged and are therefore strongly repelled by positively charged nuclei. Fermi produced neutrons through nuclear reactions like the reaction in Example 7 and then used these neutrons as bullets to initiate other reactions in various targets. Fermi immediately began to produce many new isotopes.

Fermi's desire was to create entirely new elements—elements beyond uranium on the periodic table. Fermi tried to create these new "transuranic elements" by bombarding uranium with neutrons. Fermi believed that he had created two such elements, neptunium ($^{239}_{93}$Np) and plutonium ($^{239}_{94}$Pu), through the series of reactions shown in Fig. 30-11. Although Fermi had actually failed to produce these elements, they were produced in 1940 at the University of California at Berkeley. In the years that followed, other transuranic elements were produced: americium ($Z = 95$), curium ($Z = 96$), berkelium ($Z = 97$), californium ($Z = 98$), einsteinium ($Z = 99$), fermium ($Z = 100$), and so on up to the unnamed element with atomic number 109. All the transuranic elements are unstable, though some are long lived, especially plutonium, which has a half-life of 24,000 years.

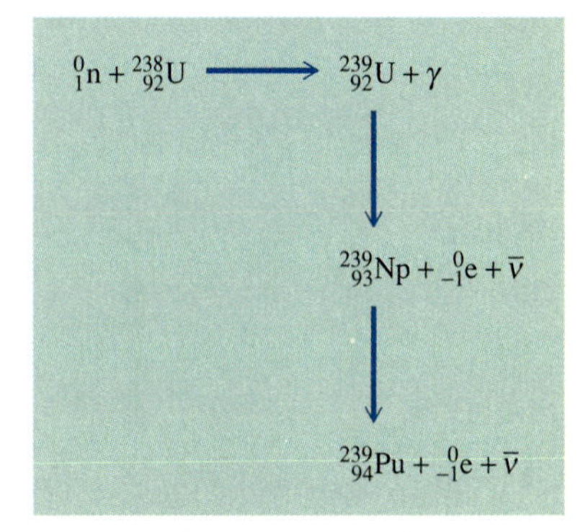

Fig. 30-11 In this nuclear reaction $^{239}_{92}$U is produced by bombardment of $^{238}_{92}$U with neutrons. A series of decays follows, producing the transuranic element $^{239}_{94}$Pu.

Fission

Without realizing it, in his uranium experiments Fermi had actually produced **nuclear fission,** a splitting of the uranium nucleus into two smaller nuclei of roughly equal size. This process was discovered in 1939 by German scientists Otto Hahn, Fritz Strassman, Lise Meitner, and Otto Frisch. Fission of the uranium nucleus is much more likely to occur in the isotope U-235 than in the more abundant isotope U-238. Fission is a two-part process (Fig. 30-12). A neutron is first absorbed by the uranium nucleus, giving it additional energy and making the nucleus highly unstable. In its excited state the uranium nucleus becomes elongated. Then the short-range strong force attracting the nucleons is no match for the long-range electrical force of repulsion, and so the nucleus splits into two smaller nuclei, along with a number of neutrons. The fission of uranium can result in several different final states:

$$^{1}_{0}n + ^{235}_{92}U \rightarrow ^{141}_{56}Ba + ^{92}_{36}Kr + 3\,^{1}_{0}n$$

$$^{1}_{0}n + ^{235}_{92}U \rightarrow ^{140}_{54}Xe + ^{94}_{38}Sr + 2\,^{1}_{0}n$$

$$^{1}_{0}n + ^{235}_{92}U \rightarrow ^{132}_{50}Sn + ^{101}_{42}Mo + 3\,^{1}_{0}n$$

All these reactions transform mass energy to kinetic energy. We can use the binding-energy curve (Fig. 30-3) to get a rough estimate of the Q-value for the fission of uranium. This graph shows that the binding energy per nucleon for U-235 is about 7.6 MeV. The average mass number of the fission fragments is about 120, for which Fig. 30-3 indicates a binding energy per nucleon of about 8.5 MeV. The larger value of the binding energy per nucleon in the system's final state means that the final state is a more tightly bound state, that is, a state of lower energy and therefore lower mass. The loss of mass energy per nucleon during the fission process equals the difference in binding energies per nucleon or roughly 8.5 MeV − 7.6 MeV = 0.9 MeV. The Q-value of the reaction, the total mass energy lost in the fission of one uranium nucleus, equals the loss in mass energy per nucleon (0.9 MeV) times the number of nucleons (235).

$$Q \approx (0.9\ \text{MeV/nucleon})(235\ \text{nucleons})$$

$$\approx 200\ \text{MeV}$$

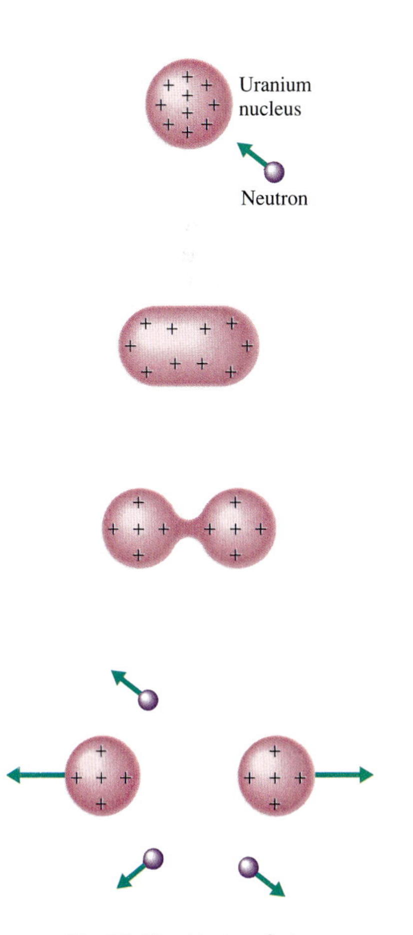

Fig. 30-12 Nuclear fission.

We can better appreciate the magnitude of the energy released in the fission of U-235 by calculating the energy liberated per kg of U-235:

$$\frac{Q}{m} = \frac{Q}{235\ \text{u}} \approx \frac{(200\ \text{MeV})(1.6 \times 10^{-13}\ \text{J/MeV})}{(235)(1.66 \times 10^{-27}\ \text{kg})}$$

$$\approx 8 \times 10^{13}\ \text{J/kg} \qquad (30\text{-}15)$$

To put this number into perspective, the heat of combustion of gasoline is 5×10^7 J/kg. Thus nuclear fission provides on the order of 10^6 times the energy of ordinary combustion. The reason for this is that fission involves nuclear binding energies, whereas combustion is an atomic process involving changes in binding energies of atomic electrons. Nuclear binding energies are on the order of 10^6 times greater than electron binding energies, and so nuclear reactions are capable of releasing 10^6 times as much energy as chemical reactions.

EXAMPLE 10 Electric Power From Fission

How long could the energy released in the fission of 1 kg of U-235 supply the electrical energy needs of a city with a population of one million if the average per capita electrical power consumption is 1 kW?

SOLUTION Power is defined as the energy supplied per unit time.

$$P = \frac{E}{t}$$

The total electrical power consumption is 10^6 kW $= 10^9$ W, and the energy supplied by fission of 1 kg of U-235 is 8×10^{13} J. Solving the preceding equation for t, we find

$$t = \frac{E}{P} = \frac{8 \times 10^{13}\ \text{J}}{10^9\ \text{W}}$$

$$= 8 \times 10^4\ \text{s} = 20\ \text{h}$$

The energy supplied by the fission of 1 kg of U-235 is sufficient to supply the electrical energy needs of a city of one million for about 20 hours. The energy supplied by burning 1 kg of gasoline would supply the city for only 0.05 s.

Not only does fission of uranium 235 liberate an enormous amount of energy, but it does so in a very short time interval—on the order of 10^{-9} s. Furthermore, each fission process typically produces 2 or 3 neutrons. Each of these neutrons is then capable of causing fission of other uranium nuclei. This suggests the possibility of a **chain reaction**—a series of fissions, with more nuclei fissioning and more energy released at each stage (Fig. 30-13). An uncontrolled chain reaction of nuclear fissions is the principle behind the "atomic" bomb. A chain reaction of nuclear fissions is also used to supply energy in a nuclear reactor, but the chain reaction is controlled. Materials are used to absorb some of the neutrons emitted through fission, so that the total number of neutrons inducing fission remains constant, and so the reactor produces heat at a constant rate. The heat is then used to operate an electric generator and to provide electrical energy.

Fusion

Fission of uranium releases energy because the binding energy per nucleon of this large nucleus is less than the binding energy per nucleon of the smaller fission fragments, as shown in the binding-energy graph (Fig. 30-3). The graph peaks at a mass number of about 60, which means that nucleons are generally more weakly bound as the mass number falls below about 60. It follows that energy can be liberated in a nu-

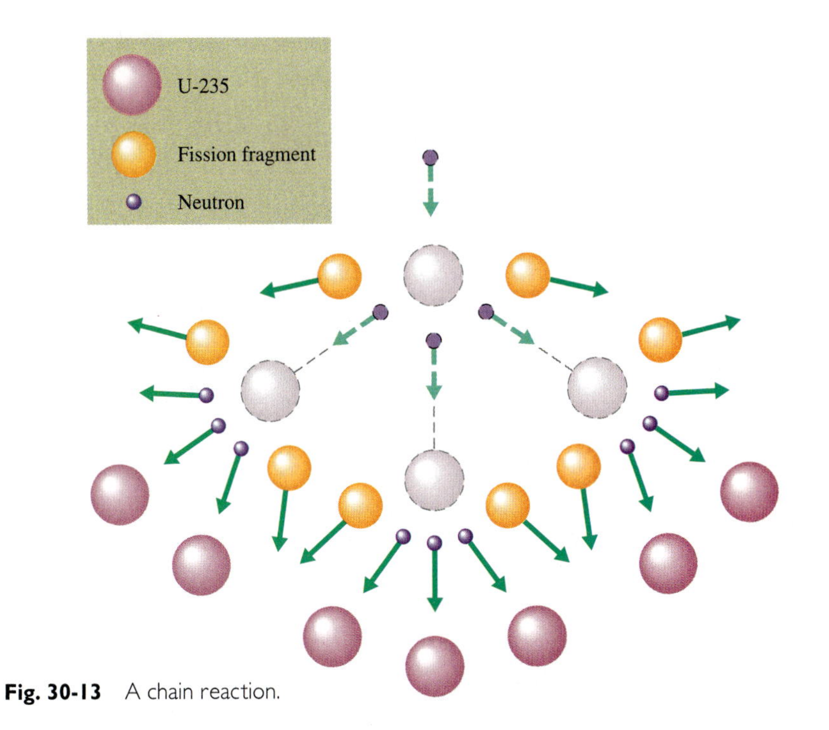

Fig. 30-13 A chain reaction.

clear reaction that fuses smaller nuclei together to form a larger nucleus. Fusion of such nuclei liberates energy because the final binding energy per nucleon is greater than the initial binding energy per nucleon. **Nuclear fusion** occurs in nature. It is the process by which the sun produces its enormous energy. Since solar energy is the ultimate source of virtually all energy on earth, nuclear fusion is the most important of all reactions in nature. The following sequence of three fusions produces solar energy.

$$^1_1\text{H} + {}^1_1\text{H} \rightarrow {}^2_1\text{H} + \text{e}^+ + \nu$$

$$^1_1\text{H} + {}^2_1\text{H} \rightarrow {}^3_2\text{He} + \gamma$$

$$^3_2\text{He} + {}^3_2\text{He} \rightarrow {}^4_2\text{He} + {}^1_1\text{H} + {}^1_1\text{H}$$

One of the simplest fusion reactions is the fusion of the two hydrogen isotopes ^1_1H and ^3_1H (tritium) to form ^4_2He.

$$^1_1\text{H} + {}^3_1\text{H} \rightarrow {}^4_2\text{He}$$

We can get a rough estimate of the energy liberated from the binding-energy graph (Fig. 30-3), which indicates that the binding energy per nucleon for ^4_2He is about 7 MeV and the binding energy per nucleon for the two hydrogen isotopes averages roughly 2 MeV. So the energy per nucleon liberated in this process is roughly 7 MeV − 2 MeV = 5 MeV, and the total energy liberated is (5 MeV/nucleon)(4 nucleons), or about 20 MeV. The energy liberated per unit mass is

$$\frac{\Delta E}{m} \approx \frac{20\text{ MeV}}{4\text{ u}} = \frac{20(1.6 \times 10^{-13}\text{ J})}{4(1.66 \times 10^{-27}\text{ kg})}$$

$$\approx 5 \times 10^{14}\text{ J/kg} \qquad (30\text{-}16)$$

This is a value typical of fusion reactions—a value even larger than the typical energy released per unit mass in fission (8×10^{13} J/kg).

EXAMPLE 11 How Long Could Solar Energy Last?

How long can the sun supply energy at its current rate of 4×10^{26} W if its mass of 2×10^{30} kg begins entirely as hydrogen and undergoes fusion?

SOLUTION We can estimate the total energy available as the product of the sun's mass and the energy per unit mass liberated in fusion, which we have estimated to be 5×10^{14} J/kg (Eq. 30-16).

$$\frac{\Delta E}{m} = 5 \times 10^{14} \text{ J/kg}$$

or

$$\begin{aligned} \Delta E &= (5 \times 10^{14} \text{ J/kg})m \\ &= (5 \times 10^{14} \text{ J/kg})(2 \times 10^{30} \text{ kg}) \\ &= 10^{45} \text{ J} \end{aligned}$$

The sun radiates energy at a rate

$$\frac{\Delta E}{\Delta t} = 4 \times 10^{26} \text{ W}$$

Thus the energy lasts for a time

$$\begin{aligned} \Delta t &= \frac{\Delta E}{4 \times 10^{26} \text{ W}} = \frac{10^{45} \text{ J}}{4 \times 10^{26} \text{ W}} \\ &= 3 \times 10^{18} \text{ s} = 8 \times 10^{10} \text{ y} \end{aligned}$$

Will the sun really last 80 billion years? A more detailed analysis of all the factors that contribute to the sun's stable radiation indicates a probable lifetime of only about 8 billion years because stability will be disrupted when hydrogen in the sun's core (about 10% of the total hydrogen) is used up. The sun's stable life is roughly half over; we expect it to last roughly another 4 billion years.

Ordinary combustion gives only about 5×10^{7} J/kg, or 10^{-7} times the energy per unit mass given by fusion. So if solar energy were provided by ordinary combustion, the sun could provide energy only for $10^{-7}(8 \times 10^{10} \text{ y}) = 8000$ y. The sun would have burned out long ago.

Unlike fission, the initiation of a fusion reaction requires very high initial kinetic energies of the fusing nuclei, since the positively charged nuclei experience a repulsive electric force. The nuclei must get close enough so that the short-range nuclear force can take effect. The maximum range of this force is only about 10^{-14} m. To get this close the nuclei must have at least enough initial kinetic energy K_i when they are far apart to provide enough electrical potential energy U_{Ef} for the nuclei to be at rest at a distance of about 10^{-14} m. Applying $U_E = q'V = q'\left(\frac{kq}{r}\right) = \frac{kqq'}{r}$ (Eqs. 18-3 and 18-7) to the fusion of ${}^1_1\text{H}$ and ${}^3_1\text{H}$, we find

$$\begin{aligned} K_i = U_{Ef} &= \frac{kqq'}{r} = \frac{ke^2}{r} \\ &= \frac{(9 \times 10^{9} \text{ N-m}^2/\text{C}^2)(1.6 \times 10^{-19} \text{ C})^2}{10^{-14} \text{ m}} \\ &= 2 \times 10^{-14} \text{ J} \end{aligned}$$

The way to provide such high energy to many nuclei is to heat the nuclei to a very high temperature. The fusion process that results is called thermonuclear fusion. If we equate the required kinetic energy to the mean kinetic energy of an atom at temperature T, we can get some idea of the temperature required for fusion. We apply Eq. 12-14:

$$\begin{aligned} K &= \tfrac{3}{2}kT \\ T &= \frac{2}{3}\left(\frac{K}{k}\right) = \frac{2}{3}\left(\frac{2 \times 10^{-14} \text{ J}}{1.38 \times 10^{-23} \text{ J/K}}\right) \\ &\approx 10^{9} \text{ K} \end{aligned}$$

Because of quantum effects, the temperature required is lower than the temperature we have calculated here—*only* about 10^7 K! Such temperatures are maintained on the sun, and so fusion can occur. A temperature of this order is also attained in the explosion of a fission bomb, which can then be used to trigger a fusion, or "hydrogen," bomb. Controlling fusion in a reactor presents enormous technical problems because of the difficulty of confining the high-temperature thermonuclear reaction. If fusion energy could be harnessed, however, deuterium could be used as a fuel, and because every 30 kg of water contains about 1 g of deuterium, the earth's oceans would provide a cheap and practically limitless supply of this fuel.

30-4 Biological Effects of Radiation

Radiation emitted by radioactive sources or by nuclear reactions can significantly affect biological matter by ionizing its atoms. **Ionizing radiation** is *any* radiation that is sufficiently energetic to ionize atoms. This includes ultraviolet radiation (UV), X rays, γ rays—all part of the electromagnetic spectrum, as well as α and β particles, neutrons, protons, and nuclei. Radiation emitted by radioactive sources, or nuclear reactions, is often quite potent in its ionizing effect because the typical energy of such radiation (on the order of MeV's) is large enough to ionize thousands of atoms. Radiation can kill cells and cause great harm to an organism—possibly radiation sickness or death, either in the short term of a few days or weeks or in the long term of years or even decades. Cancers caused by radiation may not show up for 30 years after exposure. Radiation absorbed by the reproductive organs can cause mutations, which are passed on to subsequent generations. Only rarely are such mutations beneficial.

To study the effects of radiation, we need to be able to measure the effects quantitatively. The strength of a radioactive source is indicated by its activity, typically measured in units of curies (Ci) or millicuries (mCi), where 1 Ci is defined as 3.7×10^{10} decays/s. The effect of a radioactive source on a subject exposed to it depends not only on the source's activity, but also on other factors: the kind of radiation (α, β, γ, or other), the subject's distance from the source, the presence of other materials that can absorb radiation, the length of time of the exposure, and the half-life of the source.

For example, α particles are easily absorbed by matter. Even a few centimeters of air is enough to absorb most of the α particles emitted by an α-source. Thus α particles are not likely to be damaging unless the α-source is inside the body. Unfortunately, this is exactly what happens when radioactive radon gas is inhaled. This source of α particles is a product in the natural radioactive decay series that begins with uranium 238 (Fig. 30-7). Inhaled radon gas, produced by natural sources within the earth, is estimated by the environmental protection agency to cause up to 20,000 cases of lung cancer annually in the United States.

X rays and γ rays can easily penetrate matter, and so a source of such radiation is dangerous, even when the source is outside the body. Time of exposure to a radioactive source is an obvious factor, since the longer one is exposed, the greater is the radiation that enters the body. Half-life too is an important factor because it determines how long a "hot" radioactive source continues to have a high level of activity. After, say, 20 half-lives, the activity declines to $(\frac{1}{2})^{20}$, or roughly 10^{-6} times its initial value. But the half-life could be a few seconds, or it could be many years. A radioactive source with a long half-life is more dangerous than one with a shorter half-life and the same initial activity.

The effect of radiation on a body absorbing it relates to a quantity called the **absorbed dose** D, defined as the **energy absorbed from ionizing radiation per unit mass of the absorbing body.**

$$D = \frac{E}{m} \tag{30-17}$$

The SI unit of absorbed dose is the gray (Gy), defined as one joule per kilogram.

$$1 \text{ Gy} = 1 \text{ J/kg} \tag{30-18}$$

Another commonly used unit is the rad, an acronym for radiation absorbed dose, where

$$1 \text{ rad} = 0.01 \text{ Gy} = 0.01 \text{ J/kg} \tag{30-19}$$

Absorption of different kinds of radiation causes different amounts of damage. For example, an absorbed dose of α particles equal to 1 rad does about as much damage to cells as about 10 rad of X rays. We define the **relative biological effectiveness** (RBE) of any radiation as the **number of rad of X rays that produces the same biological damage as one rad of the radiation used.** Thus α particles have an RBE of 10. Table 30-2 gives values of RBE for different types of radiation. The values depend on the energy of the radiation and the body part absorbing it. For example, neutrons are particularly damaging to the eyes, and so the RBE of neutrons is higher for the eyes than for other parts of the body.

The **equivalent dose** D_e of any absorbed radiation is defined as the **product of the absorbed dose and the relative biological effectiveness** of the kind of radiation absorbed.

$$D_e = D \times \text{RBE} \tag{30-20}$$

The SI unit of equivalent dose is the sievert (Sv), defined as the gray times RBE.

$$1 \text{ Sv} = 1 \text{ Gy} \times \text{RBE} \tag{30-21}$$

Another more commonly used unit is the rem, defined as the rad times RBE.

$$1 \text{ rem} = 1 \text{ rad} \times \text{RBE} \tag{30-22}$$

Since 1 rad = 0.01 Gy,

$$1 \text{ rem} = 0.01 \text{ Sv} \tag{30-23}$$

Table 30-2 Relative biological effectiveness (RBE)

Radiation	RBE
Photons and β particles of 30 keV or more	1.0
β particles of less than 30 keV	1.7
Neutrons and protons below 10 MeV	10 (body)
	30 (eyes)
α particles from natural radioactivity	10
Heavy-recoil nuclei	20

From Hobbie, R.: *Intermediate physics for medicine and biology*, New York, 1978, Wiley, based on James AE Jr, Wagner HN Jr, and Cooke RE: *Pediatric nuclear medicine*, Philadelphia, 1974, Saunders.

It is surprising to realize the effect that even very small amounts of ionizing radiation can have on the human body. In the early days of nuclear physics there was no realization of the damage that could be produced by radioactive sources. Becquerel, who in 1896 discovered the phenomenon of radioactivity, found a red spot on his skin beneath a vest pocket in which he had carried a sample of radium for a few hours. A painful sore developed several days after exposure and finally healed 2 months later. Pierre Curie, Marie Curie's husband, confirmed the effect Becquerel had reported by exposing his own hand to radium for 10 hours. Several days later an open wound appeared on Curie's hand. Four months later it healed. Experiments on cancerous cells soon showed that these cells that multiply so rapidly are more easily destroyed by ionizing radiation than normal cells. Radiation therapy has been used in the treatment of cancer ever since. Unfortunately, radiation can also cause cancer. Marie Curie died in 1934 of leukemia, over 30 years after her first exposure to radium, as she isolated this highly radioactive element.

Today we have a much better appreciation for the biological effects of ionizing radiation. The effects of radiation are cumulative and so depend mainly on the total equivalent dose absorbed and not so much on the interval of exposure. For example, a whole body dose of about 300 rem is usually lethal within 30 days of exposure whether the dose is absorbed in a few seconds or over several days. If, however, exposure is spread over a much longer time interval, the body's repair mechanisms have a chance to reduce the effects of radiation damage. The maximum allowable whole-body exposure for workers with occupational exposure to radiation is 5 rem per year. Absorbed doses are indicated by exposure of photographic film in radiation film badges worn by workers. Government recommendations for the general population is that whole-body absorbed doses should not exceed 0.5 rem per year. Natural background radiation, produced by cosmic rays and radioactive materials within the earth, contributes roughly 0.2 rem per year to the average individual. Variations in exposure to radon gas and variations in cosmic rays, which are much greater at high altitudes, causes even natural absorbed radiation to vary considerably among individuals. The most subtle and least understood effects of ionizing radiation are the long-term effects, such as cancer and the genetic defects passed on to future generations. Probably the best advice is to avoid any unnecessary exposure to any ionizing radiation.

EXAMPLE 12 Energy Absorbed in a Lethal Dose of Neutrons

How much energy is absorbed by a man of mass 80 kg who receives a lethal whole-body equivalent dose of 400 rem in the form of low-energy neutrons?

SOLUTION Table 30-2 shows that the RBE for whole-body absorption of low-energy neutrons is 10. Applying Eq. 30-20 ($D_e = D \times \text{RBE}$), we find the absorbed dose D.

$$D = \frac{D_e}{\text{RBE}} = \frac{400 \text{ rem}}{10}$$

$$= 40 \text{ rad} = 0.40 \text{ Gy} = 0.40 \text{ J/kg}$$

Using the definition of absorbed dose $\left(\text{Eq. 30-17: } D = \frac{E}{m}\right)$, we find the total energy absorbed by the man's body of mass 80 kg.

$$E = mD = (80 \text{ kg})(0.40 \text{ J/kg}) = 32 \text{ J}$$

To emphasize what a small amount of energy this is, we might compare it to the thermal energy required to raise the temperature of the man's body. Human tissue has a specific heat of about 3500 J/kg-C°. Thus to raise the body's temperature by 1 C° requires heat $Q = mc\,\Delta T = (80 \text{ kg})(3500 \text{ J/kg-C°})(1 \text{ C°}) = 2.8 \times 10^5$ J. This small thermal energy is roughly 10,000 times the energy deposited by the lethal dose of neutrons.

EXAMPLE 13 Energy Absorbed From Dental X rays

How much energy is deposited in the head of a dental patient who, during a dental X-ray exam, receives an equivalent dose of 0.02 rem to the head, which has a mass of 3 kg?

SOLUTION Since X rays have a RBE of 1, the absorbed dose equals the equivalent dose:

$$D = \frac{D_e}{\text{RBE}} = \frac{0.02\text{ rem}}{1} = 0.02\text{ rad} = 2 \times 10^{-4}\text{ Gy}$$

Applying the definition of absorbed dose $\left(\text{Eq. 30-17: } D = \frac{E}{m}\right)$, we find the energy deposited in the head.

$$E = mD = (3\text{ kg})(2 \times 10^{-4}\text{ Gy}) = 6 \times 10^{-4}\text{ J}$$

30-5 Elementary Particles

Elementary Particles and Fundamental Forces of the 1930s

The desire to find the basic building blocks of all matter is an enduring one. The ancient Greeks introduced the concept of an atom as the smallest, indivisible unit of matter. In the nineteenth century John Dalton proposed that each chemical element is composed of a different kind of atom, each with its own characteristic mass, and Dmitri Mendeleev later devised the periodic table of the elements, in which he arranged the elements according to their properties. Toward the end of the nineteenth century the atom began to be seen not as an indestructible unit but as something composed of smaller particles. By 1932 the electron, proton, and neutron had been discovered and were understood to be the truly "elementary particles," the smallest, indivisible pieces of matter, out of which atoms and hence all matter were made. Using quantum mechanics to describe the interaction of these particles, we obtained a very thorough understanding of the structure of atoms. Although the quantum theory of atoms would remain wonderfully successful in describing all kinds of phenomena, the picture of electrons, protons, and neutrons as the complete set of elementary particles was to dramatically change in the decades that followed. Many other supposedly elementary particles were discovered. Some of these turned out to be not at all elementary or structureless but rather composite structures, made of other particles. Of the three elementary particles of the 1930s (electrons, protons, and neutrons), only the electron is now believed to be a truly elementary particle.

By the 1930s not only were there believed to be very few kinds of elementary particles, but it was also believed that all forces observed in nature were explainable in terms of just four kinds of forces, the four **fundamental forces:**

1. **Gravitational force**
2. **Electromagnetic force**
3. **Strong nuclear force**
4. **Weak nuclear force**

The gravitational force, discovered by Newton in the seventeenth century, acts between any two masses. It is the force responsible for holding planetary systems together and of course for the earth's gravity.

The electromagnetic force acts between electric charges. It is the force responsible for holding electrons and protons together in atoms. It is also the force responsible for binding atoms together in solids. Any contact force between macroscopic bodies arises from electromagnetic interactions between the charged particles in those bodies.

The strong nuclear force acts between nucleons (neutrons and protons) but does not act on electrons. It is the force responsible for holding the nucleus together. The strong nuclear force between adjacent protons in a nucleus is much stronger than the electromagnetic force, but the strong force has a very short range. The strong force between nucleons quickly falls to zero at distances of about 10^{-14} m or greater, which accounts for the fact that the maximum diameter of a nucleus is roughly 10^{-14} m.

The weak nuclear force acts between all particles. It is the force responsible for β decay of nuclei. Like the strong nuclear force, the weak force is of very short range ($\sim 10^{-14}$ m). But, as its name implies, the weak force is much weaker than the strong nuclear force. It is even much weaker than the electromagnetic force. For this reason, weak interactions between particles are usually masked by the much stronger electromagnetic or strong interactions.

Antimatter

In 1931 Paul Dirac used theoretical arguments blending quantum mechanics and relativity to predict the existence of a new particle. The predicted particle was in all ways identical to an electron, except for its charge of $+e$, which was exactly opposite the electron's charge of $-e$. At first Dirac believed that the particle was the well-known proton, since its charge was the same as the proton's. It was soon recognized, however, that the particle predicted by Dirac's theory must have the same mass as an electron, whereas the proton is much heavier. Dirac called his particle an antielectron. This new particle was discovered in 1932 by Carl Anderson, who called the particle a **positron.** Fig. 30-14 shows one of Anderson's original photographs of the track left by a positron in a cloud chamber. At the time of his discovery, Anderson was using the cloud chamber to study cosmic rays incident on the earth from outer space. The track shown in the figure was identical to the path an electron would have made, except that it curved to the left, rather than to the right as an electron would have, which meant that its charge must be positive. Anderson concluded that he had discovered the new particle predicted by Dirac.

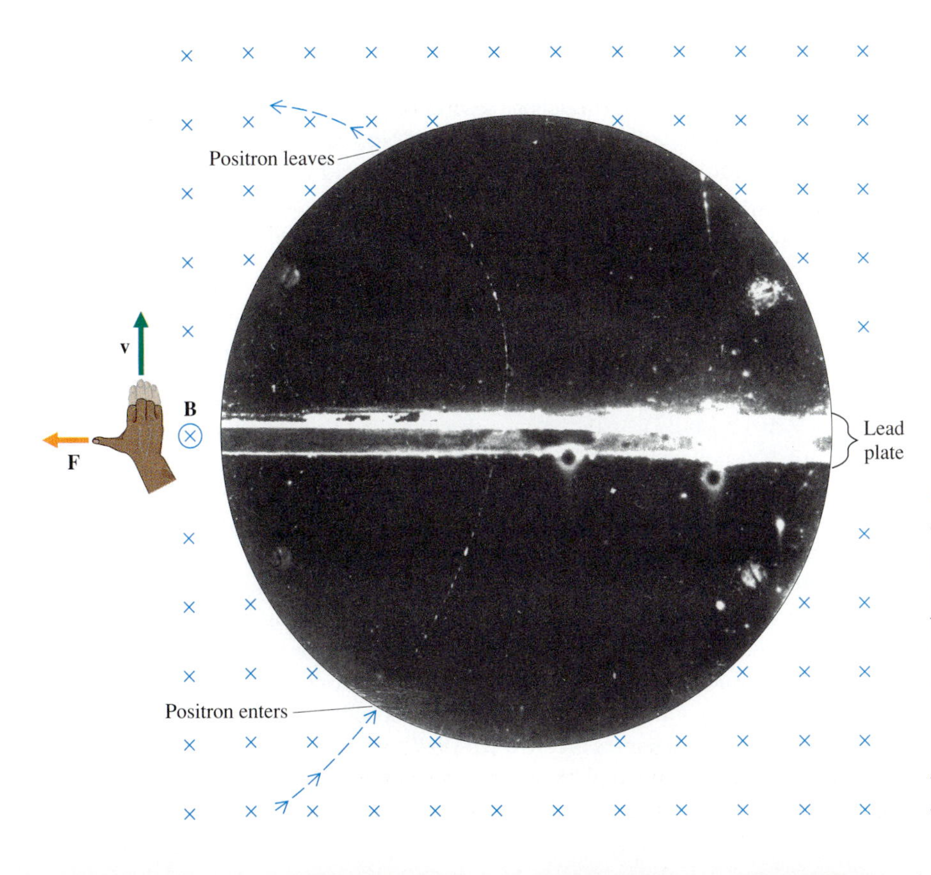

Fig. 30-14 This photo made by Anderson in 1932 shows the track of a positron in a cloud chamber. The positron enters from below, passes through a lead plate in the center, and leaves at the top of the photo. A magnetic field directed into the page exerts a magnetic force on the positron, causing its path to curve to the left. The fact that the bending is to the left indicates that the positron is positively charged.

Today we believe that **for every elementary particle there is a corresponding antiparticle.** Particles and antiparticles have the same mass (and certain other identical properties) but have opposite charges (and opposite values of certain other properties). For example, the antiproton, discovered in 1955, has the same mass as a proton but has a charge of −e. Uncharged particles such as neutrons, also have antiparticles. In this case it is not the charge but another quantity that has a different sign. There are only a very few particles that have no distinguishable antiparticles. In this case we say that the particle is its own antiparticle. The photon is one example of such a particle. The antiparticle of a photon is a photon.

Sources of Elementary Particles; Accelerators

A natural source of elementary particles are the cosmic rays that constantly bombard the earth. These rays consist mainly of high-energy protons from the sun and from interstellar space. When cosmic rays interact with the earth's atmosphere, showers of other particles are produced. The most energetic particles ever observed on earth have been cosmic rays. But the rate of production of particles produced by cosmic-ray collision is so low that it is impractical to wait for rare particles to be produced in this way.

Particle accelerators are devices producing beams of high-energy particles that mimic nature's cosmic rays but at much higher particle densities. The basic function of an accelerator is to accelerate a beam of particles—usually protons or electrons, since they are easy to produce in abundance—and then to direct the beam of high-energy particles at a target. As the particles in the beam collide with the nuclei of the target, various particles are given off and detected either in a bubble chamber or in some other kind of detector.

The maximum energy of each particle in an accelerator's beam is a very important characteristic of the accelerator because the energy determines how deeply the beam is able to penetrate matter and how effective it will be in breaking up tightly bound structures. We have seen that the binding energy per nucleon in the nucleus is roughly 8 MeV. Thus particles with kinetic energies on the order of a few MeV are energetic enough to initiate nuclear reactions. Early scattering experiments by Rutherford, Fermi, and others used for their particle beams particles with kinetic energies of a few MeV, emitted by radioactive sources. However, in order to probe the structure of the proton or to produce more exotic elementary particles, much more energy is required. Typically energy is measured in units of either gigaelectron volts (GeV) or teraelectron volts (TeV), where

$$1 \text{ GeV} = 10^9 \text{ eV} \quad \text{(gigaelectron volts, or billion electron volts)} \quad (30\text{-}24)$$

$$1 \text{ TeV} = 10^{12} \text{ eV} \quad \text{(teraelectron volts, or trillion electron volts)} \quad (30\text{-}25)$$

Fig. 30-15 Stanford Linear Accelerator.

The highest-energy accelerators in operation today are at Fermilab near Chicago (see chapter opening photo) and at CERN near Geneva, Switzerland. Both of these accelerators produce beams of protons with energies on the order of 1 TeV. The highest-energy electron beams (~50 GeV) are produced at the Stanford Linear Accelerator Center (SLAC) in California (Fig. 30-15). An enormous new accelerator had been planned for a site near Dallas, Texas. This accelerator would have used colliding beams of protons accelerated around a track 80 km long, with superconducting electromagnets to control the beams.This Superconducting Super Collider (SSC) would have achieved energies on the order of 40 TeV. Although construction had already begun, cost overruns and budgetary constraints led Congress to cancel the SSC in 1994.

Leptons, Mesons, and Baryons

Particle accelerators have produced hundreds of different kinds of subatomic particles. All but a few of these particles are unstable and quickly decay to other particles. Lifetimes vary greatly among various unstable particles. Some may live 10^{-6} s. Others decay in 10^{-20} s or less. Table 30-3 shows a few of these particles and some of their properties. For each particle there is a corresponding antiparticle with equal mass and opposite charge. The particles shown in the table are arranged in three categories or families: **leptons, mesons,** and **baryons.** Originally these names were intended to classify particles according to mass. The three names are the Greek names for lightweight particles (leptons), middleweight particles (mesons), and heavyweight particles (baryons). Classification based on mass holds only for the lightest members of each family. Thus, for example, although the electron and muon are leptons with less mass than any meson or baryon, the tau is a lepton that is heavier than many baryons.

There are more fundamental differences between the families of particles than their masses. Leptons do not experience the strong nuclear force, whereas mesons and baryons do experience the strong force. All families experience the weak nuclear force as well as the electromagnetic and gravitational forces. Only the leptons are now believed to be elementary particles. Mesons and baryons are believed to be made of other elementary particles, known as "quarks." We shall return to a discussion of quarks and see a fundamental way in which mesons and baryons differ, but first we shall describe a different class of particles known as "mediator particles."

Table 30-3 Some particles and their properties

	Particle	Symbol	Charge (e)	Rest Mass Energy (MeV)	Lifetime (s)
Leptons	Electron	e^-	−1	0.511	stable
	Muon	μ^-	−1	105.7	2.20×10^{-6}
	Tau	τ^-	−1	1784	10^{-13}
	Electron neutrino	ν_e	0	~0	stable
	Muon neutrino	ν_μ	0	~0	stable
	Tau neutrino	ν_τ	0	~0	stable
Mesons	Pion	π^+	+1	139.6	2.60×10^{-8}
		π^0	0	135.0	8.4×10^{-17}
	Kaon	K^+	+1	493.7	1.24×10^{-8}
		K^0_S	0	497.7	8.9×10^{-11}
		K^0_L	0	497.7	5.2×10^{-8}
	Eta	η^0	0	548.8	10^{-18}
Baryons	Proton	p	+1	938.3	stable
	Neutron	n	0	939.6	900
	Lambda	Λ^0	0	1116	2.6×10^{-10}
	Sigma	Σ^+	+1	1189	8.0×10^{-11}
		Σ^0	0	1192	6×10^{-20}
		Σ^-	−1	1197	1.5×10^{-10}
	Omega	Ω^-	−1	1672	8.2×10^{-11}

Mediator Particles

In Chapter 28 we learned that light has particle-like properties. Light waves are quantized, and the quantum of light is a photon. Indeed all electromagnetic waves, radio waves, microwaves, and so on are quantized. When a television transmitting antenna broadcasts a signal that is picked up by a television receiving antenna, photons are being emitted by the transmitter and absorbed by the receiver. The photons "mediate" the electromagnetic force between the electrons in the transmitting and receiving antennas. Beginning in the 1940s it began to be understood that all electromagnetic interactions, even those not involving waves, are mediated by the exchange of photons. Whenever two electric charges experience a mutual force of attraction or repulsion, it is through the absorption and emission of photons by the charges, even if the charges are at rest. We say that **the photon is the mediator of the electromagnetic force.** Presumably all the other fundamental forces have their own mediator particles.

The mediator of the gravitational force and the quantum of gravitational waves is called a **graviton,** but no one has yet detected a graviton. The weak nuclear force has three mediator particles, the $\mathbf{W}^+$, the $\mathbf{W}^-$, and the $\mathbf{Z}^0$. These particles, whose existence and masses were predicted on the basis of theoretical calculations, were discovered in 1983 in experiments performed at CERN by Carlo Rubbia and others. The mediators of the strong nuclear force are called **gluons.** Actually the strong force, which acts between nucleons, has turned out not to be a truly fundamental force after all. Instead it is the complex effect of the quarks that form the nucleons, interacting through the fundamental force known as the **color force.**

Elementary Particles and Fundamental Forces of the 1990s

As subatomic particles proliferated in the 1960s, physicists searched for underlying simplicty and order. A classification scheme for ordering particles, known as the "Eightfold Way," was proposed in 1961 by Murray Gell-Mann. This system organized the baryons and mesons into geometric patterns, based on charge and other properties, somewhat as Mendeleev's periodic table organized the elements based on their properties. And just as the periodic table led to an understanding of the structure of atoms, the Eightfold Way led to an understanding of a structure within the baryons and mesons. In 1963 Gell-Mann and George Zweig independently proposed that all baryons and mesons are composed of smaller particles that Gell-Mann called **quarks.** Today it is believed that every meson consists of some combination of two quarks and that every baryon consists of some combination of three quarks. Because the forces binding quarks together are so strong, no isolated quark has ever been observed. However, very strong indirect evidence for the existence of quarks has been found.

There are six kinds of quarks, which have been given the names **up, down, strange, charmed, beauty** (or bottom), and **truth** (or top). Each quark has an electric charge that is a fraction of the electron's charge, either $\pm\frac{1}{3}$e or $\pm\frac{2}{3}$e. In addition, each quark has another property analogous to electric charge, which has been given the name "color charge," or simply **color.** This property has nothing to do with optical color. It is just the property through which the force binding quarks together (the color force) acts. There are three distinct kinds of colors, given the names **red, green,** and **blue.**

Table 30-4 shows the leptons and quarks, which together with their antiparticles and the mediator particles, constitute the elementary particles as we know them today.

Table 30-4 Elementary particles

	Particle	Symbol	Charge (e)
Leptons	Electron	e	−1
	Muon	μ	−1
	Tau	τ	−1
	Electron neutrino	ν_e	0
	Muon neutrino	ν_μ	0
	Tau neutrino	ν_τ	0
Quarks	Up	u	$+\frac{2}{3}$
(Each comes in 3 colors: red, green, and blue)	Down	d	$-\frac{1}{3}$
	Strange	s	$-\frac{1}{3}$
	Charmed	c	$+\frac{2}{3}$
	Beauty (or bottom)	b	$-\frac{1}{3}$
	Truth (or top)	t	$+\frac{2}{3}$

Marie Curie (born Marja Sklodowska) 1867-1934

Marja (Marie) Sklodowska was born in Warsaw, Poland on November 7, 1867. Her father taught physics and mathematics, and her mother managed a private boarding school for girls. Her parents' interest in science and education must have been a significant stimulus throughout Marie's school years, which were rounded off with a high school gold medal for excellence. Her developing intellect, however, was accompanied by poor physical health.

In her youth Marie Sklodowska was keenly interested in politics and literature in addition to science. She read the works of Karl Marx and became involved in underground political movements. This background probably influenced her later disinterest in seeking financial gain from her discoveries.

At the age of 19 Marie Sklodowska began work as a governess, a lowly position that nevertheless gave her access to her employer's well-stocked scientific library. By that time, however, her dream was to escape from Poland. Her early interest in politics had led to a great desire to move to France, the "land of liberty." Her sister Bronia had studied in Paris and qualified as a medical doctor. By 1891 Marie had accumulated sufficient savings to join her sister and brother-in-law (also a doctor) and begin studies at the Sorbonne.

Marie Sklodowska met Pierre Curie shortly after passing her *licence* in both physics and mathematics at the Sorbonne. The couple were married in 1895. This initiated the most successful partnership of a married couple in the history of science. In tracing the careers of either Marie Curie or Pierre Curie, it is often difficult to separate their different contributions. When they married, Pierre was already well established in scientific research, particularly in the field of piezoelectricity (the separation of electric charge within certain crystals when they are subjected to pressure). As Marie began to develop her interest in the newly discovered phenomenon that she would later christen "radioactivity," Pierre's interests also moved to this area.

Marie Curie embarked on research toward her thesis shortly after Becqerel had discovered his "uranic rays" emanating from a compound of uranium. Becquerel's discovery is a classic example of the way in which chance and a mind able to appreciate what chance reveals to it can lead to great progress in science. Every generation of science students is told of Becquerel's photographic plate, left together with some potassium uranium sulfate in a closed drawer. When Becquerel developed the plate he was surprised to find dark patches, which he was able to show were caused by rays emanating from the uranium compound.

Marie Curie decided to undertake a precise quantitative investigation of Becquerel's uranic rays. The work was performed using an instrument devised by her husband Pierre and his brother Jacques. It was essentially a capacitor in which two metal plates 8 cm in diameter were separated by a gap of 3 cm. One of the plates was coated with a uranium compound or other substance being tested, and a potential difference of 100 volts was established between the plates. If the substance coating the plates gave off radiation, this would ionize the air between the plates and cause a small current to flow between them. This current was measured using a device based on Pierre Curie's work on piezo-electricity. It measured the amount of pressure on a crystal that was required to compensate for the current produced by the radiation. Careful measurement of this current gave a quantitative measure of how much radioactivity was being produced.

Using this apparatus, Marie Curie confirmed Becquerel's observations for uranium and found a similar effect from compounds of thorium. She reported that "thorium is even more active than uranium." Unknown to Marie Curie, the German physicist Carl Schmidt had discovered this same effect a little earlier.

The discovery that thorium oxide also emitted these new type of rays led Marie Curie to rename them as "Becquerel rays" rather than uranic rays. In the paper in which she coined this term, she also used the term "radioactive substance" for the first time. So although Marie Curie was not the first person to discover radioactivity, she did give it the general name by which it is now universally known.

Becquerel had established that the radiations he discovered were a property of uranium rather than of any other elements in his compounds or of the compounds as a whole. Marie Curie went on to quantify this observation more precisely, demonstrating that the intensity of radiation released from uranium compounds depends in a precise way on the actual number of uranium atoms in the samples. By 1898 she was able to announce that "radioactivity is an atomic property." In other words, radiation is a phenomenon released from *individual atoms*. Although it is commonplace today to associate radiation with the decay of individual atoms, this was a major clarification of the field at the time.

Together with Pierre Curie, Marie Curie then began to purify radioactive material from pitchblende, a naturally occurring uranium-containing ore. Marie noticed that the intensity of radiation released from the samples as they were purified increased to a greater extent than could be accounted for by the increasing concentration of uranium atoms alone. This led the Curies to believe that another radioactive element must be present in their samples. They called this element Polonium, after Marie's native Poland. They set about purifying the Polonium by applying laborious chemical separation techniques to large quantities of the pitchblende. This work was performed in a primitive shed and involved boiling the pitchblende, stirring it for hours with an iron rod, and constantly transferring the steadily purifying material from vessel to vessel. It was heroic physical labor as much as pioneering science, and many of the illustrations of the Curies at work depict them toiling with their pitchblende in that shed. By the end of 1898, this procedure had revealed the existence of another intensely radioactive element, christened radium.

The discovery of radium, with its very intense radioactivity, led to widespread publicity for both radioactivity and Marie and Pierre Curie. The fact that radioactive elements were a source of energy led to excited declarations in the press such as "The Curies have discovered perpetual motion!" Such claims were of course unfounded, but in 1903 the Curies and Becquerel were jointly awarded the Nobel prize in physics "in recognition of the extraordinary services they have rendered by their joint researches on the radiation phenomena." Neither Marie nor Pierre were able to attend the prize-giving ceremony, and it appears that Pierre, at least, was already suffering from radiation sickness by this time. However, his life was brought to an abrupt and tragic end not by radiation, but by a horse-drawn tram that hit him in the rue Dauphine, Paris, on April 19, 1906. He died instantly.

Marie was devastated by her husband's death, and her personality darkened markedly. She continued to work, however, and was given the professorship previously held by her husband. This made Marie Curie the first woman ever to become a professor at the Sorbonne. Her inaugural lecture attracted great media interest. She began it movingly, by repeating the final sentence that Pierre Curie had delivered in the same place; and then continued as if the flow of ideas had never been interrupted.

Assisted by a small research team, Marie Curie proceeded to define precise standards for the measurement of radioactivity and fill in various other details concerning the properties of the radioactive phenomenon. In 1911 she was awarded her second Nobel prize, this time in chemistry, in recognition of her discovery of the elements radium and polonium. She paid further tribute to her husband in her Nobel lecture, emphasizing that the award was in recognition of work they performed together "and thus pays high homage to the memory of Pierre Curie." Marie Curie was the first person to be awarded two Nobel prizes in science.

Increasing interest was by then being given to the use of radium in medicine. Throughout World War I, Marie Curie became actively involved in the development of medical units devoted to the use of radiation, both in diagnosis and treatment. She was made head of the radiological service of the Red Cross, and supervised and taught courses on the use of radiation in medicine.

Marie Curie's later years were characterized by continuing involvement in the medical applications of radiation, increasing fame, and steadily declining health. The penetrating power of radiation, which makes it of use in medical diagnosis and therapy, had also inflicted considerable damage on Marie Curie's body. For many years she had worked virtually unprotected with high-intensity radioactive sources. She had four operations between 1923 and 1930 to correct cataracts of the eyes. By 1932 there were permanent lesions on her fingers from handling radium. On June 6, 1934 she entered a sanatorium in the French Alps, where she died a few weeks later at the age of 67.

CHAPTER 30 SUMMARY

An atom consists of a central nucleus surrounded by electrons with charge −e and mass $m_e = 9.1094 \times 10^{-31}$ kg. The particles within a nucleus are called **nucleons** and are of two types: **protons,** with charge +e and mass $m_p = 1.6726 \times 10^{-27}$ kg, and **neutrons,** with zero charge and mass $m_n = 1.6750 \times 10^{-27}$ kg. An atom's **atomic number** Z indicates the number of protons in the atom's nucleus, and the neutron number N indicates the number of neutrons. The **mass number** A gives the total nucleon number and so equals the sum of Z and N

$$A = Z + N$$

Different kinds of nuclei are called **nuclides.** A specific nuclide is indicated by

$${}^{A}_{Z}\mathrm{X}$$

where X is the symbol for the element. Nuclides with the same atomic number but different mass numbers are called **isotopes.**

The approximate radius of a nucleus is given by

$$r \approx (1.2 \times 10^{-15}\ \mathrm{m})A^{1/3}$$

Nucleons are bound together by the strong nuclear force. The energy required to separate the nucleons is called the **binding energy** of the nucleus. Since bound nucleons in a nucleus have less energy than free nucleons, the nucleus has less mass than the sum of the masses of the nucleons. The difference between the total mass of the nucleons and the mass of the nucleus is the **mass defect** Δm and is related to the binding energy ΔE by

$$\Delta E = \Delta m\, c^2$$

Energy units commonly used in nuclear or elementary-particle physics are:

$$1\ \mathrm{MeV} = 10^6\ \mathrm{eV} = 1.6022 \times 10^{-13}\ \mathrm{J}$$

$$1\ \mathrm{GeV} = 10^9\ \mathrm{eV}$$

$$1\ \mathrm{TeV} = 10^{12}\ \mathrm{eV}$$

A quantity that often appears in mass-energy calculations is the product of an atomic mass unit and the speed of light squared:

$$(1\ \mathrm{u})c^2 = 931.5\ \mathrm{MeV}$$

Some nuclides are stable, whereas others spontaneously emit **alpha** (α), **beta** (β), or **gamma** (γ) radiation. Before decay a nucleus is called a "parent," and after decay it is called a "daughter." An alpha particle is a helium nucleus, ${}^{4}_{2}\mathrm{He}$, consisting of two protons and two neutrons. When a nucleus undergoes α decay, the daughter nucleus has a smaller atomic number and mass number.

α decay parent $(Z, A) \rightarrow$ daughter $(Z - 2, A - 4) + \alpha$

A nucleus undergoing β^- decay emits a β^- particle, which is an electron. The nucleus also emits an antineutrino.

β^- decay parent $(Z, A) \rightarrow$ daughter $(Z + 1, A)$ + electron + antineutrino

A nucleus undergoing β^+ decay emits a β^+ particle, which is a positron, the antiparticle of an electron. The nucleus also emits a neutrino.

β^+ decay parent $(Z, A) \rightarrow$ daughter $(Z - 1, A)$ + positron + neutrino

Gamma rays are electromagnetic radiation of very high frequency. The radiation is produced by nuclei changing from a higher-energy state to a lower-energy state.

A nuclide's **decay constant** λ is a number that indicates the fraction of parent nuclei that undergo decay during a time interval Δt

$$\frac{\Delta N}{N} = -\lambda\, \Delta t$$

The **activity** $\mathcal{A}$ of a radioactive sample is defined as the rate of decay and equals the product of the decay constant and the number of parent nuclei.

$$\mathcal{A} = \left|\frac{\Delta N}{\Delta t}\right| = \lambda N$$

Activity is measured in units of **becquerels** (Bq) or **curies** (Ci), where

$$1\ \mathrm{Bq} = 1\ \mathrm{decay/s}$$

$$1\ \mathrm{Ci} = 3.7 \times 10^{10}\ \mathrm{decays/s}$$

The number N of parent nuclei remaining at time t is related to the number N_0 at $t = 0$ by

$$N = N_0 e^{-\lambda t}$$

A sample's activity at time t is related to its initial activity $\mathcal{A}_0$ in the same way.

$$\mathcal{A} = \mathcal{A}_0 e^{-\lambda t}$$

A nuclide's **half-life** $T_{1/2}$ is the time interval during which half of the sample decays. For each successive time interval of length $T_{1/2}$, the number of parent nuclei is reduced by $\frac{1}{2}$.

$$N = (\tfrac{1}{2})^n N_0 \qquad (t = nT_{1/2})$$

A nuclide's half-life and decay constant are related by the equation

$$\lambda = \frac{\ln 2}{T_{1/2}} = \frac{0.693}{T_{1/2}}$$

Carbon dating is a technique of determining the time since the death of a specimen, such as wood or bone, which contains carbon. The method is based on the fact that living organisms contain sufficient carbon 14 to give an activity of 0.23 Bq per gram. Once dead the carbon 14 content and the activity decay, with a half-life of 5700 years.

A collision in which a nucleus is changed is called a **nuclear reaction.** The energy Q transformed from mass energy to kinetic energy in a reaction is called the **reaction energy,** or **Q-value.**

$$Q = \Delta K = -(\Delta m)c^2$$

Nuclear **fission** is the splitting of a large nucleus into two smaller fragments. Fission liberates energy. For U-235, the energy liberated per unit mass is

$$\frac{Q}{m} \approx 8 \times 10^{13} \text{ J/kg}$$

Nuclear **fusion** is the joining together of two smaller nuclei to form a larger nucleus. Fusion reactions require a temperature on the order of 10^7 K and typically liberate an energy per unit mass on the order of

$$\frac{\Delta E}{m} \approx 5 \times 10^{14} \text{ J/kg}$$

Ionizing radiation is any radiation that has sufficient energy to ionize atoms and includes electromagnetic radiation, α and β particles, neutrons, protons, and nuclei. Ionizing radiation can kill cells. The effect of radiation from a radioactive source on a subject exposed to it depends on various factors: the source's initial activity and half-life, distance from the source, time of exposure, and presence of absorbing materials. The **absorbed dose** D is the energy absorbed from ionizing radiation per unit mass of the absorbing body.

$$D = \frac{E}{M}$$

Units of absorbed dose are the gray (Gy) and the rad, where

$$1 \text{ Gy} = 1 \text{ J/kg}$$

and

$$1 \text{ rad} = 0.01 \text{ Gy} = 0.01 \text{ J/kg}$$

Absorption of different kinds of radiation causes different amounts of damage. The **relative biological effectiveness** (RBE) of any radiation is the number of rads of X rays that produces the same biological damage as one rad of the radiation used. The **equivalent dose** D_e of any absorbed radiation is the product of the absorbed dose and the radiation's RBE.

$$D_e = D \times \text{RBE}$$

Equivalent dose is measured in units of sieverts (Sv) or rem, where

$$1 \text{ Sv} = 1 \text{ Gy} \times \text{RBE}$$

and

$$1 \text{ rem} = 1 \text{ rad} \times \text{RBE} = 0.01 \text{ Sv}$$

The fundamental forces in nature are the gravitational force, the electromagnetic force, the weak interaction, and the color force. The strong nuclear force results from the color force. Elementary particles are the smallest, most basic particles, out of which other matter is built. Elementary particles are either **leptons, quarks,** or **mediator particles.** It is through the exchange of mediator particles that the fundamental forces act. Quarks are bound together by the color force. Mesons consist of two bound quarks and baryons consist of three bound quarks. Protons and neutrons are baryons. The electron is a lepton. For every particle there is a corresponding antiparticle with the same mass and opposite charge.

Questions

1 Does the density of a nucleus depend on the number of nucleons it contains?

2 How many neutrons are there in a $^{60}_{27}$Co nucleus?

3 When an $^{227}_{89}$Ac nucleus undergoes α decay, what is the mass number of the daughter nucleus?

4 After a $^{223}_{88}$Ra nucleus undergoes a series of three α decays and one β decay, what are the atomic number and mass number of the nucleus that remains?

5 What combination of α and β particles are emitted in a series of radioactive decays that transforms a $^{231}_{90}$Th nucleus into a $^{227}_{90}$Th nucleus?

6 Which is greater: (a) the mass of a $^{20}_{10}$Ne nucleus, or (b) 10 times the sum of the masses of a proton and a neutron?

7 Is the mass of a hydrogen atom greater than, less than, or equal to the mass of a proton plus the mass of an electron?

8 Which has more mass: (a) one atom of $^{32}_{16}$S, or (b) two atoms of $^{16}_{8}$O?

9 Which has more mass: (a) one atom of $^{112}_{50}$Sn, or (b) one atom of $^{54}_{24}$Cr and one atom of $^{58}_{26}$Fe?

10 One of the early mysteries of radioactivity was how it seemed to be an endless source of energy. For example, one gram of radium 226 produces heat at a constant rate of 140 calories per hour. Even after many weeks, the radium continues to supply energy at this rate. Radium 226 undergoes α decay with a half-life of 1600 years. The kinetic energy of emitted α particles produces the heat. How long would you have to wait before the heat production by a radium sample declined from 140 calories per hour to 70 calories per hour?

11 Which way would an α particle bend if it were initially moving to the right and passed through a magnetic field, directed perpendicular to the page, outward?

12 Which way would a β particle bend if it were initially moving to the left and passed through a magnetic field, directed perpendicular to the page, inward?

13 The decay constant for $^{211}_{82}Pb$ is 0.02 min^{-1}. If a sample initially contains 10^6 of these nuclei, how many can be expected to decay in 1 minute?

14 A radioactive source with a half-life of 3 hours has an initial activity of 64 mCi. How long would it take for the activity to reduce to 1 mCi?

15 Which emits more α's, β's, or γ's per second: (a) 10^{10} nuclei of an isotope with a half-life of 1 day, or (b) 10^{11} nuclei of an isotope with a half-life of 20 days?

16 In a sample of 10^{13} radium-226 nuclei, on the average 137 nuclei decay each second. In a certain 1-second interval, 120 nuclei decay. In the next second are chances greater that more than 137 nuclei will decay than that less than 137 will decay?

17 Initially sample A is a 100 mCi radioactive source with a half-life of 10 s, and sample B is a 5 mCi radioactive source with a half-life of 6 h. Which sample would be more active 1 day later?

18 An artifact from an archeological site contains 1 gram of carbon and has an activity of 0.02 Bq. Is the age of the artifact greater than 10,000 years?

19 Is the path of a neutron through a bubble chamber visible?

20 Would you expect the nuclide $^{40}_{20}Ca$ to undergo nuclear fission?

21 What nuclide could fuse with $^{14}_{7}N$ to form $^{15}_{8}O$?

22 In a controlled chain reaction within a nuclear reactor, energy is being produced at a constant rate. On the average, how many neutrons emitted per fissioning nucleus go on to cause another fission?

23 Which radiation dose would deposit more energy in your body: (a) a 1 rad dose to your hand, or (b) a 0.1 rad dose to your entire body?

24 Why could Anderson be sure that his cloud-chamber photo (Fig. 30-14) did not show an electron moving downward, rather than a positron moving upward? (HINT: The lead plate slowed the particle. What evidence is there that the particle is moving slower in the top half of the photo?)

25 The Δ^{++} particle has a charge of +2e. What is the charge of its antiparticle?

26 How many quarks are there in an α particle?

Answers to Odd-Numbered Questions

1 no; **3** 223; **5** 1 α and 2 β's; **7** less than; **9** a; **11** down; **13** 2×10^4; **15** a; **17** B; **19** no; **21** $^{1}_{1}H$; **23** b; **25** -2e

Problems (listed by sections)

30-1 Nuclear Structure

1 How many nucleons are there in 1.0 kg of matter?

2 Find the approximate radius of (a) the aluminum-27 nucleus; (b) the zinc-64 nucleus.

3 Find the approximate radius of (a) the hydrogen-1 nucleus; (b) the uranium-238 nucleus.

4 How do you represent symbolically the iron isotope that has an atomic mass of 57? Iron has an atomic number of 26.

5 How do you represent symbolically the oxygen isotope that has 10 neutrons?

6 How many neutrons are there in a $^{44}_{20}\text{Ca}$ nucleus?

7 Find the binding energy and the binding energy per nucleon of the $^{2}_{1}\text{H}$ nucleus.

8 Find the binding energy and the binding energy per nucleon of the $^{12}_{6}\text{C}$ nucleus.

9 Use the binding energy per nucleon curve (Fig. 30-3) to determine the approximate binding energy of $^{20}_{10}\text{Ne}$.

10 Use the binding energy per nucleon curve (Fig. 30-3) to determine how much energy must be supplied to separate the nucleons in a $^{200}_{80}\text{Hg}$ nucleus.

11 Given that the binding energy of the hydrogen atom is 13.6 eV, how much less is the mass of a hydrogen atom than the sum of the masses of a proton and an electron? Express the answer in atomic mass units.

12 The binding energy of the system of the earth and the moon is 3.81×10^{28} J. Find the mass defect of the earth-moon system and explain the meaning of this number.

30-2 Radioactive Decay

13 If the parent nucleus in an α decay is $^{224}_{88}\text{Ra}$, what is the daughter nucleus?

14 If the daughter nucleus in an α decay is $^{223}_{87}\text{Fr}$, what is the parent nucleus?

15 What nuclide results from the α decay of $^{242}_{94}\text{Pu}$?

16 What nuclide results from the β^- decay of $^{40}_{19}\text{K}$?

17 What nuclide results when a $^{235}_{92}\text{U}$ nucleus undergoes a succession of radioactive decays consisting of 3 α decays and 2 β^- decays?

18 How many α and β^- particles are emitted as a $^{220}_{84}\text{Po}$ nucleus undergoes a succession of radioactive decays that results in a $^{204}_{80}\text{Hg}$ nucleus?

19 Calculate the kinetic energy of the decay products in the α decay of $^{210}_{84}\text{Po}$.

★ **20** A parent nucleus, initially at rest, undergoes α decay.
(a) Use conservation of momentum to show that the ratio of the kinetic energy of the α particle to the kinetic energy of the daughter nucleus equals the ratio of the masses of the daughter nucleus and the α particle, M_D/M_α.
(b) How much of the kinetic energy calculated in Example 3 is carried by the α particle?

21 Calculate the maximum kinetic energy of an electron emitted in the β^- decay of $^{24}_{11}\text{Na}$. Assume that the daughter nucleus has negligible kinetic energy.

22 A $^{32}_{15}\text{P}$ nucleus undergoes β^- decay, emitting an electron with a kinetic energy of 1.00 MeV. How much energy is carried by the antineutrino?

★★ **23** A $^{13}_{7}\text{N}$ nucleus undergoes β^+ decay. Find the maximum kinetic energy of the emitted positron.

24 In a certain radioactive sample, one out of every 1.0×10^6 parent nuclei decays during a 10 s interval. Find the decay constant and half-life of the isotope.

25 Find the decay constant of $^{222}_{86}\text{Rn}$, which has a half-life of 3.82 days.

26 Initially a sample contains 2.0×10^{13} atoms of the radioactive isotope iodine 131, which has a half-life of 8.0 days. This isotope is used to detect thyroid disorders. How many iodine-131 atoms are present 24 days later?

27 A Geiger counter reading of a radioactive sample is initially 6800 counts per minute. The same sample gives a reading of 425 counts per minute 10.0 h later. What is the sample's half-life?

28 If a rock containing U-238 initially reads 84.0 counts per minute on a Geiger counter, what will its reading be 1 week later?

29 If a radioactive sample initially contains 8.00×10^{14} parent nuclei with a half-life of 7.00 h, how many parent nuclei remain 24.0 h later?

30 Find the activity of a sample containing 2.0×10^{13} iodine-131 nuclei ($T_{1/2} = 8.0$ days).

31 Find the activity of a sample containing 4.00×10^{-6} g of $^{60}_{27}\text{Co}$ ($T_{1/2} = 5.27$ years).

32 Find the number of $^{238}_{92}\text{U}$ nuclei in a sample with an activity of 560 Bq.

33 Find the masses of samples of the following isotopes, if each of the samples has an activity of 1.00 Ci: (a) $^{11}_{6}\text{C}$ ($T_{1/2} = 20.4$ min); (b) $^{90}_{38}\text{Sr}$ ($T_{1/2} = 29.1$ y); (c) $^{238}_{92}\text{U}$ ($T_{1/2} = 4.47 \times 10^9$ y).

34 How long would you have to wait before there was a 5.00% reduction in the rate of heat produced by a sample of $^{226}_{88}\text{Ra}$ ($T_{1/2} = 1600$ y)? See Question 10 for the historical significance of this result.

35 How long will it be before 90% of the $^{226}_{88}\text{Ra}$ nuclei now present in the earth will have decayed? Does this mean that at that time there will be only 10% as much $^{226}_{88}\text{Ra}$ in the earth as there is now?

36 An archeological specimen containing 278 g of carbon has an activity of 45.0 Bq. How old is the specimen?

37 An ancient bone contains 12.5% as much $^{14}_{6}\text{C}$ as a piece of living matter with an equal carbon content. How old is the bone?

38 It is difficult to determine with much accuracy the $^{14}_{6}\text{C}$ content of an archeological specimen if it is less than 1% of the $^{14}_{6}\text{C}$ content in a living sample containing the same mass of carbon. What is the age of a specimen for which the $^{14}_{6}\text{C}$ content has decayed to 1% of its original, living value? This is roughly the maximum age for which carbon dating works.

39 A sample of $^{232}_{90}Th$ ($T_{1/2} = 1.41 \times 10^{10}$ y) and a sample of $^{15}_{8}O$ ($T_{1/2}$ = 122 s) each have an activity of 1.00 mCi.
(a) What is the ratio of the number of $^{232}_{90}Th$ nuclei to the number of $^{15}_{8}O$ nuclei?
(b) After 24.0 h, what are the activities of the two samples?

★ **40** Initially a sample of $^{11}_{6}C$ ($T_{1/2} = 20$ min) has an activity of 3.0 mCi, and a sample of $^{13}_{7}N$ ($T_{1/2} = 10$ min) has an activity of 6.0 mCi. How long is it before the two activities are equal?

30-3 Nuclear Reactions; Fission and Fusion

41 Find the particle or nuclide denoted by X in the reaction

$$^{14}_{7}N + X \rightarrow {}^{1}_{1}H + {}^{14}_{6}C$$

42 Find the particle or nuclide denoted by X in the reaction

$$^{19}_{9}F + {}^{1}_{1}H \rightarrow {}^{4}_{2}He + X$$

43 (a) Find the Q-value of the reaction described in Problem 41.
(b) If the nitrogen target is initially at rest and X has an initial kinetic energy of 2.00 MeV, what is the final kinetic energy of the reaction products?

44 (a) Find the Q-value of the reaction described in Problem 42.
(b) If the $^{19}_{9}F$ nucleus is initially at rest and the bombarding proton has a kinetic energy of 1.00 MeV, what is the final kinetic energy of the reaction products?

45 Is the reaction $^{7}_{3}Li + {}^{1}_{1}H \rightarrow {}^{1}_{0}n + {}^{7}_{4}Be$ possible if the lithium target is at rest and the bombarding proton has a kinetic energy of 1.00 MeV?

46 Find the number of neutrons produced in the fission reaction

$$^{1}_{0}n + {}^{235}_{92}U \rightarrow {}^{136}_{54}Xe + {}^{88}_{38}Sr + \text{neutrons}$$

47 Find the fission fragment X produced in the fission reaction

$$^{1}_{0}n + {}^{235}_{92}U \rightarrow {}^{99}_{41}Nb + X + 4\,{}^{1}_{0}n$$

48 Use Appendix D to find the energy released in the fission reaction described in Problem 46.

49 Use Appendix D to find the energy released in the fission reaction

$$^{1}_{0}n + {}^{235}_{92}U \rightarrow {}^{141}_{56}Ba + {}^{92}_{36}Kr + 3\,{}^{1}_{0}n$$

50 What mass of U-235 undergoing fission would liberate energy equal to the approximate annual United States energy use of 10^{19} J?

51 The first atomic bombs each released about 8×10^{13} J of energy, the same energy that would have been released in the explosion of 20,000 tons of TNT. What mass of fissioning U-235 is required to release this much energy? (The bombs contained considerably more fissionable nuclei than the nuclei that actually fissioned.)

52 A nuclear power plant produces 10^3 MW of electrical energy. What mass of U-235 must fission each second to produce this energy if 25% of the energy liberated through fission is converted to electrical energy?

53 Use Appendix D to find the energy released in the fusion reaction

$$^{1}_{1}H + {}^{3}_{1}H \rightarrow {}^{4}_{2}He$$

54 Use Appendix D to find the energy released in the fusion reaction

$$^{3}_{2}He + {}^{3}_{2}He \rightarrow {}^{4}_{2}He + {}^{1}_{1}H + {}^{1}_{1}H$$

★**55** The relative abundance on earth of the deuterium isotope ($^{2}_{1}H$) is 0.015%.
(a) Calculate the mass of water containing 1.0 g of deuterium.
(b) What mass of deuterium undergoing fusion would liberate energy equal to the approximate annual United States energy use of 10^{19} J? Use 5×10^{14} J/kg as the energy liberated per unit mass of fusing deuterium. What volume of water contains that much deuterium?

30-4 Biological Effects of Radiation

56 During a diagnostic X-ray exam a patient absorbs 5.0×10^{-3} J of energy in tissue of mass 25 kg. Find (a) the absorbed dose; (b) the equivalent dose.

57 A woman is exposed to a source of α particles. Her body of mass 60 kg absorbs 2.0 J of energy. Find (a) the absorbed dose; (b) the equivalent dose.

58 Find the equivalent dose in rem when 2.5 rad of low-energy electrons are absorbed by the body.

59 Find the equivalent dose in rem when 3.0 rad of low-energy neutrons are absorbed by the eyes.

60 What X-ray dose would produce the same biological damage as an absorbed α-particle dose of 15 rad?

61 What absorbed dose of low-energy electrons would produce the same biological damage as an X-ray dose of 3.0 rad?

62 Find the equivalent dose of X rays that would deposit as much energy in a 75 kg body as the 120 kcal of energy derived from eating a slice of bread.

★ **63** Radiation from 1 gram of radium produces heat at a rate of 140 cal per hour in a calorimeter. If we assume that energy is absorbed at the same rate by the body of a person ingesting radium, what mass of ingested radium would produce in 1 day a lethal dose of 300 rad in a body of mass 80 kg?

64 Enough radiation is absorbed by water to raise its temperature by 10.0 C°. Find the absorbed dose.

30-5 Elementary Particles

65 The CERN accelerator straddles the French-Swiss border. Protons travel around a circular path of radius 1.0 km at approximately the speed of light. How many times per second does a proton cross the border?

66 The Fermilab accelerator beam contains 5×10^{13} protons, each with an energy of 1 TeV.

(a) Find the total energy of the beam.

(b) Find the speed of a 10^4 kg truck with kinetic energy equal to the beam's energy.

67 What is the total number of the elementary particles, including all antiparticles and the 13 mediator particles (graviton, photon, 8 gluons, and 3 weak-interaction mediators)?

68 How many elementary particles are there in a ${}^{12}_{6}C$ atom?

69 How many quarks are there in a water molecule?

70 Pi mesons are composed of up and down quarks and their antiparticles. Which quarks are in a π^+ meson?

71 Nucleons are each composed of only up and down quarks. Which quarks are in (a) a proton; (b) a neutron?

Additional Problems

★ **72** In 1.0 m^3 of water approximately what volume is occupied by nuclei?

★ **73** The isotope ${}^{90}_{38}Sr$ ($T_{1/2}$ = 29.1 y) is one of the fission fragments produced by the fission of uranium and was present in the radioactive fallout from atmospheric testing of nuclear bombs in the 1950s. Strontium is chemically similar to calcium and can contaminate milk and be deposited in bones. How long does it take for the decay of 99.0% of the strontium initially present after a bomb detonation?

★ **74** Roughly one out of every 4000 persons in the United States dies annually in traffic accidents.

(a) What is the "half-life" for death by auto accident?

(b) Suppose that the average person in the United States spends 2% of his or her life driving. If N people were constantly driving, how long would it be before traffic deaths reduced their number to $N/2$? This number is the "half-life of driving."

★ **75** According to one theory, the proton is unstable, with a half-life of approximately 10^{33} y. How many protons should decay in a 10^4 m^3 volume of water over a time interval of 1 year?

★★ **76** The prescription drug Prozac has a "biological" half life of 7 days, which means that after 7 days half of the Prozac initially in the bloodstream has been eliminated. Suppose a patient takes one 20 mg tablet daily. Since it takes 7 days for half of any initial concentration to be eliminated, the drug is entering the body faster than it is leaving, and its concentration in the bloodstream will therefore continue to increase day by day after medication begins, until the concentration is great enough that the small percentage of the drug leaving the body during one day equals the daily dose entering the body.

(a) Find the ratio of this "equilibrium" concentration to the concentration just after the first tablet.

(b) Calculate the ratio of the concentration just after the fourth tablet is taken to the concentration just after the first tablet.

Appendix A Review of Mathematics

A–1 Basic Operations
A–2 Powers of Ten and Scientific Notation
A–3 Logarithms
A–4 Algebra
A–5 Geometry
A–6 Trigonometry

This appendix contains a review of the mathematics assumed as a background for this text.

A-1 Basic Operations

The following are rules for adding, subtracting, multiplying, dividing, raising to a power, and taking roots of either numbers or algebraic quantities.

Rules

$a(b + c) = ab + ac$

$-(b + c) = -b - c$

$(a + b)(c + d) = ac + ad + bc + bd$

$(a + b)^2 = (a + b)(a + b) = a^2 + 2ab + b^2$

$(a - b)^2 = (a - b)(a - b) = a^2 - 2ab + b^2$

$$\frac{a+b}{c} = \frac{a}{c} + \frac{b}{c}$$

$$\frac{a}{b} + \frac{c}{d} = \frac{ad+bc}{bd}$$

$$\frac{a}{b} \times \frac{c}{d} = \frac{ac}{bd}$$

$$\frac{\frac{a}{b}}{\frac{c}{d}} = \frac{a}{b} \times \frac{d}{c} = \frac{ad}{bc}$$

$$\sqrt{ab} = \sqrt{a}\sqrt{b}$$

$$\sqrt{\frac{a}{b}} = \frac{\sqrt{a}}{\sqrt{b}}$$

$a^0 = 1$

$a^1 = a \qquad a^{-1} = \frac{1}{a}$

$a^2 = a \times a \qquad a^{-2} = \frac{1}{a^2}$

$a^3 = a \times a \times a \qquad a^{-3} = \frac{1}{a^3}$

$a^n = \underbrace{a \times a \times \cdots \times a}_{n \text{ factors}} \qquad a^{-n} = \frac{1}{a^n}$

Common Mistakes

$(a + b)^2 \neq a^2 + b^2$
e.g. $(3 + 2)^2 \neq 3^2 + 2^2$
$25 \neq 13$

$(a - b)^2 \neq a^2 - b^2$
e.g. $(3 - 2)^2 \neq 3^2 - 2^2$
$1 \neq 5$

$\frac{a}{b} + \frac{a}{d} \neq \frac{a}{b+d}$
e.g. $\frac{8}{2} + \frac{8}{4} \neq \frac{8}{6}$
$6 \neq \frac{4}{3}$

$\frac{a}{\frac{1}{b}} \neq \frac{a}{b}$
e.g. $\frac{4}{\frac{1}{2}} \neq \frac{4}{2}$
$8 \neq 2$

$\sqrt{a+b} \neq \sqrt{a} + \sqrt{b}$
e.g. $\sqrt{9+16} \neq \sqrt{9} + \sqrt{16}$
$5 \neq 3 + 4$

$$a^b \times a^c = a^{b+c}$$

$$\frac{a^b}{a^c} = a^{b-c}$$

$$(a^b)^c = a^{bc}$$
$$(ab)^c = a^c b^c$$

$$\left(\frac{a}{b}\right)^c = \frac{a^c}{b^c}$$

Exercises Verify the following:

1 $2(x + 3) = 2x + 6$

2 $-3(y - x) = -3y + 3x$

3 $(a + 3)^2 = a^2 + 6a + 9$

4 $(2x - 5)^2 = 4x^2 - 20x + 25$

5 $(x + 5)(y + 1) = xy + x + 5y + 5$

6 $\frac{x + 4}{2} = \frac{x}{2} + 2$

7 $\frac{2}{x} + \frac{3}{x} = \frac{5}{x}$

8 $\frac{x}{2} + \frac{x}{3} = \frac{5x}{6}$

9 $\frac{\frac{x}{4}}{\frac{2}{y}} = \frac{xy}{8}$

10 $\frac{\frac{x}{4}}{\frac{y}{2}} = \frac{x}{2y}$

11 $\sqrt{4x^2} = 2x$

12 $\sqrt{\frac{4}{x^2}} = \frac{2}{x}$

13 $2^5 = 32$

14 $2^{-6} = \frac{1}{64}$

15 $10^8 \times 10^6 = 10^{14}$

16 $\frac{10^{19}}{10^{11}} = 10^8$

17 $(10^8)^3 = 10^{24}$

18 $(2x)^3 = 8x^3$

19 $\left(\frac{4}{x}\right)^2 = \frac{16}{x^2}$

20 $\left(\frac{3x^3}{6x^2}\right)^3 = \frac{1}{8}x^3$

Powers of Ten and Scientific Notation

To express a quantity that is either very small or very large, it is convenient to express it in scientific notation, a shorthand method that uses multiples of powers of ten:

$$10^0 = 1$$

$$10^1 = 10 \qquad 10^{-1} = \frac{1}{10} = 0.1$$

$$10^2 = 10 \times 10 = 100 \qquad 10^{-2} = \frac{1}{10^2} = 0.01$$

$$10^3 = 10 \times 10 \times 10 = 1000 \qquad 10^{-3} = \frac{1}{10^3} = 0.001$$

$$\vdots \qquad \vdots$$

$$10^n = \underbrace{10 \times 10 \times \cdots \times 10}_{\text{n factors}} \qquad 10^{-n} = \frac{1}{10^n}$$

$$= \underbrace{100 \cdots 0}_{\text{n zeroes}}$$

For example, the distance from earth to the moon, in units of meters (m), is 384 000 000 m, which we express in scientific notation as 3.84×10^8 m. The exponent 8 tells us that the decimal moves 8 places to the right in going from the scientific form 3.84×10^8 to the form without powers of 10: 384 000 000.

8 places

Very small numbers are expressed by means of negative exponents. For example, the radius of a hydrogen atom is 0.000 000 000 053 m, which in scientific notation is 5.3×10^{-11} m. Notice we have shifted the decimal point 11 places in changing to scientific notation.

When we multiply or divide numbers expressed in scientific notation, we can separate the powers of ten from their multipliers, as in the following examples.

$$\begin{aligned}(4.5 \times 10^5)(2.0 \times 10^3) &= (4.5 \times 2.0) \times 10^5 \times 10^3 \\ &= 9.0 \times 10^{5+3} \\ &= 9.0 \times 10^8\end{aligned}$$

$$\frac{7.5 \times 10^{11}}{3.0 \times 10^7} = \frac{7.5}{3.0} \times \frac{10^{11}}{10^7} = 2.5 \times 10^{11-7} = 2.5 \times 10^4$$

Exercises Verify the following:

1 $3600 = 3.6 \times 10^3$

2 $0.00010 = 1.0 \times 10^{-4}$

3 $6352 = 6.352 \times 10^3$

4 $0.00295 = 2.95 \times 10^{-3}$

5 $(2 \times 10^5)(3 \times 10^{11}) = 6 \times 10^{16}$

6 $\dfrac{9 \times 10^{18}}{3 \times 10^6} = 3 \times 10^{12}$

7 $(4.17 \times 10^6)(2.00 \times 10^{12}) = 8.34 \times 10^{18}$

8 $\dfrac{5.20 \times 10^4}{4.00 \times 10^{10}} = 1.30 \times 10^{-6}$

9 $\dfrac{(2700)(3.1 \times 10^5)}{4.7 \times 10^4} = 1.8 \times 10^4$

10 $\dfrac{(6.67 \times 10^{-11})(5.43 \times 10^{17})(2.41 \times 10^6)}{(3.12 \times 10^8)^2} = 8.97 \times 10^{-4}$

A-3 Logarithms

If $y = 10^x$, we say that x is the logarithm of y to the base 10, or the **common logarithm** of y, written log y.

$$y = 10^x$$

$$\log y = x$$

For example, $$2.0 = 10^{0.30} \quad \text{and} \quad 3.0 = 10^{0.48},$$

and so $$\log 2.0 = 0.30 \quad \text{and} \quad \log 3.0 = 0.48$$

The common log of a power of 10 is easy.

$$\log 1 = 0$$

$$\log 10 = 1 \qquad \log 0.1 = -1$$

$$\log 100 = 2 \qquad \log 10^{-2} = -2$$

$$\log 10^3 = 3 \qquad \log 10^{-3} = -3$$

The following rules are often useful in working with logarithms.

$$\log (ab) = \log a + \log b$$

$$\log \frac{a}{b} = \log a - \log b$$

$$\log a^n = n \log a$$

The **natural number** e is the one number having the property that

$$e^x \approx 1 + x \text{ for any small number } x$$

For example, $e^{0.1} = 1.1$ and $e^{0.025} = 1.025$. The value of e to 3 decimal places is

$$e = 2.718$$

Physical laws are often expressed in terms of e. For example, the rate of emission by a radioactive source is proportional to e^{-ct}, where t is time and c is a constant. Since powers of e often appear in expressions of physical laws, the inverse operation, the logarithm to the base e, or **natural logarithm,** is important. We denote the natural log by ln. Thus if

$$y = e^x$$

then $$\ln y = x$$

For example, $$2.0 = e^{0.69} \quad \text{and} \quad 3.0 = e^{1.1},$$

and so $$\ln 2.0 = 0.69 \quad \text{and} \quad \ln 3.0 = 1.1$$

Of course, $$\ln e = 1, \text{ since } e = e^1$$

and $$\ln 1 = 0, \text{ since } 1 = e^0$$

The same rules listed for common logs apply to natural logs as well.

$$\ln(ab) = \ln a + \ln b$$

$$\ln \frac{a}{b} = \ln a - \ln b$$

$$\ln a^n = n \ln a$$

Scientific calculators provide values of both common and natural logs.

Exercises Verify the following:

1 $\log 10^{10} = 10$
2 $\log 10^{-23} = -23$
3 $\log 2.0 \times 10^{12} = 12.30$
4 $\log 3.0 \times 10^{-6} = -5.52$
5 $\log \frac{x^2}{y^3} = 2 \log x - 3 \log y$
6 $\ln(e^6 \times e^3) = 9$
7 $\ln(2.0 \times e^{-3t}) = 0.69 - 3t$
8 $\ln e^{4.61} = 4.61$
9 $\ln(e^7)^3 = 21$
10 $\ln \frac{e^{25}}{3.0} = 23.9$

Algebra

Proportionality

We say that two variable quantities x and y are proportional if doubling one doubles the other, or if changing one by any constant factor c changes the other by the same factor c. We denote proportionality by the symbol $\propto$. Thus

if $$y \propto x,$$

then $$x \longrightarrow cx \quad \text{means that} \quad y \longrightarrow cy$$

We can express the proportionality of x and y by means of an equation, using a constant of proportionality k, which is the ratio of the proportional quantities.

If $$\frac{y}{x} = k$$

then $$y = kx$$

For example, a circle's circumference C is proportional to its radius r. If you double the radius, the circumference doubles, or if you triple the radius, the circumference triples. The ratio of a circle's circumference to its radius is the constant 2π, where to two decimal places $\pi = 3.14$. Thus

$$C = 2\pi r$$

or $$C \simeq 6.28r$$

Exercises Verify the following:

1 If $y \propto x$ and $y = 10$ when $x = 4$, then $y = 30$ when $x = 12$
2 If $y \propto x$ and $y = 0.2$ when $x = 25$, then $y = 0.02$ when $x = 2.5$
3 If $y \propto x^2$ and $y = 5$ when $x = a$, then $y = 20$ when $x = 2a$
4 If $y \propto \frac{1}{x^2}$ and $y = 8$ when $x = \text{a}$, then $y = 2$ when $x = 2\text{a}$
5 If $y \propto \frac{1}{x^2}$ and $y = b$ when $x = a$, then $y = \frac{1}{9}b$ when $x = 3a$

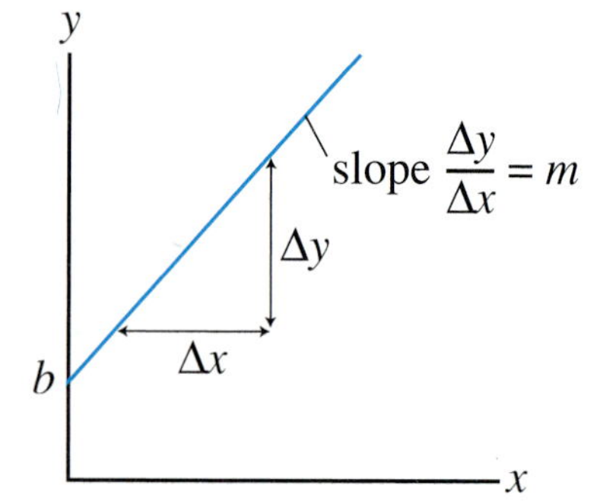

Fig. A-1

Linear Equations

The general form of a linear equation is

$$y = mx + b,$$

where m and b are constants. The graph of y vs x is a straight line of slope m and y-intercept b (Fig. A-1). For example, the graph of the equation $y = 4x - 3$ is a straight line of slope 4 and y-intercept −3.

Exercises Verify the following:

1. $y + 5x = 7$ describes a line of slope −5 and y-intercept 7.
2. $2(x - 3) = 4y$ describes a line of slope $\frac{1}{2}$ and y-intercept −1.5
3. $3(y + 2) = 6(x - 1)$ describe a line of slope 2 and y-intercept −4
4. The equation of the straight line that passes through the points $x = 0$, $y = 2$ and $x = 3$, $y = 8$ is $y = 2x + 2$

Solving an Equation for an Unknown

A single equation can be used to solve for a single unknown—either to obtain a value for the unknown or to relate it to other variables in the equation. To solve for an unknown, you must **isolate the unknown.** You can accomplish this by performing various operations: adding or subtracting terms, multiplying by factors, raising to a power, and so forth. An equation remains valid if you **perform the same operation on both sides of the equation**—that is, add or subtract the same terms from both sides, multiply both sides by the same factor, and/or raise both sides to the same power.

EXAMPLE

Solve $3x + 9 = 15$.

First subtract 9 from both sides, which gives

$$3x = 6$$

Then divide both sides by 3, which gives the solution.

$$x = 2$$

EXAMPLE

Solve the equation $y = -\frac{1}{2}gt^2$ for the unknown t. First multiply both sides by the factor $-\frac{2}{g}$, which gives

$$-\frac{2y}{g} = \left(-\frac{2}{g}\right)\left(-\frac{1}{2}gt^2\right) = t^2$$

Then take the square root of both sides.

$$t = \sqrt{\frac{-2y}{g}}$$

Exercises | **Solutions**

1 Solve for x:	a) $x + 8 = 3x$	$x = 4$
	b) $3 = \frac{2}{x} + 4$	$x = -2$
	c) $a + \frac{1}{x} = b$	$x = \frac{1}{b - a}$
	d) $y = x^3 + 4$	$x = (y - 4)^{1/3}$
2 Solve for a:	a) $v = v_0 + at$	$a = \frac{v - v_0}{t}$
	b) $x = v_0 t + \frac{1}{2}at^2$	$a = \frac{2x}{t^2} - \frac{2v_0}{t}$
3 Solve for r:	$g = \frac{GM}{r^2}$	$r = \sqrt{\frac{GM}{g}}$

Quadratic Equations

The general form of a quadratic equation is

$$ax^2 + bx + c = 0$$

where a, b, and c are constants. If $b^2 > 4ac$, there are two real solutions:

$$x = \frac{-b \pm \sqrt{b^2 - 4ac}}{2a}$$

EXAMPLE

Solve the equation $x^2 - 3x - 4 = 0$.

Applying the quadratic formula, with $a = 1$, $b = -3$, $c = -4$, we find

$$x = \frac{3 \pm \sqrt{(-3)^2 - 4(1)(-4)}}{2(1)} = \frac{3 \pm 5}{2}$$

$x = 4$ or $x = -1$

Exercises Solve for x:

	Solutions
1 $x^2 - x - 6 = 0$	$x = -2, x = 3$
2 $x^2 - 15x + 50 = 0$	$x = 10, x = 5$
3 $4x^2 - 12x + 5 = 0$	$x = 2.5, x = 0.5$
4 $3x^2 + 2x = 0$	$x = 0, x = -\frac{2}{3}$
5 $x^2 + 2x - 4 = 0$	$x = -1 + \sqrt{5}, x = -1 - \sqrt{5}$

Solving Simultaneous Linear Equations

When more than one unknown appears in an equation, additional equations must be used to solve for the unknowns. We can solve for n unknowns by using n independent*, linear equations. For example, we can solve for 2 unknowns using 2 independent, linear equations. We solve simultaneous linear equations by a variety of methods, including the substitution method, which involves solving one equation for an unknown expressed in terms of other unknowns, and then substituting the expression for that unknown into the other equations. This method is illustrated in the following example.

EXAMPLE

Solve the equations:

$$(1)\ 2x + y = 3 \qquad (2)\ 4x + y = 11$$

First we solve Eq. (1) for x in terms of y; we isolate the unknown x.

$$2x = 3 - y$$

$$x = \frac{1}{2}(3 - y)$$

Next we substitute this expression for x into Eq. (2), and proceed to solve for y.

$$4 \times \frac{1}{2}(3 - y) + y = 11$$

$$6 - 2y + y = 11$$

$$-y = 11 - 6 = 5$$

$$y = -5$$

Finally, we substitute into the last equation for x the value of y.

$$x = \frac{1}{2}(3 - y) = \frac{1}{2}[3 - (-5)] = 4$$

Thus $x = 4$, $y = -5$ are the solution to Eqs. (1) and (2), as we can verify by substituting these values back into both equations.

Exercises Solve the following simultaneous equations:

	Solutions
1 $2x + y = 4$ $x - y = -1$	$x = 1, y = 2$
2 $y = 3x + 2$ $2y = x - 4$	$x = -1.6, y = -2.8$
3 $y = -5x$ $2x + 3y = 8$	$x = -0.615, y = +3.08$
4 $x + 2y - z = 3$ $10x - y + 2z = 45$ $2x + 4y + z = 36$	$x = 3, y = 5, z = 10$

*Two linear equations in two unknowns are independent if the graphs of these equations have unequal slopes. Thus $x + y = 3$ and $x - y = 5$ are independent equations, and therefore have a unique solution: $x = 4$, $y = -1$. On the other hand, the equations $x + y = 3$ and $2x + 2y = 6$ are not independent and do not have a *unique* solution. Independence of n linear equations in n unknowns is assured if the determinant formed by the coefficients in these equations is non-zero. The n equations then have a unique solution.

A-5 Geometry

Fig. A-2 shows useful geometric relationships between angles, and Fig. A-3 gives formulas for the areas of plane figures and for the volumes of solids.

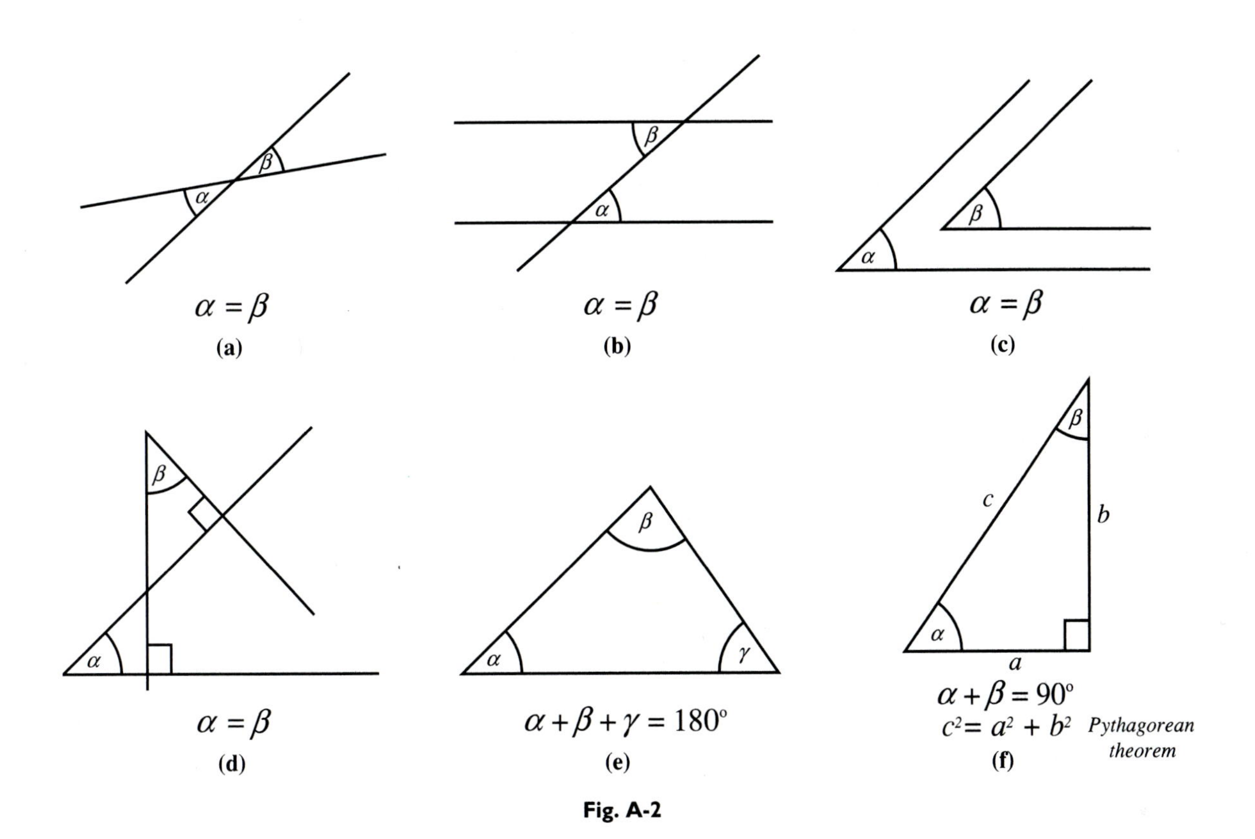

Fig. A-2

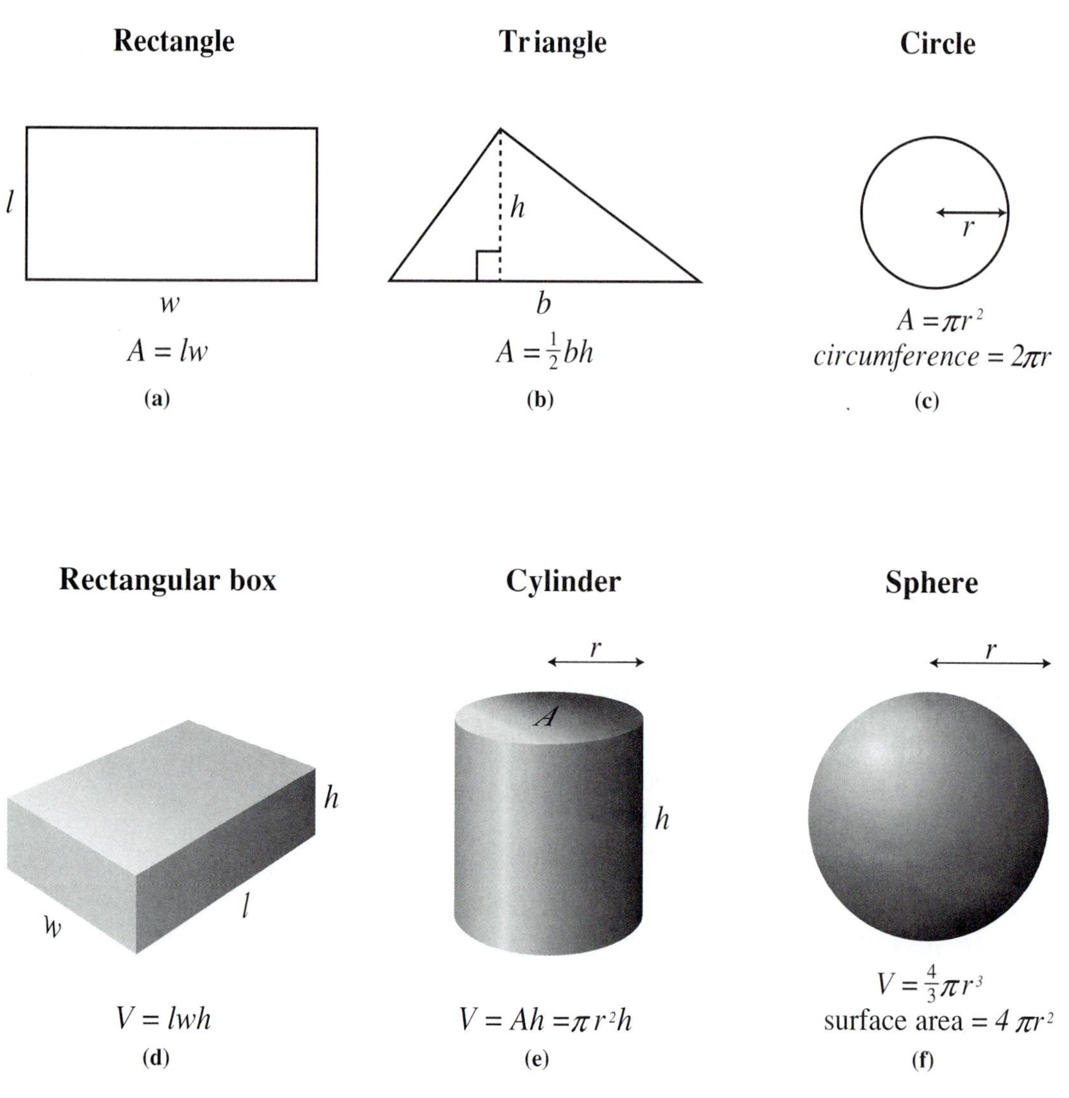
Rectangle
Triangle
Circle
l
w
$A = lw$
(a)
h
b
$A = \frac{1}{2}bh$
(b)
r
$A = \pi r^2$
circumference = $2\pi r$
(c)
Rectangular box
Cylinder
Sphere
h
w
l
$V = lwh$
(d)
r
A
h
$V = Ah = \pi r^2 h$
(e)
r
$V = \frac{4}{3}\pi r^3$
surface area = $4\pi r^2$
(f)

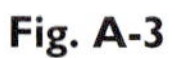
Fig. A-3

A-6 Trigonometry

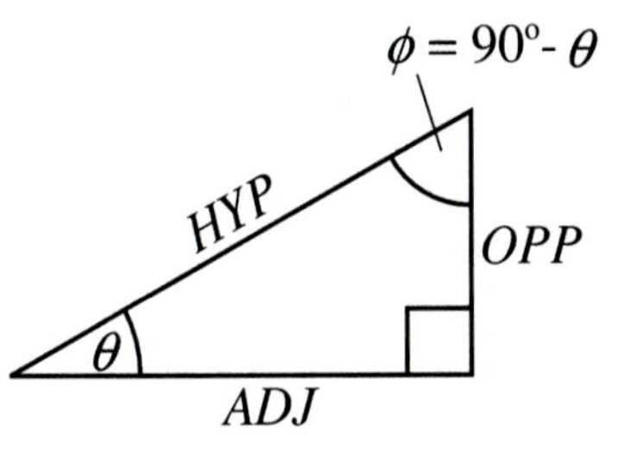

Fig. A-4

Fig. A-4 shows a right triangle, with sides labeled in reference to the angle θ. OPP is the side opposite θ, ADJ is the side adjacent to θ, and HYP is the hypotenuse.

The ratio of any two of these sides is a function only of the angle θ. The three basic functions of trigonometry are the sine, cosine, and tangent, abbreviated sin, cos, and tan respectively, and defined by the equations:

$$\sin\theta = \frac{\text{OPP}}{\text{HYP}}$$

$$\cos\theta = \frac{\text{ADJ}}{\text{HYP}}$$

$$\tan\theta = \frac{\text{OPP}}{\text{ADJ}}$$

The inverse of each of these trig functions is also given a name. These are the cosecant, secant, and cotangent, abbreviated csc, sec, and cot, where

$$\csc\theta = 1/\sin\theta$$

$$\sec\theta = 1/\cos\theta$$

$$\cot\theta = 1/\tan\theta$$

The corresponding trig functions for the angle $\phi = 90^\circ - \theta$ are similarly defined with respect to the sides opposite and adjacent to this angle. Since the opposite and adjacent sides are then interchanged, we find

$$\sin\phi = \cos\theta$$

$$\cos\phi = \sin\theta$$

$$\tan\phi = \cot\theta$$

The following trig identities are often useful:

$$\sin^2\theta + \cos^2\theta = 1 \qquad \sin(\alpha \pm \beta) = \sin\alpha\cos\beta \pm \sin\beta\cos\alpha$$

$$\tan\theta = \frac{\sin\theta}{\cos\theta} \qquad \cos(\alpha \pm \beta) = \cos\alpha\cos\beta \mp \sin\alpha\sin\beta$$

Appendix B Gauss's Law

Gauss's law is a relationship between the electric field on any closed surface and the charge enclosed by that surface. We can use Gauss's law to discover some very general properties of electric fields in and around conductors (as described in section 17-4) and to find general expressions for electric fields produced by certain symmetric configurations of charge (for example, the field of a uniform sphere of charge). Gauss's law is expressed in terms of "electric flux," a quantity related to the number of field lines passing through a surface.

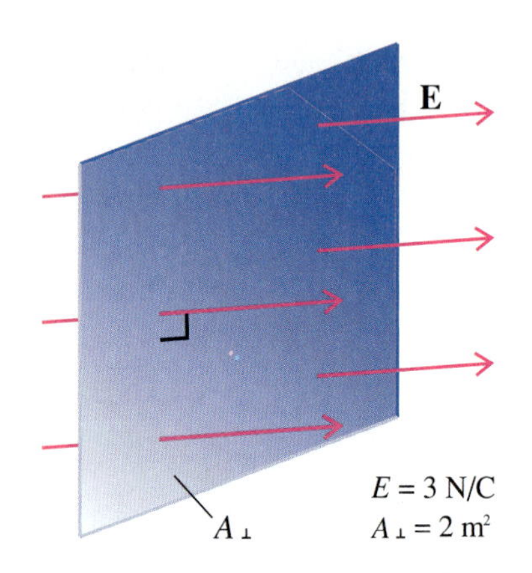

Fig. B-1 A uniform electric field of magnitude 3 N/C is directed perpendicular to a surface of area 2 m^2. With a scale factor c set equal to unity, the number of field lines passing through the surface is 6. The flux to the right is +6 N-m^2/C.

B-1 Electric Flux

We define electric flux in such a way that **the flux through any surface is proportional to the number of field lines passing through that surface.** Consider first the simplest case of a uniform electric field perpendicular to a plane surface.

Flux Through a Plane Perpendicular to a Uniform Field

Eq. 17-20 relates the number of field lines to the field's magnitude E and the perpendicular plane of area $A_\perp$, through which the field lines pass:

$$n = c\,E\,A_\perp$$

We define **electric flux,** denoted by the Greek letter Φ, to be a signed quantity, of magnitude equal to the number n when the scale factor c is set equal to 1.

$$|\Phi| = n = E\,A_\perp$$

The flux in a given direction is positive if the field points in that direction and is negative if the field is in the opposite direction:

$$\Phi = \pm E\,A_\perp \qquad \text{(B-1)}$$

In Fig. B-1, for example, 6 field lines directed to the right pass through the surface, and the flux to the right is +6 N-m^2/C. The flux to the left is − 6 N-m^2/C.

Flux of a Uniform Field Through an Arbitrary Plane Surface

Next we wish to extend the definition of electric flux through a surface in such a way that it applies to surfaces that are not perpendicular to the electric field. Again we want to define flux so that it is proportional to the number of field lines that pass through a surface. Fig. B-2 shows the same field lines passing through two plane surfaces: one of area A, whose perpendicular makes an arbitrary angle θ with the electric field, and a second of area $A_\perp$, perpendicular to the field. Because the same number of field lines pass through both surfaces, we define the flux through $A_\perp$ to be the same as the flux through A. Thus the flux to the right through A is

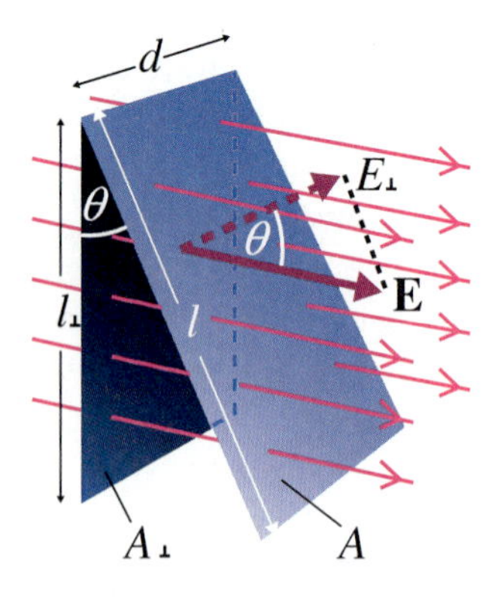

Fig. B-2 The same field lines pass through these two surfaces, and, since flux is proportional to the number of field lines, the flux through the surfaces is the same.

$$\Phi_A = \Phi_{\perp A} = E\,A_\perp$$

Fig. B-2 shows that $A_\perp$ is related to A by the equation

$$A_\perp = A\cos\theta$$

Combining the two preceding expressions, we obtain

$$\Phi_A = E\,A\cos\theta \tag{B-2}$$

We obtain a more useful equation for the flux through A if we express the flux in terms of the electric field component $E_\perp$ perpendicular to the surface. From Fig. B-2 we see that

$$E_\perp = E\cos\theta$$

Inserting this into the preceding expression for flux, we obtain

$$\Phi = E_\perp\,A \tag{B-3}$$

Suppose, for example, the electric field in Fig. B-2 has magnitude 5 N/C, the angle $\theta = 37°$, and the area $A = 2.5\ \text{m}^2$. Then $E_\perp = E\cos\theta = 5\ \text{N/C}\cos 37° = 4\ \text{N/C}$, and

$$\Phi = E_\perp A = (4\ \text{N/C})(2.5\ \text{m}^2) = 10\ \text{N-m}^2/\text{C}$$

Note that 10 field lines pass through the surface.

Flux Through a Curved Surface

We can count the field lines passing through an arbitrary curved surface by adding the number of lines passing through each small part of the surface. Accordingly, we define the flux through a curved surface as a sum of terms, each of which is the flux through a small section ΔA of the entire surface:

$$\Phi = \Sigma E_\perp\,\Delta A \tag{B-4}$$

The sections are small enough to be approximately flat and small enough that the field is nearly constant over any one section (Fig. B-3).

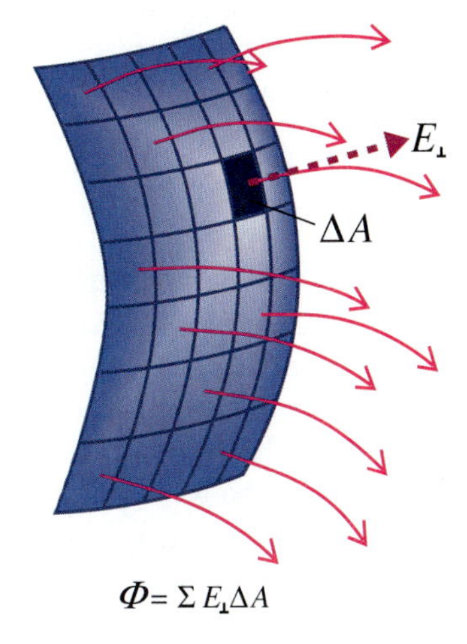

Fig. B-3 The flux of an electric field through a curved surface.

B-2 Gauss's Law and Applications

In this section we shall state Gauss's law and use it to determine electric fields. In section B-3 we derive Gauss's law from Coulomb's law.

Gauss's law is a relationship between the electric flux through a closed surface and the charge enclosed by that surface. Gauss's law states that the net outward flux of the electric field through a surface that encloses a net charge Q equals 4πk times Q.

$$\Phi = 4\pi kQ \qquad \text{(Gauss's law)} \tag{B-5}$$

Fig. B-4 shows a "Gaussian surface" that completely encloses a region of space containing a number of charges: $q_1, q_2, q_3, \ldots, q_n$. A Gaussian surface is any closed surface on which we choose to apply Gauss's law. Field lines are directed outward through some parts of the surface and therefore make a positive contribution to the outward flux; on other parts of the surface field lines are directed inward and make a negative contribution to the outward flux. When the net charge Q enclosed by the surface is positive, Gauss's law tells us that the net outward flux is positive: more field lines pass out of the surface than pass into it. When Q is negative, Gauss's law tells us that the net outward flux is negative: more field lines pass into the surface than pass out of it.

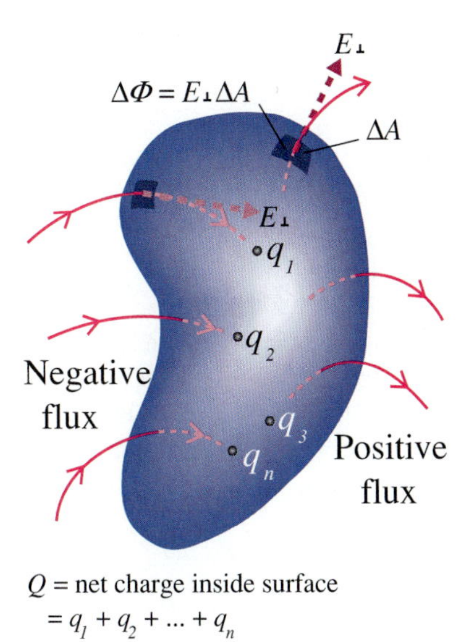

Fig. B-4 A surface encloses charges $q_1, q_2, q_3, \ldots, q_n$.

EXAMPLE 1 The Electric Field of a Uniformly Charged Infinite Sheet

Use Gauss's law to determine the magnitude and direction of the electric field produced by a positively charged infinite sheet having a uniform charge density σ.

SOLUTION We choose as our Gaussian surface a cylinder centered on the sheet of charge (Fig. B-5). By symmetry, the electric field must be everywhere perpendicular to the sheet. Therefore there is no component of **E** perpendicular to the curved part of the cylindrical surface and hence no flux through this part. The only contribution to the flux comes from the two flat ends of the cylinder, which are symmetrically placed on either side of the sheet. Because of this symmetry, the magnitude of the electric field must be the same on the two ends. The electric field is directed away from the sheet, and so there is a positive outward contribution to the flux on either side, equal to E times the area A:

$$\Phi = 2\,EA$$

The charge Q enclosed by the Gaussian surface is the charge on the circular section of the sheet, which has an area A. Thus Q equals the product of the charge density σ times A:

$$Q = \sigma A$$

Using the two equations above, we apply Gauss's law (Eq. B-5) and obtain an expression for E:

$$\Phi = 4\pi k Q$$

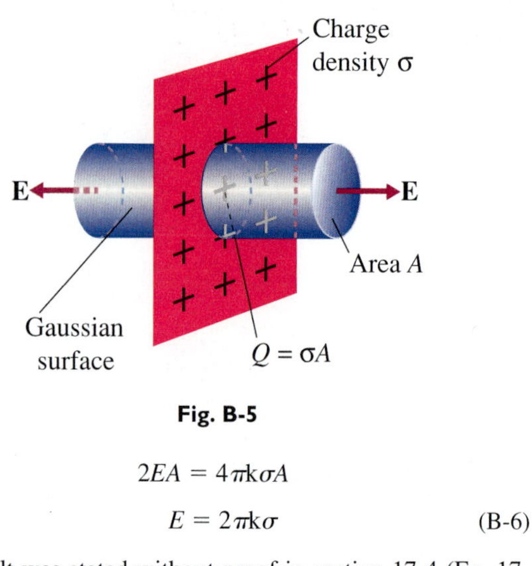

Fig. B-5

$$2EA = 4\pi k \sigma A$$

$$E = 2\pi k \sigma \qquad \text{(B-6)}$$

This result was stated without proof in section 17-4 (Eq. 17-15). The field is uniform, and hence field lines are uniformly spaced, as shown in Fig. 17-30. A negatively charged sheet having the same magnitude of charge density would produce an electric field of the same magnitude but opposite direction—that is, directed toward the negative charge.

Note that symmetry arguments were essential to the application of Gauss's law in this example. This is typical. In order for Gauss's law to provide an easy method for finding an electric field, the problem must possess sufficient symmetry so that the flux may be easily expressed in terms of E.

EXAMPLE 2 The Beginning and End of Field Lines

Prove that field lines begin only on positive charge and end only on negative charge.

SOLUTION Consider a point P at which a field line originates (Fig. B-6). Obviously there is a positive outward flux through a small Gaussian surface centered on P. According to Gauss's law, there must then be a positive charge enclosed by the surface. We can make the surface as small as we want without changing this conclusion. Thus there must be a positive point charge at P, the point where the field line originates.

Similarly, if a field line terminates at any point, one can draw a small Gaussian surface around that point. There is a negative outward flux through such a surface, and Gauss's law implies that there must be negative charge enclosed; that is, there is a negative charge at the point where the field line terminates.

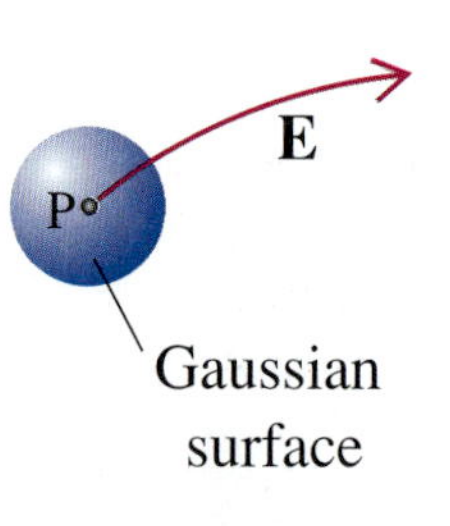

Fig. B-6

EXAMPLE 3 Charge on a Conductor

Show that all the excess charge on a charged conductor is on the surface. Assume that the distribution of charge is static.

SOLUTION As discussed in section 17-4, the free electrons in any conductor always move whenever there is an electric field inside the conductor. Thus when there is a *static* distribution of charge on a conductor, we can deduce that there is no electric field inside; in other words, the charge must arrange itself in such a way that the field everywhere inside the conductor vanishes. We apply Gauss's law over a small surface centered on an arbitrary point P inside the conductor (Fig. B-7). The field is zero on this surface, and hence Gauss's law states that the charge enclosed must also be zero. Thus the net charge at any point P inside the conductor is zero. Any excess charge must be on the conductor's surface.

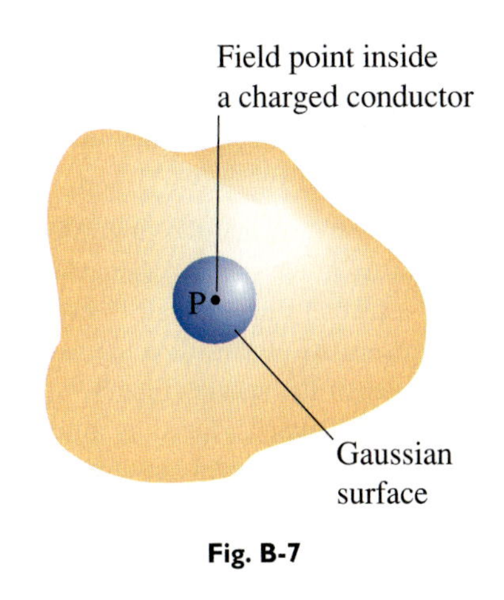

Fig. B-7

EXAMPLE 4 The Electric Field Near the Surface of a Conductor

Find a general expression for the electric field at a point just outside the surface of a conductor as a function of the charge density on the adjacent part of the surface.

SOLUTION First we note that the electric field at the surface of a conductor must be perpendicular to the surface. If this were not so, the component of the field tangent to the surface would cause free electrons to move along the surface. When excess charge is placed on a conductor, the charge very rapidly arranges itself along the surface until a static distribution is achieved—in other words, until there is no longer any field either inside the conductor or along its surface.

We apply Gauss's law to a small cylindrical surface that has one end parallel to the conducting surface and just outside it and the other end inside the conducting material (Fig. B-8). We assume the charge density σ to be positive.

$$\Phi = 4\pi kQ$$

$$EA = 4\pi k\sigma A$$

$$E = 4\pi k\sigma$$

The electric field just outside the conductor's surface is directly proportional to the charge density on that part of the surface. A negative charge density would give a field of equal magnitude but with the direction reversed. We may apply the result to either case if we express the result in terms of the magnitude of σ. We then obtain the result stated without proof in section 17-4 (Eq. 17-18):

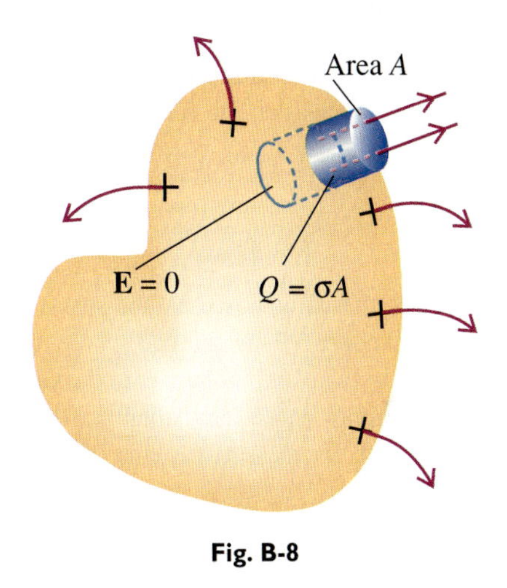

Fig. B-8

$$E = 4\pi k|\sigma|$$

EXAMPLE 5 The Electric Field of a Uniformly Charged Sphere

Prove that a uniformly charged sphere produces an electric field that, for field points outside the sphere, is the same as if the charge were concentrated at the center of the sphere. Assume the charge to be positive.

SOLUTION First we note that the field $\Delta\mathbf{E}$ produced by any charge element Δq in the sphere, when added to the field $\Delta\mathbf{E}'$ produced by a second symmetrically located charge element $\Delta q'$, gives a sum $\Delta\mathbf{E} + \Delta\mathbf{E}'$ directed radially outward away from the center of the sphere (Fig. B-9a). This means that the resultant of all fields produced by all charges within the sphere is a field directed radially outward.

E is directed radially outward

Because of the symmetry of the charge distribution, the field has the same magnitude at all field points the same distance from the center of the sphere. Thus if we apply Gauss's law to a spherical surface of radius r, centered on the sphere of charge (Fig. B-9b), we can find an expression relating the magnitude of the field to the distance r from the center of the sphere. Gauss's law states that the outward flux through the surface is given by

$$\Phi = 4\pi k Q$$

where Q is the net charge enclosed by the surface, which in this case is the total charge of the charged sphere. Because the field has a constant outward component over the Gaussian surface, we can easily apply Eq. B-3 and obtain an expression for Φ, using the equation for the surface area of a sphere ($A = 4\pi r^2$).

$$\Phi = E_{\perp} A$$

$$\Phi = E\,4\pi r^2$$

Inserting this into Gauss's law, we find

$$E\,4\pi r^2 = 4\pi k Q$$

$$E = k\frac{Q}{r^2}$$

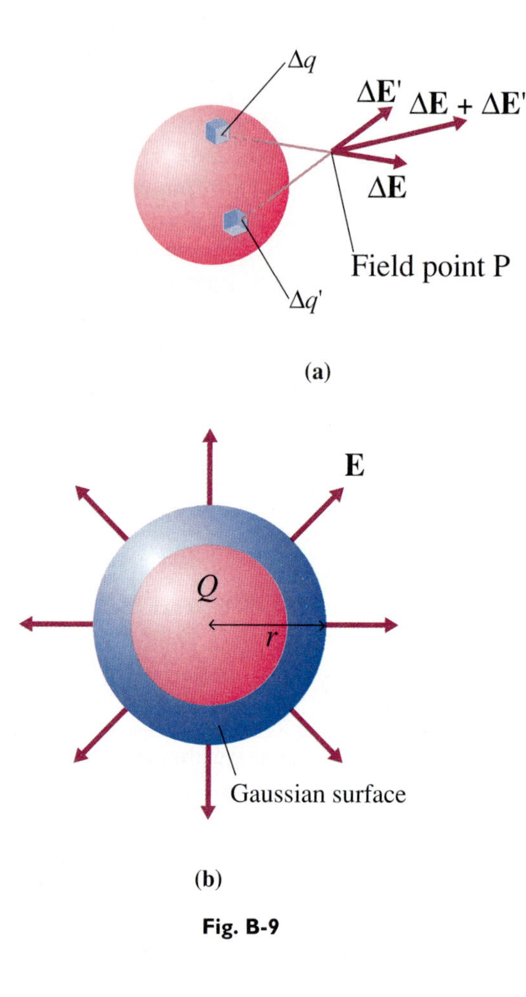

Fig. B-9

Both the magnitude and direction of the electric field outside the uniformly charged sphere are exactly the same as though that charge were concentrated as a point charge at the sphere's center.

B-3 Derivation of Gauss's Law

Gauss's law can be derived from the Coulomb expression for the electric field. We begin by considering the field of a single point charge and then consider an arbitrary field.

Field of a Single Point Charge

Suppose we calculate the total outward flux through an arbitrary closed surface when the field is produced by a single positive point charge. If the charge is inside the surface, as indicated in Fig. B-10, the net outward flux is positive. In order to calculate the flux through the arbitrary surface, we note that any field line that passes through the arbitrary surface also passes through a spherical surface centered on the charge. Hence the flux is simply the flux through the sphere, which we can easily compute using Eq. B-3, since $E_\perp$ is constant over the sphere's surface of area $4\pi r^2$:

$$\Phi = E_\perp A = \left(k\frac{q}{r^2}\right)(4\pi r^2)$$

$$\Phi = 4\pi kq \qquad (q \text{ inside any closed surface})$$

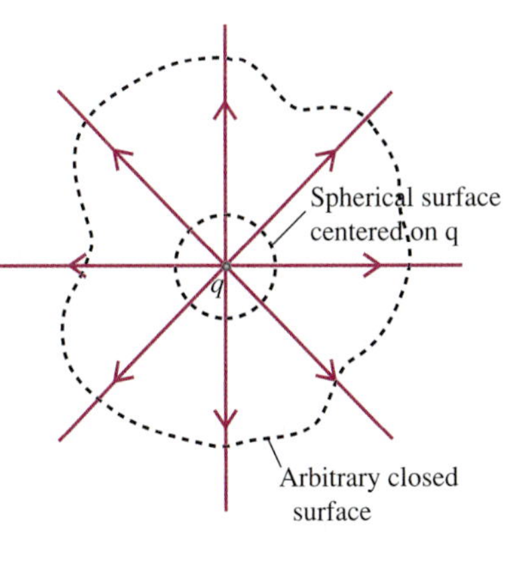

Fig. B-10 All field lines passing through the spherical surface also pass through the arbitrary surface. Therefore the flux is the same through the two surfaces.

For a negative charge, the field lines are directed inward and the outward flux is negative. Otherwise, everything is the same as for the positive charge and the preceding expression (which yields a negative value for negative q) is valid.

For a point charge outside a closed surface, it is easy to see that the flux of the field through the surface must be zero: any field line passes through the surface twice—once directed inward and once directed outward, as Fig. B-11 shows. (The charge in Fig. B-11 is positive, but, as above, our result is just as valid when q is negative.) Thus the *net* number of lines passing outward is always zero, and so the flux of the field through the surface is zero. This completes the proof of Gauss's law for the field of a single point charge.

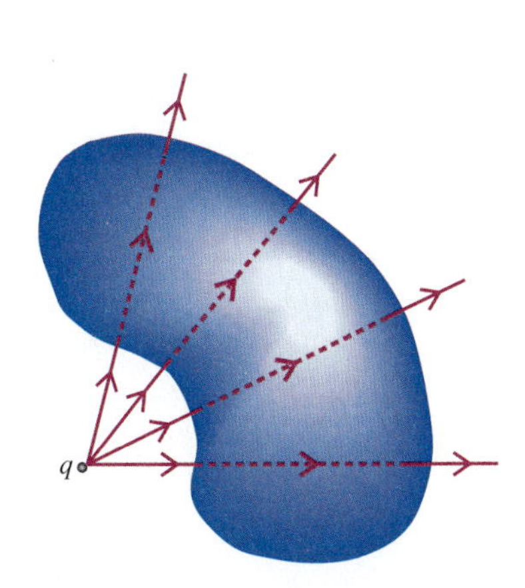

Fig. B-11 Four field lines pass into the closed surface, and four pass out. Thus the net number of field lines passing outward is zero, and so the net outward flux is zero.

Arbitrary Field

Consider the general case in which the electric field is produced by an arbitrary collection of point charges, some inside the Gaussian surface and some outside. Any charge distribution may be thought of as a collection of point charges, so this case is completely general. Suppose charge q_1 produces field $\mathbf{E}_1$, charge q_2 produces $\mathbf{E}_2$, and so forth. The total electric field at any field point is the vector sum of the fields of the point charges:

$$\mathbf{E} = \mathbf{E}_1 + \mathbf{E}_2 + \ldots$$

In this case of arbitrary charge distribution, the flux of $\mathbf{E}$ through the Gaussian surface may be expressed using Eq. B-4:

$$\Phi = \Sigma E_\perp \Delta A$$

where $E_\perp$ is the component of the resultant electric field along a direction perpendicular to the surface. This component can be expressed as the sum of the components of the single charge fields:

$$E_\perp = E_{1\perp} + E_{2\perp} + \ldots$$

Substituting this expression into the equation above, we find

$$\begin{aligned}\Phi &= \Sigma(E_{1\perp} + E_{2\perp} + \ldots)\Delta A \\ &= \Sigma E_{1\perp}\Delta A + \Sigma E_{2\perp}\Delta A + \ldots\end{aligned}$$

The quantity $\Sigma E_{1\perp}\ \Delta A$ is the flux of the field $\mathbf{E}_1$ produced by the first charge q_1, $\Sigma E_{2\perp}\ \Delta A$ is the flux of the field $\mathbf{E}_2$ of the second charge, and so on. Thus the flux of $\mathbf{E}$ is just the sum of the fluxes produced by the single-particle fields, which we label $\Phi_1, \Phi_2, \ldots$:

$$\Phi = \Phi_1 + \Phi_2 + \ldots$$

From the first part of the proof for a single point charge, we see that the flux for charges that are inside the surface is just $4\pi kq$ and the flux for charges outside is zero. In other words, only charges inside contribute to the flux; labeling these charges $q_1, q_2, \ldots, q_n$, we have

$$\begin{aligned}\Phi &= 4\pi k q_1 + 4\pi k q_2 + \cdots + 4\pi k q_n \\ &= 4\pi k(q_1 + q_2 + \cdots + q_n)\end{aligned}$$

or

$$\Phi = 4\pi k Q$$

where Q is the net charge inside: $q_1 + q_2 + \cdots + q_n$.

Appendix C Models of Electrical Conduction in Metals

Classical Model of Conduction

We shall describe a model that explains Ohm's law in terms of how the electrons in a metal respond to the presence of an electric field. This model, developed by Drude in 1900, uses classical mechanics and pictures electrons as tiny particles. Though it does not provide an adequate explanation of conduction, it is a useful starting point in attempting to relate electron motion to electric current in a metal. We shall first discuss this classical model, then point out its deficiencies, and finally outline briefly how quantum theory correctly describes the motion of conduction electrons.

In Fig. 19-33 we saw that free electrons behave much like gas molecules. Their average speed is a function of temperature and can be calculated using the formula for the rms speed of gas molecules found from kinetic theory (Eq. 12-16):

$$v_{\text{rms}} = \sqrt{3kT/m}$$

Because the electron mass is so small, this formula gives a very large thermal speed for electrons at room temperature:

$$v_{\text{rms}} = \sqrt{\frac{3(1.38 \times 10^{-23}\ \text{J/K})(300\ \text{K})}{(9.11 \times 10^{-31}\ \text{kg})}} = 1.17 \times 10^{5}\ \text{m/s}$$

The direction of electron velocity is random in the absence of an external electric field.

On the basis of this picture, the free electrons in a metal should collide with the tightly packed positive ions. A typical interionic spacing is on the order of 10^{-10} m, and one would expect this to be roughly the average distance travelled by electrons between collisions. Since the average electron speed is on the order of 10^5 m/s at room temperature, the average time τ an electron travels between collisions should be on the order of $(10^{-10}\text{ m})/(10^5\text{ m/s})$, or 10^{-15} s:

$$\tau \sim 10^{-15}\text{ s, at } T = 300\text{ K} \tag{C-1}$$

Since v_{rms} is proportional to $\sqrt{T}$, v_{rms} is greater at 300 K than at 3K by a factor of $\sqrt{100}$, or 10. Thus at T = 3K, v_{rms} is on the order of 10^4 m/s and τ is on the order of $(10^{-10}\text{ m})/(10^4\text{ m/s})$, or 10^{-14} s:

$$\tau \sim 10^{-14}\text{ s, at } T = 3\text{K} \tag{C-2}$$

Next we consider the effect of an external electric field on the conduction electrons. Each electron experiences a force $\mathbf{F} = -e\mathbf{E}$ and hence an acceleration $\mathbf{a} = \Sigma\mathbf{F}/m = -e\mathbf{E}/m$. During the time t that the field acts on an electron, the electron's velocity increases from its initial value $\mathbf{v}_0$ to $\mathbf{v}$, where

$$\mathbf{v} = \mathbf{v}_0 + \mathbf{a}t$$

or

$$\mathbf{v} = \mathbf{v}_0 - \frac{e\mathbf{E}}{m}t \tag{C-3}$$

This equation implies an electron velocity that increases in the $-\mathbf{E}$ direction as time increases, *without limit!* This infinitely increasing $\mathbf{v}$ means that when an electric field is produced in a wire, the resulting electric current will increase without limit. Of course, what we observe is a *constant* current in the wire soon after the field is turned on, with a current that is directly proportional to the field. In deriving Eq. C-3, we ignored the possibility of other forces acting on the electron. In particular, we ignored the effect of electrons colliding with the ionic lattice. Clearly it must be these collisions that prevent unlimited current growth.

Conduction electrons are accelerated by an electric field only for the brief time interval between collisions. The effect of a collision is to give an electron a new initial velocity $\mathbf{v}_0$, which is completely random in direction and which therefore shows no effect of the previous acceleration due to the field. At any instant an electron has a velocity $\mathbf{v}$ given by Eq. C-3, with t interpreted as the time since the last collision, since the effect of the field starts fresh with each collision. The average velocity $\overline{\mathbf{v}}$ of a large group of electrons is found by taking the average value of the terms in Eq. C-3. Because $\mathbf{v}_0$ is random, its average is zero, and we obtain

$$\overline{\mathbf{v}} = -\frac{e\mathbf{E}\tau}{m}$$

where τ is the average value of t, that is, the average time since the last collision (or, equivalently, the average time between collisions). The magnitude of $\overline{\mathbf{v}}$ is the average electron speed $\overline{v}$ contributing to electron current.

$$\overline{v} = \frac{eE\tau}{m}$$

Inserting this result into Eq. 19-25 ($I = |q|n\overline{v}A$), we obtain an expression for the current resulting from the motion of the free electrons:

$$I = \frac{ne^2\tau}{m}EA$$

Solving for E, we find

$$E = \frac{m}{ne^2\tau}\frac{I}{A}$$

This equation states that the electric field strength E inside the conductor is directly proportional to the current per cross sectional area $\frac{I}{A}$. Thus we have derived the microscopic form of Ohm's law, stated in chapter 19 (Eq. 19-26) as

$$E = \rho\frac{I}{A}$$

Our derivation allows us to obtain an expression for the resistivity ρ in terms of microscopic quantities. Comparing the two preceding equations, we find

$$\rho = \frac{m}{ne^2\tau} \qquad \text{(C-4)}$$

EXAMPLE Resistivity of Copper

The density of free electrons in copper is 8.5×10^{28} electrons per m^3. Use the classical theory of electron conduction to estimate the resistivity of copper at 300 K, and compare with the experimental value.

SOLUTION Inserting into Eq. C-4 the estimate of 10^{-15} s given in Eq. C-1 for the mean time τ between collisions, we obtain

$$\rho = \frac{m}{ne^2\tau} = \frac{9.1 \times 10^{-31}\text{ kg}}{(8.5 \times 10^{28}\text{ m}^{-3})(1.6 \times 10^{-19}\text{ C})^2(10^{-15}\text{ s})}$$

$$= 4 \times 10^{-7}\ \Omega\text{-m}$$

This theoretical estimate is not very close to the experimental value of $1.7 \times 10^{-8}\ \Omega$-m, given in Table 19-1. And at very low temperatures, the agreement is much worse. At $T = 3$K, $\tau \sim 10^{-14}$ s, and our classical estimate of ρ yields 4×10^{-8} Ω-m, whereas experimental values of ρ less that $10^{-12}\ \Omega$-m have been measured at this temperature for very pure copper crystals.

Quantum Model of Conduction

The discrepancy between theory and experiment in the preceding example indicates that the classical model is inadequate for a quantitative understanding of electron conduction in metals. It is necessary to employ quantum theory in order to construct a more satisfactory model. The main difference from the classical model is that in the quantum model the particle picture of electrons is replaced by a wave picture. Electron waves propagate through the medium provided by the ionic lattice. Electric current is caused by the motion of the electron waves in the direction opposite the electric field.

Resistance to the flow of these electron waves is caused not by the *existence* of the ions but rather by any *derivation from perfect regularity in the lattice*. At ordinary temperatures, the principal contribution to the resistance is from thermal vibrations of the lattice. As the ions vibrate, the perfect periodicity of the lattice is destroyed and electron wave packets are randomly scattered. Thus resistance increases with increasing temperature. At very low temperatures, electron wave packets are scattered primarily by lattice imperfections. For very pure crystals, the low-temperature resistivity may be extremely small.

Appendix D Selected Isotopes*

Atomic Number Z	Element	Symbol	Mass Number A	Atomic Mass, u	Percent Abundance, or Decay Mode if Radioactive	Half-life (If Radioactive)
0	(Neutron)	n	1	1.008 665	$\beta-$	10.3 min
1	Hydrogen	H	1	1.007 825	99.985	
	Deuterium	D	2	2.014 102	0.015	
	Tritium	T	3	3.016 050	$\beta-$	12.32 yr
2	Helium	He	3	3.016 030	0.000 14	
			4	4.002 603	≈100	
3	Lithium	Li	6	6.015 121	7.5	
			7	7.016 003	92.5	
4	Beryllium	Be	7	7.016 928	EC, γ	53.3 days
			9	9.012 183	100	
5	Boron	B	10	10.012 937	19.9	
			11	11.009 305	80.1	
6	Carbon	C	11	11.011 432	β^+, EC	20.3 min
			12	12.000 000	98.90	
			13	13.003 355	1.10	
			14	14.003 242	$\beta-$	5730 yr
7	Nitrogen	N	13	13.005 738	β^+	9.97 min
			14	14.003 074	99.63	
			15	15.000 108	0.37	
8	Oxygen	O	15	15.003 065	β^+, EC	122 s
			16	15.994 915	99.76	
			18	17.999 160	0.20	
9	Fluorine	F	19	18.998 403	100	
10	Neon	Ne	20	19.992 435	90.5	
			22	21.991 383	9.3	
11	Sodium	Na	22	21.994 434	β^+, EC, γ	2.60 yr
			23	22.989 767	100	
			24	23.990 961	$\beta-$, γ	15.0 h
12	Magnesium	Mg	24	23.985 042	78.99	
13	Aluminum	Al	27	26.981 539	100	
14	Silicon	Si	28	27.976 927	92.23	
			31	30.975 362	$\beta-$, γ	2.62 h
15	Phosphorus	P	31	30.973 762	100	
			32	31.973 907	$\beta-$	14.28 days
16	Sulfur	S	32	31 972 070	95.0	
			35	34.969 031	$\beta-$	87.2 days
17	Chlorine	Cl	35	34.968 852	75.77	
			37	36.965 903	24.23	
18	Argon	Ar	40	39.962 384	99.60	
19	Potassium	K	39	38.963 707	93.26	
			40	39.963 999	$\beta-$, EC, γ	1.26×10^9 yr
20	Calcium	Ca	40	39.962 591	96.94	
21	Scandium	Sc	45	44.955 910	100	

*Data from *Handbook of Chemistry and Physics,* 74th ed., 1993 CRC Press, Boca Raton, FL. The masses are for the neutral atom, including electrons.

Atomic Number Z	Element	Symbol	Mass Number A	Atomic Mass, u	Percent Abundance, or Decay Mode if Radioactive	Half-life (If Radioactive)
22	Titanium	Ti	48	47.947 947	73.8	
23	Vanadium	V	51	50.943 962	99.75	
24	Chromium	Cr	52	51.940 509	83.79	
25	Manganese	Mn	55	54.938 047	100	
26	Iron	Fe	56	55.934 939	91.8	
27	Cobalt	Co	59	58.933 198	100	
			60	59.933 819	$\beta-$, γ	5.271 yr
28	Nickel	Ni	58	57.935 346	68.1	
			60	59.930 788	26.2	
29	Copper	Cu	63	62.939 598	69.2	
			65	64.927 793	30.8	
30	Zinc	Zn	64	63.929 145	48.6	
			66	65.926 034	27.9	
31	Gallium	Ga	69	68.925 580	60.1	
32	Germanium	Ge	72	71.922 079	27.7	
			74	73.921 177	35.9	
33	Arsenic	As	75	74.921 594	100	
34	Selenium	Se	80	79.916 520	49.6	
35	Bromine	Br	79	78.918 336	50.69	
36	Krypton	Kr	84	83.911 507	57.0	
			89	88.917 640	$\beta-$, γ	3.15 min
			92	91.926 270	$\beta-$	1.84s
37	Rubidium	Rb	85	84.911 794	72.17	
38	Strontium	Sr	86	85.909 267	9.86	
			88	87.905 619	82.6	
			90	89.907 738	$\beta-$	29.1 yr
39	Yttrium	Y	89	88.905 849	100	
40	Zirconium	Zr	90	89.904 703	51.45	
41	Niobium	Nb	93	92.906 377	100	
42	Molybdenum	Mo	98	97.905 406	24.13	
43	Technetium	Tc	98	97.907 215	$\beta-$, γ	4.2×10^6 yr
44	Ruthenium	Ru	102	101.904 348	31.6	
45	Rhodium	Rh	103	102.905 500	100	
46	Palladium	Pd	106	105.903 478	27.33	
47	Silver	Ag	107	106.905 092	51.839	
			109	108.904 757	48.161	
48	Cadmium	Cd	114	113.903 357	28.73	
49	Indium	In	115	114.903 880	95.7; $\beta-$	4.4×10^{14} yr
50	Tin	Sn	120	119.902 200	32.59	
51	Antimony	Sb	121	120.903 821	57.3	
52	Tellurium	Te	130	129.906 229	33.9; $\beta-$	2.5×10^{21} yr
53	Iodine	I	127	126.904 473	100	
			131	130.906 114	$\beta-$, γ	8.040 days
54	Xenon	Xe	132	131.904 144	26.9	
			136	135.907 214	8.9	
55	Cesium	Cs	133	132.905 429	100	
			134	133.906 696	$\beta-$, EC, γ	2.065 yr
56	Barium	Ba	137	136.905 812	11.23	
			138	137.905 232	71.70	
			141	140.914 363	$\beta-$, γ	18.3 min
57	Lanthanum	La	139	138.906 346	99.91	
58	Cerium	Ce	140	139.905 433	88.43	

Atomic Number Z	Element	Symbol	Mass Number A	Atomic Mass, u	Percent Abundance, or Decay Mode if Radioactive	Half-life (If Radioactive)
59	Praseodymium	Pr	141	140.907 647	100	
60	Neodymium	Nd	142	141.907 719	27.13	
61	Promethium	Pm	145	144.912 743	EC, α, γ	17.7 yr
62	Samarium	Sm	152	151.919 729	26.7	
63	Europium	Eu	153	152.921 225	52.2	
64	Gadolinium	Gd	158	157.924 099	24.84	
65	Terbium	Tb	159	158.925 342	100	
66	Dysprosium	Dy	164	163.929 171	28.2	
67	Holmium	Ho	165	164.930 319	100	
68	Erbium	Er	166	165.930 290	33.6	
69	Thulium	Tm	169	168.934 212	100	
70	Ytterbium	Yb	174	173.938 859	31.8	
71	Lutetium	Lu	175	174.940 770	97.41	
72	Hafnium	Hf	180	179.946 545	35.100	
73	Tantalum	Ta	181	180.947 992	99.998	
74	Tungsten	W	184	183.950 928	30.7	
75	Rhenium	Re	187	186.955 744	62.60; $\beta-$	4.2×10^{10} yr
76	Osmium	Os	191	190.960 920	$\beta-$, γ	15.4 days
			192	191.961 467	41.0	
77	Iridium	Ir	191	190.960 584	37.3	
			193	192.962 917	62.7	
78	Platinum	Pt	195	194.964 766	33.8	
79	Gold	Au	197	196.966 543	100	
80	Mercury	Hg	202	201.970 617	29.86	
81	Thallium	Tl	205	204.974 401	70.476	
			208	207.981 988	$\beta-$, γ	3.053 min
82	Lead	Pb	206	205.974 440	24.1	
			207	206.975 872	22.1	
			208	207.976 627	52.4	
			210	209.984 163	α, $\beta-$, γ	22.6 yr
			211	210.988 735	$\beta-$, γ	36.1 min
			212	211.991 871	$\beta-$, γ	10.64 h
			214	213.999 798	$\beta-$, γ	26.8 min
83	Bismuth	Bi	209	208.980 374	100	
			211	210.987 255	α, $\beta-$, γ	2.14 min
			212	211.991 255	$\beta-$, α, γ	1.009 h
84	Polonium	Po	210	209.982 848	α, γ	138.3 days
85	Astatine	At	218	218.008 684	α, $\beta-$	1.6 s
86	Radon	Rn	222	222.017 570	α, γ	3.824 days
87	Francium	Fr	223	223.019 733	α, $\beta-$, γ	21.8 min
88	Radium	Ra	226	226.025 402	α, γ	1.60×10^{3} yr
			228	228.031 064	$\beta-$, γ	5.76 yr
89	Actinium	Ac	227	227.027 750	α, $\beta-$, γ	21.77 yr
90	Thorium	Th	228	228.028 715	α, γ	1.913 yr
			232	232.038 054	100; α, γ	1.4×10^{10} yr
			234	234.043 593	$\beta-$	24.10 days
91	Protactinium	Pa	231	231.035 880	α, γ	3.25×10^{4} yr
92	Uranium	U	232	232.037 130	α, γ	68.9 yr
			233	233.039 628	α, γ	1.592×10^{5} yr
			235	235.043 924	0.720; α, γ	7.04×10^{8} yr
			236	236.045 562	α, γ	2.34×10^{7} yr
			238	238.050 785	99.2745; α, γ	4.46×10^{9} yr
			239	239.054 289	$\beta-$, γ	23.5 min

Atomic Number Z	Element	Symbol	Mass Number A	Atomic Mass, u	Percent Abundance, or Decay Mode if Radioactive	Half-life (If Radioactive)
93	Neptunium	Np	239	239.052 933	$\beta-$, γ	2.355 days
94	Plutonium	Pu	239	239.052 157	α, γ	2.411×10^4 yr
95	Americium	Am	243	243.061 375	α, γ	7.37×10^3 yr
96	Curium	Cm	245	245.065 483	α, γ	8.5×10^3 yr
97	Berkelium	Bk	247	247.070 300	α, γ	1.4×10^3 yr
98	Californium	Cf	249	249.074 844	α, γ	351 yr
99	Einsteinium	Es	254	254.088 019	α, γ, $\beta-$	276 days
100	Fermium	Fm	253	253.085 173	EC, α, γ	3.0 days
101	Mendelevium	Md	255	255.091 081	EC, α	27 min
102	Nobelium	No	255	255.093 260	EC, α	3.1 min
103	Lawrencium	Lr	257	257.099 480	α, EC	0.65 s
104	Rutherfordium	Rf	261	261.108 690	α	1.1 min
105	Hahnium	Ha	262	262.113 760	α	34 s

Appendix E Answers to Odd-Numbered Problems

Measurements and Units

1 3×10^5 y
3 The definition would not have been consistent with earlier definitions.
5 3×10^7 volts
7 0.06 m^2
9 91.5 m
11 186,000 mi/s
13 $v = at$
15 m
17 (a) 3; (b) cannot determine; (c) 3; (d) 5; (e) 2; (f) 4
19 (a) 5780 m; (b) 2.48×10^5 s; (c) 3.82×10^{-3} kg; (d) 0.0623 m
21 477 acres
23 252 kg
25 10^{19} m^3
27 100 days
29 10^5
31 10^8 m^2
33 2000

Chapter 1

1 22 ft/s
3 (a) 3.33×10^{-6} s; (b) 2.94 s
5 3.56 s
7 9.9 s
9 0.050 m/s
11 1.280 km, 0.256 km/min
13 50.0 m/s
15 3.3 m/s, 3.0 m/s, 3.0 m/s, 3.0 m/s
17 3600 km, 27° south of east
19 0
21 230 V/m, 27° to the left of vertically upward
23 2.0 units, vertically upward
25 10.0 lb, 60.0° below the horizontal to the left
27 100 lb to the right
29 7.07 lb, 7.07 lb
31 −5.96 lb
33 (a) 7.00, 9.00; (b) 11.4, 52.1° above the positive x-axis
35 147 lb down the hill
37 10.0 lb along the positive y-axis
39 11.2 lb, 26.6° below the horizontal to the left
41 14.1 lb, 45.0° to the left of the 5.00 lb force
43 0.400 m/s, in the $+x$ direction
45 0.500 cm/s, 37.0° above the x-axis; the average speed is much greater.
47 $x = 50.0$ m; $y = -15.0$ m
49 0.11 s
51 (a) 0; (b) 100 lb to the right
53 0.469 s
55 5.66 m, 32.0° west of north
57 50 m

Chapter 2

1 10.0 m/s^2 forward
3 (a) 69.4 m/s (155 mi/h); (b) 19.3 m/s^2
5 9.00×10^4 m/s^2, west
7 10.1 cm/s, 3 cm/s^2
9 30.0 m/s^2
11 (a) 0; (b) 18.0 m
13 6.00 s, 90.0 m
15 (a) 24.3 h; (b) 8750 m/s
17 4.00 m
19 $a_x = 2.80$ m/s^2; $v_x = 17.0$ m/s
21 29.6 m
23 9.90 m/s
25 57.6 m
27 1.43 s
29 88.0 m
31 14.7 m/s, downward
33 27,400 m

35 (a) t_4; (b) t_1; (c) t_3; (d) t_5; (e) t_5; (f) t_4; (g) t_2; (h) t_4
37 Body starts from rest, accelerates at a constant rate, moves at constant speed, then accelerates at a much faster rate than the first acceleration, and then decelerates just as quickly down to the same constant speed at which it was moving before the sharp acceleration.

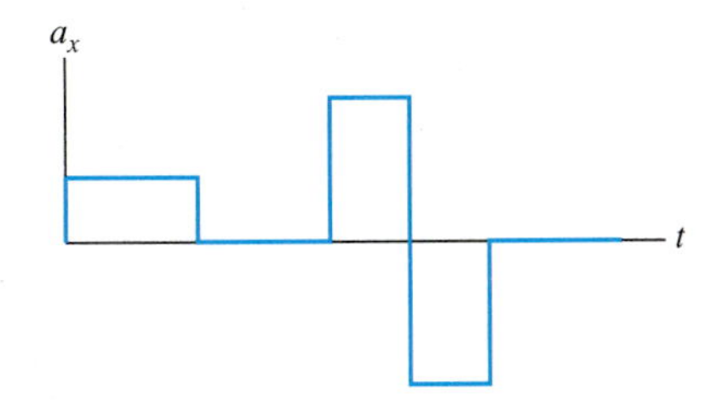

39 (a) 50.0 m; (b) 69.4 m; (c) 114 m to 140 m
41 child below, 0.461 s
45 (a) 3.00 s; (b) 5.00 s later; (c) 2.13 m/s²
47 313 m/s² toward the rear of the car
49 (a) 10.0 m/s; (b) 7.21 m/s²; (c) 10.8 m/s
51 5.00 ft
53 80.0 m/s²
55 0.221 s

Chapter 3

1 southwest (45° south of west)
3 340 m/s², 23° west of south
5 4.95 m/s², 36.0° below the negative x axis.
7 300 m/s², 46° below the horizontal
9 (a) 19.6 cm; (b) 22.6 cm
11 9.8 m/s, directed horizontally
13 2.87°
15 10.0 m/s, 26.2°, 8.08 m
17 62.6 m/s
19 27.3 m/s, 63.9° above the horizontal
21 20.0 m/s²
23 6.64 rev/s
25 1.70×10^3 m/s
27 (a) 20.0 km/h; (b) 1.20 min
29 279 m/s, 5.81° south of east
31 (a) 32.0 m/s, 38.7° south of west
33 (a) 60.0 s; (b) 33.3 m;
(c) 106 m/min, 18.5° south of east; (d) 105 m;
(e) 19.5° north of east
35 11.4 m
37 4.98 m, 8.23 m, 8.07 m
39 2.5×10^5 m/s, 2.5×10^{-10} m/s²
41 355 m/s, 0.0258 m/s²
43 (a) $v_0 = 17.5$ m/s, $\theta_0 = 52.7°$; (b) 10.6 m
45 8 m/s²
47 (a) 2.00×10^2 m/s, 36.9° south of west; (b) 60.0 m
49 (a) 4.98 m; (b) 5.96 m
51 35.0 m/s
53 5.42 m/s, 45.0° above the horizontal
55 (a) 115 m; (b) 5.00 m/s²

Chapter 4

1 1.00×10^2 kg
3 no
5 40.0 N, 30.0 N
7 94.0 N to the right
9 276 N, 188 N
11 133 m/s², directed opposite the 30.0 N force
13 1.72 m/s²
15 14.5 m/s², 22.8° above the positive x-axis
17 5.10×10^4 N
19 6750 N, directed toward the rear of the car
21 (a) a downward force of 9.80 N exerted on the book by the earth; an upward force of 9.80 N exerted on the book by the hand; an upward force of 9.80 N exerted on the earth by the book; a downward force of 9.80 N exerted on the hand by the book.
(b) 5.20 m/s², upward;
(c) 9.80 N downward, 9.80 m/s² downward.
23 972 N upward
25 The rope exerts on each team a force of 5.00×10^3 N directed toward the other team. The ground exerts a horizontal force on each team directed away from the other team, 5.10×10^3 N on B and 4.90×10^3 N on A.
27 14.7 m/s², downward
29 $T_1 = 21.7$ N; $T_2 = 17.7$ N; $T_3 = 20.0$ N
31 (a) 5.00×10^2 N; (b) 389 N, 389 N
33 80.0 N, 69.3 N
35 520 N, 640 N
37 (a) 55.0 N; (b) 3.01 m/s
39 (a) 9.80 N; (b) 0.202 s
41 2.00 m/s², down
43 (a) 4.04×10^4 N (9100 lb);
(b) 2.53×10^4 N (5700 lb);
(c) 1.50×10^3 N (340 lb)
45 210 N
47 312 N, 45.5° above the horizontal
49 time falling
51 (a) 90 cm; (b) 7.4 m/s² up
53 66.0°, 59.0°
55 $CA(v - v_w)^2$, $CA(v + v_w)^2$, CAv_w^2

Chapter 5

1 (a) 0.33; (b) 0.22
3 15 N
5 32 N
7 2°
9 160 N
11 (a) 1.0 N to the right; (b) 1.0 N to the left; (c) 9.8 N to the left
13 (a) 8.7 N; (b) 5.0 N; (c) 10.0 N; (d) 8.7 N into the plane, 5.0 N down the plane
15 34.1 m/s
17 (a) 9.8m/s^2; (b) 32 m; (c) 320 m
19 0.425
21 45.0 N
23 (a) 200 N; (b) passenger will slide along the seat.
25 700 N
27 208 N
29 4.64×10^6 m
31 (3.33 cm, 9.00 cm)
33 0.289 m
35 (a) 0.333 m to the right of A; (b) 4.00 m in the direction of the 6 N force
37 25 m west and 45 m south of the point of impact
39 (a)

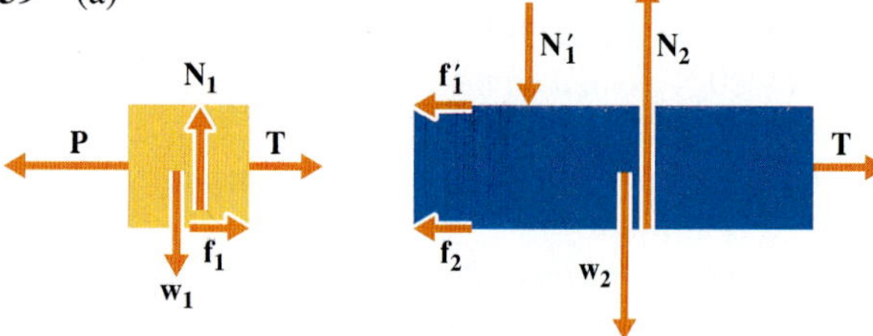

(b) coefficients of friction
41 370 N
43 3.5 m/s^2 downward
45 30 kg
47 447 N
49 4.9 m/s^2 (top), 15 m/s^2 (bottom)
53 0.279
55 3.33 cm

Chapter 6

1 0.24 y
3 8.34×10^{-13} N
5 1.74×10^{-11} N
7 9.00 kg
9 (a) 3.7×10^{-16} N; (b) 1.6×10^{-8} N
11 4.62×10^{-10} N, directed toward the center of the triangle
13 3.33 m
15 6.67×10^{-11} N-m^2/kg^2
17 (a) 24.9 N; (b) 24.9 m/s^2; (c) 12.5 m
19 1.63 m/s^2, 3.73 m/s^2
21 8.34×10^{-2} N
23 1.68×10^3 m/s, 108 min
25 1.99×10^{30} kg
27 (a) 320 N, directed away from the center of the circle; (b) 585 N
29 (a) 843 N; (b) 343 N
31 (a) 589 N, directed perpendicularly into the track; (b) 1960 N, directed 25.2° below the horizontal, extending out from the center of curvature; (c) 2600 N, directed downward; (d) 935 N, directed upward
33 3.58×10^9 kg/m^3
35 9.81 m/s^2
37 8.57 m/s; 26.4 m/s^2, directed into the wall, 21.8° below the horizontal
39 317 km
45 (a) 5.34×10^{-8}m/s^2; (b) 194 s; (c) 1.67×10^{-2} rad, or 0.957°

Chapter 7

1 (a) 225 J; (b) 2.03×10^3 J; (c) 2.00×10^5 J
3 (a) 40.0 J (b) 3.62×10^6 J
5 0
7 (a) 300 J; (b) if you lift quickly, exerting a force greater than 200 N
9 (a) 2250 J, 1590 N; (b) 2250 J, 391 N
11 (a) − 75,000 J; (b) 3750 N, opposite the direction of motion
13 (a) − 11,000 J; (b) 44 cm
15 170 m/s (380 mph)
17 11.8 m/s
19 3.60 m/s (8 mph)
21 4.00×10^5 J
23 2600 N
25 5470 m/s
27 1.66×10^4 m/s
29 (a) $-\frac{1}{2}GmM/r$; (b) -1.99×10^{11} J, 6320 m/s
31 7.50 J
33 2.94×10^3 N/m
35 9.90 cm
37 9.90 m/s
39 − 98.0 J
41 78.9 cm
43 38.7%, air resistance and friction
45 4.00×10^4 W, or 54 hp
47 165 lb

49 2.47×10^5 W, 331 hp
51 6.8×10^8 m^2
53 480 ft-lb, or 651 J
55 10.0 m/s
57 19.6 kW
59 42°
61 3890 N (875 lb); − 1200 N (− 270 lb)
63 (a) 2.27×10^3 m/s (b) 1.12×10^4 m/s (c) 24.3
65 5800 ft
67 337 N
69 4.22×10^4 m/s
71 4.36×10^4 m/s

Chapter 8

1 (a) 16.3 kg-m/s; 2640 J; (b) 1120 kg-m/s; 5600 J (c) 2.00×10^5 kg-m/s; 2.00×10^6 J
3 5.00×10^2 N
5 1.08 N-s, in the $-x$ direction
7 (a) 8.32 N-s, directed 45.0° above the horizontal; (b) 0.0202 s, 412 N
9 2.70 s
11 15.0 m/s, backward
13 2.80×10^2 m/s
15 319 kg-m/s, 1.60 s
17 (a) no; (b) 0; (c) 3.00 m/s to the left
19 19.6 J
21 5.67 m/s; 7.67 m/s
23 4.31 cm/s
25 1.50 m/s to the right; 6.75 J
27 1.88 kg-m/s, 3.65° to the left of the positive y-axis
29 (b) 0.019 N-s to the left; (c) 3.8 mm/s to the left
31 (a) $2r$ sin 50.0°; (b) 40.0°

Chapter 9

1 (a) − 22.0 rad/s; (b) 1.32 m/s
3 313 rad/s
5 22.1 rad/s
7 7.27×10^{-5} rad/s, 2.58×10^{-2} m/s^2, in a direction 40.0° N of vertically down
9 (a) clockwise; (b) -1.20×10^5 N-m
11 − 3.61 N-m
13 0, − 0.500 N-m, − 1.46 N-m
15 0, + 87.8 N-m, − 175 N-m, + 87.1 N-m
17 (a) 360 $rads/s^2$, 36 rad/s; (b) 29 m/s
19 52 revolutions
21 160 N-m
23 0.958 s, 12.0 N on the m_1 side, 15,2 N on the m_2 side
25 (a) hollow; (b) solid
27 2.65 m/s
29 (a) 5.47×10^5 J; (b) 9.77×10^3 J
31 1.81 m/s
33 (a) 2.68×10^{40} kg-m^2/s; (b) 7.05×10^{33} kg-m^2/s
35 1.47 rad/s
37 4.00 kg-m^2
39 58 W
41 200 W
43 $r_1\omega_1 = r_2\omega_2$
47 0.455 m/s^2
49 3.14 kg
51 (a) 28.6 hp; (b) 72.6 hp

Chapter 10

1 1350 N
3 332 N
5 9.00×10^2 N
7 313 N
9 (a) 1240 N; (b) 620 N, − 1610 N
11 0.397 m
13 688 N, 312 N
15 25.0 N
17 0.500 m
19 259 N + 17.5 W; 936 N, 15.7° below the horizontal
21 487 N
23 1.37 cm
25 125 m
29 tip over
31 (a) 60.9°; (b) 5.45 m/s^2
33 0.559
35 650 N, 650 N, 500 N

Chapter 11

1 (a) 0.974 N/mm^2; (b) 159 N/mm^2
3 40.4 N
5 125 cm^2
7 10.3 m
9 (a) 10.3 m; (b) 3.1 m
11 (a) 333 N; (b) 1.00×10^4 J; (c) 30.0 m; (d) 1.00×10^4 J
13 1.99 atm
15 10.1 m, 19.7 N
19 9000 N
21 1.67×10^3 kg/m^3
23 7.03×10^{-2} m^3

25 167 lb
27 1.29 m
29 0.57
31 0.022
33 1.49 mm
37 (a) − 3.85 atm; (b) air inward
39 3030 N inward
41 (a) 1.08 m/s; (b) 1.08×10^{-4} m^3/s
43 (a) 7.67 m/s, 7.67×10^{-2} m^3/s; (b) 29.4 J/kg; (c) 2260 W (3 hp)
45 (b) 1.33 m/s
47 0.246 atm
49 3.0 cm, turbulent
51 6.94 cm/s
53 2.51 m
55 15 torr
57 0.841
59 2.53 N
61 15.5 m^3
63 (a) 100 cm^3/s; (b) 29.6 cm^3/s; (c) 20.2 cm
65 0.78 mm

Chapter 12

1 32.0°F, 50.0°F, 68.0°F, 86.0°F, 104°F
3 0.110 atm
5 − 40.0°
7 1.22×10^{22}
9 2.44×10^{19}
11 (a) 2.02×10^{23}; (b) 4.54×10^5 Pa; (c) 5.19×10^5 Pa; (d) 9.73×10^5 Pa
13 0.035 atm, or 3500 Pa
15 46.9° C
17 225 K
19 (a) 1.00 kg/m^3; (b) 1.60×10^5 Pa; (c) 0; (d) 1.16×10^{22} K
21 341 K, 551 m/s
23 (a) 1; (b) no
25 0.20 atm
27 25%
29 20° C
31 5.1 cm
33 2.5 cm^3
35 0.016 mm
37 0.60 L
39 1.45 m
41 5.35×10^{-2} mm
43 (a) 28.9 atm, 1480 m/s; 49.3 atm, 1370 m/s; (b) 1400 m/s; (c) 317 K; (d) 26.1 atm, 52.2 atm
45 gap increases by 1.1×10^{-2} mm
47 10 cm
49 4.2°

Chapter 13

1 (a) 1.0×10^4 cal; (b) 1.1×10^3 cal
3 120 kcal, or 120 Cal
5 (a) 8.00×10^4 cal; (b) 6.19×10^5 cal; 43.2 min
7 0.881 cal/g-C°
9 58.8 cm^3, or 0.588 L
11 209 W/m^2
13 51.7° C
15 (a) 0.157 W; (b) 47.4 h
17 2.83×10^{23} m^2; (b) 3.96×10^{26} W; (c) 5800 K
19 827 W
21 3.0×10^3 W
23 3.3 W/m^2-C°
25 0.60 m
27 22 kcal/h
29 (a) 800 kcal; (b) 11.4 h
31 2 cm
33 2.8 min
35 66° C
37 (a) 2.0×10^{-3} cal/s; (b) 4.0×10^3 s
39 40.1° C
41 (a) $H_{rad} = 620$ W; $H_{conv} = 75$W; (c) $H_{cond} = 5.0$ W; (b) 44° C

Chapter 14

1 6230 J
3 (a) 4.62×10^5 J; (b) 3.67×10^5 J; (c) 3.67×10^5 J
5 15.0 J
7 20.0 C°
9 6.97 min
11 6.26×10^{-20} J
13 9000 J
15 (a) 1.06 J; (b) 2.10×10^5 J
17 (a) 3410 J; (b) 3410 J
19 (a) − 37 cal; (b) $Q = \Delta U = 37$ cal; $W = 0$
21 (a) 100 J; (b) 20,000 W, or 26.8 hp
23 (a) 3.4×10^7 J/L; (b) 2.4×10^6 J/km; (c) 14 km/L, or 33 mi/gal
25 0.56
27 667 J/K; $e^{4.83 \times 10^{25}}$
29 (a) 3660 J/K; (b) − 3410 J/K; (c) 250 J/K; (d) $e^{2.65 \times 10^{26}}$; $e^{-2.47 \times 10^{26}}$; $e^{1.81 \times 10^{25}}$
31 2.85 J/K
33 1500 J/K
35 (a) 2740 J; 9.12 J/K; (b) − 9.12 J/K
37 (a) $\frac{1}{6}$; (b) $\frac{1}{36}$; (c) $\frac{1}{18}$; (d) $\frac{1}{6}$
39 488 K
41 333 K
43 30

45 (a) 43 days; (b) 48 days; (c) 59 days
47 1410 Cal
49 (a) 72 cal; (b) 0.82 g
51 (a) V/V_0; (b) $(V/V_0)^2$; (c) $(V/V_0)^N$
55 (b) 8.77 W; (c) 24 W
57 28 min

Chapter 15

1 1.95×10^{-3} s
3 4 m, 4 s, 0.25 Hz
5 (a) 2.94 Hz; (b) 176 beats/min
7 (a) 3.00 m; (b) 10.0 s; (c) 0.100 Hz; (d) 1.76 m; (e) 1.76 m
9 10 cm, 0.8 rad/s, $\pi/4$ rad = 45°
11 $\pi/2$
13 π rad/s
15 -395 m/s², -790 m/s²
17 (a) 2380 N; (b) 2450 N
19 101 kg
21 893 g
23 1.73 m
25 1.38 m/s, 1.91×10^3 m/s²
27 2π s
29 9.77 m/s²
31 0.897 s
33 2.46 s
35 (a) 0.600 J; (b) 0.200 W
37 1.18 m/s; 1.06 m/s
39 (a) 0.65 kg-m²; (d) 0.72 m; (e) 1600 N
41 $2\pi\sqrt{R/g}$
43 will lose 8.64 s/day

Chapter 16

1 40.0 cm/s
3 337 m
5 1020 m
7 (a) 2.5 Hz; (b) 20 cm/s
9 $t = 20$ ms to $t = 22$ ms
11 (a) 100 m/s; (b) 8.00 cm; (c) 80.0 m/s
13 (a) 17 m to 17 mm; (b) 74 m to 7.4 cm
15 1.45 s later
17 1.79 s
19 55 m
21 74 m
23 0.560 s
25 6.88 mm
27 0
29 57.3 m/s
31 (a) 212 Hz, 189 Hz; (b) 212 Hz, 188 Hz
33 86 m/s (190 mi/h)
35 578 Hz
37 667 Hz
39 30.6 s
41 85 dB
43 6.28 W
45 133 dB
47 2.00×10^{-7} W/m²
49 5 cm, 2 Hz, 20 cm, 40 cm/s, to the right
51 $y = 1.00 \times 10^{-8} \sin(1024\pi t - 9.35x)$
53 0
55 $6 \sin(2t - 5x)$
57 2.00 cm
59 5730 Hz; 17,200 Hz; 28,700 Hz
61 260 Hz or 264 Hz, lengthen
63 440 Hz; 880 Hz; 1320 Hz
65 10 cm
67 (a) 43.0 Hz; (b) 21.5 Hz
69 402 Hz
71 (a) 3.96×10^{14} J; (b) 32, 1000
73 42.7 min
75 1.72 m
77 3.77 m/s, 474 m/s²
79 (a) 2.62×10^{-4} s; (b) 0

Chapter 17

1 (a) 1.75×10^{24}; (b) 6.02×10^{22}
3 (a) 2.89×10^{-9} N; (b) 0; (c) 3.63×10^4 N
5 2.27×10^{39}
7 (a) 3.57×10^{-4} N; (b) 3.99×10^{-9} C; (c) 2.49×10^{10}
9 (a) 1.78×10^{-16} N, vertically down; (b) 1.95×10^{14} m/s², vertically down
11 19.4 N, 21.8° above the $+x$-axis
13 5 N/C
15 D
17 1.88 N/C, 20° below the horizontal to the right
19 173 N/C
21 (a) 6.24×10^4 N/C, 72.1° above the negative x-axis; (b) 5.98×10^{12} m/s², 72.1° above the negative x-axis
23 (a) 4500 N/C, 2250 N/C; (b) 329 N/C, 795 N/C
25 1.00×10^{-8} C
27 (a) 1.60×10^{-16} N, upward; (b) negligible; (c) 2.20 mm; (d) 8.78×10^5 m/s; (e) 1.98 cm
29 -5.1×10^5 C
31 2.00×10^{-2} N/C
33 5650 N/C, directed away from slab and perpendicular to it
35 (a) 0; (b) $4\pi k\sigma$, to the right; (c) 0
37 (a) 7.96 N/C 45.0° below the negative x-axis; (b) 7.96 N/C 45.0° below the positive x-axis

39 (a) 0; (b) T, R; (c) 8.85×10^{-7} C/m^2
41 10.0 cm, 0
43 8.1×10^{-4} mm
45 0.993 cm
47 -1.00×10^{-9} C
49 2.25×10^5 N/C, to the right

Chapter 18

1 (a) 100 eV = 1.60×10^{-17} J
(b) -100 eV = -1.60×10^{-17} J
3 (a) earth; (b) 2.00×10^9 J; (c) 231 days
5 (a) 10^4 V; (b) 0; (c) 0
7 1.44×10^{-9} m
9 0
11 3110 V
13 94.1 V
15 -80.0 V
17 1530 V
19 (a) 1.00×10^3 V; (b) 8.85×10^{-10} C
21 (a) 2 μF; (b) ½ μF
23 (a) 40.0 V; (b) 3.00 μF; (c) -20.0 μC
25 (a) 3 μC, 3 V; (b) 6 μC, 3 V
27 (a) 2.33 μF; (b) 20.0 V
29 5.33 V
31 5.00×10^{-5} J
33 6.00×10^{-2}μF
35 (b) 5.00 N
37 (a) 0.443 μF/cm^2; (b) 4.43×10^{-8} C/cm^2;
(c) 1.00×10^7 V/m
39 (a) 6×10^3 V; (b) 3×10^{-9} C; (c) 4×10^{-3} N;
(d) Only a small amount of charge flows.
41 (a) 2.26×10^{-4} m^3; (b) 226 m^3
43 (a) 0.200 V; (b) 8.00×10^{-3} s; (c) 125 Hz
45 176 beats/min
47 3.0 m
49 1.41 cm/s
51 4.45 m/s
53 5.22×10^4 V
55 (a) -4.51×10^5 C; (b) -6.37×10^8 V;
(c) 1.44×10^{14} J
57 (a) 5.0×10^{-7} m; (b) 1.1×10^{-3} m^3 = 1.1 liters
59 (a) 1.0×10^9 J, 5.0×10^9 W; (b) 0.83 s; (c) 230 m^3;
(d) 10 m^3

Chapter 19

1 (a) left; (b) 16.0 A
3 2.25×10^{23}
5 438 Ω
7 1.00 Ω
9 12 A
11 (a) 260 C°; (b) 2500 C°
13 (a) + 6 A, left; (b) 1.9 Ω
15 5.00 s
17 (a) 30 V; (b) 25 V; (c) being charged
19 (a) 2.6×10^6 J; (b) 5400 J
21 (a) 144 Ω; (b) 7.20 Ω
23 340 V
25 -10 A to the right
27 (a) 2.80 min; (b) 18.8 min
29 1.9×10^{-5} m/s
31 (a) 28 min; (b) 5.8×10^{-4} m
33 (a) 15; (b) 1 in 3.4×10^{14}; (c) 1 in 2.2 million
35 (a) 8.76×10^{-2} Ω; (b) 3.59×10^{-2} Ω; (c) 0.618 V;
(d) copper
37 3.66×10^{-2} mm, 1.11 m
39 5.00×10^{-5} A
41 6.22 A

Chapter 20

1 -2 A
3 $+6$ A, down
5 6 V
7 (a) 2.0 V; (b) 1.0 W; (c) 4.3 W; (d) 3.0 W
9 (a) + 0.33 A, clockwise; (b) 2.33 V
11

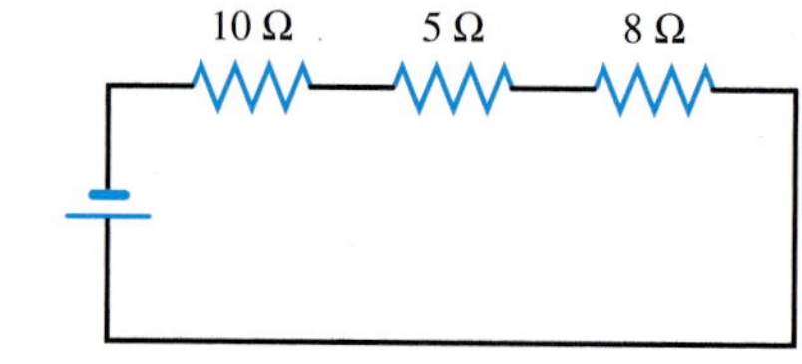

13 2.1 A
15 (a) 5.4 A; (b) 5.0 Ω
17 100 Ω
19 1.5 A
21 -5.4 V
23 (a) 0.257 A, upward; (b) 0.857 V
25 (a) 1.17 A to the left; (b) 0.081 A to the left
27 3 A, 2 A, 1 A
29 9.0 V
31 (a) 3×10^{-3} Ω, 3×10^{-3} Ω; (b) 10 μA, 1 A
33 1.00×10^4 Ω, 200 μF
35 (a) 2.0 A, down; (b) 1.0 C; (c) 1.6 A
37 4.54×10^{-5} A
39 (a) 280 μF, 3600 Ω, 1.7 mA; (b) 3.2 years
41 (a) 4700 Ω; (b) 4700 Ω-m; (c) 94,600 Ω
43 1.67 kW
45 (a) 0.012 V; (b) 24,000 V; (c) no
47 (a) 2.55×10^{-3} Ω; (b) 0.0764 V

49
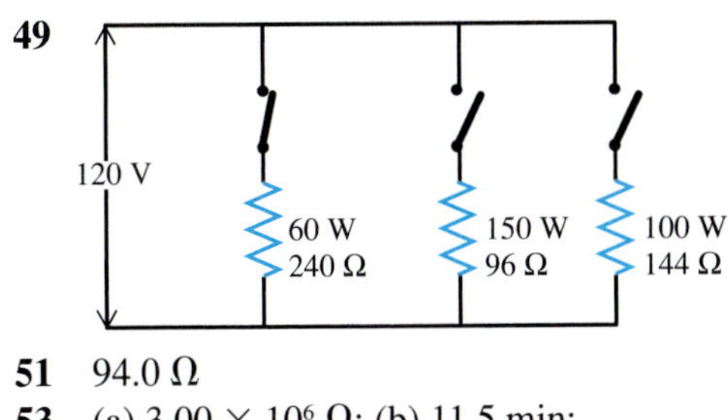

51 94.0 Ω
53 (a) 3.00×10^6 Ω; (b) 11.5 min; (c) 3.39×10^{12} Ω-m
55 (a) 100 mA; (b) 33.3 mA
57 (a) 2.55×10^{10} Ω; (b) 6370 Ω
59 (a) $\frac{1}{2} R_A (1 + \sqrt{1 + 4 R_M/R_A})$; (b) 4.50×10^6 Ω

Chapter 21

1 +, 0, −
3 2.72×10^{-17} N, directed east
5 3.13×10^7 m/s, directed in the x-z plane, 45.0° behind the + x-axis
7 (a) 0.800 N (out of the page on the right plate and into the page on the left plate); (b) 0; (c) 8.00×10^{-5} N-m
9 4.11×10^{-16} N, directed north
11 (a) 0; (b) 0.170 N, into the page; (c) 0.240 N, into the page; (d) 0; (e) 0.170 N, out of the page; (f) 0.240 N, out of the page
13 3.15 N, directed west
15 (a) 8.66 N, directed into the page; (b) 4.00×10^5 N, directed into the page (NOTE: This force would act only for a very brief interval — on the order of the time constant $RC = 10^{-9}$ s — and therefore would have very little effect.)
17 4.52 N-m
19 8.33 A
21 The proton describes a circular path of radius 3.48 cm in a horizontal plane. Viewed from above, the motion is clockwise.
23 (a) $v = E/B$; (b) 1.50×10^6 m/s; (c) they would bend upward if positively charged, downward if negatively charged; (d) they would bend downward if positively charged, upward if negatively charged.
25 30 cm
27 (a) 1.79×10^{-10} s; (b) + 1.02 cm, − 1.02 cm
29 The field in each case has a magnitude of 6.00×10^{-5} T and is perpendicular to the page. The directions are (a) inward; (b) outward; (c) inward; (d) inward.
31 2.50×10^{-3} T
33 2.14×10^{-5} T
35 (a) 8.0×10^{22} A-m², directed vertically downward; (b) 4.8×10^9 A, directed clockwise when viewed from the north pole
37 0.377 N
39 (a) 7.07×10^{-5} T, directed toward the lower left corner; (b) 5.31×10^{-4} N/m, directed away from the wire at the diagonally opposite corner
41 4.53×10^{-6} T in the x-z plane, directed 45° behind the positive x-axis.
43 (a) 7.96×10^5 A/m; (c) 2.21×10^6 W/m², 2.21×10^5 W/m², no
45 3.1×10^{-3} T
47 40 N-m
49 13 N, attractive
51 0.200 m
53 (a) 40.0 N; (b) 15.5 m/s

Chapter 22

1 (a) 0; (b) 8.0×10^{-4} T-m²; (c) 4.0×10^{-4} T-m²
3 (a) 1.25×10^{-5} T-m²; (b) 1.25×10^{-5} T-m²; (c) 0
5 (a) 0; (b) 0; (c) − 0.60 T-m²; (d) + 0.60 T-m²
7 − 30 V
9 79 A, clockwise
11 0.50 A, clockwise
13 (a) 0.313 T; (b) when it is entirely within the field, 2.50 s after it begins
15 102 V, 2.05 A
17 (a) 1.97×10^{-3} V; (b) reduces it
19 − 2.00 V
21 5.97 mA
23 (a) + 20 V; (b) − 60 V; (c) − 50 V; (d) + 48 V
25 3.00×10^3 A/s
27 0.250 J
29 (a) 9.80×10^5 J; (b) 0.0246
31 (a) 14.4 Ω; (b) 8.33 A; (c) 11.8 A; (d) 2,000 W; (e) 0
33 (a) 9.05×10^{-2} A; (b) 0.905 A; (c) 9.05 A
35 (a) 0.754 Ω; (b) 7.54 Ω; (c) 75.4 Ω
37 2.50 A, 66.3 V, 83.0 V
39 27.6 W
41 113 Hz
43 (a) 0.377 A; (b) a 5.63 mH inductor; (c) 6.67 A
45 83
47 38
49 (a) 2.0 A-m², directed outward; (b) 0
51 (a) 1.99×10^3; (b) 0.251 mm; (c) 275 Ω; (d) 3.64 ms
53 (a) 19.6 kW; (b) 240 V; (c) 81.7 A

Chapter 23

1 (a) radio waves; (b) microwaves; (c) infrared; (d) ultraviolet; (e) X rays
3 (a) 2.97 m; (b) 8.84×10^{-6} W/m^2; (c) 8.16×10^{-2} N/C
5 3.63×10^{26} W
7 408 m
9 1.67×10^{-2} s
11 8 years
13 2.00
15 750 nm
17 (a) 500 nm; (b) 333 nm
19 5.00×10^4
21 255 W/m^2
23 6.30×10^5 W/m^2
25 4.0 W/m^2; 92 W/m^2; 3.7 W/m^2
27 97.5%
29 2.40×10^{-8} s
31 22°
33 70°
35 (a) 28°, 28°, 45°; (b) no
37 0° to 48°
39 48.6°
41 from earth
43 1.50×10^8 m/s
45 684, 2.28×10^{-6} s
47 15.1°
49 1.58
51 2.3 m
53 (a) 0.33; (b) 1.77
55 (a) along a north-south axis; (b) 119 V

Chapter 24

1 C
3 (a) 7 ft; (b) 3.5 ft, no
5 7
7 (a) 2 m/s; (b) 4 m/s
9 B
11 (a) inverted; (b) yes; (c) left
15 (a) virtual, upright; (b) 1.7 cm
17 −8 mm
19 90 cm, 4.0 cm, inverted
21 10 cm, 25 cm, 25 cm
25 20 cm
27 2.00 m
29 (a) 5.0 cm; (b) 20 cm, 25 cm
31 2.8 mm, 15 mm
33 16 cm
39 −30 cm
41 0.10 d
43

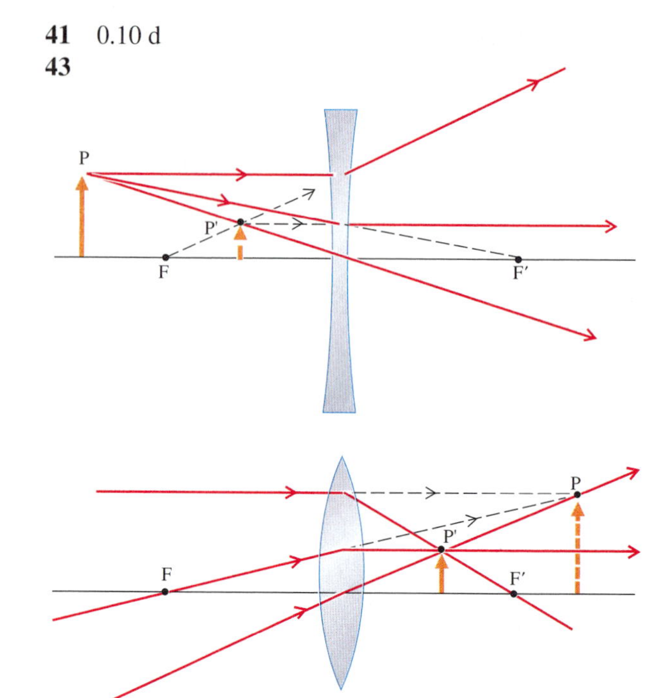

45 100 cm, −4.0
47 10 cm to the left, 5.0 cm
49 70 m
51 25 cm if camera lens is focused close-up, or any focal length from 20 cm to 25 cm for other focus settings.
53 5.6 cm to the left
55 2.00

Chapter 25

1 roughly 30 cm
3 18 mm
5 −4.0 d, −2.0 d
7 2.3 d
9 −2.0 m (virtual object)
11 5 cm to 25 cm
13 thicker at the edge
15 −2.0 d, 0
17 6.0 mm
19 5 mm
21 50 m
23 (a) 0.10 rad; (b) 0.14 rad
25 (a) 5.0; (b) 16 cm; (c) 1.5
27 (a) 0.10 rad; (b) 0.080 rad; (c) 0.20 rad; (d) 0.18 rad
29 6.3 cm
31 no
33 1.6
35 50

37 10^4
39 c, f
41 4 times
43 +46.3 cm
45 (b) 20 cm
47 85.7 cm

Chapter 26

1 (a) no; (b) yes, 0.1 N/C, directed upward
3 1.12×10^4 m
5 5.00×10^{-8} W/m^2
7 green
9 9
11 1.6 cm, orange
13 4/3
15 (a) 4450; (b) destructively
17 640 nm, 457 nm
19 3.81×10^{-2} mm
21 (a) constructively; (b) neither, there is no interference; (c) destructively
23 (a) 2.00×10^{-3} rad; (b) 4.00×10^{-3} rad
25 4.00×10^{-3} rad = 0.229°
27 (a) 14.5°, 30.0°, 48.6°, 90.0°; (b) 77.2°; (c) no direction
29 6.00×10^{-4} rad
31 6.10 cm
33 3.82 μm
35 (a) 1.82×10^{-4} rad; (b) 3.19×10^{-4} rad; (c) 6.07×10^{-5} rad; (d) 1.06×10^{-4} rad
37 580 nm, 435 nm
39 870
41 63.4°
43 89 W/m^2
45 33.7°
47 48.4°
49 546 nm
51 9
53 2.31×10^{-2} mm, 6.69 mm
55 280 m

Chapter 27

1 (a) 2.5 s; (b) 10 y
3 10.0 min
5 0.83 s
7 1.6 s
9 0.998*c*
11 (c) 0.161*c*
13 (a) 2.83 light-years; (b) 0.943 light-year
15 258 days
17 0.99995*c*
19 1.8×10^8 m/s
21 0.800*c*, toward earth
23 *c*, toward them; 0.714*c*, away from them
25 0.268*c*
27 1.1×10^{23} kg/s, a rate of nearly 2 solar masses per year
29 4.6×10^{-9} g
31 2×10^6
33 (a) 319 days; (b) 118 m/s^2. It should take considerably longer than found in (a).
35 $5.00 \times 10^{-16}\,c$

Chapter 28

1 3.11 eV to 1.78 eV
3 (a) 2.4×10^{-7} eV; (b) 1.3×10^{30}
5 1.2×10^{19}
7 (a) 7.55×10^{13}; (b) 3.18×10^{15}
9 1200 nm
11 0.34 eV
13 (a)none; (b) Na; (c) all
15 (a) 1.87 V; (b) sodium
17 (a) blue; (b) 2.63 eV
19 1.41×10^{-10} m
21 h/mc
23 100
25 (a) 1.05×10^{-27} kg-m/s, 2.21×10^{-38} kg-m/s; (b) 4770 m
27 (a) 0.100 mm; (b) 1.05×10^{-30} kg-m/s
29 6.63×10^{-26} J
31 (a) 3.87×10^{21}; (b) 0.0168 kg-m/s
33 (a) 3.46×10^{-5} m/s^2; (b) 1090 m/s

Chapter 29

1 486 nm, blue
3 122 nm
5 $n_i = 5$, $n_f = 1$
7 (a) 4.4×10^4; (b) 8.6×10^{13}
9 (a) 6.58×10^9 Hz; (b) 6.68×10^9 Hz
11 8.8×10^6 V
13 1.10×10^{-34} m
15 1.16×10^6 m/s
17 (a) 3.73×10^{-12} m; (b) 6.78×10^{-6}; (c) 147,000
19 10
21 0.031 nm
23 1.94 eV

Chapter 30

1 6.0×10^{26}
3 (a) 1.2×10^{-15} m; (b) 7.4×10^{-15} m
5 $^{18}_{8}O$
7 2.22 MeV and 1.11 MeV per nucleon
9 160 MeV
11 1.46×10^{-8} u
13 $^{220}_{86}Rn$
15 $^{238}_{92}U$
17 $^{223}_{88}Ra$
19 5.41 MeV
21 5.51 MeV
23 1.20 MeV
25 0.181 day^{-1}
27 2.50 h
29 7.43×10^{13}
31 1.68×10^{8} Bq, or 4.55 mCi
33 (a) 1.19×10^{-12} kg; (b) 7.30×10^{-6} kg; (c) 2960 kg
35 5320 y; no
37 17,100 y
39 (a) 3.64×10^{15}; (b) 1.00 mCi and 0
41 $^{0}_{1}n$
43 (a) 0.626 MeV; (b) 2.63 MeV
45 no
47 $^{133}_{51}Sb$
49 174 MeV
51 1 kg
53 19.8 MeV
55 (a) 30 kg; (b) 2×10^{4} kg; 6×10^{5} m^3
57 (a) 3.3 rad; (b) 33 rem
59 90 rem
61 1.8 rad
63 17.1 mg
65 95,000
67 61
69 54
71 (a) u, u, d; (b) u, d, d
73 193 y
75 2

Index

A

Aberration
  chromatic, 687, 688
    visual acuity and, 723, 724
  in lenses, 687–688, 690
  magnifying glass, 713
  spherical, 669, 687
    visual acuity and, 722–723
Absolute scale, 296, 297
Absolute temperature, entropy and, 350
Absorbed dose, 878, 889
Absorption
  polarization by, 748–749
  of radiation, 327–328
Acceleration
  angular, 216
    average, 216
    constant, 218
    instantaneous, 216
  average, 43, 59
    angular, 216
    on curved path, 82
    in xy plane, 65, 81
  at center of mass, 127, 131
  centripetal, 74, 217
    of earth, 151
    of moon, 145
    of satellite, 146
  circular motion and, 72–75, 370
  constant; *see* Constant acceleration
  on curved path, 65–66
  defined, 43
  of electrons
    on electric field, 439
    energy of, 783
  in equations, consistency of units and, 10
  in free fall, 49, 53–54
  friction in, 118–123
  gravitational; *see also* Gravitation
    of earth, 151
    at various points on earth, 49
  instantaneous, 44, 59
    circular motion and, 72, 73
    on curved path, 65, 82
  magnetic field and, 552–553
  in one dimension, 41–45, 60
  of particle, Newton's second law and, 92
  radial, 217
  rotational, 231
    moment of inertia and, 223
Acceleration—cont'd
  tangential, 217
    rotation about fixed axis and, 221
  zero, with no force acting on it, 90
Accelerators, 882–883
Accommodation, eye and, 700–703
Actin, 240
Action between bodies, 89
Action force, 94
Addition
  of components of vectors, 29–30
  of vectors, 24–25
Adiabatic process, 344, 358
Adjacent domains of ferromagnet, 573
Air
  dielectrics and, 476
  free fall straight line motion in, 53–54
  laser light in, 642
  refractive index of, 641
  specific heat of, 323
Air resistance
  escape velocity and, 176
  free fall and, 49, 53–54
  projectile motion and, 67
  running and, 187, 230
Airy disk, 744
Alcohols, specific heat of, 323
Algebra of vectors; *see* Vectors
Alnico, V., 573
Alpha decay, 861–862
Alpha particles, 822–823
  binding energy and, 859–860
  biological effects of, 877
  kinetic energy of, 862
  relative biological effectiveness of, 878
Alternating current circuits, 603–615, 625–626
  capacitor and, 605–608
  inductor and, 608–609
  RCL series circuit and, 609–611
  resistor and, 603–605
  resonance of, 612
  transformers and, 612–615
Alternating current source, 507–509, 512, 516
Aluminum
  atomic mass of, 856
  resistivity and temperature coefficient of, 497
Aluminum—cont'd
  specific heat of, 323
Americium, 873
Ammeter, 528
Amp; *see* Ampere
Ampere, 494
  magnetism and, 568
  volts per, 495
Ampère, André, 551
Amplitude, 366
Anderson, Carl, 881
Angle
  of incidence, 645
    critical, 649, 650
    plane mirrors and, 663
    in refraction at single spherical surface, 688–689
  phase, harmonic motion and, 367, 368
  of reflection, 645
    plane mirrors and, 663
  of refraction, 688–689
Angular acceleration, 216
  average, 216
  constant, 218
  instantaneous, 216
Angular frequency in circular motion, 369
Angular momentum
  in Bohr's model of atom, 828
    standing-wave explanation of, 834
  rotation and, 226–227, 231, 234–235
Angular velocity, 215
  average, 215
  in circular motion, 369
  instantaneous, 215
  running and, 228
Anticoronas in sky, 755
Antimatter, 881–882
Antinodes, standing waves and, 411, 415
Antiparticle, 882
Appliances
  electric circuits to, 536
  shock from, 537
Applications of physics, 3–4
Aqueous humor, 700
  light particle scattering in, 722
  refractive index of, 641
Arago study of diffraction, 741
Archimedes, 266

Archimedes' principle, 265–269, 286, 289–290
Aristotle, 56
  Newton's laws of motion and, 90
Arthroscopic surgery, 652
Astigmatism, 706, 725
Astronomical distance, 9
Atmospheric pressure, 259, 260
  with elevation, 264
Atom, 818–854
  Bohr model of, 819–830
  classical mechanics and, 835
  De Broglie wavelength and, 830–832
  early concepts of, 821
  electrons and, 821, 837, 838; *see also* Electron
  energy levels of, 828–830
    of sodium, 838
  ground state of, 837
  K shell of, 837
  L shell of, 837
  mass of selected, 856
  nucleus in, 822–825; *see also* Nucleus
  quantum mechanics and, 830–835
  spectrum of, 835–840
    Bohr model of, 819–821, 851, 852–853
    early concepts of, 821
    electrons and, 821
    nucleus and, 822–825
    Thomson's model and, 822
    visible, 838
    x-ray, 839–840, 853
  structure of, 835–840
  Thomson's model of, 822
Atomic clock, 8
Atomic energy levels, 828–830
  of sodium, 838
Atomic magnetic moments in paramagnet, 572
Atomic mass, 301
Atomic mass unit, 301, 855
Atomic number of element, 837, 855
Aurora borealis, 562, 563
Average acceleration, 43, 59
  angular, 216
  on curved path, 82
  in xy plane, 65, 81
Average collision force, 200
Average current, 494
Average density of earth, 146
Average potential difference, 474
Average power, 182, 190, 508
Average pressure in blood vessels, 283
Average speed, 19–20, 37
  center of mass, in running, 228
  of electrons, 511
Average velocity, 31–32, 40
  acceleration in one dimension and, 42, 59
  angular, 215
  constant acceleration and, 46
  instantaneous velocity and, 32–34
  linear motion and, 59
  over time interval, displacements and, 31, 35
Avogadro's number, 302
Axial tomography, computerized, 839
Axis
  optical, 666, 667
    principle rays and, 674
  of rotation
    fixed; *see* Fixed axis of rotation
    static equilibrium and, 238–239
  transmission, polarization and, 748
Axon, 532
Axon membrane, 549

B

Back emf, 597
Bacteria, magnetotactic, 574
Balance
  in measure of mass, 92
  stress and strain in, 244–247; *see also* Equilibrium
Balloons, hot-air, 300
Balmer, Johann, 820
Balmer series, 820, 831, 851
Bandwidths of light waves, 732
Barometer, 264–265
Baryons, 883
  quarks and, 884
Basal metabolic rate, 355–356
Base units, 7
  primary or secondary, 7
Batteries, 470, 503–504, 512, 516
  charging of, 517
  ideal, 506
Beats
  interference of sound waves and, 409
  superposition of waves and, 408–409, 420, 424–425
Beauty quark, 885
Becquerel, 863
Benzene, refractive index of, 641
Berkelium, 873
Bernoulli, Daniel, 277
Bernoulli's equation, 273, 276–279, 286, 291–292
  continuity equation and, 275
  Poiseuille's law compared to, 284
  steady flow and, 273–274
Beryllium, atomic mass of, 856
Beta decay, 862–863
Beta particle, relative biological effectiveness of, 878
Bimetallic strip in thermal expansion, 314
Binding energy, 858–860, 888
Biological effects of radiation, 877–880, 889, 892–893
Biomagnetism, 574
Biot, 551
Biot-Savart law, 563–564
Blackbody radiation, 799–800
Blind spot, 727
Blood flow
  heat evaporation and, 330
  Poiseuille's law and, 283
Blood pressure, 284–285
Blue color force, 885
Bodies
  action between, 89
  center of mass of; *see* Center of mass
  electrical effects on, 532–533, 534–537, 547–549
  free, action on, 102–108
  Newton's laws of motion in force acting on, 93, 102
  rigid
    center of mass of, 129
    torque and, 246
    velocity of, 215–216
  rotating; *see* Rotation
  specific heat of, 323
  symmetric, center of mass of, 129
  two or more extended, center of mass of, 129
  weight of, gravitational attraction and, 144, 158–159
Body fluids, resistivities and temperature coefficients of, 497
Bohr, Niels, 825
Bohr's model of atom, 825–828
  energy levels of, 828–830
  standing-wave explanation of angular momentum postulate of, 834
Boiling, 311
Boltzmann, Ludwig, 303, 308
Boltzmann's constant, 298, 315
  entropy and, 350
Bottom quark, 885
Bound states of electron, 826
Boyle, Robert, 299
Boyle's law, 299
Brackett series, 821, 831, 851
Brahe, Tycho, 153
Brain waves, oscilloscope measurement of, 482
Branch circuits, 536, 537
Branch points, 519
Brass, thermal expansion and, 314

Bremsstrahlung radiation, 840
Brewster's angle for glass, 751
British system of units, 7
Bubble chamber, 870
Buoyant force, 265
  fluid displacement and, 266

C

Cable, coaxial, 455
Calculation, significant figures in, 10–11
Californium, 873
Caloric theory of heat, 320–321
Calorie, 321
Calorimetry, 321–325, 334
  phase transitions in, 323
Cameras, lenses for; *see* Lenses
Capacitance, 468–475, 483, 487–489
Capacitive reactance of capacitor, 606–607
Capacitor, 468
  in alternating current circuits, 605–608
  capacitive reactance of, 606–607
  charging, 470–471
    alternating discharging and, 530–531
    in direct current circuits, 529–530
  discharging, in direct current circuits, 530–531
  electrolytic, 492
  energy stored by, 474–475
  equivalent, 471
  field energy of charged, 475
  parallel, 471–472
  parallel-plate, 469–470
  potential difference and, 470, 471
    energy storage and, 474
  in RCL circuits, 609–611
  series, 472–473
Capillarity of fluids, 270–273, 286, 290–291
  surface tension and, 272–273, 290–291
Capillary tube, 272
Capture, electron, 863
Carbon
  atomic mass of, 301, 856
  resistivity and temperature coefficient of, 497
Carbon 14, kinetic energy of beta decay of, 863
Carbon dating, 867–868
Carbon dioxide, refractive index of, 641
Carbon tetrachloride, refractive index of, 641
Carnot, Sadi, 353
Carnot cycle, 354, 358
Carnot engine, 354–355, 358
Carnot refrigerator, 364
Carnot's theorem, 354
Cartesian coordinates
  in particle trajectory, 16–17
  straight line motion and, 41
Cathode ray tube, 479, 480
Cavendish, Henry, 143
Cavendish apparatus, 143
Cavendish balance, 143
Celsius scale, 296, 297, 315
Center
  of curvature, 666
    graphical method for, 685
    principal rays and, 674
  of gravity, static equilibrium and, 242–243, 252–254
  of mass, 127–131, 132, 135–136
    coordinates of, 128
    definition of, 128–130
    derivation of dynamics of, 130–131
    energy of, in isolated system, 341, 358
    kinetic energy and, 186, 225
    motion of, 127
    running and, 186, 228, 229
    system moving as single particle in, 127
  of oscillation, 385
  of percussion, 385
Centrifugal force, 148
Centripetal acceleration, 74, 217
  of earth, 151
  of moon, 145
  of satellite, 146
Centripetal force, 125–126, 132, 134–135
Cern accelerator, 882
Certainty in measurement, 10–11
Cesium in definition of second, 8
Chain reaction, 874, 875
Characteristic x-rays, 840
Charge, electric; *see* Electric charge
Charged ring, electric field of, 440
Charging
  of batteries, 517
  of capacitor, 470–471
    in direct current circuits, 529–530
Charmed quark, 885
Chemical energy, 182
  electrical potential energy and, 504
Chromatic aberrations, 687, 688
  visual acuity and, 723, 724
Ciliary muscles, 700, 702
Circle
  circumference of, 21
  of confusion, 700
Circuit breaker, 536
Circuits
  alternating current, 603–615; *see also* Alternating current circuits
Circuits—cont'd
  direct current, 518–549; *see also* Direct current circuits
  multiloop, 526–527, 538, 544–545
  open, 519
  RC, 529–531
  RCL, 609–611
  RL, 599–601
  short, 519
Circular aperture diffraction, 744
Circular current loop, 564–565
Circular motion
  harmonic, 369–371, 383
  in plane, 72–75, 83
Circulatory system, Poiseuille's law and, 283
Circumference of circle, 21
Classical mechanics, 88
  explaining macroscopic phenomena, 835
Clausius, Rudolph, 303, 304
Clausius' statement of thermodynamics, 355, 358
Clocks, 766–767
  atomic, 8
  light, 770–773
  moving, 772–773
Cloud chamber, 870
  in discovery of positron, 881–882
Coal-powered plant, 596
Coaxial cable, 455
Coefficients
  of expansion, 311–312, 315
  of static friction, 120–121
Coherence of light sources, 733–735
Coherence length, 734
Coherent waves, 406
Coils
  in electromagnets, 572; *see also* Inductors
    electrical potential energy and, 596
  Helmholtz, 585
  magnetic forces and, 557
Cold fusion, 14–15
Collision, 204–208
  average force of, 200
  elastic
    defined, 204, 209
    in one dimension, 205–207
    in two dimensions, 207
  inelastic, 208
    completely, 205
    defined, 205, 209
  kinetic energy and, 211–212
  ordinary, 349
  reversed, 349
Collision force, average, 200

Color
in nuclear physics, 885
in visible light spectrum, 632, 721–722
Color blindness, 722
Color charge, 885
Color force, 884
Color sensitivity, 719–722
Combustion, heat from, 324
Completely inelastic collisions, 205
Components of vectors, 27–30
Compound microscope, 713, 725
Compton, Arthur, 803
Compton effect, 803–804
Compton scattering, 803
Computerized axial tomography, 839
Concave spherical mirrors, 665–669, 690, 693–694
graphical method and, 674, 675
Conducting plate, electric charge in, 442–443
Conduction, 321, 330–332, 333, 336
Conduction electrons, motion of, 510
Conductivity, thermal, 331
Conductors of electric charge, 429–430
magnetism and, 554–559, 582–584
U-shaped conductors in, 592–593
properties of, 443–444
surface of, 464
uncharged, charge distribution in, 430
Cones in retina, 719
visual acuity and, 722–723, 724
Conservation of energy; *see* Energy, conservation of
Conservative force, 180–182, 195
spring potential energy and, 178, 194
Constant
decay, 864
of proportionality, 280
Stefan-Boltzmann, 333
universal gravitational, 141
Constant $\mathcal{E}$, 590
Constant acceleration
angular, 218
average and instantaneous velocity in, 59
in free fall, 49
friction and, 118–123
linear motion at, 45–48, 118–123
universal gravitation and, 144
Constant force
gravitational, 169–173, 192–193
in path-independent work, 172–173
work done by
gravitational potential energy and, 169–170
in one dimension, 163
Constant velocity, graph of linear motion at, 54–55
Constructive interference, 406–407, 419
light waves and, 736–741, 760–762
wave-particle duality of, 804–806, 816
Contact angle of fluids, 272
Contact force, 89
Continuity equation, Bernoulli's equation and, 275
Continuous distribution of electric charge, 440–444, 447, 453–454
Convection, 321, 329–330, 333, 335
forced, 329
Convection coefficient, 329
Converging mirror, 666
Converging surfaces of lenses, 676
Conversions of units, 9
Convex spherical mirrors, 671, 690, 693–694
graphical method and, 674
linear magnification in, 672
Coordinates
Cartesian, in particle trajectory, 16–17
of center of mass, 128
Copernican system of universe, 152–153
Copernicus, Nicolaus, 1, 152–153
Copper
electric charge and, 430
magnetic field induction of, 589
resistivity and temperature coefficient of, 497
specific heat of, 323
Cornea, 699
myopia and, 703–704, 725, 727
refractive index of, 641
Coronas in sky, 450, 755
Correction of vision, 703–707
Cosmic rays, 882
Coulomb, 430
magnetism and, 568
nuclear force and, 858
Coulomb's force as conservative force, 458
Coulomb's law, 431–433, 447, 451–452
Counters of radioactive decay, 870
Critical angle of incidence, 649, 650
Critical damping, 378
Critical temperature of ferromagnet, 573
Crystalline lens, 700
refractive index of, 641
Curie, 863
Curie, Pierre and Marie, 879
Curium, 873
Current, electric, 493–517; *see also* Electric current
Curvature
center of, 666
Curvature—cont'd
center of—cont'd
graphical method for, 685
principal rays and, 674
radius of
in spherical mirrors, 671
spherical refracting surface and, 677
Curved path, acceleration on, 65–66
Cycle, 344
Carnot, 354
of motion
frequency and, 368
simple harmonic motion and, 367
Otto, 348
Cyclotron, 560
Cylinder, rotation axis of, 223

D

Dalton's law of partial pressures, 300–301, 315
Damped oscillations, 378, 384
Damping, 378
critical, 378
Daughter nucleus, 861
in decay series, 868
da Vinci, Leonardo, 94
Davisson, Clinton, 831
De Broglie, Louis, 830
De Broglie wavelength, 830–832
Decay
alpha, 861–862
beta, 862–863
gamma, 864
radioactive, 861–870, 888, 891–892
counters of, 870
rates of, 864–867
Decay constant, 864
Decay series, 868, 869
Deceleration, 42
Decibel scale, 399–403, 419, 423–424
Deep water, speed of waves in, 394
Deflection plates, vertical, 479–481
Democritus, 2
Density, 258–259, 286
average, of earth, 146
in electric charge, 441
energy, 475
of gas molecules, velocity and, 308
of nucleus, 857
Reynolds number and, 281
of water, temperature and, 314
Depth of fluid, pressure and, 261–262
De Revolutionibus, 1
Derivations, 4
Derived forces, 98
Derived units, 7
Descartes, 154, 155

Destructive interference, 406–407, 419
light waves and, 736–741, 760–762
wave-particle duality of, 804–806, 816
Detector, metal, 621
Dew, 310
Dew point, 310
Diagram, free-body, 102
Diamagnetic materials, 572
Diameter, instantaneous speed and, 21
Diamond, sparkle of, 651
Dielectric constant, 477–478
of electrolytic capacitor, 492
Dielectric strength, 476
Dielectrics, 429, 476–478, 483, 489–490
of capacitor, 492
Diffraction, 643–644, 741–748, 757, 763–764
circular aperture, 744
of electrons, 831
Fraunhofer and Fresnel, 741, 742, 744, 757
gratings of, 746–748, 758
helium-neon laser patterns of, 746–747
historical background of, 741
of light by eye, 745
light wave properties and, 735–736, 757, 759
microscope limit of, 746
number of photons in, 806
Rayleigh criterion for resolution and, 745–746
single-slit, 742–744
sodium doublet, 747
telescopes and, 717
visual acuity and, 722, 724
x-ray, 748
Diopter, 681
Dipole, electric, 466–467
Dipole moment, 448, 478
Dirac, Paul, 881
Direct current, 507
Direct current circuits, 518–549
current, potential difference, and resistance measurement in, 528–529, 538, 545–548
description of, 519
discharging capacitor in, 530–531
in electrical effects on human body, 532–533
from household electricity, 534–537, 547–549
equivalent resistance in, 523–526, 538, 541–544
Kirchhoff's rules in, 520–522, 538, 541–542
multiloop circuits in, 526–527, 538, 544–545
Direct current circuits—cont'd
potential difference and resistance measurement in, 528–529
RC circuit in, 529–531
Discharge tubes for light, 626, 635
atomic spectra and, 819
Discharging capacitor in direct current circuits, 530–531
Dispersion, 390
of light, 641–642
Displacement, 22–23, 38
addition and, 22
and average velocity over time interval, 31, 35
of particle of medium, time dependence of, 404–405, 424
relative, 76
simple harmonic motion and, 368
standing waves and, 411
total, 22–23
Distance
action on bodies at, 89
astronomical, 9
length contraction and, 776–778, 795
in equations, consistency of units and, 10
of object
spherical refracting surface and, 677
virtual objects and, 670
projectile motion and, 70–71
traveled at high speed, length contraction and, 776–778, 795
traveled by light in measurement of meter, 8
universal gravitation and, 145
Diverging mirror, 671
Diverging surfaces of lenses, 676
Division, significant figures in, 11
Domains of ferromagnet, 573
Dominoes, waves and, 387
Doppler effect, 395–399, 419, 423
electromagnetic, 397
moving source and stationary observer and, 396–397
relativistic, 794–795
stationary source and moving observer and, 395
supersonic speeds and, 398–399
Dose
absorbed, 878, 889
equivalent, 878
Double slit of Young, interference and, 737–739, 760–762
Double slit interference pattern, wave-particle duality and, 805
Down quark, 885
Dry cell battery, 503
du Fay, Charles, 428
Dust particle, uncertainties in, 833–834
Dynamics, 87
fluid, 273–279, 286, 291–292
particle, 87–138; *see also* Particle dynamics
Dyne, 93

E

Ear, sound and, 418; *see also* Sound
Earth
centripetal acceleration of, 151
electric charge on surface of, 461–462
gravitation attraction of, 144–147, 156; *see also* Gravitational potential energy
gravitational acceleration of, 151
gravitational field of, 437
heating of atmosphere of, 364
mass and average density of, 135, 146
as noninertial reference frame, 151, 160
Van Allen belts and, 562
Eclipse, solar, 659
"Eightfold Way," 884
Einstein, Albert, 765, 788–789
Einstein's postulates, 767, 793
experimental support for, 770
first postulate in, 767, 790
second postulate in, 768–769, 790
Einsteinium, 873
Elastic collisions, 205–207, 209
defined, 204
Electric charge, 427–431, 447, 452–453; *see also* Electric current
of capacitor, 470–471
in direct current circuits, 529–530
conservation of, 430
Kirchoff's rules and, 520
continuous distribution of, 440–444, 447, 452–453
density in, 441
electromagnetic force and, 880
field lines and, 444–446, 447, 454–455
historical observations of, 427–428
like or unlike, 429
moving, 509–510
negative and positive, 494
oscillation of electromagnetic waves and, 630
point source of, 437–439; *see also* Electrical potential energy, point charge of
magnetic field and, 560–562
positive, 435
electric potential and, 462
magnetic field and, 552
potential difference and, 468–469
of quark, 885

Electric charge—cont'd
in rings, electric field of, 440
in series capacitor, 472
single source of, 435–437
three, 439
two point, 431–432, 438
uniformly charged infinite plane in, 441–442
units of, 430–431
Electric current, 493–517; *see also* Electric charge
alternating, 507–509
ammeter in measuring, 528
average, 494
direct, 507
in direct current circuits, 518–549; *see also* Direct current circuits
electric power and, 500–509, 512, 516
alternating current source in, 507–509, 512, 516
batteries in, 503–504, 512, 516
emf in, 504–505
loss of, 501
voltage drop across emf source of, 506–507
induced, magnetic field and, 591, 592
instantaneous, 494
lagging behind voltage, 608
leading voltage, 606
magnetism and, 563–569
in circular loop, 564–565
conductors in, 554–559, 582–584
long-straight wire in, 566
measurement of, 528–529, 538, 545–546
moving charges and, 509–510
negative and positive, 494
Ohm's law and, 495–500, 512,514–515
macroscopic form of, 511
microscopic level of, 509–511, 516–517
in phase with voltage, 604
root-mean-square, 604
shock from, 534–537
steady-state, 519
through network, 525
time constant measure of, 600
wire carrying uniform, 511
Electric dipole, 466–467
Electric dipole moment, 448
Electric energy conversion to thermal energy, 500
Electric field, 427–456, 447, 452–453
acceleration of electron on, 439
charges in, 427–431
in charged ring, 440
continuous distribution of, 440–444, 447, 452–453
point source group of, 437–439
Electric field—cont'd
charges in—cont'd
of sheet of photocopy paper, 442
single source of, 435–437
three, 439
two, 438
uniform, 445, 446, 460–463
concept of, 434
conductor properties and, 443–444
Coulomb's law and, 431–433, 447, 451–452
definition of, 434–435
in light polarization, 731–732
lines of, 444–446, 447, 454–455
two-dimensional drawings of, 446
in sunlight, 631
uniform
dielectrics in, 478
in infinite plane, 445
potential difference in, 460–463
of sphere, 446
Electric motor
magnetic forces and, 557–559
mechanical energy and, 182
Electric potential, 457–492
capacitors and, 468–475
charging, 470–471
energy stored by, 474–475
parallel, 471–472
parallel-plate, 469–470
series, 472–473
definition of, 458–460
dielectrics and, 476–478, 483, 489–490
electrical potential energy in; *see* Electrical potential energy
in equipotential surfaces, 464
field energy and, 475
oscilloscope and, 478–482, 483, 490
point charges and, 584–585; *see also* Electrical potential energy, point charge of
several, 466–468
single, 465–466
in uniform electric field, 460–463
Electric power, 500–509, 512, 516
loss of, 501
source of
alternating current, 507–509, 512, 516
battery, 503–504, 512, 516
emf, 504–505
voltage drop across emf, 506–507
Electric storms, 455
Electrical effects on human body, 532–533
in shock from household electricity, 534–537, 547–549
Electrical forces of repulsion, 451
Electrical potential energy, 458–468, 483, 486–487
Electrical potential energy—cont'd
chemical energy and, 504
definition of, 458–460
generators and, 596
magnetic energy and, 601–603
point charge of; *see also* Electric potential, point charges and
magnetism and, 560–562, 584–585
nonuniform motion and, 562, 563
several, 466–468
single, 465–466
uniform motion and, 560–562
positive and negative charges in, 462
power loss and, 501
reference level for, 459
Electrical resistance, 495–496
equivalent, in direct current circuits, 523–526
measurement of, 528–529
temperature coefficient of, 497
units of, 495
Electrical wire; *see* Wire, electrical
Electricity, household, 532–533
shock from, 534–537, 547–549
Electrocardiogram, 533
Electroencephalogram, 482, 533
Electrolytic capacitor, 492
Electromagnet, 572
Electromagnetic Doppler effect, 397
Electromagnetic fields, defined, 551, 576, 581–582, 588
Electromagnetic force, 98
fundamental forces and, 880
proton as mediator of, 884
Electromagnetic induction, 588–627
in alternating current circuits, 603–615, 625–626
capacitor and, 605–608
inductor and, 608–609
RCL series circuit and, 609–611
resistor and, 603–605
resonance of, 612
transformers and, 612–615
Faraday's law of, 589–596, 618, 622–624
constant $\mathcal{E}$ and, 590
experiments with, 589
generators and, 594–596
Lenz's law and, 591–592
magnetic flux and, 590
motional emf and, 592–593
variable $\mathcal{E}$ and, 593
inductance in, 597–603, 618, 624
inductors and, 598–599
magnetic energy and, 601–603
RL circuits and, 599–601
self-induction and, 597–598
Michael Faraday and, 616–617

Electromagnetic radiation, 326
Compton effect and, 803
visible light spectrum and, 632
Electromagnetic waves
light as, 629–632, 656–657
intensity of wave of, 806
principle of relativity and, 769
produced by distant source, 630
Electron
acceleration of
on electric field, 439
energy of, 783
after Compton scatterings, 804
atoms and
Bohr model of, 821
configurations of, 837, 838
average speed of, 511
bound states of, 826
capture of, 863
charge of, 430
conduction, 510
energy of
accelerated, 783
after Compton scatterings, 804
excited states of, 827
magnetism and, 569, 572–573
mass of, 856
moving, 781
momentum of, after Compton scatterings, 804
moving
mass of, 781
producing current, 494
through metal, 510
orbits of, 825–826
proportional to intensity of wave, 832
transfer of, net charge in, 428
uncertainties in position of, 833–834
wave properties of, 832, 853
wavelength of, 832
in x-ray tube, 839
Electron field, 434
Electron lepton, 883
Electron microscopes, 463, 715
scanning, 719
Electron neutrino, 883
Electron volt, 462
Electron wave packets, 832
Electrostatic forces, 429
total, as conservative forces, 458
Electrostatics, defined, 427, 551, 588
Elementary particles, 880–885, 893
fundamental forces and
in 1930's, 880–881
in 1990's, 884–885
nuclear physics and, 854–893; *see also* Nuclear physics and elementary particles
Elementary particles—cont'd
sources of, 882–883
Elements
atomic number of, 837
in nuclear reactions, transuranic, 873
Emf, 504–505
back, 597
motional, 592–593
negative or positive, 590
self-induced, 597–598
source of
alternating current generator as, 595, 603
ideal, 505
power provided by, 505
resistor and, 603
voltage drop across, 506–507
Emission
of positron, 863
of radiation, 327
quantum theory and, 835
Emissivity of surface, 327
Endothermic reaction, 871
Energy, 162–198
of accelerated electron, 783
in atom
electrons and, 828–830
levels of, 828–830
binding, nuclear structure and, 858–860, 888
calorie as unit of, 321–322
center-of-mass, 341, 358
chemical, 182
concept of, 162
conservation of, 170–172
Kirchoff's rules and, 520
law of, 162
spring force in, 178–179, 194
conservative and nonconservative forces of, 180–182, 195
electrical potential; *see* Electrical potential energy
of electron after Compton scatterings, 804
graphical interpretation of, 167–168
gravitational potential, 169–177
constant force in, 169–173
internal
food and, 356–357
heat and, 321
isolated system and, 341, 358
in thermodynamic systems, 340–341
as thermodynamic variable, 339
kinetic; *see* Kinetic energy
magnetic
electromagnetic inductance and, 601–603, 618, 624
per unit volume, 602
Energy—cont'd
mass, 782
nuclear reactions and, 871
spring harmonic motion and, 372–373
mechanical; *see* Mechanical energy
of particle
in accelerator's beam, 882
in system of particles, 187–189, 196
of pendulum, 377
of photon, 800
potential; *see* Potential energy
power and, 182–184, 190, 195
quantum concepts and, 809
reaction, 871, 889
relativistic, 781–783, 795–796
rest, 782
in running, 184–187, 228–230, 235
air resistance and, 230
gravitational potential, 229
kinetic, 228
rotation and, 228–230
translational kinetic, 229
storage of
by capacitor, 474–475
for electrical power companies, 603
sun supply of, 876
thermal, 182
work and, 163–168
Energy density, 475
Engines
Carnot, thermodynamics of, 354–355
gasoline, as heat engine, 348
heat, efficiency of, 346
steam, 338
Carnot, 354
Entropy
of monatomic ideal gas, 350
thermodynamics and, 350, 358
second law of, 351–353
Equations
Bernoulli's, 273–279, 286, 291–292
consistency of units in, 10
kinematic, 47
of motion, 47
Equilibrium
rotational, 237, 238
static; *see* Static equilibrium
thermal, 296
thermodynamic system and, 339, 358
translational, 237, 238
Equipotential surfaces in electric potential, 464
Equivalent capacitor, 471
Equivalent dose, 878
Equivalent resistance
in direct current circuits, 523–526, 538, 542–544
of network, 525

Equivalent resistance—cont'd
of parallel resistors, 524, 538
of series resistors, 523, 538
Erg, 163
Escape velocity in gravitational potential energy, 176–177
Estimates, order-of-magnitude, 11–12
Eta meson, 883
"The ether," 770
Ethyl alcohol
refractive index of, 641
specific heat of, 323
Evaporation, 309
heat, 323
Excited states of electron, 827
Exponent of ten, 9
Extended source of light, 643
Eye, 699–710
accommodation in, 700–703
diffraction of light by, 745, 757, 763–764
extended objects in, 707–708
image formation defects in, 703–707
magnifier for, 710–713, 725, 728–729, 730
microscope and, 713–715, 725, 729
out-of-focus, 700, 701
photosensitive receptors of, 698
reduced, 707–708
refractive indices of, 641
sensitivity of
color, 719–722
to light, 633
structure of, 699–700, 719
telescope and, 716–718, 725, 730
vision correction in, 703–707
visual acuity of, 709–710, 722–724, 725, 728, 730
Eyeglasses
lenses for; *see* Lenses
reading, 706
Eyepiece of microscope, 713–714

F

Fahrenheit scale, 296, 297, 298
Falling, straight line motion in, 49–52, 59
Far point, 700, 704
Faraday, Michael, 589, 616–617
drawing of first generator of, 588
Faraday's law, 589–596, 618, 622–624
constant $\mathcal{E}$ and, 590
generators and, 594–596
induction experiments and, 589
inductors and, 599
Lenz's law and, 591–592
magnetic flux and, 590
motional emf and, 592–593
variable $\mathcal{E}$ and, 593
Farsightedness, 705–706
Fata morgana, 755
Fermi, Enrico, 873
Fermilab accelerator, 882
Fermium, 873
Ferromagnetic materials, 572–573
Fiber optics, 651–652
Fibrillation, ventricular, 534
Field energy of charged capacitor, 475
Field point, 435
Films, thin, interference and, 739–741
First harmonic, 415
First law of thermodynamics, 341–345, 358, 359–361
Fission, nuclear, 873–874, 875, 892
Fixed axis of rotation, 214, 221–224, 231, 233–234
moment of inertia and, 222–224
static equilibrium and, 238–239
torque and, 219, 221–222
Flow rate
in blood vessels, 283
continuity equation and, 275
defined, 274
Poiseuille's law and, 282, 283
turbulent, 280, 281
Flow tube, 275
continuity equation and, 275
Poiseuille's law and, 282
pressure at two points in fluid in, 276
Reynolds number and, 281
Fluids, 257–294
Archimedes' principle and, 265–269, 286, 289–290
Bernoulli's equation and, 273–279, 286, 291–292
capillarity and, 270–273, 286, 290–291
defined, 257
displacement of, 266
dynamics of, 273–279, 286, 291–292
Poiseuille's law and, 282–285, 286, 292–294
pressure in, 286
at rest, 258–259, 286, 288–289
properties of, 257–259
surface tension and, 270–273, 286, 290–291
viscosity of, 279–282, 282, 286
Fluorescence of atom, 839
Flux, magnetic
Faraday's law and, 590, 593
Lenz's law and, 591
Focal length, 666
linear magnification and, 684
optical power and, 681–682
telescopes and, 716–718
Focal point, 666
far point and, 704
graphical method for, 685
magnifying glass and, 712–713
principle rays and, 674
Focus, eye accommodation for, 700, 701
Fog, 310
Food
calories in, 322
combustion heat and, 324
metabolic rate and, 356–357
Foot-pound, 163
Force, 88–89, 109
action of, 94
in pairs on different bodies, 93
buoyant, 265
fluid displacement and, 266
centrifugal, 148
centripetal, 125–126, 132, 134–135
collision; *see* Collision
color, 884
conservative; *see* Conservative force
constant; *see* Constant force
contact, 89
definition of work and, 163
derived, 98
electrical
of charge, 433
of repulsion, 451
electromagnetic, 98
fundamental forces and, 880
proton as mediator of, 884
electrostatic, 429
of fluid on surface, 259
in free body's acceleration, 102
frictional; *see* Friction
fundamental, 98
color force as, 884
elementary particles and, 880–881
gravitational; *see* Gravitational force
impulsive, 200
in interaction between two bodies, 89
kinetic, 119
magnitude of
of static friction, 132
torque and, 219
Newton's laws of motion in, 93, 102–108
nonconservative, 180–182, 190, 195
normal, 119
nuclear; *see* Nuclear force
particles and, 140
elementary, 880–881
total gravitational force in, 141
pressure of gas molecules and, 303–304
reaction, 88, 94
spring; *see* Spring force

Force—cont'd
  static
    in electrostatic force, 429
    equilibrium and, 237
    of friction, 132
  surface, 118
  tension and, 99
  units of, 101
  variable
    gravitational, 173–177
    in three dimensions, 167
  viscosity and, 280
  weight on earth and, 96–97, 98
Force laws, 96–101, 109, 112–113
  gravity and, 98, 139–143, 156; *see also* Gravitational force
Forced convection, 329
Forced oscillations, 378, 384
Fovea, 700
Foveal cones in retina, 719
  visual acuity and, 722–723, 724, 725, 728
Fracture of material, 248
Frames, reference; *see* Reference frames
Franklin, Benjamin, 821
  one-fluid theory of, 428
Fraunhofer diffraction, 742, 757
  circular aperture and, 744
Free body, forces acting on, 102–108
Free fall in straight line motion, 49–52, 59, 61
Free-body diagram, 102
Frequency
  angular, in circular motion, 369, 381
  fundamental, 415
  lowest resonant, 415
  of periodic motion, 368, 381
  resonant, 379, 412–413, 612
    lowest, 415
  of sound from flute, 417
  of waves, 389–390
    periodic, 389
    spectral colors and, 632
    wave source and, 390
Frequency bandwidths of light waves, 732
Fresnel study of diffraction, 741, 742
Friction, 118–123, 133–134
  centripetal force and, 125–126
  loss of mechanical energy and, 182
  maximum force of, 119, 120
  static, 119
    coefficient of, 120–121
    magnitude of force of, 132
  work done by, 180
Frisch, Otto, 873
Fundamental forces, 98
  color force as, 884
Fundamental forces—cont'd
  elementary particles and, 880–881
Fundamental frequency, 415
Fundamental quantities, 6–7
Fusion
  cold, 14–15
  nuclear, 874–877, 892

G

G; *see* Gravitational force
Galileo Galilei, 2, 4
  Newton's laws of motion and, 90
  on scaling of structures, 247
  speed of light measurements and, 638
  straight line motion study by, 56–58
Galileo's telescope, 716
Galvani's experiment, 503
Galvanometer, 545, 583
Gamma decay, 864
Gamma ray, 326
  biological uses of, 877–880
Gap, nerve, 532
Gas
  ideal
    adiabatic or isothermal process and, 344, 358
    Carnot cycle and, 354
    kinetic energy of, 162, 163–168
    model of, 303–306, 317
    monatomic, entropy of, 350
    translational kinetic energy of, 340
  molecule of
    distribution of speeds for, 304, 305
    kinetic energy of, 162, 163–168
    Newton's laws of motion for, 306
    path of, in container, 306–308
    pressure of, against container wall, 303–304
    velocity of, in container, 307, 308
  refractive index of, 641
Gas discharge tubes, 626, 635
  atomic spectra and, 819
Gas law, ideal, 298–303, 315, 316–317
  derivation of, 306–309
Gasoline engine, 348
Gauge pressure of fluids, 265
Gauss, 552
Gay-Lussac, Joseph, 299
Geiger, Hans, 823
Geiger counter, 870
Gell-Mann, Murray, 884
Generator
  emf source and, 603
  Faraday's drawing of, 588
  Faraday's law and, 594–596
  van de Graff, 450
Geometrical optics, 662–697
  diffraction and, 643–644
  lenses and, 676–689, 690, 694–696
    aberrations in, 687–688
    derivation of single-surface refraction in, 688–689
    graphical method of, 685–686
    linear magnification of, 683–685
    negative, 681
    optical power of, 681–683
    positive, 680
    refraction at spherical surface of, 677–678, 694–695
    thin, 679–680
  plane mirrors in, 663–665, 690, 692–693
  spherical mirrors in, 665–675, 690, 693–694
    concave, 665–669
    convex, 671
    graphical method and, 674–675
    linear magnification and, 672–673
    virtual images and, 670
    virtual objects and, 670
Germanium, 497
Germer, Lester, 831
Ghost images, 656
Gilbert, William, 428
Glass
  Brewster's angle for, 751
  dielectrics and, 476
  dispersion curve for, 642
  laser light in, 642
  Pyrex, 476
  reflection and refraction of light in, 647
  resistivity and temperature coefficient of, 497
  specific heat of, 323
Glass prism, 637
  separation of colors with, 647
Glory, 755
Gluons, 884
Goggles and underwater vision, 706
Gold, 497
Graph
  energy interpretation and, 167–168
  lenses and, 685–686
  linear motion analysis and, 54–55, 62
  spherical mirrors and, 674–675
  of vectors, 23
Gratings of diffraction, 746–748, 758
Gravitation, 139–161
  force of; *see* Gravitational force
  Isaac Newton and, 154–155
  in noninertial reference frames, 147–151, 156, 160
  universal, 139–143, 156, 157–158
    constant of, 141

Gravitation—cont'd
universal—cont'd
Coulomb's law and, 432
law of, 139–143, 156; *see also* Force laws
measurement of, 143
origins of theory of, 152–153
several particles in, 141–142
Gravitational acceleration of earth, 151
Gravitational attraction of earth, 144–147, 156, 158–159
Gravitational constant, universal, 141
Gravitational field
of earth, 437
uniform, 460
Gravitational force, 96, 98, 139–161
constant, 169–173, 192–193
escape velocity and, 176–177
fundamental forces and, 880
laws of, 96, 98
measurement of, 141, 158
spring force and, 179–180
total, on given particle, 141
universal laws of, 98, 139–143, 156
variable, 173–177, 193–194
work done by, 169–170, 179–180
Gravitational potential energy, 169–177, 186, 192–194
conservation of energy in, 170–172
electrical potential energy and, 459
gravitational force in; *see* Gravitational force
in path-independent work, 172–173
running and, 186, 229
Graviton, 884
Gravity, 49, 96, 98; *see also* Gravitation
acceleration and, 151
attraction of, 144–147, 156, 158–159
center of, static equilibrium and, 242–243, 252–254
field of, 437
Gray unit, 878
Greek system of universe, 152
Green color force, 885
Grid in projectile motion recording, 16, 17
Grimaldi, Francesco, 736
Ground fault interrupter, 536
Ground state, 826
of atom, 837
Ground wire, 535
Grounding of electric charge, 534–535
appliances and, 537

H

Hahn, Otto, 873
Hale telescope, 717
Half-life, 864–867, 888
in carbon dating, 867–868
decay curve for, 865–866
Halo, 755
Hard rubber, 497
Harmonic analysis, 413–414
Harmonic motion, 366–385
damped oscillations in, 378, 384
forced oscillations in, 378, 384
mass and spring in, 371–373, 383
pendulum in, 374–377, 384
period of motion in, 371–372
resonance and, 379–380
simple, 366–368, 381, 382–383
circular and, 369–371, 383
Harmonic waves, 389
power and intensity of, 400–403
of sound, 400–403
on string, 399
Harmonics, 412–413
Hawking, Stephen, 813
Heart, electrical effects on, 533
Heat, 320–337
caloric theory of, 320–321
calorimetry and, 321–325, 334
of combustion, 324
conduction of, 321, 330–332, 333, 336
convection and, 321, 329–330, 333, 335
and internal energy, 321, 358
latent, 323
in power lines, 612
radiation and, 321, 326–328, 335
specific, 322
steady flow of, 331
waste, 346
Heat engines, 346–348, 358, 361
efficiency of, 346
Heat pump, 364
Heavy-recoil nuclei, 878
Heisenberg, Werner, 835
Heisenberg's uncertainty principle, 808, 814, 816–817
atom and, 833–834
Helium
atomic mass of, 856
refractive index of, 641
Helium-neon laser, 637–638
diffraction patterns of, 746–747
frequency bandwidth of, 732
interference and, 738–739
wave properties of light and, 734–735
Helmholtz, Hermann, 321
Helmholtz coil, 585
Henry, Joseph, 598
Henry unit, 598
Hertz, Heinrich, 630
Hertz's production, electromagnetic wave detection and, 630
Hindu system of universe, 152
Hollow cylinder, rotation axis of, 223
Hooke's law, 371
Horsepower, 183
Hot wires, 535
Hot-air balloons, 300
Hour as unit of time, 8
Household circuits, 536, 547–549
Household electricity, 532–533
electric shock from, 534–537, 547–549
Hubble space telescope, 718
Human body
electrical effects on, 532–533
shock from, 534–537, 547–549
metabolism in, 355–357, 362–363
specific heat of, 323
Human eye; *see* Eye
Humidity, 309–311, 317
Huygens, Christian, 204
Huygens-Fresnel principle, 735
in single-slit diffraction, 742
Hydroelectric power, 596
Hydrogen
atomic mass of, 856
refractive index of, 641
spectra of, 819, 820–821
structure of, 835–836
Hydrogen atom, mass of, 301
Hyperopia, 705–706

I

Ice, specific heat of, 323
Ideal battery, 506
Ideal gas; *see* Gas, ideal
Ideal gas law, 298–303, 315, 316–317
derivation of, 306–309
Ideal gas molecule, kinetic energy of, 162, 163–168
Ideal inductor, 598
in RL circuits, 599–601
Ideal solenoid, 568
Ideal source of emf, 505
Illusions with mirrors, 662
Image
in concave spherical mirror, 665–669
virtual, 667
distance of
spherical refracting surface and, 677
virtual images and, 670
of extended object in plane mirror, 664
in eye
defects in, 703–707
focusing and, 702–703
formation of, 703–707
inverted, 707

Image—cont'd
  ghost, 656
  graphical method for position and size of, 685
  in negative lenses, 681
  optical fiber transmission and, 652
  in positive lenses, 680
  in refraction at single spherical surface, 688–689
  in spherical refracting surface of thin lenses, 679–680, 695–696
  virtual, 670
    in concave spherical mirror, 667
    optical power and, 682–683
  width of, in single-slit diffraction, 744
Image point
  real, 663
  virtual, 663
Impedance in alternating current circuits, 610–611
Impulse
  defined, 200
  linear momentum and, 199–202, 210
Impulse-momentum theorem, 200
Impulsive force, 200
Incandescent light bulb, 637–638
  filament in, 337
Incidence, angle of, 645
  critical, 649, 650
  plane mirrors and, 663
  in refraction at single spherical surface, 688–689
Index of refraction, 641
  of cornea, 699
  light interference and, 739–741
Inductance, electromagnetic, 588–627, 597–603, 618, 624
  inductors in, 598–599
  magnetic energy and, 601–603
  RL circuits in, 599–601
  self-induction in, 597–598
Inductive reactance, 608
Inductors; *see also* Coils
  electromagnetic, 598–599
    in alternating current circuits, 608–609
  ideal, 598
    in RL circuits, 599–601
  in RCL circuits, 609–611
Inelastic collisions, 208
Inertia
  law of, 90
  mass as measure of, 91
  moment of
    angular momentum and, 227
    in rotation about fixed axis, 222–224, 227

Inertia—cont'd
  moment of—cont'd
    running and, 228
Inertial reference frames, 90–91, 109
  Newton's second law of motion in, 90–91
Infinite plane, uniformly charged, 441–442
Infrared radiation, 326, 633
  passing through media, 640
Instantaneous acceleration, 44, 59, 81, 82
  angular, 216
  circular motion and, 72, 73
  on curved path, 65, 82
Instantaneous current, 494
Instantaneous flux, Faraday's law and, 593
Instantaneous power, 182, 183, 190, 508
Instantaneous pressure in blood vessels, 283
Instantaneous speed, 19, 20–21, 35, 37
Instantaneous velocity, 32–34, 35, 40
  acceleration in one dimension and, 42
  angular, 215
  on graph of linear motion, 55
  linear motion and, 59
Insulation
  R-value of, 332
  short circuits and, 519
Insulators, electric charge and, 429–430
Intensity
  of light
    passing through surface, 631
    scattering and, 752
    transmitted and average, 749, 750
  of sound wave, 401
  of wave, 399–403
    electron proportional to, 832
Intensity spectrum of x-ray, 840
Interaction
  entropy statement of second law of thermodynamics and, 352–353, 358
  strong or weak, 98
  between two bodies, 89
Interference
  constructive and destructive, 406–407, 408, 419
  light waves and
    properties of, 736, 757, 760–762
    wave-particle duality of, 804–806
  wave optics and, 736–741
Interferometer of Michelson, 762, 770, 793
Internal energy
  food and, 356–357
  heat and, 321, 358
  isolated system and, 341, 358
  in thermodynamic systems, 340–341
    as variable, 339
Internal sweep, 481

Interval of time, simple harmonic motion and, 366–367
Inverted image in eye, 707
Ionizing radiation, 877–880
Iris, 700
Iron
  atomic mass of, 856
  in ferromagnetic materials, 572–573
  resistivity and temperature coefficient of, 497
  specific heat of, 323
  in strong magnetic field, 570–572
Isolated conducting plate for electric charge, 442–443
Isolated system, thermodynamics and, 351–352
Isothermal process, 344, 358
  entropy and, 350
Isotopes, 855

J

Joule, James, 321
Joule as unit of work, 163
Jupiter, gravitational attraction on, 158

K

K shell of atom, 837
Kaon meson, 883
Keck I telescope, 717
Kelvin scale, 296–297, 315
Kelvin statement of thermodynamics, 353, 358
Kepler, Johannes, 140, 153
Kepler's third law of planetary motion, 140, 153, 157
Kilogram, standard, 91
Kilowatt-hour, 183
Kinematic equations, 47
Kinetic energy, 162, 163–168, 783
  of alpha particles, 862
  in Bernoulli's equation, 277
  of beta decay of carbon, 863
  center-of-mass and, 186, 225
  collisions and, 211–212
  defined, 165
  of ideal gas molecule, 162, 163–168
  momentum and, 204–208
  nuclear reactions and, 871
  of particle of mass, 190
  of photoelectrons, 801
  relativistic correction of, 783
  rotational, 187, 224–225, 231, 233–234
  running and, 186, 187, 228
  total change in, 167
  translational, 225, 340
    in running, 229
  work and, 163–168, 190, 191–192

Kinetic force, 119
Kinetic interpretation of temperature, 304
Kinetic theory, temperature and, 295–319; *see also* Temperature
Kirchhoff's rules, 520–522, 538, 541–542
  discharging of capacitor and, 530
  first, 520
  multiloop circuits and, 527, 538
  second, 520
Kleiber's law, 355
Krypton in measurement of meter, 9

L

L shell of atom, 837
Lambda baryon, 883
Laminar flow
  fluids and, 280
  Poiseuille's law and, 282
  Reynolds number and, 281
Land, Edward, 748
Laplace, 87
Lasers
  helium-neon; *see* Helium-neon laser
  light from, 637
    in air and glass, 642
    interference in, 738–739
    polarized, 732
    reflection of, 644
    wave properties of, 734–735
  photons emitted by, 802, 815
Latent heat, 323
Laws
  Bernoulli's; *see* Bernoulli's equation
  Biot-Savart, 563–564
  Boyle's, 299
  of conservation of energy, 162
  Coulomb's, 431–433, 447, 451–452
  Dalton's, 300–301, 315
  Faraday's; *see* Faraday's law
  force, 96–101, 109, 112–113; *see also* Force
    as universal law of gravity, 139–143, 156
  gas, ideal, 298–303, 315, 316–317
    derivation of, 306–309
  Hooke's, 371
  of inertia, 90
  Kepler's third, 140, 153, 157
  Kleiber's, 355
  Lenz's, 591–592
  of Malus, 749
  Newton's; *see* Newton's laws of motion
  Ohm's; *see* Ohm's law
  physical, derivations of, 4
  Poiseuille's, 282–285, 286, 292–294
  of reflection, 645, 654, 657–659
Laws—cont'd
  of refraction or Snell's; *see* Refraction of light, laws of
  Snell's; *see* Refraction of light, laws of
  Stefan's, 327, 333
  of thermodynamics
    first, 341–345, 358, 359–361
    second; *see* Thermodynamics, second law of
  of universal gravitation, 156, 157–158; *see also* Force laws
Lead
  atomic mass of, 856
  resistivity and temperature coefficient of, 497
  specific heat of, 323
Leeuwenhoek, Anton, 713
Length
  coherence, 734
  defined, 7
  focal; *see* Focal length
  as fundamental quantity, 6
  path
    change in, 18
    in mechanical wave superposition, 406–407
    in particle trajectory, 17
  proper, 776
  wave; *see* Wavelength
Length contraction, relativity and, 776–778, 795
Lenses, 676–689, 690, 694–696
  aberrations in, 687–688
  accommodation in, 700–703
  in correcting hyperopia, 705–706
  crystalline, 700
    refractive index of, 641
  derivation of single-surface refraction in, 688–689
  graphical method of, 685–686
  linear magnification of, 683–685
  negative, 681
    graphical method for, 685
  objective, of microscope, 713–714
  optical power of, 681–683
  positive, graphical method for, 685
  refraction at spherical surface of, 677–678, 694–695
  spherical surface of, 676
  thin, 679–680
Lenz's law, 591–592
Leptons, 883, 885
Leyden jar, 490
Light, 628–661
  absorption of, in retina, 720
  beam of
    polarized, 731–732
    unpolarized, 732
Light—cont'd
  diffraction of; *see* Diffraction
  dispersion of, 641–642
  distance traveled by, in measurement of meter, 8
  electromagnetic waves and, 629–632, 656–657
  extended source of, 643
  intensity of
    passing through surface, 631
    scattering and, 752
    transmitted and average, 749, 750
  in medium, 640
  monochromatic, 635, 637
    interference and, 737
    polarization and, 731
  from moving spaceship, 780
  nature of, 632–638
  neon, 626, 635
  passing through two polarizers, 749
  point source of, 643
    plane mirrors and, 663–664
  propagation of, 638–644, 657
  quantum concepts and, 798–817; *see also* Quantum concepts
  rays of; *see* Light rays
  reflection of; *see* Reflection of light
  refraction of; *see* Refraction of light
  sources of, 633, 635–638
  speed of, 8, 638–639
    Maxwell's theory and, 629
    principle of relativity and, 767–768
  ultraviolet, passing through media, 640
  unpolarized, 749, 750
  as visible electromagnetic radiation, 326
  visible spectrum of, 632–633
    quantum theory and production of, 630
  visual sensitivity to, 633, 720–722
  wavelength of; *see* Wavelength
  waves of
    properties of, 731–736, 757, 759
    supposed trajectory of, 805
  white, 732
Light bulb filament, 337
Light clock, 770–773
Light rays, 643
  accommodation in eye to, 700, 701
  image points and
    in plane mirrors, 663–664
    in spherical concave mirrors, 668
  parallel, 643
    graphical method for, 685
    incident on spherical surface of lens, 676

Light rays—cont'd
paraxial, 668–669
on spherical refracting surface, 677, 678
principle, 674
in refraction at single spherical surface, 688–689
in single-slit diffraction, 742–744
telescopes and, 716–718, 725, 730
Lightning, 455
Limit of ratio, 20
Limiting value, 20–21
Line of action, 219
Linear accelerators, 882
Linear magnification
of lenses, 683–685
spherical mirrors and, 672–673
Linear momentum, 209
conservation of, 210–211
defined, 200
impulse and, 199–202, 210
of isolated system, 203
total, 209
Linear motion, 41–63
acceleration in, 41–48
constant, 45–48, 59, 60–61, 118–123
in one dimension, 41–45
center of mass and, 127
free fall, 49–52
in air, 53–54
friction and, 118–123
Galileo Galilei study of, 56–58
graphical analysis of, 54–55, 62
quantity of; *see* Linear momentum
Linear polarization, 731
Linear speed of rotating body, 231
Liquids, refractive indices of, 641
Locomotion, Newton's third law and, 94
Longitudinal wave, 387
Long-straight wire, electric currents in, 566
Looming, 755
Loops
in circuits, 519
magnetic forces and
circular current in, 564–565
external magnetic field in, 557–559
Lyman series, 821, 831, 851

M

Mach number, 398
Macroscopic phenomena, classical mechanics explaining, 835
Macula lutea, 723
Magic in sky, 754–755
Magnet, permanent, 569–572, 576, 587; *see also* Magnetic field
Faraday's law and, 589
Magnetic bottle, 562
Magnetic energy
electromagnetic inductance and, 601–603, 618, 624
per unit volume, 602
Magnetic field, 550–554
Biot-Savart law and, 563
of circular current loop, 564–565
defined, 551–554, 576, 581–582
historical background of study of, 550–551
Lenz's law and, 591
long-straight wire and, 566
magnetic flux and, 590
motion of point charge and, 560–562, 563
nonuniform, 562, 563
parallel wires and, 567
from permanent magnets, 569–572, 576, 587
Faraday's law and, 589
production of
by coil, 598
by electric current, 563–569, 585–586
from solenoids, 568
in sunlight, 631
torque on current loop in external, 557–559
uniform, 560–562
Magnetic field lines
closed paths of, 566
magnetic flux and, 590
Magnetic flux
Faraday's law and, 590, 593
Lenz's law and, 591
Magnetic forces on current-carrying wires, 554–559, 582–584
Magnetic materials, 572–573
Magnetic mirror, 562
Magnetic moment
generator coil and, 596
per unit volume, 570
Magnetic moment vector, 557
torque and, 558
generator coil and, 596
Magnetic poles, 550
Magnetic propulsion, 592
Magnetic quantum number, 836
Magnetic resonance imaging, 554
Magnetism, 550–587
biomagnetism in, 574
current-carrying conductors and, 554–559, 582–584
electric currents and, 563–569, 585–586
magnetic field and, 550–554, 576, 581–582, 585–586
magnetic materials in, 572–573
motion of point charge and, 560–562, 584–585
Magnetism—cont'd
permanent magnets and magnetic fields in, 569–572, 576, 587
Faraday's law and, 589
Magnetization, 570
Magnetostatics, defined, 551, 588
Magnetotactic bacteria, 574
Magnification
linear
of lenses, 683–685
spherical mirrors and, 672–673
in microscopes, 713–715, 725, 729
in telescopes, 716–718, 725, 730
Magnifier, 710–713, 725, 728–729, 730
Magnifying glass, 710–713
Magnitude of force
of static friction, 132
torque and, 219
Malus, law of, 749
Mars, gravitational attraction on, 158
Mass, 91–92, 109, 111
center of; *see* Center of mass
of earth, 135, 146
as fundamental quantity, 6
gravitational force and, 173–177, 880
harmonic motion and, 371–373, 383
simple, 366–367
of moon, 135
period of motion in, 371–372
relativistic, 781–783, 795–796
scientific standard of, 91
selected particle and atom, 856
spherical distribution of, 158–159
weight and, 92
Mass defect, nuclear structure and, 858–860
Mass energy; *see* Energy, mass
Mass number, 855
Mass spectrometer, 561
Mathematics, 4
Matter, 2
universal tendency of, to attract other matter, 139
Maximum displacement, standing waves and, 411
Maximum frictional force, 119, 120
Maximum kinetic energy of photoelectrons, 801
Maxwell, James, 303, 304, 628, 629
Maxwell distribution of molecular speeds, 304, 305
root mean square speed for, 305–306
Maxwell's equations in total internal reflection, 649
Maxwell's theory of light, 629–631
Mayer, Robert, 321
Measurement and units, 6–15
certainty in, 10–11

Measurement and units—cont'd
  fundamental quantities in, 6–7
  meter in, 8
  n-rays, polywater, and cold fusion in, 14–15
  order-of-magnitude estimates in, 11–12
  powers of ten in, 9
  second in, 7–8
  Système International, 7
  of temperature, 295–298, 315, 316
  units in
    base, 7
    conversions of, 9
    derived, 7
    names of, 9
Mechanical energy, 162, 180–181
  electric motor and, 182
  nonconservation of, 181, 182
  potential energy and
    gravitational, 170
    spring, 178, 194
  sliding down incline and, 182
  spring force in conservation of, 180–181
  total, 170, 180, 190
    conservation of, projectile motion and, 171
  wave motion and, 387
Mechanical waves, 386–426
  decibel scale for, 399–403, 423–424
  defined, 386
  description of, 386, 419, 420–422
  Doppler effect and, 395–399, 419, 423
  ear and, 418
  moving sources and observers in, 395–399
  power and intensity of, 399–403, 419, 423–424
  sound in; *see* Sound
  speed of, 392–395, 419, 422–423
  superposition of, 405–417, 420, 424–425
    beats in, 408–409, 424–425
    constructive and destructive interference in, 406, 419
    harmonic analysis in, 413–414
    harmonics in, 412–413
    path length in, 406–407
    principle of, 405
    resonant frequencies in, 412–413
    standing sound waves in, 414–417, 424–425
    standing waves on string in, 410–412
  time dependence of displacement of particle of medium in, 404–405, 424
Mechanics, classical, 88
Mediator particles, 884
Medium
  light in, 640
  sound in, speed and, 769
Meitner, Lise, 873
Membrane
  axon, 549
  nerve cell, potential difference and, 532
Mercury, 497
Mesons, 883
  quarks and, 884
Metabolic rate, basal, 355–356
Metabolism, 355–357, 362–363
  work and, 356–357
Metal
  detector of, 621
  electric charge and, 430
Metallic conductor, potential difference across, 510
Meteoroids, 175
Meter, 8
Methyl alcohol, refractive index of, 641
Michelson, Albert, 638–639
Michelson's apparatus for speed of light measurement, 639
Michelson interferometer, 762, 770, 793
Michelson-Morley experiment, 770
Microscopes, 713–715, 725, 729
  diffraction limit of, 746
  distinguishing close points with, 714–715
  electron, 463, 715
    scanning, 719
  lenses for; *see* Lenses
  as magnifiers, 710–713
Microscopic phenomena, quantum mechanics explaining, 835
Microwaves, 326
Mili, Gjon, 19
Millikan, Robert, 802
Minute as unit of time, 8
Mirages, 754–755
Mirrors
  converging, 666
  diverging, 671
  illusions with, 662
  parabolic, 669
  plane, 663–665
  spherical, 665–675
    concave, 665–669, 674, 675
    convex, 671, 672, 674
    graphical method and, 674, 675
    linear magnification in, 672
Modem, 3
Molar form of ideal gas law, 301, 302–303
Mole, 302
Molecule
  gas; *see* Gas, molecule of
  polar, 477
  as thermodynamic variable, 339
Moment
  electric dipole, 448
Moment—cont'd
  of inertia
    angular momentum and, 227
    in rotation about fixed axis, 222–224, 227
    running and, 228
Moment arm, 219
Momentum, 199–213
  angular; *see* Angular momentum
  collisions and kinetic energy in, 204–208, 211–212
  of electron after Compton scatterings, 804
  linear; *see* Linear momentum
  of photon, wavelength of light and, 814, 815–816, 830
  quantum concepts and, 807–808
  rate of change of, 209
  of system of particles, 202–204, 209, 210–211
  total, 202
Monatomic ideal gas, entropy of, 350
Monochromatic light, 635, 637
  interference and, 737
  polarization and, 731
Moon
  centripetal acceleration of, 145
  gravitational attraction on, 158
  mass of, 135
  ocean tides and, 147
  seen through telescope, 718
Motion, 1–2
  of center of mass, 127
  circular
    harmonic, 369–371, 383
    in plane, 72–75, 83
  description of, 16–40
    displacement in, 22–23
    particle trajectory in, 16–17
    problem-solving strategy in, 36
    speed in, 18–21
    vector algebra in, 23–27
    vectors and, 27–30
    velocity in, 31–34
  equations of, 47
  harmonic, 366–385; *see also* Harmonic motion
  linear; *see* Linear motion
  mass and spring, period of, 371–372
  Newton's laws of; *see* Newton's laws of motion
  in plane, 64–86
    acceleration on curved path in, 65–66, 82
    circular, 72–75, 83
    projectile, 67–72, 81, 82–83
    reference frames and relative motion, 76–80, 84

Motion—cont'd
planetary, Kepler's third law of, 140, 153
of point charge in magnetic field, 560–562, 584–585
projectile; *see* Projectile motion
quantity of; *see* Momentum
relative, 77
reference frames and, 76–80, 84
of two spacecrafts, 780
rotational; *see* Rotation, motion of
straight line; *see* Linear motion
uniform, 18, 35
Motional emf, Faraday's law and, 592–593
Mountain specter, 755
Moving charges of electric current, 509–510
Moving clocks, 772–773
Multielectrons, structure of, 837–838
Multiloop circuits, 526–527, 538, 544–545
Multimeter, 529
Multiplication
significant figures in, 11
of vectors by scalars, 27
Muon, 774
Muon lepton, 883
Muon neutrino, 883
Muscles
ciliary, 700, 702
static equilibrium and, 240–241
Muybridge, Eadweard, 16, 17
Myopia, 703–704, 725, 727
Myosin, 240

N

Names of units, 9
Nanometers, 642
Near point, 700
aging and, 706–707
Nearsightedness, 703–704, 725, 727
Negative charge, magnetic field and, 552
Negative emf, 590
Negative lenses, 681
graphical method for, 685
Negative pressure, fluids and, 273
Negative source charge, 435, 436
electric potential and, 462
Negative work
in interacting thermodynamic system, 342, 343
pressure in moving fluid and, 277
Neon lights, 626, 635
Neon tubes, 546
Neoprene rubber, 476
Neptunium, 873
decay of, 868
Nerves, electrical effects on, 532–533, 547–549
Net charge, 428
on conductor, 443
Net power, 190
Net radiation, 328
Net work
done by system over cycle, 344
kinetic energy and, 164
pressure at two points in moving fluid and, 276
of resistors, 523, 525
spring force and, 179
of system of particles, 187
Neuron, 532
Neuron pulse, 533
Neutrino, 863, 883
Neutron, 448, 854
mass of, 856
relative biological effectiveness of, 878
Neutron baryon, 883
Neutron number, 855
New Theory about Light and Colors, 154
Newton, Sir Isaac, 2, 88
on gravitation, 154–155
Newton as force of acceleration, 93
Newton's laws of motion, 88
applications of, 102–108, 113–116
center of mass and, 127
first, 90–91, 109, 111–112
inertial reference frames and, 90–91
friction and, 118–123
Galileo and Aristotle and, 90
gas molecules and, 306
linear momentum and, 199
rotation and, 214
second, 92–93, 109, 111–112
in inertial frame of reference, 148
linear momentum and, 199
pendulum motion and, 374
rotation about fixed axis and, 221
third, 93–96, 109, 111–112
Newton's rings, 760
Newton-meter, 219
Nichrome, 497
Nilsson, Lennart, 652
Nitrogen, atomic mass of, 856
Nodes, standing waves and, 411
Nonconservative force, 180–182, 195
Noninertial reference frames
earth in, 151
gravitation and, 147–151, 156, 160
Nonuniform motion, 18–20
Normal in light rays incident on object, 645
Normal force, 119
North pole, magnetic, 550
Northern lights, 562, 563
N-rays, 14–15
Nuclear fission, 873–874, 875, 892
Nuclear force, 98
fundamental forces and, 880, 881
strong, 858, 880, 881
weak, 880, 881
Nuclear fusion, 874–877, 892
Nuclear physics and elementary particles, 854–893
biological effects of radiation in, 877–880, 892–893
elementary particles in, 880–885, 893
nuclear reactions in, 871–877, 892
nuclear structure in, 854–860, 888, 890–891
radioactive decay in, 861–870, 888, 891–892
Nuclear powered plant, 596
Nuclear reactions, 871–877, 892
Nuclei; *see* Nucleus
Nucleons, 854–855
binding energy and, 860
nuclear force and, 881
Nucleus, 822–825
binding energy of, 858–860
daughter, 861
in decay series, 868
density of, 857
heavy-recoil, relative biological effectiveness of, 878
parent, 861
in decay series, 868
radioactive, activity of sample and, 866
radius of, 888
size of, 856–857
structure of
elementary particles and, 854–860, 888, 890–891
strong nuclear force and, 858
Nuclides, 855–856
stable, 857
or unstable, 872
Numerical values expressed in units, 6

O

Object
distance of; *see* Distance
virtual, 670
Objective lens of microscope, 713–714
Ocean tides, 147
Ocular of microscope, 713–714
Oersted, Hans Christian, 551
Ohm's law, 495–500, 512, 514–515
current through resistor and, 603, 605
macroscopic form of, 511
on microscopic level, 509–511
Ohmmeter, 529
Ohms, 495
Omega baryon, 883
One-loop circuit, 519

Opaque medium, 640
Open circuits, 519
Optical axis, 666, 667
principle rays and, 674
Optical fibers, 651–652
Optical instruments, eye as, 698–730; *see also* Eye
Optical power of lenses, 681–683
Optical reversibility, principle of, 666
positive lenses and, 680
Optics
fiber, 651–652
geometrical; *see* Geometrical optics
physical, 644
wave, 731–765; *see also* Wave optics
Orbit, electron, 825–826
Orbital quantum number, 836
Order-of-magnitude estimates, 11–12
Ordinary collision, 349
Oscillations
center of, 385
damped, 378, 384
electromagnetic waves and, 629–630
forced, 378, 384
physical pendulum and, 375–377
resonance and, 379, 380
simple harmonic motion and, 366–3693
Oscilloscope
electric potential and, 478–482, 483, 490
time-dependent phenomena and, 482
voltage of, 481
Otto cycle, 348
Oxygen
atomic mass of, 856
distribution of molecular speeds for, 305
Ozone, 640

P

Paper, dielectrics and, 476
Parabolic mirror, 669
Parabolic trajectory, 71
center of mass of projectile and, 130
Parallel capacitor, 471–472
Parallel light rays, 643
graphical method for, 685
incident on spherical surface of lens, 676
Parallel resistors, 524, 538
Parallel wires, magnetism and, 567
Parallel-plate capacitor, 469–470
Paramagnetic materials, 572
Paraxial rays, 668–669
on spherical refracting surface, 677, 678
Parent nucleus, 861
in decay series, 868
Parhelia, 755
Partial pressures, Dalton's law of, 300–301, 315
Particle
acceleration of, Newton's second law and, 92
alpha; *see* Alpha particles
beta, relative biological effectiveness of, 878
displacements and velocity of, 31, 35, 389, 404–405, 424
dust, uncertainties in, 833–834
elementary; *see* Elementary particles
gravitational force and, 140, 144; *see also* Gravitation
instantaneous speed of, 19, 20–21, 35
circular motion and, 72, 73
linear momentum of, 209, 210–211
in mass, 491, 856
kinetic energy of, 190
mechanical energy of, 190
mediator, 884
nuclear force and, 881
position of, at any time, 35
sedimentation rate of, 150
speed of, 35
total gravitational force on, 141
trajectory of, 17
Cartesian coordinates in, 16–17
path length in, 17
uniform speed of, 18
velocity of, circular motion and, 72, 73
wavelength of, 832
in wave-particle duality, 804–806, 816
work done on, 190
Particle accelerators, 882–883
Particle dynamics, 87–138
center of mass in, 127–131
classical mechanics in, 88
force in, 88–89; *see also* Force
centripetal, 125–126
concept of, 101
derived, 98
electromagnetic, 98
fundamental, 98
gravitational, 98
laws of, 96–101
nuclear, 98
spring, 100–101
tension, 99
friction in, 118–123
microscopic description of, 124–125
mass in, 91–92
Newton's laws of motion in
applications of, 102–108
first, 90–91
second, 92–93
third, 93–96
units in, 93
weight on earth in, 96–97
Particle dynamics—cont'd
weight on earth in,—cont'd
variation of, 98
Particle system
in energy, 187–189
momentum and, 202–204
Pascal as unit of pressure, 259
Pascal's principle, 262–263
Paschen series, 821, 831, 851
Path
curved, acceleration on, 65–66
length of
change in, 18
in mechanical wave superposition, 406–407
in particle trajectory, 17
of motion in magnetic field, 560–562, 563, 566
work done by gravity independent of, 172–173
Pauli exclusion principle, 837
Pendulum, 374–377, 381, 384
energy of, 377
physical, 375–377
simple, 374–375
small angle approximation and, 374
Percussion, center of, 385
Period
defined, 8
of motion, 366
circular, 75
Periodic table, description of, 837
Periodic waves, 389–390
Permanent magnets, 569–572, 576, 587
Faraday's law and, 589
Phase angle, harmonic motion and, 367, 368
Phasor diagram, 604, 611
Phenomena
macroscopic, classical mechanics explaining, 835
microscopic, quantum mechanics explaining, 835
oscilloscope time-dependent, 482
refraction, 754–755
of surface tension, 270–272, 286, 290–291
Philosophiae Naturalis Principia Mathematica, 2, 88
Phosphorescence of atom, 839
Photocell, 801
Photoelectric effect, 800–802
Photoelectrons, maximum kinetic energy of, 801
Photomultiplier tube, 870
Photons
blackbody radiation and, 799–800

Photons—cont'd
  Compton effect and, 803–804
  emitted by lasers, 802
  light particles and, 632
  momentum of, wavelength of light and, 830
  number of, in diffraction, 806
  photoelectric effect and, 800–802
  Planck's constant and, 799–800
  as quanta of light, 798, 800
  quantum concepts and, 799–804, 814, 815–816
  relative biological effectiveness of, 878
Photopic vision, 719–720
Photosensitive receptors of eye, 698, 719
Physical laws, derivations of, 4
Physical optics, 644
Physical pendulum, 375–377
Physics, applications of, 3–4
Pion meson, 883
Planck's constant, 799–800
Plane, motion in, 64–86
  acceleration on curved path in, 65–66
  circular, 72–75
  projectile, 67–72
  reference frames and relative, 76–80
Plane mirrors, 663–665, 690, 692–693
Plane wave, 388
Planetary motion, Kepler's third law of, 140, 153
Plutonium, 873
Point, focal; *see* Focal point
Point charge of electrical potential energy; *see* Electrical potential energy, point charge of
Point source of light, 643
  plane mirrors and, 663–664
Point source charges in electric field, 437–439
Poiseuille, Jean-Louie-Marie, 282
Poiseuille's law, 282–285, 286, 292–294
  Bernoulli's equation compared to, 284
Polar molecule, 477
Polarization
  light wave properties and, 731–732
  linear, 731
  in wave optics, 748–753, 758, 764
Polaroid sunglasses, 751
Polonium, decay of, 868
Polyethylene, 497
Polygon rule of vectors, 25
Polywater, 14–15
Porcelain, dielectrics and, 476
Position of electron
  quantum concepts and, 807–808
  uncertainties in, 833–834
Positive charge, 435
  electric potential and, 462
  magnetic field and, 552
Positive emf, 590
Positive lenses, 680
  graphical method for, 685
Positive work
  in interacting thermodynamic system, 342, 343
  pressure in moving fluid and, 277
Positron, 448
  discovery of, 881
  emitted, 863
Potassium pump, 549
Potential, electric, 457–492; *see also* Electric potential
Potential difference, 468–469
  applied across metallic conductor, 510
  average, 474
  capacitor and, 470, 471
    energy storage and, 474
  conductors and, 534–535
  in direct current circuits, 528–529
  electric shock and, 534
  Kirchoff's rules and, 520
  measurement of, 528–529, 538, 545–546
  nerve cell membrane and, 532
  of parallel resistors, 524, 538
  photoelectric effect and, 801
  power supply or battery and, 470
  transformers and, 612–615
  voltmeter in measuring, 528
Potential energy, 162
  in Bernoulli's equation, 277
  electrical, 458–468, 483, 486–487
    generators and, 596
  gravitational; *see* Gravitational potential energy
  projectile motion and, 171
  spring, 177–180
  total, 179
Pound, 93
Power
  average, 182, 190, 508
  electric; *see* Electric power
  energy and, 182–184, 190, 195
    in running, 228–230, 235
    units in, 183–184
  instantaneous, 182, 183, 190, 508
  net, 190
  optical, of lenses, 681–683
  of system of particles, 188, 196
  of ten, 9
  of waves, 399–403
Power lines
  in electric shock, 535
  heat in, 612
Power lines—cont'd
  in household circuits, 535, 536, 547–549
  transformers and, 612
Power supply, 470
Pressure
  in Bernoulli's equation, 277
  in blood vessels, 283
  in fluids, 286
    Poiseuille's law and, 282
    at rest, 258–259, 286, 288–289
  of gas, 298–303
  of gas molecules against container wall, 303–304
  in incompressible fluids with depth, 261–262
  in moving fluid, 276
  partial, Dalton's law of, 300
  saturated vapor, 309, 310
  as thermodynamic variable, 339
  at two points in fluid, 276
  vapor, 309–311, 317
Pressure gradient, 283, 284
Pressure waves, 422
Principal quantum number, 836
Principle rays, 674
Principles
  Archimedes,' 265–269, 286, 289–290
  Pascal's, 262–263
Prism, 637
  separation of colors with, 647
Problem-solving strategy in description of motion, 36
Projectile, center of mass of, 130
Projectile motion
  in plane, 67–72, 82–83
  potential energy and, 171
  recording of, 16, 17
  total mechanical energy conservation and, 171
Propagation of light, 638–644, 657
Proper length, 776
Proper time interval, 770
Proportional scaling, 247
Proportionality, constant of, 280
Proton, 448, 854
  charge of, 430
  mass of, 856
  as mediator of electromagnetic force, 884
  movement of, 491
    producing current, 494
  relative biological effectiveness of, 878
Proton baryon, 883
Pseudoforce, 148, 156
  mass and, 148
Ptolemy, 1
Ptolemaic system of universe, 152

Pulse
neuron, 533
wave, 387
Pump
heat, 364
sodium or potassium, 549
Pupil, 700
Pyrex glass, 476

Q

Quantities, fundamental, 6–7
Quantum concepts, 798–817
of nature of light, 798
photons and, 799–804, 814, 815–816
Stephen Hawking and, 813
uncertainty principle and, 807–809, 814, 816–817
wave-particle duality and, 804–806, 816
Quantum mechanics
atoms and, 830–835, 853
electrons and, 853
explaining microscopic phenomena, 835
Quantum numbers
magnetic, 836
orbital, 836
principal, 836
quantum mechanics and, 835
spin, 836
Quarks, 448, 883, 884
kinds of, 885
Quasar, 795
Quasi-static process, 339–340, 358
entropy and, 350
Quasi-stellar source, 795
Q-value of nuclear reactions, 871–872, 889, 892

R

Rad, 878
Radial acceleration, 217
Radiation, 321, 326–328, 335
absorption of, 327–328, 877–880
biological effects of, 877–880, 892–893
blackbody, 799–800
Bremsstrahlung, 840
in definition of second, 8
electromagnetic, visible light spectrum and, 632
emission of, 326–327
quantum theory and, 835
hazards of, 879
incident perpendicular to surface area, 661
infrared, 633
passing through media, 640
ionizing, 877–880
net, 328
Radiation—cont'd
reflection of, 327
scattered and polarized, 753
solar, 328, 337
ultraviolet, 633
Radiation detectors, 870
Radio waves, 326
Radioactive decay, 861–870, 888, 891–892
counters of, 870
Radioactive sample, activity of, 866–867
Radiowaves passing through media, 640
Radium
biological effects of, 879
decay of, 868
Radius
of curvature
in spherical mirrors, 671
spherical refracting surface and, 677
of nucleus, 888
Radon, decay of, 868
Rain, 310
Range, projectile motion and, 70–71
Ratio, limit of, 20
Rayleigh criterion for resolution in diffraction, 745–746
Rayleigh scattering, 752
Rays
cosmic, elementary particles and, 882
of light; *see* Light rays
RC circuit, 529–531, 538, 546–547
RCL series circuit, 609–611
Reaction
chain, 874, 875
energy of, 871, 889
force of, 88, 94
nuclear, 871–877, 892
Reading glasses, 706
Real image
point of, 663
graphical method for, 685, 686
in refraction at single spherical surface, 688–689
Real object, graphical method for, 685
Receptors, photosensitive, 698, 719
Red color force, 885
Reduced eye, 707–708
Reference frames
inertial, 90–91, 109
Newton's second law of motion in, 90–91
noninertial, 147–151, 156, 160
relative motion and, 76–80, 84
Reflection of light, 644–652, 654, 657–659
angle of, 645
plane mirrors and, 663
fiber optics and, 651
law of, 645
Reflection of light—cont'd
polarization by, 750–751
of radiation, 327
scattering and, 752
total internal, 649–651
transmittance and, 648–649
of wave, 391
Refraction of light, 644–652, 654, 657–659
angle of, in refraction at single spherical surface, 688–689
index of, 641
of cornea, 699
light interference and, 739–741
laws of, 645, 646, 654
plane mirrors and, 663
single spherical surface and, 688–689
in total internal reflection, 649, 650
in lenses
ciliary eye muscles and, 700–703
derivation of single-surface, 688–689
at spherical surface, 677–678
phenomena from, 754–755
scattering and, 752
Refrigerators, 346–348, 361
Relative biological effectiveness, 878, 889, 892–893
Relative displacement, 76
Relative humidity, 310
Relative motion, 77
reference frames and, 76–80, 84
Relative velocity, 77, 78, 79, 779–780, 795
Relativistic Doppler effect, 794–795
Relativistic mass and energy, 781–783, 795–796
Relativity, 765–797
Albert Einstein and, 765, 788–789
in length contraction, 776–778, 795
in relative velocity, 779–780, 795
in relativistic mass and energy, 781–783, 795–796
time dilation and, 770–775, 794–795
in time measurement, 766–770, 790, 793
Remanent magnetization, 573
Repulsion, electrical forces of, 451
Resistance
air; *see* Air resistance
electrical; *see* Electrical resistance
equivalent; *see* Equivalent resistance
measurement of, 528–529, 538, 545–546
transformers and, 612
Resistivity, 496–499
Resistors, 499–500
in alternating current circuits
electromagnetic induction and, 603–605
Faraday's law and, 594, 595
network of, 523

Resistors—cont'd
in RCL circuits, 609–611
in RL circuits, 599–601
Resolution
in diffraction, Rayleigh criterion for, 745–746
telescopes and, 717–718
Resonance, 379–380
of alternating current circuits, electromagnetic induction and, 612
Resonant frequency, 379, 412–413, 612
lowest, 415
Rest energy, 782
Rest mass, loss of, 782
Retina, 700, 719–722
hyperopia and, 705–706
intensity range of, 719–720
myopia and, 703–704, 725, 727
reduced eye in forming image height in, 708
structure of, 719
Reversed collision, 349
Reversibility, optical, 666
positive lenses and, 680
Reversible process, 339–340
Reynolds, Osborne, 281
Reynolds number, 281–282
Rhodopsin, 720
Richter scale, 425
Right-hand rule, 552, 555
Biot-Savart law and, 563
Rigid bodies
center of mass of, 129
rotating, torque and, 246
velocity of rotating, 215–216
Ring, Newton's, 760
RL circuits, 599–601
Rod
heat conduction through, 330–331
in retina, 719
thin, rotation axis of, 223
Roentgen, Wilhelm, 839
Root-mean-square, 508
Root-mean-square current, 604
Root-mean-square speed, 305–306
Rotation, 214–236
acceleration and, 217–218, 231
moment of inertia in, 223
angular momentum and, 226–227
in energy analysis of running, 228–230, 235
air resistance and, 230
gravitational potential energy and, 229
kinetic energy and, 228
translational kinetic energy and, 229
equilibrium and, 237, 238
fixed axis of; *see* Fixed axis of rotation
Rotation—cont'd
kinetic energy of, 187, 224–225, 231, 233–234
linear dynamics and, 231
motion of, 214–218, 231, 232
acceleration and, 216, 217–218
angular acceleration and, 216
angular velocity and, 215
velocity of rigid body and, 215–216
torque and, 216, 219–220
Rubber, 497
Running, 184–187
energy in, 184–187
rotation and, 228–230, 235
heat from muscles contracting in, 185
internal body energy in, 185–187
work done by muscles in, 184–185
Rutherford, Ernest, 822
Rutherford's scattering experiment, 822–823
R-value of insulation, 332
Rydberg constant, 820

S

St. Elmo's fire, 450
Satellite
centripetal acceleration of, 146
circular trajectory of, 74
gravitational attraction of earth and, 159
Saturated vapor, 309, 310
Saturated vapor pressure, 309, 310
Saturation in magnetization, 573
Savart, 551
Biot-Savart law and, 563–564
Scalars
multiplication of vectors by, 27
vectors and, 23
Scales
Richter, 425
for temperature measurement, 296–298, 315
Scaling, proportional, 247
Scanning electron microscope, 719
Scattering
Compton, 803
polarization by, 752–753
Rayleigh, 752
Rutherford's experiment of, 822–823
sunlight, 753, 758
Scheiner, Christoph, 707
Schroedinger, Erwin, 835
Scintillation counter, 870
Scotopic vision, 719–720
Scuba diving, 268
Second, 7–8
Second law of thermodynamics, 338, 349–355, 358, 361–362
Second law of thermodynamics—cont'd
Clausius' statement of, 355
entropy statement of, 351–353
Kelvin statement of, 353
Sedimentation rate of particles, 150
Seismic waves, 422
Self-induction, electromagnetic, 597–598
Semiconductors, 498
Sensitivity
to color, 719–722
to light, 633
spectral, of retina, 720–722
Series capacitor, 472–473
Series resistors, 523, 538
Series-parallel combinations, 524–526, 538
Shear stress, 245
laminar flow and, 280
static, fluids and, 259
Shear waves, 422
SHM; *see* Simple harmonic motion
Shock, electric, 534–537, 547–549
Short circuit, 519
SI units; *see* Système International Units
Sievert imot, 878
Sigma baryon, 883
Silicon, 497
Silver, 497
Simple harmonic motion, 366–368
circular motion and, 369–371
Single source charge in electric field, 435–437
Single-loop circuits, Kirchoff's rules in, 522, 526
Single-slit diffraction, 742–744
Single-surface refraction in lenses, 688–689
Skin, heat evaporation through, 330
Sky, magic in, 754–755
Slug, 93
Small angle approximation in pendulum motion, 374
Snell's law; *see* Refraction of light, laws of
Sodium
atomic energy levels of, 838
photoelectric effect for, 802
radioactive, sample decays in, 867
Sodium discharge tube, 635
Sodium doublet, angles of diffraction for, 747
Sodium pump, 549
Solar eclipse, 659
Solar radiation, 328, 337
Solenoids
ideal, 568
magnetic energy per unit volume and, 602
magnets and, 568–569
permanent, 571–572

Solids
refractive indices of, 641
rotation of cylindrical or spherical, 223
Sonic boom, 398–399
Sound, 386–426; *see also* Sound waves
Doppler effect of, 395–399
ear and, 418
frequency of, from flute, 417
speed of, 393–394, 419
principle of relativity and, 769
supersonic, 398–399
Sound waves; *see also* Sound
decibel scale for, 399–403
displacement of, 404–405, 424
ear and, 418
harmonic, 400–403
interference of, beats and, 409, 424–425
oscilloscope measurement of, 482
tuning fork and, 394, 414
Source charge
negative, 435, 436
positive, 435
South pole, magnetic, 550
Spacecraft
escape velocity and, 176–177
relativity and; *see* Relativity
Specific gravity, 259
Spectra
of atom; *see* Atom, spectrum of
visible, 632–633
of atom, 838
retina and, 720–722
Spectral colors, 632
Spectral lines, 819
frequency bandwidth of, 732
Spectral sensitivity of retina, 720–722
Spectrometer
mass, magnetic field and, 561
in study of retinal absorption of light, 721
Spectrum; *see* Spectra
Speed, 18–21
average, 19–20, 37
center of mass, in running, 228
of electrons, 511
dependence of mass of body on, 781
of electrons, average, 511
of flow of fluids, Reynolds number and, 281
initial, escape velocity and, 176
instantaneous, 19, 20–21, 37
of light, 8
Maxwell's theory and, 629
principle of relativity and, 767–768
nonuniform motion and, 18–20
of particle of medium, 389
problems of, 37
Speed—cont'd
of rotating body, linear, 231
of sound, 393–394, 419
principle of relativity and, 769
supersonic, 398–399
supersonic, 398–399
in Système International units, 9
uniform motion and, 18
velocity compared to, 31
of waves, 389
on string, 392–395
in water, 394
Sphere, rotation axis of, 223
Spherical aberration, 669, 687
visual acuity and, 722–723
Spherical distribution of mass, 158–159
Spherical mirrors, 665–675, 690, 693–694
concave, 665–669
graphical method and, 674, 675
convex, 671
graphical method and, 674
linear magnification in, 672
graphical method and, 674–675
linear magnification and, 672–673
virtual images and, 670
virtual objects and, 670
Spherical surface of lens, 676
refraction at, 677–678
Spherical wave, 388
diffraction of, 735
Spin quantum number, 836
Spring
force of; *see* Spring force
harmonic motion and, 371–373, 383
simple, 366–367
period of motion in, 371–372
potential energy of, 177–180
waves and, 387
Spring force, 100–101, 109
in conservation of energy, 178–179
work done by, 177–178
Stable nuclides, 857
nuclear reactions and, 872
Standard mass, 91
Standing waves, 414–417, 420
adjusting length of column of air to produce, 417
in explanation of Bohr's angular momentum postulate, 834
on string, 410–412
through pipe, 416
Stanford linear accelerator, 882
Static equilibrium, 237–256
center of gravity in, 242–243
conditions for, 237–241, 248, 250–252
stress and strain in, 244–247
Static friction; *see* Friction, static
Static potential, 532
Static shear stress, fluids and, 259
Steady flow
Bernoulli's equation and, 273–274
continuity equation and, 275
defined, 274
of heat, 331
Steady-state current, 519
Steam, specific heat of, 323
Steam engine, 338
Carnot, 354
Steel
specific heat of, 323
thermal expansion and, 314
Stefan's law, 327, 333
Stefan-Boltzmann constant, 333
Step-up transformer, 615
Stereo speakers, magnetic forces and, 556–557
Stiles-Crawford effect, 722
Straight line motion; *see* Linear motion
Straight wire, magnetism and electric currents in, 566
Strain
normal, 245
static equilibrium and, 245–247, 248, 254
Strange quark, 885
Strassman, Fritz, 873
Stress
in fluids, 258–259
normal, 245
shear, 245
laminar flow and, 280
static, in fluids, 259
static equilibrium and, 245, 248, 254
thermal, 313
Strings, waves on, 387
force of, 393
harmonics of, 399, 413
power transmitted by, 400
speed of, 392–393
standing, 410–412
transmitting force of, 393
Strong interactions, 98
Strong nuclear force, fundamental forces and, 880, 881
Structure
of atom, 835–840
nuclear
elementary particles and, 854–860, 888, 890–891
strong nuclear force, 858
Study objectives, 4–5
Subtraction of vectors, 26
Sun
energy supply of, 876

Sun—cont'd
visible-light frequencies produced by, 635
Sundogs, 755
Sunglasses, Polaroid, 751
Sunlight
electric and magnetic fields in, 631
geometric optics and, 644
passing through media, 640
scattering of, 753, 758
Super Collider, 883
Superconductivity, 502
Superconductors, 498
Superposition of waves; *see* Mechanical waves, superposition of
Superposition principle, 406
Coulomb's law and, 433
Supersonic speeds, 398–399
Surface
of conductor as equipotential surface, 464
of fluids
pressure and, 259
volume per unit time passing through, 274
Surface area, pressure and, 259
Surface charge density, 441
Surface force, 118
Surface tension, 270–273, 286, 290–291
Switch, thermostat, 314
Symmetric body, center of mass of, 129
Synapse, 532
System of particles
energy of, 187–189, 196
momentum of, 202–204, 209–210
Système International Units, 7, 9, 878

T

Tangential acceleration, 217
rotation about fixed axis and, 221
Tau lepton, 883
Tau neutrino, 883
Telephone communication, fiber optics in, 652
Telescopes, 716–718, 725, 730
as magnifiers, 710–713
Television tube, 482
Temperature, 295–319
absolute, entropy and, 350
of ferromagnet, 573
gas pressure and, 298
heat and; *see* Heat
ideal gas law and, 298–303, 306–309, 315, 316
derivation of, 306–309
ideal gas model in, 303–306, 317
kinetic interpretation of, 304
Temperature—cont'd
kinetic theory in, 303–304
measurement of, 295–298, 315, 316
nuclear reactions and, 877
saturated vapor pressure and, 310
thermal expansion in, 311–314, 317–318
as thermodynamic variable, 339
vapor pressure and humidity in, 309–311
viscosity and, 279
Temperature coefficient of resistance, 497
Tension
centripetal force and, 125–126
force laws and, 99, 109, 112–113
surface, 270–273, 286, 290–291
Terminals of power supply, 470
Tesla, 552
Thermal conductivity, 331
Thermal energy, 182
conversion of electric energy to, 500
Thermal equilibrium, 296
Thermal expansion, 311–314, 317–318
of water, 314
Thermal interaction, 296
Thermal reservoirs, 330
Thermal stress, 313
Thermodynamics, 338–365
first law of, 341–345, 358, 359–361
heat engines and refrigerators in, 346–348, 358, 361
human metabolism and, 355–357
second law of, 338, 349–355, 358, 361–362
Clausius' statement of, 355
entropy statement of, 351–353
Kelvin statement of, 353
systems of, 339–341, 358, 359
Carnot engine and, 354–355, 358
Clausius' statement of, 355, 358
expressions for work in, 343–345
interacting, 341–343, 352–353
internal energy in, 340–341, 358
isolated, 341, 351–352
Kelvin statement of, 353, 358
quasi-static or reversible process in, 339–340, 358
second law of thermodynamics and, 351–353
variables in, 339, 358
Thermometers, 295–296
Thermometric property, 296
Thermostat switch, 314
Thin films, interference and, 739–741
Thin lens equation, 679–680, 695–696
image distance and, 684–685
Thin rod, rotation axis of, 223
Thompson, Benjamin, 320
Thomson, J.J., 822
Thomson's model of atom, 822
Thorium, decay of, 868
Three charges of electric field, 439
Three-dimensional waves, 388; *see also* Sound
Thunderstorms, 455
Tides, ocean, 147
Time; *see also* Time interval
in acceleration computation, 44, 45
Cartesian coordinates and, 17
change in, uniform motion and, 18
defined, 7
dilation and, 770–775
in displacement of particle of medium, 404–405, 424
in equations, consistency of units and, 10
as fundamental quantity, 6
in measurements
Einstein's postulates for; *see* Einstein's postulates
of meter, 8
relativity and, 766–775, 790, 793, 794–795; *see also* Relativity
nature of, 766–767
quantum concepts and, 809
in RCL circuits, 609
Time constant, 530
in RL circuits, 600
Time dilation, 770–775, 794–795
Time interval; *see also* Time
average acceleration on curved path and, 65
circular motion and, 73
displacements and average velocity over, 31, 35
half-life decay curve and, 865–866
proper, 770
simple harmonic motion and, 366–367
wave properties of light and, 734–736
Titanium dioxide, 476
Tomography, computerized axial, 839
Top quark, 885
Torque
center of gravity and, 242
on current loop in external magnetic field, 557–559
rotation and, 219–220, 231, 232–233
about fixed axis, 221–222, 231, 233–234
stress and strain and, 246, 248
Torr as unit of pressure, 259
Torricelli's barometer, 264
Total internal reflection, 649–651
fiber optics and, 651
Trajectory
circular motion and, 74
instantaneous velocity and, 32–34

Trajectory—cont'd
parabolic, 71
center of mass of projectile and, 130
particle, 17
projectile motion and, 71, 130
Transformers
in alternating current circuits, 612–615
step-up, 615
Translational equilibrium, 237, 238
Translational kinetic energy, 225, 340
in running, 229
Transmission axis of Polaroid sheet, 748
Transmittance of light, 648–649
Transparence, 640
Transuranic elements in nuclear reactions, 873
Transverse wave, 387
Triple point, 296
Truth quark, 885
Tube
gas discharge, 626, 635
atomic spectra and, 819
neon, 546
photomultiplier, 870
x-ray, 839
Tungsten, 497
Tuning fork
vibrating, waves and, 414
wavelength of sound produced by, 394
Turbulent flow
fluids and, 280
Reynolds number and, 281
Twin paradox, 775
Two New Sciences, 2, 247
Two point electric charge, 431–432
electric field of, 438
Two-dimensional waves, 388
Two-loop circuit, 519

U

Ultraviolet light, 326, 633
passing through media, 640
Uncertainty in measurement, 10–11
Uncertainty principle of Heisenberg, 808
quantum concepts and, 807–809, 814, 816–817
quantum mechanics and, 833–834
Uncharged conductor, charge distribution in, 430
Underdamped system, 378
Underwater vision, 706
Uniform electric field
dielectrics in, 478
in infinite plane, 445
potential difference in, 460–463
of sphere, 446
Uniform magnetic field, 560–562
Uniform motion, 18, 35
Units, 6–15; *see also* Measurement and units
base, 7
conversions of, 9
derived, 7
of energy, calorie as, 321–322
of force, 101
names of, 9
in particle dynamics, 93
of power, 183–184
energy and, 183–184
of work, 163–164
Universal gravitation; *see* Gravitation, universal
Unpolarized light, 749, 750
Unstable nuclides, 872
Up quark, 885
Uranium
atomic mass of, 856
charge density in nucleus of, 441
decay of, 868, 869
elements beyond, 873
in nuclear fission, 873
U-shaped conductors of electric charge, 592–593

V

Vacuum, dielectrics and, 476
Vacuum wavelength, 642
Values
limiting, 20–21
numerical, expressed in units, 6
Van Allen belts, 562
Van de Graff generator, 450
Vapor, saturated, 309, 310
Vapor pressure, 309–311, 317
saturated, 309, 310
Vaporization, latent heat of, 323
Variable $\mathcal{E}$, 593
Variable force
gravitational, 173–177
in three dimensions, 167
Vector sums, 35
in electric field, 437
Vectors
addition of, 24–25, 35
by components, 29–30, 39
graphical, 38
algebra of, 23–27, 38
change in, acceleration in velocity and, 42
components of, 27–30, 39
addition by, 29–30, 39
graphical representation of, 23
instantaneous velocity and, 32–34
polygon rule of, 25
Vectors—cont'd
relative displacement of, 76, 77, 78
scalars and, 23
multiplication of, 27
subtraction of, 26, 35
graphical, 38
Velocity, 31–34, 40
acceleration and, 42, 43
constant, 45
friction in, 122, 123
angular; *see* Angular velocity
average; *see* Average velocity
center of mass and, 127
circular motion and, 72, 74, 370
constant, graph of linear motion at, 54–55
escape, in gravitational potential energy, 176–177
friction in, 122, 123
of gas molecule in container, 307, 308
instantaneous; *see* Instantaneous velocity
nonrelativistic, 779
relative, 77, 78, 79, 779–780, 795
second law of thermodynamics and, 349
speed compared to, 31
Ventricular fibrillation, 534
Vertical deflection plates, 479–481
Vertical motion in mass and spring, 373
Vessels, blood
heat evaporation and, 330
Poiseuille's law and, 283
Vibration, waves and, 413, 414
Virtual image point, 663
Virtual images, 670
in concave spherical mirror, 667
graphical method for, 685
optical power and, 682–683
in plane mirrors, 663, 665
Virtual object, 670
graphical method for, 685, 686
Viscosimeter, 279
Viscosity of fluids, 279–282, 286, 292
Poiseuille's law and, 282, 286, 292–294
Visible spectrum
of atom, 838
of light, 632–633
retina and, 720
Vision
correction of, 703–707
scotopic and photopic, 719–720
Visual acuity, 709–710, 722–724, 725, 728
factors limiting, 730
Visual sensitivity to light, 633
Vitreous humor, 700
light particle scattering in, 722
refractive index of, 641

Volt
electron, 462
per amp, 495
Volta's first battery, 503
Voltage
current in phase with, 604
current lagging behind, 608
current leading, 606
drop in
across emf source, 506–507
across resistor, 500
Kirchoff's rules and, 520–522
in transformers, 614–615
of oscilloscope, 481
Voltmeter, 481
direct current circuits and, 528
electric potential and, 459
Volume
gas pressure and, 298
magnetic energy per unit, 602
magnetic moment per unit, 570
per unit time passing through surface of fluid, 274
as thermodynamic variable, 339
Volume charge density, 441
Volume coefficient of expansion, 311–312, 315
von Guericke, Otto, 260

W

Walking, study of, 19
Waste heat, 346
Water
density of, temperature and, 314
dielectrics and, 476
diffraction of waves of, 735
latent heat of vaporization for, 323
reflection and refraction of light in, 641, 646
specific heat of, 323
vision in, 706
waves in, 388
speed of, 394, 419
Watt, 183
Wave; *see also* Wave optics
coherent, 406
description of, 387–391, 419, 421–422
exerted by one charge to another, 432
form of, at equal time intervals, 405
frequency of, 389–390
harmonic, 389, 400–403, 419
intensity of, electron proportional to, 832
length of; *see* Wavelength
longitudinal, 387
mechanical; *see* Mechanical waves
oscilloscope measurement of, 482
periodic, 389–390
Wave—cont'd
in phase, interference and, 406
phenomena of, 387
plane, 388
pressure, 422
properties of
in electrons, 853
in light rays, 731–736, 757, 759
in matter, 835
seismic, 422
shear, 422
sound; *see* Sound waves
speed of, 389, 392–395, 419, 422–423
spherical, 388
diffraction of, 735
spread of, 389–390
standing; *see* Standing waves
on string; *see* Strings, waves on
three-dimensional, 388
transverse, 387
water, 388
Wave function, 836
Wave optics, 731–765; *see also* Wave
diffraction in, 643–644, 741–748, 757, 763–764
by circular aperture, 744
Fraunhofer, 742, 744
Fresnel, 741, 742
gratings of, 746–748, 758
historical background of, 741
light wave properties and, 735–736, 757, 759
Rayleigh criterion for resolution and, 745–746
single-slit, 742–744
interference and, 736–741, 757, 760–762
light wave properties in, 731–736, 757, 759
in magic in sky, 754–755
polarization in, 748–753, 758, 764
Wave packets, 807
electron, 832
Wave pulse phenomenon, 387
Wave theory of Schroedinger, 835
Wavelength, 389, 390, 419
de Broglie, 830–832
dependence of, on index of refraction, 642
of electron and particle, 832
measurement of, 642
microscopes and, 715
momentum of photons and, 814, 815–816, 830
of sound produced by tuning fork, 394
of spectral colors, 632
in vacuum, 642
x-ray, 840
Wave-particle duality, 804–806
Weak interactions, 98
Weak nuclear force, 880, 881
Weight
centripetal force and, 125–126
force laws and, 96–97, 98, 109, 112
gravitational attraction and, 96, 144, 156, 158–159
mass and, 92
nonconservative force and, 181
true, 150
Wheatstone bridge, 529, 544
Wheel, circular motion of, 369–371
White light, 732
Wind-powered generators, 596
Wire, electrical
carrying uniform current, 511
grounding of, 535
hot, 535
magnetic forces on current-carrying, 554–559, 582–584
long-straight wires in, 566
parallel wires in, 567
Wood
resistivity and temperature coefficient of, 497
specific heat of, 323
Work
done by constant force
acting on particle, 167
gravitational potential energy and, 169–170
in one dimension, 163
done by spring force, 177–178
gravitational force and, 179–180
expressions for, 343–345
gravitational potential energy and, 169
kinetic energy and, 163–168, 190, 191–192
metabolism and, 356–357
negative
in interacting thermodynamic system, 342, 343
pressure in moving fluid and, 277
net; *see* Net work
positive
in interacting thermodynamic system, 342, 343
pressure in moving fluid and, 277
spring potential energy and, 178, 194
of system of particles, 187, 188
thermodynamics and, 356–357
units of, 163–164
Work function, photoelectric effect and, 802
Work-energy theorem, 165–166

X

X coordinate, 17
  of centers of mass, 138
X-ray, 326
  biological uses of, 877–880
  characteristic, 840
  Compton effect and, 803–804
  diffraction pattern of, 748
  intensity spectrum of, 840
X-ray—cont'd
  relative biological effectiveness of, 878, 892–893
  in spectra of atoms, 839–840, 853
  wavelength of, 840
X-ray tube, 839

Y

Y coordinate, 17
  of centers of mass, 138
Young, Thomas, 721, 737
Young's double slit, interference and, 737–739, 760–762
Young's modulus, 245

Z

Z coordinate, 17
Zero displacement, standing waves and, 411
Zweig, George, 884

Photo Credits

Introduction

I-1 English Heritage Photo Library. I-2 Smithsonian Institution. I-3 Phil A. Harrington/Peter Arnold, Inc. I-4 Far Side copyright 1990 Farworks, Inc./Dist. by Universal Press Syndicate. M-0 Berkeley Sensor and Actuator Center. M-1a, M1b Smithsonian Institution. M-1c Jody Dole/The Image Bank. M-1d National Institute of Standards and Technology Boulder Laboratories, US Department of Commerce. M-2 Patrick Watson. E1-1 Princeton Plasma Physics Laboratory. E1-2 Newsweek/Sipa Press. E1-3 Jet Propulsion Laboratory.

Chapter 1

1-0 Robert Mathena/Fundamental Photographs. 1-1 The Harold E. Edgerton 1992 Trust. 1-2 Miriam and Ira D. Wallach Division of Art, Prints and Photographs, The New York Public Library, Astor, Lenox and Tilden Foundations. 1-3 David Coletta. 1-7 After John Napier, Antiquity of Human Walking, *Scientific American,* 1967.

Chapter 2

2-0 Jet Propulsion Laboratory (photo). 2-5 James Sugar/Black Star. 2-9 Patrick Watson. 2-A Jeff Hispanski. 2-B Fundamental Photographs. Unn 2-1 Special collections, New York Public Library. 2-18 David Coletta. 2x Copyright 1994, Patricia Mitchell, Los Angeles Times.

Chapter 3

3-0 Biblioteca Ambrosiana, Milano. 3-2 Lewis Portnoy/ Spectra Action Inc. 3-6 Fundamental Photographs. 3-9 David Coletta. 3-10a Missouri Division of Tourism. 3-10b John Sohlden/Visuals Unlimited. 3-16 AIP Emilio Segrè Visual Archives. 3-18a, 3-18b David Coletta. 3-26 Anthony Neste. 3-28 David Coletta. Unn3y Steve Karp.

Chapter 4

4-0 Meteor Crater, Northern Arizona U.S.A. 4-1 David Coletta. 4-3 PASCO scientific/Central Scientific Company. 4-7a David Coletta. 4-7b American Red Cross. 4-13 Copyright by Sidney Harris: *Physics Today.* 4-14a Patrick Watson. 4-14b Vincent Coletta. 4-24 McNeil: *Biomechanics,* New York, 1975, Methuen & Co. 4-52 David Coletta.

Chapter 5

5-0 Tom Lippert. 5-4a Loubat/ Vandystadt. 5-D David Coletta. 5-9a Richard Megna/Fundamental Photographs. 5-9b The Harold E. Edgerton 1992 Trust. 5-14 Tony Stone Worldwide/J. McDermott.

Chapter 6

6-0 Erich Lessing/Art Resource, NY. 6-4 Central Scientific Company. 6-7 Jet Propulsion Laboratory. 6-9 James Sugar/Black Star. 6-10 Clyde H. Smith/ Peter Arnold, Inc. 6-14 James Sugar/Black Star. 6-Aa, 6-Ab Department Library Services, American Museum of Natural History. Unn 6-1 AIP Emilio Segrè Visual Archives. 6-20 Steward Observatory and NOAO; NASA; Jet Propulsion Laboratory. 6x, 6y, 6v, 6z Jet Propulsion Laboratory.

Chapter 7

7-0 From Early Bow Design and Construction, by Edward McEwen, Robert L. Miller and Christopher A. Bergman. Copyright June 1991 by Scientific American, Inc. All rights reserved. 7-18a New England Meteoritical Services; Jet Propulsion Laboratory. 7-18b Neg./ Trans. no. 323714 Willamete Meteorite Department of Library Services, American Museum of Natural History. 7-24 Tom Lippert. 7-28 Lewis Portnoy/Spectra Action Inc. 7-29 David Coletta. 7-32 Antoine de Saint-Exupery/Harcourt Brace & Company. 7-36 Vincent Coletta. 7-40 Illustration by Steven Simpson, *Sky & Telescope.* 7E1 David Coletta.

Chapter 8

8-0 Bonnie Sue Rauch. 8-3a, 8-3b, 8-3c, 8-3d Lewis Portnoy/ Spectra Action, Inc. 8-6 CERN Photo. 8-10 The Harold E. Edgerton 1992 Trust. 8-13 Bernice Abbott/Commerce Graphics Ltd, Inc.

Chapter 9

9-0 James Sugar/Black Star. 9-4 IBM. 9-18 Luke White. 9-19 The Harold E. Edgerton 1991 Trust.

Chapter 10

10-0 R. Krubner/H. Armstrong Roberts. 10-4a Barbara Resse, NINCDS and Raven and Johnson: *Understanding Biology,* ed 2, Dubuque, W.C. Brown. 10-15a Kenneth W. Fink, Root Resources.

Chapter 11

11-0 The Harold E. Edgerton 1992 Trust. 11-4 The Deutches Museum, Munich. 11-9c David Coletta. 11-11 James Sugar/Black Star. 11-14a National Center For Atmospheric Research. 11-14b American Red Cross. 11-22 Tom Pantages. 11-26 Patrick Watson. 11-37a Gregory Gorel/ Visuals Unlimited. 11-37b David Coletta. 11-46 David Coletta. 11-56 Warner Brothers/Courtesy of The KOBAL Collection. 11-63 Patrick Watson.

Chapter 12

12-0 Bob Firth. 12-1b Central Scientific Company. 12-1c Central Scientific Company. 12-1d Wahl Instruments, Inc. 12-2 David Coletta. 12-3 b. Taylor/H. Armstrong Roberts. 12-4 David Coletta. 12-5 H. Armstrong Roberts. 12-14 David Coletta. 12-15 Patrick Watson. 12-16 David Coletta. 12-20 Alese/Mort Pechter.

Chapter 13

13-0, 13-1 David Coletta. 13-3 Freda Greene/ Images International. 13-4 Dr. Ray Clark & M.R. Goff/Science Photo Library. 13-7 Tim Haske/Profiles West. 13-10 Tom Pantages.

Chapter 14

14-1 Science Museum London. 14-17, 14-24 David Coletta. 14-25 Michael and Cheryl Gelfand-Grant.

Chapter 15

15-0 Robert Mathena/Fundamental Photographs. 15-10 David Coletta. 15-12 James Stevenson/Science Photo Library. 15-22 Ben Rose/ The Image Bank. 15-23 Special Collections, Univ. of Washington Libraries, Neg. #3 & #4.

Chapter 16

16-0 M. Thonig/H. Armstrong Roberts. 16-1 David Coletta. 16-4 Martin Dohrn/Science Photo Library. 16-12 Visuals Unlimited. 16-21 David Coletta. 16-22 The Palm Beach Post. 16-27 David Coletta. 16-29 *PSSC Physics,* ed 2, Newton, MA, 1965, D.C. Heath and Company and Education Development Center, Inc. 16-36a Richard Megna/Fundamental Photographs. 16-39 Patrick Watson. 16-42 H. Armstrong Roberts. 16-48 U.S. Geological Survey Photographic Library.

Chapter 17

17-0 The Wellcome Trustees. 17-1 The Deutches Museum, Munich. 17-3 Central Scientific Company. 17-10a, 17-10b Harold M. Waage/ Princeton University. 17-11 David Coletta. 17-19 Harold M. Waage/Princeton University. 17-25 Westinghouse. 17-27 Harold M. Waage/Princeton University. 17-37 David Coletta. 17-43 The Magic House/St. Louis Children's Museum. 17-60 Stewart D. Halperln.

Chapter 18

18-1 Robert Holmes. 18-3 David Coletta. 18-9a(1) David Coletta. 18-9(2) Dr. Mitsuo Ohtsuki/Science Photo Library. 18-23a David Coletta. 18-23b Lawrence Livermore Laboratory. 18-26 David Coletta. 18-39 Central Scientific Company. 18-64 Tom Pantages. 18-65 Central Scientific Company.

Chapter 19

19-0 National Center for Atomspheric Research. 19-8a, 19-8b, 19-11 David Coletta. 19-A Mehau Kulyk/Science Photo Library. 19-B G. Davis/Sygma. 19-C AT&T. 19-16 David Coletta. 19-18 Smithsonian Institution. 19-20 David Coletta. 19-24(a) John Elk III. 19-24(b) General Electric. 19-24(c) Chris Boylan/Unicorn Stork Photos. 19-34, 19-36, 19-37 David Coletta. 19-42 R. Amft/H. Armstrong Roberts. 19-43 David Coletta.

Chapter 20

20-0 The Museum of Modern Art, New York/ Hillman Periodicals Fund. 20-1 Smithsonian Institution. 20-5a, 20-5b David Coletta. 20-8 Cordon Art B.V. 20-26a, 20-26b David Coletta. 20-B AL Hodgkin, AF Huxley *Journal of Physiology* 1945, 104:176. 20-32a, 20-72 David Coletta.

Chapter 21

21-4 General Electric. 21-7 Alvis Upitis/The Image Bank. 21-12 Eisbrenner Public Relations GM Electric Vehicles Impact. 21-18a David Coletta. 21-18b CERN Photo. 21-24 NASA; Eugene Lauria. 21-27b, 21-30a Fundamental Photographs/Richard Megna. 21-34 Halliday et al: *Fundamentals of Physics,* ed 4, New York, 1993, John Wiley & Sons. 21-40 Tom Hollyman/Photo Researchers, Inc. 21-A H. Vali/McGill University.

Chapter 22

22-0 Consolidated Edison of New York Inc. 22-1 Science Museum London, Paul Cox Picture Library. 22-2b, 22-3b, 22-10b David Coletta. 22-18a US Army Corps of Engineers. 22-18b Gary Gladstone/The Image Bank. 22-18c Central Hudson Gas and Electric Co. 22-18d Steve Halperin. 22-35a David Coletta. 22-35b David J. Sams/Texas Imprint. 22-35c David Coletta. Unn 22-1 AIP Emilio Segrè Visual Archives. Unn 22-2 Royal Institute of Great Britain Faraday Museum. 22-45 Patrick Watson. 22-54a Brad Jirka and St. Elmo's. 22-54b David Coletta. 22-57 Vincent Coletta.

Chapter 23

23-0 Jet Propulsion Laboratory. 23-1 Cambridge University Library/Trinity College. 23-7 Science Photo Library. 23-9a Richard Megna/ Fundamental Photographs. 23-9b, 23-10 David Coletta. 23-11 Wabash Instrument Corp./Fundamental Photographs; Bausch & Lomb. 23-15 Alese and Mort Pechter. 23-16 Carl Purcell. 23-24, 23-29, 23-33, 23-34 David Coletta. 23-37 Gem Media. 23-38b Dominique Sarraute/The Image Bank. 23-40 Circon Corporation. 23-41 Lennart Nilsson, Dell Publishing Company. 23-45 Jet Propulsion Laboratory. 23-56 Alese and Mort Pechter. 23-57 Patrick Watson. 23-61 David Coletta.

Chapter 24

24-0 Hughes Danbury Optical Systems Inc. 24-1, 24-2, 24-3B, 24-6B, 24-9, 24-10A, 24-10B, 24-11A-E, 24-17A, 24-17B, 24-19 David Coletta. 24-24 Optics Guide Three/Melles Griot. 24-26, 24-29, 24-31A-E, 24-34B(2), 24-34C(2), 24-37, 24-43, 24-46 David Coletta. 24-62 Grant Heilman/Grant Heilman Photography. 24-63 David Coletta.

Chapter 25

25-0 Lennart Nilsson, Dell Publishing Company. 25-2(1) Ralph C. Eagle, MD/Scheic Eye Institute. 25-5a, 25-5b, 25-8a, 25-8b, 25-12, 25-18(1), 25-18(2) David Coletta. 25-21 Patrick Watson. 25-22 David Coletta. 25-24 Tom Pantages. 25-25 David Coletta. 25-28 IBM Corporation, Research Division, Almaden Research Center. 25-29 AIP Emilio Segrè Visual Archives. 25-32, 25-33 Jet Propulsion Laboratory. 25-A ER Lewis, YY Zeevi, FS Werblin *Brain Res* 15:559-562, 1969. 25-38(1), 25-38(2) Alfred L. Yarbus, *Eye Movements and Vision,* 1973, Plenum Publishing Corporation. 25-39, 25-42(1) David Coletta.

Chapter 26

26-0 Peter Aprahamian/Science Photo Library. 26-8 NOAO, E.R. Degginger. 26-11 Cagnet et al: *The Atlas of Optical Phenomena.* 26-12b, 26-16 David Coletta. 26-19a, 26-19b Tom Pantages. 26-20 Phillip Rinard. 26-21a, 26-22b, 26-26, 26-27a, 26-27b, 26-27c Cagnet et al: *The Atlas of Optical Phenomena.* 26-30, 26-31, 26-32 David Coletta. 26-34 Wabash Instruments/Fundamental Photograph. 26-35 Rosalind Franklin/ Photo Researchers. 26-42a, 26-42b Polaroid Corporation. 26-46 Peter Aprahamian/Science Photo Library. 26-50 Richard Megna/Fundamental Photographs. 26-51a, 26-52 PASCO scientific. 26-C Robert Greenler/Robert Greenler Sky Photos.

Chapter 27

27-0 MGM/The KOBAL Collection. 27-1 ETH–Bibliothek. 27-2a David Coletta. 27-2b Hans Wolf/The Image Bank. 27-2c David Coletta. 27-2d Jet Propulsion Laboratory. Unn 1 The Jewish National and University Library, Albert Einstein Archives. Unn 2 Archives, California Institute of Technology. Unn X Jet Propulsion Laboratory. 27-C James Sugar/Black Star. 27-G Carl Masthay. 27-H Jet Propulsion Laboratory.

Chapter 28

28-0 Radio Corporation of America. 28-1 Freda Greene/Images International. 28-9b David Coletta. 28-12 Cagnet et al: *The Atlas of Optical Phenomena.* Unn 1 AIP Emilio Segrè Visual Archives/Physics Today Collection. Unn 2 California Institute of Technology. Unn 3 Woodfin, Camp and Associates.

Chapter 29

29-0 Dr. Mitsuo Ohtsuki/Science Photo Library. 29-2 A-C Wabash Instrument Corp./Fundamental Photographs; Bausch & Lomb. 29-3 Richard Herzberg/Prentice Hall. 29-5 AIP Emilio Segrè Visual Archives/Otto Hahn and Lawrence Badash. 29-9 AIP Emilio Segrè Visual Archives, W.F. Meggers Collection. 29-14 *PSSC Physics,* ed 2, Newton, MA, 1965, D.C. Heath & Company and Educational Development Center, Inc. 29-18 AIP Emilio Segrè Visual Archives, Francis Simon Collection. 29-19 AIP Emilio Segrè Visual Archives, W.F. Meggers Collection. 29-23a Patrick Watson. 20-24 Deutsches Museum, Munich. 29-I Hughes Aircraft Company/Electron Dynamics Division. 29-J(b) Will and Demi McIntyre/Photo Researchers. 29-R IBM. Unn 1 Tom Way/IBM Microelectronics.

Chapter 30

30-0(1), 30-0(2) Fermi National Laboratory. 30-8 Central Scientific Company. 30-10 CERN Photo. 30-14 Archives, California Institute of Technology. 30-15 Stanford Linear Accelerator Center. 30-A AIP Emilio Segrè Visual Archives, W.F. Meggers Collection.

Notation

Selected Vectors

	Velocity	Acceleration	Force	Electric field
Vector	**v**	**a**	**F**	**E**
Components	v_x, v_y	a_x, a_y	F_x, F_y	E_x, E_y
Magnitude	v	a	F	E

Other

Σ summation
$\propto$ proportional to
$\approx$ approximately equals
$>$ greater than
$<$ less than
$\bar{x}$ average value of x
Δx change in x

Greek Alphabet

Alpha	A	α	Nu	N	ν
Beta	B	β	Xi	Ξ	ξ
Gamma	Γ	γ	Omicron	O	o
Delta	Δ	δ	Pi	Π	π
Epsilon	E	ϵ	Rho	P	ρ
Zeta	Z	ζ	Sigma	Σ	σ
Eta	H	η	Tau	T	τ
Theta	Θ	θ	Upsilon	Υ	υ
Iota	I	ι	Phi	Φ	ϕ
Kappa	K	κ	Chi	X	χ
Lambda	Λ	λ	Psi	Ψ	ψ
Mu	M	μ	Omega	Ω	ω

Math Formulas

Area of a circle $= \pi r^2$
Circumference of a circle $= 2\pi r$
Surface area of a sphere $= 4\pi r^2$
Volume of a sphere $= \frac{4}{3}\pi r^3$
Volume of a cylinder $= \pi r^2 h$
Pythagoras' Theorem $c^2 = a^2 + b^2$
$\sin\theta =$ opposite side/hypotenuse
$\cos\theta =$ adjacent side/hypotenuse
$\tan\theta = \sin\theta/\cos\theta =$ opposite/adjacent
If $ax^2 + bx + c = 0$, then

$$x = \frac{-b \pm \sqrt{b^2 - 4ac}}{2a}$$

Index of Tables

2-1 Gravitational acceleration at various points on earth, 49
4-1 Systems of units for force and mass, 93
5-1 Coefficients of static friction, 120
11-1 Densities, 258
11-2 Viscosities, 281
12-1 Kelvin, Celsius, and Fahrenheit temperature scales, 297
12-2 Coefficients of thermal expansion, 312
13-1 Specific heats, 323
13-2 Thermal conductivities, 331
16-1 Speed of sound, 393
18-1 Dielectric constants and dielectric strengths, 476
19-1 Resistivities and temperature coefficients, 497
23-1 The visible spectrum, 633
23-2 Refractive indices, 641
26-1 Frequency bandwidths, 733
30-1 Masses of selected particles and atoms, 856
30-2 Relative biological effectiveness, 878
30-3 Some particles and their properties, 883
30-4 Elementary particles, 885